Many exercises are suitable for **cooperative learning**, providing opportunities to work with others.

19. If a pair of regular dice is tossed once, use the expectation formula to determine the expected sum of the numbers on the upward faces of the 2 dice.

20. Consider rolling a pair of unusual dice, for which the faces have the number of pips indicated.

Die 1: {0, 0, 0, 6, 6, 6}
Die 2: {1, 2, 3, 4, 5, 6}

a. List the sample space for the experiment.
b. Compute the probability of each possible sum of the upward faces on the dice.
c. What is the expected value of the sum of the numbers on the upward faces of the 2 dice?

21. Two dice, one labeled 1, 2, 2, 3, 3, 4 and the other labeled 1, 3, 4, 5, 6, 8, are rolled once. Use the formula for expectation to determine the expected sum of the numbers on the upward faces of the 2 dice. Dice such as these are called *Sicherman dice*.

22. Suppose you purchase a ticket for a prize and your expectation is −$1. What is the meaning of this expectation?

23. Efron's dice Suppose you are offered 1 of 2 pairs of dice, a red pair or a green pair, that are labeled as follows.

Red die 1: 0, 0, 4, 4, 4, 4
Red die 2: 2, 3, 3, 9, 10, 11
Green die 1: 3, 3, 3, 3, 3, 3
Green die 2: 0, 1, 7, 8, 8, 8

After you choose, your friend will receive the other pair. Which pair should you choose if you are going to play a game in which each of you rolls your dice and the player with the higher sum wins? Dice such as these are part of a set of 4 pairs of dice called *Efron's dice*. Which pair should you choose? Explain.

24. Lotteries The PowerBall lottery commission chooses 5 white balls from a drum containing 69 balls marked with the numbers 1 through 69, and 1 red ball from a separate drum containing 26 balls. The following table shows the approximate odds of winning certain prizes if the numbers you choose match those chosen by the lottery commission.

Match	Prize	Odds
○○○○○ + ●	Grand Prize	1 in 292,201,338.00
○○○○○	$1,000,000	1 in 11,688,053.52
○○○○ + ●	$50,000	1 in 913,129.18
○○○○	$100	1 in 36,525.17
○○○ + ●	$100	1 in 14,494.11
○○○	$7	1 in 579.76
○○ + ●	$7	1 in 701.33
○ + ●	$4	1 in 91.98
●	$4	1 in 38.32

The overall odds of winning a prize are 1 in 24.87.
The odds presented here are based on a $2 play (rounded to two decimal places).

SOURCE: http://www.powerball.com/powerball/pb_prizes.asp

Assuming the jackpot for a certain drawing is $150 million, what is your expectation for the jackpot if you purchase 1 ticket for $2? Round to the nearest cent. Assume the jackpot is not split among multiple winn

A variety of **End-of-Chapter** features help you prepare for a test.

The **Chapter Summary** reviews the major concepts discussed in the chapter. For each concept, there is a reference to a worked example illustrating how the concept is used and at least one exercise in the Chapter Review Exercises relating to that concept.

CHAPTER 11 SUMMARY

The following table summarizes essential concepts in this chapter. The references given in the right-hand column list Examples and Exercises that can be used to test your understanding of a concept.

11.1 Simple Interest	
Simple Interest Formula The simple interest formula is $I = Prt$, where I is the interest, P is the principal, r is the interest rate, and t is the time period.	See **Examples 2, 4, and 5** on pages 621 and 622, and then try Exercises 1, 2, and 3 on page 683.

Chapter Review Exercises help you review all of the concepts in the chapter. Answers to all the Chapter Review Exercises are in the answer section, along with a reference to the section from which the exercise was taken. If you miss an exercise, use that reference to review the concept.

CHAPTER 11 REVIEW EXERCISES

1. **Simple Interest** Calculate the simple interest due on a 4-month loan of $2750 if the interest rate is 6.75%.
2. **Simple Interest** Find the simple interest due on an 8-month loan of $8500 if the interest rate is 1.15% per month.
3. **Simple Interest** What is the simple interest earned in 120 days on a deposit of $4000 if the interest rate is 6.75%?
4. **Maturity Va**… simple intere… rate is 10.4%.
5. **Simple Inte**… on a 3-month… interest rate.
6. **Compound Amount** Calculate the compound amount when $3000 is deposited in an account earning 6.6% interest, compounded monthly, for 3 years.
7. **Compound Amount** What is the compound amount when $6400 is deposited in an account earning an interest rate of 6%, compounded quarterly, for 10 years?
8. **Future Value** Find the future value of $6000 earning 9% interest, compounded daily, for 3 years.

CHAPTER 11 REVIEW EXERCISES *page 683*

1. $61.88 [Sec. 11.1] 2. $782 [Sec. 11.1] 3. $90 [Sec. 11.1] 4. $7218.40 [Sec. 11.1] 5. 7.5% [Sec. 11.1] 6. $3654.90 [Sec. 11.2] 7. $11,609.72 [Sec. 11.2] 8. $7859.52 [Sec. 11.2] 9. $200.23 [Sec. 11.2] 10. $10,683.29 [Sec. 11.2] 11. a. $11,318.23 b. $3318.23 [Sec. 11.2] 12. $19,225.50 [Sec. 11.2] 13. 1.1% [Sec. 11.4] 14. $9000 [Sec. 11.4] 15. $1.59 [Sec. 11.2] 16. $43,650.68 [Sec. 11.2] 17. 6.06% [Sec. 11.2] 18. 5.4% compounded semiannually [Sec. 11.2] 19. $431.16 [Sec. 11.3] 20. $6.12 [Sec. 11.3] 21. a. $259.38 b. 12.75% [Sec. 11.3] 22. a. $36.03 b. 12.9% [Sec. 11.3] 23. $45.41 [Sec. 11.3] 24. a. $10,092.69 b. $2018.54 c. $253.01 [Sec. 11.3] 25. $664.40 [Sec. 11.3] 26. a. $540.02 b. $12,196.80 [Sec. 11.3] 27. a. Profit of $5325 b. $256.10 [Sec. 11.4] 28. 200 shares [Sec. 11.4] 29. $99,041 [Sec. 11.5] 30. a. $1659.11 b. $597,279.60 c. $341,479.60 [Sec. 11.5] 31. a. $1396.69 b. $150,665.74 [Sec. 11.5] 32. $2658.53 [Sec. 11.5] 33. $288.62 [Sec. 11.3]

The **Chapter Test** gives you a chance to practice a possible test for the chapter. Answers to all Chapter Test questions are in the answer section, along with a section reference for the question.

CHAPTER 11 TEST

1. **Simple Interest** Calculate the simple interest due on a 3-month loan of $5250 if the interest rate is 8.25%.
2. **Simple Interest** Find the simple interest earned in 180 days on a deposit of $6000 if the interest rate is 6.75%.
3. **Maturity Value** Calculate the maturity value of a simple interest, 200-day loan of $8000 if the interest rate is 9.2%.
4. **Simple Interest Rate** The simple interest charged

10. **Bonds** Suppose you purchase a $5000 bond that has a 3.8% coupon and a 10-year maturity. Calculate the total of the interest payments that you will receive.
11. **Inflation** In 2016, the median value of a single-family house was $224,000. Use an annual inflation rate of 4.3% to calculate the median value of a single family house in 2029. (*Source:* money.cnn.com)
12. **Effective Interest Rate** Calculate the effective interest rate of 6.25% compounded quarterly. Round t… the nearest hundredth of a percent.

For the Chapter Test, besides a reference to the section from which an exercise was taken, there is a reference to an example that is similar to the exercise.

CHAPTER 11 TEST *page 685*

1. $108.28 [Sec. 11.1, Example 2] 2. $202.50 [Sec. 11.1, Example 1] 3. $8408.89 [Sec. 11.1, Example 6] 4. 9% [Sec. 11.1, Example 5] 5. $7340.87 [Sec. 11.2, Check Your Progress 2] 6. $312.03 [Sec. 11.2, Example 4] 7. a. $15,331.03 b. $4831.03 [Sec. 11.1, Example 6] 8. $21,949.06 [Sec. 11.2, Example 6] 9. 1.2% [Sec. 11.4, Example 2] 10. $1900 [Sec. 11.4, Example 4] 11. $387,207.74 [Sec. 11.2, Check Your Progress 8] 12. 6.40% [Sec. 11.2, Check Your Progress 10] 13. 4.6% compounded semiannually [Sec. 11.2, Example 11] 14. $7.79 [Sec. 11.3, Example 1] 15. a. $48.56 b. 16.6% [Sec. 11.3, Example 2] 16. $56.49 [Sec. 11.3, Example 3] 17. a. Loss of $4896 b. $226.16 [Sec. 11.4, Example 3] 18. 208 shares [Sec. 11.4, Example 5] 19. a. $6985.94 b. $1397.19 c. $174.62 [Sec. 11.3, Example 4] 20. $60,083.50 [Sec. 11.5, Example 1] 21. a. $1530.69 [Sec. 11.5, Example 2a] b. $221,546.46 [Sec. 11.5, Example 4] 22. $2595.97 [Sec. 11.5, Example 2a, Example 5]

FOURTH EDITION

Mathematical Excursions

Richard N. Aufmann
Palomar College

Joanne S. Lockwood
Nashua Community College

Richard D. Nation
Palomar College

Daniel K. Clegg
Palomar College

CENGAGE Learning®

Australia • Brazil • Mexico • Singapore • United Kingdom • United States

CENGAGE Learning

***Mathematical Excursions*, Fourth Edition**
Richard N. Aufmann, Joanne S. Lockwood, Richard D. Nation, Daniel K. Clegg

Product Director: Terry Boyle

Product Manager: Rita Lombard

Content Developer: Powell Vacha

Product Assistants: Gabriela Carrascal, Abigail DeVeuve

Marketing Manager: Ana Albinson

Content Project Manager: Jennifer Risden

Art Director: Vernon Boes

Manufacturing Planner: Doug Bertke

Production Service: Graphic World Inc.

Photo Researcher: Lumina Datamatics

Text Researcher: Lumina Datamatics

Copy Editor: Graphic World Inc.

Illustrator: Graphic World Inc.

Text Designer: Hespenheide Design

Cover Designer: Hespenheide Design

Cover Image: IvanJekic/E+/Getty Images

Interior Design Image: Dudarev Mikhail/Shutterstock.com

Compositor: Graphic World Inc.

Library of Congress Control Number: 2016950658

Student Edition:
ISBN: 978-1-305-96558-4

Loose-leaf Edition:
ISBN: 978-1-337-28877-4

Cengage Learning
20 Channel Center Street
Boston, MA 02210
USA

Printed in the United States of America
Print Number: 01 Print Year: 2016

Contents

Dynamic Graphics/PictureQuest

1 Problem Solving 1

aastock/Shutterstock.com

2 Sets 47

Morphart Creation/Shutterstock.com

3 Logic 105

United Nations, New York

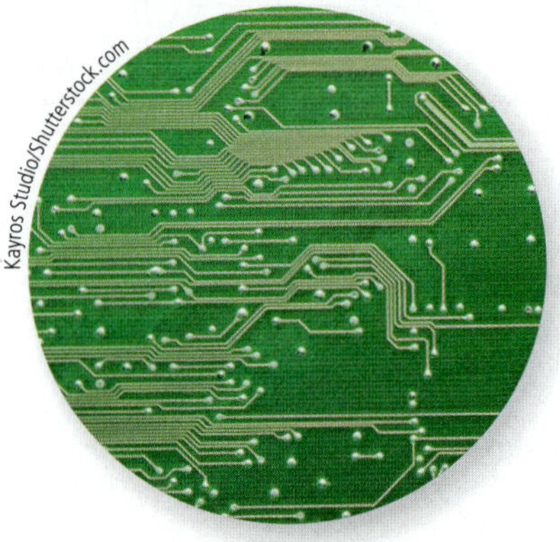

Kayros Studio/Shutterstock.com

PictureQuest

7 Measurement and Geometry 355

8 Mathematical Systems 457

Steve Allen/Stockbyte/Getty Images

9 Applications of Equations 495

cbpix/Shutterstock.com

10 Applications of Functions 555

Andy Dean Photography/Shutterstock.com

11 The Mathematics of Finance 619

nvuk/Shutterstock.com

12 Combinatorics and Probability 687

Eric Fahrner/Shutterstock.com

13 Statistics 751

Preface

Mathematical Excursions is about mathematics as a system of knowing or understanding our surroundings. It is similar to an English literature textbook, an introduction to philosophy textbook, or perhaps an introductory psychology textbook. Each of those books provides glimpses into the thoughts and perceptions of some of the world's greatest writers, philosophers, and psychologists. Reading and studying their thoughts enables us to better understand the world we inhabit.

In a similar way, *Mathematical Excursions* provides glimpses into the nature of mathematics and how it is used to understand our world. This understanding, in conjunction with other disciplines, contributes to a more complete portrait of the world. Our contention is that:

- Planning a shopping trip to several local stores, or several cities scattered across Europe, is more interesting when one has knowledge of efficient routes, which is a concept from the field of graph theory.
- Problem solving is more enjoyable after you have studied a variety of problem-solving techniques and have practiced using George Polya's four-step, problem-solving strategy.
- The challenges of sending information across the Internet are better understood by examining prime numbers.
- The perils of radioactive waste take on new meaning with knowledge of exponential functions.
- Generally, knowledge of mathematics strengthens the way we know, perceive, and understand our surroundings.

The central purpose of *Mathematical Excursions* is to explore those facets of mathematics that will strengthen your quantitative understandings of our environs. We hope you enjoy the journey.

Updates to This Edition

- Application Examples, Exercises, and Excursions have been updated to reflect recent data and trends.
- Expanded Chapter 7 with the addition of a section on measurement.
- Extension exercises have been consolidated and streamlined.

Interactive Method

The AIM FOR SUCCESS STUDENT PREFACE explains what is required of a student to be successful and how this text has been designed to foster student success. This "how to use this text" preface can be used as a lesson on the first day of class or as a project for students to complete to strengthen their study skills.

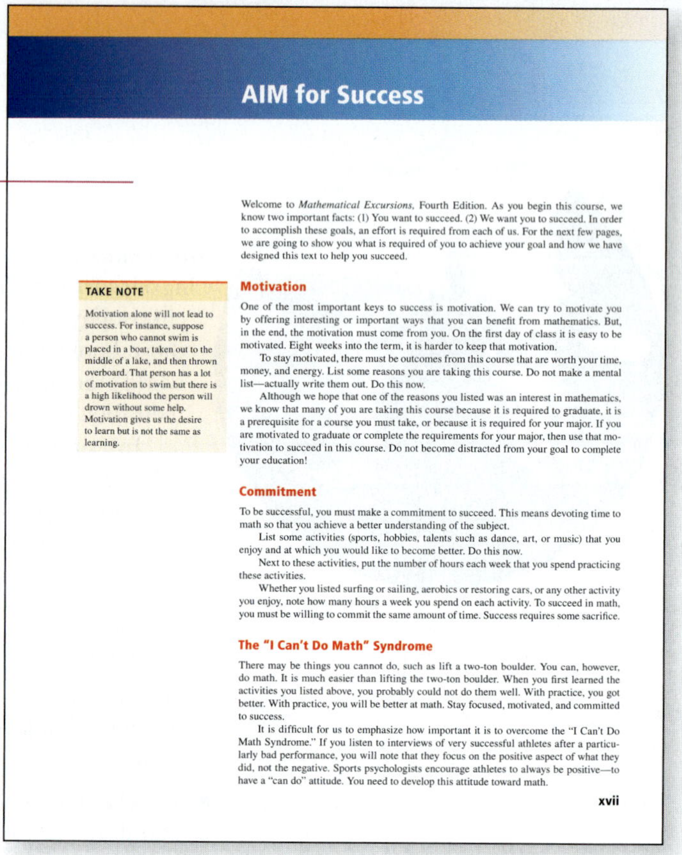

AIM for Success

Welcome to *Mathematical Excursions*, Fourth Edition. As you begin this course, we know two important facts: (1) You want to succeed. (2) We want you to succeed. In order to accomplish these goals, an effort is required from each of us. For the next few pages, we are going to show you what is required of you to achieve your goal and how we have designed this text to help you succeed.

TAKE NOTE

Motivation alone will not lead to success. For instance, suppose a person who cannot swim is placed in a boat, taken out to the middle of a lake, and then thrown overboard. That person has a lot of motivation to swim but there is a high likelihood the person will drown without some help. Motivation gives us the desire to learn but is not the same as learning.

Motivation

One of the most important keys to success is motivation. We can try to motivate you by offering interesting or important ways that you can benefit from mathematics. But, in the end, the motivation must come from you. On the first day of class it is easy to be motivated. Eight weeks into the term, it is harder to keep that motivation.

To stay motivated, there must be outcomes from this course that are worth your time, money, and energy. List some reasons you are taking this course. Do not make a mental list—actually write them out. Do this now.

Although we hope that one of the reasons you listed was an interest in mathematics, we know that many of you are taking this course because it is required to graduate, it is a prerequisite for a course you must take, or because it is required for your major. If you are motivated to graduate or complete the requirements for your major, then use that motivation to succeed in this course. Do not become distracted from your goal to complete your education!

Commitment

To be successful, you must make a commitment to succeed. This means devoting time to math so that you achieve a better understanding of the subject.

List some activities (sports, hobbies, talents such as dance, art, or music) that you enjoy and at which you would like to become better. Do this now.

Next to these activities, put the number of hours each week that you spend practicing these activities.

Whether you listed surfing or sailing, aerobics or restoring cars, or any other activity you enjoy, note how many hours a week you spend on each activity. To succeed in math, you must be willing to commit the same amount of time. Success requires some sacrifice.

The "I Can't Do Math" Syndrome

There may be things you cannot do, such as lift a two-ton boulder. You can, however, do math. It is much easier than lifting the two-ton boulder. When you first learned the activities you listed above, you probably could not do them well. With practice, you got better. With practice, you will be better at math. Stay focused, motivated, and committed to success.

It is difficult for us to emphasize how important it is to overcome the "I Can't Do Math Syndrome." If you listen to interviews of very successful athletes after a particularly bad performance, you will note that they focus on the positive aspect of what they did, not the negative. Sports psychologists encourage athletes to always be positive—to have a "can do" attitude. You need to develop this attitude toward math.

xvii

1

Problem Solving

Most occupations require good problem-solving skills. For instance, architects and engineers must solve many complicated problems as they design and construct modern buildings that are aesthetically pleasing, functional, and that meet stringent safety requirements. Two goals of this chapter are to help you become a better problem solver and to demonstrate that problem solving can be an enjoyable experience.

One problem that many have enjoyed is the Monty Hall (host of the game show *Let's Make a Deal*) problem, which is stated as follows. The grand prize in *Let's Make a Deal* is behind one of three doors. Less desirable prizes (for instance, a goat and a box of candy) are behind the other two doors. You select one of the doors, say door 1. Monty Hall reveals one of the less desirable prizes behind one of the other doors. You are then given the opportunity either to stay with your original choice or to choose the remaining closed door.

Example: You choose door 1. Monty Hall reveals a goat behind door 3. You can stay with door 1 or switch to door 2.

Marilyn vos Savant, author of the "Ask Marilyn" column featured in *Parade Magazine*, analyzed this problem,[1] claiming that you *double* your chances of winning the grand prize by switching to the other closed door. Many readers, including some mathematicians, responded with arguments that contradicted Marilyn's analysis.

What do you think? Do you have a better chance of winning the grand prize by switching to the other closed door or staying with your original choice?

Of course there is also the possibility that it does not matter, if the chances of winning are the same with either strategy.

Discuss the Monty Hall problem with some of your friends and classmates. Is everyone in agreement? Additional information on this problem is given in Exploration Exercise 54 on page 14.

1.1 Inductive and Deductive Reasoning
1.2 Problem Solving with Patterns
1.3 Problem-Solving Strategies

Marilyn vos Savant

[1] "Ask Marilyn," *Parade Magazine*, September 9, 1990, p. 15.

1

Each CHAPTER OPENER includes a list of sections that can be found within the chapter and includes an anecdote, description, or explanation that introduces the student to a topic in the chapter.

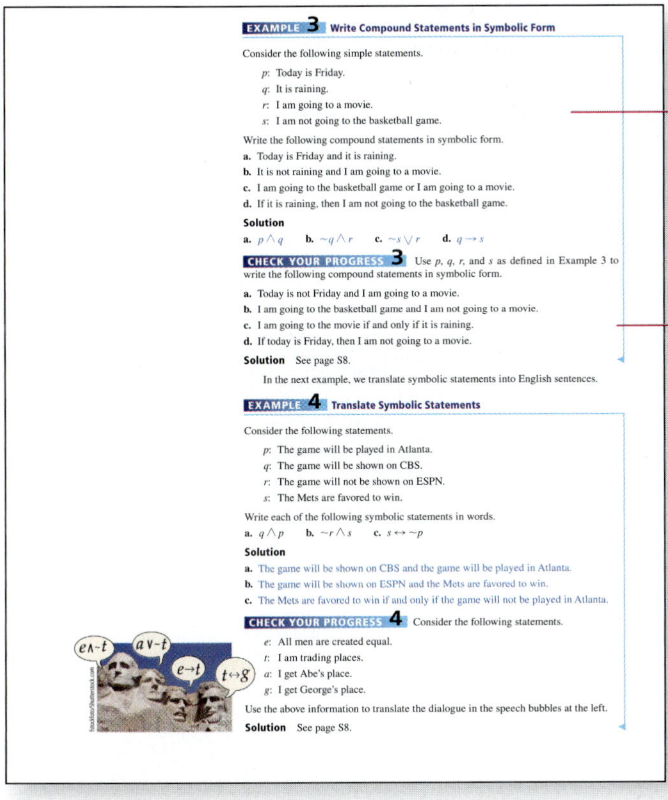
EXAMPLE 3 Write Compound Statements in Symbolic Form

Consider the following simple statements.

p: Today is Friday.
q: It is raining.
r: I am going to a movie.
s: I am not going to the basketball game.

Write the following compound statements in symbolic form.

a. Today is Friday and it is raining.
b. It is not raining and I am going to a movie.
c. I am going to the basketball game or I am going to a movie.
d. If it is raining, then I am not going to the basketball game.

Solution

a. $p \wedge q$ **b.** $\sim q \wedge r$ **c.** $\sim s \vee r$ **d.** $q \rightarrow s$

CHECK YOUR PROGRESS 3 Use p, q, r, and s as defined in Example 3 to write the following compound statements in symbolic form.

a. Today is not Friday and I am going to a movie.
b. I am going to the basketball game and I am not going to a movie.
c. I am going to the movie if and only if it is raining.
d. If today is Friday, then I am not going to a movie.

Solution See page S8.

In the next example, we translate symbolic statements into English sentences.

EXAMPLE 4 Translate Symbolic Statements

Consider the following statements.

p: The game will be played in Atlanta.
q: The game will be shown on CBS.
r: The game will not be shown on ESPN.
s: The Mets are favored to win.

Write each of the following symbolic statements in words.

a. $q \wedge p$ **b.** $\sim r \wedge s$ **c.** $s \leftrightarrow \sim p$

Solution

a. The game will be shown on CBS and the game will be played in Atlanta.
b. The game will be shown on ESPN and the Mets are favored to win.
c. The Mets are favored to win if and only if the game will not be played in Atlanta.

CHECK YOUR PROGRESS 4 Consider the following statements.

e: All men are created equal.
t: I am trading places.
a: I get Abe's place.
g: I get George's place.

Use the above information to translate the dialogue in the speech bubbles at the left.

Solution See page S8.

Each section contains a variety of WORKED EXAMPLES. Each example is given a title so that the student can see at a glance the type of problem that is being solved. Most examples include annotations that assist the student in moving from step to step, and the final answer is in color in order to be readily identifiable.

Following each worked example is a CHECK YOUR PROGRESS exercise for the student to work. By solving this exercise, the student actively practices concepts as they are presented in the text. For each Check Your Progress exercise, there is a detailed solution in the Solutions appendix.

At various places throughout the text, a QUESTION is posed about the topic that is being discussed. This question encourages students to pause, think about the current discussion, and answer the question. Students can immediately check their understanding by referring to the ANSWER to the question provided in a footnote on the same page. This feature creates another opportunity for the student to interact with the textbook.

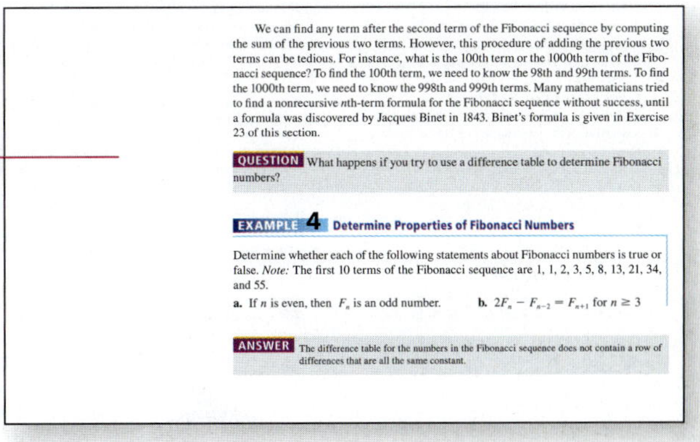
We can find any term after the second term of the Fibonacci sequence by computing the sum of the previous two terms. However, this procedure of adding the previous two terms can be tedious. For instance, what is the 100th term or the 1000th term of the Fibonacci sequence? To find the 100th term, we need to know the 98th and 99th terms. To find the 1000th term, we need to know the 998th and 999th terms. Many mathematicians tried to find a nonrecursive nth-term formula for the Fibonacci sequence without success, until a formula was discovered by Jacques Binet in 1843. Binet's formula is given in Exercise 23 of this section.

QUESTION What happens if you try to use a difference table to determine Fibonacci numbers?

EXAMPLE 4 Determine Properties of Fibonacci Numbers

Determine whether each of the following statements about Fibonacci numbers is true or false. *Note:* The first 10 terms of the Fibonacci sequence are 1, 1, 2, 3, 5, 8, 13, 21, 34, and 55.

a. If n is even, then F_n is an odd number. **b.** $2F_n - F_{n-2} = F_{n+1}$ for $n \geq 3$

ANSWER The difference table for the numbers in the Fibonacci sequence does not contain a row of differences that are all the same constant.

Each section ends with an EXCURSION along with corresponding EXCURSION EXERCISES. These activities engage students in the mathematics of the section. Some Excursions are designed as in-class cooperative learning activities that lend themselves to a hands-on approach. They can also be assigned as projects or extra credit assignments. The Excursions are a unique and important feature of this text. They provide opportunities for students to take an active role in the learning process.

EXCURSION

Earned Run Average

One measure of a pitcher's success is earned run average. **Earned run average (ERA)** is the number of earned runs a pitcher gives up for every nine innings pitched. The definition of an earned run is somewhat complicated, but basically an earned run is a run that is scored as a result of hits and base running that involves no errors on the part of the pitcher's team. If the opposing team scores a run on an error (for example, a fly ball that should have been caught in the outfield was fumbled), then that run is not an earned run.

A proportion is used to calculate a pitcher's ERA. Remember that the statistic involves the number of earned runs per *nine innings*. The answer is always rounded to the nearest hundredth. Here is an example.

During the 2015 baseball season, Clayton Kershaw gave up 55 earned runs and pitched 232.2 innings for the Los Angeles Dodgers. To calculate Clayton Kershaw's ERA, let x = the number of earned runs for every nine innings pitched. Write a proportion and then solve it for x.

$$\frac{55 \text{ earned runs}}{232.2 \text{ innings}} = \frac{x}{9 \text{ innings}}$$
$$55 \cdot 9 = 232.2 \cdot x$$
$$495 = 232.2x$$
$$\frac{495}{232.2} = \frac{232.2x}{232.2}$$
$$2.13 \approx x$$

Photo Works/Shutterstock.com
Clayton Kershaw

Clayton Kershaw's ERA for the 2015 season was 2.13.

Earned Run Average Leaders		
Major League Baseball		
Year	Player, club	ERA
2005	Roger Clemens, Houston	1.87
2006	Johan Santana, Minnesota	2.77
2007	Jake Peavy, San Diego	2.54
2008	Johan Santana, New York	2.53
2009	Zack Greinke, Kansas City	2.16
2010	Felix Hernandez, Seattle	2.27
2011	Clayton Kershaw, Los Angeles	2.28
2012	Clayton Kershaw, Los Angeles	2.53
2013	Clayton Kershaw, Los Angeles	1.83
2014	Clayton Kershaw, Los Angeles	1.77
2015	Zack Greinke, Los Angeles	1.66

EXCURSION EXERCISES

1. In 1979, his rookie year, Jeff Reardon pitched 21 innings for the New York Mets and gave up four earned runs. Calculate Reardon's ERA for 1979.
2. Roger Clemens's first year with the Boston Red Sox was 1984. During that season, he pitched 133.1 innings and gave up 64 earned runs. Calculate Clemens's ERA for 1984.
3. In 1987, Nolan Ryan had the lowest ERA of any pitcher in the major leagues. He gave up 65 earned runs and pitched 211.2 innings for the Houston Astros. Calculate Ryan's ERA for 1987.
4. During the 2015 season, Jake Arrieta of the Baltimore Orioles pitched 229 innings and had an ERA of 1.77. How many earned runs did he give up during the season?
5. Find the necessary statistics for a pitcher on your "home team," and calculate that pitcher's ERA.

EXERCISE SET 4.3

■ In the following exercises that involve weighted voting systems for voters A, B, C, ..., the systems are given in the form $\{q: w_1, w_2, w_3, w_4, \ldots, w_n\}$. The weight of voter A is w_1, the weight of voter B is w_2, the weight of voter C is w_3, and so on.

1. A weighted voting system is given by {6: 4, 3, 2, 1}.
   a. What is the quota?
   b. How many voters are in this system?
   c. What is the weight of voter B?
   d. What is the weight of the coalition {A, C}?
   e. Is {A, D} a winning coalition?
   f. Which voters are critical voters in the coalition {A, C, D}?
   g. How many coalitions can be formed?
   h. How many coalitions consist of exactly two voters?
2. A weighted voting system is given by {16: 8, 7, 4, 2, 1}.
   a. What is the quota?
   b. How many voters are in this system?
   c. What is the weight of voter C?
   d. What is the weight of the coalition {B, C}?
   e. Is {B, C, D, E} a winning coalition?
   f. Which voters are critical voters in the coalition {A, B, D}?
   g. How many coalitions can be formed?
   h. How many coalitions consist of exactly three voters?

■ In Exercises 3 to 12, calculate, if possible, the Banzhaf power index for each voter. Round to the nearest hundredth.

3. {6: 4, 3, 2}
4. {10: 7, 6, 4}
5. {10: 7, 3, 2, 1}
6. {14: 7, 5, 1, 1}
7. {19: 14, 12, 4, 3, 1}
8. {3: 1, 1, 1, 1}
9. {18: 18, 7, 3, 3, 1, 1}
10. {14: 6, 6, 4, 3, 1}
11. {80: 50, 40, 30, 25, 5}
12. {85: 55, 40, 25, 5}
13. Which, if any, of the voting systems in Exercises 3 to 12 is
    a. a dictatorship?
    b. a veto power system? *Note:* A voting system is a veto power system if any of the voters has veto power.
    c. a null system?
    d. a one-person, one-vote system?
14. Explain why it is impossible to calculate the Banzhaf power index for any voter in the null system {8: 3, 2, 1, 1}.
15. **Music Education** A music department consists of a band director and a music teacher. Decisions on motions are made by voting. If both members vote in favor of a motion, it passes. If both members vote against a motion, it fails. In the event of a tie vote, the principal of the school votes to break the tie. For this voting scheme, determine the Banzhaf power index for each department member and for the principal. *Hint:* See Example 3, page 214.
16. Four voters, A, B, C, and D, make decisions by using the voting scheme {4: 3, 1, 1, 1}, except when there is a tie. In the event of a tie, a fifth voter, E, casts a vote to break the tie. For this voting scheme, determine the Banzhaf power index for each voter, including voter E. *Hint:* See Example 3, page 214.
17. **Criminal Justice** In a criminal trial, each of the 12 jurors has one vote and all of the jurors must agree to reach a verdict. Otherwise the judge will declare a mistrial.
    a. Write the weighted voting system, in the form $\{q: w_1, w_2, w_3, w_4, \ldots, w_{12}\}$, used by these jurors.
    b. Is this weighted voting system a one-person, one-vote system?
    c. Is this weighted voting system a veto power system?
    d. Explain an easy way to determine the Banzhaf power index for each voter.
18. **Criminal Justice** In California civil court cases, each of the 12 jurors has one vote and at least 9 of the jury members must agree on the verdict.

    a. Write the weighted voting system, in the form $\{q: w_1, w_2, w_3, w_4, \ldots, w_{12}\}$, used by these jurors.
    b. Is this weighted voting system a one-person, one-vote system?
    c. Is this weighted voting system a veto power system?
    d. Explain an easy way to determine the Banzhaf power index for each voter.

The EXERCISE SETS were carefully written to provide a wide variety of exercises that range from drill and practice to interesting challenges. Exercise sets emphasize skill building, skill maintenance, concepts, and applications. Icons are used to identify various types of exercises.

 Writing exercises

 Data analysis exercises

 Graphing calculator exercises

Exercises that require the Internet

EXTENSIONS EXERCISES are placed at the end of each exercise set. These exercises are designed to extend concepts. In most cases these exercises are more challenging and require more time and effort than the preceding exercises.

18. **Fitness** An aerobic exercise instructor remembers the data given in the following table, which shows the recommended maximum exercise heart rates for individuals of the given ages.

Age (x years)	20	40	60
Maximum heart rate (y beats per minute)	170	153	136

a. Find the linear correlation coefficient for the data.
b. What is the significance of the value found in part a?
c. Find the equation of the least-squares line.
d. Use the equation from part c to predict the maximum exercise heart rate for a person who is 72.
e. Is the procedure in part d an example of interpolation or extrapolation?

EXTENSIONS

19. **Tuition** The following table shows the average annual tuition and fees at private and public 4-year colleges and universities for the school years 2009–2010 through 2014–2015. (*Source:* National Center for Education Statistics)

Four-year Colleges and Universities Tuition and Fees

Year	Private	Public
2009–2010	31,448	15,014
2010–2011	32,617	15,918
2011–2012	33,674	16,805
2012–2013	35,074	17,474
2013–2014	36,193	18,372
2014–2015	37,385	19,203

a. Using 1 for 2009–2010, 2 for 2010–2011, and so on, find the linear correlation coefficient and the equation of the least-squares line for the tuition and fees at private 4-year colleges and universities, based on the year.
b. Using 1 for 2009–2010, 2 for 2010–2011, and so on, find the linear correlation coefficient and the equation of the least-squares line for the tuition and fees at public 4-year colleges and universities, based on the year.
c. Based on the linear correlation coefficients you found in parts a and b, are the equations you wrote in parts a and b good models of the growth in tuition and fees at 4-year colleges and universities?
d. The equation of a least-squares line is written in the form $\hat{y} = ax + b$. Explain the meaning of the value of a for each equation you wrote in parts a and b.

20. Search for bivariate data (in a magazine, in a newspaper, in an almanac, or on the Internet) that can be closely modeled by a linear equation.
a. Draw a scatter diagram of the data.
b. Find the equation of the least-squares line and the linear correlation coefficient for the data.
c. Graph the least-squares line on the scatter diagram in part a.
d. Use the equation of the least-squares line to predict a range value for a specific domain value.

At the end of each chapter is a CHAPTER SUMMARY that describes the concepts presented in each section of the chapter. Each concept is paired with page numbers of examples that illustrate the concept and exercises that students can use to test their understanding of a concept.

CHAPTER 2 **SUMMARY**

The following table summarizes essential concepts in this chapter. The references given in the right-hand column list Examples and Exercises that can be used to test your understanding of a concept.

2.1 Basic Properties of Sets	
The Roster Method The roster method is used to represent a set by listing each element of the set inside a pair of braces. Commas are used to separate the elements.	See **Example 1** on page 48, and then try Exercises 1 and 2 on page 101.
Basic Number Sets Natural Numbers or Counting Numbers $N = \{1, 2, 3, 4, 5, ...\}$ Whole Numbers $W = \{0, 1, 2, 3, 4, 5, ...\}$ Integers $I = \{..., -4, -3, -2, -1, 0, 1, 2, 3, 4, ...\}$ Rational Numbers Q = the set of all terminating or repeating decimals Irrational Numbers $\mathcal{I}$ = the set of all nonterminating, nonrepeating decimals Real Numbers R = the set of all rational or irrational numbers	See **Example 3** and **Check Your Progress 3** on page 49, and then try Exercises 3 to 6 on page 101.
Set-Builder Notation Set-builder notation is used to represent a set, by describing its elements.	See **Example 5** on page 50, and then try Exercises 7 to 10 on page 101.
Cardinal Number of a Finite Set The cardinal number of a finite set is the number of elements in the set. The cardinal number of a finite set A is denoted by the notation $n(A)$.	See **Example 6** on page 51, and then try Exercises 63 to 67 on page 103.
Equal Sets and Equivalent Sets Two sets are equal if and only if they have exactly the same elements. Two sets are equivalent if and only if they have the same number of elements.	See **Example 7** on page 52, and then try Exercises 11 and 12 on page 101.

continued

CHAPTER REVIEW EXERCISES are found near the end of each chapter. These exercises were selected to help the student integrate the major topics presented in the chapter. The answers to all the Chapter Review exercises appear in the answer section along with a section reference for each exercise. These section references indicate the section or sections where a student can locate the concepts needed to solve the exercise.

CHAPTER 1 **REVIEW EXERCISES**

- In Exercises 1 to 4, determine whether the argument is an example of inductive reasoning or deductive reasoning.

1. All books written by J. K. Rowling make the best-seller list. The book *Harry Potter and the Deathly Hallows* is a J. K. Rowling book. Therefore, *Harry Potter and the Deathly Hallows* made the bestseller list.

2. Samantha got an A on each of her first four math tests, so she will get an A on the next math test.

3. We had rain each day for the last five days, so it will rain today.

4. All amoeba multiply by dividing. I have named the amoeba shown in my microscope Amelia. Therefore, Amelia multiplies by dividing.

5. Find a counterexample to show that the following conjecture is false.
Conjecture: For all numbers x, $x^4 > x$.

6. Find a counterexample to show that the following conjecture is false.
Conjecture: For all counting numbers n, $\frac{n^3 + 5n + 6}{6}$ is an even counting number.

7. Find a counterexample to show that the following conjecture is false.
Conjecture: For all numbers x, $(x + 4)^2 = x^2 + 16$.

8. Find a counterexample to show that the following conjecture is false.
Conjecture: For numbers a and b, $(a + b)^3 = a^3 + b^3$.

9. Use a difference table to predict the next term of each sequence.
a. −2, 2, 12, 28, 50, 78, ?
b. −4, −1, 14, 47, 104, 191, 314, ?

CHAPTER 1 REVIEW EXERCISES *page 41*

1. deductive [Sec. 1.1] **2.** inductive [Sec. 1.1] **3.** inductive [Sec. 1.1] **4.** deductive [Sec. 1.1] **5.** $x = 0$ provides a counterexample because $0^4 = 0$ and 0 is not greater than 0. [Sec. 1.1] **6.** $x = 4$ provides a counterexample because $\frac{(4)^3 + 5(4) + 6}{6} = 15$, which is not an even counting number. [Sec. 1.1] **7.** $x = 1$ provides a counterexample because $[(1) + 4]^2 = 25$, but $(1)^2 + 16 = 17$. [Sec. 1.1] **8.** Let $a = 1$ and $b = 1$. Then $(a + b)^3 = (1 + 1)^3 = 2^3 = 8$. However, $a^3 + b^3 = 1^3 + 1^3 = 2$. [Sec. 1.1] **9.** **a.** 112 **b.** 479 [Sec. 1.2] **10.** **a.** −72 **b.** −768 [Sec. 1.2] **11.** $a_1 = 1$, $a_2 = 12$, $a_3 = 31$, $a_4 = 58$, $a_5 = 93$, $a_{20} = 1578$ [Sec. 1.2] **12.** $a_{11} = 89$, $a_{12} = 144$ [Sec. 1.2] **13.** $a_n = 3n$ [Sec. 1.2] **14.** $a_n = n^2 + 3n + 4$ [Sec. 1.2] **15.** $a_n = n^2 + 3n + 2$ [Sec. 1.2] **16.** $a_n = 5n - 1$ [Sec. 1.2] **17.** 320 feet by 1600 feet [Sec. 1.3] **18.** $3^{15} = 14{,}348{,}907$ ways [Sec. 1.3] **19.** 48 skyboxes [Sec. 1.3] **20.** On the first trip, the rancher takes the rabbit across the river. The rancher returns alone. The rancher takes the dog across the river and returns with the rabbit. The rancher next takes the carrots across the river and returns alone. On the final trip, the rancher takes the rabbit across the river. [Sec. 1.3] **21.** \$300 [Sec. 1.3] **22.** 105 handshakes [Sec. 1.3] **23.** Answers will vary. [Sec. 1.3] **24.** Answers will vary. [Sec. 1.3] **25.** Michael: biology major; Clarissa: business major; Reggie: computer science major; Ellen: chemistry major [Sec. 1.1] **26.** Dodgers: drugstore; Pirates: supermarket; Tigers: bank; Giants: service station [Sec. 1.1] **27.** **a.** Yes. Answers will vary. **b.** No. The countries of India, Bangladesh, and Myanmar all share borders with each of the other two countries. Thus at least three colors are needed to color the map. [Sec. 1.1] **28.** **a.** The following figure shows a route that starts from North Bay and passes over each bridge once and only once. **b.** No. [Sec. 1.3]

The CHAPTER TEST exercises are designed to emulate a possible test of the material in the chapter. The answers to all the Chapter Test exercises appear in the answer section along with a section reference and an example reference for each exercise. The section references indicate the section or sections where a student can locate the concepts needed to solve the exercise, and the example references allow students to readily find an example that is similar to a given test exercise.

CHAPTER 3 TEST

1. Determine whether each sentence is a statement.
   a. Look for the cat.
   b. Clark Kent is afraid of the dark.
2. Write the negation of each statement. Start each negation with "Some," "No," or "All."
   a. Some trees are not green.
   b. No apartments are available.
3. Determine whether each statement is true or false.
   a. $5 \leq 4$
   b. $-2 \geq -2$
4. Determine the truth value of each statement given that p is true, q is false, and r is true.
   a. $(p \vee \sim q) \wedge (\sim r \wedge q)$
   b. $(r \vee \sim p) \vee [(p \vee \sim q) \leftrightarrow (q \rightarrow r)]$

■ In Exercises 5 and 6, construct a truth table for the given statement.

5. $\sim(p \wedge \sim q) \vee (q \rightarrow p)$ 6. $(r \leftrightarrow \sim q) \wedge (p \rightarrow q)$
7. Use one of De Morgan's laws to write the following in an equivalent form.

   Elle did not eat breakfast and she did not take a lunch break.
8. What is a tautology?
9. Write $p \rightarrow q$ in its equivalent disjunctive form.
10. Determine whether the given statement is true or false. Assume that x, y, and z are real numbers.
    a. $x = y$ if $|x| = |y|$. b. If $x > y$, then $xz > yz$.

CHAPTER 3 TEST *page 166*

1. a. Not a statement b. Statement [Sec. 3.1, Example 1] 2. a. All trees are green. b. Some apartments are available. [Sec. 3.1, Example 2] 3. a. False b. True [Sec. 3.1, Example 6] 4. a. False b. True [Sec. 3.3, Example 3]

p	q	5. [Sec. 3.3, Example 3]
T	T	T
T	F	T
F	T	T
F	F	T

p	q	r	6. [Sec. 3.3, Example 3]
T	T	T	F
T	T	F	T
T	F	T	F
T	F	F	F
F	T	T	F
F	T	F	T
F	F	T	T
F	F	F	F

7. It is not true that Elle ate breakfast or took a lunch break. [Sec. 3.2, Example 5]

Other Key Features

MATH MATTERS The Barber's Paradox

PictureQuest/Corbis Images

Some problems that concern sets have led to paradoxes. For instance, in 1902, the mathematician Bertrand Russell developed the following paradox. "Is the set A of all sets that are not elements of themselves an element of itself?" Both the assumption that A is an element of A and the assumption that A is not an element of A lead to a contradiction. Russell's paradox has been popularized as follows.

> The town barber shaves all males who do not shave themselves, and he shaves only those males. The town barber is a male who shaves. Who shaves the barber?

The assumption that the barber shaves himself leads to a contradiction, and the assumption that the barber does not shave himself also leads to a contradiction.

Math Matters

This feature of the text typically contains an interesting sidelight about mathematics, its history, or its applications.

HISTORICAL NOTE

Hutton Archive/Handout/Getty Images

Pythagoras (c. 580 B.C.–520 B.C.) The ancient Greek philosopher and mathematician Pythagoras (pī-thăg'ər-əs) formed a secret brotherhood that investigated topics in music, astronomy, philosophy, and mathematics. The Pythagoreans believed that the nature of the universe was directly related to mathematics and that whole numbers and the ratios formed by whole numbers could be used to describe and represent all natural events.

The Pythagoreans were particularly intrigued by the number 5 and the shape of a pentagon. They used the following figure, which is a five-pointed star inside a regular pentagon, as a secret symbol that could be used to identify other members of the brotherhood.

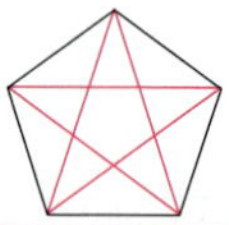

Historical Note

These margin notes provide historical background information related to the concept under discussion or vignettes of individuals who were responsible for major advancements in their fields of expertise.

POINT OF INTEREST

M.C. Escher

Waterfall by M.C. Escher

M.C. Escher (1898–1972) created many works of art that defy logic. In this lithograph, the water completes a full cycle even though the water is always traveling downward.

Point of Interest

These short margin notes provide interesting information related to the mathematical topics under discussion. Many of these are of a contemporary nature and, as such, they help students understand that math is an interesting and dynamic discipline that plays an important role in their daily lives.

TAKE NOTE

The alternative procedure for constructing a truth table, as described to the right, generally requires less writing, less time, and less effort than the truth table procedure that was used in Examples 1 and 2.

Take Note

These notes alert students to a point requiring special attention, or they are used to amplify the concepts currently being developed.

CALCULATOR NOTE

Some calculators display $\frac{7}{27}$ as 0.259259259 26. However, the last digit 6 is not correct. It is a result of the rounding process. The actual decimal representation of $\frac{7}{27}$ is the decimal 0.259259... or $0.\overline{259}$, in which the digits continue to repeat the 259 pattern forever.

Calculator Note

These notes provide information about how to use the various features of a calculator.

Instructor Resources

Annotated Instructor's Edition (ISBN 978-1-305-96559-1): The Annotated Instructor's Edition features answers to all problems in the book.

Complete Solutions Manual: This manual contains complete solutions to all the problems in the text. Available on the Instructor Companion Site.

MindTap: Through personalized paths of dynamic assignments and applications, MindTap is a digital learning solution and representation of your course that turns cookie cutter into cutting edge, apathy into engagement, and memorizers into higher-level thinkers.

The Right Content: With MindTap's carefully curated material, you get the precise content and groundbreaking tools you need for every course you teach.

Personalization: Customize every element of your course—from rearranging the learning path to inserting videos and activities.

Improved Workflow: Save time when planning lessons with all of the trusted, most current content you need in one place in MindTap.

Tracking Students' Progress in Real Time: Promote positive outcomes by tracking students in real time and tailoring your course as needed based on the analytics.

Learn more at www.cengage.com/mindtap.

Cognero (ISBN: 978-1-305-96565-2): Cengage Learning Testing Powered by Cognero is a flexible, online system that allows you to author, edit, and manage test bank content from multiple Cengage Learning solutions; create multiple test versions in an instant; and deliver tests from your LMS, your classroom, or wherever you want. Access to Cognero is available on the Instructor Companion Site.

Instructor Companion Site: This collection of book-specific lecture and class tools is available online at www.cengage.com/login. Access and download PowerPoint presentations, the solutions manual, and more.

Student Resources

Student Solutions Manual (ISBN: 978-1-305-96561-4): Go beyond the answers—see what it takes to get there and improve your grade! This manual provides worked-out, step-by-step solutions to the odd-numbered problems in the text. You'll have the information you need to truly understand how the problems are solved.

MindTap: MindTap is a digital representation of your course that provides you with the tools you need to better manage your limited time, stay organized, and be successful. You can complete assignments whenever and wherever you are ready to learn, with course material specially customized for you by your instructor and streamlined in one proven, easy-to-use interface. With an array of study tools, you'll get a true understanding of course concepts, achieve better grades, and set the groundwork for your future courses. Learn more at www.cengage.com/mindtap.

CengageBrain: Visit www.cengagebrain.com to access additional course materials and companion resources. At the CengageBrain.com home page, search for the ISBN of your title (from the back cover of your book) using the search box at the top of the page. This will take you to the product page where free companion resources can be found.

Acknowledgments

The authors would like to thank the people who have contributed many valuable suggestions over the years, including for this most recent edition.

Brenda Alberico, *College of DuPage*

Beverly R. Broomell, *Suffolk County Community College*

Duff Campbell, *Hendrix College*

Donald Cater, *Monroe Community College*

Henjin Chi, *Indiana State University*

Ivette Chuca, *El Paso Community College*

Marcella Cremer, *Richland Community College*

Amy Curry, *College of Lake County*

Houbin Fang, *Columbus State University*

Margaret Finster, *Erie Community College*

Kenny Fister, *Murray State University*

Luke Foster, *Northeastern State University*

Rita Fox, *Kalamazoo Valley Community College*

Sue Grapevine, *Northwest Iowa Community College*

Shane Griffith, *Lee University*

Elizabeth Henkle, *Longview Community College*

Charles Huelsman, *Marylhurst University*

Robert Jajcay, *Indiana State University*

Dr. Nancy R. Johnson, *Manatee Community College*

Mary Juliano, *Caldwell University*

Brian Karasek, *South Mountain Community College*

Dr. Vernon Kays, *Richland Community College*

L. Christine Kinsey, *Canisius University*

Dr. Suda Kunyosying, *Shepherd College*

Kathryn Lavelle, *Westchester Community College*

Colleen Livingston, *Bemidji University*

Roger Marty, *Cleveland State University*

Eric Matsuoka, *Leeward Community College*

Beverly Meyers, *Jefferson College*

Dr. Alec Mihailovs, *Shepherd University*

Leona Mirza, *North Park University*

Bette Nelson, *Alvin Community College*

Sara Ngo, *Anoka Ramsey Community College*

Kathleen Offenholley, *Brookdale Community College*

Kathy Pinchback, *University of Memphis*

Michael Polley, *Southeastern Community College*

Dr. Anne Quinn, *Edinboro University of Pennsylvania*

Brenda Reed, *Navarro College*

Marc Renault, *Shippensburg University*

Christopher Rider, *North Greenville University*

Cynthia Roemer, *Union County College*

Richard D. Rupp, *Del Mar College*

Sharon M. Saxton, *Cascadia Community College*

Mary Lee Seitz, *Erie Community College–City Campus*

Lars Seme, *Hendrix College*

Jaime Shinn, *Gonzaga University*

Dr. Sue Stokley, *Spartanburg Technical College*

Dr. Julie M. Theoret, *Lyndon State College*

Walter Jacob Theurer, *Fulton Montgomery Community College*

Jamie Thomas, *University of Wisconsin Colleges–Manitowoc*

William Twentyman, *ECPI College of Technology*

Nancy Vendeville, *Kalamazoo Valley Community College*

Denise A. Widup, *University of Wisconsin–Parkside*

Nancy Wilson, *Marshall University*

Jane-Marie Wright, *Suffolk Community College*

Diane Zych, *Erie Community College*

AIM for Success

Welcome to *Mathematical Excursions,* Fourth Edition. As you begin this course, we know two important facts: (1) You want to succeed. (2) We want you to succeed. In order to accomplish these goals, an effort is required from each of us. For the next few pages, we are going to show you what is required of you to achieve your goal and how we have designed this text to help you succeed.

TAKE NOTE

Motivation alone will not lead to success. For instance, suppose a person who cannot swim is placed in a boat, taken out to the middle of a lake, and then thrown overboard. That person has a lot of motivation to swim but there is a high likelihood the person will drown without some help. Motivation gives us the desire to learn but is not the same as learning.

Motivation

One of the most important keys to success is motivation. We can try to motivate you by offering interesting or important ways that you can benefit from mathematics. But, in the end, the motivation must come from you. On the first day of class it is easy to be motivated. Eight weeks into the term, it is harder to keep that motivation.

To stay motivated, there must be outcomes from this course that are worth your time, money, and energy. List some reasons you are taking this course. Do not make a mental list—actually write them out. Do this now.

Although we hope that one of the reasons you listed was an interest in mathematics, we know that many of you are taking this course because it is required to graduate, it is a prerequisite for a course you must take, or because it is required for your major. If you are motivated to graduate or complete the requirements for your major, then use that motivation to succeed in this course. Do not become distracted from your goal to complete your education!

Commitment

To be successful, you must make a commitment to succeed. This means devoting time to math so that you achieve a better understanding of the subject.

List some activities (sports, hobbies, talents such as dance, art, or music) that you enjoy and at which you would like to become better. Do this now.

Next to these activities, put the number of hours each week that you spend practicing these activities.

Whether you listed surfing or sailing, aerobics or restoring cars, or any other activity you enjoy, note how many hours a week you spend on each activity. To succeed in math, you must be willing to commit the same amount of time. Success requires some sacrifice.

The "I Can't Do Math" Syndrome

There may be things you cannot do, such as lift a two-ton boulder. You can, however, do math. It is much easier than lifting the two-ton boulder. When you first learned the activities you listed above, you probably could not do them well. With practice, you got better. With practice, you will be better at math. Stay focused, motivated, and committed to success.

It is difficult for us to emphasize how important it is to overcome the "I Can't Do Math Syndrome." If you listen to interviews of very successful athletes after a particularly bad performance, you will note that they focus on the positive aspect of what they did, not the negative. Sports psychologists encourage athletes to always be positive—to have a "can do" attitude. You need to develop this attitude toward math.

Strategies for Success

Know the Course Requirements To do your best in this course, you must know exactly what your instructor requires. Course requirements may be stated in a *syllabus*, which is a printed outline of the main topics of the course, or they may be presented orally. When they are listed in a syllabus or on other printed pages, keep them in a safe place. When they are presented orally, make sure to take complete notes. In either case, it is important that you understand them completely and follow them exactly. Be sure you know the answer to each of the following questions.

1. What is your instructor's name?
2. Where is your instructor's office?
3. At what times does your instructor hold office hours?
4. Besides the textbook, what other materials does your instructor require?
5. What is your instructor's attendance policy?
6. If you must be absent from a class meeting, what should you do before returning to class? What should you do when you return to class?
7. What is the instructor's policy regarding collection or grading of homework assignments?
8. What options are available if you are having difficulty with an assignment? Is there a math tutoring center?
9. If there is a math lab at your school, where is it located? What hours is it open?
10. What is the instructor's policy if you miss a quiz?
11. What is the instructor's policy if you miss an exam?
12. Where can you get help when studying for an exam?

Remember: Your instructor wants to see you succeed. If you need help, ask! Do not fall behind. If you were running a race and fell behind by 100 yards, you may be able to catch up, but it will require more effort than if you had not fallen behind.

TAKE NOTE

Besides time management, there must be realistic ideas of how much time is available. There are very few people who can *successfully* work full-time and go to school full-time. If you work 40 hours a week, take 15 units, spend the recommended study time given at the right, and sleep 8 hours a day, you use over 80% of the available hours in a week. That leaves less than 20% of the hours in a week for family, friends, eating, recreation, and other activities.

Time Management We know that there are demands on your time. Family, work, friends, and entertainment all compete for your time. We do not want to see you receive poor job evaluations because you are studying math. However, it is also true that we do not want to see you receive poor math test scores because you devoted too much time to work. When several competing and important tasks require your time and energy, the only way to manage the stress of being successful at both is to manage your time efficiently.

Instructors often advise students to spend twice the amount of time outside of class studying as they spend in the classroom. Time management is important if you are to accomplish this goal and succeed in school. The following activity is intended to help you structure your time more efficiently.

Take out a sheet of paper and list the names of each course you are taking this term, the number of class hours each course meets, and the number of hours you should spend outside of class studying course materials. Now create a weekly calendar with the days of the week across the top and each hour of the day in a vertical column. Fill in the calendar with the hours you are in class, the hours you spend at work, and other commitments such as sports practice, music lessons, or committee meetings. Then fill in the hours that are more flexible, such as study time, recreation, and meal times.

	Monday	Tuesday	Wednesday	Thursday	Friday	Saturday	Sunday
10–11 A.M.	History	Rev Spanish	History	Rev Span Vocab	History	Jazz Band	
11–12 P.M.	Rev History	Spanish	Study group	Spanish	Math tutor	Jazz Band	
12–1 P.M.	Math		Math		Math		Soccer

We know that many of you must work. If that is the case, realize that working 10 hours a week at a part-time job is equivalent to taking a three-unit class. If you must work, consider letting your education progress at a slower rate to allow you to be successful at both work and school. There is no rule that says you must finish school in a certain time frame.

Schedule Study Time As we encouraged you to do by filling out the time management form, schedule a certain time to study. You should think of this time like being at work or class. Reasons for "missing study time" should be as compelling as reasons for missing work or class. "I just didn't feel like it" is not a good reason to miss your scheduled study time. Although this may seem like an obvious exercise, list a few reasons you might want to study. Do this now.

Of course we have no way of knowing the reasons you listed, but from our experience one reason given quite frequently is "To pass the course." There is nothing wrong with that reason. If that is the most important reason for you to study, then use it to stay focused.

One method of keeping to a study schedule is to form a ***study group***. Look for people who are committed to learning, who pay attention in class, and who are punctual. Ask them to join your group. Choose people with similar educational goals but different methods of learning. You can gain from seeing the material from a new perspective. Limit groups to four or five people; larger groups are unwieldy.

There are many ways to conduct a study group. Begin with the following suggestions and see what works best for your group.

1. Test each other by asking questions. Each group member might bring two or three sample test questions to each meeting.
2. Practice teaching each other. Many of us who are teachers learned a lot about our subject when we had to explain it to someone else.
3. Compare class notes. You might ask other students about material in your notes that is difficult for you to understand.
4. Brainstorm test questions.
5. Set an agenda for each meeting. Set approximate time limits for each agenda item and determine a quitting time.

And now, probably the most important aspect of studying is that it should be done in relatively small chunks. If you can study only three hours a week for this course (probably not enough for most people), do it in blocks of one hour on three separate days, preferably after class. Three hours of studying on a Sunday is not as productive as three hours of paced study.

Features of This Text That Promote Success

Preparing for Class Before the class meeting in which your professor begins a new chapter, you should read the title of each section. Next, browse through the chapter material, being sure to note each word in bold type. These words indicate important concepts that you must know to learn the material. Do not worry about trying to understand all the material. Your professor is there to assist you with that endeavor. The purpose of browsing through the material is so that your brain will be prepared to accept and organize the new information when it is presented to you.

Math Is Not a Spectator Sport To learn mathematics you must be an active participant. Listening and watching your professor do mathematics is not enough. Mathematics requires that you interact with the lesson you are studying. If you have been writing down the things we have asked you to do, you were being interactive. There are other ways this textbook has been designed so that you can be an active learner.

Check Your Progress One of the key instructional features of this text is a completely worked-out example followed by a *Check Your Progress.*

EXAMPLE 8 **Applications of the Blood Transfusion Table**

Use the blood transfusion table and Figures 2.3 and 2.4 to answer the following questions.

a. Can Sue safely be given a type O+ blood transfusion?

b. Why is a person with type O− blood called a *universal donor?*

Solution

a. Sue's blood type is A−. The blood transfusion table shows that she can safely receive blood only if it is type A− or type O−. Thus it is not safe for Sue to receive type O+ blood in a blood transfusion.

b. The blood transfusion table shows that all eight blood types can safely receive type O− blood. Thus a person with type O− blood is said to be a universal donor.

page 74

Note that each Example is completely worked out and the *Check Your Progress* following the example is not. Study the worked-out example carefully by working through each step. You should do this with paper and pencil.

Now work the *Check Your Progress.* If you get stuck, refer to the page number following the word *Solution*, which directs you to the page on which the *Check Your Progress* is solved—a complete worked-out solution is provided. Try to use the given solution to get a hint for the step you are stuck on. Then try to complete your solution.

When you have completed the solution, check your work against the solution we provide.

CHECK YOUR PROGRESS 8 Use the blood transfusion table and Figures 2.3 and 2.4 to answer the following questions.

a. Is it safe for Alex to receive type A− blood in a blood transfusion?

b. What blood type do you have if you are classified as a *universal recipient?*

Solution See page S6.

page 74

Be aware that frequently there is more than one way to solve a problem. Your answer, however, should be the same as the given answer. If you have any question as to whether your method will "always work," check with your instructor or with someone in the math center.

Remember: Be an active participant in your learning process. When you are sitting in class watching and listening to an explanation, you may think that you understand. However, until you actually try to do it, you will have no confirmation of the new knowledge or skill. Most of us have had the experience of sitting in class thinking we knew how to do something only to get home and realize we didn't.

Rule Boxes Pay special attention to definitions, theorems, formulas, and procedures that are presented in a rectangular box, because they generally contain the most important concepts in each section.

Simple Interest Formula

The simple interest formula is

$$I = Prt$$

where I is the interest, P is the principal, r is the interest rate, and t is the time period.

page 620

Chapter Exercises When you have completed studying a section, do the section exercises. Math is a subject that needs to be learned in small sections and practiced continually in order to be mastered. Doing the exercises in each exercise set will help you master the problem-solving techniques necessary for success. As you work through the exercises, check your answers to the odd-numbered exercises against those in the back of the book.

Preparing for a Test There are important features of this text that can be used to prepare for a test.

- Chapter Summary
- Chapter Review Exercises
- Chapter Test

After completing a chapter, read the Chapter Summary. (See page 99 for the Chapter 2 Summary.) This summary highlights the important topics covered in each section of the chapter. Each concept is paired with page numbers of examples that illustrate the concept and exercises that will provide you with practice on the skill or technique.

Following the Chapter Summary are Chapter Review Exercises (see page 101). Doing the review exercises is an important way of testing your understanding of the chapter. The answer to each review exercise is given at the back of the book, along with, in brackets, the section reference from which the question was taken (see page A5). After checking your answers, restudy any section from which a question you missed was taken. It may be helpful to retry some of the exercises for that section to reinforce your problem-solving techniques.

Each chapter ends with a Chapter Test (see page 103). This test should be used to prepare for an exam. We suggest that you try the Chapter Test a few days before your actual exam. Take the test in a quiet place and try to complete the test in the same amount of time you will be allowed for your exam. When taking the Chapter Test, practice the strategies of successful test takers: (1) scan the entire test to get a feel for the questions; (2) read the directions carefully; (3) work the problems that are easiest for you first; and perhaps most importantly, (4) try to stay calm.

When you have completed the Chapter Test, check your answers for each exercise (see page A6). Next to each answer is, in brackets, the reference to the section from which the question was taken and an example reference for each exercise. The section references indicate the section or sections where you can locate the concepts needed to solve a given exercise, and the example reference allows you to easily find an example that is similar to the given test exercise. If you missed a question, review the material in that section and rework some of the exercises from that section. This will strengthen your ability to perform the skills in that section.

Your career goal goes here. ⟶

Is it difficult to be successful? YES! Successful music groups, artists, professional athletes, teachers, sociologists, chefs, and __________ have to work very hard to achieve their goals. They focus on their goals and ignore distractions. The things we ask you to do to achieve success take time and commitment. We are confident that if you follow our suggestions, you will succeed.

1

Problem Solving

Most occupations require good problem-solving skills. For instance, architects and engineers must solve many complicated problems as they design and construct modern buildings that are aesthetically pleasing, functional, and that meet stringent safety requirements. Two goals of this chapter are to help you become a better problem solver and to demonstrate that problem solving can be an enjoyable experience.

One problem that many have enjoyed is the Monty Hall (host of the game show *Let's Make a Deal*) problem, which is stated as follows. The grand prize in *Let's Make a Deal* is behind one of three doors. Less desirable prizes (for instance, a goat and a box of candy) are behind the other two doors. You select one of the doors, say door 1. Monty Hall reveals one of the less desirable prizes behind one of the other doors. You are then given the opportunity either to stay with your original choice or to choose the remaining closed door.

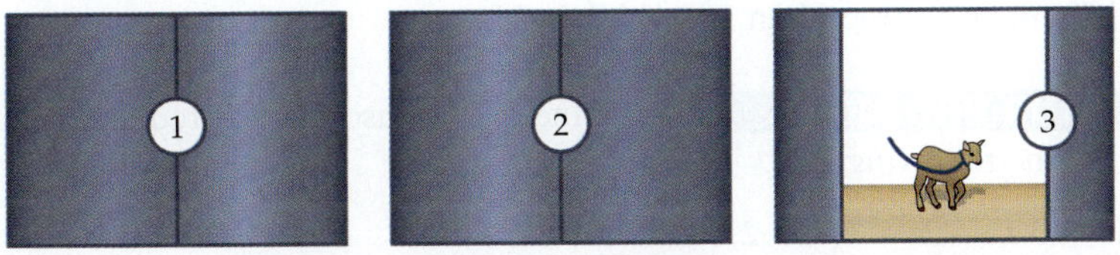

Example: You choose door 1. Monty Hall reveals a goat behind door 3. You can stay with door 1 or switch to door 2.

Marilyn vos Savant, author of the "Ask Marilyn" column featured in *Parade Magazine,* analyzed this problem,[1] claiming that you *double* your chances of winning the grand prize by switching to the other closed door. Many readers, including some mathematicians, responded with arguments that contradicted Marilyn's analysis.

Marilyn vos Savant

What do you think? Do you have a better chance of winning the grand prize by switching to the other closed door or staying with your original choice?

Of course there is also the possibility that it does not matter, if the chances of winning are the same with either strategy.

Discuss the Monty Hall problem with some of your friends and classmates. Is everyone in agreement? Additional information on this problem is given in Exploration Exercise 54 on page 14.

[1] "Ask Marilyn," *Parade Magazine*, September 9, 1990, p. 15.

SECTION 1.1 Inductive and Deductive Reasoning

Inductive Reasoning

The type of reasoning that forms a conclusion based on the examination of specific examples is called *inductive reasoning*. The conclusion formed by using inductive reasoning is a **conjecture**, since it may or may not be correct.

> **Inductive Reasoning**
>
> **Inductive reasoning** is the process of reaching a general conclusion by examining specific examples.

When you examine a list of numbers and predict the next number in the list according to some pattern you have observed, you are using inductive reasoning.

EXAMPLE 1 Use Inductive Reasoning to Predict a Number

Use inductive reasoning to predict the next number in each of the following lists.

a. 3, 6, 9, 12, 15, ? **b.** 1, 3, 6, 10, 15, ?

Solution

a. Each successive number is 3 larger than the preceding number. Thus we predict that the next number in the list is 3 larger than 15, which is 18.

b. The first two numbers differ by 2. The second and the third numbers differ by 3. It appears that the difference between any two numbers is always 1 more than the preceding difference. Since 10 and 15 differ by 5, we predict that the next number in the list will be 6 larger than 15, which is 21.

CHECK YOUR PROGRESS 1 Use inductive reasoning to predict the next number in each of the following lists.

a. 5, 10, 15, 20, 25, ? **b.** 2, 5, 10, 17, 26, ?

Solution See page S1.

Inductive reasoning is not used just to predict the next number in a list. In Example 2 we use inductive reasoning to make a conjecture about an arithmetic procedure.

EXAMPLE 2 Use Inductive Reasoning to Make a Conjecture

Consider the following procedure: Pick a number. Multiply the number by 8, add 6 to the product, divide the sum by 2, and subtract 3.

Complete the above procedure for several different numbers. Use inductive reasoning to make a conjecture about the relationship between the size of the resulting number and the size of the original number.

Solution

Suppose we pick 5 as our original number. Then the procedure would produce the following results:

Original number:	5
Multiply by 8:	$8 \times 5 = 40$
Add 6:	$40 + 6 = 46$
Divide by 2:	$46 \div 2 = 23$
Subtract 3:	$23 - 3 = 20$

> **TAKE NOTE**
>
> In Example 5, we will use a deductive method to verify that the procedure in Example 2 always yields a result that is four times the original number.

We started with 5 and followed the procedure to produce 20. Starting with 6 as our original number produces a final result of 24. Starting with 10 produces a final result of 40. Starting with 100 produces a final result of 400. In each of these cases the resulting number is four times the original number. We *conjecture* that following the given procedure produces a number that is four times the original number.

CHECK YOUR PROGRESS 2 Consider the following procedure: Pick a number. Multiply the number by 9, add 15 to the product, divide the sum by 3, and subtract 5.

Complete the above procedure for several different numbers. Use inductive reasoning to make a conjecture about the relationship between the size of the resulting number and the size of the original number.

Solution See page S1.

> **HISTORICAL NOTE**
>
> 
>
> Hulton Archive/Getty Images
>
> **Galileo Galilei** (găl′-ə-lā′ē′) entered the University of Pisa to study medicine at the age of 17, but he soon realized that he was more interested in the study of astronomy and the physical sciences. Galileo's study of pendulums assisted in the development of pendulum clocks.

Scientists often use inductive reasoning. For instance, Galileo Galilei (1564–1642) used inductive reasoning to discover that the time required for a pendulum to complete one swing, called the *period* of the pendulum, depends on the length of the pendulum. Galileo did not have a clock, so he measured the periods of pendulums in "heartbeats." The following table shows some results obtained for pendulums of various lengths. For the sake of convenience, a length of 10 inches has been designated as 1 unit.

Length of pendulum, in units	Period of pendulum, in heartbeats
1	1
4	2
9	3
16	4
25	5
36	6

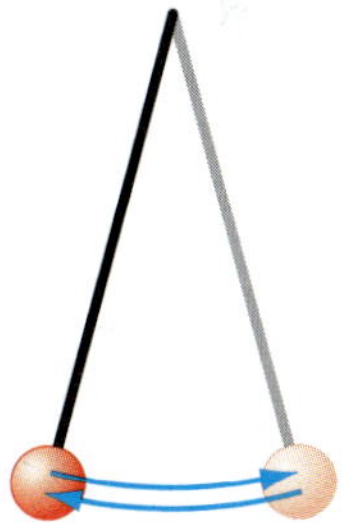

The period of a pendulum is the time it takes for the pendulum to swing from left to right and back to its original position.

EXAMPLE 3 Use Inductive Reasoning to Solve an Application

Use the data in the above table and inductive reasoning to answer each of the following questions.

a. If a pendulum has a length of 49 units, what is its period?

b. If the length of a pendulum is quadrupled, what happens to its period?

Solution

a. In the table, each pendulum has a period that is the square root of its length. Thus we conjecture that a pendulum with a length of 49 units will have a period of 7 heartbeats.

b. In the table, a pendulum with a length of 4 units has a period that is twice that of a pendulum with a length of 1 unit. A pendulum with a length of 16 units has a period that is twice that of a pendulum with a length of 4 units. It appears that quadrupling the length of a pendulum doubles its period.

Velocity of tsunami, in feet per second	Height of tsunami, in feet
6	4
9	9
12	16
15	25
18	36
21	49
24	64

CHECK YOUR PROGRESS 3 A tsunami is a sea wave produced by an underwater earthquake. The height of a tsunami as it approaches land depends on the velocity of the tsunami. Use the table at the left and inductive reasoning to answer each of the following questions.

a. What happens to the height of a tsunami when its velocity is doubled?

b. What should be the height of a tsunami if its velocity is 30 feet per second?

Solution See page S1.

Conclusions based on inductive reasoning may be incorrect. As an illustration, consider the circles shown below. For each circle, all possible line segments have been drawn to connect each dot on the circle with all the other dots on the circle.

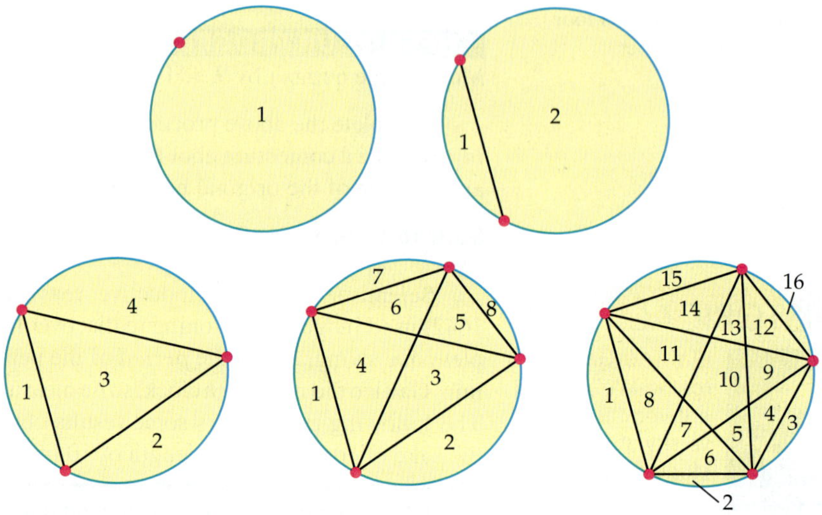

The maximum numbers of regions formed by connecting dots on a circle

TAKE NOTE

To produce the maximum number of regions, the dots on a circle must be placed so that no three line segments that connect the dots intersect at a single point.

For each circle, count the number of regions formed by the line segments that connect the dots on the circle. Your results should agree with the results in the following table.

Number of dots	1	2	3	4	5	6
Maximum number of regions	1	2	4	8	16	?

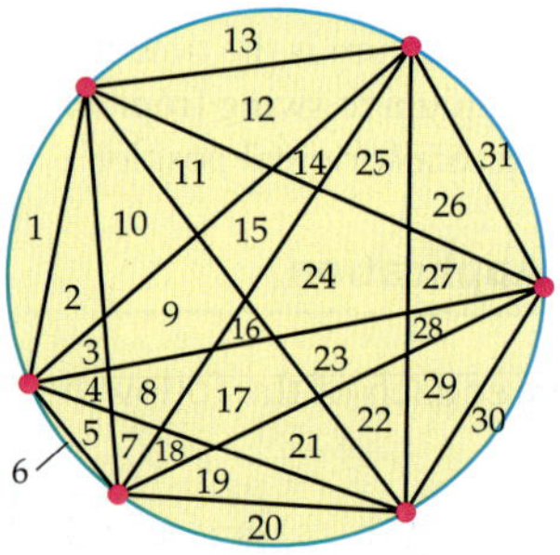

The line segments connecting six dots on a circle yield a maximum of 31 regions.

There appears to be a pattern. Each additional dot seems to double the number of regions. Guess the maximum number of regions you expect for a circle with six dots. Check your guess by counting the maximum number of regions formed by the line segments that connect six dots on a *large* circle. Your drawing will show that for six dots, the maximum number of regions is 31 (see the figure at the left), not 32 as you may have guessed. With seven dots the maximum number of regions is 57. This is a good example to keep in mind. Just because a pattern holds true for a few cases, it does not mean the pattern will continue. When you use inductive reasoning, you have no guarantee that your conclusion is correct.

Counterexamples

A statement is a true statement provided that it is true in all cases. If you can find *one case* for which a statement is not true, called a **counterexample**, then the statement is a false statement. In Example 4 we verify that each statement is a false statement by finding a counterexample for each.

EXAMPLE 4 Find a Counterexample

Verify that each of the following statements is a false statement by finding a counterexample.

For all numbers x:

a. $|x| > 0$ **b.** $x^2 > x$ **c.** $\sqrt{x^2} = x$

Solution

A statement may have many counterexamples, but we need only find one counterexample to verify that the statement is false.

a. Let $x = 0$. Then $|0| = 0$. Because 0 is not greater than 0, we have found a counterexample. Thus "for all numbers x, $|x| > 0$" is a false statement.

b. For $x = 1$ we have $1^2 = 1$. Since 1 is not greater than 1, we have found a counterexample. Thus "for all numbers x, $x^2 > x$" is a false statement.

c. Consider $x = -3$. Then $\sqrt{(-3)^2} = \sqrt{9} = 3$. Since 3 is not equal to -3, we have found a counterexample. Thus "for all numbers x, $\sqrt{x^2} = x$" is a false statement.

CHECK YOUR PROGRESS 4 Verify that each of the following statements is a false statement by finding a counterexample for each.

For all numbers x:

a. $\frac{x}{x} = 1$ **b.** $\frac{x+3}{3} = x + 1$ **c.** $\sqrt{x^2 + 16} = x + 4$

Solution See page S1.

QUESTION How many counterexamples are needed to prove that a statement is false?

Deductive Reasoning

Another type of reasoning is called *deductive reasoning*. Deductive reasoning is distinguished from inductive reasoning in that it is the process of reaching a conclusion by applying general principles and procedures.

Deductive Reasoning

Deductive reasoning is the process of reaching a conclusion by applying general assumptions, procedures, or principles.

TAKE NOTE

Example 5 is the same as Example 2, on page 2, except in Example 5 we use deductive reasoning, instead of inductive reasoning.

EXAMPLE 5 Use Deductive Reasoning to Establish a Conjecture

Use deductive reasoning to show that the following procedure produces a number that is four times the original number.

Procedure: Pick a number. Multiply the number by 8, add 6 to the product, divide the sum by 2, and subtract 3.

Solution

Let n represent the original number.

Multiply the number by 8:	$8n$
Add 6 to the product:	$8n + 6$
Divide the sum by 2:	$\frac{8n+6}{2} = 4n + 3$
Subtract 3:	$4n + 3 - 3 = 4n$

We started with n and ended with $4n$. The procedure given in this example produces a number that is four times the original number.

ANSWER One

CHECK YOUR PROGRESS 5 Use deductive reasoning to show that the following procedure produces a number that is three times the original number.

Procedure: Pick a number. Multiply the number by 6, add 10 to the product, divide the sum by 2, and subtract 5. *Hint:* Let n represent the original number.

Solution See page S1.

MATH MATTERS Deductive Reasoning in Mathematics

You may have observed that some of your math classes made extensive use of deductive reasoning to prove theorems and solve problems. The following quote by the mathematician Paul R. Halmos (1916–2006) advocates that you not limit yourself to only using deductive reasoning to prove theorems.

"Mathematics is not a deductive science—that's a cliché. When you try to prove a theorem, you don't just list the hypotheses, and then start to reason. What you do is trial and error, experimentation, guesswork."

I Want to be a Mathematician: An Automathography (1985).

Inductive Reasoning vs. Deductive Reasoning

In Example 6 we analyze arguments to determine whether they use inductive or deductive reasoning.

EXAMPLE 6 Determine Types of Reasoning

Determine whether each of the following arguments is an example of inductive reasoning or deductive reasoning.

a. During the past 10 years, a tree has produced plums every other year. Last year the tree did not produce plums, so this year the tree will produce plums.

b. All home improvements cost more than the estimate. The contractor estimated that my home improvement will cost $35,000. Thus my home improvement will cost more than $35,000.

Solution

a. This argument reaches a conclusion based on specific examples, so it is an example of inductive reasoning.

b. Because the conclusion is a specific case of a general assumption, this argument is an example of deductive reasoning.

CHECK YOUR PROGRESS 6 Determine whether each of the following arguments is an example of inductive reasoning or deductive reasoning.

a. All Gillian Flynn novels are worth reading. The novel *Gone Girl* is a Gillian Flynn novel. Thus *Gone Girl* is worth reading.

b. I know I will win a jackpot on this slot machine in the next 10 tries, because it has not paid out any money during the last 45 tries.

Solution See page S1.

Logic Puzzles

Logic puzzles, similar to the one in Example 7, can be solved by using deductive reasoning and a chart that enables us to display the given information in a visual manner.

EXAMPLE 7 Solve a Logic Puzzle

Each of four neighbors, Sean, Maria, Sarah, and Brian, has a different occupation (editor, banker, chef, or dentist). From the following clues, determine the occupation of each neighbor.

1. Maria gets home from work after the banker but before the dentist.
2. Sarah, who is the last to get home from work, is not the editor.
3. The dentist and Sarah leave for work at the same time.
4. The banker lives next door to Brian.

Solution

From clue 1, Maria is not the banker or the dentist. In the following chart, write X1 (which stands for "ruled out by clue 1") in the Banker and the Dentist columns of Maria's row.

	Editor	Banker	Chef	Dentist
Sean				
Maria		X1		X1
Sarah				
Brian				

From clue 2, Sarah is not the editor. Write X2 (ruled out by clue 2) in the Editor column of Sarah's row. We know from clue 1 that the banker is not the last to get home, and we know from clue 2 that Sarah is the last to get home; therefore, Sarah is not the banker. Write X2 in the Banker column of Sarah's row.

	Editor	Banker	Chef	Dentist
Sean				
Maria		X1		X1
Sarah	X2	X2		
Brian				

From clue 3, Sarah is not the dentist. Write X3 for this condition. There are now Xs for three of the four occupations in Sarah's row; therefore, Sarah must be the chef. Place a ✓ in that box. Since Sarah is the chef, none of the other three people can be the chef. Write X3 for these conditions. There are now Xs for three of the four occupations in Maria's row; therefore, Maria must be the editor. Insert a ✓ to indicate that Maria is the editor, and write X3 twice to indicate that neither Sean nor Brian is the editor.

	Editor	Banker	Chef	Dentist
Sean	X3		X3	
Maria	✓	X1	X3	X1
Sarah	X2	X2	✓	X3
Brian	X3		X3	

From clue 4, Brian is not the banker. Write X4 for this condition. See the following table. Since there are three Xs in the Banker column, Sean must be the banker. Place a

✓ in that box. Thus Sean cannot be the dentist. Write X4 in that box. Since there are 3 Xs in the Dentist column, Brian must be the dentist. Place a ✓ in that box.

	Editor	Banker	Chef	Dentist
Sean	X3	✓	X3	X4
Maria	✓	X1	X3	X1
Sarah	X2	X2	✓	X3
Brian	X3	X4	X3	✓

Sean is the banker, Maria is the editor, Sarah is the chef, and Brian is the dentist.

CHECK YOUR PROGRESS 7 Brianna, Ryan, Tyler, and Ashley were recently elected as the new class officers (president, vice president, secretary, treasurer) of the sophomore class at Summit College. From the following clues, determine which position each holds.

1. Ashley is younger than the president but older than the treasurer.
2. Brianna and the secretary are both the same age, and they are the youngest members of the group.
3. Tyler and the secretary are next-door neighbors.

Solution See page S1.

EXCURSION

KenKen® Puzzles: An Introduction

KenKen® is an arithmetic-based logic puzzle that was invented by the Japanese mathematics teacher Tetsuya Miyamoto in 2004. The noun "ken" has "knowledge" and "awareness" as synonyms. Hence, KenKen translates as knowledge squared, or awareness squared.

In recent years the popularity of KenKen has increased at a dramatic rate. More than a million KenKen puzzle books have been sold, and KenKen puzzles now appear in many popular newspapers, including the *New York Times* and the *Boston Globe*.

KenKen puzzles are similar to Sudoku puzzles, but they also require you to perform arithmetic to solve the puzzle.

Rules for Solving a KenKen Puzzle

For a 3 by 3 puzzle, fill in each box (square) of the grid with one of the numbers 1, 2, or 3.
For a 4 by 4 puzzle, fill in each square of the grid with one of the numbers 1, 2, 3, or 4.
For a n by n puzzle, fill in each square of the grid with one of the numbers 1, 2, 3, ..., n.
Grids range in size from a 3 by 3 up to a 9 by 9.

- Do not repeat a number in any row or column.
- The numbers in each heavily outlined set of squares, called **cages**, must combine (in some order) to produce the **target number** in the top left corner of the cage using the mathematical operation indicated.
- Cages with just one square should be filled in with the target number.
- A number can be repeated within a cage as long as it is not in the same row or column.

Here is a 4 by 4 puzzle and its solution. Properly constructed puzzles have a unique solution.

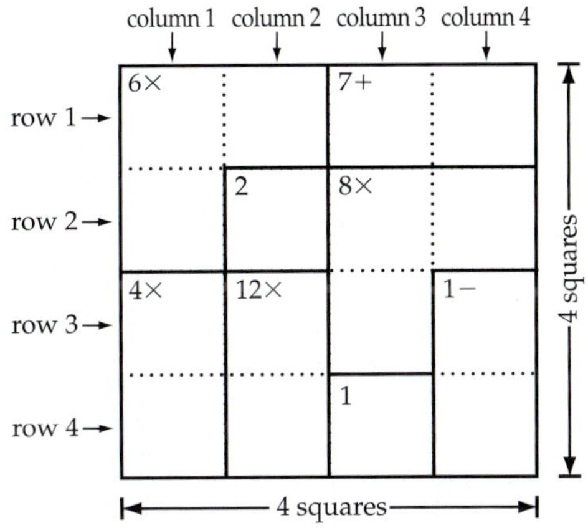

A 4 by 4 puzzle with 8 cages

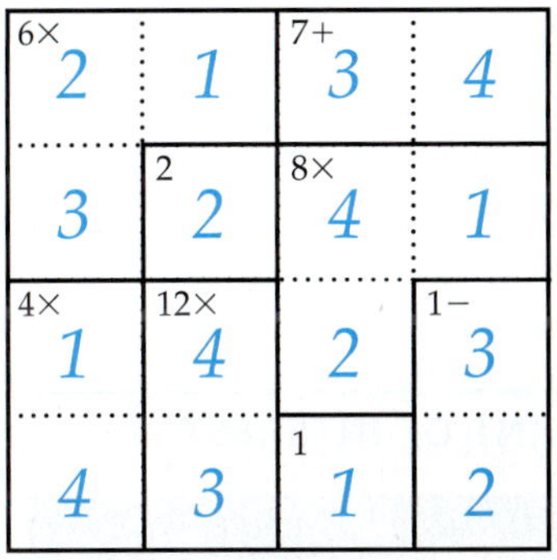

The solution to the puzzle

Basic Puzzle Solution Strategies

Single-Square Cages Fill cages that consist of a single square with the target number for that square.

Cages with Two Squares Next examine the cages with exactly two squares. Many cages that cover two squares will only have two digits that can be used to fill the cage. For instance, in a 5 by 5 puzzle, a **20×** cage with exactly two squares can only be filled with 4 and 5 or 5 and 4.

Large or Small Target Numbers Search for cages that have an unusually large or small target number. These cages generally have only a few combinations of numbers that can be used to fill the cage.

Examples:

In a 6 by 6 puzzle, a **120×** cage with exactly three squares can only be filled with 4, 5, and 6.

A **3+** cage with exactly two squares can only be filled with 1 and 2.

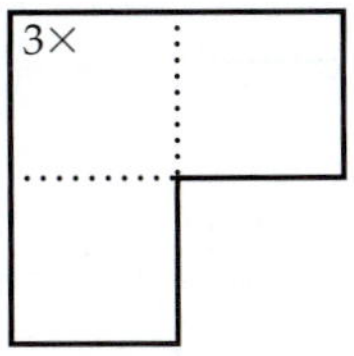

An L-shaped cage

Duplicate Digit in a Cage Consider the **3×** cage shown at the left. The digits 1, 1, and 3 produce a product of 3; however, we cannot place the two 1s in the same row or the same column. Thus the only way to fill the squares is to place the 3 in the corner of the L-shaped cage as shown below. *Remember:* A digit can occur more than once in a cage, provided that it does not appear in the same row or in the same column.

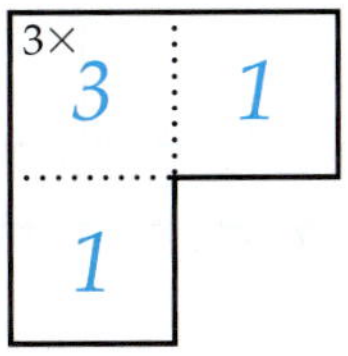

Remember the Following Rules

In an *n* by *n* puzzle, each row and column must contain every digit from 1 to *n*.

In a two-square cage that involves subtraction or division, the order of the numbers in the cage is not important. For instance, a **3−** cage with two squares could be filled with 4 and 1 or with 1 and 4. A **3÷** cage with two squares could be filled with 3 and 1 or with 1 and 3.

Make a List of Possible Digits For each cage, make a list of digits, with no regard to order, that can be used to fill the cage. See the following examples.

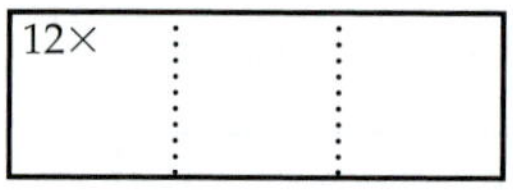

- In a 4 by 4 puzzle, this cage can only be filled with 1, 3, and 4 in some order. *Note:* 2, 2, and 3 cannot be used because the two 2s cannot be placed in the same row.

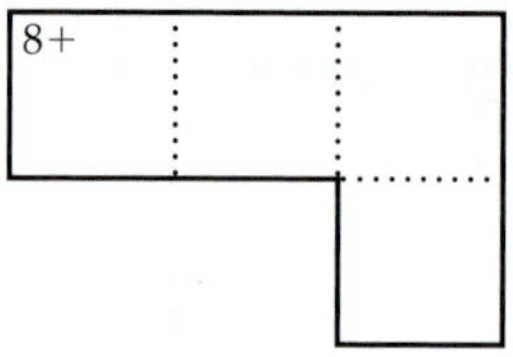

- In a 4 by 4 puzzle, this cage can be filled with 1, 1, 2, and 4, provided that the two 1s are placed in different rows. *Note:* The combination 1, 1, 3, and 3 and the combination 2, 2, 2, and 2 cannot be used because a duplicate digit would appear in the same row or column.

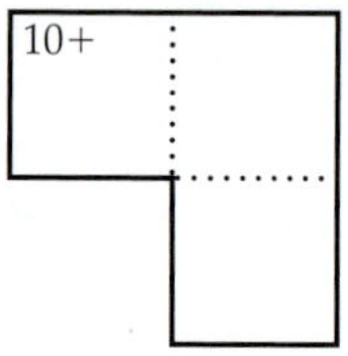

- In a 5 by 5 puzzle, this cage can be filled with the following combinations:
  - 5, 4, 1
  - 5, 3, 2
  - 4, 4, 2 provided the 2 is placed in the corner
  - 4, 3, 3 provided the 4 is placed in the corner

POINT OF INTEREST

HANDOUT/MCT/Newscom

Will Shortz, the well-known *New York Times* crossword puzzle editor, is reported to be the only person with a college degree in enigmatology, the study of puzzles.

Guess and Check In most puzzles you will reach a point where you will need to experiment. Assume that the possible digits in a particular cage are arranged in a particular manner and then see where your assumption takes you. If you find that the remaining part of a row or column cannot be filled in correctly, then you can eliminate your assumption and proceed to check out one of the remaining possible numerical arrangements for that particular cage.

It is worth noting that there are generally several different orders that can be used to fill in the squares/cages in a KenKen puzzle, even though each puzzle has a unique solution.

Many Internet sites provide additional strategies for solving KenKen puzzles and you may benefit from watching some of the video tutorials that are available online. For instance, Will Shortz has produced a video tutorial on KenKen puzzles. It is available on YouTube by searching for "Will Shortz KenKen".

INSTRUCTOR NOTE

Some students who have worked with KenKen® puzzles will be able to quickly complete these exercises. However, some students may need additional time and instructions to successfully complete them.

You may wish to provide a bonus puzzle from a newspaper or the Internet that can be used to challenge the students who have had previous experience working KenKen puzzles. The NekNek site at

http://www.mlsite.net/neknek/

has a **puzzle archive** link that provides quality puzzles in sizes from 3 by 3 up to 9 by 9.

EXCURSION EXERCISES

Solve each of the following puzzles. *Note:* The authors of this textbook are not associated with the KenKen brand. Thus the following puzzles are not official KenKen puzzles; however, each puzzle can be solved using the same techniques one would use to solve an official KenKen puzzle.

1.

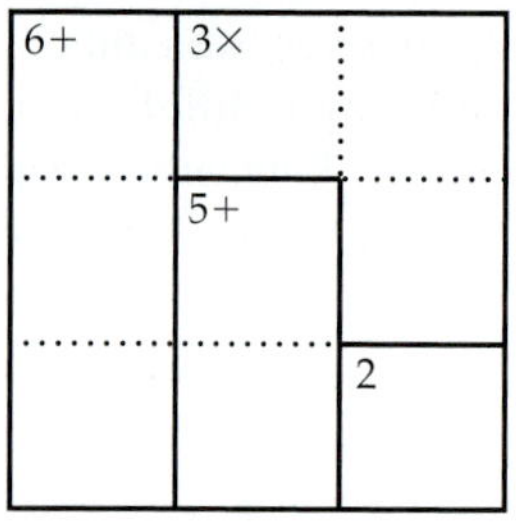

2.

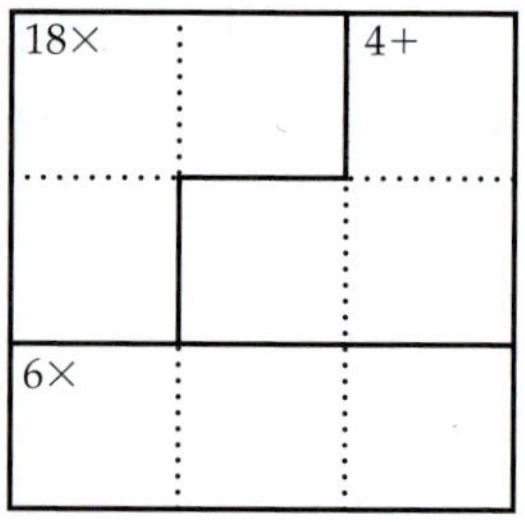

3.

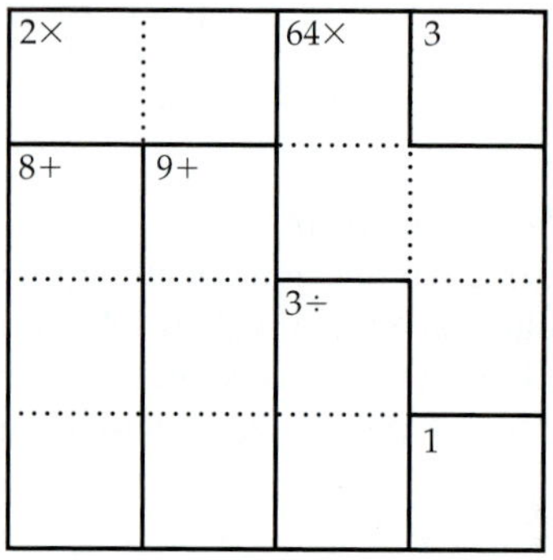

4.

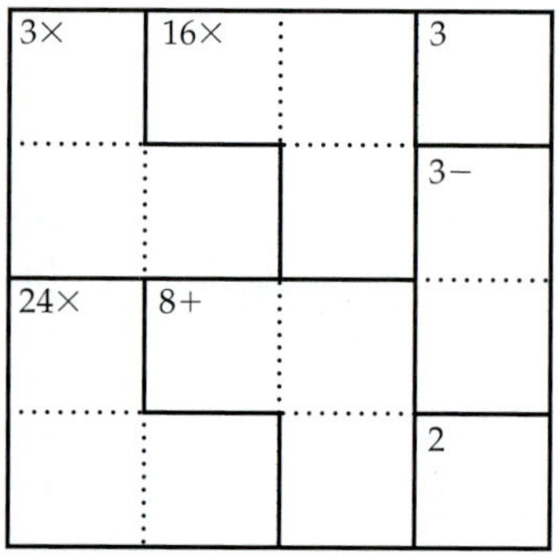

5.

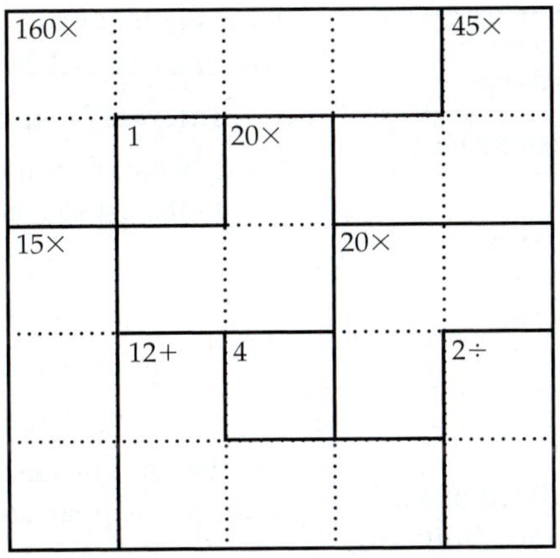

6.

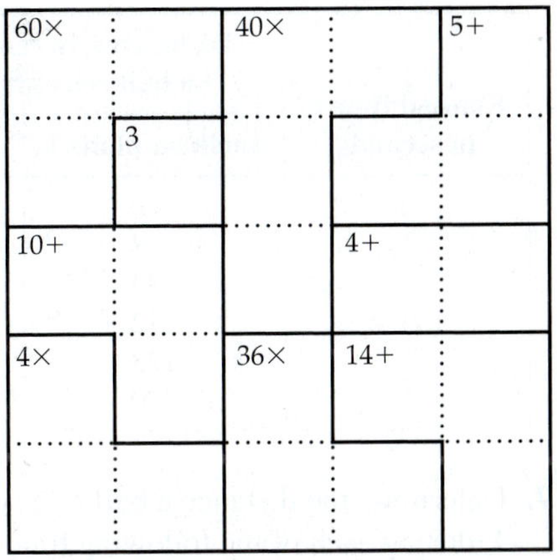

EXERCISE SET 1.1

■ In Exercises 1 to 10, use inductive reasoning to predict the next number in each list.

1. 4, 8, 12, 16, 20, 24, ?

2. 5, 11, 17, 23, 29, 35, ?

3. 3, 5, 9, 15, 23, 33, ?

4. 1, 8, 27, 64, 125, ?

5. 1, 4, 9, 16, 25, 36, 49, ?

6. 80, 70, 61, 53, 46, 40, ?

7. $\frac{3}{5}, \frac{5}{7}, \frac{7}{9}, \frac{9}{11}, \frac{11}{13}, \frac{13}{15}, ?$

8. $\frac{1}{2}, \frac{2}{3}, \frac{3}{4}, \frac{4}{5}, \frac{5}{6}, \frac{6}{7}, ?$

9. 2, 7, −3, 2, −8, −3, −13, −8, −18, ?

10. 1, 5, 12, 22, 35, ?

■ In Exercises 11 to 16, use inductive reasoning to decide whether each statement is correct. *Note:* The numbers 1, 2, 3, 4, 5, ... are called **counting numbers** or **natural numbers**. Any counting number n divided by 2 produces a remainder of 0 or 1. If $n \div 2$ has a remainder of 0, then n is an **even counting number**. If $n \div 2$ has a remainder of 1, then n is an **odd counting number**.

Even counting numbers: 2, 4, 6, 8, 10, ...

Odd counting numbers: 1, 3, 5, 7, 9, ...

11. The sum of any two even counting numbers is always an even counting number.

12. The product of an odd counting number and an even counting number is always an even counting number.

13. The product of two odd counting numbers is always an odd counting number.

14. The sum of two odd counting numbers is always an odd counting number.

15. Pick any counting number. Multiply the number by 6. Add 8 to the product. Divide the sum by 2. Subtract 4 from the quotient. The resulting number is twice the original number.

16. Pick any counting number. Multiply the number by 8. Subtract 4 from the product. Divide the difference by 2. Add 2 to the quotient. The resulting number is four times the original number.

Inclined Plane Experiments

Galileo (1564–1642) wanted to determine how the speed of a falling object changes as it falls. He conducted many free-fall experiments, but he found the motion of a falling object difficult to analyze. Eventually, he came up with the idea that it would be easier to perform experiments with a ball that rolls down a gentle incline, because the speed of the ball would be slower than the speed of a falling object. One objective of Galileo's inclined plane experiments was to determine how the speed of a ball rolling down an inclined plane changes as it rolls.

Examine the inclined plane shown below and the time–distance data in the following table.

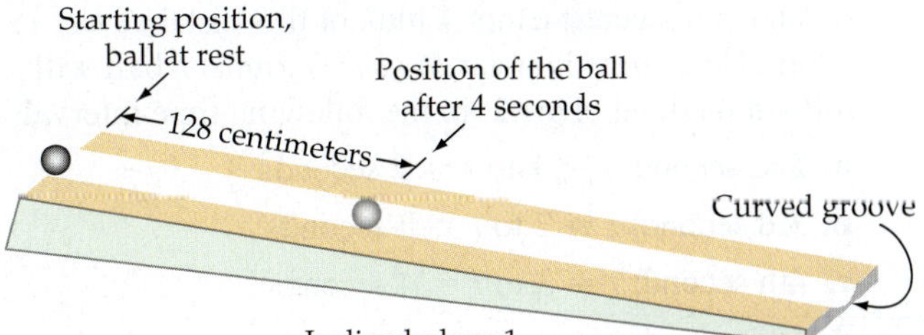

Inclined plane 1

Elapsed time t, in seconds	Distance d, in centimeters (cm), a ball rolls in t seconds (s)	
	Inclined plane 1	Inclined plane 2
0	0	0.0
1	8	6.5
2	32	26.0
3	72	58.5
4	128	104.0
5	200	162.5

17. Determine the distance a ball rolls, on inclined plane 1, during each of the following time intervals. *Hint:* To determine the distance a ball rolls in the interval $t = 1$ to $t = 2$ seconds, find the distance it rolls in 2 seconds and from this distance subtract the distance it rolls in 1 second.

a. 1st second: $t = 0$ to $t = 1$ second

b. 2nd second: $t = 1$ to $t = 2$ seconds

c. 3rd second: $t = 2$ to $t = 3$ seconds

d. 4th second: $t = 3$ to $t = 4$ seconds

e. 5th second: $t = 4$ to $t = 5$ seconds

18. Determine the distance a ball rolls, on inclined plane 2, during each of the following time intervals.

a. 1st second: $t = 0$ to $t = 1$ second

b. 2nd second: $t = 1$ to $t = 2$ seconds

c. 3rd second: $t = 2$ to $t = 3$ seconds

d. 4th second: $t = 3$ to $t = 4$ seconds

e. 5th second: $t = 4$ to $t = 5$ seconds

19. For inclined plane 1, the distance a ball rolls in the 1st second is 8 centimeters. Think of this distance as 1 unit. That is, for inclined plane 1,

$$1 \text{ unit} = 8 \text{ centimeters}$$

Determine how far *in terms of units* a ball will roll, on inclined plane 1, in the following time intervals.

a. 2nd second: $t = 1$ to $t = 2$ seconds

b. 3rd second: $t = 2$ to $t = 3$ seconds

c. 4th second: $t = 3$ to $t = 4$ seconds

d. 5th second: $t = 4$ to $t = 5$ seconds

20. For inclined plane 2, the distance a ball rolls in the 1st second is 6.5 centimeters. Think of this distance as 1 unit. Determine how far *in terms of units* a ball will roll, on inclined plane 2, in the following time intervals.

a. 2nd second: $t = 1$ to $t = 2$ seconds

b. 3rd second: $t = 2$ to $t = 3$ seconds

c. 4th second: $t = 3$ to $t = 4$ seconds

d. 5th second: $t = 4$ to $t = 5$ seconds

By using inductive reasoning and the results obtained in Exercises 19 and 20 we form the following conjecture.

Interval Distances in Units

If a ball rolls a distance of 1 unit on an inclined plane in the 1st second, then:

in the 2nd second, the ball rolls 3 units.

in the 3rd second, the ball rolls 5 units.

in the 4th second, the ball rolls 7 units.

in the 5th second, the ball rolls 9 units.

The following figure illustrates the concept presented in the above conjecture.

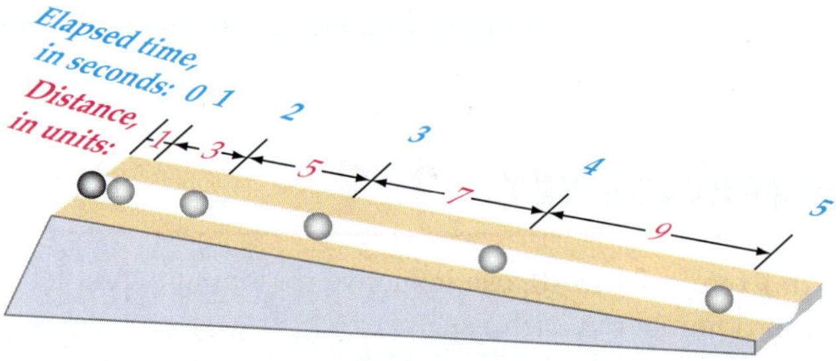

The gray circles show the position of the ball after rolling for 1 second, 2 seconds, 3 seconds, 4 seconds, and 5 seconds.

Galileo conjectured that as a ball rolls down an inclined plane, its speed is increasing, because in each successive 1-second time interval after the first 1-second time interval, it rolls *2 units* farther than it did in the preceding time interval of equal duration. This idea that the speed of a ball rolling down an inclined plane *increases uniformly* as it rolls downward was contrary to the popular belief, held in the early 1600s, that a ball rolls down an inclined plane at a constant rate.

Galileo was also interested in the relationship between the amount of time a ball rolls and the distance it rolls.

■ In Exercises 21 to 24, use inductive reasoning and the data in the inclined plane time–distance table, shown above exercise 17, to predict the answer to each question.

21. If the time a ball is allowed to roll on an inclined plane is doubled, what effect does this have on the distance the ball rolls?

22. If the time a ball is allowed to roll on an inclined plane is tripled, what effect does this have on the distance the ball rolls?

23. How far will a ball roll on inclined plane 1 in 6 seconds?

24. How far will a ball roll on inclined plane 1 in 1.5 seconds?

■ In Exercises 25 to 32, determine whether the argument is an example of inductive reasoning or deductive reasoning.

25. Emma enjoyed reading the novel *Finders Keepers* by Stephen King, so she will enjoy reading his next novel.

26. All pentagons have exactly five sides. Figure A is a pentagon. Therefore, Figure A has exactly five sides.

27. Every English setter likes to hunt. Duke is an English setter, so Duke likes to hunt.

28. Cats don't eat tomatoes. Tigger is a cat. Therefore, Tigger does not eat tomatoes.

29. A number is a *neat* number if the sum of the cubes of its digits equals the number. Therefore, 153 is a *neat* number.

30. The Atlanta Braves have won five games in a row. Therefore, the Atlanta Braves will win their next game.

31. Since

$$11 \times (1)(101) = 1111$$
$$11 \times (2)(101) = 2222$$
$$11 \times (3)(101) = 3333$$
$$11 \times (4)(101) = 4444$$
$$11 \times (5)(101) = 5555$$

we know that the product of 11 and a multiple of 101 is a number in which every digit is the same.

32. The following equations show that $n^2 - n + 11$ is a prime number for all counting numbers $n = 1, 2, 3, 4, \ldots$.

$$(1)^2 - 1 + 11 = 11 \qquad n = 1$$
$$(2)^2 - 2 + 11 = 13 \qquad n = 2$$
$$(3)^2 - 3 + 11 = 17 \qquad n = 3$$
$$(4)^2 - 4 + 11 = 23 \qquad n = 4$$

Note: A **prime number** is a counting number greater than 1 that has no counting number factors other than itself and 1. The first 15 prime numbers are 2, 3, 5, 7, 11, 13, 17, 19, 23, 29, 31, 37, 41, 43, and 47.

■ In Exercises 33 to 38, find a number that provides a counterexample to show that the given statement is false.

33. For all numbers x, $x > \frac{1}{x}$.

34. For all numbers x, $x + x > x$.

35. For all numbers x, $x^3 \geq x$.

36. For all numbers x, $|x + 3| = |x| + 3$.

37. For all numbers x, $-x < x$.

38. For all numbers x, $\frac{(x + 1)(x - 1)}{(x - 1)} = x + 1$.

■ In Exercises 39 and 40, find a pair of numbers that provides a counterexample to show that the given statement is false.

39. If the sum of two counting numbers is an even counting number, then the product of the two counting numbers is an even counting number.

40. If the product of two counting numbers is an even counting number, then both of the counting numbers are even counting numbers.

■ **Magic Squares** **A magic square of order *n*** is an arrangement of numbers in a square such that the sum of the n numbers in each row, column, and diagonal is the same number. The magic square below has order 3, and the sum of the numbers in each row, column, and diagonal is 15.

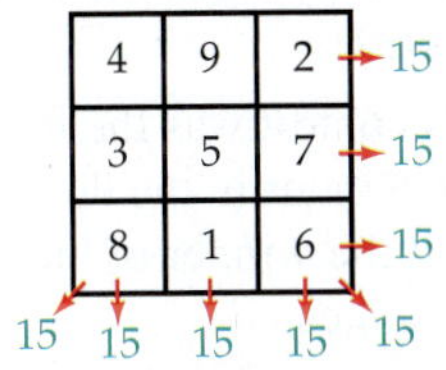

A magic square of order 3

In Exercises 41 and 42, use deductive reasoning to determine the missing numbers in each magic square.

41. A magic square of order 4

		2	13
	10	11	
	6		12
4	15		1

42. A magic square of order 5

11		7		3
4			8	16
	5	13		
10	18	1		22
23	6		2	15

43. Use deductive reasoning to show that the following procedure always produces a number that is equal to the original number.

Procedure: Pick a number. Multiply the number by 6 and add 8. Divide the sum by 2, subtract twice the original number, and subtract 4.

44. Use deductive reasoning to show that the following procedure always produces the number 5.

Procedure: Pick a number. Add 4 to the number and multiply the sum by 3. Subtract 7 and then decrease this difference by the triple of the original number.

45. Stocks Each of four siblings (Anita, Tony, Maria, and Jose) is given \$5000 to invest in the stock market. Each chooses a different stock. One chooses a utility stock, another an automotive stock, another a technology stock, and the other an oil stock. From the following clues, determine which sibling bought which stock.

a. Anita and the owner of the utility stock purchased their shares through an online brokerage, whereas Tony and the owner of the automotive stock did not.

b. The gain in value of Maria's stock is twice the gain in value of the automotive stock.

c. The technology stock is traded on NASDAQ, whereas the stock that Tony bought is traded on the New York Stock Exchange.

46. Gourmet Chefs The Changs, Steinbergs, Ontkeans, and Gonzaleses were winners in the All-State Cooking Contest. There was a winner in each of four categories: soup, entrée, salad, and dessert. From the following clues, determine in which category each family was the winner.

a. The soups were judged before the Ontkeans' winning entry.

b. This year's contest was the first for the Steinbergs and for the winner in the dessert category. The Changs and the winner of the soup category entered last year's contest.

c. The winning entrée took 2 hours to cook, whereas the Steinbergs' entrée required no cooking at all.

47. Collectibles The cities of Atlanta, Chicago, Philadelphia, and San Diego held conventions this summer for collectors of coins, stamps, comic books, and baseball cards. From the following clues, determine which collectors met in which city.

a. The comic book collectors convention was in August, as was the convention held in Chicago.

b. The baseball card collectors did not meet in Philadelphia, and the coin collectors did not meet in San Diego or Chicago.

c. The convention in Atlanta was held during the week of July 4, whereas the coin collectors convention was held the week after that.

d. The convention in Chicago had more collectors attending it than did the stamp collectors convention.

48. Map Coloring The following map shows eight states in the central time zone of the United States. Four colors have been used to color the states such that no two bordering states are the same color.

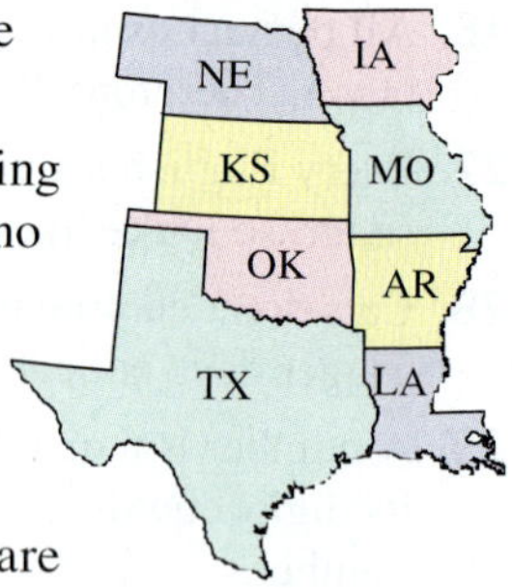

a. Can this map be colored, using only three colors, such that no two bordering states are the same color? Explain.

b. Can this map be colored, using only two colors, such that no two bordering states are the same color? Explain.

49. Driving Time You need to buy groceries at the supermarket, deposit a check at the credit union, and purchase a book at the bookstore. You can complete the errands in any order; however, you must start and end at your home. The driving time in minutes between each of these locations is given in the following figure.

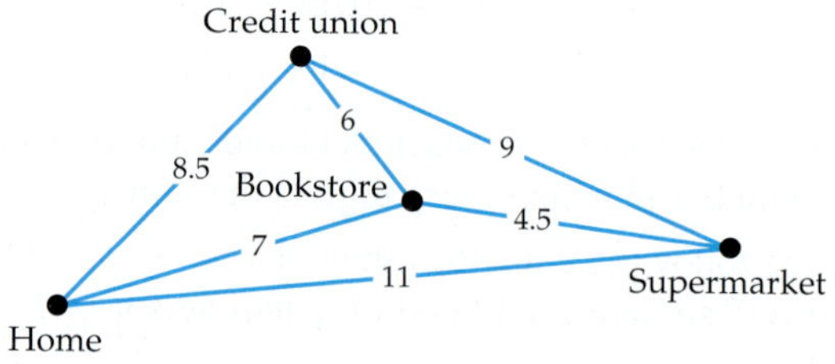

Find a route for which total driving time is less than 30 minutes.

50. Driving Time Suppose, in Exercise 49, that you need to go to the supermarket after you have completed the other two errands. What route should you take to minimize your travel time?

EXTENSIONS

51. Use inductive reasoning to predict the next letter in the following list.

O, T, T, F, F, S, S, E, ...

Hint: Look for a pattern that involves letters from words used for counting.

52. Use inductive reasoning to predict the next symbol in the following list.

Hint: Look for a pattern that involves counting numbers and symmetry about a line.

53. Counterexamples Find a counterexample to prove that the inductive argument in

a. Exercise 31 is incorrect.

b. Exercise 32 is incorrect.

54. The Monty Hall Problem Redux You can use the Internet to perform an experiment to determine the best strategy for playing the Monty Hall problem, which was stated in the Chapter 1 opener on page 1. Here is the procedure.

a. Use a search engine to find a website that provides a simulation of the Monty Hall problem. This problem is also known as the three-door problem, so search under both of these titles. Once you locate a site that provides a simulation, play the simulation 30 times using the strategy of not switching. Record the number of times you win the grand prize. Now play the simulation 30 times using the strategy of switching. How many times did you win the grand prize by not switching? How many times did you win the grand prize by switching?

b. On the basis of this experiment, which strategy seems to be the best strategy for winning the grand prize? What type of reasoning have you used?

SECTION 1.2 Problem Solving with Patterns

Terms of a Sequence

An ordered list of numbers such as

5, 14, 27, 44, 65, ...

is called a **sequence**. The numbers in a sequence that are separated by commas are the **terms** of the sequence. In the above sequence, 5 is the first term, 14 is the second term, 27 is the third term, 44 is the fourth term, and 65 is the fifth term. The three dots "..." indicate that the sequence continues beyond 65, which is the last written term. It is customary to use the subscript notation a_n to designate the ***n*th term of a sequence**. That is,

a_1 represents the first term of a sequence.

a_2 represents the second term of a sequence.

a_3 represents the third term of a sequence.

.

.

.

a_n represents the *n*th term of a sequence.

In the sequence 2, 6, 12, 20, 30, ..., $n^2 + n$, ...

$a_1 = 2$, $a_2 = 6$, $a_3 = 12$, $a_4 = 20$, $a_5 = 30$, and $a_n = n^2 + n$.

When we examine a sequence, it is natural to ask:

- What is the next term?
- What formula or rule can be used to generate the terms?

To answer these questions, we often construct a **difference table**, which shows the differences between successive terms of the sequence. The following table is a difference table for the sequence 2, 5, 8, 11, 14, ...

sequence:	2		5		8		11		14		...	
first differences:		3		3		3		3		...		(1)

Each of the numbers in row (1) of the table is the difference between the two closest numbers just above it (upper right number minus upper left number). The differences in row (1) are called the **first differences** of the sequence. In this case, the first differences are all the same. Thus, if we use the above difference table to predict the next number in the sequence, we predict that $14 + 3 = 17$ is the next term of the sequence. This prediction might be wrong; however, the pattern shown by the first differences seems to indicate that each successive term is 3 larger than the preceding term.

The following table is a difference table for the sequence 5, 14, 27, 44, 65, ...

sequence:	5		14		27		44		65		...	
first differences:		9		13		17		21		...		(1)
second differences:			4		4		4		...			(2)

In this table, the first differences are *not* all the same. In such a situation it is often helpful to compute the successive differences of the first differences. These are shown in row (2). These differences of the first differences are called the **second differences**. The differences of the second differences are called the **third differences**.

To predict the next term of a sequence, we often look for a pattern in a row of differences. For instance, in the following table, the second differences shown in blue are all the same constant, namely 4. If the pattern continues, then a 4 would also be the next second difference, and we can extend the table to the right as shown.

sequence:	5	14	27	44	65	...	
first differences:	9	13	17	21	...		(1)
second differences:	4	4	4	4	...		(2)

Now we work upward. That is, we add 4 to the first difference 21 to produce the next first difference, 25. We then add this difference to the fifth term, 65, to predict that 90 is the next term in the sequence. This process can be repeated to predict additional terms of the sequence.

sequence:	5	14	27	44	65	90	...
first differences:	9	13	17	21	25	...	(1)
second differences:	4	4	4	4	...		(2)

EXAMPLE 1 Predict the Next Term of a Sequence

Use a difference table to predict the next term in the sequence.

2, 7, 24, 59, 118, 207, ...

Solution

Construct a difference table as shown below.

sequence:	2	7	24	59	118	207	332	...
first differences:	5	17	35	59	89	125	...	(1)
second differences:	12	18	24	30	36	...		(2)
third differences:	6	6	6	6	...			(3)

The third differences, shown in blue in row (3), are all the same constant, 6. Extending row (3) so that it includes an additional 6 enables us to predict that the next second difference will be 36. Adding 36 to the first difference 89 gives us the next first difference, 125. Adding 125 to the sixth term 207 yields 332. Using the method of extending the difference table, we predict that 332 is the next term in the sequence.

CHECK YOUR PROGRESS 1 Use a difference table to predict the next term in the sequence.

1, 14, 51, 124, 245, 426, ...

Solution See page S2.

QUESTION Must the fifth term of the sequence 2, 4, 6, 8, ... be 10?

ANSWER No. The fifth term could be any number. However, if you used the method shown in Example 1, then you would predict that the fifth term is 10.

TAKE NOTE

Recall that the numbers 1, 2, 3, 4, ... are called *natural numbers* or *counting numbers*. We will often use the letter n to represent an arbitrary natural number.

*n*th-Term Formula for a Sequence

In Example 1 we used a difference table to predict the next term of a sequence. In some cases we can use patterns to predict a formula, called an ***n*th-term formula**, that generates the terms of a sequence. As an example, consider the formula $a_n = 3n^2 + n$. This formula defines a sequence and provides a method for finding any term of the sequence. For instance, if we replace n with 1, 2, 3, 4, 5, and 6, then the formula $a_n = 3n^2 + n$ generates the sequence 4, 14, 30, 52, 80, 114. To find the 40th term, replace each n with 40.

$$a_{40} = 3(40)^2 + 40 = 4840$$

In Example 2 we make use of patterns to determine an nth-term formula for a sequence given by geometric figures.

EXAMPLE 2 Find an *n*th-Term Formula

Assume the pattern shown by the square tiles in the following figures continues.

a. What is the nth-term formula for the number of tiles in the nth figure of the sequence?

b. How many tiles are in the eighth figure of the sequence?

c. Which figure will consist of exactly 320 tiles?

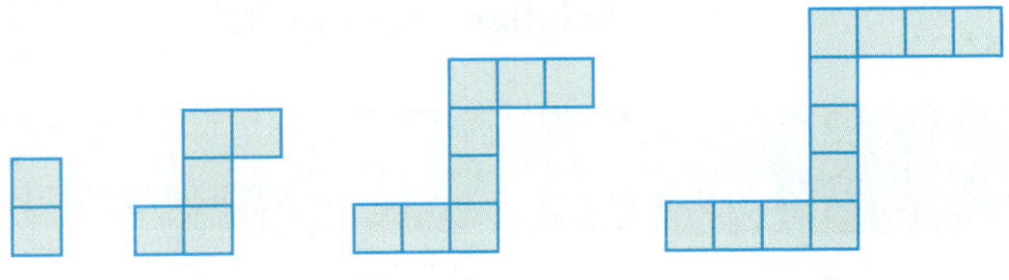

Solution

TAKE NOTE

The method of grouping used in Example 2a is not unique. Do you see a different way to group the tiles? Does your method of grouping still produce the nth-term formula $a_n = 3n - 1$?

a. Examine the figures for patterns. Note that the second figure has two tiles on each of the horizontal sections and one tile between the horizontal sections. The third figure has three tiles on each horizontal section and two tiles between the horizontal sections. The fourth figure has four tiles on each horizontal section and three tiles between the horizontal sections.

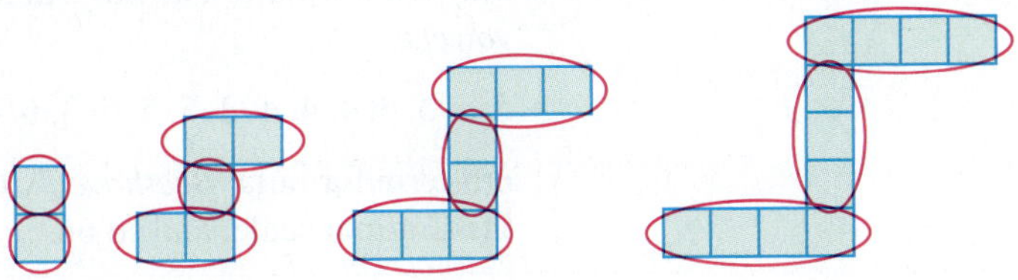

Thus the number of tiles in the nth figure is given by two groups of n plus a group of n less one. That is,

$$a_n = 2n + (n - 1)$$
$$a_n = 3n - 1$$

b. The number of tiles in the eighth figure of the sequence is $3(8) - 1 = 23$.

c. To determine which figure in the sequence will have 320 tiles, we solve the equation $3n - 1 = 320$.

$$3n - 1 = 320$$
$$3n = 321 \quad \text{• Add 1 to each side.}$$
$$n = 107 \quad \text{• Divide each side by 3.}$$

The 107th figure is composed of 320 tiles.

CHECK YOUR PROGRESS 2 Assume that the pattern shown by the square tiles in the following figure continues.

a. What is the *n*th-term formula for the number of tiles in the *n*th figure of the sequence?

b. How many tiles are in the tenth figure of the sequence?

c. Which figure will consist of exactly 419 tiles?

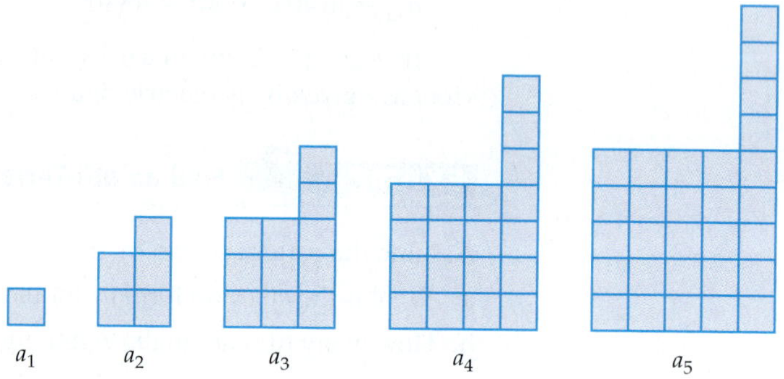

Solution See page S2.

MATH MATTERS Sequences on the Internet

If you find it difficult to determine how the terms of a sequence are being generated, you might be able to find a solution on the Internet. One resource is *The On-Line Encyclopedia of Integer Sequences*© at:

www.oeis.org

Here are two sequences from this website.

- 1, 2, 3, 5, 9, 12, 21, 22, 23, 25, 29, 31, 32, 33, 35, 39, 41, 42, 43, 45, 49, 51, 52, ...

  *n*th-term formula: *The natural numbers whose names, in English, end with vowels.*

- 5, 5, 5, 3, 4, 4, 4, 2, 5, 5, 5, 3, 6, 6, 6, 5, 10, 10, 10, 8, ...

  *n*th-term formula: *Beethoven's Fifth Symphony;* 1 stands for the first note in the C minor scale, and so on.

The Fibonacci Sequence

Leonardo of Pisa, also known as Fibonacci (fē′bə-nä′chē) (c. 1170–1250), is one of the best-known mathematicians of medieval Europe. In 1202, after a trip that took him to several Arab and Eastern countries, Fibonacci wrote the book *Liber Abaci*. In this book Fibonacci explained why the Hindu-Arabic numeration system that he had learned about during his travels was a more sophisticated and efficient system than the Roman numeration system. This book also contains a problem created by Fibonacci that concerns the birth rate of rabbits. Here is a statement of Fibonacci's rabbit problem.

Bettman/Corbis

Fibonacci

> At the beginning of a month, you are given a pair of newborn rabbits. After a month the rabbits have produced no offspring; however, every month thereafter, the pair of rabbits produces another pair of rabbits. The offspring reproduce in exactly the same manner. If none of the rabbits dies, how many pairs of rabbits will there be at the start of each succeeding month?

The solution of this problem is a sequence of numbers that we now call the **Fibonacci sequence.** The following figure shows the numbers of pairs of rabbits on the first day of each of the first six months. The larger rabbits represent mature rabbits that produce another pair of rabbits each month. The numbers in the blue region—1, 1, 2, 3, 5, 8—are the first six terms of the Fibonacci sequence.

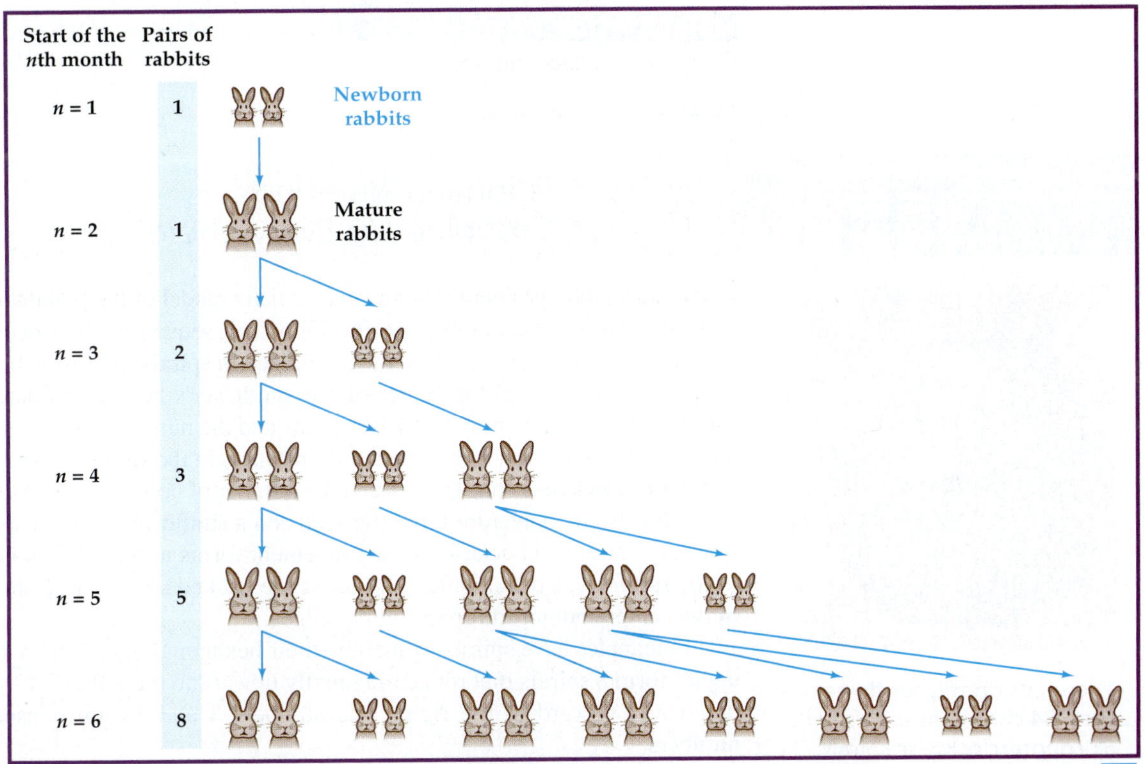

Fibonacci discovered that the number of pairs of rabbits for any month after the first two months can be determined by adding the numbers of pairs of rabbits in each of the *two previous* months. For instance, the number of pairs of rabbits at the start of the sixth month is $3 + 5 = 8$.

A **recursive definition** for a sequence is one in which each successive term of the sequence is defined by using some of the preceding terms. If we use the mathematical notation F_n to represent the nth Fibonacci number, then the numbers in the Fibonacci sequence are given by the following recursive definition.

INSTRUCTOR NOTE

To determine whether your students understand the concept of a recursive definition, ask them to find a_3, a_4, and a_5 for the sequence defined by $a_1 = 1$, $a_2 = 2$, and

$$a_n = 3a_{n-1} + 2a_{n-2}$$

for $n \geq 3$.

They should find that $a_3 = 8$, $a_4 = 28$, and $a_5 = 100$.

The Fibonacci Numbers

$F_1 = 1$, $F_2 = 1$, and $F_n = F_{n-1} + F_{n-2}$ for $n \geq 3$.

EXAMPLE 3 Find a Fibonacci Number

Use the definition of Fibonacci numbers to find the seventh and eighth Fibonacci numbers.

Solution

The first six Fibonacci numbers are 1, 1, 2, 3, 5, and 8. The seventh Fibonacci number is the sum of the two previous Fibonacci numbers. Thus,

$$\begin{aligned} F_7 &= F_6 + F_5 \\ &= 8 + 5 \\ &= 13 \end{aligned}$$

The eighth Fibonacci number is

$$F_8 = F_7 + F_6$$
$$= 13 + 8$$
$$= 21$$

CHECK YOUR PROGRESS 3 Use the definition of Fibonacci numbers to find the ninth Fibonacci number.

Solution See page S2.

MATH MATTERS Fibonacci Numbers: Cropping Up Everywhere!

Dynamic Graphics/PictureQuest

The seeds on this sunflower form 34 clockwise spirals and 55 counterclockwise spirals. The numbers 34 and 55 are consecutive Fibonacci numbers.

Fibonacci's rabbit problem is not a very realistic model of the population growth rate of rabbits; however, the numbers in the Fibonacci sequence often occur in nature. For instance, the seeds on a sunflower are arranged in spirals that curve both clockwise and counterclockwise from the center of the sunflower's head to its outer edge. In many sunflowers, the number of clockwise spirals and the number of counterclockwise spirals are consecutive Fibonacci numbers. For instance, in the sunflower shown at the left, the number of clockwise spirals is 34 and the number of counterclockwise spirals is 55.

It has been conjectured that the seeds on a sunflower grow in spirals that involve Fibonacci numbers because this arrangement forms a uniform packing. At any stage in the sunflower's development, its seeds are packed so that they are not too crowded in the center and not too sparse at the edges.

Pineapples have spirals formed by their hexagonal nubs. The nubs on many pineapples form 8 spirals that rotate diagonally upward to the left and 13 spirals that rotate diagonally upward to the right. The numbers 8 and 13 are consecutive Fibonacci numbers.

Additional information about the occurrence of Fibonacci numbers in nature can be found on the Internet.

We can find any term after the second term of the Fibonacci sequence by computing the sum of the previous two terms. However, this procedure of adding the previous two terms can be tedious. For instance, what is the 100th term or the 1000th term of the Fibonacci sequence? To find the 100th term, we need to know the 98th and 99th terms. To find the 1000th term, we need to know the 998th and 999th terms. Many mathematicians tried to find a nonrecursive nth-term formula for the Fibonacci sequence without success, until a formula was discovered by Jacques Binet in 1843. Binet's formula is given in Exercise 23 of this section.

QUESTION What happens if you try to use a difference table to determine Fibonacci numbers?

EXAMPLE 4 Determine Properties of Fibonacci Numbers

Determine whether each of the following statements about Fibonacci numbers is true or false. *Note:* The first 10 terms of the Fibonacci sequence are 1, 1, 2, 3, 5, 8, 13, 21, 34, and 55.

a. If n is even, then F_n is an odd number.

b. $2F_n - F_{n-2} = F_{n+1}$ for $n \geq 3$

ANSWER The difference table for the numbers in the Fibonacci sequence does not contain a row of differences that are all the same constant.

Solution

a. An examination of Fibonacci numbers shows that the second Fibonacci number, 1, is odd and the fourth Fibonacci number, 3, is odd, but the sixth Fibonacci number, 8, is even. Thus the statement "If n is even, then F_n is an odd number" is false.

b. Experiment to see whether $2F_n - F_{n-2} = F_{n+1}$ for several values of n. For instance, for $n = 7$, we get

$$2F_n - F_{n-2} = F_{n+1}$$
$$2F_7 - F_{7-2} = F_{7+1}$$
$$2F_7 - F_5 = F_8$$
$$2(13) - 5 = 21$$
$$26 - 5 = 21$$
$$21 = 21$$

which is true. Evaluating $2F_n - F_{n-2}$ for several additional values of n, $n \geq 3$, we find that in each case $2F_n - F_{n-2} = F_{n+1}$. Thus, by inductive reasoning, we conjecture $2F_n - F_{n-2} = F_{n+1}$ for $n \geq 3$ is a true statement. *Note:* This property of Fibonacci numbers can also be established using deductive reasoning. See Exercise 32 of this section.

TAKE NOTE

Pick any Fibonacci number larger than 1. The equation

$$2F_n - F_{n-2} = F_{n+1}$$

merely states that for numbers in the Fibonacci sequence

1, 1, 2, 3, 5, 8, 13, 21, ...

the double of a Fibonacci number, F_n, less the Fibonacci number two to its left, is the Fibonacci number just to the right of F_n.

CHECK YOUR PROGRESS 4 Determine whether each of the following statements about Fibonacci numbers is true or false.

a. $2F_n > F_{n+1}$ for $n \geq 3$ **b.** $2F_n + 4 = F_{n+3}$ for $n \geq 3$

Solution See page S2.

EXCURSION

Polygonal Numbers

The ancient Greek mathematicians were interested in the geometric shapes associated with numbers. For instance, they noticed that triangles can be constructed using 1, 3, 6, 10, or 15 dots, as shown in Figure 1.1 on page 22. They called the numbers 1, 3, 6, 10, 15, ... *triangular numbers.* The Greeks called the numbers 1, 4, 9, 16, 25, ... *square numbers* and the numbers 1, 5, 12, 22, 35, ... *pentagonal numbers.*

An nth-term formula for the triangular numbers is:

$$Triangular_n = \frac{n(n+1)}{2}$$

The square numbers have an nth-term formula of

$$Square_n = n^2$$

The nth-term formula for the pentagonal numbers is

$$Pentagonal_n = \frac{n(3n-1)}{2}$$

HISTORICAL NOTE

Hutton Archive/Handout/Getty Images

Pythagoras (c. 580 B.C.–520 B.C.) The ancient Greek philosopher and mathematician Pythagoras (pĭ-thăg′ər-əs) formed a secret brotherhood that investigated topics in music, astronomy, philosophy, and mathematics. The Pythagoreans believed that the nature of the universe was directly related to mathematics and that whole numbers and the ratios formed by whole numbers could be used to describe and represent all natural events.

The Pythagoreans were particularly intrigued by the number 5 and the shape of a pentagon. They used the following figure, which is a five-pointed star inside a regular pentagon, as a secret symbol that could be used to identify other members of the brotherhood.

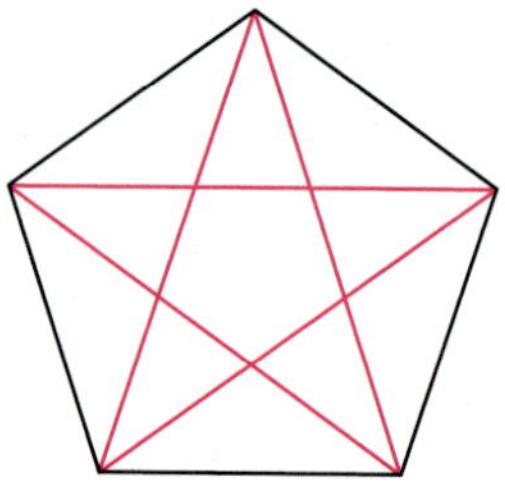

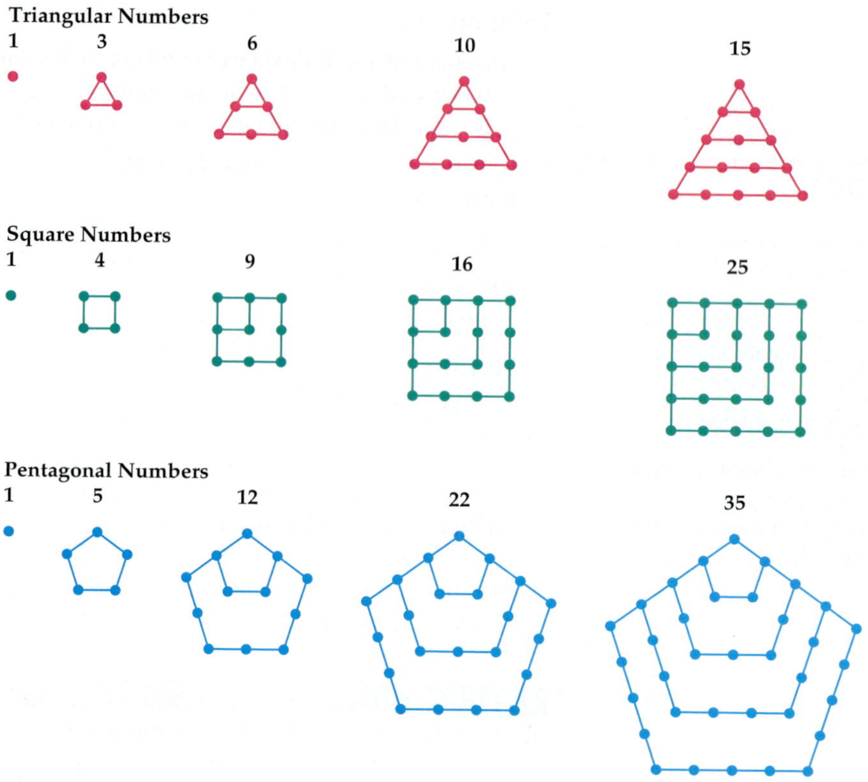

FIGURE 1.1

EXCURSION EXERCISES

1. Extend Figure 1.1 above by constructing drawings of the sixth triangular number, the sixth square number, and the sixth pentagonal number.

2. The figure below shows that the fourth triangular number, 10, added to the fifth triangular number, 15, produces the fifth square number, 25.

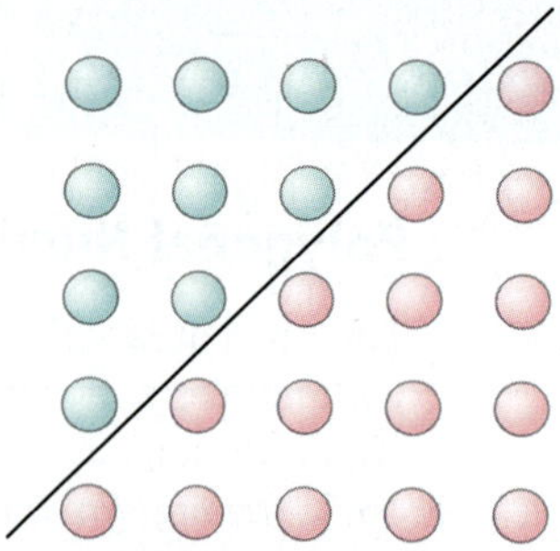

a. Use a drawing to show that the fifth triangular number added to the sixth triangular number is the sixth square number.

b. Verify that the 50th triangular number added to the 51st triangular number is the 51st square number. *Hint:* Use a numerical approach; don't use a drawing.

c. Use nth-term formulas to verify that the sum of the nth triangular number and the $(n + 1)$st triangular number is always the square number $(n + 1)^2$.

3. Construct a drawing of the fourth hexagonal number.

EXERCISE SET 1.2

■ In Exercises 1 to 6, construct a difference table to predict the next term of each sequence.

1. 1, 7, 17, 31, 49, 71, ...
2. 10, 10, 12, 16, 22, 30, ...
3. −1, 4, 21, 56, 115, 204, ...
4. 0, 10, 24, 56, 112, 190, ...
5. 9, 4, 3, 12, 37, 84, ...
6. 17, 15, 25, 53, 105, 187, ...

■ In Exercises 7 to 10, use the given nth-term formula to compute the first five terms of the sequence.

7. $a_n = \dfrac{n(2n+1)}{2}$

8. $a_n = \dfrac{n}{n+1}$

9. $a_n = 5n^2 - 3n$

10. $a_n = 2n^3 - n^2$

■ In Exercises 11 to 14, determine the nth-term formula for the number of square tiles in the nth figure.

11.

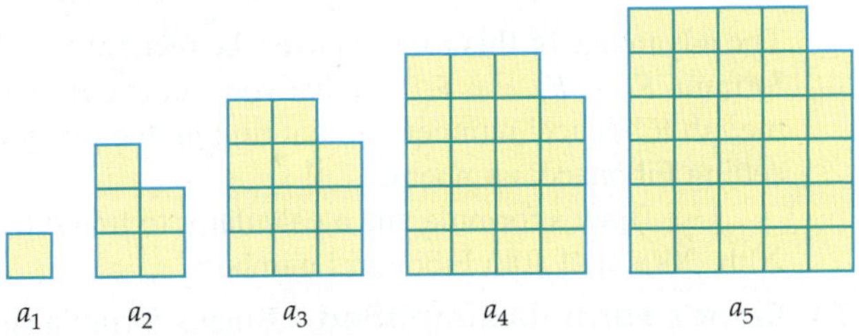

12.

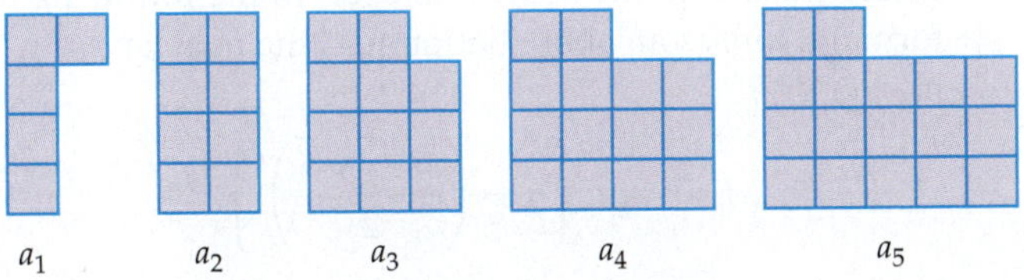

13.

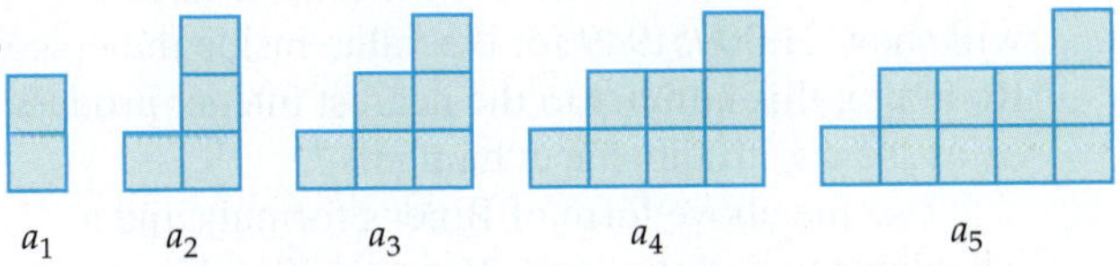

14.

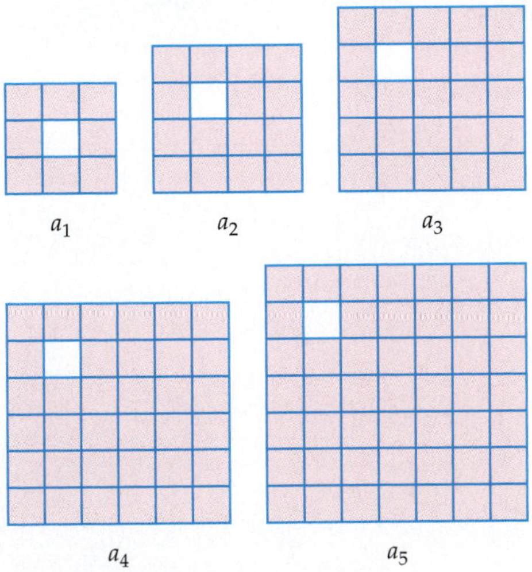

■ Cannonballs can be stacked to form a pyramid with a triangular base. Five of these pyramids are shown below. Use these figures in Exercises 15 and 16.

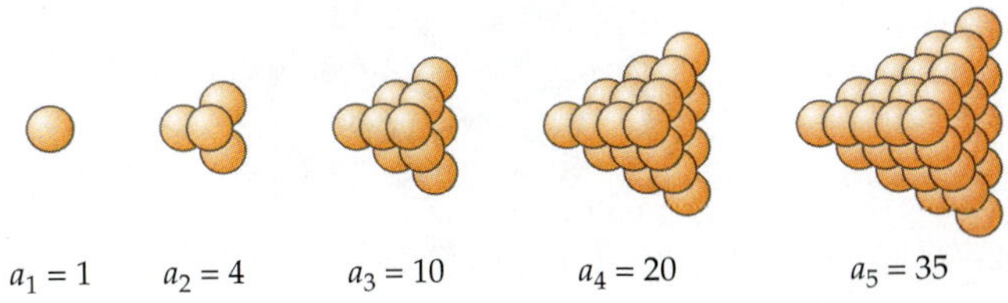

15. **a.** Use a difference table to predict the number of cannonballs in the sixth pyramid and in the seventh pyramid.

b. Write a few sentences that describe the eighth pyramid in the sequence.

16. The sequence formed by the numbers of cannonballs in the above pyramids is called the *tetrahedral sequence*. The nth-term formula for the tetrahedral sequence is

$$Tetrahedral_n = \frac{1}{6}n(n+1)(n+2)$$

Find $Tetrahedral_{10}$.

17. **Pieces vs. Cuts** One cut of a stick of licorice produces two pieces. Two cuts produce three pieces. Three cuts produce four pieces.

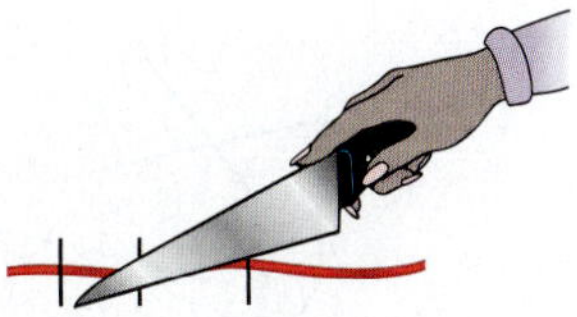

a. How many pieces are produced by five cuts and by six cuts?

b. Predict the nth-term formula for the number of pieces of licorice that are produced by n cuts of a stick of licorice.

18. Pieces vs. Cuts One straight cut across a pizza produces 2 pieces. Two cuts can produce a maximum of 4 pieces. Three cuts can produce a maximum of 7 pieces. Four cuts can produce a maximum of 11 pieces.

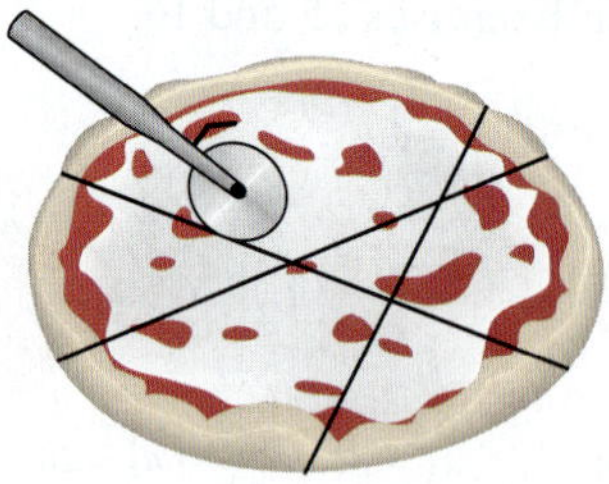

a. Use a difference table to predict the maximum number of pieces that can be produced with seven cuts.

b. How are the pizza-slicing numbers related to the triangular numbers, which are defined by

$$Triangular_n = \frac{n(n+1)}{2}?$$

19. Pieces vs. Cuts One straight cut through a thick piece of cheese produces two pieces. Two straight cuts can produce a maximum of 4 pieces. Three straight cuts can produce a maximum of 8 pieces. You might be inclined to think that every additional cut doubles the previous number of pieces. However, for four straight cuts, you will find that you get a maximum of 15 pieces. An nth-term formula for the maximum number of pieces, P_n, that can be produced by n straight cuts is

$$P_n = \frac{n^3 + 5n + 6}{6}$$

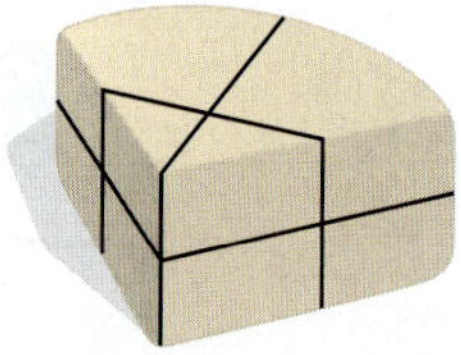

a. Use the nth-term formula to determine the maximum number of pieces that can be produced by five straight cuts.

b. What is the smallest number of straight cuts that you can use if you wish to produce at least 60 pieces? *Hint:* Use the nth-term formula and experiment with larger and larger values of n.

20. Fibonacci Properties The Fibonacci sequence has many unusual properties. Experiment to decide which of the following properties are valid. *Note:* F_n represents the nth Fibonacci number.

a. $3F_n - F_{n-2} = F_{n+2}$ for $n \geq 3$

b. $F_n F_{n+3} = F_{n+1} F_{n+2}$

c. F_{3n} is an even number.

d. $5F_n - 2F_{n-2} = F_{n+3}$ for $n \geq 3$

21. Find the third, fourth, and fifth terms of the sequence defined by $a_1 = 3$, $a_2 = 5$, and $a_n = 2a_{n-1} - a_{n-2}$ for $n \geq 3$.

22. Find the third, fourth, and fifth terms of the sequence defined by $a_1 = 2$, $a_2 = 3$, and $a_n = (-1)^n a_{n-1} + a_{n-2}$ for $n \geq 3$.

23. Binet's Formula The following formula is known as *Binet's formula* for the nth Fibonacci number.

$$F_n = \frac{1}{\sqrt{5}}\left[\left(\frac{1+\sqrt{5}}{2}\right)^n - \left(\frac{1-\sqrt{5}}{2}\right)^n\right]$$

The advantage of this formula over the recursive formula $F_n = F_{n-1} + F_{n-2}$ is that you can determine the nth Fibonacci number without finding the two preceding Fibonacci numbers.

Use Binet's formula and a calculator to find the 20th, 30th, and 40th Fibonacci numbers.

24. Binet's Formula Simplified Binet's formula (see Exercise 23) can be simplified if you round your calculator results to the nearest integer. In the following formula, *nint* is an abbreviation for "the nearest integer of."

$$f_n = nint\left\{\frac{1}{\sqrt{5}}\left(\frac{1+\sqrt{5}}{2}\right)^n\right\}$$

If you use $n = 8$ in the above formula, a calculator will show 21.00951949 for the value inside the braces. Rounding this number to the nearest integer produces 21 as the eighth Fibonacci number.

Use the above form of Binet's formula and a calculator to find the 16th, 21st, and 32nd Fibonacci numbers.

EXTENSIONS

25. A Geometric Model The ancient Greeks often discovered mathematical relationships by using geometric drawings. Study the accompanying drawing to determine what needs to be put in place of the question mark to make the equation a true statement.

$$1 + 3 + 5 + 7 + \cdots + (2n - 1) = ?$$

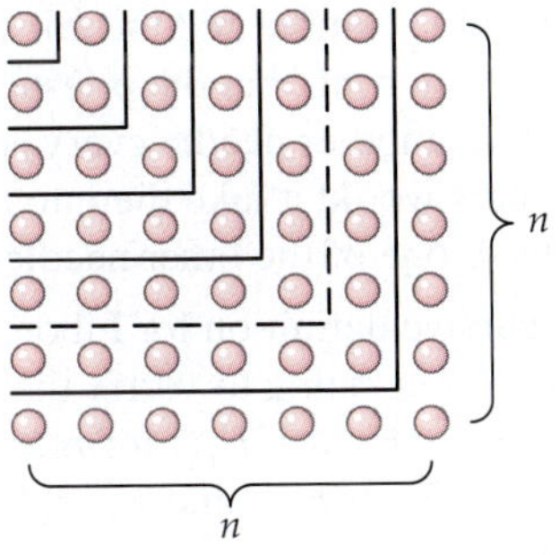

26. The nth-term formula

$$a_n = \frac{n(n-1)(n-2)(n-3)(n-4)}{4 \cdot 3 \cdot 2 \cdot 1} + 2n$$

generates 2, 4, 6, 8, 15 for $n = 1, 2, 3, 4, 5$. Make minor changes to the above formula to produce an nth-term formula (with $n = 1, 2, 3, 4$, and 5) that will generate the following finite sequences.

a. 2, 4, 6, 8, 20

b. 2, 4, 6, 8, 30

27. Fibonacci Sums Make a conjecture for each of the following sums, where F_n represents the nth Fibonacci number.

a. $F_n + 2F_{n+1} + F_{n+2} = ?$

b. $F_n + F_{n+1} + F_{n+3} = ?$

28. Fibonacci Sums Make a conjecture for each of the following sums, where F_n represents the nth Fibonacci number.

a. $F_1 + F_2 + F_3 + F_4 + \cdots + F_n = ?$

b. $F_2 + F_4 + F_6 + \cdots + F_{2n} = ?$

29. Pascal's Triangle The triangular pattern in the following figure is known as **Pascal's triangle**. Pascal's triangle has intrigued mathematicians for hundreds of years. Although it is named after the mathematician Blaise Pascal (1623–1662), there is evidence that it was first developed in China in the 1300s. The numbers in Pascal's triangle are created in the following manner. Each row begins and ends with the number 1. Any other number in a row is the sum of the two closest numbers above it. For instance, the first 10 in row 5 is the sum of the first 4 and the 6 above it in row 4.

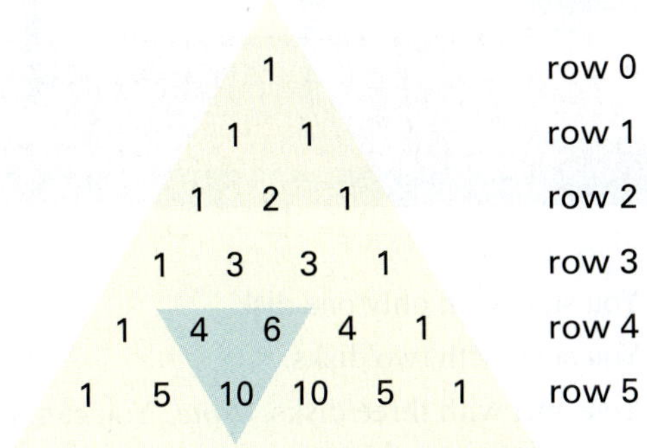

Pascal's triangle

There are many patterns that can be discovered in Pascal's triangle.

a. Find the sum of the numbers in each row, except row 0, of the portion of Pascal's triangle shown above. What pattern do you observe concerning these sums? Predict the sum of the numbers in row 9 of Pascal's triangle.

b. The numbers 1, 3, 6, 10, 15, ... , $\frac{n(n+1)}{2}$, ... are called *triangular numbers*. Where do the triangular numbers appear in Pascal's triangle?

30. A Savings Plan You save a penny on day 1. On each of the following days you save double the amount of money you saved on the previous day. How much money will you have after:

a. 5 days? **b.** 10 days? **c.** 15 days?

d. n days? *Hint:* $2^1 = 2$, $2^2 = 4$, $2^3 = 8$, $2^4 = 16$, $2^5 = 32$, ... , $2^{10} = 1024$, ... , $2^{15} = 32768$, ...

31. A Famous Puzzle The *Tower of Hanoi* is a puzzle invented by Edouard Lucas in 1883. The puzzle consists of three pegs and a number of disks of distinct diameters stacked on one of the pegs such that the largest disk is on the bottom, the next largest is placed on the largest disk, and so on as shown on page 26.

The object of the puzzle is to transfer the tower to one of the other pegs. The rules require that *only one disk be moved at a time* and that *a larger disk may not be placed on a smaller disk.* All pegs may be used.

Determine the *minimum* number of moves required to transfer all of the disks to another peg for each of the following situations.

a. You start with only one disk.

b. You start with two disks.

c. You start with three disks. (*Note:* You can use a stack of various size coins to simulate the puzzle, or you can use one of the many websites that provide a simulation of the puzzle.)

d. You start with four disks.

e. You start with five disks.

f. You start with n disks.

g. Lucas included with the Tower puzzle a legend about a tower that had 64 gold disks on one of three diamond needles. A group of priests had the task of transferring the 64 disks to one of the other needles using the same rules as the Tower of Hanoi puzzle. When they had completed the transfer, the tower would crumble and the universe would cease to exist. Assuming that the priests could transfer one disk to another needle every second, how many years would it take them to transfer all of the 64 disks to one of the other needles?

32. Use the recursive definition for Fibonacci numbers and *deductive* reasoning to verify that, for Fibonacci numbers, $2F_n - F_{n-2} = F_{n+1}$ for $n \geq 3$. *Hint:* By definition, $F_{n+1} = F_n + F_{n-1}$ and $F_n = F_{n-1} + F_{n-2}$.

SECTION 1.3 Problem-Solving Strategies

Polya's Problem-Solving Strategy

Ancient mathematicians such as Euclid and Pappus were interested in solving mathematical problems, but they were also interested in *heuristics*, the study of the methods and rules of discovery and invention. In the seventeenth century, the mathematician and philosopher René Descartes (1596–1650) contributed to the field of heuristics. He tried to develop a universal problem-solving method. Although he did not achieve this goal, he did publish some of his ideas in *Rules for the Direction of the Mind* and his better-known work *Discourse de la Methode.*

Another mathematician and philosopher, Gottfried Wilhelm Leibnitz (1646–1716), planned to write a book on heuristics titled *Art of Invention.* Of the problem-solving process, Leibnitz wrote, "Nothing is more important than to see the sources of invention which are, in my opinion, more interesting than the inventions themselves."

One of the foremost recent mathematicians to make a study of problem solving was George Polya (1887–1985). He was born in Hungary and moved to the United States in 1940. The basic problem-solving strategy that Polya advocated consisted of the following four steps.

Polya's Four-Step Problem-Solving Strategy

1. Understand the problem.
2. Devise a plan.
3. Carry out the plan.
4. Review the solution.

Polya's four steps are deceptively simple. To become a good problem solver, it helps to examine each of these steps and determine what is involved.

HISTORICAL NOTE

Chuck Painter/Stanford News Service

George Polya

After a brief stay at Brown University, George Polya (pōl'yə) moved to Stanford University in 1942 and taught there until his retirement. While at Stanford, he published 10 books and a number of articles for mathematics journals. Of the books Polya published, *How to Solve It* (1945) is one of his best known. In this book, Polya outlines a strategy for solving problems from virtually any discipline.

"A great discovery solves a great problem but there is a grain of discovery in the solution of any problem. Your problem may be modest; but if it challenges your curiosity and brings into play your inventive faculties, and if you solve it by your own means, you may experience the tension and enjoy the triumph of discovery."

Understand the Problem This part of Polya's four-step strategy is often overlooked. You must have a clear understanding of the problem. To help you focus on understanding the problem, consider the following questions.

- Can you restate the problem in your own words?
- Can you determine what is known about these types of problems?
- Is there missing information that, if known, would allow you to solve the problem?
- Is there extraneous information that is not needed to solve the problem?
- What is the goal?

Devise a Plan Successful problem solvers use a variety of techniques when they attempt to solve a problem. Here are some frequently used procedures.

- Make a list of the known information.
- Make a list of information that is needed.
- Draw a diagram.
- Make an organized list that shows all the possibilities.
- Make a table or a chart.
- Work backwards.
- Try to solve a similar but simpler problem.
- Look for a pattern.
- Write an equation. If necessary, define what each variable represents.
- Perform an experiment.
- Guess at a solution and then check your result.

Carry Out the Plan Once you have devised a plan, you must carry it out.

- Work carefully.
- Keep an accurate and neat record of all your attempts.
- Realize that some of your initial plans will not work and that you may have to devise another plan or modify your existing plan.

Review the Solution Once you have found a solution, check the solution.

- Ensure that the solution is consistent with the facts of the problem.
- Interpret the solution in the context of the problem.
- Ask yourself whether there are generalizations of the solution that could apply to other problems.

In Example 1 we apply Polya's four-step problem-solving strategy to solve a problem involving the number of routes between two points.

EXAMPLE 1 **Apply Polya's Strategy** (Solve a similar but simpler problem)

Consider the map shown in Figure 1.2. Allison wishes to walk along the streets from point A to point B. How many direct routes can Allison take?

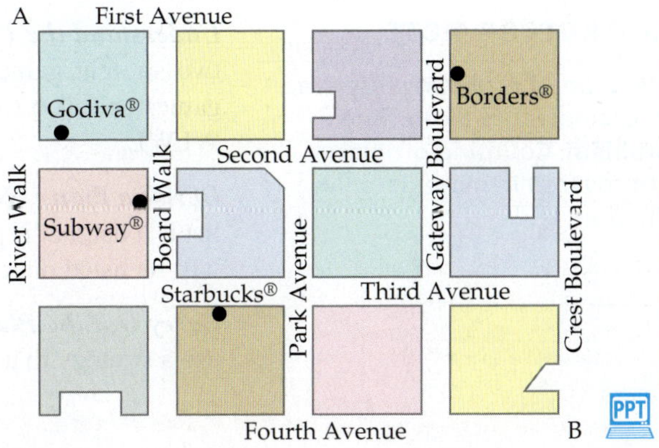

FIGURE 1.2 City map

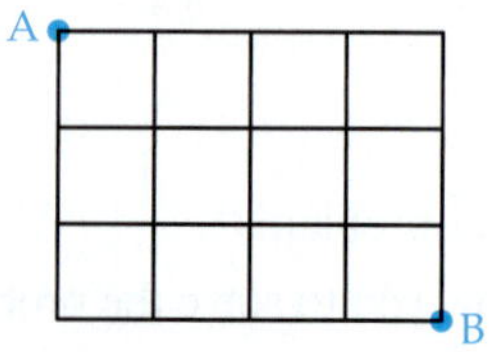

A simple diagram of the street map in Figure 1.2

Solution

Understand the Problem We would not be able to answer the question if Allison retraced her path or traveled away from point B. Thus we assume that on a direct route, she always travels along a street in a direction that gets her closer to point B.

Devise a Plan The map in Figure 1.2 has many extraneous details. Thus we make a diagram that allows us to concentrate on the essential information. See the figure at the left.

Because there are many routes, we consider the similar but simpler diagrams shown below. The number at each street intersection represents the number of routes from point A to that particular intersection.

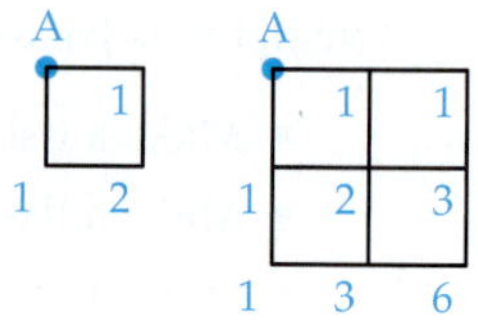

Simple street diagrams

TAKE NOTE

The strategy of working a similar but simpler problem is an important problem-solving strategy that can be used to solve many problems.

Look for patterns. It appears that the number of routes to an intersection is the *sum* of the number of routes to the adjacent intersection to its left and the number of routes to the intersection directly above. For instance, the number of routes to the intersection labeled 6 is the sum of the number of routes to the intersection to its left, which is 3, and the number of routes to the intersection directly above, which is also 3.

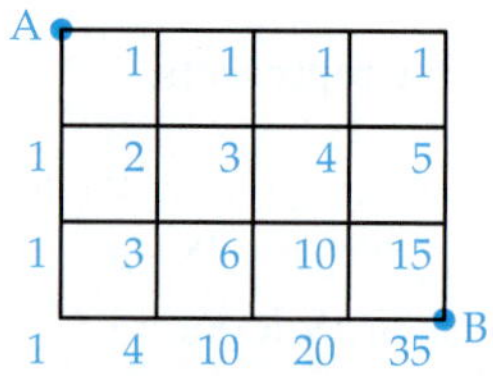

A street diagram with the number of routes to each intersection labeled

Carry Out the Plan Using the pattern discovered above, we see from the figure at the left that the number of routes from point A to point B is $20 + 15 = 35$.

Review the Solution Ask yourself whether a result of 35 seems reasonable. If you were required to draw each route, could you devise a scheme that would enable you to draw each route without missing a route or duplicating a route?

CHECK YOUR PROGRESS 1 Consider the street map in Figure 1.2 on page 27. Allison wishes to walk directly from point A to point B. How many different routes can she take if she wants to go past Starbucks on Third Avenue?

Solution See page S2.

Example 2 illustrates the technique of using an organized list.

EXAMPLE 2 Apply Polya's Strategy (Make an organized list)

A baseball team won two out of their last four games. In how many different orders could they have two wins and two losses in four games?

Solution

Understand the Problem There are many different orders. The team may have won two straight games and lost the last two (WWLL). Or maybe they lost the first two games and won the last two (LLWW). Of course there are other possibilities, such as WLWL.

INSTRUCTOR NOTE

The type of problem illustrated in Example 2 is a counting problem. Counting problems are covered in more detail in Chapter 12.

Devise a Plan We will make an *organized list* of all the possible orders. An organized list is a list that is produced using a system that ensures that each of the different orders will be listed once and only once.

Carry Out the Plan Each entry in our list must contain two Ws and two Ls. We will use a strategy that makes sure each order is considered, with no duplications. One such strategy is to always write a W unless doing so will produce too many Ws or a duplicate of one of the previous orders. If it is not possible to write a W, then and only then do we write an L. This strategy produces the six different orders shown below.

1. WWLL (Start with two wins)
2. WLWL (Start with one win)
3. WLLW
4. LWWL (Start with one loss)
5. LWLW
6. LLWW (Start with two losses)

Review the Solution We have made an organized list. The list has no duplicates and the list considers all possibilities, so we are confident that there are six different orders in which a baseball team can win exactly two out of four games.

CHECK YOUR PROGRESS 2 A true-false quiz contains five questions. In how many ways can a student answer the questions if the student answers two of the questions with "false" and the other three with "true"?

Solution See page S3.

In Example 3 we make use of several problem-solving strategies to solve a problem involving the total number of games to be played.

EXAMPLE 3 Apply Polya's Strategy (Solve a similar but simpler problem)

In a basketball league consisting of 10 teams, each team plays each of the other teams exactly three times. How many league games will be played?

Solution

Understand the Problem There are 10 teams in the league, and each team plays exactly three games against each of the other teams. The problem is to determine the total number of league games that will be played.

Devise a Plan Try the strategy of working a similar but simpler problem. Consider a league with only four teams (denoted by A, B, C, and D) in which each team plays each of the other teams only once. The diagram at the left illustrates that the games can be represented by line segments that connect the points A, B, C, and D.

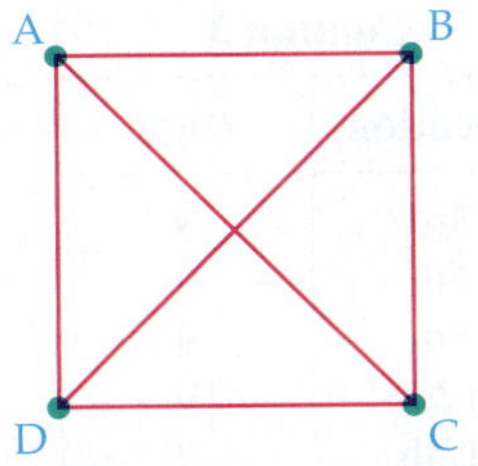

The possible pairings of a league with only four teams

Since each of the four teams will play a game against each of the other three, we might conclude that this would result in $4 \cdot 3 = 12$ games. However, the diagram shows only six line segments. It appears that our procedure has counted each game twice. For instance, when team A plays team B, team B also plays team A. To produce the correct result, we must divide our previous result, 12, by 2. Hence, four teams can play each other once in $\frac{4 \cdot 3}{2} = 6$ games.

Carry Out the Plan Using the process developed above, we see that 10 teams can play each other *once* in a total of $\frac{10 \cdot 9}{2} = 45$ games. Since each team plays each opponent exactly three times, the total number of games is $45 \cdot 3 = 135$.

AB AC AD AE AF AG AH AI AJ
BC BD BE BF BG BH BI BJ
CD CE CF CG CH CI CJ
DE DF DG DH DI DJ
EF EG EH EI EJ
FG FH FI FJ
GH GI GJ
HI HJ
IJ

An organized list of all possible games

Review the Solution We could check our work by making a diagram that includes all 10 teams represented by dots labeled A, B, C, D, E, F, G, H, I, and J. Because this diagram would be somewhat complicated, let's try the method of making an organized list. The figure at the left shows an organized list in which the notation BC represents a game between team B and team C. The notation CB is not shown because it also represents a game between team B and team C. This list shows that 45 games are required for each team to play each of the other teams once. Also notice that the first row has nine items, the second row has eight items, the third row has seven items, and so on. Thus 10 teams require

$$9 + 8 + 7 + 6 + 5 + 4 + 3 + 2 + 1 = 45$$

games if each team plays every other team once, and $45 \cdot 3 = 135$ games if each team plays exactly three games against each opponent.

CHECK YOUR PROGRESS 3 If six people greet each other at a meeting by shaking hands with one another, how many handshakes will take place?

Solution See page S3.

In Example 4 we make use of a table to solve a problem.

CALCULATOR NOTE

Some calculators display $\frac{7}{27}$ as 0.25925925926. However, the last digit 6 is not correct. It is a result of the rounding process. The actual decimal representation of $\frac{7}{27}$ is the decimal 0.259259... or $0.\overline{259}$, in which the digits continue to repeat the 259 pattern forever.

EXAMPLE 4 **Apply Polya's Strategy** (Make a table and look for a pattern)

Determine the digit 100 places to the right of the decimal point in the decimal representation $\frac{7}{27}$.

Solution

Understand the Problem Express the fraction $\frac{7}{27}$ as a decimal and look for a pattern that will enable us to determine the digit 100 places to the right of the decimal point.

Devise a Plan Dividing 27 into 7 by long division or by using a calculator produces the decimal 0.259259259... . Since the decimal representation repeats the digits 259 over and over forever, we know that the digit located 100 places to the right of the decimal point is either a 2, a 5, or a 9. A table may help us to see a pattern and enable us to determine which one of these digits is in the 100th place. Since the decimal digits repeat every three digits, we use a table with three columns.

The First 15 Decimal Digits of $\frac{7}{27}$

Column 1		Column 2		Column 3	
Location	**Digit**	**Location**	**Digit**	**Location**	**Digit**
1st	2	2nd	5	3rd	9
4th	2	5th	5	6th	9
7th	2	8th	5	9th	9
10th	2	11th	5	12th	9
13th	2	14th	5	15th	9
⋮		⋮		⋮	

Carry Out the Plan Only in column 3 is each of the decimal digit *locations* evenly divisible by 3. From this pattern we can tell that the 99th decimal digit (because 99 is evenly divisible by 3) must be a 9. Since a 2 always follows a 9 in the pattern, the 100th decimal digit must be a 2.

Review the Solution The above table illustrates additional patterns. For instance, if each of the location numbers in column 1 is divided by 3, a remainder of 1 is produced. If each of the location numbers in column 2 is divided by 3, a remainder of 2 is produced. Thus we can find the decimal digit in any location by dividing the location number by 3 and examining the remainder. For instance, to find the digit in the 3200th decimal place of $\frac{7}{27}$, merely divide 3200 by 3 and examine the remainder, which is 2. Thus, the digit 3200 places to the right of the decimal point is a 5.

CHECK YOUR PROGRESS 4 Determine the units digit (ones digit) of 4^{200}.

Solution See page S3.

Example 5 illustrates the method of working backwards. In problems in which you know a final result, this method may require the least effort.

CaseyMartin/Shutterstock.com

EXAMPLE 5 Apply Polya's Strategy (Work backwards)

In consecutive turns of a Monopoly game, Stacy first paid \$800 for a hotel. She then lost half her money when she landed on Boardwalk. Next, she collected \$200 for passing GO. She then lost half her remaining money when she landed on Illinois Avenue. Stacy now has \$2500. How much did she have just before she purchased the hotel?

Solution

Understand the Problem We need to determine the number of dollars that Stacy had just prior to her \$800 hotel purchase.

Devise a Plan We could guess and check, but we might need to make several guesses before we found the correct solution. An algebraic method might work, but setting up the necessary equation could be a challenge. Since we know the end result, let's try the method of working backwards.

Carry Out the Plan Stacy must have had \$5000 just before she landed on Illinois Avenue; \$4800 just before she passed GO; and \$9600 prior to landing on Boardwalk. This means she had \$10,400 just before she purchased the hotel.

Review the Solution To check our solution we start with \$10,400 and proceed through each of the transactions. \$10,400 less \$800 is \$9600. Half of \$9600 is \$4800. \$4800 increased by \$200 is \$5000. Half of \$5000 is \$2500.

TAKE NOTE

Example 5 can also be worked by using algebra. Let A be the amount of money Stacy had just before she purchased the hotel. Then

$$\frac{1}{2}\left[\frac{1}{2}(A - 800) + 200\right] = 2500$$
$$\frac{1}{2}(A - 800) + 200 = 5000$$
$$A - 800 + 400 = 10{,}000$$
$$A - 400 = 10{,}000$$
$$A = 10{,}400$$

CHECK YOUR PROGRESS 5 Melody picks a number. She doubles the number, squares the result, divides the square by 3, subtracts 30 from the quotient, and gets 18. What are the possible numbers that Melody could have picked? What operation does Melody perform that prevents us from knowing with 100% certainty which number she picked?

Solution See page S3.

MATH MATTERS A Mathematical Prodigy

Walter Sanders/TIME & LIFE Images/Getty Images

Carl Friedrich Gauss (1887–1985)

Carl Friedrich Gauss (gous) was a scientist and mathematician. His work encompassed several disciplines, including number theory, analysis, astronomy, and optics. He is known for having shown mathematical prowess as early as age three. It is reported that soon after Gauss entered elementary school, his teacher assigned the problem of finding the sum of the first 100 natural numbers. Gauss was able to determine the sum in a matter of a few seconds. The following solution shows the thought process he used.

Understand the Problem The sum of the first 100 natural numbers is represented by

$$1 + 2 + 3 + \cdots + 98 + 99 + 100$$

Devise a Plan Adding the first 100 natural numbers from left to right would be time consuming. Gauss considered another method. He added 1 and 100 to produce 101. He noticed that 2 and 99 have a sum of 101, and that 3 and 98 have a sum of 101. Thus the 100 numbers could be thought of as 50 pairs, each with a sum of 101. *continued*

Carry Out the Plan To find the sum of the 50 pairs, each with a sum of 101, Gauss computed $50 \cdot 101$ and arrived at 5050 as the solution.

Review the Solution Because the addends in an addition problem can be placed in any order without changing the sum, Gauss was confident that he had the correct solution.

An Extension The sum $1 + 2 + 3 + \cdots + (n - 2) + (n - 1) + n$ can be found by using the following formula.

A summation formula for the first *n* natural numbers:

$$1 + 2 + 3 + \cdots + (n - 2) + (n - 1) + n = \frac{n(n + 1)}{2}$$

Some problems can be solved by making guesses and checking. Your first few guesses may not produce a solution, but quite often they will provide additional information that will lead to a solution.

EXAMPLE 6 Apply Polya's Strategy (Guess and check)

The product of the ages, in years, of three teenagers is 4590. None of the teens are the same age. What are the ages of the teenagers?

Solution

Understand the Problem We need to determine three distinct counting numbers, from the list 13, 14, 15, 16, 17, 18, and 19, that have a product of 4590.

Devise a Plan If we represent the ages by x, y, and z, then $xyz = 4590$. We are unable to solve this equation, but we notice that 4590 ends in a zero. Hence, 4590 has a factor of 2 and a factor of 5, which means that at least one of the numbers we seek must be an even number and at least one number must have 5 as a factor. The only number in our list that has 5 as a factor is 15. Thus 15 is one of the numbers, and at least one of the other numbers must be an even number. At this point we try to solve by *guessing and checking.*

Carry Out the Plan

$15 \cdot 16 \cdot 18 = 4320$ • No. This product is too small.

$15 \cdot 16 \cdot 19 = 4560$ • No. This product is too small.

$15 \cdot 17 \cdot 18 = 4590$ • Yes. This is the correct product.

The ages of the teenagers are 15, 17, and 18.

Review the Solution Because $15 \cdot 17 \cdot 18 = 4590$ and each of the ages represents the age of a teenager, we know our solution is correct. None of the numbers 13, 14, 16, and 19 is a factor (divisor) of 4590, so there are no other solutions.

CHECK YOUR PROGRESS 6 Nothing is known about the personal life of the ancient Greek mathematician Diophantus except for the information in the following epigram. "Diophantus passed $\frac{1}{6}$ of his life in childhood, $\frac{1}{12}$ in youth, and $\frac{1}{7}$ more as a bachelor. Five years after his marriage was born a son who died four years before his father, at $\frac{1}{2}$ his father's (final) age."

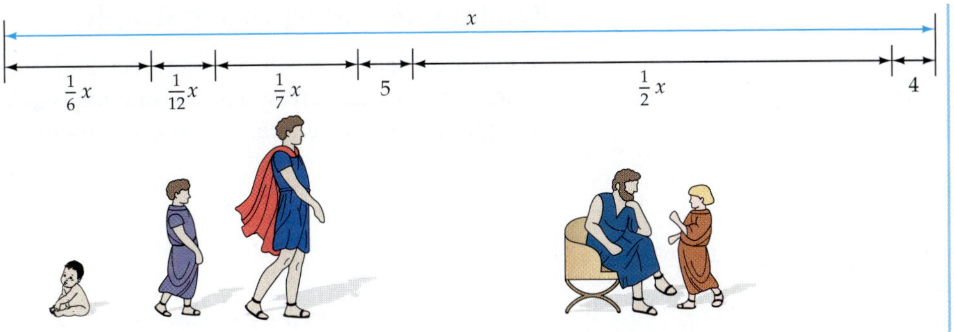

A diagram of the data, where x represents the age of Diophantus when he died

How old was Diophantus when he died? (*Hint:* Although an equation can be used to solve this problem, the method of guessing and checking will probably require less effort. Also assume that his age, when he died, is a counting number.)

Solution See page S3.

QUESTION Is the process of guessing at a solution and checking your result one of Polya's problem-solving strategies?

Some problems are deceptive. After reading one of these problems, you may think that the solution is obvious or impossible. These deceptive problems generally require that you carefully read the problem several times and that you check your solution to make sure it satisfies all the conditions of the problem.

EXAMPLE 7 Solve a Deceptive Problem

A hat and a jacket together cost \$100. The jacket costs \$90 more than the hat. What are the cost of the hat and the cost of the jacket?

Solution

Understand the Problem After reading the problem for the first time, you may think that the jacket costs \$90 and the hat costs \$10. The sum of these costs is \$100, but the cost of the jacket is only \$80 more than the cost of the hat. We need to find two dollar amounts that differ by \$90 and whose sum is \$100.

Devise a Plan Write an equation using h for the cost of the hat and $h + 90$ for the cost of the jacket.

$$h + h + 90 = 100$$

Carry Out the Plan Solve the above equation for h.

$$2h + 90 = 100 \quad \text{• Collect like terms.}$$
$$2h = 10 \quad \text{• Solve for } h.$$
$$h = 5$$

The cost of the hat is \$5 and the cost of the jacket is \$90 + \$5 = \$95.

Review the Solution The sum of the costs is \$5 + \$95 = \$100, and the cost of the jacket is \$90 more than the cost of the hat. This check confirms that the hat costs \$5 and the jacket costs \$95.

CHECK YOUR PROGRESS 7 Two U.S. coins have a total value of 35¢. One of the coins is not a quarter. What are the two coins?

Solution See page S3.

ANSWER Yes.

Reading and Interpreting Graphs

Graphs are often used to display numerical information in a visual format that allows the reader to see pertinent relationships and trends quickly. Three of the most common types of graphs are the *bar graph*, the *broken-line-graph*, and the *circle graph*.

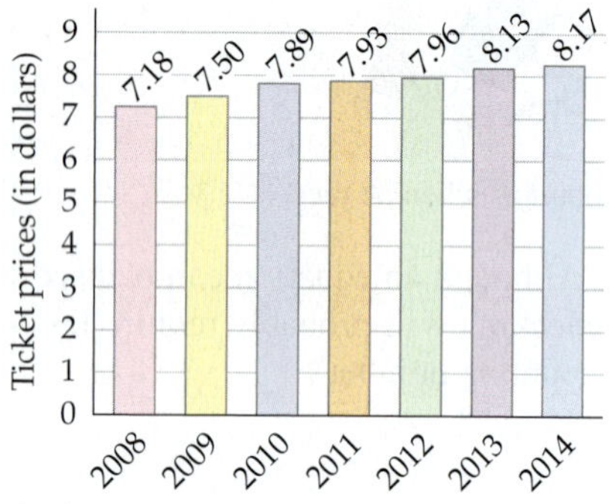

FIGURE 1.3 Average U.S. movie theatre ticket prices
SOURCE: National Association of Theatre Owners

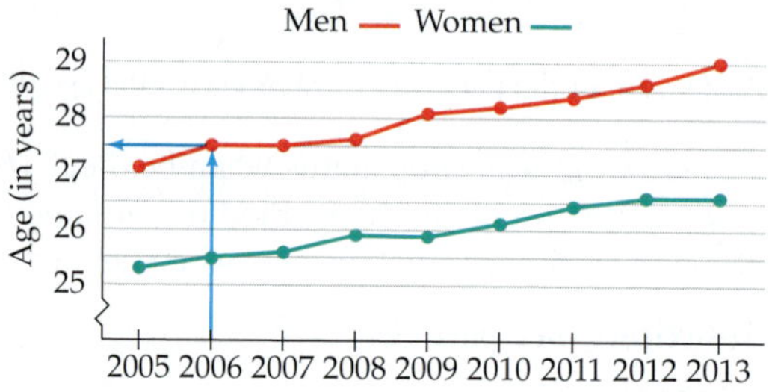

FIGURE 1.4 U.S. median age at first marriage
SOURCE: U.S. Census Bureau, U.S. Department of Commerce

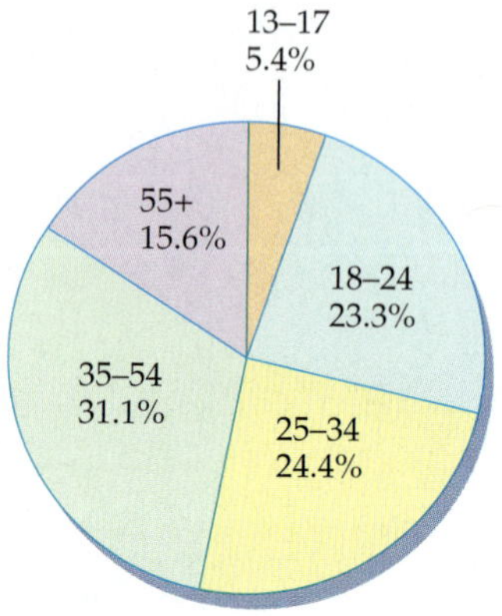

FIGURE 1.5 Classication of the 180,000,000 U.S. Facebook users by age: January 2014
SOURCE: iStrategylabs

Figure 1.3 is a bar graph that displays the average U.S. movie theatre ticket prices for the years from 2008 to 2014. The years are displayed on the horizontal axis. Each vertical bar is used to display the average ticket price for a given year. The higher the bar, the greater the average ticket price for that year.

Figure 1.4 shows two broken-line graphs. The red broken-line graph displays the median age at first marriage for men for the years from 2005 to 2013. The green broken-line graph displays the median age at first marriage for women during the same time period. The ⌇ symbol on the vertical axis indicates that the ages between 0 and 25 are not displayed. This break in the vertical axis allows the graph to be displayed in a compact form. The segments that connect points on the graph indicate trends. Increasing trends are indicated by segments that rise as they move to the right, and decreasing trends are indicated by segments that fall as they move to the right. The blue arrows in Figure 1.4 show that the median age at which men married for the first time in 2006 was 27.5 years, rounded to the nearest half of a year.

Figure 1.5 is a circle graph or pie chart that uses circular sectors to display the percentage of the 180,000,000 U.S. Facebook users in selected age groups as of January 2014.

EXAMPLE 8 Use Graphs to Solve Problems

a. Use Figure 1.3 to determine the minimum average U.S. movie theatre ticket price for the years from 2008 to 2014.

b. Use Figure 1.4 to estimate the median age at which women married for the first time in 2011. Round to the nearest half of a year.

c. Use Figure 1.5 to estimate the number of U.S. Facebook users in the 18–24 age group. Round to the nearest hundred thousand.

Solution

a. The minimum of the average ticket prices is displayed by the height of the shortest vertical bar in Figure 1.3. Thus the minimum average U.S. movie theatre ticket price for the years from 2008 to 2014 was \$7.18.

b. To estimate the median age at which women married for the first time in 2011, locate 2011 on the horizontal axis of Figure 1.4 and then move directly upward to a point on the green broken-line graph. The height of this point represents the median age at first marriage for women in 2011, and it can be estimated by moving horizontally to the vertical axis on the left. Thus the median age at first marriage for women in 2011 was 26.5 years, rounded to the nearest half of a year.

c. Figure 1.5 indicates that 23.3% of the 180,000,000 U.S. Facebook users were in the 18–24 age group.

$0.233 \cdot 180{,}000{,}000 = 41{,}940{,}000$

Thus, rounded to the nearest hundred thousand, the number of U.S. Facebook users in this age group was 41,900,000 in January 2014.

CHECK YOUR PROGRESS 8

a. Use Figure 1.3 to determine the maximum average U.S. movie theatre ticket price for the years from 2008 to 2014.

b. Use Figure 1.4 to estimate the median age at first marriage for men in 2011. Round to the nearest half of a year.

c. Use Figure 1.5 to estimate the number of U.S. Facebook users in the 25–34 age group. Round to the nearest hundred thousand.

Solution See page S4.

EXCURSION

Routes on a Probability Demonstrator

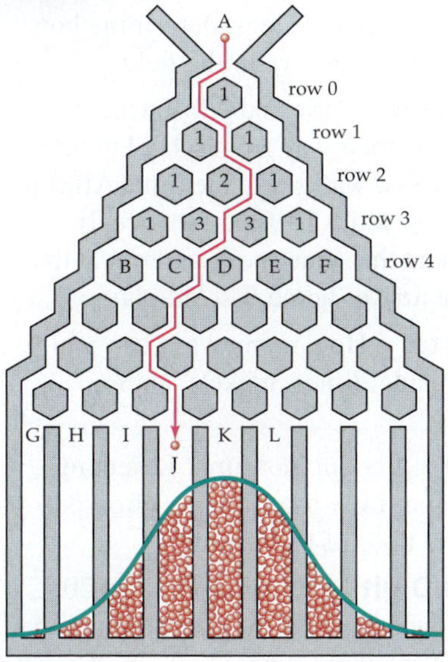

The object shown at the left is called a *Galton board* or *probability demonstrator.* It was invented by the English statistician Francis Galton (1822–1911). This particular board has 256 small red balls that are released so that they fall through an array of hexagons. The board is designed such that when a ball falls on a vertex of one of the hexagons, it is equally likely to fall to the left or to the right. As the ball continues its downward path, it strikes a vertex of a hexagon in the next row, where the process of falling to the left or to the right is repeated. After the ball passes through all the rows of hexagons, it falls into one of the bins at the bottom. In most cases the balls will form a *bell shape*, as shown by the green curve. Examine the numbers displayed in the hexagons in rows 0 through 3. Each number indicates the number of different routes that a ball can take from point A to the top of that particular hexagon.

EXCURSION EXERCISES

Use the probability demonstrator, in the left margin, to work Excursion Exercises 1 to 4.

1. How many routes can a ball take as it travels from A to B, from A to C, from A to D, from A to E, and from A to F? *Hint:* This problem is similar to Example 1 on page 27.
2. How many routes can a ball take as it travels from A to G, from A to H, from A to I, from A to J, and from A to K?
3. Explain how you know that the number of routes from A to J is the same as the number of routes from A to L.
4. Explain why the greatest number of balls tend to fall into the center bin.
5. The probability demonstrator shown to the right has nine rows of hexagons. Determine how many routes a ball can take as it travels from A to P, from A to Q, from A to R, from A to S, from A to T, and from A to U.

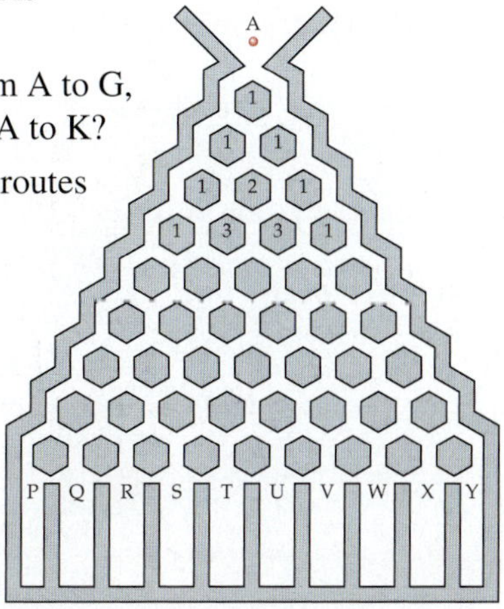

EXERCISE SET 1.3

■ Use Polya's four-step problem-solving strategy and the problem-solving procedures presented in this section to solve each of the following exercises.

1. **Number of Girls** There are 364 first-grade students in Park Elementary School. If there are 26 more girls than boys, how many girls are there?
2. **Heights of Ladders** If two ladders are placed end to end, their combined height is 31.5 feet. One ladder is 6.5 feet shorter than the other ladder. What are the heights of the two ladders?
3. **Number of Squares** How many squares are in the following figure?

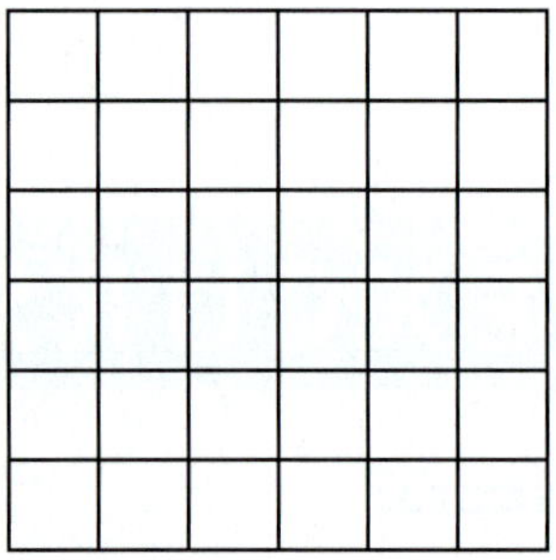

4. **Determine a Digit** What is the 44th decimal digit in the decimal representation of $\frac{1}{11}$?

$$\frac{1}{11} = 0.09090909\ldots$$

5. **Cost of a Shirt** A shirt and a tie together cost \$50. The shirt costs \$30 more than the tie. What is the cost of the shirt?
6. **Number of Games** In a basketball league consisting of 12 teams, each team plays each of the other teams exactly twice. How many league games will be played?
7. **Number of Routes** Consider the following map. Tyler wishes to walk along the streets from point A to point B. How many direct routes (no backtracking) can Tyler take?

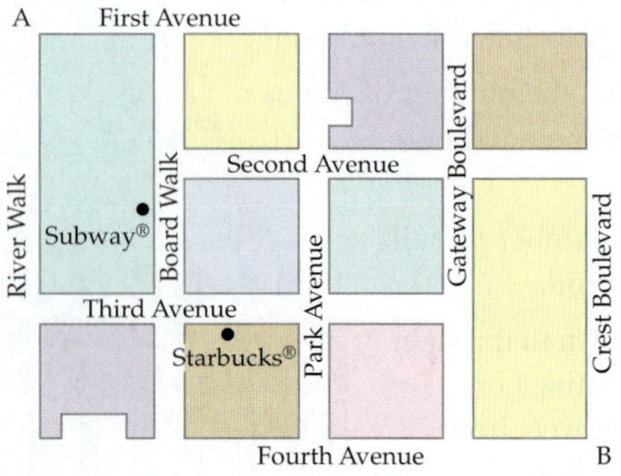

8. **Number of Routes** Use the map in Exercise 7 to answer each of the following.
   a. How many direct routes are there from A to B if you want to pass by Starbucks?
   b. How many direct routes are there from A to B if you want to stop at Subway for a sandwich?
   c. How many direct routes are there from A to B if you want to stop at Starbucks and at Subway?
9. **True-False Test** In how many ways can you answer a 12-question true-false test if you answer each question with either a "true" or a "false"?
10. **A Puzzle** A frog is at the bottom of a 17-foot well. Each time the frog leaps, it moves up 3 feet. If the frog has not reached the top of the well, then the frog slides back 1 foot before it is ready to make another leap. How many leaps will the frog need to escape the well?
11. **Number of Handshakes** If eight people greet each other at a meeting by shaking hands with one another, how many handshakes take place?
12. **Number of Line Segments** Twenty-four points are placed around a circle. A line segment is drawn between each pair of points. How many line segments are drawn?
13. **Number of Ducks and Pigs** The number of ducks and pigs in a field totals 35. The total number of legs among them is 98. Assuming each duck has exactly two legs and each pig has exactly four legs, determine how many ducks and how many pigs are in the field.
14. **Racing Strategies** Carla and Allison are sisters. They are on their way from school to home. Carla runs half the time and walks half the time. Allison runs half the distance and walks half the distance. If Carla and Allison walk at the same speed and run at the same speed, which one arrives home first? Explain.
15. **Change for a Quarter** How many ways can you make change for 25¢ using dimes, nickels, and/or pennies?
16. **Carpet for a Room** A room measures 12 feet by 15 feet. How many 3-foot by 3-foot squares of carpet are needed to cover the floor of this room?

■ **Determine the Units Digit** In Exercises 17 to 20, determine the units digit (ones digit) of the counting number represented by the exponential expression.

17. 4^{300} 18. 2^{725} 19. 3^{412} 20. 7^{146}

21. **Find Sums** Find the following sums without using a calculator. *Hint:* Apply the procedure used by Gauss. (See the Math Matters on page 31.)
   a. $1 + 2 + 3 + 4 + \cdots + 397 + 398 + 399 + 400$
   b. $1 + 2 + 3 + 4 + \cdots + 547 + 548 + 549 + 550$
   c. $2 + 4 + 6 + 8 + \cdots + 80 + 82 + 84 + 86$

22. Explain how you could modify the procedure used by Gauss (see the Math Matters on page 31) to find the following sum.

$$1 + 2 + 3 + 4 + \cdots + 62 + 63 + 64 + 65$$

23. Palindromic Numbers A **palindromic number** is a whole number that remains unchanged when its digits are written in reverse order. Find all palindromic numbers that have exactly

a. three digits and are the square of a natural number.

b. four digits and are the cube of a natural number.

24. Speed of a Car A car has an odometer reading of 15,951 miles, which is a palindromic number. (See Exercise 23.) After 2 hours of continuous driving at a constant speed, the odometer reading is the next palindromic number. How fast, in miles per hour, was the car being driven during these 2 hours?

25. A Puzzle Three volumes of the series *Mathematics: Its Content, Methods, and Meaning* are on a shelf with no space between the volumes. Each volume is 1 inch thick without its covers. Each cover is $\frac{1}{8}$ inch thick. See the following figure. A bookworm bores horizontally from the first page of Volume I to the last page of Volume III. How far does the bookworm travel?

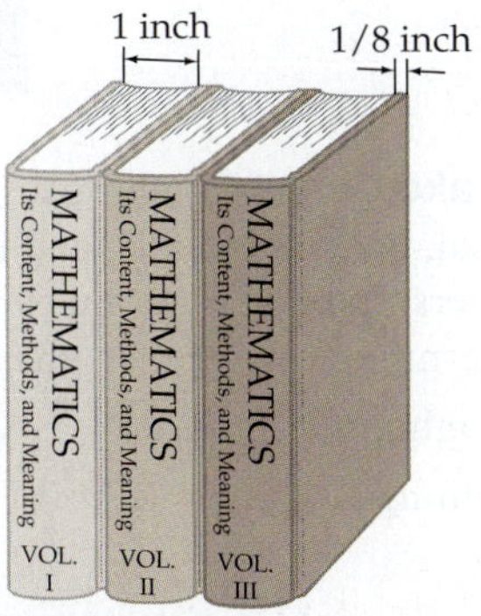

26. Connect the Dots Nine dots are arranged as shown. Is it possible to connect the nine dots with exactly four lines if you are not allowed to retrace any part of a line and you are not allowed to remove your pencil from the paper? If it can be done, demonstrate with a drawing.

27. Movie Theatre Admissions The following bar graph shows the number of U.S. and Canada movie theatre admissions for the years from 2007 to 2014.

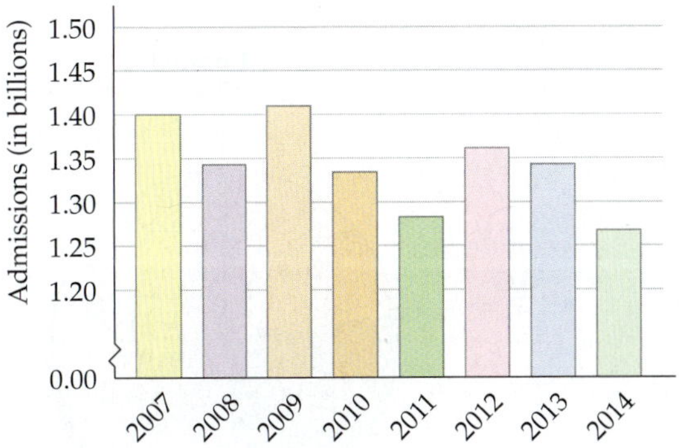

Total U.S. and Canada movie theatre admissions

SOURCE: National Association of Theatre Owners

a. Estimate the number of admissions for the year 2009. Round to the nearest tenth of a billion.

b. Which year had the least number of admissions?

c. Which year had the greatest number of admissions?

28. Box Office Revenues The following broken-line graph shows the U.S. and Canada movie theatre box office revenues, in billions of dollars, for the years from 2007 to 2014.

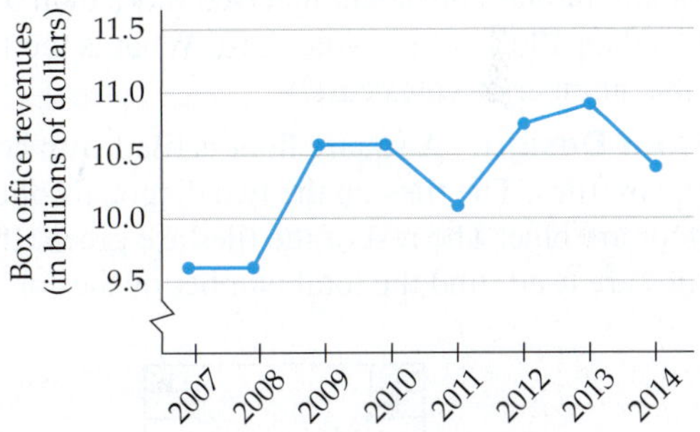

Total U.S. and Canada box office revenues

SOURCE: National Association of Theatre Owners

a. Which two years had the least box office revenues?

b. Which year had the greatest box office revenue?

c. During which two consecutive years did the box office revenues increase the most?

29. Movie Ratings and Box Office Revenue The following circle graph shows the percentage of the 10.4 billion dollar box office revenue attributed to each of the various movie ratings in 2014.

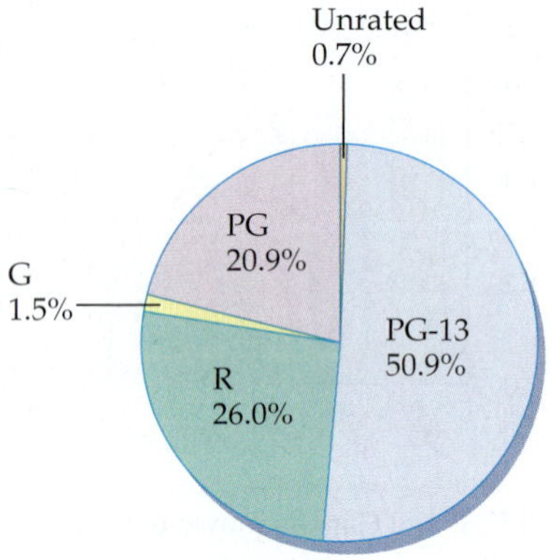

Percentage of the total box office revenue produced by each movie rating in 2014
SOURCE: Box Office Mojo

a. Which movie rating brought in the largest share of the 2014 box office revenue?

b. Determine the 2014 box office revenue produced by the PG-rated films. Round to the nearest tenth of a billion dollars.

30. Votes in an Election In a school election, one candidate for class president received more than 94%, but less than 100%, of the votes cast. What is the least possible number of votes cast?

31. Floor Design A square floor is tiled with congruent square tiles. The tiles on the two diagonals of the floor are blue. The rest of the tiles are green. If 101 blue tiles are used, find the total number of tiles on the floor.

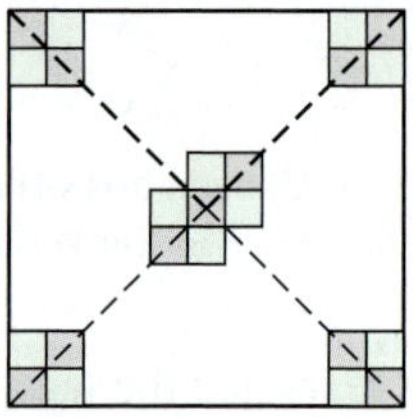

32. Number of Children How many children are there in a family wherein each girl has as many brothers as sisters, but each boy has twice as many sisters as brothers?

33. Brothers and Sisters I have two more sisters than brothers. Each of my sisters has two more sisters than brothers. How many more sisters than brothers does my youngest brother have?

34. A Coin Problem If you take 22 pennies from a pile of 57 pennies, how many pennies do you have?

35. Bacterial Growth The bacteria in a petri dish grow in a manner such that each day the number of bacteria doubles. On what day will the number of bacteria be half of the number present on the 12th day?

36. Number of River Crossings Four people on one side of a river need to cross the river in a boat that can carry a maximum load of 180 pounds. The weights of the people are 80, 100, 150, and 170 pounds.

a. Explain how the people can use the boat to get everyone to the opposite side of the river.

b. What is the minimum number of crossings that must be made by the boat?

37. Examination Scores On three examinations, Dana received scores of 82, 91, and 76. What score does Dana need on the fourth examination to raise his average to 85?

38. Puzzle from a Movie In the movie *Die Hard: With a Vengeance,* Bruce Willis and Samuel L. Jackson are given a 5-gallon jug and a 3-gallon jug and they must put *exactly* 4 gallons of water on a scale to keep a bomb from exploding. Explain how they could accomplish this feat.

Buyenlarge/Getty Images

39. Find the Fake Coin You have eight coins. They all look identical, but one is a fake and is slightly lighter than the others. Explain how you can use a balance scale to determine which coin is the fake in exactly

a. three weighings.

b. two weighings.

Problems from the Mensa Workout Mensa is a society that welcomes people from every walk of life whose IQ is in the top 2% of the population. The multiple-choice Exercises 40 to 43 are from the *Mensa Workout,* which is posted on the Internet at

www.mensa.org

40. If it were 2 hours later, it would be half as long until midnight as it would be if it were an hour later. What time is it now?

a. 18:30 **b.** 20:00 **c.** 21:00
d. 22:00 **e.** 23:30

41. Sally likes 225 but not 224; she likes 900 but not 800; she likes 144 but not 145. Which of the following does she like?

a. 1600 **b.** 1700

42. There are 1200 elephants in a herd. Some have pink and green stripes, some are all pink, and some are all blue. One third are pure pink. Is it true that 400 elephants are definitely blue?

a. Yes **b.** No

43. Following the pattern shown in the number sequence below, what is the missing number?

1 8 27 ? 125 216

a. 36 **b.** 45 **c.** 46 **d.** 64 **e.** 99

EXTENSIONS

44. Compare Exponential Expressions

a. How many times as large is $3^{(3^3)}$ than $(3^3)^3$?

b. How many times as large is $4^{(4^4)}$ than $(4^4)^4$? *Note:* Most calculators will not display the answer to this problem because it is too large. However, the answer can be determined in exponential form by applying the following properties of exponents.

$$(a^m)^n = a^{mn} \text{ and } \frac{a^m}{a^n} = a^{m-n}$$

45. A Famous Puzzle The mathematician Augustus De Morgan once wrote that he had the distinction of being x years old in the year x^2. He was 43 in the year 1849.

a. Explain why people born in the year 1980 might share the distinction of being x years old in the year x^2. *Note:* Assume x is a natural number.

b. What is the next year after 1980 for which people born in that year might be x years old in the year x^2?

46. Verify a Procedure Select a two-digit number between 50 and 100. Add 83 to your number. From this number form a new number by adding the digit in the hundreds place to the number formed by the other two digits (the digits in the tens place and the ones place). Now subtract this newly formed number from your original number. Your final result is 16. Use a deductive approach to show that the final result is always 16 regardless of which number you start with.

47. Numbering Pages How many digits does it take in total to number a book from page 1 to page 240?

48. Mini Sudoku Sudoku is a deductive reasoning, number-placement puzzle. The object in a 6 by 6 mini-Sudoku puzzle is to fill all empty squares so that the counting numbers 1 to 6 appear exactly once in each row, each column, and each of the 2 by 3 regions, which are delineated by the thick line segments. Solve the following 6 by 6 mini-Sudoku puzzle.

6	2			5	
		4	3		
	6		5		4
		1		3	
1		6	2		5
	4		1	6	

49. The Four 4s Problem The object of this exercise is to create mathematical expressions that use exactly four 4s and that simplify to a counting number from 1 to 20, inclusive. You are allowed to use the following mathematical symbols: $+$, $-$, $\times$, $\div$, $\sqrt{\ }$, (, and). For example,

$$\frac{4}{4} + \frac{4}{4} = 2,\ 4^{(4-4)} + 4 = 5, \text{ and}$$

$$4 - \sqrt{4} + 4 \times 4 = 18$$

50. A Cryptarithm The following puzzle is a famous *cryptarithm.*

```
  S E N D
+ M O R E
---------
M O N E Y
```

Each letter in the cryptarithm represents one of the digits 0 through 9. The leading digits, represented by S and M, are not zero. Determine which digit is represented by each of the letters so that the addition is correct. *Note:* A letter that is used more than once, such as M, represents the same digit in each position in which it appears.

CHAPTER 1 SUMMARY

The following table summarizes essential concepts in this chapter. The references given in the right-hand column list Examples and Exercises that can be used to test your understanding of a concept.

1.1 Inductive and Deductive Reasoning	
Inductive Reasoning Inductive reasoning is the process of reaching a general conclusion by examining specific examples. A conclusion based on inductive reasoning is called a conjecture. A conjecture may or may not be correct.	See **Examples 2 and 3** on pages 2 and 3, and then try Exercises 2, 3, and 43 on pages 41 and 44.
Deductive Reasoning Deductive reasoning is the process of reaching a conclusion by applying general assumptions, procedures, or principles.	See **Examples 6 and 7** on pages 6 and 7, and then try Exercises 1 and 25 on pages 41 and 42.
Counterexamples A statement is a true statement provided it is true in all cases. If you can find one case in which a statement is not true, called a counterexample, then the statement is a false statement.	See **Example 4** on page 4, and then try Exercises 5 and 6 on page 41.
1.2 Problem Solving with Patterns	
Sequences A sequence is an ordered list of numbers. Each number in a sequence is called a term of the sequence. The notation a_n is used to designate the nth term of a sequence. A formula that can be used to generate all the terms of a sequence is called an nth-term formula.	See **Example 2** on page 17, and then try Exercises 11 and 14 on page 41.
Difference Tables A difference table shows the differences between successive terms of a sequence, and in some cases it can be used to predict the next term in a sequence.	See **Example 1** on page 16, and then try Exercises 9 and 44 on pages 41 and 44.
Fibonacci Sequence Let F_n represent the nth Fibonacci number. Then the terms in the Fibonacci sequence are given by the recursive definition: $F_1 = 1, F_2 = 1$, and $F_n = F_{n-1} + F_{n-2}$ for $n \geq 3$	See **Example 3** on page 19, and then try Exercise 12 on page 41.
1.3 Problem-Solving Strategies	
Polya's Four-Step Problem-Solving Strategy **1.** Understand the problem. **2.** Devise a plan. **3.** Carry out the plan. **4.** Review the solution.	See **Examples 3 and 4** on pages 29 and 30, and then try Exercises 22, 23, and 33 on pages 42 and 43.
Graphs Bar graphs, circle graphs, and broken-line graphs are often used to display data in a visual format.	See **Example 8** on page 34, and then try Exercises 38 to 40 on pages 43 and 44.

CHAPTER 1 REVIEW EXERCISES

■ In Exercises 1 to 4, determine whether the argument is an example of inductive reasoning or deductive reasoning.

1. All books written by J. K. Rowling make the best-seller list. The book *Harry Potter and the Deathly Hallows* is a J. K. Rowling book. Therefore, *Harry Potter and the Deathly Hallows* made the bestseller list.
2. Samantha got an A on each of her first four math tests, so she will get an A on the next math test.
3. We had rain each day for the last five days, so it will rain today.
4. All amoeba multiply by dividing. I have named the amoeba shown in my microscope Amelia. Therefore, Amelia multiplies by dividing.
5. Find a counterexample to show that the following conjecture is false.

   Conjecture: For all numbers x, $x^4 > x$.
6. Find a counterexample to show that the following conjecture is false.

   Conjecture: For all counting numbers n, $\dfrac{n^3 + 5n + 6}{6}$ is an even counting number.
7. Find a counterexample to show that the following conjecture is false.

   Conjecture: For all numbers x, $(x + 4)^2 = x^2 + 16$.
8. Find a counterexample to show that the following conjecture is false.

   Conjecture: For numbers a and b, $(a + b)^3 = a^3 + b^3$.
9. Use a difference table to predict the next term of each sequence.

   a. $-2, 2, 12, 28, 50, 78, ?$

   b. $-4, -1, 14, 47, 104, 191, 314, ?$
10. Use a difference table to predict the next term of each sequence.

    a. $5, 6, 3, -4, -15, -30, -49, ?$

    b. $2, 0, -18, -64, -150, -288, -490, ?$
11. A sequence has an nth-term formula of

    $$a_n = 4n^2 - n - 2$$

    Use the nth-term formula to determine the first five terms of the sequence and the 20th term of the sequence.
12. The first six terms of the Fibonacci sequence are:

    1, 1, 2, 3, 5, and 8.

    Determine the 11th and 12th terms of the Fibonacci sequence.

■ In Exercises 13 to 16, determine the nth-term formula for the number of square tiles in the nth figure.

13.

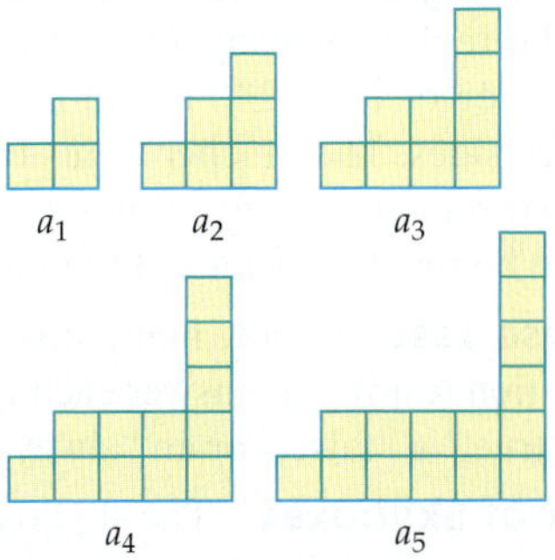

14.

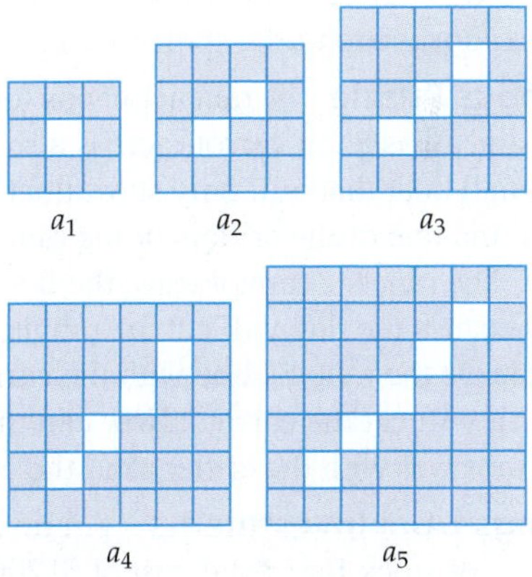

15.

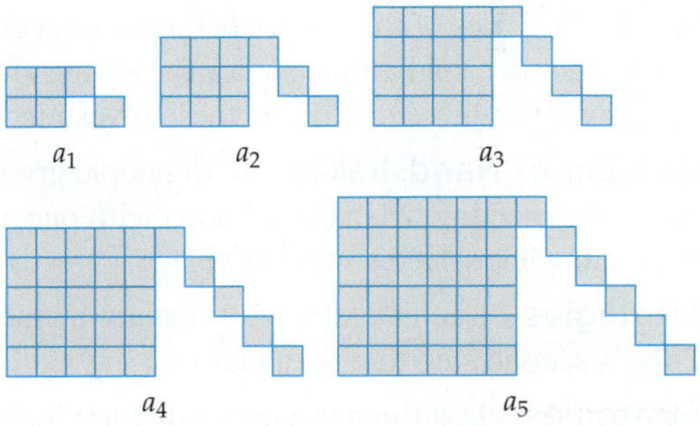

16.

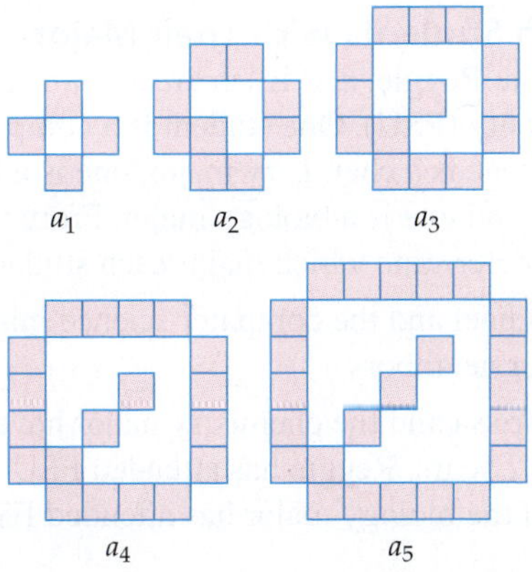

■ **Polya's Problem-Solving Strategy** In Exercises 17 to 22, solve each problem using Polya's four-step problem-solving strategy. Label your work so that each of Polya's four steps is identified.

17. Enclose a Region A rancher decides to enclose a rectangular region by using an existing fence along one side of the region and 2240 feet of new fence on the other three sides. The rancher wants the length of the rectangular region to be five times as long as its width. What will be the dimensions of the rectangular region?

18. True-False Test In how many ways can you answer a 15-question test if you answer each question with either a "true," a "false," or an "always false"?

19. Number of Skyboxes The skyboxes at a large sports arena are equally spaced around a circle. The 11th skybox is directly opposite the 35th skybox. How many skyboxes are in the sports arena?

20. A Famous Puzzle A rancher needs to get a dog, a rabbit, and a basket of carrots across a river. The rancher has a small boat that will only stay afloat carrying the rancher and one of the critters or the rancher and the carrots. The rancher cannot leave the dog alone with the rabbit because the dog will eat the rabbit. The rancher cannot leave the rabbit alone with the carrots because the rabbit will eat the carrots. How can the rancher get across the river with the critters and the carrots?

21. Earnings from Investments An investor bought 20 shares of stock for a total cost of \$1200 and then sold all the shares for \$1400. A few months later, the investor bought 25 shares of the same stock for a total cost of \$1800 and then sold all the shares for \$1900. How much money did the investor earn on these investments?

22. Number of Handshakes If 15 people greet each other at a meeting by shaking hands with one another, how many handshakes will take place?

23. Strategies List five strategies that are included in Polya's second step (devise a plan).

24. Strategies List three strategies that are included in Polya's fourth step (review the solution).

25. Match Students with Their Major Michael, Clarissa, Reggie, and Ellen are attending Florida State University (FSU). One student is a computer science major, one is a chemistry major, one is a business major, and one is a biology major. From the following clues, determine which major each student is pursuing.

a. Michael and the computer science major are next-door neighbors.

b. Clarissa and the chemistry major have attended FSU for 2 years. Reggie has attended FSU for 3 years, and the biology major has attended FSU for 4 years.

c. Ellen has attended FSU for fewer years than Michael.

d. The business major has attended FSU for 2 years.

26. Little League Baseball Each of the Little League teams in a small rural community is sponsored by a different local business. The names of the teams are the Dodgers, the Pirates, the Tigers, and the Giants. The businesses that sponsor the teams are the bank, the supermarket, the service station, and the drugstore. From the following clues, determine which business sponsors each team.

a. The Tigers and the team sponsored by the service station have winning records this season.

b. The Pirates and the team sponsored by the bank are coached by parents of the players, whereas the Giants and the team sponsored by the drugstore are coached by the director of the community center.

c. Jake is the pitcher for the team sponsored by the supermarket and coached by his father.

d. The game between the Tigers and the team sponsored by the drugstore was rained out yesterday.

27. Map Coloring The following map shows six countries in the Indian subcontinent. Four colors have been used to color the countries such that no two bordering countries are the same color.

a. Can this map be colored using only three colors, such that no two bordering countries are the same color? Explain.

b. Can this map be colored using only two colors, such that no two bordering countries are the same color? Explain.

28. Find a Route The following map shows the 10 bridges and 3 islands between the suburbs of North Bay and South Bay.

a. During your morning workout, you decide to jog over each bridge exactly once. Draw a route that you can take. Assume that you start from North Bay and that your workout concludes after you jog over the 10th bridge.

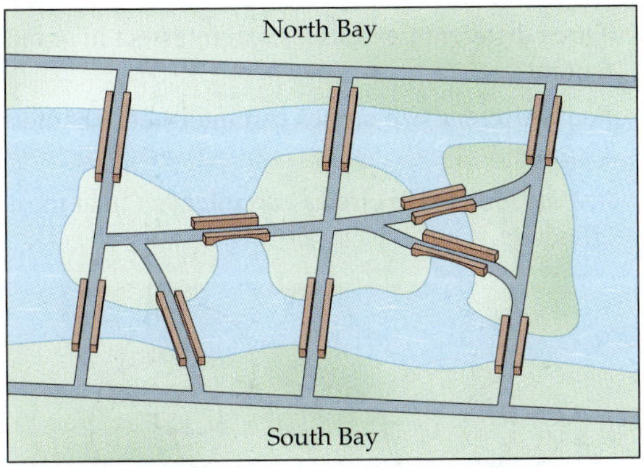

b. Assume you start your jog from South Bay. Can you find a route that crosses each bridge exactly once?

29. Areas of Rectangles Two perpendicular line segments partition the interior of a rectangle into four smaller rectangles. The areas of these smaller rectangles are x, 2, 5, and 10 square inches. Find all possible values of x.

30. Use a Pattern to Make Predictions Consider the following figures.

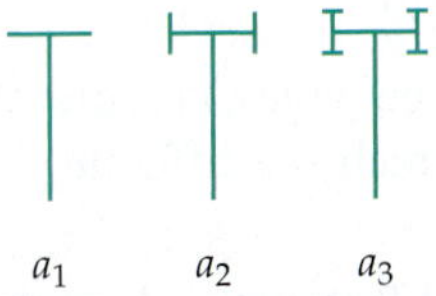

Figure a_1 consists of two line segments, and figure a_2 consists of four line segments. If the pattern of adding a smaller line segment to each end of the shortest line segments continues, how many line segments will be in

a. figure a_{10}? **b.** figure a_{30}?

31. A Cryptarithm In the following addition problem, each letter represents one of the digits 0, 1, 2, 3, 4, 5, 6, 7, 8, or 9. The leading digits represented by A and B are nonzero digits. What digit is represented by each letter?

$$\begin{array}{r} A \\ +\,BB \\ \hline ADD \end{array}$$

32. Make Change In how many different ways can change be made for a dollar using only quarters and/or nickels?

33. Counting Problem In how many different orders can a basketball team win exactly three out of their last five games?

■ **Units Digit** In Exercises 34 and 35, determine the units digit (ones digit) of the exponential expression.

34. 7^{56} **35.** 23^{85}

36. Verify a Conjecture Use deductive reasoning to show that the following procedure always produces a number that is twice the original number.

Procedure: Pick a number. Multiply the number by 4, add 12 to the product, divide the sum by 2, and subtract 6.

37. Explain why 2004 nickels are worth more than $100.

38. Gasoline Prices The following bar graph shows the average U.S. unleaded regular gasoline prices for the years from 2007 to 2013.

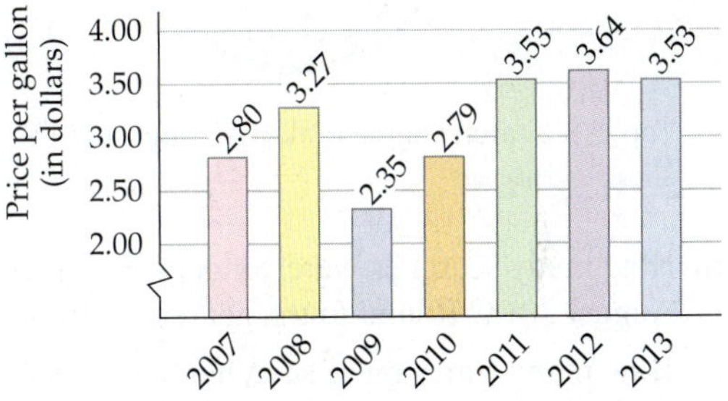

Average U.S. retail gasoline prices for unleaded regular

SOURCE: The World Almanac and Book of Facts 2015

a. What was the maximum average price per gallon during the years from 2007 to 2013?

b. During which two consecutive years did the largest price increase occur?

39. Super Bowl Ad Price The following graph shows the price for a 30-second Super Bowl ad from 2008 to 2015.

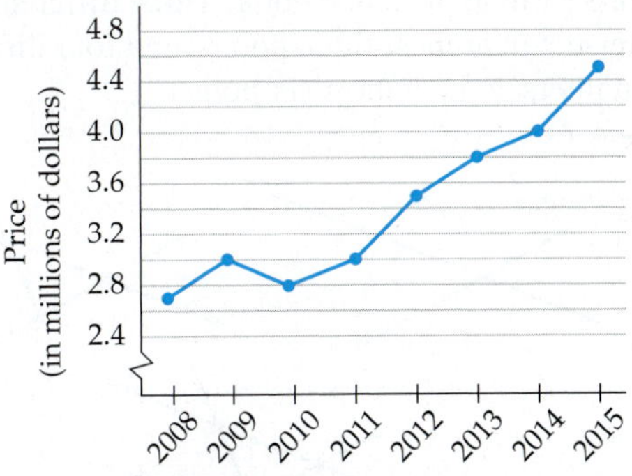

Super Bowl prices for a 30-second ad

SOURCE: Boyds Bets

a. What was the price of a 30-second Super Bowl ad in 2013? Round to the nearest tenth of a million dollars.

b. In 2015, about 118.5 million people watched the Super Bowl on television. What was the price per viewer that was paid by an advertiser that purchased a $4.5 million ad? Round to the nearest cent per viewer.

40. Search Engine Rankings The following circle graph shows the percent of the 18.6 billion U.S. searches that were conducted in August 2015 by the top five search engines.

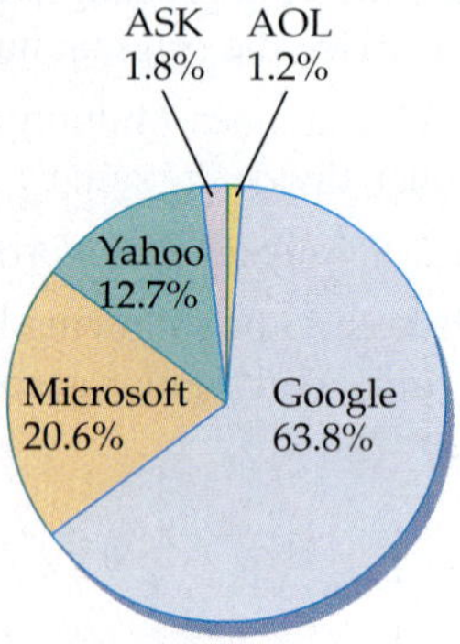

Top U.S. search engine rankings: August 2015
SOURCE: comScore

a. How many searches were conducted by Google in August 2015? Round to the nearest tenth of a billion.

b. How many times more searches were conducted by Yahoo than by ASK in August 2015? Round to the nearest tenth.

41. Palindromic Numbers Recall that palindromic numbers read the same from left to right as they read from right to left. For instance, 37,573 is a palindromic number. Find the smallest palindromic number larger than 1000 that is a multiple of 5.

42. Narcissistic Numbers A **narcissistic number** is a two-digit natural number that is equal to the sum of the squares of its digits. Find all narcissistic numbers.

43. Number of Intersections Two different lines can intersect in at most one point. Three different lines can intersect in at most three points, and four different lines can intersect in at most six points.

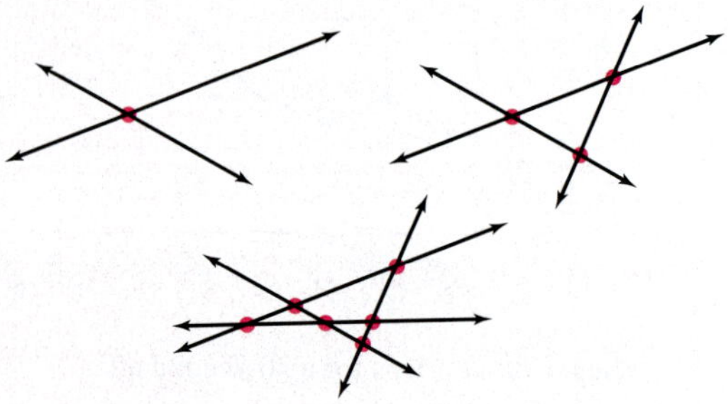

a. Determine the maximum number of intersections for five different lines.

b. Does it appear, by inductive reasoning, that the maximum number of intersection points I_n for n different lines is given by $I_n = \dfrac{n(n-1)}{2}$?

44. Number of Intersections

Two different size circles can intersect in at most 2 points.

Three different size circles can intersect in at most 6 points.

Four different size circles can intersect in at most 12 points.

Five different size circles can intersect in at most 20 points.

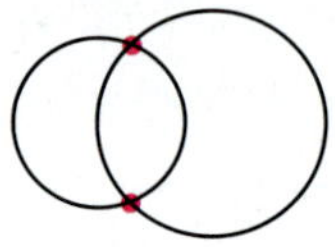

Two circles
2 intersections

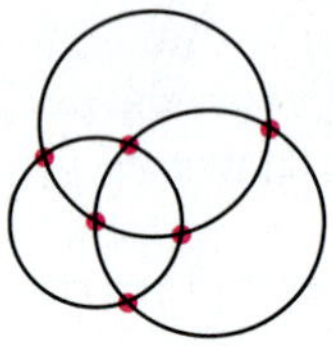

Three circles
6 intersections

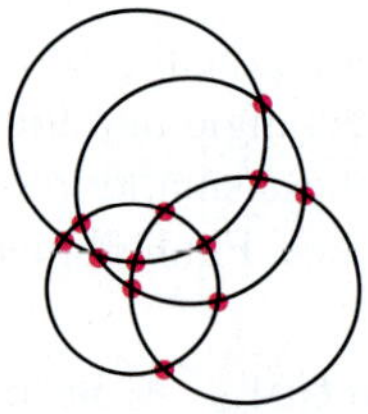

Four circles
12 intersections

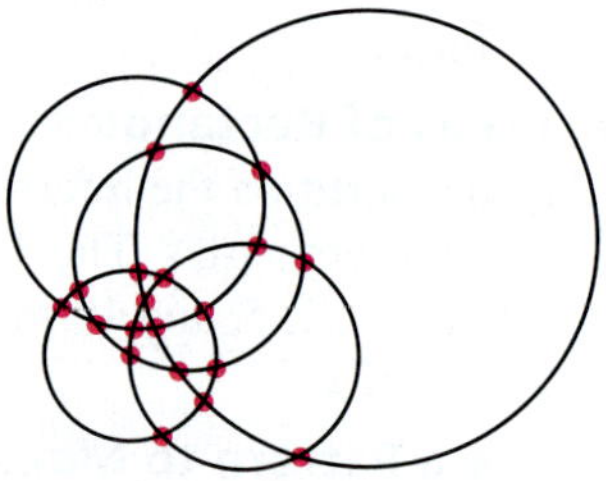

Five circles
20 intersections

Use a difference table to predict the maximum number of points in which six different size circles can intersect.

45. A Numerical Pattern A student has noticed the following pattern.

$9^1 = 9$ has 1 digit.

$9^2 = 81$ has 2 digits.

$9^3 = 729$ has 3 digits.

$\vdots$

$9^{10} = 3{,}486{,}784{,}401$ has 10 digits.

a. Find the smallest natural number n such that the number of digits in the decimal expansion of 9^n is *not* equal to n.

b. A professor indicates that you can receive five extra-credit points if you write all of the digits in the decimal expansion of $9^{(9^9)}$. Is this a worthwhile project? Explain.

CHAPTER 1 TEST

■ **Inductive vs. Deductive Reasoning** In Exercises 1 and 2, determine whether the argument is an example of inductive reasoning or deductive reasoning.

1. Two computer programs, a *bubble sort* and a *shell sort,* are used to sort data. In each of 50 experiments, the shell sort program took less time to sort the data than did the bubble sort program. Thus the shell sort program is the faster of the two sorting progams.
2. If a figure is a rectangle, then it is a parallelogram. Figure A is a rectangle. Therefore, Figure A is a parallelogram.
3. Use a difference table to predict the next term in the sequence $-1, 0, 9, 32, 75, 144, 245, \ldots$.
4. List the first 10 terms of the Fibonacci sequence.
5. In each of the following, determine the nth-term formula for the number of square tiles in the nth figure.

a.

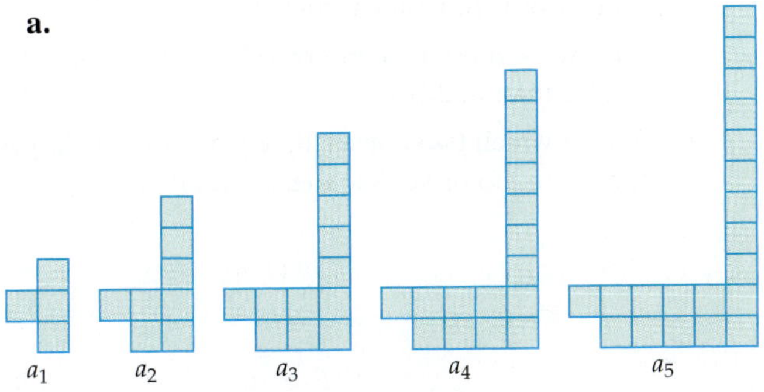

b.

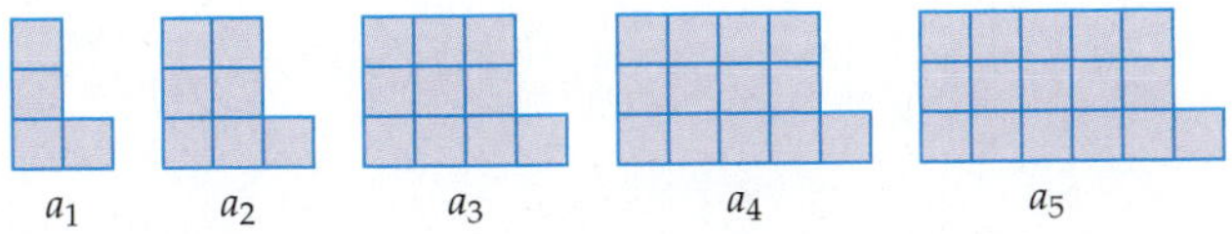

6. A sequence has an nth-term formula of

$$a_n = (-1)^n\left(\frac{n(n-1)}{2}\right)$$

Use the nth-term formula to determine the first 5 terms and the 105th term in the sequence.

7. **Terms of a Sequence** In a sequence:

$$a_1 = 3,\ a_2 = 7,\text{ and } a_n = 2a_{n-1} + a_{n-2}\text{ for } n \geq 3$$

Find a_3, a_4, and a_5.

8. **Number of Diagonals** A diagonal of a polygon is a line segment that connects nonadjacent vertices (corners) of the polygon. In the following polygons, the diagonals are shown by the blue line segments.

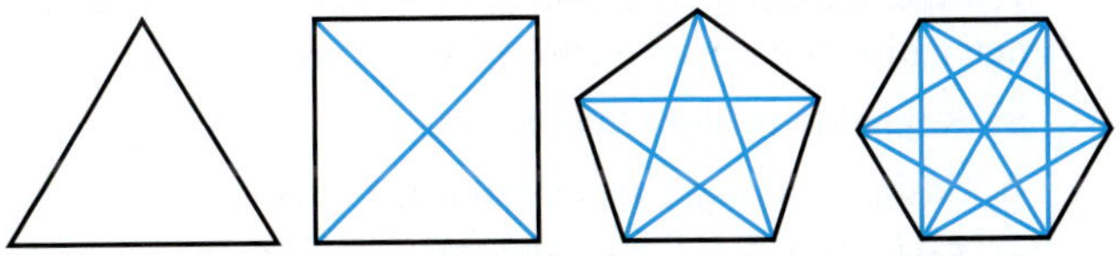

Triangle	Quadrilateral	Pentagon	Hexagon
3 sides	4 sides	5 sides	6 sides
0 diagonals	2 diagonals	5 diagonals	9 diagonals

Use a difference table to predict the number of diagonals in

a. a heptagon (a 7-sided polygon)

b. an octagon (an 8-sided polygon)

9. State the four steps of Polya's four-step problem-solving strategy.
10. **Make Change** How many different ways can change be made for a dollar using only half-dollars, quarters, and/or dimes?
11. **Counting Problem** In how many different ways can a basketball team win exactly four out of their last six games?
12. **Units Digit** What is the units digit (ones digit) of 3^{4513}?
13. **Vacation Money** Shelly has saved some money for a vacation. Shelly spends half of her vacation money on an airline ticket; she then spends \$50 for sunglasses, \$22 for a taxi, and one-third of her remaining money for a room with a view. After her sister repays her a loan of \$150, Shelly finds that she has \$326. How much vacation money did Shelly have at the start of her vacation?
14. **Number of Different Routes** How many different direct routes are there from point A to point B in the following figure?

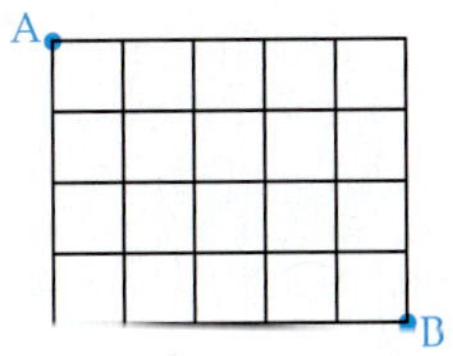

15. Number of League Games In a league of nine football teams, each team plays every other team in the league exactly once. How many league games will take place?

16. Ages of Children The four children in the Rivera family are Reynaldo, Ramiro, Shakira, and Sasha. The ages of the two teenagers are 13 and 15. The ages of the younger children are 5 and 7. From the following clues, determine the age of each of the children.

a. Reynaldo is older than Ramiro.

b. Sasha is younger than Shakira.

c. Sasha is 2 years older than Ramiro.

d. Shakira is older than Reynaldo.

17. Counterexample Find a counterexample to show that the following conjecture is false.

Conjecture: For all numbers x,

$$\frac{(x-4)(x+3)}{x-4} = x + 3.$$

18. Counterexample Find a counterexample to show that the following conjecture is false.

Conjecture: For all numbers x, $x \leq x^2$.

19. Find a Sum Find the following sum without using a calculator.

$$1 + 2 + 3 + 4 + \cdots + 497 + 498 + 499 + 500$$

20. Motor Vehicle Thefts The following graph shows the number of U.S. motor vehicle thefts for each year from 2009 to 2014.

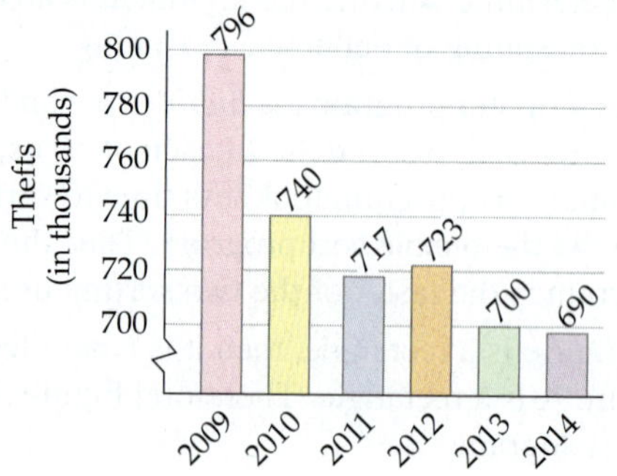

U.S. motor vehicle thefts

SOURCES: U.S. Department of Justice, Federal Bureau of Investigation, Uniform Crime Reports

a. Which one of the given years had the greatest number of U.S. motor vehicle thefts?

b. How many more U.S. motor vehicle thefts occurred in 2011 than in 2013?

c. During which two consecutive years did the largest decline in motor vehicle thefts occur?

2

Sets

In mathematics, any group or collection of objects is called a set. A simple application of sets occurs when you use a search engine (such as Google or Bing) to find a topic on the Internet. You enter a few words describing what you are searching for and click the Search button. The search engine then creates a list (set) of websites that contain a match for the words you submitted.

For instance, suppose you wish to make a cake. You search the Internet for a cake recipe and you obtain a set containing over 30 million matches. This is a very large number, so you narrow your search. One method of narrowing your search is to use the AND option found in the Advanced Search link of some search engines. An AND search is an all-words search. That is, an AND search finds only those sites that contain all of the words submitted. An AND search for "flourless chocolate cake recipe" produces a set containing 210,400 matches. This is a more reasonable number, but it is still quite large.

You narrow the search even further by using an AND search for "foolproof flourless chocolate cake recipe," which returns a few hundred matches. One of these sites provides you with a recipe that provides clear directions and has several good reviews.

Sometimes it is helpful to perform a search using the OR option. An OR search is an any-words search. That is, an OR search finds all those sites that contain any of the words you submitted.

Many additional applications of sets are given in this chapter.

Fuse/Getty Images

SECTION 2.1 Basic Properties of Sets

The Granger Collection, NY

The constellation Scorpius is a set of stars.

Sets

In an attempt to better understand the universe, ancient astronomers classified certain groups of stars as constellations. Today we still find it extremely helpful to classify items into groups that enable us to find order and meaning in our complicated world.

Any group or collection of objects is called a **set**. The objects that belong in a set are the **elements**, or **members**, of the set. For example, the set consisting of the four seasons has spring, summer, fall, and winter as its elements.

The following two methods are often used to designate a set.

- Describe the set using words.
- List the elements of the set inside a pair of braces, { }. This method is called the **roster method**. Commas are used to separate the elements.

For instance, let's use S to represent the set consisting of the four seasons. Using the roster method, we would write

$$S = \{\text{spring, summer, fall, winter}\}$$

The order in which the elements of a set are listed is not important. Thus the set consisting of the four seasons can also be written as

$$S = \{\text{winter, spring, fall, summer}\}$$

The following table gives two examples of sets, where each set is designated by a word description and also by using the roster method.

TAKE NOTE

Sets can also be designated by using *set-builder notation.* This method is described on page 50.

TABLE 2.1

Define Sets by Using a Word Description and the Roster Method

Word description	Roster method
The set of denominations of U.S. paper currency in production at this time	{\$1, \$5, \$10, \$20, \$50, \$100}
The set of states in the United States that border the Pacific Ocean	{California, Oregon, Washington, Alaska, Hawaii}

POINT OF INTEREST

US Department of the Treasury

Paper currency in denominations of \$500, \$1000, \$5000, and \$10,000 has been in circulation, but production of these bills ended in 1945. If you just happen to have some of these bills, you can still cash them for their face value.

EXAMPLE 1 Use the Roster Method to Represent a Set

Use the roster method to represent the set of the days in a week.

Solution {Sunday, Monday, Tuesday, Wednesday, Thursday, Friday, Saturday}

CHECK YOUR PROGRESS 1 Use the roster method to represent the set of months that start with the letter A.

Solution See page S4.

EXAMPLE 2 Use a Word Description to Represent a Set

Write a word description for the set

$$A = \{a, b, c, d, e, f, g, h, i, j, k, l, m, n, o, p, q, r, s, t, u, v, w, x, y, z\}$$

Solution Set A is the set of letters of the English alphabet.

CHECK YOUR PROGRESS 2 Write a word description for the set {March, May}.

Solution See page S4.

TAKE NOTE

Many sets can be described in more than one way. For instance, {Sunday, Saturday} can be described as the days of the week that begin with the letter *S*, as the days of the week that occur in a weekend, or as the first and last days of a week.

The following sets of numbers are used extensively in many areas of mathematics.

TAKE NOTE

In this chapter, the letters N, W, I, Q, $\mathcal{I}$, and R will often be used to represent the basic number sets defined at the right.

Basic Number Sets

Natural Numbers or Counting Numbers $N = \{1, 2, 3, 4, 5, \ldots\}$
Whole Numbers $W = \{0, 1, 2, 3, 4, 5, \ldots\}$
Integers $I = \{\ldots, -4, -3, -2, -1, 0, 1, 2, 3, 4, \ldots\}$
Rational Numbers Q = the set of all terminating or repeating decimals
Irrational Numbers $\mathcal{I}$ = the set of all nonterminating, nonrepeating decimals
Real Numbers R = the set of all rational or irrational numbers

The set of natural numbers is also called the set of counting numbers. The three dots ... are called an **ellipsis** and indicate that the elements of the set continue in a manner suggested by the elements that are listed.

The integers $\ldots, -4, -3, -2, -1$ are **negative integers**. The integers 1, 2, 3, 4, ... are **positive integers**. Note that the natural numbers and the positive integers are the same set of numbers. The integer zero is neither a positive nor a negative integer.

If a number in decimal form terminates or repeats a block of digits without end, then the number is a rational number. Rational numbers can also be written in the form $\frac{p}{q}$, where p and q are integers and $q \neq 0$. For example,

$$\frac{1}{4} = 0.25 \quad \text{and} \quad \frac{3}{11} = 0.\overline{27}$$

are rational numbers. The bar over the 27 means that the block of digits 27 repeats without end; that is, $0.\overline{27} = 0.27272727\ldots$.

A decimal that neither terminates nor repeats is an **irrational number**. For instance, $0.35335333533335\ldots$ is a nonterminating, nonrepeating decimal and thus is an irrational number.

Every real number is either a rational number or an irrational number.

EXAMPLE 3 Use the Roster Method to Represent a Set of Numbers

Use the roster method to write each of the given sets.

a. The set of natural numbers less than 5
b. The solution set of $x + 5 = -1$
c. The set of negative integers greater than -4

Solution

a. The set of natural numbers is given by $\{1, 2, 3, 4, 5, 6, 7, \ldots\}$. The natural numbers less than 5 are 1, 2, 3, and 4. Using the roster method, we write this set as $\{1, 2, 3, 4\}$.

b. Adding -5 to each side of the equation produces $x = -6$. The solution set of $x + 5 = -1$ is $\{-6\}$.

c. The set of negative integers greater than -4 is $\{-3, -2, -1\}$.

TAKE NOTE

1, 2, 3 is not a set. With the roster method, elements need to be placed inside braces to form a set.

CHECK YOUR PROGRESS 3 Use the roster method to write each of the given sets.

a. The set of whole numbers less than 4
b. The set of counting numbers larger than 11 and less than or equal to 19
c. The set of negative integers greater than -5 and less than 7

Solution See page S4.

Definitions Regarding Sets

A set is **well defined** if it is possible to determine whether any given item is an element of the set. For instance, the set of letters of the English alphabet is well defined. The set of *great songs* is not a well-defined set. It is not possible to determine whether any given song is an element of the set or is not an element of the set because there is no standard method for making such a judgment.

The statement "4 is an element of the set of natural numbers" can be written using mathematical notation as $4 \in N$. The symbol $\in$ is read "is an element of." To state that "-3 is not an element of the set of natural numbers," we use the "is not an element of" symbol, $\notin$, and write $-3 \notin N$.

TAKE NOTE

Recall that N denotes the set of natural numbers, I denotes the set of integers, and W denotes the set of whole numbers.

EXAMPLE 4 Apply Definitions Regarding Sets

Determine whether each statement is true or false.

a. $4 \in \{2, 3, 4, 7\}$ **b.** $-5 \in N$ **c.** $\frac{1}{2} \notin I$

d. The set of nice cars is a well-defined set.

Solution

a. Since 4 is an element of the given set, the statement is true.

b. There are no negative natural numbers, so the statement is false.

c. Since $\frac{1}{2}$ is not an integer, the statement is true.

d. The word *nice* is not precise, so the statement is false.

CHECK YOUR PROGRESS 4 Determine whether each statement is true or false.

a. $5.2 \in \{1, 2, 3, 4, 5, 6\}$ **b.** $-101 \in I$ **c.** $2.5 \notin W$

d. The set of all integers larger than π is a well-defined set.

Solution See page S4.

TAKE NOTE

Neither the set $\{0\}$ nor the set $\{\varnothing\}$ represents the empty set because each set has one element.

The **empty set**, or **null set**, is the set that contains no elements. The symbol $\varnothing$ or $\{\ \}$ is used to represent the empty set. As an example of the empty set, consider the set of natural numbers that are negative integers.

Another method of representing a set is **set-builder notation.** Set-builder notation is especially useful when describing infinite sets. For instance, in set-builder notation, the set of natural numbers greater than 7 is written as follows:

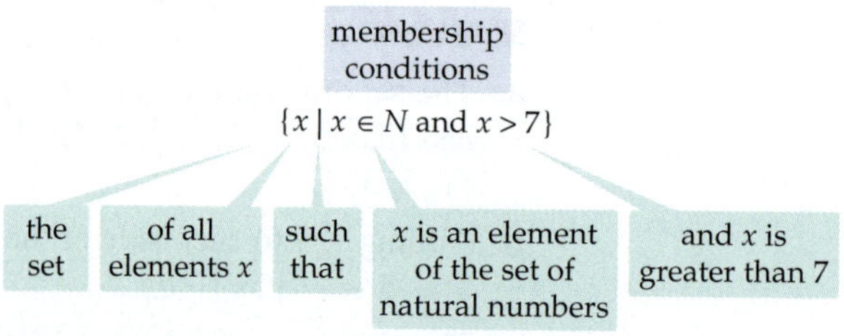

The preceding set-builder notation is read as "the set of all elements x such that x is an element of the set of natural numbers and x is greater than 7." It is impossible to list all the elements of the set, but set-builder notation defines the set by describing its elements.

EXAMPLE 5 Use Set-Builder Notation to Represent a Set

Use set-builder notation to write the following sets.

a. The set of integers greater than -3

b. The set of whole numbers less than 1000

Solution

a. $\{x \mid x \in I \text{ and } x > -3\}$ **b.** $\{x \mid x \in W \text{ and } x < 1000\}$

CHECK YOUR PROGRESS 5 Use set-builder notation to write the following sets.

a. The set of integers less than 9

b. The set of natural numbers greater than 4

Solution See page S4.

A set is **finite** if the number of elements in the set is a whole number. The **cardinal number** of a finite set is the number of elements in the set. The cardinal number of a finite set A is denoted by the notation $n(A)$. For instance, if $A = \{1, 4, 6, 9\}$, then $n(A) = 4$. In this case, A has a cardinal number of 4, which is sometimes stated as "A has a *cardinality* of 4."

TAKE NOTE

The cardinality of infinite sets is covered in Section 2.5.

EXAMPLE 6 The Cardinality of a Finite Set

Find the cardinality of each of the following sets.

a. $J = \{2, 5\}$ **b.** $S = \{3, 4, 5, 6, 7, \ldots, 31\}$ **c.** $T = \{3, 3, 7, 51\}$

Solution

a. Set J contains exactly two elements, so J has a cardinality of 2. Using mathematical notation, we state this as $n(J) = 2$.

b. Only a few elements are actually listed. The number of natural numbers from 1 to 31 is 31. If we omit the numbers 1 and 2, then the number of natural numbers from 3 to 31 must be $31 - 2 = 29$. Thus $n(S) = 29$.

c. Elements that are listed more than once are counted only once. Thus $n(T) = 3$.

CHECK YOUR PROGRESS 6 Find the cardinality of the following sets.

a. $C = \{-1, 5, 4, 11, 13\}$ **b.** $D = \{0\}$ **c.** $E = \emptyset$

Solution See page S4.

The following definitions play an important role in our work with sets.

Equal Sets

Set A is **equal** to set B, denoted by $A = B$, if and only if A and B have exactly the same elements.

For instance, $\{d, e, f\} = \{e, f, d\}$.

Equivalent Sets

Set A is **equivalent** to set B, denoted by $A \sim B$, if and only if A and B have the same number of elements. Using mathematical notation, the previous sentence can be expressed as $A \sim B$ provided $n(A) = n(B)$.

QUESTION If two sets are equal, must they also be equivalent?

ANSWER Yes. If the sets are equal, then they have exactly the same elements; therefore, they also have the same number of elements.

EXAMPLE 7 Equal Sets and Equivalent Sets

State whether each of the following pairs of sets are equal, equivalent, both, or neither.

a. {a, e, i, o, u}, {3, 7, 11, 15, 19} **b.** $\{4, -2, 7\}, \{3, 4, 7, 9\}$

Solution

a. The sets are not equal. However, each set has exactly five elements, so the sets are equivalent.

b. The first set has three elements and the second set has four elements, so the sets are not equal and are not equivalent.

CHECK YOUR PROGRESS 7 State whether each of the following pairs of sets are equal, equivalent, both, or neither.

a. $\{x \mid x \in W \text{ and } x \leq 5\}$, $\{\alpha, \beta, \Gamma, \Delta, \delta, \varepsilon\}$

b. $\{5, 10, 15, 20, 25, 30, \ldots, 80\}$, $\{x \mid x \in N \text{ and } x < 17\}$

Solution See page S4.

MATH MATTERS Georg Cantor

Science Photo Library/Photo Researchers/
Science Source

Georg Cantor

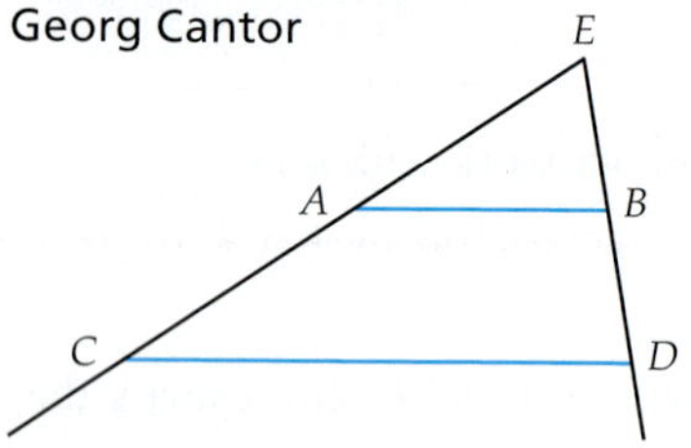

Georg Cantor (kăn′tər) (1845–1918) was a German mathematician who developed many new concepts regarding the theory of sets. Cantor studied under the famous mathematicians Karl Weierstrass and Leopold Kronecker at the University of Berlin. Although Cantor demonstrated a talent for mathematics, his professors were unaware that Cantor would produce extraordinary results that would cause a major stir in the mathematical community.

Cantor never achieved his lifelong goal of a professorship at the University of Berlin. Instead he spent his active career at the undistinguished University of Halle. It was during this period, when Cantor was between the ages of 29 and 39, that he produced his best work. Much of this work was of a controversial nature. One of the simplest of the controversial concepts concerned points on a line segment. For instance, consider the line segment $\overline{AB}$ and the line segment $\overline{CD}$ in the figure at the left. Which of these two line segments do you think contains the most points? Cantor was able to prove that they both contain the same number of points. In fact, he was able to prove that any line segment, no matter how short, contains the same number of points as a line or a plane. We will take a closer look at some of the mathematics developed by Cantor in the last section of this chapter.

EXCURSION

Fuzzy Sets

In traditional set theory, an element either belongs to a set or does not belong to the set. For instance, let $A = \{x \mid x \text{ is an even integer}\}$. Given $x = 8$, we have $x \in A$. However, if $x = 11$, then $x \notin A$. For any given integer, we can decide whether x belongs to A.

Now consider the set $B = \{x \mid x \text{ is a number close to } 10\}$. Does 8 belong to this set? Does 9.9 belong to the set? Does 10.001 belong to the set? Does 10 belong to the set? Does

−50 belong to the set? Given the imprecision of the words "close to," it is impossible to know which numbers belong to set B.

In 1965, Lotfi A. Zadeh of the University of California, Berkeley, published a paper titled *Fuzzy Sets* in which he described the mathematics of fuzzy set theory. This theory proposed that "to some degree," many of the numbers 8, 9.9, 10.001, 10, and −50 belong to set B defined in the previous paragraph. Zadeh proposed giving each element of a set a *membership grade* or *membership value*. This value is a number from 0 to 1. The closer the membership value is to 1, the greater the certainty that an element belongs to the set. The closer the membership value is to 0, the less the certainty that an element belongs to the set. Elements of fuzzy sets are written in the form (element, membership value). Here is an example of a fuzzy set.

$$C = \{(8, 0.4), (9.9, 0.9), (10.001, 0.999), (10, 1), (-50, 0)\}$$

An examination of the membership values suggests that we are certain that 10 belongs to C (membership value is 1) and we are certain that −50 does not belong to C (membership value is 0). Every other element belongs to the set "to some degree."

The concept of a fuzzy set has been used in many real-world applications. Here are a few examples.

- Control of heating and air-conditioning systems
- Compensation against vibrations in camcorders
- Voice recognition by computers
- Control of valves and dam gates at power plants
- Control of robots
- Control of subway trains
- Automatic camera focusing

HISTORICAL NOTE

Lotfi Zadeh

Lotfi Zadeh

Lotfi Zadeh's work in the area of fuzzy sets has led to a new area of mathematics called *soft computing*. On the topic of soft computing, Zadeh stated, "The essence of soft computing is that unlike the traditional, hard computing, soft computing is aimed at an accommodation with the pervasive imprecision of the real world. Thus the guiding principle of soft computing is: Exploit the tolerance for imprecision, uncertainty and partial truth to achieve tractability, robustness, low solution cost and better rapport with reality. In the final analysis, the role model for soft computing is the human mind."

EXCURSION EXERCISES

1. Mark, Erica, Larry, and Jennifer have each defined a fuzzy set to describe what they feel is a "good" grade. Each person paired the letter grades A, B, C, D, and F with a membership value. The results are as follows.

Mark: $M = \{(\text{A}, 1), (\text{B}, 0.75), (\text{C}, 0.5), (\text{D}, 0.5), (\text{F}, 0)\}$

Erica: $E = \{(\text{A}, 1), (\text{B}, 0), (\text{C}, 0), (\text{D}, 0), (\text{F}, 0)\}$

Larry: $L = \{(\text{A}, 1), (\text{B}, 1), (\text{C}, 1), (\text{D}, 1), (\text{F}, 0)\}$

Jennifer: $J = \{(\text{A}, 1), (\text{B}, 0.8), (\text{C}, 0.6), (\text{D}, 0.1), (\text{F}, 0)\}$

a. Which of the four people considers an A grade to be the only good grade?

b. Which of the four people is most likely to be satisfied with a grade of D or better?

c. Write a fuzzy set that you would use to describe the set of good grades. Consider only the letter grades A, B, C, D, and F.

2. In some fuzzy sets, membership values are given by a *membership graph* or by a *formula*. For instance, the following figure is a graph of the membership values of the fuzzy set *OLD*.

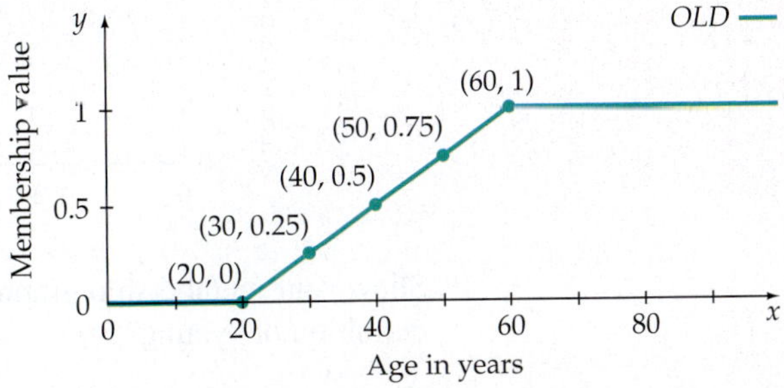

Use the membership graph of *OLD* to determine the membership value of each of the following.

a. $x = 15$ **b.** $x = 50$ **c.** $x = 65$

d. Use the graph of *OLD* to determine the age x with a membership value of 0.25.

An ordered pair (x, y) of a fuzzy set is a **crossover point** if its membership value is 0.5.

e. Find the crossover point for *OLD*.

3. The following membership graph provides a definition of real numbers x that are "about" 4.

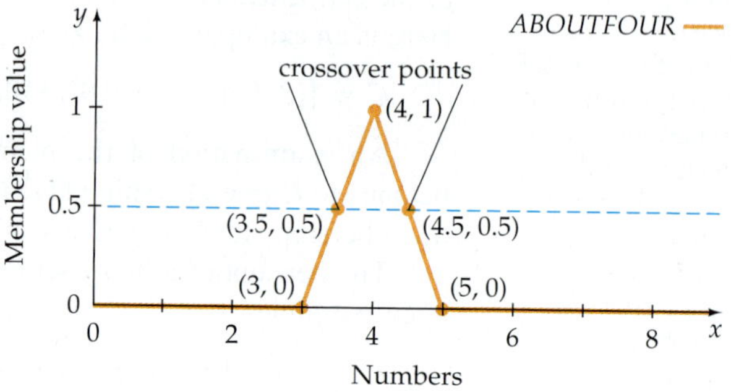

Use the graph of *ABOUTFOUR* to determine the membership value of:

a. $x = 2$ **b.** $x = 3.5$ **c.** $x = 7$

d. Use the graph of *ABOUTFOUR* to determine its crossover points.

4. The membership graphs in the following figure provide definitions of the fuzzy sets *COLD* and *WARM*.

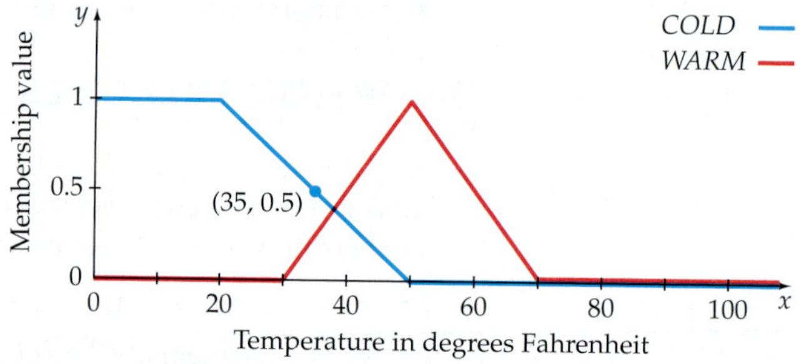

TAKE NOTE

The process of creating a fuzzy definition is called **fuzzification**. The graphs in Excursion Exercise 4 are fuzzifications of the terms *cold* and *warm*. Fuzzification is not a unique process. For instance, "cold" can be defined in many different ways.

The point (35, 0.5) on the membership graph of *COLD* indicates that the membership value for $x = 35$ is 0.5. Thus, by this definition, 35°F is 50% cold. Use the above graphs to estimate

a. the *WARM* membership value for $x = 40$.

b. the *WARM* membership value for $x = 50$.

c. the crossover points of *WARM*.

5. The membership graph in Excursion Exercise 2 shows one person's idea of what ages are "old." Use a grid similar to the following to draw a membership graph that you feel defines the concept of being "young" in terms of a person's age in years.

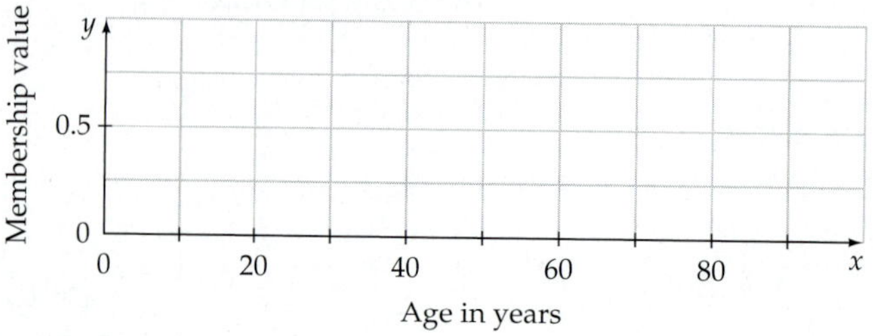

Show your membership graph to a few of your friends. Do they concur with your definition of "young"?

EXERCISE SET 2.1

■ In Exercises 1 to 14, use the roster method to write each of the given sets. For some exercises you may need to consult a reference, such as the Internet or an encyclopedia.

1. The set of U.S. coins with a value of less than 50¢
2. The set of months of the year with a name that ends with the letter y
3. The set of planets in our solar system with a name that starts with the letter M
4. The set of the seven dwarfs
5. The set of U.S. presidents who have served after Bill Clinton
6. The set of months with exactly 30 days
7. The set of negative integers greater than -6
8. The set of whole numbers less than 8
9. The set of integers x that satisfy $x - 4 = 3$
10. The set of integers x that satisfy $2x - 1 = -11$
11. The set of natural numbers x that satisfy $x + 4 = 1$
12. The set of whole numbers x that satisfy $x - 1 < 4$
13. The set of letters in the English alphabet, except the letter y, that are vowels
14. The set of natural numbers less than 5

■ In Exercises 15 to 24, write a word description of each set. There may be more than one correct description.

15. {Tuesday, Thursday}
16. {Libra, Leo}
17. {Mercury, Venus}
18. {penny, nickel, dime}
19. {1, 2, 3, 4, 5, 6, 7, 8, 9}
20. {2, 4, 6, 8}
21. $\{x \mid x \in N \text{ and } x \le 7\}$
22. $\{x \mid x \in W \text{ and } x < 5\}$
23. {1, 3, 5, 7, 9}
24. {−4, −3, −2, −1}

■ In Exercises 25 to 36, determine whether each statement is true or false. If the statement is false, give a reason.

25. $b \in \{a, b, c\}$
26. $0 \notin N$
27. $\{b\} \in \{a, b, c\}$
28. $\{1, 5, 9\} \sim \{\Psi, \Pi, \Sigma\}$
29. $\{0\} \sim \varnothing$
30. The set of large numbers is a well-defined set.
31. The set of good teachers is a well-defined set.
32. The set $\{x \mid 2 \le x \le 3\}$ is a well-defined set.
33. $\{x \mid x \in N\} = \{x \mid x \in W \text{ and } x > 0\}$
34. $0 \in \varnothing$
35. $\{2, 3, 4\} = \{2, 2, 3, 4, 4\}$
36. $\{5, 6, 7\} \sim \{9, 20, 31\}$

■ In Exercises 37 to 48, use set-builder notation to write each of the following sets.

37. {1, 2, 3, 4, 5, 6, 7, 8, 9, 10, 11, 12}
38. {45, 55, 65, 75}
39. {5, 10, 15}
40. {1, 4, 9, 16, 25, 36, 49, 64, 81}
41. {January, March, May, July, August, October, December}
42. {Iowa, Ohio, Utah}
43. {Arizona, Alabama, Arkansas, Alaska}
44. {Mexico, Canada}
45. {spring, summer}
46. {1900, 1901, 1902, 1903, 1904, ..., 1999}
47. The set of months that have less than 31 days
48. {1, 2, 3, 4, 5, ...}
49. **Average Student Debt** The following bar graph shows the average student debt in each graduating class from 2008 to 2015.

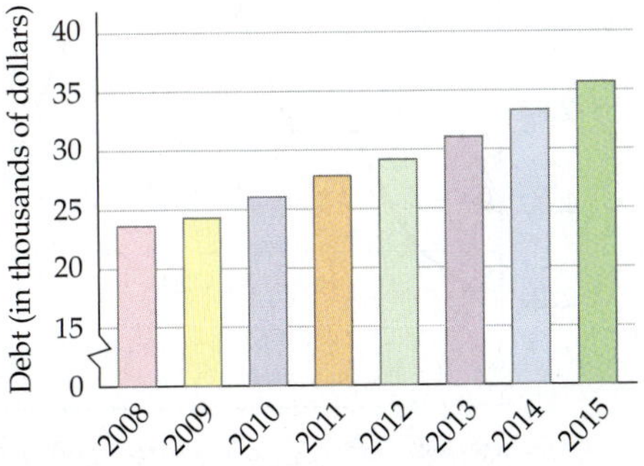

Average student debt in each graduating class
SOURCE: Mark Kantrowitz | WSJ.com

Use the information in the graph and the roster method to represent each of the following sets.

a. The set of years in which the average student debt was greater than \$30,000.
b. The set of years in which the average student debt was greater than \$20,000 but less than \$25,000.
c. The set of years in which the average student debt was less than \$30,000 but greater than \$25,000.

50. **Gold Prices** The following horizontal bar graph shows the average yearly price of gold, per troy ounce, from 2007 to 2014.

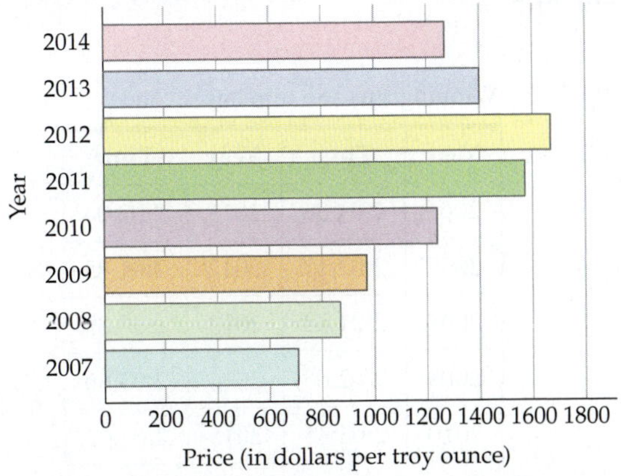

Average yearly price of gold
SOURCE: U.S. Geological Survey, U.S. Department of the Interior

Use the information in the graph and the roster method to represent each of the following sets.

a. The set of years in which the price of gold was greater than \$1500.

b. The set of years in which the price of gold was greater than \$1000 but less than \$1500.

c. The set of years in which the price of gold was less than \$1000.

51. **Gasoline Prices** The following broken line graph shows the average U.S. monthly regular gasoline prices for the months from January to November of 2015.

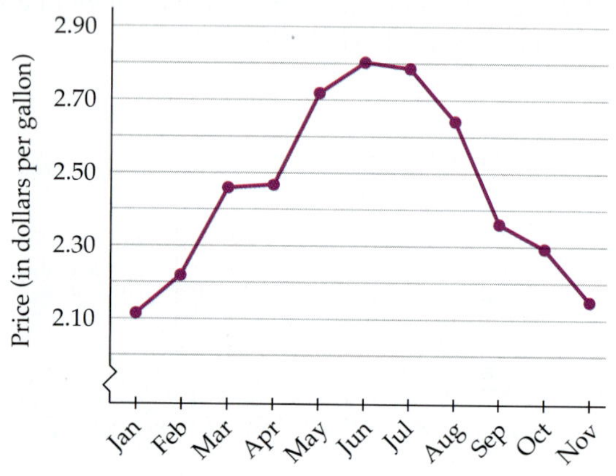

Average monthly U.S. regular gasoline prices in 2015
SOURCE: U.S. Energy Information Administration

Use the information in the above graph and the roster method to represent each of the following sets.

a. The set of months in which the price of gasoline was greater than \$2.50 per gallon.

b. The set of months in which the price of gasoline was greater than \$2.30 but less than \$2.60 per gallon.

c. The set of months in which the price of gasoline was less than \$2.20 per gallon.

52. **Winning Times** The following table shows the winning times in the Kentucky Derby from 2006 to 2015.

Winning times in minutes:seconds

Year	Time	Year	Time
2006	2:01.36	2011	2:02.04
2007	2:02.17	2012	2:01.83
2008	2:01.82	2013	2:02.89
2009	2:02.66	2014	2:03.66
2010	2:04.45	2015	2:03.02

SOURCE: http://horseracing.about.com/

Use the information in the table and the roster method to represent each of the following sets.

a. The set of years in which the winning time was greater than 2 minutes 3.00 seconds.

b. The set of years in which the winning time was less than 2 minutes 3.50 seconds but greater than 2 minutes 3.00 seconds.

c. The year in which the winning time was the least of all the given times.

■ In Exercises 53 to 62, find the cardinality of each of the following sets. For some exercises you may need to consult a reference, such as the Internet or an encyclopedia.

53. $A = \{2, 4, 6, 8, 10, 12, 14, 16, 18, 20, 22\}$

54. $B = \{7, 14, 21, 28, 35, 42, 49, 56\}$

55. $D =$ the set of all dogs that can spell "elephant"

56. $S =$ the set of all states in the United States

57. $J =$ the set of all states of the United States that border Minnesota

58. $T =$ the set of all stripes on the U.S. flag

59. $N =$ the set of all baseball teams in the National League

60. $C =$ the set of all chess pieces on a chess board at the start of a chess game

61. $\{3, 6, 9, 12, 15, \ldots, 363\}$

62. $\{7, 11, 15, 19, 23, 27, \ldots, 407\}$

■ In Exercises 63 to 70, state whether each of the given pairs of sets are equal, equivalent, both, or neither.

63. The set of U.S. senators; the set of U.S. representatives

64. The set of single-digit natural numbers; the set of pins used in a regulation bowling game

65. The set of positive whole numbers; the set of natural numbers

66. The set of single-digit natural numbers; the set of single-digit integers

67. $\{1, 2, 3\}$; $\{\text{I, II, III}\}$

68. $\{6, 8, 10, 12\}$; $\{1, 2, 3, 4\}$

69. $\{2, 5\}$; $\{0, 1\}$

70. $\{\ \}$; $\{0\}$

■ In Exercises 71 to 82, determine whether each of the sets is a well-defined set.

71. The set of good foods

72. The set of the six most heavily populated cities in the United States

73. The set of tall buildings in the city of Chicago

74. The set of states that border Colorado

75. The set of even integers

76. The set of rational numbers of the form $\frac{1}{p}$, where p is a counting number

77. The set of former presidents of the United States who are alive at the present time

78. The set of real numbers larger than 89,000

79. The set of small countries

80. The set of great cities in which to live

81. The set consisting of the best soda drinks

82. The set of fine wines

EXTENSIONS

Extension of Set-Builder Notation In the following set-builder example the variable x, which generally appears just to the left of the vertical bar, has been replaced by the algebraic expression $3x$.

$$A = \{3x \mid x \in N \text{ and } x < 5\}$$

To determine the elements of set A, first use the information to the right of the vertical bar to find that x is a natural number less than 5. Thus as x takes on the values 1, 2, 3, and 4, the algebraic expression $3x$ takes on the values 3, 6, 9, and 12. In roster form, A is written as:

$$A = \{3, 6, 9, 12\}$$

Here are two additional examples.

$$B = \{x^2 \mid x \in W \text{ and } x \le 6\} = \{0, 1, 4, 9, 16, 25, 36\}$$
$$C = \{4x - 2 \mid x \in N \text{ and } x \ge 3\} = \{10, 14, 18, 22, 26, ...\}$$

■ In Exercises 83 to 86, use the extension of set-builder notation concepts and the roster method to write each of the given sets.

83. $D = \{3x^2 - 1 \mid x \in N \text{ and } x < 5\}$

84. $E = \{2x^2 - x \mid x \in N \text{ and } x < 7\}$

85. $F = \{(-1)^x(x^3) \mid x \in N \text{ and } x \ge 2\}$

86. $G = \left\{(-1)^x\left(\frac{2}{x}\right) \middle| x \in N \text{ and } x < 7\right\}$

■ In Exercises 87 to 90, determine whether the given sets are equal.

87. $A = \{2n + 1 \mid n \in W\}$
$B = \{2n - 1 \mid n \in N\}$

88. $A = \left\{16\left(\frac{1}{2}\right)^{n-1} \middle| n \in N\right\}$
$B = \left\{16\left(\frac{1}{2}\right)^{n} \middle| n \in W\right\}$

89. $A = \{2n - 1 \mid n \in N\}$
$B = \left\{\frac{n(n+1)}{2} \middle| n \in N\right\}$

90. $A = \{3n + 1 \mid n \in W\}$
$B = \{3n - 2 \mid n \in N\}$

SECTION 2.2 Complements, Subsets, and Venn Diagrams

The Universal Set and the Complement of a Set

In complex problem-solving situations and even in routine daily activities, we need to understand the set of all elements that are under consideration. For instance, when an instructor assigns letter grades, the possible choices may include A, B, C, D, F, and I. In this case the letter H is not a consideration. When you place a telephone call, you know that the area code is given by a natural number with three digits. In this instance a rational number such as $\frac{2}{3}$ is not a consideration. The set of all elements that are being considered is called the **universal set**. We will use the letter U to denote the universal set.

The Complement of a Set

The **complement** of a set A, denoted by A', is the set of all elements of the universal set U that are not elements of A.

TAKE NOTE

To determine the complement of a set A, you must know the elements of A as well as the elements of the universal set, U, that is being used.

EXAMPLE 1 **Find the Complement of a Set**

Let $U = \{1, 2, 3, 4, 5, 6, 7, 8, 9, 10\}$, $S = \{2, 4, 6, 7\}$, and $T = \{x \mid x < 10 \text{ and } x \in \text{the odd counting numbers}\}$. Find

a. S' **b.** T'

Solution

a. The elements of the universal set are 1, 2, 3, 4, 5, 6, 7, 8, 9, and 10. From these elements we wish to exclude the elements of S, which are 2, 4, 6, and 7. Therefore $S' = \{1, 3, 5, 8, 9, 10\}$.

b. $T = \{1, 3, 5, 7, 9\}$. Excluding the elements of T from U gives us $T' = \{2, 4, 6, 8, 10\}$.

CHECK YOUR PROGRESS 1 Let $U = \{0, 2, 3, 4, 6, 7, 17\}$, $M = \{0, 4, 6, 17\}$, and $P = \{x \mid x < 7 \text{ and } x \in \text{the even natural numbers}\}$. Find

a. M' **b.** P'

Solution See page S4.

There are two fundamental results concerning the universal set and the empty set. Because the universal set contains all elements under consideration, the complement of the universal set is the empty set. Conversely, the complement of the empty set is the universal set, because the empty set has no elements and the universal set contains all the elements under consideration. Using mathematical notation, we state these fundamental results as follows:

The Complement of the Universal Set and the Complement of the Empty Set

$$U' = \varnothing \quad \text{and} \quad \varnothing' = U$$

Subsets

Consider the set of letters in the alphabet and the set of vowels {a, e, i, o, u}. Every element of the set of vowels is an element of the set of letters in the alphabet. The set of vowels is said to be a *subset* of the set of letters in the alphabet. We will often find it useful to examine subsets of a given set.

A Subset of a Set

Set A is a **subset** of set B, denoted by $A \subseteq B$, if and only if every element of A is also an element of B.

Here are two fundamental subset relationships.

Subset Relationships

$A \subseteq A$, for any set A
$\varnothing \subseteq A$, for any set A

To convince yourself that the empty set is a subset of any set, consider the following. We know that a set is a subset of a second set provided every element of the first set is an element of the second set. Pick an arbitrary set A. Because every element of the empty set (*there are none*) is an element of A, we know that $\varnothing \subseteq A$.

The notation $A \not\subseteq B$ is used to denote that A is *not* a subset of B. To show that A is not a subset of B, it is necessary to find at least one element of A that is not an element of B.

EXAMPLE 2 Apply the Definition of a Subset

Determine whether each statement is true or false.

a. $\{5, 10, 15, 20\} \subseteq \{0, 5, 10, 15, 20, 25, 30\}$

b. $W \subseteq N$

c. $\{2, 4, 6\} \subseteq \{2, 4, 6\}$

d. $\emptyset \subseteq \{1, 2, 3\}$

Solution

a. True; every element of the first set is an element of the second set.

b. False; 0 is a whole number, but 0 is not a natural number.

c. True; every set is a subset of itself.

d. True; the empty set is a subset of every set.

CHECK YOUR PROGRESS 2 Determine whether each statement is true or false.

a. $\{1, 3, 5\} \subseteq \{1, 5, 9\}$

b. The set of counting numbers is a subset of the set of natural numbers.

c. $\emptyset \subseteq U$

d. $\{-6, 0, 11\} \subseteq I$

Solution See page S4.

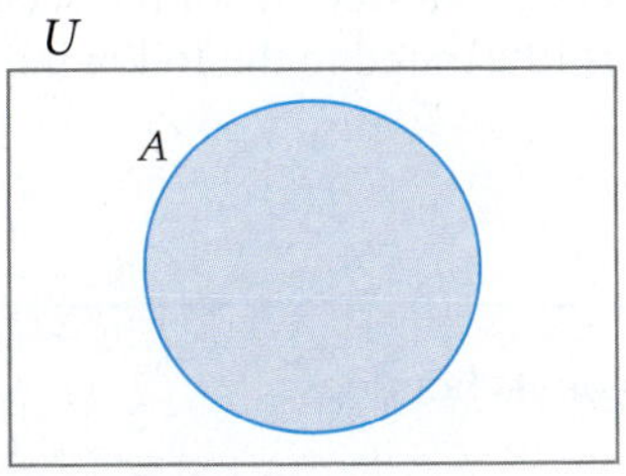

A Venn diagram

The English logician John Venn (1834–1923) developed diagrams, which we now refer to as *Venn diagrams,* that can be used to illustrate sets and relationships between sets. In a **Venn diagram**, the universal set is represented by a rectangular region and subsets of the universal set are generally represented by oval or circular regions drawn inside the rectangle. The Venn diagram at the left shows a universal set and one of its subsets, labeled as set A. The size of the circle is not a concern. The region outside of the circle, but inside of the rectangle, represents the set A'.

QUESTION What set is represented by $(A')'$?

Proper Subsets of a Set

Proper Subset

Set A is a **proper subset** of set B, denoted by $A \subset B$, if every element of A is an element of B, and $A \neq B$.

To illustrate the difference between subsets and proper subsets, consider the following two examples.

1. Let $R = \{\text{Mars, Venus}\}$ and $S = \{\text{Mars, Venus, Mercury}\}$. The first set, R, is a subset of the second set, S, because every element of R is an element of S. In addition, R is also a proper subset of S, because $R \neq S$.

ANSWER The set A' contains the elements of U that are not in A. By definition, the set $(A')'$ contains only the elements of U that are elements of A. Thus $(A')' = A$.

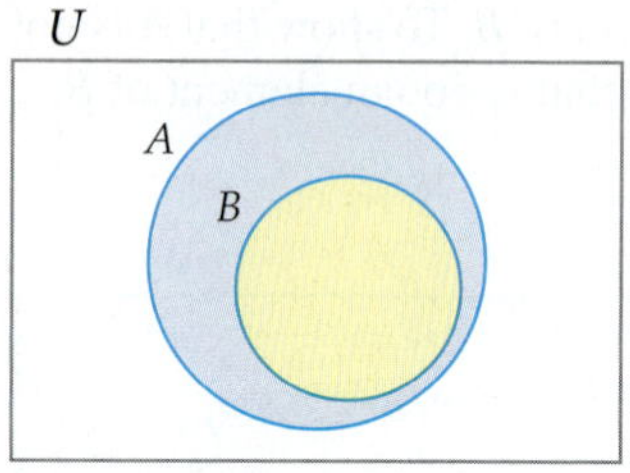

If B is a subset of A and $B \neq A$, then B is a proper subset of A.

2. Let $T = \{\text{Europe, Africa}\}$ and $V = \{\text{Africa, Europe}\}$. The first set, T, is a subset of the second set, V; however, T is *not* a proper subset of V because $T = V$.

Venn diagrams can be used to represent proper subset relationships. For instance, if a set B is a proper subset of a set A, then we illustrate this relationship in a Venn diagram by drawing a circle labeled B inside of a circle labeled A. See the Venn diagram at the left.

EXAMPLE 3 Proper Subsets

For each of the following, determine whether the first set is a proper subset of the second set.

a. {a, e, i, o, u}, {e, i, o, u, a} **b.** N, I

Solution

a. Because the sets are equal, the first set is not a proper subset of the second set.

b. Every natural number is an integer, so the set of natural numbers is a subset of the set of integers. The set of integers contains elements that are not natural numbers, such as -3. Thus the set of natural numbers is a proper subset of the set of integers.

CHECK YOUR PROGRESS 3 For each of the following, determine whether the first set is a proper subset of the second set.

a. N, W **b.** {1, 4, 5}, {5, 1, 4}

Solution See page S4.

Some applications require that we find all of the subsets of a given set. One way to find all the subsets of a given set is to use the method of making an organized list. First list the empty set, which has no elements. Next list all the sets that have exactly one element, followed by all the sets that contain exactly two elements, followed by all the sets that contain exactly three elements, and so on. This process is illustrated in the following example.

Kheng Guan Toh/Shutterstock.com

EXAMPLE 4 List All the Subsets of a Set

Set C shows the four condiments that a hot dog stand offers on its hot dogs.

C = {mustard, ketchup, onions, relish}

List all the subsets of C.

Solution

An organized list shows the following subsets.

{ }	• Subsets with 0 elements
{mustard}, {ketchup}, {onions}, {relish}	• Subsets with 1 element
{mustard, ketchup}, {mustard, onions}, {mustard, relish}, {ketchup, onions}, {ketchup, relish}, {onions, relish}	• Subsets with 2 elements
{mustard, ketchup, onions}, {mustard, ketchup, relish}, {mustard, onions, relish}, {ketchup, onions, relish}	• Subsets with 3 elements
{mustard, ketchup, onion, relish}	• Subsets with 4 elements

CHECK YOUR PROGRESS 4 List all of the subsets of {a, b, c, d, e}.

Solution See page S4.

Number of Subsets of a Set

In some applications we need to determine the number of subsets of a set *without* making an actual list of all the subsets. We know that the empty set has 0 elements and the only subset of the empty set is the empty set. A set with 1 element has 2 subsets, namely, itself and the empty set. It is easy to show that a set with 2 elements has 4 subsets and a set with 3 elements has 8 subsets. In Example 4 we found that a set with 4 elements has 16 subsets. The following diagram summarizes the above results.

# of elements in a set: n	0	1	2	3	4	5	6	...
# of subsets of the set:	1	2	4	8	16	?	?	...

It appears that if the number of elements in a set is increased by 1, then the number of subsets of the set is doubled. Thus we suspect that a set with 5 elements will have $2 \cdot 16 = 32$ subsets and a set with 6 elements will have $2 \cdot 32 = 64$ subsets. Also note that each of the numbers in the bottom row of the diagram is equal to 2^n.

$1 = 2^0 \qquad 2 = 2^1 \qquad 4 = 2^2 \qquad 8 = 2^3 \qquad \text{and} \qquad 16 = 2^4$

These observations lend support for the following theorem.

The Number of Subsets of a Set

A set with n elements has 2^n subsets.

Examples

- $\{1, 2, 3, 4, 5, 6, 7\}$ has 7 elements, so it has $2^7 = 128$ subsets.
- $\{4, 5, 6, 7, 8, \ldots, 15\}$ has 12 elements, so it has $2^{12} = 4096$ subsets.
- The empty set has 0 elements, so it has $2^0 = 1$ subset.

Consider set A with n elements. All of the 2^n subsets of A are proper subsets of A, except for A itself. Thus the number of proper subsets of A is $2^n - 1$.

The Number of Proper Subsets of a Set

A set with n elements has $2^n - 1$ subsets.

Examples

- $\{1, 3, 5, 7\}$ has 4 elements, so it has $2^4 - 1 = 16 - 1 = 15$ proper subsets.
- $\{1, 2, 3, 4, \ldots, 10\}$ has 10 elements, so it has $2^{10} - 1 = 1024 - 1 = 1023$ proper subsets.

In Example 5, we apply the formula for the number of subsets of a set to determine the number of different variations of pizzas that a restaurant can serve.

svry/Shutterstock.com

EXAMPLE 5 Pizza Variations

A restaurant sells pizzas for which you can choose from seven toppings.

a. How many different variations of pizzas can the restaurant serve?

b. What is the minimum number of toppings the restaurant must provide if it wishes to advertise that it offers over 1000 variations of its pizzas?

Solution

a. The restaurant can serve a pizza with no topping, one topping, two toppings, three toppings, and so forth, up to all seven toppings.

Let T be the set consisting of the seven toppings. The elements in each subset of T describe exactly one of the variations of toppings that the restaurant can serve. Consequently, the number of different variations of pizzas that the restaurant can serve is the same as the number of subsets of T.

Thus the restaurant can serve $2^7 = 128$ different variations of its pizzas.

b. Use the method of guessing and checking to find the smallest natural number n for which $2^n > 1000$.

$$2^8 = 256$$
$$2^9 = 512$$
$$2^{10} = 1024$$

The restaurant must provide a minimum of 10 toppings if it wishes to offer over 1000 variations of its pizzas.

CHECK YOUR PROGRESS 5 A company makes a car with 11 upgrade options.

a. How many different versions of this car can the company produce? Assume that each upgrade option is independent of the other options.

b. What is the minimum number of upgrade options the company must provide if it wishes to offer at least 8000 different versions of this car?

Solution See page S4.

MATH MATTERS The Barber's Paradox

PictureQuest/Corbis Images

Some problems that concern sets have led to paradoxes. For instance, in 1902, the mathematician Bertrand Russell developed the following paradox. "Is the set A of all sets that are not elements of themselves an element of itself?" Both the assumption that A is an element of A and the assumption that A is not an element of A lead to a contradiction. Russell's paradox has been popularized as follows.

> The town barber shaves all males who do not shave themselves, and he shaves only those males. The town barber is a male who shaves. Who shaves the barber?

The assumption that the barber shaves himself leads to a contradiction, and the assumption that the barber does not shave himself also leads to a contradiction.

EXCURSION

Subsets and Complements of Fuzzy Sets

This excursion extends the concept of fuzzy sets that was developed in the Excursion in Section 2.1. Recall that the elements of a fuzzy set are ordered pairs. For any ordered pair (x, y) of a fuzzy set, the membership value y is a real number such that $0 \leq y \leq 1$.

TAKE NOTE

A set such as {3, 5, 9} is called a *crisp set*, to distinguish it from a fuzzy set.

The set of all x-values that are being considered is called the **universal set for the fuzzy set** and it is denoted by X.

A Fuzzy Subset

If the fuzzy sets $A = \{(x_1, a_1), (x_2, a_2), (x_3, a_3), ...\}$ and $B = \{(x_1, b_1), (x_2, b_2), (x_3, b_3), ...\}$ are both defined on the universal set $X = \{x_1, x_2, x_3, ...\}$, then $A \subseteq B$ if and only if $a_i \leq b_i$ for all i.

A fuzzy set A is a **subset** of a fuzzy set B if and only if the membership value of each element of A *is less than or equal to* its corresponding membership value in set B. For instance, in Excursion Exercise 1 in Section 2.1, Mark and Erica used fuzzy sets to describe the set of good grades as follows:

Mark: $M = \{(A, 1), (B, 0.75), (C, 0.5), (D, 0.5), (F, 0)\}$

Erica: $E = \{(A, 1), (B, 0), (C, 0), (D, 0), (F, 0)\}$

In this case, fuzzy set E is a subset of fuzzy set M because each membership value of set E *is less than or equal to* its corresponding membership value in set M.

The Complement of a Fuzzy Set

Let A be the fuzzy set $\{(x_1, a_1), (x_2, a_2), (x_3, a_3), ...\}$ defined on the universal set $X = \{x_1, x_2, x_3, ...\}$. Then the **complement** of A is the fuzzy set $A' = \{(x_1, b_1), (x_2, b_2), (x_3, b_3), ...\}$, where each $b_i = 1 - a_i$.

Each element of the fuzzy set A' has a membership value that is 1 minus its membership value in the fuzzy set A. For example, the complement of

$$S = \{(\text{math}, 0.8), (\text{history}, 0.4), (\text{biology}, 0.3), (\text{art}, 0.1), (\text{music}, 0.7)\}$$

is the fuzzy set

$$S' = \{(\text{math}, 0.2), (\text{history}, 0.6), (\text{biology}, 0.7), (\text{art}, 0.9), (\text{music}, 0.3)\}.$$

The membership values in S' were calculated by subtracting the corresponding membership values in S from 1. For instance, the membership value of math in set S is 0.8. Thus the membership value of math in set S' is $1 - 0.8 = 0.2$.

EXCURSION EXERCISES

1. Let $K = \{(1, 0.4), (2, 0.6), (3, 0.8), (4, 1)\}$ and $J = \{(1, 0.3), (2, 0.6), (3, 0.5), (4, 0.1)\}$ be fuzzy sets defined on $X = \{1, 2, 3, 4\}$. Is $J \subseteq K$? Explain.
2. Consider the following membership graphs of *YOUNG* and *ADOLESCENT* defined on $X = \{x \mid 0 \leq x \leq 50\}$, where x is age in years.

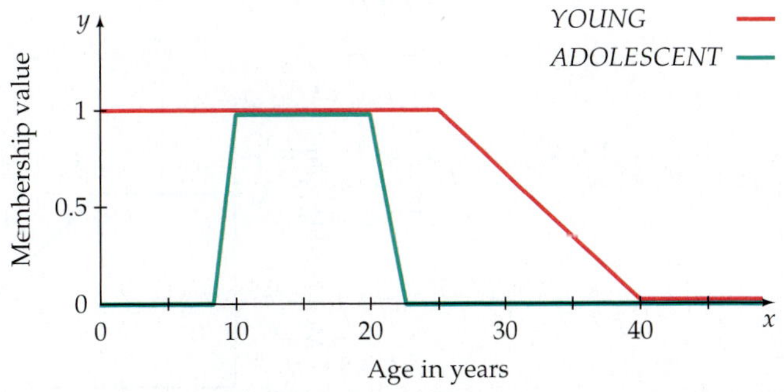

Is the fuzzy set *ADOLESCENT* a subset of the fuzzy set *YOUNG*? Explain.

3. Let the universal set be {A, B, C, D, F} and let

$$G = \{(A, 1), (B, 0.7), (C, 0.4), (D, 0.1), (F, 0)\}$$

be a fuzzy set defined by Greg to describe what he feels is a good grade. Determine G'.

4. Let $C = \{(\text{Ferrari}, 0.9), (\text{Ford Mustang}, 0.6), (\text{Dodge Neon}, 0.5), (\text{Hummer}, 0.7)\}$ be a fuzzy set defined on the universal set {Ferrari, Ford Mustang, Dodge Neon, Hummer}. Determine C'.

Consider the following membership graph.

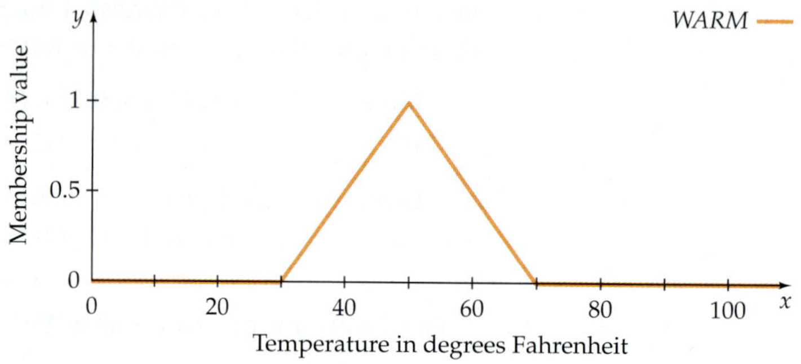

The membership graph of *WARM′* can be drawn by *reflecting* the graph of *WARM* about the graph of the line $y = 0.5$, as shown in the following figure.

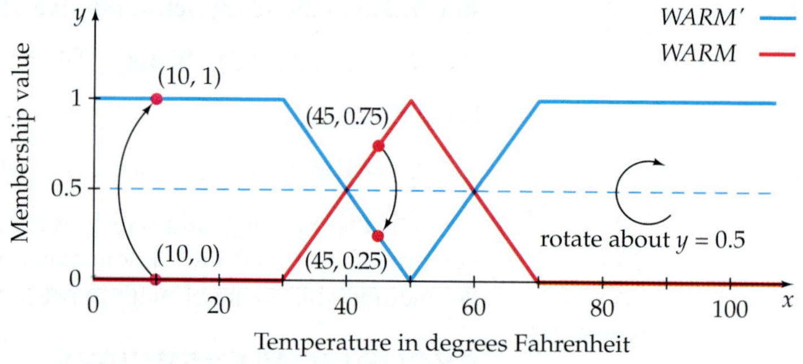

Note that when the membership graph of *WARM* is at a height of 0, the membership graph of *WARM′* is at a height of 1, and vice versa. In general, for any point (x, a) on the graph of *WARM*, there is a corresponding point $(x, 1 - a)$ on the graph of *WARM′*.

5. Use the following membership graph of *COLD* to draw the membership graph of *COLD′*.

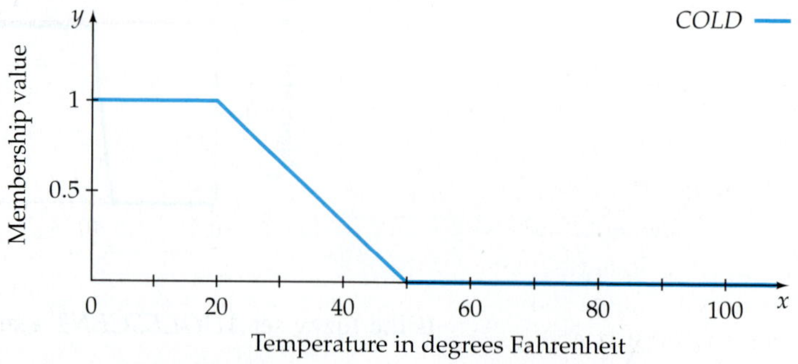

EXERCISE SET 2.2

■ In Exercises 1 to 8, find the complement of the set given that $U = \{0, 1, 2, 3, 4, 5, 6, 7, 8\}$.

1. {2, 4, 6, 7}
2. {3, 6}
3. $\varnothing$
4. {4, 5, 6, 7, 8}
5. $\{x \mid x \in W \text{ and } x < 4\}$
6. $\{x \mid x \in N \text{ and } x < 5\}$
7. The set of odd natural numbers less than 8
8. The set of even natural numbers less than 6

■ In Exercises 9 to 12, find the complement of the set given that $U = \{x \mid x \in I \text{ and } -3 \le x \le 7\}$.

9. $\{-2, 0, 2, 4, 6, 7\}$
10. $\{-3, 1, 4, 5, 6\}$
11. $\{x \mid x \in I \text{ and } -2 \le x < 3\}$
12. $\{x \mid x \in W \text{ and } x < 5\}$

■ In Exercises 13 to 22, insert either $\subseteq$ or $\not\subseteq$ in the blank space between the sets to make a true statement.

13. {a, b, c, d} {a, b, c, d, e, f, g}
14. {3, 5, 7} {3, 4, 5, 6}
15. {big, small, little} {large, petite, short}
16. {red, white, blue} {the colors in the American flag}
17. the set of integers the set of rational numbers
18. the set of real numbers the set of integers
19. $\varnothing$ {a, e, i, o, u}
20. {all sandwiches} {all hamburgers}
21. {2, 4, 6, ..., 5000} the set of even whole numbers
22. $\{x \mid x < 10 \text{ and } x \in Q\}$ the set of integers

■ In Exercises 23 to 40, let $U = \{p, q, r, s, t\}$, $D = \{p, r, s, t\}$, $E = \{q, s\}$, $F = \{p, t\}$, and $G = \{s\}$. Determine whether each statement is true or false.

23. $F \subseteq D$
24. $D \subseteq F$
25. $F \subset D$
26. $E \subset F$
27. $G \subset E$
28. $E \subset D$
29. $G' \subset D$
30. $E = F'$
31. $\varnothing \subset D$
32. $\varnothing \subset \varnothing$
33. $D' \subset E$
34. $G \in E$
35. $F \in D$
36. $G \not\subset F$
37. D has exactly eight subsets and seven proper subsets.
38. U has exactly 32 subsets.
39. F and F' each have exactly four subsets.
40. $\{0\} = \varnothing$
41. A class of 16 students has 2^{16} subsets. Use a calculator to determine how long (to the nearest hour) it would take you to write all the subsets, assuming you can write each subset in 1 second.
42. A class of 32 students has 2^{32} subsets. Use a calculator to determine how long (to the nearest year) it would take you to write all the subsets, assuming you can write each subset in 1 second.

■ In Exercises 43 to 46, list all subsets of the given set.

43. $\{\alpha, \beta\}$
44. $\{\alpha, \beta, \Gamma, \Delta\}$
45. {I, II, III}
46. $\varnothing$

■ In Exercises 47 to 54, find the number of subsets of the given set.

47. {2, 5}
48. {1, 7, 11}
49. $\{x \mid x$ is an even counting number between 7 and 21$\}$
50. $\{x \mid x$ is an odd integer between -4 and 8$\}$
51. The set of eleven players on a football team
52. The set of all letters of our alphabet
53. The set of all negative whole numbers
54. The set of all single-digit natural numbers
55. Suppose you have a nickel, two dimes, and a quarter. One of the dimes was minted in 1976, and the other one was minted in 1992.

a. Assuming you choose at least one coin, how many different sets of coins can you form?
b. Assuming you choose at least one coin, how many different sums of money can you produce?
c. Explain why the answers in part a and part b are not the same.

56. The number of subsets of a set with n elements is 2^n.

a. Use a calculator to find the exact value of 2^{18}, 2^{19}, and 2^{20}.
b. What is the largest integer power of 2 for which your calculator will display the exact value?

57. Sandwich Choices A delicatessen makes a roast-beef-on-sourdough sandwich for which you can choose from eight condiments.

Joe Gough/Shutterstock.com

a. How many different types of roast-beef-on-sourdough sandwiches can the delicatessen prepare?
b. What is the minimum number of condiments the delicatessen must have available if it wishes to offer at least 2000 different types of roast-beef-on-sourdough sandwiches?

58. Upgrade Options A company that builds homes advertises that, by choosing from its upgrade options, each of its new homes is available in 256 different variations. How many upgrade options does the company offer?

59. Omelet Choices A restaurant provides a brunch where the omelets are individually prepared. Each guest is allowed to choose from 10 different ingredients.

a. How many different types of omelets can the restaurant prepare?

b. What is the minimum number of ingredients that must be available if the restaurant wants to advertise that it offers over 4000 different omelets?

60. Truck Options A truck company makes a pickup truck with 12 upgrade options. Some of the options are air conditioning, chrome wheels, and a satellite radio.

Rob Wilson/Shutterstock.com

a. How many different versions of this truck can the company produce?

b. What is the minimum number of upgrade options the company must be able to provide if it wishes to offer at least 14,000 different versions of this truck?

61. a. Explain why $\{2\} \notin \{1, 2, 3\}$.

b. Explain why $1 \not\subseteq \{1, 2, 3\}$.

c. Consider the set $\{1, \{1\}\}$. Does this set have one or two elements? Explain.

62. a. A set has 1024 subsets. How many elements are in the set?

b. A set has 255 proper subsets. How many elements are in the set?

c. Is it possible for a set to have an odd number of subsets? Explain.

EXTENSIONS

63. Voting Coalitions Five people, designated A, B, C, D, and E, serve on a committee. To pass a motion, at least three of the committee members must vote for the motion. In such a situation, any set of three or more voters is called a **winning coalition** because if this set of people votes for a motion, the motion will pass. Any nonempty set of two or fewer voters is called a **losing coalition**.

a. List all the winning coalitions.

b. List all the losing coalitions.

64. Subsets and Pascal's Triangle Following is a list of all the subsets of {a, b, c, d}.
Subsets with

0 elements:	{ }
1 element:	{a}, {b}, {c}, {d}
2 elements:	{a, b}, {a, c}, {a, d}, {b, c}, {b, d}, {c, d}
3 elements:	{a, b, c}, {a, b, d}, {a, c, d}, {b, c, d}
4 elements:	{a, b, c, d}

There is 1 subset with zero elements, and there are 4 subsets with exactly one element, 6 subsets with exactly two elements, 4 subsets with exactly three elements, and 1 subset with exactly four elements. Note that the numbers 1, 4, 6, 4, 1 are the numbers in row 4 of Pascal's triangle, which is shown in the column to the right. Recall that the numbers in Pascal's triangle are created in the following manner. Each row begins and ends with the number 1. Any other number in a row is the sum of the two closest numbers above it. For instance, the first 10 in row 5 is the sum of the first 4 and the 6 in row 4.

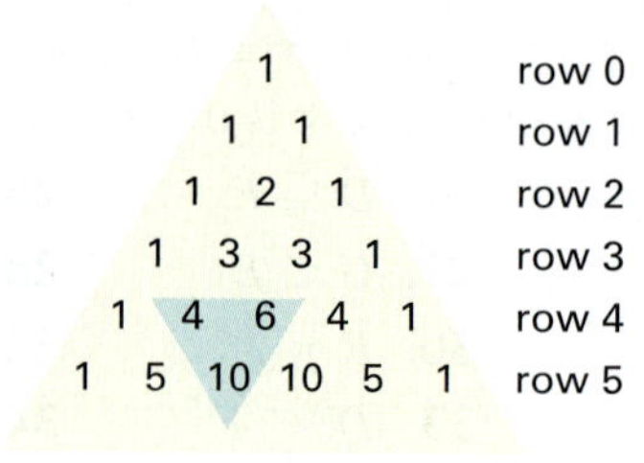

Pascal's triangle

a. Use Pascal's triangle to make a conjecture about the numbers of subsets of {a, b, c, d, e} that have: zero elements, exactly one element, exactly two elements, exactly three elements, exactly four elements, and exactly five elements. Use your work from Check Your Progress 4 on page 60 to verify that your conjecture is correct.

b. Extend Pascal's triangle to show row 6. Use row 6 of Pascal's triangle to make a conjecture about the number of subsets of {a, b, c, d, e, f} that have exactly three elements. Make a list of all the subsets of {a, b, c, d, e, f} that have exactly three elements to verify that your conjecture is correct.

SECTION 2.3 Set Operations

Intersection and Union of Sets

In Section 2.2 we defined the operation of finding the complement of a set. In this section we define the set operations *intersection* and *union*. In everyday usage, the word "intersection" refers to the *common region* where two streets cross. The intersection of two sets is defined in a similar manner.

> **Intersection of Sets**
>
> The **intersection** of sets A and B, denoted by $A \cap B$, is the set of elements common to both A and B.
>
> $$A \cap B = \{x \mid x \in A \quad \text{and} \quad x \in B\}$$

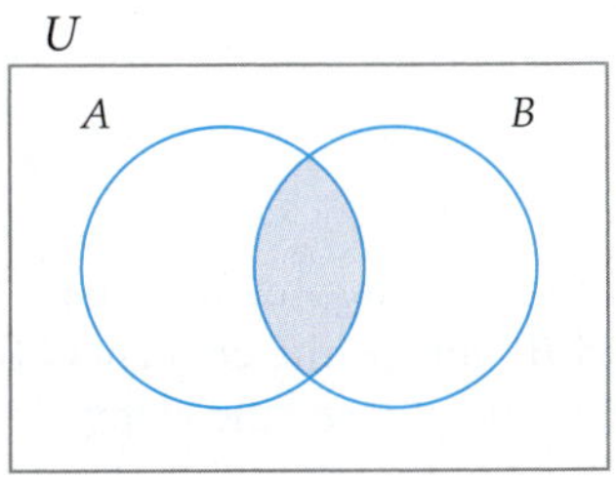

$A \cap B$

In the figure at the left, the region shown in blue represents the intersection of sets A and B.

EXAMPLE 1 Find Intersections

Let $A = \{1, 4, 5, 7\}$, $B = \{2, 3, 4, 5, 6\}$, and $C = \{3, 6, 9\}$. Find

a. $A \cap B$ **b.** $A \cap C$

Solution

a. The elements common to A and B are 4 and 5.

$$A \cap B = \{1, 4, 5, 7\} \cap \{2, 3, 4, 5, 6\}$$
$$= \{4, 5\}$$

b. Sets A and C have no common elements. Thus $A \cap C = \varnothing$.

TAKE NOTE

It is a mistake to write

$$\{1, 5, 9\} \cap \{3, 5, 9\} = 5, 9$$

The intersection of two sets is a set. Thus,

$$\{1, 5, 9\} \cap \{3, 5, 9\} = \{5, 9\}$$

CHECK YOUR PROGRESS 1 Let $D = \{0, 3, 8, 9\}$, $E = \{3, 4, 8, 9, 11\}$, and $F = \{0, 2, 6, 8\}$. Find

a. $D \cap E$ **b.** $D \cap F$

Solution See page S5.

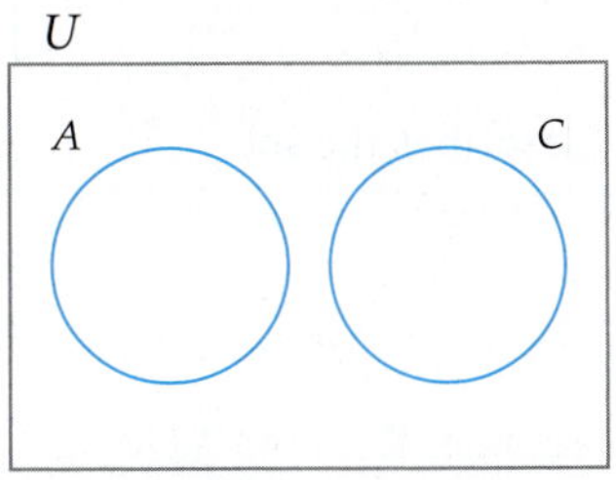

$A \cap C = \varnothing$

Two sets are **disjoint** if their intersection is the empty set. The sets A and C in Example 1b are disjoint. The Venn diagram at the left illustrates two disjoint sets.

In everyday usage, the word "union" refers to the act of uniting or joining together. The union of two sets has a similar meaning.

> **Union of Sets**
>
> The **union** of sets A and B, denoted by $A \cup B$, is the set that contains all the elements that belong to A or to B or to both.
>
> $$A \cup B = \{x \mid x \in A \quad \text{or} \quad x \in B\}$$

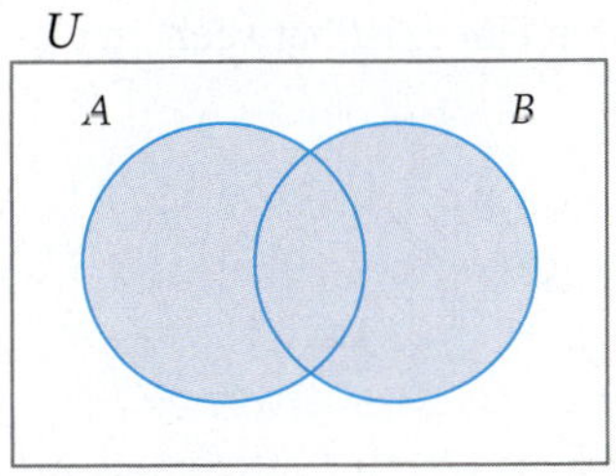

$A \cup B$

In the figure at the left, the region shown in blue represents the union of sets A and B.

EXAMPLE 2 Find Unions

Let $A = \{1, 4, 5, 7\}$, $B = \{2, 3, 4, 5, 6\}$, and $C = \{3, 6, 9\}$. Find

a. $A \cup B$ **b.** $A \cup C$

Solution

a. List all the elements of set A, which are 1, 4, 5, and 7. Then add to your list the elements of set B that have not already been listed—in this case 2, 3, and 6. Enclose all elements with a pair of braces. Thus,

$$A \cup B = \{1, 4, 5, 7\} \cup \{2, 3, 4, 5, 6\}$$
$$= \{1, 2, 3, 4, 5, 6, 7\}$$

b.
$$A \cup C = \{1, 4, 5, 7\} \cup \{3, 6, 9\}$$
$$= \{1, 3, 4, 5, 6, 7, 9\}$$

CHECK YOUR PROGRESS 2 Let $D = \{0, 4, 8, 9\}$, $E = \{1, 4, 5, 7\}$, and $F = \{2, 6, 8\}$. Find

a. $D \cup E$ **b.** $D \cup F$

Solution See page S5.

TAKE NOTE

Would you like soup or salad?

In a sentence, the word "or" means one or the other, but not both. For instance, if a menu states that you can have soup or salad with your meal, this generally means that you can have either soup or salad for the price of the meal, but not both. In this case the word "or" is said to be an *exclusive or.* In the mathematical statement "A or B," the "or" is an *inclusive or.* It means A or B, or both.

In mathematical problems that involve sets, the word "and" is interpreted to mean *intersection*. For instance, the phrase "the elements of A and B" means the elements of $A \cap B$. Similarly, the word "or" is interpreted to mean *union*. The phrase "the elements of A or B" means the elements of $A \cup B$.

EXAMPLE 3 Describe Sets

Write a sentence that describes the set.

a. $A \cup (B \cap C)$ **b.** $J \cap K'$

Solution

a. The set $A \cup (B \cap C)$ can be described as "the set of all elements that are in A, or are in B and C."

b. The set $J \cap K'$ can be described as "the set of all elements that are in J and are not in K."

CHECK YOUR PROGRESS 3 Write a sentence that describes the set.

a. $D \cap (E' \cup F)$ **b.** $L' \cup M$

Solution See page S5.

Venn Diagrams and Equality of Sets

The Venn diagram in Figure 2.1 shows the four regions formed by two intersecting sets in a universal set U. It shows the four possible relationships that can exist between an element of a universal set U and two sets A and B.

An element of U:

- may be an element of both A and B. Region i
- may be an element of A, but not B. Region ii
- may be an element of B, but not A. Region iii
- may not be an element of either A or B. Region iv

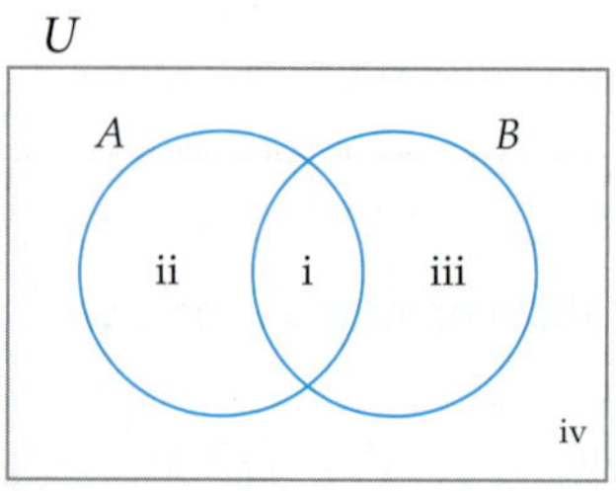

FIGURE 2.1 Venn diagram for two intersecting sets, in a universal set U.

TAKE NOTE

The sets A and B in Figure 2.1 separate the universal set, U, into four regions. These four regions can be numbered in any manner—that is, we can number any region as i, any region as ii, and so on. However, in this chapter we will use the numbering scheme shown in Figure 2.1 whenever we number the four regions formed by two intersecting sets in a universal set.

We can use Figure 2.1 to determine whether two expressions that involve two sets are equal. For instance, to determine whether $(A \cup B)'$ and $A' \cap B'$ are equal for all sets A and B, we find what region or regions each of the expressions represents in Figure 2.1.

- If both expressions are represented by the same region(s), then the expressions are equal for all sets A and B.
- If both expressions are *not* represented by the same region(s), then the expressions are *not* equal for all sets A and B.

EXAMPLE 4 Equality of Sets

Determine whether $(A \cup B)' = A' \cap B'$ for all sets A and B.

Solution

To determine the region(s) in Figure 2.1, represented by $(A \cup B)'$, first determine the region(s) that are represented by $A \cup B$.

Set	Region or regions	Venn diagram
$A \cup B$	i, ii, iii The regions obtained by joining the regions represented by A (i, ii) and the regions represented by B (i, iii)	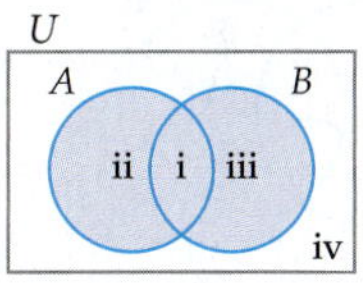
$(A \cup B)'$	iv The region in U that is not in $(A \cup B)$	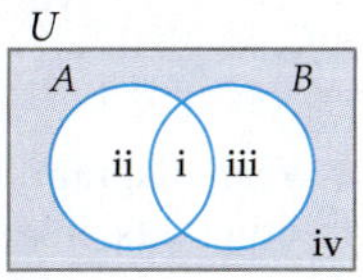

Now determine the region(s) in Figure 2.1 that are represented by $A' \cap B'$.

Set	Region or regions	Venn diagram
A'	iii, iv The regions outside of A	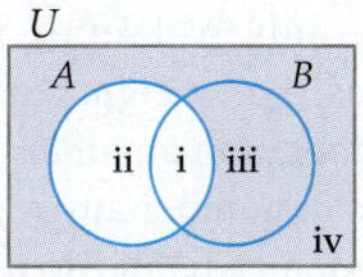
B'	ii, iv The regions outside of B	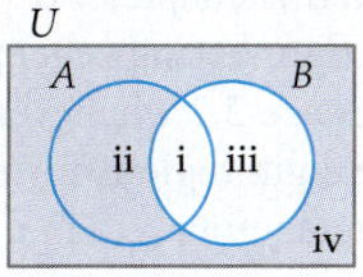
$A' \cap B'$	iv The region common to A' and B'	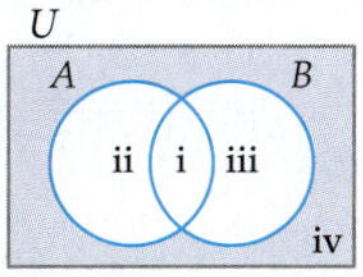

The expressions $(A \cup B)'$ and $A' \cap B'$ are both represented by region iv in Figure 2.1. Thus $(A \cup B)' = A' \cap B'$ for all sets A and B.

CHECK YOUR PROGRESS 4 Determine whether $(A \cap B)' = A' \cup B'$ for all sets A and B.

Solution See page S5.

The properties that were verified in Example 4 and Check Your Progress 4 are known as **De Morgan's laws**.

De Morgan's Laws

For all sets A and B,

$$(A \cup B)' = A' \cap B' \quad \text{and} \quad (A \cap B)' = A' \cup B'$$

De Morgan's law $(A \cup B)' = A' \cap B'$ can be stated as "the complement of the union of two sets is the intersection of the complements of the sets." De Morgan's law $(A \cap B)' = A' \cup B'$ can be stated as "the complement of the intersection of two sets is the union of the complements of the sets."

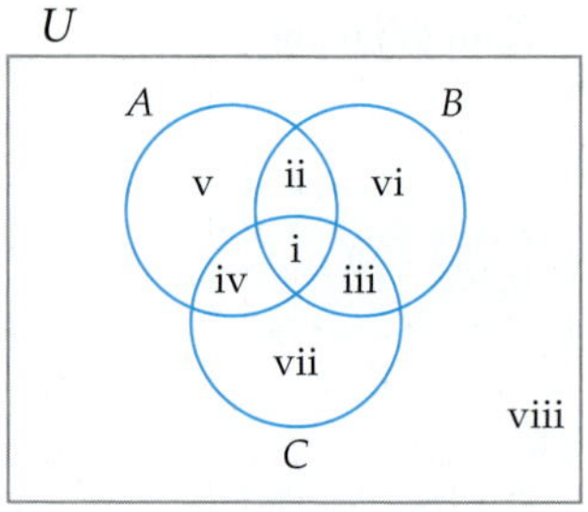

FIGURE 2.2 Venn diagram for three intersecting sets, in a universal set U

TAKE NOTE

The sets A, B, and C in Figure 2.2 separate the universal set, U, into eight regions. These eight regions can be numbered in any manner—that is, we can number any region as i, any region as ii, and so on. However, in this chapter we will use the numbering scheme shown in Figure 2.2 whenever we number the eight regions formed by three intersecting sets in a universal set.

Venn Diagrams Involving Three Sets

The Venn diagram in Figure 2.2 shows the eight regions formed by three intersecting sets in a universal set U. It shows the eight possible relationships that can exist between an element of a universal set U and three sets A, B, and C.

An element of U:

- may be an element of A, B, and C. Region i
- may be an element of A and B, but not C. Region ii
- may be an element of B and C, but not A. Region iii
- may be an element of A and C, but not B. Region iv
- may be an element of A, but not B or C. Region v
- may be an element of B, but not A or C. Region vi
- may be an element of C, but not A or B. Region vii
- may not be an element of A, B, or C. Region viii

EXAMPLE 5 Determine Regions that Represent Sets

Use Figure 2.2 to answer each of the following.

a. Which regions represent $A \cap C$?

b. Which regions represent $A \cup C$?

c. Which regions represent $A \cap B'$?

Solution

a. $A \cap C$ is represented by all the regions common to circles A and C. Thus $A \cap C$ is represented by regions i and iv.

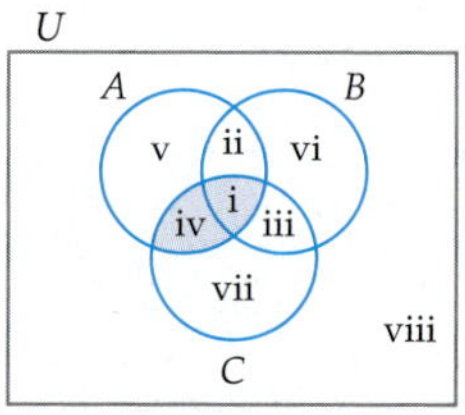

b. $A \cup C$ is represented by all the regions obtained by joining the regions in circle A (i, ii, iv, v) and the regions in circle C (i, iii, iv, vii). Thus $A \cup C$ is represented by regions i, ii, iii, iv, v, and vii.

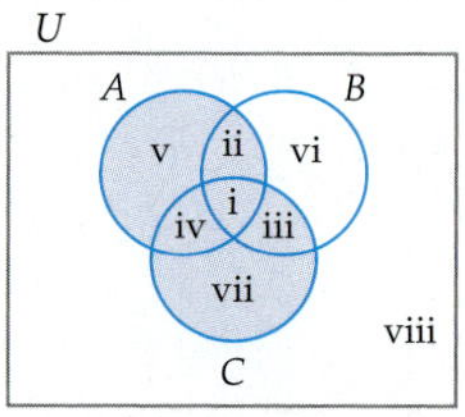

c. $A \cap B'$ is represented by all the regions common to circle A and the regions that are not in circle B. Thus $A \cap B'$ is represented by regions iv and v.

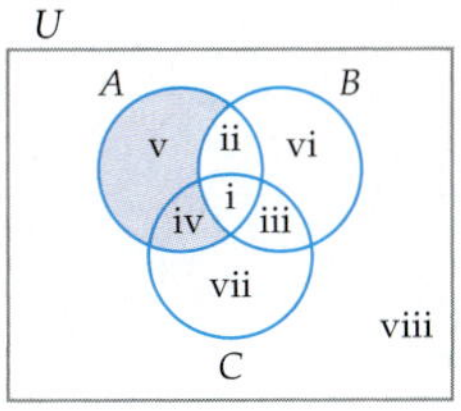

CHECK YOUR PROGRESS 5 Use Figure 2.2 to answer each of the following.

a. Which region represents $(A \cap B) \cap C$?

b. Which regions represent $A \cup B'$?

c. Which regions represent $C' \cap B$?

Solution See page S5.

In Example 6, we use Figure 2.2 to determine whether two expressions that involve three sets are equal.

EXAMPLE 6 Equality of Sets

Determine whether $A \cup (B \cap C) = (A \cup B) \cap C$ for all sets A, B, and C.

Solution

To determine the region(s) in Figure 2.2 represented by $A \cup (B \cap C)$, we join the regions in A and the regions in $B \cap C$.

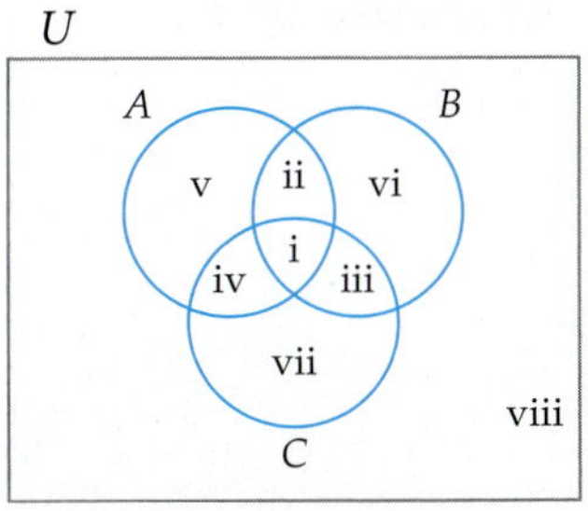

FIGURE 2.2 Displayed a second time for convenience

Set	Region or regions	Venn diagram
A	i, ii, iv, v The regions in A	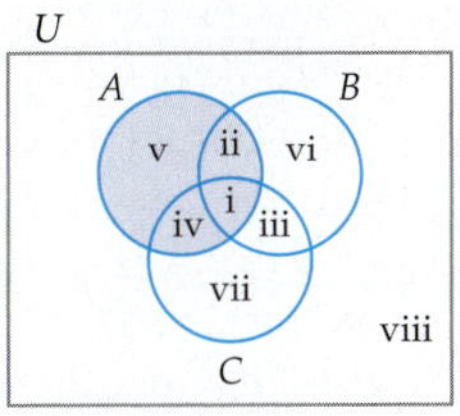
$B \cap C$	i, iii The regions common to B and C	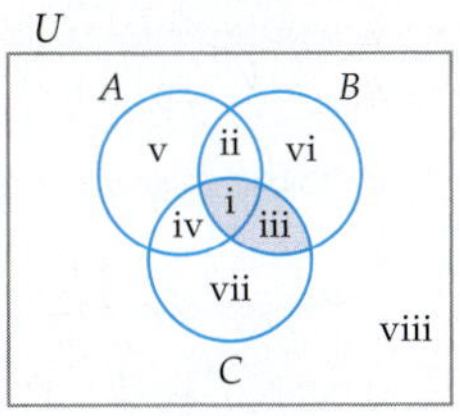
$A \cup (B \cap C)$	i, ii, iv, v, iii The regions in A joined with the regions in $B \cap C$	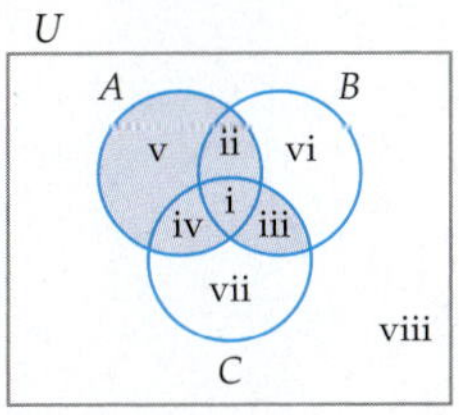

Now determine the region(s) in Figure 2.2 that are represented by $(A \cup B) \cap C$.

Set	Region or regions	Venn diagram
$A \cup B$	i, ii, iv, v, vi, iii The regions in A joined with the regions in B	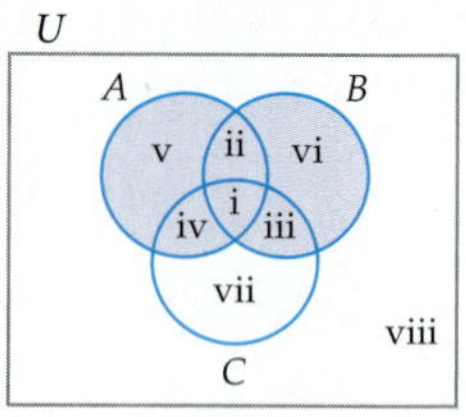
C	i, iii, iv, vii The regions in C	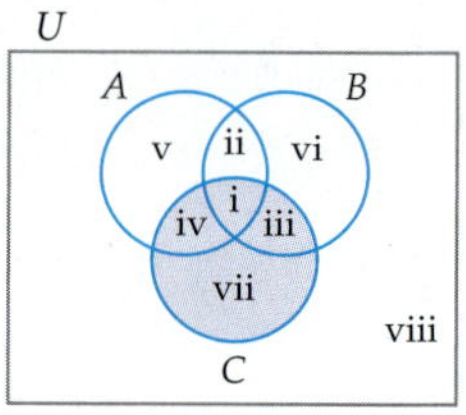
$(A \cup B) \cap C$	i, iii, iv The regions common to $A \cup B$ and C	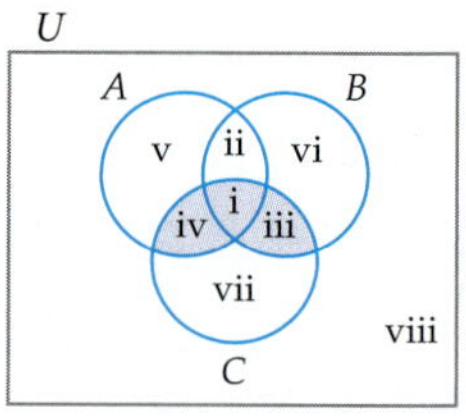

The expressions $A \cup (B \cap C)$ and $(A \cup B) \cap C$ are not represented by the same regions.

Thus $A \cup (B \cap C) \neq (A \cup B) \cap C$ for all sets A, B, and C.

CHECK YOUR PROGRESS 6 Determine whether $A \cap (B \cup C) = (A \cap B) \cup (A \cap C)$ for all sets A, B, and C.

Solution See page S6.

Venn diagrams can be used to verify each of the following properties.

Properties of Sets

For all sets A and B:

Commutative Properties

$A \cap B = B \cap A$ — Commutative property of intersection

$A \cup B = B \cup A$ — Commutative property of union

For all sets A, B, and C:

Associative Properties

$(A \cap B) \cap C = A \cap (B \cap C)$ — Associative property of intersection

$(A \cup B) \cup C = A \cup (B \cup C)$ — Associative property of union

Distributive Properties

$A \cap (B \cup C) = (A \cap B) \cup (A \cap C)$ — Distributive property of intersection over union

$A \cup (B \cap C) = (A \cup B) \cap (A \cup C)$ — Distributive property of union over intersection

QUESTION Does $(B \cup C) \cap A = (A \cap B) \cup (A \cap C)$?

Application: Blood Groups and Blood Types

HISTORICAL NOTE

Ted Spiegel/Encyclopedia/Corbis

The Nobel Prize is an award granted to people who have made significant contributions to society. Nobel Prizes are awarded annually for achievements in physics, chemistry, physiology or medicine, literature, peace, and economics. The prizes were first established in 1901 by the Swedish industrialist Alfred Nobel, who invented dynamite.

Karl Landsteiner won a Nobel Prize in 1930 for his discovery of the four different human blood groups. He discovered that the blood of each individual contains exactly one of the following combinations of antigens.

- Only A antigens (blood group A)
- Only B antigens (blood group B)
- Both A and B antigens (blood group AB)
- No A antigens and no B antigens (blood group O)

These four blood groups are represented by the Venn diagram in the left margin.

In 1941, Landsteiner and Alexander Wiener discovered that human blood may or may not contain an Rh, or rhesus, factor. Blood with this factor is called Rh-positive and is denoted by Rh+. Blood without this factor is called Rh-negative and is denoted by Rh−.

The Venn diagram in Figure 2.3 illustrates the eight blood types (A+, B+, AB+, O+, A−, B−, AB−, O−) that are possible if we consider antigens and the Rh factor.

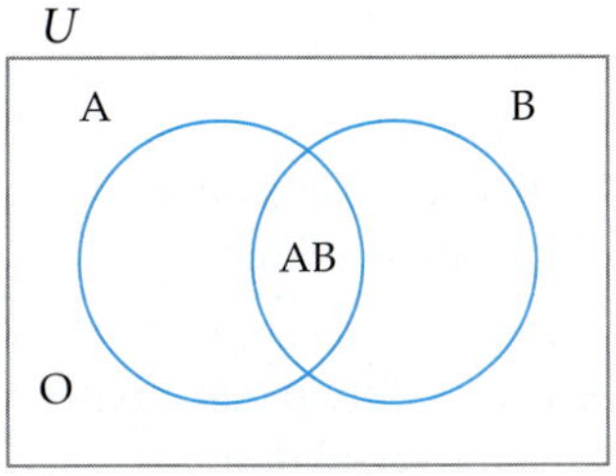

The four blood groups

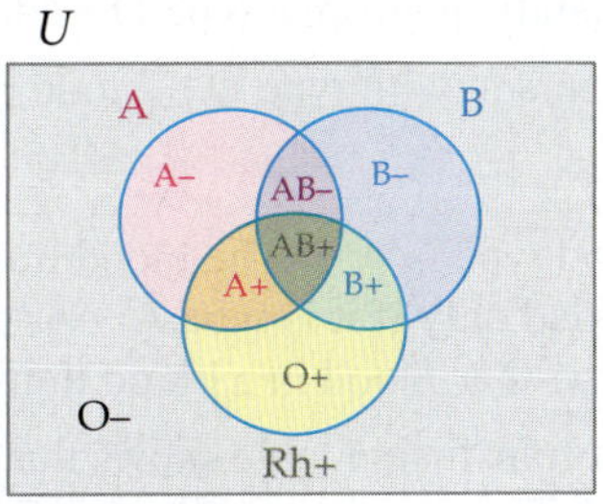

FIGURE 2.3
The eight blood types

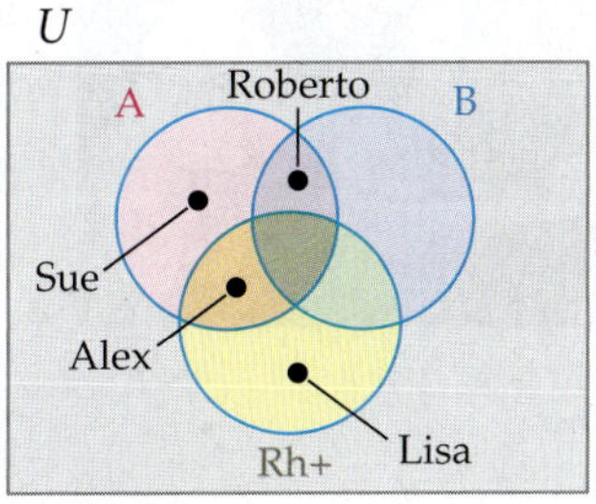

FIGURE 2.4

EXAMPLE 7 Venn Diagrams and Blood Type

Use the Venn diagrams in Figures 2.3 and 2.4 to determine the blood type of each of the following people.

a. Sue **b.** Lisa

Solution

a. Because Sue is in blood group A, not in blood group B, and not Rh+, her blood type is A−.

b. Lisa is in blood group O and she is Rh+, so her blood type is O+.

CHECK YOUR PROGRESS 7 Use the Venn diagrams in Figures 2.3 and 2.4 to determine the blood type of each of the following people.

a. Alex **b.** Roberto

Solution See page S6.

ANSWER Yes. The commutative property of intersection allows us to write $(B \cup C) \cap A$ as $A \cap (B \cup C)$ and $A \cap (B \cup C) = (A \cap B) \cup (A \cap C)$ by the distributive property of intersection over union.

The following table shows the blood types that can safely be given during a blood transfusion to persons of each of the eight blood types.

Blood Transfusion Table

Recipient blood type	Donor blood type
A+	A+, A−, O+, O−
B+	B+, B−, O+, O−
AB+	A+, A−, B+, B−, AB+, AB−, O+, O−
O+	O+, O−
A−	A−, O−
B−	B−, O−
AB−	A−, B−, AB−, O−
O−	O−

SOURCE: American Red Cross

EXAMPLE 8 Applications of the Blood Transfusion Table

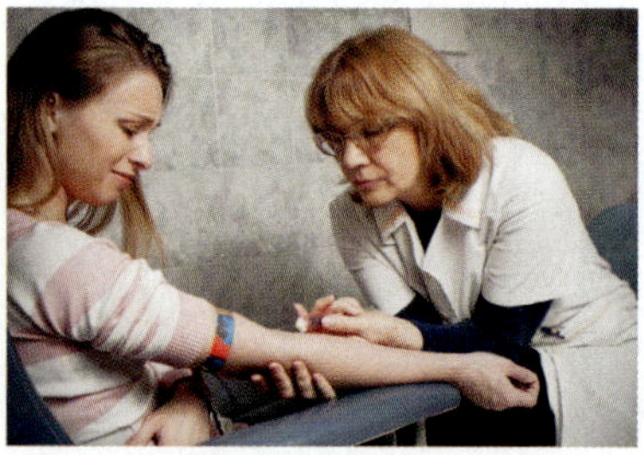

MilanMarkovic/Shutterstock.com

Use the blood transfusion table and Figures 2.3 and 2.4 to answer the following questions.

a. Can Sue safely be given a type O+ blood transfusion?

b. Why is a person with type O− blood called a *universal donor?*

Solution

a. Sue's blood type is A−. The blood transfusion table shows that she can safely receive blood only if it is type A− or type O−. Thus it is not safe for Sue to receive type O+ blood in a blood transfusion.

b. The blood transfusion table shows that all eight blood types can safely receive type O− blood. Thus a person with type O− blood is said to be a universal donor.

CHECK YOUR PROGRESS 8 Use the blood transfusion table and Figures 2.3 and 2.4 to answer the following questions.

a. Is it safe for Alex to receive type A− blood in a blood transfusion?

b. What blood type do you have if you are classified as a *universal recipient?*

Solution See page S6.

MATH MATTERS The Cantor Set

Consider the set of points formed by a line segment with a length of 1 unit. Remove the middle third of the line segment. Remove the middle third of each of the remaining 2 line segments. Remove the middle third of each of the remaining 4 line segments. Remove the middle third of each of the remaining 8 line segments. Remove the middle third of each of the remaining 16 line segments.

The first five steps in the formation of the Cantor set

The **Cantor set**, also known as **Cantor's dust**, is the set of points that *remain* after the above process is repeated infinitely many times. You might conjecture that there are no points in the Cantor set, but it can be shown that there are just as many points in the Cantor set as in the original line segment! This is remarkable because it can also be shown that the sum of the lengths of the *removed* line segments equals 1 unit, which is the length of the original line segment. You can find additional information about the remarkable properties of the Cantor set on the Internet.

EXCURSION

Union and Intersection of Fuzzy Sets

This Excursion extends the concepts of fuzzy sets that were developed in Sections 2.1 and 2.2.

There are a number of ways in which the *union of two fuzzy sets* and the *intersection of two fuzzy sets* can be defined. The definitions we will use are called the **standard union operator** and the **standard intersection operator**. These standard operators preserve many of the set relations that exist in standard set theory.

Union and Intersection of Two Fuzzy Sets

Let $A = \{(x_1, a_1), (x_2, a_2), (x_3, a_3), \ldots\}$ and $B = \{(x_1, b_1), (x_2, b_2), (x_3, b_3), \ldots\}$. Then

$$A \cup B = \{(x_1, c_1), (x_2, c_2), (x_3, c_3), \ldots\}$$

where c_i is the *maximum* of the two numbers a_i and b_i and

$$A \cap B = \{(x_1, c_1), (x_2, c_2), (x_3, c_3), \ldots\}$$

where c_i is the *minimum* of the two numbers a_i and b_i.

Each element of the fuzzy set $A \cup B$ has a membership value that is the *maximum* of its membership value in the fuzzy set A and its membership value in the fuzzy set B. Each element of the fuzzy set $A \cap B$ has a membership value that is the *minimum* of its membership value in fuzzy set A and its membership value in the fuzzy set B. In the following example, we form the union and intersection of two fuzzy sets. Let P and S be defined as follows.

Paul: $P = \{(\text{math}, 0.2), (\text{history}, 0.5), (\text{biology}, 0.7), (\text{art}, 0.8), (\text{music}, 0.9)\}$

Sally: $S = \{(\text{math}, 0.8), (\text{history}, 0.4), (\text{biology}, 0.3), (\text{art}, 0.1), (\text{music}, 0.7)\}$

Then

The maximum membership values for each of the given elements math, history, biology, art, and music

$$P \cup S = \{(\text{math}, 0.8), (\text{history}, 0.5), (\text{biology}, 0.7), (\text{art}, 0.8), (\text{music}, 0.9)\}$$

$$P \cap S = \{(\text{math}, 0.2), (\text{history}, 0.4), (\text{biology}, 0.3), (\text{art}, 0.1), (\text{music}, 0.7)\}$$

The minimum membership values for each of the given elements math, history, biology, art, and music

EXCURSION EXERCISES

In Excursion Exercise 1 of Section 2.1, we defined the following fuzzy sets.

Mark: $M = \{(A, 1), (B, 0.75), (C, 0.5), (D, 0.5), (F, 0)\}$
Erica: $E = \{(A, 1), (B, 0), (C, 0), (D, 0), (F, 0)\}$
Larry: $L = \{(A, 1), (B, 1), (C, 1), (D, 1), (F, 0)\}$
Jennifer: $J = \{(A, 1), (B, 0.8), (C, 0.6), (D, 0.1), (F, 0)\}$

Use these fuzzy sets to find each of the following.

1. $M \cup J$ 2. $M \cap J$ 3. $E \cup J'$ 4. $J \cap L'$ 5. $J \cap (M' \cup L')$

Consider the following membership graphs.

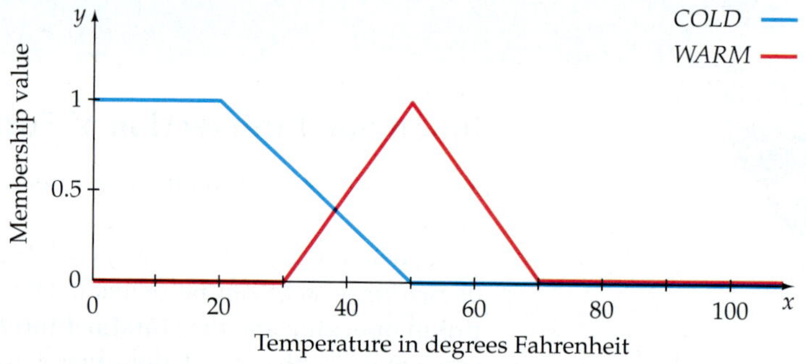

The membership graph of *COLD* ∪ *WARM* is shown in purple in the following figure. The membership graph of *COLD* ∪ *WARM* lies on either the membership graph of *COLD* or the membership graph of *WARM*, depending on which of these graphs is *higher* at any given temperature x.

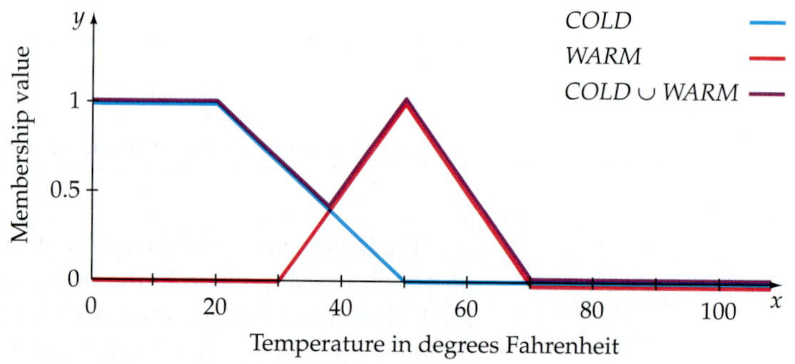

The membership graph of *COLD* ∩ *WARM* is shown in green in the following figure. The membership graph of *COLD* ∩ *WARM* lies on either the membership graph of *COLD* or the membership graph of *WARM*, depending on which of these graphs is *lower* at any given temperature x.

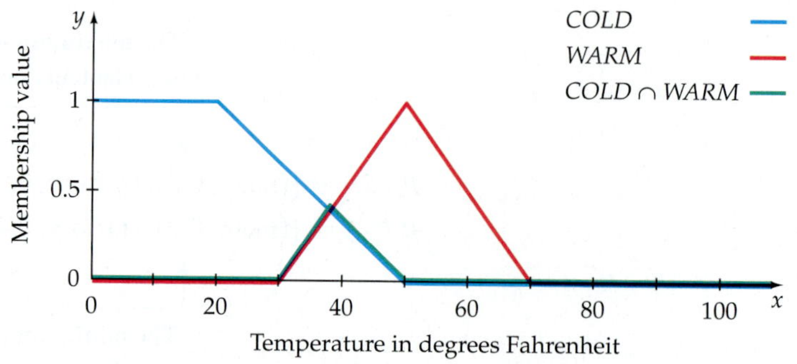

TAKE NOTE

The following lyrics from the old Scottish song *Loch Lomond* provide an easy way to remember how to draw the graph of the union or intersection of two membership graphs.

> Oh! ye'll take the high road and I'll take the low road, and I'll be in Scotland afore ye.

The graph of the union of two membership graphs takes the "high road" provided by the graphs, and the graph of the intersection of two membership graphs takes the "low road" provided by the graphs.

6. Use the following graphs to draw the membership graph of $WARM \cup HOT$.

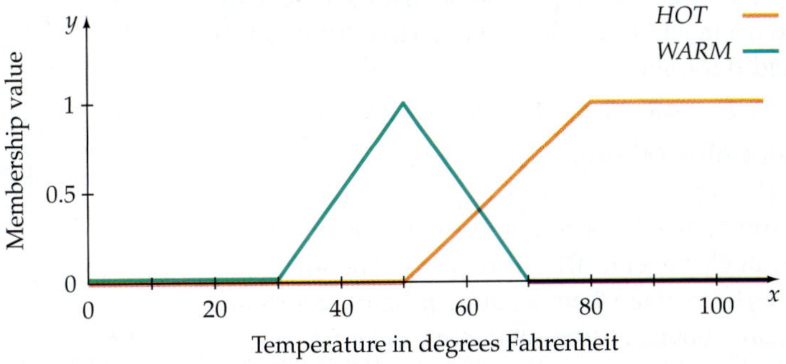

7. Let $X = \{a, b, c, d, e\}$ be the universal set. Determine whether De Morgan's law $(A \cap B)' = A' \cup B'$ holds true for the fuzzy sets
$A = \{(a, 0.3), (b, 0.8), (c, 1), (d, 0.2), (e, 0.75)\}$ and
$B = \{(a, 0.5), (b, 0.4), (c, 0.9), (d, 0.7), (e, 0.45)\}$.

EXERCISE SET 2.3

■ In Exercises 1 to 20, let $U = \{1, 2, 3, 4, 5, 6, 7, 8\}$, $A = \{2, 4, 6\}$, $B = \{1, 2, 5, 8\}$, and $C = \{1, 3, 7\}$. Find each of the following.

1. $A \cup B$

2. $A \cap B$

3. $A \cap B'$

4. $B \cap C'$

5. $(A \cup B)'$

6. $(A' \cap B)'$

7. $A \cup (B \cup C)$

8. $A \cap (B \cup C)$

9. $A \cap (B \cap C)$

10. $A' \cup (B \cap C)$

11. $B \cap (B \cup C)$

12. $A \cap A'$

13. $B \cup B'$

14. $(A \cap (B \cup C))'$

15. $(A \cup C') \cap (B \cup A')$

16. $(A \cup C') \cup (B \cup A')$

17. $(C \cup B') \cup \varnothing$

18. $(A' \cup B) \cap \varnothing$

19. $(A \cup B) \cap (B \cap C')$

20. $(B \cap A') \cup (B' \cup C)$

In Exercises 21 to 28, write a sentence that describes the given mathematical expression.

21. $L' \cup T$

22. $J' \cap K$

23. $A \cup (B' \cap C)$

24. $(A \cup B) \cap C'$

25. $T \cap (J \cup K')$

26. $(A \cap B) \cup C$

27. $(W \cap V) \cup (W \cap Z)$

28. $D \cap (E \cup F)'$

■ In Exercises 29 to 36, draw a Venn diagram to show each of the following sets.

29. $A \cap B'$

30. $(A \cap B)'$

31. $(A \cup B)'$

32. $(A' \cap B) \cup B'$

33. $A \cap (B \cup C')$

34. $A \cap (B' \cap C)$

35. $(A \cup C) \cap (B \cup C')$

36. $(A' \cap B) \cup (A \cap C')$

■ In Exercises 37 to 40, draw two Venn diagrams to determine whether the following expressions are equal for all sets A and B.

37. $A \cap B'$; $A' \cup B$

38. $A' \cap B$; $A \cup B'$

39. $A \cup (A' \cap B)$; $A \cup B$

40. $A' \cap (B \cup B')$; $A' \cup (B \cap B')$

■ In Exercises 41 to 46, draw two Venn diagrams to determine whether the following expressions are equal for all sets A, B, and C.

41. $(A \cup C) \cap B'$; $A' \cup (B \cup C)$

42. $A' \cap (B \cap C)$, $(A \cup B') \cap C$

43. $(A' \cap B) \cup C$, $(A' \cap C) \cap (A' \cap B)$

44. $A' \cup (B' \cap C)$, $(A' \cup B') \cap (A' \cup C)$

45. $((A \cup B) \cap C)'$, $(A' \cap B') \cup C'$

46. $(A \cap B) \cap C$, $((A \cup B) \cup C)'$

Additive Color Mixing Computers and televisions make use of *additive color mixing*. The following figure shows that when the *primary colors* red R, green G, and blue B are mixed together using additive color mixing, they produce white, W. Using set notation, we state this as $R \cap B \cap G = W$. The colors yellow Y, cyan C, and magenta M are called *secondary colors*. A secondary color is produced by mixing exactly two of the primary colors.

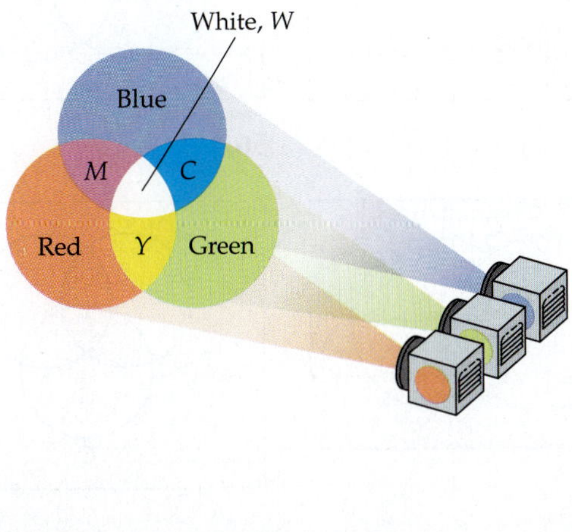

■ In Exercises 47 to 49, determine which color is represented by each of the following. Assume that the colors are being mixed using additive color mixing. (Use R for red, G for green, and B for blue.)

47. $R \cap G \cap B'$ **48.** $R \cap G' \cap B$ **49.** $R' \cap G \cap B$

Subtractive Color Mixing Artists who paint with pigments use *subtractive color mixing* to produce different colors. In a subtractive color mixing system, the primary colors are cyan C, magenta M, and yellow Y. The following figure shows that when the three primary colors are mixed in equal amounts, using subtractive color mixing, they form black, K. Using set notation, we state this as $C \cap M \cap Y = K$. In subtractive color mixing, the colors red R, blue B, and green G are the secondary colors. As mentioned previously, a secondary color is produced by mixing equal amounts of exactly two of the primary colors.

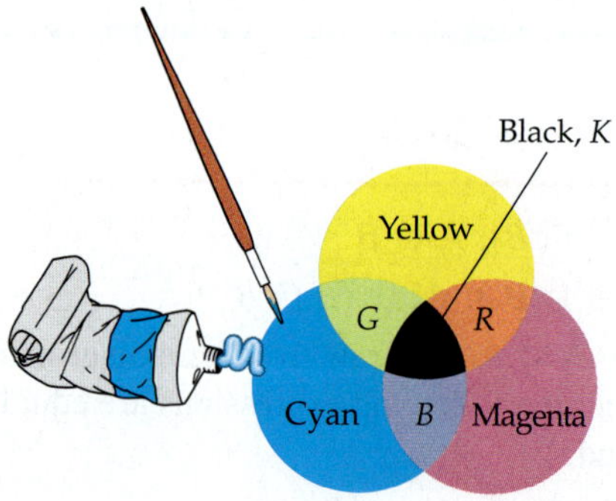

■ In Exercises 50 to 52, determine which color is represented by each of the following. Assume the colors are being mixed using subtractive color mixing. (Use C for cyan, M for magenta, and Y for yellow.)

50. $C \cap M \cap Y'$ **51.** $C' \cap M \cap Y$ **52.** $C \cap M' \cap Y$

■ In Exercises 53 to 62, use set notation to describe the shaded region. You may use any of the following symbols: A, B, C, $\cap$, $\cup$, and $'$. Keep in mind that each shaded region has more than one set description.

53.
U
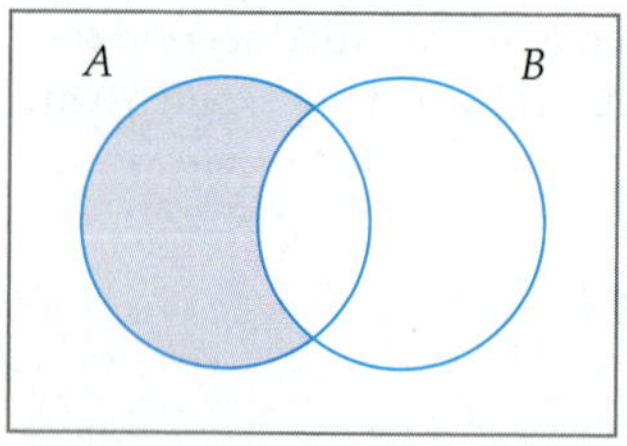

54.
U
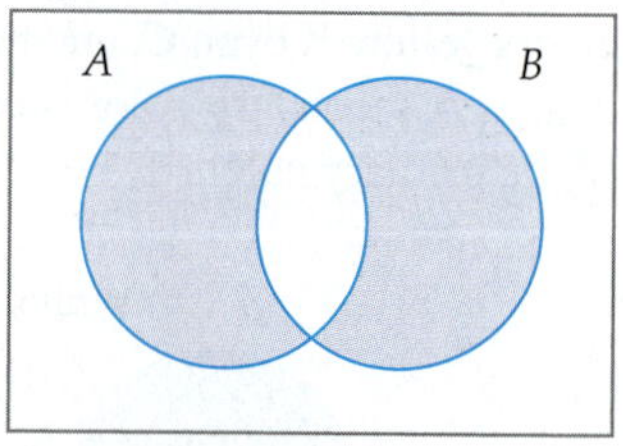

55.
U
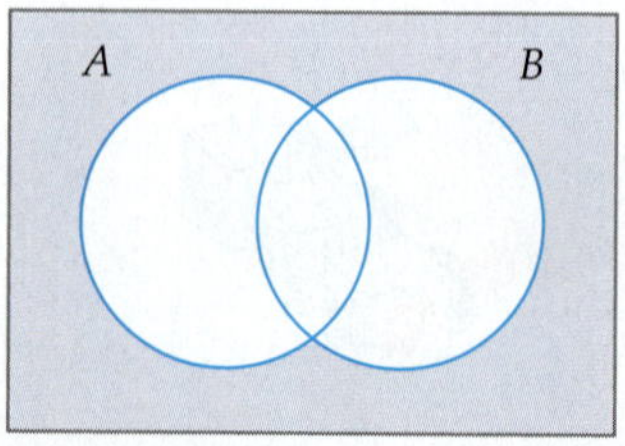

56.
U
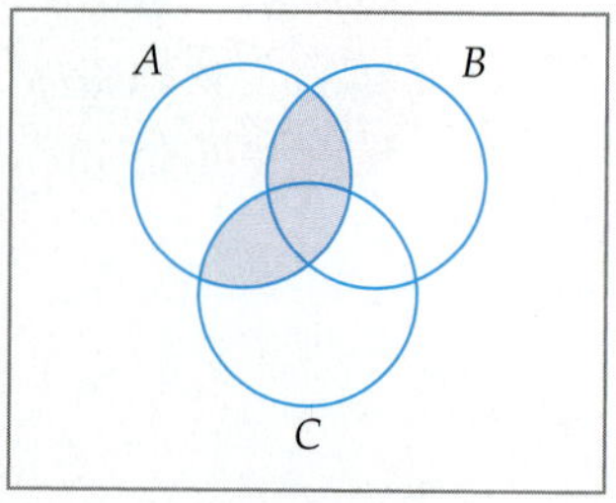

57.
U
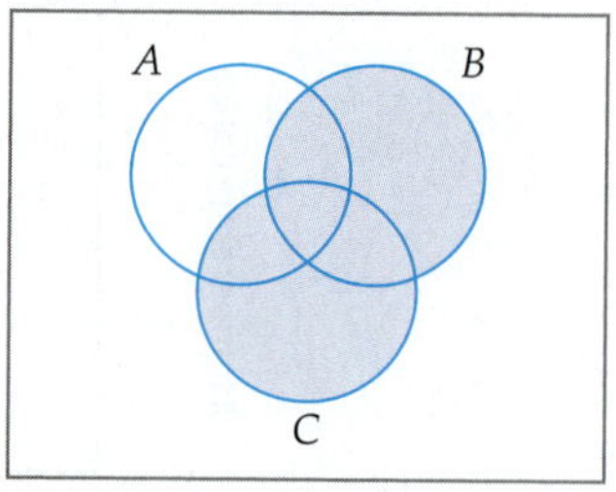

58.
U
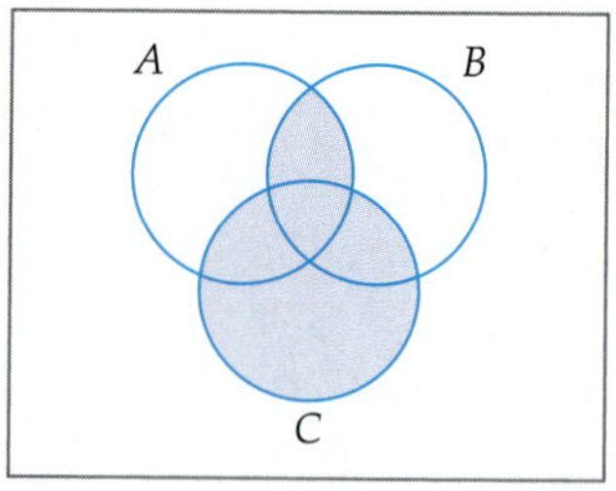

59.
U
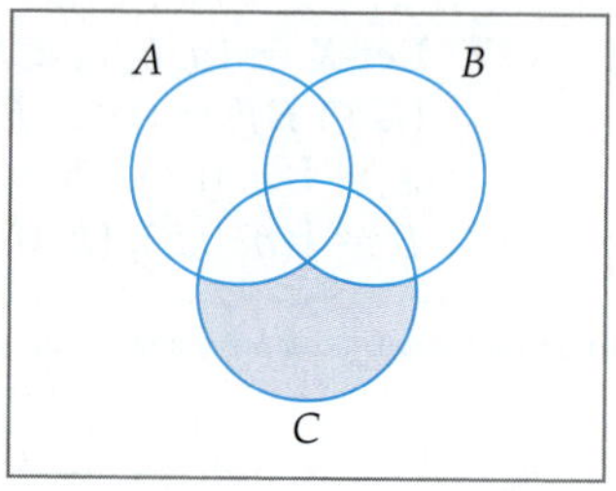

60.
U
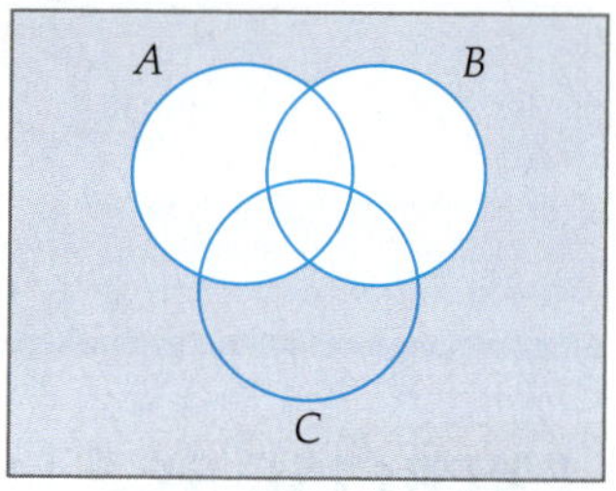

61.
U
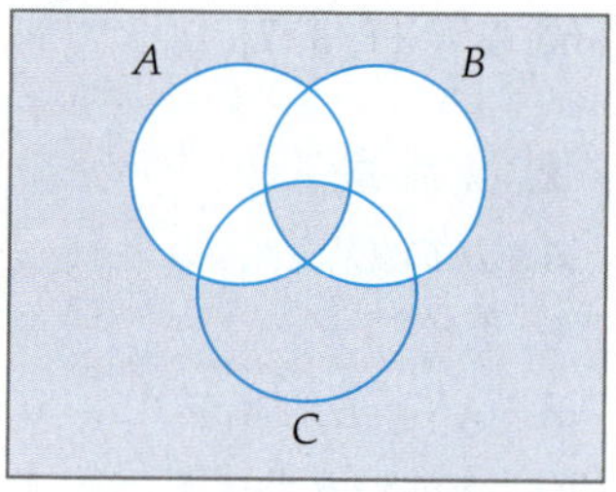

62.
U
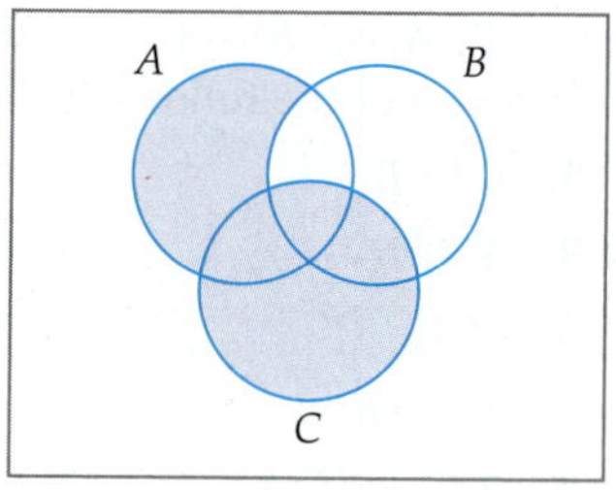

63. A Survey A special interest group plans to conduct a survey of households concerning a ban on handguns. The special interest group has decided to use the following Venn diagram to help illustrate the results of the survey. *Note:* A rifle is a gun, but it is not a handgun.

U
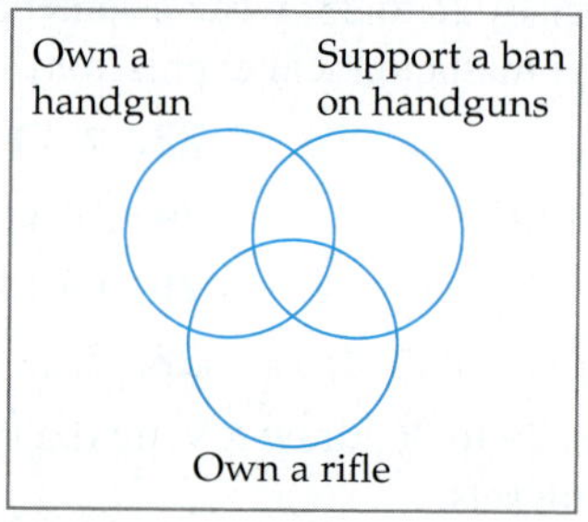

a. Shade in the regions that represent households that own a handgun and do not support the ban on handguns.

b. Shade in the region that represents households that own only a rifle and support the ban on handguns.

c. Shade in the region that represents households that do not own a gun and do not support the ban on handguns.

64. A Music Survey The administrators of an Internet music site plan to conduct a survey of college students to determine how the students acquire music. The administrators have decided to use the following Venn diagram to help tabulate the results of the survey.

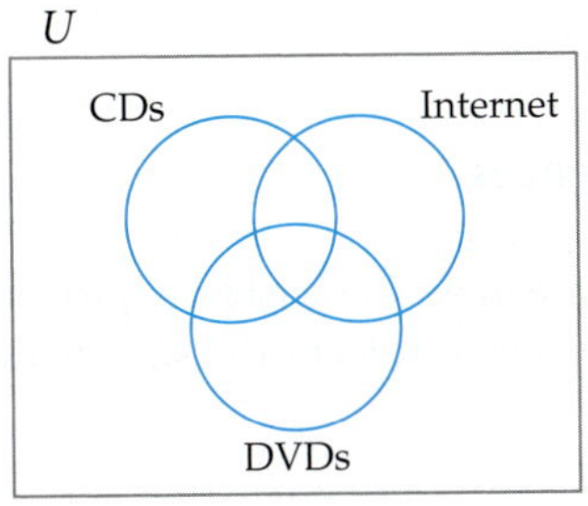

a. Shade in the region that represents students who acquire music from CDs and the Internet, but not from DVDs.

b. Shade in the regions that represent students who acquire music from CDs or the Internet.

c. Shade in the regions that represent students who acquire music from both CDs and DVDs.

■ In Exercises 65 to 68, draw a Venn diagram with each of the given elements placed in the correct region.

65. $U = \{-1, 0, 1, 2, 3, 4, 5, 6, 7, 8, 9\}$
$A = \{1, 3, 5\}$
$B = \{3, 5, 7, 8\}$
$C = \{-1, 8, 9\}$

66. $U = \{2, 4, 6, 8, 10, 12, 14\}$
$A = \{2, 10, 12\}$
$B = \{4, 8\}$
$C = \{6, 8, 10\}$

67. U = {Sue, Bob, Al, Jo, Ann, Herb, Eric, Mike, Sal}
A = {Sue, Herb}
B = {Sue, Eric, Jo, Ann}
$Rh+$ = {Eric, Sal, Al, Herb}

68. U = {Hal, Marie, Rob, Armando, Joel, Juan, Melody}
A = {Marie, Armando, Melody}
B = {Rob, Juan, Hal}
$Rh+$ = {Hal, Marie, Rob, Joel, Juan, Melody}

■ In Exercises 69 and 70, use two Venn diagrams to verify the following properties for all sets A, B, and C.

69. The associative property of intersection

$$(A \cap B) \cap C = A \cap (B \cap C)$$

70. The distributive property of intersection over union

$$A \cap (B \cup C) = (A \cap B) \cup (A \cap C)$$

EXTENSIONS

Difference of Sets Another operation that can be defined on sets A and B is the **difference of the sets**, denoted by $A - B$. Here is a formal definition of the difference of sets A and B.

$$A - B = \{x \mid x \in A \quad \text{and} \quad x \notin B\}$$

Thus $A - B$ is the set of elements that belong to A but not to B. For instance, let $A = \{1, 2, 3, 7, 8\}$ and $B = \{2, 7, 11\}$. Then $A - B = \{1, 3, 8\}$.

■ In Exercises 71 to 76, determine each difference, given that $U = \{1, 2, 3, 4, 5, 6, 7, 8, 9\}$, $A = \{2, 4, 6, 8\}$, and $B = \{2, 3, 8, 9\}$.

71. $B - A$ **72.** $A - B$ **73.** $A - B'$

74. $B' - A$ **75.** $A' - B'$ **76.** $A' - B$

77. **John Venn** Write a few paragraphs about the life of John Venn and his work in the area of mathematics.

The following Venn diagram illustrates that four sets can partition the universal set into 16 different regions.

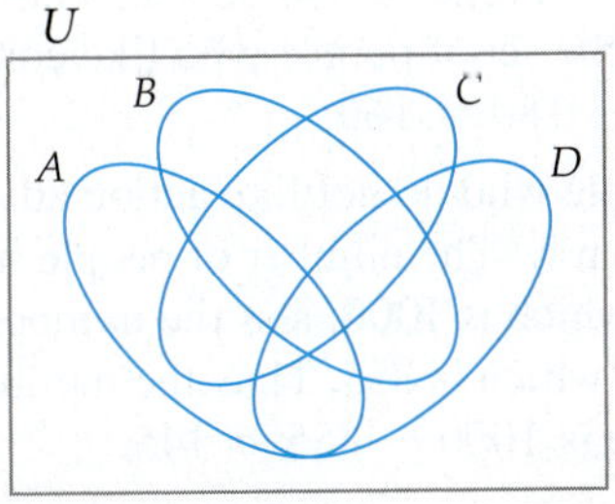

■ In Exercises 78 and 79, use a Venn diagram similar to the one at the left below to shade in the region represented by the given expression.

78. $(A \cap B) \cup (C' \cap D)$

79. $(A \cup B)' \cap (C \cap D)$

80. In an article in *New Scientist* magazine, Anthony W. F. Edwards illustrated how to construct Venn diagrams that involve many sets.[1] Search the Internet to find Edwards's method of constructing a Venn diagram for five sets and a Venn diagram for six sets. Use drawings to illustrate Edwards's method of constructing a Venn diagram for five sets and a Venn diagram for six sets. (*Source:* http://www.combinatorics.org/Surveys/ds5/VennWhatEJC.html)

[1] Anthony W. F. Edwards, "Venn diagrams for many sets," *New Scientist*, 7 January 1989, pp. 51–56.

SECTION 2.4 Applications of Sets

Surveys: An Application of Sets

Counting problems occur in many areas of applied mathematics. To solve these counting problems, we often make use of a Venn diagram and the inclusion-exclusion principle, which will be presented in this section.

GK Hart/Vikki Hart/Brand X Pictures/

EXAMPLE 1 A Survey of Preferences

A movie company is making plans for future movies it wishes to produce. The company has done a random survey of 1000 people. The results of the survey are shown below.

695 people like action adventures.

340 people like comedies.

180 people like both action adventures and comedies.

Of the people surveyed, how many people

a. like action adventures but not comedies?

b. like comedies but not action adventures?

c. do not like either of these types of movies?

Solution

A Venn diagram can be used to illustrate the results of the survey. We use two overlapping circles (see Figure 2.5). One circle represents the set of people who like action adventures and the other represents the set of people who like comedies. The region i where the circles intersect represents the set of people who like both types of movies.

We start with the information that 180 people like both types of movies and write 180 in region i. See Figure 2.6.

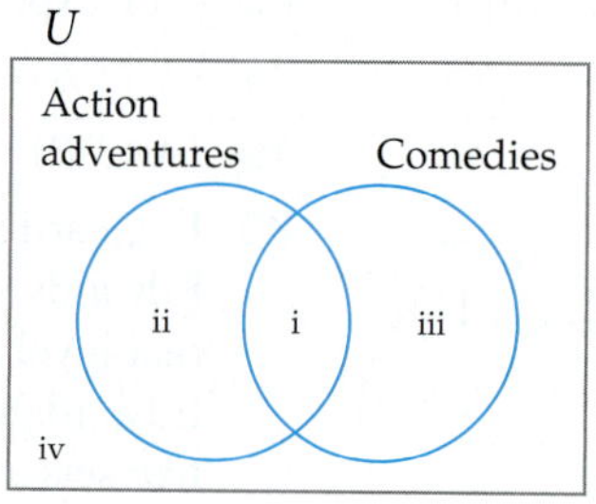

FIGURE 2.5

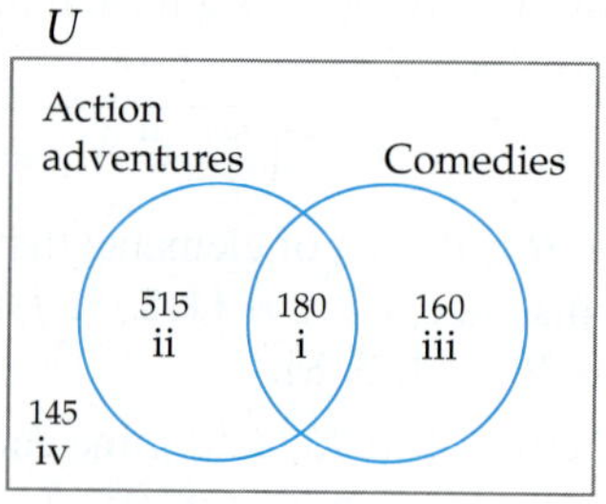

FIGURE 2.6

a. Regions i and ii have a total of 695 people. So far we have accounted for 180 of these people in region i. Thus the number of people in region ii, which is the set of people who like action adventures but do not like comedies, is $695 - 180 = 515$.

b. Regions i and iii have a total of 340 people. Thus the number of people in region iii, which is the set of people who like comedies but do not like action adventures, is $340 - 180 = 160$.

c. The number of people who do not like action adventure movies or comedies is represented by region iv. The number of people in region iv must be the total number of people, which is 1000, less the number of people accounted for in regions i, ii, and iii, which is 855. Thus the number of people who do not like either type of movie is $1000 - 855 = 145$.

CHECK YOUR PROGRESS 1 The athletic director of a school has surveyed 200 students. The survey results are shown below.

140 students like volleyball.
120 students like basketball.
85 students like both volleyball and basketball.

Of the students surveyed, how many students

a. like volleyball but not basketball?
b. like basketball but not volleyball?
c. do not like either of these sports?

Solution See page S6.

In the next example we consider a more complicated survey that involves three types of music.

EXAMPLE 2 A Music Survey

aastock/Shutterstock.com

A music teacher has surveyed 495 students. The results of the survey are listed below.

320 students like rap music.
395 students like rock music.
295 students like heavy metal music.
280 students like both rap music and rock music.
190 students like both rap music and heavy metal music.
245 students like both rock music and heavy metal music.
160 students like all three.

How many students

a. like exactly two of the three types of music?
b. like only rock music?
c. like only one of the three types of music?

Solution

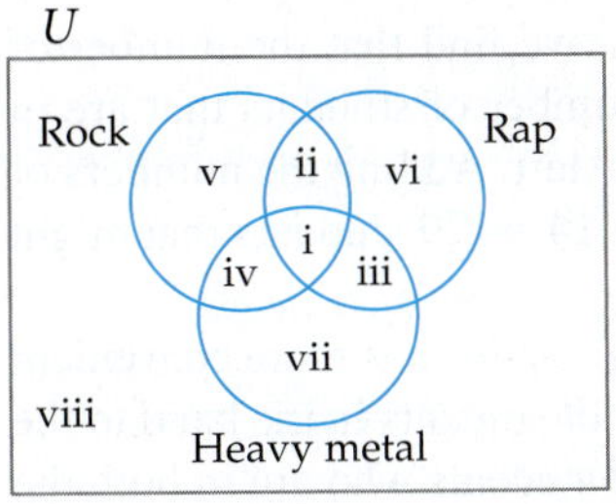

The Venn diagram at the left shows three overlapping circles. Region i represents the set of students who like all three types of music. Each of the regions v, vi, and vii represent the students who like only one type of music.

a. The survey shows that 245 students like rock and heavy metal music, so the numbers we place in regions i and iv must have a sum of 245. Since region i has 160 students, we see that region iv must have $245 - 160 = 85$ students. In a similar manner, we can determine that region ii has 120 students and region iii has 30 students. Thus $85 + 120 + 30 = 235$ students like exactly two of the three types of music.

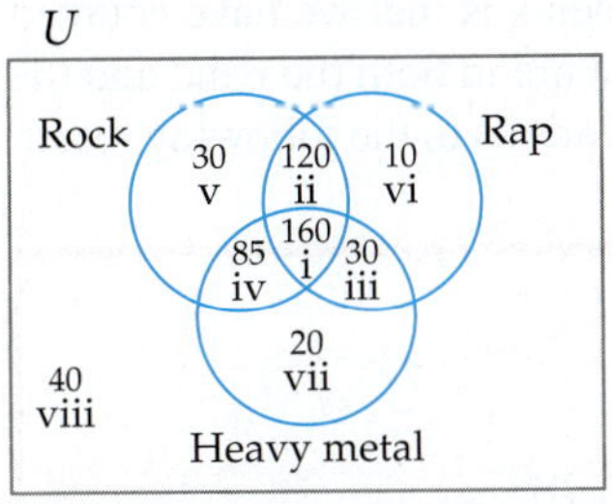

b. The sum of the students represented by regions i, ii, iv, and v must be 395. The number of students in region v must be the difference between this total and the sum of the numbers of students in region i, ii, and iv. Thus the number of students who like only rock music is $395 - (160 + 120 + 85) = 30$. See the Venn diagram at the left.

c. Using the same reasoning as in part b, we find that region vi has 10 students and region vii has 20 students. To find the number of students who like only one type of music, find the sum of the numbers of students in regions v, vi, and vii, which is $30 + 10 + 20 = 60$. See the Venn diagram at the left.

CHECK YOUR PROGRESS 2 An activities director for a cruise ship has surveyed 240 passengers. Of the 240 passengers,

135 like swimming.	80 like swimming and dancing.
150 like dancing.	40 like swimming and games.
65 like games.	25 like dancing and games.
	15 like all three activities.

How many passengers

a. like exactly two of the three types of activities?

b. like only swimming?

c. like none of these activities?

Solution See page S6.

MATH MATTERS Grace Chisholm Young (1868–1944)

Courtesy of Sylvia Wiegand

Grace Chisholm Young

Grace Chisholm Young studied mathematics at Girton College, which is part of Cambridge University. In England at that time, women were not allowed to earn a university degree, so she decided to continue her mathematical studies at the University of Göttingen in Germany, where her advisor was the renowned mathematician Felix Klein. She excelled while at Göttingen and at the age of 27 earned her doctorate in mathematics, magna cum laude. She was the first woman officially to earn a doctorate degree from a German university. Shortly after her graduation she married the mathematician William Young. Together they published several mathematical papers and books, one of which was the first textbook on set theory.

The Inclusion-Exclusion Principle

A music director wishes to take the band and the choir on a field trip. There are 65 students in the band and 30 students in the choir. The number of students in both the band and the choir is 16. How many students should the music director plan on taking on the field trip?

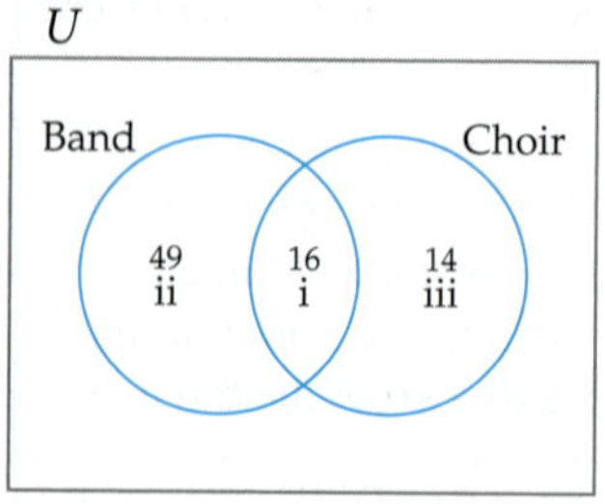

Using the process developed in the previous examples, we find that the number of students that are in only the band is $65 - 16 = 49$. The number of students that are in only the choir is $30 - 16 = 14$. See the Venn diagram at the left. Adding the numbers of students in regions i, ii, and iii gives us a total of $49 + 16 + 14 = 79$ students that might go on the field trip.

Although we can use Venn diagrams to solve counting problems, it is more convenient to make use of the following technique. First add the number of students in the band to the number of students in the choir. Then subtract the number of students who are in both the band and the choir. This technique gives us a total of $(65 + 30) - 16 = 79$ students, the same result as above. The reason we subtract the 16 students is that we have counted each of them twice. Note that first we include the students that are in both the band and the choir twice, and then we exclude them once. This procedure leads us to the following result.

TAKE NOTE

Recall that $n(A)$ represents the number of elements in set A.

The Inclusion-Exclusion Principle

For all finite sets A and B,

$$n(A \cup B) = n(A) + n(B) - n(A \cap B)$$

QUESTION What must be true of the finite sets A and B if $n(A \cup B) = n(A) + n(B)$?

EXAMPLE 3 An Application of the Inclusion-Exclusion Principle

A school finds that 430 of its students are registered in chemistry, 560 are registered in mathematics, and 225 are registered in both chemistry and mathematics. How many students are registered in chemistry or mathematics?

Solution

Let $C = \{\text{students registered in chemistry}\}$ and let $M = \{\text{students registered in mathematics}\}$.

$$\begin{aligned} n(C \cup M) &= n(C) + n(M) - n(C \cap M) \\ &= 430 + 560 - 225 \\ &= 765 \end{aligned}$$

Using the inclusion-exclusion principle, we see that 765 students are registered in chemistry or mathematics.

CHECK YOUR PROGRESS 3 A high school has 80 athletes who play basketball, 60 athletes who play soccer, and 24 athletes who play both basketball and soccer. How many athletes play either basketball or soccer?

Solution See page S7.

The inclusion-exclusion principle can be used provided we know the number of elements in any three of the four sets in the formula.

EXAMPLE 4 An Application of the Inclusion-Exclusion Principle

Given $n(A) = 15$, $n(B) = 32$, and $n(A \cup B) = 41$, find $n(A \cap B)$.

Solution

Substitute the given information in the inclusion-exclusion formula and solve for the unknown.

$$\begin{aligned} n(A \cup B) &= n(A) + n(B) - n(A \cap B) \\ 41 &= 15 + 32 - n(A \cap B) \\ 41 &= 47 - n(A \cap B) \end{aligned}$$

Thus,

$$n(A \cap B) = 47 - 41$$
$$n(A \cap B) = 6$$

CHECK YOUR PROGRESS 4 Given $n(A) = 785$, $n(B) = 162$, and $n(A \cup B) = 852$, find $n(A \cap B)$.

Solution See page S7.

ANSWER A and B must be disjoint sets.

The inclusion-exclusion formula can be adjusted and applied to problems that involve percents. In the following formula we denote "the percent in set A" by the notation $p(A)$.

The Percent Inclusion-Exclusion Formula

For all finite sets A and B,

$$p(A \cup B) = p(A) + p(B) - p(A \cap B)$$

EXAMPLE 5 An Application of the Percent Inclusion-Exclusion Formula

A blood donation organization reports that about

44% of the U.S. population has the A antigen.

15% of the U.S. population has the B antigen.

4% of the U.S. population has both the A and the B antigen.

Use the percent inclusion-exclusion formula to estimate the percent of the U.S. population that has the A antigen or the B antigen.

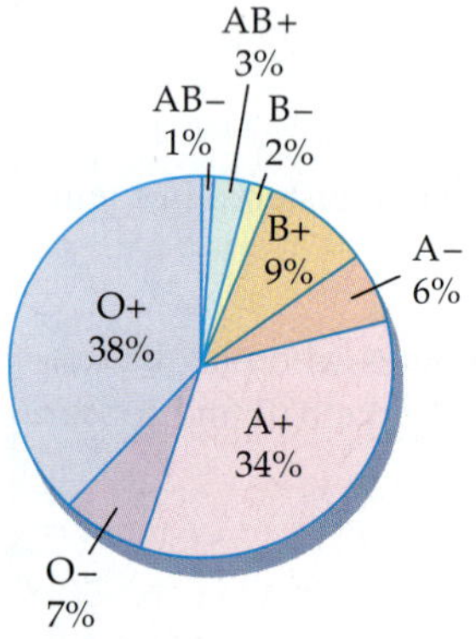

Approximate percentage of U.S. population with each blood type

Solution

We are given $p(\text{A}) = 44\%$, $p(\text{B}) = 15\%$, and $p(\text{A} \cap \text{B}) = 4\%$. Substituting in the percent inclusion-exclusion formula gives

$$\begin{aligned} p(\text{A} \cup \text{B}) &= p(\text{A}) + p(\text{B}) - p(\text{A} \cap \text{B}) \\ &= 44\% + 15\% - 4\% \\ &= 55\% \end{aligned}$$

Thus about 55% of the U.S. population has the A antigen or the B antigen.

CHECK YOUR PROGRESS 5 A blood donation organization reports that about

44% of the U.S. population has the A antigen.

84% of the U.S. population is Rh+.

91% of the U.S. population either has the A antigen or is Rh+.

Use the percent inclusion-exclusion formula to estimate the percent of the U.S. population that has the A antigen *and* is Rh+.

Solution See page S7.

In the next example, the data are provided in a table. The number in column G and row M represents the number of elements in $G \cap M$. The sum of all the numbers in column G and column B represents the number of elements in $G \cup B$.

EXAMPLE 6 A Survey Presented in Tabular Form

A survey of men M, women W, and children C concerning the use of the Internet search engines Google G, Yahoo! Y, and Bing B yielded the following results.

	Google (G)	**Yahoo! (Y)**	**Bing (B)**
Men (M)	440	310	275
Women (W)	390	280	325
Children (C)	140	410	40

Use the data in the table to find each of the following.

a. $n(W \cap Y)$ **b.** $n(G \cap C')$ **c.** $n(M \cap (G \cup B))$

Solution

a. The table shows that 280 of the women surveyed use Yahoo! as a search engine. Thus $n(W \cap Y) = 280$.

b. The set $G \cap C'$ is the set of surveyed Google users who are men or women. The number in this set is $440 + 390 = 830$.

c. The number of men in the survey that use either Google or Bing is $440 + 275 = 715$.

CHECK YOUR PROGRESS 6 Use the table in Example 6 to find each of the following.

a. $n(Y \cap C)$ **b.** $n(B \cap M')$ **c.** $n((G \cap M) \cup (G \cap W))$

Solution See page S7.

EXCURSION

Voting Systems

Jim West/Alamy Stock Photo

There are many types of voting systems. When people are asked to vote for or against a resolution, a one-person, one-vote *majority system* is often used to decide the outcome. In this type of voting, each voter receives one vote, and the resolution passes only if it receives *most* of the votes.

In any voting system, the number of votes that is required to pass a resolution is called the **quota**. A **coalition** is a set of voters each of whom votes the same way, either for or against a resolution. A **winning coalition** is a set of voters the sum of whose votes is greater than or equal to the quota. A **losing coalition** is a set of voters the sum of whose votes is less than the quota.

Sometimes you can find all the winning coalitions in a voting process by making an organized list. For instance, consider the committee consisting of Alice, Barry, Cheryl, and Dylan. To decide on any issues, they use a one-person, one-vote majority voting system. Since each of the four voters has a single vote, the quota for this majority voting system is 3. The winning coalitions consist of all subsets of the voters that have three or more people. We list these winning coalitions in the table at the left below, where A represents Alice, B represents Barry, C represents Cheryl, and D represents Dylan.

A **weighted voting system** is one in which some voters' votes carry more weight regarding the outcome of an election. As an example, consider a selection committee that consists of four people designated by A, B, C, and D. Voter A's vote has a weight of 2, and the vote of each other member of the committee has a weight of 1. The quota for this weighted voting system is 3. A winning coalition must have a weighted voting sum of at least 3. The winning coalitions are listed in the table at the right below.

Winning coalition	Sum of the votes
{A, B, C}	3
{A, B, D}	3
{A, C, D}	3
{B, C, D}	3
{A, B, C, D}	4

Winning coalition	Sum of the weighted votes
{A, B}	3
{A, C}	3
{A, D}	3
{B, C, D}	3
{A, B, C}	4
{A, B, D}	4
{A, C, D}	4
{A, B, C, D}	5

HISTORICAL NOTE

The Granger Collection, NY

An ostrakon

In ancient Greece, the citizens of Athens adopted a procedure that allowed them to vote for the expulsion of any prominent person. The purpose of this procedure, known as an *ostracism,* was to limit the political power that any one person could attain.

In an ostracism, each voter turned in a *potsherd,* a piece of pottery fragment, on which was inscribed the name of the person the voter wished to ostracize. The pottery fragments used in the voting process became known as *ostrakon.*

The person who received the majority of the votes, above some set minimum, was exiled from Athens for a period of 10 years.

A **minimal winning coalition** is a winning coalition that has no proper subset that is a winning coalition. In a minimal winning coalition each voter is said to be a **critical voter**, because if any of the voters leaves the coalition, the coalition will then become a losing coalition. In the table at the bottom right on the previous page, the minimal winning coalitions are {A, B}, {A, C}, {A, D}, and {B, C, D}. If any single voter leaves one of these coalitions, then the coalition will become a losing coalition. The coalition {A, B, C, D} is not a minimal winning coalition, because it contains at least one proper subset, for instance {A, B, C}, that is a winning coalition.

EXCURSION EXERCISES

1. A selection committee consists of Ryan, Susan, and Trevor. To decide on issues, they use a one-person, one-vote majority voting system.
   a. Find all winning coalitions.
   b. Find all losing coalitions.
2. A selection committee consists of three people designated by M, N, and P. M's vote has a weight of 3, N's vote has a weight of 2, and P's vote has a weight of 1. The quota for this weighted voting system is 4. Find all winning coalitions.
3. Determine the minimal winning coalitions for the voting system in Excursion Exercise 2.

Additional information on the applications of mathematics to voting systems is given in Chapter 4.

EXERCISE SET 2.4

■ In Exercises 1 to 10, let $U = \{$English, French, History, Math, Physics, Chemistry, Psychology, Drama$\}$, $A = \{$English, History, Psychology, Drama$\}$, $B = \{$Math, Physics, Chemistry, Psychology, Drama$\}$, and $C = \{$French, History, Chemistry$\}$.

Find each of the following.

1. $n(B \cup C)$
2. $n(A \cup B)$
3. $n(B) + n(C)$
4. $n(A) + n(B)$
5. $n[(A \cup B) \cup C]$
6. $n(A \cap B)$
7. $n(A) + n(B) + n(C)$
8. $n(A \cap B \cap C)$
9. $n[A \cup (B \cap C)]$
10. $n[A \cap (B \cup C)]$
11. Verify that for A and B as defined in Exercises 1 to 10, $n(A \cup B) = n(A) + n(B) - n(A \cap B)$.
12. Verify that for A and C as defined in Exercises 1 to 10, $n(A \cup C) = n(A) + n(C) - n(A \cap C)$.
13. Given $n(J) = 245$, $n(K) = 178$, and $n(J \cup K) = 310$, find $n(J \cap K)$.
14. Given $n(L) = 780$, $n(M) = 240$, and $n(L \cap M) = 50$, find $n(L \cup M)$.
15. Given $n(A) = 1500$, $n(A \cup B) = 2250$, and $n(A \cap B) = 310$, find $n(B)$.
16. Given $n(A) = 640$, $n(B) = 280$, and $n(A \cup B) = 765$, find $n(A \cap B)$.

■ In Exercises 17 and 18, use the given information to find the number of elements in each of the regions labeled with a question mark.

17. $n(A) = 28$, $n(B) = 31$, $n(C) = 40$, $n(A \cap B) = 15$, $n(U) = 75$

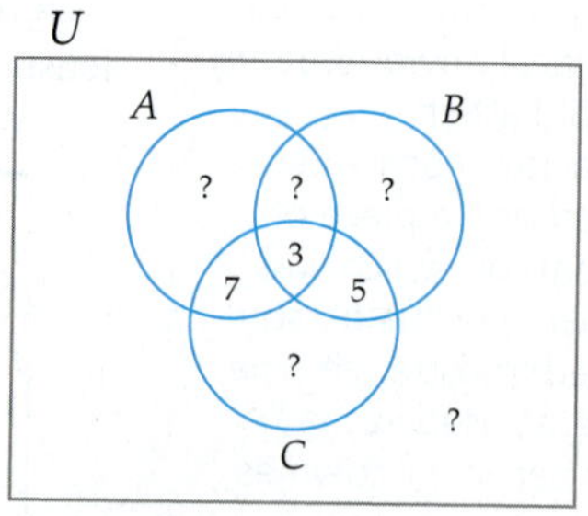

18. $n(A) = 610$, $n(B) = 440$, $n(C) = 1000$, $n(U) = 2900$

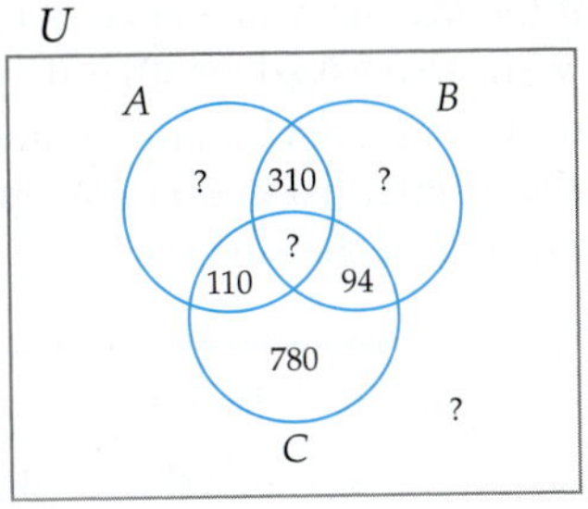

19. Investing In a survey of 600 investors, it was reported that 380 had invested in stocks, 325 had invested in bonds, and 75 had not invested in either stocks or bonds.

Rob Pitman/Shutterstock.com

a. How many investors had invested in both stocks and bonds?

b. How many investors had invested only in stocks?

20. Commuting A survey of 1500 commuters in New York City showed that 1140 take the subway, 680 take the bus, and 120 do not take either the bus or the subway.

a. How many commuters take both the bus and the subway?

b. How many commuters take only the subway?

21. First Aid Treatments A team physician has determined that of all the athletes who were treated for minor back pain, 72% responded favorably to an analgesic, 59% responded favorably to a muscle relaxant, and 44% responded favorably to both forms of treatment.

a. What percent of the athletes who were treated responded favorably to the muscle relaxant but not to the analgesic?

b. What percent of the athletes who were treated did not respond favorably to either form of treatment?

22. Gratuities The management of a hotel conducted a survey. It found that of the 2560 guests who were surveyed,

1785 tip the wait staff.
1219 tip the luggage handlers.
831 tip the maids.
275 tip the maids and the luggage handlers.
700 tip the wait staff and the maids.
755 tip the wait staff and the luggage handlers.
245 tip all three services.
210 do not tip these services.

How many of the surveyed guests tip

a. exactly two of the three services?

b. only the wait staff?

c. only one of the three services?

23. Advertising A computer company advertises its computers in *PC World*, in *PC Magazine*, and on television. A survey of 770 customers finds that the numbers of customers who are familiar with the company's computers because of the different forms of advertising are as follows.

305, *PC World*
290, *PC Magazine*
390, television
110, *PC World* and *PC Magazine*
135, *PC Magazine* and television
150, *PC World* and television
85, all three sources

How many of the surveyed customers know about the computers because of

a. exactly one of these forms of advertising?

b. exactly two of these forms of advertising?

c. *PC World* and neither of the other two forms of advertising?

24. Blood Types During one month, a blood donation center found that

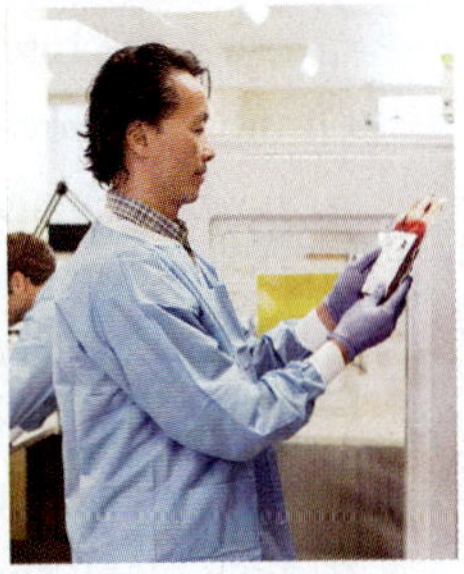

Tyler Olson/Shutterstock.com

45.8% of the donors had the A antigen.

14.2% of the donors had the B antigen.

4.1% of the donors had the A antigen and the B antigen.

84.7% of the donors were Rh+.

87.2% of the donors had the B antigen or were Rh+.

Find the percent of the donors that

a. have the A antigen or the B antigen.

b. have the B antigen and are Rh+.

25. Gun Ownership A special interest group has conducted a survey concerning a ban on handguns. *Note:* A rifle is a gun, but it is not a handgun. The survey yielded the following results for the 1000 households that responded.

271 own a handgun.

437 own a rifle.

497 supported the ban on handguns.

140 own both a handgun and a rifle.

202 own a rifle but no handgun and do not support the ban on handguns.

74 own a handgun and support the ban on handguns.

52 own both a handgun and a rifle and also support the ban on handguns.

How many of the surveyed households

a. only own a handgun and do not support the ban on handguns?

b. do not own a gun and support the ban on handguns?

c. do not own a gun and do not support the ban on handguns?

26. Acquisition of Music A survey of college students was taken to determine how the students acquired music. The survey showed the following results.

365 students acquired music from CDs.

298 students acquired music from the Internet.

268 students acquired music from DVDs.

212 students acquired music from both CDs and DVDs.

155 students acquired music from both CDs and the Internet.

36 students acquired music from DVDs, but not from CDs or the Internet.

98 students acquired music from CDs, DVDs, and the Internet.

Of those surveyed,

a. how many acquired music from CDs, but not from the Internet or DVDs?

b. how many acquired music from the Internet, but not from CDs or DVDs?

c. how many acquired music from CDs or the Internet?

d. how many acquired music from the Internet and DVDs?

27. Diets A survey was completed by individuals who were currently on the Zone diet (Z), the South Beach diet (S), or the Weight Watchers diet (W). All persons surveyed were also asked whether they were currently in an exercise program (E), taking diet pills (P), or under medical supervision (M). The following table shows the results of the survey.

		Supplements			
		E	***P***	***M***	**Totals**
Diet	***Z***	124	82	65	271
	S	101	66	51	218
	W	133	41	48	222
	Totals	358	189	164	711

Find the number of surveyed people in each of the following sets.

a. $S \cap E$

b. $Z \cup M$

c. $S' \cap (E \cup P)$

d. $(Z \cup S) \cap (M')$

e. $W' \cap (P \cup M)'$

f. $W' \cup P$

28. Financial Assistance A college study categorized its seniors (S), juniors (J), and sophomores (M) who are currently receiving financial assistance. The types of financial assistance consist of full scholarships (F), partial scholarships (P), and government loans (G). The following table shows the results of the survey.

		Financial assistance			
		F	***P***	***G***	**Totals**
Year	***S***	210	175	190	575
	J	180	162	110	452
	M	114	126	86	326
	Totals	504	463	386	1353

Find the number of students who are currently receiving financial assistance in each of the following sets.

a. $S \cap P$

b. $J \cup G$

c. $M \cup F'$

d. $S \cap (F \cup P)$

e. $J \cap (F \cup P)'$

f. $(S \cup J) \cap (F \cup P)$

EXTENSIONS

29. Given that set A has 47 elements and set B has 25 elements, determine each of the following.

a. The maximum possible number of elements in $A \cup B$

b. The minimum possible number of elements in $A \cup B$

c. The maximum possible number of elements in $A \cap B$

d. The minimum possible number of elements in $A \cap B$

30. Given that set A has 16 elements, set B has 12 elements, and set C has 7 elements, determine each of the following.

a. The maximum possible number of elements in $A \cup B \cup C$

b. The minimum possible number of elements in $A \cup B \cup C$

c. The maximum possible number of elements in $A \cap (B \cup C)$

d. The minimum possible number of elements in $A \cap (B \cup C)$

31. Search Engines The following Venn diagram displays U parceled into 16 distinct regions by four sets.

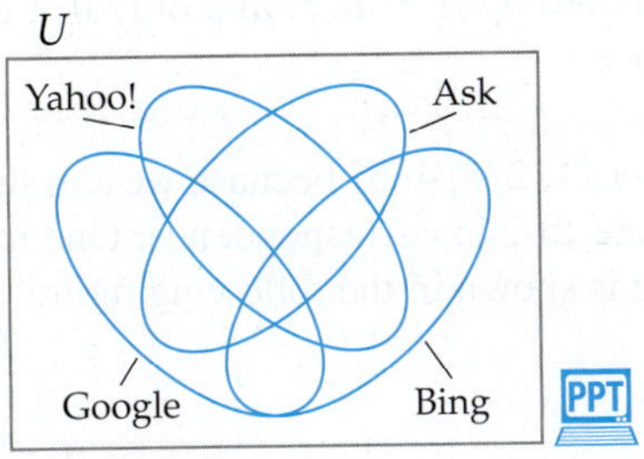

Use the preceding Venn diagram and the information below to answer the questions that follow.

A survey of 1250 Internet users shows the following results concerning the use of the search engines Google, Bing, Yahoo!, and Ask.

585 use Google.

620 use Yahoo!.

560 use Ask.

450 use Bing.

100 use only Google, Yahoo!, and Ask.

41 use only Google, Yahoo!, and Bing.

50 use only Google, Ask, and Bing.

80 use only Yahoo!, Ask, and Bing.

55 use only Google and Yahoo!.

34 use only Google and Ask.

45 use only Google and Bing.

50 use only Yahoo! and Ask.

30 use only Yahoo! and Bing.

45 use only Ask and Bing.

60 use all four.

How many of the Internet users

a. use only Google?

b. use exactly three of the four search engines?

c. do not use any of the four search engines?

32. An Inclusion-Exclusion Formula for Three Sets Exactly one of the following equations is a valid inclusion-exclusion formula for the union of three finite sets. Which equation do you think is the valid formula? *Hint:* Use the data in Example 2 on page 81 to check your choice.

a. $n(A \cup B \cup C) = n(A) + n(B) + n(C)$

b. $n(A \cup B \cup C) = n(A) + n(B) + n(C) - n(A \cap B \cap C)$

c. $n(A \cup B \cup C) = n(A) + n(B) + n(C) - n(A \cap B) - n(A \cap C) - n(B \cap C)$

d. $n(A \cup B \cup C) = n(A) + n(B) + n(C) - n(A \cap B) - n(A \cap C) - n(B \cap C) + n(A \cap B \cap C)$

SECTION 2.5 Infinite Sets

One-to-One Correspondences

Much of Georg Cantor's work with sets concerned infinite sets. Some of Cantor's work with infinite sets was so revolutionary that it was not readily accepted by his contemporaries. Today, however, his work is generally accepted, and it provides unifying ideas in several diverse areas of mathematics.

Much of Cantor's set theory is based on the simple concept of a *one-to-one correspondence.*

One-to-One Correspondence

A **one-to-one correspondence** (or 1–1 correspondence) between two sets A and B is a rule or procedure that pairs each element of A with exactly one element of B and each element of B with exactly one element of A.

Many practical problems can be solved by applying the concept of a one-to-one correspondence. For instance, consider a concert hall that has 890 seats. During a performance the manager of the concert hall observes that every person occupies exactly one seat and that every seat is occupied. Thus, without doing any counting, the manager knows that there are 890 people in attendance. During a different performance the manager notes that all but six seats are filled, and thus there are $890 - 6 = 884$ people in attendance.

Recall that two sets are equivalent if and only if they have the same number of elements. One method of showing that two sets are equivalent is to establish a one-to-one correspondence between the elements of the sets.

One-to-One Correspondence and Equivalent Sets

Two sets A and B are equivalent, denoted by $A \sim B$, if and only if A and B can be placed in a one-to-one correspondence.

Set {a, b, c, d, e} is equivalent to set {1, 2, 3, 4, 5} because we can show that the elements of each set can be placed in a one-to-one correspondence. One method of establishing this one-to-one correspondence is shown in the following figure.

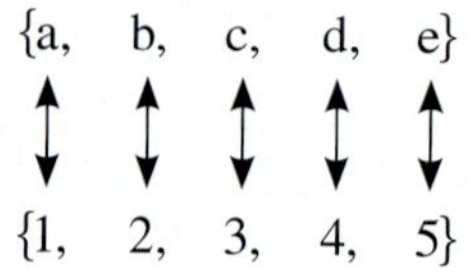

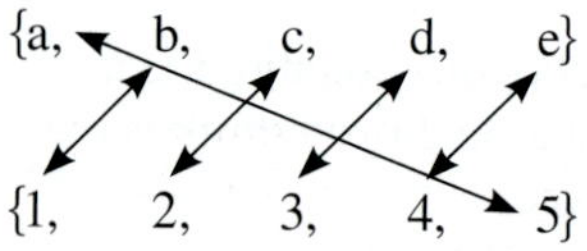

Each element of {a, b, c, d, e} has been paired with exactly one element of {1, 2, 3, 4, 5}, and each element of {1, 2, 3, 4, 5} has been paired with exactly one element of {a, b, c, d, e}. This is not the only one-to-one correspondence that we can establish. The figure at the left shows another one-to-one correspondence between the sets. In any case, we know that both sets have the same number of elements because we have established a one-to-one correspondence between the sets.

Sometimes a set is defined by including a general element. For instance, in the set $\{3, 6, 9, 12, 15, \ldots, 3n, \ldots\}$, the $3n$ (where n is a natural number) indicates that all the elements of the set are multiples of 3.

Some sets can be placed in a one-to-one correspondence with a proper subset of themselves. Example 1 illustrates this concept for the set of natural numbers.

TAKE NOTE

Many mathematicians and nonmathematicians have found the concept that $E \sim N$, as shown in Example 1, to be a bit surprising. After all, the set of natural numbers includes the even natural numbers as well as the odd natural numbers!

EXAMPLE 1 Establish a One-to-One Correspondence

Establish a one-to-one correspondence between the set of natural numbers $N = \{1, 2, 3, 4, 5, \ldots, n, \ldots\}$ and the set of even natural numbers $E = \{2, 4, 6, 8, 10, \ldots, 2n, \ldots\}$.

Solution

Write the sets so that one is aligned below the other. Draw arrows to show how you wish to pair the elements of each set. One possible method is shown in the following figure.

$$N = \{1, 2, 3, 4, \ldots, n, \ldots\}$$
$$\updownarrow \; \updownarrow \; \updownarrow \; \updownarrow \qquad \updownarrow$$
$$E = \{2, 4, 6, 8, \ldots, 2n, \ldots\}$$

In the above correspondence, each natural number $n \in N$ is paired with the even number $(2n) \in E$. The *general correspondence* $n \leftrightarrow (2n)$ enables us to determine exactly which element of E will be paired with any given element of N, and vice versa. For instance, under this correspondence, $19 \in N$ is paired with the even number $2 \cdot 19 = 38 \in E$, and $100 \in E$ is paired with the natural number $\frac{1}{2} \cdot 100 = 50 \in N$. The general correspondence $n \leftrightarrow (2n)$ establishes a one-to-one correspondence between the sets.

CHECK YOUR PROGRESS 1 Establish a one-to-one correspondence between the set of natural numbers $N = \{1, 2, 3, 4, 5, \ldots, n, \ldots\}$ and the set of odd natural numbers $D = \{1, 3, 5, 7, 9, \ldots, 2n - 1, \ldots\}$.

Solution See page S7.

Infinite Sets

Infinite Set

A set is an **infinite set** if it can be placed in a one-to-one correspondence with a proper subset of itself.

We know that the set of natural numbers N is an infinite set because in Example 1 we were able to establish a one-to-one correspondence between the elements of N and the elements of one of its proper subsets, E.

QUESTION Can the set $\{1, 2, 3\}$ be placed in a one-to-one correspondence with one of its proper subsets?

EXAMPLE 2 Verify That a Set Is an Infinite Set

Verify that $S = \{5, 10, 15, 20, \ldots, 5n, \ldots\}$ is an infinite set.

Solution

One proper subset of S is $T = \{10, 20, 30, 40, \ldots, 10n, \ldots\}$, which was produced by deleting the odd numbers in S. To establish a one-to-one correspondence between set S and set T, consider the following diagram.

$$S = \{5, 10, 15, 20, \ldots, 5n, \ldots\}$$
$$\updownarrow \; \updownarrow \; \updownarrow \; \updownarrow \qquad \updownarrow$$
$$T = \{10, 20, 30, 40, \ldots, 10n, \ldots\}$$

In the above correspondence, each $(5n) \in S$ is paired with $(10n) \in T$. The *general correspondence* $(5n) \leftrightarrow (10n)$ establishes a one-to-one correspondence between S and one of its proper subsets, namely T. Thus S is an infinite set.

TAKE NOTE

The solution shown in Example 2 is not the only way to establish that S is an infinite set. For instance, $R = \{10, 15, 20, \ldots, 5n + 5, \ldots\}$ is also a proper set of S, and the sets S and R can be placed in a one-to-one correspondence as follows.

$$S = \{5, 10, 15, 20, \ldots, 5n, \ldots\}$$
$$\updownarrow \; \updownarrow \; \updownarrow \; \updownarrow \qquad \updownarrow$$
$$R = \{10, 15, 20, 25, \ldots, 5n + 5, \ldots\}$$

This one-to-one correspondence between S and one of its proper subsets R also establishes that S is an infinite set.

ANSWER No. The set $\{1, 2, 3\}$ is a finite set with three elements. Every proper subset of $\{1, 2, 3\}$ has two or fewer elements.

CHECK YOUR PROGRESS 2 Verify that $V = \{40, 41, 42, 43, ..., 39 + n, ...\}$ is an infinite set.

Solution See page S7.

The Cardinality of Infinite Sets

The symbol $\aleph_0$ is used to represent the cardinal number for the set N of natural numbers. $\aleph$ is the first letter of the Hebrew alphabet and is pronounced *aleph*. $\aleph_0$ is read as "aleph-null." Using mathematical notation, we write this concept as $n(N) = \aleph_0$. Since $\aleph_0$ represents a cardinality larger than any finite number, it is called a **transfinite number**. Many infinite sets have a cardinality of $\aleph_0$. In Example 3, for instance, we show that the cardinality of the set of integers is $\aleph_0$ by establishing a one-to-one correspondence between the elements of the set of integers and the elements of the set of natural numbers.

EXAMPLE 3 Establish the Cardinality of the Set of Integers

Show that the set of integers $I = \{..., -5, -4, -3, -2, -1, 0, 1, 2, 3, 4, 5, ...\}$ has a cardinality of $\aleph_0$.

Solution

First we try to establish a one-to-one correspondence between I and N, with the elements in each set arranged as shown below. No general method of pairing the elements of N with the elements of I seems to emerge from this figure.

$$N = \{1, 2, 3, 4, 5, 6, 7, 8, 9, 10, 11, ...\}$$

?

$$I = \{..., -5, -4, -3, -2, -1, 0, 1, 2, 3, 4, 5, ...\}$$

If we arrange the elements of I as shown in the figure below, then *two* general correspondences, shown by the blue arrows and the red arrows, can be identified.

$$N = \{1, 2, 3, 4, 5, 6, 7, 8, 9, 10, 11, ..., 2n - 1, 2n, ...\}$$

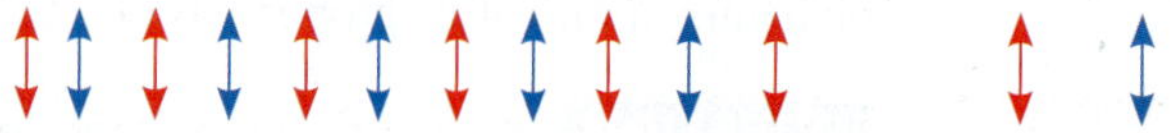

$$I = \{0, 1, -1, 2, -2, 3, -3, 4, -4, 5, -5, ..., -n + 1, n, ...\}$$

- Each even natural number $2n$ of N is paired with the integer n of I. This correspondence is shown by the blue arrows.
- Each odd natural number $2n - 1$ of N is paired with the integer $-n + 1$ of I. This correspondence is shown by the red arrows.

Together the two general correspondences $(2n) \leftrightarrow n$ and $(2n - 1) \leftrightarrow (-n + 1)$ establish a one-to-one correspondence between the elements of I and the elements of N. Thus the cardinality of the set of integers must be the same as the cardinality of the set of natural numbers, which is $\aleph_0$.

CHECK YOUR PROGRESS 3 Show that $M = \left\{\frac{1}{2}, \frac{1}{3}, \frac{1}{4}, \frac{1}{5}, ..., \frac{1}{n+1}, ...\right\}$ has a cardinality of $\aleph_0$.

Solution See page S7.

Cantor was also able to show that the set of positive rational numbers is equivalent to the set of natural numbers. Recall that a rational number is a number that can be written as a fraction $\frac{p}{q}$, where p and q are integers and $q \neq 0$. Cantor's proof used an array of rational numbers similar to the array shown on the next page.

Theorem The set $Q+$ of positive rational numbers is equivalent to the set N of natural numbers.

Proof Consider the following array of positive rational numbers.

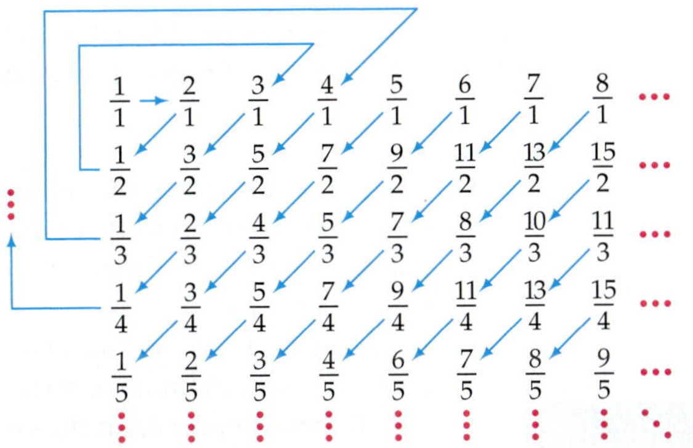

An array of all the positive rational numbers

TAKE NOTE

The rational number $\frac{2}{2}$ is not listed in the second row because $\frac{2}{2} = 1 = \frac{1}{1}$, which is already listed in the first row.

The first row of the above array contains, in order from smallest to largest, all the positive rational numbers which, *when expressed in lowest terms,* have a denominator of 1. The second row contains the positive rational numbers which, *when expressed in lowest terms,* have a denominator of 2. The third row contains the positive rational numbers which, *when expressed in lowest terms,* have a denominator of 3. This process continues indefinitely.

Cantor reasoned that every positive rational number appears once and only once in this array. Note that $\frac{3}{5}$ appears in the fifth row. In general, if $\frac{p}{q}$ is in lowest terms, then it appears in row q.

At this point, Cantor used a numbering procedure that establishes a one-to-one correspondence between the natural numbers and the positive rational numbers in the array. The numbering procedure starts in the upper left corner with $\frac{1}{1}$. Cantor considered this to be the first number in the array, so he assigned the natural number 1 to this rational number. He then moved to the right and assigned the natural number 2 to the rational number $\frac{2}{1}$. From this point on, he followed the diagonal paths shown by the arrows and assigned each number he encountered to the next consecutive natural number. When he reached the bottom of a diagonal, he moved up to the top of the array and continued to number the rational numbers in the next diagonal. The following table shows the first 10 rational numbers Cantor numbered using this scheme.

Rational number in the array	$\frac{1}{1}$	$\frac{2}{1}$	$\frac{1}{2}$	$\frac{3}{1}$	$\frac{3}{2}$	$\frac{1}{3}$	$\frac{4}{1}$	$\frac{5}{2}$	$\frac{2}{3}$	$\frac{1}{4}$
Corresponding natural number	1	2	3	4	5	6	7	8	9	10

This numbering procedure shows that each element of $Q+$ can be paired with exactly one element of N, and each element of N can be paired with exactly one element of $Q+$. Thus $Q+$ and N are equivalent sets.

The negative rational numbers $Q-$ can also be placed in a one-to-one correspondence with the set of natural numbers in a similar manner.

QUESTION Using Cantor's numbering scheme, which rational numbers in the array shown above would be assigned the natural numbers 11, 12, 13, 14, and 15?

ANSWER The rational numbers in the next diagonal, namely $\frac{5}{1}, \frac{7}{2}, \frac{4}{3}, \frac{3}{4}$, and $\frac{1}{5}$, would be assigned to the natural numbers 11, 12, 13, 14, and 15, respectively.

Countable Set

A set is a **countable set** if and only if it is a finite set or an infinite set that is equivalent to the set of natural numbers.

Every infinite set that is countable has a cardinality of $\aleph_0$. Every infinite set that we have considered up to this point is countable. You might think that all infinite sets are countable; however, Cantor was able to show that this is not the case. Consider, for example, $A = \{x \mid x \in R \text{ and } 0 < x < 1\}$ where R is the set of real numbers. To show that A is *not* a countable set, we use a *proof by contradiction,* where we assume that A is countable and then proceed until we arrive at a contradiction.

Henryk T. Kaiser/Photolibrary/Getty Images

To better understand the concept of a proof by contradiction, consider the situation in which you are at a point where a road splits into two roads. See the figure at the left. Assume you know that only one of the two roads leads to your desired destination. If you can show that one of the roads cannot get you to your destination, then you know, without ever traveling down the other road, that it is the road that leads to your destination. In the following proof, we know that either set A is a countable set or set A is not a countable set. To establish that A is *not* countable, we show that the assumption that A is countable leads to a contradiction. In other words, our assumption that A is countable must be incorrect, and we are forced to conclude that A is not countable.

Theorem The set $A = \{x \mid x \in R \text{ and } 0 < x < 1\}$ is not a countable set.

Proof by contradiction Either A is countable or A is not countable. Assume A is countable. Then we can place the elements of A, which we will represent by $a_1, a_2, a_3, a_4, \ldots,$ in a one-to-one correspondence with the elements of the natural numbers as shown below.

$$N = \{1, 2, 3, 4, \ldots, n, \ldots\}$$
$$\updownarrow \quad \updownarrow \quad \updownarrow \quad \updownarrow \qquad \updownarrow$$
$$A = \{a_1, a_2, a_3, a_4, \ldots, a_n, \ldots\}$$

For example, the numbers $a_1, a_2, a_3, a_4, \ldots, a_n, \ldots$ could be as shown below.

$$1 \leftrightarrow a_1 = 0.\boxed{3}573485\ldots$$
$$2 \leftrightarrow a_2 = 0.0\boxed{6}52891\ldots$$
$$3 \leftrightarrow a_3 = 0.68\boxed{2}3514\ldots$$
$$4 \leftrightarrow a_4 = 0.050\boxed{0}310\ldots$$
$$\vdots$$
$$n \leftrightarrow a_n = 0.3155728\ldots\boxed{5}\ldots$$
$$\vdots$$

*n*th decimal digit of a_n

INSTRUCTOR NOTE

You may wish to explain that the procedure used in the proof at the right cannot be used on the set of rational numbers between 0 and 1 because it is not possible to ensure that the number *d* created by the procedure is a rational number.

At this point we use a "diagonal technique" to construct a real number d that is greater than 0 and less than 1 and is not in the above list. We construct d by writing a decimal that *differs* from a_1 in the first decimal place, differs from a_2 in the second decimal place, differs from a_3 in the third decimal place, and, in general, differs from a_n in the nth decimal place. For instance, in the above list, a_1 has 3 as its first decimal digit. The first decimal digit of d can be any digit other than 3, say 4. The real number a_2 has 6 as its second decimal digit. The second decimal digit of d can be any digit other than 6, say 7. The real number a_3 has 2 as its third decimal digit. The third decimal digit of d can be any digit other than 2, say 3. Continue in this manner to determine the decimal digits of d. Now $d = 0.473\ldots$ must be in A because $0 < d < 1$. However, d is not in A, because d differs from each of the numbers in A in at least one decimal place.

We have reached a contradiction. Our assumption that the elements of A could be placed in a one-to-one correspondence with the elements of the natural numbers must be false. Thus A is not a countable set.

An infinite set that is not countable is said to be **uncountable**. Because the set $A = \{x | x \in R \text{ and } 0 < x < 1\}$ is uncountable, the cardinality of A is not $\aleph_0$. Cantor used the letter c, which is the first letter of the word *continuum,* to represent the cardinality of A. Cantor was also able to show that set A is equivalent to the set of all real numbers R. Thus the cardinality of R is also c. Cantor was able to prove that $c > \aleph_0$.

A Comparison of Transfinite Cardinal Numbers

$c > \aleph_0$

Up to this point, all of the infinite sets we have considered have a cardinality of either $\aleph_0$ or c. The following table lists several infinite sets and the transfinite cardinal number that is associated with each set.

The Cardinality of Some Infinite Sets

Set	Cardinal number
Natural numbers, N	$\aleph_0$
Integers, I	$\aleph_0$
Rational numbers, Q	$\aleph_0$
Irrational numbers, $\mathscr{I}$	c
Any set of the form $\{x \mid a \le x \le b\}$, where a and b are real numbers such that $a < b$	c
Real numbers, R	c

PPT

Your intuition may suggest that $\aleph_0$ and c are the only two cardinal numbers associated with infinite sets; however, this is not the case. In fact, Cantor was able to show that no matter how large the cardinal number of a set, we can find a set that has a larger cardinal number. Thus there are infinitely many transfinite numbers. Cantor's proof of this concept is now known as *Cantor's theorem.*

Cantor's Theorem

Let S be any set. The set of all subsets of S has a cardinal number that is larger than the cardinal number of S.

The set of all subsets of S is called the **power set** of S and is denoted by $P(S)$. We can see that Cantor's theorem is true for the finite set $S = \{\text{a, b, c}\}$ because the cardinality of S is 3 and S has $2^3 = 8$ subsets. The interesting part of Cantor's theorem is that it also applies to infinite sets.

Some of the following theorems can be established by using the techniques illustrated in the Excursion that follows.

Transfinite Arithmetic Theorems

- For any whole number a, $\aleph_0 + a = \aleph_0$ and $\aleph_0 - a = \aleph_0$.
- $\aleph_0 + \aleph_0 = \aleph_0$ and, in general, $\underbrace{\aleph_0 + \aleph_0 + \aleph_0 + \cdots + \aleph_0}_{\text{a finite number of aleph nulls}} = \aleph_0$.
- $c + c = c$ and, in general, $\underbrace{c + c + c + \cdots + c}_{\text{a finite number of c's}} = c$.
- $\aleph_0 + c = c$.
- $\aleph_0 c = c$.

MATH MATTERS Criticism and Praise of Cantor's Work

Georg Cantor's work in the area of infinite sets was not well received by some of his colleagues. For instance, the mathematician Leopold Kronecker tried to stop the publication of some of Cantor's work. He felt many of Cantor's theorems were ridiculous and asked, "How can one infinity be greater than another?" The following quote illustrates that Cantor was aware that his work would attract harsh criticism.

> . . . I realize that in this undertaking I place myself in a certain opposition to views widely held concerning the mathematical infinite and to opinions frequently defended on the nature of numbers.[2]

A few mathematicians were willing to show support for Cantor's work. For instance, the famous mathematician David Hilbert stated that Cantor's work was

> . . . the finest product of mathematical genius and one of the supreme achievements of purely intellectual human activity.[3]

[2] *Source:* https://en.wikipedia.org/wiki/Georg_Cantor
[3] *Source:* http://platonicrealms.com/quotes/David-Hilbert

EXCURSION

Transfinite Arithmetic

Disjoint sets are often used to explain addition. The sum 4 + 3, for example, can be deduced by selecting two disjoint sets, one with exactly four elements and one with exactly three elements. See the Venn diagram. Now form the union of the two sets. The union of the two sets has exactly seven elements; thus 4 + 3 = 7. In mathematical notation, we write

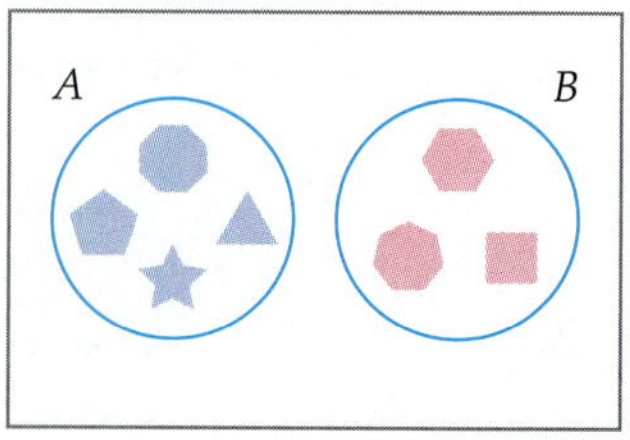

$$\begin{aligned} n(A) + n(B) &= n(A \cup B) \\ 4 + 3 &= 7 \end{aligned}$$

Cantor extended this idea to infinite sets. He reasoned that the sum $\aleph_0 + 1$ could be determined by selecting two disjoint sets, one with cardinality of $\aleph_0$ and one with cardinality of 1. In this case the set N of natural numbers and the set $Z = \{0\}$ are appropriate choices. Thus

$$\begin{aligned} n(N) + n(Z) &= n(N \cup Z) \\ &= n(W) \quad \text{• } W \text{ represents the set of whole numbers} \\ \aleph_0 \;+\; 1 &= \aleph_0 \end{aligned}$$

and, in general, for any whole number a, $\aleph_0 + a = \aleph_0$.

To find the sum $\aleph_0 + \aleph_0$, use two disjoint sets, each with cardinality of $\aleph_0$. The set E of even natural numbers and the set D of odd natural numbers satisfy the necessary conditions. Since E and D are disjoint sets, we know

$$\begin{aligned} n(E) + n(D) &= n(E \cup D) \\ &= n(N) \\ \aleph_0 \;+\; \aleph_0 &= \aleph_0 \end{aligned}$$

Thus $\aleph_0 + \aleph_0 = \aleph_0$ and, in general,

$$\underbrace{\aleph_0 + \aleph_0 + \aleph_0 + \cdots + \aleph_0}_{\text{a finite number of aleph-nulls}} = \aleph_0$$

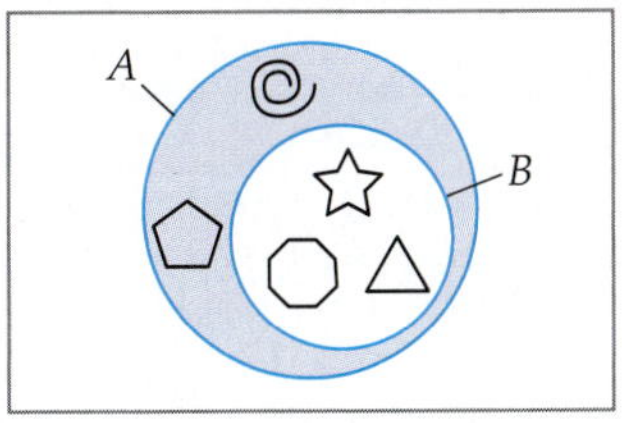

To determine a difference such as $5 - 3$ using sets, we first select a set A that has exactly five elements. We then find a subset B of this set that has exactly three elements. The difference $5 - 3$ is the cardinal number of the set $A \cap B'$, which is shown in blue in the figure at the left.

To determine $\aleph_0 - 3$, select a set with $\aleph_0$ elements, such as N, and then select a subset of this set that has exactly three elements. One such subset is $C = \{1, 2, 3\}$. The difference $\aleph_0 - 3$ is the cardinal number of the set $N \cap C' = \{4, 5, 6, 7, 8, \ldots\}$. Since $N \cap C'$ is a countably infinite set, we can conclude that $\aleph_0 - 3 = \aleph_0$. This procedure can be generalized to show that for any whole number a, $\aleph_0 - a = \aleph_0$.

EXCURSION EXERCISES

1. Use two disjoint sets to show that $\aleph_0 + 2 = \aleph_0$.
2. Use two disjoint sets other than the set of even natural numbers and the set of odd natural numbers to show that $\aleph_0 + \aleph_0 = \aleph_0$.
3. Use sets to show that $\aleph_0 - 6 = \aleph_0$.

EXERCISE SET 2.5

1. **a.** Use arrows to establish a one-to-one correspondence between $V = \{\text{a, e, i}\}$ and $M = \{3, 6, 9\}$.
   b. How many different one-to-one correspondences between V and M can be established?
2. Establish a one-to-one correspondence between the set of natural numbers $N = \{1, 2, 3, 4, 5, \ldots, n, \ldots\}$ and $F = \{5, 10, 15, 20, \ldots, 5n, \ldots\}$ by stating a general rule that can be used to pair the elements of the sets.
3. Establish a one-to-one correspondence between $D = \{1, 3, 5, \ldots, 2n - 1, \ldots\}$ and $M = \{3, 6, 9, \ldots, 3n, \ldots\}$ by stating a general rule that can be used to pair the elements of the sets.

■ In Exercises 4 to 10, state the cardinality of each set.

4. $\{2, 11, 19, 31\}$
5. $\{2, 9, 16, \ldots, 7n - 5, \ldots\}$, where n is a natural number
6. The set Q of rational numbers
7. The set R of real numbers
8. The set $\mathcal{I}$ of irrational numbers
9. $\{x \mid 5 \le x \le 9\}$
10. The set of subsets of $\{1, 5, 9, 11\}$

■ In Exercises 11 to 14, determine whether the given sets are equivalent.

11. The set of natural numbers and the set of integers

12. The set of whole numbers and the set of real numbers

13. The set of rational numbers and the set of integers

14. The set of rational numbers and the set of real numbers

■ In Exercises 15 to 18, show that the given set is an infinite set by placing it in a one-to-one correspondence with a proper subset of itself.

15. $A = \{5, 10, 15, 20, 25, 30, \ldots, 5n, \ldots\}$

16. $B = \{11, 15, 19, 23, 27, 31, \ldots, 4n + 7, \ldots\}$

17. $C = \left\{\frac{1}{2}, \frac{3}{4}, \frac{5}{6}, \frac{7}{8}, \frac{9}{10}, \ldots, \frac{2n-1}{2n}, \ldots\right\}$

18. $D = \left\{\frac{1}{2}, \frac{1}{3}, \frac{1}{4}, \frac{1}{5}, \frac{1}{6}, \ldots, \frac{1}{n+1}, \ldots\right\}$

■ In Exercises 19 to 26, show that the given set has a cardinality of $\aleph_0$ by establishing a one-to-one correspondence between the elements of the given set and the elements of N.

19. $\{50, 51, 52, 53, \ldots, n + 49, \ldots\}$

20. $\{10, 5, 0, -5, -10, -15, \ldots, -5n + 15, \ldots\}$

21. $\left\{1, \frac{1}{3}, \frac{1}{9}, \frac{1}{27}, \ldots, \frac{1}{3^{n-1}}, \ldots\right\}$

22. $\{-12, -18, -24, -30, \ldots, -6n - 6, \ldots\}$

23. $\{10, 100, 1000, \ldots, 10^n, \ldots\}$

24. $\left\{1, \frac{1}{2}, \frac{1}{4}, \frac{1}{8}, \ldots, \frac{1}{2^{n-1}}, \ldots\right\}$

25. $\{1, 8, 27, 64, \ldots, n^3, \ldots\}$

26. $\{0.1, 0.01, 0.001, 0.0001, \ldots, 10^{-n}, \ldots\}$

EXTENSIONS

27. a. Place the set $M = \{3, 6, 9, 12, 15, \ldots\}$ of positive multiples of 3 in a one-to-one correspondence with the set K of all natural numbers that are not multiples of 3. Write a sentence or two that explains the general rule you used to establish the one-to-one correspondence.

b. Use your rule to determine what number from K is paired with the number 606 from M.

c. Use your rule to determine what number from M is paired with the number 899 from K.

In the figure below, every point on line segment AB corresponds to a real number from 0 to 1 and every real number from 0 to 1 corresponds to a point on line segment AB.

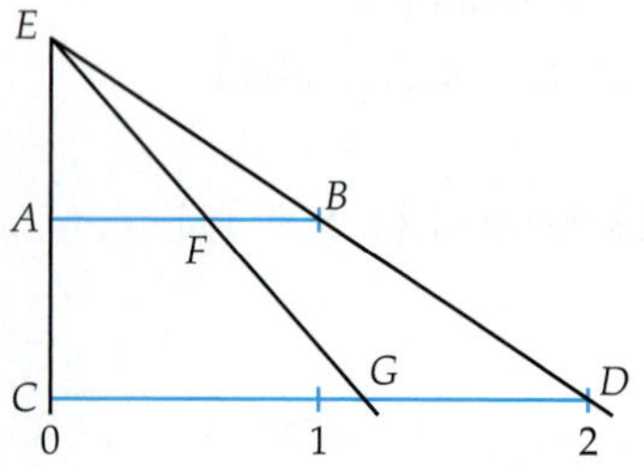

The line segment CD represents the real numbers from 0 to 2. Note that any point F on line segment AB can be paired with a unique point G on line segment CD by drawing a line from E through F. Also, any arbitrary point G on line segment CD can be paired with a unique point F on line segment AB by drawing the line EG. This geometric procedure establishes a one-to-one correspondence between the set $\{x \mid 0 \le x \le 1\}$ and the set $\{x \mid 0 \le x \le 2\}$. Thus $\{x \mid 0 \le x \le 1\} \sim \{x \mid 0 \le x \le 2\}$.

28. Draw a figure that can be used to verify each of the following.

a. $\{x \mid 0 \le x \le 1\} \sim \{x \mid 0 \le x \le 5\}$

b. $\{x \mid 2 \le x \le 5\} \sim \{x \mid 1 \le x \le 8\}$

29. Consider the semicircle with arc length π and center C and the line L_1 in the following figure. Each point on the semicircle, *other than the endpoints,* represents a unique real number between 0 and π. Each point on line L_1 represents a unique real number.

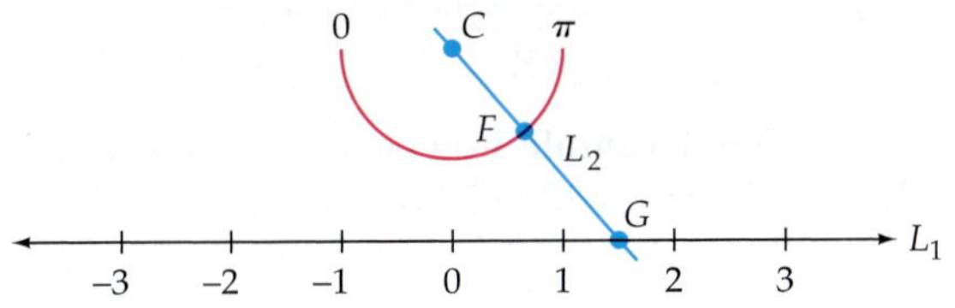

Any line through C that intersects the semicircle at a point other than one of its endpoints will intersect line L_1 at a unique point. Also, any line through C that intersects line L_1 will intersect the semicircle at a unique point that is not an endpoint of the semicircle. What can we conclude from this correspondence?

30. **The Hilbert Hotel** The Hilbert Hotel is an imaginary hotel created by the mathematician David Hilbert (1862–1943). The hotel has an infinite number of rooms. Each room is numbered with a natural number—room 1, room 2, room 3, and so on. Search the Internet for information on Hilbert's Hotel. Write a few paragraphs that explain some of the interesting questions that arise when guests arrive to stay at the hotel.

Mary Pat Campbell has written a song about a hotel with an infinite number of rooms. Her song is titled *Hotel Aleph Null—yeah.* Here are the lyrics for the chorus of her song, which is to be sung to the tune of *Hotel California* by the Eagles (*Source:* http://www.marypat.org/mathcamp/doc2001/hellrelays.html#hotel).[4]

Hotel Aleph Null—yeah
Welcome to the Hotel Aleph Null—yeah
What a lovely place (what a lovely place)
Got a lot of space
Packin' em in at the Hotel Aleph Null—yeah
Any time of year
You can find space here

[4] Reprinted by permission of Mary Pat Campbell.

CHAPTER 2 SUMMARY

The following table summarizes essential concepts in this chapter. The references given in the right-hand column list Examples and Exercises that can be used to test your understanding of a concept.

2.1 Basic Properties of Sets

The Roster Method The roster method is used to represent a set by listing each element of the set inside a pair of braces. Commas are used to separate the elements.	See **Example 1** on page 48, and then try Exercises 1 and 2 on page 101.
Basic Number Sets Natural Numbers or Counting Numbers $N = \{1, 2, 3, 4, 5, ...\}$ Whole Numbers $W = \{0, 1, 2, 3, 4, 5, ...\}$ Integers $I = \{..., -4, -3, -2, -1, 0, 1, 2, 3, 4, ...\}$ Rational Numbers Q = the set of all terminating or repeating decimals Irrational Numbers $\mathcal{J}$ = the set of all nonterminating, nonrepeating decimals Real Numbers R = the set of all rational or irrational numbers	See **Example 3** and **Check Your Progress 3** on page 49, and then try Exercises 3 to 6 on page 101.
Set-Builder Notation Set-builder notation is used to represent a set, by describing its elements.	See **Example 5** on page 50, and then try Exercises 7 to 10 on page 101.
Cardinal Number of a Finite Set The cardinal number of a finite set is the number of elements in the set. The cardinal number of a finite set A is denoted by the notation $n(A)$.	See **Example 6** on page 51, and then try Exercises 63 to 67 on page 103.
Equal Sets and Equivalent Sets Two sets are equal if and only if they have exactly the same elements. Two sets are equivalent if and only if they have the same number of elements.	See **Example 7** on page 52, and then try Exercises 11 and 12 on page 101.

continued

2.2 Complements, Subsets, and Venn Diagrams

The Universal Set and the Complement of a Set The universal set, denoted by U, is the set of all elements that are under consideration. The complement of set A, denoted by A', is the set of all elements of the universal set that are not elements of A.	See **Example 1** on page 58, and then try Exercises 22 to 24 on page 101.
Subset of a Set Set A is a subset of set B, denoted by $A \subseteq B$ if and only if every element of A is also an element of B.	See **Example 2** on page 59, and then try Exercises 25 and 26 on page 102.
Proper Subset of a Set Set A is a proper subset of set B, denoted by $A \subset B$, if every element of A is an element of B and $A \neq B$.	See **Example 3** on page 60, and then try Exercises 27 to 30 on page 102.
The Number of Subsets of a Set A set with n elements has 2^n subsets.	See **Example 5** on page 61, and then try Exercise 38 on page 102.

2.3 Set Operations

Intersection of Sets The intersection of sets A and B, denoted by $A \cap B$, is the set of elements common to both A and B. $A \cap B = \{x \mid x \in A \quad \text{and} \quad x \in B\}$ 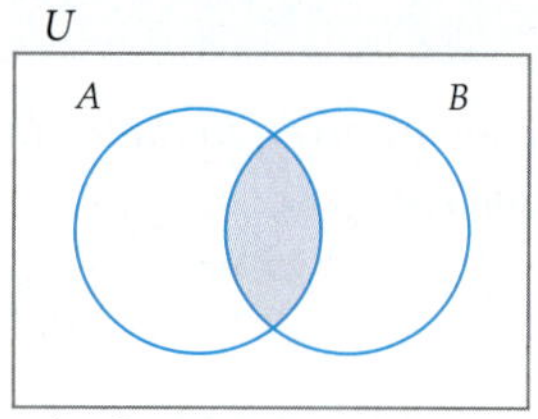 $A \cap B$	See **Example 1** on page 67, and then try Exercises 17 and 19 on page 101.
Union of Sets The union of sets A and B, denoted by $A \cup B$, is the set that contains all the elements that belong to A or to B or to both. $A \cup B = \{x \mid x \in A \quad \text{or} \quad x \in B\}$ 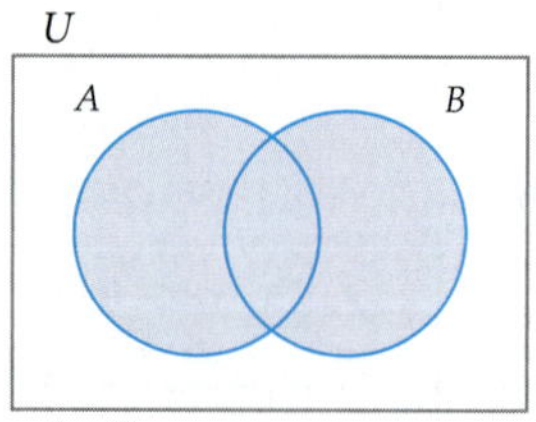 $A \cup B$	See **Example 2** on page 68, and then try Exercises 18 and 20 on page 101.
De Morgan's Laws For all sets A and B, $(A \cup B)' = A' \cap B'$ and $(A \cap B)' = A' \cup B'$	See **Example 4** and **Check Your Progress 4** on page 69, and then try Exercises 39 and 40 on page 102.

Venn Diagrams and the Equality of Set Expressions Two sets are equal if and only if they each represent the same region(s) on a Venn diagram. Venn diagrams can be used to verify each of the following properties. For all sets A, B, and C: **Commutative Properties** $A \cap B = B \cap A$ $A \cup B = B \cup A$ **Associative Properties** $(A \cap B) \cap C = A \cap (B \cap C)$ $(A \cup B) \cup C = A \cup (B \cup C)$ **Distributive Properties** $A \cap (B \cup C) = (A \cap B) \cup (A \cap C)$ $A \cup (B \cap C) = (A \cup B) \cap (A \cup C)$	See **Example 6** and **Check Your Progress 6** on page 71, and then try Exercises 45 to 48 on page 102.
2.4 Applications of Sets	
Applications Many counting problems that arise in applications involving surveys can be solved by using sets and Venn diagrams.	See **Examples 1 and 2** on page 81, and then try Exercises 53 and 54 on page 102.
The Inclusion-Exclusion Formula For all finite sets A and B, $n(A \cup B) = n(A) + n(B) - n(A \cap B)$	See **Examples 3 and 4** on page 83, and then try Exercises 55 and 56 on page 103.
2.5 Infinite Sets	
One-to-One Correspondence and Equivalent Sets Two sets A and B are equivalent, denoted by $A \sim B$, if and only if A and B can be placed in a one-to-one correspondence.	See **Examples 1 and 3** on pages 90 and 92, and then try Exercises 57 to 60 on page 103.
Infinite Set A set is an infinite set if it can be placed in a one-to-one correspondence with a proper subset of itself.	See **Example 2** on page 91, and then try Exercises 61 and 62 on page 103.

CHAPTER 2 REVIEW EXERCISES

■ In Exercises 1 to 6, use the roster method to represent each set.

1. The set of months of the year with a name that starts with the letter J
2. The set of states in the United States that do not share a common border with another state
3. The set of whole numbers less than 8
4. The set of integers that satisfy $x^2 = 64$
5. The set of natural numbers that satisfy $x + 3 \leq 7$
6. The set of counting numbers larger than -3 and less than or equal to 6

■ In Exercises 7 to 10, use set-builder notation to write each set.

7. The set of integers greater than -6
8. {April, June, September, November}
9. {Kansas, Kentucky}
10. $\{1, 8, 27, 64, 125\}$

■ In Exercises 11 and 12, state whether each of the following sets are equal, equivalent, both, or neither.

11. $\{2, 4, 6, 8\}$, $\{x \mid x \in N \text{ and } x < 5\}$
12. $\{8, 9\}$, the set of single digit whole numbers greater than 7

■ In Exercises 13 to 16, determine whether the statement is true or false.

13. $\{3\} \in \{1, 2, 3, 4\}$
14. $-11 \in I$
15. $\{a, b, c\} \sim \{1, 5, 9\}$
16. The set of small numbers is a well-defined set

■ In Exercises 17 to 24, let $U = \{2, 6, 8, 10, 12, 14, 16, 18\}$, $A = \{2, 6, 10\}$, $B = \{6, 10, 16, 18\}$, and $C = \{14, 16\}$. Find each of the following.

17. $A \cap B$
18. $A \cup B$
19. $A' \cap C$
20. $B \cup C'$
21. $A \cup (B \cap C)$
22. $(A \cup C)' \cap B'$
23. $(A \cap B')'$
24. $(A \cup B \cup C)'$

■ In Exercises 25 and 26, determine whether the first set is a subset of the second set.

25. {0, 1, 5, 9}, the set of natural numbers

26. {1, 2, 4, 8, 9.5}, the set of integers

■ In Exercises 27 to 30, determine whether the first set is a proper subset of the second set.

27. The set of natural numbers; the set of whole numbers

28. The set of integers; the set of real numbers

29. The set of counting numbers; the set of natural numbers

30. The set of real numbers; the set of rational numbers

■ In Exercises 31 to 34, list all the subsets of the given set.

31. {I, II}

32. {s, u, n}

33. {penny, nickel, dime, quarter}

34. {A, B, C, D, E}

■ In Exercises 35 to 38, find the number of subsets of the given set.

35. The set of the four musketeers

36. The set of the letters of the English alphabet

37. The set of the letters of "uncopyrightable," which is the longest English word with no repeated letters

38. The set of the seven dwarfs

■ In Exercises 39 and 40, determine whether each statement is true or false for all sets *A* and *B*.

39. $(A \cup B')' = A' \cap B$

40. $(A' \cap B')' = A \cup B$

■ In Exercises 41 to 44, draw a Venn diagram to represent the given set.

41. $A \cap B'$

42. $A' \cup B'$

43. $(A \cup B) \cup C'$

44. $A \cap (B' \cup C)$

■ In Exercises 45 to 48, draw Venn diagrams to determine whether the expressions are equal for all sets *A*, *B*, and *C*.

45. $A' \cup (B \cup C)$; $(A' \cup B) \cup (A' \cup C)$

46. $(A \cap B) \cap C'$; $(A' \cup B') \cup C$

47. $A \cap (B' \cap C)$; $(A \cup B') \cap (A \cup C)$

48. $A \cap (B \cup C)$; $A' \cap (B \cup C)$

■ In Exercises 49 and 50, use set notation to describe the shaded region.

49.

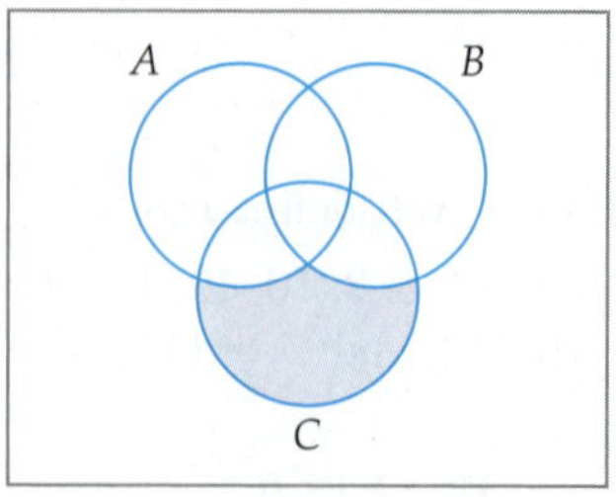

50.

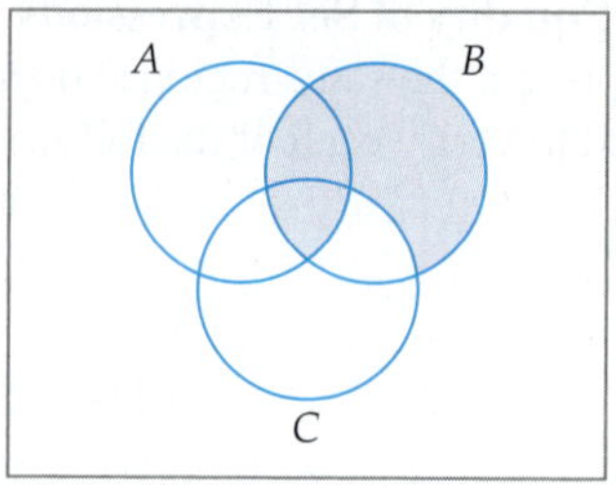

■ In Exercises 51 and 52, draw a Venn diagram with each of the given elements placed in the correct region.

51. $U = \{e, h, r, d, w, s, t\}$
$A = \{t, r, e\}$
$B = \{w, s, r, e\}$
$C' = \{s, r, d, h\}$

52. $U = \{\alpha, \beta, \Gamma, \gamma, \Delta, \delta, \varepsilon, \theta\}$
$A' = \{\beta, \Delta, \theta, \gamma\}$
$B = \{\delta, \varepsilon\}$
$C = \{\beta, \varepsilon, \Gamma\}$

53. An Exercise Survey In a survey at a health club, 208 members indicated that they enjoy aerobic exercises, 145 indicated that they enjoy weight training, 97 indicated that they enjoy both aerobics and weight training, and 135 indicated that they do not enjoy either of these types of exercise. How many members were surveyed?

Tomasz Trojanowski/Shutterstock.com

54. A Coffee Survey A gourmet coffee bar conducted a survey to determine the preferences of its customers. Of the customers surveyed,

221 like espresso.
127 like cappuccino and chocolate-flavored coffee.
182 like cappuccino.
136 like espresso and chocolate-flavored coffee.
209 like chocolate-flavored coffee.
96 like all three types of coffee.
116 like espresso and cappuccino.
82 like none of these types of coffee.

How many of the customers in the survey

a. like only chocolate-flavored coffee?

b. like cappuccino and chocolate-flavored coffee but not espresso?

c. like espresso and cappuccino but not chocolate-flavored coffee?

d. like exactly one of the three types of coffee?

■ In Exercises 55 and 56, use the inclusion-exclusion formula to answer each application.

55. On a football team, 27 of its athletes play on offense, 22 play on defense, and 43 play on offense or defense. How many of the athletes play both on offense and on defense?

56. A college finds that 625 of its students are registered in biology, 433 are registered in psychology, and 184 are registered in both biology and psychology. How many students are registered in biology or psychology?

■ In Exercises 57 to 60, establish a one-to-one correspondence between the sets.

57. $\{1, 3, 6, 10\}$, $\{1, 2, 3, 4\}$

58. $\{x \mid x > 10 \text{ and } x \in N\}$, $\{2, 4, 6, 8, \ldots, 2n, \ldots\}$

59. $\{3, 6, 9, 12, \ldots, 3n, \ldots\}$, $\{10, 100, 1000, \ldots, 10^n, \ldots\}$

60. $\{x \mid 0 \leq x \leq 1\}$, $\{x \mid 0 \leq x \leq 4\}$ (*Hint:* Use a drawing.)

■ In Exercises 61 and 62, show that the given set is an infinite set.

61. $A = \{6, 10, 14, 18, \ldots, 4n + 2, \ldots\}$

62. $B = \left\{1, \frac{1}{2}, \frac{1}{4}, \frac{1}{8}, \ldots, \frac{1}{2^{n-1}}, \ldots\right\}$

■ In Exercises 63 to 70, state the cardinality of each set.

63. $\{5, 6, 7, 8, 6\}$

64. $\{4, 6, 8, 10, 12, \ldots, 22\}$

65. $\{0, \varnothing\}$

66. The set of all states in the United States that border the Gulf of Mexico

67. The set of integers less than 1,000,000

68. The set of rational numbers between 0 and 1

69. The set of irrational numbers

70. The set of real numbers between 0 and 1

■ In Exercises 71 and 72, find each of the following, where $\aleph_0$ and c are transfinite cardinal numbers.

71. $\aleph_0 + (\aleph_0 + \aleph_0)$

72. $c + (c + c)$

CHAPTER 2 TEST

■ In Exercises 1 to 4, let $U = \{1, 2, 3, 4, 5, 6, 7, 8, 9, 10\}$, $A = \{3, 5, 7, 8\}$, $B = \{2, 3, 8, 9, 10\}$, and $C = \{1, 4, 7, 8\}$. Use the roster method to write each of the following sets.

1. $(A \cap B)'$

2. $A' \cap B$

3. $A' \cup (B \cap C')$

4. $A \cap (B' \cup C)$

■ In Exercises 5 and 6, use set-builder notation to write each of the given sets.

5. $\{0, 1, 2, 3, 4, 5, 6\}$

6. $\{-3, -2, -1, 0, 1, 2\}$

7. State the cardinality of each set.

a. The set of whole numbers less than 4

b. The set of integers

■ In Exercises 8 and 9, state whether the given sets are equal, equivalent, both, or neither.

8. a. $\{j, k, l\}$, $\{a, e, i, o, u\}$

b. $\{x \mid x \in W \text{ and } x < 4\}$, $\{9, 10, 11, 12\}$

9. a. the set of natural numbers; the set of integers

b. the set of whole numbers; the set of positive integers

10. List all of the subsets of $\{a, b, c, d\}$.

11. Determine the number of subsets of a set with 21 elements.

■ In Exercises 12 and 13, draw a Venn diagram to represent the given set.

12. $(A \cup B') \cap C$

13. $(A' \cap B) \cup (A \cap C')$

14. Use one of De Morgan's laws to write $(A \cup B)'$ as the intersection of two sets.

15. Upgrade Options An automobile company makes a sedan with nine upgrade options.

a. How many different versions of this sedan can the company produce?

b. What is the minimum number of upgrade options the company must provide if it wishes to offer at least 2500 versions of this sedan?

16. Student Demographics A college finds that 841 of its students are receiving financial aid, 525 students are business majors, and 202 students are receiving financial aid and are business majors. How many students are receiving financial aid or are business majors?

17. **Affordability of Housing** The following bar graph shows the monthly principal and interest payment needed to purchase an average-priced existing home in the United States from 2007 to 2014.

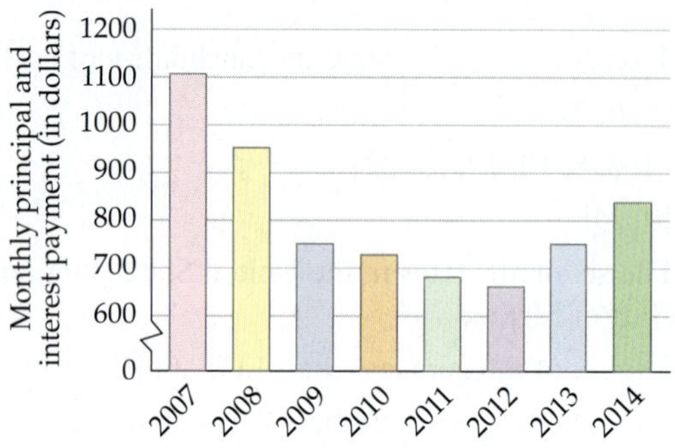

Monthly principal and interest payment for an average-priced existing home

SOURCE: National Association of Realtors as reported in the *World Almanac*, 2015

Use the information in the bar graph and the roster method to represent each of the following sets.

a. The set of years in which the monthly principal and interest payment was greater than \$800.

b. The set of months in which the monthly principal and interest payment was greater than \$700 but less than \$900.

c. The set of months in which the monthly principal and interest payment was less than \$600.

18. **A Survey** A survey of 1000 households was taken to determine how they obtained news about current events. The survey considered only television, newspapers, and the Internet as sources for news. Of the households surveyed,

724 obtained news from television.

545 obtained news from newspapers.

280 obtained news from the Internet.

412 obtained news from both television and newspapers.

185 obtained news from both television and the Internet.

105 obtained news from television, newspapers, and the Internet.

64 obtained news from the Internet but not from television or newspapers.

Of those households that were surveyed,

a. how many obtained news from television but not from newspapers or the Internet?

b. how many obtained news from newspapers but not from television or the Internet?

c. how many obtained news from television or newspapers?

d. how many did not acquire news from television, newspapers, or the Internet?

19. Show a method that can be used to establish a one-to-one correspondence between the elements of the following sets.

$$\{5, 10, 15, 20, 25, \ldots, 5n, \ldots\}, W$$

20. Verify that the following set is an infinite set by illustrating a one-to-one correspondence between the elements of the set and the elements of one of the set's proper subsets.

$$\{3, 6, 9, 12, \ldots, 3n, \ldots\}$$

3

Logic

It is not easy to summarize in a few paragraphs the subject matter known as logic. For lawyers and judges, logic is the science of correct reasoning. They often use logic to communicate more effectively, construct valid arguments, analyze legal contracts, and make decisions. Law schools consider a knowledge of logic to be one of the most important predictors of future success for their new students. A sizeable portion of the LSAT (Law School Admission Test), which is required by law school applicants as part of their admission process, concerns logical reasoning. A typical LSAT logic problem is presented in Exercise 29, page 161.

Many other professions also make extensive use of logic. For instance, programmers use logic to design computer software, electrical engineers use logic to design circuits for smart phones, and mathematicians use logic to solve problems and construct mathematical proofs.

In this chapter, you will encounter several facets of logic. Specifically, you will use logic to

- analyze information and the relationship between statements,
- determine the validity of arguments,
- determine valid conclusions based on given assumptions, and
- analyze electronic circuits.

SECTION 3.1 Logic Statements and Quantifiers

HISTORICAL NOTE

Jean Marc Keystone/Hulton Archive/Getty Images

George Boole (bo͞ol) was born in 1815 in Lincoln, England. He was raised in poverty, but he was very industrious and had learned Latin and Greek by the age of 12. Later he mastered German, French, and Italian. His first profession, at the young age of 16, was that of an assistant school teacher. At the age of 20 he started his own school.

In 1849 Boole was appointed the chairperson of mathematics at Queens College in Cork, Ireland.

Many of Boole's mathematical ideas, such as Boolean algebra, have applications in the areas of computer programming and the design of electronic circuits.

One of the first mathematicians to make a serious study of symbolic logic was Gottfried Wilhelm Leibniz (1646–1716). Leibniz tried to advance the study of logic from a merely philosophical subject to a formal mathematical subject. Leibniz never completely achieved this goal; however, several mathematicians, such as Augustus De Morgan (1806–1871) and George Boole (1815–1864), contributed to the advancement of symbolic logic as a mathematical discipline.

Boole published *The Mathematical Analysis of Logic* in 1848. In 1854 he published the more extensive work, *An Investigation of the Laws of Thought.* Concerning this document, the mathematician Bertrand Russell stated, "Pure mathematics was discovered by Boole in a work which is called *The Laws of Thought.*"

Logic Statements

Every language contains different types of sentences, such as statements, questions, and commands. For instance,

"Is the test today?" is a question.

"Go get the newspaper" is a command.

"This is a nice car" is an opinion.

"Denver is the capital of Colorado" is a statement of fact.

The symbolic logic that Boole was instrumental in creating applies only to sentences that are *statements* as defined below.

A Statement

A **statement** is a declarative sentence that is either true or false, but not both true and false.

It may not be necessary to determine whether a sentence is true to determine whether it is a statement. For instance, consider the following sentence.

Yosemite National Park is located in California.

You may not know if the sentence is true, but you do know that the sentence is either true or it is false, and that it is not both true and false. Thus, you know that the sentence is a statement.

Henry Lien/Shutterstock.com

Yosemite's Half Dome

EXAMPLE 1 Identify Statements

Determine whether each sentence is a statement.

a. Florida is a state in the United States.

b. How are you?

c. $9^9 + 2$ is a prime number.

d. $x + 1 = 5$.

Solution

a. Florida is one of the 50 states in the United States, so this sentence is true and it is a statement.

b. The sentence "How are you?" is a question; it is not a declarative sentence. Thus it is not a statement.

TAKE NOTE

The following sentence is a famous paradox:

This is a false sentence.

It is not a statement, because if we assume it to be a true sentence, then it is false, and if we assume it to be a false sentence, then it is true. Statements cannot be true and false at the same time.

c. You may not know whether $9^9 + 2$ is a prime number; however, you do know that it is a whole number larger than 1, so it is either a prime number or it is not a prime number. The sentence is either true or it is false, and it is not both true and false, so it is a statement.

d. $x + 1 = 5$ is a statement. It is known as an *open statement.* It is true for $x = 4$, and it is false for any other values of x. For any given value of x, it is true or false but not both.

CHECK YOUR PROGRESS 1 Determine whether each sentence is a statement.

a. Open the door.

b. 7055 is a large number.

c. In the year 2024, the president of the United States will be a woman.

d. $x > 3$.

Solution See page S7.

MATH MATTERS Charles Dodgson

The Granger Collection

Charles Dodgson (Lewis Carroll)

One of the best-known logicians is Charles Dodgson (1832–1898). His mathematical works include *A Syllabus of Plane Algebraical Geometry, The Fifth Book of Euclid Treated Algebraically,* and *Symbolic Logic.* Although Dodgson was a distinguished mathematician in his time, he is best known by his pen name Lewis Carroll, which he used when he published *Alice's Adventures in Wonderland* and *Through the Looking-Glass.*

Queen Victoria of the United Kingdom enjoyed *Alice's Adventures in Wonderland* to the extent that she told Dodgson she was looking forward to reading another of his books. He promptly sent her his *Syllabus of Plane Algebraical Geometry,* and it was reported that she was less than enthusiastic about the latter book.

Morphart Creation/Shutterstock.com

Simple Statements and Compound Statements

Simple Statements and Compound Statements

A **simple statement** is a statement that conveys a single idea. A **compound statement** is a statement that conveys two or more ideas.

Connecting simple statements with words and phrases such as *and, or, if ... then,* and *if and only if* creates a compound statement. For instance, "I will attend the meeting or I will go to school." is a compound statement. It is composed of the two simple statements, "I will attend the meeting." and "I will go to school." The word *or* is a connective for the two simple statements.

George Boole used symbols such as p, q, r, and s to represent simple statements and the symbols $\wedge$, $\vee$, $\sim$, $\rightarrow$, and $\leftrightarrow$ to represent connectives. See Table 3.1.

TABLE 3.1
Logic Connectives and Symbols

Statement	Connective	Symbolic form	Type of statement
not p	not	$\sim p$	negation
p and q	and	$p \wedge q$	conjunction
p or q	or	$p \vee q$	disjunction
If p, then q	If … then	$p \rightarrow q$	conditional
p if and only if q	if and only if	$p \leftrightarrow q$	biconditional

PPT

QUESTION What connective is used in a conjunction?

Truth Value and Truth Tables

The **truth value** of a simple statement is either true (T) or false (F).

The **truth value** of a compound statement depends on the truth values of its simple statements and its connectives.

A **truth table** is a table that shows the truth value of a compound statement for all possible truth values of its simple statements.

Truth Table for $\sim p$

p	$\sim p$
T	F
F	T

The *negation* of the statement "Today is Friday." is the statement "Today is not Friday." In symbolic logic, the tilde symbol $\sim$ is used to denote the negation of a statement. If a statement p is true, its negation $\sim p$ is false, and if a statement p is false, its negation $\sim p$ is true. See the table at the left. The negation of the negation of a statement is the original statement. Thus $\sim(\sim p)$ can be replaced by p in any statement.

EXAMPLE 2 Write the Negation of a Statement

Write the negation of each statement.

a. Ellie Goulding is an opera singer.

b. The dog does not need to be fed.

Solution

a. Ellie Goulding is not an opera singer.

b. The dog needs to be fed.

Luboslav Tiles/Shutterstock.com

The *Queen Mary 2*

CHECK YOUR PROGRESS 2 Write the negation of each statement.

a. The *Queen Mary 2* is the world's largest cruise ship.

b. The fire engine is not red.

Solution See page S7.

We will often find it useful to write compound statements in symbolic form.

ANSWER The connective *and*.

EXAMPLE 3 Write Compound Statements in Symbolic Form

Consider the following simple statements.

p: Today is Friday.
q: It is raining.
r: I am going to a movie.
s: I am not going to the basketball game.

Write the following compound statements in symbolic form.

a. Today is Friday and it is raining.
b. It is not raining and I am going to a movie.
c. I am going to the basketball game or I am going to a movie.
d. If it is raining, then I am not going to the basketball game.

Solution

a. $p \wedge q$ **b.** $\sim q \wedge r$ **c.** $\sim s \vee r$ **d.** $q \rightarrow s$

CHECK YOUR PROGRESS 3 Use p, q, r, and s as defined in Example 3 to write the following compound statements in symbolic form.

a. Today is not Friday and I am going to a movie.
b. I am going to the basketball game and I am not going to a movie.
c. I am going to the movie if and only if it is raining.
d. If today is Friday, then I am not going to a movie.

Solution See page S8.

In the next example, we translate symbolic statements into English sentences.

EXAMPLE 4 Translate Symbolic Statements

Consider the following statements.

p: The game will be played in Atlanta.
q: The game will be shown on CBS.
r: The game will not be shown on ESPN.
s: The Mets are favored to win.

Write each of the following symbolic statements in words.

a. $q \wedge p$ **b.** $\sim r \wedge s$ **c.** $s \leftrightarrow \sim p$

Solution

a. The game will be shown on CBS and the game will be played in Atlanta.
b. The game will be shown on ESPN and the Mets are favored to win.
c. The Mets are favored to win if and only if the game will not be played in Atlanta.

CHECK YOUR PROGRESS 4 Consider the following statements.

e: All men are created equal.
t: I am trading places.
a: I get Abe's place.
g: I get George's place.

fstockfoto/Shutterstock.com

Use the above information to translate the dialogue in the speech bubbles at the left.

Solution See page S8.

Compound Statements and Grouping Symbols

If a compound statement is written in symbolic form, then parentheses are used to indicate which simple statements are grouped together. Table 3.2 illustrates the use of parentheses to indicate groupings for some statements in symbolic form.

TABLE 3.2

Symbolic form	The parentheses indicate that:
$p \wedge (q \vee \sim r)$	q and $\sim r$ are grouped together.
$(p \wedge q) \vee r$	p and q are grouped together.
$(p \wedge \sim q) \rightarrow (r \vee s)$	p and $\sim q$ are grouped together. r and s are also grouped together.

If a compound statement is written as an English sentence, then a comma is used to indicate which simple statements are grouped together. **Statements on the same side of a comma are grouped together**. See Table 3.3.

TABLE 3.3

English sentence	The comma indicates that:
p, and q or not r.	q and $\sim r$ are grouped together because they are both on the same side of the comma.
p and q, or r.	p and q are grouped together because they are both on the same side of the comma.
If p and not q, then r or s.	p and $\sim q$ are grouped together because they are both to the left of the comma. r and s are grouped together because they are both to the right of the comma.

If a statement in symbolic form is written as an English sentence, then the simple statements that appear together in parentheses in the symbolic form will all be on the same side of the comma that appears in the English sentence.

EXAMPLE 5 Translate Compound Statements

Let p, q, and r represent the following.

p: You get a promotion.

q: You complete the training.

r: You will receive a bonus.

a. Write $(p \wedge q) \rightarrow r$ as an English sentence.

b. Write "If you do not complete the training, then you will not get a promotion and you will not receive a bonus." in symbolic form.

Solution

a. Because the p and the q statements both appear in parentheses in the symbolic form, they are placed to the left of the comma in the English sentence.

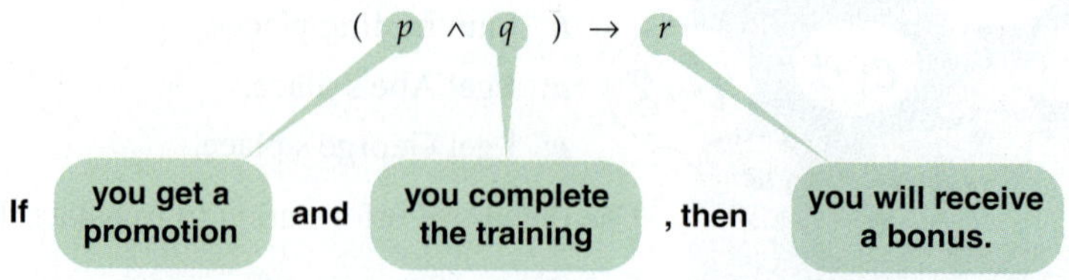

Thus the translation is: If you get a promotion and complete the training, then you will receive a bonus.

b. Because the not p and the not r statements are both to the right of the comma in the English sentence, they are grouped together in parentheses in the symbolic form.

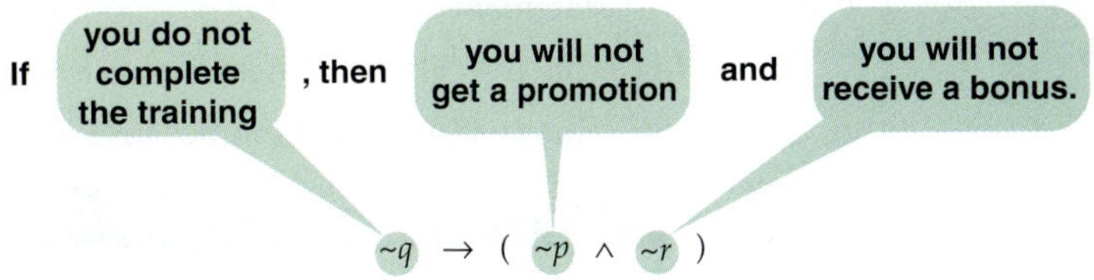

Thus the translation is: $\sim q \rightarrow (\sim p \wedge \sim r)$

Martin Lehmann/Shutterstock.com

Kesha

CHECK YOUR PROGRESS 5 Let p, q, and r represent the following.

p: Kesha's singing style is similar to Uffie's.

q: Kesha has messy hair.

r: Kesha is a rapper.

a. Write $(p \wedge q) \rightarrow r$ as an English sentence.

b. Write "If Kesha is not a rapper, then Kesha does not have messy hair and Kesha's singing style is not similar to Uffie's." in symbolic form.

Solution See page S8.

The use of parentheses in a symbolic statement may affect the meaning of the statement. For instance, $\sim(p \vee q)$ indicates the negation of the compound statement $p \vee q$. However, $\sim p \vee q$ indicates that only the p statement is negated.

The statement $\sim(p \vee q)$ is read as, "It is not true that, p or q." The statement $\sim p \vee q$ is read as, "Not p or q."

If you order cake *and* ice cream in a restaurant, the waiter will bring *both* cake and ice cream. In general, the **conjunction** $p \wedge q$ is true if both p and q are true, and the conjunction is false if either p or q is false. The truth table at the left shows the four possible cases that arise when we form a conjunction of two statements.

Truth Table for $p \wedge q$

p	q	$p \wedge q$
T	T	T
T	F	F
F	T	F
F	F	F

T: True F: False

Truth Value of a Conjunction

The conjunction $p \wedge q$ is true if and only if both p and q are true.

Sometimes the word *but* is used in place of the connective *and*. For instance, "I ride my bike to school, but I ride the bus to work," is equivalent to the conjunction, "I ride my bike to school and I ride the bus to work."

Any **disjunction** $p \vee q$ is true if p is true or q is true or both p and q are true. The truth table at the left shows that the disjunction p or q is false if both p and q are false; however, it is true in all other cases.

Truth Table for $p \vee q$

p	q	$p \vee q$
T	T	T
T	F	T
F	T	T
F	F	F

Truth Value of a Disjunction

The disjunction $p \vee q$ is true if and only if p is true, q is true, or both p and q are true.

EXAMPLE 6 Determine the Truth Value of a Statement

Determine whether each statement is true or false.

a. $7 \geq 5$.

b. 5 is a whole number and 5 is an even number.

c. 2 is a prime number and 2 is an even number.

Solution

a. $7 \geq 5$ means $7 > 5$ or $7 = 5$. Because $7 > 5$ is true, the statement $7 \geq 5$ is a true statement.

b. This is a false statement because 5 is not an even number.

c. This is a true statement because each simple statement is true.

CHECK YOUR PROGRESS 6 Determine whether each statement is true or false.

a. 21 is a rational number and 21 is a natural number.

b. $4 \leq 9$.

c. $-7 \geq -3$.

Solution See page S8.

Truth tables for the conditional and biconditional are given in Section 3.3.

Quantifiers and Negation

In a statement, the word *some* and the phrases *there exists* and *at least one* are called **existential quantifiers**. Existential quantifiers are used as prefixes to assert the existence of something.

In a statement, the words *none*, *no*, *all*, and *every* are called **universal quantifiers**. The universal quantifiers *none* and *no* deny the existence of something, whereas the universal quantifiers *all* and *every* are used to assert that every element of a given set satisfies some condition.

Recall that the negation of a false statement is a true statement and the negation of a true statement is a false statement. It is important to remember this fact when forming the negation of a quantified statement. For instance, what is the negation of the false statement, "All dogs are mean"? You may think that the negation is "No dogs are mean," but this is also a false statement. Thus the statement "No dogs are mean" is not the negation of "All dogs are mean." The negation of "All dogs are mean," which is a false statement, is in fact "Some dogs are not mean," which is a true statement. The statement "Some dogs are not mean" can also be stated as "At least one dog is not mean" or "There exists a dog that is not mean."

What is the negation of the false statement, "No doctors write in a legible manner"? Whatever the negation is, we know it must be a true statement. The negation cannot be "All doctors write in a legible manner," because this is also a false statement. The negation is "Some doctors write in a legible manner." This can also be stated as, "There exists at least one doctor who writes in a legible manner."

Table 3.4A illustrates how to write the negation of some quantified statements.

TABLE 3.4A
Quantified Statements and Their Negations

Statement	Negation
All X are Y.	Some X are not Y.
No X are Y.	Some X are Y.
Some X are not Y.	All X are Y.
Some X are Y.	No X are Y.

In Table 3.4A, the negations of the statements in the first column are shown in the second column. Also, the negation of the statements in the second column are the statements in the first column. Thus the information in Table 3.4A can be shown more compactly as in Table 3.4B.

TABLE 3.4B
Quantified Statements and Their Negations Displayed in a Compact Format

All X are Y.	negation ⟷	Some X are not Y.
No X are Y.	negation ⟷	Some X are Y.

PPT

EXAMPLE 7 Write the Negation of a Quantified Statement

Write the negation of each of the following statements.

a. Some airports are open.

b. All movies are worth the price of admission.

c. No odd numbers are divisible by 2.

Solution

a. No airports are open.

b. Some movies are not worth the price of admission.

c. Some odd numbers are divisible by 2.

CHECK YOUR PROGRESS 7 Write the negation of the following statements.

a. All bears are brown.

b. No smart phones are expensive.

c. Some vegetables are not green.

Solution See page S8.

EXCURSION

Switching Networks

Claude E. Shannon

Keystone/Getty Images

In 1939, Claude E. Shannon (1916–2001) wrote a thesis on an application of symbolic logic to *switching networks.* A switching network consists of wires and switches that can open and close. Switching networks are used in many electrical appliances, as well as in telephone equipment and computers. Figure 3.1 shows a switching network that consists of a single switch P that connects two terminals. An electric current can flow from one terminal to the other terminal provided that the switch P is in the closed position. If P is in the open position, then the current cannot flow from one terminal to the other. If a current can flow between the terminals, we say that a network is closed, and if a current cannot flow between the terminals, we say that the network is open. We designate this network by the letter P. There exists an analogy between a network P and a statement p in that a network is either open or it is closed, and a statement is either true or it is false.

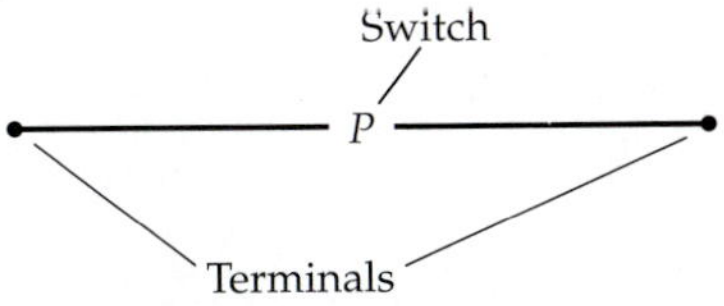

FIGURE 3.1

FIGURE 3.2 A series network

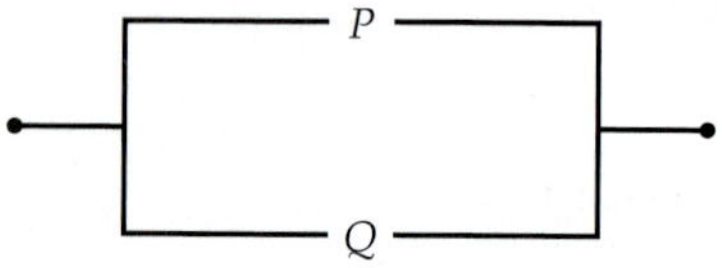

FIGURE 3.3 A parallel network

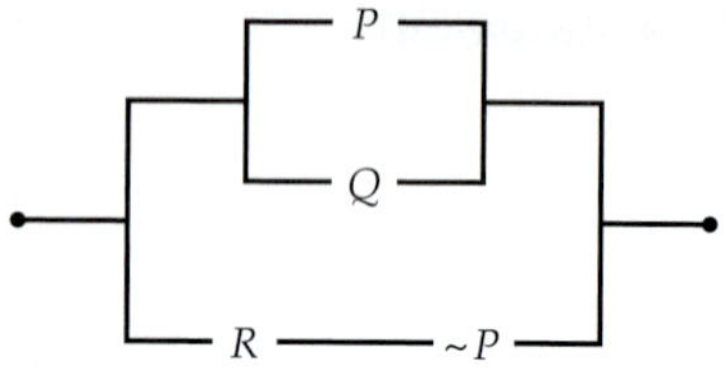

FIGURE 3.4

Figure 3.2 shows two switches P and Q connected in **series**. This series network is closed if and only if both switches are closed. We will use $P \wedge Q$ to denote this series network because it is analogous to the logic statement $p \wedge q$, which is true if and only if both p and q are true.

Figure 3.3 shows two switches P and Q connected in **parallel**. This parallel network is closed if either P or Q is closed. We will designate this parallel network by $P \vee Q$ because it is analogous to the logic statement $p \vee q$, which is true if p is true or if q is true.

Series and parallel networks can be combined to produce more complicated networks, as shown in Figure 3.4.

The network shown in Figure 3.4 is closed provided P or Q is closed or provided that both R and $\sim P$ are closed. Note that the switch $\sim P$ is closed if P is open, and $\sim P$ is open if P is closed. We use the symbolic statement $(P \vee Q) \vee (R \wedge \sim P)$ to represent this network.

If two switches are always open at the same time and always closed at the same time, then we will use the same letter to designate both switches.

EXCURSION EXERCISES

Write a symbolic statement to represent each of the networks in Excursion Exercises 1 to 6.

1.

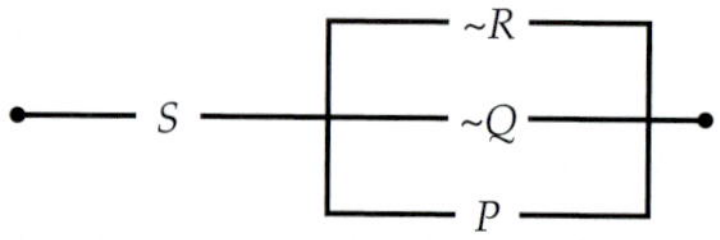

2.

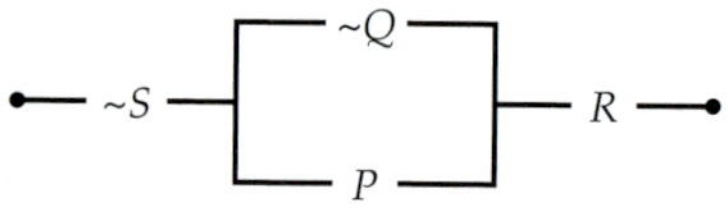

3.

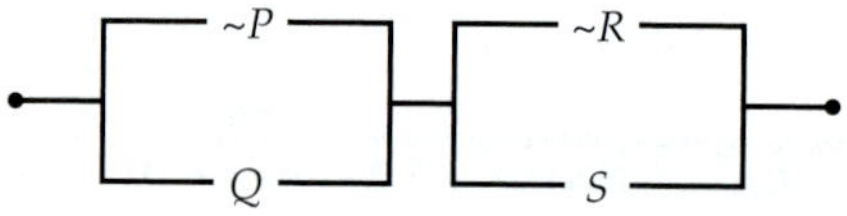

4.

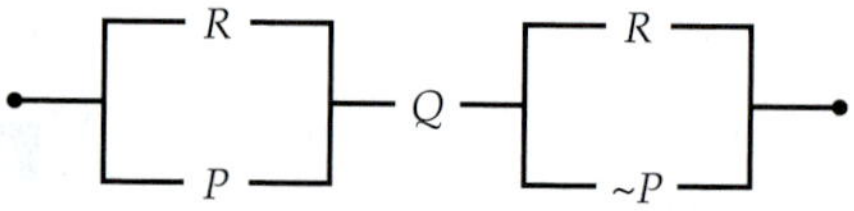

5.

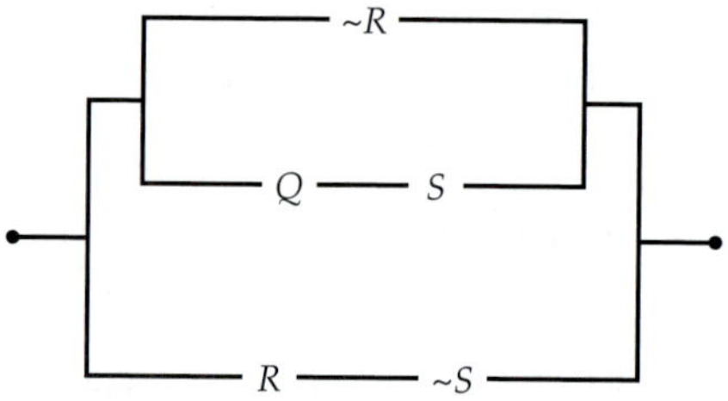

6.

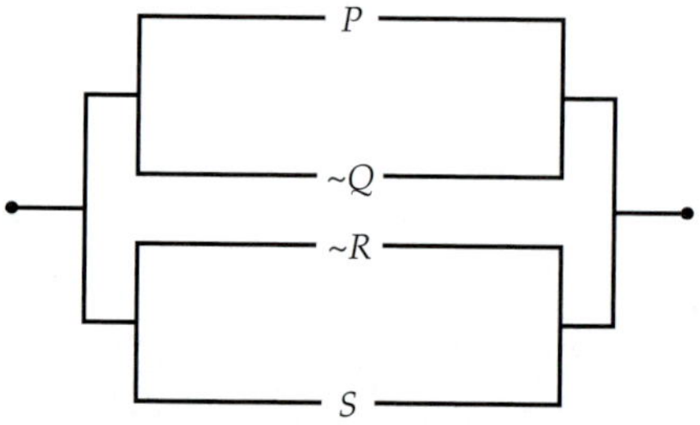

7. Which of the networks in Excursion Exercises 1 to 6 are closed networks, given that P is closed, Q is open, R is closed, and S is open?

8. Which of the networks in Excursion Exercises 1 to 6 are closed networks, given that P is open, Q is closed, R is closed, and S is closed?

In Excursion Exercises 9 to 14, draw a network to represent each statement.

9. $(\sim P \vee Q) \wedge (R \wedge P)$

10. $P \wedge [(Q \wedge \sim R) \vee R]$

11. $[\sim P \wedge Q \wedge R] \vee (P \wedge R)$

12. $(Q \vee R) \vee (S \vee \sim P)$

13. $[(\sim P \wedge R) \vee Q] \vee (\sim R)$

14. $(P \vee Q \vee R) \wedge S \wedge (\sim Q \vee R)$

Warning Circuits The circuits shown in Excursion Exercises 15 and 16 include a switching network, a warning light, and a battery. In each circuit the warning light will turn on only when the switching network is closed.

15. Consider the following circuit.

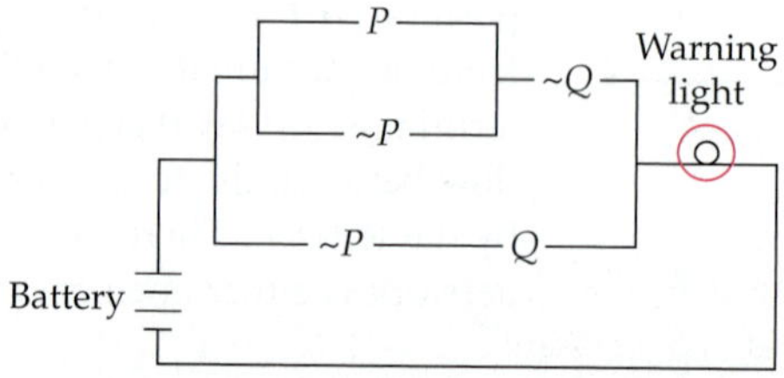

For each of the following conditions, determine whether the warning light will be on or off.

a. P is closed and Q is open. **b.** P is closed and Q is closed.

c. P is open and Q is closed. **d.** P is open and Q is open.

16. An engineer thinks that the following circuit can be used in place of the circuit shown in Excursion Exercise 15. Do you agree? Explain.

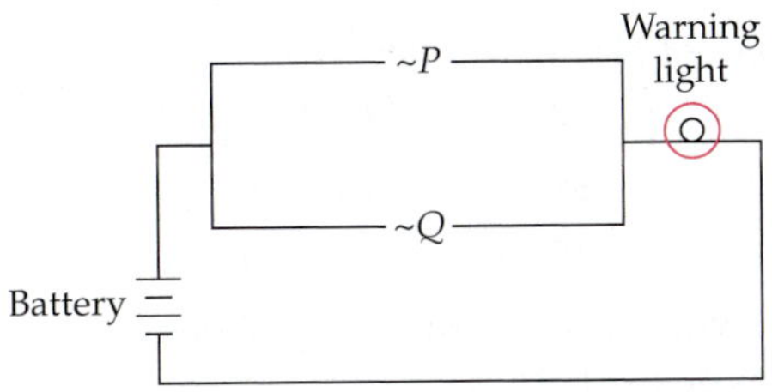

EXERCISE SET 3.1

■ In Exercises 1 to 10, determine whether each sentence is a statement.

1. *Star Wars: The Force Awakens* is the greatest movie of all time.

2. Harvey Mudd college is in Oregon.

3. The area code for Storm Lake, Iowa, is 512.

4. January 1, 2024, will be a Sunday.

5. Have a fun trip.

6. Do you like to read?

7. Mickey Mouse was the first animated character to receive a star on the Hollywood Walk of Fame.

8. Drew Brees is the starting quarterback of the Dallas Cowboys.

9. $x^2 = 25$

10. $x = x + 1$

■ In Exercises 11 to 14, determine the simple statements in each compound statement.

11. The principal will attend the class on Tuesday or Wednesday.

12. 5 is an odd number and 6 is an even number.

13. A triangle is an acute triangle if and only if it has three acute angles.

14. If this is Saturday, then tomorrow is Sunday.

■ In Exercises 15 to 18, write the negation of each statement.

15. The Giants lost the game.

16. The lunch was served at noon.

17. The game did not go into overtime.

18. The game was not shown on ABC.

■ In Exercises 19 to 26, write each sentence in symbolic form. Represent each simple statement in the sentence with the letter indicated in the parentheses. Also state whether the sentence is a conjunction, a disjunction, a negation, a conditional, or a biconditional.

19. If today is Wednesday (w), then tomorrow is Thursday (t).

20. I went to the post office (p) and the bookstore (s).

21. A triangle is an equilateral triangle (l) if and only if it is an equiangular triangle (a).

22. A number is an even number (e) if and only if it has a factor of 2 (t).

23. If it is a dog (d), it has fleas (f).

24. Polynomials that have exactly three terms (p) are called trinomials (t).

25. I will major in mathematics (m) or computer science (c).

26. All pentagons (p) have exactly five sides (s).

■ In Exercises 27 to 32, write each symbolic statement in words. Use p, q, r, s, t, and u as defined below.

p: The tour goes to Italy.

q: The tour goes to Spain.

r: We go to Venice.

s: We go to Florence.

t: The hotel fees are included.

u: The meals are not included.

27. $p \wedge \sim q$ **28.** $r \vee s$

29. $r \rightarrow \sim s$ **30.** $p \rightarrow r$

31. $s \leftrightarrow \sim r$ **32.** $\sim t \wedge u$

■ In Exercises 33 to 38, write each symbolic statement as an English sentence. Use p, q, r, s, and t as defined below.

p: Taylor Swift is a singer.

q: Taylor Swift is not a songwriter.

r: Taylor Swift is an actress.

s: Taylor Swift plays the piano.

t: Taylor Swift does not play the guitar.

33. $(p \vee r) \wedge q$

34. $\sim s \rightarrow (p \wedge \sim q)$

35. $p \rightarrow (q \wedge \sim r)$

36. $(s \wedge \sim q) \rightarrow t$

37. $(r \wedge p) \leftrightarrow q$

38. $t \leftrightarrow (\sim r \wedge \sim p)$

■ In Exercises 39 to 44, write each sentence in symbolic form. Use p, q, r, and s as defined below.

p: Stephen Curry is a football player.

q: Stephen Curry is a basketball player.

r: Stephen Curry is a rock star.

s: Stephen Curry plays for the Warriors.

39. Stephen Curry is a football player or a basketball player, and he is not a rock star.

40. Stephen Curry is a rock star, and he is not a basketball player or a football player.

41. If Stephen Curry is a basketball player and a rock star, then he is not a football player.

42. Stephen Curry is a basketball player, if and only if he is not a football player and he is not a rock star.

43. If Stephen Curry plays for the Warriors, then he is a basketball player and he is not a football player.

44. It is not true that, Stephen Curry is a football player or a rock star.

■ In Exercises 45 to 52, determine whether each statement is true or false.

45. $7 < 5$ or $3 > 1$.

46. $3 \leq 9$.

47. $(-1)^{50} = 1$ and $(-1)^{99} = -1$.

48. $7 \neq 3$ or 9 is a prime number.

49. $-5 \geq -11$.

50. $4.5 \leq 5.4$.

51. 2 is an odd number or 2 is an even number.

52. The square of any real number is a positive number.

■ In Exercises 53 to 60, write the negation of each quantified statement. Start each negation with "Some," "No," or "All."

53. Some lions are playful.

54. Some dogs are not friendly.

55. All classic movies were first produced in black and white.

56. Everybody enjoyed the dinner.

57. No even numbers are odd numbers.

58. Some actors are not rich.

59. All cars run on gasoline.

60. None of the students took my advice.

EXTENSIONS

Write Quotations in Symbolic Form In Exercises 61 to 64, translate each quotation into symbolic form. For each simple statement in the quotation, indicate what letter you used to represent the simple statement.

61. If you can count your money, you don't have a billion dollars. *J. Paul Getty*

62. If you aren't fired with enthusiasm, then you will be fired with enthusiasm. *Vince Lombardi*

63. If people concentrated on the really important things in life, there'd be a shortage of fishing poles. *Doug Larson*

64. If you're killed, you've lost a very important part of your life. *Brooke Shields*

Write Statements in Symbolic Form In Exercises 65 to 70, translate each mathematical statement into symbolic form. For each simple statement in the given statement, indicate what letter you used to represent the simple statement.

65. An angle is a right angle if and only if its measure is 90°.

66. Any angle inscribed in a semicircle is a right angle.

67. If two sides of a triangle are equal in length, then the angles opposite those sides are congruent.

68. The sum of the measures of the three angles of any triangle is 180°.

69. All squares are rectangles.

70. If the corresponding sides of two triangles are proportional, then the triangles are similar.

71. **Recreational Logic** The following diagram shows two cylindrical teapots. The yellow teapot has the same diameter as the green teapot, but it is one and one-half times as tall as the green teapot.

If the green teapot can hold a maximum of 6 cups of tea, then estimate the maximum number of cups of tea that the yellow teapot can hold. Explain your reasoning.

SECTION 3.2 Truth Tables, Equivalent Statements, and Tautologies

Truth Tables

In Section 3.1, we defined truth tables for the negation of a statement, the conjunction of two statements, and the disjunction of two statements. Each of these truth tables is shown below for review purposes.

Negation

p	$\sim p$
T	F
F	T

Conjunction

p	q	$p \wedge q$
T	T	T
T	F	F
F	T	F
F	F	F

Disjunction

p	q	$p \vee q$
T	T	T
T	F	T
F	T	T
F	F	F

p	q	**Given statement**
T	T	
T	F	
F	T	
F	F	

Standard truth table form for a given statement that involves only the two simple statements p and q

In this section, we consider methods of constructing truth tables for a statement that involves a combination of conjunctions, disjunctions, and/or negations. If the given statement involves only two simple statements, then start with a table with four rows (see the table at the left), called the **standard truth table form**, and proceed as shown in Example 1.

EXAMPLE 1 Truth Tables

a. Construct a table for $\sim(\sim p \vee q) \vee q$.

b. Use the truth table from part a to determine the truth value of $\sim(\sim p \vee q) \vee q$, given that p is true and q is false.

Solution

a. Start with the standard truth table form and then include a $\sim p$ column.

p	q	$\sim p$
T	T	F
T	F	F
F	T	T
F	F	T

Now use the truth values from the $\sim p$ and q columns to produce the truth values for $\sim p \vee q$, as shown in the rightmost column of the following table.

p	q	$\sim p$	$\sim p \vee q$
T	T	F	T
T	F	F	F
F	T	T	T
F	F	T	T

Negate the truth values in the $\sim p \vee q$ column to produce the following.

p	q	$\sim p$	$\sim p \vee q$	$\sim(\sim p \vee q)$
T	T	F	T	F
T	F	F	F	T
F	T	T	T	F
F	F	T	T	F

As our last step, we form the disjunction of $\sim(\sim p \vee q)$ with q and place the results in the rightmost column of the table. See the following table. The shaded column is the truth table for $\sim(\sim p \vee q) \vee q$.

p	q	$\sim p$	$\sim p \vee q$	$\sim(\sim p \vee q)$	$\sim(\sim p \vee q) \vee q$	
T	T	F	T	F	T	row 1
T	F	F	F	T	T	row 2
F	T	T	T	F	T	row 3
F	F	T	T	F	F	row 4

b. In row 2 of the above truth table, we see that when p is true, and q is false, the statement $\sim(\sim p \vee q) \vee q$ in the rightmost column is true.

CHECK YOUR PROGRESS 1

a. Construct a truth table for $(p \wedge \sim q) \vee (\sim p \vee q)$.

b. Use the truth table that you constructed in part a to determine the truth value of $(p \wedge \sim q) \vee (\sim p \vee q)$, given that p is true and q is false.

Solution See page S8.

p	q	r	Given statement
T	T	T	
T	T	F	
T	F	T	
T	F	F	
F	T	T	
F	T	F	
F	F	T	
F	F	F	

Standard truth table form for a statement that involves the three simple statements p, q, and r

Compound statements that involve exactly three simple statements require a standard truth table form with $2^3 = 8$ rows, as shown at the left.

EXAMPLE 2 Truth Tables

a. Construct a truth table for $(p \wedge q) \wedge (\sim r \vee q)$.

b. Use the truth table from part a to determine the truth value of $(p \wedge q) \wedge (\sim r \vee q)$, given that p is true, q is true, and r is false.

Solution

a. Using the procedures developed in Example 1, we can produce the following table. The shaded column is the truth table for $(p \wedge q) \wedge (\sim r \vee q)$. The numbers in the squares below the columns denote the order in which the columns were constructed. Each truth value in the column numbered 4 is the conjunction of the truth values to its left in the columns numbered 1 and 3.

p	q	r	$p \wedge q$	$\sim r$	$\sim r \vee q$	$(p \wedge q) \wedge (\sim r \vee q)$	
T	T	T	T	F	T	T	row 1
T	T	F	T	T	T	T	row 2
T	F	T	F	F	F	F	row 3
T	F	F	F	T	T	F	row 4
F	T	T	F	F	T	F	row 5
F	T	F	F	T	T	F	row 6
F	F	T	F	F	F	F	row 7
F	F	F	F	T	T	F	row 8
			1	2	3	4	

b. In row 2 of the above truth table, we see that $(p \wedge q) \wedge (\sim r \vee q)$ is true when p is true, q is true, and r is false.

CHECK YOUR PROGRESS 2

a. Construct a truth table for $(\sim p \wedge r) \vee (q \wedge \sim r)$.

b. Use the truth table that you constructed in part a to determine the truth value of $(\sim p \wedge r) \vee (q \wedge \sim r)$, given that p is false, q is true, and r is false.

Solution See page S8.

Alternative Method for the Construction of a Truth Table

In Example 3 we use an *alternative procedure* to construct a truth table.

TAKE NOTE

The alternative procedure for constructing a truth table, as described to the right, generally requires less writing, less time, and less effort than the truth table procedure that was used in Examples 1 and 2.

Alternative Procedure for Constructing a Truth Table

1. If the given statement has n simple statements, then start with a standard form that has 2^n rows. Enter the truth values for each simple statement and their negations.
2. Use the truth values for each simple statement and their negations to enter the truth values under each connective within a pair of grouping symbols, including parentheses ( ), brackets [ ], and braces { }. If some grouping symbols are nested inside other grouping symbols, then work from the inside out. In any situation in which grouping symbols have not been used, then we use the following **order of precedence agreement**.

   First assign truth values to negations from left to right, followed by conjunctions from left to right, followed by disjunctions from left to right, followed by conditionals from left to right, and finally by biconditionals from left to right.
3. The truth values that are entered into the column under the connective for which truth values are assigned *last*, form the truth table for the given statement.

EXAMPLE 3 **Use the Alternative Procedure to Construct a Truth Table**

Construct a truth table for $p \vee [\sim(p \wedge \sim q)]$.

Solution

Step 1: The given statement $p \vee [\sim(p \wedge \sim q)]$ has the two simple statements p and q. Thus we start with a standard form that has $2^2 = 4$ rows. In each column, enter the truth values for the statements p and $\sim q$, as shown in the columns numbered 1, 2, and 3 of the following table.

p	q	p	$\vee$	$[\sim$	$(p$	$\wedge$	$\sim q)]$
T	T	T			T		F
T	F	T			T		T
F	T	F			F		F
F	F	F			F		T
		1			2		3

Step 2: Use the truth values in columns 2 and 3 to determine the truth values to enter under the "and" connective. See column 4 in the following truth table. Now negate the truth values in column 4 to produce the truth values in column 5.

p	q	p	$\vee$	$[\sim$	$(p$	$\wedge$	$\sim q)]$
T	T	T		T	T	F	F
T	F	T		F	T	T	T
F	T	F		T	F	F	F
F	F	F		T	F	F	T
		1		5	2	4	3

Step 3: Use the truth values in the columns 1 and 5 to determine the truth values to enter under the "or" connective. See column 6 in the following table. Shaded column 6 is the truth table for $p \vee [\sim(p \wedge \sim q)]$.

p	q	p	$\vee$	$[\sim$	$(p$	$\wedge$	$\sim q)]$
T	T	T	T	T	T	F	F
T	F	T	T	F	T	T	T
F	T	F	T	T	F	F	F
F	F	F	T	T	F	F	T
		1	6	5	2	4	3

CHECK YOUR PROGRESS 3 Construct a truth table for $\sim p \vee (p \wedge q)$.

Solution See page S9.

MATH MATTERS A Three-Valued Logic

Jan Lukasiewicz (loo-kä-shā-vēch) (1878–1956) was the Polish minister of education in 1919 and served as a professor of mathematics at Warsaw University from 1920 to 1939. Most of Lukasiewicz's work was in the area of logic. He is well known for developing *Polish notation,* which was first used in logic to eliminate the need for parentheses in symbolic statements. Today *reverse Polish notation* is used by some computers and calculators to perform computations without the need to enter parentheses.

Jan Lukasiewicz was one of the first mathematicians to consider a three-valued logic in which a statement is true, false, or "somewhere between true and false." In his three-valued logic, Lukasiewicz classified the truth value of a statement as true (T), false (F), or maybe (M). The following table shows truth values for negation, conjunction, and disjunction in this three-valued logic.

p	q	Negation $\sim p$	Conjunction $p \wedge q$	Disjunction $p \vee q$
T	T	F	T	T
T	M	F	M	T
T	F	F	F	T
M	T	M	M	T
M	M	M	M	M
M	F	M	F	M
F	T	T	F	T
F	M	T	F	M
F	F	T	F	F

TAKE NOTE

In the remaining sections of this chapter, the ≡ symbol will often be used to denote that two statements are equivalent.

Equivalent Statements

Two statements are **equivalent** if they both have the same truth value for all possible truth values of their simple statements. Equivalent statements have identical truth values in the final columns of their truth tables. The notation $p \equiv q$ is used to indicate that the statements p and q are equivalent.

EXAMPLE 4 Verify That Two Statements Are Equivalent

Show that $\sim(p \vee \sim q)$ and $\sim p \wedge q$ are equivalent statements.

Solution

Construct two truth tables and compare the results. The truth tables below show that $\sim(p \vee \sim q)$ and $\sim p \wedge q$ have the same truth values for all possible truth values of their simple statements. Thus the statements are equivalent.

p	q	$\sim$	$(p$	$\vee$	$\sim q)$
T	T	F	T	T	F
T	F	F	T	T	T
F	T	T	F	F	F
F	F	F	F	T	T
		4	1	3	2

p	q	$\sim p$	$\wedge$	q
T	T	F	F	T
T	F	F	F	F
F	T	T	T	T
F	F	F	F	F
		1	3	2

identical truth values

Thus $\sim(p \vee \sim q) \equiv \sim p \wedge q$.

CHECK YOUR PROGRESS 4 Show that $p \vee (p \wedge \sim q)$ and p are equivalent.

Solution See page S9.

The truth tables in Table 3.5 show that $\sim(p \vee q)$ and $\sim p \wedge \sim q$ are equivalent statements. The truth tables in Table 3.6 show that $\sim(p \wedge q)$ and $\sim p \vee \sim q$ are equivalent statements.

TABLE 3.5

p	q	$\sim(p \vee q)$	$\sim p \wedge \sim q$
T	T	F	F
T	F	F	F
F	T	F	F
F	F	T	T

TABLE 3.6

p	q	$\sim(p \wedge q)$	$\sim p \vee \sim q$
T	T	F	F
T	F	T	T
F	T	T	T
F	F	T	T

These equivalences are known as **De Morgan's laws for statements**.

De Morgan's Laws for Statements

For any statements p and q,

$$\sim(p \vee q) \equiv \sim p \wedge \sim q$$
$$\sim(p \wedge q) \equiv \sim p \vee \sim q$$

De Morgan's laws can be used to restate certain English sentences in an equivalent form.

EXAMPLE 5 State an Equivalent Form

Use one of De Morgan's laws to restate the following sentence in an equivalent form.

It is not true that, I graduated or I got a job.

Solution

Let p represent the statement "I graduated." Let q represent the statement "I got a job." In symbolic form, the original sentence is $\sim(p \vee q)$. One of De Morgan's laws states that this is equivalent to $\sim p \wedge \sim q$. Thus a sentence that is equivalent to the original sentence is "I did not graduate and I did not get a job."

CHECK YOUR PROGRESS 5 Use one of De Morgan's laws to restate the following sentence in an equivalent form.

It is not true that, I am going to the dance and I am going to the game.

Solution See page S9.

Tautologies and Self-Contradictions

A **tautology** is a statement that is always true. A **self-contradiction** is a statement that is always false.

EXAMPLE 6 Verify Tautologies and Self-Contradictions

Show that $p \vee (\sim p \vee q)$ is a tautology.

Solution

Enter the truth values for each simple statement and its negation as shown in the columns numbered 1, 2, and 3. Use the truth values in columns 2 and 3 to determine the truth values to enter in column 4, under the "or" connective. Use the truth values in columns 1 and 4 to determine the truth values to enter in column 5, under the "or" connective.

p	q	p	$\vee$	$(\sim p$	$\vee$	$q)$
T	T	T	T	F	T	T
T	F	T	T	F	F	F
F	T	F	T	T	T	T
F	F	F	T	T	T	F
		1	5	2	4	3

Column 5 of the table shows that $p \vee (\sim p \vee q)$ is always true. Thus $p \vee (\sim p \vee q)$ is a tautology.

CHECK YOUR PROGRESS 6 Show that $p \wedge (\sim p \wedge q)$ is a self-contradiction.

Solution See page S9.

QUESTION Is the statement $x + 2 = 5$ a tautology or a self-contradiction?

ANSWER Neither. The statement is not true for all values of x, and it is not false for all values of x.

EXCURSION

Switching Networks—Part II

The Excursion in Section 3.1 introduced the application of symbolic logic to switching networks. This Excursion makes use of *closure tables* to determine under what conditions a switching network is open or closed. **In a closure table, we use a 1 to designate that a switch or switching network is closed and a 0 to indicate that it is open.**

Figure 3.5 shows a switching network that consists of the single switch P and a second network that consists of the single switch $\sim P$. The table below shows that the switching network $\sim P$ is open when P is closed and is closed when P is open.

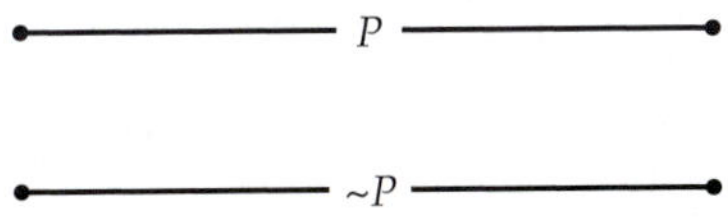

FIGURE 3.5

Negation Closure Table

P	$\sim P$
1	0
0	1

Figure 3.6 shows switches P and Q connected to form a series network. The table below shows that this series network is closed if and only if both P and Q are closed.

FIGURE 3.6 A series network

Series Network Closure Table

P	Q	$P \wedge Q$
1	1	1
1	0	0
0	1	0
0	0	0

Figure 3.7 shows switches P and Q connected to form a parallel network. The table below shows that this parallel network is closed if P is closed or if Q is closed.

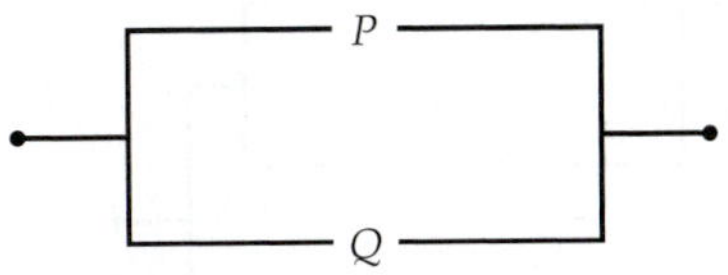

FIGURE 3.7 A parallel network

Parallel Network Closure Table

P	Q	$P \vee Q$
1	1	1
1	0	1
0	1	1
0	0	0

Now consider the network shown in Figure 3.8. To determine the required conditions under which the network is closed, we first write a symbolic statement that represents the network, and then we construct a closure table.

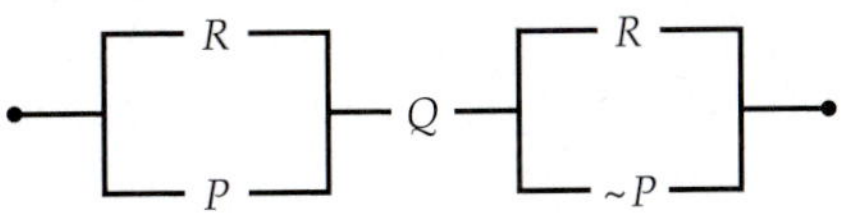

FIGURE 3.8

A symbolic statement that represents the network in Figure 3.8 is

$$[(R \vee P) \wedge Q] \wedge (R \vee \sim P)$$

The closure table for this network is shown below.

P	Q	R	$[(R$	$\vee$	$P)$	$\wedge$	$Q]$	$\wedge$	$(R$	$\vee$	$\sim P)$	
1	1	1	1	1	1	1	1	1	1	1	0	row 1
1	1	0	0	1	1	1	1	0	0	0	0	row 2
1	0	1	1	1	1	0	0	0	1	1	0	row 3
1	0	0	0	1	1	0	0	0	0	0	0	row 4
0	1	1	1	1	0	1	1	1	1	1	1	row 5
0	1	0	0	0	0	0	1	0	0	1	1	row 6
0	0	1	1	1	0	0	0	0	1	1	1	row 7
0	0	0	0	0	0	0	0	0	0	1	1	row 8
			1	6	2	7	3	9	4	8	5	

Rows 1 and 5 of the above table show that the network is closed whenever

- P is closed, Q is closed, and R is closed, or
- P is open, Q is closed, and R is closed.

Thus the switching network in Figure 3.8 is closed provided Q is closed and R is closed. The switching network is open under all other conditions.

EXCURSION EXERCISES

Construct a closure table for each of the following switching networks. Use the closure table to determine the required conditions for the network to be closed.

1.
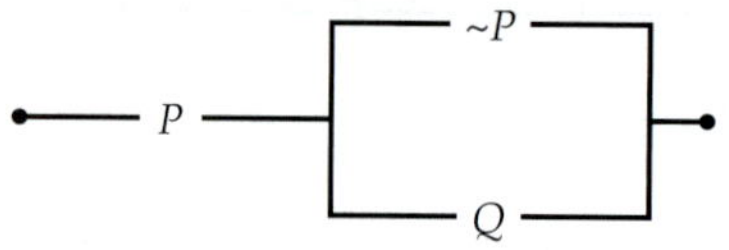

2.
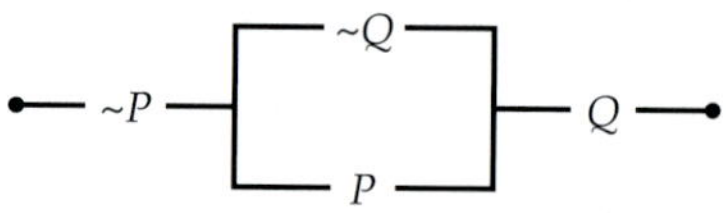

3.
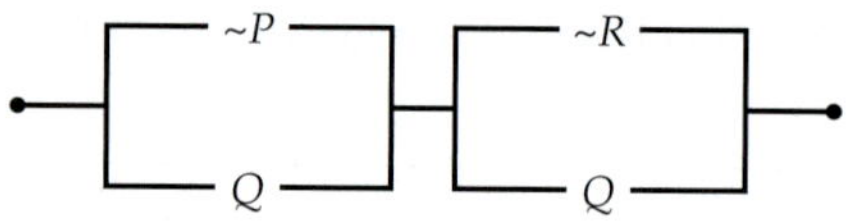

4.
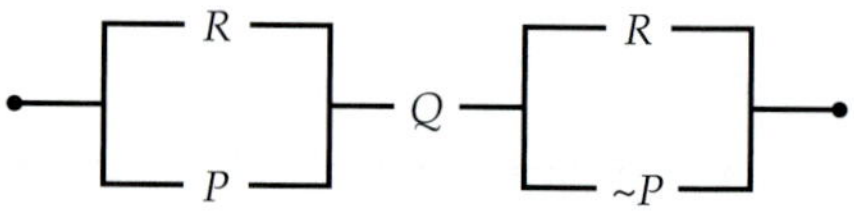

5.
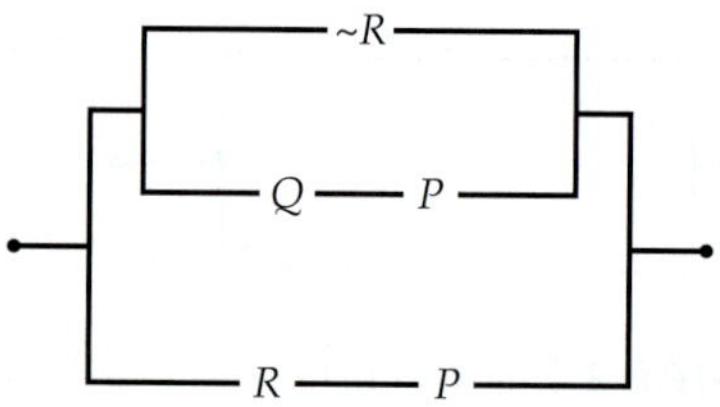

6.
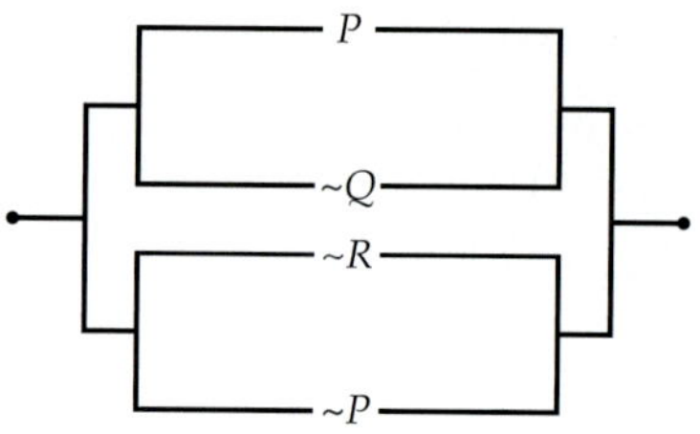

7. ***Warning Circuits***

a. The following circuit shows a switching network used in an automobile. The warning buzzer will buzz only when the switching network is closed. Construct a closure table for the switching network.

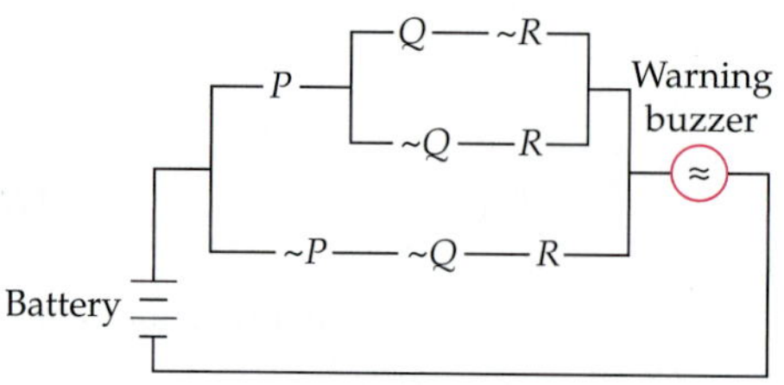

$$\{P \wedge [(Q \wedge \sim R) \vee (\sim Q \wedge R)]\} \vee [(\sim P \wedge \sim Q) \wedge R]$$

b. An engineer thinks that the following circuit can be used in place of the circuit in part a. Do you agree? *Hint:* Construct a closure table for the switching network and compare your closure table with the closure table in part a.

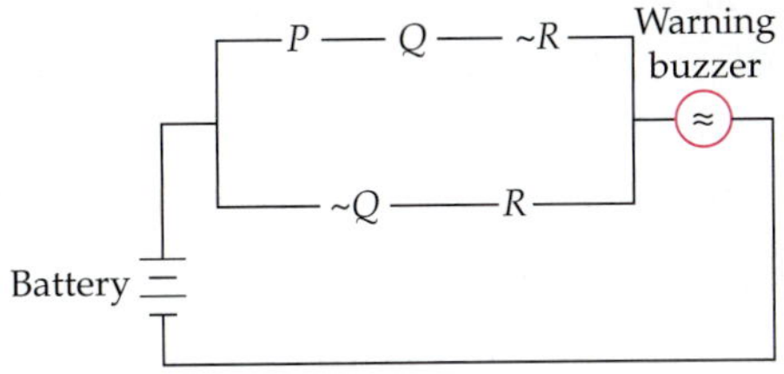

$[(P \wedge Q) \wedge {\sim}R] \vee ({\sim}Q \wedge R)$

EXERCISE SET 3.2

■ In Exercises 1 to 10, determine the truth value of the compound statement given that p is a false statement, q is a true statement, and r is a true statement.

1. $p \vee ({\sim}q \vee r)$

2. $r \wedge {\sim}(p \vee r)$

3. $(p \wedge q) \vee ({\sim}p \wedge {\sim}q)$

4. $(p \wedge q) \vee [({\sim}p \wedge {\sim}q) \vee q]$

5. $[{\sim}(p \wedge {\sim}q) \vee r] \wedge (p \wedge {\sim}r)$

6. $(p \wedge {\sim}q) \vee [(p \wedge {\sim}q) \vee r]$

7. $[(p \wedge {\sim}q) \vee {\sim}r] \wedge (q \wedge r)$

8. $({\sim}p \wedge q) \wedge [(p \wedge {\sim}q) \vee r]$

9. $[(p \wedge q) \wedge r] \vee [p \vee (q \wedge {\sim}r)]$

10. $\{[({\sim}p \wedge q) \wedge r] \vee [(p \wedge q) \wedge {\sim}r]\} \vee [p \wedge (q \wedge r)]$

11. **a.** Given that p is a false statement, what can be said about $p \wedge (q \vee r)$?

b. Explain why it is not necessary to know the truth values of q and r to determine the truth value of $p \wedge (q \vee r)$ in part a above.

12. **a.** Given that q is a true statement, what can be said about $q \vee {\sim}r$?

b. Explain why it is not necessary to know the truth value of r to determine the truth value of $q \vee {\sim}r$ in part a above.

■ In Exercises 13 to 28, construct a truth table for each compound statement.

13. ${\sim}p \vee q$

14. $(q \wedge {\sim}p) \vee {\sim}q$

15. $p \wedge {\sim}q$

16. $p \vee [{\sim}(p \wedge {\sim}q)]$

17. $(p \wedge {\sim}q) \vee [{\sim}(p \wedge q)]$

18. $(p \vee q) \wedge [{\sim}(p \vee {\sim}q)]$

19. ${\sim}(p \vee q) \wedge ({\sim}r \vee q)$

20. $[{\sim}(r \wedge {\sim}q)] \vee ({\sim}p \vee q)$

21. $(p \wedge {\sim}r) \vee [{\sim}q \vee (p \wedge r)]$

22. $[r \wedge ({\sim}p \vee q)] \wedge (r \vee {\sim}q)$

23. $[(p \wedge q) \vee (r \wedge {\sim}p)] \wedge (r \vee {\sim}q)$

24. $(p \wedge q) \wedge \{[{\sim}({\sim}p \vee r)] \wedge q\}$

25. $q \vee [{\sim}r \vee (p \wedge r)]$

26. $\{[{\sim}(p \vee {\sim}r)] \wedge {\sim}q\} \vee r$

27. $({\sim}q \wedge r) \vee [p \wedge (q \wedge {\sim}r)]$

28. ${\sim}[{\sim}p \wedge (q \wedge r)]$

■ In Exercises 29 to 36, use two truth tables to show that each of the statements are equivalent.

29. $p \vee (p \wedge r), p$

30. $q \wedge (q \vee r), q$

31. $p \wedge (q \vee r), (p \wedge q) \vee (p \wedge r)$

32. $p \vee (q \wedge r), (p \vee q) \wedge (p \vee r)$

33. $p \vee (q \wedge {\sim}p), p \vee q$

34. ${\sim}[p \vee (q \wedge r)], {\sim}p \wedge ({\sim}q \vee {\sim}r)$

35. $[(p \wedge q) \wedge r] \vee [p \wedge (q \wedge {\sim}r)], p \wedge q$

36. $[({\sim}p \wedge {\sim}q) \wedge r] \vee [(p \wedge q) \wedge {\sim}r] \vee [p \wedge (q \wedge r)],$
$(p \wedge q) \vee [({\sim}p \wedge {\sim}q) \wedge r]$

■ In Exercises 37 to 42, make use of one of De Morgan's laws to write the given statement in an equivalent form.

37. It is not true that, it rained or it snowed.

38. I did not pass the test and I did not complete the course.

39. She did not visit France and she did not visit Italy.

40. It is not true that, I bought a new car and I moved to Florida.

41. It is not true that, she received a promotion or that she received a raise.

42. It is not the case that, the students cut classes or took part in the demonstration.

■ In Exercises 43 to 48, use a truth table to determine whether the given statement is a tautology.

43. $p \vee \sim p$

44. $q \vee [\sim(q \wedge r) \wedge \sim q]$

45. $(p \vee q) \vee (\sim p \vee q)$

46. $(p \wedge q) \vee (\sim p \vee \sim q)$

47. $(\sim p \vee q) \vee (\sim q \vee r)$

48. $\sim[p \wedge (\sim p \vee q)] \vee q$

■ In Exercises 49 to 54, use a truth table to determine whether the given statement is a self-contradiction.

49. $\sim r \wedge r$

50. $\sim(p \vee \sim p)$

51. $p \wedge (\sim p \wedge q)$

52. $\sim[(p \vee q) \vee (\sim p \vee q)]$

53. $[p \wedge (\sim p \vee q)] \vee q$

54. $\sim[p \vee (\sim p \vee q)]$

55. Explain why the statement $7 \leq 8$ is a disjunction.

56. **a.** Why is the statement $5 \leq 7$ true?

b. Why is the statement $7 \leq 7$ true?

EXTENSIONS

57. How many rows are needed to construct a truth table for the statement $[p \wedge (q \vee \sim r)] \vee (s \wedge \sim t)$?

58. Explain why no truth table can have exactly 100 rows.

■ In Exercises 59 and 60, construct a truth table for the given compound statement. *Hint:* Use a table with 16 rows.

59. $[(p \wedge \sim q) \vee (q \wedge \sim r)] \wedge (r \vee \sim s)$

60. $s \wedge [\sim(\sim r \vee q) \vee \sim p]$

61. **Recreational Logic** A friend hands you the slip of paper shown below and challenges you to circle exactly four digits that have a sum of 19.

1 3 3 5 5 7 7 9 9 9

Explain how you can meet this challenge.

SECTION 3.3 The Conditional and the Biconditional

Conditional Statements

Paramount/Everett Collection

Matthew Broderick in the movie *Ferris Bueller's Day Off*

Life moves pretty fast. If you don't stop and look around once in a while, you could miss it.

The above quotation is from the movie *Ferris Bueller's Day Off.* The movie stars Matthew Broderick as Ferris, a high school senior who pretends to be ill so he can spend the day in Chicago, with his best friend Cameron and his girlfriend Sloane. The sentence, "If you don't stop and look around once in a while, you could miss it," is a *conditional statement.* **Conditional statements** can be written in *if p, then q* form or in *if p, q* form. For instance, all of the following are conditional statements.

If we order pizza, then we can have it delivered.

If you go to the movie, you will not be able to meet us for dinner.

If n is a prime number greater than 2, then n is an odd number.

In any conditional statement represented by "If p, then q" or by "If p, q," the p statement is called the **antecedent** and the q statement is called the **consequent**.

EXAMPLE 1 Identify the Antecedent and Consequent of a Conditional

Identify the antecedent and consequent in the following statements.

a. If our school was this nice, I would go there more than once a week.
—*The Basketball Diaries*

b. If you don't get in that plane, you'll regret it.
—*Casablanca*

c. If you strike me down, I shall become more powerful than you can possibly imagine.
—Obi-Wan Kenobi, Star Wars, Episode IV, *A New Hope*

Solution

a. *Antecedent:* our school was this nice
Consequent: I would go there more than once a week

b. *Antecedent:* you don't get in that plane
Consequent: you'll regret it

c. *Antecedent:* you strike me down
Consequent: I shall become more powerful than you can possibly imagine

CHECK YOUR PROGRESS 1 Identify the antecedent and consequent in each of the following conditional statements.

a. If I study for at least 6 hours, then I will get an A on the test.

b. If I get the job, I will buy a new car.

c. If you can dream it, you can do it.

Solution See page S9.

Arrow Notation

The conditional statement, "If p, then q," can be written using the **arrow notation** $p \rightarrow q$. The arrow notation $p \rightarrow q$ is read as "if p, then q" or as "p implies q."

The Truth Table for the Conditional $p \rightarrow q$

To determine the truth table for $p \rightarrow q$, consider the advertising slogan for a web authoring software product that states, "If you can use a word processor, you can create a webpage." This slogan is a conditional statement. The antecedent is p, "you can use a word processor," and the consequent is q, "you can create a webpage." Now consider the truth value of $p \rightarrow q$ for each of the following four possibilities.

TABLE 3.7

Antecedent p: you can use a word processor	Consequent q: you can create a webpage	$p \rightarrow q$	
T	T	?	row 1
T	F	?	row 2
F	T	?	row 3
F	F	?	row 4

Row 1: Antecedent T, Consequent T You can use a word processor, and you can create a webpage. In this case the truth value of the advertisement is true. To complete Table 3.7, we place a T in place of the question mark in row 1.

Row 2: Antecedent T, Consequent F You can use a word processor, but you cannot create a webpage. In this case the advertisement is false. We put an F in place of the question mark in row 2 of Table 3.7.

Row 3: Antecedent F, Consequent T You cannot use a word processor, but you can create a webpage. Because the advertisement does not make any statement about what you might or might not be able to do if you cannot use a word processor, we cannot state that the advertisement is false, and we are compelled to place a T in place of the question mark in row 3 of Table 3.7.

Row 4: Antecedent F, Consequent F You cannot use a word processor, and you cannot create a webpage. Once again we must consider the truth value in this case to be true because the advertisement does not make any statement about what you might or might not be able to do if you cannot use a word processor. We place a T in place of the question mark in row 4 of Table 3.7.

The truth table for the conditional $p \rightarrow q$ is given in Table 3.8.

TABLE 3.8
Truth Table for $p \rightarrow q$

p	q	$p \rightarrow q$
T	T	T
T	F	F
F	T	T
F	F	T

Truth Value of the Conditional $p \rightarrow q$

The conditional $p \rightarrow q$ is false if p is true and q is false. It is true in all other cases.

CALCULATOR NOTE

```
Program FACTOR

  0→dim (L1)
  Prompt N
  1→S: 2→F:0→E
  √(N)→M
  While F≤ M
  While fPart
  (N/F)=0
  E+1→E:N/F→N
  End
  If E>0
  Then
  F→L1(S)
  E→L1(S+1)
  S+2→S:0→E
  √(N)→M
  End
  If F=2
  Then
  3→F
  Else
  F+2→F
  End: End
  If N≠1
  Then
  N→L1(S)
  1→L1(S+1)
  End
  If S=1
  Then
  Disp N, "IS PRIME"
  Else
  Disp L1
```

EXAMPLE 2 Find the Truth Value of a Conditional

Determine the truth value of each of the following conditional statements.

a. If 2 is an integer, then 2 is a rational number.

b. If 3 is a negative number, then $5 > 7$.

c. If $5 > 3$, then $2 + 7 = 4$.

Solution

a. Because the consequent is true, this is a true statement.

b. Because the antecedent is false, this is a true statement.

c. Because the antecedent is true and the consequent is false, this is a false statement.

CHECK YOUR PROGRESS 2 Determine the truth value of each of the following conditional statements.

a. If $4 \geq 3$, then $2 + 5 = 6$.

b. If $5 > 9$, then $4 > 9$.

c. If Tuesday follows Monday, then April follows March.

Solution See page S9.

EXAMPLE 3 Construct a Truth Table for a Statement Involving a Conditional

Construct a truth table for $[p \wedge (q \vee {\sim}p)] \rightarrow {\sim}p$.

Solution

Enter the truth values for each simple statement and its negation as shown in columns 1, 2, 3, and 4. Use the truth values in columns 2 and 3 to determine the truth values to enter in column 5, under the "or" connective. Use the truth values in columns 1 and 5 to determine the truth values to enter in column 6 under the "and" connective. Use the truth values in columns 6 and 4 to determine the truth values to enter in column 7 under the "If . . . then" connective.

p	q	$[p$	$\wedge$	$(q$	$\vee$	$\sim p)]$	$\rightarrow$	$\sim p$
T	T	T	T	T	T	F	F	F
T	F	T	F	F	F	F	T	F
F	T	F	F	T	T	T	T	T
F	F	F	F	F	T	T	T	T
		1	6	2	5	3	7	4

CHECK YOUR PROGRESS 3 Construct a truth table for $[p \wedge (p \rightarrow q)] \rightarrow q$.

Solution See page S9.

MATH MATTERS Use Conditional Statements to Control a Calculator Program

Computer and calculator programs use conditional statements to control the flow of a program. For instance, the "If . . . Then" instruction in a TI-83 or TI-84 calculator program directs the calculator to execute a group of commands if a condition is true and to skip to the End statement if the condition is false. See the program steps below.

:If *condition*

:Then (skip to End if *condition* is false)

:*command* if *condition* is true

:End

:*commands that follow the End statement*

The TI-83/84 program FACTOR shown in the left margin of page 128, factors a natural number N into its prime factors. Note the use of the "If . . . Then" instructions highlighted in red.

An Equivalent Form of the Conditional

TABLE 3.9
Truth Table for $\sim p \vee q$

p	q	$\sim p \vee q$
T	T	T
T	F	F
F	T	T
F	F	T

The truth table for $\sim p \vee q$ is shown in Table 3.9. The truth values in this table are identical to the truth values in Table 3.8 on page 128. Hence, the conditional $p \rightarrow q$ is equivalent to the disjunction $\sim p \vee q$.

An Equivalent Form of the Conditional $p \rightarrow q$

$p \rightarrow q \equiv \sim p \vee q$

EXAMPLE 4 **Write a Conditional in Its Equivalent Disjunctive Form**

Write each of the following in its equivalent disjunctive form.

a. If I could play the guitar, I would join the band.

b. If Cam Newton cannot play, then his team will lose.

Solution

In each case we write the disjunction of the negation of the antecedent and the consequent.

a. I cannot play the guitar or I would join the band.

b. Cam Newton can play or his team will lose.

CHECK YOUR PROGRESS 4 Write each of the following in its equivalent disjunctive form.

a. If I don't move to Georgia, I will live in Houston.

b. If the number is divisible by 2, then the number is even.

Solution See page S9.

The Negation of the Conditional

Because $p \rightarrow q \equiv \sim p \vee q$, an equivalent form of $\sim(p \rightarrow q)$ is given by $\sim(\sim p \vee q)$, which, by one of De Morgan's laws, can be expressed as the conjunction $p \wedge \sim q$.

The Negation of $p \rightarrow q$

$\sim(p \rightarrow q) \equiv p \wedge \sim q$

EXAMPLE 5 Write the Negation of a Conditional Statement

Write the negation of each conditional statement.

a. If they pay me the money, I will sign the contract.

b. If the lines are parallel, then they do not intersect.

Solution

In each case, we write the conjunction of the antecedent and the negation of the consequent.

a. They paid me the money and I did not sign the contract.

b. The lines are parallel and they intersect.

CHECK YOUR PROGRESS 5 Write the negation of each conditional statement.

a. If I finish the report, I will go to the concert.

b. If the square of n is 25, then n is 5 or -5.

Solution See page S9.

The Biconditional

The statement $(p \rightarrow q) \wedge (q \rightarrow p)$ is called a **biconditional** and is denoted by $p \leftrightarrow q$, which is read as "p if and only if q."

The Biconditional $p \leftrightarrow q$

$p \leftrightarrow q \equiv [(p \rightarrow q) \wedge (q \rightarrow p)]$

EXAMPLE 6 Write Symbolic Biconditional Statements in Words

Let p, q, and r represent the following:

p: She will go on vacation.

q: She cannot take the train.

r: She cannot get a loan.

Write the following symbolic statements in words:

a. $p \leftrightarrow \sim q$ **b.** $\sim r \leftrightarrow \sim p$

Solution

a. She will go on vacation if and only if she can take the train.
b. She can get a loan if and only if she does not go on vacation.

CHECK YOUR PROGRESS 6 Use p, q, and r as defined in Example 6.

Write the following symbolic statements in words:

a. $p \leftrightarrow \sim r$ **b.** $\sim q \leftrightarrow \sim r$

Solution See page S9.

TABLE 3.10
Truth Table for $p \leftrightarrow q$

p	q	$p \leftrightarrow q$
T	T	T
T	F	F
F	T	F
F	F	T

Table 3.10 shows that $p \leftrightarrow q$ is true only when p and q have the same truth value.

EXAMPLE 7 Determine the Truth Value of a Biconditional

State whether each biconditional is true or false.

a. $x + 4 = 7$ if and only if $x = 3$. **b.** $x^2 = 36$ if and only if $x = 6$.

Solution

a. Both equations are true when $x = 3$, and both are false when $x \neq 3$. Both equations have the same truth value for any given value of x, so this is a true statement.

b. If $x = -6$, the first equation is true and the second equation is false. Thus this is a false statement.

CHECK YOUR PROGRESS 7 State whether each biconditional is true or false.

a. $x > 7$ if and only if $x > 6$. **b.** $x + 5 > 7$ if and only if $x > 2$.

Solution See page S9.

EXCURSION

Logic Gates

Modern digital computers use *gates* to process information. These gates are designed to receive two types of electronic impulses, which are generally represented as a 1 or a 0. Figure 3.9 shows a *NOT gate*. It is constructed so that a stream of impulses

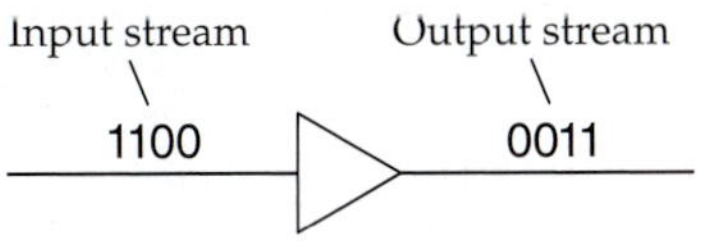

FIGURE 3.9 NOT gate

that enter the gate will exit the gate as a stream of impulses in which each 1 is converted to a 0 and each 0 is converted to a 1.

Note the similarity between the logical connective *not* and the logic gate NOT. The *not* connective converts the sequence of truth values T F to F T. The NOT gate converts the input stream 1 0 to 0 1. If the 1s are replaced with Ts and the 0s with Fs, then the NOT logic gate yields the same results as the *not* connective.

Many gates are designed so that two input streams are converted to one output stream. For instance, Figure 3.10 shows an *AND gate*. The AND gate is constructed so that a 1 is the output if and only if both input streams have a 1. In any other situation, a 0 is produced as the output.

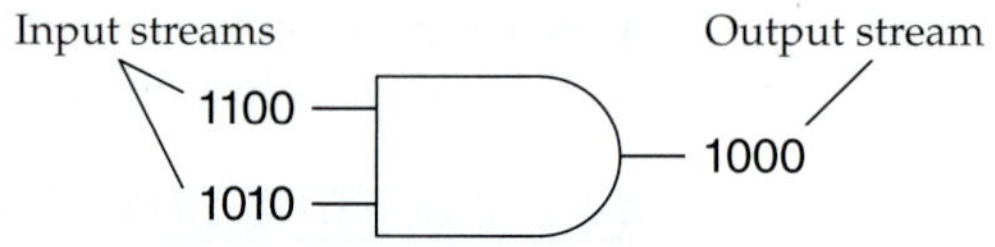

FIGURE 3.10 AND gate

Note the similarity between the logical connective *and* and the logic gate AND. The *and* connective combines the sequence of truth values T T F F with the truth values T F T F to produce T F F F. The AND gate combines the input stream 1 1 0 0 with the input stream 1 0 1 0 to produce 1 0 0 0. If the 1s are replaced with Ts and the 0s with Fs, then the AND logic gate yields the same result as the *and* connective.

The *OR gate* is constructed so that its output is a 0 if and only if both input streams have a 0. All other situations yield a 1 as the output. See Figure 3.11.

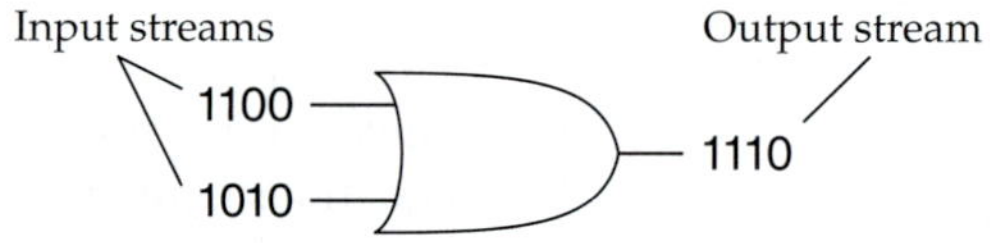

FIGURE 3.11 OR gate

Figure 3.12 shows a network that consists of a NOT gate and an AND gate.

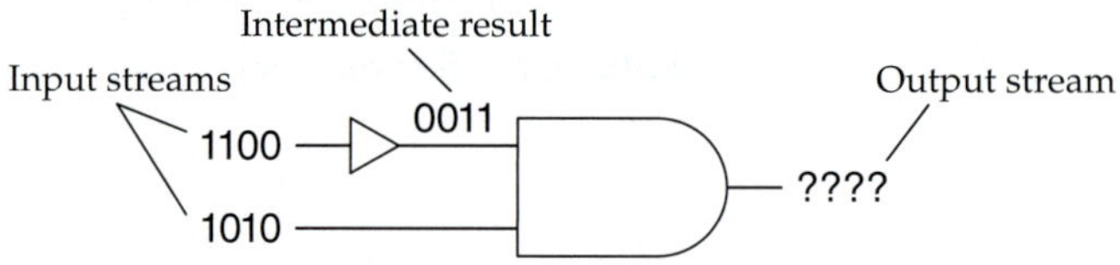

FIGURE 3.12

QUESTION What is the output stream for the network in Figure 3.12?

ANSWER 0010

EXCURSION EXERCISES

1. For each of the following, determine the output stream for the given input streams.

a.

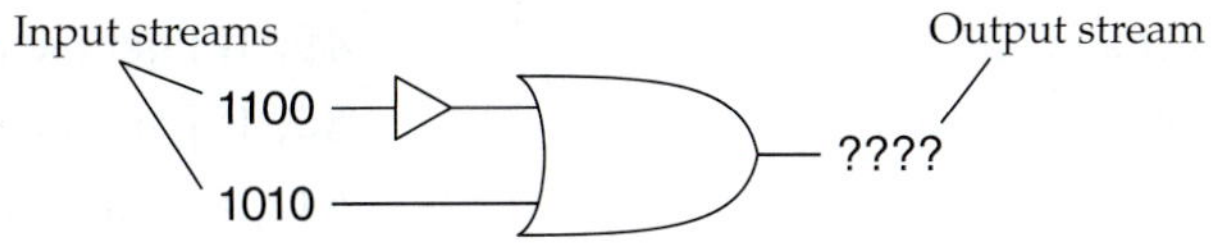

b.

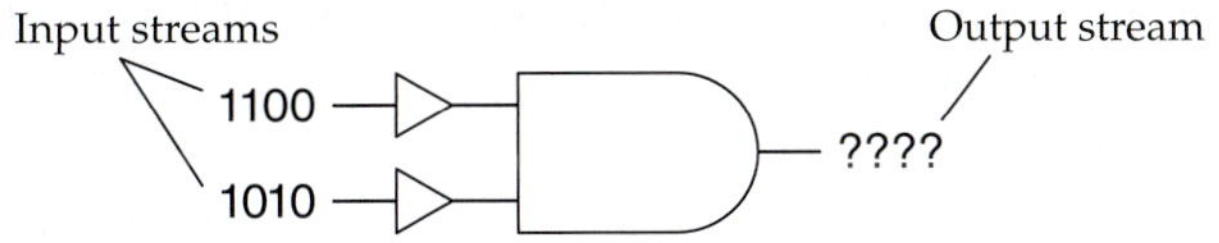

c.

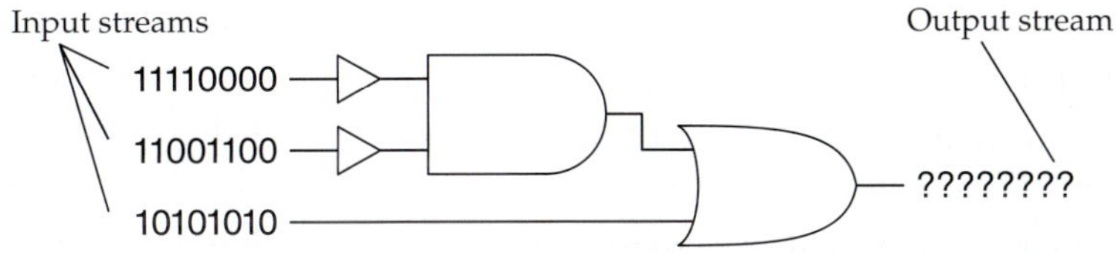

2. Construct a network using NOT, AND, and OR gates as needed that accepts the two input streams 1 1 0 0 and 1 0 1 0 and produces the output stream 0 1 1 1.

EXERCISE SET 3.3

■ In Exercises 1 to 6, identify the antecedent and the consequent of each conditional statement.

1. If I had the money, I would buy the painting.
2. If Shelly goes on the trip, she will not be able to take part in the graduation ceremony.
3. If they had a guard dog, then no one would trespass on their property.
4. If I don't get to school before 7:30, I won't be able to find a parking place.
5. If I change my major, I must reapply for admission.
6. If your blood type is type O−, then you are classified as a universal blood donor.

■ In Exercises 7 to 14, determine the truth value of the given statement.

7. If x is an even integer, then x^2 is an even integer.
8. If x is a prime number, then $x + 2$ is a prime number.
9. If all frogs can dance, then today is Monday.
10. If all cats are black, then I am a millionaire.
11. If $4 < 3$, then $7 = 8$.
12. If $x < 2$, then $x + 5 < 7$.
13. If $|x| - 6$, then $x = 6$.
14. If $\pi = 3$, then $2\pi = 6$.

■ In Exercises 15 to 24, construct a truth table for the given statement.

15. $(p \wedge \sim q) \rightarrow [\sim(p \wedge q)]$
16. $[(p \rightarrow q) \wedge p] \rightarrow p$
17. $[(p \rightarrow q) \wedge p] \rightarrow q$
18. $(\sim p \vee \sim q) \rightarrow \sim(p \wedge q)$
19. $[r \wedge (\sim p \vee q)] \rightarrow (r \vee \sim q)$
20. $[(p \rightarrow \sim r) \wedge q] \rightarrow \sim r$
21. $[(p \rightarrow q) \vee (r \wedge \sim p)] \rightarrow (r \vee \sim q)$
22. $\{p \wedge [(p \rightarrow q) \wedge (q \rightarrow r)]\} \rightarrow r$
23. $[\sim(p \rightarrow \sim r) \wedge \sim q] \rightarrow r$
24. $[p \wedge (r \rightarrow \sim q)] \rightarrow (r \vee q)$

■ In Exercises 25 to 30, write each conditional statement in its equivalent disjunctive form.

25. If she could sing, she would be perfect for the part.
26. If he does not get frustrated, he will be able to complete the job.
27. If x is an irrational number, then x is not a terminating decimal.
28. If Mr. Hyde had a brain, he would be dangerous.
29. If the fog does not lift, our flight will be cancelled.
30. If the Yankees win the pennant, Carol will be happy.

■ In Exercises 31 to 36, write the negation of each conditional statement in its equivalent conjunctive form.

31. If they offer me the contract, I will accept.
32. If I paint the house, I will get the money.
33. If pigs had wings, pigs could fly.
34. If we had a telescope, we could see that comet.
35. If she travels to Italy, she will visit her relatives.
36. If Paul could play better defense, he could be a professional basketball player.

■ In Exercises 37 to 46, state whether the given biconditional is true or false. Assume that x and y are real numbers.

37. $x^2 = 9$ if and only if $x = 3$.

38. x is a positive number if and only if $x > 0$.

39. $|x|$ is a positive number if and only if $x \neq 0$.

40. $|x + y| = x + y$ if and only if $x + y > 0$.

41. A number is a rational number if and only if the number can be written as a terminating decimal.

42. $0.\overline{3}$ is a rational number if and only if $\frac{1}{3}$ is a rational number.

43. $4 = 7$ if and only if $2 = 3$.

44. x is an even number if and only if x is not an odd number.

45. Triangle ABC is an equilateral triangle if and only if triangle ABC is an equiangular triangle.

46. Today is March 1 if and only if yesterday was February 28.

■ In Exercises 47 to 54, write each sentence in symbolic form. Use v, p, and t as defined below.

v: "I will take a vacation."

p: "I get the promotion."

t: "I will be transferred."

47. I will take a vacation if and only if I get the promotion.

48. If I do not get the promotion, then I will be transferred and I will not take a vacation.

49. If I get the promotion, I will take a vacation.

50. If I am not transferred, I will take a vacation.

51. If I am transferred, then I will not take a vacation.

52. If I will not take a vacation, then I will not be transferred and I get the promotion.

53. If I am not transferred and I get the promotion, then I will take a vacation.

54. If I get the promotion, then I will be transferred and I will take a vacation.

■ In Exercises 55 to 60, construct a truth table for each statement to determine if the statements are equivalent.

55. $p \rightarrow \sim r$, $r \vee \sim p$

56. $p \rightarrow q$, $q \rightarrow p$

57. $\sim p \rightarrow (p \vee r)$, r

58. $p \rightarrow q$, $\sim q \rightarrow \sim p$

59. $p \rightarrow (q \vee r)$, $(p \rightarrow q) \vee (p \rightarrow r)$

60. $\sim q \rightarrow p$, $p \vee q$

EXTENSIONS

The statement, "All squares are rectangles," can be written as "If a figure is a square, then it is a rectangle." In Exercises 61 to 64, write each statement given in "All *X*s are *Y*s" form in the form "If it is an *X*, then it is a *Y*."

61. All rational numbers are real numbers.

62. All whole numbers are integers.

63. All sauropods are herbivorous.

64. All paintings by Vincent van Gogh are valuable.

65. **Recreational Logic** The field of a new soccer stadium is watered by three individual sprinkler systems, as shown by the A, B, and C regions in the figure at the right. Each sprinkler system is controlled by exactly one of three on-off valves in an underground maintenance room, and each sprinkler system can be turned on without turning on the other two systems. Each of the valves is presently in the off position, and the field is dry. The valves have not been labeled, so you do not know which valve controls which sprinkler system. You want to correctly label the valves as A, B, and C. You also want to do it by making only one trip up to the field. You cannot see the field from the maintenance room, and no one is available to help you. What procedure can you use to determine how to correctly label the valves? Assume that all of the valves and all of the sprinkler systems are operating properly. Also assume that the sprinklers are either completely off or completely on. Explain your reasoning.

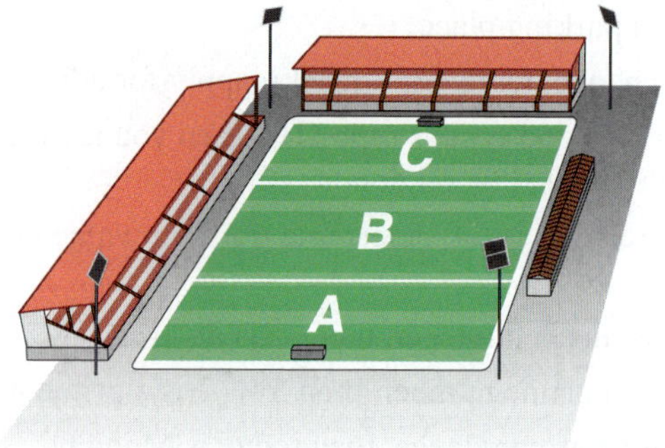

66. **A Factor Program** If you have access to a TI-83 or a TI-84 calculator, enter the program FACTOR on page 128 into the calculator and demonstrate the program to your classmates.

SECTION 3.4 The Conditional and Related Statements

Equivalent Forms of the Conditional

Every conditional statement can be stated in many equivalent forms. It is not even necessary to state the antecedent before the consequent. For instance, the conditional "If I live in Boston, then I must live in Massachusetts" can also be stated as

I must live in Massachusetts, if I live in Boston.

Table 3.11 lists some of the various forms that may be used to write a conditional statement.

TABLE 3.11
Common Forms of $p \rightarrow q$

Every conditional statement $p \rightarrow q$ can be written in the following equivalent forms.	
If p, then q.	Every p is a q.
If p, q.	q, if p.
p only if q.	q provided that p.
p implies q.	q is a necessary condition for p.
Not p or q.	p is a sufficient condition for q.

PPT

EXAMPLE 1 Write a Statement in an Equivalent Form

Write each of the following in "If p, then q" form.

a. The number is an even number provided that it is divisible by 2.

b. Today is Friday, only if yesterday was Thursday.

Solution

a. The statement, "The number is an even number provided that it is divisible by 2," is in "q provided that p" form. The antecedent is "it is divisible by 2," and the consequent is "the number is an even number." Thus its "If p, then q" form is

If it is divisible by 2, then the number is an even number.

b. The statement, "Today is Friday, only if yesterday was Thursday," is in "p only if q" form. The antecedent is "today is Friday." The consequent is "yesterday was Thursday." Its "If p, then q" form is

If today is Friday, then yesterday was Thursday.

CHECK YOUR PROGRESS 1 Write each of the following in "If p, then q" form.

a. Every square is a rectangle.

b. Being older than 30 is sufficient to show that I am at least 21.

Solution See page S9.

The Converse, the Inverse, and the Contrapositive

Every conditional statement has three related statements. They are called the *converse*, the *inverse*, and the *contrapositive*.

> **Statements Related to the Conditional Statement**
>
> The **converse** of $p \rightarrow q$ is $q \rightarrow p$.
>
> The **inverse** of $p \rightarrow q$ is $\sim p \rightarrow \sim q$.
>
> The **contrapositive** of $p \rightarrow q$ is $\sim q \rightarrow \sim p$.

The above definitions show the following:

- The converse of $p \rightarrow q$ is formed by interchanging the antecedent p with the consequent q.
- The inverse of $p \rightarrow q$ is formed by negating the antecedent p and negating the consequent q.
- The contrapositive of $p \rightarrow q$ is formed by negating both the antecedent p and the consequent q and interchanging these negated statements.

EXAMPLE 2 Write the Converse, Inverse, and Contrapositive of a Conditional

Write the converse, inverse, and contrapositive of

If I get the job, then I will rent the apartment.

Solution

Converse: If I rent the apartment, then I get the job.
Inverse: If I do not get the job, then I will not rent the apartment.
Contrapositive: If I do not rent the apartment, then I did not get the job.

CHECK YOUR PROGRESS 2 Write the converse, inverse, and contrapositive of

If we have a quiz today, then we will not have a quiz tomorrow.

Solution See page S10.

Table 3.12 shows that any conditional statement is equivalent to its contrapositive and that the converse of a conditional statement is equivalent to the inverse of the conditional statement.

TABLE 3.12
Truth Tables for Conditional and Related Statements

p	q	Conditional $p \rightarrow q$	Converse $q \rightarrow p$	Inverse $\sim p \rightarrow \sim q$	Contrapositive $\sim q \rightarrow \sim p$
T	T	T	T	T	T
T	F	F	T	T	F
F	T	T	F	F	T
F	F	T	T	T	T

$q \rightarrow p \equiv \sim p \rightarrow \sim q$

$p \rightarrow q \equiv \sim q \rightarrow \sim p$

EXAMPLE 3 Determine Whether Related Statements Are Equivalent

Determine whether the given statements are equivalent.

a. If a number ends with a 5, then the number is divisible by 5.
If a number is divisible by 5, then the number ends with a 5.

b. If two lines in a plane do not intersect, then the lines are parallel.
If two lines in a plane are not parallel, then the lines intersect.

Solution

a. The second statement is the converse of the first. The statements are not equivalent.

b. The second statement is the contrapositive of the first. The statements are equivalent.

CHECK YOUR PROGRESS 3 Determine whether the given statements are equivalent.

a. If $a = b$, then $a \cdot c = b \cdot c$.
If $a \neq b$, then $a \cdot c \neq b \cdot c$.

b. If I live in Nashville, then I live in Tennessee.
If I do not live in Tennessee, then I do not live in Nashville.

Solution See page S10.

In mathematics, it is often necessary to prove statements that are in "If p, then q" form. If a proof cannot be readily produced, mathematicians often try to prove the contrapositive "If $\sim q$, then $\sim p$." Because a conditional and its contrapositive are equivalent statements, a proof of either statement also establishes the proof of the other statement.

QUESTION A mathematician wishes to prove the following statement about the integer x.

Statement (I): If x^2 is an odd integer, then x is an odd integer.

If the mathematician is able to prove the statement, "If x is an even integer, then x^2 is an even integer," does this also prove statement (I)?

EXAMPLE 4 Use the Contrapositive to Determine a Truth Value

Write the contrapositive of each statement and use the contrapositive to determine whether the original statement is true or false.

a. If $a + b$ is not divisible by 5, then a and b are not both divisible by 5.

b. If x^3 is an odd integer, then x is an odd integer. (Assume x is an integer.)

c. If a geometric figure is not a rectangle, then it is not a square.

Solution

a. If a and b are both divisible by 5, then $a + b$ is divisible by 5. This is a true statement, so the original statement is also true.

b. If x is an even integer, then x^3 is an even integer. This is a true statement, so the original statement is also true.

c. If a geometric figure is a square, then it is a rectangle. This is a true statement, so the original statement is also true.

ANSWER Yes, because the second statement is the contrapositive of statement (I).

CHECK YOUR PROGRESS 4 Write the contrapositive of each statement and use the contrapositive to determine whether the original statement is true or false.

a. If $3 + x$ is an odd integer, then x is an even integer. (Assume x is an integer.)

b. If two triangles are not similar triangles, then they are not congruent triangles. *Note:* Similar triangles have the same shape. Congruent triangles have the same size and shape.

c. If today is not Wednesday, then tomorrow is not Thursday.

Solution See page S10.

MATH MATTERS Grace Hopper

Rear Admiral Grace Hopper

Grace Hopper (1906–1992) was a visionary in the field of computer programming. She was a mathematics professor at Vassar from 1931 to 1943, but she retired from teaching to start a career in the U.S. Navy at the age of 37.

The Navy assigned Hopper to the Bureau of Ordnance Computation at Harvard University. It was here that she was given the opportunity to program computers. It has often been reported that she was the third person to program the world's first large-scale digital computer. Grace Hopper had a passion for computers and computer programming. She wanted to develop a computer language that would be user-friendly and enable people to use computers in a more productive manner.

Grace Hopper had a long list of accomplishments. She designed some of the first computer compilers, she was one of the first to introduce English commands into computer languages, and she wrote the precursor to the computer language COBOL.

Grace Hopper retired from the Navy (for the first time) in 1966. In 1967 she was recalled to active duty and continued to serve in the Navy until 1986, at which time she was the nation's oldest active duty officer.

In 1951, the UNIVAC I computer that Grace Hopper was programming started to malfunction. The malfunction was caused by a moth that had become lodged in one of the computer's relays. Grace Hopper pasted the moth into the UNIVAC I logbook with a label that read, "computer bug." Since then computer programmers have used the word *bug* to indicate any problem associated with a computer program. Modern computers use logic gates instead of relays to process information, so actual bugs are not a problem; however, bugs such as the "Year 2000 bug" can cause serious problems.

EXCURSION

Sheffer's Stroke and the NAND Gate

In 1913, the logician Henry M. Sheffer created a connective that we now refer to as *Sheffer's stroke* (or *NAND*). This connective is often denoted by the symbol $|$. Table 3.13 shows that $p|q$ is equivalent to $\sim(p \wedge q)$. Sheffer's stroke $p|q$ is false when both p and q are true, and it is true in all other cases.

Any logic statement can be written using only Sheffer's stroke connectives. For instance, Table 3.14 shows that $p|p \equiv \sim p$ and $(p|p)|(q|q) \equiv p \vee q$.

TABLE 3.13
Sheffer's Stroke

p	q	$p \mid q$
T	T	F
T	F	T
F	T	T
F	F	T

Figure 3.13 shows a logic gate called a *NAND gate*. This gate models the Sheffer's stroke connective in that its output is 0 when both input streams are 1 and its output is 1 in all other cases.

TABLE 3.14

p	q	$p \mid p$	$(p \mid p) \mid (q \mid q)$
T	T	F	T
T	F	F	T
F	T	T	T
F	F	T	F

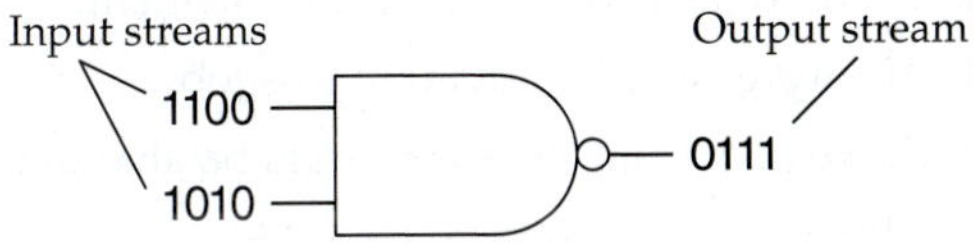

FIGURE 3.13 NAND gate

EXCURSION EXERCISES

1. a. Complete a truth table for $p \mid (q \mid q)$.

b. Use the results of Excursion Exercise 1a to determine an equivalent statement for $p \mid (q \mid q)$.

2. a. Complete a truth table for $(p \mid q) \mid (p \mid q)$.

b. Use the results of Excursion Exercise 2a to determine an equivalent statement for $(p \mid q) \mid (p \mid q)$.

3. a. Determine the output stream for the following network of NAND gates. *Note:* In a network of logic gates, a solid circle • is used to indicate a connection. A symbol such as ⮎ is used to indicate "no connection."

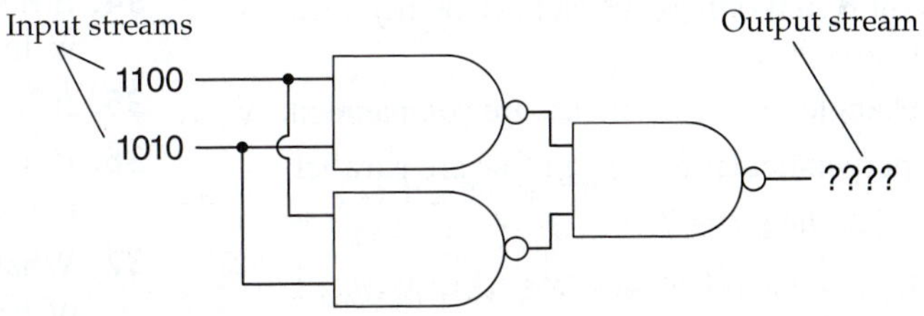

FIGURE 3.14

b. What logic gate is modeled by the network in Figure 3.14?

4. NAND gates are functionally complete in that any logic gate can be constructed using only NAND gates. Construct a network of NAND gates that would produce the same output stream as an OR gate.

EXERCISE SET 3.4

■ In Exercises 1 to 10, write each statement in "If p, then q" form.

1. We will be in good shape for the ski trip provided that we take the aerobics class.

2. We can get a dog only if we install a fence around the backyard.

3. Every odd prime number is greater than 2.

4. The triangle is a 30°-60°-90° triangle, if the length of the hypotenuse is twice the length of the shorter leg.

5. He can join the band, if he has the talent to play a keyboard.

6. Every theropod is carnivorous.

7. I will be able to prepare for the test only if I have the textbook.

8. I will be able to receive my credential provided that Education 147 is offered in the spring semester.

9. Being in excellent shape is a necessary condition for running the Boston marathon.

10. If it is an ankylosaur, it is quadrupedal.

■ In Exercises 11 to 24, write the **a.** converse, **b.** inverse, and **c.** contrapositive of the given statement.

11. If I were rich, I would quit this job.

12. If we had a car, then we would be able to take the class.

13. If she does not return soon, we will not be able to attend the party.

14. I will be in the talent show only if I can do the same comedy routine I did for the banquet.

15. Every parallelogram is a quadrilateral.

16. If you get the promotion, you will need to move to Denver.

17. I would be able to get current information about astronomy provided I had access to the Internet.

18. You need four-wheel drive to make the trip to Death Valley.

19. We will not have enough money for dinner, if we take a taxi.

20. If you are the president of the United States, then your age is at least 35.

21. She will visit Kauai only if she can extend her vacation for at least two days.

22. In a right triangle, the acute angles are complementary.

23. Two lines perpendicular to a given line are parallel.

24. If $x + 5 = 12$, then $x = 7$.

■ In Exercises 25 to 30, determine whether the given statements are equivalent.

25. If Kevin wins, we will celebrate.
If we celebrate, then Kevin will win.

26. If I save \$1000, I will go on the field trip.
If I go on the field trip, then I saved \$1000.

27. If she attends the meeting, she will make the sale.
If she does not make the sale, then she did not attend the meeting.

28. If you understand algebra, you can remember algebra.
If you do not understand algebra, you cannot remember algebra.

29. If $a > b$, then $ac > bc$.
If $a \leq b$, then $ac \leq bc$.

30. If $a < b$, then $\frac{1}{a} > \frac{1}{b}$.

If $\frac{1}{a} \leq \frac{1}{b}$, then $a \geq b$.
(Assume $a \neq 0$ and $b \neq 0$.)

■ In Exercises 31 to 36, write the contrapositive of the statement and use the contrapositive to determine whether the given statement is true or false.

31. If $3x - 7 = 11$, then $x \neq 7$.

32. If $x \neq 3$, then $5x + 7 \neq 22$.

33. If $a \neq 3$, then $|a| \neq 3$.

34. If $a + b$ is divisible by 3, then a is divisible by 3 and b is divisible by 3.

35. If $\sqrt{a + b} \neq 5$, then $a + b \neq 25$.

36. If x^2 is an even integer, then x is an even integer. (Assume x is an integer.)

37. What is the converse of the inverse of the contrapositive of $p \rightarrow q$?

38. What is the inverse of the converse of the contrapositive of $p \rightarrow q$?

EXTENSIONS

39. Give an example of a true conditional statement whose
a. converse is true. **b.** converse is false.

40. Give an example of a true conditional statement whose
a. inverse is true. **b.** inverse is false.

■ In Exercises 41 to 44, determine the original statement if the given statement is related to the original in the manner indicated.

41. *Converse:* If you can do it, you can dream it.

42. *Inverse:* If I did not have a dime, I would not spend it.

43. *Contrapositive:* If I were a singer, I would not be a dancer.

44. *Negation:* Pigs have wings and pigs cannot fly.

45. Explain why it is not possible to find an example of a true conditional statement whose contrapositive is false.

46. If a conditional statement is false, must its converse be true? Explain.

47. A Puzzle Lewis Carroll (Charles Dodgson) wrote many puzzles, many of which he recorded in his diaries. Solve the following puzzle, which appears in one of his diaries.

The Granger Collection

The Dodo says that the Hatter tells lies.

The Hatter says that the March Hare tells lies.

The March Hare says that both the Dodo and the Hatter tell lies.

Who is telling the truth?[1]

Hint: Consider the three different cases in which only one of the characters is telling the truth. In only one of these cases can all three of the statements be true.

48. Recreational Logic Consider a checkerboard with two red squares on opposite corners removed, as shown below. Determine whether it is possible to completely cover the checkerboard with 31 dominoes if each domino is placed horizontally or vertically and each domino covers exactly two squares. If it is possible, show how to do it. If it is not possible, explain why it cannot be done.

[1] This puzzle is from *Lewis Carroll's Games and Puzzles*, compiled and edited by Edward Wakeling. New York: Dover Publications, Inc., copyright 1992, p. 11, puzzle 9, "Who's Telling the Truth?"

SECTION 3.5 Symbolic Arguments

Arguments

In this section we consider methods of analyzing arguments to determine whether they are *valid* or *invalid*. For instance, consider the following argument.

> If Aristotle was human, then Aristotle was mortal. Aristotle was human. Therefore, Aristotle was mortal.

To determine whether the above argument is a valid argument, we must first define the terms *argument* and *valid argument*.

HISTORICAL NOTE

Panos Karapanagiotis/Shutterstock.com

Aristotle (ăr′ĭ-stŏt′l) (384–322 B.C.) was an ancient Greek philosopher who studied under Plato. He wrote about many subjects, including logic, biology, politics, astronomy, metaphysics, and ethics. His ideas about logic and the reasoning process have had a major impact on mathematics and philosophy.

An Argument and a Valid Argument

An **argument** consists of a set of statements called **premises** and another statement called the **conclusion**. An argument is **valid** if the conclusion is true whenever all the premises are assumed to be true. An argument is **invalid** if it is not a valid argument.

In the argument about Aristotle, the two premises and the conclusion are shown below. It is customary to place a horizontal line between the premises and the conclusion.

First Premise:	If Aristotle was human, then Aristotle was mortal.
Second Premise:	Aristotle was human.
Conclusion:	Therefore, Aristotle was mortal.

Arguments can be written in **symbolic form**. For instance, if we let h represent the statement "Aristotle was human" and m represent the statement "Aristotle was mortal," then the argument can be expressed as

$$\begin{array}{l} h \rightarrow m \\ \underline{h \qquad} \\ \therefore m \end{array}$$

The three dots $\therefore$ are a symbol for "therefore."

EXAMPLE 1 Write an Argument in Symbolic Form

Write the following argument in symbolic form.

The fish is fresh or I will not order it. The fish is fresh. Therefore I will order it.

Solution

Let f represent the statement "The fish is fresh." Let o represent the statement "I will order it." The symbolic form of the argument is

$$\begin{array}{l} f \vee \sim o \\ \underline{f \qquad} \\ \therefore o \end{array}$$

CHECK YOUR PROGRESS 1 Write the following argument in symbolic form.

If she doesn't get on the plane, she will regret it. She does not regret it. Therefore, she got on the plane.

Let p represent "She got on the plane." Let r represent "She will regret it."

Solution See page S10.

Arguments and Truth Tables

The following truth table procedure can be used to determine whether an argument is valid or invalid.

Truth Table Procedure to Determine the Validity of an Argument

1. Write the argument in symbolic form.
2. Construct a truth table that shows the truth value of each premise and the truth value of the conclusion for all combinations of truth values of the simple statements.
3. If the conclusion is true in every row of the truth table in which all the premises are true, the argument is valid. If the conclusion is false in any row in which all of the premises are true, the argument is invalid.

We will now use the above truth table procedure to determine the validity of the argument about Aristotle.

1. Once again we let h represent the statement "Aristotle was human" and m represent the statement "Aristotle was mortal." In symbolic form the argument is

$$\begin{array}{ll} h \rightarrow m & \text{First premise} \\ \underline{h \qquad} & \text{Second premise} \\ \therefore m & \text{Conclusion} \end{array}$$

2. Construct a truth table as shown below.

h	m	First premise $h \rightarrow m$	Second premise h	Conclusion m	
T	T	T	T	T	row 1
~~T~~	~~F~~	~~F~~	~~T~~	~~F~~	~~row 2~~
~~F~~	~~T~~	~~T~~	~~F~~	~~T~~	~~row 3~~
~~F~~	~~F~~	~~T~~	~~F~~	~~F~~	~~row 4~~

3. Row 1 is the only row in which all the premises are true, so it is the only row that we examine. Because the conclusion is true in row 1, the argument is valid.

In Example 2, we use the truth table method to determine the validity of a more complicated argument.

EXAMPLE 2 Determine the Validity of an Argument

Determine whether the following argument is valid or invalid.

If it rains, then the game will not be played. It is not raining. Therefore, the game will be played.

Solution

If we let r represent "it rains" and g represent "the game will be played," then the symbolic form is

$$\begin{array}{l} r \rightarrow \sim g \\ \underline{\sim r} \\ \therefore g \end{array}$$

The truth table for this argument follows.

r	g	First premise $r \rightarrow \sim g$	Second premise $\sim r$	Conclusion g	
~~T~~	~~T~~	~~F~~	~~F~~	~~T~~	~~row 1~~
~~T~~	~~F~~	~~T~~	~~F~~	~~F~~	~~row 2~~
F	T	T	T	T	row 3
F	F	T	T	F	row 4

Because the conclusion in row 4 is false and the premises are both true, the argument is invalid.

QUESTION Why do we need to examine only rows 3 and 4?

ANSWER Rows 3 and 4 are the only rows in which all of the premises are true.

CHECK YOUR PROGRESS 2 Determine the validity of the following argument.

If the stock market rises, then the bond market will fall.
The bond market did not fall.
∴ The stock market did not rise.

Let *r* represent "The stock market rises." Let *f* represent "The bond market will fall."

Solution See page S10.

The argument in Example 3 involves three statements. Thus we use a truth table with $2^3 = 8$ rows to determine the validity of the argument.

EXAMPLE 3 Determine the Validity of an Argument

Determine whether the following argument is valid or invalid.

If I am going to run the marathon, then I will buy new shoes.
If I buy new shoes, then I will not buy a television.
∴ If I buy a television, I will not run the marathon.

Solution

Label each simple statement.

m: I am going to run the marathon.
s: I will buy new shoes.
t: I will buy a television.

The symbolic form of the argument is

$$m \rightarrow s$$
$$s \rightarrow \sim t$$
$$\therefore t \rightarrow \sim m$$

The truth table for this argument follows.

m	*s*	*t*	First premise $m \rightarrow s$	Second premise $s \rightarrow \sim t$	Conclusion $t \rightarrow \sim m$	
~~T~~	~~T~~	~~T~~	~~T~~	~~F~~	~~F~~	~~row 1~~
T	T	F	T	T	T	row 2
~~T~~	~~F~~	~~T~~	~~F~~	~~T~~	~~F~~	~~row 3~~
~~T~~	~~F~~	~~F~~	~~F~~	~~T~~	~~T~~	~~row 4~~
~~F~~	~~T~~	~~T~~	~~T~~	~~F~~	~~T~~	~~row 5~~
F	T	F	T	T	T	row 6
F	F	T	T	T	T	row 7
F	F	F	T	T	T	row 8

The only rows in which both premises are true are rows 2, 6, 7, and 8. Because the conclusion is true in each of these rows, the argument is valid.

CHECK YOUR PROGRESS 3 Determine whether the following argument is valid or invalid.

If I arrive before 8 A.M., then I will make the flight.
If I make the flight, then I will give the presentation.
∴ If I arrive before 8 A.M., then I will give the presentation.

Let a represent "I arrive before 8 A.M." Let f represent "I will make the flight." Let p represent "I will give the presentation."

Solution See page S10.

Standard Forms

Arguments can be shown to be valid if they have the same symbolic form as an argument that is known to be valid. For instance, we have shown that the argument

$$\begin{array}{l} h \rightarrow m \\ \underline{h\quad\quad} \\ \therefore m \end{array}$$

is valid. This symbolic form is known as **direct reasoning**. All arguments that have this symbolic form are valid.

Table 3.15 shows four symbolic forms and the name used to identify each form. Any argument that has a symbolic form identical to one of these symbolic forms is a valid argument.

TAKE NOTE

In logic, the ability to identify standard forms of arguments is an important skill. If an argument has one of the standard forms in Table 3.15, then it is a valid argument. If an argument has one of the standard forms in Table 3.16, then it is an invalid argument. The standard forms can be thought of as laws of logic. Concerning the laws of logic, the logician Gottlob Frege (frā′gə) (1848–1925) stated, "The laws of logic are not like the laws of nature. They ... are laws of the laws of nature."

TABLE 3.15
Standard Forms of Four Valid Arguments

Direct reasoning	Contrapositive reasoning	Transitive reasoning	Disjunctive reasoning	
$p \rightarrow q$	$p \rightarrow q$	$p \rightarrow q$	$p \vee q$	$p \vee q$
p	$\sim q$	$q \rightarrow r$	$\sim p$	$\sim q$
$\therefore q$	$\therefore \sim p$	$\therefore p \rightarrow r$	$\therefore q$	$\therefore p$

PPT

Transitive reasoning can be extended to include more than two conditional premises. For instance, if the conditional premises of an argument are $p \rightarrow q$, $q \rightarrow r$, and $r \rightarrow s$, then a valid conclusion for the argument is $p \rightarrow s$.

In Example 4 we use standard forms to determine a valid conclusion for an argument.

EXAMPLE 4 Determine a Valid Conclusion for an Argument

Use a standard form from Table 3.15 to determine a valid conclusion for each argument.

a. If Kim is a lawyer (p), then she will be able to help us (q).
Kim is not able to help us ($\sim q$).
∴ ?

b. If they had a good time (g), they will return (r).
If they return (r), we will make more money (m).
∴ ?

Solution

a. The symbolic form of the premises is:

$p \rightarrow q$

$\sim q$

This matches the standard form known as *contrapositive reasoning*. Thus a valid conclusion is $\sim p$: "Kim is not a lawyer."

b. The symbolic form of the premises is:

$g \rightarrow r$

$r \rightarrow m$

This matches the standard form known as *transitive reasoning*. Thus a valid conclusion is $g \rightarrow m$: "If they had a good time, then we will make more money."

CHECK YOUR PROGRESS 4 Use a standard form from Table 3.15 to determine a valid conclusion for each argument.

a. If you can dream it (p), you can do it (q).
You can dream it (p).
$\therefore$?

b. I bought a car (c) or I bought a motorcycle (m).
I did not buy a car ($\sim c$).
$\therefore$?

Solution See page S10.

Table 3.16 shows two symbolic forms associated with invalid arguments. Any argument that has one of these symbolic forms is invalid.

TABLE 3.16
Standard Forms of Two Invalid Arguments

Fallacy of the converse	Fallacy of the inverse
$p \rightarrow q$	$p \rightarrow q$
q	$\sim p$
$\therefore p$	$\therefore \sim q$

PPT

EXAMPLE 5 **Use a Standard Form to Determine the Validity of an Argument**

Use a standard form from Table 3.15 or Table 3.16 to determine whether the following arguments are valid or invalid.

a. The program is interesting or I will watch the basketball game.
The program is not interesting.
$\therefore$ I will watch the basketball game.

b. If I have a cold, then I find it difficult to sleep.
I find it difficult to sleep.
$\therefore$ I have a cold.

Solution

a. Label the simple statements.

i: The program is interesting.

w: I will watch the basketball game.

In symbolic form the argument is

$$\begin{array}{l} i \vee w \\ \underline{\sim i \quad} \\ \therefore w \end{array}$$

This symbolic form matches one of the standard forms known as *disjunctive reasoning*. Thus the argument is valid.

b. Label the simple statements.

c: I have a cold.

s: I find it difficult to sleep.

In symbolic form the argument is

$$\begin{array}{l} c \rightarrow s \\ \underline{s \quad\quad} \\ \therefore c \end{array}$$

This symbolic form matches the standard form known as the *fallacy of the converse*. Thus the argument is invalid. Having a cold is not the only cause of sleep difficulties. For instance, you may find it difficult to sleep because you are thinking about logic.

CHECK YOUR PROGRESS 5 Use a standard form from Table 3.15 or Table 3.16 to determine whether the following arguments are valid or invalid.

a. If I go to Florida for spring break, then I will not study.
I did not go to Florida for spring break.
∴ I studied.

b. If you helped solve the crime, then you should be rewarded.
You helped solve the crime.
∴ You should be rewarded.

Solution See page S10.

POINT OF INTEREST

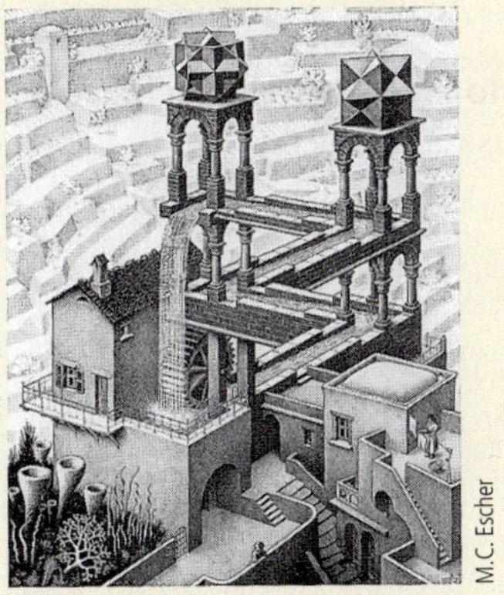

Waterfall by M.C. Escher

M.C. Escher (1898–1972) created many works of art that defy logic. In this lithograph, the water completes a full cycle even though the water is always traveling downward.

Consider an argument with the following symbolic form.

$q \rightarrow r$	Premise 1
$r \rightarrow s$	Premise 2
$\sim t \rightarrow \sim s$	Premise 3
$\underline{q \quad\quad}$	Premise 4
$\therefore t$	

To determine whether the argument is valid or invalid using a truth table would require a table with $2^4 = 16$ rows. It would be time consuming to construct such a table and, with the large number of truth values to be determined, we might make an error. Thus we consider a different approach that makes use of a sequence of valid arguments to arrive at a conclusion.

$q \rightarrow r$	Premise 1
$r \rightarrow s$	Premise 2
$\therefore q \rightarrow s$	Transitive reasoning
$q \rightarrow s$	The previous conclusion
$s \rightarrow t$	Premise 3 expressed in an equivalent form
$\therefore q \rightarrow t$	Transitive reasoning
$q \rightarrow t$	The previous conclusion
q	Premise 4
$\therefore t$	Direct reasoning

This sequence of valid arguments shows that t is a valid conclusion for the original argument. Thus the original argument is a valid argument.

EXAMPLE 6 Determine the Validity of an Argument

Determine whether the following argument is valid.

> If the movie was directed by Steven Spielberg (s), then I want to see it (w). The movie's production costs must exceed \$50 million (c) or I do not want to see it. The movie's production costs were less than \$50 million. Therefore, the movie was not directed by Steven Spielberg.

Solution

In symbolic form the argument is

$s \rightarrow w$	Premise 1
$c \vee \sim w$	Premise 2
$\sim c$	Premise 3
$\therefore \sim s$	Conclusion

Premise 2 can be written as $\sim w \vee c$, which is equivalent to $w \rightarrow c$. Applying transitive reasoning to Premise 1 and this equivalent form of Premise 2 produces

$s \rightarrow w$	Premise 1
$w \rightarrow c$	Equivalent form of Premise 2
$\therefore s \rightarrow c$	Transitive reasoning

Combining the conclusion $s \rightarrow c$ with Premise 3 gives us

$s \rightarrow c$	Conclusion from previous argument
$\sim c$	Premise 3
$\therefore \sim s$	Contrapositive reasoning

This sequence of valid arguments has produced the desired conclusion, $\sim s$. Thus the original argument is valid.

CHECK YOUR PROGRESS 6 Determine whether the following argument is valid.

> I start to fall asleep if I read a math book. I drink soda whenever I start to fall asleep. If I drink a soda, then I must eat a candy bar. Therefore, I eat a candy bar whenever I read a math book.

Hint: p whenever q is equivalent to $q \rightarrow p$.

Solution See page S11.

In the next example, we use standard forms to determine a valid conclusion for an argument with three premises.

EXAMPLE 7 Determine a Valid Conclusion for an Argument

Use all of the premises to determine a valid conclusion for the following argument.

> We will not go to Japan ($\sim j$) or we will go to Hong Kong (h). If we visit my uncle (u), then we will go to Singapore (s). If we go to Hong Kong, then we will not go to Singapore.

Solution

In symbolic form the argument is

$$\begin{array}{ll} \sim j \vee h & \text{Premise 1} \\ u \rightarrow s & \text{Premise 2} \\ \underline{h \rightarrow \sim s} & \text{Premise 3} \\ \therefore\ ? & \end{array}$$

TAKE NOTE

In Example 7 we are rewriting and reordering the statements so that transitive reasoning can be applied.

The first premise can be written as $j \rightarrow h$. The contrapositive of the second premise is $\sim s \rightarrow \sim u$. Therefore, the argument can be written as

$$\begin{array}{l} j \rightarrow h \\ \sim s \rightarrow \sim u \\ \underline{h \rightarrow \sim s} \\ \therefore\ ? \end{array}$$

Interchanging the second and third premises yields

$$\begin{array}{l} j \rightarrow h \\ h \rightarrow \sim s \\ \underline{\sim s \rightarrow \sim u} \\ \therefore\ ? \end{array}$$

An application of transitive reasoning produces

$$\begin{array}{l} j \rightarrow h \\ h \rightarrow \sim s \\ \underline{\sim s \rightarrow \sim u} \\ \therefore j \rightarrow \sim u \end{array}$$

Thus a valid conclusion for the original argument is "If we go to Japan (j), then we will not visit my uncle ($\sim u$)."

CHECK YOUR PROGRESS 7 Use all of the premises to determine a valid conclusion for the following argument.

$$\begin{array}{l} \sim m \vee t \\ t \rightarrow \sim d \\ e \vee g \\ \underline{e \rightarrow d} \\ \therefore\ ? \end{array}$$

Solution See page S11.

MATH MATTERS A Famous Puzzle: *Where Is the Missing Dollar?*

Most puzzles are designed to test your logical reasoning skills. The following puzzle is intriguing because of its simplicity; however, many people have found that they were unable to provide a satisfactory solution to the puzzle.

> Three men decide to share the cost of a hotel room. The regular room rate is \$25, but the desk clerk decides to charge them \$30 because it will be easier for each man to pay one-third of \$30 than it would be for each man to pay one-third of \$25. Each man pays \$10 and the bellhop shows them to their room.

> Shortly thereafter the desk clerk starts to feel guilty and gives the bellhop 5 one dollar bills, along with instructions to return the money to the 3 men. On the way to the room the bellhop decides to give each man \$1 and keep \$2. After all, the three men will find it difficult to split \$5 evenly.
>
> Thus each man paid \$10 dollars and received a \$1 refund. After the refund, each man has each paid \$9 for the room. The total amount the men have paid for the room is $3 \times \$9 = \$27$. The bellhop has \$2. The \$27 added to the \$2 equals \$29. *Where is the missing dollar?*
>
> A solution to this puzzle is given in the Answers to Selected Exercises Appendix, on page A9, just before the Exercise Set 3.5 answers.

INSTRUCTOR NOTE

A solution to this puzzle is also given in the Answers to All Exercises Appendix, on page A16, just before the Excursion Exercises, Section 3.5 answers.

EXCURSION

Fallacies

Any argument that is not valid is called a **fallacy**. Ancient logicians enjoyed the study of fallacies and took pride in their ability to analyze and categorize different types of fallacies. In this Excursion we consider the four fallacies known as *circulus in probando*, the fallacy of experts, the fallacy of equivocation, and the fallacy of accident.

Circulus in Probando

A fallacy of *circulus in probando* is an argument that uses a premise as the conclusion. For instance, consider the following argument.

The Chicago Bulls are the best basketball team because there is no basketball team that is better than the Chicago Bulls.

The fallacy of *circulus in probando* is also known as *circular reasoning* or *begging the question.*

Fallacy of Experts

A fallacy of experts is an argument that uses an expert (or a celebrity) to lend support to a product or an idea. Often the product or idea is outside the expert's area of expertise. The following endorsements may qualify as fallacy of experts arguments.

Jamie Lee Curtis for Activia yogurt

David Duchovny for Pedigree dog food

Kevin Bacon for the American Egg Board (What Goes Better with Eggs than Bacon?)

Fallacy of Equivocation

A fallacy of equivocation is an argument that uses a word with two interpretations in two different ways. The following argument is an example of a fallacy of equivocation.

The highway sign read $268 fine for littering,
so I decided fine, for $268, I will litter.

Fallacy of Accident

The following argument is an example of a fallacy of accident.

Everyone should visit Europe.

Therefore, prisoners on death row should be allowed to visit Europe.

Using more formal language, we can state the argument as follows.

If you are a prisoner on death row (d), then you are a person (p).

If you are a person (p), then you should be allowed to visit Europe (e).

∴ If you are a prisoner on death row, then you should be allowed to visit Europe.

The symbolic form of the argument is

$$\begin{array}{l} d \rightarrow p \\ \underline{p \rightarrow e} \\ \therefore d \rightarrow e \end{array}$$

This argument appears to be a valid argument because its symbolic form is identical to the standard form for transitive reasoning. Common sense tells us the argument is not valid, so where have we gone wrong in our analysis of the argument?

The problem occurs with the interpretation of the word "everyone." Often, when we say "everyone," we really mean "most everyone." A fallacy of accident may occur whenever we use a statement that is often true in place of a statement that is always true.

EXCURSION EXERCISES

1. Write an argument that is an example of *circulus in probando.*
2. Give an example of an argument that is a fallacy of experts.
3. Write an argument that is an example of a fallacy of equivocation.
4. Write an argument that is an example of a fallacy of accident.

5. Algebraic arguments often consist of a list of statements. In a valid algebraic argument, each statement (after the premises) can be deduced from the previous statements. The following argument that $1 = 2$ contains exactly one step that is not valid. Identify the step and explain why it is not valid.

Let

$a = b$	• Premise.
$a^2 = ab$	• Multiply each side by a.
$a^2 - b^2 = ab - b^2$	• Subtract b^2 from each side.
$(a + b)(a - b) = b(a - b)$	• Factor each side.
$a + b = b$	• Divide each side by $(a - b)$.
$b + b = b$	• Substitute b for a.
$2b = b$	• Collect like terms.
$2 = 1$	• Divide each side by b.

EXERCISE SET 3.5

■ In Exercises 1 to 8, use the indicated letters to write each argument in symbolic form.

1. If you can read this bumper sticker (r), you're too close (c). You can read the bumper sticker. Therefore, you're too close.

2. If Lois Lane marries Clark Kent (m), then Superman will get a new uniform (u). Superman does not get a new uniform. Therefore, Lois Lane did not marry Clark Kent.

3. If the price of gold rises (g), the stock market will fall (s). The price of gold did not rise. Therefore, the stock market did not fall.

4. I am going shopping (s) or I am going to the museum (m). I went to the museum. Therefore, I did not go shopping.

5. If we search the Internet (s), we will find information on logic (i). We searched the Internet. Therefore, we found information on logic.

6. If we check the sports results on ESPN (c), we will know who won the match (w). We know who won the match. Therefore, we checked the sports results on ESPN.

7. If the power goes off $(\sim p)$, then the air conditioner will not work $(\sim a)$. The air conditioner is working. Therefore, the power is not off.

8. If it snowed (s), then I did not go to my chemistry class $(\sim c)$. I went to my chemistry class. Therefore, it did not snow.

■ In Exercises 9 to 24, use a truth table to determine whether the argument is valid or invalid.

9. $p \vee \sim q$
$\underline{\sim q}$
$\therefore p$

10. $\sim p \wedge q$
$\underline{\sim p}$
$\therefore q$

11. $p \rightarrow \sim q$
$\underline{\sim q}$
$\therefore p$

12. $p \rightarrow \sim q$
$\underline{p}$
$\therefore \sim q$

13. $\sim p \rightarrow \sim q$
$\underline{\sim p}$
$\therefore \sim q$

14. $\sim p \rightarrow q$
$\underline{p}$
$\therefore \sim q$

15. $(p \rightarrow q) \wedge (\sim p \rightarrow q)$
$\underline{q}$
$\therefore p$

16. $(p \vee q) \wedge (p \wedge q)$
$\underline{p}$
$\therefore q$

17. $(p \wedge \sim q) \vee (p \rightarrow q)$
$\underline{q \vee p}$
$\therefore \sim p \wedge q$

18. $(p \wedge \sim q) \rightarrow (p \vee q)$
$\underline{q \rightarrow \sim p}$
$\therefore p \rightarrow q$

19. $(p \wedge \sim q) \vee (p \vee r)$
$\underline{r}$
$\therefore p \vee q$

20. $(p \rightarrow q) \rightarrow (r \rightarrow \sim q)$
$\underline{p}$
$\therefore \sim r$

21. $p \leftrightarrow q$
$\underline{p \rightarrow r}$
$\therefore \sim r \rightarrow \sim p$

22. $p \wedge r$
$\underline{p \rightarrow \sim q}$
$\therefore r \rightarrow q$

23. $p \wedge \sim q$
$\underline{p \leftrightarrow r}$
$\therefore q \vee r$

24. $p \rightarrow r$
$\underline{r \rightarrow q}$
$\therefore \sim p \rightarrow \sim q$

■ In Exercises 25 to 30, use the indicated letters to write the argument in symbolic form. Then use a truth table to determine whether the argument is valid or invalid.

25. If you finish your homework (h), you may attend the reception (r). You did not finish your homework. Therefore, you cannot go to the reception.

26. The X Games will be held in Oceanside (o) if and only if the city of Oceanside agrees to pay \$200,000 in prize money (a). If San Diego agrees to pay \$300,000 in prize money (s), then the city of Oceanside will not agree to pay \$200,000 in prize money. Therefore, if the X Games were held in Oceanside, then San Diego did not agree to pay \$300,000 in prize money.

27. If I can't buy the house $(\sim b)$, then at least I can dream about it (d). I can buy the house or at least I can dream about it. Therefore, I can buy the house.

28. If the winds are from the east (e), then we will not have a big surf $(\sim s)$. We do not have a big surf. Therefore, the winds are from the east.

29. If I master college algebra (c), then I will be prepared for trigonometry (t). I am prepared for trigonometry. Therefore, I mastered college algebra.

30. If it is a blot (b), then it is not a clot $(\sim c)$. If it is a zlot (z), then it is a clot. It is a blot. Therefore, it is not a zlot.

■ In Exercises 31 to 40, determine whether the argument is valid or invalid by comparing its symbolic form with the standard forms given in Tables 3.15 and 3.16. For each valid argument, state the name of its standard form.

31. If you take Art 151 in the fall, you will be eligible to take Art 152 in the spring. You were not eligible to take Art 152 in the spring. Therefore, you did not take Art 151 in the fall.

32. He will attend Stanford or Yale. He did not attend Yale. Therefore, he attended Stanford.

33. If I had a nickel for every logic problem I have solved, then I would be rich. I have not received a nickel for every logic problem I have solved. Therefore, I am not rich.

34. If it is a dog, then it has fleas. It has fleas. Therefore, it is a dog.

35. If we serve salmon, then Vicky will join us for lunch. If Vicky joins us for lunch, then Marilyn will not join us for lunch. Therefore, if we serve salmon, Marilyn will not join us for lunch.

36. If I go to college, then I will not be able to work for my Dad. I did not go to college. Therefore, I went to work for my Dad.

37. If my cat is left alone in the apartment, then she claws the sofa. Yesterday I left my cat alone in the apartment. Therefore, my cat clawed the sofa.

38. If I wish to use the new software, then I cannot continue to use this computer. I don't wish to use the new software. Therefore, I can continue to use this computer.

39. If Rita buys a new car, then she will not go on the cruise. Rita went on the cruise. Therefore, Rita did not buy a new car.

40. If Yordano Ventura pitches, then I will go to the game. I did not go to the game. Therefore, Yordano Ventura did not pitch.

■ In Exercises 41 to 46, use a sequence of valid arguments to show that each argument is valid.

41.
$$\begin{array}{l} \sim p \rightarrow r \\ r \rightarrow t \\ \underline{\sim t} \\ \therefore p \end{array}$$

42.
$$\begin{array}{l} r \rightarrow \sim s \\ s \vee \sim t \\ \underline{r} \\ \therefore \sim t \end{array}$$

43. If we sell the boat (s), then we will not go to the river $(\sim r)$. If we don't go to the river, then we will go camping (c). If we do not buy a tent $(\sim t)$, then we will not go camping. Therefore, if we sell the boat, then we will buy a tent.

44. If it is an ammonite (a), then it is from the Cretaceous period (c). If it is not from the Mesozoic era $(\sim m)$, then it is not from the Cretaceous period. If it is from the Mesozoic era, then it is at least 65 million years old (s). Therefore, if it is an ammonite, then it is at least 65 million years old.

45. If the computer is not operating $(\sim o)$, then I will not be able to finish my report $(\sim f)$. If the office is closed (c), then the computer is not operating. Therefore, if I am able to finish my report, then the office is open.

46. If he reads the manuscript (r), he will like it (l). If he likes it, he will publish it (p). If he publishes it, then you will get royalties (m). You did not get royalties. Therefore, he did not read the manuscript.

■ In Exercises 47 to 50, use all of the premises to determine a valid conclusion for the given argument.

47.
$$\begin{array}{l} \sim(p \wedge \sim q) \\ \underline{p} \\ \therefore ? \end{array}$$

48.
$$\begin{array}{l} \sim s \rightarrow q \\ \sim t \rightarrow \sim q \\ \underline{\sim t} \\ \therefore ? \end{array}$$

49. If it is a theropod, then it is not herbivorous. If it is not herbivorous, then it is not a sauropod. It is a sauropod. Therefore, ________________.

50. If you buy the car, you will need a loan. You do not need a loan or you will make monthly payments. You buy the car. Therefore, ________________.

EXTENSIONS

51. **Recreational Logic** "Are You Smarter Than a 5th Grader?" is a popular television program that requires adult contestants to answer grade-school level questions. The show is hosted by Jeff Foxworthy.

MARK BURNETT PROD/ THE KOBAL COLLECTION

Here is an arithmetic problem that many 5th grade students can solve in less than 10 seconds after the problem has been stated.

> Start with the number 32. Add 46 to the starting number and divide the result by 2. Multiply the previous result by 3 and add 27 to that product. Take the square root of the previous sum and then quadruple that result. Multiply the latest result by 5 and divide that product by 32. Multiply the latest result by 0 and then add 24 to the result. Take half of the previous result. What is the final answer?

Explain how to solve this arithmetic problem in less than 10 seconds.

52. **An Argument by Lewis Carroll** The following argument is from *Symbolic Logic* by Lewis Carroll, written in 1896. Use symbolic logic to determine whether the argument is valid or invalid.

> Babies are illogical.
>
> Nobody is despised who can manage a crocodile.
>
> Illogical persons are despised.
>
> Hence, babies cannot manage crocodiles.

SECTION 3.6 Arguments and Euler Diagrams

Arguments and Euler Diagrams

Many arguments involve sets whose elements are described using the quantifiers *all*, *some*, and *none*. The mathematician Leonhard Euler (laônhärt oi′lər) used diagrams to determine whether arguments that involved quantifiers were valid or invalid. The following figures show Euler diagrams that illustrate the four possible relationships that can exist between two sets.

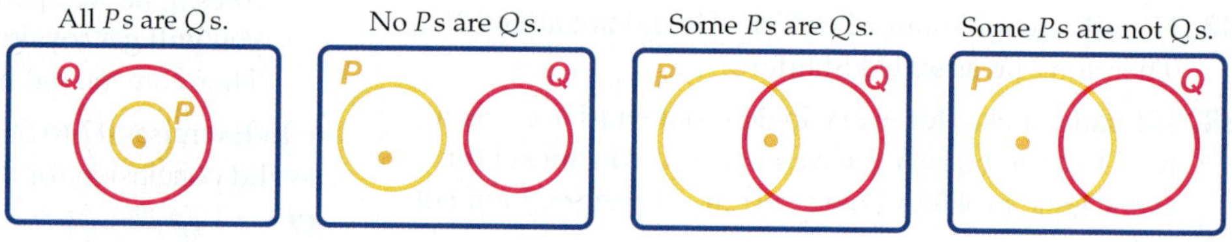

Euler diagrams

Euler used diagrams to illustrate logic concepts. Some 100 years later, John Venn extended the use of Euler's diagrams to illustrate many types of mathematics. In this section, we will construct diagrams to determine the validity of arguments. We will refer to these diagrams as Euler diagrams.

EXAMPLE 1 Use an Euler Diagram to Determine the Validity of an Argument

Use an Euler diagram to determine whether the following argument is valid or invalid.

All college courses are fun.
This course is a college course.
∴ This course is fun.

HISTORICAL NOTE

Leonhard Euler (1707–1783) Euler was an exceptionally talented Swiss mathematician. He worked in many different areas of mathematics and produced more written material about mathematics than any other mathematician. His mental computational abilities were remarkable. The French astronomer and statesman Dominque François Arago wrote,

> Euler calculated without apparent effort, as men breathe, or as eagles sustain themselves in the wind.

In 1776, Euler became blind; however, he continued to work in the disciplines of mathematics, physics, and astronomy. He even solved a problem that Newton had attempted concerning the motion of the moon. Euler performed all the necessary calculations in his head.

Solution

The first premise indicates that the set of college courses is a subset of the set of fun courses. We illustrate this subset relationship with an Euler diagram, as shown in Figure 3.15. The second premise tells us that "this course" is an element of the set of college courses. If we use c to represent "this course," then c must be placed inside the set of college courses, as shown in Figure 3.16.

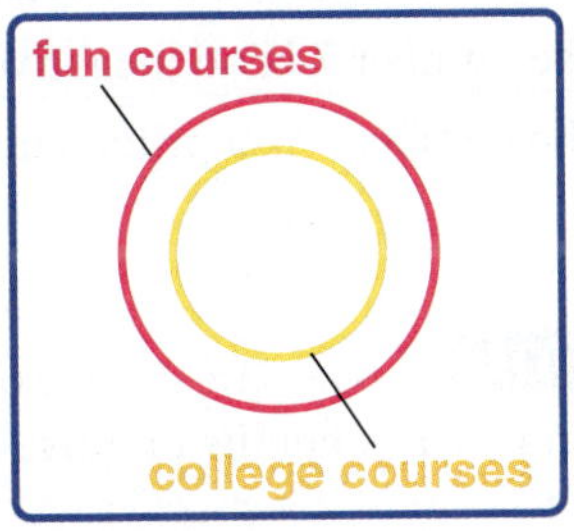

FIGURE 3.15

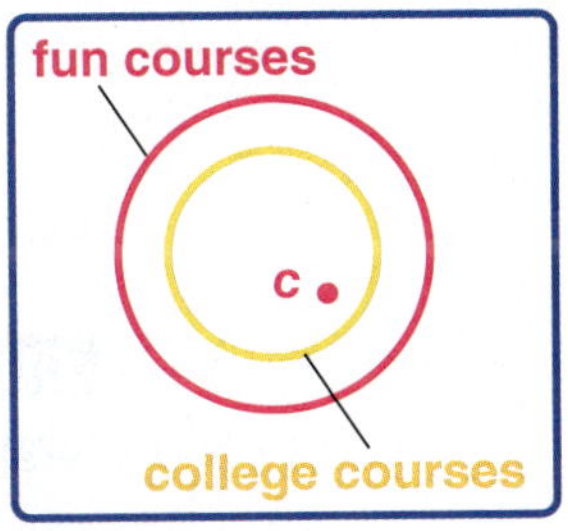

FIGURE 3.16

Figure 3.16 illustrates that c must also be an element of the set of fun courses. Thus the argument is valid.

CHECK YOUR PROGRESS 1 Use an Euler diagram to determine whether the following argument is valid or invalid.

All lawyers drive BMWs.
Susan is a lawyer.
∴ Susan drives a BMW.

Solution See page S11.

If an Euler diagram can be drawn so that the conclusion does not necessarily follow from the premises, then the argument is invalid. This concept is illustrated in the next example.

EXAMPLE 2 Use an Euler Diagram to Determine the Validity of an Argument

Use an Euler diagram to determine whether the following argument is valid or invalid.

Some Impressionist paintings are Renoirs.
Dance at Bougival is an Impressionist painting.
∴ *Dance at Bougival* is a Renoir.

Solution

The Euler diagram in Figure 3.17 illustrates the premise that some Impressionist paintings are Renoirs. Let d represent the painting *Dance at Bougival*. Figures 3.18 and 3.19 show that d can be placed in one of two regions.

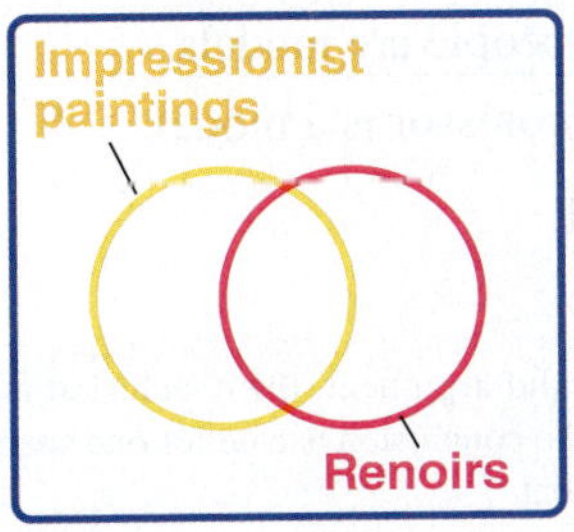

FIGURE 3.17

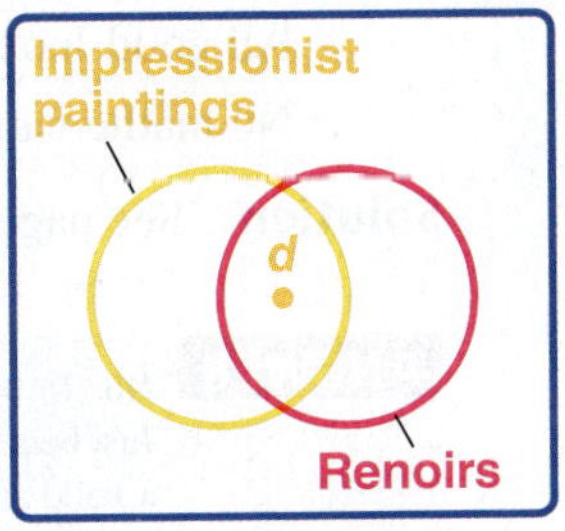

FIGURE 3.18

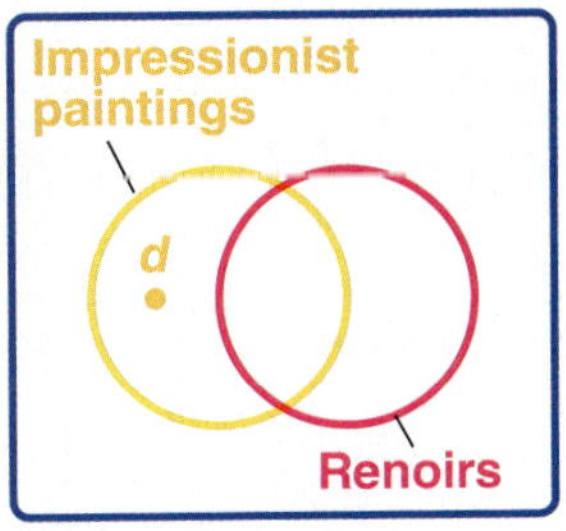

FIGURE 3.19

TAKE NOTE

Even though the conclusion in Example 2 is true, the argument is invalid.

Although Figure 3.18 supports the argument, Figure 3.19 shows that the conclusion does not necessarily follow from the premises, and thus the argument is invalid.

CHECK YOUR PROGRESS 2 Use an Euler diagram to determine whether the following argument is valid or invalid.

No prime numbers are negative.
The number 7 is not negative.
∴ The number 7 is a prime number.

Solution See page S11.

QUESTION If one particular example can be found for which the conclusion of an argument is true when its premises are true, must the argument be valid?

Some arguments can be represented by an Euler diagram that involves three sets, as shown in Example 3.

EXAMPLE 3 Use an Euler Diagram to Determine the Validity of an Argument

Use an Euler diagram to determine whether the following argument is valid or invalid.

No psychologist can juggle.
All clowns can juggle.
∴ No psychologist is a clown.

Solution

The Euler diagram in Figure 3.20 shows that the set of psychologists and the set of jugglers are disjoint sets. Figure 3.21 shows that because the set of clowns is a subset of the set of jugglers, no psychologists p are elements of the set of clowns. Thus the argument is valid.

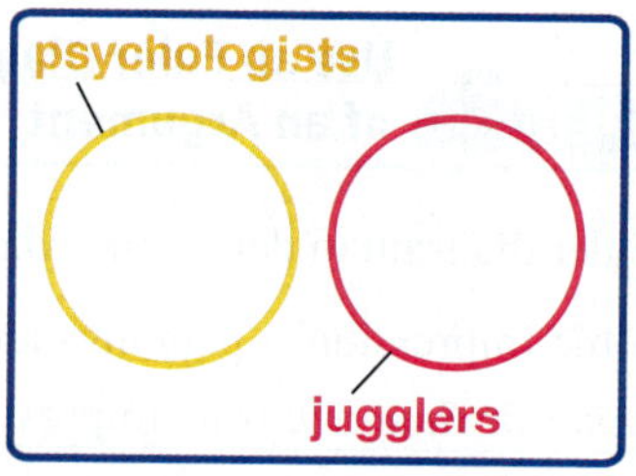

FIGURE 3.20

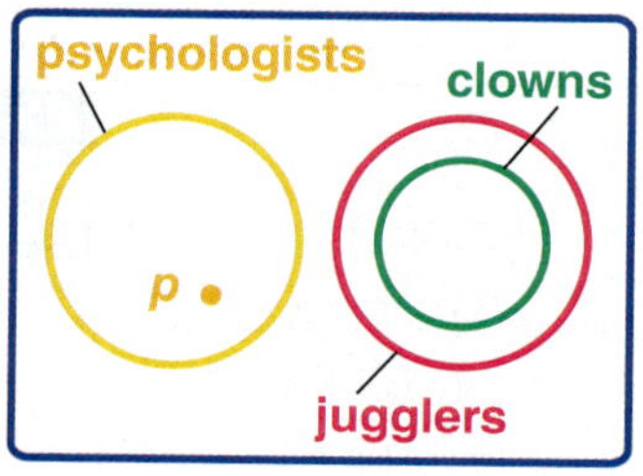

FIGURE 3.21

CHECK YOUR PROGRESS 3 Use an Euler diagram to determine whether the following argument is valid or invalid.

No mathematics professors are good-looking.
All good-looking people are models.
∴ No mathematics professor is a model.

Solution See page S11.

ANSWER No. To be a valid argument, the conclusion must be true whenever the premises are true. Just because the conclusion is true for one specific example, does not mean the argument is a valid argument.

MATH MATTERS Raymond Smullyan

Raymond Smullyan (1919–) is a concert pianist, a logician, a Taoist philosopher, a magician, a retired professor, and an author of many popular books on logic. Over a period of several years he has created many interesting logic problems. One of his logic problems is an enhancement of the classic logic puzzle that concerns two doors and two guards. One of the doors leads to heaven and the other door leads to hell. One of the guards always tells the truth and the other guard always lies. You do not know which guard always tells the truth and which guard always lies, and you are only allowed to ask one question of one of the guards. What one question should you ask that will allow you to determine which door leads to heaven?

Another logic puzzle that Raymond Smullyan created has been referred to as the "hardest logic puzzle ever." Information about this puzzle and the solution to the above puzzle concerning the two guards and the two doors can be found at: https://en.wikipedia.org/wiki/Raymond_Smullyan

Euler Diagrams and Transitive Reasoning

Example 4 uses Euler diagrams to visually illustrate transitive reasoning.

EXAMPLE 4 Use an Euler Diagram to Determine the Validity of an Argument

Use an Euler diagram to determine whether the following argument is valid or invalid.

All fried foods are greasy.
All greasy foods are delicious.
All delicious foods are healthy.
$\therefore$ All fried foods are healthy.

Solution

The figure at the left illustrates that every fried food is an element of the set of healthy foods, so the argument is valid.

TAKE NOTE

Although the conclusion in Example 4 is false, the argument in Example 4 is valid.

CHECK YOUR PROGRESS 4 Use an Euler diagram to determine whether the following argument is valid or invalid.

All squares are rhombi.
All rhombi are parallelograms.
All parallelograms are quadrilaterals.
$\therefore$ All squares are quadrilaterals.

Solution See page S11.

Using Euler Diagrams to Form Conclusions

In Example 5, we make use of an Euler diagram to determine a valid conclusion for an argument.

EXAMPLE 5 Use an Euler Diagram to Determine a Conclusion for an Argument

Use an Euler diagram and all of the premises in the following argument to determine a valid conclusion for the argument.

All *M*s are *N*s.
No *N*s are *P*s.
$\therefore$?

Solution

The first premise indicates that the set of *M*s is a subset of the set of *N*s. The second premise indicates that the set of *N*s and the set of *P*s are disjoint sets. The following Euler diagram illustrates these set relationships. An examination of the Euler diagram allows us to conclude that no *M*s are *P*s.

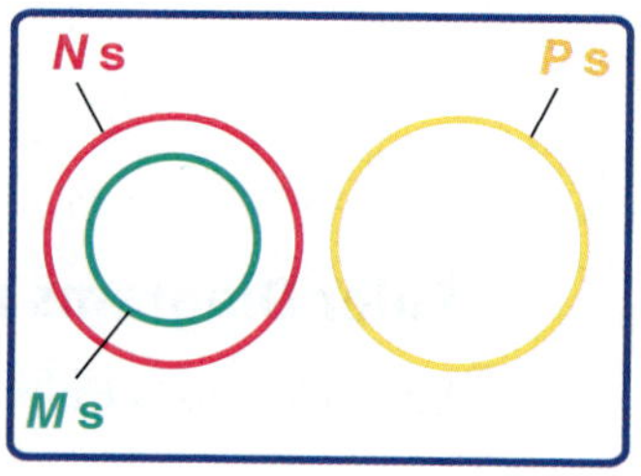

CHECK YOUR PROGRESS 5 Use an Euler diagram and all of the premises in the following argument to determine a valid conclusion for the argument.

Some rabbits are white.
All white animals like tomatoes.
$\therefore$?

Solution See page S11.

EXCURSION

Using Logic to Solve Cryptarithms

Many puzzles can be solved by making an assumption and then checking to see if the assumption is supported by the conditions (premises) associated with the puzzle. For instance, consider the following addition problem in which each letter represents a digit from 0 through 9 and different letters represent different digits.

$$\begin{array}{r} TA \\ +\ BT \\ \hline TEE \end{array}$$

TAKE NOTE

When working with cryptarithms, we assume that the leading digit of each number is a nonzero digit.

Note that the T in T E E is a carry from the middle column. Because the sum of any two single digits plus a previous carry of at most 1 is 19 or less, the T in T E E must be a 1. Replacing all the Ts with 1s produces:

```
  1A
+ B1
 1EE
```

Now B must be an 8 or a 9, because these are the only digits that could produce a carry into the leftmost column.

Case 1: Assume B is a 9. Then A must be an 8 or smaller, and A + 1 does not produce a carry into the middle column. The sum of the digits in the middle column is 10; thus E is a 0. This presents a dilemma because the units digit of A + 1 must also be a 0, which requires A to be a 9. The assumption that B is a 9 is not supported by the conditions of the problem; thus we reject the assumption that B is a 9.

Case 2: Assume B is an 8. To produce the required carry into the leftmost column, there must be a carry from the column on the right. Thus A must be a 9, and we have the result shown below.

```
  19
+ 81
 100
```

A check shows that this solution satisfies all the conditions of the problem.

EXCURSION EXERCISES

Solve the following cryptarithms. Assume that no leading digit is a 0. (*Source:* http://cryptarithms.awardspace.us/)[3]

1.
```
   SO
 + SO
  TOO
```

2.
```
   US
 + AS
  ALL
```

3.
```
   COCA
 + COLA
  OASIS
```

4.
```
     AT
   EAST
 + WEST
  SOUTH
```

[3] Copyright © 2002 by Jorge A C B Soares.

EXERCISE SET 3.6

■ In Exercises 1 to 20, use an Euler diagram to determine whether the argument is valid or invalid.

1. All frogs are poetical.
Kermit is a frog.
∴ Kermit is poetical.

2. All Oreo cookies have a filling.
All Fig Newtons have a filling.
∴ All Fig Newtons are Oreo cookies.

3. Some plants have flowers.
All things that have flowers are beautiful.
∴ Some plants are beautiful.

4. No squares are triangles.
Some triangles are equilateral.
∴ No squares are equilateral.

5. No rocker would do the mariachi.
All baseball fans do the mariachi.
∴ No rocker is a baseball fan.

6. Nuclear energy is not safe.
Some electric energy is safe.
∴ No electric energy is nuclear energy.

7. Some birds bite.
All things that bite are dangerous.
∴ Some birds are dangerous.

8. All fish can swim.
That barracuda can swim.
∴ That barracuda is a fish.

9. All men behave badly.
Some hockey players behave badly.
∴ Some hockey players are men.

10. All grass is green.
That ground cover is not green.
∴ That ground cover is not grass.

11. Most teenagers drink soda.
No CEOs drink soda.
∴ No CEO is a teenager.

12. Some students like history.
Vern is a student.
∴ Vern likes history.

13. No mathematics test is fun.
All fun things are worth your time.
∴ No mathematics test is worth your time.

14. All prudent people shun sharks.
No accountant is imprudent.
∴ No accountant fails to shun sharks.

15. All candidates without a master's degree will not be considered for the position of director.
All candidates who are not considered for the position of director should apply for the position of assistant.
∴ All candidates without a master's degree should apply for the position of assistant.

16. Some whales make good pets.
Some good pets are cute.
Some cute pets bite.
∴ Some whales bite.

17. All prime numbers are odd.
2 is a prime number.
∴ 2 is an odd number.

18. All Lewis Carroll arguments are valid.
Some valid arguments are syllogisms.
∴ Some Lewis Carroll arguments are syllogisms.

19. All aerobics classes are fun.
Jan's class is fun.
∴ Jan's class is an aerobics class.

20. No sane person takes a math class.
Some students that take a math class can juggle.
∴ No sane person can juggle.

■ In Exercises 21 to 26, use all of the premises in each argument to determine a valid conclusion for the argument.

21. All reuben sandwiches are good.
All good sandwiches have pastrami.
All sandwiches with pastrami need mustard.
∴ ?

22. All cats are strange.
Boomer is not strange.
∴ ?

23. All multiples of 11 end with a 5.
1001 is a multiple of 11.
∴ ?

24. If it isn't broken, then I do not fix it.
If I do not fix it, then I do not get paid.
∴ ?

25. Some horses are frisky.
All frisky horses are grey.
∴ ?

26. If we like to ski, then we will move to Vail.
If we move to Vail, then we will not buy a house.
If we do not buy a condo, then we will buy a house.
∴ ?

27. Examine the following three premises:

1. All people who have an Xbox play video games.
2. All people who play video games enjoy life.
3. Some mathematics professors enjoy life.

Now consider each of the following six conclusions. For each conclusion, determine whether the argument formed by the three premises and the conclusion is valid or invalid.

a. ∴ Some mathematics professors have an Xbox.
b. ∴ Some mathematics professors play video games.
c. ∴ Some people who play video games are mathematics professors.

d. ∴ Mathematics professors never play video games.

e. ∴ All people who have an Xbox enjoy life.

f. ∴ Some people who enjoy life are mathematics professors.

28. Examine the following three premises:

1. All people who drive pickup trucks like Garth Brooks.
2. All people who like Garth Brooks like country western music.
3. Some people who like heavy metal music like Garth Brooks.

Now consider each of the following five conclusions. For each conclusion, determine whether the argument formed by the three premises and the conclusion is valid or invalid.

a. ∴ Some people who like heavy metal music drive a pickup truck.

b. ∴ Some people who like heavy metal music like country western music.

c. ∴ Some people who like Garth Brooks like heavy metal music.

d. ∴ All people who drive a pickup truck like country western music.

e. ∴ People who like heavy metal music never drive a pickup truck.

EXTENSIONS

The LSAT and Logic Games

A good score on the Law School Admission Test, better known as the LSAT, is viewed by many to be the most important part of getting into a top-tier law school. Rather than testing what you've already learned, it's designed to measure and project your ability to excel in law school.

29. LSAT Practice Problem The following exercise is similar to some of the questions from the logic games section of the LSAT.

A cell phone provider sells seven different types of smart phones.

- Each phone has either a touch screen keyboard or a pushbutton keyboard.
- Each phone has a 4-inch, a 4.7-inch, or a 5.5-inch screen.
- Every phone with a 4.7-inch screen is paired with a touch screen keyboard.
- Of the seven different types of phones, most have a touch screen keyboard.
- No phone with a pushbutton keyboard is paired with a 4 inch screen.

Which one of the following statements CANNOT be true?

a. Five of the types of phones have 4.7-inch screens.

b. Five of the types of phones have 4-inch screens.

c. Four of the types of phones have a pushbutton keyboard.

d. Four of the types of phones have 5.5-inch screens.

e. Five of the types of phones have a touch screen keyboard.

30. Bilateral Diagrams Lewis Carroll (Charles Dodgson) devised a *bilateral diagram* (two-part board) to analyze syllogisms. His method has some advantages over Euler diagrams and Venn diagrams. Use a library or the Internet to find information on Carroll's method of analyzing syllogisms. Write a few paragraphs that explain his method and its advantages.

CHAPTER 3 SUMMARY

The following table summarizes essential concepts in this chapter. The references given in the right-hand column list Examples and Exercises that can be used to test your understanding of a concept.

3.1 Logic Statements and Quantifiers	
Statements A statement is a declarative sentence that is either true or false, but not both true and false. A simple statement is a statement that does not contain a connective.	See **Example 1** on page 106, and then try Exercises 1 to 6 on page 164.

continued

Compound Statements A compound statement is formed by connecting simple statements with the connectives *and, or, if . . . then,* and *if and only if.*	See **Examples 3 and 4** on page 109, and then try Exercises 7 to 10 on page 164.
Truth Values The conjunction $p \wedge q$ is true if and only if both p and q are true. The disjunction $p \vee q$ is true provide p is true, q is true, or both p and q are true.	See **Example 6** on page 111, and then try Exercises 17 to 20 on page 165.
The Negation of a Quantified Statement The information in the following table can be used to write the negation of many quantified statements. All X are Y. ←negation→ Some X are not Y. No X are Y. ←negation→ Some X are Y.	See **Example 7** on page 113, and then try Exercises 11 to 16 on pages 164 and 165.

3.2 Truth Tables, Equivalent Statements, and Tautologies

Construction of Truth Tables **1.** If the given statement has n simple statements, then start with a standard form that has 2^n rows. Enter the truth values for each simple statement and their negations. **2.** Use the truth values for each simple statement and their negations to enter the truth values under each connective within a pair of grouping symbols—parentheses ( ), brackets [ ], braces { }. If some grouping symbols are nested inside other grouping symbols, then work from the inside out. In any situation in which grouping symbols have not been used, then we use the following **order of precedence agreement**. First assign truth values to negations from left to right, followed by conjunctions from left to right, followed by disjunctions from left to right, followed by conditionals from left to right, and finally by biconditionals from left to right. **3.** The truth values that are entered into the column under the connective for which truth values are assigned *last* form the truth table for the given statement.	See **Example 3** on page 119, and then try Exercises 27 to 34 on page 165.
Equivalent Statements Two statements are equivalent if they both have the same truth value for all possible truth values of their simple statements. The notation $p \equiv q$ is used to indicate that the statements p and q are equivalent.	See **Example 4** on page 121, and then try Exercises 39 to 42 on page 165.
De Morgan's Laws for Statements For any statements p and q, $\sim(p \vee q) \equiv \sim p \wedge \sim q$ and $\sim(p \wedge q) \equiv \sim p \vee \sim q$	See **Example 5** on page 122, and then try Exercises 35 to 38 on page 165.
Tautologies and Self-Contradictions A tautology is a statement that is always true. A self-contradiction is a statement that is always false.	See **Example 6** on page 122, and then try Exercises 43 to 46 on page 165.

3.3 The Conditional and the Biconditional

Antecedent and Consequent of a Conditional In a conditional statement represented by "if p, then q" or by "if p, q," the p statement is called the antecedent and the q statement is called the consequent.	See **Example 1** on page 126, and then try Exercises 47 and 50 on page 165.
Equivalent Disjunctive Form of $p \rightarrow q$ $p \rightarrow q \equiv {\sim}p \vee q$ The conditional $p \rightarrow q$ is false when p is true and q is false. It is true in all other cases.	See **Examples 2 to 4** on pages 128 and 129, and then try Exercises 51 to 54 on page 165.
The Negation of $p \rightarrow q$ ${\sim}(p \rightarrow q) \equiv p \wedge {\sim}q$	See **Example 5** on page 130, and then try Exercises 55 to 58 on page 165.
The Biconditional $p \leftrightarrow q$ $p \leftrightarrow q \equiv [(p \rightarrow q) \wedge (q \rightarrow p)]$ The biconditional $p \leftrightarrow q$ is true only when p and q have the same truth value.	See **Example 6** on page 131, and then try Exercises 59 and 60 on page 165.

3.4 The Conditional and Related Statements

Equivalent Forms of the Conditional The conditional "if p, then q" can be stated, in English, in several equivalent forms. For example, p only if q; p implies that q; and q provided that p are all equivalent forms of if p, then q.	See **Example 1** on page 135, and then try Exercises 63 to 66 on page 165.
Statements Related to the Conditional Statement ▪ The **converse** of $p \rightarrow q$ is $q \rightarrow p$. ▪ The **inverse** of $p \rightarrow q$ is ${\sim}p \rightarrow {\sim}q$. ▪ The **contrapositive** of $p \rightarrow q$ is ${\sim}q \rightarrow {\sim}p$.	See **Examples 2 and 3** on pages 136 and 137, and then try Exercises 67 to 72 on pages 165 and 166.
A Conditional Statement and Its Contrapositive A conditional and its contrapositive are equivalent statements. Therefore, if the contrapositive of a conditional statement is a true statement, then the conditional statement must also be a true statement.	See **Example 4** on page 137, and then try Exercise 74 on page 166.

3.5 Symbolic Arguments

Valid Argument An argument consists of a set of statements called premises and another statement called the conclusion. An argument is valid if the conclusion is true whenever all the premises are assumed to be true. An argument is invalid if it is not a valid argument.	See **Examples 2 and 3** on pages 143 and 144, and then try Exercises 79 to 82 on page 166.
Symbolic Forms of Arguments **Standard Forms of Four Valid Arguments** (see table below)	See **Examples 4 to 6** on pages 145 to 148, and then try Exercises 83 to 88 on page 166.

Direct reasoning	Contrapositive reasoning	Transitive reasoning	Disjunctive reasoning	
$p \rightarrow q$	$p \rightarrow q$	$p \rightarrow q$	$p \vee q$	$p \vee q$
$\underline{p}$	$\underline{{\sim}q}$	$\underline{q \rightarrow r}$	$\underline{{\sim}p}$	$\underline{{\sim}q}$
$\therefore q$	$\therefore {\sim}p$	$\therefore p \rightarrow r$	$\therefore q$	$\therefore p$

continued

Symbolic Forms of Arguments *(continued)*

Standard Forms of Two Invalid Arguments

Fallacy of the converse	Fallacy of the inverse
$p \rightarrow q$ $\underline{q \quad}$ $\therefore p$	$p \rightarrow q$ $\underline{\sim p \quad}$ $\therefore \sim q$

3.6 Arguments and Euler Diagrams

Euler Diagrams

See **Examples 1 to 4** on pages 154 to 157, and then try Exercises 89 to 92 on page 166.

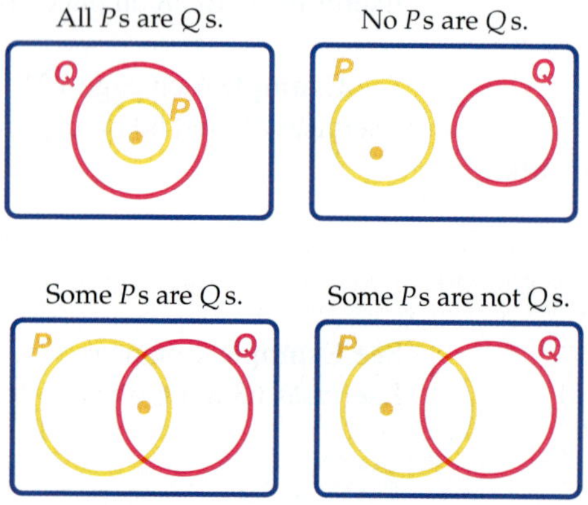

Euler diagrams can be used to determine whether arguments that involve quantifiers are valid or invalid.

Draw an Euler diagram that illustrates the conditions required by the premises of an argument.

If the conclusion of the argument must necessarily follow from all the conditions shown by the premises, then the argument is valid.

If the conclusion of the argument does not necessarily follow from the conditions shown by all the premises, then the argument is invalid.

CHAPTER 3 REVIEW EXERCISES

■ In Exercises 1 to 6, determine whether each sentence is a statement. Assume that a and b are real numbers.

1. How much is a ticket to London?

2. 91 is a prime number.

3. $a > b$

4. $a^2 \geq 0$

5. Lock the car.

6. Clark Kent is Superman.

■ In Exercises 7 to 10, write each sentence in symbolic form. Represent each simple statement of the sentence with the letter indicated in parentheses. Also state whether the sentence is a conjunction, a disjunction, a negation, a conditional, or a biconditional.

7. Today is Monday (m) and it is my birthday (b).

8. If x is divisible by 2 (d), then x is an even number (e).

9. I am going to the dance (g) if and only if I have a date (d).

10. All triangles (t) have exactly three sides (s).

■ In Exercises 11 to 16, write the negation of each quantified statement. Start each negation with "Some," "No," or "All."

11. Some dogs bite.

12. Every dessert at the Cove restaurant is good.

13. All winners receive a prize.

14. Some cameras do not use film.

15. No student finished the assignment.

16. At least one person enjoyed the story.

■ In Exercises 17 to 20, determine whether each statement is true or false.

17. $5 > 2$ or $5 = 2$.

18. $3 \neq 5$ and 7 is a prime number.

19. $4 \leq 7$.

20. $-3 < -1$.

■ In Exercises 21 to 26, determine the truth value of the statement given that p is true, q is false, and r is false.

21. $(p \wedge q) \vee (\sim p \vee q)$

22. $(p \rightarrow \sim q) \leftrightarrow \sim(p \vee q)$

23. $(p \wedge \sim q) \wedge (\sim r \vee q)$

24. $(r \wedge \sim p) \vee [(p \vee \sim q) \leftrightarrow (q \rightarrow r)]$

25. $[p \wedge (r \rightarrow q)] \rightarrow (q \vee \sim r)$

26. $(\sim q \vee \sim r) \rightarrow [(p \leftrightarrow \sim r) \wedge q]$

■ In Exercises 27 to 34, construct a truth table for the given statement.

27. $(\sim p \rightarrow q) \vee (\sim q \wedge p)$

28. $\sim p \leftrightarrow (q \vee p)$

29. $\sim(p \vee \sim q) \wedge (q \rightarrow p)$

30. $(p \leftrightarrow q) \vee (\sim q \wedge p)$

31. $(r \leftrightarrow \sim q) \vee (p \rightarrow q)$

32. $(\sim r \vee \sim q) \wedge (q \rightarrow p)$

33. $[p \leftrightarrow (q \rightarrow \sim r)] \wedge \sim q$

34. $\sim(p \wedge q) \rightarrow (\sim q \vee \sim r)$

■ In Exercises 35 to 38, make use of De Morgan's laws to write the given statement in an equivalent form.

35. It is not true that, Bob failed the English proficiency test and he registered for a speech course.

36. Ellen did not go to work this morning and she did not take her medication.

37. Wendy will go to the store this afternoon or she will not be able to prepare her fettuccine al pesto recipe.

38. Gina enjoyed the movie, but she did not enjoy the party.

■ In Exercises 39 to 42, use a truth table to show that the given pairs of statements are equivalent.

39. $\sim p \rightarrow \sim q$; $p \vee \sim q$

40. $\sim p \vee q$; $\sim(p \wedge \sim q)$

41. $p \vee (q \wedge \sim p)$; $p \vee q$

42. $p \leftrightarrow q$; $(p \wedge q) \vee (\sim p \wedge \sim q)$

■ In Exercises 43 to 46, use a truth table to determine whether the given statement is a tautology or a self-contradiction.

43. $p \wedge (q \wedge \sim p)$

44. $(p \wedge q) \vee (p \rightarrow \sim q)$

45. $[\sim(p \rightarrow q)] \leftrightarrow (p \wedge \sim q)$

46. $p \vee (p \rightarrow q)$

■ In Exercises 47 to 50, identify the antecedent and the consequent of each conditional statement.

47. If he has talent, he will succeed.

48. If I had a credential, I could get the job.

49. I will follow the exercise program provided I join the fitness club.

50. I will attend only if it is free.

■ In Exercises 51 to 54, write each conditional statement in its equivalent disjunctive form.

51. If she were tall, she would be on the volleyball team.

52. If he can stay awake, he can finish the report.

53. Rob will start, provided he is not ill.

54. Sharon will be promoted only if she closes the deal.

■ In Exercises 55 to 58, write the negation of each conditional statement in its equivalent conjunctive form.

55. If I get my paycheck, I will purchase a ticket.

56. The tomatoes will get big only if you provide them with plenty of water.

57. If you entered Cleggmore University, then you had a high score on the SAT exam.

58. If Ryan enrolls at a university, then he will enroll at Yale.

■ In Exercises 59 to 62, determine whether the given statement is true or false. Assume that x and y are real numbers.

59. $x = y$ if and only if $|x| = |y|$.

60. $x > y$ if and only if $x - y > 0$.

61. If $x^2 > 0$, then $x > 0$.

62. If $x^2 = y^2$, then $x = y$.

■ In Exercises 63 to 66, write each statement in "If p, then q" form.

63. Every nonrepeating, nonterminating decimal is an irrational number.

64. Being well known is a necessary condition for a politician.

65. I could buy the house provided that I could sell my condominium.

66. Being divisible by 9 is a sufficient condition for being divisible by 3.

■ In Exercises 67 to 72, write the **a.** converse, **b.** inverse, and **c.** contrapositive of the given statement.

67. If $x + 4 > 7$, then $x > 3$.

68. All recipes in this book can be prepared in less than 20 minutes.

69. If a and b are both divisible by 3, then $(a + b)$ is divisible by 3.

70. If you build it, they will come.

71. Every trapezoid has exactly two parallel sides.

72. If they like it, they will return.

73. What is the inverse of the contrapositive of $p \rightarrow q$?

74. Use the contrapositive of the following statement to determine whether the statement is true or false.

If today is not Monday, then yesterday was not Sunday.

■ In Exercises 75 to 78, determine the original statement if the given statement is related to the original statement in the manner indicated.

75. *Converse:* If $x > 2$, then x is an odd prime number.

76. *Negation:* The senator will attend the meeting and she will not vote on the motion.

77. *Inverse:* If their manager will not contact me, then I will not purchase any of their products.

78. *Contrapositive:* If Ginny can't rollerblade, then I can't rollerblade.

■ In Exercises 79 to 82, use a truth table to determine whether the argument is valid or invalid.

79.
$(p \wedge \sim q) \wedge (\sim p \rightarrow q)$
$\underline{p}$
$\therefore \sim q$

80.
$p \rightarrow \sim q$
$\underline{q}$
$\therefore \sim p$

81.
r
$p \rightarrow \sim r$
$\underline{\sim p \rightarrow q}$
$\therefore p \wedge q$

82.
$(p \vee \sim r) \rightarrow (q \wedge r)$
$\underline{r \wedge p}$
$\therefore p \vee q$

■ In Exercises 83 to 88, determine whether the argument is valid or invalid by comparing its symbolic form with the symbolic forms in Tables 3.15 and 3.16, pages 145 and 146.

83. We will serve either fish or chicken for lunch. We did not serve fish for lunch. Therefore, we served chicken for lunch.

84. If Mike is a CEO, then he will be able to afford to make a donation. If Mike can afford to make a donation, then he loves to ski. Therefore, if Mike does not love to ski, he is not a CEO.

85. If we wish to win the lottery, we must buy a lottery ticket. We did not win the lottery. Therefore, we did not buy a lottery ticket.

86. Robert can charge it on his MasterCard or his Visa. Robert does not use his MasterCard. Therefore, Robert charged it to his Visa.

87. If we are going to have a caesar salad, then we need to buy some eggs. We did not buy eggs. Therefore, we are not going to have a caesar salad.

88. If we serve lasagna, then Eva will not come to our dinner party. We did not serve lasagna. Therefore, Eva came to our dinner party.

■ In Exercises 89 to 92, use an Euler diagram to determine whether the argument is valid or invalid.

89.
No wizard can yodel.
<u>All lizards can yodel.</u>
∴ No wizard is a lizard.

90.
Some dogs have tails.
<u>Some dogs are big.</u>
∴ Some big dogs have tails.

91. All Italian villas are wonderful. Some wonderful villas are expensive. Therefore, some Italian villas are expensive.

92. All logicians like to sing "It's a small world after all." Some logicians have been presidential candidates. Therefore, some presidential candidates like to sing "It's a small world after all."

CHAPTER 3 TEST

1. Determine whether each sentence is a statement.

a. Look for the cat.

b. Clark Kent is afraid of the dark.

2. Write the negation of each statement. Start each negation with "Some," "No," or "All."

a. Some trees are not green.

b. No apartments are available.

3. Determine whether each statement is true or false.

a. $5 \leq 4$

b. $-2 \geq -2$

4. Determine the truth value of each statement given that p is true, q is false, and r is true.

a. $(p \vee \sim q) \wedge (\sim r \wedge q)$

b. $(r \vee \sim p) \vee [(p \vee \sim q) \leftrightarrow (q \rightarrow r)]$

■ In Exercises 5 and 6, construct a truth table for the given statement.

5. $\sim(p \wedge \sim q) \vee (q \rightarrow p)$

6. $(r \leftrightarrow \sim q) \wedge (p \rightarrow q)$

7. Use one of De Morgan's laws to write the following in an equivalent form.

Elle did not eat breakfast and she did not take a lunch break.

8. What is a tautology?

9. Write $p \rightarrow q$ in its equivalent disjunctive form.

10. Determine whether the given statement is true or false. Assume that x, y, and z are real numbers.

a. $x = y$ if $|x| = |y|$.

b. If $x > y$, then $xz > yz$.

11. Write the **a.** converse, **b.** inverse, and **c.** contrapositive of the following statement.

$$\text{If } x + 7 > 11, \text{ then } x > 4.$$

12. Write the symbolic form of direct reasoning.

13. Write the symbolic form of transitive reasoning.

14. Write the symbolic form of contrapositive reasoning.

15. Write the symbolic form of the fallacy of the inverse.

■ In Exercises 16 and 17, use a truth table to determine whether the argument is valid or invalid.

16. $(p \wedge \sim q) \wedge (\sim p \rightarrow q)$

$\underline{p \qquad\qquad\qquad\qquad}$

$\therefore \sim q$

17. r

$p \rightarrow \sim r$

$\underline{\sim p \rightarrow q}$

$\therefore p \wedge q$

■ In Exercises 18 to 22, determine whether the argument is valid or invalid. Explain how you made your decision.

18. If we wish to win the talent contest, we must practice. We did not win the contest. Therefore, we did not practice.

19. Gina will take a job in Atlanta or she will take a job in Kansas City. Gina did not take a job in Atlanta. Therefore, Gina took a job in Kansas City.

20. No wizard can glow in the dark.

Some lizards can glow in the dark.

∴ No wizard is a lizard.

21. Some novels are worth reading.

War and Peace is a novel.

∴ *War and Peace* is worth reading.

22. If I cut my night class, then I will go to the party. I went to the party. Therefore, I cut my night class.

4

Apportionment and Voting

In this chapter, we discuss two of the most fundamental principles of democracy: the right to vote and the value of that vote. The U.S. Constitution, in Article I, Section 2, states in part that

> The House of Representatives shall be composed of members chosen every second year by the people of the several states, and the electors in each state shall have the qualifications requisite for electors of the most numerous branch of the state legislature. . . . Representatives and direct taxes shall be apportioned among the several states which may be included within this union, according to their respective numbers. . . . The actual Enumeration shall be made within three years after the first meeting of the Congress of the United States, and within every subsequent term of ten years, in such manner as they shall by law direct. The number of Representatives shall not exceed one for every thirty thousand, but each state shall have at least one Representative; . . .

This article of the Constitution requires that "Representatives . . . be *apportioned* [our italics] among the several states . . . according to their respective numbers. . . ." That is, the number of representatives each state sends to Congress should be based on its population. Because populations change over time, this article also requires that the number of people within a state should be counted "within every subsequent term of ten years." This is why we have a census every 10 years.

The way representatives are *apportioned* has been a contentious issue since the founding of the United States. The first presidential veto was issued by George Washington in 1792 because he did not approve of the way the House of Representatives decided to apportion the number of representatives each state would have. Ever since that first veto, the issue of how to apportion membership in the House of Representatives among the states has been revisited many times.

Brian Lawrence/Photographer's Choice/Getty Images

SECTION 4.1 Introduction to Apportionment

HISTORICAL NOTE

The original Article I, Section 2 of the U.S. Constitution included wording as to how to count citizens. According to this article, the numbers "shall be determined by adding to the whole number of free persons, including those bound to service for a term of years, and excluding Indians not taxed, three fifths of all other Persons." The "three fifths of all other Persons" meant that a slave was counted as only $\frac{3}{5}$ of a person. This article was modified by Section 1 of the 14th Amendment to the Constitution, which, in part, states that "All persons born or naturalized in the United States, and subject to the jurisdiction thereof, are citizens of the United States and of the state wherein they reside. . . ."

The mathematical investigation into **apportionment**, which is a method of dividing a whole into various parts, has its roots in the U.S. Constitution. (See the chapter opener.) Since 1790, when the House of Representatives first attempted to apportion itself, various methods have been used to decide how many voters would be represented by each member of the House. The two competing plans in 1790 were put forward by Alexander Hamilton and Thomas Jefferson.

To illustrate how the Hamilton and Jefferson plans were used to calculate the number of representatives each state should have, we will consider the fictitious country of Andromeda, with a population of 20,000 and five states. The population of each state is given in the table at the right.

Andromeda

State	Population
Apus	11,123
Libra	879
Draco	3518
Cephus	1563
Orion	2917
Total	20,000

Andromeda's constitution calls for 25 representatives to be chosen from these states. The number of representatives is to be apportioned according to the states' respective populations.

The Hamilton Plan

Under the Hamilton plan, the total population of the country (20,000) is divided by the number of representatives (25). This gives the number of citizens represented by each representative. This number is called the *standard divisor.*

Standard Divisor

$$\text{Standard divisor} = \frac{\text{total population}}{\text{number of people to apportion}}$$

For Andromeda, we have

$$\text{Standard divisor} = \frac{\text{total population}}{\text{number of people to apportion}} = \frac{20{,}000}{25} = 800$$

QUESTION What is the meaning of the number 800 calculated above?

Now divide the population of each state by the standard divisor and round the quotient *down* to a whole number. For example, both 15.1 and 15.9 would be rounded to 15. Each whole number quotient is called a *standard quota.*

Standard Quota

The **standard quota** is the whole number part of the quotient of a population divided by the standard divisor.

TAKE NOTE

Today apportionment is applied to situations other than a population of people. For instance, *population* could refer to the number of math classes offered at a college or the number of fire stations in a city. Nonetheless, the definition of *standard divisor* is still given as though people were involved.

ANSWER It is the number of citizens represented by each representative.

State	Population	Quotient	Standard quota
Apus	11,123	$\frac{11,123}{800} \approx 13.904$	13
Libra	879	$\frac{879}{800} \approx 1.099$	1
Draco	3518	$\frac{3518}{800} \approx 4.398$	4
Cephus	1563	$\frac{1563}{800} \approx 1.954$	1
Orion	2917	$\frac{2917}{800} \approx 3.646$	3
		Total	22

From the calculations in the above table, the total number of representatives is 22, not 25 as required by Andromeda's constitution. When this happens, the Hamilton plan calls for revisiting the calculation of the quotients and assigning an additional representative to the state with the largest decimal remainder. This process is continued until the number of representatives equals the number required by the constitution. For Andromeda, we have

TAKE NOTE

Additional representatives are assigned according to the largest decimal remainders. Because the sum of the standard quotas came to only 22 representatives, we must add three more representatives. The states with the three highest decimal remainders are Cephus (1.954), Apus (13.904), and Orion (3.646). Thus each of these states gets an additional representative.

State	Population	Quotient	Standard quota	Number of representatives
Apus	11,123	$\frac{11,123}{800} \approx 13.904$	13	14
Libra	879	$\frac{879}{800} \approx 1.099$	1	1
Draco	3518	$\frac{3518}{800} \approx 4.398$	4	4
Cephus	1563	$\frac{1563}{800} \approx 1.954$	1	2
Orion	2917	$\frac{2917}{800} \approx 3.646$	3	4
		Total	22	25

The Jefferson Plan

As we saw with the Hamilton plan, dividing by the standard divisor and then rounding down does not always yield the correct number of representatives. In the previous example, we were three representatives short. The Jefferson plan attempts to overcome this difficulty by using a *modified standard divisor*. This number is chosen, by trial and error, so that the sum of the standard quotas is equal to the total number of representatives. In a specific apportionment calculation, there may be more than one number that can serve as the modified standard divisor. For instance, in the following apportionment calculation shown in the first table on the following page, we have used 740 as our modified standard divisor. However, 741 can also be used as the modified standard divisor.

TAKE NOTE

The modified standard divisor will always be smaller than the standard divisor.

State	Population	Quotient	Number of representatives
Apus	11,123	$\frac{11,123}{740} \approx 15.031$	15
Libra	879	$\frac{879}{740} \approx 1.188$	1
Draco	3518	$\frac{3518}{740} \approx 4.754$	4
Cephus	1563	$\frac{1563}{740} \approx 2.112$	2
Orion	2917	$\frac{2917}{740} \approx 3.942$	3
		Total	25

INSTRUCTOR NOTE

The most difficult part of using the Jefferson plan is finding the modified standard divisor. It is a matter of trying different divisors and then finding the sum of the whole number parts of the quotients. A divisor that yields the sum that is equal to the total number of representatives is a modified standard divisor. There is a spreadsheet program on our companion site at CengageBrain.com that will enable students to try different divisors until the correct sum of representatives is obtained. See also the Calculator Note on page 173.

The table below shows how the results of the Hamilton and Jefferson apportionment methods differ. Note that each method assigns a different number of representatives to certain states.

State	Population	Hamilton plan	Jefferson plan
Apus	11,123	14	15
Libra	879	1	1
Draco	3518	4	4
Cephus	1563	2	2
Orion	2917	4	3
	Total	25	25

Although we have applied apportionment to allocating representatives to a congress, there are many other applications of apportionment. For instance, nurses can be assigned to hospitals according to the number of patients requiring care; police officers can be assigned to precincts based on the number of reported crimes; math classes can be scheduled based on student demand for those classes.

EXAMPLE 1 Apportioning Board Members Using the Hamilton and Jefferson Methods

Ruben County

City	Population
Cardiff	7020
Solana	2430
Vista	1540
Pauma	3720
Pacific	5290

Suppose the 18 members on the board of the Ruben County environmental agency are selected according to the populations of the five cities in the county, as shown in the table at the left.

a. Use the Hamilton method to determine the number of board members each city should have.

b. Use the Jefferson method to determine the number of board members each city should have.

Solution

a. First find the total population of the cities.

$$7020 + 2430 + 1540 + 3720 + 5290 = 20{,}000$$

Now calculate the standard divisor.

$$\text{Standard divisor} = \frac{\text{population of the cities}}{\text{number of board members}} = \frac{20{,}000}{18} \approx 1111.11$$

Use the standard divisor to find the standard quota for each city.

City	Population	Quotient	Standard quota	Number of board members
Cardiff	7020	$\frac{7020}{1111.11} \approx 6.318$	6	6
Solana	2430	$\frac{2430}{1111.11} \approx 2.187$	2	2
Vista	1540	$\frac{1540}{1111.11} \approx 1.386$	1	2
Pauma	3720	$\frac{3720}{1111.11} \approx 3.348$	3	3
Pacific	5290	$\frac{5290}{1111.11} \approx 4.761$	4	5
		Total	16	18

The sum of the standard quotas is 16, so we must add 2 more members. The two cities with the largest decimal remainders are Pacific and Vista. Each of these two cities gets one additional board member. Thus the composition of the environmental board using the Hamilton method is Cardiff: 6, Solana: 2, Vista: 2, Pauma: 3, and Pacific: 5.

b. To use the Jefferson method, we must find a *modified* standard divisor that is less than the standard divisor we calculated in part a. We must do this by trial and error. For instance, if we choose 925 as the modified standard divisor, we have the following result.

City	Population	Quotient	Number of board members
Cardiff	7020	$\frac{7020}{925} \approx 7.589$	7
Solana	2430	$\frac{2430}{925} \approx 2.627$	2
Vista	1540	$\frac{1540}{925} \approx 1.665$	1
Pauma	3720	$\frac{3720}{925} \approx 4.022$	4
Pacific	5290	$\frac{5290}{925} \approx 5.719$	5
		Total	19

This result yields too many board members. Thus we must increase the modified standard divisor. By experimenting with different divisors, we find that 950 gives the correct number of board members, as shown in the table on the next page.

CALCULATOR NOTE

Using lists and the iPart function (which returns only the whole number part of a number) of a TI-83/84 calculator can be helpful when trying to find a modified standard divisor.

Press STAT ENTER to display the list editor. If a list is present in L1, use the up arrow key to highlight L1, then press CLEAR ENTER. Enter the populations of each city in L1.

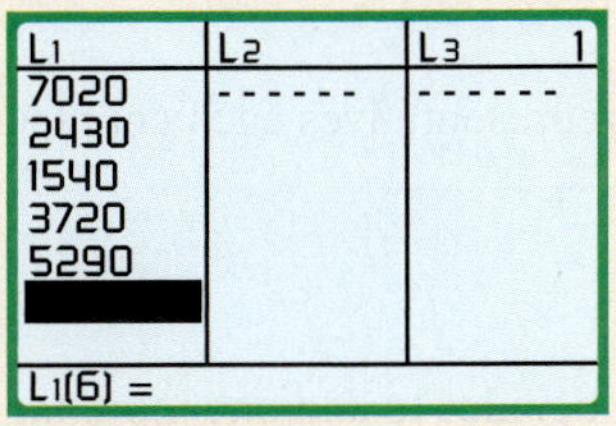

Press 2nd [QUIT]. To divide each number in L1 by a modified divisor (we are using 925 in part b of this example), enter the following.

MATH ▶ 3 2nd [L1]

÷ 925)

STO 2nd [L2] ENTER

The standard quotas are shown on the screen (and stored in the list L2). The sum of these numbers is 19, one more than the desired number of representatives.

CALCULATOR NOTE

Pressing 2nd [ENTRY] on a TI-83/84 calculator repeats the previously entered calculation, which can then be edited. This is a useful shortcut when experimenting with different modified standard divisors. For example, the Calculator Note on page 173 concludes with a calculation using 925 as a modified standard divisor. To repeat the computation using 950 as the modified standard divisor, press 2nd [ENTRY] and use the arrow keys to change 925 to 950. Then press ENTER.

City	Population	Quotient	Number of board members
Cardiff	7020	$\frac{7020}{950} \approx 7.389$	7
Solana	2430	$\frac{2430}{950} \approx 2.558$	2
Vista	1540	$\frac{1540}{950} \approx 1.621$	1
Pauma	3720	$\frac{3720}{950} \approx 3.916$	3
Pacific	5290	$\frac{5290}{950} \approx 5.568$	5
		Total	18

Thus the composition of the environmental board using the Jefferson method is Cardiff: 7, Solana: 2, Vista: 1, Pauma: 3, and Pacific: 5.

European countries' populations, 2015

Country	Population
France	66,550,000
Germany	80,850,000
Italy	61,860,000
Spain	48,150,000
Belgium	11,320,000

SOURCE: U.S. Census Bureau, International Data Base

CHECK YOUR PROGRESS 1 Suppose the 20 members of a committee from five European countries are selected according to the populations of the five countries, as shown in the table at the left.

a. Use the Hamilton method to determine the number of representatives each country should have.

b. Use the Jefferson method to determine the number of representatives each country should have.

Solution See page S12.

Suppose that the environmental agency in Example 1 decides to add one more member to the board even though the population of each city remains the same. The total number of members is now 19, and we must determine how the members of the board will be apportioned.

The standard divisor is now $\frac{20{,}000}{19} \approx 1052.63$. Using Hamilton's method, the calculations necessary to apportion the board members are shown below.

City	Population	Quotient	Standard quota	Number of board members
Cardiff	7020	$\frac{7020}{1052.63} \approx 6.669$	6	7
Solana	2430	$\frac{2430}{1052.63} \approx 2.309$	2	2
Vista	1540	$\frac{1540}{1052.63} \approx 1.463$	1	1
Pauma	3720	$\frac{3720}{1052.63} \approx 3.534$	3	4
Pacific	5290	$\frac{5290}{1052.63} \approx 5.026$	5	5
		Total	17	19

The table below summarizes the number of board members each city would have if the board consisted of 18 members (Example 1) or 19 members.

City	Hamilton apportionment with 18 board members	Hamilton apportionment with 19 board members
Cardiff	6	7
Solana	2	2
Vista	2	1
Pauma	3	4
Pacific	5	5
Total	18	19

Note that although one more board member was added, Vista lost a board member, even though the populations of the cities did not change. This is called the **Alabama paradox** and has a negative effect on fairness. In the interest of fairness, an apportionment method should not exhibit the Alabama paradox. (See the Math Matters below for other paradoxes.)

MATH MATTERS Apportionment Paradoxes

The Alabama paradox, although it was not given that name until later, was first noticed after the 1870 census. At the time, the House of Representatives had 270 seats. However, when the number of representatives in the House was increased to 280 seats, Rhode Island lost a representative.

After the 1880 census, C. W. Seaton, the chief clerk of the U.S. Census Office, calculated the number of representatives each state would have if the number were set at some number between 275 and 300. He noticed that when the number of representatives was increased from 299 to 300, Alabama lost a representative.

There are other paradoxes that involve apportionment methods. Two of them are the *population paradox* and the *new states paradox.* It is possible for the population of one state to be increasing faster than that of another state and for the state still to lose a representative. This is an example of the **population paradox**.

In 1907, when Oklahoma was added to the Union, the size of the House was increased by five representatives to accommodate Oklahoma's population. However, when the complete apportionment of the Congress was recalculated, New York lost a seat and Maine gained a seat. This is an example of the **new states paradox**.

Fairness in Apportionment

As we have seen, the choice of apportionment method affects the number of representatives a state will have. Given that fact, mathematicians and others have tried to work out an apportionment method that is fair. The difficulty lies in trying to define what is fair, so we will try to state conditions by which an apportionment plan is judged fair.

One criterion of fairness for an apportionment plan is that it should satisfy the *quota rule.*

Quota Rule

The number of representatives apportioned to a state is the standard quota or one more than the standard quota.

We can show that the Jefferson plan, for instance, does not always satisfy the quota rule by calculating the standard quota of Apus (see page 171).

$$\text{Standard quota} = \frac{\text{population of Apus}}{\text{standard divisor}} = \frac{11{,}123}{800} \approx 13$$

The standard quota of Apus is 13. However, the Jefferson plan assigns 15 representatives to that state (see page 172), two more than its standard quota. Therefore, the Jefferson method violates the quota rule.

Another measure of fairness is *average constituency.* This is the population of a state divided by the number of representatives from the state and then rounded to the nearest whole number.

Average Constituency

$$\text{Average constituency} = \frac{\text{population of a state}}{\text{number of representatives from the state}}$$

Consider the two states Hampton and Shasta in the table below.

State	Population	Representatives	Average constituency
Hampton	16,000	10	$\frac{16{,}000}{10} = 1600$
Shasta	8340	5	$\frac{8340}{5} = 1668$

Because the average constituencies are approximately equal, it seems natural to say that both states are equally represented. See the Take Note at the left.

TAKE NOTE

The idea of average constituency is an essential aspect of our democracy. To understand this, suppose state A has an average constituency of 1000 and state B has an average constituency of 10,000. When a bill is voted on in the House of Representatives, each vote has equal weight. However, a vote from a representative from state A would represent 1000 people, but a vote from a representative from state B would represent 10,000 people. Consequently, in this situation, we do not have "equal representation."

QUESTION Although the average constituencies of Hampton and Shasta are approximately equal, which state has the more favorable representation?

Now suppose that one representative will be added to one of the states. Which state is more deserving of the new representative? In other words, to be fair, which state should receive the new representative?

The changes in the average constituency are shown below.

State	Average constituency (old)	Average constituency (new)
Hampton	$\frac{16{,}000}{10} = 1600$	$\frac{16{,}000}{11} \approx 1455$
Shasta	$\frac{8340}{5} = 1668$	$\frac{8340}{6} = 1390$

From the table, there are two possibilities for adding one representative. If Hampton receives the representative, its average constituency will be 1455 and Shasta's will remain at 1668. The difference in the average constituencies is $1668 - 1455 = 213$. This difference is called the *absolute unfairness of the apportionment.*

ANSWER Because Hampton's average constituency is smaller than Shasta's, Hampton has the more favorable representation.

Absolute Unfairness of an Apportionment

The **absolute unfairness of an apportionment** is the absolute value of the difference between the average constituency of state A and the average constituency of state B.

$$|\text{Average constituency of A} - \text{average constituency of B}|$$

If Shasta receives the representative, its average constituency will be 1390 and Hampton's will remain at 1600. The absolute unfairness of apportionment is 1600 − 1390 = 210. This is summarized below.

	Hampton's average constituency	Shasta's average constituency	Absolute unfairness of apportionment
Hampton receives the new representative	1455	1668	213
Shasta receives the new representative	1600	1390	210

Because the smaller absolute unfairness of apportionment occurs if Shasta receives the new representative, it might seem that Shasta should receive the representative. However, this is not necessarily true.

To understand this concept, let's consider a somewhat different situation. Suppose an investor makes two investments, one of \$10,000 and another of \$20,000. One year later, the first investment is worth \$11,000 and the second investment is worth \$21,500. This is shown in the table below.

	Original investment	One year later	Increase
Investment A	\$10,000	\$11,000	\$1000
Investment B	\$20,000	\$21,500	\$1500

Although there is a larger increase in investment B, the increase *per dollar* of the original investment is $\frac{1500}{20{,}000} = 0.075$. On the other hand, the increase per dollar of investment A is $\frac{1000}{10{,}000} = 0.10$. Another way of saying this is that each \$1 of investment A produced a return of 10 cents (\$0.10), whereas each \$1 of investment B produced a return of 7.5 cents (\$0.075). Therefore, even though the increase in investment A was less than the increase in investment B, investment A was more productive.

A similar process is used when deciding which state should receive another representative. Rather than look at the difference in the absolute unfairness of apportionment, we determine the *relative unfairness* of adding the representative.

TAKE NOTE

The average constituency in the denominator is computed *after* the new representative is added.

Relative Unfairness of an Apportionment

The **relative unfairness of an apportionment** is the quotient of the absolute unfairness of the apportionment and the average constituency of the state receiving the new representative.

$$\frac{\text{Absolute unfairness of the apportionment}}{\text{Average constituency of the state receiving the new representative}}$$

EXAMPLE 2 Determine the Relative Unfairness of an Apportionment

Determine the relative unfairness of an apportionment that gives a new representative to Hampton rather than Shasta.

Solution

Using the table for Hampton and Shasta shown on the preceding page, we have

Relative unfairness of the apportionment

$$= \frac{\text{absolute unfairness of the apportionment}}{\text{average constituency of Hampton with a new representative}}$$

$$= \frac{213}{1455} \approx 0.146$$

The relative unfairness of the apportionment is approximately 0.146.

CHECK YOUR PROGRESS 2 Determine the relative unfairness of an apportionment that gives a new representative to Shasta rather than Hampton.

Solution See page S13.

The following principle uses the relative unfairness of an apportionment to determine how to add a representative to an existing apportionment.

Apportionment Principle

When adding a new representative to a state, the representative is assigned to the state in such a way as to give the smallest relative unfairness of apportionment.

From Example 2, the relative unfairness of adding a representative to Hampton is approximately 0.146. From Check Your Progress 2, the relative unfairness of adding a representative to Shasta is approximately 0.151. Because the smaller relative unfairness results from adding the representative to Hampton, that state should receive the new representative.

Although we have focused on assigning representatives to states, the apportionment principle can be used in many other situations.

EXAMPLE 3 Use the Apportionment Principle

The table below shows the number of paramedics and the annual number of paramedic calls for two cities. If a new paramedic is hired, use the apportionment principle to determine to which city the paramedic should be assigned.

	Paramedics	Annual paramedic calls
Tahoe	125	17,526
Erie	143	22,461

Solution

Calculate the relative unfairness of the apportionment that assigns the paramedic to Tahoe and the relative unfairness of the apportionment that assigns the paramedic to Erie. In this case, average constituency is the number of annual paramedic calls divided by the number of paramedics.

	Tahoe's annual paramedic calls per paramedic	Erie's annual paramedic calls per paramedic	Absolute unfairness of apportionment
Tahoe receives a new paramedic	$\frac{17,526}{125+1} \approx 139$	$\frac{22,461}{143} \approx 157$	$157 - 139 = 18$
Erie receives a new paramedic	$\frac{17,526}{125} \approx 140$	$\frac{22,461}{143+1} \approx 156$	$156 - 140 = 16$

If Tahoe receives the new paramedic, the relative unfairness of the apportionment is

$$\begin{aligned}&\text{Relative unfairness of the apportionment}\\&= \frac{\text{absolute unfairness of the apportionment}}{\text{Tahoe's average constituency with a new paramedic}}\\&= \frac{18}{139} \approx 0.129\end{aligned}$$

If Erie receives the new paramedic, the relative unfairness of the apportionment is

$$\begin{aligned}&\text{Relative unfairness of the apportionment}\\&= \frac{\text{absolute unfairness of the apportionment}}{\text{Erie's average constituency with a new paramedic}}\\&= \frac{16}{156} \approx 0.103\end{aligned}$$

Because the smaller relative unfairness results from adding the paramedic to Erie, that city should receive the paramedic.

CHECK YOUR PROGRESS 3 The table below shows the number of first and second grade teachers in a school district and the number of students in each of those grades. If a new teacher is hired, use the apportionment principle to determine to which grade the teacher should be assigned.

	Number of teachers	Number of students
First grade	512	12,317
Second grade	551	15,439

Solution See page S13.

Huntington-Hill Apportionment Method

As we mentioned earlier, the members of the House of Representatives are apportioned among the states every 10 years. The present method used by the House is based on the apportionment principle and is called the *method of equal proportions* or the *Huntington-Hill method.* This method has been used since 1940.

The Huntington-Hill method is implemented by calculating what is called a *Huntington-Hill number.* This number is derived from the apportionment principle. See Exercise 35, page 188.

Huntington-Hill Number

The value of $\frac{(P_A)^2}{a(a+1)}$, where P_A is the population of state A and a is the current number of representatives from state A, is called the **Huntington-Hill number** for state A.

HISTORICAL NOTE

According to the U.S. Bureau of the Census, methods of apportioning the House of Representatives have changed over time.

1790–1830: Jefferson method

1840: Webster method (See the paragraph following Exercise 26, page 187.)

1850–1900: Hamilton method

1910, 1930: Webster method

Note that 1920 is missing. In direct violation of the U.S. Constitution, the House of Representatives failed to reapportion the House in 1920.

1940–2010: Method of equal proportions or the Huntington-Hill method

All apportionment plans enacted by the House have created controversy. For instance, the constitutionality of the Huntington-Hill method was challenged by Montana in 1992 because it lost a seat to Washington after the 1990 census. For more information on this subject, see http://www.census.gov/topics/public-sector/congressional-apportionment.html.

When the Huntington-Hill method is used to apportion representatives between two states, the state with the greater Huntington-Hill number receives the next representative. This method can be extended to more than two states.

Huntington-Hill Apportionment Principle

When there is a choice of adding one representative to one of several states, the representative should be added to the state with the greatest Huntington-Hill number.

EXAMPLE 4 Use the Huntington-Hill Apportionment Principle

The table below shows the numbers of lifeguards that are assigned to three different beaches and the numbers of rescues made by lifeguards at those beaches. Use the Huntington-Hill apportionment principle to determine to which beach a new lifeguard should be assigned.

Beach	Number of lifeguards	Number of rescues
Mellon	37	1227
Donovan	51	1473
Ferris	24	889

Solution

Calculate the Huntington-Hill number for each of the beaches. In this case, the population is the number of rescues and the number of representatives is the number of lifeguards.

Mellon: $\dfrac{1227^2}{37(37+1)} \approx 1071$ Donovan: $\dfrac{1473^2}{51(51+1)} \approx 818$ Ferris: $\dfrac{889^2}{24(24+1)} \approx 1317$

Ferris has the greatest Huntington-Hill number. Thus, according to the Huntington-Hill Apportionment Principle, the new lifeguard should be assigned to Ferris.

CHECK YOUR PROGRESS 4 A university has a president's council that is composed of students from each of the undergraduate classes. If a new student representative is added to the council, use the Huntington-Hill apportionment principle to determine which class the new student council member should represent.

Class	Number of representatives	Number of students
First year	12	2015
Second year	10	1755
Third year	9	1430
Fourth year	8	1309

Solution See page S13.

TAKE NOTE

The advantage of using the Huntington-Hill apportionment principle, rather than calculating relative unfairness, occurs when there are many states that could receive the next representative. For instance, if we were to use relative unfairness to determine which of four states should receive a new representative, it would be necessary to compute the relative unfairness for every possible pairing of the states—a total of 24 computations. However, using the Huntington-Hill method, we need only calculate the Huntington-Hill number for each state—a total of four calculations. In a sense, the Huntington-Hill number provides a shortcut for applying the relative unfairness method.

Now that we have looked at various apportionment methods, it seems reasonable to ask which is the best method. Unfortunately, all apportionment methods have some flaws. This was proved by Michael Balinski and H. Peyton Young in 1982.

TAKE NOTE

When we say that an apportionment method violates a rule like the quota rule, we don't mean that every time the method is used the quota rule is violated. We mean that it is *possible* for the method to violate the quota rule.

Balinski-Young Impossibility Theorem

Any apportionment method either will violate the quota rule or will produce paradoxes such as the Alabama paradox.

The following table lists flaws that may occur in the application of different apportionment methods.

Summary of Apportionment Methods and Possible Flaws

Apportionment methods	Flaws			
	Violation of the quota rule	**Alabama paradox**	**Population paradox**	**New states paradox**
Hamilton plan	Cannot occur	May occur	May occur	May occur
Jefferson plan	May violate	Cannot occur	Cannot occur	Cannot occur
Huntington-Hill method	May violate	Cannot occur	Cannot occur	Cannot occur
Webster method*	May violate	Cannot occur	Cannot occur	Cannot occur

*The Webster method is explained in Exercise Set 4.1 after Exercise 26.

Although there is no perfect apportionment method, Balinski and Young went on to present a strong case that the Webster method (following Exercise 26 on page 187) is the system that most closely satisfies the goal of one person, one vote. However, political expediency sometimes overrules mathematical proof. Some historians have suggested that although the Huntington-Hill apportionment method was better than some of the previous methods, President Franklin Roosevelt chose this method in 1941 because it alloted one more seat to Arkansas and one less to Michigan. This essentially meant that the House of Representatives would have one more seat for the Democrats, Roosevelt's party.

EXCURSION

Apportioning the 1790 House of Representatives

The first apportionment of the House of Representatives, using the 1790 census, is given in the following table. This apportionment was calculated by using the Jefferson method. (See our companion site at CengageBrain.com for an Excel spreadsheet that will help with the computations.)

HISTORICAL NOTE

According to the U.S. Constitution, each state must have at least one representative to the House of Representatives. The remaining representatives are then assigned to the states using, since 1941, the Huntington-Hill apportionment principle.

POINT OF INTEREST

For most states, the populations shown in the table are not the actual populations of the states because each slave was counted as only three-fifths of a person. (See the Historical Note on page 170.) For instance, the actual population of Connecticut was 237,946, of which 2764 were slaves. For apportionment purposes, the number of slaves was subtracted, and then $\frac{3}{5}$ of that population was added back.

$$237{,}946 - 2764 + \frac{3}{5}(2764) \approx 236{,}841$$

Apportionment Using the Jefferson Method, 1790

State	Population	Number of representatives
Connecticut	236,841	7
Delaware	55,540	1
Georgia	70,835	2
Maryland	278,514	8
Massachusetts	475,327	14
Kentucky	68,705	2
New Hampshire	141,822	4
Vermont	85,533	2
New York	331,591	10
New Jersey	179,570	5
Pennsylvania	432,879	13
North Carolina	353,523	10
South Carolina	206,236	6
Virginia	630,560	19
Rhode Island	68,446	2

SOURCE: U.S. Census Bureau

EXCURSION EXERCISES

1. Verify this apportionment using the Jefferson method. You will have to experiment with various modified standard divisors until you reach the given representation. See the Calculator Note on page 173.
2. Find the apportionment that would have resulted if the Hamilton method had been used.
3. Give each state one representative. Use the Huntington-Hill method with $a = 1$ to determine the state that receives the next representative. The Calculator Note on page 173 will help. With the populations stored in L1, enter [2nd] L1 [x²] ÷ 2 [STO] [2nd] L2 [ENTER]. Now scroll through L2 to find the largest number.
4. Find the apportionment that would have resulted if the Huntington-Hill method (the one used for the 2010 census) had been used in 1790. See our companion site at CengageBrain.com for a spreadsheet that will help with the calculations.

EXERCISE SET 4.1

1. Explain how to calculate the standard divisor of an apportionment for a total population p with n items to apportion.

2. **Teacher Aides** A total of 25 teacher aides are to be apportioned among seven classes at a new elementary school. The enrollments in the seven classes are shown in the following table.

Class	Number of students
Kindergarten	38
First grade	39
Second grade	35
Third grade	27
Fourth grade	21
Fifth grade	31
Sixth grade	33
Total	224

   a. Determine the standard divisor. What is the meaning of the standard divisor in the context of this exercise?

   b. Use the Hamilton method to determine the number of teacher aides to be apportioned to each class.

   c. Use the Jefferson method to determine the number of teacher aides to be apportioned to each class. Is this apportionment in violation of the quota rule?

   d. How do the apportionment results produced using the Jefferson method compare with the results produced using the Hamilton method?

3. In the Hamilton apportionment method, explain how to calculate the standard quota for a particular state (group).

4. What is the quota rule?

5. **Governing Boards** The following table shows how the average constituency changes for two regional governing boards, Joshua and Salinas, when a new representative is added to each board.

	Joshua's average constituency	Salinas's average constituency
Joshua receives new board member	1215	1547
Salinas receives new board member	1498	1195

   a. Determine the relative unfairness of an apportionment that gives a new board member to Joshua rather than to Salinas. Round to the nearest thousandth.

   b. Determine the relative unfairness of an apportionment that gives a new board member to Salinas rather than to Joshua. Round to the nearest thousandth.

   c. Using the apportionment principle, determine which regional governing board should receive the new board member.

6. **Forest Rangers** The table below shows how the average constituency changes when two different national parks, Evergreen State Park and Rust Canyon Preserve, add a new forest ranger.

	Evergreen State Park's average constituency	Rust Canyon Preserve's average constituency
Evergreen receives new forest ranger	466	638
Rust Canyon receives new forest ranger	650	489

   a. Determine the relative unfairness of an apportionment that gives a new forest ranger to Evergreen rather than to Rust Canyon. Round to the nearest thousandth.

   b. Determine the relative unfairness of an apportionment that gives a new forest ranger to Rust Canyon rather than to Evergreen. Round to the nearest thousandth.

   c. Using the apportionment principle, determine which national park should receive the new forest ranger.

7. Sales Associates The table below shows the number of sales associates and the average number of customers per day at a company's two department stores. The company is planning to add a new sales associate to one of the stores. Use the apportionment principle to determine which store should receive the new employee.

Shopping mall location	Number of sales associates	Average number of customers per day
Summer Hill Galleria	587	5289
Seaside Mall Galleria	614	6215

8. Hospital Interns The table below shows the number of interns and the average number of patients admitted each day at two different hospitals. The hospital administrator is planning to add a new intern to one of the hospitals. Use the apportionment principle to determine which hospital should receive the new intern.

Hospital location	Number of interns	Average number of patients admitted per day
South Coast Hospital	128	518
Rainer Hospital	145	860

9. House of Representatives The U.S. House of Representatives currently has 435 members to represent the 308,745,538 citizens of the U.S. as determined by the 2010 census.

a. Calculate the standard divisor for the apportionment of these representatives and explain the meaning of this standard divisor in the context of this exercise.

b. According to the 2010 census, the population of Delaware was 897,934. Delaware currently has only one representative in the House of Representatives. Is Delaware currently overrepresented or underrepresented in the House of Representatives? Explain.

c. According to the 2010 census, the population of Vermont was 625,741. Vermont currently has only one representative in the House of Representatives. Is Vermont currently overrepresented or underrepresented in the House of Representatives? Explain.

10. College Enrollment The following table shows the enrollment for each of the four divisions of a college. The four divisions are liberal arts, business, humanities, and science. There are 180 new computers that are to be apportioned among the divisions based on the enrollments.

Ruth Peterkin/Shutterstock.com

Division	Enrollment
Liberal arts	3455
Business	5780
Humanities	1896
Science	4678
Total	15,809

a. What is the standard divisor for an apportionment of the computers? What is the meaning of the standard divisor in the context of this exercise?

b. Use the Hamilton method to determine the number of computers to be apportioned to each division.

c. If the computers are to be apportioned using the Jefferson method, explain why neither 86 nor 87 can be used as a modified standard divisor. Explain why 86.5 can be used as a modified standard divisor.

d. Explain why the modified standard divisor used in the Jefferson method cannot be larger than the standard divisor.

e. Use the Jefferson method to determine the number of computers to be apportioned to each division. Is this apportionment in violation of the quota rule?

f. How do the apportionment results produced using the Jefferson method compare with the results produced using the Hamilton method?

11. Medical Care A hospital district consists of six hospitals. The district administrators have decided that 48 new nurses should be apportioned based on the number of beds in each of the hospitals. The following table shows the number of beds in each hospital.

Hospital	Number of beds
Sharp	242
Palomar	356
Tri-City	308
Del Raye	190
Rancho Verde	275
Bel Aire	410
Total	1781

a. Determine the standard divisor. What is the meaning of the standard divisor in the context of this exercise?

b. Use the Hamilton method to determine the number of nurses to be apportioned to each hospital.

c. Use the Jefferson method to determine the number of nurses to be apportioned to each hospital.

d. How do the apportionment results produced using the Jefferson method compare with the results produced using the Hamilton method?

12. What is the Alabama paradox?

13. What is the population paradox?

14. What is the new states paradox?

15. What is the Balinski-Young Impossibility Theorem?

16. Apportionment of Projectors Consider the apportionment of 27 digital projectors for a school district with four campus locations labeled A, B, C, and D. The following table shows the apportionment of the projectors using the Hamilton method.

tatniz Shutterstock.com

Campus	A	B	C	D
Enrollment	840	1936	310	2744
Apportionment of 27 projectors	4	9	1	13

a. If the number of projectors to be apportioned increases from 27 to 28, what will be the apportionment if the Hamilton method is used? Will the Alabama paradox occur? Explain.

b. If the number of projectors to be apportioned using the Hamilton method increases from 28 to 29, will the Alabama paradox occur? Explain.

17. Hotel Management A company operates four resorts. The CEO of the company decides to use the Hamilton method to apportion 115 new LCD television sets to the resorts based on the number of guest rooms at each resort.

Resort	A	B	C	D
Number of guest rooms	23	256	182	301
Apportionment of 115 televisions	4	39	27	45

a. If the number of television sets to be apportioned by the Hamilton method increases from 115 to 116, will the Alabama paradox occur?

b. If the number of television sets to be apportioned by the Hamilton method increases from 116 to 117, will the Alabama paradox occur?

c. If the number of television sets to be apportioned by the Hamilton method increases from 117 to 118, will the Alabama paradox occur?

18. College Security A college apportions 40 security personnel among three education centers according to their enrollments. The following table shows the present enrollments at the three centers.

Center	A	B	C
Enrollment	356	1054	2590

a. Use the Hamilton method to apportion the security personnel.

b. After one semester, the centers have the following enrollments.

Center	A	B	C
Enrollment	370	1079	2600

Center A has an increased enrollment of 14 students, which is an increase of $\frac{14}{356} \approx 0.039 = 3.9\%$. Center B has an increased enrollment of 25 students, which is an increase of $\frac{25}{1054} \approx 0.024 = 2.4\%$.

If the security personnel are reapportioned using the Hamilton method, will the population paradox occur? Explain.

19. Management Scientific Research Corporation has offices in Boston and Chicago. The number of employees at each office is shown in the following table. There are 22 vice presidents to be apportioned between the offices.

Office	Boston	Chicago
Employees	151	1210

a. Use the Hamilton method to find each office's apportionment of vice presidents.

b. The corporation opens an additional office in San Francisco with 135 employees and decides to have a total of 24 vice presidents. If the vice presidents are reapportioned using the Hamilton method, will the new states paradox occur? Explain.

20. Education The science division of a college consists of three departments: mathematics, physics, and chemistry. The number of students enrolled in each department is shown in the following table. There are 19 clerical assistants to be apportioned among the departments.

Department	Math	Physics	Chemistry
Student enrollment	4325	520	1165

a. Use the Hamilton method to find each department's apportionment of clerical assistants.

b. The division opens a new computer science department with a student enrollment of 495. The division decides to have a total of 20 clerical assistants. If the clerical assistants are reapportioned using the Hamilton method, will the new states paradox occur? Explain.

21. Elementary School Teachers The following table shows the number of fifth and sixth grade teachers in a school district and the number of students in the two grades. The number of teachers for each of the grade levels was determined by using the Huntington-Hill apportionment method.

	Number of teachers	Number of students
Fifth grade	19	604
Sixth grade	21	698

The district has decided to hire a new teacher for either the fifth or the sixth grade.

a. Use the apportionment principle to determine to which grade the new teacher should be assigned.

b. Use the Huntington-Hill apportionment principle to determine to which grade the new teacher should be assigned. How does this result compare with the result in part a?

22. Social Workers The following table shows the number of social workers and the number of cases (the case load) handled by the social workers for two offices. The number of social workers for each office was determined by using the Huntington-Hill apportionment method.

	Number of social workers	Case load
Hill Street office	20	584
Valley office	24	712

A new social worker is to be hired for one of the offices.

a. Use the apportionment principle to determine to which office the social worker should be assigned.

b. Use the Huntington-Hill apportionment principle to determine to which office the new social worker should be assigned. How does this result compare with the result in part a?

c. The results of part b indicate that the new social worker should be assigned to the Valley office. At this moment the Hill Street office has 20 social workers and the Valley office has 25 social workers. Use the Huntington-Hill apportionment principle to determine to which office the *next* new social worker should be assigned. Assume the case loads remain the same.

23. Computer Usage The table below shows the number of computers that are assigned to four different schools and the number of students in those schools. Use the Huntington-Hill apportionment principle to determine to which school a new computer should be assigned.

School	Number of computers	Number of students
Rose	26	625
Lincoln	22	532
Midway	26	620
Valley	31	754

24. The population of Illinois increased by over 400,000 from 2000 to 2010, yet the state lost a seat in the House of Representatives. How must Illinois's population increase have compared to the changes of other states' populations?

25. The population of Louisiana increased 1.4% from 2000 to 2010 and the state lost a representative. New York's population increased at a higher rate, 2.1%, yet the state lost two representatives. Can you explain the seeming discrepancy?

26. House of Representatives Currently, the U.S. House of Representatives has 435 members who have been apportioned by the Huntington-Hill apportionment method. If the number of representatives were to be increased to 436, then, according to the 2010 census figures, North Carolina would be given the new representative. How must North Carolina's 2010 census Huntington-Hill number compare with the 2010 census Huntington-Hill numbers for the other 49 states? Explain.

The *Webster method of apportionment* is similar to the Jefferson method except that quotas are rounded up when the decimal remainder is 0.5 or greater and quotas are rounded down when the decimal remainder is less than 0.5. This method of rounding is referred to as **rounding to the nearest integer**. For instance, using the Jefferson method, a quotient of 15.91 would be rounded to 15; using the Webster method, it would be rounded to 16. A quotient of 15.49 would be rounded to 15 in both methods.

To use the Webster method you must still experiment to find a *modified standard divisor* for which the sum of the quotas rounded to the nearest integer equals the number of items to be apportioned. Although the Webster method is similar to the Jefferson method, the Webster method is generally more difficult to apply because the modified divisor may be less than, equal to, or more than the standard divisor.

A calculator can be very helpful in testing a possible modified standard divisor m when applying the Webster method of apportionment. For instance, on a TI-83/84 calculator, first store the populations in L1. Then enter

```
iPart(L1/m+.5)→L2
```

where m is the modified divisor. Then press the ENTER key. Now scroll through L2 to view the quotas rounded to the *nearest integer.* If the sum of these quotas equals the total number of items to be apportioned, then you have the Webster apportionment. If the sum of the quotas in L2 is less than the total number of items to be apportioned, try a smaller modified standard divisor. If the sum of the quotas in L2 is greater than the total number of items to be apportioned, try a larger modified standard divisor.

27. Computer Usage Use the Webster method to apportion the computers in Exercise 10, page 184. How do the apportionment results produced using the Webster method compare with the results produced using the

a. Hamilton method?

b. Jefferson method?

28. Demographics The table below shows the populations of five European countries. A committee of 20 people from these countries is to be formed using the Webster method of apportionment.

Country	Population
France	66,550,000
Germany	80,850,000
Italy	61,860,000
Spain	48,150,000
Belgium	11,320,000
Total	268,730,000

SOURCE: U.S. Census Bureau, International Data Base

a. Explain why 13,500,000 *cannot* be used as a modified standard divisor.

b. Explain why 13,750,000 *can* be used as a modified standard divisor.

c. Use the Webster apportionment method to determine the apportionment of the 20 committee members.

29. Which of the following apportionment methods can violate the quota rule?

- Hamilton method
- Jefferson method
- Webster method
- Huntington-Hill method

30. According to Michael Balinski and H. Peyton Young, which of the apportionment methods most closely satisfies the goal of one person, one vote?

31. What method is presently used to apportion the members of the U.S. House of Representatives?

EXTENSIONS

32. According to the 2010 census, what is the population of your state? How many representatives does your state have in the U.S. House of Representatives? Is your state underrepresented or overrepresented in the House of Representatives? Explain. (*Hint:* The result of Exercise 9a on page 184 shows that the ratio of representatives to citizens is about $\frac{1}{709{,}760}$.) How does your state's current number of representatives compare with the number of representatives it had after the 2000 census?

33. John Quincy Adams, the sixth president of the United States, proposed an apportionment method. Research this method, which is known as the Adams method of apportionment. Describe how this method works. Also indicate whether it satisfies the quota rule and whether it is susceptible to any paradoxes.

34. In the Huntington-Hill method of apportionment, each state is first given one representative and then additional representatives are assigned, one at a time, to the state currently having the highest Huntington-Hill number. This way of implementing the Huntington-Hill apportionment method is time consuming. Another process for implementing the Huntington-Hill apportionment method consists of using a modified divisor and a special rounding procedure that involves the *geometric mean* of two consecutive integers. Research this process for implementing the Huntington-Hill apportionment method. Apply this process to apportion 22 new security vehicles to each of the following schools, based on their student populations.

School	Number of students	Number of security vehicles
Del Mar	5230	?
Wheatly	12,375	?
West	8568	?
Mountain View	14,245	?

35. Deriving the Huntington-Hill Number The Huntington-Hill number is derived by using the apportionment principle. Let

P_A = population of state A
a = number of representatives from state A
P_B = population of state B
b = number of representatives from state B

Complete the following to derive the Huntington-Hill number.

a. Write the fraction that gives the average constituency of state A when it receives a new representative.

b. Write the fraction that gives the average constituency of state B without a new representative.

c. Express the relative unfairness of apportionment by giving state A the new representative in terms of the fractions from parts a and b.

d. Express the relative unfairness of apportionment by giving state B the new representative.

e. According to the apportionment principle, state A should receive the next representative instead of state B if the relative unfairness of giving the new representative to state A is *less than* the relative unfairness of giving the new representative to state B. Express this inequality in terms of the expressions in parts c and d.

f. Simplify the inequality and you will have the Huntington-Hill number.

SECTION 4.2 Introduction to Voting

Plurality Method of Voting

One of the most revered privileges that those of us who live in a democracy enjoy is the right to vote for our representatives. Sometimes, however, we are puzzled by the fact that the best candidate did not get elected. Unfortunately, because of the way our *plurality* voting system works, it is possible to elect someone or pass a proposition that has less than *majority support.* As we proceed through this section, we will look at the problems with plurality voting and alternatives to this system. We start with a definition.

TAKE NOTE

When an issue requires a **majority vote**, it means that more than 50% of the people voting must vote for the issue. This is not the same as a **plurality**, in which the person or issue with the most votes wins.

The Plurality Method of Voting

Each voter votes for one candidate, and the candidate with the most votes wins. The winning candidate does not have to have a majority of the votes.

EXAMPLE 1 Determine the Winner Using Plurality Voting

Fifty people were asked to rank their preferences of five varieties of chocolate candy, using 1 for their favorite and 5 for their least favorite. This type of ranking of choices is called a **preference schedule**. The results are shown in the table below.

	Rankings					
Caramel center	5	4	4	4	2	4
Vanilla center	1	5	5	5	5	5
Almond center	2	3	2	1	3	3
Toffee center	4	1	1	3	4	2
Solid chocolate	3	2	3	2	1	1
Number of voters:	17	11	9	8	3	2

TAKE NOTE

A preference schedule lists the number of people who gave a particular ranking. For example, the column shaded green means that 3 voters ranked solid chocolate first, caramel centers second, almond centers third, toffee centers fourth, and vanilla centers fifth.

According to this table, which variety of candy would win the taste test using the plurality voting system?

Solution

To answer the question, we will make a table showing the number of first-place votes for each candy.

	First-place votes
Caramel center	0
Vanilla center	17
Almond center	8
Toffee center	$11 + 9 = 20$
Solid chocolate	$3 + 2 = 5$

Because toffee centers received 20 first-place votes, this type of candy would win the plurality taste test.

CHECK YOUR PROGRESS 1 According to the table in Example 1, which variety of candy would win second place using the plurality voting system?

Solution See page S13.

Example 1 can be used to show the difference between plurality and majority. There were 20 first-place votes for toffee-centered chocolate, so it wins the taste test. However, toffee-centered chocolate was the first choice of only 40% $\left(\frac{20}{50} = 40\%\right)$ of the people voting. Thus less than half of the people voted for toffee-centered chocolate as number one, so it did not receive a majority vote.

MATH MATTERS Gubernatorial and Presidential Elections

In 1998, in a three-party race, plurality voting resulted in the election of former wrestler Jesse Ventura as governor of Minnesota, despite the fact that more than 60% of the state's voters did not vote for him. In fact, he won the governor's race with only 37% of the voters choosing him. Ventura won not because the majority of voters chose him, but because of the plurality voting method.

There are many situations that can be cited to show that plurality voting can lead to unusual results. When plurality voting is mixed with other voting methods, even the plurality winner may not win. One situation in which this can happen is in a U.S. presidential election. The president of the United States is elected not directly by popular vote but by the Electoral College.

In the 1824 presidential election, the approximate percent of the *popular* vote received by each candidate was: John Quincy Adams, 31%; Andrew Jackson, 43%; William Crawford, 13%; and Henry Clay, 13%. The vote in the Electoral College was John Quincy Adams, 84; Andrew Jackson, 99; William Crawford, 41; and Henry Clay, 37. Because none of the candidates had received 121 electoral votes (the number needed to win in 1824), by the Twelfth Amendment to the U.S. Constitution, the House of Representatives decided the election. The House elected John Quincy Adams, thereby electing a president who had less than one-third of the popular vote.

There have been three other instances when the candidate with the most popular votes in a presidential election was not elected president: 1876, Hayes versus Tilden; 1888, Harrison versus Cleveland; and 2000, Bush versus Gore.

Borda Count Method of Voting

The problem with plurality voting is that alternative choices are not considered. For instance, the result of the Minnesota governor's contest might have been quite different if voters had been asked, "Choose the candidate you prefer, but if that candidate does not receive a *majority* of the votes, which candidate would be your second choice?"

To see why this might be a reasonable alternative to plurality voting, consider the following situation. Thirty-six senators are considering an educational funding measure. Because the senate leadership wants an educational funding measure to pass, the leadership first determines that the senators prefer measure A for \$50 million over measure B for \$30 million. However, because of an unexpected dip in state revenues, measure A is removed from consideration and a new measure, C, for \$15 million, is proposed. The senate leadership determines that senators favor measure B over measure C. In summary, we have

A majority of senators favor measure A over measure B.

A majority of senators favor measure B over measure C.

From these results, it seems reasonable to think that a majority of senators would prefer measure A over measure C. However, when the senators are asked about their preferences between the two measures, measure C is preferred over measure A. To understand how this could happen, consider the preference schedule for the senators shown in the following table.

	Rankings		
Measure A: $50 million	1	3	3
Measure B: $30 million	3	1	2
Measure C: $15 million	2	2	1
Number of senators:	15	12	9

Notice that 15 senators prefer measure A over measure C, but $12 + 9 = 21$ senators, a majority of the 36 senators, prefer measure C over measure A. According to the preference schedule, if all three measures were on the ballot, A would come in first, B would come in second, and C would come in third. However, if just A and C were on the ballot, C would win over A. This paradoxical result was first discussed by Jean C. Borda in 1770.

In an attempt to remove such paradoxical results from voting, Borda proposed that voters rank their choices by giving each choice a certain number of points.

The Borda Count Method of Voting

If there are n candidates or issues in an election, each voter ranks the candidates or issues by giving n points to the voter's first choice, $n - 1$ points to the voter's second choice, and so on, with the voter's least favorite choice receiving 1 point. The candidate or issue that receives the most total points is the winner.

Applying the Borda count method to the education measures, a measure receiving a first-place vote receives 3 points. (There are three different measures.) Each measure receiving a second-place vote receives 2 points, and each measure receiving a third-place vote receives 1 point. The calculations are shown below.

		Points per vote
Measure A:	15 first-place votes:	$15 \cdot 3 = 45$
	0 second-place votes:	$0 \cdot 2 = 0$
	21 third-place votes:	$21 \cdot 1 = 21$
	Total:	66
Measure B:	12 first-place votes:	$12 \cdot 3 = 36$
	9 second-place votes:	$9 \cdot 2 = 18$
	15 third-place votes:	$15 \cdot 1 = 15$
	Total:	69
Measure C:	9 first-place votes:	$9 \cdot 3 = 27$
	27 second-place votes:	$27 \cdot 2 = 54$
	0 third-place votes:	$0 \cdot 1 = 0$
	Total:	81

Using the Borda count method, measure C is the clear winner (even though it is not the plurality winner).

HISTORICAL NOTE

The issue of whether plurality voting methods are fair has been around for more than 200 years. Jean C. Borda (1733–1799) was a member of the French Academy of Sciences when he first started thinking about the way in which people were elected to the Academy. He was concerned that the plurality method of voting might not result in the best candidate being elected. The Borda count method was born out of these concerns. It was the first attempt to mathematically quantify voting systems.

TAKE NOTE

Paradoxes occur in voting only when there are three or more candidates or issues on a ballot. If there are only two candidates in a race, then the candidate receiving the majority of the votes cast is the winner. In a two-candidate race, the majority and the plurality are the same.

POINT OF INTEREST

One way to see the difference between a plurality voting system (sometimes called a "winner-take-all" method) and the Borda count method is to consider grades earned in school. Suppose a student is going to be selected for a scholarship based on grades. If one student has 10 As and 20 Fs and another student has five As and 25 Bs, it would seem that the second student should receive the scholarship. However, if the scholarship is awarded by the plurality of As, the first student will get the scholarship. The Borda count method is closely related to the method used to calculate grade-point average (GPA).

EXAMPLE 2 Use the Borda Count Method

The members of a club are going to elect a president from four nominees using the Borda count method. If the 100 members of the club mark their ballots as shown in the table below, who will be elected president?

	Rankings					
Avalon	2	2	2	2	3	2
Branson	1	4	4	3	2	1
Columbus	3	3	1	4	1	4
Dunkirk	4	1	3	1	4	3
Number of voters:	30	24	18	12	10	6

Solution

Using the Borda count method, each first-place vote receives 4 points, each second-place vote receives 3 points, each third-place vote receives 2 points, and each last-place vote receives 1 point. The summary for each candidate is shown below.

Avalon:	0 first-place votes	$0 \cdot 4 = 0$
	90 second-place votes	$90 \cdot 3 = 270$
	10 third-place votes	$10 \cdot 2 = 20$
	0 fourth-place votes	$0 \cdot 1 = 0$
		Total 290
Branson:	36 first-place votes	$36 \cdot 4 = 144$
	10 second-place votes	$10 \cdot 3 = 30$
	12 third-place votes	$12 \cdot 2 = 24$
	42 fourth-place votes	$42 \cdot 1 = 42$
		Total 240
Columbus:	28 first-place votes	$28 \cdot 4 = 112$
	0 second-place votes	$0 \cdot 3 = 0$
	54 third-place votes	$54 \cdot 2 = 108$
	18 fourth-place votes	$18 \cdot 1 = 18$
		Total 238
Dunkirk:	36 first-place votes	$36 \cdot 4 = 144$
	0 second-place votes	$0 \cdot 3 = 0$
	24 third-place votes	$24 \cdot 2 = 48$
	40 fourth-place votes	$40 \cdot 1 = 40$
		Total 232

Avalon has the largest total score. By the Borda count method, Avalon is elected president.

TAKE NOTE

Note in Example 2 that Avalon was the winner even though that candidate did not receive any first-place votes. The Borda count method was devised to allow voters to say, "If my first choice does not win, then consider my second choice."

CHECK YOUR PROGRESS 2 The preference schedule given in Example 1 for the 50 people who were asked to rank their preferences of five varieties of chocolate candy is shown again below.

	Rankings					
Caramel center	5	4	4	4	2	4
Vanilla center	1	5	5	5	5	5
Almond center	2	3	2	1	3	3
Toffee center	4	1	1	3	4	2
Solid chocolate	3	2	3	2	1	1
Number of voters:	17	11	9	8	3	2

Determine the taste test favorite using the Borda count method.

Solution See page S14.

Plurality with Elimination

A variation of the plurality method of voting is called *plurality with elimination.* Like the Borda count method, the method of plurality with elimination considers a voter's alternate choices.

Suppose that 30 members of a regional planning board must decide where to build a new airport. The airport consultants to the regional board have recommended four different sites. The preference schedule for the board members is shown in the following table.

	Rankings			
Apple Valley	3	1	2	3
Bremerton	2	3	3	1
Coachella	1	2	4	2
Del Mar	4	4	1	4
Number of ballots:	12	11	5	2

Using the plurality with elimination method, the board members first eliminate the site with the fewest number of first-place votes. If two or more of these alternatives have the same number of first-place votes, all are eliminated unless that would eliminate all alternatives. In that case, a different method of voting must be used. From the table above, Bremerton is eliminated because it received only two first-place votes. Now a vote is retaken using the following important assumption: *Voters do not change their preferences from round to round.* This means that after Bremerton is deleted, the 12 people in the first column would adjust their preferences so that Apple Valley becomes their second choice, Coachella remains their first choice, and Del Mar becomes their third choice. For the 11 voters in the second column, Apple Valley remains their first choice, Coachella remains their second choice, and Del Mar

TAKE NOTE

When the second round of voting occurs, the two ballots that listed Bremerton as the first choice must be adjusted. The second choice on those ballots becomes the first, the third choice becomes the second, and the fourth choice becomes the third. The order of preference does not change. Similar adjustments must be made to the 12 ballots that listed Bremerton as the second choice. Because that choice is no longer available, Apple Valley becomes the second choice and Del Mar becomes the third choice. Adjustments must be made to the 11 ballots that listed Bremerton as the third choice. The fourth choice of those ballots, Del Mar, becomes the third choice. A similar adjustment must be made for the five ballots that listed Bremerton as the third choice.

POINT OF INTEREST

Plurality with elimination is used to choose the city to host the Olympic games.

A variation of this method is also used to select the Academy Award nominees and, since 2009, the winner for best picture.

Tim Boyle/Getty Images

becomes their third choice. Similar adjustments are made by the remaining voters. The new preference schedule is

	Rankings			
Apple Valley	2	1	2	2
Coachella	1	2	3	1
Del Mar	3	3	1	3
Number of ballots:	12	11	5	2

The board members now repeat the process and eliminate the site with the fewest first-place votes. In this case it is Del Mar. The new adjusted preference schedule is

	Rankings			
Apple Valley	2	1	1	2
Coachella	1	2	2	1
Number of ballots:	12	11	5	2

From this table, Apple Valley has 16 first-place votes and Coachella has 14 first-place votes. Therefore, Apple Valley is the selected site for the new airport.

EXAMPLE 3 Use the Plurality with Elimination Voting Method

A university wants to add a new sport to its existing program. To help ensure that the new sport will have student support, the students of the university are asked to rank the four sports under consideration. The results are shown in the following table.

	Rankings					
Lacrosse	3	2	3	1	1	2
Squash	2	1	4	2	3	1
Rowing	4	3	2	4	4	4
Golf	1	4	1	3	2	3
Number of ballots:	326	297	287	250	214	197

Use the plurality with elimination method to determine which of these sports should be added to the university's program.

Solution

Because rowing received no first-place votes, it is eliminated from consideration. The new preference schedule is shown below.

	Rankings					
Lacrosse	3	2	2	1	1	2
Squash	2	1	3	2	3	1
Golf	1	3	1	3	2	3
Number of ballots:	326	297	287	250	214	197

TAKE NOTE

Remember to shift the preferences at each stage of the elimination method. The 297 students who chose rowing as their third choice and golf as their fourth choice now have golf as their third choice. Check each preference schedule and update it as necessary.

From this table, lacrosse has 464 first-place votes, squash has 494 first-place votes, and golf has 613 first-place votes. Because lacrosse has the fewest first-place votes, it is eliminated. The new preference schedule is shown below.

	Rankings					
Squash	2	1	2	1	2	1
Golf	1	2	1	2	1	2
Number of ballots:	326	297	287	250	214	197

From this table, squash received 744 first-place votes and golf received 827 first-place votes. Therefore, golf is added to the sports program.

CHECK YOUR PROGRESS 3 A service club is going to sponsor a dinner to raise money for a charity. The club has decided to serve Italian, Mexican, Thai, Chinese, or Indian food. The members of the club were surveyed to determine their preferences. The results are shown in the table below.

	Rankings				
Italian	2	5	1	4	3
Mexican	1	4	5	2	1
Thai	3	1	4	5	2
Chinese	4	2	3	1	4
Indian	5	3	2	3	5
Number of ballots:	33	30	25	20	18

Use the plurality with elimination method to determine the food preference of the club members.

Solution See page S14.

HISTORICAL NOTE

Maria Nicholas Caritat (kä-re-tä) (1743–1794), the Marquis de Condorcet, was, like Borda, a member of the French Academy of Sciences. Around 1780, he showed that the Borda count method also had flaws and proposed what is now called the *Condorcet criterion*.

In addition to his work on the theory of voting, Condorcet contributed to the writing of a French constitution in 1793. He was an advocate of equal rights for women and the abolition of slavery. Condorcet was also one of the first mathematicians to try, although without much success, to use mathematics to discover principles in the social sciences.

Pairwise Comparison Voting Method

The *pairwise comparison* method of voting is sometimes referred to as the "head-to-head" method. In this method, each candidate is compared one-on-one with each of the other candidates. A candidate receives 1 point for a win, 0.5 points for a tie, and 0 points for a loss. The candidate with the greatest number of points wins the election.

A voting method that elects the candidate who wins all head-to-head matchups is said to satisfy the Condorcet criterion.

Condorcet Criterion

A candidate who wins all possible head-to-head matchups should win an election when all candidates appear on the ballot.

This is one of the *fairness criteria* that a voting method should exhibit. We will discuss other fairness criteria later in this section.

EXAMPLE 4 Use the Pairwise Comparison Voting Method

There are four proposals for the name of a new football stadium at a college: Panther Stadium, after the team mascot; Sanchez Stadium, after a large university contributor; Mosher Stadium, after a famous alumnus known for humanitarian work; and Fritz Stadium, after the college's most winning football coach. The preference schedule cast by alumni and students is shown below.

	Rankings				
Panther Stadium	2	3	1	2	4
Sanchez Stadium	1	4	2	4	3
Mosher Stadium	3	1	4	3	2
Fritz Stadium	4	2	3	1	1
Number of ballots:	752	678	599	512	487

Use the pairwise comparison voting method to determine the name of the stadium.

Solution

We will create a table to keep track of each of the head-to-head comparisons. Before we begin, note that a matchup between, say, Panther and Sanchez is the same as the matchup between Sanchez and Panther. Therefore, we will shade the duplicate cells and the cells between the same candidates. This is shown below.

versus	Panther	Sanchez	Mosher	Fritz
Panther				
Sanchez				
Mosher				
Fritz				

To complete the table, we will place the name of the winner in the cell of each head-to-head match. For instance, for the Panther–Sanchez matchup,

✓ Panther was favored over Sanchez on 678 + 599 + 512 = 1789 ballots.
Sanchez was favored over Panther on 752 + 487 = 1239 ballots.

The winner of this matchup is Panther, so that name is placed in the Panther versus Sanchez cell. Do this for each of the matchups.

✓ Panther was favored over Mosher on 752 + 599 + 512 = 1863 ballots.
Mosher was favored over Panther on 678 + 487 = 1165 ballots.

Panther was favored over Fritz on 752 + 599 = 1351 ballots.
✓ Fritz was favored over Panther on 678 + 512 + 487 = 1677 ballots.

Sanchez was favored over Mosher on 752 + 599 = 1351 ballots.
✓ Mosher was favored over Sanchez on 678 + 512 + 487 = 1677 ballots.

Sanchez was favored over Fritz on 752 + 599 = 1351 ballots.
✓ Fritz was favored over Sanchez on 678 + 512 + 487 = 1677 ballots.

Mosher was favored over Fritz on 752 + 678 = 1430 ballots.
✓ Fritz was favored over Mosher on 599 + 512 + 487 = 1598 ballots.

TAKE NOTE

Although we have shown the totals for both Panther over Sanchez and Sanchez over Panther, it is only necessary to do one of the matchups. You can then just subtract that number from the total number of ballots cast, which in this case is 3028.

Panther over Sanchez: 1789

Sanchez over Panther:
3028 − 1789 = 1239

We could have used this calculation method for all of the other matchups. For instance, in the final matchup of Mosher versus Fritz, we have

Mosher over Fritz: 1430

Fritz over Mosher:
3028 − 1430 = 1598

versus	Panther	Sanchez	Mosher	Fritz
Panther		Panther	Panther	Fritz
Sanchez			Mosher	Fritz
Mosher				Fritz
Fritz				

From the above table, Fritz has three wins, Panther has two wins, and Mosher has one win. Using pairwise comparison, Fritz Stadium is the winning name.

CHECK YOUR PROGRESS 4 One hundred restaurant critics were asked to rank their favorite restaurants from a list of four. The preference schedule for the critics is shown in the table below.

	Rankings				
Sanborn's Fine Dining	3	1	4	3	1
The Apple Inn	4	3	3	2	4
May's Steak House	2	2	1	1	3
Tory's Seafood	1	4	2	4	2
Number of ballots:	31	25	18	15	11

Use the pairwise voting method to determine the critics' favorite restaurant.

Solution See page S15.

AP Images

Kenneth J. Arrow

Fairness of Voting Methods and Arrow's Theorem

Now that we have examined various voting options, we will stop to ask which of these options is the *fairest.* To answer that question, we must first determine what we mean by fair.

In 1948, Kenneth J. Arrow was trying to develop material for his doctoral dissertation. As he studied, it occurred to him that he might be able to apply the principles of order relations to problems in social choice or voting. (An example of an order relation for real numbers is "less than.") His investigation led him to outline various criteria for a fair voting system. A paraphrasing of four fairness criteria is given below.

Fairness Criteria

1. *Majority criterion:* The candidate who receives a majority (more than 50%) of the first-place votes is the winner.
2. *Monotonicity criterion:* If candidate A wins an election, then candidate A will also win the election if the only change in the voters' preferences is that supporters of a different candidate change their votes to support candidate A.
3. *Condorcet criterion:* A candidate who wins all possible head-to-head matchups should win an election when all candidates appear on the ballot.
4. *Independence of irrelevant alternatives:* If a candidate wins an election, the winner should remain the winner in any recount in which losing candidates withdraw from the race.

There are other criteria, such as the *dictator criterion,* which we will discuss in the next section. However, what Kenneth Arrow was able to prove is that no matter what kind of voting system we devise, it is impossible for it to satisfy the fairness criteria.

POINT OF INTEREST

In the 1970s, Allan Gibbard and Mark Satterthwaite each proved a theorem (now known as the Gibbard-Satterthwaite Theorem) similar to Arrow's Impossibility Theorem. They considered elections with three or more candidates where voters rank the candidates in order of preference. The theorem states that, in essence, any reasonable method to determine the winner of the election will provide incentives to at least some of the voters to vote insincerely, meaning that a voter submits a ranking that is different from his or her true preferences.

Arrow's Impossibility Theorem

There is no voting method involving three or more choices that satisfies all four fairness criteria stated on page 197.

By Arrow's Impossibility Theorem, none of the voting methods we have discussed is fair. Not only that, but we *cannot* construct a fair voting system for three or more candidates. We will now give some examples of each of the methods we have discussed and show which of the fairness criteria are not satisfied.

EXAMPLE 5 Show That the Borda Count Method Violates the Majority Criterion

Suppose the preference schedule for three candidates, Alpha, Beta, and Gamma, is given by the table below.

	Rankings		
Alpha	1	3	3
Beta	2	1	2
Gamma	3	2	1
Number of ballots:	55	50	3

Show that using the Borda count method violates the majority criterion.

Solution

The calculations for Borda's method are shown below.

Alpha		Beta	
55 first-place votes	$55 \cdot 3 = 165$	50 first-place votes	$50 \cdot 3 = 150$
0 second-place votes	$0 \cdot 2 = 0$	58 second-place votes	$58 \cdot 2 = 116$
53 third-place votes	$53 \cdot 1 = 53$	0 third-place votes	$0 \cdot 1 = 0$
	Total 218		Total 266

Gamma	
3 first-place votes	$3 \cdot 3 = 9$
50 second-place votes	$50 \cdot 2 = 100$
55 third-place votes	$55 \cdot 1 = 55$
	Total 164

From these calculations, Beta should win the election. However, Alpha has the majority (more than 50%) of the first-place votes. This result violates the majority criterion.

CHECK YOUR PROGRESS 5 Using the table in Example 5, show that the Borda count method violates the Condorcet criterion.

Solution See page S15.

QUESTION Does the pairwise comparison voting method satisfy the Condorcet criterion?

EXAMPLE 6 Show That Plurality with Elimination Violates the Monotonicity Criterion

Suppose the preference schedule for three candidates, Alpha, Beta, and Gamma, is given by the table below.

	Rankings			
Alpha	2	3	1	1
Beta	3	1	2	3
Gamma	1	2	3	2
Number of ballots:	25	20	16	10

a. Show that, using plurality with elimination voting, Gamma wins the election.

b. Suppose that the 10 people who voted for Alpha first and Gamma second changed their votes such that they all voted for Alpha second and Gamma first. Show that, using plurality with elimination voting, Beta will now be elected.

c. Explain why this result violates the monotonicity criterion.

Solution

a. Beta received the fewest first-place votes, so Beta is eliminated. The new preference schedule is

	Rankings			
Alpha	2	2	1	1
Gamma	1	1	2	2
Number of ballots:	25	20	16	10

From this schedule, Gamma has 45 first-place votes and Alpha has 26 first-place votes, so Gamma is the winner.

b. If the 10 people who voted for Alpha first and Gamma second changed their votes such that they all voted for Alpha second and Gamma first, the preference schedule would be

	Rankings			
Alpha	2	3	1	2
Beta	3	1	2	3
Gamma	1	2	3	1
Number of ballots:	25	20	16	10

ANSWER Yes. The pairwise comparison voting method elects the person who wins all head-to-head matchups.

From this schedule, Alpha has the fewest first-place votes and is eliminated. The new preference schedule is

	Rankings			
Beta	2	1	1	2
Gamma	1	2	2	1
Number of ballots:	25	20	16	10

From this schedule, Gamma has 35 first-place votes and Beta has 36 first-place votes, so Beta is the winner.

c. This result violates the monotonicity criterion because Gamma, who won the first election, loses the second election even though Gamma received a larger number of first-place votes in the second election.

CHECK YOUR PROGRESS 6 The table below shows the preferences for three new car colors.

	Rankings		
Radiant silver	1	3	3
Electric red	2	2	1
Lightning blue	3	1	2
Number of votes:	30	27	2

Show that the Borda count method violates the independence of irrelevant alternatives criterion. (*Hint:* Reevaluate the vote if lightning blue is eliminated.)

Solution See page S15.

EXCURSION

Variations of the Borda Count Method

Sixty people were asked to select their preferences among plain iced tea, lemon-flavored iced tea, and raspberry-flavored iced tea. The preference schedule is shown in the table below.

	Rankings		
Plain iced tea	1	3	3
Lemon iced tea	2	2	1
Raspberry iced tea	3	1	2
Number of ballots:	25	20	15

POINT OF INTEREST

The Cy Young Award is given to the best pitchers in Major League Baseball. Beginning with the 1970 season, the winner has been determined using a variation of the Borda count method, where 5 points are assigned for a first-place vote, 3 points for a second-place vote, and 1 point for a third-place vote.

Scott Kane/Getty Images

Jake Arrieta of the Chicago Cubs, pictured above, won the National League Cy Young Award in 2015.

1. Using the Borda method of voting, which flavor of iced tea is preferred by this group? Which is second? Which is third?
2. Instead of using the normal Borda method, suppose the Borda method used in Exercise 1 of this Excursion assigned 1 point for first, 0 points for second, and -1 point for third place. Does this alter the preferences you found in Exercise 1?
3. Suppose the Borda method used in Exercise 1 of this Excursion assigned 10 points for first, 5 points for second, and 0 points for third place. Does this alter the preferences you found in Exercise 1?
4. Suppose the Borda method used in Exercise 1 of this Excursion assigned 20 points for first, 5 points for second, and 0 points for third place. Does this alter the preferences you found in Exercise 1?
5. Suppose the Borda method used in Exercise 1 of this Excursion assigned 25 points for first, 5 points for second, and 0 points for third place. Does this alter the preferences you found in Exercise 1?
6. Can the assignment of points for first, second, and third place change the preference order when the Borda method of voting is used?
7. Suppose the assignment of points for first, second, and third place for the Borda method of voting are consecutive integers. Can the value of the starting integer change the outcome of the preferences?

EXERCISE SET 4.2

1. What is the difference between a majority and a plurality? Is it possible to have one without the other?
2. Explain why the plurality voting system may not be the best system to use in some situations.
3. Explain how the Borda count method of voting works.
4. Explain how the plurality with elimination voting method works.
5. Explain how the pairwise comparison voting method works.
6. What does the Condorcet criterion say?
7. Is there a "best" voting method? Is one method more fair than the others?
8. Explain why, if only two candidates are running, the plurality and Borda count methods will determine the same winner.
9. **Presidential Election** The table below shows the popular vote and the Electoral College vote for the major candidates in the 2000 presidential election.

Candidate	Popular vote	Electoral College vote
George W. Bush	50,456,002	271
Al Gore	50,999,897	266
Ralph Nader	2,882,955	0

SOURCE: *Encyclopaedia Britannica* Online

a. Which candidate received the plurality of the popular vote?

b. Did any candidate receive a majority of the popular vote?

c. Who won the election?

10. Breakfast Cereal Sixteen people were asked to rank three breakfast cereals in order of preference. Their responses are given below.

Corn Flakes	3	1	1	2	3	3	2	2	1	3	1	3	1	2	1	2
Raisin Bran	1	3	2	3	1	2	1	1	2	1	2	2	3	1	3	3
Mini Wheats	2	2	3	1	2	1	3	3	3	2	3	1	2	3	2	1

If the plurality method of voting is used, which cereal is the group's first preference?

11. Cartoon Characters A kindergarten class was surveyed to determine the children's favorite cartoon characters among Dora the Explorer, SpongeBob SquarePants, and Buzz Lightyear. The students ranked the characters in order of preference; the results are shown in the preference schedule below.

	Rankings					
Dora the Explorer	1	1	2	2	3	3
SpongeBob SquarePants	2	3	1	3	1	2
Buzz Lightyear	3	2	3	1	2	1
Number of students:	6	4	6	5	6	8

a. How many students are in the class?

b. How many votes are required for a majority?

c. Using plurality voting, which character is the children's favorite?

12. Catering A 15-person committee is having lunch catered for a meeting. Three caterers, each specializing in a different cuisine, are available. In order to choose a caterer for the group, each member is asked to rank the cuisine options in order of preference. The results are given in the preference schedule below.

	Rankings				
Italian	1	1	2	3	3
Mexican	2	3	1	1	2
Japanese	3	2	3	2	1
Number of votes:	2	4	1	5	3

Using plurality voting, which caterer should be chosen?

13. Movies Fifty consumers were surveyed about their movie-watching habits. They were asked to rank the likelihood that they would participate in each listed activity. The results are summarized in the table below.

	Rankings				
Stream online	2	3	1	2	1
Go to a theater	3	1	3	1	2
Rent a Blu-ray disc or DVD	1	2	2	3	3
Number of votes:	8	13	15	7	7

Using the Borda count method of voting, which activity is the most popular choice among this group of consumers?

14. Breakfast Cereal Use the Borda count method of voting to determine the preferred breakfast cereal in Exercise 10.

15. Cartoons Use the Borda count method of voting to determine the children's favorite cartoon character in Exercise 11.

16. Catering Use the Borda count method of voting to determine which caterer the committee should hire in Exercise 12.

17. Class Election A senior high school class held an election for class president. Instead of just voting for one candidate, the students were asked to rank all four candidates in order of preference. The results are shown below.

	Rankings					
Raymond Lee	2	3	1	3	4	2
Suzanne Brewer	4	1	3	4	1	3
Elaine Garcia	1	2	2	2	3	4
Michael Turley	3	4	4	1	2	1
Number of votes:	36	53	41	27	31	45

Using the Borda count method, which student should be class president?

18. Cell Phone Usage A journalist reviewing various cellular phone services surveyed 200 customers and asked each one to rank four service providers in order of preference. The group's results are shown below.

	Rankings				
AT&T	3	4	2	3	4
Sprint	1	1	4	4	3
Verizon	2	2	1	2	1
T-Mobile	4	3	3	1	2
Number of votes:	18	38	42	63	39

Using the Borda count method, which provider is the favorite of these customers?

19. Baseball Uniforms A Little League baseball team must choose the colors for its uniforms. The coach offered four different choices, and the players ranked them in order of preference, as shown in the table below.

	Rankings			
Red and white	2	3	3	2
Green and yellow	4	1	4	1
Red and blue	3	4	2	4
Blue and white	1	2	1	3
Number of votes:	4	2	5	4

Using the plurality with elimination method, what colors should be used for the uniforms?

20. Radio Stations A number of college students were asked to rank four radio stations in order of preference. The responses are given in the table below.

	Rankings				
WNNX	3	1	1	2	4
WKLS	1	3	4	1	2
WWVV	4	2	2	3	1
WSTR	2	4	3	4	3
Number of votes:	57	72	38	61	15

Use plurality with elimination to determine the students' favorite radio station among the four.

21. Class Election Use plurality with elimination to choose the class president in Exercise 17.

22. Cell Phone Usage Use plurality with elimination to determine the preferred cellular phone service in Exercise 18.

23. Campus Club A campus club has money left over in its budget and must spend it before the school year ends. The members arrive at five different possibilities, and each member ranks them in order of preference. The results are shown in the table below.

	Rankings				
Establish a scholarship	1	2	3	3	4
Pay for several members to travel to a convention	2	1	2	1	5
Buy new computers for the club	3	3	1	4	1
Throw an end-of-year party	4	5	5	2	2
Donate to charity	5	4	4	5	3
Number of votes:	8	5	12	9	7

a. Using the plurality voting system, how should the club spend the money?

b. Use the plurality with elimination method to determine how the money should be spent.

c. Using the Borda count method of voting, how should the money be spent?

d. In your opinion, which of the previous three methods seems most appropriate in this situation? Why?

24. Recreation A company is planning its annual summer retreat and has asked its employees to rank five different choices of recreation in order of preference. The results are given in the table below.

	Rankings				
Picnic in a park	1	2	1	3	4
Water skiing at a lake	3	1	2	4	3
Amusement park	2	5	5	1	2
Riding horses at a ranch	5	4	3	5	1
Dinner cruise	4	3	4	2	5
Number of votes:	10	18	6	28	16

a. Using the plurality voting system, what activity should be planned for the retreat?

b. Use the plurality with elimination method to determine which activity should be chosen.

c. Using the Borda count method of voting, which activity should be planned?

25. X-Men Movies Fans of the *X-Men* movies have been debating on an online forum regarding which of the films is the best. To see what the overall opinion is, visitors to the site can rank four films in order of preference. The results are shown in the preference schedule below.

	Rankings			
X-Men	1	2	1	3
X2: X-Men United	4	4	2	1
X-Men: Days of Future Past	2	1	3	2
X-Men: First Class	3	3	4	4
Number of votes:	429	1137	384	582

Using pairwise comparison, which film is the favorite of the visitors to the site who voted?

26. Family Reunion The Nelson family is trying to decide where to hold a family reunion. They have asked all their family members to rank four choices in order of preference. The results are shown in the preference schedule below.

	Rankings				
Grand Canyon	3	1	2	3	1
Yosemite	1	2	3	4	4
Bryce Canyon	4	4	1	2	2
Yellowstone	2	3	4	1	3
Number of votes:	7	3	12	8	13

Use the pairwise comparison method to determine the best choice for the reunion.

27. School Mascot A new college needs to pick a mascot for its football team. The students were asked to rank four choices in order of preference; the results are tallied below.

	Rankings				
Bulldog	3	4	4	1	4
Panther	2	1	2	4	3
Hornet	4	2	1	2	2
Bobcat	1	3	3	3	1
Number of votes:	638	924	525	390	673

Using the pairwise comparison method of voting, which mascot should be chosen?

28. Election Five candidates are running for president of a charity organization. Interested persons were asked to rank the candidates in order of preference. The results are given below.

	Rankings				
P. Gibson	5	1	2	1	2
E. Yung	2	4	5	5	3
R. Allenbaugh	3	2	1	3	5
T. Meckley	4	3	4	4	1
G. DeWitte	1	5	3	2	4
Number of votes:	16	9	14	9	4

Use the pairwise comparison method to determine the president of the organization.

29. Baseball Uniforms Use the pairwise comparison method to choose the colors for the Little League uniforms in Exercise 19.

30. Radio Stations Use the pairwise comparison method to determine the favorite radio station in Exercise 20.

31. Does the winner in Exercise 11c satisfy the Condorcet criterion?

32. Does the winner in Exercise 12 satisfy the Condorcet criterion?

33. Does the winner in Exercise 23c satisfy the Condorcet criterion?

34. Does the winner in Exercise 24c satisfy the Condorcet criterion?

35. Does the winner in Exercise 17 satisfy the majority criterion?

36. Does the winner in Exercise 20 satisfy the majority criterion?

37. Election Three candidates are running for mayor. A vote was taken in which the candidates were ranked in order of preference. The results are shown in the preference schedule below.

	Rankings		
John Lorenz	1	3	3
Marcia Beasley	3	1	2
Stephen Hyde	2	2	1
Number of votes:	2691	2416	237

a. Use the Borda count method to determine the winner of the election.

b. Verify that the majority criterion has been violated.

c. Identify the candidate who wins all head-to-head comparisons.

d. Explain why the Condorcet criterion has been violated.

e. If Marcia Beasley drops out of the race for mayor (and voter preferences remain the same), determine the winner of the election again, using the Borda count method.

f. Explain why the independence of irrelevant alternatives criterion has been violated.

38. Film Competition Three films have been selected as finalists in a national student film competition. Seventeen judges have viewed each of the films and have ranked them in order of preference. The results are given in the preference schedule below.

	Rankings			
Film A	1	3	2	1
Film B	2	1	3	3
Film C	3	2	1	2
Number of votes:	4	6	5	2

a. Using the plurality with elimination method, which film should win the competition?

b. Suppose the first vote is declared invalid and a revote is taken. All of the judges' preferences remain the same except for the votes represented by the last column of the table. The judges who cast these votes both decide to switch their first place vote to film C, so their preference now is film C first, then film A, and then film B. Which film now wins using the plurality with elimination method?

c. Has the monotonicity criterion been violated?

EXTENSIONS

39. Election A campus club needs to elect four officers: a president, a vice president, a secretary, and a treasurer. The club has five volunteers. Rather than vote individually for each position, the club members will rank the candidates in order of preference. The votes will then be tallied using the Borda count method. The candidate receiving the highest number of points will be president, the candidate receiving the next highest number of points will be vice president, the candidate receiving the next highest number of points will be secretary, and the candidate receiving the next highest number of points will be treasurer. For the preference schedule shown below, determine who wins each position in the club.

	Rankings				
Cynthia	4	2	5	2	3
Andrew	2	3	1	4	5
Jen	5	1	2	3	2
Hector	1	5	4	1	4
Ahmad	3	4	3	5	1
Number of votes:	22	10	16	6	27

40. Scholarship Awards The members of a scholarship committee have ranked four finalists competing for a scholarship in order of preference. The results are shown in the preference schedule below.

	Rankings			
Francis Chandler	3	4	4	1
Michael Huck	1	2	3	4
David Chang	2	3	1	2
Stephanie Owen	4	1	2	3
Number of votes:	9	5	7	4

If you are one of the voting members and you want David Chang to win the scholarship, which voting method would you suggest that the committee use?

41. Another method of voting is to assign a "weight," or score, to each candidate rather than ranking the candidates in order. All candidates must receive a score, and two or more candidates can receive the same score from a voter. A score of 5 represents the strongest endorsement of a candidate. The scores range down to 1, which corresponds to complete disapproval of a candidate. A score of 3 represents indifference. The candidate with the most total points wins the election. The results of a sample election are given in the table.

	Weights							
Candidate A	2	1	2	5	4	2	4	5
Candidate B	5	3	5	3	2	5	3	2
Candidate C	4	5	4	1	4	3	2	1
Candidate D	1	3	2	4	5	1	3	2
Number of votes:	26	42	19	33	24	8	24	33

a. Find the winner of the election.

b. If plurality were used (assuming that a person's vote would go to the candidate that he or she gave the highest score to), verify that a different winner would result.

42. *Approval voting* is a system in which voters may vote for more than one candidate. Each vote counts equally, and the candidate with the most total votes wins the election. Many feel that this is a better system for large elections than simple plurality because it considers a voter's second choices and is a stronger measure of overall voter support for each candidate. Some organizations use approval voting to elect their officers. The United Nations uses this method to elect the secretary-general.

Suppose a math class is going to show a film involving mathematics or mathematicians on the last day of class. The options are *Stand and Deliver, Good Will Hunting, A Beautiful Mind, Proof,* and *Contact.* The students vote using approval voting. The results are as follows.

8 students vote for all five films.

8 students vote for *Good Will Hunting, A Beautiful Mind*, and *Contact.*

8 students vote for *Stand and Deliver, Good Will Hunting*, and *Contact.*

8 students vote for *A Beautiful Mind* and *Proof.*

8 students vote for *Stand and Deliver* and *Proof.*

8 students vote for *Good Will Hunting* and *Contact.*

1 student votes for *Proof.*

Which film will be chosen for the last day of class screening?

Danny E Hooks/Shutterstock.com

43. Restaurants Suppose you and three friends, David, Sara, and Cliff, are trying to decide on a pizza restaurant. You like Pizza Hut best, Round Table pizza is acceptable to you, and you definitely do not want to get pizza from Domino's. Domino's is David's favorite, and he also likes Round Table, but he won't eat pizza from Pizza Hut. Sara says she will only eat Round Table pizza. Cliff prefers Domino's, but he will also eat Round Table pizza. He doesn't like Pizza Hut. Use approval voting (see Exercise 42) to determine the pizza restaurant that the group of friends should choose.

SECTION 4.3 Weighted Voting Systems

Biased Voting Systems

A **weighted voting system** is one in which some voters have more weight on the outcome of an election. Examples of weighted voting systems are fairly common. A few examples are the stockholders of a company, the Electoral College, the United Nations Security Council, and the European Union.

MATH MATTERS The Electoral College

As mentioned in the Historical Note below, the Electoral College elects the president of the United States. The number of electors representing each state is equal to the sum of the number of senators (2) and the number of members in the House of Representatives for that state. The original intent of the framers of the Constitution was to protect the smaller states. We can verify this by computing the number of people represented by each elector. In the 2015 election, each Vermont elector represented about 209,000 people; each California elector represented about 705,000 people. To see how this gives a state with a smaller population more *power* (a word we will discuss in more detail later in this section), note that three electoral votes from Vermont represent approximately the same size population as does one electoral vote from California. Not every vote represents the same number of people.

Another peculiarity related to the Electoral College system is that it is very sensitive to small vote swings. For instance, in the 2000 election, if an additional 0.01% of the voters in Florida had cast their votes for Al Gore instead of George Bush, Gore would have won the presidential election.

HISTORICAL NOTE

The U.S. Constitution, Article 2, Section 1 states that the members of the Electoral College elect the president of the United States. The original article directed members of the College to vote for two people. However, it did not stipulate that one name was for president and the other name was for vice president. The article goes on to state that the person with the greatest number of votes becomes president and the one with the next highest number of votes becomes vice president. In 1800, Thomas Jefferson and Aaron Burr received exactly the same number of votes even though they were running on a Jefferson for president, Burr for vice president ticket. Thus the House of Representatives was asked to select the president. It took 36 different votes by the House before Jefferson was elected president. In 1804, the Twelfth Amendment to the Constitution was ratified to prevent a recurrence of the 1800 election problems.

Consider a small company with a total of 100 shares of stock and three stockholders, A, B, and C. Suppose that A owns 45 shares of the stock (which means A has 45 votes), B owns 45 shares, and C owns 10 shares. If a vote of 51 or greater is required to approve any measure before the owners, then a measure cannot be passed without two of the three owners voting for the measure. Even though C has only 10 shares, C has the same voting power as A and B.

Now suppose that a new stockholder is brought into the company and the shares of the company are redistributed so that A has 27 shares, B has 26 shares, C has 25 shares, and D has 22 shares. Note, in this case, that any two of A, B, or C can pass a measure, but D paired with any of the other shareholders cannot pass a measure. D has virtually no power even though D has only three shares less than C.

The number of votes that are required to pass a measure is called a **quota**. For the two stockholder examples above, the quota was 51. The **weight of a voter** is the number of votes controlled by the voter. In the case of the company whose stock was split A–27 shares, B–26 shares, C–25 shares, and D–22 shares, the weight of A is 27, the weight of B is 26, the weight of C is 25, and the weight of D is 22. Rather than write out in sentence form the quota and weight of each voter, we use the notation

Quota — Weights

$$\{51: \underbrace{27, 26, 25, 22}_{}\}$$

• The four numbers after the colon indicate that there are a total of four voters in this system.

This notation is very convenient. We state its more general form in the following definition.

Weighted Voting System

A weighted voting system of n voters is written $\{q: w_1, w_2, \ldots, w_n\}$, where q is the quota and w_1 through w_n represent the weights of each of the n voters.

Using this notation, we can describe various voting systems.

- **One person, one vote:** For instance, {5: 1, 1, 1, 1, 1, 1, 1, 1, 1}. In this system, each person has one vote and five votes, a majority, are required to pass a measure.
- **Dictatorship:** For instance, {20: 21, 6, 5, 4, 3}. In this system, the person with 21 votes can pass any measure. Even if the remaining four people get together, their votes do not total the quota of 20.
- **Null system:** For instance, {28: 6, 3, 5, 2}. If all the members of this system vote for a measure, the total number of votes is 16, which is less than the quota. Therefore, no measure can be passed.
- **Veto power system:** For instance, {21: 6, 5, 4, 3, 2, 1}. In this case, the sum of all the votes is 21, the quota. Therefore, if any one voter does not vote for the measure, it will fail. Each voter is said to have **veto power**. In this case, this means that even the voter with one vote can veto a measure (cause the measure not to pass). A voter has veto power whenever a measure cannot be passed without that voter's vote. If at least one voter in a voting system has veto power, the system is a veto power system.

MATH MATTERS UN Security Council: An Application of Inequalities

United Nations, New York

UN Security Council Chamber

The United Nations Security Council consists of 5 permanent members (the United States, China, France, the United Kingdom, and Russia) and 10 members that are elected by the General Assembly for a two-year term. In 2016, the 10 nonpermanent members were Angola, Egypt, Japan, Malaysia, New Zealand, Senegal, Spain, Ukraine, Uruguay, and Venezuela.

For a resolution to pass the Security Council,

1. Nine countries must vote for the resolution; and
2. If one of the five permanent members votes against the resolution, it fails.

We can express this system as a weighted voting system by describing the situation using inequalities. Assume that the nonpermanent members each have a vote with weight 1, let x be the weight of the vote of one permanent member of the Council, and let q be the quota for a vote to pass. Then, by condition 1, $q \le 5x + 4$. (The weights of the five votes of the permanent members plus the single votes of four nonpermanent members must be greater than or equal to the quota.)

By condition 2, we have $4x + 10 < q$. (If one of the permanent members opposes the resolution, it fails even if all of the nonpermanent members vote for it.)

Combining the inequalities from condition 1 and condition 2, we have

$$4x + 10 < 5x + 4$$

$$10 < x + 4 \qquad \text{• Subtract } 4x \text{ from each side.}$$

$$6 < x \qquad \text{• Subtract 4 from each side.}$$

The smallest whole number greater than 6 is 7. Therefore, the weight x of each permanent member is 7. Substituting 7 for x into $q \le 5x + 4$ and $4x + 10 < q$, we find that $q = 39$. Thus the weighted voting system of the Security Council is given by {39: 7, 7, 7, 7, 7, 1, 1, 1, 1, 1, 1, 1, 1, 1, 1}.

QUESTION Is the UN Security Council voting system a veto power system?

In a weighted voting system, a **coalition** is a set of voters each of whom votes the same way, either for or against a resolution. A **winning coalition** is a set of voters the sum of whose votes is greater than or equal to the quota. A **losing coalition** is a set of voters the sum of whose votes is less than the quota. A voter who leaves a winning coalition and thereby turns it into a losing coalition is called a **critical voter**.

As shown in the next theorem, for large numbers of voters, there are many possible coalitions.

TAKE NOTE

The number of coalitions of n voters is the number of subsets that can be formed from n voters. From Chapter 2, this is 2^n. Because a coalition must contain at least one voter, the empty set is not a possible coalition. Therefore, the number of coalitions is $2^n - 1$.

Number of Possible Coalitions of n Voters

The number of possible coalitions of n voters is $2^n - 1$.

As an example, if all electors of each state to the Electoral College cast their ballots for one candidate, then there are $2^{51} - 1 \approx 2.25 \times 10^{15}$ possible coalitions (the District of Columbia is included). The number of *winning* coalitions is far less. For instance, any coalition of 10 or fewer states cannot be a winning coalition because the largest 10 states do not have enough electoral votes to elect the president. As we proceed through this section, we will not attempt to list all the coalitions, only the winning coalitions.

EXAMPLE 1 Determine Winning Coalitions in a Weighted Voting System

Suppose that the four owners of a company, Ang, Bonhomme, Carmel, and Diaz, own, respectively, 500 shares, 375 shares, 225 shares, and 400 shares. There are a total of 1500 votes; half of this is 750, so the quota is 751. The weighted voting system for this company is {751: 500, 375, 225, 400}.

a. Determine the winning coalitions.

b. For each winning coalition, determine the critical voters.

Solution

a. A winning coalition must represent at least 751 votes. We will list these coalitions in the table below, in which we use A for Ang, B for Bonhomme, C for Carmel, and D for Diaz.

TAKE NOTE

The coalition {A, C} is not a winning coalition because the total number of votes for that coalition is 725, which is less than 751.

Winning coalition	Number of votes
{A, B}	875
{A, D}	900
{B, D}	775
{A, B, C}	1100
{A, B, D}	1275
{A, C, D}	1125
{B, C, D}	1000
{A, B, C, D}	1500

ANSWER Yes. If any of the permanent members votes against a resolution, the resolution cannot pass.

b. A voter who leaves a winning coalition and thereby creates a losing coalition is a critical voter. For instance, for the winning coalition {A, B, C}, if A leaves, the number of remaining votes is 600, which is not enough to pass a resolution. If B leaves, the number of remaining votes is 725—again, not enough to pass a resolution. If C leaves, the number of remaining votes is 875, which is greater than the quota. Therefore, A and B are critical voters for the coalition {A, B, C} and C is not a critical voter. The table below shows the critical voters for each winning coalition.

Winning coalition	Number of votes	Critical voters
{A, B}	875	A, B
{A, D}	900	A, D
{B, D}	775	B, D
{A, B, C}	1100	A, B
{A, B, D}	1275	None
{A, C, D}	1125	A, D
{B, C, D}	1000	B, D
{A, B, C, D}	1500	None

CHECK YOUR PROGRESS 1 Many countries must govern by forming coalitions from among many political parties. Suppose a country has five political parties named A, B, C, D, and E. The numbers of votes, respectively, for the five parties are 22, 18, 17, 10, and 5.

a. Determine the winning coalitions if 37 votes are required to pass a resolution.

b. For each winning coalition, determine the critical voters.

Solution See page S15.

QUESTION Is the voting system in Example 1 a dictatorship? What is the total number of possible coalitions in Example 1?

POINT OF INTEREST

John F. Banzhaf III is professor of law at George Washington University Law School. In 1965, he wrote an article called "Weighted Voting Does Not Work: A Mathematical Analysis," in which he developed the Banzhaf power index to measure voting power in systems such as the Electoral College.

In addition to his work in voting analysis, Banzhaf is credited with persuading the federal government to establish no-smoking sections on airplanes. He eventually convinced the government to ban smoking on all domestic flights.

In 2001, Banzhaf was admitted to the World Technology Network, a group whose members have included Steve Jobs, the co-founder of Apple, Inc., and Sergey Brin and Larry Page, the founders of Google.

Banzhaf Power Index

There are a number of measures of the *power* of a voter. For instance, as we saw from the Electoral College example, some electors represent fewer people and therefore their votes may have more power. As an extreme case, suppose that two electors, A and B, each represent 10 people and that a third elector, C, represents 1000 people. If a measure passes when two of the three electors vote for the measure, then A and B voting together could pass a resolution even though they represent only 20 people.

Another measure of power, called the *Banzhaf power index,* was derived by John F. Banzhaf III in 1965. The purpose of this index is to determine the power of a voter in a weighted voting system.

Banzhaf Power Index

The **Banzhaf power index** of a voter v, symbolized by $BPI(v)$, is given by

$$BPI(v) = \frac{\text{number of times voter } v \text{ is a critical voter}}{\text{number of times any voter is a critical voter}}$$

ANSWER No. There is no one shareholder who has 751 or more shares of stock. The number of possible coalitions is $2^4 - 1 = 15$.

Consider four people, A, B, C, and D, and the one-person, one-vote system given by {3: 1, 1, 1, 1}.

Winning coalition	Number of votes	Critical voters
{A, B, C}	3	A, B, C
{A, B, D}	3	A, B, D
{A, C, D}	3	A, C, D
{B, C, D}	3	B, C, D
{A, B, C, D}	4	None

To find *BPI*(A), we look under the critical voters column and find that A is a critical voter three times. The number of times any voter is a critical voter, the denominator of the Banzhaf power index, is 12. (A is a critical voter three times, B is a critical voter three times, C is a critical voter three times, and D is a critical voter three times. The sum is $3 + 3 + 3 + 3 = 12$.) Thus

$$BPI(\text{A}) = \frac{3}{12} = 0.25$$

Similarly, we can calculate the Banzhaf power index for each of the other voters.

$$BPI(\text{B}) = \frac{3}{12} = 0.25 \qquad BPI(\text{C}) = \frac{3}{12} = 0.25 \qquad BPI(\text{D}) = \frac{3}{12} = 0.25$$

In this case, each voter has the same power. This is expected in a voting system in which each voter has one vote.

Now suppose that three people, A, B, and C, belong to a dictatorship given by {3: 3, 1, 1}.

Winning coalition	Number of votes	Critical voters
{A}	3	A
{A, B}	4	A
{A, C}	4	A
{A, B, C}	5	A

The sum of the numbers of critical voters in all winning coalitions is 4. To find *BPI*(A), we look under the critical voters column and find that A is a critical voter four times. Thus

$$BPI(\text{A}) = \frac{4}{4} = 1 \qquad BPI(\text{B}) = \frac{0}{4} = 0 \qquad BPI(\text{C}) = \frac{0}{4} = 0$$

Thus A has all the power. This is expected in a dictatorship.

TAKE NOTE

The Banzhaf power index is a number between 0 and 1. If the index for voter A is less than the index for voter B, then A has less power than B. This means that A has fewer opportunities to form a winning coalition than does B.

For instance, if *BPI*(A) = 0.25 and *BPI*(B) = 0.5, then B is a critical voter twice as often as A. Thus B can enter into more winning coalitions than A. It is not true, however, that B can enter into twice the number of winning coalitions, because not all winning coalitions have critical voters.

EXAMPLE 2 Compute the BPI for a Weighted Voting System

Suppose the stock in a company is held by five people, A, B, C, D, and E. The voting system for this company is {626: 350, 300, 250, 200, 150}. Determine the Banzhaf power index for A and E.

Solution

Determine all of the winning coalitions and the critical voters in each coalition.

Winning coalition	Critical voters
{A, B}	A, B
{A, B, C}	A, B
{A, B, D}	A, B
{A, B, E}	A, B
{A, C, D}	A, C, D
{A, C, E}	A, C, E
{A, D, E}	A, D, E
{B, C, D}	B, C, D

Winning coalition	Critical voters
{B, C, E}	B, C, E
{B, D, E}	B, D, E
{A, B, C, D}	None
{A, B, C, E}	None
{A, B, D, E}	None
{A, C, D, E}	A
{B, C, D, E}	B
{A, B, C, D, E}	None

The number of times any voter is critical is 28. To find $BPI(\text{A})$, we look under the critical voters columns and find that A is a critical voter eight times. Thus

$$BPI(\text{A}) = \frac{8}{28} \approx 0.29$$

To find $BPI(\text{E})$, we look under the critical voters columns and find that E is a critical voter four times. Thus

$$BPI(\text{E}) = \frac{4}{28} \approx 0.14$$

CHECK YOUR PROGRESS 2 Suppose that a government is composed of four political parties, A, B, C, and D. The voting system for this government is {26: 18, 16, 10, 6}. Determine the Banzhaf power index for A and D.

Solution See page S16.

In many cities, the only time the mayor votes on a resolution is when there is a tie vote by the members of the city council. This is also true of the United States Senate. The vice president only votes when there is a tie vote by the senators.

In Example 3, we will calculate the Banzhaf power index for a voting system in which one voter votes only to break a tie.

EXAMPLE 3 Use the BPI to Determine a Voter's Power

Suppose a city council consists of four members, A, B, C, and D, and a mayor, M, each with an equal vote. The mayor votes only when there is a tie vote among the members of the council. In all cases, a resolution receiving three or more votes passes. Show that the Banzhaf power index for the mayor is the same as the Banzhaf power index for each city council member.

Solution

We first list all of the winning coalitions that do not include the mayor. To this list, we add the winning coalitions in which the mayor votes to break a tie.

Winning coalition (without mayor)	Critical voters
{A, B, C}	A, B, C
{A, B, D}	A, B, D
{A, C, D}	A, C, D
{B, C, D}	B, C, D
{A, B, C, D}	None

Winning coalition (mayor voting)	Critical voters
{A, B, M}	A, B, M
{A, C, M}	A, C, M
{A, D, M}	A, D, M
{B, C, M}	B, C, M
{B, D, M}	B, D, M
{C, D, M}	C, D, M

POINT OF INTEREST

In 2015, 28 countries were full members of the organization known as the European Union (EU), which at one time was referred to as the Common Market or the European Economic Community (EEC). (See Check Your Progress 3.)

By working Check Your Progress 3, you will find that the Banzhaf power index for the original EEC gave little power to Belgium and the Netherlands and no power to Luxembourg. However, there was an implicit understanding among the countries that a resolution would not pass unless all countries voted for it, thereby effectively giving each country veto power.

By examining the table, we see that A, B, C, D, and M each occur in exactly six winning coalitions. The total number of critical voters in all winning coalitions is 30. Therefore, each member of the council and the mayor have the same Banzhaf power index, which is $\frac{6}{30} = 0.2$.

CHECK YOUR PROGRESS 3 The European Economic Community (EEC) was founded in 1958 and originally consisted of Belgium, France, Germany, Italy, Luxembourg, and the Netherlands. The weighted voting system was {12: 2, 4, 4, 4, 1, 2}. Find the Banzhaf power index for each country.

Solution See page S16.

EXCURSION

Blocking Coalitions and the Banzhaf Power Index

The four members, A, B, C, and D, of an organization adopted the weighted voting system {6: 4, 3, 2, 1}. The table below shows the winning coalitions.

Winning coalition	Number of votes	Critical voters
{A, B}	7	A, B
{A, C}	6	A, C
{A, B, C}	9	A
{A, B, D}	8	A, B
{A, C, D}	7	A, C
{B, C, D}	6	B, C, D
{A, B, C, D}	10	None

Using the Banzhaf power index, we have $BPI(\text{A}) = \frac{5}{12}$.

A **blocking coalition** is a group of voters who can prevent passage of a resolution. In this case, a critical voter is one who leaves a blocking coalition, thereby producing a coalition that is no longer capable of preventing the passage of a resolution. For the voting system on the preceding page, we have

Blocking coalition	Number of votes	Number of remaining votes	Critical voters
{A, B}	7	3	A, B
{A, C}	6	4	A, C
{A, D}	5	5	A, D
{B, C}	5	5	B, C
{A, B, C}	9	1	None
{A, B, D}	8	2	A
{A, C, D}	7	3	A
{B, C, D}	6	4	B, C

If we count the number of times A is a critical voter in a winning or blocking coalition, we find what is called the *Banzhaf index*. In this case, the Banzhaf index is 10 and we write $BI(\text{A}) = 10$. Using both the winning coalition and the blocking coalition tables, we find that $BI(\text{B}) = 6$, $BI(\text{C}) = 6$, and $BI(\text{D}) = 2$. This information can be used to create an alternative definition of the Banzhaf power index.

TAKE NOTE

The Banzhaf index is always a *whole number*, whereas the Banzhaf power index is often a fraction between 0 and 1.

Banzhaf Power Index—Alternative Definition

$$BPI(\text{A}) = \frac{BI(\text{A})}{\text{sum of all Banzhaf indices for the voting system}}$$

Applying this definition to the voting system given above, we have

$$BPI(\text{A}) = \frac{BI(\text{A})}{BI(\text{A}) + BI(\text{B}) + BI(\text{C}) + BI(\text{D})} = \frac{10}{10 + 6 + 6 + 2} = \frac{10}{24} = \frac{5}{12}$$

EXCURSION EXERCISES

1. Using the data in Example 1 on page 211, list all blocking coalitions.
2. For the data in Example 1 on page 211, calculate the Banzhaf power indices for A, B, C, and D using the alternative definition.
3. Using the data in Check Your Progress 3 on page 215, list all blocking coalitions.
4. For the data in Check Your Progress 3 on page 215, calculate the Banzhaf power indices for Belgium and Luxembourg using the alternative definition.
5. Create a voting system with three members that is a dictatorship. Calculate the Banzhaf power index for each voter for this system using the alternative definition.
6. Create a voting system with four members in which one member has veto power. Calculate the Banzhaf power index for this system using the alternative definition.
7. Create a voting system with five members that satisfies the one-person, one-vote rule. Calculate the Banzhaf power index for this system using the alternative definition.

EXERCISE SET 4.3

■ In the following exercises that involve weighted voting systems for voters A, B, C, ..., the systems are given in the form $\{q: w_1, w_2, w_3, w_4, \ldots, w_n\}$. The weight of voter A is w_1, the weight of voter B is w_2, the weight of voter C is w_3, and so on.

1. A weighted voting system is given by {6: 4, 3, 2, 1}.

a. What is the quota?

b. How many voters are in this system?

c. What is the weight of voter B?

d. What is the weight of the coalition {A, C}?

e. Is {A, D} a winning coalition?

f. Which voters are critical voters in the coalition {A, C, D}?

g. How many coalitions can be formed?

h. How many coalitions consist of exactly two voters?

2. A weighted voting system is given by {16: 8, 7, 4, 2, 1}.

a. What is the quota?

b. How many voters are in this system?

c. What is the weight of voter C?

d. What is the weight of the coalition {B, C}?

e. Is {B, C, D, E} a winning coalition?

f. Which voters are critical voters in the coalition {A, B, D}?

g. How many coalitions can be formed?

h. How many coalitions consist of exactly three voters?

■ In Exercises 3 to 12, calculate, if possible, the Banzhaf power index for each voter. Round to the nearest hundredth.

3. {6: 4, 3, 2}

4. {10: 7, 6, 4}

5. {10: 7, 3, 2, 1}

6. {14: 7, 5, 1, 1}

7. {19: 14, 12, 4, 3, 1}

8. {3: 1, 1, 1, 1}

9. {18: 18, 7, 3, 3, 1, 1}

10. {14: 6, 6, 4, 3, 1}

11. {80: 50, 40, 30, 25, 5}

12. {85: 55, 40, 25, 5}

13. Which, if any, of the voting systems in Exercises 3 to 12 is

a. a dictatorship?

b. a veto power system? *Note:* A voting system is a veto power system if any of the voters has veto power.

c. a null system?

d. a one-person, one-vote system?

14. Explain why it is impossible to calculate the Banzhaf power index for any voter in the null system {8: 3, 2, 1, 1}.

15. Music Education A music department consists of a band director and a music teacher. Decisions on motions are made by voting. If both members vote in favor of a motion, it passes. If both members vote against a motion, it fails. In the event of a tie vote, the principal of the school votes to break the tie. For this voting scheme, determine the Banzhaf power index for each department member and for the principal. *Hint:* See Example 3, page 214.

16. Four voters, A, B, C, and D, make decisions by using the voting scheme {4: 3, 1, 1, 1}, except when there is a tie. In the event of a tie, a fifth voter, E, casts a vote to break the tie. For this voting scheme, determine the Banzhaf power index for each voter, including voter E. *Hint:* See Example 3, page 214.

17. Criminal Justice In a criminal trial, each of the 12 jurors has one vote and all of the jurors must agree to reach a verdict. Otherwise the judge will declare a mistrial.

a. Write the weighted voting system, in the form $\{q: w_1, w_2, w_3, w_4, \ldots, w_{12}\}$, used by these jurors.

b. Is this weighted voting system a one-person, one-vote system?

c. Is this weighted voting system a veto power system?

d. Explain an easy way to determine the Banzhaf power index for each voter.

18. Criminal Justice In California civil court cases, each of the 12 jurors has one vote and at least 9 of the jury members must agree on the verdict.

a. Write the weighted voting system, in the form $\{q: w_1, w_2, w_3, w_4, \ldots, w_{12}\}$, used by these jurors.

b. Is this weighted voting system a one-person, one-vote system?

c. Is this weighted voting system a veto power system?

d. Explain an easy way to determine the Banzhaf power index for each voter.

A voter who has a weight that is greater than or equal to the quota is called a **dictator**. In a weighted voting system, the dictator has all the power. A voter who is never a critical voter has no power and is referred to as a **dummy**. This term is not meant to be a comment on the voter's intellectual powers; it just indicates that the voter has no ability to influence an election.

■ In Exercises 19 to 22, identify any dictator and all dummies for each weighted voting system.

19. {16: 16, 5, 4, 2, 1}

20. {15: 7, 5, 3, 2}

21. {19: 12, 6, 3, 1}

22. {45: 40, 6, 2, 1}

23. Football At the beginning of each football season, the coaching staff at Vista High School must vote to decide which players to select for the team. They use the weighted voting system {4: 3, 2, 1}. In this voting system, the head coach, A, has a weight of 3, the assistant coach, B, has a weight of 2, and the junior varsity coach, C, has a weight of 1.

Kathy Burns-Millyard/Shutterstock.com

a. Compute the Banzhaf power index for each of the coaches.

b. Explain why it seems reasonable that the assistant coach and the junior varsity coach have the same Banzhaf power index in this voting system.

24. Football The head coach in Exercise 23 has decided that next year the coaching staff should use the weighted voting system {5: 4, 3, 1}. The head coach is still voter A, the assistant coach is still voter B, and the junior varsity coach is still voter C.

a. Compute the Banzhaf power index for each coach under this new system.

b. How do the Banzhaf power indices for this new voting system compare with the Banzhaf power indices in Exercise 23? Did the head coach gain any power, according to the Banzhaf power indices, with this new voting system?

25. Consider the weighted voting system {60: 4, 56, 58}.

a. Compute the Banzhaf power index for each voter in this system.

b. Voter B has a weight of 56 compared with only 4 for voter A, yet the results of part a show that voter A and voter B both have the same Banzhaf power index. Explain why it seems reasonable, in this voting system, to assign voters A and B the same Banzhaf power index.

EXTENSIONS

26. Consider the weighted voting system {17: 7, 7, 7, 2}.

a. Explain why voter D is a dummy in this system.

b. Explain an *easy* way to compute the Banzhaf power indices for this system.

27. Consider the weighted voting system $\{q: 8, 3, 3, 2\}$, with q an integer and $9 \leq q \leq 16$.

a. For what values of q is there a dummy?

b. For what values of q do all voters have the same power?

c. If a voter is a dummy for a given quota, must the voter be a dummy for all larger quotas?

28. a. In a weighted voting system, suppose that two voters have the same weight. Must they also have the same Banzhaf power index?

b. In a weighted voting system, suppose that voter A has a larger weight than voter B. Must the Banzhaf power index for voter A be larger than the Banzhaf power index for voter B?

Most voting systems are susceptible to fraudulent practices. In the following exercise, the Banzhaf power index is used to examine the power shift that occurs when a fraudulent practice is used.

29. Voter Intimidation and Tampering Consider the voting system {3: 1, 1, 1, 1, 1} with voters A, B, C, D, and E. The Banzhaf power index of each voter is 0.2.

a. Intimidation is used to force B to vote exactly as A. Thus the original {3: 1, 1, 1, 1, 1} voting system becomes

{3: 2, 0, 1, 1, 1}

The weight of A's vote is now 2 instead of 1. Does this mean that *BPI*(A) has doubled from 0.2 to 0.4? Explain.

b. Voter C tampers with the voting software so that the original {3: 1, 1, 1, 1, 1} voting system becomes

{3: 1, 1, 2, 1, 1}

The weight of C's vote is now 2 instead of 1. Does this mean that *BPI*(C) has doubled from 0.2 to 0.4? Explain.

30. **UN Security Council** The United Nations Security Council consists of 5 permanent members and 10 members who are elected for two-year terms. (See the Math Matters on page 210.) The weighted voting system of the Security Council is given by

$$\{39: 7, 7, 7, 7, 7, 1, 1, 1, 1, 1, 1, 1, 1, 1, 1\}$$

In this system, each permanent member has a voting weight of 7 and each elected member has a voting weight of 1.

United Nations, New York

a. Use the program "BPI" at the website www.math.temple.edu/~cow/bpi.html to compute the Banzhaf power index distribution for the members of the Security Council.

b. According to the Banzhaf power index distribution from part a, each permanent member has about how many times more power than an elected member?

c. Some people think that the voting system used by the United Nations Security Council gives too much power to the permanent members. A weighted voting system that would reduce the power of the permanent members is given by

$$\{30: 5, 5, 5, 5, 5, 1, 1, 1, 1, 1, 1, 1, 1, 1, 1\}$$

In this new system, each permanent member has a voting weight of 5 and each elected member has a voting weight of 1. Use the program "BPI" at www.math.temple.edu/~cow/bpi.html to compute the Banzhaf power indices for the members of the Security Council under this voting system.

d. According to the Banzhaf power indices from part c, each permanent member has about how many times more power than an elected member under this new voting system?

CHAPTER 4 SUMMARY

The following table summarizes essential concepts in this chapter. The references given in the right-hand column list Examples and Exercises that can be used to test your understanding of a concept.

4.1 Introduction to Apportionment

The Hamilton Plan To apply the Hamilton plan, first compute the standard divisor for the total population. The standard divisor is determined by dividing the total population of all the states by the number of representatives to apportion. $$\text{standard divisor} = \frac{\text{total population}}{\text{number of representatives to apportion}}$$ Then compute the standard quota for each state. The standard quota for a given state is the whole number part of the quotient of the state's population divided by the standard divisor. Initially each state is apportioned the number of representatives given by its standard quota. However, if the sum of all the standard quotas is less than the total number of representatives to apportion, then an additional representative is assigned to the state that has the largest decimal remainder of the quotients formed by the population of each state and the standard divisor. This process is continued until the total number of representatives equals the number of representatives to apportion.	See **Example 1a** on page 172, and then try Exercises 1a, 2a, and 5 to 8 on pages 222 to 224.

continued

The Hamilton plan is susceptible to three paradoxes. ▪ *The Alabama Paradox* The Alabama paradox occurs when an increase in the number of representatives to be apportioned results in a loss of a representative for a state. ▪ *The Population Paradox* The population paradox occurs when a state loses a representative to another state, even though its population is increasing at a faster rate than that of the other state. ▪ *The New States Paradox* The new states paradox occurs when the addition of a new state results in a reduction in the number of representatives of another state. The Hamilton plan does satisfy the quota rule. ▪ *The Quota Rule* The number of representatives apportioned to a state is the standard quota or more than the standard quota.	
The Jefferson Plan The Jefferson plan uses a modified standard divisor that is determined by trial and error, so that the sum of all the standard quotas equals the total number of representatives. The Jefferson plan is not susceptible to the three paradoxes that the Hamilton plan may produce, but it may violate the quota rule.	See **Example 1b** on page 172, and then try Exercises 1b, 2b, and 10 on pages 222 and 224.
Relative Unfairness of an Apportionment The average constituency of a state is defined as the quotient of the population of the state and the number of representatives from that state, rounded to the nearest whole number. One method that is used to gauge the fairness of an apportionment between two states is to examine the average constituency of the states. The state with the smaller average constituency has the more favorable representation. The absolute value of the difference between the average constituency of state A and the average constituency of state B is called the absolute unfairness of an apportionment. The relative unfairness of an apportionment is calculated when a new representative is assigned to one of two states. The relative unfairness of an apportionment that gives a new representative to state A rather than state B is the quotient of the absolute unfairness of apportionment between the two states and the average constituency of state A (the state receiving the new representative).	See **Example 2 and Check Your Progress 2** on page 178, and then try Exercises 3a and 3b on page 223.
Apportionment Principle When adding a new representative, the representative is assigned to the state in such a way as to give the smallest relative unfairness of apportionment.	See **Example 3** on page 178, and then try Exercises 3c and 4 on page 223.
The Huntington-Hill Apportionment Method The Huntington-Hill method is implemented by calculating the Huntington-Hill number and applying the Huntington-Hill apportionment principle, which states that when there is a choice of adding one representative to a state, the representative should be added to the state with the greatest Huntington-Hill number. The Huntington-Hill number for state A is $$\frac{(P_A)^2}{a(a+1)},$$ where P_A is the population of state A and a is the current number of representatives from state A.	See **Example 4** on page 180, and then try Exercise 11 on page 224.

The Huntington-Hill method is not susceptible to the three paradoxes that the Hamilton plan may produce, but it may violate the quota rule. ■ *Balinski-Young Impossibility Theorem* Any apportionment method either will violate the quota rule or will produce paradoxes such as the Alabama paradox.	
4.2 Introduction to Voting	
Plurality Method of Voting Each voter votes for one candidate, and the candidate with the most votes wins. The winning candidate does not have to have a majority of the votes.	See **Example 1** on page 189, and then try Exercises 12a and 13a on page 224.
Borda Count Method of Voting With n candidates in an election, each voter ranks the candidates by giving n points to the voter's first choice, $n - 1$ points to the voter's second choice, and so on, with the voter's least favorite choice receiving 1 point. The candidate that receives the most total points is the winner.	See **Example 2** on page 192, and then try Exercises 12b and 13b on page 224.
Plurality with Elimination Method of Voting Eliminate the candidate with the smallest number of first-place votes. Retake a vote, keeping the same ranking preferences, and eliminate the candidate with the smallest number of first-place votes. Continue until only one candidate remains.	See **Example 3** on page 194, and then try Exercises 14 and 15 on pages 224 and 225.
Pairwise Comparison Method of Voting Compare each candidate head-to-head with every other candidate. Award 1 point for a win, 0.5 points for a tie, and 0 points for a loss. The candidate with the greatest number of points wins the election.	See **Example 4** on page 196, and then try Exercises 16 and 17 on page 225.
Fairness Criteria **1.** *Majority Criterion* The candidate who receives a majority of the first-place votes is the winner. **2.** *Monotonicity Criterion* If candidate A wins an election, then candidate A will also win that election if the only change in the voter's preferences is that supporters of a different candidate change their votes to support candidate A. **3.** *Condorcet Criterion* A candidate who wins all possible head-to-head matchups should win an election when all candidates appear on the ballot. **4.** *Independence of Irrelevant Alternatives Criterion* If a candidate wins an election, the winner should remain the winner in any recount in which losing candidates withdraw from the race. ■ *Arrow's Impossibility Theorem* There is no voting method involving three or more choices that satisfies all four fairness criteria.	See **Examples 5 and 6** on pages 198 and 199, and then try Exercises 18 to 20 on page 225.

continued

4.3 Weighted Voting Systems	
Weighted Voting System A weighted voting system is one in which some voters have more weight on the outcome of an election. The number of votes required to pass a measure is called the quota. The weight of a voter is the number of votes controlled by the voter. A weighted voting system of n voters is written $\{q: w_1, w_2, \ldots, w_n\}$, where q is the quota and w_1 through w_n represent the weights of each of the n voters. In a weighted voting system, a coalition is a set of voters each of whom votes the same way, either for or against a resolution. A winning coalition is a set of voters the sum of whose votes is greater than or equal to the quota. A losing coalition is a set of voters the sum of whose votes is less than the quota. A voter who leaves a winning coalition and thereby turns it into a losing coalition is called a critical voter. The number of possible coalitions of n voters is $2^n - 1$. A voter who has a weight that is greater than or equal to the quota is called a dictator. A voter who is never a critical voter is referred to as a dummy.	See **Example 1** on page 211, and then try Exercises 21, 22, 27 and 28 on pages 225 and 226.
Banzhaf Power Index The Banzhaf power index of a voter v, symbolized by $BPI(v)$, is given by $BPI(v) = \dfrac{\text{number of times voter } v \text{ is a critical voter}}{\text{number of times any voter is a critical voter}}$	See **Examples 2 and 3** on pages 213 to 215, and then try Exercises 23 to 26 and 29 on pages 225 and 226.

CHAPTER 4 REVIEW EXERCISES

1. Education The following table shows the enrollments for the four divisions of a college. There are 50 new overhead projectors that are to be apportioned among the divisions based on the enrollments.

Division	Enrollment
Health	1280
Business	3425
Engineering	1968
Science	2936
Total	9609

a. Use the Hamilton method to determine the number of projectors to be apportioned to each division.

b. Use the Jefferson method to determine the number of projectors to be apportioned to each division.

c. Use the Webster method to determine the number of projectors to be apportioned to each division.

2. Airline Industry The following table shows the numbers of ticket agents at five airports for a small airline company. The company has hired 35 new security employees who are to be apportioned among the airports based on the number of ticket agents at each airport.

Airport	Number of ticket agents
Newark	28
Cleveland	19
Chicago	34
Philadelphia	13
Detroit	16
Total	110

a. Use the Hamilton method to apportion the new security employees among the airports.

b. Use the Jefferson method to apportion the new security employees among the airports.

c. Use the Webster method to apportion the new security employees among the airports.

3. **Airline Industry** The table below shows how the average constituency changes when two different airports, High Desert Airport and Eastlake Airport, add a new air traffic controller.

	High Desert Airport average constituency	Eastlake Airport average constituency
High Desert Airport receives new controller	297	326
Eastlake Airport receives new controller	302	253

a. Determine the relative unfairness of an apportionment that gives a new air traffic controller to High Desert Airport rather than to Eastlake Airport. Round to the nearest thousandth.

b. Determine the relative unfairness of an apportionment that gives a new air traffic controller to Eastlake Airport rather than to High Desert Airport. Round to the nearest thousandth.

c. Using the apportionment principle, determine which airport should receive the new air traffic controller.

4. **Education** The following table shows the number of English professors and the number of students taking English at two campuses of a state university. The university is planning to add a new English professor to one of the campuses. Use the apportionment principle to determine which campus should receive the new professor.

University campus	Number of English professors	Number of students taking English
Morena Valley	38	1437
West Keyes	46	1504

5. **Technology** A company has four offices. The president of the company uses the Hamilton method to apportion 66 new computer printers among the offices based on the number of employees at each office.

Office	A	B	C	D
Number of employees	19	195	308	402
Apportionment of 66 printers	1	14	22	29

a. If the number of printers to be apportioned by the Hamilton method increases from 66 to 67, will the Alabama paradox occur? Explain.

b. If the number of printers to be apportioned by the Hamilton method increases from 67 to 68, will the Alabama paradox occur? Explain.

6. **Automobile Sales** Consider the apportionment of 27 automobiles to the sales departments of a business with five regional centers labeled A, B, C, D, and E. The following table shows the Hamilton apportionment of the automobiles based on the number of sales personnel at each center.

Center	A	B	C	D	E
Number of sales personnel	31	108	70	329	49
Apportionment of 27 automobiles	2	5	3	15	2

a. If the number of automobiles to be apportioned increases from 27 to 28, what will be the apportionment if the Hamilton method is used? Will the Alabama paradox occur? Explain.

b. If the number of automobiles to be apportioned using the Hamilton method increases from 28 to 29, will the Alabama paradox occur? Explain.

7. **Music Company** MusicGalore.biz has offices in Los Angeles and Newark. The number of employees at each office is shown in the following table. There are 11 new computer file servers to be apportioned between the offices according to their numbers of employees.

Office	Los Angeles	Newark
Employees	1430	235

a. Use the Hamilton method to find each office's apportionment of file servers.

b. The corporation opens an additional office in Kansas City with 111 employees and decides to have a total of 12 file servers. If the file servers are reapportioned using the Hamilton method, will the new states paradox occur? Explain.

8. Building Inspectors A city apportions 34 building inspectors among three regions according to their populations. The following table shows the present population of each region.

Region	A	B	C
Population	14,566	3321	29,988

a. Use the Hamilton method to apportion the inspectors.

b. After a year the regions have the following populations.

Region	A	B	C
Population	15,008	3424	30,109

Region A has an increase in population of 442, which is an increase of 3.03%. Region B has an increase in population of 103, which is an increase of 3.10%. Region C has an increase in population of 121, which is an increase of 0.40%. If the inspectors are reapportioned using the Hamilton method, will the population paradox occur? Explain.

9. Is the Hamilton apportionment method susceptible to the population paradox?

10. Is the Jefferson apportionment method susceptible to the new states paradox?

11. Corporate Security The Huntington-Hill apportionment method has been used to apportion 86 security guards among three corporate office buildings according to the number of employees at each building. See the following table.

Building	Number of security guards	Number of employees
A	25	414
B	43	705
C	18	293

The corporation has decided to hire a new security guard.

a. Use the Huntington-Hill apportionment principle to determine to which building the new security guard should be assigned.

b. If another security guard is hired, bringing the total number of guards to 88, to which building should this guard be assigned?

12. Essay Contest Four finalists are competing in an essay contest. Judges have read and ranked each essay in order of preference. The results are shown in the preference schedule below.

	Rankings			
Crystal Kelley	3	2	2	1
Manuel Ortega	1	3	4	3
Peter Nisbet	2	4	1	2
Sue Toyama	4	1	3	4
Number of votes:	8	5	4	6

a. Using the plurality voting system, who is the winner of the essay contest?

b. Does this winner have a majority?

c. Use the Borda count method of voting to determine the winner of the essay contest.

13. Ski Club A campus ski club is trying to decide where to hold its winter break ski trip. The members of the club were surveyed and asked to rank five choices in order of preference. Their responses are tallied in the following table.

	Rankings				
Aspen	1	1	3	2	3
Copper Mountain	5	4	2	4	4
Powderhorn	3	2	5	1	5
Telluride	4	5	4	5	2
Vail	2	3	1	3	1
Number of votes:	14	8	11	18	12

a. Use the plurality method of voting to determine which resort the club should choose.

b. Use the Borda count method to choose the ski resort the club should visit.

14. Campus Election Four students are running for the activities director position on campus. Students were asked to rank the four candidates in order of preference. The results are shown in the table below.

	Rankings			
G. Reynolds	2	3	1	3
L. Hernandez	1	4	4	2
A. Kim	3	1	2	1
J. Schneider	4	2	3	4
Number of votes:	132	214	93	119

Use the plurality with elimination method to determine the winner of the election.

15. Consumer Preferences A group of consumers were surveyed about their favorite candy bars. Each participant was asked to rank four candy bars in order of preference. The results are given in the table below.

	Rankings				
Nestle Crunch	1	4	4	2	3
Snickers	2	1	2	4	1
Milky Way	3	2	1	1	4
Twix	4	3	3	3	2
Number of votes:	15	38	27	16	22

Use the plurality with elimination method to determine the group's favorite candy bar.

16. Use the pairwise comparison method of voting to choose the winner of the election in Exercise 14.

17. Use the pairwise comparison method of voting to choose the group's favorite candy bar in Exercise 15.

18. Homecoming Queen Three high school students are running for homecoming queen. Students at the school were allowed to rank the candidates in order of preference. The results are shown in the preference schedule below.

	Rankings		
Cynthia L.	3	2	1
Hannah A.	1	3	3
Shannon M.	2	1	2
Number of votes:	112	97	11

a. Use the Borda count method to find the winner.

b. Find a candidate who wins all head-to-head comparisons.

c. Explain why the Condorcet criterion has been violated.

d. Who wins using the plurality voting system?

e. Explain why the majority criterion has been violated.

19. Homecoming Queen In Exercise 18, suppose Cynthia L. withdraws from the homecoming queen election.

a. Assuming voter preferences between the remaining two candidates remain the same, who will be crowned homecoming queen using the Borda count method?

b. Explain why the independence of irrelevant alternatives criterion has been violated.

20. Scholarship Awards A scholarship committee must choose a winner from three finalists, Jean, Margaret, and Terry. Each member of the committee ranked the three finalists, and Margaret was selected using the plurality with elimination method. This vote was later declared invalid and a new vote was taken. All members voted using the same rankings except one, who changed her first choice from Terry to Margaret. This time Jean won the scholarship using the same voting method. Which fairness criterion was violated and why?

21. A weighted voting system for voters A, B, C, and D is given by {18: 12, 7, 6, 1}. The weight of voter A is 12, the weight of voter B is 7, the weight of voter C is 6, and the weight of voter D is 1.

a. What is the quota?

b. What is the weight of the coalition {A, C}?

c. Is {A, C} a winning coalition?

d. Which voters are critical voters in the coalition {A, C, D}?

e. How many coalitions can be formed?

f. How many coalitions consist of exactly two voters?

22. A weighted voting system for voters A, B, C, D, and E is given by {35: 29, 11, 8, 4, 2}. The weight of voter A is 29, the weight of voter B is 11, the weight of voter C is 8, the weight of voter D is 4, and the weight of voter E is 2.

a. What is the quota?

b. What is the weight of the coalition {A, D, E}?

c. Is {A, D, E} a winning coalition?

d. Which voters are critical voters in the coalition {A, C, D, E}?

e. How many coalitions can be formed?

f. How many coalitions consist of exactly two voters?

23. Calculate the Banzhaf power indices for voters A, B, and C in the weighted voting system {9: 6, 5, 3}.

24. Calculate the Banzhaf power indices for voters A, B, C, D, and E in the one-person, one-vote system {3: 1, 1, 1, 1, 1}.

25. Calculate the Banzhaf power indices for voters A, B, C, and D in the weighted voting system {31: 19, 15, 12, 10}. Round to the nearest hundredth.

26. Calculate the Banzhaf power indices for voters A, B, C, D, and E in the weighted voting system {35: 29, 11, 8, 4, 2}. Round to the nearest hundredth.

■ In Exercises 27 and 28, identify any dictator and all dummies for each weighted voting system.

27. Voters A, B, C, D, and E: {15: 15, 10, 2, 1, 1}

28. Voters A, B, C, and D: {28: 19, 6, 4, 2}

29. Four voters, A, B, C, and D, make decisions by using the weighted voting system {5: 4, 2, 1, 1}. In the event of a tie, a fifth voter, E, casts a vote to break the tie. For this voting scheme, determine the Banzhaf power index for each voter, including voter E.

CHAPTER 4 TEST

1. Postal Service The table below shows the number of mail carriers and the population for two cities, Spring Valley and Summerville. The postal service is planning to add a new mail carrier to one of the cities. Use the apportionment principle to determine which city should receive the new mail carrier.

City	Number of mail carriers	Population
Spring Valley	158	67,530
Summerville	129	53,950

2. Computer Allocation The following table shows the number of employees in the four divisions of a corporation. There are 85 new computers that are to be apportioned among the divisions based on the number of employees in each division.

Division	Number of employees
Sales	1008
Advertising	234
Service	625
Manufacturing	3114
Total	4981

a. Use the Hamilton method to determine the number of computers to be apportioned to each division.

b. Use the Jefferson method to determine the number of computers to be apportioned to each division. Does this particular apportionment violate the quota rule?

3. High School Counselors The following table shows the number of counselors and the number of students at two high schools. The current number of counselors for each school was determined using the Huntington-Hill apportionment method.

School	Number of counselors	Number of students
Cedar Falls	9	2646
Lake View	7	1984

A new counselor is to be hired for one of the schools.

a. Calculate the Huntington-Hill number for each of the schools. Round to the nearest whole number.

b. Use the Huntington-Hill apportionment principle to determine to which school the new counselor should be assigned.

4. A weighted voting system for voters A, B, C, D, and E is given by {33: 21, 14, 12, 7, 6}.

a. What is the quota?

b. What is the weight of the coalition {B, C}?

c. Is {B, C} a winning coalition?

d. Which voters are critical voters in the coalition {A, C, D}?

e. How many coalitions can be formed?

f. How many coalitions consist of exactly two voters?

5. Consumer Preference One hundred consumers ranked three brands of bottled water in order of preference. The results are shown in the preference schedule below.

	Rankings				
Arrowhead	2	3	1	3	2
Evian	3	2	2	1	1
Aquafina	1	1	3	2	3
Number of votes:	22	17	31	11	19

a. Using the plurality system of voting, which brand of bottled water is the preferred brand?

b. Does the winner have a majority?

c. Use the Borda count method of voting to determine the preferred brand of water.

6. **Executives' Preferences** A company with offices across the country will hold its annual executive meeting in one of four locations. All of the executives were asked to rank the locations in order of preference. The results are shown in the table below.

	Rankings			
New York	3	1	2	2
Dallas	2	3	1	4
Los Angeles	4	2	4	1
Atlanta	1	4	3	3
Number of votes:	19	24	7	35

Use the pairwise comparison method to determine which location should be chosen.

7. **Exam Review** A professor is preparing an extra review session the Monday before final exams and she wants as many students as possible to be able to attend. She asks all of the students to rank different times of day in order of preference. The results are given below.

	Rankings				
Morning	4	1	4	2	3
Noon	3	2	2	3	1
Afternoon	1	3	1	4	2
Evening	2	4	3	1	4
Number of votes:	12	16	9	5	13

a. Using the plurality with elimination method, for what time of day should the professor schedule the review session?

b. Use the Borda count method to determine the best time of day for the review.

8. **Budget Proposal** A committee must vote on which of three budget proposals to adopt. Each member of the committee has ranked the proposals in order of preference. The results are shown below.

	Rankings		
Proposal A	1	3	3
Proposal B	2	2	1
Proposal C	3	1	2
Number of votes:	40	9	39

a. Using the plurality system of voting, which proposal should be adopted?

b. If proposal C is found to be invalid and is eliminated, which proposal wins using the plurality system?

c. Explain why the independence of irrelevant alternatives criterion has been violated.

d. Verify that proposal B wins all head-to-head comparisons.

e. Explain why the Condorcet criterion has been violated.

9. **Drama Department** The four staff members, A, B, C, and D, of a college drama department use the weighted voting system {8: 5, 4, 3, 2} to make casting decisions for a play they are producing. In this voting system, voter A has a weight of 5, voter B has a weight of 4, voter C has a weight of 3, and voter D has a weight of 2. Compute the Banzhaf power index for each member of the department. Round each result to the nearest hundredth.

10. Three voters, A, B, and C, make decisions by using the weighted voting system {6: 5, 4, 1}. In the case of a tie, a fourth voter, D, casts a single vote to break the tie. For this voting scheme, determine the Banzhaf power index for each voter, including voter D.

5

The Mathematics of Graphs

In this chapter, you will learn how to analyze and solve a variety of problems, such as how to find the least expensive route of travel on a vacation, how to determine the most efficient order in which to run errands, and how to schedule meetings at a conference so that no one has two required meetings at the same time.

The methods we will use to study these problems can be traced back to an old recreational puzzle. In the early eighteenth century, the Pregel River in a city called Königsberg (located in modern-day Russia and now called Kaliningrad) surrounded an island before splitting in two. Seven bridges crossed the river and connected four different land areas, similar to the map drawn below.

Many citizens of the time attempted to take a stroll that would lead them across each bridge and return them to the starting point without traversing the same bridge twice. None of them could do it, no matter where they chose to start. Try it for yourself with pencil and paper. You will see that it is not easy!

In 1736 the Swiss mathematician Leonhard Euler (1707–1783) proved that it is, in fact, impossible to walk such a path. His analysis of the challenge laid the groundwork for a branch of mathematics known as *graph theory*. We will investigate how Euler approached the problem of the seven bridges of Königsberg in Section 5.1.

SECTION 5.1 Graphs and Euler Circuits

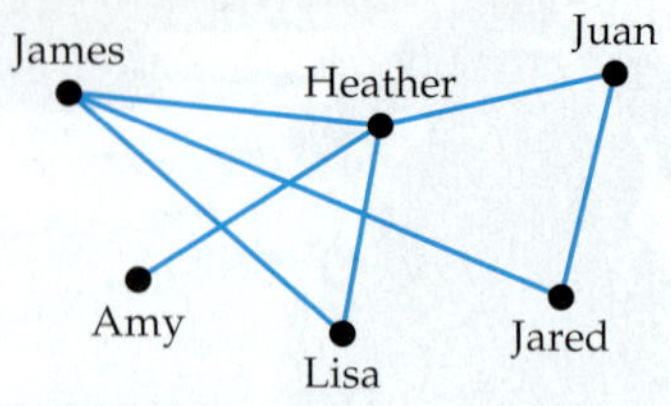

FIGURE 5.1

Introduction to Graphs

Think of all the various connections we experience in our lives—friends are connected on Facebook, cities are connected by roads, computers are connected across the Internet. A branch of mathematics called **graph theory** illustrates and analyzes connections such as these.

For example, the diagram in Figure 5.1 could represent friends that are connected on Facebook. Each dot represents a person, and a line segment connecting two dots means that those two people are friends on Facebook. This type of diagram is called a *graph*. Note that this is a different kind of graph from the graph of a function that we will discuss in Chapter 10.

TAKE NOTE

Vertices are always clearly indicated with a "dot." Edges that intersect with no marked vertex are considered to cross over each other without touching.

Graph

A **graph** is a set of points called **vertices** and line segments or curves called **edges** that connect vertices.

Graphs can be used to represent many different scenarios. For instance, the two graphs in Figure 5.2 are the same graph as in Figure 5.1 but used in different contexts. In part (a), each vertex represents a baseball team, and an edge connecting two vertices might mean that the two teams played against each other during the current season. The graph in part (b) could be used to represent the flights available on a particular airline between a selection of cities; each vertex represents a city, and an edge connecting two cities means that there is a direct flight between the two cities.

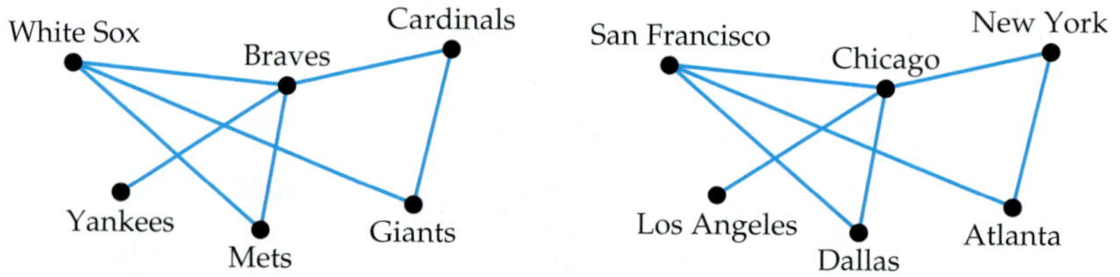

FIGURE 5.2 (a) (b)

EXAMPLE 1 Constructing a Graph

The following table lists five students at a college. An "X" indicates that the two students participate in the same study group this semester.

	Matt	Amber	Oscar	Laura	Kayla
Matt	—	X		X	
Amber	X	—	X	X	
Oscar		X	—		X
Laura	X	X		—	
Kayla			X		—

a. Draw a graph that represents this information where each vertex represents a student and an edge connects two vertices if the corresponding students study together.

b. Use your graph to answer the following questions: Which student is involved in the most study groups with the others? Which student has only one study group in common with the others? How many study groups does Laura have in common with the others?

Solution

a. We draw five vertices (in any configuration we wish) to represent the five students, and connect vertices with edges according to the table.

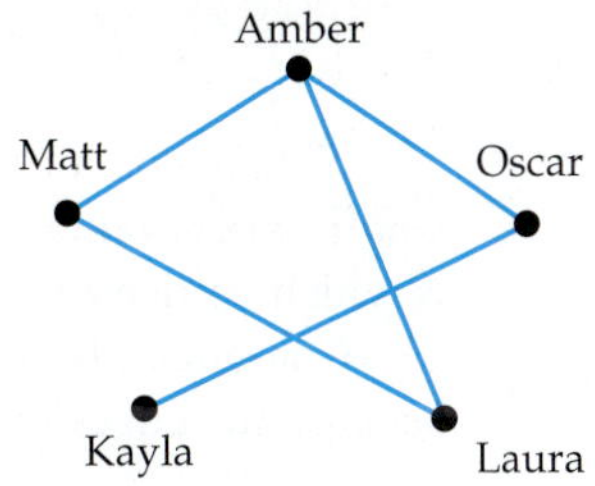

b. The vertex corresponding to Amber is connected to more edges than the others, so she is involved with more study groups (three) than the others. Kayla is the only student with one study group in common, as her vertex is the only one connected to just one edge. Laura's vertex is connected to two edges, so she shares two study groups with the others.

CHECK YOUR PROGRESS 1 The table below lists five mobile phone companies and indicates whether they have agreements to roam onto each other's networks. Draw a graph that represents this information, where each vertex represents a phone company and an edge connects two vertices if the corresponding companies have a roaming agreement. Then use the graph to answer the questions: Which phone company has roaming agreements with the most carriers? Which company can roam with only one other network?

	MobilePlus	TalkMore	SuperCell	Airwave	Lightning
MobilePlus	—	No	Yes	No	Yes
TalkMore	No	—	Yes	No	No
SuperCell	Yes	Yes	—	Yes	No
Airwave	No	No	Yes	—	Yes
Lightning	Yes	No	No	Yes	—

Solution See page S16.

In general, a graph can include vertices that are not joined to any edges, but all edges must begin and end at vertices. If two or more edges connect the same vertices, they are called **multiple edges**. If an edge begins and ends at the same vertex, it is called a **loop**.

A graph is called **connected** if any vertex can be reached from any other vertex by tracing along edges. (Essentially, the graph consists of one "piece.") A connected graph in which every possible edge is drawn between vertices (without any multiple edges) is called a **complete graph**. Several examples of graphs are shown below.

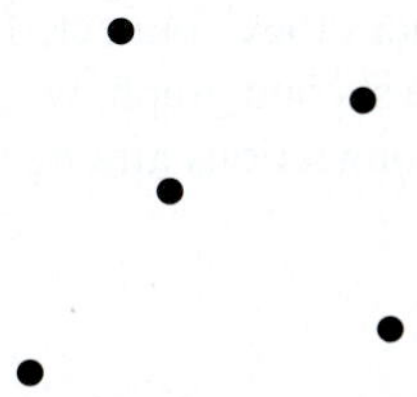

This graph has five vertices but no edges. It is not connected.

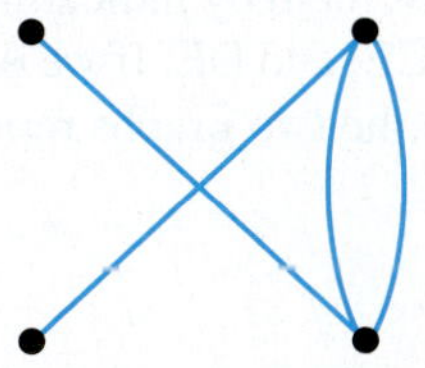

This is a connected graph that has a pair of multiple edges. Note that two edges cross in the center, but there is no vertex there. Unless a dot is drawn, the edges are considered to pass over each other without touching.

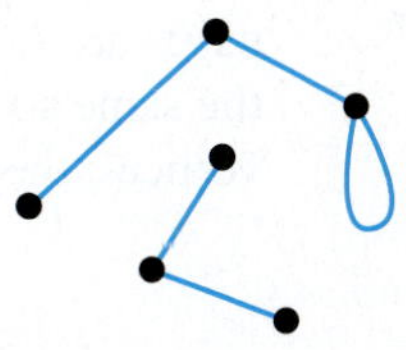

This graph is not connected; it consists of two different sections. It also contains a loop.

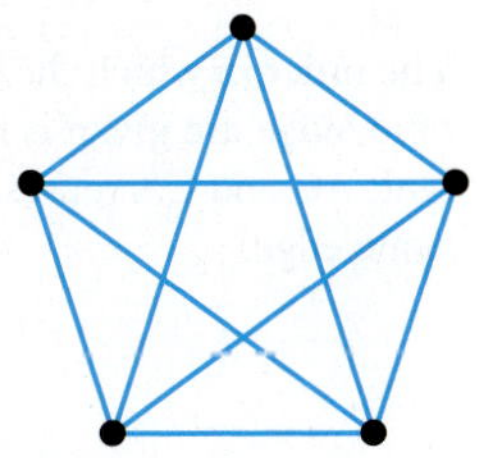

This is a complete graph with five vertices.

QUESTION Is the following graph a complete graph?

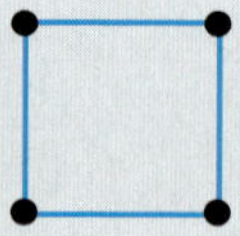

Note that it does not matter whether the edges are drawn straight or curved, and their lengths are not important. Nor is the placement of the vertices important. All that matters is which vertices are connected by edges.

Consequently, the three graphs shown below are considered **equivalent graphs** because the edges form the same connections of vertices in each graph.

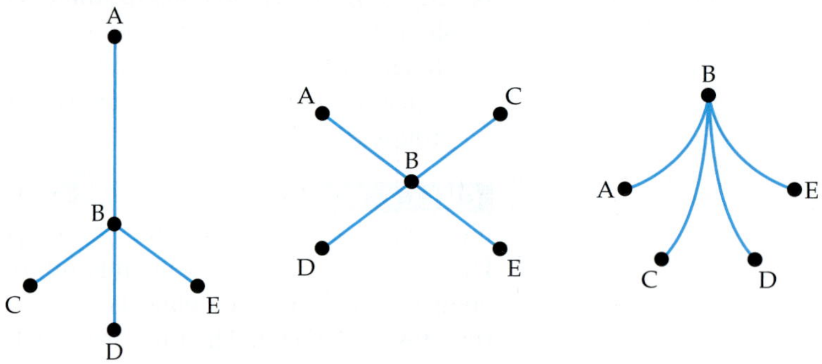

If you have difficulty seeing that these graphs are equivalent, use the labeled vertices to compare each graph. Note that in each case, vertex B has an edge connecting it to each of the other four vertices, and no other edges exist.

EXAMPLE 2 Equivalent Graphs

Determine whether the following two graphs are equivalent.

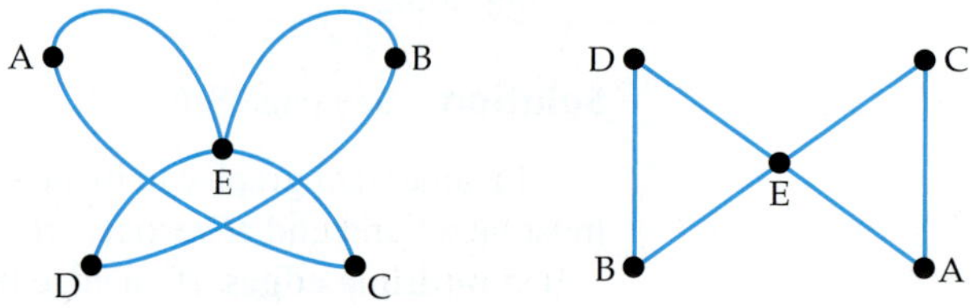

Solution

Despite the fact that the two graphs have different arrangements of vertices and edges, they are equivalent. To illustrate, we examine the edges of each graph. The first graph contains six edges; we can list them by indicating which two vertices they connect. The edges are AC, AE, BD, BE, CE, and DE. If we do the same for the second graph, we get the same six edges. Because the two graphs represent the same connections among the vertices, they are equivalent.

TAKE NOTE

The order in which the vertices of an edge are given is not important; AC and CA represent the same edge.

ANSWER No. Not every possible edge is drawn. We can add two more edges:

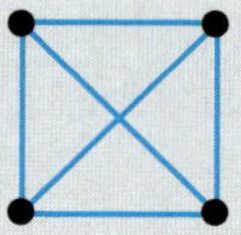

CHECK YOUR PROGRESS 2 Determine whether the following two graphs are equivalent.

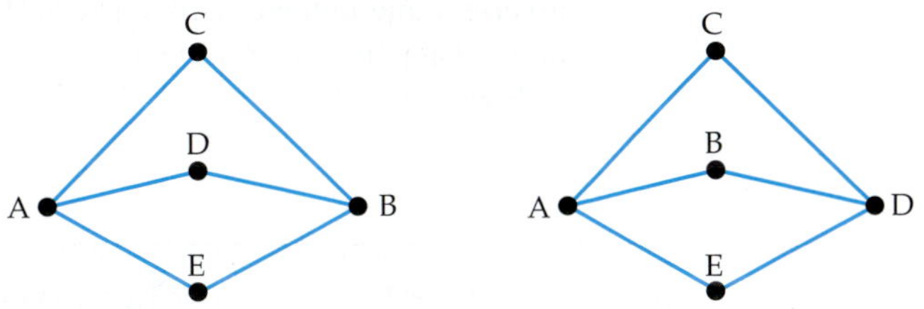

Solution See page S16.

MATH MATTERS Picture the Internet as a Graph

Graph theory is playing an ever-increasing role in the study of large and complex systems such as the collection of web pages that are published on the Internet. The image below, created by Matthew Hurst, is an illustration of various blogs that appear on the web. Each vertex represents a particular blog, and two vertices are joined by an edge if one of the corresponding blogs contains a link to the other. You can imagine how large the graph would be if all the millions of existing blogs were included.

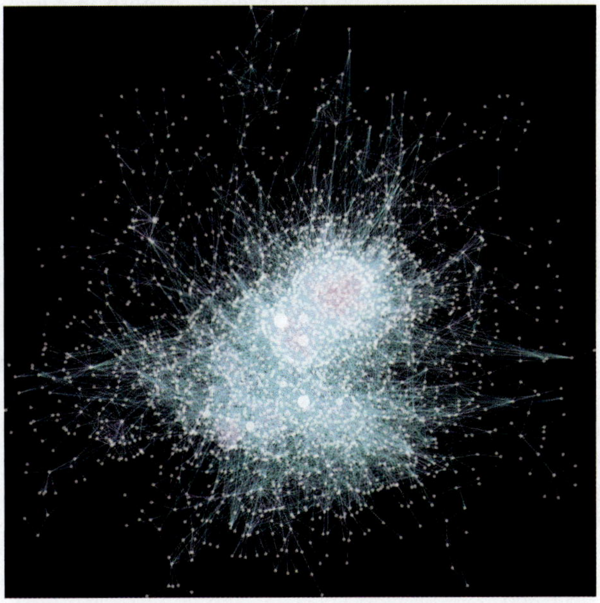

Matthew Hurst/Science Source

Another example of an extremely large graph is the graph that uses vertices to represent telephone numbers. An edge connects two vertices if a phone call was placed from one number to the other. James Abello of AT&T Shannon Laboratories analyzed a graph formed from one day's worth of calls. The graph had over 53 million vertices and 170 million edges! Interestingly, although the graph was not a connected graph (it contained 3.7 million separate components), over 80% of the vertices were part of one large connected component. Within this component, any telephone number could be linked to any other through a chain of 20 or fewer calls.

Euler Circuits

To solve the Königsberg bridges problem presented on page 229, we can represent the arrangement of land areas and bridges with a graph. Let each land area be represented by a vertex, and connect two vertices if there is a bridge spanning the corresponding land areas. Then the geographical configuration shown in Figure 5.3 becomes the graph shown in Figure 5.4.

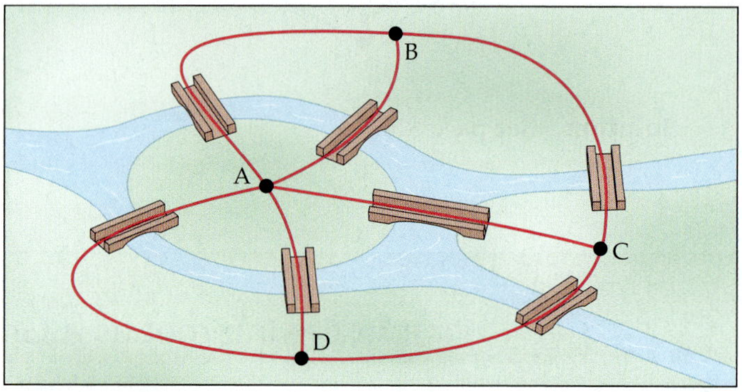

FIGURE 5.3

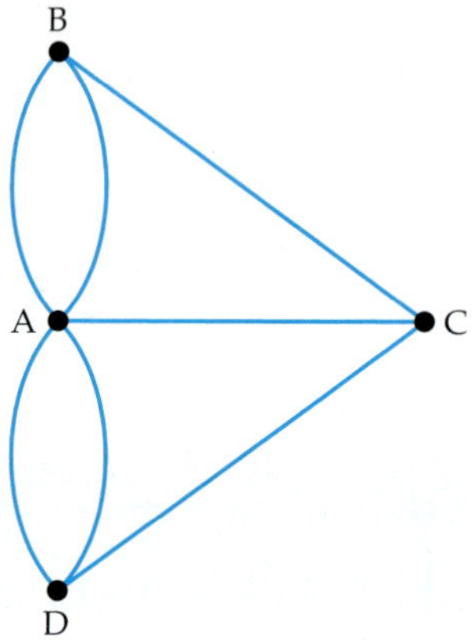

FIGURE 5.4

In terms of a graph, the original problem can be stated as follows: Can we start at any vertex, move through each edge once (but not more than once), and return to the starting vertex? Again, try it with pencil and paper. Every attempt seems to end in failure.

Before we can examine how Euler proved this task impossible, we need to establish some terminology. A **path** in a graph can be thought of as a movement from one vertex to another by traversing edges. We can refer to our movement by vertex letters. For example, in the graph in Figure 5.4, one path would be A–B–A–C.

If a path ends at the same vertex at which it started, it is considered a **closed path**, or **circuit**. For the graph in Figure 5.5, the path A–D–F–G–E–B–A is a circuit because it begins and ends at the same vertex. The path A–D–F–G–E–H is not a circuit, as the path ends at a different vertex than the one it started at.

HISTORICAL NOTE

Bettmann/Corbis

Leonhard Euler (oi′lər) (1707–1783) was one of the most prolific mathematicians of all time. He wrote hundreds of papers in almost every area of mathematics. In one of these papers, published in 1736, Euler proved that it was impossible to traverse each of the bridges of Königsberg exactly once and return to the starting point.

Although he did not present his paper in the language of graph theory, his arguments were equivalent to our discussion in this section. In fact, his proof was more general than just a proof of the Königsberg bridges problem, and it can be considered the first paper in graph theory.

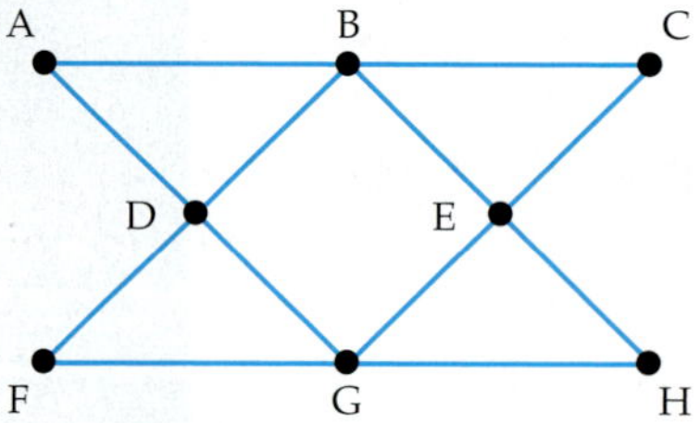

FIGURE 5.5

A circuit that uses every edge, but never uses the same edge twice, is called an **Euler circuit**. (The path may cross through vertices more than once.) The path B–D–F–G–H–E–C–B–A–D–G–E–B in Figure 5.5 is an Euler circuit. It begins and ends at the same vertex and uses each edge exactly once. (Trace the path with your pencil to verify!) The path A–B–C–E–H–G–E–B–D–A is not an Euler circuit: The path begins and ends at the same vertex but it does not use edges DF, DG, or FG. The path A–B–C–E–H–G–F–D–A–B–E–G–D–A begins and ends at A but uses edges AB and AD twice so it is not an Euler circuit.

All of this relates to the Königsberg bridges problem in the following way: Finding a path that crosses each bridge exactly once and returns to the starting point is equivalent to finding an Euler circuit for the graph in Figure 5.4.

Euler essentially proved that the graph in Figure 5.4 could not have an Euler circuit. He accomplished this by examining the number of edges that met at each vertex. The number of edges that meet at a vertex is called the **degree** of a vertex. He made the observation that in order to complete the desired path, every time you approached a vertex you would then need to leave that vertex. If you traveled through that vertex again, you would again need an approaching edge and a departing edge. Thus for an Euler circuit to exist, the degree of every vertex would have to be an even number. Furthermore, he was able to show that any graph that has even degree at every vertex must have an Euler circuit. Consequently, such graphs are called **Eulerian**.

Eulerian Graph Theorem

A connected graph is Eulerian if and only if every vertex of the graph is of even degree.

EXAMPLE 3 Identifying Eulerian Graphs

Which of the following graphs has an Euler circuit?

a.

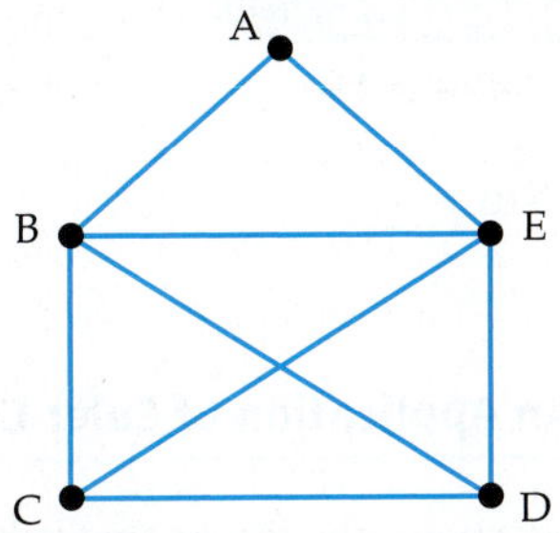

b.

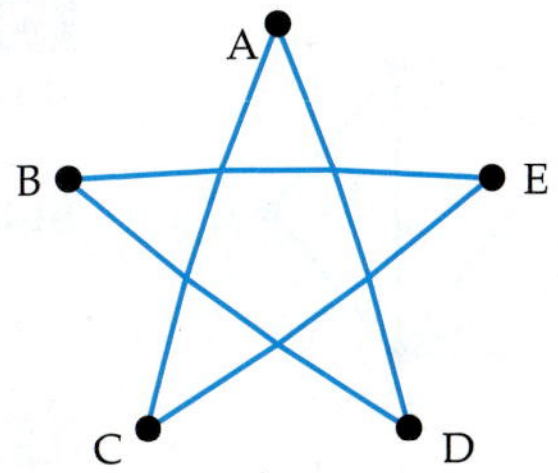

Solution

a. Vertices C and D are of odd degree. By the Eulerian graph theorem, the graph does not have an Euler circuit.

b. All vertices are of even degree. By the Eulerian graph theorem, the graph has an Euler circuit.

CHECK YOUR PROGRESS 3 Does the graph shown below have an Euler circuit?

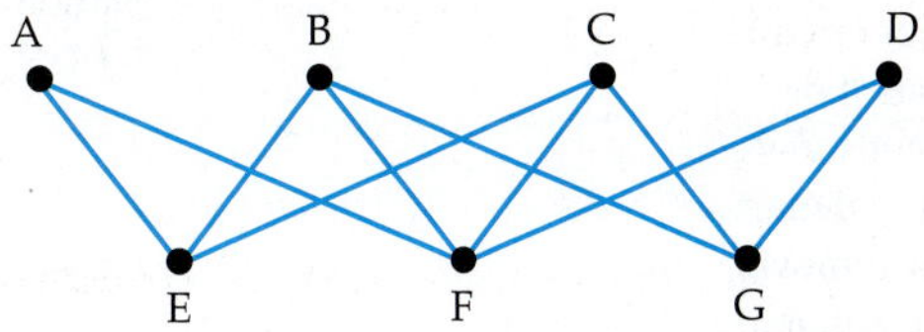

Solution See page S16.

Look again at Figure 5.4 on page 234, the graph representation of the Königsberg bridges. Because not every vertex is of even degree, we know by the Eulerian graph theorem that no Euler circuit exists. Consequently, it is not possible to begin and end at the same location near the river and cross each bridge exactly once.

The Eulerian graph theorem guarantees that when all vertices of a graph have an even degree, an Euler circuit exists, but it does not tell us how to find one. Because the graphs we will examine here are relatively small, we will rely on trial and error to find Euler circuits. There is a systematic method, called **Fleury's algorithm**, that can be used to find Euler circuits in graphs with large numbers of vertices.

TAKE NOTE

For information on and examples of finding Eulerian circuits, search for "Fleury's algorithm" on the Internet.

EXAMPLE 4 Find an Euler Circuit

Determine whether the graph shown below is Eulerian. If it is, find an Euler circuit. If it is not, explain how you know.

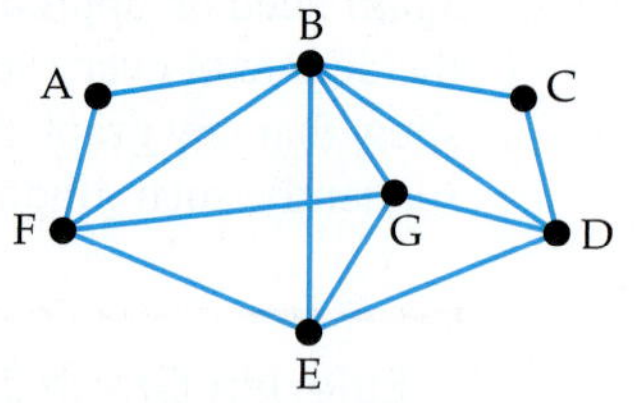

Solution

Each vertex is of even degree (2, 4, or 6), so by the Eulerian graph theorem, the graph is Eulerian. There are many possible Euler circuits in this graph. We do not have a formal method of locating one, but by trial and error, one Euler circuit is B–A–F–B–E–F–G–E–D–G–B–D–C–B.

> **TAKE NOTE**
>
> You should verify that the given path is an Euler circuit. Using your pencil, start at vertex B and trace along edges of the graph, following the vertices in order. Make sure you trace over each edge once (but none twice).

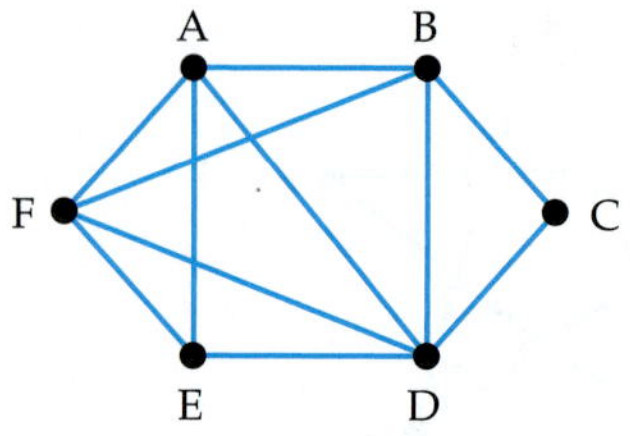

CHECK YOUR PROGRESS 4 Does the graph at the left have an Euler circuit? If so, find one. If not, explain why.

Solution See page S16.

EXAMPLE 5 An Application of Euler Circuits

The subway map below shows the tracks that subway trains traverse as well as the junctions where one can switch trains. Suppose an inspector needs to travel the full length of each track. Is it possible to plan a journey that traverses the tracks and returns to the starting point without traveling through any portion of a track more than once?

> **POINT OF INTEREST**
>
> There are a number of instances in which finding an Euler circuit has practical implications. For example, in cities where it snows in the winter, the highway department must provide snow removal for the streets. The most efficient route is an Euler circuit. In this case, the snow plow leaves the maintenance garage, travels down each street only once, and returns to the garage. The situation becomes more complicated if one-way streets are involved; nonetheless, graph theory techniques can still help find the most efficient route to follow.

Solution

We can consider the subway map a graph, with a vertex at each junction. An edge represents a track that runs between two junctions. In order to find a travel route that does not traverse the same track twice, we need to find an Euler circuit in the graph. Note, however, that the vertex representing the Civic Center junction has degree 3. Because a vertex has an odd degree, the graph cannot be Eulerian, and it is impossible for the inspector not to travel at least one track twice.

CHECK YOUR PROGRESS 5 Suppose the city of Königsberg had the arrangement of islands and bridges pictured below instead of the arrangement that we introduced previously. Would the citizens be able to complete a stroll across each bridge and return to their starting points without crossing the same bridge twice?

Solution See page S16.

Euler Paths

Perhaps the Königsberg bridges problem would have a solution if we did not need to return to the starting point. In this case, what we are looking for in Figure 5.4 on page 234 is a path (not necessarily a circuit) that uses every edge once and only once. We call such a path an **Euler path**. Euler showed that even with this relaxed condition, the bridge problem still was not solvable. The general result of his argument is given in the following theorem.

TAKE NOTE

Note that an Euler *path* does not require that we start and stop at the same vertex, whereas an Euler *circuit* does.

Euler Path Theorem

A connected graph contains an Euler path if and only if the graph has two vertices of odd degree with all other vertices of even degree. Furthermore, every Euler path must start at one of the vertices of odd degree and end at the other.

To see why this theorem is true, note that the only places at which an Euler path differs from an Euler circuit are the start and end vertices. If we never return to the starting vertex, only one edge meets there and the degree of the vertex is 1. If we do return, we cannot stop there. So we depart again, giving the vertex a degree of 3. Similarly, any return trip means that an additional two edges meet at the vertex. Thus the degree of the start vertex must be odd. By similar reasoning, the ending vertex must also be of odd degree. All other vertices, just as in the case of an Euler circuit, must be of even degree.

EXAMPLE 6 An Application of Euler Paths

A photographer would like to travel across all of the roads shown on the following map. The photographer will rent a car that need not be returned to the same city, so the trip can begin in any city. Is it possible for the photographer to design a trip that traverses all of the roads exactly once?

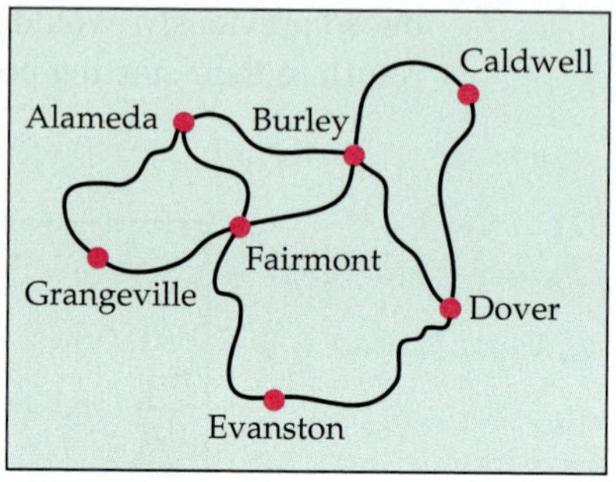

Solution

Looking at the map of roads as a graph, we see that a route that includes all of the roads but does not cover any road twice corresponds to an Euler path of the graph. Notice that only two vertices are of odd degree, the cities Alameda and Dover. Thus we know that an Euler path exists, and so it is possible for the photographer to plan a route that travels each road once. Because (abbreviating the cities) A and D are vertices of odd degree, the photographer must start at one of these cities. With a little experimentation, we find that one Euler path is A–B–C–D–B–F–A–G–F–E–D.

CHECK YOUR PROGRESS 6 A bicyclist wants to mountain bike through all the trails of a national park. A map of the park is shown below. Because the bicyclist will be dropped off in the morning by friends and picked up in the evening, she does not have a preference for where she begins and ends her ride. Is it possible for the cyclist to traverse all of the trails without repeating any portions of her trip?

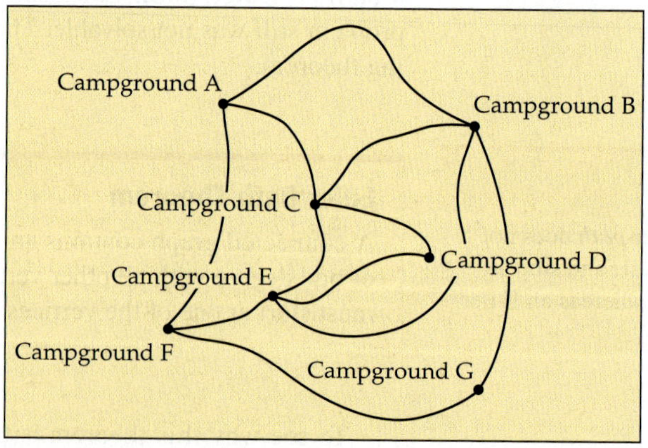

Solution See page S17.

EXAMPLE 7 An Application of Euler Paths

The following figure depicts the floor plan of an art gallery. Draw a graph that represents the floor plan, where vertices correspond to rooms and edges correspond to doorways. Is it possible to take a stroll that passes through every doorway without going through the same doorway twice? If so, does it matter whether we return to the starting point?

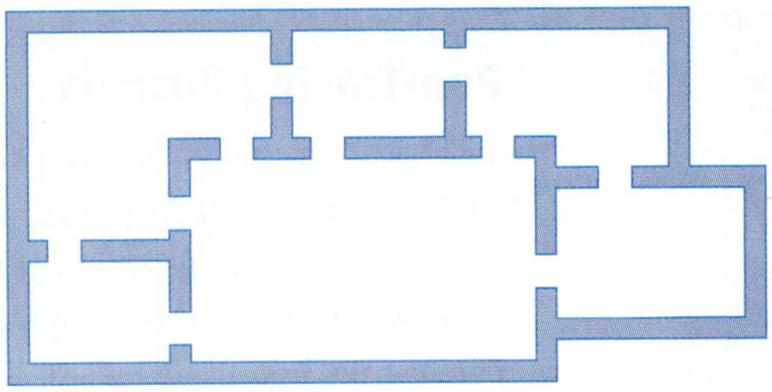

Solution

We can represent the floor plan by a graph if we let a vertex represent each room. Draw an edge between two vertices if there is a doorway between the two rooms, as shown in Figure 5.6.

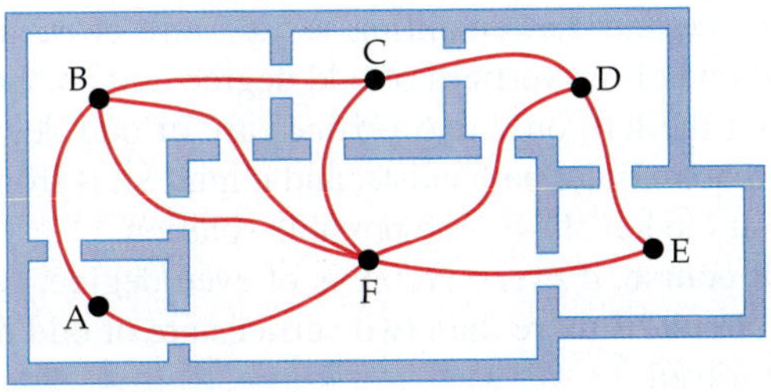

FIGURE 5.6

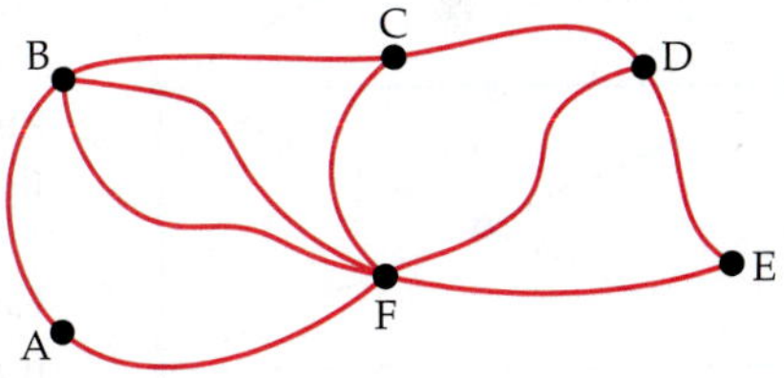

FIGURE 5.7

The graph in Figure 5.7 is equivalent to our floor plan. If we would like to tour the gallery and pass through every doorway once, we must find a path in our graph that uses every edge once (and no more). Thus we are looking for an Euler path. In the graph, two vertices are of odd degree and the others are of even degree. So we know that an Euler path exists, but not an Euler circuit. Therefore, we cannot pass through each doorway once and only once if we want to return to the starting point, but we can do it if we end up somewhere else. Furthermore, we know we must start at a vertex of odd degree—either room C or room D. By trial and error, one such path is C–B–F–B–A–F–E–D–C–F–D.

CHECK YOUR PROGRESS 7 The floor plan of a warehouse is illustrated below. Use a graph to represent the floor plan, and answer the following questions: Is it possible to walk through the warehouse so that you pass through every doorway once but not twice? Does it matter whether you return to the starting point?

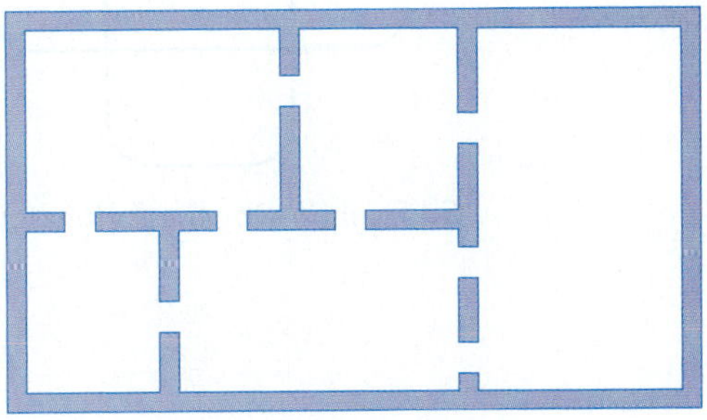

Solution See page S17.

EXCURSION

Pen-Tracing Puzzles

You may have seen puzzles like this one: Can you draw the diagram at the left without lifting your pencil from the paper and without tracing over the same segment twice?

Before reading on, try it for yourself. By trial and error, you may discover a tracing that works. Even though there are several possible tracings, you may notice that only certain starting points seem to allow a complete tracing. How do we know which point to start from? How do we even know that a solution exists?

Puzzles such as this, called "pen-tracing puzzles," can be considered problems in graph theory. If we imagine a vertex placed wherever two lines meet or cross over each other, then we have a graph. Our task is to start at a vertex and find a path that traverses every edge of the graph, without repeating any edges. In other words, we need an Euler path! (An Euler circuit would work as well.)

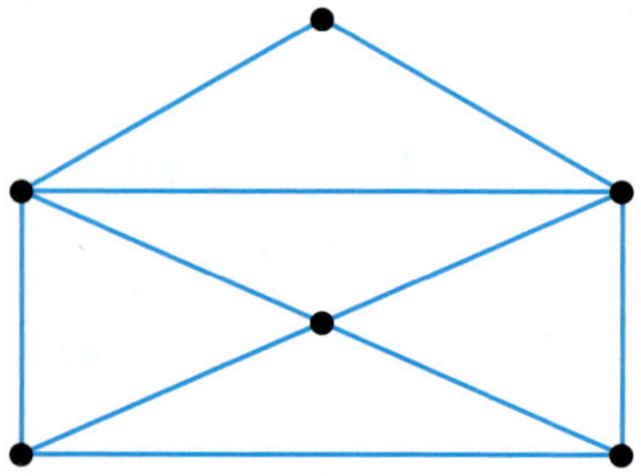

As we learned in this section, a graph has an Euler path only if two vertices are of odd degree and the remaining vertices are of even degree. Furthermore, the path must start at one of the vertices of odd degree and end at the other. In the puzzle in the upper figure at the left, only two vertices are of odd degree—the two bottom corners. So we know that an Euler path exists, and it must start from one of these two corners. The lower figure at the left shows one possible solution.

Of course, if every vertex is of even degree, we can solve the puzzle by finding an Euler circuit. If more than two vertices are of odd degree, then we know the puzzle cannot be solved.

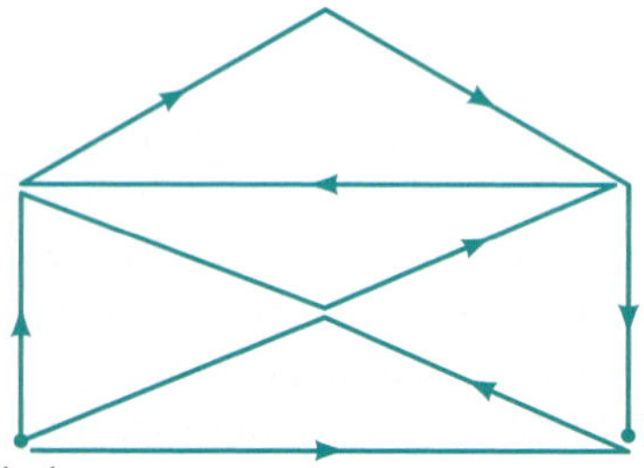

EXCURSION EXERCISES

In Exercises 1 to 4, a pen-tracing puzzle is given. See if you can find a way to trace the shape without lifting your pen and without tracing over the same segment twice.

1.

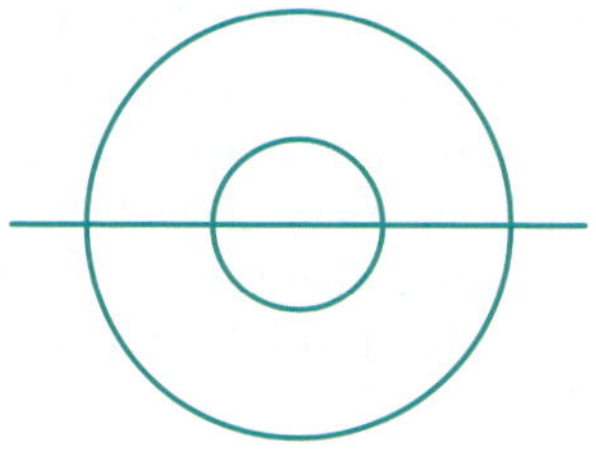

2.

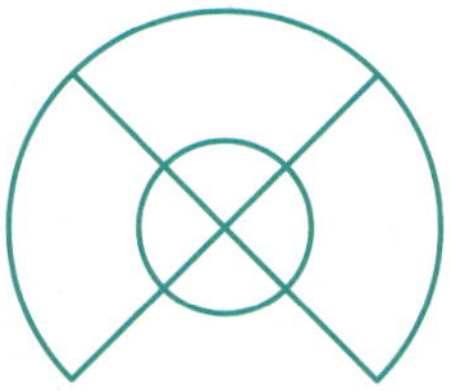

3.

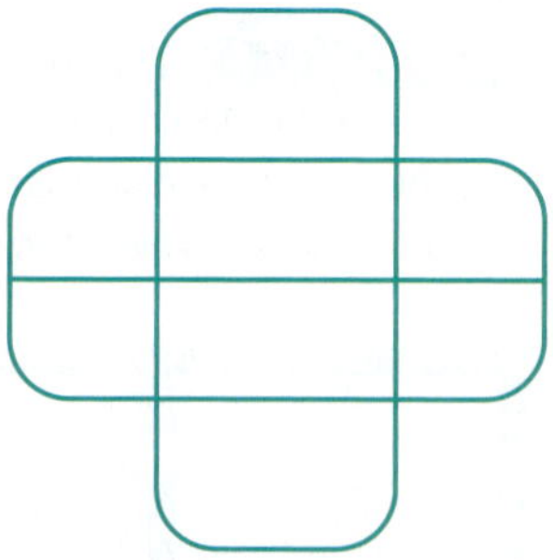

4.

5. Explain why the following pen-tracing puzzle is impossible to solve.

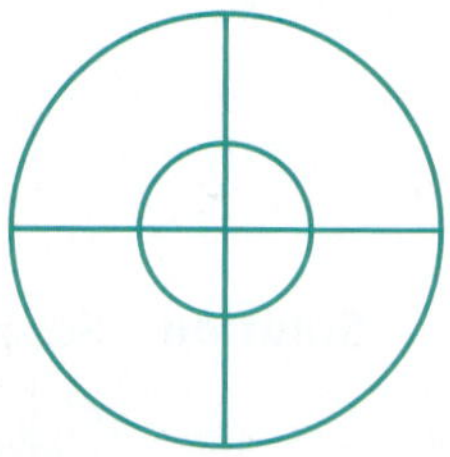

EXERCISE SET 5.1

1. **Transportation** An "X" in the table below indicates a direct train route between the corresponding cities. Draw a graph that represents this information, in which each vertex represents a city and an edge connects two vertices if there is a train route between the corresponding cities.

	Springfield	Riverside	Greenfield	Watertown	Midland	Newhope
Springfield	—		X		X	
Riverside		—		X	X	X
Greenfield	X		—	X	X	X
Watertown		X	X	—		
Midland	X	X	X		—	X
Newhope		X	X		X	—

2. **Transportation** The table below shows the nonstop flights offered by a small airline. Draw a graph that represents this information, where each vertex represents a city and an edge connects two vertices if there is a nonstop flight between the corresponding cities.

	Newport	Lancaster	Plymouth	Auburn	Dorset
Newport	—	no	yes	no	yes
Lancaster	no	—	yes	yes	no
Plymouth	yes	yes	—	yes	yes
Auburn	no	yes	yes	—	yes
Dorset	yes	no	yes	yes	—

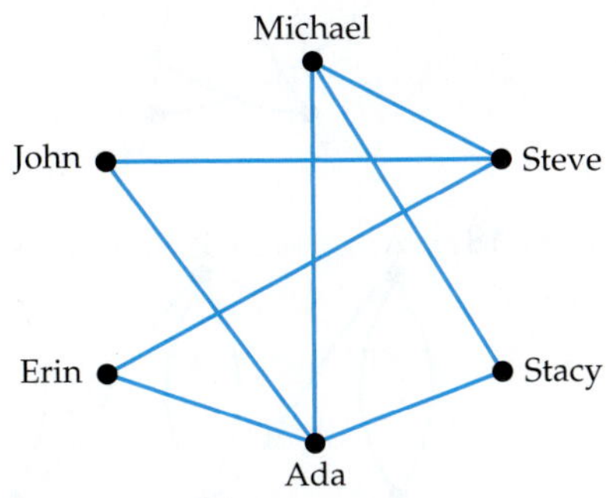

3. **Social Network** A group of friends is represented by the graph at the left. An edge connecting two names means that the two friends have spoken to each other in the last week.
   a. Have John and Stacy talked to each other in the last week?
   b. How many of the friends in this group has Steve talked to in the last week?
   c. Among this group of friends, who has talked to the most people in the last week?
   d. Why would it not make sense for this graph to contain a loop?

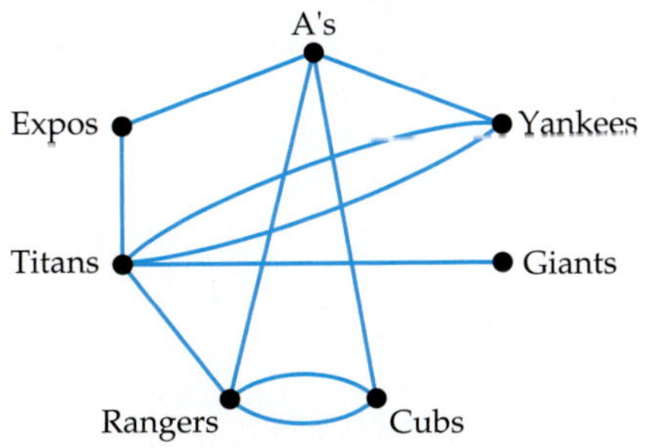

4. **Baseball** The local Little League baseball teams are represented by the graph at the left. An edge connecting two teams means that those teams have played a game against each other this season.
   a. Which team has played only one game this season?
   b. Which team has played the most games this season?
   c. Have any teams played each other twice this season?

■ In Exercises 5 to 8, determine (a) the number of edges in the graph, (b) the number of vertices in the graph, (c) the number of vertices that are of odd degree, (d) whether the graph is connected, and (e) whether the graph is a complete graph.

5.

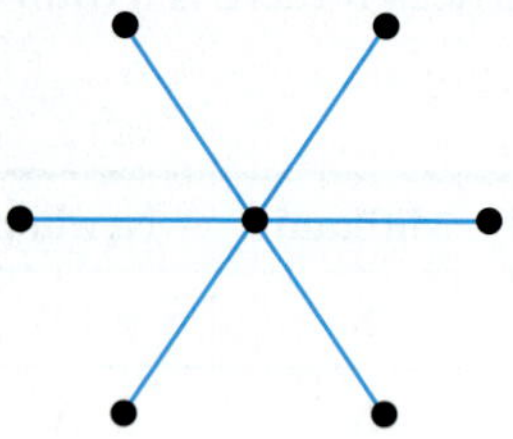

6.

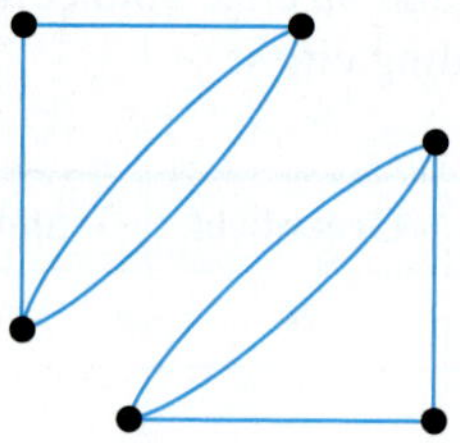

7.

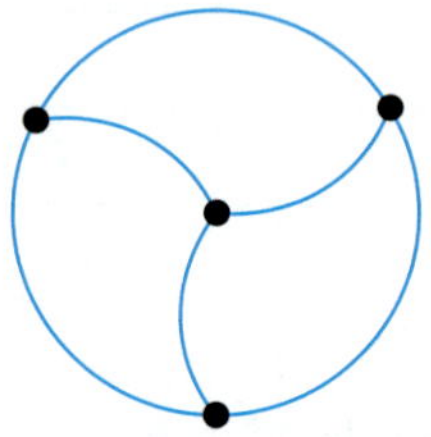

8.

■ In Exercises 9 to 12, determine whether the two graphs are equivalent.

9.

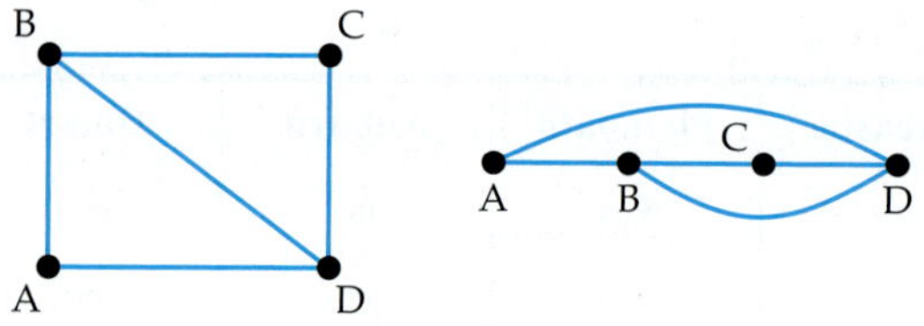

10.

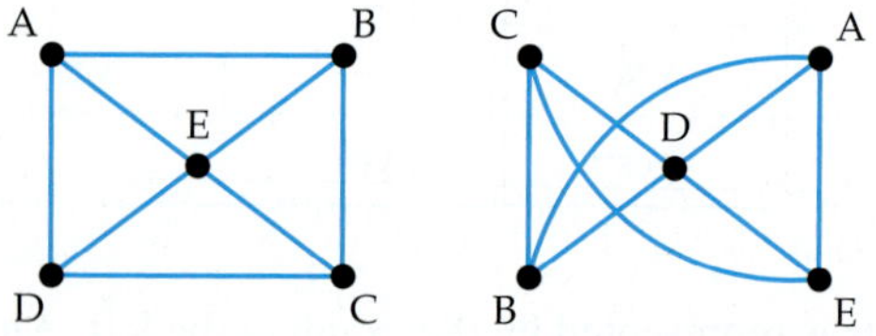

11.

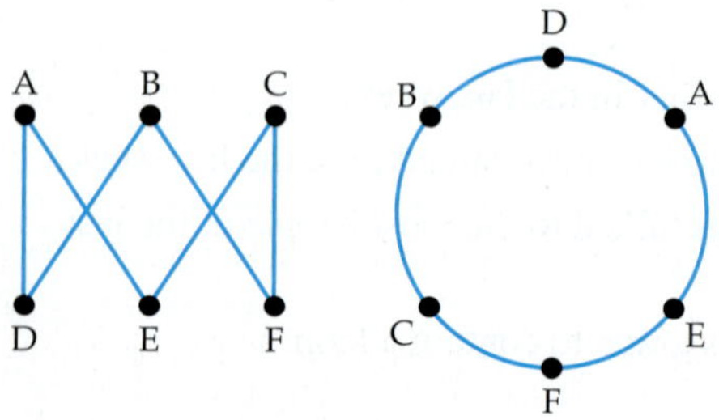

12.

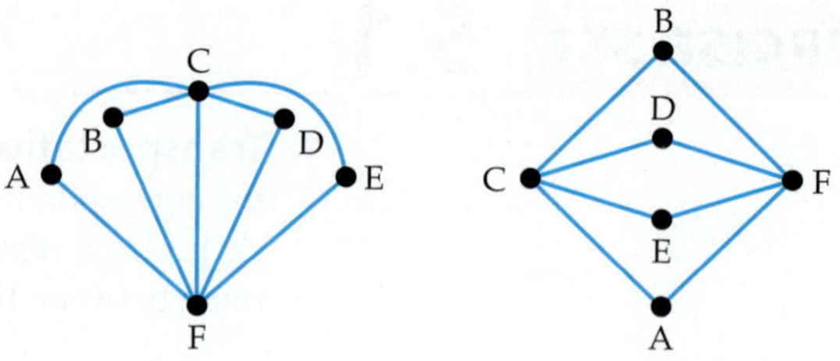

13. Explain why the following two graphs cannot be equivalent.

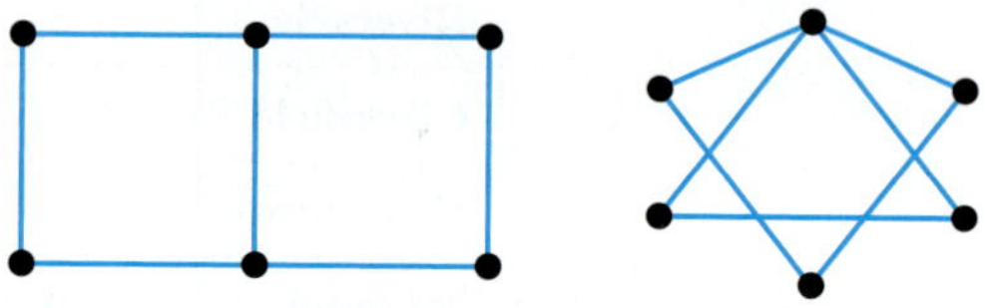

14. Label the vertices of the second graph so that it is equivalent to the first graph.

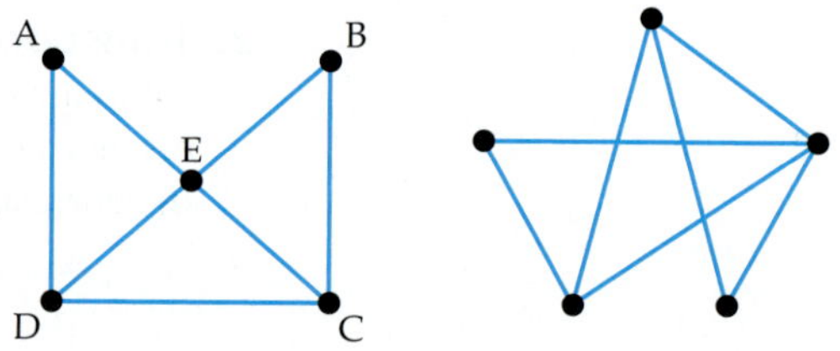

■ In Exercises 15 to 22, (a) determine whether the graph is Eulerian. If it is, find an Euler circuit. If it is not, explain why. (b) If the graph does not have an Euler circuit, does it have an Euler path? If so, find one. If not, explain why.

15.

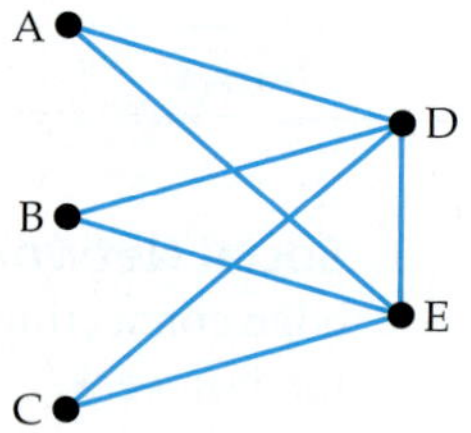

16.

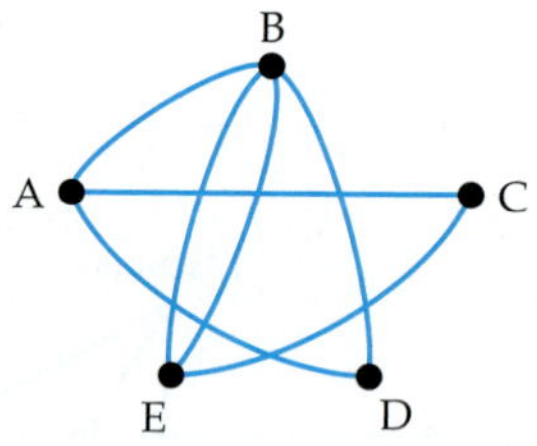

17.

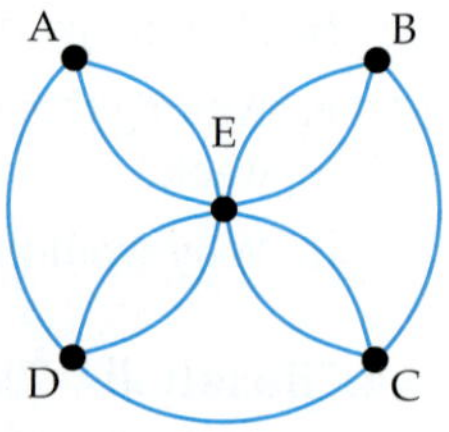

18.

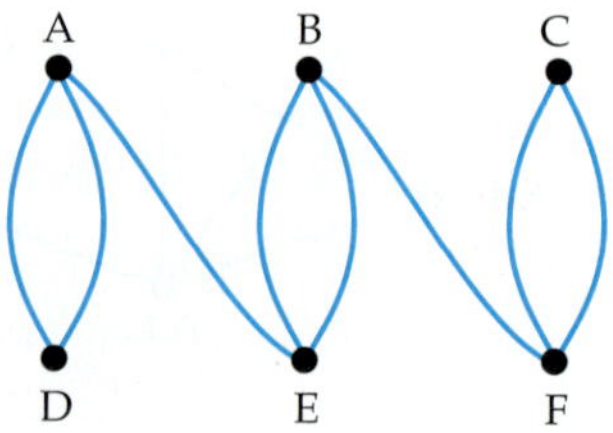

19.
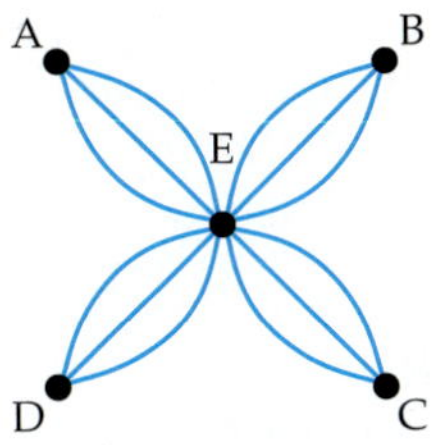

20.
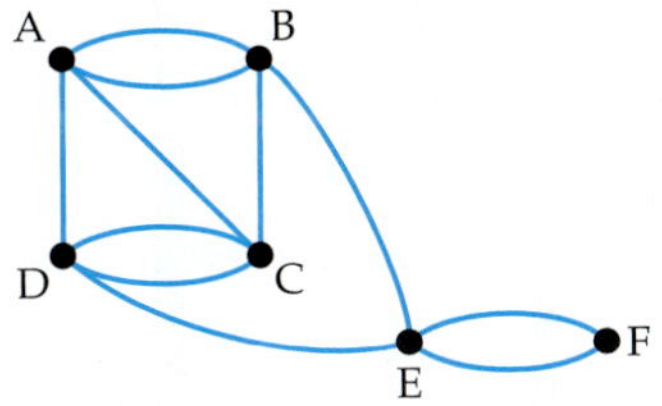

21.
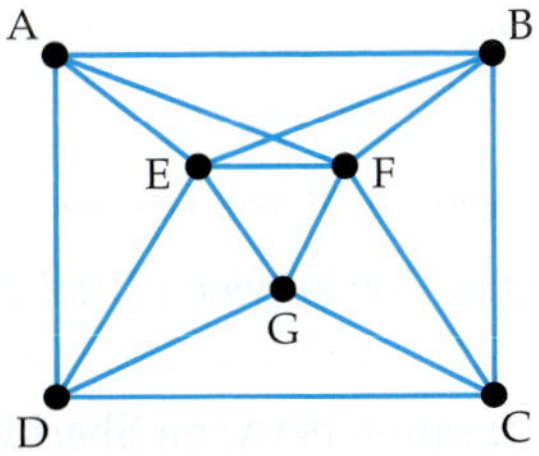

22.
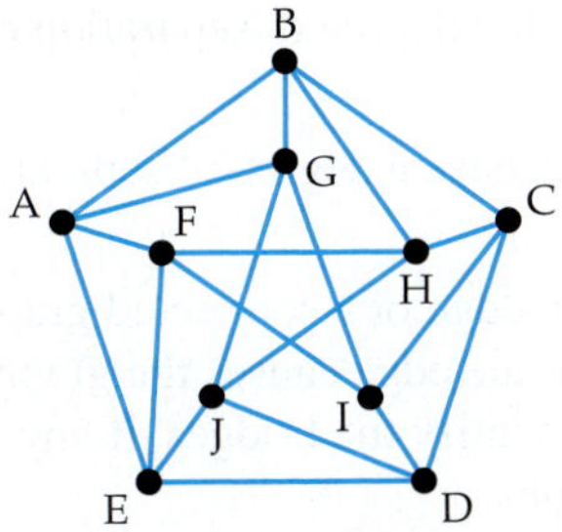

■ **Parks** In Exercises 23 and 24, a map of a park is shown with bridges connecting islands in a river to the banks.

a. Represent the map as a graph. See Figures 5.3 and 5.4 on page 234.

b. Is it possible to take a walk that crosses each bridge once and returns to the starting point without crossing any bridge twice? If not, can you do it if you do not end at the starting point? Explain how you know.

23.

24.
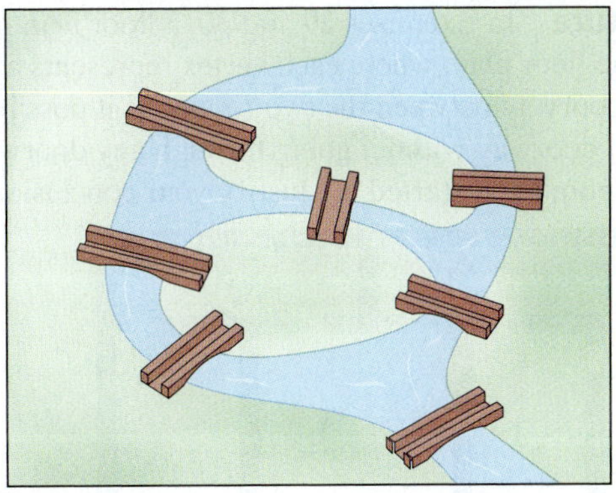

25. **Transportation** For the train routes given in Exercise 1, is it possible to travel along all of the train routes without traveling along any route twice? Explain how you reached your conclusion.

26. **Transportation** For the direct air flights given in Exercise 2, is it possible to start at one city and fly every route offered without repeating any flight if you return to the starting city? Explain how you reached your conclusion.

27. **Pets** The diagram below shows the arrangement of a Habitrail cage for a pet hamster. (Plastic tubes connect different cages.) Is it possible for a hamster to travel through every tube without going through the same tube twice? If so, find a route for the hamster to follow. Can the hamster return to its starting point without repeating any tube passages?

28. **Transportation** A subway map is shown below. Is it possible for a rider to travel the length of every subway route without repeating any segments? Justify your conclusion.

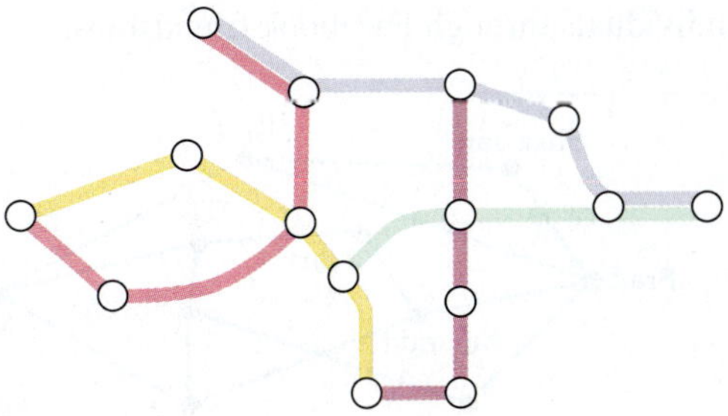

Architecture In Exercises 29 and 30, a floor plan of a museum is shown. Draw a graph that represents the floor plan, where each vertex represents a room and an edge connects two vertices if there is a doorway between the two rooms. Is it possible to walk through the museum and pass through each doorway without going through any doorway twice? Does it depend on whether you return to the room you started at? Justify your conclusion.

29.

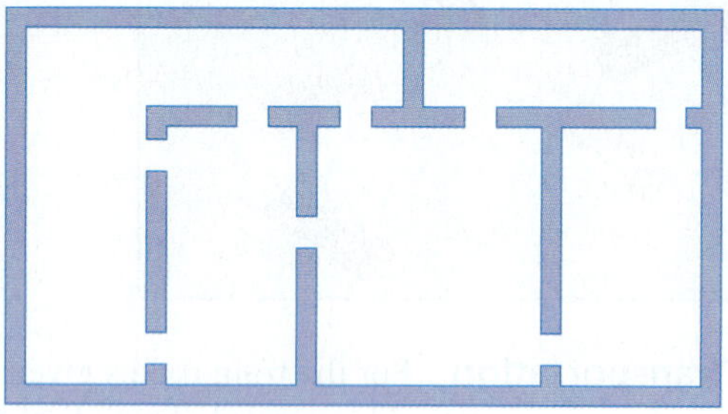

30.

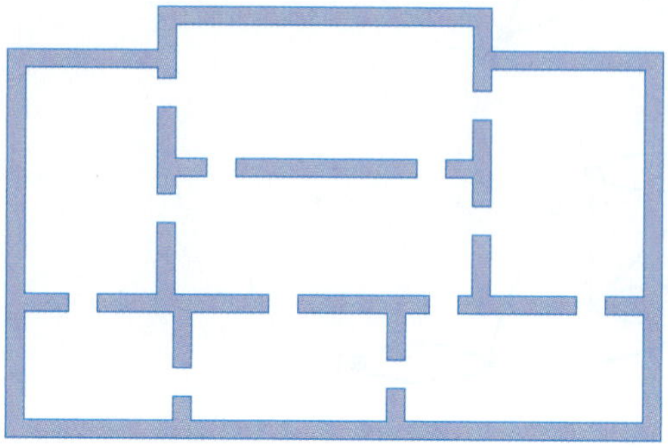

EXTENSIONS

31. **Degrees of Separation** In the graph below, an edge connects two vertices if the corresponding people have communicated by text message. Here we define the *degree of separation* between two individuals to be the minimum number of steps required to link one person to another through text message communications. This is equivalent to the (minimum) number of edges in a path connecting the corresponding vertices. For example, Karina and Lois have a degree of separation of 2.

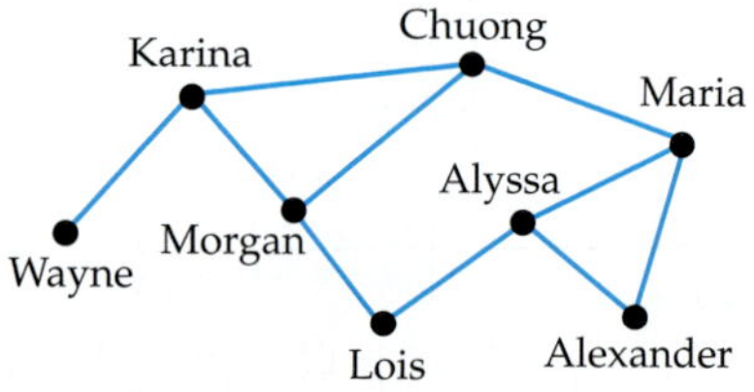

a. What is the degree of separation between Morgan and Maria?

b. What is the degree of separation between Wayne and Alexander?

c. Who has the largest degree of separation from Alyssa?

32. **Social Network** In the graph below, an edge connects two vertices if the corresponding people are friends on Facebook. Here "degree of separation" refers to the minimum number of steps required to link two individuals through Facebook friendships.

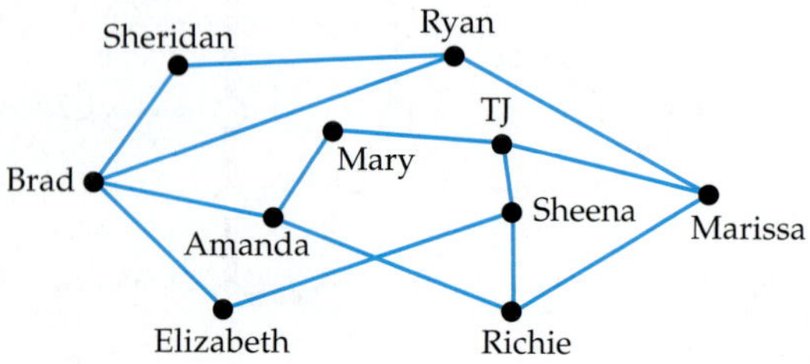

a. What is the degree of separation between Brad and Marissa?

b. What is the degree of separation between Sheridan and Sheena?

c. List the individuals with a degree of separation of two from Brad.

d. Do any two individuals have a degree of separation of more than three?

33. **Bridges of a Graph** An edge of a connected graph is called a **bridge** if deleting the edge causes the graph to no longer be connected. Identify the bridges, if any, in each of the following graphs.

a.

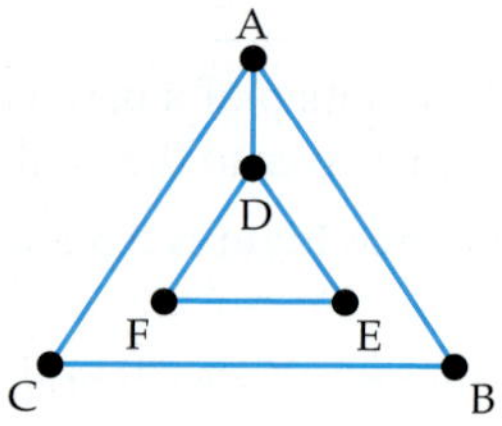

b.

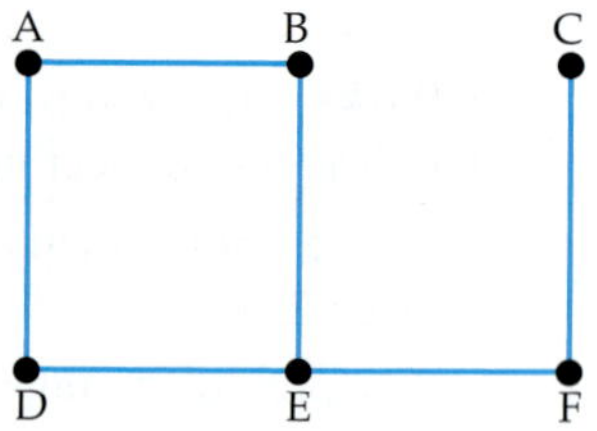

c.

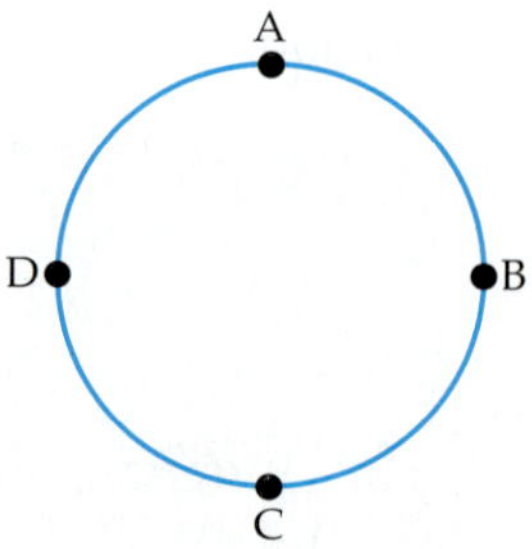

d.

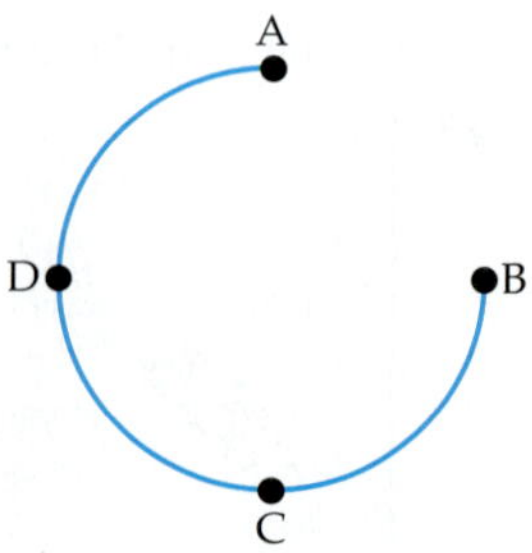

34. Travel A map of South America is shown at the left.

a. Draw a graph in which the vertices represent the 13 countries of South America and two vertices are joined by an edge if the corresponding countries share a common border.

b. Two friends are planning a driving tour of South America. They would like to drive across every border on the continent. Is it possible to plan such a route that never crosses the same border twice? What would the route correspond to on the graph?

c. Find a route the friends can follow that will start and end in Venezuela and that crosses every border while recrossing the fewest borders possible. *Hint:* On the graph, add multiple edges corresponding to border crossings that allow an Euler circuit.

SECTION 5.2 Weighted Graphs

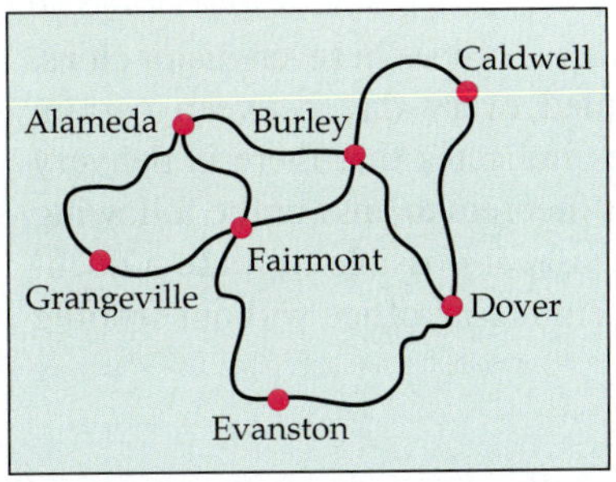

FIGURE 5.8

Hamiltonian Circuits

In Section 5.1 we looked at paths that use every edge of a graph exactly once. In some situations we may be more interested in paths that visit each vertex once, regardless of whether all edges are used or not.

For instance, Figure 5.8 shows the map of cities from Example 6 in Section 5.1. If our priority is to visit each city, we could travel along the route A–B–C–D–E–F–G–A (abbreviating the cities). This path visits each vertex once and returns to the starting vertex without visiting any vertex twice. This type of path is called a *Hamiltonian circuit.*

Hamiltonian Circuit

A **Hamiltonian circuit** is a path that begins and ends at the same vertex and passes through each vertex of a graph exactly once. A graph that contains a Hamiltonian circuit is called **Hamiltonian**.

Unfortunately we do not have a straightforward criterion to guarantee that a graph is Hamiltonian, but we do have the following helpful theorem.

HISTORICAL NOTE

Gabriel A. Dirac earned his doctorate from the University of London in 1952 and has published many papers in graph theory. He is the stepson of the famous physicist Paul Dirac, who shared the Nobel Prize in physics with Erwin Schrödinger for his work in quantum mechanics.

Dirac's Theorem

Consider a connected graph with at least three vertices and no multiple edges. Let n be the number of vertices in the graph. If every vertex has degree of at least $n/2$, then the graph must be Hamiltonian.

We must be careful, however; if our graph does not meet the requirements of this theorem, it still might be Hamiltonian. Dirac's theorem does not help us in this case.

EXAMPLE 1 Apply Dirac's Theorem

The following graph shows the available flights of a small airline. An edge between two vertices in the graph means that the airline has direct flights between the two corresponding cities. Apply Dirac's theorem to verify that the following graph is Hamiltonian.

Then find a Hamiltonian circuit. What does the Hamiltonian circuit represent in terms of flights?

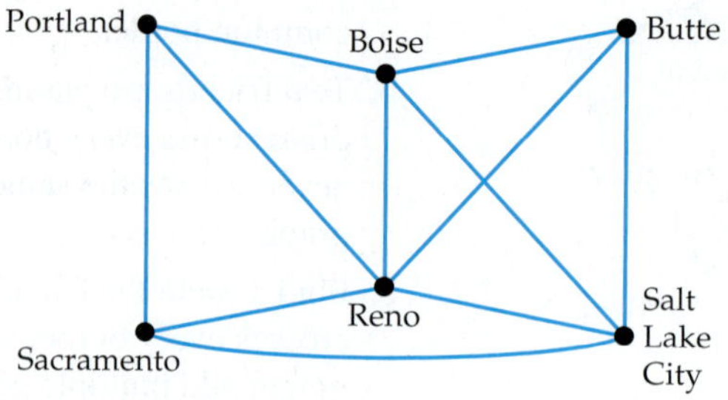

Solution

There are six vertices in the graph, so $n = 6$, and every vertex has a degree of at least $n/2 = 3$. By Dirac's theorem, the graph is Hamiltonian. This means that the graph contains a circuit that visits each vertex once and returns to the starting vertex without visiting any vertex twice. By trial and error, one Hamiltonian circuit is Portland–Boise–Butte–Salt Lake City–Reno–Sacramento–Portland, which represents a sequence of flights that visits each city and returns to the starting city without visiting any city twice.

CHECK YOUR PROGRESS 1 A large law firm has offices in seven major cities. The firm has overnight document deliveries scheduled every day between certain offices. In the graph below, an edge between vertices indicates that there is delivery service between the corresponding offices. Use Dirac's theorem to answer the following question: Using the law firm's existing delivery service, is it possible to route a document to all the offices and return the document to its originating office without sending it through the same office twice?

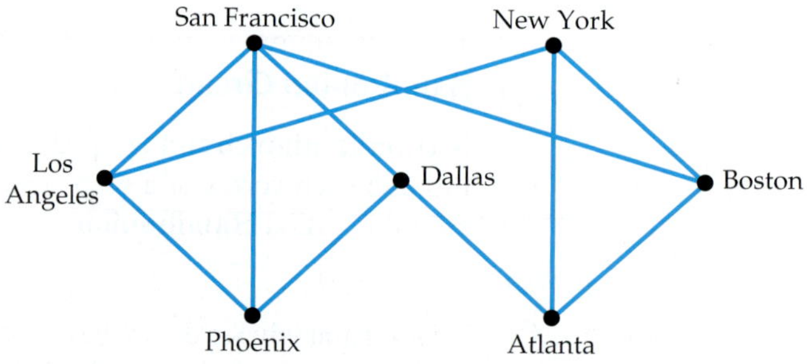

Solution See page S17.

Weighted Graphs

A Hamiltonian circuit can identify a route that visits all of the cities represented on a graph, as in Figure 5.8, but there are often a number of different paths we could use. If we are concerned with the distances we must travel between cities, chances are that some of the routes will involve a longer total distance than others. We might be interested in finding the route that minimizes the total number of miles traveled.

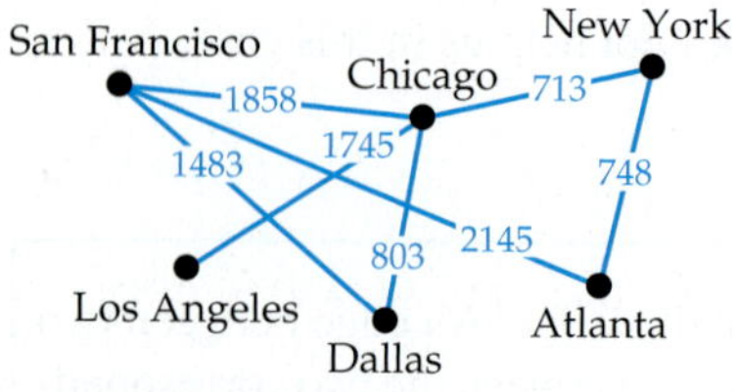

FIGURE 5.9

We can represent this situation with a *weighted graph*. A **weighted graph** is a graph in which each edge is associated with a value, called a **weight**. The value can represent any quantity we desire. In the case of distances between cities, we can label each edge with the number of miles between the corresponding cities, as in Figure 5.9. (Note that the length of an edge does not necessarily correlate to its weight.) For each Hamiltonian circuit in the weighted graph, the sum of the weights along the edges traversed gives the total distance traveled along that route. We can then compare different routes and find the one that requires the shortest total distance. This is an example of a famous problem called the *traveling salesman problem.*

EXAMPLE 2 Find Hamiltonian Circuits in a Weighted Graph

The table below lists the distances in miles between six popular cities that a particular airline flies to. Suppose a traveler would like to start in Chicago, visit the other five cities this airline flies to, and return to Chicago. Find three different routes that the traveler could follow, and find the total distance flown for each route.

	Chicago	New York	Washington, D.C.	Philadelphia	Atlanta	Dallas
Chicago	—	713	597	665	585	803
New York	713	—	No flights	No flights	748	1374
Washington, D.C.	597	No flights	—	No flights	544	1185
Philadelphia	665	No flights	No flights	—	670	1299
Atlanta	585	748	544	670	—	No flights
Dallas	803	1374	1185	1299	No flights	—

Solution

The various options will be simpler to analyze if we first organize the information in a graph. Begin by letting each city be represented by a vertex. Draw an edge between two vertices if there is a flight between the corresponding cities, and label each edge with a weight that represents the number of miles between the two cities.

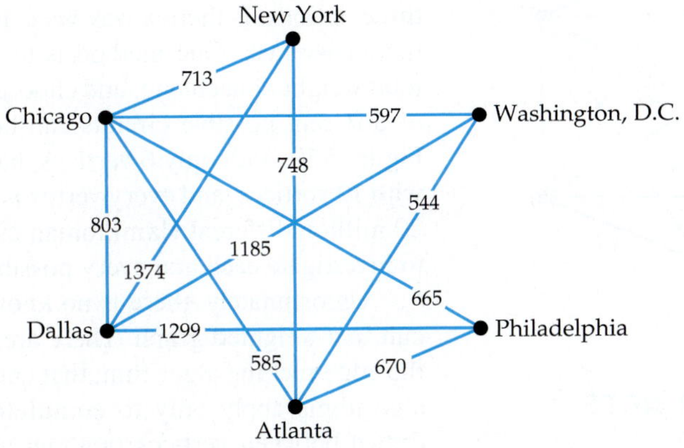

TAKE NOTE

Remember that the placement of the vertices is not important. There are many equivalent ways in which to draw the graph in Example 2.

A route that visits each city just once corresponds to a Hamiltonian circuit. Beginning at Chicago, one such circuit is Chicago–New York–Dallas–Philadelphia–Atlanta–Washington, D.C.–Chicago. By adding the weights of each edge in the circuit, we see that the total number of miles traveled is

$$713 + 1374 + 1299 + 670 + 544 + 597 = 5197$$

By trial and error, we can identify two additional routes. One is Chicago–Philadelphia–Dallas–Washington, D.C.–Atlanta–New York–Chicago. The total weight of the circuit is

$$665 + 1299 + 1185 + 544 + 748 + 713 = 5154$$

A third route is Chicago–Washington, D.C.–Dallas–New York–Atlanta–Philadelphia–Chicago. The total mileage is

$$597 + 1185 + 1374 + 748 + 670 + 665 = 5239$$

CHECK YOUR PROGRESS 2 A tourist visiting San Francisco is staying at a hotel near the Moscone Center. The tourist would like to visit five locations by bus tomorrow and then return to the hotel. The number of minutes spent traveling by bus between locations is given in the table below. (N/A in the table indicates that no convenient bus route is available.) Find two different routes for the tourist to follow and compare the total travel times.

	Moscone Center	Civic Center	Union Square	Embarcadero Plaza	Fisherman's Wharf	Coit Tower
Moscone Center	—	18	6	22	N/A	N/A
Civic Center	18	—	14	N/A	33	N/A
Union Square	6	14	—	24	28	36
Embarcadero Plaza	22	N/A	24	—	N/A	18
Fisherman's Wharf	N/A	33	28	N/A	—	14
Coit Tower	N/A	N/A	36	18	14	—

Solution See page S17.

Algorithms in Complete Graphs

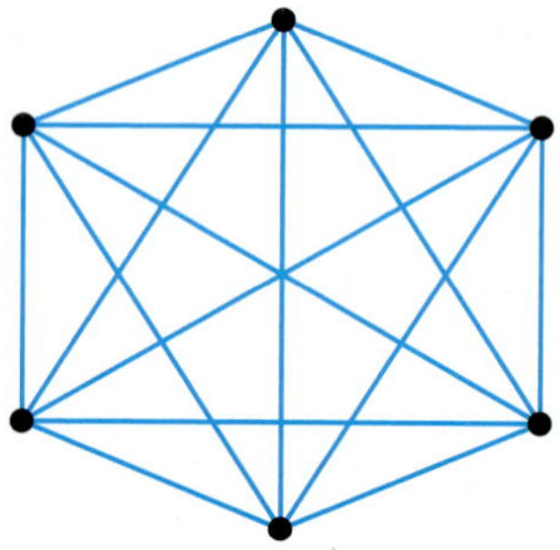

FIGURE 5.10

In Example 2, the second route we found represented the smallest total distance out of the three options. Is there a way we can find the very best route to take? It turns out that this is no easy task. One method is to list every possible Hamiltonian circuit, compute the total weight of each one, and choose the smallest total weight. Unfortunately, the number of different possible circuits can be extremely large. For instance, the graph shown in Figure 5.10, with only 6 vertices, has 60 unique Hamiltonian circuits. If we have a graph with 12 vertices, and every vertex is connected to every other by an edge, there are almost 20 million different Hamiltonian circuits! Even by using computers, it can take too long to investigate each and every possibility.

Unfortunately, there is no known shortcut for finding the optimal Hamiltonian circuit in a weighted graph. There are, however, two *algorithms*, the greedy algorithm and the edge-picking algorithm, that can be used to find a pretty good solution. Both of these algorithms apply only to **complete graphs**—graphs in which every possible edge is drawn between vertices (without any multiple edges). For instance, the graph in Figure 5.10 is a complete graph with six vertices. The circuits found by the algorithms are not guaranteed to have the smallest total weight possible, but they are often better than you would find by trial and error.

INSTRUCTOR NOTE

The number of Hamiltonian circuits in a complete graph with n vertices is $\frac{(n-1)!}{2}$.

TAKE NOTE

An algorithm is a step-by-step procedure for accomplishing a specific task.

The Greedy Algorithm

1. Choose a vertex to start at, then travel along the connected edge that has the smallest weight. (If two or more edges have the same weight, pick any one.)
2. After arriving at the next vertex, travel along the edge of smallest weight that connects to a vertex not yet visited. Continue this process until you have visited all vertices.
3. Return to the starting vertex.

The **greedy algorithm** is so called because it has us choose the "cheapest" option at every chance we get.

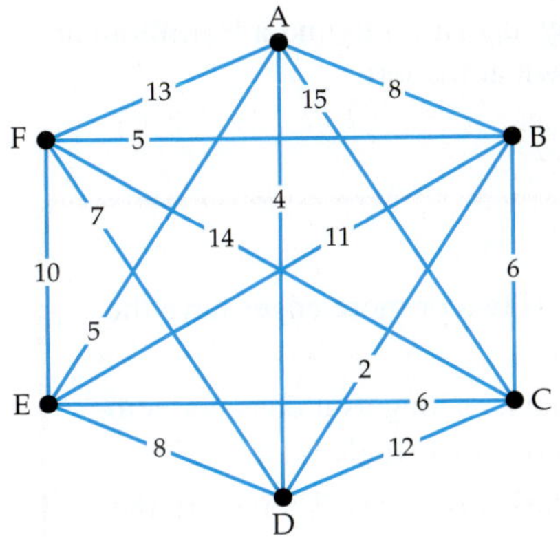

FIGURE 5.11

EXAMPLE 3 The Greedy Algorithm

Use the greedy algorithm to find a Hamiltonian circuit in the weighted graph shown in Figure 5.11. Start at vertex A.

Solution

Begin at A. The weights of the edges from A are 13, 5, 4, 15, and 8. The smallest is 4. Connect A to D.

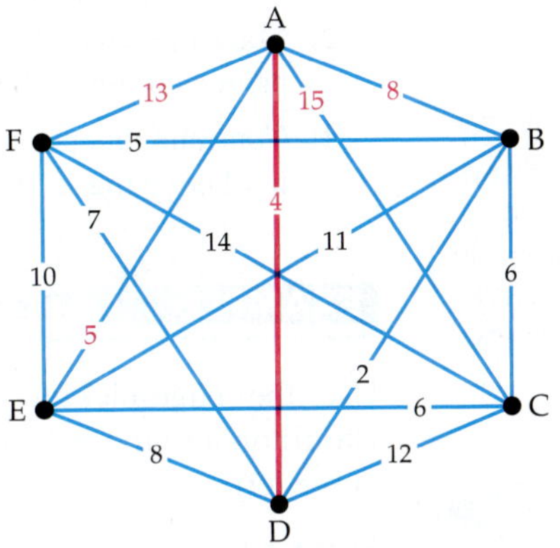

At D, the edge with the smallest weight is DB. Connect D to B.

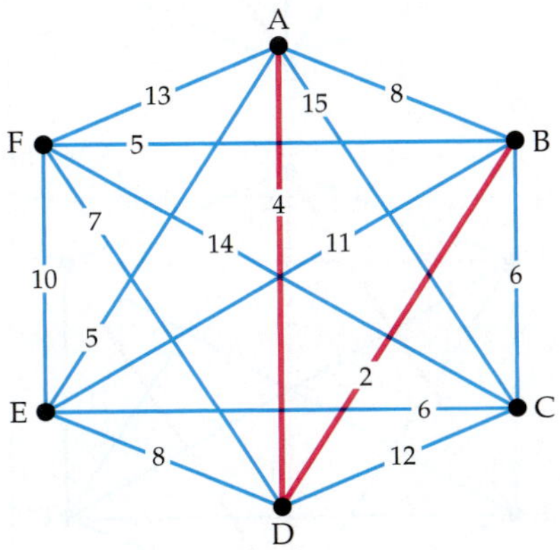

At B, the edge with the smallest weight is BF. Connect B to F.

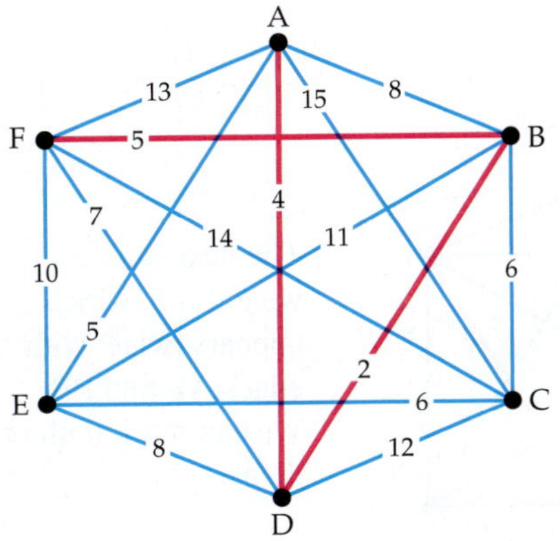

At F, the edge with the smallest weight, 7, is FD. However, D has already been visited. Choose the next smallest weight, edge FE. Connect F to E.

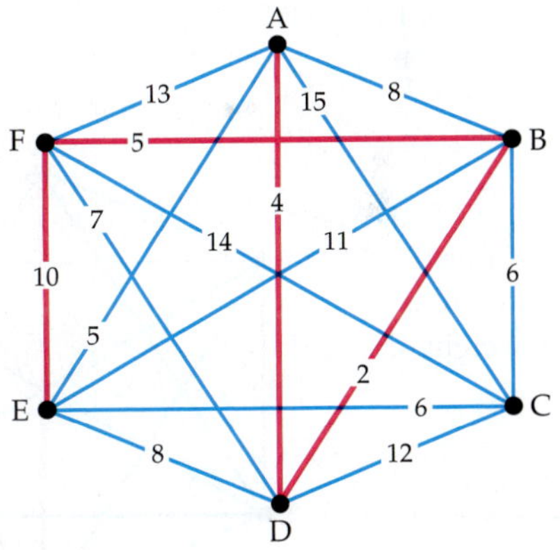

At E, the edge with the smallest weight whose vertex has not been visited is C. Connect E to C.

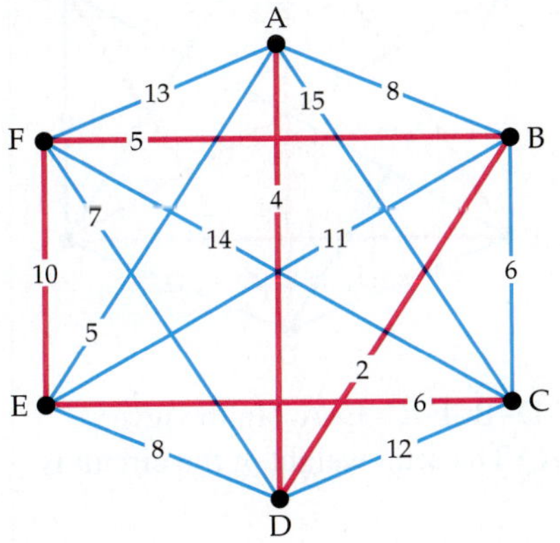

All vertices have been visited, so we are at step 3 of the algorithm. We return to the starting vertex by connecting C to A.

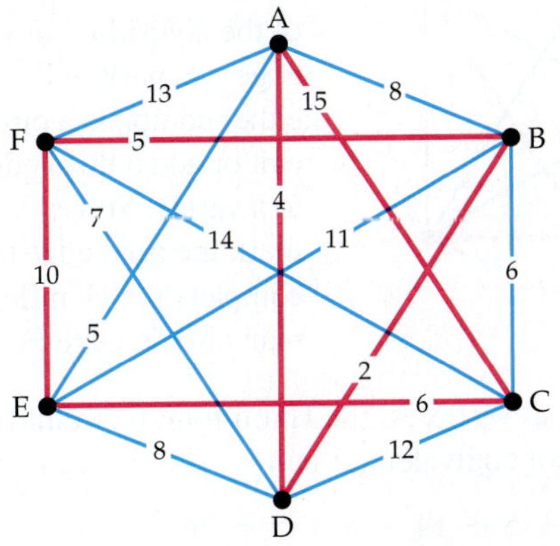

The Hamiltonian circuit is A–D–B–F–E–C–A. The weight of the circuit is

$$4 + 2 + 5 + 10 + 6 + 15 = 42$$

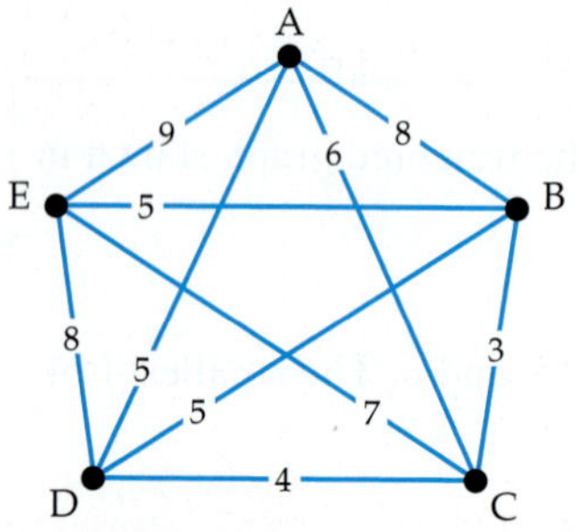

CHECK YOUR PROGRESS 3 Use the greedy algorithm to find a Hamiltonian circuit starting at vertex A in the weighted graph shown at the left.

Solution See page S17.

The Edge-Picking Algorithm

1. Mark the edge of smallest weight in the graph. (If two or more edges have the same weight, pick any one.)
2. Mark the edge of next smallest weight in the graph, as long as it does not complete a circuit and does not add a third marked edge to a single vertex.
3. Continue this process until you can no longer mark any edges. Then mark the final edge that completes the Hamiltonian circuit.

TAKE NOTE

A Hamiltonian circuit will always have exactly two edges at each vertex. In step 2 we are warned not to mark an edge that would allow three edges to meet at one vertex.

EXAMPLE 4 The Edge-Picking Algorithm

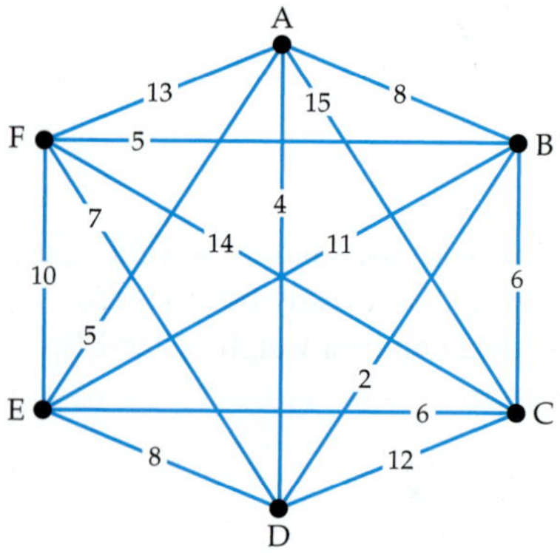

Use the edge-picking algorithm to find a Hamiltonian circuit in Figure 5.11 (reprinted at the left).

Solution

We first highlight the edge of smallest weight, namely BD with weight 2.

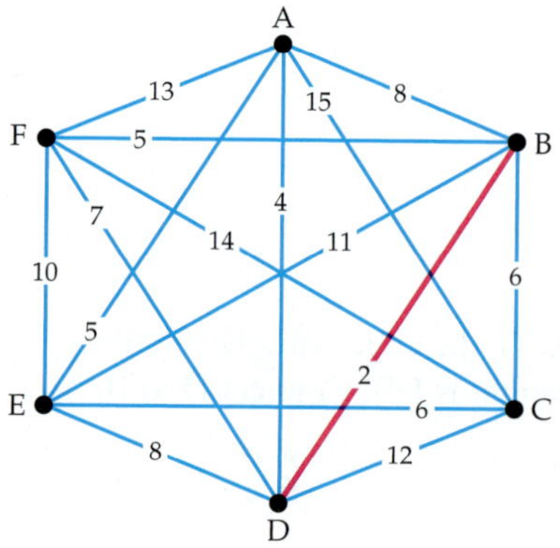

The edge of next smallest weight is AD with weight 4.

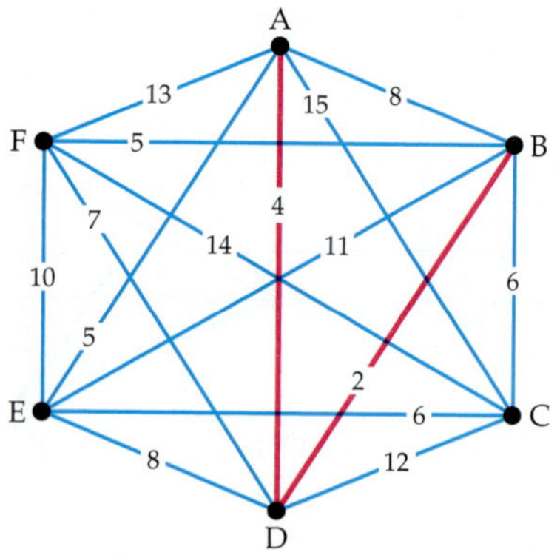

The next smallest weight is 5, which appears twice, with edges AE and FB. We can mark both of them.

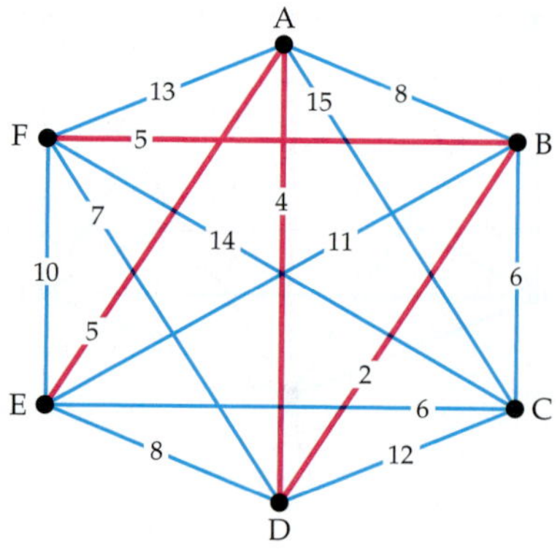

There are two edges of weight 6 (the next smallest weight), BC and EC. We cannot use BC because it would add a third marked edge to vertex B. We mark edge EC.

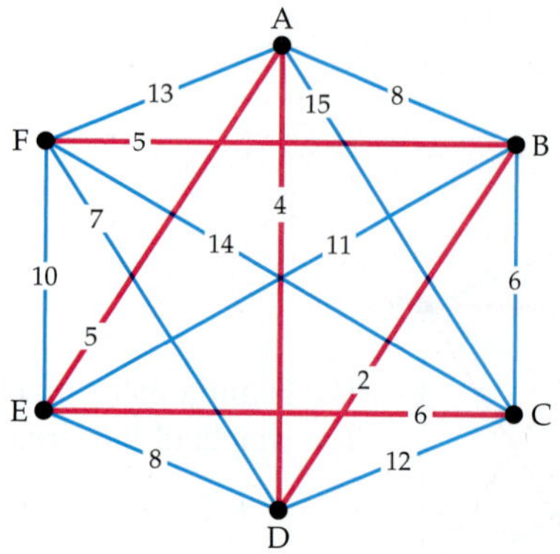

We are now at step 3 of the algorithm; any edge we mark will either complete a circuit or add a third edge to a vertex. So we mark the final edge to complete the Hamiltonian circuit, edge FC.

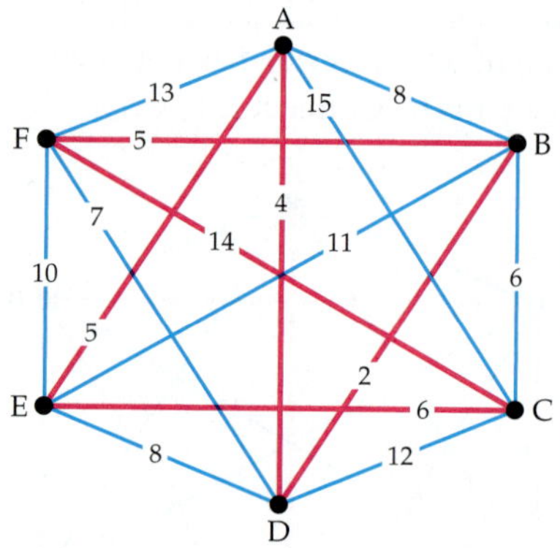

Beginning at vertex A, the Hamiltonian circuit is A–D–B–F–C–E–A. (In the reverse direction, an equivalent circuit is A–E–C–F–B–D–A.) The total weight of the circuit is

$$4 + 2 + 5 + 14 + 6 + 5 = 36$$

TAKE NOTE

A Hamiltonian circuit forms a complete loop, so we can follow along the circuit starting at any of the vertices. We can also reverse the direction in which we follow the circuit.

CHECK YOUR PROGRESS 4 Use the edge-picking algorithm to find a Hamiltonian circuit in the weighted graph in Check Your Progress 3.

Solution See page S17.

Note in Examples 3 and 4 that the two algorithms gave different Hamiltonian circuits, and in this case the edge-picking algorithm gave the more efficient route. Is this the best route? We mentioned before that the algorithms are helpful but that there is no known efficient method for finding the very best circuit. In fact, a third Hamiltonian circuit, A–D–F–B–C–E–A in Figure 5.11, has a total weight of 33, which is smaller than the weights of both routes given by the algorithms.

MATH MATTERS Computers and Traveling Salesman Problems

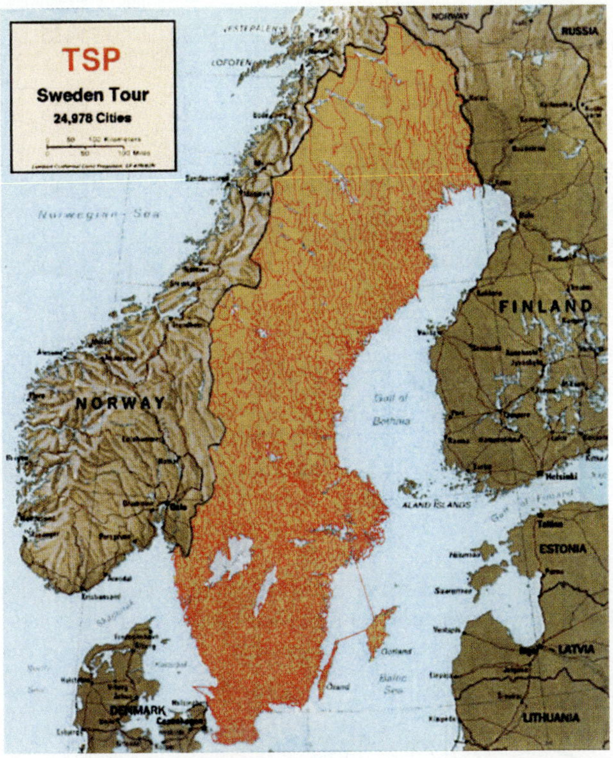

William Cook

The traveling salesman problem is a long-standing mathematical problem that has been analyzed since the 1920s by mathematicians, statisticians, and later, computer scientists. The algorithms given in this section can help us find good solutions, but they cannot guarantee that we will find the best solution.

At this point in time, the only way to find the optimal Hamiltonian circuit in a weighted graph is to find each and every possible circuit and compare the total weights of all circuits. Computers are well suited to such a task; however, as the number of vertices increases, the number of possible Hamiltonian circuits increases rapidly, and even computers are not fast enough to handle large graphs. There are so many possible circuits in a graph with a large number of vertices that finding them all could take hundreds or thousands of years on even the fastest computers.

Computer scientists continue to improve their methods and produce more sophisticated algorithms. In 2004, five researchers (David Applegate, Robert Bixby, Vasek Chvátal, William Cook, and Keld Helsgaun) were able to find the optimal route to visit all 24,978 cities in Sweden. To accomplish this task, they used a network of 96 dual-processor workstation computers. The amount of computing time required was equivalent to a single-processor computer running for 84.8 years. You can see their resulting circuit on the map.

Since that time the team has successfully solved a traveling salesman problem with 85,900 vertices arising from the connections on a computer chip.

POINT OF INTEREST

The same researchers created as a challenge a traveling salesman problem consisting of all 1,904,711 cities in the world. As of 2015, the tour found with the shortest distance was 7,515,772,212 meters long. It is not yet known if this is the optimal solution.

Applications of Weighted Graphs

In Example 2, we examined distances between cities. This is just one example of a weighted graph; the weight of an edge can be used to represent any quantity we like. For example, a traveler might be more interested in the cost of flights than the time or distance between cities. If we labeled each edge of the graph in Example 2 with the cost of traveling between the two cities, the total weight of a Hamiltonian circuit would be the total travel cost of the trip.

QUESTION Can the greedy algorithm or the edge-picking algorithm be used to identify a route to visit the cities in Example 2?

EXAMPLE 5 An Application of the Greedy and Edge-Picking Algorithms

The cost of flying between various European cities is shown in the following table. Use both the greedy algorithm and the edge-picking algorithm to find a low-cost route that visits each city just once and starts and ends in London. Which route is more economical?

	London, England	Berlin, Germany	Paris, France	Rome, Italy	Madrid, Spain	Vienna, Austria
London, England	—	\$325	\$160	\$280	\$250	\$425
Berlin, Germany	\$325	—	\$415	\$550	\$675	\$375
Paris, France	\$160	\$415	—	\$495	\$215	\$545
Rome, Italy	\$280	\$550	\$495	—	\$380	\$480
Madrid, Spain	\$250	\$675	\$215	\$380	—	\$730
Vienna, Austria	\$425	\$375	\$545	\$480	\$730	—

Solution

First we draw a weighted graph with vertices representing the cities and each edge labeled with the price of the flight between the corresponding cities.

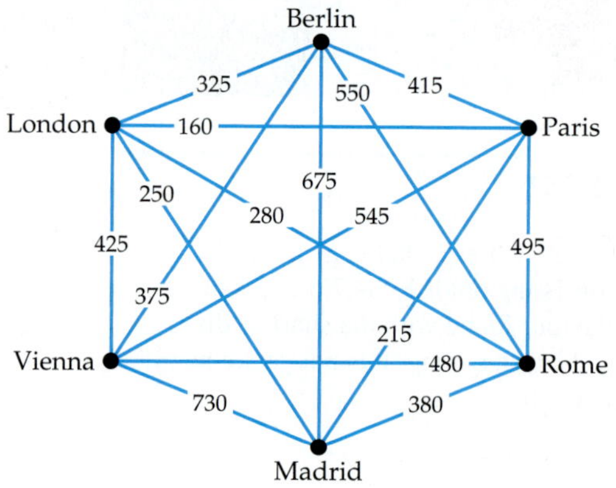

ANSWER No. Both algorithms apply only to complete graphs.

To use the greedy algorithm, start at London and travel along the edge with the smallest weight, 160, to Paris. The edge of smallest weight leaving Paris is the edge to Madrid. From Madrid, the edge of smallest weight (that we have not already traversed) is the edge to London, of weight 250. However, we cannot use this edge, because it would bring us to a city we have already seen. We can take the next-smallest-weight edge to Rome. We cannot yet return to London, so the next available edge is to Vienna, then to Berlin, and finally back to London.

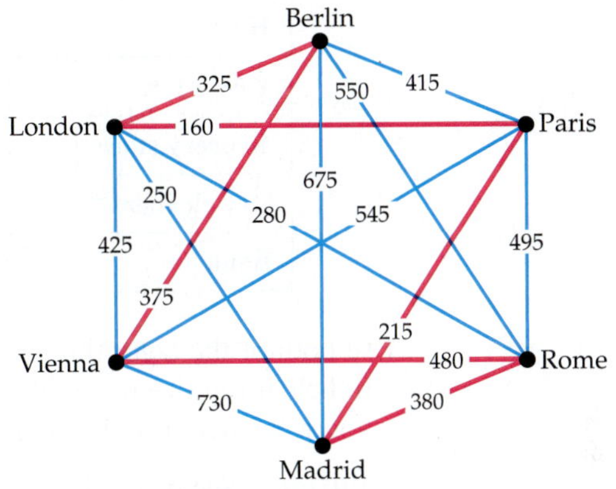

The total weight of the edges, and thus the total airfare for the trip, is

$$160 + 215 + 380 + 480 + 375 + 325 = \$1935$$

If we use the edge-picking algorithm, the edges with the smallest weights that we can highlight are London–Paris and Madrid–Paris. The edge of next smallest weight has a weight of 250, but we cannot use this edge because it would complete a circuit. We can take the edge of next smallest weight, 280, from London to Rome. We cannot take the edge of next smallest weight, 325, because it would add a third edge to the London vertex, but we can take the edge Vienna–Berlin of weight 375. We must skip the edges of weights 380, 415, and 425, but we can take the edge of weight 480, which is the Vienna–Rome edge. There are no more edges we can mark that will meet the requirements of the algorithm, so we mark the last edge to complete the circuit, Berlin–Madrid.

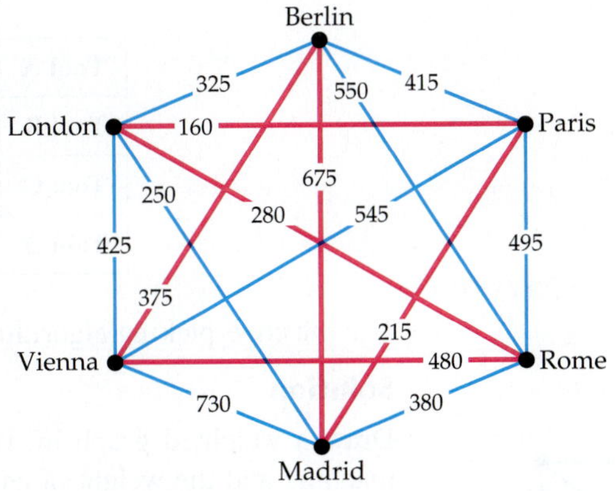

The resulting route is London–Paris–Madrid–Berlin–Vienna–Rome–London, for a total cost of

$$160 + 215 + 675 + 375 + 480 + 280 = \$2185$$

(We could also travel this route in the reverse order.)

CHECK YOUR PROGRESS 5 Susan needs to mail a package at the post office, pick up several items at the grocery store, drop off clothes at the dry cleaners, and make a deposit at her bank. The estimated driving time, in minutes, between each of these locations is given in the table below.

	Home	Post office	Grocery store	Dry cleaners	Bank
Home	—	14	12	20	23
Post office	14	—	8	12	21
Grocery store	12	8	—	17	11
Dry cleaners	20	12	17	—	18
Bank	23	21	11	18	—

Use both of the algorithms from this section to design routes for Susan to follow that will help minimize her total driving time. Assume she must start from home and return home when her errands are done.

Solution See page S17.

A wide variety of problems are actually traveling salesman problems in disguise and can be analyzed using the algorithms from this section.

EXAMPLE 6 An Application of the Edge-Picking Algorithm

A toolmaker needs to use one machine to create four different tools. The toolmaker needs to make adjustments to the machine before starting each different tool. However, since the tools have parts in common, the amount of adjustment time required depends on which tool the machine was previously used to create. The table below lists the estimated time (in minutes) required to adjust the machine from making one tool to another. The machine is currently configured for tool A and should be returned to that state when all the tools are finished.

	Tool A	Tool B	Tool C	Tool D
Tool A	—	25	6	32
Tool B	25	—	18	9
Tool C	6	18	—	15
Tool D	32	9	15	—

Use the edge-picking algorithm to determine a sequence for creating the tools.

Solution

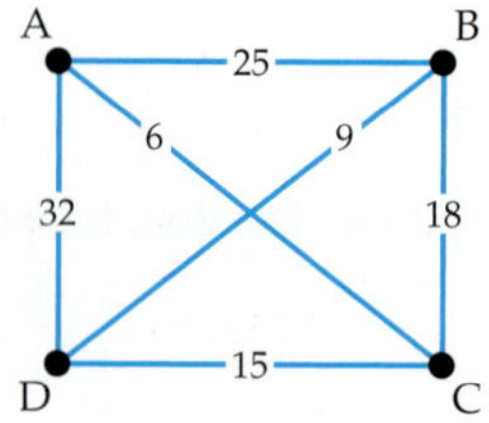

Draw a weighted graph in which each vertex represents a tool configuration of the machine and the weight of each edge is the number of minutes required to adjust the machine from one tool to another.

Using the edge-picking algorithm, we first choose edge AC, of weight 6. The next smallest weight is 9, or edge DB, followed by edge DC of weight 15. We end by completing the circuit with edge AB. The final circuit, of total weight $25 + 9 + 15 + 6 = 55$, is A–B–D–C–A. So the machine starts with tool A, is reconfigured for tool B, then for tool D, then for tool C, and finally is returned to the settings used for tool A. (Note that we can equivalently follow this sequence in the reverse order: from tool A to C, to D, to B, and back to A.)

CHECK YOUR PROGRESS 6 Businesses often network their various computers. One option is to run cables from a central hub to each computer individually; another is to connect one computer to the next, and that one to the next, and so on until you return to the first computer. Thus the computers are all connected in a large loop. Suppose a company wishes to use the latter method, and the lengths of cable (in feet) required between computers are given in the table below.

	Computer A	Computer B	Computer C	Computer D	Computer E	Computer F	Computer G
Computer A	—	43	25	6	28	30	45
Computer B	43	—	26	40	37	22	25
Computer C	25	26	—	20	52	8	50
Computer D	6	40	20	—	30	24	45
Computer E	28	37	52	30	—	49	20
Computer F	30	22	8	24	49	—	41
Computer G	45	25	50	45	20	41	—

Use the edge-picking algorithm to determine how the computers should be networked if the business wishes to use the smallest amount of cable possible.

Solution See page S18.

EXCURSION

Extending the Greedy Algorithm

When we create a Hamiltonian circuit in a graph, it is a closed loop. We can start at any vertex, follow the path, and arrive back at the starting vertex. For instance, if we use the greedy algorithm to create a Hamiltonian circuit starting at vertex A, we are actually creating a circuit that could start at any vertex in the circuit.

If we use the greedy algorithm and start from different vertices, will we always get the same result? Try it with Figure 5.12 below. If we start at vertex A, we get the circuit A–C–E–B–D–A, with a total weight of 26 (see Figure 5.13). However, if we start at vertex B, we get B–E–D–C–A–B, with a total weight of 18 (see Figure 5.14). Even though we found the second circuit by starting at B, we could traverse the same circuit beginning at A, namely A–B–E–D–C–A, or, equivalently, A–C–D–E–B–A. This circuit has a smaller total weight than the first one we found.

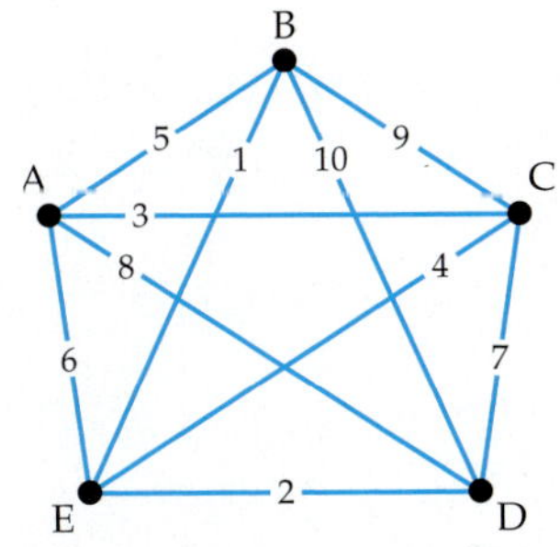

FIGURE 5.12

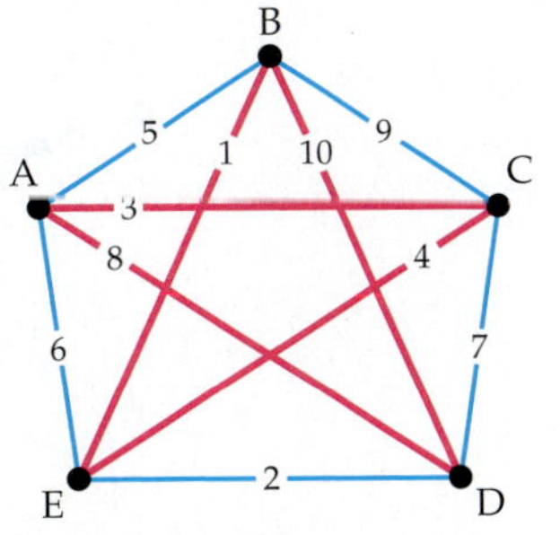

FIGURE 5.13

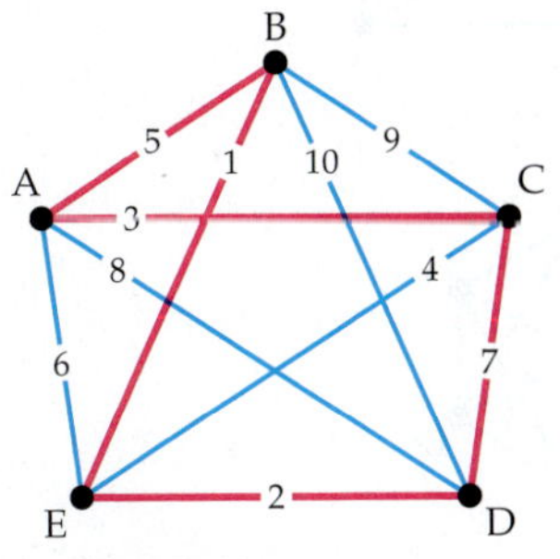

FIGURE 5.14

In other words, it may be advantageous to try the greedy algorithm starting from different vertices, even when we know that we actually want to begin the circuit at a particular vertex. To be thorough, we can extend the greedy algorithm by using it at each and every vertex, generating several different circuits. We then choose from among these the circuit with the smallest total weight and begin the circuit at the vertex we want. The following exercises ask you to apply this approach for yourself.

EXCURSION EXERCISES

1. Continue investigating Hamiltonian circuits in Figure 5.12 by using the greedy algorithm starting at vertices C, D, and E. Then compare the various Hamiltonian circuits to identify the one with the smallest total weight.
2. Use the greedy algorithm and the weighted graph below to generate a Hamiltonian circuit starting from each vertex. Then compare the different circuits to find the one of smallest total weight.

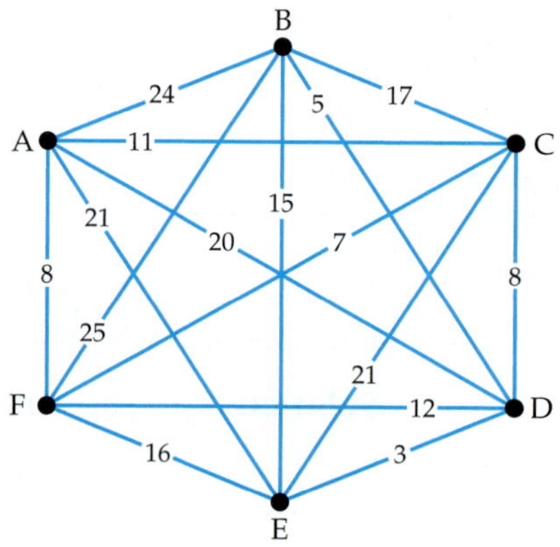

3. Use the edge-picking algorithm to find a Hamiltonian circuit in the graph in Exercise 2. How does the weight of this circuit compare with the weights of the circuits found in Exercise 2?

EXERCISE SET 5.2

■ In Exercises 1 to 4, use Dirac's theorem to verify that the graph is Hamiltonian. Then find a Hamiltonian circuit.

1. 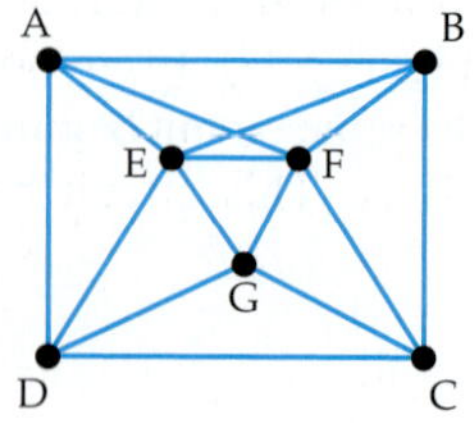

2. 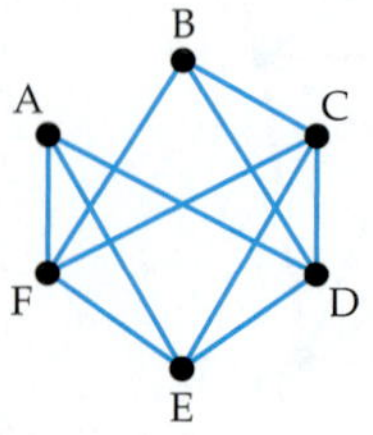

3. 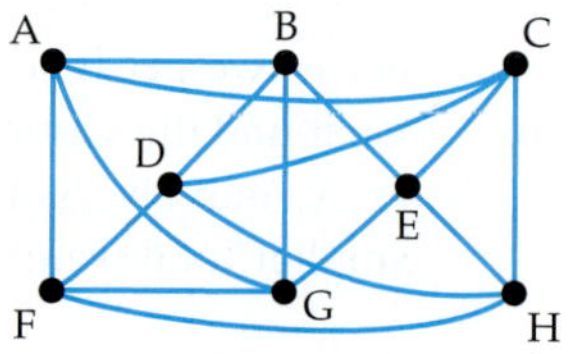

4. 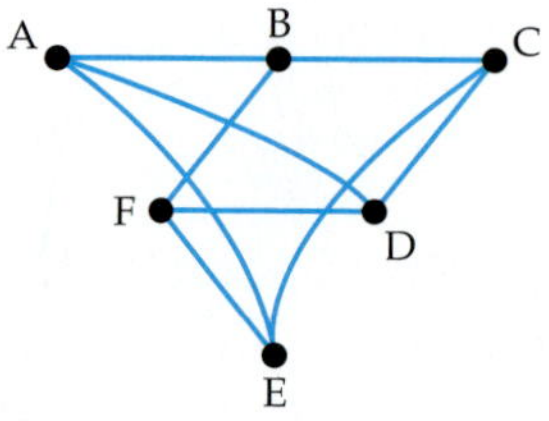

5. **Transportation** For the train routes given in Exercise 1 of Section 5.1, find a route that visits each city and returns to the starting city without visiting any city twice.

6. **Transportation** For the direct air flights given in Exercise 2 of Section 5.1, find a route that visits each city and returns to the starting city without visiting any city twice.

■ In Exercises 7 to 10, use trial and error to find two Hamiltonian circuits of different total weights, starting at vertex A in the weighted graph. Compute the total weight of each circuit.

7.

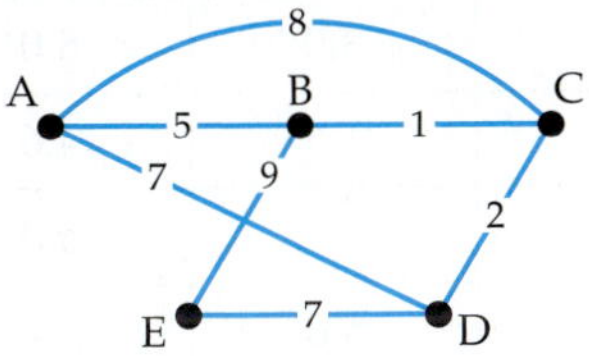

8.

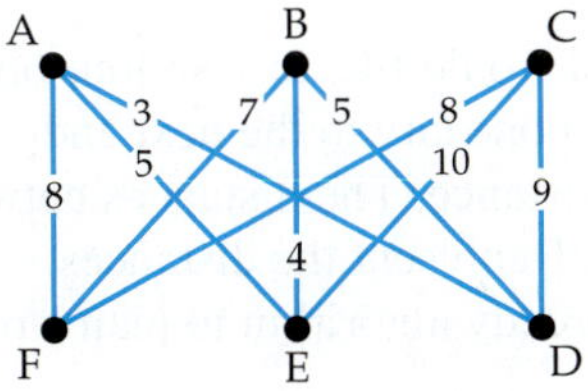

9.

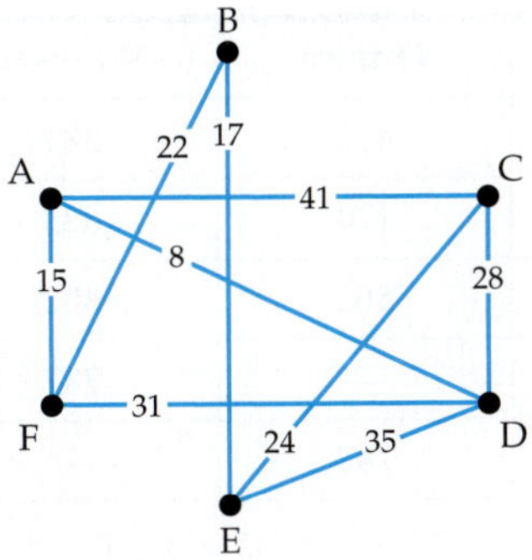

10.

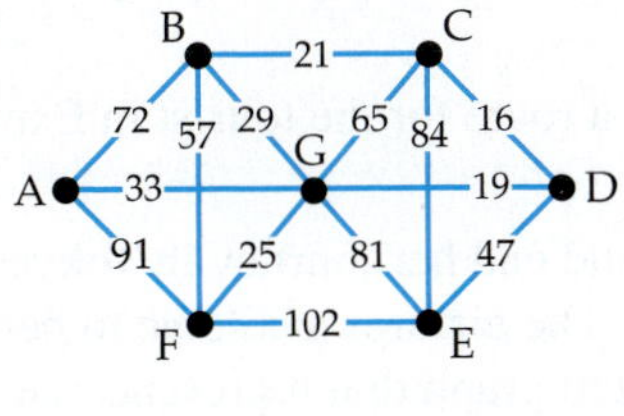

■ In Exercises 11 to 14, use the greedy algorithm to find a Hamiltonian circuit starting at vertex A in the weighted graph.

11.

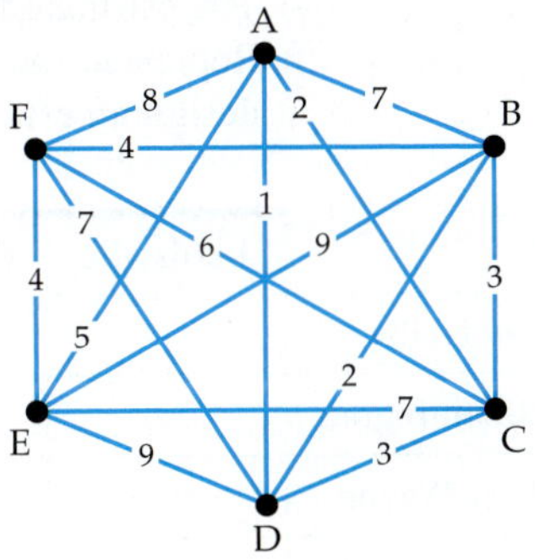

12.

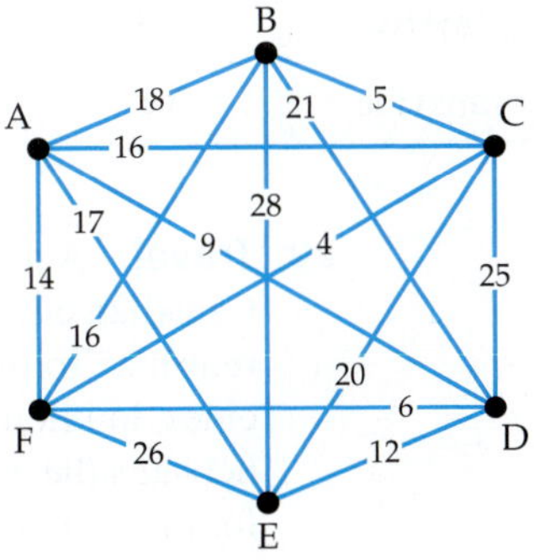

13.

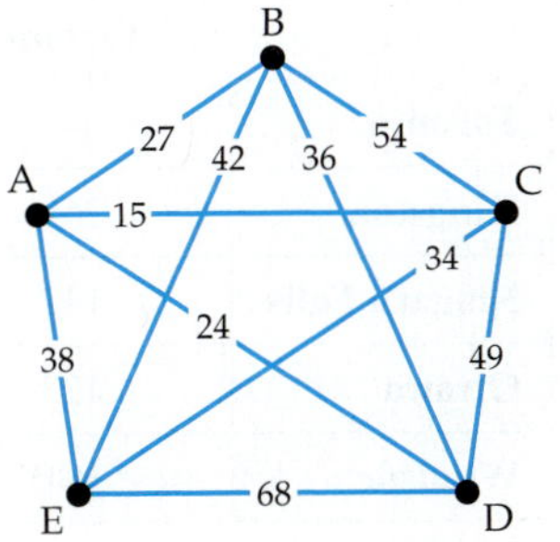

14.

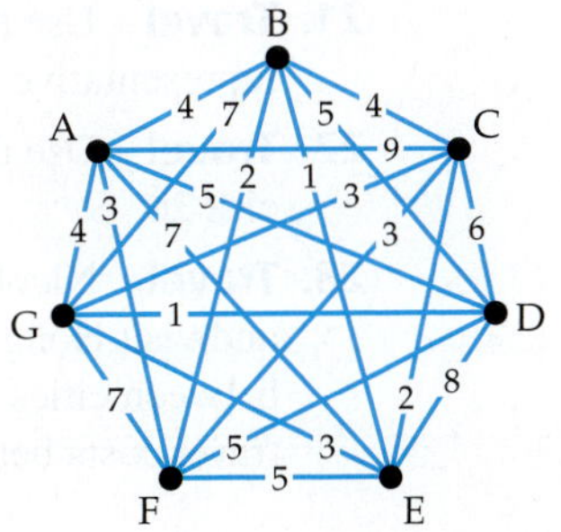

■ In Exercises 15 to 18, use the edge-picking algorithm to find a Hamiltonian circuit in the indicated graph.

15. Graph in Exercise 11

16. Graph in Exercise 12

17. Graph in Exercise 13

18. Graph in Exercise 14

19. Travel A company representative lives in Louisville, Kentucky, and needs to visit offices in five different Indiana cities over the next few days. The representative wants to drive between cities and return to Louisville at the end of the trip. The estimated driving times, in hours, between cities are given in the table below. Represent the driving times by a weighted graph. Use the greedy algorithm to design an efficient route for the representative to follow.

	Louisville	Bloomington	Fort Wayne	Indianapolis	Lafayette	Evansville
Louisville	—	3.6	6.4	3.2	4.9	3.1
Bloomington	3.6	—	4.5	1.3	2.4	3.4
Fort Wayne	6.4	4.5	—	3.3	3.0	8.0
Indianapolis	3.2	1.3	3.3	—	1.5	4.6
Lafayette	4.9	2.4	3.0	1.5	—	5.0
Evansville	3.1	3.4	8.0	4.6	5.0	—

20. Travel A tourist is staying in Toronto, Canada, and would like to visit four other Canadian cities by train. The visitor wants to go from one city to the next and return to Toronto while minimizing the total travel distance. The distances between cities, in kilometers, are given in the following table. Represent the distances between the cities using a weighted graph. Use the greedy algorithm to plan a route for the tourist.

	Toronto	Kingston	Niagara Falls	Ottawa	Windsor
Toronto	—	259	142	423	381
Kingston	259	—	397	174	623
Niagara Falls	142	397	—	562	402
Ottawa	423	174	562	—	787
Windsor	381	623	402	787	—

21. Travel Use the edge-picking algorithm to design a route for the company representative in Exercise 19.

22. Travel Use the edge-picking algorithm to design a route for the tourist in Exercise 20.

23. Travel Nicole wants to tour Asia. She will start and end her journey in Tokyo and visit Hong Kong, Bangkok, Seoul, and Beijing. The airfares available to her between cities are given in the table. Draw a weighted graph that represents the travel costs between cities and use the greedy algorithm to find a low-cost route.

	Tokyo	Hong Kong	Bangkok	Seoul	Beijing
Tokyo	—	$845	$1275	$470	$880
Hong Kong	$845	—	$320	$515	$340
Bangkok	$1275	$320	—	$520	$365
Seoul	$470	$515	$520	—	$225
Beijing	$880	$340	$365	$225	—

24. Travel The prices for traveling between five cities in Colorado by bus are given in the table below. Represent the travel costs between cities using a weighted graph. Use the greedy algorithm to find a low-cost route that starts and ends in Boulder and visits each city.

	Boulder	Denver	Colorado Springs	Grand Junction	Durango
Boulder	—	$16	$25	$49	$74
Denver	$16	—	$22	$45	$72
Colorado Springs	$25	$22	—	$58	$59
Grand Junction	$49	$45	$58	—	$32
Durango	$74	$72	$59	$32	—

25. Travel Use the edge-picking algorithm to find a low-cost route for the traveler in Exercise 23.

26. Travel Use the edge-picking algorithm to find a low-cost bus route in Exercise 24.

27. Route Planning Brian needs to visit the pet store, the shopping mall, the local farmers market, and the pharmacy. His estimated driving times (in minutes) between the locations are given in the table below. Use the greedy algorithm and the edge-picking algorithm to find two possible routes, starting and ending at home, that will help Brian minimize his total travel time.

	Home	Pet store	Shopping mall	Farmers market	Pharmacy
Home	—	18	27	15	8
Pet store	18	—	24	22	10
Shopping mall	27	24	—	20	32
Farmers market	15	22	20	—	22
Pharmacy	8	10	32	22	—

28. Route Planning A bike messenger needs to deliver packages to five different buildings and return to the courier company. The estimated biking times (in minutes) between the buildings are given in the following table. Use the greedy algorithm and the edge-picking algorithm to find two possible routes for the messenger to follow that will help minimize the total travel time.

	Courier company	Prudential building	Bank of America building	Imperial Bank building	GE Tower	Design Center
Courier company	—	10	8	15	12	17
Prudential building	10	—	10	6	9	8
Bank of America building	8	10	—	7	18	20
Imperial Bank building	15	6	7	—	22	16
GE Tower	12	9	18	22	—	5
Design Center	17	8	20	16	5	—

29. **Scheduling** A research company has a large supercomputer that is used by different teams for a variety of computational tasks. In between each task, the software must be reconfigured. The time required depends on which tasks follow which, because some settings are shared by different tasks. The times (in minutes) required to reconfigure the machine from one task to another are given in the table below. Use the greedy algorithm and the edge-picking algorithm to find time-efficient sequences in which to assign the tasks to the computer. The software configuration must start and end in the home state.

	Home state	Task A	Task B	Task C	Task D
Home state	—	35	15	40	27
Task A	35	—	30	18	25
Task B	15	30	—	35	16
Task C	40	18	35	—	32
Task D	27	25	16	32	—

30. **Computer Networks** A small office wishes to network its six computers in one large loop (see Check Your Progress 6 on page 255). The lengths of cable, in meters, required between machines are given in the table below. Use the edge-picking algorithm to find an efficient cable configuration in which to network the computers.

	Computer A	Computer B	Computer C	Computer D	Computer E	Computer F
Computer A	—	10	22	9	15	8
Computer B	10	—	12	14	16	5
Computer C	22	12	—	14	9	16
Computer D	9	14	14	—	7	15
Computer E	15	16	9	7	—	13
Computer F	8	5	16	15	13	—

EXTENSIONS

31. **Route Planning** A security officer patrolling a city neighborhood needs to drive every street each night. The officer has drawn a graph to represent the neighborhood in which the edges represent the streets and the vertices correspond to street intersections. Would the most efficient way to drive the streets correspond to an Euler circuit, a Hamiltonian circuit, or neither? (The officer must return to the starting location when finished.) Explain your answer.

32. **Route Planning** A city engineer needs to inspect the traffic signs at each street intersection of a neighborhood. The engineer has drawn a graph to represent the neighborhood, where the edges represent the streets and the vertices correspond to street intersections. Would the most efficient route to drive correspond to an Euler circuit, a Hamiltonian circuit, or neither? (The engineer must return to the starting location when finished.) Explain your answer.

33. **a.** Draw a connected graph with six vertices that has no Euler circuits and no Hamiltonian circuits.

b. Draw a graph with six vertices that has a Hamiltonian circuit but no Euler circuits.

c. Draw a graph with five vertices that has an Euler circuit but no Hamiltonian circuits.

34. **a.** Assign weights to the edges of the following complete graph so that the edge-picking algorithm gives a circuit of lower total weight than the circuit

given by the greedy algorithm. For the greedy algorithm, begin at vertex A.

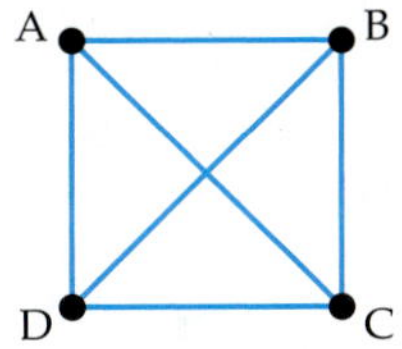

b. Assign weights to the edges of the graph so that the greedy algorithm gives a circuit of lower total weight than the circuit given by the edge-picking algorithm. For the greedy algorithm, begin at vertex A.

c. Assign weights to the edges of the graph so that there is a circuit of lower total weight than the circuits given by both the greedy algorithm (beginning at vertex A) and the edge-picking algorithm.

SECTION 5.3 Planarity and Euler's Formula

Planarity

A puzzle that was posed some time ago goes something like this: Three utility companies each need to run pipes to three houses. Can they do so without crossing over each other's pipes at any point? The puzzle is illustrated in Figure 5.15. Go ahead and try to draw pipes connecting each utility company to each house without letting any pipes cross over each other.

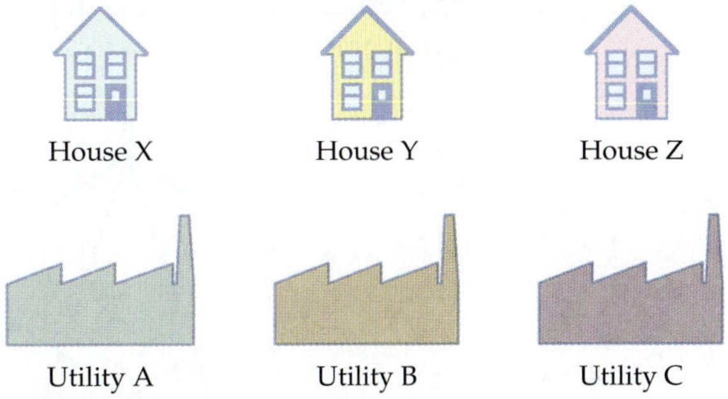

FIGURE 5.15

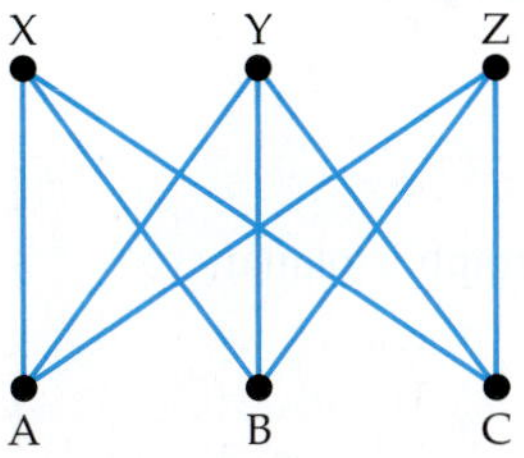

FIGURE 5.16
The Utilities Graph

One way to approach the puzzle is to express the situation in terms of a graph. Each of the houses and utility companies will be represented by a vertex, and we will draw an edge between two vertices if a pipe needs to run from one building to the other. If we were not worried about pipes crossing, we could easily draw a solution, as in Figure 5.16.

To solve the puzzle, we need to draw an equivalent graph in which no edges cross over each other. If this is possible, the graph is called a *planar graph.*

Planar Graph

A **planar graph** is a graph that can be drawn so that no edges intersect each other (except at vertices).

If the graph is drawn in such a way that no edges cross, we say that we have a **planar drawing** of the graph.

QUESTION Is this a planar drawing of a graph? Is the graph planar?

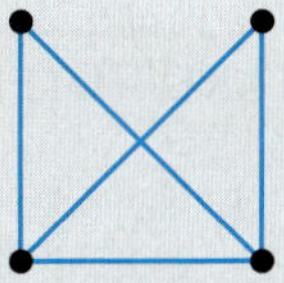

ANSWER The drawing is not planar because two edges cross. The graph is planar because we can make an equivalent planar drawing of it as shown at the right.

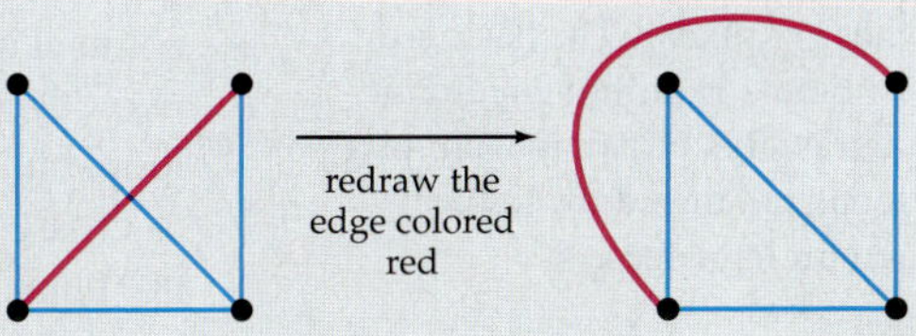

EXAMPLE 1 Identify a Planar Graph

Show that the graph below is planar.

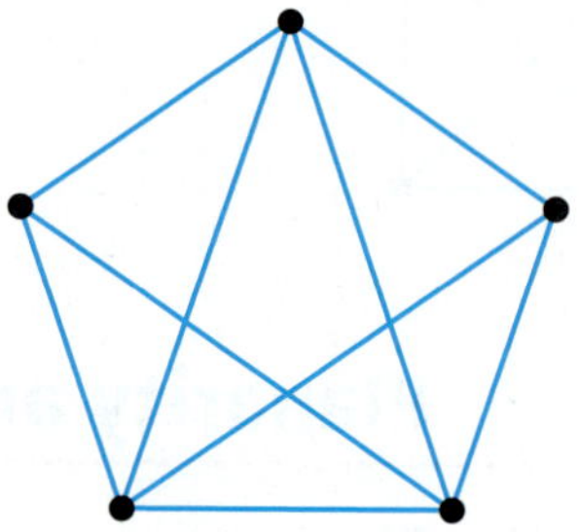

Solution

As given, the graph has several intersecting edges. However, we can redraw the graph in an equivalent form in which no edges touch except at vertices by redrawing the two red edges shown below. To verify that the second graph is equivalent to the first, we can label the vertices and check that the edges join the same vertices in each graph. Because the given graph is equivalent to a graph whose edges do not intersect, the graph is planar.

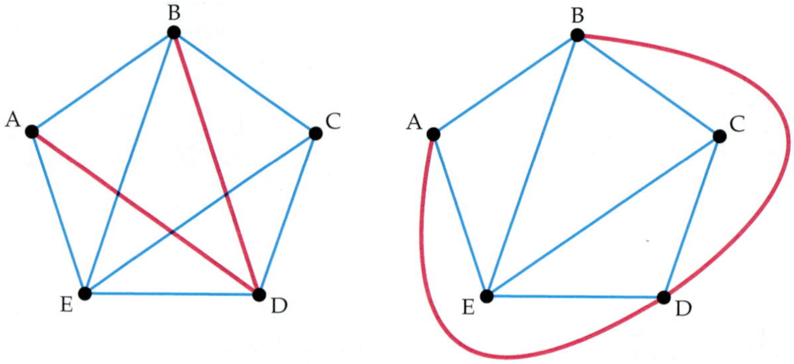

CHECK YOUR PROGRESS 1 Show that the following graph is planar.

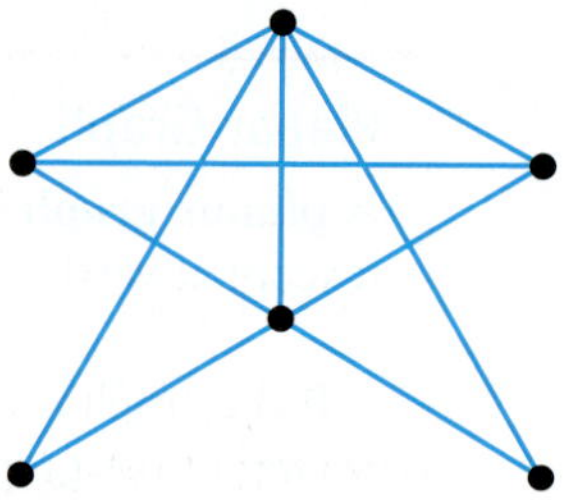

Solution See page S18.

TAKE NOTE

The strategy we used here to show that the graph in Figure 5.16 is not planar can often be used to make planar drawings of graphs that *are* planar. The basic strategy is: Find a Hamiltonian circuit, redraw this circuit in a circular loop, and then draw in the remaining edges.

To see that the graph in Figure 5.16 representing the puzzle of connecting utilities is not planar, note that the graph is Hamiltonian, and one Hamiltonian circuit is A–X–B–Y–C–Z–A. If we redraw the graph so that this circuit is drawn in a loop (see Figure 5.17 on the next page), we then need to add the edges AY, BZ, and CX. All three of these edges connect opposite vertices. We can draw only one of these edges inside the loop; otherwise two edges would cross. This means that the other two edges must be drawn outside the loop, but as you can see in Figure 5.18 on the next page, those two edges would then have to cross. Thus the graph in Figure 5.16, which we will refer to as the *Utilities Graph,* is not planar, and so the utilities puzzle is not solvable.

INSTRUCTOR NOTE

Students may find it interesting to hear that the utilities puzzle is solvable if we consider the situation on a different kind of surface instead of the flat plane (or spherical earth) we are accustomed to. In particular, the puzzle has a solution on a torus. See the *Ask Dr. Math* web page http://mathforum.org/dr.math/faq/faq.3utilities.html for an illustration and discussion.

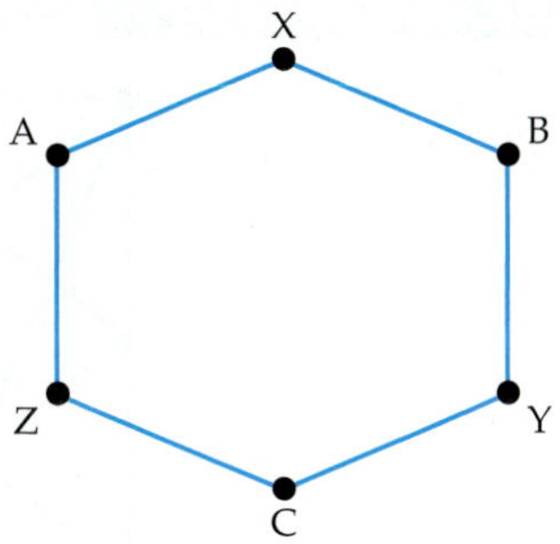

FIGURE 5.17

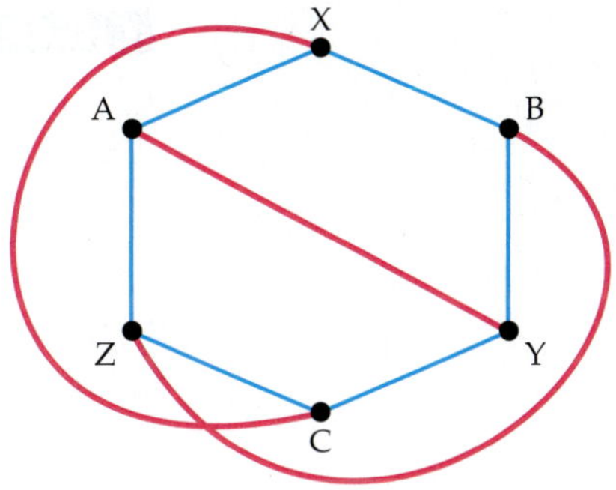

FIGURE 5.18

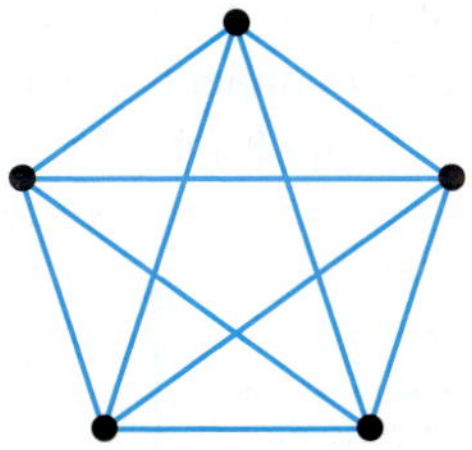

FIGURE 5.19 The complete graph K_5

One strategy we can use to show that a graph is *not* planar is to find a **subgraph**, a graph whose edges and vertices come from the given graph, that is not planar. The Utilities Graph in Figure 5.16 is a common subgraph to watch for. Another graph that is not planar is the complete graph with five vertices, denoted K_5, shown in Figure 5.19. (See Exercise 29.)

Subgraph Theorem

If a graph G has a subgraph that is not planar, then G is also not planar. In particular, if G contains the Utilities Graph or K_5 as a subgraph, G is not planar.

EXAMPLE 2 Identify a Nonplanar Graph

Show that the following graph is not planar.

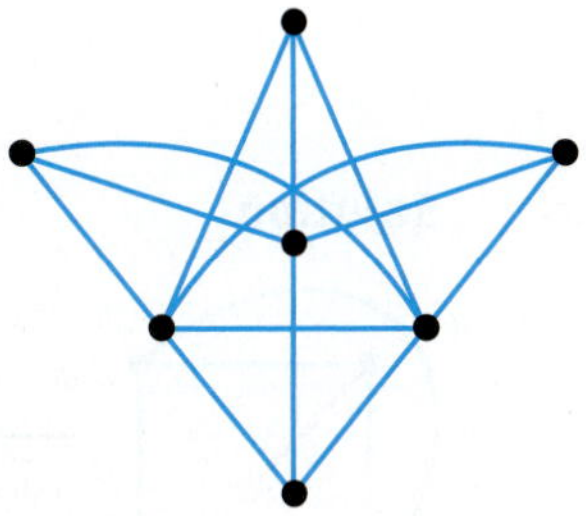

Solution

In the figure below, we have highlighted edges connecting the top six vertices. If we consider the highlighted edges and attached vertices as a subgraph, we can verify that the subgraph is the Utilities Graph. (The graph is slightly distorted compared with the version shown in Figure 5.16, but it is equivalent.) By the preceding theorem, we know that the graph is not planar.

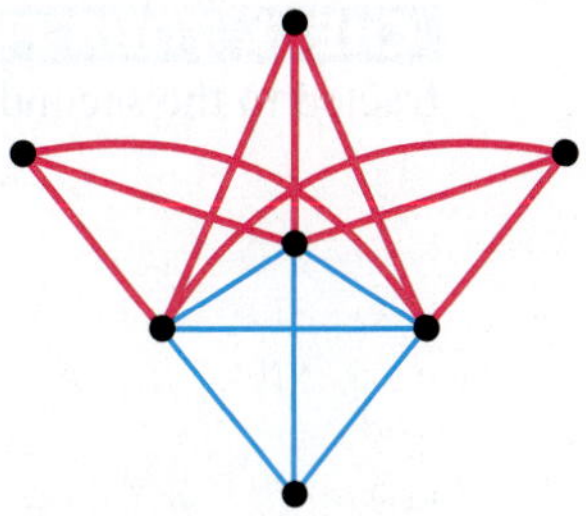

TAKE NOTE

The figure below shows how the highlighted edges of the graph in Example 2 can be transformed into the Utilities Graph.

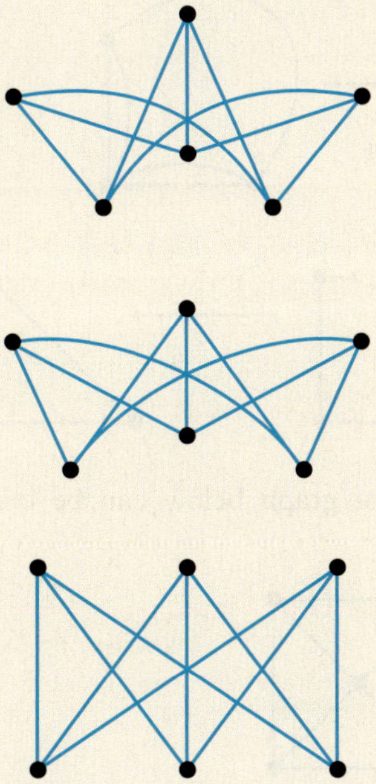

CHECK YOUR PROGRESS 2 Show that the following graph is not planar.

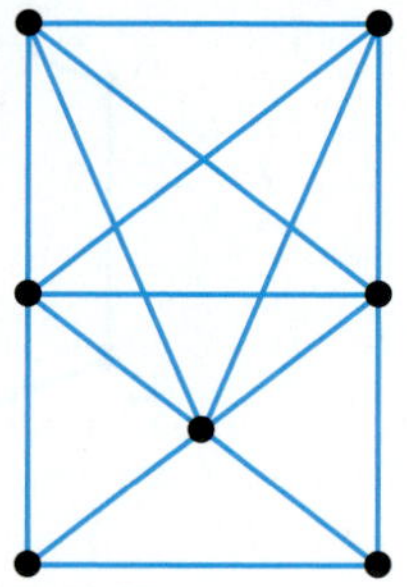

Solution See page S18.

We can expand this strategy by considering *contractions* of a subgraph. A **contraction** of a graph is formed by "shrinking" an edge until the two vertices it connects come together and blend into one. If, in the process, the graph is left with any multiple edges, we merge them into one. The process is illustrated below.

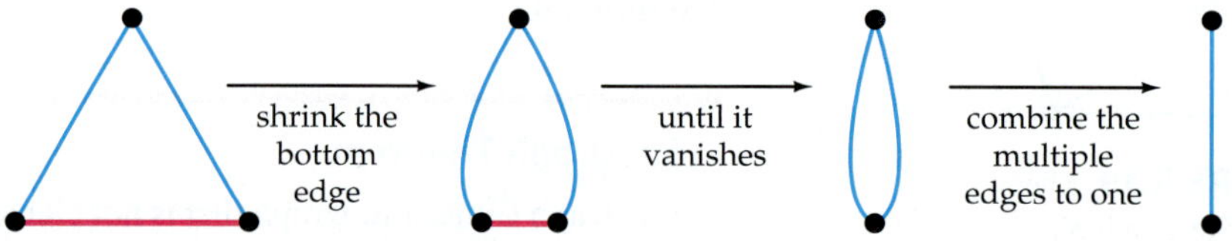

EXAMPLE 3 Contracting a Graph

Show that the first graph below can be contracted to the second graph.

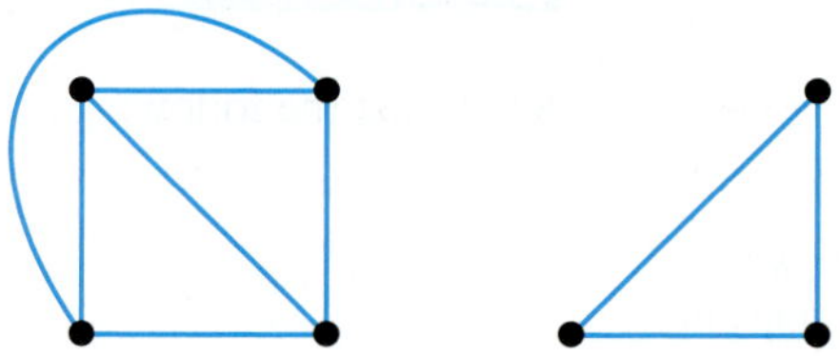

Solution

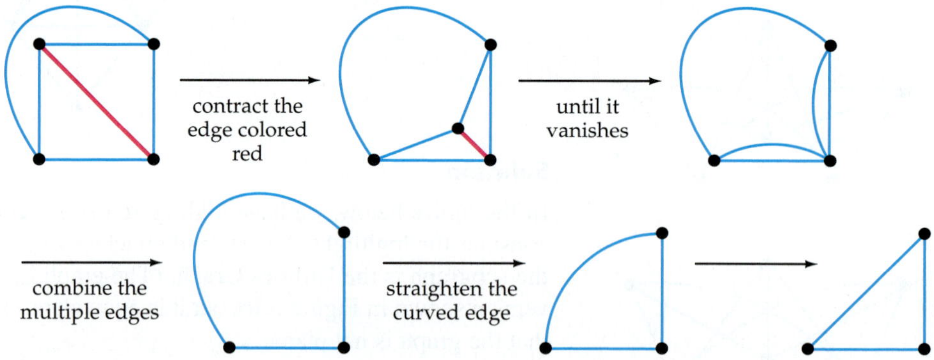

CHECK YOUR PROGRESS 3 Show that the first graph below can be contracted to the second graph.

Solution See page S18.

If we consider contractions, it turns out that the Utilities Graph and K_5 serve as building blocks for nonplanar graphs. In fact, it was proved in 1930 that any nonplanar graph will always have a subgraph that is the Utilities Graph or K_5, or a subgraph that can be contracted to the Utilities Graph or K_5. We can then expand our strategy, as given by the following theorem.

Nonplanar Graph Theorem

A graph is nonplanar if and only if it has the Utilities Graph or K_5 as a subgraph, or it has a subgraph that can be contracted to the Utilities Graph or K_5.

This gives us a definite test that will determine whether or not a graph can be drawn in such a way as to avoid crossing edges. (Note that the theorem includes the case in which the entire graph can be contracted to the Utilities Graph or K_5, as we can consider the graph a subgraph of itself.)

EXAMPLE 4 Identify a Nonplanar Graph

Show that the graph below is not planar.

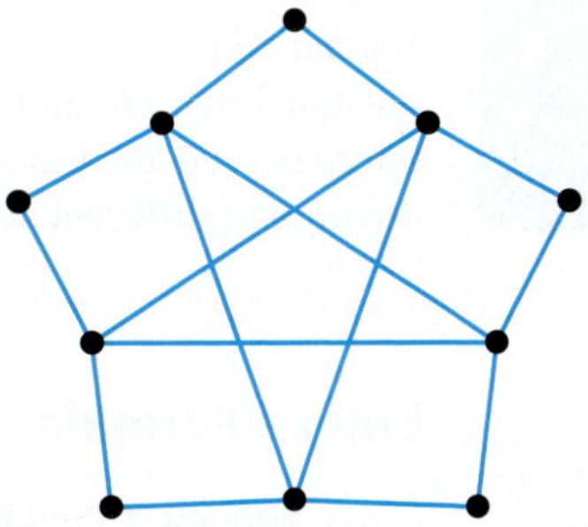

Solution

Note that the graph looks similar to K_5. In fact, we can contract some edges and make the graph look like K_5. Choose a pair of adjacent outside edges and contract one of them, as shown in the figure below.

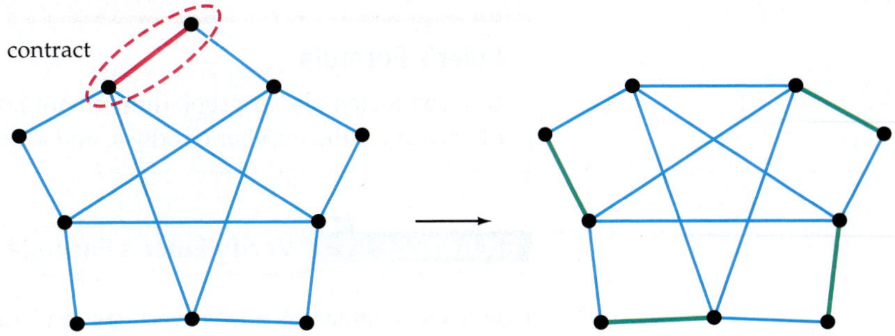

If we similarly contract the four edges colored green in the preceding figure, we arrive at the graph of K_5.

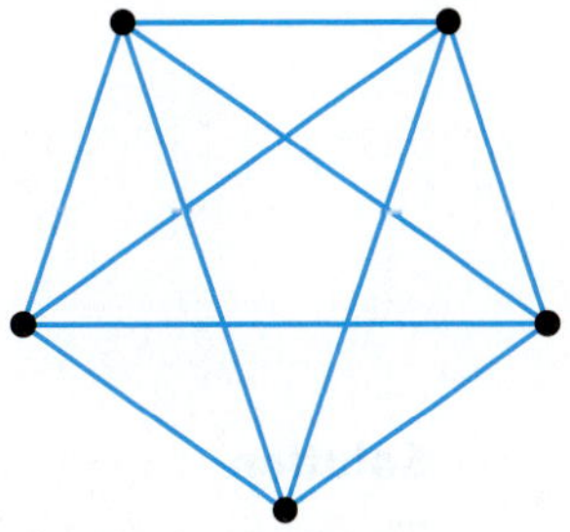

We were able to contract our graph to K_5, so by the nonplanar graph theorem, the given graph is not planar.

CHECK YOUR PROGRESS 4 Show that the graph below is not planar.

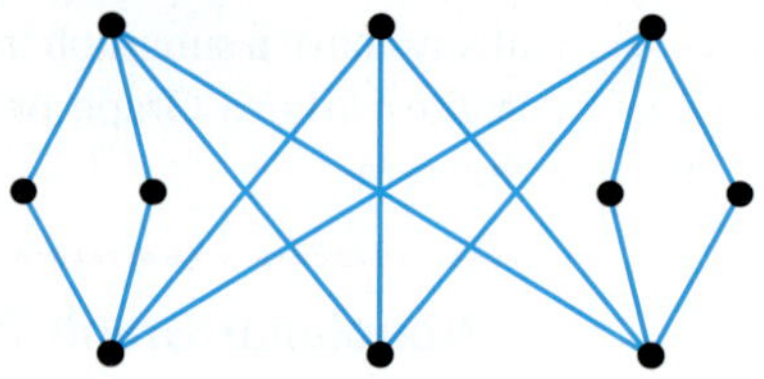

Solution See page S18.

MATH MATTERS Planar Graphs on Circuit Boards

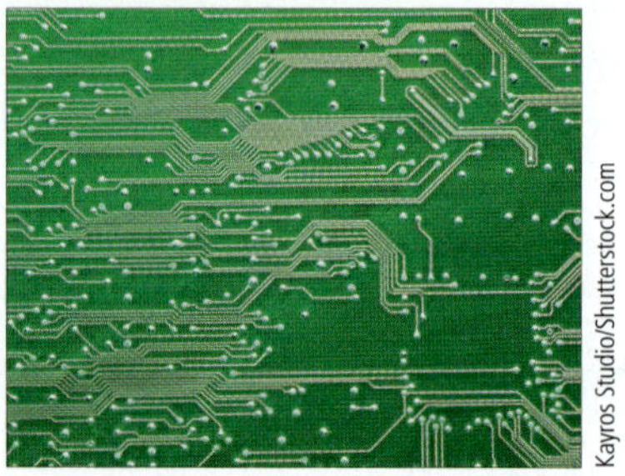

Trying to draw a graph as a planar drawing has many practical applications. For instance, the design of circuit boards used in computers and other electronic components depends on wires connecting different components without touching each other elsewhere. Circuit boards can be very complex, but in effect they require that collections of wires be arranged in a planar drawing of a planar graph. If this is impossible, special connections can be installed that allow one wire to "jump" over another without touching. Or both sides of the board can be used, or sometimes more than one board is used and the boards are then connected by wires. In effect, the graph is spread out among different surfaces.

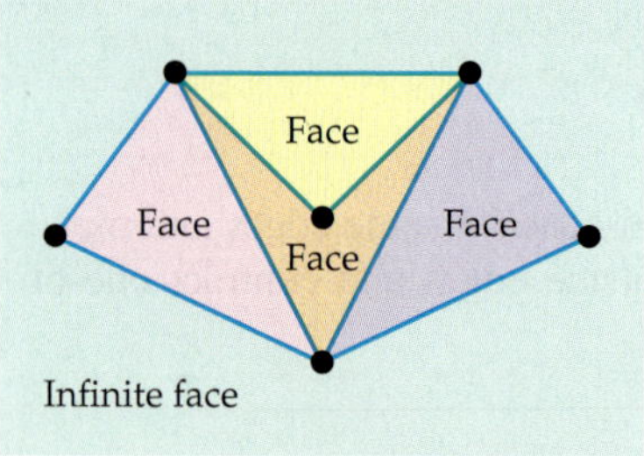

Euler's Formula

Euler noticed a connection between various features of planar graphs. In addition to edges and vertices, he looked at *faces* of a graph. In a planar drawing of a graph, the edges divide the graph into different regions called **faces**. The region surrounding the graph, or the exterior, is also considered a face, called the **infinite face**. (See the figure at the left.) The following relationship, called Euler's formula, is always true.

Euler's Formula

In a connected planar graph drawn with no intersecting edges, let v be the number of vertices, e the number of edges, and f the number of faces. Then $v + f = e + 2$.

EXAMPLE 5 Verify Euler's Formula in a Graph

Count the number of edges, vertices, and faces in the planar graph below, and then verify Euler's formula.

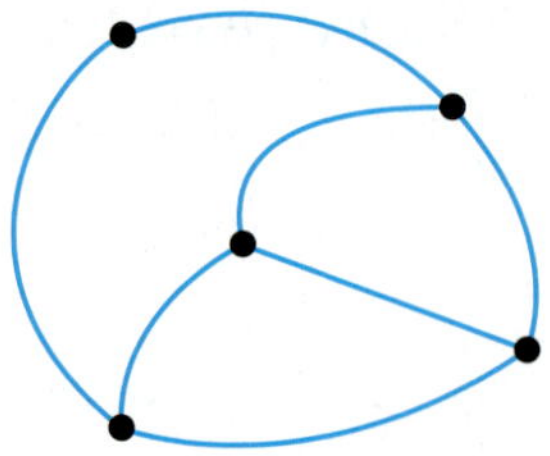

Solution

There are seven edges, five vertices, and four faces (counting the infinite face) in the graph. Thus $v + f = 5 + 4 = 9$ and $e + 2 = 7 + 2 = 9$, so $v + f = e + 2$, as Euler's Formula predicts.

CHECK YOUR PROGRESS 5 Verify Euler's formula for the planar graph below.

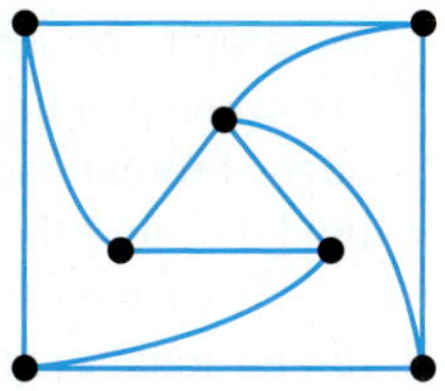

Solution See page S18.

EXCURSION

The Five Regular Convex Polyhedra

A **regular polyhedron** is a three-dimensional object in which all faces are identical. Specifically, each face is a regular polygon (meaning that each edge has the same length and each angle has the same measure). In addition, the same number of faces must meet at each vertex, or corner, of the object. Here we will discuss convex polyhedra, which means that any line joining one vertex to another is entirely contained within the object. (In other words, there are no indentations.)

It was proved long ago that only five objects fit this description—the tetrahedron, the cube, the octahedron, the dodecahedron, and the icosahedron. (See Figure 5.20 on the next page.) One way mathematicians determined that there were only five such polyhedra was to visualize these three-dimensional objects in a two-dimensional way. We will use the cube to demonstrate.

Imagine a standard cube, but with the edges made of wire and the faces empty. Now suspend the cube above a flat surface and place a light above the top face of the cube. If the light is shining downward, a shadow is created on the flat surface. This is called a *projection* of the cube.

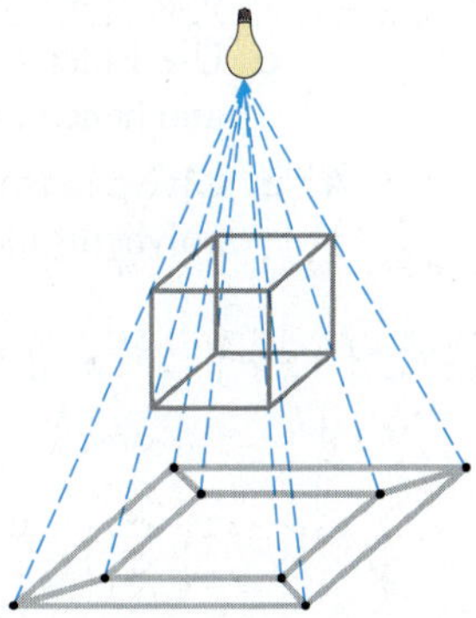

The shadow forms a graph. Note that each corner, or vertex, of the cube corresponds to a vertex of the graph, and each face of the cube corresponds to a face of the graph. The exception is the top face of the cube—its projection overlaps the entire graph. As a convention, we will identify the top face of the cube with the infinite face in the projected graph. Thus the three-dimensional cube is identified with the graph shown below.

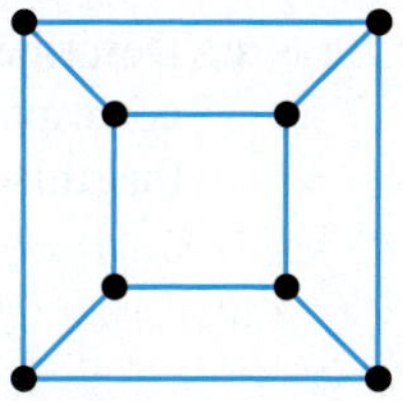

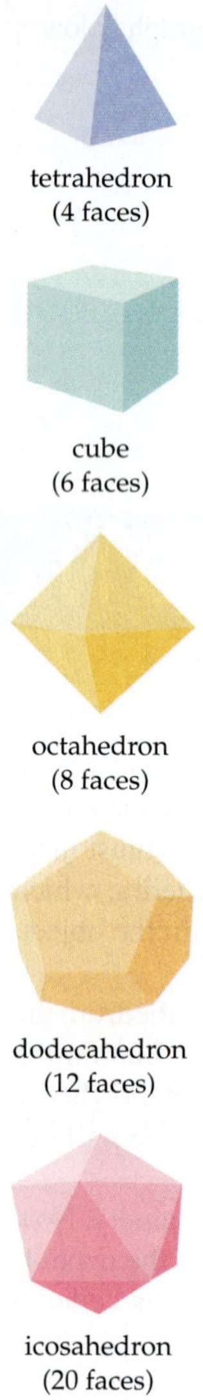

FIGURE 5.20

Because the same number of faces must meet at each vertex of a polyhedron, the same number of faces, and so the same number of edges, must meet at each vertex of the graph. We also know that each face of a polyhedron has the same shape. Although the corresponding faces of the graph are distorted, we must have the same number of edges around each face (including the infinite face). In addition, Euler's formula must hold, because the projection is always a planar graph.

Thus we can investigate different polyhedra by looking at their features in two-dimensional graphs. It turns out that only five planar graphs satisfy these requirements, and the polyhedra that these graphs are the projections of are precisely the five pictured in Figure 5.20. The following exercises ask you to investigate some of the relationships between polyhedra and their projected graphs.

EXCURSION EXERCISES

1. The tetrahedron in Figure 5.20 consists of four faces, each of which is an equilateral triangle. Draw the graph that results from a projection of the tetrahedron.
2. The following graph is the projection of one of the polyhedra in Figure 5.20. Identify the polyhedron it represents by comparing features of the graph to features of the polyhedron.

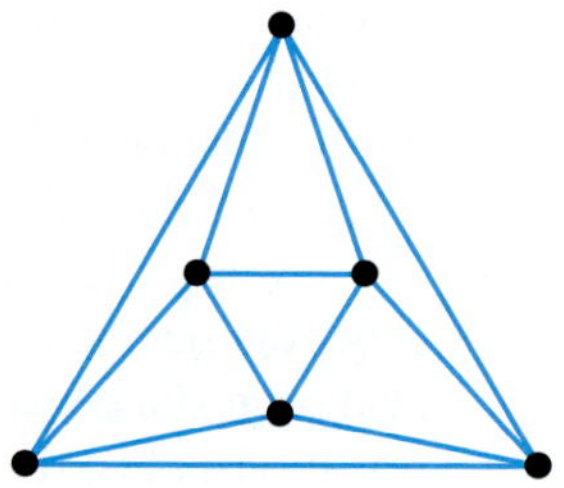

3. **a.** If we form a graph by a projection of the dodecahedron in Figure 5.20, what will the degree of each vertex be?

   b. The dodecahedron has 12 faces, each of which has 5 edges. Use this information to determine the number of edges in the projected graph.

   c. Use Euler's formula to determine the number of vertices in the projected graph, and hence in the dodecahedron.

4. **a.** Give a reason why the graph below cannot be the projection of a *regular* convex polyhedron.

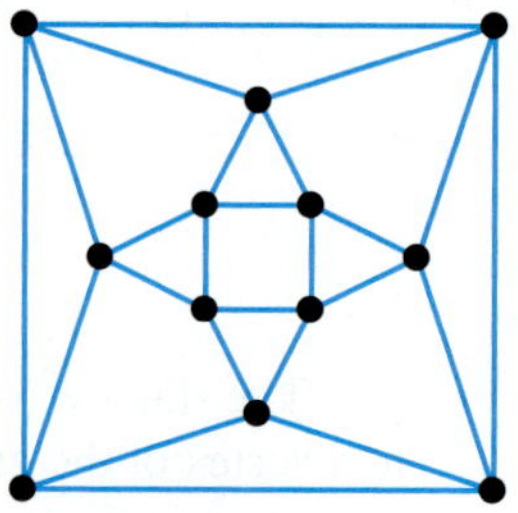

   b. If the above graph is the projection of a three-dimensional convex polyhedron, how many faces does the polyhedron have?

   c. Describe, in as much detail as you can, additional features of the polyhedron corresponding to the graph. If you are feeling adventurous, sketch a polyhedron that this graph could be a projection of!

EXERCISE SET 5.3

■ In Exercises 1 to 8, show that the graph is planar by finding a planar drawing.

1.

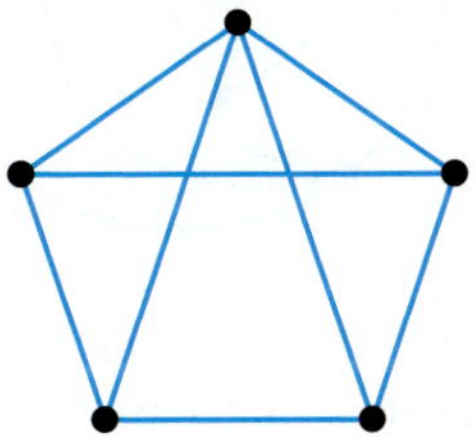

2.

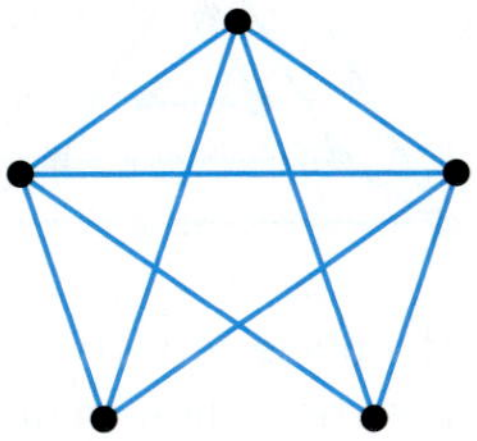

3.

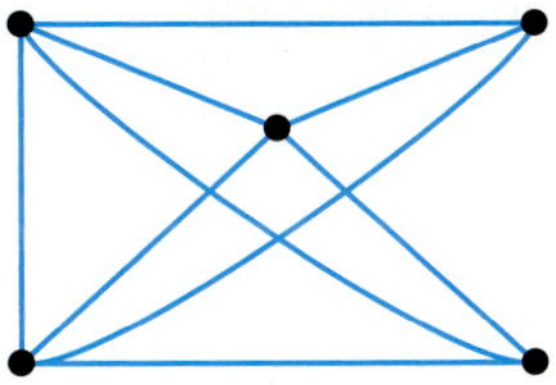

4.

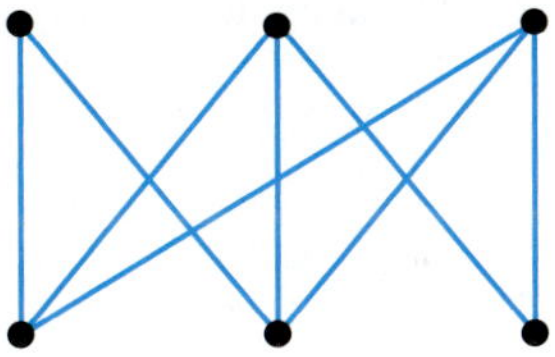

5.

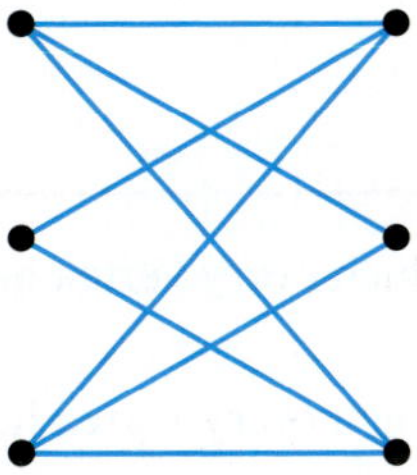

6.

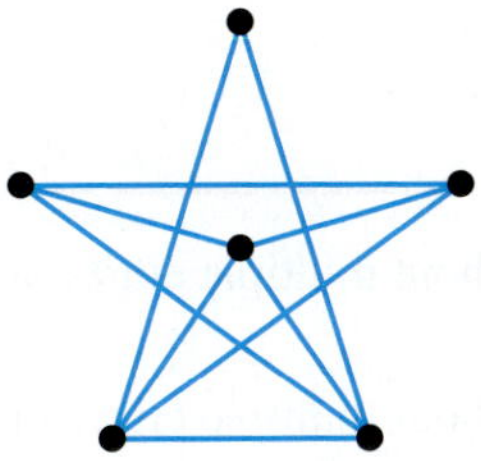

7.

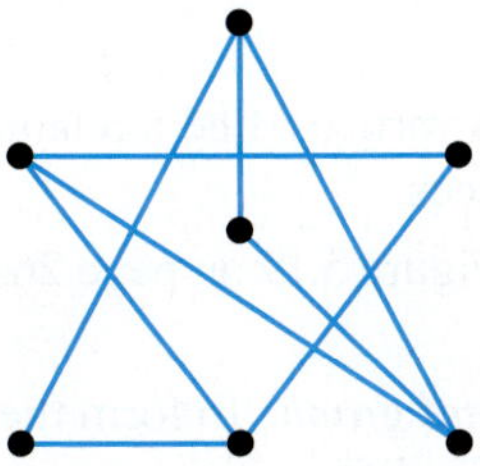

8.

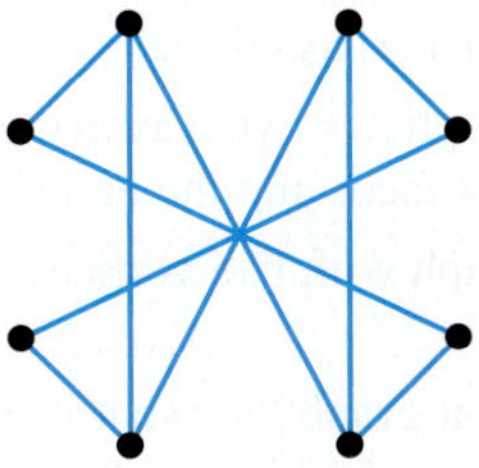

■ In Exercises 9 to 12, show that the graph is *not* planar.

9.

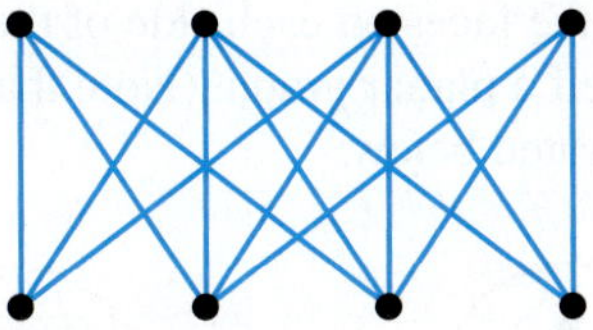

10.

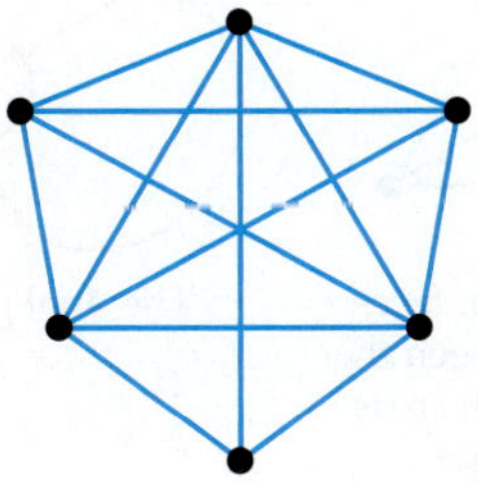

11.

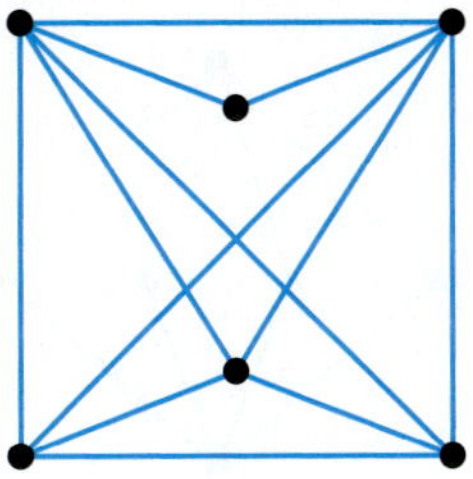

12.

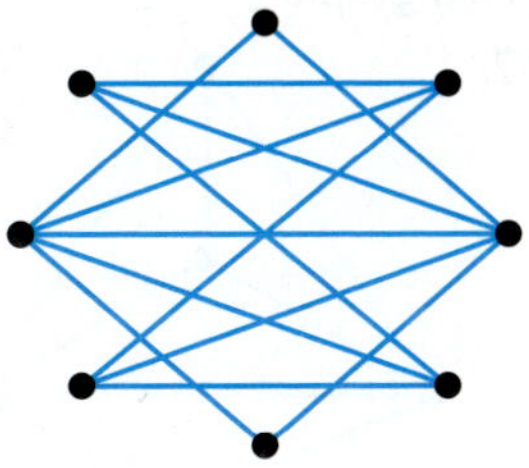

13. Show that the following graph contracts to K_5.

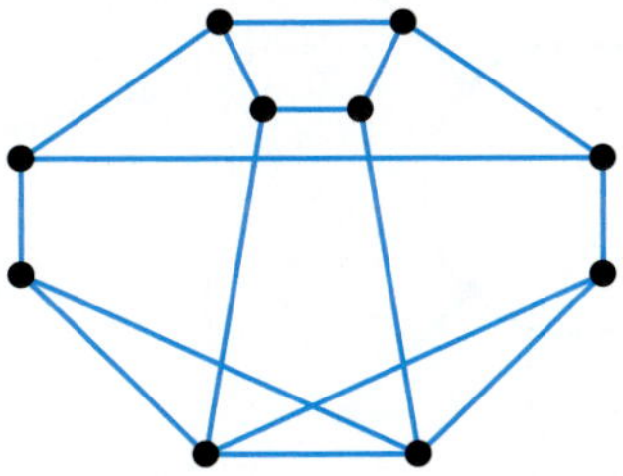

14. Show that the following graph contracts to the Utilities Graph.

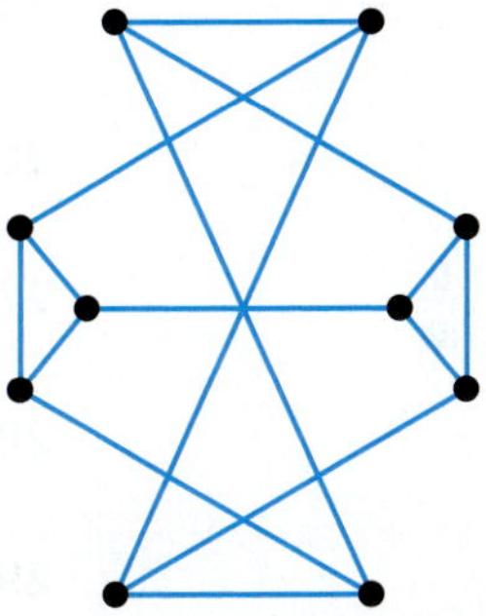

■ In Exercises 15 and 16, show that the graph is not planar by finding a subgraph whose contraction is the Utilities Graph or K_5.

15.

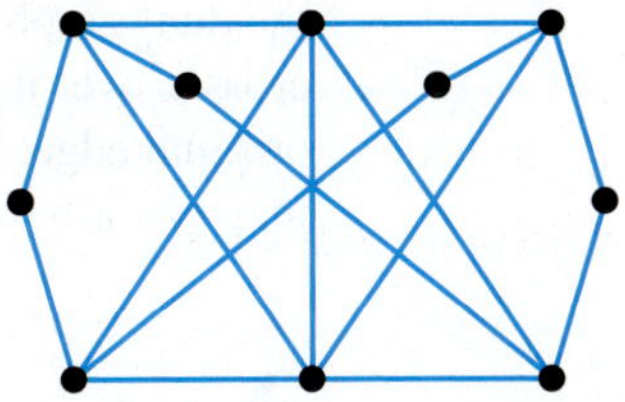

16.

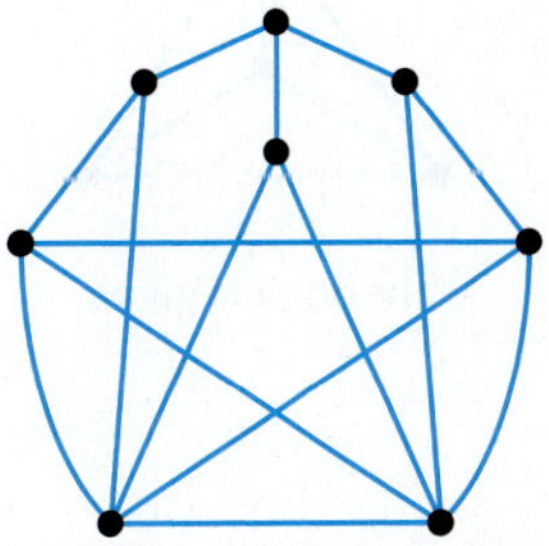

■ In Exercises 17 to 22, count the number of vertices, edges, and faces, and then verify Euler's formula for the given graph.

17.

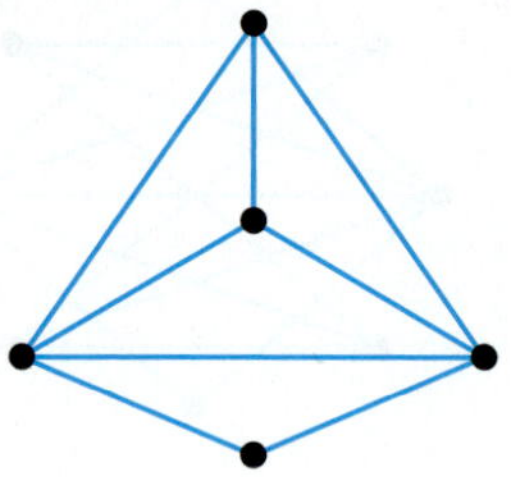

18.

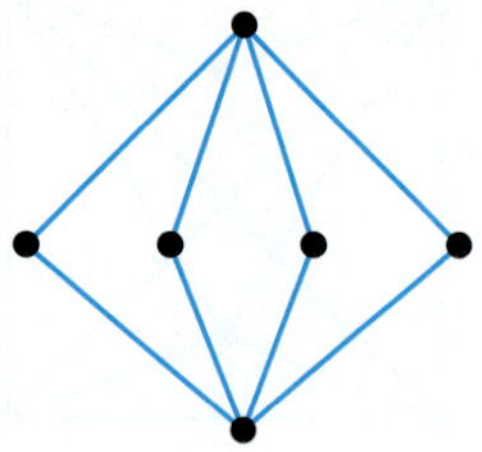

19.

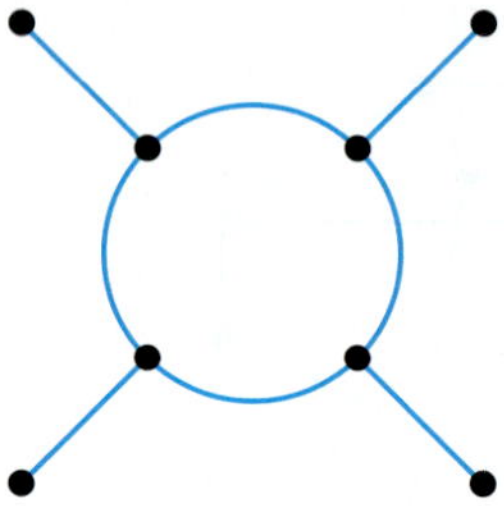

20.

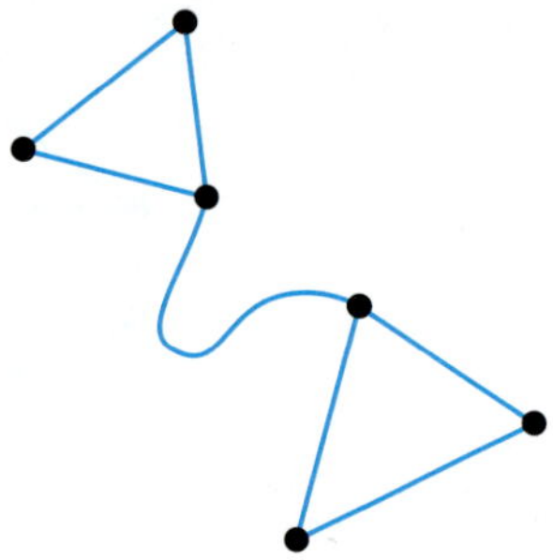

21.

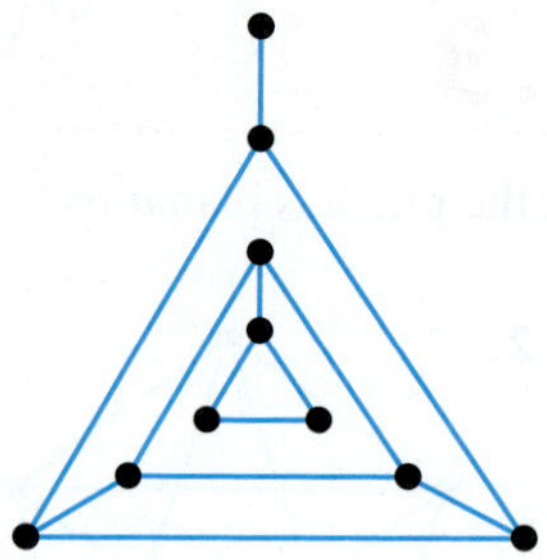

22.

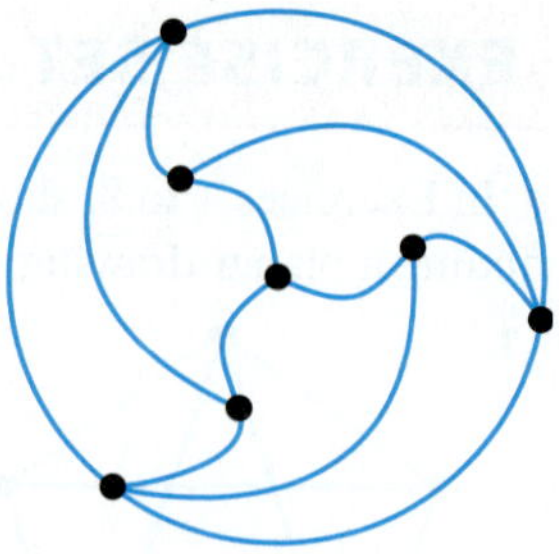

23. If a planar drawing of a graph has 15 edges and 8 vertices, how many faces does the graph have?

24. If a planar drawing of a graph has 100 vertices and 50 faces, how many edges are in the graph?

EXTENSIONS

25. Sketch a planar graph (without multiple edges or loops) in which every vertex has degree 3.

26. Sketch a planar graph (without multiple edges or loops) in which every vertex has degree 4.

27. Explain why it is not possible to draw a planar graph that contains three faces and has the same number of vertices as edges.

28. If a planar drawing of a graph has twice as many edges as vertices, find a relationship between the number of faces and the number of vertices.

29. Show that the complete graph with five vertices, K_5 (see Figure 5.19 on page 263), is not planar.

30. **Dual Graph** Every planar graph has what is called a *dual graph.* To form the dual graph, start with a planar drawing of the graph. With a different-color pen, draw a dot in each face (including the infinite face). These will be the vertices of the dual graph. Now, for every edge in the original graph, draw a new edge that crosses over it and connects the two new vertices in the faces on each side of the original edge. The resulting dual graph is always itself a planar graph. (Note that it may contain multiple edges.) The procedure is illustrated below.

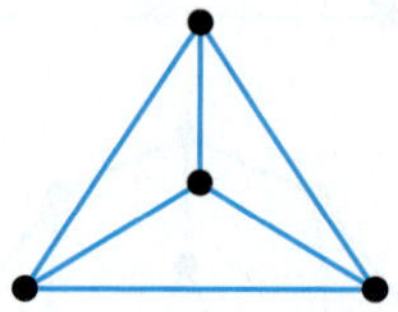

The original graph

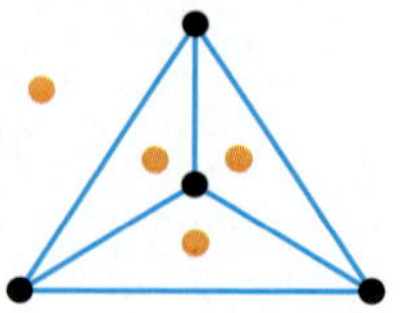

Draw dots in each face, including the infinite face.

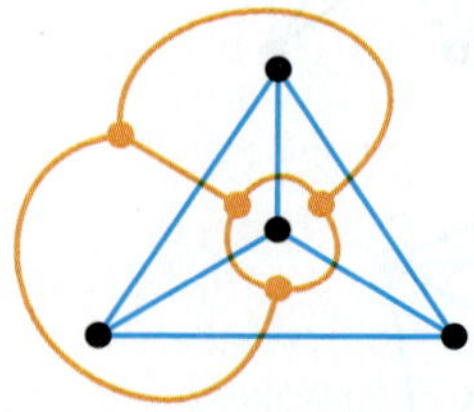

Connect adjacent faces with edges between the new vertices that cross each original edge.

The dual graph

a. Draw the dual graph of each of the planar graphs below.

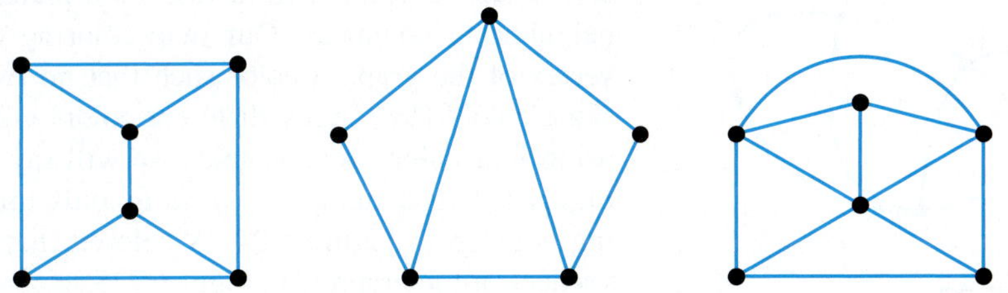

b. For each of the graphs in part a, count the number of faces, edges, and vertices in the original graph and compare your results with the numbers in the dual graph. What do you notice? Explain why this will always be true.

c. Start with your dual graph of the first graph in part a, and find *its* dual. What do you notice?

SECTION 5.4 Graph Coloring

Coloring Maps

In the mid-1800s, Francis Guthrie was trying to color a map of the counties of England. So that it would be easy to distinguish the counties, he wanted counties sharing a common border to have different colors. After several attempts, he noticed that four colors were required to color the map, but not more. This observation became known as the *four-color problem.* (It was not proved until over 100 years later; see the Math Matters on page 274.)

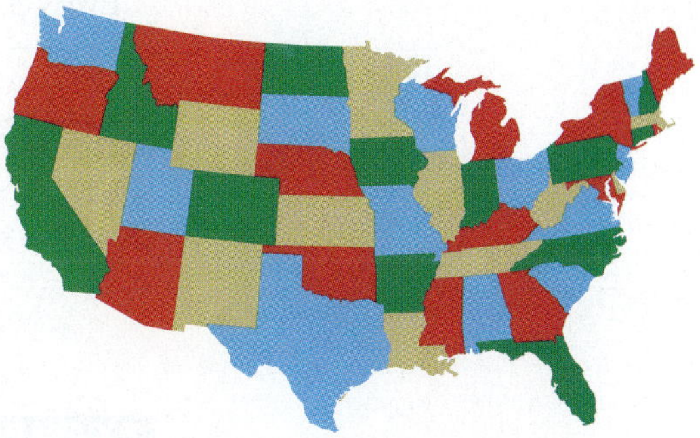

Here is a map of the contiguous states of the United States colored similarly. Note that the map has only four colors and that no two states that share a common border have the same color.

There is a connection between coloring maps and graph theory. This connection has many practical applications, from scheduling tasks, to designing computers, to playing Sudoku. Later in this section we will look more closely at some of these applications.

Suppose the map in Figure 5.21 shows the countries, labeled as letters, of a continent. We will assume that no country is split into more than one piece and countries that touch at just a corner point will not be considered neighbors. We can represent each country by a vertex, placed anywhere within the boundary of that country. We will then connect two vertices with an edge if the two corresponding countries are neighbors—that is, if they share a common boundary. The result is shown in Figure 5.22.

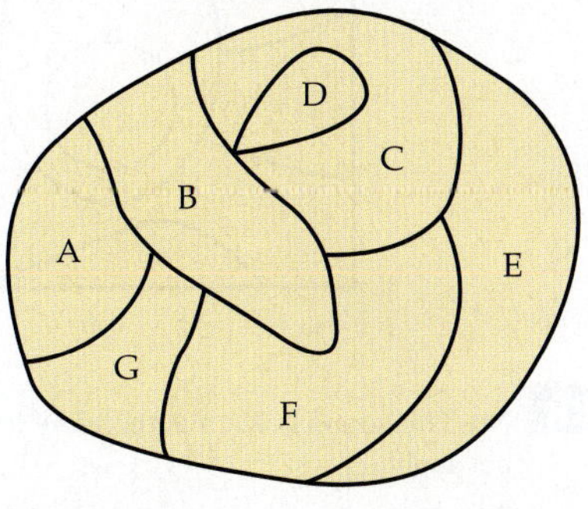

FIGURE 5.21

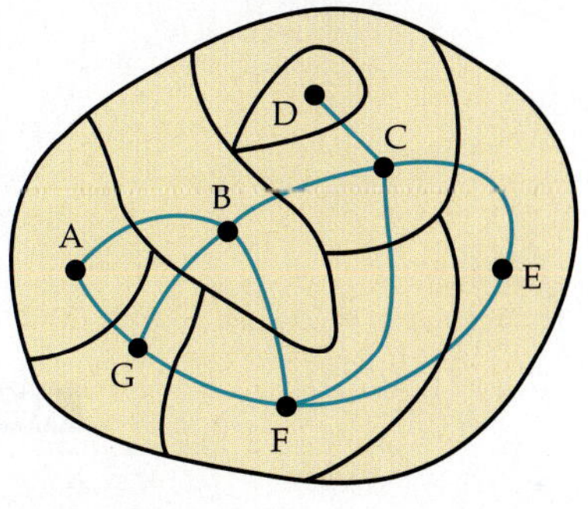

FIGURE 5.22

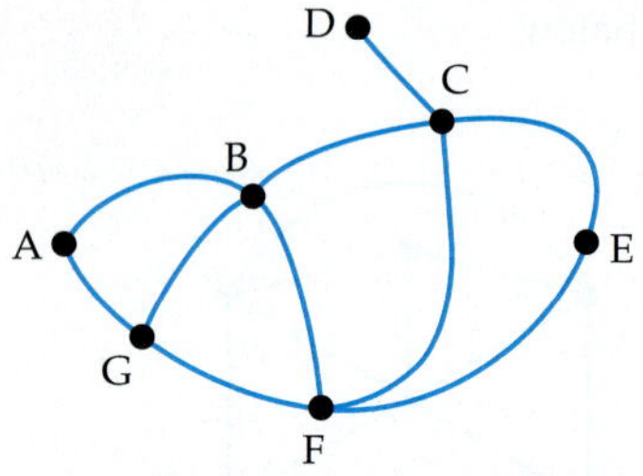

FIGURE 5.23

If we erase the boundaries of the countries, we are left with the graph in Figure 5.23. The resulting graph will always be a planar graph, because the edges simply connect neighboring countries. Our map-coloring question then becomes: Can we give each vertex of the graph a color such that no two vertices connected by an edge share the same color? How many different colors will be required? If this can be accomplished using four colors, for instance, we will say that the graph is **4-colorable**. The graph in Figure 5.23 is actually *3-colorable;* only three colors are necessary. One possible coloring is given in Figure 5.24. A colored map for Figure 5.21 based on the colors of the vertices of the graph is in Figure 5.25.

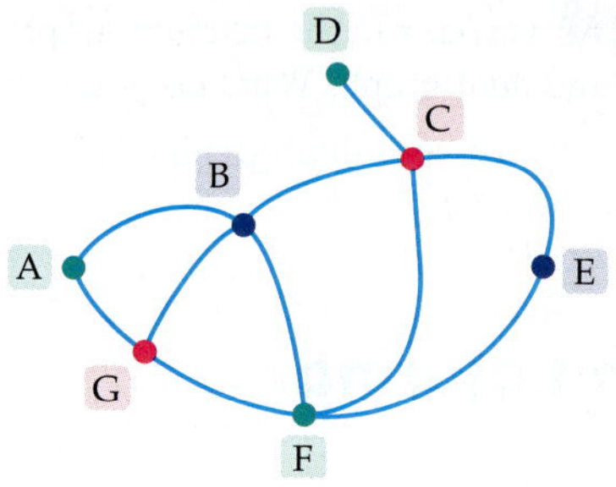

FIGURE 5.24

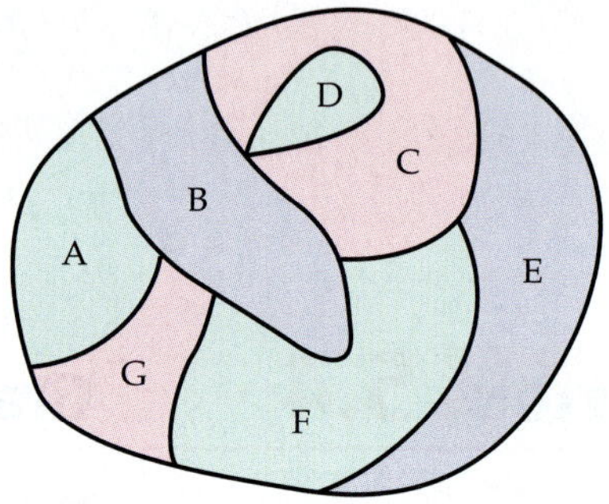

FIGURE 5.25

We can now formally state the four-color theorem.

Four-Color Theorem

Every planar graph is 4-colorable.

QUESTION The graph shown at the right requires five colors if we wish to color it such that no edge joins two vertices of the same color. Does this contradict the four-color theorem?

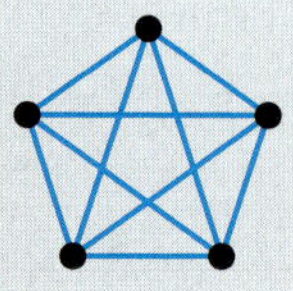

EXAMPLE 1 Using a Graph to Color a Map

The fictional map below shows the boundaries of countries on a rectangular continent. Represent the map as a graph, and find a coloring of the graph using the fewest possible number of colors. Then color the map according to the graph coloring.

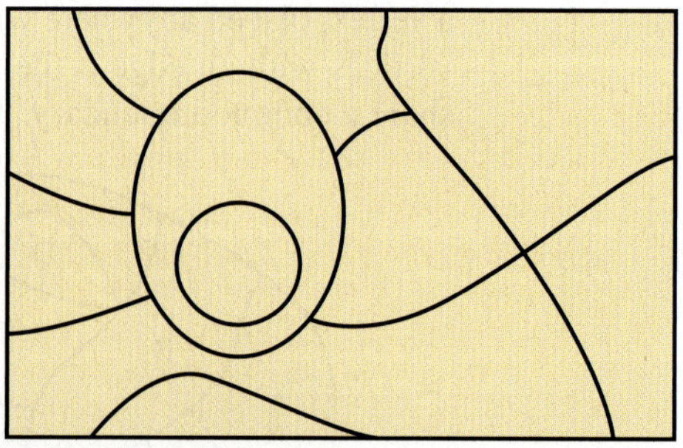

ANSWER No. This graph is K_5, and therefore it is not a planar graph, so the four-color theorem does not apply.

Solution

First draw a vertex in each country and then connect two vertices with an edge if the corresponding countries are neighbors. (See the first figure below.) Now try to color the vertices of the resulting graph so that no edge connects two vertices of the same color. We know we will need at least two colors, so one strategy is simply to pick a starting vertex, give it a color, and then assign colors to the connected vertices one by one. Try to reuse the same colors, and use a new color only when there is no other option. For this graph we will need four colors. (The four-color theorem guarantees that we will not need more than that.) To see why we will need four colors, notice that the one vertex colored green in the second figure below connects to a ring of five vertices. Three different colors are required to color the five-vertex ring, and the green vertex connects to all these, so it requires a fourth color.

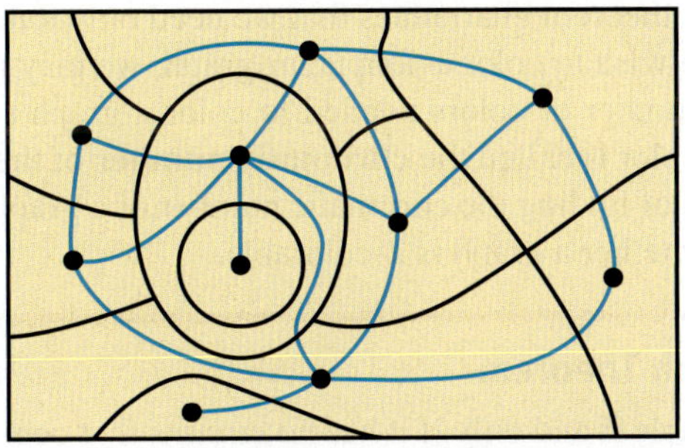

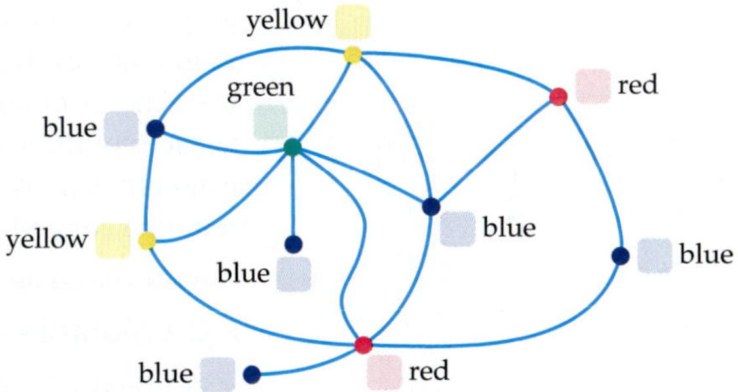

Now we color each country the same color as the coresponding vertex.

CHECK YOUR PROGRESS 1 Represent the fictional map of countries below as a graph, and determine whether the graph is 2-colorable, 3-colorable, or 4-colorable by finding a suitable coloring of the graph. Then color the map according to the graph coloring.

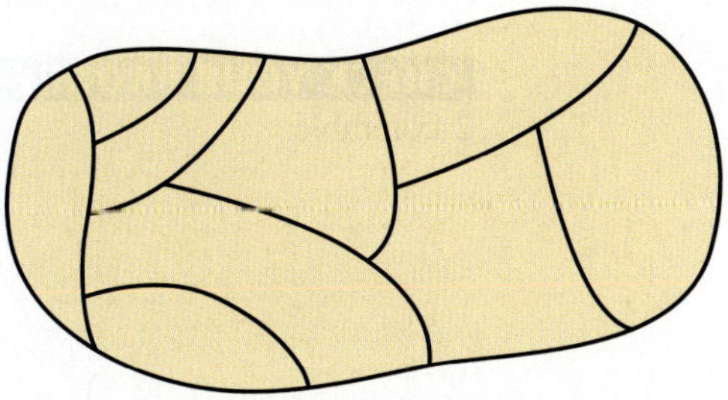

Solution See page S18.

MATH MATTERS Proving the Four-Color Theorem

The four-color theorem can be stated in a simple, short sentence, but proving it is anything but simple. The theorem was finally proved in 1976 by Wolfgang Haken and Kenneth Appel, two mathematicians at the University of Illinois. Mathematicians had long hunted for a short, elegant proof, but it turned out that the proof had to wait for the advent of computers to help sift through the many possible arrangements that can occur.

The Chromatic Number of a Graph

We mentioned previously that representing a map as a graph always results in a planar graph. The four-color theorem guarantees that we need only four colors to color a *planar* graph; however, if we wish to color a nonplanar graph, we may need more than four colors. The minimum number of colors needed to color a graph so that no edge connects vertices of the same color is called the **chromatic number** of the graph. In general, there is no efficient method of finding the chromatic number of a graph, but we do have a theorem that can tell us whether a graph is 2-colorable.

2-Colorable Graph Theorem

A graph is 2-colorable if and only if it has no circuits that consist of an odd number of vertices.

EXAMPLE 2 Determine the Chromatic Number of a Graph

Find the chromatic number of the Utilities Graph from Section 5.3.

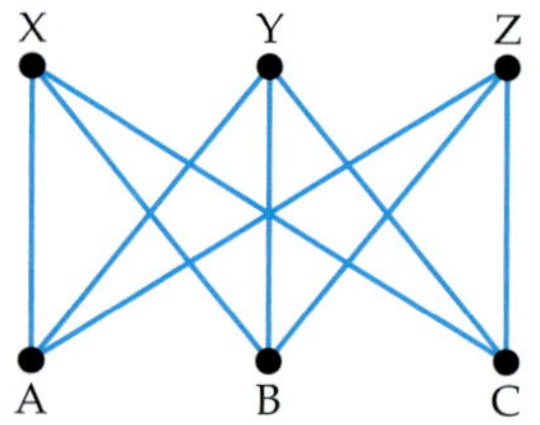

TAKE NOTE

The 2-coloring of the Utilities Graph described in Example 2 is shown below. Note that no edge connects vertices of the same color.

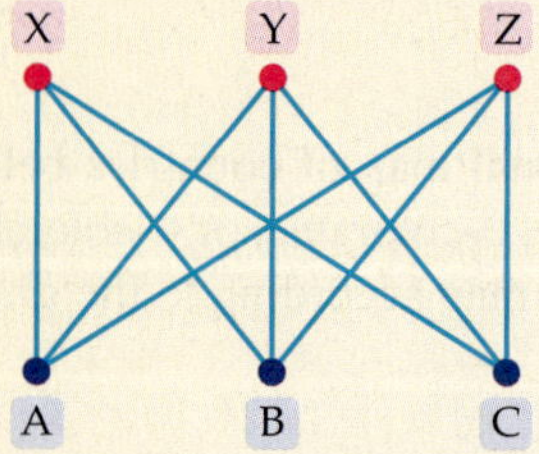

Solution

Note that the graph contains circuits such as A–Y–C–Z–B–X–A with six vertices and A–Y–B–X–A with four vertices. It seems that any circuit we find, in fact, involves an even number of vertices. It is difficult to determine whether we have looked at all possible circuits, but our observations suggest that the graph may be 2-colorable. A little trial and error confirms this if we simply color vertices A, B, and C one color and the remaining vertices another. Thus the Utilities Graph has a chromatic number of 2.

CHECK YOUR PROGRESS 2 Determine whether the following graph is 2-colorable.

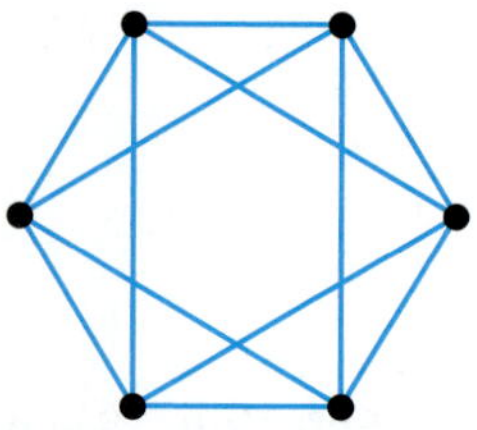

Solution See page S19.

Applications of Graph Coloring

Determining the chromatic number of a graph and finding a corresponding coloring of the graph can solve a wide assortment of practical problems. One common application is in scheduling meetings or events. This is best shown by example.

EXAMPLE 3 A Scheduling Application of Graph Coloring

Eight different school clubs want to schedule meetings on the last day of the semester. Some club members, however, belong to more than one of these clubs, so clubs that share members cannot meet at the same time. How many different time slots are required so that all members can attend all meetings? Clubs that have a member in common are indicated with an "X" in the table below.

	Ski club	Student government	Debate club	Honor society	Student newspaper	Community outreach	Campus Democrats	Campus Republicans
Ski club	—	X		X			X	X
Student government	X	—	X	X	X			
Debate club		X	—	X		X		X
Honor society	X	X	X	—	X	X		
Student newspaper		X		X	—	X	X	
Community outreach			X	X	X	—	X	X
Campus Democrats	X				X	X	—	
Campus Republicans	X		X			X		—

PPT

Solution

We can represent the given information by a graph. Each club is represented by a vertex, and an edge connects two vertices if the corresponding clubs have at least one common member.

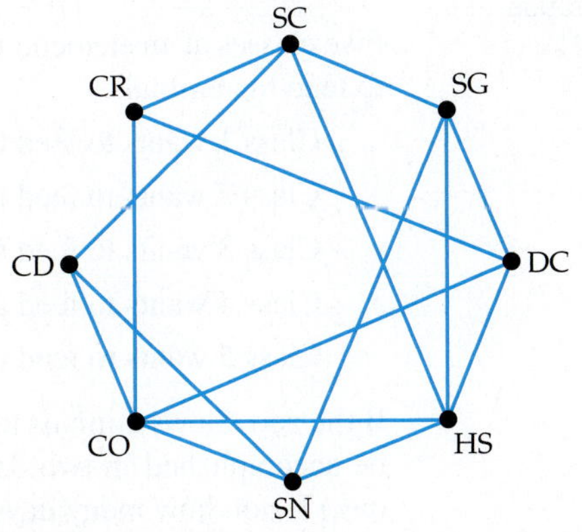

Two clubs that are connected by an edge cannot meet simultaneously. If we let a color correspond to a time slot, then we need to find a coloring of the graph that uses the fewest possible number of colors. The graph is not 2-colorable, because we can find circuits of odd length. However, by trial and error, we can find a 3-coloring. One example is shown below. Thus the chromatic number of the graph is 3, so we need three different time slots.

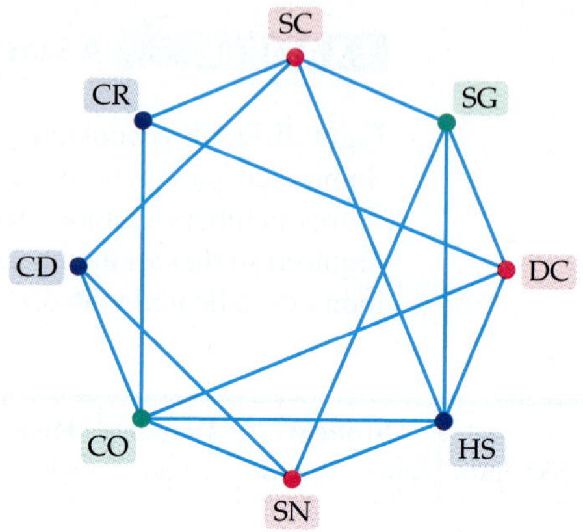

Each color corresponds to a time slot, so one scheduling is

First time slot: ski club, debate club, student newspaper

Second time slot: student government, community outreach

Third time slot: honor society, campus Democrats, campus Republicans

CHECK YOUR PROGRESS 3 Six film students have collaborated on the creation of five films.

Film A was produced by Brian, Chris, and Damon.

Film B was produced by Allison and Fernando.

Film C was produced by Damon, Erin, and Fernando.

Film D was produced by Brian and Erin.

Film E was produced by Brian, Chris, and Erin.

The college is scheduling a one-day film festival where each film will be shown once and the producers of each film will attend and participate in a discussion afterward. The college has several screening rooms available and two hours will be allotted for each film. If the showings begin at noon, create a screening schedule that allows the festival to end as early as possible while assuring that all of the producers of each film can attend that film's screening.

Solution See page S19.

EXAMPLE 4 A Scheduling Application of Graph Coloring

Five classes at an elementary school have arranged a tour at a zoo where the students get to feed the animals.

Class 1 wants to feed the elephants, giraffes, and hippos.

Class 2 wants to feed the monkeys, rhinos, and elephants.

Class 3 wants to feed the monkeys, deer, and sea lions.

Class 4 wants to feed the parrots, giraffes, and polar bears.

Class 5 wants to feed the sea lions, hippos, and polar bears.

If the zoo allows animals to be fed only once a day by one class of students, can the tour be accomplished in two days? (Assume that each class will visit the zoo only on one day.) If not, how many days will be required?

POINT OF INTEREST

Sudoku puzzles like the one shown here have become extremely popular. To complete the puzzle, one must fill each square of the grid with a digit from 1 to 9 so that each row, each column, and each 3×3 subgrid contains the digits 1 through 9 (and so no digit is repeated).

5	3			7				
6			1	9	5			
	9	8					6	
8				6				3
4			8		3			1
7				2				6
	6					2	8	
			4	1	9			5
				8			7	9

Two mathematicians, Agnes Herzberg and M. Ram Murty, used graph theory to analyze the puzzles. We can represent each of the 81 squares by a vertex; connect two vertices with an edge if the squares are in the same row, column, or subgrid. Then associate each of the numbers 1 through 9 with a different color. A solution to the puzzle corresponds to a coloring of the graph using nine colors where no edge connects vertices of the same color.

The researchers used graph theory to prove that at least eight of the nine digits must appear at the start of the puzzle in order for it to have a unique solution. It is still not known how many of the squares must be revealed initially to have a unique solution, but they found an example with only 17 given entries. They also showed that there are about 5.5 billion possible Sudoku puzzles!

Solution

No animal is listed more than twice in the tour list, so you may be tempted to say that only two days will be required. However, to get a better picture of the problem, we can represent the situation with a graph. Use a vertex to represent each class, and connect two vertices with an edge if the corresponding classes want to feed the same animal. Then we can try to find a 2-coloring of the graph, where a different color represents a different day at the zoo.

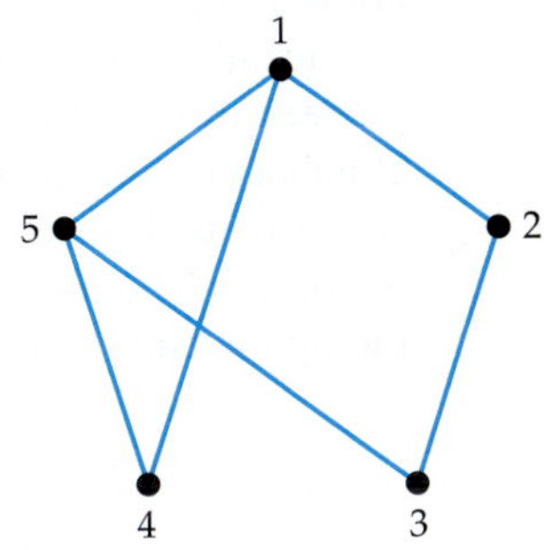

Note that the graph contains a circuit, 1–4–5–1, consisting of three vertices. This circuit will require three colors, and the remaining vertices will not require additional colors. So the chromatic number of the graph is 3; one possible coloring is given below. Using this coloring, three days are required at the zoo. On the first day classes 2 and 5, represented by the blue vertices, will visit the zoo; on the second day classes 1 and 3, represented by the red vertices, will visit the zoo; and on the third day class 4, represented by the green vertex, will visit the zoo.

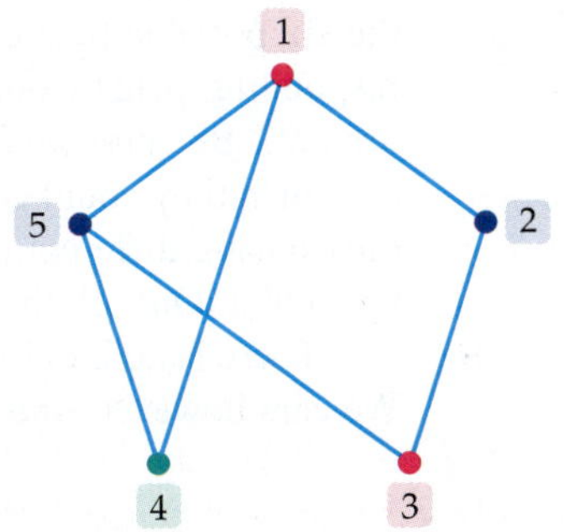

CHECK YOUR PROGRESS 4 Several delis in New York City have arranged deliveries to various buildings at lunchtime. The buildings' managements do not want more than one deli showing up at a particular building in one day, but the delis would like to deliver as often as possible. If they decide to agree on a delivery schedule, how many days will be required before each deli can return to the same building?

Deli A delivers to the Empire State Building, the Statue of Liberty, and Rockefeller Center.

Deli B delivers to the Chrysler Building, the Empire State Building, and the New York Stock Exchange.

Deli C delivers to the New York Stock Exchange, the American Stock Exchange, and the United Nations Building.

Deli D delivers to New York City Hall, the Chrysler Building, and Rockefeller Center.

Deli E delivers to Rockefeller Center, New York City Hall, and the United Nations Building.

Solution See page S19.

EXCURSION

Modeling Traffic Lights with Graphs

Have you ever watched the cycles that traffic lights go through while you were waiting for a red light to turn green? Some intersections have lights that go through several stages to allow all the different lanes of traffic to proceed safely.

Ideally, each stage of a traffic-light cycle should allow as many lanes of traffic to proceed through the intersection as possible. We can design a traffic-light cycle by modeling an intersection with a graph. Figure 5.26 shows a three-way intersection where two two-way roads meet. Each direction of traffic has turn lanes, with left-turn lights where possible. There are six different directions in which vehicles can travel, as indicated in the figure, and we have labeled each possibility with a letter.

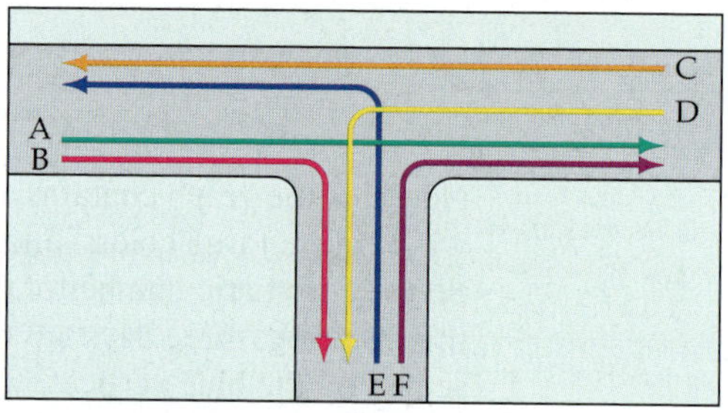

FIGURE 5.26

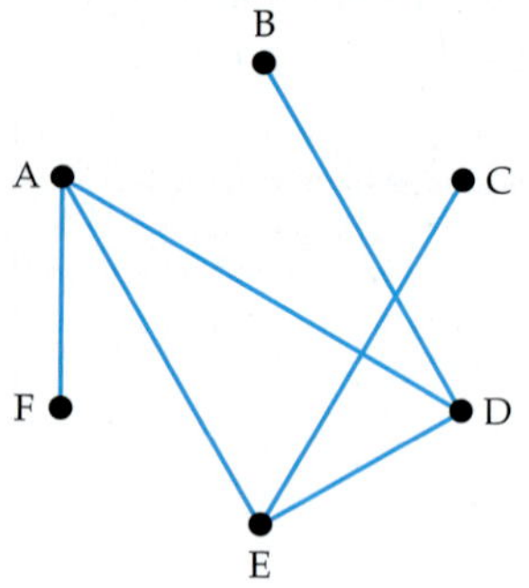

FIGURE 5.27

We can represent the traffic patterns with a graph; each vertex will represent one of the six possible traffic paths, and we will draw an edge between two vertices if the corresponding paths would allow vehicles to collide. The result is the graph shown in Figure 5.27. Because we do not want to allow vehicles to travel simultaneously along routes on which they could collide, any vertices connected by an edge can allow traffic to move only during different parts of the light cycle. We can represent each portion of the cycle by a color. Our job then is to color the graph using the fewest colors possible.

There is no 2-coloring of the graph because we have a circuit of length 3: A–D–E–A. We can, however, find a 3-coloring. One possibility is given in Figure 5.28.

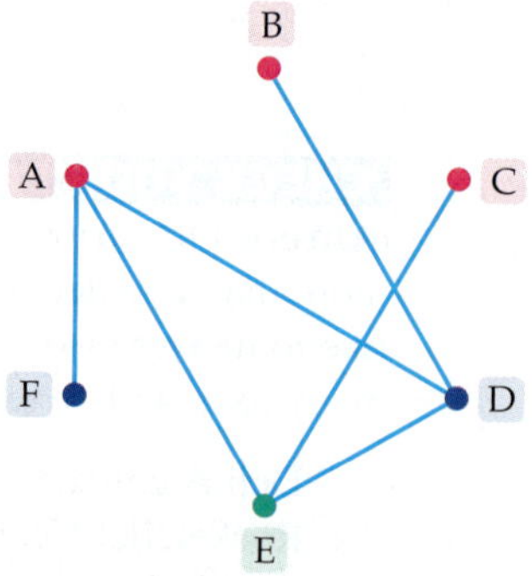

FIGURE 5.28

A 3-coloring of the graph means that the traffic lights at the intersection will have to go through a three-stage cycle. One stage will allow the traffic routes corresponding to the red vertices to proceed, the next stage will let the paths corresponding to the blue vertices to proceed, and finally, the third stage will let path E, colored green, proceed.

Although safety requires three stages for the lights, we can refine the design to allow more traffic to travel through the intersection. Note that at the third stage, only one route, path E, is scheduled to be moving. However, there is no harm in allowing path B to move

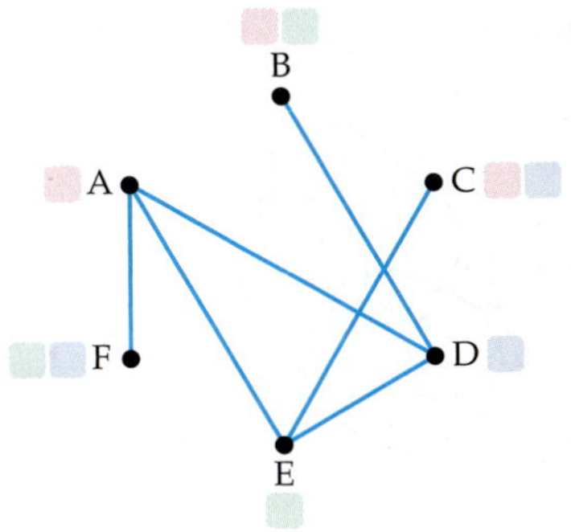

FIGURE 5.29

at the same time, since it is a right turn that doesn't conflict with route E. We could also allow path F to proceed at the same time. Adding these additional paths corresponds to adding colors to the graph in Figure 5.28. We do not want to use more than three colors, but we can add a second color to some of the vertices while maintaining the requirement that no edge can connect two vertices of the same color. The result is shown in Figure 5.29. Notice that the vertices in the triangular circuit A–D–E–A can be assigned only a single color, but the remaining vertices can accommodate two colors.

In summary, our design allows traffic paths A, B, and C to proceed during one stage of the cycle, paths C, D, and F during another, and paths B, E, and F during the third stage.

EXCURSION EXERCISES

1. A one-way road ends at a two-way street. The intersection and the different possible traffic routes are shown in the figure below. The one-way road has a left-turn light. Represent the traffic routes with a graph and use graph coloring to determine the minimum number of stages required for a light cycle.

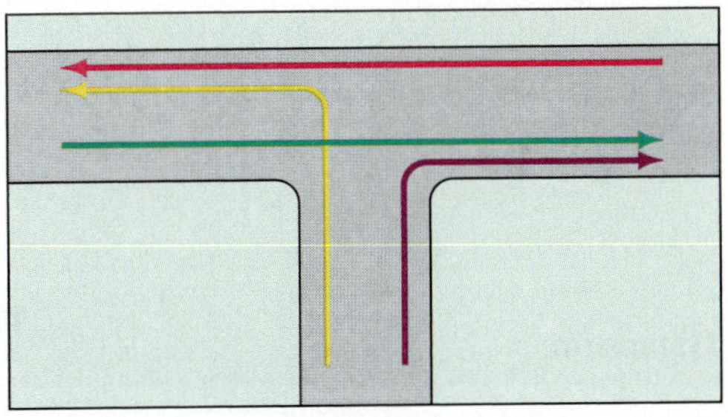

2. A one-way road intersects a two-way road in a four-way intersection. Each direction has turn lanes and left-turn lights. Represent the various traffic routes with a graph and use graph coloring to determine the minimum number of stages required for a light cycle. Then refine your design to allow as much traffic as possible to proceed at each stage of the cycle.
3. A two-way road intersects another two-way road in a four-way intersection. One road has left-turn lanes with left-turn lights, but on the other road cars are not allowed to make left turns. Represent the various traffic routes with a graph and use graph coloring to determine the minimum number of stages required for a light cycle. Then refine your design to allow as much traffic as possible to proceed at each stage of the cycle.

EXERCISE SET 5.4

■ **Map Coloring** In Exercises 1 to 4, a fictional map of the countries of a continent is given. Represent the map by a graph and find a coloring of the graph that uses the fewest possible number of colors. Then color the map according to the graph coloring you found.

1.

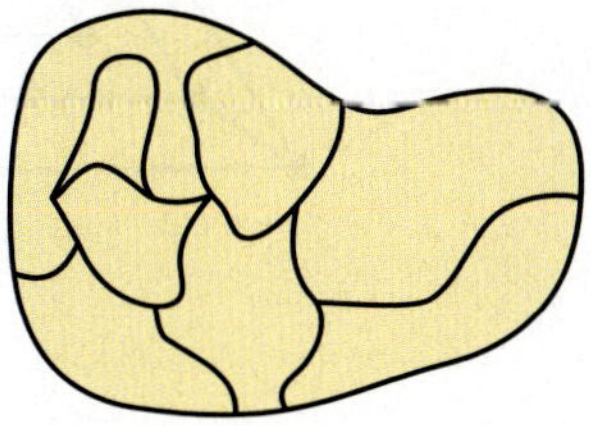

2.

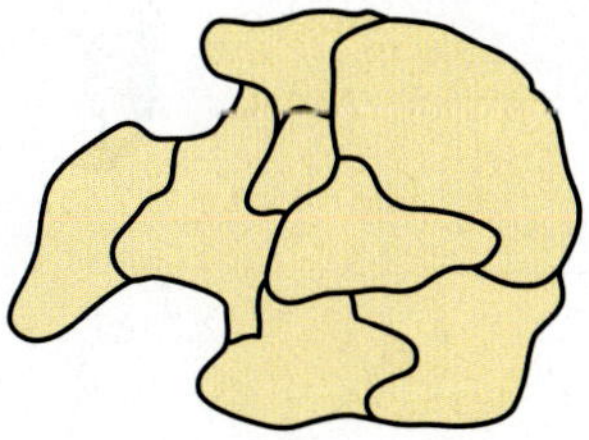

3.

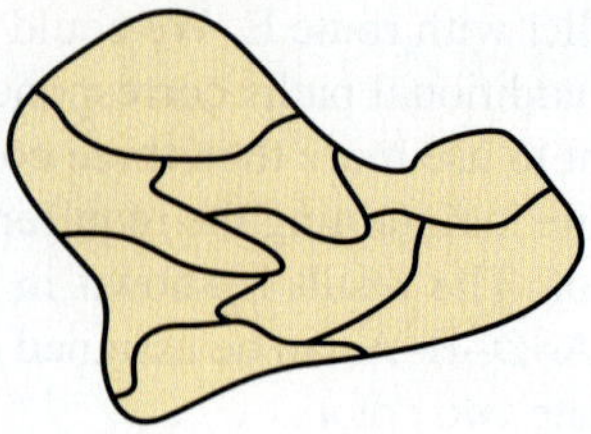

4.

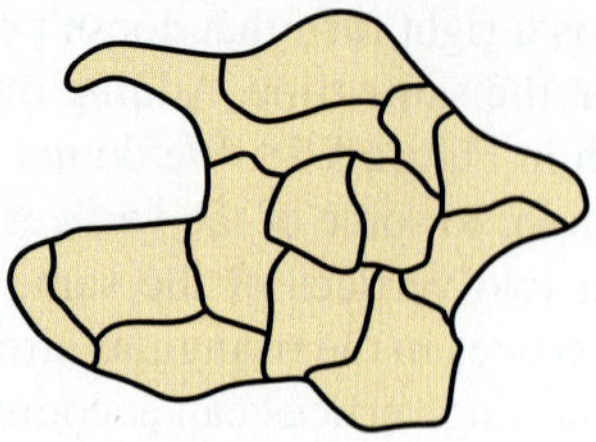

■ **Map Coloring** In Exercises 5 to 8, represent the map by a graph and find a coloring of the graph that uses the smallest possible number of colors.

5. Western portion of the United States

6. Counties of New Hampshire

7. Countries of South America

8. Provinces of South Africa

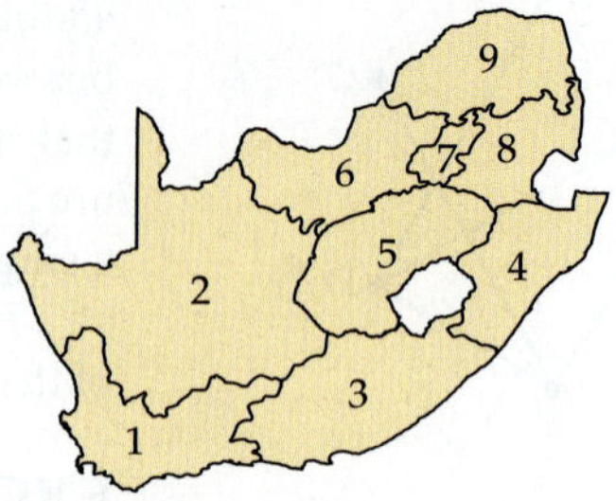

■ In Exercises 9 to 14, show that the graph is 2-colorable by finding a 2-coloring. If the graph is not 2-colorable, explain why.

9.

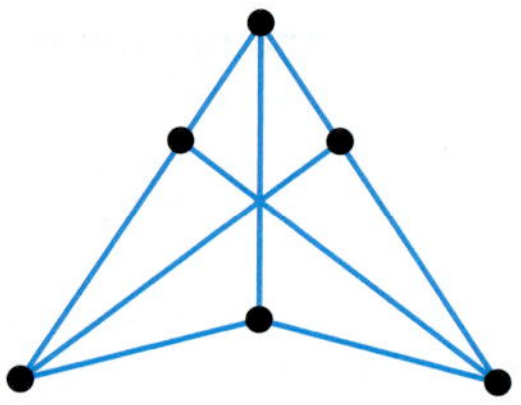

10.

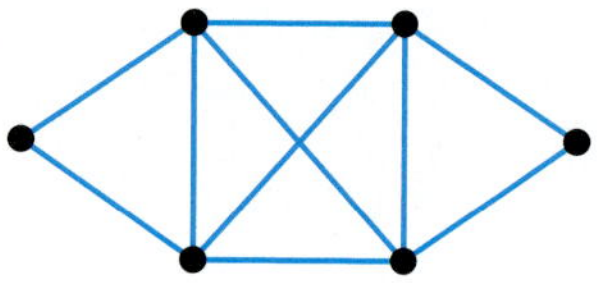

11.

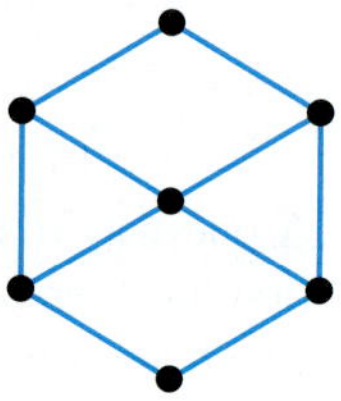

12.

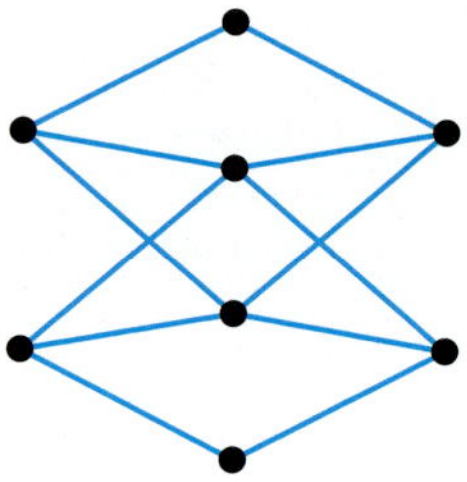

13.

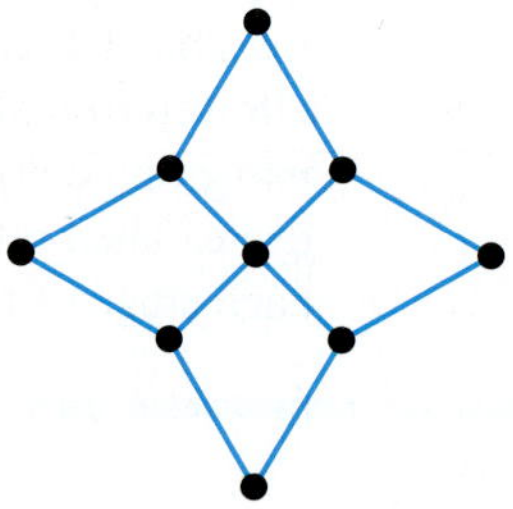

14.

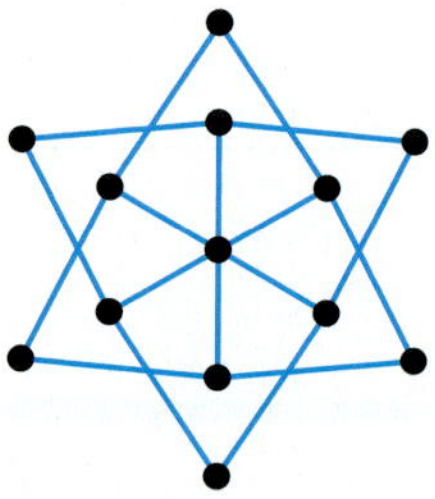

■ In Exercises 15 to 20, determine (by trial and error) the chromatic number of the graph.

15.

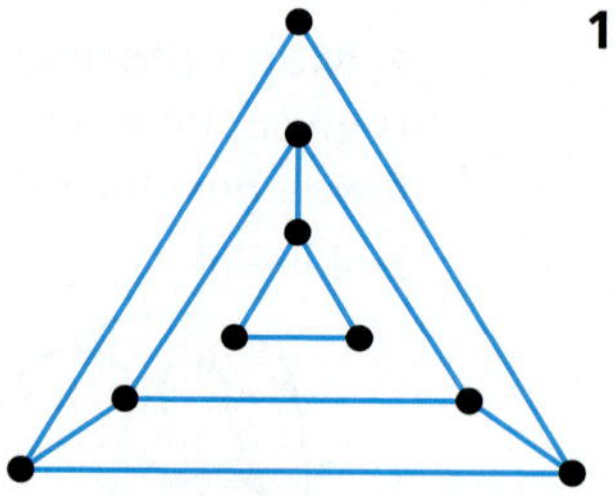

16.

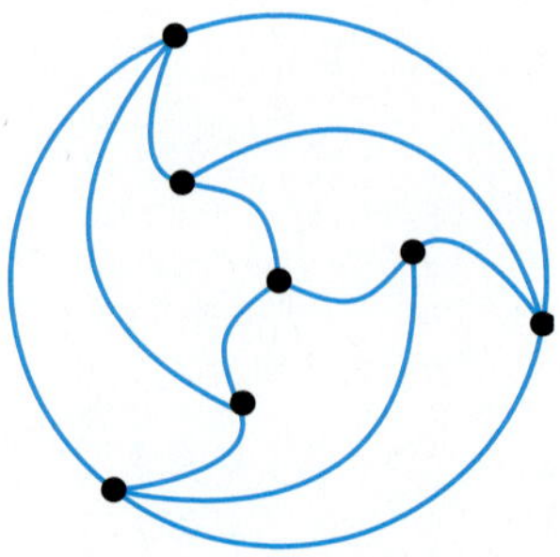

17.

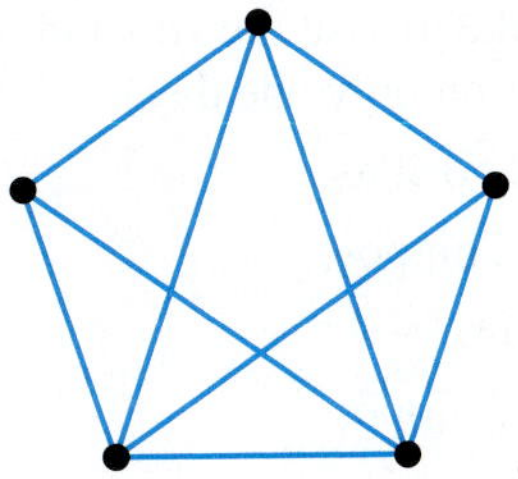

18.

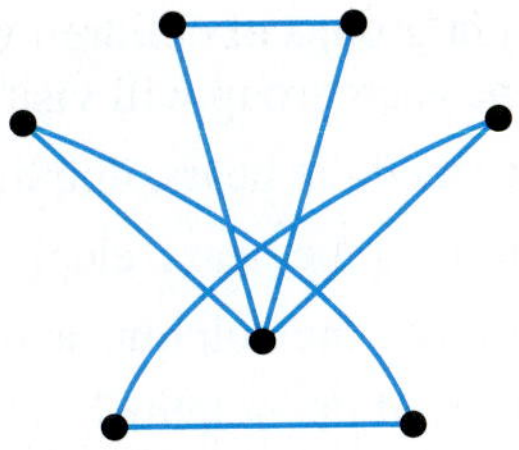

19.

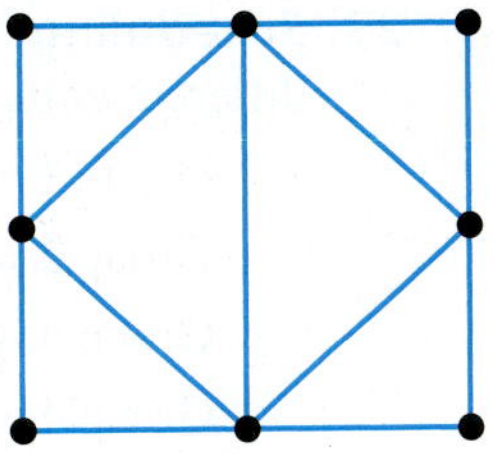

20.

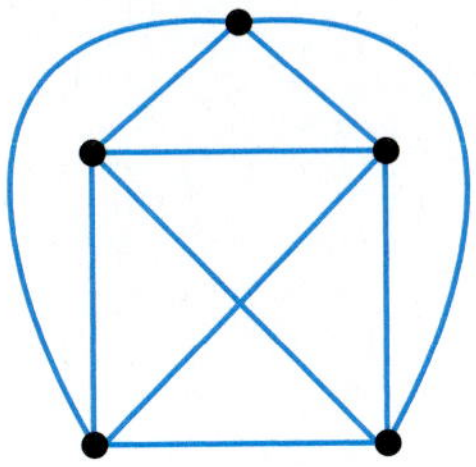

21. Scheduling Six student clubs need to hold meetings on the same day, but some students belong to more than one club. In order to avoid members missing meetings, the meetings need to be scheduled during different time slots. An "X" in the table below indicates that the two corresponding clubs share at least one member. Use graph coloring to determine the minimum number of time slots necessary to ensure that all club members can attend all meetings.

	Student newspaper	Honor society	Biology association	Gaming club	Debate team	Engineering club
Student newspaper	—		X		X	
Honor society		—	X		X	X
Biology association	X	X	—	X		
Gaming club			X	—	X	X
Debate team	X	X		X	—	
Engineering club		X		X		—

22. Scheduling Eight political committees must meet on the same day, but some members are on more than one committee. Thus any committees that have members in common cannot meet at the same time. An "X" in the following table indicates that the two corresponding committees share a member. Use graph coloring to determine the minimum number of meeting times that will be necessary so that all members can attend the appropriate meetings.

	Appropriations	Budget	Finance	Judiciary	Education	Health	Foreign affairs	Housing
Appropriations	—		X			X	X	
Budget		—		X		X		
Finance	X		—	X			X	X
Judiciary		X	X	—		X		X
Education					—		X	X
Health	X	X		X		—		
Foreign affairs	X		X		X		—	
Housing			X	X	X			—

23. Scheduling Six different groups of children would like to visit the zoo and feed different animals. (Assume each group will visit the zoo on only one day.)

Group 1 would like to feed the bears, dolphins, and gorillas.

Group 2 would like to feed the bears, elephants, and hippos.

Group 3 would like to feed the dolphins and elephants.

Group 4 would like to feed the dolphins, zebras, and hippos.

Group 5 would like to feed the bears and hippos.

Group 6 would like to feed the gorillas, hippos, and zebras.

Use graph coloring to find the minimum number of days that are required so that all groups can feed the animals they would like to feed but no animals will be fed twice on the same day. Design a schedule to accomplish this goal.

24. Scheduling Five different charity organizations send trucks on various routes to pick up donations that residents leave on their doorsteps.

Charity A covers Main St., First Ave., and State St.

Charity B covers First Ave., Second Ave., and Third Ave.

Charity C covers State St., City Dr., and Country Lane.

Charity D covers City Dr., Second Ave., and Main St.

Charity E covers Third Ave., Country Lane, and Fourth Ave.

Each charity has its truck travel down all three streets on its route on the same day, but no two charities wish to visit the same streets on the same day. Use graph coloring to design a schedule for the charities. Arrange their pickup routes so that no street is visited twice on the same day by different charities. The schedule should use the smallest possible number of days.

Creatas/PictureQuest

25. Scheduling Students in a film class have volunteered to form groups and create several short films. The class has three digital video cameras that may be checked out for one day only, and it is expected that each group will need the entire day to finish shooting. All members of each group must participate in the film they volunteered for, so a student cannot work on more than one film on any given day.

Film 1 will be made by Brian, Angela, and Kate.

Film 2 will be made by Jessica, Vince, and Brian.

Film 3 will be made by Corey, Brian, and Vince.

Film 4 will be made by Ricardo, Sarah, and Lupe.

Film 5 will be made by Sarah, Kate, and Jessica.

Film 6 will be made by Angela, Corey, and Lupe.

Use graph coloring to design a schedule for lending the cameras, using the smallest possible number of days, so that each group can shoot its film and all members can participate.

26. Animal Housing A researcher has discovered six new species of insects overseas and needs to transport them home. Some species will harm each other and so cannot be transported in the same container.

Species A cannot be housed with species C or F.

Species B cannot be housed with species D or F.

Species C cannot be housed with species A, D, or E.

Species D cannot be housed with species B, C, or F.

Species E cannot be housed with species C or F.

Species F cannot be housed with species A, B, D, or E.

Draw a graph where each vertex represents a species of insect and an edge connects two vertices if the species cannot be housed together. Then use graph coloring to determine the minimum number of containers the researcher will need to transport the insects.

EXTENSIONS

27. **Wi-Fi Stations** An office building is installing eight Wi-Fi transmitting stations. Any stations within 200 feet of each other must transmit on different channels. The engineers have a chart of the distance between each pair of stations. Suppose that they draw a graph where each vertex represents a Wi-Fi station and an edge connects two vertices if the distance between the stations is 200 feet or less. What would the chromatic number of the graph tell the engineers?

28. **Map Coloring** Draw a map of a fictional continent consisting of four countries, where the map cannot be colored with three or fewer colors without adjacent countries sharing a color.

29. If the chromatic number of a graph with five vertices is 1, what must the graph look like?

30. **Edge Coloring** In this section, we colored vertices of graphs so that no edge connected two vertices of the same color. We can also consider coloring edges, rather than vertices, so that no vertex connects two or more edges of the same color. In parts a to d, assign each edge in the graph a color so that no vertex connects two or more edges of the same color. Use the fewest number of colors possible.

a.

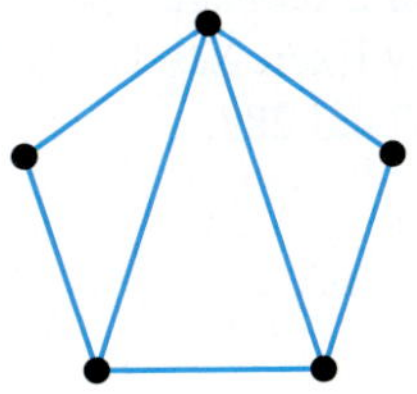

b.

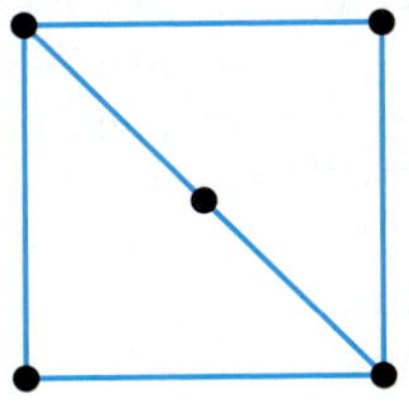

c.

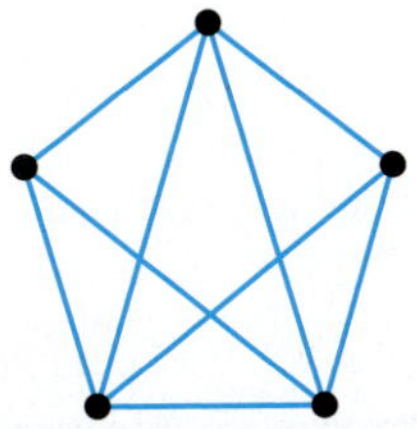

d.

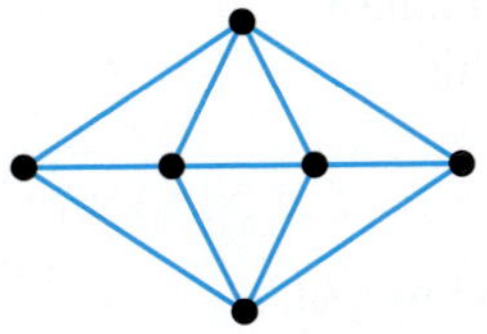

e. Explain why the number of colors required will always be at least the number of edges that meet at the vertex of highest degree in the graph.

31. **Scheduling** Edge colorings, as explained in Exercise 30, can be used to solve scheduling problems. For instance, suppose five players are competing in a tennis tournament. Each player needs to play every other player in a match (but not more than once). Each player will participate in no more than one match per day, and two matches can occur at the same time when possible. How many days will be required for the tournament? Represent the tournament as a graph, in which each vertex corresponds to a player and an edge joins two vertices if the corresponding players will compete against each other in a match. Next, color the edges, where each different color corresponds to a different day of the tournament. Because one player will not be in more than one match per day, no two edges of the same color can meet at the same vertex. If we can find an edge coloring of the graph that uses the fewest number of colors possible, it will correspond to the fewest number of days required for the tournament. Sketch a graph that represents the tournament, find an edge coloring using the fewest number of colors possible, and use your graph to design a schedule of matches for the tournament that minimizes the number of days required.

CHAPTER 5 SUMMARY

The following table summarizes essential concepts in this chapter. The references given in the right-hand column list Examples and Exercises that can be used to test your understanding of a concept.

5.1 Graphs and Euler Circuits	
Graphs A graph is a set of points called vertices and line segments or curves called edges that connect vertices. A graph is a connected graph provided any vertex can be reached from any other vertex by tracing along edges. Two graphs are equivalent graphs provided the edges form the same connections of vertices in each graph. The degree of a given vertex is equal to the number of edges that meet at the vertex.	See **Examples 1 and 2** on pages 230 and 232, and then try Exercises 1, 2, 5, and 6 on pages 285 and 286.

continued

Euler Circuits An Euler circuit is a path that uses every edge but does not use any edge more than once, and begins and ends at the same vertex. **Eulerian Graph Theorem** A connected graph is Eulerian (has an Euler circuit) if and only if every vertex of the graph is of even degree.	See **Examples 3 to 5** on pages 235 and 236, and then try part b of Exercises 7 to 10 and Exercise 11 on page 286.
Euler Paths An Euler path is a path that uses every edge but does not use any edge more than once. **Euler Path Theorem** A connected graph contains an Euler path if and only if the graph has two vertices of odd degree with all other vertices of even degree. Furthermore, every Euler path must start at one of the vertices of odd degree and end at the other.	See **Examples 6 and 7** on page 238, and then try part a of Exercises 7 to 10 on page 286.
5.2 Weighted Graphs	
Hamiltonian Circuits A Hamiltonian circuit is a path in a graph that uses each vertex exactly once and returns to the starting vertex. If a graph has a Hamiltonian circuit, the graph is said to be Hamiltonian. **Dirac's Theorem** Dirac's theorem states that in a connected graph with at least three vertices and with no multiple edges, if n is the number of vertices in the graph and every vertex has degree of at least $n/2$, then the graph must be Hamiltonian. If it is not the case that every vertex has degree of at least $n/2$, then the graph may or may not be Hamiltonian.	See **Example 1** on page 245, and then try Exercises 13 to 15 on page 287.
Weighted Graphs A weighted graph is a graph in which each edge is associated with a value, called a weight. **The Greedy Algorithm** A method of finding a Hamiltonian circuit in a complete weighted graph is given by the following greedy algorithm. **1.** Choose a vertex to start at, then travel along the connected edge that has the smallest weight. (If two or more edges have the same weight, pick any one.) **2.** After arriving at the next vertex, travel along the edge of smallest weight that connects to a vertex not yet visited. Continue this process until you have visited all vertices. **3.** Return to the starting vertex. The greedy algorithm attempts to give a circuit of minimal total weight, although it does not always succeed.	See **Examples 2 and 3** on pages 247 and 249, and then try Exercises 17, 18, and 21 on pages 287 and 288.
The Edge-Picking Algorithm Another method of finding a Hamiltonian circuit in a complete weighted graph is given by the following edge-picking algorithm. **1.** Mark the edge of smallest weight in the graph. (If two or more edges have the same weight, pick any one.) **2.** Mark the edge of next smallest weight in the graph, as long as it does not complete a circuit and does not add a third marked edge to a single vertex. **3.** Continue this process until you can no longer mark any edges. Then mark the final edge that completes the Hamiltonian circuit. The edge-picking algorithm attempts to give a circuit of minimal total weight, although it does not always succeed.	See **Examples 4 to 6** on pages 250 to 254, and then try Exercises 19, 20, and 22 on pages 287 and 288.

<table>
<tr><th colspan="2">5.3 Planarity and Euler's Formula</th></tr>
<tr><td>Planar Graph A planar graph is a graph that can be drawn so that no edges intersect each other (except at vertices).</td><td>See Example 1 on page 262, and then try Exercises 23 and 24 on page 288.</td></tr>
<tr><td>Subgraphs A subgraph of a graph is a graph whose edges and vertices come from the given graph.

The Nonplanar Graph Theorem A graph is nonplanar if and only if it has the Utilities Graph or K_5 as a subgraph, or if it has a subgraph that can be contracted to the Utilities Graph or K_5.

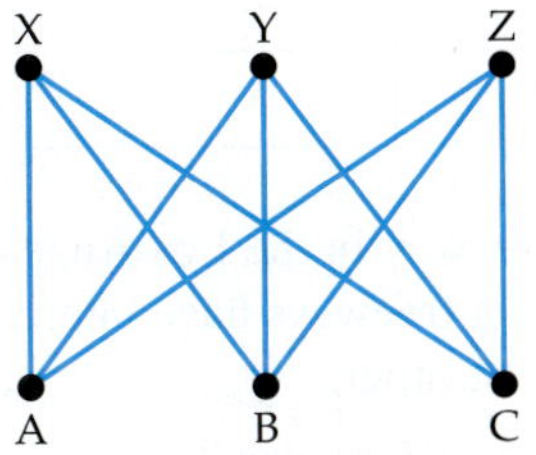

Utilities Graph

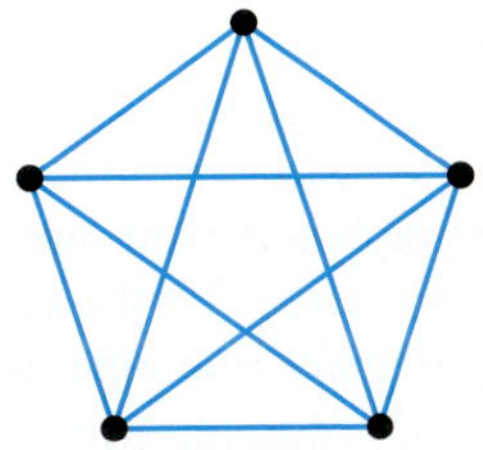
K_5 Graph</td><td>See Examples 2 and 4 on pages 263 and 265, and then try Exercises 25 and 26 on page 288.</td></tr>
<tr><td>Euler's Formula In a connected planar graph drawn with no intersecting edges, let v be the number of vertices, e the number of edges, and f the number of faces. Then $v + f = e + 2$.</td><td>See Example 5 on page 266, and then try Exercises 27 and 28 on page 288.</td></tr>
<tr><th colspan="2">5.4 Graph Coloring</th></tr>
<tr><td>Representing Maps as Graphs Draw a vertex in each region (country, state, etc.) of the map. Connect two vertices if the corresponding regions share a common border.

The Four-Color Theorem Every planar graph is 4-colorable. (In some cases less than four colors may be required. Also, if the graph is not planar, more than four colors may be necessary.)</td><td>See Example 1 on page 272, and then try Exercises 29 and 30 on page 288.</td></tr>
<tr><td>2-Colorable Graph Theorem A graph is 2-colorable if and only if it has no circuits that consist of an odd number of vertices.</td><td>See Example 2 on page 274, and then try Exercises 31 and 32 on page 289.</td></tr>
<tr><td>Applications of Graph Coloring Determining the chromatic number of a graph and finding a corresponding coloring of the graph can solve some practical applications such as scheduling meetings or events.</td><td>See Examples 3 and 4 on pages 275 and 276, and then try Exercises 33 to 35 on page 289.</td></tr>
</table>

CHAPTER 5 REVIEW EXERCISES

■ In Exercises 1 and 2, (a) determine the number of edges in the graph, (b) find the number of vertices in the graph, (c) list the degree of each vertex, and (d) determine whether the graph is connected.

1.

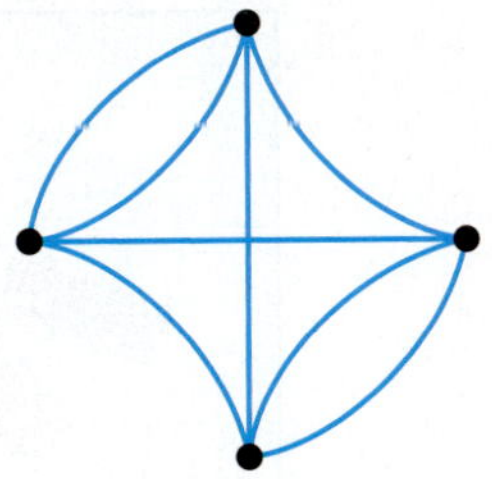

2.

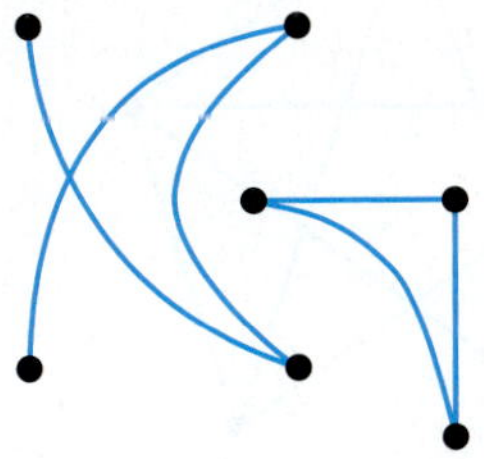

3. Soccer In the table below, an "X" indicates teams from a junior soccer league that have played each other in the current season. Draw a graph to represent the games by using a vertex to represent each team. Connect two vertices with an edge if the corresponding teams played a game against each other this season.

	Mariners	Scorpions	Pumas	Stingrays	Vipers
Mariners	—	X		X	X
Scorpions	X	—	X		
Pumas		X	—	X	X
Stingrays	X		X	—	
Vipers	X		X		—

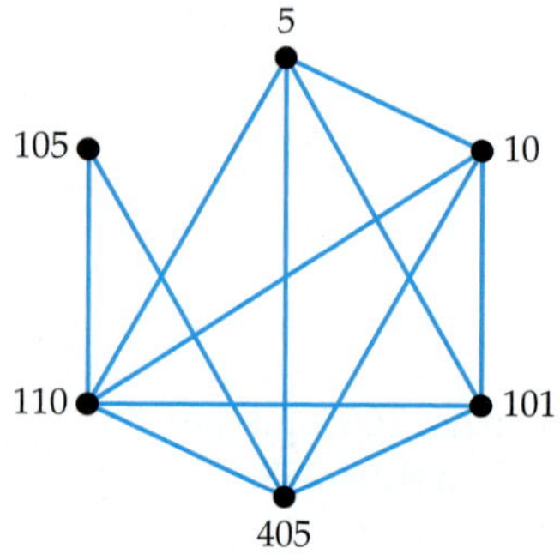

4. Each vertex in the graph at the left represents a freeway in the Los Angeles area. An edge connects two vertices if the corresponding freeways have interchanges allowing drivers to transfer from one freeway to the other.

a. Can drivers transfer from the 105 freeway to the 10 freeway?

b. Among the freeways represented, how many have interchanges with the 5 freeway?

c. Which freeways have interchanges to all the other freeways in the graph?

d. Of the freeways represented in the graph, which has the fewest interchanges to the other freeways?

■ In Exercises 5 and 6, determine whether the two graphs are equivalent.

5.

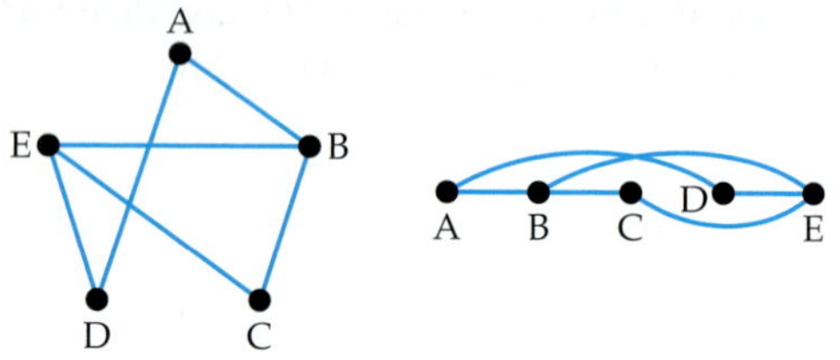

6.

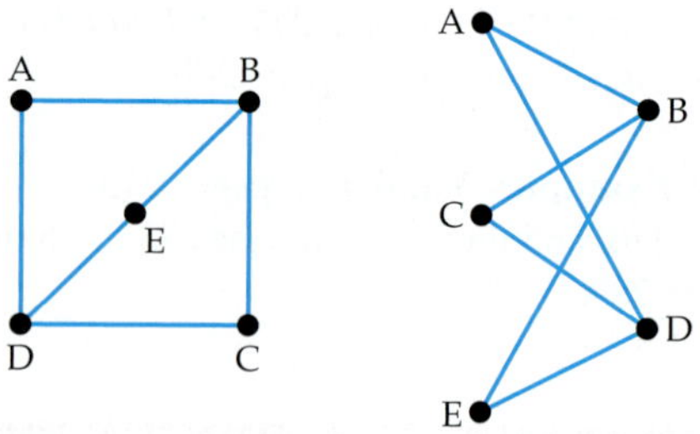

■ In Exercises 7 to 10, (a) find an Euler path if possible, and (b) find an Euler circuit if possible.

7.

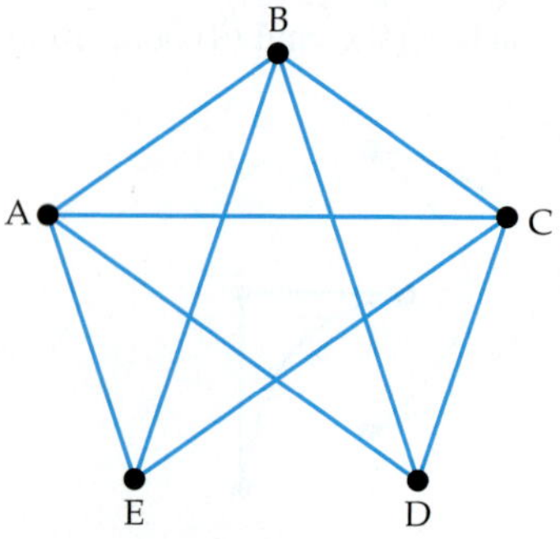

8.

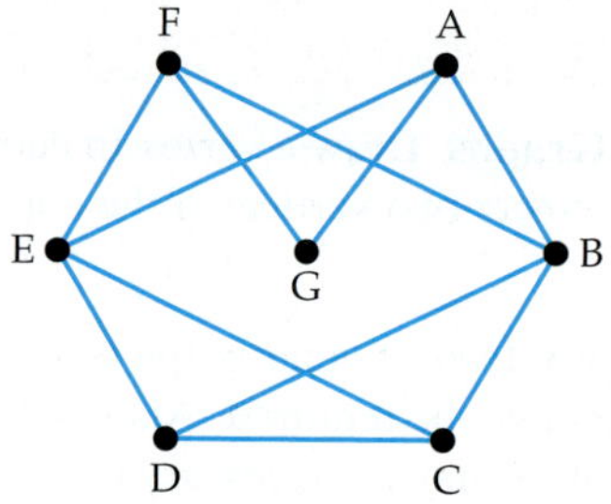

9.

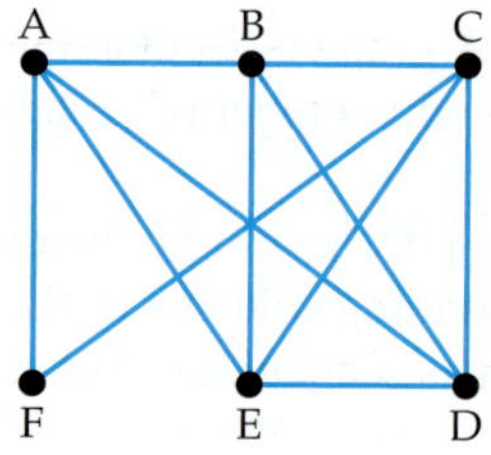

10.

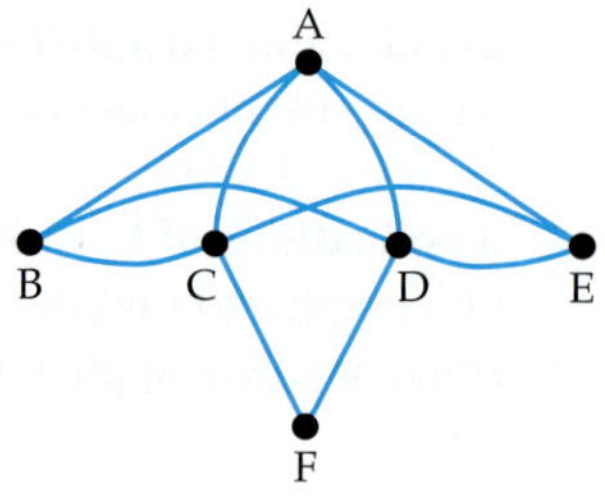

11. Parks The figure shows an arrangement of bridges connecting land areas in a park. Represent the map as a graph, and then determine whether it is possible to stroll across each bridge exactly once and return to the starting position.

12. Architecture The floor plan of a sculpture gallery is shown below. Is it possible to walk through each doorway exactly once? Is it possible to walk through each doorway exactly once and return to the starting point?

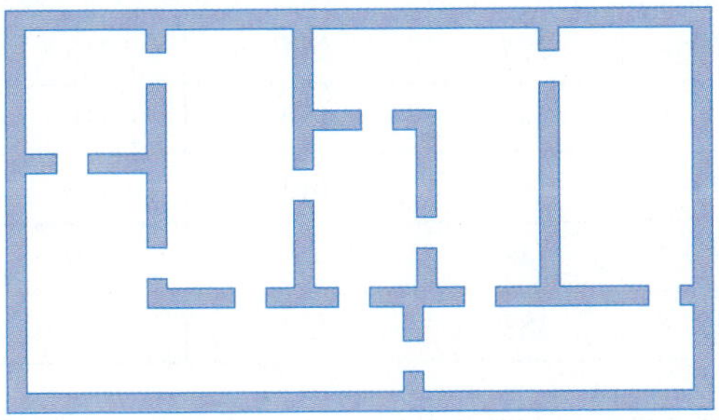

■ In Exercises 13 and 14, use Dirac's theorem to verify that the graph is Hamiltonian, and then find a Hamiltonian circuit.

13.

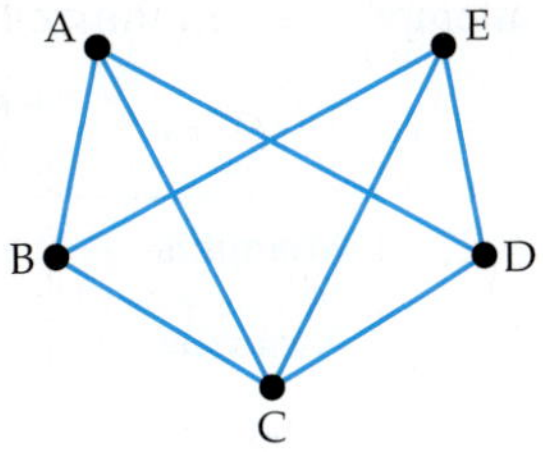

14.

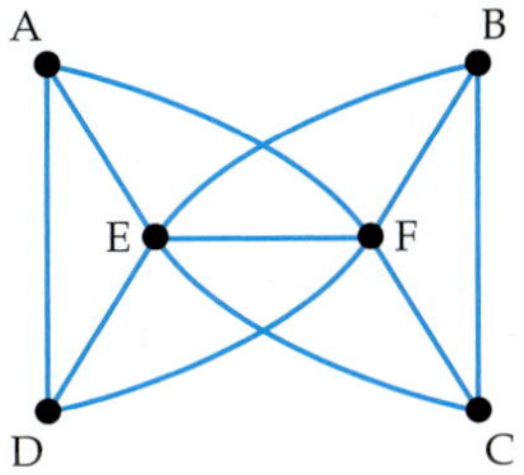

15. Travel The table below lists cities serviced by a small airline. An "X" in the table indicates a direct flight offered by the airline. Draw a graph that represents the direct flights, and use your graph to find a route that visits each city exactly once and returns to the starting city.

	Casper	Rapid City	Minneapolis	Des Moines	Topeka	Omaha	Boulder
Casper	—	X					X
Rapid City	X	—	X				
Minneapolis		X	—	X	X		X
Des Moines			X	—	X		
Topeka			X	X	—	X	X
Omaha					X	—	X
Boulder	X		X		X	X	—

16. Travel For the direct flights given in Exercise 15, find a route that travels each flight exactly once and returns to the starting city. (You may visit cities more than once.)

■ In Exercises 17 and 18, use the greedy algorithm to find a Hamiltonian circuit starting at vertex A in the weighted graph.

17.

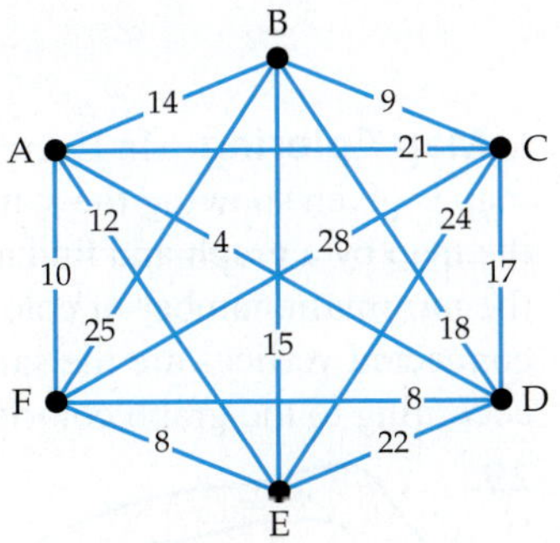

18.

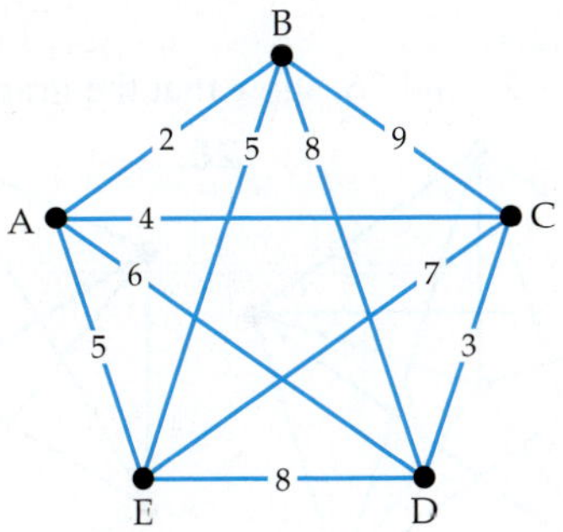

■ In Exercises 19 and 20, use the edge-picking algorithm to find a Hamiltonian circuit starting at vertex A in the weighted graph.

19. The graph in Exercise 17

20. The graph in Exercise 18

21. Efficient Route The distances, in miles, between five different cities are given in the table. Sketch a weighted graph that represents the distances, and then use the greedy algorithm to design a route that starts in Memphis, visits each city, and returns to Memphis while attempting to minimize the total distance traveled.

	Memphis	Nashville	Atlanta	Birmingham	Jackson
Memphis	—	210	394	247	213
Nashville	210	—	244	189	418
Atlanta	394	244	—	148	383
Birmingham	247	189	148	—	239
Jackson	213	418	383	239	—

22. Computer Networking A small office needs to network five computers by connecting one computer to another and forming a large loop. The length of cable needed (in feet) between pairs of machines is given in the table. Use the edge-picking algorithm to design a method to network the computers while attempting to use the smallest possible amount of cable.

	Computer A	Computer B	Computer C	Computer D	Computer E
Computer A	—	85	40	55	20
Computer B	85	—	35	40	18
Computer C	40	35	—	60	50
Computer D	55	40	60	—	30
Computer E	20	18	50	30	—

■ In Exercises 23 and 24, show that the graph is planar by finding a planar drawing.

23.

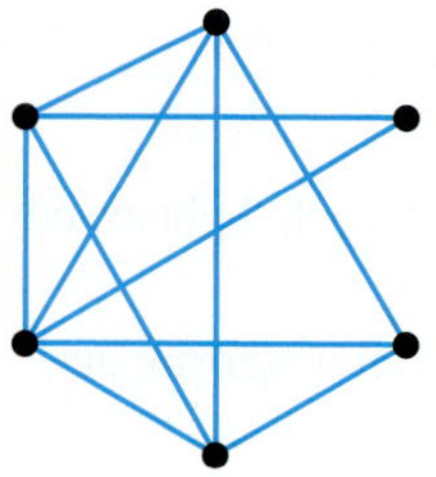

24.

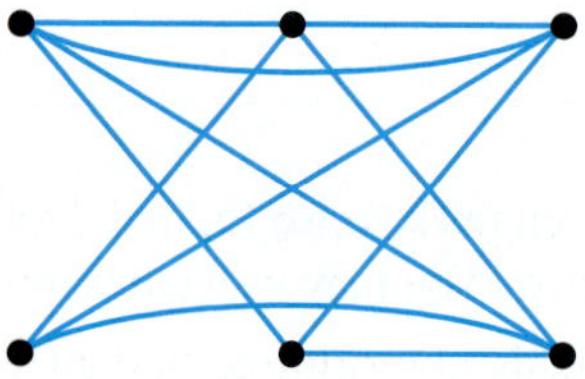

■ In Exercises 25 and 26, show that the graph is *not* planar.

25.

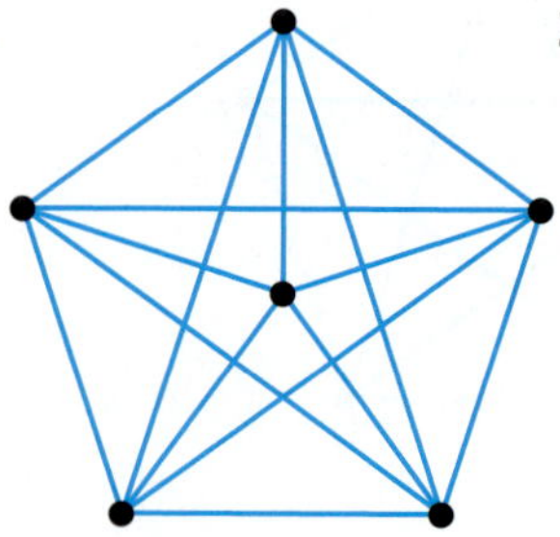

26.

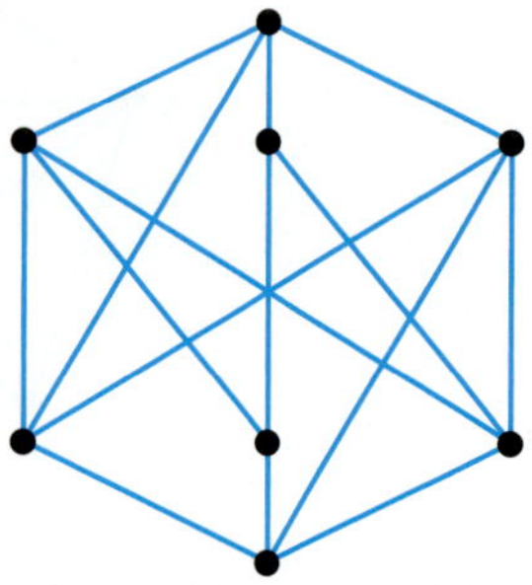

■ In Exercises 27 and 28, count the number of vertices, edges, and faces in the graph, and then verify Euler's formula.

27.

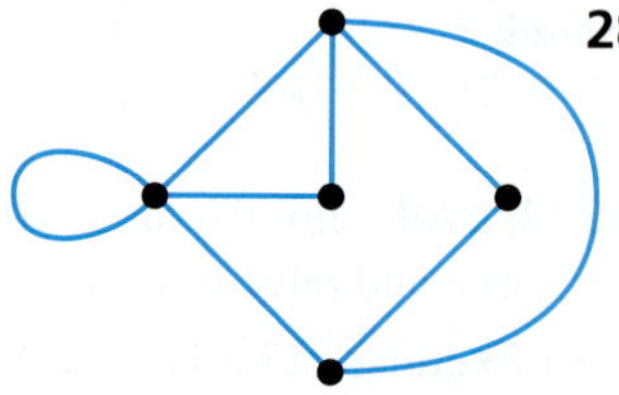

28.

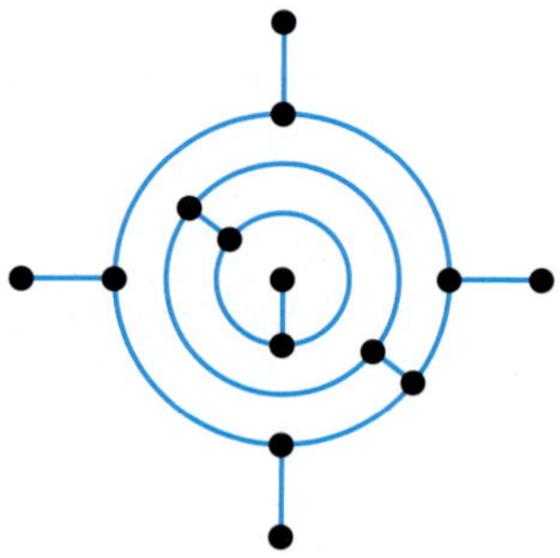

■ **Map Coloring** In Exercises 29 and 30, a fictional map is given showing the states of a country. Represent the map by a graph and find a coloring of the graph, using the minimum number of colors possible, such that no two connected vertices are the same color. Then color the map according to the graph coloring that you found.

29.

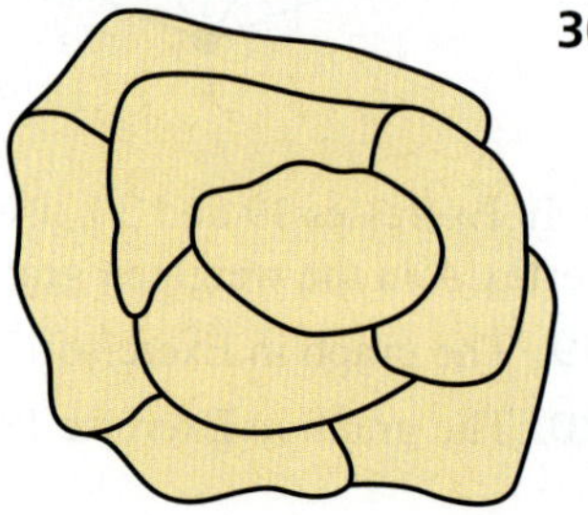

30.

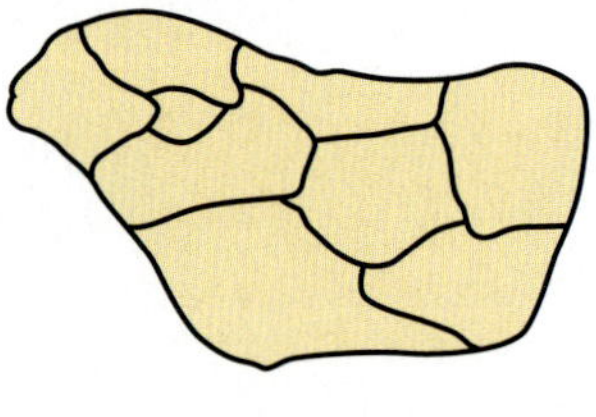

■ In Exercises 31 and 32, show that the graph is 2-colorable by finding a 2-coloring, or explain why the graph is not 2-colorable.

31.

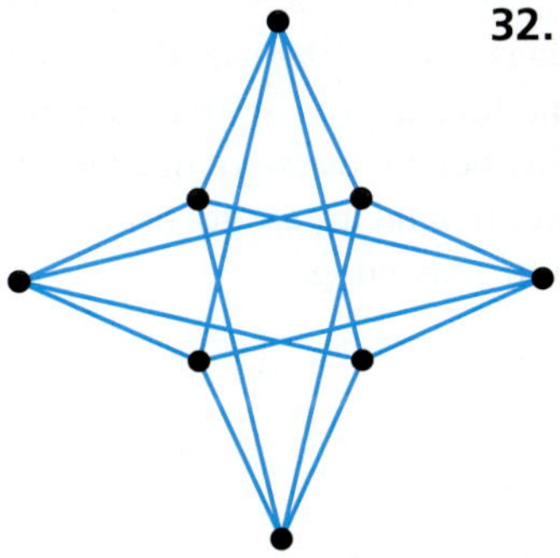

32.

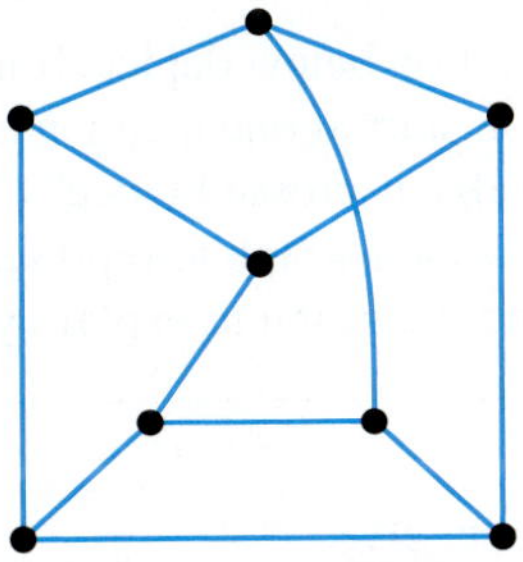

■ In Exercises 33 and 34, determine (by trial and error) the chromatic number of the graph.

33.

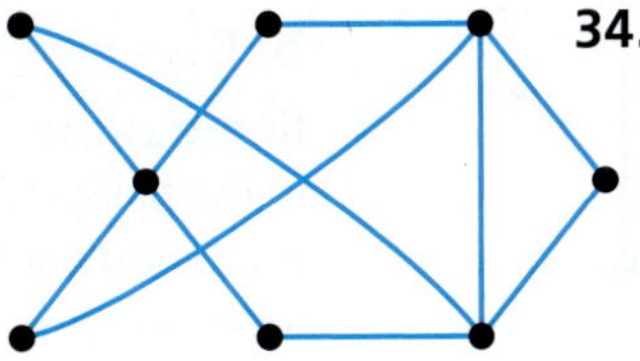

34.

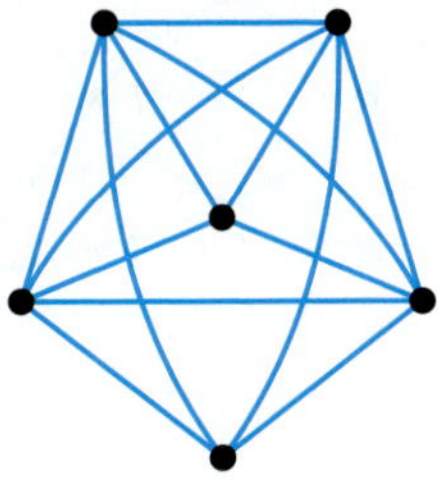

35. Scheduling A company has scheduled a retreat at a hotel resort. It needs to hold meetings the first day there, and several conference rooms are available. Some employees must attend more than one meeting, however, so the meetings cannot all be scheduled at the same time. An "X" in the table below indicates that at least one employee must attend both of the corresponding meetings, and so those two meetings must be held at different times. Draw a graph in which each vertex represents a meeting, and an edge joins two vertices if the corresponding meetings require the attendance of the same employee. Then use graph coloring to design a meeting schedule that uses the minimum number of time slots.

	Budget meeting	Marketing meeting	Executive meeting	Sales meeting	Research meeting	Planning meeting
Budget meeting	—	X	X		X	
Marketing meeting	X	—		X		
Executive meeting	X		—		X	X
Sales meeting		X		—		X
Research meeting	X		X		—	X
Planning meeting			X	X	X	—

CHAPTER 5 TEST

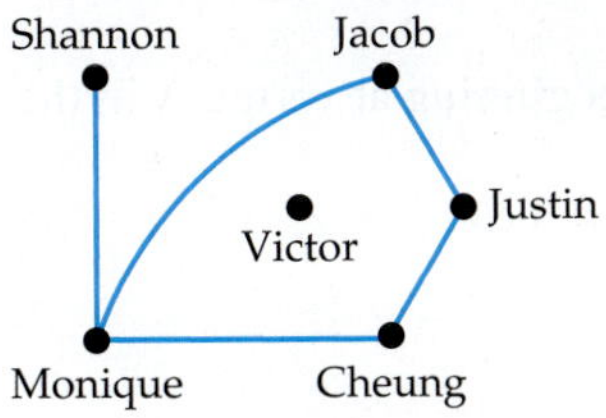

1. Social Network Each vertex in the graph at the left represents a student. An edge connects two vertices if the corresponding students have at least one class in common during the current term.

a. Is there a class that both Jacob and Cheung are enrolled in?

b. Who shares the most classes among the members of this group of students?

c. How many classes does Victor have in common with the other students?

d. Is this graph connected?

2. Determine whether the following two graphs are equivalent. Explain your reasoning.

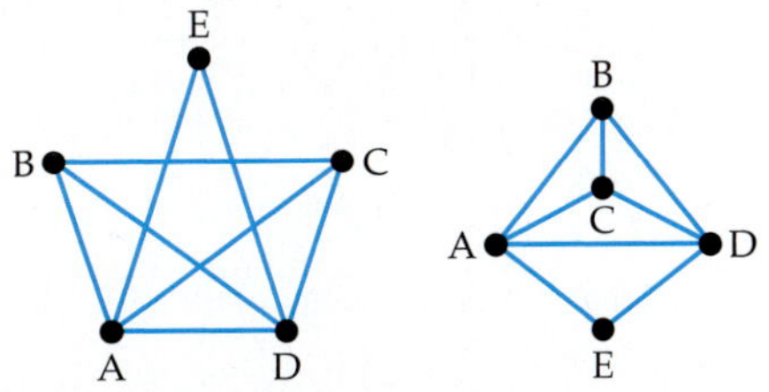

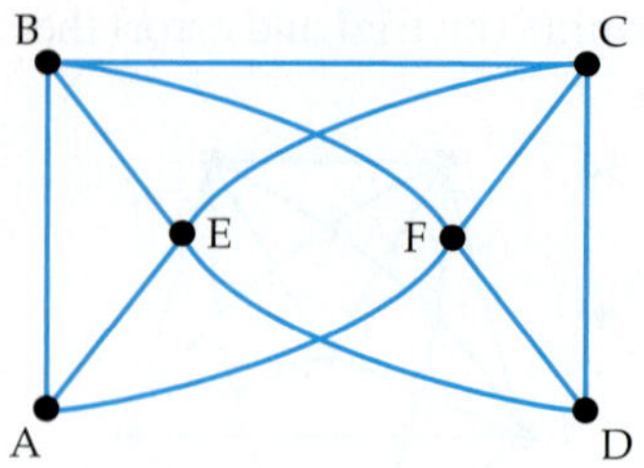

3. Answer the following questions for the graph shown at the left.

a. Is the graph Eulerian? If so, find an Euler circuit. If not, explain how you know.

b. Does the graph have an Euler path? If so, find one. If not, explain how you know.

4. Recreation The illustration below depicts bridges connecting islands in a river to the banks. Can a person start at one location and take a stroll so that each bridge is crossed once but no bridge is crossed twice? (The person does not need to return to the starting location.) Draw a graph to represent the land areas and bridges. Answer the question using the graph to explain your reasoning.

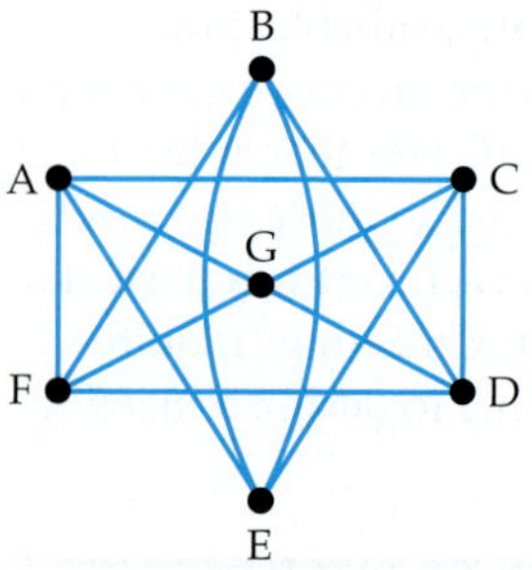

5. a. What does Dirac's theorem state? Explain how it guarantees that the graph at the left is Hamiltonian.

b. Find a Hamiltonian circuit in the graph at the left.

6. Low-Cost Route The table below shows the cost of direct train travel between various cities.

a. Draw a weighted graph that represents the train fares.

b. Use the edge-picking algorithm to find a route that begins in Angora, visits each city, and returns to Angora. What is the total cost of this route?

c. Is it possible to travel along each train route and return to the starting city without traveling any routes more than once? Explain how you know.

	Angora	Bancroft	Chester	Davenport	Elmwood
Angora	—	\$48	\$52	\$36	\$90
Bancroft	\$48	—	\$32	\$42	\$98
Chester	\$52	\$32	—	\$76	\$84
Davenport	\$36	\$42	\$76	—	\$106
Elmwood	\$90	\$98	\$84	\$106	—

7. Use the greedy algorithm to find a Hamiltonian circuit beginning at vertex A in the weighted graph shown.

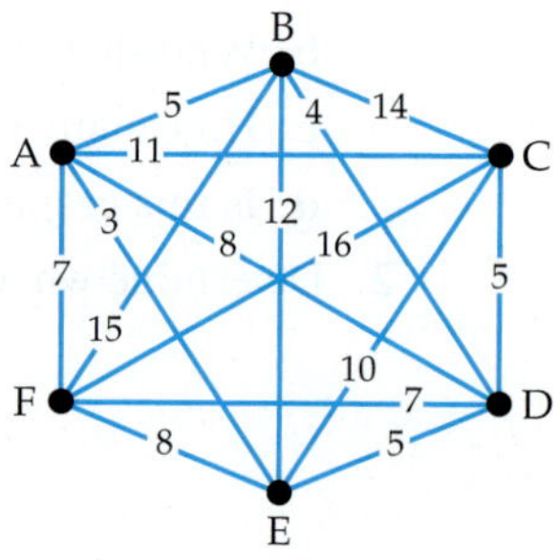

8. **a.** Sketch a planar drawing of the graph below.

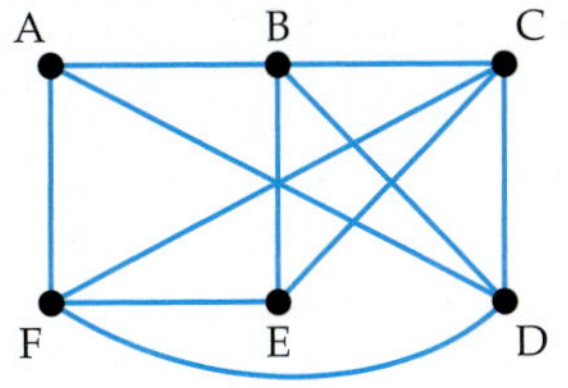

b. Show that the graph below is not planar.

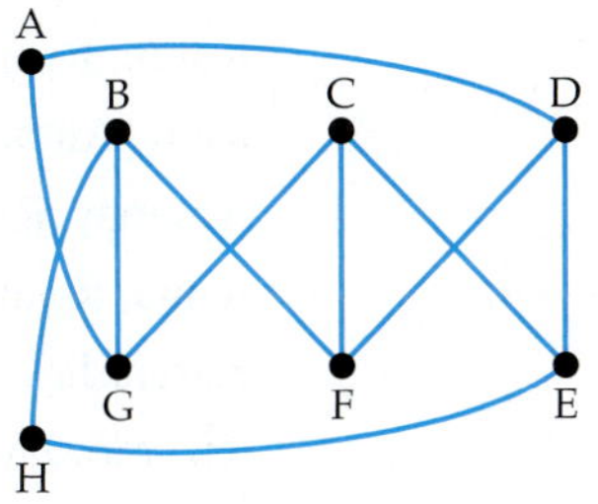

9. Answer the following questions for the graph shown below.

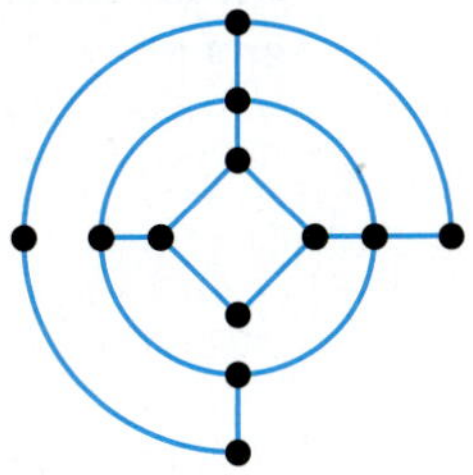

a. How many vertices are of degree 3?

b. How many faces does the graph consist of (including the infinite face)?

c. Show that Euler's formula is satisfied by this graph.

10. **Map Coloring** A fictional map of the countries of a continent is shown below.

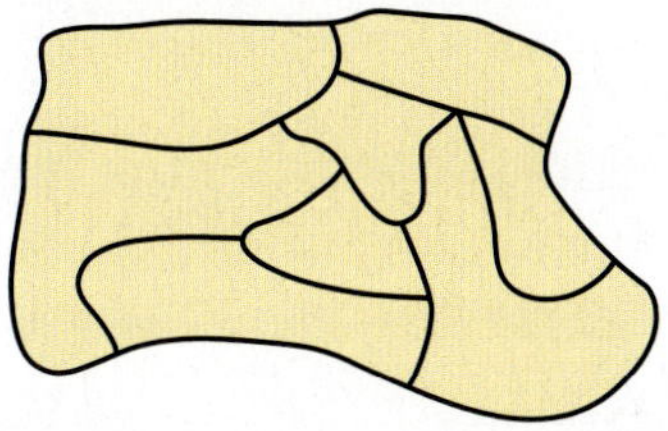

a. Represent the map by a graph in which the vertices correspond to countries and the edges indicate which countries share a common border.

b. What is the chromatic number of your graph?

c. Use your work from part b to color the countries of the map with the fewest colors possible so that no two neighboring countries share the same color.

11. For the graph shown below, find a 2-coloring of the vertices or explain why a 2-coloring is impossible.

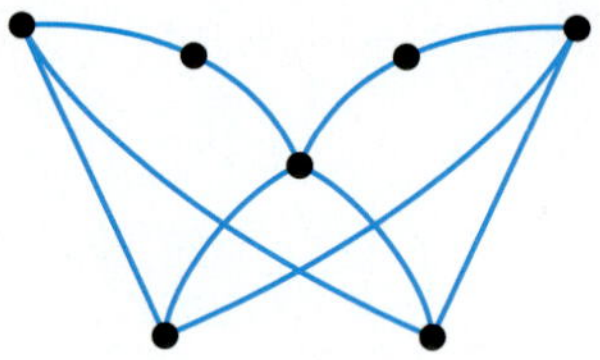

12. A group of eight friends is planning a vacation in Las Vegas, where they will split into different groups each evening to see various shows.

Karen, Ryan, and Ruby want to see Cirque du Soleil together.

Ruby, Anthony, and Jay want to see a magic show together.

Anthony, Ricardo, and Heather want to see a comedy show together.

Jenna, Ryan, and Ricardo want to see a tribute band play in concert together.

Karen, Jay, and Jenna want to see a musical together.

Ricardo, Jay, and Heather want to see a play together.

Draw a graph in which each vertex represents one of the shows, and connect vertices with an edge if at least one person wants to see both corresponding shows. Then use graph coloring to determine the fewest evenings needed so that all the friends can see the shows they would like to see, and design a schedule for the group.

6

Numeration Systems and Number Theory

We start this chapter with an examination of several numeration systems. A working knowledge of these numeration systems will enable you to better understand and appreciate the advantages of our current Hindu-Arabic numeration system.

The last two sections of this chapter cover prime numbers and topics from the field of number theory. Many of the concepts in number theory are easy to comprehend but difficult, or impossible, to prove. The mathematician Karl Friedrich Gauss (1777–1855) remarked that "it is just this which gives the higher arithmetic (number theory) that magical charm which has made it the favorite science of the greatest mathematicians, not to mention its inexhaustible wealth, wherein it so greatly surpasses other parts of mathematics." Gauss referred to mathematics as "the queen of the sciences," and he considered the field of number theory "the queen of mathematics."

There are many unsolved problems in the field of number theory. One unsolved problem, dating from the year 1742, is *Goldbach's conjecture,* which states that every even number greater than 2 can be written as the sum of two prime numbers. This conjecture has yet to be proved or disproved, despite the efforts of the world's best mathematicians. The British publishing company Faber and Faber offered a $1 million prize to anyone who could provide a proof or disproof of Goldbach's conjecture between March 20, 2000, and March 20, 2002, but the prize went unclaimed and Goldbach's conjecture remains a conjecture. The company had hoped that the prize money would entice young, mathematically talented people to work on the problem. This scenario is similar to the story line in the movie *Good Will Hunting,* in which a mathematics problem posted on a bulletin board attracts the attention of a yet-to-be-discovered math genius, played by Matt Damon.

SECTION 6.1 Early Numeration Systems

The Egyptian Numeration System

In mathematics, symbols that are used to represent numbers are called **numerals.** A number can be represented by many different numerals. For instance, the concept of "eightness" is represented by each of the following.

Hindu-Arabic: 8	Tally: 𝍸 |||	Roman: VIII
Chinese: 八	Egyptian: 𓏺𓏺𓏺𓏺𓏺𓏺𓏺𓏺	Babylonian: 𒁹𒁹𒁹𒁹𒁹𒁹𒁹𒁹

A **numeration system** consists of a set of numerals and a method of arranging the numerals to represent numbers. The numeration system that most people use today is known as the *Hindu-Arabic numeration system.* It makes use of the 10 numerals 0, 1, 2, 3, 4, 5, 6, 7, 8, and 9. Before we examine the Hindu-Arabic numeration system, it will be helpful to study the numeration systems developed by the Egyptians and the Romans.

The Egyptian numeration system uses pictorial symbols called **hieroglyphics** as numerals. The Egyptian hieroglyphic system is an **additive system** because any given number is written by using numerals whose sum equals the number. Table 6.1 gives the Egyptian hieroglyphics for powers of 10 from 1 to 1 million.

TABLE 6.1
Egyptian Hieroglyphics for Powers of 10

Hindu-Arabic numeral	Egyptian hieroglyphic	Description of hieroglyphic
1	𓏺	stroke
10	𓎆	heel bone
100	𓍢	scroll
1000	𓆼	lotus flower
10,000	𓂭	pointing finger
100,000	𓆐	fish
1,000,000	𓁨	astonished person

To write the number 300, the Egyptians wrote the scroll hieroglyphic three times: 𓍢𓍢𓍢. In the Egyptian hieroglyphic system, the order of the hieroglyphics is of no importance. Each of the following four Egyptian numerals represents 321.

𓍢𓍢𓍢𓎆𓎆𓏺, 𓎆𓎆𓍢𓏺𓍢𓍢, 𓍢𓏺𓎆𓍢𓎆𓍢, 𓏺𓎆𓎆𓍢𓍢𓍢

EXAMPLE 1 Evaluate a Numeral Written Using Egyptian Hieroglyphics

Write 𓆐𓆐𓆼𓆼𓆼𓎆𓎆𓎆𓎆𓏺𓏺 as a Hindu-Arabic numeral.

Solution

Use Table 6.1 to determine the numerical value of each hieroglyphic. Then add these values to determine the Hindu-Arabic numeral.

$$100{,}000 + 100{,}000 + 1000 + 1000 + 1000 + 10 + 10 + 10 + 10 + 1 + 1 = 203{,}042$$

CHECK YOUR PROGRESS 1 Write 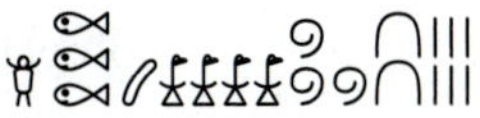 as a Hindu-Arabic numeral.

Solution See page S19.

QUESTION Do the Egyptian hieroglyphics 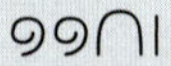 and 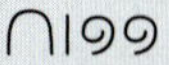 represent the same number?

EXAMPLE 2 Write a Numeral Using Egyptian Hieroglyphics

Write 2351 using Egyptian hieroglyphics.

Solution

First rewrite the Hindu-Arabic numeral as a sum of powers of ten.

$$\begin{aligned} 2351 &= 2000 + 300 + 50 + 1 \\ &= 1000 + 1000 + 100 + 100 + 100 + 10 + 10 + 10 + 10 + 10 + 1 \end{aligned}$$

Now replace each power of 10 with the Egyptian hieroglyphic, from Table 6.1, that represents the given power of ten. Omit the plus signs. Thus the Egyptian hieroglyphic for 2351 is

CHECK YOUR PROGRESS 2 Write 201,473 using Egyptian hieroglyphics.

Solution See page S19.

HISTORICAL NOTE

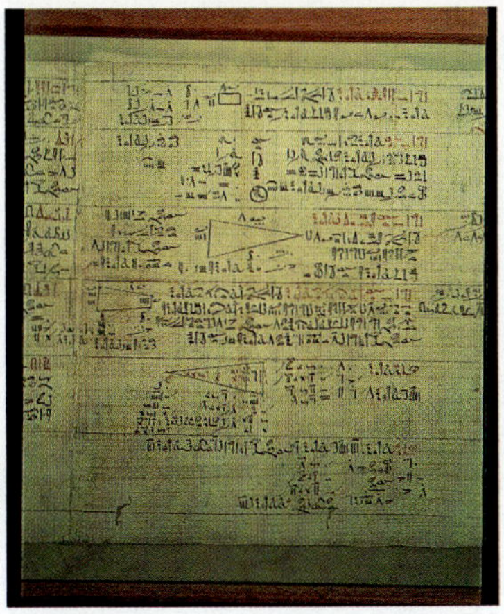

Art Resources, NY

A portion of the Rhind papyrus

The Rhind papyrus is named after Alexander Henry Rhind, who purchased the papyrus in Egypt in AD 1858. Today the Rhind papyrus is preserved in the British Museum in London.

One of the earliest written documents of mathematics is the Rhind papyrus (see the figure at the left). This tablet was found in Egypt in AD 1858, but it is estimated that the writings date back to 1650 BC. The Rhind papyrus contains 85 mathematical problems. Studying these problems has enabled mathematicians and historians to understand some of the mathematical procedures used in the early Egyptian numeration system.

The operation of addition with Egyptian hieroglyphics is a simple grouping process. In some cases the final sum can be simplified by replacing a group of hieroglyphics by a single hieroglyphic with an equivalent numerical value. This technique is illustrated in Example 3.

EXAMPLE 3 Use Egyptian Hieroglyphics to Find a Sum

Use Egyptian hieroglyphics to find 2452 + 1263.

Solution

The sum is found by combining the hieroglyphics.

$$\begin{array}{r} 2452 \\ +\,1263 \\ \hline \end{array}$$

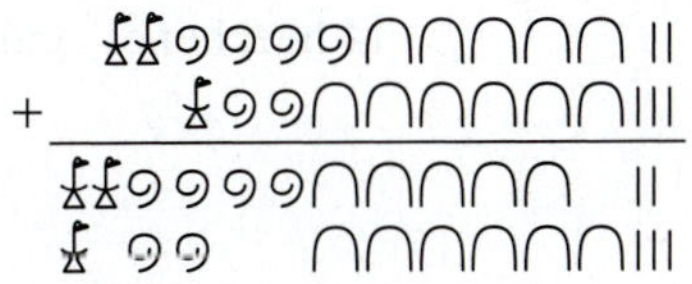

ANSWER Yes. They both represent 211. In the Egyptian numeration system, the order of the hieroglyphics is not important.

Replacing 10 heel bones with one scroll produces

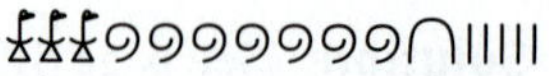 or 3715

The sum is 3715.

CHECK YOUR PROGRESS 3 Use Egyptian hieroglyphics to find 23,341 + 10,562.

Solution See page S19.

In the Egyptian numeration system, subtraction is performed by removing some of the hieroglyphics from the larger numeral. In some cases it is necessary to "borrow," as shown in the next example.

EXAMPLE 4 Use Egyptian Hieroglyphics to Find a Difference

Use Egyptian hieroglyphics to find 4345 − 3162.

Solution

Rewrite using Egyptian hieroglyphics.

4,345
− 3,162

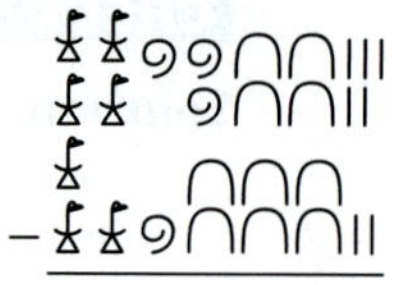

Six heel bones cannot be removed from four heel bones, so replace one scroll with 10 heel bones, resulting in a total of 14 heel bones.

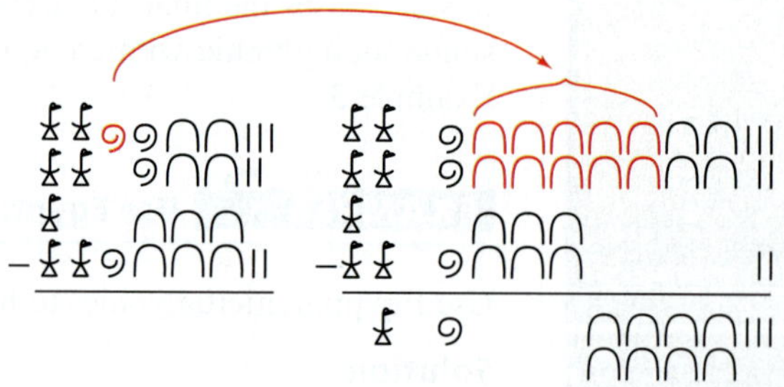

Remove 3 lotus flowers from 4 lotus flowers, 1 scroll from 2 scrolls, 6 heel bones from 14 heel bones, and 2 strokes from 5 strokes to produce:

 = 1183

CHECK YOUR PROGRESS 4 Use Egyptian hieroglyphics to find 61,432 − 45,121.

Solution See page S19.

MATH MATTERS Early Egyptian Fractions

Evidence gained from the Rhind papyrus shows that the Egyptian method of calculating with fractions was much different from the methods we use today. All Egyptian fractions (except for $\frac{2}{3}$) were represented in terms of unit fractions, which are fractions of the form $\frac{1}{n}$, for some natural number $n > 1$. The Egyptians wrote these unit fractions by placing an oval over the numeral that represented the denominator. For example,

 $= \frac{1}{3}$  $= \frac{1}{15}$

If a fraction was not a unit fraction, then the Egyptians wrote the fraction as the sum of *distinct* unit fractions. For instance,

$$\frac{2}{5} \text{ was written as the sum of } \frac{1}{3} \text{ and } \frac{1}{15}.$$

Of course, $\frac{2}{5} = \frac{1}{5} + \frac{1}{5}$, but (for some mysterious reason) the early Egyptian numeration system didn't allow repetitions. The Rhind papyrus includes a table that shows how to write fractions of the form $\frac{2}{k}$, where k is an odd number from 5 to 101, in terms of unit fractions. Some of these are listed below.

$$\frac{2}{7} = \frac{1}{4} + \frac{1}{28} \qquad \frac{2}{11} = \frac{1}{6} + \frac{1}{66} \qquad \frac{2}{19} = \frac{1}{12} + \frac{1}{76} + \frac{1}{114}$$

TABLE 6.2
Roman Numerals

Hindu-Arabic numeral	Roman numeral
1	I
5	V
10	X
50	L
100	C
500	D
1000	M

TAKE NOTE

The C used in Roman numerals is from the Latin word for "hundred," which is *centum.* The word *century,* which means a period of 100 years, is a derivation of *centum.* The M is from the Latin word for "thousand," which is *mille.* We use the word *millennium* to designate 1000 years.

The Roman Numeration System

The Roman numeration system was used in Europe during the reign of the Roman Empire. Today we still make limited use of Roman numerals on clock faces, on the cornerstones of buildings, and in numbering the volumes of periodicals and books. Table 6.2 shows the numerals used in the Roman numeration system. If the Roman numerals are listed so that each numeral has a larger value than the numeral to its right, then the value of the Roman numeral is found by adding the values of each numeral. For example,

$$\text{CLX} = 100 + 50 + 10 = 160$$

If a Roman numeral is repeated two or three times in succession, we add to determine its numerical value. For instance, XX $= 10 + 10 = 20$ and CCC $= 100 + 100 + 100 = 300$. Each of the numerals I, X, C, and M may be repeated up to three times. The numerals V, L, and D are not repeated.

Although the Roman numeration system is an additive system, it also incorporates a subtraction property. In the Roman numeration system, the value of a numeral is determined by adding the values of the numerals from left to right. However, if the value of a numeral is less than the value of the numeral to its right, the smaller value is subtracted from the next larger value. For instance, VI $= 5 + 1 = 6$; however, IV $= 5 - 1 = 4$. In the Roman numeration system, the only numerals whose values can be subtracted from the value of the numeral to the right are I, X, and C. Also, the subtraction of these values is allowed only if the value of the numeral to the right is within two rows, as shown in Table 6.2. That is, the value of the numeral to be subtracted must be *no less than* one-tenth of the value of the numeral it is to be subtracted from. For instance, XL $= 40$ and XC $= 90$, but XD does not represent 490 because the value of X is less than one-tenth the value of D. To write 490 using Roman numerals, we write CDXC.

HISTORICAL NOTE

The Roman numeration system evolved over a period of many years, and thus some Roman numerals displayed on ancient structures do not adhere to the basic rules given at the right. For instance, in the Colosseum in Rome (ca. AD 80), the numeral XXVIIII appears above archway 29 instead of the numeral XXIX.

PictureQuest

A Summary of the Basic Rules Employed in the Roman Numeration System

$$I = 1, \quad V = 5, \quad X = 10, \quad L = 50, \quad C = 100, \quad D = 500, \quad M = 1000$$

1. *Addition Property* If the numerals are listed so that each numeral has a larger value than the numeral to the right, then the value of the Roman numeral is found by adding the values of the numerals. Each of the numerals, I, X, C, and M may be repeated up to three times. The numerals V, L, and D are not repeated. If a numeral is repeated two or three times in succession, we add to determine its numerical value.
2. *Subtraction Property* The only numerals whose values can be subtracted from the value of the numeral to its right are I, X, and C. The value of the numeral to be subtracted must be no less than one-tenth of the value of the numeral to its right.

PPT

EXAMPLE 5 Evaluate a Roman Numeral

Write DCIV as a Hindu-Arabic numeral.

Solution

Because the value of D is larger than the value of C, we add their numerical values. The value of I is less than the value of V, so we subtract the smaller value from the larger value. Thus

$$\text{DCIV} = (\text{DC}) + (\text{IV}) = (500 + 100) + (5 - 1) = 600 + 4 = 604$$

CHECK YOUR PROGRESS 5 Write MCDXLV as a Hindu-Arabic numeral.

Solution See page S19.

EXAMPLE 6 Write a Hindu-Arabic Numeral as a Roman Numeral

Write 579 as a Roman numeral.

Solution

$$579 = 500 + 50 + 10 + 10 + 9$$

In Roman numerals, 9 is written as IX. Thus 579 = DLXXIX.

CHECK YOUR PROGRESS 6 Write 473 as a Roman numeral.

Solution See page S19.

In the Roman numeration system, a bar over a numeral or group of numerals is used to denote a value 1000 times the value of the numeral or group of numerals. For instance,

$$\overline{\text{V}} = 1000 \times 5 = 5000 \qquad \text{and} \qquad \overline{\text{CD}} = 1000 \times 400 = 400{,}000$$

EXAMPLE 7 Convert between Roman Numerals and Hindu-Arabic Numerals

a. Write $\overline{\text{VII}}$ as a Hindu-Arabic numeral.

b. Write 6125 as a Roman numeral.

Solution

a. $\overline{\text{VII}} = 7 \times 1000 = 7000$

b. The Roman numeral 6 is written VI, and 125 is written as CXXV. Thus in Roman numerals, 6125 is $\overline{\text{VI}}$CXXV.

TAKE NOTE

The method of writing a bar over a numeral should be used only to write Roman numerals that cannot be written using the basic rules. For instance, the Roman numeral for 2019 is MMXIX, not $\overline{\text{II}}$XIX.

CHECK YOUR PROGRESS 7

a. Write $\overline{\text{IX}}\text{L}$ as a Hindu-Arabic numeral.

b. Write 8070 as a Roman numeral.

Solution See page S19.

EXCURSION

A Rosetta Tablet for the Traditional Chinese Numeration System

Most of the knowledge we have gained about early numeration systems has been obtained from inscriptions found on ancient tablets or stones. The information provided by these inscriptions has often been difficult to interpret. For several centuries, archeologists had little success in interpreting the Egyptian hieroglyphics they had discovered. Then, in 1799, a group of French military engineers discovered a basalt stone near Rosetta in the Nile delta. This stone, which we now call the Rosetta Stone, has an inscription in three scripts: Greek, Egyptian Demotic, and Egyptian hieroglyphic. It was soon discovered that all three scripts contained the same message. The Greek script was easy to translate, and from its translation, clues were uncovered that enabled scholars to translate many of the documents that up to that time had been unreadable.

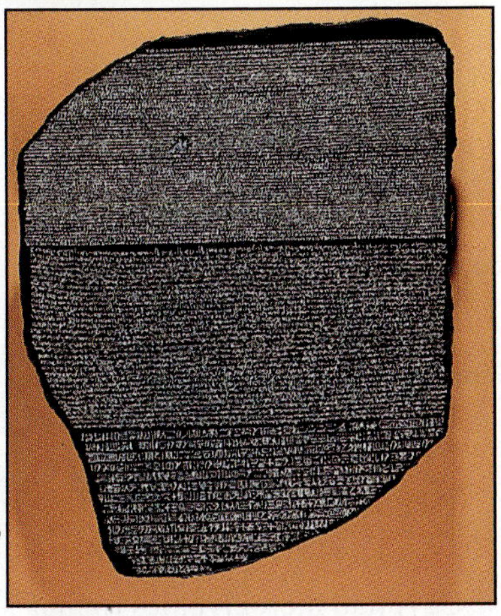

The Granger Collection, NY

The Rosetta Stone

Pretend that you are an archeologist. Your team has just discovered an old tablet that displays Roman numerals and traditional Chinese numerals. It also provides hints in the form of a crossword puzzle about the traditional Chinese numeration system. Study the inscriptions on the following tablet and then complete the Excursion Exercises that follow.

Roman:	Traditional Chinese:
I	一
II	二
III	三
IV	四
V	五
VI	六
VII	七
VIII	八
IX	九
X	十
C	百
M	千
CCCLIX	三 百 五 十 九 (written vertically)
$\overline{\text{VII}}$CCXLVI	七 千 二 百 四 十 六 (written vertically)

Clues
In the traditional Chinese numeration system:

Across
2. Numerals are arranged in a vertical ____________.
5. If a numeral is written above a power of ____, then their numerical values are multiplied.
6. There is no numeral for ______.

Down
1. The total value of a numeral is the ___ of the multiples of the powers of ten and the ones numeral.
3. If there are an odd number of numerals, then the bottom numeral represents units or _____.
4. A numeral written above a power of ten has a value from 1 to _____, inclusive.

EXCURSION EXERCISES

1. Complete the crossword puzzle shown on the previous page.
2. Write 26 as a traditional Chinese numeral.
3. Write 357 as a traditional Chinese numeral.
4. Write the Hindu-Arabic numeral given by each of the following traditional Chinese numerals.

   a. 八
百
九
十
六

   b. 二
千
四
百
六
十
五

5. **a.** How many Hindu-Arabic numerals are required to write four thousand five hundred twenty-eight?

   b. How many traditional Chinese numerals are required to write four thousand five hundred twenty-eight?

6. The traditional Chinese numeration system is no longer in use. Give a reason that may have contributed to its demise.

EXERCISE SET 6.1

■ In Exercises 1 to 12, write each Hindu-Arabic numeral using Egyptian hieroglyphics.

1. 46 **2.** 82
3. 103 **4.** 157
5. 2568 **6.** 3152
7. 23,402 **8.** 15,303
9. 65,800 **10.** 43,217
11. 1,405,203 **12.** 653,271

■ In Exercises 13 to 24, write each Egyptian numeral as a Hindu-Arabic numeral.

13. 𓆼𓆼𓍢𓎆𓎆𓎆𓏺𓏺𓏺𓏺

14. 𓆼𓍢𓍢𓍢𓎆𓏺𓏺

15. 𓍢𓍢𓍢𓍢𓎆𓎆𓏺𓏺𓏺
𓍢𓍢𓍢𓍢𓎆𓎆𓏺𓏺

16. 𓍢𓎆𓎆𓎆𓎆𓏺𓏺𓏺𓏺
𓍢𓎆𓎆𓎆𓎆𓏺𓏺𓏺

17. 𓆼𓍢𓍢𓎆𓎆𓎆𓏺𓏺

18. 𓂭𓂭 𓍢𓍢𓎆𓏺
𓆐𓆐𓂭𓆼𓆼𓍢𓎆𓎆𓏺𓏺

19. 𓆐𓆐𓂭𓂭𓆼𓎆𓏺

20. 𓂭𓂭𓂭𓆼𓆼𓍢𓍢𓍢𓎆𓎆𓏺𓏺𓏺

21. 𓂭𓂭𓂭𓆼𓆼𓆼𓍢𓍢𓍢𓎆𓎆𓎆𓏺𓏺𓏺𓏺𓏺
𓂭𓂭𓂭𓆼𓆼𓍢𓍢𓍢𓍢𓎆𓎆𓎆𓏺𓏺𓏺𓏺

22. 𓁨𓆐𓆐𓂭𓂭𓆼𓆼𓆼𓍢𓍢𓍢𓍢𓎆
𓁨𓆐𓆐𓂭 𓆼𓆼𓆼𓍢𓍢𓍢𓎆𓏺𓏺𓏺

23. 𓁨𓁨𓁨𓆐 𓆼𓍢𓍢𓏺𓏺𓏺
𓁨𓁨𓂭𓂭 𓆼𓍢𓍢𓏺𓏺𓏺

24. 𓁨𓆐𓆐𓆼𓍢𓍢
𓆐𓆐
𓁨𓆐𓆐𓍢𓍢

■ In Exercises 25 to 32, use Egyptian hieroglyphics to find each sum or difference.

25. 51 + 43 **26.** 67 + 58
27. 231 + 435 **28.** 623 + 124
29. 83 − 51 **30.** 94 − 23
31. 254 − 198 **32.** 640 − 278

■ In Exercises 33 to 46, write each Roman numeral as a Hindu-Arabic numeral.

33. CLXI **34.** LXIV
35. DCL **36.** MCX
37. MCDIX **38.** MDCCII
39. MCCXL **40.** MMDCIV
41. DCCCXL **42.** CDLV
43. $\overline{\text{IX}}$XLIV **44.** $\overline{\text{VII}}$DXVII
45. $\overline{\text{XI}}$CDLXI **46.** $\overline{\text{IV}}$CCXXI

■ In Exercises 47 to 60, write each Hindu-Arabic numeral as a Roman numeral.

47. 39 **48.** 42 **49.** 157
50. 231 **51.** 542 **52.** 783
53. 1197 **54.** 1039 **55.** 787
56. 1343 **57.** 683 **58.** 959
59. 6898 **60.** 4357

Egyptian Multiplication The Rhind papyrus contains problems that show a *doubling procedure* used by the Egyptians to find the product of two whole numbers. The following examples illustrate this doubling procedure. In the examples, we have used Hindu-Arabic numerals so that you can concentrate on the doubling procedure and not be distracted by the Egyptian hieroglyphics. The first example determines the product 5 × 27 by computing two successive doublings of 27 and then forming the sum of the blue numbers in the rows marked with a check. Note that the rows marked with a check show that one 27 is 27 and four 27s is 108. Thus five 27s is the sum of 27 and 108, or 135.

√ 1	27	double
2	54	double
√ 4	108	
5	135	← This sum is the product of 5 and 27.

In the next example, we use the Egyptian doubling procedure to find the product of 35 and 94. Because the sum of 1, 2, and 32 is 35, we add only the blue numbers in the rows marked with a check to find that 35 × 94 = 94 + 188 + 3008 = 3290.

√ 1	94	double
√ 2	188	double
4	376	double
8	752	double
16	1504	double
√ 32	3008	
35	3290	← This sum is the product of 35 and 94.

In Exercises 61 to 70, use the Egyptian doubling procedure to find each product.

61. 5 × 47

62. 7 × 58

63. 8 × 63

64. 4 × 57

65. 7 × 29

66. 9 × 33

67. 17 × 35

68. 26 × 43

69. 23 × 108

70. 72 × 215

EXTENSIONS

71. a. State a reason why you might prefer to use the Egyptian hieroglyphic numeration system rather than the Roman numeration system.

b. State a reason why you might prefer to use the Roman numeration system rather than the Egyptian hieroglyphic numeration system.

72. What is the largest number that can be written using Roman numerals without using the bar over a numeral or the subtraction property?

73. **The Ionic Greek Numeration System** The Ionic Greek numeration system assigned numerical values to the letters of the Greek alphabet. Research the Ionic Greek numeration system and write a report that explains this numeration system. Include information about some of the advantages and disadvantages of this system compared with our present Hindu-Arabic numeration system.

74. **The Method of False Position** The Rhind papyrus (see page 295) contained solutions to several mathematical problems. Some of these solutions made use of a procedure called the *method of false position*. Research the method of false position and write a report that explains this method. In your report, include a specific mathematical problem and its solution by the method of false position.

SECTION 6.2 Place-Value Systems

Expanded Form

The most common numeration system used by people today is the Hindu-Arabic numeration system. It is called the Hindu-Arabic system because it was first developed in India (around AD 800) and then refined by the Arabs. It makes use of the 10 symbols 0, 1, 2, 3, 4, 5, 6, 7, 8, and 9. The reason for the 10 symbols, called *digits,* is related to the fact that we have 10 fingers. The Hindu-Arabic numeration system is also called the *decimal system,* where the word *decimal* is a derivation of the Latin word *decem,* which means "ten."

One important feature of the Hindu-Arabic numeration system is that it is a *place-value* or *positional-value system.* This means that the numerical value of each digit in a Hindu-Arabic numeral depends on its *place* or *position* in the numeral. For instance, the 3 in 31 represents 3 tens, whereas the 3 in 53 represents 3 ones. The Hindu-Arabic numeration system is a **base ten numeration system** because the place values are the powers of 10:

$$\ldots, 10^5, 10^4, 10^3, 10^2, 10^1, 10^0$$

> **HISTORICAL NOTE**
>
> **Abu Ja'far Muhammad ibn Musa al'Khwarizmi**
> (ca. AD 790–850) Al'Khwarizmi (ăl'khwăhr'īzmee) produced two important texts. One of these texts advocated the use of the Hindu-Arabic numeration system. The 12th-century Latin translation of this book is called *Liber Algoritmi de Numero Indorum,* or *Al-Khwarizmi on the Hindu Art of Reckoning.* In Europe, the people who favored the adoption of the Hindu-Arabic numeration system became known as *algorists* because of Al'Khwarizmi's *Liber Algoritmi de Numero Indorum* text. The Europeans who opposed the Hindu-Arabic system were called *abacists.* They advocated the use of Roman numerals and often performed computations with the aid of an abacus.

The place value associated with the nth digit of a numeral (counting from right to left) is 10^{n-1}. For instance, in the numeral 7532, the 7 is the fourth digit from the right and is in the $10^{4-1} = 10^3$, or thousands, place. The numeral 2 is the first digit from the right and is in the $10^{1-1} = 10^0$, or ones, place. The *indicated sum* of each digit of a numeral multiplied by its respective place value is called the **expanded form** of the numeral.

EXAMPLE 1 Write a Numeral in Its Expanded Form

Write 4672 in expanded form.

Solution

$$\begin{aligned}4672 &= 4000 + 600 + 70 + 2\\ &= (4 \times 1000) + (6 \times 100) + (7 \times 10) + (2 \times 1)\end{aligned}$$

The above expanded form can also be written as

$$(4 \times 10^3) + (6 \times 10^2) + (7 \times 10^1) + (2 \times 10^0)$$

CHECK YOUR PROGRESS 1 Write 17,325 in expanded form.

Solution See page S19.

If a number is written in expanded form, it can be simplified to its ordinary decimal form by performing the indicated operations. The *Order of Operations Agreement* states that we should first perform the exponentiations, then perform the multiplications, and finish by performing the additions.

EXAMPLE 2 Simplify a Number Written in Expanded Form

Simplify: $(2 \times 10^3) + (7 \times 10^2) + (6 \times 10^1) + (3 \times 10^0)$

Solution

$$\begin{aligned}&(2 \times 10^3) + (7 \times 10^2) + (6 \times 10^1) + (3 \times 10^0)\\ &\quad= (2 \times 1000) + (7 \times 100) + (6 \times 10) + (3 \times 1)\\ &\quad= 2000 + 700 + 60 + 3\\ &\quad= 2763\end{aligned}$$

CHECK YOUR PROGRESS 2 Simplify:

$$(5 \times 10^4) + (9 \times 10^3) + (2 \times 10^2) + (7 \times 10^1) + (4 \times 10^0)$$

Solution See page S19.

In the next few examples, we make use of the expanded form of a numeral to compute sums and differences. An examination of these examples will help you better understand the computational algorithms used in the Hindu-Arabic numeration system.

EXAMPLE 3 Use Expanded Forms to Find a Sum

Use expanded forms of 26 and 31 to find their sum.

Solution

$$\begin{array}{rl} 26 = & (2 \times 10) + 6 \\ \underline{+\ 31} = & \underline{(3 \times 10) + 1} \\ & (5 \times 10) + 7 = 50 + 7 = 57 \end{array}$$

CHECK YOUR PROGRESS 3 Use expanded forms to find the sum of 152 and 234.

Solution See page S19.

If the expanded form of a sum contains one or more powers of 10 that have multipliers larger than 9, then we simplify by rewriting the sum with multipliers that are less than or equal to 9. This process is known as *carrying*.

EXAMPLE 4 Use Expanded Forms to Find a Sum

Use expanded forms of 85 and 57 to find their sum.

Solution

$$\begin{aligned} 85 &= (8 \times 10) + 5 \\ + 57 &= (5 \times 10) + 7 \\ \hline & (13 \times 10) + 12 \\ & (10 + 3) \times 10 + 10 + 2 \\ & 100 + 30 + 10 + 2 = 100 + 40 + 2 = 142 \end{aligned}$$

TAKE NOTE

From the expanded forms in Example 4, note that 12 is 1 ten and 2 ones. When we add columns of numbers, this is shown as "carry a 1." Because the 1 is placed in the tens column, we are actually adding 10.

$$\begin{array}{r} {}^{1} \\ 85 \\ +\ 57 \\ \hline 142 \end{array}$$

CHECK YOUR PROGRESS 4 Use expanded forms to find the sum of 147 and 329.

Solution See page S20.

In the next example, we use the expanded forms of numerals to analyze the concept of "borrowing" in a subtraction problem.

EXAMPLE 5 Use Expanded Forms to Find a Difference

Use the expanded forms of 457 and 283 to find 457 − 283.

Solution

$$\begin{aligned} 457 &= (4 \times 100) + (5 \times 10) + 7 \\ - 283 &= (2 \times 100) + (8 \times 10) + 3 \end{aligned}$$

At this point, this example is similar to Example 4 in Section 6.1. We cannot remove 8 tens from 5 tens, so 1 hundred is replaced by 10 tens.

$$\begin{aligned} 457 &= (4 \times 100) + (5 \times 10) + 7 \\ &= (3 \times 100) + (10 \times 10) + (5 \times 10) + 7 \\ &= (3 \times 100) + (15 \times 10) + 7 \end{aligned}$$

• $4 \times 100 = 3 \times 100 + 100 = 3 \times 100 + 10 \times 10$

We can now remove 8 tens from 15 tens.

$$\begin{aligned} 457 &= (3 \times 100) + (15 \times 10) + 7 \\ - 283 &= (2 \times 100) + (8 \times 10) + 3 \\ \hline &= (1 \times 100) + (7 \times 10) + 4 = 100 + 70 + 4 = 174 \end{aligned}$$

TAKE NOTE

From the expanded forms in Example 5, note that we "borrowed" 1 hundred as 10 tens. This explains how we show borrowing when numbers are subtracted using place value form.

$$\begin{array}{r} {}^{3}\ \ \ \ \\ \not{4}\,{}^{1}57 \\ -\ 2\,83 \\ \hline 1\,74 \end{array}$$

CHECK YOUR PROGRESS 5 Use expanded forms to find the difference 382 − 157.

Solution See page S20.

The Babylonian Numeration System

The Babylonian numeration system uses a base of 60. The place values in the Babylonian system are given in the following table.

TABLE 6.3
Place Values in the Babylonian Numeration System

. . .	60^3 = 216,000	60^2 = 3600	60^1 = 60	60^0 = 1

The Babylonians recorded their numerals on damp clay using a wedge-shaped stylus. A vertical wedge shape represented 1 unit and a sideways "vee" shape represented 10 units.

1
10

To represent a number smaller than 60, the Babylonians used an *additive* feature similar to that used by the Egyptians. For example, the Babylonian numeral for 32 is

For the number 60 and larger numbers, the Babylonians left a small space between groups of symbols to indicate a different place value. This procedure is illustrated in the following example.

EXAMPLE 6 **Write a Babylonian Numeral as a Hindu-Arabic Numeral**

Write  as a Hindu-Arabic numeral.

Solution

1 group of 60^2 | 31 groups of 60 | 25 ones

$$= (1 \times 60^2) + (31 \times 60) + (25 \times 1)$$
$$= 3600 + 1860 + 25$$
$$= 5485$$

CHECK YOUR PROGRESS 6 Write  as a Hindu-Arabic numeral.

Solution See page S20.

QUESTION In the Babylonian numeration system, does  = ?

ANSWER No.  = 2, whereas  = $(1 \times 60) + (1 \times 1) = 61$.

In the next example we illustrate a division process that can be used to convert Hindu-Arabic numerals to Babylonian numerals.

EXAMPLE 7 Write a Hindu-Arabic Numeral as a Babylonian Numeral

Write 8503 as a Babylonian numeral.

Solution

The Babylonian numeration system uses place values of

$$60^0, 60^1, 60^2, 60^3, \ldots .$$

Evaluating the powers produces

1, 60, 3600, 216,000, ...

The largest of these powers that is contained in 8503 is 3600. One method of finding how many groups of 3600 are in 8503 is to divide 3600 into 8503. Refer to the first division shown below. Now divide to determine how many groups of 60 are contained in the remainder 1303.

$$\begin{array}{r} 2 \\ 3600\overline{)8503} \\ \underline{7200} \\ 1303 \end{array} \qquad \begin{array}{r} 21 \\ 60\overline{)1303} \\ \underline{120} \\ 103 \\ \underline{60} \\ 43 \end{array}$$

The above computations show that 8503 consists of 2 groups of 3600 and 21 groups of 60, with 43 left over. Thus

$$8503 = (2 \times 60^2) + (21 \times 60) + (43 \times 1)$$

As a Babylonian numeral, 8503 is written

CHECK YOUR PROGRESS 7 Write 12,578 as a Babylonian numeral.

Solution See page S20.

In Example 8, we find the sum of two Babylonian numerals. If a numeral for any power of 60 is larger than 59, then simplify by decreasing that numeral by 60 and increasing the numeral in the place value to its left by 1.

EXAMPLE 8 Find the Sum of Babylonian Numerals

Find the sum of the following numerals. Write the answer as a Babylonian numeral.

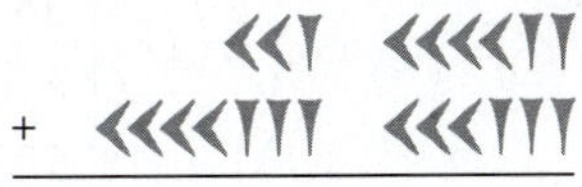

Solution

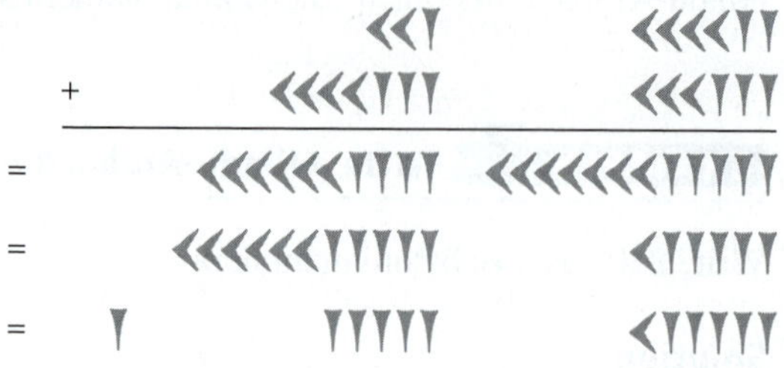

- Combine the symbols for each place value.
- Take away 60 from the ones place and add 1 to the 60s place.
- Take away 60 from the 60s place and add 1 to the 60^2 place.

CHECK YOUR PROGRESS 8 Find the sum of the following numerals. Write the answer as a Babylonian numeral.

Solution See page S20.

MATH MATTERS Zero as a Placeholder

When the Babylonian numeration system first began to develop around 1700 BC, an empty space was used to indicate that a place value was missing. This procedure of "leaving a space" can be confusing. How big is an empty space? Is that one empty space or two empty spaces? Around 300 BC, the Babylonians started to use the symbol  to indicate that a particular place value was missing. For instance,  represented $(2 \times 60^2) + (11 \times 1) = 7211$. In this case the zero placeholder indicates that there are no 60s.

The Mayan Numeration System

The Mayan civilization existed in the Yucatán area of southern Mexico and in Guatemala, Belize, and parts of El Salvador and Honduras. It started as far back as 9000 BC and reached its zenith during the period from AD 200 to AD 900. Among their many accomplishments, the Maya are best known for their complex hieroglyphic writing system, their sophisticated calendars, and their remarkable numeration system.

The Maya used three calendars—the solar calendar, the ceremonial calendar, and the Venus calendar. The solar calendar consisted of about 365.24 days. Of these, 360 days were divided into 18 months, each with 20 days. The Mayan numeration system was strongly influenced by this solar calendar, as evidenced by the use of the numbers 18 and 20 in determining place values. See Table 6.4.

TAKE NOTE

Observe that the place values used in the Mayan numeration system are not all powers of 20.

TABLE 6.4
Place Values in the Mayan Numeration System

. . .	18×20^3 = 144,000	18×20^2 = 7200	18×20^1 = 360	20^1 = 20	20^0 = 1

The Mayan numeration system was one of the first systems to use a symbol for zero as a placeholder. The Mayan numeration system used only three symbols. A dot was used to represent 1, a horizontal bar represented 5, and a conch shell represented 0. The following table shows how the Maya used a combination of these three symbols to write the whole numbers from 0 to 19. Note that each numeral contains at most four dots and at most three horizontal bars.

TABLE 6.5
Mayan Numerals

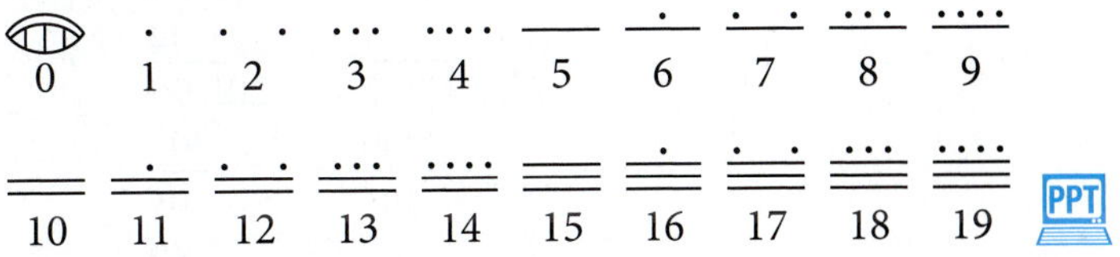

To write numbers larger than 19, the Maya used a vertical arrangement with the largest place value at the top. The following example illustrates the process of converting a Mayan numeral to a Hindu-Arabic numeral.

> **TAKE NOTE**
>
> In the Mayan numeration system,
>
> 
>
> does not represent 3 because the three dots are not all in the same row. The Mayan numeral
>
> 
>
> represents one group of 20 and two ones, or 22.

EXAMPLE 9 Write a Mayan Numeral as a Hindu-Arabic Numeral

Write each of the following as a Hindu-Arabic numeral.

a.  **b.** 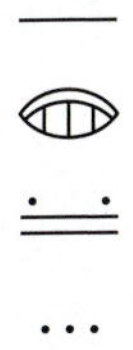

Solution

a. 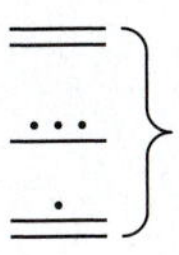

$10 \times 360 = 3600$

$8 \times 20 = 160$

$11 \times 1 = +11$

3771

b. 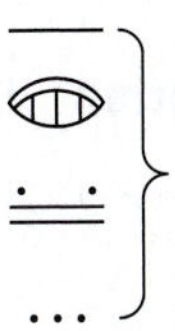

$5 \times 7200 = 36{,}000$

$0 \times 360 = 0$

$12 \times 20 = 240$

$3 \times 1 = +\ 3$

$36{,}243$

CHECK YOUR PROGRESS 9 Write each of the following as a Hindu-Arabic numeral.

a.  **b.** 

Solution See page S20.

In the next example, we illustrate how the concept of place value is used to convert Hindu-Arabic numerals to Mayan numerals.

EXAMPLE 10 Write a Hindu-Arabic Numeral as a Mayan Numeral

Write 7495 as a Mayan numeral.

Solution

The place values used in the Mayan numeration system are

$$20^0, 20^1, 18 \times 20^1, 18 \times 20^2, 18 \times 20^3, \ldots$$

or

1, 20, 360, 7200, 144,000, ...

Removing 1 group of 7200 from 7495 leaves 295. No groups of 360 can be obtained from 295, so we divide 295 by the next smaller place value of 20 to find that 295 equals 14 groups of 20 with 15 left over.

$$\begin{array}{r} 1 \\ 7200\overline{)7495} \\ \underline{7200} \\ 295 \end{array} \qquad \begin{array}{r} 14 \\ 20\overline{)295} \\ \underline{20} \\ 95 \\ \underline{80} \\ 15 \end{array}$$

Thus

$$7495 = (1 \times 7200) + (0 \times 360) + (14 \times 20) + (15 \times 1)$$

In Mayan numerals, 7495 is written as

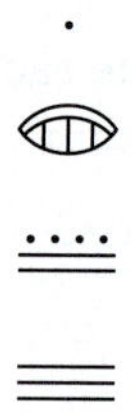

CHECK YOUR PROGRESS 10 Write 11,480 as a Mayan numeral.

Solution See page S20.

EXCURSION

Subtraction via the Nines Complement and the End-Around Carry

INSTRUCTOR NOTE

The concepts in this Excursion are extended in the Excursion in Section 6.4, page 327.

In the subtraction 5627 − 2564 = 3063, the number 5627 is called the *minuend,* 2564 is called the *subtrahend,* and 3063 is called the *difference.* In the Hindu-Arabic base ten system, subtraction can be performed by a process that involves addition and the *nines complement* of the subtrahend. The **nines complement** of a single digit n is the number $9 - n$. For instance, the nines complement of 3 is 6, the nines complement of 1 is 8, and the nines complement of 0 is 9. The nines complement of a number with more than one digit is the number that is formed by taking the nines complement of each digit. The nines complement of 25 is 74, and the nines complement of 867 is 132.

Subtraction by Using the Nines Complement and the End-Around Carry

To subtract by using the nines complement:

1. Add the nines complement of the subtrahend to the minuend.
2. Take away 1 from the leftmost digit of the sum produced in step 1 and add 1 to the units digit. This is referred to as the end-around carry procedure.

The following example illustrates the process of subtracting 2564 from 5627 by using the nines complement.

5627	Minuend
− 2564	Subtrahend
5627	Minuend
+ 7435	Replace the subtrahend with the nines complement of the subtrahend and add.
13062	
13062	Take away 1 from the leftmost digit and add 1 to the units digit. This is the end-around carry procedure.
+ 1	
3063	

Thus

$$\begin{array}{r} 5627 \\ -\ 2564 \\ \hline 3063 \end{array}$$

If the subtrahend has fewer digits than the minuend, leading zeros should be inserted in the subtrahend so that it has the same number of digits as the minuend. This process is illustrated below for 2547 − 358.

2547	Minuend
− 358	Subtrahend
2547	
− 0358	Insert a leading zero in the subtrahend.
2547	Minuend
+ 9641	Nines complement of subtrahend
12188	Add the minuend to the nines complement of the subtrahend.
12188	Take away 1 from the leftmost digit and add 1 to the units digit.
+ 1	
2189	Add 1 to the units digit of 2188.

Verify that 2189 is the correct difference.

EXCURSION EXERCISES

For Exercises 1 to 6, use the nines complement of the subtrahend and the end-around carry to find the indicated difference.

1. 724 − 351

2. 2405 − 1608

3. 91,572 − 7824

4. 214,577 − 48,231

5. 3,156,782 − 875,236

6. 54,327,105 − 7,678,235

7. Explain why the nines complement and the end-around carry procedure produce the correct answer to a subtraction problem.

EXERCISE SET 6.2

■ In Exercises 1 to 8, write each numeral in its expanded form.

1. 48

2. 93

3. 420

4. 501

5. 6803

6. 9045

7. 10,208

8. 67,482

■ In Exercises 9 to 16, simplify each expansion.

9. $(4 \times 10^2) + (5 \times 10^1) + (6 \times 10^0)$

10. $(7 \times 10^2) + (6 \times 10^1) + (3 \times 10^0)$

11. $(5 \times 10^3) + (0 \times 10^2) + (7 \times 10^1) + (6 \times 10^0)$

12. $(3 \times 10^3) + (1 \times 10^2) + (2 \times 10^1) + (8 \times 10^0)$

13. $(3 \times 10^4) + (5 \times 10^3) + (4 \times 10^2) + (0 \times 10^1) + (7 \times 10^0)$

14. $(2 \times 10^5) + (3 \times 10^4) + (0 \times 10^3) + (6 \times 10^2) + (7 \times 10^1) + (5 \times 10^0)$

15. $(6 \times 10^5) + (8 \times 10^4) + (3 \times 10^3) + (0 \times 10^2) + (4 \times 10^1) + (0 \times 10^0)$

16. $(5 \times 10^7) + (3 \times 10^6) + (0 \times 10^5) + (0 \times 10^4) + (7 \times 10^3) + (9 \times 10^2) + (0 \times 10^1) + (2 \times 10^0)$

■ In Exercises 17 to 22, use expanded forms to find each sum.

17. 35 + 41 **18.** 42 + 56

19. 257 + 138 **20.** 352 + 461

21. 1023 + 1458 **22.** 3567 + 2651

■ In Exercises 23 to 28, use expanded forms to find each difference.

23. 62 − 35 **24.** 193 − 157

25. 4725 − 1362 **26.** 85,381 − 64,156

27. 23,168 − 12,857 **28.** 59,163 − 47,956

■ In Exercises 29 to 36, write each Babylonian numeral as a Hindu-Arabic numeral.

29. 

30. 

31. 

32. 

33. 

34. 

35. 

36. 

■ In Exercises 37 to 46, write each Hindu-Arabic numeral as a Babylonian numeral.

37. 42 **38.** 57

39. 128 **40.** 540

41. 5678 **42.** 7821

43. 10,584 **44.** 12,687

45. 21,345 **46.** 24,567

■ In Exercises 47 to 52, find the sum of the Babylonian numerals. Write each answer as a Babylonian numeral.

47. 

48. 

49. 

50. 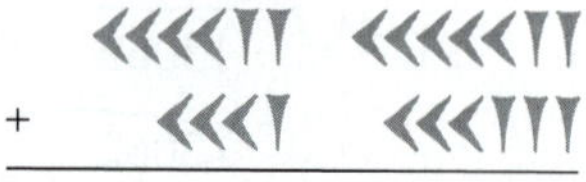

51. 

52. 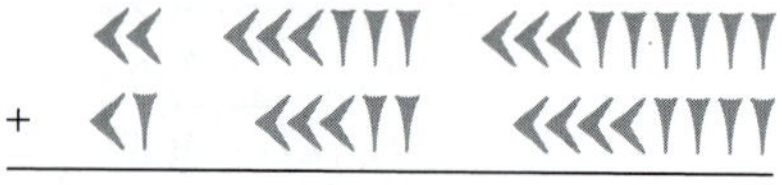

■ In Exercises 53 to 60, write each Mayan numeral as a Hindu-Arabic numeral.

53. 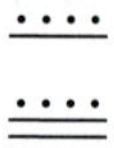

54. 

55. 

56. 

57. 

58. 

59. 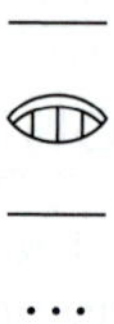

60. 

■ In Exercises 61 to 68, write each Hindu-Arabic numeral as a Mayan numeral.

61. 137 **62.** 253

63. 948 **64.** 1265

65. 1693 **66.** 2728

67. 7432 **68.** 8654

EXTENSIONS

69. a. State a reason why you might prefer to use the Babylonian numeration system instead of the Mayan numeration system.

b. State a reason why you might prefer to use the Mayan numeration system instead of the Babylonian numeration system.

70. Explain why it might be easy to mistake the number 122 for the number 4 when 122 is written as a Babylonian numeral.

71. **A Base Three Numeration System** A student has created a *base three* numeration system. The student has named this numeration system ZUT because Z, U, and T are the symbols used in this system: Z represents 0, U represents 1, and T represents 2. The place values in this system follow: ..., $3^3 = 27$, $3^2 = 9$, $3^1 = 3$, $3^0 = 1$.

Write each ZUT numeral as a Hindu-Arabic numeral.

a. TU **b.** TZT **c.** UZTT

Write each Hindu-Arabic numeral as a ZUT numeral.

d. 37 **e.** 87 **f.** 144

SECTION 6.3 Different Base Systems

Converting Non–Base Ten Numerals to Base Ten

Recall that the Hindu-Arabic numeration system is a base ten system because its place values

$$..., 10^5, 10^4, 10^3, 10^2, 10^1, 10^0$$

all have 10 as their base. The Babylonian numeration system is a base sixty system because its place values

$$..., 60^5, 60^4, 60^3, 60^2, 60^1, 60^0$$

all have 60 as their base. In general, a base b (where b is a natural number greater than 1) numeration system has place values of

$$..., b^5, b^4, b^3, b^2, b^1, b^0$$

TAKE NOTE

Recall that in the expression

$$b^n$$

b is the *base*, and n is the *exponent*.

Many people think that our base ten numeration system was chosen because it is the easiest to use, but this is not the case. In reality most people find it easier to use our base ten system only because they have had a great deal of experience with the base ten system and have not had much experience with non–base ten systems. In this section, we examine some non–base ten numeration systems. To reduce the amount of memorization that would be required to learn new symbols for each of these new systems, we will (as far as possible) make use of our familiar Hindu-Arabic symbols. For instance, if we discuss a base four numeration system that requires four basic symbols, then we will use the four Hindu-Arabic symbols 0, 1, 2, and 3 and the place values

$$..., 4^5, 4^4, 4^3, 4^2, 4^1, 4^0$$

The base eight, or **octal**, numeration system uses the Hindu-Arabic symbols 0, 1, 2, 3, 4, 5, 6, and 7, and the place values

$$..., 8^5, 8^4, 8^3, 8^2, 8^1, 8^0$$

To differentiate between bases, we will label each non–base ten numeral with a subscript that indicates the base. For instance, 23_{four} represents a base four numeral. If a numeral is written without a subscript, then it is understood that the base is ten. Thus 23 written without a subscript is understood to be the base ten numeral 23.

TAKE NOTE

Because 23_{four} is *not* equal to the base ten number 23, it is important *not* to read 23_{four} as "twenty-three." To avoid confusion, read 23_{four} as "two three base four."

To convert a non–base ten numeral to base ten, we write the numeral in its expanded form, as shown in the following example.

EXAMPLE 1 **Convert to Base Ten**

Convert 2314_{five} to base ten.

Solution

In the base five numeration system, the place values are

$$..., 5^4, 5^3, 5^2, 5^1, 5^0$$

The expanded form of 2314_{five} is

$$\begin{aligned} 2314_{five} &= (2 \times 5^3) + (3 \times 5^2) + (1 \times 5^1) + (4 \times 5^0) \\ &= (2 \times 125) + (3 \times 25) + (1 \times 5) + (4 \times 1) \\ &= 250 + 75 + 5 + 4 \\ &= 334 \end{aligned}$$

Thus $2314_{five} = 334$.

CHECK YOUR PROGRESS 1 Convert 3156_{seven} to base ten.

Solution See page S20.

QUESTION Does the notation 26_{five} make sense?

In base two, which is called the **binary numeration system,** the place values are the powers of two.

$$\ldots, 2^7, 2^6, 2^5, 2^4, 2^3, 2^2, 2^1, 2^0$$

The binary numeration system uses only the two digits 0 and 1. These *b*inary dig*its* are often called **bits**. To convert a base two numeral to base ten, write the numeral in its expanded form and then evaluate the expanded form.

POINT OF INTEREST

The base twelve numeration system is called the **duodecimal system**. A group called the Dozenal Society of America advocates the replacement of our base ten decimal system with the duodecimal system. Many members of this organization think that the duodecimal system is "a more convenient numeration system" than the base ten system.

The Dozenal Society of America has chosen a panda as its mascot because pandas have six "fingers" on each paw. The rationale for the choice of a panda is that if humans had six fingers on each hand, we would most likely have adopted a base twelve numeration system.

You can find more information about the Dozenal Society of America at www.dozenal.org

EXAMPLE 2 Convert to Base Ten

Convert 10110111_{two} to base ten.

Solution

$$\begin{aligned} 10110111_{two} &= (1 \times 2^7) + (0 \times 2^6) + (1 \times 2^5) + (1 \times 2^4) + (0 \times 2^3) \\ &\quad + (1 \times 2^2) + (1 \times 2^1) + (1 \times 2^0) \\ &= (1 \times 128) + (0 \times 64) + (1 \times 32) + (1 \times 16) + (0 \times 8) \\ &\quad + (1 \times 4) + (1 \times 2) + (1 \times 1) \\ &= 128 + 0 + 32 + 16 + 0 + 4 + 2 + 1 \\ &= 183 \end{aligned}$$

CHECK YOUR PROGRESS 2 Convert 111000101_{two} to base ten.

Solution See page S20.

The base twelve numeration system requires 12 distinct symbols. We will use the symbols 0, 1, 2, 3, 4, 5, 6, 7, 8, 9, A, and B as our base twelve numeration system symbols. The symbols 0 through 9 have their usual meaning; however, A is used to represent 10 and B to represent 11.

EXAMPLE 3 Convert to Base Ten

Convert $B37_{twelve}$ to base ten.

Solution

In the base twelve numeration system, the place values are

$$\ldots, 12^4, 12^3, 12^2, 12^1, 12^0$$

ANSWER No. The expression 26_{five} is a meaningless expression because there is no 6 in base five.

TABLE 6.6
Decimal and Hexadecimal Equivalents

Base ten decimal	Base sixteen hexadecimal
0	0
1	1
2	2
3	3
4	4
5	5
6	6
7	7
8	8
9	9
10	A
11	B
12	C
13	D
14	E
15	F

Thus

$$\begin{aligned} B37_{\text{twelve}} &= (11 \times 12^2) + (3 \times 12^1) + (7 \times 12^0) \\ &= 1584 + 36 + 7 \\ &= 1627 \end{aligned}$$

CHECK YOUR PROGRESS 3 Convert $A5B_{\text{twelve}}$ to base ten.

Solution See page S20.

Computer programmers often write programs that use the base sixteen numeration system, which is also called the **hexadecimal system**. This system uses the symbols 0, 1, 2, 3, 4, 5, 6, 7, 8, 9, A, B, C, D, E, and F. Table 6.6 shows that A represents 10, B represents 11, C represents 12, D represents 13, E represents 14, and F represents 15.

EXAMPLE 4 **Convert to Base Ten**

Convert $3E8_{\text{sixteen}}$ to base ten.

Solution

In the base sixteen numeration system, the place values are

$$\ldots, 16^4, 16^3, 16^2, 16^1, 16^0$$

Thus

$$\begin{aligned} 3E8_{\text{sixteen}} &= (3 \times 16^2) + (14 \times 16^1) + (8 \times 16^0) \\ &= 768 + 224 + 8 \\ &= 1000 \end{aligned}$$

CHECK YOUR PROGRESS 4 Convert $C24F_{\text{sixteen}}$ to base ten.

Solution See page S20.

Converting from Base Ten to Another Base

The most efficient method of converting a number written in base ten to another base makes use of a *successive division process.* For example, to convert 219 to base four, divide 219 by 4 and write the quotient 54 and the remainder 3, as shown below. Now divide the quotient 54 by the base to get a new quotient of 13 and a new remainder of 2. Continuing the process, divide the quotient 13 by 4 to get a new quotient of 3 and a remainder of 1. Because our last quotient, 3, is less than the base, 4, we stop the division process. The answer is given by the last quotient, 3, and the remainders, shown in red in the following diagram. That is, $219 = 3123_{\text{four}}$.

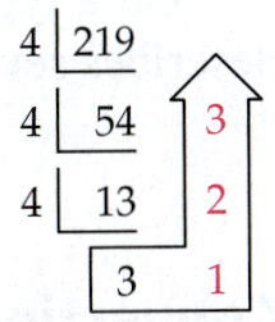

You can understand how the successive division process converts a base ten numeral to another base by analyzing the process. The first division shows there are 54 fours in 219, with **3 ones** left over. The second division shows that there are 13 sixteens (two successive divisions by 4 is the same as dividing by 16) in 219, and the remainder 2 indicates that there are **2 fours** left over. The last division shows that there are 3 sixty-fours (three successive divisions by 4 is the same as dividing by 64) in 219, and the remainder 1 indicates that there is **1 sixteen** left over. In mathematical notation these results are written as follows.

$$\begin{aligned} 219 &= (3 \times 64) + (1 \times 16) + (2 \times 4) + (3 \times 1) \\ &= (3 \times 4^3) + (1 \times 4^2) + (2 \times 4^1) + (3 \times 4^0) \\ &= 3123_{\text{four}} \end{aligned}$$

EXAMPLE 5 Convert a Base Ten Numeral to Another Base

Convert 5821 to **a.** base three and **b.** base sixteen.

Solution

a.

Divisor	Quotient	Remainder
3	5821	
3	1940	1
3	646	2
3	215	1
3	71	2
3	23	2
3	7	2
	2	1

$5821 = 21222121_{\text{three}}$

b.

Divisor	Quotient	Remainder
16	5821	
16	363	13 = D_{sixteen}
16	22	11 = B_{sixteen}
	1	6

$5821 = 16BD_{\text{sixteen}}$

CHECK YOUR PROGRESS 5 Convert 1952 to **a.** base five and **b.** base twelve.

Solution See page S20.

MATH MATTERS Music by the Numbers

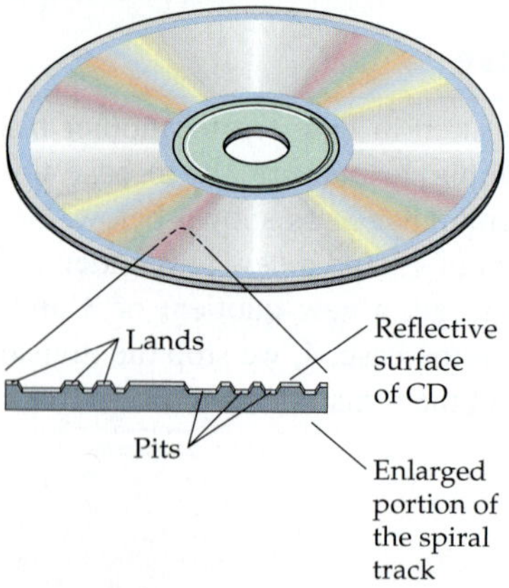

The binary numeration system is used to digitize music (as well as video). Music files played or streamed on a computer or other device are ultimately processed by the machine as a series of 1s and 0s.

For instance, when music is digitized for publication on a CD (compact disc), the sound is sampled 44,100 times per second. Each sound snapshot is converted to a 16-digit binary number representing the amplitude of the audio signal at that moment. The sequential binary numbers are encoded on the surface of the CD as flat regions called lands and small indentations called pits. (See the figure at the left.) A laser beam in a CD player tracks the spiral path, turning the reflections from the lands and lack of reflections from the pits back to a string of 0s and 1s. The player's processor then converts this binary data back to music.

On a typical CD, the spiral path that the laser follows is over 3 miles long and describes several billion binary bits!

Converting Directly between Computer Bases

Although computers compute internally by using base two (binary system), humans generally find it easier to compute with a larger base. Fortunately, there are easy conversion techniques that can be used to convert a base two numeral directly to a base eight (octal) numeral or a base sixteen (hexadecimal) numeral. Before we explain the techniques, it will help to become familiar with the information in Table 6.7, which shows the eight octal symbols and their binary equivalents.

TABLE 6.7
Octal and Binary Equivalents

Octal	Binary
0	000
1	001
2	010
3	011
4	100
5	101
6	110
7	111

To convert from octal to binary, just replace each octal numeral with its 3-bit binary equivalent.

EXAMPLE 6 Convert Directly from Base Eight to Base Two

Convert 5724_{eight} directly to binary form.

Solution

$$\begin{array}{cccc} 5 & 7 & 2 & 4_{\text{eight}} \\ \| & \| & \| & \| \\ 101 & 111 & 010 & 100_{\text{two}} \end{array}$$

$5724_{\text{eight}} = 101111010100_{\text{two}}$

CHECK YOUR PROGRESS 6 Convert 63210_{eight} directly to binary form.

Solution See page S21.

Because every group of three binary bits is equivalent to an octal symbol, we can convert from binary directly to octal by breaking a binary numeral into groups of three (from right to left) and replacing each group with its octal equivalent.

EXAMPLE 7 Convert Directly from Base Two to Base Eight

Convert 11100101_{two} directly to octal form.

Solution

Starting from the right, break the binary numeral into groups of three. Then replace each group with its octal equivalent.

This zero was inserted to make a group of three.

$$\begin{array}{ccc} 011 & 100 & 101_{\text{two}} \\ \| & \| & \| \\ 3 & 4 & 5_{\text{eight}} \end{array}$$

$11100101_{\text{two}} = 345_{\text{eight}}$

CHECK YOUR PROGRESS 7 Convert $111010011100_{\text{two}}$ directly to octal form.

Solution See page S21.

TABLE 6.8
Hexadecimal and Binary Equivalents

Hexadecimal	Binary
0	0000
1	0001
2	0010
3	0011
4	0100
5	0101
6	0110
7	0111
8	1000
9	1001
A	1010
B	1011
C	1100
D	1101
E	1110
F	1111

Table 6.8 shows the hexadecimal symbols and their binary equivalents. To convert from hexadecimal to binary, replace each hexadecimal symbol with its four-bit binary equivalent.

EXAMPLE 8 Convert Directly from Base Sixteen to Base Two

Convert BAD_{sixteen} directly to binary form.

Solution

$$\begin{array}{ccc} B & A & D_{\text{sixteen}} \\ \| & \| & \| \\ 1011 & 1010 & 1101_{\text{two}} \end{array}$$

$BAD_{\text{sixteen}} = 101110101101_{\text{two}}$

CHECK YOUR PROGRESS 8 Convert $C5A_{\text{sixteen}}$ directly to binary form.

Solution See page S21.

Because every group of four binary bits is equivalent to a hexadecimal symbol, we can convert from binary to hexadecimal by breaking the binary numeral into groups of four (from right to left) and replacing each group with its hexadecimal equivalent.

EXAMPLE 9 Convert Directly from Base Two to Base Sixteen

Convert 10110010100011_{two} directly to hexadecimal form.

Solution

Starting from the right, break the binary numeral into groups of four. Replace each group with its hexadecimal equivalent.

Insert two zeros to make a group of four.

0010	1100	1010	0011_{two}
‖	‖	‖	‖
2	C	A	3

$10110010100011_{two} = 2CA3_{sixteen}$

CHECK YOUR PROGRESS 9 Convert 101000111010010_{two} directly to hexadecimal form.

Solution See page S21.

The Double-Dabble Method

There is a shortcut that can be used to convert a base two numeral to base ten. The advantage of this shortcut, called the *double-dabble method,* is that you can start at the left of the numeral and work your way to the right without first determining the place value of each bit in the base two numeral.

INSTRUCTOR NOTE

Many students enjoy using the double-dabble method. It is best to demonstrate this method first with small base two numerals, such as 10100_{two}.

EXAMPLE 10 Apply the Double-Dabble Method

Use the double-dabble method to convert 1011001_{two} to base ten.

Solution

Start at the left with the first 1 and move to the right. Every time you pass by a 0, double your current number. Every time you pass by a 1, dabble. Dabbling is accomplished by doubling your current number and adding 1.

	$2 \cdot 1$	$2 \cdot 2 + 1$	$2 \cdot 5 + 1$	$2 \cdot 11$	$2 \cdot 22$	$2 \cdot 44 + 1$
	double	dabble	dabble	double	double	dabble
	2	5	11	22	44	89
1	0	1	1	0	0	1_{two}

As we pass by the final 1 in the units place, we dabble 44 to get 89. Thus $1011001_{two} = 89$.

CHECK YOUR PROGRESS 10 Use the double-dabble method to convert 1110010_{two} to base ten.

Solution See page S21.

EXCURSION

Information Retrieval via a Binary Search

To complete this Excursion, you must first construct a set of 31 cards that we refer to as a deck of *binary cards.* Templates and directions for constructing the cards are available on our companion site at CengageBrain.com, under the file name Binary Cards. Ask your instructor for the complete details. Use a computer to print the templates onto a medium-weight card stock similar to that used for playing cards.

INSTRUCTOR NOTE

The templates and the directions for the binary cards are available at

CengageBrain.com

Select the "Mathematical Excursions, 4th Edition" link, the "Instructors Companion Site" link and then select the "Binary Cards" link.

We are living in the information age, but information is not useful if it cannot be retrieved when you need it. The binary numeration system is vital to the retrieval of information. To illustrate the connection between retrieval of information and the binary system, examine the card in the following figure. The card is labeled with the base ten numeral 20, and the holes and notches at the top of the card represent 20 in binary notation. A hole is used to indicate a 1, and a notch is used to indicate a 0. In the figure, the card has holes in the third and fifth binary-place-value positions (counting from right to left) and notches cut out of the first, second, and fourth positions.

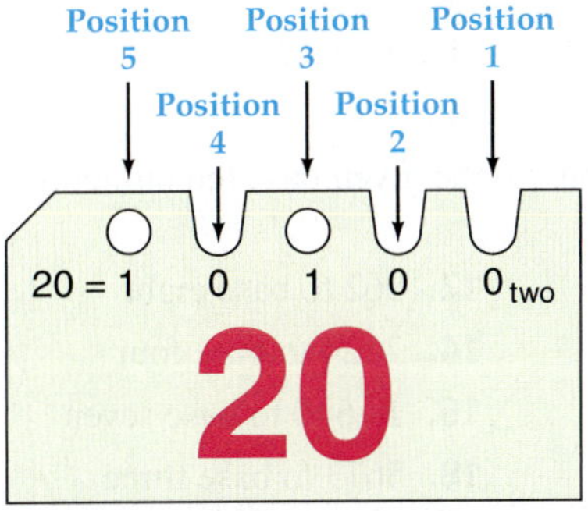

After you have constructed your deck of binary cards, take a few seconds to shuffle the deck. To find the card labeled with the numeral 20, complete the following process.

1. Use a thin dowel (or the tip of a sharp pencil) to lift out the cards that have a hole in the fifth position. *Keep* these cards and set the other cards off to the side.
2. From the cards that are *kept,* use the dowel to lift out the cards with a hole in the fourth position. Set these cards off to the side.
3. From the cards that are *kept,* use the dowel to lift out the cards that have a hole in the third position. *Keep* these cards and place the others off to the side.
4. From the cards that are *kept,* use the dowel to lift out the cards with a hole in the second position. Set these cards off to the side.
5. From the cards that are *kept,* use the dowel to lift out the card that has a hole in the first position. Set this card off to the side.

   The card that remains is the card labeled with the numeral 20. You have just completed a binary search.

EXCURSION EXERCISES

The binary numeration system can also be used to implement a *sorting* procedure. To illustrate, shuffle your deck of cards. Use the dowel to lift out the cards that have a hole in the rightmost position. Place these cards, *face up,* behind the other cards. Now use the dowel to lift out the cards that have a hole in the next position to the left. Place these cards, face up, behind the other cards. Continue this process of lifting out the cards in the next position to the left and placing them behind the other cards until you have completed the process for all five positions.

1. Examine the numerals on the cards. What do you notice about the order of the numerals? Explain why they are in this order.

2. If you wanted to sort 1000 cards from smallest to largest value by using the binary sort procedure, how many positions (where each position is either a hole or a notch) would be required at the top of each card? How many positions are needed to sort 10,000 cards?
3. Explain why the above sorting procedure cannot be implemented with base three cards.

EXERCISE SET 6.3

■ In Exercises 1 to 10, convert the given numeral to base ten.

1. 243_{five}
2. 145_{seven}
3. 67_{nine}
4. 573_{eight}
5. 3154_{six}
6. 735_{eight}
7. 13211_{four}
8. 102022_{three}
9. $\text{B}5_{\text{sixteen}}$
10. $4\text{A}_{\text{twelve}}$

■ In Exercises 11 to 20, convert the given base ten numeral to the indicated base.

11. 267 to base five
12. 362 to base eight
13. 1932 to base six
14. 2024 to base four
15. 15,306 to base nine
16. 18,640 to base seven
17. 4060 to base two
18. 5673 to base three
19. 283 to base twelve
20. 394 to base sixteen

■ In Exercises 21 to 28, use expanded forms to convert the given base two numeral to base ten.

21. 1101_{two}
22. 10101_{two}
23. 11011_{two}
24. 101101_{two}
25. 1100100_{two}
26. 11110101000_{two}
27. 10001011_{two}
28. 110110101_{two}

■ In Exercises 29 to 34, use the double-dabble method to convert the given base two numeral to base ten.

29. 101001_{two}
30. 1110100_{two}
31. 1011010_{two}
32. 10001010_{two}
33. 10100111010_{two}
34. 10000000100_{two}

■ In Exercises 35 to 46, convert the given numeral to the indicated base.

35. 34_{six} to base eight
36. 71_{eight} to base five
37. 878_{nine} to base four
38. 546_{seven} to base six
39. 1110_{two} to base five
40. 21200_{three} to base six
41. 3440_{eight} to base nine
42. 1453_{six} to base eight
43. 56_{sixteen} to base eight
44. 43_{twelve} to base six
45. $\text{A}4_{\text{twelve}}$ to base sixteen
46. $\text{C}9_{\text{sixteen}}$ to base twelve

■ In Exercises 47 to 56, convert the given numeral *directly* (without first converting to base ten) to the indicated base.

47. 352_{eight} to base two
48. $\text{A}4_{\text{sixteen}}$ to base two
49. 11001010_{two} to base eight
50. $111011100101_{\text{two}}$ to base sixteen
51. 101010001_{two} to base sixteen
52. 56721_{eight} to base two
53. $\text{BEF}3_{\text{sixteen}}$ to base two
54. $6\text{A}7\text{B}8_{\text{sixteen}}$ to base two
55. $\text{BA}5\text{CF}_{\text{sixteen}}$ to base two
56. 47134_{eight} to base two
57. **The Triple-Whipple-Zipple Method** There is a procedure that can be used to convert a base three numeral directly to base ten without using the expanded form of the numeral. Write an explanation of this procedure, which we will call the *triple-whipple-zipple* method. *Hint:* The method is an extension of the double-dabble method.
58. Determine whether the following statements are true or false.

    a. A number written in base two is divisible by 2 if and only if the number ends with a 0.

    b. In base six, the next counting number after 55_{six} is 100_{six}.

    c. In base sixteen, the next counting number after $3\text{BF}_{\text{sixteen}}$ is $3\text{C}0_{\text{sixteen}}$.

EXTENSIONS

59. **The ASCII Code** ASCII, pronounced *ask-key,* is an acronym for the American Standard Code for Information Interchange. In this code, each of the characters that can be typed on a computer keyboard is represented by a hexadecimal numeral. Research the topic of ASCII and then write a sentence or two that illustrates how to determine the ASCII code for the letter A, from an ASCII chart.

60. **The Postnet Code** The U.S. Postal Service uses a *Postnet code* to write zip codes + 4 on envelopes. The Postnet code is a bar code that is based on the binary numeration system. Postnet code is very useful because it can be read by a machine. Write a few sentences that explain how to convert a zip code + 4 to its Postnet code. What is the Postnet code for your zip code + 4?

The D'ni Numeration System In the computer game *Riven*, a D'ni numeration system is used. Although the D'ni numeration system is a base twenty-five numeration system with 25 distinct numerals, you really need to memorize only the first 5 numerals, which are shown below.

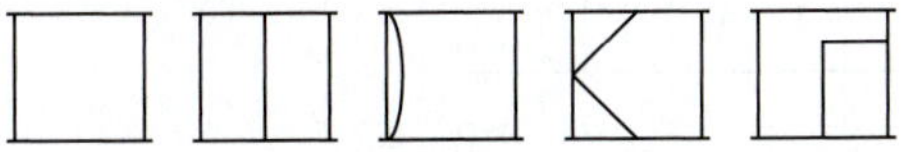

The basic D'ni numerals

If two D'ni numerals are placed side by side, then the numeral on the left is in the twenty-fives place and the numeral on the right is in the ones place. Thus  is the D'ni numeral for $(3 \times 25) + (2 \times 1) = 77$.

61. Convert the following D'ni numeral to base ten.

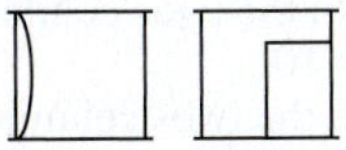

62. Convert the following D'ni numeral to base ten.

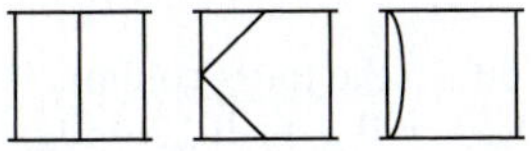

Rotating any of the D'ni numerals for 1, 2, 3, and 4 by a 90° counterclockwise rotation produces a numeral with a value five times its original value. For instance, rotating the numeral for 1 produces 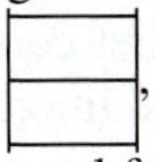, which is the D'ni numeral for 5, and rotating the numeral for 2 produces 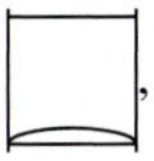, which is the D'ni numeral for 10.

63. Write the D'ni numeral for 15.

64. Write the D'ni numeral for 20.

In the D'ni numeration system explained above, many numerals are obtained by rotating a basic numeral and then overlaying it on one of the basic numerals. For instance, if you rotate the D'ni numeral for 1, you get the numeral for 5. If you then overlay the numeral for 5 on the numeral for 1, you get the numeral for $5 + 1 = 6$.

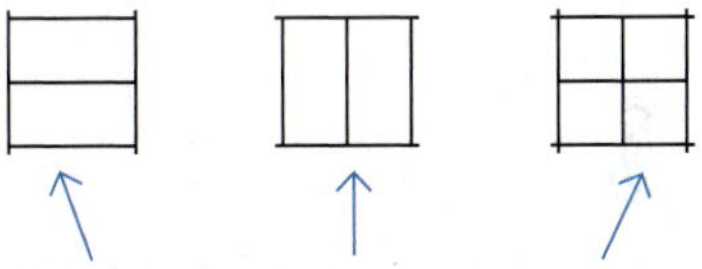

5 overlayed on 1 produces 6.

65. Write the D'ni numeral for 8.

66. Write the D'ni numeral for 22.

67. Convert the following D'ni numeral to base ten.

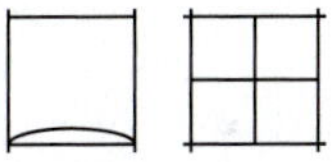

68. Convert the following D'ni numeral to base ten.

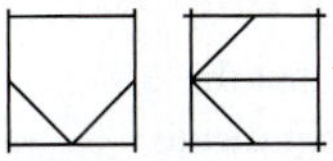

69. **a.** State one advantage of the hexadecimal numeration system over the decimal numeration system.

b. State one advantage of the decimal numeration system over the hexadecimal numeration system.

70. **a.** State one advantage of the D'ni numeration system over the decimal numeration system.

b. State one advantage of the decimal numeration system over the D'ni numeration system.

SECTION 6.4 Arithmetic in Different Bases

Addition in Different Bases

Most computers and calculators make use of the base two (binary) numeration system to perform arithmetic computations. For instance, if you use a calculator to find the sum of 9 and 5, the calculator first converts the 9 to 1001_{two} and the 5 to 101_{two}. The calculator uses electronic circuitry called *binary adders* to find the sum of 1001_{two} and 101_{two} as 1110_{two}. The calculator then converts 1110_{two} to base ten and displays the sum 14. All of

TABLE 6.9
A Binary Addition Table

(Second addend across the top; first addend down the side; sums in the body)

+	0	1
0	0	1
1	1	10

the conversions and the base two addition are done internally in a fraction of a second, which gives the user the impression that the calculator performed the addition in base ten.

The following examples illustrate how to perform arithmetic in different bases. We first consider the operation of addition in the binary numeration system. Table 6.9 is an addition table for base two. It is similar to the base ten addition table that you memorized in elementary school, except that it is much smaller because base two involves only the bits 0 and 1. The numerals shown in red in Table 6.9 illustrate that $1_{two} + 1_{two} = 10_{two}$.

EXAMPLE 1 Add Base Two Numerals

Find the sum of 11110_{two} and 1011_{two}.

Solution

Arrange the numerals vertically, keeping the bits of the same place value in the same column.

	THIRTY-TWOS	SIXTEENS	EIGHTS	FOURS	TWOS	ONES
		1	1	1	1	0_{two}
+			1	0	1	1_{two}
						1_{two}

Start by adding the bits in the ones column: $0_{two} + 1_{two} = 1_{two}$. Then move left and add the bits in the twos column. When the sum of the bits in a column exceeds 1, the addition will involve carrying, as shown below.

TAKE NOTE

In this section, assume that the small numerals, used to indicate a carry, are written in the same base as the numerals in the given problem.

	THIRTY-TWOS	SIXTEENS	EIGHTS	FOURS	TWOS	ONES
				1		
		1	1	1	1	0_{two}
+			1	0	1	1_{two}
					0	1_{two}

Add the bits in the twos column.
$1_{two} + 1_{two} = 10_{two}$
Write the 0 in the twos column and carry the 1 to the fours column.

	THIRTY-TWOS	SIXTEENS	EIGHTS	FOURS	TWOS	ONES
			1	1		
		1	1	1	1	0_{two}
+			1	0	1	1_{two}
				0	0	1_{two}

Add the bits in the fours column.
$(1_{two} + 1_{two}) + 0_{two} = 10_{two} + 0_{two} = 10_{two}$
Write the 0 in the fours column and carry the 1 to the eights column.

	THIRTY-TWOS	SIXTEENS	EIGHTS	FOURS	TWOS	ONES
	1	1	1	1		
		1	1	1	1	0_{two}
+			1	0	1	1_{two}
	1	0	1	0	0	1_{two}

Add the bits in the eights column.
$(1_{two} + 1_{two}) + 1_{two} = 10_{two} + 1_{two} = 11_{two}$
Write a 1 in the eights column and carry a 1 to the sixteens column. Continue to add the bits in each column to the left of the eights column.

The sum of 11110_{two} and 1011_{two} is 101001_{two}.

CHECK YOUR PROGRESS 1 Find the sum of 11001_{two} and 1101_{two}.

Solution See page S21.

TABLE 6.10
A Base Four Addition Table

+	0	1	2	3
0	0	1	2	3
1	1	2	3	10
2	2	3	10	11
3	3	10	11	12

There are four symbols in base four, namely 0, 1, 2, and 3. Table 6.10 shows a base four addition table that lists all the sums that can be produced by adding two base four digits. The numerals shown in red in Table 6.10 illustrate that $2_{\text{four}} + 3_{\text{four}} = 11_{\text{four}}$.

In the next example, we compute the sum of two numbers written in base four.

EXAMPLE 2 Add Base Four Numerals

Find the sum of 23_{four} and 13_{four}.

Solution

Arrange the numbers vertically, keeping the numerals of the same place value in the same column.

	SIXTEENS	FOURS	ONES
		1	
		2	3_{four}
+		1	3_{four}
			2_{four}

Add the numerals in the ones column.
Table 6.10 shows that $3_{\text{four}} + 3_{\text{four}} = 12_{\text{four}}$.
Write the 2 in the ones column and carry the 1 to the fours column.

	SIXTEENS	FOURS	ONES
	1	1	
		2	3_{four}
+		1	3_{four}
	1	0	2_{four}

Add the numerals in the fours column.
$(1_{\text{four}} + 2_{\text{four}}) + 1_{\text{four}} = 3_{\text{four}} + 1_{\text{four}} = 10_{\text{four}}$
Write the 0 in the fours column and carry the 1 to the sixteens column. Bring down the 1 that was carried to the sixteens column to form the sum 102_{four}.

The sum of 23_{four} and 13_{four} is 102_{four}.

CHECK YOUR PROGRESS 2 Find $32_{\text{four}} + 12_{\text{four}}$.

Solution See page S21.

In the previous examples, we used a table to determine the necessary sums. However, it is generally quicker to find a sum by computing the base ten sum of the numerals in each column and then converting each base ten sum back to its equivalent in the given base. The next two examples illustrate this summation technique.

EXAMPLE 3 Add Base Six Numerals

Find $25_{\text{six}} + 32_{\text{six}} + 42_{\text{six}}$.

Solution

Arrange the numbers vertically, keeping the numerals of the same place value in the same column.

	THIRTY-SIXES	SIXES	ONES
		1	
		2	5_{six}
		3	2_{six}
+		4	2_{six}
			3_{six}

Add the numerals in the ones column:
$5_{\text{six}} + 2_{\text{six}} + 2_{\text{six}} = 5 + 2 + 2 = 9$.
Convert 9 to base six. $(9 = 13_{\text{six}})$
Write the 3 in the ones column and carry the 1 to the sixes column.

$$\begin{array}{rccl} & 1 & 1 & \\ & & 2 & 5_{six} \\ & & 3 & 2_{six} \\ + & & 4 & 2_{six} \\ \hline & 1 & 4 & 3_{six} \end{array}$$

Add the numerals in the sixes column and convert the sum to base six. $1_{six} + 2_{six} + 3_{six} + 4_{six} = 1 + 2 + 3 + 4 = 10 = 14_{six}$ Write the 4 in the sixes column and carry the 1 to the thirty-sixes column. Bring down the 1 that was carried to the thirty-sixes column to form the sum 143_{six}.

$25_{six} + 32_{six} + 42_{six} = 143_{six}$

CHECK YOUR PROGRESS 3 Find $35_{seven} + 46_{seven} + 24_{seven}$.

Solution See page S21.

In the next example, we solve an addition problem that involves a base greater than ten.

EXAMPLE 4 Add Base Twelve Numerals

Find $A97_{twelve} + 8BA_{twelve}$.

Solution

ONE HUNDRED FORTY-FOURS | TWELVES | ONES

$$\begin{array}{rccl} & & 1 & \\ & A & 9 & 7_{twelve} \\ + & 8 & B & A_{twelve} \\ \hline & & & 5_{twelve} \end{array}$$

Add the numerals in the ones column.
$7_{twelve} + A_{twelve} = 7 + 10 = 17$
Convert 17 to base twelve $(17 = 15_{twelve})$.
Write the 5 in the ones column and carry the 1 to the twelves column.

$$\begin{array}{rccl} & 1 & 1 & \\ & A & 9 & 7_{twelve} \\ + & 8 & B & A_{twelve} \\ \hline & & 9 & 5_{twelve} \end{array}$$

Add the numerals in the twelves column.
$1_{twelve} + 9_{twelve} + B_{twelve} = 1 + 9 + 11 = 21 = 19_{twelve}$
Write the 9 in the twelves column and carry the 1 to the one hundred forty-fours column.

$$\begin{array}{rcccl} & 1 & 1 & 1 & \\ & & A & 9 & 7_{twelve} \\ + & & 8 & B & A_{twelve} \\ \hline & 1 & 7 & 9 & 5_{twelve} \end{array}$$

Add the numerals in the one hundred forty-fours column.
$1_{twelve} + A_{twelve} + 8_{twelve} = 1 + 10 + 8 = 19 = 17_{twelve}$
Write the 7 in the one hundred forty-fours column and carry the 1 to the one thousand seven hundred twenty-eights column. Bring down the 1 that was carried to the one thousand seven hundred twenty-eights column to form the sum 1795_{twelve}.

$A97_{twelve} + 8BA_{twelve} = 1795_{twelve}$

CHECK YOUR PROGRESS 4 Find $AC4_{sixteen} + 6E8_{sixteen}$.

Solution See page S21.

TAKE NOTE

In the following subtraction, 7 is the *minuend* and 4 is the *subtrahend.*

$7 - 4 = 3$

Subtraction in Different Bases

To subtract two numbers written in the same base, begin by arranging the numbers vertically, keeping numerals that have the same place value in the same column. It will be necessary to borrow whenever a numeral in the subtrahend is greater than its corresponding numeral in the minuend. Every number that is borrowed will be a power of the base.

EXAMPLE 5 Subtract Base Seven Numerals

Find $463_{seven} - 124_{seven}$.

Solution

Arrange the numbers vertically, keeping the numerals of the same place value in the same column.

	FORTY-NINES	SEVENS	ONES
		5 + 1	
	4	~~6~~	3_{seven}
−	1	2	4_{seven}

Because $4_{seven} > 3_{seven}$, it is necessary to borrow from the 6 in the sevens column.
(6 sevens = 5 sevens + 1 seven)

	FORTY-NINES	SEVENS	ONES
		5	10
	4	~~6~~	3_{seven}
−	1	2	4_{seven}

Borrow 1 seven from the sevens column and add $7 = 10_{seven}$ to the 3_{seven} in the ones column.

	FORTY-NINES	SEVENS	ONES
		5	13
	4	~~6~~	~~3~~$_{seven}$
−	1	2	4_{seven}
	3	3	6_{seven}

Subtract the numerals in each column. The 6_{seven} in the ones column was produced by the following arithmetic:

$$13_{seven} - 4_{seven} = 10 - 4 = 6 = 6_{seven}$$

$463_{seven} - 124_{seven} = 336_{seven}$

CHECK YOUR PROGRESS 5 Find $365_{nine} - 183_{nine}$.

Solution See page S21.

TAKE NOTE

In this section, assume that the small numerals, which are used to illustrate the borrowing process, are written in the same base as the numerals in the given subtraction problem.

TABLE 6.11
Decimal and Hexadecimal Equivalents

Base ten decimal	Base sixteen hexadecimal
0	0
1	1
2	2
3	3
4	4
5	5
6	6
7	7
8	8
9	9
10	A
11	B
12	C
13	D
14	E
15	F

EXAMPLE 6 Subtract Base Sixteen Numerals

Find $7AB_{sixteen} - 3E4_{sixteen}$.

Solution

Table 6.11 shows the hexadecimal numerals and their decimal equivalents. Because $B_{sixteen}$ is greater than $4_{sixteen}$, there is no need to borrow to find the difference in the ones column. However, $A_{sixteen}$ is less than $E_{sixteen}$, so it is necessary to borrow to find the difference in the sixteens column.

	TWO HUNDRED FIFTY-SIXES	SIXTEENS	ONES
	7	A	$B_{sixteen}$
−	3	E	$4_{sixteen}$
			$7_{sixteen}$

$$B_{sixteen} - 4_{sixteen} = 11 - 4 = 7 = 7_{sixteen}$$

	TWO HUNDRED FIFTY-SIXES	SIXTEENS	ONES
	6	10	
	~~7~~	A	$B_{sixteen}$
−	3	E	$4_{sixteen}$
			$7_{sixteen}$

Borrow 1 two hundred fifty-six from the two hundred fifty-sixes column and write 10 in the sixteens column.

	TWO HUNDRED FIFTY-SIXES	SIXTEENS	ONES
	6	1A	
	~~7~~	~~A~~	$B_{sixteen}$
−	3	E	$4_{sixteen}$
	3	C	$7_{sixteen}$

$10_{sixteen} + A_{sixteen} = 1A_{sixteen}$
The C in the sixteens column was produced by the following arithmetic:

$$1A_{sixteen} - E_{sixteen} = 26 - 14 = 12 = C_{sixteen}$$

$7AB_{sixteen} - 3E4_{sixteen} = 3C7_{sixteen}$

CHECK YOUR PROGRESS 6 Find $83A_{twelve} - 467_{twelve}$.

Solution See page S21.

Multiplication in Different Bases

To perform multiplication in bases other than base ten, it is often helpful to first write a multiplication table for the given base. Table 6.12 shows a multiplication table for base four. The numbers shown in red in the table illustrate that $2_{four} \times 3_{four} = 12_{four}$. You can verify this result by converting the numbers to base ten, multiplying in base ten, and then converting back to base four. Here is the actual arithmetic.

$$2_{four} \times 3_{four} = 2 \times 3 = 6 = 12_{four}$$

TABLE 6.12
A Base Four Multiplication Table

×	0	1	2	3
0	0	0	0	0
1	0	1	2	3
2	0	2	10	12
3	0	3	12	21

QUESTION What is $5_{six} \times 4_{six}$?

EXAMPLE 7 Multiply Base Four Numerals

Use the base four multiplication table to find $3_{four} \times 123_{four}$.

Solution

Align the numerals of the same place value in the same column. Use Table 6.12 to multiply 3_{four} times each numeral in 123_{four}. If any of these multiplications produces a two-numeral product, then write down the numeral on the right and carry the numeral on the left.

	SIXTY-FOURS	SIXTEENS	FOURS	ONES
(carry)			2	
		1	2	3_{four}
×				3_{four}
				1_{four}

$3_{four} \times 3_{four} = 21_{four}$
Write the 1 in the ones column and carry the 2.

	SIXTY-FOURS	SIXTEENS	FOURS	ONES
(carry)		2	2	
		1	2	3_{four}
×				3_{four}
			0	1_{four}

$3_{four} \times 2_{four} = 12_{four}$
$12_{four} + 2_{four}$(the carry) $= 20_{four}$
Write the 0 in the fours column and carry the 2.

	SIXTY-FOURS	SIXTEENS	FOURS	ONES
(carry)	1	2	2	
		1	2	3_{four}
×				3_{four}
	1	1	0	1_{four}

$3_{four} \times 1_{four} = 3_{four}$
$3_{four} + 2_{four}$(the carry) $= 11_{four}$
Write a 1 in the sixteens column and carry a 1 to the sixty-fours column. Bring down the 1 that was carried to the sixty-fours column to form the product 1101_{four}.

$3_{four} \times 123_{four} = 1101_{four}$

CHECK YOUR PROGRESS 7 Find $2_{four} \times 213_{four}$.

Solution See page S21.

Writing all of the entries in a multiplication table for a large base such as base twelve can be time-consuming. In such cases you may prefer to multiply in base ten and then

ANSWER $5_{six} \times 4_{six} = 20 = 32_{six}$.

convert each product back to the given base. The next example illustrates this multiplication method.

EXAMPLE 8 Multiply Base Twelve Numerals

Find $53_{\text{twelve}} \times 27_{\text{twelve}}$.

Solution

Align the numerals of the same place value in the same column. Start by multiplying each numeral of the multiplicand (53_{twelve}) by the ones numeral of the multiplier (27_{twelve}).

$$\begin{array}{c|c|c|l} & \text{ONE HUNDRED FORTY-FOURS} & \text{TWELVES} & \text{ONES} \\ & & 1 & \\ & & 5 & 3_{\text{twelve}} \\ \times & & 2 & 7_{\text{twelve}} \\ \hline & & & 9_{\text{twelve}} \end{array}$$

$$\begin{aligned} 7_{\text{twelve}} \times 3_{\text{twelve}} &= 7 \times 3 \\ &= 21 \\ &= 19_{\text{twelve}} \end{aligned}$$

Write the 9 in the ones column and carry the 1.

$$\begin{array}{c|c|c|l} & \text{ONE HUNDRED FORTY-FOURS} & \text{TWELVES} & \text{ONES} \\ & & 1 & \\ & & 5 & 3_{\text{twelve}} \\ \times & & 2 & 7_{\text{twelve}} \\ \hline & 3 & 0 & 9_{\text{twelve}} \end{array}$$

$$\begin{aligned} 7_{\text{twelve}} \times 5_{\text{twelve}} &= 7 \times 5 \\ &= 35 \\ 35 + 1\,(\text{the carry}) &= 36 \\ &= 30_{\text{twelve}} \end{aligned}$$

Write the 0 in the twelves column and write the 3 in the one hundred forty-fours column.

Now multiply each numeral of the multiplicand by the twelves numeral of the multiplier.

$$\begin{array}{c|c|c|l} & \text{ONE HUNDRED FORTY-FOURS} & \text{TWELVES} & \text{ONES} \\ & & 5 & 3_{\text{twelve}} \\ \times & & 2 & 7_{\text{twelve}} \\ \hline & 3 & 0 & 9_{\text{twelve}} \\ & & 6_{\text{twelve}} & \end{array}$$

$$\begin{aligned} 2_{\text{twelve}} \times 3_{\text{twelve}} &= 2 \times 3 \\ &= 6 \\ &= 6_{\text{twelve}} \end{aligned}$$

Write the 6 in the twelves column.

$$\begin{array}{c|c|c|c|l} & & \text{ONE HUNDRED FORTY-FOURS} & \text{TWELVES} & \text{ONES} \\ & & & 5 & 3_{\text{twelve}} \\ \times & & & 2 & 7_{\text{twelve}} \\ \hline & & 3 & 0 & 9_{\text{twelve}} \\ & & A & 6_{\text{twelve}} & \\ \hline & 1 & 1 & 6 & 9_{\text{twelve}} \end{array}$$

$$\begin{aligned} 2_{\text{twelve}} \times 5_{\text{twelve}} &= 2 \times 5 \\ &= 10 \\ &= A_{\text{twelve}} \end{aligned}$$

Write the A in the one hundred forty-fours column. Now add to produce the product 1169_{twelve}.

$53_{\text{twelve}} \times 27_{\text{twelve}} = 1169_{\text{twelve}}$

CHECK YOUR PROGRESS 8 Find $25_{\text{eight}} \times 34_{\text{eight}}$.

Solution See page S21.

MATH MATTERS The Fields Medal

The International Mathematical Union

The Fields Medal

A Nobel Prize is awarded each year in the categories of chemistry, physics, physiology, medicine, literature, and peace. However, no award is given in mathematics. Why Alfred Nobel chose not to provide an award in the category of mathematics is unclear.

The Canadian mathematician John Charles Fields (1863–1932) felt that a prestigious award should be given in the area of mathematics. Fields helped establish the Fields Medal, which was first given to Lars Valerian Ahlfors and Jesse Douglas in 1936. The International Congress of Mathematicians had planned to give two Fields Medals every four years after 1936, but because of World War II, the next Fields Medals were not given until 1950.

It was Fields' wish that the Fields Medal recognize both existing work and the promise of future achievement. Because of this concern for future achievement, the International Congress of Mathematicians decided to restrict those eligible for the Fields Medal to mathematicians under the age of 40.

Division in Different Bases

To perform a division in a base other than base ten, it is helpful to first make a list of a few multiples of the divisor. This procedure is illustrated in the following example.

INSTRUCTOR NOTE

Many students find division with non–base ten numerals to be difficult. If you are short on time, you may find that it is best to skip division with non–base ten numerals.

Instructors who decide to cover this topic may find it helpful to first review division using a base ten example.

EXAMPLE 9 Divide Base Seven Numerals

Find $253_{\text{seven}} \div 3_{\text{seven}}$.

Solution

First list a few multiples of the divisor 3_{seven}.

$3_{\text{seven}} \times 0_{\text{seven}} = 3 \times 0 = 0 = 0_{\text{seven}}$ $\quad 3_{\text{seven}} \times 4_{\text{seven}} = 3 \times 4 = 12 = 15_{\text{seven}}$

$3_{\text{seven}} \times 1_{\text{seven}} = 3 \times 1 = 3 = 3_{\text{seven}}$ $\quad 3_{\text{seven}} \times 5_{\text{seven}} = 3 \times 5 = 15 = 21_{\text{seven}}$

$3_{\text{seven}} \times 2_{\text{seven}} = 3 \times 2 = 6 = 6_{\text{seven}}$ $\quad 3_{\text{seven}} \times 6_{\text{seven}} = 3 \times 6 = 18 = 24_{\text{seven}}$

$3_{\text{seven}} \times 3_{\text{seven}} = 3 \times 3 = 9 = 12_{\text{seven}}$

Because $3_{\text{seven}} \times 6_{\text{seven}} = 24_{\text{seven}}$ is slightly less than 25_{seven}, we pick 6 as our first numeral in the quotient when dividing 25_{seven} by 3_{seven}.

```
            6
3seven ) 2 5 3seven      3seven × 6seven = 24seven
         2 4             Subtract 24seven from 25seven.
           1
```

```
quotient
      ↘
         6 3seven
3seven ) 2 5 3seven
         2 4
           1 3           Bring down the 3.
           1 2           3seven × 3seven = 12seven
             1           Subtract 12seven from 13seven.
           ↗
remainder
```

Thus $253_{\text{seven}} \div 3_{\text{seven}} = 63_{\text{seven}}$ with a remainder of 1_{seven}.

CHECK YOUR PROGRESS 9 Find $324_{five} \div 3_{five}$.

Solution See page S22.

In a base two division problem, the only multiples of the divisor used are zero times the divisor and one times the divisor.

EXAMPLE 10 Divide Base Two Numerals

Find $101011_{two} \div 11_{two}$.

Solution

The divisor is 11_{two}. The multiples of the divisor that may be needed are $11_{two} \times 0_{two} = 0_{two}$ and $11_{two} \times 1_{two} = 11_{two}$. Also note that because $10_{two} - 1_{two} = 2 - 1 = 1 = 1_{two}$, we know that

$$\begin{array}{r} 10_{two} \\ -\ \underline{1_{two}} \\ 1_{two} \end{array}$$

$$\begin{array}{r} 1\,1\,1\,0_{two} \\ 11_{two}\overline{)1\,0\,1\,0\,1\,1_{two}} \\ \underline{1\,1}\phantom{\,0\,1\,1_{two}} \\ 1\,0\,0\phantom{\,1\,1_{two}} \\ \underline{1\,1}\phantom{\,1\,1_{two}} \\ 1\,1\phantom{\,1_{two}} \\ \underline{1\,1}\phantom{\,1_{two}} \\ 0\,1\phantom{_{two}} \\ \underline{0}\phantom{_{two}} \\ 1\phantom{_{two}} \end{array}$$

Therefore, $101011_{two} \div 11_{two} = 1110_{two}$ with a remainder of 1_{two}.

CHECK YOUR PROGRESS 10 Find $1110011_{two} \div 10_{two}$.

Solution See page S22.

EXCURSION

Subtraction in Base Two via the Ones Complement and the End-Around Carry

Computers and calculators are often designed so that the number of required circuits is minimized. Instead of using separate circuits to perform addition and subtraction, engineers make use of an *end-around carry procedure* that uses addition to perform subtraction. The end-around carry procedure also makes use of the ones complement of a number. In base two, the ones complement of 0 is 1 and the ones complement of 1 is 0. Thus the ones complement of any base two number can be found by changing each 1 to a 0 and each 0 to a 1.

INSTRUCTOR NOTE

The concepts in this Excursion are an extension of the concepts in the Excursion in Section 6.2, page 308.

Subtraction Using the Ones Complement and the End-Around Carry

To subtract a base two number from a larger base two number:

1. Add the ones complement of the subtrahend to the minuend.
2. Take away 1 from the leftmost bit of the sum and add 1 to the units bit. This is referred to as the end-around carry procedure.

The following example illustrates the process of subtracting 1001_{two} from 1101_{two} using the ones complement and the end-around carry procedure.

$$\begin{array}{r} 1101_{two} \\ -\ \underline{1001_{two}} \end{array}$$ Minuend / Subtrahend

$$\begin{array}{r} 1101_{two} \\ +\ \underline{0110_{two}} \\ 10011_{two} \end{array}$$ Replace the subtrahend with the ones complement of the subtrahend and add.

$$\begin{array}{r} 10011_{two} \\ +\ \underline{1_{two}} \\ 100_{two} \end{array}$$ Take away 1 from the leftmost bit and add 1 to the ones bit. This is the end-around carry procedure.

$1101_{two} - 1001_{two} = 100_{two}$

If the subtrahend has fewer bits than the minuend, leading zeros should be inserted in the subtrahend so that it has the same number of bits as the minuend. This process is illustrated below for the subtraction $1010110_{two} - 11001_{two}$.

$$\begin{array}{r} 1010110_{two} \\ -\ \underline{11001_{two}} \end{array}$$ Minuend / Subtrahend

$$\begin{array}{r} 1010110_{two} \\ -\ \underline{0011001_{two}} \end{array}$$ Insert two leading zeros.

$$\begin{array}{r} 1010110_{two} \\ +\ \underline{1100110_{two}} \\ 10111100_{two} \end{array}$$ Ones complement of subtrahend

$$\begin{array}{r} 10111100_{two} \\ +\ \underline{1_{two}} \\ 111101_{two} \end{array}$$ Take away 1 from the leftmost bit and add 1 to the ones bit.

$1010110_{two} - 11001_{two} = 111101_{two}$

EXCURSION EXERCISES

Use the ones complement of the subtrahend and the end-around carry method to find each difference. State each answer as a base two numeral.

1. $1110_{two} - 1001_{two}$
2. $101011_{two} - 100010_{two}$
3. $101001010_{two} - 1011101_{two}$
4. $111011100110_{two} - 101010100_{two}$
5. $1111101011_{two} - 1001111_{two}$
6. $1110010101100_{two} - 100011110_{two}$

EXERCISE SET 6.4

■ In Exercises 1 to 12, find each sum in the same base as the given numerals.

1. $204_{\text{five}} + 123_{\text{five}}$
2. $323_{\text{four}} + 212_{\text{four}}$
3. $5625_{\text{seven}} + 634_{\text{seven}}$
4. $1011_{\text{two}} + 101_{\text{two}}$
5. $110101_{\text{two}} + 10011_{\text{two}}$
6. $11001010_{\text{two}} + 1100111_{\text{two}}$
7. $8B5_{\text{twelve}} + 578_{\text{twelve}}$
8. $379_{\text{sixteen}} + 856_{\text{sixteen}}$
9. $C489_{\text{sixteen}} + BAD_{\text{sixteen}}$
10. $221_{\text{three}} + 122_{\text{three}}$
11. $435_{\text{six}} + 245_{\text{six}}$
12. $5374_{\text{eight}} + 615_{\text{eight}}$

■ In Exercises 13 to 24, find each difference in the same base as the given numerals.

13. $434_{\text{five}} - 143_{\text{five}}$
14. $534_{\text{six}} - 241_{\text{six}}$
15. $7325_{\text{eight}} - 563_{\text{eight}}$
16. $6148_{\text{nine}} - 782_{\text{nine}}$
17. $11010_{\text{two}} - 1011_{\text{two}}$
18. $111001_{\text{two}} - 10101_{\text{two}}$
19. $11010100_{\text{two}} - 1011011_{\text{two}}$
20. $9C5_{\text{sixteen}} - 687_{\text{sixteen}}$
21. $43A7_{\text{twelve}} - 289_{\text{twelve}}$
22. $BAB2_{\text{twelve}} - 475_{\text{twelve}}$
23. $762_{\text{nine}} - 367_{\text{nine}}$
24. $3223_{\text{four}} - 133_{\text{four}}$

■ In Exercises 25 to 38, find each product in the same base as the given numerals.

25. $3_{\text{six}} \times 145_{\text{six}}$
26. $5_{\text{seven}} \times 542_{\text{seven}}$
27. $2_{\text{three}} \times 212_{\text{three}}$
28. $4_{\text{five}} \times 4132_{\text{five}}$
29. $5_{\text{eight}} \times 7354_{\text{eight}}$
30. $11_{\text{two}} \times 11011_{\text{two}}$
31. $10_{\text{two}} \times 101010_{\text{two}}$
32. $101_{\text{two}} \times 110100_{\text{two}}$
33. $25_{\text{eight}} \times 453_{\text{eight}}$
34. $43_{\text{six}} \times 1254_{\text{six}}$
35. $132_{\text{four}} \times 1323_{\text{four}}$
36. $43_{\text{twelve}} \times 895_{\text{twelve}}$
37. $5_{\text{sixteen}} \times BAD_{\text{sixteen}}$
38. $23_{\text{sixteen}} \times 798_{\text{sixteen}}$

■ In Exercises 39 to 49, find each quotient and remainder in the same base as the given numerals.

39. $132_{\text{four}} \div 2_{\text{four}}$
40. $124_{\text{five}} \div 2_{\text{five}}$
41. $231_{\text{four}} \div 3_{\text{four}}$
42. $672_{\text{eight}} \div 5_{\text{eight}}$
43. $5341_{\text{six}} \div 4_{\text{six}}$
44. $11011_{\text{two}} \div 10_{\text{two}}$
45. $101010_{\text{two}} \div 11_{\text{two}}$
46. $1011011_{\text{two}} \div 100_{\text{two}}$
47. $457_{\text{twelve}} \div 5_{\text{twelve}}$
48. $832_{\text{sixteen}} \div 7_{\text{sixteen}}$
49. $234_{\text{five}} \div 12_{\text{five}}$
50. If $232_x = 92$, find the base x.
51. If $143_x = 10200_{\text{three}}$, find the base x.
52. If $46_x = 101010_{\text{two}}$, find the base x.
53. Consider the addition $384 + 245$.
    a. Use base ten addition to find the sum.
    b. Convert 384 and 245 to base two.
    c. Find the base two sum of the base two numbers you found in part b.
    d. Convert the base two sum from part c to base ten.
    e. How does the answer to part a compare with the answer to part d?
54. Consider the subtraction $457 - 318$.
    a. Use base ten subtraction to find the difference.
    b. Convert 457 and 318 to base two.
    c. Find the base two difference of the base two numbers you found in part b.
    d. Convert the base two difference from part c to base ten.
    e. How does the answer to part a compare with the answer to part d?
55. Consider the multiplication 247×26.
    a. Use base ten multiplication to find the product.
    b. Convert 247 and 26 to base two.
    c. Find the base two product of the base two numbers you found in part b.
    d. Convert the base two product from part c to base ten.
    e. How does the answer to part a compare with the answer to part d?

EXTENSIONS

56. Explain the error in the following base eight subtraction.

$$\begin{array}{r} 751_{\text{eight}} \\ -\ 126_{\text{eight}} \\ \hline 625_{\text{eight}} \end{array}$$

57. Determine the base used in the following multiplication.

$$314_{\text{base } x} \times 24_{\text{base } x} = 11202_{\text{base } x}$$

58. The base ten number 12 is an even number. In base seven, 12 is written as 15_{seven}. Is 12 an odd number in base seven?

59. Explain why there is no numeration system with a base of 1.

60. A Cryptarithm In the following base four addition problem, each letter represents one of the numerals 0, 1, 2, or 3. No two different letters represent the same numeral. Determine which numeral is represented by each letter.

$$\begin{array}{rl} \text{NO} & \text{four} \\ +\ \text{AT} & \text{four} \\ \hline \text{NOT} & \text{four} \end{array}$$

61. A Cryptarithm In the following base six addition problem, each letter represents one of the numerals 0, 1, 2, 3, 4, or 5. No two different letters represent the same numeral. Determine which numeral is represented by each letter.

$$\begin{array}{rl} \text{MA} & \text{six} \\ +\ \text{AS} & \text{six} \\ \hline \text{MOM} & \text{six} \end{array}$$

62. Negative Base Numerals It is possible to use a negative number as the base of a numeration system. For instance, the negative base four numeral $32_{\text{negative four}}$ represents the number
$3 \times (-4)^1 + 2 \times (-4)^0 = -12 + 2 = -10.$

a. Convert each of the following negative base numerals to base 10:

$143_{\text{negative five}}$

$74_{\text{negative nine}}$

$10110_{\text{negative two}}$

b. Write -27 as a negative base five numeral.

c. Write 64 as a negative base three numeral.

d. Write 112 as a negative base ten numeral.

SECTION 6.5 Prime Numbers

Prime Numbers

Number theory is a mathematical discipline that is primarily concerned with the properties that are exhibited by the natural numbers. The mathematician Carl Friedrich Gauss established many theorems in number theory. As we noted in the chapter opener, Gauss called mathematics the queen of the sciences and number theory the queen of mathematics. Many topics in number theory involve the concept of a *divisor* or *factor* of a natural number.

Divisor of a Natural Number

The natural number a is a **divisor** or **factor** of the natural number b, provided there exists a natural number j such that $aj = b$.

In less formal terms, a natural number a is a divisor of the natural number b provided $b \div a$ has a remainder of 0. For instance, 10 has divisors of 1, 2, 5, and 10 because each of these numbers divides into 10 with a remainder of 0.

EXAMPLE 1 Find Divisors

Determine all of the natural number divisors of each number.

a. 6 **b.** 42 **c.** 17

Solution

a. Divide 6 by 1, 2, 3, 4, 5, and 6. The division of 6 by 1, 2, 3, and 6 each produces a natural number quotient and a remainder of 0. Thus 1, 2, 3, and 6 are divisors of 6. Dividing 6 by 4 and 6 by 5 does not produce a remainder of 0. Therefore 4 and 5 are not divisors of 6.

b. The only natural numbers from 1 to 42 that divide into 42 with a remainder of 0 are 1, 2, 3, 6, 7, 14, 21, and 42. Thus the divisors of 42 are 1, 2, 3, 6, 7, 14, 21, and 42.

c. The only natural number divisors of 17 are 1 and 17.

CHECK YOUR PROGRESS 1 Determine all of the natural number divisors of each number.

a. 9 **b.** 11 **c.** 24

Solution See page S22.

It is worth noting that every natural number greater than 1 has itself as a factor and 1 as a factor. If a natural number greater than 1 has only 1 and itself as factors, then it is a very special number known as a *prime number*.

Prime Numbers and Composite Numbers

A **prime number** is a natural number greater than 1 that has exactly two factors (divisors): itself and 1.

A **composite number** is a natural number greater than 1 that is not a prime number.

TAKE NOTE

The natural number 1 is neither a prime number nor a composite number.

The 10 smallest prime numbers are 2, 3, 5, 7, 11, 13, 17, 19, 23, and 29. Each of these numbers has only itself and 1 as factors. The 10 smallest composite numbers are 4, 6, 8, 9, 10, 12, 14, 15, 16, and 18.

EXAMPLE 2 Classify a Number as a Prime Number or a Composite Number

Determine whether each number is a prime number or a composite number.

a. 41 **b.** 51 **c.** 119

Solution

a. The only divisors of 41 and 1 are 41. Thus 41 is a prime number.

b. The divisors of 51 are 1, 3, 17, and 51. Thus 51 is a composite number.

c. The divisors of 119 are 1, 7, 17, and 119. Thus 119 is a composite number.

CHECK YOUR PROGRESS 2 Determine whether each number is a prime number or a composite number.

a. 47 **b.** 171 **c.** 91

Solution See page S22.

QUESTION Are all prime numbers odd numbers?

Divisibility Tests

To determine whether one number is divisible by a smaller number, we often apply a **divisibility test**, which is a procedure that enables one to determine whether the smaller number is a divisor of the larger number without actually dividing the smaller number

ANSWER No. The even number 2 is a prime number.

into the larger number. Table 6.13 provides divisibility tests for the numbers 2, 3, 4, 5, 6, 8, 9, 10, and 11.

TABLE 6.13
Base Ten Divisibility Tests

A number is divisible by the following divisor if:	Divisibility test	Example
2	The number is an even number.	846 is divisible by 2 because 846 is an even number.
3	The sum of the digits of the number is divisible by 3.	531 is divisible by 3 because $5 + 3 + 1 = 9$ is divisible by 3.
4	The last two digits of the number form a number that is divisible by 4.	1924 is divisible by 4 because the last two digits form the number 24, which is divisible by 4.
5	The number ends with a 0 or a 5.	8785 is divisible by 5 because it ends with 5.
6	The number is divisible by 2 and by 3.	972 is divisible by 6 because it is divisible by 2 and also by 3.
8	The last three digits of the number form a number that is divisible by 8.	19,168 is divisible by 8 because the last three digits form the number 168, which is divisible by 8.
9	The sum of the digits of the number is divisible by 9.	621,513 is divisible by 9 because the sum of the digits is 18, which is divisible by 9.
10	The last digit is 0.	970 is divisible by 10 because it ends with 0.
11	Start at one end of the number and compute the sum of every other digit. Next compute the sum of the remaining digits. If the difference of these sums is divisible by 11, then the original number is divisible by 11.	4807 is divisible by 11 because the difference of the sum of the digits shown in blue ($8 + 7 = 15$) and the sum of the remaining digits shown in red ($4 + 0 = 4$) is $15 - 4 = 11$, which is divisible by 11.

PPT

POINT OF INTEREST

Frank Nelson Cole (1861–1926) concentrated his mathematical work in the areas of number theory and group theory. He is well known for his 1903 presentation to the American Mathematical Society. Without speaking a word, he went to a chalkboard and wrote

$$2^{67} - 1 = 147573952589676412927$$

Many mathematicians considered this large number to be a prime number. Then Cole moved to a second chalkboard and computed the product

$$761838257287 \cdot 193707721$$

When the audience saw that this product was identical to the result on the first chalkboard, they gave Cole the only standing ovation ever given during an American Mathematical Society presentation. They realized that Cole had factored $2^{67} - 1$, which was a most remarkable feat, considering that no computers existed at that time.

EXAMPLE 3 Apply Divisibility Tests

Use divisibility tests to determine whether 16,278 is divisible by the following numbers.

a. 2 **b.** 3 **c.** 5 **d.** 8 **e.** 11

Solution

a. Because 16,278 is an even number, it is divisible by 2.

b. The sum of the digits of 16,278 is 24, which is divisible by 3. Therefore, 16,278 is divisible by 3.

c. The number 16,278 does not end with a 0 or a 5. Therefore, 16,278 is not divisible by 5.

d. The last three digits of 16,278 form the number 278, which is not divisible by 8. Thus 16,278 is not divisible by 8.

e. The sum of the digits with even place-value powers is $1 + 2 + 8 = 11$. The sum of the digits with odd place-value powers is $6 + 7 = 13$. The difference of these sums is $13 - 11 = 2$. This difference is not divisible by 11, so 16,278 is not divisible by 11.

CHECK YOUR PROGRESS 3 Use divisibility tests to determine whether 341,565 is divisible by each of the following numbers.

a. 3 **b.** 4 **c.** 10 **d.** 11

Solution See page S22.

Prime Factorization

The **prime factorization** of a composite number is a factorization that contains only prime numbers. Many proofs in number theory make use of the following important theorem.

> **The Fundamental Theorem of Arithmetic**
>
> Every composite number can be written as a unique product of prime numbers (disregarding the order of the factors).

To find the prime factorization of a composite number, rewrite the number as a product of two smaller natural numbers. If these smaller numbers are both prime numbers, then you are finished. If either of the smaller numbers is not a prime number, then rewrite it as a product of smaller natural numbers. Continue this procedure until all factors are primes. In Example 4 we make use of a *tree diagram* to organize the factorization process.

EXAMPLE 4 Find the Prime Factorization of a Number

Determine the prime factorization of the following numbers.

a. 84 **b.** 495 **c.** 4004

Solution

a. The following tree diagrams show two different ways of finding the prime factorization of 84, which is $2 \cdot 2 \cdot 3 \cdot 7 = 2^2 \cdot 3 \cdot 7$. Each number in the tree is equal to the product of the two smaller numbers below it. The numbers (in red) at the extreme ends of the branches are the prime factors.

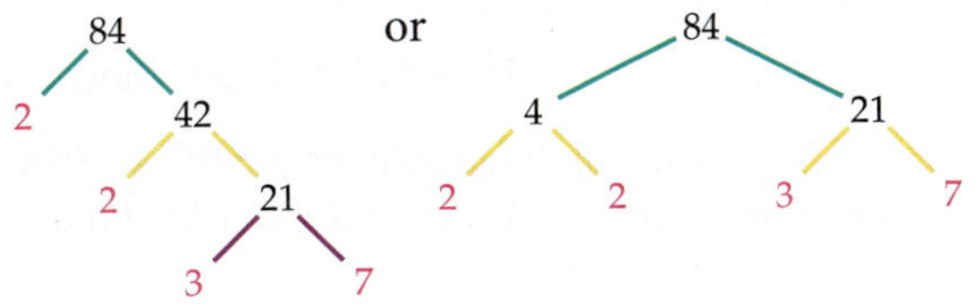

$84 = 2^2 \cdot 3 \cdot 7$

b.

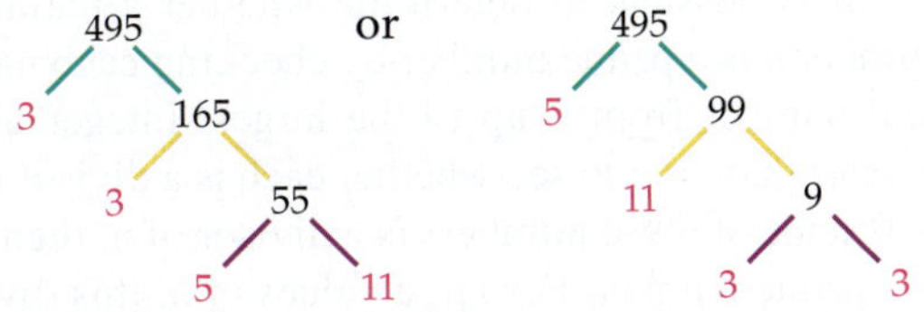

$495 = 3^2 \cdot 5 \cdot 11$

c.

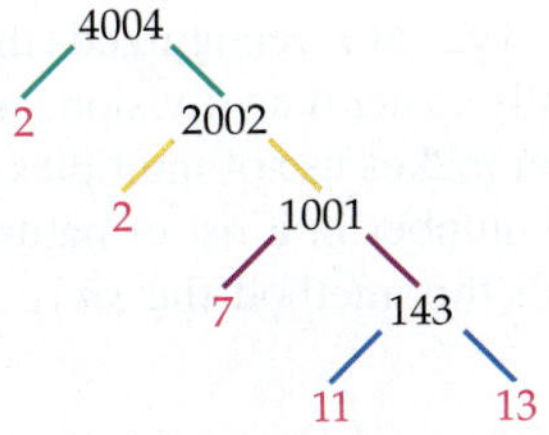

$4004 = 2^2 \cdot 7 \cdot 11 \cdot 13$

CHECK YOUR PROGRESS 4 Determine the prime factorization of the following numbers.

a. 315 **b.** 273 **c.** 1309

Solution See page S22.

TAKE NOTE

The TI-83/84 program listed on page 128 can be used to find the prime factorization of a given natural number less than 10 billion.

TAKE NOTE

The following compact division procedure can also be used to determine the prime factorization of a number.

2	4004
2	2002
7	1001
11	143
	13

In this procedure, we use only prime number divisors, and we continue until the last quotient is a prime number. The prime factorization is the indicated product of all the prime numbers, which are shown in red.

MATH MATTERS Srinivasa Ramanujan

The Granger Collection, NY

Srinivasa Ramanujan (1887–1920)

On January 16, 1913, the 26-year-old Srinivasa Ramanujan (Rä-mä'noo-jûn) sent a letter from Madras, India, to the illustrious English mathematician G.H. Hardy. The letter requested that Hardy give his opinion about several mathematical ideas that Ramanujan had developed. In the letter Ramanujan explained, "I have not trodden through the conventional regular course which is followed in a University course, but I am striking out a new path for myself." Much of the mathematics was written using unconventional terms and notation; however, Hardy recognized (after many detailed readings and with the help of other mathematicians at Cambridge University) that Ramanujan was "a mathematician of the highest quality, a man of altogether exceptional originality and power."

On March 17, 1914, Ramanujan set sail for England, where he joined Hardy in a most unusual collaboration that lasted until Ramanujan returned to India in 1919. The following famous story is often told to illustrate the remarkable mathematical genius of Ramanujan.

> After Hardy had taken a taxicab to visit Ramanujan, he made the remark that the license plate number for the taxi was "1729, a rather dull number." Ramanujan immediately responded by saying that 1729 was a most interesting number, because it is the smallest natural number that can be expressed in two different ways as the sum of two cubes.

AP/Wide World photos

$$1^3 + 12^3 = 1729 \quad \text{and} \quad 9^3 + 10^3 = 1729$$

An interesting biography of the life of Srinivasa Ramanujan is given in *The Man Who Knew Infinity: A Life of the Genius Ramanujan* by Robert Kanigel.[1]

INSTRUCTOR NOTE

The number 1729 is also a Carmichael number, a number that is not prime but satisfies Fermat's little theorem (see Exercise 35, Section 6.6). Thus 1729 provides a counterexample to the converse of Fermat's little theorem.

TAKE NOTE

To determine whether a natural number is a prime number, it is necessary to consider only divisors from 2 up to the square root of the number because every composite number n has at least one divisor less than or equal to $\sqrt{n}$. The proof of this statement is outlined in Exercise 77 of this section.

It is possible to determine whether a natural number n is a prime number by checking each natural number from 2 up to the largest integer not greater than $\sqrt{n}$ to see whether each is a divisor of n. If none of these numbers is a divisor of n, then n is a prime number. For large values of n, this division method is generally time consuming and tedious. The Greek astronomer and mathematician Eratosthenes (ca. 276–192 BC) recognized that multiplication is generally easier than division, and he devised a method that makes use of multiples to determine every prime number in a list of natural numbers. Today we call this method the *sieve of Eratosthenes*.

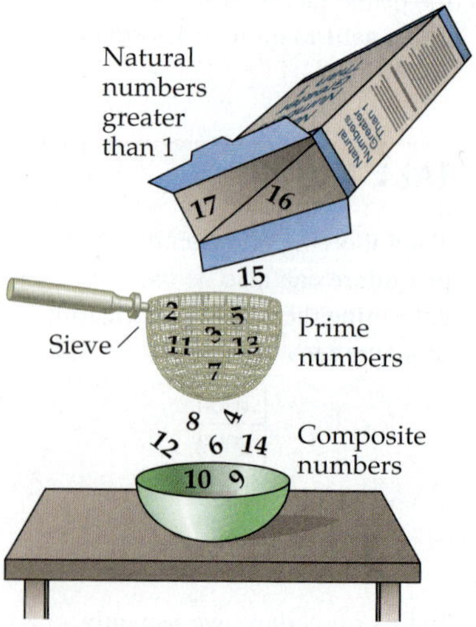

[1] Kanigel, Robert. *The Man Who Knew Infinity: A Life of the Genius Ramanujan*. New York: Simon & Schuster, 1991.

POINT OF INTEREST

In article 329 of *Disquisitiones Arithmeticae*, Gauss wrote:

> "The problem of distinguishing prime numbers from composite numbers and of resolving the latter into their prime factors is known to be one of the most important and useful in arithmetic . . . . The dignity of the science itself seems to require that every possible means be explored for the solution of a problem so elegant and so celebrated." (*Source:* Paulo Ribenboim, *The Little Book of Big Primes,* New York: Springer-Verlag, 1991)

To sift prime numbers, first make a list of consecutive natural numbers. In Table 6.14, we have listed the consecutive counting numbers from 2 to 100.

- Cross out every multiple of 2 larger than 2. The next smallest remaining number in the list is 3. Cross out every multiple of 3 larger than 3.
- Call the next smallest remaining number in the list k. Cross out every multiple of k larger than k. Repeat this step for all $k < \sqrt{100}$.

TABLE 6.14

The Sieve Method of Finding Primes

	2	3	~~4~~	5	~~6~~	7	~~8~~	~~9~~	~~10~~
11	~~12~~	13	~~14~~	~~15~~	~~16~~	17	~~18~~	19	~~20~~
~~21~~	~~22~~	23	~~24~~	~~25~~	~~26~~	~~27~~	~~28~~	29	~~30~~
31	~~32~~	~~33~~	~~34~~	~~35~~	~~36~~	37	~~38~~	~~39~~	~~40~~
41	~~42~~	43	~~44~~	~~45~~	~~46~~	47	~~48~~	~~49~~	~~50~~
~~51~~	~~52~~	53	~~54~~	~~55~~	~~56~~	~~57~~	~~58~~	59	~~60~~
61	~~62~~	~~63~~	~~64~~	~~65~~	~~66~~	67	~~68~~	~~69~~	~~70~~
71	~~72~~	73	~~74~~	~~75~~	~~76~~	~~77~~	~~78~~	79	~~80~~
~~81~~	~~82~~	83	~~84~~	~~85~~	~~86~~	~~87~~	~~88~~	89	~~90~~
~~91~~	~~92~~	~~93~~	~~94~~	~~95~~	~~96~~	97	~~98~~	~~99~~	~~100~~

PPT

The numbers in blue that are not crossed out are prime numbers. Table 6.14 shows that there are 25 prime numbers smaller than 100.

Over 2000 years ago, Euclid proved that the set of prime numbers is an infinite set. Euclid's proof is an *indirect proof* or a *proof by contradiction.* Essentially, his proof shows that for any finite list of prime numbers, we can create a number T, as described below, such that any prime factor of T can be shown to be a prime number that is not in the list. Thus there must be an infinite number of primes because it is not possible for all of the primes to be in any finite list.

INSTRUCTOR NOTE

It may be helpful to give a specific example to illustrate that

$$T = (p_1 \cdot p_2 \cdot p_3 \cdots p_r) + 1$$

is not divisible by any of the primes $p_1, p_2, p_3, \ldots, p_r$. For instance, if $T = (2 \cdot 3 \cdot 5 \cdot 7) + 1$, then T is not divisible by 2, 3, 5, or 7. In fact, T divided by any of the primes 2, 3, 5, or 7 yields a remainder of 1.

Euclid's Proof Assume that *all* of the prime numbers are contained in the list $p_1, p_2, p_3, \ldots, p_r$. Let $T = (p_1 \cdot p_2 \cdot p_3 \cdots p_r) + 1$. Either T is a prime number or T has a prime divisor. If T is a prime, then it is a prime that is not in our list and we have reached a contradiction. If T is not prime, then one of the primes $p_1, p_2, p_3, \ldots, p_r$ must be a divisor of T. However, the number T is not divisible by any of the primes $p_1, p_2, p_3, \ldots, p_r$ because each p_i divides $p_1 \cdot p_2 \cdot p_3 \cdots p_r$ but does not divide 1. Hence any prime divisor of T, say p, is a prime number that is not in the list $p_1, p_2, p_3, \ldots, p_r$. So p is yet another prime number, and $p_1, p_2, p_3, \ldots, p_r$ is not a complete list of all the prime numbers.

We conclude this section with two quotations about prime numbers. The first is by the illustrious mathematician Paul Erdös (1913–1996), and the second by the mathematics professor Don B. Zagier of the Max-Planck Institute, Bonn, Germany.

> It will be millions of years before we'll have any understanding, and even then it won't be a complete understanding, because we're up against the infinite.—
> P. Erdös, about prime numbers in *Atlantic Monthly*, November 1987, p. 74.
> *Source:* http://www.mlahanas.de/Greeks/Primes.htm.

Don B. Zagier

In a 1975 lecture, D. Zagier commented,

> There are two facts about the distribution of prime numbers of which I hope to convince you so overwhelmingly that they will be permanently engraved in your hearts. The first is that, despite their simple definition and role as the building blocks of the natural numbers, the prime numbers belong to the most arbitrary and ornery objects studied by mathematicians: they grow like weeds among the natural numbers, seeming to obey no other law than that of chance, and nobody can predict where the next one will sprout. The second fact is even more astonishing, for it states just the opposite: that the prime numbers exhibit stunning regularity, that there are laws governing their behavior, and that they obey these laws with almost military precision. (*Source:* Don B. Zagier, "The first 50 million prime numbers," *The Mathematical Intelligencer*, 1977)

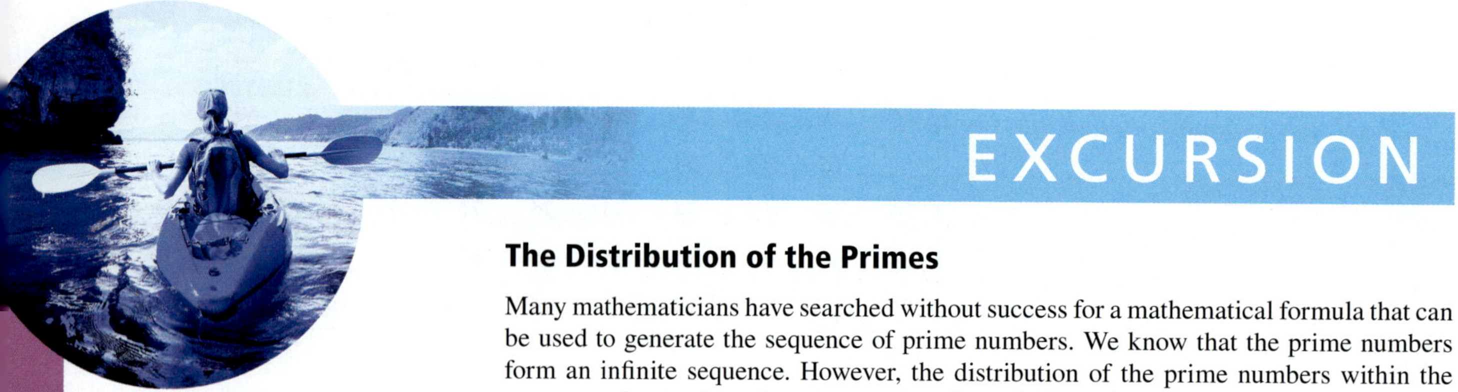

The Distribution of the Primes

Many mathematicians have searched without success for a mathematical formula that can be used to generate the sequence of prime numbers. We know that the prime numbers form an infinite sequence. However, the distribution of the prime numbers within the sequence of natural numbers is very complicated. The ratio of prime numbers to composite numbers appears to become smaller and smaller as larger and larger numbers are considered. In general, the number of consecutive composite numbers that come between two prime numbers tends to increase as the size of the numbers becomes larger; however, this increase is erratic and appears to be unpredictable.

In this Excursion, we refer to a list of two or more consecutive composite numbers as a **prime desert**. For instance, 8, 9, 10 is a prime desert because it consists of three consecutive composite numbers. The longest prime desert shown in Table 6.14 is the seven consecutive composite numbers 90, 91, 92, 93, 94, 95, and 96. A formula that involves *factorials* can be used to form prime deserts of any finite length.

n factorial

If n is a natural number, then $n!$, which is read "n factorial," is defined as

$$n! = n \cdot (n-1) \cdot \cdots \cdot 3 \cdot 2 \cdot 1$$

As an example of a factorial, consider $4! = 4 \cdot 3 \cdot 2 \cdot 1 = 24$.

The sequence

$4! + 2, 4! + 3, 4! + 4$

is a prime desert of the three composite numbers 26, 27, and 28. Figure 6.1 shows a prime desert of 10 consecutive composite numbers. Figure 6.2 shows a procedure that can be used to produce a prime desert of n composite numbers, where n is any natural number greater than 2.

$$\left.\begin{array}{l} 11! + 2 \\ 11! + 3 \\ 11! + 4 \\ 11! + 5 \\ 11! + 6 \\ \vdots \\ 11! + 10 \\ 11! + 11 \end{array}\right\} \text{A prime desert of 10 consecutive composite numbers}$$

FIGURE 6.1

$$\left.\begin{array}{l} (n+1)! + 2 \\ (n+1)! + 3 \\ (n+1)! + 4 \\ (n+1)! + 5 \\ (n+1)! + 6 \\ \vdots \\ (n+1)! + n \\ (n+1)! + (n+1) \end{array}\right\} \text{A prime desert of } n \text{ consecutive composite numbers}$$

FIGURE 6.2

A prime desert of length 1 million is shown by the sequence

$$1{,}000{,}001! + 2;\ 1{,}000{,}001! + 3;\ 1{,}000{,}001! + 4;\ \ldots\ ;\ 1{,}000{,}001! + 1{,}000{,}001$$

It appears that the distribution of prime numbers is similar to the situation wherein a mathematical gardener plants an infinite number of grass seeds on a windy day. Many of the grass seeds fall close to the gardener, but some are blown down the street and into the next neighborhood. There are gaps where no grass seeds are within 1 mile of each other. Farther down the road there are gaps where no grass seeds are within 10 miles of each other. No matter how far the gardener travels and how long it has been since the last grass seed was spotted, the gardener knows that more grass seeds will appear.

EXCURSION EXERCISES

1. Explain how you know that each of the numbers

$$1{,}000{,}001! + 2;\ 1{,}000{,}001! + 3;\ 1{,}000{,}001! + 4;\ \ldots\ ;\ 1{,}000{,}001! + 1{,}000{,}001$$

is a composite number.

2. Use factorials to generate the numbers in a prime desert of 12 consecutive composite numbers. Now use a calculator to write each number in this prime desert, as a natural number.
3. Use factorials and "..." notation to represent a prime desert of
   a. 20 consecutive composite numbers.
   b. 500,000 consecutive composite numbers.
   c. 7 billion consecutive composite numbers.

EXERCISE SET 6.5

■ In Exercises 1 to 10, determine all natural number divisors of the given number.

1. 20
2. 32
3. 65
4. 75
5. 41
6. 79
7. 110
8. 150
9. 385
10. 455

■ In Exercises 11 to 20, determine whether each number is a prime number or a composite number.

11. 21
12. 31
13. 37
14. 39
15. 101
16. 81
17. 79
18. 161
19. 203
20. 211

■ In Exercises 21 to 28, use the divisibility tests in Table 6.13 to determine whether the given number is divisible by each of the following: 2, 3, 4, 5, 6, 8, 9, and 10.

21. 210 **22.** 314 **23.** 51

24. 168 **25.** 2568 **26.** 3525

27. 4190 **28.** 6123

■ In Exercises 29 to 40, write the prime factorization of the number.

29. 18 **30.** 48 **31.** 120

32. 380 **33.** 425 **34.** 625

35. 1024 **36.** 1410 **37.** 6312

38. 3155 **39.** 18,234 **40.** 19,345

41. Use the sieve of Eratosthenes procedure to find all prime numbers from 2 to 200. *Hint:* Because $\sqrt{200} \approx 14.1$, you need to continue the sieve procedure up to $k = 13$. Note: You do not need to consider $k = 14$ because 14 is not a prime number.

42. Use your list of prime numbers from Exercise 41 to find the number of prime numbers from:

a. 2 to 50 **b.** 51 to 100

c. 101 to 150 **d.** 151 to 200

43. Twin Primes If the natural numbers n and $n + 2$ are both prime numbers, then they are said to be **twin primes**. For example, 11 and 13 are twin primes. It is not known whether the set of twin primes is an infinite set or a finite set. Use the list of primes from Exercise 41 to write all twin primes less than 200.

44. Twin Primes Find a pair of twin primes between 200 and 300. See Exercise 43.

45. Twin Primes Find a pair of twin primes between 300 and 400. See Exercise 43.

46. A Prime Triplet If the natural numbers n, $n + 2$, and $n + 4$ are all prime numbers, then they are said to be **prime triplets**. Write a few sentences that explain why the prime triplets 3, 5, and 7 are the only prime triplets.

47. Goldbach's Conjecture In 1742, Christian Goldbach conjectured that every even number greater than 2 can be written as the sum of two prime numbers. Many mathematicians have tried to prove or disprove this conjecture without succeeding. Show that *Goldbach's conjecture* is true for each of the following even numbers.

a. 24 **b.** 50

c. 86 **d.** 144

e. 210 **f.** 264

48. Perfect Squares The square of a natural number is called a **perfect square**. Pick six perfect squares. For each perfect square, determine the number of distinct natural-number factors of the perfect square. Make a conjecture about the number of distinct natural-number factors of any perfect square.

Every prime number has a divisibility test. Many of these divisibility tests are slight variations of the following divisibility test for 7.

A divisibility test for 7 To determine whether a given base ten number is divisible by 7, double the ones digit of the given number. Find the difference between this number and the number formed by omitting the ones digit from the given number. If necessary, repeat this procedure until you obtain a small final difference.

If the final difference is divisible by 7, then the given number is also divisible by 7.

If the final difference is not divisible by 7, then the given number is not divisible by 7.

Example Use the above divisibility test to determine whether 301 is divisible by 7.

Solution The double of the ones digit is 2. Subtracting 2 from 30, which is the number formed by omitting the ones digit from the original number, yields 28. Because 28 is divisible by 7, the original number 301 is divisible by 7.

■ In Exercises 49 to 56, use the divisibility test for 7, given above, to determine whether each number is divisible by 7.

49. 182 **50.** 203

51. 1001 **52.** 2403

53. 11,561 **54.** 13,842

55. 204,316 **56.** 789,327

A divisibility test for 13 To determine whether a given base ten number is divisible by 13, multiply the ones digit of the given number by 4. Find the sum of this multiple of 4 and the number formed by omitting the ones digit from the given number. If necessary, repeat this procedure until you obtain a small final sum.

If the final sum is divisible by 13, then the given number is divisible by 13.

If the final sum is not divisible by 13, then the given number is not divisible by 13.

Example Use the divisibility test for 13 to determine whether 1079 is divisible by 13.

Solution Four times the ones digit is 36. The number formed by omitting the ones digit is 107. The sum of 36 and 107 is 143. Now repeat the procedure on 143. Four times the ones digit is 12. The sum of 12 and 14, which is the number formed by omitting the ones digit, is 26. Because 26 is divisible by 13, the original number 1079 is divisible by 13.

■ In Exercises 57 to 64, use the divisibility test for 13 to determine whether each number is divisible by 13.

57. 91 **58.** 273

59. 1885 **60.** 8931

61. 14,507 **62.** 22,184

63. 13,351 **64.** 85,657

EXTENSIONS

65. Factorial Primes A prime number of the form $n! \pm 1$ is called a **factorial prime**. Recall that the notation $n!$ is called n factorial and represents the product of all natural numbers from 1 to n. For example, $4! = 4 \cdot 3 \cdot 2 \cdot 1 = 24$. Factorial primes are of interest to mathematicians because they often signal the end or the beginning of a lengthy string of consecutive composite numbers. See the Excursion on page 336.

a. Find the smallest value of n such that $n! + 1$ and $n! - 1$ are twin primes.

b. Find the smallest value of n for which $n! + 1$ is a composite number and $n! - 1$ is a prime number.

66. Primorial Primes The notation $p\#$ represents the product of all the prime numbers less than or equal to the prime number p. For instance,

$$3\# = 2 \cdot 3 = 6$$
$$5\# = 2 \cdot 3 \cdot 5 = 30$$
$$11\# = 2 \cdot 3 \cdot 5 \cdot 7 \cdot 11 = 2310$$

A **primorial prime** is a prime number of the form $p\# \pm 1$. For instance, $3\# + 1 = 2 \cdot 3 + 1 = 7$ and $3\# - 1 = 2 \cdot 3 - 1 = 5$ are both primorial primes. Large primorial primes are often examined in the search for a pair of large twin primes.

a. Find the smallest prime number p, where $p \geq 7$, such that $p\# + 1$ and $p\# - 1$ are twin primes.

b. Find the smallest prime number p, such that $p\# + 1$ is a prime number but $p\# - 1$ is a composite number.

67. A Divisibility Test for 17 Determine a divisibility test for 17. *Hint:* One divisibility test for 17 is similar to the divisibility test for 7, shown on the previous page, in that it involves the last digit of the given number and the operation of subtraction.

68. A Divisibility Test for 19 Determine a divisibility test for 19. *Hint:* One divisibility test for 19 is similar to the divisibility test for 13, shown on the previous page, in that it involves the last digit of the given number and the operation of addition.

Number of Divisors of a Composite Number The following method can be used to determine the number of divisors of a composite number. First find the prime factorization (in exponential form) of the composite number. Add 1 to each exponent in the prime factorization and then compute the product of these exponents. This product is equal to the number of divisors of the composite number. To illustrate that this procedure yields the correct result, consider the composite number 12, which has the six divisors 1, 2, 3, 4, 6, and 12. The prime factorization of 12 is $2^2 \cdot 3^1$. Adding 1 to each of the exponents produces the numbers 3 and 2. The product of 3 and 2 is 6, which agrees with the result obtained by listing all of the divisors.

■ In Exercises 69 to 74, determine the number of divisors of each composite number.

69. 60 **70.** 84

71. 297 **72.** 288

73. 360 **74.** 875

75. Kummer's Proof In the 1870s, the mathematician Eduard Kummer used a proof similar to the following to show that there exist an infinite number of prime numbers. Supply the missing reasons in parts a and b.

a. ***Proof*** Assume there exist only a finite number of prime numbers, say $p_1, p_2, p_3, \ldots, p_r$. Let $N = p_1 p_2 p_3 \cdots p_r > 2$. The natural number $N - 1$ has at least one common prime factor with N. Why?

b. Call the common prime factor from part a p_i. Now p_i divides N and p_i divides $N - 1$. Thus p_i divides their difference: $N - (N - 1) = 1$. Why?

This leads to a contradiction, because no prime number is a divisor of 1. Hence Kummer concluded that the original assumption was incorrect and there must exist an infinite number of prime numbers.

76. *Theorem* If a number of the form 111...1 is a prime number, then the number of 1s in the number is a prime number. For instance,

11 and 1,111,111,111,111,111,111

are prime numbers, and the number of 1s in each number (2 in the first number and 19 in the second number) is a prime number.

a. What is the converse of the above theorem? *Hint:* The converse of "If p then q," where p and q are statements, is "If q then p." Is the converse of a theorem always true?

b. The number $111 = 3 \cdot 37$, so 111 is not a prime number. Explain why this does not contradict the above theorem.

77. State the missing reasons in parts a, b, and c of the following proof.

Theorem Every composite number n has at least one divisor less than or equal to $\sqrt{n}$.

a. ***Proof*** Assume that a is a divisor of n. Then there exists a natural number j such that $aj = n$. Why?

b. Now a and j cannot both be greater than $\sqrt{n}$, because this would imply that $aj > \sqrt{n}\sqrt{n}$. However, $\sqrt{n}\sqrt{n}$ simplifies to n, which equals aj. What contradiction does this lead to?

c. Thus either a or j must be less than or equal to $\sqrt{n}$. Because j is also a divisor of n, the proof is complete. How do we know that j is a divisor of n?

78. **The RSA Algorithm** In 1977, Ron Rivest, Adi Shamir, and Leonard Adleman invented a method for encrypting information. Their method is known as the RSA algorithm. Today the RSA algorithm is used by VISA and MasterCard to ensure secure electronic credit card transactions. The RSA algorithm involves large prime numbers. Research the RSA algorithm and write a report about some of the reasons why this algorithm has become one of the most popular of all the encryption algorithms.

79. **Theorems and Conjectures** Use the "Conjectures and Open Problems" link on "The Prime Pages" website to determine whether each of the following statements is an established theorem or a conjecture. (*Note: n* represents a natural number).

a. There are infinitely many twin primes.

b. There are infinitely many primes of the form $n^2 + 1$.

c. There is always a prime number between n and $2n$ for $n \geq 2$.

d. There is always a prime number between n^2 and $(n + 1)^2$.

e. Every odd number greater than 5 can be written as the sum of three primes.

f. Every positive even number can be written as the difference of two primes.

SECTION 6.6 Topics from Number Theory

Perfect, Deficient, and Abundant Numbers

The ancient Greek mathematicians personified the natural numbers. For instance, they considered the odd natural numbers as male and the even natural numbers as female. They also used the concept of a *proper factor* to classify a natural number as *perfect, deficient,* or *abundant.*

Proper Factors

The **proper factors** of a natural number consist of all the natural number factors of the number other than the number itself.

For instance, the proper factors of 10 are 1, 2, and 5. The proper factors of 16 are 1, 2, 4, and 8. The proper factors of a number are also called the **proper divisors** of the number.

Perfect, Deficient, and Abundant Numbers

A natural number is

- **perfect** if it is equal to the sum of its proper factors.
- **deficient** if it is greater than the sum of its proper factors.
- **abundant** if it is less than the sum of its proper factors.

POINT OF INTEREST

Six is a number perfect in itself, and not because God created the world in 6 days; rather the contrary is true. God created the world in six days because this number is perfect, and it would remain perfect, even if the work of the six days did not exist.
—*St. Augustine* (354–430)

EXAMPLE 1 Classify a Number as Perfect, Deficient, or Abundant

Determine whether the following numbers are perfect, deficient, or abundant.

a. 6 **b.** 20 **c.** 25

Solution

a. The proper factors of 6 are 1, 2, and 3. The sum of these proper factors is $1 + 2 + 3 = 6$. Because 6 is equal to the sum of its proper divisors, 6 is a perfect number.

b. The proper factors of 20 are 1, 2, 4, 5, and 10. The sum of these proper factors is $1 + 2 + 4 + 5 + 10 = 22$. Because 20 is less than the sum of its proper factors, 20 is an abundant number.

c. The proper factors of 25 are 1 and 5. The sum of these proper factors is $1 + 5 = 6$. Because 25 is greater than the sum of its proper factors, 25 is a deficient number.

CHECK YOUR PROGRESS 1 Determine whether the following numbers are perfect, deficient, or abundant.

a. 24 **b.** 28 **c.** 35

Solution See page S22.

Mersenne Numbers and Perfect Numbers

The Granger Collection, NY

Marin Mersenne (1588–1648)

As a French monk in the religious order known as the Minims, Marin Mersenne (mər-sĕn′) devoted himself to prayer and his studies, which included topics from number theory. Mersenne took it upon himself to collect and disseminate mathematical information to scientists and mathematicians through Europe. Mersenne was particularly interested in prime numbers of the form $2^n - 1$, where n is a prime number. Today numbers of the form $2^n - 1$, where n is a prime number, are known as **Mersenne numbers**.

Some Mersenne numbers are prime and some are composite. For instance, the Mersenne numbers $2^2 - 1$ and $2^3 - 1$ are prime numbers, but the Mersenne number $2^{11} - 1 = 2047$ is not prime because $2047 = 23 \cdot 89$.

EXAMPLE 2 **Determine Whether a Mersenne Number Is a Prime Number**

Determine whether the Mersenne number $2^5 - 1$ is a prime number.

Solution

$2^5 - 1 = 31$, and 31 is a prime number. Thus $2^5 - 1$ is a Mersenne prime.

CHECK YOUR PROGRESS 2 Determine whether the Mersenne number $2^7 - 1$ is a prime number.

Solution See page S22.

The ancient Greeks knew that the first four perfect numbers were 6, 28, 496, and 8128. In fact, proposition 36 from Volume IX of Euclid's *Elements* states a procedure that uses Mersenne primes to produce a perfect number.

Euclid's Procedure for Generating a Perfect Number

If n and $2^n - 1$ are both prime numbers, then $2^{n-1}(2^n - 1)$ is a perfect number.

Euclid's procedure shows how every Mersenne prime can be used to produce a perfect number. For instance:

If $n = 2$, then $2^{2-1}(2^2 - 1) = 2(3) = 6$.
If $n = 3$, then $2^{3-1}(2^3 - 1) = 4(7) = 28$.
If $n = 5$, then $2^{5-1}(2^5 - 1) = 16(31) = 496$.
If $n = 7$, then $2^{7-1}(2^7 - 1) = 64(127) = 8128$.

The fifth perfect number was not discovered until the year 1461. It is $2^{12}(2^{13} - 1) = 33{,}550{,}336$. The sixth and seventh perfect numbers were discovered in 1588 by P. A. Cataldi. In exponential form, they are $2^{16}(2^{17} - 1)$ and $2^{18}(2^{19} - 1)$. It is interesting to observe that the ones digits of the first five perfect numbers alternate: 6, 8, 6, 8, 6. Evaluate $2^{16}(2^{17} - 1)$ to determine if this alternating pattern continues for the first six perfect numbers.

INSTRUCTOR NOTE

The sixth perfect number is 8,589,869,056. Thus the 6, 8 alternating pattern for the ones digits of perfect numbers does not hold for the first six perfect numbers.

EXAMPLE 3 Use a Given Mersenne Prime Number to Write a Perfect Number

In 1772, Leonhard Euler proved that $2^{31} - 1$ is a Mersenne prime. Use Euclid's procedure to write the perfect number associated with this prime.

Solution

Euler's procedure states that if n and $2^n - 1$ are both prime numbers, then $2^{n-1}(2^n - 1)$ is a perfect number. In this example, $n = 31$, which is a prime number. We are given that $2^{31} - 1$ is a prime number, so the perfect number we seek is $2^{30}(2^{31} - 1)$.

CHECK YOUR PROGRESS 3 In 1883, Ivan Mikheevich Pervushin proved that $2^{61} - 1$ is a Mersenne prime. Use Euclid's procedure to write the perfect number associated with this prime.

Solution See page S22.

QUESTION Must a perfect number produced by Euclid's perfect number–generating procedure be an even number?

The search for Mersenne primes still continues. As of July 2016, only 49 Mersenne primes (and 49 perfect numbers) had been discovered. The largest of these 49 Mersenne primes is $2^{74,207,281} - 1$, which has 22,338,618 digits. More information on Mersenne primes is given in Table 6.15 on page 343.

MATH MATTERS The Great Internet Mersenne Prime Search

From 1952 to 1996, large-scale computers were used to find Mersenne primes. However, during recent years, personal computers working in parallel have joined in the search. This search has been organized by an Internet organization known as the Great Internet Mersenne Prime Search (GIMPS).

Members of this group use the Internet to download a Mersenne prime program that runs on their computers. Each member is assigned a range of numbers to check for Mersenne primes. A search over a specified range of numbers can take several weeks, but because the program is designed to run in the background, you can still use your computer to perform its regular duties. As of July 2016, a total of 15 Mersenne primes had been discovered by members of the GIMPS.

One of the current goals of GIMPS is to find a prime number that has at least 100 million digits. In fact, the Electronic Frontier Foundation is offering a $150,000 reward to the first person or group to discover a 100 million–digit prime number. If you are just using your computer to run a screen saver, why not join in the search? You can get the Mersenne prime program and additional information at

http://www.mersenne.org

Who knows, maybe you will discover a new Mersenne prime and share in the reward money. Of course, we know that you really just want to have your name added to Table 6.15.

$150,000 REWARD

TO THE PERSON OR GROUP THAT FIRST DISCOVERS A PRIME NUMBER WITH AT LEAST 100 MILLION DIGITS!

Offered by the Electronic Frontier Foundation

In the last five rows of Table 6.15, a question mark has been inserted after the rank number because as of July 2016, it is not known if there are other Mersenne prime numbers larger than $2^{32582657} - 1$ but less than $2^{74207281} - 1$ that have yet to be discovered.

ANSWER Yes. The 2^{n-1} factor of $2^{n-1}(2^n - 1)$ ensures that this product will be an even number.

TABLE 6.15
Selected Known Mersenne Primes as of July, 2016 (Listed from Smallest to Largest)

Rank	Mersenne prime	Discovery date	Discoverer
1	$2^2 - 1 = 3$	BC	Ancient Greek mathematicians
2	$2^3 - 1 = 7$	BC	Ancient Greek mathematicians
3	$2^5 - 1 = 31$	BC	Ancient Greek mathematicians
4	$2^7 - 1 = 127$	BC	Ancient Greek mathematicians
5	$2^{13} - 1 = 8191$	1456	Anonymous
6	$2^{17} - 1 = 131071$	1588	Cataldi
7	$2^{19} - 1 = 524287$	1588	Cataldi
8	$2^{31} - 1 = 2147483647$	1772	Euler
9	$2^{61} - 1 = 2305843009213693951$	1883	Pervushin
⋮	⋮	⋮	⋮
40	$2^{20996011} - 1$	Nov. 2003	GIMPS / Shafer
41	$2^{24036583} - 1$	May 2004	GIMPS / Findley
42	$2^{25964951} - 1$	Feb. 2005	GIMPS / Nowak
43	$2^{30402457} - 1$	Dec. 2005	GIMPS / Cooper and Boone
44	$2^{32582657} - 1$	Sept. 2006	GIMPS / Cooper and Boone
45?	$2^{37156667} - 1$	Sept. 2008	GIMPS / Elvenich
46?	$2^{42643801} - 1$	April 2009	GIMPS / Strindmo
47?	$2^{43112609} - 1$	Aug. 2008	GIMPS / Smith
48?	$2^{57885161} - 1$	Jan. 2013	GIMPS / Cooper
49?	$2^{74207281} - 1$	Jan. 2016	GIMPS / Cooper

Source of data: The *Mersenne prime* website at: http://en.wikipedia.org/wiki/Mersenne_prime.
*GIMPS is a project involving volunteers who use computer software to search for Mersenne prime numbers. See the Math Matters on page 342.

The Number of Digits in b^x

To determine the number of digits in a Mersenne number, mathematicians make use of the following formula, which involves finding the *greatest integer* of a number. **The greatest integer** of a number k is the greatest integer less than or equal to k. For instance, the greatest integer of 5 is 5 and the greatest integer of 7.8 is 7.

The Number of Digits in b^x

The number of digits in the number b^x, where b is a natural number less than 10 and x is a natural number, is the greatest integer of $(x \log b) + 1$.

EXAMPLE 4 Determine the Number of Digits in a Mersenne Number

Find the number of digits in the Mersenne prime number $2^{19} - 1$.

CALCULATOR NOTE

To evaluate 19 log 2 on a TI-83/84 calculator, press

19 LOG 2) ENTER

On a scientific calculator, press

19 × 2 LOG =

Solution

First consider just 2^{19}. The base b is 2. The exponent x is 19.

$$\begin{aligned}(x \log b) + 1 &= (19 \log 2) + 1 \\ &\approx 5.72 + 1 \\ &= 6.72\end{aligned}$$

The greatest integer of 6.72 is 6. Thus 2^{19} has six digits. If 2^{19} were a power of 10, then $2^{19} - 1$ would have one fewer digit than 2^{19}. However, 2^{19} is not a power of 10, so $2^{19} - 1$ has the same number of digits as 2^{19}. Thus the Mersenne prime number $2^{19} - 1$ also has six digits.

CHECK YOUR PROGRESS 4 Find the number of digits in the Mersenne prime number $2^{20996011} - 1$.

Solution See page S22.

Euclid's perfect number–generating procedure produces only *even* perfect numbers. Do odd perfect numbers exist? As of July 2016, no odd perfect number had been discovered and no one had been able to prove either that odd perfect numbers exist or that they do not exist. The question of the existence of an odd perfect number is one of the major unanswered questions of number theory.

Fermat's Last Theorem

In 1637, the French mathematician Pierre de Fermat wrote in the margin of a book:

> It is impossible to divide a cube into two cubes, or a fourth power into two fourth powers, or in general any power greater than the second into two like powers, and I have a truly marvelous demonstration of it. But this margin will not contain it.

This problem, which became known as **Fermat's last theorem**, can also be stated in the following manner.

Fermat's Last Theorem

There are no natural numbers x, y, z, and n that satisfy $x^n + y^n = z^n$, where n is greater than 2.

Fermat's last theorem has attracted a great deal of attention over the last three centuries. The theorem has become so well known that it is simply called "FLT." Here are some of the reasons for its popularity:

- Very little mathematical knowledge is required to understand the statement of FLT.
- FLT is an extension of the well-known Pythagorean theorem $x^2 + y^2 = z^2$, which has several natural number solutions. Two such solutions are $3^2 + 4^2 = 5^2$ and $5^2 + 12^2 = 13^2$.
- It seems so simple. After all, while reading the text *Arithmetica* by Diophantus, Fermat wrote that he had discovered a *truly marvelous proof* of FLT. The only reason that Fermat gave for not providing his proof was that the margin of *Arithmetica* was too narrow to contain it.

Many famous mathematicians have worked on FLT. Some of these mathematicians tried to disprove FLT by searching for natural numbers x, y, z, and n that satisfied $x^n + y^n = z^n$, $n > 2$.

HISTORICAL NOTE

Bettmann/Corbis

Pierre de Fermat (1601–1665) In the mathematical community, Pierre de Fermat (fĕhr'mah) is known as the Prince of Amateurs because although he spent his professional life as a councilor and judge, he spent his leisure time working on mathematics. As an amateur mathematician, Fermat did important work in analytic geometry and calculus, but he is remembered today for his work in the area of number theory. Fermat stated theorems he had developed, but seldom did he provide the actual proofs. He did not want to waste his time showing the details required of a mathematical proof and then spend additional time defending his work once it was scrutinized by other mathematicians.

By the twentieth century, all but one of Fermat's proposed theorems had been proved by other mathematicians. The remaining unproved theorem became known as Fermat's last theorem.

EXAMPLE 5 Check a Possible Solution to Fermat's Last Theorem

Determine whether $x = 6$, $y = 8$, and $n = 3$ satisfies the equation $x^n + y^n = z^n$, where z is a natural number.

Solution

Substituting 6 for x, 8 for y, and 3 for n in $x^n + y^n = z^n$ yields

$$6^3 + 8^3 = z^3$$
$$216 + 512 = z^3$$
$$728 = z^3$$

The real solution of $z^3 = 728$ is $\sqrt[3]{728} \approx 8.99588289$, which is not a natural number. Thus $x = 6$, $y = 8$, and $n = 3$ does not satisfy the equation $x^n + y^n = z^n$, where z is a natural number.

CHECK YOUR PROGRESS 5 Determine whether $x = 9$, $y = 11$, and $n = 4$ satisfies the equation $x^n + y^n = z^n$, where z is a natural number.

Solution See page S22.

In the eighteenth century, the great mathematician Leonhard Euler was able to make some progress on a proof of FLT. He adapted a technique that he found in Fermat's notes about another problem. In this problem, Fermat gave an outline of how to prove that the special case $x^4 + y^4 = z^4$ has no natural number solutions. Using a similar procedure, Euler was able to show that $x^3 + y^3 = z^3$ also has no natural number solutions. Thus all that was left was to show that $x^n + y^n = z^n$ has no solutions with n greater than 4.

In the nineteenth century, additional work on FLT was done by Sophie Germain, Augustin Louis Cauchy, and Gabriel Lame. Each of these mathematicians produced some interesting results, but FLT still remained unsolved.

In 1983, the German mathematician Gerd Faltings used concepts from differential geometry to prove that the number of solutions to FLT must be finite. Then, in 1988, the Japanese mathematician Yoichi Miyaoka claimed he could show that the number of solutions to FLT was not only finite, but that the number of solutions was zero, and thus he had proved FLT. At first Miyaoka's work appeared to be a valid proof, but after a few weeks of examination, a flaw was discovered. Several mathematicians looked for a way to repair the flaw, but eventually they came to the conclusion that although Miyaoka had developed some interesting mathematics, he had failed to establish the validity of FLT.

In 1993, Andrew Wiles of Princeton University made a major advance toward a proof of FLT. Wiles first became familiar with FLT when he was only 10 years old. At his local public library, Wiles first learned about FLT in the book *The Last Problem* by Eric Temple Bell. In reflecting on his first thoughts about FLT, Wiles recalled

> It looked so simple, and yet all the great mathematicians in history couldn't solve it. Here was a problem that I, a ten-year-old, could understand and I knew from that moment that I would never let it go. I had to solve it.[2]

Wiles took a most unusual approach to solving FLT. Whereas most contemporary mathematicians share their ideas and coordinate their efforts, Wiles decided to work alone. After seven years of working in the attic of his home, Wiles was ready to present his work. In June of 1993, Wiles gave a series of three lectures at the Isaac Newton Institute in Cambridge, England. After showing that FLT was a corollary of his major theorem, Wiles's concluding remark was "I think I'll stop here." Many mathematicians in the audience felt that Wiles had produced a valid proof of FLT, but a formal verification by

POINT OF INTEREST

Professor Peter Goddard/Science Photo Library/Photo Researchers/Science Source

Andrew Wiles

Fermat's last theorem had been labeled by some mathematicians as the world's hardest mathematical problem. After solving Fermat's last theorem, Andrew Wiles made the following remarks: "Having solved this problem there's certainly a sense of loss, but at the same time there is this tremendous sense of freedom. I was so obsessed by this problem that for eight years I was thinking about it all the time—when I woke up in the morning to when I went to sleep at night. That's a long time to think about one thing. That particular odyssey is now over. My mind is at rest."

---

[2] Singh, Simon. *Fermat's Enigma: The Quest to Solve the World's Greatest Mathematical Problem.* New York: Walker Publishing Company, Inc., 1997, p. 6.

several mathematical referees was required before Wiles's work could be classified as an official proof. The verification process was lengthy and complex. Wiles's written work was about 200 pages in length, it covered several different areas of mathematics, and it used hundreds of sophisticated logical arguments and a great many mathematical calculations. Any mistake could result in an invalid proof. Thus it was not too surprising when a flaw was discovered in late 1993. The flaw did not necessarily imply that Wiles's proof could not be repaired, but it did indicate that it was not a valid proof in its present form.

It appeared that once again the proof of FLT had eluded a great effort by a well-known mathematician. Several months passed and Wiles was still unable to fix the flaw. It was a most depressing period for Wiles. He felt that he was close to solving one of the world's hardest mathematical problems, yet all of his creative efforts failed to turn the flawed proof into a valid proof. Several mathematicians felt that it was not possible to repair the flaw, but Wiles did not give up. Finally, in late 1994, Wiles had an insight that eventually led to a valid proof. The insight required Wiles to seek additional help from Richard Taylor, who had been a student of Wiles. On October 15, 1994, Wiles and Taylor presented to the world a proof that has now been judged to be a valid proof of FLT. Their proof is certainly not the "truly marvelous proof" that Fermat said he had discovered. But it has been deemed a wonderful proof that makes use of several new mathematical procedures and concepts.

TAKE NOTE

There are several videos, available on the Internet, that provide additional information on the search for a proof of Fermat's last theorem. One such video can be found by searching for: Catalyst-Fermat's Last Theorem (2014)-YouTube.

EXCURSION

A Sum of the Divisors Formula

Consider the numbers 10, 12, and 28 and the sums of the proper factors of these numbers.

$$10: 1 + 2 + 5 = 8 \qquad 12: 1 + 2 + 3 + 4 + 6 = 16$$
$$28: 1 + 2 + 4 + 7 + 14 = 28$$

From the above sums we see that 10 is a deficient number, 12 is an abundant number, and 28 is a perfect number. The goal of this Excursion is to find a method that will enable us to determine whether a number is deficient, abundant, or perfect without having to first find all of its proper factors and then compute their sum.

In the following example, we use the number 108 and its prime factorization $2^2 \cdot 3^3$ to illustrate that every factor of 108 can be written as a product of the powers of its prime factors. Table 6.16 includes all the proper factors of 108 (the numbers in blue) plus the factor $2^2 \cdot 3^3$, which is 108 itself. The sum of each column is shown at the bottom (the numbers in red).

TABLE 6.16
Every Factor of 108 Expressed as a Product of Powers of Its Prime Factors

	1	$1 \cdot 3$	$1 \cdot 3^2$	$1 \cdot 3^3$
	2	$2 \cdot 3$	$2 \cdot 3^2$	$2 \cdot 3^3$
	2^2	$2^2 \cdot 3$	$2^2 \cdot 3^2$	$2^2 \cdot 3^3$
Sum	7	$7 \cdot 3$	$7 \cdot 3^2$	$7 \cdot 3^3$

The sum of *all* the factors of 108 is the sum of the numbers in the bottom row.

$$\text{Sum of all factors of } 108 = 7 + 7 \cdot 3 + 7 \cdot 3^2 + 7 \cdot 3^3$$
$$= 7(1 + 3 + 3^2 + 3^3) = 7(40) = 280$$

To find the sum of just the *proper* factors, we must subtract 108 from 280, which gives us 172. Thus 108 is an abundant number.

We now look for a pattern for the sum of all the factors. Note that

Sum of left column: $1 + 2 + 2^2$ → 7

Sum of top row: $1 + 3 + 3^2 + 3^3$ → 40

$$7 \cdot 40$$

This result suggests that the sum of the factors of a number can be found by finding the sum of all the prime power factors of each prime factor and then computing the product of those sums. Because we are interested only in the sum of the proper factors, we subtract the original number. Although we have not proved this result, it is a true statement and can save much time and effort. For instance, the sum of the proper factors of 3240 can be found as follows:

$$3240 = 2^3 \cdot 3^4 \cdot 5$$

Compute the sum of *all* the prime power factors of each prime factor.

$$1 + 2 + 2^2 + 2^3 = 15 \qquad 1 + 3 + 3^2 + 3^3 + 3^4 = 121 \qquad 1 + 5 = 6$$

The sum of the proper factors of 3240 is $(15)(121)(6) - 3240 = 7650$. Thus 3240 is an abundant number.

EXCURSION EXERCISES

Use the above technique to find the sum of the proper factors of each number and then state whether the number is deficient, abundant, or perfect.

1. 200 **2.** 262 **3.** 325 **4.** 496

5. Use deductive reasoning to prove that every prime number is deficient.

6. Use inductive reasoning to decide whether every multiple of 6 greater than 6 is abundant.

EXERCISE SET 6.6

■ In Exercises 1 to 16, determine whether each number is perfect, deficient, or abundant.

1. 18 **2.** 32

3. 91 **4.** 51

5. 19 **6.** 144

7. 204 **8.** 128

9. 610 **10.** 508

11. 291 **12.** 1001

13. 176 **14.** 122

15. 260 **16.** 258

■ In Exercises 17 to 20, determine whether each Mersenne number is a prime number.

17. $2^3 - 1$ **18.** $2^5 - 1$

19. $2^7 - 1$ **20.** $2^{13} - 1$

21. In 1876, Édouard Lucas proved, without the aid of a computer, that $2^{127} - 1$ is a Mersenne prime. Use Euclid's procedure to write the perfect number associated with this prime.

22. In 1952, Raphael M. Robinson proved, with the aid of a computer, that $2^{521} - 1$ is a Mersenne prime. Use Euclid's procedure to write the perfect number associated with this prime.

■ In Exercises 23 to 30, determine the number of digits in the given Mersenne prime.

23. $2^{17} - 1$ **24.** $2^{132049} - 1$

25. $2^{1398269} - 1$ **26.** $2^{3021377} - 1$

27. $2^{6972593} - 1$ **28.** $2^{20996011} - 1$

29. $2^{37156667} - 1$ **30.** $2^{43112609} - 1$

31. Verify that $x = 9$, $y = 15$, and $n = 5$ do not yield a solution to the equation $x^n + y^n = z^n$ where z is a natural number.

32. Verify that $x = 7$, $y = 19$, and $n = 6$ do not yield a solution to the equation $x^n + y^n = z^n$ where z is a natural number.

33. Determine whether each of the following statements is a true statement, a false statement, or a conjecture.

a. If n is a prime number, then $2^n - 1$ is also a prime number.

b. Fermat's last theorem is called his last theorem because we believe that it was the last theorem he proved.

c. All perfect numbers of the form $2^{n-1}(2^n - 1)$ are even numbers.

d. Every perfect number is an even number.

34. Show that $4078^n + 3433^n = 12{,}046^n$ cannot be a solution to the equation $x^n + y^n = z^n$ where n is a natural number. *Hint:* Examine the ones digits of the powers.

35. Fermat's Little Theorem A theorem known as *Fermat's little theorem* states, "If n is a prime number and a is any natural number, then $a^n - a$ is divisible by n." Verify Fermat's little theorem for

a. $n = 7$ and $a = 12$. **b.** $n = 11$ and $a = 8$.

36. Amicable Numbers The Greeks considered the pair of numbers 220 and 284 to be *amicable* or *friendly* numbers because the sum of the proper divisors of one of the numbers is the other number.

The sum of the proper factors of 220 is

$$1 + 2 + 4 + 5 + 10 + 11 + 20 + 22 + 44 + 55 + 110 = 284$$

The sum of the proper factors of 284 is

$$1 + 2 + 4 + 71 + 142 = 220$$

Determine whether

a. 60 and 84 are amicable numbers.

b. 1184 and 1210 are amicable numbers.

37. A Sum of Cubes Property The perfect number 28 can be written as $1^3 + 3^3$. The perfect number 496 can be written as $1^3 + 3^3 + 5^3 + 7^3$. Verify that the next perfect number, 8128, can also be written as the sum of the cubes of consecutive odd natural numbers, starting with 1^3.

38. A Sum of the Digits Theorem If you sum the digits of any even perfect number (except 6), then sum the digits of the resulting number, and repeat this process until you get a single digit, that digit will be 1.

As an example, consider the perfect number 28. The sum of its digits is 10. The sum of the digits of 10 is 1.

Verify this theorem for each of the following perfect numbers.

a. 496 **b.** 8128

c. 33,550,336 **d.** 8,589,869,056

39. A Sum of Reciprocals Theorem The sum of the reciprocals of all the positive divisors of a perfect number is always 2.

Verify this theorem for each of the following perfect numbers.

a. 6 **b.** 28

EXTENSIONS

40. The Smallest Odd Abundant Number Determine the smallest odd abundant number. *Hint:* It is greater than 900 but less than 1000.

41. Fermat Numbers Numbers of the form $2^{2^n} + 1$, where n is a whole number, are called *Fermat numbers.* Fermat believed that all Fermat numbers were prime. Prove that Fermat was wrong.

42. Semiperfect Numbers Any number that is the sum of *some or all* of its proper divisors is called a **semiperfect number**. For instance, 12 is a semiperfect number because it has 1, 2, 3, 4, and 6 as proper factors, and $12 = 1 + 2 + 3 + 6$.

The first twenty-five semiperfect numbers are 6, 12, 18, 20, 24, 28, 30, 36, 40, 42, 48, 54, 56, 60, 66, 72, 78, 80, 84, 88, 90, 96, 100, 102, and 104. It has been established that every natural number multiple of a semiperfect number is semiperfect and that a semiperfect number cannot be a deficient number.

a. Use the definition of a semiperfect number to verify that 20 is a semiperfect number.

b. Explain how to verify that 200 is a semiperfect number without examining its proper factors.

43. Weird Numbers Any number that is an abundant number but not a semiperfect number (see Exercise 42) is called a **weird number**. Find the only weird number less than 100. *Hint:* The abundant numbers less than 100 are 12, 18, 20, 24, 30, 36, 40, 42, 48, 54, 56, 60, 66, 70, 72, 78, 80, 84, 88, 90, and 96.

44. A False Prediction In 1811, Peter Barlow wrote in his text *Theory of Numbers* that the eighth perfect number, $2^{30}(2^{31} - 1) =$ 2,305,843,008,139,952,128, which was discovered by Leonhard Euler in 1772, "is the greatest perfect number known at present, and probably the greatest that ever will be discovered; for as they are merely curious, without being useful, it is not likely that any person will attempt to find one beyond it." (*Source:* http://en.wikipedia.org/wiki/2147483647)

The current search for larger and larger perfect numbers shows that Barlow's prediction did not come true. Search the Internet for answers to the question "Why do people continue the search for large perfect numbers (or large prime numbers)?" Write a brief summary of your findings.

CHAPTER 6 SUMMARY

The following table summarizes essential concepts in this chapter. The references given in the right-hand column list Examples and Exercises that can be used to test your understanding of a concept.

6.1 Early Numeration Systems

The Egyptian Numeration System The Egyptian numeration system is an additive system that uses symbols called hieroglyphics as numerals.

Hindu-Arabic numeral	Egyptian hieroglyphic	Description of hieroglyphic
1	𓏺	stroke
10	𓎆	heel bone
100	𓍢	scroll
1000	𓆼	lotus flower
10,000	𓂭	pointing finger
100,000	𓆐	fish
1,000,000	𓁨	astonished person

See **Examples 1 to 4** on pages 295 to 296, and then try Exercises 1 to 4 on page 351.

The Roman Numeration System The Roman numeration system is an additive system that also incorporates a subtraction property.
Basic Rules Employed in the Roman Numeration System

$$\text{I} = 1, \ \text{V} = 5, \ \text{X} = 10, \ \text{L} = 50, \ \text{C} = 100, \ \text{D} = 500, \ \text{M} = 1000$$

1. *Addition Property* If the numerals are listed so that each numeral has a larger value than the numeral to the right, then the value of the Roman numeral is found by adding the values of the numerals. Each of the numerals, I, X, C, and M may be repeated up to three times. The numerals V, L, and D are not repeated. If a numeral is repeated two or three times in succession, we add to determine its numerical value.
2. *Subtraction Property* The only numerals whose values can be subtracted from the value of the numeral to its right are I, X, and C. The value of the numeral to be subtracted must be no less than one-tenth of the value of the numeral to its right.

See **Examples 5 to 7** on page 298, and then try Exercises 5 to 12 on page 351.

6.2 Place-Value Systems

The Hindu-Arabic Numeration System The Hindu-Arabic numeration system is a base ten place-value system. The representation of a numeral as the indicated sum of each digit of the numeral multiplied by its respective place value is called the expanded form of the numeral.

See **Examples 1 to 5** on pages 302 and 303, and then try Exercises 13 to 16 on page 351.

The Babylonian Numeration System The Babylonian numeration system is a base sixty place-value system. A vertical wedge shape represents one unit, and a sideways "vee" shape represents 10 units.

𒁹 1 𒌋 10

See **Examples 6 to 8** on pages 304 to 305, and then try Exercises 17 to 24 on page 352.

continued

The Mayan Numeration System The numerals and the place values used in the Mayan numeration system are shown below.

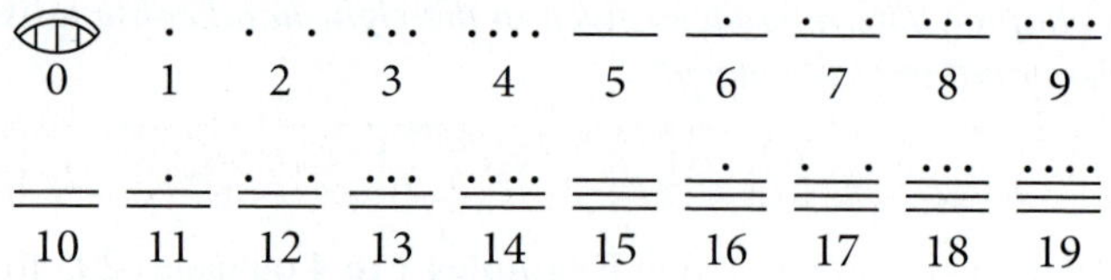

Place Values in the Mayan Numeration System

. . .	18×20^3 = 144,000	18×20^2 = 7200	18×20^1 = 360	20^1 = 20	20^0 = 1

See **Examples 9 and 10** on page 307, and then try Exercises 25 to 32 on page 352.

6.3 Different Base Systems

Convert Non–Base Ten Numerals to Base Ten In general, a base b numeration system has place values of ..., b^5, b^4, b^3, b^2, b^1, b^0. To convert a non–base ten numeral to base ten, we write the numeral in expanded form and simplify.

See **Examples 1 to 4** on pages 311 and 313, and then try Exercises 33 to 36 on page 352.

Convert from Base Ten to Another Base The method of converting a number written in base ten to another base makes use of the successive division process described on page 313.

See **Example 5** on page 314, and then try Exercises 37 to 40 on page 352.

Converting Directly between Computer Bases
To convert directly from:

- octal to binary, replace each octal numeral with its 3-bit binary equivalent.
- binary to octal, break the binary numeral into groups of three, from right to left, and replace each group with its octal equivalent.
- hexadecimal to binary, replace each hexadecimal numeral with its 4-bit binary equivalent.
- binary to hexadecimal, break the binary numeral into groups of four, from right to left, and replace each group with its hexadecimal equivalent.

See **Examples 6 to 9** on pages 315 and 316, and then try Exercises 45 to 52 on page 352.

The Double-Dabble Method The double-dabble method is a procedure that can be used to convert a base two numeral to base ten. The advantage of this method is that you can start at the left of the base two numeral and work your way to the right without first determining the place value of each bit in the base two numeral.

See **Example 10** on page 316, and then try Exercises 53 to 56 on page 352.

6.4 Arithmetic in Different Bases

Arithmetic in Different Bases This section illustrates how to perform arithmetic operations in different base systems.

See **Examples 4, 5, 8, and 9** on pages 322 to 326, and then try Exercises 57 to 64 on page 352.

6.5 Prime Numbers

Divisor of a Number The natural number a is a divisor or a factor of the natural number b provided there exists a natural number j such that $aj = b$.

See **Example 1** on page 330, and then try Exercises 65 and 66 on page 352.

Prime Numbers and Composite Numbers A prime number is a natural number greater than 1 that has exactly two factors: itself and 1. A composite number is a natural number greater than 1 that is not a prime number.	See **Example 2** on page 331, and then try Exercises 67 to 70 on page 352.
The Fundamental Theorem of Arithmetic Every composite number can be written as a unique product of prime numbers (disregarding the order of the factors).	See **Example 4** on page 333, and then try Exercises 71 to 74 on page 352.
6.6 Topics from Number Theory	
Perfect, Deficient, and Abundant Numbers A natural number is ■ **perfect** if it is equal to the sum of its proper factors. ■ **deficient** if it is greater than the sum of its proper factors. ■ **abundant** if it is less than the sum of its proper factors.	See **Example 1** on page 340, and then try Exercises 75 to 78 on page 352.
Euclid's Procedure for Generating a Perfect Number A Mersenne number is a number of the form $2^n - 1$, where n is a prime number. Some Mersenne numbers are prime and some are composite. If n is a prime number and the Mersenne number $2^n - 1$ is a prime number, then $2^{n-1}(2^n - 1)$ is a perfect number.	See **Example 3** on page 342, and then try Exercises 79 and 80 on page 352.
The Number of Digits in b^x The number of digits in the number b^x, where b is a natural number less than 10 and x is a natural number, is the greatest integer of $(x \log b) + 1$.	See **Example 4** on page 343, and then try Exercises 85 and 86 on page 352.
Fermat's Last Theorem There are no natural numbers x, y, z, and n that satisfy $x^n + y^n = z^n$ where n is greater than 2.	See **Example 5** on page 345, and then try Exercise 88 on page 352.

CHAPTER 6 REVIEW EXERCISES

1. Write 4,506,325 using Egyptian hieroglyphics.

2. Write 3,124,043 using Egyptian hieroglyphics.

3. Write the Egyptian hieroglyphic

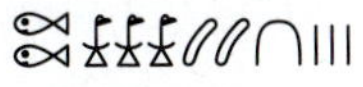

as a Hindu-Arabic numeral.

4. Write the Egyptian hieroglyphic

as a Hindu-Arabic numeral.

■ In Exercises 5 to 8, write each Roman numeral as a Hindu-Arabic numeral.

5. CCCXLIX

6. DCCLXXIV

7. $\overline{\text{IX}}$DCXL

8. $\overline{\text{XCII}}$CDXLIV

■ In Exercises 9 to 12, write each Hindu-Arabic numeral as a Roman numeral.

9. 567

10. 823

11. 2489

12. 1335

■ In Exercises 13 and 14, write each Hindu-Arabic numeral in expanded form.

13. 432

14. 456,327

■ In Exercises 15 and 16, simplify each expanded form.

15. $(5 \times 10^6) + (3 \times 10^4) + (8 \times 10^3) + (2 \times 10^2) + (4 \times 10^0)$

16. $(3 \times 10^5) + (8 \times 10^4) + (7 \times 10^3) + (9 \times 10^2) + (6 \times 10^1)$

■ In Exercises 17 to 20, write each Babylonian numeral as a Hindu-Arabic numeral.

17. 

18. 

19. 

20. 

■ In Exercises 21 to 24, write each Hindu-Arabic numeral as a Babylonian numeral.

21. 721

22. 1080

23. 12,543

24. 19,281

■ In Exercises 25 to 28, write each Mayan numeral as a Hindu-Arabic numeral.

25. 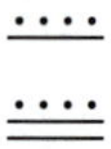

26. 

27. 

28. 

■ In Exercises 29 to 32, write each Hindu-Arabic numeral as a Mayan numeral.

29. 522

30. 346

31. 1862

32. 1987

■ In Exercises 33 to 36, convert each numeral to base ten.

33. 45_{six}

34. 172_{nine}

35. $E3_{\text{sixteen}}$

36. $1BA_{\text{twelve}}$

■ In Exercises 37 to 40, convert each base ten numeral to the indicated base.

37. 45 to base three

38. 123 to base seven

39. 862 to base eleven

40. 3021 to base twelve

■ In Exercises 41 to 44, convert each numeral to the indicated base.

41. 346_{nine} to base six

42. 1532_{six} to base eight

43. 275_{twelve} to base nine

44. $67A_{\text{sixteen}}$ to base twelve

■ In Exercises 45 to 52, convert each numeral directly (without first converting to base ten) to the indicated base.

45. 11100_{two} to base eight

46. 1010100_{two} to base eight

47. 1110001101_{two} to base sixteen

48. 11101010100_{two} to base sixteen

49. 25_{eight} to base two

50. 1472_{eight} to base two

51. $4A_{\text{sixteen}}$ to base two

52. $C72_{\text{sixteen}}$ to base two

■ In Exercises 53 to 56, use the double-dabble method to convert each base two numeral to base ten.

53. 110011010_{two}

54. 100010101_{two}

55. 10000010001_{two}

56. 11001010000_{two}

■ In Exercises 57 to 64, perform the indicated operation. Write the answers in the same base as the given numerals.

57. $235_{\text{six}} + 144_{\text{six}}$

58. $673_{\text{eight}} + 345_{\text{eight}}$

59. $672_{\text{nine}} - 135_{\text{nine}}$

60. $1332_{\text{four}} - 213_{\text{four}}$

61. $25_{\text{eight}} \times 542_{\text{eight}}$

62. $43_{\text{five}} \times 3421_{\text{five}}$

63. $1010101_{\text{two}} \div 11_{\text{two}}$

64. $321_{\text{four}} \div 12_{\text{four}}$

■ In Exercises 65 and 66, use divisibility tests to determine whether the given number is divisible by 2, 3, 4, 5, 6, 8, 9, 10, or 11.

65. 1485

66. 4268

■ In Exercises 67 to 70, determine whether the given number is a prime number or a composite number.

67. 501

68. 781

69. 689

70. 1003

■ In Exercises 71 to 74, determine the prime factorization of the given number.

71. 45

72. 54

73. 153

74. 285

■ In Exercises 75 to 78, determine whether the given number is perfect, deficient, or abundant.

75. 28

76. 81

77. 144

78. 200

■ In Exercises 79 and 80, use Euclid's perfect number–generating procedure to write the perfect number associated with the given Mersenne prime.

79. $2^{61} - 1$

80. $2^{1279} - 1$

■ In Exercises 81 to 84, use the Egyptian doubling procedure to find the given product.

81. 8×46

82. 9×57

83. 14×83

84. 21×143

85. Find the number of digits in the Mersenne number $2^{132049} - 1$.

86. Find the number of digits in the Mersenne number $2^{2976221} - 1$.

87. How many odd perfect numbers had been discovered as of July 2016?

88. Determine whether there is a natural number z such that

$$2^3 + 17^3 = z^3$$

CHAPTER 6 TEST

1. Write 3124 using Egyptian hieroglyphics.
2. Write the Egyptian hieroglyphic

as a Hindu-Arabic numeral.

3. Write the Roman numeral MCDXLVII as a Hindu-Arabic numeral.
4. Write 2609 as a Roman numeral.
5. Write 67,485 in expanded form.
6. Simplify:

$$(5 \times 10^5) + (3 \times 10^4) + (2 \times 10^2) + (8 \times 10^1) + (4 \times 10^0)$$

7. Write the Babylonian numeral

as a Hindu-Arabic numeral.

8. Write 9675 as a Babylonian numeral.
9. Write the Mayan numeral

as a Hindu-Arabic numeral.

10. Write 502 as a Mayan numeral.
11. Convert 3542_{six} to base ten.
12. Convert 2148 to **a.** base eight and **b.** base twelve.
13. Convert 4567_{eight} to binary form.
14. Convert $101010110111_{\text{two}}$ to hexadecimal form.

■ In Exercises 15 to 18, perform the indicated operation. Write the answers in the same base as the given numerals.

15. $34_{\text{five}} + 23_{\text{five}}$
16. $462_{\text{eight}} - 147_{\text{eight}}$
17. $101_{\text{two}} \times 101110_{\text{two}}$
18. $431_{\text{seven}} \div 5_{\text{seven}}$
19. Determine the prime factorization of 230.
20. Determine whether 1001 is a prime number or a composite number.
21. Use divisibility tests to determine whether 1,737,285,147 is divisible by **a.** 2, **b.** 3, or **c.** 5.
22. Use divisibility tests to determine whether 19,531,333,276 is divisible by **a.** 4, **b.** 6, or **c.** 11.
23. Determine whether 96 is perfect, deficient, or abundant.
24. Use Euclid's perfect number–generating procedure to write the perfect number associated with the Mersenne prime $2^{17} - 1$.
25. Explain how you know (without doing any arithmetic) that there is no natural number z such that

$$2^5 + 19^5 = z^5$$

7

Measurement and Geometry

Much of this chapter concerns geometric figures and their properties. Drawing complex 3D figures by hand can be tedious and time consuming. Fortunately, the advent of 3D computer drawing programs has made the process much easier.

One of the easiest 3D drawing programs to use is Google SketchUp. If you are not familiar with SketchUp, we encourage you to give it a try. A free version is available at http://sketchup.google.com/.

Even if you decide not to produce your own 3D drawings, you can still enjoy looking at some of the drawings that other SketchUp users have produced. For example, the following SketchUp drawing of a country villa incorporates many of the geometric figures presented in this chapter. After you download this drawing[1], you can use rotation and zoom tools to view the villa from any angle and distance. Use the rotation and zoom tools to check out the grand piano on the second floor.

Google SketchUp™/josh86

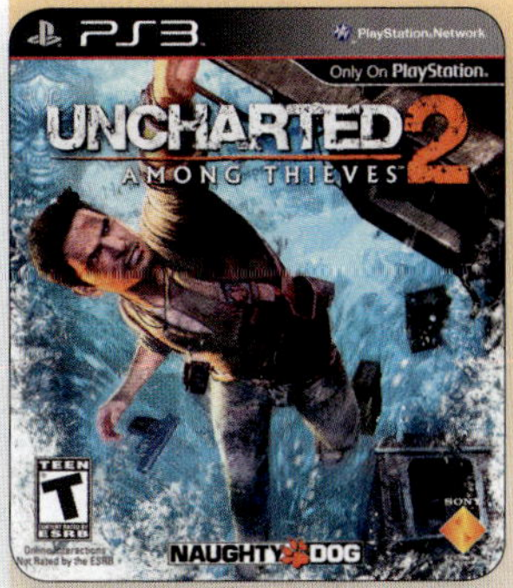

Sony Corporation of America.

3D drawing programs have been used to design video games. For instance, art director Robh Ruppel used SketchUp to design the video game *Uncharted 2: Among Thieves*. He discusses some of the details concerning the design of *Uncharted 2* in a YouTube video available at http://www.youtube.com/watch?v=8mkPRmqUlFw.

[1] The URL for the above country villa drawing is: http://sketchup.google.com/3dwarehouse/details?mid=79a4f741cbf41e86a1c299db94b0d06b&prevstart=0

SECTION 7.1 Measurement

U.S. Customary Units of Measure

Units of measure are a way to give a magnitude or size to a number. For instance, there is quite a difference between 7 feet and 7 miles. The units *feet* and *miles* are units of measure in the U.S. Customary System. In this section we are going to focus on units of length, weight, and capacity.

Here are some of the U.S. Customary units and their equivalences.

Equivalences Between Units of Length in the U.S. Customary System

12 inches (in.) = 1 foot (ft)
3 ft = 1 yard (yd)
5280 ft = 1 mile (mi)

Equivalences Between Units of Weight in the U.S. Customary System

16 ounces (oz) = 1 pound (lb)
2000 lb = 1 ton

Equivalences Between Units of Capacity in the U.S. Customary System

8 fluid ounces (fl oz) = 1 cup (c)
2 c = 1 pint (pt)
2 pt = 1 quart (qt)
4 qt = 1 gallon (gal)

These equivalences can be used to change from one unit to another unit by using a **conversion rate**. For instance, because 12 in. = 1 ft, we can write two conversion rates,

$$\frac{12\text{ in.}}{1\text{ ft}} \text{ and } \frac{1\text{ ft}}{12\text{ in.}}$$

Dimensional analysis involves using conversion rates to change from one unit of measurement to another.

EXAMPLE 1 Convert Between Units of Length

Convert 18 in. to feet.

Solution

Choose a conversion rate so that the unit in the numerator of the conversion rate is the same as the unit needed in the answer ("feet" for this problem). The unit in the denominator of the conversion rate is the same as the unit in the given measurement ("inches" for this problem). The conversion rate is $\frac{1\text{ ft}}{12\text{ in.}}$.

$$18\text{ in.} = 18\text{ in.} \times \frac{1\text{ ft}}{12\text{ in.}}$$

The unit in the numerator is the same as the unit needed in the answer.

The unit in the denominator is the same as the unit in the given measurement.

$$= \frac{18\text{ ft}}{12} = \frac{3}{2}\text{ ft}$$

$$= 1\frac{1}{2}\text{ ft}$$

$$18\text{ in.} = 1\frac{1}{2}\text{ ft}$$

CHECK YOUR PROGRESS 1

Convert 14 ft to yards.

Solution See page S23.

EXAMPLE 2 Convert Between Units of Weight

Convert $3\frac{1}{2}$ tons to pounds.

Solution

Write a conversion rate with pounds in the numerator and tons in the denominator. The conversion rate is $\frac{2000\text{ lb}}{1\text{ ton}}$.

$$3\frac{1}{2}\text{ tons} = \frac{7}{2}\cancel{\text{tons}} \times \frac{2000\text{ lb}}{1\ \cancel{\text{ton}}}$$

$$= \frac{14{,}000\text{ lb}}{2} = 7000\text{ lb}$$

$$3\frac{1}{2}\text{ tons} = 7000\text{ lb}$$

CHECK YOUR PROGRESS 2

Convert 3 lb to ounces.

Solution See page S23.

Sometimes one unit cannot be converted directly to another. For instance, consider trying to convert quarts to cups using only the capacity equivalences given earlier. In this case, it is necessary to use more than one conversion factor, as shown next.

Convert 3 qt to cups.

$$3\text{ qt} = 3\text{ qt} \times \boxed{\frac{2\text{ pt}}{1\text{ qt}}} \times \boxed{\frac{2\text{ c}}{1\text{ pt}}}$$

$$= \frac{3\ \cancel{\text{qt}}}{1} \times \frac{2\ \cancel{\text{pt}}}{1\ \cancel{\text{qt}}} \times \frac{2\text{ c}}{1\ \cancel{\text{pt}}}$$

$$= \frac{12\text{ c}}{1} = 12\text{ c}$$

$$3\text{ qt} = 12\text{ c}$$

- The direct equivalence between quarts and cups was not given earlier in the section. Use two conversion rates. First convert quarts to pints, and then convert pints to cups.

EXAMPLE 3 Convert Between Units of Capacity

Convert 42 c to quarts.

Solution

First convert cups to pints, and then convert pints to quarts.

$$42\text{ c} = 42\ \cancel{\text{c}} \times \frac{1\ \cancel{\text{pt}}}{2\ \cancel{\text{c}}} \times \frac{1\text{ qt}}{2\ \cancel{\text{pt}}}$$

$$= \frac{42\text{ qt}}{4} = 10\frac{1}{2}\text{ qt}$$

$$42\text{ c} = 10\frac{1}{2}\text{ qt}$$

CHECK YOUR PROGRESS 3

Convert 18 pt to gallons.

Solution See page S23.

The Metric System

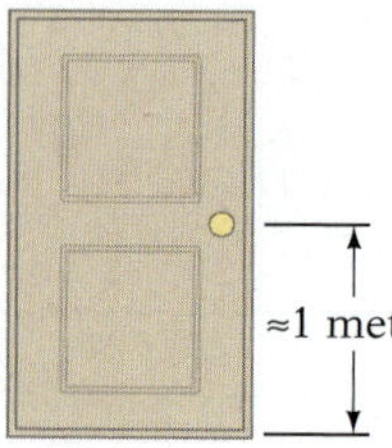

In 1789, an attempt was made to standardize units of measurement internationally in order to simplify trade and commerce between nations. A commission in France developed a system of measurement known as the **metric system**.

The basic unit of length in the metric system is the **meter**. One meter is approximately the distance from a doorknob to the floor. All units of length in the metric system are derived from the meter. Prefixes to the basic unit denote the length of each unit. For example, the prefix "centi-" means one-hundredth, so 1 centimeter is 1 one-hundredth of a meter.

Prefixes and Units of Length in the Metric System

kilo- = 1000	1 kilometer (km) = 1000 meters (m)
hecto- = 100	1 hectometer (hm) = 100 m
deca- = 10	1 decameter (dam) = 10 m
	1 meter (m) = 1 m
deci- = 0.1	1 decimeter (dm) = 0.1 m
centi- = 0.01	1 centimeter (cm) = 0.01 m
milli- = 0.001	1 millimeter (mm) = 0.001 m

Conversion between units of length in the metric system involves moving the decimal point to the right or to the left. Listing the units in order from largest to smallest will indicate how many places to move the decimal point and in which direction.

To convert 4200 cm to meters, write the units in order from largest to smallest.

km hm dam m dm cm mm

2 positions

- Converting centimeters to meters requires moving 2 positions to the left.

4200 cm = 42.00 m

2 places

- Move the decimal point the same number of places and in the same direction.

A metric measurement that involves two units is customarily written in terms of one unit. Convert the smaller unit to the larger unit, and then add.

To convert 8 km 32 m to kilometers, first convert 32 m to kilometers.

km hm dam m dm cm mm

- Converting meters to kilometers requires moving 3 positions to the left.

32 m = 0.032 km

- Move the decimal point the same number of places and in the same direction.

8 km 32 m = 8 km + 0.032 km
= 8.032 km

- Add the result to 8 km.

EXAMPLE 4 Convert Between Units of Length

Convert 0.38 m to millimeters.

Solution

0.38 m = 380 mm

CHECK YOUR PROGRESS 4

Convert 3.07 m to centimeters.

Solution See page S23.

Mass and weight are closely related. Weight is a measure of how strongly Earth is pulling on an object. Therefore, an object's weight is less in space than on Earth's surface. However, the amount of material in the object, its **mass**, remains the same. On the surface of Earth, *mass* and *weight* can be used interchangeably.

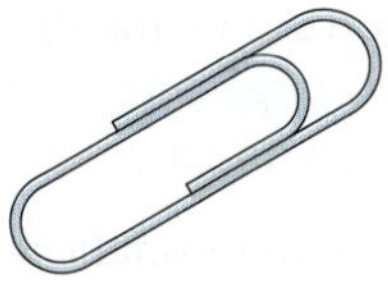

Weight ≈ 1 gram

The gram is the unit of mass in the metric system to which prefixes are added. One gram is about the weight of a paper clip.

Units of Mass in the Metric System

1 kilogram (kg) = 1000 grams (g)
1 hectogram (hg) = 100 g
1 decagram (dag) = 10 g
1 gram (g) = 1 g
1 decigram (dg) = 0.1 g
1 centigram (cg) = 0.01 g
1 milligram (mg) = 0.001 g

Conversion between units of mass in the metric system involves moving the decimal point to the right or to the left. Listing the units in order from largest to smallest will indicate how many places to move the decimal point and in which direction.

To convert 324 g to kilograms, write the units in order from largest to smallest.

kg hg dag g dg cg mg

3 positions

- Converting grams to kilograms requires moving 3 positions to the left.

324 g = 0.324 kg

3 places

- Move the decimal point the same number of places and in the same direction.

EXAMPLE 5 Convert Between Units of Mass

Convert 4.23 g to milligrams.

Solution

4.23 g = 4230 mg

CHECK YOUR PROGRESS 5

Convert 42.3 mg to grams.

Solution See page S23.

The basic unit of capacity in the metric system is the liter. One **liter** is defined as the capacity of a box that is 10 cm long on each side.

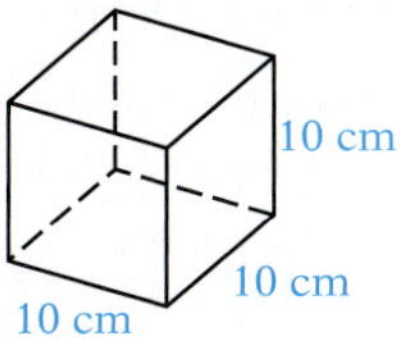

The units of capacity in the metric system have the same prefixes as the units of length.

Units of Capacity in the Metric System

1 kiloliter (kl) = 1000 L
1 hectoliter (hl) = 100 L
1 decaliter (dal) = 10 L
1 liter (L) = 1 L
1 deciliter (dl) = 0.1 L
1 centiliter (cl) = 0.01 L
1 milliliter (ml) = 0.001 L

The milliliter is equal to 1 **cubic centimeter** (cm^3). In medicine, cubic centimeter is often abbreviated cc.

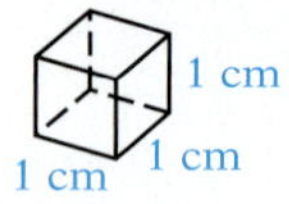

1 ml = 1 cm^3

Conversion between units of capacity in the metric system involves moving the decimal point to the right or to the left. Listing the units in order from largest to smallest will indicate how many places to move the decimal point and in which direction.

To convert 824 ml to liters, first write the units in order from largest to smallest.

kl hl dal L dl cl ml

3 positions

• Converting milliliters to liters requires moving 3 positions to the left.

824 ml = 0.824 L

3 places

• Move the decimal point the same number of places and in the same direction.

EXAMPLE 6 Convert Between Units of Capacity

Convert 4 L 32 ml to liters.

Solution

32 ml = 0.032 L

4 L 32 ml = 4 L + 0.032 L

= 4.032 L

CHECK YOUR PROGRESS 6

Convert 2 kl 167 L to liters.

Solution See page S23.

Convert Between U.S. Customary Units and Metric Units

More than 90% of the world's population uses the metric system of measurement. Therefore, converting U.S. Customary units to metric units is essential in trade and commerce—for example, in importing foreign goods and exporting domestic goods. Approximate equivalences between the two systems follow.

Units of Length	Units of Weight	Units of Capacity
1 in. = 2.54 cm	1 oz ≈ 28.35 g	1 L ≈ 1.06 qt
1 m ≈ 3.28 ft	1 lb ≈ 454 g	1 gal ≈ 3.79 L
1 m ≈ 1.09 yd	1 kg ≈ 2.2 lb	
1 mi ≈ 1.61 km		

These equivalences can be used to form conversion rates to change from one unit of measurement to another. For example, because 1 mi ≈ 1.61 km, the conversion rates $\frac{1 \text{ mi}}{1.61 \text{ km}}$ and $\frac{1.61 \text{ km}}{1 \text{ mi}}$ are both approximately equal to 1.

Convert 55 mi to kilometers.

$$55 \text{ mi} \approx 55 \text{ mi} \times \boxed{\frac{1.61 \text{ km}}{1 \text{ mi}}}$$

• The conversion rate must contain kilometers in the numerator and miles in the denominator.

$$= \frac{55 \cancel{\text{mi}}}{1} \times \frac{1.61 \text{ km}}{1 \cancel{\text{mi}}}$$

$$= \frac{88.55 \text{ km}}{1}$$

$$55 \text{ mi} \approx 88.55 \text{ km}$$

EXAMPLE 7 **Convert U.S. Customary Units to Metric Units**

The price of gasoline is \$3.89/gal. Find the cost per liter. Round to the nearest tenth of a cent.

Solution

$$\frac{\$3.89}{\text{gal}} \approx \frac{\$3.89}{\cancel{\text{gal}}} \times \frac{1\ \cancel{\text{gal}}}{3.79\ \text{L}} = \frac{\$3.89}{3.79\ \text{L}} \approx \frac{\$1.026}{1\ \text{L}}$$

$$\$3.89/\text{gal} \approx \$1.026/\text{L}$$

CHECK YOUR PROGRESS 7

The price of milk is \$3.69/gal. Find the cost per liter. Round to the nearest cent.

Solution See page S23.

Metric units are used in the United States. Cereal is sold by the gram, 35-mm film is available, and soda is sold by the liter. The same conversion rates used to convert U.S. Customary units to metric units are used to convert metric units to U.S. Customary units.

EXAMPLE 8 **Convert Metric Units to U.S. Customary Units**

Convert 200 m to feet.

Solution

$$200\ \text{m} \approx 200\ \cancel{\text{m}} \times \frac{3.28\ \text{ft}}{1\ \cancel{\text{m}}} = \frac{656\ \text{ft}}{1}$$

$$200\ \text{m} \approx 656\ \text{ft}$$

CHECK YOUR PROGRESS 8

Convert 45 cm to inches. Round to the nearest hundredth.

Solution See page S23.

EXAMPLE 9 **Convert Metric Units to U.S. Customary Units**

Convert 90 km/h to miles per hour. Round to the nearest hundredth.

Solution

$$\frac{90\ \text{km}}{\text{h}} \approx \frac{90\ \text{km}}{\text{h}} \times \frac{1\ \text{mi}}{1.61\ \text{km}} = \frac{90\ \text{mi}}{1.61\ \text{h}} \approx \frac{55.90\ \text{mi}}{1\ \text{h}}$$

$$90\ \text{km/h} \approx 55.90\ \text{mi/h}$$

CHECK YOUR PROGRESS 9

Express 75 km/h in miles per hour. Round to the nearest hundredth.

Solution See page S23.

EXCURSION

Drawing with a Straightedge and a Compass

The Elements, written by Euclid about 2300 years ago, is arguably the most influential treatise on geometry ever written. One of the most remarkable aspects of this work is that all measurements necessary for the proofs were accomplished using only a straightedge and a compass. Neither of these devices had any units written on it—Euclid used just a straight, blank piece of wood as his straightedge and a compass that could be adjusted but was not demarked by degrees.

Using these two instruments, Euclid showed how to find the midpoint of a line segment, how to construct a 90° angle, and how to draw parallel lines and many other geometric figures. For instance, here is Euclid's method for constructing a 90° angle at a point.

1. Start with a point on a line. The 90° angle will be created at this point.

2. Set the compass to a specific width and, from the point, draw two arcs that intersect the line.

3. Increase the width of the compass by some arbitrary amount. Place the point of the compass at each point where the arcs intersect the line, and draw two intersecting arcs.

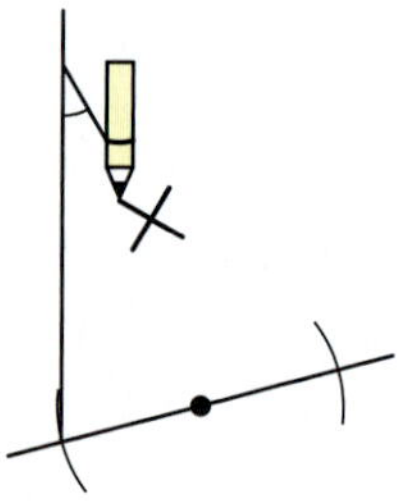

4. Draw a line through the point on the line and the intersection of the two arcs.

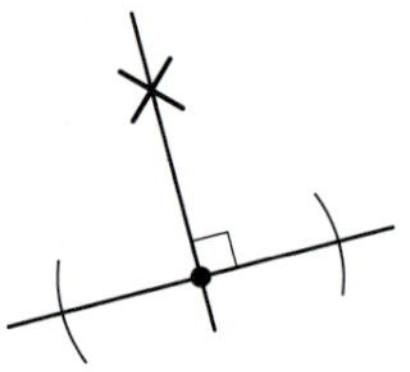

It is possible to bisect any angle using only a compass and a straightedge.

1. Start with an angle. Draw an arc through the angle with the point of the compass at the vertex of the angle.

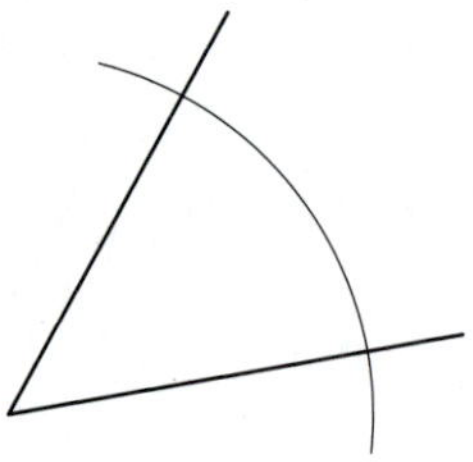

2. Draw two intersecting arcs by placing the compass at the points where the original arc intersects each side of the angle. Then draw a line between the vertex of the angle and the intersection of the two arcs.

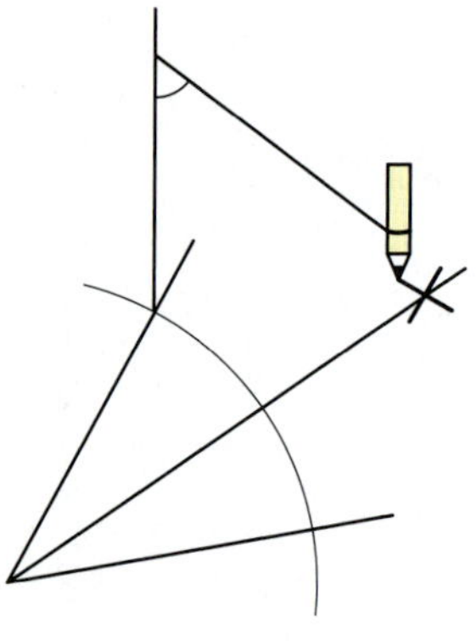

EXCURSION EXERCISES

Complete the following exercises using only a straightedge and a compass. If you get stuck, do some research on how to draw the figure with a straightedge and compass.

1. Draw the yin-and-yang symbol shown at the right. *Hint:* This symbol consists of multiple circles, each with its center on a vertical line.

2. Draw a hexagon with all sides the same length. Here is a suggestion on how to begin. First, draw two circles that are equal in size. Then add lines as shown in the diagram at the far right.

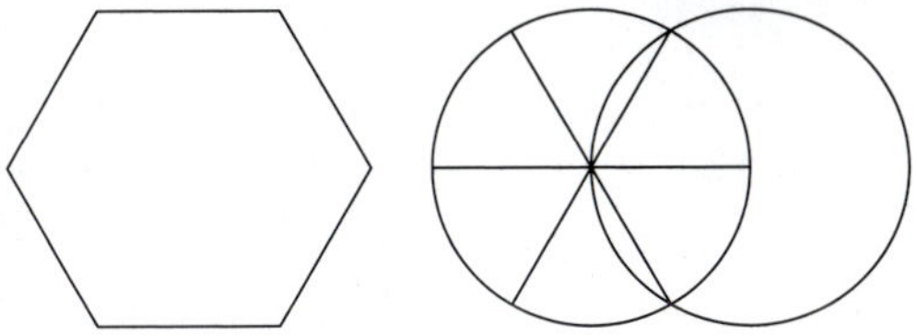

3. Find the midpoint of the line segment *AB*.

4. Draw the heart-shaped figure shown at the right. Here is a suggestion on how to begin. First, use the construction shown at the far right to draw a right triangle. Then bisect the angle and draw some circles.

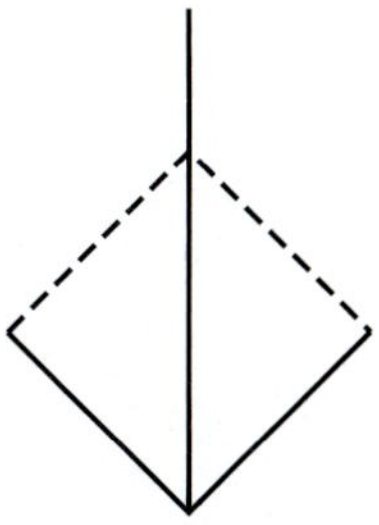

EXERCISE SET 7.1

■ For Exercises 1 to 24, convert between the two measurements.

1. 6 ft = ______ in.
2. 7920 ft = ______ mi
3. 5 yd = ______ in.
4. $\frac{1}{2}$ mi = ______ yd
5. 64 oz = ______ lb
6. 9000 lb = ______ tons
7. $1\frac{1}{2}$ lb = ______ oz
8. $\frac{4}{5}$ ton = ______ lb
9. $2\frac{1}{2}$ c = ______ fl oz
10. 10 qt = ______ gal
11. 7 gal = ______ pt
12. $1\frac{1}{2}$ qt = ______ c
13. 62 cm = ______ mm
14. 6804 m = ______ km
15. 3.21 m = ______ cm
16. 260 cm = ______ m
17. 7421 g = ______ kg
18. 43 mg = ______ g
19. 0.45 g = ______ dg
20. 0.0456 g = ______ mg
21. 7.5 ml = ______ L
22. 0.037 L = ______ ml
23. 0.435 L = ______ cm^3
24. 897 L = ______ kl

■ For Exercises 25 to 34, solve. Round to the nearest hundredth if necessary.

25. Find the weight in kilograms of a 145-pound person.
26. Find the number of liters in 14.3 gal of gasoline.
27. Express 30 mi/h in kilometers per hour.
28. Seedless watermelon costs $0.59/lb. Find the cost per kilogram.
29. Deck stain costs $32.99/gal. Find the cost per liter.
30. Find the weight, in pounds, of an 86-kilogram person.
31. Find the width, in inches, of 35-mm film.
32. Express 30 m/s in feet per second.
33. A 5-kg ham costs $10/kg. Find the cost per pound.
34. A 2.5-kg bag of grass seed costs $10.99. Find the cost per pound.

EXTENSIONS

As our scientific and technical knowledge has increased, so has our need for ever-smaller and ever-larger units of measure. The prefixes used to denote some of these units of measure are listed in the table below.

Prefixes for large units		10^n	Prefixes for small units		10^n
Name	Symbol		Name	Symbol	
yotta	Y	10^{24}	yocto	y	10^{-24}
zetta	Z	10^{21}	zepto	z	10^{-21}
exa	E	10^{18}	atto	a	10^{-18}
peta	P	10^{15}	femto	f	10^{-15}
tera	T	10^{12}	pico	p	10^{-12}
giga	G	10^{9}	nano	n	10^{-9}
mega	M	10^{6}	micro	μ	10^{-6}

These new units of measure are quickly working their way into our everyday lives. For example, it is quite easy to purchase a 1-terabyte hard drive. One terabyte (TB) is equal to 10^{12} bytes. As another example, many computers can do multiple operations in 1 nanosecond (ns). One nanosecond is equal to 10^{-9} s.

■ For Exercises 35 to 40, convert between the two units.

35. 2.3 T = ______ Y

36. 4.51 n = ______ p

37. 0.65 Z = ______ G

38. 9.46 a = ______ μ

39. 4.01 G = ______ E

40. 7.15 y = ______ f

41. The speed of light is approximately 3×10^8 m/s. What is the speed of light in Zm (zettameters) per second?

42. A light year is approximately 6,000,000,000,000,000 mi. Describe a light year using Ym (yottameters).

43. In the 1980s, it became possible to measure optical events in nanoseconds and picoseconds. Express 1 ps as a decimal.

44. A tau lepton, which is an extremely small elementary particle, has a lifetime of 3×10^{-13} s. What is the lifetime of a tau lepton in femtoseconds?

SECTION 7.2 Basic Concepts of Euclidean Geometry

Lines, Rays, Line Segments, and Angles

The word *geometry* comes from the Greek words for "earth" and "measure." Geometry has applications in such disciplines as physics, medicine, and geology. Geometry is also used in applied fields such as mechanical drawing and astronomy. Geometric forms are used in art and design. We will begin our study by introducing two basic geometric concepts: point and line.

A **point** is symbolized by drawing a dot. A **line** is determined by two distinct points and extends indefinitely in both directions, as the arrows on the line at the right indicate. This line contains points A and B and is represented by $\overleftrightarrow{AB}$. A line can also be represented by a single letter, such as ℓ.

HISTORICAL NOTE

Science Photo Library/Photo Researchers/Science Source

Geometry is one of the oldest branches of mathematics. Around 350 BC, Euclid (yoo′klĭd) of Alexandria wrote *The Elements,* which contained all of the known concepts of geometry. Euclid's contribution to geometry was to unify various concepts into a single deductive system that was based on a set of postulates.

A **ray** starts at a point and extends indefinitely in *one* direction. The point at which a ray starts is called the **endpoint** of the ray. The ray shown at the right is denoted $\overrightarrow{AB}$. Point A is the endpoint of the ray.

A **line segment** is part of a line and has two endpoints. The line segment shown at the right is denoted by $\overline{AB}$. The distance between the endpoints of $\overline{AB}$ is denoted by AB.

QUESTION Classify each diagram as a line, a ray, or a line segment.

a. 

b. 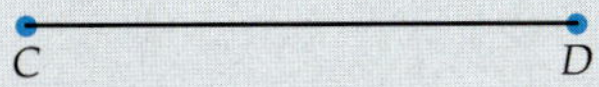

c. 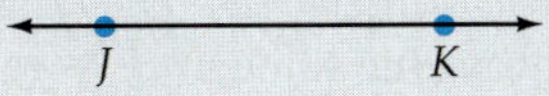

Given $AB = 22$ cm and $BC = 13$ cm in the figure at the right, then AC is the sum of the distances AB and BC.

$$AB + BC = AC$$
$$22 + 13 = AC$$
$$35 = AC$$

$AC = 35$ cm

EXAMPLE 1 Use an Equation to Find a Distance on a Line Segment

X, Y, and Z are all on line ℓ. Given $XY = 9$ m and YZ is twice XY, find XZ.

Solution

$XZ = XY + YZ$

$XZ = XY + 2(XY)$ • YZ is twice XY.

$XZ = 9 + 2(9)$ • Replace XY by 9.

$XZ = 9 + 18$ • Solve for XZ.

$XZ = 27$

$XZ = 27$ m

CHECK YOUR PROGRESS 1 A, B, and C are all on line ℓ. Given $BC = 16$ ft and $AB = \frac{1}{4}(BC)$, find AC.

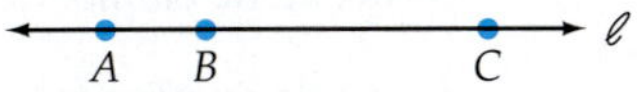

Solution See page S23.

ANSWER **a.** Ray. **b.** Line segment. **c.** Line.

In this section, we are discussing figures that lie in a plane. A **plane** is a flat surface with no thickness and no boundaries. It can be pictured as a desktop or whiteboard that extends forever. Figures that lie in a plane are called **plane figures**.

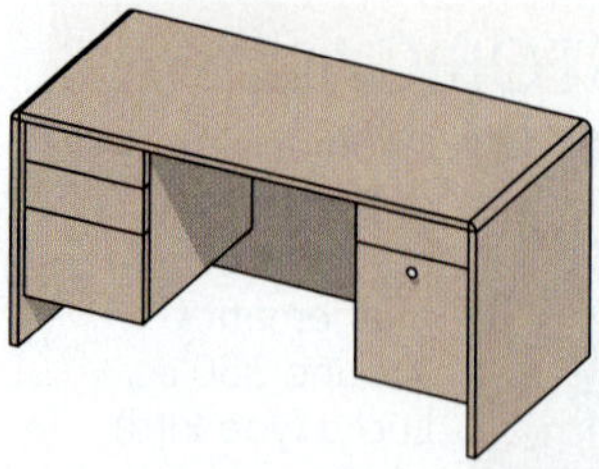

Lines in a plane can be intersecting or parallel. **Intersecting lines** cross at a point in the plane.

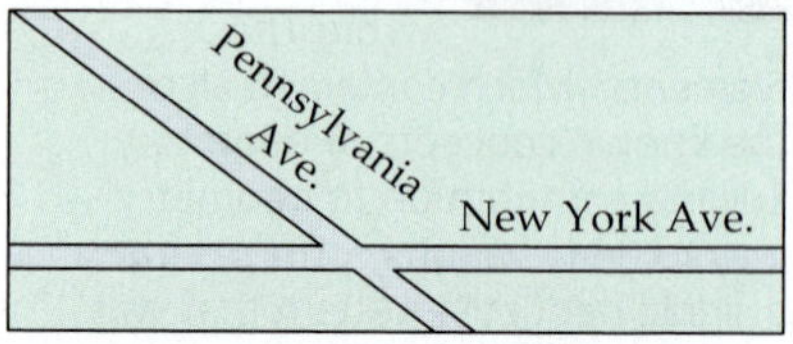

Parallel lines never intersect. The distance between them is always the same.

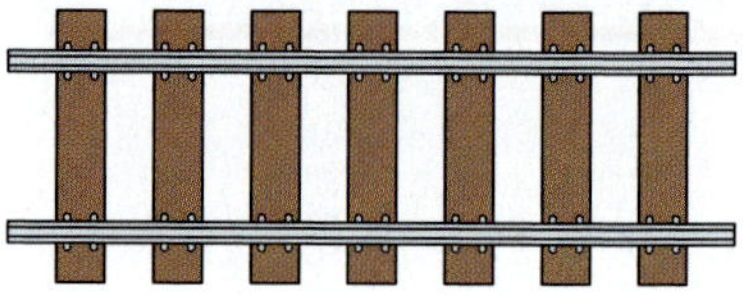

The symbol $\parallel$ means "is parallel to." In the figure at the right, $j \parallel k$ and $\overline{AB} \parallel \overline{CD}$.

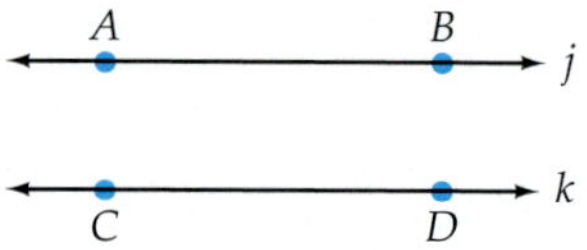

Angles

An **angle** is formed by two rays with the same endpoint. The **vertex** of the angle is the point at which the two rays meet. The rays are called the **sides** of the angle.

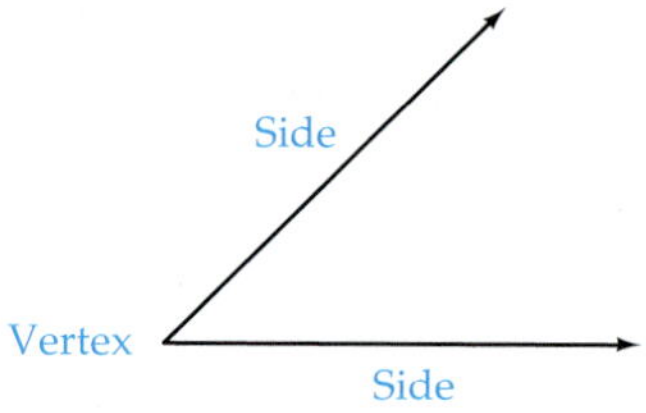

If A is a point on one ray of an angle, C is a point on the other ray, and B is the vertex, then the angle is called $\angle B$ or $\angle ABC$, where $\angle$ is the symbol for angle. Note that an angle can be named by the vertex, or by giving three points, where the second point listed is the vertex. $\angle ABC$ could also be called $\angle CBA$.

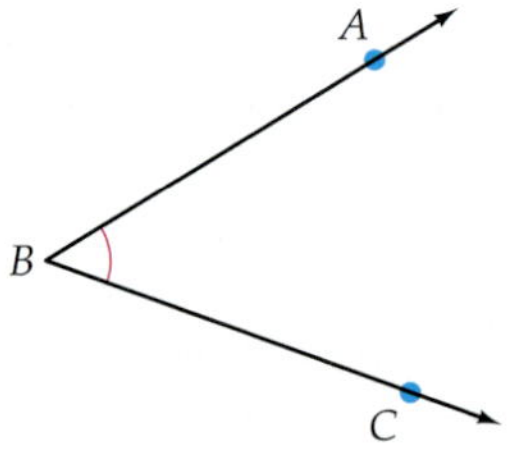

An angle can also be named by a variable written between the rays close to the vertex. In the figure at the right, $\angle x$ and $\angle QRS$ are two different names for the same angle. $\angle y$ and $\angle SRT$ are two different names for the same angle. Note that in this figure, more than two rays meet at R. In this case, the vertex alone cannot be used to name $\angle QRT$.

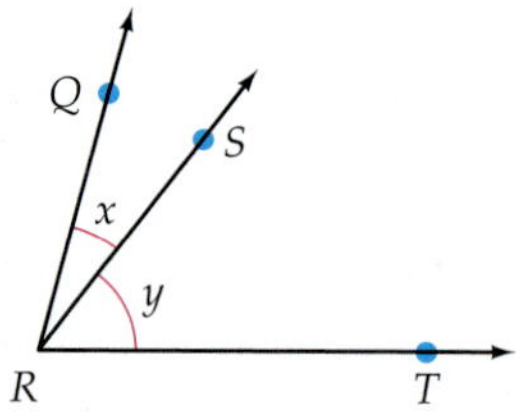

An angle is often measured in **degrees.** The symbol for degrees is a small raised circle, °. The angle formed by rotating a ray through a complete circle has a measure of 360°.

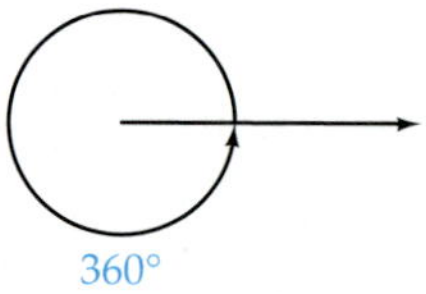

A 90° angle is called a **right angle.** The symbol ⌝ represents a right angle.

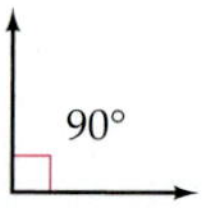

HISTORICAL NOTE

The Babylonians knew that one solar year, or the length of time it takes for the the sun to return to the same position in the cycle of seasons, as seen from Earth, was approximately 365 days. Historians suggest that the reason one complete revolution of a circle is 360° is that 360 is the closest number to 365 that is divisible by many natural numbers.

POINT OF INTEREST

Julius Elias/Shutterstock.com

The Leaning Tower of Pisa is the bell tower of the Cathedral in Pisa, Italy. The tower was designed to be vertical, but it started to lean soon after construction began in 1173. By 1990, when restoration work began, the tower was 5.5° off from the vertical. Now, post restoration, the structure leans 3.99°. (*Source:* http://en.wikipedia.org)

Perpendicular lines are intersecting lines that form right angles.

The symbol $\perp$ means "is perpendicular to." In the figure at the right, $p \perp q$ and $\overline{AB} \perp \overline{CD}$.

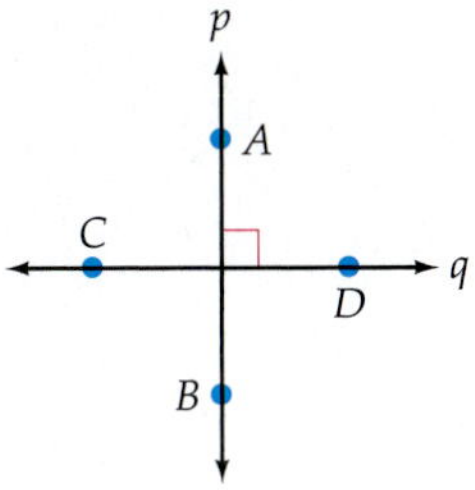

A **straight angle** is an angle whose measure is 180°. $\angle AOB$ is a straight angle.

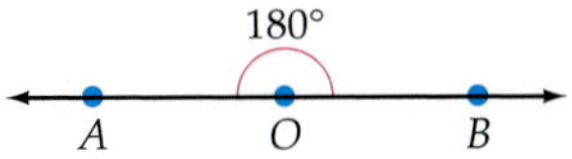

Straight angle

Complementary angles are two angles whose measures have the sum 90°.

$\angle A$ and $\angle B$ at the right are complementary angles.

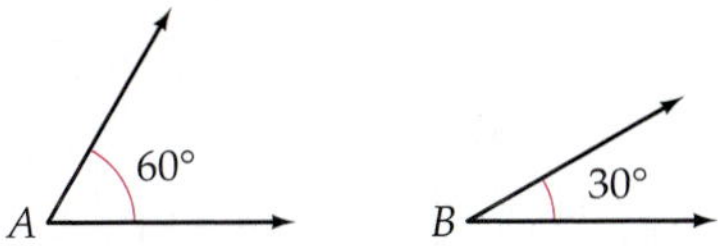

Supplementary angles are two angles whose measures have the sum 180°.

$\angle C$ and $\angle D$ at the right are supplementary angles.

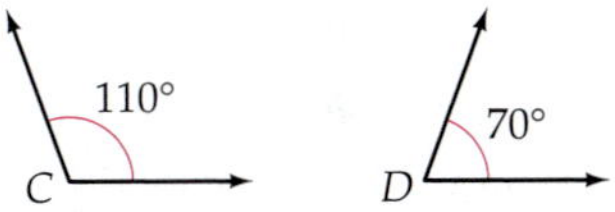

An **acute angle** is an angle whose measure is between 0° and 90°. $\angle D$ above is an acute angle. An **obtuse angle** is an angle whose measure is between 90° and 180°. $\angle C$ above is an obtuse angle.

The measure of $\angle C$ is 110°. This is often written as $m\angle C = 110°$, where m is an abbreviation for "the measure of."

EXAMPLE 2 Find the Measure of the Complement of an Angle

Find the measure of the complement of a 38° angle.

Solution

Complementary angles are two angles the sum of whose measures is 90°. To find the measure of the complement, let x represent the complement of a 38° angle. Write an equation and solve for x.

$$x + 38° = 90°$$
$$x = 52°$$

CHECK YOUR PROGRESS 2 Find the measure of the supplement of a 129° angle.

Solution See page S23.

TAKE NOTE

Co-planar means "in the same plane."

Adjacent angles are two co-planar, nonoverlapping angles that share a common vertex and a common side. In the figure at the right, $\angle DAC$ and $\angle CAB$ are adjacent angles.

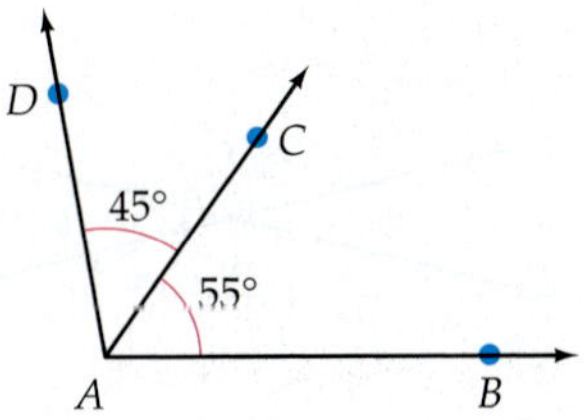

$$m\angle DAC = 45° \text{ and } m\angle CAB = 55°.$$
$$m\angle DAB = m\angle DAC + m\angle CAB$$
$$= 45° + 55° = 100°$$

In the figure at the right, $m\angle EDG = 80°$. The measure of $\angle FDG$ is three times the measure of $\angle EDF$. Find the measure of $\angle EDF$.

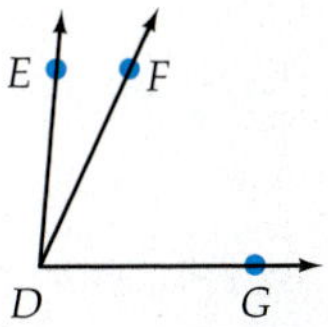

Let x = the measure of $\angle EDF$.
Then $3x$ = the measure of $\angle FDG$.
Write an equation and solve for x.

$$m\angle EDF + m\angle FDG = m\angle EDG$$
$$x + 3x = 80^\circ$$
$$4x = 80^\circ$$
$$x = 20^\circ$$

$m\angle EDF = 20^\circ$

EXAMPLE 3 Find the Measure of an Adjacent Angle

Given that $m\angle ABC$ is 84°, find the measure of $\angle x$.

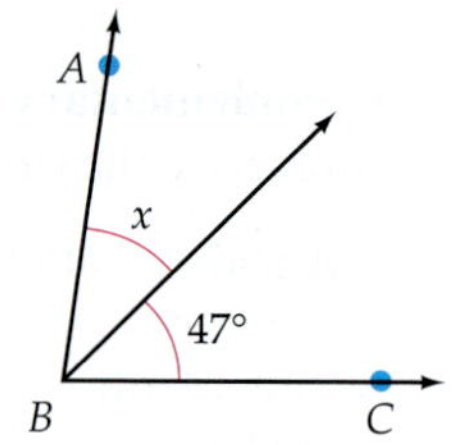

Solution

To find the measure of $\angle x$, write an equation using the fact that the sum of the measures of $\angle x$ and 47° is 84°. Solve for $m\angle x$.

$$m\angle x + 47^\circ = 84^\circ$$

$$m\angle x = 37^\circ$$

CHECK YOUR PROGRESS 3 Given that the $m\angle DEF$ is 118°, find the measure of $\angle a$.

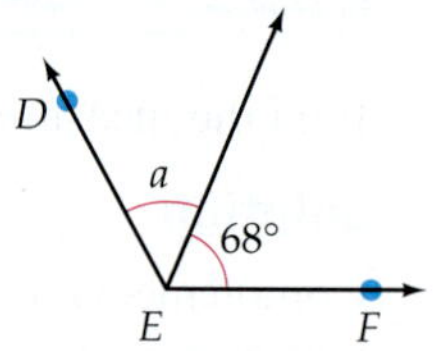

Solution See page S23.

Angles Formed by Intersecting Lines

Four angles are formed by the intersection of two lines. If the two lines are not perpendicular, then two of the angles formed are acute angles and two of the angles formed are obtuse angles. The two acute angles are always opposite each other, and the two obtuse angles are always opposite each other.

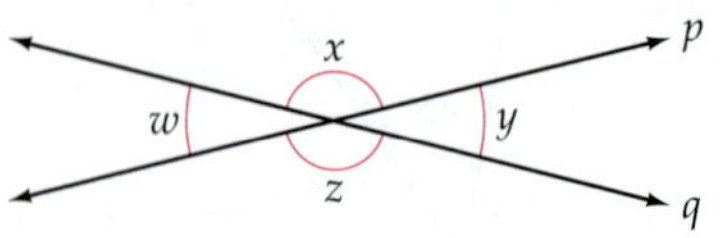

The nonadjacent angles formed by two intersecting lines are called **vertical angles.** $\angle w$ and $\angle y$ are vertical angles. $\angle x$ and $\angle z$ are vertical angles.

Vertical angles have the same measure.

$m\angle w = m\angle y$ $\qquad$ $m\angle x = m\angle z$

In the figure at the left, $\angle x$ and $\angle y$ are adjacent angles, as are $\angle y$ and $\angle z$, $\angle z$ and $\angle w$, and $\angle w$ and $\angle x$.

Adjacent angles formed by intersecting lines are supplementary angles.

$m\angle x + m\angle y = 180^\circ$ $\qquad$ $m\angle z + m\angle w = 180^\circ$

$m\angle y + m\angle z = 180^\circ$ $\qquad$ $m\angle w + m\angle x = 180^\circ$

EXAMPLE 4 Solve a Problem Involving Intersecting Lines

In the diagram at the right, $m\angle b = 115°$. Find the measures of angles a, c, and d.

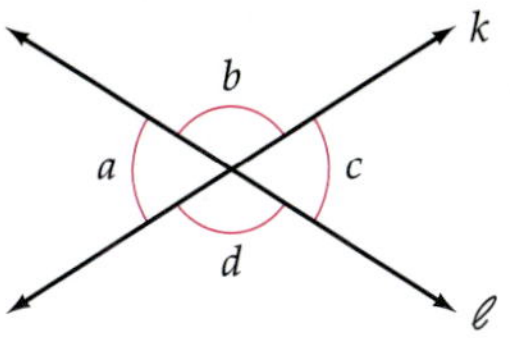

Solution

$m\angle a + m\angle b = 180°$ • $\angle a$ is supplementary to $\angle b$ because $\angle a$ and $\angle b$ are adjacent angles of intersecting lines.

$m\angle a + 115° = 180°$ • Replace $m\angle b$ with 115°.

$m\angle a = 65°$ • Subtract 115° from each side of the equation.

$m\angle c = 65°$ • $m\angle c = m\angle a$ because $\angle c$ and $\angle a$ are vertical angles.

$m\angle d = 115°$ • $m\angle d = m\angle b$ because $\angle d$ and $\angle b$ are vertical angles.

$m\angle a = 65°$, $m\angle c = 65°$, and $m\angle d = 115°$.

CHECK YOUR PROGRESS 4

In the diagram at the right, $m\angle a = 35°$. Find the measures of angles b, c, and d.

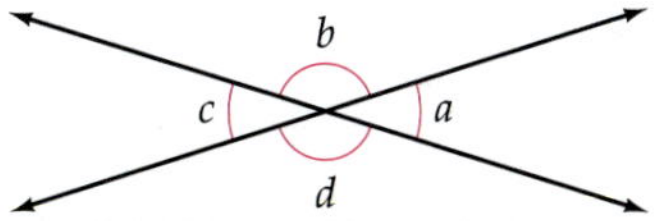

Solution See page S23.

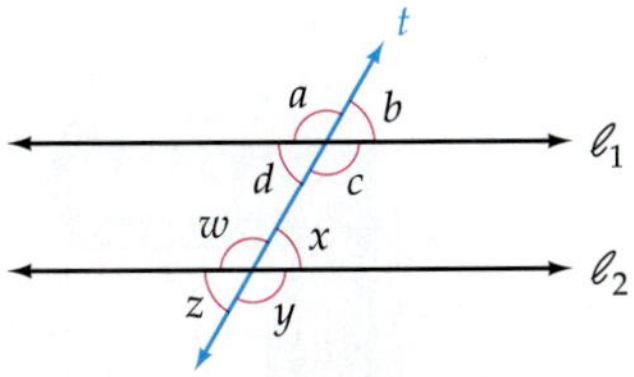

A line that intersects two other lines at different points is called a **transversal**. In the figure at the left, l_1 and l_2 are parallel lines. The eight angles formed by the transversal t that intersects the parallel lines have certain properties.

$\angle c$ and $\angle w$ are **alternate interior angles**; $\angle d$ and $\angle x$ are also alternate interior angles.

Alternate interior angles are equal.

$m\angle c = m\angle w$ $\qquad$ $m\angle d = m\angle x$

$\angle a$ and $\angle y$ are **alternate exterior angles**; $\angle b$ and $\angle z$ are also alternate exterior angles.

Alternate exterior angles are equal.

$m\angle a = m\angle y$ $\qquad$ $m\angle b = m\angle z$

$\angle a$ and $\angle w$, $\angle d$ and $\angle z$, $\angle b$ and $\angle x$, and $\angle c$ and $\angle y$ are **corresponding angles**.

Corresponding angles are equal.

$m\angle a = m\angle w$ $\qquad$ $m\angle b = m\angle x$

$m\angle d = m\angle z$ $\qquad$ $m\angle c = m\angle y$

QUESTION Which angles in the figure at the left above have the same measure as angle a? Which angles have the same measure as angle b?

POINT OF INTEREST

Many cities in the New World, unlike those in Europe, were designed using rectangular street grids. Washington, D.C. was designed this way, except that diagonal avenues were added, primarily for the purpose of enabling troop movement in the event the city required defense. As an added precaution, monuments were constructed at major intersections so that attackers would not have a straight shot down a boulevard.

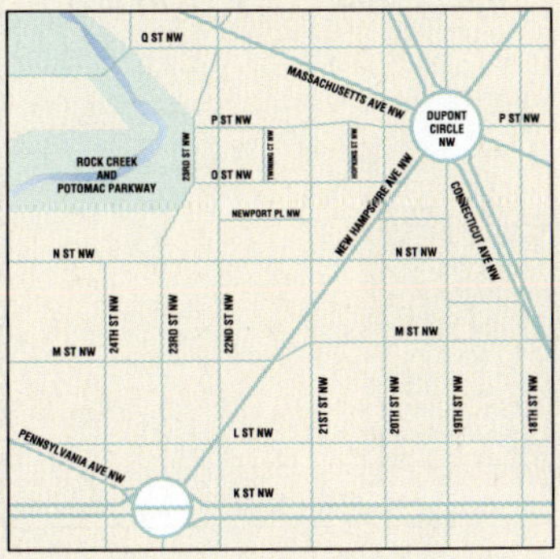

EXAMPLE 5 Solve a Problem Involving Parallel Lines Cut by a Transversal

In the diagram at the right, $\ell_1 \| \ell_2$ and $m\angle f = 58°$. Find the measures of $\angle a$, $\angle c$, and $\angle d$.

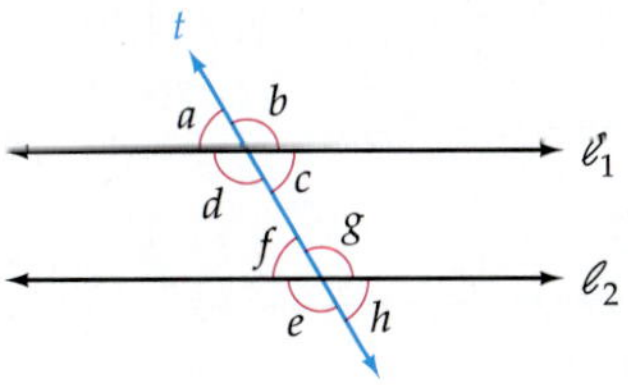

ANSWER Angles c, w, and y have the same measure as angle a. Angles d, x, and z have the same measure as angle b.

Solution

$m\angle a = m\angle f = 58°$ • $\angle a$ and $\angle f$ are corresponding angles.

$m\angle c = m\angle f = 58°$ • $\angle c$ and $\angle f$ are alternate interior angles.

$m\angle d + m\angle a = 180°$ • $\angle d$ is supplementary to $\angle a$.

$m\angle d + 58° = 180°$ • Replace $m\angle a$ with 58°.

$m\angle d = 122°$ • Subtract 58° from each side of the equation.

$m\angle a = 58°$, $m\angle c = 58°$, and $m\angle d = 122°$.

CHECK YOUR PROGRESS 5

In the diagram at the right, $\ell_1 \parallel \ell_2$ and $m\angle g = 124°$. Find the measures of $\angle b$, $\angle c$, and $\angle d$.

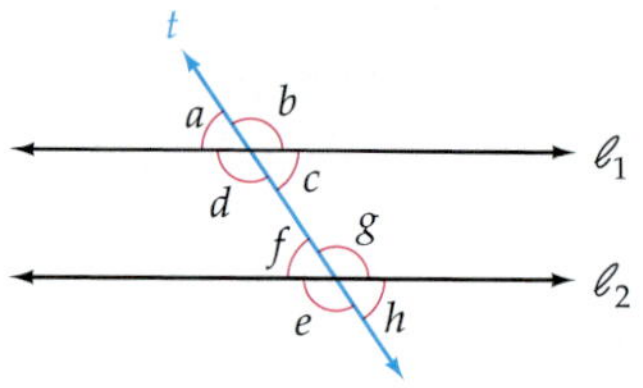

Solution See page S23.

MATH MATTERS The Principle of Reflection

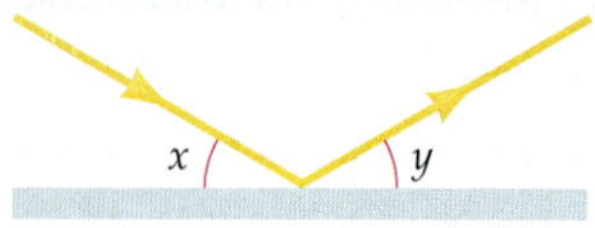

When a ray of light hits a flat surface, such as a mirror, the light is reflected at the same angle at which it hit the surface. For example, in the diagram at the left, $m\angle x = m\angle y$.

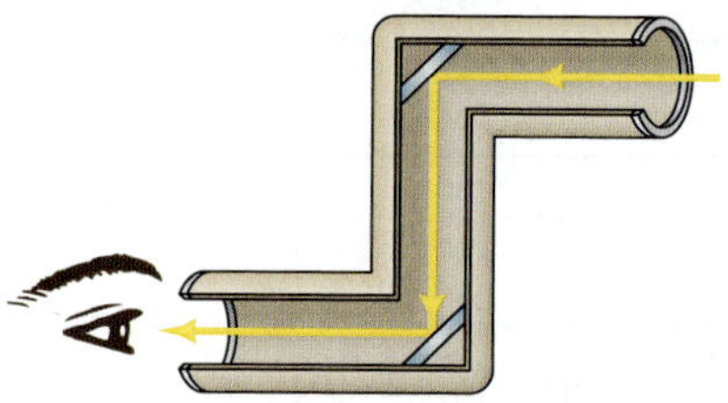

This principle of reflection is in operation in a simple periscope. In a periscope, light is reflected twice, with the result that light rays entering the periscope are parallel to the light rays at eye level.

The same principle is in operation on a billiard table. Assuming that it has no "side spin," a ball bouncing off the side of the table will bounce off at the same angle at which it hit the side. In the figure below, $m\angle w = m\angle x$ and $m\angle y = m\angle z$.

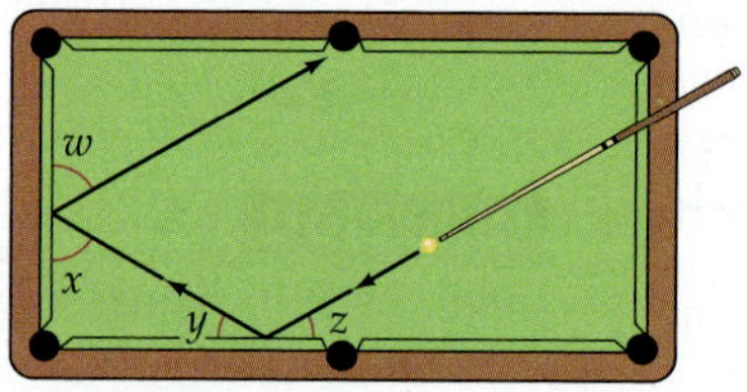

In the miniature golf shot illustrated below, $m\angle w = m\angle x$ and $m\angle y = m\angle z$.

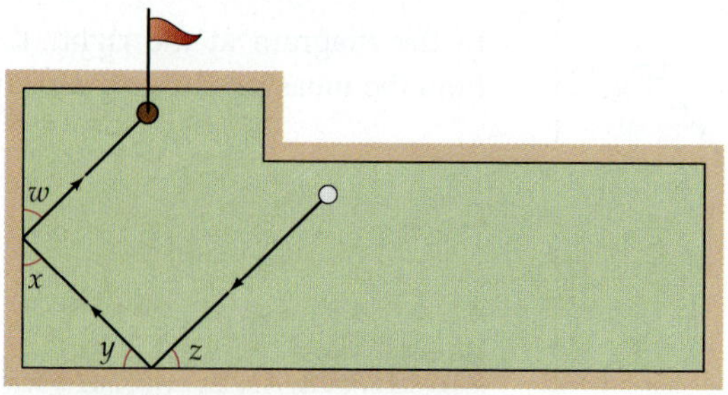

Angles of a Triangle

The figure at the right shows three intersecting lines. The plane figure formed by the line segments AB, BC, and AC is called a **triangle**.

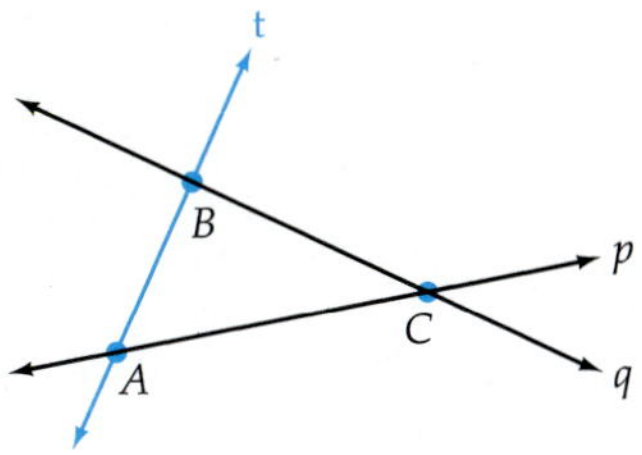

The angles within the region enclosed by the triangle are called **interior angles**. In the figure at the right, angles a, b, and c are interior angles. The sum of the measures of the interior angles of a triangle is 180°.

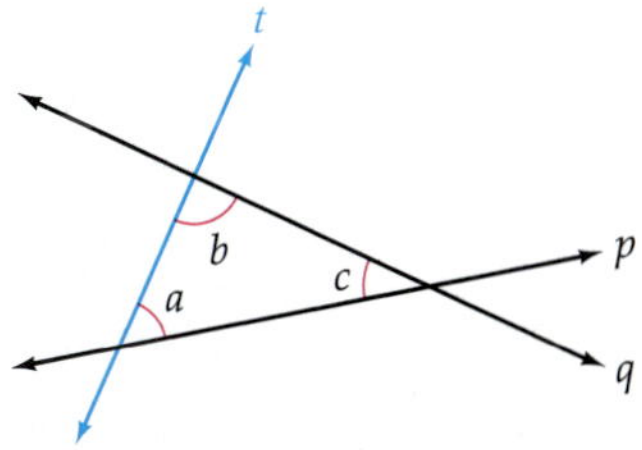

$$m\angle a + m\angle b + m\angle c = 180^\circ$$

The Sum of the Measures of the Interior Angles of a Triangle

The sum of the measures of the interior angles of a triangle is 180°.

QUESTION Can the measures of the three interior angles of a triangle be 87°, 51°, and 43°?

An **exterior angle of a triangle** is an angle that is adjacent to an interior angle of the triangle and is supplemental to the interior angle. In the figure at the right, angles m and n are exterior angles for angle a. The sum of the measure of an interior angle and one of its exterior angles is 180°.

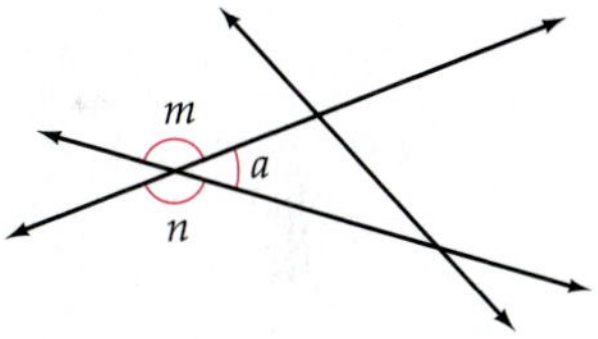

$$m\angle a + m\angle m = 180^\circ$$
$$m\angle a + m\angle n = 180^\circ$$

EXAMPLE 6 Find the Measure of the Third Angle of a Triangle

Two angles of a triangle measure 43° and 86°. Find the measure of the third angle.

Solution

Use the fact that the sum of the measures of the interior angles of a triangle is 180°. Write an equation using x to represent the measure of the third angle. Solve the equation for x.

$$x + 43^\circ + 86^\circ = 180^\circ$$
$$x + 129^\circ = 180^\circ \quad \text{• Add } 43^\circ + 86^\circ.$$
$$x = 51^\circ \quad \text{• Subtract } 129^\circ \text{ from each side of the equation.}$$

The measure of the third angle is 51°.

CHECK YOUR PROGRESS 6 One angle in a triangle is a right angle, and one angle measures 27°. Find the measure of the third angle.

Solution See page S23.

TAKE NOTE

In this text, when we refer to the angles of a triangle, we mean the interior angles of the triangle unless specifically stated otherwise.

ANSWER No, because 87° + 51° + 43° = 181°, and the sum of the measures of the three interior angles of a triangle must be 180°.

EXAMPLE 7 Solve a Problem Involving the Angles of a Triangle

In the diagram at the right, $m\angle c = 40°$ and $m\angle e = 60°$. Find the measure of $\angle d$.

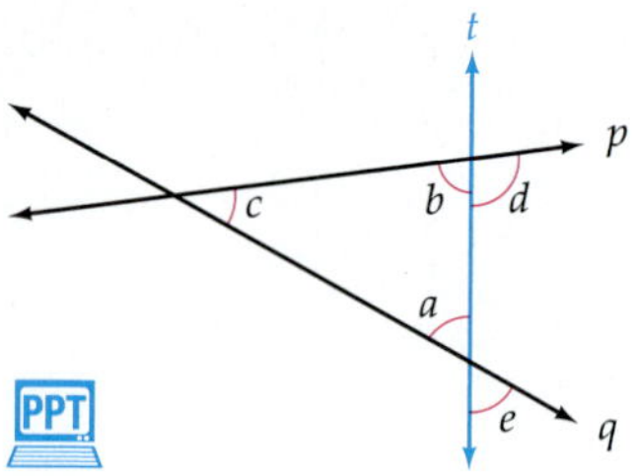

Solution

$m\angle a = m\angle e = 60°$ • $\angle a$ and $\angle e$ are vertical angles.

$m\angle c + m\angle a + m\angle b = 180°$ • The sum of the interior angles is 180°.

$40° + 60° + m\angle b = 180°$ • Replace $m\angle c$ with 40° and $m\angle a$ with 60°.

$100° + m\angle b = 180°$ • Add 40° + 60°.

$m\angle b = 80°$ • Subtract 100° from each side of the equation.

$m\angle b + m\angle d = 180°$ • $\angle b$ and $\angle d$ are supplementary angles.

$80° + m\angle d = 180°$ • Replace $m\angle b$ with 80°.

$m\angle d = 100°$ • Subtract 80° from each side of the equation.

$m\angle d = 100°$

CHECK YOUR PROGRESS 7

In the diagram at the right, $m\angle c = 35°$ and $m\angle d = 105°$. Find the measure of $\angle e$.

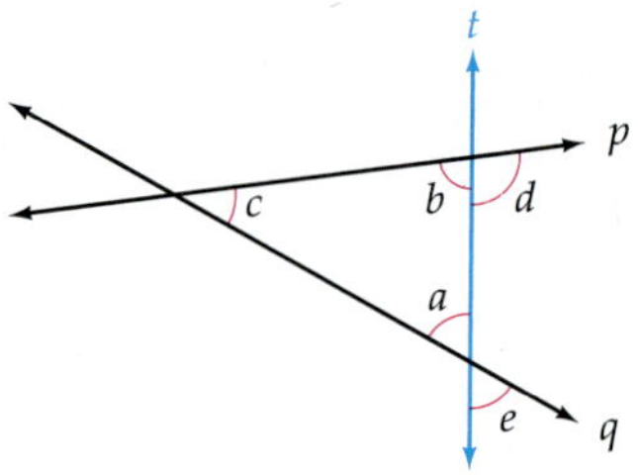

Solution See page S23.

EXCURSION

Preparing a Circle Graph

A circle graph, sometimes called a pie chart, is a circular chart divided into sectors. The ratio of the angle measure of each sector to 360 (the number of degrees in a circle) is proportional to the ratio of the number of data values represented by the sector to the total number of data values.

The circle graph at the right represents the preferences of 200 people who were asked about the temperature range they considered most comfortable for walking. The data are given in the table below.

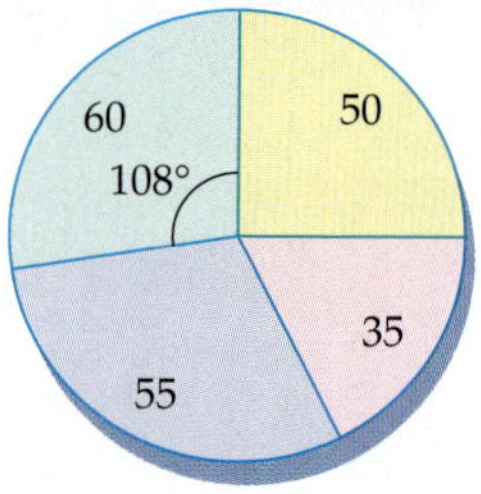

Temperature, °F	Number of people
60–64	50
65–69	55
70–74	60
75–79	35

The angle measure of the sector that represents people who chose a temperature between 70°F and 74°F is 108°. Note that $\frac{108}{360} = \frac{60}{200}$. That is, the ratio of the size of the angle, 108, to the total number of degrees in a circle, 360, is proportional to the magnitude of the data in that sector, 60, to the sum of all the data in the circle graph, 200.

Spreadsheet programs such as Excel and Numbers have built-in functions that will create a circle graph. Use such a program to prepare a circle graph for each of the exercises below.

EXCURSION EXERCISES

Prepare a circle graph for the data provided in each exercise.

1. A survey asked adults to name their favorite pizza topping. The results are shown in the table below.

Pepperoni	43%
Sausage	19%
Mushrooms	14%
Vegetables	13%
Other	7%
Onions	4%

2. A survey of children between 10 and 14 years old was conducted to determine the average amount of time they spent consuming media each day. The results are shown in the table below.

Watching TV	63 minutes
Listening to music	105 minutes
Nonschool computer use	42 minutes
Text messaging	84 minutes
Playing video games	76 minutes
Talking on cell phones	50 minutes

EXERCISE SET 7.2

1. Provide three names for the angle below.

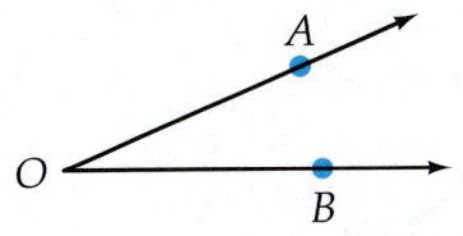

2. State the number of degrees in a full circle, a straight angle, and a right angle.
3. Find the complement of a 62° angle.
4. Find the complement of a 31° angle.
5. Find the supplement of a 162° angle.
6. Find the supplement of a 72° angle.

■ For Exercises 7 to 10, determine whether the described angle is an acute angle, is a right angle, is an obtuse angle, or does not exist.

7. The complement of an acute angle
8. The supplement of a right angle
9. The supplement of an acute angle
10. The supplement of an obtuse angle
11. Given $AB = 12$ cm, $CD = 9$ cm, and $AD = 35$ cm, find the length of $\overline{BC}$.

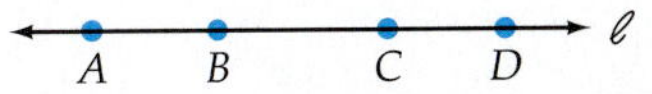

12. Given $AB = 21$ mm, $BC = 14$ mm, and $AD = 54$ mm, find the length of $\overline{CD}$.

13. Given $QR = 7$ ft and RS is three times the length of $\overline{QR}$, find the length of $\overline{QS}$.

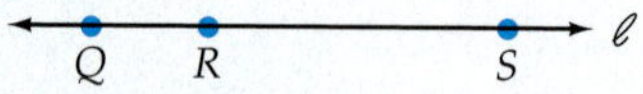

14. Given $QR = 15$ in. and RS is twice the length of $\overline{QR}$, find the length of $\overline{QS}$.

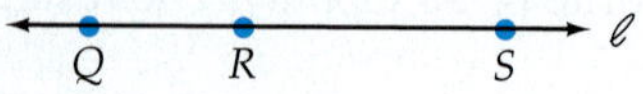

15. Given $m\angle LOM = 53°$ and $m\angle LON = 139°$, find the measure of $\angle MON$.

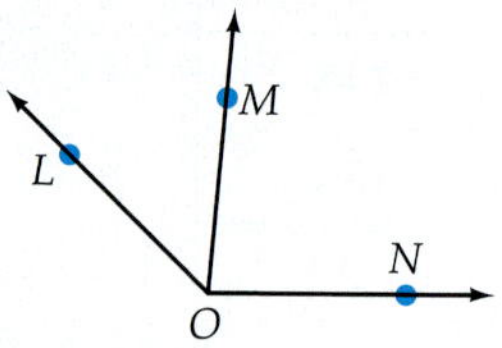

16. Given $m\angle MON = 38°$ and $m\angle LON = 85°$, find the measure of $\angle LOM$.

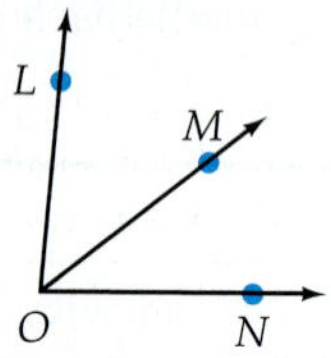

■ In Exercises 17 and 18, find the measure of $\angle x$.

17.

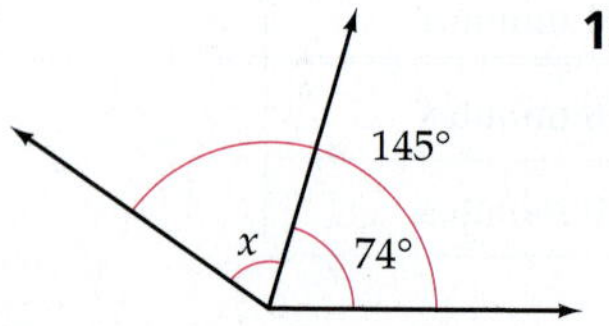

18.

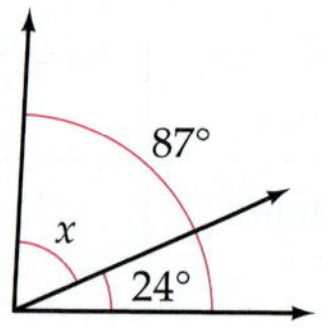

■ In Exercises 19 and 20, given that $\angle LON$ is a right angle, find the measure of $\angle x$.

19.

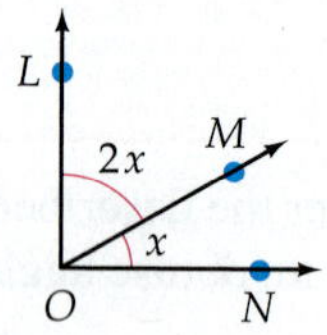

20.

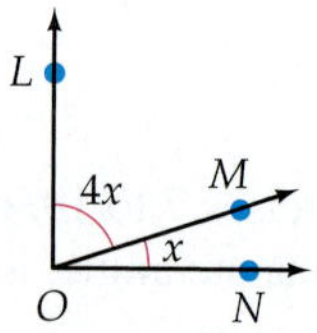

■ In Exercises 21 to 24, find the measure of $\angle a$.

21.

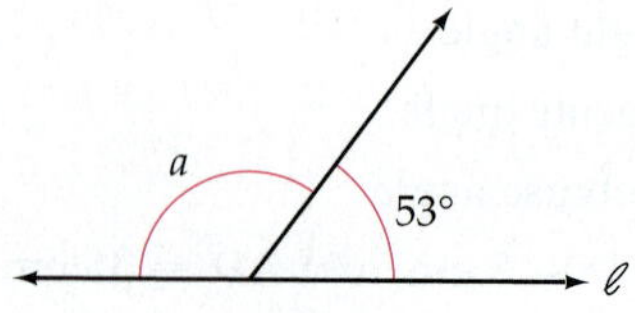

22.

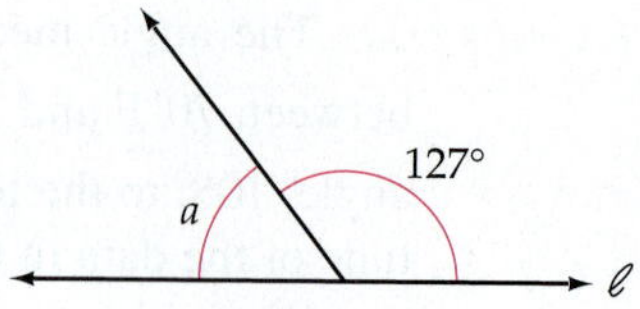

23.

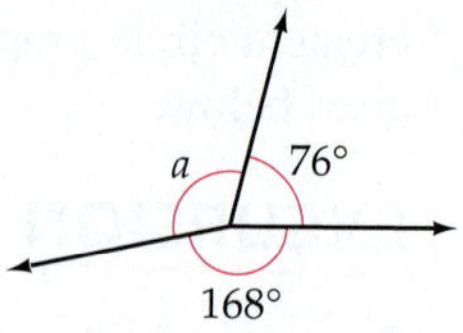

24.

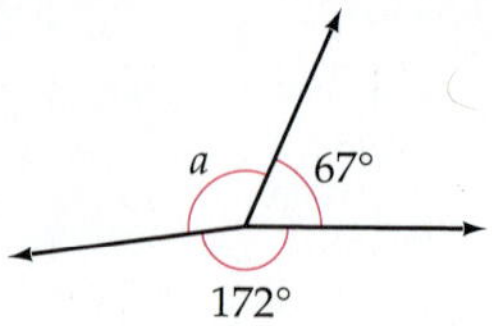

■ In Exercises 25 to 28, find the value of x.

25.

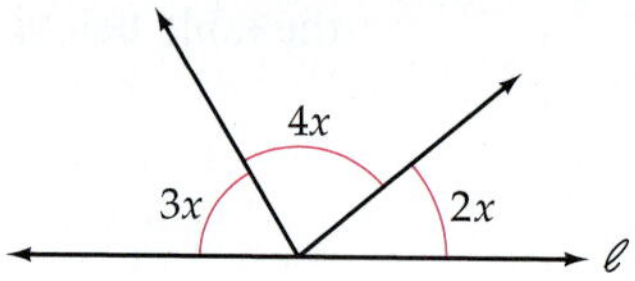

26.

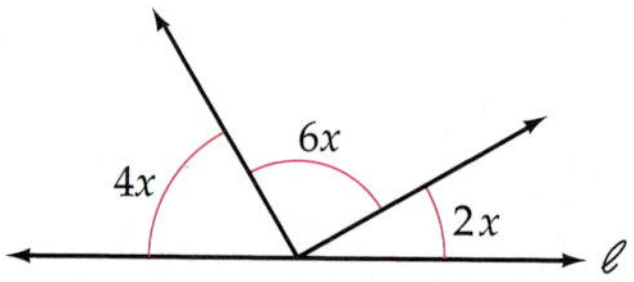

27.

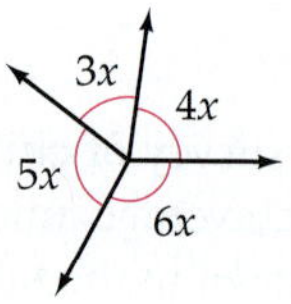

28.

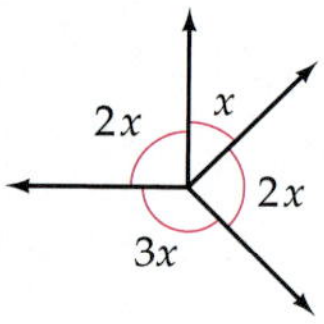

29. Given $m\angle a = 51°$, find the measure of $\angle b$.

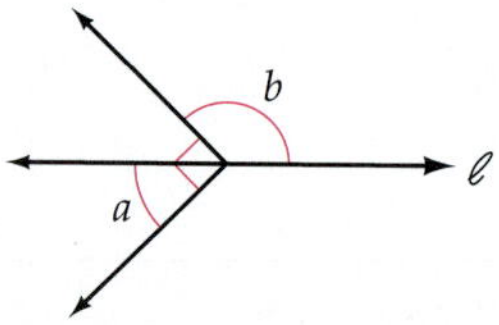

30. Given $m\angle a = 38°$, find the measure of $\angle b$.

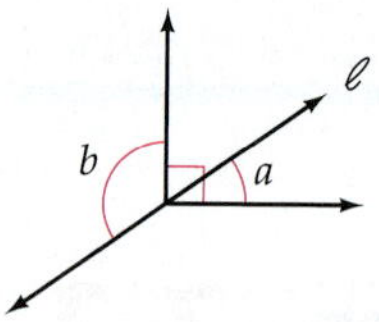

■ In Exercises 31 and 32, find the measure of $\angle x$.

31.

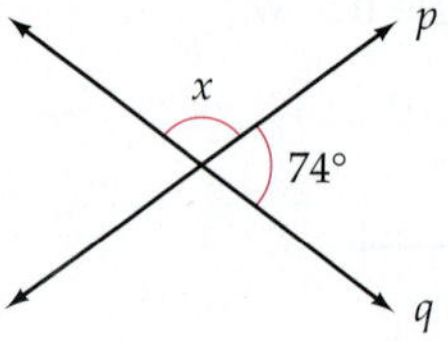

32.

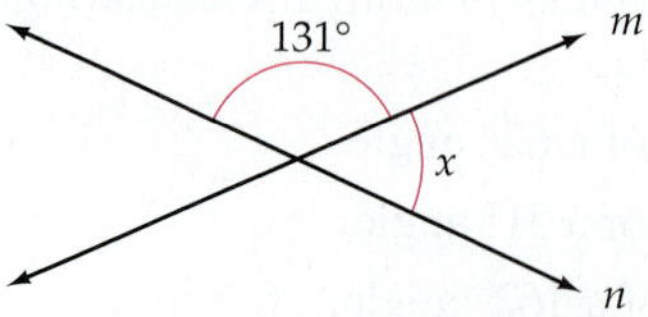

■ In Exercises 33 and 34, find the value of x.

33.

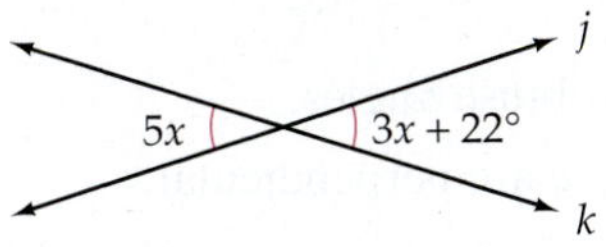

34.

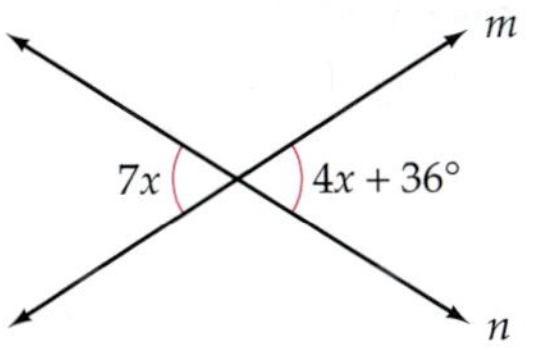

■ In Exercises 35 to 38, given that $\ell_1 \parallel \ell_2$, find the measures of angles a and b.

35.

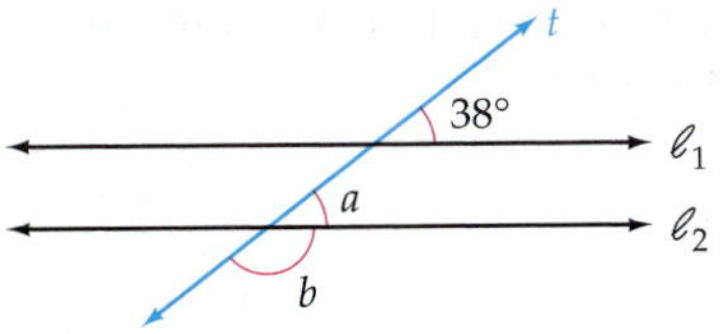

36.

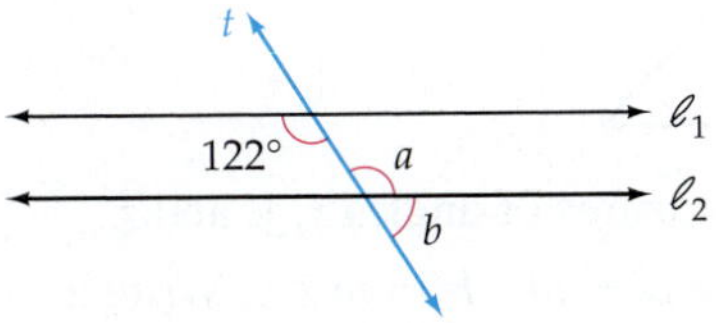

37.

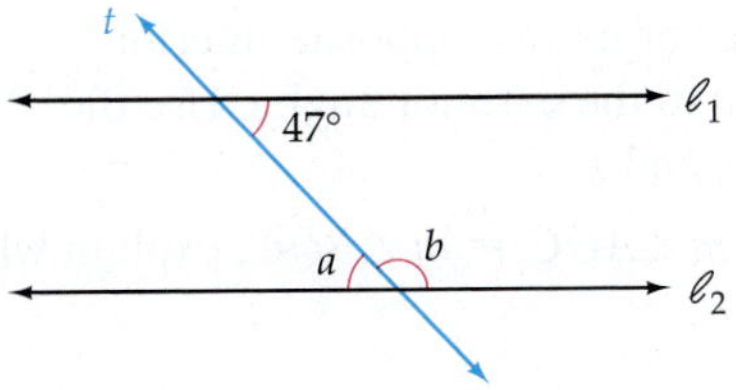

38.

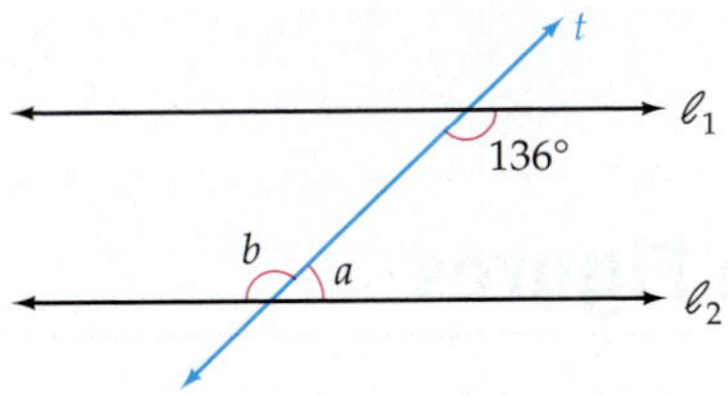

■ In Exercises 39 to 42, given that $\ell_1 \parallel \ell_2$, find x.

39.

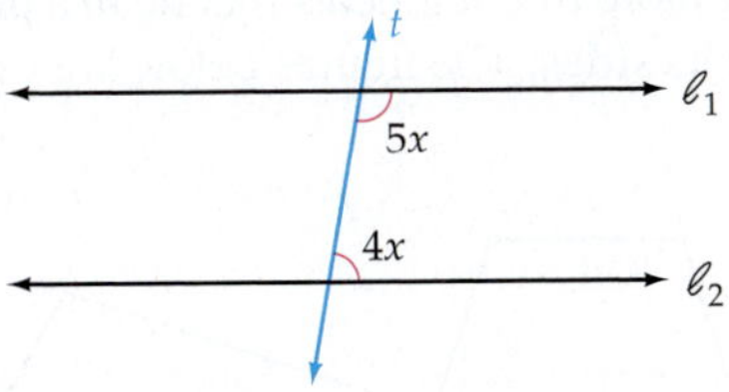

40.

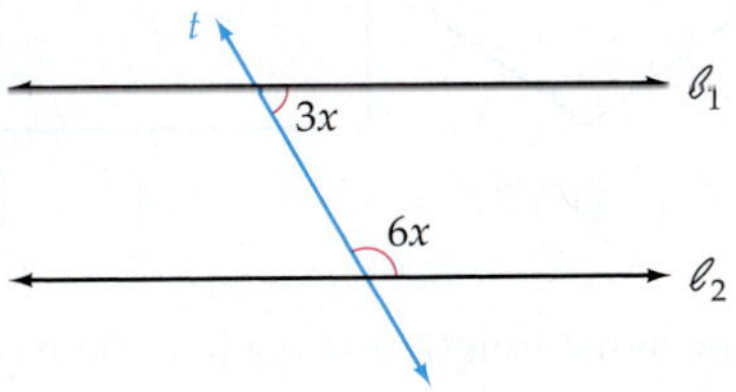

41.

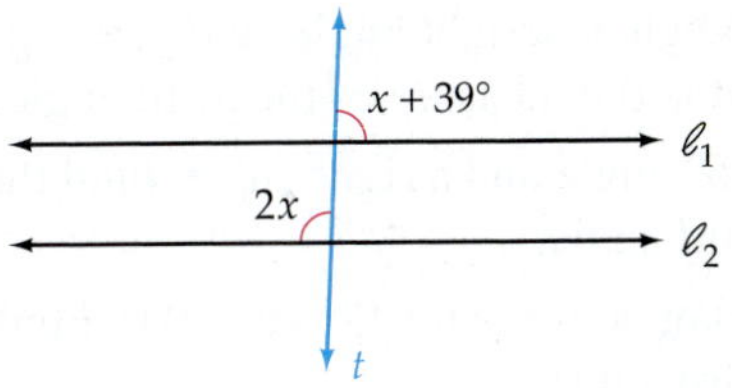

42.

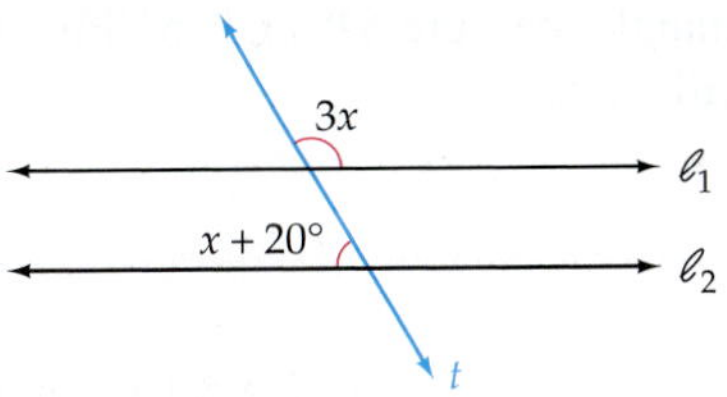

43. Given that $m\angle a = 95°$ and $m\angle b = 70°$, find the measures of angles x and y.

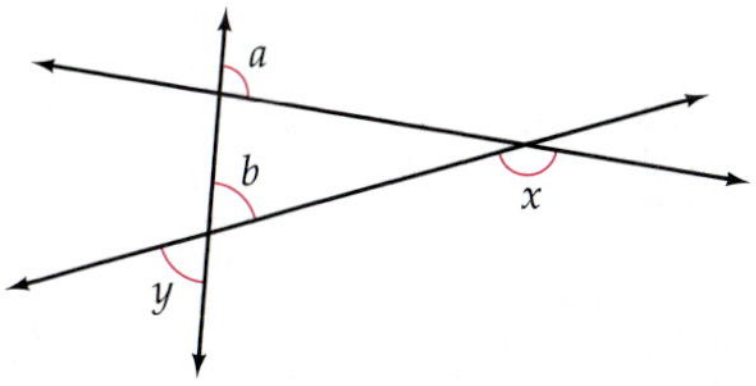

44. Given that $m\angle a = 35°$ and $m\angle b = 55°$, find the measures of angles x and y.

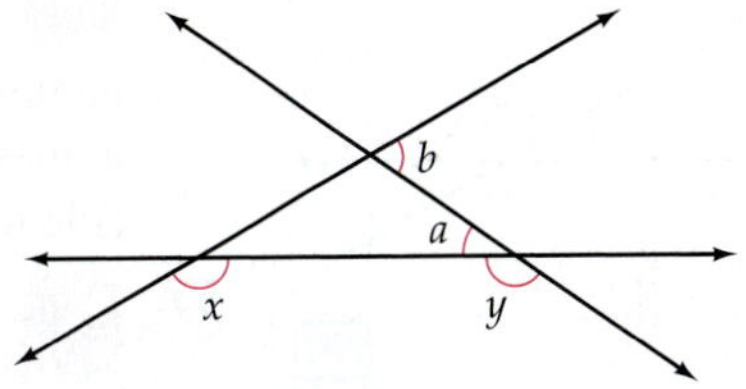

45. Given that $m\angle y = 45°$, find the measures of angles a and b.

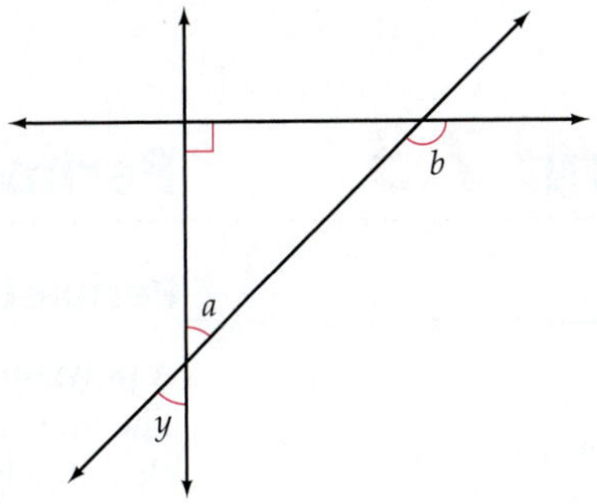

46. Given that $m\angle y = 130°$, find the measures of angles a and b.

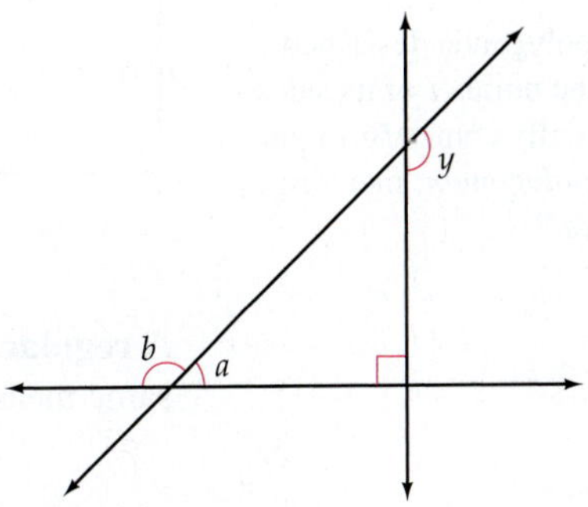

47. One angle in a triangle is a right angle, and one angle is equal to 30°. What is the measure of the third angle?

48. A triangle has a 45° angle and a right angle. Find the measure of the third angle.

49. Two angles of a triangle measure 42° and 103°. Find the measure of the third angle.

50. Two angles of a triangle measure 62° and 45°. Find the measure of the third angle.

■ For Exercises 51 to 53, determine whether the statement is true or false.

51. A triangle can have two obtuse angles.

52. The legs of a right triangle are perpendicular.

53. If the sum of two angles of a triangle is less than 90°, then the third angle is an obtuse angle.

EXTENSIONS

54. Cut out a triangle and then tear off two of the angles, as shown below. Position angle a so that it is to the left of angle b and is adjacent to angle b. Now position angle c so that it is to the right of angle b and is adjacent to angle b. Describe what you observe. What does this demonstrate?

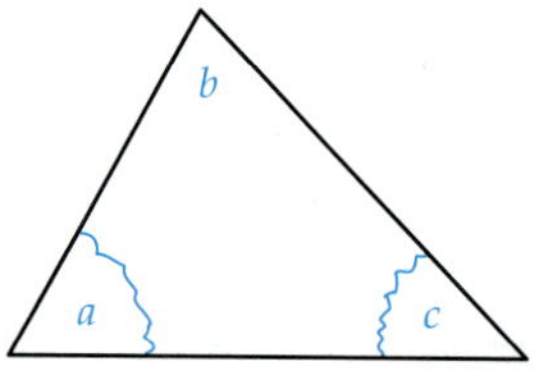

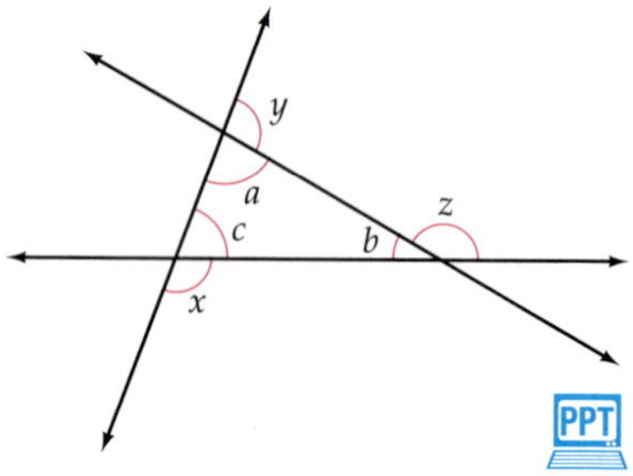

55. For the figure at the left, find the sum of the measures of angles x, y, and z.

56. For the figure at the left, explain why $m\angle a + m\angle b = m\angle x$. Write a rule that describes the relationship between the measure of an exterior angle of a triangle and the sum of the measures of its two opposite interior angles (the interior angles that are nonadjacent to the exterior angle). Use the rule to write an equation involving angles a, c, and z.

57. If $\overline{AB}$ and $\overline{CD}$ intersect at point O, and $m\angle AOC = m\angle BOC$, explain why $\overline{AB} \perp \overline{CD}$.

SECTION 7.3 Perimeter and Area of Plane Figures

Perimeter of Plane Geometric Figures

A **polygon** is a closed figure determined by three or more line segments that lie in a plane. The line segments that form the polygon are called its **sides**. The figures below are examples of polygons.

POINT OF INTEREST

Although a polygon is described in terms of the number of its sides, the word actually comes from the Latin word *polygonum*, meaning "many *angles*."

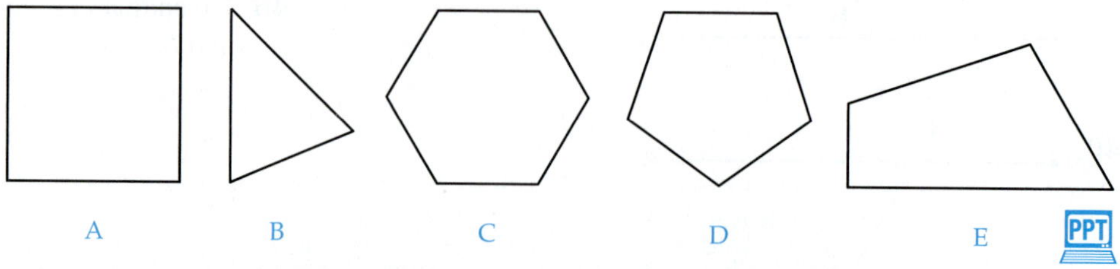

A **regular polygon** is one in which each side has the same length and each angle has the same measure. The polygons in Figures A, C, and D above are regular polygons.

AP Images

The Pentagon in Arlington, Virginia

The name of a polygon is based on the number of its sides. The table below lists the names of polygons that have from 3 to 10 sides.

Number of sides	Name of polygon
3	Triangle
4	Quadrilateral
5	Pentagon
6	Hexagon
7	Heptagon
8	Octagon
9	Nonagon
10	Decagon

Triangles and quadrilaterals are two of the most common types of polygons. Triangles are distinguished by the number of equal sides and also by the measures of their angles.

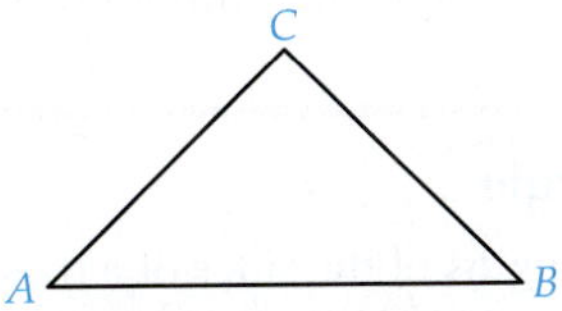

An **isosceles triangle** has exactly two sides of equal length. The angles opposite the equal sides are of equal measure.

$AC = BC$

$m\angle A = m\angle B$

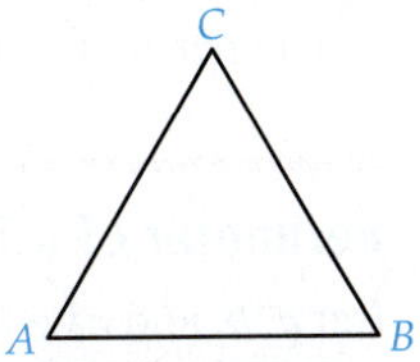

The three sides of an **equilateral triangle** are of equal length. The three angles are of equal measure.

$AB = BC = AC$

$m\angle A = m\angle B = m\angle C$

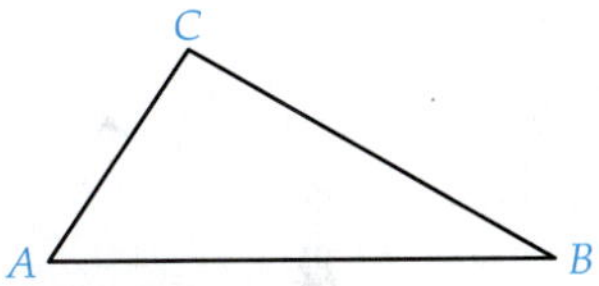

A **scalene triangle** has no two sides of equal length. No two angles are of equal measure.

PPT

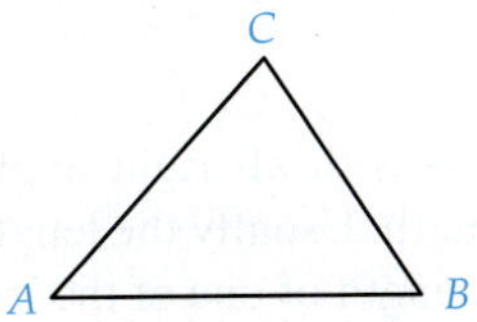

An **acute triangle** has three acute angles.

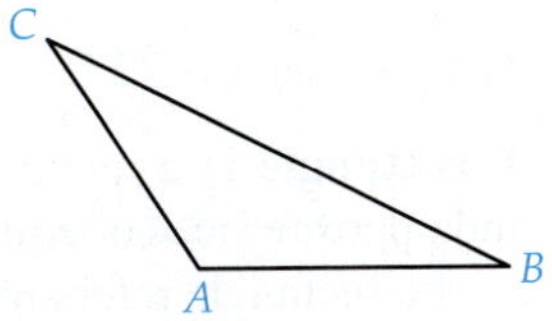

An **obtuse triangle** has one obtuse angle.

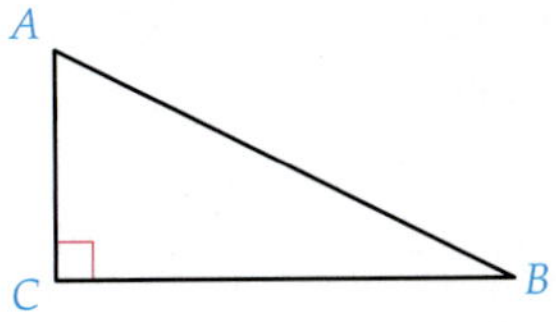

A **right triangle** has a right angle.

PPT

INSTRUCTOR NOTE

Figure 7.1 shows the relationships among different types of quadrilaterals. The description of each quadrilateral is within a drawing of that type of quadrilateral.

A **quadrilateral** is a four-sided polygon. Quadrilaterals are also distinguished by their sides and angles, as shown in Figure 7.1 on the next page. Note that a rectangle, a square, and a rhombus are different forms of a parallelogram.

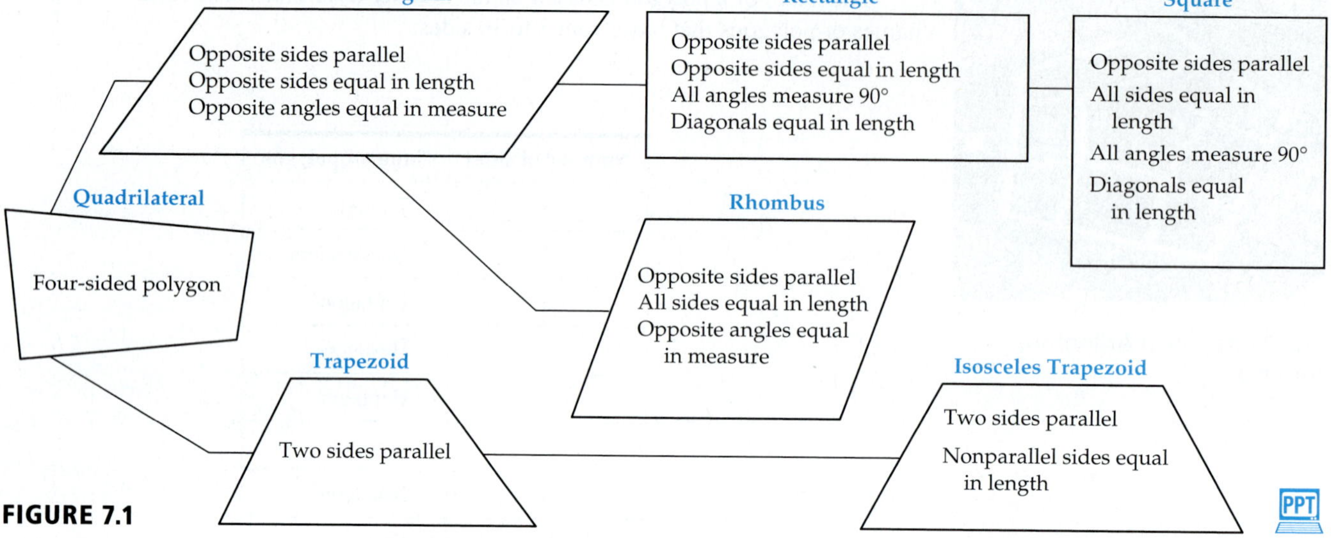

FIGURE 7.1

PPT

QUESTION **a.** What distinguishes a rectangle from other parallelograms?
b. What distinguishes a square from other rectangles?

The **perimeter** of a plane geometric figure is a measure of the distance around the figure. Perimeter is used, for example, when buying fencing for a garden or determining how much baseboard is needed for a room.

The perimeter of a triangle is the sum of the lengths of the three sides.

Perimeter of a Triangle

Let a, b, and c be the lengths of the sides of a triangle. The perimeter, P, of the triangle is given by $P = a + b + c$.

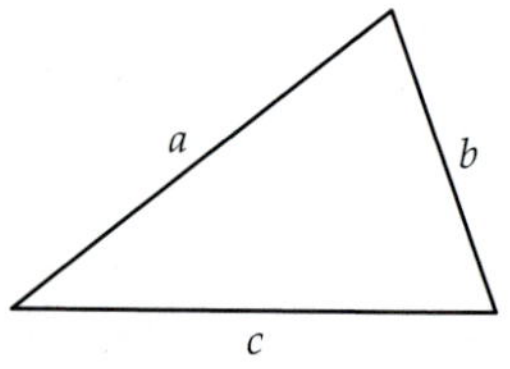

$P = a + b + c$

To find the perimeter of the triangle shown at the right, add the lengths of the three sides.

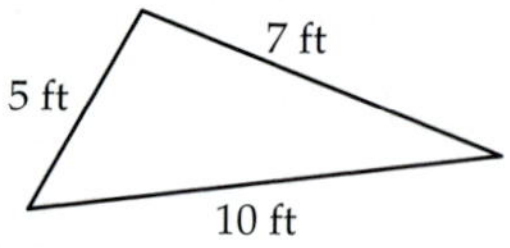

$P = 5 + 7 + 10 = 22$

The perimeter is 22 ft.

A **rectangle** is a quadrilateral with all right angles and opposite sides of equal length. Usually the length, L, of a rectangle refers to the length of one of the longer sides of the rectangle and the width, W, refers to the length of one of the shorter sides. The perimeter can then be represented as $P = L + W + L + W$.

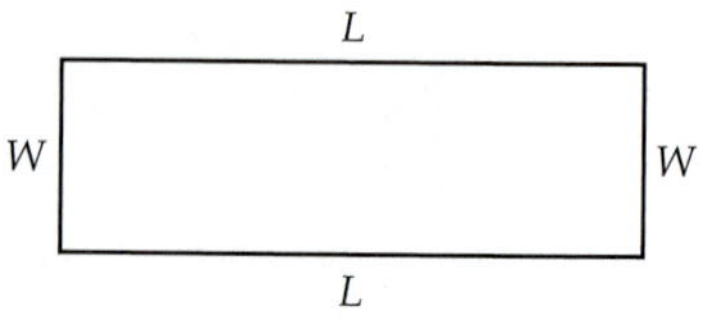

The formula for the perimeter of a rectangle is derived by combining like terms.

$P = L + W + L + W$
$P = 2L + 2W$

ANSWER **a.** In a rectangle, all angles measure 90°. **b.** In a square, all sides are equal in length.

Perimeter of a Rectangle

Let L represent the length and W the width of a rectangle. The perimeter, P, of the rectangle is given by $P = 2L + 2W$.

A **square** is a rectangle in which each side has the same length. Letting s represent the length of each side of a square, the perimeter of the square can be represented by $P = s + s + s + s$.

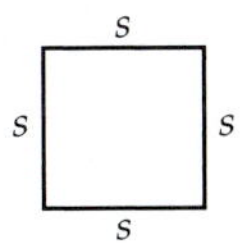

The formula for the perimeter of a square is derived by combining like terms.

$$P = s + s + s + s$$
$$P = 4s$$

Perimeter of a Square

Let s represent the length of a side of a square. The perimeter, P, of the square is given by $P = 4s$.

HISTORICAL NOTE

The Granger Collection, NY

Benjamin Banneker (băn'ĭ-kər) (1731–1806), a noted American scholar who was largely self-taught, was both a surveyor and an astronomer. As a surveyor, he was a member of the commission that defined the boundary lines and laid out the streets of the District of Columbia. (See the Point of Interest on page 369.)

Figure $ABCD$ is a parallelogram. $\overline{BC}$ is the **base** of the parallelogram. Opposite sides of a parallelogram are equal in length, so $\overline{AD}$ is the same length as $\overline{BC}$, and $\overline{AB}$ is the same length as $\overline{CD}$.

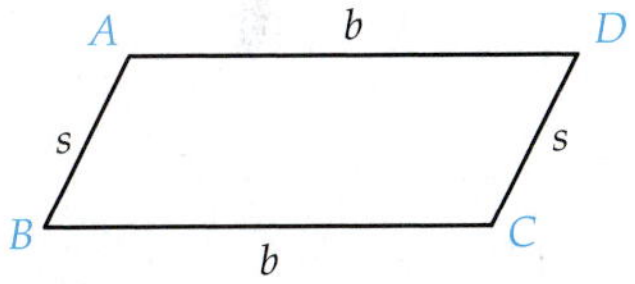

Let b represent the length of the base and s the length of an adjacent side. Then the perimeter of a parallelogram can be represented as $P = b + s + b + s$.

$$P = b + s + b + s$$

The formula for the perimeter of a parallelogram is derived by combining like terms.

$$P = 2b + 2s$$

Perimeter of a Parallelogram

Let b represent the length of the base of a parallelogram and s the length of a side adjacent to the base. The perimeter, P, of the parallelogram is given by $P = 2b + 2s$.

EXAMPLE 1 Find the Perimeter of a Rectangle

You want to trim a rectangular frame with a metal strip. The frame measures 30 in. by 20 in. Find the length of metal strip you will need to trim the frame.

Solution

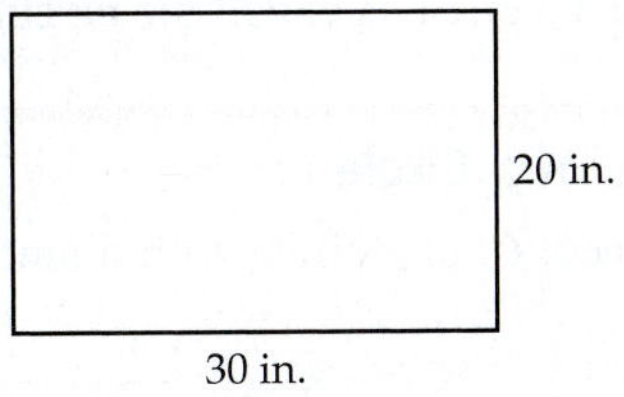

• Draw a diagram.

$P = 2L + 2W$ • Use the formula for the perimeter of a rectangle.

$P = 2(30) + 2(20)$ • The length is 30 in. Substitute 30 for L. The width is 20 in. Substitute 20 for W.

$P = 60 + 40$

$P = 100$

You will need 100 in. of the metal strip.

CHECK YOUR PROGRESS 1 Find the length of decorative molding needed to edge the top of the walls in a rectangular room that is 12 ft long and 8 ft wide.

Solution See page S24.

EXAMPLE 2 Find the Perimeter of a Square

Find the length of fencing needed to surround a square corral that measures 60 ft on each side.

Solution

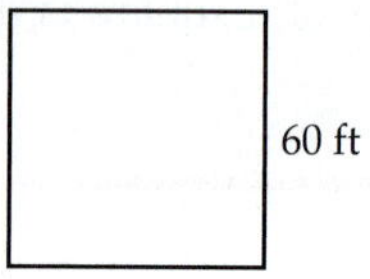

• Draw a diagram.

$P = 4s$ • Use the formula for the perimeter of a square.

$P = 4(60)$ • The length of a side is 60 ft. Substitute 60 for s.

$P = 240$

240 ft of fencing are needed.

CHECK YOUR PROGRESS 2 A homeowner plans to fence in the area around the swimming pool in the backyard. The area to be fenced in is a square measuring 24 ft on each side. How many feet of fencing should the homeowner purchase?

Solution See page S24.

POINT OF INTEREST

A **glazier** is a person who cuts, fits, and installs glass, generally in doors and windows. Of particular challenge to a glazier are intricate stained glass window designs.

Luca Moi/Shutterstock.com

A **circle** is a plane figure in which all points are the same distance from point O, called the **center** of the circle.

A **diameter** of a circle is a line segment with endpoints on the circle and passing through the center. $\overline{AB}$ is a diameter of the circle at the right. The variable d is used to designate the length of a diameter of a circle.

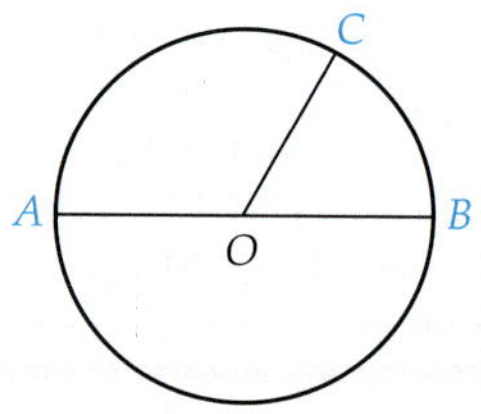

A **radius** of a circle is a line segment from the center of the circle to a point on the circle. $\overline{OC}$ is a radius of the circle at the right above. The variable r is used to designate the length of a radius of a circle.

The length of the diameter is twice the length of the radius. $d = 2r$ or $r = \dfrac{1}{2}d$

The distance around a circle is called the **circumference**.

POINT OF INTEREST

Archimedes (ar-kə-mē'dēz) (c. 287–212 BC) is the person who calculated that $\pi \approx 3\frac{1}{7}$. He actually showed that $3\frac{10}{71} < \pi < 3\frac{1}{7}$. The approximation $3\frac{10}{71}$ is a more accurate approximation of π than $3\frac{1}{7}$, but it is more difficult to use without a calculator.

Circumference of a Circle

The circumference, C, of a circle with diameter d and radius r is given by $C = \pi d$ or $C = 2\pi r$.

The formula for circumference uses the number π (pi), which is an irrational number.

The value of π can be approximated by a fraction or by a decimal. $\pi \approx 3\dfrac{1}{7} = \dfrac{22}{7}$ or $\pi \approx 3.14$

The π key on a scientific calculator gives a closer approximation of π than 3.14. A scientific calculator is used in this section to find approximate values in calculations involving π.

CALCULATOR NOTE

To approximate 16π on your calculator, enter

16 [×] [π] [=]

EXAMPLE 3 Find the Circumference of a Circle

Find the circumference of a circle with a radius of 15 cm. Round to the nearest hundredth of a centimeter.

Solution

$C = 2\pi r$ • The radius is given. Use the circumference formula that involves the radius.

$C = 2\pi(15)$ • Replace r with 15.

$C = 30\pi$ • Multiply 2 times 15.

$C \approx 94.25$ • An approximation is asked for. Use the π key on a calculator.

The circumference of the circle is approximately 94.25 cm.

CHECK YOUR PROGRESS 3 Find the circumference of a circle with a diameter of 9 km. Give the exact measure.

Solution See page S24.

EXAMPLE 4 Application of Finding the Circumference of a Circle

A bicycle tire has a diameter of 24 in. How many feet does the bicycle travel when the wheel makes 8 revolutions? Round to the nearest hundredth of a foot.

Solution

24 in. = 2 ft • The diameter is given in inches, but the answer must be expressed in feet. Convert the diameter (24 in.) to feet. There are 12 in. in 1 ft. Divide 24 by 12.

$C = \pi d$ • The diameter is given. Use the circumference formula that involves the diameter.

$C = \pi(2)$ • Replace d with 2.

$C = 2\pi$ • This is the distance traveled in 1 revolution.

$8C = 8(2\pi) = 16\pi \approx 50.27$ • Find the distance traveled in 8 revolutions.

The bicycle will travel about 50.27 ft when the wheel makes 8 revolutions.

CHECK YOUR PROGRESS 4 A tricycle tire has a diameter of 12 in. How many feet does the tricycle travel when the wheel makes 12 revolutions? Round to the nearest hundredth of a foot.

Solution See page S24.

Area of Plane Geometric Figures

Area is the amount of surface in a region. Area can be used to describe, for example, the size of a rug, a parking lot, a farm, or a national park. Area is measured in square units.

A square that measures 1 in. on each side has an area of 1 square inch, written 1 in^2.

A square that measures 1 cm on each side has an area of 1 square centimeter, written 1 cm^2.

Larger areas are often measured in square feet (ft^2), square meters (m^2), square miles (mi^2), acres (43,560 ft^2), or any other square unit.

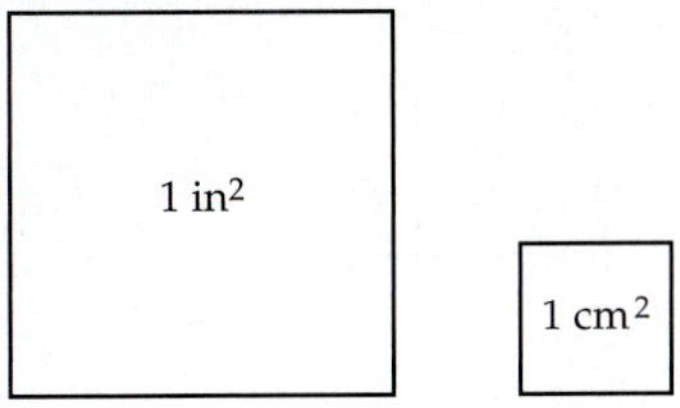

QUESTION **a.** What is the area of a square that measures 1 yd on each side?
b. What is the area of a square that measures 1 km on each side?

Area of a Rectangle

Let L represent the length and W the width of a rectangle. The area, A, of the rectangle is given by $A = LW$.

QUESTION How many squares, each 1 in. on a side, are needed to cover a rectangle that has an area of 18 in^2?

EXAMPLE 5 Find the Area of a Rectangle

How many square feet of sod are needed to cover a football field? A football field measures 360 ft by 160 ft.

Solution

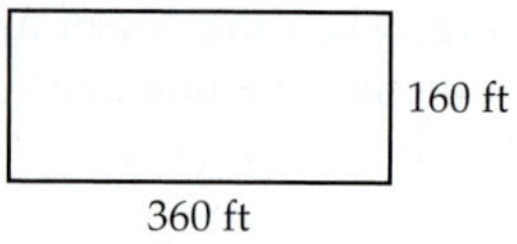

• Draw a diagram.

$A = LW$ • Use the formula for the area of a rectangle.

$A = 360(160)$ • The length is 360 ft. Substitute 360 for L. The width is 160 ft. Substitute 160 for W. Remember that LW means "L times W."

$A = 57{,}600$

57,600 ft^2 of sod is needed. • Area is measured in square units.

CHECK YOUR PROGRESS 5 Find the amount of fabric needed to make a rectangular flag that measures 308 cm by 192 cm.

Solution See page S24.

TAKE NOTE

Recall that the rules of exponents state that when multiplying variables with like bases, we add the exponents.

A square is a rectangle in which all sides are the same length. Therefore, both the length and the width of a square can be represented by s, and $A = LW = s \cdot s = s^2$.

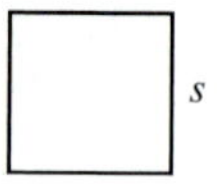

$A = s \cdot s$
$A = s^2$

Area of a Square

Let s represent the length of a side of a square. The area, A, of the square is given by $A = s^2$.

ANSWER **a.** The area is 1 square yard, written 1 yd^2. **b.** The area is 1 square kilometer, written 1 km^2.

ANSWER 18 squares, each 1 in. on a side, are needed to cover the rectangle.

EXAMPLE 6 Find the Area of a Square

A homeowner wants to carpet the family room. The floor is square and measures 6 m on each side. How much carpet should be purchased?

Solution

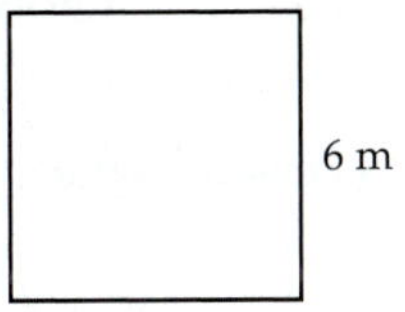

• Draw a diagram.

$A = s^2$ • Use the formula for the area of a square.

$A = 6^2$ • The length of a side is 6 m. Substitute 6 for s.

$A = 36$

36 m^2 of carpet should be purchased. • Area is measured in square units.

CHECK YOUR PROGRESS 6 Find the area of the floor of a two-car garage that is in the shape of a square that measures 24 ft on a side.

Solution See page S24.

Figure $ABCD$ is a parallelogram. $\overline{BC}$ is the **base** of the parallelogram. $\overline{AE}$, perpendicular to the base, is the **height** of the parallelogram.

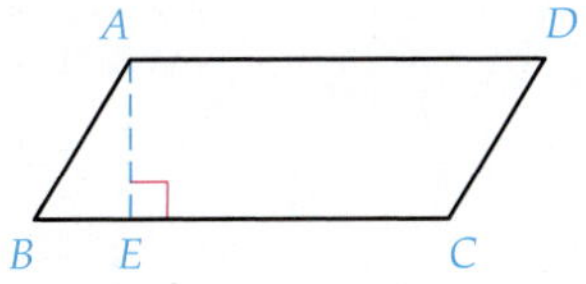

Any side of a parallelogram can be designated as the base. The corresponding height is found by drawing a line segment perpendicular to the base from the opposite side. In the figure at the right, $\overline{CD}$ is the base and $\overline{AE}$ is the height.

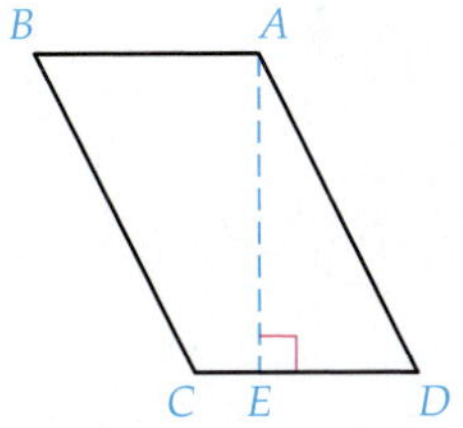

A rectangle can be formed from a parallelogram by cutting a right triangle from one end of the parallelogram and attaching it to the other end. The area of the resulting rectangle will equal the area of the original parallelogram.

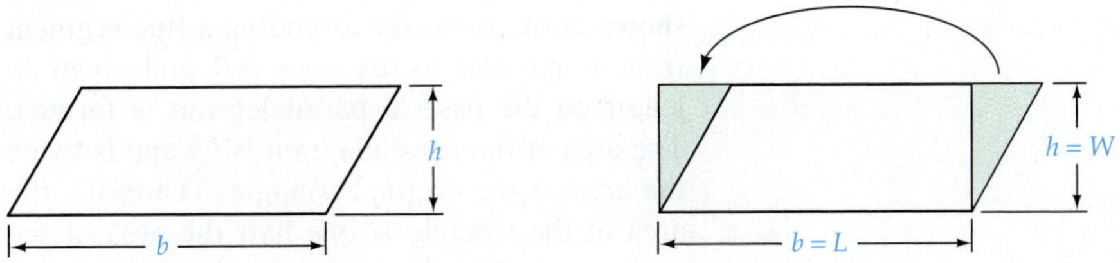

Area of a Parallelogram

Let b represent the length of the base and h the height of a parallelogram. The area, A, of the parallelogram is given by $A = bh$.

EXAMPLE 7 Find the Area of a Parallelogram

A solar panel is in the shape of a parallelogram that has a base of 2 ft and a height of 3 ft. Find the area of the solar panel.

Solution

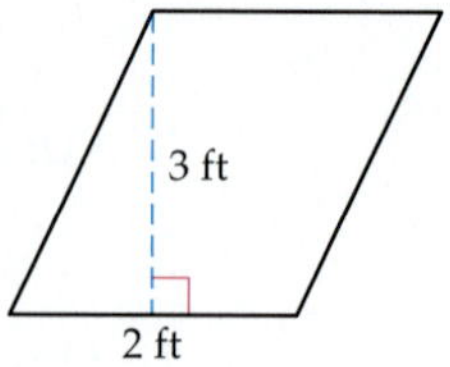

• Draw a diagram.

$A = bh$ • Use the formula for the area of a parallelogram.

$A = 2(3)$ • The base is 2 ft. Substitute 2 for b. The height is 3 ft. Substitute 3 for h. Remember that bh means "b times h."

$A = 6$

The area is 6 ft^2. • Area is measured in square units.

CHECK YOUR PROGRESS 7 A fieldstone patio is in the shape of a parallelogram that has a base measuring 14 m and a height measuring 8 m. What is the area of the patio?

Solution See page S24.

Figure ABC is a triangle. $\overline{AB}$ is the **base** of the triangle. $\overline{CD}$, perpendicular to the base, is the **height** of the triangle.

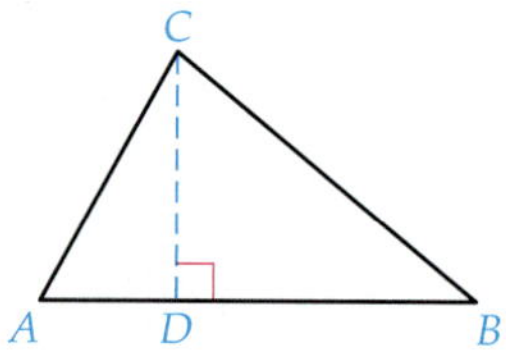

Any side of a triangle can be designated as the base. The corresponding height is found by drawing a line segment perpendicular to the base from the vertex opposite the base.

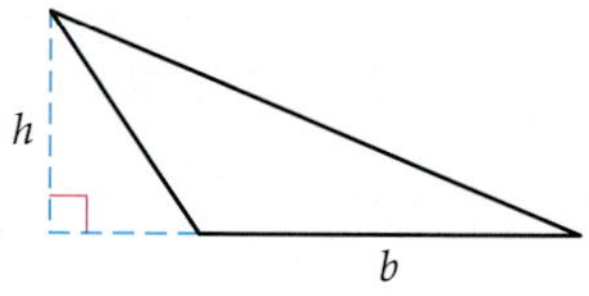

Consider triangle ABC with base b and height h, shown at the right. By extending a line segment from C parallel to the base $\overline{AB}$ and equal in length to the base, a parallelogram is formed. The area of the parallelogram is bh and is twice the area of the original triangle. Therefore, the area of the triangle is one half the area of the parallelogram, or $\frac{1}{2}bh$.

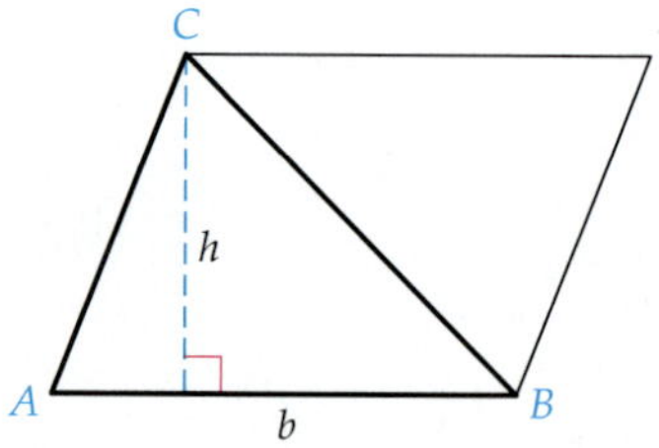

Area of a Triangle

Let b represent the length of the base and h the height of a triangle. The area, A, of the triangle is given by $A = \frac{1}{2}bh$.

EXAMPLE 8 Find the Area of a Triangle

A riveter uses metal plates that are in the shape of a triangle with a base of 12 cm and a height of 6 cm. Find the area of one metal plate.

Solution

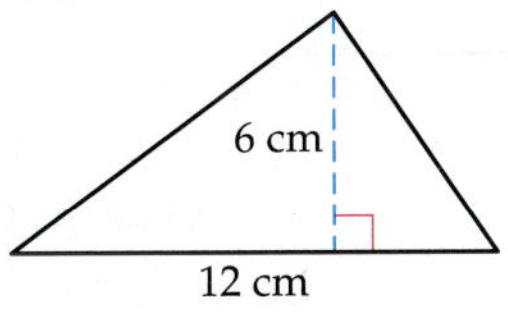

• Draw a diagram.

$A = \frac{1}{2}bh$ • Use the formula for the area of a triangle.

$A = \frac{1}{2}(12)(6)$ • The base is 12 cm. Substitute 12 for b. The height is 6 cm. Substitute 6 for h. Remember that bh means "b times h."

$A = 6(6)$

$A = 36$

The area is 36 cm². • Area is measured in square units.

CHECK YOUR PROGRESS 8 Find the amount of felt needed to make a banner that is in the shape of a triangle with a base of 18 in. and a height of 9 in.

Solution See page S24.

TAKE NOTE

The bases of a trapezoid are the parallel sides of the figure.

Figure $ABCD$ is a trapezoid. $\overline{AB}$, with length b_1, is one **base** of the trapezoid and $\overline{CD}$, with length b_2, is the other base. $\overline{AE}$, perpendicular to the two bases, is the **height**.

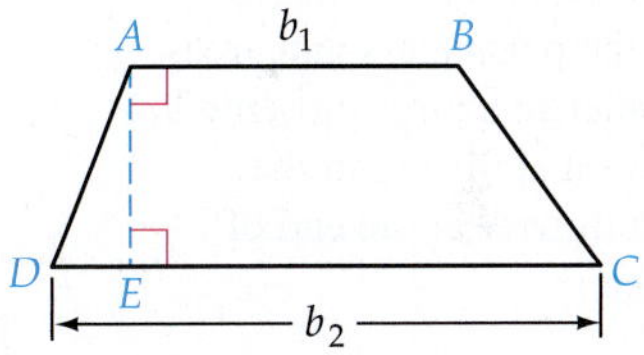

In the trapezoid at the right, the line segment $\overline{BD}$ divides the trapezoid into two triangles, ABD and BCD. In triangle ABD, b_1 is the base and h is the height. In triangle BCD, b_2 is the base and h is the height. The area of the trapezoid is the sum of the areas of the two triangles.

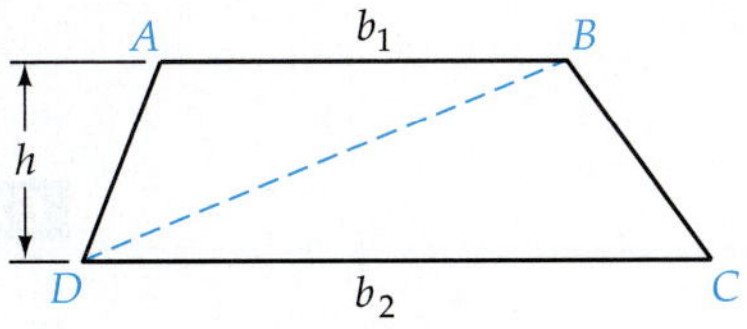

$$\text{Area of trapezoid } ABCD = \text{Area of triangle } ABD + \text{area of triangle } BCD$$
$$= \frac{1}{2}b_1h + \frac{1}{2}b_2h = \frac{1}{2}h(b_1 + b_2)$$

Area of a Trapezoid

Let b_1 and b_2 represent the lengths of the bases and h the height of a trapezoid. The area, A, of the trapezoid is given by $A = \frac{1}{2}h(b_1 + b_2)$.

EXAMPLE 9 Find the Area of a Trapezoid

A boat dock is built in the shape of a trapezoid with bases measuring 14 ft and 6 ft and a height of 7 ft. Find the area of the dock.

Solution

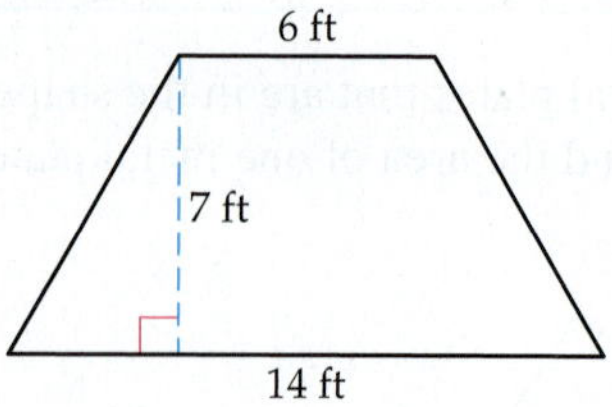

• Draw a diagram.

$A = \frac{1}{2}h(b_1 + b_2)$ • Use the formula for the area of a trapezoid.

$A = \frac{1}{2} \cdot 7(14 + 6)$ • The height is 7 ft. Substitute 7 for h. The bases measure 14 ft and 6 ft. Substitute 14 and 6 for b_1 and b_2.

$A = \frac{1}{2} \cdot 7(20)$

$A = 70$

The area is 70 ft^2. • Area is measured in square units.

CHECK YOUR PROGRESS 9 Find the area of a patio that has the shape of a trapezoid with a height of 9 ft and bases measuring 12 ft and 20 ft.

Solution See page S24.

TAKE NOTE

For your reference, all of the formulas for the perimeters and areas of the geometric figures presented in this section are listed in the Chapter Summary at the end of this chapter.

The area of a circle is the product of π and the square of the radius.

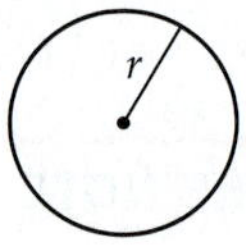

$A = \pi r^2$

The Area of a Circle

The area, A, of a circle with radius of length r is given by $A = \pi r^2$.

EXAMPLE 10 Find the Area of a Circle

Find the area of a circle with a diameter of 10 m. Round to the nearest hundredth of a square meter.

CALCULATOR NOTE

To evaluate the expression $\pi(5)^2$ on your calculator, enter

[π] [×] 5 [y^x] 2 [=]

or

[π] [×] 5 [^] 2 [ENTER]

Solution

$r = \frac{1}{2}d = \frac{1}{2}(10) = 5$ • Find the radius of the circle.

$A = \pi r^2$ • Use the formula for the area of a circle.

$A = \pi(5)^2$ • Replace r with 5.

$A = \pi(25)$ • Square 5.

$A \approx 78.54$ • An approximation is asked for. Use the π key on a calculator.

The area of the circle is approximately 78.54 m^2.

CHECK YOUR PROGRESS 10 Find the area of a circle with a diameter of 12 km. Give the exact measure.

Solution See page S24.

EXAMPLE 11 **Application of Finding the Area of a Circle**

How large a cover is needed for a circular hot tub that is 8 ft in diameter? Round to the nearest tenth of a square foot.

Solution

$r = \frac{1}{2}d = \frac{1}{2}(8) = 4$ • Find the radius of a circle with a diameter of 8 ft.

$A = \pi r^2$ • Use the formula for the area of a circle.

$A = \pi(4)^2$ • Replace r with 4.

$A = \pi(16)$ • Square 4.

$A \approx 50.3$ • Use the π key on a calculator.

The cover for the hot tub must be 50.3 ft^2.

CHECK YOUR PROGRESS 11 How much material is needed to make a circular tablecloth that is to have a diameter of 4 ft? Round to the nearest hundredth of a square foot.

Solution See page S24.

MATH MATTERS Möbius Bands

Cut out a long, narrow rectangular strip of paper.

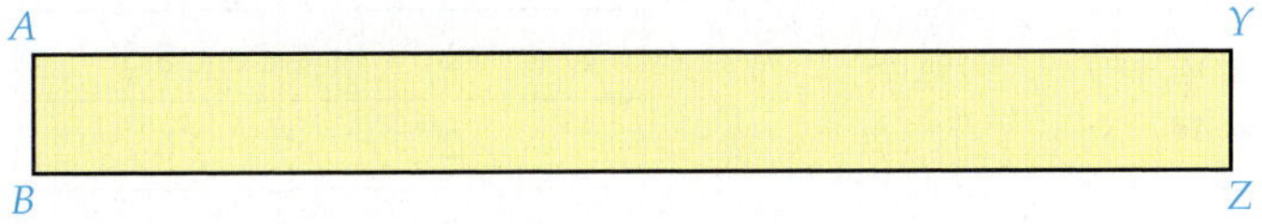

Give the strip of paper a half-twist.

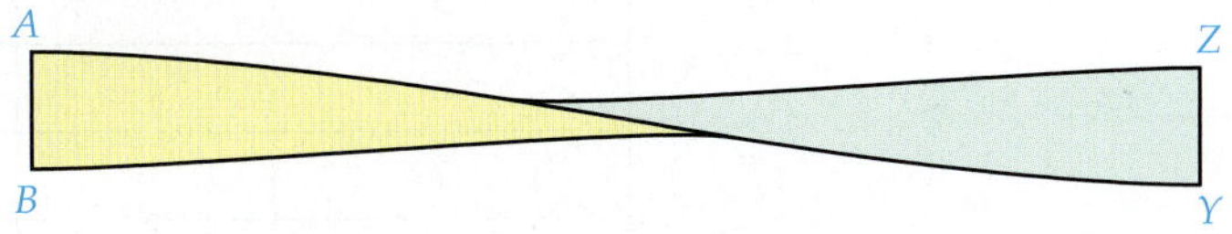

Put the ends together so that *A* meets *Z* and *B* meets *Y*. Tape the ends together. The result is a *Möbius band.* A Möbius band is also called a Möbius strip.

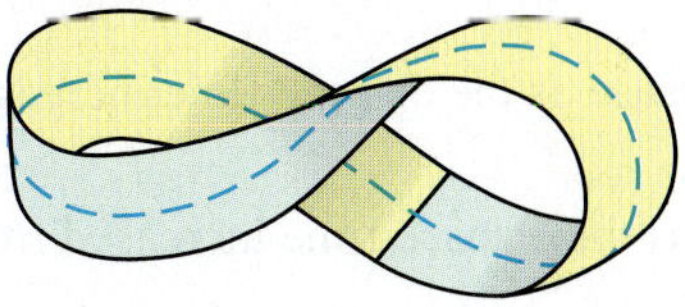

Make a Möbius band that is $1\frac{1}{2}$ in. wide. Use a pair of scissors to cut the Möbius band lengthwise down the middle, staying $\frac{3}{4}$ in. from each edge. Describe the result.

Make a Möbius band from plain white paper and then shade one side. Describe what remains unshaded on the Möbius band, and state the number of sides a Möbius band has.

EXCURSION

Perimeter and Area of a Rectangle with Changing Dimensions

A graphic artist has drawn a 5-inch by 4-inch rectangle on a computer screen. The artist is scaling the size of the rectangle in such a way that every second the upper right corner of the rectangle moves to the right 0.5 in. and downward 0.2 in.

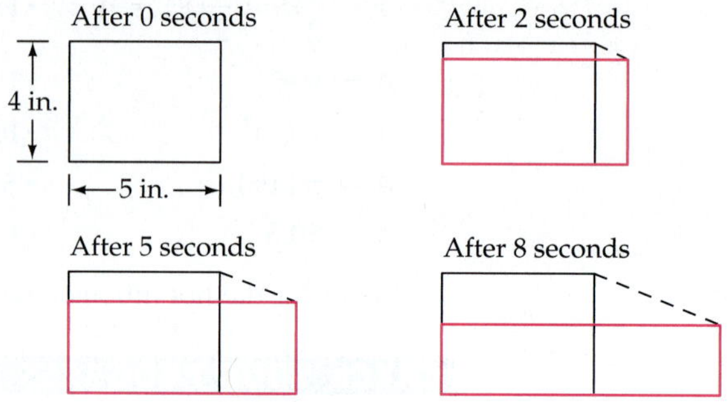

Thus after 1 s, the length of the rectangle is 5.5 in. and the width is 3.8 in. After 2 s, the length is 6 in. and the width is 3.6 in.

INSTRUCTOR NOTE

This Excursion works well as an in-class, small group activity. Students who have experience with graphic art software may more quickly grasp the concept of the changing dimensions of the rectangle, and they can explain it to other students.

EXCURSION EXERCISES

1. Complete the table below by calculating the length, width, perimeter, and area of the rectangle after the given number of seconds have elapsed.

Number of seconds elapsed	Length (in inches)	Width (in inches)	Perimeter (in inches)	Area (in square inches)
0	5	4		
1	5.5	3.8		
2	6	3.6		
3				
4				
5				
6				
7				
8				
9				
10				

Use your data in the above table to answer Exercises 2 to 6.

2. Did the perimeter of the rectangle increase or decrease as the elapsed time increased from 0 s to 5 s, in one-second increments?

3. Did the area of the rectangle increase or decrease as the elapsed time increased from 0 s to 5 s, in one-second increments?

4. Did the perimeter of the rectangle increase or decrease as the elapsed time increased from 5 s to 10 s, in one-second increments?

5. Did the area of the rectangle increase or decrease as the elapsed time increased from 5 s to 10 s, in one-second increments?

6. If the perimeter of a rectangle is increasing, does this mean that the area of the rectangle is also increasing?

EXERCISE SET 7.3

1. What is wrong with each statement?
   a. The perimeter is 40 m^2.
   b. The area is 120 ft.

■ In Exercises 2 to 8, find **(a)** the perimeter and **(b)** the area of the figure.

2. 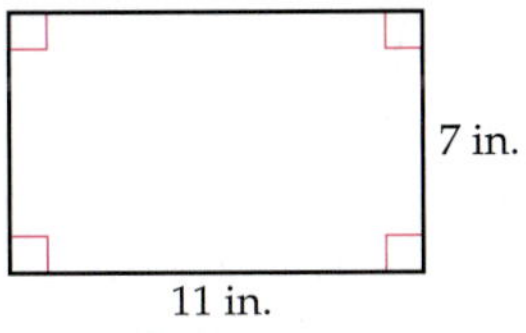

3. 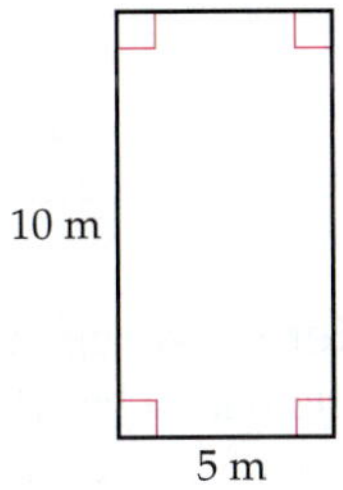

4. 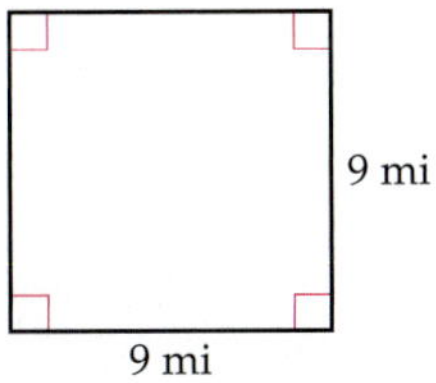

5. 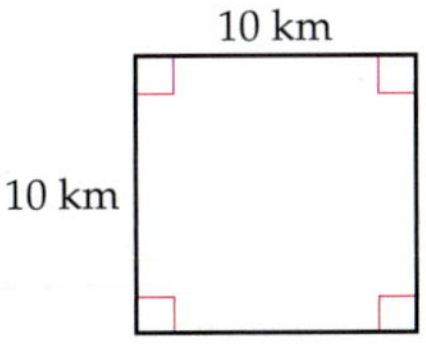

6. 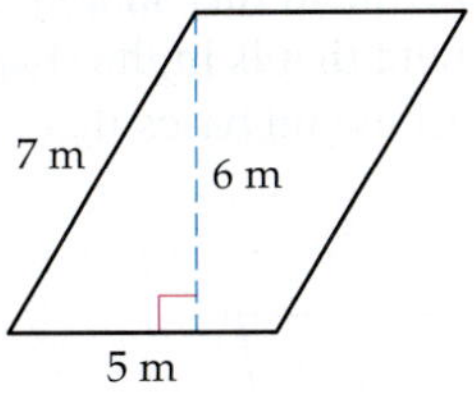

7. 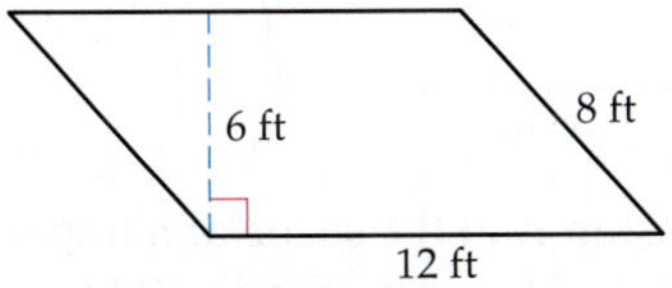

8. 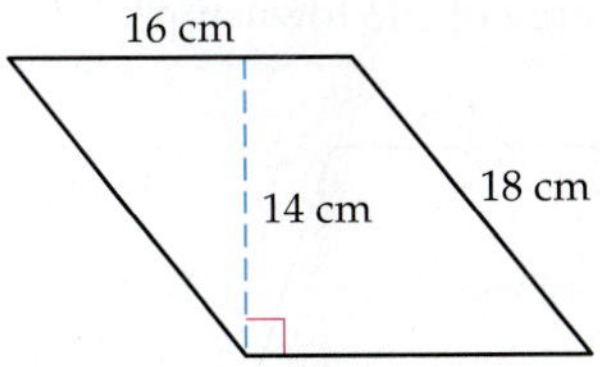

■ In Exercises 9 to 14, find **(a)** the circumference and **(b)** the area of the figure. State an exact answer and a decimal approximation rounded to the nearest hundredth.

9. 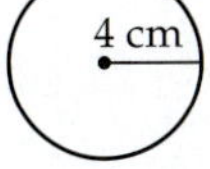

10. 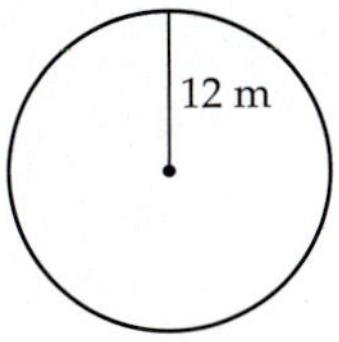

11. 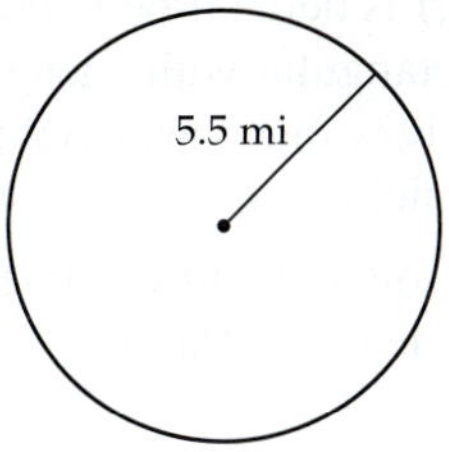

12. 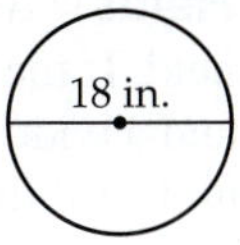

13. 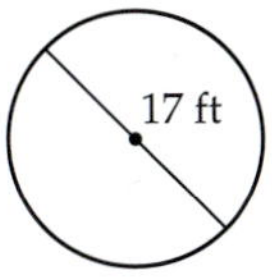

14. 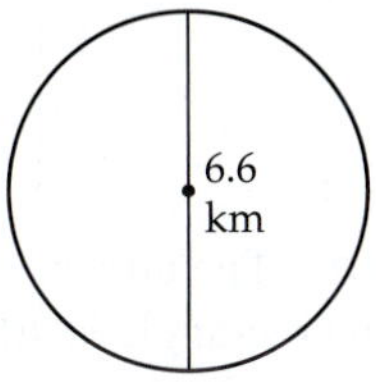

15. **Perimeter** Find the perimeter of a regular pentagon that measures 4 in. on each side.

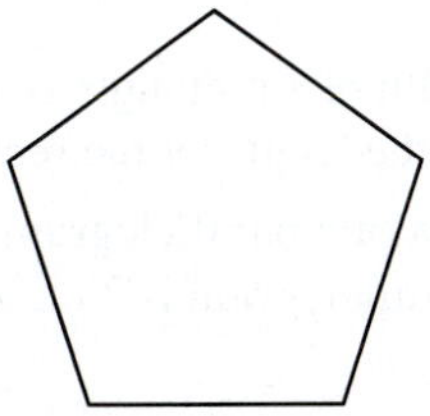

16. **Interior Decorating** Wall-to-wall carpeting is installed in a room that is 15 ft long and 10 ft wide. The edges of the carpet are held down by tack strips. How many feet of tack-strip material are needed?

17. **Cross-Country** A cross-country course is in the shape of a parallelogram with a base of length 3 mi and a side of length 2 mi. What is the total length of the cross-country course?

18. **Parks and Recreation** A rectangular playground has a length of 160 ft and a width of 120 ft. Find the length of hedge that surrounds the playground.

19. **Sewing** Bias binding is to be sewn around the edge of a rectangular tablecloth measuring 68 in. by 42 in. If the bias binding comes in packages containing 15 ft of binding, how many packages of bias binding are needed for the tablecloth?

20. **Race Tracks** The first circular dog race track opened in 1919 in Emeryville, California. The radius of the circular track was 157.64 ft. Find the circumference of the track. Use 3.14 for π. Round to the nearest foot.

21. The length of a side of a square is equal to the diameter of a circle. Which is greater, the perimeter of the square or the circumference of the circle?

22. The length of a rectangle is equal to the diameter of a circle, and the width of the rectangle is equal to the radius of the same circle. Which is greater, the perimeter of the rectangle or the circumference of the circle?

23. Construction What is the area of a square patio that measures 12 m on each side?

24. Athletic Fields Artificial turf is being used to cover a playing field. If the field is rectangular with a length of 110 yd and a width of 80 yd, how much artificial turf must be purchased to cover the field?

25. Framing The perimeter of a square picture frame is 36 in. Find the length of each side of the frame.

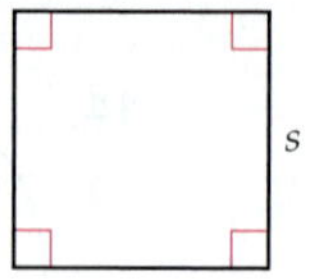

26. Area The area of a rectangle is 400 in^2. If the length of the rectangle is 40 in., what is the width?

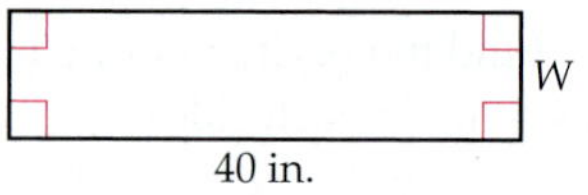

27. Area The width of a rectangle is 8 ft. If the area is 312 ft^2, what is the length of the rectangle?

28. Area The area of a parallelogram is 56 m^2. If the height of the parallelogram is 7 m, what is the length of the base?

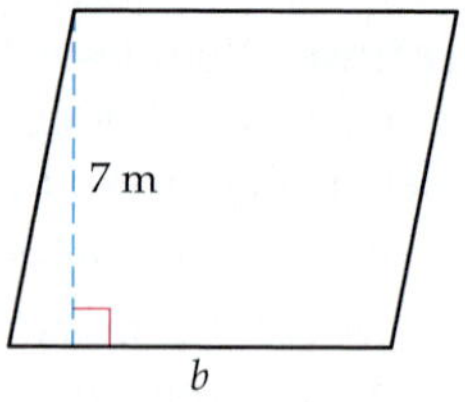

29. Storage Units You want to rent a storage unit. You estimate that you will need 175 ft^2 of floor space. You see the ad below on the Internet. You want to rent the smallest possible unit that will hold everything you want to store. Which of the six units pictured in the ad should you select?

30. Sailing A sail is in the shape of a triangle with a base of 12 m and a height of 16 m. How much canvas was needed to make the body of the sail?

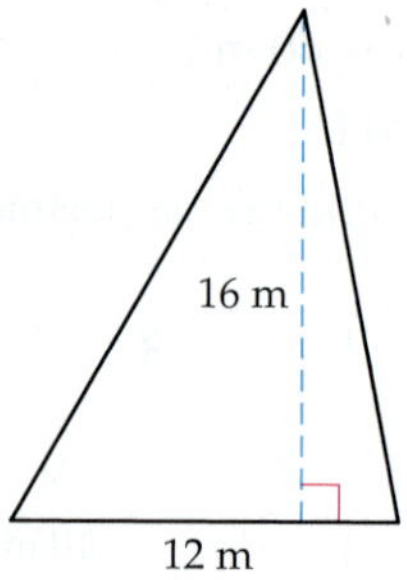

31. Gardens A vegetable garden is in the shape of a triangle with a base of 21 ft and a height of 13 ft. Find the area of the vegetable garden.

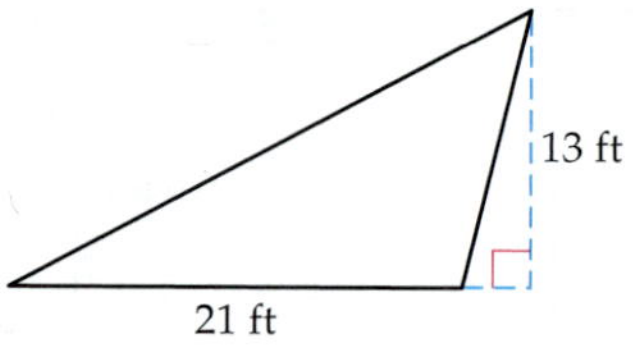

32. Athletic Fields How much artificial turf should be purchased to cover an athletic field that is in the shape of a trapezoid with a height of 15 m and bases that measure 45 m and 36 m?

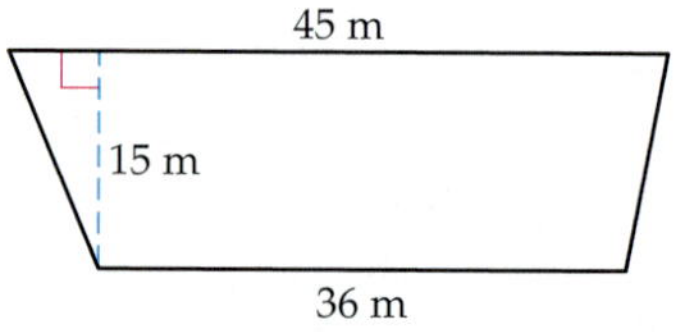

33. Land Area A township is in the shape of a trapezoid with a height of 10 km and bases measuring 9 km and 23 km. What is the land area of the township?

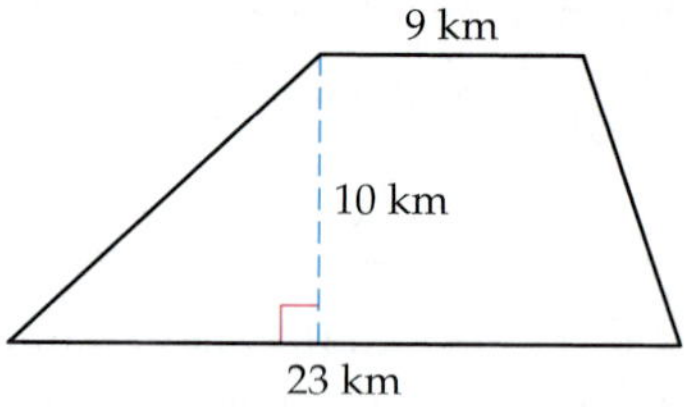

34. Parks and Recreation A city plans to plant grass seed in a public playground that has the shape of a triangle with a height of 24 m and a base of 20 m. Each bag of grass seed will seed 120 m^2. How many bags of seed should be purchased?

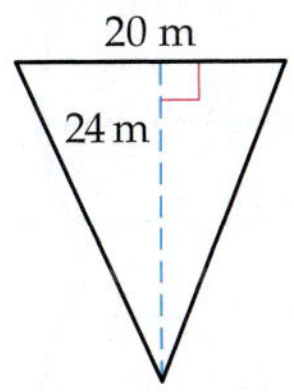

35. Home Maintenance You plan to stain the wooden deck at the back of your house. The deck is in the shape of a trapezoid with bases that measure 10 ft and 12 ft and a height of 10 ft. A quart of stain will cover 55 ft^2. How many quarts of stain should you purchase?

36. Interior Decorating A fabric wall hanging is in the shape of a triangle that has a base of 4 ft and a height of 3 ft. An additional 1 ft^2 of fabric is needed for hemming the material. How much fabric should be purchased to make the wall hanging?

37. Interior Decorating You are wallpapering two walls of a den, one measuring 10 ft by 8 ft and the other measuring 12 ft by 8 ft. The wallpaper costs $96 per roll, and each roll will cover 40 ft^2. What is the cost to wallpaper the two walls?

38. Gardens An urban renewal project involves reseeding a garden that is in the shape of a square, 80 ft on each side. Each bag of grass seed costs $12 and will seed 1500 ft^2. How much money should be budgeted for buying grass seed for the garden?

39. Carpeting You want to install wall-to-wall carpeting in the family room. The floor plan is shown below. If the cost of the carpet you would like to purchase is $38 per square yard, what is the cost of carpeting your family room? Assume that there is no waste. *Hint:* $9 \text{ ft}^2 = 1 \text{ yd}^2$.

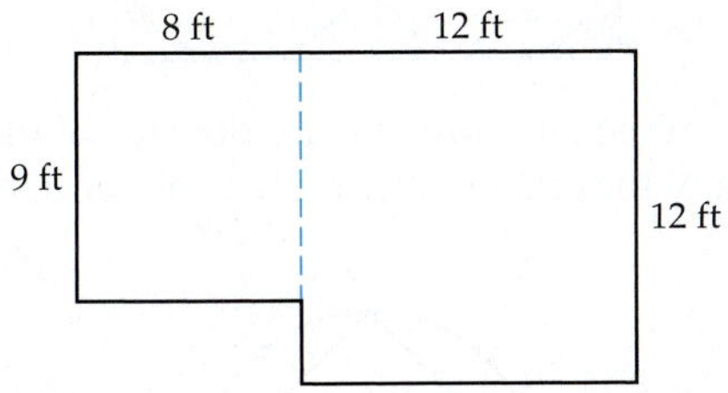

40. Interior Decorating You want to paint the rectangular walls of your bedroom. Two walls measure 16 ft by 8 ft, and the other two walls measure 12 ft by 8 ft. The paint you wish to purchase costs $28 per gallon, and each gallon will cover 400 ft^2 of wall. Find the minimum amount you will need to spend on paint.

41. Landscaping A walkway 2 m wide surrounds a rectangular plot of grass. The plot is 25 m long and 15 m wide. What is the area of the walkway?

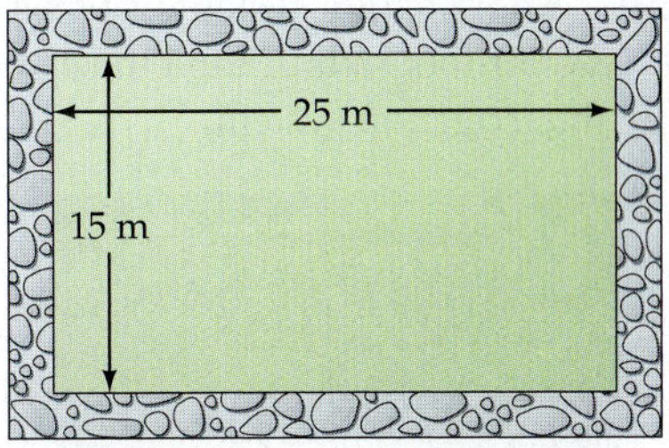

42. Draperies The material used to make pleated draperies for a window must be twice as wide as the width of the window. Draperies are being made for four windows, each 3 ft wide and 4 ft high. Because the drapes will fall slightly below the window sill and extra fabric is needed for hemming the drapes, 1 ft must be added to the height of the window. How much material must be purchased to make the drapes?

43. Carpentry Find the length of molding needed to put around a circular table that is 4.2 ft in diameter. Round to the nearest hundredth of a foot.

44. Sewing How much binding is needed to bind the edge of a circular rug that is 3 m in diameter? Round to the nearest hundredth of a meter.

45. Pulleys A pulley system is diagrammed below. If pulley B has a diameter of 16 in. and is rotating at 240 revolutions per minute, how far does a given point on the belt travel each minute that the pulley system is in operation? Assume the belt does not slip as the pulley rotates. Round to the nearest inch.

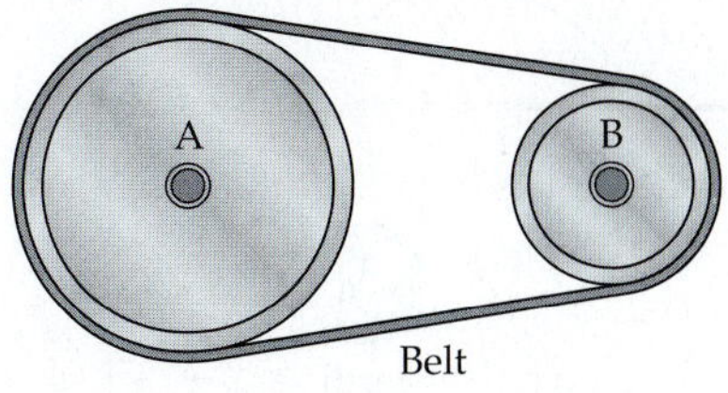

46. Bicycles A bicycle tire has a diameter of 18 in. How many feet does the bicycle travel when the wheel makes 20 revolutions? Round to the nearest hundredth of a foot.

47. Tricycles The front wheel of a tricycle has a diameter of 16 in. How many feet does the tricycle travel when the wheel makes 15 revolutions? Round to the nearest hundredth of a foot.

48. Telescopes The circular lens located on an astronomical telescope has a diameter of 24 in. Find the exact area of the lens.

49. Irrigation An irrigation system waters a circular field that has a 50-foot radius. Find the exact area watered by the irrigation system.

David Frazier/Spirit/Corbis

50. Pizza How much greater is the area of a pizza that has a radius of 10 in. than the area of a pizza that has a radius of 8 in.? Round to the nearest hundredth of a square inch.

51. Pizza A restaurant serves a small pizza that has a radius of 6 in. The restaurant's large pizza has a radius that is twice the radius of the small pizza. How much larger is the area of the large pizza? Round to the nearest hundredth of a square inch. Is the area of the large pizza more or less than twice the area of the small pizza?

52. Satellites A geostationary satellite (GEO) orbits Earth over the Equator. The orbit is circular and at a distance of 36,000 km above Earth. An orbit at this altitude allows the satellite to maintain a fixed position in relation to Earth. What is the distance traveled by a GEO satellite in one orbit around Earth? The radius of Earth is 6380 km. Round to the nearest kilometer.

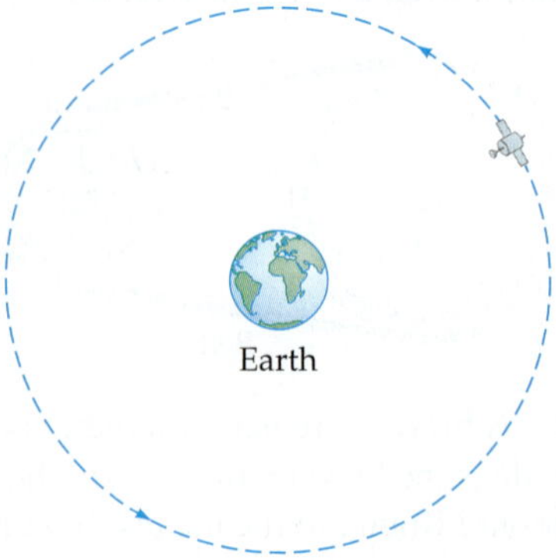

53. Lake Tahoe One way to measure the area of an irregular figure, such as a lake, is to divide the area into trapezoids that have the same height. Then measure the length of each base, calculate the area of each trapezoid, and add the areas. The following figure gives approximate dimensions for Lake Tahoe, which straddles the California and Nevada borders. Approximate the area of Lake Tahoe using the given trapezoids. Round to the nearest tenth of a square mile.

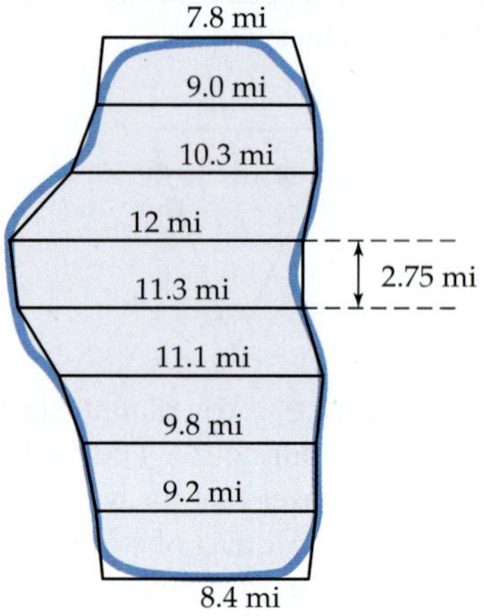

54. Ball Fields How much farther is it around the bases of a baseball diamond than around the bases of a softball diamond? *Hint:* Baseball and softball diamonds are squares.

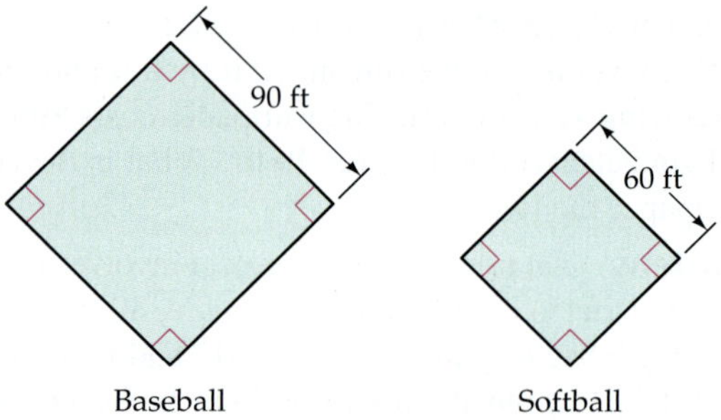

55. Area Write an expression for the area of the shaded portion of the diagram. Leave the answer in terms of π and r.

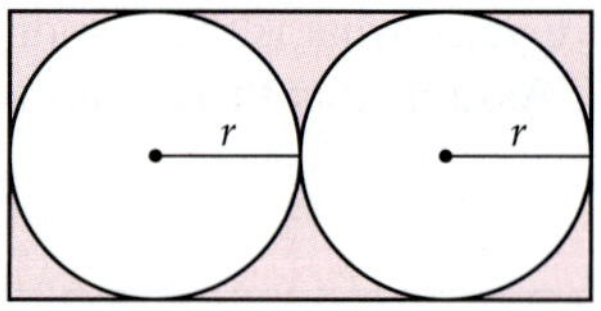

56. Area Write an expression for the area of the shaded portion of the diagram. Leave the answer in terms of π and r.

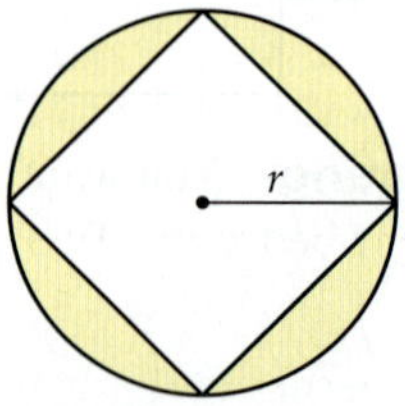

57. Area If both the length and width of a rectangle are doubled, how many times as large is the area of the resulting rectangle compared to the area of the original rectangle?

EXTENSIONS

58. A circle with radius r and circumference C is sliced into 16 identical sectors which are then arranged as shown below.

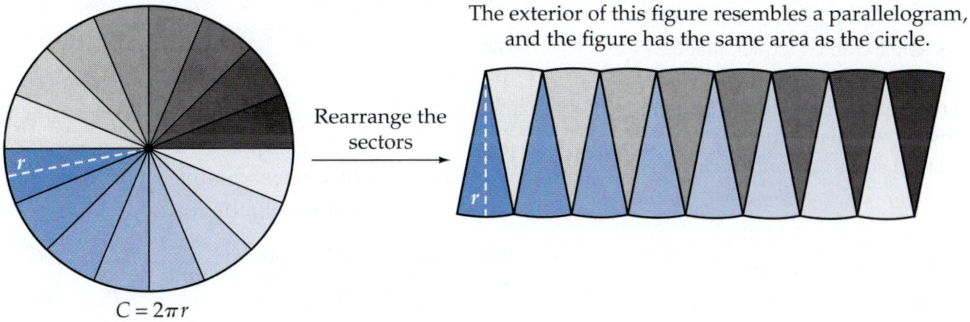

The exterior of the figure shown by the rearranged sectors resembles a parallelogram.

a. What dimension of the circle approximates the height of the parallelogram?

b. What dimension of the circle approximates the base of the parallelogram?

c. Explain how the formula for the area of a circle can be derived by using this slicing approach.

59. Heron's (or Hero's) formula is sometimes used to calculate the area of a triangle.

Heron's Formula

The area of a triangle with sides of lengths a, b, and c is given by

$$A = \sqrt{s(s - a)(s - b)(s - c)}$$

where s is the semiperimeter of the triangle:

$$s = \frac{a + b + c}{2}$$

a. Use Heron's formula to find the area of a triangle with sides that measure 4.4 in., 5.7 in., and 6.2 in. Round to the nearest tenth of a square inch.

b. Use Heron's formula to find the area of an equilateral triangle with sides that measure 8.3 cm. Round to the nearest tenth of a square centimeter.

c. Find the lengths of the sides of a triangle that has a perimeter of 12 in., given that the length of each side, in inches, is a counting number and the area of the triangle, in square inches, is also a counting number. *Hint:* All three sides are different lengths.

SECTION 7.4 Properties of Triangles

Similar Triangles

Similar objects have the same shape but not necessarily the same size. A tennis ball is similar to a basketball. A model ship is similar to an actual ship.

Similar objects have corresponding parts; for example, the rudder on the model ship corresponds to the rudder on the actual ship. The relationship between the sizes of the corresponding parts can be written as a ratio. All corresponding parts of two similar figures share the same ratio. If the rudder on the model ship is $\frac{1}{100}$ the size of the rudder on the actual ship, then the model mast is $\frac{1}{100}$ of the size of the actual mast, the width of the model is $\frac{1}{100}$ the width of the actual ship, and so on.

POINT OF INTEREST

Model trains are similar to actual trains. They come in a variety of sizes that manufacturers have agreed upon, so that, for example, an engine made by manufacturer A is able to run on a track made by manufacturer B. Listed below are three model railroad sizes by name, along with the ratio of model size to actual size.

Name	*Ratio*
Z	1:220
N	1:160
HO	1:87

Z Scale

N Scale

HO Scale

HO's ratio of 1:87 means that in every dimension, an HO scale model railroad car is $\frac{1}{87}$ the size of the real railroad car.

The two triangles *ABC* and *DEF* shown at the right are similar. Side $\overline{AB}$ corresponds to side $\overline{DE}$, side $\overline{BC}$ corresponds to side $\overline{EF}$, and side $\overline{AC}$ corresponds to side $\overline{DF}$. The ratios of the lengths of corresponding sides are equal.

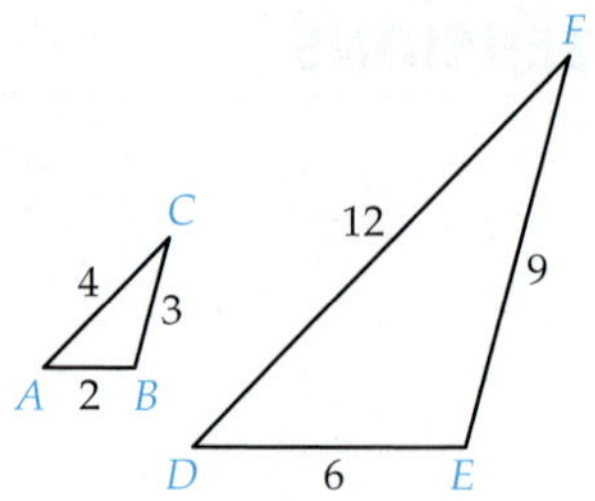

$$\frac{AB}{DE} = \frac{2}{6} = \frac{1}{3}, \quad \frac{BC}{EF} = \frac{3}{9} = \frac{1}{3}, \quad \text{and}$$

$$\frac{AC}{DF} = \frac{4}{12} = \frac{1}{3}$$

Because the ratios of corresponding sides are equal, several proportions can be formed.

$$\frac{AB}{DE} = \frac{BC}{EF}, \quad \frac{AB}{DE} = \frac{AC}{DF}, \quad \text{and} \quad \frac{BC}{EF} = \frac{AC}{DF}$$

The measures of corresponding angles in similar triangles are equal. Therefore,

$$m\angle A = m\angle D, \quad m\angle B = m\angle E, \quad \text{and}$$

$$m\angle C = m\angle F$$

Triangles *ABC* and *DEF* at the right are similar triangles. *AH* and *DK* are the heights of the triangles. The ratio of the heights of similar triangles equals the ratio of the lengths of corresponding sides.

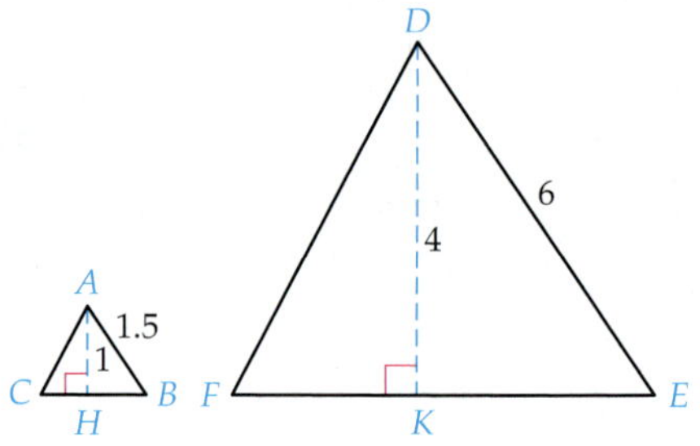

$$\text{Ratio of corresponding sides} = \frac{1.5}{6} = \frac{1}{4}$$

$$\text{Ratio of heights} = \frac{1}{4}$$

Properties of Similar Triangles

For similar triangles, the ratios of corresponding sides are equal. The ratio of corresponding heights is equal to the ratio of corresponding sides. The measures of corresponding angles are equal.

The two triangles at the right are similar triangles. Find the length of side $\overline{EF}$. Round to the nearest tenth of a meter.

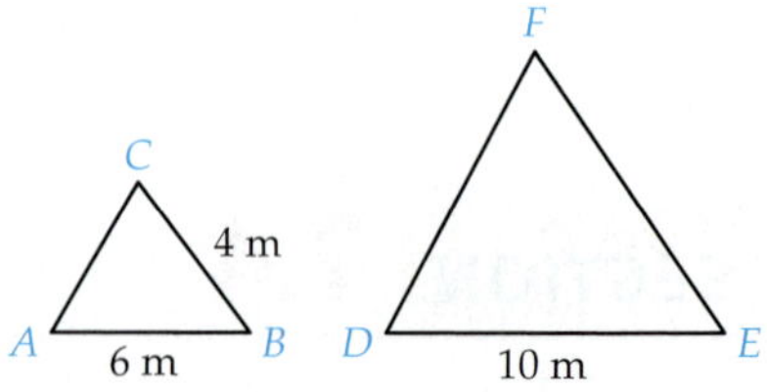

The triangles are similar, so the ratios of the lengths of corresponding sides are equal.

$$\frac{EF}{BC} = \frac{DE}{AB}$$

$$\frac{EF}{4} = \frac{10}{6}$$

$$6(EF) = 4(10)$$

$$6(EF) = 40$$

$$EF \approx 6.7$$

The length of side *EF* is approximately 6.7 m.

TAKE NOTE

The notation $\triangle ABC \sim \triangle DEF$ is used to indicate that triangle ABC is similar to triangle DEF.

INSTRUCTOR NOTE

The concept of similar triangles and the fact that the ratios of corresponding sides are equal are not obvious to students. Nonetheless, these concepts are essential in many practical applications.

You might introduce the topic of similar triangles by displaying a geometric figure on an overhead projector. Then move the projector closer to, and then farther from, the projected image to see the size of the image change. Emphasize that its shape does not change.

QUESTION What are two other proportions that can be written for the similar triangles shown in the preceding example?

EXAMPLE 1 Use Similar Triangles to Find the Unknown Height of a Triangle

Triangles ABC and DEF are similar. Find FG, the height of triangle DEF.

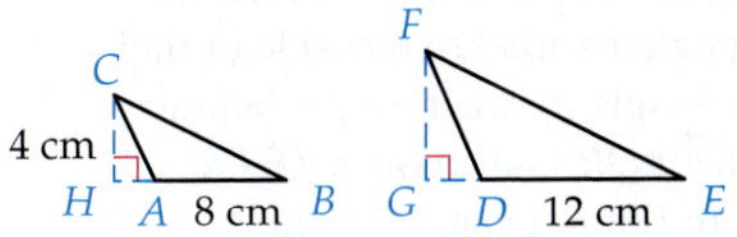

Solution

$$\frac{AB}{DE} = \frac{CH}{FG}$$ • For similar triangles, the ratio of corresponding sides equals the ratio of corresponding heights.

$$\frac{8}{12} = \frac{4}{FG}$$ • Replace AB, DE, and CH with their values.

$$8(FG) = 12(4)$$ • The cross products are equal.

$$8(FG) = 48$$

$$FG = 6$$ • Divide both sides of the equation by 8.

The height FG of triangle DEF is 6 cm.

CHECK YOUR PROGRESS 1

Triangles ABC and DEF are similar. Find FG, the height of triangle DEF.

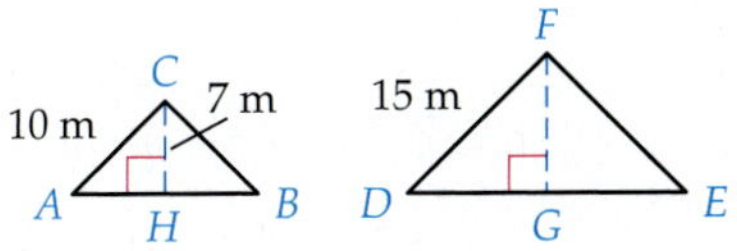

Solution See page S24.

Triangles ABC and DEF are similar triangles. Find the area of triangle ABC.

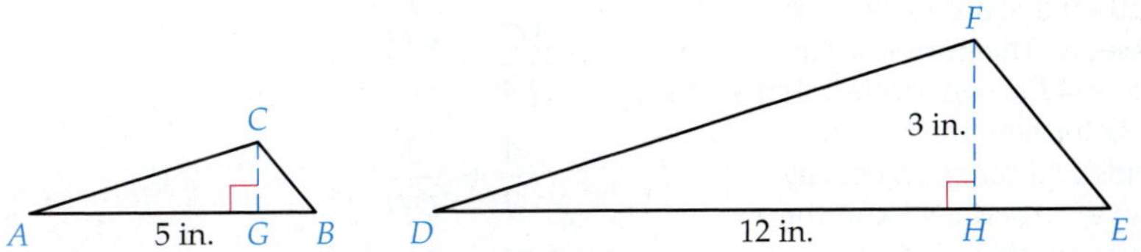

Solve a proportion to find the height of triangle ABC.

$$\frac{AB}{DE} = \frac{CG}{FH}$$

$$\frac{5}{12} = \frac{CG}{3}$$

$$12(CG) = 5(3)$$

$$12(CG) = 15$$

$$CG = 1.25$$

Use the formula for the area of a triangle. $$A = \frac{1}{2}bh$$

The base is 5 in. The height is 1.25 in. $$A = \frac{1}{2}(5)(1.25)$$

The area of triangle ABC is 3.125 in^2. $$A = 3.125$$

ANSWER In addition to $\frac{EF}{BC} = \frac{DE}{AB}$, we can write the proportions $\frac{DE}{AB} = \frac{DF}{AC}$ and $\frac{EF}{BC} = \frac{DF}{AC}$. These three proportions can also be written using the reciprocal of each fraction: $\frac{BC}{EF} = \frac{AB}{DE}$, $\frac{AB}{DE} = \frac{AC}{DF}$, and $\frac{BC}{EF} = \frac{AC}{DF}$. Also, the right and left sides of each proportion can be interchanged.

If the three angles of one triangle are equal in measure to the three angles of another triangle, then the triangles are similar.

TAKE NOTE

You can always create similar triangles by drawing a line segment inside the original triangle parallel to one side of the triangle. In the triangle below, $\overline{ST} \parallel \overline{QR}$ and triangle PST is similar to triangle PQR.

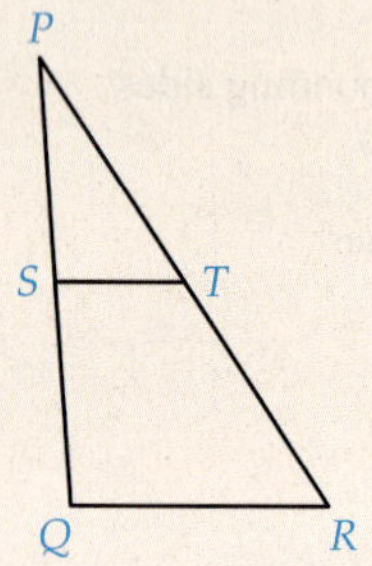

In triangle ABC at the right, line segment $\overline{DE}$ is drawn parallel to the base $\overline{AB}$. Because the measures of corresponding angles are equal, $m\angle x = m\angle r$ and $m\angle y = m\angle n$. We know that $m\angle C = m\angle C$. Thus the measures of the three angles of triangle ABC are equal, respectively, to the measures of the three angles of triangle DEC. Therefore, triangles ABC and DEC are similar triangles.

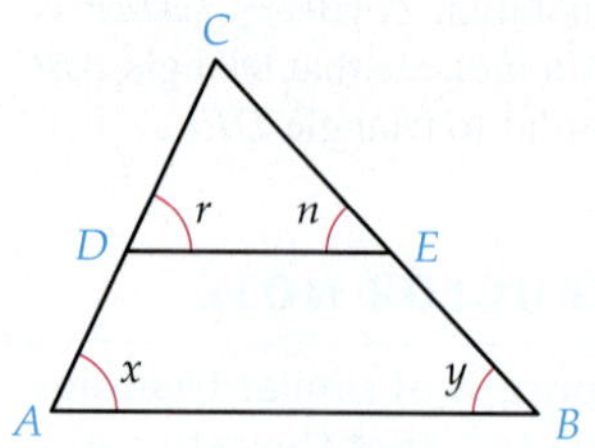

The sum of the measures of the three angles of a triangle is 180°. If two angles of one triangle are equal in measure to two angles of another triangle, then the third angles must be equal. Thus we can say that if two angles of one triangle are equal in measure to two angles of another triangle, then the two triangles are similar.

In the figure at the right, $\overline{AB}$ intersects $\overline{CD}$ at point O. Angles C and D are right angles. Find the length of $\overline{DO}$.

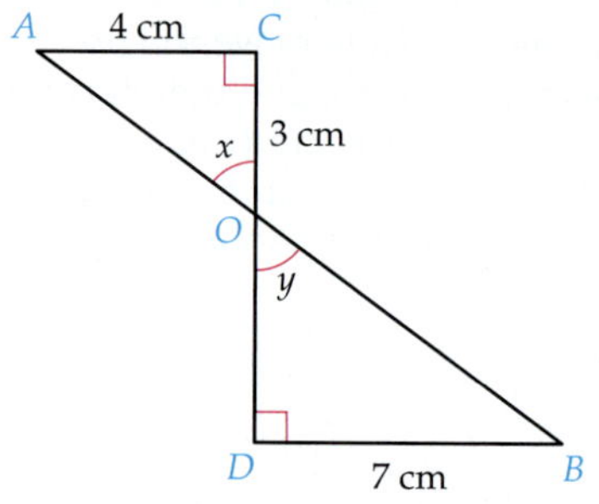

First determine whether triangles AOC and BOD are similar.

$m\angle C = m\angle D$ because they are both right angles.

$m\angle x = m\angle y$ because vertical angles have the same measure.

Because two angles of triangle AOC are equal in measure to two angles of triangle BOD, triangles AOC and BOD are similar.

HISTORICAL NOTE

Many mathematicians have studied similar objects. Thales of Miletus (ca. 624 BC–547 BC) discovered that he could determine the heights of pyramids and other objects by measuring a small object and the length of its shadow and then making use of similar triangles.

Fibonacci (1170–1250) also studied similar objects. The second section of Fibonacci's text *Liber abaci* contains practical information for merchants and surveyors. There is a chapter devoted to using similar triangles to determine the heights of tall objects.

Use a proportion to find the length of the unknown side.

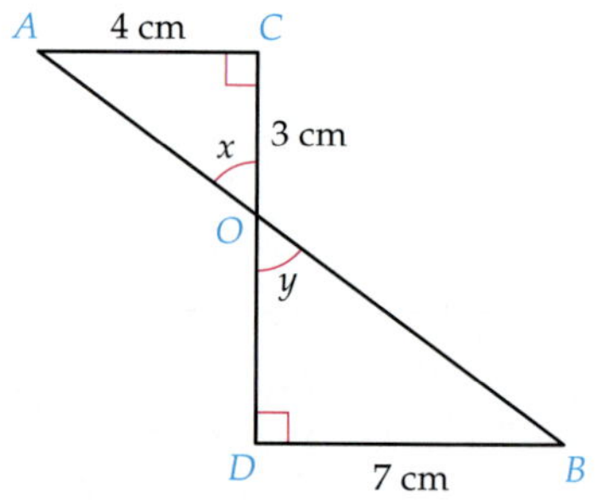

$$\frac{AC}{BD} = \frac{CO}{DO}$$

$$\frac{4}{7} = \frac{3}{DO}$$

$$4(DO) = 7(3)$$

$$4(DO) = 21$$

$$DO = 5.25$$

The length of $\overline{DO}$ is 5.25 cm.

EXAMPLE 2 Solve a Problem Involving Similar Triangles

In the figure at the right, $\angle B$ and $\angle D$ are right angles, $AB = 12$ m, $DC = 4$ m, and $AC = 18$ m. Find the length of $\overline{CO}$.

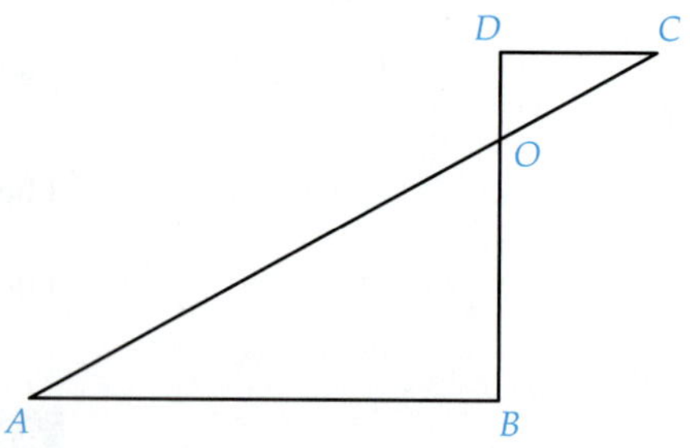

Solution

$\angle B$ and $\angle D$ are right angles. Therefore, $\angle B = \angle D$. $\angle AOB$ and $\angle COD$ are vertical angles. Therefore, $\angle AOB = \angle COD$.

Because two angles of triangle AOB are equal in measure to two angles of triangle COD, triangles AOB and COD are similar triangles.

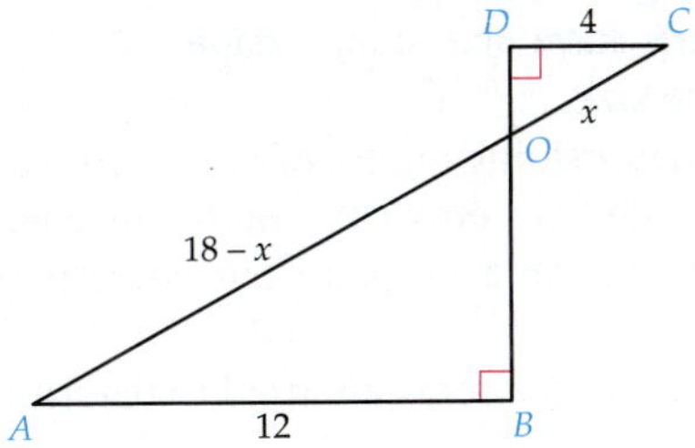

• Label the diagram using the given information. Let x represent CO. $AC = AO + CO$. Because $AC = 18$, $AO = 18 - x$.

$$\frac{DC}{BA} = \frac{CO}{AO}$$

• Triangles AOB and COD are similar triangles. The ratios of corresponding sides are equal.

$$\frac{4}{12} = \frac{x}{18 - x}$$

$$12x = 4(18 - x)$$

$$12x = 72 - 4x$$

• Use the distributive property.

$$16x = 72$$

$$x = 4.5$$

The length of $\overline{CO}$ is 4.5 m.

CHECK YOUR PROGRESS 2

In the figure at the right, $\angle A$ and $\angle D$ are right angles, $AB = 10$ cm, $CD = 4$ cm, and $DO = 3$ cm. Find the area of triangle AOB.

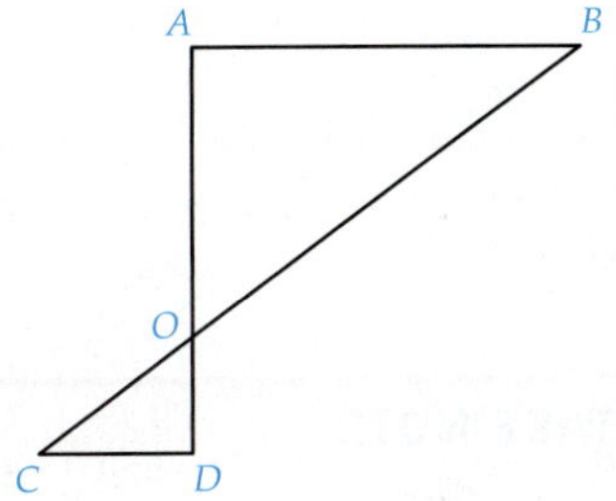

Solution See pages S24–S25.

MATH MATTERS Similar Polygons

INSTRUCTOR NOTE

The Excursion on page 400, titled "Topology: A Brief Introduction," will introduce students to the concept of studying even more basic properties of figures than their sizes and shapes.

For similar triangles, the measures of corresponding angles are equal and the ratios of the lengths of corresponding sides are equal. The same is true for similar polygons.

Quadrilaterals $ABCD$ and $LMNO$ are similar.

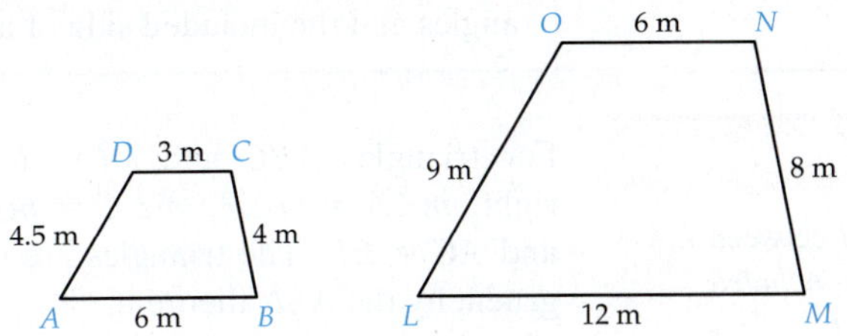

The ratio of the lengths of corresponding sides is: $\frac{AB}{LM} = \frac{6}{12} = \frac{1}{2}$

The ratio of the perimeter of $ABCD$ to the perimeter of $LMNO$ is:

$$\frac{\text{perimeter of } ABCD}{\text{perimeter of } LMNO} = \frac{17.5}{35} = \frac{1}{2}$$

Note that this ratio is the same as the ratio of corresponding sides. This is true for all similar polygons: If two polygons are similar, the ratio of their perimeters is equal to the ratio of the lengths of any pair of corresponding sides.

Congruent Triangles

TAKE NOTE

The notation $\triangle ABC \cong \triangle DEF$ is used to indicate that triangle ABC is congruent to triangle DEF.

The two triangles at the right are **congruent**. They have the same shape and the same size.

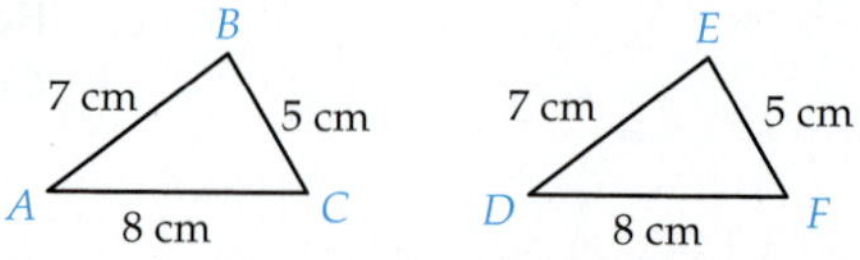

The corresponding angles of congruent triangles have the same measure and the corresponding sides are equal in length. In contrast, for similar triangles, corresponding angles have the same measure but corresponding sides are not necessarily the same length.

Three major theorems are used to determine whether two triangles are congruent.

Side-Side-Side Theorem (SSS)

If the three sides of one triangle are equal in measure to the corresponding three sides of a second triangle, the two triangles are congruent.

In the triangles at the right, $AC = DE$, $AB = EF$, and $BC = DF$. The corresponding sides of triangles ABC and DEF are equal in measure. The triangles are congruent by the SSS theorem.

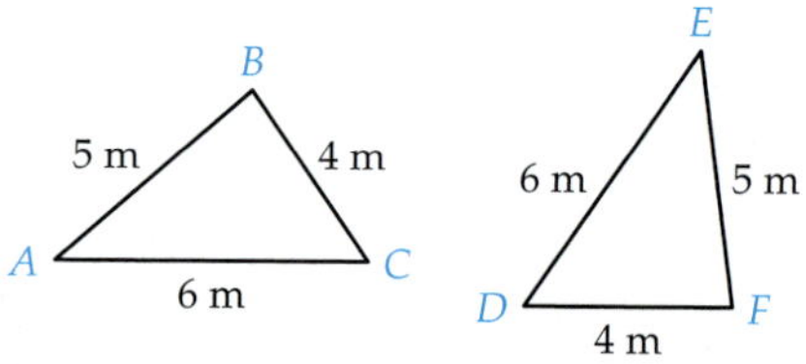

Side-Angle-Side Theorem (SAS)

If two sides and the included angle of one triangle are equal in measure to two sides and the included angle of a second triangle, the two triangles are congruent.

TAKE NOTE

$\angle A$ is *included* between sides $\overline{AB}$ and $\overline{AC}$. $\angle E$ is *included* between sides $\overline{DE}$ and $\overline{EF}$.

In the two triangles at the right, $AB = EF$, $AC = DE$, and $m\angle BAC = m\angle DEF$. The triangles are congruent by the SAS theorem.

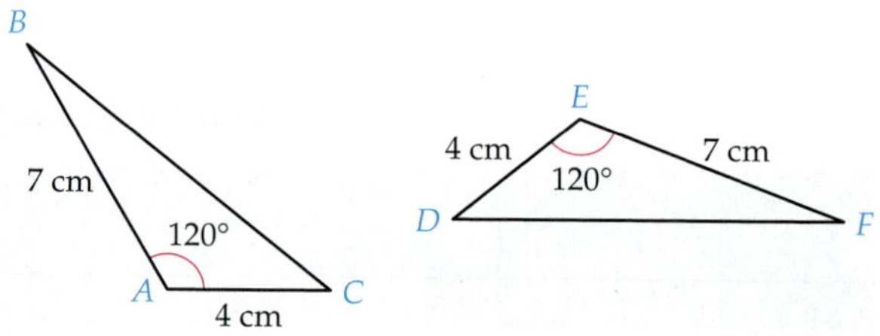

Angle-Side-Angle Theorem (ASA)

If two angles and the included side of one triangle are equal in measure to two angles and the included side of a second triangle, the two triangles are congruent.

TAKE NOTE

Side $\overline{AC}$ is *included* between $\angle A$ and $\angle C$. Side $\overline{EF}$ is *included* between $\angle E$ and $\angle F$.

For triangles ABC and DEF at the right, $m\angle A = m\angle F$, $m\angle C = m\angle E$, and $AC = EF$. The triangles are congruent by the ASA theorem.

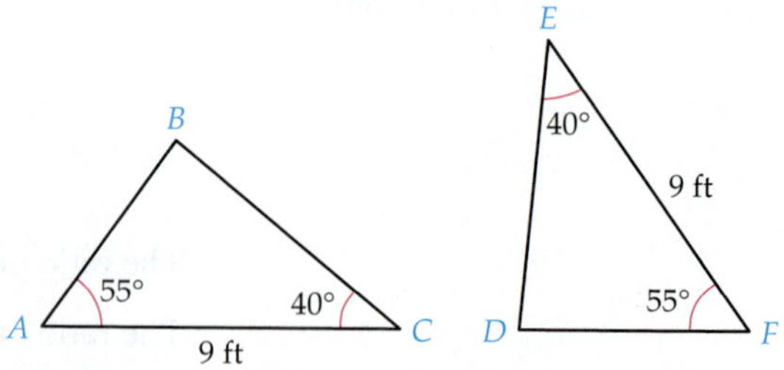

Given triangles PQR and MNO, do the conditions $m\angle P = m\angle O$, $m\angle Q = m\angle M$, and $PQ = MO$ guarantee that triangle PQR is congruent to triangle MNO?

Draw a sketch of the two triangles and determine whether one of the theorems for congruence is satisfied.

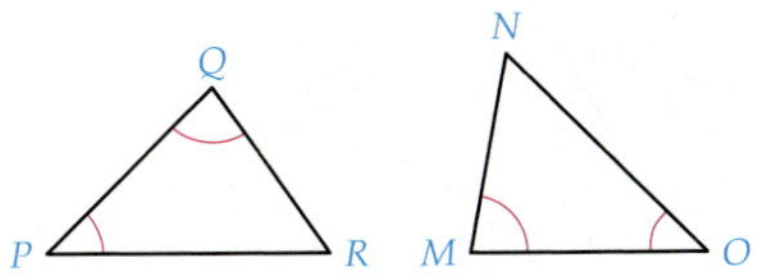

See the figures at the left. Because two angles and the included side of one triangle are equal in measure to two angles and the included side of the second triangle, the triangles are congruent by the ASA theorem.

EXAMPLE 3 Determine Whether Two Triangles Are Congruent

In the figure at the right, is triangle *ABC* congruent to triangle *DEF*?

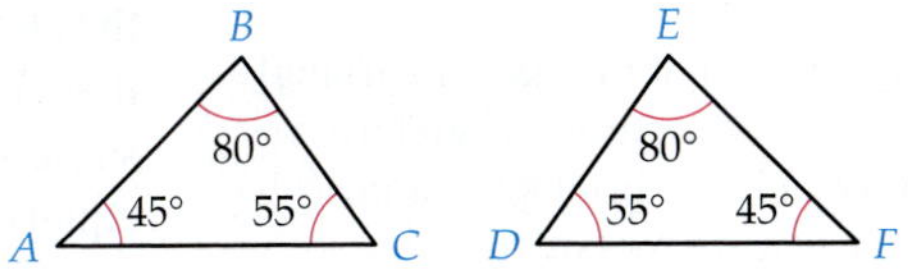

Solution

To determine whether the triangles are congruent, determine whether one of the theorems for congruence is satisfied.

The triangles do not satisfy the SSS theorem, the SAS theorem, or the ASA theorem. The triangles are not necessarily congruent.

CHECK YOUR PROGRESS 3

In the figure at the right, is triangle *PQR* congruent to triangle *MNO*?

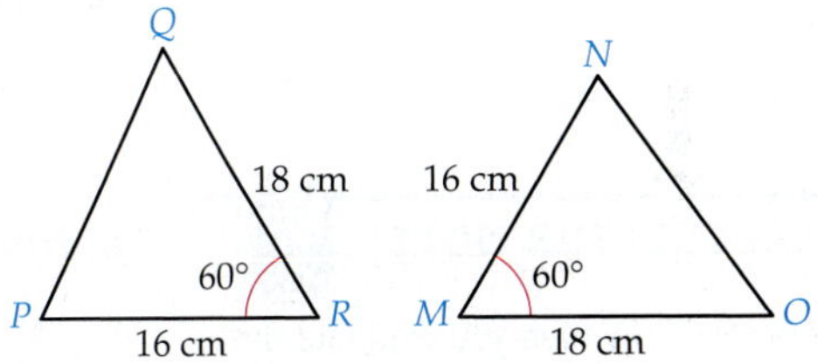

Solution See page S25.

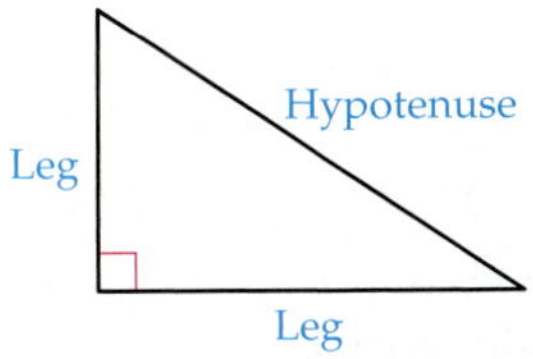

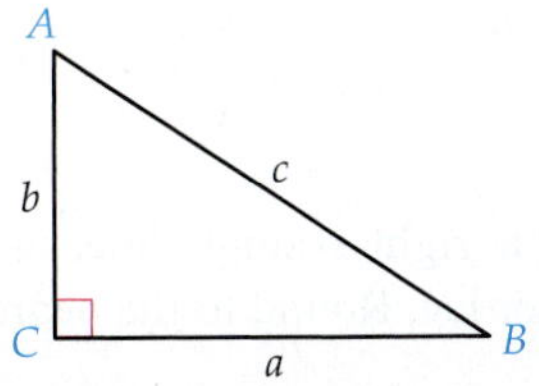

The Pythagorean Theorem

Recall that a right triangle contains one right angle. The side opposite the right angle is called the **hypotenuse**. The other two sides are called **legs**.

The angles in a right triangle are usually labeled with the capital letters *A*, *B*, and *C*, with *C* reserved for the right angle. The side opposite angle *A* is side *a*, the side opposite angle *B* is side *b*, and *c* is the hypotenuse.

The figure at the right is a right triangle with legs measuring 3 units and 4 units and a hypotenuse measuring 5 units. Each side of the triangle is also the side of a square. The number of square units in the area of the largest square is equal to the sum of the numbers of square units in the areas of the smaller squares.

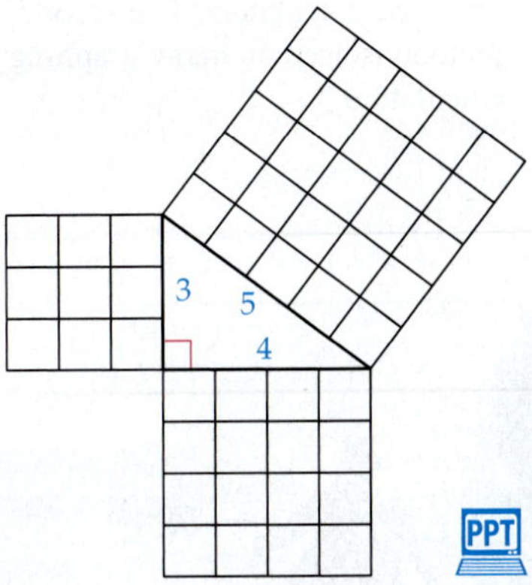

Square of the hypotenuse = Sum of the squares of the two legs

$$5^2 = 3^2 + 4^2$$
$$25 = 9 + 16$$
$$25 = 25$$

The Greek mathematician Pythagoras is generally credited with the discovery that the square of the hypotenuse of a right triangle is equal to the sum of the squares of the two legs. This is called the **Pythagorean theorem**.

The Pythagorean Theorem

If a and b are the lengths of the legs of a right triangle and c is the length of the hypotenuse, then $c^2 = a^2 + b^2$.

POINT OF INTEREST

The first known proof of the Pythagorean theorem is in a Chinese textbook that dates from 150 BC. The book is called *Nine Chapters on the Mathematical Art.* The diagram below is from that book and was used in the proof of the theorem.

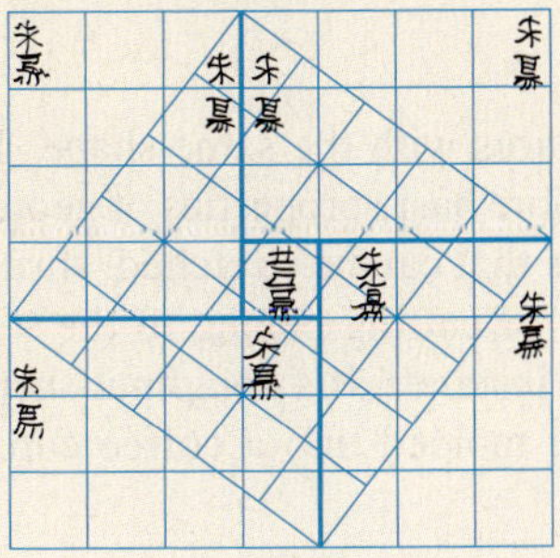

If the lengths of two sides of a right triangle are known, the Pythagorean theorem can be used to find the length of the third side.

Consider a right triangle with legs that measure 5 cm and 12 cm. Use the Pythagorean theorem, with $a = 5$ and $b = 12$, to find the length of the hypotenuse. (If you let $a = 12$ and $b = 5$, the result will be the same.) Take the square root of each side of the equation.

The length of the hypotenuse is 13 cm.

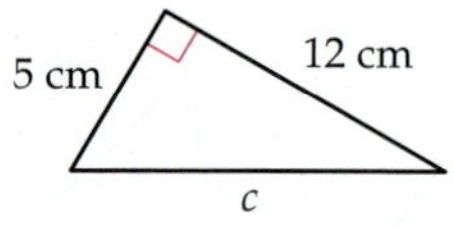

$$c^2 = a^2 + b^2$$
$$c^2 = 5^2 + 12^2$$
$$c^2 = 25 + 144$$
$$c^2 = 169$$
$$\sqrt{c^2} = \sqrt{169}$$
$$c = 13$$

TAKE NOTE

The length of the side of a triangle cannot be negative. Therefore, we take only the principal, or positive, square root of 169.

EXAMPLE 4 Determine the Length of the Unknown Side of a Right Triangle

The length of one leg of a right triangle is 8 in. The length of the hypotenuse is 12 in. Find the length of the other leg. Round to the nearest hundredth of an inch.

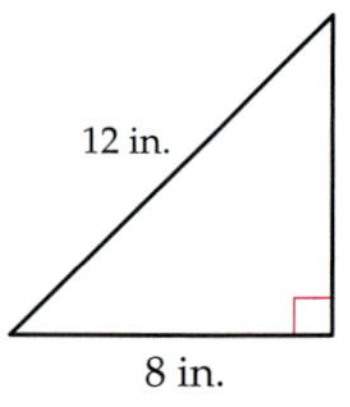

Solution

$a^2 + b^2 = c^2$	• Use the Pythagorean theorem.
$8^2 + b^2 = 12^2$	• $a = 8$, $c = 12$
$64 + b^2 = 144$	
$b^2 = 80$	• Solve for b^2. Subtract 64 from each side.
$\sqrt{b^2} = \sqrt{80}$	• Take the square root of each side of the equation.
$b \approx 8.94$	• Use a calculator to approximate $\sqrt{80}$.

The length of the other leg is approximately 8.94 in.

CALCULATOR NOTE

The way in which you evaluate the square root of a number depends on the type of calculator you have. Here are two possible keystrokes to find $\sqrt{80}$:

80 [√] [=]

or

[√] 80 [ENTER]

The first method is used on many scientific calculators. The second method is used on many graphing calculators.

CHECK YOUR PROGRESS 4 The hypotenuse of a right triangle measures 6 m, and one leg measures 2 m. Find the measure of the other leg. Round to the nearest hundredth of a meter.

Solution See page S25.

EXCURSION

Topology: A Brief Introduction

In this section, we discussed similar figures—that is, figures with the same shape. The branch of geometry called **topology** is the study of even more basic properties of geometric figures than their sizes and shapes. In topology, figures that can be stretched, shrunk, molded, or bent into the same shape without puncturing or cutting belong to the same family. They are said to be **topologically equivalent**. For instance, if a doughnut-shaped figure were made out of modeling clay, then it could be molded into a coffee cup, as shown on the next page.

INSTRUCTOR NOTE

At the website http://en.wikipedia.org/wiki/Topology, there is an award-winning animation of a coffee cup transforming into a doughnut transforming into a coffee cup.

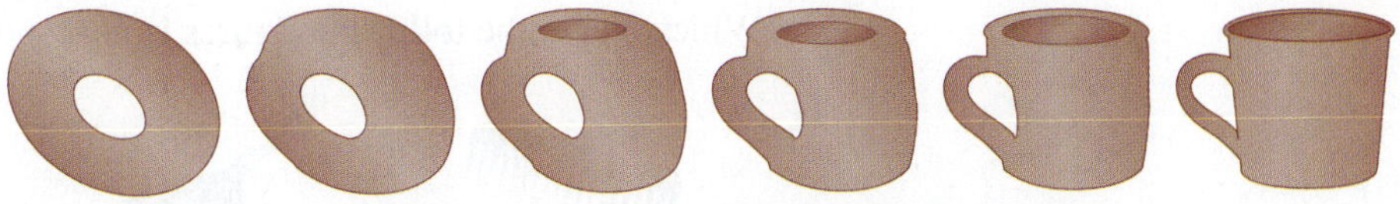

A transformation of a doughnut into a coffee cup

In topology, figures are classified according to their **genus**, where the genus is given by the number of holes in the figure. An inlet in a figure is considered to be a hole if water poured into it passes through the figure. For example, a coffee cup has a hole that is created by its handle; however, the inlet at the top of a coffee cup is not considered to be a hole, because water that is poured into this inlet does not pass through the coffee cup.

Several common geometric figures with genuses 0, 1, 2, and 3 are illustrated below. Figures with the same genus are topogically equivalent.

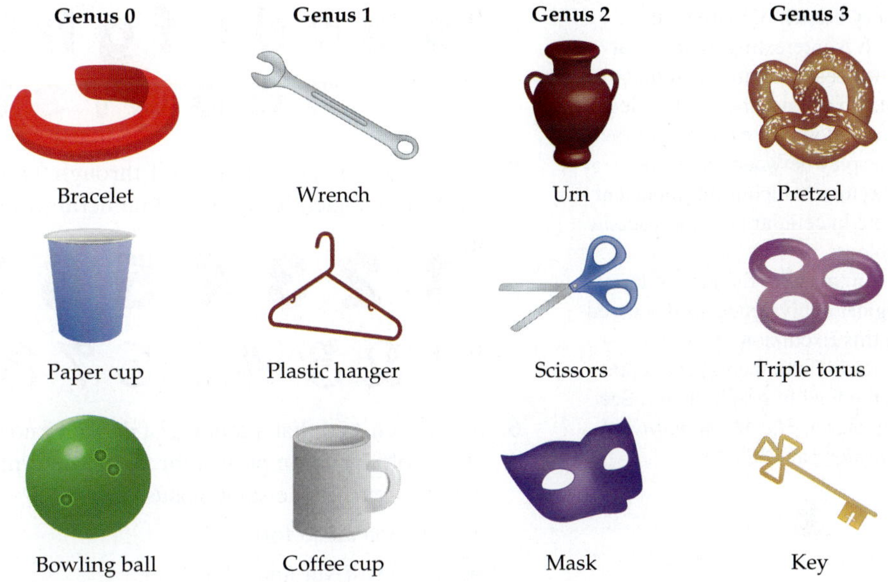

EXCURSION EXERCISES

1. Name the genus of each figure.

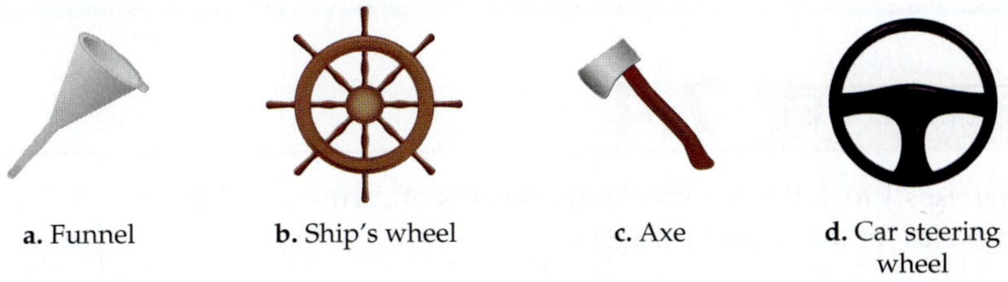

2. Which one of the following figures is not topologically equivalent to the others?

3. Which one of the following figures is not topologically equivalent to the others?

Comb Spatula Block Oar

4. In parts a and b, the letters of the alphabet are displayed using a particular font. List all the topogically equivalent letters according to their genus of 0, 1, or 2.

a. ABCDEFGHIJKLM
NOPQRSTUVWXYZ

b. a b c d e f g h i j k l m
n o p q r s t u v w x y z

5. In parts a and b, the numerals 1 through 9 are displayed using a particular font. List all the topologically equivalent numerals according to their genus of 0, 1, or 2.

a. 1 2 3 4 5 6 7 8 9

b. 1 2 3 4 5 6 7 8 9

6. It has been said that a topologist doesn't know the difference between a doughnut and a coffee cup. In parts a through f, determine whether a topologist would classify the two items in the same genus.

a. A spoon and a fork

b. A screwdriver and a hammer

c. A salt shaker and a sugar bowl

d. A bolt and its nut

e. A slice of American cheese and a slice of Swiss cheese

f. A mixing bowl and a strainer

POINT OF INTEREST

Since its early beginnings in 1736, the subject matter known as topology has expanded from a few novel ideas into a vast body of knowledge with applications in most of the sciences and nearly every branch of mathematics.

It is interesting to note that advances in an area of topology known as knot theory have led to a better understanding of the interplay between DNA and enzymes, which is an important topic in cellular and microscopic biology.

In this text, the idea of topological equivalence, as discussed in this Excursion, is used to develop fundamental concepts in the field of graph theory. See Chapter 5, *The Mathematics of Graphs*, page 229.

EXERCISE SET 7.4

■ In Exercises 1 to 4, find the ratio of the lengths of corresponding sides for the similar triangles.

1.

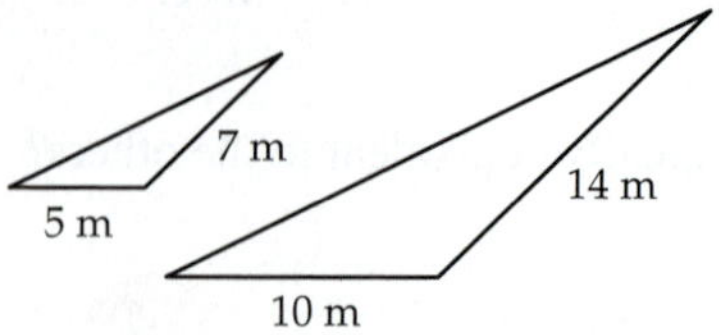

2.

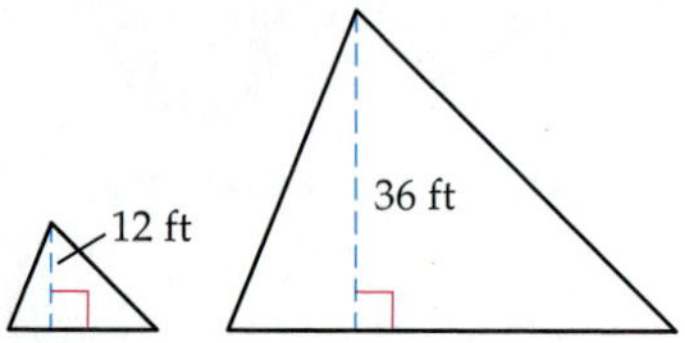

3.

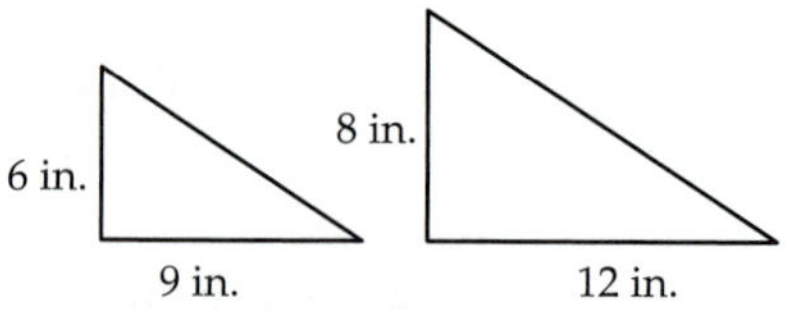

4.

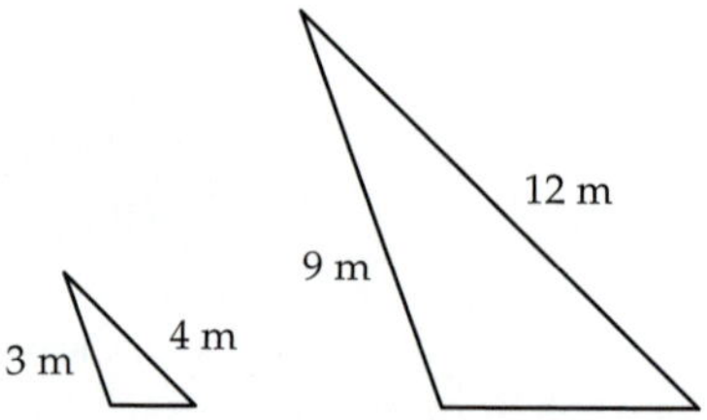

■ In Exercises 5 to 14, triangles *ABC* and *DEF* are similar triangles. Use this fact to solve each exercise. Round to the nearest tenth.

5. Find side *DE*.

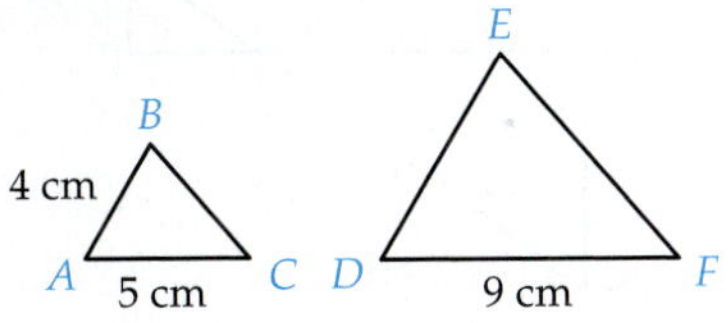

6. Find side *DE*.

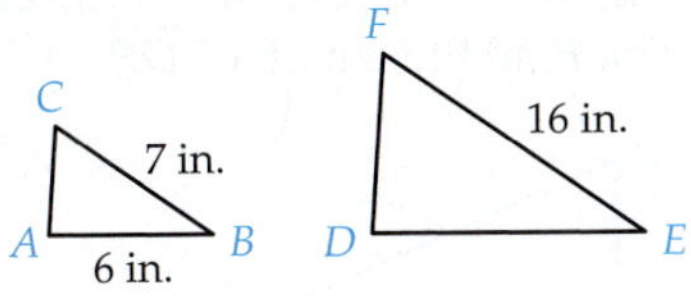

7. Find the height of triangle *DEF*.

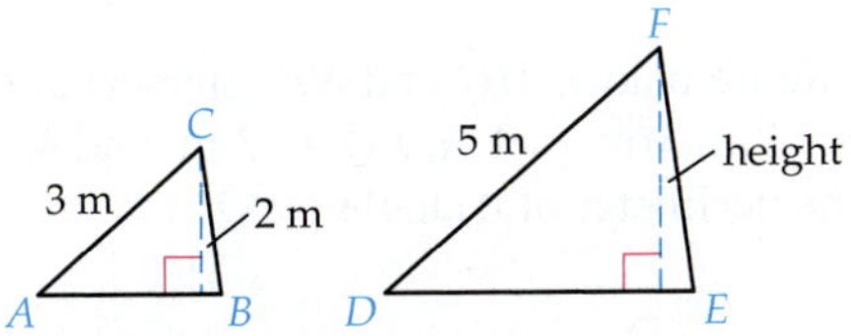

8. Find the height of triangle *ABC*.

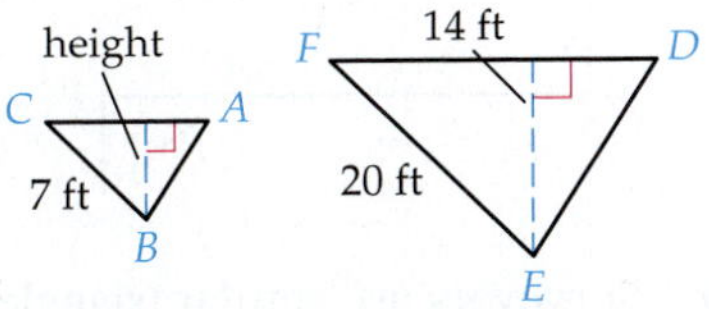

9. Find the perimeter of triangle *ABC*.

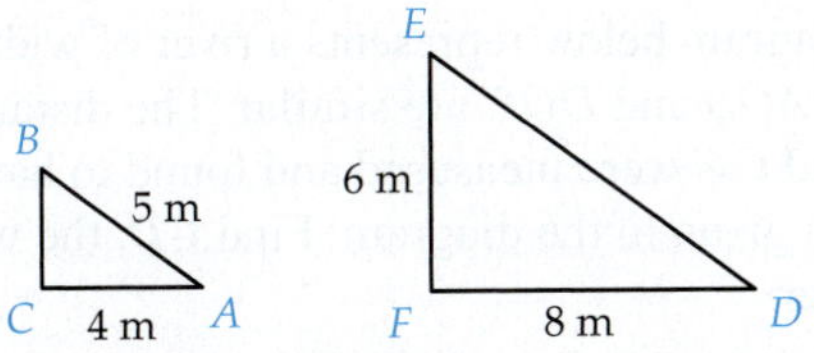

10. Find the perimeter of triangle *DEF*.

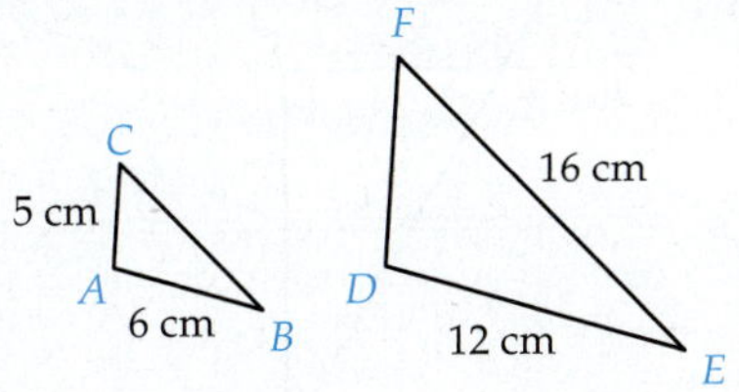

11. Find the perimeter of triangle *ABC*.

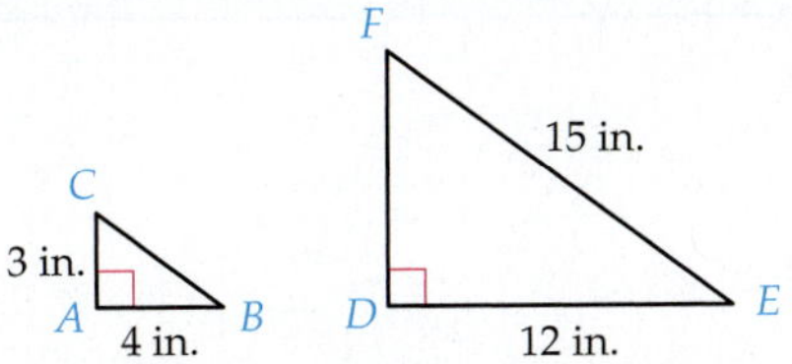

12. Find the area of triangle *DEF*.

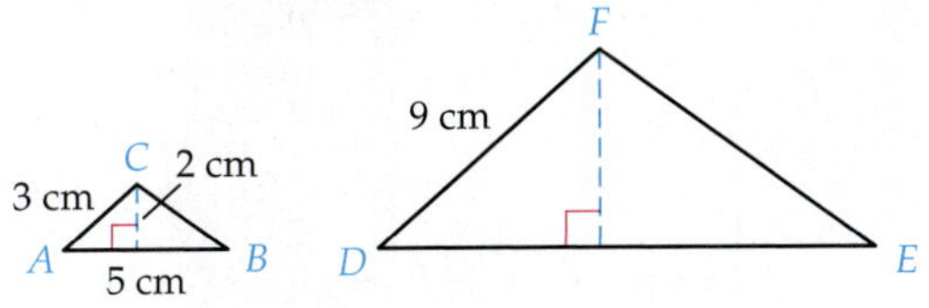

13. Find the area of triangle *ABC*.

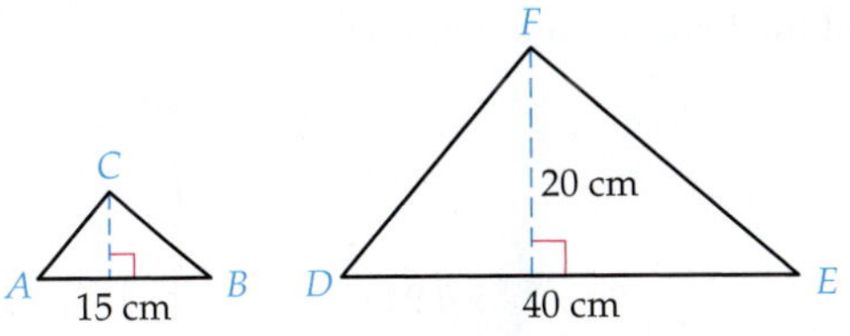

14. Find the area of triangle *DEF*.

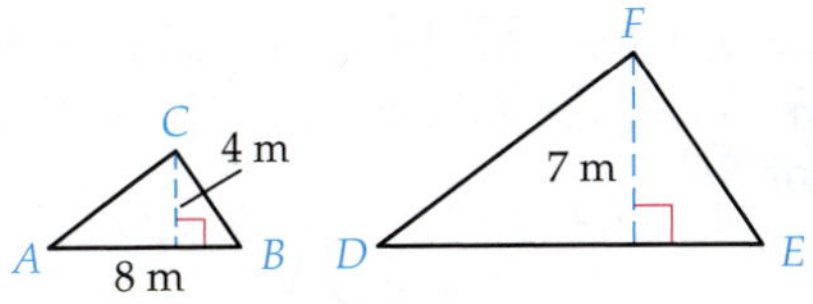

■ In Exercises 15 to 19, the given triangles are similar triangles. Use this fact to solve each exercise.

15. Find the height of the flagpole.

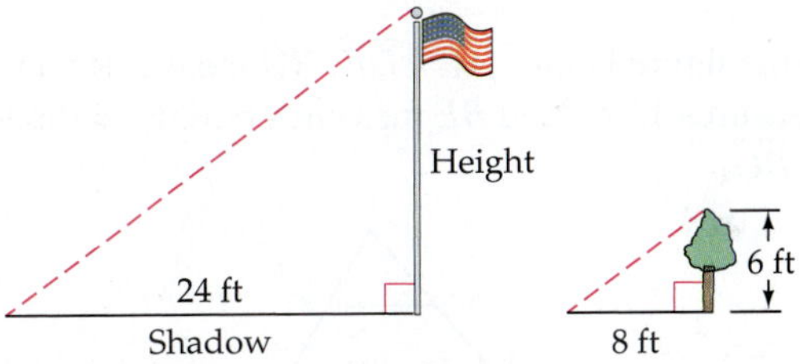

16. Find the height of the flagpole.

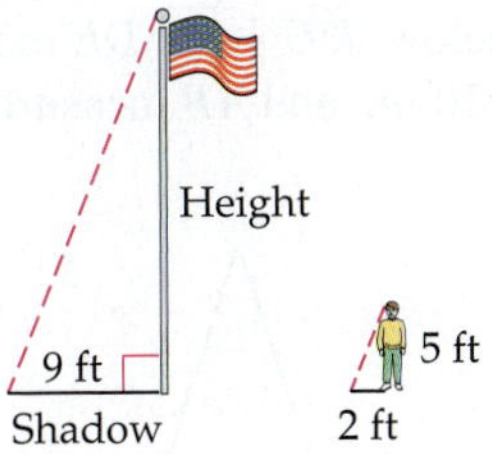

17. Find the height of the building.

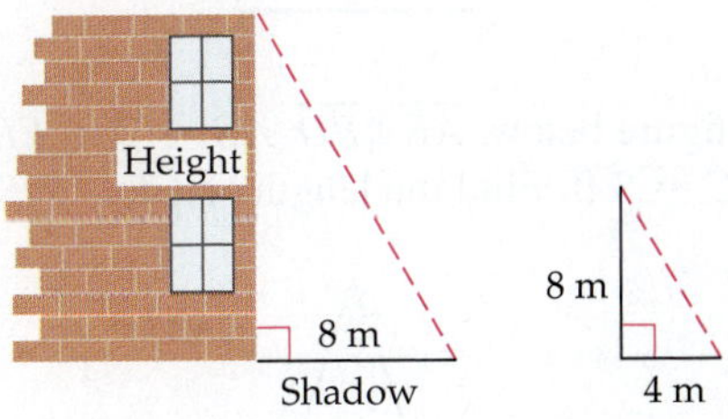

18. Find the height of the building.

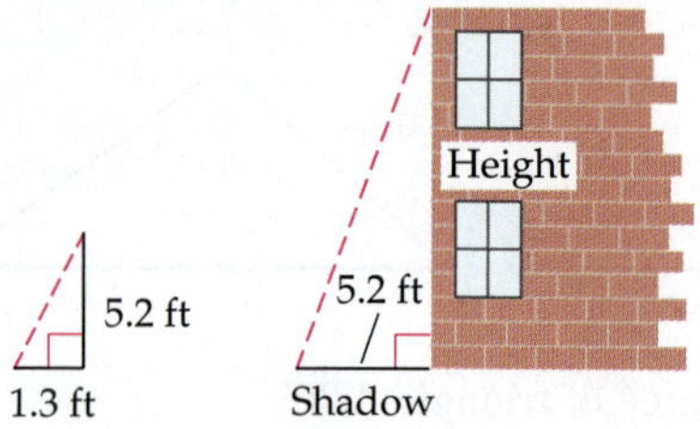

19. Find the height of the flagpole.

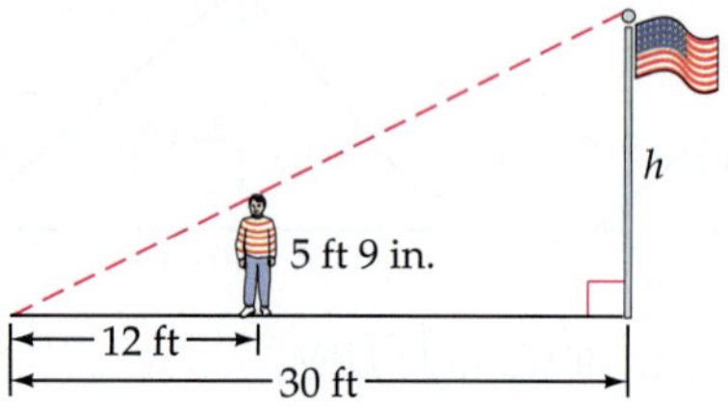

20. In the figure below, $\overline{BD} \parallel \overline{AE}$, BD measures 5 cm, AE measures 8 cm, and AC measures 10 cm. Find the length of $\overline{BC}$.

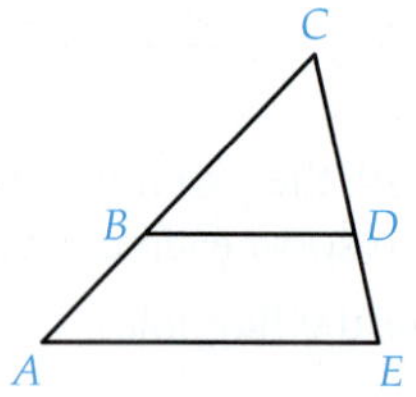

21. In the figure below, $\overline{AC} \parallel \overline{DE}$, BD measures 8 m, AD measures 12 m, and BE measures 6 m. Find the length of $\overline{BC}$.

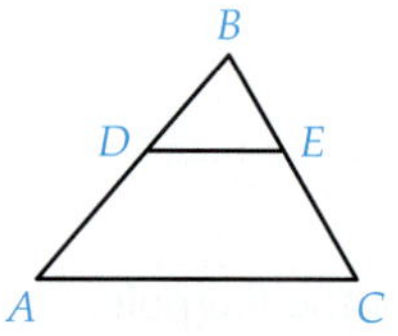

22. In the figure below, $\overline{DE} \parallel \overline{AC}$, DE measures 6 in., AC measures 10 in., and AB measures 15 in. Find the length of $\overline{DA}$.

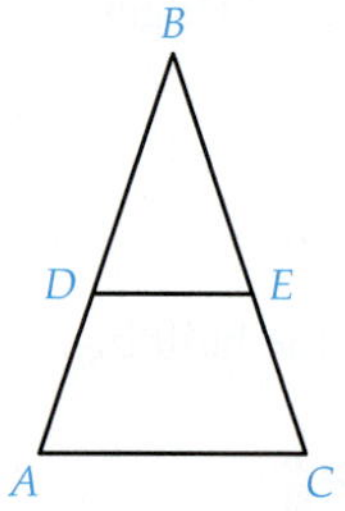

23. In the figure below, $\overline{AE} \parallel \overline{BD}$, $AB = 3$ ft, $ED = 4$ ft, and $BC = 3$ ft. Find the length of $\overline{CE}$.

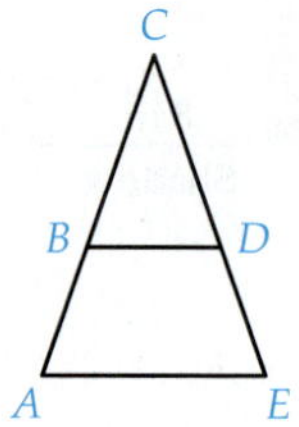

24. In the figure below, $\overline{MP}$ and $\overline{NQ}$ intersect at O, $NO = 25$ ft, $MO = 20$ ft, and $PO = 8$ ft. Find the length of $\overline{QO}$.

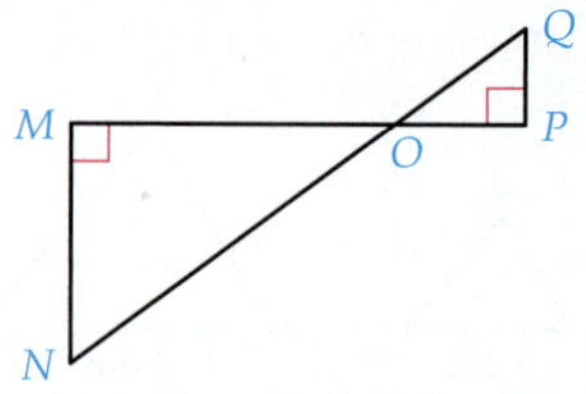

25. In the figure below, $\overline{MP}$ and $\overline{NQ}$ intersect at O, $NO = 24$ cm, $MN = 10$ cm, $MP = 39$ cm, and $QO = 12$ cm. Find the length of $\overline{OP}$.

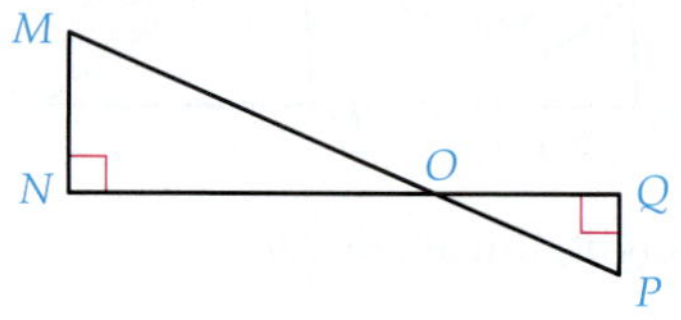

26. In the figure below, $\overline{MQ}$ and $\overline{NP}$ intersect at O, $NO = 12$ m, $MN = 9$ m, $PQ = 3$ m, and $MQ = 20$ m. Find the perimeter of triangle OPQ.

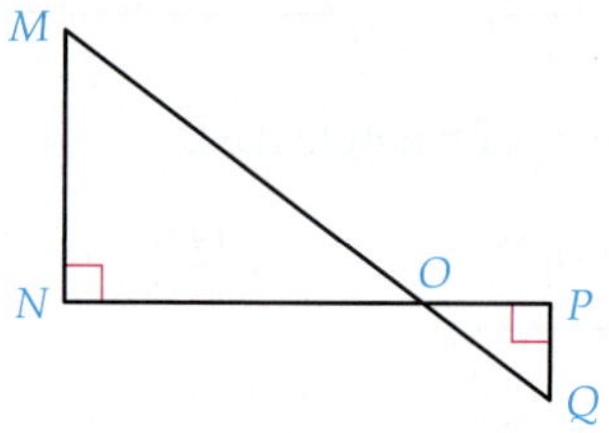

■ **Surveying** Surveyors use similar triangles to measure distances that cannot be measured directly. This is illustrated in Exercises 27 and 28.

27. The diagram below represents a river of width CD. Triangles AOB and DOC are similar. The distances AB, BO, and OC were measured and found to have the lengths given in the diagram. Find CD, the width of the river.

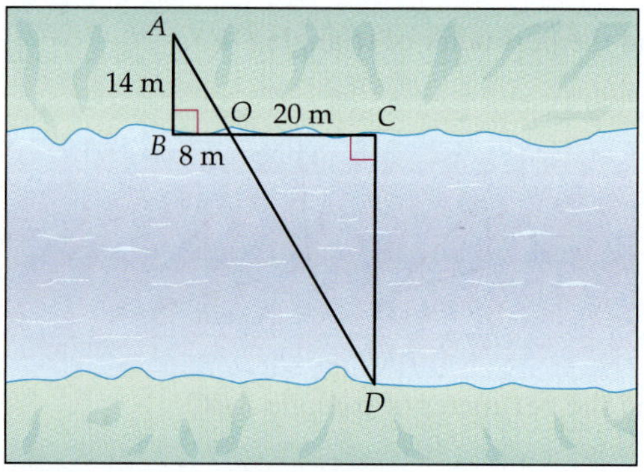

PPT

28. The diagram below shows how surveyors laid out similar triangles along the Winnepaugo River. Find the width, d, of the river.

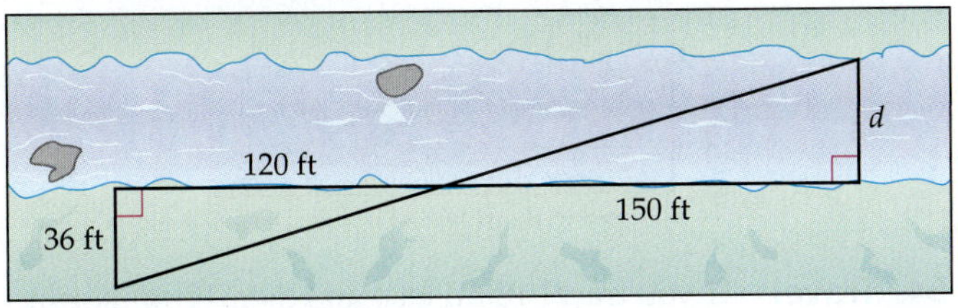

PPT

■ In Exercises 29 to 36, determine whether the two triangles are congruent. If they are congruent, state by what theorem (SSS, SAS, or ASA) they are congruent.

29.

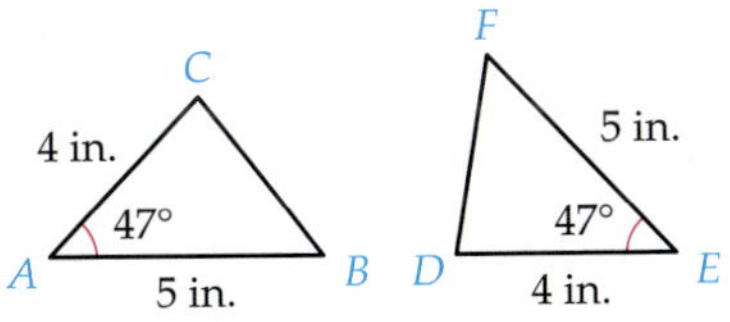

30.

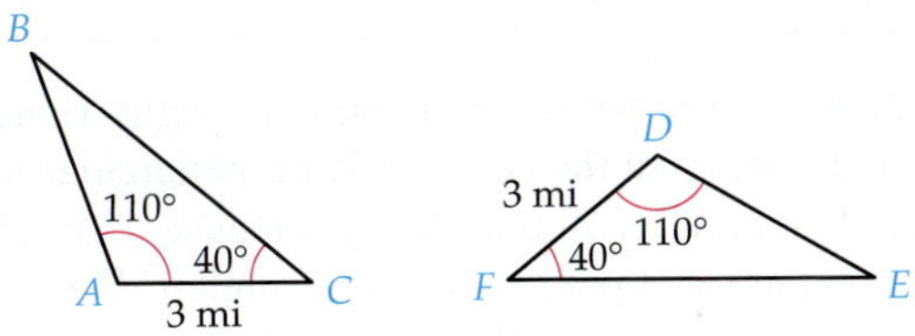

31.

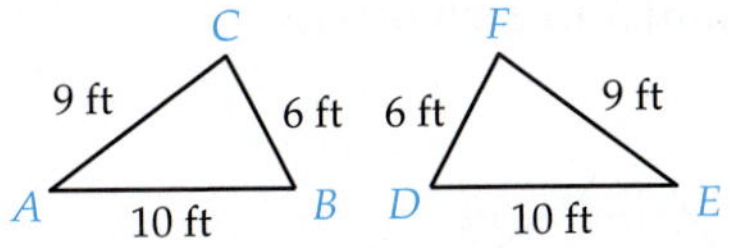

32.

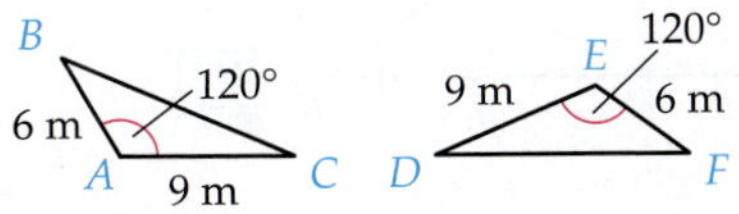

33.

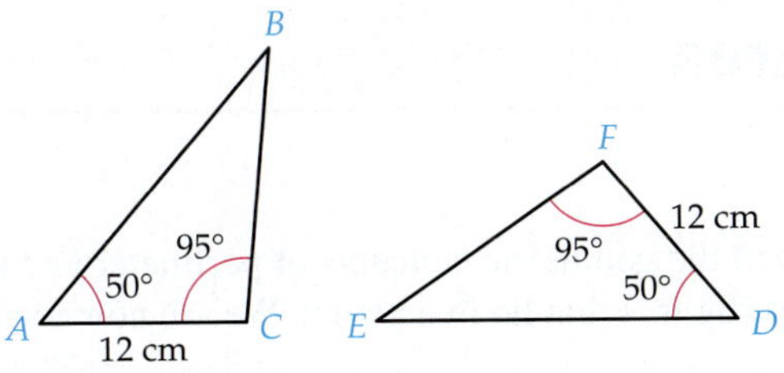

34.

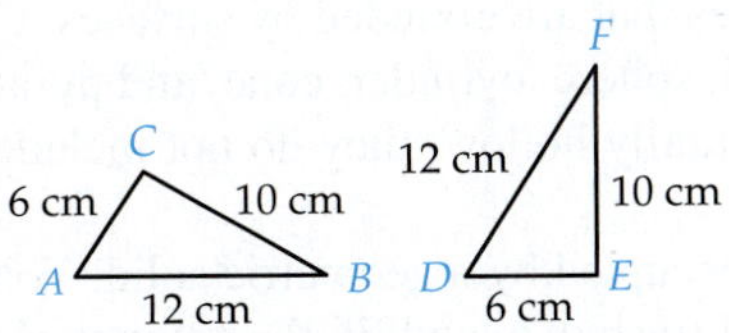

35.

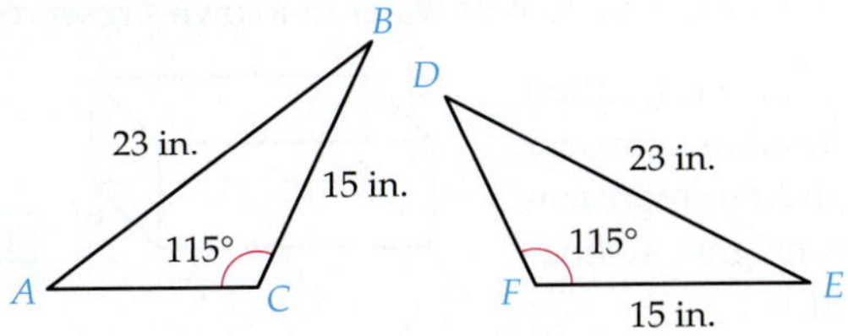

36.

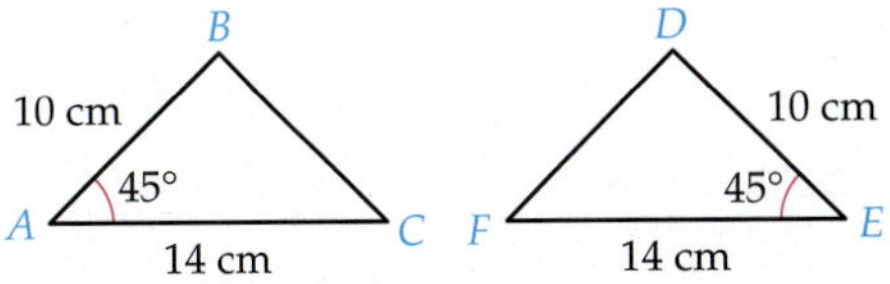

37. Given triangle ABC and triangle DEF, do the conditions $m\angle C = m\angle E$, $AC = EF$, and $BC = DE$ guarantee that triangle ABC is congruent to triangle DEF? If they are congruent, by what theorem are they congruent?

38. Given triangle PQR and triangle MNO, do the conditions $PR = NO$, $PQ = MO$, and $QR = MN$ guarantee that triangle PQR is congruent to triangle MNO? If they are congruent, by what theorem are they congruent?

39. Given triangle LMN and triangle QRS, do the conditions $m\angle M = m\angle S$, $m\angle N = m\angle Q$, and $m\angle L = m\angle R$ guarantee that triangle LMN is congruent to triangle QRS? If they are congruent, by what theorem are they congruent?

40. Given triangle DEF and triangle JKL, do the conditions $m\angle D = m\angle K$, $m\angle E = m\angle L$, and $DE = KL$ guarantee that triangle DEF is congruent to triangle JKL? If they are congruent, by what theorem are they congruent?

41. Given triangle ABC and triangle PQR, do the conditions $m\angle B = m\angle P$, $BC = PQ$, and $AC = QR$ guarantee that triangle ABC is congruent to triangle PQR? If they are congruent, by what theorem are they congruent?

42. True or false? If the ratio of the corresponding sides of two similar triangles is 1 to 1, then the two triangles are congruent.

■ In Exercises 43 to 51, find the length of the unknown side of the triangle. Round to the nearest tenth.

43.

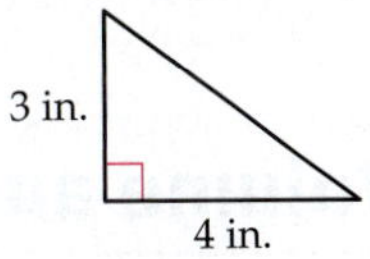

44.

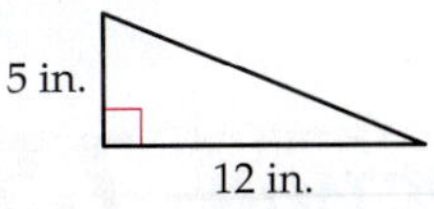

45.

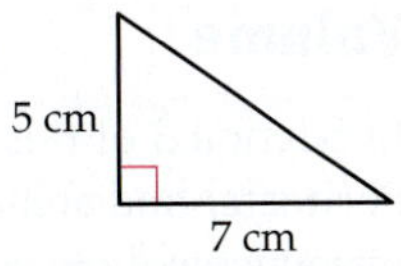

46.

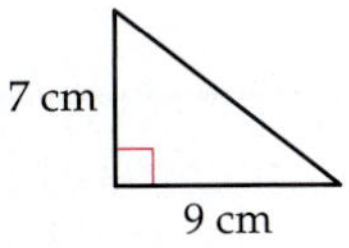

47.

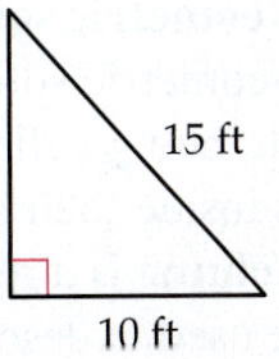

48.

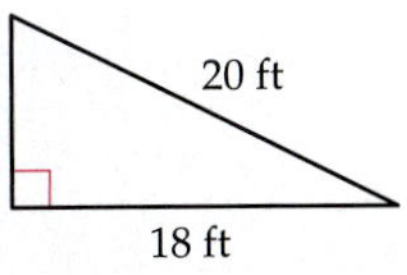

49.

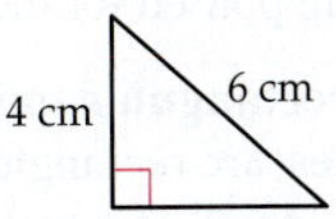

50.

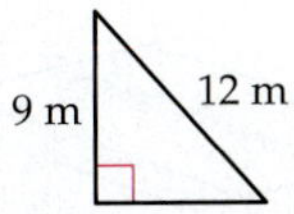

51.

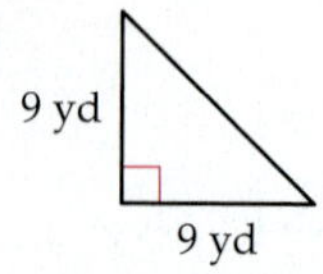

■ In Exercises 52 to 56, use the given information to solve each exercise. Round to the nearest tenth.

52. Home Maintenance A ladder 8 m long is leaning against a building. How high on the building will the ladder reach when the bottom of the ladder is 3 m from the building?

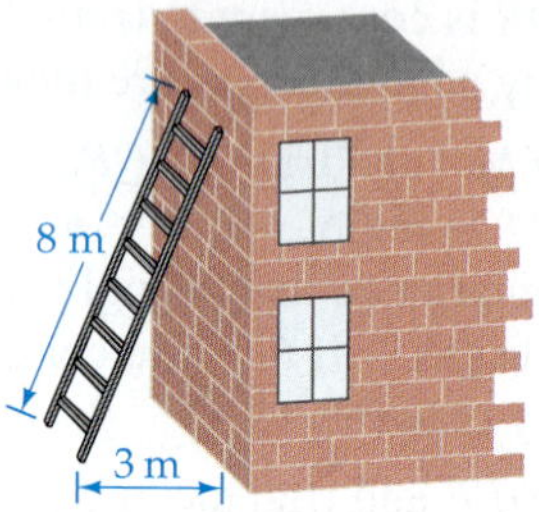

53. Mechanics Find the distance between the centers of the holes in the metal plate.

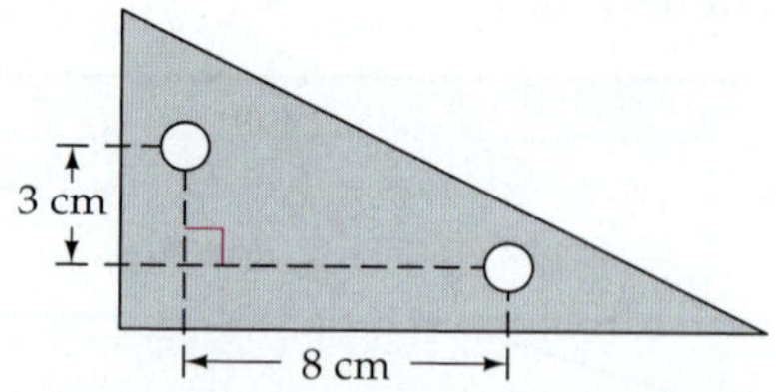

54. Travel If you travel 18 mi east and then 12 mi north, how far are you from your starting point?

55. Perimeter Find the perimeter of a right triangle with legs that measure 5 cm and 9 cm.

56. Perimeter Find the perimeter of a right triangle with legs that measure 6 in. and 8 in.

EXTENSIONS

57. Determine whether the statement is always true, sometimes true, or never true.

a. If two angles of one triangle are equal to two angles of a second triangle, then the triangles are similar triangles.

b. Two isosceles triangles are similar triangles.

c. Two equilateral triangles are similar triangles.

d. If an acute angle of a right triangle is equal to an acute angle of another right triangle, then the triangles are similar triangles.

58. In the figure below, the height of a right triangle is drawn from the right angle perpendicular to the hypotenuse. (Recall that the hypotenuse of a right triangle is the side opposite the right angle.) Verify that the two smaller triangles formed are similar to the original triangle and similar to each other.

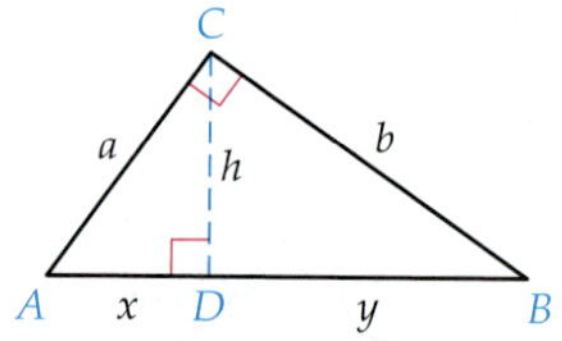

PPT

SECTION 7.5 Volume and Surface Area

Volume

In Section 3 of this chapter, we developed the geometric concepts of perimeter and area. Perimeter and area refer to plane figures (figures that lie in a plane). We are now ready to introduce *volume* of geometric solids.

Geometric solids are three-dimensional shapes that are bounded by surfaces. Common geometric solids include the rectangular solid, sphere, cylinder, cone, and pyramid. Despite being called "solids," these figures are actually hollow; they do not include the points inside their surfaces.

Volume is a measure of the amount of space occupied by a geometric solid. Volume can be used to describe, for example, the amount of trash in a landfill, the amount of concrete poured for the foundation of a house, or the amount of water in a town's reservoir.

A **rectangular solid** is one in which all six sides, called **faces**, are rectangles. The variable L is used to represent the length of a rectangular solid, W is used to represent its width, and H is used to represent its height. A shoebox is an example of a rectangular solid.

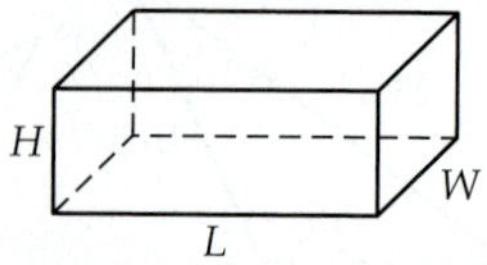

PPT

A **cube** is a special type of rectangular solid. Each of the six faces of a cube is a square. The variable s is used to represent the length of one side of a cube. A baby's block is an example of a cube.

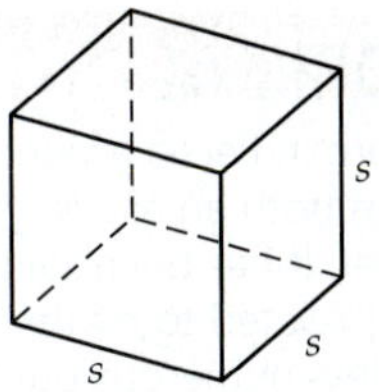

A cube that is 1 ft on each side has a volume of 1 cubic foot, which is written 1 ft^3. A cube that measures 1 cm on each side has a volume of 1 cubic centimeter, written 1 cm^3.

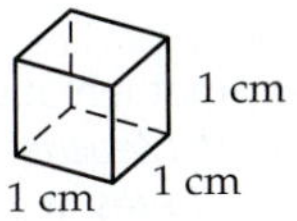

The volume of a solid is the number of cubes, each of volume 1 cubic unit, that are necessary to exactly fill the solid. The volume of the rectangular solid at the right is 24 cm^3 because it will hold exactly 24 cubes, each 1 cm on a side. Note that the volume can be found by multiplying the length times the width times the height.

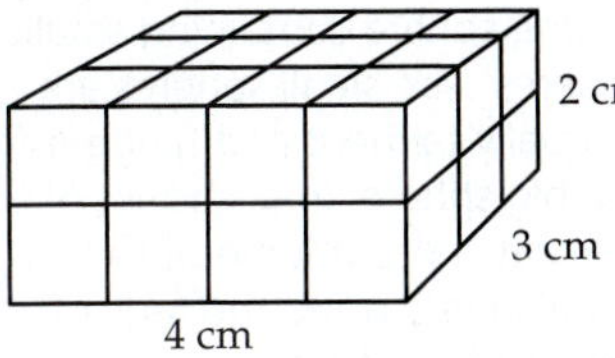

$4 \cdot 3 \cdot 2 = 24$

The volume of the solid is 24 cm^3.

Volume of a Rectangular Solid

The volume, V, of a rectangular solid with length L, width W, and height H is given by $V = LWH$.

Volume of a Cube

The volume, V, of a cube with side of length s is given by $V = s^3$.

QUESTION Which of the following are rectangular solids: juice box, baseball, can of soup, compact disc, or jewel box (plastic container) that a compact disc is packaged in?

A **sphere** is a solid in which all points are the same distance from a point O, called the **center** of the sphere. A **diameter** of a sphere is a line segment with endpoints on the sphere and passing through the center. A **radius** is a line segment from the center to a point on the sphere. $\overline{AB}$ is a diameter and $\overline{OC}$ is a radius of the sphere shown at the right. A basketball is an example of a sphere.

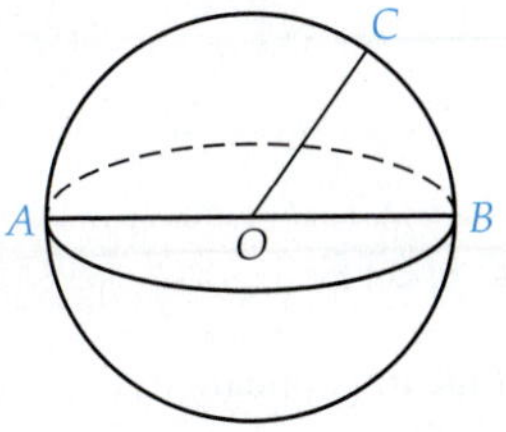

If we let d represent the length of a diameter and r represent the length of a radius, then $d = 2r$ or $r = \frac{1}{2}d$.

$$d = 2r \quad \text{or} \quad r = \frac{1}{2}d$$

Volume of a Sphere

The volume, V, of a sphere with radius of length r is given by $V = \frac{4}{3}\pi r^3$.

ANSWER A juice box and a jewel box are rectangular solids.

HISTORICAL NOTE

Al-Sijzi was born in Persia around AD 945. He was both an astrologer and a geometer. It has been said that his papers related to mathematics were less numerous but more significant than those related to astrology.

One of Al-Sijzi's mathematical works was *Book of the Measurement of Spheres by Spheres.* It contains 12 theorems regarding a large sphere containing smaller spheres. The small spheres are mutually tangent and tangent to the big sphere. In his book, Al-Sijzi calculates the volume of the large sphere using the radii of the smaller spheres.

Find the volume of a rubber ball that has a diameter of 6 in.

First find the length of a radius of the sphere. $r = \frac{1}{2}d = \frac{1}{2}(6) = 3$

Use the formula for the volume of a sphere. $V = \frac{4}{3}\pi r^3$

Replace r with 3. $V = \frac{4}{3}\pi(3)^3$

$V = \frac{4}{3}\pi(27)$

The exact volume of the rubber ball is 36π in^3. $V = 36\pi$

An approximate measure can be found by using the π key on a calculator. $V \approx 113.10$

The volume of the rubber ball is approximately 113.10 in^3.

CALCULATOR NOTE

To approximate 36π on a scientific calculator, enter

36 [×] [π] [=]

The most common cylinder, called a **right circular cylinder**, is one in which the bases are circles and are perpendicular to the height of the cylinder. The variable r is used to represent the length of the radius of a base of a cylinder, and h represents the height of the cylinder. In this text, only right circular cylinders are discussed.

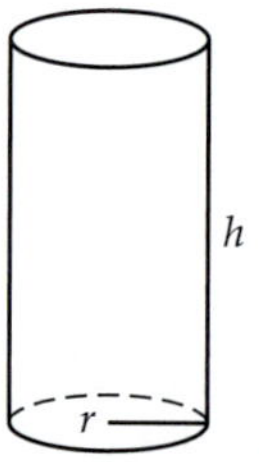

Volume of a Right Circular Cylinder

The volume, V, of a right circular cylinder is given by $V = \pi r^2 h$, where r is the radius of the base and h is the height of the cylinder.

A **right circular cone** is obtained when one base of a right circular cylinder is shrunk to a point, called the **vertex**, V. The variable r is used to represent the radius of the base of the cone, and h represents the height of the cone. The variable l is used to represent the **slant height**, which is the distance from a point on the circumference of the base to the vertex. In this text, only right circular cones are discussed. An ice cream cone is an example of a right circular cone.

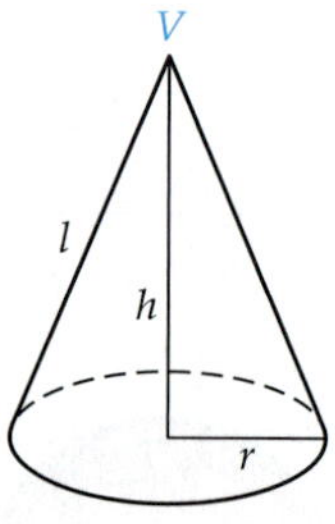

TAKE NOTE

Note that πr^2 appears in the formula for the volume of a right circular cylinder and in the formula for the volume of a right circular cone. This is because, in each case, the base of the figure is a circle.

Volume of a Right Circular Cone

The volume, V, of a right circular cone is given by $V = \frac{1}{3}\pi r^2 h$, where r is the length of a radius of the circular base and h is the height of the cone.

The base of a **regular pyramid** is a regular polygon, and the sides are isosceles triangles (two sides of the triangle are the same length). The height, h, is the distance from the vertex, V, to the base and is perpendicular to the base. The variable l is used to represent the **slant height**, which is the height of one of the isosceles triangles on the face of the pyramid. The regular square pyramid at the right has a square base. This is the only type of pyramid discussed in this text. Many Egyptian pyramids are regular square pyramids.

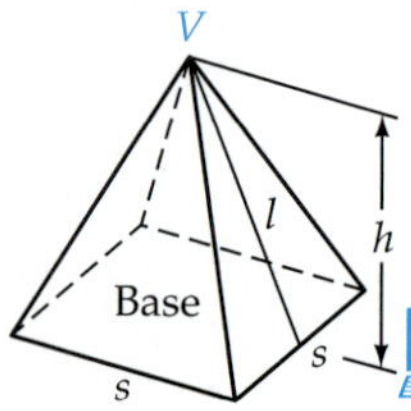

Dudarev Mikhail/Shutterstock.com

Pyramid at Giza

INSTRUCTOR NOTE

The difficulty students have distinguishing linear measure from square measure is compounded with volume measure. Ask students to give examples of things that would be measured in, for instance, feet, square feet, and cubic feet—for example, the length of a room, the area of the floor, and the volume of air in the room. Here are two more examples:

a. The distance across a lake, the area of the surface of the lake, and the volume of water in the lake

b. The length of a driveway, the area of the driveway that needs to be resealed, and the volume of asphalt used to pave the driveway

POINT OF INTEREST

A few years ago, astronomers identified a trio of supergiant stars, which were subsequently named KW Sagitarii, V354 Cephei, and KY Cygni. All three have diameters of more than 1 billion miles, or 1500 times the diameter of our sun. If one of these stars were placed in the same location as our sun, it would not only engulf Earth, but its outer boundary would extend to a point between the orbits of Jupiter and Saturn.

Volume of a Regular Square Pyramid

The volume, V, of a regular square pyramid is given by $V = \frac{1}{3}s^2h$, where s is the length of a side of the base and h is the height of the pyramid.

QUESTION Which of the following units could be used to measure the volume of a regular square pyramid?

a. ft^3 **b.** m^3 **c.** yd^2 **d.** cm^3 **e.** mi

EXAMPLE 1 Find the Volume of a Geometric Solid

Find the volume of a cube that measures 1.5 m on a side.

Solution

$V = s^3$ • Use the formula for the volume of a cube.

$V = 1.5^3$ • Replace s with 1.5.

$V = 3.375$

The volume of the cube is 3.375 m^3.

CHECK YOUR PROGRESS 1 The length of a rectangular solid is 5 m, the width is 3.2 m, and the height is 4 m. Find the volume of the solid.

Solution See page S25.

EXAMPLE 2 Find the Volume of a Geometric Solid

The radius of the base of a cone is 8 cm. The height of the cone is 12 cm. Find the volume of the cone. Round to the nearest hundredth of a cubic centimeter.

Solution

$V = \frac{1}{3}\pi r^2 h$ • Use the formula for the volume of a cone.

$V = \frac{1}{3}\pi(8)^2(12)$ • Replace r with 8 and h with 12.

$V = \frac{1}{3}\pi(64)(12)$

$V = 256\pi$ • Exact volume

$V \approx 804.25$ • Use the π key on a calculator.

The volume of the cone is approximately 804.25 cm^3.

CHECK YOUR PROGRESS 2 The length of a side of the base of a regular square pyramid is 15 m and the height of the pyramid is 25 m. Find the volume of the pyramid.

Solution See page S25.

ANSWER Volume is measured in cubic units. Therefore, the volume of a regular square pyramid could be measured in ft^3, m^3, or cm^3, but not in yd^2 or mi.

EXAMPLE 3 Find the Volume of a Geometric Solid

An oil storage tank in the shape of a cylinder is 4 m high and has a diameter of 6 m. The oil tank is two-thirds full. Find the number of cubic meters of oil in the tank. Round to the nearest hundredth of a cubic meter.

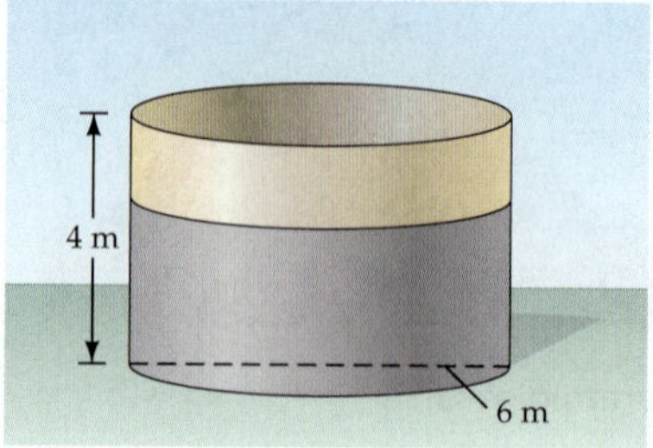

Solution

$r = \frac{1}{2}d = \frac{1}{2}(6) = 3$ • Find the radius of the base.

$V = \pi r^2 h$ • Use the formula for the volume of a cylinder.

$V = \pi(3)^2(4)$ • Replace r with 3 and h with 4.

$V = \pi(9)(4)$

$V = 36\pi$

$\frac{2}{3}(36\pi) = 24\pi$ • Multiply the volume by $\frac{2}{3}$.

≈ 75.40 • Use the π key on a calculator.

There are approximately 75.40 m^3 of oil in the storage tank.

CHECK YOUR PROGRESS 3

A silo in the shape of a cylinder is 16 ft in diameter and has a height of 30 ft. The silo is three-fourths full. Find the volume of the portion of the silo that is not being used for storage. Round to the nearest hundredth of a cubic foot.

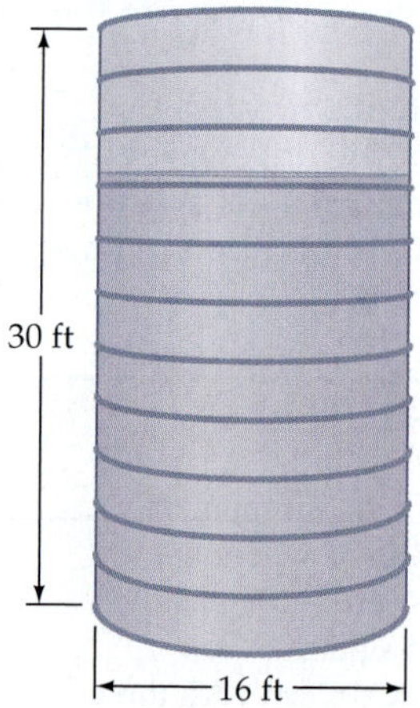

Solution See page S25.

Surface Area

The **surface area** of a solid is the total area on the surface of the solid. Suppose you want to cover a geometric solid with wallpaper. The amount of wallpaper needed is equal to the surface area of the figure.

When a rectangular solid is cut open and flattened out, each face is a rectangle. The surface area, S, of the rectangular solid is the sum of the areas of the six rectangles:

$$S = LW + LH + WH + LW + WH + LH$$

which simplifies to

$$S = 2LW + 2LH + 2WH$$

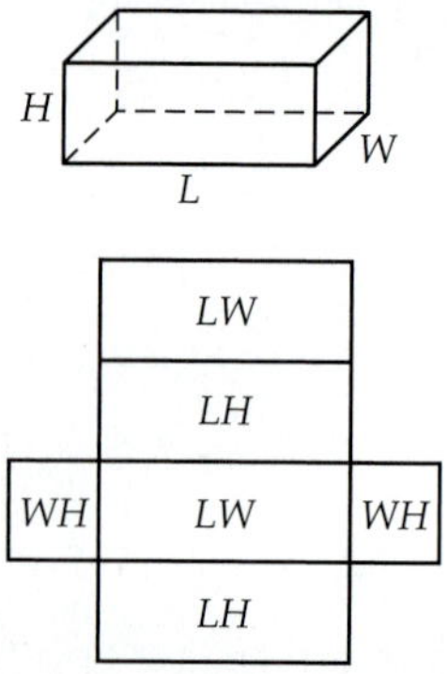

If the rectangular solid is a cube, then all three sides L, W, and H are equal. Therefore, each side can be represented by s. The surface area of a cube is

$$S = 2LW + 2LH + 2WH$$
$$S = 2 \cdot s \cdot s + 2 \cdot s \cdot s + 2 \cdot s \cdot s = 2s^2 + 2s^2 + 2s^2$$
$$S = 6s^2$$

HISTORICAL NOTE

Pappus (păp′ŭs) of Alexandria (ca. 290–350) was born in Egypt. We know the time period in which he lived because he wrote about his observation of the eclipse of the sun that took place in Alexandria on October 18, 320.

Pappus has been called the last of the great Greek geometers. His major work in geometry is *Synagoge*, or the *Mathematical Collection*. It consists of eight books. In Book V, Pappus proves that the sphere has a greater volume than any regular geometric solid with equal surface area. He also proves that if two regular solids have equal surface areas, the solid with the greater number of faces has the greater volume.

When a cylinder is cut open and flattened out, the top and bottom of the cylinder are circles. The side of the cylinder flattens out to a rectangle. The length of the rectangle is the circumference of the base, which is $2\pi r$; the width is h, the height of the cylinder. Therefore, the area of the rectangle is $2\pi rh$. The surface area, S, of the cylinder is

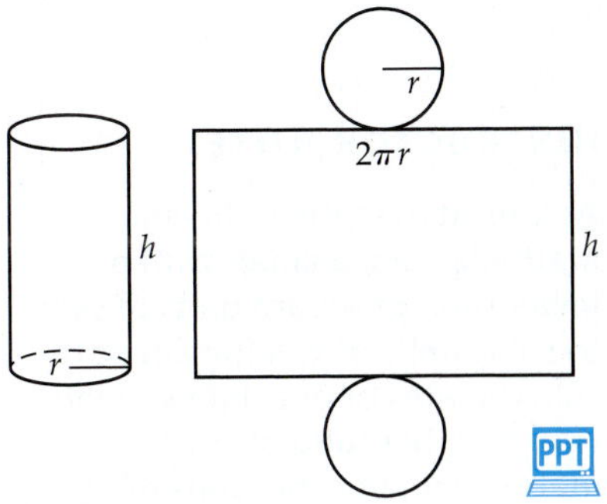

$$S = \pi r^2 + 2\pi rh + \pi r^2$$

which simplifies to

$$S = 2\pi r^2 + 2\pi rh$$

The surface area of a regular square pyramid is the area of the base plus the area of the four isosceles triangles. The length of a side of the square base is s; therefore, the area of the base is s^2. The slant height, l, is the height of each triangle, and s is the length of the base of each triangle. The surface area, S, of a regular square pyramid is

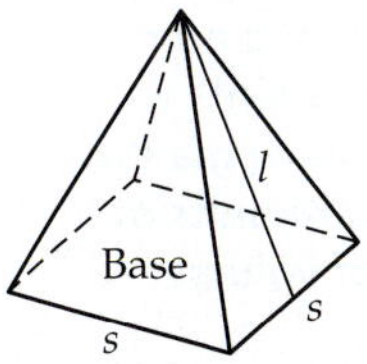

$$S = s^2 + 4\left(\frac{1}{2}sl\right)$$

which simplifies to

$$S = s^2 + 2sl$$

Formulas for the surface areas of geometric solids are given below.

Surface Areas of Geometric Solids

The surface area, S, of a **rectangular solid** with length L, width W, and height H is given by $S = 2LW + 2LH + 2WH$.

The surface area, S, of a **cube** with sides of length s is given by $S = 6s^2$.

The surface area, S, of a **sphere** with radius r is given by $S = 4\pi r^2$.

The surface area, S, of a **right circular cylinder** is given by $S = 2\pi r^2 + 2\pi rh$, where r is the radius of the base and h is the height.

The surface area, S, of a **right circular cone** is given by $S = \pi r^2 + \pi rl$, where r is the radius of the circular base and l is the slant height.

The surface area, S, of a **regular square pyramid** is given by $S = s^2 + 2sl$, where s is the length of a side of the base and l is the slant height.

QUESTION Which of the following units could be used to measure the surface area of a rectangular solid?

a. in^2 **b.** m^3 **c.** cm^2 **d.** ft^3 **e.** yd

ANSWER Surface area is measured in square units. Therefore, the surface area of a rectangular solid could be measured in in^2 or cm^2, but not in m^3, ft^3, or yd.

EXAMPLE 4 **Find the Surface Area of a Geometric Solid**

The diameter of the base of a cone is 5 m and the slant height is 4 m. Find the surface area of the cone. Round to the nearest hundredth of a square meter.

Solution

$r = \frac{1}{2}d = \frac{1}{2}(5) = 2.5$ • Find the radius of the cone.

$S = \pi r^2 + \pi rl$ • Use the formula for the surface area of a cone.

$S = \pi(2.5)^2 + \pi(2.5)(4)$ • Replace r with 2.5 and l with 4.

$S = \pi(6.25) + \pi(2.5)(4)$

$S = 6.25\pi + 10\pi$

$S = 16.25\pi$

$S \approx 51.05$

The surface area of the cone is approximately 51.05 m^2.

CHECK YOUR PROGRESS 4 The diameter of the base of a cylinder is 6 ft and the height is 8 ft. Find the surface area of the cylinder. Round to the nearest hundredth of a square foot.

Solution See page S25.

INSTRUCTOR NOTE

As indicated in the following Math Matters, animals with a lower ratio of square units of skin to cubic units of volume adapt to cold climates better than animals with a higher ratio of square units of skin to cubic units of volume. You might name four or five animals, such as the giraffe, elk, bison, sea otter, and hummingbird, and ask students to categorize them as having a low or high ratio of square units of skin to cubic units of volume.

MATH MATTERS Survival of the Fittest

hallam creations/Shutterstock.com

The ratio of an animal's surface area to the volume of its body is a crucial factor in its survival. The more square units of skin for every cubic unit of volume, the more rapidly the animal loses body heat. Therefore, animals living in a warm climate benefit from a higher ratio of surface area to volume, whereas those living in a cool climate benefit from a lower ratio.

EXCURSION

Water Displacement

A recipe for peanut butter cookies calls for 1 cup of peanut butter. Peanut butter is difficult to measure. If you have ever used a measuring cup to measure peanut butter, you know that there tend to be pockets of air at the bottom of the cup. And trying to scrape all of the peanut butter out of the cup and into the mixing bowl is a challenge.

A more convenient method of measuring 1 cup of peanut butter is to fill a 2-cup measuring cup with 1 cup of water. Then add peanut butter to the water until the water reaches the 2-cup mark. (Make sure all the peanut butter is below the top of the water.) Drain off the water, and the one cup of peanut butter drops easily into the mixing bowl.

POINT OF INTEREST

King Hiero of Syracuse commissioned a new crown from his goldsmith. When the crown was completed, Hiero suspected that the goldsmith had stolen some of the gold and replaced it with lead. Hiero asked Archimedes to determine, without defacing the surface of the crown, whether the crown was pure gold.

Archimedes knew that the first problem he had to solve was how to determine the volume of the crown. Legend has it that one day he was getting into a bathtub that was completely full of water, and the water splashed over the side. He surmised that the volume of water that poured over the side was equal to the volume of his submerged body. Supposedly, he was so excited that he jumped from the tub and went running through the streets yelling "Eureka! Eureka!," meaning "I found it! I found it!"—referring to the solution of the problem of determining the volume of the crown.

Hulton Archive/Getty Images

This method of measuring peanut butter works because when an object sinks below the surface of the water, the object displaces an amount of water that is equal to the volume of the object.

A sphere with a diameter of 4 in. is placed in a rectangular tank of water that is 6 in. long and 5 in. wide. How much does the water level rise? Round to the nearest hundredth of an inch.

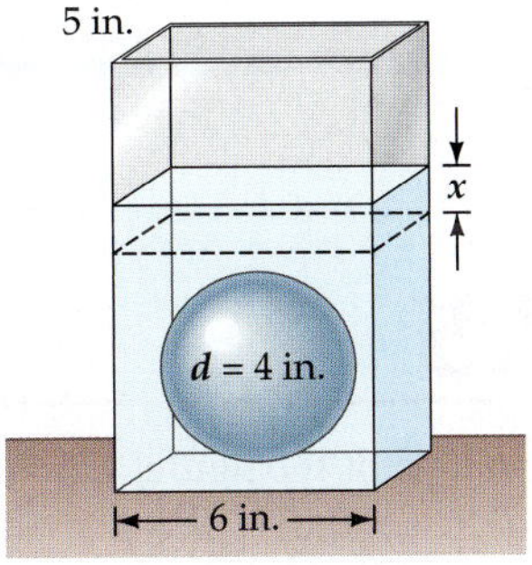

$V = \frac{4}{3}\pi r^3$ • Use the formula for the volume of a sphere.

$V = \frac{4}{3}\pi(2^3) = \frac{32}{3}\pi$ • $r = \frac{1}{2}d = \frac{1}{2}(4) = 2$

Let x represent the amount of the rise in water level. The volume of the sphere will equal the volume of the water displaced. As shown above, this volume is the volume of a rectangular solid with width 5 in., length 6 in., and height x in.

$V = LWH$ • Use the formula for the volume of a rectangular solid.

$\frac{32}{3}\pi = (6)(5)x$ • Substitute $\frac{32}{3}\pi$ for V, 6 for L, 5 for W, and x for H.

$\frac{32}{90}\pi = x$ • The exact height that the water will rise is $\frac{32}{90}\pi$.

$1.12 \approx x$ • Use a calculator to find an approximation.

The water will rise approximately 1.12 in.

EXCURSION EXERCISES

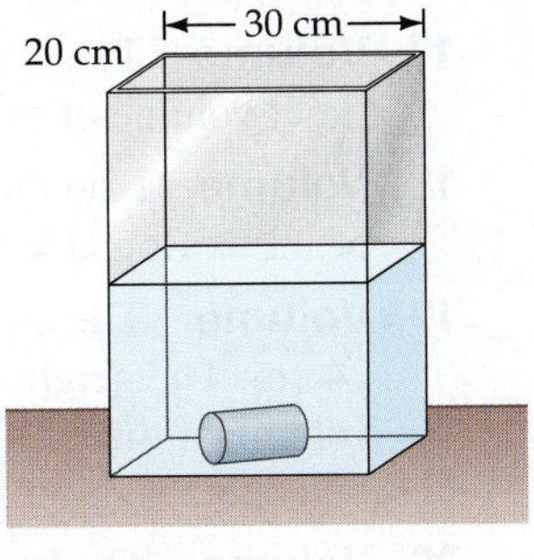

FIGURE 1

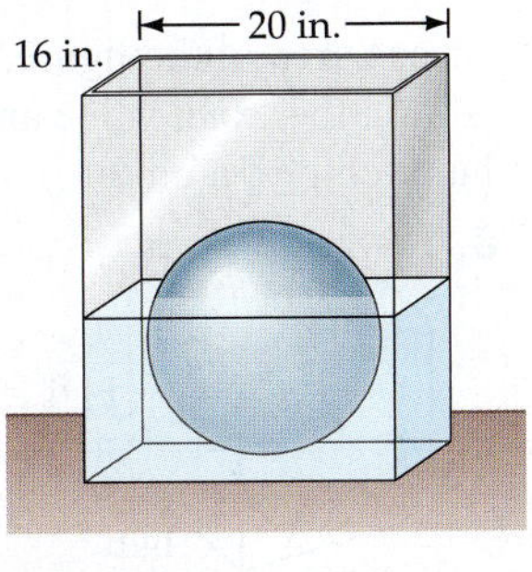

FIGURE 2

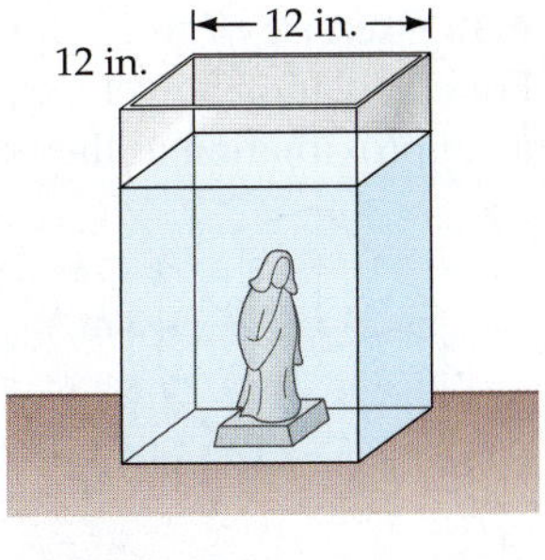

FIGURE 3

1. A cylinder with a 2-cm radius and a height of 10 cm is submerged in a tank of water that is 20 cm wide and 30 cm long (see Figure 1). How much does the water level rise? Round to the nearest hundredth of a centimeter.

2. A sphere with a radius of 6 in. is placed in a rectangular tank of water that is 16 in. wide and 20 in. long (see Figure 2). The sphere displaces water until two-thirds of the sphere, with respect to its volume, is submerged. How much does the water level rise? Round to the nearest hundredth of an inch.

3. A chemist wants to know the density of a statue that weighs 15 lb. The statue is placed in a rectangular tank of water that is 12 in. long and 12 in. wide (see Figure 3 on page 413). The water level rises 0.42 in. Find the density of the statue. Round to the nearest hundredth of a pound per cubic inch. *Hint:* Density = weight ÷ volume.

EXERCISE SET 7.5

■ In Exercises 1 to 6, find the volume of the figure. For calculations involving π, give both the exact value and an approximation to the nearest hundredth of a unit.

1.
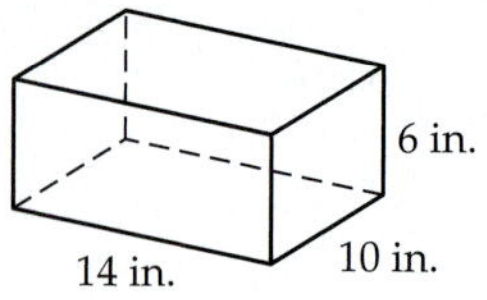

2.
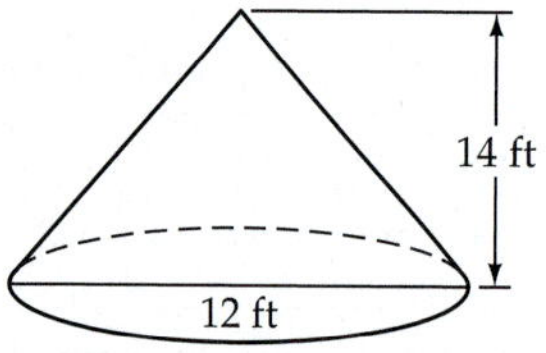

3.
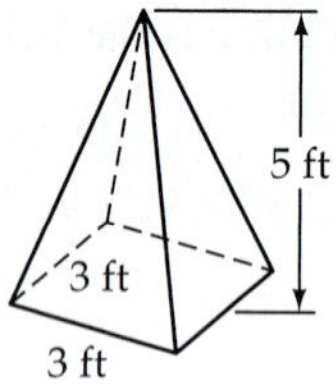

4.
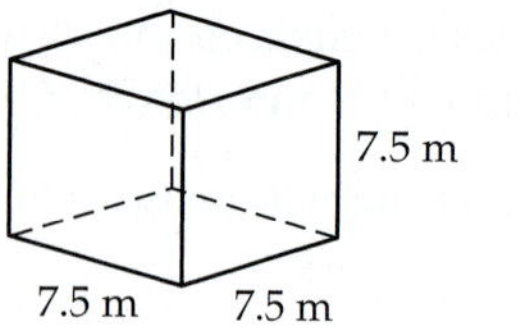

5.
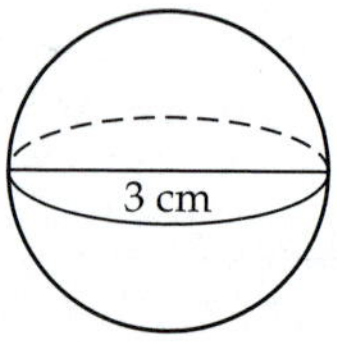

6.
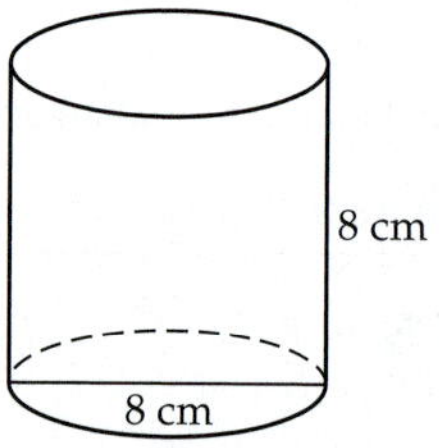

■ In Exercises 7 to 12, find the surface area of the figure. For calculations involving π, give both the exact value and an approximation to the nearest hundredth of a unit.

7.
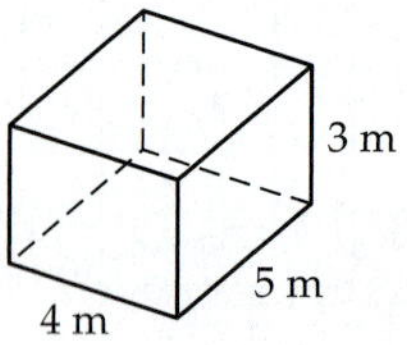

8.
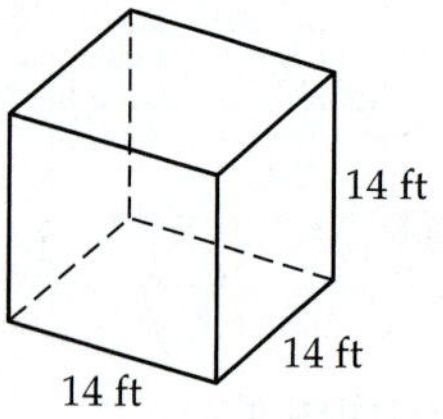

9.
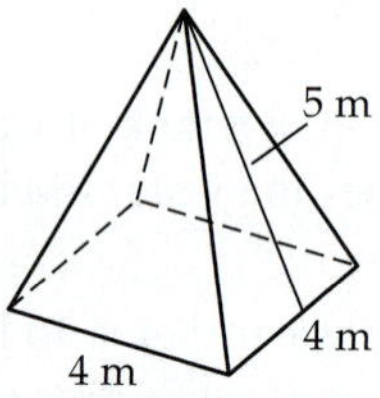

10.
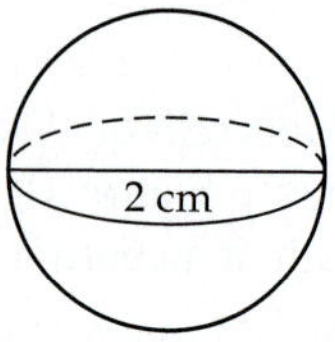

11.
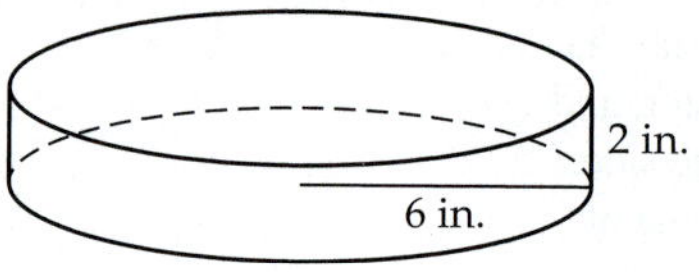

12.
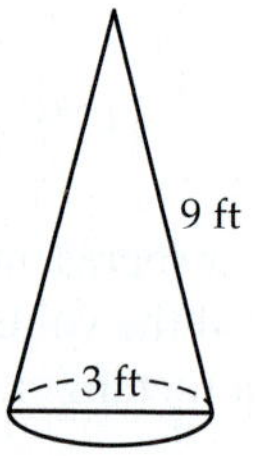

■ In Exercises 13 to 45, solve.

13. **Volume** A rectangular solid has a length of 6.8 m, a width of 2.5 m, and a height of 2 m. Find the volume of the solid.

14. **Volume** Find the volume of a rectangular solid that has a length of 4.5 ft, a width of 3 ft, and a height of 1.5 ft.

15. **Volume** Find the volume of a cube whose side measures 2.5 in.

16. **Volume** The length of a side of a cube is 7 cm. Find the volume of the cube.

17. **Volume** The diameter of a sphere is 6 ft. Find the exact volume of the sphere.

18. **Volume** Find the volume of a sphere that has a radius of 1.2 m. Round to the nearest hundredth of a cubic meter.

19. **Volume** The diameter of the base of a cylinder is 24 cm. The height of the cylinder is 18 cm. Find the volume of the cylinder. Round to the nearest hundredth of a cubic centimeter.

20. **Volume** The height of a cylinder is 7.2 m. The radius of the base is 4 m. Find the exact volume of the cylinder.

21. **Volume** The radius of the base of a cone is 5 in. The height of the cone is 9 in. Find the exact volume of the cone.

22. **Volume** The height of a cone is 15 cm. The diameter of the cone is 10 cm. Find the volume of the cone. Round to the nearest hundredth of a cubic centimeter.

23. **Volume** The length of a side of the base of a regular square pyramid is 6 in. and the height of the pyramid is 10 in. Find the volume of the pyramid.

24. **Volume** The height of a regular square pyramid is 8 m and the length of a side of the base is 9 m. What is the volume of the pyramid?

25. The length of a side of a cube is equal to the radius of a sphere. Which solid has the greater volume?

26. A sphere and a cylinder have the same radius. The height of the cylinder is equal to the radius of its base. Which solid has the greater volume?

27. **The Statue of Liberty** The index finger of the Statue of Liberty is 8 ft long. The circumference at the second joint is 3.5 ft. Use the formula for the volume of a cylinder to approximate the volume of the index finger on the Statue of Liberty. Round to the nearest hundredth of a cubic foot.

28. **Fish Hatchery** A rectangular tank at a fish hatchery is 9 m long, 3 m wide, and 1.5 m deep. Find the volume of the water in the tank when the tank is full.

Sukpaiboonwat/Shutterstock.com

29. **The Panama Canal** When the lock is full, the water in the Pedro Miguel Lock near the Pacific Ocean side of the Panama Canal fills a rectangular solid of dimensions 1000 ft long, 110 ft wide, and 43 ft deep. There are 7.48 gal of water in each cubic foot. How many gallons of water are in the lock?

Chris Jenner/Shutterstock.com

Panama Canal

30. **Surface Area** The width of a rectangular solid is 32 cm, the length is 60 cm, and the height is 14 cm. What is the surface area of the solid?

31. **Surface Area** The side of a cube measures 3.4 m. Find the surface area of the cube.

32. **Surface Area** Find the surface area of a cube with a side measuring 1.5 in.

33. **Surface Area** Find the exact surface area of a sphere with a diameter of 15 cm.

34. **Surface Area** The radius of a sphere is 2 in. Find the surface area of the sphere. Round to the nearest hundredth of a square inch.

35. **Surface Area** The radius of the base of a cylinder is 4 in. The height of the cylinder is 12 in. Find the surface area of the cylinder. Round to the nearest hundredth of a square inch.

36. **Surface Area** The diameter of the base of a cylinder is 1.8 m. The height of the cylinder is 0.7 m. Find the exact surface area of the cylinder.

37. **Surface Area** The slant height of a cone is 2.5 ft. The radius of the base is 1.5 ft. Find the exact surface area of the cone. The formula for the surface area of a cone is given on page 411.

38. **Surface Area** The diameter of the base of a cone is 21 in. The slant height is 16 in. What is the surface area of the cone? The formula for the surface area of a cone is given on page 411. Round to the nearest hundredth of a square inch.

39. **Surface Area** The length of a side of the base of a regular square pyramid is 9 in., and the pyramid's slant height is 12 in. Find the surface area of the pyramid.

40. **Surface Area** The slant height of a regular square pyramid is 18 m, and the length of a side of the base is 16 m. What is the surface area of the pyramid?

41. **Appliances** The volume of a freezer that is a rectangular solid with a length of 7 ft and a height of 3 ft is 52.5 ft^3. Find the width of the freezer.

42. **Aquariums** The length of a rectangular solid aquarium is 18 in. and the width is 12 in. If the volume of the aquarium is 1836 in^3, what is the height of the aquarium?

43. **Paint** A can of paint will cover 300 ft^2 of surface. How many cans of paint should be purchased to paint a cylinder that has a height of 30 ft and a radius of 12 ft?

44. **Ballooning** A hot air balloon is in the shape of a sphere. Approximately how much fabric was used to construct the balloon if its diameter is 32 ft? Round to the nearest square foot.

45. Surface Area The length of a side of the base of a regular square pyramid is 5 cm and the slant height of the pyramid is 8 cm. How much larger is the surface area of this pyramid than the surface area of a cone with a diameter of 5 cm and a slant height of 8 cm? Round to the nearest hundredth of a square centimeter.

■ In Exercises 46 to 51, find the volume of the figure. Round to the nearest hundredth of a unit.

46.

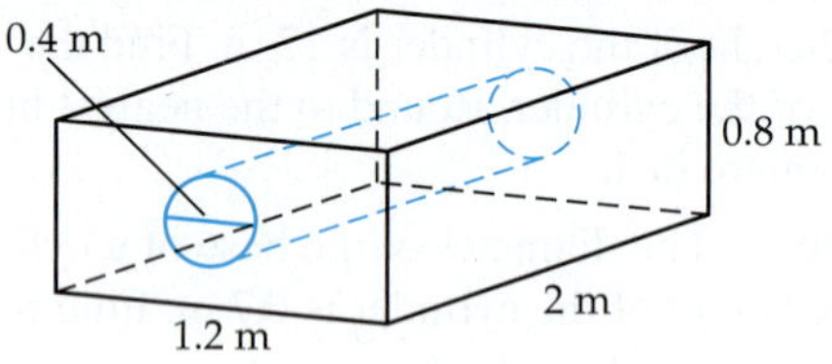

47.

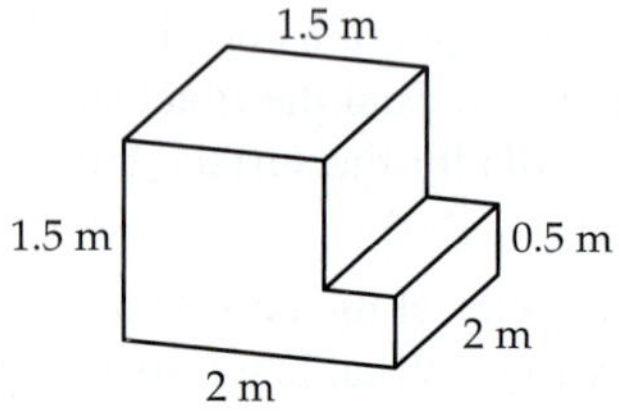

48.

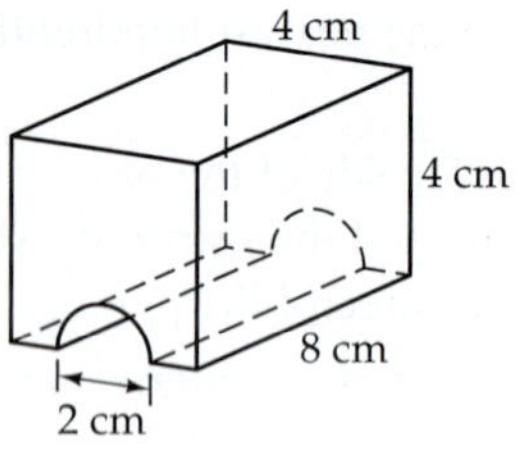

49.

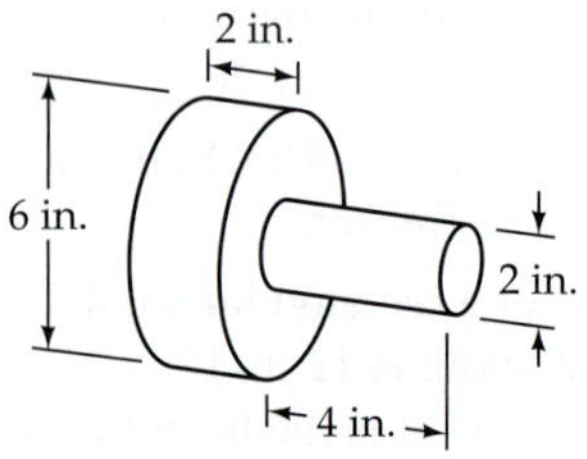

50.

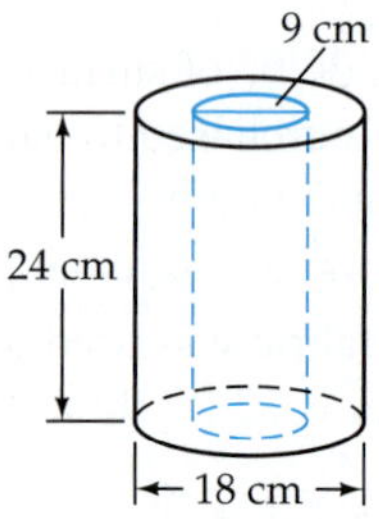

51.

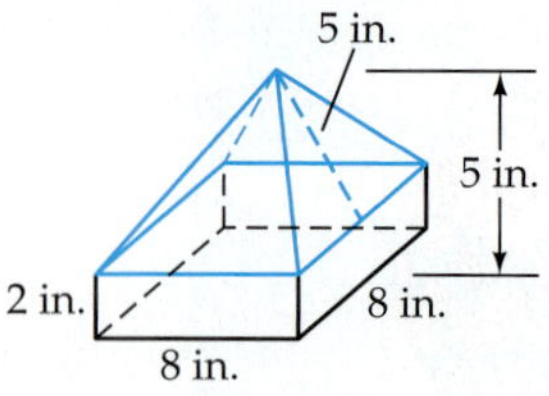

■ In Exercises 52 to 55, find the surface area of the figure. Round to the nearest hundredth of a unit.

52.

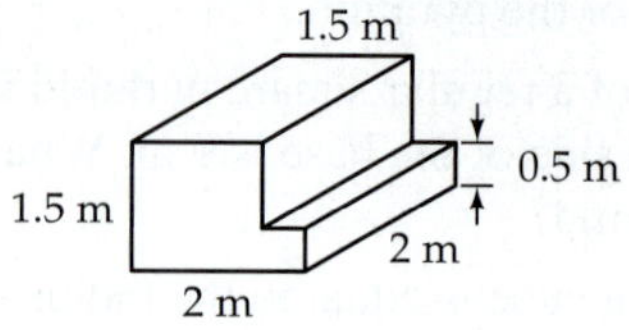

53.

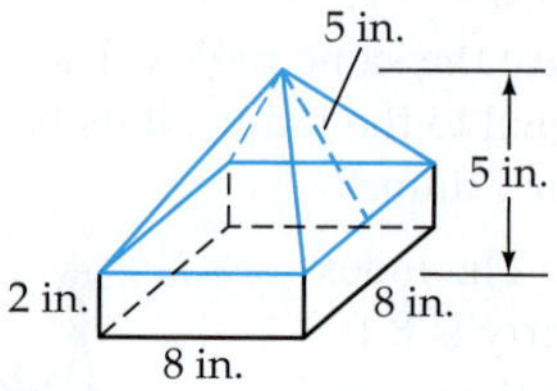

54.

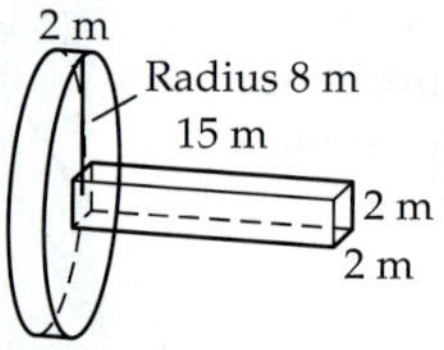

55.

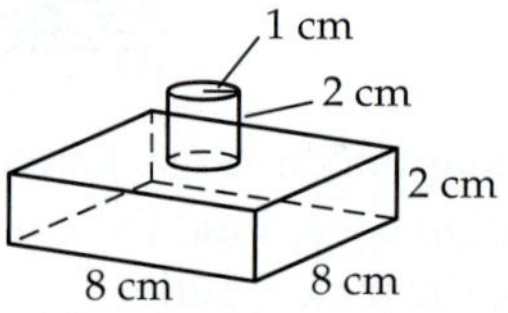

56. Oil Tanks A truck is carrying an oil tank. The tank consists of a circular cylinder with a hemisphere on each end, as shown. If the tank is half full, how many cubic feet of oil is the truck carrying? Round to the nearest hundredth of a cubic foot.

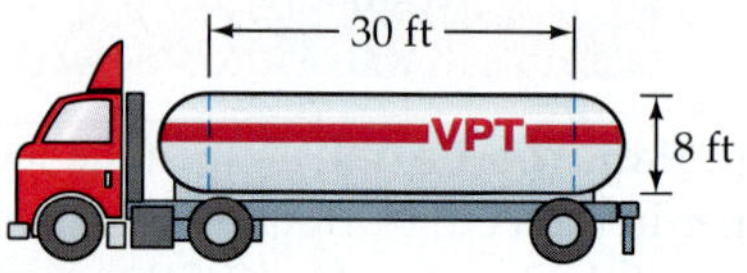

57. Swimming Pools How many liters of water are needed to fill the swimming pool shown below? (1 m^3 contains 1000 L.)

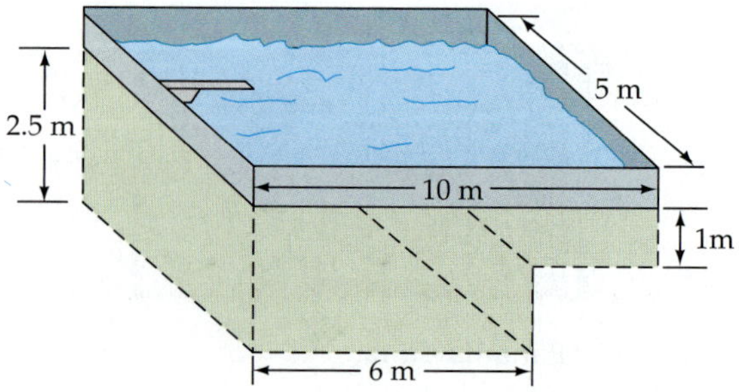

58. Metallurgy A piece of sheet metal is cut and formed into the shape shown below. Given that there are 0.24 g in 1 cm^2 of the metal, find the total number of grams of metal used. Round to the nearest hundredth of a gram.

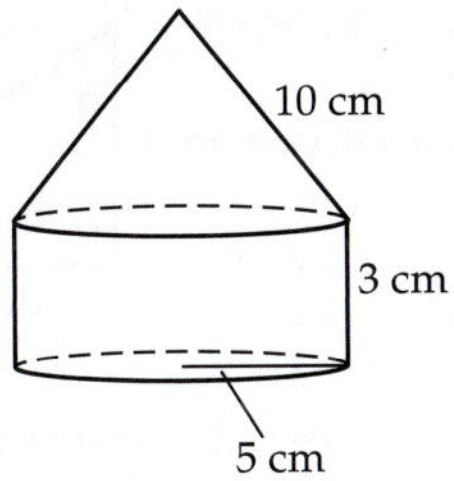

59. Gold A solid sphere of gold alloy with a radius of 0.5 cm has a value of $180. Find the value of a solid sphere of the same alloy with a radius of 1.5 cm.

60. Swimming Pools A swimming pool is built in the shape of a rectangular solid. It holds 32,000 gal of water. If the length, width, and height of the pool are each doubled, how many gallons of water will be needed to fill the pool?

EXTENSIONS

61. **a.** Draw a two-dimensional figure that can be cut out and made into a right circular cone.

b. Draw a two-dimensional figure that can be cut out and made into a regular square pyramid.

62. A sphere fits inside a cylinder as shown in the figure below. The height of the cylinder equals the diameter of the sphere. Show that the surface area of the sphere equals the surface area of the side of the cylinder.

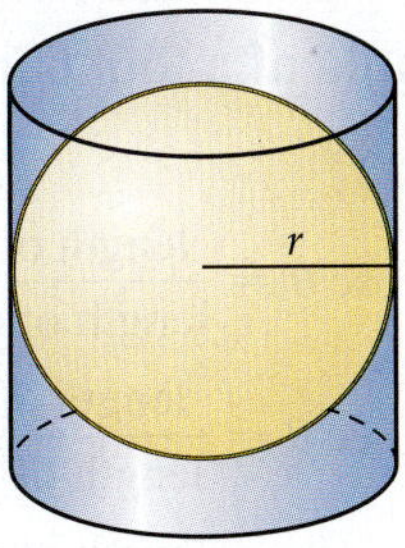

63. Determine whether the statement is always true, sometimes true, or never true.

a. The slant height of a regular square pyramid is longer than the height.

b. The slant height of a cone is shorter than the height.

c. The four triangular faces of a regular square pyramid are equilateral triangles.

64. **a.** What is the effect on the surface area of a rectangular solid of doubling the width and height?

b. What is the effect on the volume of a rectangular solid of doubling the length and width?

c. What is the effect on the volume of a cube of doubling the length of each side of the cube?

d. What is the effect on the surface area of a cylinder of doubling the radius and height?

65. Explain how you could cut through a cube so that the face of the resulting solid is

a. a square. **b.** an equilateral triangle.

c. a trapezoid. **d.** a hexagon.

SECTION 7.6 Right Triangle Trigonometry

Trigonometric Ratios of an Acute Angle

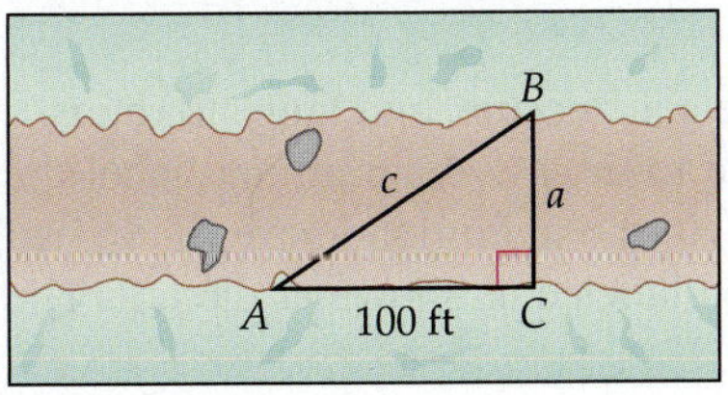

Consider the problem of engineers trying to determine the distance across a ravine in order to design a bridge. Look at the diagram at the left. It is fairly easy to measure the length of the side of the triangle that is on the land (100 ft), but the lengths of sides a and c cannot be measured easily because of the ravine.

The study of *trigonometry*, a term that comes from two Greek words meaning "triangle measurement," began about 2000 years ago, partially as a means of solving surveying problems such as the one described above. In this section, we will examine *right triangle* trigonometry—that is, trigonometry that applies only to right triangles.

When working with right triangles, it is convenient to refer to the side *opposite* an angle and the side *adjacent to* (next to) an angle. The hypotenuse of a right triangle is not adjacent to or opposite either of the acute angles of the triangle.

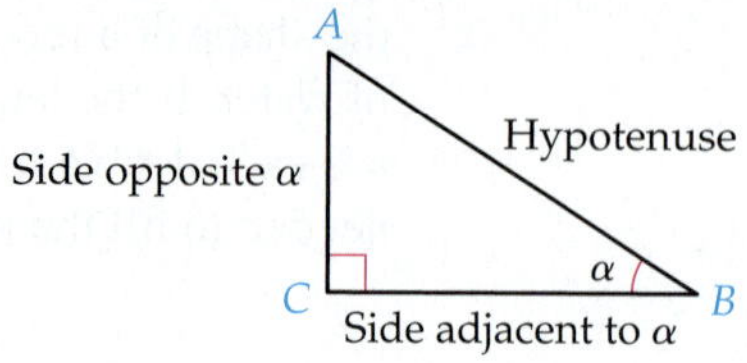

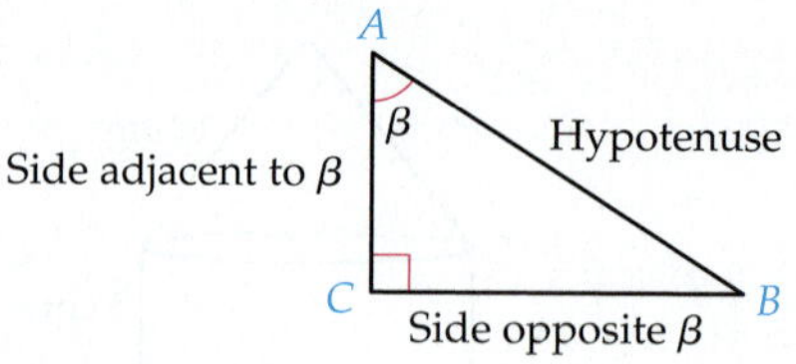

TAKE NOTE

In trigonometry, it is common practice to use Greek letters for the angles of a triangle. Here are some frequently used letters: α (alpha), β (beta), and θ (theta). The word *alphabet* is derived from the first two letters of the Greek alphabet, α and β.

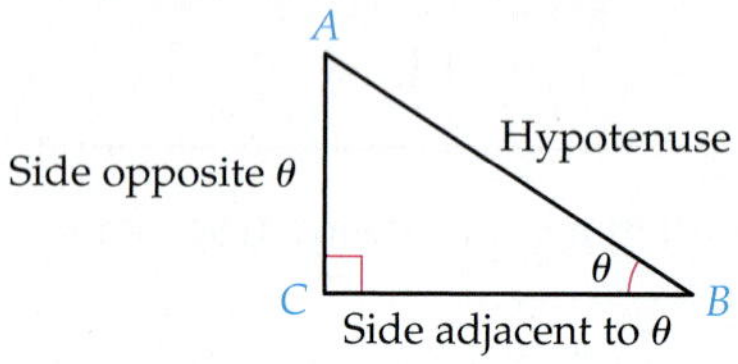

Consider the right triangle shown at the left. Six possible ratios can be formed using the lengths of the sides of the triangle.

$$\frac{\text{length of opposite side}}{\text{length of hypotenuse}} \qquad \frac{\text{length of hypotenuse}}{\text{length of opposite side}}$$

$$\frac{\text{length of adjacent side}}{\text{length of hypotenuse}} \qquad \frac{\text{length of hypotenuse}}{\text{length of adjacent side}}$$

$$\frac{\text{length of opposite side}}{\text{length of adjacent side}} \qquad \frac{\text{length of adjacent side}}{\text{length of opposite side}}$$

These ratios are called the **sine** (sin), **cosine** (cos), **tangent** (tan), **cosecant** (csc), **secant** (sec), and **cotangent** (cot) of the right triangle.

The Trigonometric Ratios of an Acute Angle of a Right Triangle

If θ is an acute angle of a right triangle ABC, then

$$\sin\theta = \frac{\text{length of opposite side}}{\text{length of hypotenuse}} \qquad \csc\theta = \frac{\text{length of hypotenuse}}{\text{length of opposite side}}$$

$$\cos\theta = \frac{\text{length of adjacent side}}{\text{length of hypotenuse}} \qquad \sec\theta = \frac{\text{length of hypotenuse}}{\text{length of adjacent side}}$$

$$\tan\theta = \frac{\text{length of opposite side}}{\text{length of adjacent side}} \qquad \cot\theta = \frac{\text{length of adjacent side}}{\text{length of opposite side}}$$

As a convenience, we will write opp, adj, and hyp as abbreviations for *the length of the* opposite side, adjacent side, and hypotenuse, respectively. Using this convention, the definitions of the trigonometric ratios are written as follows:

$$\sin\theta = \frac{\text{opp}}{\text{hyp}} \qquad \csc\theta = \frac{\text{hyp}}{\text{opp}}$$

$$\cos\theta = \frac{\text{adj}}{\text{hyp}} \qquad \sec\theta = \frac{\text{hyp}}{\text{adj}}$$

$$\tan\theta = \frac{\text{opp}}{\text{adj}} \qquad \cot\theta = \frac{\text{adj}}{\text{opp}}$$

For the remainder of this section, we will focus on the sine, cosine, and tangent ratios.

When working with trigonometric ratios, be sure to draw a diagram and label the adjacent and opposite sides of an angle. For instance, in the definition above, if we had placed θ at angle A, then the triangle would have been labeled as shown at the left. The definitions of the ratios remain the same.

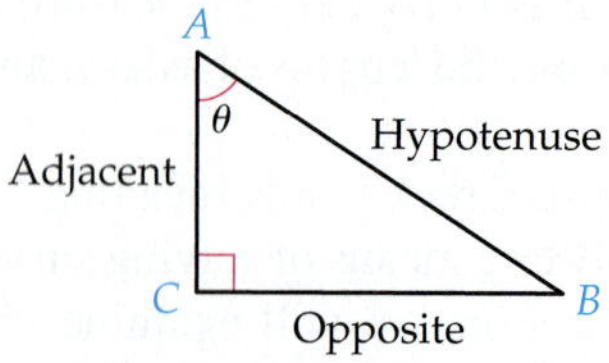

$$\sin\theta = \frac{\text{opp}}{\text{hyp}} \qquad \cos\theta = \frac{\text{adj}}{\text{hyp}} \qquad \tan\theta = \frac{\text{opp}}{\text{adj}}$$

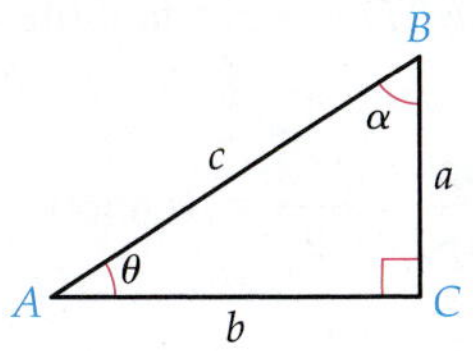

QUESTION For the right triangle shown at the left, indicate which side is

a. adjacent to $\angle A$ **b.** opposite θ **c.** adjacent to α **d.** opposite $\angle B$

EXAMPLE 1 Find the Values of Trigonometric Ratios

For the right triangle at the right, find the values of sin θ, cos θ, and tan θ.

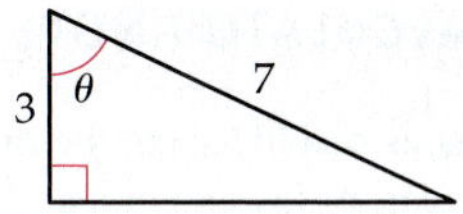

Solution

Use the Pythagorean theorem to find the length of the side opposite θ.

$$a^2 + b^2 = c^2 \qquad \text{• See the figure at the right.}$$
$$3^2 + b^2 = 7^2 \qquad \text{• } a = 3,\ c = 7$$
$$9 + b^2 = 49$$
$$b^2 = 40$$
$$b = \sqrt{40} = 2\sqrt{10}$$

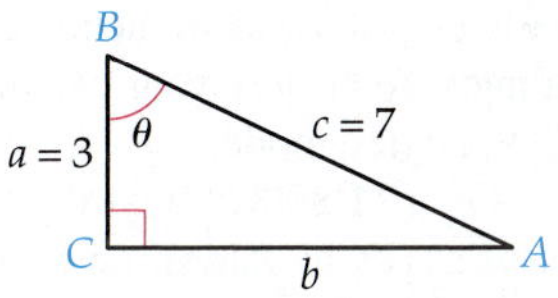

Using the definitions of the trigonometric ratios, we have

$$\sin\theta = \frac{\text{opp}}{\text{hyp}} = \frac{2\sqrt{10}}{7} \qquad \cos\theta = \frac{\text{adj}}{\text{hyp}} = \frac{3}{7} \qquad \tan\theta = \frac{\text{opp}}{\text{adj}} = \frac{2\sqrt{10}}{3}$$

INSTRUCTOR NOTE

Have students draw an isosceles right triangle similar to the one below. Then have them find the sine, cosine, and tangent of 45°.

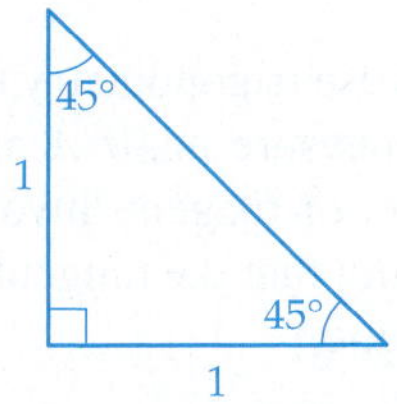

For another activity involving calculating the value of a trigonometric ratio, see the Excursion at the end of this section.

CHECK YOUR PROGRESS 1

For the right triangle at the right, find the values of sin θ, cos θ, and tan θ.

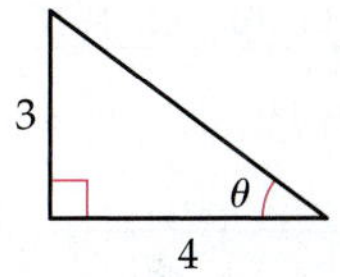

Solution See page S25.

In Example 1, we gave the exact answers. In many cases, approximate values of trigonometric ratios are given. The answers to Example 1, rounded to the nearest ten-thousandth, are

$$\sin\theta = \frac{2\sqrt{10}}{7} \approx 0.9035 \qquad \cos\theta = \frac{3}{7} \approx 0.4286 \qquad \tan\theta = \frac{2\sqrt{10}}{3} \approx 2.1082$$

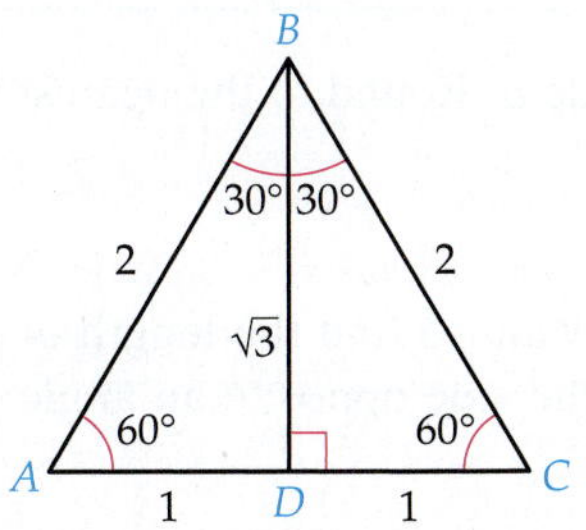

We will sometimes want to know the value of a trigonometric ratio for a given angle. Triangle ABC at the left is an equilateral triangle with sides of length 2 units and angle bisector $\overline{BD}$. Because $\overline{BD}$ bisects $\angle ABC$, the measures of $\angle ABD$ and $\angle DBC$ are both 30°. The angle bisector $\overline{BD}$ also bisects $\overline{AC}$. Therefore, $AD = 1$ and $DC = 1$. Using the Pythagorean theorem, we can find the measure of BD.

$$(DC)^2 + (BD)^2 = (BC)^2$$
$$1^2 + (BD)^2 = 2^2 \qquad \text{• } DC = 1,\ BC = 2$$
$$1 + (BD)^2 = 4 \qquad \text{• Solve for } BD.$$
$$(BD)^2 = 3$$
$$BD = \sqrt{3}$$

ANSWER **a.** b **b.** a **c.** a **d.** b

Using the definitions of the trigonometric ratios and triangle *BCD*, we can find the values of the sine, cosine, and tangent of 30° and 60°.

$$\sin 30° = \frac{\text{opp}}{\text{hyp}} = \frac{1}{2} = 0.5 \qquad \sin 60° = \frac{\text{opp}}{\text{hyp}} = \frac{\sqrt{3}}{2} \approx 0.8660$$

$$\cos 30° = \frac{\text{adj}}{\text{hyp}} = \frac{\sqrt{3}}{2} \approx 0.8660 \qquad \cos 60° = \frac{\text{adj}}{\text{hyp}} = \frac{1}{2} = 0.5$$

$$\tan 30° = \frac{\text{opp}}{\text{adj}} = \frac{1}{\sqrt{3}} \approx 0.5774 \qquad \tan 60° = \frac{\text{opp}}{\text{adj}} = \sqrt{3} \approx 1.7320$$

The properties of an equilateral triangle enabled us to calculate the values of the trigonometric ratios for 30° and 60°. Calculating values of the trigonometric ratios for most other angles, however, would be quite difficult. Fortunately, many calculators have been programmed to allow us to estimate these values.

To find tan 30° on a TI-83/84 calculator, first confirm that your calculator is in "degree mode." Press the TAN key and type in 30). Then press ENTER.

$\tan 30° \approx 0.5774$

On a scientific calculator, type in 30, and then press the TAN key.

Use a calculator to find sin 43.8° and tan 37.1° to the nearest ten-thousandth.

$\sin 43.8° \approx 0.6921$

$\tan 37.1° \approx 0.7563$

CALCULATOR NOTE

Just as distances can be measured in feet, miles, meters, and other units, angles can be measured in various units: degrees, radians, and grads. In this section, we use only degree measurements for angles, so be sure your calculator is in degree mode.

On a TI-83/84, press the MODE key to determine whether the calculator is in degree mode.

```
Normal Sci Eng
Float 0123456789
Radian Degree
Func Par Pol Seq
Connected Dot
Sequential Simul
Real a+bi re^θi
Full Horiz G-T
```

The engineers mentioned at the beginning of this section could use trigonometry to determine the distance across the ravine. Suppose the engineers measure angle *A* as 33.8°. Now they would ask, "Which trigonometric ratio, sine, cosine, or tangent, involves the side opposite angle *A* and the side adjacent to angle *A*?" Knowing that the tangent ratio is the required ratio, the engineers could write and solve the equation

$$\tan 33.8° = \frac{a}{100}$$

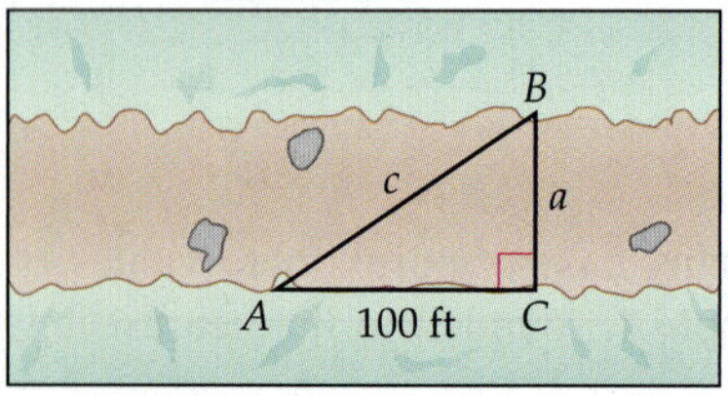

$$\tan 33.8° = \frac{a}{100}$$

$100(\tan 33.8°) = a$ • Multiply each side of the equation by 100.

$66.9 \approx a$ • Use a calculator to find tan 33.8°. Multiply the result in the display by 100.

The distance across the ravine is approximately 66.9 feet.

EXAMPLE 2 Find the Length of a Side of a Triangle

For the right triangle shown at the left, find the length of side *a*. Round to the nearest tenth of a meter.

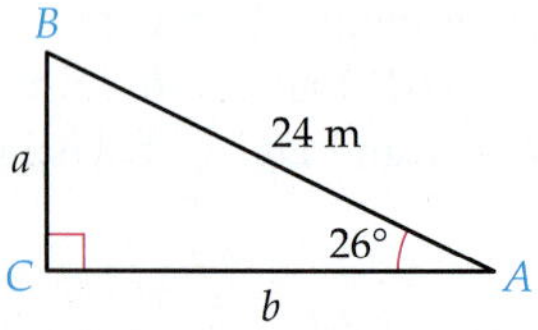

Solution

We are given the measure of ∠*A* and the hypotenuse. We want to find the length of side *a*. Side *a* is opposite ∠*A*. The sine function involves the side opposite an angle and the hypotenuse.

$$\sin A = \frac{\text{opp}}{\text{hyp}}$$

$\sin 26° = \frac{a}{24}$ • $A = 26°$, hypotenuse = 24 m

$24(\sin 26°) = a$ • Multiply each side by 24.

$10.5 \approx a$ • Use a calculator to find sin 26°. Multiply the result in the display by 24.

The length of side *a* is approximately 10.5 m.

CHECK YOUR PROGRESS 2

For the right triangle shown at the right, find the length of side a. Round to the nearest tenth of a foot.

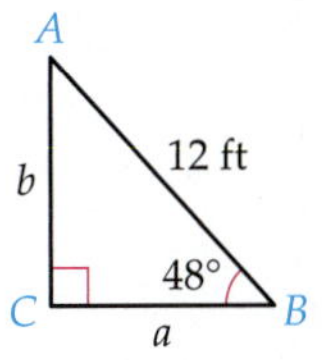

Solution See page S25.

Inverse Sine, Inverse Cosine, and Inverse Tangent

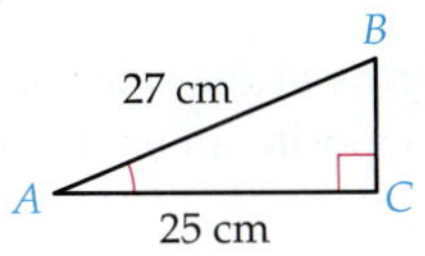

Suppose it is necessary to find the measure of $\angle A$ in the figure at the left. Because the length of the side adjacent to $\angle A$ is known and the length of the hypotenuse is known, we can write

$$\cos A = \frac{\text{adj}}{\text{hyp}}$$

$$\cos A = \frac{25}{27}$$

The solution of this equation is the angle whose cosine is $\frac{25}{27}$. This angle can be found by using the $\cos^{-1}$ key on a calculator. The expression $\cos^{-1}$ is read "the inverse cosine of."

$$\cos^{-1}\left(\frac{25}{27}\right) \approx 22.19160657$$

To the nearest tenth of a degree, the measure of $\angle A$ is 22.2°.

TAKE NOTE

The expression $\sin^{-1}(x)$ is sometimes written $\arcsin(x)$. The two expressions are equivalent. The expressions $\cos^{-1}(x)$ and $\arccos(x)$ are equivalent, as are $\tan^{-1}(x)$ and $\arctan(x)$.

Inverse Sine, Inverse Cosine, and Inverse Tangent

$\sin^{-1}(x)$ is defined as the angle whose sine is x, $0 < x < 1$.
$\cos^{-1}(x)$ is defined as the angle whose cosine is x, $0 < x < 1$.
$\tan^{-1}(x)$ is defined as the angle whose tangent is x, $x > 0$.

CALCULATOR NOTE

To find an inverse on a calculator, usually the INV or 2ND key is pressed prior to pushing the SIN, COS, or TAN key. Some calculators have $\sin^{-1}$, $\cos^{-1}$, and $\tan^{-1}$ keys. Consult the instruction manual for your calculator.

EXAMPLE 3 Evaluate an Inverse Sine Expression

Use a calculator to find $\sin^{-1}(0.9171)$. Round to the nearest tenth of a degree.

Solution

$\sin^{-1}(0.9171) \approx 66.5°$ • The calculator must be in degree mode. Press the keys for inverse sine followed by .9171). Press ENTER.

CHECK YOUR PROGRESS 3 Use a calculator to find $\tan^{-1}(0.3165)$. Round to the nearest tenth of a degree.

Solution See page S25.

TAKE NOTE

If

$\sin\theta = 0.7239,$

then

$\theta = \sin^{-1}(0.7239).$

EXAMPLE 4 Find the Measure of an Angle Using Inverse Sine

Given $\sin\theta = 0.7239$, find θ. Use a calculator. Round to the nearest tenth of a degree.

Solution

This is equivalent to finding $\sin^{-1}(0.7239)$. The calculator must be in degree mode.

$$\sin^{-1}(0.7239) \approx 46.4°$$
$$\theta \approx 46.4°$$

CHECK YOUR PROGRESS 4 Given $\tan \theta = 0.5681$, find θ. Use a calculator. Round to the nearest tenth of a degree.

Solution See page S25.

EXAMPLE 5 Find the Measure of an Angle in a Right Triangle

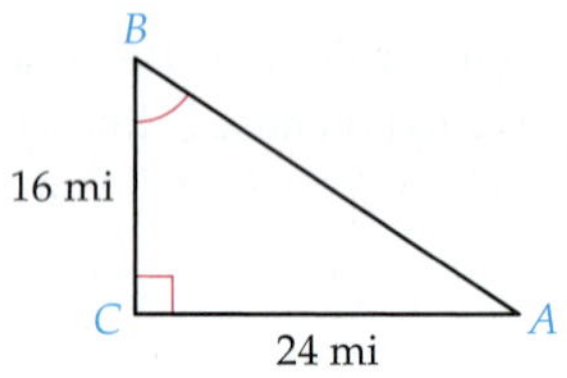

For the right triangle shown at the left, find the measure of $\angle B$. Round to the nearest tenth of a degree.

Solution

We want to find the measure of $\angle B$, and we are given the lengths of the sides opposite $\angle B$ and adjacent to $\angle B$. The tangent ratio involves the side opposite an angle and the side adjacent to that angle.

$$\tan B = \frac{\text{opp}}{\text{adj}}$$

$$\tan B = \frac{24}{16}$$

$$B = \tan^{-1}\left(\frac{24}{16}\right)$$

$$B \approx 56.3°$$

• Use the $\tan^{-1}$ key on a calculator.

The measure of $\angle B$ is approximately 56.3°.

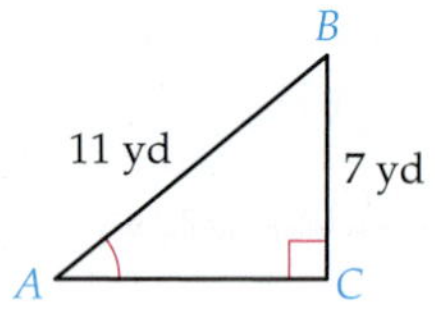

CHECK YOUR PROGRESS 5 For the right triangle shown at the left, find the measure of $\angle A$. Round to the nearest tenth of a degree.

Solution See page S25.

Angles of Elevation and Depression

One application of trigonometry, called **line-of-sight problems**, concerns an observer looking at an object.

Angles of elevation and depression are measured with respect to a horizontal line. If the object being sighted is above the observer, the acute angle formed by the line of sight and the horizontal line is an **angle of elevation**. If the object being sighted is below the observer, the acute angle formed by the line of sight and the horizontal line is an **angle of depression**.

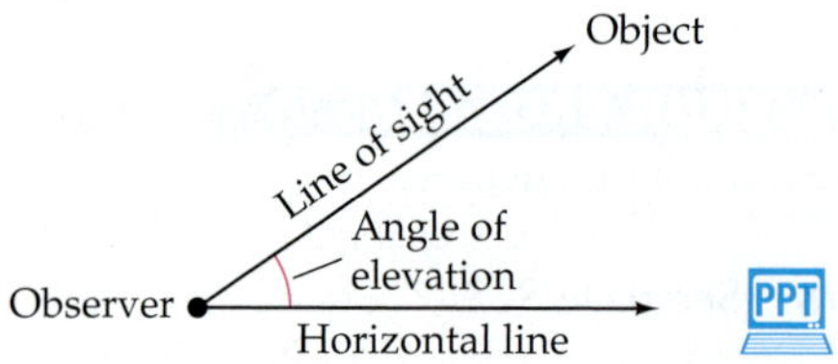

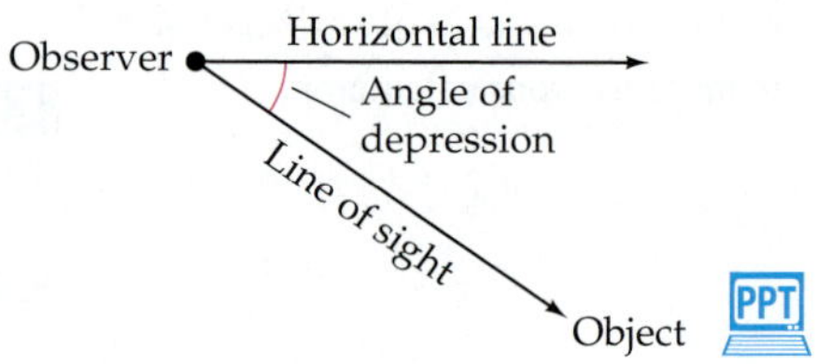

EXAMPLE 6 Solve an Angle of Elevation Problem

The angle of elevation from a point 62 ft away from the base of a flagpole to the top of the flagpole is 34°. Find the height of the flagpole. Round to the nearest tenth of a foot.

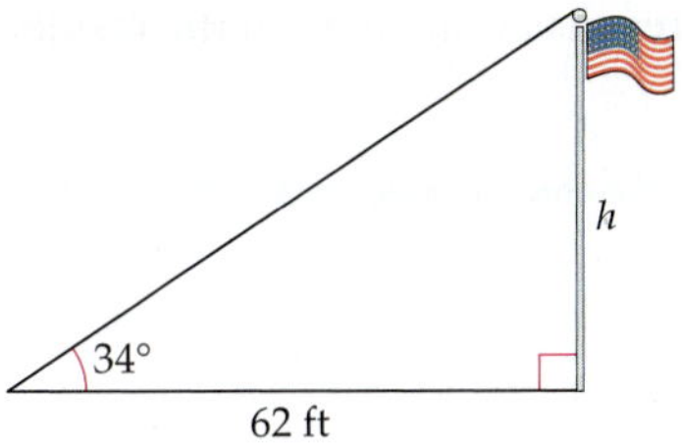

Solution

Draw a diagram. To find the height, h, write a trigonometric ratio that relates the given information and the unknown side of the triangle.

$$\tan 34^\circ = \frac{h}{62}$$

$$62(\tan 34^\circ) = h \qquad \text{• Solve for } h.$$

$$41.8 \approx h \qquad \text{• Use a calculator to find } \tan 34^\circ. \text{ Multiply the result in the display by 62.}$$

The height of the flagpole is approximately 41.8 ft.

CHECK YOUR PROGRESS 6 The angle of depression from the top of a lighthouse that is 20 m high to a boat on the water is 25°. How far is the boat from the base of the lighthouse? Round to the nearest tenth of a meter.

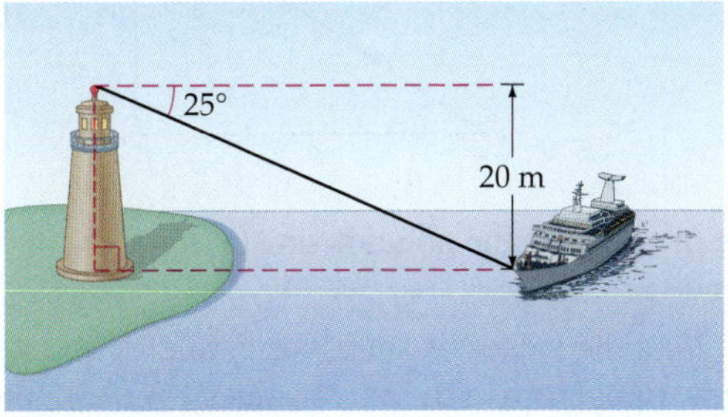

Solution See page S26.

EXCURSION

Approximating the Value of a Trigonometric Ratio

The value of a trigonometric ratio can be approximated by drawing a triangle with a given angle. To illustrate, we will choose an angle of 35°.

To find the tangent of 35° using the definitions given in this section, we can carefully construct a right triangle containing an angle of 35°. Because any two right triangles containing an angle of 35° are similar, *the value of tan 35° is the same no matter what triangle we draw.*

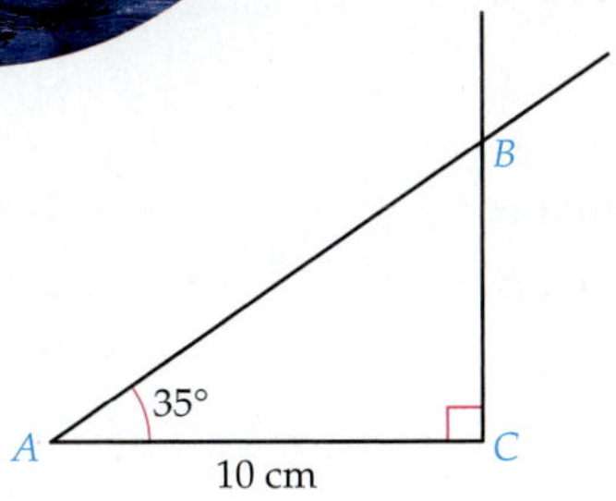

EXCURSION EXERCISES

1. Draw a horizontal line segment 10 cm long with left endpoint A and right endpoint C. See the diagram at the left.
2. Using a protractor, construct at A a 35° angle.
3. Draw at C a vertical line that intersects the terminal side of angle A at B. Your drawing should be similar to the one at the left.
4. Measure line segment BC to the nearest tenth of a centimeter.
5. Use your measurements to determine the approximate value of tan 35°.
6. Using your value for BC and the Pythagorean theorem, estimate AB. Round to the nearest centimeter.

7. Estimate sin 35° and cos 35°.

8. What are the values of sin 35°, cos 35°, and tan 35° as produced by a calculator? Round to the nearest ten-thousandth. How do these results compare with the results you obtained in Exercises 5 and 7?

EXERCISE SET 7.6

1. Use the right triangle below and sides a, b, and c to do the following:
   a. Find sin A.
   b. Find sin B.
   c. Find cos A.
   d. Find cos B.
   e. Find tan A.
   f. Find tan B.

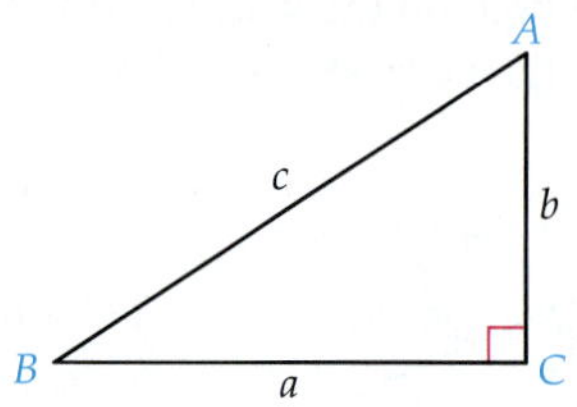

2. Explain the meaning of the notation $\sin^{-1}(x)$, $\cos^{-1}(x)$, and $\tan^{-1}(x)$.

■ In Exercises 3 to 10, find the values of sin θ, cos θ, and tan θ for the given right triangle. Give the exact values.

3.
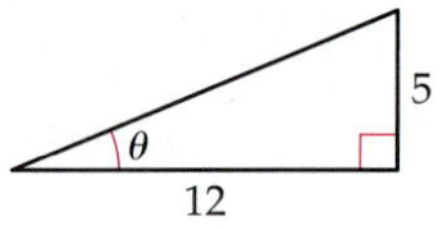

4.
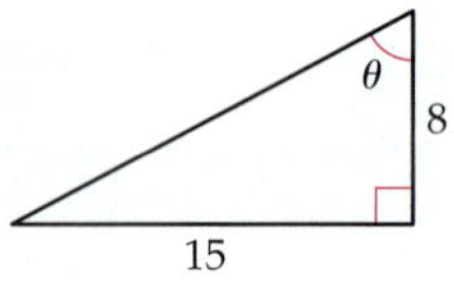

5.
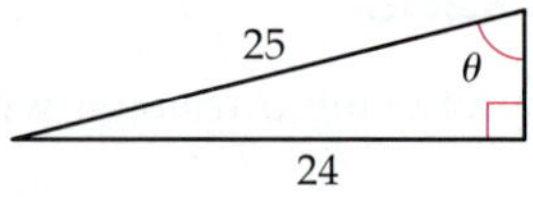

6.
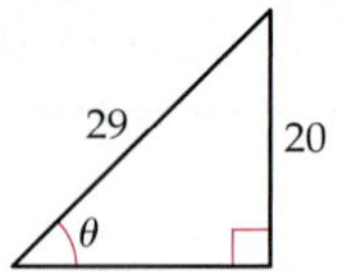

7.
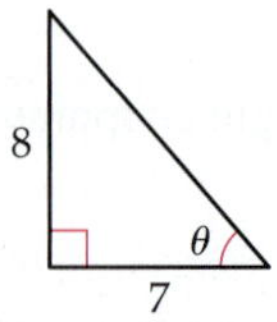

8.
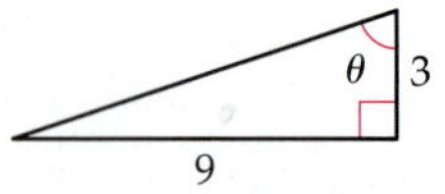

9.
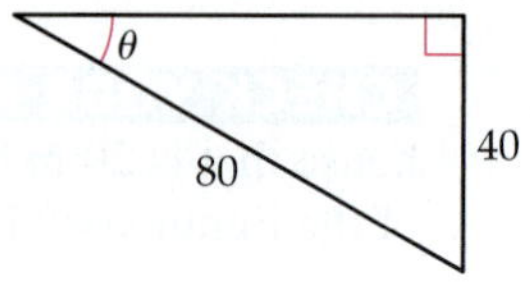

10.
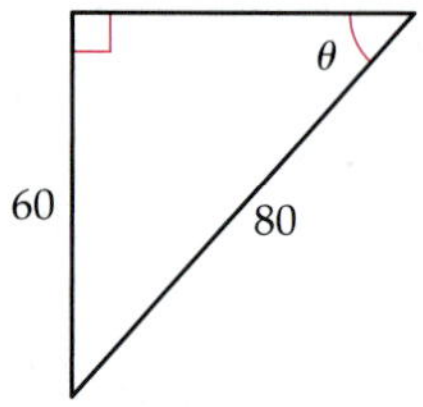

■ In Exercises 11 to 26, use a calculator to estimate the value of each of the following. Round to the nearest ten-thousandth.

11. cos 47°
12. sin 62°
13. tan 55°
14. cos 11°
15. sin 85.6°
16. cos 21.9°
17. tan 63.4°
18. sin 7.8°
19. tan 41.6°
20. cos 73°
21. sin 57.7°
22. tan 39.2°
23. sin 58.3°
24. tan 35.1°
25. cos 46.9°
26. sin 50°

■ In Exercises 27 to 42, use a calculator. Round to the nearest tenth of a degree.

27. Given $\sin\theta = 0.6239$, find θ.
28. Given $\cos\beta = 0.9516$, find β.
29. Find $\cos^{-1}(0.7536)$.
30. Find $\sin^{-1}(0.4478)$.
31. Given $\tan\alpha = 0.3899$, find α.
32. Given $\sin\beta = 0.7349$, find β.
33. Find $\tan^{-1}(0.7815)$.
34. Find $\cos^{-1}(0.6032)$.
35. Given $\cos\theta = 0.3007$, find θ.
36. Given $\tan\alpha = 1.588$, find α.
37. Find $\sin^{-1}(0.0105)$.
38. Find $\tan^{-1}(0.2438)$.
39. Given $\sin\beta = 0.9143$, find β.
40. Given $\cos\theta = 0.4756$, find θ.
41. Find $\cos^{-1}(0.8704)$.
42. Find $\sin^{-1}(0.2198)$.

■ For Exercises 43 to 56, draw a picture and label it. Then set up an equation and solve it. Show all your work. Round the measure of each angle to the nearest tenth of a degree. Round the length of a side to the nearest tenth of a unit. Assume the ground is level unless indicated otherwise.

43. Ballooning A balloon, tethered by a cable 997 ft long, was blown by a wind so that the cable made an angle of 57.6° with the ground. Find the height of the balloon off the ground.

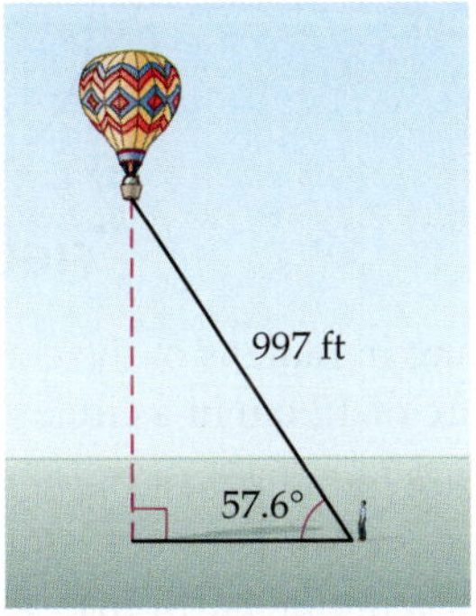

44. Roadways A road is inclined at an angle of 9.8° with the horizontal. Find the distance that one must drive on this road in order to be elevated 14.8 ft above the horizontal.

45. Home Maintenance A ladder 30.8 ft long leans against a building. If the foot of the ladder is 7.25 ft from the base of the building, find the angle the top of the ladder makes with the building.

46. Aviation A plane takes off from a field and rises at an angle of 11.4° with the horizontal. Find the height of the plane after it has traveled a distance of 1250 ft.

47. Guy Wires A guy wire whose grounded end is 16 ft from the telephone pole it supports makes an angle of 56.7° with the ground. How long is the wire?

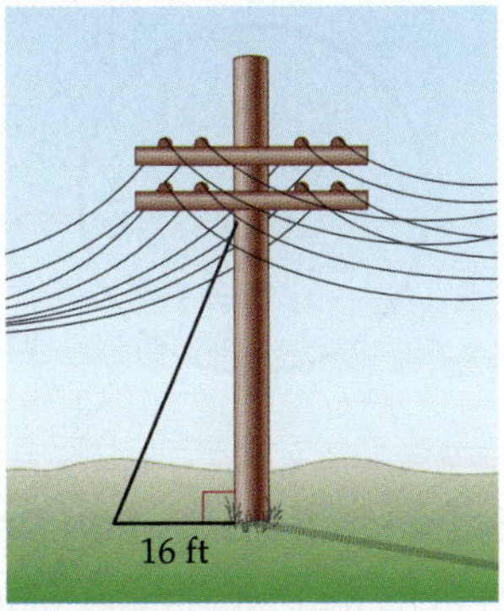

48. Angle of Depression A lighthouse built at sea level is 169 ft tall. From its top, the angle of depression to a boat below measures 25.1°. Find the distance from the boat to the foot of the lighthouse.

49. Angle of Elevation At a point 39.3 ft from the base of a tree, the angle of elevation of its top measures 53.4°. Find the height of the tree.

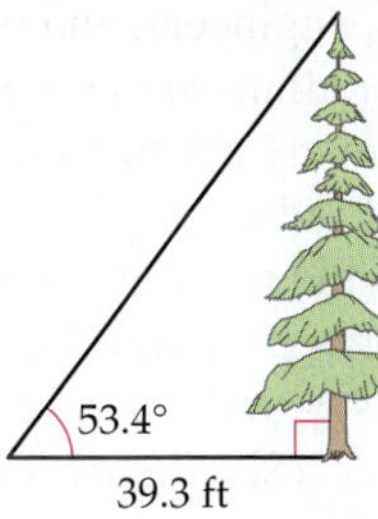

50. Angle of Depression An artillery spotter in a plane that is at an altitude of 978 ft measures the angle of depression of an enemy tank as 28.5°. How far is the enemy tank from the point on the ground directly below the spotter?

51. Home Maintenance A 15-foot ladder leans against a house. The ladder makes an angle of 65° with the ground. How far up the side of the house does the ladder reach?

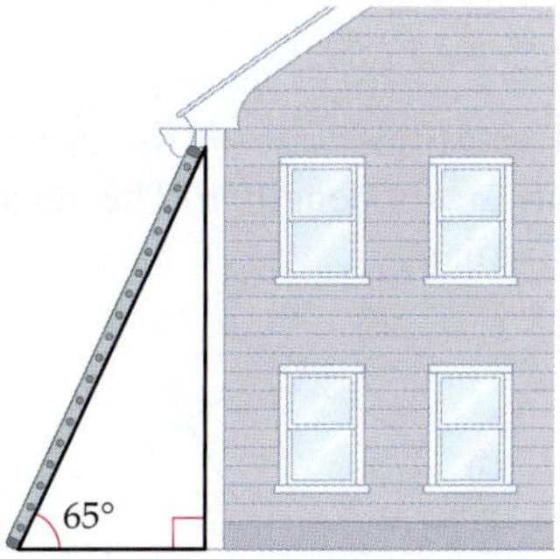

52. Angle of Elevation Find the angle of elevation of the sun when a tree 40.5 ft high casts a shadow 28.3 ft long.

53. Guy Wires A television transmitter tower is 600 ft high. If the angle between the guy wire (attached at the top) and the tower is 55.4°, how long is the guy wire?

54. Ramps A ramp used to load a racing car onto a flatbed carrier is 5.25 m long, and its upper end is 1.74 m above the lower end. Find the angle between the ramp and the road.

55. Angle of Elevation The angle of elevation of the sun is 51.3° at a time when a tree casts a shadow 23.7 yd long. Find the height of the tree.

56. Angle of Depression From the top of a building 312 ft tall, the angle of depression to a flower bed on the ground below is 12.0°. What is the distance between the base of the building and the flower bed?

EXTENSIONS

As we noted in this section, angles can also be measured in *radians*. For physicists, engineers, and other applied scientists who use calculus, radians are preferred over degrees because they simplify many calculations. To define a radian, first consider a circle of radius r and two radii $\overline{OA}$ and $\overline{OB}$. The angle θ formed by the two radii is a **central angle**. The portion of the circle between A and B is an **arc** of the circle and is written $\widehat{AB}$. We say that $\widehat{AB}$ *subtends* the angle θ. The length of the arc is s. (See Figure 1 below.)

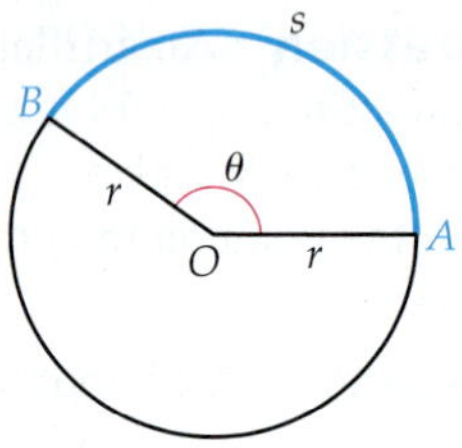

FIGURE 1

> ### Radian
>
> One **radian** is the measure of the central angle subtended by an arc of length r. The measure of θ in Figure 2 is 1 radian.

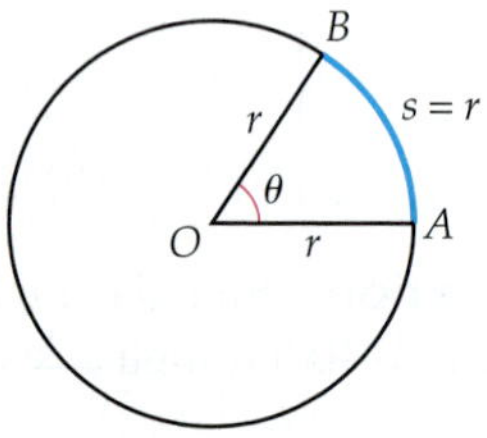

$\theta = 1$ radian

FIGURE 2

To find the radian measure of an angle subtended by an arc of length s, use the following formula.

> ### Radian Measure
>
> Given an arc of length s on a circle of radius r, the measure of the central angle subtended by the arc is $\theta = \frac{s}{r}$ radians.

For example, to find the measure in radians of the central angle subtended by an arc of 9 in. in a circle of radius 12 in., divide the length of the arc ($s = 9$ in.) by the length of the radius ($r = 12$ in.). See Figure 3.

$$\theta = \frac{9 \text{ in.}}{12 \text{ in.}} \text{ radian}$$
$$= \frac{3}{4} \text{ radian}$$

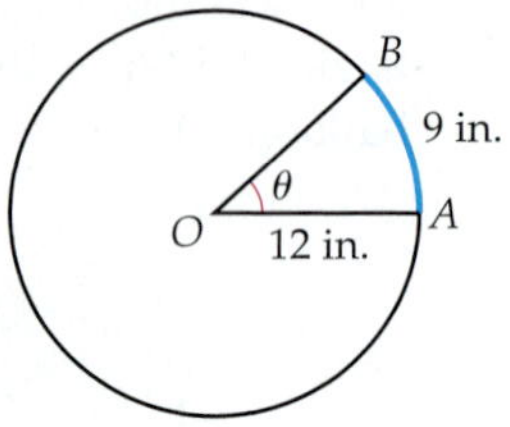

FIGURE 3

57. Find the measure in radians of the central angle subtended by an arc of 12 cm in a circle of radius 3 cm.

58. Find the measure in radians of the central angle subtended by an arc of 4 cm in a circle of radius 8 cm.

59. Find the measure in radians of the central angle subtended by an arc of 6 in. in a circle of radius 9 in.

60. Find the measure in radians of the central angle subtended by an arc of 12 ft in a circle of radius 10 ft.

Recall that the circumference of a circle is given by $C = 2\pi r$. Therefore, the radian measure of the central angle subtended by the circumference is $\theta = \frac{2\pi r}{r} = 2\pi$. In degree measure, the central angle has a measure of 360°. Thus we have 2π radians $= 360°$. Dividing each side of the equation by 2 gives π radians $= 180°$. From the last equation, we can establish the conversion factors $\frac{\pi \text{ radians}}{180°}$ and $\frac{180°}{\pi \text{ radians}}$. These conversion factors are used to convert between radians and degrees.

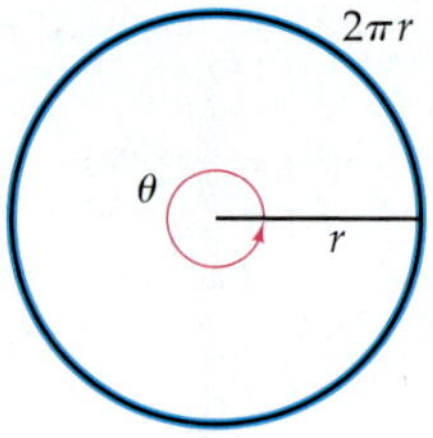

> ### Conversion Between Radians and Degrees
>
> - To convert from degrees to radians, multiply by $\frac{\pi \text{ radians}}{180°}$.
> - To convert from radians to degrees, multiply by $\frac{180°}{\pi \text{ radians}}$.

For instance, to convert 30° to radians, multiply 30° by $\frac{\pi \text{ radians}}{180°}$.

$$30° = 30°\left(\frac{\pi \text{ radians}}{180°}\right)$$
$$= \frac{\pi}{6} \text{ radian} \qquad \text{• Exact answer}$$
$$\approx 0.5236 \text{ radian} \qquad \text{• Approximate answer}$$

To convert 2 radians to degrees, multiply 2 by $\frac{180°}{\pi \text{ radians}}$.

$$2 \text{ radians} = 2\left(\frac{180°}{\pi \text{ radians}}\right)$$
$$= \left(\frac{360}{\pi}\right)° \qquad \text{• Exact answer}$$
$$\approx 114.5916° \qquad \text{• Approximate answer}$$

61. What is the measure in degrees of 1 radian?

62. Is the measure of 1 radian larger or smaller than the measure of 1°?

■ In Exercises 63 to 68, convert degree measure to radian measure. Find an exact answer and an answer rounded to the nearest ten-thousandth.

63. 45° **64.** 180° **65.** 315°

66. 90° **67.** 210° **68.** 18°

■ In Exercises 69 to 74, convert radian measure to degree measure. For Exercises 72 to 74, find an exact answer and an answer rounded to the nearest ten-thousandth.

69. $\frac{\pi}{3}$ radians **70.** $\frac{11\pi}{6}$ radians

71. $\frac{4\pi}{3}$ radians **72.** 1.2 radians

73. 3 radians **74.** 2.4 radians

SECTION 7.7 Non-Euclidean Geometry

Euclidean Geometry vs. Non-Euclidean Geometry

Some of the most popular games are based on a handful of rules that are easy to learn but still allow the game to develop into complex situations. The ancient Greek mathematician Euclid wanted to establish a geometry that was based on the *fewest* possible number of rules. He called these rules **postulates**. Euclid based his geometry on the following five postulates.

Euclid's Postulates

P1: A line segment can be drawn from any point to any other point.

P2: A line segment can be extended continuously in a straight line.

P3: A circle can be drawn with any center and any radius.

P4: All right angles have the same measure.

P5: ***The Parallel Postulate*** Through a given point not on a given line, exactly one line can be drawn parallel to the given line.

PPT

TAKE NOTE

In addition to postulates, Euclid's geometry, referred to as *Euclidean geometry*, involves definitions and some *undefined terms*. For instance, the term *point* is an undefined term in Euclidean geometry because any definition of the term *point* would require additional undefined terms. In a similar way, the term *line* is also an undefined term in Euclidean geometry.

For many centuries, the truth of these postulates was felt to be self-evident. However, a few mathematicians suspected that the fifth postulate, known as the parallel postulate, could be deduced from the other postulates. Over the years, many mathematicians tried to prove the parallel postulate, but none were successful.

Carl Friedrich Gauss (gaus′) (1777–1855) was one such mathematician. After many failed attempts to establish the parallel postulate as a theorem, Gauss came to the conclusion that the parallel postulate was an independent postulate. However, he noted that by changing this one postulate, he could create a whole new type of geometry! This is analogous to changing one of the rules of a game to create a new game.

Gauss's Alternative to the Parallel Postulate

Through a given point not on a given line, there are *at least two* lines parallel to the given line.

HISTORICAL NOTE

The Granger Collection, NY

Nikolai Lobachevsky (lō'bə-chĕf'skē) (1793–1856) was a noted Russian mathematician. Lobachevsky's concept of a non-Euclidean geometry was so revolutionary that he is called the Copernicus of geometry.

Another pioneer in **non-Euclidean geometry** (any geometry that does not include Euclid's parallel postulate) was the Russian mathematician Nikolai Lobachevsky. In a series of monthly articles that appeared in the academic journal of the University of Kazan in 1829, Lobachevsky provided a detailed investigation into the problem of the parallel postulate. He proposed that a consistent new geometry could be developed by replacing the parallel postulate with the alternative postulate, which assumes that *more than one* parallel line can be drawn through a point not on a given line. This new geometry, developed independently by both Gauss and Lobachevsky, is often called *hyperbolic geometry*.

The year 1826, in which Lobachevsky first lectured about a new non-Euclidean geometry, also marks the birth of the mathematician Bernhard Riemann. Although Riemann died of tuberculosis at age 39, he made major contributions in several areas of mathematics and physics. Riemann was the first person to consider a geometry in which the parallel postulate was replaced with the following postulate.

HISTORICAL NOTE

The Granger Collection, NY

Bernhard Riemann (rē'mən) (1826–1866). "Riemann's achievement has taught mathematicians to disbelieve in *any* geometry, or any space, as a necessary mode of human perception."[2]

Riemann's Alternative to the Parallel Postulate

Through a given point not on a given line, there exist *no* lines parallel to the given line.

Unlike the geometry developed by Lobachevsky, which was not based on a physical model, the non-Euclidean geometry of Riemann was closely associated with a sphere and the remarkable idea that because a line is an undefined term, a line on the surface of a sphere can be different from a line on a plane. It seems reasonable to suspect that "spherical lines" should retain some of the properties of lines on a plane. For example, on a plane, the shortest distance between two points is measured along the line that connects the points. The line that connects the points is an example of what is called a *geodesic*.

Geodesic

A **geodesic** is a curve C on a surface S such that for any two points on C, the portion of C between these points is the shortest path on S that joins these points.

On a sphere, the geodesic between two points is a *great circle* that connects the points.

Great Circle

A **great circle** of a sphere is a circle on the surface of the sphere whose center is at the center of the sphere. Any two points on a great circle divide the circle into two arcs. The shorter arc is the **minor arc**, and the longer arc is the **major arc**.

[2] Bell, E. T. *Men of Mathematics*. New York: Touchstone Books, Simon and Schuster, 1986.

In *Riemannian geometry*, which is also called *spherical geometry* or *elliptical geometry*, great circles, which are the geodesics of a sphere, are thought of as lines. Figure 7.2 shows a sphere and two of its great circles. Because all great circles of a sphere intersect, a sphere provides us with a model of a geometry in which there are no parallel lines.

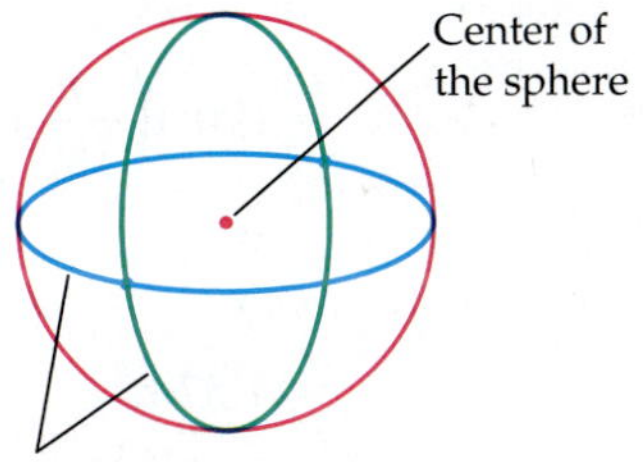

FIGURE 7.2 A sphere and its great circles serve as a physical model for Riemannian geometry.

> **TAKE NOTE**
>
> Riemannian geometry has many applications. For instance, because Earth is nearly spherical, navigation over long distances is facilitated by the use of Riemannian geometry as opposed to Euclidean geometry.

In Riemannian geometry, a triangle may have as many as three right angles. Figure 7.3 illustrates a spherical triangle with one right angle, a spherical triangle with two right angles, and a spherical triangle with three right angles.

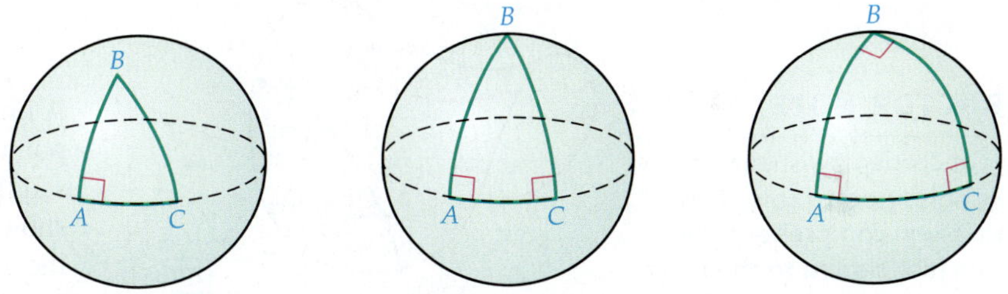

a. A spherical triangle with one right angle

b. A spherical triangle with two right angles

c. A spherical triangle with three right angles

FIGURE 7.3

The Spherical Triangle Area Formula

The area S of the spherical triangle ABC on a sphere with radius r is given by

$$S = (m\angle A + m\angle B + m\angle C - 180°)\left(\frac{\pi}{180°}\right)r^2$$

where each angle is measured in degrees.

EXAMPLE 1 Find the Area of a Spherical Triangle

Find the area of a spherical triangle with three right angles on a sphere with a radius of 1 ft. Find both the exact area and the approximate area rounded to the nearest hundredth of a square foot.

TAKE NOTE

To check the result in Example 1, use the fact that the given triangle covers $\frac{1}{8}$ of the surface of the sphere. See Figure 7.3c. The total surface area of a sphere is $4\pi r^2$. In Example 1, $r = 1$ ft, so the sphere has a surface area of 4π ft². Thus the area of the spherical triangle in Example 1 should be

$$\left(\frac{1}{8}\right)(4\pi) = \frac{\pi}{2}\text{ ft}^2$$

Solution

Apply the spherical triangle area formula.

$$S = (m\angle A + m\angle B + m\angle C - 180°)\left(\frac{\pi}{180°}\right)r^2$$

$$= (90° + 90° + 90° - 180°)\left(\frac{\pi}{180°}\right)(1)^2$$

$$= (90°)\left(\frac{\pi}{180°}\right)$$

$$= \frac{\pi}{2}\text{ ft}^2 \quad \text{• Exact area}$$

$$\approx 1.57\text{ ft}^2 \quad \text{• Approximate area}$$

CHECK YOUR PROGRESS 1 Find the area of the spherical triangle whose angles measure 200°, 90°, and 90° on a sphere with a radius of 6 in. Find both the exact area and the approximate area rounded to the nearest hundredth of a square inch.

Solution See page S26.

Mathematicians have not been able to create a three-dimensional model that *perfectly* illustrates all aspects of hyperbolic geometry. However, an infinite saddle surface can be used to visualize some of the basic aspects of hyperbolic geometry. Figure 7.4 shows a portion of an infinite saddle surface.

INSTRUCTOR NOTE

Have your students visualize a large Pringles potato chip. This will help them comprehend the saddle surface shown to the right.

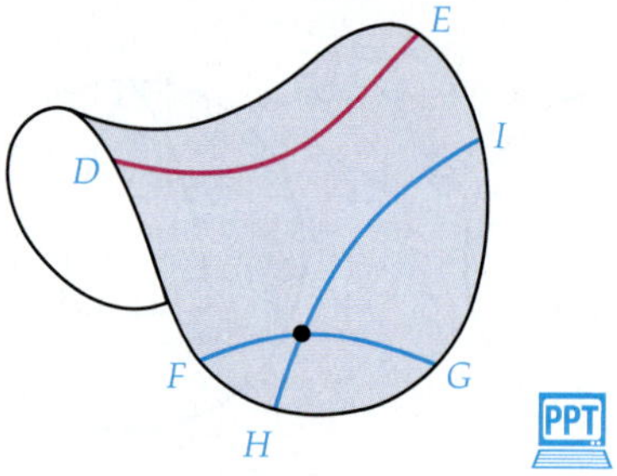

FIGURE 7.4 Portion of an infinite saddle surface

A line (geodesic) can be drawn through any two points on the saddle surface. Most lines on the saddle surface have a concave curvature, as shown by $\overleftrightarrow{DE}$, $\overleftrightarrow{FG}$, and $\overleftrightarrow{HI}$. Keep in mind that the saddle surface is an infinite surface. Figure 7.4 shows only a portion of the surface.

Parallel lines on an infinite saddle surface are defined as two lines that do not intersect. In Figure 7.4, $\overleftrightarrow{FG}$ and $\overleftrightarrow{HI}$ are *not* parallel because they intersect at a point. The lines $\overleftrightarrow{DE}$ and $\overleftrightarrow{FG}$ are parallel because they do not intersect. The lines $\overleftrightarrow{DE}$ and $\overleftrightarrow{HI}$ are also parallel lines. Figure 7.4 provides a geometric model of a hyperbolic geometry because for a given line, *more than one* parallel line exists through a point not on the given line.

Figure 7.5 shows a triangle drawn on a saddle surface. The triangle is referred to as a *hyperbolic triangle.* Due to the curvature of the sides of the hyperbolic triangle, the sum of the measures of the angles of the triangle is less than 180°. This is true for all hyperbolic triangles.

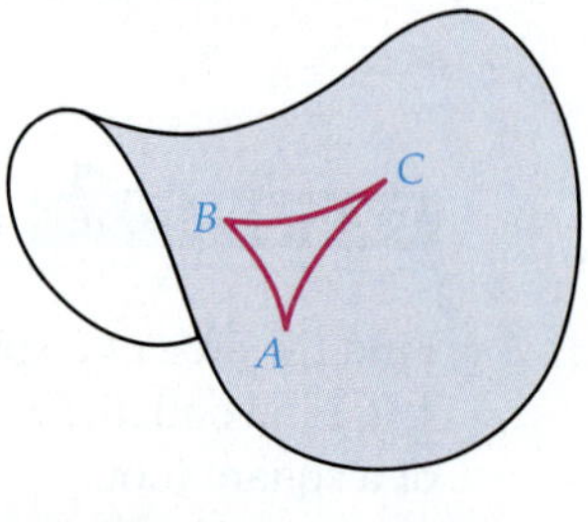

FIGURE 7.5 Hyperbolic triangle

The following chart summarizes some of the properties of plane, hyperbolic, and spherical geometries.

<table>
<tr><th>Euclidean Geometry</th><th colspan="2">Non-Euclidean Geometries</th></tr>
<tr>
<td>Euclidean or Plane Geometry (ca. 300 BC):
Through a given point not on a given line, exactly one line can be drawn parallel to the given line.
Geometry on a plane
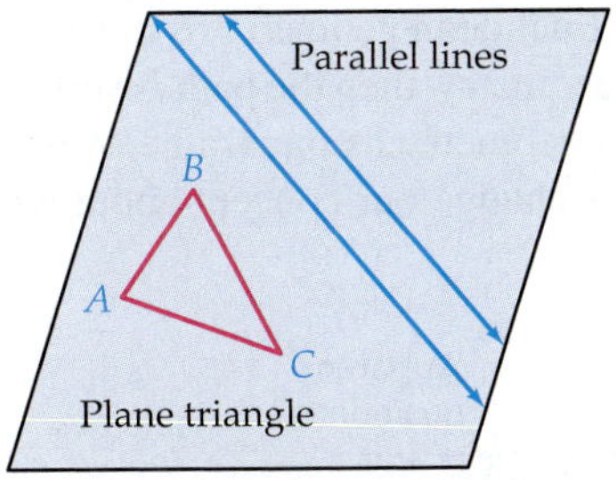

For any triangle ABC,
$m\angle A + m\angle B + m\angle C = 180°$
A triangle can have at most one right angle.
The shortest path between two points is the line segment that connects the points.</td>
<td>Lobachevskian or Hyperbolic Geometry (1826):
Through a given point not on a given line, there are at least two lines parallel to the given line.
Geometry on an infinite saddle surface
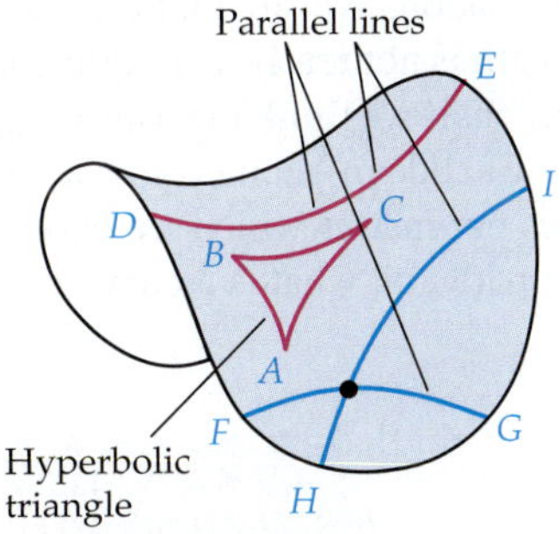

For any triangle ABC,
$m\angle A + m\angle B + m\angle C < 180°$
A triangle can have at most one right angle.
The curves shown in the above figure illustrate some of the geodesics of an infinite saddle surface.</td>
<td>Riemannian or Spherical Geometry (1855):
Through a given point not on a given line, there exist no lines parallel to the given line.
Geometry on a sphere
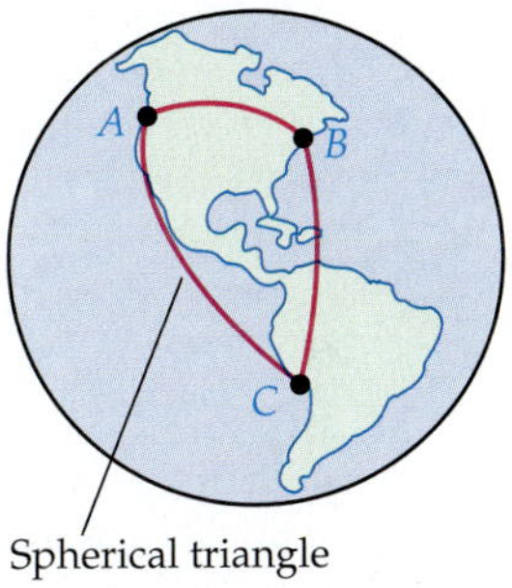

For any triangle ABC,
$180° < m\angle A + m\angle B + m\angle C < 540°$
A triangle can have one, two, or three right angles.
The shortest path between two points is the minor arc of a great circle that passes through the points.</td>
</tr>
</table>

PPT

EXAMPLE 2 Euclidean and Non-Euclidean Geometries

Determine the type of geometry (Euclidean, Riemannian, or Lobachevskian) in which two lines can intersect at a point and both of the lines can be parallel to a third line that does not pass through the intersection point.

Solution

Euclid's parallel postulate states that through a given point not on a given line, exactly one line can be drawn parallel to the given line.

Riemann's alternative to the parallel postulate states that through a given point not on a given line, there exist no lines parallel to the given line.

Gauss's alternative to the parallel postulate, which is assumed in Lobachevskian geometry, states that through a given point not on a given line, there are at least two lines parallel to the given line.

Thus, if we consider only Euclidean, Riemannian, and Lobachevskian geometries, then the condition that "two lines can intersect at a point and both of the lines can be parallel to a third line that does not pass through the intersection point" can be true only in Lobachevskian geometry.

CHECK YOUR PROGRESS 2 Determine the type of geometry in which there are no lines parallel to a given line.

Solution See page S26.

MATH MATTERS Curved Space

In 1915, Albert Einstein proposed a revolutionary theory that is now called the *general theory of relativity*. One of the major ideas of this theory is that space is curved, or "warped," by the mass of stars and planets. The greatest curvature occurs around those stars with the largest mass. Light rays in space do not travel in a straight path, but rather follow the geodesics of this curved space. Recall that the shortest path that joins two points on a given surface is on a geodesic of the surface.

It is interesting to consider the paths of light rays in space from a different perspective, in which the light rays travel along straight paths in a space that is non-Euclidean. To better understand this concept, consider an airplane that flies from Los Angeles to London. We say that the airplane flies on a straight path between the two cities. The path is not really a straight path if we use Euclidean geometry as our frame of reference, but if we think in terms of Reimannian geometry, then the path *is* straight. We can use Euclidean geometry or a non-Euclidean geometry as our frame of reference. It does not matter which we choose, but it does change our concept about what is a straight path and what is a curved path.

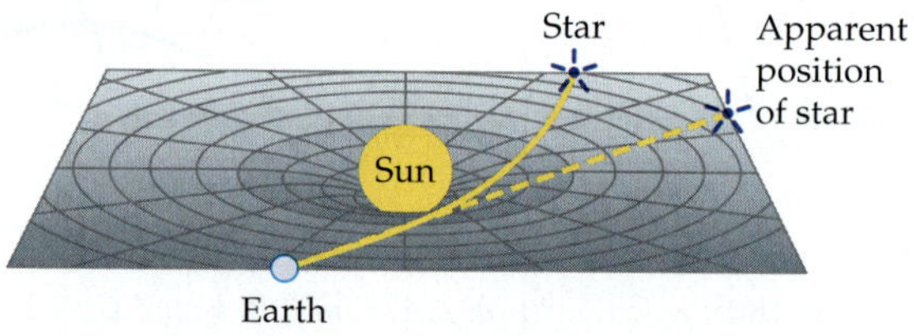

Light rays from a star follow the geodesics of space. Any light rays that pass near the sun are slightly bent. This causes some stars to appear to an observer on Earth to be in different positions than their actual positions.

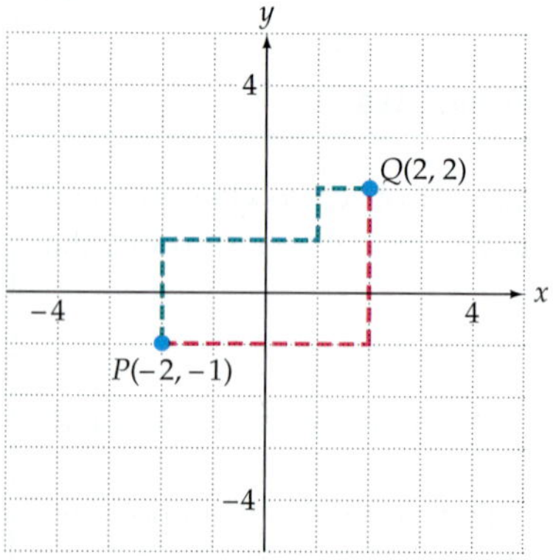

FIGURE 7.6 Two city paths from P to Q

POINT OF INTEREST

The city distance formula is also called the *Manhattan metric*. It first appeared in an article published by the mathematician Hermann Minkowski (1864–1909).

City Geometry: A Contemporary Geometry

Consider the geometric model of a city shown in Figure 7.6. In this city, all of the streets run either straight north and south or straight east and west. The distance between adjacent north–south streets is 1 block, and the distance between adjacent east–west streets is 1 block. In a city it is generally not possible to travel from P to Q along a straight path. Instead, one must travel between P and Q by traveling along the streets. As you travel from P to Q, we assume that you always travel in a direction that gets you closer to point Q. Two such paths are shown by the red and the green dashed line segments in Figure 7.6.

We will use the notation $d_C(P, Q)$ to represent the *city distance* between the points P and Q. For P and Q as shown in Figure 7.6, $d_C(P, Q) = 7$ blocks. This distance can be determined by counting the number of blocks needed to travel along the streets from P to Q or by using the following formula.

The City Distance Formula

If $P(x_1, y_1)$ and $Q(x_2, y_2)$ are two points in a city, then the **city distance** between P and Q is given by

$$d_C(P, Q) = |x_2 - x_1| + |y_2 - y_1|$$

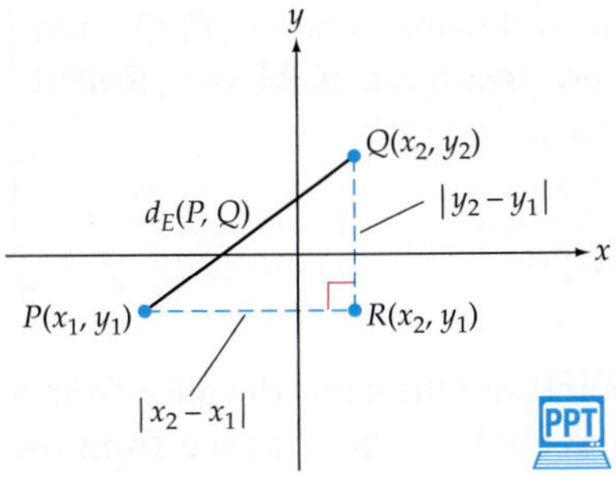

FIGURE 7.7

In Euclidean geometry, the distance between the points P and Q is defined as the length of $\overline{PQ}$. To determine a *Euclidean distance formula* for the distance between $P(x_1, y_1)$ and $Q(x_2, y_2)$, we first locate the point $R(x_2, y_1)$. See Figure 7.7.

Note that R has the same x-coordinate as Q and that R has the same y-coordinate as P. The horizontal distance between P and R is $|x_2 - x_1|$, and the vertical distance between R and Q is $|y_2 - y_1|$. Apply the Pythagorean theorem to the right triangle PRQ to produce

$$[d_E(P, Q)]^2 = |x_2 - x_1|^2 + |y_2 - y_1|^2$$

Because the square of a number cannot be negative, the absolute value signs are not necessary.

$$[d_E(P, Q)]^2 = (x_2 - x_1)^2 + (y_2 - y_1)^2$$

Take the square root of each side of the equation to produce

$$d_E(P, Q) = \sqrt{(x_2 - x_1)^2 + (y_2 - y_1)^2}$$

The Euclidean Distance Formula

If $P(x_1, y_1)$ and $Q(x_2, y_2)$ are two points in a plane, then the **Euclidean distance** between P and Q is given by

$$d_E(P, Q) = \sqrt{(x_2 - x_1)^2 + (y_2 - y_1)^2}$$

EXAMPLE 3 Find the Euclidean Distance and the City Distance Between Two Points

For each of the following, find $d_E(P, Q)$ and $d_C(P, Q)$. Assume that both $d_E(P, Q)$ and $d_C(P, Q)$ are measured in blocks. Round approximate results to the nearest tenth of a block.

a. $P(-4, -3), Q(2, -1)$ **b.** $P(2, -3), Q(-5, 4)$

Solution

a.
$$\begin{aligned} d_E(P, Q) &= \sqrt{(x_2 - x_1)^2 + (y_2 - y_1)^2} \\ &= \sqrt{[2 - (-4)]^2 + [(-1) - (-3)]^2} \\ &= \sqrt{6^2 + 2^2} \\ &= \sqrt{40} \approx 6.3 \text{ blocks} \end{aligned}$$

$$\begin{aligned} d_C(P, Q) &= |x_2 - x_1| + |y_2 - y_1| \\ &= |2 - (-4)| + |(-1) - (-3)| \\ &= |6| + |2| \\ &= 6 + 2 \\ &= 8 \text{ blocks} \end{aligned}$$

b.
$$\begin{aligned} d_E(P, Q) &= \sqrt{(x_2 - x_1)^2 + (y_2 - y_1)^2} \\ &= \sqrt{[(-5) - 2]^2 + [4 - (-3)]^2} \\ &= \sqrt{(-7)^2 + 7^2} \\ &= \sqrt{98} \approx 9.9 \text{ blocks} \end{aligned}$$

$$\begin{aligned} d_C(P, Q) &= |x_2 - x_1| + |y_2 - y_1| \\ &= |(-5) - 2| + |4 - (-3)| \\ &= |-7| + |7| \\ &= 7 + 7 \\ &= 14 \text{ blocks} \end{aligned}$$

TAKE NOTE

In Example 3 we have calculated the city distances by using the city distance formula. These distances can also be determined by counting the number of blocks needed to travel along the streets from P to Q on a rectangular coordinate grid.

CHECK YOUR PROGRESS 3 For each of the following, find $d_E(P, Q)$ and $d_C(P, Q)$. Assume that both $d_E(P, Q)$ and $d_C(P, Q)$ are measured in blocks. Round approximate results to the nearest tenth of a block.

a. $P(-1, 4), Q(3, 2)$ **b.** $P(3, -4), Q(-1, 5)$

Solution See page S26.

Recall that a circle is a plane figure in which all points are the same distance from a given center point and the length of the radius r of the circle is the distance from the center point to a point on the circle. Figure 7.8 shows a *Euclidean circle* centered at (0, 0) with a radius of 3 blocks. Figure 7.9 shows all the points in a city that are 3 blocks from the center point (0, 0). These points form a *city circle* with a radius of 3 blocks.

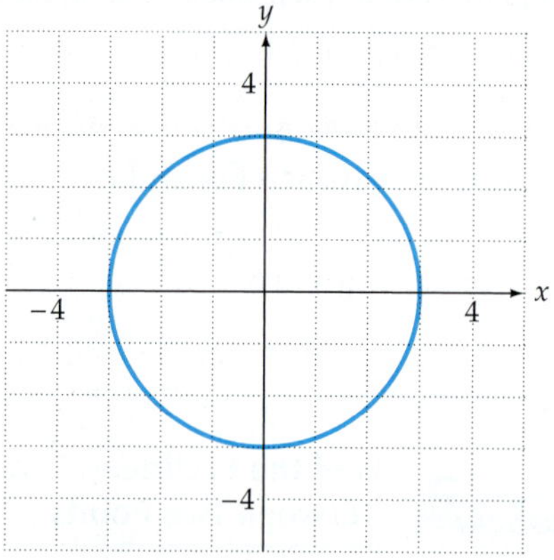

FIGURE 7.8 A Euclidean circle with center (0, 0) and a radius of 3 blocks

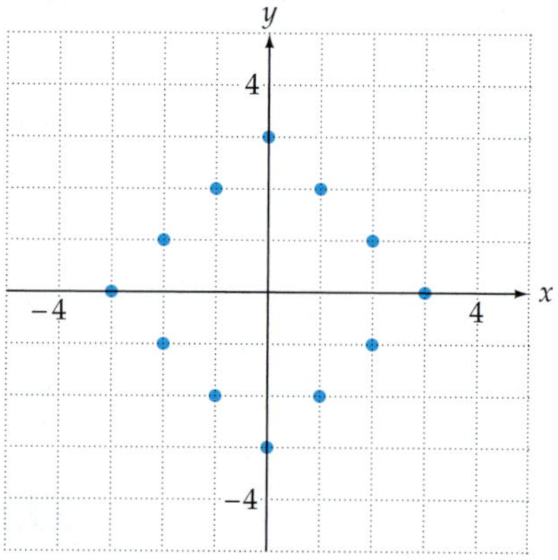

FIGURE 7.9 A city circle with center (0, 0) and a radius of 3 blocks

It is interesting to observe that the *city circle* shown in Figure 7.9 consists of just 12 points and that these points all lie on a square with vertices (3, 0), (0, 3), $(-3, 0)$, and $(0, -3)$.

EXCURSION

Finding Geodesics

Form groups of three or four students. Each group needs a roll of narrow tape or a ribbon and the two geometrical models shown at the right. The purpose of this Excursion is to use the tape (ribbon) to determine the geodesics of a surface.

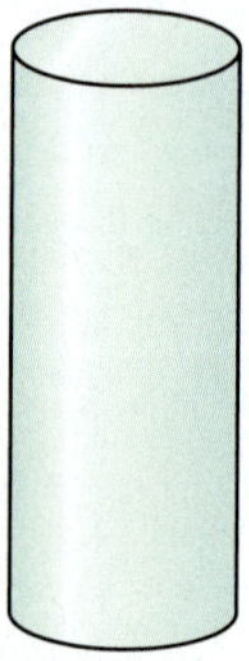

Cylinder

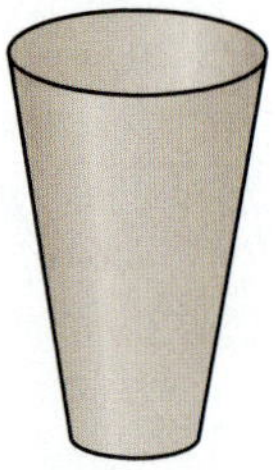

Paper cup

INSTRUCTOR NOTE

The experiments and the Excursion Exercises work best with ribbon or tape that is one-quarter inch in width.

The following three theorems can be used to determine the geodesics of a surface.

> **Geodesic Theorems**
>
> **Theorem 1** If a surface is smooth with no edges or holes, then the shortest path between any two points is on a geodesic of the surface.
>
> **Theorem 2** ***The Tape Test*** If a piece of tape is placed so that it lies flat on a smooth surface, then the center line of the tape is on a geodesic of the surface.
>
> **Theorem 3** ***Inverse of the Tape Test*** If a piece of tape does not lie flat on a smooth surface, then the center line of the tape is not on a geodesic of the surface.

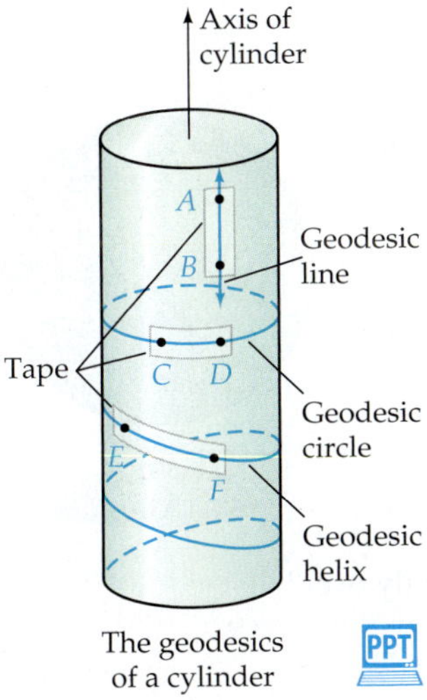

FIGURE 7.10

What are the geodesics of a cylinder? If two points A and B are as shown in Figure 7.10, then the vertical line segment between A and B is the shortest path between the points. A piece of tape can be placed so that it covers point A and point B and lies flat on the cylinder. Thus, by Theorem 2, line segment $\overline{AB}$ is on a geodesic of the cylinder.

If two points C and D are as shown in Figure 7.10, then the minor arc of a circle is the shortest path between the points. Once again we see that a piece of tape can be placed so that it covers points C and D and lies flat on the cylinder. Theorem 2 indicates that the arc $\widehat{CD}$ is on a geodesic of the cylinder.

To find a geodesic that passes through the two points E and F, start your tape at E and proceed slightly downward and to the right, toward point F. If your tape lies flat against the cylinder, you have found the geodesic for the two points. If your tape does not lie flat against the surface, then your path is not a geodesic and you need to experiment further. Eventually you will find the *circular helix* curve, shown in Figure 7.10, that allows the tape to lie flat on the cylinder.[3]

Additional experiments with the tape and the cylinder should convince you that a geodesic of a cylinder is one of the following: (a) a line parallel to the axis of the cylinder, (b) a circle with its center on the axis of the cylinder and its diameter perpendicular to the axis, or (c) a circular helix curve that has a constant slope and a center on the axis of the cylinder.

EXCURSION EXERCISES

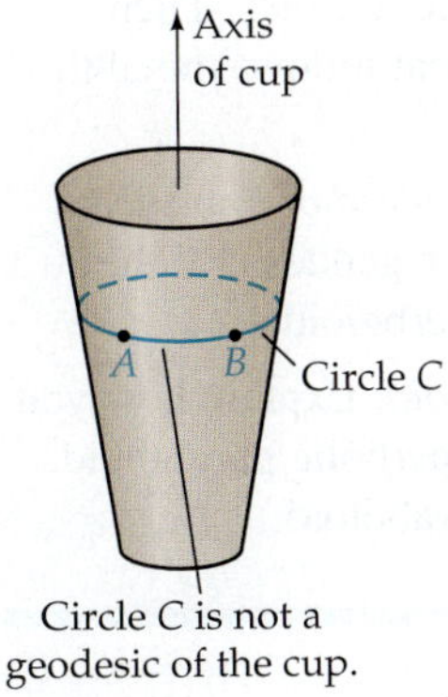

FIGURE 7.11

1. a. Place two points A and B on a paper cup so that A and B are both at the same height, as in Figure 7.11. We know circle C that passes through A and B is *not* a geodesic of the cup because a piece of tape will not lie flat when placed directly on top of circle C. Experiment with a piece of tape to determine the *actual* geodesic that passes through A and B. Make a drawing that shows this geodesic and illustrate how it differs from circle C.

b. Use a cup similar to the one in Figure 7.11 and a piece of tape to determine two other types of geodesics of the cup. Make a drawing that shows each of these two additional types of geodesics.

2. The only geodesics of a sphere are great circles. Write a sentence that explains how you can use a piece of tape to show that circle D in Figure 7.12 is not a geodesic of the sphere.

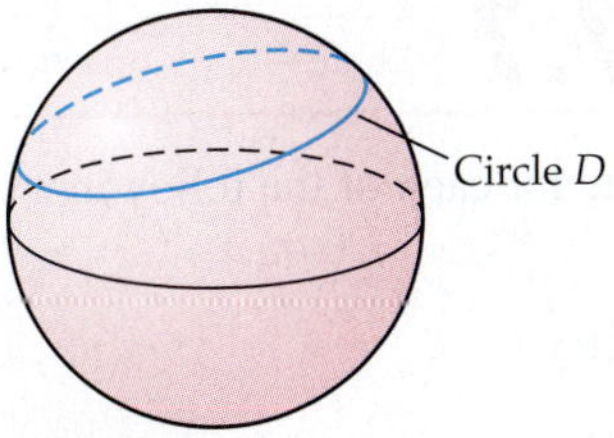

FIGURE 7.12

[3] The thread of a bolt is an example of a circular helix.

3. Write a sentence that explains how you know that circle E in Figure 7.13 is not a geodesic of the figure.

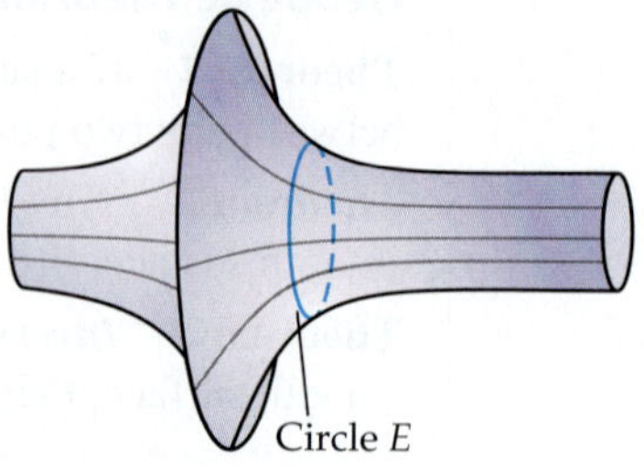

FIGURE 7.13

Excursion Exercises 4 to 6 require a world globe that shows the locations of major cities. We suggest that you use a thin ribbon, instead of tape, to determine the great circle routes in the following exercises, because tape may damage the globe.

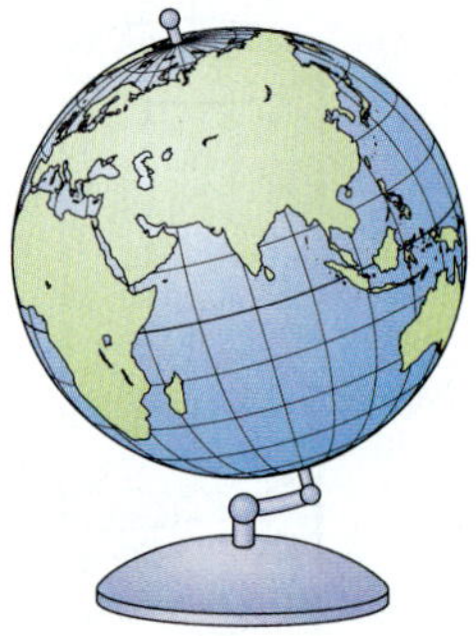

4. A pilot flies a great circle route from Miami, Florida, to Hong Kong. Which one of the following states will the plane fly over?
   - **a.** California
   - **b.** Oregon
   - **c.** Washington
   - **d.** Alaska
5. A pilot flies a great circle route from Los Angeles to London. Which one of the following cities will the plane fly over?
   - **a.** New York
   - **b.** Chicago
   - **c.** Godthaab, Greenland
   - **d.** Vancouver, Canada
6. Washington, D.C., and Seoul, Korea, both have a latitude of about 38°. How many miles (to the nearest 100 miles) will a pilot save by flying a great circle route between the cities as opposed to the route that follows the 38th parallel? *Hint:* Use a cloth measuring tape to measure the minor arc of the great circle that passes through the two cities. Use the scale on the globe to convert this distance to miles. Then use the measuring tape to determine the distance of the route that follows the 38th parallel.

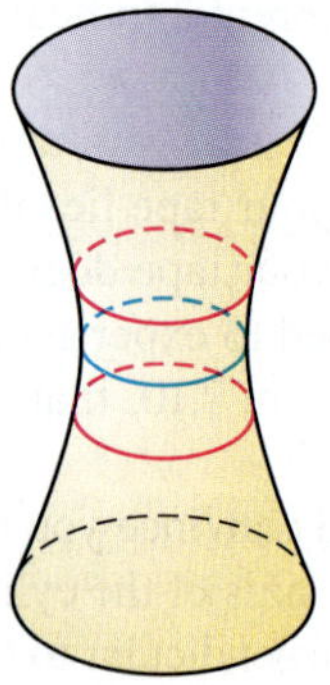

Hyperboloid of one sheet

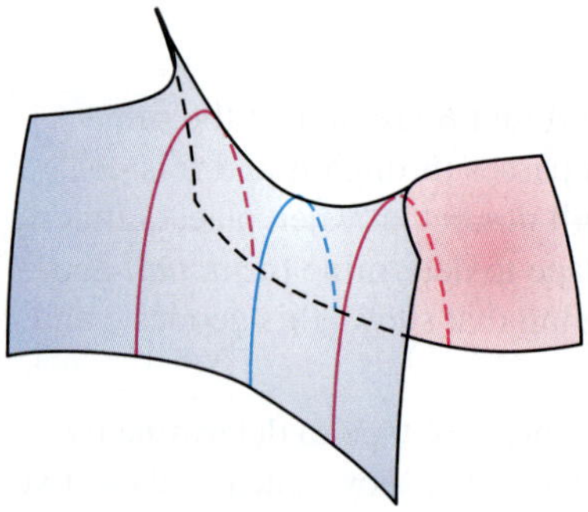

Hyperbolic paraboloid

7. **a.** The yellow surface at the left above is called a *hyperboloid of one sheet.* Explain how you can determine that the blue circle is a geodesic of the hyperboloid, but the red circles are not geodesics of the hyperboloid.

   b. The saddle surface at the left is called a *hyperbolic paraboloid.* Explain how you can determine that the blue parabola is a geodesic of the hyperbolic paraboloid, but the red parabolas are not geodesics of the hyperbolic paraboloid.

EXERCISE SET 7.7

1. State the parallel postulate for each of the following.
   - **a.** Euclidean geometry
   - **b.** Lobachevskian geometry
   - **c.** Riemannian geometry
2. Name the mathematician who is called the Copernicus of geometry.
3. Name the mathematician who was the first to consider a geometry in which Euclid's parallel postulate was replaced with "Through a given point not on a given line, there are *at least two* lines parallel to the given line."
4. What is the maximum number of right angles a triangle can have in
   - **a.** Euclidean geometry?
   - **b.** Lobachevskian geometry?
   - **c.** Riemannian geometry?

5. What name did Lobachevsky give to the geometry that he created?

6. Explain why great circles in Riemannian geometry are thought of as lines.

7. What is a geodesic?

8. In which geometry can two distinct lines be parallel to a third line but not parallel to each other?

9. What model was used in this text to illustrate hyperbolic geometry?

10. In which geometry do all perpendiculars to a given line intersect each other?

11. Find the exact area of a spherical triangle with angles of 150°, 120°, and 90° on a sphere with a radius of 1.

12. Find the area of a spherical triangle with three right angles on a sphere with a radius of 1980 mi. Round to the nearest ten thousand square miles.

City Geometry In Exercises 13 to 20, find the Euclidean distance between the points and the city distance between the points. Assume that both $d_E(P, Q)$ and $d_C(P, Q)$ are measured in blocks. Round approximate results to the nearest tenth of a block.

13. $P(-3, 1), Q(4, 1)$

14. $P(-2, 4), Q(3, -1)$

15. $P(2, -3), Q(-3, 5)$

16. $P(-2, 0), Q(3, 7)$

17. $P(-1, 4), Q(5, -2)$

18. $P(-5, 2), Q(3, -4)$

19. $P(2, 0), Q(3, -6)$

20. $P(2, -2), Q(5, -2)$

A Distance Conversion Formula The following formula can be used to convert the Euclidean distance between the points P and Q to the city distance between P and Q. In this formula, the variable m represents the slope of the line segment $\overline{PQ}$.

$$d_C(P, Q) = \frac{1 + |m|}{\sqrt{1 + m^2}} d_E(P, Q)$$

In Exercises 21 to 26, use the preceding formula to find the city distance between P and Q.

21. $d_E(P, Q) = 5$ blocks, slope of $\overline{PQ} = \frac{3}{4}$

22. $d_E(P, Q) = \sqrt{29}$ blocks, slope of $\overline{PQ} = \frac{2}{5}$

23. $d_E(P, Q) = \sqrt{13}$ blocks, slope of $\overline{PQ} = -\frac{2}{3}$

24. $d_E(P, Q) = 2\sqrt{10}$ blocks, slope of $\overline{PQ} = -3$

25. $d_E(P, Q) = \sqrt{17}$ blocks, slope of $\overline{PQ} = \frac{1}{4}$

26. $d_E(P, Q) = 4\sqrt{2}$ blocks, slope of $\overline{PQ} = -1$

27. Explain why there is no formula that can be used to convert $d_C(P, Q)$ to $d_E(P, Q)$. Assume that no additional information is given other than the value of $d_C(P, Q)$.

28. **a.** If $d_E(P, Q) = d_E(R, S)$, must $d_C(P, Q) = d_C(R, S)$? Explain.

b. If $d_C(P, Q) = d_C(R, S)$, must $d_E(P, Q) = d_E(R, S)$? Explain.

29. Plot the points in the city circle with center $(-2, -1)$ and radius $r = 2$ blocks.

30. Plot the points in the city circle with center $(1, -1)$ and radius $r = 3$ blocks.

31. Plot the points in the city circle with center (0, 0) and radius $r = 2.5$ blocks.

32. Plot the points in the city circle with center (0, 0) and radius $r = 3.5$ blocks.

33. How many points are on the city circle with center (0, 0) and radius $r = n$ blocks, where n is a natural number?

34. Which of the following city circles has the most points, a city circle with center (0, 0) and radius 4.5 blocks or a city circle with center (0, 0) and radius 5 blocks?

EXTENSIONS

Apartment Hunting in a City Use the following information to answer each of the questions in Exercises 35 and 36.

Amy and her husband Ryan are looking for an apartment located adjacent to a city street. Amy works at $P(-3, -1)$ and Ryan works at $Q(2, 3)$. Both Amy and Ryan plan to walk from their apartment to work along routes that follow the north–south and the east–west streets.

35. **a.** Plot the points where Amy and Ryan should look for an apartment if they wish the sum of the city distances they need to walk to work to be a minimum.

b. Plot the points where Amy and Ryan should look for an apartment if they wish the sum of the city distances they need to walk to work to be a minimum and they both will walk the same distance. *Hint:* Find the intersection of the city circle with center P and radius of 4.5 blocks, and the city circle with center Q and radius of 4.5 blocks.

36. **a.** Plot the points where Amy and Ryan should look for an apartment if they wish the sum of the city distances they need to walk to work to be less than or equal to 10 blocks.

b. Plot the points where Amy and Ryan should look for an apartment if they wish the sum of the city distances they need to walk to work to be less than or equal to 10 blocks and they both will walk the same distance. *Hint:* Find the intersection of the city circle with center P and radius of 5.5 blocks, and the city circle with center Q and radius of 5.5 blocks.

37. A Finite Geometry Consider a finite geometry with exactly five points: *A*, *B*, *C*, *D*, and *E*. In this geometry a line is any two of the five points. For example, the two points *A* and *B* together form the line denoted by *AB*. Parallel lines are defined as two lines that do not share a common point.

a. How many lines are in this geometry?

b. How many of the lines are parallel to line *AB*?

38. A Finite Geometry Consider a finite geometry with exactly six points: *A*, *B*, *C*, *D*, *E*, and *F*. In this geometry a line is any two of the six points. Parallel lines are defined as two lines that do not share a common point.

a. How many lines are in this geometry?

b. How many of the lines are parallel to line *AB*?

SECTION 7.8 Fractals

Fractals—Endlessly Repeated Geometric Figures

Have you ever used a computer program to enlarge a portion of a photograph? Sometimes the result is a satisfactory enlargement; however, if the photograph is enlarged too much, the image becomes blurred. For example, the photograph in Figure 7.14 below is shown at its original size. The image in Figure 7.15 is an enlarged portion of Figure 7.14, and the image in Figure 7.16 is an enlarged portion of Figure 7.15. If we continue to make enlargements of enlargements, we will produce extremely blurred images that provide little information about the original photograph.

A computer monitor displays an image using small dots called *pixels*. If a computer image is enlarged using a software program, the program must determine the color of each pixel in the enlargement. If the image file for the photograph cannot supply the needed color information for each pixel, the color of some pixels is calculated by *averaging* the numerical color values of neighboring pixels for which the image file has the color information.

Shipfactory/Shutterstock.com

FIGURE 7.14

Shipfactory/Shutterstock.com

FIGURE 7.15

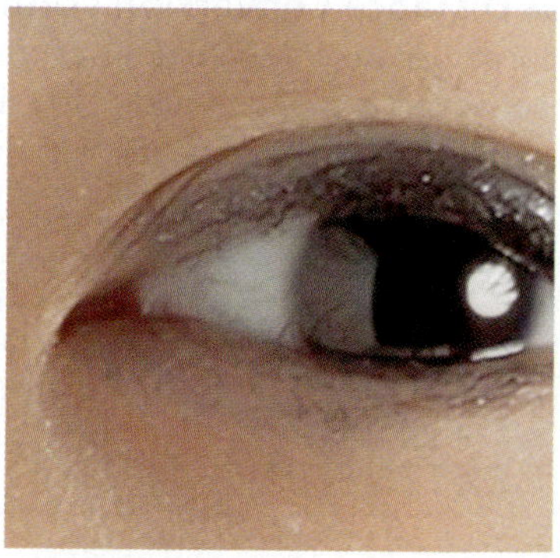

Shipfactory/Shutterstock.com

FIGURE 7.16

In the 1970s, the mathematician Benoit Mandelbrot discovered some remarkable methods that enable us to create geometric figures with a special property: if any portion of the figure is enlarged repeatedly, then additional details (not fewer details, as with the enlargement of a photograph) of the figure are displayed. Mandelbrot called these endlessly repeated geometric figures *fractals*. The fractals that we will study in this lesson can be defined as follows. A **fractal** is a geometric figure in which a self-similar motif repeats itself on an ever-diminishing scale.

Fractals are generally constructed by using **iterative processes** in which the fractal is more closely approximated as a repeated cycle of procedures is performed. For example, a fractal known as the *Koch curve* is constructed as follows.

Construction of the Koch Curve

Step 0: Start with a line segment. This initial segment is shown as stage 0 in Figure 7.17. Stage 0 of a fractal is called the **initiator** of the fractal.

Step 1: On the middle third of the line segment, draw an equilateral triangle and remove its base. The resulting curve is stage 1 in Figure 7.17. Stage 1 of a fractal is called the **generator** of the fractal.

Step 2: Replace each initiator shape (line segment, in this example) with a *scaled version* of the generator to produce the next stage of the Koch curve. The width of the scaled version of the generator is the same as the width of the line segment it replaces. Continue to repeat this step ad infinitum to create additional stages of the Koch curve.

Three applications of step 2 produce stage 2, stage 3, and stage 4 of the Koch curve, as shown in Figure 7.17.

TAKE NOTE

The line segments at any stage of the Koch curve are $\frac{1}{3}$ the length of the line segments at the previous stage. To create the next stage of the Koch curve (after stage 1), replace each line segment with a scaled version of the generator.

At any stage after stage 2, the scaled version of the generator is $\frac{1}{3}$ the size of the preceding scaled generator.

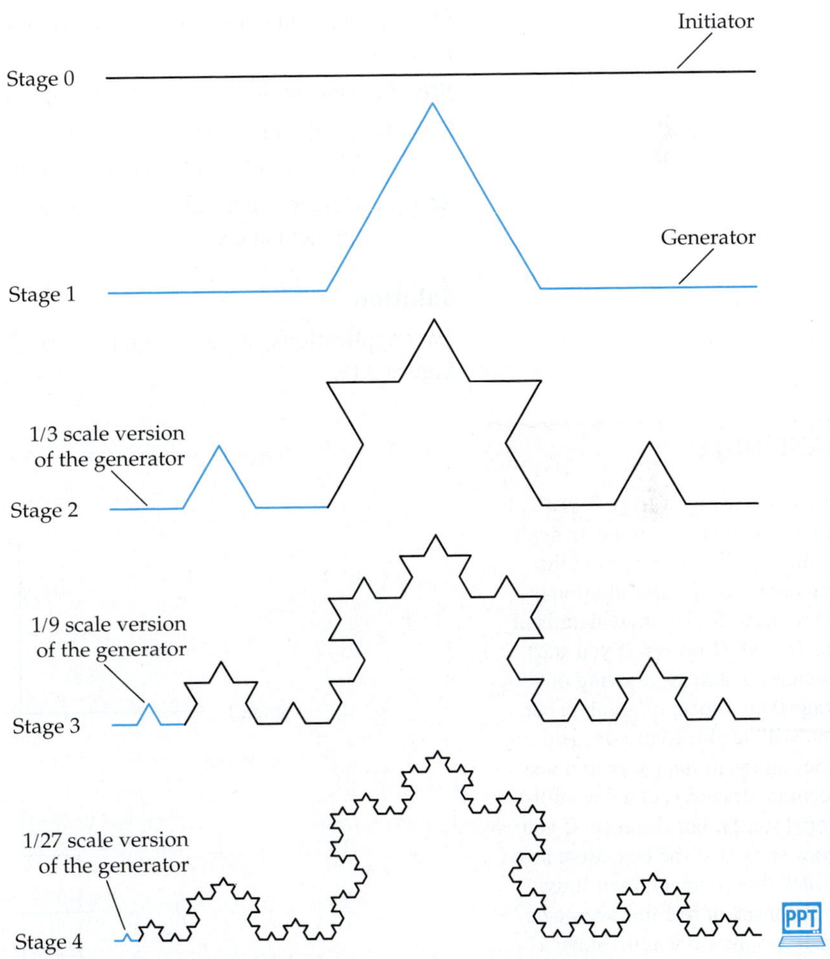

FIGURE 7.17 The first five stages of the Koch curve

HISTORICAL NOTE

The Koch curve was created by the mathematician Niels Fabian Helge von Koch (kôkh) (1870–1924). Von Koch attended Stockholm University and in 1911 became a professor of mathematics at Stockholm University. In addition to his early pioneering work with fractals, he wrote several papers on number theory.

None of the curves shown in Figure 7.17 is the Koch curve. The Koch curve is the curve that would be produced if step 2 in the above construction process were repeated ad infinitum. No one has ever seen the Koch curve, but we know that it is a very jagged curve in which the self-similar motif shown in Figure 7.17 repeats itself on an ever-diminishing scale.

POINT OF INTEREST

It can be shown that the Koch snowflake has a finite area but an infinite perimeter. Thus it has the remarkable property that you could paint the surface of the snowflake, but you could never get enough paint to paint its boundary.

The curves shown in Figure 7.18 are the first five stages of the *Koch snowflake*.

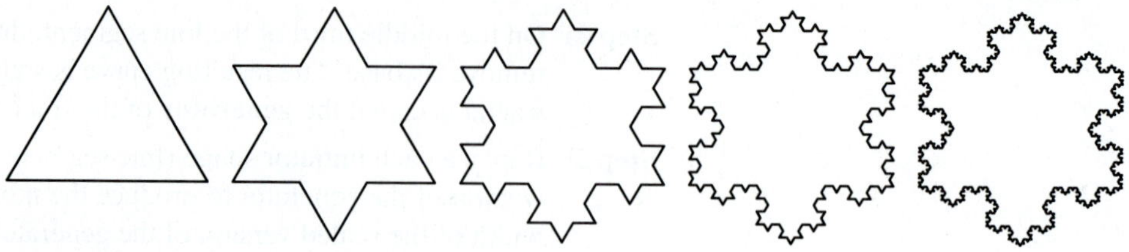

FIGURE 7.18 The first five stages of the Koch snowflake

EXAMPLE 1 Draw Stages of a Fractal

Draw stage 2 and stage 3 of the *box curve*, which is defined by the following iterative process.

Step 0: Start with a line segment as the initiator. See stage 0 in Figure 7.19.

Step 1: On the middle third of the line segment, draw a square and remove its base. This produces the generator of the box curve. See stage 1 in Figure 7.19.

Step 2: Replace each initiator shape with a scaled version of the generator to produce the next stage.

Solution

Two applications of step 2 yield stage 2 and stage 3 of the box curve, as shown in Figure 7.19.

TAKE NOTE

If your drawing of stage 0 of a fractal is small, it will be difficult to draw additional stages of the fractal, because each additional stage must display more details of the fractal. However, if you start by making a *large* drawing of stage 0 on a sheet of graph paper, you will be able to use the grid lines on the graph paper to make accurate drawings of a few additional stages. For instance, if you draw stage 0 of the box curve as a 9-inch line segment, then stage 1 will consist of five line segments, each 3 inches in length. Stage 2 will consist of 25 line segments, each 1 inch in length. The 125 line segments of stage 3 will each be $\frac{1}{3}$ of an inch in length.

Instead of erasing and drawing each new stage of the fractal on top of the previous stage, it is advantageous to draw each new stage directly below the previous stage, as shown in Figure 7.19.

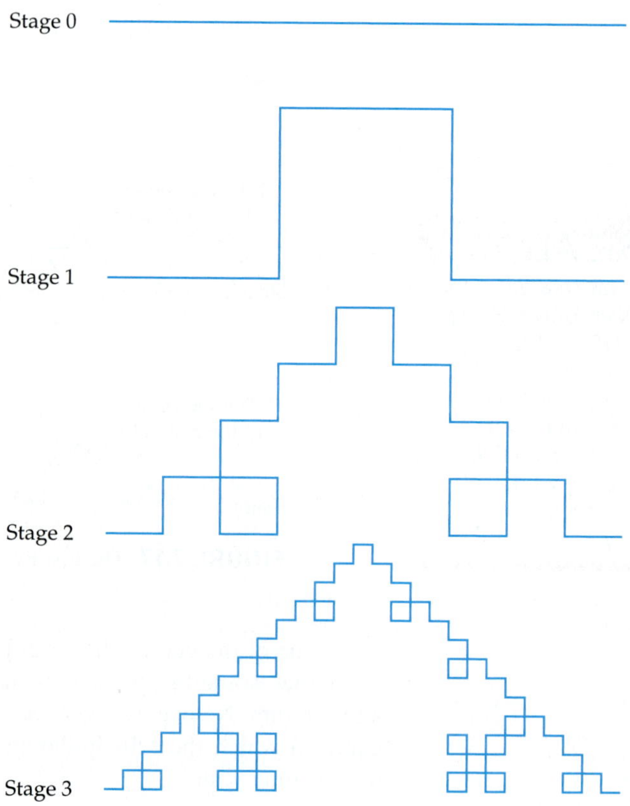

FIGURE 7.19 The first four stages of the box curve

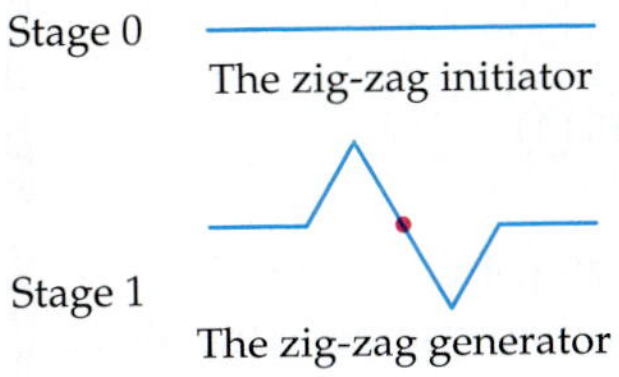

FIGURE 7.20 The initiator and generator of the zig-zag curve

CHECK YOUR PROGRESS 1 Draw stage 2 of the *zig-zag curve*, which is defined by the following iterative process.

Step 0: Start with a line segment. See stage 0 of Figure 7.20.

Step 1: Remove the middle half of the line segment and draw a zig-zag, as shown in stage 1 of Figure 7.20. Each of the six line segments in the generator is a $\frac{1}{4}$-scale replica of the initiator.

Step 2: Replace each initiator shape with the scaled version of the generator to produce the next stage. Repeat this step to produce additional stages.

Solution See page S26.

In each of the previous fractals, the initiator was a line segment. In Example 2, we use a triangle and its interior as the initiator.

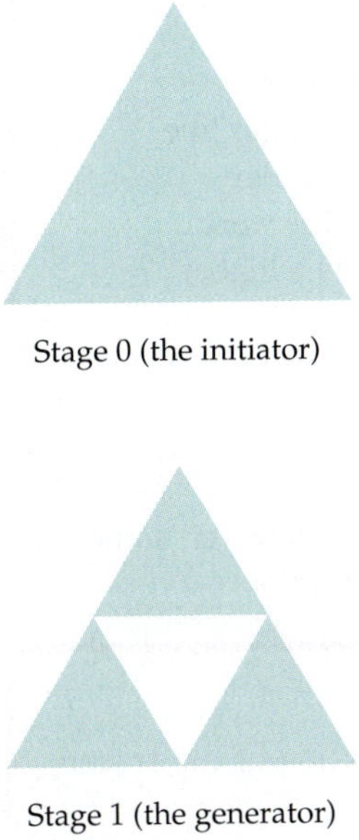

FIGURE 7.21 The initiator and generator of the Sierpinski gasket

EXAMPLE 2 Draw Stages of a Fractal

Draw stage 2 and stage 3 of the *Sierpinski gasket* (also known as the *Sierpinski triangle*), which is defined by the following iterative process.

Step 0: Start with an equilateral triangle and its interior. This is stage 0 of the Sierpinski gasket. See Figure 7.21.

Step 1: Form a new triangle by connecting the midpoints of the sides of the triangle. Remove this center triangle. The result is the three green triangles shown in stage 1 in Figure 7.21.

Step 2: Replace each initiator (green triangle) with a scaled version of the generator.

Solution

Two applications of step 2 of the above process produce stage 2 and stage 3 of the Sierpinski gasket, as shown in Figure 7.22.

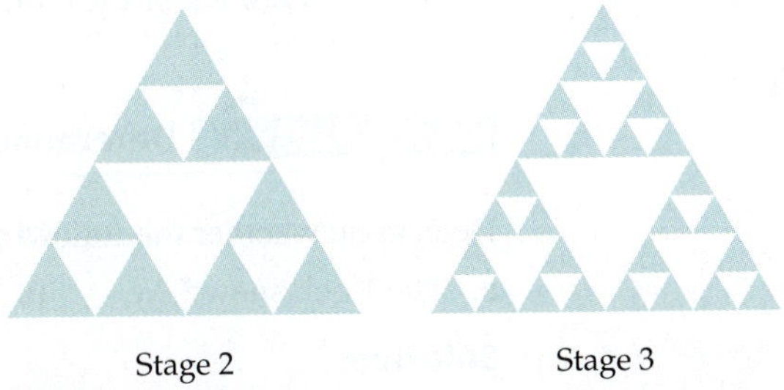

FIGURE 7.22 Stages 2 and 3 of the Sierpinski gasket

CHECK YOUR PROGRESS 2 Draw stage 2 of the *Sierpinski carpet*, which is defined by the following process.

Step 0: Start with a square and its interior. See stage 0 in Figure 7.23.

Step 1: Subdivide the square into nine smaller congruent squares and remove the center square. This yields stage 1 (the generator) shown in Figure 7.23.

Step 2: Replace each initiator (tan square) with a scaled version of the generator. Repeat this step to create additional stages of the Sierpinski carpet.

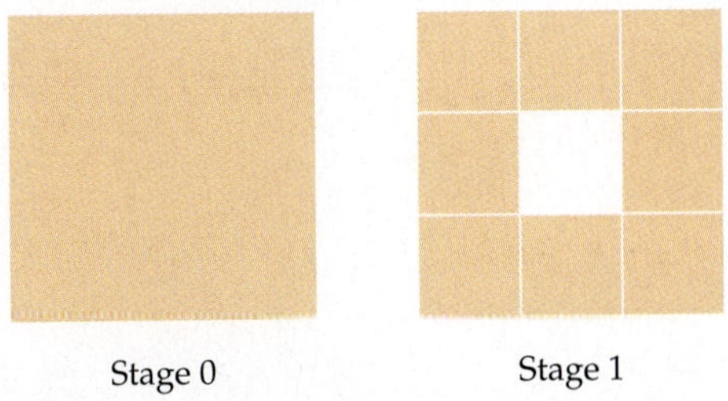

FIGURE 7.23 The initiator and generator of the Sierpinski carpet

Solution See page S26.

MATH MATTERS Benoit Mandelbrot (1924–2010)

POINT OF INTEREST

Fractal geometry is not just a chapter of mathematics, but one that helps Everyman to see the same old world differently.—Benoit Mandelbrot, IBM Research (*Source:* http://php.iupui.edu/~wijackso/fractal3.htm)

Benoit Mandelbrot is often called the father of fractal geometry. He was not the first person to create a fractal, but he was the first person to discover how some of the ideas of earlier mathematicians such as Georg Cantor, Giuseppe Peano, Helge von Koch, Waclaw Sierpinski, and Gaston Julia could be united to form a new type of geometry. Mandelbrot also recognized that many fractals share characteristics with shapes and curves found in nature. For instance, the leaves of a fern, when compared with the whole fern, are almost identical in shape, only smaller. This self-similarity characteristic is evident (to some degree) in all fractals. The following quote by Mandelbrot is from his 1983 book, *The Fractal Geometry of Nature.*[4]

> Clouds are not spheres, mountains are not cones, coastlines are not circles, and bark is not smooth, nor does lightning travel in a straight line. More generally, I claim that many patterns of Nature are so irregular and fragmented, that, compared with Euclid—a term used in this work to denote all of standard geometry—Nature exhibits not simply a higher degree but an altogether different level of complexity.

Strictly Self-Similar Fractals

All fractals show a self-similar motif on an ever-diminishing scale; however, some fractals are *strictly self-similar* fractals, according to the following definition.

Strictly Self-Similar Fractal

A fractal is said to be **strictly self-similar** if any arbitrary portion of the fractal contains a replica of the entire fractal.

EXAMPLE 3 Determine Whether a Fractal Is Strictly Self-Similar

Determine whether the following fractals are strictly self-similar.

a. The Koch snowflake **b.** The Koch curve

Solution

a. The Koch snowflake is a closed figure. Any portion of the Koch snowflake (like the portion circled in Figure 7.24) is not a closed figure. Thus the Koch snowflake is *not* a strictly self-similar fractal.

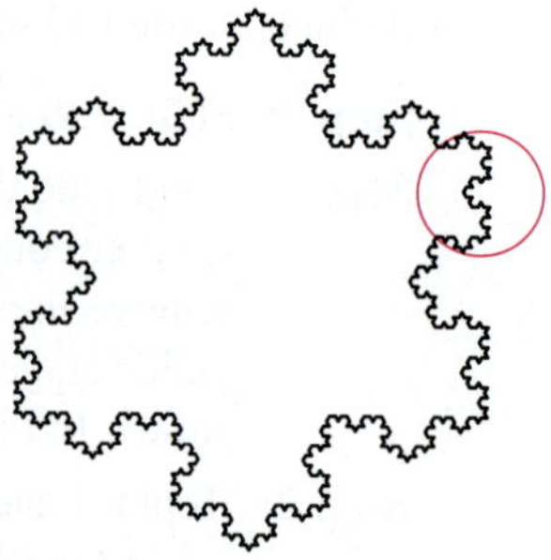

FIGURE 7.24 The portion of the Koch snowflake shown in the circle is not a replica of the entire snowflake.

[4] Mandelbrot, Benoit B. *The Fractal Geometry of Nature*. New York: W. H. Freeman and Company, 1983, p. 1.

b. Because any portion of the Koch curve replicates the entire fractal, the Koch curve is a strictly self-similar fractal. See Figure 7.25.

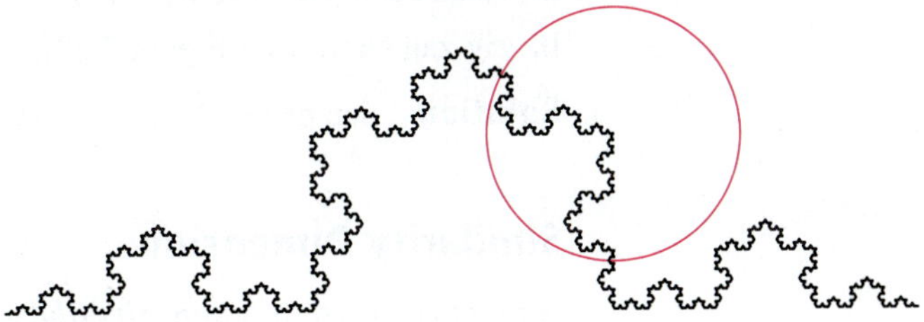

FIGURE 7.25 Any portion of the Koch curve is a replica of the entire Koch curve.

CHECK YOUR PROGRESS 3 Determine whether the following fractals are strictly self-similar.

a. The box curve (see Example 1) **b.** The Sierpinski gasket (see Example 2)

Solution See page S26.

Replacement Ratio and Scaling Ratio

Mathematicians like to assign numbers to fractals so that they can objectively compare fractals. Two numbers that are associated with many fractals are the *replacement ratio* and the *scaling ratio*.

Replacement Ratio and Scaling Ratio of a Fractal

- If the generator of a fractal consists of N replicas of the initiator, then the **replacement ratio** of the fractal is N.
- If the initiator of a fractal has linear dimensions that are r times the corresponding linear dimensions of its replicas in the generator, then the **scaling ratio** of the fractal is r.

PPT

EXAMPLE 4 **Find the Replacement Ratio and the Scaling Ratio of a Fractal**

Find the replacement ratio and the scaling ratio of the

a. box curve. **b.** Sierpinski gasket.

Solution

a. Figure 7.19 on page 440 shows that the generator of the box curve consists of five line segments and that the initiator consists of only one line segment. Thus the replacement ratio of the box curve is 5:1, or 5.

The initiator of the box curve is a line segment that is 3 times as long as the replica line segments in the generator. Thus the scaling ratio of the box curve is 3:1, or 3.

b. Figure 7.21 on page 441 shows that the generator of the Sierpinski gasket consists of three triangles and that the initiator consists of only one triangle. Thus the replacement ratio of the Sierpinski gasket is 3:1, or 3.

The initiator triangle of the Sierpinski gasket has a width that is 2 times the width of the replica triangles in the generator. Thus the scaling ratio of the Sierpinski gasket is 2:1, or 2.

CHECK YOUR PROGRESS 4 Find the replacement ratio and the scaling ratio of the

a. Koch curve (see Figure 7.17).

b. zig-zag curve (see Figure 7.20).

Solution See page S26.

Similarity Dimension

A number called the *similarity dimension* is used to quantify how densely a strictly self-similar fractal fills a region.

INSTRUCTOR NOTE

In some texts on fractals, the scaling ratio is defined as the *reciprocal* of the scaling ratio r defined on the previous page. In such texts, the similarity dimension D of a strictly self-similar fractal is given by

$$D = \frac{\log N}{\log\left(\frac{1}{r}\right)}$$

Similarity Dimension

The **similarity dimension** D of a strictly self-similar fractal is given by

$$D = \frac{\log N}{\log r}$$

where N is the replacement ratio of the fractal and r is the scaling ratio.

EXAMPLE 5 Find the Similarity Dimension of a Fractal

Find the similarity dimension, to the nearest thousandth, of the

a. Koch curve.

b. Sierpinski gasket.

Solution

TAKE NOTE

Because the Koch snowflake is not a strictly self-similar fractal, we cannot compute its similarity dimension.

a. Because the Koch curve is a strictly self-similar fractal, we can find its similarity dimension. Figure 7.17 on page 439 shows that stage 1 of the Koch curve consists of four line segments and stage 0 consists of only one line segment. Hence the replacement ratio is 4:1, or 4. The line segment in stage 0 is 3 times as long as each of the replica line segments in stage 1, so the scaling ratio is 3. Thus the Koch curve has a similarity dimension of

$$D = \frac{\log 4}{\log 3} \approx 1.262$$

b. In Example 4, we found that the Sierpinski gasket has a replacement ratio of 3 and a scaling ratio of 2. Thus the Sierpinski gasket has a similarity dimension of

$$D = \frac{\log 3}{\log 2} \approx 1.585$$

CHECK YOUR PROGRESS 5 Compute the similarity dimension, to the nearest thousandth, of the

a. box curve (see Example 1). **b.** Sierpinski carpet (see Check Your Progress 2).

Solution See page S26.

The results of Example 5 show that the Sierpinski gasket has a larger similarity dimension than the Koch curve. This means that the Sierpinski gasket fills a flat two-dimensional surface more densely than does the Koch curve.

Computers are used to generate fractals such as those shown in Figure 7.26. These fractals were *not* rendered by using an initiator and a generator, but they were rendered using iterative procedures.

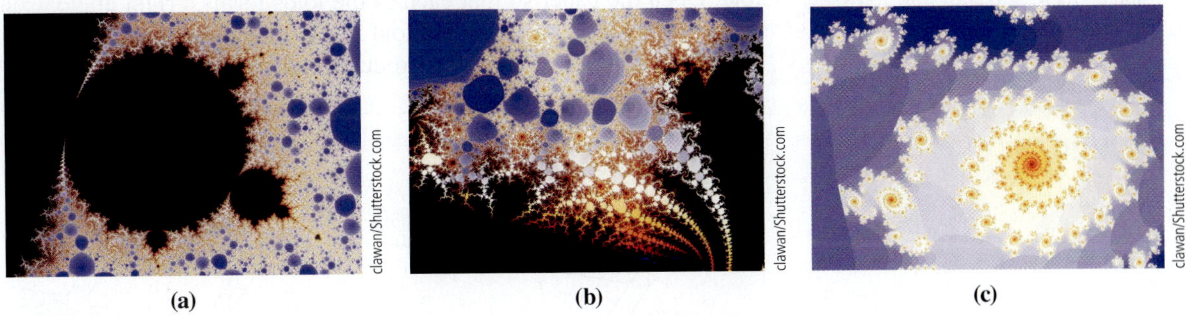

FIGURE 7.26 Computer-generated fractals

Fractals have other applications in addition to being used to produce intriguing images. For example, computer scientists have recently developed fractal image compression programs based on self-transformations of an image. An image compression program is a computer program that converts an image file to a smaller file that requires less computer memory. In some situations, these fractal compression programs outperform standardized image compression programs such as JPEG (*jay-peg*), which was developed by the Joint Photographic Experts Group.

Some cellular telephones have been manufactured with internal antennas that are fractal in design. Figure 7.27 shows a cellular telephone with an internal antenna in the shape of a stage of the Sierpinski carpet fractal. The antenna in Figure 7.28 is in the shape of a stage of the Koch curve.

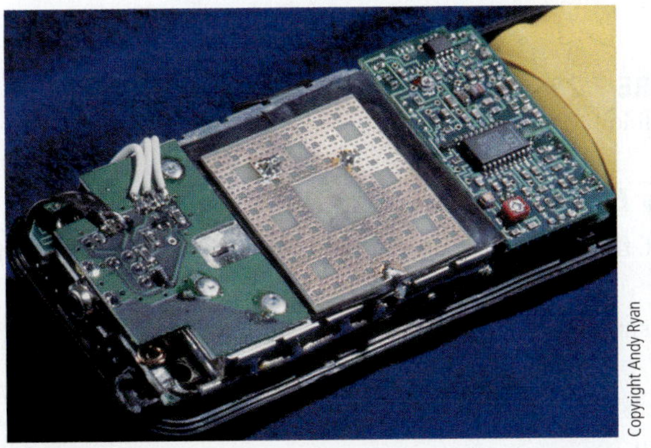

FIGURE 7.27 A Sierpinski carpet fractal antenna hidden inside a cellular telephone

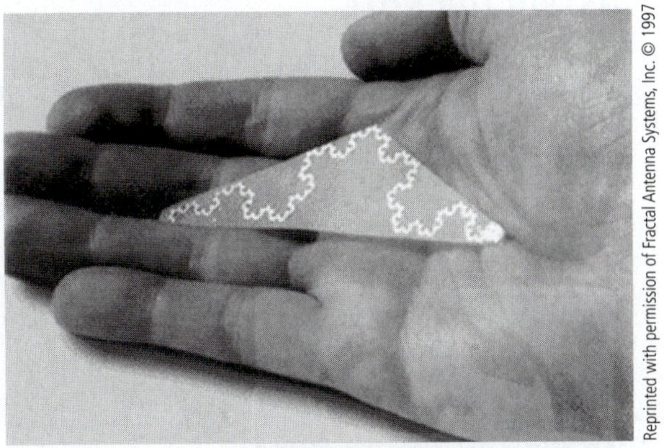

FIGURE 7.28 A Koch curve antenna designed for use in wireless communication devices

EXCURSION

The Heighway Dragon Fractal

In this Excursion, we illustrate two methods of constructing the stages of a fractal known as the *Heighway dragon.*

The Heighway Dragon via Paper Folding

The first few stages of the Heighway dragon fractal can be constructed by the repeated folding of a strip of paper. In the following discussion, we use a 1-inch-wide strip of paper

TAKE NOTE

The easiest way to construct stage 4 is to make all four folds before you open the paper. All folds must be made in the same direction.

that is 14 in. in length as stage 0. To create stage 1 of the dragon fractal, just fold the strip in half and open it so that the fold forms a right angle (see Figure 7.29). To create stage 2, fold the original strip twice. The second fold should be in the same direction as the first fold. Open the paper so that each of the folds forms a right angle. Continue the iterative process of making an additional fold in the same direction as the first fold and then forming a right angle at each fold to produce additional stages. See Figure 7.29.

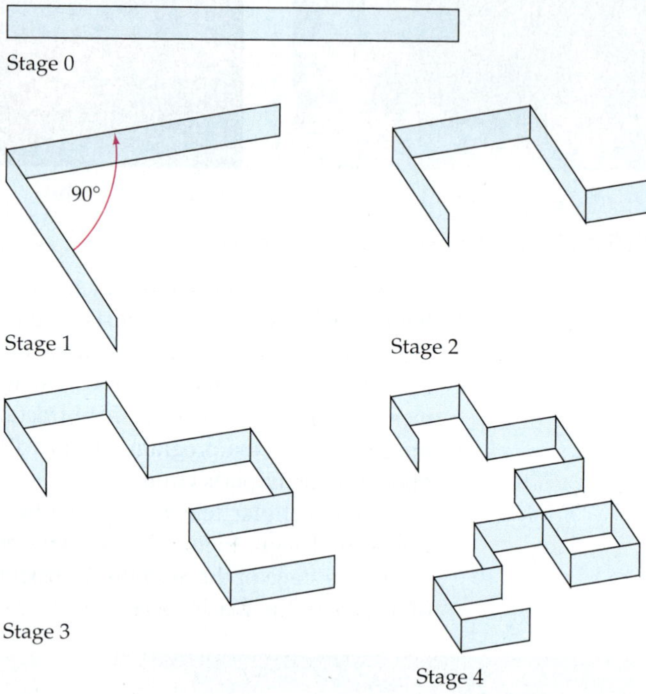

FIGURE 7.29 The first five stages of the Heighway dragon via paper folding

INSTRUCTOR NOTE

Some textbooks and Internet sites refer to the Heighway dragon fractal as the Harter-Heighway dragon.

POINT OF INTEREST

The Heighway dragon is sometimes called the "Jurassic Park fractal" because it was used in the book *Jurassic Park* to illustrate some of the surprising results that can be obtained by iterative processes.

The Heighway Dragon via the Left-Right Rule

The nth stage of the Heighway dragon can also be created by the following drawing procedure.

Step 0: Draw a small vertical line segment. Label the bottom point of this segment as vertex $v = 0$ and label the top as vertex $v = 1$.

Step 1: Use the following left-right rule to determine whether to make a left turn or a right turn.

The Left-Right Rule

At vertex v, where v is an *odd* number, go

- **right** if the remainder of v divided by 4 is 1.
- **left** if the remainder of v divided by 4 is 3.

At vertex 2, go to the right. At vertex v, where v is an *even* number greater than 2, go in the same direction in which you went at vertex $\frac{v}{2}$.

Draw another line segment of the same length as the original segment. Label the endpoint of this segment with a number that is 1 larger than the number used for the preceding vertex.

Step 2: Continue to repeat step 1 until you reach the last vertex. The last vertex of an n-stage Heighway dragon is the vertex numbered 2^n.

EXCURSION EXERCISES

1. Use a strip of paper and the folding procedure explained on page 446 to create models of the first five stages (stage 0 through stage 4) of the Heighway dragon. Explain why it would be difficult to create the 10th stage of the Heighway dragon using the paper-folding procedure.
2. Use the left-right rule to draw stage 2 of the Heighway dragon.
3. Use the left-right rule to determine the direction in which to turn at vertex 7 of the Heighway dragon.
4. Use the left-right rule to determine the direction in which to turn at vertex 50 of the Heighway dragon.
5. Use the left-right rule to determine the direction in which to turn at vertex 64 of the Heighway dragon.

EXERCISE SET 7.8

■ In Exercises 1 and 2, use an iterative process to draw stage 2 and stage 3 of the fractal with the given initiator (stage 0) and the given generator (stage 1).

1. The Cantor point set

2. Lévy's curve

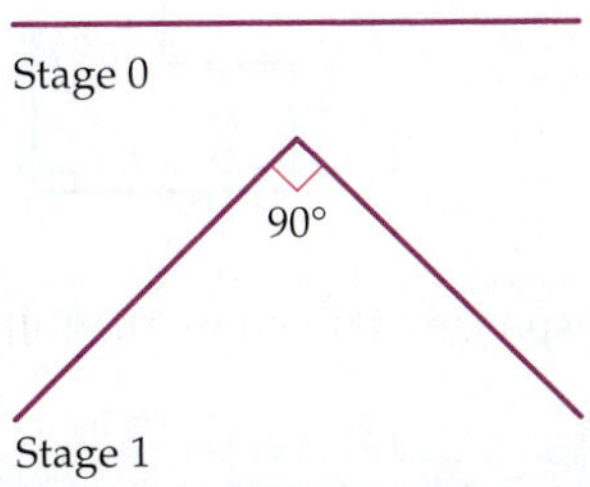

■ In Exercises 3 to 8, use an iterative process to draw stage 2 of the fractal with the given initiator (stage 0) and the given generator (stage 1).

3. The Sierpinski carpet, variation 1

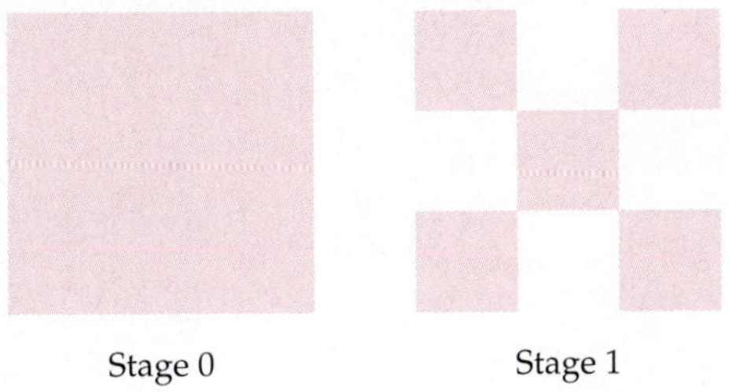

4. The Sierpinski carpet, variation 2

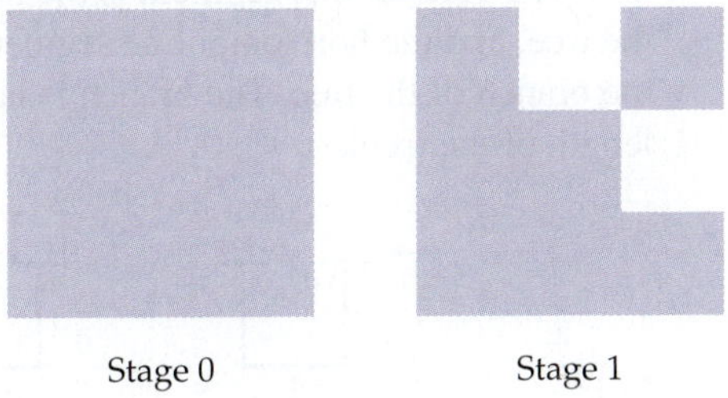

5. The river tree of Peano Cearo

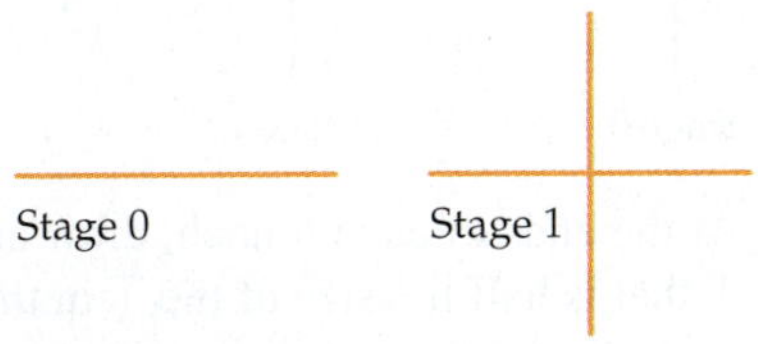

6. Minkowski's fractal

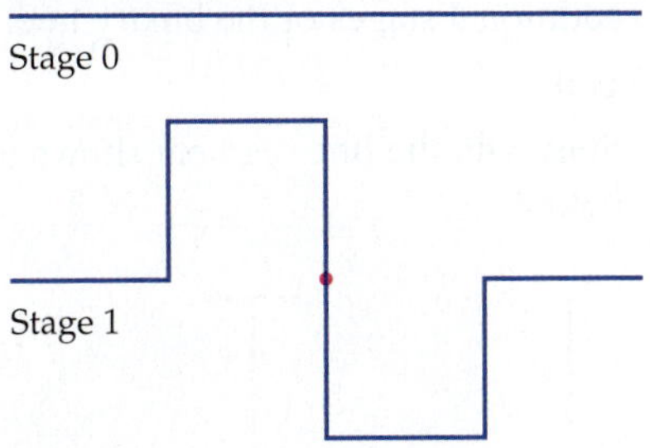

7. The square fractal

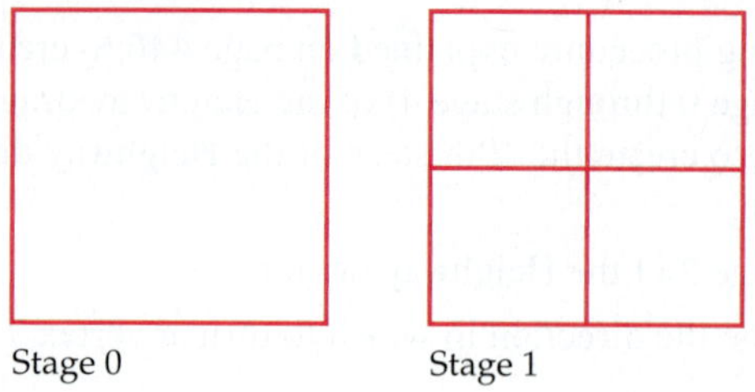

8. The cube fractal

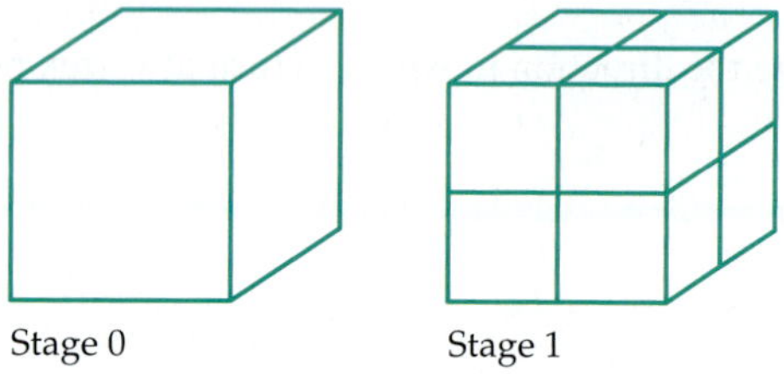

■ In Exercises 9 and 10, draw stage 3 and stage 4 of the fractal defined by the given iterative process.

9. The binary tree

Step 0: Start with a "⊤." This is stage 0 of the binary tree. The vertical line segment is the trunk of the tree, and the horizontal line segment is the branch of the tree. The branch is half the length of the trunk.

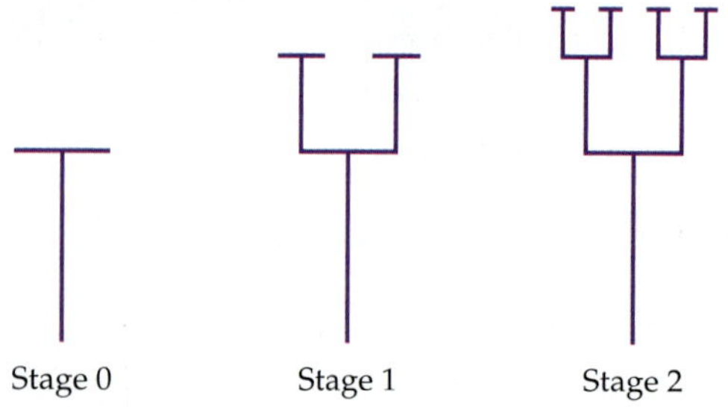

Step 1: At the ends of each branch, draw an upright ⊤ that is half the size of the ⊤ in the preceding stage.

Step 2: Continue to repeat step 1 to generate additional stages of the binary tree.

10. The I-fractal

Step 0: Start with the line segment shown as stage 0 below.

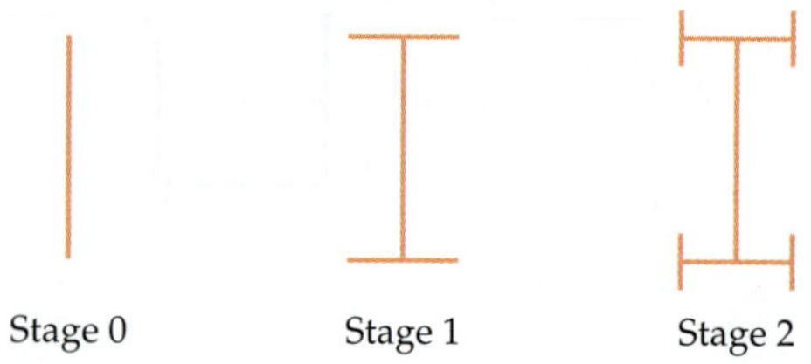

Step 1: At each end of the line segment, draw a crossbar that is half the length of the line segment it contacts. (This produces stage 1 of the I-fractal.)

Step 2: Use each crossbar from the preceding step as the connecting segment of a new "I." Attach new crossbars that are half the length of the connecting segment.

Step 3: Continue to repeat step 2 to generate additional stages of the I-fractal.

■ In Exercises 11 to 20, compute, if possible, the similarity dimension of the fractal. Round to the nearest thousandth.

11. The Cantor point set (see Exercise 1)

12. Lévy's curve (see Exercise 2)

13. The Sierpinski carpet, variation 1 (see Exercise 3)

14. The Sierpinski carpet, variation 2 (see Exercise 4)

15. The river tree of Peano Cearo (see Exercise 5)

16. Minkowski's fractal (see Exercise 6)

17. The square fractal (see Exercise 7)

18. The cube fractal (see Exercise 8)

19. The quadric Koch curve, defined by the following stages

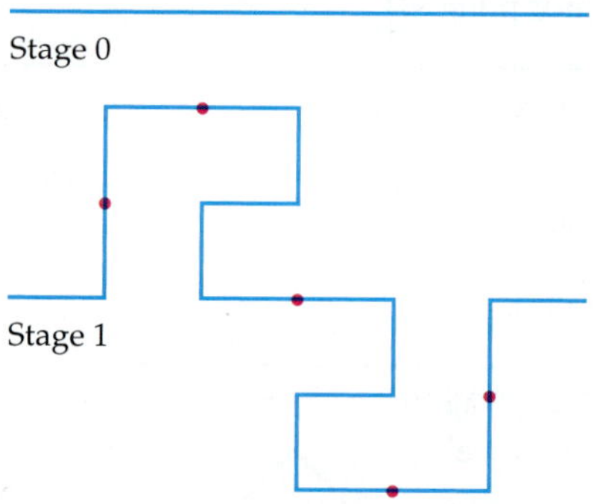

20. The Menger sponge, defined by the following stages

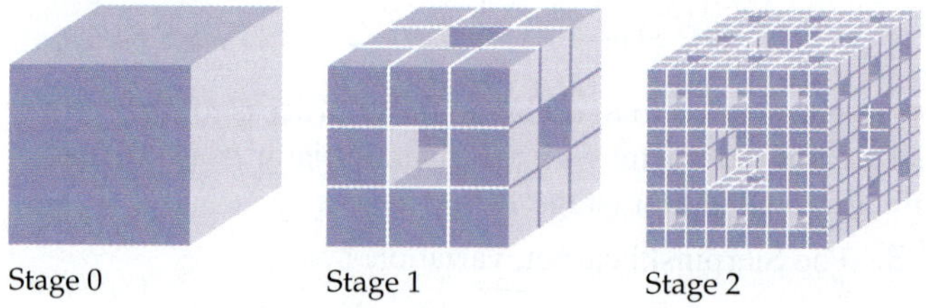

EXTENSIONS

21. Compare Similarity Dimensions

a. Rank, from largest to smallest, the similarity dimensions of the Sierpinski carpet; the Sierpinski carpet, variation 1 (see Exercise 3); and the Sierpinski carpet, variation 2 (see Exercise 4).

b. Which of the three fractals is the most dense?

22. The Peano Curve The *Peano curve* is defined by the following stages.

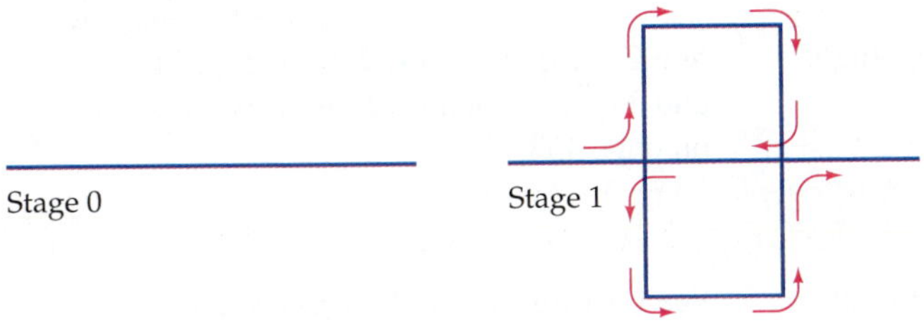

The arrows show the route used to trace the generator.

a. What is the similarity dimension of the Peano curve?

b. Explain why the Peano curve is referred to as a plane-filling curve.

23. Explain why the similarity dimension formula cannot be used to find the similarity dimension of the binary tree defined in Exercise 9.

24. Stage 0 and stage 1 of Lévy's curve (see Exercise 2) and the Heighway dragon (see page 446) are identical, but the fractals start to differ at stage 2. Make two drawings that illustrate how they differ at stage 2.

CHAPTER 7 SUMMARY

The following table summarizes essential concepts in this chapter. The references given in the right-hand column list Examples and Exercises that can be used to test your understanding of a concept.

7.1 Measurement	
U.S. Customary Units of Measure ■ Measures of length: inches, feet, yards, miles ■ Measures of weight: ounces, pounds, tons ■ Measures of capacity: fluid ounces, cups, pints, quarts, gallons	See **Examples 1 to 3** on pages 356 and 357. Then try Exercises 1 and 2 on page 453.
The Metric System The metric system uses a prefix of a base unit to denote the magnitude of a measurement. ■ Base unit of length: meter ■ Base unit of mass: gram ■ Base unit of capacity: liter Some of the prefixes used in the metric system are kilo (1000), hecto (100), deca (10), deci (0.1), centi (0.01), and milli (0.001).	See **Examples 4, 5, and 6** on pages 358–360. Then try Exercises 3, 4, and 5 on page 453.
Convert Between U.S. Customary Units and Metric Units	See **Example 7** on page 361. Then try Exercise 6 on page 453.

continued

7.2 Basic Concepts of Euclidean Geometry	
Vertical Angles The nonadjacent angles formed by two intersecting lines are called vertical angles. Vertical angles have the same measure.	See **Example 4** on page 369, and then try Exercise 14 on page 453.
Lines and Angles A line that intersects two other lines at different points is a transversal. If the lines cut by a transversal are parallel lines, several angles of equal measure are formed. ▪ Pairs of alternate interior angles have the same measure. ▪ Pairs of alternate exterior angles have the same measure. ▪ Pairs of corresponding angles have the same measure.	See **Example 5** on page 369, and then try Exercise 18 on page 453.
Triangles The sum of the measures of the interior angles of a triangle is 180°.	See **Examples 6 and 7** on pages 371 and 372, and then try Exercises 7 and 21 on page 453.
7.3 Perimeter and Area of Plane Figures	
Perimeter The perimeter of a plane geometric figure is a measure of the distance around the figure. Triangle: $P = a + b + c$ Rectangle: $P = 2L + 2W$ Square: $P = 4s$ Parallelogram: $P = 2b + 2s$ Circle: $C = \pi d$ or $C = 2\pi r$	See **Examples 1 to 3** on pages 379–381, and then try Exercises 17, 24, and 26 on pages 453 and 454.
Area Area is the amount of surface in a region. Area is measured in square units. Triangle: $A = \frac{1}{2}bh$ Rectangle: $A = LW$ Square: $A = s^2$ Parallelogram: $A = bh$ Trapezoid: $A = \frac{1}{2}h(b_1 + b_2)$ Circle: $A = \pi r^2$	See **Examples 5 to 10** on pages 382–386, and then try Exercises 15, 22, 27, and 28 on pages 453 and 454.
7.4 Properties of Triangles	
Similar Triangles The ratios of corresponding sides are equal. The ratio of corresponding heights is equal to the ratio of corresponding sides. The measures of corresponding angles are equal.	See **Examples 1 and 2** on pages 395–397, and then try Exercise 8 on page 453.
Congruent Triangles **Side-Side-Side Theorem (SSS)** If the three sides of one triangle are equal in measure to the three sides of a second triangle, the two triangles are congruent. **Side-Angle-Side Theorem (SAS)** If two sides and the included angle of one triangle are equal in measure to two sides and the included angle of a second triangle, the two triangles are congruent. **Angle-Side-Angle Theorem (ASA)** If two angles and the included side of one triangle are equal in measure to two angles and the included side of a second triangle, the two triangles are congruent.	See **Example 3** on page 399, and then try Exercise 29 on page 454.

continued

The Pythagorean Theorem If a and b are the lengths of the legs of a right triangle and c is the length of the hypotenuse, then $c^2 = a^2 + b^2$.	See **Example 4** on page 400, and then try Exercise 30 on page 454.
7.5 Volume and Surface Area	
Volume Volume is a measure of the amount of space occupied by a geometric solid. Volume is measured in cubic units. Rectangular solid: $V = LWH$ Cube: $V = s^3$ Sphere: $V = \frac{4}{3}\pi r^3$ Right circular cylinder: $V = \pi r^2 h$ Right circular cone: $V = \frac{1}{3}\pi r^2 h$ Regular square pyramid: $V = \frac{1}{3}s^2 h$	See **Examples 1, 2, and 3** on pages 409 and 410, and then try Exercises 9, 16, 20, and 23 on page 453.
Surface Area The surface area of a solid is the total area on the surface of the solid. Rectangular solid: $S = 2LW + 2LH + 2WH$ Cube: $S = 6s^2$ Sphere: $S = 4\pi r^2$ Right circular cylinder: $S = 2\pi r^2 + 2\pi rh$ Right circular cone: $S = \pi r^2 + \pi rl$ Regular square pyramid: $S = s^2 + 2sl$	See **Example 4** on page 412, and then try Exercises 11, 12, and 25 on page 453.
7.6 Right Triangle Trigonometry	
The Trigonometric Ratios of an Acute Angle of a Right Triangle If θ is an acute angle of a right triangle ABC, then $\sin\theta = \frac{\text{length of opposite side}}{\text{length of hypotenuse}}$ $\cos\theta = \frac{\text{length of adjacent side}}{\text{length of hypotenuse}}$ $\tan\theta = \frac{\text{length of opposite side}}{\text{length of adjacent side}}$ $\csc\theta = \frac{\text{length of hypotenuse}}{\text{length of opposite side}}$ $\sec\theta = \frac{\text{length of hypotenuse}}{\text{length of adjacent side}}$ $\cot\theta = \frac{\text{length of adjacent side}}{\text{length of opposite side}}$	See **Examples 1 and 2** on pages 419 and 420, and then try Exercises 31 and 32 on page 454.
Inverse Sine, Inverse Cosine, and Inverse Tangent $\sin^{-1}(x)$ is defined as the angle whose sine is x, $0 < x < 1$. $\cos^{-1}(x)$ is defined as the angle whose cosine is x, $0 < x < 1$. $\tan^{-1}(x)$ is defined as the angle whose tangent is x, $x > 0$.	See **Examples 3, 4, and 5** on pages 421 and 422, and then try Exercises 33 and 36 on page 454.

continued

Applications of Trigonometry Trigonometry is often used to solve applications that involve right triangles.	See **Example 6** on pages 422 and 423, and then try Exercises 37 and 38 on page 454.
7.7 Non-Euclidean Geometry	
Parallel Postulates **Euclidean Parallel Postulate** (*Euclidean or Plane Geometry*) Through a given point not on a given line, exactly one line can be drawn parallel to the given line. **Gauss's Alternate to the Parallel Postulate** (*Lobachevskian or Hyperbolic Geometry*) Through a given point not on a given line, there are at least two lines parallel to the given line. **Riemann's Alternative to the Parallel Postulate** (*Riemannian or Spherical Geometry*) Through a given point not on a given line, there exist no lines parallel to the given line.	See **Example 2** on page 431, and then try Exercises 42 and 43 on page 454.
The Spherical Triangle Area Formula The area S of the spherical triangle ABC on a sphere with radius r is $S = (m\angle A + m\angle B + m\angle C - 180°)\left(\frac{\pi}{180°}\right)r^2$	See **Example 1** on pages 429 and 430, and then try Exercises 44 and 45 on page 454.
The Euclidean Distance Formula and the City Distance Formula ■ The Euclidean distance between $P(x_1, y_1)$ and $Q(x_2, y_2)$ is $d_E(P, Q) = \sqrt{(x_2 - x_1)^2 + (y_2 - y_1)^2}$ ■ The city distance between $P(x_1, y_1)$ and $Q(x_2, y_2)$ is $d_C(P, Q) = \lvert x_2 - x_1\rvert + \lvert y_2 - y_1\rvert$	See **Example 3** on page 433, and then try Exercises 46 and 49 on page 454.
7.8 Fractals	
Strictly Self-Similar Fractal A fractal is a strictly self-similar fractal if any arbitrary portion of the fractal contains a replica of the entire fractal.	See **Example 3** on page 442, and then try Exercise 51 on page 454.
Replacement Ratio and Scaling Ratio of a Fractal ■ If the generator of a fractal consists of N replicas of the initiator, then the replacement ratio of the fractal is N. ■ If the initiator of a fractal has linear dimensions that are r times the corresponding linear dimensions of its replicas in the generator, then the scaling ratio of the fractal is r.	See **Example 4** on page 443, and then try Exercise 53 on page 454.
Similarity Dimension of a Fractal The similarity dimension D of a strictly self-similar fractal is $D = \frac{\log N}{\log r}$ where N is the replacement ratio of the fractal and r is the scaling ratio.	See **Example 5** on page 444, and then try Exercise 54 on page 454.

CHAPTER 7 REVIEW EXERCISES

■ For Exercises 1 to 5, convert one measurement to another.

1. 27 in. = _______ ft
2. 15 cups = _______ pints
3. 37 mm = _______ cm
4. 0.678 g = _______ mg
5. 1273 ml = _______ L
6. The price of a beverage is \$3.56 per liter. What is the price of this beverage in dollars per quart? Round to the nearest cent.
7. Given that $m\angle a = 74°$ and $m\angle b = 52°$, find the measures of angles x and y.

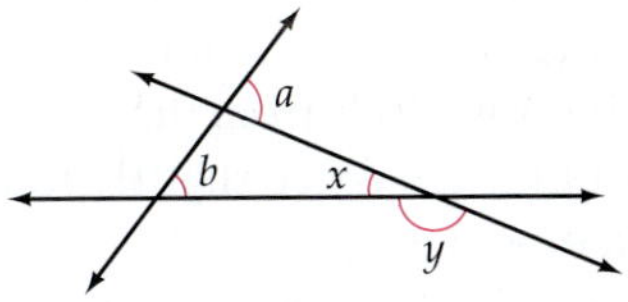

8. Triangles ABC and DEF are similar. Find the perimeter of triangle ABC.

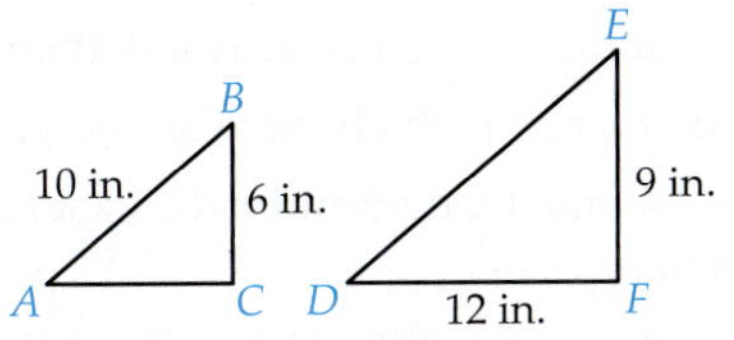

9. Find the volume of the geometric solid.

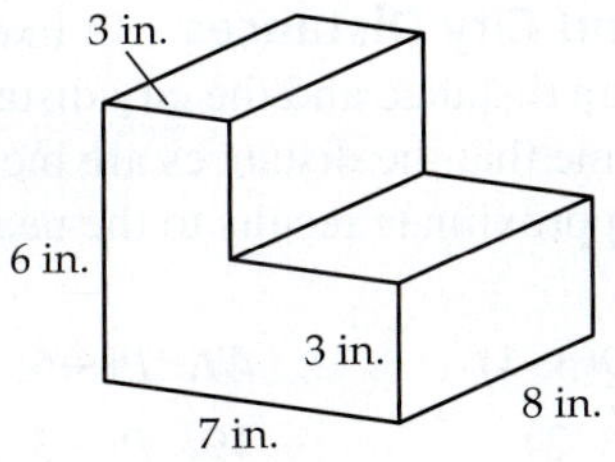

10. Find the measure of $\angle x$.

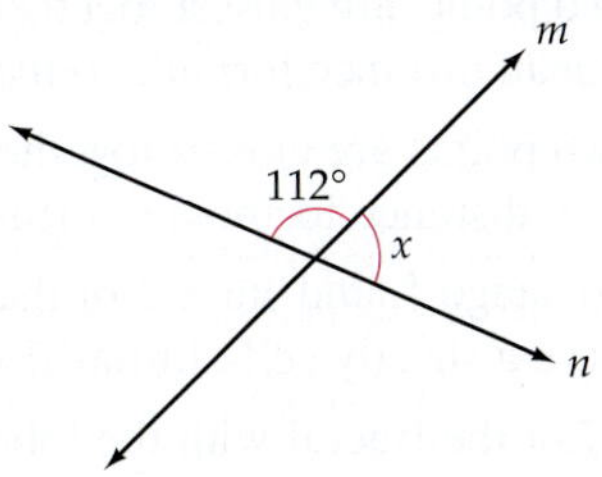

11. Find the surface area of the rectangular solid.

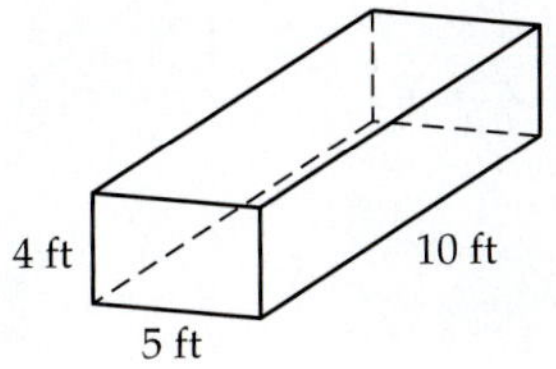

12. The length of a diameter of the base of a cylinder is 4 m, and the height of the cylinder is 8 m. Find the surface area of the cylinder. Give the exact value.
13. Given that $BC = 11$ cm and that AB is three times the length of BC, find the length of AC.

14. Given that $m\angle x = 150°$, find the measures of $\angle w$ and $\angle y$.

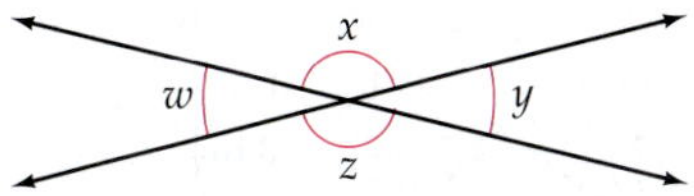

15. Find the area of a parallelogram that has a base of 6 in. and a height of 4.5 in.
16. Find the volume of the square pyramid.

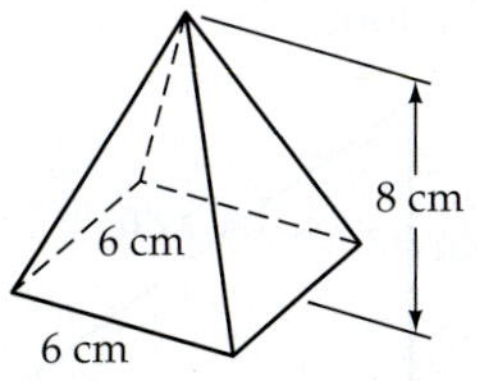

17. Find the circumference of a circle that has a diameter of 4.5 m. Round to the nearest tenth of a meter.
18. Given that $\ell_1 \parallel \ell_2$, find the measures of angles a and b.

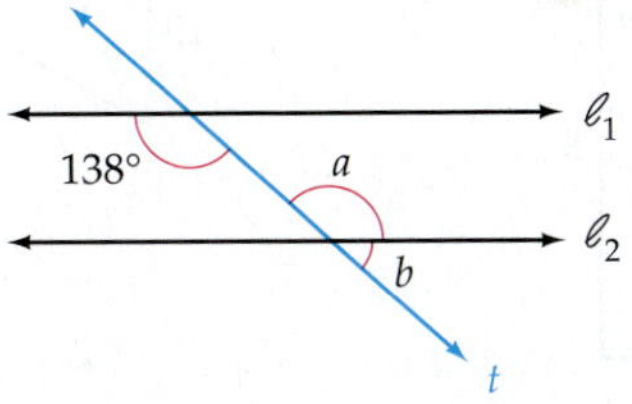

19. Find the supplement of a 32° angle.
20. Find the volume of a rectangular solid with a length of 6.5 ft, a width of 2 ft, and a height of 3 ft.
21. Two angles of a triangle measure 37° and 48°. Find the measure of the third angle.
22. The height of a triangle is 7 cm. The area of the triangle is 28 cm^2. Find the length of the base of the triangle.
23. Find the volume of a sphere that has a diameter of 12 mm. Give the exact value.
24. **Framing** The perimeter of a square picture frame is 86 cm. Find the length of each side of the frame.
25. **Paint** A can of paint will cover 200 ft^2 of surface. How many cans of paint should be purchased to paint a cylinder that has a height of 15 ft and a radius of 6 ft?

26. Parks and Recreation The length of a rectangular park is 56 yd. The width is 48 yd. How many yards of fencing are needed to surround the park?

27. Patios What is the area of a square patio that measures 9.5 m on each side?

28. Landscaping A walkway 2 m wide surrounds a rectangular plot of grass. The plot is 40 m long and 25 m wide. What is the area of the walkway?

29. Determine whether the two triangles are congruent. If they are congruent, state by what theorem they are congruent.

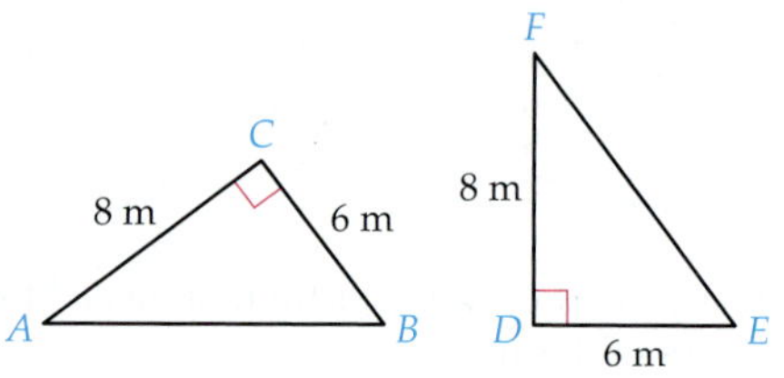

30. Find the unknown side of the triangle. Round to the nearest tenth of a foot.

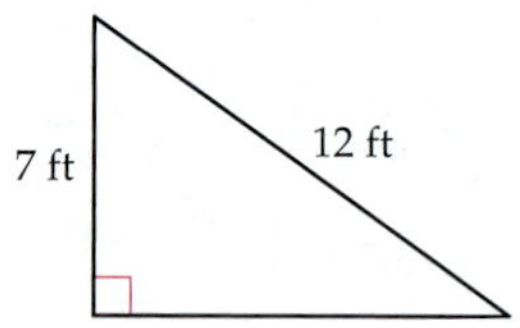

■ In Exercises 31 and 32, find the values of $\sin\theta$, $\cos\theta$, and $\tan\theta$ for the given right triangle.

31.

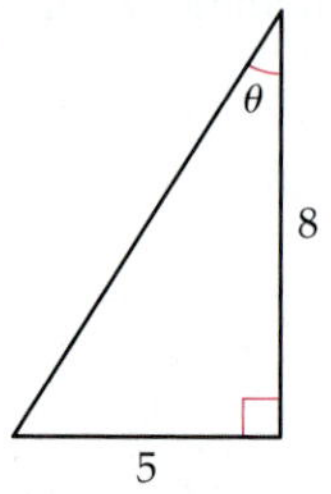

32.

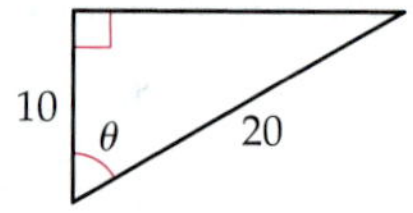

■ In Exercises 33 to 36, use a calculator. Round to the nearest tenth of a degree.

33. Find $\cos^{-1}(0.9013)$.

34. Find $\sin^{-1}(0.4871)$.

35. Given $\tan\beta = 1.364$, find β.

36. Given $\sin\theta = 0.0325$, find θ.

37. Surveying Find the distance across the marsh in the following figure. Round to the nearest tenth of a foot.

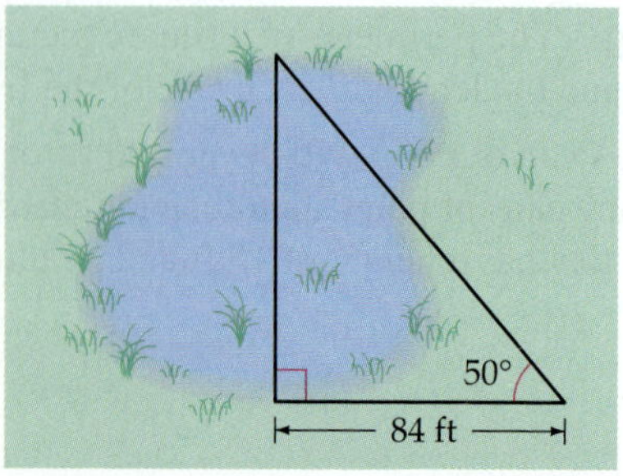

38. Angle of Depression The distance from a plane to a radar station is 200 mi, and the angle of depression is 40°. Find the number of ground miles from a point directly under the plane to the radar station. Round to the nearest tenth of a mile.

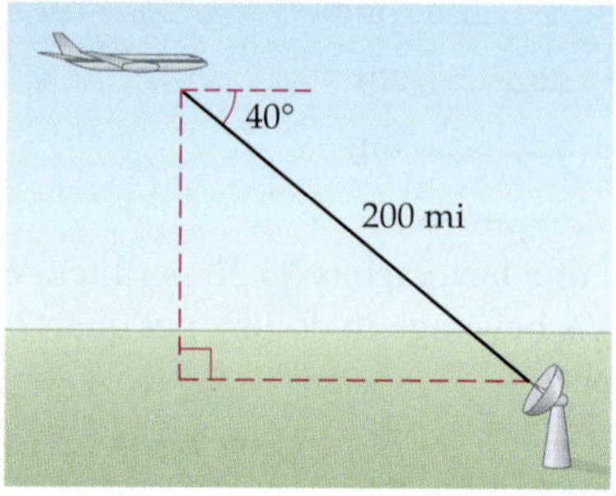

39. Angle of Elevation The angle of elevation from a point A on the ground to the top of a space shuttle is 27°. If point A is 110 ft from the base of the space shuttle, how tall is the space shuttle? Round to the nearest tenth of a foot.

40. What is another name for Riemannian geometry?

41. What is another name for Lobachevskian geometry?

42. Name a geometry in which the sum of the measures of the interior angles of a triangle is less than 180°.

43. Name a geometry in which there are no parallel lines.

■ In Exercises 44 and 45, determine the exact area of the spherical triangle.

44. Radius: 12 in.; angles: 90°, 150°, 90°

45. Radius: 5 ft; angles: 90°, 60°, 90°

■ **Euclidean and City Distances** In Exercises 46 to 49, find the Euclidean distance and the city distance between the points. Assume that the distances are measured in blocks. Round approximate results to the nearest tenth of a block.

46. $P(-1, 1)$, $Q(3, 4)$

47. $P(-5, -2)$, $Q(2, 6)$

48. $P(2, 8)$, $Q(3, 2)$

49. $P(-3, 3)$, $Q(5, -2)$

50. Consider the points $P(1, 1)$, $Q(4, 5)$, and $R(-4, 2)$.

a. Which two points are closest together if you only use the Euclidean distance formula to measure distance?

b. Which two points are closest together if you only use the city distance formula to measure distance?

51. Draw stage 0, stage 1, and stage 2 of the Koch curve. Is the Koch curve a strictly self-similar fractal?

52. Draw stage 2 of the fractal with the following initiator and generator.

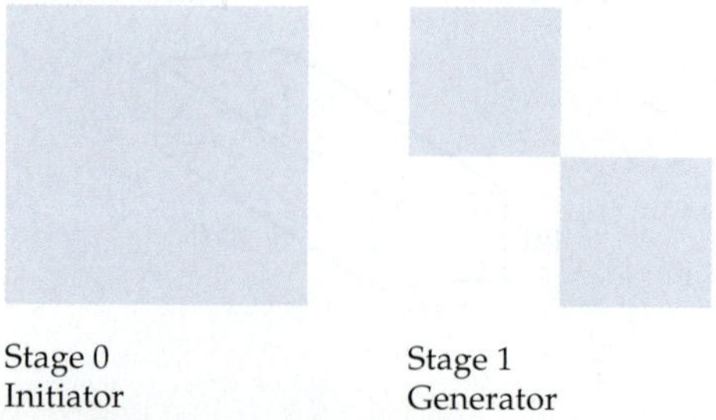

53. For the fractal defined in Exercise 52, determine the

a. relacement ratio.

b. scaling ratio.

c. similarity dimension.

54. Compute the similarity dimension of a strictly self-similar fractal with a replacement ratio of 5 and a scaling ratio of 4. Round to the nearest thousandth.

CHAPTER 7 TEST

1. Find the volume of a cylinder with a height of 6 m and a radius of 3 m. Round to the nearest tenth of a cubic meter.

2. Find the perimeter of a rectangle that has a length of 2 m and a width of 1.4 m.

3. Find the complement of a 32° angle.

4. Find the area of a circle that has a diameter of 2 m. Round to the nearest tenth of a square meter.

5. In the figure below, lines ℓ_1 and ℓ_2 are parallel. Angle x measures 30°. Find the measure of angle y.

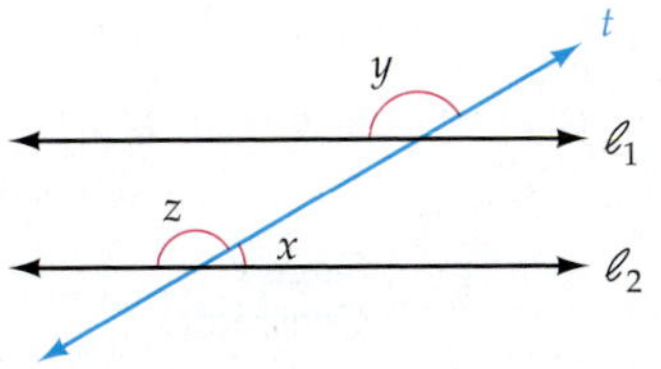

6. In the figure below, lines ℓ_1 and ℓ_2 are parallel. Angle x measures 45°. Find the measures of angles a and b.

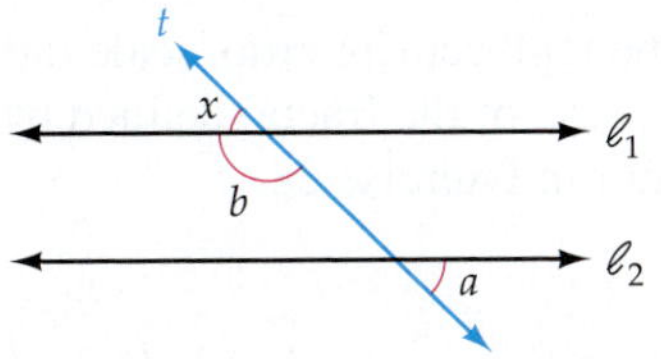

7. Convert: 1.2 m = ______ cm.

8. Find the volume of the figure. Give the exact value.

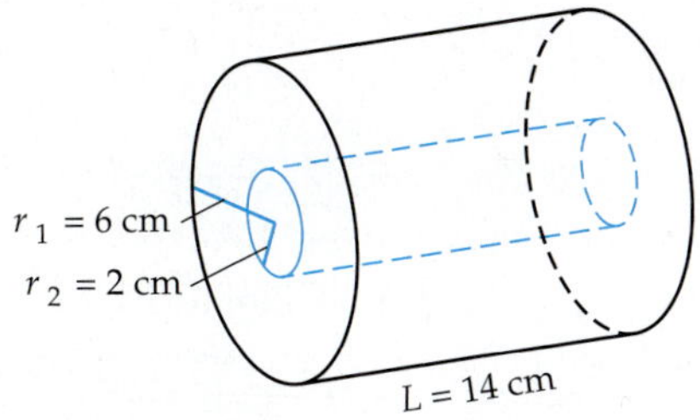

9. Triangles ABC and DEF are similar. Find side BC.

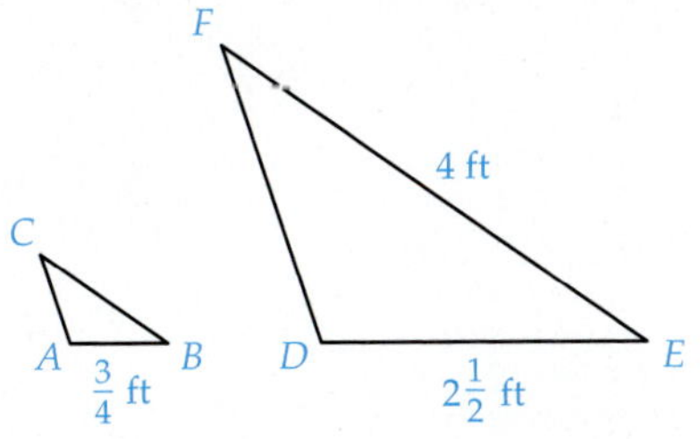

10. A right triangle has a 40° angle. Find the measures of the other two angles.

11. Find the measure of $\angle x$.

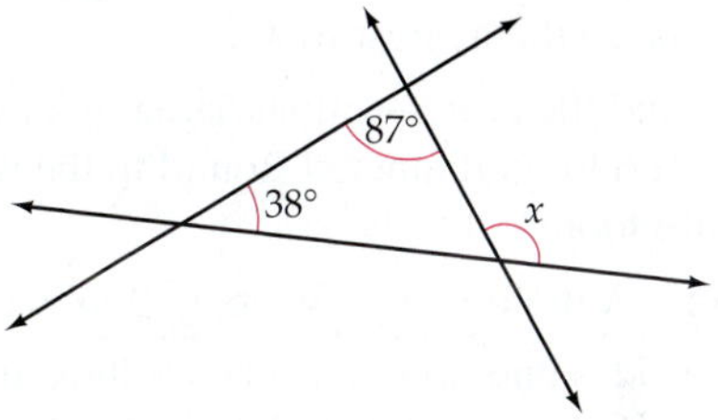

12. Find the area of the parallelogram shown below.

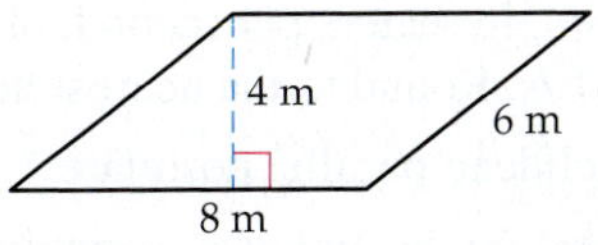

13. Surveying Find the width of the canal shown in the figure below.

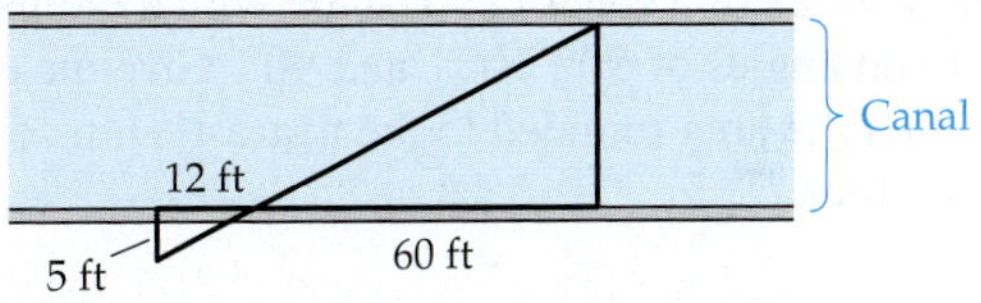

14. Pizza How much more area is in a pizza with radius 10 in. than in a pizza with radius 8 in.? Round to the nearest tenth of a square inch.

15. Determine whether the two triangles are congruent. If they are congruent, state by what theorem they are congruent.

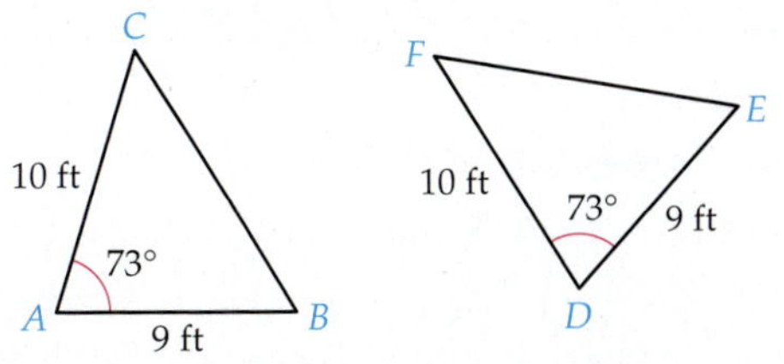

16. For the right triangle shown below, determine the length of side $\overline{BC}$. Round to the nearest tenth of a centimeter.

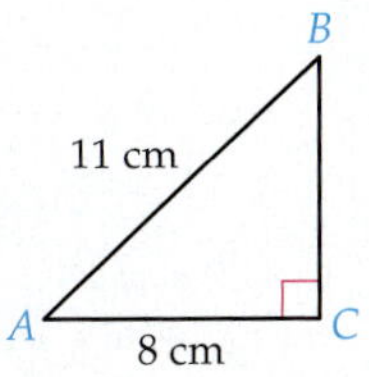

17. Find the values of sin θ, cos θ, and tan θ for the given right triangle.

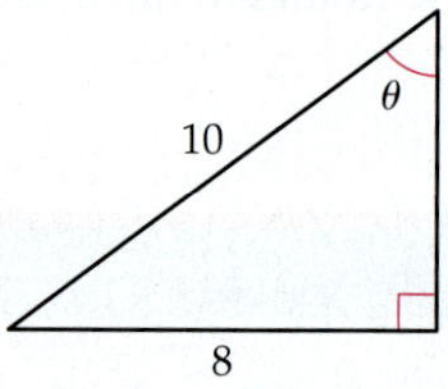

18. Angle of Elevation From a point 27 ft from the base of a Roman aqueduct, the angle of elevation to the top of the aqueduct is 78°. Find the height of the aqueduct. Round to the nearest foot.

19. Trees Find the cross-sectional area of a redwood tree that is 11 ft 6 in. in diameter. Round to the nearest tenth of a square foot.

20. Toolbox A toolbox is 14 in. long, 9 in. wide, and 8 in. high. The sides and bottom of the toolbox are $\frac{1}{2}$ in. thick. The toolbox is open at the top. Find the volume of the interior of the toolbox in cubic inches.

21. Find the speed, in meters per second, of a car that is traveling 88 ft/s. Round to the nearest hundredth.

22. State the Euclidean parallel postulate.

23. What is a great circle, and what formula is used to calculate the area of a spherical triangle formed by three arcs of three great circles?

24. Find the area of a spherical triangle with a radius of 12 ft and angles of 90°, 100°, and 90°. Give the exact area and the area rounded to the nearest tenth of a square foot.

25. City Geometry Find the Euclidean distance and the city distance between the points $P(-4, 2)$ and $Q(5, 1)$. Assume that the distances are measured in blocks. Round approximate results to the nearest tenth of a block.

26. City Geometry How many points are on the city circle with center (0, 0) and radius $r = 4$ blocks?

■ In Exercises 27 and 28, draw stage 2 of the fractal with the given initiator and generator.

27.

Stage 0
Initiator

Stage 1
Generator

28.

Stage 0
Initiator

Stage 1
Generator

29. Compute the replacement ratio, scale ratio, and similarity dimension of the fractal defined by the initiator and generator in Exercise 27.

30. Compute the replacement ratio, scale ratio, and similarity dimension of the fractal defined by the initiator and generator in Exercise 28.

8

Mathematical Systems

It's a dark, rainy night as you pull up to the drive at your home. You press a remote in your car that opens your garage door and you drive in. You press the remote again and the garage door closes.

Unbeknownst to you, there is a crook lurking on the street with a radio scanner who picks up the signal from your remote. The next time you're not home, the crook plans on using the saved signal to open your garage door and burglarize your home. However, when the crook arrives and sends the signal, the door doesn't open.

Now suppose you want to make a purchase on the Internet using a credit card. You may have noticed that the typical http:// that precedes a web address is replaced by https://. The "s" at the end indicates a *secure* website. This means that someone who may be trying to steal credit card information cannot intercept the information you send.

Although a garage door opener and a secure website may seem to be quite different, the mathematics behind each of these is similar. Both are based on modular arithmetic, one of the topics of this chapter.

Modular arithmetic, in turn, is part of a branch of mathematics called group theory, another topic in this chapter. Group theory is used in a variety of many seemingly unrelated subjects such as the structure of a diamond, wallpaper patterns, quantum physics, and the 12-tone chromatic scale in music.

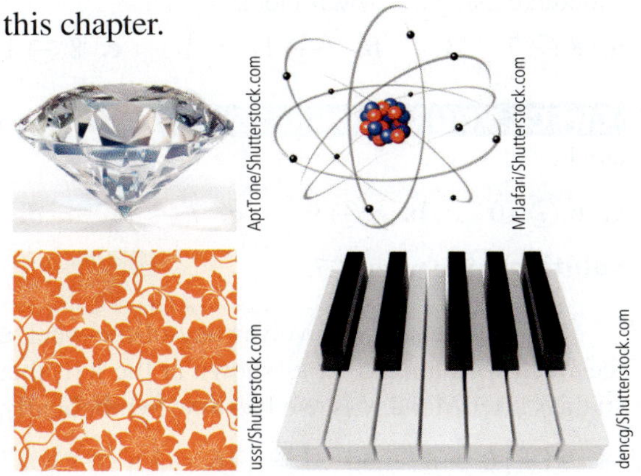

All of these are related to group theory.

SECTION 8.1 Modular Arithmetic

POINT OF INTEREST

The abbreviation A.M. comes from the Latin *ante* (before) *meridiem* (midday). The abbreviation P.M. comes from the Latin *post* (after) *meridiem* (midday).

Introduction to Modular Arithmetic

Many clocks have the familiar 12-hour design. We designate whether the time is before noon or after noon by using the abbreviations A.M. and P.M. A reference to 7:00 A.M. means 7 hours after 12:00 midnight; a reference to 7:00 P.M. means 7 hours after 12:00 noon. In both cases, once 12 is reached on the clock, we begin again with 1.

If we want to determine a time in the future or in the past, it is necessary to consider whether we have passed 12 o'clock. To determine the time 8 hours after 3 o'clock, we add 3 and 8. Because we did not pass 12 o'clock, the time is 11 o'clock (Figure 8.1A). However, to determine the time 8 hours after 9 o'clock, we must take into consideration that once we have passed 12 o'clock, we begin again with 1. Therefore, 8 hours after 9 o'clock is 5 o'clock, as shown in Figure 8.1B.

FIGURE 8.1A

FIGURE 8.1B

We will use the symbol $\oplus$ to denote addition on a 12-hour clock. Using this notation,

$$3 \oplus 8 = 11 \quad \text{and} \quad 9 \oplus 8 = 5$$

on a 12-hour clock.

We can also perform subtraction on a 12-hour clock. If the time now is 10 o'clock, then 7 hours ago the time was 3 o'clock, which is the difference between 10 and 7 $(10 - 7 = 3)$. However, if the time now is 3 o'clock, then, using Figure 8.2, we see that 7 hours ago it was 8 o'clock. If we use the symbol $\ominus$ to denote subtraction on a 12-hour clock, we can write

$$10 \ominus 7 = 3 \quad \text{and} \quad 3 \ominus 7 = 8$$

FIGURE 8.2

EXAMPLE 1 Perform Clock Arithmetic

Evaluate each of the following, where $\oplus$ and $\ominus$ indicate addition and subtraction, respectively, on a 12-hour clock.

a. $8 \oplus 7$ **b.** $7 \oplus 12$ **c.** $8 \ominus 11$ **d.** $2 \ominus 8$

Solution

Calculate using a 12-hour clock.

a. $8 \oplus 7 = 3$ **b.** $7 \oplus 12 = 7$ **c.** $8 \ominus 11 = 9$ **d.** $2 \ominus 8 = 6$

CHECK YOUR PROGRESS 1 Evaluate each of the following using a 12-hour clock.

a. $6 \oplus 10$ **b.** $5 \oplus 9$ **c.** $7 \ominus 11$ **d.** $5 \ominus 10$

Solution See page S27.

A similar example involves day-of-the-week arithmetic. If we associate each day of the week with a number, as shown at the left, then 6 days after Friday is Thursday and 16 days after Monday is Wednesday. Symbolically, we write

Monday = 1
Tuesday = 2
Wednesday = 3
Thursday = 4
Friday = 5
Saturday = 6
Sunday = 7

$$5 \boxed{+} 6 = 4 \quad \text{and} \quad 1 \boxed{+} 16 = 3$$

Note: We are using the $\boxed{+}$ symbol for days-of-the-week arithmetic to differentiate from the $\oplus$ symbol for clock arithmetic.

Another way to determine the day of the week is to note that when the sum $5 + 6 = 11$ is divided by 7, the number of days in a week, the remainder is 4, the number associated with Thursday. When $1 + 16 = 17$ is divided by 7, the remainder is 3, the number associated with Wednesday. This works because the days of the week repeat every 7 days.

The same method can be applied to 12-hour-clock arithmetic. From Example 1a, when $8 + 7 = 15$ is divided by 12, the number of hours on a 12-hour clock, the remainder is 3, the time 7 hours after 8 o'clock.

Situations such as these that repeat in cycles are represented mathematically by using **modular arithmetic**, or **arithmetic modulo *n***.

HISTORICAL NOTE

The congruent symbol consisting of three horizontal bars was introduced in print in 1801 by Carl Friedrich Gauss (1777–1855).

Modulo *n*

Two integers a and b are said to be **congruent modulo *n***, where n is a natural number, if $\frac{a-b}{n}$ is an integer. In this case, we write $a \equiv b \bmod n$. The number n is called the **modulus**. The statment $a \equiv b \bmod n$ is called a **congruence**.

INSTRUCTOR NOTE

You may wish to show students that $29 \equiv 11 \bmod 3$ also means that 29 can be arrived at by repeatedly adding 3, the modulus, to 11.

EXAMPLE 2 Determine Whether a Congruence Is True

Determine whether the congruence is true.

a. $29 \equiv 8 \bmod 3$

b. $15 \equiv 4 \bmod 6$

Solution

a. Find $\frac{29-8}{3} = \frac{21}{3} = 7$. Because 7 is an integer, $29 \equiv 8 \bmod 3$ is a true congruence.

b. Find $\frac{15-4}{6} = \frac{11}{6}$. Because $\frac{11}{6}$ is not an integer, $15 \equiv 4 \bmod 6$ is not a true congruence.

CHECK YOUR PROGRESS 2 Determine whether the congruence is true.

a. $7 \equiv 12 \bmod 5$

b. $15 \equiv 1 \bmod 8$

Solution See page S27.

HISTORICAL NOTE

The Julian calendar, introduced by Julius Caesar in 46 BC and named after him, contained 12 months and 365 days. Every fourth year, an additional day was added to the year to compensate for the difference between an ordinary year and a solar year. However, a solar year actually consists of 365.2422 days, not 365.25 days as assumed by Caesar. The difference between these numbers is 0.0078 days.

Although this is a small number, after a few centuries had passed, the ordinary year and the solar year no longer matched. If something were not done to rectify the situation, after a period of time the summer season in the northern hemisphere would be in December and the winter season would be in July.

To bring the seasons back into phase, Pope Gregory removed 10 days from October in 1582. The new calendar was called the Gregorian calendar and is the one we use today. This calendar did not win general acceptance until 1752.

For $29 \equiv 8 \bmod 3$ given in Example 2, note that $29 \div 3$ (the modulus) $= 9$ remainder 2 and that $11 \div 3$ (the modulus) $= 3$ remainder 2. Both 29 and 8 have the same remainder when divided by the modulus. This leads to an important alternate method to determine a true congruence. If $a \equiv b \bmod n$ and a and b are whole numbers, then a and b have the same remainder when divided by n.

QUESTION Using the alternate method, is $33 \equiv 49 \bmod 4$ a true congruence?

Now suppose today is Friday. To determine the day of the week 16 days from now, we observe that 14 days from now the day will be Friday, so 16 days from now the day will be Sunday. Note that the remainder when 16 is divided by 7 is 2, or, using modular notation, $16 \equiv 2 \bmod 7$. The 2 signifies 2 days after Friday, which is Sunday.

EXAMPLE 3 A Day of the Week

July 4, 2017, was a Tuesday. What day of the week is July 4, 2022?

Solution

There are 5 years between the two dates. Each year has 365 days except 2020, which has one extra day because it is a leap year. So the total number of days between the two

ANSWER Yes. $33 \div 4 = 8$ remainder 1, and $49 \div 4 = 12$ remainder 1. Both 33 and 49 have the same remainder when divided by 4.

dates is $5 \cdot 365 + 1 = 1826$. Because $1826 \div 7 = 260$ remainder 6, $1826 \equiv 6 \bmod 7$. Any multiple of 7 days past a given day will be the same day of the week. So the day of the week 1826 days after July 4, 2017, will be the same as the day 6 days after July 4, 2017. Thus July 4, 2022, will be a Monday.

CHECK YOUR PROGRESS 3 In 2016, Abraham Lincoln's birthday fell on Friday, February 12. On what day of the week does Lincoln's birthday fall in 2025?

Solution See page S27.

MATH MATTERS A Leap-Year Formula

The calculation in Example 3 required that we consider whether the intervening years contained a leap year. There is a formula, based on modular arithmetic, that can be used to determine which years are leap years.

The calendar we use today is called the Gregorian calendar. This calendar differs from the Julian calendar (see the Historical Note on page 459) in that leap years do not always occur every fourth year. Here is the rule: Let Y be the year. If $Y \equiv 0 \bmod 4$, then Y is a leap year unless $Y \equiv 0 \bmod 100$. In that case, Y is not a leap year unless $Y \equiv 0 \bmod 400$. Then Y is a leap year.

Using this rule, 2008 is a leap year because $2008 \equiv 0 \bmod 4$, and 2013 is not a leap year because $2013 \not\equiv 0 \bmod 4$. The year 1900 was *not* a leap year because $1900 \equiv 0 \bmod 100$ but $1900 \not\equiv 0 \bmod 400$. The year 2000 was a leap year because $2000 \equiv 0 \bmod 100$ and $2000 \equiv 0 \bmod 400$.

Arithmetic Operations Modulo *n*

In Example 2a we verified that $29 \equiv 8 \bmod 3$. (Both 29 and 8 have remainder 2 when divided by 3, the modulus.) There are many other numbers congruent to 8 modulo 3, but of all these, only one is a whole number less than the modulus. This number is the result when *evaluating* a modulo expression, and in this case we use an equal sign. Because $2 \equiv 8 \bmod 3$ and 2 is less than the modulus, we can write $8 \bmod 3 = 2$. In general, $m \bmod n$ becomes the remainder when m is divided by n.

Arithmetic modulo n (where n is a natural number) requires us to evaluate a modular expression after using the standard rules of arithmetic. Thus we perform the arithmetic operation and then divide by the modulus. The answer is the remainder. The result of an arithmetic operation mod n is always a whole number less than n.

TAKE NOTE

Recall that the natural numbers are the usual counting numbers: 1, 2, 3, 4, . . . The set of whole numbers consists of the natural numbers and zero.

EXAMPLE 4 Addition Modulo *n*

Evaluate: $(23 + 38) \bmod 12$

Solution

Add $23 + 38$ to produce 61. To evaluate $61 \bmod 12$, divide 61 by the modulus, 12. The answer is the remainder.

$$\begin{array}{r} 5 \\ 12\overline{)61} \\ \underline{60} \\ 1 \end{array}$$

$23 + 38 = 61$

$(23 + 38) \bmod 12 = 1$

The answer is 1.

CHECK YOUR PROGRESS 4 Evaluate: $(51 + 72) \bmod 3$

Solution See page S27.

In modular arithmetic, adding the modulus to a number does not change the equivalent value of the number. For instance,

$13 \equiv 6 \bmod 7$		$10 \equiv 1 \bmod 3$	
$20 \equiv 6 \bmod 7$	• Add 7 to 13.	$13 \equiv 1 \bmod 3$	• Add 3 to 10.
$27 \equiv 6 \bmod 7$	• Add 7 to 20.	$16 \equiv 1 \bmod 3$	• Add 3 to 13.

To understand why the value does not change, consider $7 \equiv 0 \bmod 7$ and $3 \equiv 0 \bmod 3$. That is, in mod 7 arithmetic, 7 is equivalent to 0; in mod 3 arithmetic, 3 is equivalent to 0. Just as adding 0 to a number does not change the value of the number in regular arithmetic, in modular arithmetic adding the modulus to a number does not change the value of the number. This property of modular arithmetic is sometimes used in subtraction.

It is possible to use negative numbers modulo n. For instance,

$$-2 \equiv 5 \bmod 7 \text{ because } \frac{-2-5}{7} = \frac{-7}{7} = -1, \text{ an integer.}$$

Suppose we want to find x so that $-15 \equiv x \bmod 6$. Using the definition of modulo n, we need to find x so that $\frac{-15-x}{6}$ is an integer. To do this, rewrite the expression and then try various values of x from 0 to the modulus until the value of the expression is an integer.

$$\frac{-15-x}{6} = \frac{-(15+x)}{6}$$

When $x = 0$, $\frac{-(15+0)}{6} = -\frac{15}{6}$, not an integer.

When $x = 1$, $\frac{-(15+1)}{6} = -\frac{16}{6} = -\frac{8}{3}$, not an integer.

When $x = 2$, $\frac{-(15+2)}{6} = -\frac{17}{6}$, not an integer.

When $x = 3$, $\frac{-(15+3)}{6} = -\frac{18}{6} = -3$, an integer.

$-15 \equiv 3 \bmod 6$

It may be necessary to use this idea when subtracting in modular arithmetic.

EXAMPLE 5 Subtraction Modulo *n*

Evaluate each of the following.

a. $(33 - 16) \bmod 6$ **b.** $(14 - 27) \bmod 5$

Solution

a. Subtract $33 - 16 = 17$. The result is positive. Divide the difference by the modulus, 6. The answer is the remainder.

$$\begin{array}{r} 2 \\ 6\overline{)17} \\ \underline{12} \\ 5 \end{array}$$

$(33 - 16) \bmod 6 = 5$

b. Subtract $14 - 27 = -13$. Because the answer is negative, we must find x so that $-13 \equiv x \bmod 5$. Thus we must find x so that the value of $\frac{-13-x}{5} = \frac{-(13+x)}{5}$ is an integer. Trying the whole number values of x less than 5, the modulus, we find that when $x = 2$, $\frac{-(13+2)}{5} = -\frac{15}{5} = -3$.

$(14 - 27) \bmod 5 = 2$

CHECK YOUR PROGRESS 5 Evaluate: $(21 - 43) \bmod 7$

Solution See page S27.

The methods of adding and subtracting in modular arithmetic can be used for clock arithmetic and days-of-the-week arithmetic.

EXAMPLE 6 Calculating Times

Disregarding A.M. or P.M., if it is 5 o'clock now, what time was it 57 hours ago?

Solution

The time can be determined by calculating $(5 - 57) \bmod 12$. Because $5 - 57 = -52$ is a negative number, find a whole number x less than the modulus 12, so that $-52 \equiv x \bmod 12$. This means to find x so that $\frac{-52 - x}{12} = \frac{-(52 + x)}{12}$ is an integer. Evaluating the expression for whole number values of x less than 12, we have, when $x = 8$, $\frac{-(52 + 8)}{12} = -\frac{60}{12} = -5$, an integer. Thus $(5 - 57) \bmod 12 = 8$. Therefore, if it is 5 o'clock now, 57 hours ago it was 8 o'clock.

> **TAKE NOTE**
>
> In Example 6, repeatedly adding the modulus to the difference results in the following.
>
> $-52 + 12 = -40$
>
> $-40 + 12 = -28$
>
> $-28 + 12 = -16$
>
> $-16 + 12 = -4$
>
> $-4 + 12 = 8$

CHECK YOUR PROGRESS 6 If today is Tuesday, what day of the week will it be 93 days from now?

Solution See page S27.

Problems involving multiplication can also be performed modulo n.

EXAMPLE 7 Multiplication Modulo *n*

Evaluate: $(15 \cdot 23) \bmod 11$

Solution

Find the product $15 \cdot 23$ and then divide by the modulus, 11. The answer is the remainder.

$$\begin{array}{r} 31 \\ 11\overline{)345} \\ \underline{33} \\ 15 \\ \underline{11} \\ 4 \end{array} \qquad 15 \cdot 23 = 345$$

$(15 \cdot 23) \bmod 11 = 4$

The answer is 4.

CHECK YOUR PROGRESS 7 Evaluate: $(33 \cdot 41) \bmod 17$

Solution See page S27.

Solving Congruence Equations

Solving a congruence equation means finding all whole number values of the variable for which the congruence is true.

For example, to solve $3x + 5 \equiv 3 \bmod 4$, we search for whole number values of x for which the congruence is true.

$3(0) + 5 \not\equiv 3 \bmod 4$

$3(1) + 5 \not\equiv 3 \bmod 4$

$3(2) + 5 \equiv 3 \bmod 4$ 2 is a solution.

$3(3) + 5 \not\equiv 3 \bmod 4$

$3(4) + 5 \not\equiv 3 \bmod 4$

$3(5) + 5 \not\equiv 3 \bmod 4$

$3(6) + 5 \equiv 3 \bmod 4$ 6 is a solution.

If we continued trying values, we would find that 10 and 14 are also solutions. Note that the solutions 6, 10, and 14 are all congruent to 2 modulo 4. In general, once a solution is determined, additional solutions can be found by repeatedly adding the modulus to the original solution. Thus the solutions of $3x + 5 \equiv 3 \bmod 4$ are 2, 6, 10, 14, 18, ….

When solving a congruence equation, it is necessary to check only the whole numbers less than the modulus. For the congruence equation $3x + 5 \equiv 3 \bmod 4$, we needed to check only 0, 1, 2, and 3. Each time a solution is found, additional solutions can be found by repeatedly adding the modulus to it.

A congruence equation can have more than one solution among the whole numbers less than the modulus. The next example illustrates that you must check *all* whole numbers less than the modulus.

EXAMPLE 8 Solve a Congruence Equation

Solve: $2x + 1 \equiv 3 \bmod 10$

Solution

Beginning with 0, substitute each whole number less than 10 into the congruence equation.

$x = 0$	$2(0) + 1 \not\equiv 3 \bmod 10$	Not a solution
$x = 1$	$2(1) + 1 \equiv 3 \bmod 10$	A solution
$x = 2$	$2(2) + 1 \not\equiv 3 \bmod 10$	Not a solution
$x = 3$	$2(3) + 1 \not\equiv 3 \bmod 10$	Not a solution
$x = 4$	$2(4) + 1 \not\equiv 3 \bmod 10$	Not a solution
$x = 5$	$2(5) + 1 \not\equiv 3 \bmod 10$	Not a solution
$x = 6$	$2(6) + 1 \equiv 3 \bmod 10$	A solution
$x = 7$	$2(7) + 1 \not\equiv 3 \bmod 10$	Not a solution
$x = 8$	$2(8) + 1 \not\equiv 3 \bmod 10$	Not a solution
$x = 9$	$2(9) + 1 \not\equiv 3 \bmod 10$	Not a solution

The solutions between 0 and 9 are 1 and 6; the remaining solutions are determined by repeatedly adding the modulus, 10, to these solutions. The solutions are 1, 6, 11, 16, 21, 26, ….

CHECK YOUR PROGRESS 8 Solve: $4x + 1 \equiv 5 \bmod 12$

Solution See page S27.

Not all congruence equations have a solution. For instance, $5x + 1 \equiv 3 \bmod 5$ has no solution, as shown below.

$x = 0$	$5(0) + 1 \not\equiv 3 \bmod 5$	Not a solution
$x = 1$	$5(1) + 1 \not\equiv 3 \bmod 5$	Not a solution
$x = 2$	$5(2) + 1 \not\equiv 3 \bmod 5$	Not a solution
$x = 3$	$5(3) + 1 \not\equiv 3 \bmod 5$	Not a solution
$x = 4$	$5(4) + 1 \not\equiv 3 \bmod 5$	Not a solution

Because no whole number value of x less than the modulus is a solution, there is no solution.

Additive and Multiplicative Inverses in Modular Arithmetic

Recall that if the sum of two numbers is 0, then the numbers are *additive inverses* of each other. For instance, $8 + (-8) = 0$, so 8 is the additive inverse of -8, and -8 is the additive inverse of 8.

The same concept applies in modular arithmetic. For example, $(3 + 5) \equiv 0 \bmod 8$. Thus, in mod 8 arithmetic, 3 is the additive inverse of 5, and 5 is the additive inverse of 3. **Here we consider only those whole numbers smaller than the modulus.** Note that $3 + 5 = 8$; that is, the sum of a number and its additive inverse equals the modulus. Using this fact, we can easily find the additive inverse of a number for any modulus. For instance, in mod 11 arithmetic, the additive inverse of 5 is 6 because $5 + 6 = 11$.

EXAMPLE 9 Find the Additive Inverse

Find the additive inverse of 7 in mod 16 arithmetic.

Solution

In mod 16 arithmetic, $7 + 9 = 16$, so the additive inverse of 7 is 9.

CHECK YOUR PROGRESS 9 Find the additive inverse of 6 in mod 12 arithmetic.

Solution See page S27.

If the product of two numbers is 1, then the numbers are **multiplicative inverses** of each other. For instance, $2 \cdot \frac{1}{2} = 1$, so 2 is the multiplicative inverse of $\frac{1}{2}$, and $\frac{1}{2}$ is the multiplicative inverse of 2. The same concept applies to modular arithmetic (although the multiplicative inverses will always be natural numbers). For example, in mod 7 arithmetic, 5 is the multiplicative inverse of 3 (and 3 is the multiplicative inverse of 5) because $5 \cdot 3 \equiv 1 \bmod 7$. (Here we will concern ourselves only with natural numbers less than the modulus.) To find the multiplicative inverse of a mod m, solve the modular equation $ax \equiv 1 \bmod m$ for x.

TAKE NOTE

In mod m arithmetic, every number has an additive inverse but not necessarily a multiplicative inverse. For instance, in mod 12 arithmetic, 3 does not have a multiplicative inverse. You should verify this by trying to solve the equation $3x \equiv 1 \bmod 12$ for x.

A similar situation occurs in standard arithmetic. The number 0 does not have a multiplicative inverse because there is no solution of the equation $0x = 1$.

EXAMPLE 10 Find a Multiplicative Inverse

In mod 7 arithmetic, find the multiplicative inverse of 2.

Solution

To find the multiplicative inverse of 2, solve the equation $2x \equiv 1 \bmod 7$ by trying different natural number values of x less than the modulus.

$$2x \equiv 1 \bmod 7$$

$2(1) \not\equiv 1 \bmod 7$	• Try $x = 1$.
$2(2) \not\equiv 1 \bmod 7$	• Try $x = 2$.
$2(3) \not\equiv 1 \bmod 7$	• Try $x = 3$.
$2(4) \equiv 1 \bmod 7$	• Try $x = 4$.

In mod 7 arithmetic, the multiplicative inverse of 2 is 4.

CHECK YOUR PROGRESS 10 Find the multiplicative inverse of 5 in mod 11 arithmetic.

Solution See page S27.

EXCURSION

Computing the Day of the Week

A function that is related to the modulo function is called the *floor function*. In the modulo function, we determine the remainder when one number is divided by another. In the floor function, we determine the quotient (and ignore the remainder) when one number is divided by another. The symbol for the floor function is $\lfloor\ \rfloor$. Here are some examples.

$$\left\lfloor \frac{2}{3} \right\rfloor = 0, \quad \left\lfloor \frac{10}{2} \right\rfloor = 5, \quad \left\lfloor \frac{17}{2} \right\rfloor = 8, \quad \text{and} \quad \left\lfloor \frac{2}{\sqrt{2}} \right\rfloor = 1$$

Using the floor function, we can write a formula that gives the day of the week for any date on the Gregorian calendar. The formula, known as Zeller's congruence, is given by

$$x = \left(\left\lfloor \frac{13m - 1}{5} \right\rfloor + \left\lfloor \frac{y}{4} \right\rfloor + \left\lfloor \frac{c}{4} \right\rfloor + d + y - 2c \right) \bmod 7$$

where

- d is the day of the month
- m is the month using 1 for March, 2 for April, ..., 10 for December; January and February are assigned the values 11 and 12, respectively
- y is the last two digits of the year if the month is March through December; if the month is January or February, y is the last two digits of the year *minus 1*
- c is the first two digits of the year
- x is the day of the week (using 0 for Sunday, 1 for Monday, ..., 6 for Saturday)

For example, to determine the day of the week on July 4, 1776, we have $c = 17$, $y = 76$, $d = 4$, and $m = 5$. Using these values, we can calculate x.

$$\begin{aligned} x &= \left(\left\lfloor \frac{13(5) - 1}{5} \right\rfloor + \left\lfloor \frac{76}{4} \right\rfloor + \left\lfloor \frac{17}{4} \right\rfloor + 4 + 76 - 2(17) \right) \bmod 7 \\ &= (12 + 19 + 4 + 4 + 76 - 34) \bmod 7 \\ &= 81 \bmod 7 = 4 \end{aligned}$$

Therefore, July 4, 1776, was a Thursday.

EXCURSION EXERCISES

1. Determine the day of the week on which you were born.
2. Determine the day of the week on which Abraham Lincoln's birthday (February 12) will fall in 2155.
3. Determine the day of the week on which January 1, 2050, will fall.
4. Determine the day of the week on which Valentine's Day (February 14) 1950 fell.

EXERCISE SET 8.1

■ In Exercises 1 to 16, evaluate each expression, where $\oplus$ and $\ominus$ indicate addition and subtraction, respectively, using a 12-hour clock.

Robert Brenner/PhotoEdit

1. $3 \oplus 5$
2. $6 \oplus 7$
3. $8 \oplus 4$
4. $5 \oplus 10$
5. $11 \oplus 3$
6. $8 \oplus 8$
7. $7 \oplus 9$
8. $11 \oplus 10$
9. $10 \ominus 6$
10. $2 \ominus 6$
11. $3 \ominus 8$
12. $5 \ominus 7$
13. $10 \ominus 11$
14. $5 \ominus 5$
15. $4 \ominus 9$
16. $1 \ominus 4$

■ **Military Time** In Exercises 17 to 24, evaluate each expression, where $\stackrel{+}{\triangle}$ and $\stackrel{-}{\triangle}$ indicate addition and subtraction, respectively, using military time. (Military time uses a 24-hour clock, where 2:00 A.M. is equivalent to 0200 hours and 10 P.M. is equivalent to 2200 hours.)

17. 1800 $\stackrel{+}{\triangle}$ 0900
18. 1600 $\stackrel{+}{\triangle}$ 1200
19. 0800 $\stackrel{+}{\triangle}$ 2000
20. 1300 $\stackrel{+}{\triangle}$ 1300
21. 1000 $\stackrel{-}{\triangle}$ 1400
22. 1800 $\stackrel{-}{\triangle}$ 1900
23. 0200 $\stackrel{-}{\triangle}$ 0500
24. 0600 $\stackrel{-}{\triangle}$ 2200

■ In Exercises 25 to 28, evaluate each expression, where $\boxed{+}$ and $\boxed{-}$ indicate addition and subtraction, respectively, using days-of-the-week arithmetic.

25. $6 \boxed{+} 4$
26. $3 \boxed{+} 5$
27. $2 \boxed{-} 3$
28. $3 \boxed{-} 6$

■ In Exercises 29 to 38, determine whether the congruence is true or false.

29. $5 \equiv 8 \bmod 3$
30. $11 \equiv 15 \bmod 4$
31. $5 \equiv 20 \bmod 4$
32. $7 \equiv 21 \bmod 3$
33. $21 \equiv 45 \bmod 6$
34. $18 \equiv 60 \bmod 7$
35. $88 \equiv 5 \bmod 9$
36. $72 \equiv 30 \bmod 5$
37. $100 \equiv 20 \bmod 8$
38. $25 \equiv 85 \bmod 12$

39. List five different natural numbers that are congruent to 8 modulo 6.

40. List five different natural numbers that are congruent to 10 modulo 4.

■ In Exercises 41 to 62, perform the modular arithmetic.

41. $(9 + 15) \bmod 7$
42. $(12 + 8) \bmod 5$
43. $(5 + 22) \bmod 8$
44. $(50 + 1) \bmod 15$
45. $(42 + 35) \bmod 3$
46. $(28 + 31) \bmod 4$
47. $(37 + 45) \bmod 12$
48. $(62 + 21) \bmod 2$
49. $(19 - 6) \bmod 5$
50. $(25 - 10) \bmod 4$
51. $(48 - 21) \bmod 6$
52. $(60 - 32) \bmod 9$
53. $(8 - 15) \bmod 12$
54. $(3 - 12) \bmod 4$
55. $(15 - 32) \bmod 7$
56. $(24 - 41) \bmod 8$
57. $(6 \cdot 8) \bmod 9$
58. $(5 \cdot 12) \bmod 4$
59. $(9 \cdot 15) \bmod 8$
60. $(4 \cdot 22) \bmod 3$
61. $(14 \cdot 18) \bmod 5$
62. $(26 \cdot 11) \bmod 15$

■ **Clocks and Calendars** In Exercises 63 to 70, use modular arithmetic to determine each of the following.

63. Disregarding A.M. or P.M., if it is now 7 o'clock,
 a. what time will it be 59 hours from now?
 b. what time was it 62 hours ago?

64. Disregarding A.M. or P.M., if it is now 2 o'clock,
 a. what time will it be 40 hours from now?
 b. what time was it 34 hours ago?

65. If today is Friday,
 a. what day of the week will it be 25 days from now?
 b. what day of the week was it 32 days ago?

66. If today is Wednesday,
 a. what day of the week will it be 115 days from now?
 b. what day of the week was it 81 days ago?

67. In 2015, Halloween (October 31) fell on a Saturday. On what day of the week will Halloween fall in the year 2025?

68. In 2010, April Fool's Day (April 1) fell on a Thursday. On what day of the week will April Fool's Day fall in 2021?

69. Valentine's Day (February 14) fell on a Tuesday in 2017. On what day of the week will Valentine's Day fall in 2032?

70. Cinco de Mayo (May 5) fell on a Wednesday in 2010. On what day of the week will Cinco de Mayo fall in 2028?

Jonathan Nourok/PhotoEdit

■ In Exercises 71 to 82, find all whole number solutions of the congruence equation.

71. $x \equiv 10 \bmod 3$

72. $x \equiv 12 \bmod 5$

73. $2x \equiv 12 \bmod 5$

74. $3x \equiv 8 \bmod 11$

75. $(2x + 1) \equiv 5 \bmod 4$

76. $(3x + 1) \equiv 4 \bmod 9$

77. $(2x + 3) \equiv 8 \bmod 12$

78. $(3x + 12) \equiv 7 \bmod 10$

79. $(2x + 2) \equiv 6 \bmod 4$

80. $(5x + 4) \equiv 2 \bmod 8$

81. $(4x + 6) \equiv 5 \bmod 8$

82. $(4x + 3) \equiv 3 \bmod 4$

■ In Exercises 83 to 88, find the additive inverse and the multiplicative inverse, if it exists, of the given number.

83. 4 in modulo 9 arithmetic

84. 4 in modulo 5 arithmetic

85. 7 in modulo 10 arithmetic

86. 11 in modulo 16 arithmetic

87. 3 in modulo 8 arithmetic

88. 6 in modulo 15 arithmetic

■ Modular division can be performed by considering the related multiplication problem. For instance, if $5 \div 7 = x$, then $x \cdot 7 = 5$. Similarly, the quotient $(5 \div 7) \bmod 8$ is the solution to the congruence equation $x \cdot 7 \equiv 5 \bmod 8$, which is 3. In Exercises 89 to 94, find the given quotient.

89. $(2 \div 7) \bmod 8$

90. $(4 \div 5) \bmod 8$

91. $(6 \div 4) \bmod 9$

92. $(2 \div 3) \bmod 5$

93. $(5 \div 6) \bmod 7$

94. $(3 \div 4) \bmod 7$

95. Verify that the division $5 \div 8$ has no solution in modulo 8 arithmetic.

96. Verify that the division $4 \div 4$ has more than one solution in modulo 10 arithmetic.

EXTENSIONS

97. There is only one whole number solution between 0 and 11 of the congruence equation

$$(x^2 + 3x + 7) \equiv 2 \bmod 11$$

Find the solution.

98. There are only two whole number solutions between 0 and 27 of the congruence equation

$$(x^2 + 5x + 4) \equiv 1 \bmod 27$$

Find the solutions.

99. Rolling Codes Some garage door openers are programmed using modular arithmetic. See page 457. The basic idea is that there is a computer chip in both the transmitter in your car and the receiver in your garage that are programmed exactly the same way. When the transmitter is pressed, two things happen. First, the transmitter sends a number to the receiver; second, the computer chip uses the sent number and modular arithmetic to create a new number that is sent the next time the transmitter is pressed. Here is a simple example using a modulus of 37. A real transmitter and receiver would use a number with at least 40 digits.

$x_0 = 23$ — This number is called the seed.

$x_1 = (5x_0 + 17) \bmod 37$ — This is the modular formula.

$= (5 \cdot 23 + 17) \bmod 37$ — Using this formula, a new number, 21, is calculated when the transmitter is pressed.

$= 132 \bmod 37$

$= 21$

$x_2 = (5x_1 + 17) \bmod 37$ — When the transmitter is pressed again, the last number calculated, 21, is used to find the next number.

$= (5 \cdot 21 + 17) \bmod 37$

$= 122 \bmod 37$

$= 11$

This continues each time the transmitter is pressed. A general form for this formula is written as $x_{n+1} = (5x_n + 17) \bmod 37$ and is called a recursive formula. The next number x_{n+1} is calculated using the previous number x_n.

a. Use the recursive formula $x_{n+1} = (2x_n + 13) \bmod 11$ and seed $x_0 = 3$ to find the numbers x_1 to x_9.

b. Find x_{10}. How is this number related to x_0? (Remember: The last number you calculated in part a is x_9.)

c. Why might this recursive formula not be a good one for an actual garage door opener?

100. Many people consider the 13th of the month an unlucky day when it falls on a Friday.

a. Does every year have a Friday the 13th? Explain why or why not.

b. What is the greatest number of Friday the 13ths that can occur in one year? Can you find a recent year in which this number occurred? *Hint:* A non-leap year can have two consecutive months where the 13th is a Friday.

SECTION 8.2 Applications of Modular Arithmetic

ISBN and UPC

Every book that is cataloged in the Library of Congress must have an ISBN (International Standard Book Number). This 13-digit number was created to help ensure that orders for books are filled accurately and that books are catalogued correctly.

The first three digits of an ISBN are 978 (or 979), followed by 9 digits that are divided into three groups of various lengths. These indicate the country or region, the publisher, and the title of the book. The last digit (the 13th one) is called a **check digit**.

If we label the first digit of an ISBN d_1, the second digit d_2, and so on to the 13th digit d_{13}, then the check digit is given by the following modular formula.

Formula for the ISBN Check Digit

$$d_{13} = 10 - (d_1 + 3d_2 + d_3 + 3d_4 + d_5 + 3d_6 + d_7 + 3d_8 + d_9 + 3d_{10} + d_{11} + 3d_{12}) \bmod 10$$

If $d_{13} = 10$, then the check digit is 0.

It is this check digit that is used to ensure accuracy. For instance, the ISBN for the fourth edition of the *American Heritage Dictionary* is 978-0-395-82517-4. Suppose, however, that a bookstore clerk sends an order for the *American Heritage Dictionary* and inadvertently enters the number 978-0-395-28517-4, where the clerk transposed the 8 and 2 in the five numbers that identify the book.

Correct ISBN: 978-0-395-82517-4

Incorrect ISBN: 978-0-395-28517-4

The receiving clerk calculates the check digit as follows.

$$\begin{aligned} d_{13} &= 10 - [9 + 3(7) + 8 + 3(0) + 3 + 3(9) + 5 + 3(2) + 8 + 3(5) + 1 + 3(7)] \bmod 10 \\ &= 10 - 124 \bmod 10 \\ &= 10 - 4 = 6 \end{aligned}$$

Because the check digit is 6 and not 4 as it should be, the receiving clerk knows that an incorrect ISBN has been sent. Transposition errors are among the most frequent errors that occur. The ISBN coding system will catch most of them.

EXAMPLE 1 Determine a Check Digit for an ISBN

Determine the ISBN check digit for the book *The Equation that Couldn't Be Solved* by Mario Livio. The first 12 digits of the ISBN are 978-0-7432-5820-?

Solution

$$\begin{aligned} d_{13} &= 10 - [9 + 3(7) + 8 + 3(0) + 7 + 3(4) + 3 + 3(2) + 5 + 3(8) + 2 + 3(0)] \bmod 10 \\ &= 10 - 97 \bmod 10 \\ &= 10 - 7 = 3 \end{aligned}$$

The check digit is 3.

CHECK YOUR PROGRESS 1 A purchase order for the book *The Mathematical Tourist* by Ivars Peterson includes the ISBN 978-0-716-73250-5. Determine whether this is a valid ISBN.

Solution See page S27.

Another coding scheme that is closely related to the ISBN is the UPC (Universal Product Code). This number is placed on many items and is particularly useful in grocery stores. A check-out clerk passes the product by a scanner, which reads the number from a bar code and records the price on the cash register. If the price of an item changes for a promotional sale, the price is updated in the computer, thereby relieving a clerk of having to reprice each item. In addition to pricing items, the UPC gives the store manager accurate information about inventory and the buying habits of the store's customers.

The UPC is a 12-digit number that satisfies a modular equation that is similar to the one for ISBNs. The last digit is the check digit. If we label the 12 digits of the UPC as $d_1, d_2, \ldots, d_{12}$, we can write a formula for the UPC check digit d_{12}.

Formula for the UPC Check Digit

$$d_{12} = 10 - (3d_1 + d_2 + 3d_3 + d_4 + 3d_5 + d_6 + 3d_7 + d_8 + 3d_9 + d_{10} + 3d_{11}) \bmod 10$$

If $d_{12} = 10$, then the check digit is 0.

POINT OF INTEREST

Bar codes were first patented in 1949. However, the first commercial computer did not arrive until 1951 (although there were computers for military use prior to 1951), so bar codes were not extensively used. The UPC was developed in 1973.

In addition to the ISBN and UPC, there are other coding schemes, such as EAN for European Article Numbering and SKU for Stock Keeping Unit.

EXAMPLE 2 Determine the Check Digit of a UPC

Find the check digit for the UPC of the Blu-ray Disc release of the film *Jurassic World*. The first 11 digits are 0-25192-21221-?

Solution

$$\begin{aligned} d_{12} &= 10 - [3(0) + 2 + 3(5) + 1 + 3(9) + 2 + 3(2) + 1 + 3(2) + 2 + 3(1)] \bmod 10 \\ &= 10 - 65 \bmod 10 \\ &= 10 - 5 = 5 \end{aligned}$$

The check digit is 5.

CHECK YOUR PROGRESS 2 Is 1-32342-65933-9 a valid UPC?

Solution See page S27.

The ISBN and UPC coding systems will normally catch transposition errors. There are instances, however, when they do not.

The UPC for Crisco Puritan Canola Oil with Omega-DHA is

0-51500-24275-9

Suppose, however, that the product code is written 0-51500-24725-9, where the 2 and 7 have been transposed. Calculating the check digit, we have

$$\begin{aligned} d_{12} &= 10 - [3(0) + 5 + 3(1) + 5 + 3(0) + 0 + 3(2) + 4 + 3(7) + 2 + 3(5)] \bmod 10 \\ &= 10 - 61 \bmod 10 \\ &= 10 - 1 = 9 \end{aligned}$$

The same check digit is calculated, yet the UPC has been entered incorrectly. This was an unfortunate coincidence; if any other two digits were transposed, the result would have given a different check digit and the error would have been caught. It can be shown that the ISBN and UPC coding methods will not catch a transposition error of adjancent digits a and b if $|a - b| = 5$. For the Canola Oil UPC, $|7 - 2| = 5$.

INSTRUCTOR NOTE

Ask your students to find three products at home with a UPC and then verify the validity of each UPC.

Credit Card Numbers

Companies that issue credit cards also use modular arithmetic to determine whether a credit card number is valid. This is especially important in e-commerce, where credit card information is frequently sent over the Internet. The primary coding method is based on the *Luhn algorithm,* which uses mod 10 arithmetic.

POINT OF INTEREST

Hans Peter Luhn was born in Germany and later joined IBM in the United States as a senior research engineer. He was awarded a patent in 1960 for the algorithm that bears his name. Credit card companies began using the algorithm shortly after its creation. The algorithm is in the public domain and is still widely used to identify valid card numbers.

Credit card numbers are normally 13 to 16 digits long. The first one to six digits are used to identify the card issuer. The table below shows some of the identification prefixes used by four popular card issuers.

Card issuer	Prefix	Number of digits
MasterCard	51 to 55	16
Visa	4	13 or 16
American Express	34 or 37	15
Discover	6011	16

The Luhn algorithm, used to determine whether a credit card number is valid, is calculated as follows: Beginning with the next-to-last digit (the last digit is the check digit) and reading from right to left, double every other digit. If a digit becomes a two-digit number after being doubled, treat the number as two individual digits. Now find the sum of the new list of digits; the final sum must be congruent to 0 mod 10. The Luhn algorithm is demonstrated in the next example.

EXAMPLE 3 **Determine a Valid Credit Card Number**

Determine whether 5234 8213 3410 1298 is a valid credit card number.

Solution

Highlight every other digit, beginning with the next-to-last digit and reading from right to left.

5 2 3 4 8 2 1 3 3 4 1 0 1 2 9 8

Next double each of the highlighted digits.

10 2 6 4 16 2 2 3 6 4 2 0 2 2 18 8

Finally, add all digits, treating two-digit numbers as two single digits.

$$(1 + 0) + 2 + 6 + 4 + (1 + 6) + 2 + 2 + 3 + 6 + 4 + 2 + 0 + 2 + 2 + (1 + 8) + 8 = 60$$

Because $60 \equiv 0 \bmod 10$, this is a valid credit card number.

CHECK YOUR PROGRESS 3 Is 6011 0123 9145 2317 a valid credit card number?

Solution See page S28.

Cryptology

Related to codes on books and grocery items are secret codes. These codes are used to send messages between people, companies, or nations. It is hoped that by devising a code that is difficult to break, the sender can prevent the communication from being read if it is intercepted by an unauthorized person. **Cryptology** is the study of making and breaking secret codes.

Before we discuss how messages are coded, we need to define a few terms. **Plaintext** is a message before it is coded. The line

SHE WALKS IN BEAUTY LIKE THE NIGHT

from Lord Byron's poem "She Walks in Beauty" is in plaintext. **Ciphertext** is the message after it has been written in code. The line

ODA SWHGO EJ XAWQPU HEGA PDA JECDP

is the same line of the poem in ciphertext.

The method of changing from plaintext to ciphertext is called **encryption**. The line from the poem was encrypted by substituting each letter in plaintext with the letter that is 22 letters after that letter in the alphabet. (Continue from the beginning when the end of the alphabet is reached.) This is called a *cyclical coding scheme* because each letter of the alphabet is shifted the same number of positions. The original alphabet and the substitute alphabet are shown below.

A	B	C	D	E	F	G	H	I	J	K	L	M	N	O	P	Q	R	S	T	U	V	W	X	Y	Z
↕	↕	↕	↕	↕	↕	↕	↕	↕	↕	↕	↕	↕	↕	↕	↕	↕	↕	↕	↕	↕	↕	↕	↕	↕	↕
W	X	Y	Z	A	B	C	D	E	F	G	H	I	J	K	L	M	N	O	P	Q	R	S	T	U	V

QUESTION Using the cyclical coding scheme described above where each letter is replaced by the one 22 letters after the letter, what is the plaintext word that corresponds to the ciphertext YKZA?

To **decrypt** a message means to take the ciphertext message and write it in plaintext. If a cryptologist thinks a message has been encrypted using a cyclical substitution code like the one shown above, the key to the code can be found by taking a word from the message (usually one of the longer words) and continuing the alphabet for each letter of the word. When a recognizable word appears, the key can be determined. This method is shown below using the ciphertext word XAWQPU.

X	A	W	Q	P	U	Shift four positions
Y	B	X	R	Q	V	
Z	C	Y	S	R	W	
A	D	Z	T	S	X	
B	E	A	U	T	Y	

Once a recognizable word has been found (BEAUTY), count the number of positions that the letters have been shifted (four, in this case). To decode the message, substitute the letter of the normal alphabet that comes four positions after the letter in the ciphertext.

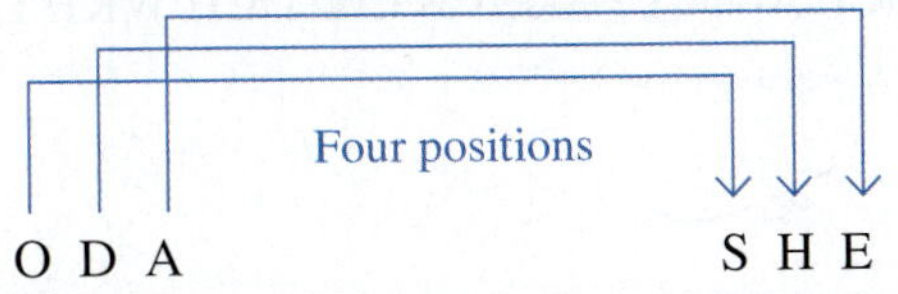

Cyclical encrypting using the alphabet is related to modular arithmetic. We begin with the normal alphabet and associate each letter with a number as shown in Table 8.1.

TABLE 8.1
Numerical Equivalents for the Letters of the Alphabet

A	B	C	D	E	F	G	H	I	J	K	L	M	N	O	P	Q	R	S	T	U	V	W	X	Y	Z
↕	↕	↕	↕	↕	↕	↕	↕	↕	↕	↕	↕	↕	↕	↕	↕	↕	↕	↕	↕	↕	↕	↕	↕	↕	↕
1	2	3	4	5	6	7	8	9	10	11	12	13	14	15	16	17	18	19	20	21	22	23	24	25	0

ANSWER CODE.

If the encrypting code is to shift each letter of the plaintext message m positions, then the corresponding letter in the ciphertext message is given by $c \equiv (p + m) \bmod 26$, where p is the numerical equivalent of the plaintext letter and c is the numerical equivalent of the ciphertext letter. The letter Z is coded as 0 because $26 \equiv 0 \bmod 26$.

Each letter in Lord Byron's poem was shifted 22 positions ($m = 22$) to the right. To code the plaintext letter S in the word SHE, we use the congruence $c \equiv (p + m) \bmod 26$.

TAKE NOTE

Because c represents a letter of the alphabet according to Table 8.1, c will always be a whole number less than or equal to 25.

$$c \equiv (p + m) \bmod 26$$
$$c \equiv (19 + 22) \bmod 26$$
$$c \equiv 41 \bmod 26$$
$$c = 15$$

• $p = 19$ (S is the 19th letter.) $m = 22$, the number of positions the letter is shifted

The 15th letter is O. Thus S is coded as O.

Once plaintext has been converted to ciphertext, there must be a method by which the person receiving the message can return the message to plaintext. For the cyclical code, the congruence is $p \equiv (c + n) \bmod 26$, where p and c are defined as before and $n = 26 - m$. The letter O in ciphertext is decoded below using the congruence $p \equiv (c + n) \bmod 26$.

$$p \equiv (c + n) \bmod 26$$
$$p \equiv (15 + 4) \bmod 26$$
$$p \equiv 19 \bmod 26$$
$$p = 19$$

• $c = 15$ (O is the 15th letter.) $n = 26 - m = 26 - 22 = 4$

The 19th letter is S. Thus O is decoded as S.

MATH MATTERS A Cipher of Caesar

Nadezhda Bolotina/Shutterstock.com

Julius Caesar allegedly used a cyclical alphabetic encrypting code to communicate with his generals. The messages were encrypted using the congruence $c \equiv (p + 3) \bmod 26$ (in modern notation). Using this coding scheme, the message BEWARE THE IDES OF MARCH would be coded as EHZDUH WKH LGHV RI PDUFK.

EXAMPLE 4 Write Messages Using Cyclical Coding

Use the cyclical alphabetic encrypting code that shifts each letter 11 positions to

a. code CATHERINE THE GREAT.

b. decode TGLY ESP EPCCTMWP.

Solution

a. The encrypting congruence is $c \equiv (p + 11) \bmod 26$. Replace p by the numerical equivalent of each letter of plaintext and determine c. The results for CATHERINE are shown below.

C	$c \equiv (3 + 11) \bmod 26 \equiv 14 \bmod 26 = 14$	Code C as N.
A	$c \equiv (1 + 11) \bmod 26 \equiv 12 \bmod 26 = 12$	Code A as L.
T	$c \equiv (20 + 11) \bmod 26 \equiv 31 \bmod 26 = 5$	Code T as E.
H	$c \equiv (8 + 11) \bmod 26 \equiv 19 \bmod 26 = 19$	Code H as S.
E	$c \equiv (5 + 11) \bmod 26 \equiv 16 \bmod 26 = 16$	Code E as P.
R	$c \equiv (18 + 11) \bmod 26 \equiv 29 \bmod 26 = 3$	Code R as C.
I	$c \equiv (9 + 11) \bmod 26 \equiv 20 \bmod 26 = 20$	Code I as T.
N	$c \equiv (14 + 11) \bmod 26 \equiv 25 \bmod 26 = 25$	Code N as Y.
E	$c \equiv (5 + 11) \bmod 26 \equiv 16 \bmod 26 = 16$	Code E as P.

Continuing, the plaintext would be coded as NLESPCTYP ESP RCPLE.

b. Because $m = 11$, $n = 26 - 11 = 15$. The ciphertext is decoded by using the congruence $p \equiv (c + 15) \bmod 26$. The results for TGLY are shown below.

T	$c \equiv (20 + 15) \bmod 26 \equiv 35 \bmod 26 = 9$	Decode T as I.
G	$c \equiv (7 + 15) \bmod 26 \equiv 22 \bmod 26 = 22$	Decode G as V.
L	$c \equiv (12 + 15) \bmod 26 \equiv 27 \bmod 26 = 1$	Decode L as A.
Y	$c \equiv (25 + 15) \bmod 26 \equiv 40 \bmod 26 = 14$	Decode Y as N.

Continuing, the ciphertext would be decoded as IVAN THE TERRIBLE.

CHECK YOUR PROGRESS 4 Use the cyclical alphabetic encrypting code that shifts each letter 17 positions to

a. encode ALPINE SKIING. **b.** decode TIFJJ TFLEKIP JBZZEX.

Solution See page S28.

The practicality of a cyclical alphabetic coding scheme is limited because it is relatively easy for a cryptologist to determine the coding scheme. (Recall the method used for the line from Lord Byron's poem.) A coding scheme that is a little more difficult to break is based on the congruence $c \equiv (ap + m) \bmod 26$, where a and 26 do not have a common factor. (For instance, a cannot be 14 because 14 and 26 have a common factor of 2.) The reason why a and 26 cannot have a common factor is related to the procedure for determining the decoding congruence. However, we will leave that discussion to other math courses.

EXAMPLE 5 Encode a Message

Use the congruence $c \equiv (5p + 2) \bmod 26$ to encode the message LASER PRINTER.

Solution

The encrypting congruence is $c \equiv (5p + 2) \bmod 26$. Replace p by the numerical equivalent of each letter from Table 8.1 and determine c. The results for LASER are shown below.

L	$c \equiv (5 \cdot 12 + 2) \bmod 26 \equiv 62 \bmod 26 = 10$	Code L as J.
A	$c \equiv (5 \cdot 1 + 2) \bmod 26 \equiv 7 \bmod 26 = 7$	Code A as G.
S	$c \equiv (5 \cdot 19 + 2) \bmod 26 \equiv 97 \bmod 26 = 19$	Code S as S.
E	$c \equiv (5 \cdot 5 + 2) \bmod 26 \equiv 27 \bmod 26 = 1$	Code E as A.
R	$c \equiv (5 \cdot 18 + 2) \bmod 26 \equiv 92 \bmod 26 = 14$	Code R as N.

Continuing, the plaintext is coded in ciphertext as JGSAN DNUTXAN.

CHECK YOUR PROGRESS 5 Use the congruence $c \equiv (3p + 1) \bmod 26$ to encode the message FLASH DRIVE.

Solution See page S28.

Decoding a message that was encrypted using the congruence $c \equiv (ap + m) \bmod n$ requires solving the congruence for p. The method relies on multiplicative inverses, which were discussed in Section 8.1.

Here we solve the congruence used in Example 5 for p.

$$c = 5p + 2$$

$$c - 2 = 5p$$
• Subtract 2 from each side of the equation.

$$21(c - 2) = 21(5p)$$
• Multiply each side of the equation by the multiplicative inverse of 5. Because $21 \cdot 5 \equiv 1 \bmod 26$, multiply each side by 21.

$$[21(c - 2)] \bmod 26 \equiv p$$

TAKE NOTE

$21(5p) \equiv p \bmod 26$ because $21 \cdot 5 = 105$ and $105 \div 26 = 4$ remainder 1.

Using this congruence equation, we can decode the ciphertext message JGSAN DNUTXAN. The method for decoding JGSAN is shown below.

J	$[21(10 - 2)] \bmod 26 \equiv 168 \bmod 26 = 12$	Decode J as L.
G	$[21(7 - 2)] \bmod 26 \equiv 105 \bmod 26 = 1$	Decode G as A.
S	$[21(19 - 2)] \bmod 26 \equiv 357 \bmod 26 = 19$	Decode S as S.
A	$[21(1 - 2)] \bmod 26 \equiv (-21) \bmod 26 = 5$	Decode A as E.
N	$[21(14 - 2)] \bmod 26 \equiv 252 \bmod 26 = 18$	Decode N as R.

Note that to decode A it was necessary to determine $(-21) \bmod 26$. Recall that this requires adding the modulus until a whole number less than 26 results. Because $(-21) + 26 = 5$, we have $(-21) \bmod 26 = 5$.

EXAMPLE 6 Decode a Message

Decode the message ACXUT CXRT, which was encrypted using the congruence $c \equiv (3p + 5) \bmod 26$.

Solution

Solve the congruence equation for p.

$$c = 3p + 5$$

$$c - 5 = 3p$$

$$9(c - 5) = 9(3p)$$
• $9(3) = 27$ and $27 \equiv 1 \bmod 26$

$$[9(c - 5)] \bmod 26 \equiv p$$

The decoding congruence is $p \equiv [9(c - 5)] \bmod 26$.

Using this congruence, we will show the details for decoding ACXUT.

A	$[9(1 - 5)] \bmod 26 \equiv (-36) \bmod 26 = 16$	Decode A as P.
C	$[9(3 - 5)] \bmod 26 \equiv (-18) \bmod 26 = 8$	Decode C as H.
X	$[9(24 - 5)] \bmod 26 \equiv 171 \bmod 26 = 15$	Decode X as O.
U	$[9(21 - 5)] \bmod 26 \equiv 144 \bmod 26 = 14$	Decode U as N.
T	$[9(20 - 5)] \bmod 26 \equiv 135 \bmod 26 = 5$	Decode T as E.

Continuing, we would decode the message as PHONE HOME.

CHECK YOUR PROGRESS 6 Decode the message IGHT OHGG, which was encrypted using the congruence $c \equiv (7p + 1) \bmod 26$.

Solution See page S28.

MATH MATTERS The Enigma Machine

markrhiggins/Shutterstock.com

During World War II, Nazi Germany encoded its transmitted messages using an Enigma machine, first patented in 1919. Each letter of the plaintext message was substituted with a different letter, but the machine changed the substitutions throughout the message so that the ciphertext appeared more random and would be harder to decipher if the message were intercepted. The Enigma machine accomplished its task by using three wheels of letters (chosen from a box of five) that rotated as letters were typed. In addition, electrical sockets corresponding to letters were connected in pairs with wires. The choice and order of the three wheels, their starting positions, and the arrangement of the electrical wires determined how the machine would encode a message. The receiver of the message needed to know the setup of the wheels and wires; with the receiver's Enigma machine configured identically, the ciphertext could be decoded. There are a staggering 150 quintillion different ways to configure the Enigma machine, making it very difficult to decode messages without knowing the setup that was used. A team of mathematicians and other code breakers, led by Alan Turing, was assembled by the British government in an attempt to decode the German messages. The team was eventually successful, and their efforts helped change the course of World War II. One aspect of the Enigma machine that aided the code breakers was the fact that the machine would never substitute a letter with itself.

As part of its *Nova* series, PBS aired an excellent account of the breaking of the Enigma codes during World War II entitled *Decoding Nazi Secrets*. Information is available at http://www.pbs.org/wgbh/nova/decoding/.

EXCURSION

Public Key Cryptography

In **public key cryptography**, there are two keys created, one for encoding a message (the public key) and one for decoding the message (the private key). One form of this scheme is known as RSA, from the first letters of the last names of Ron Rivest, Adi Shamir, and Leonard Adleman, who published the method in 1977 while professors at the Massachusetts Institute of Technology. This method uses modular arithmetic in a very unique way.

To create a coding and decoding scheme in RSA cryptography, we begin by choosing two large, distinct prime numbers p and q, say, $p = 59$ and $q = 83$. In practice, these would be prime numbers that are 200 or more digits long.

Find the product of p and q, $n = p \cdot q = 59 \cdot 83 = 4897$. The value of n is the modulus. Now find the product $z = (p - 1)(q - 1) = 58 \cdot 82 = 4756$ and randomly choose a number e, where $1 < e < z$ and e and z have no common factors. We will choose $e = 129$. Solve the congruence equation $ed \equiv 1 \text{ mod } z$ for d. For the numbers we are using, we must solve $129d \equiv 1 \text{ mod } 4756$. We won't go into the details but the solution is $d = 1401$. Recall that because $129 \cdot 1401 \equiv 1 \text{ mod } 4756$, 1401 is the multiplicative inverse of 129 mod 4756.

To receive encrypted messages, Olivia posts these values of n and e to a public key encryption service, called a certificate authority, which guarantees the integrity of her public key. If Henry wants to send Olivia a message, he codes his message as a number using a number for each letter of the alphabet. For instance, he might use $A = 11, B = 12, C = 13, \ldots$, and $Z = 36$. For Henry to send the message MATH, he would use the numbers 23 11 30 18 and code those numbers using Olivia's public key,

TAKE NOTE

The calculations at the right require a calculator that has been programmed to compute these values. These TI-83/84 programs are available on our companion site at CengageBrain.com.

POINT OF INTEREST

The Granger Collection, NYC

Sophie Germain (zhĕr-măn) (1776–1831) was born in Paris, France. Because enrollment in the university she wanted to attend was available only to men, Germain attended under the name Antoine-August Le Blanc. Eventually her ruse was discovered, but not before she came to the attention of Pierre Lagrange, one of the best mathematicians of the time. He encouraged her work and became a mentor to her.

A certain type of prime number named after Germain is called a Germain prime number. It is a prime p such that p and $2p + 1$ are both prime. For instance, 11 is a Germain prime because 11 and $2(11) + 1 = 23$ are both prime numbers. Germain primes are used in public key cryptography.

$N^{129} \bmod 4897 = M$, where N is the plaintext (23 11 30 18), and M is the ciphertext, the result of using the modular equation. This is shown below.

$23^{129} \bmod 4897 = 3065$ $\qquad$ $11^{129} \bmod 4897 = 2001$

$30^{129} \bmod 4897 = 957$ $\qquad$ $18^{129} \bmod 4897 = 2753$

Thus Henry sends Olivia the ciphertext numbers 3065, 2001, 957, and 2753.

When Olivia receives the message, she uses her private key, $M^{1401} \bmod 4897 = N$, where M is the ciphertext she received from Henry and N is the original plaintext, to decode the message. Olivia calculates

$3065^{1401} \bmod 4897 = 23$ $\qquad$ $2001^{1401} \bmod 4897 = 11$

$957^{1401} \bmod 4897 = 30$ $\qquad$ $2753^{1401} \bmod 4897 = 18$

She decodes the message as 23 11 30 18 or MATH.

EXCURSION EXERCISES

You will need the RSA program for the TI-83/84 calculators that can be downloaded from our companion site at CengageBrain.com to do these exercises. For exercises 1 to 4, use the values for n, e, and d that we used above. Code a word using $A = 11$, $B = 12$, $C = 13, \ldots,$ and $Z = 36$.

1. What is the ciphertext for the word CODE?
2. What is the ciphertext for the word HEART?
3. What is the word for a message that was received as 170, 607, 1268, 3408, 141?
4. What is the word for the message that was received as 1553, 3532, 2001, 3688, 957, 1941?
5. You can try this with your own numbers.
   - **a.** Choose two prime numbers larger than 50. These are p and q.
   - **b.** Find $n = p \cdot q$.
   - **c.** Find $z = (p - 1)(q - 1)$.
   - **d.** Choose a number e such that $1 < e < z$ and e and z have no common factors.
   - **e.** Solve $de \equiv 1 \bmod z$. You will need the program MODINV from our companion site at CengageBrain.com.

Your public encryption key is $N^e \bmod n = M$. You can give this to a friend who could then send you a message. You would decode using your private decryption key $M^d \bmod n = N$. To perform these calculations, use the RSA program above.

EXERCISE SET 8.2

■ **ISBN Numbers** In Exercises 1 to 6, determine whether the given number is a valid ISBN.

1. 978-0-281-44268-5
2. 978-1-55690-182-9
3. 978-0-671-51983-4
4. 978-0-614-35945-2
5. 978-0-143-03943-3
6. 978-0-231-10324-1

■ **ISBN Numbers** In Exercises 7 to 14, determine the correct check digit for each ISBN.

7. *The Hunger Games* by Suzanne Collins; 978-0-4390-2352-?

Ben Molyneux/
Alamy Stock Photo

8. *The Hobbit* by J.R.R. Tolkien; 978-0-345-27257-?

9. *The Girl on the Train* by Paula Hawkins; 978-0-857-52232-?

10. *Inside of a Dog: What Dogs See, Smell, and Know* by Alexandra Horowitz; 978-1-4165-8343-?

11. *David and Goliath: Underdogs, Misfits, and the Art of Battling Giants* by Malcolm Gladwell; 978-0-316-20437-?

12. *The Girl with the Dragon Tattoo* by Stieg Larsson; 978-0-307-47347-?

13. *Relativity: The Special and General Theory* by Albert Einstein; 978-0-517-88441-?

14. *Cleopatra: A Life* by Stacy Schiff; 978-0-316-00192-?

■ **UPC Codes** In Exercises 15 to 22, determine the correct check digit for the UPC.

15. 0-79893-46500-? (Organics Honey)

16. 6-53569-39973-? (Scrabble)

17. 7-14043-01126-? (Monopoly)

18. 0-32031-13439-? (Beethoven's 9th Symphony, DVD)

19. 8-88462-52148-? (Apple iPad Pro)

20. 0-33317-20666-? (TI-84 Plus CE calculator)

21. 0-41790-22106-? (Bertolli Classico olive oil)

22. 0-71818-02100-? (Guittard chocolate chips)

■ **Money Orders** Some money orders have serial numbers that consist of a 10-digit number followed by a check digit. The check digit is chosen to be the sum of the first 10 digits mod 9. In Exercises 23 to 26, determine the check digit for the given money order serial number.

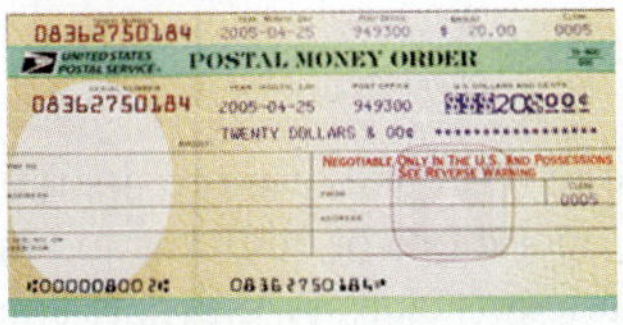

Bonnie Kamin/PhotoEdit

23. 0316615498-?

24. 5492877463-?

25. 1331497533-?

26. 3414793288-?

■ **Air Travel** Many printed airline tickets contain a 10-digit document number followed by a check digit. The check digit is chosen to be the sum of the first 10 digits mod 7. In Exercises 27 to 30, determine whether the given number is a valid document number.

27. 1182649758 2

28. 1260429984 4

29. 2026178914 5

30. 2373453867 6

■ **Credit Card Numbers** In Exercises 31 to 38, determine whether the given credit card number is a valid number.

31. 4417-5486-1785-6411

32. 5164-8295-1229-3674

33. 5591-4912-7644-1105

34. 6011-4988-1002-6487

35. 6011-0408-4977-3158

36. 4896-4198-8760-1970

37. 3715-548731-84466

38. 3401-714339-12041

■ **Encryption** In Exercises 39 to 44, encode the message by using a cyclical alphabetic encrypting code that shifts the message the stated number of positions.

39. 8 positions: THREE MUSKETEERS

40. 6 positions: FLY TONIGHT

41. 12 positions: IT'S A GIRL

42. 9 positions: MEET AT NOON

43. 3 positions: STICKS AND STONES

44. 15 positions: A STITCH IN TIME

■ **Decoding** In Exercises 45 to 48, use a cyclical alphabetic encrypting code that shifts the letters the stated number of positions to decode the encrypted message.

45. 18 positions: SYW GX WFDAYZLWFEWFL

46. 20 positions: CGUACHUNCIH LOFYM NBY QILFX

47. 15 positions: UGXTCS XC CTTS

48. 8 positions: VWJWLG QA XMZNMKB

■ **Decoding** In Exercises 49 to 52, use a cyclical alphabetic encrypting code to decode the encrypted message.

49. YVIBZM RDGG MJWDINJI

50. YBZAM HK YEBZAM

51. UDGIJCT RDDZXT

52. AOB HVS HCFDSRCSG

53. Encryption Julius Caesar supposedly used an encrypting code equivalent to the congruence $c \equiv (p + 3) \bmod 26$. Use the congruence to encrypt the message "men willingly believe what they wish."

54. Decoding Julius Caesar supposedly used an encrypting code equivalent to the congruence $c \equiv (p + 3) \bmod 26$. Use the congruence to decrypt the message WKHUH DUH QR DFFLGHQWV.

55. Encryption Use the encrypting congruence $c \equiv (3p + 2) \bmod 26$ to code the message TOWER OF LONDON.

56. Encryption Use the encrypting congruence $c \equiv (5p + 3) \bmod 26$ to code the message DAYLIGHT SAVINGS TIME.

57. Encryption Use the encrypting congruence $c \equiv (7p + 8) \bmod 26$ to code the message PARALLEL LINES.

58. Encryption Use the encrypting congruence $c \equiv (3p + 11) \bmod 26$ to code the message NONE SHALL PASS.

59. Decoding Decode the message LOFT JGMK LBS MNWMK that was encrypted using the congruence $c \equiv (3p + 4) \bmod 26$.

60. Decoding Decode the message BTYW SCRBKN UCYN that was encrypted using the congruence $c \equiv (5p + 6) \bmod 26$.

61. Decoding Decode the message SNUUHQ FM VFALHDZ that was encrypted using the congruence $c \equiv (5p + 9) \bmod 26$.

62. Decoding Decode the message GBBZ OJQBWJ TBR GJHI that was encrypted using the congruence $c \equiv (7p + 1) \bmod 26$.

EXTENSIONS

63. Money Orders Explain why the method used to determine the check digit of the money order serial numbers in Exercises 23 to 26 will not detect a transposition of digits among the first 10 digits.

64. UPC Codes Explain why the transposition of adjacent digits a and b in a UPC will go undetected only when $|a - b| = 5$.

65. Banking When banks process an electronic funds transfer (EFT), they assign a nine-digit routing number, where the ninth digit is a check digit. The check digit is chosen such that when the nine digits are multiplied in turn by the numbers 3, 7, 1, 3, 7, 1, 3, 7, and 1, the sum of the resulting products is congruent to 0 mod 10. For instance, 123456780 is a valid routing number because

$$1(3) + 2(7) + 3(1) + 4(3) + 5(7) + 6(1) + 7(3) + 8(7) + 0(1) = 150$$

and $150 \equiv 0 \bmod 10$.

a. Compute the check digit for the routing number 72859372-?

b. Verify that 584926105 is a valid routing number.

c. If a bank clerk inadvertantly types the routing number in part b as 584962105, will the computer be able to detect the error?

d. A bank employee accidentally transposed the 6 and the 1 in the routing number from part b and entered the number 584921605 into the computer system, but the computer did not detect an error. Explain why.

SECTION 8.3 Introduction to Group Theory

Introduction to Groups

An **algebraic system** is a set of elements along with one or more operations for combining the elements. The real numbers with the operations of addition and multiplication are an example of an algebraic system. Mathematicians classify this particular algebraic system as a *field.*

In the previous two sections of this chapter we discussed operations modulo n. Consider the set $\{0, 1, 2, 3, 4, 5\}$ and addition modulo 6. The set of elements is $\{0, 1, 2, 3, 4, 5\}$ and the operation is addition modulo 6. In this case there is only one operation. This is an example of an algebraic system called a *group.*

In addition to fields and groups, there are other types of algebraic systems. It is the properties of an algebraic system that distinguish it from another algebraic system. In this section, we will focus on groups.

A Group

A **group** is a set of elements, with one operation, that satisfies the following four properties:

1. The set is closed with respect to the operation.
2. The operation satisfies the associative property.
3. There is an identity element.
4. Each element has an inverse.

HISTORICAL NOTE

The Granger Collection, NY

Emmy Noether (nōh'ŭh thĭr) (1882–1935) made many contributions to mathematics and theoretical physics. One theorem she proved, sometimes referred to as Noether's theorem, relates to symmetries in physics. We will discuss symmetries of geometric figures later in this section.

Some of Noether's work led to several of the concepts of Einstein's general theory of relativity. In a letter to a colleague, Einstein described Noether's work as "penetrating, mathematical thinking."

Note from the preceding definition that a group is an algebraic system with *one* operation and that the operation must have certain characteristics. The first of these characteristics is that the set is *closed* with respect to the operation. **Closure** means that if any two elements are combined using the operation, the result must be an element of the set.

For example, the set {0, 1, 2, 3, 4, 5} with addition modulo 6 as the operation is closed. If we add two numbers of this set, modulo 6, the result is always a number of the set. For instance, $(3 + 5) \bmod 6 = 2$ and $(1 + 3) \bmod 6 = 4$.

As another example, consider the whole numbers {0, 1, 2, 3, 4, ...} with multiplication as the operation. If we multiply two whole numbers, the result is a whole number. For instance, $5 \cdot 9 = 45$ and $12 \cdot 15 = 180$. Thus the set of whole numbers is closed using multiplication as the operation. However, the set of whole numbers is not closed with division as the operation. For example, even though $36 \div 9 = 4$ (a whole number), $6 \div 4 = 1.5$, which is not a whole number. Therefore, the set of whole numbers is not closed with respect to division.

QUESTION

a. Is the set of whole numbers closed with respect to addition?

b. Is the set of whole numbers closed with respect to subtraction?

The second requirement of a group is that the operation must satisfy the associative property. Recall that the associative property of addition states that $a + (b + c) = (a + b) + c$. Addition modulo 6 is an associative operation. For instance, if we use the symbol ∇ to represent addition modulo 6, then

$$2 \nabla (5 \nabla 3) = 2 \nabla (8 \bmod 6) = 2 \nabla 2 = 4$$

and

$$(2 \nabla 5) \nabla 3 = (7 \bmod 6) \nabla 3 = 1 \nabla 3 = 4$$

Thus $2 \nabla (5 \nabla 3) = (2 \nabla 5) \nabla 3$.

Although there are some operations that do not satisfy the associative property, we will assume that an operation used in this section is associative.

The third requirement of a group is that the set must contain an *identity element*. An **identity element** is an element that, when combined with a second element using the group's operation, always returns the second element. As an illustration, if zero is added to a number, there is no change: $6 + 0 = 6$ and $0 + 1.5 = 1.5$. The number 0 is called an *additive identity*. Similarly, if we multiply any number by 1, there is no change: $\frac{1}{2} \cdot 1 = \frac{1}{2}$ and $1 \cdot 5 = 5$. The number 1 is called a *multiplicative identity*. For the set {0, 1, 2, 3, 4, 5} with addition modulo 6 as the operation, the identity element is 0. As we will soon see, an identity element does not always have to be zero or one.

The numbers 3 and -3 are called *additive inverses*. Adding these two numbers results in the additive identity: $3 + (-3) = 0$. The numbers $\frac{2}{3}$ and $\frac{3}{2}$ are called *reciprocals* or *multiplicative inverses*. Multiplying these two numbers results in the multiplicative identity: $\frac{2}{3} \cdot \frac{3}{2} = 1$. The last requirement of a group is that each element must have an *inverse*. This is a little more difficult to see for the set {0, 1, 2, 3, 4, 5} with addition modulo 6 as the operation. However, using addition modulo 6, we have

$$(0 + 0) \bmod 6 = 0 \qquad (2 + 4) \bmod 6 = 0$$
$$(1 + 5) \bmod 6 = 0 \qquad (3 + 3) \bmod 6 = 0$$

The above equations show that 0 is its own inverse, 1 and 5 are inverses, 2 and 4 are inverses, and 3 is its own inverse. Therefore, every element of this set has an inverse.

ANSWER **a.** Yes. The sum of any two whole numbers is a whole number. **b.** No. The difference between any two whole numbers is not always a whole number. For instance, $5 - 8 = -3$, but -3 is not a whole number.

We have shown that the set {0, 1, 2, 3, 4, 5} with addition modulo 6 as the operation is a group because it satisfies the four conditions stated in the definition of a group (see page 478).

EXAMPLE 1 Verify the Properties of a Group

Show that the integers with addition as the operation form a group.

Solution

We must show that the four properties of a group are satisfied.

1. The sum of two integers is always an integer. For instance,

$$-12 + 5 = -7 \qquad \text{and} \qquad -14 + (-21) = -35$$

Therefore, the integers are closed with respect to addition.

2. The associative property of addition holds true for the integers.
3. The identity element is 0, which is an integer. Therefore, the integers have an identity element for addition.
4. Each element has an inverse. If a is an integer, then $-a$ is the inverse of a.

Because each of the four conditions of a group is satisfied, the integers with addition as the operation form a group.

CHECK YOUR PROGRESS 1 Does the set {1, 2, 3} with operation multiplication modulo 4 form a group?

Solution See page S29.

HISTORICAL NOTE

Niels Henrik Abel (1802–1829) was born near Stavanger, Norway. His mathematical prowess was recognized quite early in his schooling, and he was encouraged to pursue a career in mathematics.

Abel became interested in trying to find a formula similar to the quadratic formula that could be used to solve all fifth-degree equations. In the process, he discovered that it was impossible to find such a formula. Part of Abel's proof investigated whether the equation $s_1s_2x = s_2s_1x$ was true; in other words, his conclusion depended on whether or not the commutative property held true.

Recall that the **commutative property** for an operation states that the order in which two elements are combined does not affect the result. For each of the groups discussed thus far, the operation has satisfied the commutative property. For example, the group {0, 1, 2, 3, 4, 5} with addition modulo 6 satisfies the commutative property, since, for instance, $2 + 5 = 5 + 2$. Groups in which the operation satisfies the commutative property are called **commutative groups** or **abelian groups**, after Niels Abel. The type of group we will look at next is an example of a *nonabelian group.* A **nonabelian group** is a group whose operation does not satisfy the commutative property.

Symmetry Groups

The concept of a group is very general. The elements that make up a group do not have to be numbers, and the operation does not have to be addition or multiplication. Also, a group does not have to satisfy the commutative property. We will now look at another group, called a *symmetry group,* an extension of which plays an important role in the study of atomic reactions. Symmetry groups are based on *regular polygons.* A **regular polygon** is a polygon all of whose sides have the same length and all of whose angles have the same measure.

TAKE NOTE

An **equilateral triangle** is a triangle in which all sides have the same length and each angle has a measure of 60°.

Consider two equilateral triangles, one placed inside the other, with their vertices numbered clockwise from 1 to 3. The larger triangle is the *reference triangle.* If we pick up the smaller triangle, there are several different ways in which we can set it back in its place. Each possible positioning of the inner triangle will be an element of a group. For instance, we can pick up the triangle and replace it exactly as we found it. We will call this position *I*; it will represent no change in position.

Now pick up the smaller triangle, rotate it 120° clockwise, and set it down again on top of the reference triangle. The result is shown below.

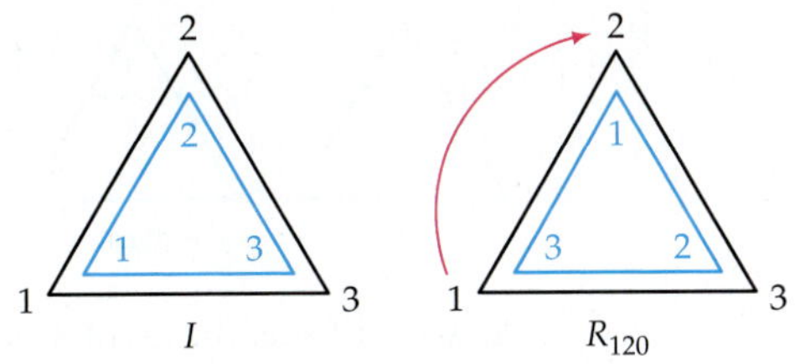

Note that the vertex originally at vertex 1 of the reference triangle is now at vertex 2, 2 is now at 3, and 3 is now at 1. We call the rotation of the triangle 120° clockwise R_{120}.

Now return the smaller triangle to its original position, where the numbers on the vertices of the triangles coincide. Consider a 240° clockwise rotation of the smaller triangle. The result of this rotation is shown below.

TAKE NOTE

We consider only rotations of the smaller triangle that result in all of its vertices coming to rest on the vertices of the reference triangle. For instance, if we rotate the smaller triangle clockwise 90°, the result is as shown below. This is not allowed.

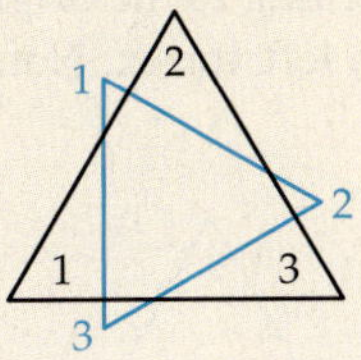

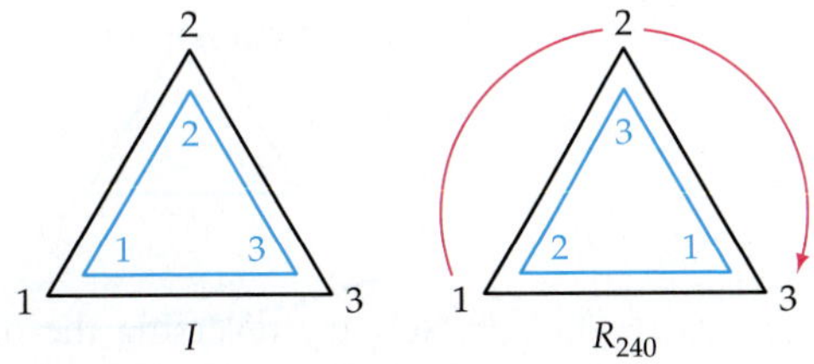

Note that the vertex originally at vertex 1 of the reference triangle is now at vertex 3, 2 is now at 1, and 3 is now at 2. We call the rotation of the triangle 240° clockwise R_{240}.

If the original triangle were rotated 360°, there would be no apparent change. The vertex at 1 would return to 1, the vertex at 2 would return to 2, and the vertex at 3 would return to 3. Because this rotation does not produce a new arrangement of the vertices, we consider it the same as the element we named *I*.

If we rotate the inner triangle *counterclockwise* 120°, the effect is the same as rotating it 240° clockwise. This rotation does not produce a different arrangement of the vertices. Similarly, a counterclockwise rotation of 240° is the same as a clockwise rotation of 120°.

In addition to rotating the triangle clockwise 120° and 240° as we did above, we could rotate the triangle about a line of symmetry that goes through a vertex before setting the triangle back down. Because there are three vertices, there are three possible results. These are shown below.

TAKE NOTE

A line of symmetry for a figure is one such that if the figure were folded along that line, the two halves of the figure would match up. See the diagram below.

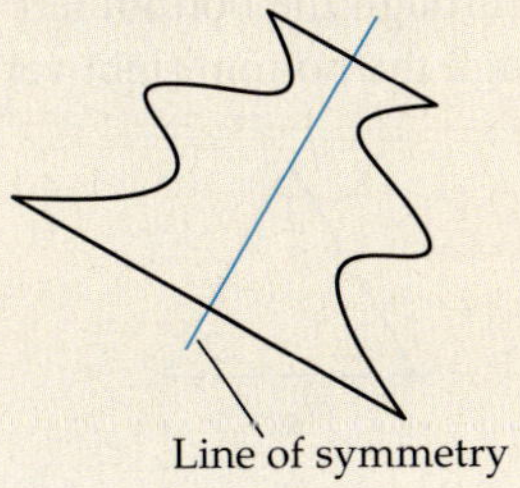

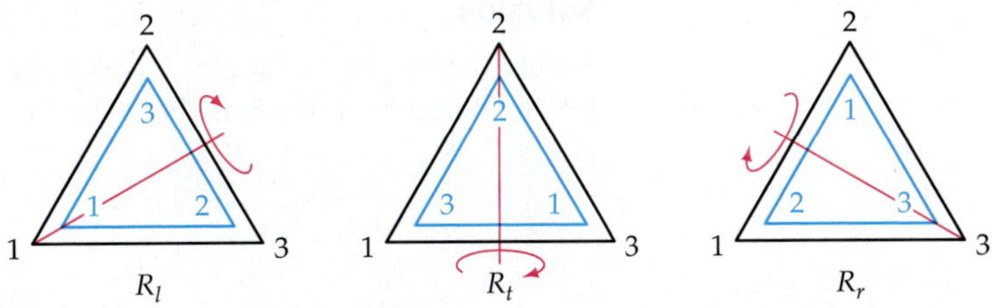

For rotation R_l, the bottom *l*eft vertex does not change, but vertices 2 and 3 are interchanged. If we rotate the triangle about the line of symmetry through the *t*op vertex, rotation R_t, vertex 2 does not change, but vertices 1 and 3 are interchanged. For rotation R_r, the bottom *r*ight vertex does not change, but vertices 1 and 2 are interchanged.

The six positions of the vertices we have seen thus far are the only possibilities: I (the triangle without any rotation), R_{120}, R_{240}, R_l, R_t, and R_r. These positions are shown below (without the reference triangle).

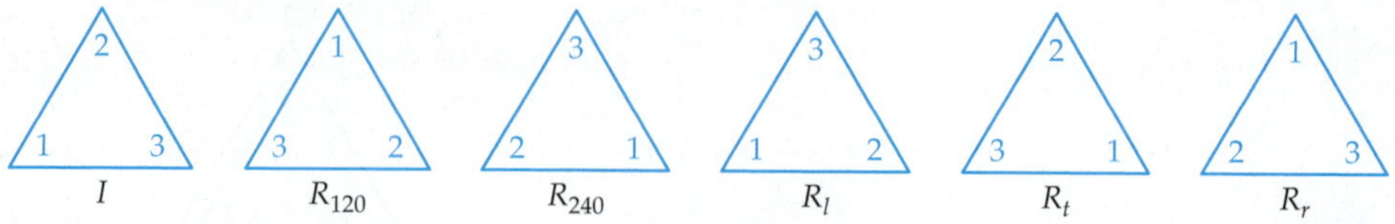

As we will see, these rotations are elements that form a group.

A group must have an operation, a method by which two elements of the group can be combined to form a third element that must also be a member of the group. (Recall that a group operation must be closed.) The operation we will use is called "followed by" and is symbolized by Δ. Next we show an example of how this operation works.

Consider $R_l \Delta R_{240}$. This means we rotate the original triangle, labeled as I, about the line of symmetry through vertex 1 "followed by" (without returning to the original position) a clockwise rotation of 240°. The result is one of the elements of the group, R_t. This operation is shown below.

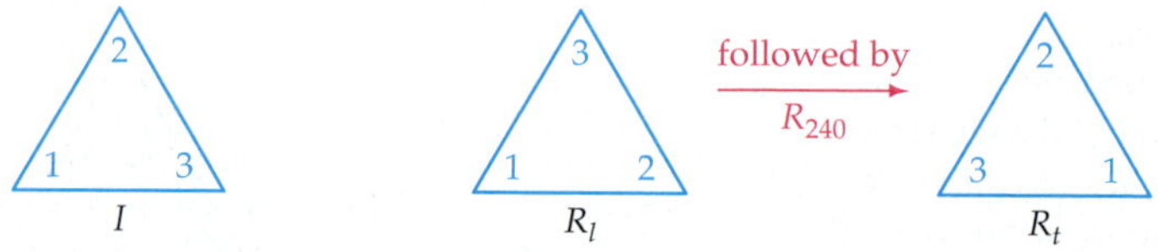

Now try reversing the operation, and consider $R_{240} \Delta R_l$. This means we rotate the original triangle, I, clockwise 240° "followed by" (without returning to the original position) a rotation about the axis of symmetry through the bottom left vertex. Note that the result is an element of the group, namely R_r, as shown below.

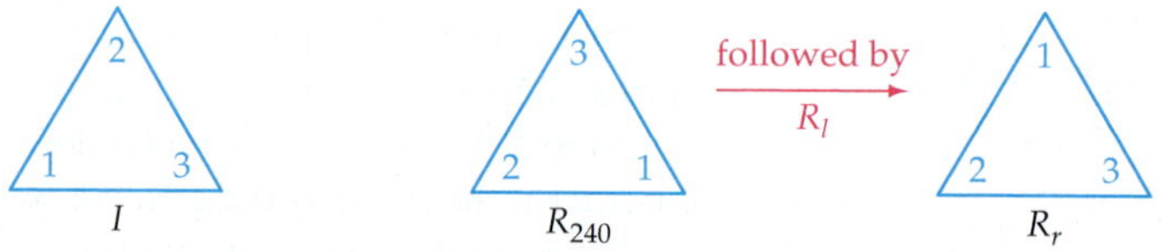

From these two examples, $R_l \Delta R_{240} = R_t$ and $R_{240} \Delta R_l = R_r$. Therefore, $R_l \Delta R_{240} \neq R_{240} \Delta_l$, which means the operation "followed by" is not commutative.

EXAMPLE 2 Perform an Operation of a Symmetry Group

Find $R_l \Delta R_r$.

Solution

Rotate the original triangle, I, about the line of symmetry through the bottom left vertex, followed by a rotation about the line of symmetry through the bottom right vertex.

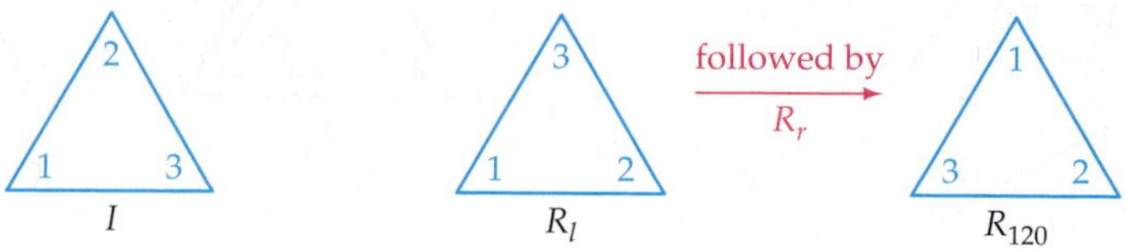

From the diagram, $R_l \Delta R_r = R_{120}$.

CHECK YOUR PROGRESS 2 Find $R_r \Delta R_{240}$.

Solution See page S29.

We have stated that the elements R_{120}, R_{240}, R_l, R_t, R_r, and I, with the operation "followed by," form a group. However, we have not demonstrated this fact, so we will do so now.

As we have seen, the set is closed with respect to the operation Δ. To show that the operation Δ is associative, think about the meaning of $x\Delta(y\Delta z)$ and $(x\Delta y)\Delta z$, where x, y, and z are elements of the group. $x\Delta(y\Delta z)$ means x, followed by the result of y followed by z; $(x\Delta y)\Delta z$ means x followed by y, followed by z. For instance,

$$R_t\Delta(R_l\Delta R_r) = R_t\Delta R_{120} = R_l \qquad \text{and} \qquad (R_t\Delta R_l)\Delta R_r = R_{120}\Delta R_r = R_l$$

All the remaining combinations of elements can be verified similarly.

The identity element is I and can be thought of as "no rotation." To see that each element has an inverse, verify the following:

$$R_{120}\Delta R_{240} = I \qquad R_l\Delta R_l = I \qquad R_t\Delta R_t = I \qquad R_r\Delta R_r = I$$

Your work should show that every element has an inverse. (The inverse of I is I.) Thus the four conditions of a group are satisfied.

To determine the outcome of the "followed by" operation on two elements of the group, we drew triangles and then rotated them as directed by the type of rotation. From a mathematical point of view, it would be nice to have a symbolic (rather than geometric) way of determining the outcome. We can do this by creating a mathematical object that describes the geometric object.

Note that any rotation of the triangle changes the position of the vertices. For R_{120}, the vertex originally at position 1 moved to position 2, the vertex at 2 moved to 3, and the vertex at 3 moved to 1. We can represent this as

$$R_{120} = \begin{pmatrix} 1 & 2 & 3 \\ 2 & 3 & 1 \end{pmatrix}$$

Similarly, for R_t, the vertex originally at position 1 moved to position 3, the vertex at 2 remained at 2, and the vertex at 3 moved to 1. This can be represented as

$$R_t = \begin{pmatrix} 1 & 2 & 3 \\ 3 & 2 & 1 \end{pmatrix}$$

The remaining four elements can be represented as

$$I = \begin{pmatrix} 1 & 2 & 3 \\ 1 & 2 & 3 \end{pmatrix}, \; R_{240} = \begin{pmatrix} 1 & 2 & 3 \\ 3 & 1 & 2 \end{pmatrix}, \; R_l = \begin{pmatrix} 1 & 2 & 3 \\ 1 & 3 & 2 \end{pmatrix}, \; R_r = \begin{pmatrix} 1 & 2 & 3 \\ 2 & 1 & 3 \end{pmatrix}$$

EXAMPLE 3 Symbolic Notation

Use symbolic notation to find $R_{120}\Delta R_t$.

Solution

To find the result of $R_{120}\Delta R_t$, we follow the movement of each vertex. The path that vertex 2 follows is highlighted.

$$R_{120}\Delta R_t = \begin{pmatrix} 1 & 2 & 3 \\ 2 & 3 & 1 \end{pmatrix} \Delta \begin{pmatrix} 1 & 2 & 3 \\ 3 & 2 & 1 \end{pmatrix}$$

- $1 \to 2 \to 2$. Thus $1 \to 2$.
- $2 \to 3 \to 1$. Thus $2 \to 1$.
- $3 \to 1 \to 3$. Thus $3 \to 3$.

$$= \begin{pmatrix} 1 & 2 & 3 \\ 2 & 1 & 3 \end{pmatrix} = R_r$$

$R_{120}\Delta R_t = R_r$

CHECK YOUR PROGRESS 3 Use symbolic notation to find $R_r\Delta R_{240}$.

Solution See page S29.

MATH MATTERS A Group of Subatomic Particles

In 1961, Murray Gell-Mann introduced what he called the *eightfold way* (after the Buddha's Eightfold Path to enlightenment and bliss) to categorize subatomic particles into eight different families, where each family was based on a certain symmetry group. Because each family was a group, each family had to satisfy the closure requirement. For that to happen, Gell-Mann's theory indicated that certain particles that had not yet been discovered must exist. Scientists began looking for these particles. One such particle, discovered in 1964, is called omega-minus. Its existence was predicted purely on the basis of the known properties of groups.

Permutation Groups

The triangular symmetry group discussed previously is an example of a special kind of group called a *permutation group*. A **permutation** is a rearrangement of objects. For instance, if we start with the arrangement of objects [♦ ♥ ♣], then one permutation of these objects is the rearrangement [♥ ♣ ♦]. If we consider each permutation of these objects as an element of a set, then the set of all possible permutations forms a group. The elements of the group are not numbers or the objects themselves, but rather the different permutations of the objects that are possible.

If we start with the numbers 1 2 3, we can represent permutations of the numbers using the same symbolic method that we used for the triangular symmetry group. For example, the permutation that rearranges 1 2 3 to 2 3 1 can be written

$$\begin{pmatrix} 1 & 2 & 3 \\ 2 & 3 & 1 \end{pmatrix}$$

which means that 1 is replaced by 2, 2 is replaced by 3, and 3 is replaced by 1.

There are only six distinct permutations of the numbers 1 2 3. We list and label them below. The identity element is named I.

$$I = \begin{pmatrix} 1 & 2 & 3 \\ 1 & 2 & 3 \end{pmatrix}, A = \begin{pmatrix} 1 & 2 & 3 \\ 2 & 3 & 1 \end{pmatrix}, B = \begin{pmatrix} 1 & 2 & 3 \\ 3 & 1 & 2 \end{pmatrix}$$

$$C = \begin{pmatrix} 1 & 2 & 3 \\ 1 & 3 & 2 \end{pmatrix}, D = \begin{pmatrix} 1 & 2 & 3 \\ 3 & 2 & 1 \end{pmatrix}, E = \begin{pmatrix} 1 & 2 & 3 \\ 2 & 1 & 3 \end{pmatrix}$$

The operation for the group is "followed by," which we will again denote by the symbol Δ. One can verify that the six elements along with this operation do indeed form a group.

POINT OF INTEREST

Bonnie Kamin, PhotoEdit

You may have played with a "15 puzzle" like the one pictured above. There is one empty slot that allows the numbered tiles to move around the board. The goal is to arrange the tiles in numerical order. The puzzle can be analyzed in the context of group theory, where each element of a group is a permutation of the tiles. If one were to remove two adjacent tiles and replace them with their positions switched, group theory could be used to prove that the puzzle becomes impossible to solve.

EXAMPLE 4 Perform an Operation in a Permutation Group

Find $B\Delta C$, where B and C are elements of the permutation group defined above.

Solution

$B\Delta C = \begin{pmatrix} 1 & 2 & 3 \\ 3 & 1 & 2 \end{pmatrix} \Delta \begin{pmatrix} 1 & 2 & 3 \\ 1 & 3 & 2 \end{pmatrix}$. In the first permutation, 1 is replaced by 3. In the second permutation, 3 is replaced by 2. When we combine these two actions, 1 is ultimately replaced by 2. Similarly, in the first permutation 2 is replaced by 1, which remains as 1 in the second permutation. 3 is replaced by 2, which is in turn replaced by 3. When these actions are combined, 3 remains as 3. The result of the operation is $\begin{pmatrix} 1 & 2 & 3 \\ 2 & 1 & 3 \end{pmatrix}$. Thus we see that $B\Delta C = E$.

CHECK YOUR PROGRESS 4 Compute $E\Delta B$, where E and B are elements of the permutation group defined previously.

Solution See page S29.

EXAMPLE 5 Inverse Element of a Permutation Group

One of the requirements of a group is that each element must have an inverse. Find the inverse of the element B of the permutation group defined previously.

Solution

Because $B = \begin{pmatrix} 1 & 2 & 3 \\ 3 & 1 & 2 \end{pmatrix}$ replaces 1 with 3, 2 with 1, and 3 with 2, its inverse must reverse these replacements. Thus we need to replace 3 with 1, 1 with 2, and 2 with 3. The inverse is the element $\begin{pmatrix} 1 & 2 & 3 \\ 2 & 3 & 1 \end{pmatrix} = A$. To verify,

$$B\Delta A = \begin{pmatrix} 1 & 2 & 3 \\ 3 & 1 & 2 \end{pmatrix}\Delta\begin{pmatrix} 1 & 2 & 3 \\ 2 & 3 & 1 \end{pmatrix} = \begin{pmatrix} 1 & 2 & 3 \\ 1 & 2 & 3 \end{pmatrix} \text{ and}$$

$$A\Delta B = \begin{pmatrix} 1 & 2 & 3 \\ 2 & 3 & 1 \end{pmatrix}\Delta\begin{pmatrix} 1 & 2 & 3 \\ 3 & 1 & 2 \end{pmatrix} = \begin{pmatrix} 1 & 2 & 3 \\ 1 & 2 & 3 \end{pmatrix}.$$

TAKE NOTE

In a nonabelian group, the order in which the two elements are combined with the group operation can affect the result. For an element to be the inverse of another element, combining the elements must give the identity element regardless of the order used.

CHECK YOUR PROGRESS 5 Find the inverse of the element D of the permutation group defined previously.

Solution See page S29.

If we consider a group with more than three numbers or objects, the number of different permutations increases dramatically. The permutation group of the five numbers 1 2 3 4 5 has 120 elements, and there are over 40,000 permutations of the numbers 1 2 3 4 5 6 7 8. In Exercise Set 8.3, you are asked to examine the permutations of 1 2 3 4.

EXCURSION

Wallpaper Groups

The symmetry group we studied based on an equilateral triangle involved rotations of the triangle that retained the position of the triangle within a reference triangle. We will now look at symmetries of an infinitely large object.

Imagine that the pattern shown in Figure 8.3 on the next page continues forever in every direction. You can think of it as an infinitely large sheet of wallpaper. Overlaying this sheet is a transparent sheet with the same pattern.

Visualize shifting the transparent sheet of wallpaper upward until the printed cats realign in what appears to be the same position. This is called a **translation**. Any such translation, in which we can shift the paper in one direction until the pattern aligns with its original appearance, is an element of a group. There are several directions in which we can shift the wallpaper (up, down, and diagonally) and an infinite number of distances we can shift the wallpaper. So the group has an infinite number of elements. Several elements are shown in Figure 8.4. Not shifting the paper at all is also considered an element;

INSTRUCTOR NOTE

To demonstrate the translations shown in Figure 8.4, make two transparencies of Figure 8.3. Overlay the transparencies, and then move one sheet in a direction until the wallpaper pattern coincides with the pattern on the second sheet.

we will call it *I*, because it is the identity element for the group. As with the symmetry groups, the operation is "followed by," denoted by Δ.

FIGURE 8.3

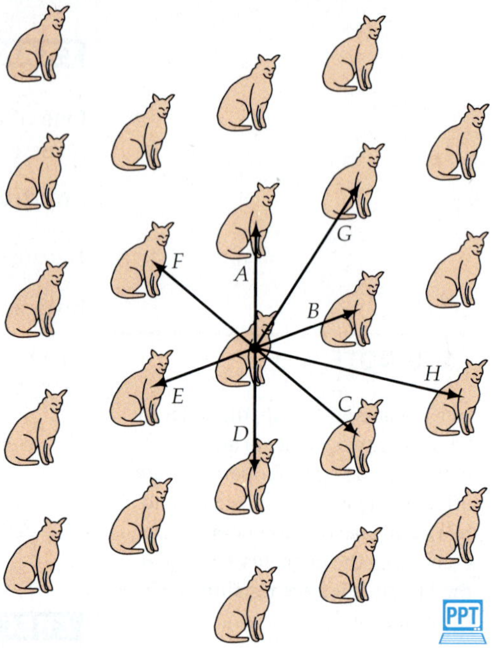

FIGURE 8.4

Some wallpaper patterns have rotational symmetries in addition to translation symmetries. The pattern in Figure 8.5 can be rotated 180° about the "×" and the wallpaper will appear to be in the exact same position. Thus the wallpaper group derived from this pattern will include this rotation as an element. Excursion Exercise 4 asks you to identify other rotational elements. Note that the group also includes translational elements.

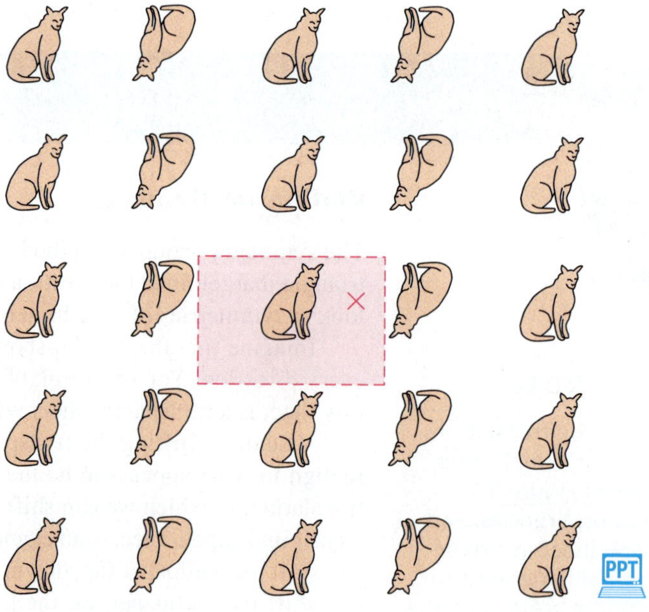

FIGURE 8.5

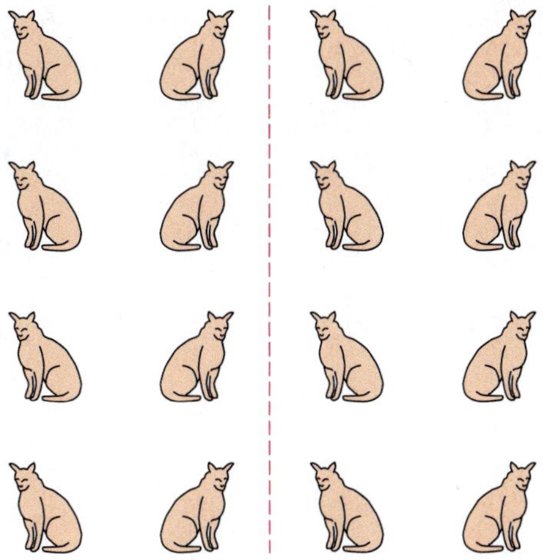

We can also consider mirror reflections. The wallpaper shown to the right can be reflected across the dashed line and its appearance will be unchanged. (There are many other vertical axes that could also be used.) So if we consider the group derived from symmetries of this wallpaper, one element will be the reflection of the entire sheet across this axis.

There is a fourth type of symmetry, called a *glide reflection*, that is a combination of a reflection and a shift. Mathematicians have determined that these are the only four types of symmetries possible and that there are only 17 distinct wallpaper groups. Of course, the visual patterns you see on actual wallpaper can appear quite varied, but if we formed groups from their symmetries, each would have to be one of the 17 known groups. (If you wish to see all 17 patterns, an excellent resource is http://www.clarku.edu/~djoyce/wallpaper/.)

EXCURSION EXERCISES

1. Explain why the properties of a group are satisfied by the symmetries of the wallpaper pattern shown in Figure 8.3. (You may assume that the operation is associative.)

2. Verify that in Figure 8.4, $A\Delta B = G$.

3. In Figure 8.6, determine where the black cat will be after applying the following translations illustrated in Figure 8.4.

a. $F\Delta(D\Delta C)$ **b.** $(C\Delta D)\Delta G$ **c.** $(A\Delta H)\Delta(E\Delta D)$

4. In Figure 8.5, there are three other locations within the dashed box about which the wallpaper can be rotated and align to its original appearance. Find the rotational center points.

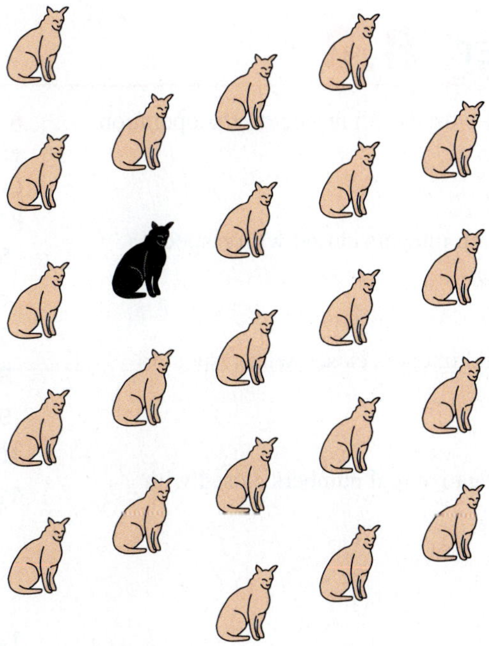

FIGURE 8.6

5. Consider the wallpaper group formed from symmetries of the pattern shown below in Figure 8.7.
   a. Identify three different elements of the group derived from translations.
   b. Find two different center points about which a 180° rotation is an element of the group.
   c. Find two different center points about which a 120° rotation is an element of the group.
   d. Identify three distinct lines that could serve as axes of reflection for elements of the group.

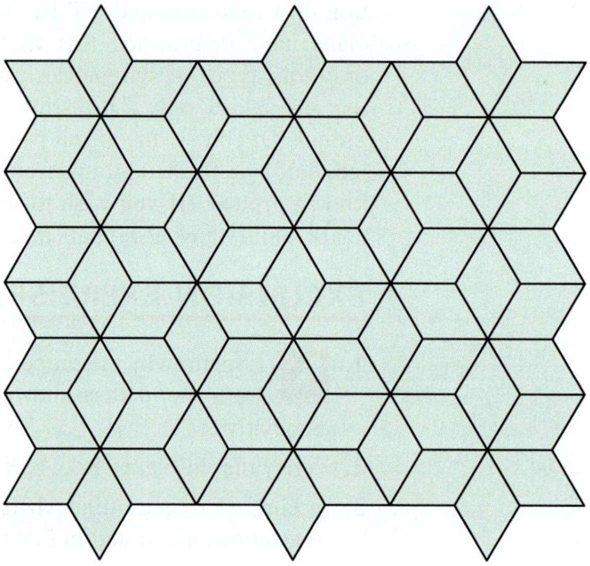

FIGURE 8.7

EXERCISE SET 8.3

1. Is the set $\{-1, 1\}$ closed with respect to the operation
   a. multiplication?
   b. addition?
2. Is the set of all even integers closed with respect to
   a. multiplication?
   b. addition?
3. Is the set of all odd integers closed with respect to
   a. multiplication?
   b. addition?
4. Is the set of all negative real numbers closed with respect to
   a. addition?
   b. subtraction?
   c. multiplication?
   d. division?

■ In Exercises 5 to 18, determine whether the set forms a group with respect to the given operation. (You may assume the operation is associative.) If the set does not form a group, determine which properties fail.

5. The even integers; addition
6. The even integers; multiplication
7. The real numbers; addition
8. The real numbers; division
9. The real numbers; multiplication
10. The real numbers *except* 0; multiplication
11. The rational numbers; addition

    (Recall that a rational number is any number that can be expressed in the form p/q, where p and q are integers, $q \neq 0$.)
12. The positive rational numbers; multiplication
13. $\{0, 1, 2, 3\}$; addition modulo 4

14. {0, 1, 2, 3, 4}; addition modulo 6

15. {0, 1, 2, 3}; multiplication modulo 4

16. {1, 2, 3, 4}; multiplication modulo 5

17. {−1, 1}; multiplication

18. {−1, 1}; division

■ Exercises 19 to 24 refer to the triangular symmetry group discussed in this section.

19. Find $R_t \Delta R_{120}$.

20. Find $R_{240} \Delta R_r$.

21. Find $R_r \Delta R_t$.

22. Find $R_t \Delta R_l$.

23. Which element is the inverse of R_{240}?

24. Which element is the inverse of R_t?

■ We can form another symmetry group by using rotations of a square rather than of a triangle. Start with a square with corners labeled 1 through 4, placed in a reference square.

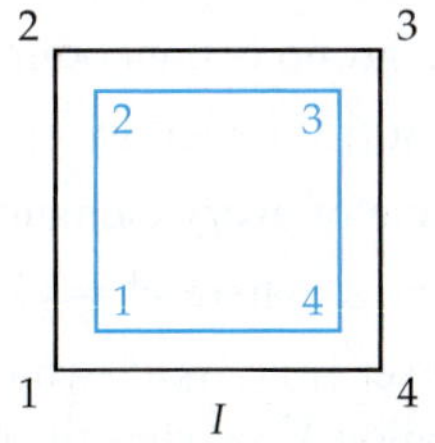

The current position of the inner square is labeled I. We can rotate the inner square clockwise 90° to give an element of the group, R_{90}. Similarly, we can rotate the square 180° or 270° to obtain the elements R_{180} and R_{270}. In addition, we can rotate the square about any of the lines of symmetry shown below.

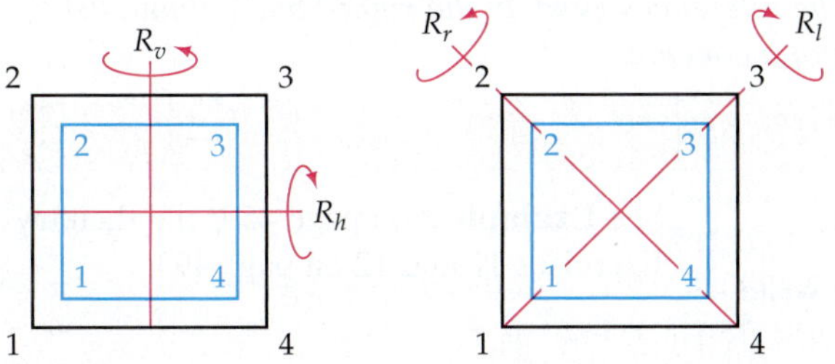

As before, the operation is "followed by," denoted by Δ. Exercises 25 to 34 refer to this symmetry group of the square.

25. Find $R_{90} \Delta R_v$.

26. Find $R_r \Delta R_{180}$.

27. Find $R_h \Delta R_{270}$.

28. Find $R_l \Delta R_h$.

29. List each of the eight elements of the group in symbolic notation. For instance, $I = \begin{pmatrix} 1 & 2 & 3 & 4 \\ 1 & 2 & 3 & 4 \end{pmatrix}$.

30. Use symbolic notation to find $R_v \Delta R_l$.

31. Use symbolic notation to find $R_h \Delta R_{90}$.

32. Use symbolic notation to find $R_{180} \Delta R_r$.

33. Find the inverse of the element R_{270}.

34. Find the inverse of the element R_v.

■ Exercises 35 to 42 refer to the group of permutations of the numbers 1 2 3 discussed in this section.

35. Find $A \Delta B$.

36. Find $B \Delta A$.

37. Find $B \Delta E$.

38. Find $D \Delta C$.

39. Find $C \Delta A$.

40. Find $B \Delta B$.

41. Find the inverse of the element A.

42. Find the inverse of the element E.

■ We can form a new permutation group by considering all permutations of the numbers 1 2 3 4. Exercises 43 to 50 refer to this group.

43. List all 24 elements of the group. Use symbolic notation, such as $I = \begin{pmatrix} 1 & 2 & 3 & 4 \\ 1 & 2 & 3 & 4 \end{pmatrix}$.

44. Find $\begin{pmatrix} 1 & 2 & 3 & 4 \\ 4 & 2 & 1 & 3 \end{pmatrix} \Delta \begin{pmatrix} 1 & 2 & 3 & 4 \\ 2 & 1 & 4 & 3 \end{pmatrix}$.

45. Find $\begin{pmatrix} 1 & 2 & 3 & 4 \\ 1 & 4 & 2 & 3 \end{pmatrix} \Delta \begin{pmatrix} 1 & 2 & 3 & 4 \\ 2 & 3 & 4 & 1 \end{pmatrix}$.

46. Find $\begin{pmatrix} 1 & 2 & 3 & 4 \\ 3 & 2 & 1 & 4 \end{pmatrix} \Delta \begin{pmatrix} 1 & 2 & 3 & 4 \\ 3 & 4 & 2 & 1 \end{pmatrix}$.

47. Find the inverse of $\begin{pmatrix} 1 & 2 & 3 & 4 \\ 2 & 4 & 1 & 3 \end{pmatrix}$.

48. Find the inverse of $\begin{pmatrix} 1 & 2 & 3 & 4 \\ 3 & 4 & 1 & 2 \end{pmatrix}$.

49. Find the inverse of $\begin{pmatrix} 1 & 2 & 3 & 4 \\ 4 & 3 & 2 & 1 \end{pmatrix}$.

50. Find the inverse of $\begin{pmatrix} 1 & 2 & 3 & 4 \\ 1 & 3 & 4 & 2 \end{pmatrix}$.

■ In Exercises 51 to 58, consider the set of elements $\{a, b, c, d\}$ with the operation ∇ as given in the table below.

∇	*a*	*b*	*c*	*d*
a	*d*	*a*	*b*	*c*
b	*a*	*b*	*c*	*d*
c	*b*	*c*	*d*	*a*
d	*c*	*d*	*a*	*b*

For example, to compute $c \nabla d$, find c in the left column and align it with d in the top row. The result is a.

51. Find $a \nabla a$.

52. Find $d \nabla c$.

53. Find $c \nabla b$.

54. Find $a \nabla d$.

55. Verify that the set with the operation ∇ is a group. (You may assume that the operation is associative.)

56. Which element of the group is the identity element?

57. Is the group commutative?

58. Identify the inverse of each element.

EXTENSIONS

59. **a.** Verify that the integers $\{1, 2, 3, 4, 5, 6\}$ with the operation multiplication modulo 7 form a group.

b. Verify that the integers $\{1, 2, 3, 4, 5\}$ with the operation multiplication modulo 6 do not form a group.

c. For which values of n are the integers $\{1, 2, \ldots, n - 1\}$ a group under the operation multiplication modulo n?

60. The set of numbers $\{2^k\}$, where k is any integer, form a group with multiplication as the operation.

a. Which element of the group is the identity element?

b. Verify that the set is closed with respect to multiplication.

c. Which element is the inverse of 2^8?

d. Which element is the inverse of $\frac{1}{32}$?

e. In general, what is the inverse of 2^k?

61. The **quaternion group** is a famous noncommutative group; we will define the group as the set $Q = \{e, r, s, t, u, v, w, x\}$ and the operation ∇ that is defined by the table to the right.

∇	e	r	s	t	u	v	w	x
e	e	r	s	t	u	v	w	x
r	r	u	t	w	v	e	x	s
s	s	x	u	r	w	t	e	v
t	t	s	v	u	x	w	r	e
u	u	v	w	x	e	r	s	t
v	v	e	x	s	r	u	t	w
w	w	t	e	v	s	x	u	r
x	x	w	r	e	t	s	v	u

a. Show that the group is noncommutative.

b. Which element is the identity element?

c. Find the inverse of every element in the group.

d. Show that $r^4 = e$, where $r^4 = r\nabla r\nabla r\nabla r$.

e. Show that the set $\{e, r, u, v\}$, which is a subset of Q, with the operation ∇ satisfies the properties of a group on its own. This is called a **subgroup** of the group.

f. Find a subgroup that has only two elements.

CHAPTER 8 SUMMARY

The following table summarizes essential concepts in this chapter. The references given in the right-hand column list Examples and Exercises that can be used to test your understanding of a concept.

8.1 Modular Arithmetic

Modulo *n* Two integers a and b are said to be congruent modulo n, where n is a natural number, if $\frac{a-b}{n}$ is an integer. In this case, we write $a \equiv b \bmod n$. The number n is called the modulus. The statement $a \equiv b \bmod n$ is called a congruence.	See **Example 2** on page 459, and then try Exercises 11 and 12 on page 493.
Arithmetic Modulo *n* After performing the operation as usual, divide the result by n; the answer is the remainder.	See **Examples 4, 5, and 7** on pages 460 to 462, and then try Exercises 15, 18, and 19 on page 493.
Solving Congruence Equations Individually check each whole number less than the modulus to see if it satisifies the congruence. Congruence equations may have more than one solution, or none at all.	See **Example 8** on page 463, and then try Exercises 26 and 27 on page 493.
Finding Inverses in Modular Arithmetic To find the additive inverse of a number modulo n, subtract the number from n. To find the multiplicative inverse (if it exists) of a number a, solve the congruence equation $ax \equiv 1 \bmod n$ for x.	See **Example 9 and 10** on page 464, and then try Exercise 29 on page 493.

8.2 Applications of Modular Arithmetic

ISBN Check Digit An ISBN is a 13-digit number used to identify a book. The 13th digit is a check digit. If we label the first digit of an ISBN d_1, the second digit d_2, and so on to the 13th digit as d_{13}, then the check digit is given by the following formula.

$$d_{13} = 10 - (d_1 + 3d_2 + d_3 + 3d_4 + d_5 + 3d_6 + d_7 + 3d_8 + d_9 + 3d_{10} + d_{11} + 3d_{12}) \bmod 10$$

If $d_{13} = 10$, then the check digit is 0.

See **Example 1** on page 468, and then try Exercise 33 on page 493.

UPC Check Digit A UPC is a 12-digit number that is used to identify a product such as a DVD, game, or grocery item. If we label the twelve digits of the UPC as $d_1, d_2, \ldots, d_{12}$, then the UPC check digit d_{12} is given by

$$d_{12} = 10 - (3d_1 + d_2 + 3d_3 + d_4 + 3d_5 + d_6 + 3d_7 + d_8 + 3d_9 + d_{10} + 3d_{11}) \bmod 10$$

If $d_{12} = 10$, then the check digit is 0.

See **Example 2** on page 469, and then try Exercise 35 on page 493.

Luhn Algorithm for Valid Credit Card Numbers The last digit of the credit card number is a check digit. Beginning with the next-to-last digit and reading from right to left, double every other digit. Treat any resulting two-digit number as two individual digits. Find the sum of the revised set of digits; the check digit is chosen such that the sum is congruent to 0 mod 10.

See **Example 3** on page 470, and then try Exercises 38 and 39 on page 493.

Cyclical Alphabetic Encryption A message can be encrypted by shifting each letter p of the plaintext message m positions through the alphabet. The encoded letter c satisfies the congruence $c \equiv (p + m) \bmod 26$, where p and c are the numerical equivalents of the plaintext and ciphertext letters, respectively. To decode the message, use the congruence $p \equiv (c + n) \bmod 26$, where $n = 26 - m$.

See **Example 4** on page 472, and then try Exercise 43 on page 493.

The $c \equiv (ap + n) \bmod 26$ Encryption Scheme Choose a natural number a that does not have any common factors with 26 and a natural number m. A plaintext letter p is encoded to a ciphertext letter c using the congruence $c \equiv (ap + n) \bmod 26$. To decode the message, solve the congruence equation $c \equiv (ap + n) \bmod 26$ for p.

See **Example 5** on page 473, and then try Exercise 46 on page 493.

8.3 Introduction to Group Theory

Properties of a Group A group is a set of objects with one operation that satisfies the following four properties.

1. The set is closed with respect to the operation.
2. The operation satisfies the associative property.
3. There is an identity element.
4. Each element has an inverse.

See **Example 1** on page 480, and then try Exercises 47, 49, and 50 on page 493.

continued

<table>
<tr>
<td>

Symmetry Group of an Equilateral Triangle The group has six elements, which consist of the different possible rotations that return the triangle to its reference triangle:

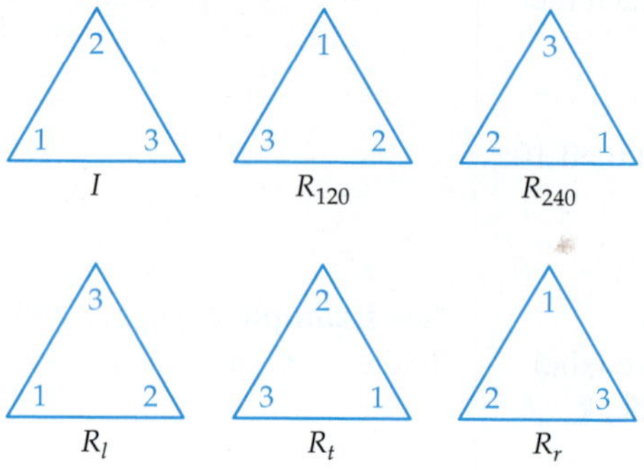

The group operation is "followed by," denoted by Δ. The operation is noncommutative.

</td>
<td>See **Example 2** on page 482, and then try Exercise 51 on page 493.</td>
</tr>
<tr>
<td>

Symbolic Notation for the Symmetry Group of an Equilateral Triangle The six elements of the group are:

$$I = \begin{pmatrix} 1 & 2 & 3 \\ 1 & 2 & 3 \end{pmatrix} \qquad R_{120} = \begin{pmatrix} 1 & 2 & 3 \\ 2 & 3 & 1 \end{pmatrix}$$

$$R_{240} = \begin{pmatrix} 1 & 2 & 3 \\ 3 & 1 & 2 \end{pmatrix} \qquad R_l = \begin{pmatrix} 1 & 2 & 3 \\ 1 & 3 & 2 \end{pmatrix}$$

$$R_t = \begin{pmatrix} 1 & 2 & 3 \\ 3 & 2 & 1 \end{pmatrix} \qquad R_r = \begin{pmatrix} 1 & 2 & 3 \\ 2 & 1 & 3 \end{pmatrix}$$

where, for instance, $\begin{pmatrix} 1 & 2 & 3 \\ 2 & 3 & 1 \end{pmatrix}$ signifies that the vertex of the equilateral triangle originally at vertex 1 of the reference triangle moves to vertex 2, the vertex at 2 moves to 3, and the vertex at 3 moves to 1.

</td>
<td>See **Example 3** on page 483, and then try Exercises 53 and 54 on page 494.</td>
</tr>
<tr>
<td>

Permutation Group of the Numbers 1 2 3 The group consists of the six permutations of the numbers 1 2 3:

$$I = \begin{pmatrix} 1 & 2 & 3 \\ 1 & 2 & 3 \end{pmatrix} \qquad A = \begin{pmatrix} 1 & 2 & 3 \\ 2 & 3 & 1 \end{pmatrix}$$

$$B = \begin{pmatrix} 1 & 2 & 3 \\ 3 & 1 & 2 \end{pmatrix} \qquad C = \begin{pmatrix} 1 & 2 & 3 \\ 1 & 3 & 2 \end{pmatrix}$$

$$D = \begin{pmatrix} 1 & 2 & 3 \\ 3 & 2 & 1 \end{pmatrix} \qquad E = \begin{pmatrix} 1 & 2 & 3 \\ 2 & 1 & 3 \end{pmatrix}$$

The group operation is "followed by," denoted by Δ.

</td>
<td>See **Examples 4 and 5** on pages 484 and 485, and then try Exercises 55, 56, and 57 on page 494.</td>
</tr>
</table>

CHAPTER 8 REVIEW EXERCISES

■ In Exercises 1 to 8, evaluate each expression, where $\oplus$ and $\ominus$ indicate addition and subtraction, respectively, using a 12-hour clock.

1. $9 \oplus 5$
2. $8 \oplus 6$
3. $10 \oplus 7$
4. $7 \oplus 11$
5. $6 \ominus 9$
6. $2 \ominus 10$
7. $3 \ominus 4$
8. $7 \ominus 12$

■ In Exercises 9 and 10, evaluate each expression, where $\boxed{+}$ and $\boxed{-}$ indicate addition and subtraction, respectively, using days-of-the-week arithmetic.

9. $5 \boxed{+} 5$
10. $2 \boxed{-} 5$

■ In Exercises 11 to 14, determine whether the congruence is true or false.

11. $17 \equiv 2 \bmod 5$
12. $14 \equiv 24 \bmod 4$
13. $35 \equiv 53 \bmod 10$
14. $12 \equiv 36 \bmod 8$

■ In Exercises 15 to 22, perform the modular arithmetic.

15. $(8 + 12) \bmod 3$
16. $(15 + 7) \bmod 6$
17. $(42 - 10) \bmod 8$
18. $(19 - 8) \bmod 4$
19. $(7 \cdot 5) \bmod 9$
20. $(12 \cdot 9) \bmod 5$
21. $(15 \cdot 10) \bmod 11$
22. $(41 \cdot 13) \bmod 8$

■ **Clocks and Calendars** In Exercises 23 and 24, use modular arithmetic to determine each of the following.

23. Disregarding A.M. or P.M., if it is now 5 o'clock,
  a. what time will it be 45 hours from now?
  b. what time was it 71 hours ago?

24. In 2015, April 15 (the day taxes are due in the United States) fell on a Wednesday. On what day of the week will April 15 fall in 2025?

■ In Exercises 25 to 28, find all whole number solutions of the congruence equation.

25. $x \equiv 7 \bmod 4$
26. $2x \equiv 5 \bmod 9$
27. $(2x + 1) \equiv 6 \bmod 5$
28. $(3x + 4) \equiv 5 \bmod 11$

■ In Exercises 29 and 30, find the additive inverse and the multiplicative inverse, if it exists, of the given number.

29. 5 in mod 7 arithmetic
30. 7 in mod 12 arithmetic

■ In Exercises 31 and 32, perform the modular division by solving a related multiplication problem.

31. $(2 \div 5) \bmod 7$
32. $(3 \div 4) \bmod 5$

■ **ISBN Numbers** In Exercises 33 and 34, determine the correct check digit for each ISBN.

33. *Oliver Twist* by Charles Dickens, 978-1-402-75425-?

34. *Oh, the Places You'll Go!* by Dr. Seuss, 978-0-679-80527-?

■ **UPC Codes** In Exercises 35 and 36, determine the correct check digit for the UPC.

35. 0-29000-07004-? (Planters Almonds)

36. 0-85391-89512-? (*Best in Show* DVD)

■ **Credit Card Numbers** In Exercises 37 to 40, determine whether the given credit card number is a valid number.

37. 5126-6993-4231-2956

38. 5383-0118-3416-5931

39. 3412-408439-82594

40. 6011-5185-8295-8328

■ **Encryption** In Exercises 41 and 42, encode the message using a cyclical alphabetic encrypting code that shifts the message the stated number of positions.

41. 7 positions: MAY THE FORCE BE WITH YOU

42. 11 positions: CANCEL ALL PLANS

■ **Encryption** In Exercises 43 and 44, use a cyclical alphabetic encrypting code to decode the encrypted message.

43. PXXM UDLT CXVXAAXF

44. HVS ROM VOG OFFWJSR

45. **Encryption** Use the encrypting congruence $c \equiv (3p + 6) \bmod 26$ to encrypt the message END OF THE LINE.

46. **Encryption** Decode the message WEU LKGGMF NHM NMGN, which was encrypted using the congruence

$$c \equiv (7p + 4) \bmod 26.$$

■ In Exercises 47 to 50, determine whether the set is a group with respect to the given operation. (You may assume the operation is associative.) If the set is not a group, determine which properties fail.

47. All rational numbers *except* 0; multiplication

48. All multiples of 3; addition

49. All negative integers; multiplication

50. $\{1, 2, 3, 4, 5, 6, 7, 8, 9, 10\}$; multiplication modulo 11

■ Exercises 51 and 52 refer to the triangular symmetry group discussed in Section 8.3.

51. Find $R_t \Delta R_r$.

52. Find $R_{240} \Delta R_t$

■ For Exercises 53 and 54, use symbolic notation, shown below, for the operations on an equilateral triangle.

$$I = \begin{pmatrix} 1 & 2 & 3 \\ 1 & 2 & 3 \end{pmatrix}, R_{120} = \begin{pmatrix} 1 & 2 & 3 \\ 2 & 3 & 1 \end{pmatrix}, R_{240} = \begin{pmatrix} 1 & 2 & 3 \\ 3 & 1 & 2 \end{pmatrix}$$

$$R_l = \begin{pmatrix} 1 & 2 & 3 \\ 1 & 3 & 2 \end{pmatrix}, R_t = \begin{pmatrix} 1 & 2 & 3 \\ 3 & 2 & 1 \end{pmatrix}, R_r = \begin{pmatrix} 1 & 2 & 3 \\ 2 & 1 & 3 \end{pmatrix}$$

53. $R_{240} \Delta R_l$

54. $R_t \Delta R_r$

■ For Exercises 55 and 56, use the elements of the permutation group on the numbers 1 2 3 shown below.

$$I = \begin{pmatrix} 1 & 2 & 3 \\ 1 & 2 & 3 \end{pmatrix}, A = \begin{pmatrix} 1 & 2 & 3 \\ 2 & 3 & 1 \end{pmatrix}, B = \begin{pmatrix} 1 & 2 & 3 \\ 3 & 1 & 2 \end{pmatrix}$$

$$C = \begin{pmatrix} 1 & 2 & 3 \\ 1 & 3 & 2 \end{pmatrix}, D = \begin{pmatrix} 1 & 2 & 3 \\ 3 & 2 & 1 \end{pmatrix}, E = \begin{pmatrix} 1 & 2 & 3 \\ 2 & 1 & 3 \end{pmatrix}$$

55. $C \Delta E$

56. $B \Delta A$

57. Find the inverse of D.

58. Find the inverse of A.

CHAPTER 8 TEST

1. Evaluate each expression, where $\oplus$ and $\ominus$ indicate addition and subtraction, respectively, using a 12-hour clock.

a. $8 \oplus 7$ **b.** $2 \ominus 9$

2. January 1, 2017, was a Sunday. What day of the week is January 1, 2026?

3. Determine whether the congruence is true or false.

a. $8 \equiv 20 \bmod 3$ **b.** $61 \equiv 38 \bmod 7$

■ In Exercises 4 to 6, perform the modular arithmetic.

4. $(25 + 9) \bmod 6$

5. $(31 - 11) \bmod 7$

6. $(5 \cdot 16) \bmod 12$

7. Scheduling Disregarding A.M. or P.M., if it is now 3 o'clock, use modular arithmetic to determine

a. what time it will be 27 hours from now.

b. what time it was 58 hours ago.

■ In Exercises 8 and 9, find all whole number solutions of the congruence equation.

8. $x \equiv 5 \bmod 9$

9. $(2x + 3) \equiv 1 \bmod 4$

10. Find the additive inverse and the multiplicative inverse, if it exists, of 5 in modulo 9 arithmetic.

11. ISBN Number Determine the correct check digit for the ISBN 978-0-739-49424-? (*Dictionary of American Slang, 4th Edition,* Barbara Ann Kipfer, Editor).

12. UPC Codes Determine the correct check digit for the UPC 6-73419-23216-? (LEGO Architecture set).

13. Credit Card Numbers Determine whether the credit card number 4232-8180-5736-4876 is a valid number.

14. Encryption Encrypt the plaintext message REPORT BACK using the cyclical alphabetic encrypting code that shifts letters 10 positions.

15. Encryption Decode the message UTSTG DPFM that was encrypted using the congruence $c \equiv (3p + 5) \bmod 26$.

16. Determine whether the set {1, 2, 3, 4} with the operation multiplication modulo 5 forms a group.

17. Determine whether the set of all odd integers with multiplication as the operation is a group. (You may assume the operation is associative.) If the set is not a group, explain which properties fail.

18. In the symmetry group of an equilateral triangle, determine the result of the operation.

a. $R_{120} \Delta R_l$ **b.** $R_t \Delta R_{240}$

19. In the permutation group of the numbers 1 2 3, determine the result of the operation.

$$\begin{pmatrix} 1 & 2 & 3 \\ 3 & 1 & 2 \end{pmatrix} \Delta \begin{pmatrix} 1 & 2 & 3 \\ 3 & 2 & 1 \end{pmatrix}$$

20. In the permutation group of the numbers 1 2 3 with operation Δ, find the inverse of $\begin{pmatrix} 1 & 2 & 3 \\ 3 & 1 & 2 \end{pmatrix}$.

9

Applications of Equations

In your study of mathematics, you have probably noticed that the problems became less concrete and more abstract. Problems that are concrete provide information pertaining to a specific instance. Abstract problems are theoretical; they are stated without reference to a specific instance. Here's an example of a concrete problem:

> If one candy bar costs 75 cents, how many candy bars can be purchased with 3 dollars?

To solve this problem, you need to calculate the number of cents in 3 dollars (multiply 3 by 100), and divide the result by the cost per candy bar (75 cents).

$$\frac{100 \cdot 3}{75} = \frac{300}{75} = 4$$

If one candy bar costs 75 cents, 4 candy bars can be purchased with 3 dollars.

Here is a related abstract problem:

If one candy bar costs c cents, how many candy bars can be purchased with d dollars?

Use the same procedure to solve the related abstract problem. Calculate the number of cents in d dollars (multiply d by 100), and divide the result by the cost per candy bar (c cents).

$$\frac{100 \cdot d}{c} = \frac{100d}{c}$$

If one candy bar costs c cents, $\frac{100d}{c}$ candy bars can be purchased with d dollars.

It is the variables in the problem above that makes it abstract. At the heart of the study of algebra is the use of variables. Variables enable us to generalize situations and state relationships among quantities. These relationships are often stated in the form of equations. In this chapter, we will be using equations to solve applications.

SECTION 9.1 First-Degree Equations and Formulas

Solving First-Degree Equations

TAKE NOTE

Recall that the order of operations agreement states that when simplifying a numerical expression you should perform the operations in the following order:

1. Perform operations inside parentheses.
2. Simplify exponential expressions.
3. Do multiplication and division from left to right.
4. Do addition and subtraction from left to right.

Suppose that the fuel economy, in miles per gallon, of a particular car traveling at a speed of v miles per hour can be calculated using the variable expression $-0.02v^2 + 1.6v + 3$, where $10 \le v \le 75$. For example, if the speed of a car is 30 miles per hour, we can calculate the fuel economy by substituting 30 for v in the variable expression and then using the order of operations agreement to evaluate the resulting numerical expression.

$$-0.02v^2 + 1.6v + 3$$

$$\begin{aligned} -0.02(30)^2 + 1.6(30) + 3 &= -0.02(900) + 1.6(30) + 3 \\ &= -18 + 48 + 3 \\ &= 33 \end{aligned}$$

The fuel economy is 33 miles per gallon.

The **terms** of a variable expression are the addends of the expression. The expression $-0.02v^2 + 1.6v + 3$ has three terms. The terms $-0.02v^2$ and $1.6v$ are **variable terms** because each contains a variable. The term 3 is a **constant term**; it does not contain a variable.

Each variable term is composed of a **numerical coefficient** and a **variable part** (the variable or variables and their exponents). For the variable term $-0.02v^2$, -0.02 is the coefficient and v^2 is the variable part.

Like terms of a variable expression are terms with the same variable part. Constant terms are also like terms. Examples of like terms are

$4x$ and $7x$

$9y$ and y

$5x^2y$ and $6x^2y$

8 and -3

POINT OF INTEREST

Albert Einstein Archive, Jerusalem, Ferdinand Schmutzer.

One of the most famous equations is $E = mc^2$. This equation, stated by Albert Einstein (īn′stīn), shows that there is a relationship between mass m and energy E. In this equation, c is the speed of light.

An **equation** expresses the equality of two mathematical expressions. Each of the following is an equation.

$8 + 5 = 13$

$4y - 6 = 10$

$x^2 - 2x + 1 = 0$

$b = 7$

Each of the equations below is a **first-degree equation in one variable**. *First degree* means that the variable has an exponent of 1.

$x + 11 = 14$

$3z + 5 = 8z$

$2(6y - 1) = 34$

TAKE NOTE

It is important to note the difference between an expression and an equation. An equation contains an equals sign; an expression doesn't.

A **solution** of an equation is a number that, when substituted for the variable, results in a true equation.

3 is a solution of the equation $x + 4 = 7$ because $3 + 4 = 7$.

9 is not a solution of the equation $x + 4 = 7$ because $9 + 4 \neq 7$.

To **solve an equation** means to find all solutions of the equation. The following properties of equations are often used to solve equations.

HISTORICAL NOTE

Finding solutions of equations has been a principal aim of mathematics for thousands of years. However, the equals sign did not occur in any text until 1557, when Robert Recorde (ca. 1510–1558) used the symbol in *The Whetstone of Witte.*

QUESTION Which of the following are first-degree equations in one variable?

a. $5y + 4 = 9 - 3(2y + 1)$ **b.** $\sqrt{x} + 9 = 16$ **c.** $p = -14$

d. $2x - 5 = x^2 - 9$ **e.** $3y + 7 = 4z - 10$

INSTRUCTOR NOTE

Students have a tendency to think of equations and expressions as the same thing. Emphasize that equations always have equals signs. Also, we *solve* equations and *simplify* expressions.

Properties of Equations

Addition Property

The same number can be added to each side of an equation without changing the solution of the equation.

$$\text{If } a = b, \text{ then } a + c = b + c.$$

Subtraction Property

The same number can be subtracted from each side of an equation without changing the solution of the equation.

$$\text{If } a = b, \text{ then } a - c = b - c.$$

Multiplication Property

Each side of an equation can be multiplied by the same *nonzero* number without changing the solution of the equation.

$$\text{If } a = b \text{ and } c \neq 0, \text{ then } ac = bc.$$

Division Property

Each side of an equation can be divided by the same *nonzero* number without changing the solution of the equation.

$$\text{If } a = b \text{ and } c \neq 0, \text{ then } \frac{a}{c} = \frac{b}{c}.$$

TAKE NOTE

In the multiplication property, it is necessary to state $c \neq 0$ so that the solutions of the equation are not changed.

For example, if $\frac{1}{2}x = 4$, then $x = 8$.

But if we multiply each side of the equation by 0, we have

$$0 \cdot \frac{1}{2}x = 0 \cdot 4$$
$$0 = 0$$

The solution $x = 8$ is lost.

In solving a first-degree equation in one variable, the goal is to rewrite the equation with the variable alone on one side of the equation and a constant term on the other side of the equation. The constant term is the solution of the equation.

For example, to solve the equation $t + 9 = -4$, use the subtraction property to subtract the constant term (9) from each side of the equation.

$$t + 9 = -4$$
$$t + 9 - 9 = -4 - 9$$
$$t = -13$$

Now the variable (t) is alone on one side of the equation and a constant term (-13) is on the other side. The solution is -13.

TAKE NOTE

You should always check the solution of an equation. The check for the example at the right is shown below.

$$\begin{array}{c|c} t + 9 & -4 \\ \hline -13 + 9 & -4 \\ -4 & -4 \end{array}$$

This is a true equation. The solution -13 checks.

ANSWER The equations in ***a*** and ***c*** are first-degree equations in one variable. The equation in ***b*** is not a first-degree equation in one variable because it contains the square root of a variable. The equation in ***d*** contains a variable with an exponent other than 1. The equation in ***e*** contains two variables.

To solve the equation $-5q = 120$, use the division property. Divide each side of the equation by the coefficient -5.

$$-5q = 120$$
$$\frac{-5q}{-5} = \frac{120}{-5}$$
$$q = -24$$

Now the variable (q) is alone on one side of the equation and a constant (-24) is on the other side. The solution is -24.

TAKE NOTE

An equation has some properties that are similar to those of a balance scale. For instance, if a balance scale is in balance and equal weights are added to each side of the scale, then the balance scale remains in balance. If an equation is true, then adding the same number to each side of the equation produces another true equation.

EXAMPLE 1 Solve a First-Degree Equation Using One of the Properties of Equations

Solve.

a. $y - 8 = 17$ **b.** $4x = -2$ **c.** $-5 = 9 + b$ **d.** $-a = -36$

Solution

a. Because 8 is subtracted from y, use the addition property to add 8 to each side of the equation.

$$y - 8 = 17$$
$$y - 8 + 8 = 17 + 8$$
$$y = 25 \quad \text{• A check will show that 25 is a solution.}$$

The solution is 25.

b. Because x is multiplied by 4, use the division property to divide each side of the equation by 4.

$$4x = -2$$
$$\frac{4x}{4} = \frac{-2}{4}$$
$$x = -\frac{1}{2} \quad \text{• A check will show that } -\tfrac{1}{2} \text{ is a solution.}$$

The solution is $-\frac{1}{2}$.

c. Because 9 is added to b, use the subtraction property to subtract 9 from each side of the equation.

$$-5 = 9 + b$$
$$-5 - 9 = 9 - 9 + b$$
$$-14 = b$$

The solution is -14.

TAKE NOTE

When the coefficient of a variable is 1 or negative 1, the 1 is usually not written; $1a$ is written as a, and $-1a$ is written as $-a$.

d. The coefficient of the variable is -1. Use the multiplication property to multiply each side of the equation by -1. (Alternatively, we can divide both sides by -1.)

$$-a = -36$$
$$-1(-1a) = -1(-36)$$
$$a = 36$$

The solution is 36.

CHECK YOUR PROGRESS 1 Solve.

a. $c - 6 = -13$ **b.** $4 = -8z$ **c.** $22 + m = -9$ **d.** $5x = 0$

Solution See page S29.

When solving more complicated first-degree equations in one variable, use the following sequence of steps.

Steps for Solving a First-Degree Equation in One Variable

1. If the equation contains fractions, multiply each side of the equation by the least common multiple (LCM) of the denominators to clear the equation of fractions.
2. Use the distributive property to remove parentheses.
3. Combine any like terms on the left side of the equation and any like terms on the right side of the equation.
4. Use the addition or subtraction property to rewrite the equation with only one variable term and only one constant term.
5. Use the multiplication or division property to rewrite the equation with the variable alone on one side of the equation and a constant term on the other side of the equation.

If one of the above steps is not needed to solve a given equation, proceed to the next step. Remember that the goal is to rewrite the equation with the variable alone on one side of the equation and a constant term on the other side of the equation.

EXAMPLE 2 Solve a First-Degree Equation Using the Properties of Equations

Solve.

a. $5x + 9 = 23 - 2x$ **b.** $8x - 3(4x - 5) = -2x + 6$

c. $\dfrac{3x}{4} - 6 = \dfrac{x}{3} - 1$

Solution

a. There are no fractions (Step 1) or parentheses (Step 2). There are no like terms on either side of the equation (Step 3). Use the addition property to rewrite the equation with only one variable term (Step 4). Add $2x$ to each side of the equation.

$$5x + 9 = 23 - 2x$$
$$5x + 2x + 9 = 23 - 2x + 2x$$
$$7x + 9 = 23$$

Use the subtraction property to rewrite the equation with only one constant term (Step 4). Subtract 9 from each side of the equation.

$$7x + 9 - 9 = 23 - 9$$
$$7x = 14$$

Use the division property to rewrite the equation with the x alone on one side of the equation (Step 5). Divide each side of the equation by 7.

$$\frac{7x}{7} = \frac{14}{7}$$
$$x = 2$$

The solution is 2.

b. There are no fractions (Step 1). Use the distributive property to remove parentheses (Step 2).

$$8x - 3(4x - 5) = -2x + 6$$
$$8x - 12x + 15 = -2x + 6$$

HISTORICAL NOTE

The letter x is used universally as the standard letter for a single unknown, which is why x-rays were so named. The scientists who discovered them did not know what they were, and so labeled them the "unknown rays," or x-rays.

Combine like terms on the left side of the equation (Step 3). Then rewrite the equation with the variable alone on one side and a constant on the other.

$$-4x + 15 = -2x + 6$$ • Combine like terms.

$$-4x + 2x + 15 = -2x + 2x + 6$$ • The addition property

$$-2x + 15 = 6$$

$$-2x + 15 - 15 = 6 - 15$$ • The subtraction property

$$-2x = -9$$

$$\frac{-2x}{-2} = \frac{-9}{-2}$$ • The division property

$$x = \frac{9}{2}$$

The solution is $\frac{9}{2}$.

TAKE NOTE

Recall that the least common multiple (LCM) of two numbers is the smallest number that both numbers divide into evenly. For the example at the right, the LCM of the denominators 4 and 3 is 12.

c. The equation contains fractions (Step 1); multiply each side of the equation by the LCM of the denominators, 12. Then rewrite the equation with the variable alone on one side and a constant on the other.

$$\frac{3x}{4} - 6 = \frac{x}{3} - 1$$

$$12\left(\frac{3x}{4} - 6\right) = 12\left(\frac{x}{3} - 1\right)$$ • The multiplication property

$$12 \cdot \frac{3x}{4} - 12 \cdot 6 = 12 \cdot \frac{x}{3} - 12 \cdot 1$$ • The distributive property

$$9x - 72 = 4x - 12$$

$$9x - 4x - 72 = 4x - 4x - 12$$ • The subtraction property

$$5x - 72 = -12$$

$$5x - 72 + 72 = -12 + 72$$ • The addition property

$$5x = 60$$

$$\frac{5x}{5} = \frac{60}{5}$$ • The division property

$$x = 12$$

The solution is 12.

CHECK YOUR PROGRESS 2 Solve.

a. $4x + 3 = 7x + 9$ **b.** $7 - (5x - 8) = 4x + 3$ **c.** $\frac{3x - 1}{4} + \frac{1}{3} = \frac{7}{3}$

Solution See page S29.

MATH MATTERS The Hubble Space Telescope

The Hubble Space Telescope was launched into orbit on April 24, 1990. Shortly thereafter, the telescope missed the stars it was targeted to photograph because it was pointing in the wrong direction. The direction was off by about one-half of 1 degree as a result of an arithmetic error—an addition instead of a subtraction.

JSC/NASA

Applications

In some applications of equations, we are given an equation that can be used to solve the problem. This is illustrated in Example 3.

EXAMPLE 3 Solve an Application

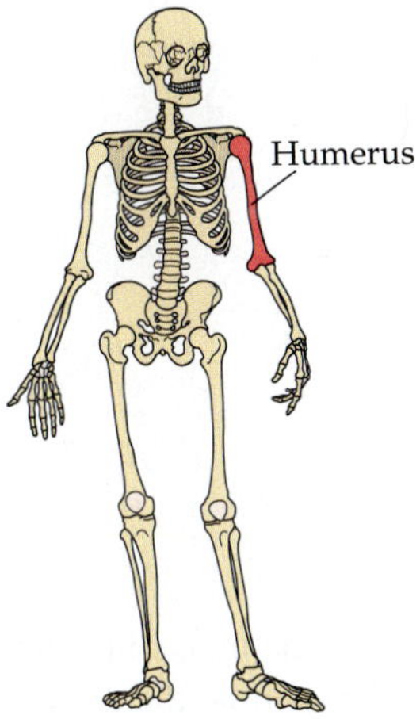

Forensic scientists have determined that the equation $H = 2.9L + 78.1$ can be used to approximate the height H, in centimeters, of an adult on the basis of the length L, in centimeters, of the adult's humerus (the bone extending from the shoulder to the elbow).

a. Use this equation to approximate the height of an adult whose humerus measures 36 cm.

b. According to this equation, what is the length of the humerus of an adult whose height is 168 cm?

Solution

a. Substitute 36 for L in the given equation. Solve the resulting equation for H.

$$\begin{aligned} H &= 2.9L + 78.1 \\ H &= 2.9(36) + 78.1 \\ H &= 104.4 + 78.1 \\ H &= 182.5 \end{aligned}$$

The adult's height is approximately 182.5 cm.

b. Substitute 168 for H in the given equation. Solve the resulting equation for L.

$$\begin{aligned} H &= 2.9L + 78.1 \\ 168 &= 2.9L + 78.1 \\ 168 - 78.1 &= 2.9L + 78.1 - 78.1 \\ 89.9 &= 2.9L \\ \frac{89.9}{2.9} &= \frac{2.9L}{2.9} \\ 31 &= L \end{aligned}$$

The length of the adult's humerus is approximately 31 cm.

vadim kozlovsky/Shutterstock.com

CHECK YOUR PROGRESS 3 The amount of garbage generated by each person living in the United States has been increasing and is approximated by the equation $P = 0.05Y - 96$, where P is the number of pounds of garbage generated per person per day and Y is the year.

a. Find the amount of garbage generated per person per day in 2015.

b. According to the equation, in what year will 5.5 lb of garbage be generated per person per day?

Solution See page S30.

In many applied problems, we are not given an equation that can be used to solve the problem. Instead, we must use the given information to write an equation whose solution answers the question stated in the problem. This is illustrated in Examples 4 and 5.

EXAMPLE 4 Solve an Application of First-Degree Equations

The cost of electricity in a certain city is \$0.16 for each of the first 300 kWh (kilowatt-hours) and \$0.26 for each kilowatt-hour over 300 kWh. Find the number of kilowatt-hours used by a family that receives a \$103.90 electric bill.

Solution

Let $k =$ the number of kilowatt-hours used by the family. Write an equation and then solve the equation for k.

> **TAKE NOTE**
>
> If the family uses 500 kWh of electricity, they are billed \$0.26/kWh for 200 kWh $(500 - 300)$. If they use 650 kWh, they are billed \$0.26/kWh for 350 kWh $(650 - 300)$. If they use k kWh, $k > 300$, they are billed \$0.26/kWh for $(k - 300)$ kWh.

[\$0.16 for each of the first 300 kWh + \$0.26 for each kilowatt-hour over 300] = [\$103.90]

$$\begin{aligned}
0.16(300) + 0.26(k - 300) &= 103.90\\
48 + 0.26k - 78 &= 103.90\\
0.26k - 30 &= 103.90\\
0.26k - 30 + 30 &= 103.90 + 30\\
0.26k &= 133.90\\
\frac{0.26k}{0.26} &= \frac{133.90}{0.26}\\
k &= 515
\end{aligned}$$

The family used 515 kWh of electricity.

CHECK YOUR PROGRESS 4 For a classified ad, an online magazine charges \$340 for the first four lines and \$76 for each additional line. Determine the number of lines that can be published in an ad for \$948.

Solution See page S30.

EXAMPLE 5 Solve an Application of First-Degree Equations

In 2014, the population of Jackson, MS was 171,200, and the population of Eugene, OR was 160,600. In recent years, Jackson's population decreased at an average rate of 600 people per year, while Eugene's increased at an average rate of 1100 people per year. If these rate changes remained stable, in what year would the populations of Jackson and Eugene be the same? Round to the nearest year.

Solution

Let $n =$ the number of years after 2014. Write an equation for when the populations would be the same and then solve the equation for n.

[The 2014 population of Jackson minus the annual decrease times n] = [The 2014 population of Eugene plus the annual increase times n]

$$\begin{aligned}
171{,}200 - 600n &= 160{,}600 + 1100n\\
171{,}200 - 600n + 600n &= 160{,}600 + 1100n + 600n\\
171{,}200 &= 160{,}600 + 1700n\\
171{,}200 - 160{,}600 &= 160{,}600 - 160{,}600 + 1700n\\
10{,}600 &= 1700n\\
\frac{10{,}600}{1700} &= \frac{1700n}{1700}\\
6 &\approx n
\end{aligned}$$

The variable n is the number of years after 2014. Add 6 to the year 2014.

$$2014 + 6 = 2020$$

To the nearest year, the populations would be the same in 2020.

> **POINT OF INTEREST**
>
> Is the population of your state increasing or decreasing? You can find out by checking a reference such as census.gov, which was the source for the data in Example 5 and Check Your Progress 5.

CHECK YOUR PROGRESS 5 In 2014, the population of Cleveland, OH, was 389,500, and the population of Tampa, FL, was 358,700. In recent years, Cleveland's population decreased at an average rate of 1600 people per year, while Tampa's increased at an average rate of 5700 people per year. If these rate changes remained stable, in what year would the populations of Cleveland and Tampa be the same? Round to the nearest year.

Solution See page S30.

Literal Equations

A **literal equation** is an equation that contains more than one variable. Examples of literal equations are:

$$2x + 3y = 6$$
$$4a - 2b + c = 0$$

A **formula** is a literal equation that states a relationship between two or more quantities in an application problem. Examples of formulas are shown below. These formulas are taken from physics, mathematics, and business.

$$\frac{1}{R_1} + \frac{1}{R_2} = \frac{1}{R}$$
$$s = a + (n - 1)d$$
$$A = P + Prt$$

QUESTION Which of the following are literal equations?

a. $5a - 3b = 7$ **b.** $a^2 + b^2 = c^2$

c. $a_1 + (n - 1)d$ **d.** $3x - 7 = 5 + 4x$

The addition, subtraction, multiplication, and division properties of equations can be used to solve some literal equations for one of the variables. In solving a literal equation for one of the variables, the goal is to rewrite the equation so that the letter being solved for is alone on one side of the equation and all numbers and other variables are on the other side. This is illustrated in Example 6.

EXAMPLE 6 Solve a Literal Equation

a. Solve $A = P(1 + i)$ for i.

b. Solve $I = \dfrac{E}{R + r}$ for R.

Solution

a. The goal is to rewrite the equation so that i is alone on one side of the equation and all other numbers and letters are on the other side. We will begin by using the distributive property on the right side of the equation.

$$A = P(1 + i)$$
$$A = P + Pi$$

ANSWER ***a*** and ***b*** are literal equations. ***c*** is not an equation. ***d*** does not have more than one variable.

Subtract P from each side of the equation.

$$A - P = P - P + Pi$$
$$A - P = Pi$$

Divide each side of the equation by P.

$$\frac{A - P}{P} = \frac{Pi}{P}$$

$$\frac{A - P}{P} = i$$

INSTRUCTOR NOTE

Some students will want to simplify the expression $\frac{A - P}{P}$. Remind them that we can divide by common *factors* only.

Many students will find it helpful if you provide additional examples of solving literal equations for an indicated variable. Here are a few possibilities.

1. Solve $PV = nRT$ for T.

   $T = \frac{PV}{nR}$

2. Solve $V = \frac{1}{3}s^2h$ for h.

   $h = \frac{3V}{s^2}$

3. Solve $S = C + rC$ for r.

   $r = \frac{S - C}{C}$

b. The goal is to rewrite the equation so that R is alone on one side of the equation and all other variables are on the other side of the equation. Because the equation contains a fraction, we will first multiply both sides of the equation by the denominator $R + r$ to clear the equation of fractions.

$$I = \frac{E}{R + r}$$

$$(R + r)I = (R + r)\frac{E}{R + r}$$

$$RI + rI = E$$

Subtract from the left side of the equation the term that does not contain a capital R.

$$RI + rI - rI = E - rI$$
$$RI = E - rI$$

Divide each side of the equation by I.

$$\frac{RI}{I} = \frac{E - rI}{I}$$

$$R = \frac{E - rI}{I}$$

CHECK YOUR PROGRESS 6

a. Solve $s = \frac{A + L}{2}$ for L.

b. Solve $L = a(1 + ct)$ for c.

Solution See page S30.

EXCURSION

Body Mass Index

Body mass index, or **BMI**, expresses the relationship between a person's height and weight. It is a measurement for gauging a person's weight-related level of risk for high blood pressure, heart disease, and diabetes. A BMI value of 25 or less indicates a very low to low risk; a BMI value of 25 to 30 indicates a low to moderate risk; a BMI of 30 or more indicates a moderate to very high risk.

The formula for body mass index is

$$B = \frac{703W}{H^2}$$

where B is the BMI, W is weight in pounds, and H is height in inches.

To determine how much a woman who is 5′4″ should weigh in order to have a BMI of 24, first convert 5′4″ to inches.

$$5'4'' = 5(12)'' + 4'' = 60'' + 4'' = 64''$$

Substitute 24 for B and 64 for H in the body mass index formula. Then solve the resulting equation for W.

$$B = \frac{703W}{H^2}$$

$$24 = \frac{703W}{64^2} \qquad \bullet\ B = 24,\ H = 64$$

$$24 = \frac{703W}{4096}$$

$$4096(24) = 4096\left(\frac{703W}{4096}\right) \qquad \bullet \text{ Multiply each side of the equation by 4096.}$$

$$98{,}304 = 703W$$

$$\frac{98{,}304}{703} = \frac{703W}{703} \qquad \bullet \text{ Divide each side of the equation by 703.}$$

$$140 \approx W$$

A woman who is 5′4″ should weigh about 140 lb in order to have a BMI of 24.

EXCURSION EXERCISES

1. Amy is 140 lb and 5′8″ tall. Calculate Amy's BMI. Round to the nearest tenth. Rank Amy as a low, moderate, or high risk for weight-related disease.
2. Roger is 5′11″. How much should he weigh in order to have a BMI of 25? Round to the nearest pound.
3. Bohdan weighs 185 lb and is 5′9″. How many pounds must Bohdan lose in order to reach a BMI of 23? Round to the nearest pound.
4. Felicia weighs 160 lb and is 5′7″. She would like to lower her BMI to 20.
   a. By how many points must Felicia lower her BMI? Round to the nearest tenth.
   b. How many pounds must Felicia lose in order to reach a BMI of 20? Round to the nearest pound.
5. **Finding BMI Using a Nomograph** Most medical professionals use a nomograph (or nomogram) to calculate the BMI of their patients. A nomograph is a diagram that shows the relationships among three or more quantities by means of scales that are arranged such that the value of one variable can be found by drawing a straight line from one scale intersecting the other scales at appropriate values. You can download a BMI nomograph from the Internet. One good source is www.pynomo.org/wiki/images/b/b4/Ex_BMI.png. On the nomograph on the next page, the line connecting a height of 5.9 ft and a weight of 150 lb intersects the BMI scale at 21, indicating that a 5.9-foot, 150-pound person has a BMI of 21.
   a. Use the nomograph to calculate Amy's BMI (see Excursion Exercise 1). Compare the results you obtained by using the nomograph with the value you obtained by using a calculator.
   b. Use the nomograph to calculate how much Roger should weigh in order to have a BMI of 25 (see Excursion Exercise 2). How does the result you obtained

by using the nomograph compare with the value you obtained by using a calculator?

c. Explain the advantages and disadvantages of using a nomograph to calculate BMI.

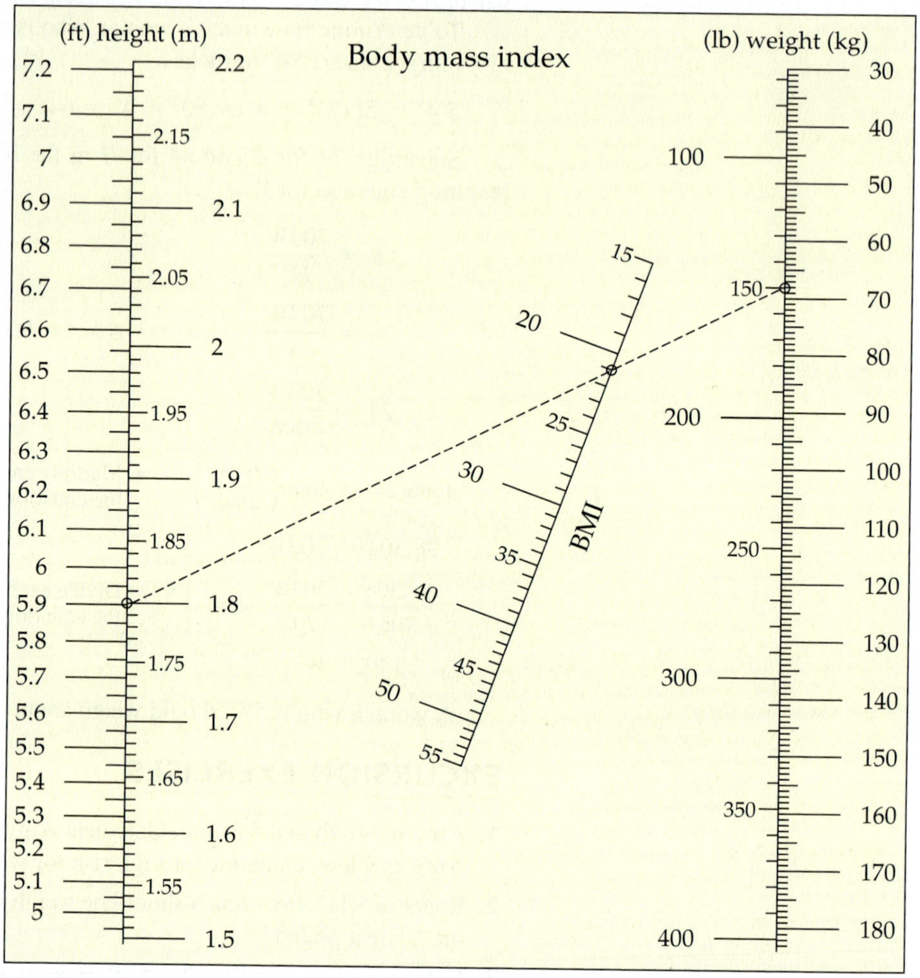

EXERCISE SET 9.1

1. What is the difference between an expression and an equation? Provide an example of each.
2. What is the solution of the equation $x = 8$? Use your answer to explain why the goal in solving an equation is to get the variable alone on one side of the equation.
3. Explain how to check the solution of an equation.

- In Exercises 4 to 41, solve the equation.

4. $x + 7 = -5$
5. $9 + b = 21$
6. $-9 = z - 8$
7. $b - 11 = 11$
8. $-3x = 150$
9. $-48 = 6z$
10. $-9a = -108$
11. $-\frac{3}{4}x = 15$
12. $\frac{5}{2}x = -10$
13. $-\frac{x}{4} = -2$
14. $\frac{2x}{5} = -8$
15. $4 - 2b = 2 - 4b$
16. $4y - 10 = 6 + 2y$
17. $5x - 3 = 9x - 7$
18. $10z + 6 = 4 + 5z$
19. $3m + 5 = 2 - 6m$
20. $6a - 1 = 2 + 2a$
21. $5x + 7 = 8x + 5$
22. $2 - 6y = 5 - 7y$
23. $4b + 15 = 3 - 2b$
24. $2(x + 1) + 5x = 23$
25. $9n - 15 = 3(2n - 1)$
26. $7a - (3a - 4) = 12$
27. $5(3 - 2y) = 3 - 4y$

28. $9 - 7x = 4(1 - 3x)$

29. $2(3b + 5) - 1 = 10b + 1$

30. $2z - 2 = 5 - (9 - 6z)$

31. $4a + 3 = 7 - (5 - 8a)$

32. $5(6 - 2x) = 2(5 - 3x)$

33. $4(3y + 1) = 2(y - 8)$

34. $2(3b - 5) = 4(6b - 2)$

35. $3(x - 4) = 1 - (2x - 7)$

36. $\frac{2y}{3} - 4 = \frac{y}{6} - 1$

37. $\frac{x}{8} + 2 = \frac{3x}{4} - 3$

38. $\frac{2x - 3}{3} + \frac{1}{2} = \frac{5}{6}$

39. $\frac{2}{3} + \frac{3x + 1}{4} = \frac{5}{3}$

40. $\frac{1}{2}(x + 4) = \frac{1}{3}(3x - 6)$

41. $\frac{3}{4}(x - 8) = \frac{1}{2}(2x + 4)$

■ **Car Payments** The monthly car payment on a 60-month car loan at a 5% rate is calculated by using the formula $P = 0.018417L$, where P is the monthly car payment and L is the loan amount. Use this formula for Exercises 42 and 43.

42. If you can afford a maximum monthly car payment of \$300, what is the maximum loan amount you can afford? Round to the nearest cent.

43. If the maximum monthly car payment you can afford is \$350, what is the maximum loan amount you can afford? Round to the nearest cent.

■ **Deep-Sea Diving** The pressure on a diver can be calculated using the formula $P = 15 + \frac{1}{2}D$, where P is the pressure in pounds per square inch and D is the depth in feet. Use this formula for Exercises 44 and 45.

44. Find the depth of a diver when the pressure on the diver is 45 lb/in^2.

45. Find the depth of a diver when the pressure on the diver is 55 lb/in^2.

■ **Foot Races** The world-record time for a 1-mile race can be approximated by $t = 16.11 - 0.0062y$, where y is the year of the race, $1950 \le y \le 2005$, and t is the time, in minutes, of the race. Use this formula for Exercises 46 and 47.

46. Approximate the year in which the first "4-minute mile" was run. The actual year was 1954.

47. In 1999, the world-record time for a 1-mile race was 3.72 min. For what year does the equation predict this record time?

■ **Black Ice** Black ice is an ice covering on roads that is especially difficult to see and therefore extremely dangerous for motorists. The distance a car traveling at 30 mph will slide after its brakes are applied is related to the outside temperature by the formula $C = \frac{1}{4}D - 45$, where C is the Celsius temperature and D is the distance, in feet, that the car will slide. Use this formula for Exercises 48 and 49.

48. Determine the distance a car will slide on black ice when the outside air temperature is -3°C.

49. How far will a car slide on black ice when the outside air temperature is -11°C?

■ **Crickets** The formula $N = 7C - 30$ approximates N, the number of times per minute a cricket chirps when the air temperature is C degrees Celsius. Use this formula for Exercises 50 and 51.

50. What is the approximate air temperature when a cricket chirps 100 times per minute? Round to the nearest tenth.

51. Determine the approximate air temperature when a cricket chirps 140 times per minute. Round to the nearest tenth.

■ **Bowling** In order to equalize all the bowlers' chances of winning, some players in a bowling league are given a handicap, or a bonus of extra points. Some leagues use the formula $H = 0.8(200 - A)$, where H is the handicap and A is the bowler's average score in past games. Use this formula for Exercises 52 and 53.

52. A bowler has a handicap of 20. What is the bowler's average score?

53. Find the average score of a bowler who has a handicap of 25.

Anton Balazh/Shutterstock.com

■ In Exercises 54 to 63, write an equation as part of solving the problem.

54. **College Tuition** The graph below shows average tuition and fees at private 4-year colleges for selected years.

a. For the 2003–04 school year, the average tuition and fees at private 4-year colleges were \$934 more than four times the average tuition and fees at public 4-year colleges. Find the average tuition and fees at public 4-year colleges for the school year 2003–04.

b. For the 2015–16 school year, the average tuition and fees at private 4-year colleges were \$4175 more than three times the average tuition and fees at public 4-year colleges. Determine the average tuition and fees at public 4-year colleges for the 2015–16 school year.

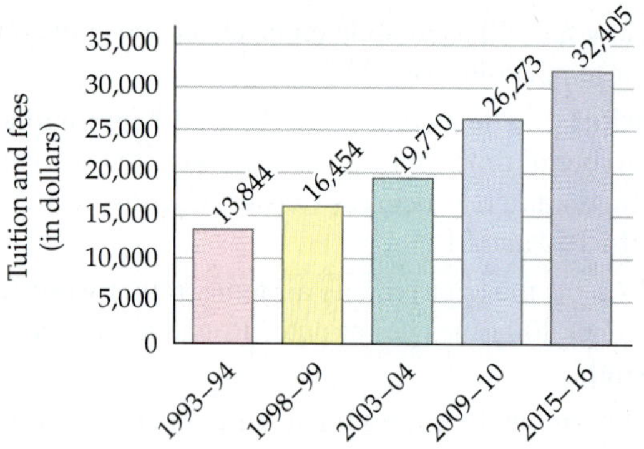

Tuition and Fees at Private 4-Year Colleges
SOURCE: The College Board

55. **Adoption** In a recent year, Americans adopted 17,438 children from foreign countries. In the graph below are the top three countries where the children were born.

a. The number of children adopted from Guatemala was 1052 less than three times the number adopted from Ethiopia. Determine the number of children adopted from Ethiopia that year.

b. The number of children adopted from China was 189 more than eight times the number adopted from Ukraine. Determine the number of children adopted from Ukraine that year.

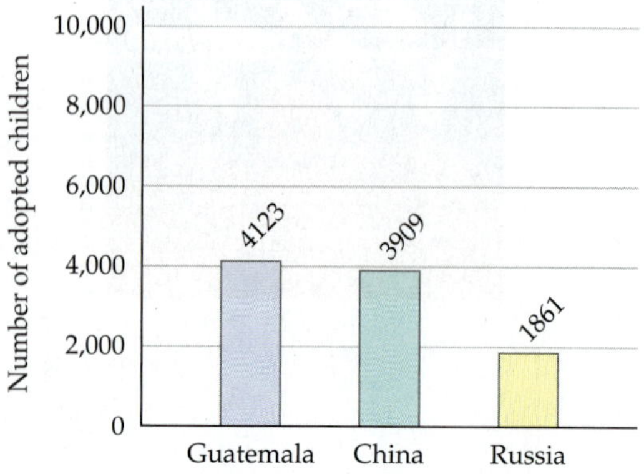

Birth Countries of Adopted American Children
SOURCE: U.S. State Department

56. **Installment Purchases** The purchase price of a large 4K LED TV, including finance charges, was \$1425. A down payment of \$300 was made, and the remainder was paid in 18 equal monthly installments. Find the monthly payment.

57. **Auto Repair** The cost to replace a water pump in a Corvette was \$355. This included \$115 for the water pump plus \$80/h for labor. How many hours of labor were required to replace the water pump?

58. **Robots** Kiva Systems, Inc., builds robots that companies can use to streamline order fulfillment operations in their warehouses. Salary and other benefits for one human warehouse worker can cost a company about \$64,000 a year, an amount that is 103 times the company's yearly maintenance and operation costs for one robot. Find the yearly costs for a robot. Round to the nearest hundred. (*Source: Boston Globe*)

59. **College Staffing** A university employs a total of 600 teaching assistants and research assistants. There are three times as many teaching assistants as research assistants. Find the number of research assistants employed by the university.

60. **Wages** A service station attendant is paid time-and-a-half for working over 40 hours per week. Last week the attendant worked 47 h and earned \$631.25. Find the attendant's regular hourly wage.

61. **Investments** An investor deposited \$5000 in two accounts. Two times the smaller deposit is \$1000 more than the larger deposit. Find the amount deposited in each account.

62. **Shipping** An overnight mail service charges \$5.60 for the first 6 oz and \$0.85 for each additional ounce or fraction of an ounce. Find the weight, in ounces, of a package that cost \$10.70 to deliver.

63. **Telecommunications** A cellular phone company charges \$59.95 per month for a service plan that includes 6 GB of data. In addition to the basic monthly rate, the company charges \$7.95 for each additional GB of data over 6 GB. A customer on this plan receives a bill for \$115.60. How much data did this customer send and receive during the month?

■ In Exercises 64 to 81, solve the formula for the indicated variable.

64. $A = \frac{1}{2}bh$; h (geometry)

65. $P = a + b + c$; b (geometry)

66. $d = rt$; t (physics)

67. $E = IR$; R (physics)

68. $PV = nRT$; R (chemistry)

69. $I = Prt$; r (business)

70. $P = 2L + 2W$; W (geometry)

71. $F = \frac{9}{5}C + 32$; C (temperature conversion)

72. $P = R - C$; C (business)

73. $A = P + Prt$; t (business)

74. $S = V_0t - 16t^2$; V_0 (physics)

75. $T = fm - gm$; f (engineering)

76. $P = \dfrac{R - C}{n}$; R (business)

77. $R = \dfrac{C - S}{t}$; S (business)

78. $V = \dfrac{1}{3}\pi r^2 h$; h (geometry)

79. $A = \dfrac{1}{2}h(b_1 + b_2)$; b_2 (geometry)

80. $a_n = a_1 + (n - 1)d$; d (mathematics)

81. $S = 2\pi r^2 + 2\pi rh$; h (geometry)

■ In Exercises 82 and 83, solve the equation for y.

82. $2x - y = 4$

83. $4x + 3y = 6$

■ In Exercises 84 and 85, solve the equation for x.

84. $ax + by + c = 0$

85. $y - y_1 = m(x - x_1)$

EXTENSIONS

■ In Exercises 86 and 87, solve the equation.

86. $3(4x + 2) = 7 - 4(1 - 3x)$

87. $4(x + 5) = 30 - (10 - 4x)$

88. Use the numbers 5, 10, and 15 to make equations by filling in the boxes: $x + \square = \square - \square$. Each equation must use all three numbers.

a. What is the largest possible solution of these equations?

b. What is the smallest possible solution of these equations?

89. Solve the equation $ax + b = cx + d$ for x. Is your solution valid for all numbers a, b, c, and d? Explain.

■ **Writing Formulas** When we know there is an explicit relationship between two quantities, often we can write a formula to express the relationship.

For example, suppose that a toll of \$3.75 is collected from each vehicle that crosses a particular bridge. Let A be the total amount of money collected, and let c be the number of vehicles that cross the bridge on a given day. Then,

$$A = \$3.75c$$

is a formula that expresses the total amount of money collected from vehicles on any given day.

In Exercises 90 to 93, write a formula for the situation. Include as part of your answer a list of variables that were used, and state what each variable represents.

90. Write a formula to represent the total cost to rent a copier from a company that charges \$325 per month plus \$0.08 per copy made.

91. Suppose you buy a used car with 30,000 mi on it. You expect to drive the car about 750 mi per month. Write a formula to represent the total number of miles the car has been driven after you have owned it for m months.

92. A parking garage charges \$7.50 for the first hour and \$5.25 for each additional hour. Write a formula to represent the parking charge for parking in this garage for h hours. Assume h is a counting number greater than 1.

93. Write a formula to represent the total cost to rent a car from a company that rents cars for \$29.95 per day plus 50¢ for every mile driven over 100 mi. Assume the car will be driven more than 100 mi.

SECTION 9.2 Rate, Ratio, and Proportion

Rates

The word *rate* is used frequently in our everyday lives. It is used in such contexts as unemployment rate, tax rate, interest rate, hourly rate, infant mortality rate, school dropout rate, inflation rate, and postage rate.

A **rate** is a comparison of two quantities and can be written as a fraction. For instance, if a car travels 135 mi on 6 gal of gas, then the miles-to-gallons rate is written

$$\frac{135 \text{ mi}}{6 \text{ gal}}$$

Note that the units (miles and gallons) are written as part of the rate.

POINT OF INTEREST

Unit rates are used in a wide variety of situations. One unit rate you may not be familiar with is used in the airline industry to describe air circulation: cubic feet of air per minute per person. Typical rates are: economy class, 7 ft^3/min/person; first class, 50 ft^3/minute/person; cockpit, 150 ft^3/min/person.

A **unit rate** is a rate in which the number in the denominator is 1. To find a unit rate, divide the number in the numerator of the rate by the number in the denominator of the rate. For the preceding example,

$$135 \div 6 = 22.5$$

The unit rate is $\frac{22.5 \text{ mi}}{1 \text{ gal}}$.

This rate can be written 22.5 mi/gal or 22.5 miles per gallon, where the word *per* has the meaning "for every."

Unit rates make comparisons easier. For example, if you travel 37 mph and I travel 43 mph, we know that I am traveling faster than you are. It is more difficult to compare speeds if we are told that you are traveling $\frac{111 \text{ mi}}{3 \text{ h}}$ and I am traveling $\frac{172 \text{ mi}}{4 \text{ h}}$.

EXAMPLE 1 Calculate a Unit Rate

A dental hygienist earns $1304 for working a 40-hour week. What is the hygienist's hourly rate of pay?

Solution

The hygienist's rate of pay is $\frac{\$1304}{40 \text{ h}}$.

To find the hourly rate of pay, divide 1304 by 40.

$$1304 \div 40 = 32.6$$

$$\frac{\$1304}{40 \text{ h}} = \frac{\$32.60}{1 \text{ h}} = \$32.60/\text{h}$$

The hygienist's hourly rate of pay is $32.60/h.

CHECK YOUR PROGRESS 1 You pay $6.75 for 1.5 lb of hamburger. What is the cost per pound?

Solution See page S30.

EXAMPLE 2 Solve an Application of Unit Rates

A teacher earns a salary of $53,280 per year. Currently the school year consists of 180 days. If the school year were extended to 220 days, as is proposed in some states, what annual salary should the teacher be paid if the salary is based on the number of days worked per year?

Solution

Find the current salary per day.

$$\frac{\$53{,}280}{180 \text{ days}} = \frac{\$296}{1 \text{ day}} = \$296/\text{day}$$

Multiply the salary per day by the number of days in the proposed school year.

$$\frac{\$296}{1 \text{ day}} \cdot 220 \text{ days} = \$296(220) = \$65{,}120$$

The teacher's annual salary should be $65,120.

CHECK YOUR PROGRESS 2 In January 2016, the federal minimum wage was $7.25/h, and the minimum wage in California was $10.00. How much greater is an employee's pay for working 35 h and earning the California minimum wage rather than the federal minimum wage?

Solution See page S30.

Monkey Business Images/Shutterstock.com

Another application of unit rate is the unit price information that grocery stores are required to provide to customers. The **unit price** of a product is its cost per unit of measure.

Suppose that the price of a 2-pound box of spaghetti is \$2.79. The unit price of the spaghetti is the cost per pound. To find the unit price, write the rate as a unit rate.

The numerator is the price and the denominator is the quantity. Divide the number in the numerator by the number in the denominator.

$$\frac{\$2.79}{2 \text{ lb}} = \frac{\$1.395}{1 \text{ lb}}$$

The unit price of the spaghetti is \$1.395/lb.

Unit pricing is used by consumers to answer the question "Which is the better buy?" The answer is that the product with the lower unit price is the more economical purchase.

EXAMPLE 3 Determine the More Economical Purchase

Which is the more economical purchase, an 18-ounce jar of peanut butter priced at \$2.69 or a 28-ounce jar of peanut butter priced at \$3.99?

Solution

Find the unit price for each item.

$$\frac{\$2.69}{18 \text{ oz}} \approx \frac{\$0.149}{1 \text{ oz}} \qquad \frac{\$3.99}{28 \text{ oz}} \approx \frac{\$0.143}{1 \text{ oz}}$$

Compare the two prices per ounce.

$$\$0.149 > \$0.143$$

The item with the lower unit price is the more economical purchase.
The more economical purchase is the 28-ounce jar priced at \$3.99.

CHECK YOUR PROGRESS 3 Which is the more economical purchase, 32 oz of detergent for \$6.29 or 48 oz of detergent for \$8.29?

Solution See page S30.

POINT OF INTEREST

According to the Centers for Disease Control and Prevention, in a recent year the teen birth rate in the United States declined to 24.2 live births per 1000 females age 15 to 19.

Rates such as crime statistics or data on fatalities are often written as rates per hundred, per thousand, per hundred thousand, or per million. For example, the table below shows bicycle deaths per million people in a recent year in the states with the highest rates. (*Source:* www.nhtsa.dot.gov)

Rates of Bicycle Fatalities (Deaths per Million People)	
Florida	6.8
Arizona	4.7
California	3.7
Oklahoma	3.4
South Carolina	3.1

The rates in this table are easier to read than they would be if they were unit rates. Consider that the bicycle fatalities in South Carolina would be written as 0.0000031 as a unit rate. It is easier to understand that 3.1 out of every million people living in South Carolina die in bicycle accidents.

QUESTION In the table on page 511, what does the rate given for Florida mean?

Ragnarock/Shutterstock.com

Another application of rates is in the area of international trade. Suppose a company in France purchases a shipment of sneakers from an American company. The French company must exchange euros, which is France's currency, for U.S. dollars in order to pay for the order. The number of euros that are equivalent to one U.S. dollar is called the *exchange rate*. The table below shows the exchange rates per U.S. dollar for three foreign countries and the European Union on January 19, 2016. Use this table for Example 4 and Check Your Progress 4.

Exchange Rates per U.S. Dollar	
British pound	0.7058
Canadian dollar	1.4574
Japanese yen	117.44
Euro	0.9156

EXAMPLE 4 Solve an Application Using Exchange Rates

a. How many Japanese yen are needed to pay for an order costing \$15,000?

b. Find the number of British pounds that would be exchanged for \$5000.

Solution

a. Multiply the number of yen per \$1 by 15,000.

$$15{,}000(117.44) = 1{,}761{,}600$$

1,761,600 yen are needed to pay for an order costing \$15,000.

b. Multiply the number of pounds per \$1 by 5000.

$$5000(0.7058) = 3529$$

3529 British pounds would be exchanged for \$5000.

CHECK YOUR PROGRESS 4

a. How many Canadian dollars would be needed to pay for an order costing \$20,000?

b. Find the number of euros that would be exchanged for \$25,000.

Solution See page S30.

POINT OF INTEREST

It is believed that billiards was invented in France during the reign of Louis XI (1423–1483). In the United States, the standard billiard table is 4 ft 6 in. by 9 ft. This is a ratio of 1:2. The same ratio holds for carom and snooker tables, which are 5 ft by 10 ft.

Ratios

A **ratio** is the comparison of two quantities that have the same units. A ratio can be written in three different ways:

1. As a fraction $\frac{2}{3}$
2. As two numbers separated by a colon (:) 2:3
3. As two numbers separated by the word *to* 2 to 3

Although units, such as hours, miles, or dollars, are written as part of a rate, units are not written as part of a ratio.

ANSWER Florida's rate of 6.8 means that 6.8 out of every million people living in Florida die in bicycle accidents.

According to the U.S. Bureau of Labor Statistics, there are 123 million married people in the United States, and 82 million of these people are in the labor force. The ratio of the number of married people in the labor force to the total number of married people in the country is calculated below. Note that the ratio is written in simplest form.

$$\frac{82{,}000{,}000}{123{,}000{,}000} = \frac{2}{3} \quad \text{or} \quad 2:3 \quad \text{or} \quad 2 \text{ to } 3$$

The ratio 2 to 3 tells us that 2 out of every 3 married people in the United States are part of the labor force.

Given that 82 million of the 123 million married people in the country work in the labor force, we can calculate the number of married people who do not work in the labor force.

$$123 \text{ million} - 82 \text{ million} = 41 \text{ million}$$

The ratio of the number of married people who are not in the labor force to the number of married people who are is:

$$\frac{41{,}000{,}000}{82{,}000{,}000} = \frac{1}{2} \quad \text{or} \quad 1:2 \quad \text{or} \quad 1 \text{ to } 2$$

The ratio 1 to 2 tells us that for every 1 married person who is not in the labor force, there are 2 married people who are in the labor force.

POINT OF INTEREST

Ratios have applications to many disciplines. Investors talk of price–earnings ratios. Accountants use the current ratio, which is the ratio of current assets to current liabilities. Metallurgists use ratios to make various grades of steel.

EXAMPLE 5 Determine a Ratio in Simplest Form

AVAVA/Shutterstock.com

A survey revealed that, on average, eighth graders watch approximately 21 h of television each week. Find the ratio, as a fraction in simplest form, of the number of hours spent watching television to the total number of hours in a week.

Solution

A ratio is the comparison of two quantities with the same units. In this problem we are given both hours and weeks. We must first convert 1 week to hours.

$$\frac{24 \text{ h}}{1 \text{ day}} \cdot 7 \text{ days} = (24 \text{ h})(7) = 168 \text{ h}$$

Write in simplest form the ratio of the number of hours spent watching television to the number of hours in 1 week.

$$\frac{21 \text{ h}}{1 \text{ wk}} = \frac{21 \text{ h}}{168 \text{ h}} = \frac{21}{168} = \frac{1}{8}$$

The ratio is $\frac{1}{8}$.

CHECK YOUR PROGRESS 5 According to the National Low Income Housing Coalition, a minimum wage worker living in Georgia must work 72 h/wk, 52 wk/yr, to afford the rent on an average one-bedroom apartment and be within the federal standard of not paying more than 30% of income for housing. Find the ratio, as a fraction in simplest form, of the number of hours this worker must spend working per week to the total number of hours in a week.

Solution See page S31.

A **unit ratio** is a ratio in which the number in the denominator is 1. One situation in which a unit ratio is used is student–faculty ratios. The table on page 514 shows the number of full-time men and women undergraduates, as well as the number of full-time faculty, at two universities in the Pacific 10. Use this table for Example 6 and Check Your Progress 6. (*Source:* National Center for Education Statistics, nces.ed.gov)

University	Men	Women	Faculty
Oregon State University	10,018	8459	1471
University of Oregon	8849	9824	1364

EXAMPLE 6 Determine a Unit Ratio

Calculate the student–faculty ratio at Oregon State University. Round to the nearest whole number. Write the ratio using the word *to*.

Solution

Add the number of male undergraduates and the number of female undergraduates to determine the total number of students.

$$10{,}018 + 8459 = 18{,}477$$

Write the ratio of the total number of students to the number of faculty. Divide the numerator and denominator by the denominator. Then round the numerator to the nearest whole number.

$$\frac{18{,}477}{1471} \approx \frac{12.56}{1} \approx \frac{13}{1}$$

The ratio is approximately 13 to 1.

CHECK YOUR PROGRESS 6

Calculate the student–faculty ratio at the University of Oregon. Round to the nearest whole number. Write the ratio using the word *to*.

Solution See page S31.

HISTORICAL NOTE

Proportions were studied by the earliest mathematicians. Clay tablets uncovered by archeologists show evidence of the use of proportions in Egyptian and Babylonian cultures dating from 1800 BC.

Proportions

Now that you have an understanding of rates and ratios, you are ready to work with proportions. A **proportion** is an equation that states the equality of two rates or ratios. The following are examples of proportions.

$$\frac{250 \text{ mi}}{5 \text{ h}} = \frac{50 \text{ mi}}{1 \text{ h}} \qquad \frac{3}{6} = \frac{1}{2}$$

The first example above is the equality of two rates. Note that the units in the numerators (miles) are the same and the units in the denominators (hours) are the same. The second example is the equality of two ratios. Remember that units are not written as part of a ratio.

The definition of a proportion can be stated as follows: If $\frac{a}{b}$ and $\frac{c}{d}$ are equal ratios or rates, then $\frac{a}{b} = \frac{c}{d}$ is a proportion.

Each of the four members in a proportion is called a **term**. Each term is numbered as shown below.

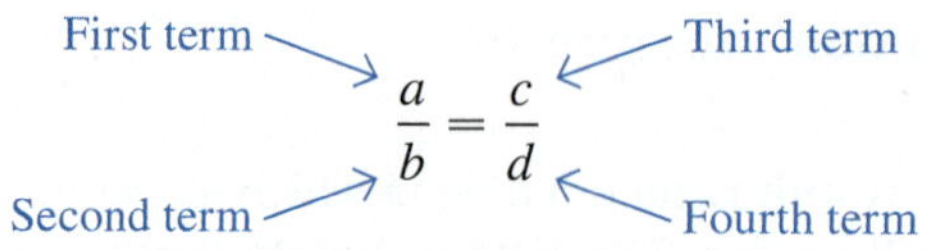

The second and third terms of the proportion are called the **means** and the first and fourth terms are called the **extremes**.

If we multiply both sides of the proportion by the product of the denominators, we obtain the following result.

$$\frac{a}{b} = \frac{c}{d}$$

$$bd\left(\frac{a}{b}\right) = bd\left(\frac{c}{d}\right)$$

$$ad = bc$$

Note that ad is the product of the extremes and bc is the product of the means. In any proportion, the product of the means equals the product of the extremes. This is sometimes phrased, "the cross products are equal."

In the proportion $\frac{3}{4} = \frac{9}{12}$, the cross products are equal.

$$\frac{3}{4} \quad \frac{9}{12} \qquad 4 \cdot 9 = 36 \longleftarrow \text{Product of the means}$$

$$3 \cdot 12 = 36 \longleftarrow \text{Product of the extremes}$$

QUESTION For the proportion $\frac{5}{8} = \frac{10}{16}$, **a.** name the first and third terms, **b.** write the product of the means, and **c.** write the product of the extremes.

Sometimes one of the terms in a proportion is unknown. In this case, it is necessary to solve the proportion for the unknown number. The **cross-products method**, which is based on the fact that the product of the means equals the product of the extremes, can be used to solve the proportion. Remember that the cross-products method is just a short cut for multiplying each side of the equation by the least common multiple of the denominators.

Cross-Products Method of Solving a Proportion

If $\frac{a}{b} = \frac{c}{d}$, then $ad = bc$.

EXAMPLE 7 Solve a Proportion

Solve: $\frac{8}{5} = \frac{n}{6}$

Solution

Use the cross-products method of solving a proportion: the product of the means equals the product of the extremes. Then solve the resulting equation for n.

$$\frac{8}{5} = \frac{n}{6}$$

$$8 \cdot 6 = 5 \cdot n$$

$$48 = 5n$$

$$\frac{48}{5} = \frac{5n}{5}$$

$$9.6 = n$$

The solution is 9.6.

TAKE NOTE

Be sure to check the solution.

$$\frac{8}{5} = \frac{9.6}{6}$$

$$8 \cdot 6 = 5 \cdot 9.6$$

$$48 = 48$$

The solution checks.

CHECK YOUR PROGRESS 7 Solve: $\frac{42}{x} = \frac{5}{8}$

Solution See page S31.

ANSWER **a.** The first term is 5. The third term is 10. **b.** The product of the means is 8(10) = 80. **c.** The product of the extremes is 5(16) = 80.

Proportions are useful for solving a wide variety of application problems. Remember that when we use the given information to write a proportion involving two rates, the units in the numerators of the rates need to be the same and the units in the denominators of the rates need to be the same. It is helpful to keep in mind that when we write a proportion, we are stating that two rates or ratios are equal.

EXAMPLE 8 **Solve an Application Using a Proportion**

If you travel 290 mi in your car on 15 gal of gasoline, how far can you travel in your car on 12 gal of gasoline under similar driving conditions?

Solution

Let $x =$ the unknown number of miles.
Write a proportion and then solve the proportion for x.

$$\frac{290\text{ mi}}{15\text{ gal}} = \frac{x\text{ mi}}{12\text{ gal}}$$
• The unit miles is in the numerators. The unit gallons is in the denominators.

$$\frac{290}{15} = \frac{x}{12}$$

$$290 \cdot 12 = 15 \cdot x$$
• Use the cross-products method of solving a proportion.

$$3480 = 15x$$

$$232 = x$$

You can travel 232 mi on 12 gal of gasoline.

TAKE NOTE

We have written a proportion with the unit "miles" in the numerators and the unit "gallons" in the denominators. It would also be correct to have "gallons" in the numerators and "miles" in the denominators.

CHECK YOUR PROGRESS 8 On a map, a distance of 2 cm represents 15 km. What is the distance between two cities that are 7 cm apart on the map?

Solution See page S31.

EXAMPLE 9 **Solve an Application Using a Proportion**

The table below shows three of the universities in the Big Ten Conference and their student–faculty ratios as of 2014. (*Source: U.S. News & World Report*) There are approximately 28,100 full-time undergraduate students at Purdue University. Approximate the number of faculty at Purdue University.

University	Student-faculty ratio
Michigan State University	17 to 1
University of Illinois	18 to 1
Purdue University	13 to 1

Solution

Let $F =$ the number of faculty members.
Write a proportion and then solve the proportion for F.

$$\frac{13\text{ students}}{1\text{ faculty}} = \frac{28{,}100\text{ students}}{F\text{ faculty}}$$

$$13 \cdot F = 1(28{,}100)$$

$$13F = 28{,}100$$

$$\frac{13F}{13} = \frac{28{,}100}{13}$$

$$F \approx 2162$$

There are approximately 2162 faculty members at Purdue University.

TAKE NOTE

Student–faculty ratios are rounded to the nearest whole number, so they are approximations. When we use an approximate ratio in a proportion, the solution will be an approximation.

CHECK YOUR PROGRESS 9 The profits of a firm are shared by its two partners in the ratio 7:5. If the partner receiving the larger amount of this year's profits receives $84,000, what amount does the other partner receive?

Solution See page S31.

EXAMPLE 10 Solve an Application Using a Proportion

The dosage for a certain medication is 2 oz for every 50 lb of body weight. How many ounces of this medication are required for a person who weighs 175 lb?

Solution

Let $n =$ the number of ounces required for a person who weighs 175 lb. Write and solve a proportion. One rate is 2 oz per 50 lb of body weight.

$$\frac{2 \text{ oz}}{50 \text{ lb}} = \frac{n \text{ oz}}{175 \text{ lb}}$$

$$2(175) = 50 \cdot n$$

$$350 = 50n$$

$$\frac{350}{50} = \frac{50n}{50}$$

$$7 = n$$

A 175-pound person requires 7 oz of the medication.

CHECK YOUR PROGRESS 10

In 2013, the number of deaths from fire in the United States was 11.0 deaths per million people. How many people died from fire in the United States in 2013? Use a figure of 318 million for the population of the United States. Round your answer to the nearest thousand. (*Source:* U.S. Fire Administration, www.usfa.dhs.gov)

Solution See page S31.

MATH MATTERS Scale Models for Special Effects

Olivier LeClerc/Gamma Presse

When you see an exploding spacecraft or a sprinting dinosaur in a film, you are experiencing the work of special effects artists. These professionals often create physical scale models of buildings, vehicles, or creatures that appear full size when we witness them in a film. Artists also create three-dimensional computer models, sometimes by using 3D scanners with physical models, to design computer-generated imagery for films. Whether working in the physical or digital realm, the artists use ratios and proportions to determine the correct sizes and dimensions of these models.

EXCURSION

Earned Run Average

One measure of a pitcher's success is earned run average. **Earned run average (ERA)** is the number of earned runs a pitcher gives up for every nine innings pitched. The definition of an earned run is somewhat complicated, but basically an earned run is a run that is scored as a result of hits and base running that involves no errors on the part of the pitcher's team. If the opposing team scores a run on an error (for example, a fly ball that should have been caught in the outfield was fumbled), then that run is not an earned run.

A proportion is used to calculate a pitcher's ERA. Remember that the statistic involves the number of earned runs per *nine innings*. The answer is always rounded to the nearest hundredth. Here is an example.

Earned Run Average Leaders		
Major League Baseball		
Year	Player, club	ERA
2005	Roger Clemens, Houston	1.87
2006	Johan Santana, Minnesota	2.77
2007	Jake Peavy, San Diego	2.54
2008	Johan Santana, New York	2.53
2009	Zack Greinke, Kansas City	2.16
2010	Felix Hernandez, Seattle	2.27
2011	Clayton Kershaw, Los Angeles	2.28
2012	Clayton Kershaw, Los Angeles	2.53
2013	Clayton Kershaw, Los Angeles	1.83
2014	Clayton Kershaw, Los Angeles	1.77
2015	Zack Greinke, Los Angeles	1.66

During the 2015 baseball season, Clayton Kershaw gave up 55 earned runs and pitched 232.2 innings for the Los Angeles Dodgers. To calculate Clayton Kershaw's ERA, let $x =$ the number of earned runs for every nine innings pitched. Write a proportion and then solve it for x.

$$\frac{55 \text{ earned runs}}{232.2 \text{ innings}} = \frac{x}{9 \text{ innings}}$$
$$55 \cdot 9 = 232.2 \cdot x$$
$$495 = 232.2x$$
$$\frac{495}{232.2} = \frac{232.2x}{232.2}$$
$$2.13 \approx x$$

Photo Works/Shutterstock.com

Clayton Kershaw

Clayton Kershaw's ERA for the 2015 season was 2.13.

EXCURSION EXERCISES

1. In 1979, his rookie year, Jeff Reardon pitched 21 innings for the New York Mets and gave up four earned runs. Calculate Reardon's ERA for 1979.
2. Roger Clemens's first year with the Boston Red Sox was 1984. During that season, he pitched 133.1 innings and gave up 64 earned runs. Calculate Clemens's ERA for 1984.
3. In 1987, Nolan Ryan had the lowest ERA of any pitcher in the major leagues. He gave up 65 earned runs and pitched 211.2 innings for the Houston Astros. Calculate Ryan's ERA for 1987.
4. During the 2015 season, Jake Arrieta of the Baltimore Orioles pitched 229 innings and had an ERA of 1.77. How many earned runs did he give up during the season?
5. Find the necessary statistics for a pitcher on your "home team," and calculate that pitcher's ERA.

EXERCISE SET 9.2

1. Provide two examples of situations in which unit rates are used.
2. Provide two examples of situations in which ratios are used.

■ In Exercises 3 to 8, write the expression as a unit rate.

3. 582 mi in 12 h
4. 138 mi on 6 gal of gasoline
5. 544 words typed in 8 min
6. 100 m in 8 s
7. $9100 for 350 shares of stock
8. 1000 ft^2 of wall covered with 2.5 gal of paint
9. **Wages** A machinist earns $682.50 for working a 35-hour week. What is the machinist's hourly rate of pay?

10. **Space Vehicles** The Space Shuttle's solid rocket boosters are a pair of rockets used during the first 2 min of powered flight. Each booster burns 680,400 kg of propellant in 2.5 min. How much propellant does each booster burn in 1 min?

11. **Photography** During filming, an IMAX camera uses 65-mm film at a rate of 5.6 ft/s.

    a. At what rate per minute does the camera go through film?

    b. How quickly does the camera use a 500-foot roll of 65-mm film? Round to the nearest second.

12. **Consumerism** Which is the more economical purchase, a 30-ounce jar of mayonnaise for $4.29 or a 48-ounce jar of mayonnaise for $6.29?

13. **Consumerism** Which is the more economical purchase, an 18-ounce box of corn flakes for $3.49 or a 24-ounce box of corn flakes for $3.89?

14. **Wages** You have a choice of receiving a wage of $34,000/year, $2840/month, $650 per week, or $16.50/h. Which pay choice would you take? Assume a 40-hour work week and 52 weeks of work per year.

Blank Archives/Getty Images

Babe Ruth

15. **Baseball** Baseball statisticians calculate a hitter's at-bats per home run by dividing the number of times the player has been at bat by the number of home runs the player has hit.

    a. Calculate the at-bats per home run for each player in the table below. Round to the nearest tenth.

    b. Which player has the lowest rate of at-bats per home run? Which player has the second lowest rate?

    c. Why is this rate used for comparison rather than the number of home runs a player has hit?

Year	Baseball player	Number of times at bat	Number of home runs hit	Number of at-bats per home run
1921	Babe Ruth	540	59	
1927	Babe Ruth	540	60	
1932	Jimmie Foxx	585	58	
1938	Hank Greenberg	556	58	
1961	Roger Maris	590	61	
1961	Mickey Mantle	514	54	
1964	Willie Mays	558	52	
1998	Mark McGwire	509	70	
1998	Sammy Sosa	643	66	
2001	Barry Bonds	476	73	
2002	Alex Rodriguez	624	57	
2006	Ryan Howard	581	58	
2013	Chris Davis	584	53	

PPT

16. **Population Density** The table below shows the population and area of three countries. The population density of a country is the number of people per square mile.

a. Which country has the lowest population density?

b. How many more people per square mile are there in India than in the United States? Round to the nearest whole number.

Country	Population	Area (in square miles)
Australia	22,751,000	2,938,000
India	1,251,696,000	1,146,000
United States	321,369,000	3,535,000

17. **E-mail** The Radicati Group compiled the following estimates on consumer use of e-mail worldwide.

a. Complete the last column of the table below by calculating the estimated number of messages per day that each user receives. Round to the nearest tenth.

b. The predicted number of messages per person per day in 2019 is how many times the estimated number in 2015? Round to the nearest hundredth.

Year	Number of users (in millions)	Messages per day (in billions)	Messages per person per day
2015	2586	205.6	
2017	2760	225.3	
2019	2943	246.5	

PPT

Exchange Rates The table below shows the exchange rates per U.S. dollar for four foreign countries on January 27, 2016. Use this table for Exercises 18 to 21.

Exchange Rates per U.S. Dollar	
Australian dollar	1.4216
Danish krone	6.8628
Indian rupee	67.8697
Mexican peso	18.4398

18. How many Danish kroner are equivalent to $5000?

19. Find the number of Indian rupees that would be exchanged for $45,000.

20. Find the cost, in Mexican pesos, of an order of American computer hardware costing $35,000.

21. Calculate the cost, in Australian dollars, of an American car costing $29,000.

Real Estate Dean Baker, co-director at the Center for Economic and Policy Research in Washington, D.C., suggests that the buy-versus-rent question can be answered using the price-to-rent ratio. Find two houses of similar size and quality in comparable neighborhoods, one for sale and the other for rent. Divide the price of the house for sale by the total cost of the rental for 1 year. If the quotient is higher than 20, renting might be the better option. If the quotient is below 15, buying might be the better option.

22. A house in San Diego, California, is priced at $530,000. The rent on a comparable house is $1800 per month. Find the price-to-rent ratio. Round to the nearest tenth. Does the ratio suggest that you buy or rent a home in San Diego?

23. A house in Orlando, Florida, is priced at $155,000. The rent on a comparable house is $1150 per month. Find the price-to-rent ratio. Round to the nearest tenth. Does the ratio suggest that you buy or rent a home in Orlando?

Student–Faculty Ratios The table below shows the number of full-time men and women undergraduates and the number of full-time faculty at several universities in the Big East. Use this table for Exercises 24 and 25. Round ratios to the nearest whole number. (*Source:* National Center for Education Statistics, nces.ed.gov)

University	Men	Women	Faculty
Georgetown University	3244	3982	1350
Syracuse University	6487	8045	1078
University of Connecticut	8851	8826	2007
West Virginia University	11,434	9429	2044

24. Calculate the student–faculty ratio at Syracuse University. Write the ratio using a colon and using the word *to*. What does this ratio mean?

25. Which school listed has the lowest student–faculty ratio?

In Exercises 26 to 37, solve the proportion. Round to the nearest hundredth.

26. $\frac{3}{8} = \frac{x}{12}$ **27.** $\frac{3}{y} = \frac{7}{40}$ **28.** $\frac{7}{12} = \frac{25}{d}$

29. $\frac{16}{d} = \frac{25}{40}$ **30.** $\frac{15}{45} = \frac{72}{c}$ **31.** $\frac{120}{c} = \frac{144}{25}$

32. $\frac{65}{20} = \frac{14}{a}$ **33.** $\frac{4}{a} = \frac{9}{5}$ **34.** $\frac{0.5}{2.3} = \frac{b}{20}$

35. $\frac{1.2}{2.8} = \frac{b}{32}$ **36.** $\frac{0.7}{1.2} = \frac{6.4}{x}$ **37.** $\frac{2.5}{0.6} = \frac{165}{x}$

38. Gravity The ratio of weight on the moon to weight on Earth is 1:6. How much would a 174-pound person weigh on the moon?

39. Management A management consulting firm recommends that the ratio of middle-management salaries to management trainee salaries be 5:4. Using this recommendation, what is the annual middle-management salary if the annual management trainee salary is $52,000?

40. Medication The dosage of a cold medication is 2 mg for every 80 lb of body weight. How many milligrams of this medication are required for a person who weighs 220 lb?

41. Fuel Consumption If your car can travel 70.5 mi on 3 gal of gasoline, how far can the car travel on 14 gal of gasoline under similar driving conditions?

42. Scale Drawings The scale on the architectural plans for a new house is 1 in. equals 3 ft. Find the length and width of a room that measures 5 in. by 8 in. on the drawing.

43. Scale Drawings The scale on a map is 1.25 in. equals 10 mi. Find the distance between two cities that are 2 in. apart on the map.

44. Art Leonardo da Vinci measured various distances on the human body in order to make accurate drawings. He determined that generally the ratio of the kneeling height of a person to the standing height of that person was $\frac{3}{4}$. Using this ratio, determine the standing height of a person who has a kneeling height of 48 in.

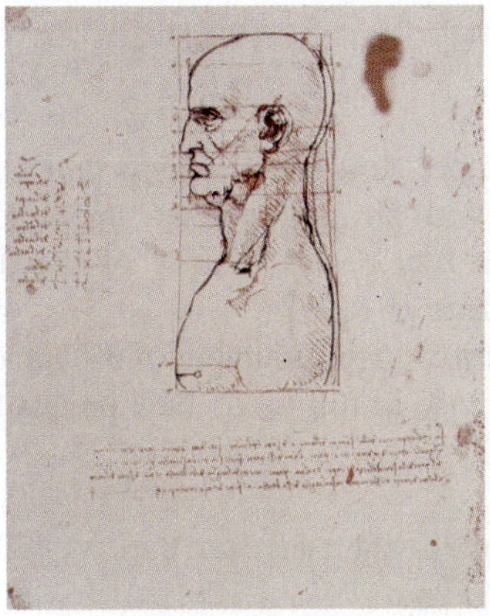

Cameraphoto Arte, Venice/Art Resources, NY

45. Art In one of Leonardo da Vinci's notebooks, he wrote that ". . . from the top to the bottom of the chin is the sixth part of a face, and it is the fifty-fourth part of the man." Suppose the distance from the top to the bottom of the chin of a person is 1.25 in. Using da Vinci's measurements, find the height of the person.

46. Elections A pre-election survey showed that two out of every three eligible voters would cast ballots in the county election. There are 240,000 eligible voters in the county. How many people are expected to vote in the election?

47. Food Waste One study estimated that in the U.S., the average family of four wastes $590 worth of food each year. Estimate the cost of food wasted by **a.** the average family of three and **b.** the average family of five.

48. Lotteries Three people put their money together to buy lottery tickets. The first person put in $25, the second person put in $30, and the third person put in $35. One of their tickets was a winning ticket. If they won $4.5 million, what was the first person's share of the winnings?

49. Nutrition A pancake 4 in. in diameter contains 5 g of fat. How many grams of fat are in a pancake 6 in. in diameter? Explain how you arrived at your answer.

Michael C. Gray/Shutterstock.com

EXTENSIONS

■ In Exercises 50 and 51, assume each denominator is a non-zero real number.

50. Determine whether the statement is true or false.

a. The quotient $a \div b$ is a ratio.

b. If $\frac{a}{b} = \frac{c}{d}$, then $\frac{b}{a} = \frac{d}{c}$.

c. If $\frac{a}{b} = \frac{c}{d}$, then $\frac{a}{c} = \frac{b}{d}$.

d. If $\frac{a}{b} = \frac{c}{d}$, then $\frac{a}{d} = \frac{c}{b}$.

51. If $\frac{a}{b} = \frac{c}{d}$, show that $\frac{a+b}{b} = \frac{c+d}{d}$.

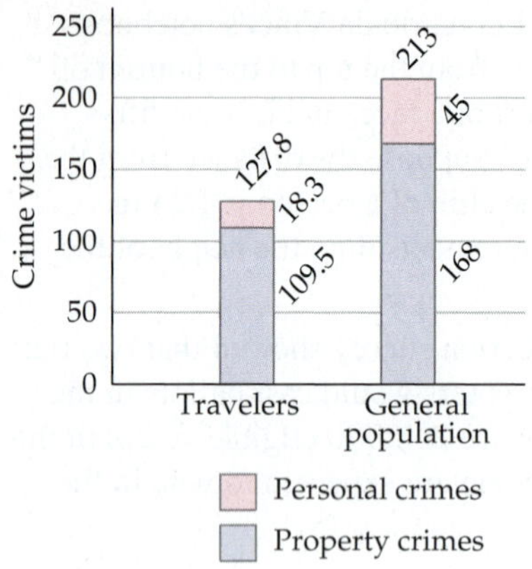

Crime Victims per 1000 Adults per Year

SOURCE: Travel Industry Association of America

52. **Crime Rates** According to a recent study, the crime rate against travelers in the United States is lower than that against the general population. The article reporting this study included a bar graph similar to the one at the left.

a. Why are the figures reported based on crime victims per 1000 adults per year?

b. Use the given figures to write the proportion

$$\frac{\text{personal crimes against travelers}}{\text{property crimes against travelers}} = \frac{\text{personal crimes against general population}}{\text{property crimes against general population}}$$

Is the proportion true?

c. Why might the crime rate against travelers be lower than that against the general population?

53. **The House of Representatives** The U.S. House of Representatives has a total of 435 members. These members represent the 50 states in proportion to each state's population. As stated in Article XIV, Section 2, of the Constitution of the United States, "Representatives shall be apportioned among the several states according to their respective numbers, counting the whole number of persons in each state."

Official White House Photo by Lawrence Jackson

U.S. House of Representatives

a. Find the population of each state according to the 2010 U.S. Census. Based on the state populations, determine how many representatives each state should elect to Congress.

b. Compare your list against the actual number of representatives that each state has.

SECTION 9.3 Percent

Percents

An understanding of percent is vital to comprehending the events that take place in our world today. We are constantly confronted with phrases such as "unemployment of 7%," "annual inflation of 4%," "6% increase in fuel prices," "25% of the daily minimum requirement," and "increase in tuition and fees of 10%."

Percent means "for every 100." Therefore, unemployment of 5% means that 5 out of every 100 people are unemployed. An increase in tuition of 10% means that tuition has gone up $10 for every $100 it cost previously.

POINT OF INTEREST

Of all the errors made on federal income tax returns, the four most common errors account for 76% of the mistakes. These errors include an omitted entry (30.7%), an incorrect entry (19.1%), an error in mathematics (17.4%), and an entry on the wrong line (8.8%).

QUESTION When adults were asked to name their favorite cookie, 52% said chocolate chip. What does this statistic mean? (*Source:* WEAREVER)

A percent is a ratio of a number to 100. Thus $\frac{1}{100} = 1\%$, $\frac{50}{100} = 50\%$, and $\frac{99}{100} = 99\%$. Because $1\% = \frac{1}{100}$ and $\frac{1}{100} = 0.01$, we can also write 1% as 0.01.

$$\mathbf{1\%} = \frac{1}{100} = \mathbf{0.01}$$

The equivalence $1\% = 0.01$ is used to write a percent as a decimal or to write a decimal as a percent.

ANSWER 52 out of every 100 people surveyed responded that their favorite cookie was chocolate chip. (In the same survey, the following responses were also given: oatmeal raisin, 10%; peanut butter, 9%; oatmeal, 7%; sugar, 4%; molasses, 4%; chocolate chip oatmeal, 3%.)

To write 17% as a decimal:

$$17\% = 17(1\%) = 17(0.01) = 0.17$$

Note that this is the same as removing the percent sign and moving the decimal point two places to the left.

To write 0.17 as a percent:

$$0.17 = 17(0.01) = 17(1\%) = 17\%$$

Note that this is the same as moving the decimal point two places to the right and writing a percent sign at the right of the number.

EXAMPLE 1 Write a Percent as a Decimal

Write the percent as a decimal.

a. 24% **b.** 183% **c.** 6.5% **d.** 0.9%

Solution

To write a percent as a decimal, remove the percent sign and move the decimal point two places to the left.

a. 24% = 0.24

b. 183% = 1.83

c. 6.5% = 0.065

d. 0.9% = 0.009

CHECK YOUR PROGRESS 1 Write the percent as a decimal.

a. 74% **b.** 152% **c.** 8.3% **d.** 0.6%

Solution See page S31.

EXAMPLE 2 Write a Decimal as a Percent

Write the decimal as a percent.

a. 0.62 **b.** 1.5 **c.** 0.059 **d.** 0.008

Solution

To write a decimal as a percent, move the decimal point two places to the right and write a percent sign.

a. 0.62 = 62%

b. 1.5 = 150%

c. 0.059 = 5.9%

d. 0.008 = 0.8%

CHECK YOUR PROGRESS 2 Write the decimal as a percent.

a. 0.3 **b.** 1.65 **c.** 0.072 **d.** 0.004

Solution See page S31.

POINT OF INTEREST

The National Safety Council estimates that 27% of car crashes in a recent year were attributable to cell phone use and texting.

The equivalence $1\% = \frac{1}{100}$ is used to write a percent as a fraction.

To write 16% as a fraction:

$$16\% = 16(1\%) = 16\left(\frac{1}{100}\right) = \frac{16}{100} = \frac{4}{25}$$

Note that this is the same as removing the percent sign and multiplying by $\frac{1}{100}$. The fraction is written in simplest form.

EXAMPLE 3 Write a Percent as a Fraction

Write the percent as a fraction.

a. 25% **b.** 120% **c.** 7.5% **d.** $33\frac{1}{3}\%$

Solution

To write a percent as a fraction, remove the percent sign and multiply by $\frac{1}{100}$. Then write the fraction in simplest form.

a. $25\% = 25\left(\frac{1}{100}\right) = \frac{25}{100} = \frac{1}{4}$

b. $120\% = 120\left(\frac{1}{100}\right) = \frac{120}{100} = 1\frac{20}{100} = 1\frac{1}{5}$

c. $7.5\% = 7.5\left(\frac{1}{100}\right) = \frac{7.5}{100} = \frac{75}{1000} = \frac{3}{40}$

d. $33\frac{1}{3}\% = \frac{100}{3}\% = \frac{100}{3}\left(\frac{1}{100}\right) = \frac{1}{3}$

INSTRUCTOR NOTE

Example 3d and Check Your Progress 3d are difficult for students. Here is another in-class example to use.

Write $12\frac{1}{2}\%$ as a fraction.

Solution:

$$12\frac{1}{2}\% = \frac{25}{2}\%$$
$$= \left(\frac{25}{2}\right)\left(\frac{1}{100}\right)$$
$$= \frac{1}{8}$$

CHECK YOUR PROGRESS 3 Write the percent as a fraction.

a. 8% **b.** 180% **c.** 2.5% **d.** $66\frac{2}{3}\%$

Solution See page S31.

TAKE NOTE

To write a fraction as a decimal, divide the number in the numerator by the number in the denominator. For example,

$$\frac{4}{5} = 4 \div 5 = 0.8.$$

To write a fraction as a percent, first write the fraction as a decimal. Then write the decimal as a percent.

EXAMPLE 4 Write a Fraction as a Percent

Write the fraction as a percent.

a. $\frac{3}{4}$ **b.** $\frac{5}{8}$ **c.** $\frac{1}{6}$ **d.** $1\frac{1}{2}$

Solution

To write a fraction as a percent, write the fraction as a decimal. Then write the decimal as a percent.

a. $\frac{3}{4} = 0.75 = 75\%$

b. $\frac{5}{8} = 0.625 = 62.5\%$

c. $\frac{1}{6} = 0.1\overline{6} = 16.\overline{6}\%$

d. $1\frac{1}{2} = 1.5 = 150\%$

CHECK YOUR PROGRESS 4 Write the fraction as a percent.

a. $\frac{1}{4}$ **b.** $\frac{3}{8}$ **c.** $\frac{5}{6}$ **d.** $1\frac{2}{3}$

Solution See page S31.

MATH MATTERS College Graduates' Job Expectations

The table below compares the expectations of 2015 college graduates with the realities of those who graduated in 2013 or 2014. (*Source:* Accenture)

Expectations of 2015 graduates	Realities of 2013–2014 graduates
80% believed their education prepared them well	64% felt their education prepared them well
72% completed an internship, apprenticeship, or co-op	47% found a job as a result of an internship, apprenticeship, or co-op
82% considered job availability before selecting a major	64% are working in their chosen field
85% expect to earn more than $25,000 per year	59% earn more than $25,000 per year

Percent Problems: The Proportion Method

INSTRUCTOR NOTE

This objective explains the proportion method of solving the basic percent equation. Coverage of this material is optional.

Finding the solution of an application problem involving percent generally requires writing and solving an equation. Two methods of writing the equation will be developed in this section—the *proportion method* and the *basic percent equation*. We will present the proportion method first.

The proportion method of solving a percent problem is based on writing two ratios. One ratio is the percent ratio, written $\frac{\text{percent}}{100}$. The second ratio is the amount-to-base ratio, written $\frac{\text{amount}}{\text{base}}$, where the *base* is the number that the percentage will be taken of, and the *amount* is the result after the percentage is taken. These two ratios form the proportion used to solve percent problems.

The Proportion Used to Solve Percent Problems

$$\frac{\text{percent}}{100} = \frac{\text{amount}}{\text{base}}$$

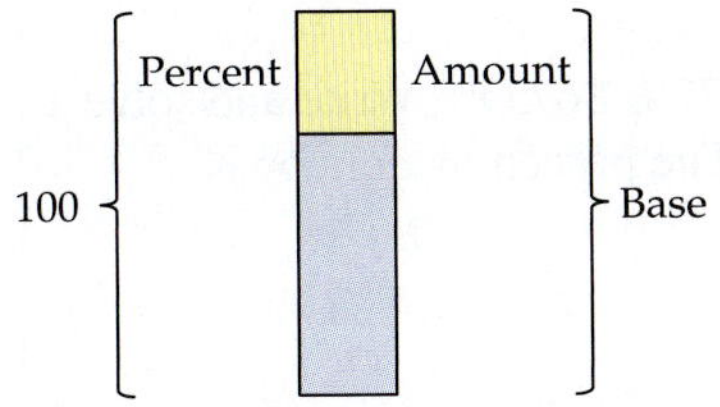

Diagram of the Proportion Method of Solving Percent Problems

The proportion method can be illustrated by a diagram. The rectangle at the left is divided into two parts. On the left, the whole rectangle is represented by 100 and the part by percent. On the right, the whole rectangle is represented by the base and the part by the amount. The ratio of percent to 100 is equal to the ratio of the amount to the base.

When solving a percent problem, first identify the percent, the base, and the amount. It is helpful to know that the base usually follows the phrase "percent of."

QUESTION In the statement "15% of 40 is 6," which number is the percent? Which number is the base? Which number is the amount?

ANSWER The percent is 15. The base is 40. (It follows the phrase "percent of.") The amount is 6.

INSTRUCTOR NOTE

Some students find it easier to remember the proportion method by using the equation

$$\frac{is}{of} = \frac{n}{100}$$

EXAMPLE 5 **Solve a Percent Problem for the Base Using the Proportion Method**

The average size of a new single-family house in 2014 was 2690 ft^2. This is 158% of the average size of a new house in 1980. What was the average size of a new house in 1980? Round to the nearest whole number.

Solution

We want to answer the question "158% of what number is 2690?" Write and solve a proportion. The percent is 158%. The amount is 2690. The base is the average size of a new house in 1980.

$$\frac{\text{percent}}{100} = \frac{\text{amount}}{\text{base}}$$

$$\frac{158}{100} = \frac{2690}{B}$$

$$158 \cdot B = 100(2690)$$

$$158B = 269{,}000$$

$$\frac{158B}{158} = \frac{269{,}000}{158}$$

$$B \approx 1703$$

The average size of a new house in 1980 was 1703 ft^2.

CHECK YOUR PROGRESS 5 A used Toyota Corolla was purchased for $12,950. This is 70% of the cost when new. What was the cost of the Toyota Corolla when it was new?

Solution See page S31.

POINT OF INTEREST

According to the U.S. Department of Agriculture, of the 430 billion pounds of food produced annually in the United States, about 133 billion pounds are wasted. This is approximately 31% of all the food produced in the United States.

EXAMPLE 6 **Solve a Percent Problem for the Percent Using the Proportion Method**

According to the Bureau of Labor Statistics, in a recent year the average American family had an income of $66,877 and spent $6759 on food. What percent of the family income was spent on food? Round to the nearest percent.

Solution

We want to answer the question "What percent of $66,877 is $6759?" Write and solve a proportion. The base is $66,877. The amount is $6759. The percent is unknown.

$$\frac{\text{percent}}{100} = \frac{\text{amount}}{\text{base}}$$

$$\frac{p}{100} = \frac{6759}{66{,}877}$$

$$p \cdot 66{,}877 = 100(6759)$$

$$66{,}877p = 675{,}900$$

$$\frac{66{,}877p}{66{,}877} = \frac{675{,}900}{66{,}877}$$

$$p \approx 10$$

Ten percent of the family income was spent on food.

CHECK YOUR PROGRESS 6

An estimated 43.5 million adults in the United States are caretakers for an older friend or relative. Of these adults, 18.705 million said they feel they did not have a choice in this role. What percent of the adult caretakers in the United States feel they did not have a choice in this role? (*Source:* TIME, February 1, 2010)

Solution See page S31.

EXAMPLE 7 Solve a Percent Problem for the Amount Using the Proportion Method

Thirty-two percent of the world population of 7.3 billion people do not have access to improved sanitation facilities. How many people worldwide do not have access to improved sanitation facilities? (*Source:* World Health Organization, Fact Sheet No. 392, June 2015)

Solution

We want to answer the question, "32% of 7.3 billion is what number?" Write and solve a proportion. The percent is 32%. The base is 7.3 billion. The amount is the number of people who do not have access to improved sanitation facilities.

$$\frac{\text{percent}}{100} = \frac{\text{amount}}{\text{base}}$$

$$\frac{32}{100} = \frac{A}{7.3}$$

$$32(7.3) = 100(A)$$

$$233.6 = 100A$$

$$\frac{233.6}{100} = \frac{100A}{100}$$

$$2.336 = A$$

About 2.34 billion people worldwide do not have access to improved sanitation facilities.

CHECK YOUR PROGRESS 7 A General Motors buyer incentive program offered a 3.5% rebate on the selling price of a new car. What rebate would a customer receive who purchased a $32,500 car under this program?

Solution See page S32.

Percent Problems: The Basic Percent Equation

A second method of solving a percent problem is to use the basic percent equation.

The Basic Percent Equation

$PB = A$, where P is the percent, B is the base, and A is the amount.

When solving a percent problem using the proportion method, we have to first identify the percent, the base, and the amount. The same is true when solving percent problems using the basic percent equation. Remember that the base usually follows the phrase "percent of."

When using the basic percent equation, the percent must be written as a decimal or a fraction. This is illustrated in Example 8.

EXAMPLE 8 Solve a Percent Problem for the Amount Using the Basic Percent Equation

A real estate broker receives a commission of 3% of the selling price of a house. Find the amount the broker receives on the sale of a $275,000 home.

Solution

We want to answer the question "3% of $275,000 is what number?" Use the basic percent equation. The percent is $3\% = 0.03$. The base is 275,000. The amount is the amount the broker receives on the sale of the home.

$$PB = A$$
$$0.03(275{,}000) = A$$
$$8250 = A$$

The real estate broker receives a commission of $8250 on the sale.

CHECK YOUR PROGRESS 8 New Hampshire public school teachers contribute 5% of their wages to the New Hampshire Retirement System. What amount is contributed during one year by a teacher whose annual salary is $46,875?

Solution See page S32.

EXAMPLE 9 Solve a Percent Problem for the Base Using the Basic Percent Equation

An investor received a payment of $480, which was 12% of the value of the investment. Find the value of the investment.

Solution

We want to answer the question "12% of what number is 480?" Use the basic percent equation. The percent is $12\% = 0.12$. The amount is 480. The base is the value of the investment.

$$PB = A$$
$$0.12B = 480$$
$$\frac{0.12B}{0.12} = \frac{480}{0.12}$$
$$B = 4000$$

The value of the investment is $4000.

CHECK YOUR PROGRESS 9 A real estate broker receives a commission of 3% of the selling price of a house. If the broker receives a commission of $14,370 on the sale of a home, what was the selling price of the home?

Solution See page S32.

EXAMPLE 10 Solve a Percent Problem for the Percent Using the Basic Percent Equation

If you answer 96 questions correctly on a 120-question exam, what percent of the questions did you answer correctly?

Solution

We want to answer the question "What percent of 120 questions is 96 questions?" Use the basic percent equation. The base is 120. The amount is 96. The percent is unknown.

$$PB = A$$
$$P \cdot 120 = 96$$
$$\frac{P \cdot 120}{120} = \frac{96}{120}$$
$$P = 0.8$$
$$P = 80\%$$

You answered 80% of the questions correctly.

CHECK YOUR PROGRESS 10 If you answer 63 questions correctly on a 90-question exam, what percent of the questions did you answer correctly?

Solution See page S32.

cappi thompson/Shutterstock.com

The table below shows the average cost in the United States for five of the most popular home remodeling projects and the average percent of that cost recouped when the home is sold. Use this table for Example 11 and Check Your Progress 11. (*Source:* cgi.money.cnn.com)

Home remodeling project	Average cost	Percent recouped
Addition to the master suite	$94,331	72%
Major kitchen remodeling	$54,241	80%
Home office remodeling	$20,057	63%
Bathroom remodeling	$12,918	85%
Basement remodeling	$56,724	79%

EXAMPLE 11 Solve an Application Using the Basic Percent Equation

Find the difference between the cost of remodeling the basement of your home and the amount by which the remodeling increases the sale price of your home.

Solution

The cost of remodeling the basement is $56,724, and the sale price increases by 79% of that amount. We need to find the difference between $56,724 and 79% of $56,724.

Use the basic percent equation to find 79% of $56,724. The percent is 79% = 0.79. The base is 56,724. The amount is unknown.

$$PB = A$$
$$0.79(56{,}724) = A$$
$$44{,}811.96 = A$$

Subtract 44,811.96 (the amount of the cost that is recouped when the home is sold) from 56,724 (the cost of remodeling the basement).

$$56{,}724 - 44{,}811.96 = 11{,}912.04$$

The difference between the cost of remodeling the basement and the increase in the value of your home is $11,912.04.

POINT OF INTEREST

According to Sallie Mae's *How America Pays for College* report, 54% of college students lived at home in 2014. That's up from 43% in 2010.

CHECK YOUR PROGRESS 11

Find the difference between the cost of a major kitchen remodeling in your home and the amount by which the remodeling increases the sale price of your home.

Solution See page S32.

Percent Increase

When a family moves from one part of the country to another, they are concerned about the difference in the cost of living. Will food, housing, and gasoline cost more in that part of the country? Will they need a larger salary in order to make ends meet?

We can use one number to represent the increased cost of living from one city to another so that no matter what salary you make, you can determine how much you will need to earn in order to maintain the same standard of living. That one number is a percent.

For example, look at the information in the table below. (*Source:* http://cgi.money.cnn.com/tools/costofliving/)

If you live in	and are moving to	you will need to make this percent of your current salary
Cincinnati, Ohio	San Francisco, California	191
St. Louis, Missouri	Boston, Massachusetts	153
Denver, Colorado	New York, New York	207

A family in Cincinnati living on \$60,000 per year would need 191% of their current income to maintain the same standard of living in San Francisco. Likewise, a family living on \$150,000 per year would need 191% of their current income.

$$60{,}000(1.91) = 114{,}600 \qquad 150{,}000(1.91) = 286{,}500$$

The family from Cincinnati living on \$60,000 would need an annual income of \$114,600 in San Francisco to maintain their standard of living. The family living on \$150,000 would need an annual income of \$286,500 in San Francisco to maintain their standard of living. No matter what a family's present income, they can use 191% to determine their necessary comparable income.

QUESTION How much would a family in Denver, Colorado, living on \$55,000 per year need in New York City to maintain a comparable lifestyle? Use the table above.

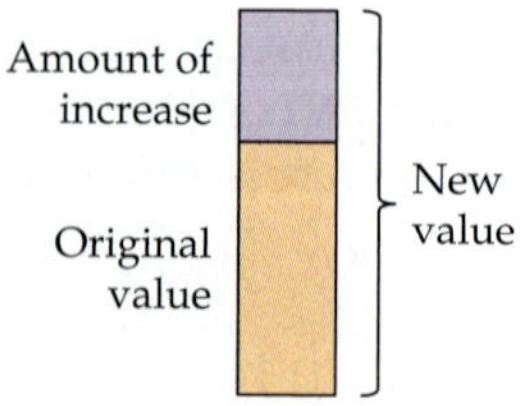

The cost of living in San Francisco is 191% of the cost of living in Cincinnati; this means that a family moving from Cincinnati to San Francisco will see a 91% *increase* in their cost of living. **Percent increase** is used to show how much a quantity has increased over its original value. Statements that illustrate the use of percent increase include "sales volume increased by 11% over last year's sales volume" and "employees received an 8% pay increase."

The **federal debt** is the amount the government owes after borrowing the money it needs to pay for its expenses. It is considered a good measure of how much of the

ANSWER In New York City, the family would need \$55,000(2.07) = \$113,850 per year to maintain a comparable lifestyle.

government's spending is financed by debt as opposed to taxation. The graph below shows the federal debt at the end of the fiscal years 1995, 2000, 2005, 2010, and 2015. A fiscal year is the 12-month period that the annual budget spans, from October 1 to September 30. Use the graph for Example 12 and Check Your Progress 12.

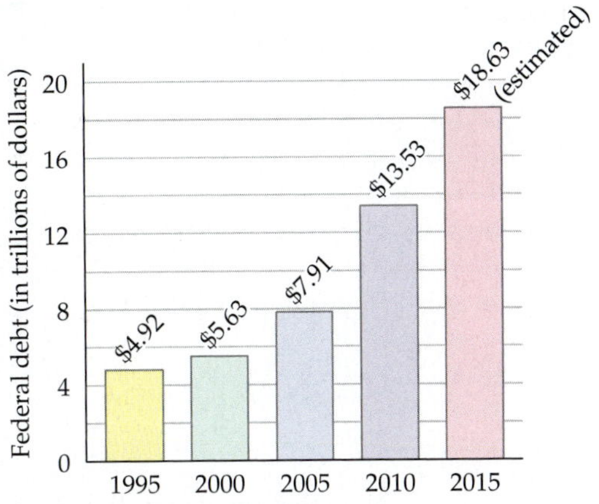

SOURCE: www.whitehouse.gov

EXAMPLE 12 Solve an Application Involving Percent Increase

Find the percent increase in the federal debt from 2005 to 2010. Round to the nearest tenth of a percent.

Solution

Calculate the amount of increase in the federal debt from 2005 to 2010.

$$13.53 - 7.91 = 5.62$$

We will use the basic percent equation. (The proportion method could also be used.) The base is the debt in 2005. The amount is the amount of increase in the debt. The percent is unknown.

$$PB = A$$
$$P \cdot 7.91 = 5.62$$
$$\frac{P \cdot 7.91}{7.91} = \frac{5.62}{7.91}$$
$$P \approx 0.710$$

The percent increase in the federal debt from 2005 to 2010 was 71.0%.

HISTORICAL NOTE

The largest percent increase, for a single day, in the Dow Jones Industrial Average occurred on March 15, 1933. The Dow gained approximately 15% of its value.

CHECK YOUR PROGRESS 12

Find the percent increase in the federal debt from 1995 to 2015. Round to the nearest tenth of a percent.

Solution See page S32.

Notice in Example 12 that the percent increase is a measure of the *amount of increase* over an *original value*. Therefore, in the basic percent equation, the amount A is the *amount of increase* and the base B is the *original value*, in this case the debt in 2005.

Percent Decrease

The federal debt is not the same as the federal deficit. The **federal deficit** is the amount by which government spending exceeds the federal budget. The table below shows projected federal deficits. (*Source:* www.usgovernmentspending.com)

Year	Federal deficit (in billions of dollars)
2010	\$1294
2011	\$1300
2012	\$1087
2013	\$680
2014	\$485
2015	\$439

Note that the deficit listed for 2013 is less than the deficit listed for 2012. This decrease can be expressed as a percent. First find the amount of decrease in the deficit from 2012 to 2013.

$$1087 - 680 = 407$$

We will use the basic percent equation to find the percent. The base is the deficit in 2012. The amount is the amount of decrease.

$$\begin{aligned} PB &= A \\ P \cdot 1087 &= 407 \\ \frac{P \cdot 1087}{1087} &= \frac{407}{1087} \\ P &\approx 0.374 \end{aligned}$$

The federal deficit decreased by 37.4% from 2012 to 2013.

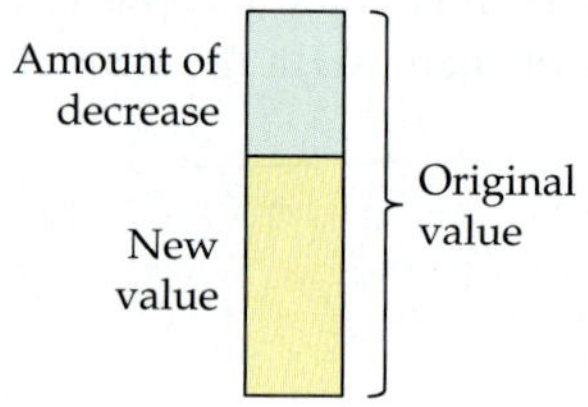

The percent used to measure the decrease in the federal deficit is a *percent decrease.* **Percent decrease** is used to show how much a quantity has decreased from its original value. Statements that illustrate the use of percent decrease include "the president's approval rating has decreased 9% over last month" and "there has been a 15% decrease in the number of industrial accidents."

Note in the deficit example above that the percent decrease is a measure of the *amount of decrease* over an *original value.* Therefore, in the basic percent equation, the amount A is the *amount of decrease* and the base B is the *original value,* in this case the deficit in 2012.

EXAMPLE 13 Solve an Application Involving Percent Decrease

According to the National Highway Traffic Safety Administration, there were 4668 deaths from motorcycle accidents in 2013 while there were 4986 deaths in 2012. This decrease reverses a rising trend from previous years. Find the percent decrease in deaths due to motorcycle accidents from 2012 to 2013. Round to the nearest tenth of a percent.

Solution

First find the amount of decrease.

$$4986 - 4668 = 318$$

We will use the basic percent equation to find the percent decrease. The base is the number of deaths in 2012 (4986). The amount is the decrease in the number of deaths (318).

$$PB = A$$
$$P \cdot 4986 = 318$$
$$\frac{P \cdot 4986}{4986} = \frac{318}{4986}$$
$$P \approx 0.064$$

The percent decrease in the number of deaths from motorcycle accidents is 6.4%.

CHECK YOUR PROGRESS 13

Find the percent decrease in the federal deficit from 2013 to 2014. Use the table on page 532. Round to the nearest tenth of a percent.

Solution See page S32.

EXCURSION

Federal Income Tax

Income taxes are the chief source of revenue for the federal government. If you are employed, your employer probably withholds some money from each of your paychecks for federal income tax. At the end of each year, your employer sends you a **Wage and Tax Statement Form (W-2 form)**, which states the amount of money you earned that year and how much was withheld for taxes.

Every employee is required by law to prepare an income tax return by April 15 of each year and send it to the Internal Revenue Service (IRS). On the income tax return, you must report your total income, or **gross income**. Then you subtract from the gross income any adjustments (such as deductions for charitable contributions or exemptions for people who are dependent on your income) to determine your **adjusted gross income**. You use your adjusted gross income and either a tax table or a tax rate schedule to determine your **tax liability**, or the amount of income you owe to the federal government.

After calculating your tax liability, compare it with the amount withheld for federal income tax, as shown on your W-2 form. If the tax liability is less than the amount withheld, you are entitled to a tax refund. If the tax liability is greater than the amount withheld, you owe the IRS money; you have a **balance due**.

The 2015 Tax Rate Schedules table is shown on page 534. To use this table for the exercises that follow, first classify the taxpayer as single, married filing jointly, or married filing separately. Then determine into which range the adjusted gross income falls. Then perform the calculations shown to the right of that range to determine the tax liability.

For example, consider a taxpayer who is single and has an adjusted gross income of \$48,720. To find this taxpayer's tax liability, use the portion of the table headed "Section A" for taxpayers whose filing status is single.

An income of \$48,720 falls in the range \$37,450 to \$90,750. The tax is \$5,156.25 + 25% of the amount over \$37,450. Find the amount over \$37,450.

$$\$48{,}720 - 37{,}450 = \$11{,}270$$

Calculate the tax liability:

$$\begin{aligned}\$5{,}156.25 + 25\%(\$11{,}270) &= \$5{,}156.25 + 0.25(\$11{,}270)\\ &= \$5{,}156.25 + \$2817.50\\ &= \$7973.75\end{aligned}$$

The taxpayer's liability is $7973.75.

2015 Tax Rate Schedules

Section A—If your filing status is **Single**

If your taxable income is: Over—	But not over—	The tax is:	of the amount over—
$0	$9,225	... 10%	$0
9,225	37,450	$922.50 + 15%	9,225
37,450	90,750	5,156.25 + 25%	37,450
90,750	189,300	18,481.25 + 28%	90,750
189,300	411,500	46,075.25 + 33%	189,300
411,500	413,200	119,401.25 + 35%	411,500
413,200	...	119,996.25 + 39.6%	413,200

Section B—If your filing status is **Married filing jointly** or **Qualifying widow(er)**

If your taxable income is: Over—	But not over—	The tax is:	of the amount over—
$0	$18,450	... 10%	$0
18,450	74,900	$1,845.00 + 15%	18,450
74,900	151,200	10,312.50 + 25%	74,900
151,200	230,450	29,387.50 + 28%	151,200
230,450	411,500	51,577.50 + 33%	230,450
411,500	464,850	111,324.00 + 35%	411,500
464,850	...	129,996.50 + 39.6%	464,850

Section C—If your filing status is **Married filing separately**

If your taxable income is: Over—	But not over—	The tax is:	of the amount over—
$0	$9,225	... 10%	$0
9,225	37,450	$922.50 + 15%	9,225
37,450	75,600	5,156.25 + 25%	37,450
75,600	115,225	14,693.75 + 28%	75,600
115,225	205,750	25,788.75 + 33%	115,225
205,750	232,425	55,662.00 + 35%	205,750
232,425	...	64,998.25 + 39.6%	232,425

PPT

EXCURSION EXERCISES

Use the 2015 Tax Rate Schedules to solve Exercises 1 to 6.

1. Joseph Abruzzio is married and filing separately. He has an adjusted gross income of $63,850. Find Joseph's tax liability.

2. Angela Lopez is single and has an adjusted gross income of $31,680. Find Angela's tax liability.

3. Dee Pinckney is married and filing jointly. She has an adjusted gross income of \$58,120. The W-2 form shows the amount withheld as \$7124. Find Dee's tax liability and determine her tax refund or balance due.

4. Jeremy Littlefield is single and has an adjusted gross income of \$152,600. His W-2 form lists the amount withheld as \$36,500. Find Jeremy's tax liability and determine his tax refund or balance due.

5. Does a taxpayer in the 33% tax bracket pay 33% of his or her earnings in income tax? Explain your answer.

6. In the table for single taxpayers, how were the figures \$922.50 and \$5156.25 arrived at?

EXERCISE SET 9.3

1. Name three situations in which percent is used.

2. Multiplying a number by 300% is the same as multiplying it by what whole number?

Complete the table of equivalent fractions, decimals, and percents.

	Fraction	Decimal	Percent
3.	$\frac{1}{2}$		
4.		0.75	
5.			40%
6.	$\frac{3}{8}$		
7.		0.7	
8.			56.25%
9.	$\frac{11}{20}$		
10.		0.52	
11.			15.625%
12.	$\frac{9}{50}$		

13. **e-Filed Tax Returns** The IRS reported that as of April 17, 2015, it had received 132 million tax returns for 2014. Of these, 90% were filed electronically. How many of the returns were filed electronically? Round to the nearest million.

14. **Credit Cards** A credit card company offers an annual 2% cash-back rebate on all gasoline purchases. If a family spent \$6200 on gasoline purchases over the course of a year, what was the family's rebate at the end of the year?

15. **Charitable Contributions** During a recent year, charitable contributions in the United States totaled \$358 billion. The graph at the right shows to whom this money was donated. Determine how much money was donated to educational organizations. (*Source:* Giving USA Foundation)

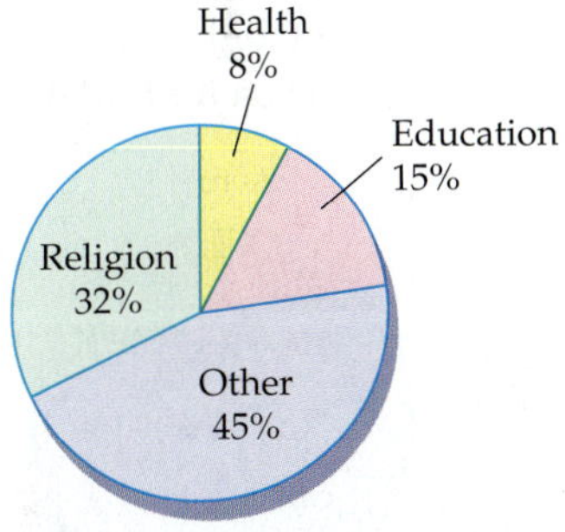

Recipients of Charitable Contributions in the United States

16. **Television** A survey by the *Boston Globe* questioned elementary and middle-school students about television. Sixty-eight students, or 42.5% of those surveyed, said that they had a television in their bedroom at home. How many students were included in the survey?

17. **Motorists** A survey of 1236 adults nationwide asked, "What irks you most about the actions of other motorists?" The response "tailgaters" was given by 293 people. What percent of those surveyed were most irked by tailgaters? Round to the nearest tenth of a percent. (*Source:* Reuters/Zogby)

18. **Wind Energy** In a recent year, wind machines in the United States generated 181.7 billion kWh of electricity, enough to serve over 16 million households. The nation's total electricity production that year was 4094 billion kWh. (*Source:* Energy Information Administration) What percent of the total energy production was generated by wind machines? Round to the nearest tenth of a percent.

TranceDrumer/Shutterstock.com

19. **Mining** During 1 year, approximately 2,240,000 oz of gold went into the manufacturing of electronic equipment in the United States. This is 16% of all the gold mined in the United States that year. How many ounces of gold were mined in the United States that year?

20. **Time Management** The two circle graphs show how surveyed employees actually spend their time and how they would prefer to spend their time. Assume that employees have 112 hours a week that are not spent sleeping. Round answers to the nearest tenth of an hour. (*Source: Wall Street Journal* Supplement from *Families and Work Institute*)

a. What is the actual number of hours per week that employees spend with family and friends?

b. What is the number of hours that employees would prefer to spend on their jobs or careers?

c. What is the difference between the number of hours an employee would prefer to spend on him- or herself and the actual amount of time the employee spends on him- or herself?

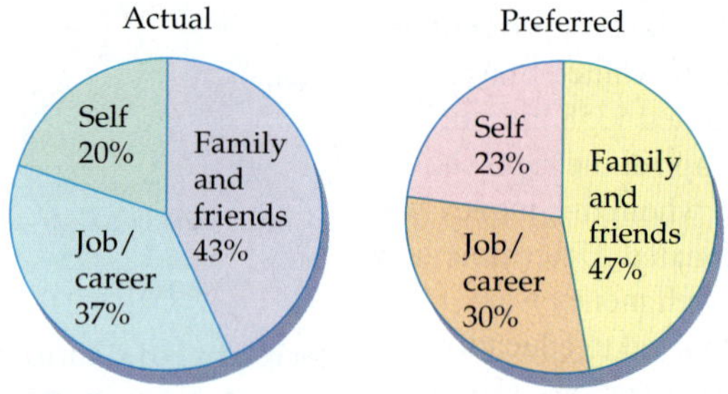

21. **Taxes** A TurboTax online survey asked people how they planned to use their tax refunds. Seven hundred forty people, or 22% of the respondents, said they would save the money. How many people responded to the survey?

22. **Diabetes** Approximately 9.3% of the American population has diabetes. Within this group, 21.0 million are diagnosed, while 8.1 million are undiagnosed. (*Source:* Centers for Disease Control and Prevention) What percent of Americans with diabetes have not been diagnosed with the disease? Round to the nearest tenth of a percent.

23. **Education** Of the 78 million baby boomers living in the United States, 45 million have some college experience but no college degree. (*Sources:* The National Center for Education Statistics; U.S. Census Bureau; *McCook Daily Gazette*) What percent of the baby boomers living in the United States have some college experience but have not earned a college degree? Round to the nearest tenth of a percent.

24. **Telecommunications** The number of Internet users worldwide went from 0.4 billion in 2000 to 3.2 billion in 2015. (*Source:* International Telecommunication Union) Find the percent increase in the number of Internet users from 2000 to 2015.

25. **Demographics** The graph below shows the projected growth of the number of Americans aged 85 and older.

a. What is the percent increase in the population of this age group from 1995 to 2030?

b. What is the percent increase in the population of this age group from 2030 to 2050?

c. What is the percent increase in the population of this age group from 1995 to 2050?

d. How many times larger is the population in 2050 than in 1995? How could you determine this number from the answer to part c?

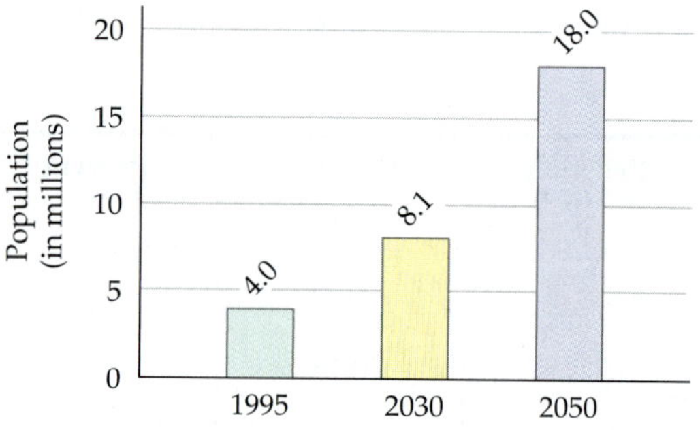

Projected Growth (in millions) of the Population of Americans Aged 85 and Older
SOURCE: U.S. Census Bureau

26. **Auto Sales** U.S. auto sales increased from 16.5 million in 2014 to 17.5 million in 2015. (*Source: Los Angeles Times*) Find the percent increase in auto sales from 2014 to 2015. Round to the nearest tenth of a percent.

27. **Cable TV** In 2006, 65.4 million people subscribed to cable television. In 2013, that number had decreased to 54.4 million. (*Source:* Federal Communications Commission) Find the percent decrease in the number of cable TV subscribers from 2006 to 2013. Round to the nearest tenth of a percent.

28. **Consumption of Eggs** During the last 50 years, the consumption of eggs in the United States has decreased by 17%. Fifty years ago, the average consumption was 307 eggs per person per year. What is the average consumption of eggs today?

29. **Millionaire Households** The following table shows the estimated number of millionaire households (households with a net worth of at least $1 million, not including primary residence) in the United States for selected years. (*Source:* Spectrem Group)

a. What is the percent increase in the estimated number of millionaire households from 2000 to 2007? Round to the nearest tenth of a percent.

b. Find the percent decrease in the estimated number of millionaire households from 2007 to 2008. Round to the nearest tenth of a percent.

Year	Millionaire households
2000	6,300,000
2007	9,200,000
2008	6,700,000
2013	9,600,000

30. **The Military** The graph below shows the number of active-duty U.S. military personnel, in thousands, in 1990 and 2014. Which branch of the military had the greatest percent decrease in personnel from 1990 to 2014? What was the percent decrease for this branch of the service? Round to the nearest tenth of a percent.

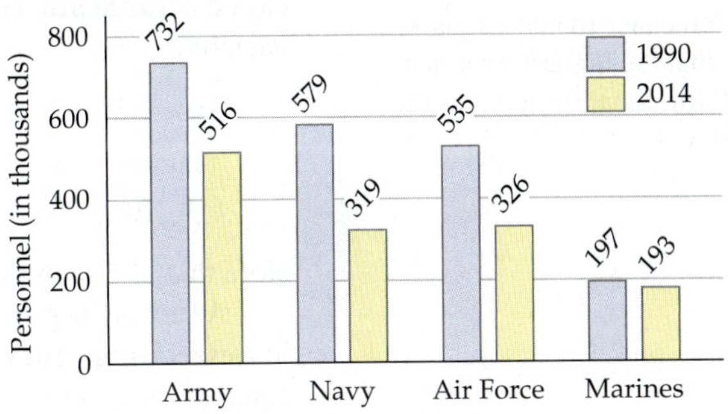

Number of Active-Duty U.S. Military Personnel
SOURCE: Department of Defense

EXTENSIONS

31. **Salaries** Your employer agrees to give you a 5% raise after 1 year on the job, a 6% raise the next year, and a 7% raise the following year. Is your salary after the third year greater than, less than, or the same as it would be if you had received a 6% raise each year?

32. **Work Habits** Approximately 73% of Americans who work in large offices work on weekends, either at home or in the office. The table below shows the average number of hours these workers report they work on a weekend. Approximately what percent of Americans who work in large offices work 11 or more hours on weekends?

Number of hours worked on weekends	Percent
0–1	3%
2–5	32%
6–10	42%
11 or more	23%

33. **Nielsen Ratings** Nielsen Media Research surveys television viewers to determine the numbers of people watching particular shows. They estimated that for the 2015–2016 television season, there were an estimated 113.3 million U.S. households with televisions. Each **rating point** represents 1% of that number, or 1,133,000. Therefore, for instance, if *60 Minutes* received a rating of 5.8, then 5.8% of all U.S. households with televisions, or (0.058)(113,300,000) = 6,571,400 households, were tuned to that program.

A rating point does not mean that 1,133,000 people are watching a program. A rating point refers to the number of households with television sets tuned to that program; there may be more than one person watching a television set in the household.

Nielsen Media Research also describes a program's share of the market. **Share** is the percent of households with television sets in use that are tuned to a program. Suppose that the same week that *60 Minutes* received 5.8 rating points, the show received a share of 11%. This would mean that 11% of all households with a television *turned on* were tuned to *60 Minutes*, whereas 5.8% of all households with a television were tuned to the program.

a. If *NCIS* received a Nielsen rating of 10.1 and a share of 17, how many TV households watched the program that week? How many TV households were watching television during that hour? Round to the nearest hundred thousand.

b. Suppose that *The Big Bang Theory* received a rating of 5.6 and a share of 11. How many TV households watched the program that week? How many TV households were watching television during that hour? Round to the nearest hundred thousand.

c. Suppose that *Modern Family* received a rating of 7.5 during a week in which 19,781,000 people were watching the show. Find the average number of people per TV household who watched the program. Round to the nearest tenth.

SECTION 9.4 Second-Degree Equations

Second-Degree Equations in Standard Form

POINT OF INTEREST

The word *quadratic* comes from the Latin word *quadratus*, which means "to make square." Note that the highest term in a quadratic equation contains the variable squared.

In Section 9.1, we introduced first-degree equations in one variable. A **second-degree equation in one variable** is an equation that can be written in the form $ax^2 + bx + c = 0$, where a and b are coefficients, c is a constant, and $a \neq 0$. An equation of this form is also called a **quadratic equation**. Here are three examples of second-degree equations in one variable.

$4x^2 - 7x + 1 = 0$ $\quad a = 4, b = -7, c = 1$

$3z^2 - 6 = 0$ $\quad a = 3, b = 0, c = -6$

$t^2 + 10t = 0$ $\quad a = 1, b = 10, c = 0$

Note that although the value of a cannot be 0, the value of b or c can be 0.

A second-degree equation is in **standard form** when the expression $ax^2 + bx + c$ is in **descending order** (the exponents on the variables decrease from left to right) and set equal to zero. For instance, $2x^2 + 8x - 3 = 0$ is written in standard form; $x^2 = 4x - 8$ is not in standard form.

QUESTION Which of the following are second-degree equations written in standard form?

a. $3y^2 + 5y - 2 = 0$ **b.** $8p - 4p^2 + 7 = 0$ **c.** $z^3 - 6z + 9 = 0$
d. $4r^2 + r - 1 = 6$ **e.** $v^2 - 16 = 0$

EXAMPLE 1 Write a Quadratic Equation in Standard Form

Write the equation $x^2 = 3x - 8$ in standard form.

Solution

Subtract $3x$ from each side of the equation.

$$x^2 = 3x - 8$$
$$x^2 - 3x = 3x - 3x - 8$$
$$x^2 - 3x = -8$$

Then add 8 to each side of the equation.

$$x^2 - 3x + 8 = -8 + 8$$
$$x^2 - 3x + 8 = 0$$

CHECK YOUR PROGRESS 1 Write $2s^2 = 6 - 4s$ in standard form.

Solution See page S32.

Solving Second-Degree Equations by Factoring

Recall that the multiplication property of zero states that the product of a number and zero is zero.

If a is a real number, then $a \cdot 0 = 0$.

ANSWER The equations in **a** and **e** are second-degree equations in standard form. The equation in **b** is not in standard form because $8p - 4p^2 + 7$ is not written in descending order. The equation in **c** is not a second-degree equation because there is an exponent of 3 on the variable. The equation in **d** is not in standard form because the expression on the left side is not set equal to 0.

Consider the equation $a \cdot b = 0$. If this is a true equation, then either $a = 0$ or $b = 0$. This is summarized in the principle of zero products.

Principle of Zero Products

If the product of two factors is zero, then at least one of the factors must be zero.

If $ab = 0$, then $a = 0$ or $b = 0$.

The principle of zero products is often used to solve equations. This is illustrated in Example 2.

TAKE NOTE

Note that both 4 and -6 check as solutions. The equation $(x - 4)(x + 6) = 0$ has two solutions.

EXAMPLE 2 Solve an Equation Using the Principle of Zero Products

Solve: $(x - 4)(x + 6) = 0$

Solution

In the expression $(x - 4)(x + 6)$, we are multiplying two numbers. Because their product is 0, one of the numbers must be equal to zero. The number $x - 4 = 0$ or the number $x + 6 = 0$. Solve each of these equations for x.

$$(x - 4)(x + 6) = 0$$

$$x - 4 = 0 \qquad x + 6 = 0$$

$$x = 4 \qquad x = -6$$

Check:

$(x - 4)(x + 6) = 0$		$(x - 4)(x + 6) = 0$	
$(4 - 4)(4 + 6)$	0	$(-6 - 4)(-6 + 6)$	0
$(0)(10)$	0	$(-10)(0)$	0
$0 = 0$		$0 = 0$	

The solutions are 4 and -6.

CHECK YOUR PROGRESS 2 Solve: $(n + 5)(2n - 3) = 0$

Solution See page S32.

HISTORICAL NOTE

Chu Shih-chieh, a Chinese mathematician who lived around 1300, wrote the book *Ssu-yuan yu-chien,* which dealt with equations of degree as high as 14.

A second-degree equation can be solved by using the principle of zero products when the expression $ax^2 + bx + c$ is factorable. This is illustrated in Example 3.

EXAMPLE 3 Solve a Quadratic Equation by Factoring

Solve: $2x^2 + x = 6$

Solution

In order to use the principle of zero products to solve a second-degree equation, the equation must be in standard form. Subtract 6 from each side of the given equation.

$$2x^2 + x = 6$$

$$2x^2 + x - 6 = 6 - 6$$

$$2x^2 + x - 6 = 0$$

Factor $2x^2 + x - 6$.

$$(2x - 3)(x + 2) = 0$$

CALCULATOR NOTE

These solutions can be checked on a scientific calculator. Many calculators require the following keystrokes.

2 [×] [(] 1.5 [INV] [x²] [)] [+] 1.5 [=]

and

2 [×] [(] 2 [+/−] [INV] [x²] [)] [+] 2 [+/−] [=]

On a graphing calculator, enter

2 [×] 1.5 [x²] [+] 1.5 [ENTER]

and

2 [×] [(] [(−)] 2 [)] [x²] [+] [(−)] 2 [ENTER]

In each case, the display is 6.

Use the principle of zero products. Set each factor equal to zero. Then solve each equation for x.

$$2x - 3 = 0 \qquad\qquad x + 2 = 0$$
$$2x = 3 \qquad\qquad x = -2$$
$$x = \frac{3}{2}$$

Check:

$2x^2 + x = 6$		$2x^2 + x = 6$	
$2\left(\frac{3}{2}\right)^2 + \frac{3}{2}$	6	$2(-2)^2 + (-2)$	6
$2\left(\frac{9}{4}\right) + \frac{3}{2}$	6	$2(4) + (-2)$	6
$\frac{9}{2} + \frac{3}{2}$	6	$8 + (-2)$	6
$6 = 6$		$6 = 6$	

The solutions are $\frac{3}{2}$ and -2.

CHECK YOUR PROGRESS 3 Solve: $2x^2 = x + 1$

Solution See page S32.

Note from Example 3 the steps involved in solving a second-degree equation by factoring. These are outlined below.

Steps in Solving a Second-Degree Equation by Factoring

1. Write the equation in standard form.
2. Factor the expression $ax^2 + bx + c$.
3. Use the principle of zero products to set each factor of the polynomial equal to zero.
4. Solve each of the resulting equations for the variable.

Solving Second-Degree Equations by Using the Quadratic Formula

When using only integers, not all second-degree equations can be solved by factoring. Any equation that cannot be solved easily by factoring can be solved by using the *quadratic formula,* which is given below.

HISTORICAL NOTE

The quadratic formula is one of the oldest formulas in mathematics. It is not known where it was first derived or by whom. However, there is evidence of its use as early as 2000 BC in ancient Babylonia (the area of the Middle East now known as Iraq). Hindu mathematicians were using it over a thousand years ago.

The Quadratic Formula

The solutions of the equation $ax^2 + bx + c = 0$, $a \neq 0$, are

$$x = \frac{-b + \sqrt{b^2 - 4ac}}{2a} \quad \text{and} \quad x = \frac{-b - \sqrt{b^2 - 4ac}}{2a}$$

The quadratic formula is frequently written in the form

$$x = \frac{-b \pm \sqrt{b^2 - 4ac}}{2a}$$

To use the quadratic formula, first write the second-degree equation in standard form. Determine the values of a, b, and c. Substitute the values of a, b, and c into the quadratic formula. Then evaluate the resulting expression.

EXAMPLE 4 Solve a Quadratic Equation by Using the Quadratic Formula

Solve the equation $2x^2 = 4x - 1$ by using the quadratic formula. Give exact solutions and approximate solutions to the nearest thousandth.

Solution

Write the equation in standard form by subtracting $4x$ from each side of the equation and adding 1 to each side of the equation. Then determine the values of a, b, and c.

$$2x^2 = 4x - 1$$
$$2x^2 - 4x + 1 = 0$$
$$a = 2, b = -4, c = 1$$

Substitute the values of a, b, and c into the quadratic formula. Then evaluate the resulting expression.

$$x = \frac{-b \pm \sqrt{b^2 - 4ac}}{2a}$$
$$x = \frac{-(-4) \pm \sqrt{(-4)^2 - 4(2)(1)}}{2(2)} = \frac{4 \pm \sqrt{16 - 8}}{4}$$
$$= \frac{4 \pm \sqrt{8}}{4} = \frac{4 \pm 2\sqrt{2}}{4} = \frac{2(2 \pm \sqrt{2})}{2(2)} = \frac{2 \pm \sqrt{2}}{2}$$

The exact solutions are $\frac{2 + \sqrt{2}}{2}$ and $\frac{2 - \sqrt{2}}{2}$.

$$\frac{2 + \sqrt{2}}{2} \approx 1.707 \qquad \frac{2 - \sqrt{2}}{2} \approx 0.293$$

To the nearest thousandth, the solutions are 1.707 and 0.293.

CHECK YOUR PROGRESS 4 Solve the equation $2x^2 = 8x - 5$ by using the quadratic formula. Give exact solutions and approximate solutions to the nearest thousandth.

Solution See page S33.

CALCULATOR NOTE

To find the decimal approximation of $\frac{2 + \sqrt{2}}{2}$ on a scientific calculator, use the following keystrokes.

(2 + 2 √) ÷ 2 =

Note that parentheses are used to ensure that the entire numerator is divided by the denominator.

On a graphing calculator, enter

(2 + 2nd √ 2)) ÷ 2 ENTER

INSTRUCTOR NOTE

It may help to give examples of just determining a, b, and c for different equations. Here are some suggestions.

$2x = x^2 - 3$

$4x - 3x^2 + 3 = 0$

$2x^2 - x = 0$

Students will ask whether it matters to which side of the equation a term is moved. To convince these students that it does not matter, have them solve $3x = x^2 - 4$ by rewriting it as

$x^2 - 3x - 4 = 0$

and as

$-x^2 + 3x + 4 = 0$

One of the difficulties that students have with the quadratic formula is correct substitution. Have them first make sure that the equation is in standard form. Then circling a, b, and c may help. Remind them that the sign that precedes a number is the sign of the number. For Example 4, we have

$\overset{a}{(2)}x^2 \overset{b}{(-4)}x \overset{c}{(+1)} = 0$

The exact solutions to Example 4 are irrational numbers. It is also possible for a quadratic equation to have no real number solutions. This is illustrated in Example 5.

EXAMPLE 5 Solve a Quadratic Equation by Using the Quadratic Formula

Solve by using the quadratic formula: $t^2 + 7 = 3t$

Solution

Write the equation in standard form by subtracting $3t$ from each side of the equation. Then determine the values of a, b, and c.

$$t^2 + 7 = 3t$$
$$t^2 - 3t + 7 = 0$$
$$a = 1, b = -3, c = 7$$

Substitute the values of a, b, and c into the quadratic formula. Then evaluate the resulting expression.

$$t = \frac{-b \pm \sqrt{b^2 - 4ac}}{2a}$$

$$t = \frac{-(-3) \pm \sqrt{(-3)^2 - 4(1)(7)}}{2(1)}$$

$$= \frac{3 \pm \sqrt{9 - 28}}{2} = \frac{3 \pm \sqrt{-19}}{2}$$

TAKE NOTE

The square root of a negative number is not a real number because there is no real number that, when squared, equals a negative number. Therefore, $\sqrt{-19}$ is not a real number.

$\sqrt{-19}$ is not a real number.

The equation has no real number solutions.

CHECK YOUR PROGRESS 5 Solve by using the quadratic formula: $z^2 = -6 - 2z$

Solution See page S33.

MATH MATTERS The Discriminant

In Example 5, the second-degree equation has no real number solutions. In the quadratic formula, the quantity $b^2 - 4ac$ under the radical sign is called the **discriminant**. When a, b, and c are real numbers, the discriminant determines whether or not a quadratic equation has real number solutions.

The Effect of the Discriminant on the Solutions of a Second-Degree Equation

1. If $b^2 - 4ac \geq 0$, the equation has real number solutions.
2. If $b^2 - 4ac < 0$, the equation has no real number solutions.

For example, for the equation $x^2 - 4x - 5 = 0$, $a = 1$, $b = -4$, and $c = -5$.

$$b^2 - 4ac = (-4)^2 - 4(1)(-5) = 16 + 20 = 36$$
$$36 > 0$$

The discriminant is greater than 0. The equation has real number solutions.

Applications of Second-Degree Equations

Second-degree equations have many applications to the real world. Examples 6 and 7 illustrate two such applications.

AP Images/ROB GRIFFITHER

EXAMPLE 6 Solve an Application of Quadratic Equations by Factoring

An arrow is projected straight up into the air with an initial velocity of 48 ft/s. At what times will the arrow be 32 ft above the ground? Use the equation $h = 48t - 16t^2$, where h is the height, in feet, above the ground after t seconds.

Solution

We are asked to find the times when the arrow will be 32 ft above the ground, so we are given a value for h. Substitute 32 for h in the given equation and solve for t.

$$h = 48t - 16t^2$$
$$32 = 48t - 16t^2$$

> **TAKE NOTE**
>
> It would also be correct to subtract 32 from each side of the equation. However, many people prefer to have the coefficient of the squared term positive.

This is a second-degree equation. Write the equation in standard form by adding $16t^2$ to each side of the equation and subtracting $48t$ from each side of the equation.

$$16t^2 - 48t + 32 = 0$$
$$16(t^2 - 3t + 2) = 0$$
$$t^2 - 3t + 2 = 0 \quad \text{• Divide each side of the equation by 16.}$$
$$(t - 1)(t - 2) = 0$$
$$t - 1 = 0 \qquad t - 2 = 0$$
$$t = 1 \qquad t = 2$$

The arrow will be 32 ft above the ground 1 s after its release and 2 s after its release.

CHECK YOUR PROGRESS 6 An object is projected straight up into the air with an initial velocity of 64 ft/s. At what times will the object be on the ground? Use the equation $h = 64t - 16t^2$, where h is the height, in feet, above the ground after t seconds.

Solution See page S33.

EXAMPLE 7 Solve an Application of Quadratic Equations by Using the Quadratic Formula

A baseball player hits a ball. The height of the ball above the ground after t seconds can be approximated by the equation $h = -16t^2 + 75t + 5$. When will the ball hit the ground? Round to the nearest hundredth of a second.

Solution

We are asked to determine the number of seconds from the time the ball is hit until it is on the ground. When the ball is on the ground, its height above the ground is 0 ft. Substitute 0 for h and solve for t.

$$h = -16t^2 + 75t + 5$$
$$0 = -16t^2 + 75t + 5$$
$$16t^2 - 75t - 5 = 0$$

> **TAKE NOTE**
>
> The time until the ball hits the ground cannot be a negative number. Therefore, -0.07 s is not a solution of this application.

This is a second-degree equation. It is not easily factored. Use the quadratic formula to solve for t.

$$a = 16, b = -75, c = -5$$
$$t = \frac{-b \pm \sqrt{b^2 - 4ac}}{2a}$$
$$t = \frac{-(-75) \pm \sqrt{(-75)^2 - 4(16)(-5)}}{2(16)} = \frac{75 \pm \sqrt{5945}}{32}$$
$$t = \frac{75 + \sqrt{5945}}{32} \approx 4.75 \qquad t = \frac{75 - \sqrt{5945}}{32} \approx -0.07$$

The ball strikes the ground 4.75 s after the baseball player hits it.

CHECK YOUR PROGRESS 7 A basketball player shoots at a basket that is 25 ft away. The height h, in feet, of the ball above the ground after t seconds is given by $h = -16t^2 + 19t + 3.5$. How many seconds after the ball is released does it hit the basket? *Note:* The basket is 10 ft off the ground. Round to the nearest hundredth.

Solution See page S33.

EXCURSION

The Sum and Product of the Solutions of a Quadratic Equation

The solutions of the equation $x^2 + 3x - 10 = 0$ are -5 and 2.

$$x^2 + 3x - 10 = 0$$
$$(x + 5)(x - 2) = 0$$
$$x + 5 = 0 \qquad x - 2 = 0$$
$$x = -5 \qquad x = 2$$

TAKE NOTE

Look closely at the example at the right, in which -5 and 2 are solutions of the quadratic equation $(x + 5)(x - 2) = 0$. Using variables, we can state that if s_1 and s_2 are solutions of a quadratic equation, then the quadratic equation can be written in the form $(x - s_1)(x - s_2) = 0$.

Note that the sum of the solutions is equal to $-b$, the opposite of the coefficient of x.

$$-5 + 2 = -3$$

The product of the solutions is equal to c, the constant term.

$$-5(2) = -10$$

This illustrates the following theorem regarding the solutions of a quadratic equation.

The Sum and Product of the Solutions of a Quadratic Equation

If s_1 and s_2 are the solutions of a quadratic equation of the form $ax^2 + bx + c = 0$, $a \neq 0$, then

the sum of the solutions $s_1 + s_2 = -\frac{b}{a}$, and

the product of the solutions $s_1 s_2 = \frac{c}{a}$.

In this section, the method we used to check the solutions of a quadratic equation was to substitute the solutions back into the original equation. An alternative method is to use the sum and product of the solutions.

For example, let's check that -2 and 6 are the solutions of the equation $x^2 - 4x - 12 = 0$. For this equation, $a = 1$, $b = -4$, and $c = -12$. Let $s_1 = -2$ and $s_2 = 6$.

TAKE NOTE

The result is the same if we let $s_1 = 6$ and $s_2 = -2$.

$$\begin{array}{c|c} \multicolumn{2}{c}{s_1 + s_2 = -\frac{b}{a}} \\ \hline -2 + 6 & -\frac{-4}{1} \\ \multicolumn{2}{c}{4 = 4} \end{array} \qquad \begin{array}{c|c} \multicolumn{2}{c}{s_1 s_2 = \frac{c}{a}} \\ \hline -2(6) & \frac{-12}{1} \\ \multicolumn{2}{c}{-12 = -12} \end{array}$$

The solutions check.

In Example 4, we found that the exact solutions of the equation $2x^2 = 4x - 1$ are $\frac{2 + \sqrt{2}}{2}$ and $\frac{2 - \sqrt{2}}{2}$. Use the sum and product of the solutions to check these solutions.

Write the equation in standard form. Then determine the values of a, b, and c.

$$2x^2 = 4x - 1$$
$$2x^2 - 4x + 1 = 0$$
$$a = 2, b = -4, c = 1$$

Let $s_1 = \frac{2+\sqrt{2}}{2}$ and $s_2 = \frac{2-\sqrt{2}}{2}$.

$s_1 + s_2 = -\frac{b}{a}$		$s_1 s_2 = \frac{c}{a}$	
$\frac{2+\sqrt{2}}{2} + \frac{2-\sqrt{2}}{2}$	$-\frac{-4}{2}$	$\left(\frac{2+\sqrt{2}}{2}\right)\left(\frac{2-\sqrt{2}}{2}\right)$	$\frac{1}{2}$
$\frac{2+\sqrt{2}+2-\sqrt{2}}{2}$	2	$\frac{4-2}{4}$	$\frac{1}{2}$
$\frac{4}{2}$	2	$\frac{2}{4}$	$\frac{1}{2}$
$2 = 2$		$\frac{1}{2} = \frac{1}{2}$	

The solutions check.

If we divide both sides of the equation $ax^2 + bx + c = 0$, $a \neq 0$, by a, the result is the equation

$$x^2 + \frac{b}{a}x + \frac{c}{a} = 0$$

Using this model and the sum and products of the solutions of a quadratic equation, we can find a quadratic equation given its solutions. The method is given below.

A Quadratic Equation with Solutions s_1 and s_2

A quadratic equation with solutions s_1 and s_2 is

$$x^2 - (s_1 + s_2)x + s_1 s_2 = 0$$

To write a quadratic equation that has solutions $\frac{2}{3}$ and 1, let $s_1 = \frac{2}{3}$ and $s_2 = 1$. Substitute these values into the equation $x^2 - (s_1 + s_2)x + s_1 s_2 = 0$ and simplify.

$$x^2 - (s_1 + s_2)x + s_1 s_2 = 0$$

$$x^2 - \left(\frac{2}{3} + 1\right)x + \left(\frac{2}{3} \cdot 1\right) = 0$$

$$x^2 - \frac{5}{3}x + \frac{2}{3} = 0$$

Assuming we want a, b, and c to be integers, then we can multiply each side of the equation by 3 to clear fractions.

$$3\left(x^2 - \frac{5}{3}x + \frac{2}{3}\right) = 3(0)$$

$$3x^2 - 5x + 2 = 0$$

A quadratic equation with solutions $\frac{2}{3}$ and 1 is $3x^2 - 5x + 2 = 0$.

EXCURSION EXERCISES

In Exercises 1 to 8, solve the equation and then check the solutions using the sum and product of the solutions.

1. $x^2 - 10 = 3x$

2. $x^2 + 16 = 8x$

3. $3x^2 + 5x = 12$

4. $3x^2 + 8x = 3$

5. $x^2 = 6x + 3$
6. $x^2 = 2x + 5$
7. $4x + 1 = 4x^2$
8. $x + 1 = x^2$

In Exercises 9 to 16, write a quadratic equation that has integer coefficients and has the given pair of solutions.

9. -1 and 6
10. -5 and -4
11. 3 and $\frac{1}{2}$
12. $-\frac{3}{4}$ and 2
13. $\frac{1}{4}$ and $-\frac{3}{2}$
14. $\frac{2}{3}$ and $-\frac{2}{3}$
15. $2 + \sqrt{2}$ and $2 - \sqrt{2}$
16. $1 + \sqrt{3}$ and $1 - \sqrt{3}$

EXERCISE SET 9.4

■ In Exercises 1 to 24, first try to solve the equation by factoring. If you are unable to solve the equation by factoring, solve the equation by using the quadratic formula. For equations with solutions that are irrational numbers, give exact solutions and approximate solutions to the nearest thousandth.

1. $r^2 - 3r = 10$
2. $p^2 + 5p = 6$
3. $t^2 = t + 1$
4. $u^2 = u + 3$
5. $y^2 - 6y = 4$
6. $w^2 + 4w = 2$
7. $9z^2 - 18z = 0$
8. $4y^2 + 20y = 0$
9. $z^2 = z + 4$
10. $r^2 = r - 1$
11. $2s^2 = 4s + 5$
12. $3u^2 = 6u + 1$
13. $r^2 = 4r + 7$
14. $s^2 + 6s = 1$
15. $2x^2 = 9x + 18$
16. $3y^2 = 4y + 4$
17. $6x - 11 = x^2$
18. $-8y - 17 = y^2$
19. $4 - 15u = 4u^2$
20. $3 - 2y = 8y^2$
21. $6y^2 - 4 = 5y$
22. $6v^2 - 3 = 7v$
23. $y - 2 = y^2 - y - 6$
24. $8s - 11 = s^2 - 6s + 8$

25. Write a second-degree equation that you can solve by factoring.

26. Write a second-degree equation that you can solve by using the quadratic formula but not by factoring.

■ In Exercises 27 to 42, round answers to nearest hundredth where appropriate.

Paul Yates/Shutterstock.com

27. Golf The height h, in feet, of a golf ball t seconds after it has been hit is given by the equation $h = -16t^2 + 60t$. How many seconds after the ball is hit will the height of the ball be 36 ft?

28. Geometry The area A, in square meters, of a rectangle with a perimeter of 100 meters is given by the equation $A = 50w - w^2$, where w is the width of the rectangle in meters. What is the width of a rectangle if its area is 400 m^2?

29. Mathematics In the diagram below, the total number of circles T when there are n rows is given by $T = 0.5n^2 + 0.5n$. Verify the formula for the four figures shown. Determine the number of rows when the total number of circles is 55.

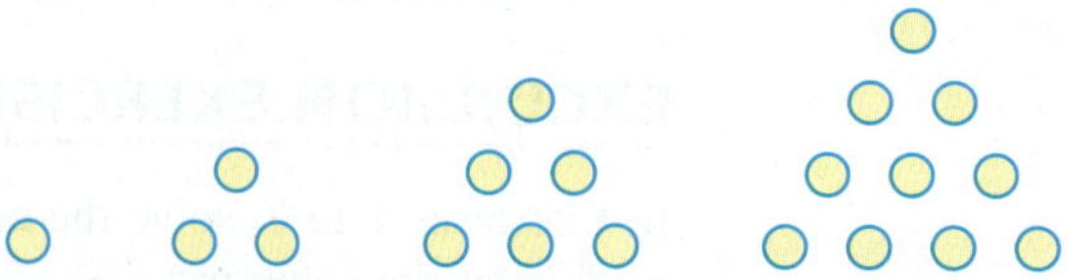

30. Astronautics If an astronaut on the moon throws a ball upward with an initial velocity of 6 m per second, its approximate height h, in meters, after t seconds is given by the equation $h = -0.8t^2 + 6t$. How long after it is released will the ball be 8 m above the surface of the moon?

31. Demography The equation $y = 0.03x^2 + 0.36x + 34.6$ describes the number of people y, in millions, aged 65 and older in the United States in year x, where $x = 0$ corresponds to the year 2000. Approximate the year in which there will be 50 million people aged 65 and older in the United States.

32. Alzheimer's The equation $y = 0.002x^2 + 0.05x + 2$ describes the number of Americans y, in millions, with Alzheimer's in year x, where $x = 0$ corresponds to the year 1980. Find the year in which 15 million Americans are expected to have Alzheimer's.

33. Stopping Distance When a driver decides to stop a car, it takes time first for the driver to react and put a foot on the brake, and then it takes additional time for the car to slow down. The total distance traveled during this period of time is called the *stopping distance* of the car. For some cars, the stopping distance d, in feet, is given by the equation $d = 0.05r^2 + r$, where r is the speed of the car in miles per hour.

a. Find the distance needed to stop a car traveling at 60 mph.

b. If skid marks at an accident site are 75 ft long, how fast was the car traveling?

34. Cliff Divers At La Quebrada in Acapulco, Mexico, cliff divers dive from a rock cliff that is 27 m above the water. The equation $h = -x^2 + 2x + 27$ describes the height h, in meters, of the diver above the water when the diver is x feet away from the cliff. Note that x is the horizontal distance between the diver's present position and the diver's "take off" point.

a. When the diver enters the water, how far is the diver from the cliff?

b. Does the diver ever reach a height of 28 m above the water? If so, how far is the diver from the cliff at that time?

c. Does the diver ever reach a height of 30 m above the water? If so, how far is the diver from the cliff at that time?

35. Football The hang time of a football that is kicked on the opening kickoff is given by $s = -16t^2 + 88t + 1$, where s is the height in feet of the football t seconds after leaving the kicker's foot. What is the hang time of a kickoff that hits the ground without being caught?

36. Softball In a slow pitch softball game, the height of a ball thrown by a pitcher can be approximated by the equation $h = -16t^2 + 24t + 4$, where h is the height, in feet, of the ball and t is the time, in seconds, since it was released by the pitcher. If a batter hits the ball when it is 2 ft off the ground, for how many seconds has the ball been in the air?

37. Fire Science The path of water from a hose on a fire tugboat can be approximated by the equation $y = -0.005x^2 + 1.2x + 10$, where y is the height, in feet, of the water above the ocean when the water is x feet from the tugboat. When the water from the hose is 5 ft above the ocean, at what distance from the tugboat is it?

German Autobahn System

38. Stopping Distance In Germany there are no speed limits on some portions of the autobahn (the highway). Other portions have a speed limit of 180 kph (approximately 112 mph). The distance d, in meters, required to stop a car traveling v kilometers per hour is $d = 0.0056v^2 + 0.14v$. Approximate the maximum speed a driver can be going and still be able to stop within 150 m.

39. Soccer A penalty kick in soccer is made from a penalty mark that is 36 ft from a goal that is 8 ft high. A possible equation for the flight of a penalty kick is $h = -0.002x^2 + 0.36x$, where h is the height, in feet, of the ball x feet from the goal. Assuming that the flight of the kick is toward the goal and that it is not touched by the goalie, will the ball land in the net?

40. Springboard Diving An event in the Summer Olympics is 10-meter springboard diving. In this event, the height h, in meters, of a diver above the water t seconds after jumping is given by the equation $h = -4.9t^2 + 7.8t + 10$. What is the height above the water of a diver after 2 s?

41. Fountains The Water Arc is a fountain that shoots water across the Chicago River from a water cannon. The path of the water can be approximated by the equation $h = -0.006x^2 + 1.2x + 10$, where x is the horizontal distance, in feet, from the cannon and h is the height, in feet, of the water above the river. On one particular day, some people were walking along the opposite side of the river from the Water Arc when a pulse of water was shot in their direction. If the distance from the Water Arc to the people was 220 ft, did they get wet from the cannon's water?

42. Model Rockets A model rocket is launched with an initial velocity of 200 ft per second. The height h, in feet, of the rocket t seconds after the launch is given by $h = -16t^2 + 200t$. How many seconds after the launch will the rocket be 300 ft above the ground?

43. First-Class Postage The graph below shows the cost for a first-class postage stamp from the 1950s to 2015. A second-degree equation that approximately models these data is $y = 0.00657x^2 - 0.330x - 0.0633$, $x \geq 50$, where $x = 50$ corresponds to the year 1950 and y is the cost in cents of a first-class stamp. Using the model equation, determine what the model predicts the cost of a first-class stamp will be in the year 2030. Round to the nearest cent.

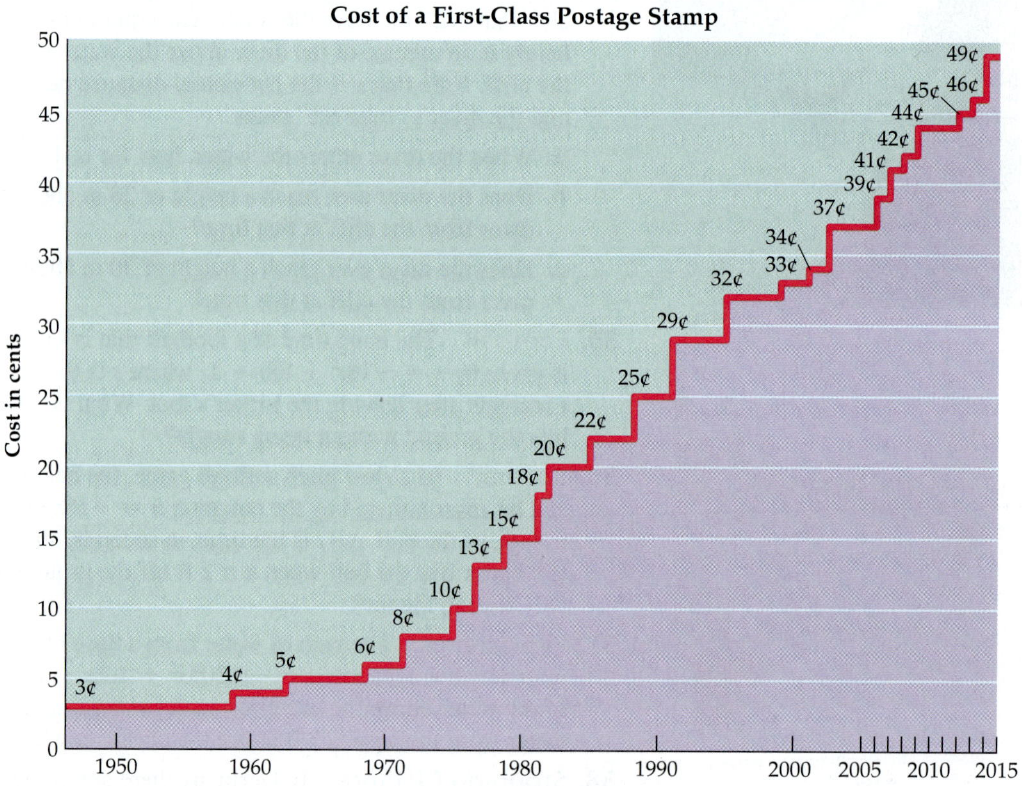

EXTENSIONS

44. Show that the solutions of the equation $ax^2 + bx = 0$, $a \neq 0$, are 0 and $-\frac{b}{a}$.

45. In a second-degree equation in standard form, why is the expression $ax^2 + bx + c$ factorable over the integers only when the discriminant is a perfect square? (See Math Matters, page 542.)

■ In Exercises 46 to 49, solve the equation for x.

46. $x^2 + 16ax + 48a^2 = 0$

47. $x^2 - 8bx + 15b^2 = 0$

48. $3x^2 - 4cx + c^2 = 0$

49. $2x^2 - xy - 3y^2 = 0$

50. Show that the equation $x^2 + bx - 1 = 0$ always has real number solutions, regardless of the value of b.

CHAPTER 9 SUMMARY

The following table summarizes essential concepts in this chapter. The references given in the right-hand column list Examples and Exercises that can be used to test your understanding of a concept.

9.1 First-Degree Equations and Formulas	
Steps for Solving a First-Degree Equation in One Variable **1.** If the equation contains fractions, multiply each side of the equation by the least common multiple (LCM) of the denominators to clear the equation of fractions. **2.** Use the distributive property to remove parentheses. **3.** Combine any like terms on the left side of the equation and any like terms on the right side of the equation. **4.** Use the addition or subtraction property to rewrite the equation with only one variable term and only one constant term. **5.** Use the multiplication or division property to rewrite the equation with the variable alone on one side of the equation and a constant term on the other side of the equation.	See **Example 2** on page 499, and then try Exercise 3 on page 550.
Applications of Solving First-Degree Equations The solution to an application problem may require solving a first-degree equation.	See **Examples 3, 4, and 5** on pages 501 and 502, and then try Exercises 11, 13, and 14 on page 550.
Solve a Literal Equation for One of the Variables The goal is to rewrite the equation so that the letter being solved for is alone on one side of the equation, and all numbers and other variables are on the other side.	See **Example 6** on page 503, and then try Exercises 9 and 10 on page 550.
9.2 Rate, Ratio, and Proportion	
Calculate a Unit Rate Divide the number in the numerator of the rate by the number in the denominator of the rate.	See **Example 1** on page 510, and then try Exercise 15 on page 550.
Write a Ratio A ratio can be written in three different ways: as a fraction, as two numbers separated by a colon (:), or as two numbers separated by the word *to*. Although units, such as hours, miles, or dollars, are written as part of a rate, units are not written as part of a ratio.	See **Example 5** on page 513, and then try Exercise 16 on page 550.
Cross-Products Method of Solving a Proportion If $\frac{a}{b} = \frac{c}{d}$, then $ad = bc$.	See **Examples 7 and 8** on pages 515 and 516, and then try Exercises 4 and 21 on pages 550 and 551.
9.3 Percent	
The Proportion Used to Solve Percent Problems $$\frac{\text{percent}}{100} = \frac{\text{amount}}{\text{base}}$$	See **Examples 5, 6, and 7** on pages 526 and 527, and then try Exercises 24 and 25 on page 552.
The Basic Percent Equation $PB = A$, where P is the percent, B is the base, and A is the amount.	See **Examples 8, 9, and 10** on pages 528 and 529, and then try Exercises 28 and 29 on page 552.

continued

Percent Increase and Percent Decrease The percent increase is the amount of increase divided by the original value, expressed as a percent. The percent decrease is the amount of decrease divided by the original value, expressed as a percent.	See **Examples 12 and 13** on pages 531 and 532, and then try Exercises 17, 26, and 27 on pages 551 and 552.
9.4 Second-Degree Equations	
Steps in Solving a Second-Degree Equation by Factoring **1.** Write the equation in standard form. **2.** Factor the polynomial $ax^2 + bx + c$. **3.** Use the principle of zero products to set each factor of the polynomial equal to zero. **4.** Solve each of the resulting equations for the variable.	See **Example 3** on page 539, and then try Exercise 6 on page 550.
The Quadratic Formula The solutions of the equation $ax^2 + bx + c = 0, a \neq 0$, are $x = \frac{-b \pm \sqrt{b^2 - 4ac}}{2a}$.	See **Examples 4 and 5** on page 541, and then try Exercises 7 and 8 on page 550.
Applications of Solving Second-Degree Equations The solution to an application problem may require solving a second-degree equation.	See **Examples 6 and 7** on pages 542 and 543, and then try Exercises 31 and 32 on page 552.

CHAPTER 9 REVIEW EXERCISES

■ In Exercises 1 to 8, solve the equation.

1. $5x + 3 = 10x - 17$

2. $3x + \frac{1}{8} = \frac{1}{2}$

3. $6x + 3(2x - 1) = -27$

4. $\frac{5}{12} = \frac{n}{8}$

5. $4y^2 + 9 = 0$

6. $x^2 - x = 30$

7. $x^2 = 4x - 1$

8. $x + 3 = x^2$

■ In Exercises 9 and 10, solve the formula for the given variable.

9. $4x + 3y = 12; y$

10. $f = v + at; t$

11. Meteorology In June, the temperature at various elevations of the Grand Canyon can be approximated by the equation $T = -0.005x + 113.25$, where T is the temperature in degrees Fahrenheit and x is the elevation (distance above sea level) in feet. Use this equation to find the elevation at Inner Gorge, the bottom of the canyon, where the temperature is 101°F.

Bryant Jayme/Shutterstock.com

The Grand Canyon

12. Falling Objects Find the time that it takes for the velocity of a falling object to increase from 4 ft/s to 100 ft/s. Use the equation $v = v_0 + 32t$, where v is the final velocity of the falling object, v_0 is the initial velocity, and t is the time it takes for the object to fall.

13. Chemistry A chemist mixes 100 g of water at 80°C with 50 g of water at 20°C. Use the formula $m_1(T_1 - T) = m_2(T - T_2)$ to find the final temperature of the water after mixing. In this equation, m_1 is the quantity of water at the hotter temperature, T_1 is the temperature of the hotter water, m_2 is the quantity of water at the cooler temperature, T_2 is the temperature of the cooler water, and T is the final temperature of water after mixing.

14. Telemarketing At a telemarketing firm, an employee is paid \$12 an hour plus \$0.75 for each call completed. During an 8-hour day, the employee's compensation was \$159.75. How many calls did the employee complete?

15. Fuel Consumption An automobile was driven 326.6 mi on 11.5 gal of gasoline. Find the number of miles driven per gallon of gas.

16. Real Estate A house with an original value of \$280,000 increased in value to \$350,000 in 5 years. Write, as a fraction in simplest form, the ratio of the increase in value to the original value of the house.

17. **Social Media** In July 2012, Instagram announced 80 million users of its photo sharing app. In September 2015, they announced that the number of users had reached 400 million. Find the percent increase in the number of Instagram users during that time period. Round to the nearest percent.

18. **City Populations** The table below shows the population and area of the five most populous cities in the United States.

a. The cities are listed in the table according to population, from largest to smallest. Rank the cities according to population density, from largest to smallest.

b. How many more people per square mile are there in New York than in Houston? Round to the nearest whole number.

City	Population	Area (in square miles)
New York	8,400,000	321.8
Los Angeles	3,900,000	467.4
Chicago	2,900,000	228.469
Houston	2,300,000	594.03
Phoenix	1,600,000	136

19. **Student–Faculty Ratios** The table below shows the number of full-time men and women undergraduates, as well as the number of full-time faculty, at five colleges in Arizona. In parts a, b, and c, round ratios to the nearest whole number. (*Source:* National Center for Education Statistics, nces.ed.gov)

University	Men	Women	Faculty
Arizona State University	20,309	15,955	2018
Embry-Riddle Aeronautical University	1441	428	93
Northern Arizona University	8215	10,958	1055
Prescott College	156	215	56
University of Arizona	14,054	15,475	2343

a. Calculate the student–faculty ratio at Prescott College. Write the ratio using a colon and using the word *to*. What does this ratio mean?

b. Which school listed has the lowest student–faculty ratio? The highest?

c. Which schools listed have the same student–faculty ratio?

20. **Advertising** The Randolph Company spent $350,000 for advertising last year. Department A and Department B share the cost of advertising in the ratio 3 : 7. Find the amount allocated to each department.

21. **Gardening** Three tablespoons of a liquid plant fertilizer are to be added to every 4 gal of water. How many tablespoons of fertilizer are required for 10 gal of water?

22. **Federal Expenditures** The table below shows how each dollar of projected spending by the federal government for a recent year was distributed. (*Source:* Congressional Budget Office) Of the items listed, defense is the only discretionary spending by the federal government; all other items are fixed expenditures. The government predicted total expenses of $3.6 trillion for the year.

a. Is more or less than one-fifth of federal spending spent on health care?

b. Find the ratio of the fixed expenditures to the discretionary spending.

c. Find the amount of the budget to be spent on fixed expenditures.

d. Find the amount of the budget to be spent on Social Security.

How Your Federal Tax Dollar Is Spent	
Health care	21 cents
Social Security	20 cents
Defense	20 cents
Other social aid	14 cents
Remaining government agencies and programs	19 cents
Interest on national debt	6 cents

23. **Demographics** According to the U.S. Bureau of the Census, the population of males and females in the United States in 2025 and 2050 is projected to be as shown in the table below.

Year	Males	Females
2025	164,119,000	170,931,000
2050	193,234,000	200,696,000

a. What percent of the projected population in 2025 is female? Round to the nearest tenth of a percent.

b. Does the percent of the projected population that is female in 2050 differ by more or less than 1% from the percent that is female in 2025?

24. **Diet** Americans consume 7 billion hot dogs from Memorial Day through Labor Day. This is 35% of the hot dogs consumed annually in the United States. (*Source:* National Hot Dog & Sausage Council: American Meat Institute) How many hot dogs do Americans consume annually?

25. **Boston Marathon** In the 2015 Boston Marathon, 27,167 runners started the race and 26,598 finished. What percent of the runners who started the course finished the race? Round to the nearest tenth of a percent.

26. **Nutrition** The table below shows the fat, saturated fat, cholesterol, and calorie content of a 90-gram ground-beef burger and a 90-gram soy burger.

a. Compared to the beef burger, by what percent is the fat content decreased in the soy burger?

b. What is the percent decrease in cholesterol in the soy burger compared to the beef burger?

c. Calculate the percent decrease in calories in the soy burger compared to the beef burger.

	Beef burger	Soy burger
Fat	24 g	4 g
Saturated fat	10 g	1.5 g
Cholesterol	75 mg	0 mg
Calories	280	140

27. **The Military** On Veterans Day in 2000, there were 26.6 million U.S. veterans. By Veterans Day in 2010, the number of U.S. veterans had dropped to 23.1 million. (*Source:* Department of Veterans Affairs) Find the percent decrease in the number of veterans from 2000 to 2010. Round to the nearest tenth of a percent.

28. **Police Officers** The graph below shows the causes of death for all police officers killed in the line of duty during a recent year. What percent of the deaths were due to auto accidents? Round to the nearest percent.

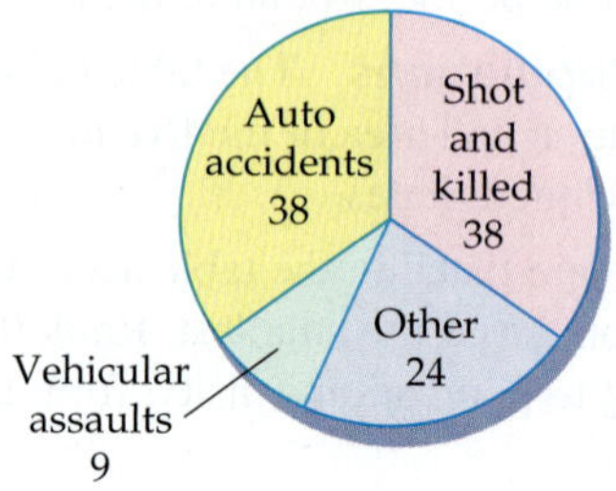

Causes of Death for Police Officers Killed in the Line of Duty

SOURCE: www.policespecial.com

29. **Retirement Programs** Massachusetts teachers enrolled in the Retirement-Plus savings program contribute 11% of their salaries to the program. What amount is contributed during 1 year by a member of this program who earns an annual salary of $64,000?

30. **Vacation Days** In Italy, workers take an average of 42 vacation days per year. This number is 3 more than three times the average number of vacation days that workers take each year in the United States. (*Source:* World Tourism Organization) On average, how many vacation days do U.S. workers take per year?

31. **Model Rockets** A small rocket is shot from the edge of a cliff. The height h, in meters, of the rocket above the cliff is given by $h = 30t - 5t^2$, where t is the time in seconds after the rocket is shot. Find the times at which the rocket is 25 m above the cliff.

32. **Sports** The height h, in feet, of a ball t seconds after being thrown from a height of 6 ft is given by the equation $h = -16t^2 + 32t + 6$. After how many seconds is the ball 18 ft above the ground? Round to the nearest tenth.

CHAPTER 9 TEST

■ In Exercises 1 to 5, solve the equation.

1. $\frac{x}{4} - 3 = \frac{1}{2}$

2. $x + 5(3x - 20) = 10(x - 4)$

3. $\frac{7}{16} = \frac{x}{12}$

4. $x^2 = 12x - 27$

5. $3x^2 - 4x = 1$

■ In Exercises 6 and 7, solve the formula for the given variable.

6. $x - 2y = 15$; y

7. $C = \frac{5}{9}(F - 32)$; F

Lee Prince/Shutterstock.com

Old Faithful

8. Geysers Old Faithful is a geyser in Yellowstone National Park. It is so named because of its regular eruptions for the past 100 years. An equation that can predict the approximate time until the next eruption is $T = 12.4L + 32$, where T is the time, in minutes, until the next eruption and L is the duration, in minutes, of the last eruption. Use this equation to determine the duration of the last eruption when the time between two eruptions is 63 min.

9. Energy The cost of electricity in a certain city is \$0.16 for each of the first 300 kWh and \$0.20 for each kilowatt-hour over 300 kWh. Find the number of kilowatt-hours used by a family with a \$74.25 electric bill.

10. Rate of Speed You drive 246.6 mi in 4.5 h. Find your average rate in miles per hour.

11. Parks The table below lists the largest city parks in the United States. The land acreage of Griffith Park in Los Angeles is 3 acres more than five times the acreage of New York's Central Park. What is the acreage of Central Park?

City park	Land acreage
Cullen Park (Houston)	10,534
Fairmount Park (Philadelphia)	8,700
Griffith Park (Los Angeles)	4,218
Eagle Creek Park (Indianapolis)	3,800
Pelham Bay Park (Bronx, NY)	2,764
Mission Bay Park (San Diego)	2,300

12. Baseball The table below shows six Major League lifetime record holders for batting. (*Source: Information Please Almanac*)

Baseball player	Number of times at bat	Number of home runs hit	Number of at-bats per home run
Ty Cobb	11,429	4,191	
Billy Hamilton	6,284	2,163	
Rogers Hornsby	8,137	2,930	
Joe Jackson	4,981	1,774	
Tris Speaker	10,195	3,514	
Ted Williams	7,706	2,654	

a. Calculate the number of at-bats per home run for each player in the table. Round to the nearest thousandth.

b. The players are listed in the table alphabetically. Rank the players according to the number of at-bats per home run, starting with the best rate.

13. Social Media In 2015, Twitter announced that they had 320 million monthly active users, and 80% of users were active on mobile devices. How many monthly active users were active on mobile devices in 2015?

14. Partnerships The two partners in a partnership share the profits of their business in the ratio 5 : 3. Last year the profits were \$360,000. Find the amount received by each partner.

15. Gardening The directions on a bag of plant food recommend 0.5 lb for every 50 ft^2 of lawn. How many pounds of plant food should be used on a lawn that measures 275 ft^2?

16. Crime Rates The table below lists U.S. cities with populations over 100,000 that had a high number of violent crimes per 1000 residents per year. Violent crimes include murder, rape, aggravated assault, and robbery. (*Source:* FBI Uniform Crime Reports)

City	Violent crimes per 1000 people
Baltimore, Maryland	13.39
St. Louis, Missouri	16.79
Detroit, Michigan	19.89
Memphis, Tennessee	17.41
Oakland, California	16.85

a. Which city has the highest rate of violent crimes?

b. The population of Baltimore is approximately 624,000. Estimate the number of violent crimes committed in that city. Round to the nearest whole number.

17. Pets During a recent year, nearly 1.2 million dogs or litters were registered with the American Kennel Club. The most popular breed was the Labrador retriever, with 172,841 registered. What percent of the registrations were Labrador retrievers? Round to the nearest tenth of a percent. (*Source:* American Kennel Club)

18. Smartphone Sales Apple Inc. reported that it sold 74,779,000 iPhones during its first quarter of 2016. This is up from sales of 48,046,000 the previous quarter. Find the percent increase in iPhones sold in the first quarter of 2016 as compared to the previous quarter. Round to the nearest tenth of a percent.

19. Compact Disc Sales In 2005, the total number of music CDs sold was 598.9 million, while in 2015, it was 125.6 million. (*Source:* Soundscan)

a. Find the percent decrease in the number of CDs sold from 2005 to 2015. Round to the nearest tenth of a percent.

b. If the percent decrease in the number of CDs sold from 2015 to 2025 is the same as it was from 2005 to 2015, how many CDs will be sold in 2025?

20. Shot Put The equation $h = -16t^2 + 28t + 6$, where $0 \leq t \leq 1.943$, can be used to find the height h, in feet, of a shot t seconds after a shot putter has released it. After how many seconds is the shot 10 ft above the ground? Round to the nearest tenth of a second.

10

Applications of Functions

The following photo shows some parabolic water arcs in a spring garden exhibit at the Bellagio Hotel in Las Vegas. It is amazing that the free flowing water in these arcs does not disperse as, for example, in water coming out of a garden hose. The cross-section of each of these arcs measures about an inch in diameter and very little, if any, water drips from the arcs.

This water effect is achieved by using laminar flow nozzles. These nozzles organize the turbulent water flow in a pipe so that all the water is traveling at the same speed and in the same direction.

A quadratic function, one of the topics of this chapter, can be used to model the path the water takes in one of these arcs. See Exercise 43, page 591.

Dick Nation

SECTION 10.1 Rectangular Coordinates and Functions

Introduction to Rectangular Coordinate Systems

When archeologists excavate a site, a *coordinate grid* is laid over the site so that records can be kept of exactly where each item was found. The grid below is from an archeological dig at Poggio Colla, a site in the Mugello area about 20 miles northeast of Florence, Italy.

David Lyons/Alamy Stock Photo

In mathematics we encounter a similar problem, that of locating a point in a plane. One way to solve the problem is to use a *rectangular coordinate system*.

A **rectangular coordinate system** is formed by two number lines, one horizontal and one vertical, that intersect at the zero point of each line. The point of intersection is called the **origin**. The two number lines are called the **coordinate axes**, or simply the **axes**. Frequently, the horizontal axis is labeled the x-axis and the vertical axis is labeled the y-axis. In this case, the axes form what is called the ***xy*-plane**.

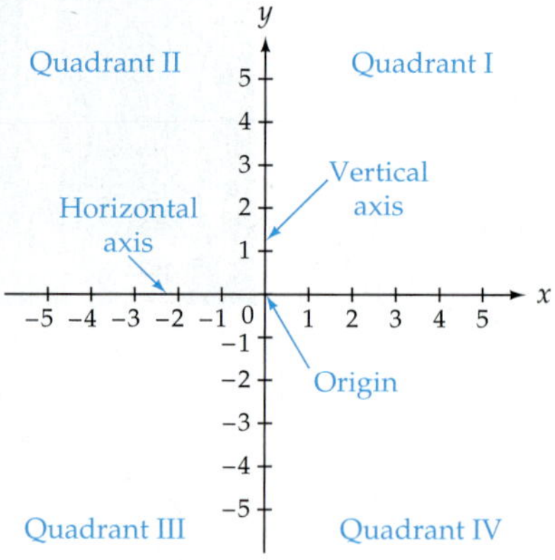

The two axes divide the plane into four regions called **quadrants**, which are numbered counterclockwise, using Roman numerals, from I to IV, starting at the upper right.

Each point in the plane can be identified by a pair of numbers called an **ordered pair**. The first number of the ordered pair measures a horizontal change from the y-axis and is called the **abscissa**, or ***x*-coordinate**. The second number of the ordered pair measures a vertical change from the x-axis and is called the **ordinate**, or ***y*-coordinate**. The ordered pair (x, y) associated with a point is also called the **coordinates** of the point.

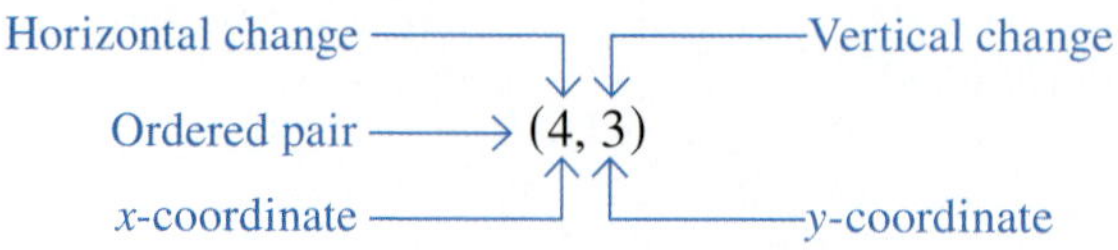

HISTORICAL NOTE

The concept of a coordinate system developed over time, culminating in 1637 with the publication of *Discourse on the Method for Rightly Directing One's Reason and Searching for Truth in the Sciences* by René Descartes (day-KART) (1596–1650) and *Introduction to Plane and Solid Loci* by Pierre de Fermat (fĕr-MA) (1601–1665). Of the two mathematicians, Descartes is usually given more credit for developing the concept of a coordinate system. In fact, he became so famous in Le Haye, the town in which he was born, that the town was renamed Le Haye–Descartes.

POINT OF INTEREST

The word *abscissa* has the same root as the word *scissors*. When open, a pair of scissors looks like an x.

To **graph**, or **plot**, a point means to place a dot at the coordinates of the point. For example, to graph the ordered pair (4, 3), start at the origin. Move 4 units to the right and then 3 units up. Draw a dot. To graph $(-3, -4)$, start at the origin. Move 3 units left and then 4 units down. Draw a dot.

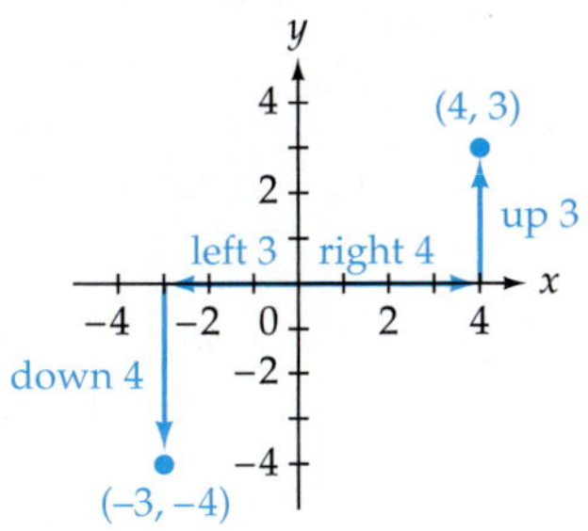

The **graph of an ordered pair** is the dot drawn at the coordinates of the point in the plane. The graphs of the ordered pairs (4, 3) and $(-3, -4)$ are shown at the right.

The graphs of the points whose coordinates are (2, 3) and (3, 2) are shown at the right. Note that they are different points. The order in which the numbers in an ordered pair are listed is important.

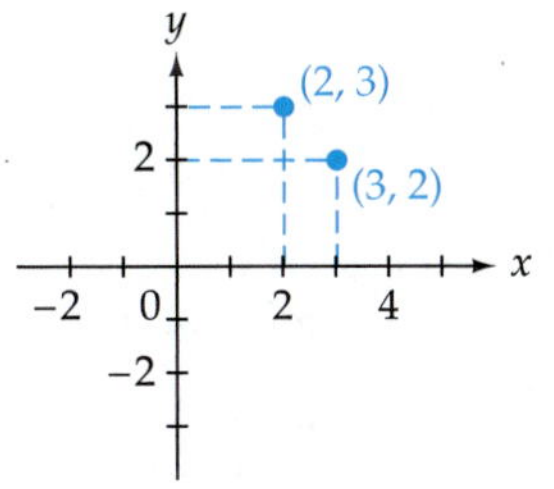

If the axes are labeled with letters other than x or y, then we refer to the ordered pair using the given labels. For instance, if the horizontal axis is labeled t and the vertical axis is labeled d, then the ordered pairs are written as (t, d). We sometimes refer to the first number in an ordered pair as the **first coordinate** of the ordered pair and to the second number as the **second coordinate** of the ordered pair.

One purpose of a coordinate system is to draw a graph of the solutions of an **equation in two variables**. Examples of equations in two variables are shown at the right.

$$y = 3x - 2$$
$$x^2 + y^2 = 25$$
$$s = t^2 - 4t + 1$$

A **solution of an equation in two variables** is an ordered pair that makes the equation a true statement. For instance, as shown below, (2, 4) is a solution of $y = 3x - 2$ but $(3, -1)$ is not a solution of the equation.

y	$3x - 2$	
4	$3(2) - 2$	• $x = 2, y = 4$
4	$6 - 2$	
$4 = 4$		• Checks.

y	$3x - 2$	
-1	$3(3) - 2$	• $x = 3, y = -1$
-1	$9 - 2$	
$-1 \neq 7$		• Does not check.

QUESTION Is $(-2, 1)$ a solution of $y = 3x + 7$?

The **graph of an equation in two variables** is a drawing of all the ordered-pair solutions of the equation. To create a graph of an equation, find some ordered-pair solutions of the equation, plot the corresponding points, and then connect the points with a smooth curve.

EXAMPLE 1 Graph an Equation in Two Variables

Graph $y = 3x - 2$.

Solution

To find ordered-pair solutions of the equation $y = 3x - 2$, select various values of x and calculate the corresponding values of y. Plot the ordered pairs. After the ordered pairs have been graphed, draw a smooth curve through the points. It is convenient to keep track of the solutions in a table.

ANSWER Yes, because $1 = 3(-2) + 7$.

TAKE NOTE

An *ordered pair* is a pair of coordinates, and the *order* in which the coordinates are listed matters. This concept is *very* important.

TAKE NOTE

An ordered pair is of the form (x, y). For the ordered pair (2, 4), 2 is the x value and 4 is the y value. Substitute 2 for x and 4 for y.

HISTORICAL NOTE

The Granger Collection, NY

Maria Graëtana Agnesi (an'yayzee) (1718–1799) was probably the first woman to write a calculus text. A report on the text made by a committee of the Académie des Sciences in Paris stated: "It took much skill and sagacity to reduce, as the author has done, to almost uniform methods these discoveries scattered among the works of modern mathematicians and often presented by methods very different from each other. Order, clarity and precision reign in all parts of this work.... We regard it as the most complete and best made treatise."

There is a graph named after Agnesi called the Witch of Agnesi. This graph came by its name because of an incorrect translation from Italian to English of a work by Agnesi. There is also a crater on Venus named after Agnesi called the Crater of Agnesi.

When choosing values of x, we often choose integer values because the resulting ordered pairs are easier to graph.

x	$y = 3x - 2$	(x, y)
-2	$3(-2) - 2 = -8$	$(-2, -8)$
-1	$3(-1) - 2 = -5$	$(-1, -5)$
0	$3(0) - 2 = -2$	$(0, -2)$
1	$3(1) - 2 = 1$	$(1, 1)$
2	$3(2) - 2 = 4$	$(2, 4)$
3	$3(3) - 2 = 7$	$(3, 7)$

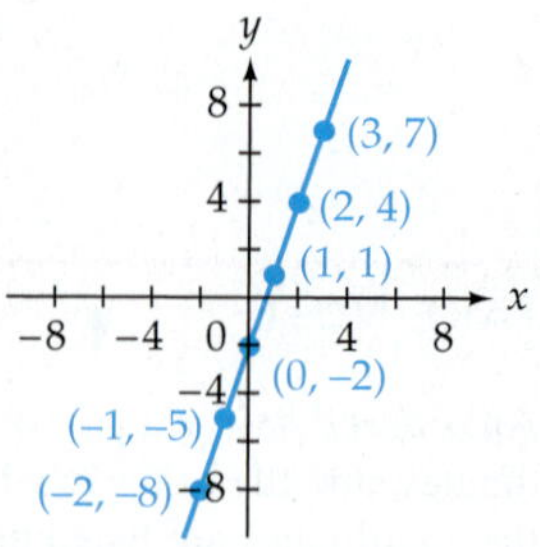

CHECK YOUR PROGRESS 1 Graph $y = -2x + 3$.

Solution See page S33.

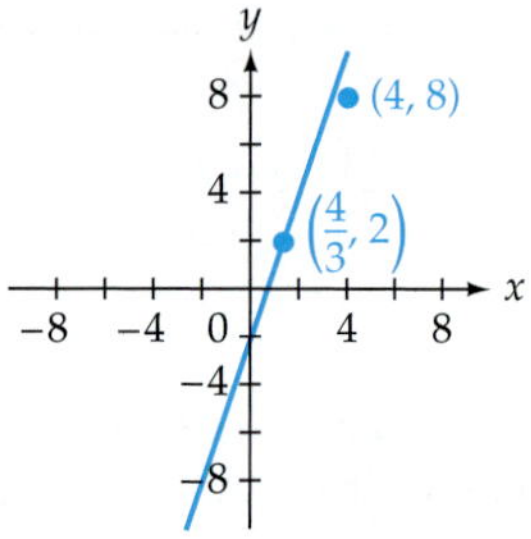

The graph of $y = 3x - 2$ is shown again at the left. Note that the ordered pair $\left(\frac{4}{3}, 2\right)$ is a solution of the equation and is a point on the graph. The ordered pair $(4, 8)$ is *not* a solution of the equation and is *not* a point on the graph. Every ordered-pair solution of the equation is a point on the graph, and every point on the graph is an ordered-pair solution of the equation.

EXAMPLE 2 Graph an Equation in Two Variables

Graph $y = x^2 + 4x$.

Solution

Select various values of x and calculate the corresponding values of y. Plot the ordered pairs. After the ordered pairs have been graphed, draw a smooth curve through the points. Here is a table showing ordered-pair solutions.

TAKE NOTE

As this example shows, it may be necessary to graph quite a number of points before a reasonably accurate graph can be drawn.

x	$y = x^2 + 4x$	(x, y)
-5	$(-5)^2 + 4(-5) = 5$	$(-5, 5)$
-4	$(-4)^2 + 4(-4) = 0$	$(-4, 0)$
-3	$(-3)^2 + 4(-3) = -3$	$(-3, -3)$
-2	$(-2)^2 + 4(-2) = -4$	$(-2, -4)$
-1	$(-1)^2 + 4(-1) = -3$	$(-1, -3)$
0	$(0)^2 + 4(0) = 0$	$(0, 0)$
1	$(1)^2 + 4(1) = 5$	$(1, 5)$

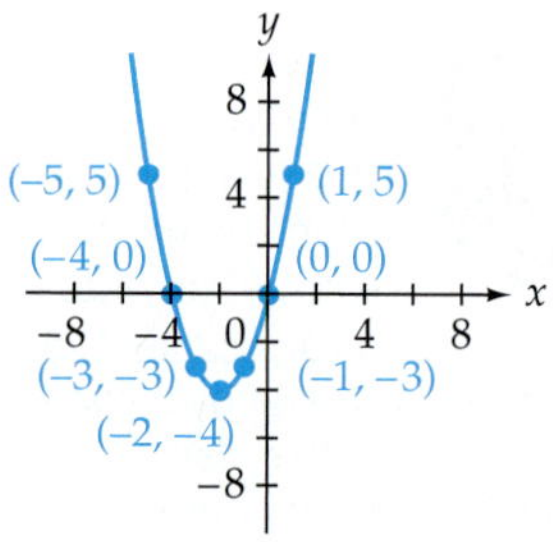

CHECK YOUR PROGRESS 2 Graph $y = -x^2 + 1$.

Solution See page S33.

MATH MATTERS Smartphone Coordinate Systems

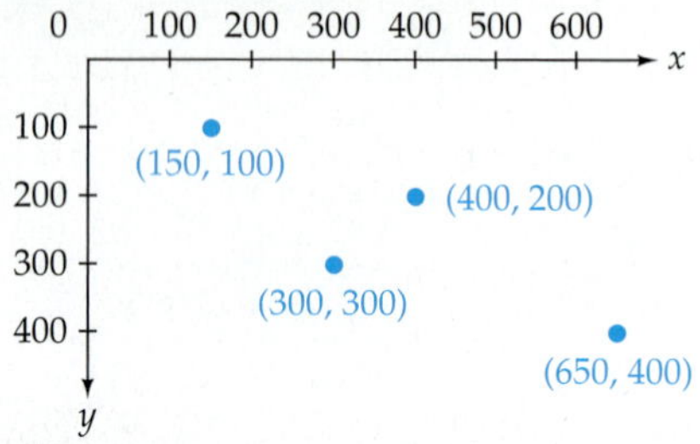

The standard coordinate system on a cell phone is quite different from the xy-coordinate system discussed above. On a cell phone grid, the origin $(0, 0)$ represents the top left point of the screen, as shown at the left. The points $(150, 100)$, $(300, 300)$, $(400, 200)$, and $(650, 400)$ are shown on the graph. Note that positive y values are plotted below the x-axis in this coordinate system.

Introduction to Functions

> **TAKE NOTE**
>
> A car that uses 1 gal of gas to travel 25 mi uses $\frac{1}{25} = 0.04$ gal to travel 1 mi.

An important part of mathematics is the study of the relationship between two known quantities. For instance, as a car is driven, the fuel in the gas tank is burned. There is a correspondence between the number of gallons of fuel used and the number of miles traveled. If a car gets 25 mi/gal, then the car consumes on average 0.04 gal of fuel for each mile driven. If, for the sake of simplicity, we assume that the car always consumes 0.04 gal of gasoline for each mile driven, then the equation $g = 0.04d$ defines how the number of gallons used, g, depends on the number of miles driven, d.

Distance traveled (in miles), d	25	50	100	250	300
Fuel used (in gallons), g	1	2	4	10	12

The ordered pairs in this table are only some of the possible ordered pairs. Other possibilities are (90, 3.6), (125, 5), and (235, 9.4). If all of the ordered pairs of the equation were plotted, the graph would be a portion of a line, as shown below. Note that all the ordered pairs of the equation lie on the same line.

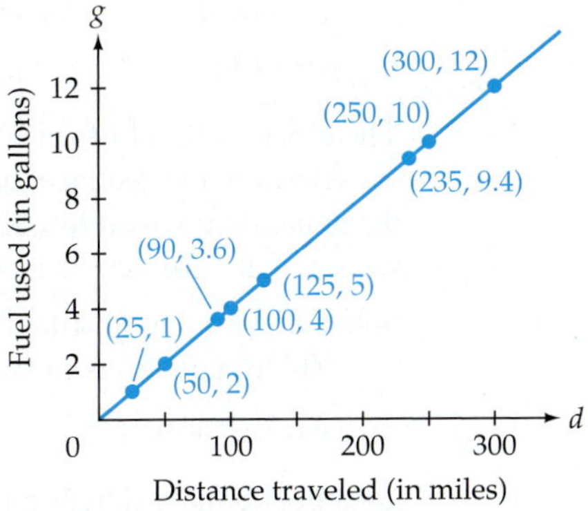

$g = 0.04d,\ d \geq 0$

QUESTION What is the meaning of the ordered pair (125, 5)?

The ordered pairs, the graph, and the equation are all different ways of expressing the correspondence between the two variables. This correspondence, which pairs the number of miles driven with the number of gallons of fuel used, is called a *function*.

Here are some additional examples of functions, along with a specific example of each.

To each real number 5	there corresponds ⟶	its square 25
To each score on an exam 87	there corresponds ⟶	a grade B
To each student Alexander Sterling	there corresponds ⟶	a student identification number S18723519

ANSWER A car that gets 25 mi per gallon can travel 125 mi on 5 gal of fuel.

Note that each result within a correspondence is *unique*. For instance, for the real number 5, there is *exactly one* square, 25. With this in mind, we now state the definition of a function.

Function, Domain, and Range

A **function** is a set of ordered pairs formed by a relationship between two sets that uniquely associates the members of one set, called the **domain**, with the members of the other set, called the **range**. The domain is the set of all first coordinates of the ordered pairs of the function. The range is the set of all second coordinates of the ordered pairs of the function.

Test score	Grade
90–100	A
80–89	B
70–79	C
60–69	D
0–59	F

Consider the function that pairs a test score with a letter grade, as shown at the left. The domain is the set of whole numbers from 0 to 100. The range is the set {A, B, C, D, F}.

A function is frequently defined by an equation in two variables. For instance, when gravity is the only force acting on a falling body, a function that describes the distance s, in feet, an object will fall in t seconds is given by $s = 16t^2$.

Given a value of t (time), the value of s (the distance the object falls) can be found. For instance, given $t = 3$,

$s = 16t^2$

$s = 16(3)^2$ • Replace t by 3.

$s = 16(9)$ • Simplify.

$s = 144$

The object falls 144 ft in 3 s.

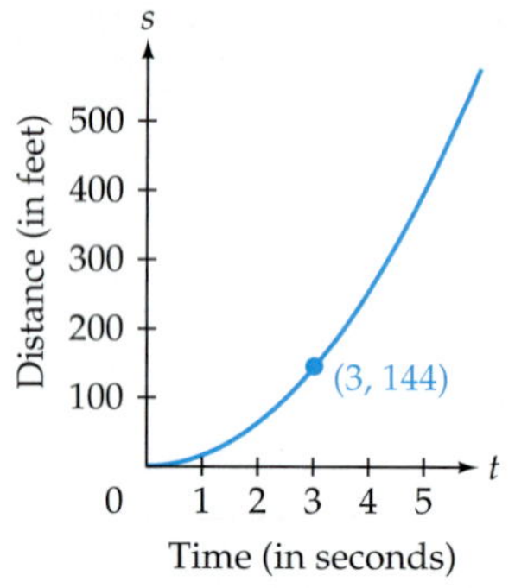

Because the distance the object falls *depends on* how long it has been falling, s is called the **dependent variable** and t is called the **independent variable**. For the equation $s = 16t^2$, we say that "distance is a function of time." A graph of the function is shown at the left. Some of the ordered pairs of this function are (3, 144), (1, 16), (0, 0), and $\left(\frac{1}{4}, 1\right)$.

Not all equations in two variables define a function. For instance,

$y^2 = x^2 + 9$

does not define a function because

$5^2 = 4^2 + 9$ and $(-5)^2 = 4^2 + 9$

The ordered pairs (4, 5) and (4, −5) both satisfy the equation. Consequently, there are two ordered pairs with the same first coordinate, 4, but *different* second coordinates, 5 and −5. By definition, the equation does not define a function. The phrase "y is a function of x," or a similar phrase with different variables, is used to describe those equations in two variables that define functions.

Function notation is frequently used for equations that define functions. Just as the letter x is commonly used as a variable, the letter f is commonly used to name a function.

TAKE NOTE

The notation $f(x)$ does *not* mean "f times x." The letter f stands for the name of the function, and $f(x)$ is the value of the function at x.

To describe the relationship between a number and its square using function notation, we can write $f(x) = x^2$. The symbol $f(x)$ is read "the value of f at x" or "f of x." The symbol $f(x)$ is the **value of the function** and represents the value of the dependent variable for a given value of the independent variable. We will often write $y = f(x)$ to emphasize the relationship between the independent variable, x, and the dependent variable, y. Remember: y and $f(x)$ are different symbols for the same number. Also, the *name* of the function is f; the *value* of the function at x is $f(x)$.

The process of finding $f(x)$ for a given value of x is called **evaluating the function**. For instance, to evaluate $f(x) = x^2$ when $x = 4$, replace x by 4 and simplify.

$f(x) = x^2$

$f(4) = 4^2 = 16$ • Replace x by 4. Then simplify.

The *value* of the function is 16 when $x = 4$. This means that an ordered pair of the function is (4, 16).

QUESTION Is $(3, -9)$ an ordered pair of $f(x) = x^2$?

EXAMPLE 3 Evaluate a Function

Evaluate $s(t) = 2t^2 - 3t + 1$ when $t = -2$.

Solution

$$s(t) = 2t^2 - 3t + 1$$
$$s(-2) = 2(-2)^2 - 3(-2) + 1 \quad \text{• Replace } t \text{ by } -2\text{. Then simplify.}$$
$$= 15$$

The value of the function is 15 when $t = -2$.

CHECK YOUR PROGRESS 3 Evaluate $f(z) = z^2 - z$ when $z = -3$.

Solution See page S33.

TAKE NOTE

To evaluate a function, you can use open parentheses in place of the variable. For instance,

$$s(t) = 2t^2 - 3t + 1$$
$$s(\;) = 2(\;)^2 - 3(\;) + 1$$

To evaluate the function, fill in each set of parentheses with the same number and then use the order of operations agreement to evaluate the numerical expression on the right side of the equation. To review the order of operations agreement, see the Take Note on page 496.

Any letter or combination of letters can be used to name a function. In the next example, the letters *SA* are used to name a *s*urface *a*rea function.

EXAMPLE 4 Application of Evaluating a Function

The surface area of a cube (the sum of the areas of each of the six faces) is given by $SA(s) = 6s^2$, where $SA(s)$ is the surface area of the cube and s is the length of one side of the cube. Find the surface area of a cube that has a side of length 10 cm.

Solution

$$SA(s) = 6s^2$$
$$SA(10) = 6(10)^2 \quad \text{• Replace } s \text{ by } 10.$$
$$= 6(100) \quad \text{• Simplify.}$$
$$= 600$$

The surface area of the cube is 600 cm^2.

CHECK YOUR PROGRESS 4 A **diagonal** of a polygon is a line segment from one vertex to a nonadjacent vertex, as shown at the left. The total number of diagonals of a polygon is given by $N(s) = \dfrac{s^2 - 3s}{2}$, where $N(s)$ is the total number of diagonals and s is the number of sides of the polygon. Find the total number of diagonals of a polygon with 12 sides.

Solution See page S34.

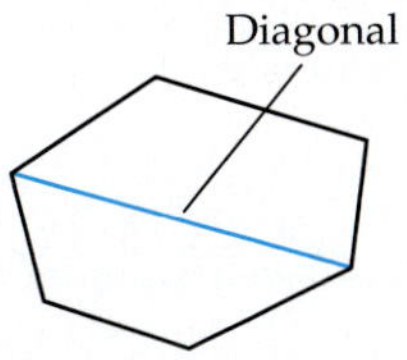

One of the diagonals of a hexagon

Graphs of Functions

Often the graph of a function can be drawn by finding ordered pairs of the function, plotting the points corresponding to the ordered pairs, and then connecting the points with a smooth curve.

ANSWER No. $f(3) = 3^2 = 9 \neq -9$.

For example, to graph $f(x) = x^3 + 1$, select several values of x and evaluate the function at each value. Recall that $f(x)$ and y are different symbols for the same quantity.

TAKE NOTE

We are basically creating the graph of an equation in two variables, as we did earlier. The only difference is the use of function notation.

x	$f(x) = x^3 + 1$	(x, y)
-2	$f(-2) = (-2)^3 + 1 = -7$	$(-2, -7)$
-1	$f(-1) = (-1)^3 + 1 = 0$	$(-1, 0)$
0	$f(0) = (0)^3 + 1 = 1$	$(0, 1)$
1	$f(1) = (1)^3 + 1 = 2$	$(1, 2)$
2	$f(2) = (2)^3 + 1 = 9$	$(2, 9)$

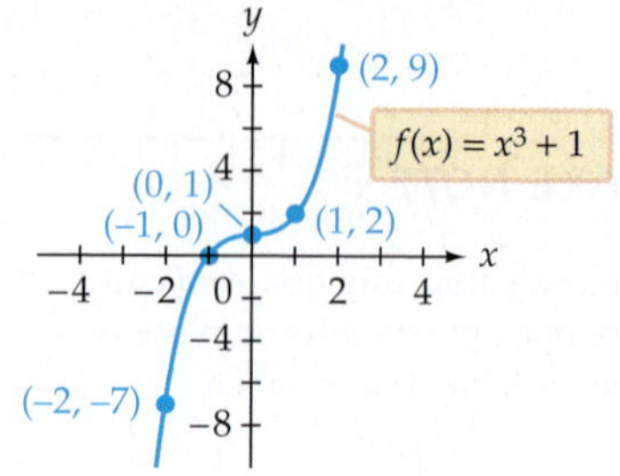

Plot the ordered pairs and draw a smooth curve through the points.

EXAMPLE 5 Graph a Function

INSTRUCTOR NOTE

For the graph of $h(x) = x^2 - 3$ at the right, ask students the following questions.

1. Are there other values of y for which there are two different x-coordinates? If so, name one.
2. Are there values of y for which there is one unique x-coordinate? If so, name it.
3. Are there values of y for which there is no x-coordinate? If so, name one.

Graph $h(x) = x^2 - 3$.

Solution

x	$h(x) = x^2 - 3$	(x, y)
-3	$h(-3) = (-3)^2 - 3 = 6$	$(-3, 6)$
-2	$h(-2) = (-2)^2 - 3 = 1$	$(-2, 1)$
-1	$h(-1) = (-1)^2 - 3 = -2$	$(-1, -2)$
0	$h(0) = (0)^2 - 3 = -3$	$(0, -3)$
1	$h(1) = (1)^2 - 3 = -2$	$(1, -2)$
2	$h(2) = (2)^2 - 3 = 1$	$(2, 1)$
3	$h(3) = (3)^2 - 3 = 6$	$(3, 6)$

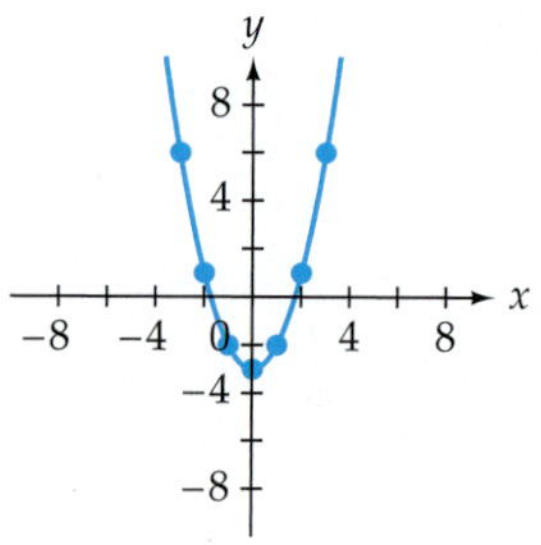

Plot the ordered pairs and draw a smooth curve through the points.

CHECK YOUR PROGRESS 5 Graph $f(x) = 2 - \frac{3}{4}x$.

Solution See page S34.

EXCURSION

Dilations of a Geometric Figure

A **dilation** of a geometric figure changes the size of the figure by either enlarging it or reducing it. This is accomplished by multiplying the coordinates of the figure by a positive number called the **dilation constant**. When the dilation constant is greater than 1, the geometric figure is enlarged. When the dilation constant is between 0 and 1, the

> **POINT OF INTEREST**
>
> Photocopy machines have reduction and enlargement features that function essentially as constants of dilation. The numbers are usually expressed as a percent. A copier selection of 50% reduces the size of the object being copied by 50%. A copier selection of 125% increases the size of the object being copied by 25%.

geometric figure is reduced. Examples of enlarging and reducing a geometric figure are shown below.

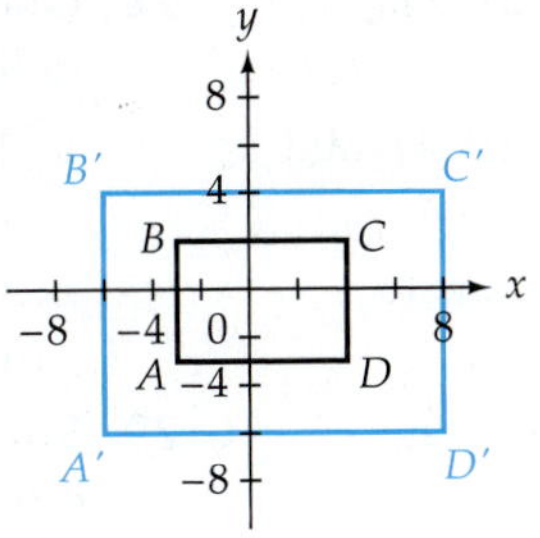

$ABCD$ was enlarged by multiplying its x- and y-coordinates by 2. The result is $A'B'C'D'$.

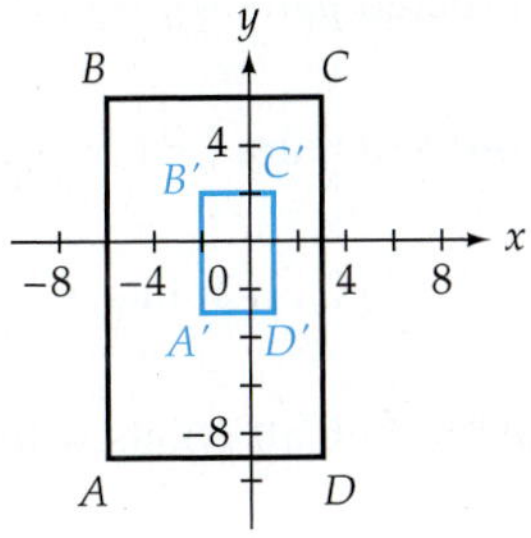

$ABCD$ was reduced by multiplying its x- and y-coordinates by $\frac{1}{3}$. The result is $A'B'C'D'$.

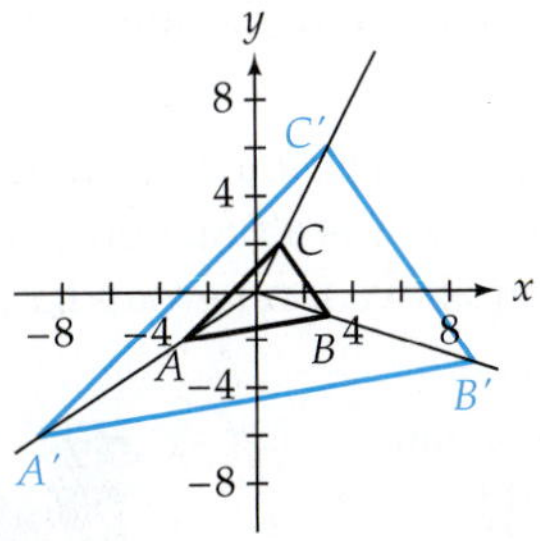

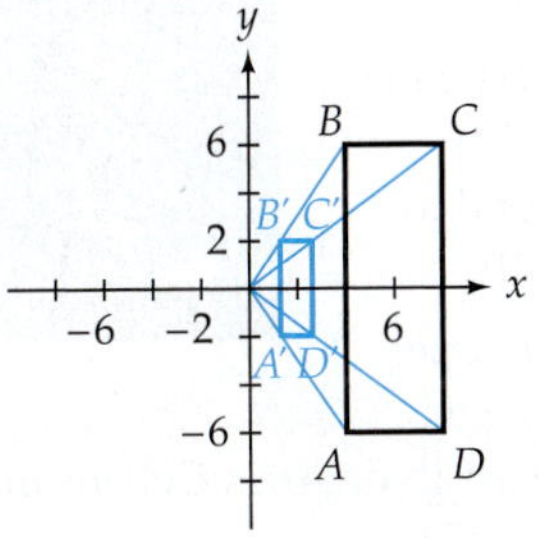

When each of the coordinates of a figure is multiplied by the same number in order to produce a dilation, the **center of dilation** will be the origin of the coordinate system. For triangle ABC at the left, a constant of dilation of 3 was used to produce triangle $A'B'C'$. Note that lines through the corresponding vertices of the two triangles intersect at the origin, the center of dilation.

For rectangle $ABCD$ at the left, a constant of dilation of $\frac{1}{3}$ was used to produce rectangle $A'B'C'D'$. Lines through the corresponding vertices of the two rectangles intersect at the origin, the center of dilation.

For a figure in space, or one not oriented in the xy-plane, the center of dilation can be any point. See Exercises 3 and 4 below.

EXCURSION EXERCISES

1. A dilation is performed on the figure with vertices $A(-2, 0)$, $B(2, 0)$, $C(4, -2)$, $D(2, -4)$, and $E(-2, -4)$.

a. Draw the original figure and a new figure using 2 as the dilation constant.

b. Draw the original figure and a new figure using $\frac{1}{2}$ as the dilation constant.

2. Because each of the coordinates of a geometric figure is multiplied by a number, the lengths of the sides of the figure will change. It is possible to show that the lengths change by a factor equal to the constant of dilation. In this exercise, you will examine the effect of a dilation on the angles of a geometric figure. Draw some figures and then draw a dilation of each figure using the origin as the center of dilation. Using a protractor, determine whether the measures of the angles of the dilated figure are different from the measures of the corresponding angles of the original figure.

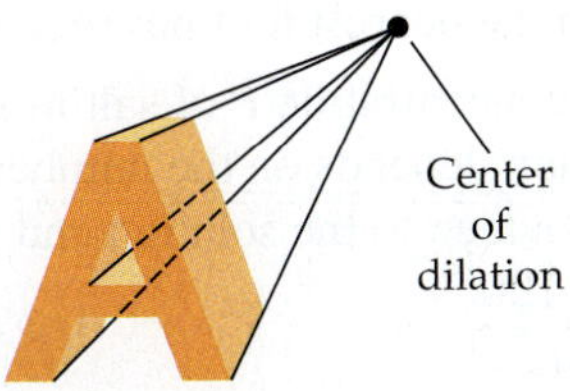

3. Graphic artists use centers of dilation to create three-dimensional (3D) effects. Consider the block letter A shown at the left. Draw another block letter A by changing the center of dilation to see how it affects the 3D look of the letter. You determine how thick to draw the letter. To enhance the 3D look, make the face of the letter darker than the rest of it. Programs such as PowerPoint use these methods to create various shading options for design elements in a presentation.

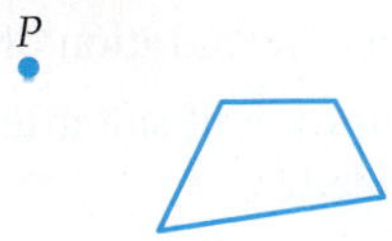

4. Draw an enlargement and a reduction of the quadrilateral at the left for the given center of dilation P.

5. On a blank piece of paper, draw a rectangle 4 in. by 6 in. with the center of the rectangle in the center of the paper. Make various photocopies of the rectangle using the reduction and enlargement settings on a copy machine. Where is the center of dilation for the copy machine?

EXERCISE SET 10.1

1. Graph the ordered pairs $(0, -1)$, $(2, 0)$, $(3, 2)$, and $(-1, 4)$.
2. Graph the ordered pairs $(-1, -3)$, $(0, -4)$, $(0, 4)$, and $(3, -2)$.
3. Draw a line through all points with an x-coordinate of 2.
4. Draw a line through all points with an x-coordinate of -3.
5. Draw a line through all points with a y-coordinate of -3.
6. Draw a line through all points with a y-coordinate of 4.
7. Graph the ordered-pair solutions of $y = x^2$ when $x = -2, -1, 0, 1$, and 2.
8. Graph the ordered-pair solutions of $y = -x^2 + 1$ when $x = -2, -1, 0, 1$, and 2.
9. Graph the ordered-pair solutions of $y = |x + 1|$ when $x = -5, -3, 0, 3$, and 5.
10. Graph the ordered-pair solutions of $y = -2|x|$ when $x = -3, -1, 0, 1$, and 3.
11. Graph the ordered-pair solutions of $y = x^3 - 2$ when $x = -1, 0, 1$, and 2.
12. Graph the ordered-pair solutions of $y = -x^3 + 1$ when $x = -1, 0$, and 1.

■ In Exercises 13 to 20, graph each equation.

13. $y = 2x - 1$
14. $y = -3x + 2$
15. $y = \frac{2}{3}x + 1$
16. $y = -\frac{x}{2} - 3$
17. $y = 2x^2 - 1$
18. $y = -3x^2 + 2$
19. $y = |x - 1|$
20. $y = |x - 3|$

■ In Exercises 21 to 26, evaluate the function for the given value.

21. $f(x) = 2x + 7; x = -2$
22. $y(x) = 1 - 3x; x = -4$
23. $f(t) = t^2 - t - 3; t = 3$
24. $P(n) = n^2 - 4n - 7; n = -3$
25. $T(p) = \frac{p^2}{p - 2}; p = 0$
26. $s(t) = \frac{4t}{t^2 + 2}; t = 2$

27. **Geometry** The perimeter P of a square is a function of the length s of one of its sides and is given by $P(s) = 4s$.
    a. Find the perimeter of a square whose side is 4 m.
    b. Find the perimeter of a square whose side is 5 ft.

28. **Geometry** The area of a circle is a function of its radius and is given by $A(r) = \pi r^2$.
    a. Find the area of a circle whose radius is 3 in. Round to the nearest tenth of a square inch.
    b. Find the area of a circle whose radius is 12 cm. Round to the nearest tenth of a square centimeter.

29. **Sports** The height h, in feet, of a ball that is released 4 ft above the ground with an initial upward velocity of 80 ft/s is a function of the time t, in seconds, the ball is in the air and is given by

$$h(t) = -16t^2 + 80t + 4, \ 0 \le t \le 5.04$$

    a. Find the height of the ball above the ground 2 s after it is released.
    b. Find the height of the ball above the ground 4 s after it is released.

30. **Forestry** The distance d, in miles, a forest fire ranger can see from an observation tower is a function of the height h, in feet, of the tower above level ground and is given by $d(h) = 1.5\sqrt{h}$.
    a. Find the distance a ranger can see whose eye level is 20 ft above level ground. Round to the nearest tenth of a mile.
    b. Find the distance a ranger can see whose eye level is 35 ft above level ground. Round to the nearest tenth of a mile.

Russell Illig/Photodisc Green/Getty Images

31. **Sound** The speed s, in feet per second, of sound in air depends on the temperature t of the air in degrees Celsius and is given by $s(t) = \frac{1087\sqrt{t + 273}}{16.52}$.
    a. What is the speed of sound in air when the temperature is 0°C (the temperature at which water freezes)? Round to the nearest foot per second.
    b. What is the speed of sound in air when the temperature is 25°C? Round to the nearest foot per second.

32. **Mixtures** The percent concentration P of salt in a particular salt water solution depends on the number of grams x of salt that are added to the solution and is given by $P(x) = \frac{100x + 100}{x + 10}$.
    a. What is the original percent concentration of salt?
    b. What is the percent concentration of salt after 5 more grams of salt are added?

33. Pendulums The time T, in seconds, it takes a pendulum to make one swing depends on the length of the pendulum and is given by $T(L) = 2\pi\sqrt{\frac{L}{32}}$, where L is the length of the pendulum in feet.

a. Find the time it takes the pendulum to make one swing if the length of the pendulum is 3 ft. Round to the nearest hundredth of a second.

b. Find the time it takes the pendulum to make one swing if the length of the pendulum is 9 in. Round to the nearest tenth of a second.

34. Botany Botanists have determined that some species of weed grow in a circular pattern. For one such species, the area A, in square meters, can be approximated by the function $A(t) = 0.005\pi t^2$, where t is the time, in days, after the growth of the weed first can be observed. Find the area this weed will cover 100 days after the growth is first observed. Round to the nearest square meter.

Malgorzata Kistryn/Shutterstock.com

■ In Exercises 35 to 50, graph the function.

35. $f(x) = 2x - 5$

36. $f(x) = -2x + 4$

37. $f(x) = -x + 4$

38. $f(x) = 3x - 1$

39. $g(x) = \frac{2}{3}x + 2$

40. $h(x) = \frac{5}{2}x - 1$

41. $F(x) = -\frac{1}{2}x + 3$

42. $F(x) = -\frac{3}{4}x - 1$

43. $f(x) = x^2 - 1$

44. $f(x) = x^2 + 2$

45. $f(x) = -x^2 + 4$

46. $f(x) = -2x^2 + 5$

47. $g(x) = x^2 - 4x$

48. $h(x) = x^2 + 4x$

49. $P(x) = x^2 - x - 6$

50. $P(x) = x^2 - 2x - 3$

EXTENSIONS

51. Suppose f is a function. Is it possible to have $f(2) = 4$ and $f(2) = 7$? Explain your answer.

52. Suppose f is a function and $f(a) = 4$ and $f(b) = 4$. Does this mean that $a = b$?

53. If $f(x) = 2x + 5$ and $f(a) = 9$, find a.

54. If $f(x) = x^2$ and $f(a) = 9$, find a.

55. Let $f(a, b) =$ the sum of a and b.

Let $g(a, b) =$ the product of a and b.

Find $f(2, 5) + g(2, 5)$.

56. Consider the function given by $M(x, y) = \frac{x + y}{2} + \frac{|x - y|}{2}$.

a. Complete the following table.

x	y	$M(x, y) = \frac{x + y}{2} + \frac{\lvert x - y\rvert}{2}$
−5	11	$M(-5, 11) = \frac{-5 + 11}{2} + \frac{\lvert -5 - 11\rvert}{2} = 11$
10	8	
−3	−1	
12	−13	
−11	15	

b. Extend the table by choosing some additional values of x and y.

c. How is the value of the function related to the values of x and y? *Hint:* For $x = -5$ and $y = 11$, the value of the function was 11, the value of y.

d. The function $M(x, y)$ is sometimes referred to as the *maximum function*. Why is this a good name for this function?

e. Create a *minimum function*—that is, a function that yields the minimum of two numbers x and y. *Hint:* The function is similar in form to the maximum function.

SECTION 10.2 Properties of Linear Functions

Intercepts

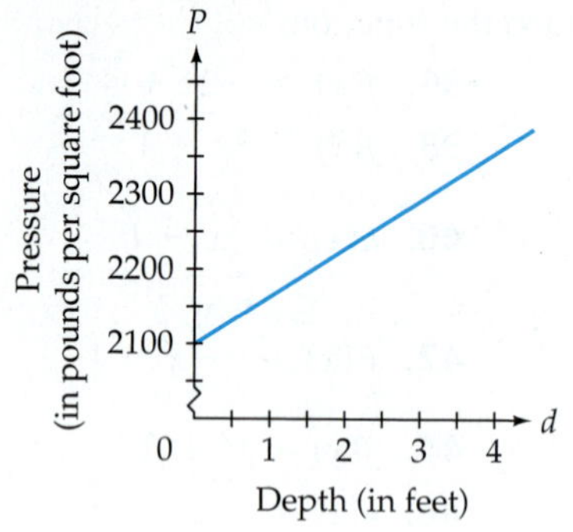

The graph at the left shows the pressure on a diver as the diver descends in the ocean. The equation of this graph can be represented by $P(d) = 64d + 2100$, where $P(d)$ is the pressure, in pounds per square foot, on a diver d feet below the surface of the ocean. By evaluating the function for various values of d, we can determine the pressure on the diver at those depths. For instance, when $d = 2$, we have

$$P(d) = 64d + 2100$$
$$P(2) = 64(2) + 2100$$
$$= 128 + 2100 = 2228$$

The pressure on a diver 2 ft below the ocean's surface is 2228 lb/sq ft.

The function $P(d) = 64d + 2100$ is an example of a *linear function.*

Linear Function

A **linear function** is one that can be written in the form $f(x) = mx + b$, where m is the coefficient of x and b is a constant.

For the linear function $P(d) = 64d + 2100$, $m = 64$ and $b = 2100$.

Here are some other examples of linear functions.

$f(x) = 2x + 5$ • $m = 2, b = 5$

$g(t) = \frac{2}{3}t - 1$ • $m = \frac{2}{3}, b = -1$

$v(s) = -2s$ • $m = -2, b = 0$

$h(x) = 3$ • $m = 0, b = 3$

$f(x) = 2 - 4x$ • $m = -4, b = 2$

Note that different variables can be used to designate a linear function.

QUESTION Which of the following are linear functions?

a. $f(x) = 2x^2 + 5$ **b.** $g(x) = 1 - 3x$

Consider the linear function $f(x) = 2x + 4$. The graph of the function is shown below, along with a table listing some of its ordered pairs.

TAKE NOTE

Note that the graph of a *linear* function is a straight *line*. Observe that when the graph crosses the x-axis, the y-coordinate is 0. When the graph crosses the y-axis, the x-coordinate is 0. The table confirms these observations.

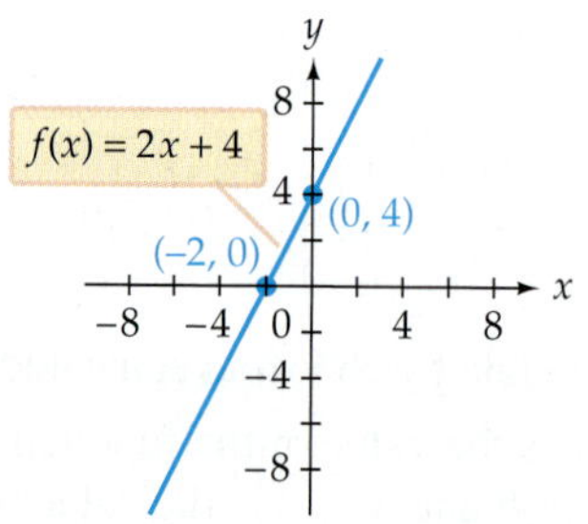

x	$f(x) = 2x + 4$	(x, y)
−3	$f(-3) = 2(-3) + 4 = -2$	$(-3, -2)$
−2	$f(-2) = 2(-2) + 4 = 0$	$(-2, 0)$
−1	$f(-1) = 2(-1) + 4 = 2$	$(-1, 2)$
0	$f(0) = 2(0) + 4 = 4$	$(0, 4)$
1	$f(1) = 2(1) + 4 = 6$	$(1, 6)$

ANSWER **a.** Because $f(x) = 2x^2 + 5$ has an x^2 term, f is not a linear function. **b.** Because $g(x) = 1 - 3x$ can be written in the form $f(x) = mx + b$ as $g(x) = -3x + 1$ ($m = -3$ and $b = 1$), g is a linear function.

From the table and the graph on the facing page, we can see that when $x = -2$, $y = 0$, and the graph crosses the x-axis at $(-2, 0)$. The point $(-2, 0)$ is called the **x-intercept** of the graph. When $x = 0$, $y = 4$, and the graph crosses the y-axis at $(0, 4)$. The point $(0, 4)$ is called the **y-intercept** of the graph.

EXAMPLE 1 Find the x- and y-intercepts of a Graph

Find the x- and y-intercepts of the graph of $g(x) = -3x + 2$.

Solution

When a graph crosses the x-axis, the y-coordinate of the point is 0. Therefore, to find the x-intercept, replace $g(x)$ by 0 and solve the equation for x. [Recall that $g(x)$ is another name for y.]

$$g(x) = -3x + 2$$
$$0 = -3x + 2 \quad \text{• Replace } g(x) \text{ by } 0.$$
$$-2 = -3x$$
$$\frac{2}{3} = x$$

The x-intercept is $\left(\frac{2}{3}, 0\right)$.

When a graph crosses the y-axis, the x-coordinate of the point is 0. Therefore, to find the y-intercept, evaluate the function when x is 0.

$$g(x) = -3x + 2$$
$$g(0) = -3(0) + 2 \quad \text{• Evaluate } g(x) \text{ when } x = 0.$$
$$= 2$$

The y-intercept is $(0, 2)$.

CHECK YOUR PROGRESS 1 Find the x- and y-intercepts of the graph of $f(x) = \frac{1}{2}x + 3$.

Solution See page S34.

TAKE NOTE

To find the y-intercept of $y = mx + b$ [we have replaced $f(x)$ by y], let $x = 0$. Then

$$y = mx + b$$
$$y = m(0) + b$$
$$= b$$

The y-intercept is $(0, b)$. This result is shown at the right.

In Example 1, note that the y-coordinate of the y-intercept of $g(x) = -3x + 2$ has the same value as b in the equation $f(x) = mx + b$. This is always true.

y-intercept

The y-intercept of the graph of $f(x) = mx + b$ is $(0, b)$.

TAKE NOTE

We are working with the function $P(d) = 64d + 2100$. Therefore, the intercept on the horizontal axis of a graph of the function is a d-intercept rather than an x-intercept, and the intercept on the vertical axis is a P-intercept rather than a y-intercept.

If we evaluate the linear function that models pressure on a diver, $P(d) = 64d + 2100$, at 0, we have

$$P(d) = 64d + 2100$$
$$P(0) = 64(0) + 2100 = 2100$$

In this case, the P-intercept (the intercept on the vertical axis) is $(0, 2100)$. In the context of the application, this means that the pressure on a diver 0 ft below the ocean's surface is 2100 lb/sq ft. Another way of saying "zero feet below the ocean's surface" is "at sea level." Thus the pressure on the diver, or anyone else for that matter, at sea level is 2100 lb/sq ft.

Both the x- and y-intercept can have a special meaning in the context of an application problem. This is demonstrated in the next example.

EXAMPLE 2 Application of the Intercepts of a Linear Function

After a parachute is deployed, a function that models the height of the parachutist above the ground is $f(t) = -10t + 2800$, where $f(t)$ is the height, in feet, of the parachutist t seconds after the parachute is deployed. Find the intercepts on the vertical and horizontal axes and explain what they mean in the context of the problem.

Solution

To find the intercept on the vertical axis, evaluate the function when t is 0.

$$f(t) = -10t + 2800$$
$$f(0) = -10(0) + 2800 = 2800$$

The intercept on the vertical axis is (0, 2800). This means that the parachutist is 2800 ft above the ground when the parachute is deployed.

To find the intercept on the horizontal axis, set $f(t) = 0$ and solve for t.

$$f(t) = -10t + 2800$$
$$0 = -10t + 2800$$
$$-2800 = -10t$$
$$280 = t$$

The intercept on the horizontal axis is (280, 0). This means that the parachutist reaches the ground 280 s after the parachute is deployed. Note that the parachutist reaches the ground when $f(t) = 0$.

CHECK YOUR PROGRESS 2 A function that models the descent of a certain small airplane is given by $g(t) = -20t + 8000$, where $g(t)$ is the height, in feet, of the airplane t seconds after it begins its descent. Find the intercepts on the vertical and horizontal axes, and explain what they mean in the context of the problem.

Solution See page S34.

Slope of a Line

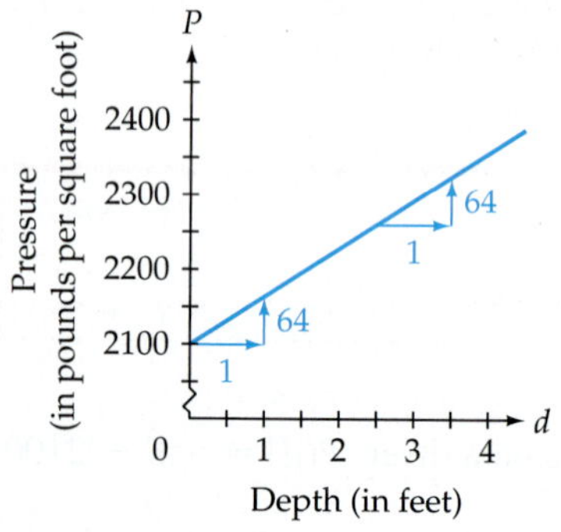

Consider again the linear function $P(d) = 64d + 2100$, which models the pressure on a diver as the diver descends below the ocean's surface. From the graph at the left, you can see that when the depth of the diver increases by 1 ft, the pressure on the diver increases by 64 lb/ft^2. This can be verified algebraically.

$P(0) = 64(0) + 2100 = 2100$ • Pressure at sea level

$P(1) = 64(1) + 2100 = 2164$ • Pressure after descending 1 ft

$2164 - 2100 = 64$ • Change in pressure

If we choose two other depths that differ by 1 ft, such as 2.5 ft and 3.5 ft (see the graph at the left), the change in pressure is the same.

$P(2.5) = 64(2.5) + 2100 = 2260$ • Pressure at 2.5 ft below the surface

$P(3.5) = 64(3.5) + 2100 = 2324$ • Pressure at 3.5 ft below the surface

$2324 - 2260 = 64$ • Change in pressure

The **slope** of a line is the change in the vertical direction caused by one unit of change in the horizontal direction. For $P(d) = 64d + 2100$, the slope is 64. In the context of the problem, the slope means that the pressure on a diver increases by 64 lb/ft^2 for each additional foot the diver descends. Note that the slope (64) has the same value as the coefficient of d in $P(d) = 64d + 2100$. This connection between the slope and the coefficient of the variable in a linear function always holds.

Slope

For a linear function given by $f(x) = mx + b$, the slope of the graph of the function is m, the coefficient of the x-variable.

QUESTION 1 What is the slope of each of the following?

a. $y = -2x + 3$ **b.** $f(x) = x + 4$ **c.** $g(x) = 3 - 4x$

d. $y = \frac{1}{2}x - 5$

The slope of a line can be calculated by using the coordinates of any two distinct points on the line and the following formula.

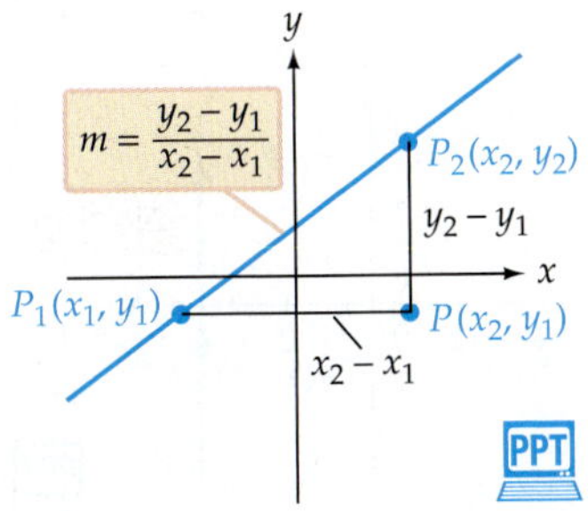

Slope of a Line

Let (x_1, y_1) and (x_2, y_2) be two points on a nonvertical line. Then the **slope** of the line through the two points is the ratio of the change in the y-coordinates to the change in the x-coordinates.

$$m = \frac{\text{change in } y}{\text{change in } x} = \frac{y_2 - y_1}{x_2 - x_1}, \; x_1 \neq x_2$$

QUESTION 2 Why is the restriction $x_1 \neq x_2$ required in the definition of slope?

EXAMPLE 3 Find the Slope of a Line between Two Points

Find the slope of the line between the two points.

a. $(-4, -3)$ and $(-1, 1)$ **b.** $(-2, 3)$ and $(1, -3)$

c. $(-1, -3)$ and $(4, -3)$ **d.** $(4, 3)$ and $(4, -1)$

Solution

a. $(x_1, y_1) = (-4, -3)$, $(x_2, y_2) = (-1, 1)$

$$m = \frac{y_2 - y_1}{x_2 - x_1} = \frac{1 - (-3)}{-1 - (-4)} = \frac{4}{3}$$

The slope is $\frac{4}{3}$. A *positive* slope indicates that the line slopes *upward* to the right. For this particular line, the value of y *increases* by $\frac{4}{3}$ when x increases by 1.

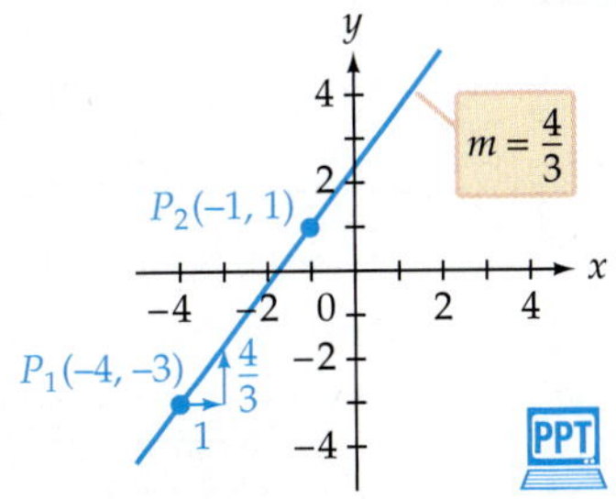

ANSWER 1 **a.** 2 **b.** 1 **c.** −4 **d.** $\frac{1}{2}$

ANSWER 2 If $x_1 = x_2$, then the difference $x_2 - x_1 = 0$. This would make the denominator 0, and division by 0 is not defined.

TAKE NOTE

When we talk about y values increasing (as in part a) or decreasing (as in part b), we always mean as we move from left to right on the coordinate grid.

b. $(x_1, y_1) = (-2, 3), (x_2, y_2) = (1, -3)$

$$m = \frac{y_2 - y_1}{x_2 - x_1} = \frac{-3 - 3}{1 - (-2)} = \frac{-6}{3} = -2$$

The slope is -2. A *negative* slope indicates that the line slopes *downward* to the right. For this particular line, the value of y *decreases* by 2 when x increases by 1.

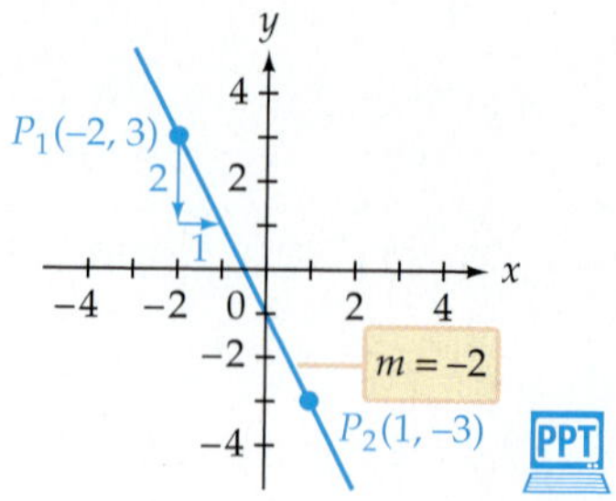

c. $(x_1, y_1) = (-1, -3), (x_2, y_2) = (4, -3)$

$$m = \frac{y_2 - y_1}{x_2 - x_1} = \frac{-3 - (-3)}{4 - (-1)} = \frac{0}{5} = 0$$

The slope is 0. A *zero* slope indicates that the line is *horizontal*. For this particular line, the value of y *stays the same* when x increases by any amount.

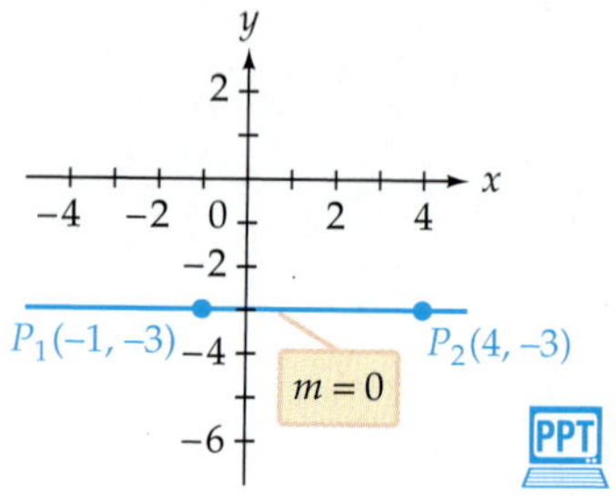

TAKE NOTE

A horizontal line has zero slope. A line that has no slope, or whose slope is undefined, is a vertical line. Note that when $y_1 = y_2$ in the formula for slope, the slope of the line through the two points is zero. When $x_1 = x_2$, the slope is undefined.

d. $(x_1, y_1) = (4, 3), (x_2, y_2) = (4, -1)$

$$m = \frac{y_2 - y_1}{x_2 - x_1} = \frac{-1 - 3}{4 - 4} = \frac{-4}{0}$$

Division by 0 is undefined.

If the denominator of the slope formula is zero, the line has *no slope*. Sometimes we say that the slope of a vertical line is *undefined*.

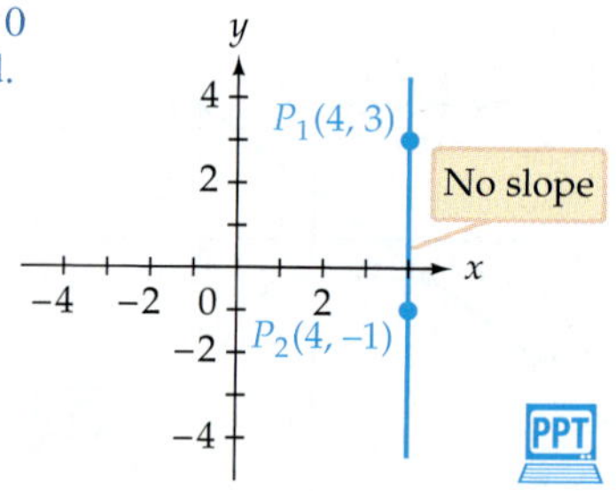

CHECK YOUR PROGRESS 3 Find the slope of the line between the two points.

a. $(-6, 5)$ and $(4, -5)$ **b.** $(-5, 0)$ and $(-5, 7)$

c. $(-7, -2)$ and $(8, 8)$ **d.** $(-6, 7)$ and $(1, 7)$

Solution See page S34.

The graph below shows the distance traveled by a jogger who is jogging at a rate of 6 mph. The table shows the distances traveled after various intervals of time. These values have been plotted on the graph, and a line has been drawn through the points.

Time, *t*, in hours	0.5	1	1.5	2	2.5	3
Distance, *d*, in miles	3	6	9	12	15	18

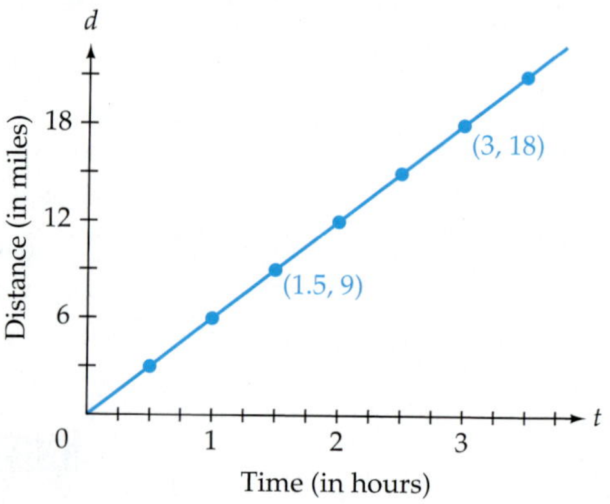

The slope of the line can be calculated using two of the points on the graph.

$$m = \frac{\text{change in } d}{\text{change in } t} = \frac{18 \text{ mi} - 9 \text{ mi}}{3 \text{ h} - 1.5 \text{ h}}$$

$$= \frac{9 \text{ mi}}{1.5 \text{ h}} = 6 \text{ mph}$$

Note that the slope of the line is the same as the speed of the jogger.

The speed of the jogger is given in miles *per* hour. Any real-world concept that can be expressed using the word *per* is described mathematically as slope. Example 4 is an illustration of this concept.

EXAMPLE 4 Application of Slope

The graph at the right shows the temperature T, in degrees Celsius, at x kilometers above sea level. Write a sentence that explains the meaning of the slope in the context of this application.

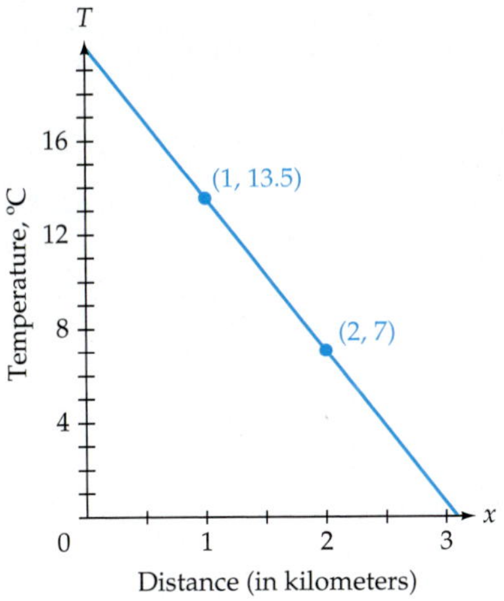

Solution

$$m = \frac{\text{change in } T}{\text{change in } x} = \frac{7°\text{C} - 13.5°\text{C}}{2 \text{ km} - 1 \text{ km}} = \frac{-6.5°\text{C}}{1 \text{ km}} = -6.5°\text{C/km}$$

The slope is -6.5°C/km. The slope means that the temperature is decreasing (because the slope is negative) 6.5°C for each 1-kilometer increase in height above sea level.

CHECK YOUR PROGRESS 4

The graph at the right shows the distance that a homing pigeon has flown after various intervals of time. What is the meaning of the slope in the context of this application?

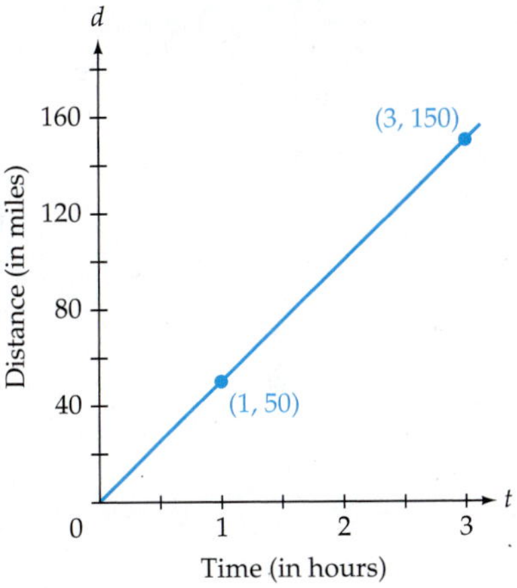

Solution See page S34.

MATH MATTERS

Julius Elias/Shutterstock.com

Galileo Galilei (găl-ĭ'lā-ē) (1564–1642) was one of the most influential scientists of his time. In addition to inventing the telescope, with which he discovered the moons of Jupiter, Galileo successfully argued that Aristotle's assertion that heavy objects drop at a greater velocity than lighter ones was incorrect. According to legend, Galileo went to the top of the Leaning Tower of Pisa and dropped two balls at the same time, one weighing twice the other. His assistant, standing on the ground, observed that both balls reached the ground at the same time.

There is no historical evidence that Galileo actually performed this experiment, but he did correctly reason that if Aristotle's assertion were true, then balls of different weights should roll down a ramp at different speeds. Galileo carried out this experiment and was able to show that, in fact, balls of different weights reached the end of the ramp at the same time. Galileo was not able to determine why this happened, and it took Isaac Newton, born the same year that Galileo died, to formulate the first theory of gravity.

Slope–Intercept Form of a Straight Line

The value of the slope of a line gives the change in y for a 1-*unit* change in x. For instance, a slope of -3 means that y changes by -3 as x changes by 1; a slope of $\frac{4}{3}$ means that y changes by $\frac{4}{3}$ as x changes by 1. Because it is difficult to graph a change of $\frac{4}{3}$, it is easier to think of a fractional slope in terms of integer changes in x and y. As shown below, for a slope of $\frac{4}{3}$ we have

$$m = \frac{\text{change in } y}{\text{change in } x} = \frac{4}{3}$$

That is, for a slope of $\frac{4}{3}$, y changes by 4 as x changes by 3.

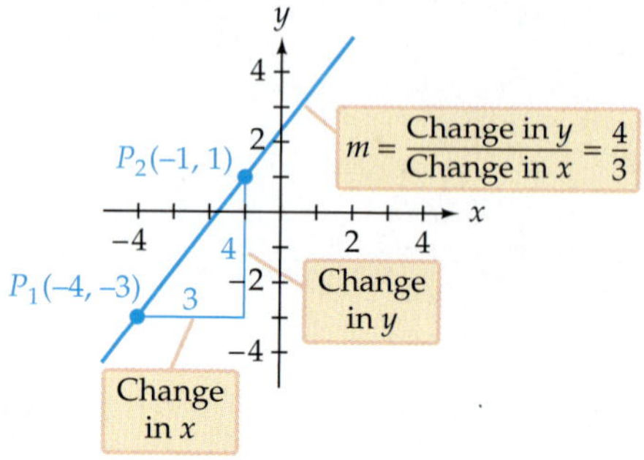

EXAMPLE 5 Graph a Line Given a Point on the Line and the Slope

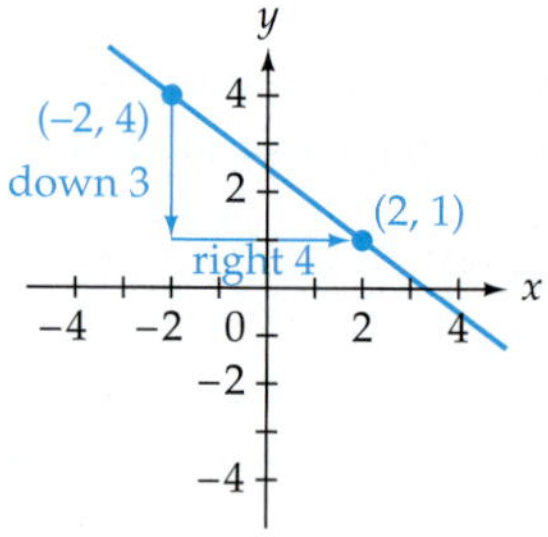

Draw the line that passes through $(-2, 4)$ and has slope $-\frac{3}{4}$.

Solution

Place a dot at $(-2, 4)$ and then rewrite $-\frac{3}{4}$ as $\frac{-3}{4}$. Starting from $(-2, 4)$, move 3 units down (the change in y) and then 4 units to the right (the change in x). Place a dot at that location and then draw a line through the two points. See the graph at the left.

CHECK YOUR PROGRESS 5 Draw the line that passes through $(2, 4)$ and has slope -1.

Solution See page S34.

Because the slope and y-intercept can be determined directly from the equation $f(x) = mx + b$, this equation is called the *slope–intercept form* of a straight line.

Slope–Intercept Form of the Equation of a Line

The graph of $f(x) = mx + b$ is a straight line with slope m and y-intercept $(0, b)$.

When a function is written in this form, it is possible to create a quick graph of the function.

EXAMPLE 6 Graph a Linear Function Using the Slope and *y*-intercept

Graph $f(x) = -\frac{2}{3}x + 4$ by using the slope and y-intercept.

Solution

From the equation, the slope is $-\frac{2}{3}$ and the y-intercept is $(0, 4)$. Place a dot at the y-intercept. We can write the slope as $m = -\frac{2}{3} = \frac{-2}{3}$. Starting from the y-intercept, move 2 units down and 3 units to the right and place another dot. Now draw a line through the two points.

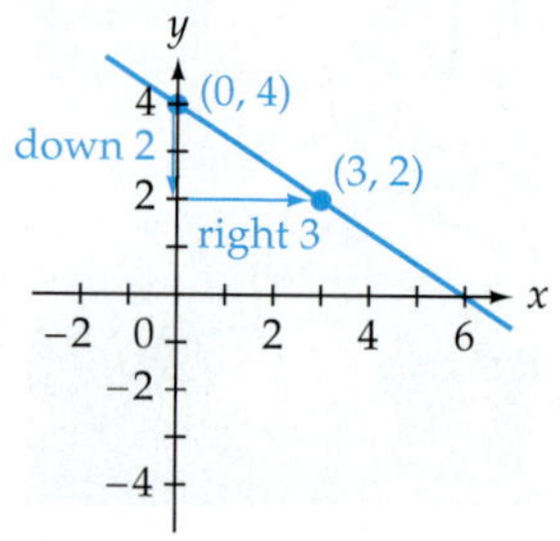

CHECK YOUR PROGRESS 6 Graph $y = \frac{3}{4}x - 5$ by using the slope and y-intercept.

Solution See page S34.

EXCURSION

Negative Velocity

We can expand the concept of velocity to include negative velocity. Suppose a car travels in a straight line starting at a given point. If the car is moving to the right, then we say that its velocity is positive. If the car is moving to the left, then we say that its velocity is negative. For instance, a velocity of -45 mph means the car is moving to the left at 45 mph.

If we were to graph the motion of an object on a distance–time graph, a positive velocity would be indicated by a positive slope and a negative velocity by a negative slope.

EXCURSION EXERCISES

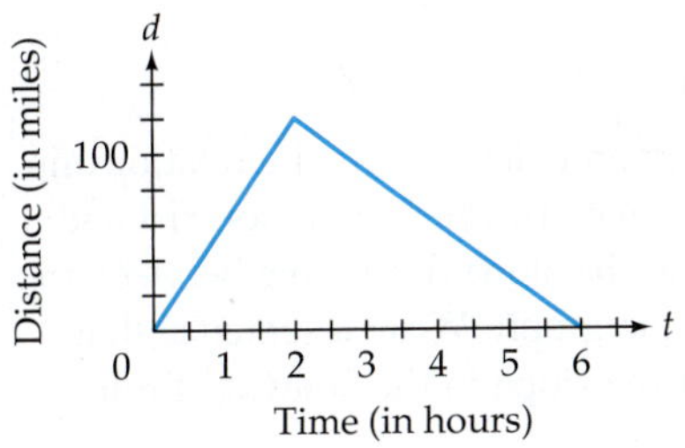

The graph at the left represents a car traveling on a straight road. Answer the following questions on the basis of this graph.

1. Between what two times is the car moving to the right?
2. Between what two times does the car have a positive velocity?
3. Between what two times is the car moving to the left?
4. Between what two times does the car have a negative velocity?
5. After 2 h, how far is the car from its starting position?
6. How long after the car leaves its starting position does it return to its starting position?
7. What is the velocity of the car during its first 2 h of travel?
8. What is the velocity of the car during its last 4 h of travel?

The graph below represents another car traveling on a straight road, but this car's motion is a little more complicated. Use this graph for the questions below.

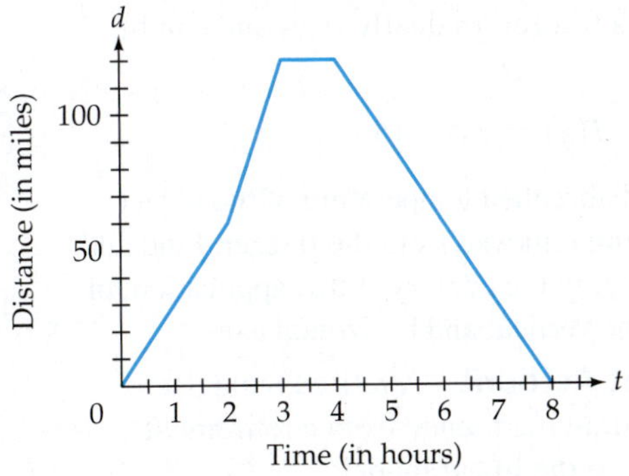

9. What is the slope of the line between hours 3 and 4?
10. What is the velocity of the car between hours 3 and 4? Is the car moving?
11. During which of the following intervals of time is the absolute value of the velocity greatest: 0 to 2 h, 2 to 3 h, 3 to 4 h, or 4 to 8 h? (Recall that the absolute value of a real number a is the distance between a and 0 on the number line.)

EXERCISE SET 10.2

■ In Exercises 1 to 14, find the x- and y-intercepts of the graph of the equation.

1. $f(x) = 3x - 6$
2. $f(x) = 2x + 8$
3. $y = \frac{2}{3}x - 4$
4. $y = -\frac{3}{4}x + 6$
5. $y = -x - 4$
6. $y = -\frac{x}{2} + 1$
7. $3x + 4y = 12$
8. $5x - 2y = 10$
9. $2x - 3y = 9$
10. $4x + 3y = 8$
11. $\frac{x}{2} + \frac{y}{3} = 1$
12. $\frac{x}{3} - \frac{y}{2} = 1$
13. $x - \frac{y}{2} = 1$
14. $-\frac{x}{4} + \frac{y}{3} = 1$

15. **Crickets** There is a relationship between the number of times a cricket chirps per minute and the air temperature. A linear model of this relationship is given by

$$f(x) = 7x - 30$$

where x is the temperature in degrees Celsius and $f(x)$ is the number of chirps per minute. Find and discuss the meaning of the x-intercept in the context of this application.

16. **Travel** An approximate linear model that gives the remaining distance, in miles, a plane must travel from Los Angeles to Paris is given by

$$s(t) = 6000 - 500t$$

where $s(t)$ is the remaining distance t hours after the flight begins. Find and discuss the meaning, in the context of this application, of the intercepts on the vertical and horizontal axes.

17. **Refrigeration** The temperature of an object taken from a freezer gradually rises and can be modeled by

$$T(x) = 3x - 15$$

where $T(x)$ is the Fahrenheit temperature of the object x minutes after being removed from the freezer. Find and discuss the meaning, in the context of this application, of the intercepts on the vertical and horizontal axes.

18. **Retirement Account** A retired biologist begins withdrawing money from a retirement account according to the linear model

$$A(t) = 100{,}000 - 2500t$$

where $A(t)$ is the amount, in dollars, remaining in the account t months after withdrawals begin. Find and discuss the meaning, in the context of this application, of the intercepts on the vertical and horizontal axes.

■ In Exercises 19 to 34, find the slope of the line containing the two points.

19. (1, 3), (3, 1)
20. (2, 3), (5, 1)
21. (−1, 4), (2, 5)
22. (3, −2), (1, 4)
23. (−1, 3), (−4, 5)
24. (−1, −2), (−3, 2)
25. (0, 3), (4, 0)
26. (−2, 0), (0, 3)
27. (2, 4), (2, −2)
28. (4, 1), (4, −3)
29. (2, 5), (−3, −2)
30. (4, 1), (−1, −2)
31. (2, 3), (−1, 3)
32. (3, 4), (0, 4)
33. (0, 4), (−2, 5)
34. (−2, −5), (−4, −1)

35. **Travel** The graph below shows the relationship between the distance traveled by a motorist and the time of travel. Find the slope of the line between the two points shown on the graph. Write a sentence that states the meaning of the slope in the context of this application.

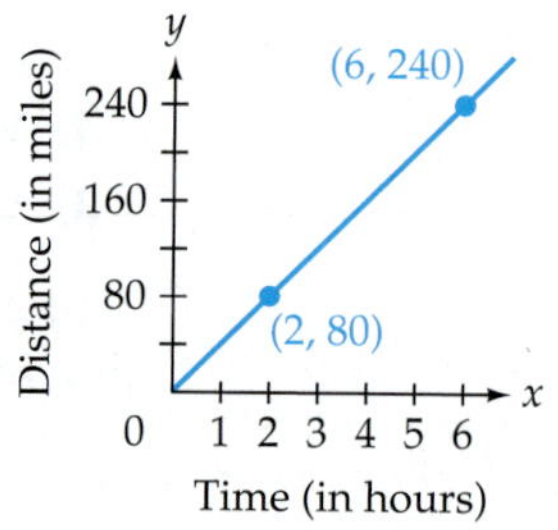

36. **Depreciation** The graph below shows the relationship between the value of a building and the depreciation allowed for income tax purposes. Find the slope of the line between the two points shown on the graph. Write a sentence that states the meaning of the slope in the context of this application.

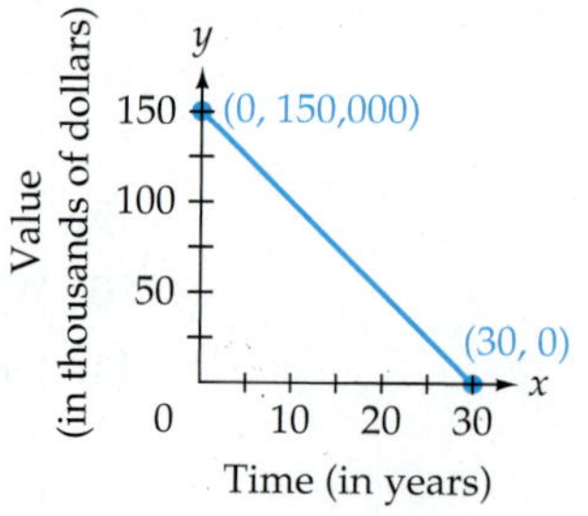

37. Aviation The graph below shows the height of a plane above an airport during its 30-minute descent from cruising altitude to landing. Find the slope of the line. Write a sentence that explains the meaning of the slope.

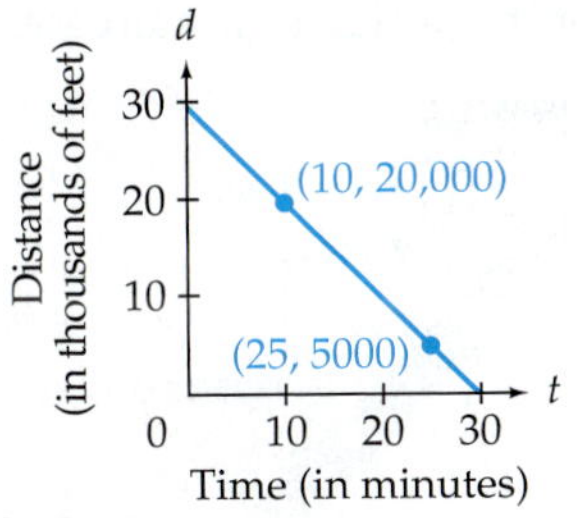

38. Mortgages The graph below shows the relationship between the monthly payment on a mortgage and the amount of the mortgage. Find the slope of the line between the two points shown on the graph. Write a sentence that states the meaning of the slope in the context of this application.

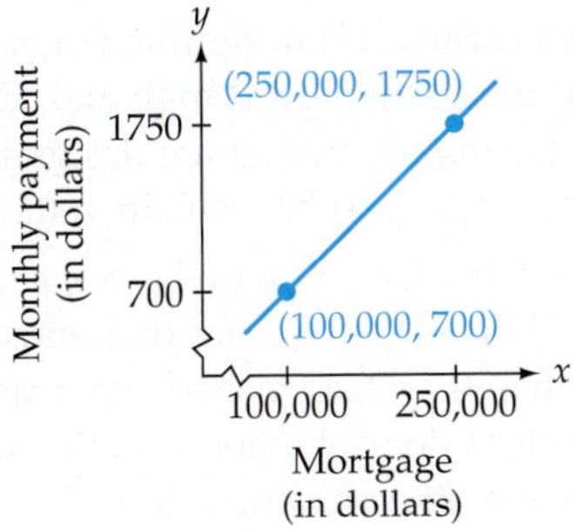

39. Panama Canal Ships in the Panama Canal are lowered through a series of locks. A ship is lowered as the water in a lock is discharged. The graph below shows the number of gallons of water N remaining in a lock t minutes after the valves are opened to discharge the water. Find the slope of the line. Write a sentence that explains the meaning of the slope.

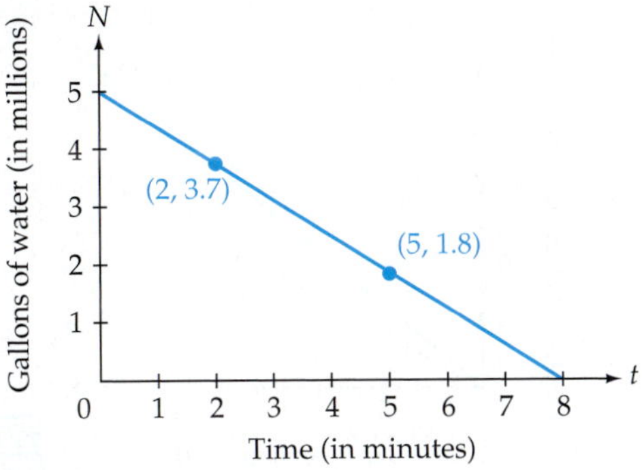

40. Graph the line that passes through the point $(-1, -3)$ and has slope $\frac{4}{3}$.

41. Graph the line that passes through the point $(-2, -3)$ and has slope $\frac{5}{4}$.

42. Graph the line that passes through the point $(-3, 0)$ and has slope -3.

43. Graph the line that passes through the point $(2, 0)$ and has slope -1.

■ In Exercises 44 to 49, graph using the slope and y-intercept.

44. $f(x) = \frac{1}{2}x + 2$

45. $f(x) = \frac{2}{3}x - 3$

46. $f(x) = -\frac{3}{2}x$

47. $f(x) = \frac{3}{4}x$

48. $f(x) = \frac{1}{3}x - 1$

49. $f(x) = -\frac{3}{2}x + 6$

EXTENSIONS

50. Jogging Lois and Tanya start from the same place at the same time on a straight jogging course and jog in the same direction. Lois is jogging at 9 km/h, and Tanya is jogging at 6 km/h. The graphs at the right show the distance each jogger has traveled after x hours and the distance between the joggers after x hours. Which graph represents the distance Lois has traveled in x hours? Which graph represents the distance Tanya has traveled in x hours? Which graph represents the distance between Lois and Tanya after x hours?

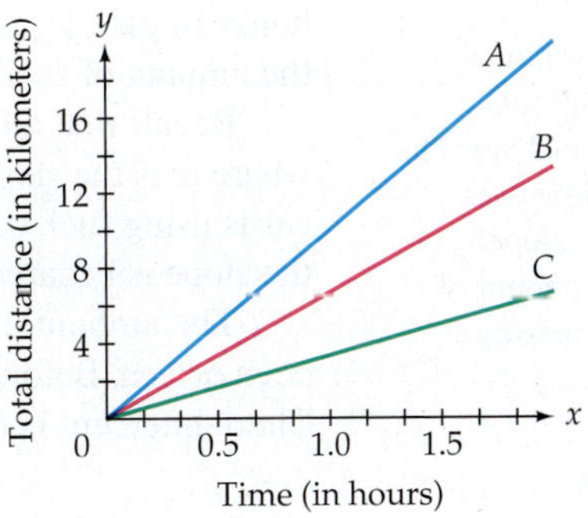

51. Chemistry A chemist is filling two cans from a faucet that releases water at a constant rate. Can 1 has a diameter of 20 mm and can 2 has a diameter of 30 mm.

a. In the following graph, which line represents the depth of the water in can 1 after x seconds?

b. Use the graph to estimate the difference in the depths of the water in the two cans after 15 s.

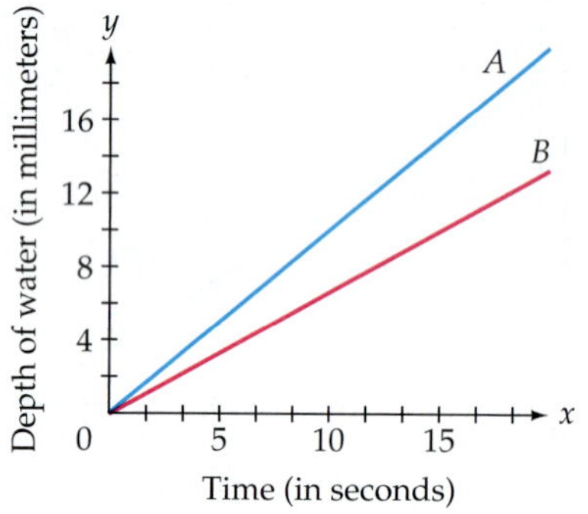

52. If (2, 3) are the coordinates of a point on a line that has slope 2, what is the y-coordinate of the point on the line at which $x = 4$?

53. If $(-1, 2)$ are the coordinates of a point on a line that has slope -3, what is the y-coordinate of the point on the line at which $x = 1$?

54. What effect does increasing the coefficient of x have on the graph of $y = mx + b$?

55. What effect does decreasing the coefficient of x have on the graph of $y = mx + b$?

56. What effect does increasing the constant term have on the graph of $y = mx + b$?

57. What effect does decreasing the constant term have on the graph of $y = mx + b$?

58. Construction When you climb a staircase, the flat part of a stair that you step on is called the *tread* of the stair. The *riser* is the vertical part of the stair. The slope of a staircase is the ratio of the length of the riser to the length of the tread. Because the design of a staircase may affect safety, most cities have building codes that give rules for the design of a staircase.

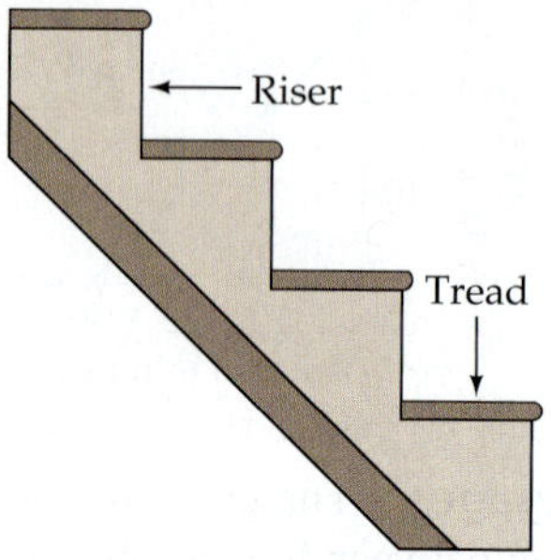

a. The traditional design of a staircase calls for a 9-inch tread and an 8.25-inch riser. What is the slope of this staircase?

b. A newer design for a staircase uses an 11-inch tread and a 7-inch riser. What is the slope of this staircase?

c. An architect is designing a house with a staircase that is 8 ft high and 12 ft long. Is the architect using the traditional design in part a or the newer design in part b? Explain your answer.

d. Staircases that have a slope between 0.5 and 0.7 are usually considered safer than those with a slope greater than 0.7. Design a safe staircase that goes from the first floor of a house to the second floor, which is 9 ft above the first floor.

e. Measure the tread and riser for three staircases you encounter. Do these staircases match the traditional design in part a or the newer design in part b?

SECTION 10.3 Finding Linear Models

Finding Linear Models

TAKE NOTE

When creating a linear model, the slope is the quantity that is expressed using the word *per*. The car discussed at the right uses 0.04 gal *per* mile. The slope is negative because the amount of fuel in the tank is decreasing.

Suppose that a car burns 0.04 gal of gas per mile driven and that the fuel tank, which holds 18 gal of gas, is full. Using this information, we can determine a linear model for the amount of fuel remaining in the gas tank after driving x miles.

Recall that a linear function is one that can be written in the form $f(x) = mx + b$, where m is the slope of the line and b is the y-intercept. The slope is the rate at which the car is using fuel, 0.04 gal per mile. Because the amount of fuel in the tank is decreasing, the slope is negative, and we have $m = -0.04$.

The amount of fuel left in the tank depends on the number of miles, x, the car has been driven. Before the car starts (that is, when $x = 0$), there are 18 gal of gas in the tank. The y-intercept is (0, 18).

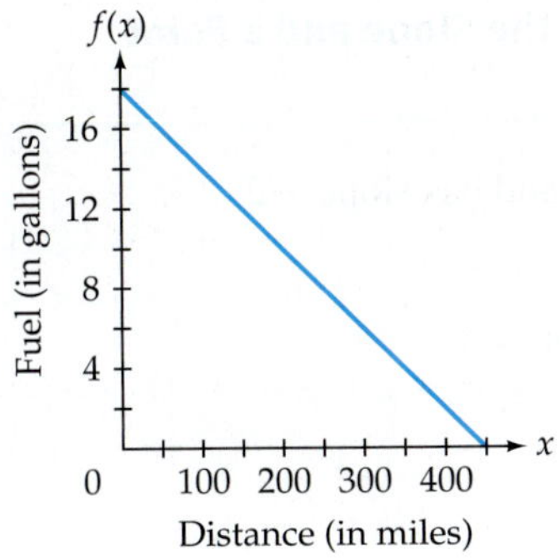

Using this information, we can create the linear function.

$f(x) = mx + b$

$f(x) = -0.04x + 18$ • Replace m by -0.04 and b by 18.

In the linear function $f(x) = -0.04x + 18$, $f(x)$ represents the amount of fuel, in gallons, remaining in the tank after driving x miles. The graph of the function is shown at the left.

The x-intercept of a graph is the point at which $f(x) = 0$. For this application, $f(x) = 0$ when 0 gal of fuel remain in the tank. Thus replacing $f(x)$ by 0 in $f(x) = -0.04x + 18$ and solving for x will give the number of miles the car can be driven before running out of gas.

$f(x) = -0.04x + 18$

$0 = -0.04x + 18$ • Replace $f(x)$ by 0.

$-18 = -0.04x$

$450 = x$

The car can travel 450 mi before running out of gas.

EXAMPLE 1 Application of Finding a Linear Model Given the Slope and y-intercept

Suppose a 20-gallon gas tank contains 2 gal of gas when a motorist decides to fill the tank. If the gas pump fills the tank at a rate of 0.1 gal per second, find a linear function that models the amount of fuel in the tank t seconds after fueling begins.

Solution

When fueling begins, at $t = 0$, there are 2 gal of gas in the tank. Therefore, the y-intercept is $(0, 2)$. The slope is the rate at which fuel is being added to the tank. Because the amount of fuel in the tank is increasing, the slope is positive and we have $m = 0.1$ gal/s. To find the linear function, replace m and b by their respective values.

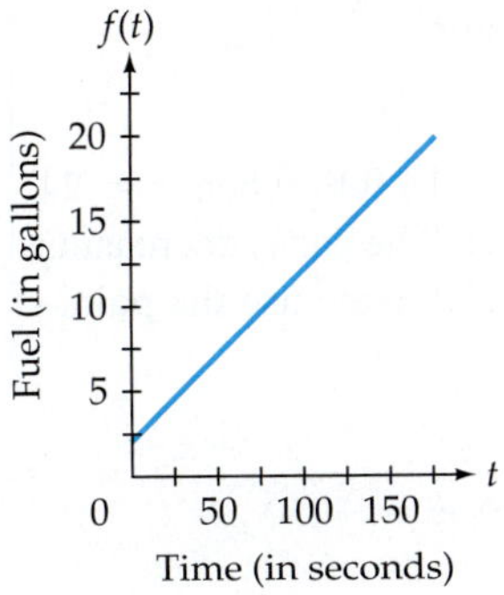

$f(t) = mt + b$

$f(t) = 0.1t + 2$ • Replace m by 0.1 and b by 2.

The linear function is $f(t) = 0.1t + 2$, where $f(t)$ is the number of gallons of fuel in the tank t seconds after fueling begins.

CHECK YOUR PROGRESS 1 The boiling point of water at sea level is 100°C. The boiling point decreases 3.5°C per 1-kilometer increase in altitude. Find a linear function that gives the boiling point of water as a function of altitude.

Solution See page S35.

For each of the previous examples, we determined the y-intercept from the given information. This information enabled us to determine b for the linear function $f(x) = mx + b$. In some cases, a point other than the y-intercept is given. In such a case, the *point–slope formula* is used to find the equation of the line.

TAKE NOTE

Using parentheses may help when substituting into the point–slope formula.

$$y - y_1 = m(x - x_1)$$
$$\downarrow \quad \downarrow \quad \downarrow$$
$$y - (\ ) = (\ )[x - (\ )]$$

Point–Slope Formula of a Straight Line

Let (x_1, y_1) be a point on a line and let m be the slope of the line. Then the equation of the line can be found using the point–slope formula

$$y - y_1 = m(x - x_1)$$

EXAMPLE 2 **Find the Equation of a Line Given the Slope and a Point on the Line**

Find the equation of the line that passes through $(1, -3)$ and has slope -2.

Solution

$$y - y_1 = m(x - x_1) \quad \bullet \text{ Use the point–slope formula.}$$
$$y - (-3) = -2(x - 1) \quad \bullet\ m = -2, (x_1, y_1) = (1, -3)$$
$$y + 3 = -2x + 2$$
$$y = -2x - 1$$

Note that we wrote the equation of the line as $y = -2x - 1$. We could also write the equation in function notation as $f(x) = -2x - 1$.

TAKE NOTE

Recall that $f(x)$ and y are different symbols for the same quantity, the value of the function at x.

CHECK YOUR PROGRESS 2 Find the equation of the line that passes through $(-2, 2)$ and has slope $-\frac{1}{2}$.

Solution See page S35.

INSTRUCTOR NOTE

Once the applications of finding a linear model have been presented, students are prepared to work on the Excursion on page 580, A Linear Business Model.

EXAMPLE 3 **Application of Finding a Linear Model Given a Point and the Slope**

Based on data from the *Kelley Blue Book*, the value of a certain car decreases approximately \$250/month. If the value of the car 2 years after it was purchased was \$14,000, find a linear function that models the value of the car after x months of ownership. Use this function to find the value of the car after 3 years of ownership.

Solution

Let V represent the value of the car after x months. Then $V = 14{,}000$ when $x = 24$ (2 years is 24 months). A solution of the equation is $(24, 14{,}000)$. The car is decreasing in value at a rate of \$250 per month. Therefore, the slope is -250. Now use the point–slope formula to find the linear equation that models the function.

$$V - V_1 = m(x - x_1)$$
$$V - 14{,}000 = -250(x - 24) \quad \bullet\ x_1 = 24, V_1 = 14{,}000, m = -250$$
$$V - 14{,}000 = -250x + 6000$$
$$V = -250x + 20{,}000$$

A linear function that models the value of the car after x months of ownership is $V(x) = -250x + 20{,}000$.

To find the value of the car after 3 years (36 months), evaluate the function when $x = 36$.

$$V(x) = -250x + 20{,}000$$
$$V(36) = -250(36) + 20{,}000 = 11{,}000$$

The value of the car is \$11,000 after 3 years of ownership.

CHECK YOUR PROGRESS 3 Whales, dolphins, and porpoises communicate using high-pitched sounds that travel through the water. The speed at which sound travels depends on many factors, one of which is the depth of the water. At approximately 1000 m below sea level, the speed of sound is 1480 m/s. The speed of sound increases at a constant rate of 0.017 m/s for each additional meter below 1000 m. Write a linear function for the speed of sound in terms of the number of meters below sea level. Use the function to approximate the speed of sound 2500 m below sea level.

Solution See page S35.

The next example shows how to find the equation of a line given two points on the line.

TAKE NOTE

There are many ways to find the equation of a line. However, in every case, there must be enough information to determine a point on the line and the slope of the line. When you are doing problems of this type, look for different ways that information may be presented. For instance, in Example 4, even though the slope of the line is not given, knowing two points enables us to find the slope.

INSTRUCTOR NOTE

You might have students use the point (3, 2) in the point–slope formula at the right and verify that the same equation results.

EXAMPLE 4 Find the Equation of a Line Given Two Points on the Line

Find the equation of the line that passes through $P_1(6, -4)$ and $P_2(3, 2)$.

Solution

Find the slope of the line between the two points.

$$m = \frac{y_2 - y_1}{x_2 - x_1} = \frac{2 - (-4)}{3 - 6} = \frac{6}{-3} = -2$$

Use the point–slope formula to find the equation of the line. We will use the point $(6, -4)$. [We could use the point (3, 2); the equation of the line would be the same.]

$$y - y_1 = m(x - x_1)$$

$$y - (-4) = -2(x - 6) \quad \bullet\ m = -2, x_1 = 6, y_1 = -4$$

$$y + 4 = -2x + 12$$

$$y = -2x + 8$$

CHECK YOUR PROGRESS 4 Find the equation of the line that passes through $P_1(-2, 3)$ and $P_2(4, 1)$.

Solution See page S35.

MATH MATTERS Perspective: Using Straight Lines in Art

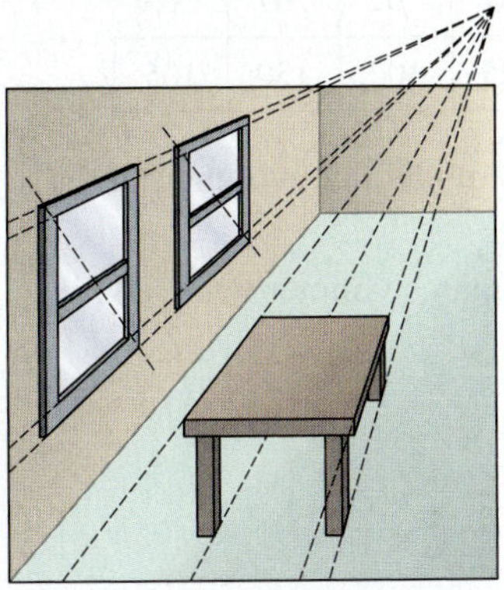

Many paintings we see today have a three-dimensional quality to them, even though they are painted on a flat surface. This was not always the case. It wasn't until the Renaissance that artists started to paint "in perspective." Using lines is one way to create this perspective. Here is a simple example.

Draw a dot, called the *vanishing point*, and a rectangle on a piece of paper. Draw windows as shown. To keep the perspective accurate, the lines through opposite corners of the windows should be parallel. A table in proper perspective is created in the same way.

This method of creating perspective was employed by Leonardo da Vinci. Use the Internet to find and print a copy of his painting *The Last Supper*. Using a ruler, see whether you can find the vanishing point by drawing two lines along the top edges of the windows on the sides of the painting.

EXAMPLE 5 Find a Linear Model

Sodium thiosulfate is used by photographers to develop some types of film. The amount of this chemical that will dissolve in water depends on the temperature of the water. The table below gives the number of grams of sodium thiosulfate that will dissolve in 100 ml of water at various temperatures.

Temperature, x, in degrees Celsius	20	35	50	60	75	90	100
Sodium thiosulfate dissolved, y, in grams	50	80	120	145	175	205	230

a. Use the points (20, 50) and (100, 230) to find a linear model that will predict the number of grams of sodium thiosulfate that will dissolve in water at a temperature of x degrees Celsius.

b. How many grams of sodium thiosulfate does the model predict will dissolve in 100 ml of water when the temperature is 70°C? Round to the nearest tenth of a gram.

Solution

a. Find the slope of the line between the two points.

$$\text{Slope} = \frac{230 - 50}{100 - 20} = \frac{180}{80} = 2.25$$

Use the point–slope formula to find the equation of the line.

$$y - y_1 = m(x - x_1)$$
$$y - 50 = 2.25(x - 20)$$
$$y - 50 = 2.25x - 45$$
$$y = 2.25x + 5$$

The linear model is $y = 2.25x + 5$.

b. Evaluate the linear model when $x = 70$.

$$y = 2.25x + 5$$
$$y = 2.25(70) + 5$$
$$y = 162.5$$

The model predicts that 162.5 g of sodium thiosulfate will dissolve in 100 ml of water at a temperature of 70°C.

CHECK YOUR PROGRESS 5 The heights and weights of women swimmers on a college swim team are given in the table below.

Height, x, in inches	68	64	65	67	62	67	65
Weight, y, in pounds	132	108	108	125	102	130	105

a. Use the points (68, 132) and (62, 102) to find a linear model that will predict the weight, in pounds, of a swimmer given her height.

b. Use your linear model to estimate the weight of a woman swimmer who is 63 in. tall. Round to the nearest pound.

Solution See page S35.

EXCURSION

A Linear Business Model

Two people decide to open a business reconditioning toner cartridges for copy machines. They rent a building for \$7000/year and estimate that building maintenance, taxes, and insurance will cost \$6500/year. Each person wants to make \$12/h in the first year and will work 10 h per day for 260 days of the year. Assume that it costs \$28 to restore a cartridge and that the restored cartridge can be sold for \$45.

EXCURSION EXERCISES

1. Write a linear function for the total cost C to operate the business and restore n cartridges during the first year, not including the hourly wage the owners wish to earn.
2. Write a linear function for the total revenue R the business will earn during the first year by selling n cartridges.

3. How many cartridges must the business restore and sell annually to break even, not including the hourly wage the owners wish to earn?
4. How many cartridges must the business restore and sell annually for the owners to pay all expenses and earn the hourly wage they desire?
5. Suppose the entrepreneurs are successful in their business and are restoring and selling 25 cartridges each day of the 260 days the business is open. What will be their hourly wage for the year if the profit is shared equally?
6. As the company becomes successful and is selling and restoring 25 cartridges each day of the 260 days it is open, the entrepreneurs decide to hire a part-time employee. The employee works 4 h per day for 260 days and is paid \$8/h. How many additional cartridges must be restored and sold each year just to cover the cost of the new employee? You can ignore employee costs such as social security, worker's compensation, and other benefits.
7. Suppose the company decides that it could increase its business by advertising. Answer Exercises 1, 2, 3, and 5 if the owners decide to spend \$400/month on advertising.

EXERCISE SET 10.3

■ In Exercises 1 to 8, find the equation of the line that passes through the given point and has the given slope.

1. $(0, 5), m = 2$
2. $(2, 3), m = \frac{1}{2}$
3. $(-1, 7), m = -3$
4. $(0, 0), m = \frac{1}{2}$
5. $(3, 5), m = -\frac{2}{3}$
6. $(0, -3), m = -1$
7. $(-2, -3), m = 0$
8. $(4, -5), m = -2$

■ In Exercises 9 to 16, find the equation of the line that passes through the given points.

9. $(0, 2), (3, 5)$
10. $(0, -3), (-4, 5)$
11. $(0, 3), (2, 0)$
12. $(-2, -3), (-1, -2)$
13. $(2, 0), (0, -1)$
14. $(3, -4), (-2, -4)$
15. $(-2, 5), (2, -5)$
16. $(2, 1), (-2, -3)$

17. **Telecommunications** A cellular phone company offers several different service options. One option, for people who plan on using the phone only in emergencies, costs the user \$4.95/month plus \$0.59/min for each minute the phone is used. Write a linear function for the monthly cost of the phone in terms of the number of minutes the phone is used. Use the function to find the cost of using the cellular phone for 13 min in 1 month.

18. **Hotel Industry** The operator of a hotel estimates that 500 rooms per night will be rented if the room rate per night is \$100. For each \$10 increase in the price of a room, six fewer rooms per night will be rented. Determine a linear function that will predict the number of rooms that will be rented per night for a given price per room. Use this model to predict the number of rooms that will be rented if the room rate is \$150/night.

19. **Construction** A general building contractor estimates that the cost to build a new home is \$30,000 plus \$85 for each square foot of floor space in the house. Determine a linear function that gives the cost of building a house that contains x square feet of floor space. Use this model to determine the cost to build a house that contains 1800 ft^2 of floor space.

20. **Compensation** An account executive receives a base salary plus a commission. On \$20,000 in monthly sales, an account executive would receive compensation of \$2700. On \$50,000 in monthly sales, an account executive would receive compensation of \$4500. Determine a linear function that yields the compensation of a sales executive for x dollars in monthly sales. Use this model to determine the compensation of an account executive who has \$85,000 in monthly sales.

21. **Car Sales** A manufacturer of economy cars has determined that 50,000 cars per month can be sold at a price of \$18,000 per car. At a price of \$17,500, the number of cars sold per month would increase to 55,000. Determine a linear function that predicts the number of cars that will be sold at a price of x dollars. Use this model to predict the number of cars that will be sold at a price of \$17,000.

22. **Calculator Sales** A manufacturer of graphing calculators has determined that 10,000 calculators per week will be sold at a price of \$95 per calculator. At a price of \$90, it is estimated that 12,000 calculators will be sold. Determine a linear function that predicts the number of calculators that will be sold per week at a price of x dollars. Use this model to predict the number of calculators that will be sold at a price of \$75.

23. Test Scores The data in the table below show five students' reading test grade and final exam grade in a history class.

Reading test score, x	8.5	9.4	10.0	11.4	12.0
History final exam grade, y	64	68	76	87	92

a. Use the points (8.5, 64) and (10.0, 76) to find a linear model that will predict a student's final exam grade given his or her reading test score.

b. Use your linear model to estimate a student's final exam grade in the history class given that the student's reading test grade was 10.5. Round to the nearest whole number.

24. Stress A research hospital did a study on the relationship between stress and diastolic blood pressure. The results from eight patients are given in the table below. Blood pressure values are measured in milliliters of mercury.

Stress test score, x	55	62	58	78	92	88	75	80
Blood pressure, y	70	85	72	85	96	90	82	85

a. Use the points (55, 70) and (92, 96) to find a linear model that will predict the blood pressure of a patient for a given stress test score. Round the slope to the nearest ten-thousandth.

b. Use your linear model to estimate the diastolic blood pressure of a person whose stress test score was 85. Round to the nearest whole number.

25. Sports The data in the table below show the amount of water, in milliliters, a professional tennis player loses for various times, in minutes, of play during a tennis match.

Time of workout (in minutes), x	10	20	30	40	50	60
Water lost (in milliliters), y	600	900	1200	1500	2000	2300

a. Use the points (10, 600) and (60, 2300) to find a linear model that will predict the water lost by a player who has been playing for x minutes.

b. Use your linear model to estimate the amount of water lost after playing a tennis match for 25 min. Round to the nearest milliliter.

26. Fuel Efficiency An automotive engineer studied the relationship between the speed of a car and the number of miles traveled per gallon of fuel consumed at that speed. The results of the study are shown in the table below.

Speed (in miles per hour), x	40	25	30	50	60	80	55	35	45
Consumption (in miles traveled per gallon), y	26	27	28	24	22	21	23	27	25

a. Use the points (25, 27) and (60, 22) to find a linear model that will predict the fuel consumption of a car traveling at x miles per hour.

b. Use your linear model to estimate the expected number of miles traveled per gallon of fuel consumed for a car traveling at 65 mph. Round to the nearest mile per gallon.

27. **Meteorology** A meteorologist studied the maximum temperatures at various latitudes for January of a certain year. The results of the study are shown in the table below.

Latitude (in °N), x	22	30	36	42	56	51	48
Maximum temperature (in °F), y	80	65	47	54	21	44	52

a. Use the points (22, 80) and (56, 21) to find a linear model that will predict the maximum temperature at a latitude of x degrees north of the Equator.

b. Use your linear model to estimate the expected maximum temperature in January at a latitude of 45°N. Round to the nearest degree.

28. **Zoology** A zoologist studied the running speeds of animals in terms of the animals' body lengths. The results of the study are shown in the table below.

Body length (in centimeters), x	1	9	15	16	24	25	60
Running speed (in meters per second), y	1	2.5	7.5	5	7.4	7.6	20

a. Use the points (9, 2.5) and (60, 20) to find a linear model that will predict the running speed of an animal with a body length of x centimeters.

b. Use your linear model to estimate the expected running speed of a deer mouse, whose body length is 10 cm. Round to the nearest tenth of a meter per second.

EXTENSIONS

29. A line contains the points $(4, -1)$ and $(2, 1)$. Find the coordinates of three other points on the line.

30. If f is a linear function for which $f(1) = 3$ and $f(-1) = 5$, find $f(4)$.

31. The ordered pairs $(0, 1)$, $(4, 9)$, and $(3, n)$ are solutions of the same linear equation. Find n.

32. The ordered pairs $(2, 2)$, $(-1, 5)$, and $(3, n)$ are solutions of the same linear equation. Find n.

33. Is there a linear function that contains the ordered pairs $(2, 4)$, $(-1, -5)$, and $(0, 2)$? If so, find the function and explain why there is such a function. If not, explain why there is no such function.

34. Is there a linear function that contains the ordered pairs $(5, 1)$, $(4, 2)$, and $(0, 6)$? If so, find the function and explain why there is such a function. If not, explain why there is no such function.

35. **Travel** Assume that the maximum speed your car will travel varies linearly with the steepness of the hill it is climbing or descending. If the hill is 5° up, your car can travel 77 km/h. If the hill is 2° down $(-2°)$, your car can travel 154 km/h. When your car's top speed is 99 km/h, how steep is the hill? State your answer in degrees, and note whether the car is climbing or descending.

SECTION 10.4 Quadratic Functions

Properties of Quadratic Functions

Recall that a linear function is a function of the form $f(x) = mx + b$. The graph of a linear function is a straight line with slope m and y-intercept $(0, b)$.

A **quadratic function** in a single variable x is a function of the form $f(x) = ax^2 + bx + c$, $a \neq 0$. Examples of quadratic functions are given below.

$f(x) = x^2 - 3x + 1$ • $a = 1, b = -3, c = 1$

$g(t) = -2t^2 - 4$ • $a = -2, b = 0, c = -4$

$h(p) = 4 - 2p - p^2$ • $a = -1, b = -2, c = 4$

$f(x) = 2x^2 + 6x$ • $a = 2, b = 6, c = 0$

POINT OF INTEREST

The suspension cables of some bridges, such as the Golden Gate Bridge, have the shape of a parabola.

Hulton Archive/Archive Photos/Getty Images

The photo above shows the roadway of the bridge being assembled in sections and attached to suspender ropes. The bridge was opened to vehicles on May 28, 1937.

TAKE NOTE

The axis of symmetry is a vertical line. The vertex of the parabola is the point that lies on the axis of symmetry.

The graph of a quadratic function in a single variable x is a **parabola**. The graphs of two quadratic functions are shown below.

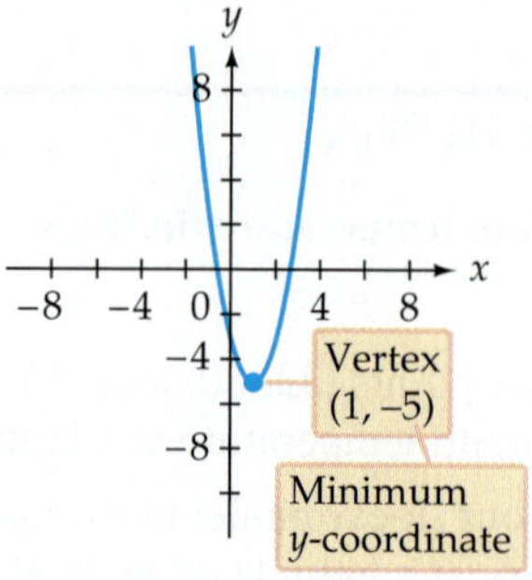

$a > 0$

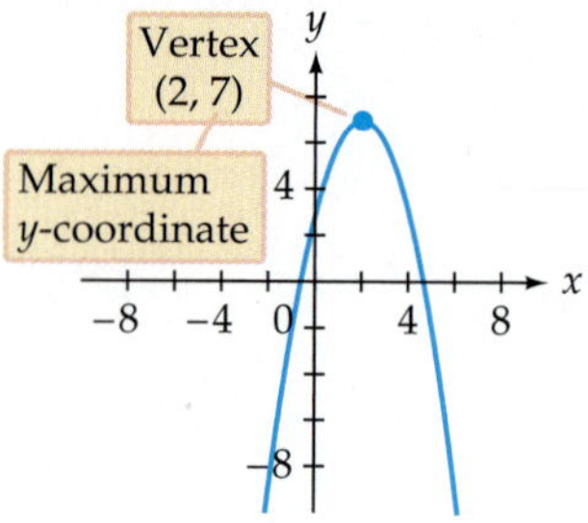

$a < 0$

For the figure on the left above, $f(x) = 2x^2 - 4x - 3$. The value of a is *positive* $(a = 2)$ and the graph opens up. For the figure on the right above, $f(x) = -x^2 + 4x + 3$. The value of a is *negative* $(a = -1)$ and the graph opens down. The point at which the graph of a parabola has a minimum or a maximum is called the *vertex* of the parabola. The **vertex** of a parabola is the point with the smallest y-coordinate when $a > 0$ and the point with largest y-coordinate when $a < 0$.

The **axis of symmetry** of the graph of a quadratic function is a vertical line that passes through the vertex of the parabola. To understand the concept of the axis of symmetry of a graph, think of folding the graph along that line. The two portions of the graph will match up.

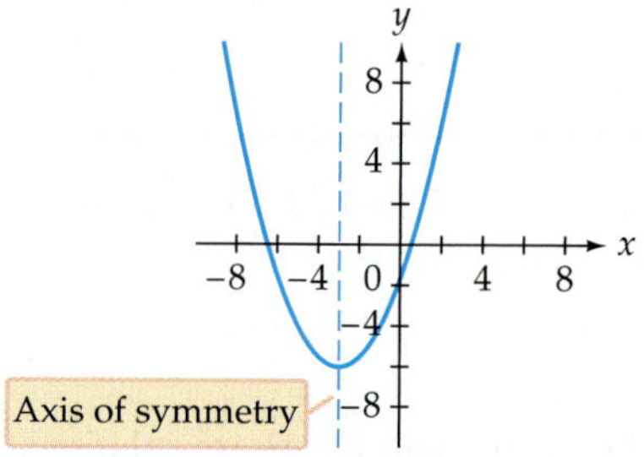

The following formula enables us to determine the vertex of a parabola.

Vertex of a Parabola

Let $f(x) = ax^2 + bx + c$ be the equation of a parabola. The coordinates of the vertex are

$$\left(-\frac{b}{2a}, f\left(-\frac{b}{2a}\right)\right)$$

EXAMPLE 1 Find the Vertex of a Parabola

Find the vertex of the parabola whose equation is $y = -3x^2 + 6x + 1$.

Solution

$$x = -\frac{b}{2a} = -\frac{6}{2(-3)} = 1$$

- Find the x-coordinate of the vertex. $a = -3$, $b = 6$

$$y = -3x^2 + 6x + 1$$
$$y = -3(1)^2 + 6(1) + 1$$
$$y = 4$$

- Find the y-coordinate of the vertex by replacing x by 1 and solving for y.

The vertex is $(1, 4)$.

- The x-coordinate of the vertex is 1. The y-coordinate of the vertex is 4.

CHECK YOUR PROGRESS 1 Find the vertex of the parabola whose equation is $y = x^2 - 2$.

Solution See page S35.

MATH MATTERS Paraboloids

The movie *Contact* was based on a novel by astronomer Carl Sagan. In the movie, Jodie Foster plays an astronomer who is searching for extraterrestrial intelligence. One scene from the movie takes place at the Very Large Array (VLA) in New Mexico. The VLA consists of 27 large radio telescopes whose dishes are paraboloids, the 3D version of a parabola. A parabolic shape is used because of the following reflective property: When the parallel rays of light, or radio waves, strike the surface of a parabolic mirror whose axis of symmetry is parallel to these rays, they are reflected to the same point.

Jeffrey M. Frank/Shutterstock.com

cbpix/Shutterstock.com

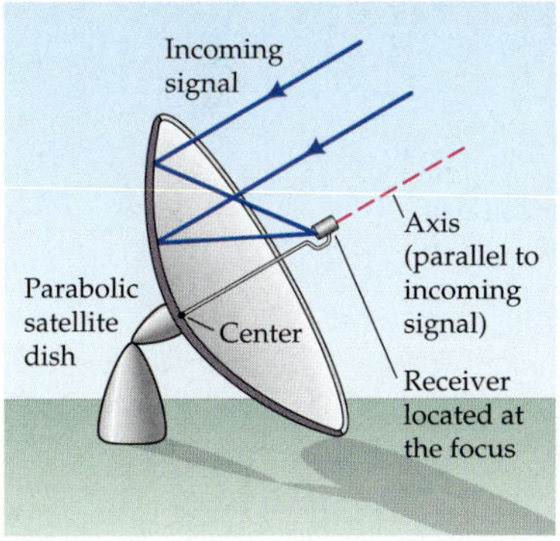

The photos above show the layout of the radio telescopes of the VLA and a more detailed picture of one of the telescopes. The figure at the far right shows the reflective property of a parabola. Note that all the incoming rays are reflected to the focus.

The reflective property of a parabola is also used in optical telescopes and headlights on a car. In the case of headlights, the bulb is placed at the focus and the light is reflected along parallel rays from the reflective surface of the headlight, thereby making a more concentrated beam of light.

x-Intercepts of Parabolas

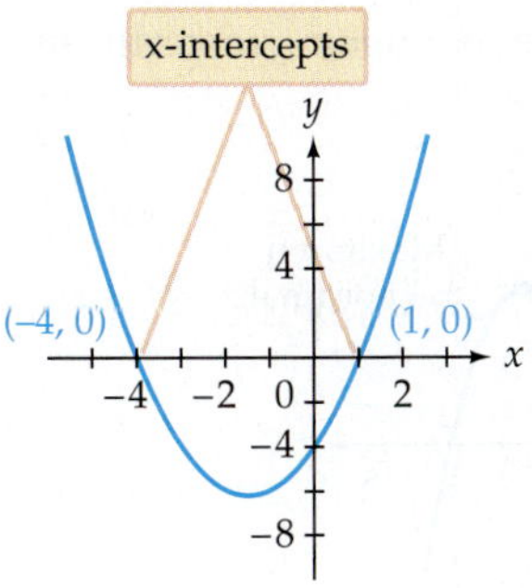

Recall that a point at which a graph crosses the x- or y-axis is called an *intercept* of the graph. The x-intercepts of the graph of an equation can be found by setting $y = 0$.

The graph of $y = x^2 + 3x - 4$ is shown at the left. The points whose coordinates are $(-4, 0)$ and $(1, 0)$ are x-intercepts of the graph. We can algebraically determine the x-intercepts by solving an equation.

EXAMPLE 2 **Find the *x*-intercepts of a Parabola**

Find the x-intercepts of the graph of the parabola given by the equation.

a. $y = 4x^2 + 4x + 1$

b. $y = x^2 + 2x - 2$

Solution

a. $y = 4x^2 + 4x + 1$

$0 = 4x^2 + 4x + 1$ • Let $y = 0$.

$0 = (2x + 1)(2x + 1)$ • Solve for x by factoring.

$2x + 1 = 0 \qquad 2x + 1 = 0$ • Set each factor equal to zero.

$x = -\frac{1}{2} \qquad x = -\frac{1}{2}$ • Solve each equation for x.

TAKE NOTE

When a quadratic equation has two solutions that are the same number, the solution is called a **double root** of the equation.

The x-intercept is $\left(-\frac{1}{2}, 0\right)$.

When a quadratic equation has a double root, the graph of the quadratic equation just touches the x-axis and is said to be *tangent* to the x-axis. See the graph at the right.

Some graphs of parabolas do not have any x-intercepts.

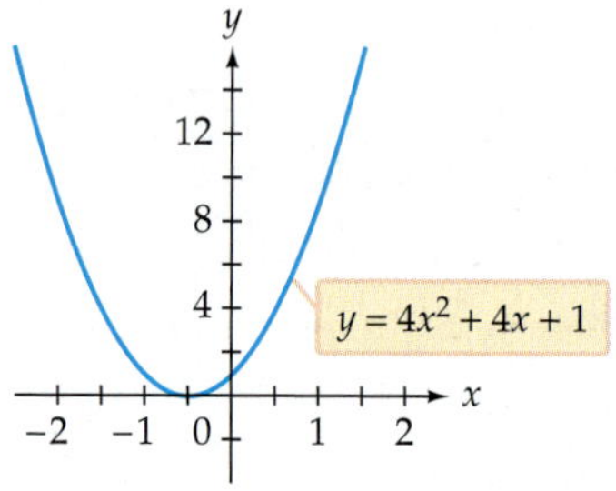

b. $y = x^2 + 2x - 2$

$0 = x^2 + 2x - 2$ • Let $y = 0$. Try to solve the quadratic equation by factoring (as we did in part a). If that cannot be done easily, then use the quadratic formula.

$x = \frac{-b \pm \sqrt{b^2 - 4ac}}{2a}$ • The trinomial $x^2 + 2x - 2$ is nonfactorable over the integers. Use the quadratic formula to solve for x. $a = 1, b = 2, c = -2$

$= \frac{-(2) \pm \sqrt{(2)^2 - 4(1)(-2)}}{2(1)}$

$= \frac{-2 \pm \sqrt{4 + 8}}{2} = \frac{-2 \pm \sqrt{12}}{2}$

$= \frac{-2 \pm 2\sqrt{3}}{2} = -1 \pm \sqrt{3}$

The intercepts are $\left(-1 + \sqrt{3}, 0\right)$ and $\left(-1 - \sqrt{3}, 0\right)$.

CHECK YOUR PROGRESS 2 Find the x-intercepts of the graph of the parabola given by the equation.

a. $y = 2x^2 - 5x + 2$ **b.** $y = x^2 + 4x + 4$

Solution See page S35.

POINT OF INTEREST

Calculus is a branch of mathematics that can be used to calculate the maxima or minima of functions other than quadratic functions. Such calculations are very important in applied mathematics. For instance, an automotive engineer wants to design a car whose shape will *minimize* the effect of air flow. Another engineer tries to *maximize* the efficiency of a car's engine. Similarly, an economist may try to determine which business practices will *minimize* cost and *maximize* profit.

Minimum and Maximum of a Quadratic Function

Note that for the graphs below, when $a > 0$, the vertex is the point with the minimum y-coordinate. When $a < 0$, the vertex is the point with the maximum y-coordinate.

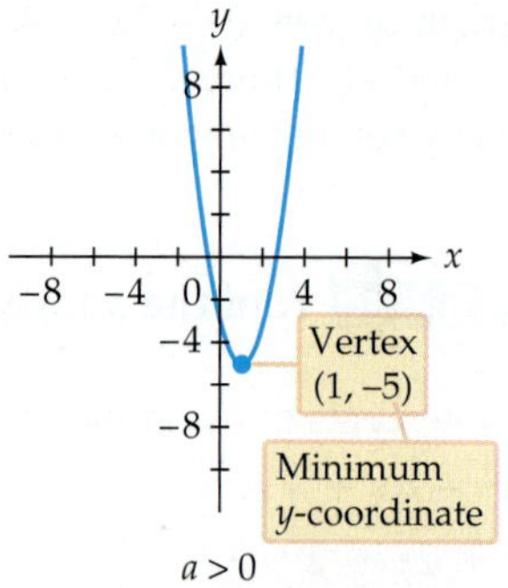

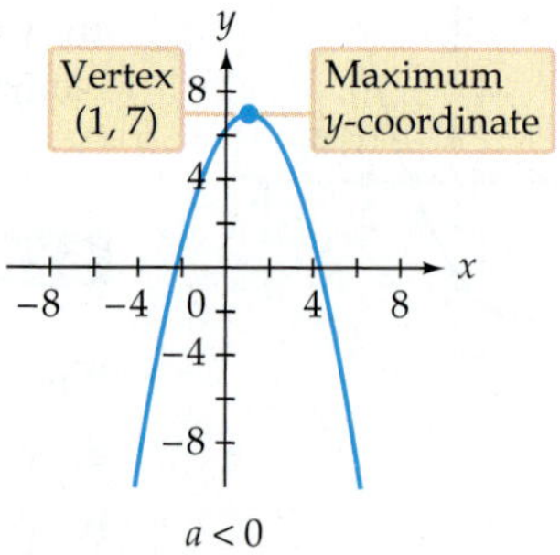

Finding the minimum or maximum value of a quadratic function is a matter of finding the vertex of the graph of the function.

INSTRUCTOR NOTE

Students often have difficulty distinguishing between the maximum (or minimum) *value of the function* and the *value of x* that produces the maximum (or minimum).

EXAMPLE 3 Find the Minimum or Maximum Value of a Quadratic Function

Find the maximum value of $f(x) = -2x^2 + 4x + 3$.

Solution

$$x = -\frac{b}{2a} = -\frac{4}{2(-2)} = 1$$

• Find the x-coordinate of the vertex. $a = -2$, $b = 4$

$$f(x) = -2x^2 + 4x + 3$$
$$f(1) = -2(1)^2 + 4(1) + 3$$

• Find the y-coordinate of the vertex by replacing x by 1 and evaluating.

$$f(1) = 5$$

• The vertex is $(1, 5)$.

The maximum value of the function is 5, the y-coordinate of the vertex.

CHECK YOUR PROGRESS 3 Find the minimum value of $f(x) = 2x^2 - 3x + 1$.

Solution See page S35.

QUESTION The vertex of a parabola that opens up is $(-4, 7)$. What is the minimum value of the function?

Applications of Quadratic Functions

EXAMPLE 4 Application of Finding the Minimum of a Quadratic Function

A mining company has determined that the cost c, in dollars per ton, of mining a mineral is given by $c(x) = 0.2x^2 - 2x + 12$, where x is the number of tons of the mineral mined. Find the number of tons of the mineral that should be mined to minimize the cost. What is the minimum cost?

Solution

To find the number of tons of the mineral that should be mined to minimize the cost and to find the minimum cost, find the x- and y-coordinates of the vertex of the graph of $c(x) = 0.2x^2 - 2x + 12$.

$$x = -\frac{b}{2a} = -\frac{-2}{2(0.2)} = 5$$

• Find the x-coordinate of the vertex. $a = 0.2$, $b = -2$

To minimize the cost, 5 tons of the mineral should be mined.

$$c(x) = 0.2x^2 - 2x + 12$$

• Find the y-coordinate of the vertex by replacing x by 5 and evaluating.

$$c(5) = 0.2(5)^2 - 2(5) + 12$$
$$c(5) = 7$$

The minimum cost per ton is \$7.

CHECK YOUR PROGRESS 4 Find two numbers whose difference is 10 and whose product is a minimum. What is the minimum product of the two numbers?

Solution See page S36.

ANSWER The minimum value of the function is 7, the y-coordinate of the vertex.

EXAMPLE 5 **Application of Finding the Maximum of a Quadratic Function**

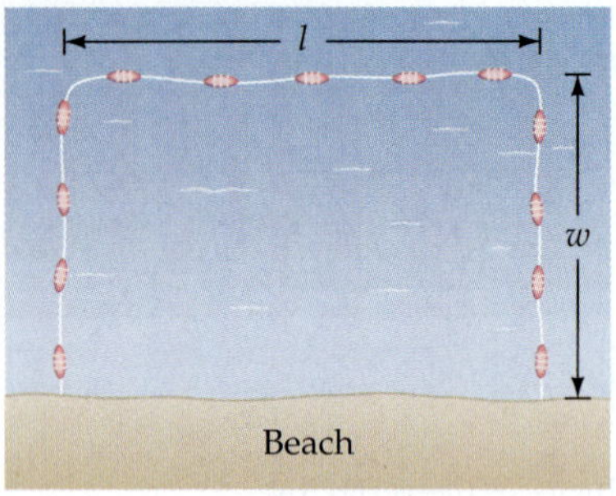

A lifeguard has 600 ft of rope with buoys attached with which to lay out a rectangular swimming area on a lake. If the beach forms one side of the rectangle, find the dimensions of the rectangle that will enclose the greatest swimming area.

Solution

Let l represent the length of the rectangle, let w represent the width, and let A (which is unknown) represent the area. See the figure at the left. Use these variables to write expressions for the perimeter and area of the rectangle.

Perimeter: $w + l + w = 600$ • There are 600 ft of rope.

$2w + l = 600$

Area: $A = lw$

The goal is to maximize A. To do this, first write A in terms of a single variable. This can be accomplished by solving $2w + l = 600$ for l and then substituting into $A = lw$.

$2w + l = 600$ • Solve for l.

$l = -2w + 600$

$A = lw$

$= (-2w + 600)w$ • Substitute $-2w + 600$ for l.

$A = -2w^2 + 600w$ • Multiply. This is now a quadratic equation. $a = -2, b = 600$

Find the w-coordinate of the vertex.

$$w = -\frac{b}{2a} = -\frac{600}{2(-2)} = 150$$ • Find the w-coordinate of the vertex. $a = -2, b = 600$

The width is 150 ft. To find l, replace w by 150 in $l = -2w + 600$ and solve for l.

$l = -2w + 600$

$l = -2(150) + 600 = -300 + 600 = 300$

The length is 300 ft. The dimensions of the rectangle with maximum area are 150 ft by 300 ft.

CHECK YOUR PROGRESS 5 A mason is forming a rectangular floor for a storage shed. The perimeter of the rectangle is 44 ft. What dimensions will give the floor a maximum area?

Solution See page S36.

EXCURSION

Reflective Properties of a Parabola

The shape of the graph of a parabola is based on the following geometric definition of a parabola.

Parabola

A parabola is the set of points in the plane that are equidistant from a fixed line (the **directrix**) and a fixed point (the **focus**) not on the line.

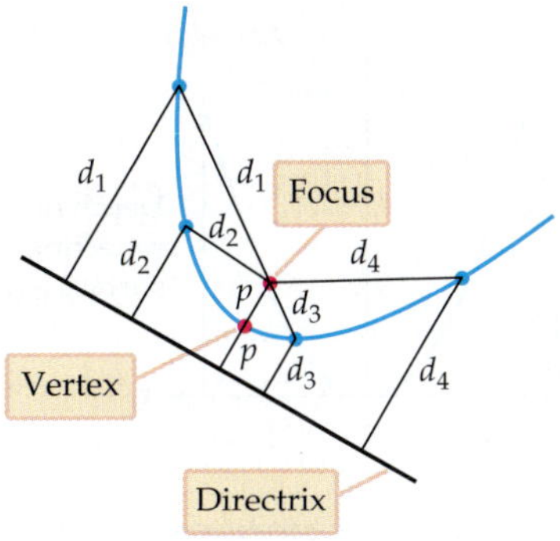

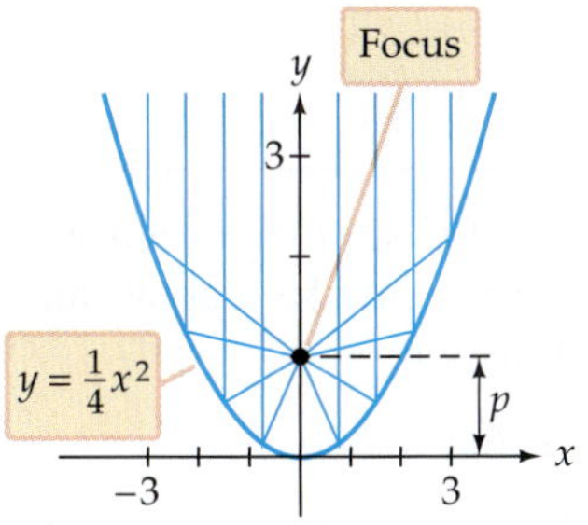

This geometric definition of a parabola is illustrated in the figure at the left. Basically, for a point to be on a parabola, the distance from the point to the focus must equal the distance from the point to the directrix. Note also that the vertex is halfway between the focus and the directrix. This distance is traditionally labeled p.

The general form of the equation of a parabola that opens up with vertex at the origin can be written in terms of the distance p between the vertex and focus as $\mathbf{y = \frac{1}{4p}x^2}$. For this equation, the coordinates of the focus are $(0, p)$. For instance, to find the coordinates of the focus for $y = \frac{1}{4}x^2$, let $\frac{1}{4p} = \frac{1}{4}$ and solve for p.

$$\frac{1}{4p} = \frac{1}{4}$$

$$1 = \frac{4p}{4}$$ • Multiply each side of the equation by $4p$.

$$1 = p$$ • Simplify the right side of the equation.

The coordinates of the focus are $(0, 1)$.

EXCURSION EXERCISES

1. Find the coordinates of the focus for the parabola whose equation is $y = 0.4x^2$.
2. Optical telescopes work on the same principle as radio telescopes (see Math Matters, page 585) except that light hits a mirror that has been shaped into a paraboloid. The light is reflected to the focus, where another mirror reflects it through a lens to the observer. See the diagram below.

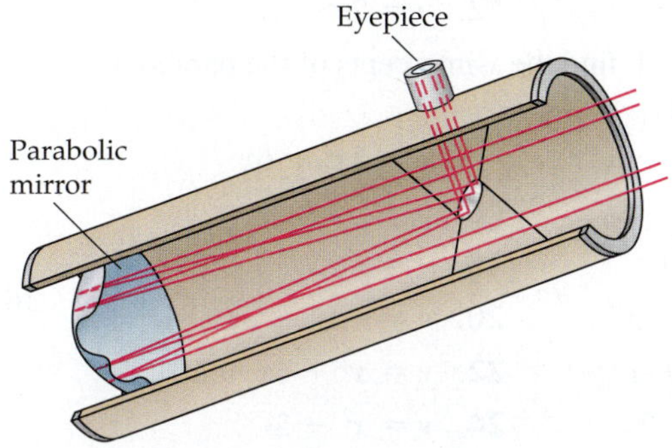

The telescope at the Palomar Observatory in California has a parabolic mirror. The circle at the top of the parabolic mirror has a 200-inch diameter. An equation that approximates the parabolic cross-section of the surface of the mirror is $y = \frac{1}{2639}x^2$, where x and y are measured in inches. How far is the focus from the vertex of the mirror?

Bill W. Marsh/Photo Researchers/Science Source

Palomar Observatory with the shutters open

If a point on a parabola whose vertex is at the origin is known, then the equation of the parabola can be found. For instance, if $(4, 1)$ is a point on a parabola with vertex at the origin, then we can find the equation as follows:

$$y = \frac{1}{4p}x^2$$ • Begin with the general form of the equation of a parabola.

$$1 = \frac{1}{4p}(4)^2$$ • The known point is $(4, 1)$. Replace x by 4 and y by 1.

$$1 = \frac{4}{p}$$ • Solve for p.

$$p = 4$$ • $p = 4$ in the equation $y = \frac{1}{4p}x^2$.

The equation of the parabola is $y = \frac{1}{16}x^2$.

3. Find a flashlight and measure the diameter of its lens cover and the depth of the reflecting parabolic surface. See the diagram at the right.

a. If coordinate axes are set up as shown, find the equation of the parabola.

b. Find the location of the focus. Explain why the filament of the light bulb should be placed at this point.

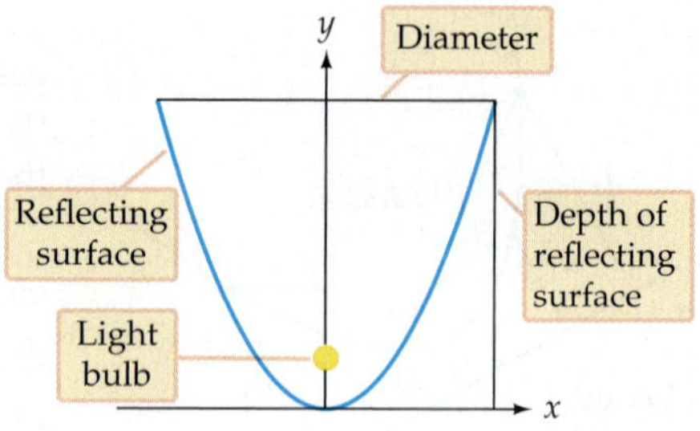

EXERCISE SET 10.4

■ In Exercises 1 to 12, find the vertex of the graph of the equation.

1. $y = x^2 - 2$ **2.** $y = x^2 + 2$

3. $y = -x^2 - 1$ **4.** $y = -x^2 + 3$

5. $y = -\frac{1}{2}x^2 + 2$ **6.** $y = \frac{1}{2}x^2$

7. $y = 2x^2 - 1$ **8.** $y = x^2 - 2x$

9. $y = x^2 - x - 2$ **10.** $y = x^2 - 3x + 2$

11. $y = 2x^2 - x - 5$ **12.** $y = 2x^2 - x - 3$

■ In Exercises 13 to 24, find the x-intercepts of the parabola given by the equation.

13. $y = 2x^2 - 4x$ **14.** $y = 3x^2 + 6x$

15. $y = 4x^2 + 11x + 6$ **16.** $y = x^2 - 9$

17. $y = x^2 + 2x - 1$ **18.** $y = x^2 + 4x - 3$

19. $y = -x^2 - 4x - 5$ **20.** $y = -x^2 - 2x + 1$

21. $y = -x^2 + 4x + 1$ **22.** $y = x^2 + 6x + 10$

23. $y = 2x^2 - 5x - 3$ **24.** $y = x^2 - 2$

■ In Exercises 25 to 32, find the minimum or maximum value of the quadratic function. State whether the value is a minimum or a maximum.

25. $f(x) = x^2 - 2x + 3$

26. $f(x) = 2x^2 + 4x$

27. $f(x) = -2x^2 + 4x - 5$

28. $f(x) = 3x^2 + 3x - 2$

29. $f(x) = x^2 - 5x + 3$

30. $f(x) = 2x^2 - 3x$

31. $f(x) = -x^2 - x + 2$

32. $f(x) = -3x^2 + 4x - 2$

33. The graph of which of the following equations is a parabola with the largest minimum value?

a. $y = x^2 - 2x - 3$

b. $y = x^2 - 10x + 20$

c. $y = 3x^2 - 1$

34. Sports The height s, in feet, of a ball thrown upward at an initial speed of 64 ft/s from a cliff 50 ft above an ocean beach is given by the function

$$s(t) = -16t^2 + 64t + 50$$

where t is the time in seconds after the ball is released. Find the maximum height above the beach that the ball will attain.

35. Sports The height s, in feet, of a ball thrown upward at an initial speed of 80 ft/s from a platform 50 ft high is given by

$$s(t) = -16t^2 + 80t + 50$$

where t is the time in seconds after the ball is released. Find the maximum height above the ground that the ball will attain.

36. Business A tour operator believes that the profit P, in dollars, from selling x tickets is given by

$$P(x) = 40x - 0.25x^2$$

Using this model, what is the maximum profit the tour operator can expect?

37. Manufacturing A manufacturer of camera lenses estimates that the average monthly cost C of producing camera lenses is given by the function

$$C(x) = 0.1x^2 - 20x + 2000$$

where x is the number of lenses produced each month. Find the number of lenses the company should produce to minimize the average cost.

38. Water Treatment A pool is treated with a chemical to reduce the number of algae. The number of algae in the pool t days after the treatment can be approximated by the function

$$A(t) = 40t^2 - 400t + 500$$

How many days after treatment will the pool have the least number of algae?

39. Civil Engineering The suspension cable that supports a footbridge hangs in the shape of a parabola. The height h, in feet, of the cable above the bridge is given by the function

$$h(x) = 0.25x^2 - 0.8x + 25,\ 0 < x < 3.2$$

where x is the distance in feet measured from the left tower toward the right tower. What is the minimum height of the cable above the bridge?

40. Annual Income The net annual income I, in dollars, of a family physician can be modeled by the equation

$$I(x) = -290(x - 48)^2 + 148{,}000$$

where x is the age of the physician and $27 \le x \le 70$. Find (a) the age at which the physician's income will be a maximum and (b) the maximum income.

41. Pitching Karen is throwing an orange to her brother Saul, who is standing on the balcony of their home. The height h, in feet, of the orange above the ground t seconds after it is thrown is given by

$$h(t) = -16t^2 + 32t + 4,\ 0 \le t \le 2.118$$

If Saul's outstretched arms are 18 ft above the ground, will the orange ever be high enough so that he can catch it?

42. Football Some football fields are built in a parabolic-mound shape so that water will drain off the field. A model for the shape of such a field is given by

$$h(x) = -0.00023475x^2 + 0.0375x$$

where h is the height of the field in feet at a distance of x feet from the sideline and $0 \le x \le 159.744$. What is the maximum height of the field? Round to the nearest tenth of a foot.

43. Bellagio Water Feature A water feature at the Bellagio Hotel in Las Vegas shoots water in a parabolic path from a nozzle at ground level. See the Chapter Opener on page 555. The height h, in feet, of the water above the ground, x feet to the right of the nozzle, is given by

$$h(x) = -\frac{1}{9}x^2 + \frac{8}{3}x,\ 0 \le x \le 24$$

a. What is the height of the water 6 ft to the right of the nozzle?

b. What is the height of the water 15 ft to the right of the nozzle?

c. What is the maximum height the water attains?

44. Stopping Distance On wet concrete, the stopping distance s, in feet, of a car traveling v miles per hour is given by

$$s(v) = 0.055v^2 + 1.1v,\ s \ge 0$$

At what maximum speed could a car be traveling on wet concrete and still stop at a stop sign 44 ft away?

45. Fuel Efficiency The fuel efficiency of an average car is given by the equation

$$E(v) = -0.018v^2 + 1.476v + 3.4,\ E \ge 0$$

where E is the fuel efficiency in miles per gallon and v is the speed of the car in miles per hour.

a. What speed will yield the maximum fuel efficiency?

b. What is the maximum fuel efficiency?

46. Ranching A rancher has 200 ft of fencing with which to build a rectangular corral alongside an existing fence. Determine the dimensions of the corral that will maximize the enclosed area.

EXTENSIONS

47. To prepare astronauts for the experience of zero gravity (technically, microgravity) in space, the National Aeronautics and Space Administration (NASA) uses a specially designed jet. A pilot accelerates the jet upward to an altitude of approximately 9000 m and then reduces power. At that time, the plane continues upward, noses over, and begins to descend until the pilot increases power. The maneuver is then repeated. The figure at the right shows one such maneuver.

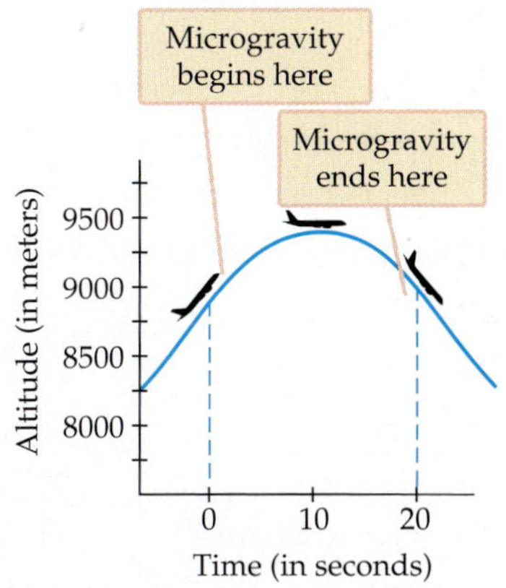

The altitude $A(t)$, in meters, of the jet t seconds after power has been reduced can be approximated by $A(t) = -4.9t^2 + 90t + 9000$. If the pilot increases power when the plane descends to 9000 m, thereby ending microgravity, for how long do the astronauts experience microgravity during one of these maneuvers? Round to the nearest tenth of a second.

SECTION 10.5 Exponential Functions

Introduction to Exponential Functions

Mobile data traffic includes cell phone calls, text messages, Internet browsing, email, and a host of other activities that are performed on a smartphone or tablet device. The graph in Figure 10.1 estimates the number of exabytes (1 exabyte equals 10^{18} bytes, or 1 million terabytes) of data that will be consumed by these devices for the years shown.

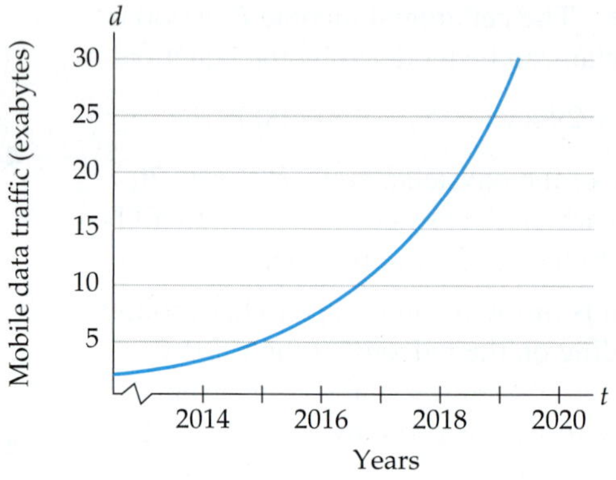

FIGURE 10.1

The graph in Figure 10.1 is a model of *exponential growth*. From the graph, we can predict that approximately 25 exabytes of data will be consumed in 2019. That's enough to fill 25 million 1-terabyte hard drives!

When light enters water, the intensity of the light decreases with the depth of the water. The graph in Figure 10.2 shows a model, for Lake Michigan, of the decrease in the percent of light available as the depth of the water increases. This model is also an exponential function. In this case y is decreasing (decaying), and the model is an *exponential decay function.*

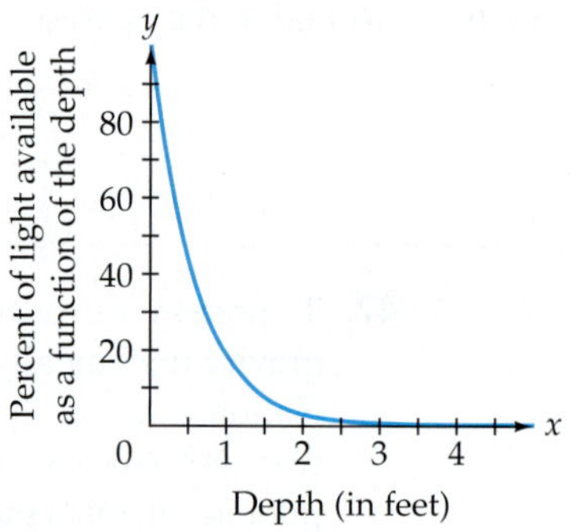

FIGURE 10.2

> **Exponential Functions**
>
> The exponential function with base b is defined by $f(x) = b^x$, where $b > 0$, $b \neq 1$, and x is any real number.

The base b of $f(x) = b^x$ must be positive. If the base were a negative number, the value of the function would not be a real number for some values of x. For instance, if $b = -4$

and $x = \frac{1}{2}$, then $f\left(\frac{1}{2}\right) = (-4)^{1/2} = \sqrt{-4}$, which is not a real number. Also, $b \neq 1$ because if $b = 1$, then $b^x = 1^x = 1$, a constant function.

Below we evaluate $f(x) = 2^x$ for $x = 3$ and $x = -2$.

$$f(x) = 2^x$$
$$f(3) = 2^3 = 8$$
$$f(-2) = 2^{-2} = \frac{1}{2^2} = \frac{1}{4}$$

To evaluate the exponential function $f(x) = 2^x$ for an irrational number such as $x = \sqrt{2}$, we use a rational approximation of $\sqrt{2}$ (for instance, 1.4142) and a calculator to obtain an approximation of the function.

$$f(\sqrt{2}) = 2^{\sqrt{2}} \approx 2^{1.4142} \approx 2.6651$$

EXAMPLE 1 Evaluate an Exponential Function

Evaluate $f(x) = 3^x$ at $x = 2$, $x = -4$, and $x = \pi$. Round approximate results to the nearest hundred thousandth.

Solution

$$f(2) = 3^2 = 9$$
$$f(-4) = 3^{-4} = \frac{1}{3^4} = \frac{1}{81}$$
$$f(\pi) = 3^{\pi} \approx 3^{3.1415927} \approx 31.54428$$

CHECK YOUR PROGRESS 1 Evaluate $g(x) = \left(\frac{1}{2}\right)^x$ when $x = 3$, $x = -1$, and $x = \sqrt{3}$. Round approximate results to the nearest thousandth.

Solution See page S36.

CALCULATOR NOTE

To evaluate $f(\pi)$, we used a calculator. For a scientific calculator, enter

3 [y^x] π [=]

For the TI-83/84 graphing calculator, enter

3 [^] [2nd] π [ENTER]

Graphs of Exponential Functions

The graph of $f(x) = 2^x$ is shown below. The coordinates of some of the points on the graph are given in the table.

INSTRUCTOR NOTE

It is important for students to distinguish between $P(x) = x^2$ and $f(x) = 2^x$. The first is a polynomial function; the second is an exponential function. Polynomial functions have a variable base and a constant exponent, while exponential functions are characterized by a constant base and a variable exponent. You might graph $P(x) = x^2$ and $f(x) = 2^x$ on the same coordinate system so that students can see the difference between these graphs.

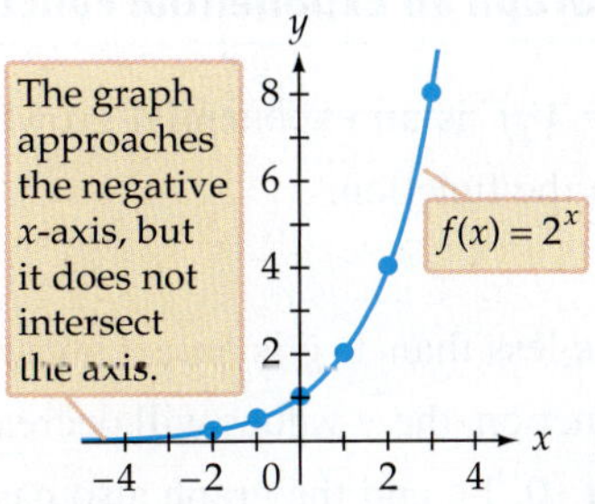

FIGURE 10.3

x	$f(x) = 2^x$	(x, y)
−2	$f(-2) = 2^{-2} = \frac{1}{4}$	$\left(-2, \frac{1}{4}\right)$
−1	$f(-1) = 2^{-1} = \frac{1}{2}$	$\left(-1, \frac{1}{2}\right)$
0	$f(0) = 2^0 = 1$	(0, 1)
1	$f(1) = 2^1 = 2$	(1, 2)
2	$f(2) = 2^2 = 4$	(2, 4)
3	$f(3) = 2^3 = 8$	(3, 8)

INSTRUCTOR NOTE

The Excursion on page 598 can be used to help your students understand how rapidly the value of an exponential growth function increases.

Observe that the values of y *increase* as x increases. This is an exponential growth function. This is typical of the graph of all exponential functions for which the base is *greater than* 1. For the function $f(x) = 2^x$, $b = 2$, which is greater than 1.

Now consider the graph of an exponential function for which the base is between 0 and 1. The graph of $f(x) = \left(\frac{1}{2}\right)^x$ is shown below. The coordinates of some of the points on the graph are given in the table.

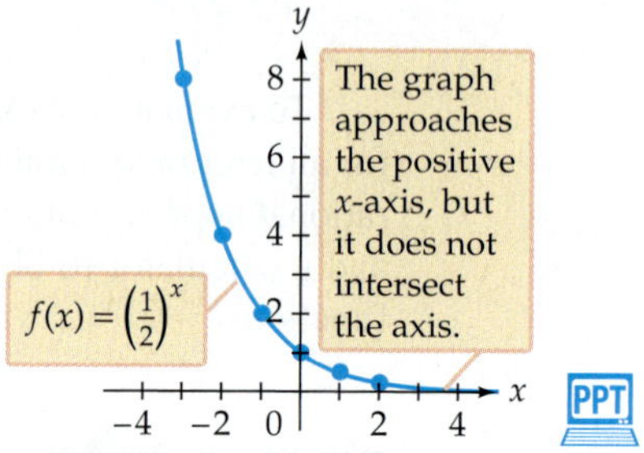

FIGURE 10.4

x	$f(x) = \left(\frac{1}{2}\right)^x$	(x, y)
-3	$f(-3) = \left(\frac{1}{2}\right)^{-3} = 8$	$(-3, 8)$
-2	$f(-2) = \left(\frac{1}{2}\right)^{-2} = 4$	$(-2, 4)$
-1	$f(-1) = \left(\frac{1}{2}\right)^{-1} = 2$	$(-1, 2)$
0	$f(0) = \left(\frac{1}{2}\right)^{0} = 1$	$(0, 1)$
1	$f(1) = \left(\frac{1}{2}\right)^{1} = \frac{1}{2}$	$\left(1, \frac{1}{2}\right)$
2	$f(2) = k\left(\frac{1}{2}\right)^{2} = \frac{1}{4}$	$\left(2, \frac{1}{4}\right)$

Observe that the values of y *decrease* as x increases. This is an exponential decay function. This is typical of the graph of all exponential functions for which the positive base is *less than* 1. For the function $f(x) = \left(\frac{1}{2}\right)^x$, $b = \frac{1}{2}$, which is **between 0 and 1**.

QUESTION Is $f(x) = 0.25^x$ an exponential growth or exponential decay function?

EXAMPLE 2 Graph an Exponential Function

State whether $g(x) = \left(\frac{3}{4}\right)^x$ is an exponential growth function or an exponential decay function. Then graph the function.

Solution

Because the base $\frac{3}{4}$ is less than 1, g is an exponential decay function. Because it is an exponential decay function, the y-values will decrease as x increases. The y-intercept of the graph is the point (0, 1), and the graph also passes through $\left(1, \frac{3}{4}\right)$. Plot a few additional points. Then draw a smooth curve through the points.

ANSWER The base is 0.25, which is less than 1. It is an exponential decay function.

x	$g(x) = \left(\frac{3}{4}\right)^x$	(x, y)
−3	$g(-3) = \left(\frac{3}{4}\right)^{-3} = \frac{64}{27}$	$\left(-3, \frac{64}{27}\right)$
−2	$g(-2) = \left(\frac{3}{4}\right)^{-2} = \frac{16}{9}$	$\left(-2, \frac{16}{9}\right)$
−1	$g(-1) = \left(\frac{3}{4}\right)^{-1} = \frac{4}{3}$	$\left(-1, \frac{4}{3}\right)$
0	$g(0) = \left(\frac{3}{4}\right)^{0} = 1$	$(0, 1)$
1	$g(1) = \left(\frac{3}{4}\right)^{1} = \frac{3}{4}$	$\left(1, \frac{3}{4}\right)$
2	$g(2) = \left(\frac{3}{4}\right)^{2} = \frac{9}{16}$	$\left(2, \frac{9}{16}\right)$
3	$g(3) = \left(\frac{3}{4}\right)^{3} = \frac{27}{64}$	$\left(3, \frac{27}{64}\right)$

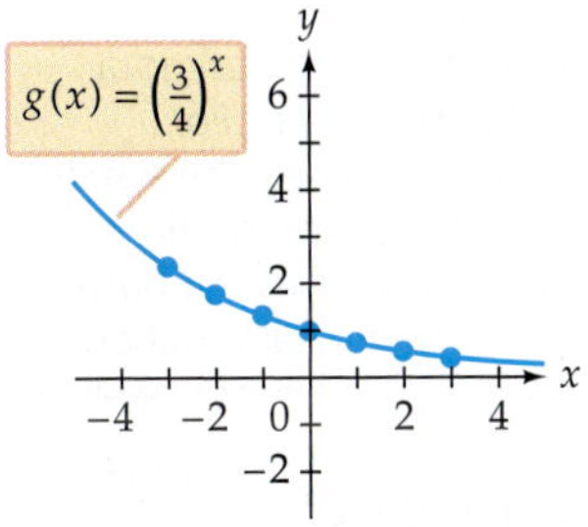

CHECK YOUR PROGRESS 2 State whether $f(x) = \left(\frac{3}{2}\right)^x$ is an exponential growth function or an exponential decay function. Then graph the function.

Solution See page S36.

The Natural Exponential Function

The irrational number π is often used in applications that involve circles. Another irrational number, denoted by the letter e, is useful in applications that involve growth or decay.

The Number *e*

The number e is defined as the number that

$$\left(1 + \frac{1}{n}\right)^n$$

approaches as n gets larger and larger.

n	$\left(1 + \frac{1}{n}\right)^n$
10	2.59374246
100	2.70481383
1000	2.71692393
10,000	2.71814593
100,000	2.71826824
1,000,000	2.71828047

The letter e was chosen in honor of the Swiss mathematician Leonhard Euler. He was able to compute the value of e to several decimal places by evaluating $\left(1 + \frac{1}{n}\right)^n$ for large values of n, as shown at the left. The value of e accurate to eight decimal places is 2.71828183.

The Natural Exponential Function

For all real numbers x, the function defined by $f(x) = e^x$ is called the **natural exponential function**.

A calculator with an e^x key can be used to evaluate e^x for specific values of x. For instance,

$$e^2 \approx 7.389056, \qquad e^{3.5} \approx 33.115452, \qquad \text{and} \qquad e^{-1.4} \approx 0.246597$$

HISTORICAL NOTE

Leonhard Euler (oi′lər) (1707–1783) Some mathematicians consider Euler to be the greatest mathematician of all time. He certainly was the most prolific writer of mathematics of all time. He was the first to introduce many of the mathematical notations that we use today. For instance, he introduced the symbol π for pi, the function notation $f(x)$, and the letter e as the base of the natural exponential function.

TAKE NOTE

In the figure below, compare the graph of $f(x) = e^x$ with the graphs of $g(x) = 2^x$ and $h(x) = 3^x$. Because $2 < e < 3$, the graph of $f(x) = e^x$ lies between the graphs of g and h.

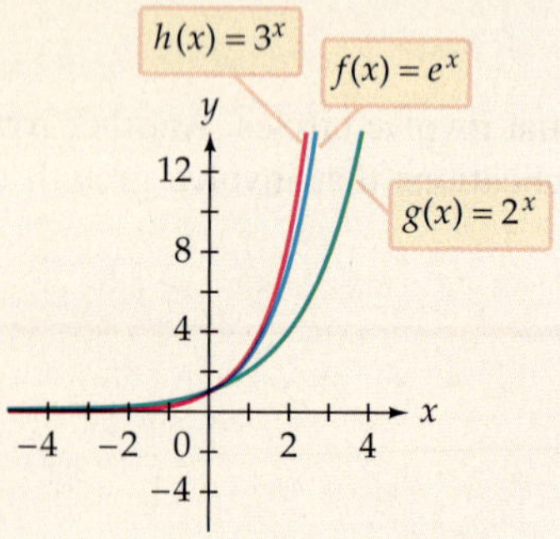

The graph of the natural exponential function can be constructed by plotting a few points or by using a graphing calculator.

EXAMPLE 3 Graph a Natural Exponential Function

Graph $f(x) = e^x$.

Solution

Use a calculator to find range values for a few domain values. The range values in the table below have been rounded to the nearest tenth.

x	−2	−1	0	1	2
$f(x) = e^x$	0.1	0.4	1.0	2.7	7.4

Plot the points given in the table and then connect the points with a smooth curve. Because $e > 1$, as x increases, e^x increases. Thus the values of y increase as x increases. As x decreases, e^x becomes closer to zero. For instance, when $x = -5$, $e^{-5} \approx 0.0067$. Thus as x decreases, the graph gets closer and closer to the x-axis. The y-intercept is (0, 1).

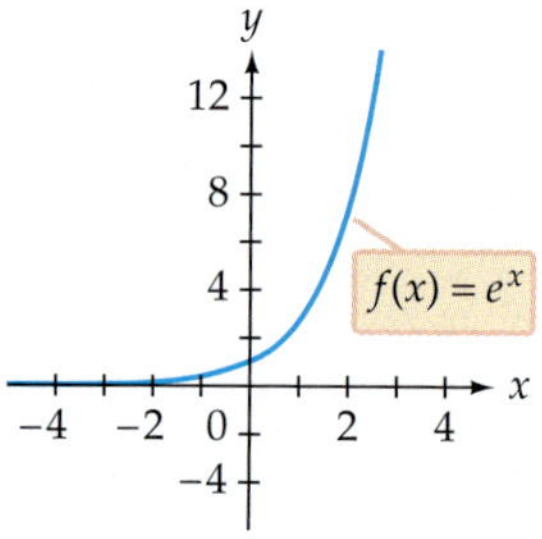

The graph at the right was produced on a TI-83/84 graphing calculator by entering e^x in the Y= menu.

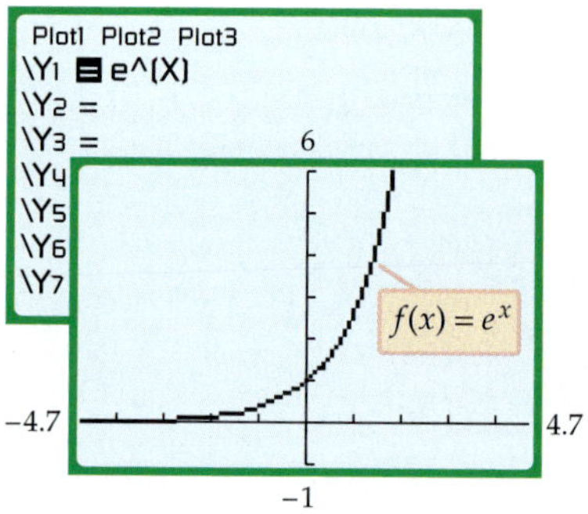

CHECK YOUR PROGRESS 3 Graph $f(x) = e^{-x} + 2$.

Solution See page S36.

Applications of Exponential Functions

Many applications can be effectively modeled by an exponential function. For instance, when money is deposited into a compound interest account, the value of the money can be represented by an exponential growth function. When physicians use a test that involves a radioactive element, the amount of radioactivity remaining in the patient's body can be modeled by an exponential decay function.

TAKE NOTE

Here is an application of exponential functions. Assume you have a long strip of paper that is $\frac{1}{16}$ in. thick and that can be folded in half infinitely many times. After how many folds would the paper be as high as the Empire State Building (1250 ft)?

INSTRUCTOR NOTE

The problem above states that the height of the Empire State Building is 1250 ft, which is 15,000 in. After being folded in half only 18 times, the paper is 16,384 in. high, which is taller than the Empire State Building.

EXAMPLE 4 Application of an Exponential Function

When an amount of money P is placed in an account that earns compound interest, the value A of the money after t years is given by the compound interest formula

$$A = P\left(1 + \frac{r}{n}\right)^{nt}$$

where r is the annual interest rate as a decimal and n is the number of compounding periods per year. Suppose \$500 is placed in an account that earns 8% interest compounded daily. Find the value of the investment after 5 years.

Solution

Use the compound interest formula. Because interest is compounded daily, $n = 365$.

$$\begin{aligned} A &= P\left(1 + \frac{r}{n}\right)^{nt} \\ &= 500\left(1 + \frac{0.08}{365}\right)^{365(5)} && \bullet\ P = 500, r = 0.08, n = 365, t = 5 \\ &\approx 500(1.491759) \approx 745.88 \end{aligned}$$

After 5 years, there is \$745.88 in the account.

CHECK YOUR PROGRESS 4 The radioactive isotope iodine-131 is used to monitor thyroid activity. The number of grams N of iodine-131 in the body t hours after an injection is given by $N(t) = 1.5\left(\frac{1}{2}\right)^{t/193.7}$. Find the number of grams of the isotope in the body 24 h after an injection. Round to the nearest ten-thousandth of a gram.

Solution See page S36.

The next example is based on Newton's law of cooling. This exponential function can be used to model the temperature of something that is being cooled.

EXAMPLE 5 Application of an Exponential Function

A cup of coffee is heated to 160°F and placed in a room that maintains a temperature of 70°F. The temperature of the coffee after t minutes is given by $T(t) = 70 + 90e^{-0.0485t}$. Find the temperature of the coffee 20 min after it is placed in the room. Round to the nearest degree.

Solution

Evaluate the function $T(t) = 70 + 90e^{-0.0485t}$ for $t = 20$.

$$\begin{aligned} T(t) &= 70 + 90e^{-0.0485t} \\ T(20) &= 70 + 90e^{-0.0485(20)} && \bullet\ \text{Substitute 20 for } t. \\ &\approx 70 + 34.1 \\ &\approx 104.1 \end{aligned}$$

After 20 min, the temperature of the coffee is about 104°F.

CHECK YOUR PROGRESS 5 The function $A(t) = 200e^{-0.014t}$ gives the amount of aspirin, in milligrams, in a patient's bloodstream t minutes after the aspirin has been administered. Find the amount of aspirin in the patient's bloodstream after 45 min. Round to the nearest milligram.

Solution See page S37.

MATH MATTERS Marie Curie

Courtesy of the History Factory

In 1903, Marie Sklodowska-Curie (1867–1934) became the first woman to receive a Nobel prize in physics. She was awarded the prize along with her husband, Pierre Curie, and Henri Becquerel for their discovery of radioactivity. In fact, Marie Curie coined the word *radioactivity*.

In 1911, Marie Curie became the first person to win a second Nobel prize, this time alone and in chemistry, for the isolation of radium. She also discovered the element polonium, which she named after Poland, her birthplace. The radioactive phenomena that she studied are modeled by exponential decay functions.

The stamp at the left was printed in 1938 to commemorate Pierre and Marie Curie. When the stamp was issued, there was a surcharge added to the regular price of the stamp. The additional revenue was donated to cancer research. Marie Curie died in 1934 from leukemia, which was caused by her constant exposure to radiation from the radioactive elements she studied.

In 1935, one of Marie Curie's daughters, Irene Joloit-Curie, won a Nobel Prize in chemistry.

EXCURSION

Chess and Exponential Functions

According to legend, when Sissa Ben Dahir of India invented the game of chess, King Shirham was so impressed with the game that he summoned the game's inventor and offered him the reward of his choosing. Sissa Ben Dahir pointed to the chessboard and requested, for his reward, one grain of wheat on the first square, two grains of wheat on the second square, four grains on the third square, eight grains on the fourth square, and so on for all 64 squares on the chessboard. The king considered this a very modest reward and said he would grant the inventor's wish.

EXCURSION EXERCISES

1. This portion of this Excursion will enable you to find a formula for the total amount of wheat on the first n squares of the chessboard. You may want to use the following chart as you answer the questions. It may help you see a pattern.

Square number, n	1	2	3	4	5	6	7
Number of grains of wheat on square n	1	2	4				
Total number of grains of wheat on squares 1 through n	1	$1 + 2 =$	$3 + 4 =$				

a. How many grains of wheat are on each of the first seven squares?

What is the total number (the sum) of grains of wheat on the first

b. two squares? **c.** three squares?

d. four squares? **e.** five squares?

f. six squares? **g.** seven squares?

2. Use inductive reasoning to find a function that gives the total number (the sum) of grains of wheat on the first n squares of the chessboard. Test your function to ensure that it works for parts b through g.
3. If all 64 squares of the chessboard are piled with wheat as requested by Sissa Ben Dahir, how many grains of wheat are on the board?
4. A grain of wheat weighs approximately 0.000008 kg. Find the total weight of the wheat requested by Sissa Ben Dahir.
5. In a recent year, a total of 6.5×10^8 metric tons of wheat were produced in the world. At this level, how many years of wheat production would be required to fill the request of Sissa Ben Dahir? *Hint:* One metric ton equals 1000 kg.

EXERCISE SET 10.5

1. Given $f(x) = 3^x$, evaluate:
   a. $f(2)$ b. $f(0)$ c. $f(-2)$
2. Given $H(x) = 2^x$, evaluate:
   a. $H(-3)$ b. $H(0)$ c. $H(2)$
3. Given $g(x) = 2^{x+1}$, evaluate:
   a. $g(3)$ b. $g(1)$ c. $g(-3)$
4. Given $F(x) = 3^{x-2}$, evaluate:
   a. $F(-4)$ b. $F(-1)$ c. $F(0)$
5. Given $G(r) = \left(\frac{1}{2}\right)^{2r}$, evaluate:
   a. $G(0)$ b. $G\left(\frac{3}{2}\right)$ c. $G(-2)$
6. Given $R(t) = \left(\frac{1}{3}\right)^{3t}$, evaluate:
   a. $R\left(-\frac{1}{3}\right)$ b. $R(1)$ c. $R(-2)$
7. Given $h(x) = e^{x/2}$, evaluate the following. Round to the nearest ten-thousandth.
   a. $h(4)$ b. $h(-2)$ c. $h\left(\frac{1}{2}\right)$
8. Given $f(x) = e^{2x}$, evaluate the following. Round to the nearest ten-thousandth.
   a. $f(-2)$ b. $f\left(-\frac{2}{3}\right)$ c. $f(2)$
9. Given $H(x) = e^{-x+3}$, evaluate the following. Round to the nearest ten-thousandth.
   a. $H(-1)$ b. $H(3)$ c. $H(5)$
10. Given $g(x) = e^{-x/2}$, evaluate the following. Round to the nearest ten-thousandth.
    a. $g(-3)$ b. $g(4)$ c. $g\left(\frac{1}{2}\right)$
11. Given $F(x) = 2^{x^2}$, evaluate the following. Round to the nearest ten-thousandth.
    a. $F(2)$ b. $F(-2)$ c. $F\left(\frac{3}{4}\right)$
12. Given $Q(x) = 2^{-x^2}$, evaluate:
    a. $Q(3)$ b. $Q(-1)$ c. $Q(-2)$
13. Given $f(x) = e^{\frac{-x^2}{2}}$, evaluate the following. Round to the nearest ten-thousandth.
    a. $f(-2)$ b. $f(2)$ c. $f(-3)$
14. Given $h(x) = e^{-2x} + 1$, evaluate the following. Round to the nearest ten-thousandth.
    a. $h(-1)$ b. $h(3)$ c. $h(-2)$

In Exercises 15 to 24, graph the function.

15. $f(x) = 2^x + 1$
16. $f(x) = 3^x - 2$
17. $g(x) = 3^{x/2}$
18. $h(x) = 2^{-x/2}$
19. $f(x) = 2^{x+3}$
20. $g(x) = 4^{-x} + 1$
21. $H(x) = 2^{2x}$
22. $F(x) = 2^{-x}$
23. $f(x) = e^{-x}$
24. $y(x) = e^{2x}$

Investments In Exercises 25 and 26, use the compound interest formula $A = P\left(1 + \frac{r}{n}\right)^{nt}$, where P is the amount deposited, A is the value of the money after t years, r is the annual interest rate as a decimal, and n is the number of compounding periods per year.

25. A computer network specialist deposits \$2500 into a retirement account that earns 7.5% annual interest, compounded daily. What is the value of the investment after 20 years?
26. A \$10,000 certificate of deposit (CD) earns 5% annual interest, compounded daily. What is the value of the investment after 20 years?
27. **Investments** Some banks use continuous compounding of an amount invested. In this case, the equation that models the value of an initial investment of P dollars in t years at an annual interest rate of r is given by $A = Pe^{rt}$. Using this equation, find the value in 5 years of an investment of \$2500 that earns 5% annual interest.

28. Isotopes An isotope of technetium is used to prepare images of internal body organs. This isotope has a half-life (time required for half the material to erode) of approximately 6 h. If a patient is injected with 30 mg of this isotope, what will be the technetium level in the patient after 3 h? Use the function $A(t) = 30\left(\frac{1}{2}\right)^{t/6}$, where A is the technetium level, in milligrams, in the patient after t hours. Round to the nearest tenth of a microgram.

29. Isotopes Iodine-131 is an isotope that is used to study the functioning of the thyroid gland. This isotope has a half-life (time required for half the material to erode) of approximately 8 days. If a patient is given an injection that contains 8 micrograms of iodine-131, what will be the iodine level in the patient after 5 days? Use the function $A(t) = 8\left(\frac{1}{2}\right)^{t/8}$, where A is the amount of the isotope, in micrograms, in the patient after t days. Round to the nearest tenth of a microgram.

30. Welding The percent of correct welds that a student can make will increase with practice and can be approximated by the function $P(t) = 100[1 - (0.75)^t]$, where P is the percent of correct welds and t is the number of weeks of practice. Find the percent of correct welds that a student will make after 4 weeks of practice. Round to the nearest percent.

31. Music The "concert A" note on a piano is the first A below middle C. When that key is struck, the string associated with the key vibrates 440 times per second. The next A above concert A vibrates twice as fast. An exponential function with a base of 2 is used to determine the frequency of the 11 notes between the two As. Find this function. *Hint:* The function is of the form $f(x) = k \cdot 2^{(cx)}$, where k and c are constants. Also $f(0) = 440$ and $f(12) = 880$.

32. Atmospheric Pressure Atmospheric pressure changes as you rise above Earth's surface. At an altitude of h kilometers, where $0 < h < 80$, the pressure P in newtons per square centimeter is approximately modeled by the function

$$P(h) = 10.13e^{-0.116h}$$

a. What is the approximate pressure at 40 km above Earth?

b. What is the approximate pressure on Earth's surface?

c. Does atmospheric pressure increase or decrease as you rise above Earth's surface?

EXTENSIONS

33. Car Payments The formula used to calculate a monthly car lease or payment is given by $P = \dfrac{Ar(1+r)^n - Vr}{(1+r)^n - 1}$, where P is the monthly payment, A is the amount of the loan, r is the *monthly* interest rate as a decimal, n is the number of months of the loan or lease, and V is the residual value of the car at the end of the lease. For a car purchase, $V = 0$.

a. Suppose you lease a car for 5 years. Find the monthly payment if the lease amount is \$10,000, the residual value is \$6000, and the annual interest rate is 6%.

b. Suppose you purchase a car and secure a 5-year loan for \$10,000 at an annual interest rate of 6%. Find the monthly payment.

c. Why are the answers to parts a and b different?

The total amount C that has been repaid on a loan or lease is given by $C = \dfrac{(P - Ar)[(1+r)^n - 1]}{r}$.

d. Using the lease payment in part a, find the total amount that will be repaid in 5 years. How much remains to be paid?

e. Using the monthly payment in part b, find the total amount that will be repaid in 5 years. How much remains to be paid?

f. Explain why the answers to parts d and e make sense.

SECTION 10.6 Logarithmic Functions

Introduction to Logarithmic Functions

Suppose a bacteria colony that originally contained 1000 bacteria doubles in size every hour. The table at the right shows the number of bacteria in the colony after 1, 2, and 3 h.

Time (in hours)	Number of bacteria
0	1000
1	2000
2	4000
3	8000

The exponential function $A = 1000(2^t)$, where A is the number of bacteria in the colony at time t, is a model of the growth of the colony. For instance, when $t = 3$ h, we have

$$A = 1000(2^t)$$
$$A = 1000(2^3) \quad \text{• Replace } t \text{ by } 3.$$
$$A = 1000(8) = 8000$$

After 3 h there are 8000 bacteria in the colony.

Now we ask, "How long will it take for there to be 32,000 bacteria in the colony?" To answer the question, we must solve the *exponential equation* $32{,}000 = 1000(2^t)$. By trial and error, we find that when $t = 5$,

$$A = 1000(2^t)$$
$$A = 1000(2^5) \quad \text{• Replace } t \text{ by } 5.$$
$$A = 1000(32) = 32{,}000$$

After 5 h there will be 32,000 bacteria in the colony.

Now suppose we want to know how long it will be before the colony reaches 50,000 bacteria. To answer this question, we must find t such that $50{,}000 = 1000(2^t)$. Using trial and error again, we find that

$$1000(2^5) = 32{,}000 \qquad \text{and} \qquad 1000(2^6) = 64{,}000$$

Because 50,000 is between 32,000 and 64,000, we conclude that t is between 5 and 6 h. If we try $t = 5.5$ (halfway between 5 and 6), we get

$$A = 1000(2^t)$$
$$A = 1000(2^{5.5}) \quad \text{• Replace } t \text{ by } 5.5.$$
$$A \approx 1000(45.25) = 45{,}250$$

In 5.5 h, there are approximately 45,250 bacteria in the colony. Because this is less than 50,000, the value of t must be a little greater than 5.5.

We could continue to use trial and error to find the correct value of t, but it would be more efficient if we could just solve the exponential equation $50{,}000 = 1000(2^t)$ for t. If we follow the procedures for solving equations discussed earlier in the text, we have

$$50{,}000 = 1000(2^t)$$
$$50 = 2^t \quad \text{• Divide each side of the equation by 1000.}$$

To proceed to the next step, it would be helpful to have a function that would find the power of 2 that produces 50.

Around the mid-16th century, mathematicians created such a function, which we now call a *logarithmic function*. We write the solution of $50 = 2^t$ as $t = \log_2 50$. This is read "t equals the logarithm base 2 of 50" and it means "t equals the power of 2 that produces 50." When logarithms were first introduced, tables were used to find a numerical value of t. Today, a calculator is used. Using a calculator, we can approximate the value of t as 5.644. This means that $2^{5.644} \approx 50$.

The equivalence of the expressions $50 = 2^t$ and $t = \log_2 50$ are described in the following definition of logarithm.

CALCULATOR NOTE

Using a calculator, we can verify that $2^{5.644} \approx 50$. On a graphing calculator, press 2 [^] 5.644 [ENTER].

TAKE NOTE

Recall that when we tried $t = 5.5$, we stated that the actual value of t must be greater than 5.5. Note that 5.644 is a little greater than 5.5.

Logarithm

For $b > 0$, $b \neq 1$, $y = \log_b x$ is equivalent to $x = b^y$. Read $\log_b x$ as "the logarithm of x, base b" or "log base b of x."

TAKE NOTE

The idea of a function that performs the opposite of a given function occurs frequently. For instance, the opposite of a "doubling" function, one that doubles a given number, is a "halving" function, one that takes one-half of a given number. We could write $f(x) = 2x$ for the doubling function and $g(x) = \frac{1}{2}x$ for the halving function. We call these functions *inverse functions* of one another. In a similar manner, exponential and logarithmic functions are inverses of one another.

For every exponential equation there is a corresponding logarithmic equation, and for every logarithmic equation there is a corresponding exponential equation. Here are some examples.

Exponential equation	Logarithmic equation
$2^5 = 32$	$\log_2 32 = 5$
$3^2 = 9$	$\log_3 9 = 2$
$5^{-2} = \frac{1}{25}$	$\log_5 \frac{1}{25} = -2$
$7^0 = 1$	$\log_7 1 = 0$

QUESTION Which of the following is the logarithmic form of $4^3 = 64$?

a. $\log_4 3 = 64$ **b.** $\log_3 4 = 64$ **c.** $\log_4 64 = 3$

HISTORICAL NOTE

Logarithms were developed independently by Jobst Burgi (1552–1632) and John Napier (1550–1617) as a means of simplifying the calculations of astronomers. The idea was to devise a method by which two numbers could be multiplied by performing additions. Napier is usually given credit for logarithms because he published his results first.

In Napier's original work, the logarithm of 10,000,000 was 0. After this work was published, Napier, in discussions with Henry Briggs (1561–1631), decided that tables of logarithms would be easier to use if the logarithm of 1 were 0. Napier died before new tables could be prepared, and Briggs took on the task. His table consisted of logarithms accurate to 30 decimal places, all accomplished without a calculator!

The equation $y = \log_b x$ is the logarithmic form of $b^y = x$, and the equation $b^y = x$ is the *exponential form* of $y = \log_b x$. These two forms state exactly the same relationship between x and y.

EXAMPLE 1 Write a Logarithmic Equation in Exponential Form and an Exponential Equation in Logarithmic Form

a. Write $2 = \log_{10}(x + 5)$ in exponential form.

b. Write $2^{3x} = 64$ in logarithmic form.

Solution

Use the definition of logarithm: $y = \log_b x$ is equivalent to $b^y = x$.

a. $2 = \log_{10}(x + 5)$ is equivalent to $10^2 = x + 5$.

b. $2^{3x} = 64$ is equivalent to $\log_2 64 = 3x$.

CHECK YOUR PROGRESS 1

a. Write $\log_2(4x) = 10$ in exponential form.

b. Write $10^3 = 2x$ in logarithmic form.

Solution See page S37.

The relationship between the exponential and logarithmic forms can be used to evaluate some logarithms. The solutions to these types of problems are based on the equality of exponents property.

Equality of Exponents Property

If $b > 0$ and $b^x = b^y$, then $x = y$.

ANSWER **c.** $\log_4 64 = 3$ is equivalent to $4^3 = 64$.

INSTRUCTOR NOTE

Some students see the definition of logarithm as artificial. The following may help.

Defining a logarithm as the inverse of the exponential function is similar to defining square root as the inverse of square.

"If $49 = x^2$, what is x?" The answer is "the square root of 49."

Now consider the question "If $8 = 2^x$, what is x?" The answer is "the logarithm, base 2, of 8."

This analogy can be extended to motivate the need for tables or calculators.

"If $19 = x^2$, what is x?"

"If $21 = 10^x$, what is x?"

EXAMPLE 2 Evaluate Logarithmic Expressions

Evaluate the logarithms.

a. $\log_8 64$ **b.** $\log_2\left(\frac{1}{8}\right)$

Solution

a. $\log_8 64 = x$ • Write an equation.

$8^x = 64$ • Write the equation in its equivalent exponential form.

$8^x = 8^2$ • Write 64 in exponential form using 8 as the base.

$x = 2$ • Solve for x using the equality of exponents property.

$\log_8 64 = 2$

b. $\log_2\left(\frac{1}{8}\right) = x$ • Write an equation.

$2^x = \frac{1}{8}$ • Write the equation in its equivalent exponential form.

$2^x = 2^{-3}$ • Write $\frac{1}{8}$ in exponential form using 2 as the base.

$x = -3$ • Solve for x using the equality of exponents property.

$\log_2\left(\frac{1}{8}\right) = -3$

CHECK YOUR PROGRESS 2 Evaluate the logarithms.

a. $\log_{10} 0.001$ **b.** $\log_5 125$

Solution See page S37.

EXAMPLE 3 Solve a Logarithmic Equation

Solve: $\log_3 x = 2$

Solution

$\log_3 x = 2$

$3^2 = x$ • Write the equation in its equivalent exponential form.

$9 = x$ • Simplify the exponential expression.

CHECK YOUR PROGRESS 3 Solve: $\log_2 x = 6$

Solution See page S37.

Not all logarithms can be evaluated by rewriting the logarithm in its equivalent exponential form and using the equality of exponents property. For instance, if we tried to evaluate $\log_{10} 18$, it would be necessary to solve the equivalent exponential equation $10^x = 18$. The difficulty here is trying to rewrite 18 in exponential form with 10 as a base.

Common and Natural Logarithms

Two of the most frequently used logarithmic functions are *common logarithms,* which have base 10, and *natural logarithms,* which have base e (the base of the natural exponential function).

Common and Natural Logarithms

The function defined by $f(x) = \log_{10} x$ is called the **common logarithmic function**. It is customarily written without the base as $f(x) = \log x$.

The function defined by $f(x) = \log_e x$ is called the **natural logarithmic function**. It is customarily written as $f(x) = \ln x$.

Most scientific or graphing calculators have a [log] key for evaluating common logarithms and an [ln] key to evaluate natural logarithms. For instance,

$\log 24 \approx 1.3802112$ and $\ln 81 \approx 4.3944492$

EXAMPLE 4 Solve Common and Natural Logarithmic Equations

Solve each of the following equations. Round to the nearest thousandth.

a. $\log x = -1.5$ **b.** $\ln x = 3$

Solution

a. $\log x = -1.5$

$10^{-1.5} = x$ • Write the equation in its equivalent exponential form.

$0.032 \approx x$ • Use a calculator to approximate the value of the exponential expression.

b. $\ln x = 3$

$e^3 = x$ • Write the equation in its equivalent exponential form.

$20.086 \approx x$ • Use a calculator to approximate the value of the exponential expression.

CHECK YOUR PROGRESS 4 Solve each of the following equations. Round to the nearest thousandth.

a. $\log x = -2.1$ **b.** $\ln x = 2$

Solution See page S37.

MATH MATTERS Zipf's Law

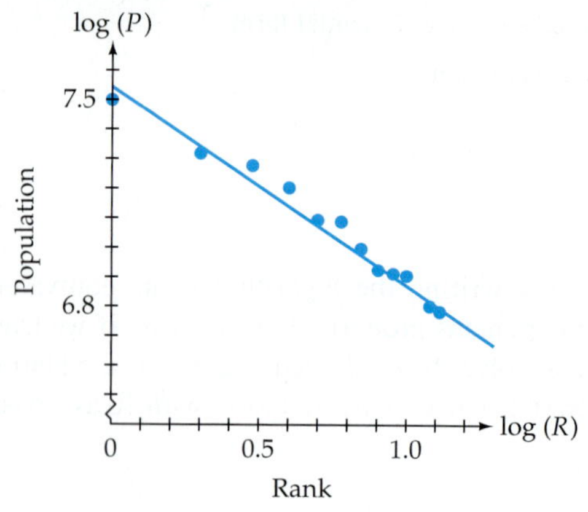

George Zipf (1902–1950) was a lecturer in German and philology at Harvard University. Philology is the study of the change in language over time. Zipf's law, as originally formulated, referred to the frequency of occurrence of a word in a book, magazine article, or other written material, and its rank. The word used most frequently had rank 1, the next most frequently used word had rank 2, and so on. Zipf hypothesized that if the x-axis were the logarithm of a word's rank and the y-axis were the logarithm of the frequency of the word, then the graph of rank versus frequency would lie on an approximately straight line.

Zipf's law has been extended to demographics. For instance, let the population of a U.S. state be P and its rank in population be R. The graph of the points $(\log R, \log P)$ lies on an approximately straight line. The graph at the left shows Zipf's law as applied to the populations and ranks of the 12 most populated states.

Recently, Zipf's law has been applied to website traffic, where websites are ranked according to the number of hits they receive. Assumptions about web traffic are used by engineers and programmers who study ways to make the Internet more efficient.

Graphs of Logarithmic Functions

The graph of a logarithmic function can be drawn by first rewriting the function in its exponential form. This procedure is illustrated in Example 5.

EXAMPLE 5 Graph a Logarithmic Function

Graph $f(x) = \log_3 x$.

Solution

To graph $f(x) = \log_3 x$, first write the equation in the form $y = \log_3 x$. Then write the equivalent exponential equation $x = 3^y$. Because this equation is solved for x, choose values of y and calculate the corresponding values of x, as shown in the table below.

$x = 3^y$	$\frac{1}{9}$	$\frac{1}{3}$	1	3	9
y	−2	−1	0	1	2

Plot the ordered pairs and connect the points with a smooth curve.

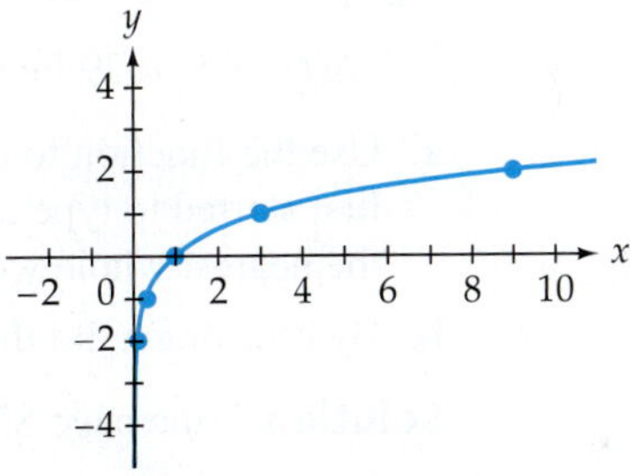

CHECK YOUR PROGRESS 5 Graph $f(x) = \log_5 x$.

Solution See page S37.

The graphs of $y = \log x$ and $y = \ln x$ can be drawn on a graphing calculator by using the [log] and [ln] keys. The graphs are shown below for a TI-83/84 calculator.

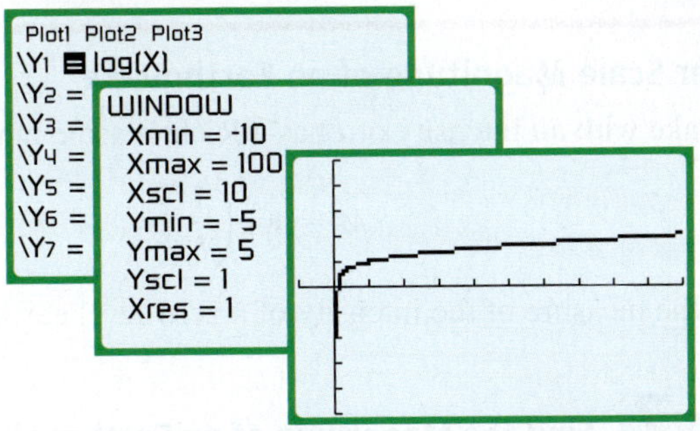

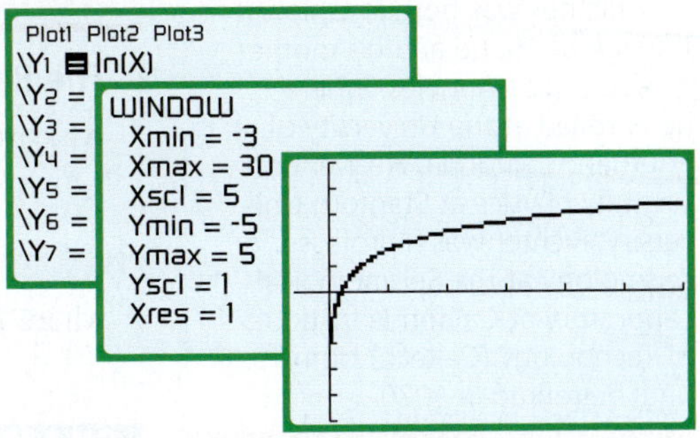

Applications of Logarithmic Functions

EXAMPLE 6 Application of a Logarithmic Function

During the 1980s and 1990s, the average time T of a major league baseball game tended to increase each year. If the year 1981 is represented by $x = 1$, then the function

$$T(x) = 149.57 + 7.63 \ln x$$

approximates the average time T, in minutes, of a major league baseball game for the years $x = 1$ to $x = 19$.

a. Use the function to determine the average time of a major league baseball game during the 1981 season and during the 1999 season. Round to the nearest hundredth of a minute.

b. By how much did the average time of a major league baseball game increase from 1981 to 1999?

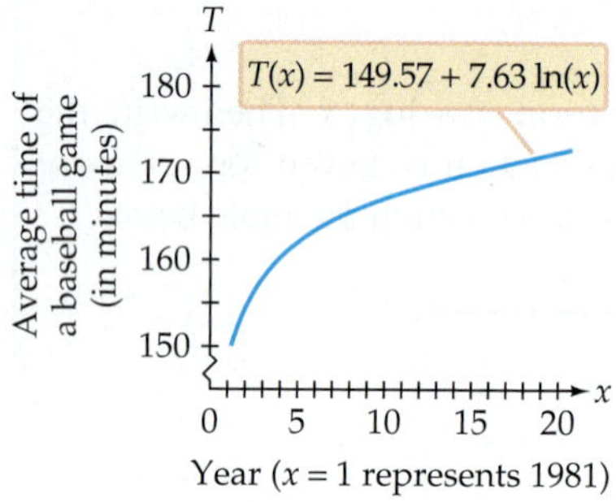

Solution

a. The year 1981 is represented by $x = 1$ and the year 1999 by $x = 19$.

$T(1) = 149.57 + 7.63\ln(1) = 149.57$

In 1981, the average time of a major league baseball game was about 149.57 min.

$T(19) = 149.57 + 7.63\ln(19) \approx 172.04$

In 1999, the average time of a major league baseball game was about 172.04 min.

b. $T(19) - T(1) \approx 172.04 - 149.57 \approx 22.47$

From 1981 to 1999, the average time of a major league baseball game increased by about 22.47 min.

CHECK YOUR PROGRESS 6 The following function models the average typing speed S, in words per minute, of a student who has been typing for t months.

$$S(t) = 5 + 29\ln(t + 1),\ 0 \le t \le 9$$

a. Use the function to determine the student's average typing speed when the student first started to type and the student's average typing speed after 3 months. Round to the nearest whole word per minute.

b. By how much did the typing speed increase during the 3 months?

Solution See page S37.

HISTORICAL NOTE

AP Images

The Richter scale was created by the seismologist Charles Francis Richter (rĭk′tər) (1900–1985) in 1935.

Richter was born in Ohio. At the age of 16, he and his mother moved to Los Angeles, where he enrolled at the University of Southern California. He went on to study physics at Stanford University. Richter was a professor of seismology at the Seismological Laboratory at California Institute of Technology (Caltech) from 1936 until he retired in 1970.

Logarithmic functions are often used to convert very large or very small numbers into numbers that are easier to comprehend. For instance, the *Richter scale*, which measures the magnitude of an earthquake, uses a logarithmic function to scale the intensity of an earthquake's shock waves I into a number M, which for most earthquakes is in the range of 0 to 10. The intensity I of an earthquake is often given in terms of the constant I_0, where I_0 is the intensity of the smallest earthquake (called a **zero-level earthquake**) that can be measured on a seismograph near the earthquake's epicenter. The following formula is used to compute the Richter scale magnitude of an earthquake.

The Richter Scale Magnitude of an Earthquake

An earthquake with an intensity of I has a Richter scale magnitude of

$$M = \log\left(\frac{I}{I_0}\right)$$

where I_0 is the measure of the intensity of a zero-level earthquake.

EXAMPLE 7 Find the Magnitude of an Earthquake

Find the Richter scale magnitude of the 2009 earthquake near the Samoa Islands, which had an intensity of $I = 125{,}892{,}541I_0$. Round to the nearest tenth.

Solution

$$M = \log\left(\frac{I}{I_0}\right) = \log\left(\frac{125{,}892{,}541I_0}{I_0}\right) = \log(125{,}892{,}541) \approx 8.1$$

The 2009 earthquake near the Samoa Islands had a Richter scale magnitude of 8.1.

TAKE NOTE

Note in Example 7 that we did not need to know the value of I_0 to determine the Richter scale magnitude of the quake.

CHECK YOUR PROGRESS 7 What is the Richter scale magnitude of an earthquake whose intensity is twice that of the Samoa Islands' earthquake in Example 7?

Solution See page S37.

If you know the Richter scale magnitude of an earthquake, then you can determine the intensity of the earthquake.

EXAMPLE 8 Find the Intensity of an Earthquake

Find the intensity of the June 15, 2010, Southern California earthquake, which measured 5.7 on the Richter scale. Round to the nearest whole number.

Solution

$$\log\left(\frac{I}{I_0}\right) = 5.7$$

$$\frac{I}{I_0} = 10^{5.7} \quad \text{• Write in exponential form.}$$

$$I = 10^{5.7}I_0 \quad \text{• Solve for } I.$$

$$I \approx 501{,}187I_0$$

The earthquake that struck Southern California on June 15, 2010, had an intensity that was approximately 501,187 times the intensity of a zero-level earthquake.

CHECK YOUR PROGRESS 8 On August 5, 2010, an earthquake measuring 4.8 on the Richter scale struck Wyoming. Find the intensity of the earthquake. Round to the nearest whole number.

Solution See page S37.

POINT OF INTEREST

The pH scale was created by the Danish biochemist Søren Sørensen (sû′rn-sn) in 1909 to measure the acidity of water used in the brewing of beer. pH is an abbreviation for *pondus hydrogenii,* which translates as "potential hydrogen."

Logarithmic scales are also used in chemistry. Chemists use logarithms to determine the pH of a liquid, which is a measure of the liquid's **acidity** or **alkalinity**. (You may have tested the pH of a swimming pool or an aquarium.) Pure water, which is considered neutral, has a pH of 7.0. The pH scale ranges from 0 to 14, with 0 corresponding to the most acidic solutions and 14 to the most alkaline. Lemon juice has a pH of about 2, whereas household ammonia measures about 11.

Specifically, the acidity of a solution is a function of the hydronium-ion concentration of the solution. Because the hydronium-ion concentration of a solution can be very small (with values as low as 0.00000001 mole per liter), pH measures the acidity or alkalinity of a solution using a logarithmic scale.

TAKE NOTE

One mole is equivalent to 6.022×10^{23} ions.

The pH of a Solution

The pH of a solution with a hydronium-ion concentration of H^+ moles per liter is given by

$$\text{pH} = -\log[H^+]$$

EXAMPLE 9 Calculate the pH of a Liquid

Find the pH of each liquid. Round to the nearest tenth.

a. Orange juice containing an H^+ concentration of 2.8×10^{-4} mole per liter

b. Milk containing an H^+ concentration of 3.97×10^{-7} mole per liter

c. A baking soda solution containing an H^+ concentration of 3.98×10^{-9} mole per liter

Solution

a. $pH = -\log[H^+]$

$pH = -\log(2.8 \times 10^{-4}) \approx 3.6$

The orange juice has a pH of about 3.6.

b. $pH = -\log[H^+]$

$pH = -\log(3.97 \times 10^{-7}) \approx 6.4$

The milk has a pH of about 6.4.

c. $pH = -\log[H^+]$

$pH = -\log(3.98 \times 10^{-9}) \approx 8.4$

The baking soda solution has a pH of about 8.4.

CHECK YOUR PROGRESS 9 Find the pH of each liquid. Round to the nearest tenth.

a. A cleaning solution containing an H^+ concentration of 2.41×10^{-13} mole per liter

b. A cola soft drink containing an H^+ concentration of 5.07×10^{-4} mole per liter

c. Rainwater containing an H^+ concentration of 6.31×10^{-5} mole per liter

Solution See page S37.

The following figure illustrates the pH scale, along with the corresponding hydronium-ion concentrations. A solution with a pH less than 7 is an **acid**, and a solution with a pH greater than 7 is an **alkaline solution**, or a **base**. Because the scale is logarithmic, a solution with a pH of 5 is 10 times more acidic than a solution with a pH of 6. From Example 9, we see that the orange juice and milk are acids, whereas the baking soda solution is a base.

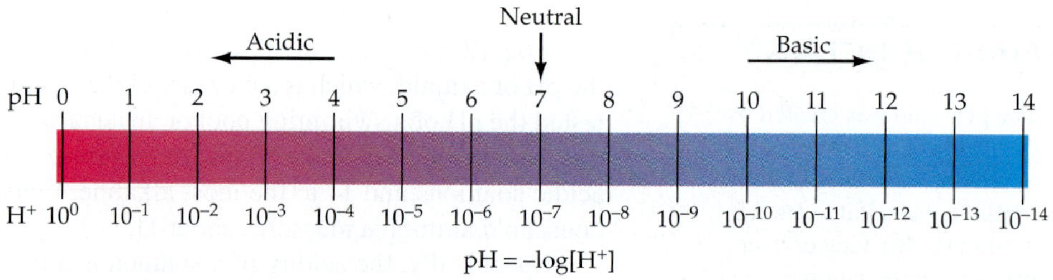

The figure above shows how the pH function scales small numbers on the H^+ axis into larger and more convenient numbers on the pH axis.

EXAMPLE 10 Find the Hydronium-Ion Concentration of a Liquid

A sample of blood has a pH of 7.3. Find the hydronium-ion concentration of the blood. Remember that the hydronium-ion concentration is measured in moles per liter.

Solution

$$pH = -\log[H^+]$$

$$7.3 = -\log[H^+]$$ • Substitute 7.3 for pH.

$$-7.3 = \log[H^+]$$ • Multiply both sides by -1.

$$10^{-7.3} = H^+$$ • Change to exponential form.

$$5.0 \times 10^{-8} \approx H^+$$ • Use a calculator to evaluate $10^{-7.3}$ and write the answer in scientific notation.

The hydronium-ion concentration of the blood is about 5.0×10^{-8} mole per liter.

CHECK YOUR PROGRESS 10 The water in the Great Salt Lake in Utah has a pH of 10.0. Find the hydronium-ion concentration of the water.

Solution See page S37.

EXCURSION

Benford's Law

Research by psychologists has shown that humans are very inept at trying to come up with random numbers, that is, a list of numbers with no discernible pattern. Forensic accounting makes use of accounting principles and this human weakness to investigate fraud. One mathematical tool used to examine whether data are fraudulent is **Benford's Law**.

This law was first discovered by the mathematician Simon Newcomb in 1881 and was then rediscovered by the physicist Frank Benford in 1938. Benford's law states that the probability P that the digit d is the first digit of a number selected at random from a wide range of numbers is given by

$$P(d) = \log\left(1 + \frac{1}{d}\right)$$

POINT OF INTEREST

Benford's law has been used to identify fraudulent accountants. In most cases these accountants are unaware of Benford's law and have replaced valid numbers with numbers selected at random. Their numbers do not conform to Benford's law. Hence, an audit is warranted.

d	$P(d) = \log(1 + 1/d)$
1	0.301
2	0.176
3	0.125
4	______
5	______
6	______
7	______
8	______
9	______

EXCURSION EXERCISES

1. Use Benford's law to complete the table and bar graph shown below.

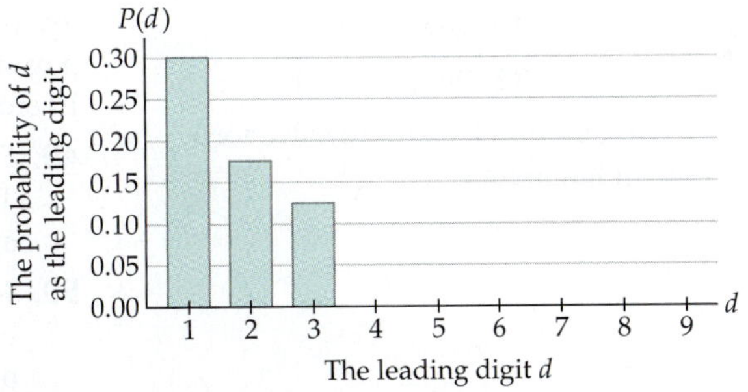

Benford's law applies to most data with a wide range. For instance, it applies to

- the populations of the cities in the United States
- the numbers of dollars in the savings accounts at your local bank
- the numbers of miles driven during a month by each person in a state
- the areas of lakes
- the sizes of files on a personal computer

2. Use Benford's law to find the probability that in a U.S. city selected at random, the number of telephones in that city will be a number starting with a 6.

3. Use Benford's law to find how many times as many purchases you have made for dollar amounts that start with a 1 than for dollar amounts that start with a 9.

4. Explain why Benford's law would not apply to the set of telephone numbers in a small city such as Le Mars, Iowa.

5. Explain why Benford's law would not apply to the set of all the ages, in years, of the students at a local high school.

EXERCISE SET 10.6

In Exercises 1 to 8, write the exponential equation in logarithmic form.

1. $7^2 = 49$

2. $10^3 = 1000$

3. $5^4 = 625$

4. $2^{-3} = \frac{1}{8}$

5. $10^{-4} = 0.0001$

6. $3^5 = 729$

7. $10^y = x$

8. $e^y = x$

In Exercises 9 to 16, write the logarithmic equation in exponential form.

9. $\log_3 81 = 4$

10. $\log_2 16 = 4$

11. $\log_5 125 = 3$

12. $\log_4 64 = 3$

13. $\log_4 \frac{1}{16} = -2$

14. $\log_2 \frac{1}{16} = -4$

15. $\ln x = y$

16. $\log x = y$

In Exercises 17 to 24, evaluate the logarithm.

17. $\log_3 81$

18. $\log_7 49$

19. $\log 100$

20. $\log 0.001$

21. $\log_3 \frac{1}{9}$

22. $\log_7 \frac{1}{7}$

23. $\log_2 64$

24. $\log 0.01$

In Exercises 25 to 32, solve the equation for x.

25. $\log_3 x = 2$

26. $\log_5 x = 1$

27. $\log_7 x = -1$

28. $\log_8 x = -2$

29. $\log_3 x = -2$

30. $\log_5 x = 3$

31. $\log_4 x = 0$

32. $\log_4 x = -1$

In Exercises 33 to 40, use a calculator to solve for x. Round to the nearest hundredth.

33. $\log x = 2.5$

34. $\log x = 3.2$

35. $\ln x = 2$

36. $\ln x = 4$

37. $\log x = 0.35$

38. $\log x = 0.127$

39. $\ln x = \frac{8}{3}$

40. $\ln x = \frac{1}{2}$

In Exercises 41 to 46, graph the function.

41. $g(x) = \log_2 x$

42. $g(x) = \log_4 x$

43. $f(x) = \log_3(2x - 1)$

44. $f(x) = -\log_2 x$

45. $f(x) = \log_2(x - 1)$

46. $f(x) = \log_3(x - 2)$

Light The percent of light that will pass through a material is given by the formula $\log P = -kd$, where P is the percent of light passing through the material, k is a constant that depends on the material, and d is the thickness of the material in centimeters. Use this formula for Exercises 47 and 48.

47. The constant k for a piece of opaque glass that is 0.5 cm thick is 0.2. Find the percent of light that will pass through the glass. Round to the nearest percent.

48. The constant k for a piece of tinted glass is 0.5. How thick is a piece of this glass that allows 60% of the light incident to the glass to pass through it? Round to the nearest hundredth of a centimeter.

Philip Meyer/Shutterstock.com

Sound The number of decibels, D, of a sound can be given by the equation $D = 10(\log I + 16)$, where I is the power of the sound measured in watts. Use this formula for Exercises 49 and 50. Round to the nearest decibel.

49. Find the number of decibels of normal conversation. The power of the sound of normal conversation is approximately 3.2×10^{-10} watts.

50. The loudest sound made by an animal is made by the blue whale and can be heard over 500 mi away. The power of the sound is 630 watts. Find the number of decibels of the sound emitted by the blue whale.

pH of a Solution For Exercises 51 and 52, use the equation

$$\text{pH} = -\log(\text{H}^+)$$

where H^+ is the hydronium-ion concentration of a solution. Round to the nearest hundredth.

51. Find the pH of the digestive solution of the stomach, for which the hydronium-ion concentration is 0.045 mole per liter.

52. Find the pH of a morphine solution used to relieve pain, for which the hydronium-ion concentration is 3.2×10^{-10} mole per liter.

Earthquakes For Exercises 53 to 57, use the Richter scale equation $M = \log \frac{I}{I_0}$, where M is the magnitude of an earthquake, I is the intensity of the shock waves, and I_0 is the measure of the intensity of a zero-level earthquake.

53. On February 27, 2010, an earthquake struck offshore Maule, Chile. The earthquake had an intensity of $I = 630{,}957{,}345I_0$. Find the Richter scale magnitude of the earthquake. Round to the nearest tenth.

54. The earthquake on May 12, 2008, in Eastern Sichuan in China had an intensity of $I = 79{,}432{,}823I_0$. Find the Richter scale magnitude of the earthquake. Round to the nearest tenth.

55. An earthquake in Japan on March 2, 1933, measured 8.9 on the Richter scale. Find the intensity of the earthquake in terms of I_0. Round to the nearest whole number.

56. An earthquake that occurred in Alaska in 1964 measured 9.2 on the Richter scale. Find the intensity of the earthquake in terms of I_0. Round to the nearest whole number.

57. How many times as strong is an earthquake whose magnitude is 8 than one whose magnitude is 6?

Astronomy Astronomers use the distance modulus function $M(r) = 5 \log r - 5$, where M is the distance modulus and r is the distance of a star from Earth in parsecs. (One parsec is approximately 1.92×10^{13} mi, or approximately 20 trillion miles.) Use this function for Exercises 58 and 59. Round to the nearest tenth.

58. The distance modulus of the star Betelgeuse is 5.89. How many parsecs from Earth is this star?

59. The distance modulus of Alpha Centauri is -1.11. How many parsecs from Earth is this star?

EXTENSIONS

60. As mentioned in this section, the main motivation for the development of logarithms was to aid astronomers and other scientists with arithmetic calculations. The idea was to enable scientists to multiply large numbers by adding the logarithms of the numbers. Division of large numbers was accomplished by subtracting the logarithms of the numbers. Follow through the exercises below to see how this was accomplished. For simplicity, we will use small numbers (2 and 3) to illustrate the procedure.

a. Write each of the equations $\log 2 = 0.30103$ and $\log 3 = 0.47712$ in exponential form.

b. Replace 2 and 3 in $x = 2 \cdot 3$ with the exponential expressions from part a. (In this exercise, we know that x is 6. However, if the two numbers were very large, the value of x would not be obvious.)

c. Simplify the exponential expression. Recall that to multiply two exponential expressions with the same base, we *add* the exponents.

d. If you completed part c correctly, you should have

$$x = 10^{0.77815}$$

Write this expression in logarithmic form.

e. Using a calculator, verify that the solution of the equation you created in part d is 6. *Note:* When tables of logarithms were used, a scientist would have had to look through the table to find 0.77815 and observe that it was the logarithm of 6.

61. Replace the denominator 2 and the numerator 3 in $x = \frac{3}{2}$ by the exponential expressions from part a of Exercise 60. Simplify the expression by *subtracting* the exponents. Write the answer in logarithmic form and verify that $x = 1.5$.

62. Write a few sentences explaining how adding the logarithms of two numbers can be used to find the product of the two numbers.

63. Write a few sentences explaining how subtracting the logarithms of two numbers can be used to find the quotient of the two numbers.

Earthquakes Seismologists generally determine the Richter scale magnitude of an earthquake by examining a *seismogram*, an example of which is shown below.

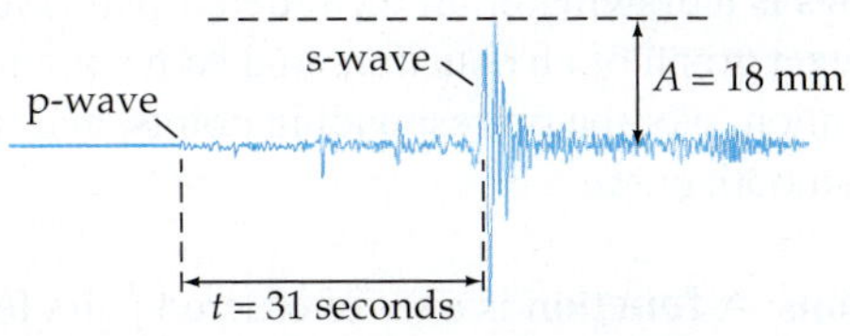

The magnitude of an earthquake cannot be determined just by examining the amplitude of a seismogram because this amplitude decreases as the distance between the epicenter of the earthquake and the observation station increases. To account for the distance between the epicenter and the observation station, a seismologist examines a seismogram for both small waves, called *p-waves*, and larger waves, called *s-waves*. The Richter scale magnitude M of the earthquake is a function of both the amplitude A of the s-waves and the time t between the occurrence of the s-waves and the occurrence of the p-waves. In the 1950s, Charles Richter developed the Amplitude-Time-Difference Formula shown below to determine the magnitude M of an earthquake from the data in a seismogram.

The Amplitude-Time-Difference Formula

The Richter scale magnitude of an earthquake is given by

$$M = \log A + 3 \log 8t - 2.92$$

where A is the amplitude, in millimeters, of the s-waves on a seismogram and t is the time, in seconds, between the s-waves and the p-waves.

64. Find the Richter scale magnitude of the earthquake that produced the seismogram shown on the preceding page. Round to the nearest tenth.

65. Find the Richter scale magnitude of the earthquake that produced the seismograph shown below. Round to the nearest tenth.

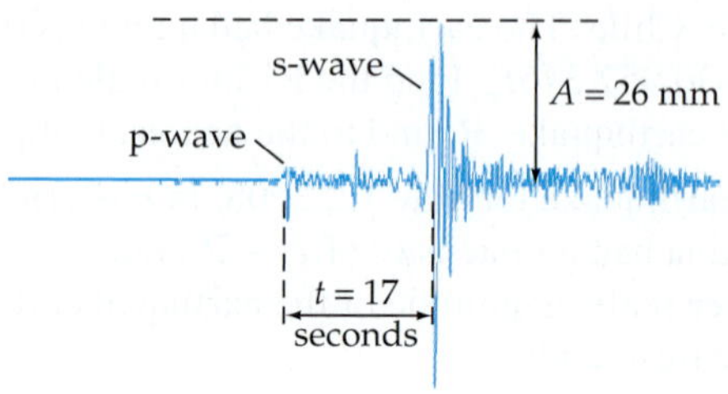

CHAPTER 10 SUMMARY

The following table summarizes essential concepts in this chapter. The references given in the right-hand column list Examples and Exercises that can be used to test your understanding of a concept.

10.1 Rectangular Coordinates and Functions	
Rectangular Coordinate System A rectangular coordinate system is formed by two number lines, one horizontal and one vertical, that intersect at the zero point of each line. The point of intersection is called the **origin**. The two number lines are the **axes**. Each point in the plane can be identified by a pair of numbers called an **ordered pair**. The first number of the ordered pair measures a horizontal change from the y-axis. The second number measures a vertical change from the x-axis. The ordered pair (x, y) associated with a point is also called the **coordinates** of the point. A **solution of an equation in two variables** is an ordered pair that makes the equation a true statement. The **graph of an equation in two variables** is a drawing of all the ordered-pair solutions of the equation. To create a graph of an equation, find some ordered-pair solutions of the equation, plot the corresponding points, and then connect the points with a smooth curve.	See **Examples 1 and 2** on pages 557 and 558, and then try Exercises 1 to 8 on page 614.
Functions A **function** is a set of ordered pairs formed by a relationship between two sets that uniquely associates the members of one set, called the **domain**, with the members of the other set, called the **range**. Function notation is frequently used for equations that define functions. The symbol $f(x)$ is read "the value of f at x" or "f of x." The variable y and the symbol $f(x)$ are different symbols for the same number. The name of the function is f; the value of the function at x is $f(x)$. The process of finding $f(x)$ for a given value of x is called **evaluating the function**.	See **Examples 3, 4, and 5** on pages 561 and 562, and then try Exercises 9, 10, 17, 18, and 24 on pages 614 and 615 .

continued

10.2 Properties of Linear Functions

Linear Function A linear function is one that can be written in the form $f(x) = mx + b$, where m is the coefficient of x and b is a constant.

The point at which the graph of a linear function crosses the x-axis is called the ***x*-intercept** of the graph. The point at which the graph crosses the y-axis is called the ***y*-intercept** of the graph. The y-intercept of the graph of $f(x) = mx + b$ is $(0, b)$.

See **Examples 1 and 2** on pages 567 and 568, and then try Exercises 27, 29, and 31 on page 615.

Slope of a Line Let (x_1, y_1) and (x_2, y_2) be two points on a vertical line. Then the slope of the line through the two points is the ratio of the change in the y-coordinates to the change in the x-coordinates.

$$m = \frac{y_2 - y_1}{x_2 - x_1}, x_1 \neq x_2$$

See **Examples 3 and 4** on pages 569 and 571, and then try Exercises 32, 34, and 36 on page 615.

Slope–Intercept Form of the Equation of a Line The graph of $f(x) = mx + b$ is a straight line with slope m and y-intercept $(0, b)$.

See **Example 6** on page 572, and then try Exercises 39, 41, and 43 on page 615.

10.3 Finding Linear Models

Finding Linear Models The solution to an application problem may require finding a linear model given the slope and the y-intercept, given the slope and a point other than the y-intercept, or given two points.

When given the slope and a point other than the y-intercept, use the **Point-Slope Formula of a Straight Line** to find the equation of the line:

$$y - y_1 = m(x - x_1),$$

where (x_1, y_1) is a point on the line and m is the slope.

See **Examples 1, 3, 4, and 5** on pages 577–580, and then try Exercises 44 and 45 on page 615.

10.4 Quadratic Functions

Properties of Quadratic Functions A quadratic function in a single variable x is a function of the form $f(x) = ax^2 + bx + c$, $a \neq 0$. The graph of a quadratic function is a **parabola**. When the value of a is positive, the graph opens up. When the value of a is negative, the graph opens down.

The **vertex** of a parabola is the point with the smallest y-coordinate when $a > 0$ and the point with the largest y-coordinate when $a < 0$.

The coordinates of the vertex are $\left(-\frac{b}{a}, f\left(-\frac{b}{a}\right)\right)$.

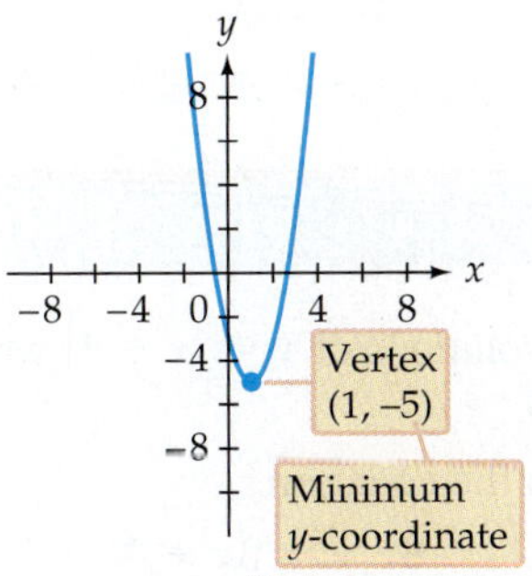

$a > 0$

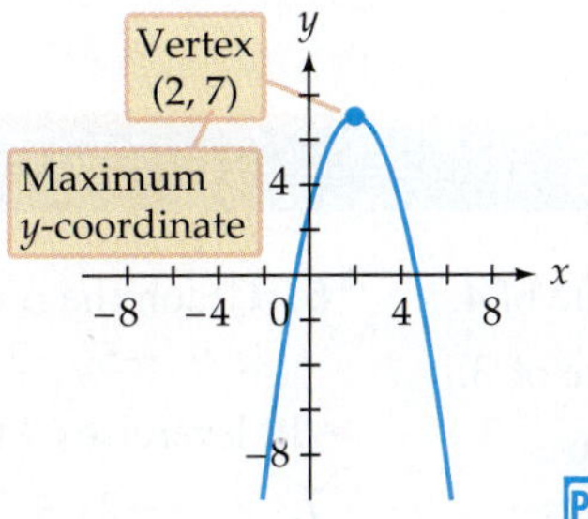

$a < 0$

The **axis of symmetry** is a vertical line that passes through the vertex of the parabola.

See **Example 1** on page 584, and then try Exercises 46 and 48 on page 616.

continued

x-Intercepts of a Parabola The point at which a graph crosses the x-axis is an x-intercept of the graph. The x-intercepts of the graph of an equation can be found by letting $y = 0$.	See **Example 2** on pages 585 and 586, and then try Exercises 50 and 52 on page 616.
Minimum and Maximum of a Quadratic Function Finding the minimum or maximum value of a quadratic function of the form $f(x) = ax^2 + bx + c$ is a matter of finding the vertex of the graph of the function. When $a > 0$, the minimum value is the y-coordinate of the vertex. When $a < 0$, the maximum value is the y-coordinate of the vertex.	See **Examples 3, 4, and 5** on pages 587 and 588, and then try Exercises 54, 56, and 58 on page 616.
10.5 Exponential Functions	
Exponential Functions The exponential function is defined by $f(x) = b^x$, where b is the base, $b > 0$, $b \neq 1$, and x is any real number. An exponential growth function is one in which $b > 1$. An exponential decay function is one in which b is between 0 and 1. For all real numbers x, the function defined by $f(x) = e^x$ is the **natural exponential function**.	See **Examples 1, 2, and 3** on pages 593–596, and then try Exercises 13, 14, 15, and 21 on pages 614 and 615.
Applications of Exponential Functions Some applications can be modeled by an exponential function.	See **Examples 4 and 5** on page 597, and then try Exercises 60, 61, and 63 on page 616.
10.6 Logarithmic Functions	
Logarithmic Functions For $b > 0$, $b \neq 1$, $y = \log_b x$ is equivalent to $x = b^y$. The relationship between the exponential and logarithmic forms can be used to evaluate some logarithms. The solutions to some of these types of problems are based on the **equality of exponents property**: If $b > 0$ and $b^x = b^y$, then $x = y$.	See **Examples 1, 2, and 3** on pages 602 and 603, and then try Exercises 64, 66, and 68 on page 616.
Common and Natural Logarithms The function defined by $f(x) = \log_{10} x$ is the common logarithmic function. It is customarily written without the base as $f(x) = \log x$. The function defined by $f(x) = \log_e x$ is the natural logarithmic function. It is customarily written as $f(x) = \ln x$.	See **Example 4** on page 604, and then try Exercises 70 and 71 on page 616.
Applications of Logarithmic Functions Some applications can be modeled by a logarithmic function.	See **Examples 6 to 10** on pages 605–608, and then try Exercises 72 and 73 on page 616.

CHAPTER 10 REVIEW EXERCISES

1. Draw a line through all points with an x-coordinate of 4.
2. Draw a line through all points with a y-coordinate of 3.
3. Graph the ordered-pair solutions of $y = 2x^2$ when $x = -2, -1, 0, 1$, and 2.
4. Graph the ordered-pair solutions of $y = 2x^2 - 5$ when $x = -2, -1, 0, 1$, and 2.
5. Graph the ordered-pair solutions of $y = -2x + 1$ when $x = -2, -1, 0, 1$, and 2.
6. Graph the ordered-pair solutions of $y = |x + 1|$ when $x = -5, -3, 0, 3$, and 5.

■ In Exercises 7 to 16, graph the equation.

7. $y = -2x + 1$
8. $y = 3x + 2$
9. $f(x) = x^2 + 2$
10. $f(x) = x^2 - 3x + 1$
11. $y = |x + 4|$
12. $f(x) = 2|x| - 1$
13. $f(x) = 2^x - 3$
14. $f(x) = 3^{-x+2}$
15. $f(x) = e^{0.5x}$
16. $f(x) = \log_2(x - 2)$

■ In Exercises 17 to 23, evaluate the function for the given value.

17. $f(x) = 4x - 5;\ x = -2$

18. $g(x) = 2x^2 - x - 2;\ x = 3$

19. $s(t) = \dfrac{4}{3t - 5};\ t = -1$

20. $R(s) = s^3 - 2s^2 + s - 3;\ s = -2$

21. $f(x) = 2^{x-3};\ x = 5$

22. $g(x) = \left(\dfrac{2}{3}\right)^x;\ x = 2$

23. $T(r) = 2e^r + 1;\ r = 2$. Round to the nearest hundredth.

24. Geometry The volume of a sphere is a function of its radius and is given by $V(r) = \dfrac{4\pi r^3}{3}$, where r is the radius of the sphere.

a. Find the volume of a sphere whose radius is 3 in. Round to the nearest tenth of a cubic inch.

b. Find the volume of a sphere whose radius is 12 cm. Round to the nearest tenth of a cubic centimeter.

25. Sports The height h of a ball that is released 5 ft above the ground with an initial upward velocity of 64 ft/s is a function of the time t, in seconds, the ball is in the air and is given by $h(t) = -16t^2 + 64t + 5$.

a. Find the height of the ball above the ground 2 s after it is released.

b. Find the height of the ball above the ground 4 s after it is released.

26. Mixtures The percent concentration P of sugar in a water solution depends on the amount of sugar that is added to the solution and is given by $P(x) = \dfrac{100x + 100}{x + 20}$, where x is the number of grams of sugar added.

a. What is the original percent concentration of sugar?

b. What is the percent concentration of sugar after 10 more grams of sugar are added? Round to the nearest tenth of a percent.

■ In Exercises 27 to 30, find the x- and y-intercepts of the graph of the function.

27. $f(x) = 2x + 10$

28. $f(x) = \dfrac{3}{4}x - 9$

29. $3x - 5y = 15$

30. $4x + 3y = 24$

31. Depreciation The accountant for a small business uses the model $V(t) = 25{,}000 - 5000t$ to approximate the value, $V(t)$, of a delivery van t years after its purchase. Find and discuss the meaning of the intercepts, on the vertical and horizontal axes, in the context of this application.

■ In Exercises 32 to 35, find the slope of the line containing the given points.

32. $(3, 2), (2, -3)$

33. $(-1, 4), (-3, -1)$

34. $(2, -5), (-4, -5)$

35. $(5, 2), (5, 7)$

36. Fuel Consumption The graph below shows how the amount of gas in the tank of a car decreases as the car is driven. Find the slope of the line. Write a sentence that explains the meaning of the slope.

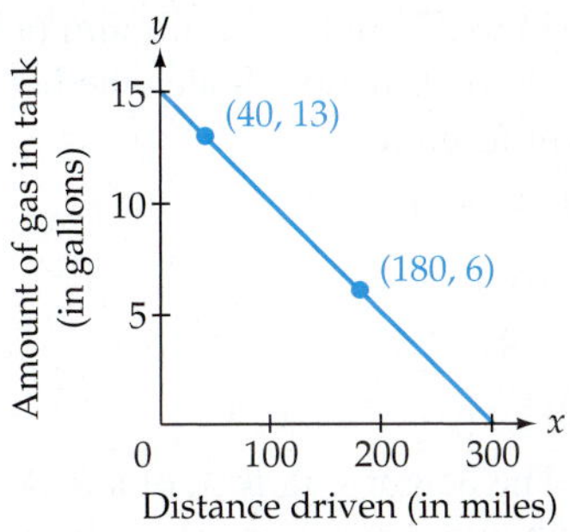

37. Graph the line that passes through the point $(3, -2)$ and has slope -2.

38. Graph the line that passes through the point $(-1, -3)$ and has slope $\frac{3}{4}$.

■ In Exercises 39 to 42, find the equation of the line that contains the given point and has the given slope.

39. $(-2, 3),\ m = 2$

40. $(1, -4),\ m = 1$

41. $(-3, 1),\ m = \dfrac{2}{3}$

42. $(4, 1),\ m = \dfrac{1}{4}$

43. Graph $f(x) = \frac{3}{2}x - 1$ using the slope and y-intercept.

44. Interior Decorating A dentist's office is being recarpeted. The cost to install the new carpet is \$100 plus \$12 per square foot of carpeting.

a. Determine a linear function for the cost to carpet the office.

b. Use this function to determine the cost to carpet 288 ft^2 of floor space.

45. Gasoline Sales The manager of Valley Gas Mart has determined that 10,000 gal of regular unleaded gasoline can be sold each week if the price is the same as that of Western QuickMart, a gas station across the street. If the manager increases the price \$0.02 above Western QuickMart's price, the manager will sell 500 less gallons per week. If the manager decreases the price \$0.02 below that of Western QuickMart, 500 more gallons of gasoline per week will be sold.

a. Determine a linear function that will predict the number of gallons of gas per week that Valley Gas Mart can sell as a function of the price relative to that of Western QuickMart.

b. Use the model to predict the number of gallons of gasoline Valley Gas Mart will sell if its price is \$0.03 below that of Western QuickMart.

■ In Exercises 46 to 49, find the vertex of the graph of the function.

46. $y = x^2 + 2x + 4$

47. $y = -2x^2 - 6x + 1$

48. $f(x) = -3x^2 + 6x - 1$

49. $f(x) = x^2 + 5x - 1$

■ In Exercises 50 to 53, find the x-intercepts of the parabola given by the equation.

50. $y = x^2 + x - 20$

51. $y = x^2 + 2x - 1$

52. $f(x) = 2x^2 + 9x + 4$

53. $f(x) = x^2 + 4x + 6$

■ In Exercises 54 to 57, find the minimum or maximum value of the quadratic function. State whether the value is a maximum or a minimum.

54. $y = -x^2 + 4x + 1$

55. $y = 2x^2 + 6x - 3$

56. $f(x) = x^2 - 4x - 1$

57. $f(x) = -2x^2 + 3x - 1$

58. Physics The height s, in feet, of a rock thrown upward at an initial speed of 80 ft/s from a cliff 25 ft above an ocean beach is given by the function $s(t) = -16t^2 + 80t + 25$, where t is the time in seconds after the rock is released. Find the maximum height above the beach that the rock will attain.

59. Manufacturing A manufacturer of DVDs estimates that the average daily cost C of producing DVDs is given by $C(x) = 0.01x^2 - 40x + 50{,}000$, where x is the number of DVDs produced each day. Find the number of DVDs the company should produce in order to minimize the average daily cost.

60. Investments A \$5000 certificate of deposit (CD) earns 6% annual interest compounded daily. What is the value of the investment after 15 years? Use the compound interest formula $A = P\left(1 + \frac{r}{n}\right)^{nt}$, where P is the amount deposited, A is the value of the money after t years, r is the annual interest rate as a decimal, and n is the number of compounding periods per year.

61. Isotopes An isotope of technetium has a half-life of approximately 6 h. If a patient is injected with 10 mg of this isotope, what will be the technetium level in the patient after 2 h? Use the function $A(t) = 10\left(\frac{1}{2}\right)^{t/6}$, where A is the technetium level, in milligrams, in the patient after t hours. Round to the nearest hundredth of a milligram.

62. Isotopes Iodine-131 has a half-life of approximately 8 days. If a patient is given an injection that contains 8 micrograms of iodine-131, what will be the iodine level in the patient after 10 days? Use the function $A(t) = 8\left(\frac{1}{2}\right)^{t/8}$, where A is the amount of iodine-131, in micrograms, in the patient after t days. Round to the nearest hundredth of a milligram.

63. Golf A golf ball is dropped from a height of 6 ft. On each successive bounce, the ball rebounds to a height that is $\frac{2}{3}$ of the previous height.

a. Find an exponential model for the height of the ball after the nth bounce.

b. What is the height of the ball after the fifth bounce? Round to the nearest hundredth of a foot.

■ In Exercises 64 to 67, evaluate the logarithm.

64. $\log_3 243$

65. $\log_2 \frac{1}{16}$

66. $\log_4 \frac{1}{4}$

67. $\log_2 64$

■ In Exercises 68 to 71, solve for x. Round to the nearest ten-thousandth.

68. $\log_4 x = 3$

69. $\log_3 x = \frac{1}{3}$

70. $\ln x = 2.5$

71. $\log x = 2.4$

72. Astronomy Use the distance modulus function $M(r) = 5 \log r - 5$, where M is the distance modulus and r is the distance of a star from Earth in parsecs, to find the distance in parsecs to a star whose distance modulus is 3.2. Round to the nearest tenth of a parsec.

73. Sound The number of decibels, D, of a sound can be given by the function $D(I) = 10(\log I + 16)$, where I is the power of the sound measured in watts. The pain threshold for a sound for most humans is approximately 0.01 watt. Find the number of decibels of this sound.

CHAPTER 10 TEST

■ In Exercises 1 and 2, evaluate the function for the given value of the independent variable.

1. $s(t) = -3t^2 + 4t - 1;\ t = -2$

2. $f(x) = 3^{x-4};\ x = 2$

3. Evaluate: $\log_5 125$

4. Solve for x: $\log_6 x = 2$

■ In Exercises 5 to 8, graph the function.

5. $f(x) = 2x - 3$

6. $f(x) = x^2 + 2x - 3$

7. $f(x) = 2^x - 5$

8. $f(x) = \log_3(x - 1)$

9. Find the slope of the line that passes through $(3, -1)$ and $(-2, -4)$.

10. Find the equation of the line that passes through $(3, 5)$ and has slope $\frac{2}{3}$.

11. Find the vertex of the graph of $f(x) = x^2 + 6x - 1$.

12. Find the x-intercepts of the parabola given by the equation $y = x^2 + 2x - 8$.

13. Find the minimum or maximum value of the quadratic function $y = -x^2 - 3x + 10$ and state whether the value is a minimum or a maximum.

14. **Travel** The distance d, in miles, of a small plane from its final destination is given by $d(t) = 250 - 100t$, where t is the time, in hours, remaining for the flight. Find and discuss the meaning of the intercepts of the graph of the function.

15. **Sports** The height h, in feet, of a ball that is thrown straight up and released 8 ft above the ground is given by $h(t) = -16t^2 + 48t + 8$, where t is the time in seconds. Find the maximum height the ball attains.

16. **Isotopes** A radioactive isotope has a half-life of 5 h. If a chemist has a 10-gram sample of this isotope, what amount will remain after 8 h? Use the function $A(t) = 10\left(\frac{1}{2}\right)^{t/5}$, where A is the amount of the isotope, in grams, remaining after t hours. Round to the nearest hundredth of a gram.

17. **Earthquakes** Two earthquakes struck Papua, Indonesia, in February 2004. One had a magnitude of 7.0 on the Richter scale, and the second had a magnitude of 7.3 on the Richter scale. How many times as great was the intensity of the second earthquake than that of the first? Round to the nearest tenth.

18. **Farming** The manager of an orange grove has determined that when there are 320 trees per acre, the average yield per tree is 260 lb of oranges. If the number of trees is increased to 330 trees per acre, the average yield per tree decreases to 245 lb. Find a linear model for the average yield per tree as a function of the number of trees per acre.

19. A traffic engineer gathered data on the number of cars that traveled a rural highway and the number of accidents on the highway. The results are recorded in the table below.

Number of cars, x	2456	3421	2590	3346	3915	2894
Number of accidents, y	74	133	91	107	117	104

a. Use the points (2456, 74) and (3915, 117) to find a linear model that predicts the number of accidents that will occur given the number of cars on the road. Round the slope and y-intercept to the nearest thousandth.

b. Use your linear model to predict the number of accidents that will occur when there are 2600 cars traveling the road. Round to the nearest whole number.

20. **Light** The equation $\log P = -kd$ gives the relationship between the percent P, as a decimal, of light passing through a substance of thickness d. The value of k for a swimming pool is approximately 0.05. At what depth, in meters, will the percent of light be 75% of the light at the surface of the pool? Round to the nearest tenth.

11

The Mathematics of Finance

We interact, on a daily basis, with people who want to separate us from our money. Advertisers tempt us to spend our hard-earned cash with images of cars, furniture, cruises, fast food, technology, and a million other products and services. At the same time, we are encouraged by banks and investment companies to save money for retirement or educational expenses by investing in retirement plans or college savings plans.

There are a multitude of ways to pay for both our purchases and our investments. We can pay for purchases by using credit in the form of bank loans, credit card loans, and mortgages. We can invest money through employer-sponsored 401(k) plans, state-sponsored college savings plans, IRAs (individual retirement accounts), mutual funds, and other investment vehicles. To assess these offers of financial assistance, we use a branch of mathematics called finance. The mathematics of finance is used by bankers, financial planners, hedge fund managers, stockbrokers, and hopefully you, after you finish this chapter.

Many of us are wary of investing our money in the stock market because of the risk involved and because the stock market seems very complicated. To give you a quick preview of how mathematics can help you make informed decisions, imagine that you have decided to hire a financial planner to assist you with your investment planning. Some good information about certified financial planners—individuals who have passed a series of tests on financial planning—can be found on the website www.cfp.net.

Financial planners charge fees for their services. It is very important to understand how these fees are structured because this will impact how the value of your investment grows. Let's assume that you begin with \$1000 and are committed to adding \$1000 each year to your investment. You want to know how much money you will have earned by the end of 25 years. This amount will depend on both the interest rate you expect to earn and the fee (normally a percentage of the value of your investment portfolio) charged by your financial advisor. For our example, we will assume that your investment earns 5% interest per year, on average. If your financial advisor charges an annual fee of 1% of your investment portfolio, then at the end of 25 years, your investment will be worth about \$44,000. If your advisor charges a fee of 0.25% per year, then after 25 years your investment will be worth about \$49,000, which is \$5000 more! Learning about the mathematics of finance is well worth the effort.

SECTION 11.1 Simple Interest

Simple Interest

HISTORICAL NOTE

The earliest loans date back to 3000 B.C., and interest on those loans may have extended over generations, not 4 or 5 years, as is the case for today's typical car loan. One of the first written records of an interest rate occurs in the Code of Hammurabi. Hammurabi ruled Babylon from 1795 to 1750 B.C. He is known for being the first ruler to write a set of laws that defined people's rights. In this Code, he allowed annual interest rates to be as high as 30%.

When you deposit money in a bank—for example, in a savings account—you are permitting the bank to use your money. The bank may lend the deposited money to customers to buy cars or make renovations on their homes. The bank pays you for the privilege of using your money. The amount paid to you is called **interest**. If you are the one borrowing money from a bank, the amount you pay for the privilege of using that money is also called interest.

The amount deposited in a bank or borrowed from a bank is called the **principal**. The amount of interest paid is usually given as a percent of the principal. The percent used to determine the amount of interest is called the **interest rate**. If you deposit $1000 in a savings account paying 5% interest per year, $1000 is the principal and the annual interest rate is 5%.

Interest paid on the original principal is called **simple interest**. The formula used to calculate simple interest is given below.

Simple Interest Formula

The simple interest formula is

$$I = Prt$$

where I is the interest, P is the principal, r is the interest rate, and t is the time period.

INSTRUCTOR NOTE

Emphasize that the simple interest formula requires that the interest rate and the time have comparable units. If the *annual* interest rate is given, then the time must be in *years*. If a *monthly* interest rate is given (as on most credit cards), then the time must be in *months*.

In the simple interest formula, the time t is expressed in the same units as the rate. For example, if the rate is given as an annual interest rate, then the time is measured in years; if the rate is given as a monthly interest rate, then the time must be expressed in months.

Interest rates are most commonly expressed as annual interest rates. Therefore, unless stated otherwise, we will assume the interest rate is an annual interest rate.

Interest rates are generally given as percents. Before performing calculations involving an interest rate, write the interest rate as a decimal.

EXAMPLE 1 Calculate Simple Interest

Calculate the simple interest earned in 1 year on a deposit of $1000 if the interest rate is 5%.

Solution

Use the simple interest formula. Substitute the following values into the formula: $P = 1000$, $r = 5\% = 0.05$, and $t = 1$.

$$I = Prt$$
$$I = 1000(0.05)(1)$$
$$I = 50$$

The simple interest earned is $50.

CHECK YOUR PROGRESS 1 Calculate the simple interest earned in 1 year on a deposit of $500 if the interest rate is 4%.

Solution See page S37.

EXAMPLE 2 Calculate Simple Interest

Calculate the simple interest due on a 3-month loan of \$2000 if the interest rate is 6.5%.

Solution

Use the simple interest formula. Substitute the values $P = 2000$ and $r = 6.5\% = 0.065$ into the formula. Because the interest rate is an annual rate, the time must be measured in years: $t = \frac{3 \text{ months}}{1 \text{ year}} = \frac{3 \text{ months}}{12 \text{ months}} = \frac{3}{12}$.

$$I = Prt$$

$$I = 2000(0.065)\left(\frac{3}{12}\right)$$

$$I = 32.5$$

The simple interest due is \$32.50.

CHECK YOUR PROGRESS 2 Calculate the simple interest due on a 4-month loan of \$1500 if the interest rate is 5.25%.

Solution See page S37.

EXAMPLE 3 Calculate Simple Interest

Calculate the simple interest due on a 2-month loan of \$500 if the interest rate is 1.5% per month.

Solution

Use the simple interest formula. Substitute the values $P = 500$ and $r = 1.5\% = 0.015$ into the formula. Because the interest rate is *per month,* the time period of the loan is expressed as the number of months: $t = 2$.

$$I = Prt$$

$$I = 500(0.015)(2)$$

$$I = 15$$

The simple interest due is \$15.

CHECK YOUR PROGRESS 3 Calculate the simple interest due on a 5-month loan of \$700 if the interest rate is 1.25% per month.

Solution See page S38.

POINT OF INTEREST

A PIRG (Public Interest Research Group) survey found that 29% of credit reports contained errors that could result in the denial of a loan. This is why financial advisors recommend that consumers check their credit ratings.

Remember that in the simple interest formula, time t is measured in the same period as the interest rate. Therefore, if the time period of a loan with an annual interest rate is given in days, it is necessary to convert the time period of the loan to a fractional part of a year. There are two methods for converting time from days to years: the exact method and the ordinary method. Using the exact method, the number of days of the loan is divided by 365, the number of days in a year.

$$\textbf{Exact method: } t = \frac{\text{number of days}}{365}$$

The ordinary method is based on there being an average of 30 days in a month and 12 months in a year ($30 \cdot 12 = 360$). Using this method, the number of days of the loan is divided by 360.

$$\textbf{Ordinary method: } t = \frac{\text{number of days}}{360}$$

The ordinary method is used by most businesses. Therefore, unless otherwise stated, the ordinary method will be used in this text.

EXAMPLE 4 Calculate Simple Interest

Calculate the simple interest due on a 45-day loan of \$3500 if the annual interest rate is 8%.

Solution

Use the simple interest formula. Substitute the following values into the formula: $P = 3500$, $r = 8\% = 0.08$, and $t = \frac{\text{number of days}}{360} = \frac{45}{360}$.

$$I = Prt$$

$$I = 3500(0.08)\left(\frac{45}{360}\right)$$

$$I = 35$$

The simple interest due is \$35.

CHECK YOUR PROGRESS 4 Calculate the simple interest due on a 120-day loan of \$7000 if the annual interest rate is 5.25%.

Solution See page S38.

HISTORICAL NOTE

Some historians think that the dollar sign evolved from the Mexican or Spanish letters P and S, for pesos or piastres or pieces of eight. The S gradually came to be written over the P. The \$ symbol was widely used before the adoption of the U.S. dollar in 1785.

The simple interest formula can be used to find the interest rate on a loan when the interest, principal, and time period of the loan are known. An example is given below.

EXAMPLE 5 Calculate the Simple Interest Rate

The simple interest charged on a 6-month loan of \$3000 is \$150. Find the simple interest rate.

Solution

Use the simple interest formula. Solve the equation for r.

$$I = Prt$$

$$150 = 3000(r)\left(\frac{6}{12}\right)$$

$$150 = 1500r \qquad \bullet\ 3000\left(\frac{6}{12}\right) = 1500$$

$$0.10 = r \qquad \bullet\ \text{Divide each side of the equation by 1500.}$$

$$r = 10\% \qquad \bullet\ \text{Write the decimal as a percent.}$$

The simple interest rate on the loan is 10%.

CHECK YOUR PROGRESS 5 The simple interest charged on a 6-month loan of \$12,000 is \$462. Find the simple interest rate.

Solution See page S38.

MATH MATTERS Is the American Dream Dead?

This chapter is about money, a topic of great importance to all of us. The possibility of going "from rags to riches" is a dream that has long attracted people from all over the world to the United States. Because the United States has such a large middle class, the leap from poverty to wealth, or even from the middle class to the upper class, is much easier to make than in almost any other country in the world. But many feel that the "American Dream" of starting out with nothing and working one's way to the top has become unattainable. Do you think this is true? Has the gap between the rich and the poor in the United States become insurmountable? Furthermore, do you think it is the government's responsibility to restore the middle class and decrease the income gap between the rich and the poor? Sixty-six percent of all Americans believe the government should be doing more in this regard. The breakdown by political affiliation is shown in the figure below. (*Source:* www.pbs.org (http://www.pbs.org/newshour/making-sense/why-half-of-u-s-adults-no-longer-believe-in-the-american-dream/)

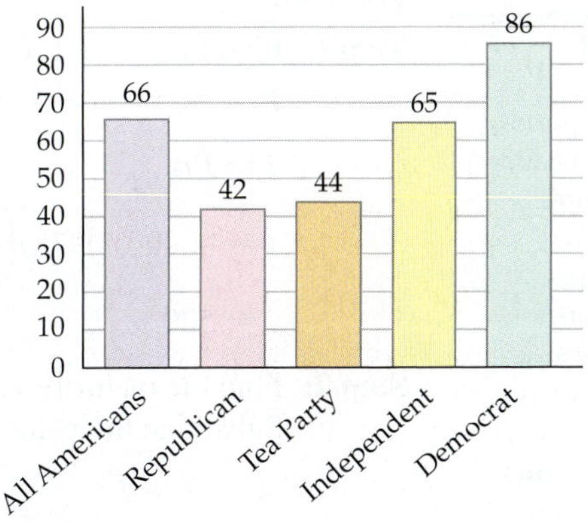

FIGURE 11.1 The percent of Americans who agree that the government should do more to reduce the gap between the rich and the poor, by political affiliation.

Future Value and Maturity Value

When you borrow money, the total amount to be repaid to the lender is the sum of the principal and interest. This sum is calculated using the following future value or maturity value formula for simple interest.

Future Value or Maturity Value Formula for Simple Interest

The future or maturity value formula for simple interest is

$$A = P + I$$

where A is the amount after the interest, I, has been added to the principal, P.

This formula can be used for loans or investments. When used for a loan, A is the total amount to be repaid to the lender; this sum is called the **maturity value** of the loan. In Example 5, the simple interest charged on the loan of \$3000 was \$150. The maturity value of the loan is therefore \$3000 + \$150 = \$3150.

POINT OF INTEREST

You may have heard the term *legal tender.* Section 102 of the Coinage Act of 1965 states, in part, that

> All coins and currencies of the United States, regardless of when coins are issued, shall be legal tender for all debts, public and private, public charges, taxes, duties, and dues.

This means that when you offer U.S. currency to a creditor, you have made a valid and legal offer of payment of your debt.

For an investment, such as a deposit in a bank savings account, A is the total amount on deposit after the interest earned has been added to the principal. This sum is called the **future value** of the investment.

QUESTION Is the stated sum a maturity value or a future value?

a. The sum of the principal and the interest on an investment

b. The sum of the principal and the interest on a loan

EXAMPLE 6 Calculate a Maturity Value

Calculate the maturity value of a simple interest, 8-month loan of \$8000 if the interest rate is 9.75%.

Solution

Step 1: Find the interest. Use the simple interest formula. Substitute the values $P = 8000$, $r = 9.75\% = 0.0975$, and $t = \frac{8}{12}$ into the formula.

$$I = Prt$$

$$I = 8000(0.0975)\left(\frac{8}{12}\right)$$

$$I = 520$$

Step 2: Find the maturity value. Use the maturity value formula for simple interest. Substitute the values $P = 8000$ and $I = 520$ into the formula.

$$A = P + I$$

$$A = 8000 + 520$$

$$A = 8520$$

The maturity value of the loan is \$8520.

POINT OF INTEREST

There are three major categories of loans used to pay for education: (1) student loans (for example, Stafford and Perkins loans), (2) parent loans (for example, PLUS loans), and (3) private student loans, which are also called alternative student loans. There is also a fourth type, called a consolidation loan, that allows borrowers to combine all their loans into one loan with a single payment.

CHECK YOUR PROGRESS 6 Calculate the maturity value of a simple interest, 9-month loan of \$4000 if the interest rate is 8.75%.

Solution See page S38.

Recall that the simple interest formula states that $I = Prt$. We can substitute Prt for I in the future or maturity value formula, as follows.

$$A = P + I$$

$$A = P + Prt$$

$$\mathbf{A = P(1 + rt)} \qquad \text{• Factor } P \text{ from each term.}$$

In the final equation, A is the future value of an investment or the maturity value of a loan, P is the principal, r is the interest rate, and t is the time period.

We used the formula $A = P + I$ in Example 6. The formula $A = P(1 + rt)$ is used in Examples 7 and 8. Note that two steps were required to find the solution in Example 6, but only one step is required in Examples 7 and 8.

ANSWER **a.** Future value **b.** Maturity value

EXAMPLE 7 Calculate the Maturity Value Using $A = P(1 + rt)$

Calculate the maturity value of a simple interest, 3-month loan of \$3800. The interest rate is 6%.

Solution

Substitute the following values into the formula $A = P(1 + rt)$: $P = 3800$, $r = 6\% = 0.06$, and $t = \frac{3}{12}$.

$$A = P(1 + rt)$$
$$A = 3800\left[1 + 0.06\left(\frac{3}{12}\right)\right]$$
$$A = 3800(1 + 0.015)$$
$$A = 3800(1.015)$$
$$A = 3857$$

The maturity value of the loan is \$3857.

CHECK YOUR PROGRESS 7 Calculate the maturity value of a simple interest, 1-year loan of \$6700. The interest rate is 8.9%.

Solution See page S38.

EXAMPLE 8 Calculate the Future Value Using $A = P(1 + rt)$

Find the future value after 1 year of \$850 in an account earning 8.2% simple interest.

Solution

Because $t = 1$, $rt = r(1) = r$. Therefore, $1 + rt = 1 + r = 1 + 0.082 = 1.082$.

$$A = P(1 + rt)$$
$$A = 850(1.082)$$
$$A = 919.7$$

The future value of the account after 1 year is \$919.70.

CHECK YOUR PROGRESS 8 Find the future value after 1 year of \$680 in an account earning 6.4% simple interest.

Solution See page S38.

POINT OF INTEREST

The bill of largest value ever printed in the United States was a \$100,000 bill. The statesman featured on the bill was Woodrow Wilson, the 28th president of the United States. The bill was used for transactions between Federal Reserve Banks only.

B Christopher/Alamy Stock Photo

Recall that the formula $A = P + I$ states that A is the amount after the interest has been added to the principal. Subtracting P from each side of this equation yields the following formula.

$$I = A - P$$

This formula states that the amount of interest paid is equal to the total amount minus the principal. This formula is used in Example 9.

EXAMPLE 9 Calculate the Simple Interest Rate

The maturity value of a 3-month loan of \$4000 is \$4085. What is the simple interest rate?

Solution

First find the amount of interest paid. Subtract the principal from the maturity value.

$$I = A - P$$
$$I = 4085 - 4000$$
$$I = 85$$

Find the simple interest rate by solving the simple interest formula for r.

$$I = Prt$$

$$85 = 4000(r)\left(\frac{3}{12}\right)$$

$$85 = 1000r \qquad \bullet\ 4000\left(\frac{3}{12}\right) = 1000$$

$$0.085 = r \qquad \bullet\ \text{Divide each side of the equation by 1000.}$$

$$r = 8.5\% \qquad \bullet\ \text{Write the decimal as a percent.}$$

The simple interest rate on the loan is 8.5%.

CHECK YOUR PROGRESS 9 The maturity value of a 4-month loan of \$9000 is \$9240. What is the simple interest rate?

Solution See page S38.

EXCURSION

Interest on a Car Loan

If you have a car loan, there is a good chance that each loan payment includes principal (the amount that goes toward paying off the original loan) and interest (the amount you are charged for borrowing), which is based on the outstanding balance. The *outstanding balance* refers to the portion of the original debt that has not yet been repaid. Because you pay some part of the principal each month, the outstanding balance decreases each month.

Here is an example. Suppose you obtain a 4-year car loan of \$10,000 at an annual interest rate of 5%. The monthly payment on the loan is \$230.29. Later in the chapter, we will show how that monthly payment is calculated.

One month after you obtain the loan, when your first payment is due, you still owe a total of \$10,000 *plus interest.* The amount of interest owed can be calculated using the formula $I = Prt$. The principal is \$10,000, the interest rate is $\frac{0.05}{12}$ (the annual interest rate, expressed as a decimal, divided by the number of months in a year), and the time is 1 month.

$$\begin{aligned} I &= Prt \\ &= 10{,}000\left(\frac{0.05}{12}\right)(1) \\ &\approx 41.67 \end{aligned}$$

The interest you owe for the first month is \$41.67. By subtracting this amount from the monthly payment of \$230.29, you can calculate the amount of the first payment that goes toward repaying the principal:

$$\$230.29 - \$41.67 = \$188.62$$

So, with the first payment, you have repaid \$188.62 of the original \$10,000 loan. You now owe $\$10{,}000 - \$188.62 = \$9811.38$.

After two months, you owe \$9811.38 plus the interest due on that amount. Performing a similar calculation, we have

$$I = Prt = 9811.38\left(\frac{0.05}{12}\right)(1) \approx 40.88$$

Note that the interest you owe for the second month is less than the interest you owed for the first month. This is because you have already paid back some of the loan. This time, the amount that goes toward reducing the principal is:

$$\$230.29 - \$40.88 = \$189.41$$

After the second payment is made, you have repaid \$189.41 of the \$9811.38 outstanding loan balance. You now owe $\$9811.38 - \$189.41 = \$9621.97$.

Notice that the amount you paid toward reducing the principal is greater than the amount you paid toward the principal in the first month. Every month, more and more of your payment will go toward reducing the principal, and less and less will go toward interest charges.

EXCURSION EXERCISES

1. Find the amount of interest you owe for the third month.
2. How much of your payment goes toward reducing the principal in the third month?
3. How much do you owe after making the payment for the third month?
4. Find the amount of interest you owe for the fourth month.
5. How much of your payment goes toward reducing the principal in the fourth month?
6. How much do you owe after making the payment for the fourth month?
7. Your last payment of \$230.29 includes the remaining balance of the loan plus the interest due on that remaining balance. Find the remaining balance of the loan for the last payment.

EXERCISE SET 11.1

1. Explain how to convert a number of months to a fractional part of a year.
2. Explain how to convert a number of days to a fractional part of a year.
3. Explain what each variable in the simple interest formula represents.

■ In Exercises 4 to 17, calculate the simple interest earned. Round to the nearest cent.

4. $P = \$2000$, $r = 6\%$, $t = 1$ year
5. $P = \$8000$, $r = 7\%$, $t = 1$ year
6. $P = \$3000$, $r = 5.5\%$, $t = 6$ months
7. $P = \$7000$, $r = 6.5\%$, $t = 6$ months
8. $P = \$4200$, $r = 8.5\%$, $t = 3$ months
9. $P - \$9000$, $r = 6.75\%$, $t = 4$ months
10. $P = \$12{,}000$, $r = 7.8\%$, $t = 45$ days
11. $P = \$3000$, $r = 9.6\%$, $t = 21$ days
12. $P = \$4000$, $r = 8.4\%$, $t = 33$ days
13. $P = \$7000$, $r = 7.2\%$, $t = 114$ days
14. $P = \$800$, $r = 1.5\%$ monthly, $t = 3$ months
15. $P = \$2000$, $r = 1.25\%$ monthly, $t = 5$ months
16. $P = \$3500$, $r = 1.8\%$ monthly, $t = 4$ months
17. $P = \$1600$, $r = 1.75\%$ monthly, $t = 6$ months

■ In Exercises 18 to 23, use the formula $A = P(1 + rt)$ to calculate the maturity value of the simple interest loan.

18. $P = \$8500$, $r = 6.8\%$, $t = 6$ months
19. $P = \$15{,}000$, $r = 8.9\%$, $t = 6$ months
20. $P = \$4600$, $r = 9.75\%$, $t = 4$ months
21. $P = \$7200$, $r = 7.95\%$, $t = 4$ months
22. $P = \$13{,}000$, $r = 1.4\%$ monthly, $t = 3$ months
23. $P = \$2800$, $r = 9.2\%$, $t = 3$ months

■ In Exercises 24 to 29, calculate the simple interest rate.

24. $P = \$8000$, $I = \$500$, $t = 1$ year
25. $P = \$1600$, $I = \$120$, $t = 1$ year
26. $P = \$4000$, $I = \$190$, $t = 6$ months
27. $P = \$2000$, $I = \$80$, $t = 6$ months
28. $P = \$500$, $I = \$10.25$, $t = 3$ months
29. $P = \$1200$, $I = \$37.20$, $t = 4$ months

30. Simple Interest Calculate the simple interest earned in 1 year on a deposit of \$1900 if the interest rate is 8%.

31. Simple Interest You deposit \$1500 in an account earning 5.2% interest. Calculate the simple interest earned in 6 months.

32. Simple Interest Calculate the simple interest due on a 2-month loan of \$800 if the interest rate is 1.5% per month.

33. Simple Interest Calculate the simple interest due on a 45-day loan of \$1600 if the interest rate is 9%.

34. Simple Interest Calculate the simple interest due on a 150-day loan of \$4800 if the interest rate is 7.25%.

35. Maturity Value Calculate the maturity value of a simple interest, 8-month loan of \$7000 if the interest rate is 8.7%.

36. Maturity Value Calculate the maturity value of a simple interest, 10-month loan of \$6600 if the interest rate is 9.75%.

37. Maturity Value Calculate the maturity value of a simple interest, 1-year loan of \$5200. The interest rate is 5.1%.

38. Future Value You deposit \$880 in an account paying 9.2% simple interest. Find the future value of the investment after 1 year.

39. Future Value You deposit \$750 in an account paying 7.3% simple interest. Find the future value of the investment after 1 year.

40. Simple Interest Rate The simple interest charged on a 6-month loan of \$6000 is \$270. Find the simple interest rate.

41. Simple Interest Rate The simple interest charged on a 6-month loan of \$18,000 is \$918. Find the simple interest rate.

42. Simple Interest Rate The maturity value of a 4-month loan of \$3000 is \$3097. Find the simple interest rate.

43. Simple Interest Rate Find the simple interest rate on a 3-month loan of \$5000 if the maturity value of the loan is \$5125.

44. Late Payments Your property tax bill is \$1200. The county charges a penalty of 11% simple interest for late payments. How much do you owe if you pay the bill 2 months past the due date?

45. Late Payments Your electric bill is \$132. You are charged 9% simple interest for late payments. How much do you owe if you pay the bill 1 month past the due date?

46. Certificate of Deposit If you withdraw part of your money from a certificate of deposit before the date of maturity, you must pay an interest penalty. Suppose you invested \$5000 in a 1-year certificate of deposit paying 8.5% interest. After 6 months, you decide to withdraw \$2000. Your interest penalty is 3 months simple interest on the \$2000. What interest penalty do you pay?

47. Maturity Value \$10,000 is borrowed for 140 days at an 8% interest rate. Calculate the maturity value by the exact method and by the ordinary method. Which method yields the greater maturity value? Who benefits from using the ordinary method rather than the exact method, the borrower or the lender?

EXTENSIONS

48. Interest has been described as a rental fee for money. Explain why this is an apt description of interest.

49. On July 31, at 4 P.M., you open a money market account that pays 5% interest, compounded daily, and you deposit \$500 in the account. Your deposit is credited as of August 1. At the beginning of September, you receive a statement from the bank that shows that during the month of August, you received \$2.15 in interest. The interest has been added to your account, bringing the total deposit to \$502.15. At the beginning of October, you receive a statement from the bank that shows that during the month of September, you received \$2.09 in interest on the \$502.15 on deposit. Why did you receive less interest during the second month, when there was more money on deposit?

SECTION 11.2 Compound Interest

Compound Interest

Simple interest is generally used for loans of 1 year or less. For loans of more than 1 year, the interest paid on the money borrowed is called *compound interest.* **Compound interest** is interest calculated not only on the original principal, but also on any interest that has already been earned.

POINT OF INTEREST

Warren Buffet, one of the richest men in the world, is sometimes referred to as "the Oracle of Omaha," a nod to his hometown of Omaha, Nebraska. When asked about the secret to his success, Buffet replied, "My wealth has come from a combination of living in America, some lucky genes, and compound interest." Compound interest is the topic of this section.

Lacy O'Toole/CNBC/NBCU/Getty Images

To illustrate compound interest, suppose you deposit \$1000 in a savings account earning 5% interest, compounded annually (once a year).

During the first year, the interest earned is calculated as follows.

$$I = Prt$$
$$I = \$1000(0.05)(1) = \$50$$

At the end of the first year, the total amount in the account is

$$A = P + I$$
$$A = \$1000 + \$50 = \$1050$$

During the second year, the interest earned is calculated using the amount in the account at the end of the first year.

$$I = Prt$$
$$I = \$1050(0.05)(1) = \$52.50$$

Note that the interest earned during the second year (\$52.50) is greater than the interest earned during the first year (\$50). This is because the interest earned during the first year was added to the original principal, and the interest for the second year was calculated using this sum. If the account earned simple interest rather than compound interest, the interest earned each year would be the same (\$50).

At the end of the second year, the total amount in the account is the sum of the amount in the account at the end of the first year and the interest earned during the second year.

$$A = P + I$$
$$A = \$1050 + 52.50 = \$1102.50$$

The interest earned during the third year is calculated using the amount in the account at the end of the second year (\$1102.50).

$$I = Prt$$
$$I = \$1102.50(0.05)(1) = \$55.125 \approx \$55.13$$

The interest earned each year keeps increasing. This is the effect of compound interest.

In this example, the interest is compounded annually. However, compound interest can be compounded semiannually (twice a year), quarterly (four times a year), monthly, or daily. The frequency with which the interest is compounded is called the **compounding period**.

If, in the preceding example, interest is compounded quarterly rather than annually, then the first interest payment on the \$1000 in the account occurs after 3 months $\left(t = \frac{3}{12} = \frac{1}{4};\ \text{3 months is one-quarter of a year}\right)$. That interest is then added to the account, and the interest earned for the second quarter is calculated using that sum.

End of 1st quarter: $I = Prt = \$1000(0.05)\left(\frac{3}{12}\right) = \$12.50$

$A = P + I = \$1000 + \$12.50 = \$1012.50$

End of 2nd quarter: $I = Prt = \$1012.50(0.05)\left(\frac{3}{12}\right) = \$12.65625 \approx \$12.66$

$A = P + I = \$1012.50 + \$12.66 = \$1025.16$

End of 3rd quarter: $I = Prt = \$1025.16(0.05)\left(\frac{3}{12}\right) = \$12.8145 \approx \$12.81$

$A = P + I = \$1025.16 + \$12.81 = \$1037.97$

End of 4th quarter: $I = Prt = \$1037.97(0.05)\left(\frac{3}{12}\right) = \$12.974625 \approx \$12.97$

$A = P + I = \$1037.97 + \$12.97 = \$1050.94$

The total amount in the account at the end of the first year is \$1050.94.

POINT OF INTEREST

The U.S. Bureau of Engraving and Printing is responsible for printing our currency. In addition to printing \$1, \$2, \$5, \$10, \$20, \$50, and \$100 bills, the Bureau also used to print \$500, \$1000, \$5000, and \$10,000 bills. Production of these large-denomination bills was stopped during World War II. In 1969, the Federal Reserve Board stopped distribution of these large-denomination bills. Below is an image of a \$5000 note.

PAUL J. RICHARDS/AAFP/Getty Images

When the interest is compounded quarterly, the account earns more interest ($50.94) than when the interest is compounded annually ($50). In general, an increase in the number of compounding periods results in an increase in the interest earned by an account.

In the example above, the formulas $I = Prt$ and $A = P + I$ were used to show the amount of interest added to the account each quarter. The formula $A = P(1 + rt)$ can be used to calculate A at the end of each quarter. For example, the amount in the account at the end of the first quarter is

$$A = P(1 + rt)$$
$$A = 1000\left[1 + 0.05\left(\frac{3}{12}\right)\right]$$
$$A = 1000(1.0125)$$
$$A = 1012.50$$

This amount, $1012.50, is the same as the amount calculated on the preceding page using the formula $A = P + I$ to find the amount at the end of the first quarter.

The formula $A = P(1 + rt)$ is used in Example 1.

EXAMPLE 1 Calculate Future Value

You deposit $500 in an account earning 6% interest, compounded semiannually. How much is in the account at the end of 1 year?

Solution

The interest is compounded every 6 months. Calculate the amount in the account after the first 6 months. $t = \frac{6}{12}$.

$$A = P(1 + rt)$$
$$A = 500\left[1 + 0.06\left(\frac{6}{12}\right)\right]$$
$$A = 515$$

Calculate the amount in the account after the second 6 months.

$$A = P(1 + rt)$$
$$A = 515\left[1 + 0.06\left(\frac{6}{12}\right)\right]$$
$$A = 530.45$$

The total amount in the account at the end of 1 year is $530.45.

CHECK YOUR PROGRESS 1 You deposit $2000 in an account earning 4% interest, compounded monthly. How much is in the account at the end of 6 months?

Solution See page S38.

POINT OF INTEREST

The top seven sources of financial news for professionals in the field of finance are listed below.

- MarketWatch News Viewer
- Boomerang Portal
- Reuters
- Forbes
- *The Wall Street Journal*
- *The Financial Times*
- CNBC

(*Source:* www.investopedia.com)

In calculations that involve compound interest, the sum of the principal and the interest that has been added to it is called the **compound amount**. In Example 1, the compound amount is $530.45.

The calculations necessary to determine compound interest and compound amounts can be simplified by using a formula. Consider an amount P deposited into an account paying an annual interest rate r, compounded annually.

The interest earned during the first year is

$$I = Prt$$
$$I = Pr(1) \quad \bullet\ t = 1.$$
$$I = Pr$$

The compound amount A in the account after 1 year is the sum of the original principal and the interest earned during the first year:

$A = P + I$

$A = P + Pr$

$A = P(1 + r)$ • Factor P from each term.

During the second year, the interest is calculated on the compound amount at the end of the first year, $P(1 + r)$.

$I = Prt$

$I = P(1 + r)r(1)$ • Replace P with $P(1 + r)$; $t = 1$.

$I = P(1 + r)r$

The compound amount A in the account after 2 years is the sum of the compound amount at the end of the first year and the interest earned during the second year:

$A = P + I$

$A = P(1 + r) + P(1 + r)r$ • Replace P with $P(1 + r)$ and I with $P(1 + r)r$.

$A = 1[P(1 + r)] + [P(1 + r)]r$

$A = P(1 + r)(1 + r)$ • Factor $P(1 + r)$ from each term.

$A = P(1 + r)^2$ • Write $(1 + r)(1 + r)$ as $(1 + r)^2$.

During the third year, the interest is calculated on the compound amount at the end of the second year, $P(1 + r)^2$.

$I = Prt$

$I = P(1 + r)^2r(1)$ • Replace P with $P(1 + r)^2$; $t = 1$.

$I = P(1 + r)^2r$

The compound amount A in the account after 3 years is the sum of the compound amount at the end of the second year and the interest earned during the third year:

$A = P + I$

$A = P(1 + r)^2 + P(1 + r)^2r$ • Replace P with $P(1 + r)^2$ and I with $P(1 + r)^2r$.

$A = 1[P(1 + r)^2] + [P(1 + r)^2]r$

$A = P(1 + r)^2(1 + r)$ • Factor $P(1 + r)^2$ from each term.

$A = P(1 + r)^3$ • Write $(1 + r)^2(1 + r)$ as $(1 + r)^3$.

Note that the compound amount at the end of each year is the previous year's compound amount multiplied by $(1 + r)$. The exponent on $(1 + r)$ is equal to the number of compounding periods. Generalizing from this, we can state that the compound amount after n years is $A = P(1 + r)^n$.

In deriving this equation, interest was compounded annually; therefore, $t = 1$. Applying a similar argument for more frequent compounding periods, we derive the following compound amount formula. This formula enables us to calculate the compound amount for any number of compounding periods per year.

Compound Amount Formula

The compound amount formula is

$$A = P\left(1 + \frac{r}{n}\right)^{nt}$$

where A is the compound amount, P is the amount of money deposited, r is the annual interest rate, n is the number of compounding periods per year, and t is the number of years.

POINT OF INTEREST

It is believed that U.S. currency is green because at the time of the introduction of the smaller-size bills in 1929, green pigment was readily available in large quantities. The color was resistant to chemical and physical changes, and green was psychologically associated with strong, stable government credit. (*Source*: www.moneyfactory.gov)

Alex Wong/Getty Images

Values of n (Number of Compounding Periods per Year)

If interest is compounded	then n =
annually	1
semiannually	2
quarterly	4
monthly	12
daily	360

To illustrate how to determine the values of the variables in this formula, consider depositing \$5000 in an account earning 6% interest, compounded quarterly, for a period of 3 years.

The amount deposited is P:	$P = 5000$
The annual interest rate is 6%:	$r = 6\% = 0.06$
When interest is compounded quarterly, there are four compounding periods per year:	$n = 4$
The time is 3 years:	$t = 3$

Recall that compound interest can be compounded annually (once a year), semiannually (twice a year), quarterly (four times a year), monthly, or daily. The possible values of n (the number of compounding periods per year) are recorded in the table at the left.

PPT

QUESTION What is the value of n when interest is compounded monthly?

Recall that the future value of an investment is the value of the investment after the original principal has been invested for a period of time. In other words, it is the principal plus the interest earned. Therefore, it is the compound amount A in the compound amount formula.

EXAMPLE 2 Calculate the Compound Amount

Calculate the compound amount when \$10,000 is deposited in an account earning 8% interest, compounded semiannually, for 4 years.

Solution

Use the compound amount formula.

$P = 10{,}000,\ r = 8\% = 0.08,\ n = 2,\ t = 4$

$$A = P\left(1 + \frac{r}{n}\right)^{nt}$$
$$A = 10{,}000\left(1 + \frac{0.08}{2}\right)^{2\cdot 4}$$
$$A = 10{,}000(1 + 0.04)^{8}$$
$$A = 10{,}000(1.04)^{8}$$
$$A \approx 10{,}000(1.368569)$$
$$A \approx 13{,}685.69$$

The compound amount after 4 years is approximately \$13,685.69.

CHECK YOUR PROGRESS 2 Calculate the compound amount when \$4000 is deposited in an account earning 6% interest, compounded monthly, for 2 years.

Solution See page S39.

TAKE NOTE

When using the compound amount formula, write the interest rate r as a decimal.

CALCULATOR NOTE

When using a scientific calculator to solve the compound amount formula for A, use the keystroking sequence

1 [+] r [÷] n [=] [y^x] [(] n [x] t [)] [x] P [=]

When using a graphing calculator, use the sequence

P [(] 1 [+] r [÷] n [)] [^] [(] n [x] t [)] [ENTER]

ANSWER $n = 12$.

EXAMPLE 3 Calculate Future Value

Calculate the future value of \$5000 earning 9% interest, compounded daily, for 3 years.

Solution

Use the compound amount formula.
$P = 5000,\ r = 9\% = 0.09,\ n = 360,\ t = 3$

$$A = P\left(1 + \frac{r}{n}\right)^{nt}$$
$$A = 5000\left(1 + \frac{0.09}{360}\right)^{360 \cdot 3}$$
$$A = 5000(1 + 0.00025)^{1080}$$
$$A = 5000(1.00025)^{1080}$$
$$A \approx 5000(1.3099202)$$
$$A \approx 6549.60$$

The future value after 3 years is approximately \$6549.60.

INSTRUCTOR NOTE

You might have students assume that they invest \$2500 in an IRA today and that they don't add any additional money to the account before they retire. Have them find the value of the account in 40 or 45 years. Here are a few examples: After 40 years, compounded monthly, at a 7% annual interest rate, the value will be \$40,778.53. After 45 years, compounded monthly, at an 8% annual interest rate, the value will be \$90,409.00.

CHECK YOUR PROGRESS 3 Calculate the future value of \$2500 earning 9% interest, compounded daily, for 4 years.

Solution See page S39.

The formula $I = A - P$ was used in Section 11.1 to find the interest earned on an investment or the interest paid on a loan. This same formula is used for compound interest. It is used in Example 4 to find the interest earned on an investment.

POINT OF INTEREST

The majority of Americans say their children should start saving for retirement earlier than they did. The figure below shows the ages at which members of each generation started saving for retirement, and the amounts they are saving. (*Source:* money.cnn.com)

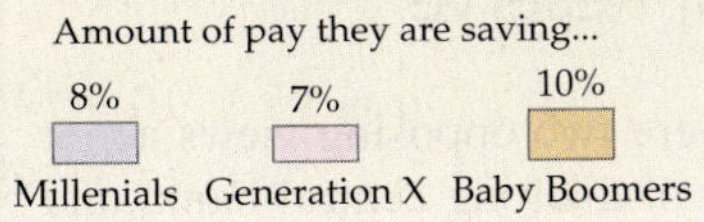

EXAMPLE 4 Calculate Compound Interest

How much interest is earned in 2 years on \$4000 deposited in an account paying 6% interest, compounded quarterly?

Solution

Calculate the compound amount. Use the compound amount formula.
$P = 4000,\ r = 6\% = 0.06,\ n = 4,\ t = 2$

$$A = P\left(1 + \frac{r}{n}\right)^{nt}$$
$$A = 4000\left(1 + \frac{0.06}{4}\right)^{4 \cdot 2}$$
$$A = 4000(1 + 0.015)^{8}$$
$$A = 4000(1.015)^{8}$$
$$A \approx 4000(1.1264926)$$
$$A \approx 4505.97$$

Calculate the interest earned. Use the formula $I = A - P$.

$$I = A - P$$
$$I = 4505.97 - 4000$$
$$I = 505.97$$

The amount of interest earned is approximately \$505.97.

CHECK YOUR PROGRESS 4 How much interest is earned in 6 years on \$8000 deposited in an account paying 9% interest, compounded monthly?

Solution See page S39.

An alternative to using the compound amount formula is to use a calculator that has finance functions built into it. The TI-83/84 graphing calculator is used in Example 5.

EXAMPLE 5 Calculate Compound Interest

Use the finance feature of a calculator to determine the compound amount when \$2000 is deposited in an account earning an interest rate of 12%, compounded quarterly, for 10 years.

Solution

On a TI-83/84 calculator, press [APPS] [ENTER]. The [APPS] key has red lettering on a black key.

Press [ENTER] to select 1: TVM Solver. TVM stands for time value of money.

N is the number of compounding periods, or $n \cdot t$ in the compound amount formula.

After N = , enter 4×10.

After I% = , enter 12.

After PV = , enter -2000. (See the note below.)

After PMT = , enter 0. Press [ENTER] twice.

After P/Y = , enter 4.

After C/Y = , enter 4.

Use the up arrow key to place the cursor at FV = .

Press [ALPHA] [Solve]. The [ALPHA] key has white lettering on a green key.

The solution is displayed to the right of FV = .

The compound amount is \$6524.08.

Note: For most financial calculators and financial computer programs such as Excel, money that is paid out (such as the \$2000 that is being deposited in this example) is entered as a negative number.

CHECK YOUR PROGRESS 5 Use the finance feature of a calculator to determine the compound amount when \$3500 is deposited in an account earning an interest rate of 6%, compounded semiannually, for 5 years.

Solution See page S39.

CALCULATOR NOTE

On the TI-83/84 calculator screen below, the variable N is the number of compounding periods, I% is the interest rate, PV is the present value of the money, PMT is the payment, P/Y is the number of payments per year, C/Y is the number of compounding periods per year, and FV is the future value of the money. TVM represents the time value of money.

Present value is discussed a little later in this section. Enter the principal for the present value amount.

"Solve" is located above the [ENTER] key on the calculator.

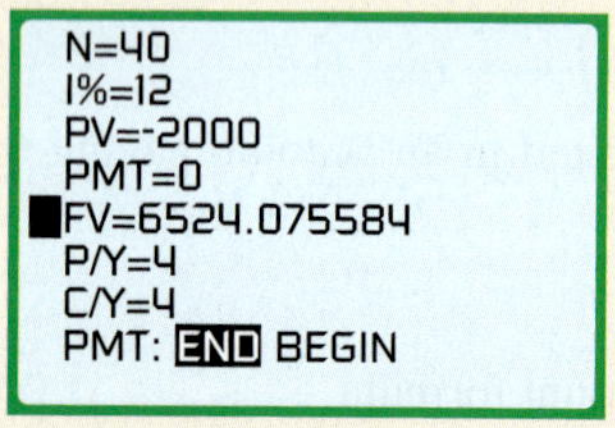

TI-83/84 calculator screen display

MATH MATTERS The Origins of the Federal Reserve

Joey Kotfica/Getty Images

The Federal Reserve Bank of San Francisco

In the late 1700s, before the Civil War began, there were two opposing views about banks in the United States. **Federalists** were in favor of a strong central bank, and **anti-Federalists** preferred to leave banks in the hands of the individual states. In 1791, the First Bank of the United States was created with a 20-year charter. Its main backer, Alexander Hamilton, was killed in a famous duel with Aaron Burr (who at the time was Thomas Jefferson's vice-president). When the charter expired, so did the first central bank. In 1816, the Second Bank of the United States was created, also with a 20-year charter, but President Andrew Jackson vetoed its renewal in 1832. The United States then entered the free banking era, which lasted until 1913, at which time the Third Bank of the United States was established. This bank was called the Federal Reserve System, and it is still the central bank of the United States today.

Present Value

The **present value** of an investment is the original principal invested, or the value of the investment before it earns any interest. Therefore, it is the principal, P, in the compound amount formula. Present value is used to determine how much money must be invested today in order for an investment to have a specific value at a future date.

The formula for the present value of an investment is found by solving the compound amount formula for P.

$$A = P\left(1+\frac{r}{n}\right)^{nt}$$

$$\frac{A}{\left(1+\frac{r}{n}\right)^{nt}} = \frac{P\left(1+\frac{r}{n}\right)^{nt}}{\left(1+\frac{r}{n}\right)^{nt}}$$

• Divide each side of the equation by $\left(1+\frac{r}{n}\right)^{nt}$.

$$\frac{A}{\left(1+\frac{r}{n}\right)^{nt}} = P$$

Present Value Formula

The present value formula is

$$P = \frac{A}{\left(1+\frac{r}{n}\right)^{nt}}$$

where P is the original principal invested, A is the compound amount, r is the annual interest rate, n is the number of compounding periods per year, and t is the number of years.

The present value formula is used in Example 6.

EXAMPLE 6 Calculate Present Value

How much money should be invested in an account that earns 8% interest, compounded quarterly, in order to have \$30,000 in 5 years?

Solution

Use the present value formula.
$A = 30{,}000$, $r = 8\% = 0.08$, $n = 4$, $t = 5$

$$P = \frac{A}{\left(1+\frac{r}{n}\right)^{nt}}$$

$$P = \frac{30{,}000}{\left(1+\frac{0.08}{4}\right)^{4\cdot 5}} = \frac{30{,}000}{1.02^{20}} \approx \frac{30{,}000}{1.485947396}$$

$$P \approx 20{,}189.14$$

\$20,189.14 should be invested in the account in order to have \$30,000 in 5 years.

CALCULATOR NOTE

To calculate P in the present value formula using a calculator, you can first calculate $\left(1+\frac{r}{n}\right)^{nt}$.

Next use the [1/x] key or the [x^{-1}] key. This will place the value of $\left(1+\frac{r}{n}\right)^{nt}$ in the denominator.

Then multiply by the value of A.

CHECK YOUR PROGRESS 6 How much money should be invested in an account that earns 9% interest, compounded semiannually, in order to have \$20,000 in 5 years?

Solution See page S39.

EXAMPLE 7 Calculate Present Value

Use the finance feature of a calculator to determine how much money should be invested in an account that earns 7% interest, compounded monthly, in order to have $50,000 in 10 years.

Solution

On a TI-83/84 calculator, press APPS ENTER.

Press ENTER to select 1 TVM Solver.

After N = , enter 12 × 10.

After I% = , enter 7. Press ENTER twice.

After PMT = , enter 0.

After FV = , enter 50000.

After P/Y = , enter 12. • Number of payments per year

After C/Y = , enter 12. • Number of compounding periods per year

Use the up arrow key to place the cursor at PV = .

Press ALPHA [Solve].

The solution is displayed to the right of PV = .

$24,879.81 should be invested in the account in order to have $50,000 in 10 years.

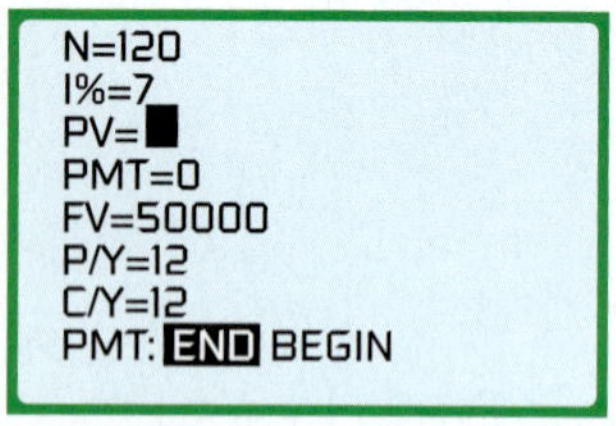

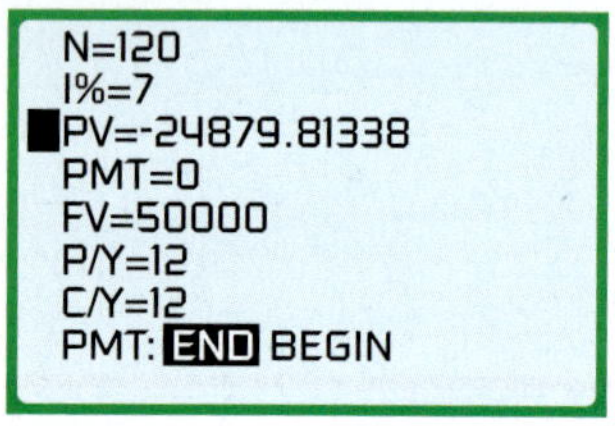

TI-83/84 calculator screen displays

CHECK YOUR PROGRESS 7 Use the finance feature of a calculator to determine how much money should be invested in an account that earns 6% interest, compounded daily, in order to have $25,000 in 15 years.

Solution See page S39.

Inflation

We have discussed compound interest and its effect on the growth of an investment. After your money has been invested for a period of time in an account that pays interest, you will have more money than you originally deposited. But does that mean you will be able to buy more with the compound amount than you were able to buy with the original investment at the time you deposited the money? The answer is not necessarily, and the reason is the effect of inflation.

Suppose the price of a large-screen Smart LED TV is $1500. You have enough money to purchase the TV, but decide instead to invest the $1500 in an account paying 6% interest, compounded monthly. After 1 year, the compound amount is $1592.52. But during that same year, the rate of inflation was 7%. The TV now costs

$$\begin{aligned}\$1500 \text{ plus } 7\% \text{ of } \$1500 &= \$1500 + 0.07(\$1500)\\ &= \$1500 + \$105\\ &= \$1605\end{aligned}$$

Because $\$1592.52 < \$1605$, you have actually lost purchasing power. At the beginning of the year, you had enough money to buy the large-screen LED TV; at the end of the year, the compound amount is not enough to pay for that same TV. Your money has actually lost value because it can buy less now than it could 1 year ago.

Inflation is an economic condition during which there are increases in the costs of goods and services. Inflation is expressed as a percent; for example, we speak of an annual inflation rate of 7%.

To calculate the effects of inflation, we use the same procedure we used to calculate compound amount. This process is illustrated in Example 8. Although inflation rates vary dramatically, in this section we will assume constant annual inflation rates, and we will use annual compounding in solving inflation problems. In other words, $n = 1$ for these exercises.

POINT OF INTEREST

The average life of a dollar bill differs from the average lives of other denominations of U.S. currency. See the list below. (*Source:* www.federalreserve.gov)

$1	70 months
$5	66 months
$10	54 months
$20	95 months
$50	102 months
$100	180 months

INSTRUCTOR NOTE

Many students will be interested in the Excursion on page 641. The topic is the Consumer Price Index, which measures the effect of inflation on the cost of goods and services.

EXAMPLE 8 Calculate the Effect of Inflation on Salary

Suppose your annual salary today is \$35,000. You want to know what an equivalent salary will be in 20 years—that is, a salary that will have the same purchasing power. Assume a 6% inflation rate.

Solution

Use the compound amount formula given on page 631, with $P = 35{,}000$, $r = 6\% = 0.06$, and $t = 20$. The inflation rate is an annual rate, so $n = 1$.

$$A = P\left(1 + \frac{r}{n}\right)^{nt}$$

$$A = 35{,}000\left(1 + \frac{0.06}{1}\right)^{1\cdot 20}$$

$$A = 35{,}000(1.06)^{20}$$

$$A \approx 35{,}000(3.20713547)$$

$$A \approx 112{,}249.74$$

Twenty years from now, you need to earn an annual salary of approximately \$112,249.74 in order to have the same purchasing power.

CHECK YOUR PROGRESS 8 Assume that the average new car sticker price in 2013 was \$28,000. Use an annual inflation rate of 5% to estimate the average new car sticker price in 2030.

Solution See page S39.

POINT OF INTEREST

Inflation rates can change dramatically over time. The table below shows the five lowest and the five highest inflation rates for the past 30 years. (*Source:* inflationdata.com)

The Five Lowest Inflation Rates

Inflation rate	Year
−0.34%	2009
0.12%	2015
1.47%	2013
1.55%	1998
1.59%	2002

The Five Highest Inflation Rates

Inflation rate	Year
13.58%	1980
10.35%	1981
6.16%	1982
5.39%	1990
4.83%	1989

The present value formula can be used to determine the effect of inflation on the future purchasing power of a given amount of money. Substitute the inflation rate for the interest rate in the present value formula. The compounding period is 1 year. Again we will assume a constant rate of inflation.

EXAMPLE 9 Calculate the Effect of Inflation on Future Purchasing Power

Suppose you purchase an insurance policy in 2015 that will provide you with \$250,000 when you retire in 2050. Assuming an annual inflation rate of 8%, what will be the purchasing power of the \$250,000 in 2050?

Solution

Use the present value formula. $A = 250{,}000$, $r = 8\% = 0.08$, $t = 35$. The inflation rate is an annual rate, so $n = 1$.

$$P = \frac{A}{\left(1 + \frac{r}{n}\right)^{nt}}$$

$$P = \frac{250{,}000}{\left(1 + \frac{0.08}{1}\right)^{1\cdot 35}}$$

$$P = \frac{250{,}000}{(1 + 0.08)^{35}}$$

$$P \approx \frac{250{,}000}{14.785344}$$

$$P \approx 16{,}908.64$$

Assuming an annual inflation rate of 8%, the purchasing power of \$250,000 will be about \$16,908.64 in 2050.

CHECK YOUR PROGRESS 9 Suppose you purchase an insurance policy in 2015 that will provide you with $500,000 when you retire in 40 years. Assuming an annual inflation rate of 7%, what will be the purchasing power of half a million dollars in 2055?

Solution See page S40.

MATH MATTERS The Rule of 72

The **Rule of 72** states that the number of years necessary for prices to double is approximately equal to 72 divided by the annual inflation rate.

$$\text{Years to double} = \frac{72}{\text{annual inflation rate}}$$

For example, at an annual inflation rate of 6%, prices will double in approximately 12 years.

$$\text{Years to double} = \frac{72}{\text{annual inflation rate}} = \frac{72}{6} = 12$$

Effective Interest Rate

When interest is compounded, the annual rate of interest is called the **nominal rate**. The **effective rate** is the simple interest rate that would yield the same amount of interest after 1 year. When a bank advertises a "7% annual interest rate compounded daily and yielding 7.25%," the nominal interest rate is 7% and the effective rate is 7.25%.

QUESTION A bank offers a savings account that pays 2.75% annual interest, compounded daily and yielding 2.79%. What is the effective rate on this account? What is the nominal rate?

POINT OF INTEREST

According to the Federal Reserve Board, the value of U.S. currency in circulation increased 1013% from 1980 to 2015, to $1.39 trillion. (*Source:* www.federalreserve.gov)

Consider $100 deposited at 6%, compounded monthly, for 1 year.

The future value after 1 year is $106.17.

$$A = P\left(1 + \frac{r}{n}\right)^{nt}$$
$$A = 100\left(1 + \frac{0.06}{12}\right)^{12\cdot 1}$$
$$A = 100(1 + 0.005)^{12}$$
$$A \approx 106.17$$

The interest earned in 1 year is $6.17.

$$I = A - P$$
$$I = 106.17 - 100$$
$$I = 6.17$$

ANSWER The effective rate is 2.79%. The nominal rate is 2.75%.

Now consider \$100 deposited at an annual simple interest rate of 6.17%.

The interest earned in 1 year is \$6.17.

$$I = Prt$$
$$I = 100(0.0617)(1)$$
$$I = 6.17$$

The interest earned on \$100 is the same when it is deposited at 6% compounded monthly as when it is deposited at an annual simple interest rate of 6.17%. 6.17% is the effective annual rate of a deposit that earns 6% compounded monthly.

In this example \$100 was used as the principal. When we use \$100 for P, we multiply the interest rate by 100. Remember that the interest rate is written as a decimal in the equation $I = Prt$, and a decimal is written as a percent by multiplying by 100. Therefore, when $P = 100$, the digits in the interest earned on the investment (\$6.17) are the same as the digits in the effective annual rate (6.17%).

EXAMPLE 10 Calculate the Effective Interest Rate

A credit union offers a certificate of deposit at an annual interest rate of 3%, compounded monthly. Find the effective rate. Round to the nearest hundredth of a percent.

TAKE NOTE

To compare two investments or loan agreements, calculate the effective interest rate of each. When you are investing your money, you want the highest effective rate so that your money will earn more interest. If you are borrowing money, you want the lowest effective rate so that you will pay less interest on the loan.

Solution

Use the compound amount formula to find the future value of \$100 after 1 year.
$P = 100$, $r = 3\% = 0.03$, $n = 12$, $t = 1$

$$A = P\left(1 + \frac{r}{n}\right)^{nt}$$
$$A = 100\left(1 + \frac{0.03}{12}\right)^{12\cdot 1}$$
$$A = 100(1 + 0.0025)^{12}$$
$$A = 100(1.0025)^{12}$$
$$A \approx 100(1.030415957)$$
$$A \approx 103.04$$

Find the interest earned on the \$100.

$$I = A - P$$
$$I = 103.04 - 100$$
$$I = 3.04$$

The effective interest rate is 3.04%.

CHECK YOUR PROGRESS 10 A bank offers a certificate of deposit at an annual interest rate of 4%, compounded quarterly. Find the effective rate. Round to the nearest hundredth of a percent.

Solution See page S40.

To compare two investments or loan agreements, we could calculate the effective annual rate of each. However, a shorter method involves comparing the compound amounts of each. Because the value of

$$\left(1 + \frac{r}{n}\right)^{nt}$$

is the compound amount of \$1, we can compare the value of

$$\left(1+\frac{r}{n}\right)^{nt}$$

for each alternative.

EXAMPLE 11 Compare Annual Yields

One bank advertises an interest rate of 5.5%, compounded quarterly, on a certificate of deposit. Another bank advertises an interest rate of 5.25%, compounded monthly. Which investment has the higher annual yield?

Solution

Calculate $\left(1+\frac{r}{n}\right)^{nt}$ for each investment.

$$\left(1+\frac{r}{n}\right)^{nt} = \left(1+\frac{0.055}{4}\right)^{4\cdot 1} \approx 1.0561448 \qquad \left(1+\frac{r}{n}\right)^{nt} = \left(1+\frac{0.0525}{12}\right)^{12\cdot 1} \approx 1.0537819$$

Compare the two compound amounts.

$$1.0561448 > 1.0537819$$

An investment that earns 5.5% compounded quarterly has a higher annual yield than an investment that earns 5.25% compounded monthly.

CHECK YOUR PROGRESS 11 Which investment has the higher annual yield, one that earns 5% compounded quarterly or one that earns 5.25% compounded semiannually?

Solution See page S40.

MATH MATTERS Saving for Retirement

The graph below shows the results of a survey in which workers were asked to estimate the amount of money they would need to retire comfortably.

Amount of Savings Workers Think They Need for Retirement (Percent of workers, by year)

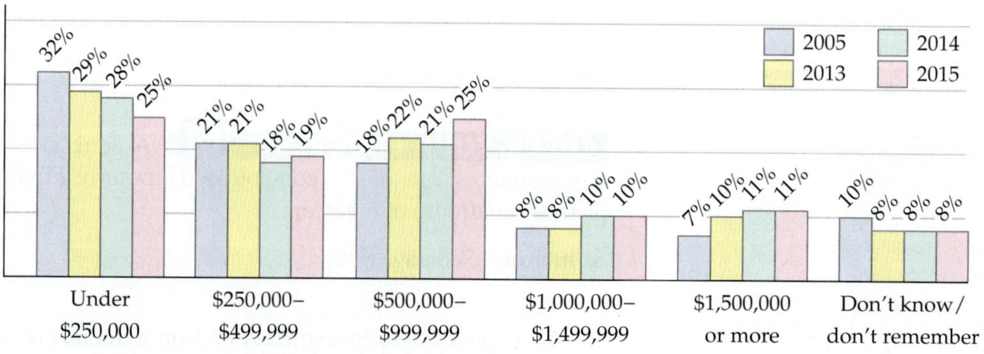

Source: Employee Benefit Research Institute and Greenwald & Associates, 2005–2015 Retirement Confidence Surveys

EXCURSION

Consumer Price Index

An **index number** measures the change in a quantity, such as cost, over a period of time. One of the most widely used indexes is the Consumer Price Index (CPI). The CPI, which includes the selling prices of key consumer goods and services, indicates the relative change in the price of these items over time. It measures the effect of inflation on the cost of goods and services.

The main components of the Consumer Price Index, shown below, are the costs of housing, food, transportation, medical care, clothing, recreation, and education.

TAKE NOTE

The category "Recreation" includes television sets, cable TV, pets and pet products, sports equipment, and admission tickets.

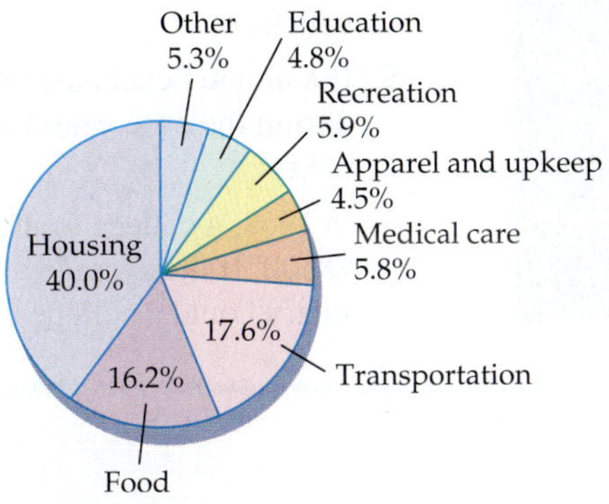

Components of the Consumer Price Index
SOURCE: U.S. Bureau of Labor Statistics

TAKE NOTE

You can obtain current and historical data on the Consumer Price Index by visiting the website of the Bureau of Labor Statistics at www.bls.gov.

The CPI is a measure of the cost of living for consumers. The government publishes monthly and annual figures on the Consumer Price Index.

The Consumer Price Index has a base period, 1982–1984, from which to make comparisons. The CPI for the base period is 100. The CPI for October 2010 was 218.9. This means that $100 in the period 1982–1984 had the same purchasing power as $218.90 in October of 2010.

An index number is actually a percent written without a percent sign. The CPI of 218.9 for October 2010 means that the average cost of consumer goods at that time was 218.9% of their cost in the 1982–1984 period.

The table below gives the CPI for various products in December 2015.

Product	CPI
All items	236.525
Food	247.903
Housing	282.394
Apparel	122.792
Transportation	294.081
Medical care	481.983
Recreation	115.626
Education	244.777

The Consumer Price Index, December 2015
SOURCE: U.S. Bureau of Labor Statistics

EXCURSION EXERCISES

Solve the following.

Andersen Ross/Getty Images

1. The CPI for 2010 was 218.056. What percent of the base period prices were the consumer prices in 2010?
2. The CPI for 2014 was 236.736. The percent increase in consumer prices from 2014 to 2015 was 0.8%. Find the CPI for 2015. Round to the nearest hundredth.
3. The CPI for 2012 was 229.939. The CPI for 2015 was 237.017. Find the percent increase in consumer prices for this time period. Round to the nearest hundredth of a percent.
4. Of the items listed in the table on the preceding page, are there any items that cost about three times as much in 2015 as they cost during the base period? If so, which ones?
5. If a movie ticket cost \$9 in 2015, what would a comparable movie ticket have cost during the base years? Recall that the category "Recreation" includes admission tickets.
6. In 1968, a college graduate could expect to earn an average annual starting salary of \$8500. If the average annual inflation rate from 1968 to 2016 was 4.1%, what starting salary could a college graduate expect to earn in 2016?

EXERCISE SET 11.2

■ In Exercises 1 to 8, calculate the compound amount. Use the compound amount formula.

1. $P = \$1200$, $r = 7\%$ compounded semiannually, $t = 12$ years
2. $P = \$3500$, $r = 8\%$ compounded semiannually, $t = 14$ years
3. $P = \$500$, $r = 9\%$ compounded quarterly, $t = 6$ years
4. $P = \$7000$, $r = 4\%$ compounded quarterly, $t = 9$ years
5. $P = \$8500$, $r = 9\%$ compounded monthly, $t = 10$ years
6. $P = \$6400$, $r = 6\%$ compounded monthly, $t = 3$ years
7. $P = \$9600$, $r = 9\%$ compounded daily, $t = 3$ years
8. $P = \$1700$, $r = 9\%$ compounded daily, $t = 5$ years

■ In Exercises 9 to 14, calculate the compound amount. Use a calculator with a financial mode.

9. $P = \$1600$, $r = 8\%$ compounded quarterly, $t = 10$ years
10. $P = \$4200$, $r = 6\%$ compounded semiannually, $t = 8$ years
11. $P = \$3000$, $r = 5\%$ compounded monthly, $t = 5$ years
12. $P = \$9800$, $r = 10\%$ compounded quarterly, $t = 4$ years
13. $P = \$1700$, $r = 9\%$ compounded semiannually, $t = 3$ years
14. $P = \$8600$, $r = 3\%$ compounded semiannually, $t = 5$ years

■ In Exercises 15 to 20, calculate the future value.

15. $P = \$7500$, $r = 6\%$ compounded monthly, $t = 5$ years
16. $P = \$1800$, $r = 9.5\%$ compounded annually, $t = 10$ years
17. $P = \$4600$, $r = 5\%$ compounded semiannually, $t = 12$ years
18. $P = \$9000$, $r = 5.5\%$ compounded quarterly, $t = 3$ years
19. $P = \$22{,}000$, $r = 9\%$ compounded monthly, $t = 7$ years
20. $P = \$5200$, $r = 8.1\%$ compounded daily, $t = 9$ years

■ In Exercises 21 to 26, calculate the present value.

21. $A = \$25{,}000$, $r = 10\%$ compounded quarterly, $t = 12$ years
22. $A = \$20{,}000$, $r = 6\%$ compounded monthly, $t = 5$ years
23. $A = \$40{,}000$, $r = 7.5\%$ compounded annually, $t = 35$ years
24. $A = \$10{,}000$, $r = 4\%$ compounded semiannually, $t = 30$ years
25. $A = \$15{,}000$, $r = 8\%$ compounded quarterly, $t = 5$ years
26. $A = \$50{,}000$, $r = 3\%$ compounded monthly, $t = 5$ years

27. **Compound Amount** Calculate the compound amount when \$8000 is deposited in an account earning 8% interest, compounded quarterly, for 5 years.

28. **Compound Amount** Calculate the compound amount when \$3000 is deposited in an account earning 10% interest, compounded semiannually, for 3 years.

29. **Compound Amount** If you leave \$2500 in an account earning 9% interest, compounded daily, how much money will be in the account after 4 years?

30. **Compound Amount** What is the compound amount when \$1500 is deposited in an account earning an interest rate of 6%, compounded monthly, for 2 years?

31. **Future Value** What is the future value of \$4000 earning 6% interest, compounded monthly, for 6 years?

32. **Future Value** Calculate the future value of \$8000 earning 8% interest, compounded quarterly, for 10 years.

33. **Compound Interest** How much interest is earned in 3 years on \$2000 deposited in an account paying 6% interest, compounded quarterly?

34. **Compound Interest** How much interest is earned in 5 years on \$8500 deposited in an account paying 9% interest, compounded semiannually?

35. **Compound Interest** Calculate the amount of interest earned in 8 years on \$15,000 deposited in an account paying 10% interest, compounded quarterly.

36. **Compound Interest** Calculate the amount of interest earned in 6 years on \$20,000 deposited in an account paying 4% interest, compounded monthly.

37. **Compound Interest** How much money should be invested in an account that earns 6% interest, compounded monthly, in order to have \$15,000 in 5 years?

38. **Compound Interest** How much money should be invested in an account that earns 7% interest, compounded quarterly, in order to have \$10,000 in 5 years?

39. **Compound Interest** \$1000 is deposited for 5 years in an account that earns 9% interest.
    - **a.** Calculate the simple interest earned.
    - **b.** Calculate the interest earned if interest is compounded daily.
    - **c.** How much more interest is earned on the account when the interest is compounded daily?

40. **Compound Interest** \$10,000 is deposited for 2 years in an account that earns 5% interest.
    - **a.** Calculate the simple interest earned.
    - **b.** Calculate the interest earned if interest is compounded daily.
    - **c.** How much more interest is earned on the account when the interest is compounded daily?

41. **Future Value** \$15,000 is deposited for 4 years in an account earning 8% interest.
    - **a.** Calculate the future value of the investment if interest is compounded semiannually.
    - **b.** Calculate the future value if interest is compounded quarterly.
    - **c.** How much greater is the future value of the investment when the interest is compounded quarterly?

42. **Future Value** \$25,000 is deposited for 3 years in an account earning 6% interest.
    - **a.** Calculate the future value of the investment if interest is compounded annually.
    - **b.** Calculate the future value if interest is compounded semiannually.
    - **c.** How much greater is the future value of the investment when the interest is compounded semiannually?

43. **Compound Interest** \$10,000 is deposited for 2 years in an account earning 8% interest.
    - **a.** Calculate the interest earned if interest is compounded semiannually.
    - **b.** Calculate the interest earned if interest is compounded quarterly.
    - **c.** How much more interest is earned on the account when the interest is compounded quarterly?

44. **Compound Interest** \$20,000 is deposited for 5 years in an account earning 6% interest.
    - **a.** Calculate the interest earned if interest is compounded annually.
    - **b.** Calculate the interest earned if interest is compounded semiannually.
    - **c.** How much more interest is earned on the account when the interest is compounded semiannually?

45. **Loans** To help pay your college expenses, you borrow \$7000 and agree to repay the loan at the end of 5 years at 8% interest, compounded quarterly.
    - **a.** What is the maturity value of the loan?
    - **b.** How much interest are you paying on the loan?

46. **Loans** You borrow \$6000 to help pay your college expenses. You agree to repay the loan at the end of 5 years at 10% interest, compounded quarterly.
    - **a.** What is the maturity value of the loan?
    - **b.** How much interest are you paying on the loan?

47. Present Value A couple plans to save for their child's college education. What principal must be deposited by the parents when their child is born in order to have \$40,000 when the child reaches the age of 18? Assume the money earns 8% interest, compounded quarterly.

Andresr/Shutterstock.com

48. Present Value A couple plans to invest money for their child's college education. What principal must be deposited by the parents when their child turns 10 in order to have \$30,000 when the child reaches the age of 18? Assume that the money earns 8% interest, compounded quarterly.

49. Present Value You want to retire in 30 years with \$1,000,000 in investments.

a. How much money would you have to invest today at 9% interest, compounded daily, in order to have \$1,000,000 in 30 years?

b. How much will the \$1,000,000 generate in interest each year if it is invested at 9% interest, compounded daily?

50. Present Value You want to retire in 40 years with \$1,000,000 in investments.

a. How much money must you invest today at 8.1% interest, compounded daily, in order to have \$1,000,000 in 40 years?

b. How much will the \$1,000,000 generate in interest each year if it is invested at 8.1% interest, compounded daily?

51. Compound Amount You deposit \$5000 in a two-year certificate of deposit (CD) earning 3.1% interest, compounded daily. At the end of the 2 years, you reinvest the compound amount in another 2-year CD. The interest rate on the second CD is 2.2%, compounded daily. What is the compound amount when the second CD matures?

52. Compound Amount You deposit \$7500 in a 2-year certificate of deposit (CD) earning 2.4% interest, compounded daily. At the end of the 2 years, you reinvest the compound amount plus an additional \$7500 in another 2-year CD. The interest rate on the second CD is 2.9%, compounded daily. What is the compound amount when the second CD matures?

53. Inflation The average monthly rent for a three-bedroom apartment in Denver, Colorado, is \$1249. Using an annual inflation rate of 7%, find the average monthly rent in 15 years.

54. Inflation The average cost of housing in Greenville, North Carolina, is \$140,300. Using an annual inflation rate of 7%, find the average cost of housing in 10 years.

55. Inflation Suppose your salary in 2015 is \$40,000. Assuming an annual inflation rate of 7%, what salary do you need to earn in 2020 in order to have the same purchasing power?

56. Inflation Suppose your salary in 2015 is \$50,000. Assuming an annual inflation rate of 6%, what salary do you need to earn in 2025 in order to have the same purchasing power?

57. Inflation In 2014, you purchase an insurance policy that will provide you with \$125,000 when you retire in 2054. Assuming an annual inflation rate of 6%, what will be the purchasing power of the \$125,000 in 2054?

58. Inflation You purchase an insurance policy in the year 2015 that will provide you with \$250,000 when you retire in 25 years. Assuming an annual inflation rate of 8%, what will be the purchasing power of the quarter-of-a-million dollars in 2040?

59. Inflation A retired couple have a fixed income of \$3500 per month. Assuming an annual inflation rate of 7%, what is the purchasing power of their monthly income in 5 years?

60. Inflation A retired couple have a fixed income of \$46,000 per year. Assuming an annual inflation rate of 6%, what is the purchasing power of their annual income in 10 years?

■ In Exercises 61 to 68, calculate the effective annual rate for an investment that earns the given rate of return. Round to the nearest hundredth of a percent.

61. 7.2% interest compounded quarterly

62. 8.4% interest compounded quarterly

63. 7.5% interest compounded monthly

64. 6.9% interest compounded monthly

65. 8.1% interest compounded daily

66. 6.3% interest compounded daily

67. 5.94% interest compounded monthly

68. 6.27% interest compounded monthly

■ **Inflation** In Exercises 69 to 76, you are given the 2009 price of an item. Use an inflation rate of 6% to calculate its price in 2014, 2019, and 2029. Round to the nearest cent.

69. Gasoline: \$3.00 per gallon

70. Milk: \$3.35 per gallon

71. Loaf of bread: \$2.69

72. Sunday newspaper: \$2.25

73. Ticket to a movie: \$9

74. Paperback novel: \$12.00

75. House: \$275,000

76. Car: \$24,000

■ In Exercises 77 to 82, calculate the purchasing power using an annual inflation rate of 7%. Round to the nearest cent.

77. \$50,000 in 10 years

78. \$25,000 in 8 years

79. \$100,000 in 20 years

80. \$30,000 in 15 years

81. \$75,000 in 5 years

82. \$20,000 in 25 years

83. a. Complete the table.

Nominal rate	Effective rate
4% annual compounding	
4% semiannual compounding	
4% quarterly compounding	
4% monthly compounding	
4% daily compounding	

b. As the number of compounding periods increases, does the effective rate increase or decrease?

84. Effective Interest Rate Beth Chipman has money in a savings account that earns an annual interest rate of 3%, compounded quarterly. What is the effective rate of interest on Beth's account? Round to the nearest hundredth of a percent.

85. Effective Interest Rate Blake Hamilton has money in a savings account that earns an annual interest rate of 3%, compounded monthly. What is the effective rate of interest on Blake's savings? Round to the nearest hundredth of a percent.

86. Annual Yield One bank advertises an interest rate of 6.6%, compounded quarterly, on a certificate of deposit. Another bank advertises an interest rate of 6.25%, compounded monthly. Which investment has the higher annual yield?

87. Annual Yield Which has the higher annual yield, 6% compounded quarterly or 6.25% compounded semiannually?

88. Annual Yield Which investment has the higher annual yield, one earning 7.8% compounded monthly or one earning 7.5% compounded daily?

89. Annual Yield One bank advertises an interest rate of 5.8%, compounded quarterly, on a certificate of deposit. Another bank advertises an interest rate of 5.6%, compounded monthly. Which investment has a higher annual yield?

EXTENSIONS

90. Continuous Compounding In our discussion of compound interest, we used annual, semiannual, monthly, quarterly, and daily compounding periods. When interest is compounded daily, it is compounded 360 times a year. If interest were compounded twice daily, it would be compounded $360(2) = 720$ times a year. If interest were compounded four times a day, it would be compounded $360(4) = 1440$ times a year. Remember that the more frequent the compounding period, the more interest earned on the account. Therefore, if interest is compounded more frequently than daily, an investment will earn even more interest than it would if interest were compounded daily.

Some banking institutions advertise **continuous compounding**, which means that the number of compounding periods per year gets very, very large. When compounding continuously, instead of using the compound amount formula $A = P\left(1 + \frac{r}{n}\right)^{nt}$, we use the following formula.

$$A = Pe^{rt}$$

In this formula, A is the compound amount when P dollars are deposited at an annual interest rate of r percent compounded continuously for t years. The number e is approximately equal to 2.7182818.

The number e is found in many real-world applications. It is an irrational number, so its decimal representation never terminates or repeats. Calculators have an $\boxed{e^x}$ key for evaluating exponential expressions in which e is the base.

To calculate the compound amount when \$10,000 is invested for 5 years at an interest rate of 10%, compounded continuously, use the formula for continuous compounding. Substitute the following values into the formula: $P = 10{,}000$, $r = 10\% = 0.10$, and $t = 5$.

$$A = Pe^{rt}$$
$$A \approx 10{,}000(2.7182818)^{0.10(5)}$$
$$A \approx 16{,}487.21$$

The compound amount is \$16,487.21.

In the following exercises, calculate the compound interest when interest is compounded continuously.

a. $P = \$5000, r = 8\%, t = 6$ years

b. $P = \$8000, r = 7\%, t = 15$ years

c. $P = \$12{,}000, r = 9\%, t = 10$ years

d. $P = \$7000, r = 6\%, t = 8$ years

e. $P = \$3000, r = 7.5\%, t = 4$ years

f. $P = \$9000, r = 8.6\%, t = 5$ years

Solve the following exercises.

g. Calculate the compound amount when \$2500 is deposited in an account earning 11% interest, compounded continuously, for 12 years.

h. What is the future value of \$15,000 earning 9.5% interest, compounded continuously, for 7 years?

i. How much interest is earned in 9 years on \$6000 deposited in an account paying 10% interest, compounded continuously?

j. \$25,000 is deposited for 10 years in an account that earns 8% interest. Calculate the future value of the investment if interest is compounded quarterly and if interest is compounded continuously. How much greater is the future value of the investment when interest is compounded continuously?

SECTION 11.3 Credit Cards and Consumer Loans

Credit Cards

When a customer uses a credit card to make a purchase, the customer is actually receiving a loan. Therefore, there is frequently an added cost to the consumer who purchases on credit. This added cost may be in the form of an annual fee or interest charges on purchases. A **finance charge** is an amount paid in excess of the cash price; it is the cost to the customer for the use of credit.

Most credit card companies issue monthly bills. The due date on the bill is usually 1 month after the billing date (the date the bill is prepared and sent to the customer). If the bill is paid in full by the due date, the customer pays no finance charge. If the bill is not paid in full by the due date, a finance charge is added to the next bill.

Suppose a credit card billing date is the 10th day of each month. If a credit card purchase is made on April 15, then May 10 is the billing date (the 10th day of the month following April). The due date is June 10 (one month from the billing date). If the bill is paid in full before June 10, no finance charge is added. However, if the bill is not paid in full, interest charges on the outstanding balance will start to accrue (be added) on June 10, and any purchase made after June 10 will immediately start accruing interest.

The most common method of determining finance charges is the **average daily balance method**. Interest charges are based on the credit card's average daily balance, which is calculated by dividing the sum of the total amounts owed each day of the month by the number of days in the billing period.

Nerthuz/Shutterstock.com

Average Daily Balance

$$\text{Average daily balance} = \frac{\text{sum of the total amounts owed each day of the month}}{\text{number of days in the billing period}}$$

POINT OF INTEREST

The average American has 2.24 credit cards. There are about 180 million credit card holders in the United States. Of these, 58% pay the full balance on the card each month. For the remaining card holders, the average credit card balance is \$6600.

An example of calculating the average daily balance follows.

Suppose an unpaid bill for \$315 had a due date of April 10. A purchase of \$28 was made on April 12, and \$123 was charged on April 24. A payment of \$50 was made on April 15. The next billing date is May 10. The interest on the average daily balance is 1.5% per month. Find the finance charge on the May 10 bill.

Solution

To find the finance charge, first prepare a table showing the unpaid balance for each purchase, the number of days the balance is owed, and the product of these numbers. A negative sign in the Payments or Purchases column of the table indicates that a payment was made on that date.

Date	Payments or purchases	Balance each day	Number of days until balance changes	Unpaid balance times number of days
April 10–11		\$315	2	\$630
April 12–14	\$28	\$343	3	\$1029
April 15–23	−\$50	\$293	9	\$2637
April 24–May 9	\$123	\$416	16	\$6656
Total				\$10,952

The sum of the total amounts owed each day of the month is \$10,952.

Find the average daily balance.

$$\text{Average daily balance} = \frac{\text{sum of the total amounts owed each day of the month}}{\text{number of days in the billing period}}$$

$$= \frac{10{,}952}{30} \approx 365.07$$

Find the finance charge.

$$I = Prt$$
$$I = 365.07(0.015)(1)$$
$$I \approx 5.48$$

The finance charge on the May 10 bill is \$5.48.

EXAMPLE 1 Calculate Interest on a Credit Card Bill

An unpaid bill for \$620 had a due date of March 10. A purchase of \$214 was made on March 15, and \$67 was charged on March 30. A payment of \$200 was made on March 22. The interest on the average daily balance is 1.5% per month. Find the finance charge on the April 10 bill.

Solution

First calculate the sum of the total amounts owed each day of the month.

Date	Payments or purchases	Balance each day	Number of days until balance changes	Unpaid balance times number of days
March 10–11		\$620	5	\$3100
March 15–21	\$214	\$834	7	\$5838
March 22–29	\$200	\$634	8	\$5072
March 30–April 9	\$67	\$701	11	\$7711
Total				\$21,721

The sum of the total amounts owed each day of the month is \$21,721.

POINT OF INTEREST

In 1950, Frank McNamara issued to 200 friends a card that could be used to pay for food at various restaurants in New York. The card, called a Diner's Card, spawned the credit card industry.

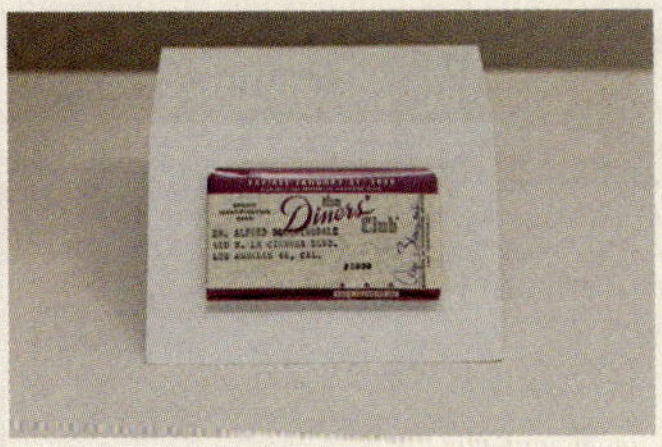

B Christopher/Alamy Stock Photo

Find the average daily balance.

$$\text{Average daily balance} = \frac{\text{sum of the total amounts owed each day of the month}}{\text{number of days in the billing period}}$$

$$= \frac{21{,}721}{31} \approx \$700.68$$

Find the finance charge.

$I = Prt$

$I = 700.68(0.015)(1)$

$I \approx 10.51$

The finance charge on the April 10 bill is \$10.51.

CHECK YOUR PROGRESS 1 A bill for \$1024 was due on July 1. Purchases of \$315 were made on July 7, and \$410 was charged on July 22. A payment of \$400 was made on July 15. The interest on the average daily balance is 1.2% per month. Find the finance charge on the August 1 bill.

Solution See page S40.

MATH MATTERS Credit Card Debt

The graph below shows how long it would take you to pay off a credit card debt of \$5000 if you paid the minimum monthly payment* plus an additional amount each month. For instance, the bar above 50 shows that it would take approximately 57 months to repay the credit card balance if you paid the minimum monthly payment plus \$50. The number inside the bar shows approximately how much interest you would pay over the repayment period.

If you have credit card debt and want to determine how long it will take you to pay off the debt, enter "debt payoff calculator" in a search engine on the Internet. Find a calculator that will calculate how long it will take to pay off the debt and the total amount of interest you will pay.

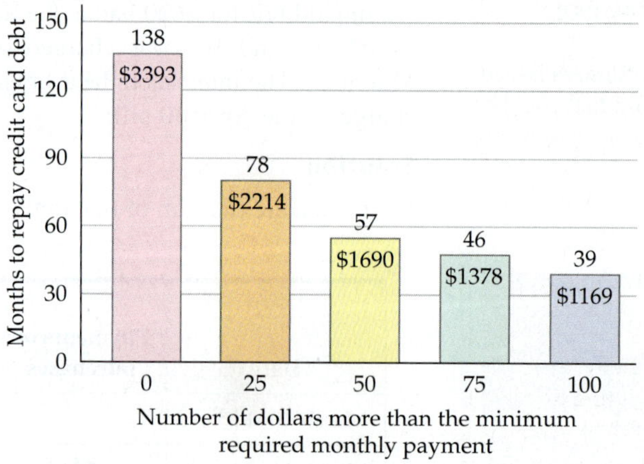

SOURCE: CardWeb.com

Note from the graph that if you paid \$100 more than the minimum due each month, the debt would be repaid approximately 99 months sooner and you would save approximately \$2224 in interest charges.

*For this illustration, we are assuming that the minimum monthly payment is 2% of the credit card balance plus the interest accrued for one month. The monthly interest rate on the credit card balance is 1.5%.

Annual Percentage Rate

TAKE NOTE

Note that both the effective interest rate discussed in Section 11.2 and the annual percentage rate discussed here reflect the cost of a loan on a yearly basis. Both are useful to the consumer interested in comparing loan offers.

Federal law, in the form of the Truth in Lending Act, requires that credit customers be made aware of the cost of credit. This law, passed by Congress in 1969, requires that a business issuing credit inform the consumer of all details of a credit transaction, including the true annual interest rate. The **true annual interest rate**, also called the **annual percentage rate (APR)** or **annual percentage yield (APY)**, is the effective annual interest rate on which credit payments are based.

The idea behind the APR is that interest is owed only on the *unpaid balance* of the loan. For instance, suppose you decide to borrow \$2400 from a bank that advertises a 10% *simple* interest rate. You want a 6-month loan and agree to repay the loan in six equal monthly payments. The simple interest due on the loan is

$$I = Prt$$
$$I = \$2400(0.10)\left(\frac{6}{12}\right)$$
$$I = \$120$$

The total amount to be repaid to the bank is

$$A = P + I$$
$$A = \$2400 + \$120$$
$$A = \$2520$$

The amount of each monthly payment is

$$\text{Monthly payment} = \frac{2520}{6} = \$420$$

CALCULATOR NOTE

You can calculate APR using the Finance option on a TI-83/84 by pressing APPS and then selecting Finance. Press ENTER. Input the values. Note on the screen below that the payment is entered as a negative number. Amounts paid out are normally entered as negative numbers.

```
N=6
I%=0
PV=2400
PMT=-420
FV=0
P/Y=12
C/Y=12
PMT: END BEGIN
```

Move the cursor to I% and then press ALPHA [SOLVE]. The actual APR is then displayed next to I%.

```
N=6
I%=16.94488071
PV=2400
PMT=-420
FV=0
P/Y=12
C/Y=12
PMT: END BEGIN
```

The APR is approximately 16.9%.

During the first month, you still owe a total of \$2400. The interest on that amount is

$$I = Prt$$
$$I = \$2400(0.10)\left(\frac{1}{12}\right)$$
$$I = \$20$$

At the end of the first month, of the \$420 payment you make, \$20 is the interest payment and \$400 is applied to reducing the loan. Therefore, during the second month, you owe \$2400 − \$400 = \$2000. The interest on that amount is

$$I = Prt$$
$$I = \$2000(0.10)\left(\frac{1}{12}\right) \approx 16.667$$
$$I = \$16.67$$

At the end of the second month, of the \$420 payment you make, \$16.67 is the interest payment and \$403.33 is applied to reducing the loan. Therefore, during the third month, you owe \$2000 − \$403.33 = \$1596.67.

The point of these calculations is to demonstrate that each month, the amount you owe is decreasing, and not by a constant amount. From our calculations, the loan decreased by \$400 the first month and by \$403.33 the second month. Similar calculations were made in the car loan example presented in the Excursion at the end of Section 11.1.

The Truth in Lending Act stipulates that the interest rate for a loan be calculated only on the amount owed at a particular time, not on the original amount borrowed. All loans must be stated according to this standard, thereby making it possible for a consumer to compare different loans.

We can use the following formula to estimate the annual percentage rate (APR) on a simple interest rate installment loan.

Approximate Annual Percentage Rate (APR) Formula for a Simple Interest Rate Loan

The annual percentage rate (APR) of a simple interest rate loan can be approximated by

$$\text{APR} \approx \frac{2nr}{n+1}$$

where n is the number of payments and r is the simple interest rate.

For the loan described on the preceding page, $n = 6$ and $r = 10\% = 0.10$.

$$\begin{aligned}\text{APR} &\approx \frac{2nr}{n+1}\\ &\approx \frac{2(6)(0.10)}{6+1} = \frac{1.2}{7} \approx 0.171\end{aligned}$$

The annual percentage rate on the loan is approximately 17.1%. Recall that the simple interest rate was 10%, much less than the actual rate. The Truth in Lending Act provides the consumer with a standard interest rate, APR, so that it is possible to compare loans. The 10% simple interest loan described above is equivalent to an APR loan of about 17%.

EXAMPLE 2 Calculate a Finance Charge and an APR

You purchase a refrigerator for \$675. You pay 20% down and agree to repay the balance in 12 equal monthly payments. The finance charge on the balance is 9% simple interest.

a. Find the finance charge.

b. Estimate the annual percentage rate. Round to the nearest tenth of a percent.

Solution

a. To find the finance charge, first calculate the down payment.

$$\begin{aligned}\text{Down payment} &= \text{percent down} \times \text{purchase price}\\ &= 0.20 \times 675 = 135\end{aligned}$$

$$\begin{aligned}\text{Amount financed} &= \text{purchase price} - \text{down payment}\\ &= 675 - 135 = 540\end{aligned}$$

Calculate the interest owed on the loan.

$$\begin{aligned}\text{Interest owed} &= \text{finance rate} \times \text{amount financed}\\ &= 0.09 \times 540 = 48.60\end{aligned}$$

The finance charge is \$48.60.

b. Use the APR formula to estimate the annual percentage rate.

$$\begin{aligned}\text{APR} &\approx \frac{2nr}{n+1}\\ &\approx \frac{2(12)(0.09)}{12+1} = \frac{2.16}{13} \approx 0.166\end{aligned}$$

The annual percentage rate is approximately 16.6%.

CHECK YOUR PROGRESS 2 You purchase a washing machine and dryer for \$750. You pay 20% down and agree to repay the balance in 12 equal monthly payments. The finance charge on the balance is 8% simple interest.

a. Find the finance charge.

b. Estimate the annual percentage rate. Round to the nearest tenth of a percent.

Solution See page S40.

Consumer Loans: Calculating Monthly Payments

The stated interest rate for most consumer loans, such as a car loan, is normally the annual percentage rate, APR, as required by the Truth in Lending Act. The payment amount for these loans is given by the following formula.

Payment Formula for an APR Loan

The payment for a loan based on APR is given by

$$PMT = A\left(\frac{\frac{r}{n}}{1-\left(1+\frac{r}{n}\right)^{-nt}}\right)$$

where PMT is the payment, A is the loan amount, r is the annual interest rate, n is the number of payments per year, and t is the number of years.

QUESTION For a 4-year loan repaid on a monthly basis, what is the value of nt in the formula above?

The payment formula given above is used to calculate monthly payments on most consumer loans. In Example 3 we calculate the monthly payment for a new refrigerator, and in Example 4 we calculate the monthly payment for a car loan.

CALCULATOR NOTE

You can calculate the monthly payment using the Finance option on a TI-83/84 calculator. Press APPS ENTER ENTER. Input the known values. Typically, financial calculations use PV (present value) for the loan amount and FV (future value) for the amount owed at the end of the loan period, usually 0. P/Y = 12 and C/Y = 12 mean that payments and interest are calculated monthly (12 times a year). Now place the cursor at PMT= and press ALPHA [SOLVE].

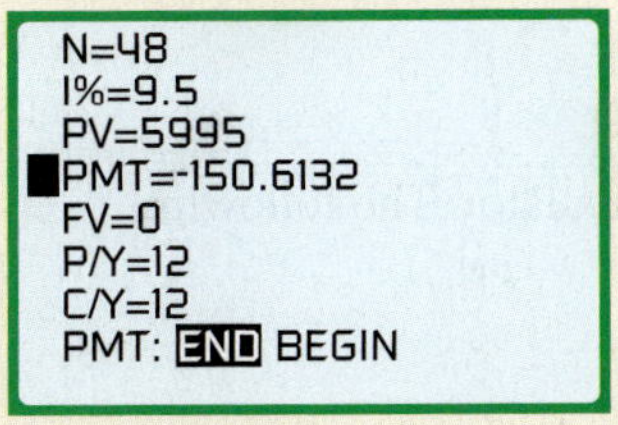

The monthly payment is \$150.61.

EXAMPLE 3 Calculate a Monthly Payment

Integrated Technologies is offering anyone who purchases a refrigerator an annual interest rate of 9.5% for 4 years. If Andrea Smyer purchases a luxury model refrigerator for \$5995 from Integrated Technologies, find her monthly payment.

Solution

To calculate the monthly payment, you will need a calculator. The following keystrokes will work on most scientific calculators. You could also use a calculator with a finance app, as shown in the Calculator Note at the left.

First calculate $\frac{r}{n}$ and store the result.

$$\frac{r}{n} = \frac{0.095}{12} \approx 0.00791667$$

Keystrokes: 0.095 ÷ 12 ≈ 0.00791667 STO

ANSWER nt = (number of payments per year)(number of years) = (12)(4) = 48.

Calculate the monthly payment. For a 4-year loan, $nt = 12(4) = 48$.

$$PMT = A\left(\frac{\frac{r}{n}}{1-\left(1+\frac{r}{n}\right)^{-nt}}\right)$$

$$= 5995\left(\frac{0.095/12}{1-(1+0.095/12)^{-48}}\right) \approx 150.61$$

Keystrokes: 5995 [×] [RCL] [=] [÷] [(] 1 [−] [(] 1 [+] [RCL] [)] [y^x] 48 [+/−] [)] [=]

The monthly payment is $150.61.

CHECK YOUR PROGRESS 3 Carlos Menton purchases a new laptop computer from Knox Computer Solutions for $1499. If the sales tax is 4.25% of the purchase price and Carlos finances the total cost, including sales tax, for 3 years at an annual interest rate of 8.4%, find the monthly payment.

Solution See page S41.

EXAMPLE 4 Calculate a Car Payment

A web page designer purchases a car for $18,395.

a. If the sales tax is 6.5% of the purchase price, find the amount of the sales tax.

b. If the car license fee is 1.2% of the purchase price, find the amount of the license fee.

c. If the designer makes a $2500 down payment, find the amount of the loan the designer needs.

d. Assuming the designer gets the loan in part c at an annual interest rate of 7.5% for 4 years, determine the monthly car payment.

Solution

a. Sales tax $= 0.065(18{,}395) = 1195.675$

The sales tax is $1195.68.

b. License fee $= 0.012(18{,}395) = 220.74$

The license fee is $220.74.

c. Loan amount = purchase price + sales tax + license fee − down payment

$= 18{,}395 + 1195.68 + 220.74 - 2500$

$= 17{,}311.42$

The loan amount is $17,311.42.

d. To calculate the monthly payment, you will need a calculator. The following keystrokes will work on most scientific calculators.

First calculate $\frac{r}{n}$ and store the result.

$$\frac{r}{n} = \frac{0.075}{12} = 0.00625$$

Keystrokes: 0.075 [÷] 12 = 0.00625 [STO]

CALCULATOR NOTE

A typical TI-83/84 screen for the calculation in Example 4 is shown below.

```
N=48
I%=7.5
PV=17311.42
PMT=-418.5711
FV=0
P/Y=12
C/Y=12
PMT: END BEGIN
```

The monthly payment is \$418.57.

Calculate the monthly payment. For a 4-year loan, $nt = 12(4) = 48$.

$$PMT = A\left(\frac{\frac{r}{n}}{1-\left(1+\frac{r}{n}\right)^{-nt}}\right)$$

$$= 17{,}311.42\left(\frac{0.00625}{1-(1+0.00625)^{-48}}\right) \approx 418.57$$

Keystrokes:

17311.42 [x] [RCL] [=] [÷] [(] 1 [−] [(] 1 [+] [RCL] [)] [y^x] 48 [+/−] [)] [=]

The monthly payment is \$418.57.

CHECK YOUR PROGRESS 4 A school superintendent purchases a new sedan for \$26,788.

a. If the sales tax is 5.25% of the purchase price, find the amount of the sales tax.

b. The superintendent makes a \$2500 down payment and the license fee is \$145. Find the amount the superintendent must finance.

c. Assuming the superintendent gets the loan in part b at an annual interest rate of 8.1% for 5 years, determine the superintendent's monthly car payment.

Solution See page S41.

MATH MATTERS Payday Loans

mikeledray/Shutterstock.com

An ad reads,

Get cash until payday! Loans of \$100 or more available.

These ads refer to *payday loans*, which go by a variety of names, such as cash advance loans, check advance loans, post-dated check loans, or deferred deposit check loans. These types of loans are offered by finance companies and check-cashing companies.

Typically a borrower writes a personal check payable to the lender for the amount borrowed plus a *service fee*. The company gives the borrower the amount of the check minus the fee, which is normally a percent of the amount borrowed. The amount borrowed is usually repaid after payday, normally within a few weeks.

Under the Truth in Lending Act, the cost of a payday loan must be disclosed. Along with other information, the borrower must receive, in writing, a statement of the APR for such a loan. To understand just how expensive these loans can be, suppose a borrower receives a loan for \$100 for 2 weeks and pays a fee of \$10. The APR for this loan can be calculated using the formula in this section or a graphing calculator. Screens for a TI-83/84 calculator are shown below.

```
N=1
I%=0
PV=100
PMT=-110
FV=0
P/Y=26
C/Y=26
PMT: END BEGIN
```

N = 1 (number of payments)
I% is unknown.
PV = 100 (amount borrowed)
PMT = −110 (the payment)
FV = 0 (no money is owed after all the payments)
P/Y = 26 (There are 26 2-week periods in 1 year.)
C/Y = 26

TAKE NOTE

Some states have enacted laws that regulate the amount a person can borrow on a payday loan and the maximum fee that can be charged. In California, for instance, the maximum amount that can be borrowed at one time is \$300, and the maximum fee is \$45. The borrower receives \$265. The APR on the loan is 460% for a two-week loan!

continued

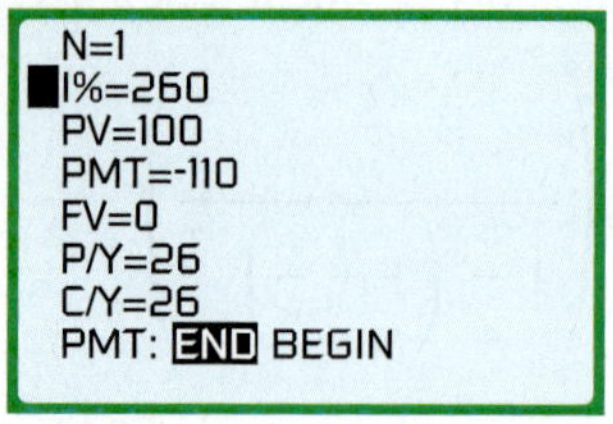

Place the cursor at I%=. Press ALPHA [SOLVE].

The annual interest rate is 260%.

To give you an idea of the enormity of a 260% APR, if the loan on the refrigerator in Example 3, page 651, were based on a 260% interest rate, the monthly payment on the refrigerator would be $1299.02!

Consumer Loans: Calculating Loan Payoffs

Sometimes a consumer wants to pay off a loan before the end of the loan term. For instance, suppose you have a five-year car loan but would like to purchase a new car after owning your car for 4 years. Because there is still 1 year remaining on the loan, you must pay off the remaining loan amount before purchasing another car.

This is not as simple as just multiplying the monthly car payment by 12 to arrive at the payoff amount. The reason, as we mentioned earlier, is that each payment includes both interest and principal. By solving the payment formula for an APR loan (see page 651) for A, the amount of the loan, we can calculate the payoff amount, which is just the remaining principal.

TAKE NOTE

The APR loan payoff formula applies only to those situations in which all regular payments that have come due have been paid on time and no extra money has been paid toward the principal.

APR Loan Payoff Formula

The payoff amount for a loan based on APR is given by

$$A = PMT\left(\frac{1 - \left(1 + \frac{r}{n}\right)^{-U}}{\frac{r}{n}}\right)$$

where A is the loan payoff, PMT is the payment, r is the annual interest rate, n is the number of payments per year, and U is the number of *remaining* (or *unpaid*) payments.

EXAMPLE 5 Calculate a Payoff Amount

Allison Werke wants to pay off the loan on her jet ski that she has owned for 18 months. Allison's monthly payment is $284.67 on a 2-year loan at an annual percentage rate of 8.7%. Find the payoff amount.

Solution

Because Allison has owned the jet ski for 18 months of a 24-month (two-year) loan, she has six payments remaining. Thus $U = 6$, the number of unpaid or remaining payments. Here are the keystrokes to find the loan payoff.

Calculate $\frac{r}{n}$ and store the result.

$$\frac{r}{n} = \frac{0.087}{12} = 0.00725$$

Keystrokes: 0.087 [÷] 12 [=] 0.00725 [STO]

Use the APR loan payoff formula.

$$A = PMT\left(\frac{1-\left(1+\frac{r}{n}\right)^{-U}}{\frac{r}{n}}\right)$$

$$= 284.67\left(\frac{1-(1+0.00725)^{-6}}{0.00725}\right) \approx 1665.50$$

Keystrokes: 284.67 [x] [(] 1 [−] [(] 1 [+] [RCL] [)] [y^x] 6 [+/−] [)] [÷] [RCL] [=]

The loan payoff is $1665.50.

CHECK YOUR PROGRESS 5 Aaron Jefferson has a 5-year car loan based on an annual percentage rate of 8.4%. The monthly payment is $592.57. After 3 years, Aaron decides to purchase a new car and must pay off his car loan. Find the payoff amount.

Solution See page S41.

Student Loans

The student loan program that benefits many students today was part Title IV of the Higher Education Act of 1965. In 1988, Congress renamed the student loan program after Senator Robert T. Stafford, who authored a number of bills in support of students seeking post-secondary education. Stafford loans, as they are now called, are available to students through the Department of Education. Because the U.S. government guarantees the repayment of these loans to lenders, the interest rate on Stafford loans is usually lower than the rate on private student loans.

There are basically two types of Stafford loans: subsidized and unsubsidized. For a subsidized loan, the interest on the loan is paid by the government while the student is in school. For unsubsidized loans, the student is responsible for all the interest on the loan while in school. In 2011, Congress eliminated subsidized loans for students in any graduate school program.

There are many different repayment plans for student loans. We will look at payment plans that are fixed for the entire term of the loan. For a summary of all payment plans, see https://studentloans.gov.

EXAMPLE 6 Calculate the Monthly Payment of a Subsidized Student Loan

Lois Wellington received a 9-year subsidized student loan of $32,000 at an annual interest rate of 4.5%. Determine her monthly payment on the loan after she graduates in 3 years.

Solution

Because Lois has a subsidized loan, she does not have to pay the interest that would normally accrue on a loan during the time she does not make payments. Therefore, her payment can be calculated using the payment formula for an APR loan. The payment can be calculated using the keystrokes shown in Example 4. For this example, we will use Excel's PMT function.

	A	B
1	Present value	32000
2	Annual interest rate as a decimal	0.045
3	Number of years to repay loan	9
4	Monthly payment	−$360.88

Each cell in a spreadsheet is defined by its column and row. Present value is in cell A1; 32000 is in cell B1. Calculation of the monthly payment is based on the PMT function. Here is the format.

= PMT(rate, nper, pv, [fv], [type])

The equals sign is used to tell Excel that a formula follows: *PMT* is the name of the function, *rate* is the interest rate per period, *nper* is the total number of payments, *pv* is the present value (amount of the loan), [*fv*] is the future value of the loan, and [*type*] indicates whether the payment is due at the end of a period (type = 0) or the beginning of a period (type = 1). The brackets around *fv* and *type* indicate that these values are optional. If values are not given, Excel assumes they are 0.

For this example, the PMT function is

= PMT(B2/12,B3*12,B1,0,0)

TAKE NOTE

The negative sign in front of the monthly payment is used to show that Lois must make a payment—money is subtracted from her account.

Note that the *cell locations* are placed in the formula, not the actual values. Also note that because B2 is the annual interest rate, we must divide by 12 to get the monthly interest rate; because B3 is the number of years of the loan, we must multiply by 12 to get the number of monthly payments.

The monthly payment is calculated in B4. Lois's monthly payment is $360.88.

CHECK YOUR PROGRESS 6 Hudson Zavello received a 10-year subsidized student loan of $25,000 at an annual interest rate of 4.9%. Determine his monthly payment after he graduates in 2 years.

Solution See page S41.

For a non-subsidized loan, interest on the loan accrues from the beginning of the loan. This interest is added to the loan amount and then the monthly payment is calculated. To emphasize the difference, we are going to rework Example 6 assuming a non-subsidized loan.

EXAMPLE 7 Calculate the Monthly Payment of a Non-subsidized Student Loan

Lois Wellington received a 9-year non-subsidized student loan of $32,000 at an annual interest rate of 4.5%. Determine her monthly payment on the loan after she graduates in 3 years.

Solution

Because this is a non-subsidized loan, Lois owes interest on the loan for the 3 years she is not making payments. This is a simple interest loan, so we use the simple interest rate formula.

$$I = Prt \quad \text{• Simple interest rate formula}$$
$$= 32{,}000(0.045)(3) \quad \text{• } P = 32{,}000,\ r = 0.045,\ t = 3$$
$$= 4320$$

When Lois begins making monthly payments, she owes $4320 in interest plus the $32,000 she borrowed, for a total of $36,320. To calculate her monthly payment, we will use the spreadsheet in Example 6. Lois's monthly payment is $409.60.

	A	B
1	Present value	36320
2	Annual interest rate as a decimal	0.045
3	Number of years to repay loan	9
4	Monthly payment	−$409.60

Note that this payment is almost $50 more per month than that of the subsidized loan.

CHECK YOUR PROGRESS 7 Hudson Zavello received a 10-year non-subsidized student loan of $25,000 at an annual interest rate of 4.9%. Determine his monthly payment after he graduates in 2 years.

Solution See page S41.

Examples 6 and 7 illustrate some of the power of using a spreadsheet. Because the calculation is based on the number in the cell, changing that number automatically recalculates any value that depends on that number.

EXCURSION

Car Leases

Leasing a car may result in lower monthly car payments. However, at the end of the lease term, you do not own the car. Ownership of the car reverts to the dealer, who can then sell it as a used car and realize the profit from the sale.

The value of the car at the end of the lease term is called the **residual value** of the car. The residual value of a car is frequently based on a percent of the manufacturer's suggested retail price (MSRP) and normally varies between 40% and 60% of the MSRP, depending on the type of lease.

For instance, suppose the MSRP of a car is \$18,500, and the residual value is 45% of the MSRP. Then

$$\begin{aligned}\text{Residual value} &= 0.45 \cdot 18{,}500 \\ &= 8325\end{aligned}$$

The residual value is \$8325. This is the amount the dealer thinks the car will be worth at the end of the lease period. The person leasing the car, the lessee, usually has the option of purchasing the car at that price at the end of the lease.

In addition to the residual value of the car, the monthly lease payment for a car takes into consideration *net capitalized cost, the money factor, average monthly finance charge,* and *average monthly depreciation.* Each of these terms is defined below.

Net capitalized cost = negotiated price − down payment − trade-in value

$$\textbf{Money factor} = \frac{\text{annual interest rate}}{24}$$

Average monthly finance charge
= (net capitalized cost + residual value) × money factor

$$\textbf{Average monthly depreciation} = \frac{\text{net capitalized cost} - \text{residual value}}{\text{term of the lease in months}}$$

Using these definitions, we have the following formula for a monthly lease payment.

Monthly Lease Payment Formula

The monthly lease payment formula is given by $P = F + D$, where P is the monthly lease payment, F is the average monthly finance charge, and D is the average monthly depreciation of the car.

Here is an example of calculating a monthly lease payment for a car.

The director of human resources for a company decides to lease a car for 30 months. Suppose the annual interest rate is 8.4%, the negotiated price is \$29,500, there is no trade-in, and the down payment is \$5000. Find the monthly lease payment. Assume that the residual value is 55% of the MSRP of \$33,400.

Solution

$$\begin{aligned}\text{Net capitalized cost} &= \text{negotiated price} - \text{down payment} - \text{trade-in value} \\ &= 29{,}500 - 5000 - 0 = 24{,}500\end{aligned}$$

$$\text{Residual value} = 0.55(33{,}400) = 18{,}370$$

$$\text{Money factor} = \frac{\text{Annual interest rate}}{24} = \frac{0.084}{24} = 0.0035$$

Average monthly finance charge

$= \text{(net capitalized cost + residual value)} \times \text{money factor}$

$= (24{,}500 + 18{,}370) \times 0.0035$

≈ 150.05

$$\text{Average monthly depreciation} = \frac{\text{net capitalized cost} - \text{residual value}}{\text{term of the lease in months}}$$

$$= \frac{24{,}500 - 18{,}370}{30}$$

$$\approx 204.33$$

Monthly lease payment

$= \text{average monthly finance charge} + \text{average monthly depreciation}$

$= 150.05 + 204.33$

$= 354.38$

The monthly lease payment is $354.38.

TAKE NOTE

When a person purchases a car, any state sales tax must be paid at the time of the purchase. However, with a lease, you make a state sales tax payment each month.

Suppose, for instance, that in this example the state sales tax is 6% of the monthly lease payment. Then,

Total monthly lease payment

$= 354.38 + 0.06(354.38)$

≈ 375.64

EXCURSION EXERCISES

1. Suppose you decide to obtain a 4-year lease for a car and negotiate a selling price of $28,990, including license fees. The trade-in value of your old car is $3850. If you make a down payment of $2400, the money factor is 0.0027, and the residual value is $15,000, find each of the following.
   - **a.** The net capitalized cost
   - **b.** The average monthly finance charge
   - **c.** The average monthly depreciation
   - **d.** The monthly lease payment
2. Marcia Scripps obtains a 5-year lease for a Ford pickup and negotiates a selling price of $37,115, including license fees. The trade-in value of her old car is $2950. Assuming that she makes a down payment of $3000, the money factor is 0.0035, and the residual value is $16,500, find each of the following.
   - **a.** The net capitalized cost
   - **b.** The average monthly finance charge
   - **c.** The average monthly depreciation
   - **d.** The monthly lease payment
3. Jorge Cruz obtains a 3-year lease for an economy sedan and negotiates a selling price of $22,100. The annual interest rate is 8.1%, the residual value is $15,000, and Jorge makes a down payment of $1000. Find each of the following.
   - **a.** The net capitalized cost
   - **b.** The money factor
   - **c.** The average monthly finance charge
   - **d.** The average monthly depreciation
   - **e.** The monthly lease payment
4. Suppose you obtain a 5-year lease for a Porsche and negotiate a selling price of $165,000. The annual interest rate is 8.4%, the residual value is $85,000, and you make a down payment of $5000. Find each of the following.
   - **a.** The net capitalized cost
   - **b.** The money factor
   - **c.** The average monthly finance charge
   - **d.** The average monthly depreciation
   - **e.** The monthly lease payment

5. Find the monthly lease payment for a car for which the negotiated price is \$31,900, the annual interest rate is 8%, the length of the lease is 5 years, and the residual value is 40% of the MSRP of \$33,395. There is no down payment or trade-in.

6. Find the monthly lease payment for a car for which the negotiated price is \$32,450, the annual interest rate is 3%, the length of the lease is 3 years, and the residual value is 35% of the MSRP of \$34,990. There is no down payment or trade-in.

EXERCISE SET 11.3

■ In Exercises 1 to 4, calculate the finance charge for a credit card that has the given average daily balance and interest rate.

1. Average daily balance: \$118.72; monthly interest rate: 1.25%
2. Average daily balance: \$391.64; monthly interest rate: 1.75%
3. Average daily balance: \$10,154.87; monthly interest rate: 1.5%
4. Average daily balance: \$20,346.91; monthly interest rate: 1.25%
5. **Average Daily Balance** A credit card account had a \$244 balance on March 5. A purchase of \$152 was made on March 12, and a payment of \$100 was made on March 28. Find the average daily balance if the billing date is April 5.
6. **Average Daily Balance** A credit card account has a \$768 balance on April 1. A purchase of \$316 was made on April 5, and a payment of \$200 was made on April 18. Find the average daily balance if the new billing date is May 1.
7. **Finance Charges** A charge account had a balance of \$944 on May 5. A purchase of \$255 was made on May 17, and a payment of \$150 was made on May 20. The interest on the average daily balance is 1.5% per month. Find the finance charge on the June 5 bill.
8. **Finance Charges** A charge account had a balance of \$655 on June 1. A purchase of \$98 was made on June 17, and a payment of \$250 was made on June 15. The interest on the average daily balance is 1.2% per month. Find the finance charge on the July 1 bill.
9. **Finance Charges** On August 10, a credit card account had a balance of \$345. A purchase of \$56 was made on August 15, and \$157 was charged on August 27. A payment of \$75 was made on August 15. The interest on the average daily balance is 1.25% per month. Find the finance charge on the September 10 bill.
10. **Finance Charges** On May 1, a credit card account had a balance of \$189. Purchases of \$213 were made on May 5, and \$102 was charged on May 21. A payment of \$150 was made on May 25. The interest on the average daily balance is 1.5% per month. Find the finance charge on the June 1 bill.

In Exercises 11 and 12, you may want to use the spreadsheet program available on our companion site at CengageBrain.com. This spreadsheet automates the finance charge procedure shown in this section.

11. **Finance Charges** The activity date, company, and amount for a credit card bill are shown below. The due date of the bill is September 15. On August 15, there was an unpaid balance of \$1236.43. Find the finance charge if the interest rate is 1.5% per month.

Activity date	Company	Amount
August 15	Unpaid balance	1236.43
August 16	Veterinary clinic	125.00
August 17	Gasoline	23.56
August 18	Olive's restaurant	53.45
August 20	Seaside market	41.36
August 22	Monterey Hotel	223.65
August 25	Airline tickets	310.00
August 30	Bike 101	23.26
September 1	Trattoria Maria	36.45
September 12	Seaside Market	41.25
September 13	Credit card payment	−1345.00

12. Finance Charges The activity date, company, and amount for a credit card bill are shown below. The due date of the bill is July 10. On June 10, there was an unpaid balance of $987.81. Find the finance charge if the interest rate is 1.8% per month.

Activity date	Company	Amount
June 10	Unpaid balance	987.81
June 11	Jan's Surf Shop	156.33
June 12	Albertson's	45.61
June 15	The Down Shoppe	59.84
June 16	News Mart	18.54
June 20	Cardiff Delicatessen	23.09
June 22	The Olde Golf Mart	126.92
June 28	Lee's Hawaiian Restaurant	41.78
June 30	City Food Drive	100.00
July 2	Credit card payment	−1000.00
July 8	Safeway Stores	161.38

Use a calculator for Exercises 13 to 39.

■ In Exercises 13 to 16, use the approximate annual percentage rate formula.

13. APR Chuong Ngo borrows $2500 from a bank that advertises a 9% simple interest rate and repays the loan in three equal monthly payments. Estimate the APR. Round to the nearest tenth of a percent.

14. APR Charles Ferrara borrows $4000 from a bank that advertises an 8% simple interest rate. If he repays the loan in six equal monthly payments, estimate the APR. Round to the nearest tenth of a percent.

15. APR Kelly Ang buys a computer system for $2400 and makes a 15% down payment. If Kelly agrees to repay the balance in 24 equal monthly payments at an annual simple interest rate of 10%, estimate the APR for Kelly's loan.

16. APR Jill Richards purchases a stereo system for $1500. She makes a 20% down payment and agrees to repay the balance in 12 equal payments. If the finance charge on the balance is 7% simple interest, estimate the APR. Round to the nearest tenth of a percent.

17. Monthly Payments Arrowood's Camera Store advertises a Canon Power Shot S95 camera for $400, including taxes. If you finance the purchase of this camera for 1 year at an annual percentage rate of 6.9%, find the monthly payment.

18. Monthly Payments Optics Mart offers a Meade ETX-LS 6 telescope for $1249, including taxes. If you finance the purchase of this telescope for 2 years at an annual percentage rate of 7.2%, what is the monthly payment?

19. Buying on Credit A surf shop offers a Channel Islands 6′10 MBM surfboard for $649. The sales tax is 7.25% of the purchase price.

a. What is the total cost, including sales tax?

b. If you make a down payment of 25% of the total cost, find the down payment.

c. Assuming you finance the remaining cost at an annual interest rate of 5.7% for 6 months, find the monthly payment.

20. Buying on Credit Waterworld marina offers a motorboat with a mercury engine for $38,250. The sales tax is 6.5% of the purchase price.

a. What is the total cost, including sales tax?

b. If you make a down payment of 20% of the total cost, find the down payment.

c. Assuming you finance the remaining cost at an annual interest rate of 5.7% for 3 years, find the monthly payment.

21. Buying on Credit After becoming a commercial pilot, Lorna Kao decides to purchase a 1978 Cessna 182 for $64,995. Assuming the sales tax is 5.5% of the purchase price, find each of the following.

a. What is the total cost, including sales tax?

b. If Lorna makes a down payment of 20% of the total cost, find the down payment.

c. Assuming Lorna finances the remaining cost at an annual interest rate of 7.15% for 10 years, find the monthly payment.

Paul Bowen/Science Faction/ Getty Images

22. Buying on Credit Donald Savchenko purchased new living room furniture for $2488. Assuming the sales tax is 7.75% of the purchase price, find each of the following.

a. What is the total cost, including sales tax?

b. If Donald makes a down payment of 15% of the total cost, find the down payment.

c. Assuming that Donald finances the remaining cost at an annual interest rate of 8.16% for 2 years, find the monthly payment.

23. **Car Payments** Luis Mahla purchases a used Porsche Boxster for $42,600 and finances the entire amount at an annual interest rate of 5.7% for 5 years. Find the monthly payment. Assume the sales tax is 6% of the purchase price and the license fee is 1% of the purchase price.

24. **Car Payments** Suppose you negotiate a selling price of $26,995 for a Ford Explorer. You make a down payment of 10% of the selling price and finance the remaining balance for 3 years at an annual interest rate of 7.5%. The sales tax is 7.5% of the selling price, and the license fee is 0.9% of the selling price. Find the monthly payment.

25. **Car Payments** Margaret Hsi purchases a classic car for $24,500. She makes a down payment of $3000 and finances the remaining amount for 4 years at an annual interest rate of 8.5%. The sales tax is 5.5% of the selling price and the license fee is $331. Find the monthly payment.

26. **Car Payments** Chris Schmaltz purchases a Toyota Avalon for $34,119. Chris makes a down payment of $5000 and finances the remaining amount for 5 years at an annual interest rate of 7.6%. The sales tax is 6.25% of the selling price, and the license fee is $429. Find the monthly payment.

27. **Car Payments** Suppose you purchase a car for a total price of $25,445, including taxes and license fee, and finance that amount for 4 years at an annual interest rate of 8%.
   a. Find the monthly payment.
   b. What is the total amount of interest paid over the term of the loan?

28. **Car Payments** Adele Paolo purchased an SUV for a total price of $21,425, including taxes and license fee, and financed that amount for 5 years at an annual interest rate of 7.8%.
   a. Find the monthly payment.
   b. What is the total amount of interest paid over the term of the loan?

29. **Loan Payoffs** Angela Montery has a 5-year car loan for a Jeep Wrangler at an annual interest rate of 6.3% and a monthly payment of $603.50. After 3 years, Angela decides to purchase a new car. What is the payoff on Angela's loan?

30. **Loan Payoffs** Suppose you have a 4-year car loan at an annual interest rate of 7.2% and a monthly payment of $587.21. After $2\frac{1}{2}$ years, you decide to purchase a new car. What is the payoff on your loan?

31. **Loan Payoffs** Suppose you have a 4-year car loan at an annual interest rate of 8.9% and a monthly payment of $303.52. After 3 years, you decide to purchase a new car. What is the payoff on your loan?

32. **Loan Payoffs** Ming Li has a 3-year car loan at an annual interest rate of 9.3% and a monthly payment of $453.68. After 1 year, Ming decides to purchase a new car. What is the payoff on his loan?

33. **Student Loans** Samuel Ng received a 5-year subsidized student loan of $12,000 at an annual interest rate of 5%. What are Samuel's monthly loan payments for this loan when he graduates from college in 2 years?

34. **Student Loans** Melissa Hernandez received an 8-year subsidized student loan of $21,000 at an annual interest rate of 4.1%. What are Melissa's monthly loan payments for this loan when she graduates in 1 year?

35. **Student Loans** Angelica Reardon received a 5-year non-subsidized student loan of $17,000 at an annual interest rate of 6.2%. What are Angelica's monthly loan payments for this loan after she graduates in 4 years?

36. **Student Loans** Jeffery Wei received a 6-year non-subsidized student loan of $30,000 at an annual interest rate of 5.2%. What are Jeffery's monthly loan payments for this loan after he graduates in 4 years?

EXTENSIONS

37. **Car Trade-Ins** You may have heard advertisements from car dealerships that say something like "Bring in your car, paid for or not, and we'll take it as a trade-in for a new car." The advertisement does not go on to say that you have to pay off the remaining loan balance or that balance gets added to the price of the new car.
   a. Suppose you are making payments of $235.73 per month on a 4-year car loan that has an annual interest rate of 8.4%. After making payments for 3 years, you decide to purchase a new car. What is the loan payoff?
   b. You negotiate a price, including taxes, of $18,234 for the new car. What is the actual amount you owe for the new car when the loan payoff is included?
   c. If you finance the amount in part b for 4 years at an annual interest rate of 8.4%, what is the new monthly payment?

38. Credit Card Debt The APR loan payoff formula can be used to determine how many months it will take to pay off a credit card debt if the minimum monthly payment is made each month. For instance, suppose that you have a credit card bill of \$620.50, the minimum payment is \$13, and the interest rate is 18% per year. Using a graphing calculator, we can determine the number of months, n, it will take to pay off the debt. Enter the values shown on the TI-83/84 calculator screen below. Move the cursor to N= and press ALPHA [SOLVE]. It will take over 84 months (or approximately 7 years) to pay off the credit card debt, assuming you do not make additional purchases.

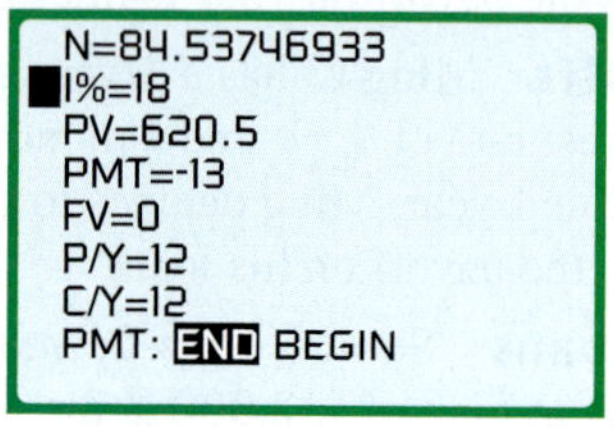

a. Find the number of months it will take to pay off a credit card debt of \$1283.34 if the minimum payment is \$27 and the annual interest rate is 19.6%. Round to the nearest month.

b. How much interest will be paid on the credit card debt in part a?

c. If you have credit card debt, determine how many months it would take to pay off your debt by making the minimum monthly payments. How much interest would you pay? You may want to use a debt payoff calculator, as mentioned in the Math Matters on page 648, to determine the answers.

39. Student Loans Suppose a graduate student receives a non-subsidized student loan of \$12,000 for each of the 4 years the student pursues a PhD. If the annual interest rate is 5% and the student has a 10-year repayment program, what are the student's monthly payments on the loans after graduation?

SECTION 11.4 Stocks, Bonds, and Mutual Funds

HISTORICAL NOTE

Topham/The Image Works

Have you heard the tongue-in-cheek advice for making a fortune in the stock market that goes something like "Buy low, sell high"? The phrase is adapted from a comment made by Hetty Green (1834–1916), who was also known as the Witch of Wall Street. She turned a \$1 million inheritance into \$100 million and became the richest woman in the world at that time. In today's terms, that \$100 million would be worth over \$2 billion. When asked how she became so successful, she replied, "There is no secret in fortune making. All you have to do is buy cheap and sell dear, act with thrift and shrewdness, and be persistent."

Stocks

Stocks, bonds, and mutual funds are investment vehicles, but they differ in nature.

When owners of a company want to raise money, generally to expand their business, they may decide to sell part of the company to investors. An investor who purchases a part of the company is said to own *stock* in the company. Stock is measured in shares; a **share of stock** in a company is a certificate that indicates partial ownership in the company. The owners of the certificates are called **stockholders** or **shareholders**. As owners, the stockholders share in the profits or losses of the corporation.

A company may distribute profits to its shareholders in the form of **dividends**. A dividend is usually expressed as a per-share amount—for example, \$0.07 per share.

EXAMPLE 1 Calculate Dividends Paid to a Stockholder

A stock pays an annual dividend of \$0.84 per share. Calculate the dividends paid to a shareholder who has 200 shares of the company's stock.

Solution

$(\$0.84 \text{ per share}) \times (200 \text{ shares}) = \$168$

The shareholder receives \$168 in dividends.

CHECK YOUR PROGRESS 1 A stock pays an annual dividend of \$0.72 per share. Calculate the dividends paid to a shareholder who has 550 shares of the company's stock.

Solution See page S41.

The **dividend yield**, which is used to compare companies' dividends, is the amount of the dividend divided by the stock price and is expressed as a percent. Determining a dividend yield is similar to calculating the simple interest rate earned on an investment. You can think of the dividend as the interest earned, the stock price as the principal, and the yield as the interest rate.

POINT OF INTEREST

Usually the dividends earned on shares of stock are sent to shareholders. However, in a **dividend reinvestment plan**, or DRIP, the dividends earned are applied toward the purchase of more shares.

EXAMPLE 2 Calculate a Dividend Yield

A stock pays an annual dividend of \$1.75 per share. The stock is trading at \$70. Find the dividend yield.

Solution

$$I = Prt$$

$$1.75 = 70r(1)$$ • Let I = annual dividend and P = the stock price. The time is 1 year.

$$1.75 = 70r$$

$$0.025 = r$$ • Divide each side of the equation by 70.

The dividend yield is 2.5%.

CHECK YOUR PROGRESS 2 A stock pays an annual dividend of \$0.82 per share. The stock is trading at \$51.25. Find the dividend yield.

Solution See page S41.

The **market value** of a share of stock is the price for which a stockholder is willing to sell a share of the stock and a buyer is willing to purchase it. Shares are always sold to the highest bidder. A **brokerage firm** is a dealer of stocks that acts as your agent when you want to buy or sell shares of stock. The **brokers** in the firm charge commissions for their service. Most trading of stocks happens on a stock exchange. **Stock exchanges** are businesses whose purpose it is to bring together buyers and sellers of stock. The largest stock exchange in the United States is the New York Stock Exchange. Shares of stock are also bought and sold through the National Association of Securities Dealers Automated Quotation System, which is commonly referred to as the NASDAQ. Every working day, each stock exchange provides financial institutions, Internet website hosts, newspapers, and other publications with data on the trading activity of all the stocks traded on that exchange. Table 11.1 shows a portion of a stock table.

TABLE 11.1
Selected Stocks from the Dow Jones Industrial Average for a Day in 2015

Name	Symbol	Open	High	Low	Close	Net chg	% Chg	Volume	52-week high	52-week low	Div	Yield	PE	YTD %chg
Apple	APPL	113.84	115.01	113.61	115.01	1.73	1.51	41654100	134.54	92.	2.08	2.14	10.05	8.48
Boeing	BA	141.13	146.45	141.05	145.41	5.09	3.5	6976300	158.83	115.02	4.36	3.63	15.8	0.56
International Business Machines	IBM	140.42	143.72	140.30	142.75	3.14	2.2	5583200	176.3	118.	5.20	4.17	9.16	3.59
Johnson & Johnson	JNJ	97.134	99.08	96.42	98.802	1.89	1.91	9887900	105.49	81.79	3.00	2.87	18.88	−3.97
General Electric	GE	28.74	29.52	28.70	29.36	0.72	2.47	81476400	31.49	19.37	0.92	3.16	16.97	−6.1
Procter & Gamble	PG	73.22	74.96	72.97	74.20	1.25	1.68	14080900	86.78	65.02	2.65	3.25	27.17	−6.09

The meaning of each column head is given below. The information refers to the first company in the table, Apple.

Open The number 113.84 in the "Open" column indicates that the opening price of the stock was \$113.84. This means that in the first trade of the day, the price of a share of Apple stock was \$113.84.

High/low On the day in question, the highest price paid for a share of Apple stock was $115.01, and the lowest price paid was $113.61.

Close The next number, 115.01, indicates that the closing price of the stock was $115.01. This means that in the final trade of the day, before the market closed, the price of a share of Apple stock was $115.01.

Net chg The number 1.73 indicates that the stock's closing price was $1.73 higher than it was on the previous trading day.

%Chg 1.51 represents the percent change in the price from the previous day's closing price to today's closing price. A positive number indicates a percent increase in price. A negative number indicates a percent decrease in price.

Vol Vol refers to the volume of shares sold. 41,654,100 shares of Apple stock were sold on the day in question.

52-week high/low The next two numbers show that, in the last 52 weeks, the highest price a share of Apple stock sold for was $134.54, and the lowest price was $92.00.

Div The number 2.08 means that the company is currently paying an annual dividend of $2.08 per share of stock.

Yield The current yield on the company's stock is 2.14%. The dividend of $2.08 is 2.14% of the current purchase price of a share of the stock.

PE The heading PE refers to the price-to-earnings ratio, the purchase price per share divided by the earnings per share.

YTD %chg The number 8.48 in the last column indicates that the price of a share of Apple stock has increased 8.48% thus far in the given calendar year.

POINT OF INTEREST

According to the Board of Governors of the Federal Reserve System, the percent of Americans who own stock decreased from 2007 to 2015. The table below gives the percents of stock owned, by annual income and by age group, for both years.

	% in 2007	% in 2015
All adults	65	55
$75K and up	90	88
$30K–$75K	72	56
Less than $30K	28	21
18–34 years	52	49
35–54 years	73	58
55 and older	65	57

EXAMPLE 3 Calculate Profits or Losses and Expenses in Selling Stock

Suppose you owned 500 shares of stock in General Electric. You purchased the shares at a price of $22.08 per share and sold them at the closing price of the stock given in Table 11.1.

a. Ignoring dividends, what was your profit or loss on the sale of the stock?

b. If your broker charges 2.4% of the total sale price, what was the broker's commission?

Solution

a. From Table 11.1, the selling price per share was $29.36.

The selling price per share is greater than the purchase price per share.

You made a profit on the sale of the stock.

$$\begin{aligned}\text{Profit} &= \text{selling price} - \text{purchase price}\\ &= 500(\$29.36) - 500(\$22.08)\\ &= \$14{,}680 - \$11{,}040\\ &= \$3640\end{aligned}$$

The profit on the sale of the stock was $3640.

b.

$$\begin{aligned}\text{Commission} &= 2.4\%(\text{selling price})\\ &= 0.024(\$14{,}680)\\ &= \$352.32\end{aligned}$$

The broker's commission was $352.32.

CHECK YOUR PROGRESS 3 Use Table 11.1. Suppose you bought 300 shares of General Electric at the 52-week low and sold the shares at the 52-week high.

a. Ignoring dividends, what was your profit or loss on the sale of the stock?

b. If your broker charges 2.1% of the total sale price, what was the broker's commission? Round to the nearest cent.

Solution See page S42.

Bonds

INSTRUCTOR NOTE

You may want to assign the Excursion on page 667. It describes Treasuries, which are bills, notes, and bonds issued by the U.S. government.

When a corporation issues stock, it is *selling* part of the company to the stockholders. When it issues a **bond**, the corporation is *borrowing* money from the bondholders; a **bondholder** lends money to a corporation. Corporations, the federal government, government agencies, states, and cities all issue bonds. These entities need money to operate—for example, to fund the federal deficit, repair roads, or build a new factory—so they borrow money from the public by issuing bonds.

Bonds are usually issued in units of \$1000. The price paid for the bond is the **face value**. The issuer promises to repay the bondholder on a particular day, called the **maturity date**, at a given rate of interest, called the **coupon**.

POINT OF INTEREST

Municipal bonds are issued by states, cities, counties, and other governments to raise money to build schools, highways, sewer systems, hospitals, and other projects for the public good. The income from many municipal bonds is exempt from federal and/or state taxes.

Assume that a bond with a \$1000 face value has a 5% coupon and a 10-year maturity date. The bondholder collects interest payments of \$50 in each of those 10 years. The payments are calculated using the simple interest formula, as shown below.

$$I = Prt$$
$$I = 1000(0.05)(1)$$
$$I = 50$$

At the end of the 10-year period, the bondholder receives from the issuer the \$1000 face value of the bond.

EXAMPLE 4 Calculate Interest Payments on a Bond

A bond with a \$10,000 face value has a 3% coupon and a 5-year maturity date. Calculate the total of the interest payments paid to the bondholder.

Solution

Use the simple interest formula to find the annual interest payments. Substitute the following values into the formula: $P = 10{,}000$, $r = 3\% = 0.03$, and $t = 1$.

$$I = Prt$$
$$I = 10{,}000(0.03)(1)$$
$$I = 300$$

Multiply the annual interest payment by the term of the bond (5 years).

$$300(5) = 1500$$

The total of the interest payments paid to the bondholder is \$1500.

CHECK YOUR PROGRESS 4 A bond has a \$15,000 face value, a 4-year maturity, and a 3.5% coupon. What is the total of the interest payments paid to the bondholder?

Solution See page S42.

A key difference between stocks and bonds is that stocks make no promises about dividends or returns, whereas the issuer of a bond guarantees that, provided the issuer remains solvent, it will pay back the face value of the bond plus interest.

Mutual Funds

An **investment trust** is a company whose assets are stocks and bonds. These companies do not manufacture a product but instead purchase stocks and bonds with the hope that their value will increase. A **mutual fund** is an example of an investment trust.

When investors purchase shares in a mutual fund, they are adding their money to a pool along with many other investors. The investments within a mutual fund are called the fund's **portfolio**. The investors in a mutual fund share the fund's profits or losses from the investments in the portfolio.

An advantage of owning shares in a mutual fund is that your money is managed by full-time professionals whose job it is to research and evaluate stocks; you own stocks without having to choose which individual stocks to buy or decide when to sell them. Another advantage is that by owning shares in the fund, you have purchased shares of stock in many different companies. This diversification helps to reduce some of the risks of investing.

Because a mutual fund owns many different stocks, each share of the fund represents a fractional interest in each stock. Each day, the value of a share in the fund, called the **net asset value of the fund**, or **NAV**, depends on the performance of the stocks in the fund. It is calculated by the following formula.

Net Asset Value of a Mutual Fund

The net asset value (NAV) of a mutual fund is given by

$$NAV = \frac{A - L}{N}$$

where A is the total fund assets, L is the total fund liabilities, and N is the number of shares outstanding.

EXAMPLE 5 Calculate the Net Asset Value of, and the Number of Shares Purchased in, a Mutual Fund

A mutual fund has $600 million worth of stock, $5 million worth of bonds, and $1 million in cash. The fund's total liabilities amount to $2 million. There are 25 million shares outstanding. You invest $15,000 in this fund.

a. Calculate the NAV.

b. How many shares are you purchasing?

Solution

a. $NAV = \frac{A - L}{N}$

$= \frac{606 \text{ million} - 2 \text{ million}}{25 \text{ million}}$

• $A = 600$ million $+ 5$ million $+ 1$ million $= 606$ million, $L = 2$ million, $N = 25$ million

$= 24.16$

The NAV of the fund is $24.16.

b. $\frac{15{,}000}{24.16} \approx 620$

• Divide the amount invested by the cost per share of the fund. Round down to the nearest whole number.

You are purchasing 620 shares of the mutual fund.

CHECK YOUR PROGRESS 5 A mutual fund has $750 million worth of stock, $750,000 in cash, and $1,500,000 in other assets. The fund's total liabilities amount to $1,500,000. There are 20 million shares outstanding. You invest $10,000 in this fund.

a. Calculate the NAV.

b. How many shares are you purchasing?

Solution See page S42.

MATH MATTERS Growth of Mutual Funds

Where do Americans invest their money? You might correctly assume that the largest number of Americans invest in real estate, as every homeowner is considered to have an investment in real estate. But more and more Americans are investing their money in the stock market, and many of them are doing so by purchasing shares in mutual funds. The table below shows data on mutual funds for selected years from 1980 to 2015. (*Source:* Investment Company Institute)

Year	Total number of U.S. households (millions)	Total value of mutual funds owned by U.S. households (millions of $)	Percent of U.S. households owning mutual funds
1980	80.8	4.6	5.7
1990	93.3	23.4	25.1
2000	106.4	48.6	45.7
2010	117.5	53.2	45.3
2015	124.6	53.6	43.0

EXCURSION

Treasury Bills

The bonds issued by the United States government are called Treasuries. Some investors prefer to invest in Treasuries, rather than the stock market, because their investment is backed by the federal government. For this reason, Treasuries are considered the safest of all investments. They are grouped into three categories.

U.S. Treasury bills have maturities of under 1 year.

U.S. Treasury notes have maturities ranging from 2 to 10 years.

U.S. Treasury bonds have maturities ranging from 10 to 30 years.

This Excursion will focus on Treasury bills.

The **face value** of a Treasury bill is the amount of money received by the investor on the maturity date of the bill. Treasury bills are sold on a **discount basis**; that is, the interest on the bill is computed and subtracted from the face value to determine the bill's cost.

TAKE NOTE

You can obtain more information about the various Treasury securities at http://www.publicdebt.treas.gov. Using this site, investors can buy securities directly from the government, thereby avoiding the service fee charged by banks and brokerage firms. This website also provides information about Treasury bills. In the past, paper bills were issued, but now all Treasury bills are issued and held electronically.

Suppose a company invests in a \$50,000 United States Treasury bill at 3.35% interest for 28 days. The bank through which the bill is purchased charges a service fee of \$15. What is the cost of the Treasury bill?

To find the cost, first find the interest. Use the simple interest formula.

$$
\begin{aligned}
I &= Prt \\
&= 50{,}000(0.0335)\left(\frac{28}{360}\right) \\
&\approx 130.28
\end{aligned}
$$

The interest earned is \$130.28.

Find the cost of the Treasury bill.

$$
\begin{aligned}
\text{Cost} &= (\text{face value} - \text{interest}) + \text{service fee} \\
&= (50{,}000 - 130.28) + 15 \\
&= 49{,}869.72 + 15 \\
&= 49{,}884.72
\end{aligned}
$$

The cost of the Treasury bill is \$49,884.72.

EXCURSION EXERCISES

1. The face value of a Treasury bill is \$30,000. The interest rate is 2.32% and the bill matures in 182 days. The bank through which the bill is purchased charges a service fee of \$15. What is the cost of the Treasury bill?
2. The face value of a 91-day Treasury bill is \$20,000. The interest rate is 2.96%. The purchaser buys the bill through Treasury Direct and pays no service fee. Calculate the cost of the Treasury bill.
3. A company invests in a 29-day, \$60,000 United States Treasury bill at 2.28% interest. The bank through which the bill is purchased charges a service fee of \$35. Calculate the cost of the Treasury bill.
4. A \$40,000 United States Treasury bill, purchased at 1.96% interest, matures in 92 days. The purchaser is charged a service fee of \$20. What is the cost of the Treasury bill?

EXERCISE SET 11.4

1. **Annual Dividends** A stock pays an annual dividend of \$1.02 per share. Calculate the dividends paid to a shareholder who has 375 shares of the company's stock.
2. **Annual Dividends** A stock pays an annual dividend of \$0.58 per share. Calculate the dividends paid to a shareholder who has 1500 shares of the company's stock.
3. **Annual Dividends** Calculate the dividends paid to a shareholder who has 850 shares of a stock that is paying an annual dividend of \$0.63 per share.
4. **Annual Dividends** Calculate the dividends paid to a shareholder who has 400 shares of a stock that is paying an annual dividend of \$0.91 per share.
5. **Dividend Yield** Find the dividend yield for a stock that pays an annual dividend of \$1.24 per share and has a current price of \$49.375. Round to the nearest hundredth of a percent.
6. **Dividend Yield** The Blackburn Computer Company has declared an annual dividend of \$0.50 per share. The stock is trading at \$40 per share. Find the dividend yield.
7. **Dividend Yield** A stock that pays an annual dividend of \$0.58 per share has a current price of \$31.75. Find the dividend yield. Round to the nearest hundredth of a percent.
8. **Dividend Yield** The Moreau Corporation is paying an annual dividend of \$0.65 per share. If the price of a share of the stock is \$81.25, what is the dividend yield on the stock?

Use the partial stock table shown below for Exercises 9 to 16. Round dollar amounts to the nearest cent when necessary.

TABLE 11.1
Selected Stocks from the Dow Jones Industrial Average for a Day in 2015

Name	Symbol	Open	High	Low	Close	Net chg	% Chg	Volume	52-week high	52-week low	Div	Yield	PE	YTD %chg
Apple	APPL	113.84	115.01	113.61	115.01	1.73	1.51	41654100	134.54	92.	2.08	2.14	10.05	8.48
Boeing	BA	141.13	146.45	141.05	145.41	5.09	3.5	6976300	158.83	115.02	4.36	3.63	15.8	0.56
International Business Machines	IBM	140.42	143.72	140.30	142.75	3.14	2.2	5583200	176.3	118.	5.20	4.17	9.16	3.59
Johnson & Johnson	JNJ	97.134	99.08	96.42	98.802	1.89	1.91	9887900	105.49	81.79	3.00	2.87	18.88	−3.97
General Electric	GE	28.74	29.52	28.70	29.36	0.72	2.47	81476400	31.49	19.37	0.92	3.16	16.97	−6.1
Procter & Gamble	PG	73.22	74.96	72.97	74.20	1.25	1.68	14080900	86.78	65.02	2.65	3.25	27.17	−6.09

9. Stock Tables For Boeing (BA):

a. What is the difference between the highest and lowest prices paid for this stock during the last 52 weeks?

b. Suppose that you own 750 shares of this stock. What dividend do you receive this year?

c. How many shares of this stock were sold during the trading day?

d. Did the price of a share of this stock increase or decrease during the day shown in the table?

e. What was the price of a share of this stock at the start of the trading day?

10. Stock Tables For International Business Machines (IBM):

a. What is the difference between the highest and lowest prices paid for this stock during the last 52 weeks?

b. Suppose that you own 750 shares of this stock. What dividend do you receive this year?

c. How many shares of this stock were sold during the trading day?

d. Did the price of a share of this stock increase or decrease during the day shown in the table?

e. What was the price of a share of this stock at the start of the trading day?

11. Stock Purchases At the closing price per share of Johnson & Johnson (JNJ), how many shares of the stock can you purchase for $5000?

12. Stock Purchases At the closing price per share of General Electric (GE), how many shares of the stock can you purchase for $2500?

13. Stock Sale Suppose that you owned 1000 shares of stock in Procter & Gamble (PG). You purchased the shares at a price of $48.96 per share and sold them at the closing price of the stock given in the table.

a. Ignoring dividends, what was your profit or loss on the sale of the stock?

b. If your broker charges 1.9% of the total sale price, what was the broker's commission?

14. Stock Sale Gary Walters owned 400 shares of stock in General Electric (GE). He purchased the shares at a price of $38.06 per share and sold them at the closing price of the stock given in the table.

a. Ignoring dividends, what was Gary's profit or loss on the sale of the stock?

b. If his broker charges 2.5% of the total sale price, what was the broker's commission?

15. Stock Sale Michele Desjardins bought 800 shares of Apple (APPL) at the 52-week low and sold the shares at the 52-week high shown in the table.

a. Ignoring dividends, what was Michele's profit or loss on the sale of the stock?

b. If her broker charges 2.3% of the total sale price, what was the broker's commission?

16. Stock Sale Suppose that you bought 1200 shares of IBM at the 52 week low and sold the shares at the 52-week high shown in the table.

a. Ignoring dividends, what was your profit or loss on the sale of the stock?

b. If your broker charges 2.25% of the total sale price, what was the broker's commission?

17. **Bonds** A bond with a face value of $6000 and a 4.2% coupon has a 5-year maturity. Find the annual interest paid to the bondholder.

18. **Bonds** The face value on a bond is $15,000. It has a 10-year maturity and a 3.75% coupon. What is the annual interest paid to the bondholder?

19. **Bonds** A bond with an $8000 face value has a 3.5% coupon and a 3-year maturity. What is the total of the interest payments paid to the bondholder?

20. **Bonds** A bond has a $12,000 face value, an 8-year maturity, and a 2.95% coupon. Find the total of the interest payments paid to the bondholder.

21. **Mutual Funds** A mutual fund has total assets of $50,000,000 and total liabilities of $5,000,000. There are 2,000,000 shares outstanding. Find the net asset value of the mutual fund.

22. **Mutual Funds** A mutual fund has total assets of $25,000,000 and total liabilities of $250,000. There are 1,500,000 shares outstanding. Find the net asset value of the mutual fund.

23. **Mutual Funds** A mutual fund has total assets of $15 million and total liabilities of $1 million. There are 2 million shares outstanding. You invest $5000 in this fund. How many shares are you purchasing?

24. **Mutual Funds** A mutual fund has total assets of $12 million and total liabilities of $2 million. There are 1 million shares outstanding. You invest $2500 in this fund. How many shares are you purchasing?

25. **Mutual Funds** A mutual fund has $500 million worth of stock, $500,000 in cash, and $1 million in other assets. The fund's total liabilities amount to $2 million. There are 10 million shares outstanding. You invest $12,000 in this fund. How many shares are you purchasing?

26. **Mutual Funds** A mutual fund has $250 million worth of stock, $10 million worth of bonds, and $1 million in cash. The fund's total liabilities amount to $1 million. There are 13 million shares outstanding. You invest $10,000 in this fund. How many shares are you purchasing?

EXTENSIONS

27. **Load and No-Load Funds** All mutual funds carry fees. One type of fee is called a "load." This is an additional fee that generally is paid at the time you invest your money in the mutual fund. A no-load mutual fund does not charge this up-front fee.

    Suppose you invested $2500 in a 4% load mutual fund 2 years ago. The 4% fee was paid out of the $2500 invested. The fund has earned 8% during each of the past 2 years. There was a management fee of 0.015% charged at the end of each year. A friend of yours invested $2500 2 years ago in a no-load fund that has earned 6% during each of the past 2 years. This fund charged a management fee of 0.15% at the end of each year. Find the difference between the values of the two investments now.

28. **Investing in the Stock Market** (This activity assumes that the class has been divided into groups of three or four students.) Imagine that your group has $10,000 to invest in each of 10 stocks. Use the stock table in today's paper or the Internet to determine the price you would pay per share. Determine the number of shares of each stock you will purchase. Check the value of each stock every business day for the next four weeks. Assume that you sell your shares at the end of the fourth week. Calculate the group's profit or loss over the 4-week period. Compare your profits or losses with those of the other groups in your class.

29. Find a mutual fund table in a daily newspaper. You can find one in the same section where the stock tables are printed. Explain the meaning of the heading of each column in the table.

SECTION 11.5 Home Ownership

Initial Expenses

POINT OF INTEREST

The U.S. Census Bureau keeps a vast selection of statistics. Two categories for which data are collected are the prices of existing homes and the prices of new homes. Recent statistics showed that the median price of an existing home was $210,000, while the median price of a new home was $283,000.

When you purchase a home, you generally make a down payment and finance the remainder of the purchase price with a loan obtained through a bank or savings and loan association. The amount of the down payment can vary, but it is normally between 10% and 30% of the selling price. The **mortgage** is the amount that is borrowed to buy the real estate. The amount of the mortgage is the difference between the selling price and the down payment.

Mortgage = selling price − down payment

This formula is used to find the amount of the mortgage. For example, suppose you buy a $240,000 home with a down payment of 25%. First find the down payment by computing 25% of the purchase price.

$$\text{Down payment} = 25\% \text{ of } 240{,}000 = 0.25(240{,}000) = 60{,}000$$

Then find the mortgage by subtracting the down payment from the selling price.

$$\begin{aligned}\text{Mortgage} &= \text{selling price} - \text{down payment}\\ &= 240{,}000 - 60{,}000\\ &= 180{,}000\end{aligned}$$

The mortgage is $180,000.

The down payment is generally the largest initial expense in purchasing a home, but there are other expenses associated with the purchase. These payments are due at the closing, when the sale of the house is finalized, and are called **closing costs**. The bank may charge fees for attorneys, credit reports, loan processing, and title searches. There may also be a **loan origination fee**. This fee is usually expressed in **points**. One point is equal to 1% of the mortgage.

Suppose you purchase a home and obtain a loan for $180,000. The bank charges a fee of 1.5 points. To find the charge for points, multiply the loan amount by 1.5%.

TAKE NOTE

1.5 points means 1.5%.
1.5% = 0.015

$$\begin{aligned}\text{Points} &= 1.5\% \text{ of } 180{,}000 = 0.015(180{,}000)\\ &= 2700\end{aligned}$$

The charge for points is $2700.

EXAMPLE 1 Calculate a Down Payment and the Closing Costs

The purchase price of a home is $392,000. A down payment of 20% is made. The bank charges $450 in fees plus $2\frac{1}{2}$ points. Find the total of the down payment and the closing costs.

Andy Dean Photography/Shutterstock.com

Solution

First find the down payment.

$$\begin{aligned}\text{Down payment} &= 20\% \text{ of } 392{,}000 = 0.20(392{,}000)\\ &= 78{,}400\end{aligned}$$

The down payment is $78,400.

Next find the mortgage.

$$\begin{aligned}\text{Mortgage} &= \text{selling price} - \text{down payment}\\ &= 392{,}000 - 78{,}400\\ &= 313{,}600\end{aligned}$$

The mortgage is $313,600.

Then, calculate the charge for points.

$$\begin{aligned}\text{Points} &= 2\frac{1}{2}\% \text{ of } 313{,}600 = 0.025(313{,}600) \qquad \bullet\ 2\frac{1}{2}\% = 2.5\% = 0.025\\ &= 7840\end{aligned}$$

The charge for points is $7840.

Finally, find the sum of the down payment and the closing costs.

$$78{,}400 + 450 + 7840 = 86{,}690$$

The total of the down payment and the closing costs is $86,690.

CHECK YOUR PROGRESS 1 The purchase price of a home is $410,000. A down payment of 25% is made. The bank charges $375 in fees plus 1.75 points. Find the total of the down payment and the closing costs.

Solution See page S42.

Mortgages

When a bank agrees to provide you with a mortgage, you agree to pay off that loan in monthly payments. If you fail to make the payments, the bank has the right to **foreclose**, which means that the bank takes possession of the property and has the right to sell it.

There are many types of mortgages available to home buyers today, so the terms of mortgages differ considerably. Some mortgages are **adjustable rate mortgages (ARMs)**. The interest rate charged on an ARM is adjusted periodically to more closely reflect current interest rates. The mortgage agreement specifies exactly how often and by how much the interest rate can change.

A **fixed-rate mortgage**, or **conventional mortgage**, is one in which the interest rate charged on the loan remains the same throughout the life of the mortgage. For a fixed-rate mortgage, the amount of the monthly payment also remains unchanged throughout the term of the loan.

The term of a mortgage can vary. Terms of 15, 20, 25, and 30 years are most common.

The monthly payment on a mortgage is the **mortgage payment**. The amount of the mortgage payment depends on the amount of the mortgage, the interest rate on the loan, and the term of the loan. This payment is calculated by using the payment formula for an APR loan given in Section 11.3. We will restate the formula here.

INSTRUCTOR NOTE

Many students will have heard of equity loans but do not understand what they are. You may want to discuss the following with them.

Home equity is the difference between the fair market value of your home (the price at which it would probably sell if you were to put it on the market today) and the amount you still owe on the mortgage. For example, if the market value of your home is \$125,000 and you still owe \$75,000 on the mortgage, your home equity is \$125,000 − \$75,000 = \$50,000.

Many banks allow homeowners to borrow on their equity. The bank charges fees to originate the loan and charges you interest, and you pay back the loan as you would any other loan, normally with monthly payments. In a home equity loan, your house serves as collateral, or security, for the loan.

Mortgage Payment Formula

The mortgage payment for a mortgage is given by

$$PMT = A\left(\frac{\frac{r}{n}}{1-\left(1+\frac{r}{n}\right)^{-nt}}\right)$$

where PMT is the monthly mortgage payment, A is the amount of the mortgage, r is the annual interest rate, n is the number of payments per year, and t is the number of years.

POINT OF INTEREST

The national home ownership rate in the United States in 2016 was approximately 63%, a rate comparable to the rate in 1990. The highest rate on record was about 69%, which occurred during the housing boom of 2006.

EXAMPLE 2 Calculate a Mortgage Payment

Suppose Allison Sommerset purchases a condominium and secures a loan of \$134,000 for 30 years at an annual interest rate of 6.5%.

a. Find the monthly mortgage payment.

b. What is the total of the payments over the life of the loan?

c. Find the amount of interest paid on the loan over the 30 years.

Solution

a. First calculate $\frac{r}{n}$ and store the result.

$$\frac{r}{n} = \frac{0.065}{12} \approx 0.00541667$$

Keystrokes: 0.065 [÷] 12 ≈ 0.00541667 [STO]

Calculate the monthly payment. For a 30-year loan, $nt = 12(30) = 360$.

$$PMT = A\left(\frac{\frac{r}{n}}{1-\left(1+\frac{r}{n}\right)^{-nt}}\right)$$

$$= 134{,}000\left(\frac{0.065/12}{1-(1+0.065/12)^{-360}}\right) \approx 846.97$$

Keystrokes:

134000 [x] [RCL] [=] [÷] [(] 1 [−] [(] 1 [+] [RCL] [)] [y^x] 360 [+/−] [)] [=]

The monthly mortgage payment is \$846.97.

b. To determine the total of the payments, multiply the number of payments (360) by the monthly payment ($846.97).

$$846.97(360) = 304{,}909.20$$

The total of the payments over the life of the loan is $304,909.20.

c. To determine the amount of interest paid, subtract the mortgage from the total of the payments.

$$304{,}909.20 - 134{,}000 = 170{,}909.20$$

The amount of interest paid over the life of the loan is $170,909.20.

CHECK YOUR PROGRESS 2 Suppose Antonio Scarletti purchases a home and secures a loan of $223,000 for 25 years at an annual interest rate of 7%.

a. Find the monthly mortgage payment.

b. What is the total of the payments over the life of the loan?

c. Find the amount of interest paid on the loan over the 25 years.

Solution See page S42.

A portion of a mortgage payment pays the current interest owed on the loan, and the remaining portion of the mortgage payment is used to reduce the principal owed on the loan. This process of paying off the principal and the interest, which is similar to paying a car loan, is called **amortizing the loan.**

In Example 2, the mortgage payment on a $134,000 mortgage at 6.5% for 30 years was $846.97. The amount of the first payment that is interest and the amount that is applied to the principal can be calculated using the simple interest formula.

$$\begin{aligned} I &= Prt \\ &= 134{,}000(0.065)\left(\frac{1}{12}\right) \\ &\approx 725.83 \end{aligned}$$

- $P = 134{,}000$, the current loan amount; $r = 0.065$; $t = \frac{1}{12}$
- Round to the nearest cent.

Of the $846.97 mortgage payment, $725.83 is an interest payment. The remainder is applied toward reducing the principal.

Principal reduction = $846.97 − $725.83 = $121.14

After the first month's mortgage payment, the balance on the loan (the amount that remains to be paid) is calculated by subtracting the principal paid on the mortgage from the mortgage.

Loan balance after first month = $134,000 − $121.14 = $133,878.86

The portion of the second mortgage payment that is applied to interest and the portion that is applied to the principal can be calculated in the same manner. In the calculation, the figure used for the principal, P, is the current balance on the loan, $133,878.86.

$$\begin{aligned} I &= Prt \\ &= 133{,}878.86(0.065)\left(\frac{1}{12}\right) \\ &\approx 725.18 \end{aligned}$$

- $P = 133{,}878.86$, the current loan amount; $r = 0.065$; $t = \frac{1}{12}$
- Round to the nearest cent.

Principal reduction = $846.97 − $725.18 = $121.79

Of the second mortgage payment, $725.18 is an interest payment and $121.79 is a payment toward the principal.

Loan balance after second month = $133,878.86 − $121.79 = $133,757.07

The interest payment, principal payment, and balance on the loan can be calculated in this manner for all of the mortgage payments throughout the life of the loan—all 360 of them! Alternatively, a computer can be programmed to make these calculations and print out the information. The printout is called an **amortization schedule**. It lists, for each mortgage payment, the payment number, the interest payment, the amount applied toward the principal, and the resulting balance to be paid.

Each month, the amount of the mortgage payment that is an interest payment decreases and the amount applied toward the principal increases. This is because you are paying interest on a decreasing balance each month. Mortgage payments early in the life of a mortgage are largely interest payments; mortgage payments late in the life of a mortgage are largely payments toward the principal.

The partial amortization schedule below shows the breakdown of the monthly payments for the first 12 months of the loan discussed in Example 2.

TAKE NOTE

You can find a spreadsheet on our companion site at CengageBrain.com, that will enable you to obtain a complete amortization schedule for any fixed-rate loan. Because car loans and home loans are calculated using the same formula, the spreadsheet can also be used to show the principal and interest payments on a car loan.

Amortization Schedule			
Loan Amount	$134,000.00		
Interest Rate	6.50%		
Term of Loan	30		
Monthly Payment	$846.97		
Month	**Amount of interest**	**Amount of principal**	**New loan amount**
1	$725.83	$121.14	$133,878.86
2	$725.18	$121.79	$133,757.07
3	$724.52	$122.45	$133,634.61
4	$723.85	$123.12	$133,511.50
5	$723.19	$123.78	$133,387.71
6	$722.52	$124.45	$133,263.26
7	$721.84	$125.13	$133,138.13
8	$721.16	$125.81	$133,012.32
9	$720.48	$126.49	$132,885.84
10	$719.80	$127.17	$132,758.66
11	$719.11	$127.86	$132,630.80
12	$718.42	$128.55	$132,502.25

QUESTION Using the amortization schedule above, how much of the loan has been paid off after 1 year?

ANSWER After 1 year (12 months), the loan amount is $132,502.25. The original loan was $134,000. The amount that has been paid off is

$$\$134{,}000 - \$132{,}502.25 = \$1497.75.$$

POINT OF INTEREST

According to the Joint Center for Housing Studies at Harvard University, consumers in the United States spend an average of 49% of their income on housing costs, which include costs associated with their residence, utilities, household supplies, and home furnishings.

EXAMPLE 3 Calculate Principal and Interest for a Mortgage Payment

You purchase a condominium for $98,750 and obtain a 30-year, fixed-rate mortgage at 7.25%. After paying a down payment of 20%, how much of the second payment is interest and how much is applied toward the principal?

Solution

First find the down payment by multiplying the percent of the purchase price that is the down payment by the purchase price.

$0.20(98{,}750) = 19{,}750$

The down payment is \$19,750.

Find the mortgage by subtracting the down payment from the purchase price.

$98{,}750 - 19{,}750 = 79{,}000$

The mortgage is \$79,000.

Calculate $\frac{r}{n}$ and store the result.

$$\frac{r}{n} = \frac{\text{annual interest rate}}{\text{number of payments per year}} = \frac{0.0725}{12} \approx 0.00604167$$

Keystrokes: 0.0725 [÷] 12 ≈ 0.00604167 [STO]

Calculate the monthly payment. For a 30-year loan, $nt = 12(30) = 360$.

$$PMT = A\left(\frac{\frac{r}{n}}{1 - \left(1 + \frac{r}{n}\right)^{-nt}}\right)$$

$$= 79{,}000\left(\frac{0.0725/12}{1 - (1 + 0.0725/12)^{-360}}\right) \approx 538.92$$

Keystrokes:

79000 [x] [RCL] [=] [÷] [(] 1 [-] [(] 1 [+] [RCL] [)] [y^x] 360 [+/-] [)] [=]

The monthly payment is \$538.92.

Find the amount of interest paid on the first mortgage payment by using the simple interest formula.

$I = Prt$

$= 79{,}000(0.0725)\left(\frac{1}{12}\right)$ • $P = 79{,}000$, the current loan amount; $r = 0.0725$; $t = \frac{1}{12}$

≈ 477.29 • Round to the nearest cent.

Find the principal paid on the first mortgage payment by subtracting the interest paid from the monthly mortgage payment.

$538.92 - 477.29 = 61.63$

Calculate the balance on the loan after the first mortgage payment by subtracting the principal paid from the mortgage.

$79{,}000 - 61.63 = 78{,}938.37$

Find the amount of interest paid on the second mortgage payment.

$I = Prt$

$= 78{,}938.37(0.0725)\left(\frac{1}{12}\right)$ • $P = 78{,}938.37$, the current loan amount; $r = 0.0725$; $t = \frac{1}{12}$

≈ 476.92 • Round to the nearest cent.

The interest paid on the second payment was \$476.92.

Find the principal paid on the second mortgage payment.

$538.92 - 476.92 = 62.00$

The principal paid on the second payment was \$62.

CHECK YOUR PROGRESS 3 You purchase a home for \$295,000. You obtain a 30-year conventional mortgage at 6.75% after paying a down payment of 25% of the purchase price. Of the first month's payment, how much is interest and how much is applied toward the principal?

Solution See page S42.

When a home is sold before the term of the loan has expired, the homeowner must pay the lender the remaining balance on the loan. To calculate that balance, we can use the APR loan payoff formula from Section 11.3.

APR Loan Payoff Formula

The payoff amount for a mortgage is given by

$$A = PMT\left(\frac{1-\left(1+\frac{r}{n}\right)^{-U}}{\frac{r}{n}}\right)$$

where A is the loan payoff, PMT is the mortgage payment, r is the annual interest rate, n is the number of payments per year, and U is the number of *remaining* (or *unpaid*) payments.

PPT

EXAMPLE 4 Calculate a Mortgage Payoff

A homeowner has a monthly mortgage payment of \$645.32 on a 30-year loan at an annual interest rate of 7.2%. After making payments for 5 years, the homeowner decides to sell the house. What is the payoff for the mortgage?

Solution

Use the APR loan payoff formula. The homeowner has been making payments for 5 years, or 60 months. There are 360 months in a 30-year loan, so there are $360 - 60 = 300$ unpaid or remaining payments; $U = 300$.

$$A = PMT\left(\frac{1-\left(1+\frac{r}{n}\right)^{-U}}{\frac{r}{n}}\right)$$

$$= 645.32\left(\frac{1-(1+0.006)^{-300}}{0.006}\right)$$

• $PMT = 645.32$; $\frac{r}{n} = \frac{0.072}{12} = 0.006$; $U = 300$, the number of remaining payments

$$\approx 89{,}679.01$$

```
N=300
I%=7.2
PV=89679.0079
PMT=-645.32
FV=0
P/Y=12
C/Y=12
PMT: END BEGIN
```

Here are the keystrokes to compute the payoff on a scientific calculator. The same calculation using a graphing calculator is shown at the left.

Calculate i: 0.072 [÷] 12 [=] 0.006 [STO]

Calculate the payoff: 645.32 [×] [(] 1 [−] [(] 1 [+] [RCL] [)] [y^x] 300 [+/−] [)] [÷] [RCL] [=]

The loan payoff is \$89,679.01.

CHECK YOUR PROGRESS 4 Ava Rivera has a monthly mortgage payment of $846.82 on her condo. After making payments for 4 years, she decides to sell the condo. If she has a 25-year loan at an annual interest rate of 6.9%, what is the payoff for the mortgage?

Solution See page S42.

MATH MATTERS Biweekly and Two-Step Mortgages

A variation of the fixed-rate mortgage is the *biweekly mortgage.* Borrowers make payments on a 30-year loan, but they pay half of a monthly payment every 2 weeks, which adds up to 26 half-payments a year, or 13 monthly payments. By making one extra monthly payment each year, the homeowner can pay off the mortgage in about $17\frac{1}{2}$ years.

Another type of mortgage is the *two-step mortgage.* Its name is derived from the fact that the life of the loan has two stages. The first step is a low fixed rate for the first 7 years of the loan, and the second step is a different, and probably higher, fixed rate for the remaining 23 years of the loan. This loan is appealing to those homeowners who do not anticipate owning the home beyond the initial low-interest-rate period; they do not need to worry about the increased interest rate during the second step.

Ongoing Expenses

In addition to a monthly mortgage payment, there are other ongoing expenses associated with home ownership. Among these expenses are the costs of insurance, property tax, and utilities such as heat, electricity, and water.

Services such as schools, police and fire protection, road maintenance, and recreational services, which are provided by cities and counties, are financed by the revenue received from taxes levied on real property, or property taxes. Property tax is normally an annual expense that can be paid on a monthly, quarterly, semiannual, or annual basis.

Homeowners who obtain a mortgage must carry fire insurance. This insurance guarantees that the lender will be repaid in the event of a fire.

EXAMPLE 5 Calculate a Total Monthly Payment

A homeowner has a monthly mortgage payment of $1145.60 and an annual property tax bill of $1074. The annual fire insurance premium is $600. Find the total monthly payment for the mortgage, property tax, and fire insurance.

Solution

Find the monthly property tax bill by dividing the annual property tax bill by 12.

$$1074 \div 12 = 89.50$$

The monthly property tax bill is $89.50.

Find the monthly fire insurance bill by dividing the annual fire insurance bill by 12.

$$600 \div 12 = 50$$

The monthly fire insurance bill is $50.

Find the sum of the mortgage payment, the monthly property tax bill, and the monthly fire insurance bill.

$$1145.60 + 89.50 + 50.00 = 1285.10$$

The monthly payment for the mortgage, property tax, and fire insurance is $1285.10.

POINT OF INTEREST

The home ownership rate in the United States for the fourth quarter of 2015 was 63.8%. The following list gives home ownership rates by region during the same quarter.

Northeast: 61.6%

Midwest: 68.1%

South: 65.3%

West: 59.0%

It is interesting to note that the home ownership rate in the United States in 1950 was 55.0%, significantly lower than it is today. (*Source:* U.S. Bureau of the Census)

CHECK YOUR PROGRESS 5 A homeowner has a monthly mortgage payment of \$1492.89, an annual property tax bill of \$2332.80, and an annual fire insurance premium of \$450. Find the total monthly payment for the mortgage, property tax bill, and fire insurance.

Solution See page S43.

EXCURSION

Home Ownership Issues

There are a number of issues that a person must think about when purchasing a home. One such issue is the difference between the interest rate on which the loan payment is based and the APR. For instance, a bank may offer a loan at an annual interest rate of 6.5%, but then go on to say that the APR is 7.1%.

The discrepancy is a result of the Truth in Lending Act. This act requires that the APR be based on all loan fees. This includes points and other fees associated with the purchase. To calculate the APR, a computer or financial calculator is necessary.

Suppose that you decide to purchase a home and you secure a 30-year, \$285,000 loan at an annual interest rate of 6.5%.

EXCURSION EXERCISES

1. Calculate the monthly payment for the loan.
2. If points are 1.5% of the loan amount, find the fee for points.
3. Add the fee for points to the loan amount. This is the modified mortgage on which the APR is calculated.
4. Using the result from Excursion Exercise 3 as the mortgage and the monthly payment from Excursion Exercise 1, determine the interest rate. (This is where the financial or graphing calculator is necessary. See page 650 for details.) The result is the APR required by the Truth in Lending Act. For this example we have included only points. In most situations, other fees would be included as well.

Another issue to research when purchasing a home is that of points and mortgage interest rates. Usually, higher points are associated with a lower mortgage interest rate. The question for the homebuyer is: Should I pay higher points for a lower mortgage interest rate, or pay lower points for a higher mortgage interest rate? The answer to that question depends on many factors, one of which is the amount of time the homeowner plans on staying in the home.

Consider two typical situations for a 30-year, \$100,000 mortgage. Option 1 offers an annual mortgage interest rate of 7.25% and a loan origination fee of 1.5 points. Option 2 offers an annual interest rate of 7% and a loan origination fee of 2 points.

5. Calculate the monthly payments for Option 1 and Option 2.
6. Calculate the loan origination fees for Option 1 and Option 2.
7. What is the total amount paid, including points, after 2 years for each option?
8. What is the total amount paid, including points, after 3 years for each option?
9. Which option is more cost effective if you stay in the home for 2 years or less? Which option is more cost effective if you stay in the home for 3 years or more? Explain your answer.

EXERCISE SET 11.5

1. **Mortgages** You buy a $258,000 home with a down payment of 25%. Find the amount of the down payment and the mortgage amount.
2. **Mortgages** Greg Walz purchases a home for $325,000 with a down payment of 10%. Find the amount of the down payment and the mortgage amount.
3. **Points** Clarrisa Madison purchases a home and secures a loan of $250,000. The bank charges a fee of 2.25 points. Find the charge for points.
4. **Points** Jerome Thurber purchases a home and secures a loan of $170,000. The bank charges a fee of $2\frac{3}{4}$ points. Find the charge for points.
5. **Closing Costs** The purchase price of a home is $309,000. A down payment of 30% is made. The bank charges $350 in fees plus 3 points. Find the total of the down payment and the closing costs.
6. **Closing Costs** The purchase price of a home is $243,000. A down payment of 20% is made. The bank charges $425 in fees plus 4 points. Find the total of the down payment and the closing costs.
7. **Closing Costs** The purchase price of a condominium is $121,500. A down payment of 25% is made. The bank charges $725 in fees plus $3\frac{1}{2}$ points. Find the total of the down payment and the closing costs.
8. **Closing Costs** The purchase price of a manufactured home is $159,000. A down payment of 20% is made. The bank charges $815 in fees plus 1.75 points. Find the total of the down payment and the closing costs.
9. **Mortgage Payments** Find the mortgage payment for a 25-year loan of $129,000 at an annual interest rate of 7.75%.
10. **Mortgage Payments** Find the mortgage payment for a 30-year loan of $245,000 at an annual interest rate of 6.5%.
11. **Mortgage Payments** Find the mortgage payment for a 15-year loan of $223,500 at an annual interest rate of 8.15%.
12. **Mortgage Payments** Find the mortgage payment for a 20-year loan of $149,900 at an annual interest rate of 8.5%.
13. **Mortgage Payments** Leigh King purchased a townhouse and obtained a 30-year loan of $152,000 at an annual interest rate of 7.75%.
    a. What is the mortgage payment?
    b. What is the total of the payments over the life of the loan?
    c. Find the amount of interest paid on the mortgage loan over the 30 years.
14. **Mortgage Payments** Richard Miyashiro purchased a condominium and obtained a 25-year loan of $199,000 at an annual interest rate of 8.25%.
    a. What is the mortgage payment?
    b. What is the total of the payments over the life of the loan?
    c. Find the amount of interest paid on the mortgage loan over the 25 years.
15. **Interest Paid** Ira Patton purchased a home and obtained a 15-year loan of $219,990 at an annual interest rate of 8.7%. Find the amount of interest paid on the loan over the 15 years.
16. **Interest Paid** Leona Jefferson purchased a home and obtained a 30-year loan of $437,750 at an annual interest rate of 7.5%. Find the amount of interest paid on the loan over the 30 years.
17. **Principal and Interest** Marcel Thiessen purchased a home for $208,500 and obtained a 15-year, fixed-rate mortgage at 9% after paying a down payment of 10%. Of the first month's mortgage payment, how much is interest and how much is applied to the principal?
18. **Principal and Interest** You purchase a condominium for $173,000. You obtain a 30-year, fixed-rate mortgage loan at 12% after paying a down payment of 25%. Of the second month's mortgage payment, how much is interest and how much is applied to the principal?
19. **Principal and Interest** You purchase a cottage for $185,000. You obtain a 20-year, fixed-rate mortgage loan at 12.5% after paying a down payment of 30%. Of the second month's mortgage payment, how much is interest and how much is applied to the principal?

Rafal Mielczarek/Shutterstock.com

20. **Principal and Interest** Fay Nguyen purchased a second home for $183,000 and obtained a 25-year, fixed-rate mortgage loan at 9.25% after paying a down payment of 30%. Of the second month's mortgage payment, how much is interest and how much is applied to the principal?
21. **Loan Payoffs** After making payments of $913.10 for 6 years on your 30-year loan at 8.5%, you decide to sell your home. What is the loan payoff?

22. **Loan Payoffs** Christopher Chamberlain has a 25-year mortgage loan at an annual interest rate of 7.75%. After making payments of $1011.56 for $3\frac{1}{2}$ years, Christopher decides to sell his home. What is the loan payoff?

23. **Loan Payoffs** Iris Chung has a 15-year mortgage loan at an annual interest rate of 7.25%. After making payments of $672.39 for 4 years, Iris decides to sell her home. What is the loan payoff?

24. **Loan Payoffs** After making payments of $736.98 for 10 years on your 30-year loan at 6.75%, you decide to sell your home. What is the loan payoff?

25. **Total Monthly Payment** A homeowner has a mortgage payment of $996.60, an annual property tax bill of $594, and an annual fire insurance premium of $300. Find the total monthly payment for the mortgage, property tax, and fire insurance.

26. **Total Monthly Payment** Malcolm Rothschild has a mortgage payment of $1753.46, an annual property tax bill of $1023, and an annual fire insurance premium of $780. Find the total monthly payment for the mortgage, property tax, and fire insurance.

27. **Total Monthly Payment** Baka Onegin obtains a 25-year mortgage loan of $259,500 at an annual interest rate of 7.15%. Her annual property tax bill is $1320 and her annual fire insurance premium is $642. Find the total monthly payment for the mortgage, property tax, and fire insurance.

28. **Total Monthly Payment** Suppose you obtain a 20-year mortgage loan of $198,000 at an annual interest rate of 8.4%. The annual property tax bill is $972 and the annual fire insurance premium is $486. Find the total monthly payment for the mortgage, property tax, and fire insurance.

29. **Mortgage Loans** Consider a mortgage loan of $150,000 at an annual interest rate of 8.125%.
    a. How much greater is the mortgage payment if the term is 15 years rather than 30 years?
    b. How much less is the amount of interest paid over the life of the 15-year loan than over the life of the 30-year loan?

30. **Mortgage Loans** Consider a mortgage loan of $359,960 at an annual interest rate of 7.875%.
    a. How much greater is the mortgage payment if the term is 15 years rather than 30 years?
    b. How much less is the amount of interest paid over the life of the 15-year loan than over the life of the 30-year loan?

31. **Mortgage Loans** The Mendez family is considering a mortgage loan of $349,500 at an annual interest rate of 6.75%.
    a. How much greater is their mortgage payment if the term is 20 years rather than 30 years?
    b. How much less is the amount of interest paid over the life of the 20-year loan than over the life of the 30-year loan?

32. **Mortgage Loans** Herbert Bloom is considering a mortgage loan of $322,495 at an annual interest rate of 7.5%.
    a. How much greater is his mortgage payment if the term is 20 years rather than 30 years?
    b. How much less is the amount of interest paid over the life of the 20-year loan than over the life of the 30-year loan?

33. **Affordability** A couple has saved $25,000 for a down payment on a home. Their bank requires a minimum down payment of 20%. What is the maximum price they can offer for a house in order to have enough money for the down payment?

34. **Affordability** You have saved $18,000 for a down payment on a house. Your bank requires a minimum down payment of 15%. What is the maximum price you can offer for a home in order to have enough money for the down payment?

35. **Affordability** You have saved $39,400 to make a down payment and pay the closing costs on your future home. Your bank informs you that a 15% down payment is required and that the closing costs should be $380 plus 4 points. What is the maximum price you can offer for a home in order to have enough money for the down payment and the closing costs?

EXTENSIONS

36. **Amortization Schedules** Suppose you have a 30-year mortgage loan for $119,500 at an annual interest rate of 8.25%. For which monthly payment does the amount of principal paid first exceed the amount of interest paid? For this exercise, you will need a spreadsheet program for producing amortization schedules. You can find one on our companion site at CengageBrain.com.

37. **Amortization Schedules** Does changing the amount of the loan in Exercise 36 change the number of the monthly payment for which the amount of principal paid first exceeds the amount of interest paid? For this exercise, you will need a spreadsheet program for producing amortization schedules. You can find one on our companion site at CengageBrain.com.

38. **Amortization Schedules** Does changing the interest rate of the loan in Exercise 36 change the number of the monthly payment for which the amount of principal paid first exceeds the amount of interest paid? For this exercise, you will need a spreadsheet program for producing amortization schedules. You can find one on our companion site at CengageBrain.com.

39. **Buying and Selling a Home** Suppose you buy a house for \$208,750, make a down payment that is 30% of the purchase price, and secure a 30-year loan for the balance at an annual interest rate of 7.75%. The points on the loan are 1.5% and there are additional lender fees of \$825.

   a. How much is due at closing? Note that the down payment is due at closing.

   b. After 5 years, you decide to sell your house. What is the loan payoff?

   c. Because of inflation, you are able to sell your house for \$248,000. Assuming that the selling fees are 6% of the selling price, what are the proceeds of the sale after deducting selling fees? Do not include the interest paid on the mortgage. Remember to consider the loan payoff.

   d. The percent return on an investment equals $\dfrac{\text{proceeds from sale}}{\text{total closing costs}} \times 100$. Find the percent return on your investment. Round to the nearest percent.

CHAPTER 11 SUMMARY

The following table summarizes essential concepts in this chapter. The references given in the right-hand column list Examples and Exercises that can be used to test your understanding of a concept.

11.1 Simple Interest

Simple Interest Formula The simple interest formula is $I = Prt$, where I is the interest, P is the principal, r is the interest rate, and t is the time period.	See **Examples 2, 4, and 5** on pages 621 and 622, and then try Exercises 1, 2, and 3 on page 683.
Future Value or Maturity Value Formula for Simple Interest The future or maturity value formula for simple interest is $$A = P + I$$ where A is the amount after the interest, I, has been added to the principal, P. The future value or maturity value formula can also be written $A = P(1 + rt)$, where A is the future or maturity value, P is the principal, r is the interest rate, and t is the time period.	See **Examples 6 and 7** on pages 624 and 625, and then try Exercise 4 on page 683.

11.2 Compound Interest

Compound Amount Formula The compound amount formula is $$A = P\left(1 + \frac{r}{n}\right)^{nt}$$ where A is the compound amount, P is the amount of money deposited, r is the annual interest rate, n is the number of compounding periods per year, and t is the number of years.	See **Examples 2, 3, and 4** on pages 632 and 633, and then try Exercises 7, 8, and 9 on page 683.
Present Value Formula The present value formula is $$P = \frac{A}{\left(1 + \frac{r}{n}\right)^{nt}}$$ where P is the original principal invested, A is the compound amount, r is the annual interest rate, n is the number of compounding periods per year, and t is the number of years.	See **Example 6** on page 635, and then try Exercises 10 and 12 on pages 683 and 684.

continued

Inflation To calculate the effects of inflation, use the same procedure used to calculate compound amount. Assume a constant annual inflation rate, and use $n = 1$ in the calculation.	See **Examples 8 and 9** on page 637, and then try Exercises 15 and 16 on page 684.
Effective Interest Rate When interest is compounded, the annual rate of interest is the nominal rate. The effective rate is the simple interest rate that would yield the same amount of interest after 1 year.	See **Examples 10 and 11** on pages 639 and 640, and then try Exercises 17 and 18 on page 684.
11.3 Credit Cards and Consumer Loans	
Average Daily Balance Average daily balance = $\dfrac{\text{sum of the total amounts owed each day of the month}}{\text{number of days in the billing cycle}}$	See **Example 1** on page 647, and then try Exercises 19 and 20 on page 684.
Approximate Annual Percentage Rate (APR) Formula for a Simple Interest Loan The annual percentage rate (APR) of a simple interest rate loan can be approximated by $\text{APR} \approx \dfrac{2nr}{n+1}$ where n is the number of payments and r is the simple interest rate.	See **Example 2** on page 650, and then try Exercises 21 and 22 on page 684.
Payment Formula for an APR Loan The payment for a loan based on APR is given by $PMT = A\left(\dfrac{\frac{r}{n}}{1-\left(1+\frac{r}{n}\right)^{-nt}}\right)$ where PMT is the payment, A is the loan amount, r is the annual interest rate, n is the number of payments per year, and t is the number of years.	See **Examples 3 and 4** on pages 651 and 652, and then try Exercises 23, 24, and 25 on page 684.
APR Loan Payoff Formula The payoff amount for a loan based on APR is given by $A = PMT\left(\dfrac{1-\left(1+\frac{r}{n}\right)^{-U}}{\frac{r}{n}}\right)$ where A is the loan payoff, PMT is the payment, r is the annual interest rate, n is the number of payments per year, and U is the number of *remaining* (or *unpaid*) payments.	See **Example 5** on page 654, and then try Exercise 26 on page 684.
11.4 Stocks, Bonds, and Mutual Funds	
Stocks A company may distribute profits to its shareholders in the form of dividends. A dividend is usually expressed as a per-share amount. The dividend yield is the amount of the dividend divided by the stock price and is expressed as a percent.	See **Examples 1 and 2** on pages 662 and 663, and then try Exercise 13 on page 684.

Bonds The price paid for a bond is the face value. The issuer promises to repay the bondholder on a particular day, called the maturity date, at a given interest rate, called the coupon.	See **Example 4** on page 665, and then try Exercise 14 on page 684.
Net Asset Value of a Mutual Fund The net asset value of a mutual fund is given by $NAV = \frac{A - L}{N}$ where A is the total fund assets, L is the total fund liabilities, and N is the number of shares outstanding.	See **Example 5** on page 666, and then try Exercise 28 on page 684.
11.5 Home Ownership	
Mortgage Payment Formula The mortgage payment for a mortgage is given by $PMT = A\left(\frac{\frac{r}{n}}{1 - \left(1 + \frac{r}{n}\right)^{-nt}}\right)$ where PMT is the monthly mortgage payment, A is the amount of the mortgage, r is the annual interest rate, n is the number of payments per year, and t is the number of years.	See **Examples 2 and 3** on pages 672–673 and 674–675, and then try Exercise 30 on page 685.
APR Loan Payoff Formula The payoff amount for a mortgage is given by $A = PMT\left(\frac{1 - \left(1 + \frac{r}{n}\right)^{-U}}{\frac{r}{n}}\right)$ where A is the loan payoff, PMT is the mortgage payment, r is the annual interest rate, n is the number of payments per year, and U is the number of *remaining* (or *unpaid*) payments.	See **Example 4** on page 676, and then try Exercise 26 on page 684.

CHAPTER 11 REVIEW EXERCISES

1. **Simple Interest** Calculate the simple interest due on a 4-month loan of $2750 if the interest rate is 6.75%.
2. **Simple Interest** Find the simple interest due on an 8-month loan of $8500 if the interest rate is 1.15% per month.
3. **Simple Interest** What is the simple interest earned in 120 days on a deposit of $4000 if the interest rate is 6.75%?
4. **Maturity Value** Calculate the maturity value of a simple interest, 108-day loan of $7000 if the interest rate is 10.4%.
5. **Simple Interest Rate** The simple interest charged on a 3-month loan of $6800 is $127.50. Find the simple interest rate.
6. **Compound Amount** Calculate the compound amount when $3000 is deposited in an account earning 6.6% interest, compounded monthly, for 3 years.
7. **Compound Amount** What is the compound amount when $6400 is deposited in an account earning an interest rate of 6%, compounded quarterly, for 10 years?
8. **Future Value** Find the future value of $6000 earning 9% interest, compounded daily, for 3 years.
9. **Compound Interest** Calculate the amount of interest earned in 4 years on $600 deposited in an account paying 7.2% interest, compounded daily.
10. **Present Value** How much money should be invested in an account that earns 8% interest, compounded semiannually, in order to have $18,500 in 7 years?

11. Loans To help pay your college expenses, you borrow \$8000 and agree to repay the loan at the end of 5 years at 7% interest, compounded quarterly.

a. What is the maturity value of the loan?

b. How much interest are you paying on the loan?

12. Present Value A couple plans to save for their child's college education. What principal must be deposited by the parents when their child is born in order to have \$80,000 when the child reaches the age of 18? Assume the money earns 8% interest, compounded quarterly.

13. Dividend Yield A stock pays an annual dividend of \$0.66 per share. The stock is trading at \$60. Find the dividend yield.

14. Bonds A bond with a \$20,000 face value has a 4.5% coupon and a 10-year maturity. Calculate the total of the interest payments paid to the bondholder.

15. Inflation In 2011, the price of 1 lb of baking potatoes was \$0.89. Use an annual inflation rate of 6% to calculate the price of 1 lb of baking potatoes in 2021. Round to the nearest cent.

16. Inflation You purchase a bond that will provide you with \$75,000 in 8 years. Assuming an annual inflation rate of 7%, what will be the purchasing power of the \$75,000 in 8 years?

17. Effective Interest Rate Calculate the effective interest rate of 5.90% compounded monthly. Round to the nearest hundredth of a percent.

18. Annual Yield Which has the higher annual yield, 5.2% compounded quarterly or 5.4% compounded semiannually?

19. Average Daily Balance A credit card account had a \$423.35 balance on March 11. A purchase of \$145.50 was made on March 18, and a payment of \$250 was made on March 29. Find the average daily balance if the billing date is April 11.

20. Finance Charges On September 10, a credit card account had a balance of \$450. A purchase of \$47 was made on September 20, and \$157 was charged on September 25. A payment of \$175 was made on September 28. The interest on the average daily balance is 1.25% per month. Find the finance charge on the October 10 bill.

21. APR Arlene McDonald borrows \$1500 from a bank that advertises a 7.5% simple interest rate and repays the loan in six equal monthly payments.

a. Find the monthly payment.

b. Estimate the APR. Round to the nearest tenth of a percent.

22. APR Suppose you purchase a laptop computer for \$449, make a 10% down payment, and agree to repay the balance in 12 equal monthly payments. The finance charge on the balance is 7% simple interest.

a. Find the monthly payment.

b. Estimate the APR. Round to the nearest tenth.

23. Monthly Payments Photo Experts offers a Nikon camera for \$999, including taxes. If you finance the purchase of this camera for 2 years at an annual interest rate of 8.5%, find the monthly payment.

24. Monthly Payments Abeni Silver purchases a Samsung 4K Ultra HD TV for \$9499. The sales tax is 6.25% of the purchase price.

a. What is the total cost, including sales tax?

b. If Abeni makes a down payment of 20% of the total cost, find the down payment.

c. Assuming that Abeni finances the remaining cost at an annual interest rate of 8% for 3 years, find the monthly payment.

25. Car Payments Suppose that you decide to purchase a new car. You go to a credit union to get preapproval for your loan. The credit union offers you an annual interest rate of 3.25% for 3 years. The purchase price of the car you select is \$28,450, including taxes, and you make a 20% down payment. What is your monthly payment?

26. Loan Payoffs Dasan Houston obtains a \$28,000, 5-year loan for a hybrid car at an annual interest rate of 5.9%.

a. Find the monthly payment.

b. After 3 years, Dasan decides to purchase a new car. What is the payoff on his loan?

27. Stock Sale Suppose that you purchased 500 shares of stock at a price of \$28.75 per share and sold them for \$39.40 per share.

a. Ignoring dividends, what was your profit or loss on the sale of the stock?

b. If your broker charges 1.3% of the total sale price, what was the broker's commission?

28. Mutual Funds A mutual fund has total assets of \$34 million and total liabilities of \$4 million. There are 2 million shares outstanding. You invest \$3000 in this fund. How many shares are you purchasing?

29. Closing Costs The purchase price of a seaside cottage is \$459,000. A down payment of 20% is made. The bank charges \$815 in fees plus 1.75 points. Find the total of the down payment and the closing costs.

30. **Mortgage Payments** Suppose you purchase a condominium and obtain a 30-year loan of $255,800 at an annual interest rate of 6.75%.

 a. What is the mortgage payment?

 b. What is the total of the payments over the life of the loan?

 c. Find the amount of interest paid on the mortgage loan over the 30 years.

31. **Mortgage Payments and Loan Payoffs** Garth Santacruz purchased a condominium and obtained a 25-year loan of $189,000 at an annual interest rate of 7.5%.

 a. What is the mortgage payment?

 b. After making payments for 10 years, Garth decides to sell his home. What is the loan payoff?

32. **Total Monthly Payments** Geneva Goldberg obtains a 15-year loan of $278,950 at an annual interest rate of 7%. Her annual property tax bill is $1134 and her annual fire insurance premium is $681. Find the total monthly payment for the mortgage, property tax, and fire insurance.

33. **Student Loans** A student receives a non-subsidized Stafford loan of $17,000 at an annual interest rate of 4.1% for 6 years. What are the monthly payments on the loan when the student graduates 2 years later?

CHAPTER 11 TEST

1. **Simple Interest** Calculate the simple interest due on a 3-month loan of $5250 if the interest rate is 8.25%.

2. **Simple Interest** Find the simple interest earned in 180 days on a deposit of $6000 if the interest rate is 6.75%.

3. **Maturity Value** Calculate the maturity value of a simple interest, 200-day loan of $8000 if the interest rate is 9.2%.

4. **Simple Interest Rate** The simple interest charged on a 2-month loan of $7600 is $114. Find the simple interest rate.

5. **Compound Amount** What is the compound amount when $4200 is deposited in an account earning an interest rate of 7%, compounded monthly, for 8 years?

6. **Compound Interest** Calculate the amount of interest earned in 3 years on $1500 deposited in an account paying 6.3% interest, compounded daily.

7. **Maturity Value** To help pay for a new truck, you borrow $10,500 and agree to repay the loan in 4 years at 9.5% interest, compounded monthly.

 a. What is the maturity value of the loan?

 b. How much interest are you paying on the loan?

8. **Present Value** A young couple wants to save money to buy a house. What principal must be deposited by the couple in order to have $30,000 in 5 years? Assume the money earns 6.25% interest, compounded daily.

9. **Dividend Yield** A stock that has a market value of $40 pays an annual dividend of $0.48 per share. Find the dividend yield.

10. **Bonds** Suppose you purchase a $5000 bond that has a 3.8% coupon and a 10-year maturity. Calculate the total of the interest payments that you will receive.

11. **Inflation** In 2016, the median value of a single-family house was $224,000. Use an annual inflation rate of 4.3% to calculate the median value of a single family house in 2029. (*Source:* money.cnn.com)

12. **Effective Interest Rate** Calculate the effective interest rate of 6.25% compounded quarterly. Round to the nearest hundredth of a percent.

13. **Annual Yield** Which has the higher annual yield, 4.4% compounded monthly or 4.6% compounded semiannually?

14. **Finance Charges** On October 15, a credit card account had a balance of $515. A purchase of $75 was made on October 20, and a payment of $250 was made on October 28. The interest on the average daily balance is 1.8% per month. Find the finance charge on the November 15 bill.

15. **APR** Suppose that you purchase a 2-in-1 laptop computer for $629, make a 15% down payment, and agree to repay the balance in 12 equal monthly payments. The finance charge on the balance is 9% simple interest.

 a. Find the monthly payment.

 b. Estimate the APR. Round to the nearest tenth of a percent.

16. **Monthly Payments** Technology Pro offers a new computer for $1899, including taxes. If you finance the purchase of this computer for 3 years at an annual percentage rate of 4.5%, find your monthly payment.

17. Stock Sale Suppose you purchased 800 shares of stock at a price of \$31.82 per share and sold them for \$25.70 per share.

a. Ignoring dividends, what was your profit or loss on the sale of the stock?

b. If your broker charges 1.1% of the total sale price, what was the broker's commission?

18. Mutual Funds A mutual fund has total assets of \$42 million and total liabilities of \$6 million. There are 3 million shares outstanding. You invest \$2500 in this fund. How many shares are you purchasing?

19. Monthly Payments Kalani Canfield purchases a deluxe hot tub for \$6575. The sales tax is 6.25% of the purchase price.

a. What is the total cost, including sales tax?

b. If Kalani makes a down payment of 20% of the total cost, find the down payment.

c. Assuming Kalani finances the remaining cost at an annual interest rate of 7.8% for 3 years, find the monthly payment.

20. Closing Costs The purchase price of a house is \$262,250. A down payment of 20% is made. The bank charges \$815 in fees plus 3.25 points. Find the total of the down payment and the closing costs.

21. Mortgage Payments and Loan Payoffs Bernard Mason purchased a house and obtained a 30-year loan of \$236,000 at an annual interest rate of 6.75%.

a. What is the mortgage payment?

b. After making payments for 5 years, Bernard decides to sell his home. What is the loan payoff?

22. Total Monthly Payment Zelda MacPherson obtains a 20-year loan of \$312,000 at an annual interest rate of 7.25%. Her annual property tax bill is \$1044 and her annual fire insurance premium is \$516. Find the total monthly payment for the mortgage, property tax, and fire insurance.

12

Combinatorics and Probability

The PowerBall lottery combines 44 states, the District of Columbia, and the Virgin Islands into a single lottery. To win the grand prize in this lottery, a player must select the five numbers from 1 to 69 and the one number from 1 to 26 that are chosen by the lottery commission. There are 292,201,338 different ways in which a player can make that selection. Thus a person's chance of winning the PowerBall lottery grand prize is 1 in 292,201,338. To put this in perspective, a person is about 600 times more likely to be struck by lightning than to win the PowerBall lottery grand prize.

When no person selects the winning lottery numbers, the grand prize increases. As of May 2016, the largest PowerBall jackpot was approximately \$1.59 billion, which was won on January 14, 2016, and shared among three winning tickets.

This chapter focuses on the mathematics necessary to calculate the number of ways in which certain events, such as selecting a winning lottery number, can happen. This is part of the study of combinatorics. Once the number of ways in which an event can occur has been calculated, it is possible to determine the probability of that event. Probability is another topic covered in this chapter.

Alex Grimm GRI/JV/REUTERS

SECTION 12.1 The Counting Principle

Counting by Making a List

Tatiana Popova/Shutterstock.com

Combinatorics is the study of counting the different outcomes of some task. For example, if a coin is flipped, the side facing upward will be a head or a tail. The outcomes can be listed as {H, T}. There are two possible outcomes.

If a regular six-sided die is rolled, the possible outcomes are , , , , , . The outcomes can also be listed as {1, 2, 3, 4, 5, 6}. There are six possible outcomes.

EXAMPLE 1 Counting by Forming a List

List and then count the number of different outcomes that are possible when one letter from the word *Tennessee* is chosen.

Solution

The possible outcomes are {T, e, n, s}. There are four possible outcomes.

CHECK YOUR PROGRESS 1 List and then count the number of different outcomes that are possible when one letter is chosen from the word *Mississippi.*

Solution See page S43.

In combinatorics, an **experiment** is an activity with an observable outcome. The set of all possible outcomes of an experiment is called the **sample space** of the experiment. Flipping a coin, rolling a die, and choosing a letter from the word *Tennessee* are experiments. The sample spaces are {H, T}, {1, 2, 3, 4, 5, 6}, and {T, e, n, s}, respectively.

An **event** is one or more of the possible outcomes of an experiment. Flipping a coin and having a head show on the upward face, rolling a 5 when a die is tossed, and choosing a T from one of the letters in the word *Tennessee* are all examples of events. An event is a *subset* of the sample space.

EXAMPLE 2 Listing the Elements of an Event

One number is chosen from the sample space

$$S = \{1, 2, 3, 4, 5, 6, 7, 8, 9, 10, 11, 12, 13, 14, 15, 16, 17, 18, 19, 20\}$$

List the elements in the following events.

a. The number is even.

b. The number is divisible by 5.

c. The number is a prime number.

Solution

a. {2, 4, 6, 8, 10, 12, 14, 16, 18, 20}

b. {5, 10, 15, 20}

c. {2, 3, 5, 7, 11, 13, 17, 19}

CHECK YOUR PROGRESS 2 One digit is chosen from the digits 0 through 9. The sample space S is {0, 1, 2, 3, 4, 5, 6, 7, 8, 9}. List the elements in the following events.

a. The number is odd.

b. The number is divisible by 3.

c. The number is greater than 7.

Solution See page S43.

Counting by Making a Table

Each of the experiments given above illustrates a *single-stage experiment.* A **single-stage experiment** is an experiment for which there is a single outcome. Experiments that have two, three, or more stages are called **multi-stage experiments**. To count the number of outcomes of such an experiment, a systematic procedure is helpful. Using a table to record results is one such procedure.

Consider the two-stage experiment of rolling two dice, one red and one green. How many different outcomes are possible? To determine the number of outcomes, make a table with the different outcomes of rolling the red die across the top and the different outcomes of rolling the green die down the side.

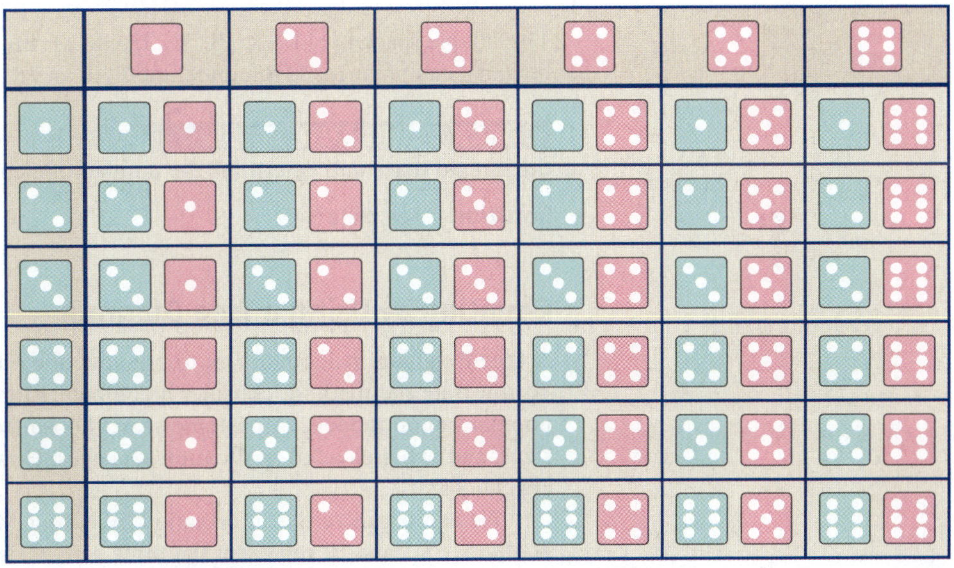

Outcomes of Rolling Two Dice

PPT

By counting the number of entries in the diagram above, we see that there are 36 different outcomes of the experiment of rolling two dice. The sample space is

From the table, several different events can be discussed.

- The sum of the pips (dots) on the upward faces is 7. There are six outcomes of this event. They are .
- The sum of the pips on the upward faces is 11. There are two outcomes of this event. They are 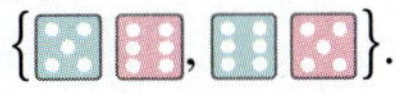.
- The numbers of pips on the upward faces are equal. There are six outcomes of this event. They are 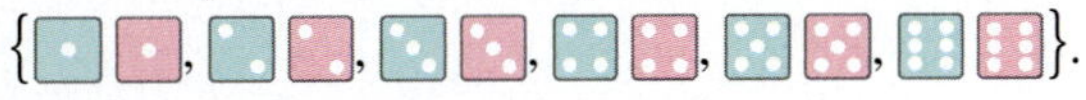.

EXAMPLE 3 **Counting Using a Table**

Two-digit numbers are formed from the digits 1, 3, and 8. Find the sample space and determine the number of elements in the sample space.

Solution

Use a table to list all the different two-digit numbers that can be formed by using the digits 1, 3, and 8.

	1	3	8
1	11	13	18
3	31	33	38
8	81	83	88

The sample space is {11, 13, 18, 31, 33, 38, 81, 83, 88}. There are nine two-digit numbers that can be formed from the digits 1, 3, and 8.

CHECK YOUR PROGRESS 3 A die is tossed and then a coin is flipped. Find the sample space and determine the number of elements in the sample space.

Solution See page S43.

Counting by Using a Tree Diagram

A **tree diagram** is another way to organize the outcomes of a multi-stage experiment. To illustrate the method, consider a computer store offering special prices on its most popular laptop models. A customer can choose from two sizes of RAM, three screen sizes, and two preloaded application packages. How many different laptops can customers choose?

We can organize the information by letting M_1 and M_2 represent the two sizes of RAM; S_1, S_2, and S_3 represent the three screen sizes; and A_1 and A_2 represent the two application packages (see Figure 12.1).

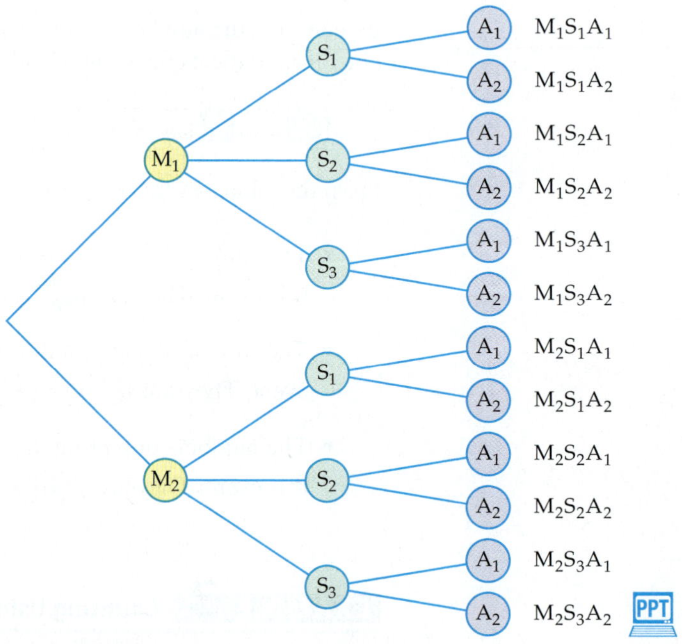

FIGURE 12.1 There are 12 possible laptops.

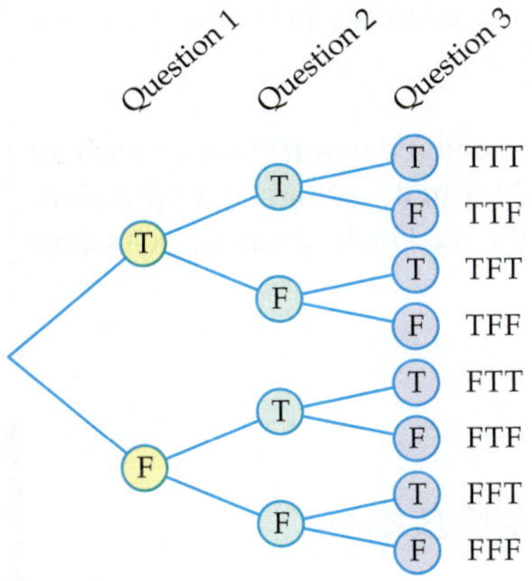

FIGURE 12.2

INSTRUCTOR NOTE

To convince students that explicitly listing all possible outcomes of an experiment can easily get out of hand, suggest that they draw a tree diagram for all 10 questions in Example 4. They should quickly see that it is not a reasonable request.

EXAMPLE 4 Counting Using a Tree Diagram

A true/false test consists of 10 questions. Draw a tree diagram to show the number of ways to answer the first three questions.

Solution

See the tree diagram in Figure 12.2. There are eight possible ways to answer the first three questions.

CHECK YOUR PROGRESS 4 Draw a tree diagram to determine the sample space for Example 3.

Solution See page S43.

The Counting Principle

For each of the previous problems, the possible outcomes were listed and then counted to determine the number of different outcomes. However, it is not always possible or practical to list and count outcomes. For example, the number of different five-card poker hands that can be drawn from a standard deck of 52 playing cards is 2,598,960. Trying to create a list of these hands would be quite time consuming.

Consider again the problem of selecting a laptop. By using a tree diagram, we listed the 12 possible laptops. Another way to arrive at this result is to find the product of the numbers of choices available for RAM sizes, screen sizes, and application packages.

$$\begin{bmatrix}\text{number of}\\ \text{RAM sizes}\end{bmatrix} \times \begin{bmatrix}\text{number of}\\ \text{screen sizes}\end{bmatrix} \times \begin{bmatrix}\text{number of}\\ \text{application packages}\end{bmatrix} = \begin{bmatrix}\text{number of}\\ \text{laptops}\end{bmatrix}$$

$$2 \times 3 \times 2 = 12$$

For the example of tossing two dice, there were 36 possible outcomes. We can arrive at this result without listing the outcomes by finding the product of the number of possible outcomes of rolling the red die and the number of possible outcomes of rolling the green die.

$$\begin{bmatrix}\text{outcomes}\\ \text{of red die}\end{bmatrix} \times \begin{bmatrix}\text{outcomes}\\ \text{of green die}\end{bmatrix} = \begin{bmatrix}\text{number of}\\ \text{outcomes}\end{bmatrix}$$

$$6 \times 6 = 36$$

This method of determining the number of outcomes of a multi-stage experiment without listing them is called the **counting principle**.

Counting Principle

Let E be a multi-stage experiment. If $n_1, n_2, n_3, \ldots, n_k$ are the number of possible outcomes of each of the k stages of E, then there are $n_1 \cdot n_2 \cdot n_3 \cdot \cdots \cdot n_k$ possible outcomes for E.

EXAMPLE 5 Counting by Using the Counting Principle

In horse racing, betting on a *trifecta* refers to choosing the exact order of the first three horses across the finish line. If there are eight horses in a race, how many trifectas are possible, assuming there are no ties?

Solution

Any one of the eight horses can be first, so $n_1 = 8$. Because a horse cannot finish both first and second, there are seven horses that can finish second; thus $n_2 = 7$. Similarly, there are six horses that can finish third; $n_3 = 6$. By the counting principle, there are $8 \cdot 7 \cdot 6 = 336$ possible trifectas.

CHECK YOUR PROGRESS 5 Nine runners are entered in a 100-meter dash for which a gold, silver, and bronze medal will be awarded for first, second, and third place finishes, respectively. In how many possible ways can the medals be awarded? (Assume that there are no ties.)

Solution See page S43.

MATH MATTERS Coding Characters for Web Pages

So that any web page will look the same on computers all over the world, standards must be adopted and adhered to by web page developers. Developers must be able to code the English alphabet, punctuation marks, special characters such as $, ¢, and %, and all the numerals. Developers also need to be able to display letters in the Cyrillic alphabet, such as Ж and Й; letters in the French alphabet, like ç; the Japanese symbol あ; and a host of other characters and special symbols.

The standard default coding system endorsed by the World Wide Web Consortium (W3C) is Unicode Transformation Format-8 (UTF-8). This coding system uses one to four bytes to represent each character.

Here are some of the representations of characters used in UTF-8.

Character	UTF-8
¢	11000010 101000010
A	01000001
Ж	11010000 10010110
あ	11100011 10000001 10000010

TAKE NOTE

A *byte* is a sequence of eight digits using just 0 and 1. For instance, 01100101 is a byte.

We can use the counting techniques discussed in this chapter to count the number of unique characters that can be created with UTF-8. There are over one million possible characters.

Browsers use other character sets as well, such as ISO Latin 1. You can see which character set your browser is using by selecting Preferences and viewing the font characteristics.

Counting With and Without Replacement

Consider an experiment in which three balls colored red, blue, and green are placed in a box. A person reaches into the box and repeatedly pulls out a colored ball, keeping note of the color picked. The sequence of colors will depend on whether the balls are returned to the box after each pick. We say that the experiment can be performed *with replacement* or *without replacement*.

Consider the following two situations.

1. How many four-digit numbers can be formed from the digits 1 through 9 if no digit can be repeated?
2. How many four-digit numbers can be formed from the digits 1 through 9 if a digit can be used repeatedly?

In the first case, there are nine choices for the first digit ($n_1 = 9$). Because a digit cannot be repeated, the first digit chosen cannot be used again. Thus there are only eight choices for the second digit ($n_2 = 8$). Because neither of the first two digits can be used as the third digit, there are only seven choices for the third digit ($n_3 = 7$). Similarly, there are six choices for the fourth digit ($n_4 = 6$). By the counting principle, there are $9 \cdot 8 \cdot 7 \cdot 6 = 3024$ four-digit numbers in which no digit is repeated.

In the second case, there are nine choices for the first digit ($n_1 = 9$). Because a digit can be used repeatedly, the first digit chosen can be used again. Thus there are nine choices for the second digit ($n_2 = 9$) and, similarly, nine choices for the third and fourth digits ($n_3 = 9, n_4 = 9$). By the counting principle, there are $9 \cdot 9 \cdot 9 \cdot 9 = 6561$ four-digit numbers when digits can be used repeatedly.

The set of four-digit numbers created without replacement includes numbers such as 3867, 7941, and 9128. In these numbers, no digit is repeated. However, numbers such as 6465, 9911, and 2222, each of which contains at least one repeated digit, can be created only with replacement.

QUESTION Does a multi-stage experiment performed with replacement generally have more or fewer possible outcomes than the same experiment performed without replacement?

EXAMPLE 6 Counting With and Without Replacement

From the letters a, b, c, d, and e, how many four-letter groups can be formed if

a. a letter can be used more than once?

b. each letter can be used exactly once?

Solution

a. Because each letter can be repeated, there are $5 \cdot 5 \cdot 5 \cdot 5 = 625$ possible four-letter groups.

b. Because each letter can be used only once, there are $5 \cdot 4 \cdot 3 \cdot 2 = 120$ four-letter groups in which no letter is repeated.

CHECK YOUR PROGRESS 6 In how many ways can three awards be given to five students if

a. each student may receive more than one award?

b. each student may receive no more than one award?

Solution See page S43.

ANSWER More. Each stage of an experiment (after the first) performed without replacement will have fewer possible outcomes than the preceding stage. Performed with replacement, each stage of the experiment has the same number of outcomes.

EXCURSION

Decision Trees

Decision trees are tree diagrams that are used to solve problems that involve many choices. To illustrate, suppose we are given eight coins, one of which is counterfeit and slightly heavier than the other seven. Using a balance scale, we must find the counterfeit coin.

INSTRUCTOR NOTE

There is a much more difficult version of this puzzle in which it is not known whether the counterfeit coin is heavier or lighter than the other coins. See http://mathforum.org/library/drmath/view/57947.html.

Designate the coins as c_1, c_2, c_3, c_4, c_5, c_6, c_7, and c_8. One way to determine the counterfeit coin is to weigh coins in pairs. This method is illustrated by the decision tree in Figure 12.3. In this case, it would take from one to four weighings to determine the counterfeit coin.

A second method is to divide the coins into two groups of four coins each, place each group on a pan of the balance scale, and take the coins from the side that goes down. Now divide these four coins into two groups of two coins each and weigh them on the balance scale. Again keep the coins from the heavier side. Weighing one of these coins against the other will reveal the counterfeit coin. The decision tree for this procedure is shown in Figure 12.4. Using this method, the counterfeit coin will be found in three weighings.

A third method is even more efficient. Divide the coins into three groups. Two of the groups contain three coins, and the third group contains two coins. Place each of the three-coin groups on the balance scale. If they balance, the counterfeit coin is in the third group. Placing a coin from the third group on each of the balance pans will determine the counterfeit coin. If the three coin groups do not balance, then take two of the coins from the pan that goes down and weigh them against each other. If these balance, the third coin is the counterfeit. If not, the counterfeit is the coin on the pan that goes down. This method requires only two weighings and is shown in the decision tree in Figure 12.5.

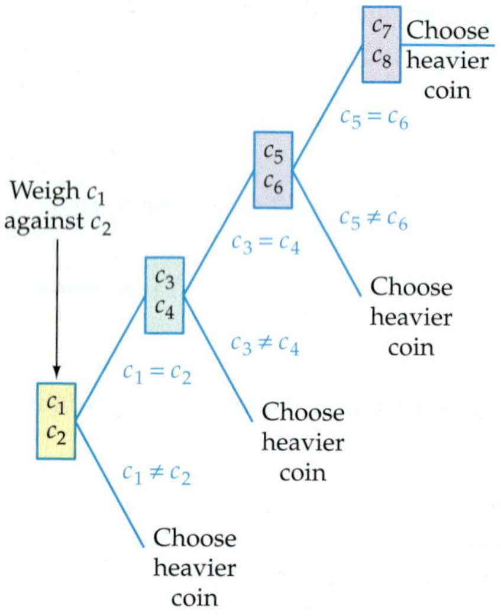

FIGURE 12.3 Using this method, it may take up to four weighings to determine which is the counterfeit coin.

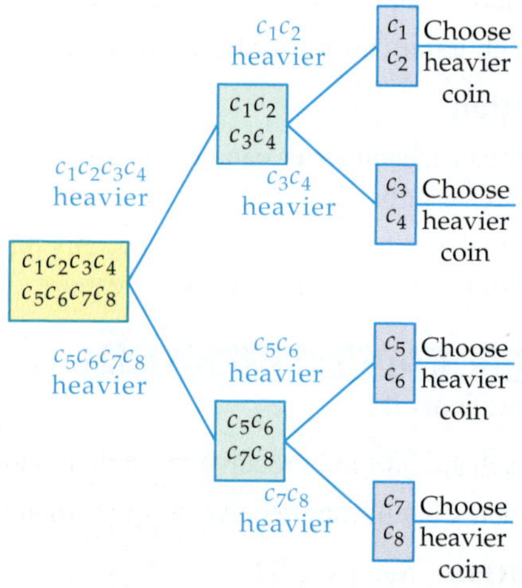

FIGURE 12.4 Using this method, it will always take three weighings to determine which is the counterfeit coin.

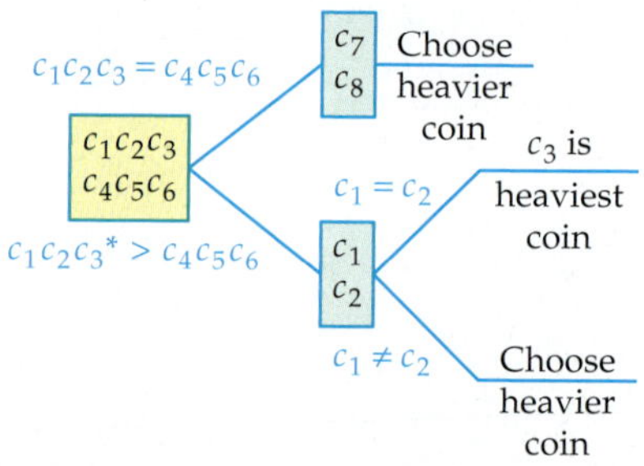

*We are assuming that $c_1c_2c_3$ is heavier than $c_4c_5c_6$. If $c_4c_5c_6$ is heavier, use those coins in the final weighing.

FIGURE 12.5 Using this method, it will take only two weighings to determine which is the counterfeit coin.

EXCURSION EXERCISES

For each of the following problems, draw a decision tree and determine the minimum number of weighings necessary to identify the counterfeit coin.

1. In a stack of 12 identical-looking coins, one is counterfeit and is lighter than the remaining 11 coins.
2. In a stack of 13 identical-looking coins, one is counterfeit and is heavier than the remaining 12 coins.

EXERCISE SET 12.1

■ In Exercises 1 to 10, list the elements of the sample space defined by each experiment.

1. Select an even single-digit whole number.
2. Select an odd single-digit whole number.
3. Select one day from the days of the week.
4. Select one month from the months of the year.
5. Toss a coin twice.
6. Toss a coin three times.
7. Roll a single die and then toss a coin.
8. Toss a coin and then choose a digit from the digits 1 through 4.
9. Choose a complete dinner from a dinner menu that allows a customer to choose from two salads, three entrees, and two desserts.
10. Choose a car during a new car promotion that allows a buyer to choose from three body styles, two radios, and two interior color schemes.

■ For Exercises 11 and 12, list the sample space of paths that start at A and pass through each vertex of the figure exactly once.

11. Square $ABCD$

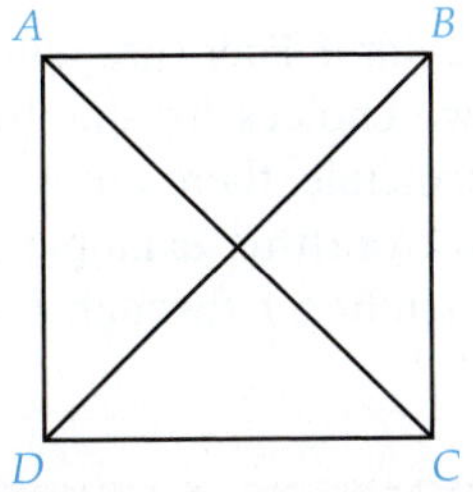

12. Pentagon $ABCDE$

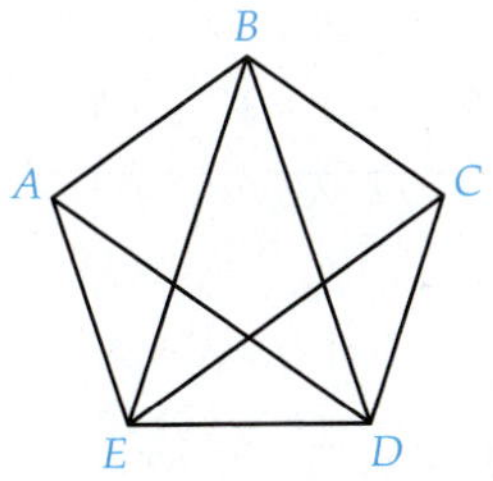

■ In Exercises 13 to 18, use the counting principle to determine the number of elements in the sample space.

13. Two digits are selected without replacement from the digits 1, 2, 3, and 4.
14. Two digits are selected with replacement from the digits 1, 2, 3, and 4.
15. The possible ways to complete a multiple-choice test consisting of 20 questions, with each question having four possible answers (a, b, c, or d)
16. The possible ways to complete a true–false examination consisting of 25 questions
17. The possible four-digit telephone number extensions that can be formed if 0, 8, and 9 are excluded as the first digit
18. The possible six-character passwords that can be formed using a letter from a to h and five numbers from 1 to 9. Assume a letter or number cannot be used more than once in any six-character password.

■ In Exercises 19 to 24, use the following experiment. Two-digit numbers are formed, with replacement, from the digits 0–9.

19. How many two-digit numbers are possible?
20. How many two-digit even numbers are possible?
21. How many numbers are divisible by 5?
22. How many numbers are divisible by 3?
23. How many numbers are greater than 37?
24. How many numbers are less than 59?

■ In Exercises 25 to 28, use the following experiment. Four cards labeled A, B, C, and D are randomly placed in four boxes labeled A, B, C, and D. Each box receives exactly one card.

25. In how many ways can the cards be placed in the boxes?
26. Count the number of elements in the event that no box contains a card with the same letter as the box.
27. Count the number of elements in the event that *at least* one card is placed in the box with the corresponding letter.
28. If you add the answer for Exercise 26 and the answer for Exercise 27, is the sum the answer for Exercise 25? Why or why not?

■ **Lotteries** In Exercises 29 to 32, use the following experiment. A state lottery game consists of choosing one card from each of the four suits in a standard deck of playing cards. (There are 13 cards in each suit.)

29. Count the number of elements in the sample space.
30. Count the number of elements in the event that an ace, a king, a queen, and a jack are chosen.
31. Count the number of ways in which four aces can be chosen.
32. Count the number of ways in which four cards, each of a different face value, can be chosen.
33. Write a lesson that you could use to explain the meanings of the words *experiment, sample space,* and *event* as they apply to combinatorics.
34. Explain how a tree diagram is used to count the number of ways an experiment can be performed.

EXTENSIONS

35. Computer Programming A main component of any computer programming language is its method of repeating a series of computations. Each programming language has its own syntax for performing those "loops." In one programming language, BASIC, the structure is similar to the display below.

```
FOR I = 1 TO 10 ——— Start with I = 1.
    FOR J = 1 TO 15 — Start with J = 1.
        SUM = I + J
    NEXT J ← Increase J by 1 until J > 15.
NEXT I ←—— Increase I by 1 until I > 10.
END
```

The program repeats each loop until the index variables, I and J in this case, exceed a certain value. After this program is executed, how many times will the instruction SUM = I + J have been executed? What is the final value of SUM?

36. Review the rules of the game of checkers, and make a tree diagram that shows all of the first two moves that are possible by one player of a checker game. Assume that no moves are blocked by the opponent's checkers. (*Hint:* It may help to number the squares of the checkerboard.)

SECTION 12.2 Permutations and Combinations

Factorial

Suppose four different colored squares are arranged in a row. One possibility is shown below.

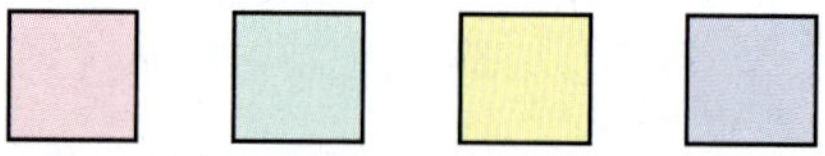

How many different ways are there to order the colors? There are four choices for the first square, three choices for the second square, two choices for the third square, and one choice for the fourth square. By the counting principle, there are $4 \cdot 3 \cdot 2 \cdot 1 = 24$ different arrangements of the four squares. Note from this example that the number of arrangements equals the product of the natural numbers n through 1, where n is the number of objects. This product is called a *factorial.*

n Factorial

***n* factorial** is the product of the natural numbers n through 1 and is symbolized by $\boldsymbol{n!}$.

$$n! = n \cdot (n-1) \cdot (n-2) \cdot \cdots \cdot 3 \cdot 2 \cdot 1$$

CALCULATOR NOTE

The factorial of a number becomes quite large for even relatively small numbers. For instance, 58! is the approximate number of atoms in the known universe. The number 70! is greater than 10 with 100 zeros after it. This number is larger than most scientific calculators can handle.

Here are some examples:

$$5! = 5 \cdot 4 \cdot 3 \cdot 2 \cdot 1 = 120$$

$$8! = 8 \cdot 7 \cdot 6 \cdot 5 \cdot 4 \cdot 3 \cdot 2 \cdot 1 = 40{,}320$$

$$1! = 1$$

On some occasions it will be necessary to use 0! (zero factorial). Because it is impossible to define zero factorial in terms of a product of natural numbers, a standard definition is used.

Zero Factorial

$$0! = 1$$

A factorial can be written in terms of smaller factorials. This is useful when calculating large factorials. For example,

$$10! = 10 \cdot 9!$$
$$10! = 10 \cdot 9 \cdot 8!$$
$$10! = 10 \cdot 9 \cdot 8 \cdot 7!$$

EXAMPLE 1 Simplify Factorials

Evaluate: **a.** $5! - 3!$ **b.** $\frac{9!}{6!}$

Solution

a. $5! - 3! = (5 \cdot 4 \cdot 3 \cdot 2 \cdot 1) - (3 \cdot 2 \cdot 1) = 120 - 6 = 114$

b. $\frac{9!}{6!} = \frac{9 \cdot 8 \cdot 7 \cdot \cancel{6!}}{\cancel{6!}} = 9 \cdot 8 \cdot 7 = 504$

CHECK YOUR PROGRESS 1 Evaluate: **a.** $7! + 4!$ **b.** $\frac{8!}{4!}$

Solution See page S43.

Permutations

Determining the number of possible ordered arrangements of a group of distinct objects, as we did with the squares earlier, is one application of the counting principle. Each arrangement of this type is called a *permutation*.

Permutation

A **permutation** is an arrangement of objects in a definite order.

For example, abc and cba are two different permutations of the letters a, b, and c. As a second example, 122 and 212 are two different permutations of one 1 and two 2s.

The counting principle is used to count the number of different permutations of any set of objects. We will begin our discussion using sets of *distinct* objects; that is, sets in which no two objects are the same. For instance, the objects a, b, c, d are distinct, whereas the objects □, ☆, ○, ☆ are not.

Most music players allow the user to create a playlist, which is a list of songs that can be played on the device. Many of these players have a *shuffle* feature that plays the songs in a playlist in a different order each time.

Suppose a playlist consists of two songs, a rock song and a reggae song. If the shuffle feature is used, the songs could be played in two orders:

rock then reggae   or   reggae then rock

Thus there are two choices for the first song (rock or reggae) and there is one choice for the second song (whichever song was not played first). By the counting principle, there are $2 \cdot 1 = 2! = 2$ permutations or orders in which to play the songs.

With three songs in a playlist, one rock, one reggae, and one country, there are three choices for the first song, two choices for the second song, and one choice for the third song. By the counting principle, there are $3 \cdot 2 \cdot 1 = 3! = 6$ permutations in which the three songs could be played.

Permutation 1	Permutation 2	Permutation 3	Permutation 4	Permutation 5	Permutation 6
Rock	Rock	Reggae	Reggae	Country	Country
Reggae	Country	Rock	Country	Rock	Reggae
Country	Reggae	Country	Rock	Reggae	Rock

POINT OF INTEREST

If a playlist has 10 songs, there are $10! = 3{,}628{,}800$ possible permutations of the songs in shuffle mode. If each song were 2 min long, it would take over 120,000 h to listen to every permutation.

With four songs in a playlist, there are $4 \cdot 3 \cdot 2 \cdot 1 = 4! = 24$ orders in which the songs could be played. In general, if there are n songs in a playlist, then there are $n!$ permutations, or orders, in which the songs could be played.

Suppose now that you have a playlist that consists of eight songs but you have time to listen to only three of the songs. In shuffle mode, any one of the eight songs could play first, then any one of the seven remaining songs could play second, and then any one of the remaining six songs could play third. By the counting principle, there are $8 \cdot 7 \cdot 6 = 336$ permutations in which the songs could be played.

The following formula can be used to determine the number of permutations of n distinct objects (the songs in the example above), of which k are selected.

Permutation Formula for Distinct Objects

The number of permutations of n distinct objects selected k at a time is

$$P(n, k) = \frac{n!}{(n-k)!}$$

Applying this formula to the situation above in which there were eight songs ($n = 8$), of which only three songs ($k = 3$) could be played, we have

$$P(8, 3) = \frac{8!}{(8-3)!} = \frac{8!}{5!} = \frac{8 \cdot 7 \cdot 6 \cdot 5!}{5!} = 8 \cdot 7 \cdot 6 = 336$$

There are 336 permutations of playing the songs. This is the same answer we obtained using the counting principle.

EXAMPLE 2 Counting Permutations

A university tennis team consists of six players who are ranked from 1 through 6. If a tennis coach has 10 players from which to choose, how many different ranked tennis teams can the coach select?

Solution

Because the players on the tennis team are ranked from 1 through 6, a team with player A in position 1 is different from a team with player A in position 2. Therefore, the number of different teams is the number of permutations of 10 players selected 6 at a time.

$$P(10, 6) = \frac{10!}{(10-6)!} = \frac{10!}{4!} = \frac{10 \cdot 9 \cdot 8 \cdot 7 \cdot 6 \cdot 5 \cdot 4!}{4!}$$
$$= 10 \cdot 9 \cdot 8 \cdot 7 \cdot 6 \cdot 5 = 151{,}200$$

There are 151,200 possible tennis teams.

CHECK YOUR PROGRESS 2 A college golf team consists of five players who are ranked from 1 through 5. If a golf coach has eight players from which to choose, how many different ranked golf teams can the coach select?

Solution See page S43.

EXAMPLE 3 Counting Permutations

In 2015, 18 horses were entered in the Kentucky Derby. How many different finishes of first through fourth place were possible?

Solution

Because the order in which the horses finish the race is important, the number of possible finishes of first through fourth place is $P(18, 4)$.

$$P(18, 4) = \frac{18!}{(18-4)!} = \frac{18!}{14!} = \frac{18 \cdot 17 \cdot 16 \cdot 15 \cdot 14!}{14!}$$
$$= 18 \cdot 17 \cdot 16 \cdot 15 = 73{,}440$$

There were 73,440 possible finishes of first through fourth places.

CHECK YOUR PROGRESS 3 There were 43 cars entered in the 2015 Daytona 500 NASCAR race. In how many different ways could the first, second, and third place prizes have been awarded?

Solution See page S43.

MATH MATTERS How Many Shuffles?

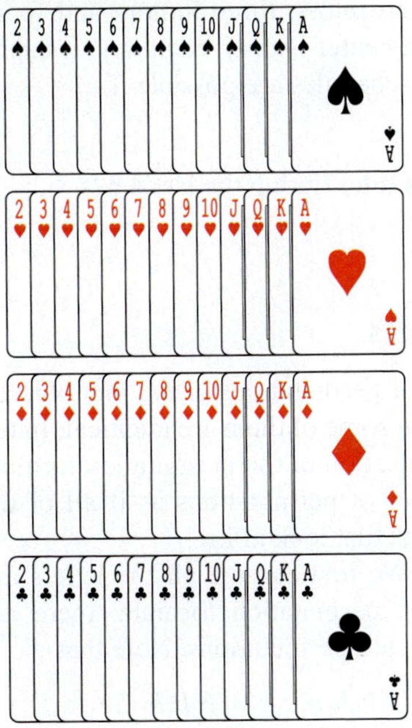

A standard deck of playing cards

A standard deck of playing cards consists of 52 different cards divided into four suits: spades (♠), hearts (♥), diamonds (♦), and clubs (♣). Each shuffle of the deck results in a new arrangement of the cards. We could say that each shuffle results in a new permutation of the cards. There are $P(52, 52) = 52! \approx 8 \times 10^{67}$ (that's 8 with 67 zeros after it) possible arrangements.

Suppose a deck has each of the four suits arranged in order from 2 through ace. How many shuffles are necessary to achieve a randomly ordered deck in which any card is equally likely to occur in any position in the deck?

Two mathematicians, Dave Bayer of Columbia University and Persi Diaconis of Harvard University, have shown that seven shuffles are enough. Their proof has many applications to complicated counting problems. One problem in particular is that of analyzing speech patterns. Solving this problem is critical to enabling computers to interpret human speech.

Applying Several Counting Techniques

The permutation formula is derived from the counting principle. This formula is a convenient way of expressing the number of ways in which the items in an ordered list can be arranged. Both the permutation formula and the counting principle are needed to solve some counting problems.

EXAMPLE 4 Counting Using Several Methods

Five women and four men are to be seated in a row of nine chairs. How many different seating arrangements are possible if

a. there are no restrictions on the seating arrangements?

b. the women sit together and the men sit together?

Solution

Because seating arrangements have a definite order, they are permutations.

a. If there are no restrictions on the seating arrangements, then the number of seating arrangements is $P(9, 9)$.

$$P(9, 9) = \frac{9!}{(9 - 9)!} = \frac{9!}{0!} = 9! = 362{,}880$$

There are 362,880 seating arrangements.

b. This is a multi-stage experiment, so both the permutation formula and the counting principle will be used. There are 5! ways to arrange the women and 4! ways to arrange the men. We must also consider that either the women or the men could be seated at the beginning of the row. There are two ways to do this. By the counting principle, there are $2 \cdot 5! \cdot 4!$ ways to seat the women together and the men together.

$$2 \cdot 5! \cdot 4! = 5760$$

There are 5760 arrangements in which women sit together and men sit together.

CHECK YOUR PROGRESS 4 There are seven tutors, three juniors and four seniors, who must be assigned to the 7 h that a math center is open each day. If each tutor works 1 h per day, how many different tutoring schedules are possible if

a. there are no restrictions?

b. the juniors tutor during the first 3 h and the seniors tutor during the last 4 h?

Solution See page S43.

Permutations of Indistinguishable Objects

Up to this point we have been counting the number of permutations of *distinct* objects. We now look at the situation of arranging objects when some of them are identical. In the case of identical or indistinguishable objects, a modification of the permutation formula is necessary. The general idea is to count the number of permutations as if all of the objects were distinct and then remove the permutations that look alike.

Consider the permutations of the letters *bbbcc*. We first assume all the letters are different by labeling them as $b_1b_2b_3c_1c_2$. Using the permutation formula, there are $5! = 120$ permutations. Now we need to remove repeated permutations. Note that

$$b_1b_2b_3c_1c_2 \quad b_1b_3b_2c_1c_2 \quad b_2b_1b_3c_1c_2 \quad b_2b_3b_1c_1c_2 \quad b_3b_2b_1c_1c_2 \quad b_3b_1b_2c_1c_2$$

are all distinct permutations that end with c_1c_2. However, if we replace each b_1, b_2, and b_3 with b, all six of these permutations will look the same. Thus there are $3! = 6$ times too many arrangements for each arrangement of c_1 and c_2. A similar argument applies to the c's. There are $2! = 2$ times too many permutations of the c's for each arrangement of b's.

Combining the results above, the number of permutations of *bbbcc* is

$$\frac{5!}{3! \cdot 2!} = \frac{5 \cdot 4 \cdot 3 \cdot 2 \cdot 1}{(3 \cdot 2 \cdot 1) \cdot (2 \cdot 1)} = 10$$

There are 10 distinct permutations of *bbbcc*.

Permutations of Objects, Some of Which Are Identical

The number of distinguishable permutations of n objects of r different types, where k_1 identical objects are of one type, k_2 are of another, and so on, is given by

$$\frac{n!}{k_1! \cdot k_2! \cdot \cdots \cdot k_r!}$$

where $k_1 + k_2 + \cdots + k_r = n$.

EXAMPLE 5 Permutations of Identical Objects

A password requires 7 characters. If a person who lives at 155 Nunn Road wants a password to be an arrangement of the characters 155NUNN, how many different passwords are possible?

Solution

We are looking for the number of permutations of the characters 155NUNN. With $n = 7$ (number of characters), $k_1 = 1$ (number of 1's), $k_2 = 2$ (number of 5's), $k_3 = 3$ (number of N's), and $k_4 = 1$ (number of U's) we have

$$\frac{7!}{1! \cdot 2! \cdot 3! \cdot 1!} = \frac{7 \cdot 6 \cdot 5 \cdot 4 \cdot 3!}{2 \cdot 3!} = 420$$

There are 420 possible passwords.

CHECK YOUR PROGRESS 5 Eight coins—3 pennies, 2 nickels, and 3 dimes—are placed in a single stack. How many different stacks are possible if

a. there are no restrictions on the placement of the coins?

b. the dimes must stay together?

Solution See page S43.

INSTRUCTOR NOTE

Students often confuse permutations and combinations. Emphasize that if the particular order of objects makes a difference, then we are counting permutations. You can also point out that there are more permutations of objects than combinations.

Combinations

For some arrangements of objects, the order in which the objects are arranged is important. These are permutations. If a telephone extension is 2537, then the digits must be dialed in exactly that order. On the other hand, if you were to receive a \$1 bill, a \$5 bill, and a \$10 bill, you would have \$16 regardless of the order in which you received the bills. A **combination** is a collection of objects in which the order of the objects is not important. The three-letter sequences acb and bca are *different* permutations but the *same* combination.

QUESTION From a group of 45 applicants, five identical scholarships will be awarded. Is the number of ways in which the scholarships can be awarded determined by permutations or combinations?

ANSWER Combinations. The order in which the scholarship winners are chosen is not important.

The formula for finding the number of combinations is derived in much the same manner as the formula for finding the number of permutations of identical objects. Consider the problem of finding the number of possible combinations when choosing three letters from the letters a, b, c, d, and e, without replacement. For each choice of three letters, there are 3! permutations. For example, choosing the letters a, d, and e gives the following six permutations.

ade aed dea dae ead eda

Because there are six permutations and each permutation represents the *same* combination, the number of permutations is six times the number of combinations. This is true each time three letters are selected. Therefore, to find the number of combinations of five objects chosen three at a time, divide the number of permutations by 3! = 6. The number of combinations of five objects chosen three at a time is

$$\frac{P(5,3)}{3!} = \frac{5!}{3! \cdot (5-3)!} = \frac{5!}{3! \cdot 2!} = \frac{5 \cdot 4 \cdot 3!}{3! \cdot 2!} = \frac{5 \cdot 4}{2 \cdot 1} = 10$$

CALCULATOR NOTE

Some calculators can compute permutations and combinations directly. For instance, on a TI-83/84 calculator, enter 11 nCr 5 for $C(11, 5)$. The nCr operation is accessible in the probability menu after pressing the MATH key.

Combination Formula

The number of combinations of n objects chosen k at a time is

$$C(n,k) = \frac{P(n,k)}{k!} = \frac{n!}{k! \cdot (n-k)!}$$

In Section 12.1, we stated that there were 2,598,960 possible 5-card poker hands. This number was calculated using the combination formula. Because the 5-card hand ace of hearts, king of diamonds, queen of clubs, jack of spades, 10 of hearts is exactly the same as the 5-card hand king of diamonds, jack of spades, queen of clubs, 10 of hearts, ace of hearts, the order of the cards is not important, and therefore the number of hands is a combination. The number of different 5-card poker hands is the combination of 52 cards chosen 5 at a time, which is given by $C(52, 5)$.

$$C(52,5) = \frac{52!}{5! \cdot (52-5)!} = \frac{52!}{5! \cdot 47!} = \frac{52 \cdot 51 \cdot 50 \cdot 49 \cdot 48 \cdot 47!}{5! \cdot 47!}$$
$$= \frac{52 \cdot 51 \cdot 50 \cdot 49 \cdot 48}{5 \cdot 4 \cdot 3 \cdot 2 \cdot 1} = 2{,}598{,}960$$

EXAMPLE 6 Counting Using the Combination Formula

An emergency room at a hospital has 11 nurses on staff. Each night a team of 5 nurses is on duty. In how many different ways can the team of 5 nurses be chosen?

Solution

This is a combination problem, because the order in which the nurses are chosen is not important. The 5 nurses N_1, N_2, N_3, N_4, N_5 are the same as the 5 nurses N_3, N_5, N_1, N_2, N_4.

$$C(11,5) = \frac{11!}{5! \cdot (11-5)!} = \frac{11!}{5! \cdot 6!} = \frac{11 \cdot 10 \cdot 9 \cdot 8 \cdot 7 \cdot 6!}{5! \cdot 6!}$$
$$= \frac{11 \cdot 10 \cdot 9 \cdot 8 \cdot 7}{5 \cdot 4 \cdot 3 \cdot 2 \cdot 1} = 462$$

There are 462 possible teams of 5 nurses.

CHECK YOUR PROGRESS 6 A restaurant employs 16 waiters and waitresses. In how many ways can a group of 9 waiters and waitresses be chosen for the lunch shift?

Solution See page S44.

EXAMPLE 7 **Counting Using the Combination Formula and the Counting Principle**

A committee of 5 is chosen from 5 mathematicians and 6 economists. How many different committees are possible if the committee must include 2 mathematicians and 3 economists?

Solution

Because a committee of professors A, B, C, D, and E is exactly the same as a committee of professors B, D, E, A, and C, choosing a committee is an example of choosing a combination. There are 5 mathematicians from whom 2 are chosen, which is equivalent to $C(5, 2)$ combinations. There are 6 economists from whom 3 are chosen, which is equivalent to $C(6, 3)$ combinations. Therefore, by the counting principle, there are $C(5, 2) \cdot C(6, 3)$ ways to choose 2 mathematicians and 3 economists.

$$C(5, 2) \cdot C(6, 3) = \frac{5!}{2! \cdot 3!} \cdot \frac{6!}{3! \cdot 3!} = 10 \cdot 20 = 200$$

There are 200 possible committees consisting of 2 mathematicians and 3 economists.

CHECK YOUR PROGRESS 7 An IRS auditor randomly chooses 5 tax returns to audit from a stack of 10 tax returns, 4 of which are from corporations and 6 of which are from individuals. In how many different ways can the auditor choose the tax returns if the auditor wants to include 3 corporate and 2 individual returns?

Solution See page S44.

MATH MATTERS Buying Every Possible Lottery Ticket

A lottery prize in Pennsylvania reached \$65 million. A resident of the state suggested that it might be worth buying a ticket for every possible combination of numbers. To win the \$65 million, a player has to correctly select 6 of 50 numbers. Each ticket costs \$1. Because the order in which the numbers are drawn is not important, the number of different possible tickets is $C(50, 6) = 15{,}890{,}700$. Thus it would cost the resident \$15,890,700 to purchase tickets for every possible combination of numbers.

It might seem that an approximately \$16 million investment to win \$65 million is reasonable. Unfortunately, when prize levels reach such lofty heights, many more people play the lottery. This increases the chances that more than one person will select the winning combination of numbers. In fact, eight people chose the winning numbers, and each received approximately \$8 million. Now the \$16 million investment does not look very appealing.

A standard deck of playing cards contains 4 suits: spades, hearts, diamonds, and clubs. Each suit has 13 cards: 2 through 10, jack, queen, king, and ace. See the Math Matters on page 699.

EXAMPLE 8 **Counting Problems with Cards**

From a standard deck of playing cards, 5 cards are chosen. How many 5-card combinations contain

a. 2 kings and 3 queens?

b. 5 hearts?

c. 5 cards of the same suit?

Solution

a. There are $C(4, 2)$ ways of choosing 2 kings from 4 kings and $C(4, 3)$ ways of choosing 3 queens from 4 queens. By the counting principle, there are $C(4, 2) \cdot C(4, 3)$ ways of choosing 2 kings and 3 queens.

$$C(4, 2) \cdot C(4, 3) = \frac{4!}{2! \cdot 2!} \cdot \frac{4!}{3! \cdot 1!} = 6 \cdot 4 = 24$$

There are 24 ways of choosing 2 kings and 3 queens.

b. There are $C(13, 5)$ ways of choosing 5 hearts from 13 hearts.

$$C(13, 5) = \frac{13!}{5! \cdot 8!} = 1287$$

There are 1287 ways to choose 5 hearts from 13 hearts.

Five cards of the same suit is called a *flush* in poker.

c. From part b, there are also 1287 ways of choosing 5 spades from 13 spades, 5 clubs from 13 clubs, or 5 diamonds from 13 diamonds. Because there are 4 suits from which to choose and $C(13, 5)$ ways of choosing 5 cards from a suit, by the counting principle there are $4 \cdot C(13, 5)$ ways to choose 5 cards of the same suit.

$$4 \cdot C(13, 5) = 4 \cdot 1287 = 5148$$

There are 5148 ways of choosing 5 cards of the same suit from a standard deck of playing cards.

CHECK YOUR PROGRESS 8 Five cards are chosen from a standard deck of playing cards. How many 5-card combinations contain 4 cards of the same suit?

Solution See page S44.

EXCURSION

Choosing Numbers in Keno

A popular gambling game called keno, first introduced in China over 2000 years ago, is played in many casinos. In keno, there are 80 balls numbered from 1 to 80. The casino randomly chooses 20 balls from the 80 balls. These are "lucky balls" because if a gambler chooses some of the numbers on these balls, there is a possibility of winning money. The amount that is won depends on the number of lucky numbers the gambler has selected. The number of ways in which a casino can choose 20 balls from 80 is

$$C(80, 20) = \frac{80!}{20! \cdot 60!} \approx 3{,}535{,}000{,}000{,}000{,}000{,}000$$

A keno card is used to mark the numbers chosen.

Once the casino chooses the 20 lucky balls, the remaining 60 balls are unlucky for the gambler. A gambler who chooses 5 numbers will have from 0 to 5 lucky numbers.

Let's consider the case in which 2 of the 5 numbers chosen by the gambler are lucky numbers. Because 5 numbers were chosen, there must be 3 unlucky numbers among the 5 numbers. The number of ways of choosing 2 lucky numbers from 20 lucky numbers is $C(20, 2)$. The number of ways of choosing 3 unlucky numbers from 60 unlucky numbers is $C(60, 3)$. By the counting principle, there are $C(20, 2) \cdot C(60, 3) = 190 \cdot 34{,}220 = 6{,}501{,}800$ ways to choose 2 lucky and 3 unlucky numbers.

EXCURSION EXERCISES

For each of the following exercises, assume that a gambler playing keno has randomly chosen 4 numbers.

1. In how many ways can the gambler choose no lucky numbers?

2. In how many ways can the gambler choose exactly 1 lucky number?

3. In how many ways can the gambler choose exactly 2 lucky numbers?

4. In how many ways can the gambler choose exactly 3 lucky numbers?

5. In how many ways can the gambler choose 4 lucky numbers?

EXERCISE SET 12.2

■ In Exercises 1 to 30, evaluate the expression.

1. $8!$

2. $5!$

3. $9! - 5!$

4. $(9 - 5)!$

5. $(8 - 3)!$

6. $8! - 3!$

7. $P(8, 5)$

8. $P(7, 2)$

9. $P(9, 7)$

10. $P(10, 5)$

11. $P(8, 0)$

12. $P(7, 0)$

13. $P(8, 8)$

14. $P(10, 10)$

15. $P(8, 2) \cdot P(5, 3)$

16. $\dfrac{P(10, 4)}{P(8, 4)}$

17. $\dfrac{P(6, 0)}{P(6, 6)}$

18. $\dfrac{P(6, 3) \cdot P(5, 2)}{P(4, 3)}$

19. $C(9, 2)$

20. $C(8, 6)$

21. $C(12, 0)$

22. $C(11, 11)$

23. $C(6, 2) \cdot C(7, 3)$

24. $C(8, 5) \cdot C(9, 4)$

25. $\dfrac{C(10, 4) \cdot C(5, 2)}{C(15, 6)}$

26. $\dfrac{C(4, 3) \cdot C(5, 2)}{C(9, 5)}$

27. $\dfrac{C(9, 7) \cdot C(5, 3)}{C(14, 10)}$

28. $3! \cdot C(8, 5)$

29. $4! \cdot C(10, 3)$

30. $5! \cdot C(18, 0)$

■ In Exercises 31 to 34, how many combinations are possible? Assume that the items are distinct.

31. 7 items chosen 5 at a time

32. 8 items chosen 3 at a time

33. 12 items chosen 7 at a time

34. 11 items chosen 11 at a time

35. Is it possible to calculate $C(7, 9)$? Think of your answer in terms of 7 items chosen 9 at a time.

36. Is it possible to calculate $C(n, k)$ where $k > n$? See Exercise 35 for some help.

37. Music Downloads A student downloaded 5 music files to a portable music player. In how many different orders can the songs be played?

38. Elections The board of directors of a corporation must select a president, a secretary, and a treasurer. In how many possible ways can this be accomplished if there are 20 members on the board of directors?

39. Elections A committee of 16 students must select a president, a vice-president, a secretary, and a treasurer. In how many possible ways can this be accomplished?

40. The Olympics A gold, a silver, and a bronze medal are awarded in an Olympic event. In how many possible ways can the medals be awarded for a 200-meter sprint in which there are 9 runners?

Everett Collection Inc/Alamy Stock Photo

41. Music Festival Six country music bands and 3 rock bands are signed up to perform at an all-day festival. How many different orders can the bands play in if

a. there are no restrictions on the order?

b. all the bands of each type must perform in a row?

42. Radio Show Five rock songs and 6 rap songs are on a disc jockey's playlist for a radio show. How many different orders can the songs be played in if

a. there are no restrictions on the order?

b. all the rap songs are played consecutively?

43. Passwords A password requires 8 characters. If a soccer player whose jersey number is 77 wants his password to be an arrangement of the numbers and letters in SOCCER77, how many passwords are possible?

44. Gardening A gardener is planting a row of tulip bulbs. She has a mix of bulbs for 10 yellow tulips and 6 red tulips. How many color arrangements are possible for the row of tulips?

45. Firefighters At a certain fire station, one team of firefighters consists of 8 firefighters. If there are 24 firefighters qualified for a team, how many different teams are possible?

M. Norris/Shutterstock.com

46. Platoons A typical platoon consists of 20 soldiers. If there are 30 soldiers available to create a platoon, how many different platoons are possible?

47. Exam Questions A professor gives his students 7 essay questions to prepare for an exam. Only 3 of the questions will actually appear on the exam. How many different exams are possible?

48. Test Banks A math quiz is generated by randomly choosing 5 questions from a test bank consisting of 50 questions. How many different quizzes are possible?

49. Committee Selection A committee of 6 people is chosen from 8 women and 8 men. How many different committees are possible that consist of 3 women and 3 men?

50. Quality Control In a shipment of 20 smartphones, 2 are defective. How many ways can a quality control inspector randomly test 5 smartphones, of which 2 are defective?

51. Football In the National Football Conference (NFC), the NFC East division has 4 teams. During the regular season, each NFC East team plays each of the other NFC East teams twice. How many games are played between teams of the NFC East during a regular season?

52. Basketball In the Eastern Conference of the National Basketball Association (NBA), the Atlantic, Southeast, and Central Divisions have 5 teams each. During the regular season, each team plays every other team in its own division four times a year, plays six of the teams from the other two divisions four times a year, plays the remaining four teams from the other two divisions three times a year, and plays each team in the Western conference (which has 15 teams) twice. How many games are played by a team in the Eastern Conference during the regular season?

53. Geometry A hexagon is a 6-sided plane figure. A diagonal is a line segment connecting any 2 nonadjacent vertices. How many diagonals are possible?

54. Geometry Seven distinct points are drawn on a circle. How many different triangles can be drawn in which each vertex of the triangle is at 1 of the 7 points?

55. Softball Eighteen people decide to play softball. In how many ways can the 18 people be divided into 2 teams of 9 people?

56. Bowling Fifteen people decide to form a bowling league. In how many ways can the 15 people be divided into 3 teams of 5 people each?

57. Signal Flags The Coast Guard uses signal flags as a method of communicating between ships. If 4 different flags are available and the order in which the flags are raised is important, how many different signals are possible? Assume that 4 flags are raised.

58. Pizza Toppings A restaurant offers a special pizza with any 5 toppings. If the restaurant has 12 toppings from which to choose, how many different special pizzas are possible?

59. Letter Arrangements How many different letter arrangements are possible using all the letters of the word *committee*?

60. Color Arrangements A set of 12 plates came with 4 red, 4 green, and 4 yellow plates, but 2 red plates have been broken. The remaining plates are stacked on a shelf. How many different color arrangements are possible for the stack of plates?

61. Coin Tosses Ten identical coins are tossed. How many possible arrangements of the coins include 5 heads and 5 tails?

62. Coin Tosses Twelve identical coins are tossed. How many possible arrangements of the coins include 8 heads and 4 tails?

63. Concerts Three groups will perform at a choral concert. One group will sing 3 pieces, one group will sing 4 pieces, and one group will sing 2 pieces. In how many possible orders can the 9 pieces be performed, assuming each group performs all its songs consecutively?

64. Morse Code In 1835, Samuel Morse, a professor of art at New York University, devised a code that could be transmitted over a wire by using an electric current. This was the invention of the telegraph. The code used is called Morse code. It consists of a dot or a dash or a combination of up to 5 dots and/or dashes. For instance, the letter c is represented by — • — •, and • — represents the letter a. The numeral 0 is represented by 5 dashes, — — — — —. How many different symbols can be represented using Morse code?

■ Exercises 65 to 70 refer to a standard deck of playing cards. Assume that 5 cards are randomly chosen from the deck.

65. How many hands contain 4 aces?

66. How many hands contain 2 aces and 2 kings?

67. How many hands contain exactly 3 jacks?

68. How many hands contain exactly 3 jacks and 2 queens?

69. How many hands contain exactly two 7s?

70. How many hands contain exactly two 7s and two 8s?

71. Write a few sentences that explain the difference between a permutation and a combination of distinct objects. Give examples of each.

72. Francis Bacon, a contemporary of William Shakespeare, invented a cipher (a secret code) based on permutations of the letters *a* and *b*. Write an essay on Bacon's method and the intended use of his scheme.

EXTENSIONS

73. Hamiltonian Circuits In Chapter 5, we examined the problem of determining various paths through a network (or graph). One particular path, called a *Hamiltonian circuit*, visited each vertex (dot) of a graph exactly once and returned to the starting vertex. A Hamiltonian circuit is shown in green in the following network. The number associated with each line (edge) of the graph indicates the distance between two vertices. (The segments are not drawn to scale.) In most applications, the object is to find the *shortest* path. The green path shown here is the shortest path that visits each vertex exactly once and returns to the starting vertex.

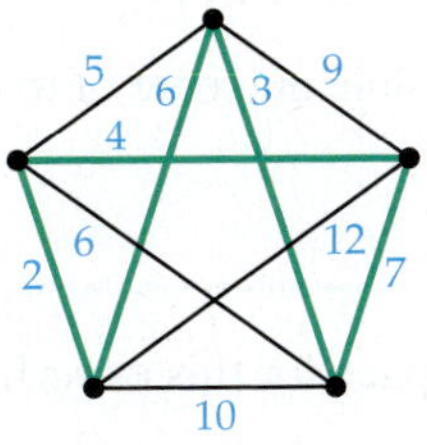

One method of searching for the shortest Hamiltonian circuit is by trial and error. For a network with a small number of vertices, this procedure will work fine. As the number of vertices increases, the likelihood of finding the shortest path using trial-and-error becomes extremely remote.

The counting principle can be used to find the number of possible paths through a network in which every pair of vertices possible is connected by a line. For the complete network with 5 vertices shown at the top of the next column, beginning at the home vertex labeled *H*, there are 4 choices for the next vertex to visit, 3 choices for the next, 2 choices for the next, and finally 1 choice that returns to *H*. By the counting principle, there are $4 \cdot 3 \cdot 2 \cdot 1 = 24$ possible circuits. Traveling circuit *HABCDH* is the same as traveling the circuit in the reverse order, *HDCBAH*. Thus there are $\frac{24}{2} = 12$ possible circuits. In general, there are $\frac{(n-1)!}{2}$ possible Hamiltonian circuits through a network of n vertices, where $n \geq 3$.

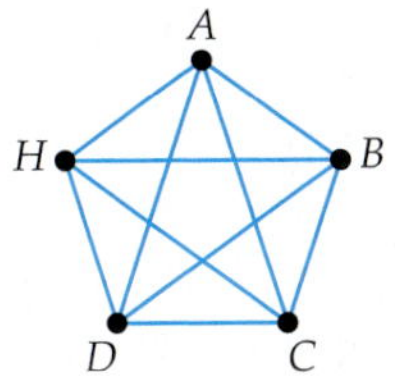

a. A network has 8 vertices, and each vertex is connected to the other vertices. How many Hamiltonian circuits are possible?

b. For the network below, find the number of possible Hamiltonian circuits.

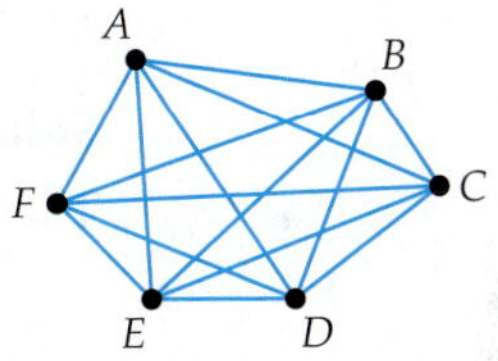

c. Suppose a network has 20 vertices (with every pair of vertices connected by a line segment) and a computer is available that can analyze 1 million paths per second. How many years would it take this computer to find all the possible Hamiltonian circuits of this network? (Assume that a year is 365 days.)

d. Suppose a network has 40 vertices (with every pair of vertices connected by a line segment) and a computer is available that can analyze 1 trillion paths per second. How many years would it take this computer to find all the possible Hamiltonian circuits? Assume a year is 365 days.

SECTION 12.3 Probability and Odds

Introduction to Probability

In California, the likelihood of selecting the winning lottery numbers in the Super Lotto Plus game is approximately 1 chance in 41,000,000. In contrast, the likelihood of being struck by lightning is about 1 chance in 500,000. Comparing the likelihood of winning the Super Lotto Plus to the likelihood of being struck by lightning, you are about 80 times more likely to be struck by lightning than to pick the winning California lottery numbers.

POINT OF INTEREST

Improbable and *unlikely* are not the same as *impossible*. Winning a lottery is an unlikely event, but not an impossible event, as one Texas native—who has won that state's lottery four times in 17 years—knows. Some mathmaticians estimate the probability of her four wins to be as low as 1 chance in 18,000,000,000,000,000,000,000,000.

The likelihood of the occurrence of a particular event is described by a number between 0 and 1. (You can think of this as a percentage between 0% and 100%, inclusive.) This number is called the **probability** of the event. An event that is not very likely has a probability close to 0; an event that is very likely has a probability close to 1 (100%). For instance, the probability of being struck by lightning is close to 0. However, if you randomly choose a basketball player from the National Basketball Association, it is very likely that the player is over 6 feet tall, so the probability is close to 1.

Because any event has from 0% to 100% chance of occurring, probabilities are always between 0 and 1, inclusive. If an event *must* occur, its probability is 1. If an event *cannot* occur, its probability is 0.

Probabilities can be calculated by considering the outcomes of experiments. Here are some examples of experiments.

- Flip a coin and observe the outcome as a head or a tail.
- Select a company and observe its annual profit.
- Record the time a person spends at the checkout line in a supermarket.

The sample space of an experiment is the set of all possible outcomes of the experiment. For example, consider tossing a coin three times and observing the outcome as a head or a tail. Using H for head and T for tail, the sample space is

$S = \{HHH, HHT, HTH, HTT, THH, THT, TTH, TTT\}$

Note that the sample space consists of *every* possible outcome of tossing three coins.

EXAMPLE 1 Find a Sample Space

A single die is rolled once. What is the sample space for this experiment?

Solution

The sample space is the set of possible outcomes of the experiment.

$S =$ 

CHECK YOUR PROGRESS 1 A coin is tossed twice. What is the sample space for this experiment?

Solution See page S44.

Formally, an event is a subset of a sample space. Using the sample space of Example 1, here are some possible events:

- There are an even number of pips (dots) facing up. The event is

  $E_1 =$ 
- The number of pips facing up is greater than 4. The event is $E_2 =$ 
- The number of pips facing up is less than 20. The event is

  $E_3 =$ 

  Because the number of pips facing up is always less than 20, this event will always occur. The event and the sample space are the same.
- The number of pips facing up is greater than 15. The event is $E_4 = \varnothing$, the empty set. This is an impossible event; the number of pips facing up cannot be greater than 15.

Outcomes of some experiments are **equally likely**, which means that the chance of any one outcome is just as likely as the chance of any other. For instance, if 4 balls of the same size but different colors—red, blue, green, and white—are placed in a box and a blindfolded person chooses 1 ball, the chance of choosing a green ball is the same as the chance of choosing any other color ball.

TAKE NOTE

As an example of an experiment whose outcomes are not equally likely, consider tossing a thumbtack and recording whether it lands with the point up or on its side. There are only two possible outcomes for this experiment, but the outcomes are not equally likely.

In the case of equally likely outcomes, the probability of an event is based on the number of elements in the event and the number of elements in the sample space. We will use $n(E)$ to denote the number of elements in the event E and $n(S)$ to denote the number of elements in the sample space S.

Probability of an Event

For an experiment with sample space S of *equally likely outcomes,* the probability $P(E)$ of an event E is given by

$$P(E) = \frac{n(E)}{n(S)} = \frac{\text{number of elements in } E}{\text{total number of elements in sample space } S}$$

Because each outcome of rolling a fair die is equally likely, the probability of the events E_1 through E_4 described on page 708 can be determined from the formula for the probability of an event.

$$P(E_1) = \frac{3}{6} \begin{matrix}\leftarrow \text{Number of elements in } E_1 \\ \leftarrow \text{Number of elements in the sample space}\end{matrix}$$
$$= \frac{1}{2}$$

The probability of rolling an even number of pips on a single roll of one die is $\frac{1}{2}$ (or 50%).

$$P(E_2) = \frac{2}{6} \begin{matrix}\leftarrow \text{Number of elements in } E_2 \\ \leftarrow \text{Number of elements in the sample space}\end{matrix}$$
$$= \frac{1}{3}$$

The probability of rolling a number greater than 4 on a single roll of one die is $\frac{1}{3}$.

$$P(E_3) = \frac{6}{6} \begin{matrix}\leftarrow \text{Number of elements in } E_3 \\ \leftarrow \text{Number of elements in the sample space}\end{matrix}$$
$$= 1$$

The probability of rolling a number less than 20 on a single roll of one die is 1 (or 100%). Recall that the probability of any event that is certain to occur is 1.

$$P(E_4) = \frac{0}{6} \begin{matrix}\leftarrow \text{Number of elements in } E_4 \\ \leftarrow \text{Number of elements in the sample space}\end{matrix}$$
$$= 0$$

The probability of rolling a number greater than 15 on a single roll of one die is 0. (It is not possible to roll any number greater than 6.)

EXAMPLE 2 Probability of Equally Likely Outcomes

A fair coin—one for which it is equally likely that heads or tails will result from a single toss—is tossed 3 times. What is the probability that 2 heads and 1 tail are tossed?

Solution

Determine the number of elements in the sample space. The sample space must include every possible toss of a head or a tail (in order) in 3 tosses of the coin.

$S = \{\text{HHH, HHT, HTH, HTT, THH, THT, TTH, TTT}\}$

TAKE NOTE

In Example 2, we calculated the probability as $\frac{3}{8}$. However, we could have expressed the probability as a decimal, 0.375, or as a percent, 37.5%. A probability can always be expressed as a fraction, a decimal, or a percent.

The elements in the event are $E = \{HHT, HTH, THH\}$.

$$P(E) = \frac{n(E)}{n(S)} = \frac{3}{8}$$

The probability is $\frac{3}{8}$.

CHECK YOUR PROGRESS 2 If a fair die is rolled once, what is the probability that an odd number will show on the upward face?

Solution See page S44.

QUESTION Is it possible that the probability of some event could be 1.23?

FIGURE 12.6 Outcomes of the roll of 2 dice

EXAMPLE 3 Calculating Probabilities with Dice

Two fair dice are tossed. What is the probability that the sum of the pips on the upward faces of the 2 dice equals 8?

Solution

The dice must be considered as distinct, so there are 36 possible outcomes. ( and  are considered different outcomes.) Therefore, $n(S) = 36$. The sample space is shown at the left. (See also Section 12.1.) Let E represent the event that the sum of the pips on the upward faces is 8. These outcomes are circled in Figure 12.6. By counting the number of circled pairs, $n(E) = 5$.

$$P(E) = \frac{n(E)}{n(S)} = \frac{5}{36}$$

The probability that the sum of the pips is 8 is $\frac{5}{36}$.

CHECK YOUR PROGRESS 3 Two fair dice are tossed. What is the probability that the sum of the pips on the upward faces of the 2 dice equals 7?

Solution See page S44.

Empirical Probability

Probabilities such as those calculated in the preceding examples are sometimes referred to as **theoretical probabilities**. In Example 2, we assumed that, in theory, we had a perfectly balanced coin, and we calculated the probability based on the fact that each outcome was equally likely. Similarly, we assumed the dice in Example 3 were equally likely to land with any of the 6 faces upward.

When a probability is based on data gathered from an experiment, it is called an **experimental** or **empirical probability**. For instance, if we tossed a thumbtack 100 times and recorded the number of times it landed "point up," the results might be as shown in the table at the left. From this experiment, the empirical probability of "point up" is

Point up	15
Side	85
Total	100

$$P(\text{point up}) = \frac{15}{100} = 0.15$$

Empirical Probability of an Event

If an experiment is performed repeatedly and the occurrence of the event E is observed, the probability $P(E)$ of the event E is given by

$$P(E) = \frac{\text{number of times event } E \text{ occurred}}{\text{number of times the experiment was performed}}$$

ANSWER No. All probabilities must be between 0 and 1, inclusive.

EXAMPLE 4 Calculate an Empirical Probability

A survey showed the following information on the ages and party affiliations of registered voters in a certain city. If one voter is chosen at random from this survey, what is the probability that the voter is a Republican?

Age	Republican	Democrat	Independent	Other	Total
18–28	205	432	98	112	847
29–38	311	301	109	83	804
39–49	250	251	150	122	773
≥50	272	283	142	107	804
Total	1038	1267	499	424	3228

Solution

Let R be the event that a Republican is selected. Then

$$P(R) = \frac{1038}{3228} \begin{array}{l} \leftarrow \text{Number of Republicans in the survey} \\ \leftarrow \text{Total number of people surveyed} \end{array}$$

$$\approx 0.32$$

The probability that the selected person is a Republican is approximately 0.32.

CHECK YOUR PROGRESS 4 Using the data from Example 4, what is the probability that a randomly selected person is between the ages of 39 and 49?

Solution See page S44.

HISTORICAL NOTE

TopFoto/The Image Works

Gregor Mendel (mĕn′dl) (1822–1884) was an Augustinian monk and teacher. His study of the transmission of traits from one generation to another was meant to confirm the prevailing theory of the day that environment influenced the traits of a plant. However, his research seemed to suggest that plants have certain "determiners" that dictate the characteristics of the next generation. Thus began the study of heredity. It took over 30 years for Mendel's conclusions to be accepted in the scientific community.

Application to Genetics

Completed in April 2003, the Human Genome Project was a 13-year-long project designed to completely map the genetic make-up of *Homo sapiens*. Researchers hope to use this information to treat and prevent certain hereditary diseases.

The concept behind this project began with Gregor Mendel and his work on flower color and how it was transmitted from generation to generation. From his studies, Mendel concluded that flower color seems to be predictable in future generations by making certain assumptions about a plant's color "determiner." He concluded that red was a *dominant* determiner of color and that white was a *recessive* determiner. Today, geneticists talk about the *gene* for flower color and the *allele* of the gene. A gene consists of two dominant alleles (two red), a dominant and a recessive allele (red and white), or two recessive alleles (two white). Because red is the dominant allele, a flower will be white only if no dominant allele is present.

Later work by Reginald Punnett (1875–1967) showed how to determine the probability of certain flower colors by using a **Punnett square**. Using a capital letter for a dominant allele (R for red) and the corresponding lower-case letter for a recessive allele (r for white), Punnett arranged the alleles of the parents in a square. A parent could be RR, Rr, or rr.

Suppose that the genotype (genetic composition) of one parent is rr and the genotype of the other is Rr. The first parent is represented in the left column of the square, and the second parent is represented in the top row. The genotypes of the offspring are shown in the body of the table and are the result of combining one allele from each parent.

Parents	R	r
r	Rr	rr
r	Rr	rr

Because each of the genotypes of the offspring is equally likely, the probability that a flower will be red is $\frac{1}{2}$ (two Rr genotypes of the four possible genotypes), and the probability that a flower will be white is $\frac{1}{2}$ (two rr genotypes of the four possible genotypes).

EXAMPLE 5 Probability Using a Punnett Square

A child will have cystic fibrosis if the child inherits the recessive gene from both parents. Using F for the normal allele and f for the mutant recessive allele, suppose a parent who is Ff (said to be a *carrier*) and a parent who is FF (does not have the mutant allele) decide to have a child.

a. What is the probability that the child will have cystic fibrosis? (To have the disease, the child must be ff.)

b. What is the probability that the child will be a carrier?

Solution

Make a Punnett square.

Parents	F	F
F	FF	FF
f	Ff	Ff

a. To have the disease, the child must be ff. From the table, there is no combination of the alleles that will produce ff. Therefore, the child cannot have the disease, and the probability is 0.

b. To be a carrier, exactly one allele must be f. From the table, there are two cases out of four of a genotype with one f. The probability that the child will be a carrier is $\frac{2}{4} = \frac{1}{2}$.

CHECK YOUR PROGRESS 5 For a certain type of hamster, the color cinnamon, C, is dominant and the color white, c, is recessive. If both parents are Cc, what is the probability that an offspring will be white?

Solution See page S44.

TAKE NOTE

The probability that a parent will pass a certain genetic characteristic on to a child is one-half. However, the probability that a parent has a certain genetic characteristic is not one-half. For instance, the probability of passing on the allele for cystic fibrosis by a parent who has the mutant allele is 0.5. However, the probability that a person randomly selected from the population has the mutant allele is less than 0.025. We will discuss this further in Section 12.5.

Calculating Odds

Statistics kept by some fantasy sports enthusiasts show that the *odds in favor* of a professional basketball player making a free throw are 9 to 2. These odds indicate that out of every 11 (9 + 2) free throws, a player is expected to make 9 baskets.

A **favorable outcome** of an experiment is one that satisfies some event. For instance, in the case of free-throw attempts by professional basketball players, you can assume that out of 11 free-throw attempts, there will be 9 favorable outcomes (9 baskets). The opposite event, a missed basket, is an **unfavorable outcome**. Odds are frequently expressed in terms of favorable and unfavorable outcomes.

Odds of an Event

Let E be an event in a sample space of equally likely outcomes. Then

$$\textbf{Odds in favor of } E = \frac{\text{number of favorable outcomes}}{\text{number of unfavorable outcomes}}$$

$$\textbf{Odds against } E = \frac{\text{number of unfavorable outcomes}}{\text{number of favorable outcomes}}$$

INSTRUCTOR NOTE

Make sure students understand that a fraction such as $\frac{1}{4}$, when being used to express odds, has a different meaning than it does when being used to express a probability.

When the odds of an event are written in fractional form, the fraction bar is read as the word *to*. Thus odds of $\frac{3}{2}$ are read as "3 to 2." We can also write odds of $\frac{3}{2}$ as 3:2. This form is also read as "3 to 2."

EXAMPLE 6 Calculate Odds

If a pair of fair dice is rolled once, what are the odds in favor of rolling a sum of 7?

Solution

Let E be the event of rolling a sum of 7. From Figure 12.6 on page 710, the 6 favorable outcomes are E = {⚀⚅, ⚁⚄, ⚂⚃, ⚃⚂, ⚄⚁, ⚅⚀}. The unfavorable outcomes are the remaining 30 possibilities. (Because there are 36 possible outcomes when tossing 2 dice and 6 of them are favorable, there are $36 - 6 = 30$ unfavorable outcomes.)

$$\text{Odds in favor of } E = \frac{\text{number of favorable outcomes}}{\text{number of unfavorable outcomes}} = \frac{6}{30} = \frac{1}{5}$$

The odds in favor of rolling a sum of 7 are 1 to 5.

CHECK YOUR PROGRESS 6 If 3 red, 4 white, and 5 blue balls are placed in a box and 1 ball is randomly selected from the box, what are the odds against the ball being blue?

Solution See page S44.

Elsa/Getty Images Sport/Getty Images

When Triple Crown winner American Pharoah ran in the 2015 Breeder's Cup, bettors estimated that the horse's *odds against* winning the race were 3 to 5. This means bettors expected that American Pharoah would *lose* 3 races and win 2.

POINT OF INTEREST

The Triple Crown is awarded to a horse that wins the Kentucky Derby, the Preakness, and the Belmont Stakes in a single year.

Odds express the likelihood of an event and are therefore related to probability. When the odds of an event are known, the probability of the event can be determined. Conversely, when the probability of an event is known, the odds of the event can be determined.

The Relationship between Odds and Probability

1. Suppose E is an event in a sample space and that the *odds in favor* of E are $\frac{a}{b}$. Then $P(E) = \frac{a}{a+b}$.
2. Suppose E is an event in a sample space. Then the *odds in favor* of E are $\frac{P(E)}{1 - P(E)}$.

EXAMPLE 7 Determine Probability from Odds

In 2010, the racehorse Drosselmeyer won the Belmont Stakes, beating the favorite Ice Box. The odds against Drosselmeyer winning the race were 12 to 1. What was the probability of Drosselmeyer winning the race?

Solution

Because the odds *against* Drosselmeyer winning the race were 12 to 1, the odds *in favor* of Drosselmeyer winning the race were 1 to 12, or, as a fraction, $\frac{1}{12}$. Now use the formula for calculating the probability of an event when the odds in favor are known.

$$P(E) = \frac{a}{a+b}$$

$$P(E) = \frac{1}{1+12} = \frac{1}{13} \qquad \bullet\ a = 1,\ b = 12$$

The probability of Drosselmeyer winning the race was $\frac{1}{13}$, or about 7.7%.

CHECK YOUR PROGRESS 7 A report issued by the Southern California Earthquake Data Center estimates that the probability of an earthquake of magnitude 6.7 or greater within 30 years in the San Francisco Bay Area is about 60%. What are the odds in favor of such an earthquake occurring in that region in the next 30 years?

Solution See page S44.

MATH MATTERS The Birth of Probability Theory

Pierre Fermat and Blaise Pascal are generally considered to be the founders of the study of probability. For many historians, the starting point of the formal discussion of probability is contained in a letter from Pascal to Fermat written in July 1654.

> The Chevalier de Mere said to me that he found a falsehood in the theory of numbers for the following reason. If one wants to throw a six with a single die, there is an advantage in undertaking to do it in four throws, as the odds are 671 to 625. If one throws two sixes with a pair of dice, there is a disadvantage in having only 24 throws. However, 24 to 36 (which is the number of cases for two dice) is 4 to 6 (which is the number of cases on one die). This is the great "scandal" which makes him proclaim loftily that the theorems are not constant and Arithmetic is self-contradictory.

Basically (although it certainly isn't obvious from this letter), the Chevalier was claiming that there is a better than 50% chance of tossing a 6 in 4 rolls of a single die. By his reasoning, he concluded that there should be a better than 50% chance of rolling a pair of 6s in 24 tosses of 2 dice. However, the Chevalier tested his theory and found that 25 tosses were needed. From his tests, he concluded a "falsehood in the theory of numbers." In this letter, Pascal was mocking the inability of the Chevalier to correctly determine the probabilities involved.

EXCURSION

The Value of Pi by Simulation

A **simulation** is an activity designed to approximate a given situation. For instance, pilots fly a simulator to practice maneuvers that would be dangerous to try during actual flight. This Excursion uses a simulation to approximate the value of π.

This Excursion will work better if four or five people work together. Get 25 toothpicks, and then tape several sheets of blank paper together to form a large rectangular shape. Use a large ruler to draw parallel lines on the paper such that the distance between the lines is the length of one toothpick. Then drop all of the toothpicks from approximately knee height onto the paper. (Make sure that none of the toothpicks land off the paper.) Record the number of toothpicks that cross any of the parallel lines. Repeat this process 10 times for a total of 250 toothpicks dropped.[1]

[1] See our companion site at CengageBrain.com for a computer program that can be used to simulate the dropping of the toothpicks. This program will enable you to perform the experiment many thousands of times.

EXCURSION EXERCISES

1. Using the recorded data, calculate the empirical probability that a dropped toothpick will cross a line.
2. The theoretical probability that a toothpick will cross a line is $\frac{2}{\pi}$. Use a calculator to find the value of $\frac{2}{\pi}$ rounded to four decimal places.
3. The experimental value you calculated in Exercise 1 should be close to the theoretical value you calculated in Exercise 2 (at least it should be if you dropped a toothpick around 1000 times). Calculate the percent error between the experimental value and the theoretical value. (The percent error can be determined by finding the difference between the two values and dividing it by the theoretical value.)
4. Explain how performing this experiment thousands of times could give you an approximate value for π.
5. Using your data, what approximate value for π do you calculate?
6. This experiment is based on a famous 18th-century problem called the *Buffon needle problem*. Look up "Buffon needle problem" on the Internet and write a short essay about this problem and how it applies to this Excursion.

EXERCISE SET 12.3

■ In Exercises 1 to 6, list the elements of the sample space for each experiment.

1. A coin is flipped 3 times.
2. An even number between 1 and 11 is selected at random.
3. One day in the first 2 weeks of November is selected.
4. A current U.S. coin is selected from a piggy bank.
5. A state is selected from the U.S. states whose names begin with the letter A.
6. A month is selected from the months that have exactly 30 days.

■ In Exercises 7 to 15, assume that it is equally likely for a child to be born a boy or a girl and that the Lin family is planning on having 3 children.

7. List the elements of the sample space for the genders of the 3 children.
8. List the elements of the event that the Lins have 2 boys and 1 girl.
9. List the elements of the event that the Lins have at least 2 girls.
10. List the elements of the event that the Lins have no girls.
11. List the elements of the event that the Lins have at least 1 girl.
12. Compute the probability that the Lins will have 2 boys and 1 girl.
13. Compute the probability that the Lins will have at least 2 girls.
14. Compute the probability that the Lins will have no girls.
15. Compute the probability that the Lins will have at least 1 girl.

■ In Exercises 16 to 18, a coin is tossed 4 times. Assuming the coin is equally likely to land on heads or tails, compute the probability of each event occurring.

16. 2 heads and 2 tails
17. 1 head and 3 tails
18. All 4 coin tosses are identical.

■ In Exercises 19 to 22, a dodecahedral die (one with 12 sides numbered from 1 to 12) is tossed once. Find each of the following probabilities.

grum_1/Shutterstock.com

19. The number on the upward face is 12.
20. The number on the upward face is not 10.
21. The number on the upward face is divisible by 4.
22. The number on the upward face is less than 5 or greater than 9.

■ In Exercises 23 to 32, 2 regular 6-sided dice are tossed. Compute the probability that the sum of the pips on the upward faces of the 2 dice is each of the following. (See Figure 12.6 on page 710 for the sample space of this experiment.)

23. 6

24. 11

25. 2

26. 12

27. 1

28. 14

29. At least 10

30. At most 5

31. An even number

32. An odd number

33. If 2 dice are rolled, compute the probability of rolling doubles (both dice show the same number of pips).

34. If 2 dice are rolled, compute the probability of *not* rolling doubles.

■ In Exercises 35 to 38, a card is selected at random from a standard deck of playing cards.

35. Compute the probability that the card is a 9.

36. Compute the probability that the card is a face card (jack, queen, or king).

37. Compute the probability that the card is between 5 and 9, inclusive.

38. Compute the probability that the card is between 3 and 6, inclusive.

■ **Voter Characteristics** In Exercises 39 to 44, use the data given in Example 4, page 711, to compute the probability that a randomly chosen voter from the survey will satisfy the following. Round to the nearest hundredth.

39. The voter is a Democrat.

40. The voter is not a Republican.

41. The voter is 50 years old or older.

42. The voter is under 39 years old.

43. The voter is between 39 and 49 and is registered as an Independent.

44. The voter is under 29 and is registered as a Democrat.

■ **Education Levels** In Exercises 45 to 48, a survey asked 850 respondents about their highest levels of completed education. The results are given in the following table.

Education completed	Number of respondents
No high school diploma	52
High school diploma	234
Associate's degree or 2 years of college	274
Bachelor's degree	187
Master's degree	67
Ph.D. or professional degree	36

If a respondent from the survey is selected at random, compute the probability of each of the following.

45. The respondent did not complete high school.

46. The respondent has an associate's degree or 2 years of college (but not more).

47. The respondent has a Ph.D. or professional degree.

48. The respondent has a degree beyond a bachelor's degree.

■ **Annual Salaries** A random survey asked respondents about their current annual salaries. The results are given in the following table. Use the table for Exercises 49 to 52.

Salary range	Number of respondents
Below \$18,000	24
\$18,000–\$27,999	41
\$28,000–\$35,999	52
\$36,000–\$45,999	58
\$46,000–\$59,999	43
\$60,000–\$79,999	39
\$80,000–\$99,999	22
\$100,000 or more	14

If a respondent from the survey is selected at random, compute the probability of the following.

49. The respondent earns from \$36,000 to \$45,999 annually.

50. The respondent earns from \$60,000 to \$79,999 per year.

51. The respondent earns at least \$80,000 per year.

52. The respondent earns less than \$36,000 annually.

53. **Genotypes** The following Punnett square for flower color shows two parents of genotype Rr, where R corresponds to the dominant red flower allele and r represents the recessive white flower allele. (See Example 5.)

Parents	R	r
R	RR	Rr
r	Rr	rr

What is the probability that the offspring of these parents will have white flowers?

54. **Genotypes** One parent plant with red flowers has genotype RR and the other with white flowers has genotype rr, where R is the dominant allele for a red flower and r is the recessive allele for a white flower. Compute the probability of one of the offspring having white flowers. *Hint:* Draw a Punnett square.

55. Genotypes The eye color of mice is determined by a dominant allele E, corresponding to black eyes, and a recessive allele e, corresponding to red eyes. If two mice, one of genotype EE and the other of genotype ee, have offspring, compute the probability of one of the offspring having red eyes. *Hint:* Draw a Punnett square.

56. Genotypes The height of a certain plant is determined by a dominant allele T corresponding to tall plants and a recessive allele t corresponding to short (or dwarf) plants. If both parent plants have genotype Tt, compute the probability that the offspring plants will be tall. *Hint:* Draw a Punnett square.

57. Explain the difference between the probability of an event and the odds of the same event.

58. Give an example of an event that has probability 0 and one that has probability 1.

■ In Exercises 59 to 64, the odds in favor of an event are given. Compute the probability of the event.

59. 1 to 2 **60.** 1 to 4

61. 3 to 7 **62.** 3 to 5

63. 8 to 5 **64.** 11 to 9

■ In Exercises 65 to 70, the probability of an event is given. Find the odds in favor of the event.

65. 0.2 **66.** 0.6

67. 0.375 **68.** 0.28

69. 0.55 **70.** 0.81

71. Game Shows The game board for the television show "Jeopardy" is divided into 6 categories, with each category containing 5 answers. In the Double Jeopardy round, there are 2 hidden Daily Double squares. What are the odds in favor of choosing a Daily Double square on the first turn?

72. Game Shows The spinner for the television game show "Wheel of Fortune" is shown below. On a single spin of the wheel, what are the odds against stopping on $300?

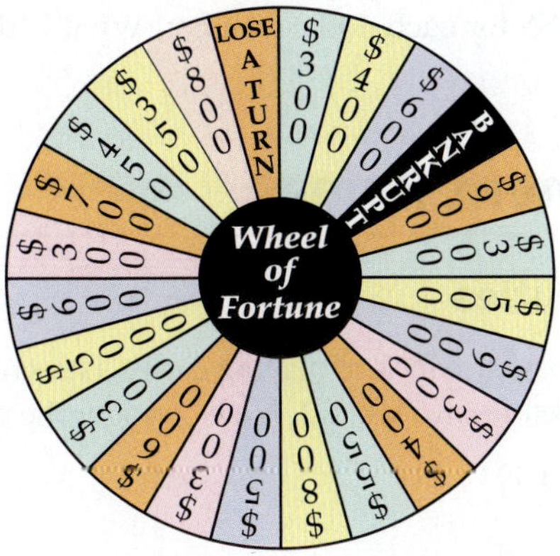

© Cengage Learning

73. If a single fair die is rolled, what are the odds in favor of rolling an even number?

74. If a card is randomly pulled from a standard deck of playing cards, what are the odds in favor of pulling a heart?

75. A coin is tossed 4 times. What are the odds against the coin showing heads all 4 times?

Earthquakes For Exercises 76 to 78, use the table, which is based on information from an article discussing the probabilities of earthquakes occurring in California.

Earthquake of magnitude . . .	Will occur in California within . . .	Probability or odds
7.5 or greater	next 30 years	Probability: 46%
7.5 or greater	any one year	Odds against: 48 to 49
6.7 or greater	next 30 years	Odds in favor: 332 to 1

SOURCE: http://www.digitjournal.com

76. What are the odds in favor of an earthquake of magnitude 7.5 or greater occurring in California within the next 30 years?

77. What is the probability of an earthquake of magnitude 7.5 or greater occurring in California within any 1 year?

78. What is the probability of an earthquake of magnitude 6.7 or greater occurring in California within the next 30 years?

79. Football A bookmaker has placed 8 to 3 odds *against* a particular football team winning its next game. What is the probability, in the bookmaker's view, of the team winning?

80. Candy Colors A snack-size bag of M&Ms candies contains 12 red candies, 12 blue, 7 green, 13 brown, 3 orange, and 10 yellow. If a candy is randomly picked from the bag, compute

Rachel Epstein/The Image Works

a. the odds of getting a green M&M.

b. the probability of getting a green M&M.

EXTENSIONS

81. If 4 cards labeled A, B, C, and D are randomly placed in 4 boxes also labeled A, B, C, and D, 1 to each box, find the probability that no card will be in a box with the same letter.

82. Determine the probability that if 10 coins are tossed, 5 heads and 5 tails will result.

83. In a family of 3 children, all of whom are girls, a family member new to probability reasons that the probability that each child would be a girl is 0.5. Therefore, the probability that the family would have 3 girls is $0.5 + 0.5 + 0.5 = 1.5$. Explain why this reasoning is not valid.

■ In Exercises 84 and 85, a hand of 5 cards is dealt from a standard deck of playing cards. You may want to review the material on combinations before doing these exercises.

84. Find the probability that the hand will contain all 4 aces.

85. Find the probability that the hand will contain 3 jacks and 2 queens.

Roulette Exercises 86 to 91 use the casino game roulette. Roulette is played by spinning a wheel with 38 numbered slots. The numbers 1 through 36 appear on the wheel, half of them colored black and half colored red. Two slots, numbered 0 and 00, are colored green. A ball is placed on the spinning wheel and allowed to come to rest in one of the slots. Bets are placed on where the ball will land.

nvuk/Shutterstock.com

86. You can place a bet that the ball will stop in a black slot. If you win, the casino will pay you \$1 for each dollar you bet. What is the probability of winning this bet?

87. You can bet that the ball will land on an odd number. If you win, the casino will pay you \$1 for each dollar you bet. What is the probability of winning this bet?

88. You can bet that the ball will land on any number from 1 to 12. If you win, the casino will pay you \$2 for each dollar you bet. What is the probability of winning this bet?

89. You can bet that the ball will land on any particular number. If you win, the casino will pay you \$35 for each dollar you bet. What is the probability of winning this bet?

90. You can bet that the ball will land on one of 0 or 00. If you win, the casino will pay you \$17 for each dollar you bet. What is the probability of winning this bet?

91. You can bet that the ball will land on certain groups of 6 numbers (such as 1 to 6). If you win, the casino will pay you \$5 for each dollar you bet. What is the probability of winning this bet?

SECTION 12.4 Addition and Complement Rules

The Addition Rule for Probabilities

Suppose you draw a single card from a standard deck of playing cards. The sample space S consists of the 52 cards of the deck. Therefore, $n(S) = 52$. Now consider the events

$$E_1 = \text{a 4 is drawn} = \{\spadesuit 4, \heartsuit 4, \diamondsuit 4, \clubsuit 4\}$$

$$E_2 = \text{a spade is drawn}$$
$$= \{\spadesuit A, \spadesuit 2, \spadesuit 3, \spadesuit 4, \spadesuit 5, \spadesuit 6, \spadesuit 7, \spadesuit 8, \spadesuit 9, \spadesuit 10, \spadesuit J, \spadesuit Q, \spadesuit K\}$$

It is possible, on one draw, to satisfy the conditions of both events: the ♠4 could be drawn. This card is an element of both E_1 and E_2.

Now consider the events

E_3 = a 5 is drawn = {♠5, ♥5, ♦5, ♣5}

E_4 = a king is drawn = {♠K, ♥K, ♦K, ♣K}

In this case, it is not possible to draw one card that satisfies the conditions of both events. There are no elements common to both sets. Two events that cannot both occur at the same time are called **mutually exclusive events**. The events E_3 and E_4 are mutually exclusive events, whereas E_1 and E_2 are not.

Mutually Exclusive Events

Two events A and B are mutually exclusive if they cannot occur at the same time. That is, A and B are mutually exclusive when $A \cap B = \varnothing$.

QUESTION A die is rolled once. Let E be the event that an even number is rolled and let O be the event that an odd number is rolled. Are the events E and O mutually exclusive?

The probability of either of two mutually exclusive events occurring can be determined by adding the probabilities of the individual events.

Probability of Mutually Exclusive Events

If A and B are two mutually exclusive events, then the probability of A or B occurring is

$$P(A \text{ or } B) = P(A) + P(B)$$

EXAMPLE 1 Probability of Mutually Exclusive Events

Suppose a single card is drawn from a standard deck of playing cards. Find the probability of drawing a 5 or a king.

Solution

Let A = {♠5, ♥5, ♦5, ♣5} and B = {♠K, ♥K, ♦K, ♣K}. There are 52 cards in a standard deck of playing cards; thus $n(S) = 52$. Because the events are mutually exclusive, we can use the formula for the probability of mutually exclusive events.

$P(A \text{ or } B) = P(A) + P(B)$ • Formula for the probability of mutually exclusive events

$= \frac{1}{13} + \frac{1}{13} = \frac{2}{13}$ • $P(A) = \frac{4}{52} = \frac{1}{13}$, $P(B) = \frac{4}{52} = \frac{1}{13}$

The probability of drawing a 5 or a king is $\frac{2}{13}$.

CHECK YOUR PROGRESS 1 Two fair dice are tossed once. What is the probability of rolling a 7 or an 11? For the sample space for this experiment, see page 689.

Solution See page S44.

ANSWER Yes. It is not possible to roll an even number and an odd number on a single roll of the die.

Consider the experiment of rolling two dice. Let A be the event of rolling a sum of 8 and let B be the event of rolling a double (the same number on both dice).

$A = \{$  $\}$

$B = \{$ 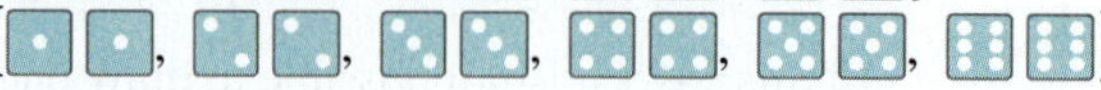 $\}$

> **TAKE NOTE**
>
> The $P(A \text{ and } B)$ term in the Addition Rule for Probabilities is subtracted to compensate for the overcounting of the first two terms of the formula. If two events are mutually exclusive, then $A \cap B = \varnothing$. Therefore, $n(A \cap B) = 0$ and $P(A \text{ and } B) = \frac{n(A \cap B)}{n(S)} = 0$. For mutually exclusive events, the Addition Rule for Probabilities is the same as the formula for the probability of mutually exclusive events.

These events are *not* mutually exclusive because it is possible to satisfy the conditions of each event on one toss of the dice—a  could be rolled. Therefore, $P(A \text{ or } B)$, the probability of a sum of 8 or a double, cannot be calculated using the formula for the probability of mutually exclusive events. However, a modification of that formula can be used.

> **Addition Rule for Probabilities**
>
> If A and B are two events in a sample space S, then
>
> $$P(A \text{ or } B) = P(A) + P(B) - P(A \text{ and } B)$$

> **TAKE NOTE**
>
> Recall that the probability of an event A is $P(A) = \frac{n(A)}{n(S)}$. Therefore, $P(A \text{ and } B) = \frac{n(A \cap B)}{n(S)}$.

Using this formula with

$A = \{$ 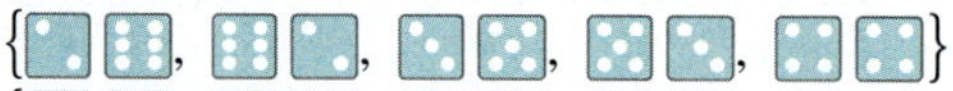 $\}$

$B = \{$ 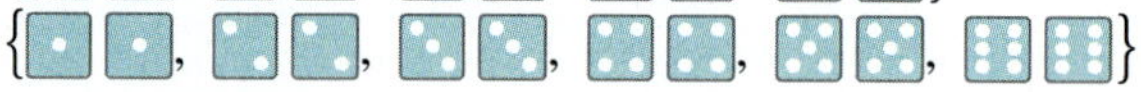 $\}$

$A \cap B = \{$  $\}$

the probability of A or B can be calculated.

$$P(A \text{ or } B) = P(A) + P(B) - P(A \text{ and } B)$$

$$= \frac{5}{36} + \frac{6}{36} - \frac{1}{36} \qquad \bullet\ P(A) = \frac{5}{36},\ P(B) = \frac{6}{36},\ P(A \cap B) = \frac{1}{36}$$

$$= \frac{10}{36} = \frac{5}{18}$$

On a single roll of two dice, the probability of rolling a sum of 8 or a double is $\frac{5}{18}$.

EXAMPLE 2 **Use the Addition Rule for Probabilities**

The table at the left shows data from an experiment conducted to test the effectiveness of a flu vaccine. If one person is selected from this population, what is the probability that the person was vaccinated or contracted the flu?

	F	No F	Total
V	21	198	219
No V	76	195	271
Total	97	393	490

V: Vaccinated
F: Contracted the flu

Solution

Let $V = \{\text{people who were vaccinated}\}$ and $F = \{\text{people who contracted the flu}\}$. These events are not mutually exclusive because there are 21 people who were vaccinated and who contracted the flu. The sample space S consists of the 490 people who participated in the experiment. From the table, $n(V) = 219$, $n(F) = 97$, and $n(V \text{ and } F) = 21$.

$$P(V \text{ or } F) = P(V) + P(F) - P(V \text{ and } F)$$

$$= \frac{219}{490} + \frac{97}{490} - \frac{21}{490}$$

$$= \frac{295}{490} \approx 0.602$$

The probability of selecting a person who was vaccinated or who contracted the flu is approximately 60.2%.

CHECK YOUR PROGRESS 2 The data in the table on the next page show the starting salaries of college graduates with selected degrees. If one person is chosen from this population, what is the probability that the person has a degree in business or has a starting salary between \$20,000 and \$24,999?

	Degree			
Salary (in \$)	**Engineering**	**Business**	**Chemistry**	**Psychology**
Less than 20,000	0	4	1	12
20,000–24,999	4	16	3	16
25,000–29,999	7	21	5	15
30,000–34,999	12	35	5	7
35,000 or more	12	22	4	5

Solution See page S45.

The Complement of an Event

Consider the experiment of tossing a single die once. The sample space is

$S = \{$⚀, ⚁, ⚂, ⚃, ⚄, ⚅$\}$

Now consider the event $E = \{$⚂$\}$, that is, the event of tossing a ⚂. The probability of E is

$$P(E) = \frac{1}{6} \begin{array}{l} \leftarrow \text{Number of elements in } E \\ \leftarrow \text{Number of elements in the sample space} \end{array}$$

The **complement** of an event E, symbolized by E^c, is the "opposite" event of E. The complement includes all those outcomes of S that are not in E and excludes the outcomes in E. For the event E above, E^c is the event of not tossing a ⚂. Thus

$E^c = \{$⚀, ⚁, ⚃, ⚄, ⚅$\}$

Note that because E and E^c are opposite events, they are mutually exclusive, and their union is the entire sample space S. Thus $P(E) + P(E^c) = P(S)$. But $P(S) = 1$, so $P(E^c) = 1 - P(E)$.

Probability of the Complement of an Event

If E is an event and E^c is the complement of the event, then

$$P(E^c) = 1 - P(E)$$

Continuing our example, the probability of not tossing a ⚂ is given by

$P(\text{not a ⚂}) = 1 - P(\text{⚂})$

$\quad = 1 - \frac{1}{6} = \frac{5}{6}$

You can also verify the probability of E^c directly:

$$P(\text{not a ⚂}) = \frac{5}{6} \begin{array}{l} \leftarrow \text{Number of elements in } E^c \\ \leftarrow \text{Number of elements in the sample space} \end{array}$$

INSTRUCTOR NOTE

It may be helpful to have students practice formulating a probability in terms of the complement of an event. Ask the students how they might use the complement to determine the probability of the following situations:

- A player does not roll doubles in the game of Monopoly.
- A randomly selected student on campus is enrolled in at least 2 classes.
- A randomly selected student in the class has received at least 1 traffic violation.

EXAMPLE 3 Find a Probability by the Complement Rule

The probability of tossing a sum of 11 on the toss of 2 dice is $\frac{1}{18}$. What is the probability of not tossing a sum of 11 on the toss of 2 dice?

Solution

Use the formula for the probability of the complement of an event.
Let $E = \{\text{toss a sum of } 11\}$. Then $E^c = \{\text{toss a sum that is not } 11\}$.

$$P(E^c) = 1 - P(E)$$

$$= 1 - \frac{1}{18} = \frac{17}{18} \qquad \bullet\ P(E) = \frac{1}{18}$$

The probability of not tossing a sum of 11 is $\frac{17}{18}$.

CHECK YOUR PROGRESS 3 The probability that a person has type A blood is 34%. What is the probability that a person does not have type A blood?

Solution See page S45.

TAKE NOTE

The phrase "at least one" means 1 or more. Tossing a coin 3 times and asking the probability of getting at least 1 head means calculating the probability of getting 1, 2, or 3 heads.

Suppose we toss a coin 3 times and want to calculate the probability of having heads occur *at least once*. We could list all the possibilities of tossing 3 coins, as shown below, and then find the ones that contain at least 1 head.

$$\underbrace{\{\text{HHH, HHT, HTH, HTT, THH, THT, TTH}}_{\text{at least one head}}\text{, TTT}\}$$

The probability of at least 1 head is $\frac{7}{8}$.

Another way to calculate this result is to use the formula for the probability of the complement of an event. Let $E = \{\text{at least 1 head}\}$. From the list above, note that E contains every outcome except TTT (no heads). Thus $E^c = \{\text{TTT}\}$ and we have

$$P(E) = 1 - P(E^c)$$

$$= 1 - \frac{1}{8} = \frac{7}{8} \qquad \bullet\ P(E^c) = \frac{n(E^c)}{n(S)} = \frac{1}{8}$$

This is the same result that we calculated above. As we will see, sometimes working with a complement is much less work than proceeding directly.

Combinatoric Formulas and Probability

In many cases, the principles of counting that were discussed in Sections 12.1 and 12.2 are part of the process of calculating a probability.

EXAMPLE 4 Find a Probability Using the Complement Rule

A die is tossed 4 times. What is the probability that a ⚅ will show on the upward face at least once?

Solution

Let $E = \{\text{at least one } 6\}$. Then $E^c = \{\text{no 6s}\}$. To calculate the number of elements in the sample space (all possible outcomes of tossing a die 4 times) and the number of items in E^c, we will use the counting principle.

Because on each toss of the die there are 6 possible outcomes,

$$n(S) = 6 \cdot 6 \cdot 6 \cdot 6 = 1296$$

On each toss of the die there are 5 numbers that are not 6s. Therefore,

$$n(E^c) = 5 \cdot 5 \cdot 5 \cdot 5 = 625$$

$$\begin{aligned} P(E) &= 1 - P(E^c) \\ &= 1 - \frac{625}{1296} = \frac{671}{1296} \\ &\approx 0.518 \end{aligned}$$

When a die is tossed 4 times, the probability that a  will show on the upward face at least once is approximately 0.518.

CHECK YOUR PROGRESS 4 A pair of dice is rolled 3 times. What is the probability that a sum of 7 will occur at least once?

Solution See page S45.

MATH MATTERS The Monty Hall Problem

A famous probability puzzle began with the game show *Let's Make a Deal,* of which Monty Hall was the host, and goes something like the following. Suppose you appear on the show and are shown three closed doors. Behind one of the doors is the grand prize; behind the other two doors are less desirable prizes (like a goat!). If you select the door hiding the grand prize, you win that prize. The probability of randomly choosing the grand prize, of course, is 1/3. After you choose a door, the show's host (who knows which door the grand prize is hiding behind) does not immediately open it. Instead, he opens one of the other two doors and reveals a goat. Obviously you are relieved that you did not choose that particular door, but he then asks if you would like to switch your choice to the third door. Should you stay with your original choice, or switch? Most people would say at first that it makes no difference. However, computer simulations that play the game over and over have shown that you *should* switch. In fact, you double your chances of winning the grand prize if you give up your first choice! See Exercise 54 on page 726 for a mathematical investigation, and then try a simulation of your own with Exercises 56 and 57.

In the next example, we use the combination formula $C(n, r) = \frac{n!}{r!(n-r)!}$ to find the number of ways in which r objects can be chosen from n objects.

EXAMPLE 5 Find a Probability Using the Combination Formula

Suppose a manufacturing process for tableware produces 40 dinner plates, of which 3 are defective. If 5 plates are randomly selected from the 40, what is the probability that at least one is defective?

Solution

Let E = {at least one plate is defective}. It is easier to work with the complement event, E^c = {no plates are defective}. To calculate the number of elements in the sample space (all possible outcomes of choosing 5 plates from 40), use the combination formula with $n = 40$ (the total number of plates) and $r = 5$ (the number of plates that are chosen). Then

$$\begin{aligned} n(S) = C(40, 5) &= \frac{40!}{5!(40-5)!} \qquad \bullet\ n = 40, r = 5 \\ &= \frac{40!}{5!\,35!} = 658{,}008 \end{aligned}$$

To find the number of outcomes that contain no defective plates, all of the plates chosen must come from the 37 nondefective plates. Therefore, we must calculate the number of ways in which 5 objects can be chosen from 37. Thus $n = 37$ (the number of nondefective plates) and $r = 5$ (the number of plates chosen).

$$n(E^c) = C(37, 5) = \frac{37!}{5!(37-5)!} \qquad \bullet\ n = 37, r = 5$$

$$= \frac{37!}{5!32!} = 435{,}897$$

$$P(E) = 1 - P(E^c)$$

$$= 1 - \frac{435{,}897}{658{,}008} = \frac{222{,}111}{658{,}008}$$

$$\approx 0.338$$

The probability is approximately 0.338, or 33.8%.

CHECK YOUR PROGRESS 5 The winner of a contest will be blindfolded and then allowed to reach into a hat containing 31 \$1 bills and 4 \$100 bills. The winner can remove 4 bills from the hat and keep the money. Find the probability that the winner will pull out at least 1 \$100 bill.

Solution See page S45.

EXCURSION

Keno Revisited

In Section 12.2 (see pages 704–705), we looked at the popular casino game keno, in which a player chooses numbers from 1 to 80 and hopes that the casino will draw balls with the same numbers.

A player can choose only 1 number or as many as 20. The casino will then pick 20 numbered balls from the 80 possible; if enough of the player's numbers match the lucky numbers the casino chooses, the player wins money. The amount won varies according to how many numbers were chosen and how many are "lucky."

EXCURSION EXERCISES

1. A gambler playing keno randomly chooses 5 numbers. What is the probability that the gambler will match at least 1 lucky number?
2. If 5 numbers are chosen, compute the probability of matching fewer than 5 lucky numbers.
3. If the keno player chooses 15 numbers, bets \$1, and matches 13 of the lucky numbers, the gambler will be paid \$12,000. What is the probability of this occurring?
4. If the keno player chooses 15 numbers and matches 5 or 6 of the lucky numbers, the gambler gets the bet back but is not paid any extra. What is the probability of this occurring?
5. Some casinos will let you choose up to 20 numbers. In this case, if you don't match any of the lucky numbers, the casino pays you! Although this may seem unusual, it is actually more difficult not to match any of the lucky numbers than it is to match a few of them. Compute the probability of not matching any of the lucky numbers at all, and compare it to the probability of matching 5 lucky numbers.
6. If 20 numbers are chosen, find the probability of matching *at least one* lucky number.

EXERCISE SET 12.4

1. What are mutually exclusive events? How do you calculate the probability of mutually exclusive events?
2. Give an example of two mutually exclusive events and an example of two events that are not mutually exclusive.

■ In Exercises 3 to 6, first verify that the compound event consists of two mutually exclusive events, and then compute the probability of the compound event occurring.

3. A single card is drawn from a standard deck of playing cards. Find the probability of drawing a 4 or an ace.
4. A single card is drawn from a standard deck of playing cards. Find the probability of drawing a heart or a club.
5. Two dice are rolled. Find the probability of rolling a 2 or a 10.
6. Two dice are rolled. Find the probability of rolling a 7 or an 8.
7. If $P(A) = 0.2$, $P(B) = 0.5$, and $P(A \text{ and } B) = 0.1$, find $P(A \text{ or } B)$.
8. If $P(A) = 0.6$, $P(B) = 0.4$, and $P(A \text{ and } B) = 0.2$, find $P(A \text{ or } B)$.
9. If $P(A) = 0.3$, $P(B) = 0.8$, and $P(A \text{ or } B) = 0.9$, find $P(A \text{ and } B)$.
10. If $P(A) = 0.7$, $P(A \text{ and } B) = 0.4$, and $P(A \text{ or } B) = 0.8$, find $P(B)$.

■ In Exercises 11 to 14, suppose you ask a friend to randomly choose an integer between 1 and 10, inclusive.

11. What is the probability that the number will be more than 6 or odd?
12. What is the probability that the number will be less than 5 or even?
13. What is the probability that the number will be even or prime?
14. What is the probability that the number will be prime or greater than 7?

■ In Exercises 15 to 20, two dice are rolled. Determine the probability of each of the following. ("Doubles" means that both dice show the same number.)

15. Rolling a 6 or doubles
16. Rolling a 7 or doubles
17. Rolling an even number or doubles
18. Rolling a number greater than 7 or doubles
19. Rolling an odd number or a number less than 6
20. Rolling an even number or a number greater than 9

■ In Exercises 21 to 26, a single card is drawn from a standard deck. Find the probability of each of the following events.

21. Drawing an 8 or a spade
22. Drawing an ace or a red card
23. Drawing a jack or a face card
24. Drawing a red card or a face card
25. Drawing a diamond or a black card
26. Drawing a spade or a red card

■ **Employment** In Exercises 27 to 30, use the data in the table below, which shows the employment status of individuals in a particular town by age group.

Age	Full-time	Part-time	Unemployed
0–17	24	164	371
18–25	185	203	148
26–34	348	67	27
35–49	581	179	104
≥50	443	162	173

27. If a person is randomly chosen from the town's population, what is the probability that the person is aged 26 to 34 or is employed part-time?
28. If a person is randomly chosen from the town's population, what is the probability that the person is at least 50 years old or unemployed?
29. If a person is randomly chosen from the town's population, what is the probability that the person is under 18 or employed part-time?
30. If a person is randomly chosen from the town's population, what is the probability that the person is 18 or older or employed full-time?
31. **Contests** If the probability of winning a particular contest is 0.04, what is the probability of not winning the contest?
32. **Weather** Suppose the probability that it will rain tomorrow is 0.38. What is the probability that it will not rain tomorrow?

Professional Sports The National Collegiate Athletic Association (NCAA) keeps statistics on the numbers of seniors playing on men's NCAA teams who are drafted to play on professional teams. Exercises 33 and 34 use some of these statistics.

33. There is a 1 in 75 chance that a senior on an NCAA basketball team will be drafted by a National Basketball Association (NBA) team. What is the probability that a senior NCAA basketball player will not be drafted to play on an NBA team?
34. The odds in favor of a senior on an NCAA ice hockey team being drafted by a National Hockey League (NHL) team are 1 to 25. What is the probability that a senior NCAA ice hockey player will not be drafted to play on an NHL team?

■ In Exercises 35 to 46, use the formula for the probability of the complement of an event.

35. Two dice are tossed. What is the probability of not tossing a 7?

36. Two dice are tossed. What is the probability of not getting doubles?

37. Two dice are tossed. What is the probability of getting a sum of at least 4?

38. Two dice are tossed. What is the probability of getting a sum of at most 11?

39. A single card is drawn from a deck. What is the probability of not drawing an ace?

40. A single card is drawn from a deck. What is the probability of not drawing a face card?

41. A coin is flipped 4 times. What is the probability of getting at least 1 tail?

42. A coin is flipped 4 times. What is the probability of getting at least 2 heads?

43. A single die is rolled 3 times. What is the probability that a 1 will show on the upward face at least once?

44. A single die is rolled 4 times. Find the probability that a 5 will be rolled at least once.

45. A pair of dice is rolled 3 times. What is the probability that a sum of 8 on the 2 dice will occur at least once?

46. A pair of dice is rolled 4 times. Compute the probability that a sum of 11 on the 2 dice will occur at least once.

47. A magician shuffles a standard deck of playing cards and allows an audience member to pull out a card, look at it, and replace it in the deck. Two additional people do the same. Find the probability that of the 3 cards drawn, at least 1 is a face card.

48. If a person draws 3 cards from a standard deck (without replacing them), what is the probability that at least 1 of the cards is a face card?

49. **E-Readers** An electronics store receives a shipment of 30 e-readers. Unbeknownst to the store, 4 of the e-readers are defective. If the store sells 12 of these e-readers the first day, what is the probability that at least 1 of the 12 buyers will get a defective e-reader?

AP Images/Michael Probst

50. **Blu-ray Players** An electronics store currently has 28 new Blu-ray players in stock, of which 5 are defective. If customers buy 3 Blu-ray players, what is the probability that at least 1 of them will be defective?

51. **Prize Drawing** Three employees of a restaurant each contributed 1 business card for a random drawing to win a prize. Forty-two business cards were received in all, and 3 cards will be drawn for prizes. Determine the probability that at least 1 of the restaurant employees will win a prize.

52. **Coins** A bag contains 44 U.S. quarters and 6 Canadian quarters. (The coins are identical in size.) If 5 quarters are randomly picked from the bag, what is the probability of getting at least 1 Canadian quarter?

EXTENSIONS

53. **Blackjack** In the game blackjack, a player is dealt 2 cards from a standard deck of playing cards. The player has a blackjack if one card is an ace and the other card is a 10, a jack, a queen, or a king. In some casinos, blackjack is played with more than 1 standard deck of playing cards. Does using more than 1 deck of cards change the probability of getting a blackjack?

54. **Monty Hall Problem** The *Monty Hall problem* is described in the Math Matters on page 723. The question arises, "If the contestant changes his or her original choice of door, what is the probability of choosing the door hiding the grand prize?" To answer this question, complete the following.

a. What is the probability that the contestant will choose the grand prize on the first try?

b. What is the probability that the contestant will not choose the grand prize on the first try?

c. What is the probability that the person will choose the grand prize by switching?

55. **Door Codes** A planned community has 300 homes, each with an automatic garage door opener operated by a code of 8 numbers. The homeowner sets each of the 8 numbers to be 0 or 1. For example, a door opener code might be 01101001. Assuming all the homes in the community are sold, what is the probability that at least 2 homeowners will set their door openers to use the same code and will therefore be able to open each other's garage doors?

56. Monty Hall Problem When someone is first presented with the Monty Hall problem mentioned in Exercise 54, there is a tendency for that person to think that switching his or her choice of door does not make any difference. You can actually simulate the Monty Hall game using playing cards and show that a player is twice as likely to win by switching from the first choice. To do this you will need at least 2 people and 3 cards, say the ace of spades to represent the grand prize, and the 2 of hearts and the 2 of diamonds, which represent the other 2 prizes. Shuffle the cards and place them face down on the table. Choose 1 of the cards. The other player picks up the remaining 2 cards, removes 1 of the cards that is not the ace (it is possible that both cards will not be the ace), and places the other card face down on the table. Now you may either stay with your original selection or change to the remaining card. Perform this experiment 30 times staying with your original selection and 30 times switching to the other card. Keep a record of how many times staying with your original selection resulted in selecting the ace and how many times switching cards resulted in choosing the ace. On the basis of your results, is staying or switching the better strategy?

57. Monty Hall Problem The benefit of switching doors in the Monty Hall problem is even more dramatic if there are more hidden prizes from which to choose. Instead of using 3 cards as in Exercise 56, use 5 cards, the four 2s and the ace of spades. Shuffle them well, place them face down on a table, and choose 1 card. Another person then picks up the remaining 4 cards, removes three 2s, and places the remaining card face down on the table. Now you may either stay with your original selection or switch to the remaining card. Perform this experiment 30 times staying with your original selection and 30 switching to the other card. Keep a record of how many times staying with your original selection resulted in selecting the ace and how many times changing resulted in choosing the ace. On the basis of your results, is staying or switching the better strategy? About how many more times did switching result in choosing the ace than did staying with your original choice?

58. Poker In 5-card stud poker, a hand containing 5 cards of the same suit from a standard deck is called a flush. In this exercise, you will compute the probability of getting a flush.

a. How many 5-card poker hands are possible?

b. How many 5-card poker hands are possible that contain all spades?

c. What is the probability of getting a 5-card poker hand containing all spades?

d. What is the probability of getting a 5-card poker hand containing all hearts? All diamonds? All clubs?

e. Are the events of getting 5 spades, 5 hearts, 5 diamonds, or 5 clubs mutually exclusive?

f. What is the probability of getting a flush in 5-card stud poker?

SECTION 12.5 Conditional Probability

Conditional Probability

	F	No F	Total
V	21	198	219
No V	76	195	271
Total	97	393	490

V: Vaccinated
F: Contracted the flu

In the preceding section, we discussed the effectiveness of a flu vaccination in preventing the onset of the flu. The table from that discussion is shown again at the left. From the table, we can calculate the probability that one person randomly selected from this population will contract the flu.

$$P(F) = \frac{n(F)}{n(S)} = \frac{97}{490} \approx 0.198$$

• $n(F) = 97, n(S) = 490$ (S denotes the sample space)

Now consider a slightly different situation. We could ask, "What is the probability that a person randomly chosen from this population will contract the flu *given* that the person received the flu vaccination?"

In this case, we know that the person received a flu vaccination, and we want to determine the probability that the person will contract the flu. Therefore, the only part of the table that is of concern to us is the top row. In this case, we have

	F	No F	Total
V	21	198	219

$$P(F \text{ given } V) = \frac{21}{219} \approx 0.096$$

Thus the probability that an individual will contract the flu given that the individual has been vaccinated is about 0.096.

The probability of an event B occurring given that some other event A has already occurred is called a **conditional probability** and is denoted $P(B|A)$.

Conditional Probability Formula

If A and B are two events in a sample space S, then the conditional probability of B given that A has occurred is

$$P(B|A) = \frac{P(A \text{ and } B)}{P(A)}$$

The symbol $P(B|A)$ is read "the probability of B given A."

To see how this formula applies to the flu data on the preceding page, let

$S = \{\text{all people participating in the test}\}$

$F = \{\text{people who contracted the flu}\}$

$V = \{\text{people who were vaccinated}\}$

Then $F \cap V = \{\text{people who contracted the flu } \textit{and} \text{ were vaccinated}\}$.

$$P(F|V) = \frac{P(F \text{ and } V)}{P(V)} = \frac{\frac{21}{490}}{\frac{219}{490}}$$

- $P(F \text{ and } V) = \dfrac{n(F \cap V)}{n(S)} = \dfrac{21}{490}$
- $P(V) = \dfrac{n(V)}{n(S)} = \dfrac{219}{490}$

$$= \frac{21}{219} \approx 0.096$$

The probability that one person selected from this population contracted the flu given that the person received the vaccination, $P(F|V)$, is about 0.096. Our answer agrees with the calculation we performed directly from the table, but the Conditional Probability Formula enables us to find conditional probabilities even when we cannot compute them directly.

QUESTION In the preceding example, what is the interpretation of $P(V|F)$?

EXAMPLE 1 Determine a Conditional Probability

The data in the table below show the results of a survey used to determine the number of adults who received financial help from their parents for certain purchases.

Age	College tuition	Buy a car	Buy a house	Total
18–28	405	253	261	919
29–39	389	219	392	1000
40–49	291	146	245	682
50–59	150	71	112	333
≥ 60	62	15	98	175
Total	1297	704	1108	3109

ANSWER $P(V|F)$ is the probability that a person selected from this population has been vaccinated, given that the person contracted the flu.

If one person is selected from this survey, what is the probability that the person received financial help to purchase a home, given that the person is between the ages of 29 and 39?

Solution

Let $B = \{$adults who received financial help for a home purchase$\}$ and $A = \{$adults between 29 and 39$\}$. From the table, $n(A \cap B) = 392$, $n(A) = 1000$, and $n(S) = 3109$. Using the Conditional Probability Formula, we have

$$P(B|A) = \frac{P(A \text{ and } B)}{P(A)} = \frac{\frac{392}{3109}}{\frac{1000}{3109}}$$

- $P(A \text{ and } B) = \frac{n(A \cap B)}{n(S)} = \frac{392}{3109}$
- $P(A) = \frac{n(A)}{n(S)} = \frac{1000}{3109}$

$$= \frac{392}{1000} = 0.392$$

The probability that a person received financial help to purchase a home, given that the person is between the ages of 29 and 39, is 0.392.

CHECK YOUR PROGRESS 1 Two dice are tossed, one after the other. What is the probability that the result is a sum of 6, given that the first toss is not a 3?

Solution See page S45.

MATH MATTERS Bayes' Theorem and Bertrand's Box Paradox

HISTORICAL NOTE

Thomas Bayes was born in London and became a Nonconformist minister who also dabbled in mathematics. Of the mathematical papers he wrote, only one was published in his lifetime, and that paper was published anonymously. His work on probability was not discovered until after his death.

Bayes' Theorem, named after Thomas Bayes (1702–1761), gives a method of computing conditional probabilities by knowing the reverse conditional probabilities. Consider the following problem, known as Bertrand's Box Paradox. Three boxes each contain 2 coins. One box has 2 gold coins, another has 2 silver coins, and the third box has 1 gold coin and 1 silver coin. A box is chosen at random, and a coin is pulled out (without looking at the other coin). If the coin that is taken out is gold, what is the probability that the other coin in the same box is also gold?

Many people initially guess that the probability is $\frac{1}{2}$. To check, let B_1 be the event that the first box (with 2 gold coins) is chosen, B_2 the event that the second box (with 2 silver coins) is chosen, and B_3 the event that the third box (with 1 silver and 1 gold coin) is chosen. In addition, let G represent the event that a gold coin is pulled from a box, and S the event that a silver coin is chosen. The conditional probability of pulling a gold coin from a box, given that we know which box was chosen, is simple to compute. What we want to find, however, is the reverse: the conditional probability that we have chosen the first box, given that a gold coin was chosen, $P(B_1|G)$. Bayes' Theorem gives us a way to find this probability. In this case, the theorem states

$$P(B_1|G) = \frac{P(G|B_1)P(B_1)}{P(G|B_1)P(B_1) + P(G|B_2)P(B_2) + P(G|B_3)P(B_3)}$$

$$= \frac{1 \cdot \frac{1}{3}}{1 \cdot \frac{1}{3} + 0 \cdot \frac{1}{3} + \frac{1}{2} \cdot \frac{1}{3}} = \frac{2}{3}$$

Thus, there is actually a $\frac{2}{3}$ chance that if a gold coin is pulled from a box, the other coin in the box is also gold.

Product Rule for Probabilities

Suppose that 2 cards are drawn, without replacement, from a standard deck of playing cards. Let A be the event that an ace is drawn on the first draw and B the event that an ace is drawn on the second draw. Then the probability that an ace is drawn on the first *and* second draws is $P(A \text{ and } B)$. To find this probability, we can solve the conditional probability formula for $P(A \text{ and } B)$.

$$\frac{P(A \text{ and } B)}{P(A)} = P(B|A)$$

$$P(A) \cdot \frac{P(A \text{ and } B)}{P(A)} = P(A) \cdot P(B|A) \qquad \text{• Multiply each side by } P(A).$$

$$P(A \text{ and } B) = P(A) \cdot P(B|A)$$

This is called the Product Rule for Probabilities.

Product Rule for Probabilities

If A and B are two events in a sample space S, then

$$P(A \text{ and } B) = P(A) \cdot P(B|A)$$

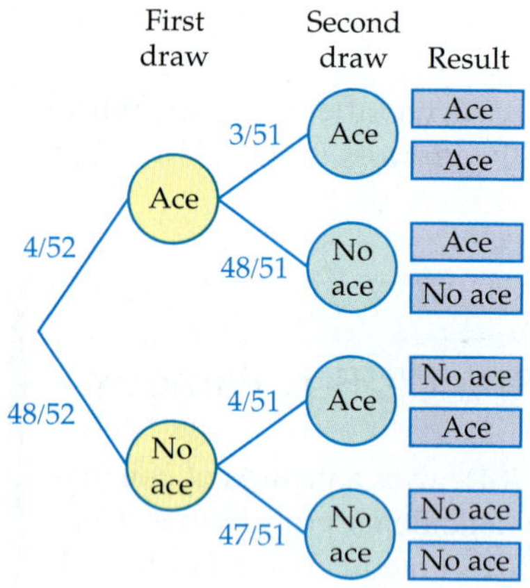

For the problem just given, $P(A \text{ and } B)$ is the product of $P(A)$, the probability that the first card drawn is an ace, and $P(B|A)$, the probability of an ace on the second draw *given* that the first card drawn was an ace.

The tree diagram at the left shows the possible outcomes of drawing 2 cards from a deck without replacement. On the first draw, there are 4 aces in the deck of 52 cards. Therefore, $P(A) = \frac{4}{52} = \frac{1}{13}$. On the second draw, there are only 51 cards remaining and only 3 aces (an ace was drawn on the first draw). Therefore, $P(B|A) = \frac{3}{51} = \frac{1}{17}$. Putting these calculations together, we have

$$P(A \cap B) = P(A) \cdot P(B|A)$$
$$= \frac{1}{13} \cdot \frac{1}{17} = \frac{1}{221}$$

The probability of drawing an ace on the first and second draws is $\frac{1}{221}$.

The Product Rule for Probabilities can be extended to more than two events. The probability that a certain sequence of events will occur in succession is the product of the probabilities of each of the events *given* that the preceding events have occurred.

Probability of Successive Events

The probability of two or more events occurring in succession is the product of the conditional probabilities of each of the events.

TAKE NOTE

In part a, there are originally 12 balls in the box. After a red ball is selected, there are only 11 balls remaining, of which 3 are white. Thus $P(B|A) = \frac{3}{11}$. After a red ball and a white ball are selected, there are 10 balls left, of which 5 are green. Thus $P(C|A \text{ and } B) = \frac{5}{10}$.

In part b, we have a similar situation. However, after a white ball is selected, there are 11 balls remaining, of which only 2 are white. Therefore, $P(B|A) = \frac{2}{11}$.

EXAMPLE 2 Find the Probability of Successive Events

A box contains 4 red, 3 white, and 5 green balls. Suppose that 3 balls are randomly selected from the box in succession, without replacement.

a. What is the probability that first a red, then a white, and then a green ball are selected?

b. What is the probability that 2 white balls followed by 1 green ball are selected?

Solution

a. Let $A = \{\text{a red ball is selected first}\}$, $B = \{\text{a white ball is selected second}\}$, and $C = \{\text{a green ball is selected third}\}$. Then

$$P(A \text{ followed by } B \text{ followed by } C) = P(A) \cdot P(B|A) \cdot P(C|A \text{ and } B)$$
$$= \frac{4}{12} \cdot \frac{3}{11} \cdot \frac{5}{10}$$
$$= \frac{1}{22}$$

The probability of choosing a red, then a white, then a green ball is $\frac{1}{22}$.

b. Let $A = \{\text{a white ball is selected first}\}$, $B = \{\text{a white ball is selected second}\}$, and $C = \{\text{a green ball is selected third}\}$. Then

$$\begin{aligned} P(A \text{ followed by } B \text{ followed by } C) &= P(A) \cdot P(B \mid A) \cdot P(C \mid A \text{ and } B) \\ &= \frac{3}{12} \cdot \frac{2}{11} \cdot \frac{5}{10} \\ &= \frac{1}{44} \end{aligned}$$

The probability of choosing 2 white balls followed by 1 green ball is $\frac{1}{44}$.

CHECK YOUR PROGRESS 2 A standard deck of playing cards is shuffled and 3 cards are dealt. Find the probability that the cards dealt are a spade followed by a heart followed by another spade.

Solution See page S45.

Independent Events

Earlier in this section we considered the probability of drawing 2 aces in a row from a standard deck of playing cards. Because the cards were drawn without replacement, the probability of an ace on the second draw *depended* on the result of the first draw.

Now consider the case of tossing a coin twice. The outcome of the first coin toss has no effect on the outcome of the second toss. So the probability of the coin flipping to a head or a tail on the second toss is not affected by the result of the first toss. When the outcome of a first event does not affect the outcome of a second event, the events are called *independent*.

Independent Events

If A and B are two events in a sample space and $P(B \mid A) = P(B)$, then A and B are called **independent events**.

For a mathematical verification, consider tossing a coin twice. We can compute the probability that the second toss comes up heads, given that the first coin toss came up heads. If A is the event of a head on the first toss, then $A = \{\text{HH, HT}\}$. Let B be the event of a head on the second toss. Then $B = \{\text{HH, TH}\}$. The sample space is $S = \{\text{HH, HT, TH, TT}\}$. The conditional probability $P(B \mid A)$ (the probability of a head on the second toss given a head on the first toss) is

$$P(B \mid A) = \frac{P(A \text{ and } B)}{P(A)} = \frac{\frac{1}{4}}{\frac{1}{2}} = \frac{1}{2}$$

- $P(A \text{ and } B) = \dfrac{n(A \cap B)}{n(S)} = \dfrac{1}{4}$
- $P(A) = \dfrac{n(A)}{n(S)} = \dfrac{2}{4} = \dfrac{1}{2}$

Thus $P(B \mid A) = \frac{1}{2}$. Note, however, that $P(B) = \frac{n(B)}{n(S)} = \frac{2}{4} = \frac{1}{2}$. Therefore, in the case of tossing a coin twice, the probability of the second event does not depend on the outcome of the first event, and we have $P(B \mid A) = P(B)$.

In general, this result enables us to simplify the product rule when two events are independent; the probability of two independent events occurring in succession is simply the product of the probabilities of the individual events.

TAKE NOTE

The product rule for independent events can be extended to more than 2 events. If E_1, E_2, E_3, and E_4 are independent events, then the probability that all 4 events will occur is $P(E_1) \cdot P(E_2) \cdot P(E_3) \cdot P(E_4)$.

Product Rule for Independent Events

If A and B are two independent events from the sample space S, then $P(A \text{ and } B) = P(A) \cdot P(B)$.

EXAMPLE 3 Find the Probability of Independent Events

A pair of dice is tossed twice. What is the probability that the first roll is a sum of 7 and the second roll is a sum of 11?

TAKE NOTE

See page 689 for all the possible outcomes of the roll of 2 dice.

Solution

The rolls of a pair of dice are independent; the probability of a sum of 11 on the second roll does not depend on the outcome of the first roll. Let $A = \{\text{sum of 7 on the first roll}\}$ and $B = \{\text{sum of 11 on the second roll}\}$. Then

$$P(A \text{ and } B) = P(A) \cdot P(B) = \frac{6}{36} \cdot \frac{2}{36} = \frac{1}{108}$$

CHECK YOUR PROGRESS 3 A coin is tossed 3 times. What is the probability that heads appears on all 3 tosses?

Solution See page S45.

Applications of Conditional Probability

Conditional probability is used in many real-world situations, such as to determine the efficacy of a drug test, to verify the accuracy of genetic testing, and to analyze evidence in legal proceedings.

EXAMPLE 4 Drug Testing and Conditional Probability

Suppose a company claims it has a test that is 95% effective in determining whether an athlete is using a steroid. That is, if an athlete is using a steroid, the test will be positive 95% of the time. In the case of a negative result, the company says its test is 97% accurate. That is, even if an athlete is not using steroids, it is possible that the test will be positive in 3% of the cases. Such an occurrence is called a **false positive**. Suppose this test is given to a group of athletes in which 10% of the athletes are using steroids. What is the probability that a randomly chosen athlete actually uses steroids, given that the athlete's test is positive?

Solution

Let S be the event that an athlete uses steroids and let T be the event that the test is positive. Then the probability we wish to determine is $P(S \mid T)$. Using the Conditional Probability Formula, we have

$$P(S \mid T) = \frac{P(S \text{ and } T)}{P(T)}$$

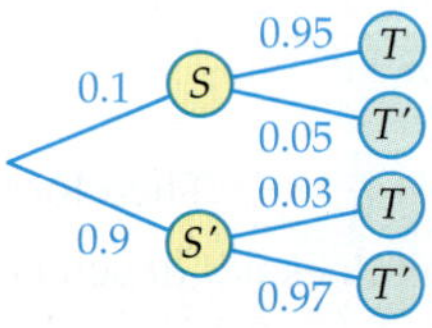

A tree diagram, shown at the left, can be used to calculate this probability. A positive test result can occur in two ways: either an athlete using steroids correctly tests positive, or an athlete not using steroids incorrectly tests positive. The probability of a positive test result, $P(T)$, corresponds to an athlete following path ST or path $S'T$ in the tree diagram. (S' symbolizes no steroid use and T' symbolizes a negative result.) $P(S \text{ and } T)$, the probability of using steroids and getting a positive test result, is path ST. Thus,

$$P(S \mid T) = \frac{P(S \text{ and } T)}{P(T)}$$

$$= \frac{(0.1)(0.95)}{(0.1)(0.95) + (0.9)(0.03)} \approx 0.779$$

Given that an athlete tests positive, the probability that the athlete actually uses steroids is approximately 77.9%.

POINT OF INTEREST

The calculation in Example 4 is actually an illustration of Bayes' Theorem, described in the Math Matters on page 729.

This result is probably lower than you might expect considering that the manufacturer claims its test is 95% accurate! In fact, the prevalence of false-positive test results can be much more dramatic in cases of conditions that are present in a small percentage of the population. (See Exercise 61, page 736.)

CHECK YOUR PROGRESS 4 A pharmaceutical company has a test that is 95% effective in determining whether a person has a certain genetic defect. However, the test may give a false positive result in 4% of cases. Suppose this particular genetic defect occurs in 2% of the population. Given that a person tests positive, what is the probability that the person actually has the defect?

Solution See page S45.

EXCURSION

Sharing Birthdays

Have you ever been introduced to someone at a party or other social gathering and discovered that you share the same birthday? It seems like an amazing coincidence, but exactly how rare is this coincidence?

As an example, suppose 4 people have gathered for a dinner party. We can determine the probability that at least 2 of the guests have the same birthday. (For simplicity, we will ignore the February 29th birthday from leap years.) Let E be the event that at least 2 people share a birthday. In this case, it is easier to look at the complement E^C, the event that no one shares the same birthday.

If we start with 1 of the guests, then the second guest cannot share the same birthday, so that person has 364 possible dates for his or her birthday from a total of 365. Thus the conditional probability that the second guest has a different birthday, given that we know the first person's birthday, is $\frac{364}{365}$. Similarly, the third person has 363 possible birthday dates that do not coincide with those of the first two guests. So the conditional probability that the third person does not share a birthday with either of the first 2 guests, given that we know the birthdays of the first 2 people, is $\frac{363}{365}$. The probability of the fourth guest having a distinct birthday is, similarly, $\frac{362}{365}$. We can use the Product Rule for Probabilities to find the probability that all of these conditions are met; that is, none of the 4 guests share a birthday.

Elena Elisseeva/Shutterstock.com

$$P(E^c) = \frac{364}{365} \cdot \frac{363}{365} \cdot \frac{362}{365} \approx 0.984$$

Then $P(E) = 1 - P(E^c) \approx 0.016$, so there is about a 1.6% chance that in a group of 4 people, 2 or more will have the same birthday.

It would require 366 people gathered together to *guarantee* that 2 people in the group will have the same birthday. But how many people would be necessary to guarantee that the chance of at least 2 of them sharing a birthday is at least 50/50? Make a guess before you proceed through the exercises. The results may surprise you!

INSTRUCTOR NOTE

You can introduce this Excursion by first asking the class how likely they think it is that 2 of the students in the class have the same birthday. Then ask each student when his or her birthday is and determine if any of the students share a birthday. (If you have a large number of students, you can proceed month by month.) If 2 or more students do have a common birthday, it can serve as evidence that perhaps a shared birthday is not such a rare occurrence in a relatively small group of people.

EXCURSION EXERCISES

1. If 8 people are present at a meeting, find the probability that at least 2 share a common birthday.
2. Compute the probability that at least 2 people among a group of 15 have the same birthday.

3. If 23 people are in attendance at a party, what is the probability that at least 2 share a birthday?

4. In a group of 40 people, what would you estimate to be the probability that at least 2 people share a birthday? If you have the patience, compute the probability to check your guess.

EXERCISE SET 12.5

1. What is a conditional probability?

2. Explain the difference between independent events and dependent events.

■ In Exercises 3 to 6, compute the conditional probabilities $P(A \mid B)$ and $P(B \mid A)$.

3. $P(A) = 0.7$, $P(B) = 0.4$, $P(A \text{ and } B) = 0.25$

4. $P(A) = 0.45$, $P(B) = 0.8$, $P(A \text{ and } B) = 0.3$

5. $P(A) = 0.61$, $P(B) = 0.18$, $P(A \text{ and } B) = 0.07$

6. $P(A) = 0.2$, $P(B) = 0.5$, $P(A \text{ and } B) = 0.2$

■ **Employment** In Exercises 7 to 10, use the data in the table below, which shows the employment status of individuals in a particular town by age group.

	Full-time	Part-time	Unemployed
0–17	24	164	371
18–25	185	203	148
26–34	348	67	27
35–49	581	179	104
≥50	443	162	173

7. If a person in this town is selected at random, find the probability that the individual is employed part-time, given that he or she is between the ages of 35 and 49.

8. If a person in the town is randomly selected, what is the probability that the individual is unemployed, given that he or she is 50 years old or older?

9. A person from the town is randomly selected; what is the probability that the individual is employed full-time, given that he or she is between 18 and 49 years of age?

10. A person from the town is randomly selected; what is the probability that the individual is employed part-time, given that he or she is at least 35 years old?

■ **Video Games** In Exercises 11 to 14, use the data in the following table, which shows the results of a survey of 2000 gamers concerning their favorite home video game systems, organized by age group. If a survey participant is selected at random, determine the probability of each of the following. Round to the nearest hundredth.

	Nintendo Wii U	Microsoft Xbox One	Sega Genesis	Sony Playstation 4
0–12	63	84	55	51
13–18	105	139	92	113
19–24	248	217	83	169
≥25	191	166	88	136

11. The participant prefers the Nintendo Wii U system.

12. The participant prefers the Xbox One, given that the person is between the ages of 13 and 18.

13. The participant prefers Sega Genesis, given that the person is between the ages of 13 and 24.

14. The participant is under 12 years of age, given that the person prefers the Playstation 4 system.

15. A pair of dice is tossed. Find the probability that the sum on the 2 dice is 8, given that the sum is even.

16. A pair of dice is tossed. Find the probability that the sum on the 2 dice is 12, given that doubles are rolled.

17. A pair of dice is tossed. What is the probability that doubles are rolled, given that the sum on the 2 dice is less than 7?

18. A pair of dice is tossed. What is the probability that the sum on the 2 dice is 8, given that the sum is more than 6?

19. What is the probability of drawing 2 cards in succession (without replacement) from a standard deck and having them both be face cards? Round to the nearest thousandth.

20. Two cards are drawn from a standard deck without replacement. Find the probability that both cards are hearts. Round to the nearest thousandth.

21. Two cards are drawn from a standard deck without replacement. What is the probability that the first card is a spade and the second card is red? Round to the nearest thousandth.

22. Two cards are drawn from a standard deck without replacement. What is the probability that the first card is a king and the second card is not? Round to the nearest thousandth.

■ **Candy Colors** In Exercises 23 to 26, a snack-size bag of M&Ms candies is opened. Inside, there are 12 red candies, 12 blue, 7 green, 13 brown, 3 orange, and 10 yellow. Three candies are pulled from the bag in succession, without replacement.

23. Determine the probability that the first candy drawn is blue, the second is red, and the third is green.

24. Determine the probability that the first candy drawn is brown, the second is orange, and the third is yellow.

25. What is the probability that the first two candies drawn are green and the third is red?

26. What is the probability that the first candy drawn is orange, the second is blue, and the third is orange?

■ In Exercises 27 to 30, 3 cards are dealt from a shuffled standard deck of playing cards.

27. Find the probability that the first card dealt is red, the second is black, and the third is red.

28. Find the probability that the first 2 cards dealt are clubs and the third is a spade.

29. What is the probability that the 3 cards dealt are, in order, an ace, a face card, and an 8? (A face card is a jack, queen, or king.)

30. What is the probability that the 3 cards dealt are, in order, a red card, a club, and another red card?

■ **Student Attendance** In Exercises 31 to 34, the probability that a student enrolled at a local high school will be absent on a particular day is 0.04, assuming that the student was in attendance the previous school day. However, if a student is absent, the probability that he or she will be absent again the following day is 0.11. For each exercise, assume that the student was in attendance the previous day.

31. What is the probability that a student will be absent 3 days in a row?

32. What is the probability that a student will be absent 2 days in a row but then show up on the third day?

33. Find the probability that a student will be absent, attend the next day, but then be absent again the third day.

34. Find the probability that a student will be absent 4 days in a row.

■ In Exercises 35 to 38, determine whether the events are independent.

35. A single die is rolled and then rolled a second time.

36. Numbered balls are pulled from a bin one by one to determine the winning lottery numbers.

37. Numbers are written on slips of paper in a hat; 1 person pulls out a slip of paper without replacing it, then a second person pulls out a slip of paper.

38. In order to determine who goes first in a game, 1 person picks a number between 1 and 10, then a second person picks a number from the remaining 9 numbers.

■ In Exercises 39 to 44, a pair of dice is tossed twice.

39. Find the probability that both rolls give a sum of 8.

40. Find the probability that the first roll is a sum of 6 and the second roll is a sum of 12.

41. Find the probability that the first roll is a total of at least 10 and the second roll is a total of at least 11.

42. Find the probability that both rolls result in doubles.

43. Find the probability that both rolls give even sums.

44. Find the probability that both rolls give at most a sum of 4.

45. A fair coin is tossed 4 times in succession. Find the probability of getting 2 heads followed by 2 tails.

46. **Monopoly** In the game of Monopoly, a player is sent to jail if he or she rolls doubles with a pair of dice 3 times in a row. What is the probability of rolling doubles 3 times in succession?

CaseyMartin/Shutterstock.com

47. Find the probability of tossing a pair of dice 3 times in succession and getting a sum of at least 10 on all 3 tosses.

48. Find the probability of tossing a pair of dice 3 times in succession and getting a sum of at most 3 on all 3 tosses.

■ In Exercises 49 to 54, a card is drawn from a standard deck and replaced. After the deck is shuffled, another card is pulled.

49. What is the probability that both cards pulled are aces?

50. What is the probability that both cards pulled are face cards?

51. What is the probability that the first card drawn is a spade and the second card is a diamond?

52. What is the probability that the first card drawn is an ace and the second card is not an ace?

53. Find the probability that the first card drawn is a heart and the second card is a spade.

54. Find the probability that the first card drawn is a face card and the second card is black.

55. A standard deck of playing cards is shuffled, and 3 people each choose a card. Find the probability that the first 2 cards chosen are diamonds and the third card is black if

a. the cards are chosen *with* replacement.

b. the cards are chosen *without* replacement.

56. A standard deck of playing cards is shuffled and 3 people each choose a card. Find the probability that all 3 cards are face cards if

a. the cards are chosen *with* replacement.

b. the cards are chosen *without* replacement.

57. A bag contains 5 red marbles, 4 green marbles, and 8 blue marbles. Find the probability of pulling 2 red marbles followed by a green marble if the marbles are pulled from the bag

a. with replacement.

b. without replacement.

58. A box contains 3 medium t-shirts, 5 large t-shirts, and 4 extra-large t-shirts. If someone randomly chooses 3 t-shirts from the box, find the probability that the first t-shirt is large, the second is medium, and the third is large if the shirts are chosen

a. with replacement. **b.** without replacement.

59. Drug Testing A company that performs drug testing guarantees that its test determines a positive result with 97% accuracy. However, the test also gives 6% false positives. If 5% of those being tested actually have the drug present in their bloodstream, find the probability that a person testing positive has actually been using drugs. Round to the nearest thousandth.

60. Genetic Testing A test for a genetic disorder can detect the disorder with 94% accuracy. However, the test will incorrectly report positive results for 3% of those without the disorder. If 12% of the population has the disorder, find the probability that a person testing positive actually has the genetic disorder. Round to the nearest thousandth.

61. Disease Testing A pharmaceutical company has developed a test for a rare disease that is present in 0.5% of the population. The test is 98% accurate in determining a positive result, and the chance of a false positive is 4%. What is the probability that someone who tests positive actually has the disease? Round to the nearest thousandth.

EXTENSIONS

62. Suppose you are standing at a street corner and flip a coin to decide whether you will go north or south from your current position. When you reach the next intersection, you repeat the procedure. (This problem is a simplified version of what is called a *random walk* problem. Problems of this type are important in economics, physics, chemistry, biology, and other disciplines.)

a. After performing this experiment 3 times, what is the probability that you will be 3 blocks north of your original position?

b. After performing this experiment 4 times, what is the probability that you will be 2 blocks north of your original position?

c. After performing this experiment 4 times, what is the probability that you will be back at your original position?

63. Monty Hall Problem This is another explanation, using conditional probability, of the Monty Hall problem discussed in Exercise 54 of the last section. Let the door chosen by the contestant be labeled 1 and the other two doors be labeled 2 and 3. In the following exercises, we will use A to represent the event that the grand prize is behind door 1, B to represent the event that the prize is behind door 2, and C to represent the event that the prize is behind door 3. We will use $\overline{A}$ to represent that Monty Hall opens door 1, $\overline{B}$ to represent that he opens door 2, and $\overline{C}$ to represent that he opens door 3.

a. What is the probability that Monty Hall opens door 2 given that the grand prize is behind door 1? This is $P(\overline{B}\,|\,A)$.

b. What is the probability that Monty Hall opens door 2 given that the grand prize is behind door 2? This is $P(\overline{B}\,|\,B)$.

c. What is the probability that Monty Hall opens door 2 given that the grand prize is behind door 3? This is $P(\overline{B}\,|\,C)$.

d. The probability that the grand prize is behind door 1 given that door 2 is opened (you have not switched choices) is given by Bayes' Theorem in the following form.

$$P(A\,|\,\overline{B}) = \frac{P(\overline{B}\,|\,A)P(A)}{P(\overline{B}\,|\,A)P(A) + P(\overline{B}\,|\,B)P(B) + P(\overline{B}\,|\,C)P(C)}$$

What is the probability?

e. Find the probability of choosing the grand prize if you switch doors. That is, find

$$P(C\,|\,\overline{B}) = \frac{P(\overline{B}\,|\,C)P(C)}{P(\overline{B}\,|\,A)P(A) + P(\overline{B}\,|\,B)P(B) + P(\overline{B}\,|\,C)P(C)}$$

f. Is switching the better strategy?

SECTION 12.6 Expectation

Expectation

Suppose a barrel contains a large number of balls, half of which have the number 1000 painted on them and the other half of which have the number 500 painted on them. As the grand prize winner of a contest, you get to reach into the barrel (blindfolded, of course) and select 10 balls. Your prize is the sum of the numbers on the balls in cash.

If you are very lucky, all of the balls will have 1000 painted on them, and you will win \$10,000. If you are very unlucky, all of the balls will have 500 painted on them, and you will win \$5000. Most likely, however, approximately one-half of the balls will have 1000 painted on them and one-half will have 500 painted on them. The amount of your winnings in this case will be $5(1000) + 5(500)$, or \$7500. Because 10 balls are drawn, your amount of winnings per ball is $\frac{\$7500}{10} = \$750$.

The number \$750 is called the *expected value* or *expectation* of the game. You cannot win \$750 on one draw, but if given the opportunity to draw many times, you will win, on average, \$750 per ball.

QUESTION Can an expectation be negative?

We can also calculate expectation by using probabilities. For the game above, one-half of the balls have the number 1000 painted on them and one-half have the number 500 painted on them. Therefore, $P(1000) = \frac{1}{2}$ and $P(500) = \frac{1}{2}$. The expectation is calculated as follows.

Expectation

$$= (\text{probability of winning } \$1000) \cdot \$1000 + (\text{probability of winning } \$500) \cdot \$500$$

$$= P(1000) \cdot \$1000 + P(500) \cdot \$500$$

$$= \frac{1}{2} \cdot \$1000 + \frac{1}{2} \cdot \$500 = \$500 + \$250 = \$750$$

The general result for experiments with numerical outcomes follows.

INSTRUCTOR NOTE

You can point out to students that expectation is like a weighted average. We are finding the average value of the outcomes, but each outcome is weighted by the probability of its occurrence.

Expectation

Let $S_1, S_2, S_3, \ldots, S_n$ be the possible numerical outcomes of an experiment, and let $P(S_1), P(S_2), P(S_3), \ldots, P(S_n)$ be the probabilities of those outcomes. Then the **expectation** of the experiment is

$$P(S_1) \cdot S_1 + P(S_2) \cdot S_2 + P(S_3) \cdot S_3 + \cdots + P(S_n) \cdot S_n$$

That is, to find the expectation of an experiment, multiply the probability of each outcome of the experiment by the outcome and then add the results.

Photodisc/Getty Images

EXAMPLE 1 Expectation in Gambling

One of the wagers in roulette is to place a bet on 1 of the numbers from 0 to 36 or on 00. If that number comes up, the player wins 35 times the amount bet (and keeps the original bet). Suppose a player bets \$1 on a number. What is the player's expectation?

ANSWER Yes. For instance, if the expected value of a gambling game is negative, it simply means that, in the long run, a person will lose that amount of money, on average, on each play.

Solution

Let S_1 be the event that the player's number comes up and the player wins \$35. Because there are 38 numbers from which to choose, $P(S_1) = \frac{1}{38}$. Let S_2 be the event that the player's number does not come up and the player therefore loses \$1. Then $P(S_2) = 1 - \frac{1}{38} = \frac{37}{38}$.

$$\begin{aligned}\text{Expectation} &= P(S_1) \cdot S_1 + P(S_2) \cdot S_2 \\ &= \frac{1}{38}(35) + \frac{37}{38}(-1) = -\frac{1}{19} \\ &\approx -0.053\end{aligned}$$

• The amount the player can win is entered as a positive number. The amount that can be lost is entered as a negative number.

The player's expectation is approximately $-\$0.053$. This means that, on average, the player will lose about \$0.05 every time this bet is made.

TAKE NOTE

In Example 1, suppose the player bets \$5 instead of \$1. The payoff is then \$175 $(35 \cdot 5)$ if the player wins and $-\$5$ if the player loses. The expectation is $\frac{1}{38}(175) + \frac{37}{38}(-5) = -\frac{5}{19}$. Note that the bet is 5 times greater and the expectation, $-\frac{5}{19}$, is 5 times the expectation when \$1 is bet. Thus a player who makes \$5 bets can expect to lose 5 times as much money as a player who makes \$1 bets.

CHECK YOUR PROGRESS 1 In roulette it is possible to place a wager that 1 of the numbers between 1 and 12 (inclusive) will come up. If it does, the player wins twice the amount bet. Suppose a player bets \$5 that a number between 1 and 12 will come up. What is the player's expectation?

Solution See page S46.

In Example 1, the fact that the player is losing approximately 5 cents on each dollar bet means that the casino's expectation is positive 5 cents; it is earning (on average) 5 cents for every dollar spent making that particular bet at the roulette wheel. An individual player may get lucky, but over time the casino can plan on a predictable profit.

MATH MATTERS A Bargain at Any Price?

HISTORICAL NOTE

Daniel Bernoulli (bər-nōō′lē) (1700–1782) was the son of Jean Bernoulli I and the nephew of Jacques Bernoulli. For a time, he was a professor in St. Petersburg, Russia, where he collaborated with Leonhard Euler. There he wrote a paper on probability and expectation in which he discussed the game described at the right, now known as the St. Petersburg Paradox.

The Swiss mathematician Daniel Bernoulli discussed the following game in a paper he published in 1738. A fair coin is repeatedly tossed until the coin comes up tails. Let $n =$ the number of total coin flips. When the coin comes up tails, you are paid 2^n dollars. Thus, if the first flip is tails, you are paid \$2. If the coin comes up heads 5 times in a row and the sixth flip comes up tails, you are paid $2^6 = \$64$. How much would you pay to play such a game? \$5? \$20? The game becomes interesting when we compute the expected value:

$$\begin{aligned}&\text{Expectation}\\ &= P(\mathrm{T}) \cdot 2^1 + P(\mathrm{HT}) \cdot 2^2 + P(\mathrm{HHT}) \cdot 2^3 + P(\mathrm{HHHT}) \cdot 2^4 + \cdots \\ &= \frac{1}{2} \cdot 2 + \frac{1}{4} \cdot 4 + \frac{1}{8} \cdot 8 + \frac{1}{16} \cdot 16 + \cdots \\ &= 1 + 1 + 1 + 1 + \cdots\end{aligned}$$

There is no maximum number of times the coin can be flipped, and the expectation is infinite! Theoretically, it would be worthwhile to play no matter how high the fee is.

Business Applications of Expectation

When an insurance company sells a life insurance policy, the premium (the cost to purchase the policy) is based to a large extent on the probability that the insured person will outlive the term of the policy. Such probabilities are found in *mortality tables,* which give the probability that a person of a certain age will live 1 more year. The insurance company wants to know its expectation on a policy—that is, how much it will have to pay out, on average, for each policy it writes.

HISTORICAL NOTE

Edmond Halley (hăl'ē) (1656–1742), of Halley's comet fame, also created for the city of Breslau, Germany, one of the first mortality tables, which he published in 1693. This was one of the first attempts to relate mortality and age in a population.

HISTORICAL NOTE

Pierre Simon Laplace (lə-plās') (1749–1827) made many important contributions to mathematics and astronomy. In 1812, he published his *Thérie Analytique des Probabilitiés*, in which he extended the theories of probability beyond analyzing games of chance and applied them to many practical and scientific problems. In his book he states, "The most important questions of life are indeed, for the most part, really only problems of probability."

EXAMPLE 2 Expectation in Insurance

According to mortality tables published in the National Vital Statistics Report, the probability that a 21-year-old will die within 1 year is approximately 0.000962. Suppose that the premium for a 1-year, \$25,000 life insurance policy for a 21-year-old is \$32. What is the insurance company's expectation for this policy?

Solution

Let S_1 be the event that the person dies within 1 year. Then $P(S_1) = 0.000962$, and the company must pay out \$25,000. Because the company charged \$32 for the policy, the company's actual loss is \$24,968. Let S_2 be the event that the policy holder does not die during the year of the policy. Then $P(S_2) = 0.999038$, and the company keeps the premium of \$32. The expectation is

$$\begin{aligned}\text{Expectation} &= P(S_1) \cdot S_1 + P(S_2) \cdot S_2 \\ &= 0.000962(-24{,}968) + 0.999038(32) \\ &= 7.95\end{aligned}$$

• The amount the company pays out is entered as a negative number. The amount the company receives is entered as a positive number.

The company's expectation is \$7.95, so the company earns, on average, \$7.95 for each policy sold.

CHECK YOUR PROGRESS 2 The probability that an 18-year-old will die within 1 year is approximately 0.000753. Suppose that the premium for a 1-year, \$10,000 life insurance policy for an 18-year-old is \$45. What is the insurance company's expectation for this policy?

Solution See page S46.

Expectation is also used when a company bids on a project. The company must try to predict the costs and amount of work involved if it is to make a bid that will ensure a profit. At the same time, if the bid is too high, the client may reject the offer. Because it is impossible to predict in advance the exact requirements of a job, probabilities can be used to analyze the likelihood of making a profit.

EXAMPLE 3 Expected Company Profits

Suppose a software company bids on a project to update the database program for an accounting firm. The software company assesses its potential profit as shown in the following table.

Profit/Loss	Probability
\$75,000	0.10
\$50,000	0.25
\$20,000	0.50
−\$10,000	0.10
−\$25,000	0.05

What is the profit expectation for the company?

Profit/Loss	Probability
\$500,000	0.05
\$250,000	0.30
\$150,000	0.35
−\$100,000	0.20
−\$350,000	0.10

Solution

The company's expected profit is

$$\begin{aligned}&\text{Expectation}\\&= 0.10(75{,}000) + 0.25(50{,}000) + 0.50(20{,}000) + 0.10(-10{,}000) + 0.05(-25{,}000)\\&= 7500 + 12{,}500 + 10{,}000 - 1000 - 1250 = 27{,}750\end{aligned}$$

The company's expected profit is \$27,750.

CHECK YOUR PROGRESS 3 A road construction company bids on a project to build a new freeway. The company estimates its potential profit as shown in the table at the left. What is the profit expectation for the company?

Solution See page S46.

EXCURSION

Chuck-a-luck

Chuck-a-luck is a game of chance in which 3 dice in a cage are tumbled. Bets are placed on the values that the 3 dice will show. The table below shows the different bets and the amounts won if \$1 is wagered.

Chuck Schwartz/Visuals Unlimited

Numbers bet	Bet on a number from 1 through 6	
	if 1 die matches	Pays \$1
	if 2 dice match	Pays \$2
	if all 3 dice match	Pays \$10
Field bet	Bet that the sum of the numbers showing on all 3 dice will be 5, 6, 7, 8, 13, 14, 15, or 16	Pays \$1
Over 10	Bet that the sum of the numbers showing on all 3 dice will be more than 10	Pays \$1
Under 11	Bet that the sum of the numbers showing on all 3 dice will be less than 11	Pays \$1

HISTORICAL NOTE

Chuck-a-luck, also known (sometimes with slight variations) as "Bird Cage," "Sic Bo," or "Sweat," is one of the oldest dice games. It was probably the most popular dice game played by the soldiers of the Civil War. The term *tinhorn gambler* is derived from the practice of using a metal chute in place of a cage to tumble the dice.

EXCURSION EXERCISES

1. Compute the probability that if a bet is placed on the number 5, all 3 dice will show a 5.
2. What is the probability that *exactly* 1 die will show a 5?
3. What is the probability that 2 (but not 3) dice will show a 5?
4. Find the probability that none of the dice will show a 5.
5. What is the expectation for a \$1 bet placed on the number 5?
6. Determine the expectation for wagering \$1 on the Over 10 bet.
7. Determine the expectation for wagering \$1 on the Under 11 bet.
8. Which bet is most favorable for the player? Which is most favorable for the casino? (The expectation for a \$1 Field bet is −4 cents.)

EXERCISE SET 12.6

1. The outcomes of an experiment and the probability of each outcome are given in the table below. Compute the expectation for this experiment.

Outcome	Probability
30	0.15
40	0.2
50	0.4
60	0.05
70	0.2

2. The outcomes of an experiment and the probability of each outcome are given in the table below. Compute the expectation for this experiment.

Outcome	Probability
5	0.4
6	0.3
7	0.1
8	0.08
9	0.07
10	0.05

3. **Roulette** One of the wagers in the game of roulette is to place a bet that the ball will land on a black number. (Eighteen of the numbers are black, 18 are red, and 2 are green.) If the ball lands on a black number, the player wins the amount bet. If a player bets $1, find the player's expectation.

4. **Roulette** One of the wagers in roulette is to bet that the ball will stop on a number that is a multiple of 3. (Both 0 and 00 are not included.) If the ball stops on such a number, the player wins double the amount bet. If a player bets $1, compute the player's expectation.

■ **Casino Games** Many casinos have a game called the Big Six Money Wheel, which has 54 slots in which are displayed a Joker, the casino logo, and various dollar amounts, as shown in the table at the top of the next column. Players may bet on the Joker, the casino logo, or one or more dollar denominations. The wheel is spun and if the wheel stops on the same place as the player's bet, the player wins that amount for each dollar bet. Exercises 5 to 8 use this game.

Denomination	Number of slots
$40 (Joker)	1
$40 (Casino logo)	1
$20	2
$10	4
$5	7
$2	15
$1	24

Medioimages/Photodisc/Getty Images

5. If a player bets $1 on the Joker denomination, find the player's expectation.

6. If a player bets $1 on the $20 denomination, find the player's expectation.

7. If a player bets $1 on the $5 denomination, find the player's expectation.

8. If a player bets $1 on the $2 denomination, find the player's expectation.

Life Insurance Exercises 9 to 14 use data taken from mortality tables published in the National Vital Statistics Report.

9. The probability that a 22-year-old female in the United States will die within 1 year is approximately 0.000487. If an insurance company sells a 1-year, $25,000 life insurance policy to such a person for $75, what is the company's expectation?

10. The probability that a 28-year-old male in the United States will die within 1 year is approximately 0.001362. If an insurance company sells a 1-year, $20,000 life insurance policy to such a person for $155, what is the company's expectation?

11. The probability that an 80-year-old male in the United States will die within 1 year is approximately 0.070471. If an insurance company sells a 1-year, $10,000 life insurance policy to such a person for $495, what is the company's expectation?

12. The probability that an 80-year-old female in the United States will die within 1 year is approximately 0.050409. If an insurance company sells a 1-year, $15,000 life insurance policy to such a person for $860, what is the company's expectation?

13. The probability that a 30-year-old male in the United States will die within 1 year is about 0.001406. An insurance company is preparing to sell a 30-year-old male a 1-year, $30,000 life insurance policy. How much should it charge for its premium in order to have a positive expectation for the policy?

14. The probability that a 25-year-old female in the United States will die within 1 year is about 0.000509. An insurance company is preparing to sell a 25-year-old female a 1-year, $75,000 life insurance policy. How much should it charge for its premium in order to have a positive expectation for the policy?

15. Construction A construction company has been hired to build a custom home. The builder estimates the probabilities of potential profit (or loss) as shown in the table below. What is the profit expectation for the company?

Profit/Loss	Probability
$100,000	0.10
$60,000	0.40
$30,000	0.25
$0	0.15
−$20,000	0.08
−$40,000	0.02

Marquisphoto/Shutterstock.com

16. Painting A professional painter has been hired to paint a commercial building for $18,000. From this fee, the painter must buy supplies and pay employees. The painter estimates the potential profit as shown in the table below. What is the profit expectation for the painter?

Profit/Loss	Probability
$10,000	0.15
$8,000	0.35
$5,000	0.2
$3,000	0.2
$1,000	0.1

17. Design Consultant A consultant has been hired to redesign a company's production facility. The consultant estimates the probabilities of her potential profit as shown in the table below. What is her profit expectation?

Profit/Loss	Probability
$40,000	0.05
$30,000	0.2
$20,000	0.5
$10,000	0.2
$5,000	0.05

18. Office Rentals A real estate company has purchased an office building with the intention of renting office space to small businesses. The company estimates the probabilities of potential profit (or loss) as shown in the table below. What is the profit expectation for the company?

Profit/Loss	Probability
$700,000	0.15
$400,000	0.25
$200,000	0.25
$50,000	0.20
−$100,000	0.10
−$250,000	0.05

EXTENSIONS

19. If a pair of regular dice is tossed once, use the expectation formula to determine the expected sum of the numbers on the upward faces of the 2 dice.

20. Consider rolling a pair of unusual dice, for which the faces have the number of pips indicated.

Die 1: {0, 0, 0, 6, 6, 6}

Die 2: {1, 2, 3, 4, 5, 6}

a. List the sample space for the experiment.

b. Compute the probability of each possible sum of the upward faces on the dice.

c. What is the expected value of the sum of the numbers on the upward faces of the 2 dice?

21. Two dice, one labeled 1, 2, 2, 3, 3, 4 and the other labeled 1, 3, 4, 5, 6, 8, are rolled once. Use the formula for expectation to determine the expected sum of the numbers on the upward faces of the 2 dice. Dice such as these are called *Sicherman dice.*

22. Suppose you purchase a ticket for a prize and your expectation is −$1. What is the meaning of this expectation?

23. **Efron's dice** Suppose you are offered 1 of 2 pairs of dice, a red pair or a green pair, that are labeled as follows.

Red die 1: 0, 0, 4, 4, 4, 4

Red die 2: 2, 3, 3, 9, 10, 11

Green die 1: 3, 3, 3, 3, 3, 3

Green die 2: 0, 1, 7, 8, 8, 8

After you choose, your friend will receive the other pair. Which pair should you choose if you are going to play a game in which each of you rolls your dice and the player with the higher sum wins? Dice such as these are part of a set of 4 pairs of dice called *Efron's dice.* Which pair should you choose? Explain.

24. **Lotteries** The PowerBall lottery commission chooses 5 white balls from a drum containing 69 balls marked with the numbers 1 through 69, and 1 red ball from a separate drum containing 26 balls. The following table shows the approximate odds of winning certain prizes if the numbers you choose match those chosen by the lottery commission.

Match	Prize	Odds
○○○○○ + ●	Grand Prize	1 in 292,201,338.00
○○○○○	$1,000,000	1 in 11,688,053.52
○○○○ + ●	$50,000	1 in 913,129.18
○○○○	$100	1 in 36,525.17
○○○ + ●	$100	1 in 14,494.11
○○○	$7	1 in 579.76
○○ + ●	$7	1 in 701.33
○ + ●	$4	1 in 91.98
●	$4	1 in 38.32

The overall odds of winning a prize are 1 in 24.87.
The odds presented here are based on a $2 play
(rounded to two decimal places).

SOURCE: http://www.powerball.com/powerball/pb_prizes.asp

Assuming the jackpot for a certain drawing is $150 million, what is your expectation for the jackpot if you purchase 1 ticket for $2? Round to the nearest cent. Assume the jackpot is not split among multiple winners.

CHAPTER 12 SUMMARY

The following table summarizes essential concepts in this chapter. The references given in the right-hand column list Examples and Exercises that can be used to test your understanding of a concept.

12.1 The Counting Principle	
Sample Spaces An experiment is an activity with an observable outcome. The sample space of an experiment is the set of all possible outcomes. A table or a tree diagram can be used to list all the outcomes in the sample space of a multi-stage experiment.	See **Examples 3 and 4** on pages 689 to 691, and then try Exercises 3 and 4 on page 746.
The Counting Principle Let E be a multi-stage experiment. If $n_1, n_2, n_3, \ldots, n_k$ are the numbers of possible outcomes of each of the k stages of E, then there are $n_1, n_2, n_3, \ldots, n_k$ possible outcomes for E.	See **Example 5** on pages 691 to 692, and then try Exercises 5 and 7 on page 746.

continued

Counting With and Without Replacement Some multi-stage experiments involve repeatedly choosing an element from a given set. If an element is returned to the set and can be chosen again, the experiment is performed *with replacement.* If an element is not returned to the set and cannot be chosen again, the experiment is performed *without replacement.*	See **Example 6** on page 693, and then try Exercises 1 and 2 on page 746.
12.2 Permutations and Combinations	
***n* factorial** *n* factorial is the product of the natural numbers *n* through 1. $n! = (n-1)\cdot(n-2)\cdot\cdots\cdot 3\cdot 2\cdot 1$ $0! = 1$	See **Example 1** on page 697, and then try Exercises 9 to 11 on page 746.
Permutation Formula for Distinct Objects A permutation is an arrangement of objects in a definite order. The number of permutations of *n* distinct objects selected *k* at a time is $P(n, k) = \dfrac{n!}{(n-k)!}$	See **Examples 2 and 3** on pages 698 and 699, and then try Exercises 15 and 20 on pages 746 and 747.
Permutations of Objects, Some of Which Are Identical The number of permutations of *n* objects of *r* different types, where k_1 identical objects are of one type, k_2 are of another, and so on, is $\dfrac{n!}{k_1!\cdot k_2!\cdot\cdots\cdot k_r!}$, where $k_1 + k_2 + \cdots + k_r = n$.	See **Example 5** on page 701, and then try Exercise 18 on page 746.
Combination Formula A combination is a collection of objects for which the order is not important. The number of combinations of *n* objects chosen *k* at a time is $C(n, k) = \dfrac{P(n, k)}{k!} = \dfrac{n!}{k!\cdot(n-k)!}$	See **Example 6** on page 702, and then try Exercises 21 and 22 on page 747.
Applying Several Counting Techniques Some counting problems require using the counting principle along with a permutation or combination formula.	See **Examples 4, 7, and 8** on pages 700, 703, and 704, and then try Exercises 23 to 25 on page 747.
12.3 Probability and Odds	
Theoretical Probability of an Event For an experiment with sample space *S* of equally likely outcomes, the theoretical probability *P*(*E*) of an event *E* is given by $P(E) = \dfrac{n(E)}{n(S)} = \dfrac{\text{number of elements in } E}{\text{number of elements in } S}$	See **Examples 2 and 3** on pages 709 and 710, and then try Exercises 27 and 31 on page 747.
Empirical Probability of an Event If an experiment is performed repeatedly and the occurrence of the event *E* is observed, the empirical probability *P*(*E*) of the event *E* is given by $P(E) = \dfrac{\text{number of times event } E \text{ occurred}}{\text{number of times the experiment was performed}}$	See **Example 4** on page 711, and then try Exercise 29 on page 747.
Punnet Square Using a capital letter to represent a dominant allele and a lower-case letter to represent a recessive allele, a Punnet square shows all possible genotypes of the offspring of two parents with given genotypes.	See **Example 5** on page 712, and then try Exercise 44 on page 747.

Odds of an Event Let E be an event in a sample space of equally likely outcomes. Then $$\text{Odds in favor of } E = \frac{\text{number of favorable outcomes}}{\text{number of unfavorable outcomes}}$$ $$\text{Odds against } E = \frac{\text{number of unfavorable outcomes}}{\text{number of favorable outcomes}}$$	See **Example 6** on page 713, and then try Exercises 41 and 42 on page 747.
Relationship between Odds and Probability If the odds in favor of event E are $\frac{a}{b}$, then $P(E) = \frac{a}{a+b}$. This relationship is also expressed by the equation $$\text{Odds in favor of } E = \frac{P(E)}{1 - P(E)}$$	See **Example 7** on page 713, and then try Exercise 43 on page 747.
12.4 Addition and Complement Rules	
Probability of Mutually Exclusive Events Two events A and B are mutually exclusive if they cannot occur at the same time. In this case, the probability of A or B occurring is $$P(A \text{ or } B) = P(A) + P(B)$$	See **Example 1** on page 719, and then try Exercises 27 and 31 on page 747.
Addition Rule for Probabilities If A and B are two events in a sample space S, then $$P(A \text{ or } B) = P(A) + P(B) - P(A \text{ and } B)$$	See **Example 2** on page 720, and then try Exercise 38 on page 747.
Probability of the Complement of an Event The complement of an event E in a sample space S includes all those outcomes of S that are not in E and excludes the outcomes in E. The symbol for the complement of E is E^c. To find the probability of E^c, use the relationship $$P(E^c) = 1 - P(E)$$	See **Example 3** on page 722, and then try Exercise 39 on page 747.
Using Combinatorics Formulas to Find Probabilities In many cases, the process of finding a probability involves using the counting principle and/or a permutation or combination formula.	See **Examples 4 and 5** on pages 722 to 724, and then try Exercises 60 and 63 on page 748.
12.5 Conditional Probability	
Conditional Probability Formula If A and B are two events in a sample space S, then the conditional probability of B given that A has occurred is $$P(B \mid A) = \frac{P(A \text{ and } B)}{P(A)}$$	See **Example 1** on pages 728 to 729, and then try Exercises 35 and 36 on page 747.
Product Rule for Probabilities If A and B are two events in a sample space S, then $$P(A \text{ and } B) = P(A) \cdot P(B \mid A)$$ **Probability of Successive Events** The probability of two or more events occurring in succession is the product of the conditional probabilities of each of the events.	See **Example 2** on pages 730 to 731, and then try Exercises 51 and 57 on page 748.

continued

Product Rule for Independent Events If A and B are two events in a sample space S and $P(B \mid A) = P(B)$, then A and B are independent events. In this case $$P(A \text{ and } B) = P(A) \cdot P(B)$$	See **Example 3** on page 732, and then try Exercise 39 on page 747.
Using Several Formulas to Find Probabilities The process of finding a conditional probability may involve using other probability formulas.	See **Example 4** on page 732, and then try Exercise 61 on page 748.
12.6 Expectation	
Expectation If $S_1, S_2, S_3, \dots, S_n$ are the possible numerical outcomes of an experiment and $P(S_1), P(S_2), P(S_3), \dots, P(S_n)$ are the probabilities of those outcomes, then the expectation of the experiment is $$P(S_1) \cdot S_1 + P(S_2) \cdot S_2 + P(S_3) \cdot S_3 + \cdots + P(S_n) \cdot S_n$$	See **Example 1** on pages 737 to 738, and then try Exercises 64 and 65 on page 748.
Business Applications of Expectation Insurance companies can use expectation to help determine life insurance premiums. Other businesses can use expectation to determine potential profits.	See **Examples 2 and 3** on pages 739 to 740, and then try Exercises 68 and 70 on page 748.

CHAPTER 12 REVIEW EXERCISES

■ In Exercises 1 and 2, list the elements of the sample space for the given experiment.

1. Two-digit numbers are formed, with replacement, from the digits 1, 2, and 3.

2. Two-digit numbers are formed, without replacement, from the digits 2, 6, and 8.

3. Use a tree diagram to list all possible outcomes that result from tossing 4 coins.

4. Use a table to list all possible 2-character codes that can be formed from one of the digits 7, 8, or 9 followed by one of the letters A or B.

5. An athletic shoe store sells running shoes in 3 styles that come in 4 colors. Each color comes in 6 sizes. How many distinct shoes are available?

6. The combination for a lock to a bicycle chain contains 4 numbers chosen from the numbers 0 through 9. How many different lock combinations are possible? Assume that a number can be used more than once.

7. **Serial Numbers** In the 1970s and 1980s, the Conn music company assigned the first 4 characters in serial numbers of instruments in the following way. The first character is one of the letters G or H and indicates the decade in which the instrument was made: "G" for 1970s and "H" for 1980s. The second character indicates the month of the year in which the instrument was made: "A" for January, "B" for February, and so on. The third character is a number from 0 to 9 indicating the year of the decade in which the instrument was made, and the fourth character is a number from 0 to 9 indicating the type of instrument: 1 = cornet, 2 = trumpet, 3 = alto horn, 4 = French horn, 5 = mellophone, 6 = valve trombone, 7 = slide trombone, 8 = euphonium, 9 = tuba, 0 = sousaphone. How many 4-character sequences are possible?

8. **Codes** A *biquinary code* is a code that consists of 2 different binary digits (a binary digit is a 0 or a 1) followed by 5 binary digits for which there are no restrictions. How many biquinary codes are possible?

■ In Exercises 9 to 14, evaluate each expression.

9. $7!$

10. $8! - 4!$

11. $\dfrac{9!}{2!\,3!\,4!}$

12. $P(10, 6)$

13. $P(8, 3)$

14. $\dfrac{C(6, 2) \cdot C(8, 3)}{C(14, 5)}$

15. In how many different ways can 7 people arrange themselves in a line to receive service from a bank teller?

16. A matching test has 7 definitions that are to be paired with 7 words. Assuming each word corresponds to exactly 1 definition, how many different matches are possible by random matching?

17. A matching test has 7 definitions to be matched with 5 words. Assuming each word corresponds to exactly 1 definition, how many different matches are possible by random matching?

18. How many distinct arrangements are possible using the letters of the word *letter?*

19. Twelve identical coins are tossed. How many distinct arrangements are possible consisting of 4 heads and 8 tails?

20. Work Shifts Three positions are open at a manufacturing plant: the day shift, the swing shift, and the night shift. In how many different ways can 5 people be assigned to the 3 shifts?

21. A professor assigns 25 homework problems, of which 10 will be graded. How many different sets of 10 problems can the professor choose to grade?

22. Stock Portfolios A stockbroker recommends 11 stocks to a client. If the client will invest in 3 of the stocks, how many different 3-stock portfolios can be selected?

23. Quality Control A quality control inspector receives a shipment of 15 computer monitors, of which 3 are defective. If the inspector randomly chooses 5 monitors, how many different sets can be formed that consist of 3 nondefective monitors and 2 defective monitors?

24. In how many ways can 9 people be seated in 9 chairs if 2 of the people refuse to sit next to each other?

25. How many 5-card poker hands consist of 4 of a kind (4 aces, 4 kings, 4 queens, and so on)?

26. If it is equally likely that a child will be born a boy or a girl, compute the probability that a family of 4 children will have 1 boy and 3 girls.

27. If a coin is tossed 3 times, what is the probability of getting 1 head and 2 tails?

28. A large company currently employs 5739 men and 7290 women. If an employee is selected at random, what is the probability that the employee is a woman?

■ **Enrollment** In Exercises 29 and 30, use the table below, which shows the number of students at a university who are currently in each class level.

Class level	Number of students
First year	642
Sophomore	549
Junior	483
Senior	445
Graduate student	376

29. If a student is selected at random, what is the probability that the student is an upper-division undergraduate student (junior or senior)?

30. If a student is selected at random, what is the probability that the student is not a graduate student?

■ In Exercises 31 to 36, a pair of dice is tossed.

31. Find the probability that the sum of the pips on the 2 upward faces is 9.

32. Find the probability that the sum on the 2 dice is not 11.

33. Find the probability that the sum on the 2 dice is at least 10.

34. Find the probability that the sum on the 2 dice is an even number or a number less than 5.

35. What is the probability that the sum on the 2 dice is 9, given that the sum is odd?

36. What is the probability that the sum on the 2 dice is 8, given that doubles were rolled?

■ In Exercises 37 to 40, a single card is selected from a standard deck of playing cards.

37. What is the probability that the card is a heart or a black card?

38. What is the probability that the card is a heart or a jack?

39. What is the probability that the card is not a 3?

40. What is the probability that the card is red, given that it is not a club?

41. If a pair of dice is rolled, what are the odds in favor of getting a sum of 6?

42. If 1 card is drawn from a standard deck of playing cards, what are the odds that the card is a heart?

43. If the odds against an event occurring are 4 to 5, compute the probability of the event occurring.

44. Genotypes The hair length of a particular rodent is determined by a dominant allele H, corresponding to long hair, and a recessive allele h, corresponding to short hair. Draw a Punnett square for parents of genotypes Hh and hh, and compute the probability that the offspring of the parents will have short hair.

45. Two cards are drawn, without replacement, from a standard deck of playing cards. The probability that exactly 1 card is an ace is 0.145. The probability that exactly 1 card is a face card (jack, queen, or king) is 0.362, and the probability that a selection of 2 cards will contain an ace or a face card is 0.471. Find the probability that the 2 cards are an ace and a face card.

46. Surveys A recent survey asked 1000 people whether they liked cheese-flavored corn chips (642 people), jalapeño-flavored chips (487 people), or both (302 people). If 1 person is chosen from this survey, what is the probability that the person does not like either of the 2 flavors?

■ In Exercises 47 to 51, a box contains 24 different colored chips that are identical in size. Five are black, 4 are red, 8 are white, and 7 are yellow.

47. If a chip is selected at random, what is the probability that the chip will be yellow or white?

48. If a chip is selected at random, what are the odds in favor of getting a red chip?

49. If a chip is selected at random, find the probability that the chip is yellow given that it is not white.

50. If 5 chips are randomly chosen, without replacement, what is the probability that none of them is red?

51. If 3 chips are chosen without replacement, find the probability that the first one is yellow, the second is white, and the third is yellow.

■ **Voting** In Exercises 52 to 56, use the table below, which shows the number of voters in a city who voted for or against a proposition (or abstained from voting) according to their party affiliations. Round answers to the nearest hundredth.

	For	Against	Abstained
Democrat	8452	2527	894
Republican	2593	5370	1041
Independent	1225	712	686

52. If a voter is chosen at random, compute the probability that the person voted against the proposition.

53. If a voter is chosen at random, compute the probability that the person is a Democrat or an Independent.

54. If a voter is randomly chosen, what is the probability that the person abstained from voting on the proposition and is not a Republican?

55. A voter is randomly selected. What is the probability that the individual voted for the proposition, given that the voter is a registered Independent?

56. A voter is randomly selected. What is the probability that the individual is registered as a Democrat, given that the person voted against the proposition?

57. A single die is rolled 3 times in succession. What is the probability that each roll gives a 6?

58. A single die is rolled 5 times in succession. Find the probability that a 6 will be rolled at least once.

59. A single die is rolled 5 times in succession. Find the probability that exactly 2 of the rolls give a 6.

60. A person draws a card from a standard deck and replaces it; she then does this 3 more times. What is the probability that she drew a spade at least once?

61. **Disease Testing** A veterinarian uses a test to determine whether a dog has a disease that affects 7% of the dog population. The test correctly gives a positive result for 98% of dogs that have the disease, but gives false positives for 4% of dogs that do not have the disease. If a dog tests positive, what is the probability that the dog has the disease?

62. **Weather** Suppose that in your area in the wintertime, if it rains one day, there is a 65% chance that it will rain the next day. If it does not rain on a given day, there is only a 15% chance that it will rain the following day. What is the probability that, if it didn't rain today, it will rain the next 2 days but not the following 2?

63. **Batteries** About 1.2% of AA batteries produced by a particular manufacturer are defective. If a consumer buys a box of 12 of these batteries, what is the probability that at least 1 battery is defective?

64. Suppose it costs $4 to play a game in which a single die is rolled and you win the amount of dollars that the die shows. What is the expectation for the game?

65. I will flip 2 coins. If both coins come up tails, I will pay you $5. If one shows heads and one shows tails, you will pay me $2. If both coins come up heads, we will call it a draw. What is your expectation for this game?

66. **Raffle Tickets** For a fundraiser, an elementary school is selling 800 raffle tickets for $1 each. From these, 5 tickets will be drawn. One of the winners gets $200 and the others each get $75. If you buy 1 raffle ticket, what is your expectation?

67. If a pair of dice is rolled 65 times, on how many rolls can we expect to get a total of 4?

68. **Life Insurance** The probability that a 29-year-old female in the United States will die within 1 year is approximately 0.000595. If an insurance company sells a 1-year, $40,000 life insurance policy to a 29-year-old for $320, what is the company's expectation?

69. **Life Insurance** The probability that a 19-year-old male in the United States will die within 1 year is approximately 0.001188. If an insurance company sells a 1-year, $25,000 life insurance policy to a 19-year-old for $795, what is the company's expectation?

70. **Construction** A construction company has bid on a building renovation project. The company estimates the probabilities of potential profit (or loss) as shown in the table below. What is the profit expectation for the company?

Profit/loss	Probability
$25,000	0.20
$15,000	0.25
$10,000	0.20
$5,000	0.15
$0	0.10
−$5,000	0.10

CHAPTER 12 TEST

1. **Driver's License Numbers** A certain driver's license number begins with one of the letters A, D, G, or K and is followed by one of the digits 2, 3, or 4. List the elements in the sample space of the first two digits of this driver's license number.
2. **Computer Systems** A computer system can be configured using 1 of 3 processors of different speeds, 1 of 4 disk drives of different sizes, 1 of 3 monitors, and 1 of 2 graphics cards. How many different computer systems are possible?
3. **Transistors** In a very simple computer chip, 10 wires go from one transistor to a second transistor. For the computer to function, instructions must be sent between these 2 transistors. In how many ways can 4 different instructions be sent between the 2 transistors if no 2 instructions can be sent along the same wire?
4. A matching test asks students to match 10 words with 15 definitions. Assuming each word corresponds to exactly 1 definition, how many different matches are possible?
5. **Softball** How many games are necessary in a softball league consisting of 8 teams if each team must play each of the other teams once?
6. A coin and a regular 6-sided die are tossed together once. What is the probability that the coin shows a head or the die has a 5 on the upward face?
7. A person draws 4 cards in succession from a standard deck of playing cards, replacing each card before drawing the next. What is the probability that the person draws at least 1 ace?
8. What is the probability of drawing 2 cards in succession (without replacement) from a standard deck of playing cards and having them both be hearts?
9. Four cards are drawn from a standard deck of playing cards without replacement. What is the probability that none of them are 9s?
10. Three coins are tossed once. What are the odds in favor of the coins showing all heads?
11. **Disease Testing** A new medical test can determine whether a human has a disease that affects 5% of the population. The test correctly gives a positive result for 99% of people who have the disease, but it gives a false positive for 3% of people who do not have the disease. If a person tests positive, what is the probability that the person has the disease?
12. **Advertising** The table below shows the number of men and the number of women who responded either positively or negatively to a new commercial. If 1 person is chosen from this group, find the probability that the person is a woman, given that the person responded negatively. Round to the nearest thousandth.

	Positive	Negative	Total
Men	684	736	1420
Women	753	642	1395
Total	1437	1378	2815

13. **Genotypes** Straight or curly hair for a hamster is determined by a dominant allele S, corresponding to straight hair, and a recessive allele s, which gives curly hair. If 1 parent is of genotype Ss and the other is of genotype ss, compute the probability that the offspring of the parents will have curly hair.
14. **Inventories** A software company is preparing a bid to create a new inventory program for an auto parts company. The software company estimates the probabilities of potential profit (or loss) as shown in the table below. What is the profit expectation for the company?

Profit/loss	Probability
\$75,000	0.18
\$50,000	0.36
\$25,000	0.31
\$0	0.08
−\$10,000	0.05
−\$20,000	0.02

13

Statistics

The U.S. government collects data on the population of the United States. It then issues *statistical* reports that indicate changes and trends in the U.S. population. For instance, according to *The World Factbook*, published by the Central Intelligence Agency (CIA), in 2015 there were approximately 105 males for every 100 females between the ages of 15 and 24. However, in the category of people 65 years old and older, there were approximately 79 men for every 100 women. See the graph below.

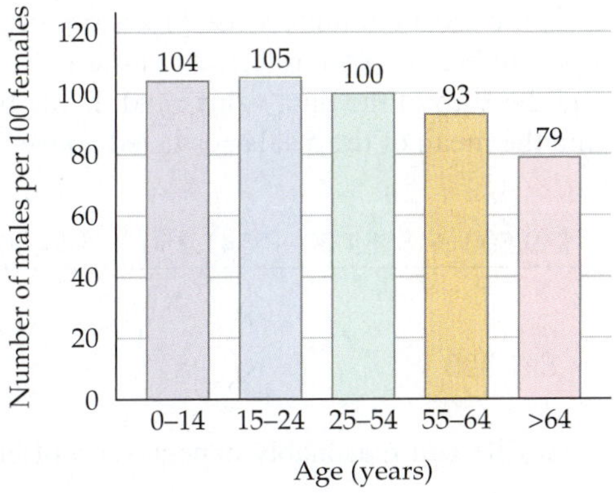

SOURCE: CIA, *The World Factbook*

Here are some other statistics from *The World Factbook*:

- There are 2.45 physicians per 1000 people in the United States.
- In 1910, the mean annual family income in the United States was $687. In 2015, the mean annual family income was approximately $68,500.

However, the *median* annual family income was approximately $51,900. The difference between the mean and the median is one of the topics of this chapter.

SECTION 13.1 Measures of Central Tendency

The Arithmetic Mean

Statistics involves the collection, organization, summarization, presentation, and interpretation of data. The branch of statistics that involves the collection, organization, summarization, and presentation of data is called **descriptive statistics**. The branch that interprets and draws conclusions from the data is called **inferential statistics**.

One of the most basic statistical concepts involves finding *measures of central tendency* of a set of numerical data. It is often helpful to find numerical values that locate, in some sense, the *center* of a set of data. Suppose Elle is a senior at a university. In a few months she plans to graduate and start a career as a landscape architect. A survey of five landscape architects from last year's senior class shows that they received job offers with the following yearly salaries.

\$43,750 \$39,500 \$38,000 \$41,250 \$44,000

Before Elle interviews for a job, she wishes to determine an *average* of these 5 salaries. This average should be a "central" number around which the salaries cluster. We will consider three types of averages, known as the *arithmetic mean*, the *median*, and the *mode*. Each of these averages is a **measure of central tendency** for the numerical data.

The *arithmetic mean* is the most commonly used measure of central tendency. The arithmetic mean of a set of numbers is often referred to as simply the *mean*. To find the mean for a set of data, find the sum of the data values and divide by the number of data values. For instance, to find the mean of the 5 salaries listed above, Elle would divide the sum of the salaries by 5.

$$\text{Mean} = \frac{\$43{,}750 + \$39{,}500 + \$38{,}000 + \$41{,}250 + \$44{,}000}{5}$$

$$= \frac{\$206{,}500}{5} = \$41{,}300$$

The mean suggests that Elle can reasonably expect a job offer at a salary of about \$41,300.

In statistics it is often necessary to find the sum of a set of numbers. The traditional symbol used to indicate a summation is the Greek letter *sigma*, Σ. Thus the notation Σx, called **summation notation**, denotes the sum of all the numbers in a given set. We can define the mean using summation notation.

Mean

The **mean** of n numbers is the sum of the numbers divided by n.

$$\text{Mean} = \frac{\Sigma x}{n}$$

Statisticians often collect data from small portions of a large group in order to determine information about the group. In such situations the entire group under consideration is known as the **population**, and any subset of the population is called a **sample**. It is traditional to denote the mean of a *sample* by $\bar{x}$ (which is read as "x bar") and to denote the mean of a *population* by the Greek letter μ (lowercase mu).

EXAMPLE 1 Find a Mean

Six friends in a biology class of 20 students received test grades of

92, 84, 65, 76, 88, and 90

Find the mean of these test scores.

Solution

The 6 friends are a sample of the population of 20 students. Use $\bar{x}$ to represent the mean.

$$\bar{x} = \frac{\Sigma x}{n} = \frac{92 + 84 + 65 + 76 + 88 + 90}{6} = \frac{495}{6} = 82.5$$

The mean of these test scores is 82.5.

CHECK YOUR PROGRESS 1 A doctor ordered 4 separate blood tests to measure a patient's total blood cholesterol levels. The test results were

245, 235, 220, and 210

Find the mean of the blood cholesterol levels.

Solution See page S46.

The Median

Another type of average is the *median*. Essentially, the median is the *middle number* or the *mean of the two middle numbers* in a list of numbers that have been arranged in numerical order from smallest to largest or largest to smallest. Any list of numbers that is arranged in numerical order from smallest to largest or largest to smallest is a **ranked list**.

POINT OF INTEREST

Peter French/Pacific Stock/Design Pics/Superstock

The average price of the homes in a neighborhood is often stated in terms of the median price of the homes that have been sold over a given time period. The median price, rather than the mean, is used because it is easy to calculate and less sensitive to extreme prices.

Median

The median of a ranked list of n numbers is:

- the middle number if n is odd.
- the mean of the two middle numbers if n is even.

EXAMPLE 2 Find a Median

Find the median of the data in the following lists.

a. 4, 8, 1, 14, 9, 21, 12 **b.** 46, 23, 92, 89, 77, 108

Solution

a. The list 4, 8, 1, 14, 9, 21, 12 contains 7 numbers. The median of a list with an odd number of entries is found by ranking the numbers and finding the middle number. Ranking the numbers from smallest to largest gives

1, 4, 8, 9, 12, 14, 21

The middle number is 9. Thus 9 is the median.

b. The list 46, 23, 92, 89, 77, 108 contains 6 numbers. The median of a list of data with an even number of entries is found by ranking the numbers and computing the mean of the two middle numbers. Ranking the numbers from smallest to largest gives

23, 46, 77, 89, 92, 108

The two middle numbers are 77 and 89. The mean of 77 and 89 is 83. Thus 83 is the median of the data.

CHECK YOUR PROGRESS 2 Find the median of the data in the following lists.

a. 14, 27, 3, 82, 64, 34, 8, 51

b. 21.3, 37.4, 11.6, 82.5, 17.2

Solution See page S46.

QUESTION The median of the ranked list 3, 4, 7, 11, 17, 29, 37 is 11. If the maximum value 37 is increased to 55, what effect will this have on the median?

The Mode

A third type of average is the *mode*.

Mode

The **mode** of a list of numbers is the number that occurs most frequently.

Some lists of numbers do not have a mode. For instance, in the list 1, 6, 8, 10, 32, 15, 49, each number occurs exactly once. Because no number occurs more often than the other numbers, there is no mode.

A list of numerical data can have more than one mode. For instance, in the list 4, 2, 6, 2, 7, 9, 2, 4, 9, 8, 9, 7, the number 2 occurs three times and the number 9 occurs three times. Each of the other numbers occurs less than three times. Thus 2 and 9 are both modes for the data.

EXAMPLE 3 Find a Mode

Find the mode of the data in the following lists.

a. 18, 15, 21, 16, 15, 14, 15, 21 **b.** 2, 5, 8, 9, 11, 4, 7, 23

Solution

a. In the list 18, 15, 21, 16, 15, 14, 15, 21, the number 15 occurs more often than the other numbers. Thus 15 is the mode.

b. Each number in the list 2, 5, 8, 9, 11, 4, 7, 23 occurs only once. Because no number occurs more often than the others, there is no mode.

CHECK YOUR PROGRESS 3 Find the mode of the data in the following lists.

a. 3, 3, 3, 3, 3, 4, 4, 5, 5, 5, 8 **b.** 12, 34, 12, 71, 48, 93, 71

Solution See page S46.

POINT OF INTEREST

Daniel Acker/Bloomberg/Getty Images

For professional sports teams, the salaries of a few very highly paid players can lead to large differences between the mean and the median salary. In 2015, the median Major League Baseball (MLB) salary was about $1.65 million. The mean salary was $4.2 million, about 2.5 times the median. (*Source:* sports.yahoo.com)

The mean, the median, and the mode are all averages; however, they are generally not equal. The mean of a set of data is the most sensitive of the averages. A change in any of the numbers changes the mean, and the mean can be changed drastically by changing an extreme value.

In contrast, the median and the mode of a set of data are usually not changed by changing an extreme value.

When a data set has one or more extreme values that are very different from the majority of data values, the mean will not necessarily be a good indicator of an average value. In the following example, we compare the mean, median, and mode for the salaries of 5 employees of a small company.

Salaries: $370,000 $60,000 $36,000 $20,000 $20,000

The sum of the 5 salaries is $506,000. Hence the mean is

$$\frac{506{,}000}{5} = 101{,}200$$

ANSWER The median will remain the same because 11 will still be the middle number in the ranked list.

The median is the middle number, \$36,000. Because the \$20,000 salary occurs the most, the mode is \$20,000. The data contain one extreme value that is much larger than the other values. This extreme value makes the mean considerably larger than the median. Most of the employees of this company would probably agree that the median of \$36,000 better represents the average of the salaries than does either the mean or the mode.

MATH MATTERS Average Rate for a Round Trip

Suppose you average 60 mph on a one-way trip of 60 mi. On the return trip you average 30 mph. You might be tempted to think that the average of 60 mph and 30 mph, which is 45 mph, is the average rate for the entire trip. However, this is not the case. Because you were traveling more slowly on the return trip, the return trip took longer than the original trip to your destination. More time was spent traveling at the slower speed. Thus the average rate for the round trip is less than the average (mean) of 60 mph and 30 mph.

To find the actual average rate for the round trip, use the formula

$$\text{Average rate} = \frac{\text{total distance}}{\text{total time}}$$

The total round-trip distance is 120 mi. The time spent traveling to your destination was 1 h, and the time spent on the return trip was 2 h. The total time for the round trip was 3 h. Thus,

$$\text{Average rate} = \frac{\text{total distance}}{\text{total time}} = \frac{120}{3} = 40 \text{ mph}$$

The Weighted Mean

A value called the *weighted mean* is often used when some data values are more important than others. For instance, many professors determine a student's course grade from the student's tests and the final examination. Consider the situation in which a professor counts the final examination score as 2 test scores. To find the weighted mean of the student's scores, the professor first assigns a weight to each score. In this case the professor could assign each of the test scores a weight of 1 and the final exam score a weight of 2. A student with test scores of 65, 70, and 75 and a final examination score of 90 has a weighted mean of

$$\frac{(65 \times 1) + (70 \times 1) + (75 \times 1) + (90 \times 2)}{5} = \frac{390}{5} = 78$$

Note that the numerator of the weighted mean above is the sum of the products of each test score and its corresponding weight. The number 5 in the denominator is the sum of all the weights $(1 + 1 + 1 + 2 = 5)$. The procedure for finding the weighted mean can be generalized as follows.

The Weighted Mean

The **weighted mean** of the n numbers $x_1, x_2, x_3, \ldots, x_n$ with the respective assigned weights $w_1, w_2, w_3, \ldots, w_n$ is

$$\text{Weighted mean} = \frac{\Sigma(x \cdot w)}{\Sigma w}$$

where $\Sigma(x \cdot w)$ is the sum of the products formed by multiplying each number by its assigned weight, and Σw is the sum of all the weights.

Many colleges use the 4-point grading system:

$A = 4, B = 3, C = 2, D = 1, F = 0$

A student's grade point average (GPA) is calculated as a weighted mean, where the student's grade in each course is given a weight equal to the number of units (or credits) that course is worth. Use this 4-point grading system for Example 4 and Check Your Progress 4.

TABLE 13.1
Dillon's Grades, Fall Semester

Course	Course grade	Course units
English	B	4
History	A	3
Chemistry	D	3
Algebra	C	4

EXAMPLE 4 Find a Weighted Mean

Table 13.1 shows Dillon's fall semester course grades. Use the weighted mean formula to find Dillon's GPA for the fall semester.

Solution

The B is worth 3 points, with a weight of 4; the A is worth 4 points with a weight of 3; the D is worth 1 point, with a weight of 3; and the C is worth 2 points, with a weight of 4. The sum of all the weights is 4 + 3 + 3 + 4, or 14.

$$\text{Weighted mean} = \frac{(3 \times 4) + (4 \times 3) + (1 \times 3) + (2 \times 4)}{14}$$

$$= \frac{35}{14} = 2.5$$

Dillon's GPA for the fall semester is 2.5.

TABLE 13.2
Janet's Grades, Spring Semester

Course	Course grade	Course units
Biology	A	4
Statistics	B	3
Business	C	3
Psychology	F	2
CAD	B	2

CHECK YOUR PROGRESS 4 Table 13.2 shows Janet's spring semester course grades. Use the weighted mean formula to find Janet's GPA for the spring semester. Round to the nearest hundredth.

Solution See page S46.

Data that have not been organized or manipulated in any manner are called **raw data**. A large collection of raw data may not provide much readily observable information. A **frequency distribution**, which is a table that lists observed events and the frequency of occurrence of each observed event, is often used to organize raw data. For instance, consider the following table, which lists the number of laptop computers owned by families in each of 40 homes in a subdivision.

TABLE 13.3
Number of Laptop Computers per Household

2	0	3	1	2	1	0	4
2	1	1	7	2	0	1	1
0	2	2	1	3	2	2	1
1	4	2	5	2	3	1	2
2	1	2	1	5	0	2	5

The frequency distribution in Table 13.4 on the next page was constructed using the data from Table 13.3. The first column of the frequency distribution consists of the numbers 0, 1, 2, 3, 4, 5, 6, and 7. The corresponding frequency of occurrence, *f*, of each of the numbers in the first column is listed in the second column.

TABLE 13.4
A Frequency Distribution for Table 13.3

Observed event Number of laptop computers, x	*Frequency* Number of households, f, with x laptop computers
0	5
1	12
2	14
3	3
4	2
5	3
6	0
7	1
	40 total

This row indicates that there are 14 households with 2 laptop computers.

The formula for a weighted mean can be used to find the mean of the data in a frequency distribution. The only change is that the weights $w_1, w_2, w_3, \ldots, w_n$ are replaced with the frequencies $f_1, f_2, f_3, \ldots, f_n$. This procedure is illustrated in the next example.

EXAMPLE 5 Find the Mean of Data Displayed in a Frequency Distribution

Find the mean of the data in Table 13.4.

Solution

The numbers in the right-hand column of Table 13.4 are the frequencies f for the numbers in the first column. The sum of all the frequencies is 40.

$$\text{Mean} = \frac{\Sigma(x \cdot f)}{\Sigma f}$$

$$= \frac{(0 \cdot 5) + (1 \cdot 12) + (2 \cdot 14) + (3 \cdot 3) + (4 \cdot 2) + (5 \cdot 3) + (6 \cdot 0) + (7 \cdot 1)}{40}$$

$$= \frac{79}{40}$$

$$= 1.975$$

The mean number of laptop computers per household for the homes in the subdivision is 1.975.

CHECK YOUR PROGRESS 5 A housing division consists of 45 homes. The following frequency distribution shows the number of homes in the subdivision that are two-bedroom homes, the number that are three-bedroom homes, the number that are four-bedroom homes, and the number that are five-bedroom homes. Find the mean number of bedrooms for the 45 homes.

Observed event Number of bedrooms, x	*Frequency* Number of homes with x bedrooms
2	5
3	25
4	10
5	5
	45 total

Solution See page S46.

EXCURSION

Linear Interpolation and Animation

Linear interpolation is a method used to find a particular number between two given numbers. For instance, if a table lists the two entries 0.3156 and 0.8248, then the value exactly halfway between the numbers is the mean of the numbers, which is 0.5702. To find the number that is 0.2 of the way from 0.3156 to 0.8248, compute 0.2 times the difference between the numbers and, because the first number is smaller than the second number, add this result to the smaller number.

$0.8248 - 0.3156 = 0.5092$ ⟵ Difference between the table entries

$0.2 \cdot (0.5902) = 0.10184$ ⟵ 0.2 of the above difference

$0.3156 + 0.10184 = 0.41744$ ⟵ Interpolated result, which is 0.2 of the way between the two table entries

The above linear interpolation process can be used to find an intermediate number that is any specified fraction of the difference between two given numbers.

EXCURSION EXERCISES

1. Use linear interpolation to find the number that is 0.7 of the way from 1.856 to 1.972.

2. Use linear interpolation to find the number that is 0.3 of the way from 0.8765 to 0.8652. Note that because 0.8765 is larger than 0.8652, three-tenths of the difference between 0.8765 and 0.8652 must be subtracted from 0.8765 to find the desired number.

3. A calculator shows that $\sqrt{2} \approx 1.414$ and $\sqrt{3} \approx 1.732$. Use linear interpolation to estimate $\sqrt{2.4}$. *Hint:* Find the number that is 0.4 of the difference between 1.414 and 1.732 and add this number to the smaller number, 1.414. Round your estimate to the nearest thousandth.

4. We know that $2^1 = 2$ and $2^2 = 4$. Use linear interpolation to estimate $2^{1.2}$.

5. At the present time, a football player weighs 325 lb. There are 90 days until the player needs to report to spring training at a weight of 290 lb. The player wants to lose weight at a constant rate. That is, the player wants to lose the same amount of weight each day of the 90 days. What weight, to the nearest tenth of a pound, should the player attain in 25 days?

Graphic artists use computer drawing programs, such as Adobe Illustrator, to draw the intermediate frames of an animation. For instance, in the following figure, the artist drew the small green apple on the left and the large ripe apple on the right. The drawing program used interpolation procedures to draw the five apples between the two apples drawn by the artist.

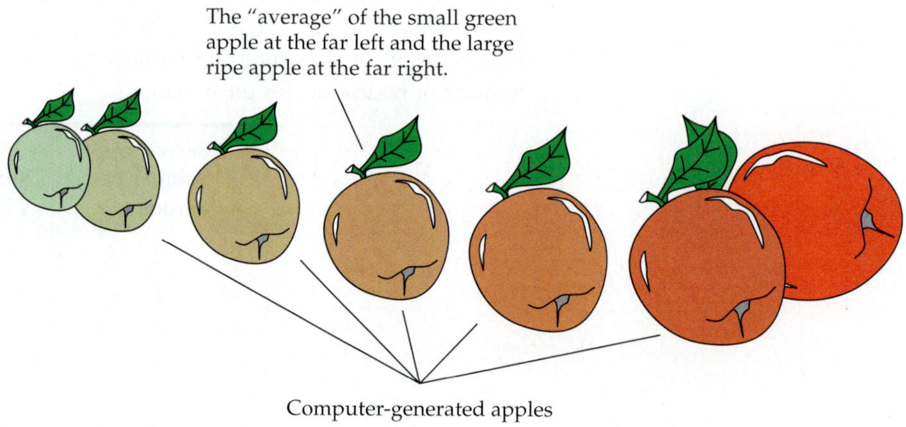

EXERCISE SET 13.1

■ In Exercises 1 to 10, find the mean, median, and mode(s), if any, for the given data. Round noninteger means to the nearest tenth.

1. 2, 7, 5, 7, 14

2. 8, 3, 3, 17, 9, 22, 19

3. 11, 8, 2, 5, 17, 39, 52, 42

4. 101, 88, 74, 60, 12, 94, 74, 85

5. 2.1, 4.6, 8.2, 3.4, 5.6, 8.0, 9.4, 12.2, 56.1, 78.2

6. 5, 5, 5, 5, 5, 5, 5, 5, 5, 5, 5, 5

7. 255, 178, 192, 145, 202, 188, 178, 201

8. 118, 105, 110, 118, 134, 155, 166, 166, 118

9. −12, −8, −5, −5, −3, 0, 4, 9, 21

10. −8.5, −2.2, 4.1, 4.1, 6.4, 8.3, 9.7

11. a. If exactly one number in a set of data is changed, will this necessarily change the mean of the set? Explain.

b. If exactly one number in a set of data is changed, will this necessarily change the median of the set? Explain.

12. If a set of data has a mode, then *must* the mode be one of the numbers in the set? Explain.

13. Academy Awards The following table displays the ages of female actors when they starred in their Oscar-winning Best Actor performances.

Ages of Best Female Actor Award Recipients, Academy Awards, 1980–2015

41	33	31	74	33	49	38	61	21	41	26	80
42	29	33	36	45	49	39	34	26	25	33	35
35	28	30	29	61	32	33	45	66	25	46	55

Find the mean and the median for the data in the table. Round to the nearest tenth.

14. Academy Awards The following table displays the ages of male actors when they starred in their Oscar-winning Best Actor performances.

Ages of Best Male Actor Award Recipients, Academy Awards, 1980–2015

40	42	37	76	39	53	45	36	62	43	51	32
42	54	52	37	38	32	45	60	46	40	36	47
29	43	37	38	45	50	48	60	43	58	46	33

Find the mean and the median for the data in the table. Round to the nearest tenth.

15. Dental Schools Dental schools provide urban statistics to their students.

a. Use the following data to decide which of the two cities you would pick to set up your practice in.

Cloverdale: Population, 18,250
Median price of a home, $167,000
Dentists, 12; median age, 49
Mean number of patients, 1294.5

Barnbridge: Population, 27,840
Median price of a home, $204,400
Dentists, 17.5; median age, 53
Mean number of patients, 1148.7

b. Explain how you made your decision.

16. Expense Reports A salesperson records the following daily expenditures during a 10-day trip.

$185.34	$234.55	$211.86	$147.65	$205.60
$216.74	$1345.75	$184.16	$320.45	$88.12

In your opinion, does the mean or the median of the expenditures best represent the salesperson's average daily expenditure? Explain your reasoning.

Grade Point Average In some 4.0 grading systems, a student's grade point average (GPA) is calculated by assigning letter grades the following numerical values.

A = 4.00	B− = 2.67	D+ = 1.33
A− = 3.67	C+ = 2.33	D = 1.00
B+ = 3.33	C = 2.00	D− = 0.67
B = 3.00	C− = 1.67	F = 0.00

■ In Exercises 17 to 20, use the above grading system to find each student's GPA. Round to the nearest hundredth.

17. Jerry's Grades, Fall Semester

Course	Course grade	Course units
English	A	3
Anthropology	A	3
Chemistry	B	4
French	C+	3
Theatre	B−	2

18. Rhonda's Grades, Spring Semester

Course	Course grade	Course units
English	C	3
History	D+	3
Computer science	B+	2
Calculus	B−	3
Photography	A−	1

19. Tessa's cumulative GPA for 3 semesters was 3.24 for 46 course units. Her fourth semester GPA was 3.86 for 12 course units. What is Tessa's cumulative GPA for all 4 semesters?

20. Richard's cumulative GPA for 3 semesters was 2.0 for 42 credits. His fourth semester GPA was 4.0 for 14 course units. What is Richard's cumulative GPA for all 4 semesters?

21. Calculate a Course Grade A professor grades students on 5 tests, a project, and a final examination. Each test counts as 10% of the course grade. The project counts as 20% of the course grade. The final examination counts as 30% of the course grade. Samantha has test scores of 70, 65, 82, 94, and 85. Samantha's project score is 92. Her final examination score is 80. Use the weighted mean formula to find Samantha's average for the course. *Hint:* The sum of all the weights is 100% = 1.

22. Calculate a Course Grade A professor grades students on 4 tests, a term paper, and a final examination. Each test counts as 15% of the course grade. The term paper counts as 20% of the course grade. The final examination counts as 20% of the course grade. Alan has test scores of 80, 78, 92, and 84. Alan received an 84 on his term paper. His final examination score was 88. Use the weighted mean formula to find Alan's average for the course. *Hint:* The sum of all the weights is 100% = 1.

Baseball In baseball, a batter's *slugging average*, which measures the batter's power as a hitter, is a type of weighted mean. If s, d, t, and h represent the numbers of singles, doubles, triples, and home runs, respectively, that a player achieves in n times at bat, then the player's slugging average is $\dfrac{s + 2d + 3t + 4h}{n}$.

Transcendental Graphics/Getty Images

■ In Exercises 23 to 26, find the player's slugging average for the season or seasons described. Slugging averages are given to the nearest thousandth.

23. Babe Ruth, in his first season with the New York Yankees (1920), was at bat 458 times and achieved 73 singles, 36 doubles, 9 triples, and 54 home runs. In this season, Babe Ruth achieved his highest slugging average, which stood as a major league record until 2001.

24. Babe Ruth, over his 22-year career, was at bat 8399 times and hit 1517 singles, 506 doubles, 136 triples, and 714 home runs.

25. Albert Pujols, in his 2006 season with the St. Louis Cardinals, was at bat 535 times and achieved 94 singles, 33 doubles, 1 triple, and 49 home runs.

26. Albert Pujols, during 10 years with the St. Louis Cardinals (2001–2010), was at bat 5733 times and hit 1051 singles, 426 doubles, 15 triples, and 408 home runs.

■ In Exercises 27 to 30, find the mean, the median, and all modes for the data in the given frequency distribution.

27. Points Scored by Lynn

Points scored in a basketball game	Frequency
2	6
4	5
5	6
9	3
10	1
14	2
19	1

28. Mystic Pizza Company

Hourly pay rates for employees	Frequency
$8.00	14
$11.50	9
$14.00	8
$16.00	5
$19.00	2
$22.50	1
$35.00	1

29. Quiz Scores

Scores on a biology quiz	Frequency
2	1
4	2
6	7
7	12
8	10
9	4
10	3

30. Ages of Science Fair Contestants

Age	Frequency
7	3
8	4
9	6
10	15
11	11
12	7
13	1

Meteorology In Exercises 31 to 34, use the following information about another measure of central tendency for a set of data, called the *midrange*. The **midrange** is defined as the value that is halfway between the minimum data value and the maximum data value. That is,

$$\text{Midrange} = \frac{\text{minimum value} + \text{maximum value}}{2}$$

The midrange is often stated as the *average* of a set of data in situations in which there are a large amount of data and the data are constantly changing. Many weather reports state the average daily temperature of a city as the midrange of the temperatures achieved during that day. For instance, if the minimum daily temperature of a city was 60° and the maximum daily temperature was 90°, then the midrange of the temperatures is $\frac{60° + 90°}{2} = 75°$.

31. Find the midrange of the following daily temperatures, which were recorded at 3-hour intervals.

52°, 65°, 71°, 74°, 76°, 75°, 68°, 57°, 54°

32. Find the midrange of the following daily temperatures, which were recorded at three-hour intervals.

−6°, 4°, 14°, 21°, 25°, 26°, 18°, 12°, 2°

33. During a 24-hour period on January 23–24, 1916, the temperature in Browning, Montana, decreased from a high of 44°F to a low of −56°F. Find the midrange of the temperatures during this 24-hour period. (*Source:* National Oceanic and Atmospheric Administration)

34. During a 2-minute period on January 22, 1943, the temperature in Spearfish, South Dakota, increased from a low of −4°F to a high of 45°F. Find the midrange of the temperatures during this 2-minute period. (*Source:* National Oceanic and Atmospheric Administration)

35. Test Scores After 6 biology tests, Ruben has a mean score of 78. What score does Ruben need on the next test to raise his average (mean) to 80?

36. Test Scores After 4 algebra tests, Alisa has a mean score of 82. One more 100-point test is to be given in this class. All of the test scores are of equal importance. Is it possible for Alisa to raise her average (mean) to 90? Explain.

37. Baseball For the first half of a baseball season, a player had 92 hits out of 274 times at bat. The player's batting average was $\frac{92}{274} \approx 0.336$. During the second half of the season, the player had 60 hits out of 282 times at bat. The player's batting average was $\frac{60}{282} \approx 0.213$.

a. What is the average (mean) of 0.336 and 0.213?

b. What is the player's batting average for the complete season?

c. Does the answer in part a equal the average in part b?

38. Commuting Times Mark averaged 60 mph during the 30-mile trip to college. Because of heavy traffic he was able to average only 40 mph during the return trip. What was Mark's average speed for the round trip?

EXTENSIONS

Consider the data in the following table.

Summary of Yards Gained in Two Football Games

	Game 1	Game 2	Combined statistics for both games
Warren	12 yds on 4 carries Average: 3 yds/carry	78 yds on 16 carries Average: 4.875 yds/carry	90 yds on 20 carries Average: 4.5 yds/carry
Barry	120 yds on 30 carries Average: 4 yds/carry	100 yds on 20 carries Average: 5 yds/carry	220 yds on 50 carries Average: 4.4 yds/carry

- In the first game, Barry has the better average.
- In the second game, Barry has the better average.
- If the statistics for the games are combined, Warren has the better average.

You may be surprised by the above results. After all, how can it be that Barry has the better average in game 1 and game 2, but he does not have the better average for both games? In statistics, an example such as this is known as a **Simpson's paradox**.

■ Form groups of three or four students to work Exercises 39 and 40.

39. Consider the following data.

Batting Statistics for Two Baseball Players

	First month	Second month	Both months
Dawn	2 hits; 5 at-bats Average: ?	19 hits; 49 at-bats Average: ?	? hits; ? at-bats Average: ?
Joanne	29 hits; 73 at-bats Average: ?	31 hits; 80 at-bats Average: ?	? hits; ? at-bats Average: ?

Is this an example of a Simpson's paradox? Explain.

40. Consider the following data.

Test Scores for Two Students

	English	History	English and history combined
Wendy	84, 65, 72, 91, 99, 84 Average: ?	66, 84, 75, 77, 94, 96, 81 Average: ?	Average: ?
Sarah	90, 74 Average: ?	68, 78, 98, 76, 68, 92, 88, 86 Average: ?	Average: ?

Is this an example of a Simpson's paradox? Explain.

SECTION 13.2 Measures of Dispersion

The Range

TABLE 13.5
Soda Dispensed (ounces)

Machine 1	Machine 2
9.52	8.01
6.41	7.99
10.07	7.95
5.85	8.03
8.15	8.02
$\bar{x} = 8.0$	$\bar{x} = 8.0$

In the preceding section we introduced three types of average values for a data set—the mean, the median, and the mode. Some characteristics of a set of data may not be evident from an examination of averages. For instance, consider a soft-drink dispensing machine that should dispense 8 oz of your selection into a cup. Table 13.5 shows data for two of these machines.

The mean data value for each machine is 8 oz. However, look at the variation in data values for Machine 1. The quantity of soda dispensed is very inconsistent—in some cases the soda overflows the cup, and in other cases too little soda is dispensed. The machine obviously needs adjustment. Machine 2, on the other hand, is working just fine. The quantity dispensed is very consistent, with little variation

This example shows that average values do not reflect the *spread* or *dispersion* of data. To measure the spread or dispersion of data, we must introduce statistical values known as the *range* and the *standard deviation.*

Range

The **range** of a set of data values is the difference between the greatest data value and the least data value.

EXAMPLE 1 Find a Range

Find the range of the numbers of ounces dispensed by Machine 1 in Table 13.5.

Solution

The greatest number of ounces dispensed is 10.07 and the least is 5.85. The range of the numbers of ounces dispensed is $10.07 - 5.85 = 4.22$ oz.

CHECK YOUR PROGRESS 1 Find the range of the numbers of ounces dispensed by Machine 2 in Table 13.5.

Solution See page S46.

MATH MATTERS A World Record Range

Robert Wadlow

The tallest man for whom there is irrefutable evidence was Robert Pershing Wadlow. On June 27, 1940, Wadlow was 8 ft 11.1 in. tall. The shortest man for whom there is reliable evidence is Chandra Bahadur Dangi. On February 26, 2012, he was 21.5 in. tall. (*Source:* Guinness World Records) The range of the heights of these men is $107.1 - 21.5 = 85.6$ in.

TABLE 13.6
Machine 2: Deviations from the Mean

x	$x - \bar{x}$
8.01	$8.01 - 8 = 0.01$
7.99	$7.99 - 8 = -0.01$
7.95	$7.95 - 8 = -0.05$
8.03	$8.03 - 8 = 0.03$
8.02	$8.02 - 8 = 0.02$
Sum of deviations $= 0$	

The Standard Deviation

The range of a set of data is easy to compute, but it can be deceiving. The range is a measure that depends only on the two most extreme values, and as such it is very sensitive. A measure of dispersion that is less sensitive to extreme values is the *standard deviation.* The standard deviation of a set of numerical data makes use of the amount by which each individual data value deviates from the mean. These deviations, represented by $(x - \bar{x})$, are positive when the data value x is greater than the mean $\bar{x}$ and are negative when x is less than the mean $\bar{x}$. The sum of all the deviations $(x - \bar{x})$ is 0 for all sets of data. This is shown in Table 13.6 for the Machine 2 data of Table 13.5.

Because the sum of all the deviations of the data values from the mean is *always* 0, we cannot use the sum of the deviations as a measure of dispersion for a set of data. Instead, the standard deviation uses the sum of the *squares* of the deviations.

TAKE NOTE

You may question why a denominator of $n - 1$ is used instead of n when we compute a sample standard deviation. The reason is that a sample standard deviation is often used to estimate the population standard deviation, and it can be shown mathematically that the use of $n - 1$ tends to yield better estimates.

Standard Deviations for Populations and Samples

If $x_1, x_2, x_3, \ldots, x_n$ is a *population* of n numbers with a mean of μ, then the **standard deviation** of the population is $\sigma = \sqrt{\frac{\Sigma(x-\mu)^2}{n}}$ (1).

If $x_1, x_2, x_3, \ldots, x_n$ is a *sample* of n numbers with a mean of $\bar{x}$, then the **standard deviation** of the sample is $s = \sqrt{\frac{\Sigma(x-\bar{x})^2}{n-1}}$ (2).

PPT

Most statistical applications involve a sample rather than a population, which is the complete set of data values. Sample standard deviations are designated by the lowercase letter s. In those cases in which we *do* work with a population, we designate the standard deviation of the population by σ, which is the lowercase Greek letter sigma. We can use the following procedure to calculate the standard deviation of n numbers.

Procedure for Computing a Standard Deviation

1. Determine the mean of the n numbers.
2. For each number, calculate the deviation (difference) between the number and the mean of the numbers.
3. Calculate the square of each deviation and find the sum of these squared deviations.
4. If the data is a *population*, then divide the sum by n. If the data is a *sample*, then divide the sum by $n - 1$.
5. Find the square root of the quotient in Step 4.

PPT

EXAMPLE 2 Find the Standard Deviation

The following numbers were obtained by sampling a population.

2, 4, 7, 12, 15

Find the standard deviation of the sample.

Solution

Step 1: The mean of the numbers is

$$\bar{x} = \frac{2 + 4 + 7 + 12 + 15}{5} = \frac{40}{5} = 8$$

Step 2: For each number, calculate the deviation between the number and the mean.

x	$x - \bar{x}$
2	$2 - 8 = -6$
4	$4 - 8 = -4$
7	$7 - 8 = -1$
12	$12 - 8 = 4$
15	$15 - 8 = 7$

TAKE NOTE

Because the sum of the deviations is always 0, you can use this as a means to check your arithmetic. That is, if your deviations from the mean do not add to 0, then you know you have made an error.

Step 3: Calculate the square of each deviation in Step 2, and find the sum of these squared deviations.

x	$x - \bar{x}$	$(x - \bar{x})^2$
2	$2 - 8 = -6$	$(-6)^2 = 36$
4	$4 - 8 = -4$	$(-4)^2 = 16$
7	$7 - 8 = -1$	$(-1)^2 = 1$
12	$12 - 8 = 4$	$4^2 = 16$
15	$15 - 8 = 7$	$7^2 = 49$
		118 ← Sum of the squared deviations

Step 4: Because we have a sample of $n = 5$ values, divide the sum 118 by $n - 1$, which is 4.

$$\frac{118}{4} = 29.5$$

Step 5: The standard deviation of the sample is $s = \sqrt{29.5}$. To the nearest hundredth, the standard deviation is $s = 5.43$.

CHECK YOUR PROGRESS 2 A student has the following quiz scores: 5, 8, 16, 17, 18, 20. Find the standard deviation for this population of quiz scores.

Solution See page S46.

In the next example we use standard deviations to determine which company produces batteries that are most consistent with regard to their life expectancy.

EXAMPLE 3 Use Standard Deviations

A consumer group has tested a sample of 8 size-D batteries from each of 3 companies. The results of the tests are shown in the following table. According to these tests, which company produces batteries for which the values representing hours of constant use have the smallest standard deviation?

Company	Hours of constant use per battery
EverSoBright	6.2, 6.4, 7.1, 5.9, 8.3, 5.3, 7.5, 9.3
Dependable	6.8, 6.2, 7.2, 5.9, 7.0, 7.4, 7.3, 8.2
Beacon	6.1, 6.6, 7.3, 5.7, 7.1, 7.6, 7.1, 8.5

Solution

The mean for each sample of batteries is 7 h.
The batteries from EverSoBright have a standard deviation of

$$s_1 = \sqrt{\frac{(6.2 - 7)^2 + (6.4 - 7)^2 + \cdots + (9.3 - 7)^2}{7}}$$
$$= \sqrt{\frac{12.34}{7}} \approx 1.328 \text{ h}$$

The batteries from Dependable have a standard deviation of

$$s_2 = \sqrt{\frac{(6.8 - 7)^2 + (6.2 - 7)^2 + \cdots + (8.2 - 7)^2}{7}}$$
$$= \sqrt{\frac{3.62}{7}} \approx 0.719 \text{ h}$$

The batteries from Beacon have a standard deviation of

$$s_3 = \sqrt{\frac{(6.1 - 7)^2 + (6.6 - 7)^2 + \cdots + (8.5 - 7)^2}{7}}$$
$$= \sqrt{\frac{5.38}{7}} \approx 0.877 \text{ h}$$

The batteries from Dependable have the smallest standard deviation. According to these results, the Dependable company produces the most consistent batteries with regard to life expectancy under constant use.

INSTRUCTOR NOTE

You may wish to show your students the formulas

$$\sigma = \sqrt{\frac{\Sigma(x^2)}{n} - \mu^2} \text{ and}$$

$$s = \sqrt{\frac{n(\Sigma x^2) - (\Sigma x)^2}{n(n-1)}}$$

These formulas provide a more efficient method of computing, by hand, the standard deviation of $x_1, x_2, \ldots, x_n$ for large values of n.

If your students have access to graphing calculators, the procedure shown in Example 4 will enable them to quickly and effortlessly find the standard deviation of a set of data.

CHECK YOUR PROGRESS 3 A consumer testing agency has tested the strengths of 3 brands of $\frac{1}{8}$-inch rope. The results of the tests are shown in the following table. According to the sample test results, which company produces $\frac{1}{8}$-inch rope for which the breaking point has the smallest standard deviation?

Company	Breaking point of $\frac{1}{8}$-inch rope in pounds
Trustworthy	122, 141, 151, 114, 108, 149, 125
Brand X	128, 127, 148, 164, 97, 109, 137
NeverSnap	112, 121, 138, 131, 134, 139, 135

Solution See page S47.

Many calculators have built-in statistics features for calculating the mean and standard deviation of a set of numbers. The next example illustrates these features on a TI-83/84 graphing calculator.

EXAMPLE 4 Use a Calculator to Find the Mean and Standard Deviation

Use a graphing calculator to find the mean and standard deviation of the times in the following table. Because the table contains all the winning times for this race (up to the year 2012), the data set is a population.

Olympic Women's 400-Meter Dash Results, in Seconds, 1969–2012

52.0	51.08	49.29	48.88	48.83	48.65	48.83	48.25	49.11	49.41	49.62	49.55

Solution

On a TI-83/84 calculator, press STAT ENTER and then enter the above times into list L1. Press STAT ▻ ENTER ENTER. The calculator displays the mean and standard deviations shown below. Because we are working with a population, we are interested in the population standard deviation. From the calculator screen, $\bar{x} \approx 49.458$ s and $\sigma x \approx 1.021$ s.

TAKE NOTE

Because the calculations of the population mean and the sample mean are the same, a graphing calculator uses the same symbol $\bar{x}$ for both. The symbols for the population standard deviation, σx, and the sample standard deviation, sx, are different.

TI-83/84 Display of List 1

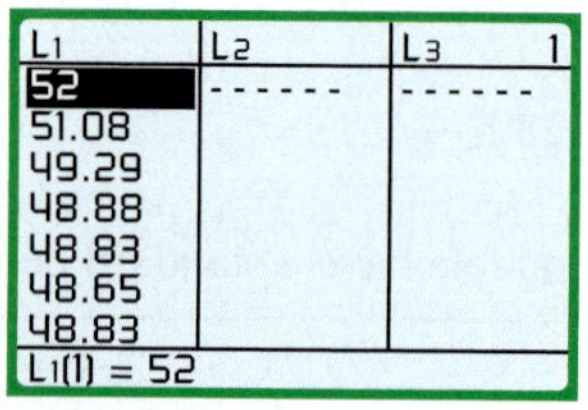

TI-83/84 Display of $\bar{x}$, s, and σ

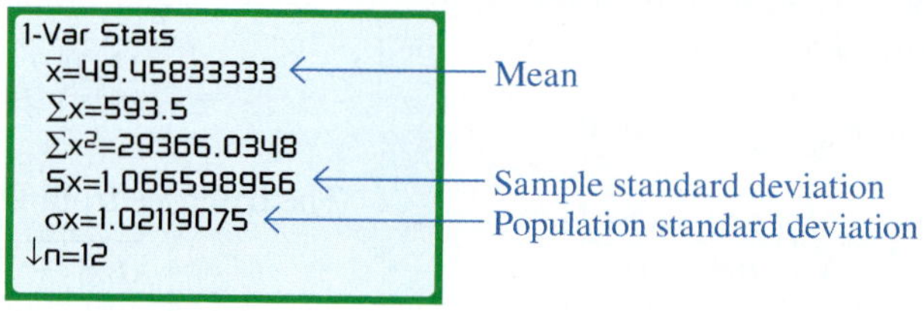

CHECK YOUR PROGRESS 4 Use a graphing calculator to find the mean and the population standard deviation of the race times in the following table.

Olympic Men's 400-Meter Dash Results, in Seconds, 1900–2012

49.4	49.2	53.2	50.0	48.2	49.6	47.6	47.8	46.2
46.5	46.2	45.9	46.7	44.9	45.1	43.8	44.66	44.26
44.60	44.27	43.87	43.50	43.49	43.84	44.00	43.75	43.94

Solution See page S47.

The Variance

A statistic known as the *variance* is also used as a measure of dispersion. The **variance** for a given set of data is the square of the standard deviation of the data. The following chart shows the mathematical notations that are used to denote standard deviations and variances.

Notations for Standard Deviation and Variance

σ is the standard deviation of a population.

σ^2 is the variance of a population.

s is the standard deviation of a sample.

s^2 is the variance of a sample.

EXAMPLE 5 Find the Variance

Find the variance for the sample given in Example 2.

Solution

In Example 2, we found $s = \sqrt{29.5}$. The variance is the square of the standard deviation. Thus the variance is $s^2 = (\sqrt{29.5})^2 = 29.5$.

CHECK YOUR PROGRESS 5 Find the variance for the population given in Check Your Progress 2.

Solution See page S47.

QUESTION Can the variance of a data set be less than the standard deviation of the data set?

Although the variance of a set of data is an important measure of dispersion, it has a disadvantage that is not shared by the standard deviation: the variance does not have the same unit of measure as the original data. For instance, if a set of data consists of times measured in hours, then the variance of the data will be measured in *square* hours. The standard deviation of this data set is the square root of the variance, and as such it is measured in hours, which is a more intuitive unit of measure.

ANSWER Yes. The variance is less than the standard deviation whenever the standard deviation is less than 1.

EXCURSION

A Geometric View of Variance and Standard Deviation[1]

The following geometric explanation of the variance and standard deviation of a set of data is designed to provide you with a deeper understanding of these important concepts.

Consider the data $x_1, x_2, \ldots, x_n$, which are arranged in ascending order. The average, or mean, of these data is

$$\mu = \frac{\Sigma x_i}{n}$$

and the variance is

$$\sigma^2 = \frac{\Sigma(x_i - \mu)^2}{n}$$

In the last formula, each term $(x_i - \mu)^2$ can be pictured as the area of a square whose sides are of length $|x_i - \mu|$, the distance between the ith data value and the mean. We will refer to these squares as *tiles*, denoting by T_i the area of the tile associated with the data value x_i. Thus $\sigma^2 = \frac{\Sigma T_i}{n}$, which means that the variance may be thought of as the *area of the average-sized tile* and the standard deviation as the length of a side of this average-sized tile. By drawing the tiles associated with a data set, as shown on the next page, you can

TAKE NOTE

Up to this point we have used $\mu = \frac{\Sigma x}{n}$ as the formula for the mean. However, many statistics texts use the formula $\mu = \frac{\Sigma x_i}{n}$ for the mean. Letting the subscript i vary from 1 to n helps us to remember that we are finding the sum of all the numbers $x_1, x_2, x_3, \ldots, x_n$.

[1] Adapted with permission from "Chebyshev's Theorem: A Geometric Approach," *The College Mathematics Journal,* Vol. 26, No. 2, March 1995. Article by Pat Touhey, College Misericordia, Dallas, PA 18612.

visually estimate an average-sized tile, and thus you can roughly approximate the variance and standard deviation.

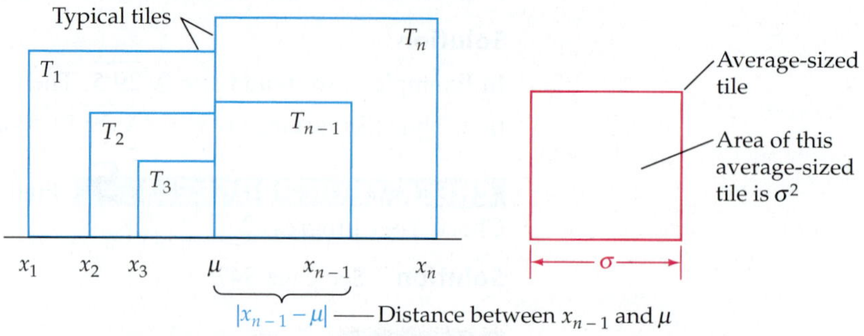

A typical data set, with its associated tiles and average-sized tile

These geometric representations of variance and standard deviation enable us to visualize how these values are used as measures of the dispersion of a set of data. If all of the data are bunched up near the mean, it is clear that the average-sized tile will be small and, consequently, so will its side length, which represents the standard deviation. But if even a small portion of the data lies far from the mean, the average-sized tile may be rather large, and thus its side length will also be large.

EXCURSION EXERCISES

1. This exercise makes use of the geometric procedure just explained to calculate the variance and standard deviation of the population 2, 5, 7, 11, 15. The following figure shows the given set of data labeled on a number line, along with its mean, which is 8.

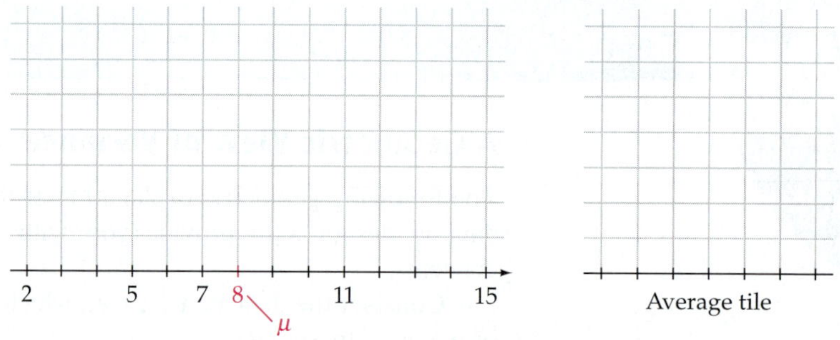

a. Draw the tile associated with each of the five data values 2, 5, 7, 11, and 15.

b. Label each tile with its area.

c. Find the sum of the areas of all the tiles.

d. Find the average (mean) of the areas of all 5 tiles.

e. To the right of the above number line, draw a tile whose area is the average found in part d.

f. What is the variance of the data? What geometric figure represents the variance?

g. What is the standard deviation of the data? What geometric figure represents the standard deviation?

2. **a.** to **g.** Repeat all of the steps described in Excursion Exercise 1 for the data set

6, 8, 9, 11, 16

h. Which of the data sets in these two Excursion exercises has the larger mean? Which data set has the larger standard deviation?

EXERCISE SET 13.2

1. **Meteorology** During a 12-hour period on December 24, 1924, the temperature in Fairfield, Montana, dropped from a high of 63°F to a low of −21°F. What was the range of the temperatures during this period? (*Source:* National Oceanic and Atmospheric Administration)

2. **Meteorology** During a 2-hour period on January 12, 1911, the temperature in Rapid City, South Dakota, dropped from a high of 49°F to a low of −13°F. What was the range of the temperatures during this period? (*Source:* National Oceanic and Atmospheric Administration)

■ In Exercises 3 to 12, find the range, the standard deviation, and the variance for the given *samples*. Round noninteger results to the nearest tenth.

3. 1, 2, 5, 7, 8, 19, 22
4. 3, 4, 7, 11, 12, 12, 15, 16
5. 2.1, 3.0, 1.9, 1.5, 4.8
6. 5.2, 11.7, 19.1, 3.7, 8.2, 16.3
7. 48, 91, 87, 93, 59, 68, 92, 100, 81
8. 93, 67, 49, 55, 92, 87, 77, 66, 73, 96, 54
9. 4, 4, 4, 4, 4, 4, 4, 4, 4, 4, 4, 4, 4, 4, 4, 4, 4
10. 8, 6, 8, 6, 8, 6, 8, 6, 8, 6, 8, 6, 8
11. −8, −5, −12, −1, 4, 7, 11
12. −23, −17, −19, −5, −4, −11, −31

13. **Mountain Climbing** A mountain climber plans to buy some rope to use as a lifeline. Which of the following would be the better choice? Explain your choice.

    Rope A: Mean breaking strength: 500 lb; standard deviation of 100 lb

    Rope B: Mean breaking strength: 500 lb; standard deviation of 10 lb

14. **Lotteries** Which would you expect to be larger: the standard deviation of 5 random numbers picked from 1 to 47 in the California Super Lotto, or the standard deviation of 5 random numbers picked from 1 to 69 in the multistate PowerBall lottery?

15. **Weights of Students** Which would you expect to be the larger standard deviation: the standard deviation of the weights of 25 students in a first-grade class, or the standard deviation of the weights of 25 students in a college statistics course?

16. Evaluate the accuracy of the following statement: When the mean of a data set is large, the standard deviation will be large.

17. **Fuel Efficiency** The fuel efficiency, in miles per gallon, of 10 small utility trucks was measured. The results are recorded in the table below.

Fuel Efficiency (mpg)

22	25	23	27	15	24	24	32	23	22	25	22

Find the mean and sample standard deviation of these data. Round to the nearest hundredth.

18. **Waiting Times** A customer at a specialty coffee shop observed the amount of time, in minutes, that each of 20 customers spent waiting to receive an order. The results are recorded in the table below.

Time (min) to receive order

3.2	4.0	3.8	2.4	4.7	5.1	4.6	3.5	3.5	6.2
3.5	4.9	4.5	5.0	2.8	3.5	2.2	3.9	5.3	2.9

Find the mean and sample standard deviation of these data. Round to the nearest hundredth.

19. **Fast-food Calories** A survey of 10 fast-food restaurants noted the number of calories in a mid-sized hamburger. The results are given in the table below.

Calories in a mid-sized hamburger

514	507	502	498	496	506	458	478	463	514

Find the mean and sample standard deviation of these data. Round to the nearest hundredth.

20. **Energy Drinks** A survey of 16 energy drinks noted the caffeine concentration of each drink in milligrams per ounce. The results are given in the table below.

Concentration of caffeine (mg/oz)

9.1	7.5	7.8	8.9	9.0	8.2	9.1	8.7
9.0	7.7	8.8	8.9	9.0	9.1	8.2	8.9

Find the mean and sample standard deviation of these data. Round to the nearest hundredth.

21. **Weekly Commute Times** A survey of 15 large cities noted the average weekly commute times, in hours, of the residents of each city. The results are recorded in the table below.

Weekly commute time (h)

4.5	4.0	5.8	5.4	4.7
4.0	3.6	3.9	4.7	3.7
4.6	3.4	3.5	3.9	4.4

Find the mean and sample standard deviation of these data. Round to the nearest hundredth.

22. Biology Some studies show that the mean normal human body temperature is actually somewhat lower than the commonly given value of 98.6°F. This is reflected in the following data set of body temperatures.

Body Temperatures (°F) of 30 Healthy Adults

97.1	97.8	98.0	98.7	99.5	96.3
98.4	98.5	98.0	100.8	98.6	98.2
99.0	99.3	98.8	97.6	97.4	99.0
97.4	96.4	98.0	98.1	97.8	98.5
98.7	98.8	98.2	97.6	98.2	98.8

a. Find the mean and *sample* standard deviation of the body temperatures. Round each result to the nearest hundredth.

b. Are there any temperatures in the data set that do not lie within 2 standard deviations of the mean? If so, list them.

23. Recording Industry The table below shows a random sample of the lengths of songs in a playlist.

Lengths of Songs (minutes:seconds)

3:42	3:40	3:50	3:17	3:15	3:37
2:27	3:01	3:47	3:49	4:02	3:30

a. Find the mean and *sample* standard deviation of the song lengths, in seconds. Round each result to the nearest second.

b. Are there any song lengths in the data set that do not lie within 1 standard deviation of the mean? If so, list them.

EXTENSIONS

24. Pick 5 numbers and compute the *population* standard deviation of the numbers.

a. Add a nonzero constant c to each of your original numbers, and compute the standard deviation of this new population.

b. Use the results of part a and inductive reasoning to state what happens to the standard deviation of a population when a nonzero constant c is added to each data item.

25. Pick 6 numbers and compute the *population* standard deviation of the numbers.

a. Double each of your original numbers and compute the standard deviation of this new population.

b. Use the results of part a and inductive reasoning to state what happens to the standard deviation of a population when each data item is multiplied by a positive constant k.

26. a. All of the numbers in a sample are the same number. What is the standard deviation of the sample?

b. If the standard deviation of a sample is 0, must all of the numbers in the sample be the same number?

c. If two samples both have the same standard deviation, are the samples necessarily identical?

27. Under what condition would the variance of a sample be equal to the standard deviation of the sample?

SECTION 13.3 Measures of Relative Position

z-Scores

Consider an Internet site that offers movie downloads. Based on data kept by the site, an estimate of the mean time to download a certain movie is 12 min, with a standard deviation of 4 min. When you download this movie, the download takes 20 min, which you think is an unusually long time for the download. On the other hand, when your friend downloads the movie, the download takes only 6 min, and your friend is pleasantly surprised at how quickly she receives the movie. In each case, a data value far from the mean is unexpected.

The graph below shows the download times for this movie using two different measures: the number of *minutes* a download time is from the mean and the number of *standard deviations* the download time is from the mean.

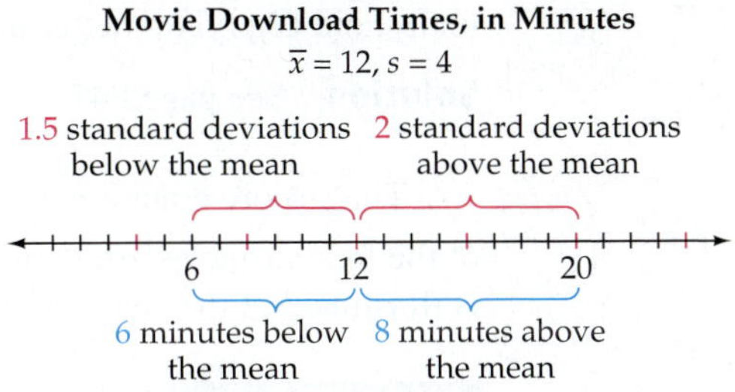

The number of standard deviations between a data value and the mean is known as the data value's *z-score* or *standard score*.

z-Score

The **z-score** for a given data value x is the number of standard deviations that x is above or below the mean of the data. The following formulas show how to calculate the z-score for a data value x in a population and in a sample.

$$\text{Population: } z_x = \frac{x - \mu}{\sigma} \qquad \text{Sample: } z_x = \frac{x - \bar{x}}{s}$$

QUESTION What does a z-score of 3 for a data value x represent? What does a z-score of -1 for a data value x represent?

In the next example, we use a student's z-scores for two tests to determine how well the student did on each test in comparison to the other students.

EXAMPLE 1 Compare z-Scores

Raul has taken two tests in his chemistry class. He scored 72 on the first test, for which the mean of all scores was 65 and the standard deviation was 8. He received a 60 on a second test, for which the mean of all scores was 45 and the standard deviation was 12. In comparison to the other students, did Raul do better on the first test or the second test?

Solution

Find the z-score for each test.

$$z_{72} = \frac{72 - 65}{8} = 0.875 \qquad z_{60} = \frac{60 - 45}{12} = 1.25$$

Raul scored 0.875 standard deviation above the mean on the first test and 1.25 standard deviations above the mean on the second test. These z-scores indicate that, in comparison to his classmates, Raul scored better on the second test than he did on the first test.

ANSWER A z-score of 3 for a data value x means that x is 3 standard deviations above the mean. A z-score of -1 for a data value x means that x is 1 standard deviation below the mean.

CHECK YOUR PROGRESS 1 Cheryl has taken two quizzes in her history class. She scored 15 on the first quiz, for which the mean of all scores was 12 and the standard deviation was 2.4. Her score on the second quiz, for which the mean of all scores was 11 and the standard deviation was 2.0, was 14. In comparison to her classmates, did Cheryl do better on the first quiz or the second quiz?

Solution See page S47.

The z-score equation $z_x = \frac{x - \bar{x}}{s}$ involves four variables. If the values of any three of the four variables are known, you can solve for the unknown variable. This procedure is illustrated in the next example.

EXAMPLE 2 Use z-Scores

A consumer group tested a sample of 100 light bulbs. It found that the mean life expectancy of the bulbs was 842 h, with a standard deviation of 90. One particular light bulb from the DuraBright Company had a z-score of 1.2. What was the life span of this light bulb?

Solution

Substitute the given values into the z-score equation and solve for x.

$$z_x = \frac{x - \bar{x}}{s}$$

$$1.2 = \frac{x - 842}{90} \qquad \bullet\ z_x = 1.2, \bar{x} = 842, s = 90$$

$$108 = x - 842 \qquad \bullet\ \text{Solve for } x.$$

$$950 = x$$

The light bulb had a life span of 950 h.

CHECK YOUR PROGRESS 2 Roland received a score of 70 on a test for which the mean score was 65.5. Roland has learned that the z-score for his test is 0.6. What is the standard deviation for this set of test scores?

Solution See page S47.

Percentiles

Most standardized examinations provide scores in terms of *percentiles*, which are defined as follows:

pth Percentile

A value x is called the **pth percentile** of a data set provided $p\%$ of the data values are less than x.

EXAMPLE 3 Using Percentiles

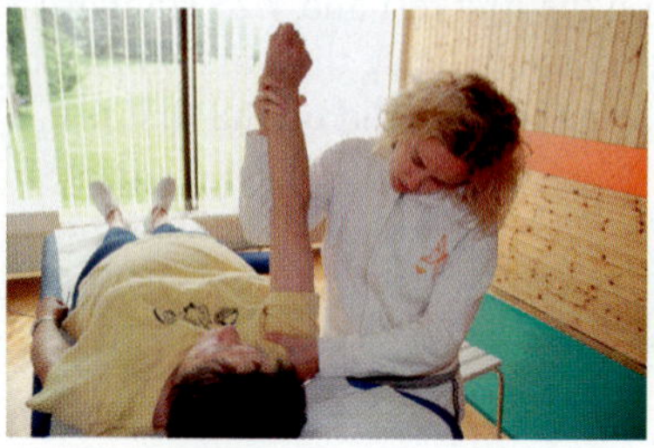
Eric Fahrner/Shutterstock.com

In a recent year, the median annual salary for a physical therapist was \$74,480. If the 90th percentile for the annual salary of a physical therapist was \$105,900, find the percent of physical therapists whose annual salary was

a. more than \$74,480.

b. less than \$105,900.

c. between \$74,480 and \$105,900.

Solution

a. By definition, the median is the 50th percentile. Therefore, 50% of the physical therapists earned more than \$74,480 per year.

b. Because \$105,900 is the 90th percentile, 90% of all physical therapists made less than \$105,900.

c. From parts a and b, $90\% - 50\% =$ 40% of the physical therapists earned between \$74,480 and \$105,900.

CHECK YOUR PROGRESS 3 The median annual salary for a police dispatcher in a large city was \$44,528. If the 25th percentile for the annual salary of a police dispatcher was \$32,761, find the percent of police dispatchers whose annual salaries were

a. less than \$44,528.

b. more than \$32,761.

c. between \$32,761 and \$44,528.

Solution See page S47.

The following formula can be used to find the percentile that corresponds to a particular data value in a set of data.

Percentile for a Given Data Value

Given a set of data and a data value x,

$$\text{Percentile of score } x = \frac{\text{number of data values less than } x}{\text{total number of data values}} \cdot 100$$

EXAMPLE 4 Find a Percentile

On a reading examination given to 900 students, Elaine's score of 602 was higher than the scores of 576 of the students who took the examination. What is the percentile for Elaine's score?

Solution

$$\begin{aligned}\text{Percentile} &= \frac{\text{number of data values less than 602}}{\text{total number of data values}} \cdot 100 \\ &= \frac{576}{900} \cdot 100 \\ &= 64\end{aligned}$$

Elaine's score of 602 places her at the 64th percentile.

CHECK YOUR PROGRESS 4 On an examination given to 8600 students, Hal's score of 405 was higher than the scores of 3952 of the students who took the examination. What is the percentile for Hal's score?

Solution See page S47.

MATH MATTERS Standardized Tests and Percentiles

Standardized tests, such as the Scholastic Assessment Test (SAT), are designed to measure all students by a *single* standard. The SAT is used by many colleges as part of their admissions criteria. The SAT I is a 3-hour examination that measures verbal and mathematical reasoning skills. Scores on each portion of the test range from 200 to 800 points. SAT scores are generally reported in points and percentiles. Sometimes students are confused by the percentile score. For instance, if a student scores 650 points on the mathematics portion of the SAT and is told that this score is in the 85th percentile, this *does not* indicate that the student answered 85% of the questions correctly. An 85th percentile score means that the student scored higher than 85% of the students who took the test. Consequently, the student scored lower than 15% (100% − 85%) of the students who took the test.

Quartiles

The three numbers Q_1, Q_2, and Q_3 that partition a ranked data set into four (approximately) equal groups are called the **quartiles** of the data. For instance, for the data set below, the values $Q_1 = 11$, $Q_2 = 29$, and $Q_3 = 104$ are the quartiles of the data.

2, 5, 5, 8, 11, 12, 19, 22, 23, 29, 31, 45, 83, 91, 104, 159, 181, 312, 354

(arrows: 11 ↑ Q_1; 29 ↑ Q_2; 104 ↑ Q_3)

The quartile Q_1 is called the *first quartile.* The quartile Q_2 is called the *second quartile.* It is the median of the data. The quartile Q_3 is called the *third quartile.* The following method of finding quartiles makes use of medians.

The Median Procedure for Finding Quartiles

1. Rank the data.
2. Find the median of the data. This is the second quartile, Q_2.
3. The first quartile, Q_1, is the median of the data values less than Q_2. The third quartile, Q_3, is the median of the data values greater than Q_2.

EXAMPLE 5 Use Medians to Find the Quartiles of a Data Set

The following table lists the calories per 100 milliliters of 25 popular sodas. Find the quartiles for the data.

Calories, per 100 milliliters, of Selected Sodas

43	37	42	40	53	62	36	32	50	49
26	53	73	48	45	39	45	48	40	56
41	36	58	42	39					

Solution

Step 1: Rank the data as shown in the following table.

1) 26	**2)** 32	**3)** 36	**4)** 36	**5)** 37	**6)** 39	**7)** 39	**8)** 40	**9)** 40
10) 41	**11)** 42	**12)** 42	**13)** 43	**14)** 45	**15)** 45	**16)** 48	**17)** 48	**18)** 49
19) 50	**20)** 53	**21)** 53	**22)** 56	**23)** 58	**24)** 62	**25)** 73		

Step 2: The median of these 25 data values has a rank of 13. Thus the median is 43. The second quartile Q_2 is the median of the data, so $Q_2 = 43$.

Step 3: There are 12 data values less than the median and 12 data values greater than the median. The first quartile is the median of the data values less than the median. Thus Q_1 is the mean of the data values with ranks of 6 and 7.

$$Q_1 = \frac{39 + 39}{2} = 39$$

The third quartile is the median of the data values greater than the median. Thus Q_3 is the mean of the data values with ranks of 19 and 20.

$$Q_3 = \frac{50 + 53}{2} = 51.5$$

CHECK YOUR PROGRESS 5 The following table lists the weights, in ounces, of 15 avocados in a random sample. Find the quartiles for the data.

Weights, in ounces, of Avocados

12.4	10.8	14.2	7.5	10.2	11.4	12.6	12.8	13.1	15.6
9.8	11.4	12.2	16.4	14.5					

Solution See page S47.

Box-and-Whisker Plots

A **box-and-whisker plot** (sometimes called a **box plot**) is often used to provide a visual summary of a set of data. A box-and-whisker plot shows the median, the first and third quartiles, and the minimum and maximum values of a data set. See the figure below.

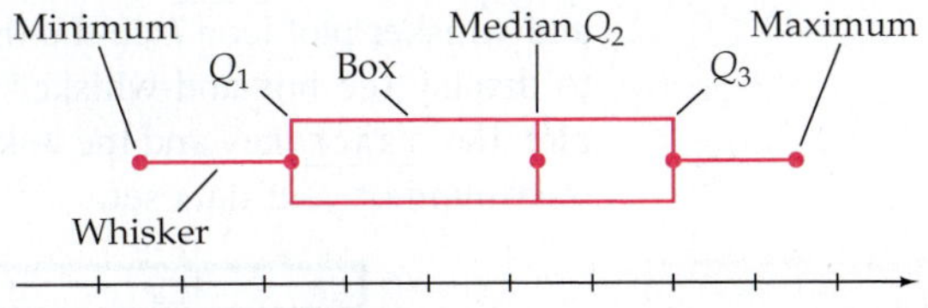

A box-and-whisker plot

TAKE NOTE

If a set of data is not ranked, then you need to rank the data so that you can easily determine the median, the first and third quartiles, and the minimum and maximum.

Construction of a Box-and-Whisker Plot

1. Draw a horizontal scale that extends from the minimum data value to the maximum data value.
2. Above the scale, draw a rectangle (box) with its left side at Q_1 and its right side at Q_3.
3. Draw a vertical line segment across the rectangle at the median, Q_2.
4. Draw a horizontal line segment, called a whisker, that extends from Q_1 to the minimum and another whisker that extends from Q_3 to the maximum.

HISTORICAL NOTE

Bettmann/Getty Images

John W. Tukey (1915–2000) made many important contributions to the field of statistics during his career at Princeton University and Bell Labs. He invented box plots when he was working on exploratory data analysis.

EXAMPLE 6 Construct a Box-and-Whisker Plot

Construct a box-and-whisker plot for the data set in Example 5.

Solution

For the data set in Example 5, we determined that $Q_1 = 39$, $Q_2 = 43$, and $Q_3 = 51.5$. The minimum data value for the data set is 26, and the maximum data value is 73. Thus the box-and-whisker plot is as shown below.

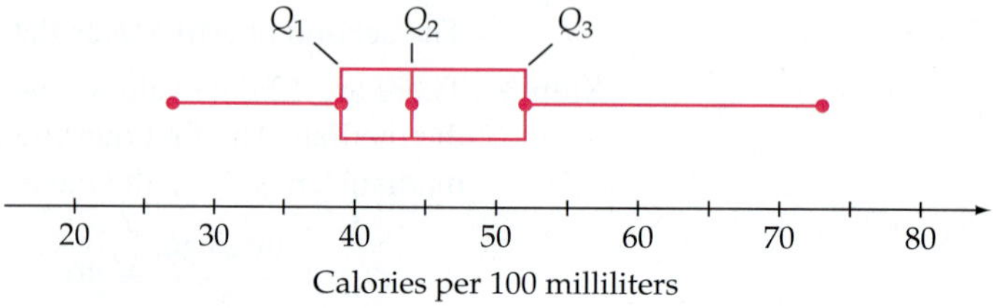

CHECK YOUR PROGRESS 6 Construct a box-and-whisker plot for the following data.

Number of Rooms Occupied in a Resort during an 18-Day Period

86	77	58	45	94	96	83	76	75
65	68	72	78	85	87	92	55	61

Solution See page S47.

Box plots are popular because they are easy to construct and they illustrate several important features of a data set in a simple diagram. Note from the box plot in Example 6 that we can easily estimate

- the quartiles of the data.
- the range of the data.
- the position of the middle half of the data as shown by the length of the box.

Some graphing calculators can be used to produce box-and-whisker plots. For instance, on a TI-83/84, you enter the data into a list, as shown on the first screen in Figure 13.1. The WINDOW menu is used to enter appropriate boundaries that contain all the data. Use the key sequence 2nd [STAT PLOT] ENTER and choose from the Type menu the box-and-whisker plot icon (see the third screen in Figure 13.1). The GRAPH key is then used to display the box-and-whisker plot. After the calculator displays the box-and-whisker plot, the TRACE key and the ⊳ key enable you to view Q_1, Q_2, Q_3, and the minimum and maximum of your data set.

TAKE NOTE

The following data were used to produce the box plot shown in Figure 13.1.

21.2, 20.5, 17.0, 16.8, 16.8, 16.5, 16.2, 14.0, 13.7, 13.3, 13.1, 13.0, 12.4, 12.1, 12.0

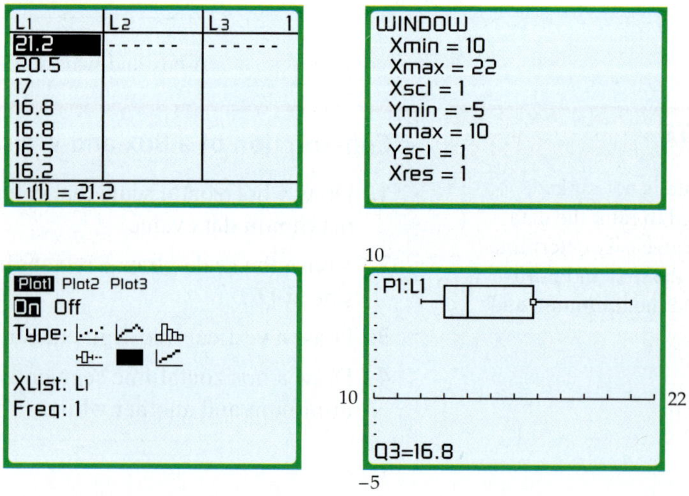

FIGURE 13.1 TI-83/84 screen displays

EXCURSION

Stem-and-Leaf Diagrams

The relative position of each data value in a small set of data can be graphically displayed by using a *stem-and-leaf diagram.* For instance, consider the following history test scores:

65, 72, 96, 86, 43, 61, 75, 86, 49, 68, 98, 74, 84, 78, 85, 75, 86, 73

A Stem-and-Leaf Diagram of a Set of History Test Scores

Stems	Leaves
4	3 9
5	
6	1 5 8
7	2 3 4 5 5 8
8	4 5 6 6 6
9	6 8

Legend: 8|6 represents 86

In the stem-and-leaf diagram at the left, we have organized the history test scores by placing all of the scores that are in the 40s in the top row, the scores that are in the 50s in the second row, the scores that are in the 60s in the third row, and so on. The tens digits of the scores have been placed to the left of the vertical line. In this diagram, they are referred to as *stems.* The ones digits of the test scores have been placed in the proper row to the right of the vertical line. In this diagram, they are the *leaves.* It is now easy to make observations about the distribution of the scores. Only two of the scores are in the 90s. Six of the scores are in the 70s, and none of the scores are in the 50s. The lowest score is 43, and the highest is 98.

Steps in Construction of a Stem-and-Leaf Diagram

1. Determine the stems and list them in a column from smallest to largest or largest to smallest.
2. List the remaining digit of each stem as a leaf to the right of the stem.
3. Include a *legend* that explains the meaning of the stems and the leaves. Include a title for the diagram.

The choice of how many leading digits to use as the stem will depend on the particular data set. For instance, consider the following data set, in which a travel agent has recorded the amount spent by customers for a cruise.

Amount Spent for a Cruise

\$3600	\$4700	\$7200	\$2100	\$5700	\$4400	\$9400
\$6200	\$5900	\$2100	\$4100	\$5200	\$7300	\$6200
\$3800	\$4900	\$5400	\$5400	\$3100	\$3100	\$4500
\$4500	\$2900	\$3700	\$3700	\$4800	\$4800	\$2400

One method of choosing the stems is to let each thousands digit be a stem and each hundreds digit be a leaf. If the stems and leaves are assigned in this manner, then the notation 2|1, with a stem of 2 and a leaf of 1, represents a cost of \$2100, and 5|4 represents a cost of \$5400. A stem-and-leaf diagram can now be constructed by writing all of the stems in a column from smallest to largest to the left of a vertical line and writing the corresponding leaves to the right of the line. See the diagram below.

Amount Spent for a Cruise

Stems	Leaves
2	1 1 4 9
3	1 1 6 7 7 8
4	1 4 5 5 7 8 8 9
5	2 4 4 7 9
6	2 2
7	2 3
8	
9	4

Legend:
7|3 represents \$7300

Sometimes two sets of data can be compared by using a *back-to-back stem-and-leaf diagram*, in which common stems are listed in the middle column of the diagram. Leaves from one data set are displayed to the right of the stems, and leaves from the other data set are displayed to the left. For instance, the back-to-back stem-and-leaf diagram below shows the test scores for two classes that took the same test. It is easy to see that the 8 A.M. class did better on the test because it had more scores in the 80s and 90s and fewer scores in the 40s, 50s, and 60s. The number of scores in the 70s was the same for both classes.

Biology Test Scores

8 A.M. class		10 A.M. class
2	4	5 8
7	5	6 7 9 9
5 8	6	2 3 4 8
1 2 3 3 3 7 8	7	1 3 3 5 5 6 8
4 4 5 5 6 8 8 9	8	2 3 6 6 6
2 4 5 5 8	9	4 5

Legend: 3|7 represents 73

Legend: 8|2 represents 82

EXCURSION EXERCISES

1. The following table lists the ages of customers who purchased a cruise. Construct a stem-and-leaf diagram for the data.

Ages of Customers Who Purchased a Cruise

32	45	66	21	62	68	72
61	55	23	38	44	77	64
46	50	33	35	42	45	51
51	28	40	41	52	52	33

2. Two groups of people were part of a test to determine how long, in seconds, it took to solve a logic problem when exposed to different ambient noise levels. Group 1 was given the problem in a room where a constant decibel (dB) level of 65 dB was maintained. For Group 2, the decibel level was maintained at 30 dB. The results are shown in the stem-and-leaf diagram below.

Group 1	Stem	Group 2
	3	3
4	4	2 2 6 8
8 3 1	5	0 4 4 4 5 8
6 2 2 2 1	6	2 3 5
5 3 2 1	7	4
6 1	8	

Legend: 8|5 represents 58 seconds

Legend: 5|8 represents 58 seconds

a. How many people in Group 1 required more than 60 s to solve the problem?

b. How many people in Group 2 required more than 60 s to solve the problem?

c. By just looking at the stem-and-leaf diagram, which group appears to have the larger mean time to solve the problem?

3. The exercise heart rate, in beats per minute (bpm), of 20 people was tested before and after a 10-week training program. The results are recorded in the table below.

Before	128	128	131	151	141	139	128	139	161	156
	136	134	134	136	116	174	158	148	156	144
After	125	107	121	140	150	149	126	119	134	138
	164	140	134	129	123	133	139	117	128	139

Draw a back-to-back stem-and-leaf diagram for these data.

EXERCISE SET 13.3

■ In Exercises 1 to 4, round each z-score to the nearest hundredth.

1. A data set has a mean of $\bar{x} = 75$ and a standard deviation of 11.5. Find the z-score for each of the following.

a. $x = 85$ **b.** $x = 95$

c. $x = 50$ **d.** $x = 75$

2. A data set has a mean of $\bar{x} = 212$ and a standard deviation of 40. Find the z-score for each of the following.

a. $x = 200$ **b.** $x = 224$

c. $x = 300$ **d.** $x = 100$

3. A data set has a mean of $\bar{x} = 6.8$ and a standard deviation of 1.9. Find the z-score for each of the following.

a. $x = 6.2$ **b.** $x = 7.2$

c. $x = 9.0$ **d.** $x = 5.0$

4. A data set has a mean of $\bar{x} = 4010$ and a standard deviation of 115. Find the z-score for each of the following.

a. $x = 3840$ **b.** $x = 4200$

c. $x = 4300$ **d.** $x = 4030$

5. Blood Pressure A blood pressure test was given to 450 women ages 20 to 36. It showed that their mean systolic blood pressure was 119.4 mm Hg, with a standard deviation of 13.2 mm Hg.

a. Determine the z-score, to the nearest hundredth, for a woman who had a systolic blood pressure reading of 110.5 mm Hg.

b. The z-score for one woman was 2.15. What was her systolic blood pressure reading?

6. Fruit Juice A random sample of 1000 oranges showed that the mean amount of juice per orange was 7.4 fluid ounces, with a standard deviation of 1.1 fluid ounces.

a. Determine the z-score, to the nearest hundredth, of an orange that produced 6.6 fluid ounces of juice.

b. The z-score for one orange was 3.15. How much juice was produced by this orange? Round to the nearest tenth of a fluid ounce.

7. Cholesterol A test involving 380 men ages 20 to 24 found that their blood cholesterol levels had a mean of 182 mg/dl and a standard deviation of 44.2 mg/dl.

a. Determine the z-score, to the nearest hundredth, for one of the men who had a blood cholesterol level of 214 mg/dl.

b. The z-score for one man was -1.58. What was his blood cholesterol level? Round to the nearest hundredth.

8. Tire Wear A random sample of 80 tires showed that the mean mileage per tire was 41,700 mi, with a standard deviation of 4300 mi.

a. Determine the z-score, to the nearest hundredth, for a tire that provided 46,300 mi of wear.

b. The z-score for one tire was -2.44. What mileage did this tire provide? Round your result to the nearest hundred miles.

9. Test Scores Which of the following three test scores is the highest relative score?

a. A score of 65 on a test with a mean of 72 and a standard deviation of 8.2

b. A score of 102 on a test with a mean of 130 and a standard deviation of 18.5

c. A score of 605 on a test with a mean of 720 and a standard deviation of 116.4

10. Physical Fitness Which of the following fitness scores is the highest relative score?

a. A score of 42 on a test with a mean of 31 and a standard deviation of 6.5

b. A score of 1140 on a test with a mean of 1080 and a standard deviation of 68.2

c. A score of 4710 on a test with a mean of 3960 and a standard deviation of 560.4

11. Reading Test On a reading test, Shaylen's score of 455 was higher than the scores of 4256 of the 7210 students who took the test. Find the percentile, rounded to the nearest percent, for Shaylen's score.

12. Placement Exams On a placement examination, Rick scored lower than 1210 of the 12,860 students who took the exam. Find the percentile, rounded to the nearest percent, for Rick's score.

13. Test Scores Kevin scored at the 65th percentile on a test given to 9840 students. How many students scored lower than Kevin?

14. Test Scores Rene scored at the 84th percentile on a test given to 12,600 students. How many students scored higher than Rene?

15. Median Income In 2015, the median family income in the United States was \$66,650. (*Source:* U.S. Census Bureau) If the 90th percentile for the 2015 median four-person family income was \$178,500, find the percentage of families whose 2015 income was

a. more than \$66,650. **b.** more than \$178,500.

c. between \$66,650 and \$178,500.

16. Monthly Rents A recent survey by the U.S. Census Bureau determined that the median monthly housing rent was \$708. If the first quartile for monthly housing rent was \$570, find the percent of monthly housing rents that were

a. more than \$570. **b.** less than \$708.

c. between \$570 and \$708.

17. Commute to School A survey was given to 18 students. One question asked about the one-way distance the student had to travel to attend college. The results, in miles, are shown in the following table. Use the median procedure for finding quartiles to find the first, second, and third quartiles for the data.

Miles Traveled to Attend College								
12	18	4	5	26	41	1	8	10
10	3	28	32	10	85	7	5	15

18. Prescriptions The following table shows the number of prescriptions a doctor wrote each day for a 36-day period. Use the median procedure for finding quartiles to find the first, second, and third quartiles for the data.

Number of Prescriptions Written per Day					
8	12	14	10	9	16
7	14	10	7	11	16
11	12	8	14	13	10
9	14	15	12	10	8
10	14	8	7	12	15
14	10	9	15	10	12

19. Home Sales The accompanying table shows the median selling prices of existing single-family homes in the United States in the four regions of the country for an 11-year period. Prices have been rounded to the nearest hundred. Draw a box-and-whisker plot of the data for each of the four regions. Write a few sentences that explain any differences you found.

Median Prices of Homes Sold in the United States over an 11-year Period

Year	Northwest	Midwest	South	West
1	227,400	169,700	148,000	196,400
2	246,400	172,600	155,400	213,600
3	264,300	178,000	163,400	238,500
4	264,500	184,300	168,100	260,900
5	315,800	205,000	181,100	283,100
6	343,800	216,900	197,300	332,600
7	346,000	213,500	208,200	337,700
8	320,200	208,600	217,700	330,900
9	343,600	198,900	203,700	294,800
10	302,500	189,200	194,800	263,700
11	335,500	197,600	196,000	259,700

SOURCE: U.S. Census Bureau

20. The table below shows the heights, in inches, of 15 randomly selected National Basketball Association (NBA) players and 15 randomly selected Division I National Collegiate Athletic Association (NCAA) players.

NBA	84	76	79	75	81	81	76	85
	78	79	78	78	84	75	76	
NCAA	78	73	73	78	77	76	75	74
	74	81	75	78	78	79	73	

Using the same scale, draw a box-and-whisker plot for each of the two data sets, placing the second plot below the first. Write a valid conclusion based on the data.

21. The table below shows the numbers of bushels of barley cultivated per acre for 12 one-acre plots of land for two different strains of barley, PHT-34 and CBX-21.

PHT-34	CBX-21
43	56
49	47
47	44
38	45
47	46
45	50
50	48
46	60
46	53
46	50
45	49
43	52

Using the same scale, draw a box-and-whisker plot for each of the two data sets, placing the PHT-34 plot below the CBX-21 plot. Write a valid conclusion based on the data.

22. The blood lead concentrations, in micrograms per deciliter (μg/dL), of 20 children from two different neighborhoods were measured. The results are recorded in the table.

Neighborhood 1	3.97	3.91	3.98	3.70	4.13	3.97	4.01	3.88	4.11	3.70
	3.96	3.77	4.30	4.08	4.12	4.93	3.93	3.94	3.85	3.83
Neighborhood 2	4.31	4.22	3.78	4.10	4.34	4.20	4.35	4.20	4.01	4.04
	4.28	4.12	4.59	4.12	4.01	3.85	3.96	4.28	4.39	4.13

Using the same scale, draw a box-and-whisker plot for each of the two data sets, placing the second plot below the first. Considering that high blood lead concentrations are harmful to humans, in which of the two neighborhoods would you prefer to live?

EXTENSIONS

23. **a.** The population 3, 4, 9, 14, and 20 has a mean of 10 and a standard deviation of 6.356. The z-scores for each of the five data values are $z_3 \approx -1.101$, $z_4 \approx -0.944$, $z_9 \approx -0.157$, $z_{14} \approx 0.629$, and $z_{20} \approx 1.573$. Find the mean and the standard deviation of these z-scores.

b. The population 2, 6, 12, 17, 22, and 25 has a mean of 14 and a standard deviation of 8.226. The z-scores for each of the six data values are $z_2 \approx -1.459$, $z_6 \approx -0.973$, $z_{12} \approx -0.243$, $z_{17} \approx 0.365$, $z_{22} \approx 0.973$, and $z_{25} \approx 1.337$. Find the mean and the standard deviation of these z-scores.

c. Use the results of part a and part b to make a conjecture about the mean and standard deviation of the z-scores for any set of population data.

24. For each of the following, determine whether the statement is true or false.

a. For any given set of data, the median of the data equals the mean of Q_1 and Q_3.

b. For any given set of data, $Q_3 - Q_2 = Q_2 - Q_1$.

c. A z-score for a given data value x in a set of data can be a negative number.

d. If a student answers 75% of the questions on a test correctly, then the student's score on the test will place the student at the 75th percentile.

SECTION 13.4 Normal Distributions

Frequency Distributions and Histograms

Large sets of data are often displayed using a *grouped frequency distribution* or a *histogram.* For instance, consider the following situation. An Internet service provider (ISP) has installed new computers. To estimate the new download times its subscribers will experience, the ISP surveyed 1000 of its subscribers to determine the time required for each subscriber to download a particular file from an Internet site. The results of that survey are summarized in Table 13.7 on the next page.

TABLE 13.7
A Grouped Frequency Distribution with 12 Classes

Download time (in seconds)	Number of subscribers
0–5	6
5–10	17
10–15	43
15–20	92
20–25	151
25–30	192
30–35	190
35–40	149
40–45	90
45–50	45
50–55	15
55–60	10

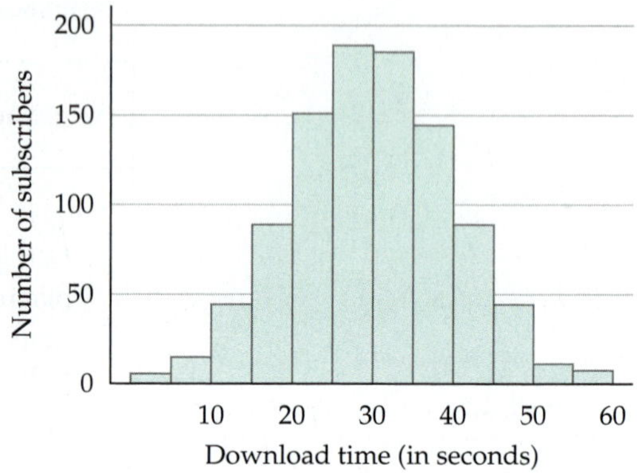

FIGURE 13.2 A histogram for the frequency distribution in Table 13.7

Table 13.7 is called a **grouped frequency distribution**. It shows how often (frequently) certain events occurred. Each interval, 0–5, 5–10, and so on, is called a **class**. This distribution has 12 classes. For the 10–15 class, 10 is the **lower class boundary** and 15 is the **upper class boundary**. Any data value that lies on a common boundary is assigned to the higher class. The *graph* of a frequency distribution is called a **histogram**. A histogram provides a pictorial view of how the data are distributed. In Figure 13.2, the height of each bar of the histogram indicates how many subscribers experienced the download times shown by the class on the base of the bar.

Examine the distribution in Table 13.8 below. It shows the *percent* of subscribers that are in each class, as opposed to the frequency distribution in Table 13.7, which shows the *number* of customers in each class. The type of frequency distribution that lists the *percent* of data in each class is called a **relative frequency distribution**. The **relative frequency histogram** in Figure 13.3 was drawn by using the data in the relative frequency distribution. It shows the *percent* of subscribers along its vertical axis.

TABLE 13.8
A Relative Frequency Distribution

Download time (in seconds)	Percent of subscribers
0–5	0.6
5–10	1.7
10–15	4.3
15–20	9.2
20–25	15.1
25–30	19.2
30–35	19.0
35–40	14.9
40–45	9.0
45–50	4.5
50–55	1.5
55–60	1.0

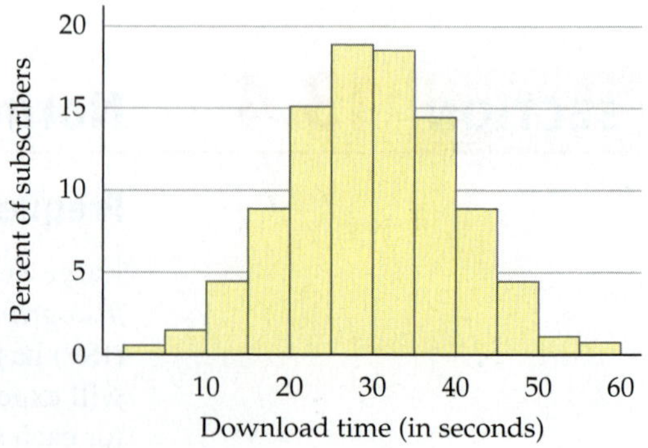

FIGURE 13.3 A relative frequency histogram

One advantage of using a relative frequency distribution instead of a grouped frequency distribution is that there is a direct correspondence between the percent values of the relative frequency distribution and probabilities. For instance, in the relative frequency distribution in Table 13.8, the percent of the data that lies between 35 s and 40 s is 14.9%. Thus, if a subscriber is chosen at random, the probability that the subscriber will require at least 35 s but less than 40 s to download the file is 0.149.

EXAMPLE 1 Use a Relative Frequency Distribution

Use the relative frequency distribution in Table 13.8 to determine the

a. *percent* of subscribers who required at least 25 s to download the file.

b. *probability* that a subscriber chosen at random will require at least 5 s but less than 20 s to download the file.

Solution

a. The percent of data in all the classes with a lower boundary of 25 s or more is the sum of the percents printed in red in Table 13.9 below. Thus the percent of subscribers who required at least 25 s to download the file is 69.1%.

TABLE 13.9

Download time (in seconds)	Percent of subscribers	
0–5	0.6	
5–10	1.7	Sum is 15.2%
10–15	4.3	
15–20	9.2	
20–25	15.1	
25–30	19.2	Sum is 69.1%
30–35	19.0	
35–40	14.9	
40–45	9.0	
45–50	4.5	
50–55	1.5	
55–60	1.0	

b. The percent of data in all the classes with a lower boundary of 5 s and an upper boundary of 20 s is the sum of the percents printed in blue in Table 13.9 above. Thus the percent of subscribers who required at least 5 s but less than 20 s to download the file is 15.2%. The probability that a subscriber chosen at random will require at least 5 s but less than 20 s to download the file is 0.152.

CHECK YOUR PROGRESS 1 Use the relative frequency distribution in Table 13.8 to determine the

a. *percent* of subscribers who required less than 25 s to download the file.

b. *probability* that a subscriber chosen at random will require at least 10 s but less than 30 s to download the file.

Solution See page S47.

Normal Distributions and the Empirical Rule

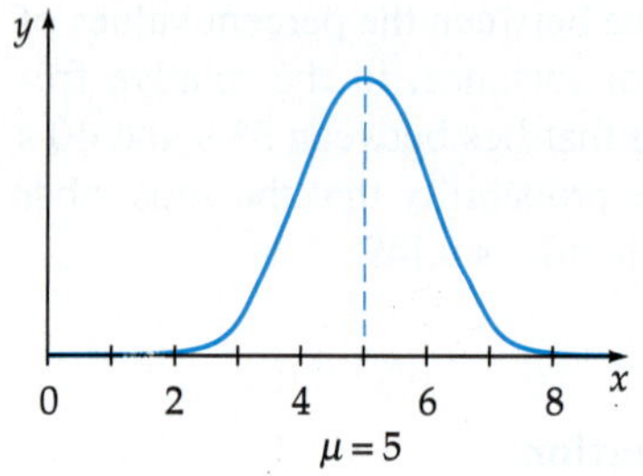

One of the most important statistical distributions of data is known as a *normal distribution*. This distribution occurs in a variety of applications. Types of data that may demonstrate a normal distribution include the lengths of leaves on a tree, the weights of newborns in a hospital, the lengths of time of a student's trip from home to school over a period of months, the SAT scores of a large group of students, and the life spans of light bulbs.

A **normal distribution** forms a bell-shaped curve that is symmetric about a vertical line through the mean of the data. A graph of a normal distribution with a mean of 5 is shown at the left.

Properties of a Normal Distribution

Every normal distribution has the following properties.

- The graph is symmetric about a vertical line through the mean of the distribution.
- The mean, median, and mode are equal.
- The y-value of each point on the curve is the *percent* (expressed as a decimal) of the data at the corresponding x-value.
- Areas under the curve that are symmetric about the mean are equal.
- The total area under the curve is 1.

PPT

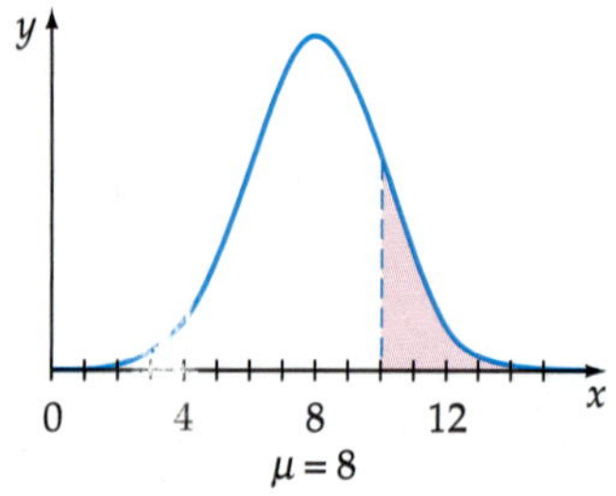

QUESTION What is the area under the curve to the right of the mean for a normal distribution?

In the normal distribution shown at the left, the area of the shaded region is 0.159 units. This region represents the fact that 15.9% of the data values are greater than or equal to 10. Because the area under the curve is 1, the unshaded region under the curve has area 1 − 0.159, or 0.841, representing the fact that 84.1% of the data are less than 10.

The following rule, called the Empirical Rule, describes the percents of data that lie within 1, 2, and 3 standard deviations of the mean in a normal distribution.

INSTRUCTOR NOTE

Remind your students that the Empirical Rule can be applied only to normal distributions, and it can be applied only to data values that are 1, 2, or 3 standard deviations above or below the mean.

Empirical Rule for a Normal Distribution

In a normal distribution, approximately

- 68% of the data lie within 1 standard deviation of the mean.
- 95% of the data lie within 2 standard deviations of the mean.
- 99.7% of the data lie within 3 standard deviations of the mean.

A normal distribution

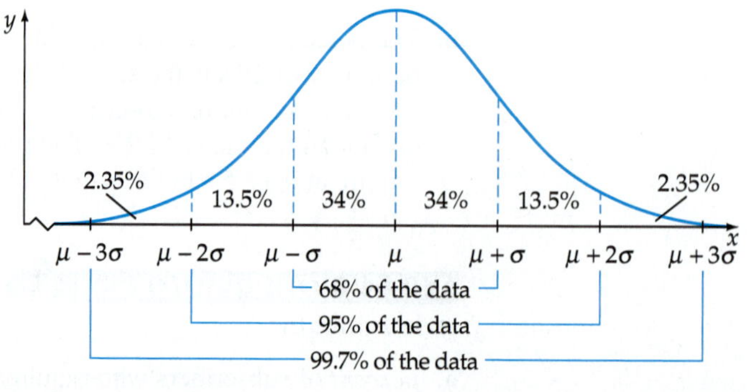

PPT

ANSWER Because a normal distribution is symmetric about the mean, the area under the curve to the right of the mean is one-half the total area. The total area under a normal distribution is 1, so the area under the curve to the right of the mean is 0.5.

EXAMPLE 2 Use the Empirical Rule to Solve an Application

A survey of 1000 U.S. gas stations found that the price charged for a gallon of regular gas could be closely approximated by a normal distribution with a mean of \$3.10 and a standard deviation of \$0.18. How many of the stations charge

a. between \$2.74 and \$3.46 for a gallon of regular gas?

b. less than \$3.28 for a gallon of regular gas?

c. more than \$3.46 for a gallon of regular gas?

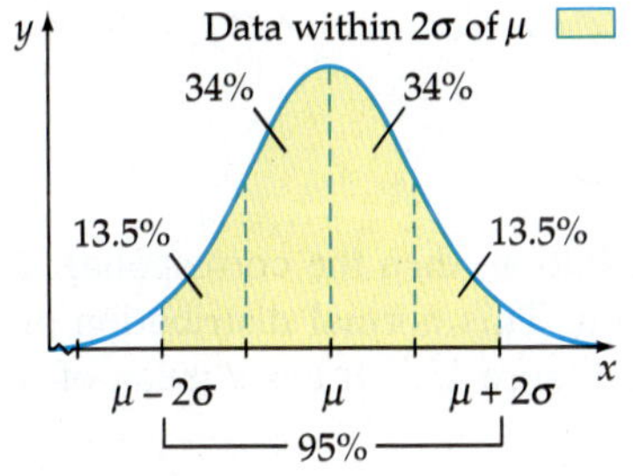

FIGURE 13.4

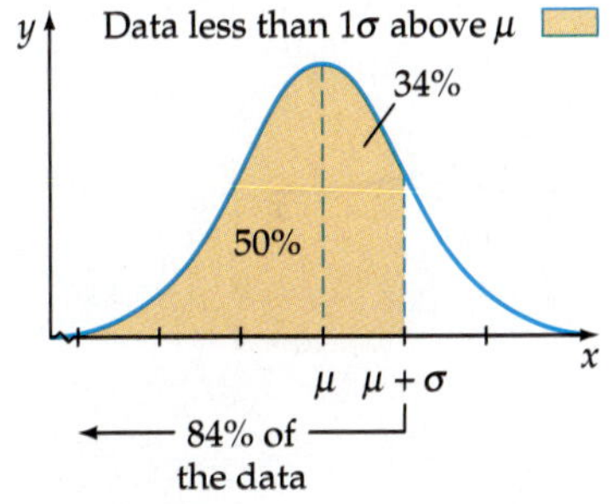

FIGURE 13.5

Solution

a. The \$2.74 per gallon price is 2 standard deviations below the mean. The \$3.46 price is 2 standard deviations above the mean. In a normal distribution, 95% of all data lie within 2 standard deviations of the mean. See Figure 13.4. Therefore approximately

$$(95\%)(1000) = (0.95)(1000) = 950$$

of the stations charge between \$2.74 and \$3.46 for a gallon of regular gas.

b. The \$3.28 price is 1 standard deviation above the mean. See Figure 13.5. In a normal distribution, 34% of all data lie between the mean and 1 standard deviation above the mean. Thus, approximately

$$(34\%)(1000) = (0.34)(1000) = 340$$

of the stations charge between \$3.10 and \$3.28 for a gallon of regular gasoline. Half of the 1000 stations, or 500 stations, charge less than the mean. Therefore about $340 + 500 = 840$ of the stations charge less than \$3.28 for a gallon of regular gas.

c. The \$3.46 price is 2 standard deviations above the mean. In a normal distribution, 95% of all data are within 2 standard deviations of the mean. This means that the other 5% of the data will lie either more than 2 standard deviations above the mean or more than 2 standard deviations below the mean. We are interested only in the data that are more than 2 standard deviations above the mean, which is $\frac{1}{2}$ of 5%, or 2.5%, of the data. See Figure 13.6. Thus about $(2.5\%)(1000) = (0.025)(1000) = 25$ of the stations charge more than \$3.46 for a gallon of regular gas.

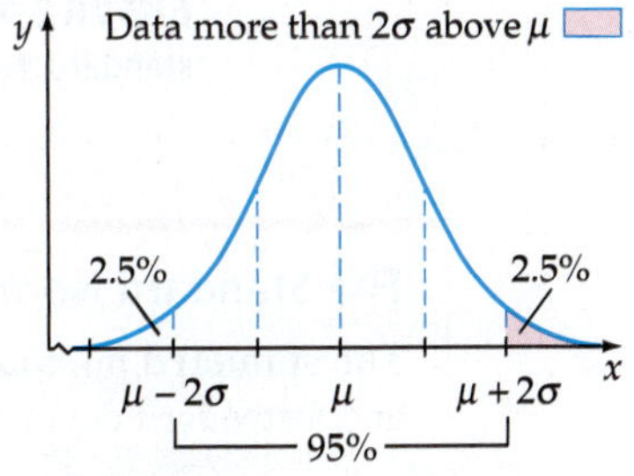

FIGURE 13.6

CHECK YOUR PROGRESS 2 A vegetable distributor knows that during the month of August, the weights of its tomatoes are normally distributed with a mean of 0.61 lb and a standard deviation of 0.15 lb.

a. What percent of the tomatoes weigh less than 0.76 lb?

b. In a shipment of 6000 tomatoes, how many tomatoes can be expected to weigh more than 0.31 lb?

c. In a shipment of 4500 tomatoes, how many tomatoes can be expected to weigh from 0.31 lb to 0.91 lb?

Solution See pages S47–S48.

QUESTION Can the Empirical Rule be applied to all data sets?

The Standard Normal Distribution

It is often helpful to convert data values x to z-scores, as we did in the previous section by using the z-score formulas:

$$z_x = \frac{x - \mu}{\sigma} \quad \text{or} \quad z_x = \frac{x - \overline{x}}{s}$$

If the original distribution of x values is a normal distribution, then the corresponding distribution of z-scores will also be a normal distribution. This normal distribution of z-scores is called the *standard normal distribution*. See Figure 13.7. It has a mean of 0 and a standard deviation of 1.

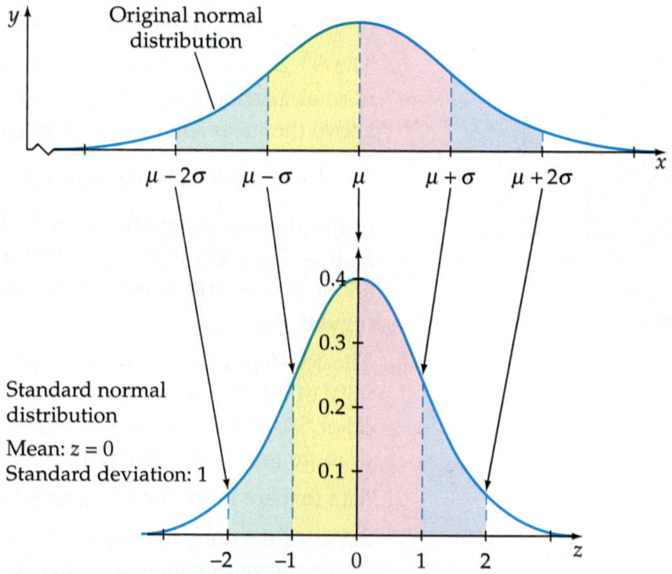

FIGURE 13.7 Conversion of a normal distribution to the standard normal distribution

The Standard Normal Distribution

The **standard normal distribution** is the normal distribution that has a mean of 0 and a standard deviation of 1.

Tables and calculators are often used to determine the area under a portion of the standard normal curve. We will refer to this type of area as an *area of the standard normal distribution*. Table 13.10 gives the approximate areas of the standard normal distribution between the mean 0 and z standard deviations from the mean. See Figure 13.8. Table 13.10 indicates that the area A of the standard normal distribution from the mean 0 up to $z = 1.34$ is 0.410 square unit.

ANSWER No. The Empirical Rule can only be applied to normal distributions.

TABLE 13.10
Area Under the Standard Normal Curve

z	A	z	A	z	A	z	A	z	A	z	A
0.00	0.000	0.56	0.212	1.12	0.369	1.68	0.454	2.24	0.487	2.80	0.497
0.01	0.004	0.57	0.216	1.13	0.371	1.69	0.454	2.25	0.488	2.81	0.498
0.02	0.008	0.58	0.219	1.14	0.373	1.70	0.455	2.26	0.488	2.82	0.498
0.03	0.012	0.59	0.222	1.15	0.375	1.71	0.456	2.27	0.488	2.83	0.498
0.04	0.016	0.60	0.226	1.16	0.377	1.72	0.457	2.28	0.489	2.84	0.498
0.05	0.020	0.61	0.229	1.17	0.379	1.73	0.458	2.29	0.489	2.85	0.498
0.06	0.024	0.62	0.232	1.18	0.381	1.74	0.459	2.30	0.489	2.86	0.498
0.07	0.028	0.63	0.236	1.19	0.383	1.75	0.460	2.31	0.490	2.87	0.498
0.08	0.032	0.64	0.239	1.20	0.385	1.76	0.461	2.32	0.490	2.88	0.498
0.09	0.036	0.65	0.242	1.21	0.387	1.77	0.462	2.33	0.490	2.89	0.498
0.10	0.040	0.66	0.245	1.22	0.389	1.78	0.462	2.34	0.490	2.90	0.498
0.11	0.044	0.67	0.249	1.23	0.391	1.79	0.463	2.35	0.491	2.91	0.498
0.12	0.048	0.68	0.252	1.24	0.393	1.80	0.464	2.36	0.491	2.92	0.498
0.13	0.052	0.69	0.255	1.25	0.394	1.81	0.465	2.37	0.491	2.93	0.498
0.14	0.056	0.70	0.258	1.26	0.396	1.82	0.466	2.38	0.491	2.94	0.498
0.15	0.060	0.71	0.261	1.27	0.398	1.83	0.466	2.39	0.492	2.95	0.498
0.16	0.064	0.72	0.264	1.28	0.400	1.84	0.467	2.40	0.492	2.96	0.498
0.17	0.067	0.73	0.267	1.29	0.401	1.85	0.468	2.41	0.492	2.97	0.499
0.18	0.071	0.74	0.270	1.30	0.403	1.86	0.469	2.42	0.492	2.98	0.499
0.19	0.075	0.75	0.273	1.31	0.405	1.87	0.469	2.43	0.492	2.99	0.499
0.20	0.079	0.76	0.276	1.32	0.407	1.88	0.470	2.44	0.493	3.00	0.499
0.21	0.083	0.77	0.279	1.33	0.408	1.89	0.471	2.45	0.493	3.01	0.499
0.22	0.087	0.78	0.282	1.34	0.410	1.90	0.471	2.46	0.493	3.02	0.499
0.23	0.091	0.79	0.285	1.35	0.411	1.91	0.472	2.47	0.493	3.03	0.499
0.24	0.095	0.80	0.288	1.36	0.413	1.92	0.473	2.48	0.493	3.04	0.499
0.25	0.099	0.81	0.291	1.37	0.415	1.93	0.473	2.49	0.494	3.05	0.499
0.26	0.103	0.82	0.294	1.38	0.416	1.94	0.474	2.50	0.494	3.06	0.499
0.27	0.106	0.83	0.297	1.39	0.418	1.95	0.474	2.51	0.494	3.07	0.499
0.28	0.110	0.84	0.300	1.40	0.419	1.96	0.475	2.52	0.494	3.08	0.499
0.29	0.114	0.85	0.302	1.41	0.421	1.97	0.476	2.53	0.494	3.09	0.499
0.30	0.118	0.86	0.305	1.42	0.422	1.98	0.476	2.54	0.494	3.10	0.499
0.31	0.122	0.87	0.308	1.43	0.424	1.99	0.477	2.55	0.495	3.11	0.499
0.32	0.126	0.88	0.311	1.44	0.425	2.00	0.477	2.56	0.495	3.12	0.499
0.33	0.129	0.89	0.313	1.45	0.426	2.01	0.478	2.57	0.495	3.13	0.499
0.34	0.133	0.90	0.316	1.46	0.428	2.02	0.478	2.58	0.495	3.14	0.499
0.35	0.137	0.91	0.319	1.47	0.429	2.03	0.479	2.59	0.495	3.15	0.499
0.36	0.141	0.92	0.321	1.48	0.431	2.04	0.479	2.60	0.495	3.16	0.499
0.37	0.144	0.93	0.324	1.49	0.432	2.05	0.480	2.61	0.495	3.17	0.499
0.38	0.148	0.94	0.326	1.50	0.433	2.06	0.480	2.62	0.496	3.18	0.499
0.39	0.152	0.95	0.329	1.51	0.434	2.07	0.481	2.63	0.496	3.19	0.499
0.40	0.155	0.96	0.331	1.52	0.436	2.08	0.481	2.64	0.496	3.20	0.499
0.41	0.159	0.97	0.334	1.53	0.437	2.09	0.482	2.65	0.496	3.21	0.499
0.42	0.163	0.98	0.336	1.54	0.438	2.10	0.482	2.66	0.496	3.22	0.499
0.43	0.166	0.99	0.339	1.55	0.439	2.11	0.483	2.67	0.496	3.23	0.499
0.44	0.170	1.00	0.341	1.56	0.441	2.12	0.483	2.68	0.496	3.24	0.499
0.45	0.174	1.01	0.344	1.57	0.442	2.13	0.483	2.69	0.496	3.25	0.499
0.46	0.177	1.02	0.346	1.58	0.443	2.14	0.484	2.70	0.497	3.26	0.499
0.47	0.181	1.03	0.348	1.59	0.444	2.15	0.484	2.71	0.497	3.27	0.499
0.48	0.184	1.04	0.351	1.60	0.445	2.16	0.485	2.72	0.497	3.28	0.499
0.49	0.188	1.05	0.353	1.61	0.446	2.17	0.485	2.73	0.497	3.29	0.499
0.50	0.191	1.06	0.355	1.62	0.447	2.18	0.485	2.74	0.497	3.30	0.500
0.51	0.195	1.07	0.358	1.63	0.448	2.19	0.486	2.75	0.497	3.31	0.500
0.52	0.198	1.08	0.360	1.64	0.449	2.20	0.486	2.76	0.497	3.32	0.500
0.53	0.202	1.09	0.362	1.65	0.451	2.21	0.486	2.77	0.497	3.33	0.500
0.54	0.205	1.10	0.364	1.66	0.452	2.22	0.487	2.78	0.497		
0.55	0.209	1.11	0.367	1.67	0.453	2.23	0.487	2.79	0.497		

TAKE NOTE

In Table 13.10, the entries below A give the area under the standard normal curve from 0 to the z-value listed to the left of the A value. The A values have been rounded to the nearest thousandth.

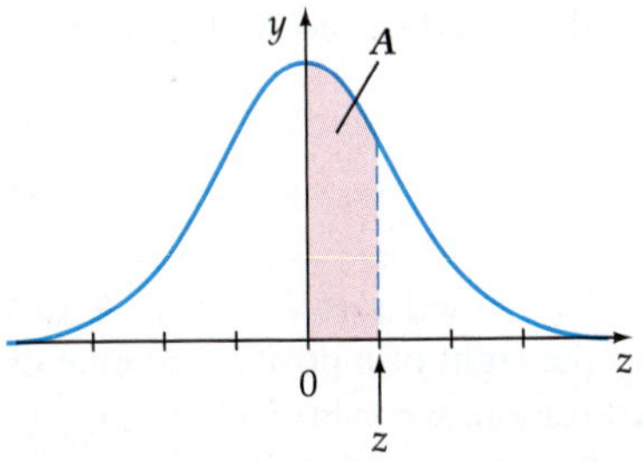

FIGURE 13.8 A is the area of the shaded region.

TAKE NOTE

To find the A value for the z-score 0.382, round 0.382 to the nearest hundredth to produce 0.38. Then use Table 13.10 to find that $A \approx 0.148$.

Because the standard normal distribution is symmetrical about the mean of 0, we can also use Table 13.10 to find the area of a region that is located to the left of the mean. This process is explained in Example 3.

EXAMPLE 3 Use Symmetry to Determine an Area

Find the area of the standard normal distribution between $z = -1.44$ and $z = 0$.

Solution

Because the standard normal distribution is symmetrical about the center line $z = 0$, the area of the standard normal distribution between $z = -1.44$ and $z = 0$ is equal to the area between $z = 0$ and $z = 1.44$. See Figure 13.9. The entry in Table 13.10 associated with $z = 1.44$ is 0.425. Thus the area of the standard normal distribution between $z = -1.44$ and $z = 0$ is 0.425 square unit.

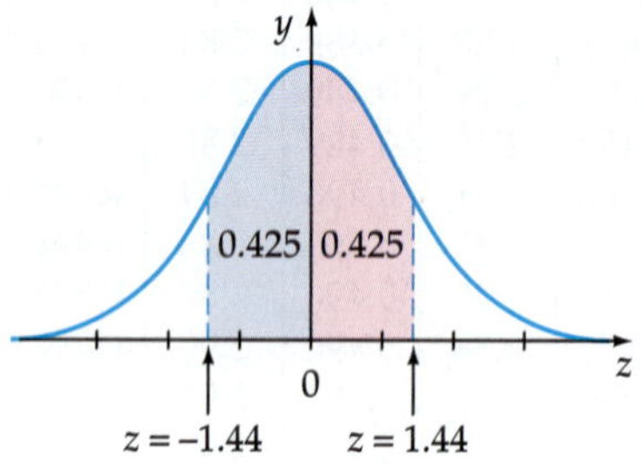

FIGURE 13.9 Symmetrical region

CHECK YOUR PROGRESS 3 Find the area of the standard normal distribution between $z = -0.67$ and $z = 0$.

Solution See page S48.

In Figure 13.10, the region to the right of $z = 0.82$ is called a *tail region*. A **tail region** is a region of the standard normal distribution to the right of a positive z-value or to the left of a negative z-value. To find the area of a tail region, we subtract the entry in Table 13.10 from 0.500. This procedure is illustrated in the next example.

EXAMPLE 4 Find the Area of a Tail Region

Find the area of the standard normal distribution to the right of $z = 0.82$.

Solution

Table 13.10 indicates that the area from $z = 0$ to $z = 0.82$ is 0.294 square unit. The area to the right of $z = 0$ is 0.500 square unit. Thus the area to the right of $z = 0.82$ is $0.500 - 0.294 = 0.206$ square unit. See Figure 13.10.

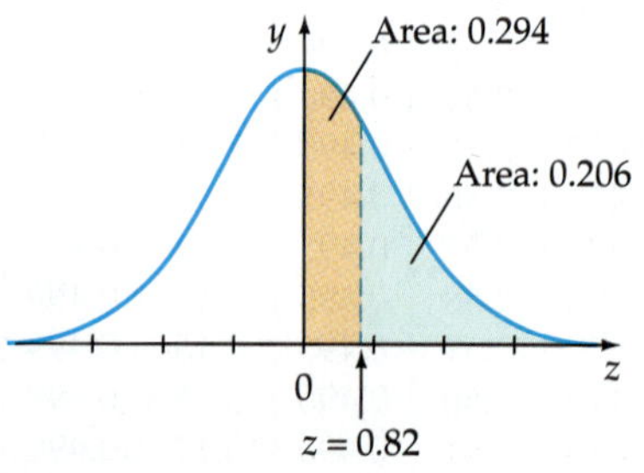

FIGURE 13.10 Area of a tail region

CHECK YOUR PROGRESS 4 Find the area of the standard normal distribution to the left of $z = -1.47$.

Solution See page S48.

The Standard Normal Distribution, Areas, Percentages, and Probabilities

In the standard normal distribution, the area of the distribution from $z = a$ to $z = b$ represents

- the *percentage* of z-values that lie in the interval from a to b.
- the *probability* that z lies in the interval from a to b.

Because the area of a portion of the standard normal distribution can be interpreted as a percentage of the data or as a probability that the variable lies in a particular interval, we can use the standard normal distribution to solve many application problems.

EXAMPLE 5 Solve an Application

A soda machine dispenses soda into 12-ounce cups. Tests show that the actual amount of soda dispensed is normally distributed, with a mean of 11.5 oz and a standard deviation of 0.2 oz.

a. What percent of cups will receive less than 11.25 oz of soda?

b. What percent of cups will receive between 11.2 oz and 11.55 oz of soda?

c. If a cup is filled at random, what is the probability that the machine will overflow the cup?

Solution

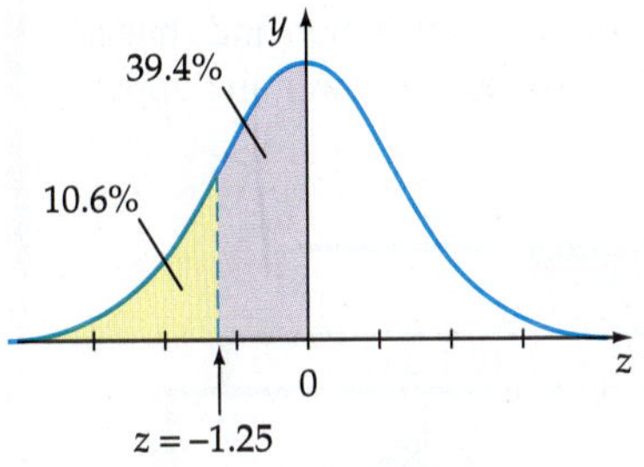

FIGURE 13.11 Portion of data to the left of $z = -1.25$

a. Recall that the formula for the z-score for a data value x is

$$z_x = \frac{x - \bar{x}}{s}$$

The z-score for 11.25 oz is

$$z_{11.25} = \frac{11.25 - 11.5}{0.2} = -1.25$$

Table 13.10 indicates that 0.394 (39.4%) of the data in a normal distribution are between $z = 0$ and $z = 1.25$. Because the data are normally distributed, 39.4% of the data is also between $z = 0$ and $z = -1.25$. The percent of data to the left of $z = -1.25$ is 50% − 39.4% = 10.6%. See Figure 13.11. Thus 10.6% of the cups filled by the soda machine will receive less than 11.25 oz of soda.

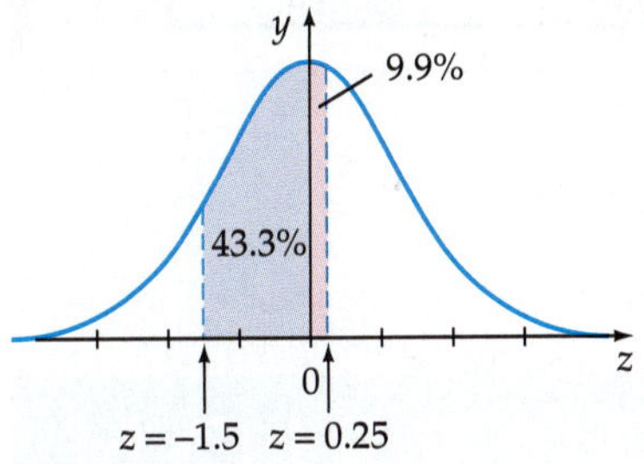

FIGURE 13.12 Portion of data between two z-scores

b. The z-score for 11.55 ounces is

$$z_{11.55} = \frac{11.55 - 11.5}{0.2} = 0.25$$

Table 13.10 indicates that 0.099 (9.9%) of the data in a normal distribution is between $z = 0$ and $z = 0.25$.

The z-score for 11.2 oz is

$$z_{11.2} = \frac{11.2 - 11.5}{0.2} = -1.5$$

Table 13.10 indicates that 0.433 (43.3%) of the data in a normal distribution are between $z = 0$ and $z = 1.5$. Because the data are normally distributed, 43.3% of the data are also between $z = 0$ and $z = -1.5$. See Figure 13.12. Thus the percent of the cups that the vending machine will fill with between 11.2 oz and 11.55 oz of soda is 43.3% + 9.9% = 53.2%.

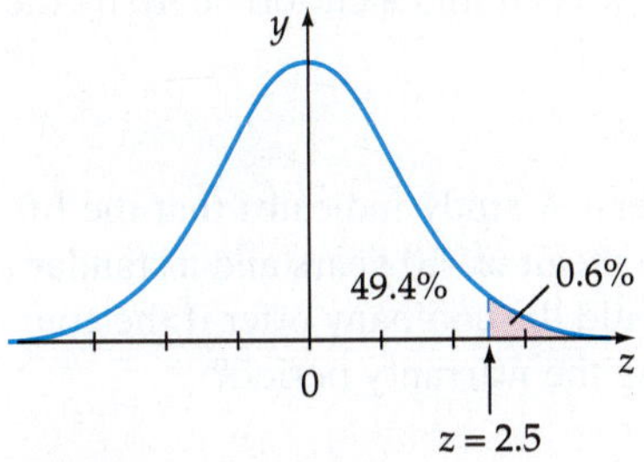

FIGURE 13.13 Portion of data to the right of $z = 2.5$

c. A cup will overflow if it receives more than 12 oz of soda. The z-score for 12 oz is

$$z_{12} = \frac{12 - 11.5}{0.2} = 2.5$$

Table 13.10 indicates that 0.494 (49.4%) of the data in the standard normal distribution are between $z = 0$ and $z = 2.5$. The percent of data to the right of $z = 2.5$ is determined by subtracting 49.4% from 50%. See Figure 13.13. Thus 0.6% of the time the machine produces an overflow, and the probability that a cup filled at random will overflow is 0.006.

CHECK YOUR PROGRESS 5 A study shows that the lengths of the careers of professional football players are nearly normally distributed, with a mean of 6.1 years and a standard deviation of 1.8 years.

a. What percent of professional football players have a career of more than 9 years?

b. If a professional football player is chosen at random, what is the probability that the player will have a career of between 3 and 4 years?

Solution See page S48.

MATH MATTERS Find the Area of a Portion of the Standard Normal Distribution by Using a Calculator

Some calculators can be used to find the area of a portion of the standard normal distribution. For instance, the TI-83/84 screen displays below indicate that the area of the standard normal distribution from a lower bound of $z = 0$ to an upper bound of $z = 1.34$ is about 0.409877 square unit. This is a more accurate value than the entry given in Table 13.10, which is 0.410.

Select the `normalcdf(` function from the `DISTR` menu. Enter your lower bound, followed by a comma and your upper bound. Press `ENTER`.

The `ShadeNorm` instruction under `DRAW` in the `DISTR` menu draws the standard normal distribution and shades the region between the lower and upper bounds.

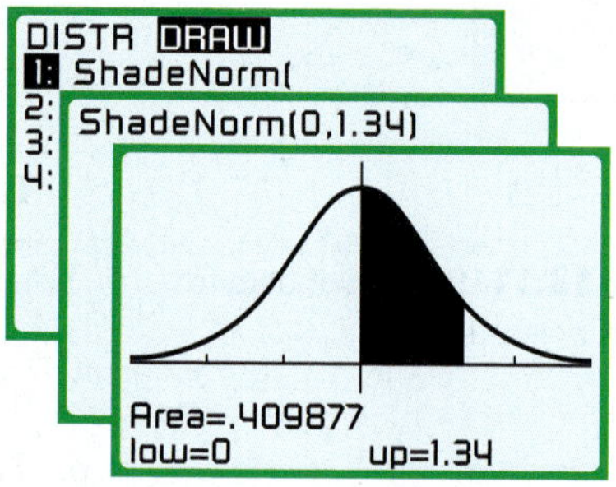

EXCURSION

Cut-Off Scores

A *cut-off score* is a score that separates data into two groups such that the data in one group satisfy a certain requirement and the data in the other group do not satisfy the requirement. If the data are normally distributed, then we can find a cut-off score by the method shown in the following example.

EXAMPLE

The OnTheGo company manufactures laptop computers. A study indicates that the life spans of its computers are normally distributed, with a mean of 4.0 years and a standard deviation of 1.2 years. How long a warranty period should the company offer if the company wishes less than 4% of its computers to fail during the warranty period?

Solution

Figure 13.14 shows a standard normal distribution with 4% of the data to the left of some unknown z-score and 46% of the data to the right of the z-score but to the left of the mean of 0. Using Table 13.10, we find that the z-score associated with an area of $A = 0.46$ is 1.75. Our unknown z-score is to the left of 0, so it must be negative. Thus $z_x = -1.75$. If we let x represent the time in years that a computer is in use, then x is related to the z-scores by the formula

$$z_x = \frac{x - \bar{x}}{s}$$

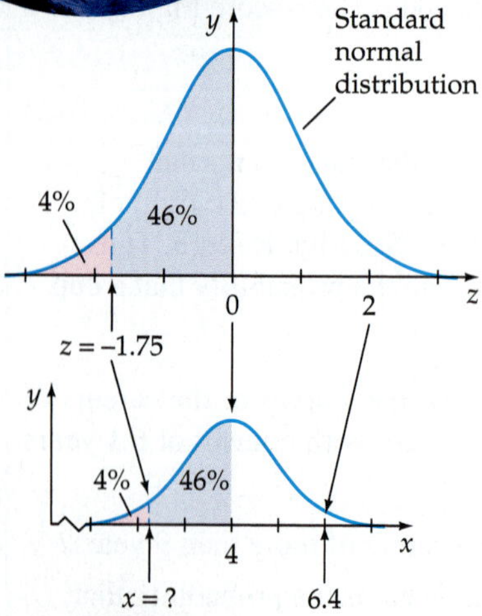

FIGURE 13.14 Finding a cut-off score

Solving for x with $\bar{x} = 4.0$, $s = 1.2$, and $z = -1.75$ gives us

$$-1.75 = \frac{x - 4.0}{1.2}$$
$$(-1.75)(1.2) = x - 4.0$$
$$x = 4.0 - 2.1$$
$$x = 1.9 \qquad \text{The cut-off score}$$

Hence the company can provide a 1.9-year warranty and expect less than 4% of its computers to fail during the warranty period.

EXCURSION EXERCISES

1. A professor finds that the grades in a large class are normally distributed. The mean of the grades is 64, and the standard deviation is 10. If the professor decides to give an A grade to the students in the top 9% of the class, what is the cut-off score for an A?
2. The results of a statewide examination of the reading skills of sixth-grade students are normally distributed, with a mean score of 104 and a standard deviation of 16. The students in the top 10% are to receive an award, and those in the bottom 14% will be required to take a special reading class.
   a. What score does a student need in order to receive an award?
   b. What is the cut-off score that will be used to determine whether a student will be required to take the special reading class?
3. A secondary school system finds that the 440-yard-dash times of its female students are normally distributed, with an average time of 72 s and a standard deviation of 5.5 s. What time does a runner need in order to be in the 9% of runners with the best times? Round to the nearest hundredth of a second.

EXERCISE SET 13.4

1. **Boys' Heights** Humans are, on average, taller today than they were 200 years ago. Today, the mean height of 14-year-old boys is about 65 in. Use the following relative frequency distribution of heights of a group of 14-year-old boys from the 19th century to answer the following questions.

Heights of a Group of 19th-Century Boys, Age 14

Height (in inches)	Percent of boys
Under 50	0.2
50–54	7.0
55–59	46.0
60–64	41.0
65–69	5.8

SOURCE: *Journal of the Anthropological Institute of Great Britain and Ireland*

a. What percent of the group of 19th-century boys was at least 65 in. tall?

b. What is the probability that one of the 19th-century boys selected at random was at least 55 in. tall but less than 65 in. tall?

2. **Biology** A biologist measured the lengths of hundreds of cuckoo bird eggs. Use the relative frequency distribution below to answer the questions that follow.

Lengths of Cuckoo Bird Eggs

Length (in millimeters)	Percent of eggs
18.75–19.75	0.8
19.75–20.75	4.0
20.75–21.75	17.3
21.75–22.75	37.9
22.75–23.75	28.5
23.75–24.75	10.7
24.75–25.75	0.8

SOURCE: Biometrika

a. What percent of the group of eggs was less than 21.75 mm long?

b. What is the probability that one of the eggs selected at random was at least 20.75 mm long but less than 24.75 mm long?

■ In Exercises 3 to 8, use the Empirical Rule to answer each question.

3. In a normal distribution, what percent of the data lie

a. within 2 standard deviations of the mean?

b. more than 1 standard deviation above the mean?

c. between 1 standard deviation below the mean and 2 standard deviations above the mean?

4. In a normal distribution, what percent of the data lie

a. within 3 standard deviations of the mean?

b. more than 2 standard deviations below the mean?

c. between 2 standard deviations below the mean and 3 standard deviations above the mean?

5. Shipping During 1 week, an overnight delivery company found that the weights of its parcels were normally distributed, with a mean of 24 oz and a standard deviation of 6 oz.

a. What percent of the parcels weighed between 12 oz and 30 oz?

b. What percent of the parcels weighed more than 42 oz?

6. Baseball A baseball franchise finds that the attendance at its home games is normally distributed, with a mean of 16,000 and a standard deviation of 4000.

a. What percent of the home games have an attendance between 12,000 and 20,000 people?

b. What percent of the home games have an attendance of fewer than 8000 people?

7. Traffic A highway study of 8000 vehicles that passed by a checkpoint found that their speeds were normally distributed, with a mean of 61 mph and a standard deviation of 7 mph.

a. How many of the vehicles had a speed of more than 68 mph?

b. How many of the vehicles had a speed of less than 40 mph?

8. Women's Heights A survey of 1000 women ages 20 to 30 found that their heights were normally distributed, with a mean of 65 in. and a standard deviation of 2.5 in.

a. How many of the women have a height that is within 1 standard deviation of the mean?

b. How many of the women have a height that is between 60 in. and 70 in.?

■ In Exercises 9 to 16, find the area, to the nearest thousandth, of the standard normal distribution between the given z-scores.

9. $z = 0$ and $z = 1.5$

10. $z = 0$ and $z = 1.9$

11. $z = 0$ and $z = -1.85$

12. $z = 0$ and $z = -2.3$

13. $z = 1$ and $z = 1.9$

14. $z = 0.7$ and $z = 1.92$

15. $z = -1.47$ and $z = 1.64$

16. $z = -0.44$ and $z = 1.82$

■ In Exercises 17 to 24, find the area, to the nearest thousandth, of the indicated region of the standard normal distribution.

17. The region where $z > 1.3$

18. The region where $z > 1.92$

19. The region where $z < -2.22$

20. The region where $z < -0.38$

21. The region where $z > -1.45$

22. The region where $z < 1.82$

23. The region where $z < 2.71$

24. The region where $z < 1.92$

■ In Exercises 25 to 30, find the z-score, to the nearest hundredth, that satisfies the given condition.

25. 0.200 square unit of the area of the standard normal distribution is to the right of z.

26. 0.227 square unit of the area of the standard normal distribution is to the right of z.

27. 0.184 square unit of the area of the standard normal distribution is to the left of z.

28. 0.330 square unit of the area of the standard normal distribution is to the left of z.

29. 0.363 square unit of the area of the standard normal distribution is to the right of z.

30. 0.440 square unit of the area of the standard normal distribution is to the left of z.

■ In Exercises 31 to 40, answer each question. Round z-scores to the nearest hundredth and then find the required A values using Table 13.10 on page 787.

31. Cholesterol Levels The cholesterol levels of a group of young women at a university are normally distributed, with a mean of 185 and a standard deviation of 39. What percent of the young women have a cholesterol level

a. greater than 219?

b. between 190 and 225?

32. Biology A biologist found the wingspans of a group of monarch butterflies to be normally distributed, with a mean of 52.2 mm and a standard deviation of 2.3 mm. What percent of the butterflies had a wingspan

a. less than 48.5 mm?

b. between 50 and 55 mm?

33. Light Bulbs A manufacturer of light bulbs finds that one light bulb model has a mean life span of 1025 h with a standard deviation of 87 h. What percent of these light bulbs will last

a. at least 950 h?

b. between 800 and 900 h?

34. Heart Rates The resting heart rates of a group of healthy adult men were found to have a mean of 73.4 beats per minute, with a standard deviation of 5.9 beats per minute. What percent of these men had a resting heart rate of

a. greater than 80 beats per minute?

b. between 70 and 85 beats per minute?

35. Cereal Weight The weights of all the boxes of corn flakes filled by a machine are normally distributed, with a mean weight of 14.5 oz and a standard deviation of 0.4 oz. What percent of the boxes will

a. weigh less than 14 oz?

b. weigh between 13.5 oz and 15.5 oz?

36. Telephone Calls A telephone company has found that the lengths of its long distance telephone calls are normally distributed, with a mean of 225 s and a standard deviation of 55 s. What percent of its long distance calls are

a. longer than 340 s?

b. between 200 and 300 s?

37. Rope Strength The breaking point of a particular type of rope is normally distributed, with a mean of 350 lb and a standard deviation of 24 lb. What is the probability that a piece of this rope chosen at random will have a breaking point of

a. less than 320 lb?

b. between 340 and 370 lb?

38. Tire Mileage The mileage for WearEver tires is normally distributed, with a mean of 48,000 mi and a standard deviation of 7400 mi. What is the probability that the WearEver tires you purchase will provide a mileage of

a. more than 60,000 mi?

b. between 40,000 and 50,000 mi?

39. Grocery Store Lines The amount of time customers spend waiting in line at a grocery store is normally distributed, with a mean of 2.5 min and a standard deviation of 0.75 min. Find the probability that the time a customer spends waiting is

a. less than 3 min.

b. less than 1 min.

40. IQ Tests A psychologist finds that the intelligence quotients of a group of patients are normally distributed, with a mean of 102 and a standard deviation of 16. Find the percent of the patients with IQs

a. above 114.

b. between 90 and 118.

41. Heights Consider the data set of the heights of all babies born in the United States during a particular year. Do you think this data set is nearly normally distributed? Explain.

42. Weights Consider the data set of the weights of all Valencia oranges grown in California during a particular year. Do you think this data set is nearly normally distributed? Explain.

EXTENSIONS

■ In Exercises 43 to 49, determine whether the given statement is true or false.

43. The standard normal distribution has a mean of 0.

44. Every normal distribution is a bell-shaped distribution.

45. In a normal distribution, the mean, the median, and the mode of the distribution are all located at the center of the distribution.

46. The mean of a normal distribution is always larger than the standard deviation of the distribution.

47. The standard deviation of the standard normal distribution is 1.

48. If a data value x from a normal distribution is positive, then its z-score also must be positive.

49. All normal distributions have a mean of 0.

50. a. Make a sketch of two normal distributions that have the same standard deviation but different means.

b. Make a sketch of two normal distributions that have the same mean but different standard deviations.

51. Determine the approximate z-scores for the first quartile and the third quartile of the standard normal distribution.

SECTION 13.5 Linear Regression and Correlation

Linear Regression

When performing research studies, scientists often wish to know whether two variables are related. If the variables are determined to be related, a scientist may then wish to find an equation that can be used to model the relationship. For instance, a geologist might want to know whether there is a relationship between the duration of an eruption of a geyser and the time between eruptions. A first step in this determination is to collect some data. Data involving two variables are called **bivariate data**. Table 13.11 gives bivariate data showing the time between two eruptions and the duration of the second eruption for 10 eruptions of the geyser Old Faithful.

TABLE 13.11

Time between eruptions (in seconds), x	272	227	237	238	203	270	218	226	250	245
Duration of eruption (in seconds), y	89	79	83	82	81	85	78	81	85	79

Once the data are collected, a **scatter diagram** or **scatter plot** can be drawn, as shown in Figure 13.15.

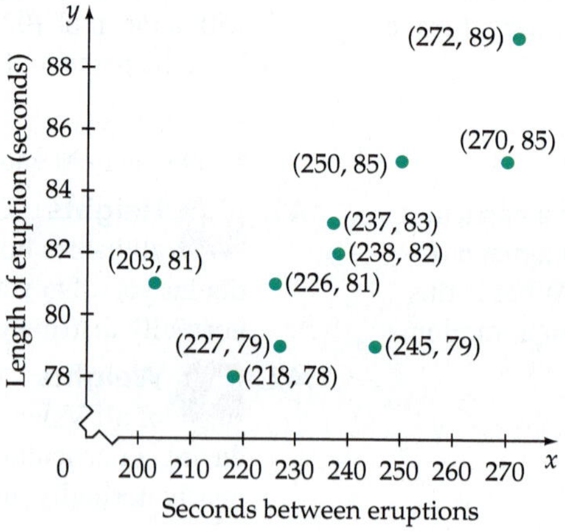

FIGURE 13.15

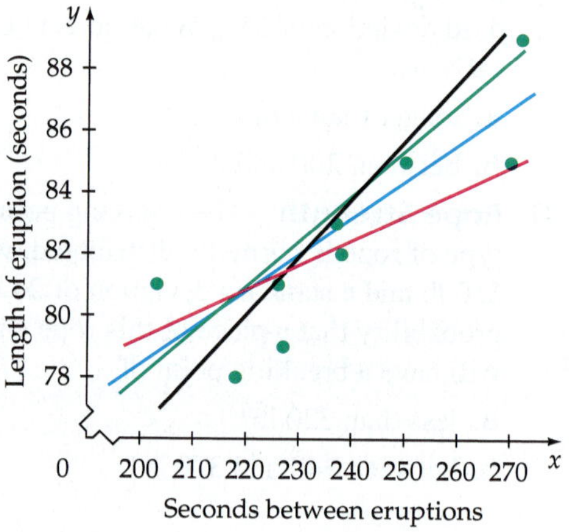

FIGURE 13.16

One way for the geologist to create a model of the relationship between the time between two eruptions and the duration of the second eruption is to find a line that *approximates* the data points plotted in the scatter plot. There are many such lines that can be drawn, as shown in Figure 13.16.

Of all the possible lines that can be drawn, the one that is usually of most interest is called the *line of best fit* or the *least-squares regression line*. The least-squares regression line is the line that fits the data better than any other line that might be drawn. The least-squares regression line is defined as follows.

The Least-Squares Regression Line

The **least-squares regression line** for a set of bivariate data is the line that minimizes the sum of the squares of the vertical deviations from each data point to the line.

In this definition, the phrase "minimizes the sum of the squares of the vertical deviations" is somewhat daunting. Referring to Figure 13.17, it means that of all the lines possible, the linear equation that minimizes the sum

$$d_1^2 + d_2^2 + d_3^2 + d_4^2 + d_5^2 + d_6^2 + d_7^2 + d_8^2 + d_9^2 + d_{10}^2$$

is the equation of the line of best fit. In this expression, each d_n represents the distance from data point n to the line.

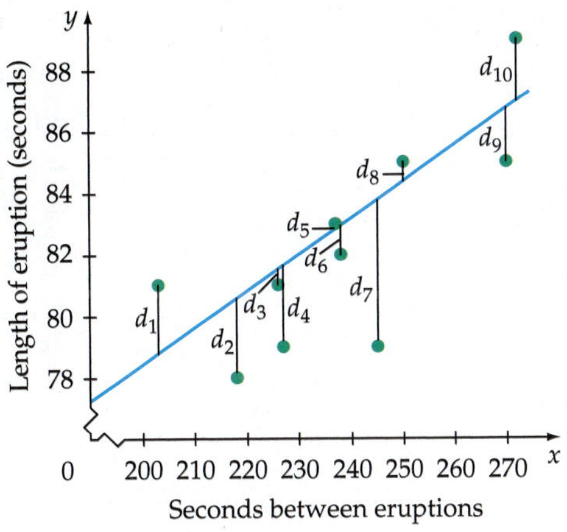

FIGURE 13.17

Applying some techniques from calculus, it is possible to find a formula for the least-squares line.

> **TAKE NOTE**
>
> It is traditional to use the symbol $\hat{y}$ (pronounced "y-hat") in place of y in the equation of the least-squares line. This helps distinguish between the line's y-values (the values calculated from the equation) and the data's y-values (the values given in the table).

The Formula for the Least-Squares Line

The equation of the least-squares line for the n ordered pairs

$$(x_1, y_1), (x_2, y_2), (x_3, y_3), \ldots, (x_n, y_n)$$

is $\hat{y} = ax + b$, where

$$a = \frac{n\Sigma xy - (\Sigma x)(\Sigma y)}{n\Sigma x^2 - (\Sigma x)^2} \quad \text{and} \quad b = \bar{y} - a\bar{x}$$

PPT

To apply this formula to the data for Old Faithful, we first find the value of each summation.

$$\Sigma x = 2386 \qquad \Sigma y = 822 \qquad \Sigma x^2 = 573{,}560 \qquad \Sigma xy = 196{,}636$$

> **TAKE NOTE**
>
> There are other lines, such as the median–median line, that can be used to approximate a set of data.

Next, we use these values to find the value of a.

$$a = \frac{n\Sigma xy - (\Sigma x)(\Sigma y)}{n\Sigma x^2 - (\Sigma x)^2}$$

$$a = \frac{(10)(196{,}636) - (2386)(822)}{(10)(573{,}560) - (2386)^2} \approx 0.1189559666$$

We then find the values of $\bar{x}$ and $\bar{y}$,

$$\bar{x} = \frac{\Sigma x}{n} = \frac{2386}{10} = 238.6 \quad \text{and} \quad \bar{y} = \frac{\Sigma y}{n} = \frac{822}{10} = 82.2$$

and use them to find the y-intercept, b.

$$\begin{aligned} b &= \bar{y} - a\bar{x} \\ &\approx 82.2 - 0.1189559666(238.6) \\ &= 53.81710637 \end{aligned}$$

The regression equation is $\hat{y} = 0.1189559666x + 53.81710637$. The graph of the regression equation and a scatter plot of the data are shown below.

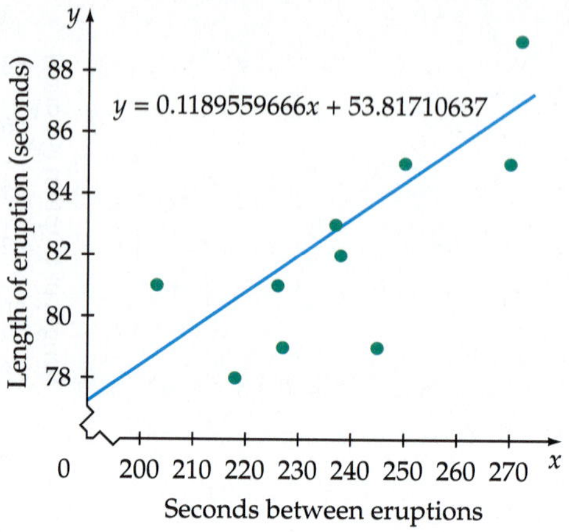

We can now use the regression equation to estimate the duration of an eruption given the time between eruptions. For instance, if the time between two eruptions is 200 seconds, then the estimated duration of the second eruption is

$$\begin{aligned}\hat{y} &= 0.1189559666x + 53.81710637\\ &= 0.1189559666(200) + 53.81710637\\ &\approx 78\end{aligned}$$

The approximate duration of the eruption is 78 seconds.

As our example demonstrates, it can be challenging to calculate all of the values needed to find a regression line. Fortunately, many computer programs and calculators can perform these calculations. The following example shows the use of a TI-84 to find the regression line for the Old Faithful data.

Enter the data from Table 13.11 into L1 and L2, as shown at the right.

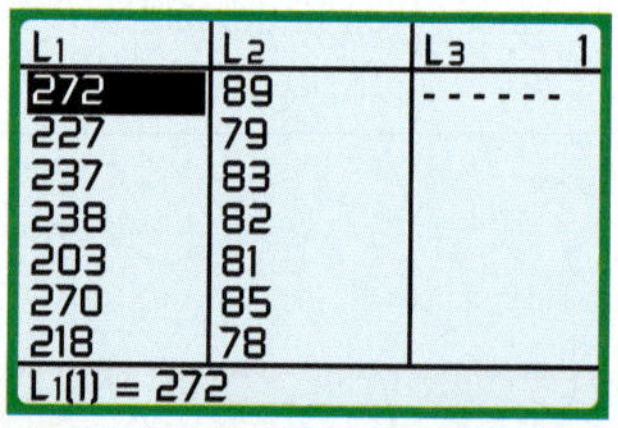

Press the STAT key, tab to CALC, and scroll to 4:. Then press ENTER.

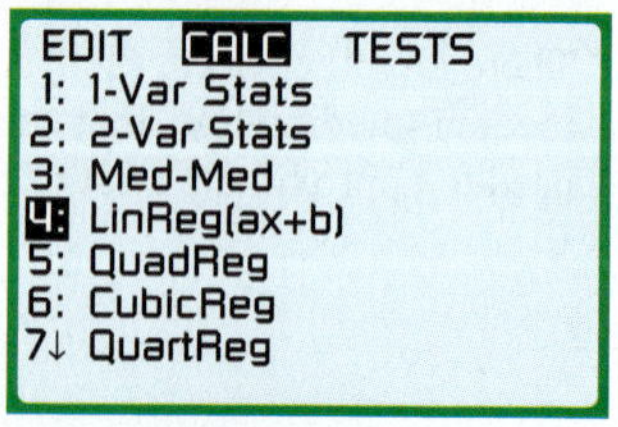

Scroll to the Store RegEQ line. Press the VARS key and scroll to Y-VARS. Press ENTER twice.

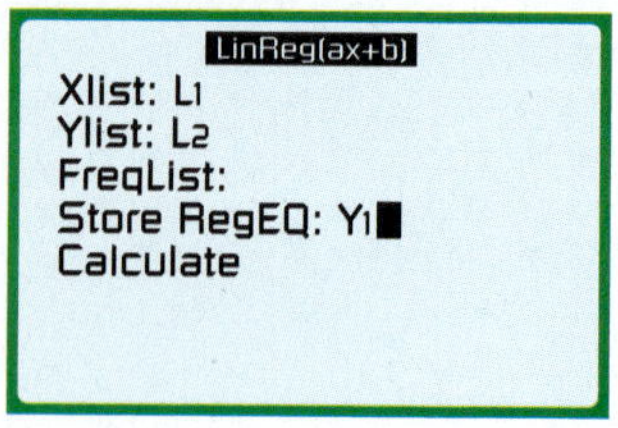

Press ENTER twice. The slope a and y-intercept b of the least-squares line are shown. You will see two additional values, r^2 and r, displayed on the screen. We will discuss the meanings of these values later.

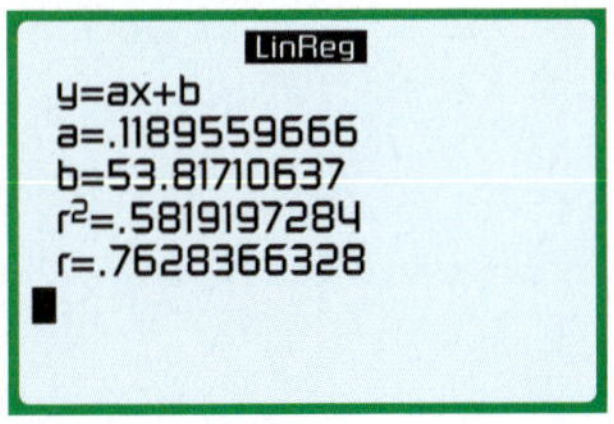

The equation for the regression line is stored in Y1. Using 200 seconds as the time between eruptions, as we did above, we can calculate the expected duration of the eruption as follows.

Press the VARS key and scroll to Y-VARS. Press ENTER twice. Now enter "(200)" and press ENTER. The predicted duration of the eruption is approximately 78 seconds.

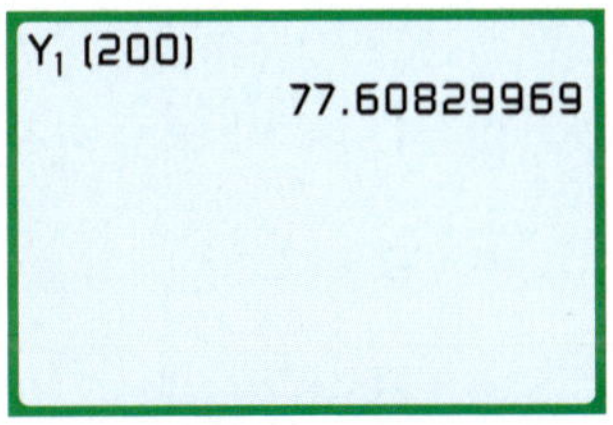

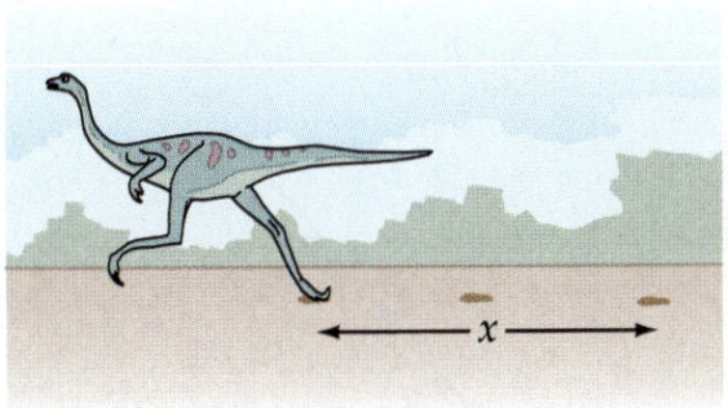

Here is an additional example of calculating regression lines. Professor R. McNeill Alexander wanted to determine whether the *stride length* of a dinosaur, as shown by its fossilized footprints, could be used to estimate the speed of the dinosaur. Stride length for an animal is defined as the distance x from a particular point on a footprint to that same point on the next footprint of the same foot. (See the figure at the left.) Because dinosaurs are extinct, Alexander and fellow scientist A. S. Jayes carried out experiments with many types of animals, including adult men, dogs, camels, ostriches, and elephants. Some of the results from these experiments are recorded in Table 13.12. These data will be used in the examples that follow.

TABLE 13.12
Speeds for Selected Stride Lengths

a. Adult men

Stride length (m)	2.5	3.0	3.3	3.5	3.8	4.0	4.2	4.5
Speed (m/s)	3.4	4.9	5.5	6.6	7.0	7.7	8.3	8.7

b. Dogs

Stride length (m)	1.5	1.7	2.0	2.4	2.7	3.0	3.2	3.5
Speed (m/s)	3.7	4.4	4.8	7.1	7.7	9.1	8.8	9.9

c. Camels

Stride length (m)	2.5	3.0	3.2	3.4	3.5	3.8	4.0	4.2
Speed (m/s)	2.3	3.9	4.4	5.0	5.5	6.2	7.1	7.6

EXAMPLE 1 Find the Equation of a Least-Squares Line

Find the equation of the least-squares line for the ordered pairs in Table 13.12a.

Solution

Enter the data into a calculator or software program that supports regresson equations. Here are the results using a TI-84 calculator.

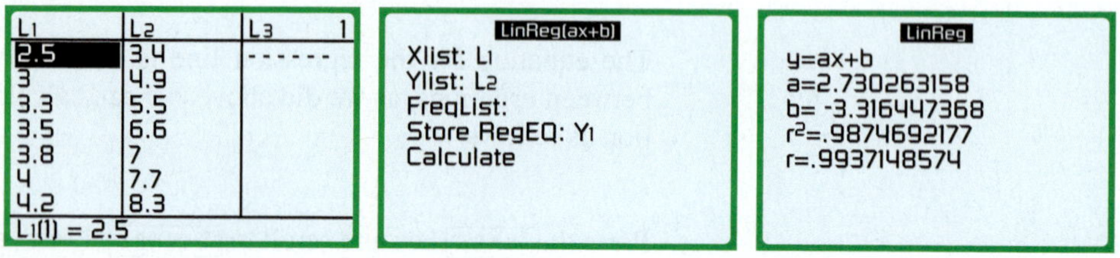

The regression equation is $\hat{y} = 2.730263158x - 3.316447368$.

CHECK YOUR PROGRESS 1 Find the equation of the least-squares line for the stride length and speed of camels given in Table 13.12c.

Solution See page S48.

It can be proved that for any set of ordered pairs, the graph of the ordered pair $(\bar{x}, \bar{y})$ is a point on the least-squares line that models the set. This fact can serve as a check. If you have calculated the least-squares line for a set of ordered pairs and you find that the point $(\bar{x}, \bar{y})$ does not lie on your line, then you know that you have made an error.

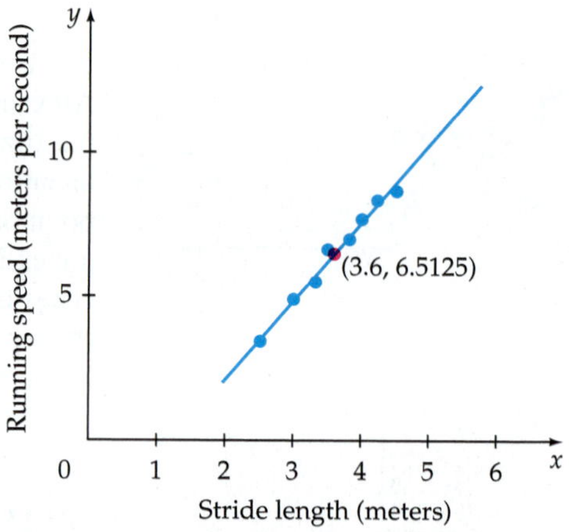

Once the equation of the least-squares line is found, it can be used to make predictions. This procedure is illustrated in the next example.

EXAMPLE 2 Use a Least-Squares Line to Make a Prediction

Use the equation of the least-squares line from Example 1 to predict the average speed of an adult man for each of the following stride lengths. Round your results to the nearest tenth of a meter per second.

a. 2.8 m **b.** 4.8 m

Solution

a. From Example 1, the regression equation is $\hat{y} = 2.730263158x - 3.316447368$. Substitute 2.8 for x and evaluate the resulting expression.

$$\begin{aligned}\hat{y} &= 2.730263158x - 3.316447368\\ &= 2.730263158(2.8) - 3.316447368 \approx 4.328\end{aligned}$$

The predicted average speed of an adult man with a stride length of 2.8 m is 4.3 m/s.

b. From Example 1, the regression equation is $\hat{y} = 2.730263158x - 3.316447368$. Substitute 4.8 for x and evaluate the resulting expression.

$$\begin{aligned}\hat{y} &= 2.730263158x - 3.316447368\\ &= 2.730263158(4.8) - 3.316447368 \approx 9.789\end{aligned}$$

The predicted average speed of an adult man with a stride length of 4.8 m is 9.8 m/s.

CHECK YOUR PROGRESS 2 Use the equation of the least-squares line from Check Your Progress 1 to predict the average speed of a camel for each of the following stride lengths. Round your results to the nearest tenth of a meter per second.

a. 2.7 m

b. 4.5 m

Solution See page S48.

TAKE NOTE

Sometimes values predicted by extrapolation are not reasonable. For instance, if we wish to predict the speed of a man with a stride length of $x = 20$ m, the least-squares equation $\hat{y} = 2.730263158x - 3.316447368$ gives us a speed of about 51.3 m/s. Because the maximum stride length of adult men is considerably less than 20 m, we should not trust this prediction.

The procedure in Example 2a made use of an equation to determine a point between given data points. This procedure is referred to as **interpolation**. In Example 2b, an equation was used to determine a point to the right of the given data points. The process of using an equation to determine a point to the right or left of given data points is referred to as **extrapolation**. See Figure 13.18.

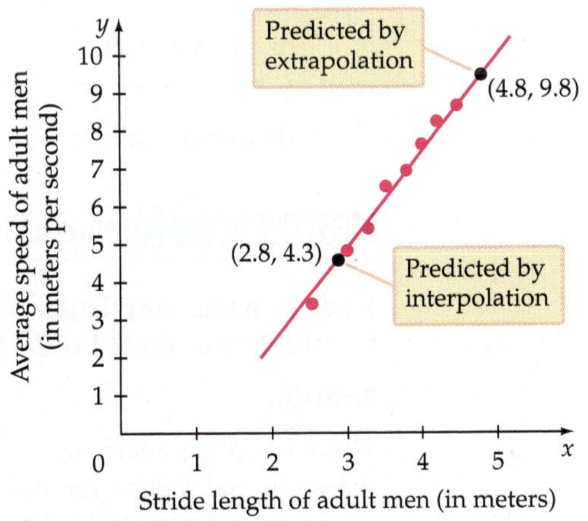

FIGURE 13.18 Interpolation and extrapolation

Linear Correlation Coefficient

To determine the strength of a linear relationship between two variables, statisticians use a statistic called the *linear correlation coefficient*, which is denoted by the variable r and is defined as follows.

HISTORICAL NOTE

Karl Pearson (pîr'sən) spent most of his career as a mathematics professor at University College, London. Some of his major contributions concerned the development of statistical procedures such as regression analysis and correlation. He was particularly interested in applying these statistical concepts to the study of heredity. The term *standard deviation* was invented by Pearson, and because of his work in the area of correlation, the formal name given to the linear correlation coefficient is the *Pearson product moment coefficient of correlation*. Pearson was a co-founder of the statistical journal *Biometrika*.

Linear Correlation Coefficient

For the n ordered pairs $(x_1, y_1), (x_2, y_2), (x_3, y_3), \ldots, (x_n, y_n)$, the **linear correlation coefficient** r is given by

$$r = \frac{n(\Sigma xy) - (\Sigma x)(\Sigma y)}{\sqrt{n(\Sigma x^2) - (\Sigma x)^2} \cdot \sqrt{n(\Sigma y^2) - (\Sigma y)^2}}$$

PPT

If the linear correlation coefficient r is positive, the relationship between the variables has a **positive correlation**. In this case, if one variable increases, the other variable also tends to increase. If r is negative, the linear relationship between the variables has a **negative correlation**. In this case, if one variable increases, the other variable tends to decrease.

Figure 13.19 on the next page shows some scatter diagrams along with the type of linear correlation that exists between the x and y variables. The closer $|r|$ is to 1, the stronger the linear relationship is between the variables.

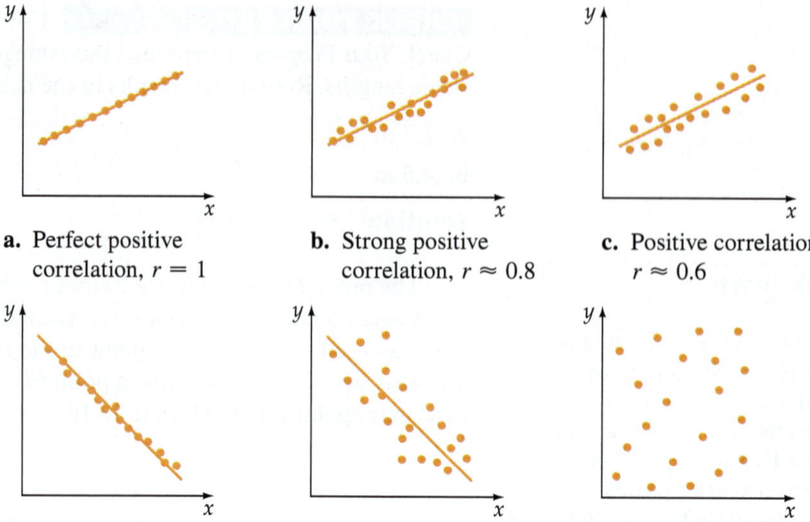

a. Perfect positive correlation, $r = 1$
b. Strong positive correlation, $r \approx 0.8$
c. Positive correlation, $r \approx 0.6$
d. Strong negative correlation, $r \approx -0.9$
e. Negative correlation, $r \approx -0.5$
f. Little or no linear correlation, $r \approx 0$

FIGURE 13.19 Linear correlation

EXAMPLE 3 Find a Linear Correlation Coefficient

Find the linear correlation coefficient for stride length versus speed of an adult man. Use the data in Table 13.12a. Round your result to the nearest hundredth.

Solution

The correlation coefficient r is displayed on the same screen as the values for the regression equation. The screen from Example 1 is shown again at the right.

The linear correlation coefficient, rounded to the nearest hundredth, is 0.99.

CHECK YOUR PROGRESS 3 Find the linear correlation coefficient for stride length versus speed of a camel as given in Table 13.12c. Round your result to the nearest hundredth.

Solution See page S48.

QUESTION What is the significance of the fact that the linear correlation coefficient is positive in Example 3?

The linear correlation coefficient indicates the strength of a linear relationship between two variables; however, it does not indicate the presence of a *cause-and-effect relationship*. For instance, the data in Table 13.13 show the hours per week that a student spent playing pool and the student's weekly algebra test scores for those same weeks.

TABLE 13.13
Algebra Test Scores vs. Hours Spent Playing Pool

Hours per week spent playing pool	4	5	7	8	10
Weekly algebra test score	52	60	72	79	83

ANSWER It indicates a positive correlation between a man's stride length and his speed. That is, as a man's stride length increases, his speed also increases.

The linear correlation coefficient for the ordered pairs in the table is $r \approx 0.98$. Thus there is a strong positive linear relationship between the student's algebra test scores and the time the student spent playing pool. This does not mean that the higher algebra test scores were caused by the increased time spent playing pool. The fact that the student's test scores increased with the increase in the time spent playing pool could be due to many other factors, or it could just be a coincidence.

In your work with applications that involve the linear correlation coefficient *r*, it is important to remember the following properties of *r*.

Properties of the Linear Correlation Coefficient

1. The linear correlation coefficient *r* is always a real number between -1 and 1, inclusive. In the case in which
   - all of the ordered pairs lie on a line with positive slope, *r* is 1.
   - all of the ordered pairs lie on a line with negative slope, *r* is -1.
2. For any set of ordered pairs, the linear correlation coefficient *r* and the slope of the least-squares line both have the same sign.
3. Interchanging the variables in the ordered pairs does not change the value of *r*. Thus the value of *r* for the ordered pairs $(x_1, y_1), (x_2, y_2), \ldots, (x_n, y_n)$ is the same as the value of *r* for the ordered pairs $(y_1, x_1), (y_2, x_2), \ldots, (y_n, x_n)$.
4. The value of *r* does not depend on the units used. You can change the units of a variable from, for example, feet to inches, and the value of *r* will remain the same.

EXCURSION

Exponential Regression

Earlier in this chapter we examined linear regression models. In some cases, an exponential function may more closely model a set of data. For example, suppose a diamond merchant has determined the values of several white diamonds that have different weights, measured in carats, but are similar in quality. See the table below.

4.00 ct	3.00 ct	2.00 ct	1.75 ct	1.50 ct	1.25 ct	1.00 ct	0.75 ct	0.50 ct
$14,500	$10,700	$7900	$7300	$6700	$6200	$5800	$5000	$4600

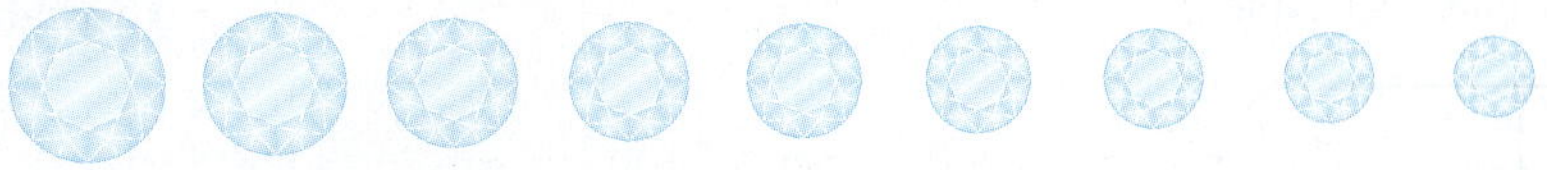

TAKE NOTE

The value of a diamond is generally determined by its color, cut, clarity, and carat weight. These characteristics of a diamond are known as the four c's. In the example at the right we have assumed that the color, cut, and clarity of all of the diamonds are similar. This assumption enables us to model the value of each diamond as a function of just its carat weight.

We can use the data in the table to determine an exponential growth function that models the values of the diamonds as a function of their weights, and then use the model to predict the value of a 3.5-carat diamond of similar quality. Using a graphing calculator, the exponential regression equation is $y \approx 4067.6(1.3816)^x$, where *x* is the carat weight of the diamond and *y* is the value of the diamond.

```
ExpReg
 y = a*b^x
 a = 4067.641145
 b = 1.381644186
 r2 = .994881215
 r = .9974373238
```

POINT OF INTEREST

The Hope Diamond, shown below, is the world's largest deep-blue diamond. It weighs 45.52 carats. We should not expect the function $y \approx 4067.6(1.3816)^x$ to yield an accurate value of the Hope Diamond because the Hope Diamond is not the same type of diamond as the diamonds in the example, and its weight is much larger.

The Hope Diamond is on display at the Smithsonian Museum of Natural History in Washington, D.C.

To use the regression equation to predict the value of a 3.5-carat diamond of similar quality, substitute 3.5 for x and evaluate.

$$y \approx 4067.6(1.3816)^x$$
$$y \approx 4067.6(1.3816)^{3.5}$$
$$y \approx 12{,}609$$

According to the modeling function, the value of a 3.5-carat diamond of similar quality is $12,609.

Note on the calculator screen on the preceding page that r^2, the coefficient of determination, is about 0.9949, which is very close to 1. This indicates that the equation $y \approx 4067.6(1.3816)^x$ provides a good fit for the data.

EXCURSION EXERCISE

1. The following table shows Earth's atmospheric pressure P at an altitude of a kilometers. Find an exponential function that models the atmospheric pressure as a function of altitude. Use the function to estimate, to the nearest tenth, the atmospheric pressure at an altitude of 11 km.

Altitude, *a*, in kilometers above sea level	Atmospheric pressure, *P*, in newtons per square centimeter
0	10.3
2	8.0
4	6.4
6	5.1
8	4.0
10	3.2
12	2.5
14	2.0
16	1.6
18	1.3

EXERCISE SET 13.5

1. Which of the scatter diagrams below suggests the

a. strongest positive linear correlation between the x and y variables?

b. strongest negative linear correlation between the x and y variables?

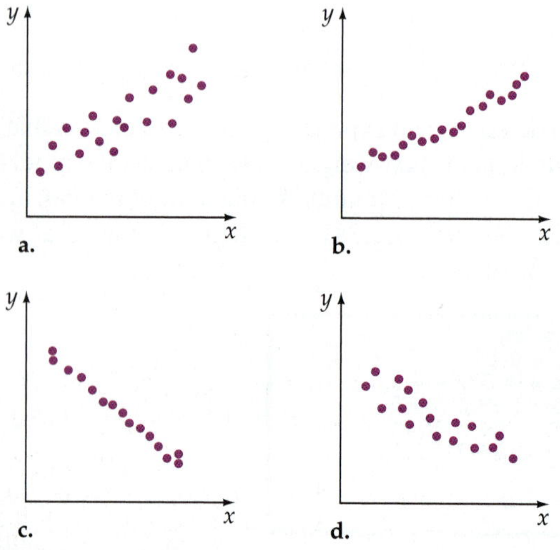

2. Which of the scatter diagrams below suggests

a. a nearly perfect positive linear correlation between the x and y variables?

b. little or no linear correlation between the x and y variables?

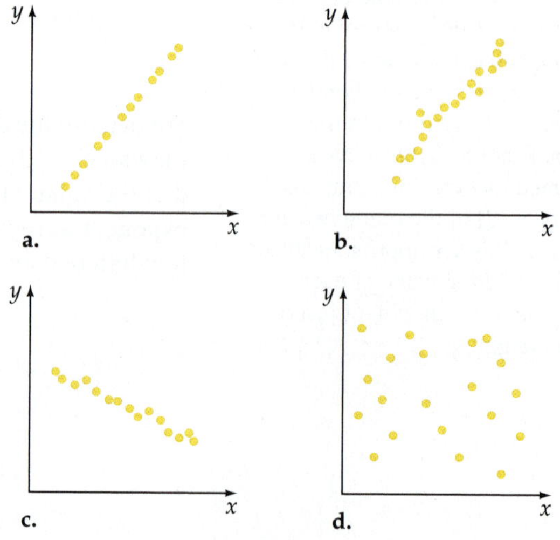

3. Given the bivariate data:

x	1	2	3	5	6
y	7	5	3	2	1

a. Draw a scatter diagram for the data.

b. Find n, Σx, Σy, Σx^2, $(\Sigma x)^2$, and Σxy.

c. Find a, the slope of the least-squares line, and b, the y-intercept of the least-squares line.

d. Draw the least-squares line on the scatter diagram from part a.

e. Is the point $(\overline{x}, \overline{y})$ on the least-squares line?

f. Use the equation of the least-squares line to predict the value of y when $x = 3.4$.

g. Find, to the nearest hundredth, the linear correlation coefficient.

4. Given the bivariate data:

x	3	4	5	6	7
y	2	3	3	5	5

a. Draw a scatter diagram for the data.

b. Find n, Σx, Σy, Σx^2, $(\Sigma x)^2$, and Σxy.

c. Find a, the slope of the least-squares line, and b, the y-intercept of the least-squares line.

d. Draw the least-squares line on the scatter diagram from part a.

e. Is the point $(\overline{x}, \overline{y})$ on the least-squares line?

f. Use the equation of the least-squares line to predict the value of y when $x = 7.3$.

g. Find, to the nearest hundredth, the linear correlation coefficient.

■ In Exercises 5 to 10, find the equation of the least-squares line and the linear correlation coefficient for the given data. Round the constants, a, b, and r to the nearest hundredth.

5. $\{(2, 6), (3, 6), (4, 8), (6, 11), (8, 18)\}$

6. $\{(2, -3), (3, -4), (4, -9), (5, -10), (7, -12)\}$

7. $\{(-3, 11.8), (-1, 9.5), (0, 8.6), (2, 8.7), (5, 5.4)\}$

8. $\{(-7, -11.7), (-5, -9.8), (-3, -8.1), (1, -5.9), (2, -5.7)\}$

9. $\{(1, 4.1), (2, 6.0), (4, 8.2), (6, 11.5), (8, 16.2)\}$

10. $\{(2, 5), (3, 7), (4, 8), (6, 11), (8, 18), (9, 21)\}$

In Exercises 11 to 18, use the statistics features of a graphing calculator.

11. **Value of a Corvette** The following table gives retail values of a 2010 Corvette for various odometer readings. (*Source:* Kelley Blue Book website)

Odometer reading	Retail value
13,000	\$52,275
18,000	\$51,525
20,000	\$51,200
25,000	\$50,275
29,000	\$49,625
32,000	\$49,075

a. Find the equation of the least-squares line for the data. Round constants to the nearest thousandth.

b. Use the equation from part a to predict the retail price of a 2010 Corvette with an odometer reading of 30,000.

c. Find the linear correlation coefficient for these data.

d. What is the significance of the fact that the linear correlation coefficient is negative for these data?

12. **Paleontology** The following table shows the length, in centimeters, of the humerus and the total wingspan, in centimeters, of several pterosaurs, which are extinct flying reptiles. (*Source:* Southwest Educational Development Laboratory)

Pterosaur Data

Humerus	Wingspan	Humerus	Wingspan
24	600	20	500
32	750	27	570
22	430	15	300
17	370	15	310
13	270	9	240
4.4	68	4.4	55
3.2	53	2.9	50
1.5	24		

a. Find the equation of the least-squares line for the data. Round constants to the nearest hundredth.

b. Use the equation from part a to determine, to the nearest centimeter, the projected wingspan of a pterosaur if its humerus is 54 cm.

13. **Health** The U.S. Centers for Disease Control and Prevention (CDC) use a measure called body mass index (BMI) to determine whether a person is overweight. A BMI between 25.0 and 29.9 is considered overweight, and a BMI of 30.0 or more is considered obese. The following table shows the percents of U.S. males 18 years old or older who were obese in the years indicated, judging on the basis of BMI. (*Source:* Centers for Disease Control and Prevention)

Year	Percent obese
2003	22.7
2004	23.9
2005	24.9
2006	25.3
2007	26.5
2008	26.6
2009	27.6

a. Using 3 for 2003, 4 for 2004, and so on, find the equation of the least-squares line for the data.

b. Use the equation from part a to predict the percent of overweight males in 2015.

14. **Health** The U.S. Centers for Disease Control and Prevention (CDC) use a measure called body mass index (BMI) to determine whether a person is overweight. A BMI between 25.0 and 29.9 is considered overweight, and a BMI of 30.0 or more is considered obese. The following table shows the percents of U.S. females 18 years old or older who were overweight in the years indicated, judging on the basis of BMI. (*Source:* Centers for Disease Control and Prevention)

Year	Percent obese
2003	23.3
2004	23.7
2005	24.3
2006	25.6
2007	25.2
2008	27.6
2009	26.8

a. Using 3 for 2003, 4 for 2004, and so on, find the equation of the least-squares line for the data.

b. Use the equation from part a to predict the percent of overweight females in 2015.

15. **Wireless Phone** The following table shows the approximate numbers of wireless telephone subscriptions in the United States for recent years.

U.S. Wireless Telephone Subscriptions

Year	2005	2006	2007	2008	2009	2010
Subscriptions, in millions	194	220	243	263	277	293

SOURCE: CTIA Semi-Annual Wireless Survey, Midyear 2010

a. Find the linear correlation coefficient for the data.

b. On the basis of the value of the linear correlation coefficient, would you conclude, at the $|r| > 0.9$ level, that the data can be reasonably modeled by a linear equation? Explain.

16. **Life Expectancy** The average remaining lifetimes for men of various ages in the United States are given in the following table. (*Source:* National Institutes of Health)

Average Remaining Lifetimes for Men

Age	Years	Age	Years
0	74.9	65	16.8
15	60.6	75	10.2
35	42.0		

Use the linear correlation coefficient to determine whether there is a strong correlation, at the level $|r| > 0.9$, between a man's age and the average remaining lifetime of that man.

17. **Life Expectancy** The average remaining lifetimes for women of various ages in the United States are given in the following table. (*Source:* National Institutes of Health)

Average Remaining Lifetimes for Women

Age	Years	Age	Years
0	79.9	65	19.5
15	65.6	75	12.1
35	46.2		

a. Find the equation of the least-squares line for the data.

b. Use the equation from part a to estimate the remaining lifetime of a woman of age 25.

c. Is the procedure in part b an example of interpolation or extrapolation?

18. **Fitness** An aerobic exercise instructor remembers the data given in the following table, which shows the recommended maximum exercise heart rates for individuals of the given ages.

Age (x years)	20	40	60
Maximum heart rate (y beats per minute)	170	153	136

a. Find the linear correlation coefficient for the data.

b. What is the significance of the value found in part a?

c. Find the equation of the least-squares line.

d. Use the equation from part c to predict the maximum exercise heart rate for a person who is 72.

e. Is the procedure in part d an example of interpolation or extrapolation?

EXTENSIONS

19. **Tuition** The following table shows the average annual tuition and fees at private and public 4-year colleges and universities for the school years 2009–2010 through 2014–2015. (*Source:* National Center for Education Statistics)

Four-year Colleges and Universities Tuition and Fees

Year	Private	Public
2009–2010	31,448	15,014
2010–2011	32,617	15,918
2011–2012	33,674	16,805
2012–2013	35,074	17,474
2013–2014	36,193	18,372
2014–2015	37,385	19,203

a. Using 1 for 2009–2010, 2 for 2010–2011, and so on, find the linear correlation coefficient and the equation of the least-squares line for the tuition and fees at private 4-year colleges and universities, based on the year.

b. Using 1 for 2009–2010, 2 for 2010–2011, and so on, find the linear correlation coefficient and the equation of the least-squares line for the tuition and fees at public 4-year colleges and universities, based on the year.

c. Based on the linear correlation coefficients you found in parts a and b, are the equations you wrote in parts a and b good models of the growth in tuition and fees at 4-year colleges and universities?

d. The equation of a least-squares line is written in the form $\hat{y} = ax + b$. Explain the meaning of the value of a for each equation you wrote in parts a and b.

20. Search for bivariate data (in a magazine, in a newspaper, in an almanac, or on the Internet) that can be closely modeled by a linear equation.

a. Draw a scatter diagram of the data.

b. Find the equation of the least-squares line and the linear correlation coefficient for the data.

c. Graph the least-squares line on the scatter diagram in part a.

d. Use the equation of the least-squares line to predict a range value for a specific domain value.

CHAPTER 13 SUMMARY

The following table summarizes essential concepts in this chapter. The references given in the right-hand column list Examples and Exercises that can be used to test your understanding of a concept.

13.1 Measures of Central Tendency	
Mean, Median, and Mode The *mean* of n numbers is the sum of the numbers divided by n. The *median* of a ranked list of n numbers is the middle number if n is odd, or the mean of the two middle numbers if n is even. The *mode* of a list of numbers is the number that occurs most frequently.	See **Examples 1, 2, and 3** on pages 752 to 754, and then try Exercise 1 on page 808.

continued

Weighted Mean The formula for the weighted mean of the n numbers $x_1, x_2, x_3, \ldots, x_n$ is $$\text{Weighted mean} = \frac{\Sigma(x \cdot w)}{\Sigma w}$$ where $\Sigma(x \cdot w)$ is the sum of the products formed by multiplying each number by its assigned weight, and Σw is the sum of all the weights.	See **Example 4** on page 756, and then try Exercise 7 on page 808.
13.2 Measures of Dispersion	
Range The range of a set of data values is the difference between the greatest data value and the least data value.	See **Example 1** on page 762, and then try Exercise 5 on page 808.
Standard Deviation and Variance If $x_1, x_2, x_3, \ldots, x_n$ is a *population* of n numbers with mean μ, then the standard deviation of the population is $$\sigma = \sqrt{\frac{\Sigma(x - \mu)^2}{n}}, \text{ and the variance is } \frac{\Sigma(x - \mu)^2}{n}.$$ If $x_1, x_2, x_3, \ldots, x_n$ is a *sample* of n numbers with mean $\bar{x}$, then the standard deviation of the population is $$s = \sqrt{\frac{\Sigma(x - \bar{x})^2}{n - 1}}, \text{ and the variance is } \frac{\Sigma(x - \bar{x})^2}{n - 1}.$$	See **Examples 2 and 5** on pages 764 and 767, and then try Exercise 9 on page 808.
13.3 Measures of Relative Position	
z-score The z-score for a given data value x is the number of standard deviations that x is above or below the mean. z-score for a population data value: $z_x = \frac{x - \mu}{\sigma}$ z-score for a sample data value: $z_x = \frac{x - \bar{x}}{s}$	See **Example 1** on page 771, and then try Exercises 8a and 12 on page 808.
Percentiles A value x is called the pth percentile of a data set provided $p\%$ of the data values are less than x. Given a set of data and a data value x, $$\text{Percentile score of } x = \frac{\text{number of data values less than } x}{\text{total number of data values}} \cdot 100$$	See **Example 4** on page 773, and then try Exercise 8b on page 808.
Quartiles and Box-and-Whisker Plots The quartiles of a data set are the three numbers Q_1, Q_2, and Q_3 that partition the ranked data into four (approximately) equal groups. Q_2 is the median of the data, Q_1 is the median of the data values less than Q_2, and Q_3 is the median of the data values greater than Q_2. A box-and-whisker plot is a display used to show the quartiles and the maximum and minimum values of a data set.	See **Examples 5 and 6** on pages 774 to 776, and then try Exercise 13 on page 808.

13.4 Normal Distributions

Frequency Distributions A frequency distribution displays a data set by dividing the data into intervals, or classes, and listing the number of data values that fall into each interval. A relative freqency distribution lists the percent of data in each interval.	See **Example 1** on page 783, and then try Exercise 15 on page 809.
Normal Distributions and the Empirical Rule A normal distribution of data is a bell-shaped curve that is symmetric about a vertical line through the mean. The y-value of each point on the curve is the *percent* (expressed as a decimal) of the data at the corresponding x-value. The total area under the curve is 1. The Empirical Rule for a normal distribution states that approximately 68% of the data lie within 1 standard deviation of the mean, 95% of the data lie within 2 standard deviations of the mean, and 99.7% of the data lie within 3 standard deviations of the mean.	See **Example 2** on page 785, and then try Exercise 20 on page 809.
Using the Standard Normal Distribution The standard normal distribution is the normal distribution that has a mean of 0 and a standard deviation of 1. Any normal distribution can be converted into the standard normal distribution by converting data values to their z-scores. Then the percent of data values that lie in a given interval can be found as the area under the standard normal curve between the z-scores of the endpoints of the given interval. Table 13.10 on page 787 gives the areas under the standard normal curve for z-scores between 0 and 3.33.	See **Example 5** on page 789, and then try Exercise 21 on page 809.

13.5 Linear Regression and Correlation

Least-Squares Line Bivariate data are data given as ordered pairs. The least-squares regression line, or least-squares line, for a set of bivariate data is the line that minimizes the sum of the squares of the vertical deviations from each data point to the line. The equation of the least-squares line for the n ordered pairs $(x_1, y_1), (x_2, y_2), (x_3, y_3), \ldots, (x_n, y_n)$ is $\hat{y} = ax + b$, where $a = \dfrac{n\Sigma xy - (\Sigma x)(\Sigma y)}{n(\Sigma x^2) - (\Sigma x)^2}$ and $b = \bar{y} - a\bar{x}$. The equation of the least-squares line can be used to predict the value of one variable when the value of the other variable is known.	See **Examples 1 and 2** on page 798, and then try Exercises 24b and 24c on page 810.
Linear Correlation Coefficient The linear correlation coefficient r measures the strength of a linear relationship between two variables. The closer $\lvert r\rvert$ is to 1, the stronger the linear relationship is between the variables. For the n ordered pairs $(x_1, y_1), (x_2, y_2), (x_3, y_3), \ldots, (x_n, y_n)$, the linear correlation coefficient is $r = \dfrac{n\Sigma xy - (\Sigma x)(\Sigma y)}{\sqrt{n\Sigma x^2 - (\Sigma x)^2} \cdot \sqrt{n\Sigma y^2 - (\Sigma y)^2}}$	See **Example 3** on page 800, and then try Exercise 24a on page 810.

CHAPTER 13 REVIEW EXERCISES

1. Find the mean, median, mode, range, population variance, and population standard deviation for the following data. Round noninteger values to the nearest tenth.

   12, 17, 14, 12, 8, 19, 21

2. A set of data has a mean of 16, a median of 15, and a mode of 14. Which of these numbers must be a value in the data set?

3. Write a set of data with five data values for which the mean, median, and mode are all 55.

4. State whether the mean, median, or mode is being used.

   a. In 2002, there were as many people aged 25 and younger in the world as there were people aged 25 and older.

   b. The majority of full-time students carry a load of 15 credit hours per semester.

   c. The average annual return on an investment is 6.5%.

5. **Bridges** The lengths of cantilever bridges in the United States are shown below. Find the mean, median, mode, and range of the data.

   Bridge Length (in feet)

   Baton Rouge (Louisiana), 1235
   Commodore John Barry (Pennsylvania), 1644
   Greater New Orleans (Louisiana), 1576
   Longview (Washington), 1200
   Patapsco River (Maryland), 1200
   Queensboro (New York), 1182
   Tappan Zee (New York), 1212
   Transbay Bridge (California), 1400

6. **Average Speed** Cleone traveled 45 mi to her sister's house in 1 h. The return trip took 1.5 h. What was Cleone's average rate for the entire trip?

7. **Grade Point Average** In a 4.0 grading system, each letter grade has the following numerical value.

A = 4.00	B− = 2.67	D+ = 1.33
A− = 3.67	C+ = 2.33	D = 1.00
B+ = 3.33	C = 2.00	D− = 0.67
B = 3.00	C− = 1.67	F = 0.00

   Use the weighted mean formula to find the grade point average for a student with the following grades. Round to the nearest hundredth.

Course	Credits	Grade
Mathematics	3	A
English	3	C+
Computers	2	B−
Biology	4	B
Art	1	A

8. **Test Scores** A teacher finds that the test scores of a group of 40 students have a mean of 72 and a standard deviation of 8.

   a. If Ann has a test score of 82, what is Ann's z-score?

   b. Ann's score is higher than that of 35 of the 40 students who took the test. Find the percentile, rounded to the nearest percent, for Ann's score.

9. **Airline Industry** An airline recorded the times it took for a ground crew to unload the baggage from an airplane. The recorded times, in minutes, were 12, 18, 20, 14, and 16. Find the *sample* standard deviation and the variance of these times. Round your results to the nearest hundredth of a minute.

10. **Ticket Prices** The following table gives the average annual admission prices to U.S. movie theaters for the years 2006 to 2015.

Average Annual Admission Price, 2006–2015

$6.55	$6.88	$7.18	$7.50	$7.89
$7.93	$7.96	$8.15	$8.17	$8.12

Source: Theatrical Market Statistics, 2015, Motion Picture Association of America

   Find the mean, median, and standard deviation for this *sample* of admission prices. Round to the nearest cent.

11. **Test Scores** One student received test scores of 85, 92, 86, and 89. A second student received scores of 90, 97, 91, and 94 (exactly 5 points more on each test).

    a. What is the relationship between the means of the 2 students' test scores?

    b. What is the relationship between the standard deviations of the 2 students' test scores?

12. A *population* data set has a mean of 81 and a standard deviation of 5.2. Find the z-scores for each of the following. Round to the nearest hundredth.

    a. $x = 72$ **b.** $x = 84$

13. **Cholesterol Levels** The cholesterol levels for 10 adults are shown below. Draw a box-and-whisker plot of the data.

Cholesterol Levels

310	185	254	221	170
214	172	208	164	182

14. Test Scores The following histogram shows the distribution of the test scores for a history test.

a. How many students scored at least 84 on the test?

b. How many students took the test?

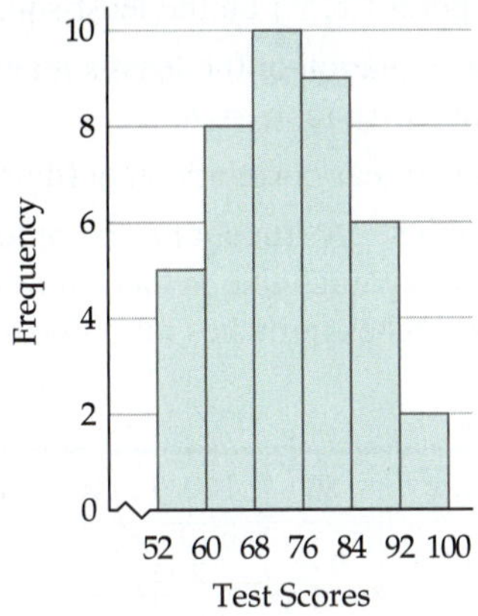

15. Teacher Salaries Use the following relative frequency distribution to determine the

a. *percent* of the states that paid an average teacher salary of at least \$48,000.

b. *probability, as a decimal,* that a state selected at random paid an average teacher salary of at least \$56,000 but less than \$72,000.

Average Salaries of Public School Teachers, 2014–2015

Average salary, s	Number of states	Relative frequency
$\$40,000 \le s < \$44,000$	2	4%
$\$44,000 \le s < \$48,000$	9	18%
$\$48,000 \le s < \$52,000$	14	28%
$\$52,000 \le s < \$56,000$	4	8%
$\$56,000 \le s < \$60,000$	9	18%
$\$60,000 \le s < \$64,000$	2	4%
$\$64,000 \le s < \$68,000$	4	8%
$\$68,000 \le s < \$72,000$	2	4%
$\$72,000 \le s < \$76,000$	3	6%
$\$76,000 \le s < \$80,000$	1	2%

SOURCE: National Education Association

16. Greenhouse Gas Emissions The table below shows annual greenhouse gas emissions, in tons of carbon dioxide (CO_2), by vehicle fuel efficiency rating in miles per gallon (mpg).

MPG rating	15	20	25	30	35	40	45	50
Tons of CO_2	12	9	7.2	6	5.1	4.5	4	4.8

SOURCE: fueleconomy.gov

Is there a linear relation, at the $|r| > 0.9$ level, between vehicle fuel efficiency and greenhouse gas emissions?

17. Alternative Fuels Alternative fuel vehicles that run on nonpetroleum-based fuels cannot refuel at traditional gas stations. Use the table of the numbers of alternative fuel stations in the United States to answer the questions below.

Alternative Fuel Stations in the U.S.

Year	Number
2011	10,071
2012	20,498
2013	27,159
2014	36,805
2015	39,963

SOURCE: U.S. Department of Energy

a. Using 11 for 2011, 12 for 2012, and so on, find the equation of the least-squares line for the data.

b. Use your equation from part a to predict the number of alternative fuel stations in the United States in 2015.

18. Test Scores A professor gave a final examination to 110 students. Eighteen students had scores that were more than 1 standard deviation above the mean. With this information, can you conclude that 18 of the students had scores that were more than 1 standard deviation below the mean? Explain.

19. Waiting Time The amount of time customers spend waiting in line at the ticket counter of an amusement park is normally distributed, with a mean of 6.5 min and a standard deviation of 1 min. Find the probability that the time a customer will spend waiting is:

a. less than 8 min. **b.** less than 6 min.

20. Pet Food The weights of all the sacks of dog food filled by a machine are normally distributed, with an average weight of 50 lb and a standard deviation of 0.5 lb. What percent of the sacks will

a. weigh less than 49.5 lb?

b. weigh between 49 and 51 lb?

21. Telecommunication A telephone manufacturer finds that the life spans of its telephones are normally distributed, with a mean of 6.5 years and a standard deviation of 0.5 year.

a. What percent of its telephones will last at least 7.25 years?

b. What percent of its telephones will last between 5.8 years and 6.8 years?

c. What percent of its telephones will last less than 6.9 years?

22. Astronomy The following table gives the distances, in millions of miles, of Earth from the sun at selected times during the year.

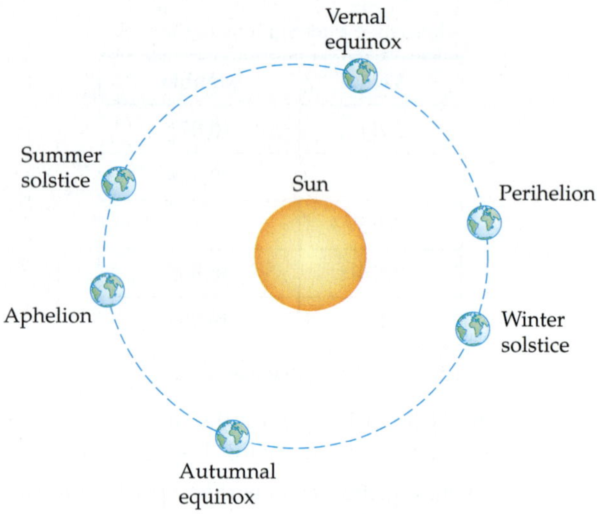

Position	Distance (millions of miles)
Perihelion	91.4
Vernal equinox	92.6
Summer solstice	94.5
Aphelion	94.6
Autumnal equinox	94.3
Winter solstice	91.5

On the basis of these data, what is the mean distance of Earth from the sun?

23. Given the bivariate data

x	10	12	14	15	16
y	8	7	5	4	1

a. Draw a scatter diagram for the data.

b. Find n, Σx, Σy, Σx^2, $(\Sigma x)^2$, and Σxy.

c. Find a, the slope of the least-squares regression line, and b, the y-intercept of the least-squares line.

d. Draw the least-squares line on the scatter diagram from part a.

e. Is the point $(\overline{x}, \overline{y})$ on the least-squares line?

f. Use the equation of the least-squares line to predict the value of y for $x = 8$.

g. Find the linear correlation coefficient.

24. Physics A student has recorded the data in the following table, which shows the distance a spring stretches in inches for a given weight in pounds.

Weight, x	80	100	110	150	170
Distance, y	6.2	7.4	8.3	11.1	12.7

a. Find the linear correlation coefficient.

b. Find the equation of the least-squares line.

c. Use the equation of the least-squares line from part b to predict the distance a weight of 195 lb will stretch the spring.

25. Internet A test of an Internet service provider showed the following download times (in seconds) for files of various sizes (in megabytes).

a. Find the equation of the least-squares line for these data.

b. On the basis of the value of the linear correlation coefficient, is a linear model of these data a reasonable model?

c. Use the equation of the least-squares line from part a to predict the expected download time of a file that is 100 megabytes in size.

26. Blood alcohol content (BAC) is measured in grams of alcohol per deciliter of blood. For instance, a BAC reading of 0.08% (a level that is considered legally intoxicated in most states) means that one deciliter of the person's blood contains 0.08 gram of alcohol. A toxicologist recorded the times elapsed, in hours, until the blood alcohol levels of eight adults who had consumed various amounts of alcohol were less than 0.005%. The results are given in the table.

BAC (%)	0.01	0.05	0.08	0.02	0.03	0.16	0.1	0.04
Time (h) for BAC level to reach 0.005%	0.8	3.5	5.4	0.9	2.1	9.2	6.4	2.1

a. Find the regression equation and linear correlation coefficient for these data.

b. Use the equation to predict the time it would take for a person with a BAC of 0.06% to reach a BAC level of less than 0.005%. Round to the nearest tenth.

CHAPTER 13 TEST

1. Find the mean, median, and mode for the following data. Round noninteger values to the nearest tenth.

3, 7, 11, 12, 7, 9, 15

2. Grade Point Average Use the 4.0 grading system:

$$A = 4, B = 3, C = 2, D = 1, F = 0$$

A student's grade point average (GPA) is calculated as a weighted mean, where the student's grade in each course is given a weight equal to the number of units that course is worth. Find Justin's GPA for the fall semester. Round to the nearest hundredth.

Justin's Grades, Fall Semester

Course	Course grade	Course units
Algebra	A	3
English	B	3
Biology	C	4
History	C	3
Psychology	A	2

3. Find the range, standard deviation, and variance for the following sample data.

7, 11, 12, 15, 31, 22

4. A *sample* data set has a mean of $\bar{x} = 65$ and a standard deviation of 10.2. Find the *z*-scores for each of the following. Round to the nearest hundredth.

a. $x = 77$ **b.** $x = 60$

5. Basketball Draw a box-and-whisker plot for the following data.

Points Scored by Top 20 Women's National Basketball Association Players in a Recent Year

729	716	702	656	618	606	575
555	548	535	528	522	521	509
484	479	470	465	453	422	

SOURCE: Women's National Basketball Association

6. Movie Attendance Use the following relative frequency distribution to estimate the *percent* of the movie attendees who were

a. at least 40 years of age.

b. at least 18 but less than 40 years of age.

Movie Attendance by Age Group

Age group	Percent of total yearly admissions
2–11	15%
12–17	12%
18–24	12%
25–39	23%
40–49	15%
50–59	11%
≥60	11%

7. During 1 month, an overnight delivery company found that the weights of its parcels were normally distributed, with a mean of 34 oz and a standard deviation of 10 oz. Use the Empirical Rule to determine

a. the percent of the parcels that weighed between 34 oz and 54 oz.

b. the percent of the parcels that weighed less than 24 oz.

8. Box Weights The weights of all the boxes of cake mix filled by a machine are normally distributed, with a mean weight of 18.0 oz and a standard deviation of 0.8 oz. What percent of the boxes will

a. weigh less than 17 oz?

b. weigh between 18.4 and 19.0 oz?

9. A psychologist wants to determine whether there is a relationship between how long it takes a subject to complete a manual task and the number of hours of sleep the subject had the night before. The results from a study of 10 people are given in the following table.

Hours of sleep	6.2	8.1	7.5	8.4	5.0	6.2	4.8	8.0	3.8	5.9
Minutes to complete task	9.0	8.6	8.4	8.6	10.0	9.3	9.9	8.9	10.4	9.1

a. Find the linear correlation coefficient for the data.

b. On the basis of your answer to part a, is there a strong linear relationship, at the $|r| > 0.9$ level, between hours of sleep and minutes to complete a task?

10. Nutrition The following table shows the percent of water and the number of calories in various canned soups to which 100 g of water are added.

a. Find the equation of the least-squares line for the data. Round constants to the nearest hundredth.

b. Use the equation in part a to find the expected number of calories in a soup that is 89% water. Round to the nearest whole number.

Solutions to Check Your Progress Problems

CHAPTER 1

SECTION 1.1

Check your progress 1, *page 2*

a. Each successive number is 5 larger than the preceding number. Thus we predict that the next number in the list is 5 larger than 25, which is 30.

b. The first two numbers differ by 3. The second and third numbers differ by 5. It appears that the difference between any two numbers is always 2 more than the preceding difference. Thus we predict that the next number will be 11 more than 26, which is 37.

Check your progress 2, *page 3*

If the original number is 2, then $\dfrac{2 \times 9 + 15}{3} - 5 = 6$, which is three times the original number.

If the original number is 7, then $\dfrac{7 \times 9 + 15}{3} - 5 = 21$, which is three times the original number.

If the original number is -12, then $\dfrac{-12 \times 9 + 15}{3} - 5 = -36$, which is three times the original number.

It appears, by inductive reasoning, that the procedure produces a number that is three times the original number.

Check your progress 3, *page 3*

a. It appears that when the velocity of a tsunami is doubled, its height is quadrupled.

b. A tsunami with a velocity of 30 feet per second will have a height that is four times that of a tsunami with a speed of 15 feet per second. Thus, we predict a height of $4 \times 25 = 100$ feet for a tsunami with a velocity of 30 feet per second.

Check your progress 4, *page 5*

a. Let $x = 0$. Then $\dfrac{x}{x} \neq 1$, because division by 0 is undefined.

b. Let $x = 1$. Then $\dfrac{x+3}{3} = \dfrac{1+3}{3} = \dfrac{4}{3}$, whereas $x + 1 = 1 + 1 = 2$.

c. Let $x = 3$. Then $\sqrt{x^2 + 16} = \sqrt{3^2 + 16} = \sqrt{25} = 5$, whereas $x + 4 = 3 + 4 = 7$.

Check your progress 5, *page 6*

Let n represent the original number.

Multiply the number by 6:	$6n$
Add 10 to the product:	$6n + 10$
Divide the sum by 2:	$\dfrac{6n + 10}{2} = 3n + 5$
Subtract 5:	$3n + 5 - 5 = 3n$

The procedure always produces a number that is three times the original number.

Check your progress 6, *page 6*

a. The conclusion is a specific case of a general assumption, so the argument is an example of deductive reasoning.

b. The argument reaches a conclusion based on specific examples, so the argument is an example of inductive reasoning.

Check your progress 7, *page 8*

From clue 1, we know that Ashley is not the president or the treasurer. In the following chart, write X1 (which stands for "ruled out by clue 1") in the President and Treasurer columns of Ashley's row.

	Pres.	V. P.	Sec.	Treas.
Brianna				
Ryan				
Tyler				
Ashley	X1			X1

From clue 2, Brianna is not the secretary. We know from clue 1 that the president is not the youngest, and we know from clue 2 that Brianna and the secretary are the youngest members of the group. Thus Brianna is not the president. In the chart, write X2 for these two conditions. Also we know from clues 1 and 2 that Ashley is not the secretary, because she is older than the treasurer. Write an X2 in the Secretary column of Ashley's row.

	Pres.	V. P.	Sec.	Treas.
Brianna	X2		X2	
Ryan				
Tyler				
Ashley	X1		X2	X1

At this point we see that Ashley must be the vice president and that none of the other members is the vice president. Thus we can update the chart as shown below.

	Pres.	V. P.	Sec.	Treas.
Brianna	X2	X2	X2	
Ryan		X2		
Tyler		X2		
Ashley	X1	√	X2	X1

Now we can see that Brianna must be the treasurer and that neither Ryan nor Tyler is the treasurer. Update the chart as shown below.

	Pres.	V. P.	Sec.	Treas.
Brianna	X2	X2	X2	√
Ryan		X2		X2
Tyler		X2		X2
Ashley	X1	√	X2	X1

From clue 3, we know that Tyler is not the secretary. Thus we can conclude that Tyler is the president and Ryan must be the secretary. See the chart below.

	Pres.	V. P.	Sec.	Treas.
Brianna	X2	X2	X2	√
Ryan	X3	X2	√	X2
Tyler	√	X2	X3	X2
Ashley	X1	√	X2	X1

Tyler is the president, Ashley is the vice president, Ryan is the secretary, and Brianna is the treasurer.

SECTION 1.2

Check your progress 1, *page 16*

sequence:	1	14	51	124	245	426	679	...	
first differences:		13	37	73	121	181	253	...	(1)
second differences:			24	36	48	60	72	...	(2)
third differences:				12	12	12	12	...	(3)

Using the method of extending the difference table, we predict that 679 is the next term in the sequence.

Check your progress 2, *page 18*

a. Each figure after the first figure consists of a square region and a "tail." The number of tiles in the square region is n^2, and the number of tiles in the tail is $n - 1$. Thus the nth-term formula for the number of tiles in the nth figure is $a_n = n^2 + n - 1$.

b. Let $n = 10$. Then $n^2 + n - 1 = (10)^2 + (10) - 1 = 109$.

c.

$n^2 + n - 1 = 419$ • A quadratic equation.

$n^2 + n - 420 = 0$ • Write in standard form.

$(n + 21)(n - 20) = 0$ • Factor.

$n + 21 = 0$ or $n - 20 = 0$ • Set each factor equal to 0.

$n = -21$ $\quad n = 20$

We consider only the positive result. The 20th figure will consist of 419 tiles.

Check your progress 3, *page 20*

$$F_9 = F_8 + F_7$$
$$= 21 + 13$$
$$= 34$$

Check your progress 4, *page 21*

a. The inequality $2F_n > F_{n+1}$ is true for $n = 3, 4, 5, 6, \ldots, 10$. Thus, by inductive reasoning, we conjecture that the statement is a true statement.

b. The equality $2F_n + 4 = F_{n+3}$ is not true for $n = 4$, because $2F_4 + 4 = 2(3) + 4 = 10$ and $F_{4+3} = F_7 = 13$. Thus the statement is a false statement.

SECTION 1.3

Check your progress 1, *page 28*

Understand the Problem In order to go past Starbucks, Allison must walk along Third Avenue from Boardwalk to Park Avenue.

Devise a Plan Label each intersection that Allison can pass through with the number of routes to that intersection. If she can reach an intersection from two different routes, then the number of routes to that intersection is the sum of the numbers of routes to the two adjacent intersections.

Carry Out the Plan The following figure shows the number of routes to each of the intersections that Allison could pass through. Thus there are a total of nine routes that Allison can take if she wishes to walk directly from point A to point B and pass by Starbucks.

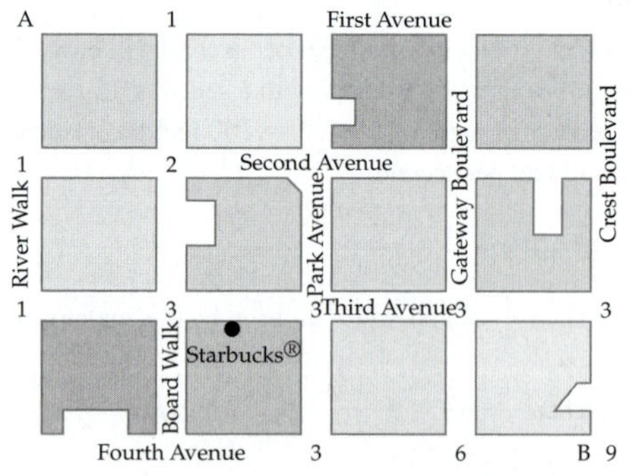

Review the Solution The total of nine routes seems reasonable. We know from Example 1 that if Allison can take any route, the total number of routes is 35. Requiring Allison to go past Starbucks eliminates several routes.

Check your progress 2, *page 29*

Understand the Problem There are several ways to answer the questions so that two answers are "false" and three answers are "true." One way is TTTFF and another is FFTTT.

Devise a Plan Make an organized list. Try the strategy of listing a T unless doing so will produce too many Ts or a duplicate of one of the previous orders in your list.

Carry Out the Plan (Start with three Ts in a row.)

TTTFF	(1)
TTFTF	(2)
TTFFT	(3)
TFTTF	(4)
TFTFT	(5)
TFFTT	(6)
FTTTF	(7)
FTTFT	(8)
FTFTT	(9)
FFTTT	(10)

Review the Solution Each entry in the list has two Fs and three Ts. Since the list is complete and has no duplications, we know that there are 10 ways for a student to mark two questions with "false" and the other three with "true."

Check your progress 3, *page 30*

Understand the Problem There are six people, and each person shakes hands with each of the other people.

Devise a Plan Each person will shake hands with five other people (a person won't shake his or her own hand; that would be silly). Since there are six people, we could multiply 6 times 5 to get the total number of handshakes. However, this procedure would count each handshake exactly twice, so we must divide this product by 2 for the actual answer.

Carry Out the Plan 6 times 5 is 30. 30 divided by 2 is 15.

Review the Solution Denote the people by the letters A, B, C, D, E, and F. Make an organized list. Remember that AB and BA represent the same people shaking hands, so do not list both AB and BA.

AB	AC	AD	AE	AF
BC	BD	BE	BF	
CD	CE	CF		
DE	DF			
EF				

The method of making an organized list verifies that if six people shake hands with each other, there will be a total of 15 handshakes.

Check your progress 4, *page 30*

Understand the Problem We need to find the ones digit of 4^{200}.

Devise a Plan Compute a few powers of 4 to see if there are any patterns. $4^1 = 4$, $4^2 = 16$, $4^3 = 64$, and $4^4 = 256$. It appears that the units digit (ones digit) of 4^{200} must be either a 4 or a 6.

Carry Out the Plan If the exponent n is an even counting number, then 4^n has a ones digit of 6. If the exponent n is an odd counting number, then 4^n has a ones digit of 4. Because 200 is an even counting number, we conjecture that 4^{200} has a ones digit of 6.

Review the Solution You could try to check the answer by using a calculator, but you would find that 4^{200} is too large to be displayed. Thus we need to rely on the patterns we have observed to conclude that 6 is indeed the ones digit of 4^{200}.

Check your progress 5, *page 31*

Understand the Problem We are asked to find the possible numbers that Melody could have started with.

Devise a Plan Work backward from 18 and do the inverse of each operation that Melody performed.

Carry Out the Plan To get 18, Melody subtracted 30 from a number, so that number was $18 + 30 = 48$. To get 48, she divided a number by 3, so that number was $48 \times 3 = 144$. To get 144, she squared a number. She could have squared either 12 or -12 to produce 144. If the number she squared was 12, then she must have doubled 6 to get 12. If the number she squared was -12, then the number she doubled was -6.

Review the Solution We can check by starting with 6 or -6. If we do exactly as Melody did, we end up with 18. The operation that prevents us from knowing with 100% certainty which number she started with is the squaring operation. We have no way of knowing whether the number she squared was a positive number or a negative number.

Check your progress 6, *page 32*

Understand the Problem We need to find Diophantus's age when he died.

Devise a Plan Read the hint and then look for clues that will help you make an educated guess. You know from the given information that Diophantus's age must be divisible by 6, 12, 7, and 2. Find a number divisible by all of these numbers and check to see if it is a possible solution to the problem.

Carry Out the Plan All multiples of 12 are divisible by 6 and 2, but the smallest multiple of 12 that is divisible by 7 is $12 \times 7 = 84$. Thus we conjecture that Diophantus's age when he died was $x = 84$ years.

If $x = 84$, then $\frac{1}{6}x = 14$, $\frac{1}{12}x = 7$, $\frac{1}{7}x = 12$, and $\frac{1}{2}x = 42$. Then $\frac{1}{6}x + \frac{1}{12}x + \frac{1}{7}x + 5 + \frac{1}{2}x + 4 = 14 + 7 + 12 + 5 + 42 + 4 = 84$.

It seems that 84 years is a correct solution to the problem.

Review the Solution After 84, the next multiple of 12 that is divisible by 7 is 168. The number 168 also satisfies all the conditions of the problem, but it is unlikely that Diophantus died at the age of 168 years or at any age older than 168 years. Hence the only reasonable solution is 84 years.

Check your progress 7, *page 33*

Understand the Problem We need to determine two U.S. coins that have a total value of 35¢, given that one of the coins is not a quarter.

Devise a Plan Experiment with different coins to try to produce 35¢. After a few attempts, you should conclude that one of the coins must be a quarter. Consider that the problem may be a *deceptive problem.*

Carry Out the Plan A total of 35¢ can be produced by using a dime and a quarter. One of the coins is a quarter, but it is also true that *one of the coins, the dime, is not a quarter.*

Review the Solution A dime and a quarter satisfy all the conditions of the problem. No other combination of coins satisfies the conditions of the problem. Thus the only solution is a dime and a quarter.

Check your progress 8, *page 35*

a. The maximum of the average yearly ticket prices is displayed by the tallest vertical bar in Figure 1.3. Thus the maximum of the average yearly U.S. movie theatre ticket prices for the years from 2008 to 2014 was \$8.17 in 2014.

b. To estimate the median age at which men married for the first time in 2011, locate 2011 on the horizontal axis of Figure 1.4 and then move directly upward to the point on the red broken-line graph. The height of this point represents the median age at first marriage for men in 2011, and it can be estimated by moving horizontally to the vertical axis. Thus the median age at first marriage for men in 2011 was 28.5 years, rounded to the nearest half of a year.

c. Figure 1.5 indicates that 24.4% of the 180,000,000 U.S. Facebook users were in the 25–34 age group in January of 2014.

$$0.244 \cdot 180{,}000{,}000 = 43{,}920{,}000$$

Thus, rounded to the nearest hundred thousand, the number of U.S. Facebook users in this age group was 43,900,000 in January of 2014.

CHAPTER 2

SECTION 2.1

Check your progress 1, *page 48*

The only months that start with the letter A are April and August. When we use the roster method, the set is given by {April, August}.

Check your progress 2, *page 48*

The set {March, May} is the set of all months that start with the letter M. The set can also be described as the third and fifth months of the year.

Check your progress 3, *page 49*

a. $\{0, 1, 2, 3\}$ **b.** $\{12, 13, 14, 15, 16, 17, 18, 19\}$

c. $\{-4, -3, -2, -1\}$

Check your progress 4, *page 50*

a. False **b.** True **c.** True **d.** True

Check your progress 5, *page 51*

a. $\{x \mid x \in I \text{ and } x < 9\}$ **b.** $\{x \mid x \in N \text{ and } x > 4\}$

Check your progress 6, *page 51*

a. $n(C) = 5$ **b.** $n(D) = 1$ **c.** $n(E) = 0$

Check your progress 7, *page 52*

a. The sets are not equal but they both contain six elements. Thus the sets are equivalent.

b. The sets are not equal but they both contain 16 elements. Thus the sets are equivalent.

SECTION 2.2

Check your progress 1, *page 58*

a. $M = \{0, 4, 6, 17\}$. The set of elements in $U = \{0, 2, 3, 4, 6, 7, 17\}$ but not in M is $M' = \{2, 3, 7\}$.

b. $P = \{2, 4, 6\}$. The set of elements in $U = \{0, 2, 3, 4, 6, 7, 17\}$ but not in P is $P' = \{0, 3, 7, 17\}$.

Check your progress 2, *page 59*

a. False. The number 3 is an element of the first set but not an element of the second set. Therefore, the first set is not a subset of the second set.

b. True. The set of counting numbers is the same set as the set of natural numbers, and every set is a subset of itself.

c. True. The empty set is a subset of every set.

d. True. Each element of the first set is an integer.

Check your progress 3, *page 60*

a. Yes, because every natural number is a whole number, and the whole numbers include 0, which is not a natural number.

b. The first set is not a proper subset of the second set because the sets are equal.

Check your progress 4, *page 60*

Subsets with zero elements: { }

Subsets with one element: $\{a\}, \{b\}, \{c\}, \{d\}, \{e\}$

Subsets with two elements: $\{a, b\}, \{a, c\}, \{a, d\}, \{a, e\}, \{b, c\}, \{b, d\}, \{b, e\}, \{c, d\}, \{c, e\}, \{d, e\}$

Subsets with three elements: $\{a, b, c\}, \{a, b, d\}, \{a, b, e\}, \{a, c, d\}, \{a, c, e\}, \{a, d, e\}, \{b, c, d\}, \{b, c, e\}, \{b, d, e\}, \{c, d, e\}$

Subsets with four elements: $\{a, b, c, d\}, \{a, b, c, e\}, \{a, b, d, e\}, \{a, c, d, e\}, \{b, c, d, e\}$

Subsets with five elements: $\{a, b, c, d, e\}$

Check your progress 5, *page 62*

a. The company offers 11 upgrade options. Each option is independent of the other options. Thus the company can produce

$$2^{11} = 2048$$

different versions of the car.

b. Use the method of guessing and checking to find the smallest natural number n for which $2^n > 8000$.

$$2^{11} = 2048$$

$$2^{12} = 4096$$

$$2^{13} = 8192$$

The company must provide a minimum of 13 upgrade options if it wishes to offer at least 8000 different versions of the car.

SECTION 2.3

Check your progress 1, *page 67*

a. $D \cap E = \{0, 3, 8, 9\} \cap \{3, 4, 8, 9, 11\}$
$= \{3, 8, 9\}$

b. $D \cap F = \{0, 3, 8, 9\} \cap \{0, 2, 6, 8\}$
$= \{0, 8\}$

Check your progress 2, *page 68*

a. $D \cup E = \{0, 4, 8, 9\} \cup \{1, 4, 5, 7\}$
$= \{0, 1, 4, 5, 7, 8, 9\}$

b. $D \cup F = \{0, 4, 8, 9\} \cup \{2, 6, 8\}$
$= \{0, 2, 4, 6, 8, 9\}$

Check your progress 3, *page 68*

a. The set $D \cap (E' \cup F)$ can be described as "the set of all elements that are in *D*, and in *F* or not *E*."

b. The set $L' \cup M$ can be described as "the set of all elements that are in *M* or are not in *L*."

Check your progress 4, *page 69*

To determine the region(s) represented by $(A \cap B)'$, first determine the region(s) in Figure 2.1 that are represented by $A \cap B$.

Set	Region or regions	Venn diagram
$A \cap B$	i The region common to *A* and *B*	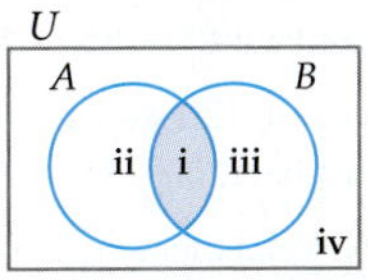
$(A \cap B)'$	ii, iii, iv The regions in *U* that are not in $A \cap B$	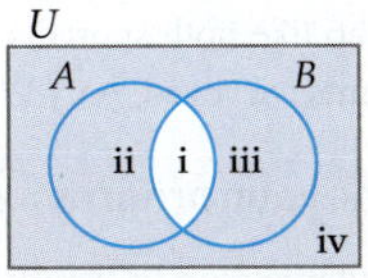

Now determine the region(s) in Figure 2.1 that are represented by $A' \cup B'$.

Set	Region or regions	Venn diagram
A'	iii, iv The regions outside of *A*	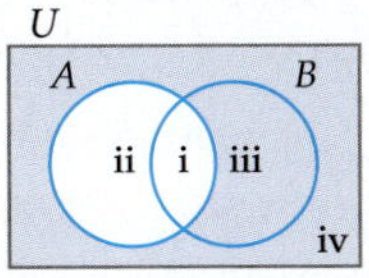
B'	ii, iv The regions outside of *B*	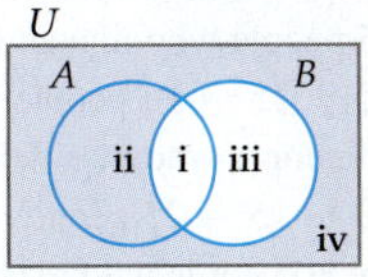
$A' \cup B'$	ii, iii, iv The regions formed by joining the regions in A' and the regions in B'	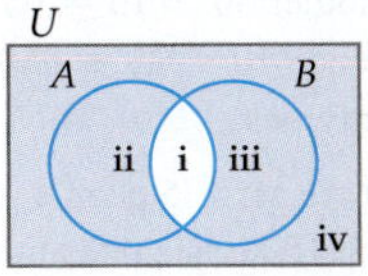

The expressions $(A \cap B)'$ and $A' \cup B'$ are both represented by regions ii, iii, and iv. Thus $(A \cap B)' = A' \cup B'$ for all sets *A* and *B*.

Check your progress 5, *page 71*

The following solutions reference the regions shown in Figure 2.2, which is displayed below for convenience.

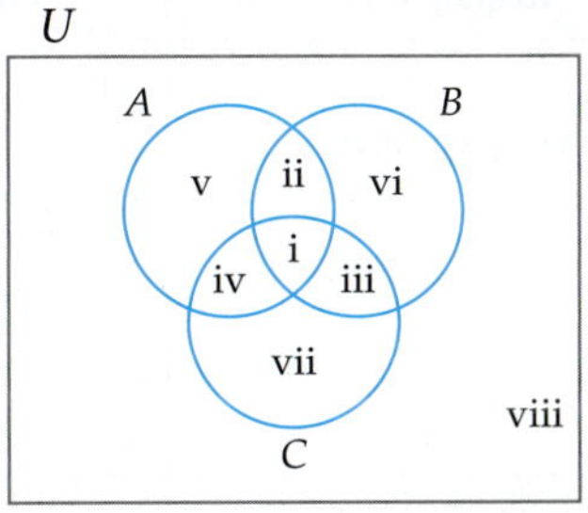

a. $A \cap B$ is represented by the regions i and ii. *C* is represented by regions i, iii, iv, and vii. $(A \cap B) \cap C$ is represented by the regions that are common to $A \cap B$ and *C*.

Thus $(A \cap B) \cap C$ is represented by region i.

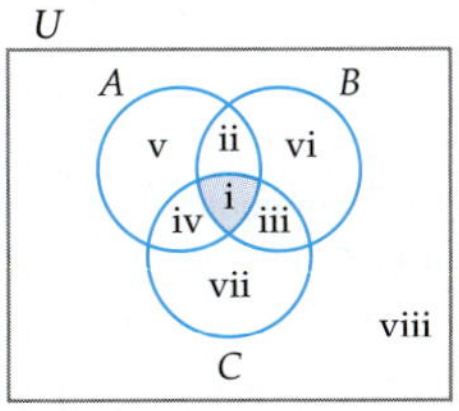

b. *A* is represented by the regions i, ii, iv, and v. B' is represented by the regions outside of *B*: iv, v, vii, and viii.

$A \cup B'$ is represented by the regions formed by joining the regions in *A* and the regions in B'.

Thus $A \cup B'$ is represented by the regions i, ii, iv, v, vii, and viii.

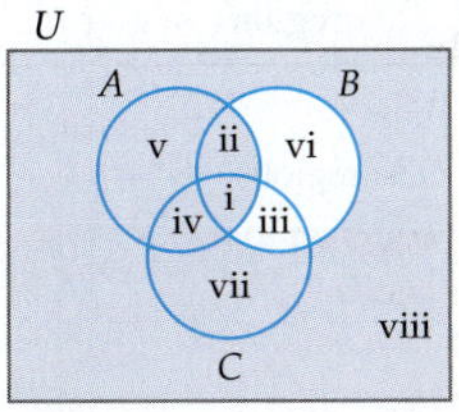

c. C' is represented by the regions outside of circle *C*: ii, v, vi, and viii.

B is represented by all the regions inside circle *B*: i, ii, iii, and vi.

$C' \cap B$ is represented by the regions that are common to C' and *B*.

Thus $C' \cap B$ is represented by regions ii and vi.

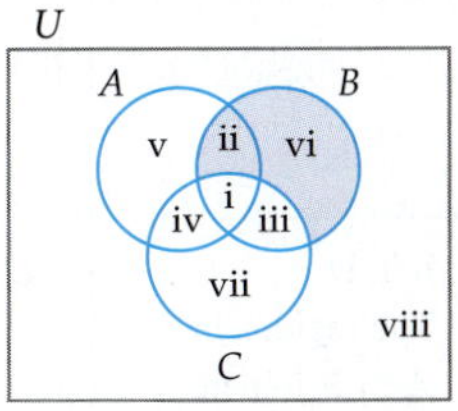

Check your progress 6, *page 72*

Determine the regions represented by $A \cap (B \cup C)$.

Set	Region or regions	Venn diagram
A	i, ii, iv, v The regions in A	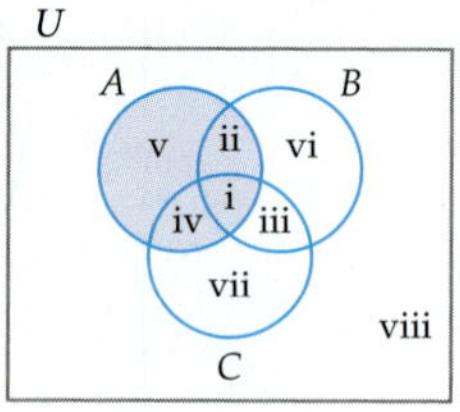
$B \cup C$	i, ii, iii, iv, vi, vii The regions in B joined with the regions in C	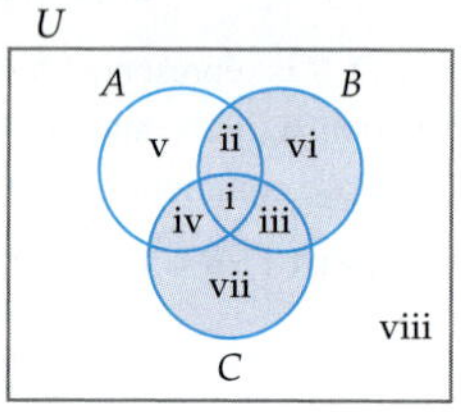
$A \cap (B \cup C)$	i, ii, iv The regions common to A and $B \cup C$	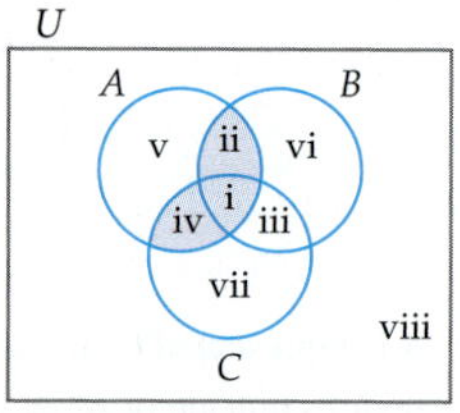

Now determine the regions represented by $(A \cap B) \cup (A \cap C)$.

Set	Region or regions	Venn diagram
$A \cap B$	i, ii The regions common to A and B	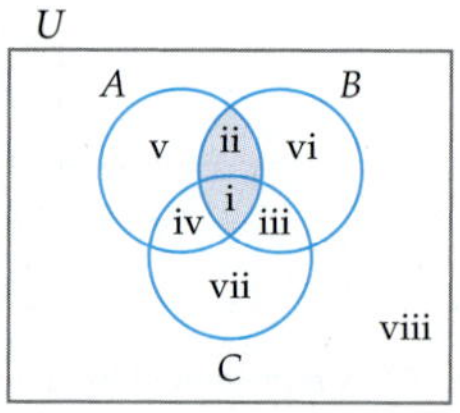
$A \cap C$	i, iv The regions common to A and C	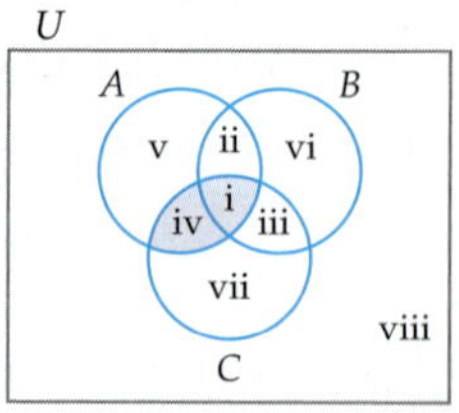
$(A \cap B) \cup (A \cap C)$	i, ii, iv The regions in $A \cap B$ joined with the regions in $A \cap C$	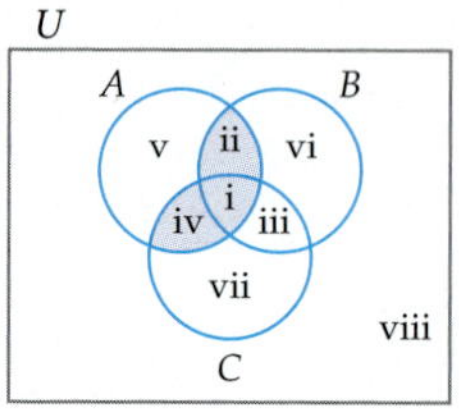

The expressions $A \cap (B \cup C)$ and $(A \cap B) \cup (A \cap C)$ are both represented by the regions i, ii, and iv. Thus $A \cap (B \cup C) = (A \cap B) \cup (A \cap C)$ for all sets A and B.

Check your progress 7, *page 73*

a. Because Alex is in blood group A, not in blood group B, and is Rh+, his blood type is A+.

b. Roberto is in both blood group A and blood group B. Roberto is not Rh+. Thus Roberto's blood type is AB−.

Check your progress 8, *page 74*

a. Alex's blood type is A+. The blood transfusion table shows that a person with blood type A+ can safely receive type A− blood.

b. The blood transfusion table shows that a person with type AB+ blood can safely receive each of the eight different types of blood. Thus a person with AB+ blood is classified as a universal recipient.

SECTION 2.4

Check your progress 1, *page 81*

The intersection of the two sets includes the 85 students who like both volleyball and basketball.

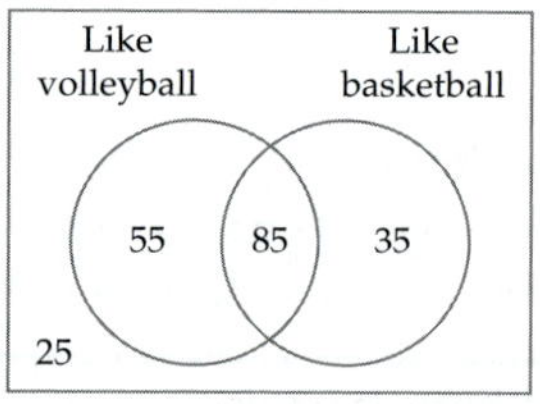

a. Because 140 students like volleyball and 85 like both sports, there must be $140 - 85 = 55$ students who like only volleyball.

b. Because 120 students like basketball and 85 like both sports, there must be $120 - 85 = 35$ students who like only basketball.

c. The Venn diagram shows that the number of students who like only volleyball plus the number who like only basketball plus the number who like both sports is $55 + 35 + 85 = 175$. Thus of the 200 students surveyed, only $200 - 175 = 25$ do not like either of the sports.

Check your progress 2, *page 82*

The intersection of the three sets includes the 15 people who like all three activities.

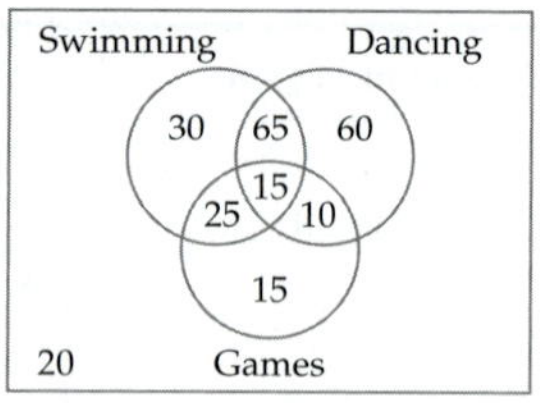

a. There are 25 people who like dancing and games. This includes the 15 people who like all three activities. Thus there must be another $25 - 15 = 10$ people who like only dancing and games. There are 40 people who like swimming and games. Thus there must be another $40 - 15 = 25$ people who like only swimming and games. There are 80 people who like swimming and dancing. Thus there must be another $80 - 15 = 65$ people who like only swimming and dancing. Hence $10 + 25 + 65 = 100$ people who like exactly two of the three activities.

b. There are 135 people who like swimming. We have determined that 15 people like all three activities, 25 like only swimming and games, and 65 like only swimming and dancing. This means that $135 - (15 + 25 + 65) = 30$ people like only swimming.

c. There are a total of 240 passengers surveyed. The Venn diagram shows that $15 + 25 + 10 + 15 + 30 + 65 + 60 = 220$ passengers like at least one of the activities. Thus $240 - 220 = 20$ passengers like none of the activities.

Check your progress 3, *page 83*

Let $B = \{$the set of students who play basketball$\}$.

Let $S = \{$the set of students who play soccer$\}$.

$$\begin{aligned} n(B \cup S) &= n(B) + n(S) - n(B \cap S) \\ &= 80 + 60 - 24 \\ &= 116 \end{aligned}$$

Using the inclusion-exclusion principle, we see that 116 students play either basketball or soccer.

Check your progress 4, *page 83*

$$\begin{aligned} n(A \cup B) &= n(A) + n(B) - n(A \cap B) \\ 852 &= 785 + 162 - n(A \cap B) \\ 852 &= 947 - n(A \cap B) \\ n(A \cap B) &= 947 - 852 \\ n(A \cap B) &= 95 \end{aligned}$$

Check your progress 5, *page 84*

$$\begin{aligned} p(\text{A} \cup \text{Rh+}) &= p(\text{A}) + p(\text{Rh+}) - p(\text{A} \cap \text{Rh+}) \\ 91\% &= 44\% + 84\% - p(\text{A} \cap \text{Rh+}) \\ 91\% &= 128\% - p(\text{A} \cap \text{Rh+}) \\ p(\text{A} \cap \text{Rh+}) &= 128\% - 91\% \\ p(\text{A} \cap \text{Rh+}) &= 37\% \end{aligned}$$

About 37% of the U.S. population has the A antigen and is Rh+.

Check your progress 6, *page 85*

a. The number 410 appears in both the column labeled "Yahoo!" and the row labeled "children." The table shows that 410 children surveyed use Yahoo! as a search engine. Thus,

$$n(Y \cap C) = 410$$

b. The set $B \cap M'$ is the set of surveyed Bing users who are women or children. Thus,

$$\begin{aligned} n(B \cap M') &= 325 + 40 \\ &= 365 \end{aligned}$$

c. The set $G \cap M$ represents the set of surveyed Google users who are men. The table shows that this set includes 440 people. The set $G \cap W$ represents the set of surveyed Google users who are women. The table shows that this set includes 390 people. Thus $n((G \cap M) \cup (G \cap W)) = 440 + 390 = 830$.

SECTION 2.5

Check your progress 1, *page 91*

Write the sets so that one is aligned below the other. Draw arrows to show how you wish to pair the elements of each set. One possible method is shown in the following figure.

$$\begin{array}{l} N = \{1, 2, 3, 4, \ldots, \quad n, \quad \ldots\} \\ \quad\;\;\, \updownarrow\; \updownarrow\; \updownarrow\; \updownarrow \qquad\quad \updownarrow \\ D = \{1, 3, 5, 7, \ldots, 2n - 1, \ldots\} \end{array}$$

In the preceding correspondence, each natural number $n \in N$ is paired with the odd number $(2n - 1) \in D$. The *general correspondence* $n \leftrightarrow (2n - 1)$ enables us to determine exactly which element of D will be paired with any given element of N, and vice versa. For instance, under this correspondence, $8 \in N$ is paired with the odd number $2 \cdot 8 - 1 = 15$, and $21 \in D$ is paired with the natural number $\frac{21 + 1}{2} = 11$. The general correspondence $n \leftrightarrow (2n - 1)$ establishes a one-to-one correspondence between the sets.

Check your progress 2, *page 92*

One proper subset of V is $P = \{41, 42, 43, 44, \ldots, 40 + n \ldots\}$, which was produced by deleting 40 from set V. To establish a one-to-one correspondence between V and P, consider the following diagram.

$$\begin{array}{l} V = \{40, 41, 42, 43, \ldots, 39 + n, \ldots\} \\ \qquad\;\; \updownarrow \quad \updownarrow \quad \updownarrow \quad \updownarrow \qquad\quad \updownarrow \\ P = \{41, 42, 43, 44, \ldots, 40 + n, \ldots\} \end{array}$$

In the above correspondence, each element of the form $39 + n$ from set V is paired with an element of the form $40 + n$ from set P. The general correspondence $(39 + n) \leftrightarrow (40 + n)$ establishes a one-to-one correspondence between V and P. Because V can be placed in a one-to-one correspondence with a proper subset of itself, V is an infinite set.

Check your progress 3, *page 92*

The following figure shows that we can establish a one-to-one correspondence between M and the set of natural numbers N by pairing $\frac{1}{n+1}$ of set M with n of set N.

$$\begin{array}{l} M = \left\{\frac{1}{2}, \frac{1}{3}, \frac{1}{4}, \frac{1}{5}, \ldots, \frac{1}{n+1}, \ldots\right\} \\ \qquad\;\; \updownarrow \;\; \updownarrow \;\; \updownarrow \;\; \updownarrow \qquad\quad \updownarrow \\ N = \{\,1,\; 2,\; 3,\; 4,\; \ldots,\quad n, \quad \ldots\} \end{array}$$

Thus the cardinality of M must be the same as the cardinality of N, which is $\aleph_0$.

CHAPTER 3

SECTION 3.1

Check your progress 1, *page 107*

a. The sentence "Open the door" is a command. It is not a statement.

b. The word *large* is not a precise term. It is not possible to determine whether the sentence "7055 is a large number" is true or false, and thus the sentence is not a statement.

c. You may not know whether the given sentence is true or false, but you know that the sentence is either true or false and that it is not both true and false. Thus the sentence is a statement.

d. The sentence $x > 3$ is a statement because for any given value of x, the inequality $x > 3$ is true or false, but not both.

Check your progress 2, *page 108*

a. The *Queen Mary* 2 is not the world's largest cruise ship.

b. The fire engine is red.

Check your progress 3, *page 109*

a. $\sim p \wedge r$

b. $\sim s \wedge \sim r$

c. $r \leftrightarrow q$

d. $p \rightarrow \sim r$

Check your progress 4, *page 109*

$e \wedge \sim t$: All men are created equal and I am not trading places.

$a \vee \sim t$: I get Abe's place or I am not trading places.

$e \rightarrow t$: If all men are created equal, then I am trading places.

$t \leftrightarrow g$: I am trading places if and only if I get George's place.

Check your progress 5, *page 111*

a. If Kesha's singing style is similar to Uffie's and Kesha has messy hair, then Kesha is a rapper.

b. $\sim r \rightarrow (\sim q \wedge \sim p)$

Check your progress 6, *page 112*

a. True. A conjunction of two statements is true provided that both statements are true.

b. True. A disjunction of two statements is true provided that at least one statement is true.

c. False. If both statements of a disjunction are false, then the disjunction is false.

Check your progress 7, *page 113*

a. Some bears are not brown.

b. Some smartphones are expensive.

c. All vegetables are green.

SECTION 3.2

Check your progress 1, *page 118*

a.

p	q	$\sim p$	$\sim q$	$p \wedge \sim q$	$\sim p \vee q$	$(p \wedge \sim q) \vee (\sim p \vee q)$	
T	T	F	F	F	T	T	row 1
T	F	F	T	T	F	T	row 2
F	T	T	F	F	T	T	row 3
F	F	T	T	F	T	T	row 4
		1	2	3	4	5	

b. p is true and q is false in row 2 of the above truth table. The truth value of $(p \wedge \sim q) \vee (\sim p \vee q)$ in row 2 is T (true).

Check your progress 2, *page 119*

a.

p	q	r	$\sim p$	$\sim r$	$\sim p \wedge r$	$q \wedge \sim r$	$(\sim p \wedge r) \vee (q \wedge \sim r)$	
T	T	T	F	F	F	F	F	row 1
T	T	F	F	T	F	T	T	row 2
T	F	T	F	F	F	F	F	row 3
T	F	F	F	T	F	F	F	row 4
F	T	T	T	F	T	F	T	row 5
F	T	F	T	T	F	T	T	row 6
F	F	T	T	F	T	F	T	row 7
F	F	F	T	T	F	F	F	row 8
			1	2	3	4	5	

b. p is false, q is true, and r is false in row 6 of the above truth table. The truth value of $(\sim p \wedge r) \vee (q \wedge \sim r)$ in row 6 is T (true).

Check your progress 3, *page 120*

The given statement has two simple statements. Thus we use a standard form that has $2^2 = 4$ rows.

Step 1 Enter the truth values for each simple statement and their negations. See columns 1, 2, and 3 in the table below.

Step 2 Use the truth values in columns 2 and 3 to determine the truth values to enter under the "and" connective. See column 4 in the table below.

Step 3 Use the truth values in columns 1 and 4 to determine the truth values to enter under the "or" connective. See column 5 in the table below.

p	q	$\sim p$	$\vee$	$(p$	$\wedge$	$q)$
T	T	F	T	T	T	T
T	F	F	F	T	F	F
F	T	T	T	F	F	T
F	F	T	T	F	F	F
		1	5	2	4	3

The truth table for $\sim p \vee (p \wedge q)$ is displayed in column 5.

Check your progress 4, *page 121*

p	q	p	$\vee$	$(p$	$\wedge$	$\sim q)$
T	T	T	T	T	F	F
T	F	T	T	T	T	T
F	T	F	F	F	F	F
F	F	F	F	F	F	T
		1	5	2	4	3

The truth values in column 1 are the same as those in column 5. Thus the above truth table shows that $p \equiv p \vee (p \wedge \sim q)$.

Check your progress 5, *page 122*

Let d represent "I am going to the dance." Let g represent "I am going to the game." The original sentence in symbolic form is $\sim(d \wedge g)$. Applying one of De Morgan's laws, we find that $\sim(d \wedge g) \equiv \sim d \vee \sim g$. Thus an equivalent form of "It is not true that, I am going to the dance and I am going to the game" is "I am not going to the dance or I am not going to the game."

Check your progress 6, *page 122*

The following truth table shows that $p \wedge (\sim p \wedge q)$ is always false. Thus $p \wedge (\sim p \wedge q)$ is a self-contradiction.

p	q	p	$\wedge$	$(\sim p$	$\wedge$	$q)$
T	T	T	F	F	F	T
T	F	T	F	F	F	F
F	T	F	F	T	T	T
F	F	F	F	T	F	F
		1	5	2	4	3

SECTION 3.3

Check your progress 1, *page 127*

a. *Antecedent:* I study for at least 6 hours

Consequent: I will get an A on the test

b. *Antecedent:* I get the job

Consequent: I will buy a new car

c. *Antecedent:* you can dream it

Consequent: you can do it

Check your progress 2, *page 128*

a. Because the antecedent is true and the consequent is false, the statement is a false statement.

b. Because the antecedent is false, the statement is a true statement.

c. Because the consequent is true, the statement is a true statement.

Check your progress 3, *page 129*

p	q	$[p$	$\wedge$	$(p$	$\rightarrow$	$q)]$	$\rightarrow$	q
T	T	T	T	T	T	T	T	T
T	F	T	F	T	F	F	T	F
F	T	F	F	F	T	T	T	T
F	F	F	F	F	T	F	T	F
		1	6	2	5	3	7	4

Check your progress 4, *page 130*

a. I will move to Georgia or I will live in Houston.

b. The number is not divisible by 2 or the number is even.

Check your progress 5, *page 130*

a. I finished the report and I did not go to the concert.

b. The square of n is 25 and n is not 5 or -5.

Check your progress 6, *page 131*

a. She will go on vacation if and only if she can get a loan.

b. She can take the train if and only if she can get a loan.

Check your progress 7, *page 131*

a. Let $x = 6.5$. Then the first inequality of the biconditional is false, and the second inequality of the biconditional is true. Thus the given biconditional statement is false.

b. Both inequalities of the biconditional are true for $x > 2$, and both inequalities are false for $x \leq 2$. Because both inequalities have the same truth value for any real number x, the given biconditional is true.

SECTION 3.4

Check your progress 1, *page 135*

a. If a geometric figure is a square, then it is a rectangle.

b. If I am older than 30, then I am at least 21.

Check your progress 2, *page 136*

Converse: If we are not going to have a quiz tomorrow, then we will have a quiz today.

Inverse: If we don't have a quiz today, then we will have a quiz tomorrow.

Contrapositive: If we have a quiz tomorrow, then we will not have a quiz today.

Check your progress 3, *page 137*

a. The second statement is the inverse of the first statement. Thus the statements are not equivalent. This can also be demonstrated by the fact that the first statement is true for $c = 0$ and the second statement is false for $c = 0$.

b. The second statement is the contrapositive of the first statement. Thus the statements are equivalent.

Check your progress 4, *page 138*

a. *Contrapositive:* If x is an odd integer, then $3 + x$ is an even integer. The contrapositive is true and so the original statement is also true.

b. *Contrapositive:* If two triangles are congruent triangles, then the two triangles are similar triangles. The contrapositive is true and so the original statement is also true.

c. *Contrapositive:* If tomorrow is Thursday, then today is Wednesday. The contrapositive is true and so the original statement is also true.

SECTION 3.5

Check your progress 1, *page 142*

Let p represent the statement "She got on the plane." Let r represent the statement "She will regret it." Then the symbolic form of the argument is

$$\begin{array}{l} \sim p \rightarrow r \\ \underline{\sim r \quad\quad} \\ \therefore\ p \end{array}$$

Check your progress 2, *page 144*

Let r represent the statement "The stock market rises." Let f represent the statement "The bond market will fall." Then the symbolic form of the argument is

$$\begin{array}{l} r \rightarrow f \\ \underline{\sim f \quad\quad} \\ \therefore\ \sim r \end{array}$$

The truth table for this argument is as follows:

		First premise	Second premise	Conclusion	
r	f	$r \rightarrow f$	$\sim f$	$\sim r$	
T	T	T	F	F	row 1
T	F	F	T	F	row 2
F	T	T	F	T	row 3
F	F	T	T	T	row 4

Row 4 is the only row in which all the premises are true, so it is the only row that we examine. Because the conclusion is true in row 4, the argument is valid.

Check your progress 3, *page 145*

Let a represent the statement "I arrive before 8 A.M." Let f represent the statement "I will make the flight." Let p represent the statement "I will give the presentation." Then the symbolic form of the argument is

$$\begin{array}{l} a \rightarrow f \\ \underline{f \rightarrow p} \\ \therefore\ a \rightarrow p \end{array}$$

The truth table for this argument is as follows:

			First premise	Second premise	Conclusion	
a	f	p	$a \rightarrow f$	$f \rightarrow p$	$a \rightarrow p$	
T	T	T	T	T	T	row 1
T	T	F	T	F	F	row 2
T	F	T	F	T	T	row 3
T	F	F	F	T	F	row 4
F	T	T	T	T	T	row 5
F	T	F	T	F	T	row 6
F	F	T	T	T	T	row 7
F	F	F	T	T	T	row 8

The only rows in which all the premises are true are rows 1, 5, 7, and 8. In each of these rows the conclusion is also true. Thus the argument is a valid argument.

Check your progress 4, *page 146*

a. The symbolic form of the premises matches the direct reasoning standard form. Thus a valid conclusion is q: You can do it.

b. The symbolic form of the premises matches one of the disjunctive reasoning standard forms. Thus a valid conclusion is m: I bought a motorcycle.

Check your progress 5, *page 147*

a. Let f represent "I go to Florida for spring break." Let $\sim s$ represent "I will not study." Then the symbolic form of the argument is

$$\begin{array}{l} f \rightarrow \sim s \\ \underline{\sim f \quad\quad} \\ \therefore\ s \end{array}$$

This argument has the form of the fallacy of the inverse. Thus the argument is invalid.

b. Let s represent "You helped solve the crime." Let r represent "You should be rewarded." Then the symbolic form of the argument is

$$\begin{array}{l} s \rightarrow r \\ \underline{s \quad\quad} \\ \therefore\ r \end{array}$$

This matches the symbolic form of a direct reasoning argument. Thus the argument is valid.

Check your progress 6, *page 148*

Let r represent "I read a math book." Let f represent "I start to fall asleep." Let d represent "I drink a soda." Let e represent "I eat a candy bar." Then the symbolic form of the argument is

$$\begin{array}{l} r \to f \\ f \to d \\ \underline{d \to e} \\ \therefore r \to e \end{array}$$

The argument has the form of transitive reasoning extended to include three conditional premises. Thus the argument is valid.

Check your progress 7, *page 149*

We are given the following premises:

$$\begin{array}{l} \sim m \vee t \\ t \to \sim d \\ e \vee g \\ \underline{e \to d} \\ \therefore ? \end{array}$$

The first premise can be written as $m \to t$, the third premise can be written as $\sim e \to g$, and the fourth premise can be written as $\sim d \to \sim e$. Thus the argument can be expressed in the following equivalent form.

$$\begin{array}{l} m \to t \\ t \to \sim d \\ \sim e \to g \\ \underline{\sim d \to \sim e} \\ \therefore ? \end{array}$$

If we switch the order of the third and fourth premises, then we have the following equivalent form.

$$\begin{array}{l} m \to t \\ t \to \sim d \\ \sim d \to \sim e \\ \underline{\sim e \to g} \\ \therefore ? \end{array}$$

An application of transitive reasoning extended to include four premises produces $m \to g$ as a valid conclusion for the argument. *Note:* Although $m \to \sim e$ is also a valid conclusion for the argument, we do not list it as our answer because it can be obtained without using all of the given premises.

SECTION 3.6

Check your progress 1, *page 155*

The following Euler diagram shows that the argument is valid.

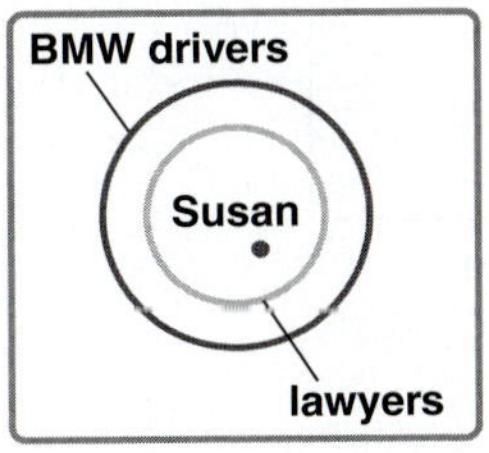

Check your progress 2, *page 156*

From the given premises we can conclude that 7 may or may not be a prime number. Thus the argument is invalid.

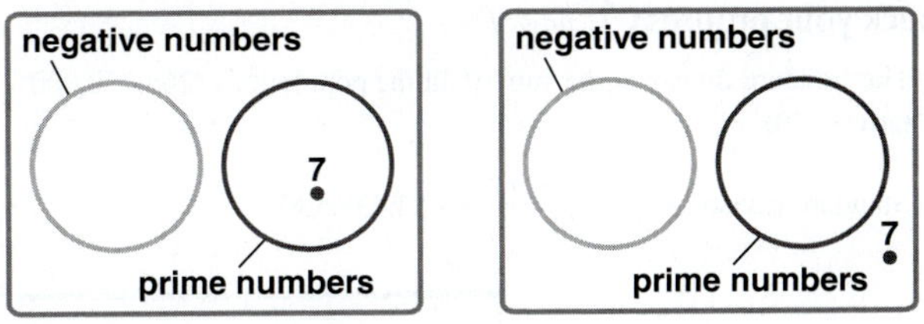

Check your progress 3, *page 156*

From the given premises we can construct two possible Euler diagrams.

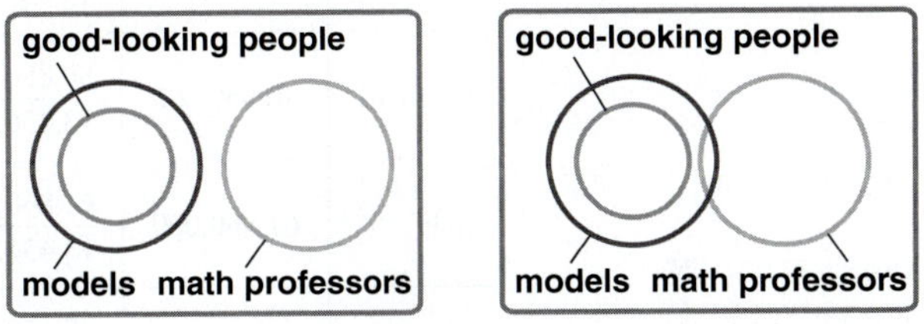

From the rightmost Euler diagram we can determine that the argument is invalid.

Check your progress 4, *page 157*

The following Euler diagram illustrates that all squares are quadrilaterals, so the argument is a valid argument.

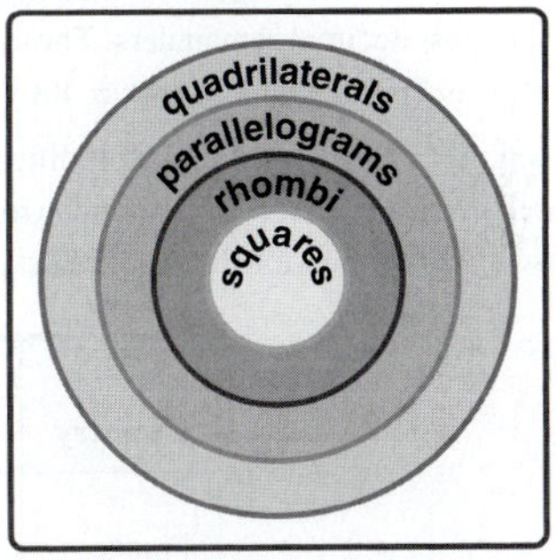

Check your progress 5, *page 158*

The following Euler diagrams illustrate two possible cases. In both cases we see that all white rabbits like tomatoes.

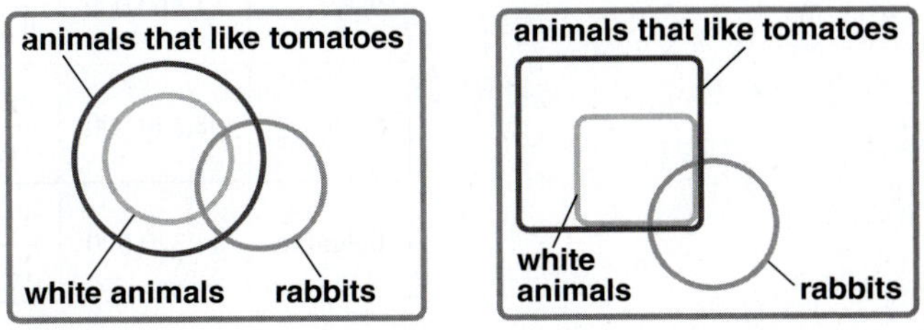

CHAPTER 4

SECTION 4.1

Check your progress 1, *page 174*

a. The standard divisor is the sum of all the populations (268,730,000) divided by the number of representatives (20).

$$\text{Standard divisor} = \frac{268{,}730{,}000}{20} = 13{,}436{,}500$$

Country	Population	Quotient	Standard quota	Number of representatives
France	66,550,000	$\frac{66{,}550{,}000}{13{,}436{,}500} \approx 4.953$	4	5
Germany	80,850,000	$\frac{80{,}850{,}000}{13{,}436{,}500} \approx 6.017$	6	6
Italy	61,860,000	$\frac{61{,}860{,}000}{13{,}436{,}500} \approx 4.604$	4	5
Spain	48,150,000	$\frac{48{,}150{,}000}{13{,}436{,}500} \approx 3.584$	3	3
Belgium	11,320,000	$\frac{11{,}320{,}000}{13{,}436{,}500} \approx 0.842$	0	1
		Total	17	20

Because the sum of the standard quotas is 17 and not 20, we add one representative to each of the three countries with the largest decimal remainders. These are France, Belgium, and Italy. Thus the composition of the committee is France: 5, Germany: 6, Italy: 5, Spain: 3, and Belgium: 1.

b. To use the Jefferson method, we must find a modified divisor such that the sum of the standard quotas is 20. This modified divisor is found by trial and error but is always less than or equal to the standard divisor. We are using 12,000,000 for the modified standard divisor.

Country	Population	Quotient	Number of representatives
France	66,550,000	$\frac{66{,}550{,}000}{12{,}000{,}000} \approx 5.546$	5
Germany	80,850,000	$\frac{80{,}850{,}000}{12{,}000{,}000} \approx 6.738$	6
Italy	61,860,000	$\frac{61{,}860{,}000}{12{,}000{,}000} = 5.155$	5
Spain	48,150,000	$\frac{48{,}150{,}000}{12{,}000{,}000} \approx 4.013$	4
Belgium	11,320,000	$\frac{11{,}320{,}000}{12{,}000{,}000} \approx 0.943$	0
		Total	20

Thus the composition of the committee is France: 5, Germany: 6, Italy: 5, Spain: 4, and Belgium: 0.

Check your progress 2, *page 178*

$$\text{Relative unfairness of the apportionment} = \frac{\text{absolute unfairness of the apportionment}}{\text{average constituency of Shasta with a new representative}}$$

$$= \frac{210}{1390} \approx 0.151$$

The relative unfairness of the apportionment is approximately 0.151.

Check your progress 3, *page 179*

Calculate the relative unfairness of the apportionment that assigns the teacher to the first grade and the relative unfairness of the apportionment that assigns the teacher to the second grade. In this case, the average constituency is the number of students divided by the number of teachers.

	First grade number of students per teacher	Second grade number of students per teacher	Absolute unfairness of apportionment
First grade receives teacher	$\frac{12{,}317}{512 + 1} \approx 24$	$\frac{15{,}439}{551} \approx 28$	$28 - 24 = 4$
Second grade receives teacher	$\frac{12{,}317}{512} \approx 24$	$\frac{15{,}439}{551 + 1} \approx 28$	$28 - 24 = 4$

If the first grade receives the new teacher, then the relative unfairness of the apportionment is

$$\text{Relative unfairness of the apportionment} = \frac{\text{absolute unfairness of the apportionment}}{\text{first grade's average constituency with a new teacher}}$$

$$= \frac{4}{24} \approx 0.167$$

If the second grade receives the new teacher, then the relative unfairness of the apportionment is

$$\text{Relative unfairness of the apportionment} = \frac{\text{absolute unfairness of the apportionment}}{\text{second grade's average constituency with a new teacher}}$$

$$= \frac{4}{28} \approx 0.143$$

Because the smaller relative unfairness results from adding the teacher to the second grade, that class should receive the new teacher.

Check your progress 4, *page 180*

Calculate the Huntington-Hill number for each of the classes. In this case, the population is the number of students.

First year:

$$\frac{2015^2}{12(12 + 1)} \approx 26{,}027$$

Second year:

$$\frac{1755^2}{10(10 + 1)} \approx 28{,}000$$

Third year:

$$\frac{1430^2}{9(9 + 1)} \approx 22{,}721$$

Fourth year:

$$\frac{1309^2}{8(8 + 1)} \approx 23{,}798$$

Because the second-year class has the greatest Huntington-Hill number, the new representative should represent the second-year class.

SECTION 4.2

Check your progress 1, *page 190*

To answer the question, we will make a table showing the number of second-place votes for each candy.

	Second-place votes
Caramel center	3
Vanilla center	0
Almond center	$17 + 9 = 26$
Toffee center	2
Solid chocolate	$11 + 8 = 19$

The largest number of second-place votes (26) were for almond centers. Almond centers would win second place using the plurality voting system.

Check your progress 2, *page 193*

Using the Borda count method, each first-place vote receives 5 points, each second-place vote receives 4 points, each third-place vote receives 3 points, each fourth-place vote receives 2 points, and each last-place vote receives 1 point. The summaries for the five varieties are as follows.

Caramel:

0 first-place votes	$0 \cdot 5 = 0$
3 second-place votes	$3 \cdot 4 = 12$
0 third-place votes	$0 \cdot 3 = 0$
30 fourth-place votes	$30 \cdot 2 = 60$
17 fifth-place votes	$17 \cdot 1 = 17$
	Total 89

Vanilla:

17 first-place votes	$17 \cdot 5 = 85$
0 second-place votes	$0 \cdot 4 = 0$
0 third-place votes	$0 \cdot 3 = 0$
0 fourth-place votes	$0 \cdot 2 = 0$
33 fifth-place votes	$33 \cdot 1 = 33$
	Total 118

Almond:

8 first-place votes	$8 \cdot 5 = 40$
26 second-place votes	$26 \cdot 4 = 104$
16 third-place votes	$16 \cdot 3 = 48$
0 fourth-place votes	$0 \cdot 2 = 0$
0 fifth-place votes	$0 \cdot 1 = 0$
	Total 192

Toffee:

20 first-place votes	$20 \cdot 5 = 100$
2 second-place votes	$2 \cdot 4 = 8$
8 third-place votes	$8 \cdot 3 = 24$
20 fourth-place votes	$20 \cdot 2 = 40$
0 fifth-place votes	$0 \cdot 1 = 0$
	Total 172

Chocolate:

5 first-place votes	$5 \cdot 5 = 25$
19 second-place votes	$19 \cdot 4 = 76$
26 third-place votes	$26 \cdot 3 = 78$
0 fourth-place votes	$0 \cdot 2 = 0$
0 fifth-place votes	$0 \cdot 1 = 0$
	Total 179

Using the Borda count method, almond centers is the first choice.

Check your progress 3, *page 195*

	Rankings				
Italian	2	5	1	4	3
Mexican	1	4	5	2	1
Thai	3	1	4	5	2
Chinese	4	2	3	1	4
Indian	5	3	2	3	5
Number of ballots:	33	30	25	20	18

Indian food received no first place votes, so it is eliminated.

	Rankings				
Italian	2	4	1	3	3
Mexican	1	3	4	2	1
Thai	3	1	3	4	2
Chinese	4	2	2	1	4
Number of ballots:	33	30	25	20	18

In this ranking, Chinese food received the fewest first-place votes, so it is eliminated.

	Rankings				
Italian	2	3	1	2	3
Mexican	1	2	3	1	1
Thai	3	1	2	3	2
Number of ballots:	33	30	25	20	18

In this ranking, Italian food received the fewest first-place votes, so it is eliminated.

	Rankings				
Mexican	1	2	2	1	1
Thai	2	1	1	2	2
Number of ballots:	33	30	25	20	18

In this ranking, Thai food received the fewest first-place votes, so it is eliminated. The preference for the banquet food is Mexican.

Check your progress 4, *page 197*

Do a head-to-head comparison for each of the restaurants and enter each winner in the table below. For instance, in the Sanborn's versus Apple Inn comparison, Sanborn's was favored by 31 + 25 + 11 = 67 critics. In the Apple Inn versus Sanborn's comparison, Apple Inn was favored by 18 + 15 = 33 critics. Therefore, Sanborn's wins this head-to-head match. The completed table is shown below.

versus	Sanborn's	Apple Inn	May's	Tory's
Sanborn's		Sanborn's	May's	Sanborn's
Apple Inn			May's	Tory's
May's				May's
Tory's				

From the table, May's has the most points, so it is the critics' choice.

Check your progress 5, *page 198*

Do a head-to-head comparison for each of the candidates and enter each winner in the table below.

versus	Alpha	Beta	Gamma
Alpha		Alpha	Alpha
Beta			Beta
Gamma			

From this table, Alpha is the winner. However, using the Borda count method (See Example 5), Beta is the winner. Thus the Borda count method violates the Condorcet criterion.

Check your progress 6, *page 200*

	Rankings		
Radiant silver	1	3	3
Electric red	2	2	1
Lightning blue	3	1	2
Number of votes:	30	27	2

Using the Borda count method, we have

Silver:

30 first-place votes	30 · 3 =	90
0 second-place votes	0 · 2 =	0
29 third-place votes	29 · 1 =	29
	Total	119

Red:

2 first-place votes	2 · 3 =	6
57 second-place votes	57 · 2 =	114
0 third-place votes	0 · 1 =	0
	Total	120

Blue:

27 first-place votes	27 · 3 =	81
2 second-place votes	2 · 2 =	4
30 third-place votes	30 · 1 =	30
	Total	115

Using this method, electric red is the preferred color. Now suppose we eliminate the third-place choice (lightning blue). This gives the following table.

	Rankings		
Radiant silver	1	2	2
Electric red	2	1	1
Number of votes:	30	27	2

Recalculating the results, we have

Silver:

30 first-place votes	30 · 2 =	60
29 second-place votes	29 · 1 =	29
	Total	89

Red:

29 first-place votes	29 · 2 =	58
30 second-place votes	30 · 1 =	30
	Total	88

Now radiant silver is the preferred color. By deleting an alternative, the result of the voting changed. This violates the irrelevant alternatives criterion.

SECTION 4.3

Check your progress 1, *page 212*

a. and **b.**

Winning coalition	Number of votes	Critical voters
{A, B}	40	A, B
{A, C}	39	A, C
{A, B, C}	57	A
{A, B, D}	50	A, B
{A, B, E}	45	A, B
{A, C, D}	49	A, C
{A, C, E}	44	A, C
{A, D, E}	37	A, D, E
{B, C, D}	45	B, C, D
{B, C, E}	40	B, C, E
{A, B, C, D}	67	None
{A, B, C, E}	62	None
{A, B, D, E}	55	A
{A, C, D, E}	54	A
{B, C, D, E}	50	B, C
{A, B, C, D, E}	72	None

Check your progress 2, *page 214*

Winning coalition	Number of votes	Critical voters
{A, B}	34	A, B
{A, C}	28	A, C
{B, C}	26	B, C
{A, B, C}	44	None
{A, B, D}	40	A, B
{A, C, D}	34	A, C
{B, C, D}	32	B, C
{A, B, C, D}	50	None

The number of times any voter is critical is 12.

$$BPI(\text{A}) = \frac{4}{12} = \frac{1}{3}$$

$$BPI(\text{D}) = \frac{0}{12} = 0$$

Check your progress 3, *page 215*

The countries are represented as follows: B, Belgium; F, France; G, Germany; I, Italy; L, Luxembourg; and N, Netherlands.

Winning coalition	Number of votes	Critical voters
{F, G, I}	12	F, G, I
{B, F, G, I}	14	F, G, I
{B, F, G, I, L}	15	F, G, I
{B, F, G, I, N}	16	None
{B, F, G, I, L, N}	17	None
{B, F, G, N}	12	B, F, G, N
{B, F, I, N}	12	B, F, I, N
{B, G, I, N}	12	B, G, I, N
{B, G, I, N, L}	13	B, G, I, N
{B, F, I, N, L}	13	B, F, I, N
{B, F, G, N, L}	13	B, F, G, N
{F, G, I, L}	13	F, G, I
{F, G, I, L, N}	15	F, G, I
{F, G, I, N}	14	F, G, I

The number of times all votes are critical is 42. The BPIs of the nations are:

$$BPI(\text{B}) = \frac{6}{42} = \frac{1}{7}$$

$$BPI(\text{F}) = \frac{10}{42} = \frac{5}{21}$$

$$BPI(\text{G}) = \frac{10}{42} = \frac{5}{21}$$

$$BPI(\text{I}) = \frac{10}{42} = \frac{5}{21}$$

$$BPI(\text{L}) = \frac{0}{42} = 0$$

$$BPI(\text{N}) = \frac{6}{42} = \frac{1}{7}$$

CHAPTER 5

SECTION 5.1

Check your progress 1, *page 231*

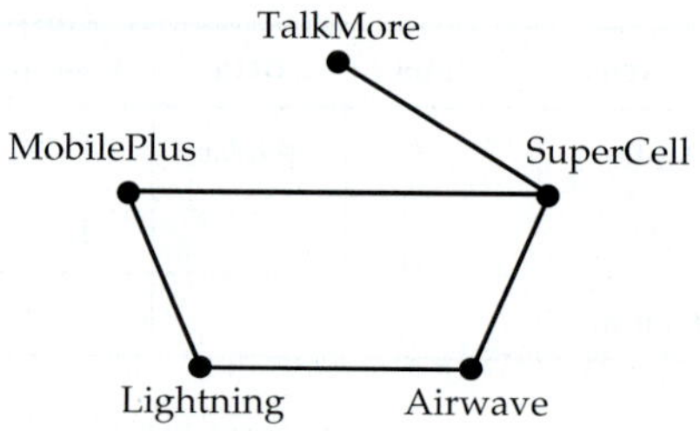

The vertex corresponding to SuperCell is connected to more edges than the others, so SuperCell has roaming agreements with the most carriers. TalkMore can roam with only one network because the corresponding vertex is connected to only one edge.

Check your progress 2, *page 233*

Because the second graph has edge AB and the first graph does not, the two graphs are not equivalent.

Check your progress 3, *page 235*

Vertices B, C, E, and G are of odd degree. By the Eulerian Graph Theorem, the graph does not have an Euler circuit.

Check your progress 4, *page 236*

One vertex in the graph is of degree 3, and another is of degree 5. Because not all vertices are of even degree, the graph is not Eulerian.

Check your progress 5, *page 237*

Represent the land areas and bridges with a graph, as we did for the Königsberg bridges earlier in the section. The vertices of the resulting graph, shown in the second figure, all have even degree. Thus we know that the graph has an Euler circuit. An Euler circuit corresponds to a stroll that crosses each bridge and returns to the starting point without crossing any bridge twice.

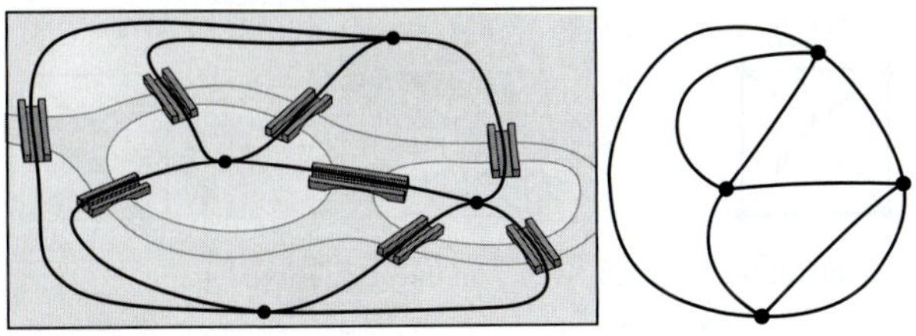

Check your progress 6, *page 238*

Consider the campground map as a graph. A route through all the trails that does not repeat any trails corresponds to an Euler path. Because only two vertices (A and F) are of odd degree, we know that an Euler path exists. Furthermore, the path must begin at A and end at F or begin at F and end at A. By trial and error, one Euler path is A–B–C–D–E–B–G–F–E–C–A–F.

Check your progress 7, *page 239*

Represent the floor plan with a graph, as in Example 7.

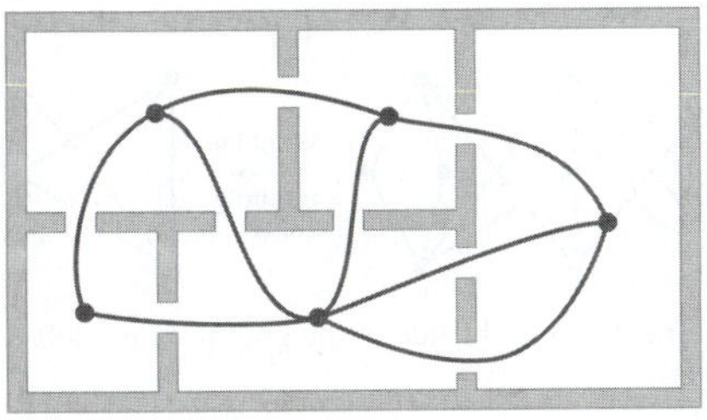

A stroll passing through each doorway just once corresponds to an Euler circuit or path. Because four vertices are of odd degree, no Euler circuit or path exists, so it is not possible to take such a stroll.

SECTION 5.2

Check your progress 1, *page 246*

The graph has seven vertices, so $n = 7$ and $n/2 = 3.5$. Several vertices are of degree less than $n/2$, so Dirac's theorem does not apply. Still, a routing for the document may be possible. By trial and error, one such route is Los Angeles–New York–Boston–Atlanta–Dallas–Phoenix–San Francisco–Los Angeles.

Check your progress 2, *page 248*

Draw a graph in which the vertices represent locations and the edges indicate available bus routes between locations. Each edge should be given a weight corresponding to the number of minutes for the bus ride.

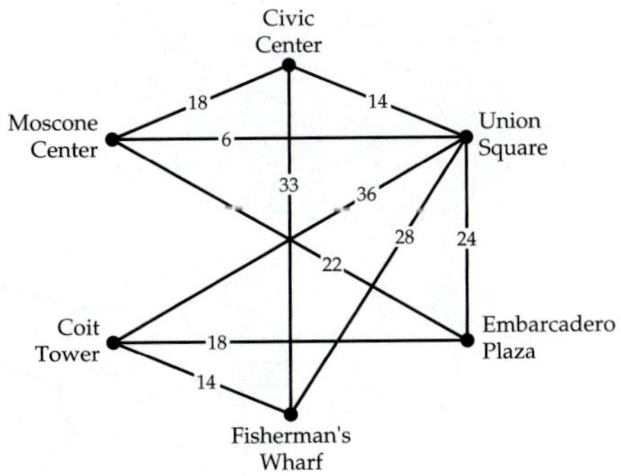

A route that visits each location and returns to the Moscone Center corresponds to a Hamiltonian circuit. Using the graph, we find that one such route is Moscone Center–Civic Center–Union Square–Fisherman's Wharf–Coit Tower–Embarcadero Plaza–Moscone Center, with a total weight of $18 + 14 + 28 + 14 + 18 + 22 = 114$. Another route is Moscone Center–Union Square–Embarcadero Plaza–Coit Tower–Fisherman's Wharf–Civic Center–Moscone Center, with a total weight of $6 + 24 + 18 + 14 + 33 + 18 = 113$. The travel time is one minute less for the second route.

Check your progress 3, *page 250*

Starting at vertex A, the edge of smallest weight is the edge to D, with weight 5. From D, take the edge of weight 4 to C, and then the edge of weight 3 to B. From B, the edge of least weight to a vertex not yet visited is the edge to vertex E (with weight 5). This is the last vertex, so we return to A along the edge of weight 9. Thus the Hamiltonian circuit is A–D–C–B–E–A, with a total weight of 26.

Check your progress 4, *page 251*

The smallest weight appearing in the graph is 3, so we mark edge BC. The next smallest weight is 4, on edge CD. Three edges have weight 5, but we cannot mark edge BD, because it would complete a circuit. We can, however, mark edge AD. The next valid edge of smallest weight is BE, also of weight 5. No more edges can be marked without completing a circuit or adding a third edge to a vertex, so we mark the final edge, AE, to complete the Hamiltonian circuit. In this case, the edge-picking algorithm generated the same circuit as the greedy algorithm did in Check Your Progress 3.

Check your progress 5, *page 254*

Represent the time between locations with a weighted graph.

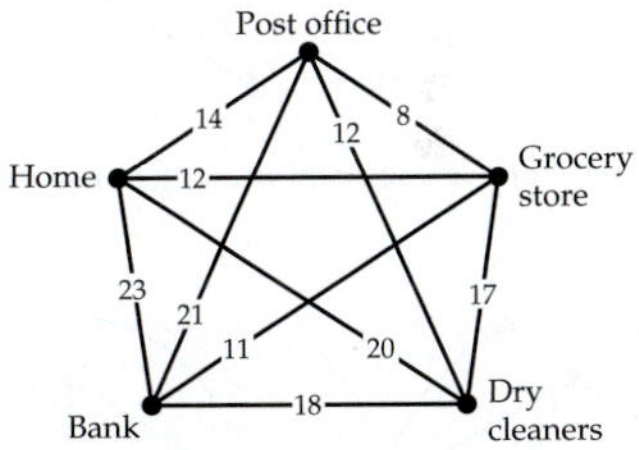

Starting at the home vertex and using the greedy algorithm, we first use the edge to the grocery store (of weight 12) followed by the edge of weight 8 to the post office and then the edge of weight 12 to the dry cleaners. The edge of next smallest weight is to the grocery store, but that vertex has already been visited, so we take the edge to the bank, with weight 18. All vertices have now been visited, so we select the last edge, of weight 23, to return home. The total weight is 73, corresponding to a total driving time of 73 minutes.

For the edge-picking algorithm, we first select the edge of weight 8, followed by the edge of weight 11. Two edges have weight 12, but one adds a third edge to the grocery store vertex, so we must choose the edge from the post office to the dry cleaners. The next smallest weight is 14, but that edge would add a third edge to a vertex, as would the edge of weight 17. The edge of weight 18 would complete a circuit too early, so the next edge we can select is that of weight 20, the edge from home to the dry cleaners. The final step is to select the edge from home to the bank to complete the circuit. The resulting route is home–dry cleaners–post office–grocery store–bank–home (we could travel the same route in the reverse order) with a total travel time of 74 minutes.

Check your progress 6, *page 255*

Represent the computer network by a graph in which the weights of the edges indicate the distances between computers.

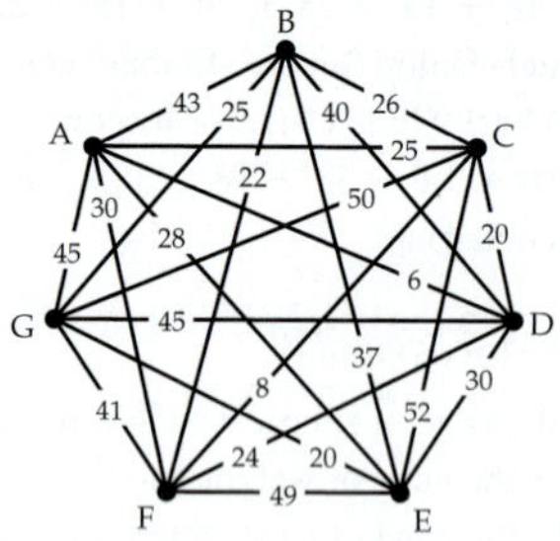

The edges with the smallest weights, which can all be chosen, are those of weights 6, 8, 20, 20, and 22. The edge of next smallest weight, 24, cannot be selected. There are two edges of weight 25; edge AC would add a third edge to vertex C, but edge BG can be chosen. All that remains is to complete the circuit with edge AE. The computers should be networked in this order: A, D, C, F, B, G, E, and back to A.

SECTION 5.3

Check your progress 1, *page 262*

First redraw the highlighted edge as shown below.

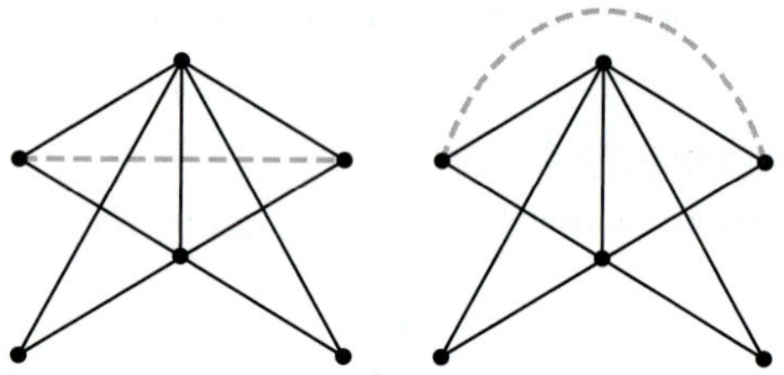

Now redraw the two lower vertices and the edges that meet there, as shown below.

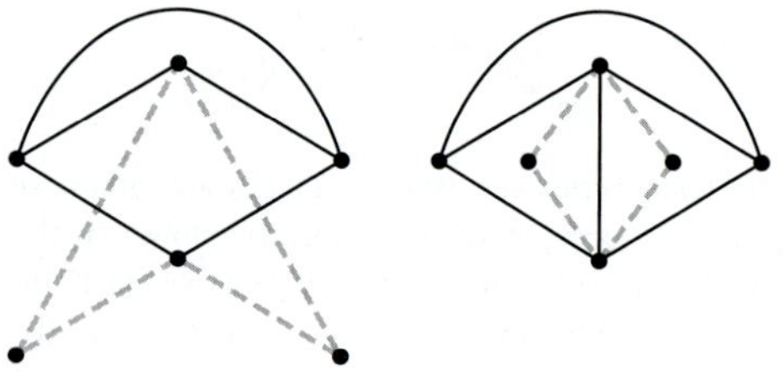

The result is a graph with no intersecting edges. Therefore, the graph is planar.

Check your progress 2, *page 264*

The highlighted edges in the graph, considered as a subgraph, form the graph K_5. (It is upside down and slightly distorted compared with the version shown in Figure 5.19.)

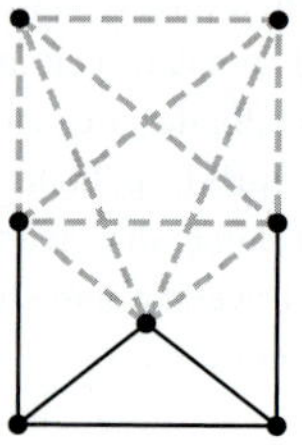

Check your progress 3, *page 264*

First contract the highlighted edge and combine the multiple edges as shown below.

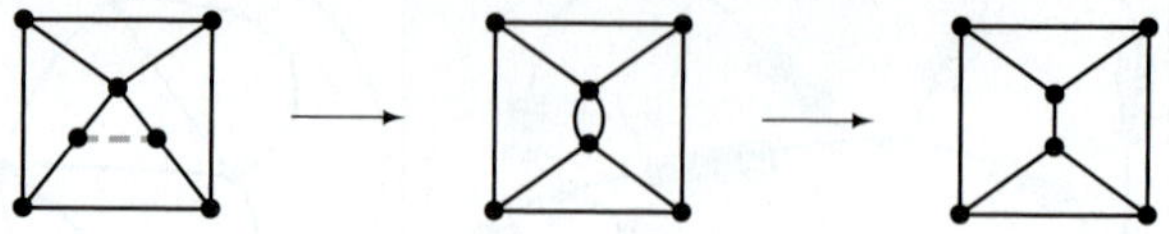

Then contract the highlighted edge in the center of the graph.

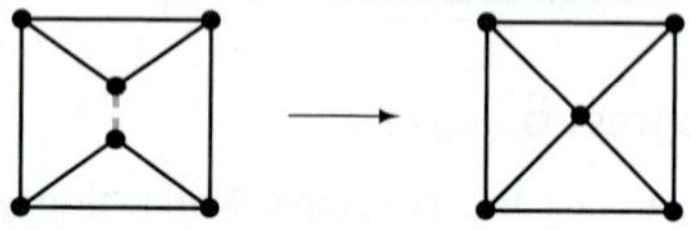

Check your progress 4, *page 266*

The graph looks similar to the Utilities Graph. Contract edges as shown below, and combine the resulting multiple edges.

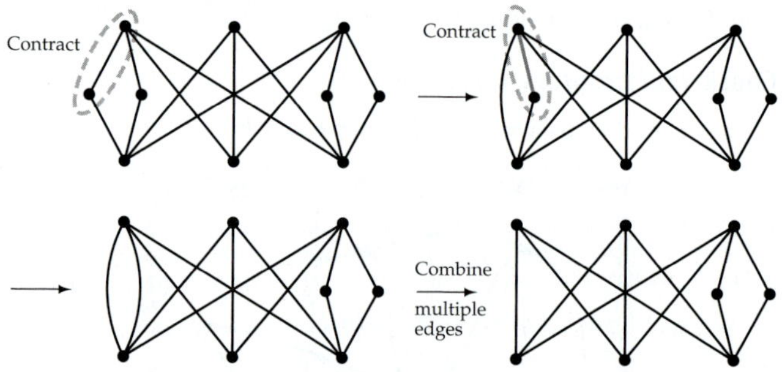

If we do the same on the right side of the graph, we are left with the Utilities Graph.

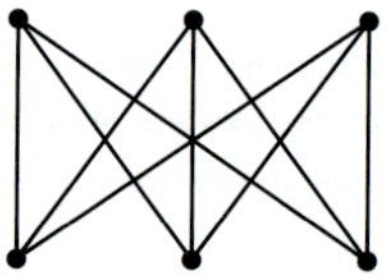

Therefore, the graph is not planar.

Check your progress 5, *page 267*

There are 11 edges in the graph, seven vertices, and six faces (including the infinite face). Then $v + f = 7 + 6 = 13$ and $e + 2 = 11 + 2 = 13$, so $v + f = e + 2$.

SECTION 5.4

Check your progress 1, *page 273*

Draw a graph on the map as in Example 1. More than two colors are required to color the resulting graph, but, by experimenting, the graph can be colored with three colors. Thus the graph is 3-colorable.

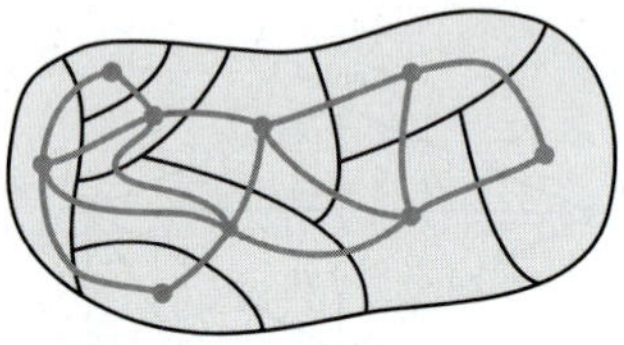

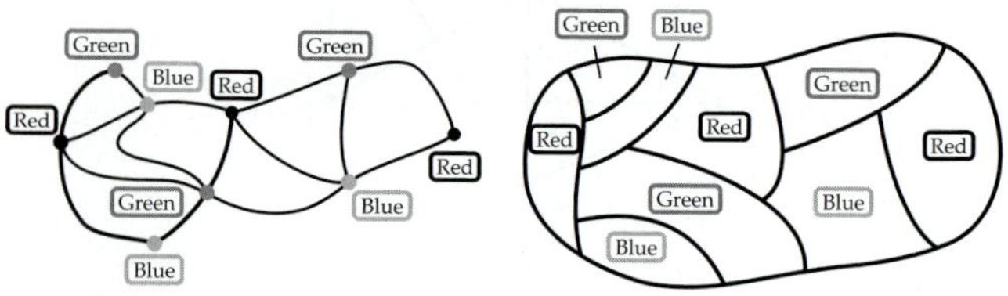

Check your progress 2, *page 274*

There are several locations in the graph at which three edges form a triangle. Because a triangle is a circuit with an odd number of vertices, the graph is not 2-colorable.

Check your progress 3, *page 276*

Draw a graph in which each vertex corresponds to a film and an edge joins two vertices if one person needs to attend both of the corresponding films. We can use colors to represent the different times at which the films can be viewed. No two vertices connected by an edge can share the same color, because that would mean one person would have to attend two films at the same time.

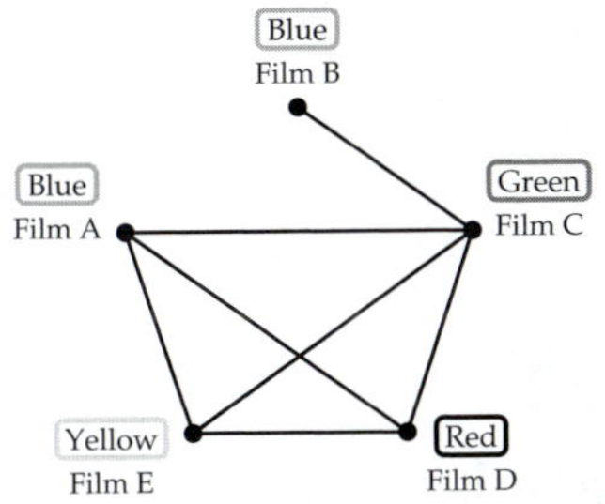

It is not possible to color the vertices with only three colors; one possible 4-coloring is shown. This means that four different time slots will be required to show the films, and the earliest that the festival can end is 8:00 P.M. A schedule can be set using the coloring in the graph. From 12 to 2, the films labeled blue, film A and film B, can be shown in two different rooms. The remaining films are represented by unique colors and so will require their own viewing times. Film C can be shown from 2 to 4, film D from 4 to 6, and film E from 6 to 8.

Check your progress 4, *page 277*

Draw a graph in which each vertex represents a deli and an edge connects two vertices if the corresponding delis deliver to a common building. Try to color the vertices using the fewest number of colors possible; each color can correspond to a day of the week that the delis can deliver.

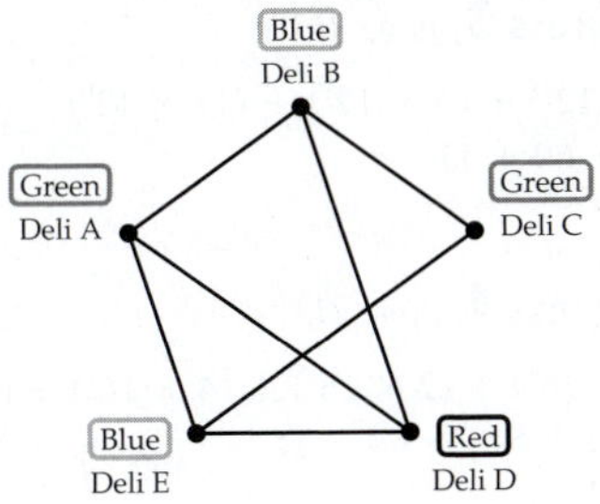

As shown, a 3-coloring is possible (but a 2-coloring is not). Therefore three different delivery days will be necessary—delis A and C deliver on one day, delis B and E on another day, and deli D on a third day.

CHAPTER 6

SECTION 6.1

Check your progress 1, *page 295*

$(1 \times 1{,}000{,}000) + (3 \times 100{,}000) + (1 \times 10{,}000) + (4 \times 1000)$
$+ (3 \times 100) + (2 \times 10) + (6 \times 1) = 1{,}314{,}326$

Check your progress 2, *page 295*

Check your progress 3, *page 296*

23,341
+ 10,562

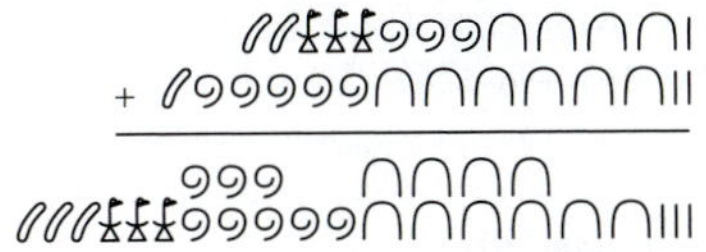

Replace 10 heel bones with one scroll and then add to produce:

which is 33,903.

Check your progress 4, *page 296*

61,432
− 45,121

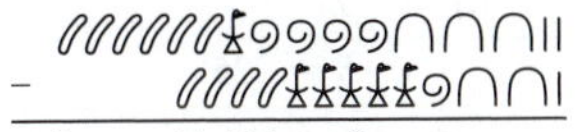

Replace one pointing finger with 10 lotus flowers and then subtract to produce:

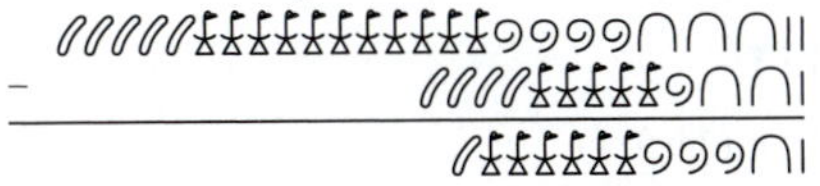

which is 16,311.

Check your progress 5, *page 298*

$\text{MCDXLV} = \text{M} + (\text{CD}) + (\text{XL}) + \text{V}$
$= 1000 + 400 + 40 + 5 = 1445$

Check your progress 6, *page 298*

$473 = 400 + 70 + 3 = \text{CD} + \text{LXX} + \text{III} = \text{CDLXXIII}$

Check your progress 7, *page 299*

a. $\overline{\text{IX}}\text{L} = \overline{\text{IX}} + \text{L} = (1000 \times 9) + 50 = 9050$

b. $8070 = 8000 + 70 = \overline{\text{VIII}} + \text{LXX} = \overline{\text{VIII}}\text{LXX}$

SECTION 6.2

Check your progress 1, *page 302*

$17{,}325 = 10{,}000 + 7000 + 300 + 20 + 5$
$= (1 \times 10{,}000) + (7 \times 1000) + (3 \times 100) + (2 \times 10) + 5$
$= (1 \times 10^4) + (7 \times 10^3) + (3 \times 10^2) + (2 \times 10^1)$
$+ (5 \times 10^0)$

Check your progress 2, *page 302*

$(5 \times 10^4) + (9 \times 10^3) + (2 \times 10^2) + (7 \times 10^1) + (4 \times 10^0)$
$= (5 \times 10{,}000) + (9 \times 1000) + (2 \times 100)$
$+ (7 \times 10) + (4 \times 1)$
$= 50{,}000 + 9000 + 200 + 70 + 4$
$= 59{,}274$

Check your progress 3, *page 303*

$152 = (1 \times 100) + (5 \times 10) + 2$
$+ 234 = (2 \times 100) + (3 \times 10) + 4$
$(3 \times 100) + (8 \times 10) + 6 = 386$

Check your progress 4, *page 303*

$$\begin{array}{rl} 147 = & (1 \times 100) + (4 \times 10) + \ \ 7 \\ \underline{+\ 329} = & \underline{(3 \times 100) + (2 \times 10) + \ \ 9} \\ & (4 \times 100) + (6 \times 10) + 16 \end{array}$$

Replace 16 with $(1 \times 10) + 6$

$$\begin{aligned} &= (4 \times 100) + (6 \times 10) + (1 \times 10) + 6 \\ &= (4 \times 100) + (7 \times 10) + 6 \\ &= 476 \end{aligned}$$

Check your progress 5, *page 303*

$$\begin{array}{rl} 382 = & (3 \times 100) + (8 \times 10) + 2 \\ \underline{-\ 157} = & \underline{(1 \times 100) + (5 \times 10) + 7} \end{array}$$

Because $7 > 2$, it is necessary to borrow by rewriting (8×10) as $(7 \times 10) + 10$.

$$\begin{array}{rl} 382 = & (3 \times 100) + (7 \times 10) + 12 \\ \underline{-\ 157} = & \underline{(1 \times 100) + (5 \times 10) + \ \ 7} \\ & (2 \times 100) + (2 \times 10) + \ \ 5 = 225 \end{array}$$

Check your progress 6, *page 304*

$= (21 \times 60^2) + (5 \times 60) + (34 \times 1)$
$= 75{,}600 + 300 + 34 = 75{,}934$

Check your progress 7, *page 305*

$$\begin{array}{r} 3 \\ 3600\overline{)12578} \\ \underline{10800} \\ 1778 \end{array} \qquad \begin{array}{r} 29 \\ 60\overline{)1778} \\ \underline{120} \\ 578 \\ \underline{540} \\ 38 \end{array}$$

Thus $12{,}578 = (3 \times 60^2) + (29 \times 60) + (38 \times 1) =$

Check your progress 8, *page 306*

Combine the symbols for each place value.

Replace ten 𒁹s in the ones place with a 𒌋.

Take away 60 from the ones place and add 1 to the 60s place.

Take away 60 from the 60s place and add 1 to the 60^2 place.

Thus

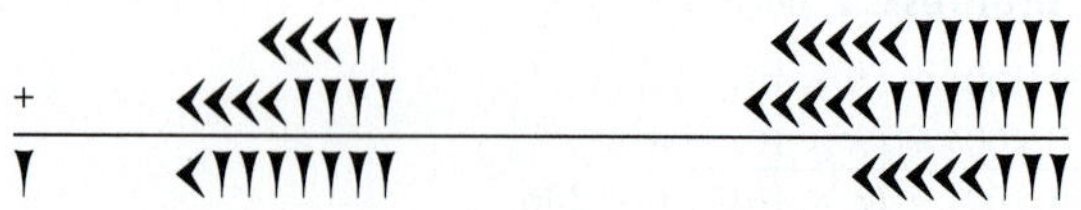

Check your progress 9, *page 307*

a. $(16 \times 360) + (0 \times 20) + (1 \times 1) = 5761$

b. $(9 \times 7200) + (1 \times 360) + (10 \times 20) + (4 \times 1) = 65{,}364$

Check your progress 10, *page 308*

$$\begin{array}{r} 1 \\ 7200\overline{)11480} \\ \underline{7200} \\ 4280 \end{array} \qquad \begin{array}{r} 11 \\ 360\overline{)4280} \\ \underline{360} \\ 680 \\ \underline{360} \\ 320 \end{array} \qquad \begin{array}{r} 16 \\ 20\overline{)320} \\ \underline{20} \\ 120 \\ \underline{120} \\ 0 \end{array}$$

Thus $11{,}480 = (1 \times 7200) + (11 \times 360) + (16 \times 20) + (0 \times 1)$.
In Mayan numerals this is

SECTION 6.3

Check your progress 1, *page 312*

$$\begin{aligned} 3156_{seven} &= (3 \times 7^3) + (1 \times 7^2) + (5 \times 7^1) + (6 \times 7^0) \\ &= (3 \times 343) + (1 \times 49) + (5 \times 7) + (6 \times 1) \\ &= 1029 + 49 + 35 + 6 \\ &= 1119 \end{aligned}$$

Check your progress 2, *page 312*

$$\begin{aligned} 111000101_{two} &= (1 \times 2^8) + (1 \times 2^7) + (1 \times 2^6) + (0 \times 2^5) \\ &\quad + (0 \times 2^4) + (0 \times 2^3) + (1 \times 2^2) \\ &\quad + (0 \times 2^1) + (1 \times 2^0) \\ &= (1 \times 256) + (1 \times 128) + (1 \times 64) + (0 \times 32) \\ &\quad + (0 \times 16) + (0 \times 8) + (1 \times 4) \\ &\quad + (0 \times 2) + (1 \times 1) \\ &= 256 + 128 + 64 + 0 + 0 + 0 + 4 + 0 + 1 \\ &= 453 \end{aligned}$$

Check your progress 3, *page 313*

$$\begin{aligned} A5B_{twelve} &= (10 \times 12^2) + (5 \times 12^1) + (11 \times 12^0) \\ &= 1440 + 60 + 11 \\ &= 1511 \end{aligned}$$

Check your progress 4, *page 313*

$$\begin{aligned} C24F_{sixteen} &= (12 \times 16^3) + (2 \times 16^2) + (4 \times 16^1) + (15 \times 16^0) \\ &= 49{,}152 + 512 + 64 + 15 \\ &= 49{,}743 \end{aligned}$$

Check your progress 5, *page 314*

a.

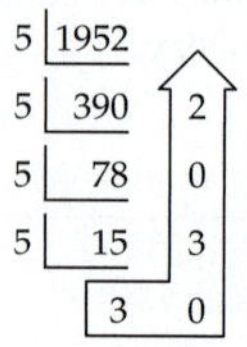

$1952 = 30302_{five}$

b.

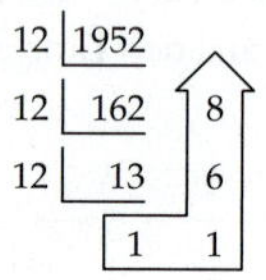

$1952 = 1168_{twelve}$

Check your progress 6, *page 315*

$$\begin{array}{ccccc} 6 & 3 & 2 & 1 & 0_{eight} \\ \Vert & \Vert & \Vert & \Vert & \Vert \\ 110 & 011 & 010 & 001 & 000_{two} \end{array}$$

$63210_{eight} = 110011010001000_{two}$

Check your progress 7, *page 315*

$$\begin{array}{cccc} 111 & 010 & 011 & 100_{two} \\ \Vert & \Vert & \Vert & \Vert \\ 7 & 2 & 3 & 4_{eight} \end{array}$$

$111010011100_{two} = 7234_{eight}$

Check your progress 8, *page 315*

$$\begin{array}{ccc} C & 5 & A_{sixteen} \\ \Vert & \Vert & \Vert \\ 1100 & 0101 & 1010_{two} \end{array}$$

$C5A_{sixteen} = 110001011010_{two}$

Check your progress 9, *page 316*

Insert a zero to make a group of four.

$$\begin{array}{cccc} 0101 & 0001 & 1101 & 0010_{two} \\ \Vert & \Vert & \Vert & \Vert \\ 5 & 1 & D & 2_{sixteen} \end{array}$$

$101000111010010_{two} = 51D2_{sixteen}$

Check your progress 10, *page 316*

	dabble 3	dabble 7	double 14	double 28	dabble 57	double 114
1	1	1	0	0	1	0_{two}

$1110010_{two} = 114$

SECTION 6.4

Check your progress 1, *page 320*

$$\begin{array}{r} {\scriptstyle 1\,1\;\;\;1\;\;\;\;} \\ 11001_{two} \\ +\;\;\;1101_{two} \\ \hline 100110_{two} \end{array}$$

Check your progress 2, *page 321*

$$\begin{array}{r} {\scriptstyle 1\,1\;\;\;\;} \\ 32_{four} \\ +\;12_{four} \\ \hline 110_{four} \end{array}$$

Check your progress 3, *page 322*

$$\begin{array}{r} {\scriptstyle 1\,2\;\;\;\;\;\;} \\ 35_{seven} \\ 46_{seven} \\ +\;24_{seven} \\ \hline 141_{seven} \end{array}$$

Check your progress 4, *page 322*

$$\begin{array}{r} {\scriptstyle 1\,1\;\;\;\;\;\;\;\;\;\;\;} \\ AC4_{sixteen} \\ +\;\;6E8_{sixteen} \\ \hline 11AC_{sixteen} \end{array}$$

Check your progress 5, *page 323*

$$\begin{array}{r} 2+1\;\;\;\;\;\;\;\; \\ \not{3}65_{nine} \\ -\;183_{nine} \\ \hline \end{array} \qquad \begin{array}{r} 10\;\;\;\;\;\; \\ 265_{nine} \\ -\;183_{nine} \\ \hline \end{array} \qquad \begin{array}{r} 16\;\;\;\;\;\; \\ 2\not{6}5_{nine} \\ -\;183_{nine} \\ \hline 172_{nine} \end{array}$$

Because $8_{nine} > 6_{nine}$, it is necessary to borrow from the 3 in the first column at the left.

Borrow 1 nine from the first column and add $9 = 10_{nine}$ to the 6 in the middle column.

$16_{nine} - 8_{nine} = 15 - 8 = 7 = 7_{nine}$

Check your progress 6, *page 324*

$$\begin{array}{rccc} & 7 & 10 & \\ & \not{8} & 3 & A_{twelve} \\ - & 4 & 6 & 7_{twelve} \\ \hline & 3 & 9 & 3_{twelve} \end{array}$$

- $A_{twelve} - 7_{twelve} = 10 - 7 = 3 = 3_{twelve}$
- $10_{twelve} + 3_{twelve} = 13_{twelve} = 15$ $\quad 15 - 6 = 9 = 9_{twelve}$
- $7_{twelve} - 4_{twelve} = 3_{twelve}$

Check your progress 7, *page 324*

$$\begin{array}{rcccc} & & & 1 & \\ & & 2 & 1 & 3_{four} \\ \times & & & & 2_{four} \\ \hline & 1 & 0 & 3 & 2_{four} \end{array}$$

- $2_{four} \times 3_{four} = 12_{four}$
- $2_{four} \times 1_{four} + 1_{four} = 3_{four}$
- $2_{four} \times 2_{four} = 10_{four}$

Check your progress 8, *page 325*

$$\begin{array}{rccc} & & 2 & \\ & & 3 & 4_{eight} \\ \times & & 2 & 5_{eight} \\ \hline & 2 & 1 & 4_{eight} \end{array}$$

- $5_{eight} \times 4_{eight} = 20 = 24_{eight}$
- $5_{eight} \times 3_{eight} + 2_{eight} = 15 + 2 = 17 = 21_{eight}$

$$\begin{array}{rcccc} & & & 1 & \\ & & & 3 & 4_{eight} \\ & \times & & 2 & 5_{eight} \\ \hline & & 2 & 1 & 4_{eight} \\ & & 7 & 0_{eight} & \\ \hline & 1 & 1 & 1 & 4_{eight} \end{array}$$

- $2_{eight} \times 4_{eight} = 8 = 10_{eight}$
- $2_{eight} \times 3_{eight} + 1_{eight} = 6 + 1 = 7 = 7_{eight}$

Check your progress 9, *page 327*

First list a few multiples of 3_{five}.

$3_{\text{five}} \times 0_{\text{five}} = 0_{\text{five}}$

$3_{\text{five}} \times 1_{\text{five}} = 3_{\text{five}}$

$3_{\text{five}} \times 2_{\text{five}} = 11_{\text{five}}$

$3_{\text{five}} \times 3_{\text{five}} = 14_{\text{five}}$

$3_{\text{five}} \times 4_{\text{five}} = 22_{\text{five}}$

```
          1                    10                   104
3five)324five        3five)324five        3five)324five
      3                    3                    3
      -                    -                    -
      2                    2                    2
                           0                    0
                           -                    -
                           24                   24
                                                22
                                                --
                                                 2
```

Thus $324_{\text{five}} \div 3_{\text{five}} = 104_{\text{five}}$ with a remainder of 2_{five}.

Check your progress 10, *page 327*

The divisor is 10_{two}. The multiples of the divisor are $10_{\text{two}} \times 0_{\text{two}} = 0_{\text{two}}$ and $10_{\text{two}} \times 1_{\text{two}} = 10_{\text{two}}$.

```
          111001two
10two)1110011two
      10
      --
       11
       10
       --
        10
        10
        --
         00
          0
         --
          01
           0
          --
           11
           10
           --
            1
```

Thus $1110011_{\text{two}} \div 10_{\text{two}} = 111001_{\text{two}}$ with a remainder of 1_{two}.

SECTION 6.5

Check your progress 1, *page 331*

a. Divide 9 by 1, 2, 3, ..., 9 to determine that the only natural number divisors of 9 are 1, 3, and 9.

b. Divide 11 by 1, 2, 3, ..., 11 to determine that the only natural number divisors of 11 are 1 and 11.

c. Divide 24 by 1, 2, 3, ..., 24 to determine that the only natural number divisors of 24 are 1, 2, 3, 4, 6, 8, 12, and 24.

Check your progress 2, *page 331*

a. The only divisors of 47 are 1 and 47. Thus 47 is a prime number.

b. 171 is divisible by 3, 9, 19, and 57. Thus 171 is a composite number.

c. The divisors of 91 are 1, 7, 13, and 91. Thus 91 is a composite number.

Check your progress 3, *page 332*

a. The sum of the digits of 341,565 is 24; therefore, 341,565 is divisible by 3.

b. The number 341,565 is not divisible by 4 because the number formed by the last two digits, 65, is not divisible by 4.

c. The number 341,565 is not divisible by 10 because it does not end in 0.

d. The sum of the digits with even place-value powers is 14. The sum of the digits with odd place-value powers is 10. The difference of these sums is 4. Thus 341,565 is not divisible by 11.

Check your progress 4, *page 333*

a.

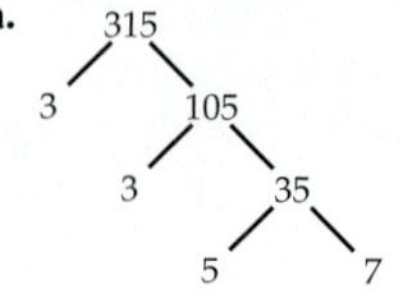

$315 = 3^2 \cdot 5 \cdot 7$

b.

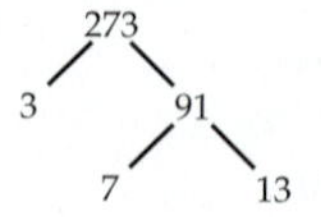

$273 = 3 \cdot 7 \cdot 13$

c.

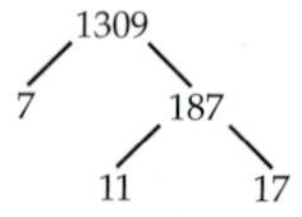

$1309 = 7 \cdot 11 \cdot 17$

SECTION 6.6

Check your progress 1, *page 341*

a. The proper factors of 24 are 1, 2, 3, 4, 6, 8, and 12. The sum of these proper factors is 36. Because 24 is less than the sum of its proper factors, 24 is an abundant number.

b. The proper factors of 28 are 1, 2, 4, 7, and 14. The sum of these proper factors is 28. Because 28 equals the sum of its proper factors, 28 is a perfect number.

c. The proper factors of 35 are 1, 5, and 7. The sum of these proper factors is 13. Because 35 is larger than the sum of its proper factors, 35 is a deficient number.

Check your progress 2, *page 341*

$2^7 - 1 = 127$, which is a prime number.

Check your progress 3, *page 342*

The exponent $n = 61$ is a prime number, and we are given that $2^{61} - 1$ is a prime number, so the perfect number we seek is $2^{60}(2^{61} - 1)$.

Check your progress 4, *page 344*

First consider $2^{20996011}$. The base b is 2. The exponent x is 20,996,011.

$$\begin{aligned}(x \log b) + 1 &= (20{,}996{,}011 \log 2) + 1\\ &\approx 6{,}320{,}429.1 + 1\\ &= 6{,}320{,}430.1\end{aligned}$$

The greatest integer of 6,320,430.1 is 6,320,430. Thus $2^{20996011}$ has 6,320,430 digits. The number $2^{20996011}$ is not a power of 10, so the Mersenne prime number $2^{20996011} - 1$ also has 6,320,430 digits.

Check your progress 5, *page 345*

Substituting 9 for x, 11 for y, and 4 for n in $x^n + y^n = z^n$ yields

$$\begin{aligned}9^4 + 11^4 &= z^4\\ 6561 + 14{,}641 &= z^4\\ 21{,}202 &= z^4\end{aligned}$$

The real solution of $z^4 = 21{,}202$ is $\sqrt[4]{21{,}202} \approx 12.066858$, which is not a natural number. Thus $x = 9$, $y = 11$, and $n = 4$ do not satisfy the equation $x^n + y^n = z^n$ where z is a natural number.

CHAPTER 7

SECTION 7.1

Check your progress 1, *page 357*

$$14 \text{ ft} = 14 \cancel{\text{ft}} \times \frac{1 \text{ yd}}{3 \cancel{\text{ft}}} = \frac{14 \text{ yd}}{3} = 4\frac{2}{3} \text{ yd}$$

Check your progress 2, *page 357*

$$3 \text{ lb} = 3 \cancel{\text{lb}} \times \frac{16 \text{ oz}}{1 \cancel{\text{lb}}} = 48 \text{ oz}$$

Check your progress 3, *page 357*

First convert pints to quarts, and then convert quarts to gallons.

$$18 \text{ pt} = 18 \cancel{\text{pt}} \times \frac{1 \cancel{\text{qt}}}{2 \cancel{\text{pt}}} \times \frac{1 \text{ gal}}{4 \cancel{\text{qt}}}$$

$$= \frac{18 \text{ gal}}{8} = 2\frac{1}{4} \text{ gal}$$

Check your progress 4, *page 358*

3.07 m = 307 cm

Check your progress 5, *page 359*

42.3 mg = 0.0423 g

Check your progress 6, *page 360*

2 kl = 2000 L

2 kl 167 L = 2000 L + 167 L

= 2167 L

Check your progress 7, *page 361*

$$\frac{\$3.69}{\text{gal}} \approx \frac{\$3.69}{\cancel{\text{gal}}} \times \frac{1 \cancel{\text{gal}}}{3.79 \text{ L}}$$

$$= \frac{\$3.69}{3.79 \text{ L}} \approx \frac{\$.97}{\text{L}}$$

\$3.69/gal ≈ \$.97/L

Check your progress 8, *page 361*

$$45 \text{ cm} = \frac{45 \cancel{\text{cm}}}{1} \times \frac{1 \text{ in.}}{2.54 \cancel{\text{cm}}}$$

$$= \frac{45 \text{ in.}}{2.54} \approx 17.72 \text{ in.}$$

45 cm ≈ 17.72 in.

Check your progress 9, *page 361*

$$\frac{75 \text{ km}}{\text{h}} \approx \frac{75 \cancel{\text{km}}}{\text{h}} \times \frac{1 \text{ mi}}{1.61 \cancel{\text{km}}}$$

$$= \frac{75 \text{ mi}}{1.61 \text{ h}} \approx 46.58 \text{ mi/h}$$

75 km/h ≈ 46.58 mi/h

SECTION 7.2

Check your progress 1, *page 365*

$$AB + BC = AC$$

$$\frac{1}{4}(BC) + BC = AC$$

$$\frac{1}{4}(16) + 16 = AC$$

$$4 + 16 = AC$$

$$20 = AC$$

$AC = 20$ ft

Check your progress 2, *page 367*

Supplementary angles are two angles the sum of whose measures is 180°. To find the supplement, let x represent the supplement of a 129° angle.

$$x + 129 = 180$$

$$x = 51$$

The supplement of a 129° angle is a 51° angle.

Check your progress 3, *page 368*

$$m\angle a + 68^\circ = 118^\circ$$

$$m\angle a = 50^\circ$$

Check your progress 4, *page 369*

$$m\angle b + m\angle a = 180^\circ$$

$$m\angle b + 35^\circ = 180^\circ$$

$$m\angle b = 145^\circ$$

$$m\angle c = m\angle a = 35^\circ$$

$$m\angle d = m\angle b = 145^\circ$$

$m\angle b = 145^\circ$, $m\angle c = 35^\circ$, and $m\angle d = 145^\circ$.

Check your progress 5, *page 370*

$$m\angle b = m\angle g = 124^\circ$$

$$m\angle d = m\angle g = 124^\circ$$

$$m\angle c + m\angle b = 180^\circ$$

$$m\angle c + 124^\circ = 180^\circ$$

$$m\angle c = 56^\circ$$

$m\angle b = 124^\circ$, $m\angle c = 56^\circ$, and $m\angle d = 124^\circ$.

Check your progress 6, *page 371*

Let x represent the measure of the third angle.

$$x + 90^\circ + 27^\circ = 180^\circ$$

$$x + 117^\circ = 180^\circ$$

$$x = 63^\circ$$

The measure of the third angle is 63°.

Check your progress 7, *page 372*

$$m\angle b + m\angle d = 180^\circ$$

$$m\angle b + 105^\circ = 180^\circ$$

$$m\angle b = 75^\circ$$

$$m\angle a + m\angle b + m\angle c = 180^\circ$$

$$m\angle a + 75^\circ + 35^\circ = 180^\circ$$

$$m\angle a + 110^\circ = 180^\circ$$

$$m\angle a = 70^\circ$$

$$m\angle e = m\angle a = 70^\circ$$

SECTION 7.3

Check your progress 1, *page 380*

$P = 2L + 2W$
$P = 2(12) + 2(8)$
$P = 24 + 16$
$P = 40$

You will need 40 ft of molding to edge the top of the walls.

Check your progress 2, *page 380*

$P = 4s$
$P = 4(24)$
$P = 96$

The homeowner should purchase 96 ft of fencing.

Check your progress 3, *page 381*

$C = \pi d$
$C = 9\pi$

The circumference of the circle is 9π km.

Check your progress 4, *page 381*

12 in. = 1 ft

$C = \pi d$
$C = \pi(1)$
$C = \pi$

$12C = 12\pi \approx 37.70$

The tricycle travels approximately 37.70 ft when the wheel makes 12 revolutions.

Check your progress 5, *page 382*

$A = LW$
$A = 308(192)$
$A = 59{,}136$

59,136 cm^2 of fabric is needed.

Check your progress 6, *page 383*

$A = s^2$
$A = 24^2$
$A = 576$

The area of the floor is 576 ft^2.

Check your progress 7, *page 384*

$A = bh$
$A = 14(8)$
$A = 112$

The area of the patio is 112 m^2.

Check your progress 8, *page 385*

$$A = \frac{1}{2}bh$$

$$A = \frac{1}{2}(18)(9)$$

$$A = 9(9)$$

$$A = 81$$

81 in^2 of felt is needed.

Check your progress 9, *page 386*

$$A = \frac{1}{2}h(b_1 + b_2)$$

$$A = \frac{1}{2} \cdot 9(12 + 20)$$

$$A = \frac{1}{2} \cdot 9(32)$$

$$A = \frac{9}{2} \cdot (32)$$

$$A = 144$$

The area of the patio is 144 ft^2.

Check your progress 10, *page 386*

$$r = \frac{1}{2}d = \frac{1}{2}(12) = 6$$

$A = \pi r^2$
$A = \pi(6)^2$
$A = 36\pi$

The area of the circle is 36π km^2.

Check your progress 11, *page 387*

$$r = \frac{1}{2}d = \frac{1}{2}(4) = 2$$

$A = \pi r^2$
$A = \pi(2)^2$
$A = \pi(4)$
$A \approx 12.57$

Approximately 12.57 ft^2 of material is needed.

SECTION 7.4

Check your progress 1, *page 395*

$$\frac{AC}{DF} = \frac{CH}{FG}$$

$$\frac{10}{15} = \frac{7}{FG}$$

$$10(FG) = (15)7$$

$$10(FG) = 105$$

$$FG = 10.5$$

The height FG of triangle DEF is 10.5 m.

Check your progress 2, *page 397*

$\angle A$ and $\angle D$ are right angles. Therefore, $\angle A = \angle D$. $\angle AOB$ and $\angle COD$ are vertical angles. Therefore, $\angle AOB = \angle COD$. Because two angles of triangle AOB are equal in measure to two angles of triangle DOC, triangles AOB and DOC are similar triangles.

$$\frac{AO}{DO} = \frac{AB}{DC}$$

$$\frac{AO}{3} = \frac{10}{4}$$

$$4(AO) = 3(10)$$

$$4(AO) = 30$$

$$AO = 7.5$$

$A = \frac{1}{2}bh$

$A = \frac{1}{2}(10)(7.5)$

$A = 5(7.5)$

$A = 37.5$

The area of triangle AOB is 37.5 cm^2.

Check your progress 3, *page 399*

Because two sides and the included angle of one triangle are equal in measure to two sides and the included angle of the second triangle, the triangles are congruent by the SAS theorem.

Check your progress 4, *page 400*

$a^2 + b^2 = c^2$ • Use the Pythagorean theorem.

$2^2 + b^2 = 6^2$ • $a = 2, c = 6$

$4 + b^2 = 36$

$b^2 = 32$ • Solve for b^2. Subtract 4 from each side.

$\sqrt{b^2} = \sqrt{32}$ • Take the square root of each side of the equation.

$b \approx 5.66$ • Use a calculator to approximate $\sqrt{32}$.

The length of the other leg is approximately 5.66 m.

SECTION 7.5

Check your progress 1, *page 409*

$V = LWH$

$V = 5(3.2)(4)$

$V = 64$

The volume of the solid is 64 m^3.

Check your progress 2, *page 409*

$V = \frac{1}{3}s^2h$

$V = \frac{1}{3}(15)^2(25)$

$V = \frac{1}{3}(225)(25)$

$V = 1875$

The volume of the pyramid is 1875 m^3.

Check your progress 3, *page 410*

$r = \frac{1}{2}d = \frac{1}{2}(16) = 8$

$V = \pi r^2 h$

$V = \pi(8)^2(30)$

$V = \pi(64)(30)$

$V = 1920\pi$

$\frac{1}{4}(1920\pi) = 480\pi$

≈ 1507.96

Approximately 1507.96 ft^3 are not being used for storage.

Check your progress 4, *page 412*

$r = \frac{1}{2}d = \frac{1}{2}(6) = 3$

$S = 2\pi r^2 + 2\pi rh$

$S = 2\pi(3)^2 + 2\pi(3)(8)$

$S = 2\pi(9) + 2\pi(3)(8)$

$S = 18\pi + 48\pi$

$S = 66\pi$

$S \approx 207.35$

The surface area of the cylinder is approximately 207.35 ft^2.

SECTION 7.6

Check your progress 1, *page 419*

Use the Pythagorean theorem to find the length of the hypotenuse.

$a^2 + b^2 = c^2$

$3^2 + 4^2 = c^2$

$9 + 16 = c^2$

$25 = c^2$

$\sqrt{25} = \sqrt{c^2}$

$5 = c$

$\sin\theta = \frac{\text{opp}}{\text{hyp}} = \frac{3}{5},$

$\cos\theta = \frac{\text{adj}}{\text{hyp}} = \frac{4}{5},$

$\tan\theta = \frac{\text{opp}}{\text{adj}} = \frac{3}{4}$

Check your progress 2, *page 421*

We are given the measure of $\angle B$ and the hypotenuse. We want to find the length of side a. The cosine function involves the side adjacent and the hypotenuse.

$$\cos B = \frac{\text{adj}}{\text{hyp}}$$

$$\cos 48^\circ = \frac{a}{12}$$

$$12(\cos 48^\circ) = a$$

$$8.0 \approx a$$

The length of side a is approximately 8.0 ft.

Check your progress 3, *page 421*

$\tan^{-1}(0.3165) \approx 17.6^\circ$

Check your progress 4, *page 422*

$\theta \approx \tan^{-1}(0.5681)$

$\theta \approx 29.6^\circ$

Check your progress 5, *page 422*

We want to find the measure of $\angle A$, and we are given the length of the side opposite $\angle A$ and the hypotenuse. The sine function involves the side opposite an angle and the hypotenuse.

$$\sin A = \frac{\text{opp}}{\text{hyp}}$$

$$\sin A = \frac{7}{11}$$

$$A = \sin^{-1}\frac{7}{11}$$

$$A \approx 39.5^\circ$$

The measure of $\angle A$ is approximately 39.5°.

Check your progress 6, *page 423*

Let d be the distance from the base of the lighthouse to the boat.

$$\tan 25° = \frac{20}{d}$$
$$d(\tan 25°) = 20$$
$$d = \frac{20}{\tan 25°}$$
$$d \approx 42.9$$

The boat is approximately 42.9 m from the base of the lighthouse.

SECTION 7.7

Check your progress 1, *page 430*

$$S = (m\angle A + m\angle B + m\angle C - 180°)\left(\frac{\pi}{180°}\right)r^2$$
$$= (200° + 90° + 90° - 180°)\left(\frac{\pi}{180°}\right)(6)^2$$
$$= (200°)\left(\frac{\pi}{180°}\right)(36)$$
$$= 40\pi \text{ in}^2 \quad \text{• Exact area}$$
$$\approx 125.66 \text{ in}^2 \quad \text{• Approximate area}$$

Check your progress 2, *page 431*

In Example 2, we observed that only Riemannian geometry has the property that there exist no lines parallel to a given line. Thus, in this Check Your Progress, the type of geometry must be Riemannian.

Check your progress 3, *page 434*

a. $d_E(P, Q) = \sqrt{(x_2 - x_1)^2 + (y_2 - y_1)^2}$
$= \sqrt{[3 - (-1)]^2 + [2 - 4]^2}$
$= \sqrt{4^2 + (-2)^2}$
$= \sqrt{20} \approx 4.5$ blocks

$d_C(P, Q) = |x_2 - x_1| + |y_2 - y_1|$
$= |3 - (-1)| + |2 - 4|$
$= |4| + |-2|$
$= 4 + 2$
$= 6$ blocks

b. $d_E(P, Q) = \sqrt{(x_2 - x_1)^2 + (y_2 - y_1)^2}$
$= \sqrt{[(-1) - 3]^2 + [5 - (-4)]^2}$
$= \sqrt{(-4)^2 + 9^2}$
$= \sqrt{97} \approx 9.8$ blocks

$d_C(P, Q) = |x_2 - x_1| + |y_2 - y_1|$
$= |(-1) - 3| + |5 - (-4)|$
$= |-4| + 9$
$= 4 + 9$
$= 13$ blocks

SECTION 7.8

Check your progress 1, *page 441*

Replace each line segment with a scaled version of the generator. As you move from left to right, your first zig should be to the left.

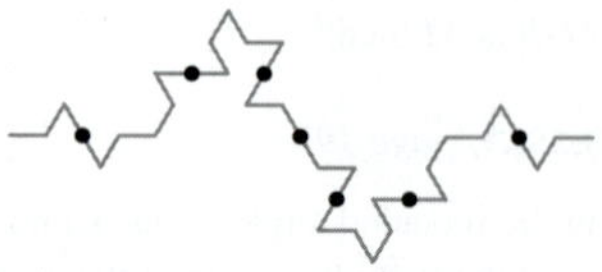

Stage 2 of the zig-zag curve

Check your progress 2, *page 441*

Replace each square with a scaled version of the generator.

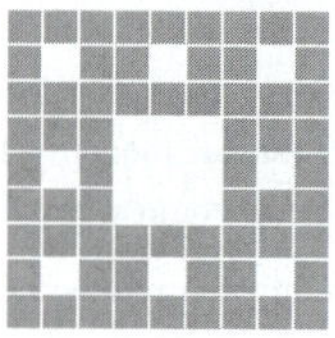

Stage 2 of the Sierpinski carpet

Check your progress 3, *page 443*

a. Any portion of the box curve replicates the entire fractal, so the box curve is a strictly self-similar fractal.

b. Any portion of the Sierpinski gasket replicates the entire fractal, so the Sierpinski gasket is a strictly self-similar fractal.

Check your progress 4, *page 444*

a. The generator of the Koch curve consists of four line segments, and the initiator consists of only one line segment. Thus the replacement ratio of the Koch curve is 4:1, or 4. The initiator of the Koch curve is a line segment that is 3 times as long as the replica line segments in the generator. Thus the scaling ratio of the Koch curve is 3:1, or 3.

b. The generator of the zig-zag curve consists of six line segments, and the initiator consists of only one line segment. Thus the replacement ratio of the zig-zag curve is 6:1, or 6. The initiator of the zig-zag curve is a line segment that is 4 times as long as the replica line segments in the generator. Thus the scaling ratio of the zig-zag curve is 4:1, or 4.

Check your progress 5, *page 444*

a. In Example 4, we determined that the replacement ratio of the box curve is 5 and the scaling ratio of the box curve is 3. Thus the similarity dimension of the box curve is $D = \frac{\log 5}{\log 3} \approx 1.465$.

b. The replacement ratio of the Sierpinski carpet is 8, and the scaling ratio of the Sierpinski carpet is 3. Thus the similarity dimension of the Sierpinski carpet is $D = \frac{\log 8}{\log 3} \approx 1.893$.

CHAPTER 8

SECTION 8.1

Check your progress 1, *page 458*

a. $6 \oplus 10 = 4$

b. $5 \oplus 9 = 2$

c. $7 \ominus 11 = 8$

d. $5 \ominus 10 = 7$

Check your progress 2, *page 459*

a. Find $\frac{7 - 12}{5} = \frac{-5}{5} = -1$. Because -1 is an integer, $7 \equiv 12 \bmod 5$ is a true congruence.

b. Find $\frac{15 - 1}{8} = \frac{14}{8} = \frac{7}{4}$. Because $\frac{7}{4}$ is not an integer, $15 \equiv 1 \bmod 8$ is not a true congruence.

Check your progress 3, *page 460*

The years 2016, 2020, and 2024 are leap years, so there are 3 years between the two dates with 366 days and 6 years with 365 days. The total number of days between the dates is $3 \cdot 366 + 6 \cdot 365 = 3288$. $3288 \div 7 = 469$ remainder 5, so $3288 \equiv 5 \bmod 7$. The day of the week 3288 days after Friday, February 12, 2016, will be the same as the day 5 days later, a Wednesday.

Check your progress 4, *page 460*

$51 + 72 = 123$, and $123 \div 3 = 41$ remainder 0, so $(51 + 72) \bmod 3 = 0$.

Check your progress 5, *page 462*

$21 - 43 = -22$, a negative number, so we must find x so that $-22 \equiv x \bmod 7$. Thus we must find x so that $\frac{-22 - x}{7} = \frac{-(22 + x)}{7}$ is an integer. Trying the whole number values of x less than 7, we find that when $x = 6$, $\frac{-(22 + 6)}{7} = \frac{-28}{7} = -4$. $(21 - 43) \bmod 7 = 6$.

Check your progress 6, *page 462*

Tuesday corresponds to 2 (see the chart on page 458), so the day of the week 93 days from now is represented by $(2 + 93) \bmod 7$. Because $95 \div 7 = 13$ remainder 4, $(2 + 93) \bmod 7 = 4$, which corresponds to Thursday.

Check your progress 7, *page 462*

$33 \cdot 41 = 1353$ and $1353 \div 17 = 79$ remainder 10, so $(33 \cdot 41) \bmod 17 = 10$.

Check your progress 8, *page 463*

Substitute each whole number from 0 to 11 into the congruence.

$4(0) + 1 \not\equiv 5 \bmod 12$	Not a solution
$4(1) + 1 \equiv 5 \bmod 12$	1 is a solution.
$4(2) + 1 \not\equiv 5 \bmod 12$	Not a solution
$4(3) + 1 \not\equiv 5 \bmod 12$	Not a solution
$4(4) + 1 \equiv 5 \bmod 12$	4 is a solution.
$4(5) + 1 \not\equiv 5 \bmod 12$	Not a solution
$4(6) + 1 \not\equiv 5 \bmod 12$	Not a solution
$4(7) + 1 \equiv 5 \bmod 12$	7 is a solution.
$4(8) + 1 \not\equiv 5 \bmod 12$	Not a solution
$4(9) + 1 \not\equiv 5 \bmod 12$	Not a solution
$4(10) + 1 \equiv 5 \bmod 12$	10 is a solution.
$4(11) + 1 \not\equiv 5 \bmod 12$	Not a solution

The solutions from 0 to 11 are 1, 4, 7, and 10. The remaining solutions are obtained by repeatedly adding the modulus 12 to these solutions. So the solutions are 1, 4, 7, 10, 13, 16, 19, 22, ...

Check your progress 9, *page 464*

In mod 12 arithmetic, $6 + 6 = 12$, so the additive inverse of 6 is 6.

Check your progress 10, *page 465*

Solve the congruence equation $5x \equiv 1 \bmod 11$ by substituting whole number values of x less than the modulus.

$5(1) \not\equiv 1 \bmod 11$
$5(2) \not\equiv 1 \bmod 11$
$5(3) \not\equiv 1 \bmod 11$
$5(4) \not\equiv 1 \bmod 11$
$5(5) \not\equiv 1 \bmod 11$
$5(6) \not\equiv 1 \bmod 11$
$5(7) \not\equiv 1 \bmod 11$
$5(8) \not\equiv 1 \bmod 11$
$5(9) \equiv 1 \bmod 11$

In mod 11 arithmetic, the multiplicative inverse of 5 is 9.

SECTION 8.2

Check your progress 1, *page 468*

Use the formula for the ISBN check digit.

$$\begin{aligned} d_{13} &= 10 - [9 + 3(7) + 8 + 3(0) + 7 + 3(1) + 6 + 3(7) + 3 + 3(2) + 5 + 3(0)] \bmod 10 \\ &= 10 - 89 \bmod 10 \\ &= 10 - 9 = 1 \end{aligned}$$

Because the check digit does not match, the ISBN is invalid.

Check your progress 2, *page 469*

Use the formula for the UPC check digit.

$$\begin{aligned} d_{12} &= 10 - [3(1) + 3 + 3(2) + 3 + 3(4) + 2 + 3(6) + 5 + 3(9) + 3 + 3(3)] \bmod 10 \\ &= 10 - 91 \bmod 10 \\ &= 10 - 1 = 9 \end{aligned}$$

Because the check digit matches, the UPC is valid.

Check your progress 3, *page 470*

Highlight every other digit, reading from right to left:

6 0 1 1 0 1 2 3 9 1 4 5 2 3 1 7

Double the highlighted digits:

12 0 2 1 0 1 4 3 18 1 8 5 4 3 2 7

Add all the digits, treating two-digit numbers as two single digits:

$(1 + 2) + 0 + 2 + 1 + 0 + 1 + 4 + 3 + (1 + 8) + 1 + 8 + 5 + 4 + 3 + 2 + 7 = 53$

Because $53 \not\equiv 0 \bmod 10$, this is not a valid credit card number.

Check your progress 4, *page 473*

a. The encrypting congruence is $c \equiv (p + 17) \bmod 26$.

A	$c \equiv (1 + 17) \bmod 26 \equiv 18 \bmod 26 = 18$	Code A as R.
L	$c \equiv (12 + 17) \bmod 26 \equiv 29 \bmod 26 = 3$	Code L as C.
P	$c \equiv (16 + 17) \bmod 26 \equiv 33 \bmod 26 = 7$	Code P as G.
I	$c \equiv (9 + 17) \bmod 26 \equiv 26 \bmod 26 = 0$	Code I as Z.
N	$c \equiv (14 + 17) \bmod 26 \equiv 31 \bmod 26 = 5$	Code N as E.
E	$c \equiv (5 + 17) \bmod 26 \equiv 22 \bmod 26 = 22$	Code E as V.
S	$c \equiv (19 + 17) \bmod 26 \equiv 36 \bmod 26 = 10$	Code S as J.
K	$c \equiv (11 + 17) \bmod 26 \equiv 28 \bmod 26 = 2$	Code K as B.
G	$c \equiv (7 + 17) \bmod 26 \equiv 24 \bmod 26 = 24$	Code G as X.

Thus the plaintext ALPINE SKIING is coded as RCGZEV JBZZEX.

b. To decode, because $m = 17$, $n = 26 - 17 = 9$, and the decoding congruence is $p \equiv (c + 9) \bmod 26$.

T	$c \equiv (20 + 9) \bmod 26 \equiv 29 \bmod 26 = 3$	Decode T as C.
I	$c \equiv (9 + 9) \bmod 26 \equiv 18 \bmod 26 = 18$	Decode I as R.
F	$c \equiv (6 + 9) \bmod 26 \equiv 15 \bmod 26 = 15$	Decode F as O.
J	$c \equiv (10 + 9) \bmod 26 \equiv 19 \bmod 26 = 19$	Decode J as S.

Continuing, the ciphertext TIFJJ TFLEKIP JBZZEX decodes as CROSS COUNTRY SKIING.

Check your progress 5, *page 474*

The encrypting congruence is $c \equiv (3p + 1) \bmod 26$.

F	$c \equiv (3 \cdot 6 + 1) \bmod 26 \equiv 19 \bmod 26 = 19$	Code F as S.
L	$c \equiv (3 \cdot 12 + 1) \bmod 26 \equiv 37 \bmod 26 = 11$	Code L as K.
A	$c \equiv (3 \cdot 1 + 1) \bmod 26 \equiv 4 \bmod 26 = 4$	Code A as D.
S	$c \equiv (3 \cdot 19 + 1) \bmod 26 \equiv 58 \bmod 26 = 6$	Code S as F.
H	$c \equiv (3 \cdot 8 + 1) \bmod 26 \equiv 25 \bmod 26 = 25$	Code H as Y.

Continuing, the plaintext FLASH DRIVE is coded as SKDFY MCBOP.

Check your progress 6, *page 474*

Solve the congruence equation $c \equiv (7p + 1) \bmod 26$ for p.

$$c = 7p + 1$$
$$c - 1 = 7p$$
• Subtract 1 from each side of the equation.
$$15(c - 1) = 15(7p)$$
• Multiply each side of the equation by the multiplicative inverse of 7. Because $7 \cdot 15 \equiv 1 \bmod 26$, multiply each side by 15.

$[15(c - 1)] \bmod 26 \equiv p$

The decoding congruence is $p \equiv [15(c - 1)] \bmod 26$.

I	$p \equiv [15(9 - 1)] \bmod 26 \equiv 120 \bmod 26 = 16$	Decode I as P.
G	$p \equiv [15(7 - 1)] \bmod 26 \equiv 90 \bmod 26 = 12$	Decode G as L.
H	$p \equiv [15(8 - 1)] \bmod 26 \equiv 105 \bmod 26 = 1$	Decode H as A.
T	$p \equiv [15(20 - 1)] \bmod 26 \equiv 285 \bmod 26 = 25$	Decode T as Y.

Continuing, the ciphertext IGHT OHGG decodes as PLAY BALL.

SECTION 8.3

Check your progress 1, *page 480*

1. Determine whether the operation is closed by finding all possible products.

   $(1 \cdot 1) \bmod 4 \equiv 1 \bmod 4 = 1$
   $(1 \cdot 2) \bmod 4 \equiv 2 \bmod 4 = 2$
   $(1 \cdot 3) \bmod 4 \equiv 3 \bmod 4 = 3$
   $(2 \cdot 2) \bmod 4 \equiv 4 \bmod 4 = 0$
   $(2 \cdot 3) \bmod 4 \equiv 6 \bmod 4 = 2$
   $(3 \cdot 3) \bmod 4 \equiv 9 \bmod 4 = 1$

   The result of each product is in the set. The operation is closed.
2. Modulo operations satisfy the associative property.
3. 1 is the identity element.
4. Determine whether each element has an inverse. From the above calculation, 1 and 3 have an inverse. However, $(2 \cdot 2) \bmod 4 \equiv 0 \bmod 4 = 0 \neq 1$. Therefore, 2 does not have an inverse.

   The set and operation do not form a group.

Check your progress 2, *page 482*

Rotate the original triangle, I, about the line of symmetry through the bottom right vertex, followed by a clockwise rotation of 240°.

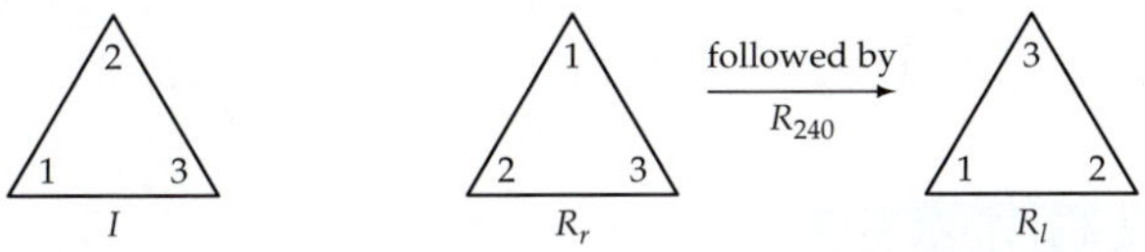

Therefore, $R_r \Delta R_{240} = R_l$.

Check your progress 3, *page 483*

$$R_r \Delta R_{240} = \begin{pmatrix} 1 & 2 & 3 \\ 2 & 1 & 3 \end{pmatrix} \Delta \begin{pmatrix} 1 & 2 & 3 \\ 3 & 1 & 2 \end{pmatrix}$$
$$= \begin{pmatrix} 1 & 2 & 3 \\ 1 & 3 & 2 \end{pmatrix} = R_l$$

- $1 \to 2 \to 1$. Thus $1 \to 1$.
- $2 \to 1 \to 3$. Thus $2 \to 3$.
- $3 \to 3 \to 2$. Thus $3 \to 2$.

Check your progress 4, *page 485*

$E \Delta B = \begin{pmatrix} 1 & 2 & 3 \\ 2 & 1 & 3 \end{pmatrix} \Delta \begin{pmatrix} 1 & 2 & 3 \\ 3 & 1 & 2 \end{pmatrix}$. 1 is replaced by 2, which is then replaced by 1 in the second permutation. Thus 1 remains as 1. 2 is replaced by 1, which is then replaced by 3, so ultimately, 2 is replaced by 3. Finally, 3 remains as 3 in the first permutation but is replaced by 2 in the second, so ultimately 3 is replaced by 2.

The result is $\begin{pmatrix} 1 & 2 & 3 \\ 1 & 3 & 2 \end{pmatrix}$, which is C. Thus $E \Delta B = C$.

Check your progress 5, *page 485*

$D = \begin{pmatrix} 1 & 2 & 3 \\ 3 & 2 & 1 \end{pmatrix}$ replaces 1 with 3, 2 with 2, and 3 with 1. Reversing these, we need to replace 3 with 1, leave 2 alone, and replace 1 with 3. This is the element $\begin{pmatrix} 1 & 2 & 3 \\ 3 & 2 & 1 \end{pmatrix}$, which is D again.

Thus D is its own inverse.

CHAPTER 9

SECTION 9.1

Check your progress 1, *page 498*

a.
$$\begin{aligned} c - 6 &= -13 \\ c - 6 + 6 &= -13 + 6 \\ c &= -7 \end{aligned}$$

The solution is -7.

b.
$$\begin{aligned} 4 &= -8z \\ \frac{4}{-8} &= \frac{-8z}{-8} \\ -\frac{1}{2} &= z \end{aligned}$$

The solution is $-\frac{1}{2}$.

c.
$$\begin{aligned} 22 + m &= -9 \\ 22 - 22 + m &= -9 - 22 \\ m &= -31 \end{aligned}$$

The solution is -31.

d.
$$\begin{aligned} 5x &= 0 \\ \frac{5x}{5} &= \frac{0}{5} \\ x &= 0 \end{aligned}$$

The solution is 0.

Check your progress 2, *page 500*

a.
$$\begin{aligned} 4x + 3 &= 7x + 9 \\ 4x - 7x + 3 &= 7x - 7x + 9 \\ -3x + 3 &= 9 \\ -3x + 3 - 3 &= 9 - 3 \\ -3x &= 6 \\ \frac{-3x}{-3} &= \frac{6}{-3} \\ x &= -2 \end{aligned}$$

The solution is -2.

b.
$$\begin{aligned} 7 - (5x - 8) &= 4x + 3 \\ 7 - 5x + 8 &= 4x + 3 \\ 15 - 5x &= 4x + 3 \\ 15 - 5x - 4x &= 4x - 4x + 3 \\ 15 - 9x &= 3 \\ 15 - 15 - 9x &= 3 - 15 \\ -9x &= -12 \\ \frac{-9x}{-9} &= \frac{-12}{-9} \\ x &= \frac{4}{3} \end{aligned}$$

The solution is $\frac{4}{3}$.

c.

$$\frac{3x-1}{4}+\frac{1}{3}=\frac{7}{3}$$
$$12\left(\frac{3x-1}{4}+\frac{1}{3}\right)=12\left(\frac{7}{3}\right)$$
$$12\cdot\frac{3x-1}{4}+12\cdot\frac{1}{3}=12\cdot\frac{7}{3}$$
$$9x-3+4=28$$
$$9x+1=28$$
$$9x+1-1=28-1$$
$$9x=27$$
$$\frac{9x}{9}=\frac{27}{9}$$
$$x=3$$

The solution is 3.

Check your progress 3, *page 501*

a. $P=0.05Y-96$

$P=0.05(2015)-96$

$P=100.75-96$

$P=4.75$

The amount of garbage was about 4.75 lb/day.

b.

$$P=0.05Y-96$$
$$5.5=0.05Y-96$$
$$5.5+96=0.05Y-96+96$$
$$101.5=0.05Y$$
$$\frac{101.5}{0.05}=\frac{0.05Y}{0.05}$$
$$2030=Y$$

The year will be 2030.

Check your progress 4, *page 502*

$340 for the first four lines + $76 for each additional line	=	**$948**

Let $L=$ the number of lines in the ad.

$$340+76(L-4)=948$$
$$340+76L-304=948$$
$$36+76L=948$$
$$36-36+76L=948-36$$
$$76L=912$$
$$\frac{76L}{76}=\frac{912}{76}$$
$$L=12$$

The ad can contain 12 lines.

Check your progress 5, *page 503*

Let $n=$ the number of years after 2014.

The 2014 population of Cleveland minus an annual decrease times *n*	=	**The 2014 population of Tampa plus the annual increase times *n***

$$389{,}500-1600n=358{,}700+5700n$$
$$389{,}500-1600n+1600n=358{,}700+5700n+1600n$$
$$389{,}500=358{,}700+7300n$$
$$389{,}500-358{,}700=358{,}700-358{,}700+7300n$$
$$30{,}800=7300n$$
$$\frac{30{,}800}{7300}=\frac{7300n}{7300}$$
$$4\approx n$$

$2014+4=2018$

The populations would be the same in 2018.

Check your progress 6, *page 504*

a.

$$s=\frac{A+L}{2}$$
$$2\cdot s=2\cdot\frac{A+L}{2}$$
$$2s=A+L$$
$$2s-A=A-A+L$$
$$2s-A=L$$

b.

$$L=a(1+ct)$$
$$\frac{L}{a}=\frac{a(1+ct)}{a}$$
$$\frac{L}{a}=1+ct$$
$$\frac{L}{a}-1=1-1+ct$$
$$\frac{L}{a}-1=ct$$
$$\frac{\frac{L}{a}-1}{t}=\frac{ct}{t}$$
$$\frac{\frac{L}{a}-1}{t}=c$$
$$\left(\frac{L}{a}-1\right)\left(\frac{1}{t}\right)=c$$
$$\frac{L}{at}-\frac{1}{t}=c$$

SECTION 9.2

Check your progress 1, *page 510*

$6.75\div1.5=4.5$

$$\frac{\$6.75}{1.5\text{ lb}}=\frac{\$4.50}{1\text{ lb}}=\$4.50/\text{pound}$$

The hamburger costs $4.50/lb.

Check your progress 2, *page 510*

Find the difference in the hourly wage.

$10.00 − $7.25 = $2.75

Multiply the difference in the hourly wage by 35.

$2.75(35) = $96.25

An employee's pay for working 35 h and earning the California minimum wage is $96.25 greater.

Check your progress 3, *page 511*

$$\frac{\$6.29}{32\text{ oz}}\approx\frac{\$0.197}{1\text{ oz}}\qquad\frac{\$8.29}{48\text{ oz}}\approx\frac{\$0.173}{1\text{ oz}}$$

$0.197 > $0.173

The more economical purchase is 48 oz of detergent for $8.29.

Check your progress 4, *page 512*

a. 20,000(1.4574) = 29,148

29,148 Canadian dollars would be needed to pay for an order costing $20,000.

b. 25,000(0.9156) = 22,890

22,890 euros would be exchanged for $25,000.

Check your progress 5, *page 513*

$$\frac{24\text{ h}}{1\text{ day}} \cdot 7\text{ days} = (24\text{ h})(7) = 168\text{ h}$$

$$\frac{72\text{ h}}{1\text{ wk}} = \frac{72\text{ h}}{168\text{ h}} = \frac{72}{168} = \frac{3}{7}$$

The ratio is $\frac{3}{7}$.

Check your progress 6, *page 514*

$$8849 + 9824 = 18{,}673$$

$$\frac{18{,}673}{1364} \approx \frac{13.69}{1} \approx \frac{14}{1}$$

The ratio is 14 to 1.

Check your progress 7, *page 515*

$$\begin{aligned} \frac{42}{x} &= \frac{5}{8} \\ 42 \cdot 8 &= x \cdot 5 \\ 336 &= 5x \\ \frac{336}{5} &= \frac{5x}{5} \\ 67.2 &= x \end{aligned}$$

The solution is 67.2.

Check your progress 8, *page 516*

$$\begin{aligned} \frac{15\text{ km}}{2\text{ cm}} &= \frac{x\text{ km}}{7\text{ cm}} \\ \frac{15}{2} &= \frac{x}{7} \\ 15 \cdot 7 &= 2 \cdot x \\ 105 &= 2x \\ \frac{105}{2} &= \frac{2x}{2} \\ 52.5 &= x \end{aligned}$$

The distance between the two cities is 52.5 km.

Check your progress 9, *page 517*

$$\begin{aligned} \frac{7}{5} &= \frac{\$84{,}000}{x\text{ dollars}} \\ \frac{7}{5} &= \frac{84{,}000}{x} \\ 7 \cdot x &= 5 \cdot 84{,}000 \\ 7x &= 420{,}000 \\ \frac{7x}{7} &= \frac{420{,}000}{7} \\ x &= 60{,}000 \end{aligned}$$

The other partner receives \$60,000.

Check your progress 10, *page 517*

$$\begin{aligned} \frac{11.0\text{ deaths}}{1{,}000{,}000\text{ people}} &= \frac{d\text{ deaths}}{318{,}000{,}000\text{ people}} \\ 11.0(318{,}000{,}000) &= 1{,}000{,}000 \cdot d \\ 3{,}498{,}000{,}000 &= 1{,}000{,}000d \\ \frac{3{,}498{,}000{,}000}{1{,}000{,}000} &= \frac{1{,}000{,}000d}{1{,}000{,}000} \\ 3498 &= d \end{aligned}$$

Approximately 3500 people died from fire in the United States in 2013.

SECTION 9.3

Check your progress 1, *page 523*

a. $74\% = 0.74$

b. $152\% = 1.52$

c. $8.3\% = 0.083$

d. $0.6\% = 0.006$

Check your progress 2, *page 523*

a. $0.3 = 30\%$

b. $1.65 = 165\%$

c. $0.072 = 7.2\%$

d. $0.004 = 0.4\%$

Check your progress 3, *page 524*

a. $8\% = 8\left(\frac{1}{100}\right) = \frac{8}{100} = \frac{2}{25}$

b. $180\% = 180\left(\frac{1}{100}\right) = \frac{180}{100} = 1\frac{80}{100} = 1\frac{4}{5}$

c. $2.5\% = 2.5\left(\frac{1}{100}\right) = \frac{2.5}{100} = \frac{25}{1000} = \frac{1}{40}$

d. $66\frac{2}{3}\% = \frac{200}{3}\% = \frac{200}{3}\left(\frac{1}{100}\right) = \frac{2}{3}$

Check your progress 4, *page 524*

a. $\frac{1}{4} = 0.25 = 25\%$

b. $\frac{3}{8} = 0.375 = 37.5\%$

c. $\frac{5}{6} = 0.83\overline{3} = 83.\overline{3}\%$

d. $1\frac{2}{3} = 1.66\overline{6} = 166.\overline{6}\%$

Check your progress 5, *page 526*

$$\begin{aligned} \frac{\text{percent}}{100} &= \frac{\text{amount}}{\text{base}} \\ \frac{70}{100} &= \frac{12{,}950}{B} \\ 70 \cdot B &= 100(12{,}950) \\ 70B &= 1{,}295{,}000 \\ \frac{70B}{70} &= \frac{1{,}295{,}000}{70} \\ B &= 18{,}500 \end{aligned}$$

The Corolla cost \$18,500 when it was new.

Check your progress 6, *page 527*

$$\begin{aligned} \frac{\text{percent}}{100} &= \frac{\text{amount}}{\text{base}} \\ \frac{p}{100} &= \frac{18{,}705{,}000}{43{,}500{,}000} \\ p \cdot 43{,}500{,}000 &= 100(18{,}705{,}000) \\ 43{,}500{,}000p &= 1{,}870{,}500{,}000 \\ \frac{43{,}500{,}000p}{43{,}500{,}000} &= \frac{1{,}870{,}500{,}000}{43{,}500{,}000} \\ p &= 43 \end{aligned}$$

43% of the adult caretakers felt that they did not have a choice in this role.

Check your progress 7, *page 527*

$$\frac{\text{percent}}{100} = \frac{\text{amount}}{\text{base}}$$
$$\frac{3.5}{100} = \frac{A}{32{,}500}$$
$$3.5(32{,}500) = 100(A)$$
$$113{,}750 = 100A$$
$$\frac{113{,}750}{100} = \frac{100A}{100}$$
$$1137.5 = A$$

The customer would receive a rebate of $1137.50.

Check your progress 8, *page 528*

$$PB = A$$
$$0.05(46{,}875) = A$$
$$2343.75 = A$$

The teacher contributes $2343.75.

Check your progress 9, *page 528*

$$PB = A$$
$$0.03B = 14{,}370$$
$$\frac{0.03B}{0.03} = \frac{14{,}370}{0.03}$$
$$B = 479{,}000$$

The selling price of the home was $479,000.

Check your progress 10, *page 529*

$$PB = A$$
$$P \cdot 90 = 63$$
$$\frac{P \cdot 90}{90} = \frac{63}{90}$$
$$P = 0.7$$
$$P = 70\%$$

You answered 70% of the questions correctly.

Check your progress 11, *page 530*

$$PB = A$$
$$0.80(54{,}241) = A$$
$$43{,}392.80 = A$$
$$54{,}241 - 43{,}392.80 = 10{,}848.20$$

The difference between the cost of the remodeling and the increase in value of your home is $10,848.20.

Check your progress 12, *page 531*

$$18.63 - 4.92 = 13.71$$
$$PB = A$$
$$P \cdot 4.92 = 13.71$$
$$\frac{P \cdot 4.92}{4.92} = \frac{13.71}{4.92}$$
$$P \approx 2.787$$

The percent increase in the federal debt from 1995 to 2015 was 278.7%.

Check your progress 13, *page 533*

$$680 - 485 = 195$$
$$PB = A$$
$$P \cdot 680 = 195$$
$$\frac{P \cdot 680}{680} = \frac{195}{680}$$
$$P \approx 0.287$$

The federal deficit decreased by 28.7% from 2013 to 2014.

SECTION 9.4

Check your progress 1, *page 538*

$$2s^2 = 6 - 4s$$
$$2s^2 + 4s = 6 - 4s + 4s$$
$$2s^2 + 4s = 6$$
$$2s^2 + 4s - 6 = 6 - 6$$
$$2s^2 + 4s - 6 = 0$$

Check your progress 2, *page 539*

$$(n + 5)(2n - 3) = 0$$
$$n + 5 = 0 \qquad 2n - 3 = 0$$
$$n = -5 \qquad 2n = 3$$
$$n = \frac{3}{2}$$

Check:

$$\begin{array}{r|l} \multicolumn{2}{c}{(n + 5)(2n - 3) = 0} \\ \hline (-5 + 5)[2(-5) - 3] & 0 \\ 0(-13) & 0 \\ 0 = & 0 \end{array} \qquad \begin{array}{r|l} \multicolumn{2}{c}{(n + 5)(2n - 3) = 0} \\ \hline \left(\frac{3}{2} + 5\right)\left(2 \cdot \frac{3}{2} - 3\right) & 0 \\ \frac{13}{2}(3 - 3) & 0 \\ 0 = & 0 \end{array}$$

The solutions are -5 and $\frac{3}{2}$.

Check your progress 3, *page 540*

$$2x^2 = x + 1$$
$$2x^2 - x = x - x + 1$$
$$2x^2 - x = 1$$
$$2x^2 - x - 1 = 1 - 1$$
$$2x^2 - x - 1 = 0$$
$$(2x + 1)(x - 1) = 0$$
$$2x + 1 = 0 \qquad x - 1 = 0$$
$$2x = -1 \qquad x = 1$$
$$x = -\frac{1}{2}$$

Check:

$$\begin{array}{r|l} \multicolumn{2}{c}{2x^2 = x + 1} \\ \hline 2\left(-\frac{1}{2}\right)^2 & -\frac{1}{2} + 1 \\ 2\left(\frac{1}{4}\right) & \frac{1}{2} \\ \frac{1}{2} = & \frac{1}{2} \end{array} \qquad \begin{array}{r|l} \multicolumn{2}{c}{2x^2 = x + 1} \\ \hline 2(1)^2 & 1 + 1 \\ 2(1) & 2 \\ 2 = & 2 \end{array}$$

The solutions are $-\frac{1}{2}$ and 1.

Check your progress 4, *page 541*

$$2x^2 = 8x - 5$$
$$2x^2 - 8x + 5 = 0$$

$a = 2, b = -8, c = 5$

$$x = \frac{-b \pm \sqrt{b^2 - 4ac}}{2a}$$
$$x = \frac{-(-8) \pm \sqrt{(-8)^2 - 4(2)(5)}}{2(2)} = \frac{8 \pm \sqrt{64 - 40}}{4}$$
$$= \frac{8 \pm \sqrt{24}}{4} = \frac{8 \pm 2\sqrt{6}}{4} = \frac{2(4 \pm \sqrt{6})}{2(2)} = \frac{4 \pm \sqrt{6}}{2}$$

The exact solutions are $\frac{4 + \sqrt{6}}{2}$ and $\frac{4 - \sqrt{6}}{2}$.

$$\frac{4 + \sqrt{6}}{2} \approx 3.225 \qquad \frac{4 - \sqrt{6}}{2} \approx 0.775$$

To the nearest thousandth, the solutions are 3.225 and 0.775.

Check your progress 5, *page 542*

$$z^2 = -6 - 2z$$
$$z^2 + 2z + 6 = 0$$

$a = 1, b = 2, c = 6$

$$z = \frac{-b \pm \sqrt{b^2 - 4ac}}{2a}$$
$$z = \frac{-(2) \pm \sqrt{(2)^2 - 4(1)(6)}}{2(1)}$$
$$= \frac{-2 \pm \sqrt{4 - 24}}{2} = \frac{-2 \pm \sqrt{-20}}{2}$$

$\sqrt{-20}$ is not a real number.

The equation has no real number solutions.

Check your progress 6, *page 543*

$$h = 64t - 16t^2$$
$$0 = 64t - 16t^2$$
$$16t^2 - 64t = 0$$
$$16t(t - 4) = 0$$

$16t = 0 \qquad t - 4 = 0$

$t = 0 \qquad t = 4$

The object will be on the ground at 0 s and after 4 s.

Check your progress 7, *page 543*

$$h = -16t^2 + 19t + 3.5$$
$$10 = -16t^2 + 19t + 3.5$$
$$16t^2 - 19t + 3.5 = 0$$

$a = 16, b = -19, c = 3.5$

$$t = \frac{-b \pm \sqrt{b^2 - 4ac}}{2a}$$
$$t = \frac{-(-19) \pm \sqrt{(-19)^2 - 4(16)(3.5)}}{2(16)} = \frac{19 \pm \sqrt{137}}{32}$$
$$t = \frac{19 + \sqrt{137}}{32} \approx 0.96 \qquad t = \frac{19 - \sqrt{137}}{32} \approx 0.23$$

The solution $t \approx 0.23$ s is not reasonable because we know from experience that the ball cannot reach the basket in 0.23 s. The ball hits the basket 0.96 s after the ball is released.

CHAPTER 10

SECTION 10.1

Check your progress 1, *page 558*

x	$y = -2x + 3$	(x, y)
-2	$-2(-2) + 3 = 7$	$(-2, 7)$
-1	$-2(-1) + 3 = 5$	$(-1, 5)$
0	$-2(0) + 3 = 3$	$(0, 3)$
1	$-2(1) + 3 = 1$	$(1, 1)$
2	$-2(2) + 3 = -1$	$(2, -1)$
3	$-2(3) + 3 = -3$	$(3, -3)$

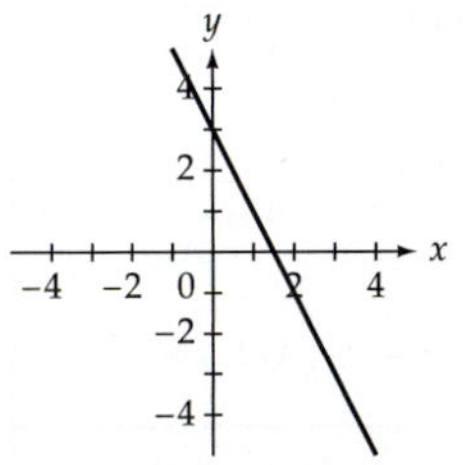

Check your progress 2, *page 558*

x	$y = -x^2 + 1$	(x, y)
-3	$-(-3)^2 + 1 = -8$	$(-3, -8)$
-2	$-(-2)^2 + 1 = -3$	$(-2, -3)$
-1	$-(-1)^2 + 1 = 0$	$(-1, 0)$
0	$-(0)^2 + 1 = 1$	$(0, 1)$
1	$-(1)^2 + 1 = 0$	$(1, 0)$
2	$-(2)^2 + 1 = -3$	$(2, -3)$
3	$-(3)^2 + 1 = -8$	$(3, -8)$

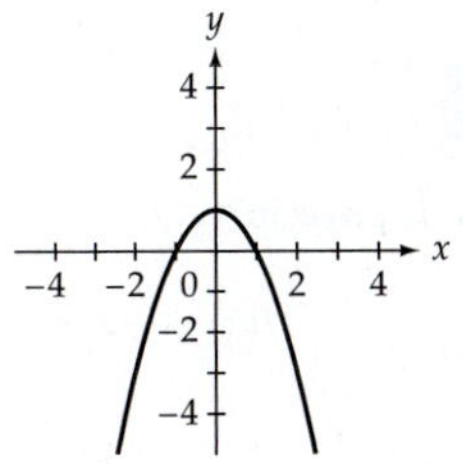

Check your progress 3, *page 561*

$$f(z) = z^2 - z$$
$$f(-3) = (-3)^2 - (-3)$$
$$= 12$$

The value of the function is 12 when $z = -3$.

Check your progress 4, *page 561*

$$N(s) = \frac{s^2 - 3s}{2}$$
$$N(12) = \frac{(12)^2 - 3(12)}{2}$$
$$= \frac{144 - 36}{2}$$
$$= 54$$

A polygon with 12 sides has 54 diagonals.

Check your progress 5, *page 562*

x	$f(x) = 2 - \frac{3}{4}x$	(x, y)
-3	$f(-3) = 2 - \frac{3}{4}(-3) = 4\frac{1}{4}$	$\left(-3, 4\frac{1}{4}\right)$
-2	$f(-2) = 2 - \frac{3}{4}(-2) = 3\frac{1}{2}$	$\left(-2, 3\frac{1}{2}\right)$
-1	$f(-1) = 2 - \frac{3}{4}(-1) = 2\frac{3}{4}$	$\left(-1, 2\frac{3}{4}\right)$
0	$f(0) = 2 - \frac{3}{4}(0) = 2$	$(0, 2)$
1	$f(1) = 2 - \frac{3}{4}(1) = 1\frac{1}{4}$	$\left(1, 1\frac{1}{4}\right)$
2	$f(2) = 2 - \frac{3}{4}(2) = \frac{1}{2}$	$\left(2, \frac{1}{2}\right)$
3	$f(3) = 2 - \frac{3}{4}(3) = -\frac{1}{4}$	$\left(3, -\frac{1}{4}\right)$

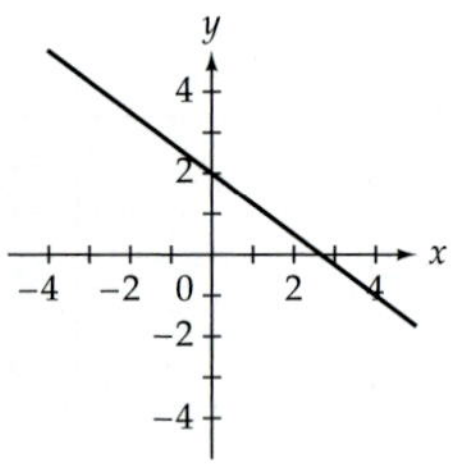

SECTION 10.2

Check your progress 1, *page 567*

$$f(x) = \frac{1}{2}x + 3$$
$$0 = \frac{1}{2}x + 3$$
$$-3 = \frac{1}{2}x$$
$$-6 = x$$

The x-intercept is $(-6, 0)$.

$$f(x) = \frac{1}{2}x + 3$$
$$f(0) = \frac{1}{2}(0) + 3$$
$$= 3$$

The y-intercept is $(0, 3)$.

Check your progress 2, *page 568*

$$g(t) = -20t + 8000$$
$$g(0) = -20(0) + 8000 = 8000$$

The intercept on the vertical axis is $(0, 8000)$. This means that the plane is at an altitude of 8000 ft when it begins its descent.

$$g(t) = -20t + 8000$$
$$0 = -20t + 8000$$
$$-8000 = -20t$$
$$400 = t$$

The intercept on the horizontal axis is $(400, 0)$. This means that the plane reaches the ground 400 s after beginning its descent.

Check your progress 3, *page 570*

a. $(x_1, y_1) = (-6, 5), (x_2, y_2) = (4, -5)$

$$m = \frac{y_2 - y_1}{x_2 - x_1} = \frac{-5 - 5}{4 - (-6)} = \frac{-10}{10} = -1$$

The slope is -1.

b. $(x_1, y_1) = (-5, 0), (x_2, y_2) = (-5, 7)$

$$m = \frac{y_2 - y_1}{x_2 - x_1} = \frac{7 - 0}{-5 - (-5)} = \frac{7}{0}$$

The slope is undefined.

c. $(x_1, y_1) = (-7, -2), (x_2, y_2) = (8, 8)$

$$m = \frac{y_2 - y_1}{x_2 - x_1} = \frac{8 - (-2)}{8 - (-7)} = \frac{10}{15} = \frac{2}{3}$$

The slope is $\frac{2}{3}$.

d. $(x_1, y_1) = (-6, 7), (x_2, y_2) = (1, 7)$

$$m = \frac{y_2 - y_1}{x_2 - x_1} = \frac{7 - 7}{1 - (-6)} = \frac{0}{7} = 0$$

The slope is 0.

Check your progress 4, *page 571*

The slope is 50 mph. The slope means that the homing pigeon can travel 50 mi for each hour of flight.

Check your progress 5, *page 572*

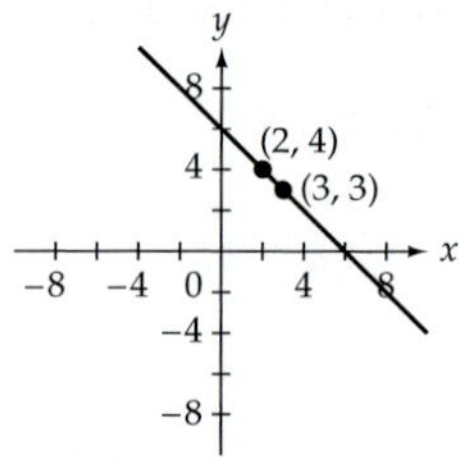

Check your progress 6, *page 573*

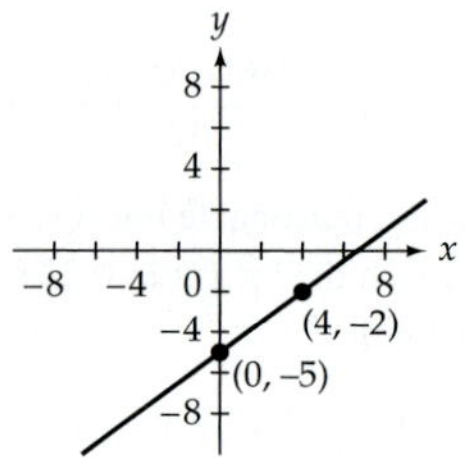

SECTION 10.3

Check your progress 1, *page 577*

$f(a) = ma + b$
$f(a) = -3.5a + 100$

The linear function is $f(a) = -3.5a + 100$, where $f(a)$ is the boiling point of water at an altitude of a kilometers above sea level.

Check your progress 2, *page 578*

$$y - y_1 = m(x - x_1)$$
$$y - 2 = -\frac{1}{2}[x - (-2)]$$
$$y - 2 = -\frac{1}{2}x - 1$$
$$y = -\frac{1}{2}x + 1$$

Check your progress 3, *page 578*

Let S represent the speed of sound at a depth of x meters. Then $S = 1480$ when $x = 1000$. A solution of the equation is (1000, 1480). The speed of sound is increasing at a rate of 0.017 m/s. Therefore, the slope is 0.017. Now use the point–slope formula to find the linear equation that models the function.

$$S - S_1 = m(x - x_1)$$
$$S - 1480 = 0.017(x - 1000)$$
$$S - 1480 = 0.017x - 17$$
$$S = 0.017x + 1463$$

A linear function that models the speed of sound at a depth of x meters is $S(x) = 0.017x + 1463$.

To find the speed of sound 2500 m below sea level, evaluate the function when $x = 2500$.

$$S(x) = 0.017x + 1463$$
$$S(2500) = 0.017(2500) + 1463$$
$$S(2500) = 42.5 + 1463 = 1505.5$$

The speed of sound 2500 m below sea level is 1505.5 m/s.

Check your progress 4, *page 579*

$$m = \frac{y_2 - y_1}{x_2 - x_1} = \frac{1 - 3}{4 - (-2)} = \frac{-2}{6} = -\frac{1}{3}$$
$$y - y_1 = m(x - x_1)$$
$$y - 3 = -\frac{1}{3}[x - (-2)]$$
$$y - 3 = -\frac{1}{3}x - \frac{2}{3}$$
$$y = -\frac{1}{3}x + \frac{7}{3}$$

Check your progress 5, *page 580*

a. Find the slope of the line between the two points.

$$\text{Slope} = \frac{102 - 132}{62 - 68} = \frac{-30}{-6} = 5$$

Use the point-slope formula to find the equation of the line.

$$y - y_1 = m(x - x_1)$$
$$y - 132 = 5(x - 68)$$
$$y - 132 = 5x - 340$$
$$y = 5x - 208$$

The linear model is $y = 5x - 208$.

b. Evaluate the linear model when $x = 63$.

$$y = 5x - 208$$
$$y = 5(63) - 208$$
$$y = 107$$

The estimated weight of a swimmer who is 63 in. tall is 107 lb.

SECTION 10.4

Check your progress 1, *page 585*

$$a = 1, b = 0; \; -\frac{b}{2a} = -\frac{0}{2(1)} = 0$$
$$y = x^2 - 2$$
$$y = (0)^2 - 2$$
$$y = -2$$

The vertex is $(0, -2)$.

Check your progress 2, *page 586*

a.
$$y = 2x^2 - 5x + 2$$
$$0 = 2x^2 - 5x + 2$$
$$0 = (2x - 1)(x - 2)$$
$$2x - 1 = 0 \qquad x - 2 = 0$$
$$x = \frac{1}{2} \qquad x = 2$$

The x-intercepts are $\left(\frac{1}{2}, 0\right)$ and $(2, 0)$.

b.
$$y = x^2 + 4x + 4$$
$$0 = x^2 + 4x + 4$$
$$0 = (x + 2)(x + 2)$$
$$x + 2 = 0 \qquad x + 2 = 0$$
$$x = -2 \qquad x = -2$$

The x-intercept is $(-2, 0)$.

Check your progress 3, *page 587*

$$a = 2, b = -3; \; -\frac{b}{2a} = -\frac{-3}{2(2)} = \frac{3}{4}$$
$$f(x) = 2x^2 - 3x + 1$$
$$f\left(\frac{3}{4}\right) = 2\left(\frac{3}{4}\right)^2 - 3\left(\frac{3}{4}\right) + 1$$
$$f\left(\frac{3}{4}\right) = -\frac{1}{8}$$

The vertex is $\left(\frac{3}{4}, -\frac{1}{8}\right)$. The minimum value of the function is $-\frac{1}{8}$, the y-coordinate of the vertex.

Check your progress 4, *page 587*

Let x represent one number. Because the difference between the two numbers is 10, $x + 10$ represents the other number. [*Note:* $(x + 10) - (x) = 10$.] Then their product is represented by $x(x + 10) = x^2 + 10x$.

To find one of the two numbers, find the x-coordinate of the vertex of $f(x) = x^2 + 10x$.

$$f(x) = x^2 + 10x$$
$$x = -\frac{b}{2a} = -\frac{10}{2(1)} = -5$$

To find the other number, replace x in $x + 10$ by the x-coordinate of the vertex and evaluate.

$$x + 10$$
$$-5 + 10 = 5$$

The numbers are -5 and 5.

To find the minimum product, evalute the function at the x-coordinate of the vertex.

$$\begin{aligned} f(x) &= x^2 + 10x \\ f(-5) &= (-5)^2 + 10(-5) \\ &= 25 - 50 \\ &= -25 \end{aligned}$$

The minimum product of the two numbers is -25.

Check your progress 5, *page 588*

Perimeter: $$\begin{aligned} w + l + w + l &= 44 \\ 2w + 2l &= 44 \\ w + l &= 22 \\ l &= -w + 22 \end{aligned}$$

Area: $$\begin{aligned} A &= lw \\ &= (-w + 22)w \\ &= -w^2 + 22w \end{aligned}$$

$$w = -\frac{b}{2a} = -\frac{22}{2(-1)} = 11$$

The width is 11 ft.

$$l = -w + 22$$
$$l = -(11) + 22 = 11$$

The length is 11 ft.

The dimensions of the rectangle with maximum area are 11 ft by 11 ft.

SECTION 10.5

Check your progress 1, *page 593*

$$g(x) = \left(\frac{1}{2}\right)^x$$
$$g(3) = \left(\frac{1}{2}\right)^3 = \frac{1}{8}$$
$$g(-1) = \left(\frac{1}{2}\right)^{-1} = \frac{1}{\frac{1}{2}} = 2$$
$$g(\sqrt{3}) = \left(\frac{1}{2}\right)^{\sqrt{3}} \approx \left(\frac{1}{2}\right)^{1.732} \approx 0.301$$

Check your progress 2, *page 595*

Because the base $\frac{3}{2}$ is greater than 1, f is an exponential growth function.

x	$f(x) = \left(\frac{3}{2}\right)^x$	(x, y)
-3	$f(-3) = \left(\frac{3}{2}\right)^{-3} = \frac{8}{27}$	$\left(-3, \frac{8}{27}\right)$
-2	$f(-2) = \left(\frac{3}{2}\right)^{-2} = \frac{4}{9}$	$\left(-2, \frac{4}{9}\right)$
-1	$f(-1) = \left(\frac{3}{2}\right)^{-1} = \frac{2}{3}$	$\left(-1, \frac{2}{3}\right)$
0	$f(0) = \left(\frac{3}{2}\right)^0 = 1$	$(0, 1)$
1	$f(1) = \left(\frac{3}{2}\right)^1 = \frac{3}{2}$	$\left(1, \frac{3}{2}\right)$
2	$f(2) = \left(\frac{3}{2}\right)^2 = \frac{9}{4}$	$\left(2, \frac{9}{4}\right)$
3	$f(3) = \left(\frac{3}{2}\right)^3 = \frac{27}{8}$	$\left(3, \frac{27}{8}\right)$

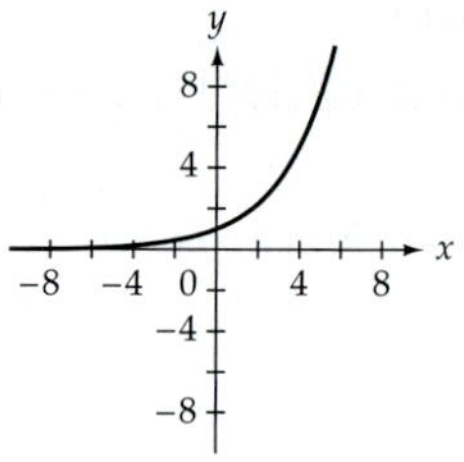

Check your progress 3, *page 596*

x	-2	-1	0	1	2
$f(x) = e^{-x} + 2$	9.4	4.7	3	2.4	2.1

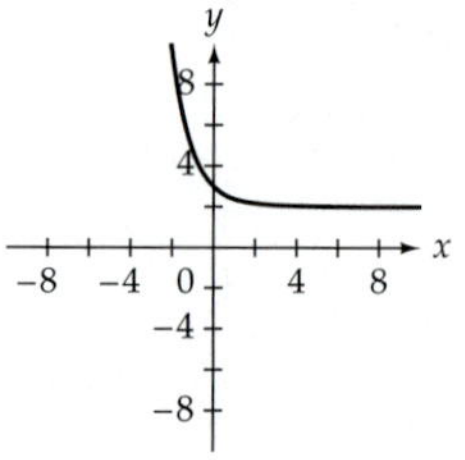

Check your progress 4, *page 597*

$$N(t) = 1.5\left(\tfrac{1}{2}\right)^{t/193.7}$$
$$\begin{aligned} N(24) &= 1.5\left(\tfrac{1}{2}\right)^{24/193.7} \\ &\approx 1.5(0.9177) \approx 1.3766 \end{aligned}$$

After 24 h, there are approximately 1.3766 g of the isotope in the body.

Check your progress 5, *page 597*

$$A(t) = 200e^{-0.014t}$$
$$A(45) = 200e^{-0.014(45)}$$
$$\approx 107$$

After 45 min, there are approximately 107 mg of aspirin in the patient's bloodstream.

SECTION 10.6

Check your progress 1, *page 602*

a. $2^{10} = 4x$

b. $\log_{10} 2x = 3$

Check your progress 2, *page 603*

a. $\log_{10} 0.001 = x$
$10^x = 0.001$
$10^x = 10^{-3}$
$x = -3$
$\log_{10} 0.001 = -3$

b. $\log_5 125 = x$
$5^x = 125$
$5^x = 5^3$
$x = 3$
$\log_5 125 = 3$

Check your progress 3, *page 603*

$\log_2 x = 6$
$2^6 = x$
$64 = x$

Check your progress 4, *page 604*

a. $\log x = -2.1$
$10^{-2.1} = x$
$0.008 \approx x$

b. $\ln x = 2$
$e^2 = x$
$7.389 \approx x$

Check your progress 5, *page 605*

$y = \log_5 x$
$5^y = x$

$x = 5^y$	$\frac{1}{25}$	$\frac{1}{5}$	1	5	25
y	−2	−1	0	1	2

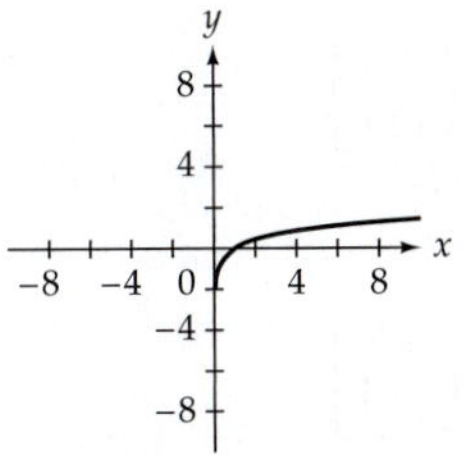

Check your progress 6, *page 606*

a. $S(0) = 5 + 29\ln(0 + 1) = 5$

The average typing speed when the student first started to type was 5 words/min.

$S(3) = 5 + 29\ln(3 + 1) \approx 45$

The average typing speed after 3 months was about 45 words/min.

b. $S(3) - S(0) = 45 - 5 = 40$

The typing speed increased by 40 words/min during the 3 months.

Check your progress 7, *page 607*

$I = 2 \cdot (125{,}892{,}541 I_0) = 251{,}785{,}082 I_0$

$$M = \log\left(\frac{I}{I_0}\right) = \log\left(\frac{251{,}785{,}082 I_0}{I_0}\right) = \log(251{,}785{,}082) \approx 8.4$$

The Richter scale magnitude of an earthquake whose intensity is twice that of the Samoa Islands earthquake is 8.4.

Check your progress 8, *page 607*

$$\log\left(\frac{I}{I_0}\right) = 4.8$$
$$\frac{I}{I_0} = 10^{4.8}$$
$$I = 10^{4.8} I_0$$
$$I \approx 63{,}096 I_0$$

The August 5, 2010, earthquake had an intensity that was approximately 63,096 times the intensity of a zero-level earthquake.

Check your progress 9, *page 608*

a. $\text{pH} = -\log[\text{H}^+] = -\log(2.41 \times 10^{-13}) \approx 12.6$

The cleaning solution has a pH of 12.6.

b. $\text{pH} = -\log[\text{H}^+] = -\log(5.07 \times 10^{-4}) \approx 3.3$

The cola soft drink has a pH of 3.3.

c. $\text{pH} = -\log[\text{H}^+] = -\log(6.31 \times 10^{-5}) \approx 4.2$

The rainwater has a pH of 4.2.

Check your progress 10, *page 609*

$$\text{pH} = -\log[\text{H}^+]$$
$$10.0 = -\log[\text{H}^+]$$
$$-10.0 = \log[\text{H}^+]$$
$$10^{-10.0} = \text{H}^+$$
$$1.0 \times 10^{-10} = \text{H}^+$$

The hydronium-ion concentration of the water in the Great Salt Lake in Utah is 1.0×10^{-10} mole per liter.

CHAPTER 11

SECTION 11.1

Check your progress 1, *page 620*

$P = 500,\ r = 4\% = 0.04,\ t = 1$

$I = Prt$
$I = 500(0.04)(1)$
$I = 20$

The simple interest earned is \$20.

Check your progress 2, *page 621*

$P = 1500,\ r = 5.25\% = 0.0525$

$$t = \frac{4 \text{ months}}{1 \text{ year}} = \frac{4 \text{ months}}{12 \text{ months}} = \frac{4}{12}$$

$I = Prt$

$$I = 1500(0.0525)\left(\frac{4}{12}\right)$$

$I = 26.25$

The simple interest due is \$26.25.

Check your progress 3, *page 621*

$P = 700,\ r = 1.25\% = 0.0125,\ t = 5$

$$I = Prt$$
$$I = 700(0.0125)(5)$$
$$I = 43.75$$

The simple interest due is $43.75.

Check your progress 4, *page 622*

$P = 7000,\ r = 5.25\% = 0.0525$

$$t = \frac{\text{number of days}}{360} = \frac{120}{360}$$

$$I = Prt$$
$$I = 7000(0.0525)\left(\frac{120}{360}\right)$$
$$I = 122.5$$

The simple interest due is $122.50.

Check your progress 5, *page 622*

$$I = Prt$$
$$462 = 12{,}000(r)\left(\frac{6}{12}\right)$$
$$462 = 6000r$$
$$0.077 = r$$
$$r = 7.7\%$$

The simple interest rate on the loan is 7.7%.

Check your progress 6, *page 624*

Find the interest.

$P = 4000,\ r = 8.75\% = 0.0875,\ t = \frac{9}{12}$

$$I = Prt$$
$$I = 4000(0.0875)\left(\frac{9}{12}\right)$$
$$I = 262.50$$

Find the maturity value.

$$A = P + I$$
$$A = 4000 + 262.50$$
$$A = 4262.50$$

The maturity value of the loan is $4262.50.

Check your progress 7, *page 625*

$P = 6700,\ r = 8.9\% = 0.089,\ t = 1$

$$A = P(1 + rt)$$
$$A = 6700[1 + 0.089(1)]$$
$$A = 6700(1 + 0.089)$$
$$A = 6700(1.089)$$
$$A = 7296.30$$

The maturity value of the loan is $7296.30.

Check your progress 8, *page 625*

$P = 680,\ r = 6.4\% = 0.064,\ t = 1$

$$A = P(1 + rt)$$
$$A = 680[1 + 0.064(1)]$$
$$A = 680(1 + 0.064)$$
$$A = 680(1.064)$$
$$A = 723.52$$

After 1 year, $723.52 is in the account.

Check your progress 9, *page 626*

$$I = A - P$$
$$I = 9240 - 9000$$
$$I = 240$$

$$I = Prt$$
$$240 = 9000(r)\left(\frac{4}{12}\right)$$
$$240 = 3000r$$
$$0.08 = r$$
$$r = 8\%$$

The simple interest rate on the loan is 8%.

SECTION 11.2

Check your progress 1, *page 630*

$$A = P(1 + rt)$$
$$A = 2000\left[1 + 0.04\left(\frac{1}{12}\right)\right]$$
$$A \approx 2006.67$$

$$A = P(1 + rt)$$
$$A \approx 2006.67\left[1 + 0.04\left(\frac{1}{12}\right)\right]$$
$$A \approx 2013.36$$

$$A = P(1 + rt)$$
$$A = 2013.36\left[1 + 0.04\left(\frac{1}{12}\right)\right]$$
$$A \approx 2020.07$$

$$A = P(1 + rt)$$
$$A = 2020.07\left[1 + 0.04\left(\frac{1}{12}\right)\right]$$
$$A \approx 2026.80$$

$$A = P(1 + rt)$$
$$A = 2026.80\left[1 + 0.04\left(\frac{1}{12}\right)\right]$$
$$A \approx 2033.56$$

$$A = P(1 + rt)$$
$$A = 2033.56\left[1 + 0.04\left(\frac{1}{12}\right)\right]$$
$$A \approx 2040.34$$

The total amount in the account at the end of 6 months is $2040.34.

Check your progress 2, *page 632*

Use the compound amount formula.

$P = 4000, r = 6\% = 0.06, n = 12, t = 2$

$$A = P\left(1 + \frac{r}{n}\right)^{nt}$$
$$A = 4000\left(1 + \frac{0.06}{12}\right)^{12 \cdot 2}$$
$$A = 4000(1 + 0.005)^{24}$$
$$A = 4000(1.005)^{24}$$
$$A \approx 4000(1.127160)$$
$$A \approx 4508.64$$

The compound amount after 2 years is approximately \$4508.64.

Check your progress 3, *page 633*

Use the compound amount formula.

$P = 2500, r = 9\% = 0.09, n = 360, t = 4$

$$A = 2500\left(1 + \frac{0.09}{360}\right)^{360 \cdot 4}$$
$$A = 2500(1 + 0.00025)^{1440}$$
$$A = 2500(1.00025)^{1440}$$
$$A \approx 2500(1.4332649)$$
$$A \approx 3583.16$$

The future value after 4 years is approximately \$3583.16.

Check your progress 4, *page 633*

Calculate the compound amount. Use the compound amount formula.

$P = 8000, r = 9\% = 0.09, n = 12, t = 6$

$$A = P\left(1 + \frac{r}{n}\right)^{nt}$$
$$A = 8000\left(1 + \frac{0.09}{12}\right)^{12 \cdot 6}$$
$$A = 8000(1 + 0.0075)^{72}$$
$$A = 8000(1.0075)^{72}$$
$$A \approx 8000(1.7125527)$$
$$A \approx 13{,}700.42$$

$$I = A - P$$
$$I = 13{,}700.42 - 8000$$
$$I = 5700.42$$

The amount of interest earned is approximately \$5700.42.

Check your progress 5, *page 634*

The following solution utilizes the finance feature of a TI-83/84 calculator.

Press [APPS] [ENTER].

Press [ENTER] to select 1: TVM Solver.

N is the number of compounding periods, or $n \cdot t$ in the compound amount formula.

After N =, enter 2 × 5.
After I% =, enter 6.
After PV =, enter −3500.
After PMT =, enter 0. Press [ENTER] twice.
After P/Y =, enter 2.
After C/Y =, enter 2.
Use the up arrow key to place the cursor at FV =.

Press [ALPHA] [Solve].

The solution is displayed to the right of FV =.

The compound amount is \$4703.71.

Check your progress 6, *page 635*

Use the present value formula.

$A = 20{,}000, r = 9\% = 0.09, n = 2, t = 5$

$$P = \frac{A}{\left(1 + \frac{r}{n}\right)^{nt}}$$
$$P = \frac{20{,}000}{\left(1 + \frac{0.09}{2}\right)^{2 \cdot 5}}$$
$$P = \frac{20{,}000}{(1 + 0.045)^{10}}$$
$$P \approx \frac{20{,}000}{1.552969}$$
$$P \approx 12{,}878.55$$

\$12,878.55 should be invested in the account.

Check your progress 7, *page 636*

The following solution utilizes the finance feature of a TI-83/84 calculator.

Press [APPS] [ENTER].

Press [ENTER] to select 1: TVM Solver.

After N =, enter 360 × 15.
After I% =, enter 6. Press [ENTER] twice.
After PMT =, enter 0.
After FV =, enter 25000.
After P/Y =, enter 360.
After C/Y =, enter 360.
Use the up arrow key to place the cursor at PV =.

Press [ALPHA] [Solve].

The solution is displayed to the right of PV =.

\$10,165.00 should be invested in the account.

Check your progress 8, *page 637*

Use the compound amount formula.

$P = 28{,}000, r = 5\% = 0.05, t = 17$

The inflation rate is an annual rate, so $n = 1$.

$$A = P\left(1 + \frac{r}{n}\right)^{nt}$$
$$A = 28{,}000\left(1 + \frac{0.05}{1}\right)^{1 \cdot 17}$$
$$A = 28{,}000(1 + 0.05)^{17}$$
$$A = 28{,}000(1.05)^{17}$$
$$A \approx 28{,}000(2.2920183)$$
$$A \approx 64{,}176.51$$

The average new car sticker price in 2030 will be approximately \$64,176.51.

Check your progress 9, *page 638*

Use the present value formula.

$A = 500{,}000,\ r = 7\% = 0.07,\ t = 40$

The inflation rate is an annual rate, so $n = 1$.

$$P = \frac{A}{\left(1 + \frac{r}{n}\right)^{nt}}$$
$$P = \frac{500{,}000}{\left(1 + \frac{0.07}{1}\right)^{1\cdot 40}}$$
$$P = \frac{500{,}000}{(1 + 0.07)^{40}}$$
$$P \approx \frac{500{,}000}{14.9744578}$$
$$P \approx 33{,}390.19$$

In 2055, the purchasing power of \$500,000 will be approximately \$33,390.19.

Check your progress 10, *page 639*

$P = 100,\ r = 4\% = 0.04,\ n = 4,\ t = 1$

$$A = P\left(1 + \frac{r}{n}\right)^{nt}$$
$$A = 100\left(1 + \frac{0.04}{4}\right)^{4\cdot 1}$$
$$A = 100(1 + 0.01)^4$$
$$A = 100(1.01)^4$$
$$A \approx 100(1.040604)$$
$$A \approx 104.06$$

$$I = A - P$$
$$I = 104.06 - 100$$
$$I = 4.06$$

The effective interest rate is 4.06%.

Check your progress 11, *page 640*

$$\left(1 + \frac{r}{n}\right)^{nt} = \left(1 + \frac{0.05}{4}\right)^{4\cdot 1} \approx 1.050945$$

$$\left(1 + \frac{r}{n}\right)^{nt} = \left(1 + \frac{0.0525}{2}\right)^{2\cdot 1} \approx 1.053189$$

An investment that earns 5.25% compounded semiannually has a higher annual yield than an investment that earns 5% compounded quarterly.

SECTION 11.3

Check your progress 1, *page 648*

Date	Payments or purchases	Balance each day	Number of days until balance changes	Unpaid balance times number of days
July 1–6		\$1024	6	\$6144
July 7–14	\$315	\$1339	8	\$10,712
July 15–21	−\$400	\$939	7	\$6573
July 22–31	\$410	\$1349	10	\$13,490
Total				\$36,919

$$\text{Average daily balance} = \frac{\text{sum of the total amounts owed each day of the month}}{\text{number of days in the billing period}}$$
$$= \frac{36{,}919}{31} \approx \$1190.94$$

$$I = Prt$$
$$I = 1190.94(0.012)(1)$$
$$I \approx 14.29$$

The finance charge on the August 1 bill is \$14.29.

Check your progress 2, *page 651*

a. Down payment = Percent down × purchase price

$= 0.20 \times 750 = 150$

Amount financed = purchase price − down payment

$= 750 - 150 = 600$

Interest owed = finance rate × amount financed

$= 0.08 \times 600 = 48$

The finance charge is \$48.

b. $$\text{APR} \approx \frac{2nr}{n+1} \approx \frac{2(12)(0.08)}{12+1} \approx \frac{1.92}{13} \approx 0.148$$

The annual percentage rate is approximately 14.8%.

Check your progress 3, *page 652*

Sales tax amount = sales tax rate × purchase price
= 0.0425 × 1499 ≈ 63.71

Amount financed = purchase price + sales tax amount
= 1499 + 63.71 = 1562.71

$$\frac{r}{n} = \frac{0.084}{12} = 0.007$$
$$nt = 12(3) = 36$$

$$PMT = A\left(\frac{\frac{r}{n}}{1 - \left(1 + \frac{r}{n}\right)^{-nt}}\right)$$

$$PMT = 1562.71\left(\frac{0.007}{1 - (1 + 0.007)^{-36}}\right)$$

$$PMT \approx 49.26$$

The monthly payment is \$49.26.

Check your progress 4, *page 653*

a. Sales tax = 0.0525(26,788) = 1406.37

b. Loan amount
= purchase price + sales tax + license fee − down payment
= 26,788 + 1406.37 + 145 − 2500
= 25,839.37

c. $\frac{r}{n} = \frac{0.081}{12} = 0.00675$

$nt = 12(5) = 60$

$$PMT = A\left(\frac{\frac{r}{n}}{1 - \left(1 + \frac{r}{n}\right)^{-nt}}\right)$$

$$PMT = 25{,}839.37\left(\frac{0.00675}{1 - (1 + 0.00675)^{-60}}\right)$$

$$PMT \approx 525.17$$

The monthly payment is \$525.17.

Check your progress 5, *page 655*

Because Aaron has owned the car for 36 months of a 60-month (5-year) loan, he has 24 payments remaining. Thus $U = 24$, the number of unpaid or remaining payments.

$$\frac{r}{n} = \frac{0.084}{12} = 0.007$$

$$A = PMT\left(\frac{1 - \left(1 + \frac{r}{n}\right)^{-U}}{\frac{r}{n}}\right)$$

$$A = 592.57\left(\frac{1 - (1 + 0.007)^{-24}}{0.007}\right)$$

$$A \approx 13{,}049.34$$

The loan payoff is \$13,049.34.

Check your progress 6, *page 656*

Using the spreadsheet from Example 6, we have

	A	B
1	Present value	25000
2	Annual interest rate as a decimal	0.049
3	Number of years to repay loan	10
4	Monthly payment	−\$263.94

His monthly payment is \$263.94.

Check your progress 7, *page 656*

First, calculate the interest owed on the loan for the 2 years before payments begin.

$I = Prt$ • Simple interest rate formula
$= 25{,}000(0.049)(2)$ • $P = 25{,}000$, $r = 0.049$, $t = 2$
$= 2450$

Add this amount to the amount borrowed, \$25,000, to determine the present value on which the loan payment will be calculated.

Present value = 25,000 + 2450 = 27,450.

Using the spreadsheet from Example 7, we have

	A	B
1	Present value	27450
2	Annual interest rate as a decimal	0.049
3	Number of years to repay loan	10
4	Monthly payment	−\$289.81

Hudson's monthly payment is \$289.81.

SECTION 11.4

Check your progress 1, *page 662*

(\$0.72 per share) × (550 shares) = \$396

The shareholder receives \$396 in dividends.

Check your progress 2, *page 663*

$I = Prt$
$0.82 = 51.25r(1)$ • Let I = annual dividend and P = the stock price. The time is 1 year.
$0.82 = 51.25r$
$0.016 = r$ • Divide each side of the equation by 51.25.

The dividend yield is 1.6%.

Check your progress 3, *page 665*

a. From Table 11.1, the selling price per share was \$31.49. The purchase price per share was \$19.37. The selling price per share is greater than the purchase price per share. You made a profit on the sale of the stock.

$$\begin{aligned}\text{Profit} &= \text{selling price} - \text{purchase price}\\ &= 300(\$31.49) - 300(\$19.37)\\ &= \$9447 - \$5811\\ &= \$3636\end{aligned}$$

The profit on the sale of the stock was \$3636.

b. $$\begin{aligned}\text{Commission} &= 2.1\%(\text{selling price})\\ &= 0.021(\$9447)\\ &\approx \$198.387\end{aligned}$$

The broker's commission was \$198.39.

Check your progress 4, *page 665*

Use the simple interest formula to find the annual interest payments. Substitute the following values into the formula: $P = 15{,}000$, $r = 3.5\% = 0.035$, and $t = 1$.

$$I = Prt$$
$$I = 15{,}000(0.035)(1)$$
$$I = 525$$

Multiply the annual interest payment by the term of the bond.

$$525(4) = 2100$$

The total of the interest payments paid to the bondholder is \$2100.

Check your progress 5, *page 667*

a. $$\begin{aligned}A - L &= (750 \text{ million} + 0.75 \text{ million} + 1.5 \text{ million}) - 1.5 \text{ million}\\ &= 750.75 \text{ million}\end{aligned}$$

$$N = 20 \text{ million}$$

$$\text{NAV} = \frac{A - L}{N} = \frac{750.75 \text{ million}}{20 \text{ million}} = 37.5375$$

The NAV of the fund is \$37.5375.

b. $$\frac{10{,}000}{37.5375} \approx 266$$ • Divide the amount invested by the cost per share of the fund. Round down to the nearest whole number.

You are purchasing 266 shares of the mutual fund.

SECTION 11.5

Check your progress 1, *page 671*

$$\begin{aligned}\text{Down payment} &= 25\% \text{ of } 410{,}000 = 0.25(410{,}000)\\ &= 102{,}500\end{aligned}$$

$$\begin{aligned}\text{Mortgage} &= \text{selling price} - \text{down payment}\\ &= 410{,}000 - 102{,}500\\ &= 307{,}500\end{aligned}$$

$$\begin{aligned}\text{Points} &= 1.75\% \text{ of } 307{,}500 = 0.0175(307{,}500)\\ &= 5381.25\end{aligned}$$

$$\text{Total} = 102{,}500 + 375 + 5381.25 = 108{,}256.25$$

The total of the down payment and the closing costs is \$108,256.25.

Check your progress 2, *page 673*

a. $$\frac{r}{n} = \frac{0.07}{12} \approx 0.00583333$$

$$nt = 12(25) = 300$$

$$PMT = A\left(\frac{\frac{r}{n}}{1 - \left(1 + \frac{r}{n}\right)^{-nt}}\right)$$

$$PMT \approx 223{,}000\left(\frac{0.00583333}{1 - (1 + 0.00583333)^{-300}}\right)$$

$$PMT \approx 1576.12$$

The monthly payment is \$1576.12.

b. $\text{Total} = 1576.12(300) = 472{,}836$

The total of the payments over the life of the loan is \$472,836.

c. $\text{Interest} = 472{,}836 - 223{,}000 = 249{,}836$

The amount of interest paid over the life of the loan is \$249,836.

Check your progress 3, *page 676*

$$\text{Down payment} = 0.25(295{,}000) = 73{,}750$$
$$\text{Mortgage} = 295{,}000 - 73{,}750 = 221{,}250$$

$$\frac{r}{n} = \frac{0.0675}{12} = 0.005625$$

$$nt = 12(30) = 360$$

$$PMT = A\left(\frac{\frac{r}{n}}{1 - \left(1 + \frac{r}{n}\right)^{-nt}}\right)$$

$$PMT = 221{,}250\left(\frac{0.005625}{1 - (1 + 0.005625)^{-360}}\right)$$

$$PMT \approx 1435.02$$

The monthly payment is \$1435.02.

$$\begin{aligned}I &= Prt\\ &= 221{,}250(0.0675)\left(\frac{1}{12}\right)\\ &\approx 1244.53\end{aligned}$$

The interest paid on the first payment is \$1244.53.

$$\text{Principal} = 1435.02 - 1244.53 = 190.49$$

The principal paid on the first payment is \$190.49.

Check your progress 4, *page 677*

Use the APR loan payoff formula. The homeowner has been making payments for 4 years, or 48 months. There are 300 payments in a 25-year loan, so there are $300 - 48 = 252$ unpaid or remaining payments; $U = 252$.

$$\frac{r}{n} = \frac{0.069}{12} = 0.00575$$

$$A = PMT\left(\frac{1 - \left(1 + \frac{r}{n}\right)^{-U}}{\frac{r}{n}}\right)$$

$$A = 846.82\left(\frac{1 - (1 + 0.00575)^{-252}}{0.00575}\right)$$

$$A \approx 112{,}548.79$$

The mortgage payoff is \$112,548.79.

Check your progress 5, *page 678*

Monthly property tax = 2332.80 ÷ 12 = 194.40
Monthly fire insurance = 450 ÷ 12 = 37.50
Total monthly payment = 1492.89 + 194.40 + 37.50 = 1724.79

The total monthly payment for mortgage, property tax, and fire insurance is $1724.79.

CHAPTER 12

SECTION 12.1

Check your progress 1, *page 688*

The possible outcomes are {M, i, s, p}. There are 4 possible outcomes.

Check your progress 2, *page 688*

a. {1, 3, 5, 7, 9} **b.** {0, 3, 6, 9} **c.** {8, 9}

Check your progress 3, *page 690*

	H	T
1	1H	1T
2	2H	2T
3	3H	3T
4	4H	4T
5	5H	5T
6	6H	6T

The sample space has 12 elements:

{1H, 1T, 2H, 2T, 3H, 3T, 4H, 4T, 5H, 5T, 6H, 6T}

Check your progress 4, *page 691*

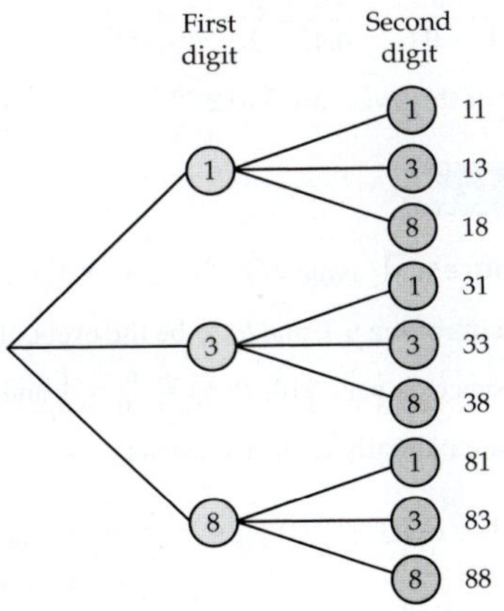

Check your progress 5, *page 692*

Any of the 9 runners could win the gold medal, so $n_1 = 9$. That leaves $n_2 = 8$ runners that could win silver, and $n_3 = 7$ possibilities for bronze. By the counting principle, there are 9 · 8 · 7 = 504 possible ways the medals can be awarded.

Check your progress 6, *page 693*

a. Because a student can receive more than 1 award and each award has 5 possible destinations, there are 5 · 5 · 5 = 125 ways for the awards to be given.

b. Once a student has an award, that student cannot receive another award, so there are 5 · 4 · 3 = 60 different ways for the awards to be given.

SECTION 12.2

Check your progress 1, *page 697*

a. $7! + 4! = (7 \cdot 6 \cdot 5 \cdot 4 \cdot 3 \cdot 2 \cdot 1) + (4 \cdot 3 \cdot 2 \cdot 1)$
$= 5040 + 24 = 5064$

b. $\dfrac{8!}{4!} = \dfrac{8 \cdot 7 \cdot 6 \cdot 5 \cdot \cancel{4!}}{\cancel{4!}} = 8 \cdot 7 \cdot 6 \cdot 5 = 1680$

Check your progress 2, *page 699*

Because the players are ranked, the number of different golf teams possible is the number of permutations of 8 players selected 5 at a time.

$$P(8, 5) = \frac{8!}{(8-5)!} = \frac{8!}{3!} = \frac{8 \cdot 7 \cdot 6 \cdot 5 \cdot 4 \cdot \cancel{3!}}{\cancel{3!}}$$
$$= 8 \cdot 7 \cdot 6 \cdot 5 \cdot 4 = 6720$$

There are 6720 possible golf teams.

Check your progress 3, *page 699*

The order in which the cars finish is important, so the number of ways to place first, second, and third is

$$P(43, 3) = \frac{43!}{(43-3)!} = \frac{43 \cdot 42 \cdot 41 \cdot \cancel{40!}}{\cancel{40!}} = 74{,}046$$

There are 74,046 different ways to award the first, second, and third place prizes.

Check your progress 4, *page 700*

a. With no restrictions, there are 7 tutors available for 7 h, so the number of schedules is

$$P(7, 7) = \frac{7!}{(7-7)!} = \frac{7!}{0!} = 7! = 5040$$

There are 5040 possible schedules.

b. This is a multistage experiment; there are 3! ways to schedule the juniors and 4! ways to schedule the seniors. By the counting principle, the number of different tutoring schedules is 3! · 4! = 6 · 24 = 144.

There are 144 tutor schedules.

Check your progress 5, *page 701*

a. With $n = 8$ coins and $k_1 = 3$ (number of pennies), $k_2 = 2$ (number of nickels), and $k_3 = 3$ (number of dimes), the number of different possible stacks is

$$\frac{8!}{3! \cdot 2! \cdot 3!} = \frac{8 \cdot 7 \cdot 6 \cdot 5 \cdot 4 \cdot 3!}{3! \cdot 2! \cdot 3!} = \frac{8 \cdot 7 \cdot 6 \cdot 5 \cdot 4}{6 \cdot 2} = 560$$

There are 560 possible stacks.

b. Not including the dimes, there are $\frac{5!}{3! \cdot 2!} = 10$ ways to stack the pennies and nickels. The dimes are identical, so there is only 1 way to arrange the dimes together, but there are 6 different locations in the stack of pennies and nickels into which the dimes could be placed. By the counting principle, the total number of ways in which the stack of coins can be arranged if the dimes are together is 10 · 6 = 60.

Check your progress 6, *page 702*

The order in which the waiters and waitresses are chosen is not important, so the number of ways to choose 9 people from 16 is

$$C(16, 9) = \frac{16!}{9! \cdot (16-9)!} = \frac{16!}{9! \cdot 7!}$$
$$= \frac{16 \cdot 15 \cdot 14 \cdot 13 \cdot 12 \cdot 11 \cdot 10 \cdot 9!}{9! \cdot 7!}$$
$$= \frac{16 \cdot 15 \cdot 14 \cdot 13 \cdot 12 \cdot 11 \cdot 10}{7 \cdot 6 \cdot 5 \cdot 4 \cdot 3 \cdot 2 \cdot 1} = 11{,}440$$

There are 11,440 possible groups of 9 waiters and waitresses.

Check your progress 7, *page 703*

There are $C(4, 3)$ ways for the auditor to choose 3 corporate tax returns and $C(6, 2)$ ways to choose 2 individual tax returns. By the counting principle, the total number of ways in which the auditor can choose the returns is

$$C(4, 3) \cdot C(6, 2) = \frac{4!}{3! \cdot 1!} \cdot \frac{6!}{2! \cdot 4!} = 4 \cdot 15 = 60$$

There are 60 ways to choose the tax returns.

Check your progress 8, *page 704*

For any single suit, there are $C(13, 4)$ ways of choosing 4 cards. That leaves $52 - 13 = 39$ cards of other suits from which to choose the fifth card. In addition, there are 4 different suits we could start with. By the counting principle, the number of 5-card combinations containing 4 cards of the same suit is

$$4 \cdot C(13, 4) \cdot 39 = 4 \cdot \frac{13!}{4! \cdot 9!} \cdot 39 = 4 \cdot 715 \cdot 39 = 111{,}540$$

There are 111,540 five-card combinations containing 4 cards of the same suit.

SECTION 12.3

Check your progress 1, *page 708*

$S = \{HH, HT, TH, TT\}$

Check your progress 2, *page 710*

The sample space for rolling a single die is $S = \{1, 2, 3, 4, 5, 6\}$. The elements in the event that an odd number is rolled are $E = \{1, 3, 5\}$. Then

$$P(E) = \frac{n(E)}{n(S)} = \frac{3}{6} = \frac{1}{2}$$

The probability that an odd number will be rolled is $\frac{1}{2}$.

Check your progress 3, *page 710*

The sample space is shown in Figure 12.6 on page 710.

Let E be the event that the sum of the pips on the upward faces is 7; the elements of this event are

$E = \{$ 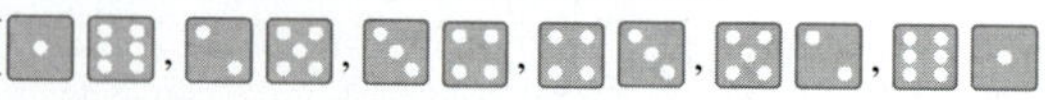 $\}$.

Then the probability of rolling a 7 is

$$P(E) = \frac{n(E)}{n(S)} = \frac{6}{36} = \frac{1}{6}$$

The probability that the sum is 7 is $\frac{1}{6}$.

Check your progress 4, *page 711*

Let E be the event that a person between the ages of 39 and 49 is selected. Then

$$P(E) = \frac{773}{3228} \approx 0.24$$

The probability the selected person is between the ages of 39 and 49 is approximately 0.24.

Check your progress 5, *page 712*

Make a Punnett square.

Parents	C	c
C	CC	Cc
c	Cc	cc

To be white, the offspring must be cc. From the table, only 1 of the 4 possible genotypes is cc, so the probability that an offspring will be white is $\frac{1}{4}$.

Check your progress 6, *page 713*

Let E be the event of selecting a blue ball. Because there are 5 blue balls, there are 5 favorable outcomes, leaving 7 unfavorable outcomes.

$$\text{Odds against } E = \frac{\text{number of unfavorable outcomes}}{\text{number of favorable outcomes}} = \frac{7}{5}$$

The odds against selecting a blue ball from the box are 7 to 5.

Check your progress 7, *page 714*

Let E represent the event of an earthquake of magnitude 6.7 or greater in the Bay Area in the next 30 years. Then the probability of E, $P(E)$, is 0.6. The odds in favor of this event are given by

$$\text{Odds in favor} = \frac{P(E)}{1 - P(E)}$$
$$= \frac{0.6}{1 - 0.6} = \frac{0.6}{0.4} = \frac{3}{2}$$

The odds in favor of this event are 3 to 2.

SECTION 12.4

Check your progress 1, *page 719*

Let A be the event of rolling a 7, and let B be the event of rolling an 11. From the sample space on page 710, $P(A) = \frac{6}{36} = \frac{1}{6}$ and $P(B) = \frac{2}{36} = \frac{1}{18}$. Because A and B are mutually exclusive events,

$$P(A \text{ or } B) = P(A) + P(B) = \frac{1}{6} + \frac{1}{18} = \frac{4}{18} = \frac{2}{9}$$

The probability of rolling a 7 or an 11 is $\frac{2}{9}$.

Check your progress 2, *pages 720–721*

Let $A = \{\text{people with a degree in business}\}$ and $B = \{\text{people with a starting salary between \$20,000 and \$24,999}\}$. Then, from the table, $n(A) = 4 + 16 + 21 + 35 + 22 = 98$, $n(B) = 4 + 16 + 3 + 16 = 39$, and $n(A \text{ and } B) = 16$. The total number of people represented in the table is 206.

$$P(A \text{ or } B) = P(A) + P(B) - P(A \text{ and } B)$$
$$= \frac{98}{206} + \frac{39}{206} - \frac{16}{206} = \frac{121}{206} \approx 0.587$$

The probability of choosing a person who has a degree in business or a starting salary between \$20,000 and \$24,999 is about 58.7%.

Check your progress 3, *page 722*

If E is the event that a person has type A blood, then E^C is the event that the person does not have type A blood, and

$$P(E^c) = 1 - P(E) = 1 - 0.34 = 0.66$$

The probability that a person does not have type A blood is 66%.

Check your progress 4, *page 723*

Let $E = \{\text{at least 1 roll of sum 7}\}$; then $E^c = \{\text{no sum of 7 is rolled}\}$. Using Figure 12.6 on page 710, there are 36 possibilities for each toss of the dice. Thus $n(S) = 36 \cdot 36 \cdot 36 = 46{,}656$. For each roll of the dice, there are 30 numbers that do not total 7, so $n(E^c) = 30 \cdot 30 \cdot 30 = 27{,}000$. Then

$$P(E) = 1 - P(E^c) = 1 - \frac{27{,}000}{46{,}656} = \frac{19{,}656}{46{,}656} \approx 0.421$$

There is about a 42.1% chance of rolling a sum of 7 at least once.

Check your progress 5, *page 724*

Let $E = \{\text{at least one \$100 bill}\}$; then $E^c = \{\text{no \$100 bills}\}$. The number of elements in the sample space is the number of ways we can choose 4 bills from 35:

$$n(S) = C(35, 4) = \frac{35!}{4!\,(35-4)!} = \frac{35!}{4!\,31!} = 52{,}360$$

To count the number of ways not to choose any \$100 bills, we need to compute the number of ways we can choose 4 \$1 bills from the 31 \$1 bills available.

$$n(E^c) = C(31, 4) = \frac{31!}{4!\,(31-4)!} = \frac{31!}{4!\,27!} = 31{,}465$$
$$P(E) = 1 - P(E^c) = 1 - \frac{n(E^c)}{n(S)}$$
$$= 1 - \frac{31{,}465}{52{,}360} = \frac{20{,}895}{52{,}360} \approx 0.399$$

The probability of pulling out at least one \$100 bill is about 39.9%.

SECTION 12.5

Check your progress 1, *page 729*

Let $B = \{\text{the sum is 6}\}$ and $A = \{\text{the first die is not a 3}\}$. From Figure 12.6 on page 710, there are 4 possible rolls of the dice for which the first die is not a 3 and the sum is 6. So $P(A \text{ and } B) = \frac{4}{36} = \frac{1}{9}$.

There are 30 possibilities for which the first die is not a 3, so $P(A) = \frac{30}{36} = \frac{5}{6}$. Then

$$P(B \mid A) = \frac{P(A \text{ and } B)}{P(A)} = \frac{\frac{1}{9}}{\frac{5}{6}} = \frac{2}{15}$$

The probability of rolling a 6 given that the first toss is not a 3 is $\frac{2}{15}$.

Check your progress 2, *page 731*

Let $A = \{\text{a spade is dealt first}\}$, $B = \{\text{a heart is dealt second}\}$, and $C = \{\text{a spade is dealt third}\}$. Then

$$P(A \text{ and } B \text{ and } C) = P(A) \cdot P(B \mid A) \cdot P(C \mid A \text{ and } B)$$
$$= \frac{13}{52} \cdot \frac{13}{51} \cdot \frac{12}{50} = \frac{13}{850}$$

The probability is $\frac{13}{850}$, or about 0.015.

Check your progress 3, *page 732*

Each coin toss is independent of the others because the probability of getting heads on any toss is not affected by the results of the other coin tosses. Let $E_1 = \{\text{heads on the first toss}\}$, $E_2 = \{\text{heads on the second toss}\}$, and $E_3 = \{\text{heads on the third toss}\}$. The events are independent, and the probability of flipping heads is $\frac{1}{2}$, so

$$P(E_1 \text{ and } E_2 \text{ and } E_3) = P(E_1) \cdot P(E_2) \cdot P(E_3) = \frac{1}{2} \cdot \frac{1}{2} \cdot \frac{1}{2} = \frac{1}{8}$$

Check your progress 4, *page 733*

Let D be the event that a person has the genetic defect, and let T be the event that the test for the defect is positive. We are asked for $P(D \mid T)$, which can be calculated by

$$P(D \mid T) = \frac{P(D \text{ and } T)}{P(T)}$$

A tree diagram will help us compute the needed probabilities.

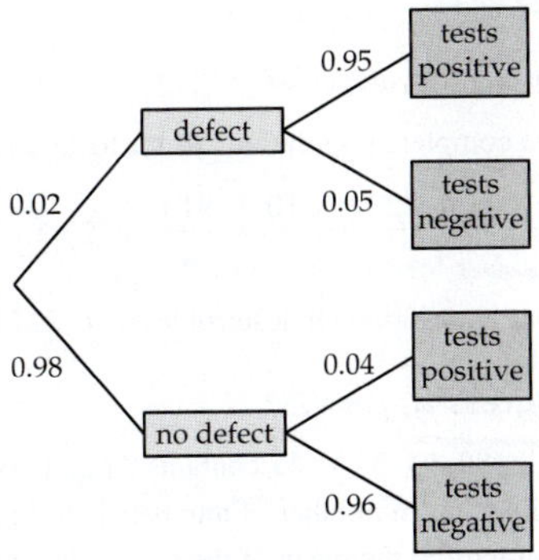

From the diagram, $P(D \text{ and } T) = P(D) \cdot P(T \mid D) = (0.02)(0.95)$. To compute $P(T)$, we need to combine two branches from the diagram, one corresponding to a correct positive test result when the person has the defect, and one corresponding to a false positive result when the person does not have the defect: $P(T) = (0.02)(0.95) + (0.98)(0.04)$. Then

$$P(D \mid T) = \frac{P(D \text{ and } T)}{P(T)} = \frac{(0.02)(0.95)}{(0.02)(0.95) + (0.98)(0.04)} \sim 0.326$$

There is only a 32.6% chance that a person who tests positive actually has the defect.

SECTION 12.6

Check your progress 1, *page 738*

Let S_1 be the event that the roulette ball lands on a number from 1 to 12, in which case the player wins \$10. There are 38 possible numbers, so $P(S_1) = \frac{12}{38}$. Let S_2 be the event that a number from 1 to 12 does not come up, in which case the player loses \$5. Then $P(S_2) = 1 - \frac{12}{38} = \frac{26}{38}$.

$$\text{Expectation} = P(S_1) \cdot S_1 + P(S_2) \cdot S_2$$
$$= \frac{12}{38}(10) + \frac{26}{38}(-5) = -\frac{5}{19} \approx -0.263$$

The player's expectation is about −\$0.263.

Check your progress 2, *page 739*

Let S_1 be the event that the person will die within 1 year. Then $P(S_1) = 0.000753$, and the company must pay out \$10,000. Because the company received a premium of \$45, the actual loss is \$9955. Let S_2 be the event that the policy holder does not die during the year of the policy. Then $P(S_2) = 0.999247$ and the company keeps the premium. The expectation is

$$\text{Expectation} = P(S_1) \cdot S_1 + P(S_2) \cdot S_2$$
$$= 0.000753(-9955) + 0.999247(45)$$
$$= 37.47$$

The company's expectation is \$37.47.

Check your progress 3, *page 740*

$$\text{Expectation} = 0.05(500{,}000) + 0.30(250{,}000) + 0.35(150{,}000)$$
$$+ 0.20(-100{,}000) + 0.10(-350{,}000)$$
$$= 97{,}500$$

The company's profit expectation is \$97,500.

CHAPTER 13

SECTION 13.1

Check your progress 1, *page 753*

The four tests are a complete population. Use μ to represent the mean.

$$\mu = \frac{\Sigma x}{n} = \frac{245 + 235 + 220 + 210}{4} = \frac{910}{4} = 227.5$$

The mean of the patient's blood cholesterol levels is 227.5.

Check your progress 2, *page 753*

a. The list 14, 27, 3, 82, 64, 34, 8, 51 contains 8 numbers. The median of a list of data with an even number of numbers is found by ranking the numbers and computing the mean of the two middle numbers. Ranking the numbers from smallest to largest gives 3, 8, 14, 27, 34, 51, 64, 82. The two middle numbers are 27 and 34. The mean of 27 and 34 is 30.5. Thus 30.5 is the median of the data.

b. The list 21.3, 37.4, 11.6, 82.5, 17.2 contains 5 numbers. The median of a list of data with an odd number of numbers is found by ranking the numbers and finding the middle number. Ranking the numbers from smallest to largest gives 11.6, 17.2, 21.3, 37.4, 82.5. The middle number is 21.3. Thus 21.3 is the median.

Check your progress 3, *page 754*

a. In the list 3, 3, 3, 3, 3, 4, 4, 5, 5, 5, 8, the number 3 occurs more often than the other numbers. Thus 3 is the mode.

b. In the list 12, 34, 12, 71, 48, 93, 71, the numbers 12 and 71 both occur twice and the other numbers occur only once. Thus 12 and 71 are both modes for the data.

Check your progress 4, *page 756*

The A is worth 4 points, with a weight of 4; the B in Statistics is worth 3 points, with a weight of 3; the C is worth 2 points, with a weight of 3; the F is worth 0 points, with a weight of 2; and the B in CAD is worth 3 points, with a weight of 2. The sum of all the weights is 4 + 3 + 3 + 2 + 2, or 14.

$$\text{Weighted mean} = \frac{(4 \times 4) + (3 \times 3) + (2 \times 3) + (0 \times 2) + (3 \times 2)}{14}$$
$$= \frac{37}{14} \approx 2.64$$

Janet's GPA for the spring semester is approximately 2.64.

Check your progress 5, *page 757*

$$\text{Mean} = \frac{\Sigma(x \cdot f)}{\Sigma f}$$
$$= \frac{(2 \cdot 5) + (3 \cdot 25) + (4 \cdot 10) + (5 \cdot 5)}{45}$$
$$= \frac{150}{45}$$
$$= 3\frac{1}{3}$$

The mean number of bedrooms per household for the homes in the subdivision is $3\frac{1}{3}$.

SECTION 13.2

Check your progress 1, *page 763*

The greatest number of ounces dispensed is 8.03, and the least is 7.95. The range of the number of ounces is 8.03 − 7.95 = 0.08 oz.

Check your progress 2, *page 764*

$$\mu = \frac{5 + 8 + 16 + 17 + 18 + 20}{6} = \frac{84}{6} = 14$$

x	$x - \mu$	$(x - \mu)^2$
5	$5 - 14 = -9$	$(-9)^2 = 81$
8	$8 - 14 = -6$	$(-6)^2 = 36$
16	$16 - 14 = 2$	$2^2 = 4$
17	$17 - 14 = 3$	$3^2 = 9$
18	$18 - 14 = 4$	$4^2 = 16$
20	$20 - 14 = 6$	$6^2 = 36$
		Sum: 182

$$\sigma = \sqrt{\frac{\Sigma(x - \mu)^2}{n}} = \sqrt{\frac{182}{6}} \approx \sqrt{30.33} \approx 5.51$$

The standard deviation for this population is approximately 5.51.

Check your progress 3, *page 765*

The mean for each sample of rope is 130 lb.

The rope from Trustworthy has a breaking point standard deviation of

$$s_1 = \sqrt{\frac{(122-130)^2 + (141-130)^2 + \cdots + (125-130)^2}{6}}$$

$$= \sqrt{\frac{1752}{6}} \approx 17.1 \text{ lb}$$

The rope from Brand X has a breaking point standard deviation of

$$s_2 = \sqrt{\frac{(128-130)^2 + (127-130)^2 + \cdots + (137-130)^2}{6}}$$

$$= \sqrt{\frac{3072}{6}} \approx 22.6 \text{ lb}$$

The rope from NeverSnap has a breaking point standard deviation of

$$s_3 = \sqrt{\frac{(112-130)^2 + (121-130)^2 + \cdots + (135-130)^2}{6}}$$

$$= \sqrt{\frac{592}{6}} \approx 9.9 \text{ lb}$$

The breaking point values of the rope from NeverSnap have the lowest standard deviation.

Check your progress 4, *page 766*

The mean is approximately 46.092.

The population standard deviation is approximately 2.476.

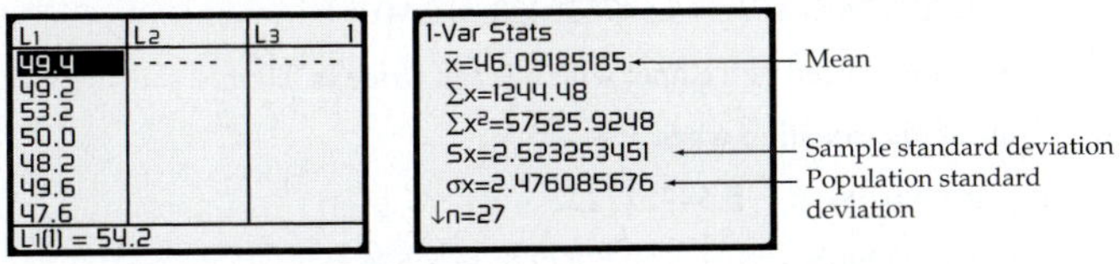

Check your progress 5, *page 767*

In Check Your Progress 2, we found $\sigma \approx \sqrt{30.33}$. Variance is the square of the standard deviation. Thus the variance is $\sigma^2 \approx (\sqrt{30.33})^2 = 30.33$.

SECTION 13.3

Check your progress 1, *page 772*

$$z_{15} = \frac{15-12}{2.4} = 1.25 \qquad z_{14} = \frac{14-11}{2.0} = 1.5$$

These z-scores indicate that in comparison to her classmates, Cheryl did better on the second quiz than she did on the first quiz.

Check your progress 2, *page 772*

$$z_x = \frac{x-\mu}{\sigma}$$

$$0.6 = \frac{70-65.5}{\sigma}$$

$$\sigma = \frac{4.5}{0.6} = 7.5$$

The standard deviation for this set of test scores is 7.5.

Check your progress 3, *page 773*

a. By definition, the median is the 50th percentile. Therefore, 50% of the police dispatchers earned less than \$44,528 per year.

b. Because \$32,761 is in the 25th percentile, 100% − 25% = 75% of all police dispatchers made more than \$32,761.

c. From parts a and b, 50% − 25% = 25% of the police dispatchers earned between \$32,761 and \$44,528.

Check your progress 4, *page 773*

$$\text{Percentile} = \frac{\text{number of data values less than 405}}{\text{total number of data values}} \cdot 100$$

$$= \frac{3952}{8600} \cdot 100$$

$$\approx 46$$

Hal's score of 405 places him at the 46th percentile.

Check your progress 5, *page 775*

Rank the data.

7.5 9.8 10.2 10.8 11.4 11.4 12.2 12.4 12.6 12.8 13.1 14.2 14.5 15.6 16.4

The median of these 15 data values has a rank of 8. Thus the median is 12.4. The second quartile, Q_2, is the median of the data, so $Q_2 = 12.4$.

The first quartile is the median of the seven values less than Q_2. Thus Q_1 has a rank of 4, so $Q_1 = 10.8$.

The third quartile is the median of the values greater than Q_2. Thus Q_3 has a rank of 12, so $Q_3 = 14.2$.

Check your progress 6, *page 776*

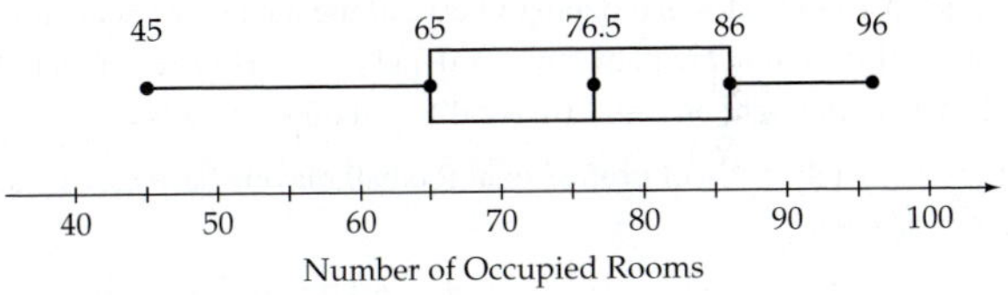

SECTION 13.4

Check your progress 1, *page 783*

a. The percent of data in all classes with an upper bound of 25 s or less is the sum of the percents for the first five classes in Table 13.8. Thus the percent of subscribers who required less than 25 s to download the file is 30.9%.

b. The percent of data in all the classes with a lower bound of at least 10 s and an upper bound of 30 s or less is the sum of the percents in the third through sixth classes in Table 13.8. Thus the percent of subscribers who required from 10 s to 30 s to download the file is 47.8%. The probability that a subscriber chosen at random will require from 10 s to 30 s to download the file is 0.478.

Check your progress 2, *page 785*

a. 0.76 lb is 1 standard deviation above the mean of 0.61 lb. In a normal distribution, 34% of all data lie between the mean and 1 standard deviation above the mean, and 50% of all data lie below the mean. Thus 34% + 50% = 84% of the tomatoes weigh less than 0.76 lb.

b. 0.31 lb is 2 standard deviations below the mean of 0.61 lb. In a normal distribution, 47.5% of all data lie between the mean and 2 standard deviations below the mean, and 50% of all data lie above the mean. This gives a total of 47.5% + 50% = 97.5% of the tomatoes that weigh more than 0.31 lb. Therefore

$$(97.5\%)(6000) = (0.975)(6000) = 5850$$

of the tomatoes can be expected to weigh more than 0.31 lb.

c. 0.31 lb is 2 standard deviations below the mean of 0.61 lb and 0.91 lb is 2 standard deviations above the mean of 0.61 lb. In a normal distribution, 95% of all data lie within 2 standard deviations of the mean. Therefore

$$(95\%)(4500) = (0.95)(4500) = 4275$$

of the tomatoes can be expected to weigh from 0.31 lb to 0.91 lb.

Check your progress 3, *page 788*

The area of the standard normal distribution between $z = -0.67$ and $z = 0$ is equal to the area between $z = 0$ and $z = 0.67$. The entry in Table 13.10 associated with $z = 0.67$ is 0.249. Thus the area of the standard normal distribution between $z = -0.67$ and $z = 0$ is 0.249 square unit.

Check your progress 4, *page 788*

Table 13.10 indicates that the area from $z = 0$ to $z = -1.47$ is 0.429 square unit. The area to the left of $z = 0$ is 0.500 square unit. Thus the area to the left of $z = -1.47$ is $0.500 - 0.429 = 0.071$ square unit.

Check your progress 5, *page 789*

Round z-scores to the nearest hundredth so you can use Table 13.10.

a. $z_9 = \dfrac{9 - 6.1}{1.8} \approx 1.61$

Table 13.10 indicates that 0.446 (44.6%) of the data in the standard normal distribution are between $z = 0$ and $z = 1.61$. The percent of the data to the right of $z = 1.61$ is 50% − 44.6% = 5.4%.

Approximately 5.4% of professional football players have careers of more than 9 years.

b. $z_3 = \dfrac{3 - 6.1}{1.8} \approx -1.72$ $\qquad z_4 = \dfrac{4 - 6.1}{1.8} \approx -1.17$

From Table 13.10:

$A_{1.72} = 0.457$ $\qquad A_{1.17} = 0.379$

$0.457 - 0.379 = 0.078$

The probability that a professional football player chosen at random will have a career of between 3 and 4 years is about 0.078.

SECTION 13.5

Check your progress 1, *page 798*

Here are the results using a TI-84 calculator.

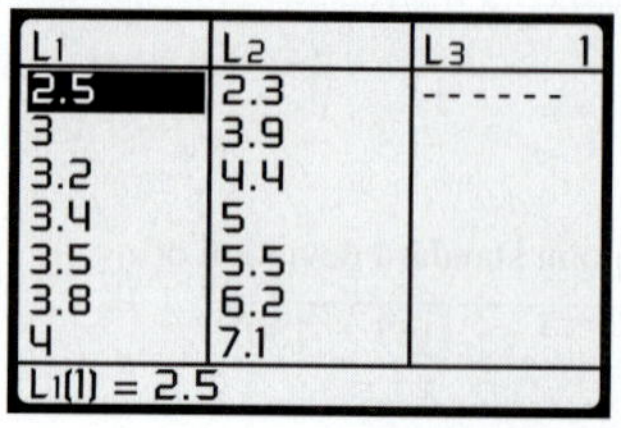

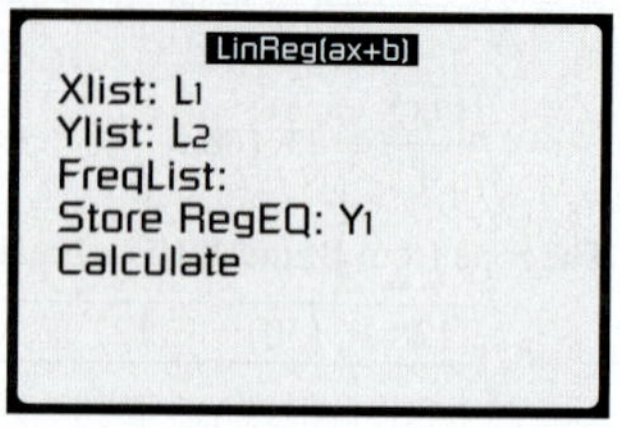

The regression equation is $\hat{y} \approx 3.12962963x - 5.547222222$.

Check your progress 2, *page 799*

Use the regression equation from Check Your Progress 1.

a. Evaluate the equation when $x = 2.7$ m.

$$\hat{y} \approx 3.12962963x - 5.547222222$$
$$\approx 3.12962963(2.7) - 5.547222222 \approx 2.903$$

The average speed of a camel with a 2.7 m stride is 2.9 m/s.

b. Evaluate the equation when $x = 4.5$ m.

$$\hat{y} \approx 3.12962963x - 5.547222222$$
$$\approx 3.12962963(4.5) - 5.547222222 \approx 8.536$$

The average speed of a camel with a 4.5 m stride is 8.5 m/s.

Check your progress 3, *page 800*

The TI-84 screenshot from Check Your Progress 1 is repeated here.

Rounded to the nearest hundredth, the linear correlation coefficient is 1.00.

Answers to All Exercises

CHAPTER 1

EXCURSION EXERCISES, SECTION 1.1 *page 10*

1.

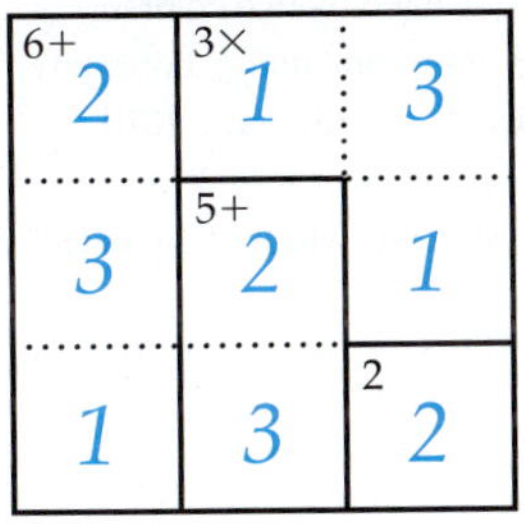

2.

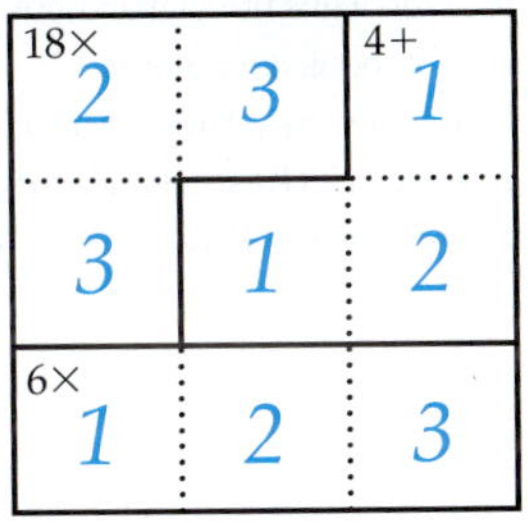

3.

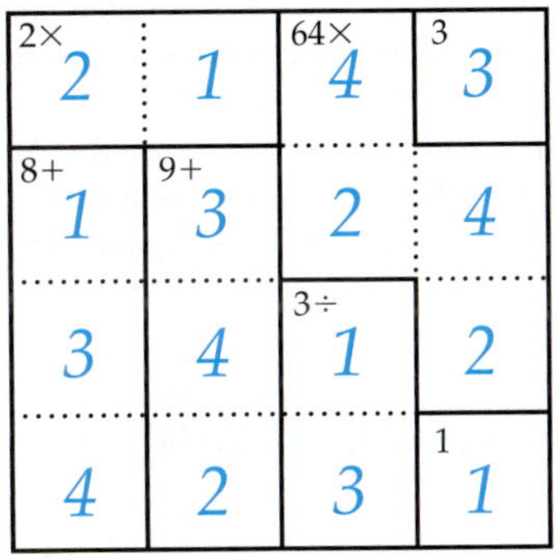

4.

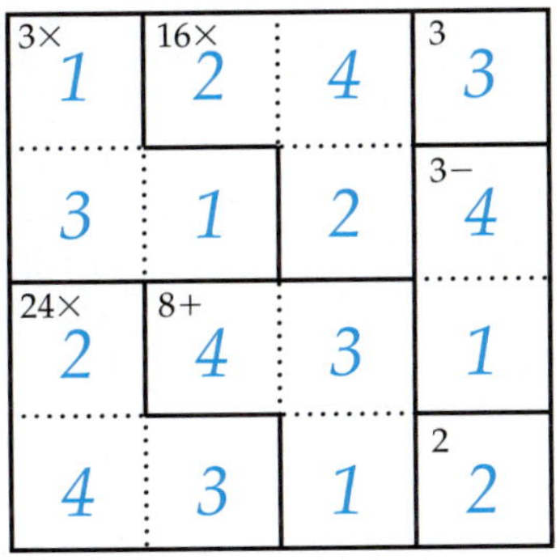

5.

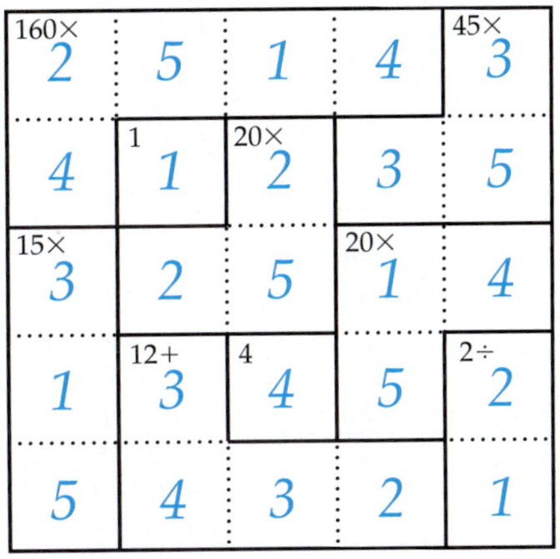

6.

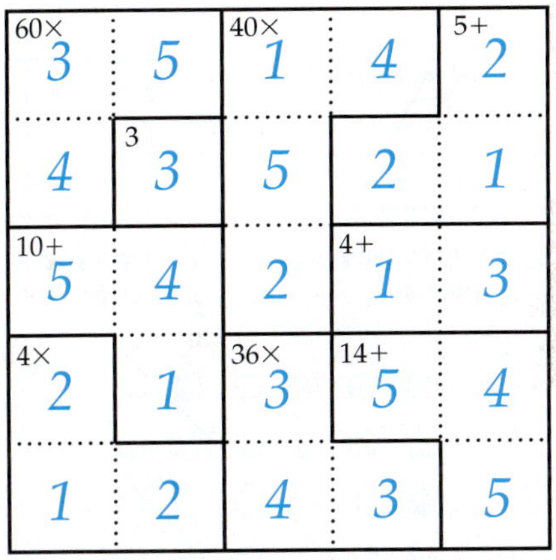

EXERCISE SET 1.1 *page 11*

1. 28 **2.** 41 **3.** 45 **4.** 216 **5.** 64 **6.** 35 **7.** $\frac{15}{17}$ **8.** $\frac{7}{8}$ **9.** −13 **10.** 51 **11.** correct
12. correct **13.** correct **14.** incorrect **15.** incorrect **16.** correct **17. a.** 8 cm **b.** 24 cm **c.** 40 cm
d. 56 cm **e.** 72 cm **18. a.** 6.5 cm **b.** 19.5 cm **c.** 32.5 cm **d.** 45.5 cm **e.** 58.5 cm **19. a.** 3 units
b. 5 units **c.** 7 units **d.** 9 units **20. a.** 3 units **b.** 5 units **c.** 7 units **d.** 9 units
21. The distance is quadrupled. **22.** The distance is multiplied by a factor of 9. **23.** 288 cm **24.** 18 cm **25.** inductive
26. deductive **27.** deductive **28.** deductive **29.** deductive **30.** inductive **31.** inductive **32.** inductive

In Exercises 33 to 38, only one possible answer is given. Your answers may vary from the given answers.

33. $x = \frac{1}{2}$ **34.** $x = -3$ **35.** $x = \frac{1}{2}$ **36.** $x = -2$ **37.** $x = -3$ **38.** $x = 1$

39. Consider 1 and 3. $1 + 3 = 4$, which is an even counting number, however, $1 \cdot 3 = 3$, which is an odd counting number.

40. Consider 2 and 3. $2 \cdot 3 = 6$, which is an even counting number; however, 3 is an odd counting number.

41.

16	3	2	13
5	10	11	8
9	6	7	12
4	15	14	1

42.

11	24	7	20	3
4	12	25	8	16
17	5	13	21	9
10	18	1	14	22
23	6	19	2	15

43. n

$6n + 8$

$\frac{6n + 8}{2} = 3n + 4$

$3n + 4 - 2n = n + 4$

$n + 4 - 4 = n$

44. n

$n + 4$

$3(n + 4) = 3n + 12$

$3n + 12 - 7 = 3n + 5$

$3n + 5 - 3n = 5$

45. Maria: the utility stock; Jose: the automotive stock; Anita: the technology stock; Tony: the oil stock **46.** Changs: entrée; Steinbergs: salad; Ontkeans: dessert; Gonzaleses: soup **47.** Atlanta: stamps; Chicago: baseball cards; Philadelphia: coins; San Diego: comic books **48. a.** Yes. Change the color of Iowa to yellow and Oklahoma to blue. **b.** No. One possible explanation: Oklahoma, Arkansas, and Louisiana must each have a color different from the color of Texas, and they cannot all be the same color. Thus the map cannot be colored using only two colors. **49.** Home, bookstore, supermarket, credit union, home; or home, credit union, supermarket, bookstore, home **50.** Home, credit union, bookstore, supermarket, home **51.** N, because the first letter of Nine is N. **52.** A 6 preceded by a backwards 6, that is, ∂6. **53. a.** 1010 is a multiple of 101 because $10 \times 101 = 1010$, but 11×1010 equals 11,110, and the digits of 11,110 are not all the same. **b.** For $n = 11$, $n^2 - n + 11 = (11)^2 - 11 + 11 = 11^2$, which is not a prime number. **54. a.** Answers will vary. **b.** Most students will find, by inductive reasoning, that the best strategy for winning the grand prize is by switching.

EXCURSION EXERCISES, SECTION 1.2 *page 22*

1.

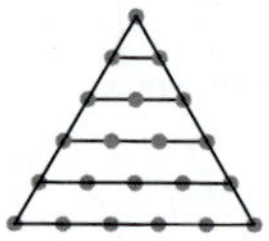

The sixth triangular number is 21.

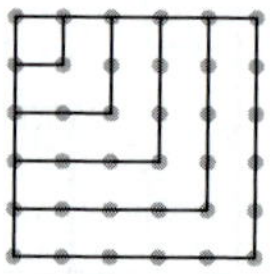

The sixth square number is 36.

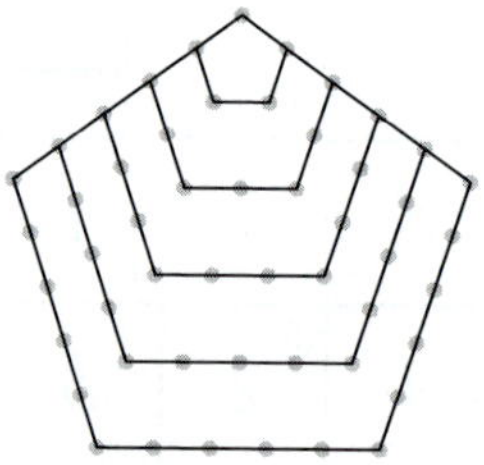

The sixth pentagonal number is 51.

2. a.

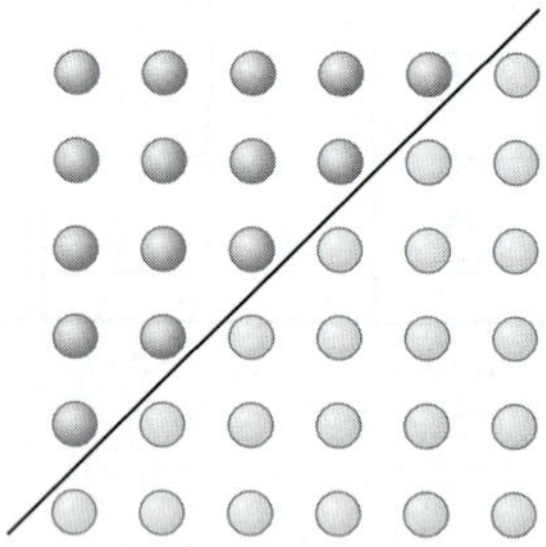

The fifth triangular number is 15. The sixth triangular number is 21. $15 + 21 = 36$, which is the sixth square number.

b. $1275 + 1326 = 2601 = 51^2$

c. $\frac{n(n+1)}{2} + \frac{(n+1)(n+2)}{2} = \frac{2n^2 + 4n + 2}{2}$

$= n^2 + 2n + 1$

$= (n + 1)^2$

3.

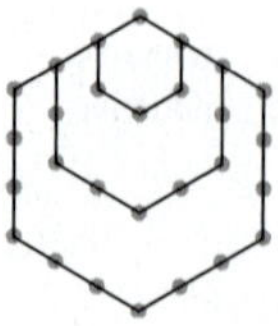

The fourth hexagonal number is 28.

EXERCISE SET 1.2 page 23

1. 97 **2.** 40 **3.** 329 **4.** 280 **5.** 159 **6.** 305 **7.** $\frac{3}{2}, 5, \frac{21}{2}, 18, \frac{55}{2}$ **8.** $\frac{1}{2}, \frac{2}{3}, \frac{3}{4}, \frac{4}{5}, \frac{5}{6}$
9. 2, 14, 36, 68, 110 **10.** 1, 12, 45, 112, 225 **11.** $a_n = n^2 + n - 1$ **12.** $a_n = 3n + 2$ **13.** $a_n = 2n$
14. $a_n = (n + 2)^2 - 1 = n^2 + 4n + 3$ **15. a.** There are 56 cannonballs in the sixth pyramid and 84 cannonballs in the seventh pyramid. **b.** The eighth pyramid has eight levels of cannonballs. The number of cannonballs in the nth level is given by the nth number in the sequence 1, 3, 6, 10, 15, 21, 28, 36. Thus the total number of cannonballs in the eighth pyramid is $1 + 3 + 6 + 10 + 15 + 21 + 28 + 36 = 120$. **16.** $Tet_{10} = 220$
17. a. Five cuts produce six pieces and six cuts produce seven pieces. **b.** $a_n = n + 1$ **18. a.** 29 pieces **b.** The nth pizza-slicing number is one more than the nth triangular number. **19. a.** 26 pieces **b.** 7 cuts **20. a.** valid **b.** not valid **c.** valid **d.** valid **21.** $a_3 = 7, a_4 = 9, a_5 = 11$ **22.** $a_3 = -1, a_4 = 2, a_5 = -3$ **23.** $F_{20} = 6765, F_{30} = 832{,}040, F_{40} = 102{,}334{,}155$ **24.** $F_{16} = 987, F_{21} = 10{,}946, F_{32} = 2{,}178{,}309$ **25.** n^2 **26. a.** $a_n = 2\left[\frac{n(n-1)(n-2)(n-3)(n-4)}{4 \cdot 3 \cdot 2 \cdot 1}\right] + 2n$
b. $a_n = 4\left[\frac{n(n-1)(n-2)(n-3)(n-4)}{4 \cdot 3 \cdot 2 \cdot 1}\right] + 2n$ **27. a.** $F_n + 2F_{n+1} + F_{n+2} = F_{n+4}$ **b.** $F_n + F_{n+1} + F_{n+3} = F_{n+4}$
28. a. $F_1 + F_2 + F_3 + F_4 + \cdots + F_n = F_{n+2} - 1$ **b.** $F_2 + F_4 + F_6 + \cdots + F_{2n} = F_{2n+1} - 1$ **29. a.** The sum of the numbers in row 1 is 2. The sum of the numbers in row 2 is 4. The sum of the numbers in row 3 is 8. The sum of the numbers in row 4 is 16. The sum of the numbers in row 5 is 32. It appears that the sum of the numbers in the nth row, where $n \geq 1$, is 2^n. If we use this pattern, then the predicted sum of the numbers in row 9 is $2^9 = 512$. **b.** The triangular numbers appear in the third diagonals. **30. a.** 31 cents **b.** 1023 cents = $10.23 **c.** 32,767 cents = $327.67 **d.** $2^n - 1$ cents **31. a.** 1 move **b.** 3 moves **c.** 7 moves **d.** 15 moves **e.** 31 moves **f.** $2^n - 1$ moves **g.** about 5.85×10^{11} years

32.
$$\begin{aligned} F_{n+1} &= F_n + F_{n-1} \\ &= (F_{n-1} + F_{n-2}) + F_{n-1} \\ &= 2F_{n-1} + F_{n-2} \\ &= 2F_{n-1} + 2F_{n-2} - F_{n-2} \\ &= 2(F_{n-1} + F_{n-2}) - F_{n-2} \\ &= 2F_n - F_{n-2} \end{aligned}$$

EXCURSION EXERCISES, SECTION 1.3 page 35

1. 1, 4, 6, 4, 1 **2.** A to G: 1 route; A to H: 8 routes; A to I: 28 routes; A to J: 56 routes; A to K: 70 routes **3.** The figure is symmetrical about a vertical line from A to K. Since J is the same distance to the left of the line AK as L is to the right of AK, the same number of routes lead from A to J as lead from A to L. **4.** More routes lead to the center bin than to any of the other bins. **5.** A to P: 1 route; A to Q: 9 routes; A to R: 36 routes; A to S: 84 routes; A to T: 126 routes; A to U: 126 routes

EXERCISE SET 1.3 page 36

1. 195 girls **2.** 12.5 feet and 19 feet **3.** 91 squares **4.** 9 **5.** $40 **6.** 132 games **7.** 18 direct routes **8. a.** 4 direct routes **b.** 3 direct routes **c.** 2 direct routes **9.** $2^{12} = 4096$ ways **10.** 8 leaps **11.** 28 handshakes **12.** 276 line segments **13.** 21 ducks, 14 pigs **14.** Carla arrives home first because she spends more time running than Allison. **15.** 12 ways **16.** 20 carpet squares **17.** 6 **18.** 2 **19.** 1 **20.** 9 **21. a.** 80,200 **b.** 151,525 **c.** 1892 **22.** One method is to apply the procedure by Gauss to find the sum of the numbers from 1 to 64 and then add 65 to this total.
23. a. 121, 484, and 676 **b.** 1331 **24.** 55 miles per hour **25.** $1\frac{1}{2}$ inches

26. Yes, it is possible, if you "think outside the box."

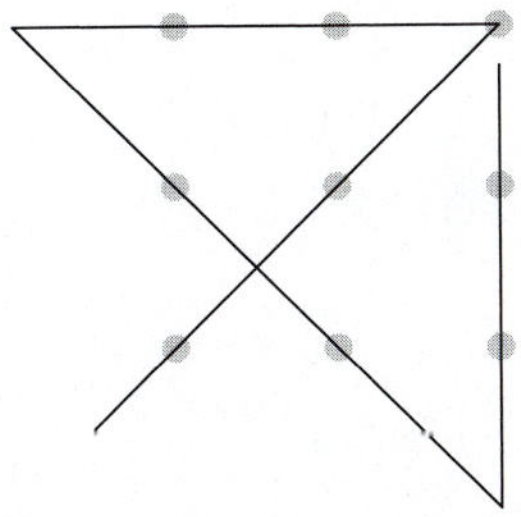

27. a. 1.4 billion admissions **b.** 2014 **c.** 2009 **28. a.** 2007 and 2008 **b.** 2013 **c.** 2008 and 2009 **29. a.** PG-13 **b.** $2.2 billion **30.** 17 votes **31.** 2601 tiles **32.** seven children **33.** Four more sisters than brothers **34.** 22 pennies **35.** the 11th day **36. a.** Let A, B, C, and D represent the four people of weights 80, 100, 150, and 170 pounds, respectively. A and B make the first trip across, then A comes back alone. C crosses the river and B comes back alone. C is now on the opposite bank. Repeat this procedure to get D to the opposite bank. Then one more trip will get both A and B to the opposite bank. **b.** The minimum number of crossings is nine. **37.** 91 **38.** Fill the 5-gallon jug. Pour from it into the 3-gallon jug until the 3-gallon jug is full. Now empty the 3-gallon jug. You now have 2 gallons in the 5-gallon jug. Pour these 2 gallons into the 3-gallon jug. Fill the 5-gallon jug again and pour from it into the 3-gallon jug. At this point there are already 2 gallons in the 3-gallon jug, so only 1 more gallon will fit. Thus, when the 3-gallon jug is full, the 5-gallon jug will contain exactly 4 gallons.

39. a. Place four coins on the left balance pan and the other four on the right balance pan. The pan that is the higher contains the fake coin. Take the four coins from the higher pan and use the balance scale to compare the weight of two of these coins to the weight of the other two. The pan that is the higher contains the fake coin. Take the two coins from the higher pan and use the balance scale to compare the weight of one of these coins to the weight of the other. The pan that is the higher contains the fake coin. This procedure enables you to determine the fake coin in three weighings. **b.** Place three of the coins on one of the balance pans and another three coins on the other. If the pans balance, then the fake coin is one of the two remaining coins. You can use the balance scale to determine which of the remaining coins is the fake coin because it will be lighter than the other coin. If the three coins on the left pan do not balance with the three coins on the right pan, then the fake coin must be one of the three coins on the higher pan. Pick any two coins from these three and place one on each balance pan. If these two coins do not balance, then the one that is the higher is the fake. If the coins balance, then the third coin (the one that you did not place on the balance pan) is the fake. In any case, this procedure enables you to determine the fake coin in two weighings. **40.** c. 21:00. 2 hours later (23:00, 1 hour before midnight) is half the time until midnight as 1 hour later (22:00, 2 hours before midnight). **41.** a. 1600. Sally likes perfect squares. **42.** b. No. The other 800 elephants can be any mix of all blue and pink and green stripes. **43.** d. 64. Each number is the cube of a term in the sequence 1, 2, 3, 4, 5, 6. **44. a.** $3^{18} = 387{,}420{,}489$ times as large **b.** 4^{240} times as large **45. a.** People born in 1980 will be 45 in 2025 ($2025 = 45^2$). **b.** 2070, because people born in 2070 will be 46 in 2116 ($2116 = 46^2$). **46.** Adding 83 is the same as adding 100 and subtracting 17. Thus after you add 83, you will have a number that has 1 as the hundreds digit. The number formed by the tens digit and the units digit will be 17 less than your original number. After you add the hundreds digit, 1, to the other two digits of this new number, you will have a number that is 16 less than your original number. If you subtract this number from your original number, you must get 16. **47.** 612 digits **48.**

6	2	3	4	5	1
5	1	4	3	2	6
3	6	2	5	1	4
4	5	1	6	3	2
1	3	6	2	4	5
2	4	5	1	6	3

49. Answers will vary. **50.** M = 1, S = 9, E = 5, N = 6, D = 7, O = 0, R = 8, Y = 2

CHAPTER 1 REVIEW EXERCISES *page 41*

1. deductive [Sec. 1.1] **2.** inductive [Sec. 1.1] **3.** inductive [Sec. 1.1] **4.** deductive [Sec. 1.1] **5.** $x = 0$ provides a counterexample because $0^4 = 0$ and 0 is not greater than 0. [Sec. 1.1] **6.** $x = 4$ provides a counterexample because $\frac{(4)^3 + 5(4) + 6}{6} = 15$, which is not an even counting number. [Sec. 1.1] **7.** $x = 1$ provides a counterexample because $[(1) + 4]^2 = 25$, but $(1)^2 + 16 = 17$. [Sec. 1.1] **8.** Let $a = 1$ and $b = 1$. Then $(a + b)^3 = (1 + 1)^3 = 2^3 = 8$. However, $a^3 + b^3 = 1^3 + 1^3 = 2$. [Sec. 1.1] **9. a.** 112 **b.** 479 [Sec. 1.2] **10. a.** −72 **b.** −768 [Sec. 1.2] **11.** $a_1 = 1$, $a_2 = 12$, $a_3 = 31$, $a_4 = 58$, $a_5 = 93$, $a_{20} = 1578$ [Sec. 1.2] **12.** $a_{11} = 89$, $a_{12} = 144$ [Sec. 1.2] **13.** $a_n = 3n$ [Sec. 1.2] **14.** $a_n = n^2 + 3n + 4$ [Sec. 1.2] **15.** $a_n = n^2 + 3n + 2$ [Sec. 1.2] **16.** $a_n = 5n - 1$ [Sec. 1.2] **17.** 320 feet by 1600 feet [Sec. 1.3] **18.** $3^{15} = 14{,}348{,}907$ ways [Sec. 1.3] **19.** 48 skyboxes [Sec. 1.3] **20.** On the first trip, the rancher takes the rabbit across the river. The rancher returns alone. The rancher takes the dog across the river and returns with the rabbit. The rancher next takes the carrots across the river and returns alone. On the final trip, the rancher takes the rabbit across the river. [Sec. 1.3] **21.** $300 [Sec. 1.3] **22.** 105 handshakes [Sec. 1.3] **23.** Answers will vary. [Sec. 1.3] **24.** Answers will vary. [Sec. 1.3] **25.** Michael: biology major; Clarissa: business major; Reggie: computer science major; Ellen: chemistry major [Sec. 1.1] **26.** Dodgers: drugstore; Pirates: supermarket; Tigers: bank; Giants: service station [Sec. 1.1] **27. a.** Yes. Answers will vary. **b.** No. The countries of India, Bangladesh, and Myanmar all share borders with each of the other two countries. Thus at least three colors are needed to color the map. [Sec. 1.1] **28. a.** The following figure shows a route that starts from North Bay and passes over each bridge once and only once. **b.** No. [Sec. 1.3]

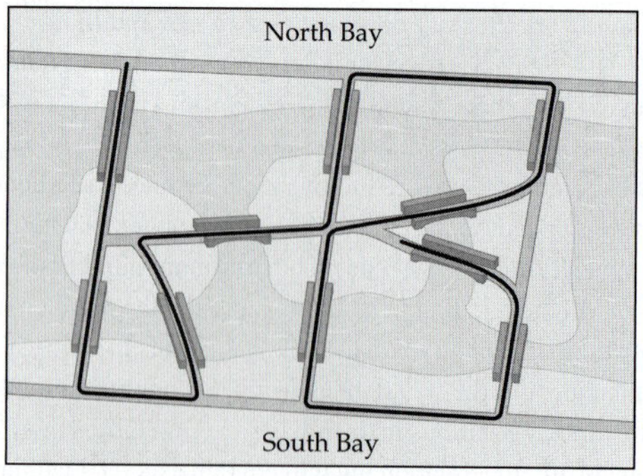

29. 1 square inch; 4 square inches; 25 square inches [Sec. 1.3] **30. a.** $2^{10} = 1024$ **b.** $2^{30} = 1{,}073{,}741{,}824$ [Sec. 1.2] **31.** A represents 1, B represents 9, and D represents 0. [Sec. 1.3] **32.** 5 ways [Sec. 1.3] **33.** 10 different orders [Sec. 1.3] **34.** 1 [Sec. 1.3] **35.** 3 [Sec. 1.3]

36. n

$4n$

$4n + 12$

$\frac{4n + 12}{2} = 2n + 6$

$2n + 6 - 6 = 2n$ [Sec. 1.1]

37. Each nickel is worth 5 cents. Thus 2004 nickels are worth $2004 \times 5 = 10{,}020$ cents, or \$100.20. [Sec. 1.3]

38. a. \$3.64 per gallon in 2012 **b.** 2010 to 2011 [Sec. 1.3] **39. a.** \$3.8 million **b.** 4 cents per viewer [Sec. 1.3] **40. a.** 11.9 billion searches **b.** 7.1 [Sec. 1.3] **41.** 5005 [Sec. 1.3] **42.** There are no narcissistic numbers. [Sec. 1.3] **43. a.** 10 intersections **b.** Yes. [Sec. 1.2] **44.** 30 points [Sec. 1.2] **45. a.** 22 **b.** No. $9^9 = 387{,}420{,}489$. Thus $9^{(9^9)}$ is the product of 387,420,489 nines. At one multiplication per second, this computation would take about 12.3 years. [Sec. 1.1/1.3]

CHAPTER 1 TEST *page 45*

1. inductive [Sec. 1.1, Example 6] **2.** deductive [Sec. 1.1, Check Your Progress 6] **3.** 384 [Sec. 1.2, Example 1] **4.** 1, 1, 2, 3, 5, 8, 13, 21, 34, 55 [Sec. 1.2, Example 3] **5. a.** $a_n = 4n$ **b.** $a_n = 3n + 1$ [Sec. 1.2, Example 2a] **6.** $a_1 = 0, a_2 = 1, a_3 = -3, a_4 = 6, a_5 = -10$, and $a_{105} = -5460$ [Sec. 1.2, Example 2b] **7.** $a_3 = 17, a_4 = 41, a_5 = 99$ [Sec. 1.2, Example 2a] **8. a.** 14 diagonals **b.** 20 diagonals [Sec. 1.2, Check Your Progress 1] **9.** Understand the problem. Devise a plan. Carry out the plan. Review the solution. [Sec. 1.3, Example 1] **10.** 6 ways [Sec. 1.3, Example 2] **11.** 15 ways [Sec. 1.3, Example 2] **12.** 3 [Sec. 1.3, Check Your Progress 4] **13.** \$672 [Sec. 1.3, Example 5] **14.** 126 routes [Sec. 1.3, Example 1] **15.** 36 games [Sec. 1.3, Example 3] **16.** Reynaldo is 13, Ramiro is 5, Shakira is 15, and Sasha is 7. [Sec. 1.1, Example 7] **17.** $x = 4$ provides a counterexample because division by zero is undefined. [Sec. 1.1, Example 4] **18.** $\frac{1}{2}$ provides a counterexample because $\frac{1}{2} > \left(\frac{1}{2}\right)^2$. [Sec. 1.1, Example 4] **19.** 125,250 [Sec. 1.2, Math Matters] **20. a.** 2009 **b.** 17,000 motor vehicle thefts **c.** 2009 to 2010 [Sec. 1.3, Example 8a]

CHAPTER 2

EXCURSION EXERCISES, SECTION 2.1 *page 53*

1. a. Erica **b.** Larry **c.** Answers will vary. **2. a.** 0 **b.** 0.75 **c.** 1 **d.** 30 **e.** (40, 0.5) **3. a.** 0 **b.** 0.5 **c.** 0 **d.** (3.5, 0.5) and (4.5, 0.5) **4. a.** 0.5 **b.** 1 **c.** (40, 0.5) and (60, 0.5) **5.** Answers will vary.

EXERCISE SET 2.1 *page 55*

1. {penny, nickel, dime, quarter} **2.** {January, February, May, July} **3.** {Mercury, Mars} **4.** {Bashful, Dopey, Doc, Grumpy, Happy, Sleepy, Sneezy} **5.** The answer in the year 2016: {George W. Bush, Barack Obama} **6.** {April, June, September, November} **7.** $\{-5, -4, -3, -2, -1\}$ **8.** $\{0, 1, 2, 3, 4, 5, 6, 7\}$ **9.** $\{7\}$ **10.** $\{-5\}$ **11.** $\{\ \}$ **12.** $\{0, 1, 2, 3, 4\}$ **13.** {a, e, i, o, u} **14.** $\{1, 2, 3, 4\}$

In Exercises 15 to 24, only one possible answer is given.

15. The set of days of the week that begin with the letter T **16.** The set of the signs of the zodiac that begin with the letter L **17.** The set consisting of the two planets in our solar system that are closest to the sun **18.** The set of U.S. coins with a value less than 25¢ **19.** The set of single-digit natural numbers **20.** The set of even natural numbers less than 9 **21.** The set of natural numbers less than or equal to 7 **22.** The set of whole numbers less than 5 **23.** The set of odd natural numbers less than 10 **24.** The set of negative integers greater than -5 **25.** True **26.** True **27.** False; $b \in \{a, b, c\}$, but $\{b\}$ is not an element of $\{a, b, c\}$. **28.** True **29.** False; $\{0\}$ has one element, whereas $\emptyset$ has no elements. **30.** False; the word "large" is a relative term. **31.** False; the word "good" is subjective. **32.** True **33.** True **34.** False; the empty set has no elements. **35.** True **36.** True

In Exercises 37 to 48, only one possible answer is given.

37. $\{x \mid x \in N \text{ and } x < 13\}$ **38.** $\{x \mid x$ is a natural number that ends with a 5, and x is greater than 40 but less than 80$\}$ **39.** $\{x \mid x$ is a multiple of 5 and $4 < x < 16\}$ **40.** $\{x \mid x$ is a positive square number less than or equal to 81$\}$ **41.** $\{x \mid x$ is the name of a month that has 31 days$\}$ **42.** $\{x \mid x$ is the name of a U.S. state with exactly four letters$\}$ **43.** $\{x \mid x$ is the name of a U.S. state that begins with the letter A$\}$ **44.** $\{x \mid x$ is a country that shares a boundary with the United States$\}$ **45.** $\{x \mid x$ is a season that starts with the letter s$\}$ **46.** $\{x \mid x \in N \text{ and } 1900 \le x \le 1999\}$ **47.** {February, April, June, September, November}

48. $\{x|x \in N\}$ **49. a.** {2013, 2014, 2015} **b.** {2008, 2009} **c.** {2010, 2011, 2012} **50. a.** {2011, 2012} **b.** {2010, 2013, 2014} **c.** {2007, 2008, 2009} **51. a.** {May, June, July, August} **b.** {March, April, September} **c.** {January, November} **52. a.** {2010, 2014, 2015} **b.** {2015} **c.** {2006} **53.** 11 **54.** 8 **55.** 0 **56.** 50 **57.** 4 **58.** 13 **59.** 16 **60.** 32 **61.** 121 **62.** 101 **63.** Neither **64.** Neither **65.** Both **66.** Neither **67.** Equivalent **68.** Equivalent **69.** Equivalent **70.** Neither **71.** Not well defined **72.** Well defined **73.** Not well defined **74.** Well defined **75.** Well defined **76.** Well defined **77.** Well defined **78.** Well defined **79.** Not well defined **80.** Not well defined **81.** Not well defined **82.** Not well defined **83.** {2, 11, 26, 47} **84.** {1, 6, 15, 28, 45, 66} **85.** {8, −27, 64, −125, 216, −343, ...} **86.** $\left\{-2, 1, -\frac{2}{3}, \frac{1}{2}, -\frac{2}{5}, \frac{1}{3}\right\}$ **87.** $A = B$ **88.** $A = B$ **89.** $A \neq B$ **90.** $A = B$

EXCURSION EXERCISES, SECTION 2.2 *page 63*

1. Yes, because the membership value of each element of *J* is less than or equal to its membership value in set *K*.
2. Yes, because the graph of the fuzzy set *ADOLESCENT* is always below or at the same height as the graph of the fuzzy set *YOUNG*.
3. $G' = \{(\text{A}, 0), (\text{B}, 0.3), (\text{C}, 0.6), (\text{D}, 0.9), (\text{F}, 1)\}$ **4.** $C' = \{(\text{Ferrari}, 0.1), (\text{Ford Mustang}, 0.4), (\text{Dodge Neon}, 0.5), (\text{Hummer}, 0.3)\}$
5.

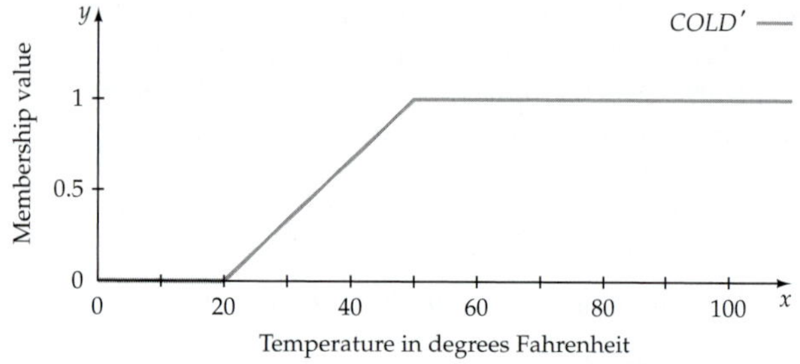

EXERCISE SET 2.2 *page 65*

1. {0, 1, 3, 5, 8} **2.** {0, 1, 2, 4, 5, 7, 8} **3.** $U = \{0, 1, 2, 3, 4, 5, 6, 7, 8\}$ **4.** {0, 1, 2, 3} **5.** {4, 5, 6, 7, 8} **6.** {0, 5, 6, 7, 8} **7.** {0, 2, 4, 6, 8} **8.** {0, 1, 3, 5, 6, 7, 8} **9.** {−3, −1, 1, 3, 5} **10.** {−2, −1, 0, 2, 3, 7} **11.** {−3, 3, 4, 5, 6, 7} **12.** {−3, −2, −1, 5, 6, 7} **13.** $\subseteq$ **14.** $\not\subseteq$ **15.** $\not\subseteq$ **16.** $\subseteq$ **17.** $\subseteq$ **18.** $\not\subseteq$ **19.** $\subseteq$ **20.** $\not\subseteq$ **21.** $\subseteq$ **22.** $\not\subseteq$ **23.** True **24.** False **25.** True **26.** False **27.** True **28.** False **29.** False **30.** False **31.** True **32.** False **33.** True **34.** False **35.** False **36.** True **37.** False **38.** True **39.** False **40.** False **41.** 18 hours **42.** 136 years **43.** $\emptyset, \{\alpha\}, \{\beta\}, \{\alpha, \beta\}$
44. $\emptyset, \{\alpha\}, \{\beta\}, \{\Gamma\}, \{\Delta\}, \{\alpha, \beta\}, \{\alpha, \Gamma\}, \{\alpha, \Delta\}, \{\beta, \Gamma\}, \{\beta, \Delta\}, \{\Gamma, \Delta\}, \{\alpha, \beta, \Gamma\}, \{\alpha, \beta, \Delta\}, \{\alpha, \Gamma, \Delta\}, \{\beta, \Gamma, \Delta\}, \{\alpha, \beta, \Gamma, \Delta\}$
45. $\emptyset$, {I}, {II}, {III}, {I, II}, {I, III}, {II, III}, {I, II, III} **46.** $\emptyset$ **47.** 4 subsets **48.** 8 subsets **49.** 128 subsets **50.** 64 subsets **51.** 2048 subsets **52.** 67,108,864 subsets **53.** 1 subset **54.** 512 subsets **55. a.** 15 different sets **b.** 10 different sums **c.** Two different sets of coins have the same value.
56. a. $2^{18} = 262{,}144$
$2^{19} = 524{,}288$
$2^{20} = 1{,}048{,}576$
b. Answers will vary. 2^{33} for the TI-83/84 calculators.
57. a. 256 sandwiches **b.** 11 condiments **58.** 8 **59. a.** 1024 omelets **b.** 12 ingredients **60. a.** 4096 different trucks **b.** 14 upgrade options **61. a.** {1, 2, 3} has only three elements, namely 1, 2, and 3. Because {2} is not equal to 1, 2, or 3, $\{2\} \notin \{1, 2, 3\}$. **b.** 1 is not a set, so it cannot be a subset. **c.** The given set has the elements 1 and {1}. Because $1 \neq \{1\}$, there are exactly two elements in {1, {1}}.
62. a. 10 elements **b.** 8 elements **c.** Yes. The empty set has exactly one subset. **63. a.** {A, B, C}, {A, B, D}, {A, B, E}, {A, C, D}, {A, C, E}, {A, D, E}, {B, C, D}, {B, C, E}, {B, D, E}, {C, D, E}, {A, B, C, D}, {A, B, C, E}, {A, B, D, E}, {A, C, D, E}, {B, C, D, E}, {A, B, C, D, E} **b.** {A}, {B}, {C}, {D}, {E}, {A, B}, {A, C}, {A, D}, {A, E}, {B, C}, {B, D}, {B, E}, {C, D}, {C, E}, {D, E} **64. a.** 1, 5, 10, 10, 5, 1 **b.** Row 6: 1, 6, 15, 20, 15, 6, 1. The 20 subsets of {a, b, c, d, e, f} that have exactly three elements are {a, b, c}, {a, b, d}, {a, b, e}, {a, b, f}, {a, c, d}, {a, c, e}, {a, c, f}, {a, d, e}, {a, d, f}, {a, e, f}, {b, c, d}, {b, c, e}, {b, c, f}, {b, d, e}, {b, d, f}, {b, e, f}, {c, d, e}, {c, d, f}, {c, e, f}, {d, e, f}

EXCURSION EXERCISES, SECTION 2.3 *page 76*

1. $M \cup J = \{(\text{A}, 1), (\text{B}, 0.8), (\text{C}, 0.6), (\text{D}, 0.5), (\text{F}, 0)\}$ **2.** $M \cap J = \{(\text{A}, 1), (\text{B}, 0.75), (\text{C}, 0.5), (\text{D}, 0.1), (\text{F}, 0)\}$
3. $E \cup J' = \{(\text{A}, 1), (\text{B}, 0.2), (\text{C}, 0.4), (\text{D}, 0.9), (\text{F}, 1)\}$ **4.** $J \cap L' = \{(\text{A}, 0), (\text{B}, 0), (\text{C}, 0), (\text{D}, 0), (\text{F}, 0)\}$
5. $J \cap (M' \cup L') = \{(\text{A}, 0), (\text{B}, 0.25), (\text{C}, 0.5), (\text{D}, 0.1), (\text{F}, 0)\}$

6.
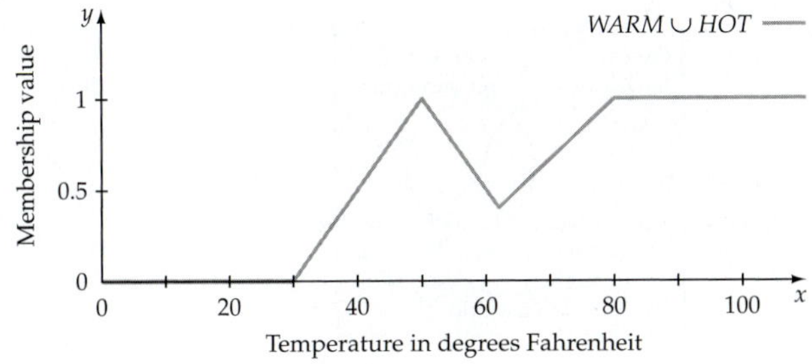

7. Yes.

EXERCISE SET 2.3 *page 77*

1. {1, 2, 4, 5, 6, 8} **2.** {2} **3.** {4, 6} **4.** {2, 5, 8} **5.** {3, 7} **6.** {2, 3, 4, 6, 7} **7.** $U = \{1, 2, 3, 4, 5, 6, 7, 8\}$ **8.** {2} **9.** $\varnothing$ **10.** $A' = \{1, 3, 5, 7, 8\}$ **11.** $B = \{1, 2, 5, 8\}$ **12.** $\varnothing$ **13.** $U = \{1, 2, 3, 4, 5, 6, 7, 8\}$ **14.** {1, 3, 4, 5, 6, 7, 8} **15.** {2, 5, 8} **16.** $U = \{1, 2, 3, 4, 5, 6, 7, 8\}$ **17.** {1, 3, 4, 6, 7} **18.** $\varnothing$ **19.** {2, 5, 8} **20.** {1, 3, 4, 5, 6, 7, 8}

In Exercises 21 to 28, one possible answer is given. Your answers may vary from the given answers.

21. The set of all elements that are not in *L* or are in *T* **22.** The set of all elements that are in *K* and are not in *J* **23.** The set of all elements that are in *A*, or are in *C* but not in *B* **24.** The set of all elements that are in either *A* or *B*, and are not in *C* **25.** The set of all elements that are in *T*, and are also in *J* or not in *K* **26.** The set of all elements that are in both *A* and *B*, or are in *C* **27.** The set of all elements that are in both *W* and *V*, or are in both *W* and *Z* **28.** The set of all elements that are in *D*, but are not in either *E* or *F*

29.
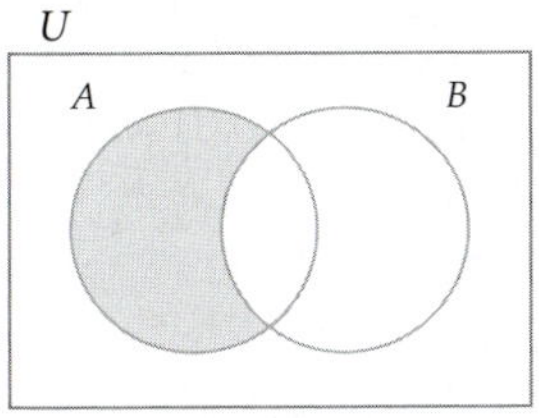

30.
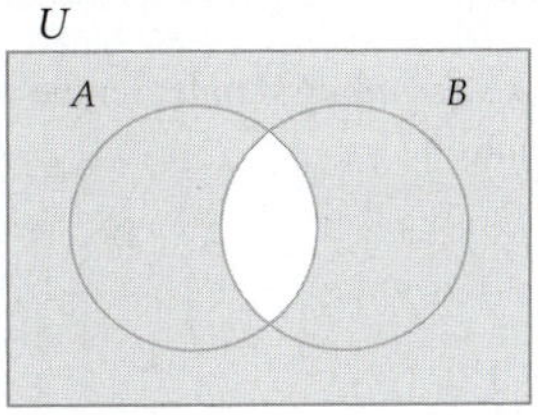

31.
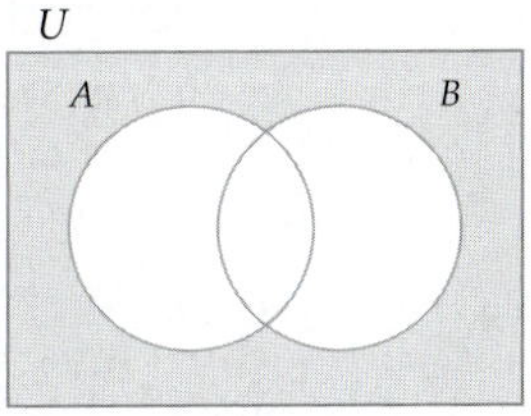

32.
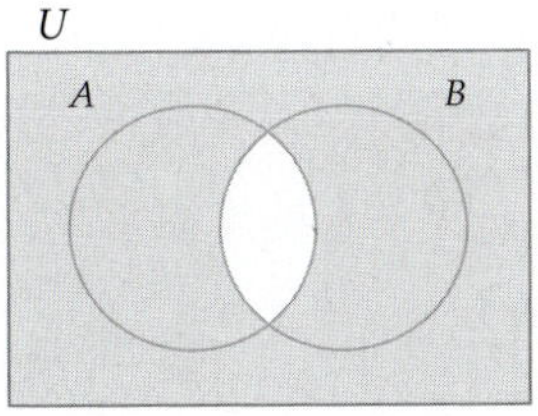

33.
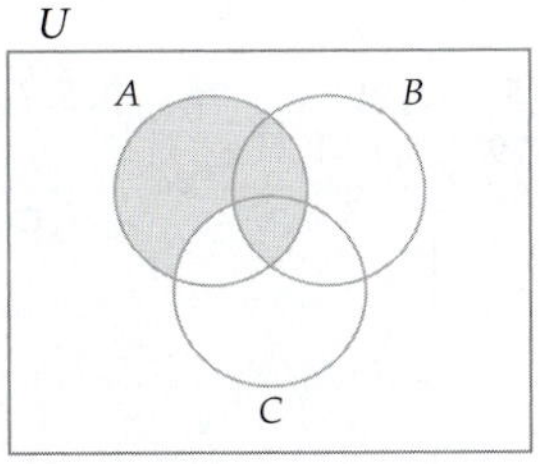

34.
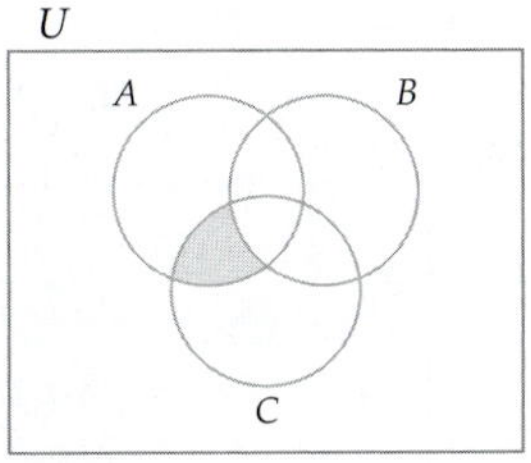

35.
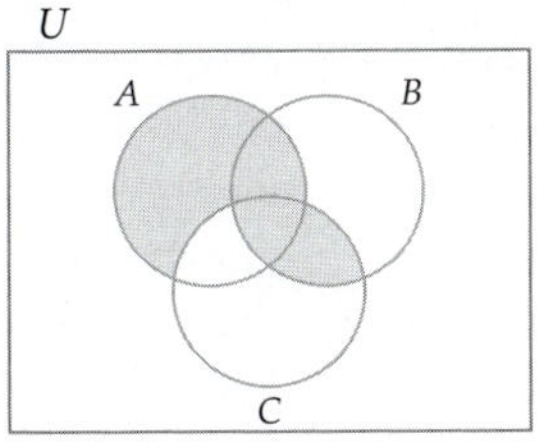

36.
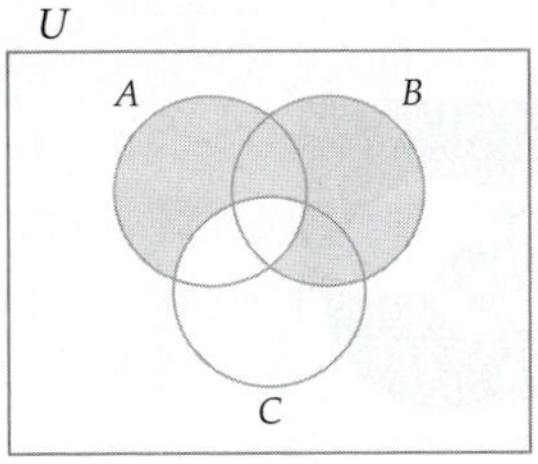

37. Not equal **38.** Not equal **39.** Equal **40.** Equal **41.** Not equal **42.** Not equal **43.** Not equal **44.** Equal **45.** Equal **46.** Not equal **47.** Yellow **48.** Magenta **49.** Cyan **50.** Blue **51.** Red **52.** Green

In Exercises 53 to 62, one possible answer is given. Your answers may vary from the given answers.

53. $A \cap B'$ **54.** $(A \cup B) \cap (A \cap B)'$ **55.** $(A \cup B)'$ **56.** $A \cap (B \cup C)$ **57.** $B \cup C$ **58.** $C \cup (A \cap B)$ **59.** $C \cap (A \cup B)'$ **60.** $(A \cup B)'$ **61.** $(A \cup B)' \cup (A \cap B \cap C)$ **62.** $C \cup (A \cap B')$

63. **a.**

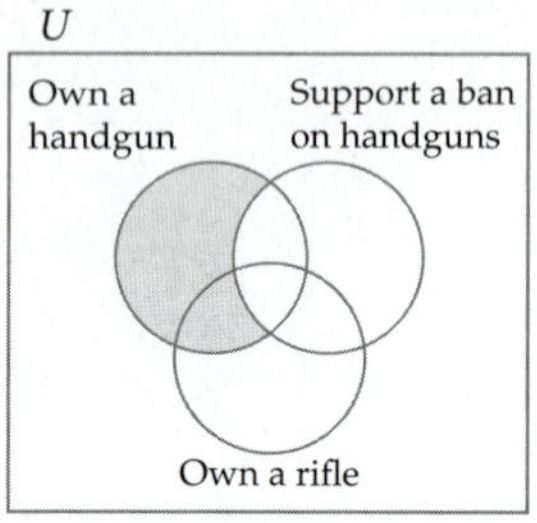

b.

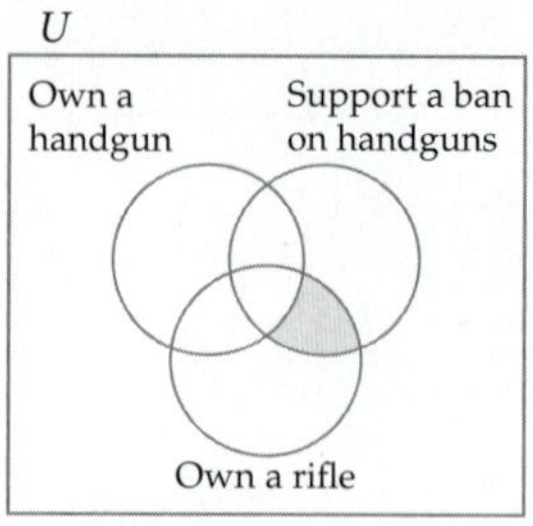

c.

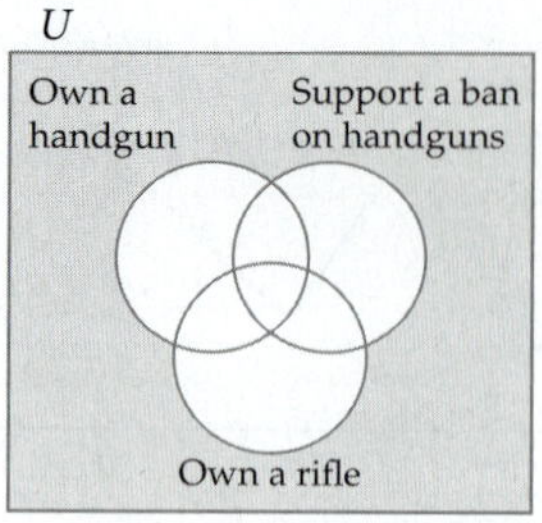

64. **a.**

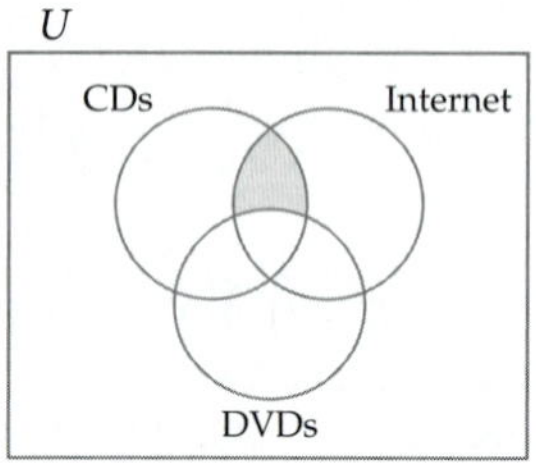

b.

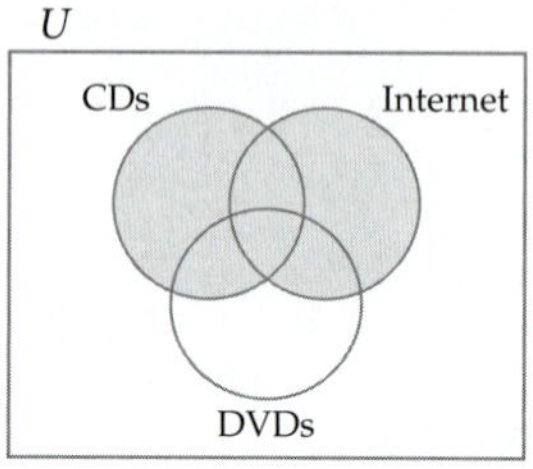

c.

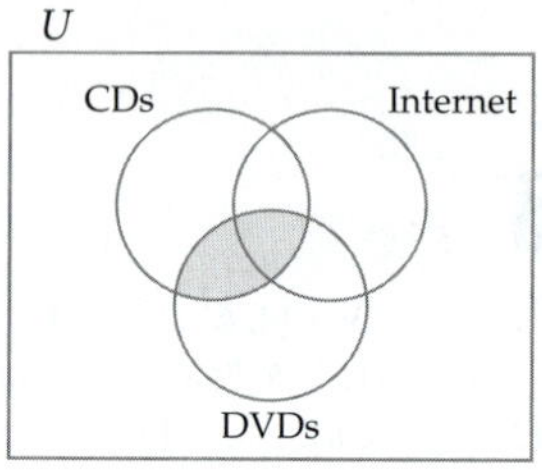

65.

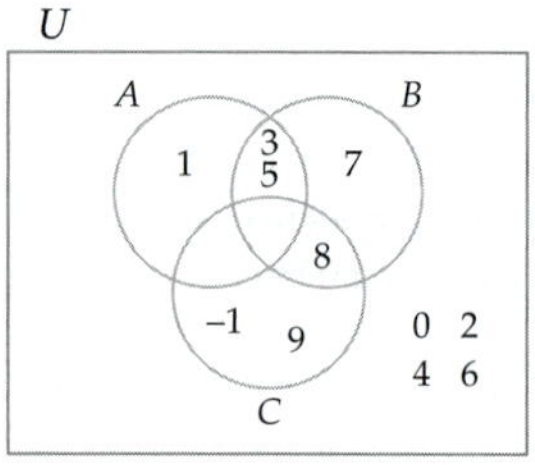

66.

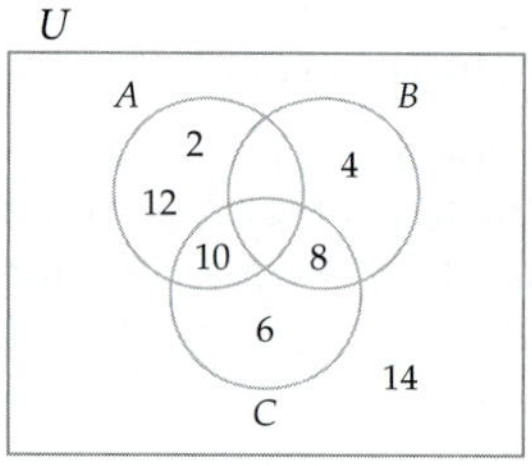

67.

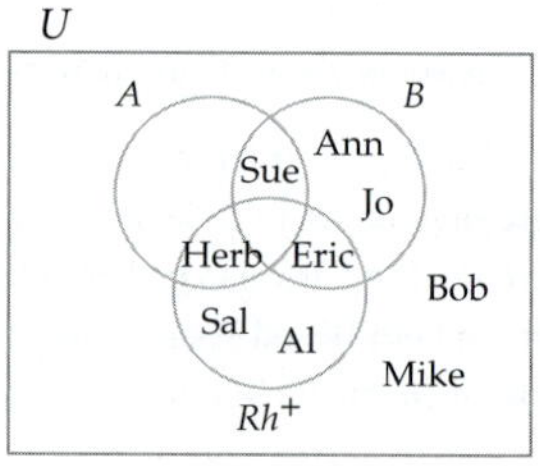

68.

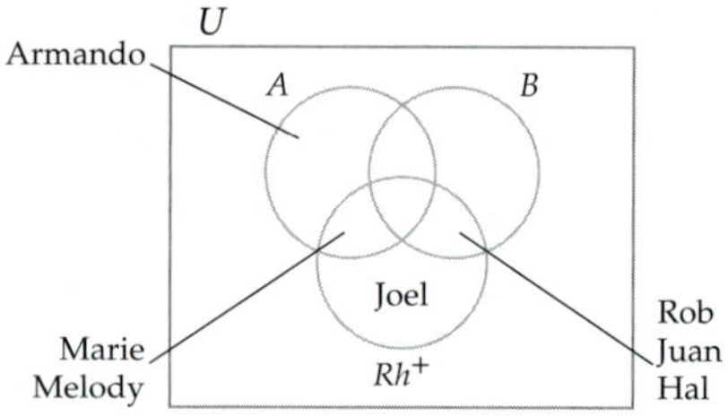

See the *Instructor's Solutions Manual* for the verifications for Exercises 69 and 70.

71. {3, 9} **72.** {4, 6} **73.** {2, 8} **74.** {1, 5, 7} **75.** {3, 9} **76.** {1, 5, 7}

77. Responses will vary.

78.

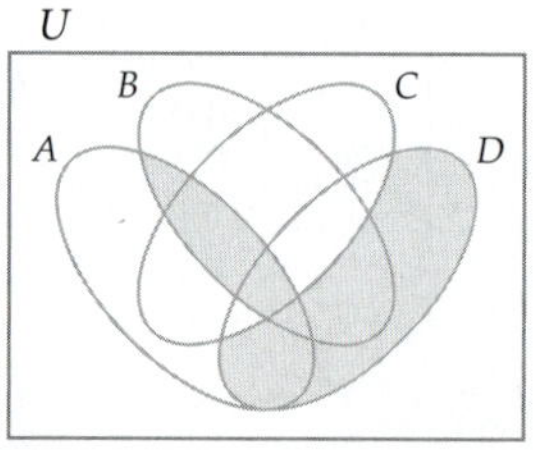

79.

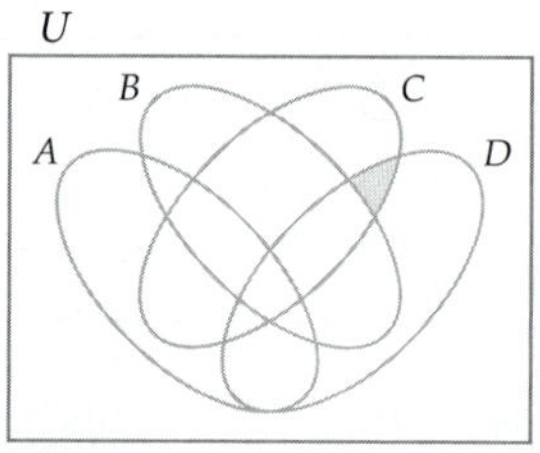

80.

A Venn diagram for five sets

A Venn diagram for six sets

SOURCE: http://www.combinatorics.org/Surveys/ds5/VennWhatEJC.html

EXCURSION EXERCISES, SECTION 2.4 page 86

1. a. {Ryan, Susan}, {Ryan, Trevor}, {Susan, Trevor}, {Ryan, Susan, Trevor} **b.** ∅, {Ryan}, {Susan}, {Trevor} **2.** {M, N}, {M, P}, {M, N, P}
3. {M, N}, {M, P}

EXERCISE SET 2.4 page 86

1. 7 **2.** 7 **3.** 8 **4.** 9 **5.** 8 **6.** 2 **7.** 12 **8.** 0 **9.** 5 **10.** 3
11. $n(A \cup B) = 7$; $n(A) + n(B) - n(A \cap B) = 4 + 5 - 2 = 7$ **12.** $n(A \cup C) = 6$; $n(A) + n(C) - n(A \cap C) = 4 + 3 - 1 = 6$
13. 113 **14.** 970
15. 1060 **16.** 155
17.

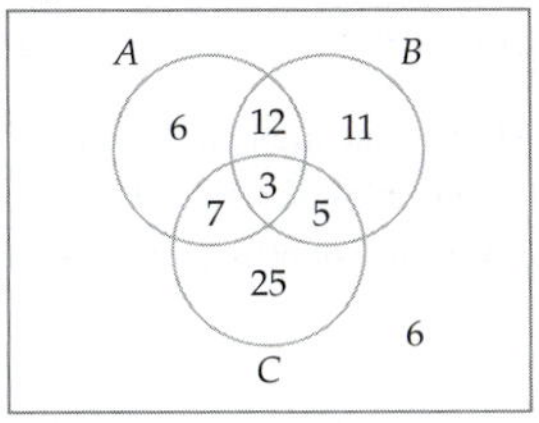

18.

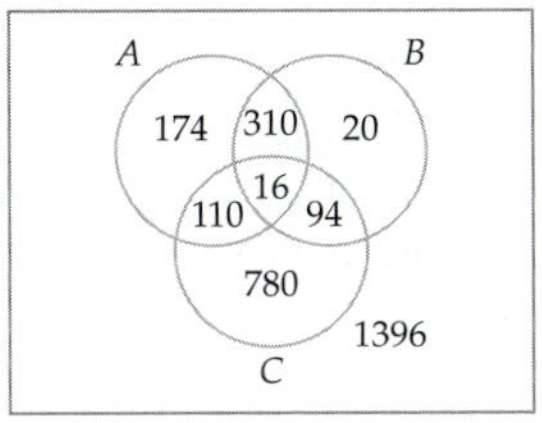

19. a. 180 investors **b.** 200 investors **20. a.** 440 commuters **b.** 700 commuters **21. a.** 15% **b.** 13%
22. a. 995 guests **b.** 575 guests **c.** 1110 guests **23. a.** 450 customers **b.** 140 customers **c.** 130 customers
24. a. 55.9% **b.** 11.7% **25. a.** 109 households **b.** 328 households **c.** 104 households **26. a.** 96 students
b. 123 students **c.** 508 students **d.** 118 students **27. a.** 101 people **b.** 370 people **c.** 380 people
d. 373 people **e.** 225 people **f.** 530 people **28. a.** 175 students **b.** 728 students **c.** 963 students
d. 385 students **e.** 110 students **f.** 727 students **29. a.** 72 elements **b.** 47 elements **c.** 25 elements
d. 0 elements **30. a.** 35 elements **b.** 16 elements **c.** 16 elements **d.** 0 elements **31. a.** 200 users
b. 271 users **c.** 16 users **32.** d

EXCURSION EXERCISES, SECTION 2.5 page 97

1. Let $C = \{3, 4, 5, 6, \dots, n + 2, \dots\}$. C and N have the same cardinality because the elements of C can be paired with the elements of N using the general correspondence $(n + 2) \leftrightarrow n$.
Let $D = \{1, 2\}$. Because $C \cup D = N$, we can establish the following equations.
$n(C) + n(D) = n(N)$
$\aleph_0 + 2 = \aleph_0$

2. Let W be the set of whole numbers and let B be the set of negative integers. Then
$n(W) + n(B) = n(W \cup B)$
$\aleph_0 + \aleph_0 = n(I)$
$\aleph_0 + \aleph_0 = \aleph_0$

3. Let $C = \{1, 2, 3, 4, 5, 6\}$. Then
$n(N) - n(C) = n(N \cap C')$
$\aleph_0 - 6 = n(\{7, 8, 9, 10, \dots\})$
$\aleph_0 - 6 = \aleph_0$

EXERCISE SET 2.5 page 97

1. a. One possible one-to-one correspondence between V and M is given by
$V = \{a, e, i\}$
↕ ↕ ↕
$M = \{3, 6, 9\}$
b. 6 **2.** Pair n of N with $5n$ of F.

3. Pair $(2n - 1)$ of D with $(3n)$ of M. **4.** 4 **5.** $\aleph_0$ **6.** $\aleph_0$ **7.** c **8.** c **9.** c **10.** 16
11. Equivalent **12.** Not equivalent **13.** Equivalent **14.** Not equivalent **15.** Let $S = \{10, 20, 30, \dots, 10n, \dots\}$. Then S is a proper subset of A. A rule for a one-to-one correspondence between A and S is $(5n) \leftrightarrow (10n)$. Because A can be placed in a one-to-one correspondence with a proper subset of itself, A is an infinite set. **16.** Let $F = \{15, 19, 23, 27, \dots, 4n + 11, \dots\}$. Then F is a proper subset of B. A rule for a one-to-one correspondence between B and F is $(4n + 7) \leftrightarrow (4n + 11)$. Because B can be placed in a one-to-one correspondence with a proper subset of itself, B is an infinite set.

17. Let $R = \left\{\frac{3}{4}, \frac{5}{6}, \frac{7}{8}, \ldots, \frac{2n+1}{2n+2}, \ldots\right\}$. Then R is a proper subset of C. A rule for a one-to-one correspondence between C and R is $\left(\frac{2n-1}{2n}\right) \leftrightarrow \left(\frac{2n+1}{2n+2}\right)$. Because C can be placed in a one-to-one correspondence with a proper subset of itself, C is an infinite set.

18. Let $H = \left\{\frac{1}{3}, \frac{1}{4}, \frac{1}{5}, \frac{1}{6}, \ldots, \frac{1}{n+2}, \ldots\right\}$. Then H is a proper subset of D. A rule for a one-to-one correspondence between D and H is $\left(\frac{1}{n+1}\right) \leftrightarrow \left(\frac{1}{n+2}\right)$. Because D can be placed in a one-to-one correspondence with a proper subset of itself, D is an infinite set.

In Exercises 19 to 26, let $N = \{1, 2, 3, 4, \ldots, n, \ldots\}$. Then a one-to-one correspondence between the given sets and the set of natural numbers N is given by the following general correspondences.

19. $(n + 49) \leftrightarrow n$ **20.** $(-5n + 15) \leftrightarrow n$ **21.** $\left(\frac{1}{3^{n-1}}\right) \leftrightarrow n$ **22.** $(-6n - 6) \leftrightarrow n$ **23.** $(10^n) \leftrightarrow n$ **24.** $\left(\frac{1}{2^{n-1}}\right) \leftrightarrow n$ **25.** $(n^3) \leftrightarrow n$ **26.** $(10^{-n}) \leftrightarrow n$ **27. a.** There are several ways to establish the one-to-one correspondence. Here is one method: For any natural number n, the two natural numbers preceding $3n$ are not multiples of 3. Pair these two numbers, $3n - 2$ and $3n - 1$, with the multiples of 3 given by $6n - 3$ and $6n$, respectively. Using the two general correspondences $(6n - 3) \leftrightarrow (3n - 2)$ and $(6n) \leftrightarrow (3n - 1)$ (as shown below), we can establish a one-to-one correspondence between the multiples of 3 (set M) and the set K of all natural numbers that are not multiples of 3.

$M = \{3, 6, 9, 12, 15, 18, \ldots, 6n - 3, \quad 6n, \quad \ldots\}$

$\qquad\qquad\qquad\qquad\qquad\qquad\quad \updownarrow \qquad\quad \updownarrow$

$K = \{1, 2, 4, 5, 7, 8, \ldots, 3n - 2, 3n - 1, \ldots\}$

The following answers in parts b and c were produced by using the correspondences established in part a. **b.** 302 **c.** 1800

28. a. In the following figure, the line from E that passes through $\overline{AB}$ and $\overline{CD}$ illustrates a method of establishing a one-to-one correspondence between the sets $\{x \mid 0 \le x \le 1\}$ and $\{x \mid 0 \le x \le 5\}$.

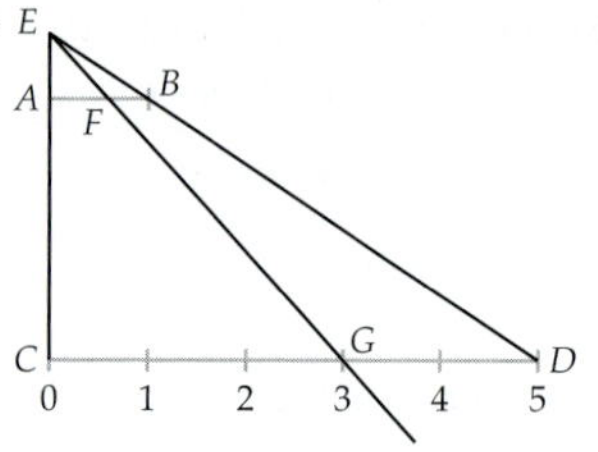

b. In the following figure, the line from E that passes through $\overline{AB}$ and $\overline{CD}$ illustrates a method of establishing a one-to-one correspondence between the sets $\{x \mid 2 \le x \le 5\}$ and $\{x \mid 1 \le x \le 8\}$.

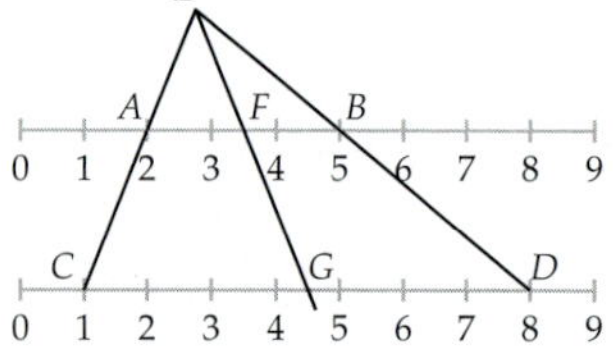

29. The set of real numbers x such that $0 < x < \pi$ is equivalent to the set of all real numbers. **30.** Written responses will vary.

CHAPTER 2 REVIEW EXERCISES *page 101*

1. {January, June, July} [Sec. 2.1] **2.** {Alaska, Hawaii} [Sec. 2.1] **3.** {0, 1, 2, 3, 4, 5, 6, 7} [Sec. 2.1] **4.** $\{-8, 8\}$ [Sec. 2.1] **5.** {1, 2, 3, 4} [Sec. 2.1] **6.** {1, 2, 3, 4, 5, 6} [Sec. 2.1] **7.** $\{x \mid x \in I \text{ and } x > -6\}$ [Sec. 2.1] **8.** $\{x \mid x$ is the name of a month with exactly 30 days$\}$ [Sec. 2.1] **9.** $\{x \mid x$ is the name of a U.S. state that begins with the letter K$\}$ [Sec. 2.1] **10.** $\{x^3 \mid x = 1, 2, 3, 4, \text{ or } 5\}$ [Sec. 2.1] **11.** Equivalent [Sec. 2.1] **12.** Both equal and equivalent [Sec. 2.1] **13.** False [Sec. 2.1] **14.** True [Sec. 2.1] **15.** True [Sec. 2.1] **16.** False [Sec. 2.1] **17.** {6, 10} [Sec. 2.3] **18.** {2, 6, 10, 16, 18} [Sec. 2.3] **19.** $C = \{14, 16\}$ [Sec. 2.3] **20.** {2, 6, 8, 10, 12, 16, 18} [Sec. 2.3] **21.** {2, 6, 10, 16} [Sec. 2.3] **22.** {8, 12} [Sec. 2.3] **23.** {6, 8, 10, 12, 14, 16, 18} [Sec. 2.3] **24.** {8, 12} [Sec. 2.3] **25.** No [Sec. 2.2] **26.** No [Sec. 2.2] **27.** Proper subset [Sec. 2.2] **28.** Proper subset [Sec. 2.2] **29.** Not a proper subset [Sec. 2.2] **30.** Not a proper subset [Sec. 2.2] **31.** ∅, {I}, {II}, {I, II} [Sec. 2.2] **32.** ∅, {s}, {u}, {n}, {s, u}, {s, n}, {u, n}, {s, u, n} [Sec. 2.2] **33.** ∅, {penny}, {nickel}, {dime}, {quarter}, {penny, nickel}, {penny, dime}, {penny, quarter}, {nickel, dime}, {nickel, quarter}, {dime, quarter}, {penny, nickel, dime}, {penny, nickel, quarter}, {penny, dime, quarter}, {nickel, dime, quarter}, {penny, nickel, dime, quarter} [Sec. 2.2] **34.** ∅, {A}, {B}, {C}, {D}, {E}, {A, B}, {A, C}, {A, D}, {A, E}, {B, C}, {B, D}, {B, E}, {C, D}, {C, E}, {D, E}, {A, B, C}, {A, B, D}, {A, B, E}, {A, C, D}, {A, C, E}, {A, D, E}, {B, C, D}, {B, C, E}, {B, D, E}, {C, D, E}, {A, B, C, D}, {A, B, C, E}, {A, B, D, E}, {A, C, D, E}, {B, C, D, E}, {A, B, C, D, E} [Sec. 2.2] **35.** $2^4 = 16$ subsets [Sec. 2.2] **36.** $2^{26} = 67{,}108{,}864$ subsets [Sec. 2.2] **37.** $2^{15} = 32{,}768$ subsets [Sec. 2.2]

38. $2^7 = 128$ subsets [Sec. 2.2] **39.** True [Sec. 2.3] **40.** True [Sec. 2.3]

41. U [Sec. 2.3]

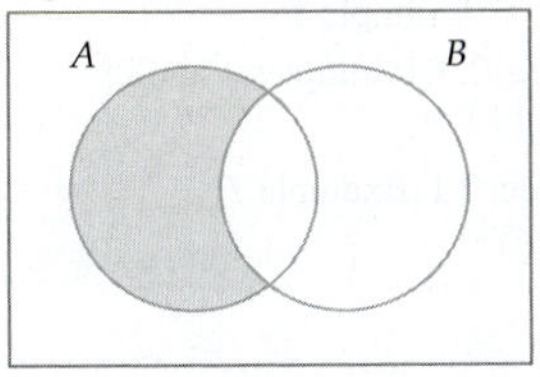

42. U [Sec. 2.3]

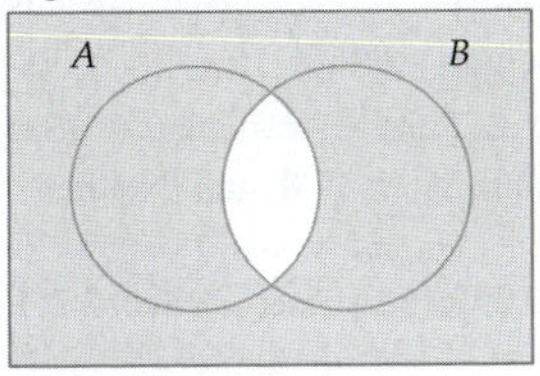

43. U [Sec. 2.3]

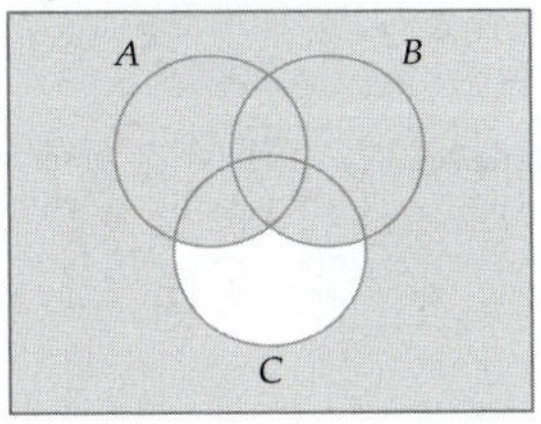

44. U [Sec. 2.3]

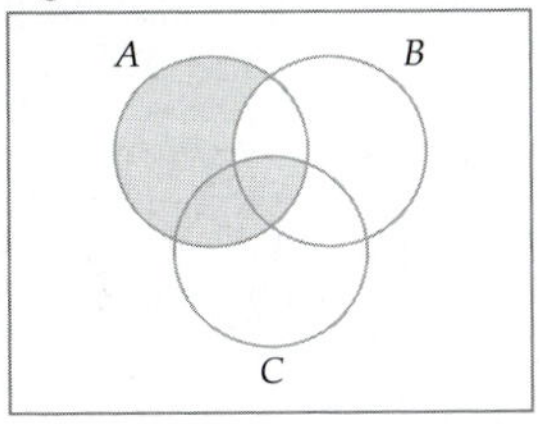

45. Equal [Sec. 2.3]

46. Not equal [Sec. 2.3] **47.** Not equal [Sec. 2.3] **48.** Not equal [Sec. 2.3]

49. $(A \cup B)' \cap C$ or $C \cap (A' \cap B')$ [Sec. 2.3] **50.** $(A \cap B) \cup (B \cap C')$ [Sec. 2.3]

51. U [Sec. 2.3]

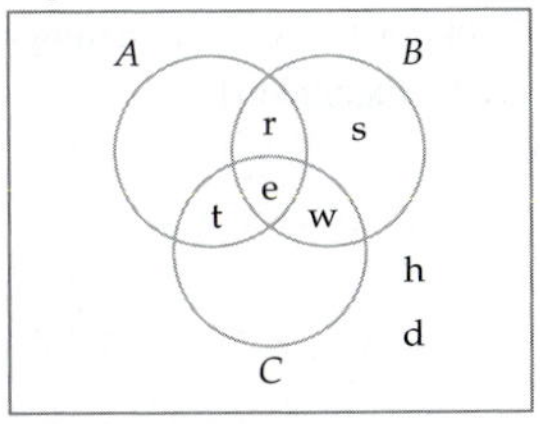

52. U [Sec. 2.3]

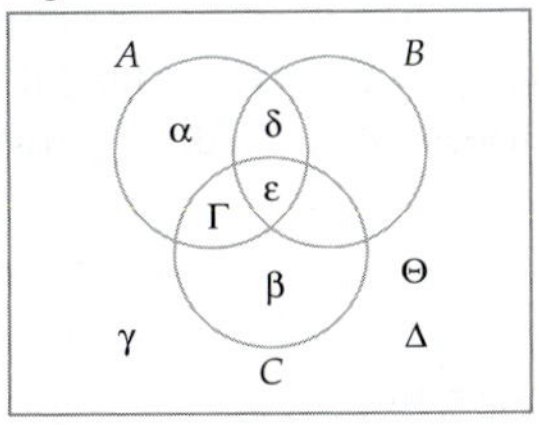

53. 391 members [Sec. 2.4]

54. a. 42 customers **b.** 31 customers **c.** 20 customers **d.** 142 customers [Sec. 2.4] **55.** 6 athletes [Sec. 2.4]

56. 874 students [Sec. 2.4]

57. One possible one-to-one correspondence between $\{1, 3, 6, 10\}$ and $\{1, 2, 3, 4\}$ is given by

$\{1, 3, 6, 10\}$
$\updownarrow\ \updownarrow\ \updownarrow\ \updownarrow$
$\{1, 2, 3, 4\}$ [Sec. 2.5]

58. $\{x \mid x > 10 \text{ and } x \in N\} = \{11, 12, 13, 14, \ldots, n + 10, \ldots\}$
Thus a one-to-one correspondence between the sets is given by

$\{11, 12, 13, 14, \ldots, n + 10, \ldots\}$
$\updownarrow\ \updownarrow\ \updownarrow\ \updownarrow\ \ \updownarrow$
$\{2, 4, 6, 8, \ldots, 2n, \ldots\}$ [Sec. 2.5]

59. One possible one-to-one correspondence between the sets is given by

$\{3, 6, 9, \ldots, 3n, \ldots\}$
$\updownarrow\ \updownarrow\ \updownarrow\ \ \updownarrow$
$\{10, 100, 1000, \ldots, 10^n, \ldots\}$ [Sec. 2.5]

60. In the following figure, the line from E that passes through $\overline{AB}$ and $\overline{CD}$ illustrates a method of establishing a one-to-one correspondence between $\{x \mid 0 \le x \le 1\}$ and $\{x \mid 0 \le x \le 4\}$.

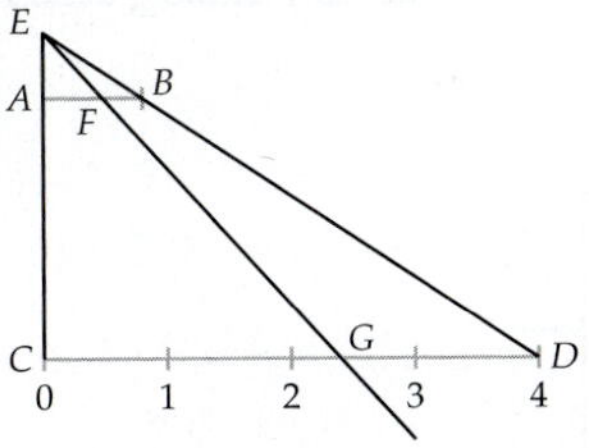

[Sec. 2.5]

61. A proper subset of A is $S = \{10, 14, 18, \ldots, 4n + 6, \ldots\}$. A one-to-one correspondence between A and S is given by

$A = \{6, 10, 14, 18, \ldots, 4n + 2, \ldots\}$
$\updownarrow\ \updownarrow\ \updownarrow\ \updownarrow\ \ \updownarrow$
$S = \{10, 14, 18, 22, \ldots, 4n + 6, \ldots\}$

Because A can be placed in a one-to-one correspondence with a proper subset of itself, A is an infinite set. [Sec. 2.5]

62. A proper subset of B is $T = \left\{\frac{1}{2}, \frac{1}{4}, \frac{1}{8}, \frac{1}{16}, \ldots, \frac{1}{2^n}, \ldots\right\}$. A one-to-one correspondence between B and T is given by

$$B = \left\{1, \frac{1}{2}, \frac{1}{4}, \frac{1}{8}, \ldots, \frac{1}{2^{n-1}}, \ldots\right\}$$
$$\updownarrow\ \updownarrow\ \updownarrow\ \updownarrow\ \ \updownarrow$$
$$T = \left\{\frac{1}{2}, \frac{1}{4}, \frac{1}{8}, \frac{1}{16}, \ldots, \frac{1}{2^n}, \ldots\right\}$$

Because B can be placed in a one-to-one correspondence with a proper subset of itself, B is an infinite set. [Sec. 2.5]

63. 5 [Sec. 2.1] **64.** 10 [Sec. 2.1] **65.** 2 [Sec. 2.1] **66.** 5 [Sec. 2.1] **67.** $\aleph_0$ [Sec. 2.5]

68. $\aleph_0$ [Sec. 2.5] **69.** c [Sec. 2.5] **70.** c [Sec. 2.5] **71.** $\aleph_0$ [Sec. 2.5] **72.** c [Sec. 2.5]

CHAPTER 2 TEST page 103

1. {1, 2, 4, 5, 6, 7, 9, 10} [Sec. 2.2, Example 1; Sec. 2.3, Example 1] **2.** {2, 9, 10} [Sec. 2.2, Example 1; Sec. 2.3, Example 1]
3. {1, 2, 3, 4, 6, 9, 10} [Sec. 2.2, Example 1; Sec. 2.3, Examples 1 and 2] **4.** {5, 7, 8} [Sec. 2.2, Example 1; Sec. 2.3, Examples 1 and 2]
5. $\{x \mid x \in W \text{ and } x < 7\}$ [Sec. 2.1, Example 5] **6.** $\{x \mid x \in I \text{ and } -3 \le x \le 2\}$ [Sec. 2.1, Example 5]
7. a. 4 **b.** $\aleph_0$ [Sec. 2.1, Example 6; Sec. 2.5, Example 3] **8. a.** Neither **b.** Equivalent [Sec. 2.1, Example 7]
9. a. Equivalent **b.** Equivalent [Sec. 2.5, Example 3]
10. { }
{a}, {b}, {c}, {d}
{a, b}, {a, c}, {a, d}, {b, c}, {b, d}, {c, d}
{a, b, c}, {a, b, d}, {a, c, d}, {b, c, d}
{a, b, c, d} [Sec. 2.2, Example 4]
11. $2^{21} = 2{,}097{,}152$ subsets [Sec. 2.2, Example 5]
12. 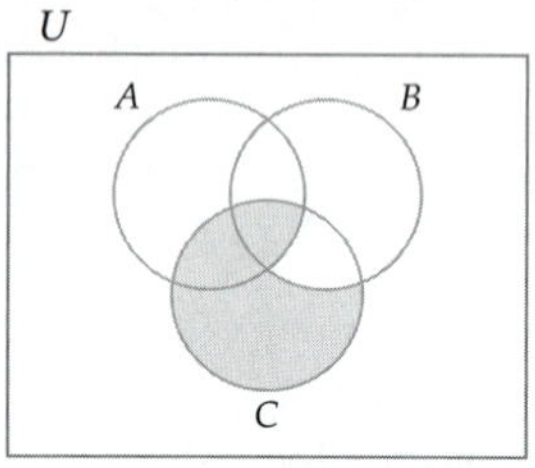

[Sec. 2.3, Example 6]
13. 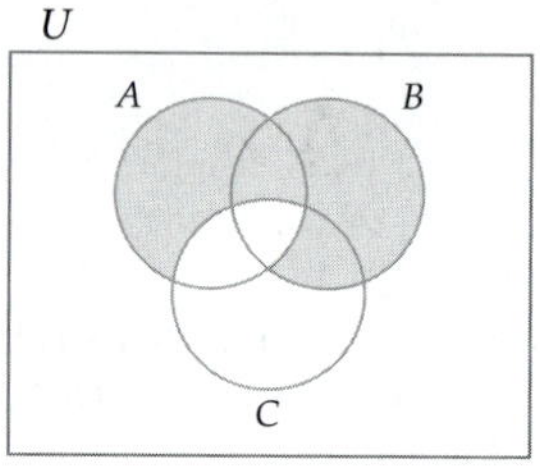

[Sec. 2.3, Example 6]
14. $A' \cap B'$ [Sec. 2.3, Example 4 and Check Your Progress 4] **15. a.** $2^9 = 512$ different versions **b.** 12 options [Sec. 2.2, Example 5]
16. 1164 [Sec. 2.4, Example 3] **17. a.** {2007, 2008, 2014} **b.** {2009, 2010, 2013, 2014} **c.** ∅ [Sec. 2.1 Example 1]
18. a. 232 households **b.** 102 households **c.** 857 households **d.** 79 households [Sec. 2.4, Example 2]
19. One possible method: {5, 10, 15, 20, 25, ..., 5n, ...}
↕ ↕ ↕ ↕ ↕ ↕
{0, 1, 2, 3, 4, ..., $n - 1$, ...}
$(5n) \leftrightarrow (n - 1)$ [Sec. 2.5, Example 1]
20. One possible method: {3, 6, 9, 12, ..., 3n, ...}
↕ ↕ ↕ ↕ ↕
{6, 12, 18, 24, ..., 6n, ...}
$(3n) \leftrightarrow (6n)$ [Sec. 2.5, Example 2]

CHAPTER 3

EXCURSION EXERCISES, SECTION 3.1 page 114

1. $S \wedge (\sim R \vee \sim Q \vee P)$ **2.** $\sim S \wedge (\sim Q \vee P) \wedge R$ **3.** $(\sim P \vee Q) \wedge (\sim R \vee S)$ **4.** $(R \vee P) \wedge Q \wedge (R \vee \sim P)$
5. $[\sim R \vee (Q \wedge S)] \vee (R \wedge \sim S)$ **6.** $(P \vee \sim Q) \vee (\sim R \vee S)$ **7.** 2, 5, 6 **8.** 3, 4, 5, 6
9. 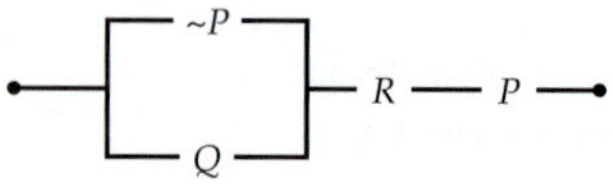

10. 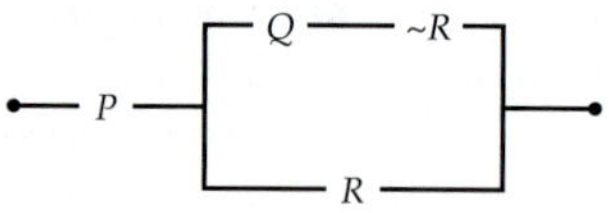

11. 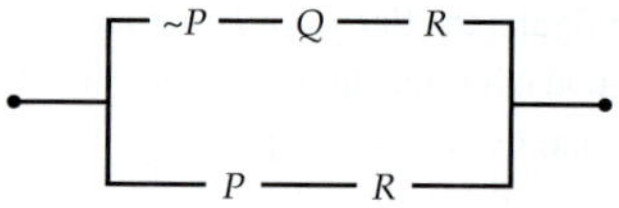

12. 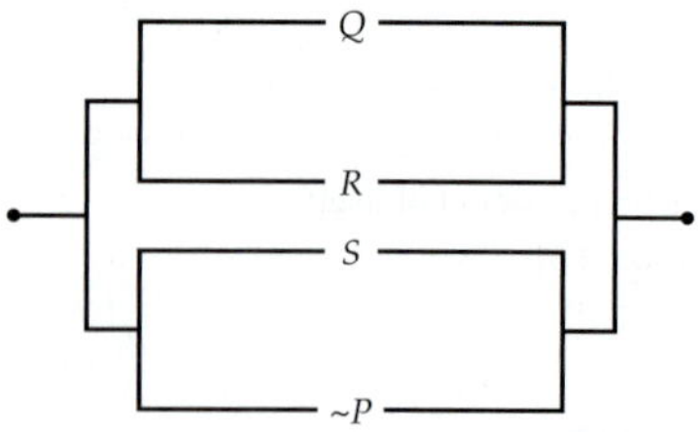

13. 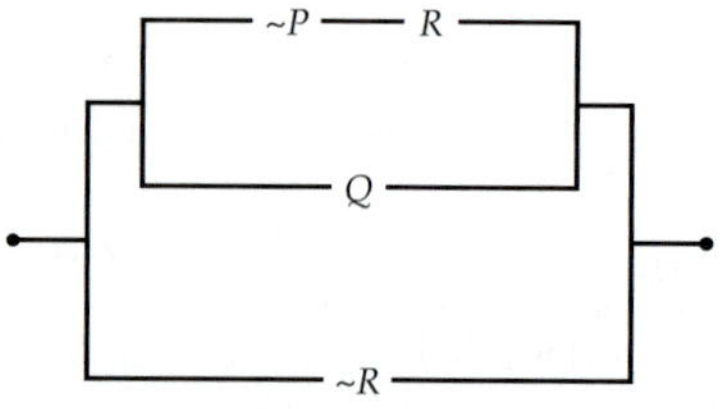

14. 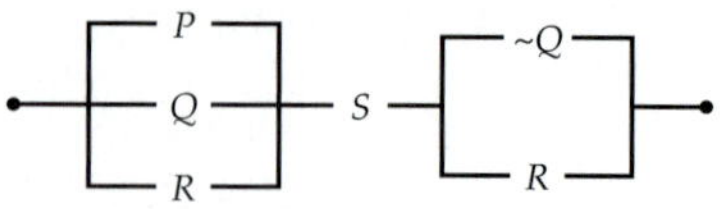

15. a. On **b.** Off **c.** On **d.** On
16. Both of the circuits are closed under exactly the same conditions, and both of the circuits are open under exactly the same conditions. Thus the circuit in Exercise 16 can be used in place of the circuit in Exercise 15.

EXERCISE SET 3.1 *page 115*

1. Not a statement **2.** Statement **3.** Statement **4.** Statement **5.** Not a statement **6.** Not a statement **7.** Statement **8.** Statement **9.** Statement **10.** Statement **11.** One simple statement is "The principal will attend the class on Tuesday." The other simple statement is "The principal will attend the class on Wednesday." **12.** One simple statement is "5 is an odd number." The other simple statement is "6 is an even number." **13.** One simple statement is "A triangle is an acute triangle." The other simple statement is "It has three acute angles." **14.** One simple statement is "This is Saturday." The other simple statement is "Tomorrow is Sunday." **15.** The Giants did not lose the game. **16.** The lunch was not served at noon. **17.** The game went into overtime. **18.** The game was shown on ABC. **19.** $w \rightarrow t$, conditional **20.** $p \wedge s$, conjunction **21.** $l \leftrightarrow a$, biconditional **22.** $e \leftrightarrow t$, biconditional **23.** $d \rightarrow f$, conditional **24.** $p \rightarrow t$, conditional **25.** $m \vee c$, disjunction **26.** $p \rightarrow s$, conditional **27.** The tour goes to Italy and the tour does not go to Spain. **28.** We will go to Venice or we will go to Florence. **29.** If we go to Venice, then we will not go to Florence. **30.** If the tour goes to Italy, then we will go to Venice. **31.** We will go to Florence if and only if we do not go to Venice. **32.** The hotel fees are not included and the meals are not included. **33.** Taylor Swift is a singer or she is an actress, and she is not a songwriter. **34.** If Taylor Swift does not play the piano, then she is a singer and she is a songwriter. **35.** If Taylor Swift is a singer, then she is not a songwriter and she is not an actress. **36.** If Taylor Swift plays the piano and she is a songwriter, then she does not play the guitar. **37.** Taylor Swift is an actress and a singer, if and only if she is not a songwriter. **38.** Taylor Swift does not play the guitar, if and only if she is not an actress and she is not a singer. **39.** $(p \vee q) \wedge \sim r$ **40.** $r \wedge (\sim q \vee p)$ **41.** $(q \wedge r) \rightarrow \sim p$ **42.** $q \leftrightarrow (\sim p \wedge \sim r)$ **43.** $s \rightarrow (q \wedge \sim p)$ **44.** $\sim(p \vee r)$ **45.** True **46.** True **47.** True **48.** True **49.** True **50.** True **51.** True **52.** False **53.** No lions are playful. **54.** All dogs are friendly. **55.** Some classic movies were not first produced in black and white. **56.** Some people did not enjoy the dinner. **57.** Some even numbers are odd numbers. **58.** All actors are rich. **59.** Some cars do not run on gasoline. **60.** Some of the students took my advice. **61.** $p \rightarrow q$, where p represents "you can count your money" and q represents "you don't have a billion dollars." **62.** $e \rightarrow f$, where e represents "you aren't fired with enthusiasm" and f represents "you will be fired with enthusiasm." **63.** $p \rightarrow q$, where p represents "people concentrated on the really important things in life" and q represents "there'd be a shortage of fishing poles." **64.** $k \rightarrow l$, where k represents "you're killed" and l represents "you've lost a very important part of your life." **65.** $p \leftrightarrow q$, where p represents "an angle is a right angle" and q represents "its measure is 90°." **66.** $i \rightarrow r$, where i represents "an angle is inscribed in a semicircle" and r represents "the angle is a right angle." **67.** $p \rightarrow q$, where p represents "two sides of a triangle are equal in length" and q represents "the angles opposite those sides are congruent." **68.** $t \rightarrow s$, where t represents "this sum is the sum of the measures of the three angles of a triangle" and s represents "this sum equals 180°." **69.** $p \rightarrow q$, where p represents "it is a square" and q represents "it is a rectangle." **70.** $p \rightarrow s$, where p represents "the corresponding sides of two triangles are proportional" and s represents "the triangles are similar." **71.** 6 cups. In these teapots the tea level cannot rise above its spout opening, because any extra tea will flow out the spout. Because both spout openings are at the same height, the maximum number of cups they can hold is the same.

EXCURSION EXERCISES, SECTION 3.2 *page 124*

The closure tables for Excursion Exercises 1 to 7 are shown below.

P	*Q*	1.	2.
1	1	1	0
1	0	0	0
0	1	0	0
0	0	0	0

P	*Q*	*R*	3.	4.	5.	6.	7.a.	7.b.
1	1	1	1	1	1	1	0	0
1	1	0	1	0	1	1	1	1
1	0	1	0	0	1	1	1	1
1	0	0	0	0	1	1	0	0
0	1	1	1	1	0	1	0	0
0	1	0	1	0	1	1	0	0
0	0	1	0	0	0	1	1	1
0	0	0	1	0	1	1	0	0

1. The network is closed provided P and Q are both closed. **2.** The network is never closed. **3.** The network is closed if Q is closed *or* if both P and R are open. **4.** The network is closed provided that Q and R are both closed. **5.** The network is closed provided P is closed *or* R is open. **6.** The network is always closed. **7.** The closure table for the switching network in **7.b.** is the same as the closure table for the switching network in **7.a.** This means that both of the circuits are closed under exactly the same conditions and both circuits are open under exactly the same conditions. Thus the circuit in **7.b.** can be used in place of the circuit in **7.a.**

EXERCISE SET 3.2 *page 125*

1. True **2.** False **3.** False **4.** True **5.** False **6.** True **7.** False **8.** True **9.** False
10. True **11. a.** If p is false, then $p \wedge (q \vee r)$ must be a false statement. **b.** For a conjunctive statement to be true, it is necessary that both components of the statement be true. Because it is given that one of the components (p) is false, $p \wedge (q \vee r)$ must be a false statement.
12. a. It is a true statement. **b.** A disjunction, of two statements, is a true statement if at least one of the statements is a true statement.

p	q	13.	14.	15.	16.	17.	18.
T	T	T	F	F	T	F	F
T	F	F	T	T	T	T	F
F	T	T	T	F	T	T	T
F	F	T	T	F	T	T	F

p	q	r	19.	20.	21.	22.	23.	24.	25.	26.	27.	28.
T	T	T	F	T	T	T	T	F	T	T	F	T
T	T	F	F	T	T	F	F	T	T	F	T	T
T	F	T	F	F	T	F	F	F	T	T	T	T
T	F	F	F	T	T	F	F	F	T	F	F	T
F	T	T	F	T	F	T	T	F	T	T	F	F
F	T	F	F	T	F	F	F	F	T	F	F	T
F	F	T	F	T	T	T	T	F	F	T	T	T
F	F	F	T	T	T	F	F	F	T	F	F	T

See the *Instructor's Solutions Manual* for the solutions to Exercises 29 to 36. **37.** It did not rain and it did not snow.
38. It is not true that, I passed the test or completed the course. **39.** She did not visit either France or Italy.
40. I did not buy a new car or I did not move to Florida. **41.** She did not get a promotion and she did not receive a raise.
42. The students did not cut classes and did not take part in the demonstration. **43.** Tautology **44.** Tautology **45.** Tautology
46. Tautology **47.** Tautology **48.** Tautology **49.** Self-contradiction **50.** Self-contradiction **51.** Self-contradiction
52. Self-contradiction **53.** Not a self-contradiction **54.** Self-contradiction **55.** The symbol $\leq$ means "less than or equal to."
56. a. The given statement is true because $5 < 7$. **b.** The given statement is true because $7 = 7$. **57.** $2^5 = 32$
58. There is no whole number n for which $2^n = 100$. **59.** F F F T T T F T F F F T F F F F
60. F F F F T F F F T F T F T F T F **61.** Circle the 1 and the three 9s. Then invert the paper so that the digits are upside down and hand it back to your friend.

EXCURSION EXERCISES, SECTION 3.3 *page 133*

1. a. 1011 **b.** 0001 **c.** 10101011 **2.** Either

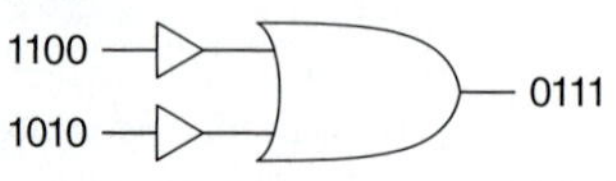

Or

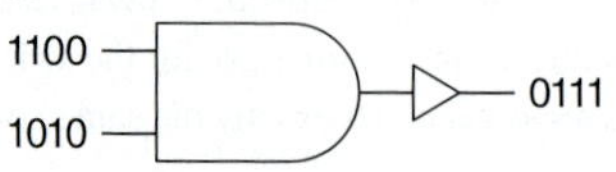

EXERCISE SET 3.3 *page 133*

1. *Antecedent:* I had the money
Consequent: I would buy the painting

2. *Antecedent:* Shelly goes on the trip
Consequent: she will not be able to take part in the graduation ceremony

3. *Antecedent:* they had a guard dog
Consequent: no one would trespass on their property
4. *Antecedent:* I don't get to school before 7:30
Consequent: I won't be able to find a parking place
5. *Antecedent:* I change my major
Consequent: I must reapply for admission
6. *Antecedent:* your blood type is O−
Consequent: you are classified as a universal donor
7. True **8.** False **9.** True **10.** True **11.** True **12.** True **13.** False **14.** True

p	q	**15.**	**16.**	**17.**	**18.**
T	T	T	T	T	T
T	F	T	T	T	T
F	T	T	T	T	T
F	F	T	T	T	T

p	q	r	**19.**	**20.**	**21.**	**22.**	**23.**	**24.**
T	T	T	T	T	T	T	T	T
T	T	F	T	T	F	T	T	T
T	F	T	T	T	T	T	T	T
T	F	F	T	T	T	T	T	F
F	T	T	T	F	T	T	T	T
F	T	F	T	T	F	T	T	T
F	F	T	T	T	T	T	T	T
F	F	F	T	T	T	T	T	T

25. She cannot sing or she would be perfect for the part. **26.** He will get frustrated or he will be able to complete the job. **27.** Either x is not an irrational number or x is not a terminating decimal. **28.** Mr. Hyde does not have a brain or he would be dangerous. **29.** The fog must lift or our flight will be cancelled. **30.** The Yankees won't win the pennant or else Carol would be happy. **31.** They offered me the contract and I didn't accept. **32.** I painted the house and I did not get the money. **33.** Pigs have wings and they still can't fly. **34.** We had a telescope and we could not see the comet. **35.** She traveled to Italy and she didn't visit her relatives. **36.** Paul could play better defense and he could not be a professional basketball player. **37.** False **38.** True **39.** True **40.** False **41.** False **42.** True **43.** True **44.** False **45.** True **46.** False **47.** $v \leftrightarrow p$ **48.** $\sim p \rightarrow (t \wedge \sim v)$ **49.** $p \rightarrow v$ **50.** $\sim t \rightarrow v$ **51.** $t \rightarrow \sim v$ **52.** $\sim v \rightarrow (\sim t \wedge p)$ **53.** $(\sim t \wedge p) \rightarrow v$ **54.** $p \rightarrow (t \wedge v)$ **55.** Not equivalent **56.** Not equivalent **57.** Not equivalent **58.** Equivalent **59.** Equivalent **60.** Equivalent **61.** If a number is a rational number, then it is a real number. **62.** If a number is a whole number, then it is an integer. **63.** If an animal is a sauropod, then it is herbivorous. **64.** If a painting was painted by Vincent van Gogh, then the painting is valuable. **65.** Turn two of the valves on. After one minute, turn off one of these valves. When you get up to the field, the sprinklers will be running on the region that is controlled by the valve that is still in the on position. The region that is wet but not receiving any water is controlled by the valve you turned off. The region that is completely dry is the region that is controlled by the valve that you left in the off position. **66.** Student demonstration

EXCURSION EXERCISES, SECTION 3.4 *page 139*

1. a.

p	q	p	|	$(q \mid q)$
T	T	T	T	F
T	F	T	F	T
F	T	F	T	F
F	F	F	T	T

b. $p \mid (q \mid q) \equiv p \rightarrow q$

2. a.

p	q	$(p \mid q)$	|	$(p \mid q)$
T	T	F	T	F
T	F	T	F	T
F	T	T	F	T
F	F	T	F	T

b. $(p|q)|(p|q) \equiv p \wedge q$ **3. a.** 1000 **b.** AND gate **4.**

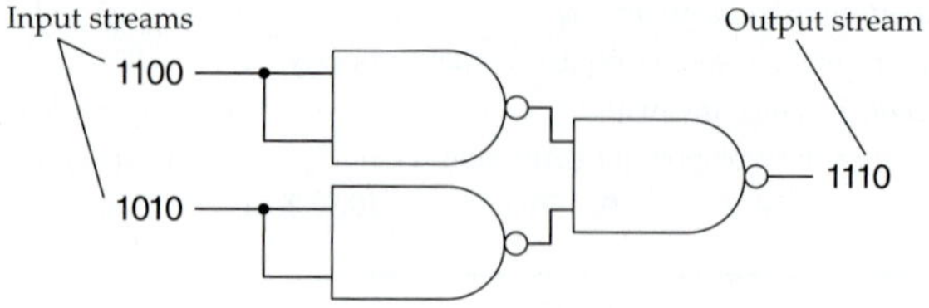

EXERCISE SET 3.4 *page 139*

1. If we take the aerobics class, then we will be in good shape for the ski trip. **2.** If we get a dog, then we will install a fence around the backyard. **3.** If the number is an odd prime number, then it is greater than 2. **4.** If the length of the hypotenuse is twice the length of the shorter leg, then the triangle is a 30°-60°-90° triangle. **5.** If he has the talent to play a keyboard, then he can join the band. **6.** If the dinosaur is a theropod, then it is carnivorous. **7.** If I was able to prepare for the test, then I had the textbook. **8.** If Education 147 is offered in the spring semester, then I will be able to receive my credential. **9.** If you ran the Boston marathon, then you are in excellent shape. **10.** If it is an ankylosaur, then it is quadrupedal. **11. a.** If I quit this job, then I am rich. **b.** If I were not rich, then I would not quit this job. **c.** If I would not quit this job, then I would not be rich. **12. a.** If we were able to take the class, then we had a car. **b.** If we did not have a car, then we were not able to take the class. **c.** If we were not able to take the class, then we did not have a car. **13. a.** If we are not able to attend the party, then she did not return soon. **b.** If she returns soon, then we will be able to attend the party. **c.** If we are able to attend the party, then she returned soon. **14. a.** If I can do the same comedy routine that I did for the banquet, then I will be in the talent show. **b.** If I am not in the talent show, then I was not allowed to do the same comedy routine that I did for the banquet. **c.** If I cannot do the same comedy routine that I did for the banquet, then I will not be in the talent show. **15. a.** If a figure is a quadrilateral, then it is a parallelogram. **b.** If a figure is not a parallelogram, then it is not a quadrilateral. **c.** If a figure is not a quadrilateral, then it is not a parallelogram. **16. a.** If you need to move to Denver, then you got the promotion. **b.** If you did not get the promotion, then you will not need to move to Denver. **c.** If you do not need to move to Denver, then you did not get the promotion. **17. a.** If I am able to get current information about astronomy, then I have access to the Internet. **b.** If I do not have access to the Internet, then I will not be able to get current information about astronomy. **c.** If I am not able to get current information about astronomy, then I don't have access to the Internet. **18. a.** If you need four-wheel drive, then you are going on a trip to Death Valley. **b.** If you are not going on a trip to Death Valley, then you will not need four-wheel drive. **c.** If you do not need four-wheel drive, then you are not going on a trip to Death Valley. **19. a.** If we don't have enough money for dinner, then we took a taxi. **b.** If we did not take a taxi, then we will have enough money for dinner. **c.** If we have enough money for dinner, then we did not take a taxi. **20. a.** If your age is at least 35, then you are the president of the United States. **b.** If you are not the president of the United States, then your age is less than 35. **c.** If your age is less than 35, then you are not the president of the United States. **21. a.** If she can extend her vacation for at least two days, then she will visit Kauai. **b.** If she does not visit Kauai, then she could not extend her vacation for at least two days. **c.** If she cannot extend her vacation for at least two days, then she will not visit Kauai. **22. a.** If the acute angles of a triangle are complementary, then the triangle is a right triangle. **b.** If a triangle is not a right triangle, then the acute angles are not complementary. **c.** If the acute angles of a triangle are not complementary, then the triangle is not a right triangle. **23. a.** If two lines are parallel, then the two lines are perpendicular to a given line. **b.** If two lines are not perpendicular to a given line, then the two lines are not parallel. **c.** If two lines are not parallel, then the two lines are not both perpendicular to a given line. **24. a.** If $x = 7$, then $x + 5 = 12$. **b.** If $x + 5 \neq 12$, then $x \neq 7$. **c.** If $x \neq 7$, then $x + 5 \neq 12$. **25.** Not equivalent **26.** Not equivalent **27.** Equivalent **28.** Not equivalent **29.** Not equivalent **30.** Equivalent **31.** If $x = 7$, then $3x - 7 \neq 11$. The original statement is true. **32.** If $5x + 7 = 22$, then $x = 3$. The original statement is true. **33.** If $|a| = 3$, then $a = 3$. The original statement is false. **34.** If a is not divisible by 3 or b is not divisible by 3, then $a + b$ is not divisible by 3. The original statement is false. **35.** If $a + b = 25$, then $\sqrt{a + b} = 5$. The original statement is true. **36.** If x is an odd integer, then x^2 is an odd integer. The original statement is true. **37.** $p \rightarrow q$ **38.** $p \rightarrow q$ **39. a.** and **b.** Answers will vary. **40. a.** and **b.** Answers will vary. **41.** If you can dream it, then you can do it. **42.** If I had a dime, I would spend it. **43.** If I were a dancer, then I would not be a singer. **44.** Pigs do not have wings or pigs could fly. This answer can also be written as "If pigs have wings, then pigs can fly." **45.** A conditional statement and its contrapositive are equivalent. They always have the same truth values. **46.** Yes. The conditional $p \rightarrow q$ is false only when p is true (T) and q is false (F). Under these conditions the converse $q \rightarrow p$ would be of the form F $\rightarrow$ T, which is true. **47.** The Hatter is telling the truth. **48.** It is not possible. Each of the 31 dominoes must cover one red square and one black square, but the checkerboard has 30 red squares and 32 black squares.

Solution to the *Where is the Missing Dollar?* puzzle in the Math Matters on page 150:
There is no missing dollar. The desk clerk has \$25. The men received a total of \$3 from the refund and the bellhop has \$2 of the refund. \$25 + \$3 + \$2 = \$30, which is the original total the men paid. *Note:* In the puzzle it states that each man paid \$9 for the room, but this is not correct. Each man paid \$9 for his share of the room and his share of the tip, if you can call it a tip, that the bellhop pocketed. Thus the men spent a total of 3 × \$9 = \$27 for the room and the tip to the bellhop. The remaining \$3 was given back to them by the bellhop.

EXCURSION EXERCISES, SECTION 3.5 *page 151*

1. to 4. Answers will vary. **5.** The step in which each side is divided by $(a - b)$ is not valid because $a - b = 0$, and division by zero is an undefined operation.

EXERCISE SET 3.5 page 152

1. $\begin{array}{l} r \to c \\ \underline{r} \\ \therefore c \end{array}$
2. $\begin{array}{l} m \to u \\ \underline{\sim u} \\ \therefore \sim m \end{array}$
3. $\begin{array}{l} g \to s \\ \underline{\sim g} \\ \therefore \sim s \end{array}$
4. $\begin{array}{l} s \vee m \\ \underline{m} \\ \therefore \sim s \end{array}$
5. $\begin{array}{l} s \to i \\ \underline{s} \\ \therefore i \end{array}$
6. $\begin{array}{l} c \to w \\ \underline{w} \\ \therefore c \end{array}$
7. $\begin{array}{l} \sim p \to \sim a \\ \underline{a} \\ \therefore p \end{array}$
8. $\begin{array}{l} s \to \sim c \\ \underline{c} \\ \therefore \sim s \end{array}$

9. Invalid **10.** Valid **11.** Invalid **12.** Valid **13.** Valid **14.** Invalid **15.** Invalid **16.** Valid **17.** Invalid **18.** Invalid **19.** Invalid **20.** Invalid **21.** Valid **22.** Invalid **23.** Valid **24.** Invalid

25. $\begin{array}{l} h \to r \\ \underline{\sim h} \\ \therefore \sim r \end{array}$ Invalid
26. $\begin{array}{l} o \leftrightarrow a \\ \underline{s \to \sim a} \\ \therefore o \to \sim s \end{array}$ Valid
27. $\begin{array}{l} \sim b \to d \\ \underline{b \vee d} \\ \therefore b \end{array}$ Invalid
28. $\begin{array}{l} e \to \sim s \\ \underline{\sim s} \\ \therefore e \end{array}$ Invalid
29. $\begin{array}{l} c \to t \\ \underline{t} \\ \therefore c \end{array}$ Invalid
30. $\begin{array}{l} b \to \sim c \\ z \to c \\ \underline{b} \\ \therefore \sim z \end{array}$ Valid

31. Valid argument, contrapositive reasoning **32.** Valid argument, disjunctive reasoning **33.** Invalid argument, fallacy of the inverse **34.** Invalid argument, fallacy of the converse **35.** Valid argument, transitive reasoning **36.** Invalid argument, fallacy of the inverse **37.** Valid argument, direct reasoning **38.** Invalid argument, fallacy of the inverse **39.** Valid argument, contrapositive reasoning **40.** Valid argument, contrapositive reasoning

See the *Instructor's Solutions Manual* for the solutions to Exercises 41 to 46. **47.** q **48.** s **49.** It is not a theropod. **50.** You will make monthly payments. **51.** 12. Any number multiplied by 0 produces 0. Thus the only arithmetic you need to perform are the last two operations. **52.** Valid

EXCURSION EXERCISES, SECTION 3.6 page 159

1. T = 1, S = 5, O = 0 **2.** U = 8, S = 5, A = 1, L = 0 **3.** C = 8, O = 1, A = 6, L = 0, S = 2, I = 9 **4.** A = 3, T = 6, E = 7, S = 1, W = 4, O = 2, U = 0, H = 8

EXERCISE SET 3.6 page 159

1. Valid **2.** Invalid **3.** Valid **4.** Invalid **5.** Valid **6.** Invalid **7.** Valid **8.** Invalid **9.** Invalid **10.** Valid **11.** Invalid **12.** Invalid **13.** Invalid **14.** Valid **15.** Valid **16.** Invalid **17.** Valid **18.** Invalid **19.** Invalid **20.** Invalid **21.** All reuben sandwiches need mustard. **22.** Boomer is not a cat. **23.** 1001 ends with a 5. **24.** If it isn't broken, then I do not get paid. **25.** Some horses are grey. **26.** If we like to ski, then we will buy a condo. **27. a.** Invalid **b.** Invalid **c.** Invalid **d.** Invalid **e.** Valid **f.** Valid **28. a.** Invalid **b.** Valid **c.** Valid **d.** Valid **e.** Invalid

29. c cannot be true, because most of the seven different types of phones, have a touch screen keyboard.

30. The following diagram shows a bilateral diagram as devised by Charles Dodgson. Note that the diagram has four regions, much like a Venn diagram for two sets, but in the bilateral diagram each region is a square and all of the regions are the same size. The upper-left region marked xy corresponds to the $X \cap Y$ region of a Venn diagram. The fact that the regions in a bilateral diagram are all large square regions allowed Dodgson to place red and grey counters in the squares or on the boundaries between the squares. A red (dark) counter indicated the existence of that type of element. A grey (light) counter was used to indicate that a particular type of element did not exist. For instance, the bilateral diagram on the right below indicates the existence of an xy element and the nonexistence of an xy' element. These bilateral diagrams, along with the red and grey counters, allowed Dodgson to solve many logic problems in a visual manner. He also devised a trilateral diagram that could be used to solve logic problems involving three sets of elements. (*Source:* http://durendal.org/lcsl052.html)

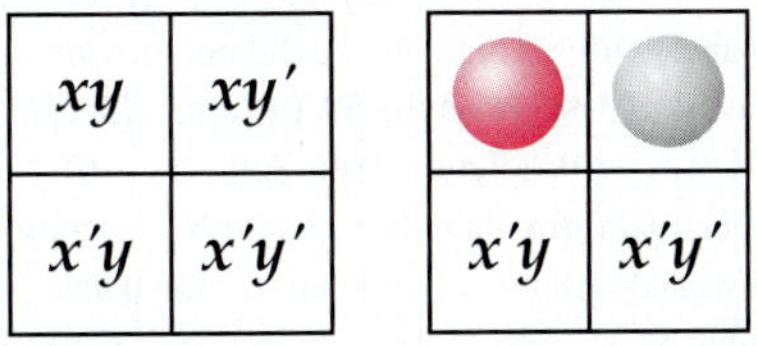

A bilateral diagram

CHAPTER 3 REVIEW EXERCISES page 164

1. Not a statement [Sec. 3.1] **2.** Statement [Sec. 3.1] **3.** Statement [Sec. 3.1] **4.** Statement [Sec. 3.1] **5.** Not a statement [Sec. 3.1] **6.** Statement [Sec. 3.1] **7.** $m \wedge b$, conjunction [Sec. 3.1] **8.** $d \to e$, conditional [Sec. 3.1]

9. $g \leftrightarrow d$, biconditional [Sec. 3.1] **10.** $t \rightarrow s$, conditional [Sec. 3.1] **11.** No dogs bite. [Sec. 3.1] **12.** Some desserts at the Cove restaurant are not good. [Sec. 3.1] **13.** Some winners do not receive a prize. [Sec. 3.1] **14.** All cameras use film. [Sec. 3.1] **15.** Some students finished the assignment. [Sec. 3.1] **16.** Nobody enjoyed the story. [Sec. 3.1] **17.** True [Sec. 3.1] **18.** True [Sec. 3.1] **19.** True [Sec. 3.1] **20.** True [Sec. 3.1] **21.** False [Sec. 3.2] **22.** False [Sec. 3.2/3.3] **23.** True [Sec. 3.2] **24.** True [Sec. 3.2/3.3] **25.** True [Sec. 3.2/3.3] **26.** False [Sec. 3.2/3.3]

p	*q*	**27.** [Sec. 3.2/3.3]	**28.** [Sec. 3.2/3.3]	**29.** [Sec. 3.2/3.3]	**30.** [Sec. 3.2/3.3]
T	T	T	F	F	T
T	F	T	F	F	T
F	T	T	T	F	F
F	F	F	F	F	T

p	*q*	*r*	**31.** [Sec. 3.2/3.3]	**32.** [Sec. 3.2/3.3]	**33.** [Sec. 3.2/3.3]	**34.** [Sec. 3.2/3.3]
T	T	T	T	F	F	T
T	T	F	T	T	F	T
T	F	T	T	T	T	T
T	F	F	F	T	T	T
F	T	T	T	F	F	F
F	T	F	T	F	F	T
F	F	T	T	T	F	T
F	F	F	T	T	F	T

35. Bob passed the English proficiency test or he did not register for a speech course. [Sec. 3.2] **36.** It is not true that, Ellen went to work this morning or she took her medication. [Sec. 3.2] **37.** It is not the case that, Wendy will not go to the store this afternoon and she will be able to prepare her fettuccine al pesto recipe. [Sec. 3.2] **38.** It is not the case that, Gina did not enjoy the movie or she enjoyed the party. [Sec. 3.2/3.3]
See the *Instructor's Solutions Manual* for solutions to Exercises 39 to 42.
43. Self-contradiction [Sec. 3.2] **44.** Tautology [Sec. 3.2/3.3] **45.** Tautology [Sec. 3.2/3.3] **46.** Tautology [Sec. 3.2/3.3]
47. *Antecedent:* he has talent
Consequent: he will succeed [Sec. 3.3]
48. *Antecedent:* I had a credential
Consequent: I could get the job [Sec. 3.3]
49. *Antecedent:* I join the fitness club
Consequent: I will follow the exercise program [Sec. 3.3]
50. *Antecedent:* I will attend
Consequent: it is free [Sec. 3.3]
51. She is not tall or she would be on the volleyball team. [Sec. 3.3] **52.** He cannot stay awake or he would finish the report. [Sec. 3.3] **53.** Rob is ill or he would start. [Sec. 3.3] **54.** Sharon will not be promoted or she closes the deal. [Sec. 3.3] **55.** I get my paycheck and I do not purchase a ticket. [Sec. 3.3] **56.** The tomatoes will get big and you did not provide them with plenty of water. [Sec. 3.3] **57.** You entered Cleggmore University and you did not have a high score on the SAT exam. [Sec. 3.3] **58.** Ryan enrolled at a university and he did not enroll at Yale. [Sec. 3.3] **59.** False [Sec. 3.3] **60.** True [Sec. 3.3] **61.** False [Sec. 3.3] **62.** False [Sec. 3.3] **63.** If a real number has a nonrepeating, nonterminating decimal form, then the real number is irrational. [Sec. 3.4] **64.** If you are a politician, then you are well known. [Sec. 3.4] **65.** If I can sell my condominium, then I can buy the house. [Sec. 3.4] **66.** If a number is divisible by 9, then the number is divisible by 3. [Sec. 3.4] **67. a.** *Converse:* If $x > 3$, then $x + 4 > 7$. **b.** *Inverse:* If $x + 4 \leq 7$, then $x \leq 3$. **c.** *Contrapositive:* If $x \leq 3$, then $x + 4 \leq 7$. [Sec. 3.4] **68. a.** *Converse:* If a recipe can be prepared in less than 20 minutes, then the recipe is in this book. **b.** *Inverse:* If a recipe is not in this book, then the recipe cannot be prepared in less than 20 minutes. **c.** *Contrapositive:* If a recipe cannot be prepared in less than 20 minutes, then the recipe is not in this book. [Sec. 3.4] **69. a.** *Converse:* If $(a + b)$ is divisible by 3, then a and b are both divisible by 3. **b.** *Inverse:* If a and b are not both divisible by 3, then $(a + b)$ is not divisible by 3. **c.** *Contrapositive:* If $(a + b)$ is not divisible by 3, then a and b are not both divisible by 3. [Sec. 3.4]

70. a. *Converse:* If they come, then you built it. **b.** *Inverse:* If you do not build it, then they will not come. **c.** *Contrapositive:* If they do not come, then you did not build it. [Sec. 3.4] **71. a.** *Converse:* If it has exactly two parallel sides, then it is a trapezoid. **b.** *Inverse:* If it is not a trapezoid, then it does not have exactly two parallel sides. **c.** *Contrapositive:* If it does not have exactly two parallel sides, then it is not a trapezoid. [Sec. 3.4] **72. a.** *Converse:* If they returned, then they liked it. **b.** *Inverse:* If they do not like it, then they will not return. **c.** *Contrapositive:* If they do not return, then they did not like it. [Sec. 3.4] **73.** $q \rightarrow p$, the converse of the original statement [Sec. 3.4] **74.** True **75.** If x is an odd prime number, then $x > 2$. [Sec. 3.4] **76.** If the senator attends the meeting, then she will vote on the motion. [Sec. 3.4] **77.** If their manager contacts me, then I will purchase some of their products. [Sec. 3.4] **78.** If I can rollerblade, then Ginny can rollerblade. [Sec. 3.4] **79.** Valid [Sec. 3.5] **80.** Valid [Sec. 3.5] **81.** Invalid [Sec. 3.5] **82.** Valid [Sec. 3.5] **83.** Valid argument, disjunctive reasoning [Sec. 3.5] **84.** Valid argument, transitive reasoning [Sec. 3.5] **85.** Invalid argument, fallacy of the inverse [Sec. 3.5] **86.** Valid argument, disjunctive reasoning [Sec. 3.5] **87.** Valid argument, contrapositive reasoning [Sec. 3.5] **88.** Invalid argument, fallacy of the inverse [Sec. 3.5] **89.** Valid [Sec. 3.6] **90.** Invalid [Sec. 3.6] **91.** Invalid [Sec. 3.6] **92.** Valid [Sec. 3.6]

CHAPTER 3 TEST *page 166*

1. a. Not a statement **b.** Statement [Sec. 3.1, Example 1] **2. a.** All trees are green. **b.** Some apartments are available. [Sec. 3.1, Example 2] **3. a.** False **b.** True [Sec. 3.1, Example 6] **4. a.** False **b.** True [Sec. 3.3, Example 3]

p	q	**5. [Sec. 3.3, Example 3]**
T	T	T
T	F	T
F	T	T
F	F	T

p	q	r	**6. [Sec. 3.3, Example 3]**
T	T	T	F
T	T	F	T
T	F	T	F
T	F	F	F
F	T	T	F
F	T	F	T
F	F	T	T
F	F	F	F

7. It is not true that Elle ate breakfast or took a lunch break. [Sec. 3.2, Example 5]

8. A tautology is a statement that is always true. [Sec. 3.2, Example 6] **9.** $\sim p \vee q$ [Sec. 3.3, Example 4] **10. a.** False **b.** False [Sec. 3.3, Example 2] **11. a.** *Converse:* If $x > 4$, then $x + 7 > 11$. **b.** *Inverse:* If $x + 7 \leq 11$, then $x \leq 4$. **c.** *Contrapositive:* If $x \leq 4$, then $x + 7 \leq 11$. [Sec. 3.4, Example 2]

12. $p \rightarrow q$
$\underline{p}$
$\therefore q$ [Sec. 3.5, Table 3.15]

13. $p \rightarrow q$
$\underline{q \rightarrow r}$
$\therefore p \rightarrow r$ [Sec. 3.5, Table 3.15]

14. $p \rightarrow q$
$\underline{\sim q}$
$\therefore \sim p$ [Sec. 3.5, Table 3.15]

15. $p \rightarrow q$
$\underline{\sim p}$
$\therefore \sim q$ [Sec. 3.5, Table 3.16]

16. Valid [Sec. 3.5, Example 2]

17. Invalid [Sec. 3.5, Example 3] **18.** Invalid argument; the argument is a fallacy of the inverse. [Sec. 3.5, Table 3.16] **19.** Valid argument; disjunctive reasoning [Sec. 3.5, Table 3.15] **20.** Invalid argument, as shown by an Euler diagram [Sec. 3.6, Example 2] **21.** Invalid argument, as shown by an Euler diagram [Sec. 3.6, Example 2] **22.** Invalid argument; the argument is a fallacy of the converse. [Sec. 3.5, Table 3.16]

CHAPTER 4

EXCURSION EXERCISES, SECTION 4.1 page 182

2. Connecticut: 7; Delaware: 2; Georgia: 2; Maryland: 8; Massachusetts: 14; Kentucky: 2; New Hampshire: 4; Vermont: 2; New York: 10; New Jersey: 5; Pennsylvania: 13; North Carolina: 10; South Carolina: 6; Virginia: 18; Rhode Island: 2 **3.** Virginia **4.** Connecticut: 7; Delaware: 2; Georgia: 2; Maryland: 8; Massachusetts: 14; Kentucky: 2; New Hampshire: 4; Vermont: 3; New York: 10; New Jersey: 5; Pennsylvania: 12; North Carolina: 10; South Carolina: 6; Virginia: 18; Rhode Island: 2

EXERCISE SET 4.1 page 183

1. To calculate the standard divisor, divide the total population p by the number of items to apportion n. **2. a.** 8.96. There is one teacher aid for every 8.96 students. **b.** Kindergarten: 4; first: 4; second: 4; third: 3; fourth: 2; fifth: 4; sixth: 4 **c.** Kindergarten: 4; first: 5; second: 4; third: 3; fourth: 2; fifth: 3; sixth: 4. No **d.** Using the Jefferson method, the first grade receives one more teacher's aide and the fifth grade receives one less teacher's aide compared to the outcome with the Hamilton method. **3.** The standard quota for a state is the whole number part of the quotient of the state's population divided by the standard divisor. **4.** The quota rule stipulates that the number of representatives apportioned to a state must be equal to or one more than the standard quota. **5. a.** 0.273 **b.** 0.254 **c.** Salinas **6. a.** 0.369 **b.** 0.329 **c.** Rust Canyon **7.** Seaside Mall **8.** Rainer **9. a.** 709,760; There is one representative for every 709,760 citizens in the U.S. **b.** Underrepresented. The average constituency is greater than the standard divisor. **c.** Overrepresented. The average constituency is less than the standard divisor. **10. a.** 87.8. There is one new computer for every 88 students. **b.** Liberal arts: 39; business: 66; humanities: 22; science: 53 **c.** Using 86 as a modified standard divisor yields an apportionment of 183, which is greater than the number of available new computers. Using 87 yields an apportionment of 179, which is less than the number of available new computers. Using 86.5 yields an apportionment that is equal to the number of available new computers. **d.** Using a number larger than the standard divisor results in an apportionment that is less than the number of available new computers. **e.** Liberal arts: 39; business: 66; humanities: 21; science: 54. No **f.** Using the Jefferson method, the humanities division receives one less computer and the science division receives one more computer when compared with the Hamilton method. **11. a.** 37.10. There is one new nurse for every 37.10 beds. **b.** Sharp: 7; Palomar: 10; Tri-City: 8; Del Raye: 5; Rancho Verde: 7; Bel Aire: 11 **c.** Sharp: 7; Palomar: 10; Tri-City: 8; Del Raye: 5; Rancho Verde: 7; Bel Aire: 11 **d.** They are identical. **12.** The Alabama paradox occurs when adding one representative to the number of representatives to be apportioned results in a state losing a representative even though the populations do not change. **13.** The population paradox occurs when the population of one state is increasing faster than that of another state, yet the first state still loses a representative. **14.** The new states paradox occurs when a new state is added along with the appropriate number of representatives to account for its population, but after reapportionment, another state ends up losing a representative. **15.** The Balinski-Young Impossibility Theorem states that any apportionment method either will violate the quota rule or will produce paradoxes such as the Alabama paradox. **16. a.** A: 4; B: 9; C: 2; D: 13. No. None of the campus locations lost a projector. **b.** A: 4; B: 10; C: 1; D: 14. Yes. Increasing the number of projectors to be apportioned from 28 to 29 causes campus C to lose a projector. **17. a.** Yes **b.** No **c.** Yes **18. a.** A: 4; B: 10; C: 26 **b.** Yes. The enrollment of center A increased at a greater rate than that of center B, yet after reappportionment, center A lost a security person and center B gained one. **19. a.** Boston: 2; Chicago: 20 **b.** Yes. Chicago lost a vice president while Boston gained one. **20. a.** Math: 14; physics: 1; chemistry: 4 **b.** Yes. After adding a new department and adjusting the number of clerical assistants to account for the increased total enrollment, reapportionment causes the math department to lose an assistant even though its population has not changed. **21. a.** Sixth grade **b.** Sixth grade. Same **22. a.** Valley office **b.** Valley office. Same **c.** Hill Street office **23.** Valley **24.** Other states must have had proportionally larger increases in population. **25.** New York's population is much larger than Louisiana's, so New York has many more representatives, and a percentage change in its population adds or subtracts more representatives than for smaller states. **26.** North Carolina's 2010 census Huntington-Hill number must be larger than the numbers of all other states because, using the Huntington-Hill apportionment method, a new representative is awarded to the state with the highest Huntington-Hill number. **27. a.** They are the same. **b.** Using the Jefferson method, the humanities division gets one less computer and the sciences division gets one more computer compared with using the Webster method. **28. a.** Using 13,500,000 as a modified standard divisor yields 21 committee members. **b.** Using 13,750,000 as a modified standard divisor yields 20 committee members. **c.** France: 5; Germany: 6; Italy: 4; Spain: 4; Belgium: 1 **29.** The Jefferson and Webster methods **30.** The Webster method **31.** The Huntington-Hill method **32.** Answers will vary. **33.** Calculate each state's standard quota using the standard divisor, and always round up to the next highest whole number. Modify the standard divisor until the sum of the states' calculated quotas equals the number of representatives to be apportioned. The Adams method violates the quota rule but is not susceptible to any paradoxes. **34.** Del Mar: 3; Wheatly: 7; West: 5; Mountain View: 7

35. a. $\frac{P_A}{a+1}$ **b.** $\frac{P_B}{b}$ **c.** $\frac{\frac{P_B}{b} - \frac{P_A}{a+1}}{\frac{P_A}{a+1}}$ **d.** $\frac{\frac{P_A}{a} - \frac{P_B}{b+1}}{\frac{P_B}{b+1}}$ **e.** $\frac{\frac{P_A}{a} - \frac{P_B}{b+1}}{\frac{P_B}{b+1}} < \frac{\frac{P_B}{b} - \frac{P_A}{a+1}}{\frac{P_A}{a+1}}$ **f.** $\frac{(P_B)^2}{b(b+1)} < \frac{(P_A)^2}{a(a+1)}$

EXCURSION EXERCISES, SECTION 4.2 page 201

1. Lemon, raspberry, plain **2.** No **3.** No **4.** Yes **5.** Yes **6.** Yes **7.** No

EXERCISE SET 4.2 *page 201*

1. A majority means that a choice receives more than 50% of the votes. A plurality means that the choice with the most votes wins. It is possible to have a plurality without a majority when there are more than two choices. **2.** If there are more than two choices, the winning choice may not have majority support. **3.** If there are n choices in an election, each voter ranks the choices by giving n points to the voter's first choice, $n - 1$ points to the voter's second choice, and so on, with the voter's least favorite choice receiving 1 point. The choice with the most points is the winner. **4.** Voters rank their preferences for all choices. The choice with the fewest first-place votes is eliminated, and preferences are renumbered assuming that voters do not change their preferences for the remaining choices when one choice is removed. This process of eliminating the choice of lowest preference is continued until the winner is determined. **5.** In the pairwise comparison voting method, each choice is compared one-on-one with each of the other choices. A choice receives 1 point for a win, 0.5 points for a tie, and 0 points for a loss. The choice with the greatest number of points is the winner. **6.** The Condorcet criterion states that a candidate who wins all possible head-to-head matchups should win an election when all candidates appear on the ballot.
7. No; no **8.** If there are only two candidates, the plurality winner will have a majority of the first-place votes. Using the Borda count method, two points are assigned to the first choice and one point is assigned to the second choice. **9. a.** Al Gore **b.** No **c.** George W. Bush **10.** There is a tie between Corn Flakes and Raisin Bran. **11. a.** 35 **b.** 18 **c.** Buzz Lightyear **12.** There is a tie between Italian and Mexican. **13.** Stream online **14.** There is a tie between Corn Flakes and Raisin Bran. **15.** SpongeBob SquarePants **16.** Mexican **17.** Elaine Garcia **18.** Verizon **19.** Blue and white **20.** WKLS **21.** Raymond Lee **22.** Verizon **23. a.** Buy new computers for the club. **b.** Pay for several members to travel to a convention. **c.** Pay for several members to travel to a convention. **d.** Answers will vary. **24. a.** Amusement park **b.** Amusement park **c.** There is a tie between picnic in a park and amusement park. **25.** *X-Men: Days of Future Past* **26.** Grand Canyon **27.** Hornet **28.** There is a three-way tie among P. Gibson, R. Allenbaugh, and G. DeWitte. **29.** Blue and white **30.** There is a three-way tie among WKLS, WNNX, and WWVV. **31.** No **32.** No **33.** Yes **34.** No **35.** No **36.** No **37. a.** Stephen Hyde **b.** Stephen Hyde received the smallest number of first-place votes. **c.** John Lorenz **d.** The candidate who wins all head-to-head matches does not win the election. **e.** John Lorenz **f.** The candidate winning the original election (Stephen Hyde) did not remain the winner in a recount after a losing candidate withdrew from the race. **38. a.** Film A **b.** Film B **c.** The monotonicity criterion has not been violated; Film A won the first vote. **39.** Ahmad is president, Jen is vice president, Andrew is secretary, and Hector is treasurer. **40.** The Borda count method **41. a.** Candidate B **b.** Candidate A **42.** There is a tie between *Good Will Hunting* and *Contact.* **43.** Round Table

EXCURSION EXERCISES, SECTION 4.3 *page 216*

1. {A, B}, {A, D}, {B, D}, {A, B, C}, {A, B, D}, {A, C, D}, {B, C, D} **2.** $BPI(\text{A}) = \frac{1}{3}$, $BPI(\text{B}) = \frac{1}{3}$, $BPI(\text{C}) = 0$, $BPI(\text{D}) = \frac{1}{3}$

3. {B, F}, {B, G}, {B, I}, {F, G}, {F, I}, {F, N}, {G, I}, {G, N}, {I, N}, {B, F, G}, {B, F, I}, {B, F, L}, {B, F, N}, {B, G, I}, {B, G, L}, {B, G, N}, {B, I, L}, {B, I, N}, {F, G, I}, {F, G, L}, {F, G, N}, {F, I, L}, {F, I, N}, {F, L, N}, {G, I, L}, {G, I, N}, {G, L, N}, {I, L, N}, {B, F, G, I}, {B, F, G, L}, {B, F, G, N}, {B, F, I, L}, {B, F, I, N}, {B, F, L, N}, {B, G, I, L}, {B, G, I, N}, {B, G, L, N}, {B, I, L, N}, {F, G, I, L}, {F, G, I, N}, {F, G, L, N}, {F, I, L, N}, {G, I, L, N}, {B, F, G, I, L}, {B, F, G, I, N}, {B, F, G, L, N}, {B, F, I, L, N}, {B, G, I, L, N}, {F, G, I, L, N}

4. $BPI(\text{Belgium}) = \frac{1}{7}$; $BPI(\text{Luxembourg}) = 0$ **5.** Answers will vary. **6.** Answers will vary. **7.** Answers will vary.

EXERCISE SET 4.3 *page 217*

1. a. 6 **b.** 4 **c.** 3 **d.** 6 **e.** No **f.** A and C **g.** 15 **h.** 6 **2. a.** 16 **b.** 5 **c.** 4 **d.** 11 **e.** No **f.** A, B, and D **g.** 31 **h.** 10 **3.** 0.60, 0.20, 0.20 **4.** 0.33, 0.33, 0.33 **5.** 0.50, 0.30, 0.10, 0.10 **6.** 0.25, 0.25, 0.25, 0.25 **7.** 0.36, 0.28, 0.20, 0.12, 0.04 **8.** 0.25, 0.25, 0.25, 0.25 **9.** 1.00, 0.00, 0.00, 0.00, 0.00, 0.00 **10.** 0.27, 0.27, 0.18, 0.18, 0.09 **11.** 0.44, 0.20, 0.20, 0.12, 0.04 **12.** 0.5, 0.3, 0.1, 0.1 **13. a.** Exercise 9 **b.** Exercises 3, 5, 6, 9, and 12 **c.** None **d.** Exercise 8 **14.** There are no winning coalitions. **15.** 0.33, 0.33, 0.33 **16.** 0.5, 0.125, 0.125, 0.125, 0.125 **17. a.** {12: 1, 1, 1, 1, 1, 1, 1, 1, 1, 1, 1, 1} **b.** Yes **c.** Yes **d.** Divide the voting power, 1, by the quota, 12. **18. a.** {9: 1, 1, 1, 1, 1, 1, 1, 1, 1, 1, 1, 1} **b.** Yes **c.** No **d.** Take the inverse of the total number of voters. **19.** Dictator: A; dummies: B, C, D, E **20.** Dummy: D **21.** None **22.** Dummies: C, D **23. a.** 0.60, 0.20, 0.20 **b.** Answers will vary. **24. a.** 0.60, 0.20, 0.20 **b.** They are the same. No **25. a.** 0.33, 0.33, 0.33 **b.** This system has the same effect as a one-person, one-vote system. **26. a.** Voter D is never a critical voter in a winning coalition. **b.** Because D is a dummy voter, $BPI(\text{D}) = 0$. The remaining votes all have equal weight, so their BPIs are the same. The sum of the BPIs must be 1, so each remaining voter's BPI will be 1 divided by 3. **27. a.** 11 and 14 **b.** 15 and 16 **c.** No

28. a. Yes **b.** No **29. a.** No. $BPI(\text{A})$ in this fraudulent system is 0.5. **b.** No. $BPI(\text{C})$ in this fraudulent system is $\frac{10}{26} \approx 0.385$.

30. a. $BPI(7) \approx 0.1669$, $BPI(1) \approx 0.0165$ **b.** About 10 times **c.** $BPI(5) \approx 0.142$, $BPI(1) \approx 0.029$ **d.** About 5 times

CHAPTER 4 REVIEW EXERCISES *page 222*

1. a. Health: 7; business: 18; engineering: 10; science: 15 **b.** Health: 6; business: 18; engineering: 10; science: 16 **c.** Health: 7; business: 18; engineering: 10; science: 15 [Sec. 4.1] **2. a.** Newark: 9; Cleveland: 6; Chicago: 11; Philadelphia: 4; Detroit: 5 **b.** Newark: 9; Cleveland: 6; Chicago: 11; Philadelphia: 4; Detroit: 5 **c.** Newark: 9; Cleveland: 6; Chicago: 11; Philadelphia: 4; Detroit: 5 [Sec. 4.1] **3. a.** 0.098 **b.** 0.194 **c.** High Desert [Sec. 4.1] **4.** Morena Valley [Sec. 4.1] **5. a.** No. None of the offices loses a new printer. **b.** Yes. Office A drops from two new printers to only one new printer. [Sec. 4.1] **6. a.** A: 2; B: 5; C: 3; D: 16; E: 2. No. None of the centers loses an automobile. **b.** No. None of the centers loses an automobile. [Sec. 4.1] **7. a.** Los Angeles: 9; Newark: 2 **b.** Yes. Newark loses a computer file server. [Sec. 4.1] **8. a.** A: 10; B: 3; C: 21 **b.** Yes. The population of region B grew at a higher rate than the population of region A, yet region B lost an inspector to region A. [Sec. 4.1] **9.** Yes [Sec. 4.1] **10.** No [Sec. 4.1] **11. a.** A **b.** B [Sec. 4.1] **12. a.** Manuel Ortega **b.** No **c.** Crystal Kelley [Sec. 4.2] **13. a.** Vail **b.** Aspen [Sec. 4.2] **14.** A. Kim [Sec. 4.2] **15.** Snickers [Sec. 4.2] **16.** A. Kim [Sec. 4.2] **17.** Snickers [Sec. 4.2] **18. a.** Shannon M. **b.** Hannah A. **c.** Hannah A. won all head-to-head comparisons but lost the overall election. **d.** Hannah A. **e.** Hannah A. received a majority of the first-place votes but lost the overall election. [Sec. 4.2] **19. a.** Hannah A. **b.** Cynthia L., a losing candidate, withdrew from the race and caused a change in the overall winner of the election. [Sec. 4.2] **20.** The monotonicity criterion was violated because the only change was that the supporter of a losing candidate changed his or her vote to support the original winner, but the original winner did not win the second vote. [Sec. 4.2] **21. a.** 18 **b.** 18 **c.** Yes **d.** A and C **e.** 15 **f.** 6 [Sec. 4.3] **22. a.** 35 **b.** 35 **c.** Yes **d.** A **e.** 31 **f.** 10 [Sec. 4.3] **23.** 0.60, 0.20, 0.20 [Sec. 4.3] **24.** 0.20, 0.20, 0.20, 0.20, 0.20 [Sec. 4.3] **25.** 0.42, 0.25, 0.25, 0.08 [Sec. 4.3] **26.** 0.62, 0.14, 0.14, 0.05, 0.05 [Sec. 4.3] **27.** Dictator: A; dummies: B, C, D, E [Sec. 4.3] **28.** Dummy: D [Sec. 4.3] **29.** 0.50, 0.125, 0.125, 0.125, 0.125 [Sec. 4.3]

CHAPTER 4 TEST *page 226*

1. Spring Valley [Sec. 4.1, Example 3] **2. a.** Sales: 17; advertising: 4; service: 11; manufacturing: 53 **b.** Sales: 17; advertising: 4; service: 10; manufacturing: 54; no [Sec. 4.1, Examples 1 and 2] **3. a.** Cedar Falls ≈ 77,792; Lake View ≈ 70,290 **b.** Cedar Falls [Sec. 4.1, Example 4] **4. a.** 33 **b.** 26 **c.** No **d.** A and C **e.** 31 **f.** 10 [Sec. 4.3, Example 1] **5. a.** Aquafina **b.** No **c.** Evian [Sec. 4.2, Examples 1 and 2] **6.** New York [Sec. 4.2, Example 4] **7. a.** Afternoon **b.** Noon [Sec. 4.2, Examples 2 and 3] **8. a.** Proposal A **b.** Proposal B **c.** Eliminating a losing choice changed the outcome of the vote. **e.** Proposal B won all head-to-head comparisons but lost the vote when all the choices were on the ballot. [Sec. 4.2, Examples 1, 2, and 4] **9.** 0.42, 0.25, 0.25, 0.08 [Sec. 4.3, Example 2] **10.** 0.40, 0.20, 0.20, 0.20 [Sec. 4.3, Example 3]

CHAPTER 5

EXCURSION EXERCISES, SECTION 5.1 *page 240*

1. One answer is

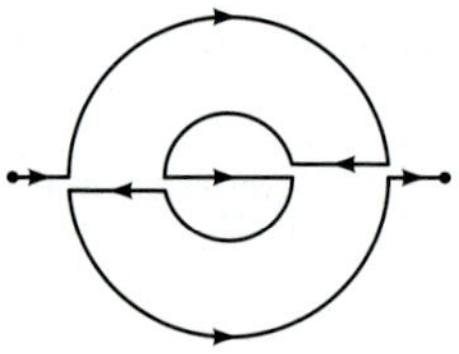

2. One answer is

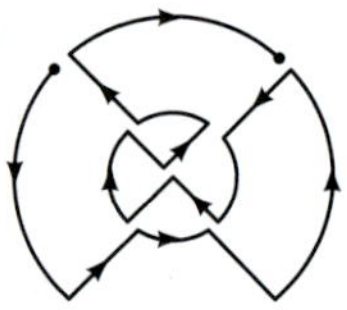

3. One answer is

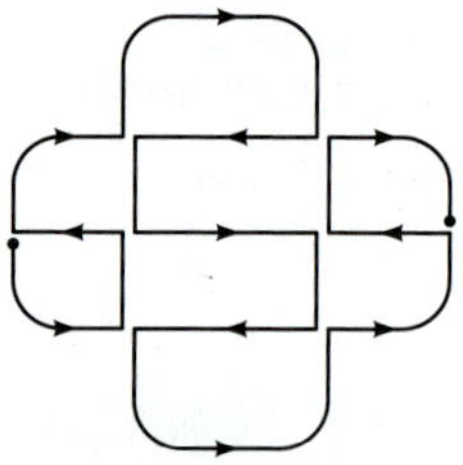

4. One answer is

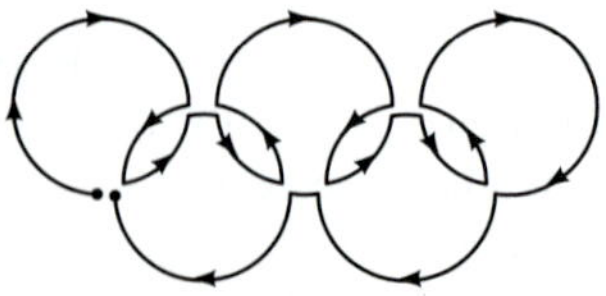

5. More than two vertices are of odd degree.

EXERCISE SET 5.1 *page 241*

1.

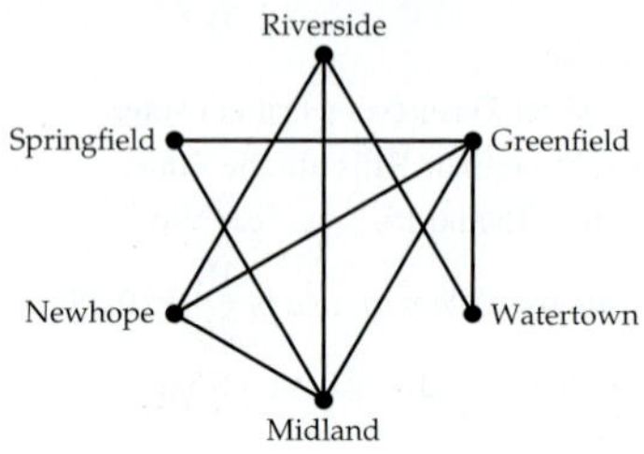

2.

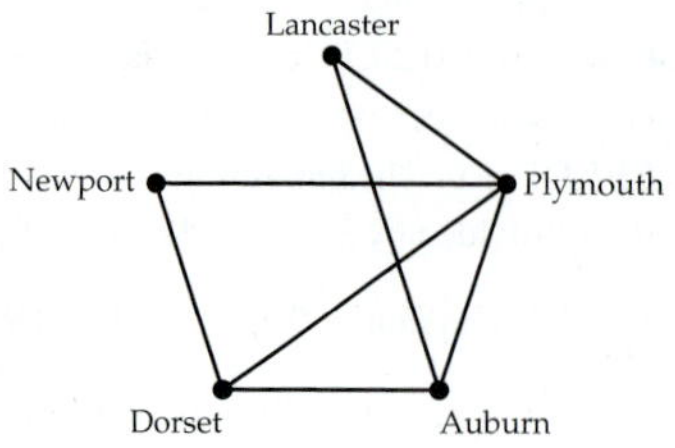

3. a. No **b.** 3 **c.** Ada **d.** A loop would correspond to a friend speaking to himself or herself.
4. a. Giants **b.** Titans **c.** Yes **5. a.** 6 **b.** 7 **c.** 6 **d.** Yes **e.** No
6. a. 8 **b.** 6 **c.** 4 **d.** No **e.** No **7. a.** 6 **b.** 4 **c.** 4 **d.** Yes **e.** Yes
8. a. 6 **b.** 2 **c.** 2 **d.** Yes **e.** No **9.** Equivalent **10.** Not equivalent **11.** Not equivalent
12. Not equivalent **13.** The graph on the right has a vertex of degree 4, and the graph on the left does not.
14.

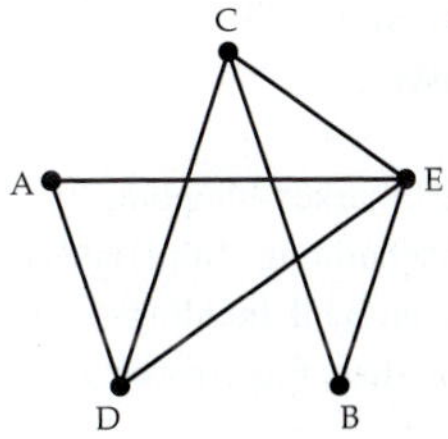

or

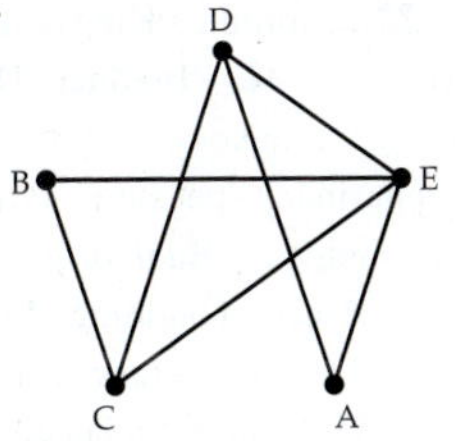

15. a. D–A–E–B–D–C–E–D

16. a. Not Eulerian **b.** A–B–D–A–C–E–B–E **17. a.** Not Eulerian **b.** A–E–A–D–E–D–C–E–C–B–E–B
18. a. Not Eulerian **b.** No **19. a.** Not Eulerian **b.** No **20. a.** A–B–A–D–C–D–E–F–E–B–C–A
21. a. Not Eulerian **b.** E–A–D–E–G–D–C–G–F–C–B–F–A–B–E–F
22. A–B–C–D–E–A–G–B–H–C–I–D–J–E–F–H–J–G–I–F–A
23. a.

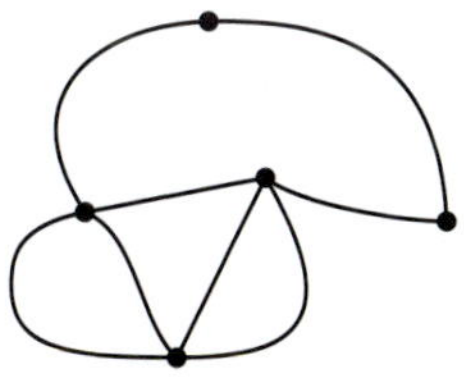

b. Yes **24. a.**

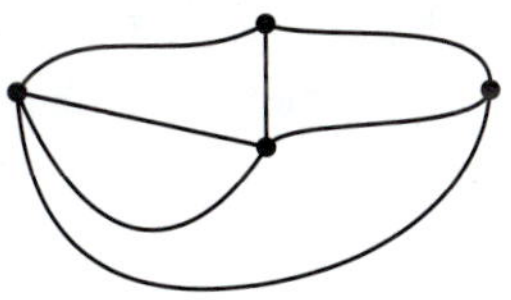

b. No; yes

25. Yes **26.** No **27.** Yes, but the hamster cannot return to its starting point.
28. Yes (exactly two vertices are of odd degree) **29.**

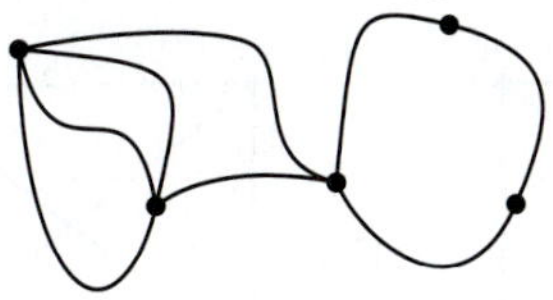

Yes. You can always return to the starting room.

30.

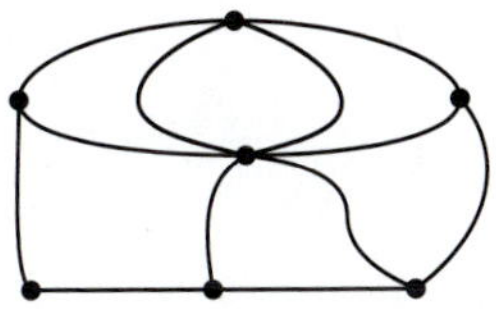

Not possible

31. a. 2 **b.** 4 **c.** Wayne
32. a. 2 **b.** 3 **c.** Marissa, Maty, Richie, Sheena **d.** No
33. a. AD **b.** EF, CF **c.** none **d.** BC, CD, AD

34. a.

b. Not possible; an Euler path
c. Venezuela–Guyana–Suriname–French Guiana–Brazil–Suriname–Guyana–Brazil–Venezuela–Colombia–Brazil–Peru–Colombia–Peru–Bolivia–Brazil–Paraguay–Bolivia–Paraguay–Argentina–Brazil–Uruguay–Argentina–Bolivia–Chile–Argentina–Chile–Peru–Ecuador–Colombia–Venezuela

EXCURSION EXERCISES, SECTION 5.2 *page 256*

1. C–A–B–E–D–C, total weight 18; D–E–B–A–C–D, total weight 18; E–B–A–C–D–E, total weight 18
2. A–F–C–D–E–B–A, total weight 65; B–D–E–F–C–A–B, total weight 66; C–F–A–D–E–B–C, total weight 70; D–E–B–C–F–A–D, total weight 70; E–D–B–C–F–A–E, total weight 61; F–C–D–E–B–A–F, total weight 65
3. A–F–C–B–D–E–A, total weight 61

EXERCISE SET 5.2 *page 256*

1. A–B–C–D–E–G–F–A **2.** A–D–B–F–C–E–A **3.** A–B–E–C–H–D–F–G–A
4. A–B–C–D–F–E–A **5.** Springfield–Greenfield–Watertown–Riverside–Newhope–Midland–Springfield
6. Newport–Plymouth–Lancaster–Auburn–Dorset–Newport **7.** A–B–E–D–C–A, total weight 31; A–D–E–B–C–A, total weight 32

8. A–F–B–D–C–E–A, total weight 44; A–D–C–E–B–F–A, total weight 41 **9.** A–D–C–E–B–F–A, total weight 114; A–C–D–E–B–F–A, total weight 158 **10.** A–G–B–C–D–E–F–A, total weight 339; A–F–G–E–D–C–B–A, total weight 353 **11.** A–D–B–C–F–E–A **12.** A–D–F–C–B–E–A **13.** A–C–E–B–D–A **14.** A–F–B–E–C–G–D–A **15.** A–D–B–F–E–C–A **16.** A–D–F–C–B–E–A **17.** A–C–E–B–D–A **18.** A–F–B–E–C–G–D–A **19.** Louisville–Evansville–Bloomington–Indianapolis–Lafayette–Fort Wayne–Louisville **20.** Toronto–Niagara Falls–Kingston–Ottawa–Windsor–Toronto **21.** Louisville–Evansville–Bloomington–Indianapolis–Lafayette–Fort Wayne–Louisville **22.** Toronto–Kingston–Ottawa–Windsor–Niagara Falls–Toronto **23.** Tokyo–Seoul–Beijing–Hong Kong–Bangkok–Tokyo **24.** Boulder–Denver–Colorado Springs–Grand Junction–Durango–Boulder **25.** Tokyo–Seoul–Beijing–Hong Kong–Bangkok–Tokyo **26.** Boulder–Denver–Colorado Springs–Durango–Grand Junction– Boulder **27.** Home–pharmacy–pet store–farmers market–shopping mall–home; home–pharmacy–pet store–shopping mall–farmers market–home **28.** Courier company–Bank of America building–Imperial Bank building–Prudential building–GE Tower–Design Center–courier company; courier company–Bank of America building–Imperial Bank building–Prudential building–Design Center–GE Tower–courier company **29.** Home state–task B–task D–task A–task C–home state. Edge-picking algorithm gives the same sequence. **30.** A–D–E–C–B–F–A **31.** Euler circuit **32.** Hamiltonian circuit

33. a.

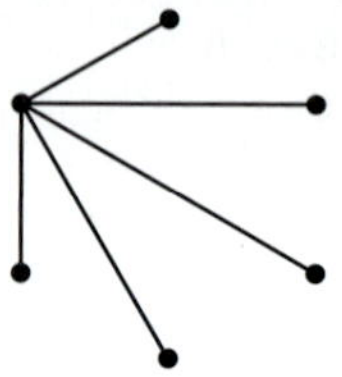

(many possible answers)

b.

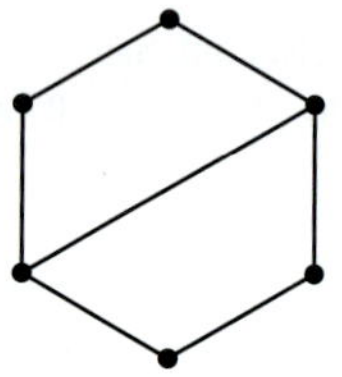

(many possible answers)

c.

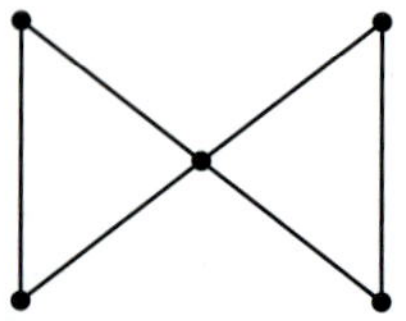

(many possible answers)

34. a.

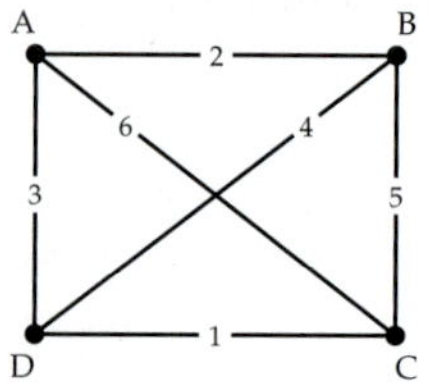

b.

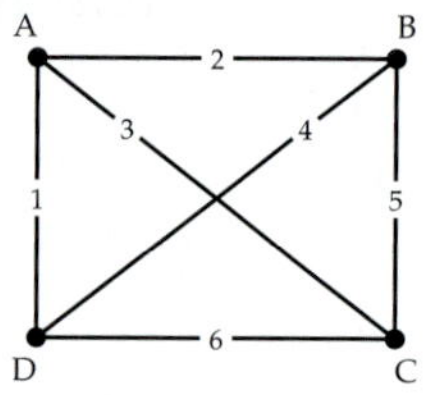

c.

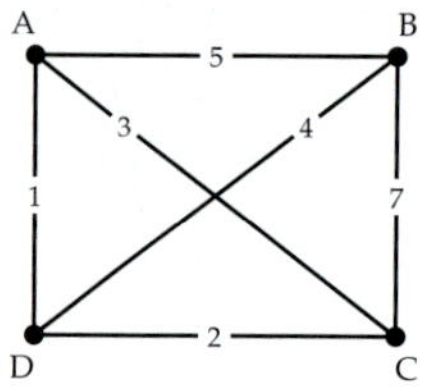

EXCURSION EXERCISES, SECTION 5.3 *page 268*

1.

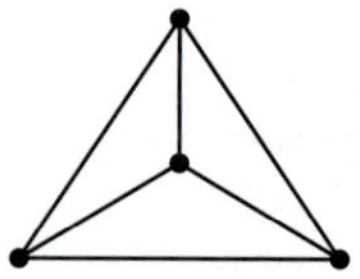

2. Octahedron **3. a.** 3 **b.** 30 **c.** 20

4. a. Not all faces of the graph have the same number of edges; thus not all faces of the corresponding polyhedron are the same shape. **b.** 14 **c.** Answers will vary.

EXERCISE SET 5.3 *page 269*

1.

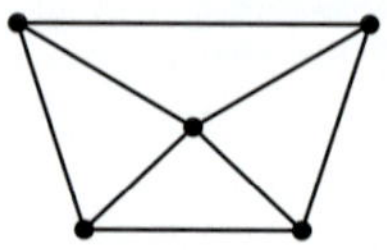

2.

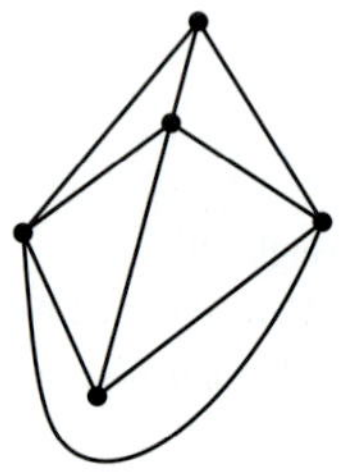

3.

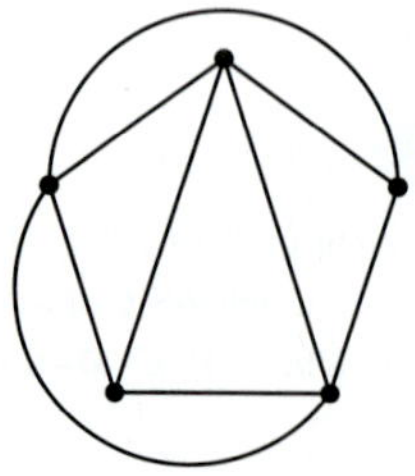

4.

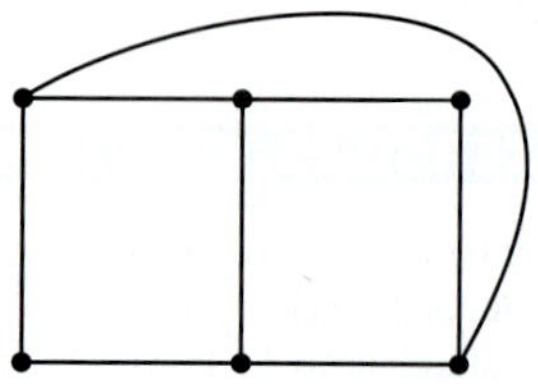

5.

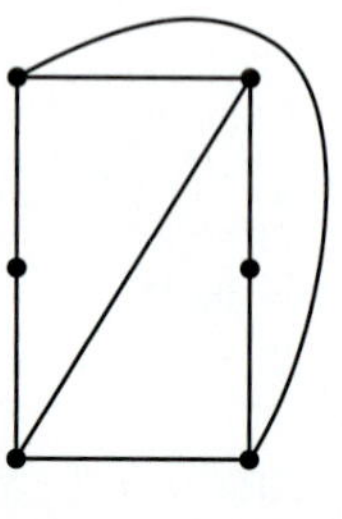

6.

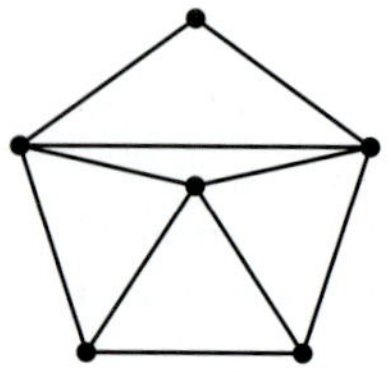

7.

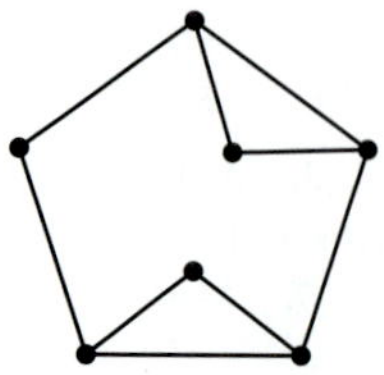

8.

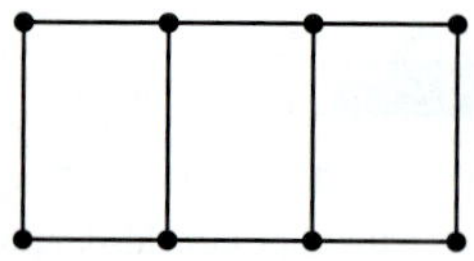

See the *Instructor's Solutions Manual* for the solutions to Exercises 9 to 16. **17.** 5 faces, 5 vertices, 8 edges
18. 4 faces, 6 vertices, 8 edges **19.** 2 faces, 8 vertices, 8 edges **20.** 3 faces, 6 vertices, 7 edges
21. 5 faces, 10 vertices, 13 edges **22.** 7 faces, 7 vertices, 12 edges **23.** 9 **24.** 148

25.

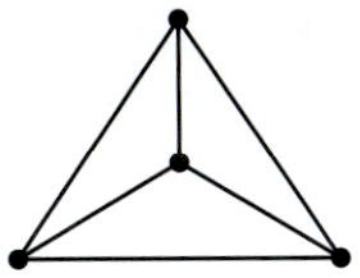

26.

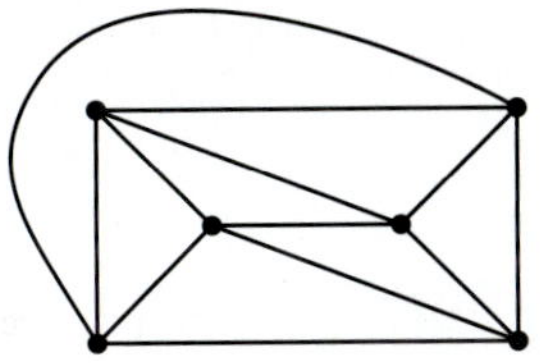

27. Euler's formula cannot be satisfied. **28.** There are two more faces than vertices.

29. With the labeling shown in the first figure, A–B–C–D–E–A is a Hamiltonian circuit, as shown in the second figure. Of the remaining five edges to be drawn, only two can be drawn inside the circuit without intersections, and it is impossible to draw the other three edges outside the circuit without intersections, as shown in the third figure. (Edge CE cannot be drawn.)

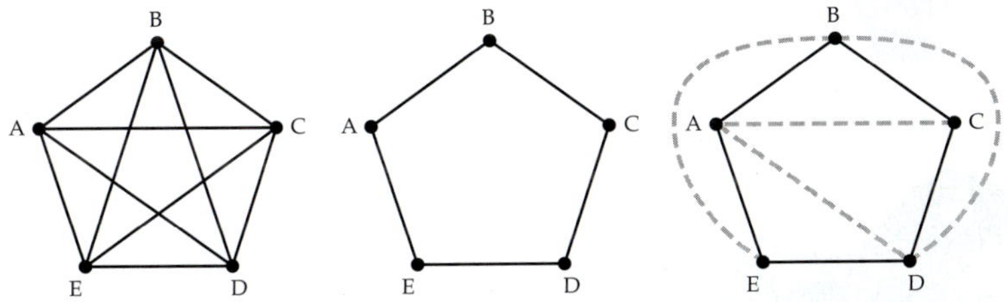

30. a.

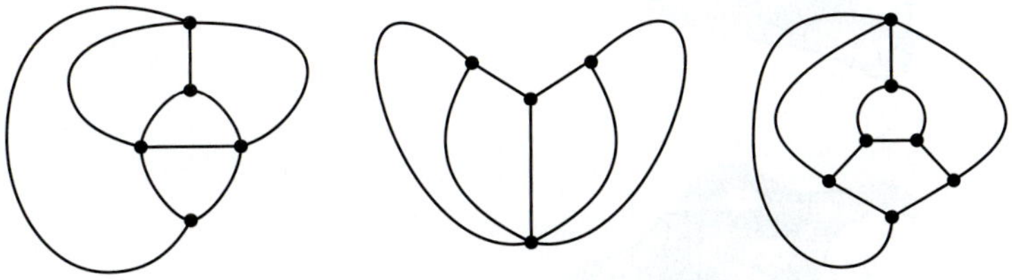

b. The number of vertices in the dual graph is equal to the number of faces in the original graph, and the number of faces in the dual graph is equal to the number of vertices in the original graph.

c. The dual of the dual is equivalent to the original graph.

EXCURSION EXERCISES, SECTION 5.4 *page 279*

1.

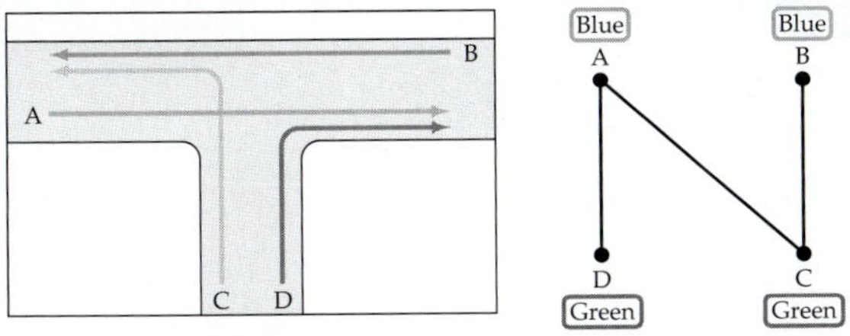

Two stages are required.

2.

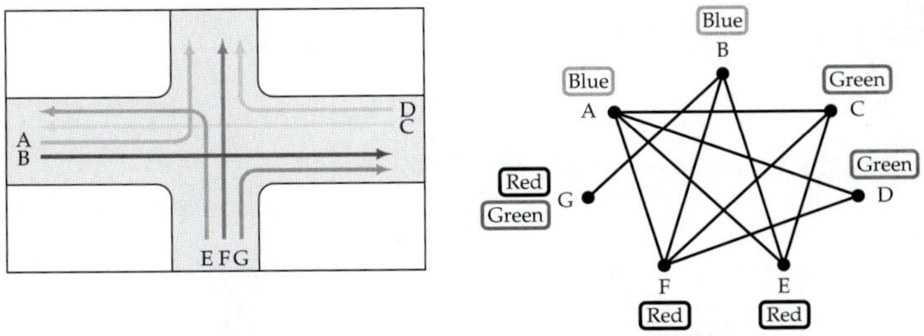

Three stages are required.

3.

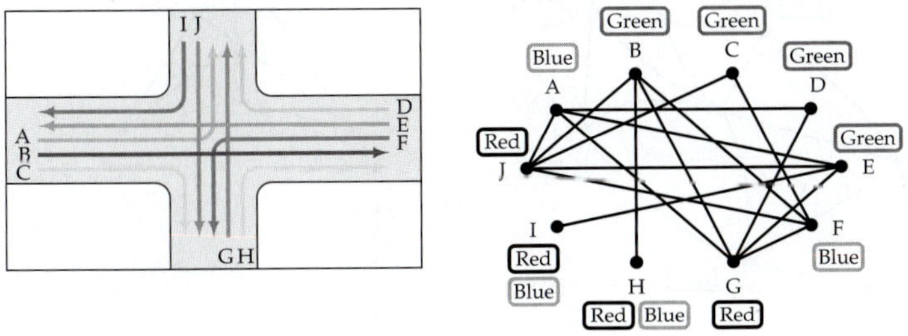

Three stages are required.

EXERCISE SET 5.4 *page 279*

1. Requires three colors

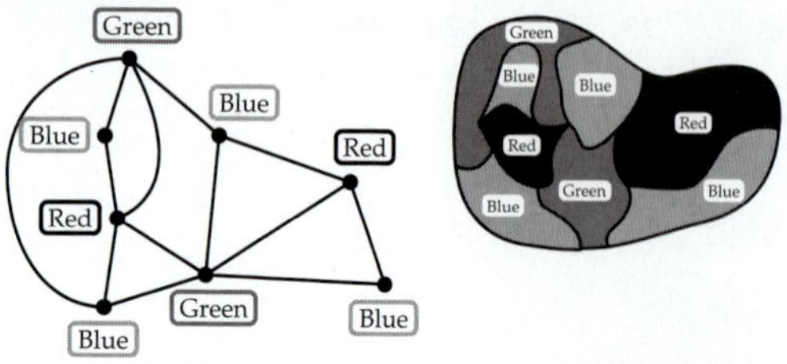

2. Requires four colors

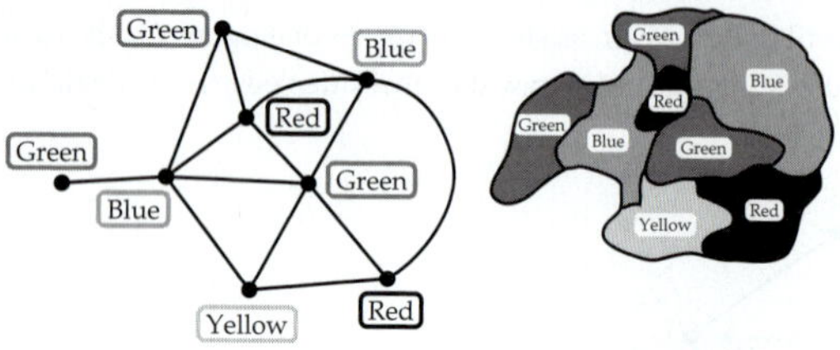

3. Requires three colors

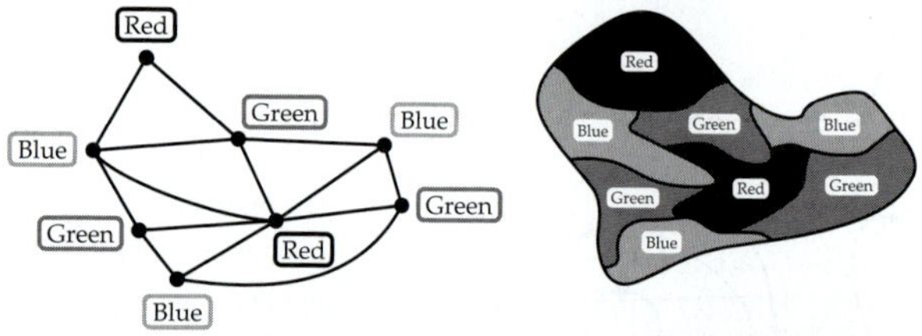

4. Requires four colors

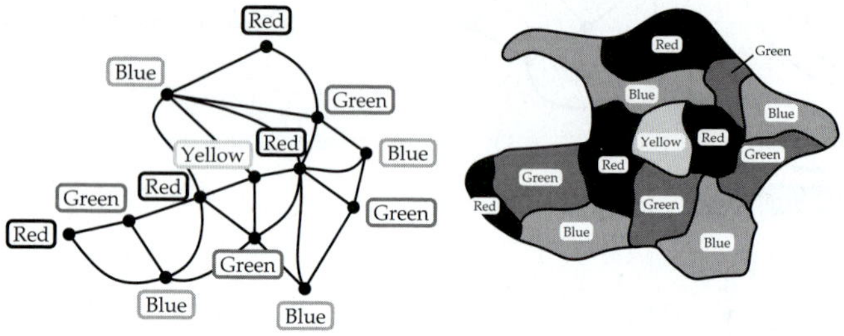

5.

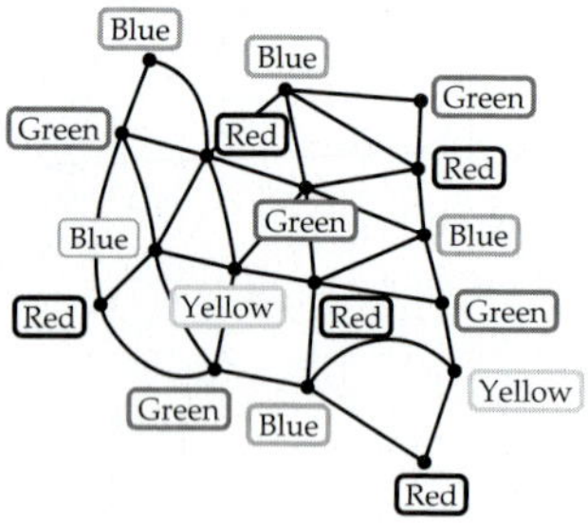

6.

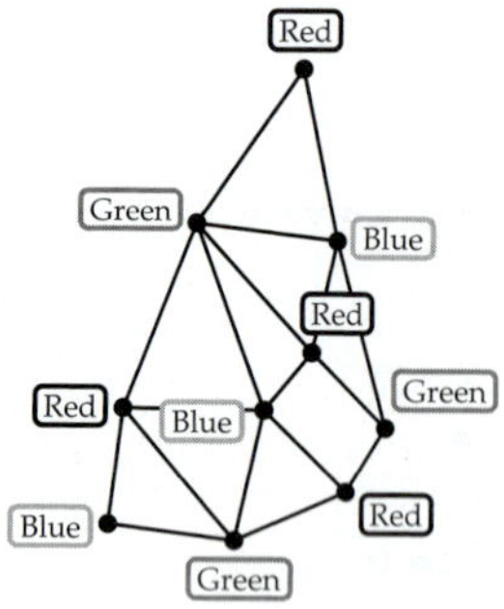

7.

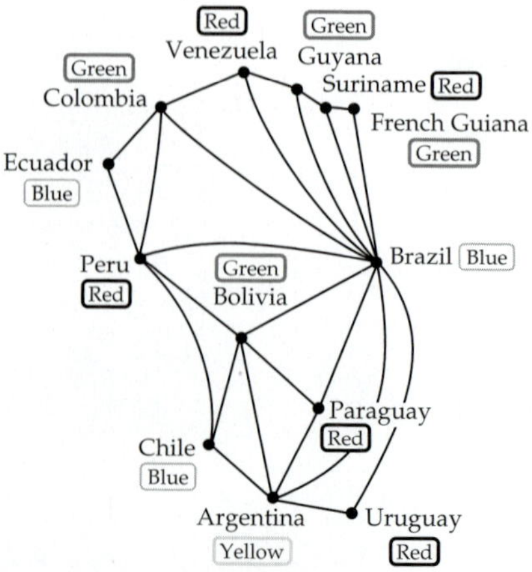

8.

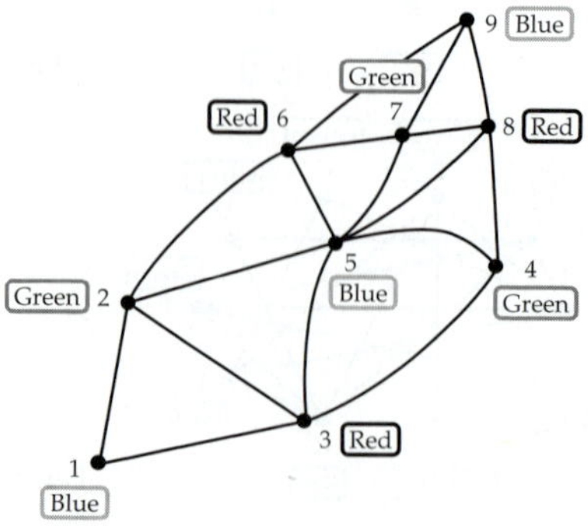

9.

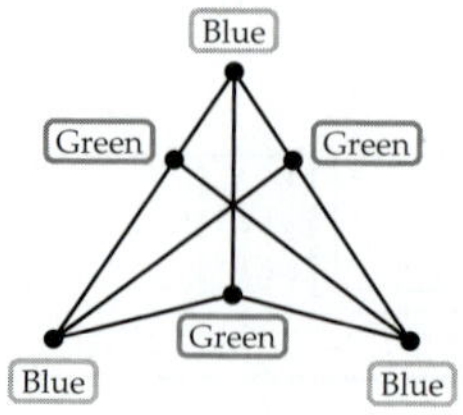

10. Not 2-colorable **11.** Not 2-colorable

12.

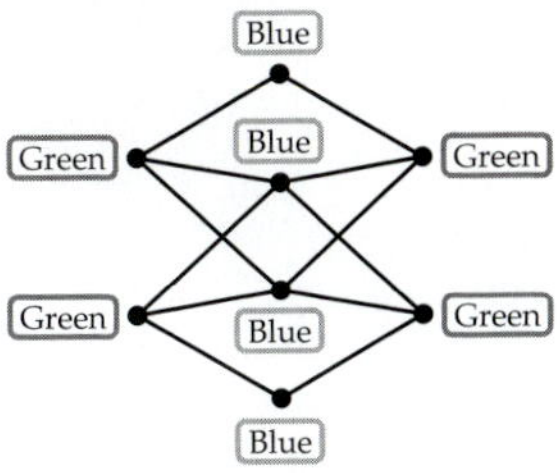

13.

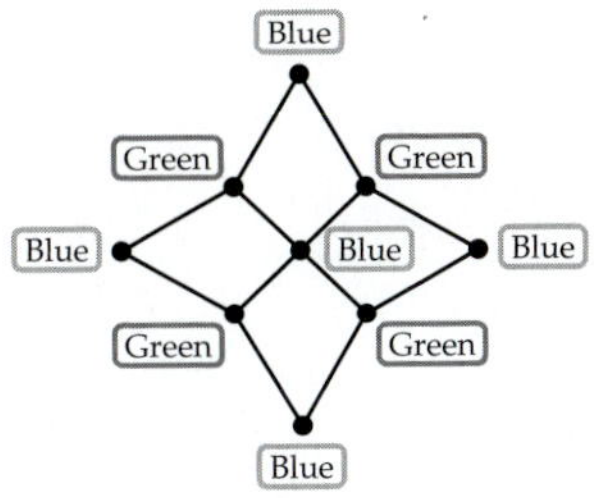

14.

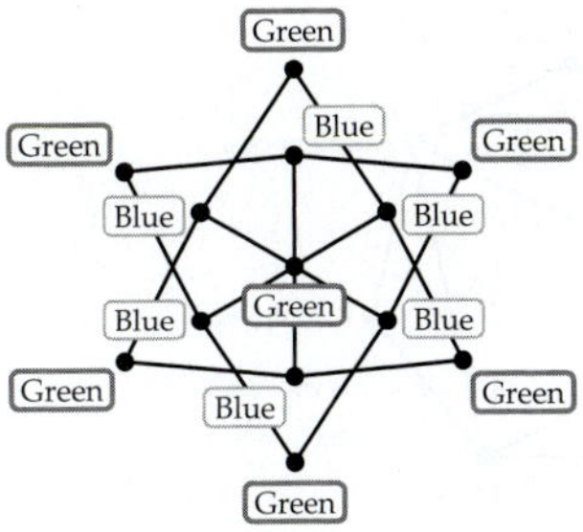

15. 3

16. 4 **17.** 4 **18.** 3 **19.** 3 **20.** 5 **21.** Two time slots **22.** Three meeting times

23. Five days; one possible schedule: group 1, group 2, groups 3 and 5, group 4, group 6

24. Three days; one possible schedule: charities A and E, charities B and C, charity D **25.** Three days: films 1 and 4, films 2 and 6, films 3 and 5

26. 3 **27.** The minimum number of transmitting channels needed

28.

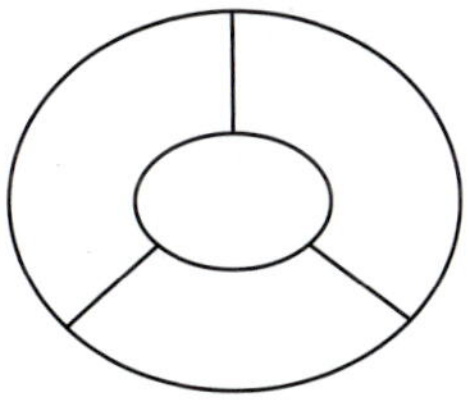

29. Disconnected vertices with no edges

30. a.

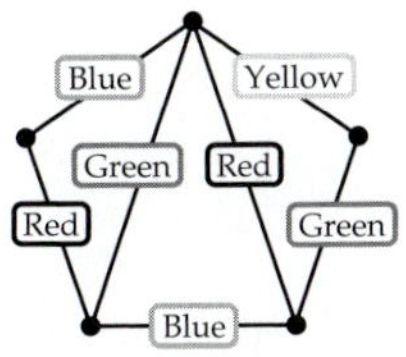

b.

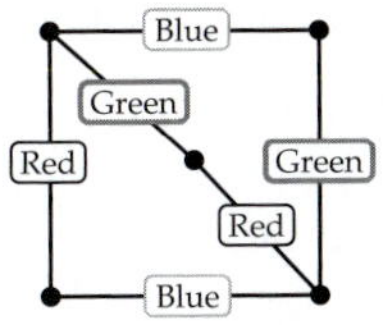

c.

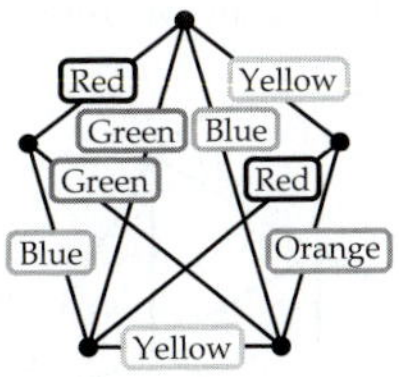

d.

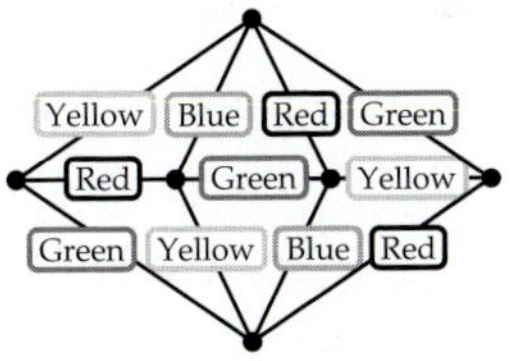

e. Answers will vary.

31.

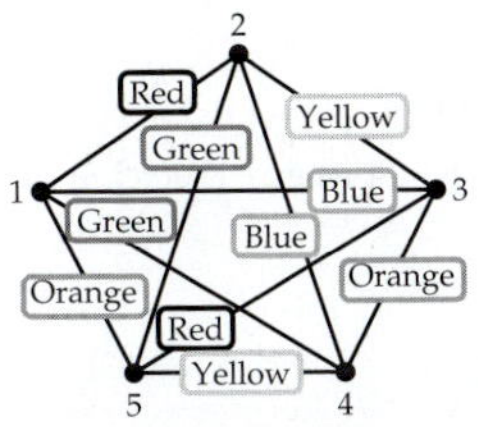

Day 1: 1 vs. 3, 2 vs. 4;
day 2: 1 vs. 4, 2 vs. 5;
day 3: 1 vs. 2, 3 vs. 5;
day 4: 2 vs. 3, 4 vs. 5;
day 5: 1 vs. 5, 3 vs. 4

CHAPTER 5 REVIEW EXERCISES *page 285*

1. a. 8 **b.** 4 **c.** All vertices have degree 4. **d.** Yes [Sec. 5.1] **2. a.** 6 **b.** 7 **c.** 1, 1, 2, 2, 2, 2, 2 **d.** No [Sec. 5.1] **3.**

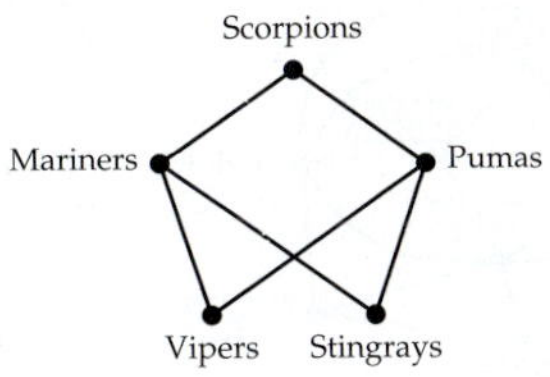

[Sec. 5.1]

4. a. No **b.** 4 **c.** 110, 405 **d.** 105 [Sec. 5.1] **5.** Equivalent [Sec. 5.1] **6.** Equivalent [Sec. 5.1]

7. a. E–A–B–C–D–B–E–C–A–D **b.** Not possible [Sec. 5.1] **8. a.** Not possible **b.** Not possible [Sec. 5.1]

9. a. and b. F–A–E–C–B–A–D–B–E–D–C–F [Sec. 5.1] **10. a.** B–A–E–C–A–D–F–C–B–D–E

b. Not possible [Sec. 5.1] **11.** Yes

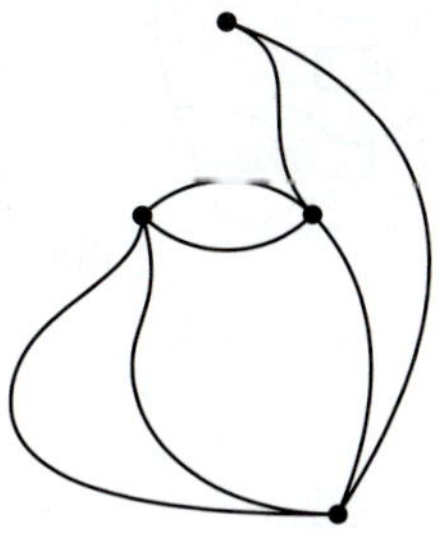

[Sec. 5.1]

12. Yes; no [Sec. 5.1]

13. A–B–C–E–D–A [Sec. 5.2] **14.** A–D–F–B–C–E–A [Sec. 5.2]

15.

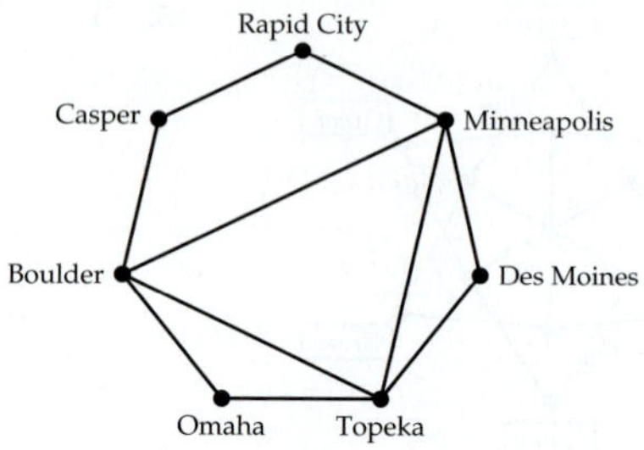

Casper–Rapid City–Minneapolis–Des Moines–Topeka–Omaha–Boulder–Casper [Sec. 5.2]

16. Casper–Boulder–Topeka–Minneapolis–Boulder–Omaha–Topeka–Des Moines–Minneapolis–Rapid City–Casper [Sec. 5.1]

17. A–D–F–E–B–C–A [Sec. 5.2] **18.** A–B–E–C–D–A [Sec. 5.2]

19. A–D–F–E–C–B–A [Sec. 5.2] **20.** A–B–E–D–C–A [Sec. 5.2]

21.

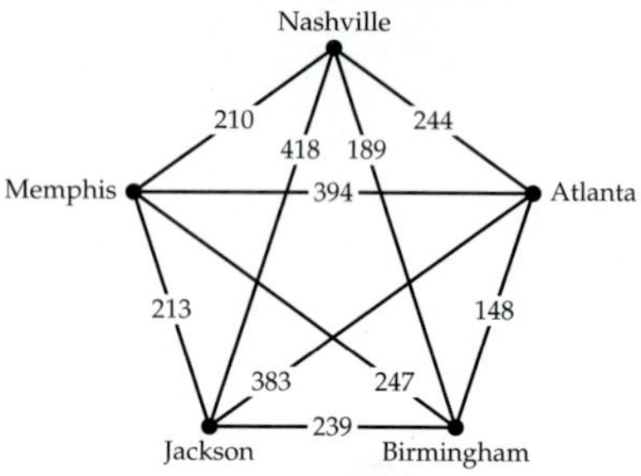

Memphis–Nashville–Birmingham–Atlanta–Jackson–Memphis [Sec. 5.2]

22. A–E–B–C–D–A [Sec. 5.2] **23.**

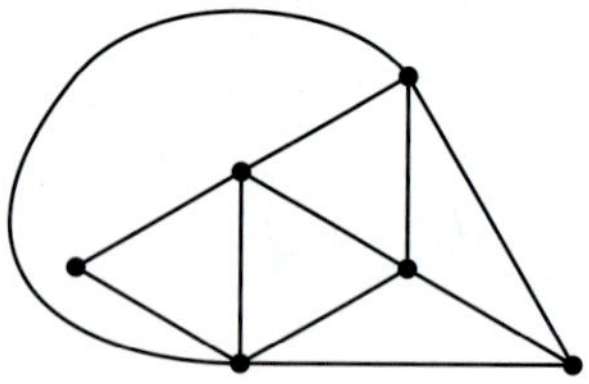

[Sec. 5.3]

24.

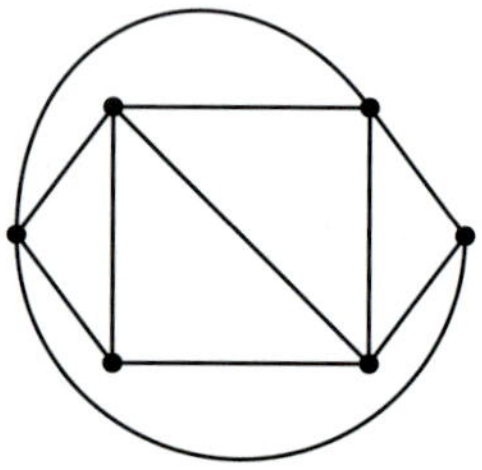

[Sec. 5.3]

See the *Instructor's Solutions Manual* for the solutions to Exercises 25 and 26.

27. 5 vertices, 8 edges, 5 faces [Sec. 5.3] **28.** 14 vertices, 16 edges, 4 faces [Sec. 5.3]

29. Requires four colors [Sec. 5.4]

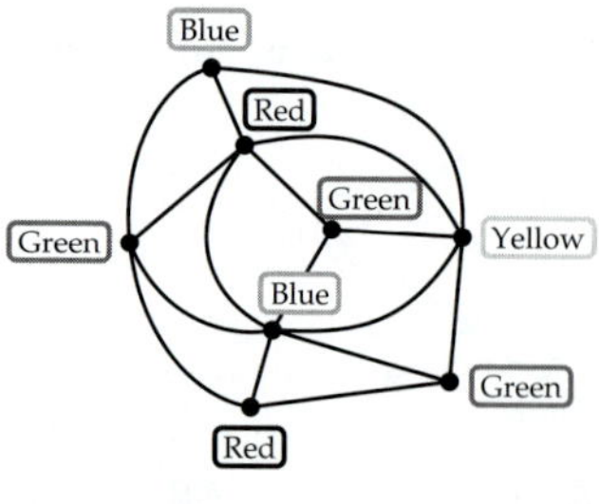

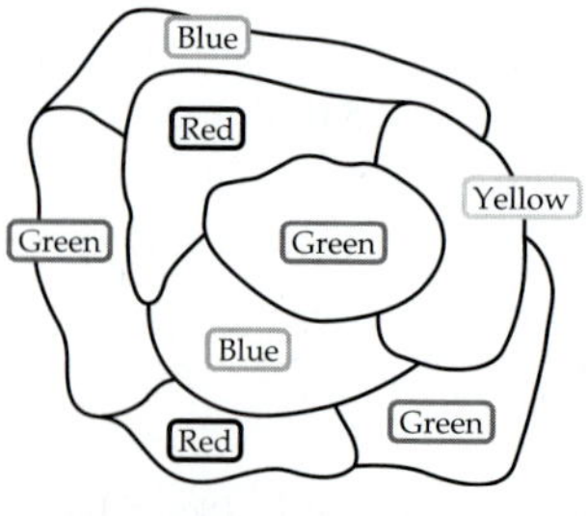

30. Requires four colors

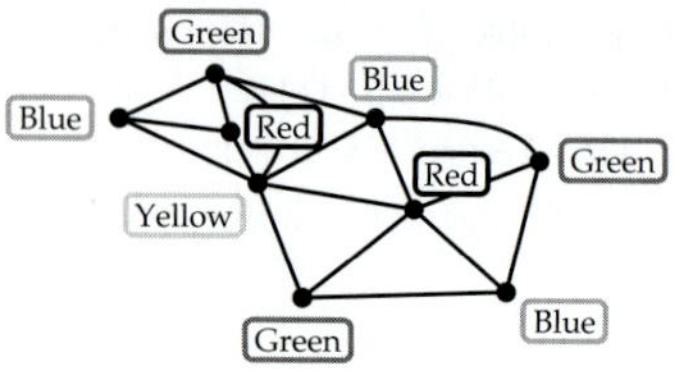

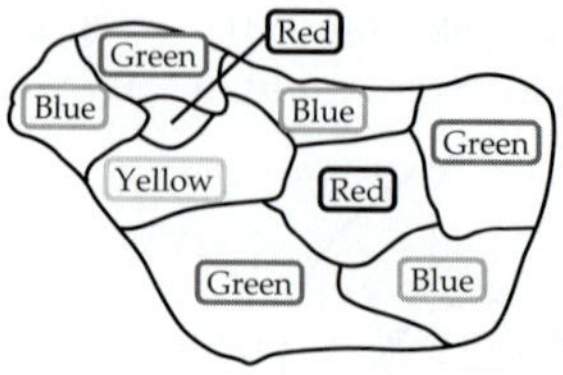

[Sec. 5.4]

31. 2-colorable

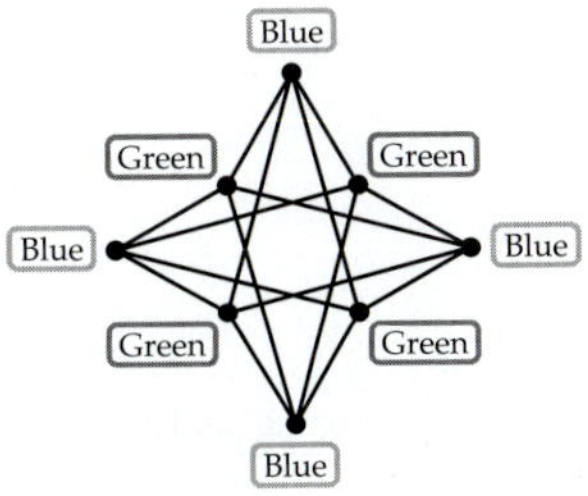

[Sec. 5.4]

32. Not 2-colorable [Sec. 5.4]
33. 3 [Sec. 5.4]
34. 5 [Sec. 5.4]

35. Three time slots: budget and planning, marketing and executive, sales and research

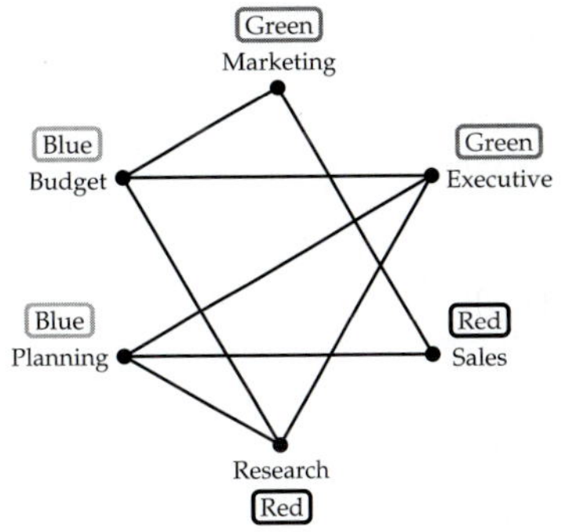

[Sec. 5.4]

CHAPTER 5 TEST *page 289*

1. a. No **b.** Monique **c.** 0 **d.** No [Sec. 5.1, Example 1] **2.** Equivalent [Sec. 5.1, Example 2]

3. a. No **b.** A–B–E–A–F–D–C–F–B–C–E–D [Sec. 5.1, Examples 4 and 6]

4. No

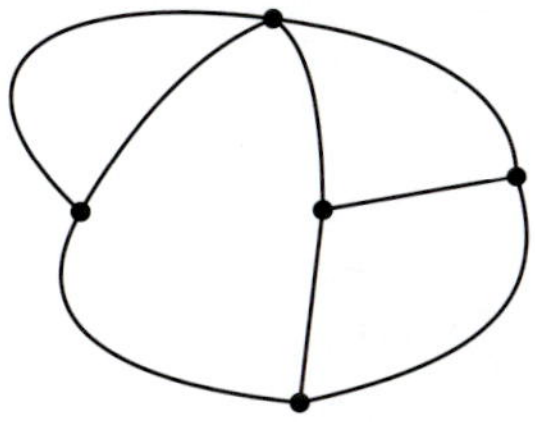

[Sec. 5.1, Check Your Progress 5 and Example 6]

5. a. See page 245.
b. A–G–C–D–F–B–E–A [Sec. 5.2, Example 1]

6. a.

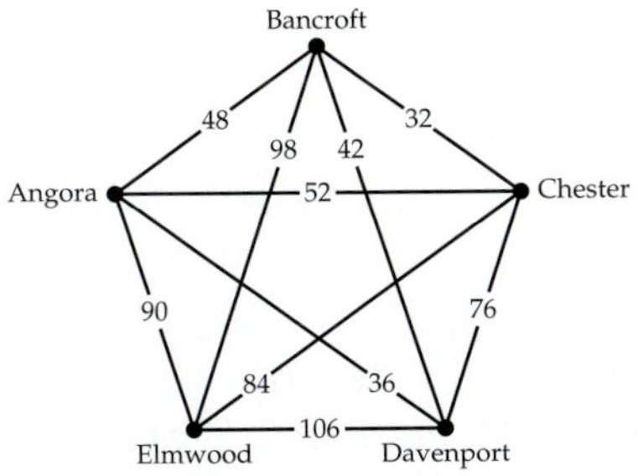

b. Angora–Elmwood–Chester–Bancroft–Davenport–Angora; $284
c. Yes [Sec. 5.2, Examples 2 and 4]

7. A–E–D–B–C–F–A [Sec. 5.2, Example 3]

8. a.

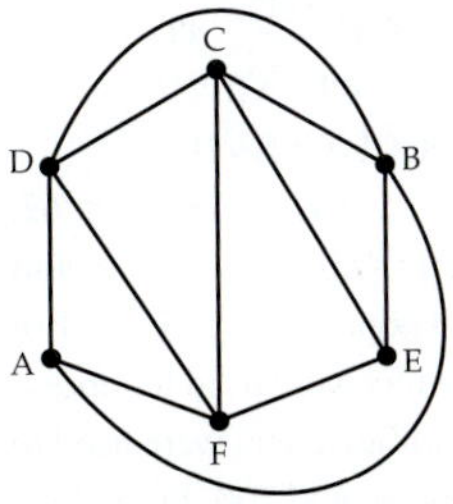

[Sec. 5.3, Example 1]

See the *Instructor's Solutions Manual* for the solution to Exercise 8, part b.

9. a. 6 **b.** 7
c. $v = 12$, $f = 7$, $e = 17$; $12 + 7 = 17 + 2$ [Sec. 5.3, Example 5]

10. a.

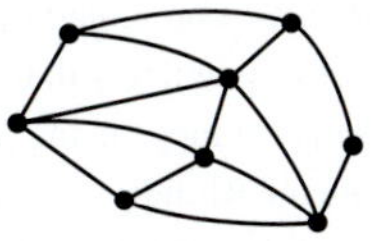

b. 3 **c.**

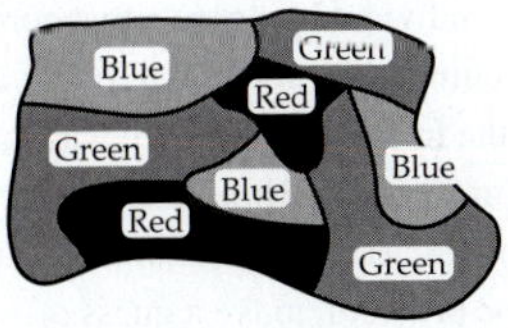

[Sec. 5.4, Example 1]

11. 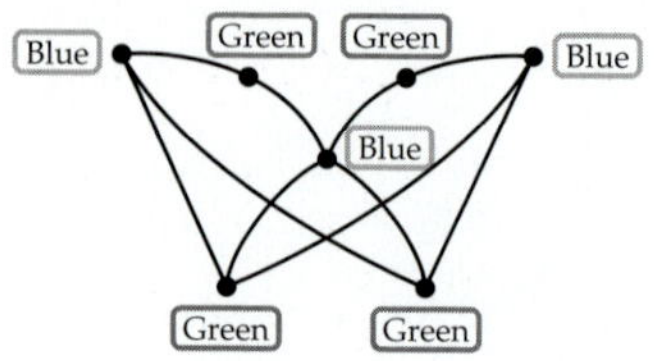

[Sec. 5.4, Example 2]

12. Three evenings: Cirque du Soleil and a play on one evening; a magic show and a tribute band concert on another evening; and a comedy show and a musical on a third evening [Sec. 5.4, Example 4]

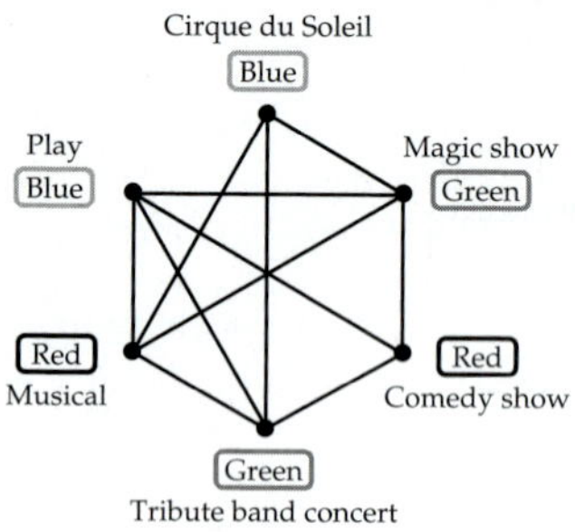

CHAPTER 6

EXCURSION EXERCISES, SECTION 6.1 *page 300*

1. 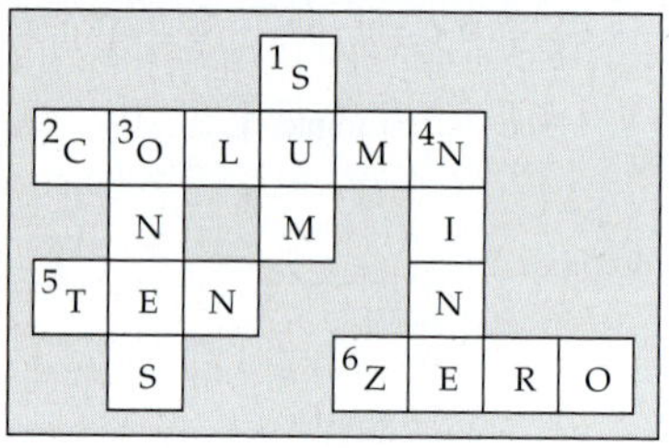

2. 二十六 **3.** 三百五十七 **4. a.** 896 **b.** 2465 **5. a.** 4 **b.** 7 **6.** Answers will vary.

EXERCISE SET 6.1 *page 300*

1.  **2.**  **3.**  **4.** 
5.  **6.**  **7.** 
8.  **9.** 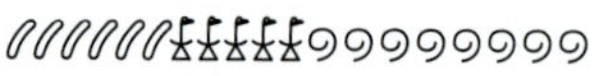 **10.** 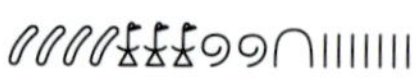
11.  **12.**  **13.** 2134 **14.** 1312 **15.** 845
16. 287 **17.** 1232 **18.** 232,333 **19.** 221,011 **20.** 32,323 **21.** 65,769 **22.** 2,436,723 **23.** 5,122,406
24. 2,601,400 **25.** 94 **26.** 125 **27.** 666 **28.** 747 **29.** 32 **30.** 71 **31.** 56 **32.** 362
33. 161 **34.** 64 **35.** 650 **36.** 1110 **37.** 1409 **38.** 1702 **39.** 1240 **40.** 2604 **41.** 840
42. 455 **43.** 9044 **44.** 7517 **45.** 11,461 **46.** 4221 **47.** XXXIX **48.** XLII **49.** CLVII
50. CCXXXI **51.** DXLII **52.** DCCLXXXIII **53.** MCXCVII **54.** MXXXIX **55.** DCCLXXXVII
56. MCCCXLIII **57.** DCLXXXIII **58.** CMLIX **59.** $\overline{\text{VI}}$DCCCXCVIII **60.** $\overline{\text{IV}}$CCCLVII **61.** 235 **62.** 406
63. 504 **64.** 228 **65.** 203 **66.** 297 **67.** 595 **68.** 1118 **69.** 2484 **70.** 15,480 **71. a.** and **b.** Answers will vary. **72.** MMMDCCCLXXXVIII = 3888 **73.** The Ionic Greek numeration system gained popularity in Greco-Roman times, starting around 100 BC. It consisted of 24 Greek letters and 3 letters taken from the Phoenician alphabet. The first nine letters were used to represent the counting numbers from 1 to 9. The next nine letters were used to represent the multiples of 10 from 10 to 90. The third set of nine letters were used to represent the multiples of 100 from 100 to 900. For example, α (alpha) represented 1, λ (lambda) represented 30, and ϕ (phi) represented 500. In the Ionic Greek numeration system, the value of a numeral was the sum of each of the individual numerals that comprised it. For instance, 531 was represented by $\alpha\lambda\phi$ $(500 + 30 + 1)$. A prime symbol was used to multiply the value of a numeral by 1000. One of the advantages of the Ionic Greek system was that large numbers could be written using very few symbols. A disadvantage of the Ionic Greek system was that it was difficult to perform arithmetic using Ionic numerals. **74.** The method of false position is used in Problems 24 to 29 of the Rhind papyrus. Ahmes, the scribe who wrote the Rhind papyrus, illustrates the method of false position in the solution of Problem 24. Problem 24 asks the question, "a quantity added to one-quarter of the quantity is 15; what is the quantity?" In modern notation the problem is $x + \frac{x}{4} = 15$. To solve the problem, make a guess of, say, $x = 4$. Then $x + \frac{x}{4} = 4 + \frac{4}{4} = 5$. This result is only one-third the desired result, so triple your guess to produce the correct solution $x = 12$. This method of making a guess, checking it, and adjusting the original guess became known as the method of false position.

EXCURSION EXERCISES, SECTION 6.2 *page 309*

1. 373 **2.** 797 **3.** 83,748 **4.** 166,346 **5.** 2,281,546 **6.** 46,648,870

7. The nines complement and the end-around carry procedure produce the correct answer to a subtraction problem because applying these processes **is equivalent to adding and subtracting the same number from a given difference**.

Example: $641 - 235 = 641 + 999 - 235 - 999$

$= 641 + (999 - 235) - 999$ • Add 999 and subtract 999.

$= (641 + 764) - 999$ • $999 - 235$ is 764, which can also be obtained by taking the nines complement of 235.

$= 1405 - 1000 + 1$ • From $641 + 764 = 1405$, subtract 999 by subtracting 1000 and adding 1. This is the end-around carry procedure.

$= 406$

Adding the nines complement of 235 to 641 produces a result that is 999 larger than $641 - 235$. Applying the end-around carry procedure decreases the previous sum by 999. Thus adding the nines complement and performing the end-around carry nullify each other and produce the correct difference.

EXERCISE SET 6.2 *page 309*

1. $(4 \times 10^1) + (8 \times 10^0)$ **2.** $(9 \times 10^1) + (3 \times 10^0)$ **3.** $(4 \times 10^2) + (2 \times 10^1) + (0 \times 10^0)$ **4.** $(5 \times 10^2) + (0 \times 10^1) + (1 \times 10^0)$ **5.** $(6 \times 10^3) + (8 \times 10^2) + (0 \times 10^1) + (3 \times 10^0)$ **6.** $(9 \times 10^3) + (0 \times 10^2) + (4 \times 10^1) + (5 \times 10^0)$ **7.** $(1 \times 10^4) + (0 \times 10^3) + (2 \times 10^2) + (0 \times 10^1) + (8 \times 10^0)$ **8.** $(6 \times 10^4) + (7 \times 10^3) + (4 \times 10^2) + (8 \times 10^1) + (2 \times 10^0)$ **9.** 456 **10.** 763 **11.** 5076 **12.** 3128 **13.** 35,407 **14.** 230,675 **15.** 683,040 **16.** 53,007,902 **17.** 76 **18.** 98 **19.** 395 **20.** 813 **21.** 2481 **22.** 6218 **23.** 27 **24.** 36 **25.** 3363 **26.** 21,225 **27.** 10,311 **28.** 11,207 **29.** 23 **30.** 45 **31.** 97 **32.** 746 **33.** 72,133 **34.** 76,272 **35.** 2,171,466 **36.** 4,575,694 **37.**  **38.** 

39.  **40.**  **41.**  **42.** 

43.  **44.**  **45.** 

46.  **47.**  **48.**  **49.** 

50.  **51.**  **52.** 

53. 194 **54.** 302 **55.** 1803 **56.** 3100 **57.** 14,492 **58.** 2420 **59.** 36,103 **60.** 57,706

61. 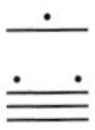 **62.**  **63.**  **64.**  **65.**  **66.**  **67.** 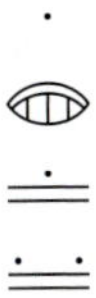 **68.** 

69. a. and **b.** Answers will vary. **70.** If the space between a first and second pair of wedge-shaped numerals is too small, then the numeral will appear as four wedges, which is the Babylonian numeral for 4. **71. a.** 7 **b.** 20 **c.** 35 **d.** UUZU **e.** UZZTZ **f.** UTUZZ

EXCURSION EXERCISES, SECTION 6.3 *page 317*

1. The cards are in numeric order from smallest, $00001_{\text{two}} = 1$, to largest, $11111_{\text{two}} = 31$.

- The first implementation of the sorting procedure places all odd numerals behind the even numerals. Thus the base two numerals of the form xxxx1 are behind the base two numerals of the form xxxx0, where x represents either a 0 or a 1.
- The second implementation of the sorting procedure places the base two numerals of the form xxx1x behind the base two numerals of the form xxx0x. Thus, from front to back, the base two numerals are now in the order xxx00, xxx01, xxx10, xxx11.
- The third implementation of the sorting procedure places the base two numerals of the form xx1xx behind the base two numerals of the form xx0xx. Thus the order of the cards from front to back is now given by the base two numerals xx000, xx001, xx010, xx011, xx100, xx101, xx110, xx111.
- The fourth implementation of the sorting procedure places the base two numerals of the form x1xxx behind the base two numerals of the form x0xxx. Thus the order of the cards from front to back is now given by the base two numerals x0000, x0001, x0010, x0011, x0100, x0101, x0110, x0111, x1000, x1001, x1010, x1011, x1100, x1101, x1110, x1111.
- The fifth implementation of the sorting procedure places the base two numerals of the form 1xxxx behind the base two numerals of the form 0xxxx. Thus the order of the cards from front to back is now given by the base two numerals 00001, 00010, 00011, 00100, 00101, 00110, 00111, 01000, 01001, 01010, 01011, 01100, 01101, 01110, 01111, 10000, 10001, 10010, 10011, 10100, 10101, 10110, 10111, 11000, 11001, 11010, 11011, 11100, 11101, 11110, 11111. In terms of base ten numerals, the cards are now arranged in the order 1, 2, 3, 4, ..., 31.

2. 10; 14 **3.** A base three number system requires three numerals, such as 0, 1, and 2. A notch in a card can be used to represent a 0, and a hole in a card can be used to represent a 1, as in base two, but there is no convenient method that can be used to represent the numeral 2.

EXERCISE SET 6.3 *page 318*

1. 73 **2.** 82 **3.** 61 **4.** 379 **5.** 718 **6.** 477 **7.** 485 **8.** 305 **9.** 181 **10.** 58 **11.** 2032_{five} **12.** 552_{eight} **13.** 12540_{six} **14.** 133220_{four} **15.** 22886_{nine} **16.** 105226_{seven} **17.** $111111011100_{\text{two}}$ **18.** 21210010_{three} **19.** $1B7_{\text{twelve}}$ **20.** $18A_{\text{sixteen}}$ **21.** 13 **22.** 21 **23.** 27 **24.** 45 **25.** 100 **26.** 1960 **27.** 139 **28.** 437 **29.** 41 **30.** 116 **31.** 90 **32.** 138 **33.** 1338 **34.** 1028 **35.** 26_{eight} **36.** 212_{five} **37.** 23033_{four} **38.** 1143_{six} **39.** 24_{five} **40.** 543_{six} **41.** 2446_{nine} **42.** 611_{eight} **43.** 126_{eight} **44.** 123_{six} **45.** $7C_{\text{sixteen}}$ **46.** 149_{twelve} **47.** 11101010_{two} **48.** 10100100_{two} **49.** 312_{eight} **50.** $EE5_{\text{sixteen}}$ **51.** 151_{sixteen} **52.** $101110111010001_{\text{two}}$ **53.** $1011111011110011_{\text{two}}$ **54.** $110101001111011000_{\text{two}}$ **55.** $1011101001011100111_{\text{two}}$ **56.** $100111001011100_{\text{two}}$ **57.** Answers will vary.

58. a. true **b.** true **c.** true **59.** The chart below shows the American Standard Code for Information Interchange (ASCII) character set. Many modern computers make use of ASCII to represent internally the characters that can be typed on a computer keyboard. The chart for coding ASCII characters uses the hexadecimal numeration system. For instance, the capital letter A is shown in row 4 and column 1. Thus the ASCII code for the capital letter A is 41_{sixteen}, which is 65 in base ten and 01000001 as an 8-bit binary numeral. The lower-case letter m is shown in row 6 and column D. The ASCII code for a lower-case m is $6D_{\text{sixteen}}$, which is 109 in base ten and 01101101 as an 8-bit binary numeral. (*Source:* Wikipedia)

	0	1	2	3	4	5	6	7	8	9	A	B	C	D	E	F
0	NUL	SOH	STX	ETX	EOT	ENQ	ACK	BEL	BS	HT	LF	VT	FF	CR	SO	SI
1	DLE	DC1	DC2	DC3	DC4	NAK	SYN	ETB	CAN	EM	SUB	ESC	FS	GS	RS	US
2	SPC	!	"	#	$	%	&	'	(	)	*	+	,	–	.	/
3	0	1	2	3	4	5	6	7	8	9	:	;	<	=	>	?
4	@	A	B	C	D	E	F	G	H	I	J	K	L	M	N	O
5	P	Q	R	S	T	U	V	W	X	Y	Z	[	\	]	^	_
6	`	a	b	c	d	e	f	g	h	i	j	k	l	m	n	o
7	p	q	r	s	t	u	v	w	x	y	z	{	|	}	~	DEL

60. All Postnet codes consist of tall bars and short bars that can be read by a machine. Every Postnet code for a zip code + 4 starts and ends with a tall bar. Each digit in a zip code +4 is represented by five bars, of which two are tall bars and three are short bars. The following table shows how each digit is represented. (*Source:* Wikipedia)

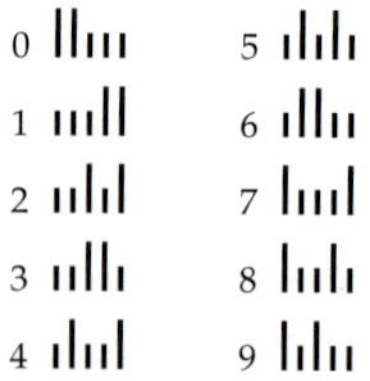

Postnet Code

For every digit except 0, each bar position has a weight, or place value, assigned to it. From left to right, the weights are 7, 4, 2, 1, and 0. A tall bar can be thought of as a 1 and a short bar as a 0. For example, the digit 3 is

$$(0 \times 7) + (0 \times 4) + (1 \times 2) + (1 \times 1) + (0 \times 0) = 2 + 1 = 3$$

Postnet code also utilizes a "checksum" digit at the end. This checksum digit is chosen such that the sum of all the digits will be a multiple of 10. If a machine makes a mistake in reading one of the digits, the checksum will alert the operator to that fact.

61. 54 **62.** 702 **63.**  **64.** 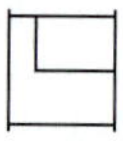 **65.** 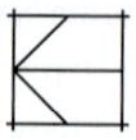

66. 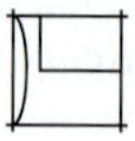 **67.** 256 **68.** 383 **69. a.** and **b.** Answers will vary. **70. a.** and **b.** Answers will vary.

EXCURSION EXERCISES, SECTION 6.4 *page 328*

1. 101_{two} **2.** 1001_{two} **3.** 11101101_{two} **4.** $110110010010_{\text{two}}$ **5.** 1110011100_{two} **6.** $1101110001110_{\text{two}}$

EXERCISE SET 6.4 page 329

1. 332_{five} 2. 1201_{four} 3. 6562_{seven} 4. 10000_{two} 5. 1001000_{two} 6. 100110001_{two} 7. 1271_{twelve} 8. $BCF_{sixteen}$ 9. $D036_{sixteen}$ 10. 1120_{three} 11. 1124_{six} 12. 6211_{eight} 13. 241_{five} 14. 253_{six} 15. 6542_{eight} 16. 5256_{nine} 17. 1111_{two} 18. 100100_{two} 19. 1111001_{two} 20. $33E_{sixteen}$ 21. $411A_{twelve}$ 22. $B639_{twelve}$ 23. 384_{nine} 24. 3030_{four} 25. 523_{six} 26. 4003_{seven} 27. 1201_{three} 28. 32133_{five} 29. 45234_{eight} 30. 1010001_{two} 31. 1010100_{two} 32. 100000100_{two} 33. 14207_{eight} 34. 104130_{six} 35. 321222_{four} 36. 31403_{twelve} 37. $3A61_{sixteen}$ 38. $109C8_{sixteen}$ 39. 33_{four}; remainder 0_{four} 40. 34_{five}; remainder of 1_{five} 41. Quotient 33_{four}; remainder 0_{four} 42. Quotient 130_{eight}; remainder 2_{eight} 43. Quotient 1223_{six}; remainder 1_{six} 44. Quotient 1101_{two}; remainder 1_{two} 45. Quotient 1110_{two}; remainder 0_{two} 46. Quotient 10110_{two}; remainder 11_{two} 47. Quotient $A8_{twelve}$; remainder 3_{twelve} 48. Quotient $12B_{sixteen}$; remainder $5_{sixteen}$ 49. Quotient 14_{five}; remainder 11_{five} 50. Six 51. Eight 52. Nine 53. a. 629 b. $384 = 110000000_{two}$; $245 = 11110101_{two}$ c. 1001110101_{two} d. 629 e. Same 54. a. 139 b. $457 = 111001001_{two}$; $318 = 100111110_{two}$ c. 10001011_{two} d. 139 e. Same 55. a. 6422 b. $247 = 11110111_{two}$; $26 = 11010_{two}$ c. 1100100010110_{two} d. 6422 e. Same 56. If you borrow in a base eight numeration system, you must borrow a power of 8. In the subtraction, a group of 10, which is not a power of 8, was borrowed to produce the incorrect result 625_{eight}. 57. Base seven 58. No. 12 is a multiple of 2 and thus it is an even number regardless of the base in which it is written. 59. In a base one numeration system, 0 would be the only numeral, and the place values would be $1^0, 1^1, 1^2, 1^3, \ldots$, each of which equals 1. Thus 0 is the only number you could write using a base one numeration system. 60. N = 1, A = 3, T = 2, and O = 0 61. M = 1, A = 4, S = 3, and O = 0 62. a. 8; −59; 18 b. $1413_{negative\ five}$ c. $11101_{negative\ three}$ d. $292_{negative\ ten}$

EXCURSION EXERCISES, SECTION 6.5 page 337

1. The first number in the list is divisible by 2, the second is divisible by 3, the third is divisible by 4, and so on, and the last number is divisible by 1,000,001. Thus each number in the list is a composite number. 2. One solution is provided by 13! + 2, 13! + 3, 13! + 4, ..., 13! + 13. 6,227,020,802; 6,227,020,803; 6,227,020,804; 6,227,020,805; 6,227,020,806; 6,227,020,807; 6,227,020,808; 6,227,020,809; 6,227,020,810; 6,227,020,811; 6,227,020,812; 6,227,020,813.
3. a. 21! + 2, 21! + 3, 21! + 4, ..., 21! + 21 b. 500,001! + 2, 500,001! + 3, 500,001! + 4, ..., 500,001! + 500,001 c. 7,000,000,001! + 2, 7,000,000,001! + 3, 7,000,000,001! + 4, ..., 7,000,000,001! + 7,000,000,001

EXERCISE SET 6.5 page 337

1. 1, 2, 4, 5, 10, 20 2. 1, 2, 4, 8, 16, 32 3. 1, 5, 13, 65 4. 1, 3, 5, 15, 25, 75 5. 1, 41 6. 1, 79 7. 1, 2, 5, 10, 11, 22, 55, 110 8. 1, 2, 3, 5, 6, 10, 15, 25, 30, 50, 75, 150 9. 1, 5, 7, 11, 35, 55, 77, 385 10. 1, 5, 7, 13, 35, 65, 91, 455 11. Composite 12. Prime 13. Prime 14. Composite 15. Prime 16. Composite 17. Prime 18. Composite 19. Composite 20. Prime 21. 2, 3, 5, 6, and 10 22. 2 23. 3 24. 2, 3, 4, 6, and 8 25. 2, 3, 4, 6, and 8 26. 3 and 5 27. 2, 5, and 10 28. 3 29. $2 \cdot 3^2$ 30. $2^4 \cdot 3$ 31. $2^3 \cdot 3 \cdot 5$ 32. $2^2 \cdot 5 \cdot 19$ 33. $5^2 \cdot 17$ 34. 5^4 35. 2^{10} 36. $2 \cdot 3 \cdot 5 \cdot 47$ 37. $2^3 \cdot 3 \cdot 263$ 38. $5 \cdot 631$ 39. $2 \cdot 3^2 \cdot 1013$ 40. $5 \cdot 53 \cdot 73$ 41. 2, 3, 5, 7, 11, 13, 17, 19, 23, 29, 31, 37, 41, 43, 47, 53, 59, 61, 67, 71, 73, 79, 83, 89, 97, 101, 103, 107, 109, 113, 127, 131, 137, 139, 149, 151, 157, 163, 167, 173, 179, 181, 191, 193, 197, 199 42. a. 15 b. 10 c. 10 d. 11 43. 3 and 5, 5 and 7, 11 and 13, 17 and 19, 29 and 31, 41 and 43, 59 and 61, 71 and 73, 101 and 103, 107 and 109, 137 and 139, 149 and 151, 179 and 181, 191 and 193, 197 and 199 44. 227 and 229, or 239 and 241, or 269 and 271, or 281 and 283 45. 311 and 313, or 347 and 349 46. One of *every* three consecutive odd natural numbers is divisible by 3. If n is a natural number greater than 3, one of the numbers n, $n + 2$, or $n + 4$ is divisible by 3 and hence at least one of the numbers n, $n + 2$, or $n + 4$ is not a prime number. Thus 3, 5, and 7 are the only prime triplets. In Exercise 47, parts a to f, only one possible sum is given. 47. a. 24 = 5 + 19 b. 50 = 3 + 47 c. 86 = 3 + 83 d. 144 = 5 + 139 e. 210 = 11 + 199 f. 264 = 7 + 257 48. Conjecture: Every perfect square has an odd number of distinct natural number factors. 49. Yes 50. Yes 51. Yes 52. No 53. No 54. No 55. Yes 56. Yes 57. Yes 58. Yes 59. Yes 60. Yes 61. No 62. No 63. Yes 64. Yes 65. a. $n = 3$ b. $n = 4$ 66. a. $p = 11$ b. $p = 7$ 67. To determine whether a given number is divisible by 17, multiply the ones digit of the given number by 5. Find the difference between this result and the number formed by omitting the ones digit from the given number. Keep repeating this procedure until you obtain a small final difference. If the final difference is divisible by 17, then the given number is divisible by 17. If the final difference is not divisible by 17, then the given number is not divisible by 17. 68. To determine whether a given number is divisible by 19, multiply the ones digit of the given number by 2. Find the sum of this result and the number formed by omitting the ones digit from the given number. Keep repeating this procedure until you obtain a small final sum. If the final sum is divisible by 19, then the given number is divisible by 19. If the final sum is not divisible by 19, then the given number is not divisible by 19. 69. 12 70. 12 71. 8 72. 18 73. 24 74. 8 75. a. N is a product of all the prime numbers. Thus $N - 1$, which must have at least one prime factor, must share at least one prime factor with the number N. b. The distributive property of multiplication over addition (subtraction) 76. a. If a number consists of k 1s, where k is a prime number, then the number is a prime number. This statement is not true because 111 consists of exactly three 1s and 111 is not a prime number.

b. The theorem stated in Exercise 76 applies only to *prime numbers* of the form 111 . . . 1. **77. a.** Definition of a divisor **b.** $aj > aj$ **c.** Because $aj = n$, where a and j are both natural numbers, both a and j are by definition divisors of n. **78.** RSA is a powerful method of encryption. It allows one to communicate securely through an insecure channel such as the Internet. The RSA encryption algorithm is a fairly simple mathematical procedure that involves large prime numbers. Computers can easily implement the encryption algorithm. One important feature of the algorithm is that anyone (the public) can use the algorithm to encrypt a message, but only those people with the "private key" can decrypt a message. The RSA algorithm is considered very secure because in order to crack a coded message, one must factor a very large number. By present techniques this is extremely time-consuming, even with modern high-speed computers. **79. a.** to **f.** Each statement is a conjecture (as of July 2016).

EXCURSION EXERCISES, SECTION 6.6 *page 347*

1. The sum of the proper factors of 200 is 265. Thus 200 is an abundant number. **2.** The sum of the proper factors of 262 is 134. Thus 262 is a deficient number. **3.** The sum of the proper factors of 325 is 109. Thus 325 is a deficient number. **4.** The sum of the proper factors of 496 is 496. Thus 496 is a perfect number. **5.** The only divisors of a prime number are 1 and the number itself. Thus the only proper factor of a prime number is 1. Hence every prime number is deficient. **6.** Because we cannot find a multiple of 6 greater than 6 that is *not* abundant, we conjecture, using inductive reasoning, that every multiple of 6 greater than 6 is an abundant number.

EXERCISE SET 6.6 *page 347*

1. Abundant **2.** Deficient **3.** Deficient **4.** Deficient **5.** Deficient **6.** Abundant **7.** Abundant **8.** Deficient **9.** Deficient **10.** Deficient **11.** Deficient **12.** Deficient **13.** Abundant **14.** Deficient **15.** Abundant **16.** Abundant **17.** Prime **18.** Prime **19.** Prime **20.** Prime **21.** $2^{126}(2^{127} - 1)$ **22.** $2^{520}(2^{521} - 1)$ **23.** 6 **24.** 39,751 **25.** 420,921 **26.** 909,526 **27.** 2,098,960 **28.** 6,320,430 **29.** 11,185,272 **30.** 12,978,189 **31.** $9^5 + 15^5 = 818{,}424$. Because $15^5 < 818{,}424$ and $16^5 > 818{,}424$, we know there is no natural number z such that $z^5 = 9^5 + 15^5$. **32.** $7^6 + 19^6 = 47{,}163{,}530$. Because $19^6 < 47{,}163{,}530$ and $20^6 > 47{,}163{,}530$, we know there is no natural number z such that $z^6 = 7^6 + 19^6$. **33. a.** False. For instance, if $n = 11$, then $2^{11} - 1 = 2047 = 23 \cdot 89$. **b.** False. Fermat's last theorem was the last of Fermat's theories (conjectures) that other mathematicians were able to establish. **c.** True **d.** Conjecture **34.** For any natural number n, the number 4078^n has a ones digit of 8, 4, 2, or 6. For any natural number n, the number 3433^n has a ones digit of 3, 9, 7, or 1. Thus, for any natural number n, the number $4078^n + 3433^n$ has an *odd* ones digit. For any natural number n, the number $12{,}046^n$ has a ones digit of 6. Therefore, it is not possible for $4078^n + 3433^n$ to equal $12{,}046^n$ for any natural number n. **35. a.** $12^7 - 12 = 35{,}831{,}796$, which is divisible by 7. **b.** $8^{11} - 8 = 8{,}589{,}934{,}584$, which is divisible by 11. **36. a.** No **b.** Yes **37.** $8128 = 1^3 + 3^3 + 5^3 + \cdots + 13^3 + 15^3$. See the *Instructor's Solutions Manual* for the verifications in Exercises 38 and 39. **40.** 945 **41.** The first five Fermat numbers formed using $n = 0, 1, 2, 3$, and 4 are all prime numbers. In 1732, Euler discovered that the sixth Fermat number, 4,294,967,297, formed using $n = 5$, is not a prime number because it is divisible by 641. **42. a.** $20 = 1 + 4 + 5 + 10$ **b.** 200 is a natural number multiple of the semiperfect number 20. Thus 200 is also a semiperfect number. **43.** 70 **44.** Answers will vary.

CHAPTER 6 REVIEW EXERCISES *page 351*

1.  [Sec. 6.1] **2.**  [Sec. 6.1] **3.** 223,013 [Sec. 6.1] **4.** 221,354 [Sec. 6.1] **5.** 349 [Sec. 6.1] **6.** 774 [Sec. 6.1] **7.** 9640 [Sec. 6.1] **8.** 92,444 [Sec. 6.1] **9.** DLXVII [Sec. 6.1] **10.** DCCCXXIII [Sec. 6.1] **11.** MMCDLXXXIX [Sec. 6.1] **12.** MCCCXXXV [Sec. 6.1] **13.** $(4 \times 10^2) + (3 \times 10^1) + (2 \times 10^0)$ [Sec. 6.2] **14.** $(4 \times 10^5) + (5 \times 10^4) + (6 \times 10^3) + (3 \times 10^2) + (2 \times 10^1) + (7 \times 10^0)$ [Sec. 6.2] **15.** 5,038,204 [Sec. 6.2] **16.** 387,960 [Sec. 6.2] **17.** 801 [Sec. 6.2] **18.** 1603 [Sec. 6.2] **19.** 76,441 [Sec. 6.2] **20.** 87,393 [Sec. 6.2] **21.**  [Sec. 6.2] **22.**  [Sec. 6.2] **23.**  [Sec. 6.2] **24.**  [Sec. 6.2] **25.** 194 [Sec. 6.2] **26.** 267 [Sec. 6.2] **27.** 2178 [Sec. 6.2] **28.** 6580 [Sec. 6.2] **29.**  [Sec. 6.2] **30.**  [Sec. 6.2] **31.**  [Sec. 6.2] **32.**  [Sec. 6.2] **33.** 29 [Sec. 6.3] **34.** 146 [Sec. 6.3] **35.** 227 [Sec. 6.3] **36.** 286 [Sec. 6.3] **37.** 1200_{three} [Sec. 6.3] **38.** 234_{seven} [Sec. 6.3] **39.** 714_{eleven} [Sec. 6.3] **40.** $18B9_{twelve}$ [Sec. 6.3] **41.** 1153_{six} [Sec. 6.3] **42.** 640_{eight} [Sec. 6.3] **43.** 458_{nine} [Sec. 6.3] **44.** $B62_{twelve}$ [Sec. 6.3] **45.** 34_{eight} [Sec. 6.3] **46.** 124_{eight} [Sec. 6.3] **47.** $38D_{sixteen}$ [Sec. 6.3] **48.** $754_{sixteen}$ [Sec. 6.3] **49.** 10101_{two} [Sec. 6.3] **50.** 1100111010_{two} [Sec. 6.3] **51.** 1001010_{two} [Sec. 6.3] **52.** 110001110010_{two} [Sec. 6.3] **53.** 410 [Sec. 6.3] **54.** 277 [Sec. 6.3] **55.** 1041 [Sec. 6.3] **56.** 1616 [Sec. 6.3] **57.** 423_{six} [Sec. 6.4] **58.** 1240_{eight} [Sec. 6.4] **59.** 536_{nine} [Sec. 6.4] **60.** 1113_{four} [Sec. 6.4] **61.** 16412_{eight} [Sec. 6.4] **62.** 324203_{five} [Sec. 6.4] **63.** Quotient 11100_{two}; remainder 1_{two} [Sec. 6.4] **64.** Quotient 21_{four}; remainder 3_{four} [Sec. 6.4] **65.** 3, 5, 9, and 11 [Sec. 6.5] **66.** 2, 4, and 11 [Sec. 6.5] **67.** Composite [Sec. 6.5] **68.** Composite [Sec. 6.5] **69.** Composite [Sec. 6.5]

70. Composite [Sec. 6.5] **71.** $3^2 \cdot 5$ [Sec. 6.5] **72.** $2 \cdot 3^3$ [Sec. 6.5] **73.** $3^2 \cdot 17$ [Sec. 6.5] **74.** $3 \cdot 5 \cdot 19$ [Sec. 6.5] **75.** Perfect [Sec. 6.6] **76.** Deficient [Sec. 6.6] **77.** Abundant [Sec. 6.6] **78.** Abundant [Sec. 6.6] **79.** $2^{60}(2^{61} - 1)$ [Sec. 6.6] **80.** $2^{1278}(2^{1279} - 1)$ [Sec. 6.6] **81.** 368 [Sec. 6.1] **82.** 513 [Sec. 6.1] **83.** 1162 [Sec. 6.1] **84.** 3003 [Sec. 6.1] **85.** 39,751 [Sec. 6.6] **86.** 895,932 [Sec. 6.6] **87.** Zero [Sec. 6.6] **88.** No [Sec. 6.6]

CHAPTER 6 TEST *page 353*

1. 𓆼𓆼𓆼𓍢∩∩|||| [Sec. 6.1, Example 1] **2.** 4263 [Sec. 6.1, Example 2] **3.** 1447 [Sec. 6.1, Example 5] **4.** MMDCIX [Sec. 6.1, Example 6] **5.** $(6 \times 10^4) + (7 \times 10^3) + (4 \times 10^2) + (8 \times 10^1) + (5 \times 10^0)$ [Sec. 6.2, Example 1] **6.** 530,284 [Sec. 6.2, Example 2] **7.** 37,274 [Sec. 6.2, Example 6] **8.** 𒁹𒁹 𒌋𒌋𒌋𒌋𒁹 𒌋𒁹𒁹𒁹𒁹𒁹 [Sec. 6.2, Example 7] **9.** 1305 [Sec. 6.2, Example 9] **10.** [Sec. 6.2, Example 10]

·

· ·
—

· ·

11. 854 [Sec. 6.3, Example 1] **12. a.** 4144_{eight} **b.** $12B0_{\text{twelve}}$ [Sec. 6.3, Example 5] **13.** $100101110111_{\text{two}}$ [Sec. 6.3, Example 6] **14.** $AB7_{\text{sixteen}}$ [Sec. 6.3, Example 9] **15.** 112_{five} [Sec. 6.4, Example 2] **16.** 313_{eight} [Sec. 6.4, Example 5] **17.** 11100110_{two} [Sec. 6.4, Example 7] **18.** Quotient 61_{seven}; remainder 3_{seven} [Sec. 6.4, Example 9] **19.** $2 \cdot 5 \cdot 23$ [Sec. 6.5, Example 4] **20.** Composite [Sec. 6.5, Example 2] **21. a.** No **b.** Yes **c.** No [Sec. 6.5, Example 3] **22. a.** Yes **b.** No **c.** No [Sec. 6.5, Example 3] **23.** Abundant [Sec. 6.6, Example 1] **24.** $2^{16}(2^{17} - 1)$ [Sec. 6.6, Example 3] **25.** Since Fermat's Last Theorem has recently been proven, we know that z is not a natural number. [Sec. 6.6 FLT]

CHAPTER 7

EXERCISE SET 7.1 *page 363*

1. 72 in. **2.** $1\frac{1}{2}$ mi **3.** 180 in. **4.** 880 yd **5.** 4 lb **6.** $4\frac{1}{2}$ tons **7.** 24 oz **8.** 1600 lb **9.** 20 fl oz **10.** $2\frac{1}{2}$ gal **11.** 56 pt **12.** 6 c **13.** 620 mm **14.** 6.804 km **15.** 321 cm **16.** 2.60 m **17.** 7.421 kg **18.** 0.043 g **19.** 4.5 dg **20.** 45.6 mg **21.** 0.0075 L **22.** 37 ml **23.** 435 cm^3 **24.** 0.897 kl **25.** 65.91 kg **26.** 54.20 L **27.** 48.3 km/h **28.** \$1.30/kg **29.** \$8.70/L **30.** 189.2 lb **31.** 1.38 in. **32.** 98.4 ft/s **33.** \$4.55/lb **34.** \$2.00/lb **35.** 2.3×10^{-12} Y **36.** 4.51×10^3 p **37.** 6.5×10^{11} G **38.** $9.46 \times 10^{-12}\ \mu$ **39.** 4.01×10^{-9} E **40.** 7.15×10^{-9} f **41.** 3×10^{-13} Zm/s **42.** Approximately 9.66×10^{-6} Ym **43.** 0.000000000001 s **44.** 3×10^2 fs

EXCURSION EXERCISES, SECTION 7.2 *page 373*

1.

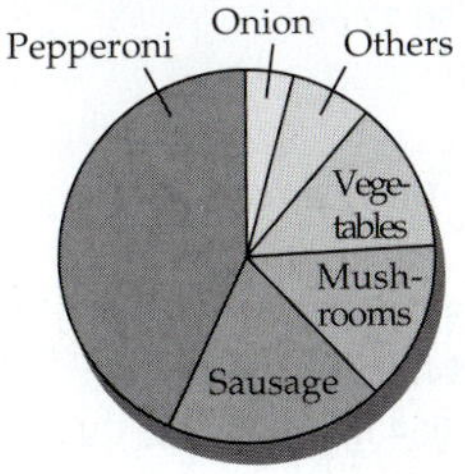

2.

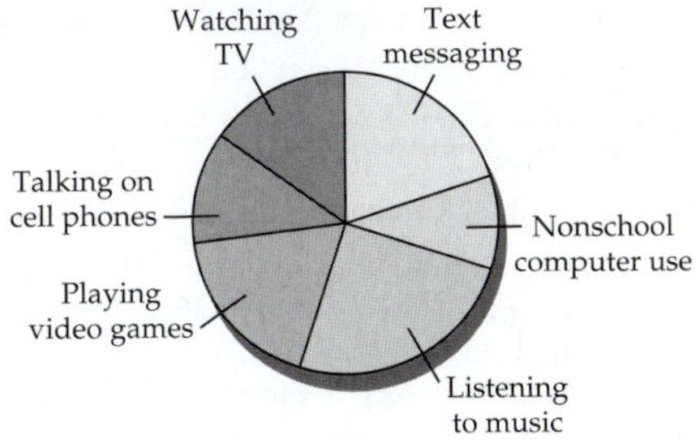

EXERCISE SET 7.2 *page 373*

1. $\angle O$, $\angle AOB$, and $\angle BOA$ **2.** 360°; 180°; 90° **3.** A 28° angle **4.** A 59° angle **5.** An 18° angle **6.** A 108° angle **7.** An acute angle **8.** A right angle **9.** An obtuse angle **10.** An acute angle **11.** 14 cm **12.** 19 mm **13.** 28 ft **14.** 45 in. **15.** 86° **16.** 47° **17.** 71° **18.** 63° **19.** 30° **20.** 18° **21.** 127° **22.** 53° **23.** 116° **24.** 121° **25.** 20° **26.** 15° **27.** 20° **28.** 45° **29.** 141° **30.** 128° **31.** 106° **32.** 49° **33.** 11° **34.** 12° **35.** $m\angle a = 38°$, $m\angle b = 142°$ **36.** $m\angle a = 122°$, $m\angle b = 58°$ **37.** $m\angle a = 47°$, $m\angle b = 133°$ **38.** $m\angle a = 44°$, $m\angle b = 136°$ **39.** 20° **40.** 20° **41.** 47° **42.** 40° **43.** $m\angle x = 155°$, $m\angle y = 70°$ **44.** $m\angle x = 160°$, $m\angle y = 145°$ **45.** $m\angle a = 45°$, $m\angle b = 135°$ **46.** $m\angle a = 40°$, $m\angle b = 140°$ **47.** 60° **48.** 45° **49.** 35° **50.** 73° **51.** False **52.** True **53.** True

54. The three angles form a straight angle. The sum of the measures of the angles of a triangle is 180°. **55.** 360° **56.** The sum of the measures of the interior angles of a triangle is 180°; therefore, $m\angle a + m\angle b + m\angle c = 180°$. The sum of the measures of an interior and exterior angle is 180°; therefore, $m\angle c + m\angle x = 180°$. Solving this equation for $m\angle c$, $m\angle c = 180° - m\angle x$. Substitute $180° - m\angle x$ for $m\angle c$ in the equation $m\angle a + m\angle b + m\angle c = 180°$: $m\angle a + m\angle b + 180° - m\angle x = 180°$. Add $m\angle x$ to each side of the equation, and subtract 180° from each side of the equation: $m\angle a + m\angle b = m\angle x$. The measure of an exterior angle of a triangle is equal to the sum of the measures of the two opposite interior angles: $m\angle a + m\angle c = m\angle z$. **57.** $\angle AOC$ and $\angle BOC$ are supplementary angles; therefore, $m\angle AOC + m\angle BOC = 180°$. Because $m\angle AOC = m\angle BOC$, by substitution, $m\angle AOC + m\angle AOC = 180°$. Therefore, $2(m\angle AOC) = 180°$, and $m\angle AOC = 90°$. Hence $\overline{AB} \perp \overline{CD}$.

EXCURSION EXERCISES, SECTION 7.3 *page 388*

1.

Number of seconds elapsed	Length (in inches)	Width (in inches)	Perimeter (in inches)	Area (in square inches)
0	5	4	18	20
1	5.5	3.8	18.6	20.9
2	6	3.6	19.2	21.6
3	6.5	3.4	19.8	22.1
4	7	3.2	20.4	22.4
5	7.5	3	21	22.5
6	8	2.8	21.6	22.4
7	8.5	2.6	22.2	22.1
8	9	2.4	22.8	21.6
9	9.5	2.2	23.4	20.9
10	10	2	24	20

2. Increased **3.** Increased **4.** Increased **5.** Decreased **6.** No

EXERCISE SET 7.3 *page 389*

1. a. Perimeter is not measured in square units. **b.** Area is measured in square units. **2. a.** 36 in. **b.** 77 in^2 **3. a.** 30 m **b.** 50 m^2 **4. a.** 36 mi **b.** 81 mi^2 **5. a.** 40 km **b.** 100 km^2 **6. a.** 24 m **b.** 30 m^2 **7. a.** 40 ft **b.** 72 ft^2 **8. a.** 68 cm **b.** 224 cm^2 **9. a.** 8π cm; 25.13 cm **b.** 16π cm^2; 50.27 cm^2 **10. a.** 24π m; 75.40 m **b.** 144π m^2; 452.39 m^2 **11. a.** 11π mi; 34.56 mi **b.** 30.25π mi^2; 95.03 mi^2 **12. a.** 18π in.; 56.55 in. **b.** 81π in^2; 254.47 in^2 **13. a.** 17π ft; 53.41 ft **b.** 72.25π ft^2; 226.98 ft^2 **14. a.** 6.6π km; 20.73 km **b.** 10.89π km^2; 34.21 km^2 **15.** 20 in. **16.** 50 ft **17.** 10 mi **18.** 560 ft **19.** 2 packages **20.** 990 ft **21.** Perimeter of the square **22.** Circumference of the circle **23.** 144 m^2 **24.** 8800 yd^2 **25.** 9 in. **26.** 10 in. **27.** 39 ft **28.** 8 m **29.** 10×20 unit **30.** 96 m^2 **31.** 136.5 ft^2 **32.** 607.5 m^2 **33.** 160 km^2 **34.** 2 bags **35.** 2 qt **36.** 7 ft^2 **37.** $480 **38.** $60 **39.** $912 **40.** $56 **41.** 176 m^2 **42.** 120 ft^2 **43.** 13.19 ft **44.** 9.42 m **45.** 12,064 in. **46.** 94.25 ft **47.** 62.83 ft **48.** 144π in^2 **49.** 2500π ft^2 **50.** 113.10 in^2 **51.** 339.29 in^2 larger; more than twice the size **52.** 266,281 km **53.** 222.2 mi^2 **54.** 120 ft **55.** $8r^2 - 2\pi r^2$ **56.** $\pi r^2 - 2r^2$ **57.** 4 times as large **58. a.** r **b.** $\frac{1}{2}C = \frac{1}{2}(2\pi r) = \pi r$ **c.** $A = bh = (\pi r)r = \pi r^2$ **59.** 12.1 in^2 **b.** 29.8 cm^2 **c.** 3 in., 4 in., 5 in.

EXCURSION EXERCISES, SECTION 7.4 *page 401*

1. a. 1 **b.** 8 **c.** 0 **d.** 3 **2.** Nail **3.** Spatula **4. a.** Genus 0: C, E, F, G, H, I, J, K, L, M, N, S, T, U, V, W, X, Y, Z; genus 1: A, D, O, P, Q, R; genus 2: B **b.** Genus 0: c, f, h, i, j, k, l, m, n, r, s, t, u, v, w, x, y, z; genus 1: a, b, d, e, g, o, p, q; genus 2: none **5. a.** Genus 0: 1, 2, 3, 5, 7; genus 1: 4, 6, 9; genus 2: 8 **b.** Genus 0: 6, 9; genus 1: 1, 4, 5, 7, 8; genus 2: 2, 3 **6. a.** Yes **b.** Yes **c.** No **d.** No **e.** No **f.** No

EXERCISE SET 7.4 *page 402*

1. $\frac{1}{2}$ **2.** $\frac{1}{3}$ **3.** $\frac{3}{4}$ **4.** $\frac{1}{3}$ **5.** 7.2 cm **6.** 13.7 in. **7.** 3.3 m **8.** 4.9 ft **9.** 12 m **10.** 38 cm **11.** 12 in. **12.** 45 cm^2 **13.** 56.3 cm^2 **14.** 49 m^2 **15.** 18 ft **16.** 22.5 ft **17.** 16 m **18.** 20.8 ft **19.** $14\frac{3}{8}$ ft **20.** 6.25 cm **21.** 15 m **22.** 6 in. **23.** 8 ft **24.** 10 ft **25.** 13 cm **26.** 12 m **27.** 35 m **28.** 45 ft **29.** Yes, SAS theorem **30.** Yes, ASA theorem **31.** Yes, SSS theorem **32.** Yes, SAS theorem **33.** Yes, ASA theorem **34.** Yes, SSS theorem **35.** No **36.** Yes, SAS theorem **37.** Yes, SAS theorem **38.** Yes, SSS theorem **39.** No **40.** Yes, ASA theorem **41.** No **42.** True **43.** 5 in. **44.** 13 in. **45.** 8.6 cm **46.** 11.4 cm **47.** 11.2 ft **48.** 8.7 ft **49.** 4.5 cm **50.** 7.9 m **51.** 12.7 yd **52.** 7.4 m **53.** 8.5 cm **54.** 21.6 mi **55.** 24.3 cm **56.** 24 in. **57. a.** Always true **b.** Sometimes true **c.** Always true **d.** Always true
58. Triangle ABC is similar to triangle ACD because both have a right angle and $m\angle A = m\angle A$. The measures of two angles of one triangle are equal to the measures of two angles of the other triangle.
Triangle ABC is similar to triangle CBD because both have a right angle and $m\angle B = m\angle B$. The measures of two angles of one triangle are equal to the measures of two angles of the other triangle.
Triangles ACD and BCD are right triangles because both have a right angle. $\angle A$ is complementary to $\angle B$, $\angle ACD$ is complementary to $\angle CBD$, $\angle ACD$ is complementary to $\angle A$, and $\angle BCD$ is complementary to $\angle B$. Therefore, $m\angle A = m\angle BCD$ and $m\angle B = m\angle ACD$. The measures of two angles of one triangle are equal to the measures of two angles of the other triangle. Triangles ACD and CBD are similar triangles.

EXCURSION EXERCISES, SECTION 7.5 *page 413*

1. 0.21 cm **2.** 1.88 in. **3.** 0.25 lb/in^3

EXERCISE SET 7.5 *page 414*

1. 840 in^3 **2.** 168π ft^3; 527.79 ft^3 **3.** 15 ft^3 **4.** 421.875 m^3 **5.** 4.5π cm^3; 14.14 cm^3 **6.** 128π cm^3; 402.12 cm^3 **7.** 94 m^2 **8.** 1176 ft^2 **9.** 56 m^2 **10.** 4π cm^2; 12.57 cm^2 **11.** 96π in^2; 301.59 in^2 **12.** 15.75π ft^2; 49.48 ft^2 **13.** 34 m^3 **14.** 20.25 ft^3 **15.** 15.625 in^3 **16.** 343 cm^3 **17.** 36π ft^3 **18.** 7.24 m^3 **19.** 8143.01 cm^3 **20.** 115.2π m^3 **21.** 75π in^3 **22.** 392.70 cm^3 **23.** 120 in^3 **24.** 216 m^3 **25.** Sphere **26.** Sphere **27.** 7.80 ft^3 **28.** 40.5 m^3 **29.** 35,380,400 gal **30.** 6416 cm^2 **31.** 69.36 m^2 **32.** 13.5 in^2 **33.** 225π cm^2 **34.** 50.27 in^2 **35.** 402.12 in^2 **36.** 2.88π m^2 **37.** 6π ft^2 **38.** 874.15 in^2 **39.** 297 in^2 **40.** 832 m^2 **41.** 2.5 ft **42.** 8.5 in. **43.** 11 cans **44.** 3217 ft^2 **45.** 22.53 cm^2 **46.** 1.67 m^3 **47.** 5 m^3 **48.** 115.43 cm^3 **49.** 69.12 in^3 **50.** 4580.44 cm^3 **51.** 192 in^3 **52.** 19 m^2 **53.** 208 in^2 **54.** 622.65 m^2 **55.** 204.57 cm^2 **56.** 888.02 ft^3 **57.** 95,000 L **58.** 79.17 g **59.** \$4860 **60.** 256,000 gal

61. a. Drawings will vary. For example:

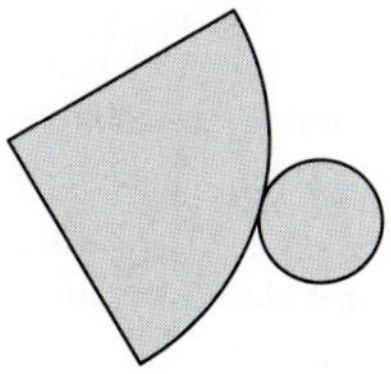

b. Drawings will vary. For example:

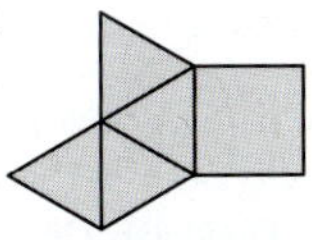

62. Surface area of the sphere $= 4\pi r^2$. Surface area of the side of the cylinder $= 2\pi rh = 2\pi r(2r) = 4\pi r^2$. **63. a.** Always true **b.** Never true **c.** Sometimes true **64. a.** Doubled + $4WH$ **b.** Quadrupled **c.** 8 times as large **d.** Quadrupled
65. a. For example, make a cut perpendicular to the top and bottom faces and parallel to two of the sides. **b.** For example, beginning at an edge that is perpendicular to the bottom face, cut at an angle through to the bottom face. **c.** For example, beginning at the top face at a distance d from the vertex, cut at an angle to the bottom face, ending at a distance greater than d from the opposite vertex. **d.** For example, beginning on the top face at a distance d from a vertex, cut across the cube to a point just above the opposite vertex.

EXCURSION EXERCISES, SECTION 7.6 *page 423*

Drawings should be similar to that shown in the text. Approximate values will vary. As produced by a calculator, $\sin 35° = 0.5736$, $\cos 35° = 0.8192$, and $\tan 35° = 0.7002$.

EXERCISE SET 7.6 page 424

1. a. $\frac{a}{c}$ **b.** $\frac{b}{c}$ **c.** $\frac{b}{c}$ **d.** $\frac{a}{c}$ **e.** $\frac{a}{b}$ **f.** $\frac{b}{a}$ **2.** $\sin^{-1}(x)$ means the angle whose sine is x. $\cos^{-1}(x)$ means the angle whose cosine is x. $\tan^{-1}(x)$ means the angle whose tangent is x. **3.** $\sin\theta = \frac{5}{13}, \cos\theta = \frac{12}{13}, \tan\theta = \frac{5}{12}$
4. $\sin\theta = \frac{15}{17}, \cos\theta = \frac{8}{17}, \tan\theta = \frac{15}{8}$ **5.** $\sin\theta = \frac{24}{25}, \cos\theta = \frac{7}{25}, \tan\theta = \frac{24}{7}$ **6.** $\sin\theta = \frac{20}{29}, \cos\theta = \frac{21}{29}, \tan\theta = \frac{20}{21}$
7. $\sin\theta = \frac{8}{\sqrt{113}}, \cos\theta = \frac{7}{\sqrt{113}}, \tan\theta = \frac{8}{7}$ **8.** $\sin\theta = \frac{3}{\sqrt{10}}, \cos\theta = \frac{1}{\sqrt{10}}, \tan\theta = 3$ **9.** $\sin\theta = \frac{1}{2}, \cos\theta = \frac{\sqrt{3}}{2}, \tan\theta = \frac{1}{\sqrt{3}}$
10. $\sin\theta = \frac{3}{4}, \cos\theta = \frac{\sqrt{7}}{4}, \tan\theta = \frac{3}{\sqrt{7}}$ **11.** 0.6820 **12.** 0.8829 **13.** 1.4281 **14.** 0.9816 **15.** 0.9971
16. 0.9278 **17.** 1.9970 **18.** 0.1357 **19.** 0.8878 **20.** 0.2924 **21.** 0.8453 **22.** 0.8156 **23.** 0.8508
24. 0.7028 **25.** 0.6833 **26.** 0.7660 **27.** 38.6° **28.** 17.9° **29.** 41.1° **30.** 26.6° **31.** 21.3°
32. 47.3° **33.** 38.0° **34.** 52.9° **35.** 72.5° **36.** 57.8° **37.** 0.6° **38.** 13.7° **39.** 66.1°
40. 61.6° **41.** 29.5° **42.** 12.7° **43.** 841.8 ft **44.** 87.0 ft **45.** 13.6° **46.** 247.1 ft **47.** 29.1 ft
48. 360.8 ft **49.** 52.9 ft **50.** 1801.3 ft **51.** 13.6 ft **52.** 55.1° **53.** 1056.6 ft **54.** 19.4° **55.** 29.6 yd
56. 1467.8 ft **57.** 4 radians **58.** $\frac{1}{2}$ radian **59.** $\frac{2}{3}$ radian **60.** 1.2 radians **61.** $\left(\frac{180}{\pi}\right)^\circ$ **62.** Larger than
63. $\frac{\pi}{4}$ radian; 0.7854 radian **64.** π radians; 3.1416 radians **65.** $\frac{7\pi}{4}$ radians; 5.4978 radians **66.** $\frac{\pi}{2}$ radians; 1.5708 radians
67. $\frac{7\pi}{6}$ radians; 3.6652 radians **68.** $\frac{\pi}{10}$ radian; 0.3142 radian **69.** 60° **70.** 330° **71.** 240° **72.** $\left(\frac{216}{\pi}\right)^\circ$; 68.7549°
73. $\left(\frac{540}{\pi}\right)^\circ$; 171.8873° **74.** $\left(\frac{432}{\pi}\right)^\circ$; 137.5099°

EXCURSION EXERCISES, SECTION 7.7 page 435

1. a.

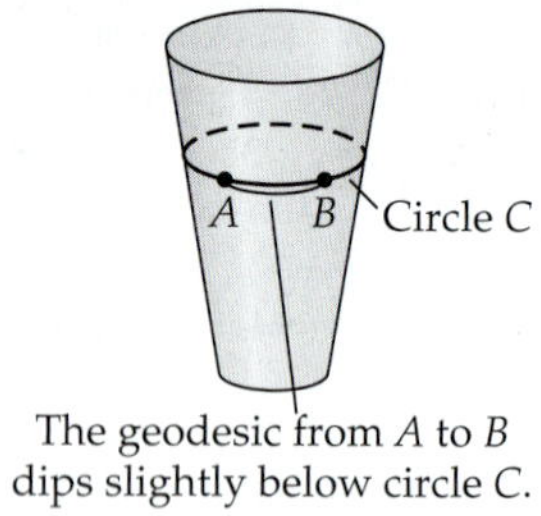

b.

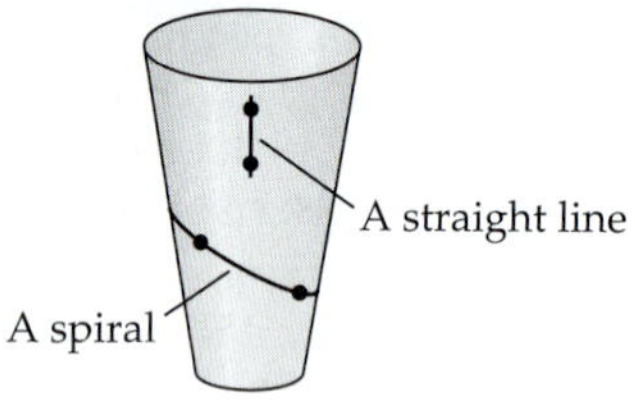

2. A piece of tape placed directly over circle D does not lie flat. Thus by the inverse of the tape test theorem, circle D is not a geodesic of the sphere.
3. A piece of tape placed directly over circle E will not lie flat. Thus by the inverse of the tape test theorem, circle E is not a geodesic of the figure.
4. d. Alaska **5. c.** Godthaab, Greenland **6.** 1800 mi **7. a.** A piece of tape placed directly over the blue circle will lie flat. A piece of tape placed directly over either of the red circles will not lie flat. **b.** A piece of tape placed directly over the blue parabola will lie flat. A piece of tape placed directly over either of the red parabolas will not lie flat.

EXERCISE SET 7.7 page 436

1. a. Through a given point not on a given line, exactly one line can be drawn parallel to the given line. **b.** Through a given point not on a given line, there are at least two lines parallel to the given line. **c.** Through a given point not on a given line, there exist no lines parallel to the given line.
2. Nikolai Lobachevsky **3.** Carl Friedrich Gauss **4. a.** One **b.** One **c.** Three **5.** Imaginary geometry
6. The great circles of a sphere are the geodesics of the sphere, just as lines are the geodesics of a plane. **7.** A geodesic is a curve on a surface such that for any two points of the curve, the portion of the curve between the points is the shortest path on the surface that joins these points. **8.** Lobachevskian or hyperbolic geometry **9.** An infinite saddle surface **10.** Riemannian geometry **11.** π square units **12.** 6,160,000 mi^2
13. $d_E(P, Q) = \sqrt{49} = 7$ blocks, $d_C(P, Q) = 7$ blocks **14.** $d_E(P, Q) = \sqrt{50} \approx 7.1$ blocks, $d_C(P, Q) = 10$ blocks
15. $d_E(P, Q) = \sqrt{89} \approx 9.4$ blocks, $d_C(P, Q) = 13$ blocks **16.** $d_E(P, Q) = \sqrt{74} \approx 8.6$ blocks, $d_C(P, Q) = 12$ blocks
17. $d_E(P, Q) = \sqrt{72} \approx 8.5$ blocks, $d_C(P, Q) = 12$ blocks **18.** $d_E(P, Q) = \sqrt{100} = 10$ blocks, $d_C(P, Q) = 14$ blocks
19. $d_E(P, Q) = \sqrt{37} \approx 6.1$ blocks, $d_C(P, Q) = 7$ blocks **20.** $d_E(P, Q) = \sqrt{9} = 3$ blocks, $d_C(P, Q) = 3$ blocks

21. $d_C(P, Q) = 7$ blocks **22.** $d_C(P, Q) = 7$ blocks **23.** $d_C(P, Q) = 5$ blocks **24.** $d_C(P, Q) = 8$ blocks
25. $d_C(P, Q) = 5$ blocks **26.** $d_C(P, Q) = 8$ blocks **27.** A city distance may be associated with more than one Euclidean distance. For example, if $P = (0, 0)$ and $Q = (2, 0)$, then the city distance between the points is 2 blocks and the Euclidean distance is also 2 blocks. However, if $P = (0, 0)$ and $Q = (1, 1)$, then the city distance between the points is still 2 blocks, but the Euclidean distance is $\sqrt{2}$ blocks. **28. a.** No, not necessarily. For instance, if $P = (0, 0)$, $Q = (3, 4)$, $R = (1, 0)$, and $S = (6, 0)$, then $d_E(P, Q) = d_E(R, S) = 5$ blocks, but $d_C(P, Q) \neq d_C(R, S)$. **b.** No, not necessarily. For instance, if $P = (0, 0)$, $Q = (1, 1)$, $R = (2, 0)$, and $S = (4, 0)$, then $d_C(P, Q) = d_C(R, S) = 2$ blocks, but $d_E(P, Q) \neq d_E(R, S)$.

29.
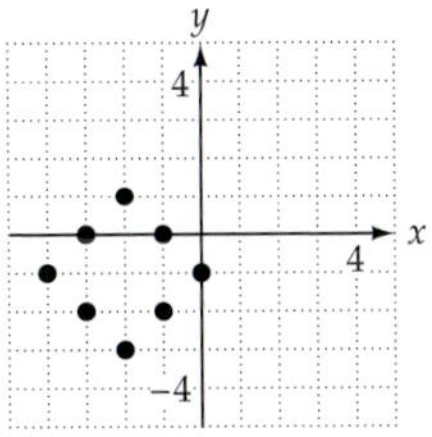

30.
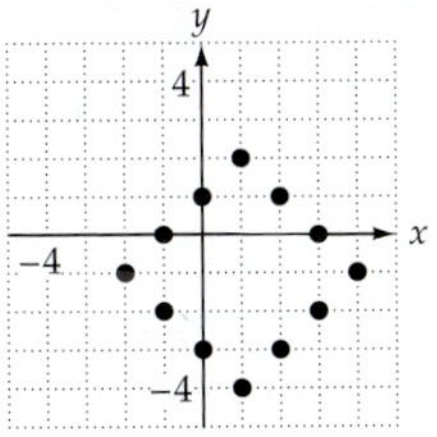

31.
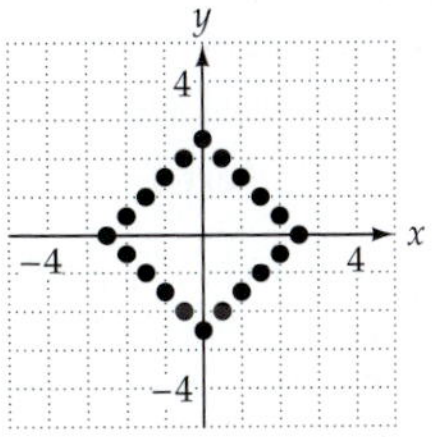

32.
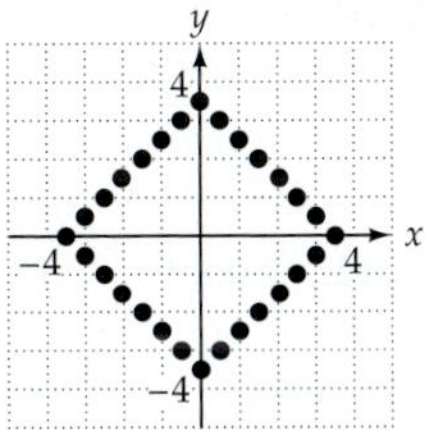

33. $4n$ **34.** The city circle with a radius of 4.5 blocks has the most points.

35. a.
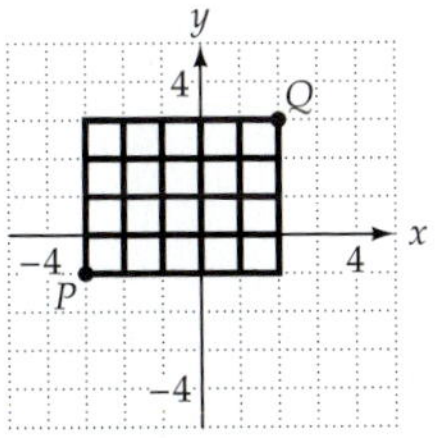

b.
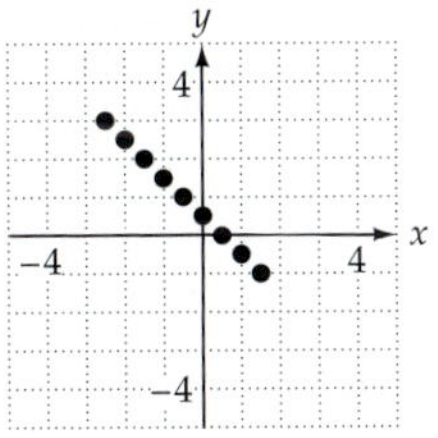

36. a.
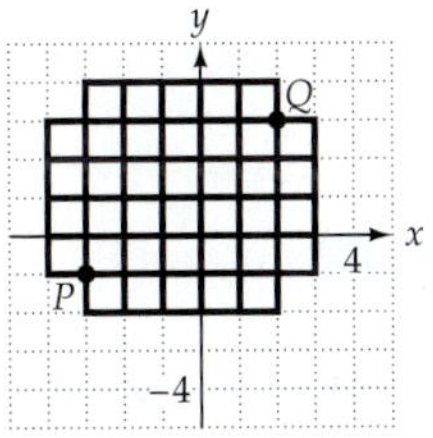

b.
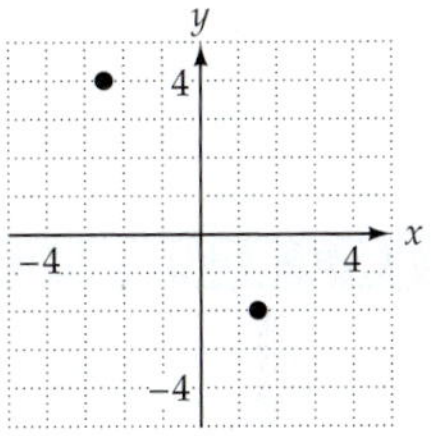

37. a. 10 **b.** 3 **38. a.** 15 **b.** 6

EXCURSION EXERCISES, SECTION 7.8 *page 447*

1. The folding procedure would require you to fold the paper 10 times. This is not humanly possible with a small piece of paper. (Try it yourself. Were you able to make more than six folds?)

2.
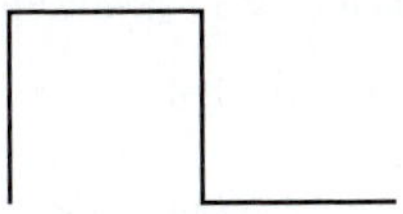

3. Left **4.** Right **5.** Right

EXERCISE SET 7.8 *page 447*

1.
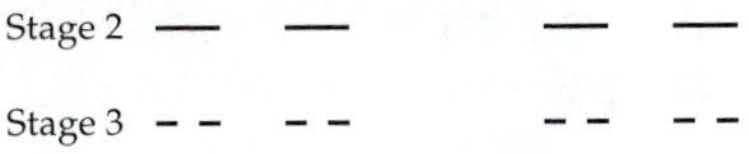

2.

Stage 2 Stage 3

3.
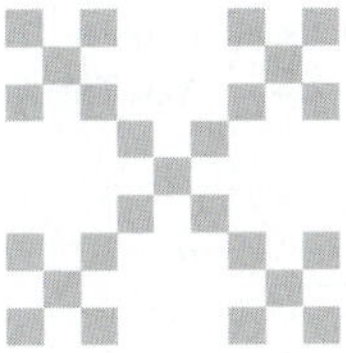
Stage 2

4.
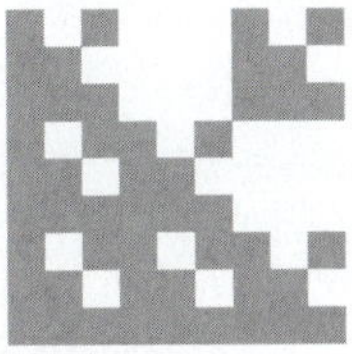
Stage 2

5.
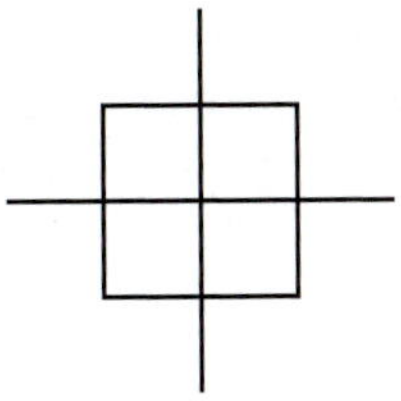
Stage 2

6.
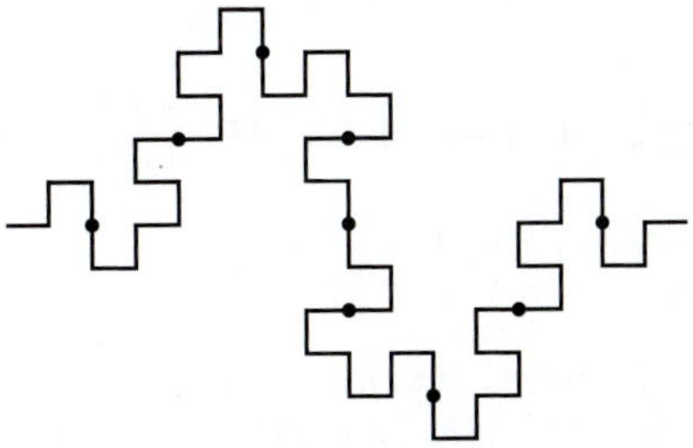
Stage 2

7.
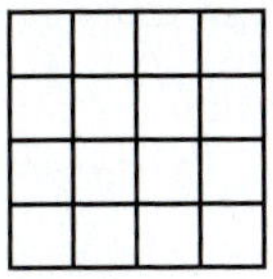
Stage 2

8. 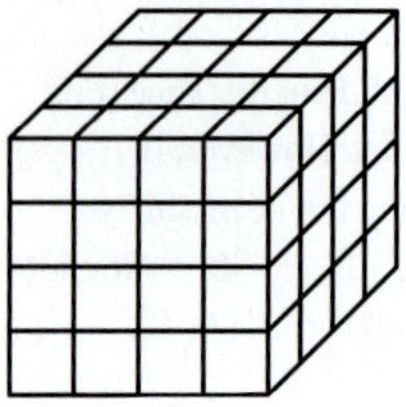

Stage 2

9. 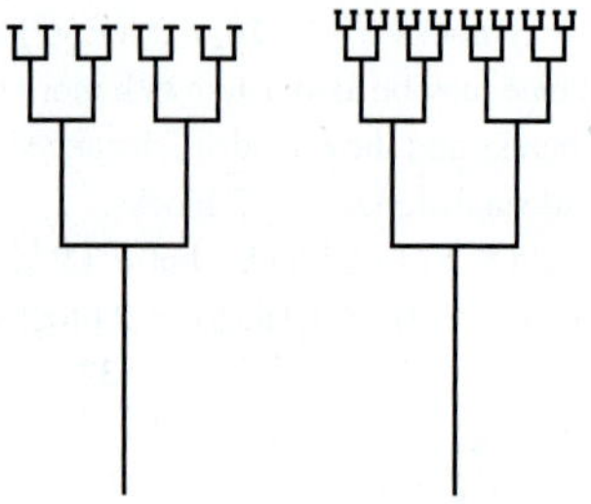

Stage 3 Stage 4

10. 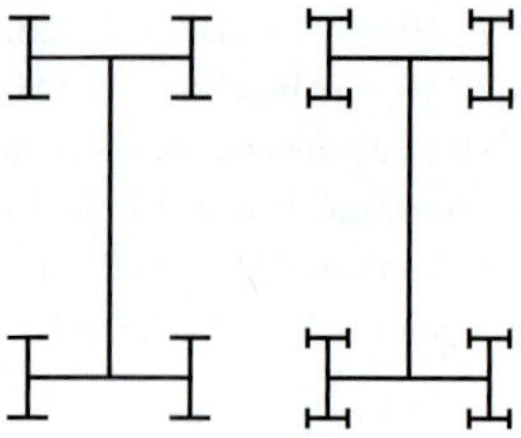

Stage 3 Stage 4

11. 0.631 **12.** 2.000 **13.** 1.465 **14.** 1.771 **15.** 2.000 **16.** 1.500 **17.** 2.000
18. 3.000 **19.** 1.613 **20.** 2.727 **21. a.** Sierpinski carpet, 1.893; variation 2, 1.771; variation 1, 1.465
b. The Sierpinski carpet **22. a.** 2 **b.** The Peano curve completely covers a square region in a plane.
23. The binary tree fractal is not a strictly self-similar fractal. **24.** 

Stage 2 of Lévy's curve Stage 2 of the Heighway dragon

CHAPTER 7 REVIEW EXERCISES *page 453*

1. $2\frac{1}{4}$ [Sec. 7.1] **2.** $7\frac{1}{2}$ [Sec. 7.1] **3.** 3.7 [Sec. 7.1] **4.** 678 [Sec. 7.1] **5.** 1.273 [Sec. 7.1]
6. \$3.36 per qt [Sec. 7.1] **7.** $m\angle x = 22°$; $m\angle y = 158°$ [Sec. 7.2] **8.** 24 in. [Sec. 7.4] **9.** 240 in^3 [Sec. 7.5]
10. 68° [Sec. 7.2] **11.** 220 ft^2 [Sec. 7.5] **12.** 40π m^2 [Sec. 7.5] **13.** 44 cm [Sec. 7.2]
14. $m\angle w = 30°$; $m\angle y = 30°$ [Sec. 7.2] **15.** 27 in^2 [Sec. 7.3] **16.** 96 cm^3 [Sec. 7.5] **17.** 14.1 m [Sec. 7.3]
18. $m\angle a = 138°$; $m\angle b = 42°$ [Sec. 7.2] **19.** A 148° angle [Sec. 7.2] **20.** 39 ft^3 [Sec. 7.5] **21.** 95° [Sec. 7.2]
22. 8 cm [Sec. 7.3] **23.** 288π mm^3 [Sec. 7.5] **24.** 21.5 cm [Sec. 7.3] **25.** 4 cans [Sec. 7.5] **26.** 208 yd [Sec. 7.3]
27. 90.25 m^2 [Sec. 7.3] **28.** 276 m^2 [Sec. 7.3] **29.** The triangles are congruent by the SAS theorem. [Sec. 7.4]
30. 9.7 ft [Sec. 7.4] **31.** $\sin\theta = \frac{5\sqrt{89}}{89}$, $\cos\theta = \frac{8\sqrt{89}}{89}$, $\tan\theta = \frac{5}{8}$ [Sec. 7.6] **32.** $\sin\theta = \frac{\sqrt{3}}{2}$, $\cos\theta = \frac{1}{2}$, $\tan\theta = \sqrt{3}$ [Sec. 7.6]
33. 25.7° [Sec. 7.6] **34.** 29.2° [Sec. 7.6] **35.** 53.8° [Sec. 7.6] **36.** 1.9° [Sec. 7.6] **37.** 100.1 ft [Sec. 7.6]
38. 153.2 mi [Sec. 7.6] **39.** 56.0 ft [Sec. 7.6] **40.** Spherical geometry or elliptical geometry [Sec. 7.7]
41. Hyperbolic geometry [Sec. 7.7] **42.** Lobachevskian or hyperbolic geometry [Sec. 7.7] **43.** Riemannian or spherical geometry [Sec. 7.7]
44. 120π in^2 [Sec. 7.7] **45.** $\frac{25\pi}{3}$ ft^2 [Sec. 7.7] **46.** $d_E(P, Q) = 5$ blocks, $d_C(P, Q) = 7$ blocks [Sec. 7.7]
47. $d_E(P, Q) = \sqrt{113} \approx 10.6$ blocks, $d_C(P, Q) = 15$ blocks [Sec. 7.7] **48.** $d_E(P, Q) = \sqrt{37} \approx 6.1$ blocks, $d_C(P, Q) = 7$ blocks [Sec. 7.7]
49. $d_E(P, Q) = \sqrt{89} \approx 9.4$ blocks, $d_C(P, Q) = 13$ blocks [Sec. 7.7] **50. a.** P and Q **b.** P and R [Sec. 7.7]
51. 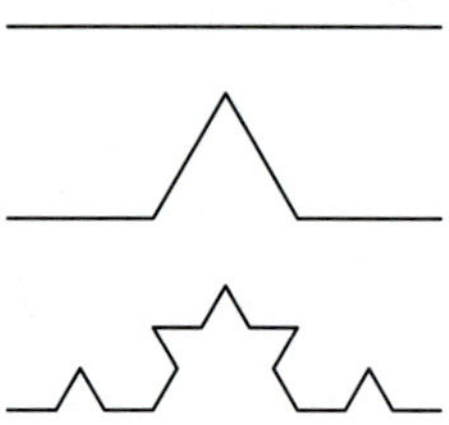

Stage 0
Stage 1
Stage 2

Yes. The Koch curve is a strictly self-similar fractal. [Sec. 7.8]

52. 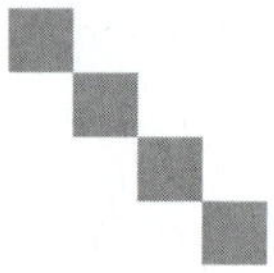

Stage 2 [Sec. 7.8]

53. a. 2 **b.** 2 **c.** $D = \frac{\log 2}{\log 2} = 1$ [Sec. 7.8] **54.** $\frac{\log 5}{\log 4} \approx 1.161$ [Sec. 7.8]

CHAPTER 7 TEST *page 455*

1. 169.6 m^3 [Sec. 7.5, Example 3] **2.** 6.8 m [Sec. 7.3, Example 1] **3.** A 58° angle [Sec. 7.2, Example 2]
4. 3.1 m^2 [Sec. 7.3, Example 10] **5.** 150° [Sec. 7.2, Example 5] **6.** $m\angle a = 45°$; $m\angle b = 135°$ [Sec. 7.2, Example 5]
7. 1200 cm [Sec. 7.1, Example 4] **8.** 448π cm^3 [Sec. 7.5, Example 3] **9.** $1\frac{1}{5}$ ft [Sec. 7.4, Example 2]
10. 90° and 50° [Sec. 7.2, Example 8] **11.** 125° [Sec. 7.2, Example 7] **12.** 32 m^2 [Sec. 7.3, Example 7]
13. 25 ft [Sec. 7.4, Example 2] **14.** 113.1 in^2 [Sec. 7.3, Example 11] **15.** The triangles are congruent by the SAS theorem.
[Sec. 7.4, Example 3] **16.** 7.5 cm [Sec. 7.4, Example 4] **17.** $\sin\theta = \frac{4}{5}$, $\cos\theta = \frac{3}{5}$, $\tan\theta = \frac{4}{3}$ [Sec. 7.6, Example 1]

18. 127 ft [Sec. 7.6, Example 6] **19.** 103.9 ft^2 [Sec. 7.3, Example 11] **20.** 780 in^3 [Sec. 7.5, Check Your Progress 1] **21.** 26.82 m/s [Sec. 7.1, Example 9] **22.** Through a given point not on a given line, exactly one line can be drawn parallel to the given line. [Sec. 7.7, Example 2] **23.** A great circle of a sphere is a circle on the surface of the sphere whose center is at the center of the sphere.

$S = (m\angle A + m\angle B + m\angle C - 180°)\left(\frac{\pi}{180°}\right)r^2$ [Sec. 7.7, Example 1] **24.** 80π $ft^2 \approx 251.3$ ft^2 [Sec. 7.7, Example 1]

25. $d_E(P, Q) = \sqrt{82} \approx 9.1$ blocks, $d_C(P, Q) = 10$ blocks [Sec. 7.7, Example 3] **26.** 16 [Sec. 7.7, Example 3]

27. 

Stage 2 [Sec. 7.8, Examples 1 and 2]

28. 

Stage 2 [Sec. 7.8, Examples 1 and 2]

29. Replacement ratio: 2; scale ratio: 2; similarity dimension: 1 [Sec. 7.8, Examples 4 and 5]

30. Replacement ratio: 3; scale ratio: 2; similarity dimension: $\frac{\log 3}{\log 2} \approx 1.585$ [Sec. 7.8, Examples 4 and 5]

CHAPTER 8

EXCURSION EXERCISES, SECTION 8.1 *page 465*

1. Answers will vary. **2.** Wednesday **3.** Saturday **4.** Tuesday

EXERCISE SET 8.1 *page 466*

1. 8 **2.** 1 **3.** 12 **4.** 3 **5.** 2 **6.** 4 **7.** 4 **8.** 9 **9.** 4 **10.** 8 **11.** 7 **12.** 10 **13.** 11 **14.** 12 **15.** 7 **16.** 9 **17.** 0300 **18.** 0400 **19.** 0400 **20.** 0200 **21.** 2000 **22.** 2300 **23.** 2100 **24.** 0800 **25.** 3 **26.** 1 **27.** 6 **28.** 4 **29.** True **30.** True **31.** False **32.** False **33.** True **34.** True **35.** False **36.** False **37.** True **38.** True **39.** Possible answers are 2, 8, 14, 20, 26, 32, 38, ... **40.** Possible answers are 2, 6, 10, 14, 18, 22, 26, 30, ... **41.** 3 **42.** 0 **43.** 3 **44.** 6 **45.** 2 **46.** 3 **47.** 10 **48.** 1 **49.** 3 **50.** 3 **51.** 3 **52.** 1 **53.** 5 **54.** 3 **55.** 4 **56.** 7 **57.** 3 **58.** 0 **59.** 7 **60.** 1 **61.** 2 **62.** 1 **63. a.** 6 o'clock **b.** 5 o'clock **64. a.** 6 o'clock **b.** 4 o'clock **65. a.** Tuesday **b.** Monday **66. a.** Saturday **b.** Saturday **67.** Friday **68.** Thursday **69.** Saturday **70.** Friday **71.** 1, 4, 7, 10, 13, 16, ... **72.** 2, 7, 12, 17, 22, 27, ... **73.** 1, 6, 11, 16, 21, 26, ... **74.** 10, 21, 32, 43, 54, 65, ... **75.** 0, 2, 4, 6, 8, 10, 12, ... **76.** 1, 4, 7, 10, 13, 16, ... **77.** No solutions **78.** 5, 15, 25, 35, 45, 55, ... **79.** 0, 2, 4, 6, 8, 10, 12, ... **80.** 6, 14, 22, 30, 38, 46, ... **81.** No solutions **82.** All whole numbers **83.** 5, 7 **84.** 1, 4 **85.** 3, 3 **86.** 5, 3 **87.** 5, 3 **88.** 9, none **89.** 6 **90.** 4 **91.** 6 **92.** 4 **93.** 2 **94.** 6 **97.** 4 **98.** 3, 19 **99. a.** 8, 7, 5, 1, 4, 10, 0, 2, 6 **b.** 3; it equals x_0. **c.** The sequence of numbers repeats after every 10 presses. **100. a.** Yes; answers will vary. **b.** 3; 1998

EXCURSION EXERCISES, SECTION 8.2 *page 476*

1. 4870, 607, 141, 3532 **2.** 2753, 3532, 2001, 1268, 957 **3.** WORLD **4.** BEAUTY **5.** Answers will vary.

EXERCISE SET 8.2 *page 476*

1. No **2.** Yes **3.** Yes **4.** No **5.** Yes **6.** No **7.** 8 **8.** 7 **9.** 0 **10.** 1 **11.** 8 **12.** 9 **13.** 6 **14.** 2 **15.** 5 **16.** 5 **17.** 7 **18.** 3 **19.** 2 **20.** 7 **21.** 0 **22.** 6 **23.** 7 **24.** 1 **25.** 3 **26.** 4 **27.** Yes **28.** No **29.** Yes **30.** Yes **31.** Yes **32.** Yes **33.** No **34.** No **35.** No **36.** Yes **37.** Yes **38.** No **39.** BPZMM UCASMBMMZA **40.** LRE ZUTOMNZ **41.** UF'E M SUDX **42.** VNNC JC WXXW **43.** VWLFNV DQG VWRQHV **44.** P HIXIRW XC IXBT **45.** AGE OF ENLIGHTENMENT **46.** IMAGINATION RULES THE WORLD **47.** FRIEND IN NEED **48.** NOBODY IS PERFECT **49.** DANGER WILL ROBINSON **50.** FIGHT OR FLIGHT **51.** FORTUNE COOKIE **52.** MAN THE TORPEDOES **53.** PHQ ZLOOLQJOB EHOLHYH ZKDW WKHB ZLVK **54.** THERE ARE NO ACCIDENTS **55.** JUSQD UT LURNUR **56.** WHXKVLQY THIVULT YVPB **57.** PODONNQN NSBQK **58.** ADAZ PINUU GNPP **59.** TURN BACK THE CLOCK **60.** THIS MORTAL COIL **61.** BARREL OF MONKEYS **62.** LOOK BEFORE YOU LEAP **63.** Because the check digit is simply the sum of the first 10 digits mod 9, the same digits in a different order will give the same sum and hence the same check digit. **64.** Every other digit is multiplied by 3 before the sum mod 10 is computed. Two adjacent digits a and b could be transposed and still give the same sum mod 10 if $3a + b \equiv a + 3b$ mod 10, which is satisfied when $a \equiv b$ mod 5, or, equivalently, $|a - b| = 5$. **65. a.** 1 **c.** Yes **d.** The digits 6 and 1 are multiplied by 1 and 3, so the situation is identical to that in Exercise 64.

EXCURSION EXERCISES, SECTION 8.3 *page 487*

1. Any translation followed by another is itself a translation, so the set of translations is closed. Translations are associative, the identity element is I, and any translation can be reversed, so each element has an inverse.

3. a. to **c.**

4.

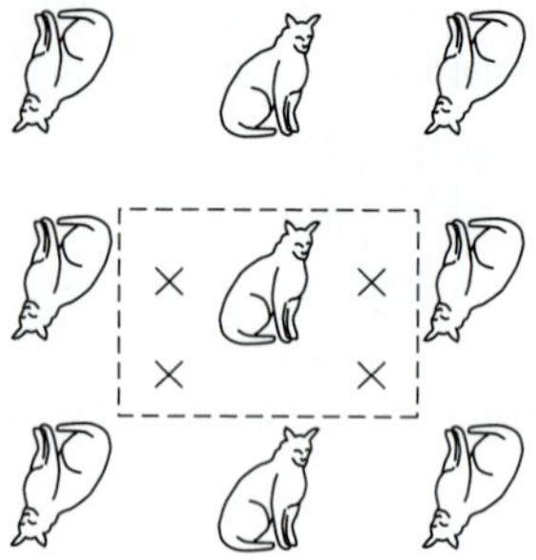

5. a.

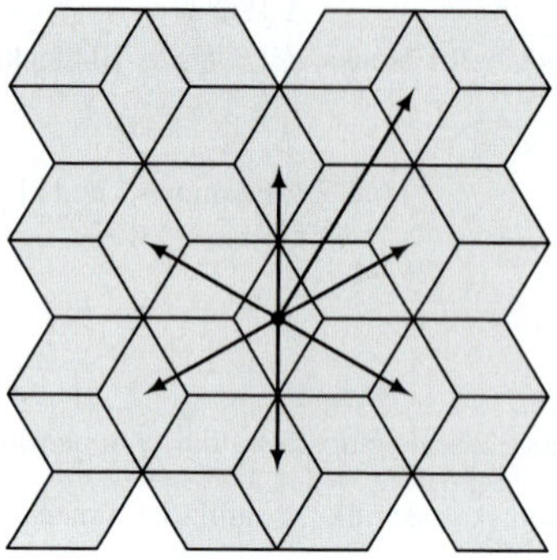

b.

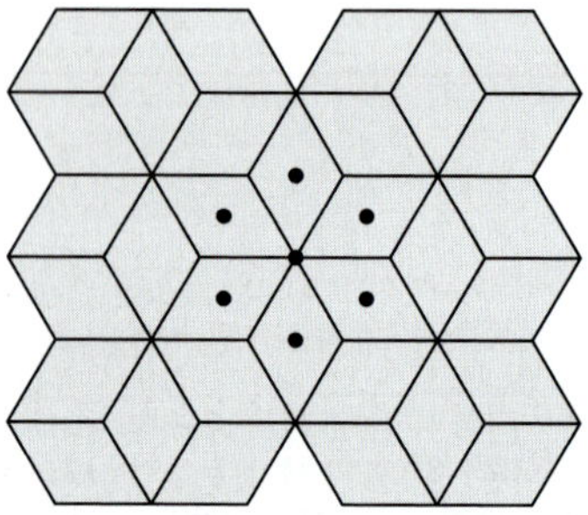

c.

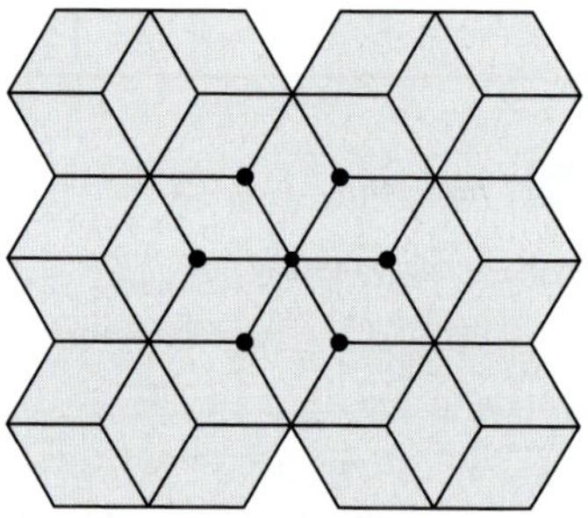

d.

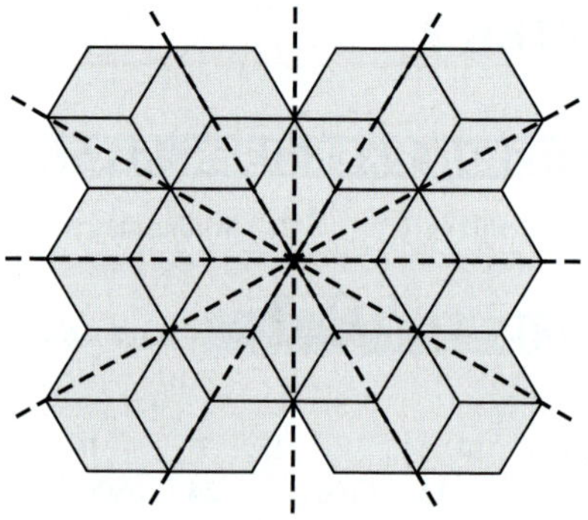

EXERCISE SET 8.3 *page 488*

1. a. Yes **b.** No **2. a.** Yes **b.** Yes **3. a.** Yes **b.** No **4. a.** Yes **b.** No **c.** No **d.** No **5.** Yes **6.** No; properties 3 and 4 fail. **7.** Yes **8.** No; properties 1, 2, and 4 fail. **9.** No; property 4 fails. **10.** Yes **11.** Yes **12.** Yes **13.** Yes **14.** No; properties 1 and 4 fail. **15.** No; property 4 fails. **16.** Yes **17.** Yes **18.** Yes **19.** R_l **20.** R_t **21.** R_{120} **22.** R_{120} **23.** R_{120} **24.** R_t **25.** R_r **26.** R_l **27.** R_l **28.** R_{90}

29. $I = \begin{pmatrix}1&2&3&4\\1&2&3&4\end{pmatrix}$, $R_{90} = \begin{pmatrix}1&2&3&4\\2&3&4&1\end{pmatrix}$, $R_{180} = \begin{pmatrix}1&2&3&4\\3&4&1&2\end{pmatrix}$, $R_{270} = \begin{pmatrix}1&2&3&4\\4&1&2&3\end{pmatrix}$, $R_v = \begin{pmatrix}1&2&3&4\\4&3&2&1\end{pmatrix}$, $R_h = \begin{pmatrix}1&2&3&4\\2&1&4&3\end{pmatrix}$, $R_r = \begin{pmatrix}1&2&3&4\\3&2&1&4\end{pmatrix}$, $R_l = \begin{pmatrix}1&2&3&4\\1&4&3&2\end{pmatrix}$ **30.** R_{90} **31.** R_r **32.** R_l **33.** R_{90} **34.** R_v **35.** I **36.** I **37.** D **38.** A **39.** E **40.** A **41.** B **42.** E

43. $\begin{pmatrix}1&2&3&4\\1&2&3&4\end{pmatrix}$, $\begin{pmatrix}1&2&3&4\\1&2&4&3\end{pmatrix}$, $\begin{pmatrix}1&2&3&4\\1&3&2&4\end{pmatrix}$, $\begin{pmatrix}1&2&3&4\\1&3&4&2\end{pmatrix}$, $\begin{pmatrix}1&2&3&4\\1&4&2&3\end{pmatrix}$, $\begin{pmatrix}1&2&3&4\\1&4&3&2\end{pmatrix}$, $\begin{pmatrix}1&2&3&4\\2&1&3&4\end{pmatrix}$, $\begin{pmatrix}1&2&3&4\\2&1&4&3\end{pmatrix}$, $\begin{pmatrix}1&2&3&4\\2&3&1&4\end{pmatrix}$, $\begin{pmatrix}1&2&3&4\\2&3&4&1\end{pmatrix}$, $\begin{pmatrix}1&2&3&4\\2&4&1&3\end{pmatrix}$, $\begin{pmatrix}1&2&3&4\\2&4&3&1\end{pmatrix}$, $\begin{pmatrix}1&2&3&4\\3&1&2&4\end{pmatrix}$, $\begin{pmatrix}1&2&3&4\\3&1&4&2\end{pmatrix}$, $\begin{pmatrix}1&2&3&4\\3&2&1&4\end{pmatrix}$, $\begin{pmatrix}1&2&3&4\\3&2&4&1\end{pmatrix}$, $\begin{pmatrix}1&2&3&4\\3&4&1&2\end{pmatrix}$, $\begin{pmatrix}1&2&3&4\\3&4&2&1\end{pmatrix}$, $\begin{pmatrix}1&2&3&4\\4&1&2&3\end{pmatrix}$, $\begin{pmatrix}1&2&3&4\\4&1&3&2\end{pmatrix}$, $\begin{pmatrix}1&2&3&4\\4&2&1&3\end{pmatrix}$, $\begin{pmatrix}1&2&3&4\\4&2&3&1\end{pmatrix}$, $\begin{pmatrix}1&2&3&4\\4&3&1&2\end{pmatrix}$, $\begin{pmatrix}1&2&3&4\\4&3&2&1\end{pmatrix}$ **44.** $\begin{pmatrix}1&2&3&4\\3&1&2&4\end{pmatrix}$ **45.** $\begin{pmatrix}1&2&3&4\\2&1&3&4\end{pmatrix}$ **46.** $\begin{pmatrix}1&2&3&4\\2&4&3&1\end{pmatrix}$ **47.** $\begin{pmatrix}1&2&3&4\\3&1&4&2\end{pmatrix}$ **48.** $\begin{pmatrix}1&2&3&4\\3&4&1&2\end{pmatrix}$ **49.** $\begin{pmatrix}1&2&3&4\\4&3&2&1\end{pmatrix}$ **50.** $\begin{pmatrix}1&2&3&4\\1&4&2&3\end{pmatrix}$ **51.** d **52.** a **53.** c **54.** c **55.** Answers will vary. **56.** b **57.** Yes **58.** a and c are inverses of each other; b and d are their own inverses. **59. a.** and **b.** Answers will vary. **c.** Values of n that are prime **60. a.** 2^0 **b.** Answers will vary. **c.** 2^{-8} **d.** 2^5 **e.** 2^{-k} **61. a.** One possible answer is $r \nabla s \neq s \nabla r$. **b.** e **c.** e and u are their own inverses; r and v, s and w, and t and x are inverses of each other. **d.** and **e.** Answers will vary. **f.** $\{e, u\}$

CHAPTER 8 REVIEW EXERCISES page 493

1. 2 [Sec. 8.1] **2.** 2 [Sec. 8.1] **3.** 5 [Sec. 8.1] **4.** 6 [Sec. 8.1] **5.** 9 [Sec. 8.1] **6.** 4 [Sec. 8.1] **7.** 11 [Sec. 8.1] **8.** 7 [Sec. 8.1] **9.** 3 [Sec. 8.1] **10.** 4 [Sec. 8.1] **11.** True [Sec. 8.1] **12.** False [Sec. 8.1] **13.** False [Sec. 8.1] **14.** True [Sec. 8.1] **15.** 2 [Sec. 8.1] **16.** 4 [Sec. 8.1] **17.** 0 [Sec. 8.1] **18.** 3 [Sec. 8.1] **19.** 8 [Sec. 8.1] **20.** 3 [Sec. 8.1] **21.** 7 [Sec. 8.1] **22.** 5 [Sec. 8.1] **23. a.** 2 o'clock **b.** 6 o'clock [Sec. 8.1] **24.** Tuesday [Sec. 8.1] **25.** 3, 7, 11, 15, 19, 23, ... [Sec. 8.1] **26.** 7, 16, 25, 34, 43, 52, ... [Sec. 8.1] **27.** 0, 5, 10, 15, 20, 25, 30, ... [Sec. 8.1] **28.** 4, 15, 26, 37, 48, 59, 70, ... [Sec. 8.1] **29.** 2, 3 [Sec. 8.1] **30.** 5, 7 [Sec. 8.1] **31.** 6 [Sec. 8.1] **32.** 2 [Sec. 8.1] **33.** 8 [Sec. 8.2] **34.** 4 [Sec. 8.2] **35.** 2 [Sec. 8.2] **36.** 1 [Sec. 8.2] **37.** No [Sec. 8.2] **38.** Yes [Sec. 8.2] **39.** No [Sec. 8.2] **40.** No [Sec. 8.2] **41.** THF AOL MVYJL IL DPAO FVB [Sec. 8.2] **42.** NLYNPW LWW AWLYD [Sec. 8.2] **43.** GOOD LUCK TOMORROW [Sec. 8.2] **44.** THE DAY HAS ARRIVED [Sec. 8.2] **45.** UVR YX NDU PGVU [Sec. 8.2] **46.** YOU PASSED THE TEST [Sec. 8.2] **47.** Yes [Sec. 8.3] **48.** Yes [Sec. 8.3] **49.** No, properties 1, 3, and 4 fail. [Sec. 8.3] **50.** Yes [Sec. 8.3] **51.** R_{240} [Sec. 8.3] **52.** R_l [Sec. 8.3] **53.** R_r [Sec. 8.3] **54.** R_{240} [Sec. 8.3] **55.** A [Sec. 8.3] **56.** I [Sec. 8.3] **57.** D [Sec. 8.3] **58.** B [Sec. 8.3]

CHAPTER 8 TEST page 494

1. a. 3 **b.** 5 [Sec. 8.1, Example 1] **2.** Thursday [Sec. 8.1, Example 3] **3. a.** True **b.** False [Sec. 8.1, Example 2] **4.** 4 [Sec. 8.1, Example 4] **5.** 6 [Sec. 8.1, Example 5] **6.** 8 [Sec. 8.1, Example 7] **7. a.** 6 o'clock **b.** 5 o'clock [Sec. 8.1, Example 6] **8.** 5, 14, 23, 32, 41, 50, ... [Sec. 8.1, Example 8] **9.** 1, 3, 5, 7, 9, 11, ... [Sec. 8.1, Example 8] **10.** 4, 2 [Sec. 8.1, Example 9 and Example 10] **11.** 0 [Sec. 8.2, Example 1] **12.** 6 [Sec. 8.2, Example 2] **13.** Yes [Sec. 8.2, Example 3] **14.** BOZYBD LKMU [Sec. 8.2, Example 4] **15.** NEVER QUIT [Sec. 8.2, Example 4] **16.** Yes [Sec. 8.3, Check Your Progress 1] **17.** No, property 4 fails; many elements do not have an inverse. [Sec. 8.3, Example 1] **18. a.** R_t **b.** R_r [Sec. 8.3, Example 2] **19.** $\begin{pmatrix} 1 & 2 & 3 \\ 1 & 3 & 2 \end{pmatrix}$ [Sec. 8.3, Example 4] **20.** $\begin{pmatrix} 1 & 2 & 3 \\ 2 & 3 & 1 \end{pmatrix}$ [Sec. 8.3, Example 5]

CHAPTER 9

EXCURSION EXERCISES, SECTION 9.1 page 505

1. 21.3; low risk **2.** 179 lb **3.** 29 lb **4. a.** 5.1 points **b.** 32 lb **5. a.–b.** Answers will vary. Values found using the nomograph should be close to those found using a calculator. **c.** A disadvantage is that the nomograph provides a visual estimate rather than an exact value. An advantage is that the nomograph gives a quick and relatively accurate estimate without having to use a formula.

EXERCISE SET 9.1 page 506

1. An equation expresses the equality of two mathematical expressions. An equation contains an equals sign. An expression does not. **2.** The solution is 8. Explanations will vary. **3.** Substitute the solution back into the original equation and confirm the equality. **4.** -12 **5.** 12 **6.** -1 **7.** 22 **8.** -50 **9.** -8 **10.** 12 **11.** -20 **12.** -4 **13.** 8 **14.** -20 **15.** -1 **16.** 8 **17.** 1 **18.** $-\frac{2}{5}$ **19.** $-\frac{1}{3}$ **20.** $\frac{3}{4}$ **21.** $\frac{2}{3}$ **22.** 3 **23.** -2 **24.** 3 **25.** 4 **26.** 2 **27.** 2 **28.** -1 **29.** 2 **30.** $\frac{1}{2}$ **31.** $\frac{1}{4}$ **32.** 5 **33.** -2 **34.** $-\frac{1}{9}$ **35.** 4 **36.** 6 **37.** 8 **38.** 2 **39.** 1 **40.** 8 **41.** -32 **42.** \$16,289.30 **43.** \$19,004.18 **44.** 60 ft **45.** 80 ft **46.** 1953 **47.** 1998 **48.** 168 ft **49.** 136 ft **50.** 18.6°C **51.** 24.3°C **52.** 175 **53.** 168.75 **54. a.** \$4694 **b.** \$9410 **55. a.** 1725 children **b.** 465 children **56.** \$62.50 **57.** 3 h **58.** \$600 **59.** 150 research assistants **60.** \$12.50 **61.** \$2000 and \$3000 **62.** More than 11 oz but not over 12 oz **63.** 13 GB **64.** $h = \frac{2A}{b}$ **65.** $b = P - a - c$ **66.** $t = \frac{d}{r}$ **67.** $R = \frac{E}{I}$ **68.** $R = \frac{PV}{nT}$ **69.** $r = \frac{I}{Pt}$ **70.** $W = \frac{P - 2L}{2}$ **71.** $C = \frac{5}{9}(F - 32)$ **72.** $C = R - P$ **73.** $t = \frac{A - P}{Pr}$ **74.** $V_0 = \frac{S + 16t^2}{t}$ **75.** $f = \frac{T + gm}{m}$ **76.** $R = Pn + C$ **77.** $S = C - Rt$ **78.** $h = \frac{3V}{\pi r^2}$ **79.** $b_2 = \frac{2A}{h} - b_1$ **80.** $d = \frac{a_n - a_1}{n - 1}$ **81.** $h = \frac{S}{2\pi r} - r$ **82.** $y = 2x - 4$ **83.** $y = 2 - \frac{4}{3}x$ **84.** $x = \frac{-by - c}{a}$ **85.** $x = \frac{y - y_1}{m} + x_1$ **86.** No solution **87.** Every real number is a solution.

88. a. 0 **b.** −20 **89.** $x = \frac{d - b}{a - c}$. $a \neq c$ or the denominator equals zero and the expression is undefined.

90. $T = 325m + 0.08x$, where T is the total cost, m is the number of months, and x is the number of copies.

91. $T = 750m + 30{,}000$, where T is the total number of miles driven and m is the number of months you have owned the car.

92. $C = 7.50 + 5.25(h - 1)$, where C is the parking charge and h is the number of hours you have parked.

93. $C = 29.95d + 0.50(m - 100)$, where C is the total cost, d is the number of days, and m is the number of miles driven.

EXCURSION EXERCISES, SECTION 9.2 *page 518*

1. 1.71 **2.** 4.33 **3.** 2.77 **4.** 45 **5.** Answers will vary.

EXERCISE SET 9.2 *page 518*

1. Examples will vary. **2.** Examples will vary. **3.** 48.5 mph **4.** 23 mi/gal **5.** 68 words/min **6.** 12.5 m/s **7.** \$26 per share **8.** 400 ft^2/gal **9.** \$19.50/hr **10.** 272,160 kg **11. a.** 336 ft/min **b.** 89 s **12.** A 48-ounce jar of mayonnaise for \$6.29 **13.** 24 ounces for \$3.89 **14.** \$16.50/h **15. a.** 9.2, 9, 10.1, 9.6, 9.7, 9.5, 10.7, 7.3, 9.7, 6.5, 10.9, 10.0, 11.0 **b.** Barry Bonds; Mark McGwire **c.** Explanations may vary. **16. a.** Australia **b.** 1001 more people per square mile **17. a.** 79.5, 81.6, 83.8 **b.** 1.05 times greater **18.** 34,314 Danish kroner **19.** 3,054,136.5 Indian rupees **20.** 645,393 Mexican pesos **21.** 41,226.4 Australian dollars **22.** 24.5; rent **23.** 11.2; buy **24.** 13 : 1 or 13 to 1. There is one faculty member for every 13 students at Syracuse University. **25.** Georgetown University **26.** 4.5 **27.** 17.14 **28.** 42.86 **29.** 25.6 **30.** 216 **31.** 20.83 **32.** 4.31 **33.** 2.22 **34.** 4.35 **35.** 13.71 **36.** 10.97 **37.** 39.6 **38.** 29 lb **39.** \$65,000 **40.** 5.5 mg **41.** 329 mi **42.** 24 ft; 15 ft **43.** 16 mi **44.** 64 in. **45.** 67.5 in. **46.** 160,000 people **47. a.** \$442.50 **b.** \$737.50 **48.** \$1,250,000 **49.** 11.25 g. Explanations will vary. **50. a.** True **b.** True **c.** True **d.** False

51.
$$\frac{a}{b} = \frac{c}{d}$$
$$\frac{a}{b} + 1 = \frac{c}{d} + 1$$
$$\frac{a}{b} + \frac{b}{b} = \frac{c}{d} + \frac{d}{d}$$
$$\frac{a + b}{b} = \frac{c + d}{d}$$

52. a. To make comparisons easier **b.** $\frac{18.3}{109.5} = \frac{45}{168}$; no **c.** Answers will vary. **53. a.** and **b.** The number of seats per state is AL 7, AK 1, AZ 8, AR 4, CA 53, CO 7, CT 5, DE 1, DC 3, FL 25, GA 13, HI 2, ID 2, IL 19, IN 9, IA 5, KS 4, KY 6, LA 7, ME 2, MD 8, MA 10, MI 15, MN 8, MS 4, MO 9, MT 1, NE 3, NV 3, NH 2, NJ 13, NM 3, NY 29, NC 13, ND 1, OH 18, OK 5, OR 5, PA 19, RI 2, SC 6, SD 1, TN 9, UT 3, VT 1, VA 11, WA 9, WV 3, WI 8, WY 1. Students will find that the calculated number of representatives per state does not match the actual number of representatives in the following states: AZ, FL, GA, IL, IA, LA, MA, MI, MN, MO, NV, NJ, NY, OH, PA, SC, TX, UT, WA.

EXCURSION EXERCISES, SECTION 9.3 *page 534*

1. \$11,756.25 **2.** \$4290.75 **3.** \$7795.50; balance due of \$671.50 **4.** \$35,799.25; tax refund of \$700.75

5. No. The taxpayer pays 33% on the amount earned over a given figure.

6. \$922.50 = 10% of \$9225; \$5156.25 = (10% of \$9225) + [15% of (\$37,450 − \$9225)]

EXERCISE SET 9.3 *page 535*

1. Answers will vary. **2.** 3 **3.** 0.5; 50% **4.** $\frac{3}{4}$; 75% **5.** $\frac{2}{5}$; 0.4 **6.** 0.375; 37.5% **7.** $\frac{7}{10}$; 70% **8.** $\frac{9}{16}$; 0.5625 **9.** 0.55; 55% **10.** $\frac{13}{25}$; 52% **11.** $\frac{5}{32}$; 0.15625 **12.** 0.18; 18% **13.** 119 million returns **14.** \$124 **15.** \$53.7 billion **16.** 160 students **17.** 23.7% **18.** 4.4% **19.** 14 million oz **20. a.** 48.2 h **b.** 33.6 h **c.** 3.4 h **21.** 3364 people **22.** 27.8% **23.** 57.7% **24.** 700% **25. a.** 102.5% **b.** 122.2% **c.** 350% **d.** It is 4.5 times larger. Convert the percent to a decimal and add 1. **26.** 6.1% **27.** 16.8% **28.** 255 eggs per person per year **29. a.** 46.0% **b.** 27.2% **30.** Navy; 44.9% **31.** Less than **32.** 16.79% **33. a.** 11,400,000 TV households; 67,100,000 TV households **b.** 6,300,000 TV households; 57,300,000 TV households **c.** 2.3 people

EXCURSION EXERCISES, SECTION 9.4 *page 545*

1. −2 and 5 **2.** 4 **3.** -3 and $\frac{4}{3}$ **4.** -3 and $\frac{1}{3}$ **5.** $3 + 2\sqrt{3}$ and $3 - 2\sqrt{3}$ **6.** $1 + \sqrt{6}$ and $1 - \sqrt{6}$
7. $\frac{1+\sqrt{2}}{2}$ and $\frac{1-\sqrt{2}}{2}$ **8.** $\frac{1+\sqrt{5}}{2}$ and $\frac{1-\sqrt{5}}{2}$ **9.** $x^2 - 5x - 6 = 0$ **10.** $x^2 + 9x + 20 = 0$ **11.** $2x^2 - 7x + 3 = 0$
12. $4x^2 - 5x - 6 = 0$ **13.** $8x^2 + 10x - 3 = 0$ **14.** $9x^2 - 4 = 0$ **15.** $x^2 - 4x + 2 = 0$ **16.** $x^2 - 2x - 2 = 0$

EXERCISE SET 9.4 *page 546*

1. −2 and 5 **2.** −6 and 1 **3.** $\frac{1+\sqrt{5}}{2}$ and $\frac{1-\sqrt{5}}{2}$; −0.618 and 1.618 **4.** $\frac{1+\sqrt{13}}{2}$ and $\frac{1-\sqrt{13}}{2}$; −1.303 and 2.303
5. $3 + \sqrt{13}$ and $3 - \sqrt{13}$; −0.606 and 6.606 **6.** $-2 + \sqrt{6}$ and $-2 - \sqrt{6}$; −4.449 and 0.449 **7.** 0 and 2 **8.** 0 and −5
9. $\frac{1+\sqrt{17}}{2}$ and $\frac{1-\sqrt{17}}{2}$; −1.562 and 2.562 **10.** No real number solutions **11.** $\frac{2+\sqrt{14}}{2}$ and $\frac{2-\sqrt{14}}{2}$; −0.871 and 2.871
12. $\frac{3+2\sqrt{3}}{3}$ and $\frac{3-2\sqrt{3}}{3}$; −0.155 and 2.155 **13.** $2 + \sqrt{11}$ and $2 - \sqrt{11}$; −1.317 and 5.317
14. $-3 + \sqrt{10}$ and $-3 - \sqrt{10}$; −6.162 and 0.162 **15.** $-\frac{3}{2}$ and 6 **16.** $-\frac{2}{3}$ and 2 **17.** No real number solutions
18. No real number solutions **19.** −4 and $\frac{1}{4}$ **20.** $-\frac{3}{4}$ and $\frac{1}{2}$ **21.** $-\frac{1}{2}$ and $\frac{4}{3}$ **22.** $-\frac{1}{3}$ and $\frac{3}{2}$
23. $1 + \sqrt{5}$ and $1 - \sqrt{5}$; −1.236 and 3.236 **24.** $7 + \sqrt{30}$ and $7 - \sqrt{30}$; 1.523 and 12.477
25. Answers will vary. **26.** Answers will vary. **27.** 0.75 s and 3 s **28.** 10 m
29. $T = 0.5(1)^2 + 0.5(1) = 1$; $T = 0.5(2)^2 + 0.5(2) = 3$; $T = 0.5(3)^2 + 0.5(3) = 6$; $T = 0.5(4)^2 + 0.5(4) = 10$; 10 rows
30. After 1.73 s and after 5.77 s **31.** 2017 **32.** 2049 **33. a.** 240 ft **b.** 30 mph
34. a. 6.29 ft **b.** Yes; 1 ft **c.** No **35.** 5.51 s **36.** 1.58 s **37.** 244.10 ft
38. 151.64 km/h **39.** No **40.** 6 m **41.** No **42.** 1.74 s and 10.76 s **43.** 68 cents
44.
$$ax^2 + bx = 0$$
$$x(ax + b) = 0$$
$$x = 0 \qquad ax + b = 0$$
$$ax = -b$$
$$x = -\frac{b}{a}$$
45. If the discriminant is not a perfect square, the radical expression in the quadratic formula will not simplify to a whole number.
46. $-12a$ and $-4a$ **47.** $3b$ and $5b$ **48.** c and $\frac{c}{3}$ **49.** $-y$ and $\frac{3}{2}y$
50. $b^2 - 4ac = b^2 - 4(1)(-1) = b^2 + 4$. If $b^2 - 4ac \geq 0$, then the equation has real number solutions. Because $b^2 \geq 0$ for all real numbers b, $b^2 + 4 \geq 4$ for all real numbers b. Therefore, the equation has real number solutions regardless of the value of b.

CHAPTER 9 REVIEW EXERCISES *page 550*

1. 4 [Sec. 9.1] **2.** $\frac{1}{8}$ [Sec. 9.1] **3.** −2 [Sec. 9.1] **4.** $\frac{10}{3}$ [Sec. 9.2] **5.** No real number solutions [Sec. 9.4]
6. −5 and 6 [Sec. 9.4] **7.** $2 + \sqrt{3}$ and $2 - \sqrt{3}$ [Sec. 9.4] **8.** $\frac{1+\sqrt{13}}{2}$ and $\frac{1-\sqrt{13}}{2}$ [Sec. 9.4] **9.** $y = -\frac{4}{3}x + 4$ [Sec. 9.1]
10. $t = \frac{f-v}{a}$ [Sec. 9.1] **11.** 2450 ft [Sec. 9.1] **12.** 3 s [Sec. 9.1] **13.** 60°C [Sec. 9.1] **14.** 85 calls [Sec. 9.1]
15. 28.4 mi/gal [Sec. 9.2] **16.** $\frac{1}{4}$ [Sec. 9.2] **17.** 400% [Sec. 9.3] **18. a.** New York, Chicago, Phoenix, Los Angeles, Houston
b. 22,231 more people per square mile [Sec. 9.2] **19. a.** 7 : 1, 7 to 1. There are 7 students for each faculty member at the university.
b. Prescott College, Embry-Riddle Aeronautical University **c.** Arizona State University and Northern Arizona University [Sec. 9.2]
20. Department A: \$105,000; Department B: \$245,000 [Sec. 9.2] **21.** 7.5 tablespoons [Sec. 9.2] **22. a.** More than one-fifth
b. 4:1 **c.** \$2.88 trillion **d.** \$0.72 trillion or \$720 billion [Sec. 9.2] **23. a.** 51.0% **b.** Less than [Sec. 9.3]
24. 20 billion hot dogs [Sec. 9.3] **25.** 97.9% [Sec. 9.3] **26. a.** $83.\overline{3}\%$ **b.** 100% **c.** 50% [Sec. 9.3]
27. 13.2% [Sec. 9.3] **28.** 35% [Sec. 9.3] **29.** \$7040 [Sec. 9.3] **30.** 13 days [Sec. 9.1] **31.** 1 s and 5 s [Sec. 9.4]
32. 0.5 s and 1.5 s [Sec. 9.4]

CHAPTER 9 TEST page 552

1. 14 [Sec. 9.1, Example 2c] **2.** 10 [Sec. 9.1, Example 2b] **3.** $\frac{21}{4}$ [Sec. 9.2, Example 7] **4.** 3 and 9 [Sec. 9.4, Example 3]
5. $\frac{2+\sqrt{7}}{3}$ and $\frac{2-\sqrt{7}}{3}$ [Sec. 9.4, Example 4] **6.** $y = \frac{1}{2}x - \frac{15}{2}$ [Sec. 9.1, Example 6a] **7.** $F = \frac{9}{5}C + 32$ [Sec. 9.1, Example 6a]
8. 2.5 min [Sec. 9.1, Example 3] **9.** 431.25 kWh [Sec. 9.1, Example 4] **10.** 54.8 mph [Sec. 9.2, Example 1]
11. 843 acres [Sec. 9.1, Check Your Progress 4] **12. a.** 2.727, 2.905, 2.777, 2.808, 2.901, 2.904
b. Ty Cobb, Rogers Hornsby, Joe Jackson, Tris Speaker, Ted Williams, Billy Hamilton [Sec. 9.2, Example 1]
13. 256 million [Sec. 9.3, Example 7] **14.** \$225,000 and \$135,000 [Sec. 9.2, Check Your Progress 9]
15. 2.75 lb [Sec. 9.2, Example 10] **16. a.** Detroit, Michigan **b.** 8355 violent crimes [Sec. 9.2, Example 4]
17. 14.4% [Sec. 9.3, Check Your Progress 6] **18.** 55.6% [Sec. 9.3, Example 12] **19. a.** 79.0% **b.** 26.4 million [Sec. 9.3, Example 13]
20. 0.2 s and 1.6 s [Sec. 9.4, Example 7]

CHAPTER 10

EXCURSION EXERCISES, SECTION 10.1 page 593

1. a.

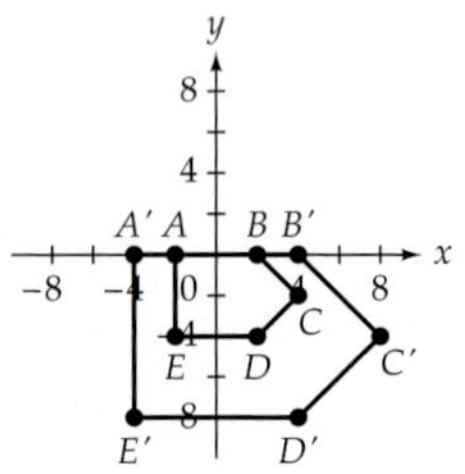

b.

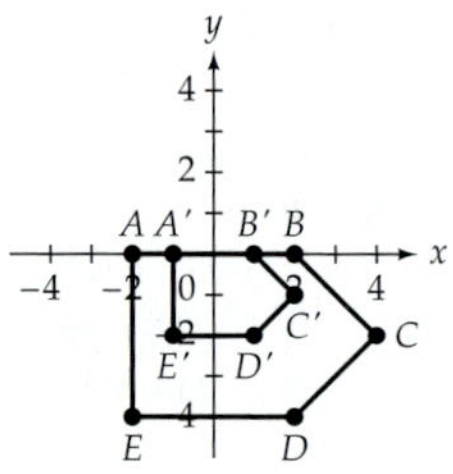

2. Drawings will vary. The angles of the dilated figure have the same measures as the corresponding angles of the original figure.

3. Drawings will vary. **4.** Drawings will vary.
5. The center of dilation is the center of the edge of the paper that is placed next to the guide marks on the copier.

EXERCISE SET 10.1 page 594

1.

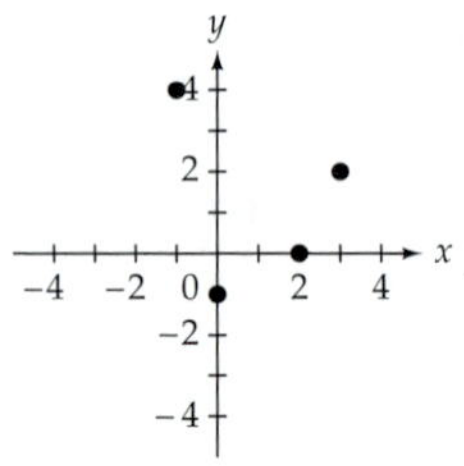

2.

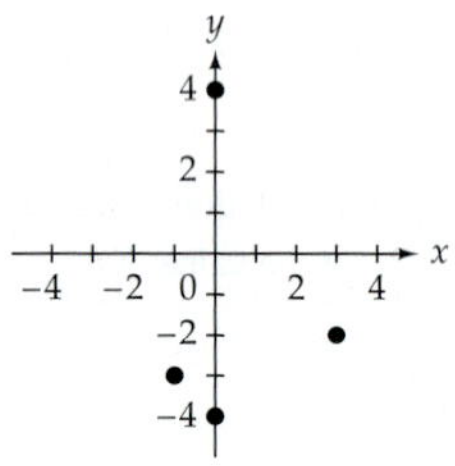

3.

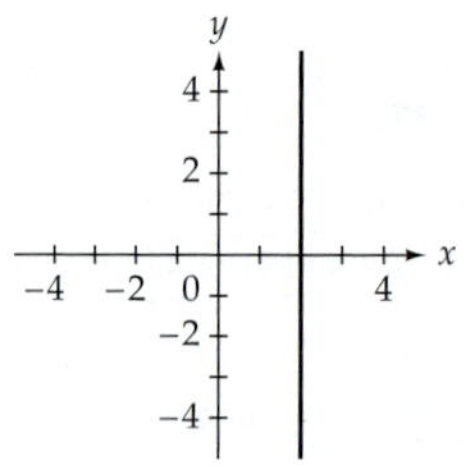

4.

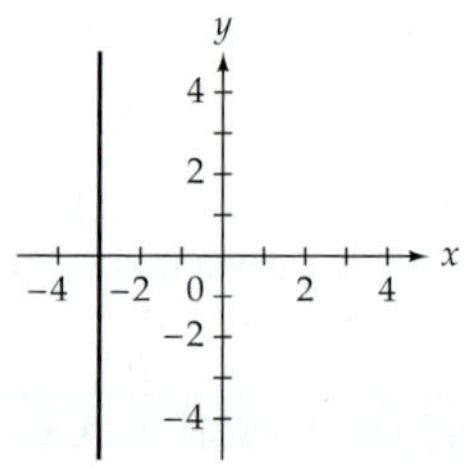

5.

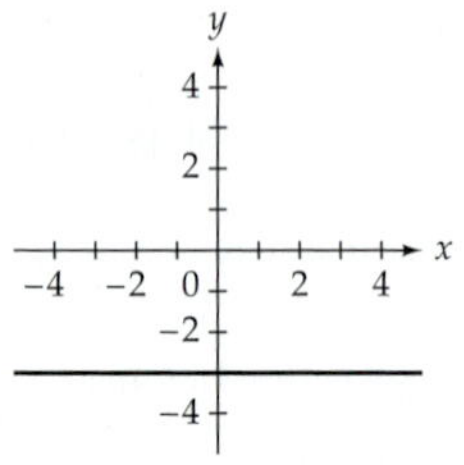

6.

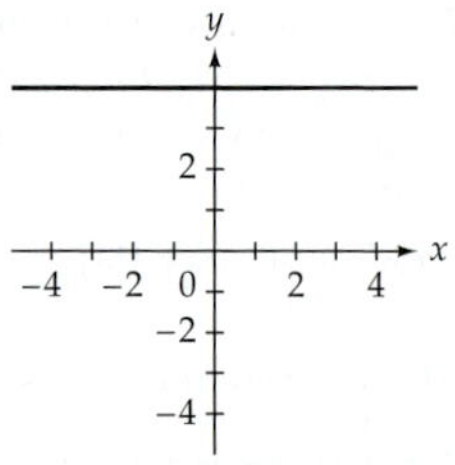

7.

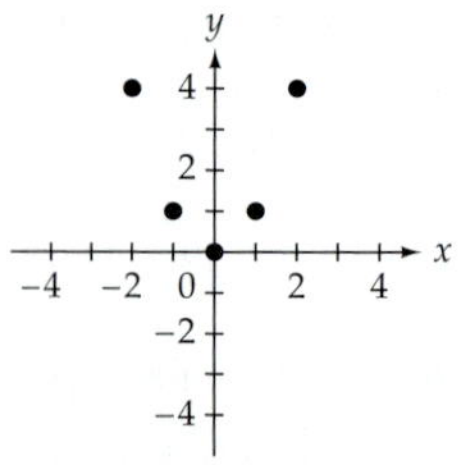

8.

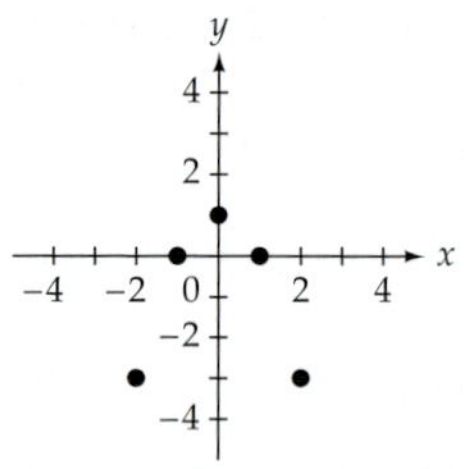

9.

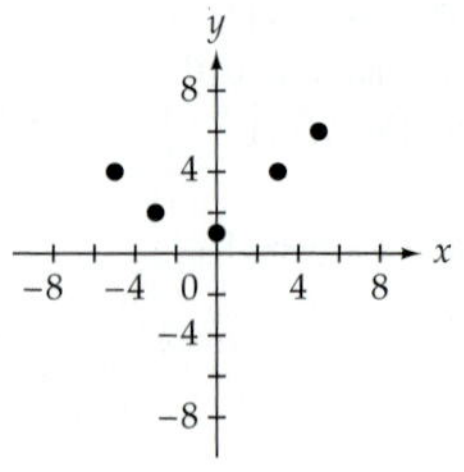

10.

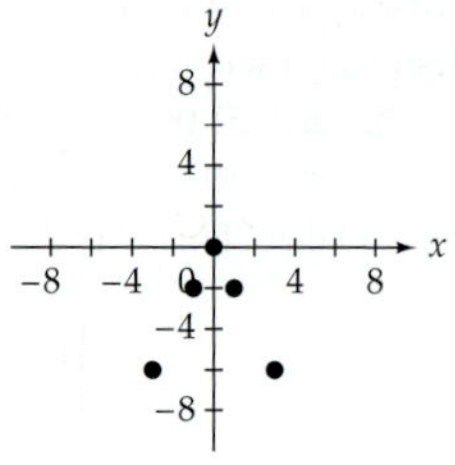

11.

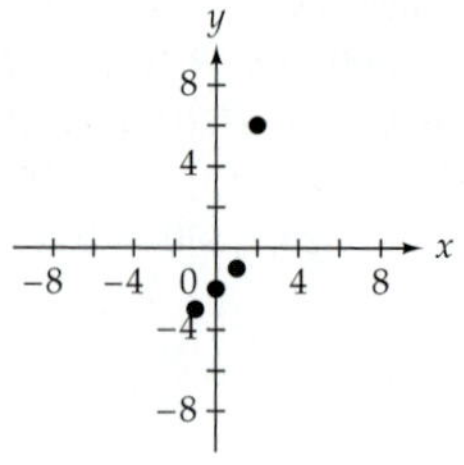

12.

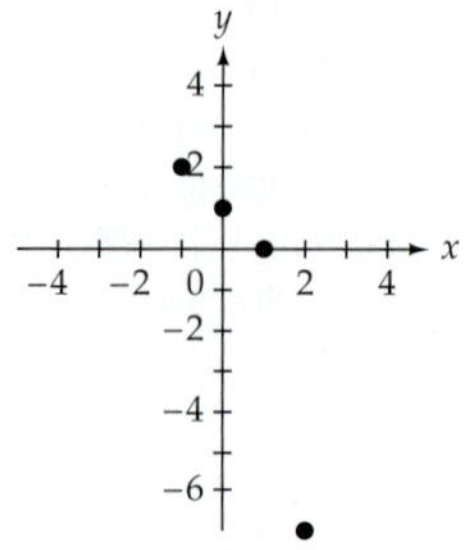

13. 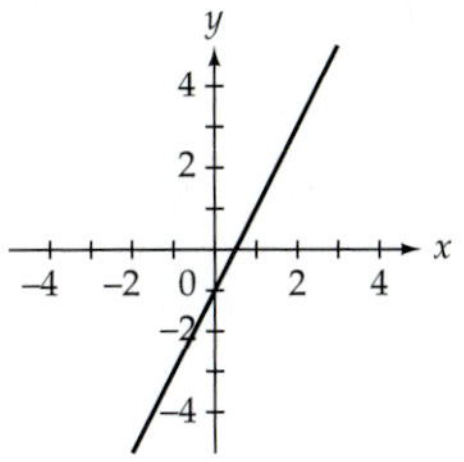

14. 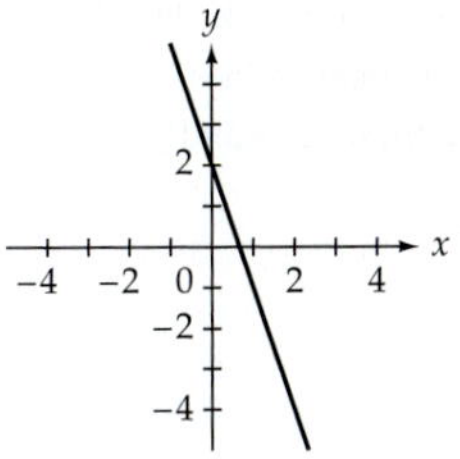

15. 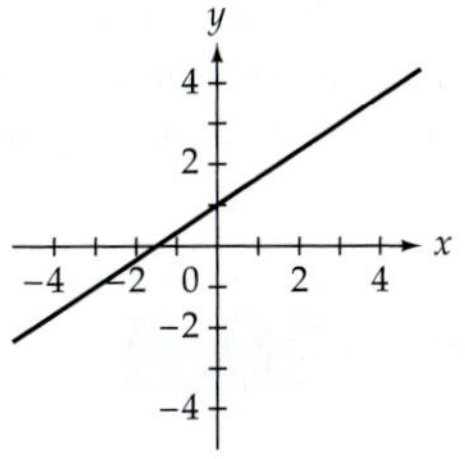

16. 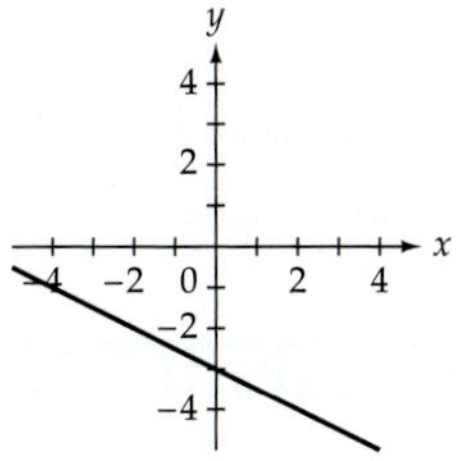

17. 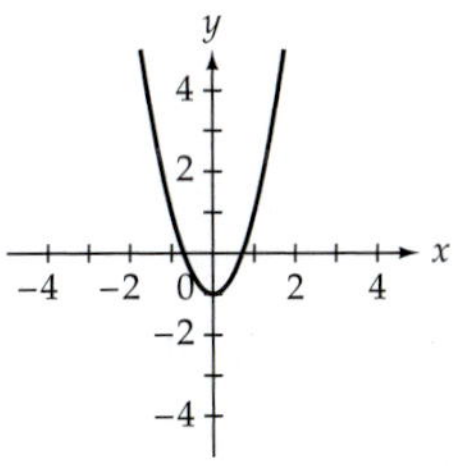

18. 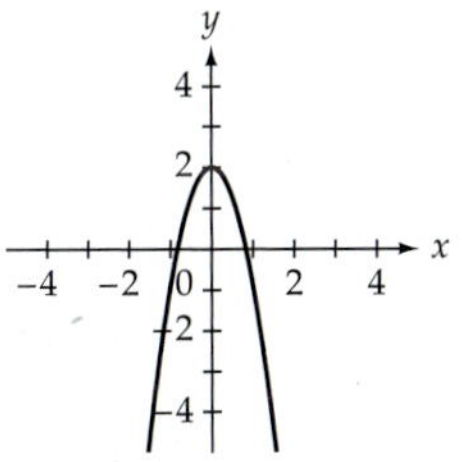

19. 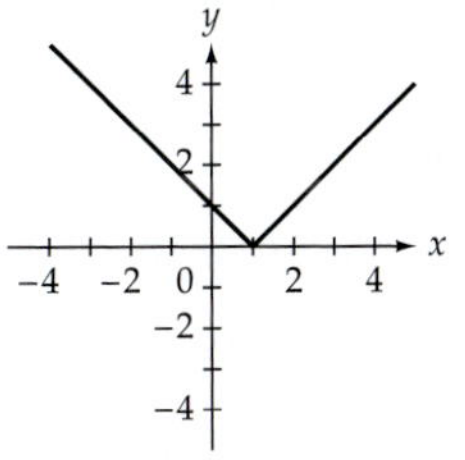

20. 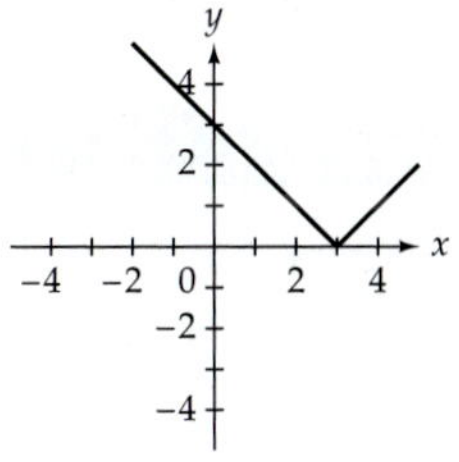

21. 3 **22.** 13 **23.** 3 **24.** 14 **25.** 0 **26.** $\frac{4}{3}$ **27. a.** 16 m **b.** 20 ft

28. a. 28.3 in^2 **b.** 452.4 cm^2 **29. a.** 100 ft **b.** 68 ft **30. a.** 6.7 mi **b.** 8.9 mi

31. a. 1087 ft/s **b.** 1136 ft/s **32. a.** 10% **b.** 40% **33. a.** 1.92 s **b.** 1.0 s **34.** 157 m^2

35. 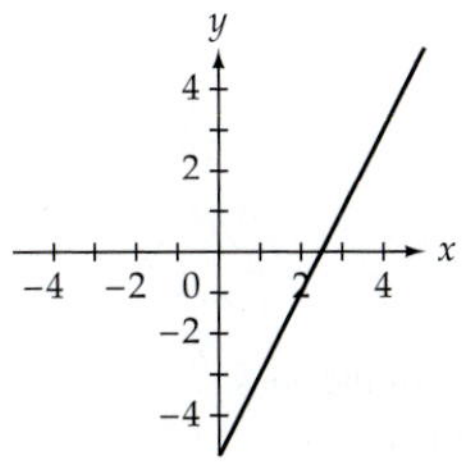

36. 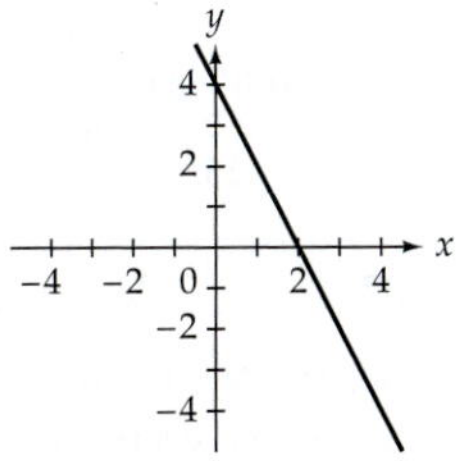

37. 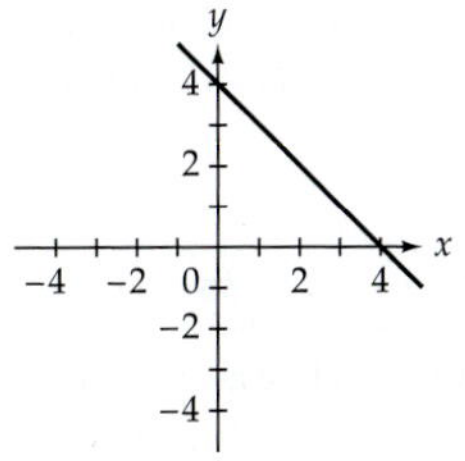

38. 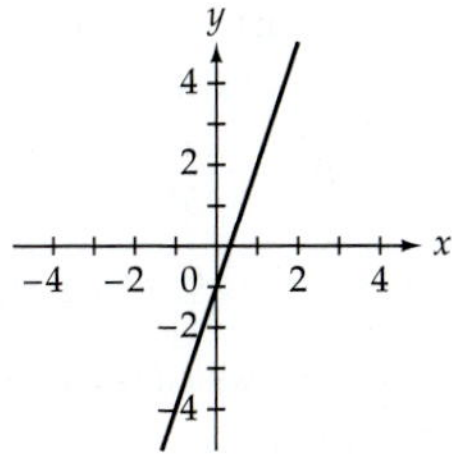

39. 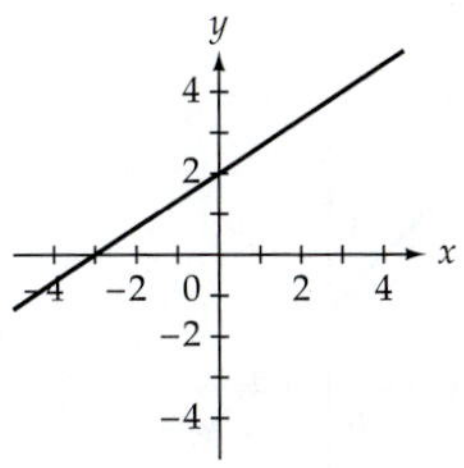

40. 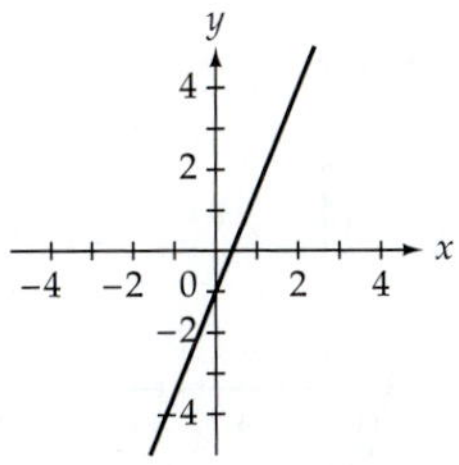

41. 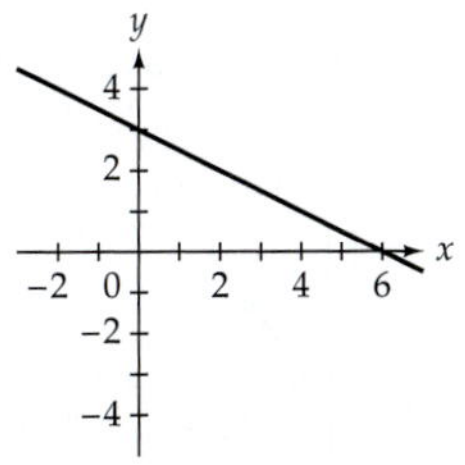

42. 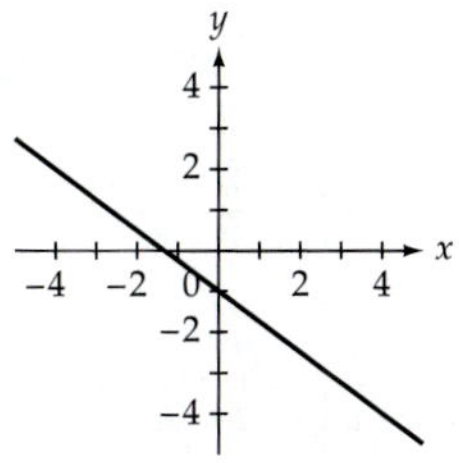

43. 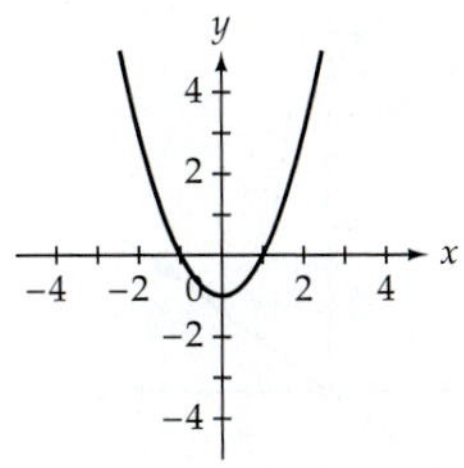

44. 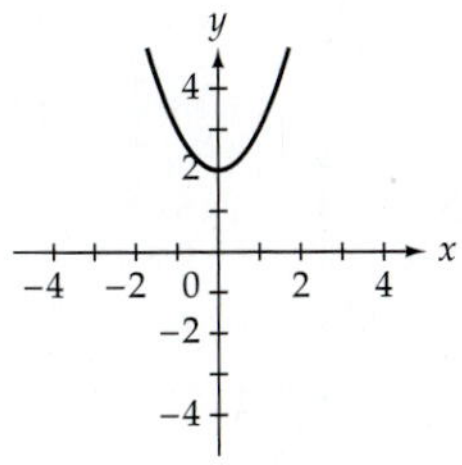

45. 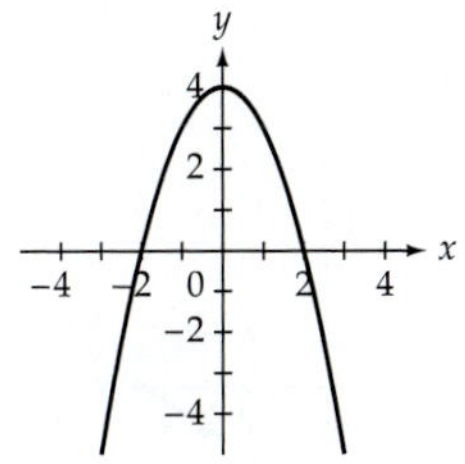

46. 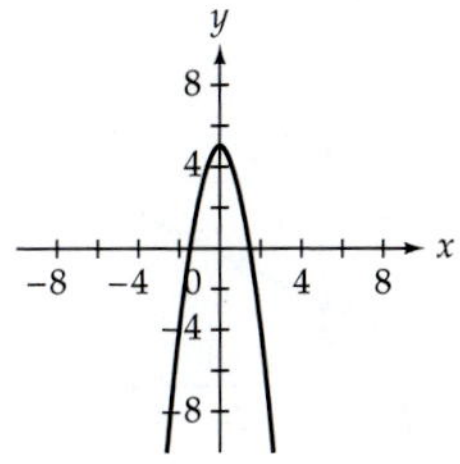

47. 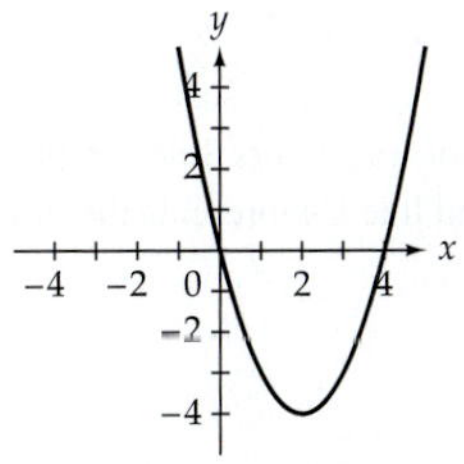

48. 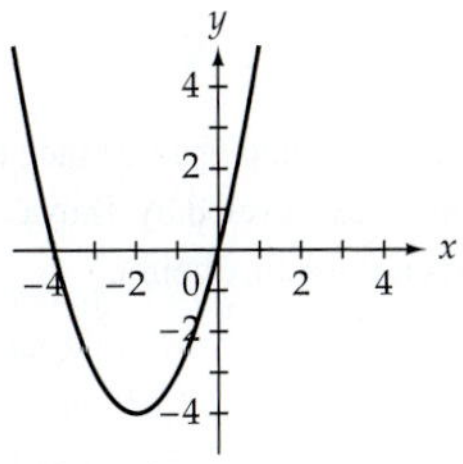

49. 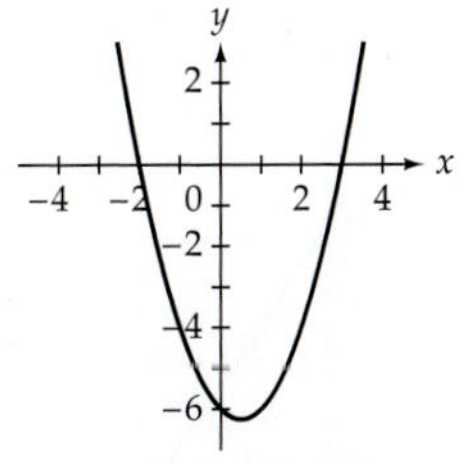

50. 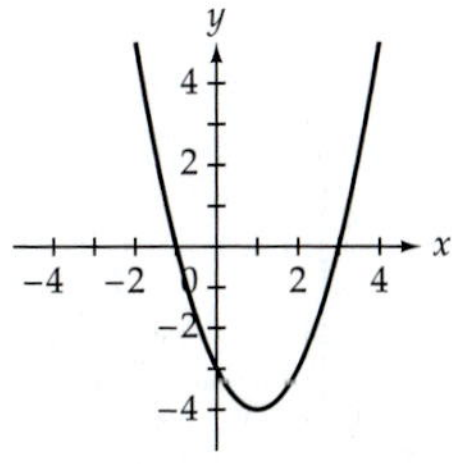

51. No. A function cannot have different elements in the range corresponding to one element in the domain.

52. No. Explanations will vary. For example, if $f(x) = x^2$, $f(2) = 4$ and $f(-2) = 4$, but $2 \neq -2$.

53. 2 **54.** $a = 3$ or -3 **55.** 17 **56. a.** $M(10, 8) = 10$; $M(-3, -1) = -1$; $M(12, -13) = 12$; $M(-11, 15) = 15$
b. Answers will vary. **c.** The value of the function is equal to the larger of the two input values.
d. It is a good name for the function because the value of the function is equal to the maximum of x and y.
e. Answers will vary. For example, $M(x, y) = \frac{x + y}{2} - \frac{|x - y|}{2}$.

EXCURSION EXERCISES, SECTION 10.2 *page 604*

1. 0 to 2 h **2.** 0 to 2 h **3.** 2 to 6 h **4.** 2 to 6 h **5.** 120 mi **6.** 6 h
7. 60 mph **8.** −30 mph **9.** 0 **10.** 0 mph; no **11.** 2 to 3 h

EXERCISE SET 10.2 *page 605*

1. $(2, 0), (0, -6)$ **2.** $(-4, 0), (0, 8)$ **3.** $(6, 0), (0, -4)$ **4.** $(8, 0), (0, 6)$ **5.** $(-4, 0), (0, -4)$
6. $(2, 0), (0, 1)$ **7.** $(4, 0), (0, 3)$ **8.** $(2, 0), (0, -5)$ **9.** $\left(\frac{9}{2}, 0\right), (0, -3)$ **10.** $(2, 0), \left(0, \frac{8}{3}\right)$ **11.** $(2, 0), (0, 3)$
12. $(3, 0), (0, -2)$ **13.** $(1, 0), (0, -2)$ **14.** $(-4, 0), (0, 3)$ **15.** $\left(\frac{30}{7}, 0\right)$; at $\frac{30}{7}$°C the cricket stops chirping.
16. The intercept on the horizontal axis, $(12, 0)$, means that after flying for 12 h, the remaining distance is 0 mi.
The intercept on the vertical axis, $(0, 6000)$, means that before the trip begins, the remaining distance is 6000 mi.
17. The intercept on the vertical axis is $(0, -15)$. This means that the temperature of the object is −15°F before it is removed from the freezer.
The intercept on the horizontal axis is $(5, 0)$. This means that it takes 5 min for the temperature of the object to reach 0°F.
18. The intercept on the horizontal axis, $(40, 0)$, means that 40 months after withdrawals begin, there is \$0 remaining in the account.
The intercept on the vertical axis, $(0, 100{,}000)$, means that before withdrawals begin, there is \$100,000 in the account.
19. -1 **20.** $-\frac{2}{3}$ **21.** $\frac{1}{3}$ **22.** -3 **23.** $-\frac{2}{3}$ **24.** -2 **25.** $-\frac{3}{4}$ **26.** $\frac{3}{2}$
27. Undefined **28.** Undefined **29.** $\frac{7}{5}$ **30.** $\frac{3}{5}$ **31.** 0 **32.** 0 **33.** $-\frac{1}{2}$ **34.** -2
35. The slope is 40, which means the motorist was traveling at 40 mph. **36.** The slope is −5000, which means the value of the house decreases \$5000 each year. **37.** The slope is −1000, which means that the height of the plane is decreasing at a rate of 1000 ft/min.
38. The slope is 0.007, which means the payment per dollar of mortgage is \$0.007.
39. The slope is $-\frac{19}{30}$, which means that the water in the lock decreases by 0.63 million gal/min.

40.
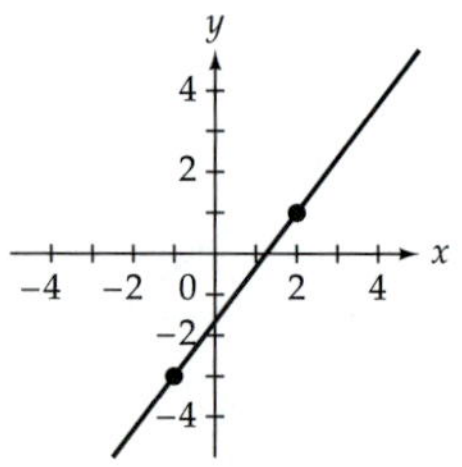

41.
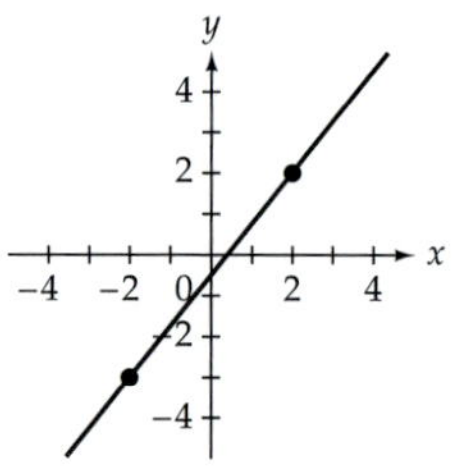

42.
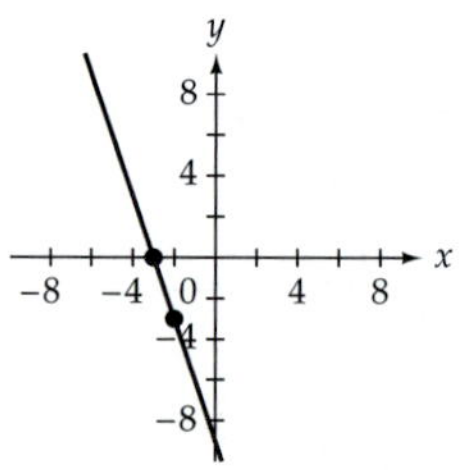

43.
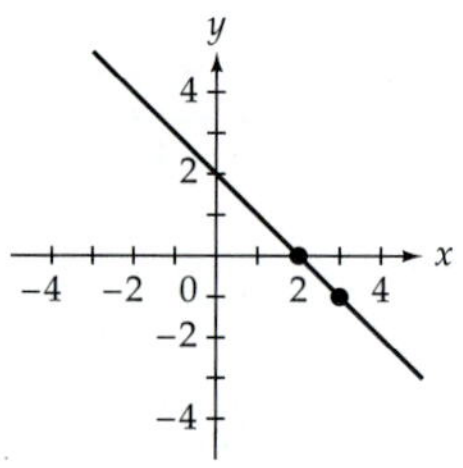

44.
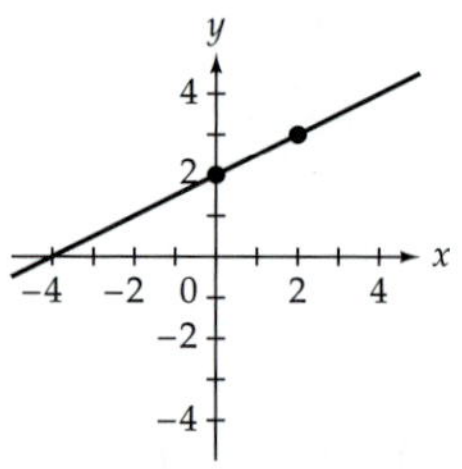

45.
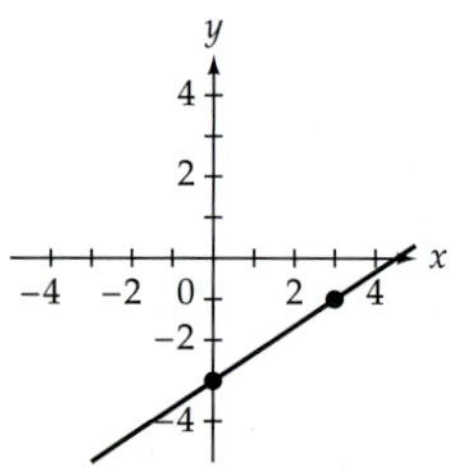

46.
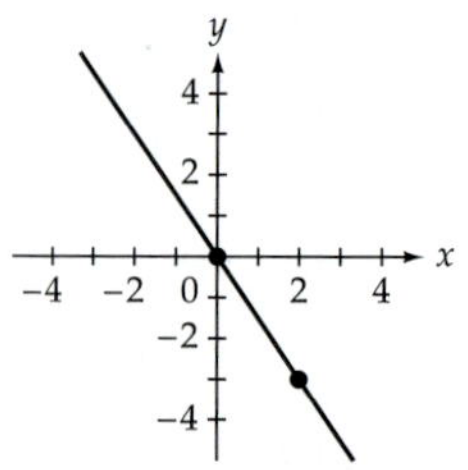

47.
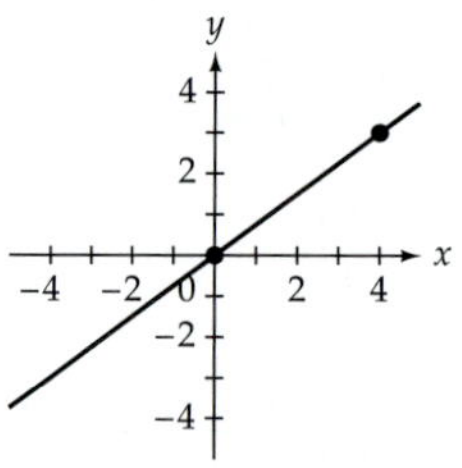

48.
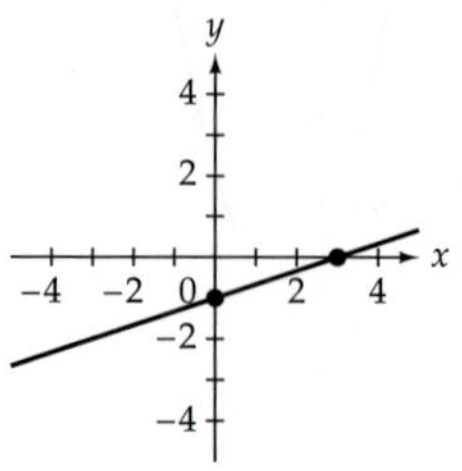

49.
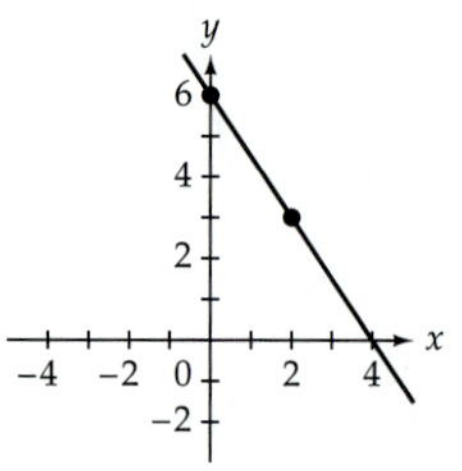

50. Line A represents the distance traveled by Lois in x hours, line B represents the distance traveled by Tanya in x hours, and line C represents the distance between them in x hours.

51. a. Line A represents can 1. **b.** Approximately 5 mm **52.** 7 **53.** −4
54. It rotates the line counterclockwise. **55.** It rotates the line clockwise.
56. It raises the line on the rectangular coordinate system. **57.** It lowers the line on the rectangular coordinate system.
58. a. $0.91\overline{6}$ **b.** $0.\overline{63}$ **c.** New design. $0.\overline{6}$ is closer to $0.\overline{63}$ than to $0.91\overline{6}$. **d.** Designs will vary. **e.** Answers will vary.

EXCURSION EXERCISES, SECTION 10.3 *page 614*

1. $C = 28n + 13{,}500$ **2.** $R = 45n$ **3.** Approximately 794 cartridges **4.** Approximately 4465 cartridges
5. Approximately $18.65/h **6.** Approximately 489 additional cartridges
7. $C = 28n + 18{,}300$; $R = 45n$; approximately 1076 cartridges; approximately $17.73/h

EXERCISE SET 10.3 *page 615*

1. $y = 2x + 5$ **2.** $y = \frac{1}{2}x + 2$ **3.** $y = -3x + 4$ **4.** $y = \frac{1}{2}x$ **5.** $y = -\frac{2}{3}x + 7$ **6.** $y = -x - 3$
7. $y = -3$ **8.** $y = -2x + 3$ **9.** $y = x + 2$ **10.** $y = -2x - 3$ **11.** $y = -\frac{3}{2}x + 3$ **12.** $y = x - 1$
13. $y = \frac{1}{2}x - 1$ **14.** $y = -4$ **15.** $y = -\frac{5}{2}x$ **16.** $y = x - 1$ **17.** $f(x) = 0.59x + 4.95$; $12.62
18. $R(x) = -\frac{3}{5}x + 560$; 470 rooms **19.** $C(x) = 85x + 30{,}000$; $183,000 **20.** $C(x) = 0.06x + 1500$; $6600
21. $N(x) = -10x + 230{,}000$; 60,000 cars **22.** $S(x) = -400x + 48{,}000$; 18,000 calculators **23. a.** $y = 8x - 4$ **b.** 80
24. a. $y = 0.7027x + 31.352$ **b.** 91 ml of mercury **25. a.** $y = 34x + 260$ **b.** 1110 ml
26. a. $y = -0.143x + 30.571$ **b.** 21 mi/gal **27. a.** $y = -1.735x + 118.176$ **b.** 40°F
28. a. $y = 0.343x - 0.588$ **b.** 2.8 m/s **29.** Answers will vary. For example, (0, 3), (1, 2), and (3, 0).
30. 0 **31.** 7 **32.** 1 **33.** No. The three points do not lie on a straight line.
34. $y = -x + 6$. The slope of the line between any two of the points is −1. **35.** 3°. The car is climbing.

EXCURSION EXERCISES, SECTION 10.4 *page 624*

1. (0, 0.625) **2.** 659.75 in. **3. a.** Answers will vary. **b.** Answers will vary. The filament of the light bulb should be placed at the focus so that all light hitting the paraboloid will be bounced off to form a strong beam of light from the flashlight.

EXERCISE SET 10.4 *page 625*

1. (0, −2) **2.** (0, 2) **3.** (0, −1) **4.** (0, 3) **5.** (0, 2) **6.** (0, 0) **7.** (0, −1) **8.** (1, −1)
9. $\left(\frac{1}{2}, -\frac{9}{4}\right)$ **10.** $\left(\frac{3}{2}, -\frac{1}{4}\right)$ **11.** $\left(\frac{1}{4}, -\frac{41}{8}\right)$ **12.** $\left(\frac{1}{4}, -\frac{25}{8}\right)$ **13.** (0, 0), (2, 0) **14.** (0, 0), (−2, 0)
15. $(-2, 0), \left(-\frac{3}{4}, 0\right)$ **16.** (3, 0), (−3, 0) **17.** $(-1 + \sqrt{2}, 0), (-1 - \sqrt{2}, 0)$ **18.** $(-2 + \sqrt{7}, 0), (-2 - \sqrt{7}, 0)$
19. None **20.** $(-1 + \sqrt{2}, 0), (-1 - \sqrt{2}, 0)$ **21.** $(2 + \sqrt{5}, 0), (2 - \sqrt{5}, 0)$ **22.** None
23. $\left(-\frac{1}{2}, 0\right), (3, 0)$ **24.** $(\sqrt{2}, 0), (-\sqrt{2}, 0)$ **25.** Minimum: 2 **26.** Minimum: −2 **27.** Maximum: −3
28. Minimum: $-\frac{11}{4}$ **29.** Minimum: −3.25 **30.** Minimum: $-\frac{9}{8}$ **31.** Maximum: $\frac{9}{4}$ **32.** Maximum: $-\frac{2}{3}$
33. c **34.** 114 ft **35.** 150 ft **36.** $1600 **37.** 100 lenses **38.** 5 days **39.** 24.36 ft
40. a. 48 years **b.** $148,000 **41.** Yes **42.** 1.5 feet **43. a.** 12 ft **b.** 15 ft **c.** 16 ft
44. 20 mph **45. a.** 41 mph **b.** 33.658 mi/gal **46.** Length: 100 ft; width: 50 ft **47.** 18.4 s

EXCURSION EXERCISES, SECTION 10.5 *page 636*

1. a. 1, 2, 4, 8, 16, 32, 64 **b.** 3 **c.** 7 **d.** 15 **e.** 31 **f.** 63 **g.** 127 **2.** Answers will vary. For example, $f(n) = 2^{n-1} + (2^{n-1} - 1)$ or $f(n) = 2(2^{n-1}) - 1$. **3.** Approximately 18,000,000,000,000,000,000 grains of wheat
4. Approximately 144 trillion kilograms **5.** 222 years

EXERCISE SET 10.5 page 637

1. a. 9 **b.** 1 **c.** $\frac{1}{9}$ **2. a.** $\frac{1}{8}$ **b.** 1 **c.** 4 **3. a.** 16 **b.** 4 **c.** $\frac{1}{4}$ **4. a.** $\frac{1}{729}$ **b.** $\frac{1}{27}$ **c.** $\frac{1}{9}$ **5. a.** 1 **b.** $\frac{1}{8}$ **c.** 16 **6. a.** 3 **b.** $\frac{1}{27}$ **c.** 729 **7. a.** 7.3891 **b.** 0.3679 **c.** 1.2840 **8. a.** 0.0183 **b.** 0.2636 **c.** 54.5982 **9. a.** 54.5982 **b.** 1 **c.** 0.1353 **10. a.** 4.4817 **b.** 0.1353 **c.** 0.7788 **11. a.** 16 **b.** 16 **c.** 1.4768 **12. a.** $\frac{1}{512}$ **b.** $\frac{1}{2}$ **c.** $\frac{1}{16}$ **13. a.** 0.1353 **b.** 0.1353 **c.** 0.0111 **14. a.** 8.3891 **b.** 1.0025 **c.** 55.5982

15.

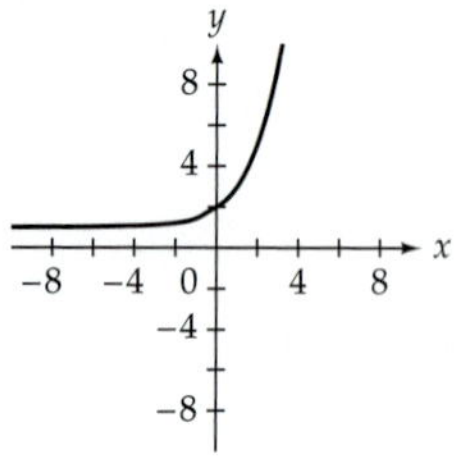

16.

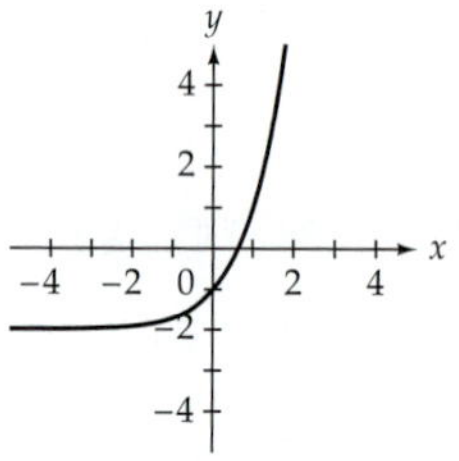

17.

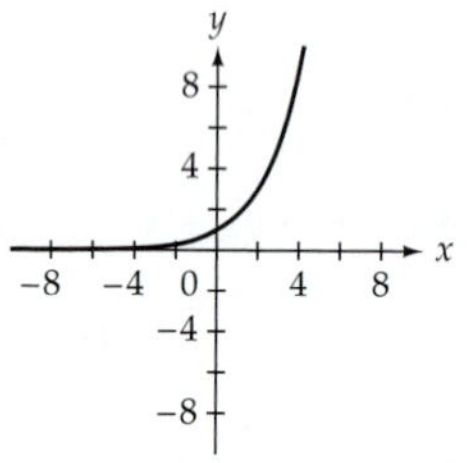

18.

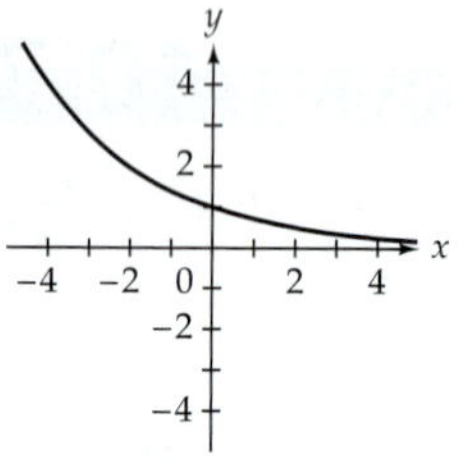

19.

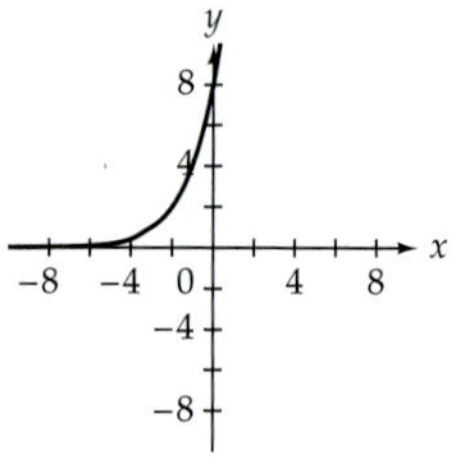

20.

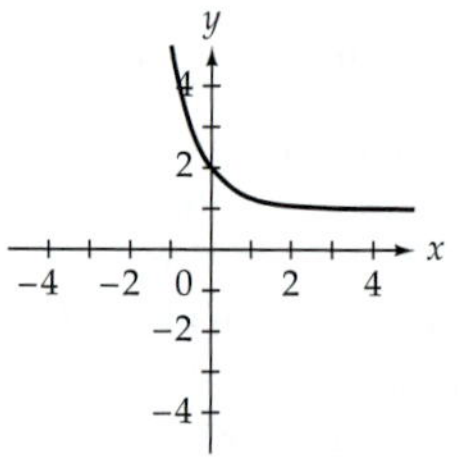

21.

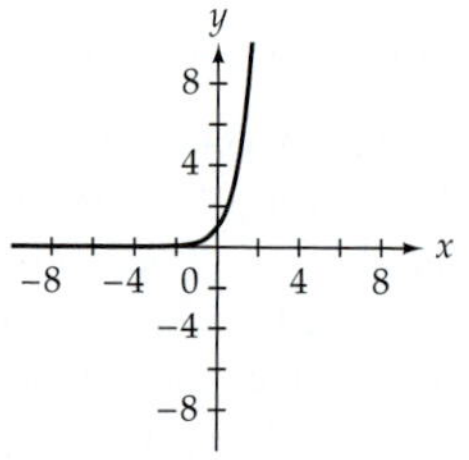

22.

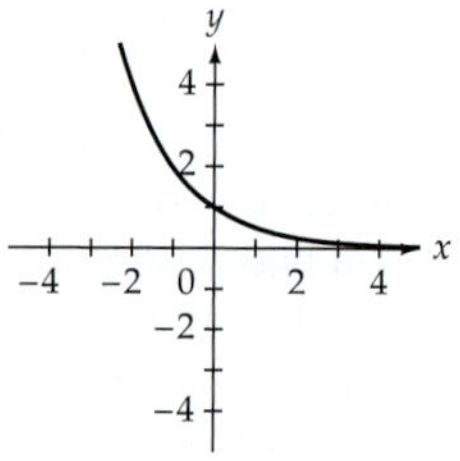

23.

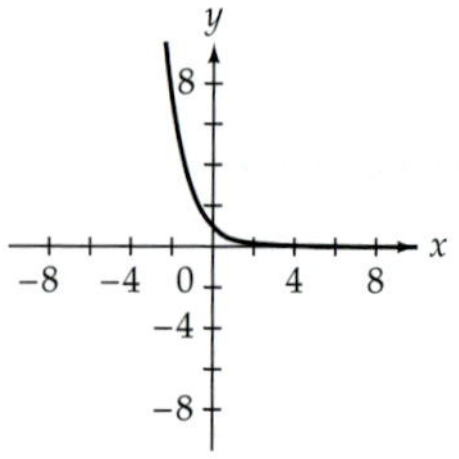

24.

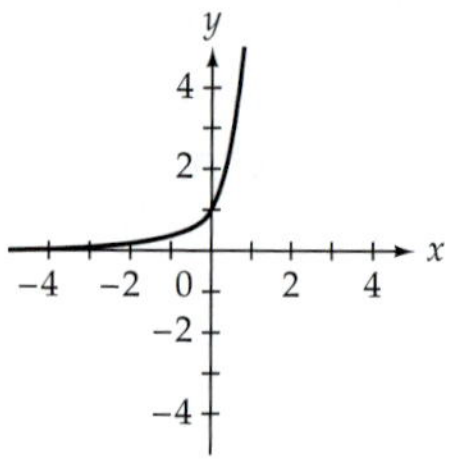

25. \$11,202.50
26. \$27,180.96
27. \$3210.06
28. 21.2 mg
29. 5.2 micrograms
30. 68%
31. $f(n) = 440(2^{n/12})$

32. a. 0.098 newton/cm² **b.** 10.13 newtons/cm² **c.** Decreases
33. a. \$107.33 **b.** \$193.33 **c.** There is no residual value on the car when it is purchased. **d.** \$3999.92; \$6000.08 **e.** \$10,000.14; approximately \$0 **f.** The \$6000 remaining on the leased car is its residual value.

EXCURSION EXERCISES, SECTION 10.6 page 649

1.

d	$P(d) = \log(1 + 1/d)$
1	0.301
2	0.176
3	0.125
4	0.097
5	0.079
6	0.067
7	0.058
8	0.051
9	0.046

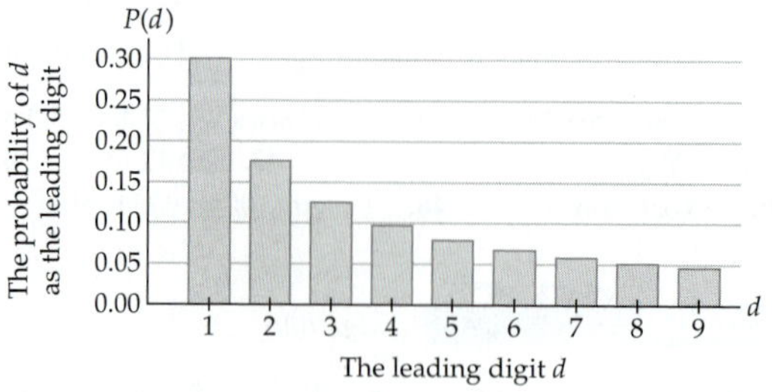

2. 6.7% **3.** About 6.5 times **4.** In a small city, not all four-digit numbers are used. Benford's law applies only to data with a wide range.
5. There is not a wide range of values.

EXERCISE SET 10.6 *page 650*

1. $\log_7 49 = 2$ **2.** $\log 1000 = 3$ **3.** $\log_5 625 = 4$ **4.** $\log_2 \frac{1}{8} = -3$ **5.** $\log 0.0001 = -4$ **6.** $\log_3 729 = 5$
7. $\log x = y$ **8.** $\ln x = y$ **9.** $3^4 = 81$ **10.** $2^4 = 16$ **11.** $5^3 = 125$ **12.** $4^3 = 64$ **13.** $4^{-2} = \frac{1}{16}$
14. $2^{-4} = \frac{1}{16}$ **15.** $e^y = x$ **16.** $10^y = x$ **17.** 4 **18.** 2 **19.** 2 **20.** −3 **21.** −2 **22.** −1
23. 6 **24.** −2 **25.** 9 **26.** 5 **27.** $\frac{1}{7}$ **28.** $\frac{1}{64}$ **29.** $\frac{1}{9}$ **30.** 125 **31.** 1 **32.** $\frac{1}{4}$
33. 316.23 **34.** 1584.89 **35.** 7.39 **36.** 54.60 **37.** 2.24 **38.** 1.34 **39.** 14.39 **40.** 1.65

41.
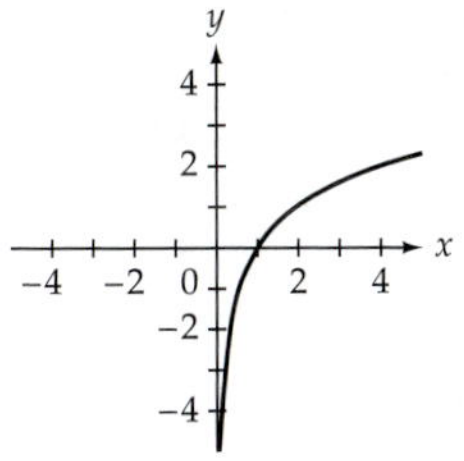

42.
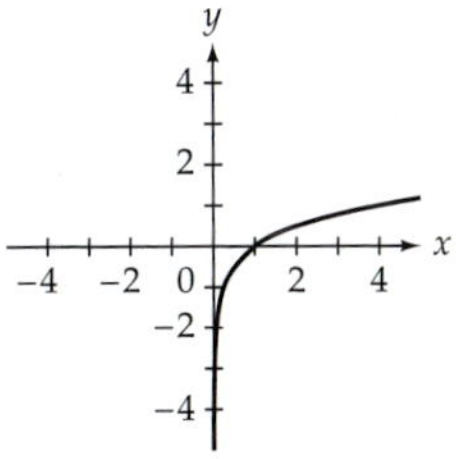

43.
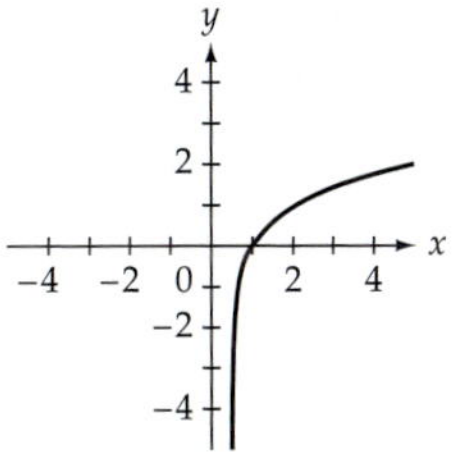

44.
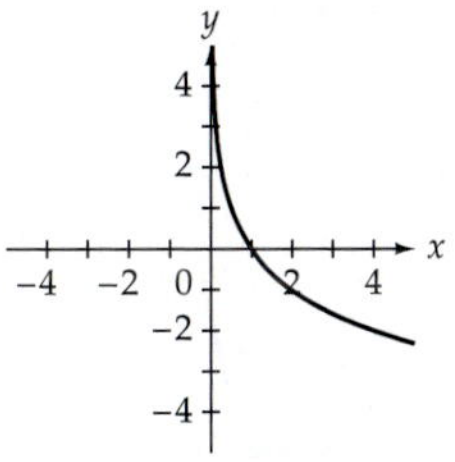

45.
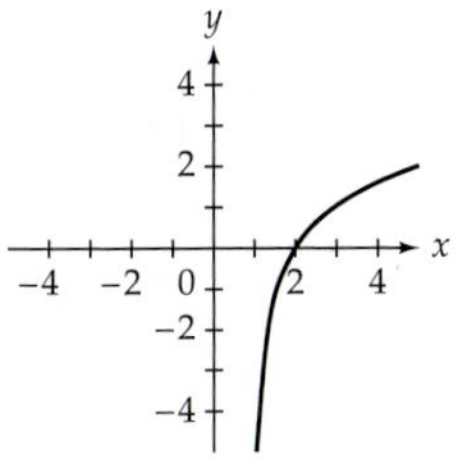

46.
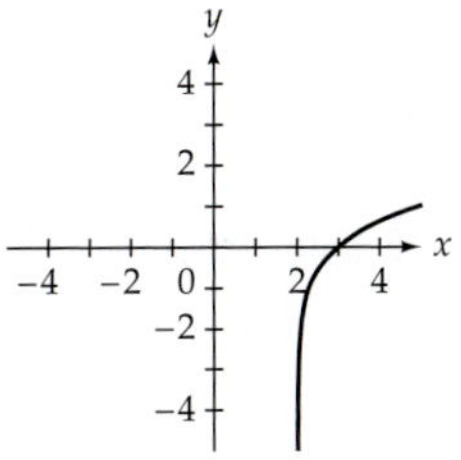

47. 79% **48.** 0.44 cm **49.** 65 decibels

50. 188 decibels **51.** 1.35 **52.** 9.49 **53.** 8.8 **54.** 7.9 **55.** $794{,}328{,}235I_0$ **56.** $1{,}584{,}893{,}192I_0$
57. 100 times as strong **58.** 150.7 parsecs **59.** 6.0 parsecs **60. a.** $10^{0.30103} = 2$; $10^{0.47712} = 3$ **b.** $x = 10^{0.30103} \cdot 10^{0.47712}$
c. $x = 10^{0.77815}$ **d.** $\log x = 0.77815$ **e.** $10^{0.77815} \approx 6$ **61.** $x = \frac{10^{0.47712}}{10^{0.30103}} = 10^{0.17609} \approx 1.5$; $\log x = 0.17609$
62. Answers will vary. **63.** Answers will vary. **64.** 5.5 **65.** 4.9

CHAPTER 10 REVIEW EXERCISES *page 655*

1.
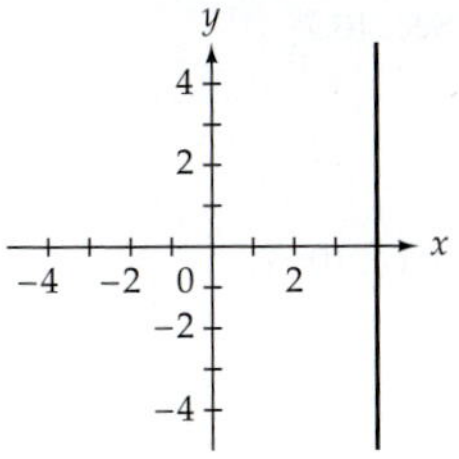

[Sec. 10.1]

2.
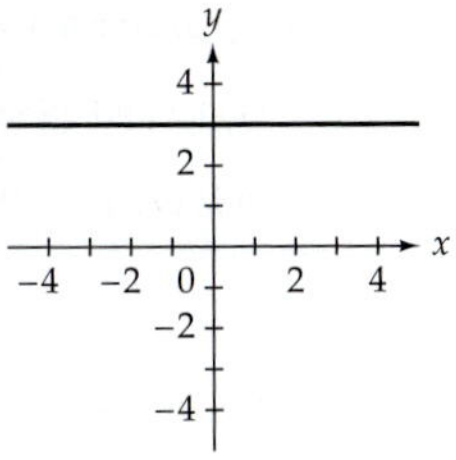

[Sec. 10.1]

3.
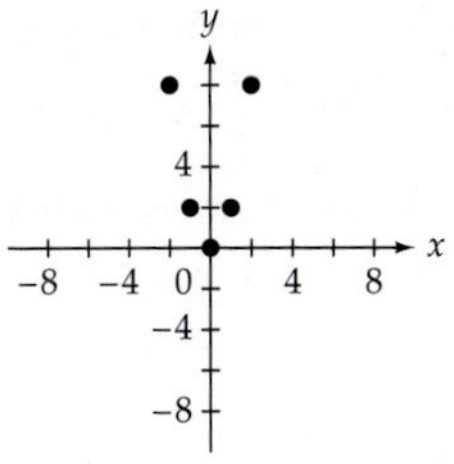

[Sec. 10.1]

4.
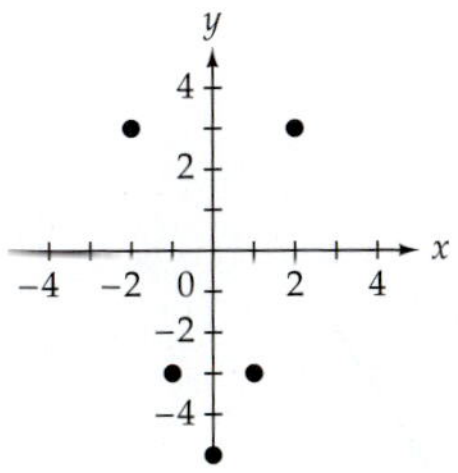

[Sec. 10.1]

5.
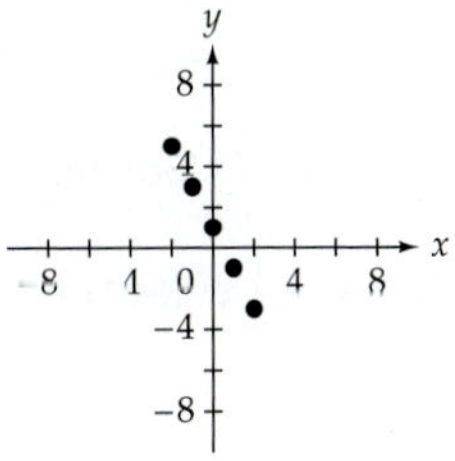

[Sec. 10.1]

6.
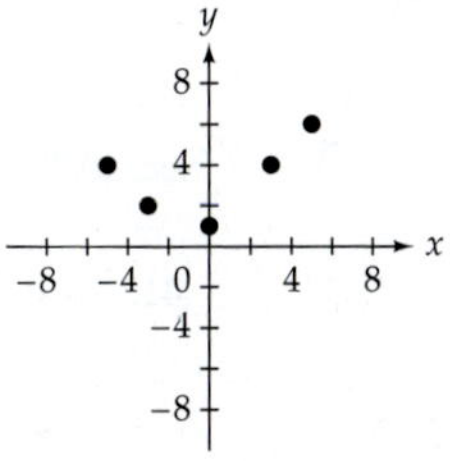

[Sec. 10.1]

7.
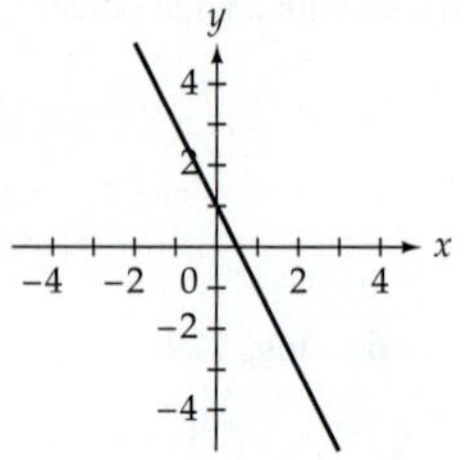

[Sec. 10.1]

8.
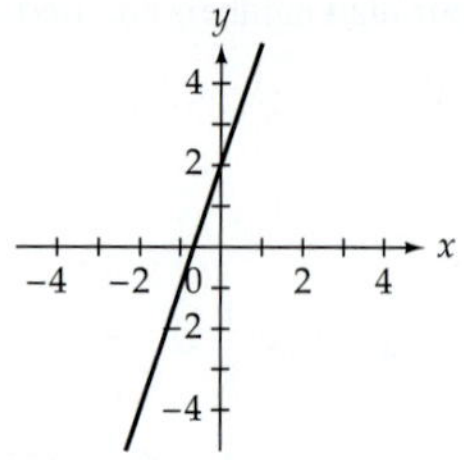

[Sec. 10.1]

9.
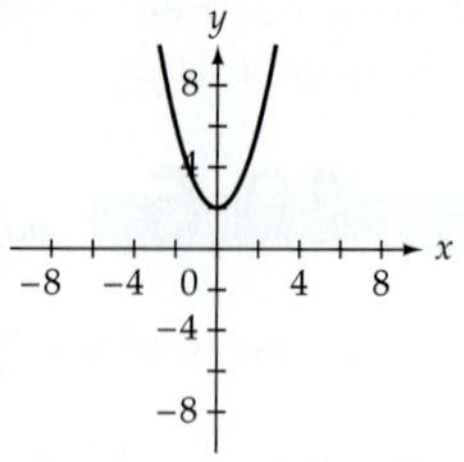

[Sec. 10.1]

10.
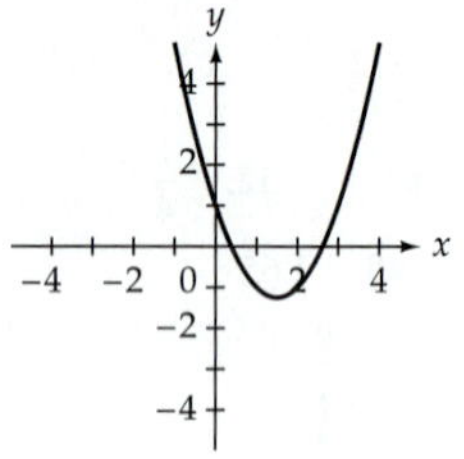

[Sec. 10.1]

11.
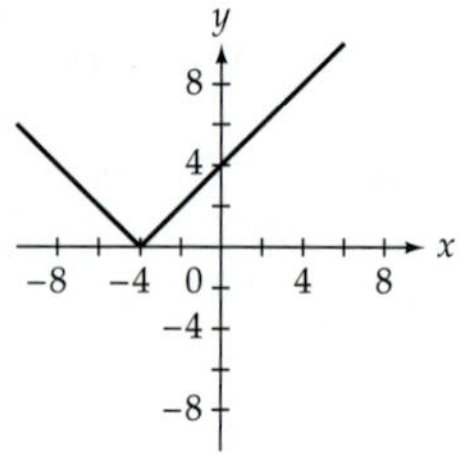

[Sec. 10.1]

12.
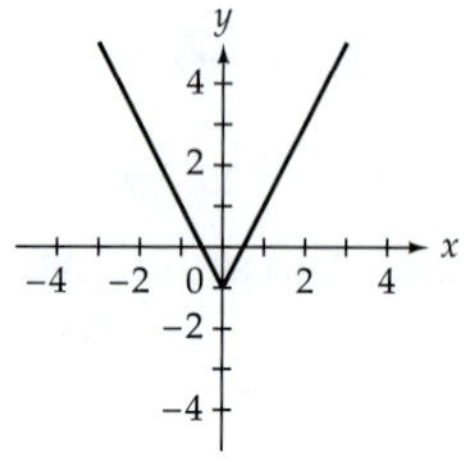

[Sec. 10.1]

13.
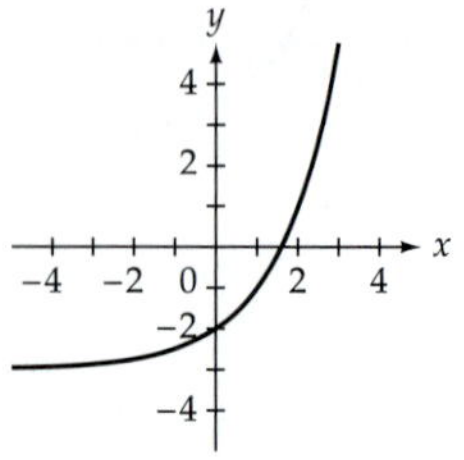

[Sec. 10.5]

14.
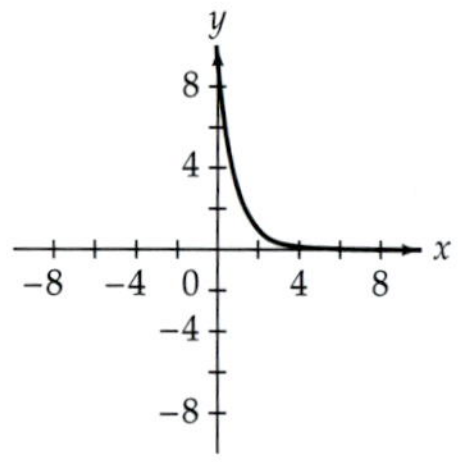

[Sec. 10.5]

15.
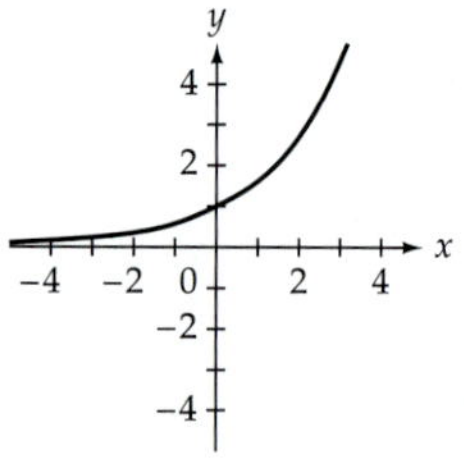

[Sec. 10.5]

16.
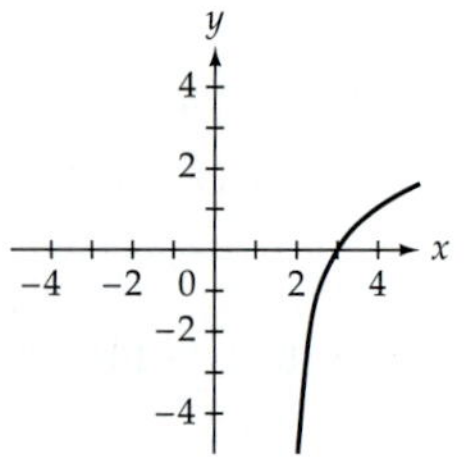

[Sec. 10.6]

17. -13 [Sec. 10.1] **18.** 13 [Sec. 10.1] **19.** $-\frac{1}{2}$ [Sec. 10.1]

20. -21 [Sec. 10.1] **21.** 4 [Sec. 10.5] **22.** $\frac{4}{9}$ [Sec. 10.5] **23.** 15.78 [Sec. 10.5] **24. a.** 113.1 in^3
b. 7238.2 cm^3 [Sec. 10.1] **25. a.** 69 ft **b.** 5 ft [Sec. 10.1] **26. a.** 5% **b.** 36.7% [Sec. 10.1]
27. $(-5, 0), (0, 10)$ [Sec. 10.2] **28.** $(12, 0), (0, -9)$ [Sec. 10.2] **29.** $(5, 0), (0, -3)$ [Sec. 10.2]
30. $(6, 0), (0, 8)$ [Sec. 10.2] **31.** The intercept on the vertical axis is $(0, 25{,}000)$. This means that the value of the delivery van was $25,000 when it was new. The intercept on the horizontal axis is $(5, 0)$. This means that after 5 years the delivery van will be worth $0. [Sec. 10.2]
32. 5 [Sec. 10.2] **33.** $\frac{5}{2}$ [Sec. 10.2] **34.** 0 [Sec. 10.2] **35.** Undefined [Sec. 10.2]
36. The slope is -0.05, which means that 0.05 gal of fuel is used for each mile the car is driven. [Sec. 10.2]

37.
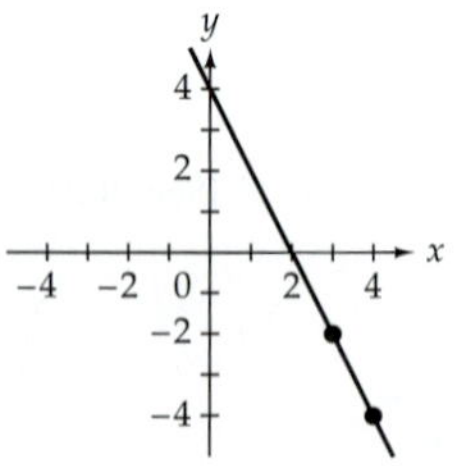

[Sec. 10.2]

38.
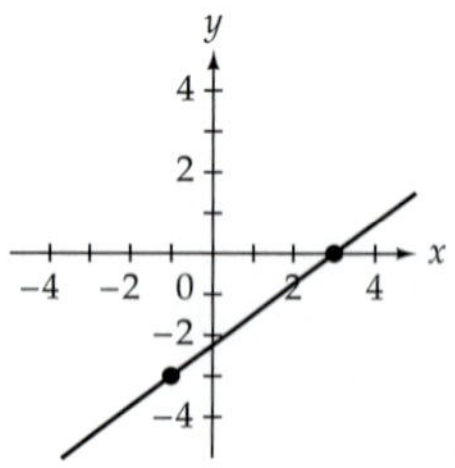

[Sec. 10.2]

39. $y = 2x + 7$ [Sec. 10.3]

40. $y = x - 5$ [Sec. 10.3] **41.** $y = \frac{2}{3}x + 3$ [Sec. 10.3] **42.** $y = \frac{1}{4}x$ [Sec. 10.3] **43.**
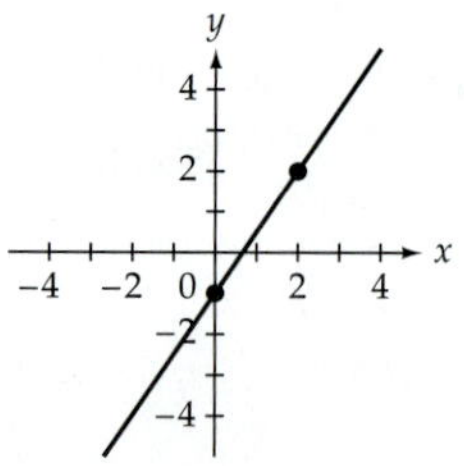

[Sec. 10.2]

44. a. $f(x) = 12x + 100$ **b.** \$3556 [Sec. 10.3] **45. a.** $A(p) = -25{,}000x + 10{,}000$ **b.** 10,750 gal [Sec. 10.3]
46. $(-1, 3)$ [Sec. 10.4] **47.** $\left(-\frac{3}{2}, \frac{11}{2}\right)$ [Sec. 10.4] **48.** $(1, 2)$ [Sec. 10.4] **49.** $(-2.5, -7.25)$ [Sec. 10.4]
50. $(4, 0), (-5, 0)$ [Sec. 10.4] **51.** $(-1 + \sqrt{2}, 0), (-1 - \sqrt{2}, 0)$ [Sec. 10.4] **52.** $\left(-\frac{1}{2}, 0\right), (-4, 0)$ [Sec. 10.4]
53. None [Sec. 10.4] **54.** 5, maximum [Sec. 10.4] **55.** $-\frac{15}{2}$, minimum [Sec. 10.4] **56.** -5, minimum [Sec. 10.4]
57. $\frac{1}{8}$, maximum [Sec. 10.4] **58.** 125 ft [Sec. 10.4] **59.** 2000 DVDs [Sec. 10.4] **60.** \$12,297.11 [Sec. 10.5]
61. 7.94 mg [Sec. 10.5] **62.** 3.36 micrograms [Sec. 10.5] **63. a.** $H(n) = 6\left(\frac{2}{3}\right)^n$ **b.** 0.79 ft [Sec. 10.5] **64.** 5 [Sec. 10.6]
65. -4 [Sec. 10.6] **66.** -1 [Sec. 10.6] **67.** 6 [Sec. 10.6] **68.** 64 [Sec. 10.6] **69.** 1.4422 [Sec. 10.6]
70. 12.1825 [Sec. 10.6] **71.** 251.1886 [Sec. 10.6] **72.** 43.7 parsecs [Sec. 10.6] **73.** 140 decibels [Sec. 10.6]

CHAPTER 10 TEST *page 657*

1. -21 [Sec. 10.1, Example 3] **2.** $\frac{1}{9}$ [Sec. 10.5, Example 1] **3.** 3 [Sec. 10.6, Example 2a] **4.** 36 [Sec. 10.6, Example 3]

5.
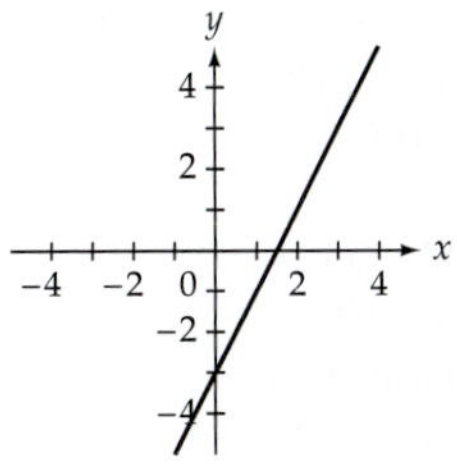
[Sec. 10.1, Check Your Progress 5]

6.
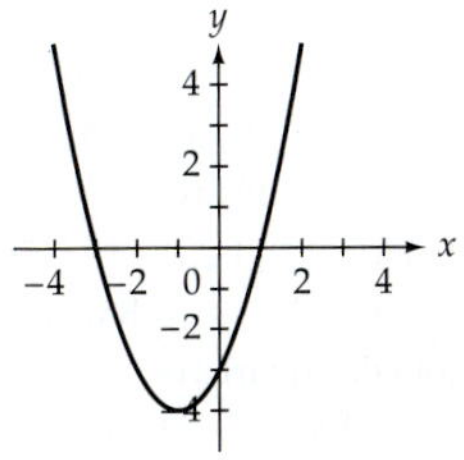
[Sec. 10.1, Example 5]

7.
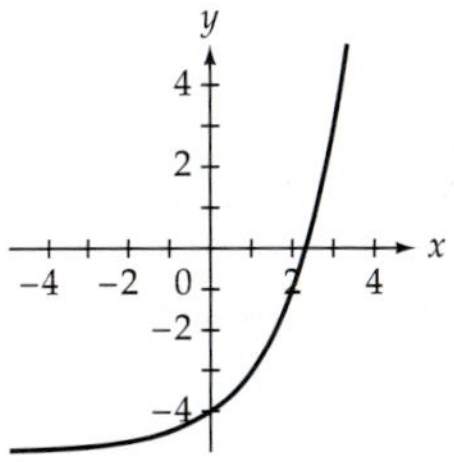
[Sec. 10.5, Example 2]

8.
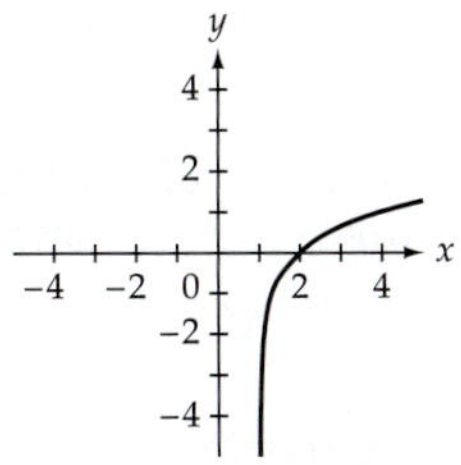
[Sec. 10.6, Example 5]

9. $\frac{3}{5}$ [Sec. 10.2, Example 3a] **10.** $y = \frac{2}{3}x + 3$ [Sec. 10.3, Check Your Progress 2]

11. $(-3, -10)$ [Sec. 10.4, Example 1] **12.** $(-4, 0), (2, 0)$ [Sec. 10.4, Check Your Progress 2a]
13. $\frac{49}{4}$, maximum [Sec. 10.4, Example 3] **14.** The vertical intercept is $(0, 250)$. This means that the plane starts 250 mi from its destination. The horizontal intercept is $(2.5, 0)$. This means that it takes the plane 2.5 h to reach its destination. [Sec. 10.2, Example 2]
15. 44 ft [Sec. 10.4, Example 4] **16.** 3.30 g [Sec. 10.5, Check Your Progress 4] **17.** 2.0 times greater [Sec. 10.6, Check Your Progress 7]
18. $y = -1.5x + 740$ [Sec. 10.3, Example 4] **19. a.** $y = 0.029x + 2.776$ **b.** 78 accidents [Sec. 10.3, Example 5]
20. 2.5 m [Sec. 10.6, Example 6a]

CHAPTER 11

EXCURSION EXERCISES, SECTION 11.1 *page 627*

1. \$40.09 **2.** \$190.20 **3.** \$9431.77 **4.** \$39.30 **5.** \$190.99 **6.** \$9240.78 **7.** \$227.50

EXERCISE SET 11.1 *page 627*

1. Divide the number of months by 12. **2.** Using the ordinary method, divide the number of days by 360.
3. I is the interest, P is the principal, r is the interest rate, and t is the time period. **4.** \$120 **5.** \$560 **6.** \$82.50 **7.** \$227.50
8. \$89.25 **9.** \$202.50 **10.** \$117 **11.** \$16.80 **12.** \$30.80 **13.** \$159.60 **14.** \$36 **15.** \$125
16. \$252 **17.** \$168 **18.** \$8789 **19.** \$15,667.50 **20.** \$4749.50 **21.** \$7390.80 **22.** \$13,546
23. \$2864.40 **24.** 6.25% **25.** 7.5% **26.** 9.5% **27.** 8% **28.** 8.2% **29.** 9.3% **30.** \$152 **31.** \$39
32. \$24 **33.** \$18 **34.** \$145 **35.** \$7406 **36.** \$7136.25 **37.** \$5465.20 **38.** \$960.96 **39.** \$804.75

40. 9% **41.** 10.2% **42.** 9.7% **43.** 10% **44.** $1222 **45.** $132.99 **46.** $42.50
47. The ordinary method. The lender benefits. **48.** Answers will vary. **49.** There are fewer days in September than there are in August.

EXCURSION EXERCISES, SECTION 11.2 *page 642*

1. 218.056% **2.** 238.63 **3.** 3.08% **4.** Yes; housing and transportation **5.** $7.78 **6.** $58,486.27

EXERCISE SET 11.2 *page 642*

1. $2739.99 **2.** $10,495.46 **3.** $852.88 **4.** $10,015.38 **5.** $20,836.54 **6.** $7658.76 **7.** $12,575.23
8. $2665.98 **9.** $3532.86 **10.** $6739.77 **11.** $3850.08 **12.** $14,548.16 **13.** $2213.84 **14.** $9980.65
15. $10,116.38 **16.** $4460.81 **17.** $8320.14 **18.** $10,602.61 **19.** $41,210.44 **20.** $10,778.75 **21.** $7641.78
22. $14,827.44 **23.** $3182.47 **24.** $3047.82 **25.** $10,094.57 **26.** $43,043.46 **27.** $11,887.58
28. $4020.29 **29.** $3583.16 **30.** $1690.74 **31.** $5728.18 **32.** $17,664.32 **33.** $391.24 **34.** $4700.24
35. $18,056.35 **36.** $5414.84 **37.** $11,120.58 **38.** $7068.25 **39. a.** $450 **b.** $568.22 **c.** $118.22
40. a. $1000 **b.** $1051.63 **c.** $51.63 **41. a.** $20,528.54 **b.** $20,591.79 **c.** $63.25
42. a. $29,775.40 **b.** $29,851.31 **c.** $75.91 **43. a.** $1698.59 **b.** $1716.59 **c.** $18.00
44. a. $6764.51 **b.** $6878.33 **c.** $113.82 **45. a.** $10,401.63 **b.** $3401.63 **46. a.** $9831.70 **b.** $3831.70
47. $9612.75 **48.** $15,919.00 **49. a.** $67,228.19 **b.** $94,161.98 **50. a.** $39,178.17 **b.** $84,361.02
51. $5559.09 **52.** $16,286.48 **53.** $3446.03 **54.** $275,991.34 **55.** $56,102.07 **56.** $89,542.38 **57.** $12,152.77
58. $36,504.48 **59.** $2495.45 **60.** $25,686.16 **61.** 7.40% **62.** 8.67% **63.** 7.76% **64.** 7.12%
65. 8.44% **66.** 6.50% **67.** 6.10% **68.** 6.45% **69.** $4.01, $5.37, $9.62 **70.** $4.48, $6.00, $10.74
71. $3.60, $4.82, $8.63 **72.** $3.01, $4.03, $7.22 **73.** $12.04, $16.12, $28.86 **74.** $16.06, $21.49, $38.49
75. $368,012.03, $492,483.12, $881,962.25 **76.** $32,117.41, $42,980.34, $76,971.25 **77.** $25,417.46 **78.** $14,550.23
79. $25,841.90 **80.** $10,873.38 **81.** $53,473.96 **82.** $3684.98 **83. a.** 4.0%, 4.04%, 4.06%, 4.07%, 4.08%
b. increase **84.** 3.03% **85.** 3.04% **86.** 6.6% compounded quarterly **87.** 6.25% compounded semiannually
88. 7.8% compounded monthly **89.** 5.8% compounded quarterly **90. a.** $3080.37 **b.** $14,861.21 **c.** $17,515.24
d. $4312.52 **e.** $1049.58 **f.** $4835.32 **g.** $9358.55 **h.** $29,167.36 **i.** $8757.62 **j.** $437.53

EXCURSION EXERCISES, SECTION 11.3 *page 658*

1. a. $22,740 **b.** $101.90 **c.** $161.25 **d.** $263.15 **2. a.** $31,165 **b.** $166.83 **c.** $244.42 **d.** $411.25
3. a. $21,100 **b.** 0.003375 **c.** $121.84 **d.** $169.44 **e.** $291.28 **4. a.** $160,000 **b.** 0.0035 **c.** $857.50
d. $1250 **e.** $2107.50 **5.** $459.89 **6.** $617.08

EXERCISE SET 11.3 *page 659*

1. $1.48 **2.** $6.85 **3.** $152.32 **4.** $254.34 **5.** $335.87 **6.** $955.20 **7.** $15.34
8. $6.81 **9.** $5.00 **10.** $5.65 **11.** $26.87 **12.** $20.37 **13.** 13.5% **14.** 13.7%
15. 19.2% **16.** 12.9% **17.** $34.59 **18.** $56.03 **19. a.** $696.05 **b.** $174.01 **c.** $88.46
20. a. $40,736.25 **b.** $8147.25 **c.** $987.00 **21. a.** $68,569.73 **b.** $13,713.95 **c.** $641.17
22. a. $2680.82 **b.** $402.12 **c.** $103.23 **23.** $874.88 **24.** $826.28 **25.** $571.31 **26.** $636.32
27. a. $621.19 **b.** $4372.12 **28. a.** $432.37 **b.** $4517.20 **29.** $13,575.25 **30.** $9990.66
31. $3472.57 **32.** $9900.81 **33.** $226.45 **34.** $256.95 **35.** $412.14 **36.** $587.01
37. a. $2704.15 **b.** $20,938.15 **c.** $515.10 **38. a.** 92 months **b.** $1200.66 **c.** Answers will vary.
39. $572.75

EXCURSION EXERCISES, SECTION 11.4 *page 668*

1. $29,663.13 **2.** $19,850.36 **3.** $59,924.80 **4.** $39,819.64

EXERCISE SET 11.4 *page 668*

1. $382.50 **2.** $870.00 **3.** $535.50 **4.** $364.00 **5.** 2.51% **6.** 1.25% **7.** 1.83%
8. 0.8% **9. a.** $43.81 **b.** $3270.00 **c.** 6,976,300 shares **d.** increase **e.** $141.13 **10. a.** $58.30
b. $3900.00 **c.** 5,583,200 shares **d.** increase **e.** $140.42 **11.** 50 shares **12.** 85 shares
13. a. Profit of $25,240.00 **b.** $1409.80 **14. a.** Loss of $3480 **b.** $293.60 **15. a.** Profit of $34,032
b. $2475.54 **16. a.** Profit of $69,960 **b.** $4760.10 **17.** $252 **18.** $562.50 **19.** $840 **20.** $2832
21. $22.50 **22.** $16.50 **23.** 714 shares **24.** 250 shares **25.** 240 shares **26.** 500 shares
27. The no-load fund's value ($2800.58) is $2.06 greater than the load fund's value ($2798.52). **28.** Answers will vary. **29.** Answers will vary.

EXCURSION EXERCISES, SECTION 11.5 *page 678*

1. \$1801.39 **2.** \$4275 **3.** \$289,275 **4.** 6.36% **5.** Option 1: \$682.18; Option 2: \$665.30
6. Option 1: \$1500; Option 2: \$2000 **7.** Option 1: \$17,872.32; Option 2: \$17,967.20 **8.** Option 1: \$26,058.48; Option 2: \$25,950.80
9. Option 1 is more cost effective if you stay in the home for 2 years or less. Option 2 is more cost effective if you stay in the home for 3 years or more. Explanations will vary.

EXERCISE SET 11.5 *page 679*

1. \$64,500; \$193,500 **2.** \$32,500; \$292,500 **3.** \$5625 **4.** \$4675 **5.** \$99,539 **6.** \$56,801
7. \$34,289.38 **8.** \$34,841 **9.** \$974.37 **10.** \$1548.57 **11.** \$2155.28 **12.** \$1300.87
13. a. \$1088.95 **b.** \$392,022 **c.** \$240,022 **14. a.** \$1569.02 **b.** \$470,706 **c.** \$271,706
15. \$174,606 **16.** \$664,141.60 **17.** Interest: \$1407.38; principal: \$495.89 **18.** Interest: \$1297.13; principal: \$37.50
19. Interest: \$1347.68; principal: \$123.62 **20.** Interest: \$986.59; principal: \$110.43 **21.** \$112,025.49 **22.** \$126,874.00
23. \$61,039.75 **24.** \$96,924.63 **25.** \$1071.10 **26.** \$1903.71 **27.** \$2022.50 **28.** \$1827.28
29. a. \$330.57 **b.** \$140,972.40 **30. a.** \$804.08 **b.** \$325,058.40 **31. a.** \$390.62 **b.** \$178,273.20
32. a. \$343.07 **b.** \$188,254.80 **33.** \$125,000 **34.** \$120,000 **35.** \$212,065 **36.** 260th payment **37.** No
38. Yes. If the interest rate is lower, it will take fewer months. **39. a.** \$65,641.88 **b.** \$138,596.60 **c.** \$28,881.52 **d.** 44%

CHAPTER 11 REVIEW EXERCISES *page 683*

1. \$61.88 [Sec. 11.1] **2.** \$782 [Sec. 11.1] **3.** \$90 [Sec. 11.1] **4.** \$7218.40 [Sec. 11.1] **5.** 7.5% [Sec. 11.1]
6. \$3654.90 [Sec. 11.2] **7.** \$11,609.72 [Sec. 11.2] **8.** \$7859.52 [Sec. 11.2] **9.** \$200.23 [Sec. 11.2]
10. \$10,683.29 [Sec. 11.2] **11. a.** \$11,318.23 **b.** \$3318.23 [Sec. 11.2] **12.** \$19,225.50 [Sec. 11.2]
13. 1.1% [Sec. 11.4] **14.** \$9000 [Sec. 11.4] **15.** \$1.59 [Sec. 11.2] **16.** \$43,650.68 [Sec. 11.2]
17. 6.06% [Sec. 11.2] **18.** 5.4% compounded semiannually [Sec. 11.2] **19.** \$431.16 [Sec. 11.3]
20. \$6.12 [Sec. 11.3] **21. a.** \$259.38 **b.** 12.75% [Sec. 11.3] **22. a.** \$36.03 **b.** 12.9% [Sec. 11.3]
23. \$45.41 [Sec. 11.3] **24. a.** \$10,092.69 **b.** \$2018.54 **c.** \$253.01 [Sec. 11.3] **25.** \$664.40 [Sec. 11.3]
26. a. \$540.02 **b.** \$12,196.80 [Sec. 11.3] **27. a.** Profit of \$5325 **b.** \$256.10 [Sec. 11.4]
28. 200 shares [Sec. 11.4] **29.** \$99,041 [Sec. 11.5] **30. a.** \$1659.11 **b.** \$597,279.60 **c.** \$341,479.60 [Sec. 11.5]
31. a. \$1396.69 **b.** \$150,665.74 [Sec. 11.5] **32.** \$2658.53 [Sec. 11.5] **33.** \$288.62 [Sec. 11.3]

CHAPTER 11 TEST *page 685*

1. \$108.28 [Sec. 11.1, Example 2] **2.** \$202.50 [Sec. 11.1, Example 1] **3.** \$8408.89 [Sec. 11.1, Example 6]
4. 9% [Sec. 11.1, Example 5] **5.** \$7340.87 [Sec. 11.2, Check Your Progress 2] **6.** \$312.03 [Sec. 11.2, Example 4]
7. a. \$15,331.03 **b.** \$4831.03 [Sec. 11.1, Example 6] **8.** \$21,949.06 [Sec. 11.2, Example 6] **9.** 1.2% [Sec. 11.4, Example 2]
10. \$1900 [Sec. 11.4, Example 4] **11.** \$387,207.74 [Sec. 11.2, Check Your Progress 8] **12.** 6.40% [Sec. 11.2, Check Your Progress 10]
13. 4.6% compounded semiannually [Sec. 11.2, Example 11] **14.** \$7.79 [Sec. 11.3, Example 1] **15. a.** \$48.56
b. 16.6% [Sec. 11.3, Example 2] **16.** \$56.49 [Sec. 11.3, Example 3] **17. a.** Loss of \$4896 **b.** \$226.16 [Sec. 11.4, Example 3]
18. 208 shares [Sec. 11.4, Example 5] **19. a.** \$6985.94 **b.** \$1397.19 **c.** \$174.62 [Sec. 11.3, Example 4]
20. \$60,083.50 [Sec. 11.5, Example 1] **21. a.** \$1530.69 [Sec. 11.5, Example 2a] **b.** \$221,546.46 [Sec. 11.5, Example 4]
22. \$2595.97 [Sec. 11.5, Example 2a, Example 5]

CHAPTER 12

EXCURSION EXERCISES, SECTION 12.1 *page 694*

1. Three weighings **2.** Three weighings

EXERCISE SET 12.1 *page 695*

1. {0, 2, 4, 6, 8} **2.** {1, 3, 5, 7, 9} **3.** {Monday, Tuesday, Wednesday, Thursday, Friday, Saturday, Sunday}
4. {January, February, March, April, May, June, July, August, September, October, November, December}
5. {HH, TT, HT, TH} **6.** {HHH, HHT, HTH, THH, HTT, THT, TTH, TTT}
7. {1H, 2H, 3H, 4H, 5H, 6H, 1T, 2T, 3T, 4T, 5T, 6T} **8.** {H1, H2, H3, H4, T1, T2, T3, T4}
9. $\{S_1E_1D_1, S_1E_1D_2, S_1E_2D_1, S_1E_2D_2, S_1E_3D_1, S_1E_3D_2, S_2E_1D_1, S_2E_1D_2, S_2E_2D_1, S_2E_2D_2, S_2E_3D_1, S_2E_3D_2\}$
10. $\{B_1R_1C_1, B_1R_1C_2, B_1R_2C_1, B_1R_2C_2, B_2R_1C_1, B_2R_1C_2, B_2R_2C_1, B_2R_2C_2, B_3R_1C_1, B_3R_1C_2, B_3R_2C_1, B_3R_2C_2\}$

11. {ABCD, ABDC, ACBD, ACDB, ADBC, ADCB} **12.** {ABCDE, ABCED, ABDCE, ABDEC, ABECD, ABEDC, ADBCE, ADBEC, ADCBE, ADCEB, ADEBC, ADECB, AEBCD, AEBDC, AECBD, AECDB, AEDBC, AEDCB}
13. 12 **14.** 16 **15.** 4^{20} **16.** 2^{25} **17.** 7000 **18.** 725,760 **19.** 90 **20.** 45 **21.** 18
22. 30 **23.** 62 **24.** 49 **25.** 24 **26.** 9 **27.** 15 **28.** Yes; see the *Instructor's Solutions Manual.*
29. 13^4 **30.** 24 **31.** 1 **32.** 17,160 **33.** Answers will vary. **34.** Answers will vary. **35.** 150; 25
36. In the tree diagram, 9–13 signifies moving a checker from square 9 to square 13.

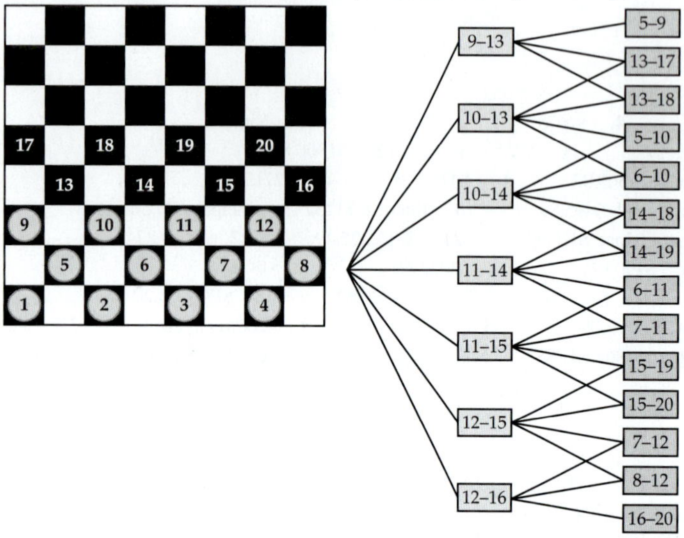

EXCURSION EXERCISES, SECTION 12.2 *page 705*

1. 487,635 **2.** 684,400 **3.** 336,300 **4.** 68,400 **5.** 4845

EXERCISE SET 12.2 *page 705*

1. 40,320 **2.** 120 **3.** 362,760 **4.** 24 **5.** 120 **6.** 40,314 **7.** 6720 **8.** 42 **9.** 181,440
10. 30,240 **11.** 1 **12.** 1 **13.** 40,320 **14.** 3,628,800 **15.** 3360 **16.** 3 **17.** $\frac{1}{720}$ **18.** 100
19. 36 **20.** 28 **21.** 1 **22.** 1 **23.** 525 **24.** 7056 **25.** $\frac{60}{143}$ **26.** $\frac{20}{63}$ **27.** $\frac{360}{1001}$
28. 336 **29.** 2880 **30.** 120 **31.** 21 **32.** 56 **33.** 792 **34.** 1 **35.** No **36.** No
37. 120 **38.** 6840 **39.** 43,680 **40.** 504 **41. a.** 362,880 **b.** 8640 **42. a.** 39,916,800 **b.** 518,400
43. 10,080 **44.** 8008 **45.** 735,471 **46.** 30,045,015 **47.** 35 **48.** 2,118,760 **49.** 3136
50. 816 **51.** 12 **52.** 82 **53.** 9 **54.** 35 **55.** 48,620 **56.** 756,756 **57.** 24 **58.** 792
59. 45,360 **60.** 3150 **61.** 252 **62.** 495 **63.** 1728 **64.** 62 **65.** 48 **66.** 1728
67. 4512 **68.** 24 **69.** 103,776 **70.** 1584 **71.** Answers will vary. **72.** Answers will vary.
73. a. 2520 **b.** 60 **c.** About 1929 years **d.** About 3.23×10^{26} years

EXCURSION EXERCISES, SECTION 12.3 *page 715*

1. Answers will vary. **2.** 0.6366 **3–6.** Answers will vary.

EXERCISE SET 12.3 *page 715*

1. {HHH, HHT, HTH, HTT, THH, THT, TTH, TTT} **2.** {2, 4, 6, 8, 10}
3. {Nov. 1, Nov. 2, Nov. 3, Nov. 4, Nov. 5, Nov. 6, Nov. 7, Nov. 8, Nov. 9, Nov. 10, Nov. 11, Nov. 12, Nov. 13, Nov. 14}
4. {penny, nickel, dime, quarter, half-dollar, dollar} **5.** {Alaska, Alabama, Arizona, Arkansas}
6. {April, June, September, November} **7.** {BBB, BBG, BGB, BGG, GBB, GBG, GGB, GGG} **8.** {BBG, BGB, GBB}
9. {BGG, GBG, GGB, GGG} **10.** {BBB} **11.** {BBG, BGB, BGG, GBB, GBG, GGB, GGG}
12. $\frac{3}{8}$ **13.** $\frac{1}{2}$ **14.** $\frac{1}{8}$ **15.** $\frac{7}{8}$ **16.** $\frac{3}{8}$ **17.** $\frac{1}{4}$ **18.** $\frac{1}{8}$ **19.** $\frac{1}{12}$ **20.** $\frac{11}{12}$
21. $\frac{1}{4}$ **22.** $\frac{7}{12}$ **23.** $\frac{5}{36}$ **24.** $\frac{1}{18}$ **25.** $\frac{1}{36}$ **26.** $\frac{1}{36}$ **27.** 0 **28.** 0 **29.** $\frac{1}{6}$

30. $\frac{5}{18}$ **31.** $\frac{1}{2}$ **32.** $\frac{1}{2}$ **33.** $\frac{1}{6}$ **34.** $\frac{5}{6}$ **35.** $\frac{1}{13}$ **36.** $\frac{3}{13}$ **37.** $\frac{5}{13}$ **38.** $\frac{4}{13}$ **39.** $\frac{1267}{3228} \approx 0.39$ **40.** $\frac{2190}{3228} \approx 0.68$ **41.** $\frac{804}{3228} \approx 0.25$ **42.** $\frac{1651}{3228} \approx 0.51$ **43.** $\frac{150}{3228} \approx 0.05$ **44.** $\frac{432}{3228} \approx 0.13$ **45.** $\frac{26}{425}$ **46.** $\frac{137}{425}$ **47.** $\frac{18}{425}$ **48.** $\frac{103}{850}$ **49.** $\frac{58}{293}$ **50.** $\frac{39}{293}$ **51.** $\frac{36}{293}$ **52.** $\frac{117}{293}$ **53.** $\frac{1}{4}$ **54.** 0 **55.** 0 **56.** $\frac{3}{4}$ **57.** Answers will vary. **58.** Answers will vary. **59.** $\frac{1}{3}$ **60.** $\frac{1}{5}$ **61.** $\frac{3}{10}$ **62.** $\frac{3}{8}$ **63.** $\frac{8}{13}$ **64.** $\frac{11}{20}$ **65.** 1 to 4 **66.** 3 to 2 **67.** 3 to 5 **68.** 7 to 18 **69.** 11 to 9 **70.** 81 to 19 **71.** 1 to 14 **72.** 3 to 1 **73.** 1 to 1 **74.** 1 to 3 **75.** 15 to 1 **76.** 23 to 27 **77.** $\frac{49}{97}$ **78.** $\frac{332}{333}$ **79.** $\frac{3}{11}$ **80. a.** 7 to 50 **b.** $\frac{7}{57}$ **81.** $\frac{3}{8}$ **82.** $\frac{63}{256}$ **83.** Answers will vary. **84.** $\frac{1}{54{,}145}$ **85.** $\frac{1}{108{,}290}$ **86.** $\frac{9}{19}$ **87.** $\frac{9}{19}$ **88.** $\frac{6}{19}$ **89.** $\frac{1}{38}$ **90.** $\frac{1}{19}$ **91.** $\frac{3}{19}$

EXCURSION EXERCISES, SECTION 12.4 *page 724*

1. 77.3% **2.** 99.9% **3.** 0.00000207% **4.** 26.2% **5.** Probability of not matching any lucky numbers: 0.12%; Probability of matching 5 lucky numbers: 23.3% **6.** 99.9%

EXERCISE SET 12.4 *page 725*

1. Answers will vary. **2.** Answers will vary. **3.** $\frac{2}{13}$ **4.** $\frac{1}{2}$ **5.** $\frac{1}{9}$ **6.** $\frac{11}{36}$ **7.** 0.6 **8.** 0.8 **9.** 0.2 **10.** 0.5 **11.** $\frac{7}{10}$ **12.** $\frac{7}{10}$ **13.** $\frac{4}{5}$ **14.** $\frac{7}{10}$ **15.** $\frac{5}{18}$ **16.** $\frac{1}{3}$ **17.** $\frac{1}{2}$ **18.** $\frac{1}{2}$ **19.** $\frac{11}{18}$ **20.** $\frac{5}{9}$ **21.** $\frac{4}{13}$ **22.** $\frac{7}{13}$ **23.** $\frac{3}{13}$ **24.** $\frac{8}{13}$ **25.** $\frac{3}{4}$ **26.** $\frac{3}{4}$ **27.** $\frac{1150}{3179}$ **28.** $\frac{84}{187}$ **29.** $\frac{1170}{3179}$ **30.** $\frac{2644}{3179}$ **31.** 0.96 **32.** 0.62 **33.** $\frac{74}{75}$, or about 98.7% **34.** $\frac{25}{26}$, or about 96.2% **35.** $\frac{5}{6}$ **36.** $\frac{5}{6}$ **37.** $\frac{11}{12}$ **38.** $\frac{35}{36}$ **39.** $\frac{12}{13}$ **40.** $\frac{10}{13}$ **41.** $\frac{15}{16}$ **42.** $\frac{11}{16}$ **43.** 42.1% **44.** 51.8% **45.** 36.1% **46.** 20.4% **47.** 54.5% **48.** 55.3% **49.** 88.8% **50.** 45.9% **51.** 20.4% **52.** 48.7% **53.** No **54. a.** $\frac{1}{3}$ **b.** $\frac{2}{3}$ **c.** $\frac{2}{3}$ **55.** 100% **56.** and **57.** Answers will vary. **58. a.** 2,598,960 **b.** 1287 **c.** 0.000495 **d.** 0.000495 **e.** Yes **f.** 0.00198

EXCURSION EXERCISES, SECTION 12.5 *page 733*

1. 7.43% **2.** 25.3% **3.** 50.7% **4.** 89.1%

EXERCISE SET 12.5 *page 734*

1. Answers will vary. **2.** Answers will vary. **3.** $P(A \mid B) = 0.625$; $P(B \mid A) \approx 0.357$ **4.** $P(A \mid B) = 0.375$; $P(B \mid A) \approx 0.667$ **5.** $P(A \mid B) \approx 0.389$; $P(B \mid A) \approx 0.115$ **6.** $P(A \mid B) = 0.4$; $P(B \mid A) = 1$ **7.** $\frac{179}{864}$ **8.** $\frac{173}{778}$ **9.** $\frac{557}{921}$ **10.** $\frac{341}{1642}$ **11.** 0.30 **12.** 0.31 **13.** 0.15 **14.** 0.11 **15.** $\frac{5}{18}$ **16.** $\frac{1}{6}$ **17.** $\frac{1}{5}$ **18.** $\frac{5}{21}$ **19.** 0.050 **20.** 0.059 **21.** 0.127 **22.** 0.072 **23.** $\frac{6}{1045}$ **24.** $\frac{13}{5852}$ **25.** $\frac{3}{1045}$ **26.** $\frac{3}{7315}$ **27.** $\frac{13}{102}$ **28.** $\frac{13}{850}$ **29.** $\frac{8}{5525}$ **30.** $\frac{13}{204}$ **31.** 0.000484 **32.** 0.003916 **33.** 0.001424 **34.** 0.00005324 **35.** Independent **36.** Not independent **37.** Not independent **38.** Not independent **39.** $\frac{25}{1296}$ **40.** $\frac{5}{1296}$ **41.** $\frac{1}{72}$ **42.** $\frac{1}{36}$ **43.** $\frac{1}{4}$ **44.** $\frac{1}{36}$ **45.** $\frac{1}{16}$ **46.** $\frac{1}{216}$ **47.** $\frac{1}{216}$ **48.** $\frac{1}{1728}$ **49.** $\frac{1}{169}$ **50.** $\frac{9}{169}$ **51.** $\frac{1}{16}$ **52.** $\frac{12}{169}$ **53.** $\frac{1}{16}$ **54.** $\frac{3}{26}$ **55. a.** $\frac{1}{32}$ **b.** $\frac{13}{425}$ **56. a.** $\frac{27}{2197}$ **b.** $\frac{11}{1105}$

57. a. $\frac{100}{4913}$ **b.** $\frac{1}{51}$ **58. a.** $\frac{25}{576}$ **b.** $\frac{1}{22}$ **59.** 0.46 **60.** 0.81 **61.** 0.11
62. a. $\frac{1}{8}$ **b.** $\frac{1}{4}$ **c.** $\frac{3}{8}$ **63. a.** $\frac{1}{2}$ **b.** 0 **c.** 1 **d.** $\frac{1}{3}$ **e.** $\frac{2}{3}$ **f.** Yes

EXCURSION EXERCISES, SECTION 12.6 *page 740*

1. $\frac{1}{216}$ **2.** $\frac{25}{72}$ **3.** $\frac{5}{72}$ **4.** $\frac{125}{216}$ **5.** −5 cents **6.** 0 **7.** 0 **8.** Under 11 or over 10; numbers bet

EXERCISE SET 12.6 *page 741*

1. 49.5 **2.** 6.27 **3.** −5 cents **4.** −5 cents **5.** −24 cents **6.** −22 cents **7.** −22 cents
8. −17 cents **9.** $62.83 **10.** $127.76 **11.** −$209.71 **12.** $103.87 **13.** More than $42.18
14. More than $38.18 **15.** $39,100 **16.** $6000 **17.** $20,250 **18.** $242,500 **19.** 7
20. a. {01, 02, 03, 04, 05, 06, 61, 62, 63, 64, 65, 66} **b.** All sums have a probability of $\frac{1}{12}$. **c.** 6.5
21. 7 **22.** Answers will vary. **23.** Red dice **24.** −$1.09

CHAPTER 12 REVIEW EXERCISES *page 746*

1. {11, 12, 13, 21, 22, 23, 31, 32, 33} [Sec. 12.1] **2.** {26, 28, 62, 68, 82, 86} [Sec. 12.1] **3.** {HHHH, HHHT, HHTH, HHTT, HTHH, HTHT, HTTH, HTTT, THHH, THHT, THTH, THTT, TTHH, TTHT, TTTH, TTTT} [Sec. 12.1]
4. {7A, 8A, 9A, 7B, 8B, 9B} [Sec. 12.1] **5.** 72 [Sec. 12.1] **6.** 10,000 [Sec. 12.1] **7.** 2400 [Sec. 12.1] **8.** 64 [Sec. 12.1]
9. 5040 [Sec. 12.2] **10.** 40,296 [Sec. 12.2] **11.** 1260 [Sec. 12.2] **12.** 151,200 [Sec. 12.2] **13.** 336 [Sec. 12.2]
14. $\frac{60}{143}$ [Sec. 12.2] **15.** 5040 [Sec. 12.2] **16.** 5040 [Sec. 12.2] **17.** 2520 [Sec. 12.2] **18.** 180 [Sec. 12.2]
19. 495 [Sec. 12.2] **20.** 60 [Sec. 12.2] **21.** 3,268,760 [Sec. 12.2] **22.** 165 [Sec. 12.2] **23.** 660 [Sec. 12.2]
24. 282,240 [Sec. 12.2] **25.** 624 [Sec. 12.2] **26.** $\frac{1}{4}$ [Sec. 12.3] **27.** $\frac{3}{8}$ [Sec. 12.3] **28.** 0.56 [Sec. 12.3]
29. 0.37 [Sec. 12.3] **30.** 0.85 [Sec. 12.3] **31.** $\frac{1}{9}$ [Sec. 12.3] **32.** $\frac{17}{18}$ [Sec. 12.4] **33.** $\frac{1}{6}$ [Sec. 12.4]
34. $\frac{5}{9}$ [Sec. 12.4] **35.** $\frac{2}{9}$ [Sec. 12.5] **36.** $\frac{1}{6}$ [Sec. 12.5] **37.** $\frac{3}{4}$ [Sec. 12.4] **38.** $\frac{4}{13}$ [Sec. 12.4] **39.** $\frac{12}{13}$ [Sec. 12.4]
40. $\frac{2}{3}$ [Sec. 12.5] **41.** 5 to 31 [Sec. 12.3] **42.** 1 to 3 [Sec. 12.3] **43.** $\frac{5}{9}$ [Sec. 12.3] **44.** $\frac{1}{2}$ [Sec. 12.3]
45. 0.036 [Sec. 12.4] **46.** $\frac{173}{1000}$ [Sec. 12.4] **47.** $\frac{5}{8}$ [Sec. 12.3] **48.** 1 to 5 [Sec. 12.3] **49.** $\frac{7}{16}$ [Sec. 12.5]
50. $\frac{646}{1771}$ [Sec. 12.5] **51.** $\frac{7}{253}$ [Sec. 12.4] **52.** 0.37 [Sec. 12.3] **53.** 0.62 [Sec. 12.4] **54.** 0.07 [Sec. 12.3]
55. 0.47 [Sec. 12.5] **56.** 0.29 [Sec. 12.5] **57.** $\frac{1}{216}$ [Sec. 12.5] **58.** 0.60 [Sec. 12.4] **59.** 0.16 [Sec. 12.5]
60. $\frac{175}{256}$ [Sec. 12.4] **61.** 0.648 [Sec. 12.5] **62.** 0.029 [Sec. 12.5] **63.** 0.135 [Sec. 12.4]
64. −50 cents [Sec. 12.6] **65.** 25 cents [Sec. 12.6] **66.** −37 cents [Sec. 12.6] **67.** About 5.4 [Sec. 12.6]
68. $296.20 [Sec. 12.6] **69.** $765.30 [Sec. 12.6] **70.** $11,000 [Sec. 12.6]

CHAPTER 12 TEST *page 749*

1. {A2, D2, G2, K2, A3, D3, G3, K3, A4, D4, G4, K4} [Sec. 12.1, Example 3 and Check Your Progress 4] **2.** 72 [Sec. 12.1, Example 5]
3. 5040 [Sec. 12.1, Example 5; Sec 12.2, Example 2] **4.** About 1.09×10^{10} [Sec. 12.2, Example 2] **5.** 28 [Sec. 12.2, Example 6]
6. $\frac{7}{12}$ [Sec. 12.3, Example 3] **7.** 27.4% [Sec. 12.4, Example 4] **8.** $\frac{1}{17}$ [Sec. 12.3, Example 3]
9. 0.72 [Sec. 12.4, Check Your Progress 3] **10.** 1 to 7 [Sec. 12.3, Example 6] **11.** 0.635 [Sec. 12.5, Example 4]
12. 0.466 [Sec. 12.5, Example 1] **13.** $\frac{1}{2}$ [Sec. 12.3, Example 5] **14.** $38,350 [Sec. 12.6, Example 3]

CHAPTER 13

EXCURSION EXERCISES, SECTION 13.1 *page 758*

1. 1.9372 **2.** 0.87311 **3.** 1.541 **4.** 2.4 **5.** 315.3 lb

EXERCISE SET 13.1 *page 759*

1. 7; 7; 7 **2.** 11.6; 9; 3 **3.** 22; 14; no mode **4.** 73.5; 79.5; 74 **5.** 18.8; 8.1; no mode **6.** 5; 5; 5 **7.** 192.4; 190; 178 **8.** 132.2; 118; 118 **9.** 0.1; −3; −5 **10.** 3.1; 4.1; 4.1 **11. a.** Yes. The mean is computed by using the sum of all the data. **b.** No. The median is not affected unless the middle value, or one of the two middle values, in a data set is changed.
12. Yes. The mode is the number (or numbers) in a data set that occurs most frequently. **13.** ≈40.0 years; 35 years
14. ≈44.9 years; 43 years **15. a.** Answers will vary. **b.** Answers will vary.
16. Answers will vary. However, the median is more representative of most of the expenditures. **17.** 3.22
18. 2.36 **19.** 3.37 **20.** 2.5 **21.** 82 **22.** 84.5 **23.** 0.847 **24.** 0.690
25. 0.671 **26.** 0.624 **27.** ≈6.1 points; 5 points; 2 points and 5 points **28.** ≈\$12.58; \$11.50; \$8.00
29. ≈7.2; 7; 7 **30.** ≈10.1 years; 10 years; 10 years **31.** 64° **32.** 10° **33.** −6°F
34. 20.5°F **35.** 92 **36.** No. Alisa needs a score of 122 to raise her average to 90, but the next test only counts as 100 points.
37. a. ≈0.275 **b.** ≈0.273 **c.** No **38.** 48 mph
39. Yes. Joanne has a smaller average for the first month and the second month, but she has a larger average for both months.
40. No. Wendy has a better average in English and in history, and she also has a better average for English and history combined.

EXCURSION EXERCISES, SECTION 13.2 *page 768*

1. a. and **b.**

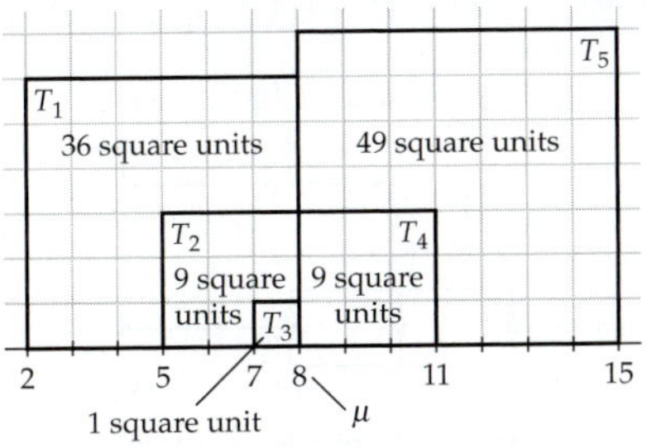

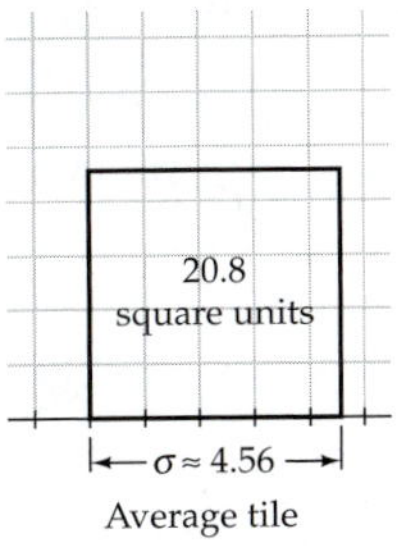

c. 104 square units
d. 20.8 square units
e. See the figure at the left.
f. The variance is 20.8. It is the area of the average tile shown at the left.
g. The standard deviation is $\sqrt{20.8} \approx 4.56$. It is the width of the average tile.

2. a. and **b.**

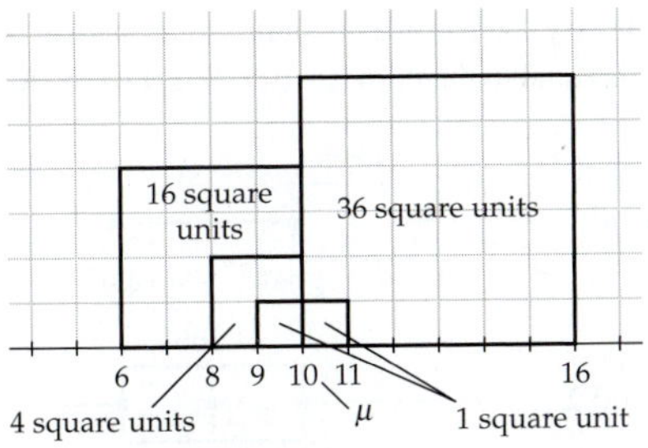

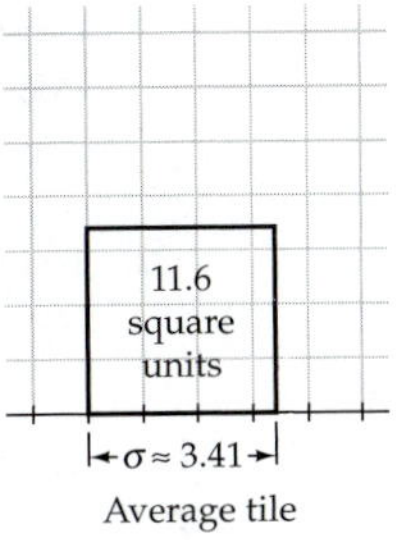

c. 58 square units
d. 11.6 square units
e. See the figure at the left.
f. The variance is 11.6. It is the area of the average tile shown at the left.
g. The standard deviation is $\sqrt{11.6} \approx 3.41$. It is the width of the average tile.
h. The data set in Excursion Exercise 2 has the larger mean, but the data set in Excursion Exercise 1 has the larger standard deviation.

EXERCISE SET 13.2 *page 769*

1. 84°F **2.** 62°F **3.** 21; 8.2; 67.1 **4.** 13; 4.8; 23.4 **5.** 3.3; 1.3; 1.7 **6.** 15.4; 6.1; 37.7
7. 52; 17.7; 311.6 **8.** 47; 16.9; 284.5 **9.** 0; 0; 0 **10.** 2; 1.0; 1.1 **11.** 23; 8.3; 69.6 **12.** 27; 9.8; 95.6
13. Opinions will vary. However, many climbers would consider rope B to be safer because of its smaller standard deviation in breaking strength.
14. The PowerBall lottery because the range is greater. **15.** The students in the college statistics course because the range of weights is greater.
16. The statement is not true. Consider, for example, the data sets 98, 99, 100, 101, 102 and 2, 4, 6, 8, 10. **17.** 23.67 mpg; 3.92 mpg
18. 3.98 min; 1.05 min **19.** 493.6 cal; 20.30 cal **20.** 8.62 mg/oz; 0.55 mg/oz **21.** 4.27 h, 0.69 h
22. a. 98.25°F; 0.90°F **b.** Yes; 96.3°F, 96.4°F, 100.8°F **23. a.** 210 s, or 3 min 30 s; 26 s **b.** Yes; 2:27, 3:01, 4:02
24. a. Answers will vary. **b.** The population standard deviation remains the same.
25. a. Answers will vary. **b.** The standard deviation of the new data is *k* times the standard deviation of the original data.
26. a. 0 **b.** Yes **c.** No **27.** If the variance is 0 or 1

EXCURSION EXERCISES, SECTION 13.3

page 778

1. Ages of Customers Who Purchased a Cruise

Stem	Leaves
2	1 3 8
3	2 3 3 5 8
4	0 1 2 4 5 5 6
5	0 1 1 2 2 5
6	1 2 4 6 8
7	2 7

Legend:
5|2 represents 52 years old

2. a. 11 people **b.** 4 people **c.** Group 1

3.

Before	Stem	After
	10	7
6	11	7 9
8 8 8	12	1 3 5 6 8 9
9 9 6 6 4 4 1	13	3 4 4 8 9 9
8 4 1	14	0 0 9
8 6 6 1	15	0
1	16	4
4	17	

Legend:
8|12 represents
128 beats per minute

Legend:
13|3 represents
133 beats per minute

EXERCISE SET 13.3

page 779

1. a. ≈ 0.87 **b.** ≈ 1.74 **c.** ≈ -2.17 **d.** 0.0 **2. a.** -0.3 **b.** 0.3 **c.** 2.2 **d.** -2.8
3. a. ≈ -0.32 **b.** ≈ 0.21 **c.** ≈ 1.16 **d.** ≈ -0.95 **4. a.** ≈ -1.48 **b.** ≈ 1.65 **c.** ≈ 2.52 **d.** ≈ 0.17
5. a. ≈ -0.67 **b.** 147.78 mm Hg **6. a.** ≈ -0.73 **b.** ≈ 10.9 fluid ounces **7. a.** ≈ 0.72 **b.** ≈ 112.16 mg/dl
8. a. ≈ 1.07 **b.** $\approx 31{,}200$ miles **9.** The score in part a. **10.** The score in part a. **11.** ≈ 59th percentile
12. ≈ 91st percentile **13.** 6396 students **14.** 2016 students **15. a.** 50% **b.** 10% **c.** 40%
16. a. 75% **b.** 50% **c.** 25% **17.** $Q_1 = 5$, $Q_2 = 10$, $Q_3 = 26$ **18.** $Q_1 = 9$, $Q_2 = 11$, $Q_3 = 14$

19.

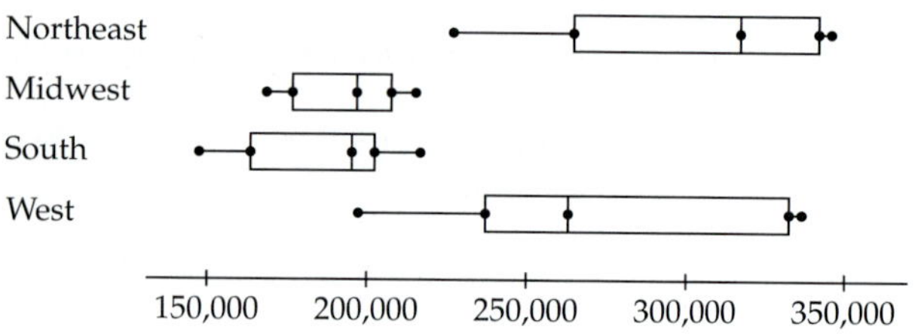

Answers will vary. Here are some possibilities. The region with the lowest median was the South; the region with the highest median was the Northeast. The range of prices was greatest for the West.

20.

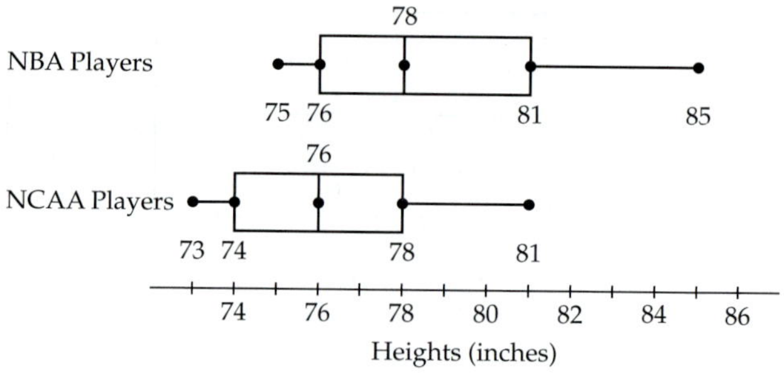

Answers will vary. Here is one possibility: The median height of an NBA player is equal to the third-quartile height of an NCAA player.

21.

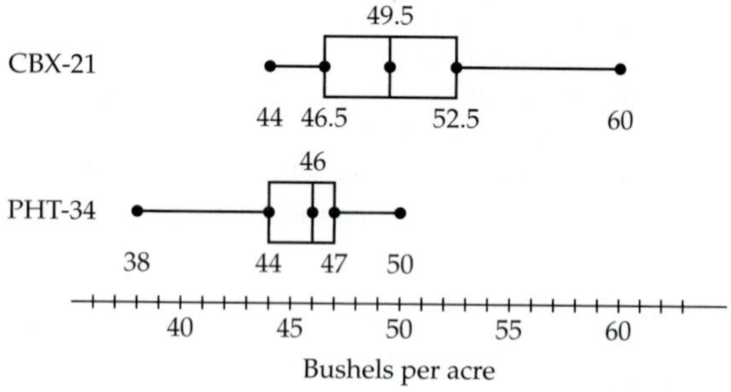

Answers will vary. Here is one possibility: The maximum number of bushels cultivated per acre for PHT-34 is approximately equal to the median number of bushels cultivated per acre for CBX-21.

22.

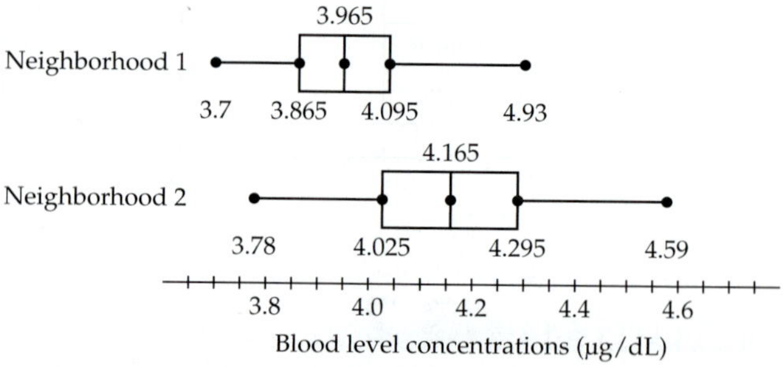

Answers will vary. Here is one possibility: The median blood lead concentration of children in Neighborhood 1 is less than the median blood lead concentration of children in Neighborhood 2, making Neighborhood 1 the healthier environment in which to live.

23. a. $\mu = 0, \sigma = 1$ **b.** $\mu = 0, \sigma = 1$ **c.** $\mu = 0, \sigma = 1$ **24. a.** False **b.** False **c.** True **d.** False

EXCURSION EXERCISES, SECTION 13.4

page 791

1. 77.4 **2. a.** At least 124.48 **b.** 86.72 **3.** 64.63 s or less

EXERCISE SET 13.4 page 791

1. a. 5.8% **b.** 0.87 **2. a.** 22.1% **b.** 0.944 **3. a.** 95% **b.** 16% **c.** 81.5% **4. a.** 99.7% **b.** 2.5% **c.** 97.35% **5. a.** 81.5% **b.** 0.15% **6. a.** 68% **b.** 2.5% **7. a.** 1280 vehicles **b.** 12 vehicles **8. a.** 680 women **b.** 950 women **9.** 0.433 square unit **10.** 0.471 square unit **11.** 0.468 square unit **12.** 0.489 square unit **13.** 0.130 square unit **14.** 0.215 square unit **15.** 0.878 square unit **16.** 0.636 square unit **17.** 0.097 square unit **18.** 0.027 square unit **19.** 0.013 square unit **20.** 0.352 square unit **21.** 0.926 square unit **22.** 0.966 square unit **23.** 0.997 square unit **24.** 0.973 square unit **25.** $z = 0.84$ **26.** $z = 0.75$ **27.** $z = -0.90$ **28.** $z = -0.44$ **29.** $z = 0.35$ **30.** $z = -0.15$ **31. a.** 19.2% **b.** 29.6% **32. a.** 5.4% **b.** 72% **33. a.** 80.5% **b.** 7% **34. a.** 13.1% **b.** 69.5% **35. a.** 10.6% **b.** 98.8% **36. a.** 1.8% **b.** 58.7% **37. a.** 0.106 **b.** 0.460 **38. a.** 0.053 **b.** 0.466 **39. a.** 0.749 **b.** 0.023 **40. a.** 22.7% **b.** 61.4% **41.** Answers will vary. **42.** Answers will vary. **43.** True **44.** True **45.** True **46.** False **47.** True **48.** False **49.** False **50. a.** and **b.** Answers will vary. **51.** ≈ -0.67 and 0.67

EXCURSION EXERCISES, SECTION 13.5 page 802

1. $\hat{y} \approx 10.14681746(0.8910371309)^x$; 2.9 newtons/cm^2

EXERCISE SET 13.5 page 802

1. a. b **b.** c **2. a.** a **b.** d

3. a.

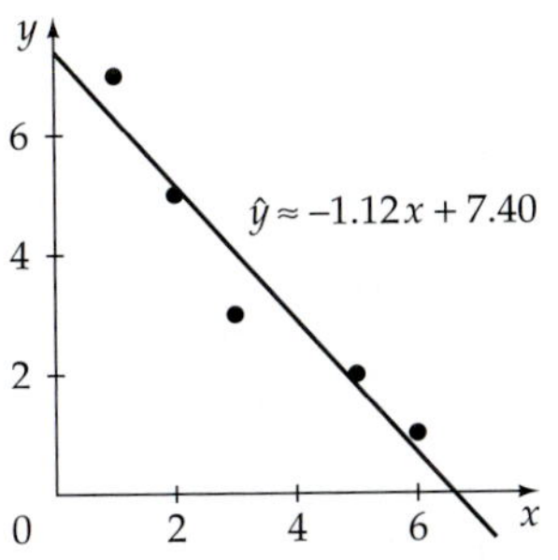

b. $n = 5$, $\Sigma x = 17$, $\Sigma y = 18$, $\Sigma x^2 = 75$, $(\Sigma x)^2 = 289$, $\Sigma xy = 42$
c. $a = -\dfrac{48}{43} \approx -1.12$, $b = \dfrac{318}{43} \approx 7.40$
d. See the graph in part a.
e. Yes
f. $y \approx 3.6$
g. $r \approx -0.96$

4. a.

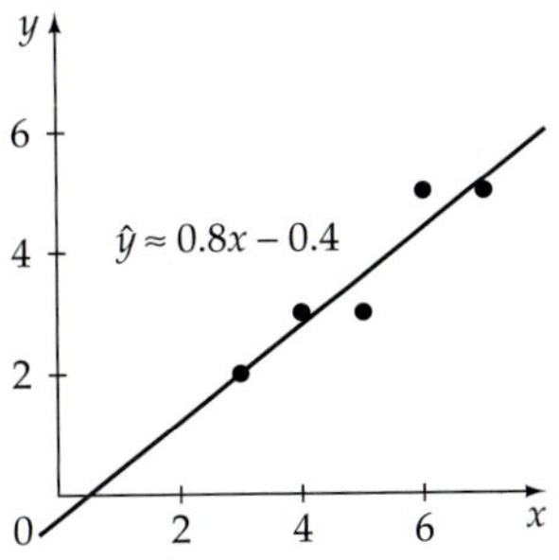

b. $n = 5$, $\Sigma x = 25$, $\Sigma y = 18$, $\Sigma x^2 = 135$, $(\Sigma x)^2 = 625$, $\Sigma xy = 98$
c. $a = 0.8$, $b = -0.4$
d. See the graph in part a.
e. Yes
f. $y = 5.44$
g. $r \approx 0.94$

5. $\hat{y} \approx 2.01x + 0.56$; $r \approx 0.96$

6. $\hat{y} \approx -1.92x + 0.46$; $r \approx -0.94$ **7.** $\hat{y} \approx -0.72x + 9.23$; $r \approx -0.96$ **8.** $\hat{y} \approx 0.66x - 6.66$; $r \approx 0.99$ **9.** $\hat{y} \approx 1.66x + 2.25$; $r \approx 0.99$ **10.** $\hat{y} \approx 2.25x - 0.36$; $r \approx 0.98$ **11. a.** $\hat{y} \approx -0.170x + 54{,}545.585$ **b.** \$49,444 **c.** $r \approx -0.999$ **d.** As the number of miles the car is driven increases, the value of the car decreases. **12. a.** $\hat{y} \approx 23.56x - 24.43$ **b.** $\approx$1248 cm **13. a.** $\hat{y} \approx 0.775x + 20.707$ **b.** $\approx$32.3% **14. a.** $\hat{y} \approx 0.686x + 21.1$ **b.** $\approx$ 31.4% **15. a.** $r \approx 0.99$ **b.** Yes, at least for the years 2005 to 2010. The correlation coefficient is very close to 1, which indicates a strong correlation. **16.** Yes. The correlation coefficient is very close to -1, which indicates a strong linear correlation. **17. a.** $\hat{y} \approx -0.91x + 79.21$ **b.** $\approx$56 years **c.** Interpolation **18. a.** $r = -1.00$ **b.** The graph of the least-squares line passes through all three data points. **c.** $\hat{y} = -0.85x + 187$ **d.** $\approx$126 beats per minute **e.** Extrapolation **19. a.** $r = 0.999525765$; $y = 1194.657x + 30{,}217.2$ **b.** $r = 0.9993403551$; $y = 827.8857x + 14{,}233.4$ **c.** Yes **d.** The value of a indicates the approximate yearly increase in tuition and fees. For instance, the number 827.8857 in part b means that tuition and fees at public universities increased by about \$828 each year. **20. a.** to **d.** Answers will vary.

CHAPTER 13 REVIEW EXERCISES page 808

1. 14.7; 14; 12; 9; 17.6; 4.2 [Sec. 13.1/13.2] **2.** The mode [Sec. 13.1] **3.** Answers will vary. [Sec. 13.1] **4. a.** Median **b.** Mode **c.** Mean [Sec. 13.1] **5.** 1331.125 ft; 1223.5 ft; 1200 ft; 462 ft [Sec. 13.1/13.2] **6.** 36 mph [Sec. 13.1] **7.** $\approx$3.20 [Sec. 13.1] **8. a.** 1.25 **b.** The 88th percentile [Sec. 13.3] **9.** $\approx$3.16; $\approx$10.00 [Sec. 13.2]

10. \$7.63; \$7.91; \$0.58 [Sec. 13.1/13.2] **b.** They are the same. [Sec. 13.1/13.2]

11. a. The second student's mean is 5 points higher than the first student's mean.

12. a. ≈ -1.73 **b.** ≈ 0.58 [Sec. 13.3]

13.

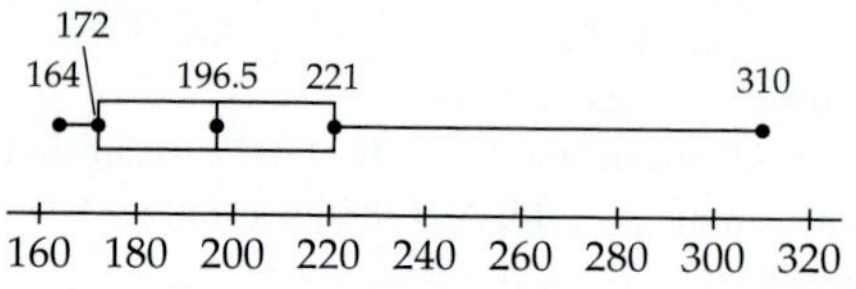

[Sec. 13.3]

14. a. 8 **b.** 40 [Sec. 13.4]

15. a. 78% **b.** 0.34 [Sec. 13.4]

16. Yes. $r \approx -0.9004$, which indicates a strong linear correlation. [Sec. 13.5]

17. a. $\hat{y} \approx 7609.1x - 72{,}019.1$ **b.** 42,117 stations [Sec. 13.5]

18. No. No information is given about how the scores are distributed below the mean. [Sec. 13.4]

19. a. 0.933 **b.** 0.309 [Sec. 13.4] **20. a.** 16% **b.** 95% [Sec. 13.4]

21. a. 6.7% **b.** 64.5% **c.** 78.8% [Sec. 13.4] **22.** 93.15 million mi [Sec. 13.1]

23. a.

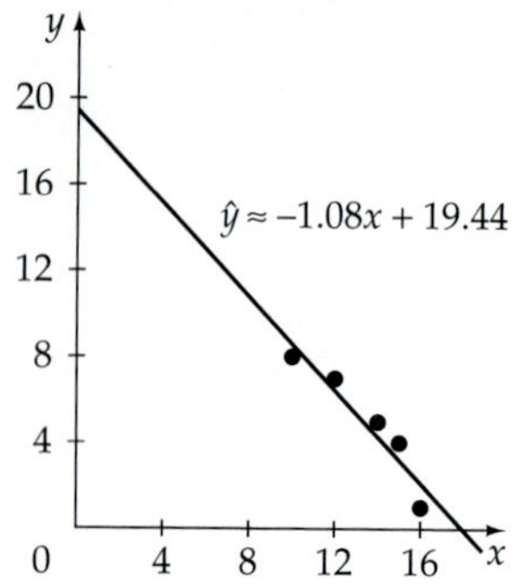

b. $n = 5$, $\Sigma x = 67$, $\Sigma y = 25$, $\Sigma x^2 = 921$, $(\Sigma x)^2 = 4489$, $\Sigma xy = 310$
c. $a \approx -1.08$, $b \approx 19.44$
d. See the graph in part a.
e. Yes
f. ≈ 10.8
g. $r \approx -0.95$ [Sec. 13.5]

24. a. $r \approx 0.999$ **b.** $\hat{y} \approx 0.07x + 0.29$ **c.** 13.94 in. [Sec. 13.5]

25. a. $\hat{y} \approx 0.018x + 0.0005$ **b.** Yes. $r \approx 0.999$, which is very close to 1. **c.** 1.80 s [Sec. 13.5]

26. a. $\hat{y} = 58.89921372x + 0.1924231594$; $r \approx 0.9911051145$ **b.** 3.7 h [Sec. 13.5]

CHAPTER 13 TEST *page 811*

1. 9.1; 9; 7 [Sec. 13.1, Examples 1, 2, and 3] **2.** 2.87 [Sec. 13.1, Example 4]

3. 24; ≈ 8.76; ≈ 76.7 [Sec. 13.2, Examples 1, 2, and 3] **4. a.** ≈ 1.18 **b.** ≈ -0.49 [Sec. 13.3, Example 1]

5.

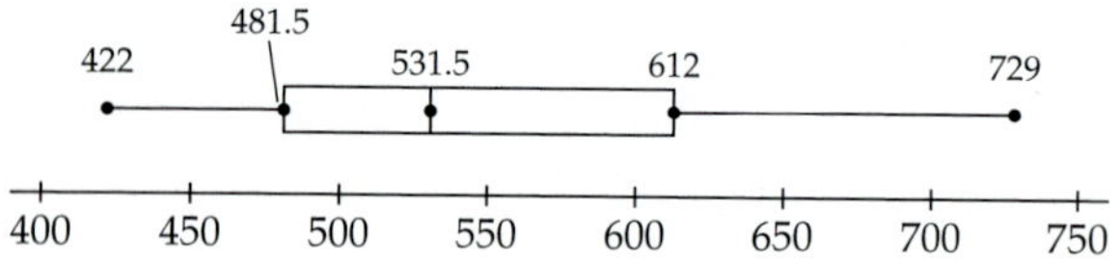

[Sec. 13.3, Examples 5 and 6]

6. a. 37% **b.** 35% [Sec. 13.4, Example 1]

7. a. 47.5% **b.** 16% [Sec. 13.4, Example 2] **8. a.** 10.6% **b.** 20.3% [Sec. 13.4, Example 4]

9. a. $r = -0.9308039961$ **b.** Yes. $|-0.9308039961| > 0.9$ [Sec. 13.5, Example 3]

10. a. $\hat{y} \approx -7.98x + 767.12$ **b.** 57 calories [Sec. 13.5, Examples 1 and 2]

Index of Applications

Index

Math Through the Ages

10,000 BC–0

10,000 BC Agricultural villages are evident in many parts of the world.

6300 BC The techniques for creating pottery are known.

6200 BC The first evidence of writing appears.

3400 BC The first symbols for numbers are used.

3250 BC Construction of the great pyramid Cheops begins.

Alexander Ryabintsev/Shutterstock.com

3000 BC Babylonians use a place-value numeration system.

2200 BC Approximate time of the Trojan War.

1850 BC The Pythagorean theorem is known by Babylonians.

1700 BC The approximate year the Rhind papyrus is written. It shows early algorithms for multiplication and division.

776 BC The first Olympic games are held.

Denis Kornilov/Shutterstock.com

700 BC A symbol for zero is introduced as a place holder. It is not considered a number.

575 BC Thales of Miletus, sometimes known as the first mathematician, uses deductive reasoning to prove theorems in geometry.

475 BC Babylonians use geometry to study astronomy.

450 BC The Greeks begin to use written numerals.

387 BC Plato founds his Academy.

300 BC Euclid writes *Elements*, the first systematic development of geometry.

290 BC Aristarchus uses geometry to calculate the distance to the moon.

260 BC Archimedes establishes pi as approximately equal to 22/7.

259 BC The construction of the Great Wall of China begins.

235 BC Eratosthenes estimates the circumference of Earth and develops a method for finding prime numbers.

225 BC Apollonius writes a treatise on conics and first uses the words *parabola*, *ellipse*, and *hyperbola*.

69 BC Cleopatra is born.

0–1499

79 Vesuvius erupts, destroying Pompeii and Herculaneum.

240 Mayan civilization uses a base 20 numeration system.

250 Diophantus writes *Arithmetica*.

400 Hypatia is the first recorded female mathematician.

476 The fall of Rome.

594 Decimal notation is used in India.

628 Brahmagupta gives rules for using zero and negative numbers in computations.

700 Mayan mathematicians introduce zero into their numeration system.

800 Charlemagne is crowned Holy Roman Emperor.

810 Al-Khwarizmi, also known as the father of algebra, writes a treatise on solving equations. From that treatise, the word *algebra* is derived.

875 Approximate date the first printed book is produced in China.

1202 Fibonacci writes *The Book of the Abacus* and helps popularize the Hindu-Arabic number system in Europe.

1206 The mechanical clock is invented.

1275 Yang Hui gives the first account of what becomes known as "Pascal's Triangle."

1290 Marco Polo travels the world.

1445 The approximate year in which the Gutenberg printing press is invented.

1482 Euclid's *Elements* becomes the first mathematics book to be printed.

1492 Christopher Columbus discovers America.

1492 Leonardo Da Vinci paints the Mona Lisa.

Classic Image/Alamy Stock Photo

1500–1699

1545 Cardan publishes *Ars Magna*, which gives the first formula to solve any cubic equation.

1564 Shakespeare is born.

1585 Stevin writes *De Thiende*, which gives a treatment of decimal fractions.

1591 Letters are first used to represent variables.

1614 Johannes Kepler publishes his laws on elliptical orbits of the planets.

1614 John Napier publishes his work on logarithms.

1617 Henry Briggs introduces logarithms base 10.

1620 Pilgrims land on Plymouth Rock.

1620 First American Indian reservation established in Connecticut.

1620 Gunter makes the first primitive slide rule.

1647 Fermat states that $x^n + y^n = z^n$ has no integer solutions for $n > 2$. This becomes known as Fermat's last theorem.

1665 Isaac Newton develops his infinitesimal calculus.

1687 Isaac Newton publishes *Philosophiæ Naturalis Principia Mathematica*, which is regarded as one of the best scientific papers ever written.

1692 Salem witch trials are held.

1698 The first steam engine is patented.

URRRA/Shutterstock.com

1700–1799

1706 William Jones introduces the symbol π to represent the ratio of the circumference to the diameter of a circle. π

1707 Abraham De Moivre uses trigonometric functions to represent complex numbers.

1709 Bartolomeo Christofori invents the piano.

1727 Leonhard Euler introduces the symbol e for the base of the natural logarithms.

1731 Benjamin Banneker, the first African American mathematician and scientist, is born. He helped survey Washington, D.C.

1735 Leonhard Euler introduces the notation $f(x)$.

1736 Leonhard Euler creates introductory graph theory and uses it to solve the Königsberg bridges problem.

1748 Maria Gaëtana Agnesi writes the most respected calculus text in Italy.

1755 Samuel Johnson publishes the first dictionary of the English language.

1761 Pi is proved to be an irrational number.

1776 The Declaration of Independence is signed.

1781 Caroline Herschel and her brother William discover Uranus.

1785 Marie Jean Condorcet publishes a treatise on voting theory.

1786 Caroline Herschel becomes the first woman to discover a comet.

1789 George Washington becomes the first president of the United States.

1796 Edward Jenner develops a smallpox vaccination.

1799 The Rosetta Stone is discovered.

1799 Carl Friedrich Gauss proves the Fundamental Theorem of Algebra.